MW00837091

HANDBOOK OF
MECHANICAL ENGINEERING
CALCULATIONS

Other McGraw-Hill Handbooks of Interest

Avallone, Baumeister • MARKS' STANDARD HANDBOOK FOR MECHANICAL
ENGINEERS

Bleier • FAN HANDBOOK

Brady et al. • MATERIALS HANDBOOK

Brink • HANDBOOK OF FLUID SEALING

Chironis, Sclater • MECHANISMS AND MECHANICAL DEVICES SOURCEBOOK

Czernik • GASKET HANDBOOK

Harris • SHOCK AND VIBRATION HANDBOOK

Hicks • STANDARD HANDBOOK OF ENGINEERING CALCULATIONS

Lingaiah • MACHINE DESIGN DATA HANDBOOK

Parmley • STANDARD HANDBOOK OF FASTENING AND JOINING

Rothbart • MECHANICAL DESIGN HANDBOOK

Shigley, Mischke • STANDARD HANDBOOK OF MACHINE DESIGN

Suchy • DIE DESIGN HANDBOOK

Walsh • MCGRAW-HILL MACHINING AND METALWORKING HANDBOOK

Walsh • ELECTROMECHANICAL DESIGN HANDBOOK

HANDBOOK OF MECHANICAL ENGINEERING CALCULATIONS

Tyler G. Hicks, M.E., P.E. Editor

International Engineering Associates
Member: American Society of Mechanical Engineers
United States Naval Institute
American Merchant Marine Museum Foundation

McGRAW-HILL

New York   San Francisco   Washington, D.C.   Auckland   Bogotá
Caracas   Lisbon   London   Madrid   Mexico City   Milan
Montreal   New Delhi   San Juan   Singapore
Sydney   Tokyo   Toronto

Library of Congress Cataloging-in-Publication Data

Handbook of mechanical engineering calculations / Tyler G. Hicks.
   editor.
     p.   cm.
   Includes index.
   ISBN 0-07-028813-5
   1. Mechanical engineering—Handbooks, manuals, etc. I. Hicks,
Tyler Gregory
TJ151.H327 1997
621—dc21   97-34001
                                            CIP

McGraw-Hill

*A Division of The **McGraw·Hill** Companies*

  3 4 5 6 7 8 9 0 DOC/DOC 9 0 2 1 0 9 8

ISBN 0-07-028813-5

*The sponsoring editor for this book was Harold B. Crawford, the editing
supervisor was Reina Zatylny, and the production supervisor was Clare
Stanley. It was set in Times Roman by Pro-Image Corporation.*

Printed and bound by R. R. Donnelley & Sons Company.

This book is printed on acid-free paper.

This handbook is dedicated to all engineers everywhere—and especially mechanical engineers, of whom the editor is one—with the hope that the calculation procedures presented will save time and energy for them. The editor thoroughly enjoyed the monumental task of preparing this handbook and he hopes its users will derive pleasure from the time-savings the book delivers. It is the further hope of the editor that the calculation procedures herein will help produce better designs for use worldwide.

Tyler G. Hicks, M.E.; P.E.

CONTENTS

CONTRIBUTORS AND ADVISERS

In preparing the various sections of this handbook, the following individuals either contributed sections, or portions of sections, or advised the editor or contributors, or both, on the optimum content of specific sections. The affiliations shown are those prevailing at the time of the preparation of the contributed material or the recommendations as to section content.

In choosing the procedures and worked-out problems, these specialists used a number of guidelines, including: (1) What are the most common applied problems that must be solved in this discipline? (2) What are the most accurate methods for solving these problems? (3) What other problems might be met in this discipline? When the answers to these and other related questions were obtained, the procedures and worked-out problems were chosen. Thus, the handbook represents a cross section of the thinking of a large number of experienced practicing engineers, project directors, and educators.

To those who might claim that the use of step-by-step solution procedures and worked-out examples makes engineering "too easy," the editor points out that for many years engineering educators have recognized the importance and value of problem solving in the development of engineering judgment and experience. Problems courses have been popular in numerous engineering schools for many years and are still given in many schools. However, with the greater emphasis on engineering science in most engineering schools, there is less time for the problems courses. The result is that many of today's graduates can benefit from a more extensive study of specific problem-solving procedures.

EDMUND B. BESSELIEVRE, P.E., *Consultant,* Forrest & Cotton, Inc.

ROBERT L. DAVIDSON, *Consulting Engineer*

STEPHEN M. EBER, P.E., Ebasco Services, Inc.

GERALD M. EISENBERG, *Project Engineering Administrator,* American Society of Mechanical Engineers

V. GANAPATHY, *Heat Transfer Specialist,* ABCO Industries, Inc.

CHARLES F. HAFER, P.E.

GREGORY T. HICKS, R.A., Gregory T. Hicks and Associates, Architects

S. DAVID HICKS, International Engineering Associates

TYLER G. HICKS, P.E., International Engineering Associates

EDGAR J. KATES, P.E., *Consulting Engineer*

MAX KURTZ, P.E., *Consulting Engineer*

JOSEPH LETO, P.E., *Consulting Engineer*

JEROME F. MUELLER, P.E., Mueller Engineering Corp.

GEORGE M. MUSCHAMP, *Consulting Engineer,* Honeywell, Inc.

RUFUS OLDENBERGER, *Professor,* Purdue University

JOHN S. REARICK, P.E., *Consulting Engineer*

RAYMOND J. ROARK, *Professor,* University of Wisconsin

LYMAN F. SCHEEL, *Consulting Engineer*

B. G. A. SKROTZKI, P.E., *Power* Magazine

S. W. SPIELVOGEL, *Piping Engineering Consultant*

KEVIN D. WILLS, M.S.E., P.E., *Consulting Engineer,* Stanley Consultants, Inc.

As the handbook user will see, the editor also drew on the published works of numerous engineers and scientists appearing in many technical magazines and journals. Each of these individuals is cited in the calculation procedure in which their work appears. Their position title and affiliation are given as of the time of original publication of the cited procedure or data.

PREFACE

This handbook presents thousands of worked-out calculation procedures (some 5000 procedures when the related calculations are included) covering the major areas of mechanical engineering. The content of the handbook is completely up-to-date and it covers current mechanical engineering practice throughout the world, presented in both the United States Customary System (USCS) and the International System (SI) units. Hence, the calculation procedures presented in this handbook are applicable in every country of the world.

The purpose of this handbook is to provide mechanical engineers—and all other engineers and technical personnel—with specific step-by-step calculation procedures for the most common design and operating problems met in daily practice. While specialists in a given discipline may know of and use more advanced methods, the procedures given in this handbook will produce safe and usable results for the majority of situations met by practicing mechanical engineers.

The handbook is useful to a variety of users—mechanical engineers, electrical engineers, civil engineers, chemical engineers, engineering students in every discipline, professional-engineer's license candidates, civil-service examination candidates, design engineers, drafters, estimators, large and small engineering firms—and many others who must make mechanical engineering calculations of many types. A major feature of this handbook is that the user can enter at any topic he or she needs guidance on—without the necessity of reviewing introductory material. Each procedure is a complete unit in itself, giving the handbook user the steps to follow, along with a worked-out example showing how to perform a specific calculation. Millions of users throughout the world find this approach to engineering calculations a valid way to save time and money in performing routine—and many non-routine—design calculations.

While some people may quarrel with direct solution of engineering design procedures, there continues to be a growing need for such information. Why? For several reasons, namely (1) today's engineering graduates have an excellent grounding in fundamentals. But their academic schedules are so full of basics that there is little time for practical applications. Thus, the new engineering graduate, when asked to choose a pump for—let's say—an industrial-plant process-water-supply system, will be puzzled as to how to apply his or her basic knowledge. This handbook shows the exact steps for the new engineer (or experienced engineer with a short memory) to follow to choose such a pump. (2) The availability of specialized computer programs for designing certain parts of an overall system—say steam piping—leaves the user of such programs almost helpless when faced with the design of other types of piping. The procedures in this handbook allow such people to refresh their memories and skills so they can perform the needed calculation tasks with assurance and competency. (3) No matter how "pure" an engineering design may be, there is still the economic urgency to "get it out by 4:00 P.M. on a Friday afternoon, either on-budget, or under-budget." This handbook allows engineers to achieve this important economic goal. Further, in today's litigious society, if the design is not completed on time and on budget, the engineer may wind up in court trying to explain why project targets were not met. The direct calculation procedures and solutions given in this handbook help every engineer avoid the misery, harassment, fear, and economic losses created by schedule and cost lawsuits.

To cover the enormous field that mechanical engineering encompasses, the handbook is divided into four major parts—Power Generation, Plant and Facilities Engineering, Environmental Control, and Design Engineering. Each part of the handbook presents a comprehensive coverage of its portion of the overall field of mechanical engineering.

In selecting procedures for the handbook the editor chose those that combine modern developments with new calculation methods. Thus, in the area of power generation there are many calculations directed at gas turbines. Why? Because in recent years the gas turbine has become the predominant prime mover in both power-plant expansion and in new generating capacity. The heat-recovery steam generator (HRSG)—unheard of just a few years ago—today accounts for almost as much steam-generating capacity as the conventional fossil-fuel-fired boiler. Hence, a number of the procedures in the power generation portion of this handbook cover HRSGs—their selection, sizing, and efficiency. And since HRSGs are almost always associated with the exhaust of a gas turbine, the two deserve—and receive—many detailed calculation procedures in this modern handbook. Cogeneration of power—popular throughout the world as an energy-conservation technique—is also the subject of numerous computations in this part of the handbook.

Another example of the developments covered in this handbook is the introduction of the aero-derivative gas turbine for power-plant expansion, topping service, repowering, and modernization. These highly efficient gas turbines save construction time, reduce space requirements, shorten environmental approval time, and allow greater fuel flexibility. All these topics are covered in this handbook.

What engineers formerly called Plant Engineering, is now more often termed Facilities Engineering. In this part of the handbook, important procedures are given for selecting, designing, and sizing piping and pumping systems, air and gas compressors, vacuum systems, materials handling, heat exchangers, and refrigeration machines and systems. Important new topics covered in this part of the handbook include non-metallic piping systems, non-polluting heat insulation for facilities piping, alternative refrigerants and refrigeration systems, and a variety of materials-handling systems. Using the calculation methods given in this part of the handbook the engineer can design a modern plant that will be environmentally friendly—as the environmentalists say.

Since the environment impacts everyone's life, the third part of this handbook interfaces the mechanical engineer with environmental rules and regulations. And since the formation of the Environmental Protection Agency (EPA) the rules and regulations governing engineering designs have grown more stringent every year. These rules and regulations impact every aspect of engineering design—from wastewater treatment and control to boiler-plant emissions to indoor air quality in offices and factories. Each of these—and many other—topics are presented in calculation-procedure form in this handbook.

Today's engineer often finds that environmental issues are stronger than economic ones—especially in the power-generation field. Since power plants are a prime source of atmospheric, stream, and land pollution, environmental considerations become primary in the calculation and design of new, rehabilitated, and expanded systems.

Design engineering—formerly termed Machine Design—is also a critical part of today's life. Conventional shafts, gears, power transmissions, couplings, clutches, springs, brakes, etc. must be designed from both a safety and an environmental standpoint. Other developments, including computer control of metalworking, robotics, and non-metallics replacing metals in some designs, all make the design engineer's job tougher. Calculation procedures given in this part of the handbook are designed to give the engineer better control of his or her designs.

The *step-by-step practical and applied calculation procedures* in this handbook are arranged so they can be followed by anyone with an engineering or scientific background. Each worked-out procedure presents fully explained and illustrated steps for solving design problems in industrial, commercial, marine, power-generation, municipal, military, petrochemical, manufacturing, metalworking, government, academic, license-examination, and a variety of other real-world situations. For any applied problem all the handbook user need do is place his or her calculation sheets alongside this handbook and follow the step-by-step procedure line for line to obtain the desired solution for the actual real-life problem. By following the calculation procedures in this handbook, the engineer, scientist, or technician will obtain accurate results in minimum time with least effort.

Four factors have changed mechanical-engineering calculations in recent years. These factors intrude on the conventional approaches to analyses and designs. Calculation procedures presented in this handbook take these four factors into consideration and show the engineer how to cope with them. The four factors are:

1. *The environment:* Engineering both produces (to some extent), and cleans up, the environment. So the mechanical engineer must prepare environmentally clean designs while meeting budget constraints. These requirements change many steps in today's calculation procedures. Thousands of environmental regulations from federal, state, county, and city regulators impact every stage of engineering design. To help all engineers cope better with keeping current with today's regulations, this handbook includes pertinent environmental data that will help the engineer tailor a design to include environmental considerations that make it more acceptable to regulators having control over the project.

2. *The economy:* In years past there was enormous growth in certain engineering-based industries. For example, the electric-power industry historically grew 10 percent per year. But growth slowed as more energy-efficient devices were introduced. So today the mechanical engineer is faced with increasing energy capacity at a slower rate while spending less for more environmentally gentle equipment. These requirements change the entire approach to calculations for new designs. This handbook emphasizes the new considerations the economy has introduced.

3. *The environmental lobby:* With no major wars taking place now, and none predicted for the near-term future, the "cause freaks" have jumped on environmental issues as their battle cry. The result is that we find that many environmentally acceptable engineering designs are not acceptable to the environmental lobby. The "new" engineer must satisfy these people—if his or her design is to be accepted by the environmental lobby in the community in which the design will be located, or which it serves. Crafting such designs can be an extremely challenging task for any engineer.

4. *The regulators:* Today's engineer must recognize many regulatory agencies—Environmental Protection Agency (EPA), Occupational Health and Safety

Agency (OSHA), Coastal Commission, plus hundreds of Acts passed by Congress. Loaded on top of these federal regulations and acts are State Acts passed by one or more of the 50 states. The result? A nightmare of regulations governing every aspect of engineering design and manufacture. Many of the regulations and Acts are reflected in the calculation procedures presented in this handbook.

Thousands of computer programs are available for solving common repetitive engineering design problems. Prices of such programs range from a few hundred dollars to several thousand dollars, depending on the program and its developer.

Engineers using such programs often find the data-entry time requirements are excessive for quick one-off-type calculations. When typical design calculations are needed, most engineers turn today to either their laptop computer with its built-in electronic calculator, or to their separate electronic calculator, and perform the necessary steps to provide the solution desired. But where repetitive calculations are required, the computer program will save time and energy in the usual medium-size or large engineering office. Small engineering offices generally resort to manual calculation for even repetitive procedures because the investment for one or more programs is difficult to justify.

Even where computer programs are used extensively, careful engineers still insist on manually checking results on a random basis to be certain the program is accurate. This checking can be speeded by using any of the calculation procedures in this handbook. Many engineers remark to the editor that they feel safer, knowing they have manually checked and verified the results of a computer-program calculation. With liability for engineering designs extending beyond the lifetime of the designer, every engineer seeks the "security blanket" provided by the manual verification of the results furnished by a computer program run on a laptop, desktop, or main-frame computer. This handbook gives its user the tools needed for manual verification of some 5000 calculation procedures.

Thus, this modern handbook is designed to answer the four needs above, plus those of engineers new to applied calculation procedures, those of engineers who've been away from calculations for a few years because of other assignments, plus all others needing proficiency in mechanical engineering calculations. The editor hopes that these needs are adequately met. He welcomes comments—and corrections—from users of the handbook. And the editor will personally answer every letter or fax sent to him in care of the publisher.

In a handbook of this size—some 1500 pages—with nearly half of every page comprised of mathematical material, errors can occur. For this reason, the editor asks each user of the handbook to call to his attention any errors that are found. The errors will be corrected in the next printing of the handbook.

Further, if a user feels that one or more important calculation procedures have been excluded from this handbook, the editor would like to have these called to his attention. And if a reader would like to submit a favorite calculation procedure for possible inclusion in the next edition, the editor will be glad to receive the procedure and evaluate it. All accepted procedures will be fully acknowledged as to source and contributor in the next edition. To have a procedure considered, just send the name of the procedure to the editor in care of the publisher. The editor will respond, indicating if he is interested in seeing the full procedure. Please do *not* send the full procedure unless requested to do so by the editor.

Lastly, thank you for using this handbook. I hope it helps you in all aspects of modern mechanical engineering practice.

Tyler G. Hicks, P.E.

ACKNOWLEDGMENTS

The contributors and advisers consulted hundreds of sources when preparing the material for inclusion in this handbook. Besides using the books and other publications listed as references in the Bibliography at the back of this handbook, the editor, contributors and advisers consulted and drew material from technical magazines and journals, trade-association standards, engineering and scientific papers, industrial and engineering catalogs, and a variety of similar publications. Most of these are noted in appropriate places throughout the handbook. Additional acknowledgments, listed in the order received, are given below.

Data and charts credited to the Hydraulic Institute are reprinted from the *Hydraulic Institute Standards,* copyright by the Hydraulic Institute, and from the *Pipe Friction Manual,* copyright by the Hydraulic Institute. Data on diesel engine cooling systems are reprinted from *Marine Diesel Standard Practices,* copyright by the Diesel Engine Manufacturers Association. Data on minimum requirements for plumbing are drawn from the *American Standard National Plumbing Code* with permission of the publisher, The American Society of Mechanical Engineers.

Specific firms, trade associations, and publications that were extremely helpful in supplying data for various sections of the handbook include Martin Marietta Corporation; *Electronic Design* magazine; Dresser Industries Inc.—Dresser Industrial Valve and Instrument Division; Ingersoll-Rand Company; Anaconda American Brass Company; Waterloo Register Division—Dynamics Corporation of America; ITT Hammel-Dahl; *Mechanical Engineering,* a monthly publication of The American Society of Mechanical Engineers; McQuay, Inc.; The G. C. Breidert Co.; Modine Manufacturing Company; Rubber Manufacturers Association; Condenser Service & Engineering Co., Inc.; Armstrong Machine Works; American Air Filter Company; Crane Company; *Machine Design* magazine; The RAND Corporation; Texas Instruments Incorporated; McGraw-Hill Publications Company, McGraw-Hill, Inc.; Morse Chain Company; Grinnell Corporation; General Electric Company; The B. F. Goodrich Company; American Standard Inc.; the American Society of Heating, Refrigerating and Air-Conditioning Engineers; International Engineering Associates; Taylor Instrument Process Control Division of Sybron Corporation; Clark-Reliance Corporation; American Society for Testing and Materials; Acoustical and Insulating Materials Association; W. S. Dickey Clay Manufacturing Co.; Flexonics Division, Universal Oil Products Co.; Dunham-Bush, Inc.; Carrier Air Conditioning Company; National Industrial Leather Association; Worthington Corporation; Goulds Pumps, Inc. Illustrations and problems credited to Carrier Air Conditioning Company are copyrighted by Carrier Air Conditioning Company.

Individuals who were helpful to the editor of this handbook at one or more times before, and during its preparation, include Lyman F. Scheel, Consulting Engineer; Jack Jaklitsch, Editor, *Mechanical Engineering;* Spencer A. Tucker, Martin & Tucker; Paul V. DeLuca, Porta Systems Corp.; Professor Steven Edelglass, Cooper Union; Professor William Vopat, Cooper Union; Professor Theodore Baumeister, Columbia University; Frederick S. Merritt, Consulting Engineer; James J. O'Connor, Editor, *Power* magazine; Nathan R. Grossner, Consulting Engineer; Nicholas P. Chironis, *Product Engineering;* Franklin D. Yeaple, *Product Engineering* and *Design Engineering;* John D. Constance, Consulting Engineer; John R. Miller, Texas Instruments Incorporated; Rupert Le Grand, *American Machinst;* Ronald G. Kogan, United Computing Systems, Inc.; Al Brons, Flexonics Div., Universal Oil Products Company; Carl W. MacPhee, ASHRAE *Guide and Data Book;* Frank P. Anderson, Secretary, Hydraulic Institute; Joseph Mittleman, *Electronics;* Cheryl A. Shaver, E.E., who was a major help in metricating several sections of the handbook; Thomas F. Epley, Editorial Director, U.S. Naval Institute Press; Janet Eyler, *Electronics;* Charles R. Hafer, P.E.; Calvin S. Cronan, *Chemical Engineering* Magazine; Nicholas Chopey, Executive Editor, *Chemical Engineering* Magazine; Joseph C. McCabe, Editor-Publisher, *Combustion* Magazine; Francis J. Lavoie, Managing Editor, *Machine Design* Magazine; Donald E. Fink, Managing Editor—Technical, *Aviation Week & Space Technology* Magazine; Richard J. Zanetti, Editor-in-Chief, *Chemical Engineering* Magazine; Robert G. Schwieger, Editorial Director, *Power* Magazine; Barbara LoSchiavo, Editorial Support, *Machine Design* Magazine; Michael G. Ivanovich, Editor, *Heating/Piping/Air Conditioning* magazine; Calmac Manufacturing Corp. whose *Ice Bank* is a registered trademark of that corporation; Joseph Leto, P.E., Consulting Engineer; Gerald M. Eisenberg, Project Engineering Administrator, American Society of Mechanical Engineers, who contributed a number of new procedures and ideas; Stephen M. Eber, P.E., Ebasco Services, Inc., who also contributed a number of new procedures and ideas; Jerome Mueller, P.E., Mueller Engineering Corporation, who was most helpful with thoughts on applied calculation procedures; V. Ganapathy, Heat Transfer Specialist, ABCO Industries, Inc.; Kevin D. Wills, M.S.E., P.E., Consulting Engineer, Stanley Consultants, Inc.; Joseph B. Shanley, Mechanical Engineer, who metricated many illustrations and procedures; and numerous working engineers and scientists in firms and universities in the United States and abroad.

Completion of this handbook would not have been possible without the enormous help of my wife, Mary Shanley Hicks, a publishing professional, who worked thousands of hours on the computer and at many other tasks to put the gigantic manuscript into publishable condition. She—more than the editor—deserves whatever accolades this handbook earns. And she has the editor's grateful thanks for keeping him on the job until the manuscript was finished.

HOW TO USE THIS HANDBOOK

There are two ways to enter this handbook to obtain the maximum benefit from the time invested. The first entry is through the index; the second is through the table of contents of the section covering the discipline, or related discipline, concerned. Each method is discussed in detail below.

Index. Great care and considerable time were expended on preparation of the index of this handbook so that it would be of maximum use to every reader. As a general guide, enter the index using the generic term for the type of calculation procedure being considered. Thus, for the design of a piping system, enter at *piping*. From here, progress to the specific type of piping being considered—such as *steam transmission line*. Once the page number or numbers of the appropriate calculation procedure are determined, turn to them to find the step-by-step instructions and worked-out example that can be followed to solve the problem quickly and accurately.

Contents. The contents of each section lists the titles of the calculation procedures contained in that section. Where extensive use of any section is contemplated, the editor suggests that the reader might benefit from an occasional glance at the table of contents of that section. Such a glance will give the user of this handbook an understanding of the breadth and coverage of a given section, or a series of sections. Then, when he or she turns to this handbook for assistance, the reader will be able more rapidly to find the calculation procedure he or she seeks.

Calculation Procedures. Each calculation procedure is a unit in itself. However, any given calculation procedure will contain subprocedures that might be useful to the reader. Thus, a calculation procedure on pump selection will contain subprocedures on pipe friction loss, pump static and dynamic heads, etc. Should the reader of this handbook wish to make a computation using any of such subprocedures, he or she will find the worked-out steps that are presented both useful and precise. Hence, the handbook contains numerous valuable procedures that are useful in solving a variety of applied engineering problems.

One other important point that should be noted about the calculation procedures presented in this handbook is that many of the calculation procedures are equally applicable in a variety of disciplines. Thus, a piping-system selection procedure can be used for mechanical-, civil-, chemical-, electrical-, and nuclear-engineering activities, as well as some others. Hence, the reader might consider a temporary neutrality for his or her particular specialty when using the handbook because the calculation procedures are designed for universal use.

Any of the calculation procedures presented can be programmed on a computer. Such programming permits rapid solution of a variety of design problems. With the growing use of low-cost time sharing, more engineering design problems are being solved using a remote terminal in the engineering office. The editor hopes that engineers throughout the world will make greater use of work stations and portable computers in solving applied engineering problems. This modern equipment promises greater speed and accuracy for nearly all the complex design problems that must be solved in today's world of engineering.

To make the calculation procedures more amenable to computer solution (while maintaining ease of solution with a handheld calculator), a number of the algorithms in the handbook have been revised to permit faster programming in a computer environment. Likewise, all the calculation procedures in this handbook have their algorithms in programmable form. This enhances ease of solution for any method used—work station, portable computer, or calculator.

SI Usage. The technical and scientific community throughout the world accepts the SI (System International) for use in both applied and theoretical calculations. With such widespread acceptance of SI, every engineer must become proficient in the use of this system of units if he or she is to remain up-to-date. For this reason, every calculation procedure in this handbook is given in both the United States Customary System (USCS) and SI. This will help all engineers become proficient in using both systems of units. In this handbook the USCS unit is generally given first, followed by the SI value in parentheses or brackets. Thus, if the USCS unit is 10 ft, it will be expressed as 10 ft (3 m).

Engineers accustomed to working in USCS are often timid about using SI. There really aren't any sound reasons for these fears. SI is a logical, easily understood, and readily manipulated group of units. Most engineers grow to prefer SI, once they become familiar with it and overcome their fears. This handbook should do much to "convert" USCS-user engineers to SI because it presents all calculation procedures in both the known and unknown units.

Overseas engineers who must work in USCS because they have a job requiring its usage will find the dual-unit presentation of calculation procedures most helpful. Knowing SI, they can easily convert to USCS because all procedures, tables, and illustrations are presented in dual units.

Learning SI. An efficient way for the USCS-conversant engineer to learn SI is to follow these steps:

1. List the units of measurement commonly used in your daily work.
2. Insert, opposite each USCS unit, the usual SI unit used; Table 1 shows a variety of commonly used quantities and the corresponding SI units.

TABLE 1   Commonly Used USCS and SI Units*

USCS unit	SI unit	SI symbol	Conversion factor—multiply USCS unit by this factor to obtain the SI unit
square feet	square meters	m^2	0.0929
cubic feet	cubic meters	m^3	0.2831
pounds per square inch	kilopascal	kPa	6.894
pound force	newton	N	4.448
foot pound torque	newton-meter	$N \cdot m$	1.356
Btu per pound	kilojoule per kilogram	kJ/kg	2.326
gallons per minute	liters per second	L/s	0.06309
Btu per cubic foot	kilojoule per cubic meter	kJ/m^3	37.26

*Because of space limitations this table is abbreviated. For a typical engineering practice an actual table would be many times this length.

3. Find, from a table of conversion factors, such as Table 2, the value to use to convert the USCS unit to SI, and insert it in your list. (Most engineers prefer a conversion factor that can be used as a multiplier of the USCS unit to give the SI unit.)
4. Apply the conversion factors whenever you have an opportunity. Think in terms of SI when you encounter a USCS unit.
5. Recognize—here and now—that the most difficult aspect of SI is becoming comfortable with the names and magnitude of the units. Numerical conversion is simple, once you've set up *your own* conversion table. So think pascal whenever you encounter pounds per square inch pressure, newton whenever you deal with a force in pounds, etc.

SI Table for a Mechanical Engineer.   Let's say you're a mechanical engineer and you wish to construct a conversion table and SI literacy document for yourself. List the units you commonly meet in your daily work; Table 1 is the list compiled by one mechanical engineer. Next, list the SI unit equivalent for the USCS unit. Obtain the equivalent from Table 2. Then, using Table 2 again, insert the conversion multiplier in Table 1.

Keep Table 1 handy at your desk and add new units to it as you encounter them in your work. Over a period of time you will build a personal conversion table that will be valuable to you whenever you must use SI units. Further, since *you* compiled the table, it will have a familiar and nonfrightening look, which will give you greater confidence in using SI.

Units Used.   In preparing the calculation procedures in this handbook, the editors and contributors used standard SI units throughout. In a few cases, however, certain units are still in a state of development. For example, the unit *tonne* is used

TABLE 2   Typical Conversion Table*

To convert from	To	Multiply by	
square feet	square meters	9.290304	E − 02
foot per second squared	meter per second squared	3.048	E − 01
cubic feet	cubic meters	2.831685	E − 02
pound per cubic inch	kilogram per cubic meter	2.767990	E + 04
gallon per minute	liters per second	6.309	E − 02
pound per square inch	kilopascal	6.894757	
pound force	newton	4.448222	
British thermal unit per square foot	joule per square meter	1.135653	E + 04
British thermal unit per hour	Watt	2.930711	E − 01
British thermal unit per cubic foot	kilojoule per cubic meter	3.725697	E + 01
foot-pound torque	newton-meter	1.355818	
British thermal unit per pound	joule per kilogram	2.326	E + 03

Note: The E indicates an exponent, as in scientific notation, followed by a positive or negative number, representing the power of 10 by which the given conversion factor is to be multiplied before use. Thus, for the square feet conversion factor, $9.290304 \times \frac{1}{100} = 0.09290304$, the factor to be used to convert square feet to square meters. For a positive exponent, as in converting British thermal units per cubic foot to kilojoule per cubic meter, $3.725697 \times 10 = 37.25697$.

Where a conversion factor cannot be found, simply use the dimensional substitution. Thus, to convert pounds per cubic inch to kilograms per cubic meter, find 1 lb = 0.4535924 kg, and 1 in^3 = 0.00001638706 m^3. Then, 1 lb/in^3 = 0.4535924 kg/0.00001638706 m^3 = 27,67990, or 2.767990 E + 04.

*This table contains only selected values. See the U.S. Department of the Interior *Metric Manual,* or National Bureau of Standards, *The International System of Units (SI),* both available from the U.S. Government Printing Office (GPO), for far more comprehensive listings of conversion factors.

in certain industries, such as metalworking. This unit is therefore used in the metalworking section of this handbook because it represents current practice. However, only a few SI units are still under development. Hence, users of this handbook face little difficulty from this situation.

Computer-aided Calculations. Widespread availability of personal computers (both laptop and desktop), programmable pocket calculators and low-cost work stations allows engineers and designers to save thousands of hours of calculation time. Yet each calculation procedure must be programmed, unless the engineer is willing to use off-the-shelf software. The editor—observing thousands of engineers over the years—detects reluctance among technical personnel to use untested and unproven software programs in their daily calculations. Hence, the tested and proven procedures in this handbook form excellent programming input for computers, programmable pocket calculators, work stations, and mainframes.

A variety of software application programs can be used to put the procedures in this handbook on computer. Typical of these are MathSoft, Algor, and similar programs.

There are a number of advantages for the engineer who programs his or her own calculation procedures, namely: (1) The engineer knows, understands, and approves *every* step in the procedure; (2) there are *no* questionable, unknown, or legally worrisome steps in the procedure; (3) the engineer has complete faith in the result because he or she knows every component of it; and (4) if a variation of the procedure is desired, it is relatively easy for the engineer to make the needed changes in the program, using this handbook as the source of the steps and equations to apply.

Modern computer equipment provides greater speed and accuracy for almost all complex design calculations. The editor hopes that engineers throughout the world will make greater use of available computing equipment in solving applied engineering problems. Becoming computer literate is a necessity for every engineer, no matter which field he or she chooses as a specialty. The procedures in this handbook simplify every engineer's task of becoming computer literate because the steps given comprise—to a great extent—the steps in the computer program that can be written.

HANDBOOK OF
MECHANICAL ENGINEERING
CALCULATIONS

P · A · R · T 1

POWER GENERATION

SECTION 1
MODERN POWER-PLANT CYCLES AND EQUIPMENT

Cycle Analyses

CHOOSING BEST OPTION FOR BOOSTING COMBINED-CYCLE PLANT OUTPUT

Select the best option to boost the output of a 230-MW facility based on a 155-MW natural-gas-fired gas turbine (GT) featuring a dry low NO_x combustor (Fig. 1). The plant has a heat-recovery steam generator (HRSG) which is a triple-pressure design with an integral deaerator. A reheat condensing steam turbine (ST) is used and it is coupled to a cooling-tower/surface-condenser heat sink turbine inlet. Steam conditions are 1450-lb/in^2 (gage)/1000°F (9991-kPa/538°C). Unit ratings are for operation at International Standard Organization (ISO) conditions. Evaluate the various technologies considered for summer peaking conditions with a dry bulb (DB) temperature of 95°F and 60 percent RH (relative humidity) (35°C and 60 percent RH). The plant heat sink is a four-cell, counterflow, mechanical-draft cooling tower optimized to achieve a steam-turbine exhaust pressure of 3.75 inHg absolute (9.5 cmHg) for all alternatives considered in this evaluation. Base circulating-water system includes a surface condenser and two 50 percent-capacity pumps. Water-treatment, consumption, and disposal-related O&M (operating & maintenance)

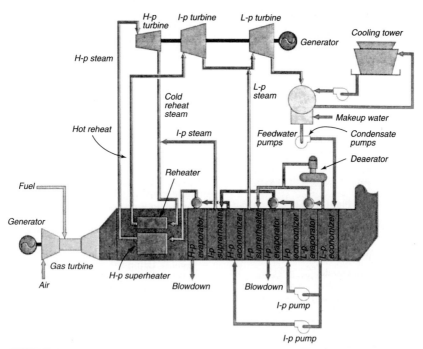

FIGURE 1 155-MW natural-gas-fired gas turbine featuring a dry low NO_x combustor (*Power*).

costs for the zero-discharge facility are assumed to be $3/1000 gal ($3/3.8 m³) of raw water, $6/1000 gal ($6/3.8 m³) of treated demineralized water, and $5/1000 gal ($5/3.8 m³) of water disposal. The plant is configured to burn liquid distillate as a backup fuel.

Calculation Procedure:

1. *List the options available for boosting output*

Seven options can be developed for boosting the output of this theoretical reference plant. Although plant-specific issues will have a significant effect on selecting an option, comparing performance based on a reference plant, Fig. 1, can be helpful. Table 1 shows the various options available in this study for boosting output. The comparisons shown in this procedure illustrate the characteristics, advantages, and disadvantages of the major power augmentation technologies now in use.

Amidst the many advantages of gas turbine (GT) combined cycles (CC) popular today from various standpoints (lower investment than for new greenfield plants, reduced environmental impact, and faster installation and startup), one drawback is that the achievable output decreases significantly as the ambient inlet air temperature increases. The lower density of warm air reduces mass flow through the GT. And, unfortunately, hot weather typically corresponds to peak power loads in many areas. So the need to meet peak-load and power-sales contract requirements causes many power engineers and developers to compensate for ambient-temperature-output loss.

The three most common methods of increasing output include: (1) injecting water or steam into the GT, (2) precooling GT inlet air, and/or (3) supplementary firing of the heat-recovery steam generator (HRSG). All three options require significant capital outlays and affect other performance parameters. Further, the options

TABLE 1 Performance Summary for Enhanced-Output Options

Measured change from base case	Case 1 Evap. cooler	Case 2 Mech. chiller	Case 3 Absorp. chiller	Case 4 Steam injection	Case 5 Water injection	Case 6[1] Supp.-fired HRSG	Case 7[2] Supp-fired HRSG
GT output, MW	5.8	20.2	20.2	21.8	15.5	0	0
ST output, MW	0.9	2.4	−2.1	−13	3.7	8	35
Plant aux. load, MW	0.05	4.5	0.7	400	0.2	0.4	1
Net plant output, MW	6.65	18.1	17.4	8.4	19	7.6	34
Net heat rate, Btu/kWh[3]	15	55	70	270	435	90	320
Incremental costs							
Change in total water cost, $/h	15	35	35	115	85	35	155
Change in wastewater cost, $/h	1	17	17	2	1	1	30
Change in capital cost/ net output, $/kW	180	165	230	75	15	70	450

[1]Partial supplementary firing.
[2]Full supplementary firing.
[3]Based on lower heating value of fuel.

may uniquely impact the operation and/or selection of other components, including boiler feedwater and condensate pumps, valves, steam turbine/generators, condensers, cooling towers, and emissions-control systems.

2. *Evaluate and analyze inlet-air precooling*
Evaporative cooling, Case 1, Table 1, boosts GT output by increasing the density and mass flow of the air entering the unit. Water sprayed into the inlet-air stream cools the air to a point near the ambient wet-bulb temperature. At reference conditions of 95°F (35°C) DB and 60 percent RH, an 85 percent effective evaporative cooler can alter the inlet-air temperature and moisture content to 85°F (29°C) and 92 percent RH, respectively, using conventional humidity chart calculations, page 16.79. This boosts the output of both the GT and—because of energy added to the GT exhaust—the steam turbine/generator. Overall, plant output for Case 1 is increased by 5.8 MW GT output + 0.9 MW ST output—plant auxiliary load of 0.9 MW = 6.65 MW, or 3.3 percent. The CC heat rate is improved 0.2 percent, or 15 Btu/kWh (14.2 kJ/kWh). The total installed cost for the evaporative cooling system, based on estimates provided by contractors and staff, is $1.2-million. The incremental cost is $1,200,000/6650 kW = $180.45/kW for this ambient condition.

The effectiveness of the same system operating in less-humid conditions—say 95°F DB (35°C) and 40 percent RH—is much greater. In this case, the same evaporative cooler can reduce inlet-air temperature to 75°F DB (23.9°C) by increasing RH to 88 percent. Here, CC output is increased by 7 percent, heat rate is improved (reduced) by 1.9 percent, and the incremental installed cost is $85/kW, computed as above. As you can clearly see, the effectiveness of evaporative cooling is directly related to reduced RH.

Water-treatment requirements must also be recognized for this Case, No. 1. Because demineralized water degrades the integrity of evaporative-cooler film media, manufacturers may suggest that only raw or filtered water be used for cooling purposes. However, both GT and evaporative-cooler suppliers specify limits for turbidity, pH, hardness, and sodium (Na) and potassium (K) concentrations in the injected water. Thus, a nominal increase in water-treatment costs can be expected. In particular, the cooling water requires periodic blowdown to limit solids buildup and system scaling. Overall, the evaporation process can significantly increase a plant's makeup-water feed rate, treatment, and blowdown requirements. Compared to the base case, water supply costs increase by $15/h of operation for the first approach, and $20/h for the second, lower RH mode. Disposal of evaporative-cooler blowdown costs $1/h in the first mode, $2/h in the second. Evaporative cooling has little or no effect on the design of the steam turbine.

3. *Evaluate the economics of inlet-air chilling*
The effectiveness of evaporative cooling is limited by the RH of the ambient air. Further, the inlet air cannot be cooled below the wet-bulb (WB) temperature of the inlet air. Thus, chillers may be used for further cooling of the inlet air below the wet-bulb temperature. To achieve this goal, industrial-grade mechanical or absorption air-conditioning systems are used, Fig. 2. Both consist of a cooling medium (water or a refrigerant), an energy source to drive the chiller, a heat exchanger for extracting heat from the inlet air, and a heat-rejection system.

A mechanical chilling system, Case 2, Table 1, is based on a compressor-driven unit. The compressor is the most expensive part of the system and consumes a significant amount of energy. In general, chillers rated above 12-million Btu/h (3.5 MW) (1000 tons of refrigeration) (3500 kW) employ centrifugal compressors. Units smaller than this may use either screw-type or reciprocating compressors. Overall,

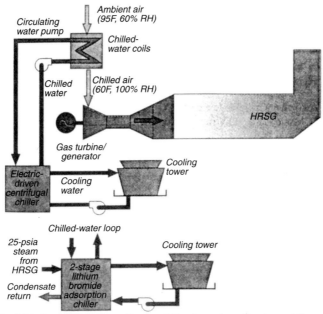

FIGURE 2 Inlet-air chilling using either centrifugal or absorption-type chillers, boosts the achieveable mass flow and power output during warm weather (*Power*).

compressor-based chillers are highly reliable and can handle rapid load changes without difficulty.

A centrifugal-compressor-based chiller can easily reduce the temperature of the GT inlet air from 95°F (35°C) to 60°F (15.6°C) DB—a level that is generally accepted as a safe lower limit for preventing icing on compressor inlet blades—and achieve 100 percent RH. This increases plant output by 20.2 MW for GT + 2.4 MW for ST − 4.5 MW plant auxiliary load = 18.1 MW, or 8.9 percent. But it degrades the net CC heat rate by 0.8 percent and results in a 1.5-in-(3.8-cm)-H_2O inlet-air pressure drop because of heat-exchanger equipment located in the inlet-air stream.

Cooling requirements of the chilling system increase the plant's required circulating water flow by 12,500 gal/min (47.3 m^3/min). Combined with the need for increased steam condensing capacity, use of a chiller may necessitate a heat sink 25 percent larger than the base case. The total installed cost for the mechanical chilling system for Case 2 is $3-million, or about $3,000,000/18,100 kW = $165.75/kW of added output. Again, costs come from contractor and staff studies.

Raw-water consumption increase the plant's overall O&M costs by $35/h when the chiller is operating. Disposal of additional cooling-tower blowdown costs $17/h. The compressor used in Case 2 consumes about 4 MW of auxiliary power to handle the plant's 68-million Btu/h (19.9 MW) cooling load.

4. *Analyze an absorption chilling system*

Absorption chilling systems are somewhat more complex than mechanical chillers. They use steam or hot water as the cooling motive force. To achieve the same inlet-air conditions as the mechanical chiller (60°F DB, 100 percent RH) (15.6°C, 100 percent RH), an absorption chiller requires about 111,400 lb/h (50,576 kg/h) of 10.3-lb/in^2 (gage) (70.9-kPa) saturated steam, or 6830 gal/min (25.9 m^3/min) of 370°F (188°C) hot water.

Cost-effective supply of this steam or hot water requires a redesign of the reference plant. Steam is extracted from the low-pressure (l-p) steam turbine at 20.3 lb/in^2 (gage) (139.9 kPa) and attemperated until it is saturated. In this case, the absorption chiller increases plant output by 8.7 percent or 17.4 MW but degrades the plant's heat rate by 1 percent.

Although the capacity of the absorption cooling system's cooling-water loop must be twice that of the mechanical chiller's, the size of the plant's overall heat sink is identical—25 percent larger than the base case—because the steam extracted from the l-p turbine reduces the required cooling capacity. Note that this also reduces steam-turbine output by 2 MW compared to the mechanical chiller, but has less effect on overall plant output.

Cost estimates summarized in Table 1 show that the absorption chilling system required here costs about $4-million, or about $230/kW of added output. Compared to the base case, raw-water consumption increases O&M costs by $35/h when the chiller is operating. Disposal of additional cooling-water blowdown adds $17/h.

Compared to mechanical chillers, absorption units may not handle load changes as well; therefore they may not be acceptable for cycling or load-following operation. When forced to operate below their rated capacity, absorption chillers suffer a loss in efficiency and reportedly require more operator attention than mechanical systems.

Refrigerant issues affect the comparison between mechanical and absorption chilling. Mechanical chillers use either halogenated or nonhalogenated fluorocarbons at this time. Halogenated fluorocarbons, preferred by industry because they reduce the compressor load compared to nonhalogenated materials, will be phased out by the end of the decade because of environmental considerations (destruction of the ozone layer). Use of nonhalogenated refrigerants is expected to increase both the cost and parasitic power consumption for mechanical systems, at least in the near term. However, absorption chillers using either ammonia or lithium bromide will be unaffected by the new environmental regulations.

Off-peak thermal storage is one way to mitigate the impact of inlet-air chilling's major drawback: high parasitic power consumption. A portion of the plant's electrical or thermal output is used to make ice or cool water during off-peak hours. During peak hours, the chilling system is turned off and the stored ice and/or cold water is used to chill the turbine inlet air. A major advantage is that plants can maximize their output during periods of peak demand when capacity payments are at the highest level. Thermal storage and its equipment requirements are analyzed elsewhere in this handbook—namely at page 18.70.

5. *Compare steam and water injection alternatives*

Injecting steam or water into a GT's combustor can significantly increase power output, but either approach also degrades overall CC efficiency. With steam injection, steam extracted from the bottoming cycle is typically injected directly into the GT's combustor, Fig. 3. For advanced GTs, the steam source may be extracted from either the high-pressure (h-p) turbine exhaust, an h-p extraction, or the heat recovery steam generator's (HRSG) h-p section.

Cycle economics and plant-specific considerations determine the steam extraction point. For example, advanced, large-frame GTs require steam pressures of 410 to 435 lb/in^2 (gage) (2825 to 2997 kPa). This is typically higher than the economically optimal range of h-p steam turbine exhaust pressures of 285 to 395 lb/in^2

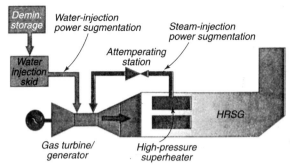

FIGURE 3   Water or steam injection can be used for both power augmentation and NO_x control (*Power*).

(gage) (1964 to 2722 kPa). Thus, steam must be supplied from either the HRSG or an h-p turbine extraction ahead of the reheat section.

Based on installed-cost considerations alone, extracting steam from the HRSG is favored for peaking service and may be accomplished without altering the reheat steam turbine. But if a plant operates in the steam-injection mode for extended periods, extracting steam from the turbine or increasing the h-p turbine exhaust pressure becomes more cost-effective.

Injecting steam from the HRSG superheat section into the GT increases unit output by 21.8 MS, Case 4 Table 1, but decreases the steam turbine/generator's output by about 12.8 MW. Net gain to the CC is 8.4 MW. But CC plant heat rate also suffers by 4 percent, or 270 Btu/kWh (256.5 kJ/kWh).

Because the steam-injection system requires makeup water as pure as boiler feedwater, some means to treat up to 350 gal/min (22.1 L/s) of additional water is necessary. A dual-train demineralizer this size could cost up to $1.5-million. However, treated water could also be bought from a third party and stored. Or portable treatment equipment could be rented during peak periods to reduce capital costs. For the latter case, the average expected cost for raw and treated water is about $130/h of operation.

This analysis assumes that steam- or water-injection equipment is already in place for NO_x control during distillate-fuel firing. Thus, no additional capital cost is incurred.

When water injection is used for power augmentation or NO_x control, the recommended water quality may be no more than filtered raw water in some cases, provided the source meets pH, turbidity, and hardness requirements. Thus, water-treatment costs may be negligible. Water injection, Case 5 Table 1, can increase the GT output by 15.5 MW.

In Case 5, the bottoming cycle benefits from increased GT-exhaust mass flow, increasing steam turbine/generator output by about 3.7 MW. Overall, the CC output increases by 9.4 percent or 19 MW, but the net plant heat rate suffers by 6.4 percent, or 435 Btu/kWh (413.3 kJ/kWh). Given the higher increase in the net plant heat rate and lower operating expenses, water injection is preferred over steam injection in this case.

6.   Evaluate supplementary-fired HRSG for this plant

The amount of excess O_2 in a GT exhaust gas generally permits the efficient firing of gaseous and liquid fuels upstream of the HRSG, thereby increasing the output from the steam bottoming cycle. For this study, two types of supplementary firing are considered—(1) partial supplementary firing, Case 6 Table 1, and (2) full supplementary firing, Case 7 Table 1.

There are three main drawbacks to supplementary firing for peak power enhancement, including 910 lower cycle efficiency, (2) higher NO_x and CO emissions, (3) higher costs for the larger plant equipment required.

For this plant, each 100-million Btu/h (29.3 MW) of added supplementary firing capacity increases the net plant output by 5.5 percent, but increases the heat rate by 2 percent. The installed cost for supplementary firing can be significant because all the following equipment is affected: (1) boiler feed pumps, (2) condensate pumps, (3) steam turbine/generator, (4) steam and water piping and valves, and (5) selective-catalytic reduction (SCR) system. Thus, a plant designed for supplementary firing to meet peak-load requirements will operate in an inefficient, off-design condition for most of the year.

7. Compare the options studied and evaluate results

Comparing the results in Table 1 shows that mechanical chilling, Case 2, gives the largest increase in plant output for the least penalty on plant heat rate—*i.e.*, 18.1 MW output for a net heat rate increase of 55 Btu/kWh (52.3 kJ/kWh). However, this option has the highest estimated installed cost ($3-million), and has a relatively high incremental installed cost.

Water injection, Case 5 Table 1, has the dual advantage of high added net output and low installed cost for plants already equipped with water-injection skids for NO_x control during distillate-fuel firing. Steam injection, Case 4 Table 1, has a significantly higher installed cost because of water-treatment requirements.

Supplementary firing, Cases 6 and 7 Table 1, proves to be more acceptable for plants requiring extended periods of increased output, not just seasonal peaking.

This calculation procedure is the work of M. Boswell, R. Tawney, and R. Narula, all of Bechtel Corporation, as reported in *Power* magazine, where it was edited by Steven Collins. SI values were added by the editor of this handbook.

Related Calculations. Use of gas turbines for expanding plant capacity or for repowering older stations is a popular option today. GT capacity can be installed quickly and economically, compared to conventional steam turbines and boilers. Further, the GT is environmentally acceptable in most areas. So long as there is a supply of combustible gas, the GT is a viable alternative that should be considered in all plant expansion and repowering today, and especially where environmental conditions are critical.

SELECTING GAS-TURBINE HEAT-RECOVERY BOILERS

Choose a suitable heat-recovery boiler equipped with an evaporator and economizer to serve a gas turbine in a manufacturing plant where the gas flow rate is 150,000 lb/h (68,040 kg/h) at 950°F (510°C) and which will generate steam at 205 lb/in^2 (gage) (1413.5 kPa). Feedwater enters the boiler at 227°F (108.3°C). Determine if supplementary firing of the exhaust is required to generate the needed steam. Use an approach temperature of 20°F (36°C) between the feedwater and the water leaving the economizer.

Calculation Procedure:

1. Determine the critical gas inlet-temperature

Turbine exhaust gas (TEG) typically leaves a gas turbine at 900–1000°F (482–538°C) and has about 13 to 16 percent free oxygen. If steam is injected into the gas turbine for NO_x control, the oxygen content will decrease by 2 to 5 percent by volume. To evaluate whether supplementary firing of the exhaust is required to generate needed steam, a knowledge of the temperature profiles in the boiler is needed.

Prepare a gas/steam profile for this heat-recovery boiler as shown in Fig. 4. TEG enters on the left at 950°F (510°C). Steam generated in the boiler at 205 lb/in² (gage) (1413.5 kPa) has a temperature of 390°F (198.9°C), from steam tables. For steam to be generated in the boiler, two conditions must be met: (1) The "pinch point" temperature must be greater than the saturated steam temperature of 390°F (198.9°C), and (2) the temperature of the saturated steam leaving the boiler economizer must be greater than that of the feedwater. The pinch point occurs somewhere along the TEG temperature line, Fig. 4, which starts at the inlet temperature of 950°F (510°C) and ends at the boiler gas outlet temperature, which is to be determined by calculation. A pinch-point temperature will be assumed during the calculation and its suitability determined.

To determine the critical gas inlet-temperature, T_1, get from the steam tables the properties of the steam generated by this boiler: t_s = 390°F (198.9°C); h_f, heat of saturated liquid = 364 Btu/lb (846.7 kJ/kg); h_s, total heat of saturated vapor = 1199.6 Btu/lb (2790.3 kJ/kg); h_w, heat of saturated liquid of feedwater leaving the economizer at 370°F (187.8°C) = 342 Btu/lb (795.5 kJ/kg); and h_f, heat of saturated liquid of the feedwater at 227°F (108.3°C) = 196.3 Btu/lb (456.6 kJ/kg).

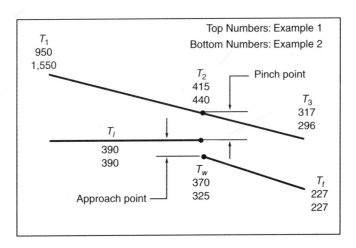

950°F (510°C)	1550°F (843°C)	390°F (199°C)	390°F (199°C)
415°F (213°C)	440°F (227°C)	370°F (188°C)	325°F (163°C)
317°F (158°C)	296°F (147°C)	227°F (108°C)	227°F (108°C)

FIGURE 4 Gas/steam profile and data (*Chemical Engineering*).

Writing an energy balance across the evaporator neglecting heat and blowdown losses, we get: $(T_1 - T_2)/(T_1 - T_3) = (h_s - h_w)/(h_s - h_f) = X$, where T_1 = gas temperature in boiler, °F (°C); T_2 = pinch-point gas temperature, °F (°C); T_3 = outlet gas temperature for TEG, °F (°C); enthalpy, h, values as listed above; X = ratio of temperature or enthalpy differences. Substituting, $X = (1199.6 - 342)/(1199.9 - 196.3) = 0.855$, using enthalpy values as given above.

The critical gas inlet-temperature, $T_{1c} = (t_s - Xt_f)/(1 - X)$, where t_s = temperature of saturated steam, °F (°C); t_f = temperature of feedwater, °F (°C); other symbols as before. Using the values determined above, $T_{1c} = [390 - (0.855)(227)]/(1 - 0.855) = 1351$°F (732.8°C).

2. Determine the system pinch point and gas/steam profile

Up to a gas inlet temperature of approximately 1351°F (732.8°C), the pinch point can be arbitrarily selected. Beyond this, the feedwater inlet temperature limits the temperature profile. Let's then select a pinch point of 25°F (13.9°C), Fig. 4. Then, T_2, the gas-turbine gas temperature at the pinch point, °F (°C) = t_f + pinch-point temperature difference, or 390°F + 25°F = 415°F (212.8°C).

Setting up an energy balance across the evaporator, assuming a heat loss of 2 percent and a blowdown of 3 percent, leads to: $Q_{evap} = W_e (1 - \text{heat loss})$(TEG heat capacity, Btu/°F) $(T_1 - T_2)$, where W_e = TEG flow, lb/h; heat capacity of TEG = 0.27 Btu/°F; T_1 = TEG inlet temperature, °F (°C). Substituting, Q_{evap} = 150,000(0.98)(0.27)(950 − 415) = 21.23 × 10⁶ Btu/h (6.22 MW).

The rate of steam generation, $W_s = Q_{evap}/[(h_s - h_w) + \text{blowdown percent} \times (h_l - h_w)]$, where the symbols are as given earlier. Substituting, W_s = 21.23 × 10⁶/[(1199.6 − 342) + 0.03 × (364 − 342)] = 24,736 lb/h (11,230 kg/h).

Determine the boiler economizer duty from $Q_{econ} = (1 + \text{blowdown})(W_s) (h_w - h_f)$, where symbols are as before. Substituting, Q_{econ} = 1.03(24,736)(342 − 196.3) = 3.71 × 10⁶ Btu/h (1.09 MW).

The gas exit-temperature, $T_3 = T_2 - Q_{econ}/\text{TEG gas flow, lb/h}$)(1 − heat loss)(heat capacity, Btu/lb °F). Since all values are known, T_3 = 415 − 3.71 × 10⁶/(150,000 × 0.98 × 0.27) = 317°F (158°C). Figure 4 shows the temperature profile for this installation.

Related Calculations. Use this procedure for heat-recovery boilers fired by gas-turbine exhaust in any industry or utility application. Such boilers may be unfired, supplementary fired, or exhaust fired, depending on steam requirements.

Typically, the gas pressure drop across the boiler system ranges from 6 to 12 in (15.2 to 30.5 cm) of water. There is an important tradeoff: a lower pressure drop means the gas-turbine power output will be higher, while the boiler surface and the capital cost will be higher, and vice versa. Generally, a lower gas pressure drop offers a quick payback time.

If ΔP_e is the additional gas pressure in the system, the power, kW, consumed in overcoming this loss can be shown approximately from $P = 5 \times 10^{-8} (W_e \Delta P_e T /E$, where E = efficiency of compression).

To show the application of this equation and the related payback period, assume W_e = 150,000 lb/g (68,100 kg/h), T = 1000°R (average gas temperature in the boiler, ΔP_e = 4 in water (10.2 cm), and E = 0.7. Then $P = 5 \times 10^{-8}$ (150,000 × 4 × 1000/0.7) = 42 kW.

If the gas turbine output is 4000 kW, nearly 1 percent of the power is lost due to the 4-in (10.2-cm) pressure drop. If electricity costs 7 cent/kWh, and the gas turbine runs 8000 h/yr, the annual loss will be 8000 × 0.07 × 42 = \$23,520. If the incremental cost of a boiler having a 4-in (10.2-cm) lower pressure drop is, say \$22,000, the payback period is about one year.

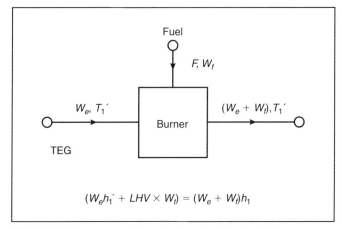

FIGURE 5 Gas/steam profile for fired mode (*Chemical Engineering*).

If steam requirements are not stated for a particular gas inlet condition, and maximum steaming rate is desired, a boiler can be designed with a low pinch point, a large evaporator, and an economizer. Check the economizer for steaming. Such a choice results in a low gas exit temperature and a high steam flow.

Then, the incremental boiler cost must be evaluated against the additional steam flow and gas-pressure drop. For example, Boiler A generates 24,000 lb/h (10,896 kg/h), while Boiler B provides 25,000 lb/h (11,350 kg/h) for the same gas pressure-drop but costs $30,000 more. Is Boiler B worth the extra expense?

To answer this question, look at the annual differential gain in steam flow. Assuming steam costs $3.50/1000 lb (3.50/454 kg), the annual differential gain in steam flow = $1000 \times 3.5 \times 8000/1000 = \$28,000$. Thus, the simple payback is about a year ($30,000 vs $28,000), which is attractive. You must, however, be certain you assess payback time against the actual amount of time the boiler will operate. If the boiler is likely to be used for only half this period, then the payback time is actually two years.

The general procedure presented here can be used for any type industry using gas-turbine heat-recovery boilers—chemical, petroleum, power, textile, food, etc. This procedure is the work of V. Ganapathy, Heat-Transfer Specialist, ABCO Industries, Inc., and was presented in *Chemical Engineering* magazine.

When supplementary fuel is added to the turbine exhaust gas before it enters the boiler, or between boiler surfaces, to increase steam production, one has to perform an energy balance around the burner, Fig. 5, to evaluate accurately the gas temperature increase that can be obtained.

V. Ganapathy, cited above, has a computer program he developed to speed this calculation.

GAS-TURBINE CYCLE EFFICIENCY ANALYSIS AND OUTPUT DETERMINATION

A gas turbine consisting of a compressor, combustor, and an expander has air entering at 60°F (15.6°C) and 14.0 lb/in² (abs) (96.5 kPa). Inlet air is compressed to 56 lb/in² (abs) (385.8 kPa); the isentropic efficiency of the compressor is 82 percent. Sufficient fuel is injected to give the mixture of fuel vapor and air a heating value of 200 Btu/lb (466 kJ/kg). Assume complete combustion of the fuel. The expander reduces the flow pressure to 14.9 lb/in² (abs), with an engine efficiency of 85 percent. Assuming that the combustion products have the same thermodynamic properties as air, $c_p = 0.24$, and is constant. The isentropic exponent may be taken as 1.4. (*a*) Find the temperature after compression, after combustion, and at the exhaust. (*b*) Determine the Btu/lb (kJ/kg) of air supplied, the work delivered by the expander, the net work produced by the gas turbine, and its thermal efficiency.

Calculation Procedure:

1. Plot the ideal and actual cycles
Draw the ideal cycle as 1-2-3-4-1, Figs. 6 and 7. Actual compression takes place along 1-2′. Actual heat added lies along 2′-3′. The ideal expansion process path is 3′-4′. Ideal work = c_p (ideal temperature difference). Actual work = c_p (actual temperature difference).

2. Find the temperature after compression
Use the relation $(T_2/T_1) = (P_2/P_1)^{(k-1)/k}$, where T_1 = entering air temperature, °R; T_2 = temperature after adiabatic compression, °R; P_1 = entering air pressure, in units given above; P_2 = pressure after compression, in units given above; k = isentropic exponent = 1.4. With an entering air temperature, T_1 of 60°F (15.6°C), or 60 + 460 = 520°R, and using the data given, $T_2 = 520[(56/14)]^{(1.4-1)/1.4} = 772.7°R$, or 772.7 − 520 = 252.7°F (122.6°C).

(*a*) Here we have isentropic compression in the compressor with an efficiency of 85 percent. Using the equation, Efficiency, isentropic = $(c_p)(T_2 - T_1)/(c_p)(T_{2'} - T_1)$, and solve for $T_{2'}$, the temperature after isentropic compression. Solving, $T_{2'} = 0.82 = 0.24(772.7 - 520)/0.24(T_{2'} - 520) = 828.4°R$, or 368°F. This is the temperature after compression.

3. Determine the temperature after combustion
To find the temperature after combustion, use the relation Heating value of fuel = $Q = c_p(T_{3'} - T_{2'})$, where $T_{3'}$ = temperature after combustion, °R. Substituting, $200 = 0.24(T_{3'} - 828)$. Solving, $T_{3'} = 1661.3°R$; 1201.3°F (649.6°C).

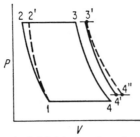

FIGURE 6 Ideal gas-turbine cycle, 1-2-3-4-1. Actual compression takes place along 1-2′; actual heat added 2′-3′; ideal expansion 3′-4′.

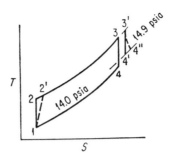

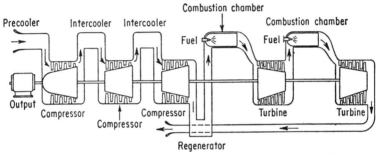

FIGURE 7   Ideal gas-turbine cycle T-S diagram with the same processes as in Fig. 6; complete-cycle gas turbine shown below the T-S diagram.

4.   Find the temperature at the exhaust of the gas turbine

Using an approach similar to that above, determine T_4 from $(T_{4'}/T_{3'}) = [(P_{4'}/P_{3'})]^{k-1/k}$. Substituting and solving for $T_{4'} = 1661[(14.9/56)]^{(1.4-1)/1.4} = 1137.9°R$, or $677.8°F$ ($358.8°C$).

Now use the equation for gas-turbine efficiency, namely, Turbine efficiency = $c_p(T_{3'} - T_{4''})/c_p (T_{3'} - T_{4'}) = 0.85$, and solve for $T_{4''}$, the temperature after expansion, at the exhaust. Substituting as earlier, $T_{4''} = 1218.2°R$, $758.2°F$ ($403.4°C$). This is the temperature after expansion, *i.e.*, at the exhaust of the gas turbine.

5.   Determine the work of compression, expander work, and thermal efficiency

(*b*) The work of compression = $c_p(T_{2'} - T_1) = 0.24(828 - 520) = 74.16$ Btu (78.23 J).

The work delivered by the expander = $c_p(T_{2'} - T_1) = 0.24 (1661 - 1218) = 106.32$ Btu (112.16 J).

The net work = $106.3 - 74.2 = 32.1$ Btu (33.86 J). Then, the thermal efficiency = net work/heat supplied = $32.1/200 = 0.1605$, 16.6 percent thermal efficiency.

Related Calculations.   With the widespread use today of gas turbines in a variety of cycles in industrial and central-station plants, it is important that an engineer be able to analyze this important prime mover. Because gas turbines can be quickly installed and easily hooked to heat-recovery steam generators (HRSG), they are more popular than ever before in history.

Further, as aircraft engines become larger—such as those for the Boeing 777 and the Airbus 340—the power output of aeroderivative machines increases at little cost to the power industry. The result is further application of gas turbines for topping, expansion, cogeneration and a variety of other key services throughout the world of power generation and energy conservation.

With further refinement in gas-turbine cycles, specific fuel consumption, Fig. 8, declines. Thus, the complete cycle gas turbine has the lowest specific fuel consumption, with the regenerative cycle a close second in the 6-to-1 compression-ratio range.

Two recent developments in gas-turbine plants promise much for the future. The first of these developments is the single-shaft combined-cycle gas and steam turbine, Fig. 9. In this cycle, the gas turbine exhausts into a heat-recovery steam generator (HRSG) that supplies steam to the turbine. This cycle is the most significant electric generating system available today. Further, its capital costs are significantly lower than competing nuclear, fossil-fired steam, and renewable-energy stations. Other advantages include low air emissions, low water consumption, smaller space requirements, and a reduced physical profile, Fig. 10. All these advantages are important in today's strict permitting and siting processes.

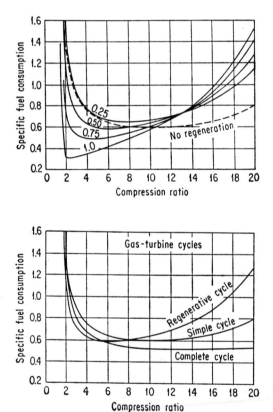

FIGURE 8 With further gas-turbine cycle refinement, the specific fuel consumption declines. These curves are based on assumed efficiencies with $T_3 = 1400$ F (760 C).

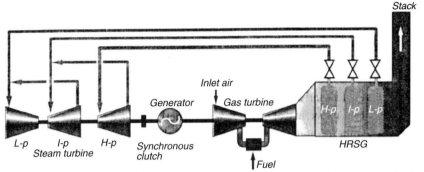

FIGURE 9 Single-shaft combined-cycle technology can reduce costs and increase thermal efficiency over multi-shaft arrangements. This concept is popular in Europe (*Power*).

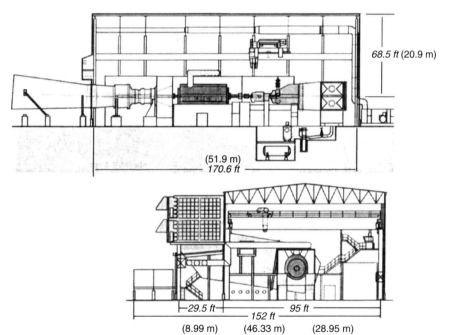

FIGURE 10 Steam turbine, electric generator, and gas turbine fit into one compact building when all three machines are arranged on a single shaft. Net result: Reduced site footprint and civil-engineering work (*Power*).

Having the gas turbine, steam turbine, and generator all on one shaft simplifies plant design and operation, and may lower first costs. When used for large reheat cycles, as shown here, separate high-pressure (h-p), intermediate-pressure (i-p), and low-pressure (l-p) turbine elements are all on the same shaft as the gas turbine and generator. Modern high-technology combined-cycle single-shaft units deliver a simple-cycle net efficiency of 38.5 percent for a combine-cycle net efficiency of 58 percent on a lower heating value (LHV) basis.

The second important gas-turbine development worth noting is the dual-fueled turbine located at the intersection of both gas and oil pipelines. Being able to use either fuel gives the gas turbine greater opportunity to increase its economy by switching to the lowest-cost fuel whenever necessary. Further developments along these lines is expected in the future.

The data in the last three paragraphs and the two illustrations are from *Power* magazine.

DETERMINING BEST-RELATIVE-VALUE OF INDUSTRIAL GAS TURBINES USING A LIFE-CYCLE COST MODEL

An industrial application requires a 21-MW continuous electrical output year-round. Five different gas turbines are under consideration. Determine which of these five turbines is the best choice, using a suitable life-cycle cost analysis.

Calculation Procedure:

1. Assemble the cost data for each gas turbine being considered
Assemble the cost data as shown below for each of the five gas turbines identified by the letters A through E. Contact the gas-turbine manufacturers for the initial cost, $/kW, thermal efficiency, availability, fuel consumption, generator efficiency, and maintenance cost, $/kWh. List these data as shown below.

The loan period, years, will be the same for all the gas turbines being considered, and is based on an equipment life-expectancy of 20 years. Interest rate on the capital investment for each turbine will vary, depending on the amount invested and the way in which the loan must be repaid and will be provided by the accounting department of the firm considering gas-turbine purchase.

Equipment Attributes for Typical Candidates*

Parameter	Gas-turbine candidates				
	A	B	C	D	E
Initial cost, $/kW	205	320	275	320	200
Thermal efficiency, %	32.5	35.5	34.0	36.5	30.0
Loan period, yr	20	20	20	20	20
Availability	0.96	0.94	0.95	0.94	0.96
Fuel cost, $/million Btu	4	4	4	4	4
Interes, %	6.5	8.0	7.0	8.5	7.5
Generator efficiency, %	98.0	98.5	98.5	98.0	98.5
Maintenance cost, $/kWh	0.004	0.005	0.005	0.005	0.004

*Assuming an equipment life of 20 years, an output of 21 MW.

2. Select a life-cycle cost model for the gas turbines being considered
A popular and widely used life-cycle cost model for gas turbines has three parts: (1) the annual investment cost, C_p; (2) annual fuel cost, C_f; (3) annual maintenance cost, C_m. Summing these three annual costs, all of which are expressed in mils/kWh, gives C_T, the life-cycle cost model. The equations for each of the three components are given below, along with the life-cycle working model, C_T:

The life-cycle cost model (C_T) consists of annual investment cost (C_p) + annual fuel cost (C_f) + annual maintenance cost (C_m). Equations for these values are:

$$C_p = \frac{l\{i/[1-(1-i)^{-n}]\}}{(A)(kW)(8760)(G)}$$

where l = initial capital cost of equipment, dollars
$\quad i$ = interest rate
$\quad n$ = number of payment periods
$\quad A$ = availability (expressed as decimal)
$\quad kW$ = kilowatts of electricity produced
$\quad 8760$ = total hours in year
$\quad G$ = efficiency of electric generator

$$C_f = E(293)$$

where E = thermal efficiency of gas turbine
$\quad 293$ = conversion of Btu to kWh

$$C_m = M/kW$$

where M = maintenance cost, dollars per operating (fired) hour.

Thus, the life-cycle working model can be expressed as

$$C_T = \frac{l\{i/[1-(1-i)^{-n}]\}}{(A)(kW)(8760)(G)} + F/E(293) + M/kW$$

where F = fuel cost, dollars per million Btu (higher heating value)

To evaluate the comparative capital cost of a gas-turbine electrical generating package the above model uses the capital-recovery factor technique. This approach spreads the initial investment and interest costs for the repayment period into an equal annual expense using the time value of money. The approach also allows for the comparison of other periodic expenses, like fuel and maintenance costs.

3. Perform the computation for each of the gas turbines being considered
Using the compiled data shown above, compute the values for C_p, C_f, and C_m, and sum the results. List for each of the units as shown below.

Results from Cost Model

Unit	Mils/kWh produced
A	48.3
B	47.5
C	48.3
D	46.6
E	51.9

4. Analyze the findings of the life-cycle model
Note that the initial investment cost for the turbines being considered ranges between $200 and $320/kW. On a $/kW basis, only unit E at the $200 level, would be considered. However, the life-cycle cost model, above, shows the cost per kWh

produced for each of the gas-turbine units being considered. This gives a much different perspective of the units.

From a life-cycle standpoint, the choice of unit E over unit D would result in an added expenditure of about $975,000 annually during the life span of the equipment, found from $[(51.9 - 46.6)/1000](8760 \text{ hr/yr})(21,000 \text{ kW}) = \$974,988$; this was rounded to $975,000. Since the difference in the initial cost between units D and E is $6,720,000 - \$4,200,000 = \$2,520,000$, this cost difference will be recovered in $2,520,000/974,988 = 2.58$ years, or about one-eighth of the 20-year life span of the equipment.

Also, note that the 20-year differential in cost/kWh produced between units D and E is equivalent to over 4.6 times the initial equipment cost of unit E. When considering the values output of a life-cycle model, remember that such values are only as valid as the data input. So take precautions to input both reasonable and accurate data to the life-cycle cost model. Be careful in attempting to distinguish model outputs that vary less than 0.5 mil from one another.

Since the predictions of this life-cycle cost model cannot be compared to actual measurements at this time, a potential shortcoming of the model lies with the validity of the data and assumptions used for input. For this reason, the model is best applied to establish comparisons to differentiate between several pieces of competing equipment.

Related Calculations. The first gas turbines to enter industrial service in the early 1950s represented a blend of steam-turbine and aerothermodynamic design. In the late 1950s/early-1960s, lightweight industrial gas turbines derived directly from aircraft engines were introduced into electric power generation, pipeline compression, industrial power generation, and a variety of other applications. These machines had performance characteristics similar to their steam-turbine counterparts, namely pressure ratios of about 12:1, firing temperatures of 1200–1500°F (649–816°C), and thermal efficiencies in the 23–27 percent range.

In the 1970s, a new breed of aeroderivative gas turbines entered industrial service. These units, with simple-cycle thermal efficiencies in the 32–37 percent bracket, represented a new technological approach to aerothermodynamic design.

Today, these second-generation units are joined by hybrid designs that incorporate some of the aeroderivative design advances but still maintain the basic structural concepts of the heavy-frame machines. These hybrid units are not approaching the simple-cycle thermal-efficiency levels reached by some of the early second-generation aeroderivative units first earmarked for industrial use.

Traditionally, the major focus has been on first cost of industrial gas-turbine units, not on operating cost. Experience with higher-technology equipment, however, reveals that a low first cost does not mean a lower total cost during the expected life of the equipment. Conversely, reliable, high-quality equipment with demonstrated availability will be remembered long after the emotional distress associated with high initial cost is forgotten.

The life-cycle cost model presented here uses 10 independent variables. A single-point solution can easily be obtained, but multiple solutions require repeated calculations. Although curves depicting simultaneous variations in all variables would be difficult to interpret, simplified diagrams can be constructed to illustrate the relative importance of different variables.

Thus, the simplified diagrams shown in Fig. 11, all plot production cost, mils/kWh, versus investment cost. All the plots are based on continuous operation of 8760 h/yr at 21-MW capacity with an equipment life expectancy of 20 years.

The curves shown depict the variation in production cost of electricity as a function of initial investment cost for various levels of thermal efficiency, loan

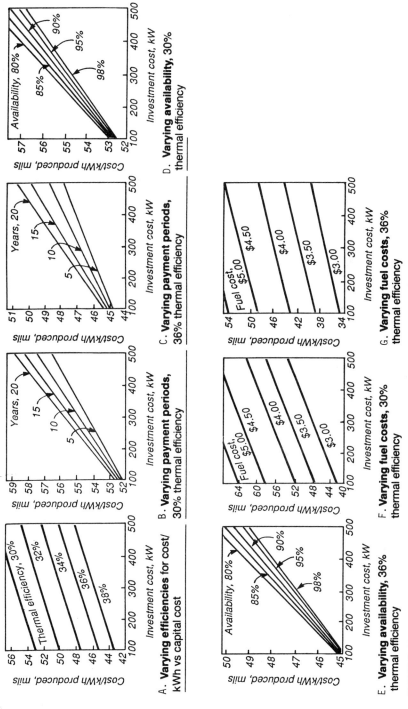

FIGURE 11 Economic study plots for life-cycle costs (*Power*).

A. **Varying efficiencies for cost/kWh vs capital cost**

B. **Varying payment periods, 30% thermal efficiency**

C. **Varying payment periods, 36% thermal efficiency**

D. **Varying availability, 30% thermal efficiency**

E. **Varying availability, 36% thermal efficiency**

F. **Varying fuel costs, 30% thermal efficiency**

G. **Varying fuel costs, 36% thermal efficiency**

repayment period, gas-turbine availability, and fuel cost. Each of these factors is an element in the life-cycle cost model presented here.

This procedure is the work of R. B. Spector, General Electric Co., as reported in *Power* magazine.

TUBE BUNDLE VIBRATION AND NOISE DETERMINATION IN HRSGs

A tubular air heater 11.7 ft (3.57 m) wide, 12.5 ft (3.81 m) deep and 13.5 ft (4.11 m) high is used in a boiler plant. Carbon steel tubes 2 in (5.08 cm) in outer diameter and 0.08 in (0.20 cm) thick are used in inline fashion with a traverse pitch of 3.5 in (8.89 cm) and a longitudinal pitch of 3 in (7.62 m). There are 40 tubes wide and 60 tubes deep in the heater; 300,000 lb (136,200 kg) of air flows across the tubes at an average temperature of 219°F (103.9°C). The tubes are fixed at both ends. Tube mass per unit length = 1.67 lb/ft (2.49 kg/m). Check this air heater for possible tube vibration problems.

Calculation Procedure:

1. Determine the mode of vibration for the tube bundle
Whenever a fluid flows across a tube bundle such as boiler tubes in an evaporator, economizer, HRSG, superheater, or air heater, vortices are formed and shed in the wake beyond the tubes. This shedding on alternate sides of the tubes causes a harmonically varying force on the tubes perpendicular to the normal flow of the fluid. It is a self-excited vibration. If the frequency of the Von Karman vortices, as they are termed, coincides with the natural frequency of vibration of the tubes, then resonance occurs and the tubes vibrate, leading to possible damage of the tubes.

Vortex shedding is most prevalent in the range of Reynolds numbers from 300 to 200,000, the range in which most boilers operate. Another problem encountered with vortex shedding is acoustic vibration, which is normal to both the fluid flow and tube length observed in only gases and vapors. This occurs when the vortex shedding frequency is close to the acoustic frequency. Excessive noise is generated, leading to large gas pressure drops and bundle and casing damage. The starting point in the evaluation for noise and vibration is the estimation of various frequencies.

Use the listing of *C* values shown below to determine the mode of vibration. Note that *C* is a factor determined by the end conditions of the tube bundle.

	Mode of vibration		
End conditions	1	2	3
Both ends clamped	22.37	61.67	120.9
One end clamped, one end hinged	15.42	49.97	104.2
Both hinged	9.87	39.48	88.8

Since the tubes are fixed at both ends, *i.e.*, clamped, select the mode of vibration as 1, with *C* = 22.37. For most situations, Mode 1 is the most important case.

2. Find the natural frequency of the tube bundle
Use the relation, $f_n = 90C[d_o^4 - d_i^4]/(L^2 - M^{0.5})$. Substituting, with $C = 22.37$, $f_n = (90)(22.37)[2^4 - 1.84^4]^{0.5}/(13.5^2 - 1.67^{0.5}) = 18.2$ cycles per second (cps). In Mode 2, $f_n = 50.2$, as $C = 61.67$.

3. Compute the vortex shedding frequency
To compute the vortex shedding frequency we must know several factors, the first of which is the Strouhl Number, S. Using Fig. 12 with a transverse pitch/diameter of 1.75 and a longitudinal pitch diameter of 1.5 we find $S = 0.33$. Then, the air density $= 40/(460 - 219) = 0.059$ lb/ft^3 (0.95 kg/m^3); free gas area $= 40(3.5 - 2)(13.5/12) = 67.5$ ft^2 (6.3 m^2); gas velocity, $V = 300,000/(67.5)(0.059)(3600) = 21$ ft/s (6.4 m/s).
   Use the relation, $f_c = 12(S)(V)/d_o = 12(0.33)(21)/2 = 41.6$ cps, where $f_c =$ vortex shedding frequency, cps.

4. Determine the acoustic frequency
As with vortex frequency, we must first determine several variables, namely: absolute temperature $=$ °R $= 219 + 460 = 679$°R; sonic velocity, $V_s = 49(679)^{0.5} = 1277$ ft/s (389.2 m/s); wave length, $\lambda = 2(w)/n$, where $w =$ width of tube bank, ft (m); $n =$ mode of vibration $= 1$ for this tube bank; then $\lambda = 2(11.7)/1 = 23.4$ ft (7.13 m).
   The acoustic frequency, $f_a = (V_s)/\lambda$, where $V_s =$ velocity of sound at the gas temperature in the duct or shell, ft/s (m/s); $V_s = [(g)(\rho)(RT)]^{0.5}$, where $R =$ gas constant $= 1546/$molecular weight of the gas; $T =$ gas temperature, °R; $\rho =$ ratio of gas specific heats, typically 1.4 for common flue gases; the molecular weight $= 29$. Simplifying, we get $V_s = 49(T)^{0.5}$, as shown above. Substituting, $f_a = 1277/23.4 = 54.5$ cps. For $n = 2$; $f_a = 54.4(2) = 109$ cps. The results for Modes 1 and 2 are summarized in the tabulation below.

Mode of vibration	1	2
n		
f_n, cps	18.2	50.2
f_c, cps	41.6	41.6
f_a (without baffles)	54.5	109
f_a (with baffles)	109	218

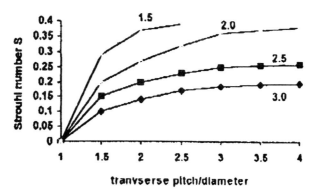

FIGURE 12   Strouhl number, S, for inline tube banks. Each curve represents a different longitudinal pitch/diameter ratio (Chen).

The tube natural frequency and the vortex shedding frequency are far apart. Hence, the tube bundle vibration problem is unlikely to occur. However, the vortex shedding and acoustic frequencies are close. If the air flow increases slightly, the two frequencies will be close. By inserting a baffle in the tube bundle (dividing the ductwork into two along the gas flow direction) we can double the acoustic frequency as the width of the gas path is now halved. This increases the difference between vortex shedding and acoustic frequencies and prevents noise problems.

Noise problems arise when the acoustic and vortex shedding frequencies are close—usually within 20 percent. Tube bundle vibration problems arise when the vortex shedding frequency and natural frequency of the bundle are close—within 20 percent. Potential noise problems must also be considered at various turndown conditions of the equipment.

Related Calculations. For a thorough analysis of a plant or its components, evaluate the performance of heat-transfer equipment as a function of load. Analyze at various loads the possible vibration problems that might occur. At low loads in the above case, tube bundle vibration is likely, while at high loads acoustic vibration is likely without baffles. Hence, a wide range of performance must be reviewed before finalizing any tube bundle design, Fig. 13.

This procedure is the work of V. Ganapathy, Heat Transfer Specialist, ABCO Industries, Inc.

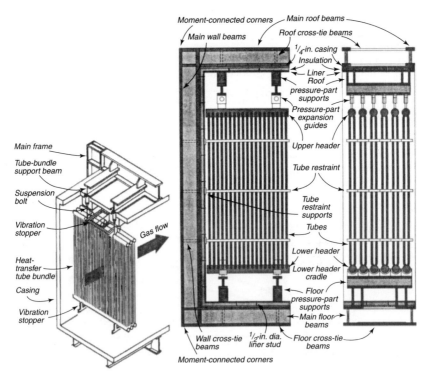

FIGURE 13 Tube bundles in HRSGs require appropriate support mechanisms; thermal cycling in combined-cycle units makes this consideration even more important (*Power*).

DETERMINING OXYGEN AND FUEL INPUT IN GAS-TURBINE PLANTS

In a gas-turbine HRSG (heat-recovery steam generator) it is desired to raise the temperature of 150,000 lb/h (68,100 kg/h) of exhaust gases from 950°F (510°C) to 1575°F (857.2°C) in order to nearly double the output of the HRSG. If the exhaust gases contain 15 percent oxygen by volume, determine the fuel input and oxygen consumed, using the gas specific-heat method.

Calculation Procedure:

1. *Determine the air equivalent in the exhaust gases*
In gas-turbine based cogeneration/combined-cycle projects the HRSG may be fired to generate more steam than that produced by the gas-turbine exhaust gases. Typically, the gas-turbine exhaust gas contains 14 to 15 percent oxygen by volume. So the question arises: How much fuel can be fired to generate more steam? Would the oxygen in the exhaust gases run out if we fired to a desired temperature? These questions are addressed in this procedure.

   If 0 percent oxygen is available in W_g lb/h (kg/h) of exhaust gases, the air-equivalent W_a in lb/h (kg/h) is given by: $W_a = 100(W_g)(32O_x)/[23(100)(29.5)] = 0.0417\ W_g(O)$. In this relation, we are converting the oxygen from a volume basis to a weight basis by multiplying by its molecular weight of 32 and dividing by the molecular weight of the exhaust gases, namely 29.5. Then multiplying by (100/23) gives the air equivalent as air contains 23 percent by weight of oxygen.

2. *Relate the air required with the fuel fired using the MM Btu (kJ) method*
Each MM Btu (kJ) of fuel fired (HHV basis) requires a certain amount of air, A. If Q = amount of fuel fired in the turbine exhaust gases on a LHV basis (calculations for turbine exhaust gases fuel input are done on a low-heating-value basis) then the fuel fired in lb/h (kJ/h) = $W_f = Q/LHV$.

   The heat input on an HHV basis = $W_f(HHV)/(10^6) = (Q/LHV)(HHV)/10^6$ Btu/h (kJ/h). Air required lb/h (kg/h) = $(Q/LHV)(HHV)(A)$, using the MM Btu, where A = amount of air required, lb (kg) per MM Btu (kJ) fired. The above quantity = air available in the exhaust gases, $W_a = 0.0417\ W_g(O)$.

3. *Simplify the gas relations further*
From the data in step 2, $(Q/LHV)(HHV)(A)/10^6 = 0.0417\ W_g(O)$. For natural gas and fuel oils it can be shown that $(LHV/A_xHHV) = 0.00124$. For example, LHV of methane = 21,520 Btu/lb (50,055.5 kJ/kg); HHV = 23,879 Btu/lb (55,542.6 kJ/kg), and A = 730 lb (331.4 kg). Hence, $(LHV/A_xHHV) = 21,520/(730 \times 23,879) = 0.00124$. By substituting in the equation in step 1, we have $Q = 58.4 (W_g)(O)$. This is an important equation because it relates the oxygen consumption from the exhaust gases to the burner fuel consumption.

4. *Find the fuel input to the HRSG*
The fuel input is given by $W_g + h_{g1} + Q = (W_g + W_f)(h_{g2})$, where h_{g1} and h_{g2} are the enthalpies of the exhaust gas before and after the fuel burner; W_f = fuel input, lb/h (kg/h); Q = fuel input in Btu/h (kJ/h).

The relation above requires enthalpies of the gases before and after the burner, which entails detailed combustion calculations. However, considering that the mass of fuel is a small fraction of the total gas flow through the HRSG, the fuel flow can be neglected. Using a specific heat for the gases of 0.31 Btu/lb °F (1297.9 J/kg K), we have, $Q = 150,000(0.31)(1575 - 950) = 29 \times 10^6$ Btu/h (8.49 kW).

The percent of oxygen by volume, $O = (29 \times 10^6)/(58.4 \times 150,000) = 3.32$ percent. That is, only 3.32 percent oxygen by volume is consumed and we still have $15.00 - 3.32 = 11.68$ percent left in the flue gases. Thus, more fuel can be fired and the gases will not run out of oxygen for combustion.

Typically, the final oxygen content of the gases can go as low as 2 to 3 percent using 3 percent final oxygen, the amount of fuel that can be fired = $(150,000)(58.4)(15 - 3) = 105$ MM Btu/h (110.8 MMJ/h). It can be shown through an HRSG simulation program (contact the author for more information) that all of the fuel energy goes into steam. Thus, if the unfired HRSG were generating 23,000 lb/h (10,442 kg/h) of steam with an energy absorption of 23 MM Btu/h (24.3 MM J/h), approximately, the amount of steam that can be generated by firing fuel in the HRSG = $23 + 105 = 128$ MM Btu/h (135 MM J/h), or 128,000 lb/h (58,112 kg/h) of steam. This is close to a firing temperature of 3000 to 3100°F (1648 to 1704°C).

Related Calculations. Using the methods given elsewhere in this handbook, one may make detailed combustion calculations and obtain a flue-gas analysis after combustion. Then compute the enthalpies of the exhaust gas before and after the burner. Using this approach, you can check the burner duty more accurately than using the gas specific-heat method presented above. This procedure is the work of V. Ganapathy, Heat Transfer Specialist, ABCO Industries, Inc.

Power magazine recently commented on the place of gas turbines in today's modern power cycles thus: Using an HRSG with a gas turbine enhances the overall efficiency of the cycle by recovering heat in the gas-turbine's hot exhaust gases. The recovered heat can be used to generate steam in the HRSG for either (1) injection back into the gas turbine, Fig. 14, (2) use in district heating or an industrial process, (3) driving a steam turbine-generator in a combined-cycle arrangement, or (4) any combination of the first three.

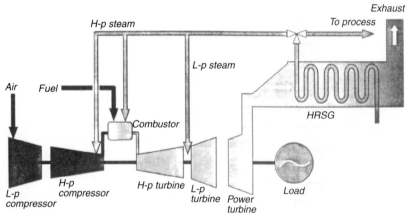

FIGURE 14 Steam injection systems offer substantial improvement in both capacity and efficiency (*Power*).

Steam injection into the gas turbine has many benefits, including: (1) achievable output is increased by 25 percent or more, depending on the gas-turbine design, (2) part-load gas-turbine efficiency can be significantly improved, (3) gas-fired NO_x emissions can be markedly reduced—up to the 15–45 ppm range in many cases, (4) operating flexibility is improved for cogeneration plants because electrical and thermal outputs can be balanced to optimize overall plant efficiency and profitability.

Combined-cycle gas-turbine plants are inherently more efficient than simple-cycle plants employing steam injection. Further, combined-cycle plants may also be considered more adaptable to cogeneration compared to steam-injected gas turbines. The reason for this is that the maximum achievable electrical output decreases significantly for steam-injected units in the cogeneration mode because less steam is available for use in the gas turbine. In contrast, the impact of cogeneration on electrical output is much less for combined-cycle plants.

Repowering in the utility industry can use any of several plant-revitalization schemes. One of the most common repowering options employed or considered today by utilities consists of replacing an aging steam generator with a gas-turbine/generator and HRSG, Fig. 15. It is estimated that within the next few years, more than 3500 utility power plants will have reached their 30th birthdays. A significant number of these facilities—more than 20 GW of capacity by some estimates—are candidates for repowering, an option that can cut emissions and boost plant efficiency, reliability, output, and service life.

And repowering often proves to be more economical, per cost of kilowatt generated, compared to other options for adding capacity. Further, compared to building a new power plant, the permitting process for repowering its typically much shorter and less complex. The HRSG will often have a separate firing capability such as that discussed in this calculation procedure.

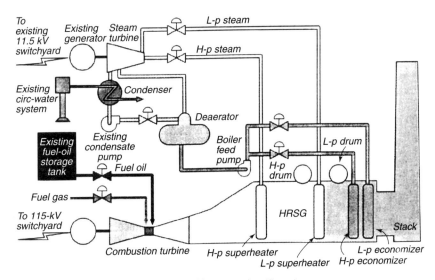

FIGURE 15 HRSG and gas turbine used in repowering (*Power*).

These comments from *Power* magazine were prepared by Steven Collins, Assistant Editor of the publication.

HEAT-RECOVERY STEAM GENERATOR (HRSG) SIMULATION

A gas turbine exhausts 140,000 lb/h (63,560 kg/h) of gas at 980°F (526.7°C) to an HRSG generating saturated steam at 200 lb/in² (gage) (1378 kPa). Determine the steam-generation and design-temperature profiles if the feedwater temperature is 230°F (110°C) and blowdown = 5 percent. The average gas-turbine exhaust gas specific heat is 0.27 Btu/lb °F (1.13 kJ/kg °C) at the evaporator and 0.253 Btu/lb °F (1.06 kJ/kg °C) at the economizer. Use a 20°F (11.1°C) pinch point, 15°F (8.3°C) approach point and 1 percent heat loss. Evaluate the evaporator duty, steam flow, economizer duty, and exit-gas temperature for normal load conditions. Then determine how the HRSG off-design temperature profile changes when the gas-turbine exhaust-gas flow becomes 165,000 lb/h (74,910 kg/h) at 880°F (471°C) with the HRSG generating 150-lb/in² (gage) (1033.5 kPa) steam with the feedwater temperature remaining the same.

Calculation Procedure:

1. Compute the evaporator duty and steam flow

Engineers should be able to predict both the design and off-design performance of an HRSG, such as that in Fig. 16, under different conditions of exhaust flow, temperature, and auxiliary firing without delving into the mechanical design aspects of tube size, length, or fin configuration. This procedure shows how to make such predictions for HRSGs of various sizes by using simulation techniques.

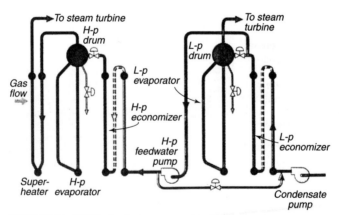

FIGURE 16 HRSG circuit shown is used by at least one manufacturer to prevent steaming in the economizer during startup and low-load operation (*Power*).

HRSGs operate at different exhaust-gas conditions. For example, variations in ambient temperature or gas-turbine load affect exhaust-gas flow and temperature. This, in turn, affects HRSG performance, temperature profiles, efficiency, and steam generation. The tool consultants use for evaluating HRSG performance under different operating conditions is simulation. With this tool you can: (1) Predict off-design performance of an HRSG; (2) Predict auxiliary fuel consumption for periods when the gas-turbine exhaust-gas flow is insufficient to generate the required steam flow; (3) Evaluate options for improving an HRSG system; (4) Evaluate field data for validating an HRSG design; (5) Evaluate different HRSG configurations for maximizing efficiency.

In this HRSG, using steam-table data, the saturation temperature of 200-lb/in^2 (gage) (1378-kPa) steam = 388°F (197.8°C). The gas temperature leaving the evaporator with the 20°F (11.1°C) pinch point = 388 + 20 = 408°F (208.9°C). Water temperature entering the evaporator = saturated-steam temperature − the approach point temperature difference, or 388 − 15 = 373°F (189.4°C).

Then, the energy absorbed by the evaporator, Q_1 = (gas flow, lb/h)(1.0 − heat loss)(gas specific heat, Btu/lb °F)(gas-turbine exhaust gas HRSG entering temperature, °F − gas temperature leaving evaporator, °F). Or, Q_1 = (140,000)(0.99)(0.27)(980 − 408) = 21.4 MM Btu/h (6.26 MW). The enthalpy absorbed by the steam in the evaporator, Btu/lb (kJ/kg) = (enthalpy of the saturated steam in the HRSG outlet − enthalpy of the feedwater entering the evaporator at 373°F) + (blowdown percentage)(enthalpy of the saturated liquid of the outlet steam − enthalpy of the water entering the evaporator, all in Btu/lb). Or, enthalpy absorbed in the evaporator = (1199.3 − 345) + (0.05)(362.2 − 345) = 855.2 Btu/lb (1992.6 kJ/kg). The quantity of steam generated = (Q_1, energy absorbed by the evaporator, Btu/h)/(enthalpy absorbed by the steam in the evaporator, Btu/lb) = (21.4 × 10^6)/855.2 = 25,023 lb/h (11,360 kg/h).

2. Determine the economizer duty and exit gas temperature

The economizer duty = (steam generated, lb/h)(enthalpy of water entering the economizer, Btu/lb − enthalpy − enthalpy of the feedwater at 230°F, Btu/lb)(1 + blowdown percentage) = (25,023)(345 − 198.5)(1.05) = 3.849 MM Btu/h (1.12 MW).

The gas temperature drop through the economizer = (economizer duty)/(gas flow rate, lb/h)(1 − heat loss percentage)(specific heat of gas, Btu/lb °F) = (3.849 × 10^6)/(140,000)(0.99)(0.253) = 109.8°F (60.9°C). Hence, the exit-gas temperature from the economizer = (steam saturation temperature, °F − exit-gas temperature from the economizer, °F) = (408 − 109) = 299°F (148.3°C).

3. Calculate the constant K for evaporator performance

In simulating evaporator performance the constant K_1 is used to compute revised performance under differing flow conditions. In equation form, K_1 = ln[(temperature of gas-turbine exhaust gas entering the HRSG, °F − HRSG saturated steam temperature, °F)/(gas temperature leaving the evaporator, °F − HRSG saturate steam temperature, °F)]/(gas flow, lb/h). Substituting, K_1 = ln[(980 − 388)/(408 − 388)]/140,000 = 387.6, where the temperatures used reflect design condition.

4. Compute the revised evaporator performance

Under the revised performance conditions, using the given data and the above value of K_1 and solving for T_{g2}, the evaporator exit gas temperature, ln[(880 − 366)/(T_{g2} − 366)] = 387.6(165,000)$^{-0.4}$; T_{g2} = 388°F (197.8°C). Then, the evaporator duty, using the same equation as in step 1 above = (165,000)(0.99)(0.27)(880 − 388) = 21.7 MM Btu/h (6.36 MW).

In this calculation, we assumed that the exhaust-gas analysis had not changed. If there are changes in the exhaust-gas analysis, then the gas properties must be evaluated and corrections made for variations in the exhaust-gas temperature. See *Waste Heat Boiler Deskbook* by V. Ganapathy for ways to do this.

5. Find the assumed duty, Q_a, for the economizer
Let the economizer leaving-water temperature = 360°F (182.2°C). The enthalpy of the feedwater = 332 Btu/lb (773.6 kJ/kg); saturated-steam enthalpy = 1195.7 Btu/lb (2785.9 kJ/kg); saturated liquid enthalpy = 338.5 Btu/lb (788.7 kJ/kg). Then, the steam flow, as before, = $(21.5 \times 10^6)/[(1195.7 − 332) + 0.05 (338.5 − 332)]$ = 25,115.7 lb/h (11,043 kg/h). Then, the assumed duty for the economizer, Q_a = (25,115.7)(1.05)(332 − 198.5) = 3.52 MM Btu/h (1.03 MW).

6. Determine the UA value for the economizer in both design and off-design conditions
For the design conditions, $UA = Q/(\Delta T)$, where Q = economizer duty from step 2, above; ΔT = design temperature conditions from the earlier data in this procedure. Solving, $UA = (3.84 \times 10^6)/\{[(299 − 230) − (408 − 373)]/\ln(69/35)\}$ = 76,800 Btu/h °F (40.5 kW). For off-design conditions, UA = (UA at design conditions)(gas flow at off-design/gas flow at design conditions)$^{0.65}$ = $(76,800)(165,000/140,000)^{0.65}$ = 85,456 Btu/h F (45.1 kW).

7. Calculate the economizer duty
The energy transferred = $Q_t = (UA)(\Delta T)$. Based on 360°F (182.2°C) water leaving the economizer, Q_a = 3.52 MM Btu/h (1.03 MW). Solving for t_{g2} as before = 382 − $[(3.52 \times 10^6)/(165,000)(0.9)(0.253)]$ = 388 − 85 = 303°F (150.6°C). Then, $\Delta T = [(303 − 230) − (388 − 360)]/\ln(73/28)$ = 47°F (26.1°C). The energy transferred = $Q_t = (UA)(\Delta T)$ = (85,456)(47) = 4.01 MM Btu/h (1.18 MW).

Since the assumed and transferred duty do not match, *i.e.,* 3.52 MM Btu/h vs. 4.01 MM Btu/h, another iteration is required. Continued iteration will show that when $Q_a = Q_t$ = 3.55 MM Btu/h (1.04 MW), and the temperature of the water leaving the economizer = 366°F (185.6°C) (saturation) and exit-gas temperature = 301°F (149.4°C), the amount of steam generated = 25,310 lb/h (11,491 kg/h).

Related Calculations.   Studying the effect of gas inlet temperature and gas flows on HRSG performance will show that at lower steam generation rates or at lower pressures that the economizer water temperature approaches saturation temperature, a situation called "steaming" in the economizer. This steaming condition should be avoided by generating more steam by increasing the inlet gas temperature or through supplementary firing, or by reducing exhaust-gas flow.

Supplementary firing in an HRSG also improves the efficiency of the HRSG in two ways: (1) The economizer acts as a bigger heat sink as more steam and hence more feedwater flows through the economizer. This reduces the exit gas temperature. So with a higher gas inlet temperature to the HRSG we have a lower exit gas temperature, thanks to the economizer. (2) Additional fuel burned in the HRSG reduces the excess air as more air is not added; instead, the excess oxygen is used. In conventional boilers we know that the higher the excess air, the lower the boiler efficiency. Similarly, in the HRSG, the efficiency increases with more supplementary firing. HRSGs used in combined-cycle steam cycles, Fig. 17, may use multiple pressure levels, gas-turbine steam injection, reheat, selective-catalytic-reduction (SCR) elements for NO_x control, and feedwater heating. Such HRSGs require ex-

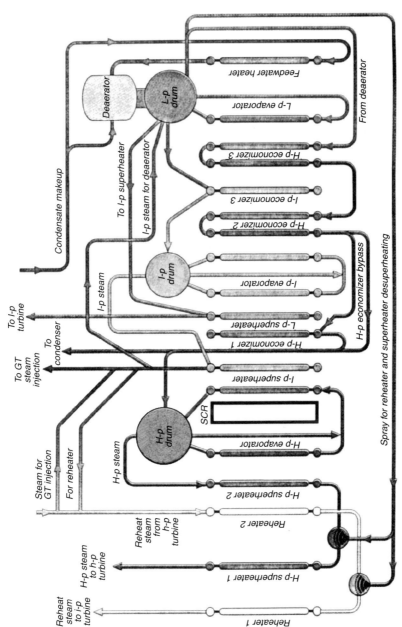

FIGURE 17 HRSGs in combined-cycle steam cycles are somewhat more involved when multiple pressure levels, gas-turbine steam injection, reheat, SCR, and feedwater heating are used (*Power*).

TABLE 2 HRSG Performance in Fired Mode

Item	Case 1	Case 2	Case 3
Gas flow, lb/h (kg/h)	150,000 (68,100)	150,000 (68,100)	150,000 (68,100)
Inlet gas temp, °F (°C)	900 (482.2)	900 (482.2)	900 (482.2)
Firing temperature, °F (°C)	900 (482.2)	1290 (698.9)	1715 (935.0)
Burner duty, MM Btu/h (LHV)*	0 (0)	17.3 (5.06)	37.6 (11.01)
Steam flow, lb/h (kg/hr)	22,780 (10,342)	40,000 (18,160)	60,000 (27,240)
Steam pressure, lb/in² (gage) (kPa)	200 (1378.0)	200 (1378.0)	200 (1378.0)
Feed water temperature, °F (°C)	240 (115.6)	240 (115.6)	240 (115.6)
Exit gas temperature, °F (°C)	327 (163.9)	315 (157.2)	310 (154.4)
System efficiency, %	68.7	79.2	84.90
Steam duty, MM Btu/h (MW)	22.67 (6.64)	39.90 (11.69)	59.90 (17.55)

*(MW)

1.32

tensive analysis to determine the best arrangement of the various heat-absorbing surfaces.

For example, an HRSG generates 22,780 lb/h (10.342 kg/h) of steam in the unfired mode. The various parameters are shown in Table 2. Studying this table shows that as the steam generation rate increases, more and more of the fuel energy goes into making steam. Fuel utilization is typically 100 percent in an HRSG. The ASME efficiency is also shown in the table.

This simulation was done using the HRSG simulation software developed by the author, V. Ganapathy, Heat Transfer Specialist, ABCO Industries, Inc.

PREDICTING HEAT-RECOVERY STEAM GENERATOR (HRSG) TEMPERATURE PROFILES

A gas turbine exhausts 150,000 lb/h (68,100 kg/h) of gas at 900°F (482.2°C) to an HRSG generating steam at 450 lb/in^2 (gage) (3100.5 kPa) and 650°F (343.3°C). Feedwater temperature to the HRSG is 240°F (115.6°C) and blowdown is 2 percent. Exhaust gas analysis by percent volume is: CO_2 = 3; H_2O = 7; O_2 = 15. Determine the steam generation and temperature profiles with a 7-lb/in^2 (48.2-kPa) pressure drop in the superheater, giving an evaporator pressure of 450 + 7 = 457 lb/in^2 (gage) (3148.7 kPa) for a saturation temperature of the steam of 460°F (237.8°C). There is a heat loss of 1 percent in the HRSG. Find the ASME efficiency for this HRSG unit.

Calculation Procedure:

1. Select the pinch and approach points for the HRSG

Gas turbine heat recovery steam generators (HRSGs) are widely used in cogeneration and combined-cycle plants. Unlike conventionally fired steam generators where the rate of steam generation is predetermined and can be achieved, steam-flow determination in an HRSG requires an analysis of the gas/steam temperature profiles. This requirement is mainly because we are starting at a much lower gas temperature—900 to 1100°F—(482.2 to 593.3°C) at the HRSG inlet, compared to 3000 to 3400°F (1648.9 to 1871.1°C) in a conventionally fired boiler. As a result, the exit gas temperature from an HRSG cannot be assumed. It is a function of the operating steam pressure, steam temperature, and pinch and approach points used, Fig. 18.

Typically, the pinch and approach points range from 10 to 30°F (5.56 to 16.6°C). Higher values may be used if less steam generation is required. In this case, we will use 20°F (11.1°C) pinch point (= $T_{g3} - t_s$) and 10°F (5.56°C) approach (= $t_s - t_{w2}$). Hence, the gas temperature leaving the evaporator = 460 + 20 = 480°F (248.9°C), and the water temperature leaving the economizer = 460 − 10 = 450°F (232.2°C).

2. Compute the steam generation rate

The energy transferred to the superheater and evaporator = $Q_1 + Q_2$ = (rate of gas flow, lb/h)(gas specific heat, Btu/lb °F)(entering gas temperature, °F − temperature of gas leaving evaporator, °F)(1.0 percent heat loss) = (150,000)(0.267)(900 − 480)(0.99) = 16.65 MM Btu/h (4.879 MW).

The enthalpy absorbed by the steam in the evaporator and superheater = (enthalpy of the superheated steam at 450 lb/in^2 (gage) and 650°F − enthalpy of the

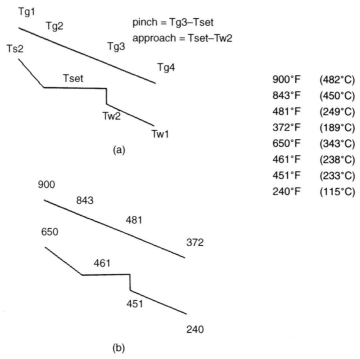

pinch = Tg3–Tset
approach = Tset–Tw2

900°F	(482°C)
843°F	(450°C)
481°F	(249°C)
372°F	(189°C)
650°F	(343°C)
461°F	(238°C)
451°F	(233°C)
240°F	(115°C)

FIGURE 18   Gas/steam temperature profiles.

water entering the evaporator at 450°F) + (blowdown percentage)(enthalpy of the saturated liquid at the superheated condition − enthalpy of the water entering the evaporator, all expressed in Btu/lb). Or, enthalpy absorbed in the evaporator and superheater = (1330.8 − 431.2) + (0.02)(442.3 − 431.2) = 899.8 Btu/lb (2096.5 kJ/kg).

To compute the steam generation rate, set up the energy balance, $899.8(W_s)$ = 16.65 MM Btu/h, where W_s = steam generation rate.

3. Calculate the energy absorbed by the superheater and the exit gas temperature

Q_1, the energy absorbed by the superheater = (steam generation rate, lb/h)(enthalpy of superheated steam, Btu/lb − enthalpy of saturated steam at the superheater pressure, Btu/lb) = (18,502)(1330.8 − 1204.4) = 2.338 MM Btu/h (0.685 MW).

The superheater gas-temperature drop = (Q_1)/(rate of gas-turbine exhaust-gas flow, lb/h)(1.0 − heat loss)(gas specific heat) = (2,338,000)/(150,000)(0.99)(0.273) = 57.67°F, say 58°F (32.0°C). Hence, the superheater exit gas temperature = 900 − 58 = 842°F (450°C). In this calculation the exhaust-gas specific heat is taken as 0.273 because the gas temperature in the superheater is different from the inlet gas temperature.

4. Compute the energy absorbed by the evaporator

The total energy absorbed by the superheater and evaporator, from the above, is 16.65 MM Btu/h (4.878 MW). Hence, the evaporator duty = Q_2 = 16.65 − 2.34 = 14.31 MM Btu/h (4.19 MW).

5. *Determine the economizer duty and exit-gas temperature*

The economizer duty, Q_3 = (rate of steam generation, lb/h)(1 + blowdown expressed as a decimal)(enthalpy of water leaving the economizer − enthalpy of feedwater at 240°F) = (18,502)(1.02)(431.2 − 209.6) = 4.182 MM Btu/h (1.225 MW).

The HRSG exit gas temperature = (480, the exit gas temperature at the evaporator computed in step 1, above) − (economizer duty)/(gas-turbine exhaust-gas flow, lb/h)(1.0 − heat loss)(exhaust gas specific heat) = 371.73°F (188.9°C); round to 372°F (188.9°C). Note that you must compute the gas specific heat at the average gas temperature of each of the heat-transmission surfaces.

6. *Compute the ASME HRSG efficiency*

The ASME Power Test Code PTC 4.4 defines the efficiency of an HRSG as: E = efficiency = (energy absorbed by the steam and fluids)/[gas flow $\times$ inlet enthalpy + fuel input to HRSG on LHV basis]. In the above case, E = (16.65 + 4.182)(10^6)/(150,000 $\times$ 220) = 0.63, or 63 percent. In this computation, 220 Btu/lb (512.6 kJ/kg) is the enthalpy of the exhaust gas at 900°F (482.2°C) and (16.65 + 4.182) is the total energy absorbed by the steam in MM Btu/h (MW).

Related Calculations. Note that the exit gas temperature is high. Further, without having done this analytical mathematical analysis, the results could not have been guessed correctly. Minor variations in the efficiency will result if one assumes different pinch and approach points. Hence, it is obvious that one cannot assume a value for the exit gas temperature—say 300°F (148.9°C)—and compute the steam generation.

The gas/steam temperature profile is also dependent on the steam pressure and steam temperature. The higher the steam temperature, the lower the steam generation rate and the higher the exit gas temperature. Arbitrary assumption of the exit gas temperature or pinch point can lead to temperature cross situations. Table 3 shows the exit gas temperatures for several different steam parameters. From the table, it can be seen that the higher the steam pressure, the higher the saturation temperature, and hence, the higher the exit gas temperature. Also, the higher the steam temperature, the higher the exit gas temperature. This results from the reduced steam generation, resulting in a smaller heat sink at the economizer.

This procedure is the work of V. Ganapathy, Heat Transfer Specialist, ABCO Industries, who is the author of several works listed in the references for this section.

TABLE 3 HRSG Exit Gas Temperatures Versus Steam Parameters*

Pressure lb/in² (gage) (kPa)	Steam temp °F (°C)	Saturation temp °F (°C)	Exit gas °F (°C)
100 (689)	sat (170)	338 (170)	300 (149)
150 (1034)	sat (186)	366 (186)	313 (156)
250 (1723)	sat (208)	406 (208)	332 (167)
400 (2756)	sat (231)	448 (231)	353 (178)
400 (2756)	600 (316)	450 (232)	367 (186)
600 (4134)	sat (254)	490 (254)	373 (189)
600 (4134)	750 (399)	492 (256)	398 (203)

*Pinch point = 20°F (11.1°C); approach = 15°F (8.3°C); gas inlet temperature = 900°F (482.2°C); blowdown = 0; feedwater temperature = 230°F (110°C).

STEAM TURBOGENERATOR EFFICIENCY AND STEAM RATE

A 20,000-kW turbogenerator is supplied with steam at 300 lb/in² (abs) (2067.0 kPa) and a temperature of 650°F (343.3°C). The backpressure is 1 in (2.54 cm) Hg absolute. At best efficiency, the steam rate is 10 lb (25.4 kg) per kWh. (*a*) What is the combined thermal efficiency (CTE) of this unit? (*b*) What is the combined engine efficiency (CEE)? (*c*) What is the ideal steam rate?

Calculation Procedure:

1. Determine the combined thermal efficiency
(*a*) Combined thermal efficiency, CTE = $(3413/w_r)(1/[h_1 - h_2])$, where w_r = combined steam rate, lb/kWh (kg/kWh); h_1 = enthalpy of steam at throttle pressure and temperature, Btu/lb (kJ/kg); h_2 = enthalpy of steam at the turbine backpressure, Btu/lb (kJ/kg). Using the steam tables and Mollier chart and substituting in this equation, CTE = $(3413/10)(1/[1340.6 - 47.06])$ = 0.2638, or 26.38 percent.

2. Find the combined engine efficiency
(*b*) Combined engine efficiency, CEE = $(w_i)/(w_e)$ = (weight of steam used by ideal engine, lb/kWh)/(weight of steam used by actual engine, lb/kWh). The weights of steam used may also be expressed as Btu/lb (kJ/kg). Thus, for the ideal engine, the value is 3413 Btu/lb (7952.3 kJ/kg). For the actual turbine, $h_1 - h_{2x}$ is used, h_{2x} is the enthalpy of the wet steam at exhaust conditions; h_1 is as before.

Since the steam expands isentropically into the wet region below the dome of the *T-S* diagram, Fig. 19, we must first determine the quality of the steam at point 2 either from a *T-S* diagram or Mollier chart or by calculation. By calculation using the method of mixtures and the entropy at each point: $S_1 = S_2 = 0.0914 + (x_2)(1.9451)$. Then $x_2 = (1.6508 - 0.0914)/1.9451 = 0.80$, or 80 percent quality. Substituting and summing, using steam-table values, $h_{2x} = 47.06 + 0.8(1047.8) = 885.3$ Btu/lb (2062.7 kJ/kg).

(*c*) To find the CEE we first must obtain the ideal steam rate, $w_i = 3413/(h_1 - h_{2x}) = 3413/(1340.6 - 885.3) = 7.496$ lb/kWh (3.4 kg/kWh).

Now, CEE = (7.496/10)(100) = 74.96 percent. This value is excellent for such a plant and is in a range being achieved today.

Related Calculations. Use this approach to analyze the efficiency of any turbogenerator used in central-station, industrial, marine, and other plants.

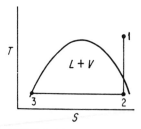

FIGURE 19 *T-S* diagrams for steam turbine.

TURBOGENERATOR REHEAT-REGENERATIVE CYCLE ALTERNATIVES ANALYSIS

A turbogenerator operates on the reheating-regenerative cycle with one stage of reheat and one regenerative feedwater heater. Throttle steam at 400 lb/in² (abs) (2756.0 kPa) and 700°F (371.1°C) is used. Exhaust at 2-in (5.08-cm) Hg is taken from the turbine at a pressure of 63 lb/in² (abs) (434 kPa) for both reheating and feedwater heating with reheat to 700°F (371.1°C). For an ideal turbine working under these conditions, find: (a) Percentage of throttle steam bled for feedwater heating; (b) Heat converted to work per pound (kg) of throttle steam; (3) Heat supplied per pound (kg) of throttle steam; (d) Ideal thermal efficiency; (e) Other ways to heat feedwater and increase the turbogenerator output. Figure 20 shows the layout of the cycle being considered, along with a Mollier chart of the steam conditions.

Calculation Procedure:

1. Using the steam tables and Mollier chart, list the pertinent steam conditions
Using the subscript 1 for throttle conditions, list the key values for the cycle thus:

$$P_1 = 400 \text{ lb/in}^2 \text{ (abs) (2756.0 kPa)}$$
$$t_1 = 700°F \text{ (371.1°C)}$$
$$H_1 = 1362.2 \text{ Btu/lb (3173.9 kJ/kg)}$$
$$S_1 = 1.6396$$
$$H_2 = 1178 \text{ Btu/lb (2744.7 kJ/kg)}$$
$$H_3 = 1380.1 \text{ Btu/lb (3215.6 kJ/kg)}$$
$$H_4 = 1035.8 \text{ (2413.4 kJ/kg)}$$
$$H_5 = 69.1 \text{ Btu/lb (161.0 kJ/kg)}$$
$$H_6 = 265.27 \text{ Btu/lb (618.07 kJ/kg)}$$

2. Determine the percentage of throttle steam bled for feedwater heating
(a) Set up the ratio for the feedwater heater of (heat added in the feedwater heater)/(heat supplied to the heater)(100). Or, using the enthalpy data from step 1 above, $(H_6 - H_5)/(H_2 - H_5)(100) = (265.26 - 69.1)/(1178 - 69.1)(100) = 17.69$ percent of the throttle steam is bled for feedwater heating.

3. Find the heat converted to work per pound (kg) of throttle steam
(b) The heat converted to work is the enthalpy difference between the throttle steam and the bleed steam at point 2 plus the enthalpy difference between points 3 and 4 times the percentage of throttle flow between these points. In equation form, heat converted to work $= H_1 - H_2 + (1.00 - 0.1769)(H_3 - H_4) = (1362.2 - 1178) + (0.0823)(1380.1 - 1035.8) = 467.55$ Btu/lb (1089.39 kJ/kg).

4. Calculate the heat supplied per pound (kg) of throttle steam
(c) The heat supplied per pound (kg) of throttle steam $= (H_1 - H_6) + (H_3 - H_2) = (1362.3 - 265.27) + (1380.1 - 1178) = 1299.13$ Btu/lb (3026.97 kJ/kg).

400 psia (2756 kPa)

63 psia (434 kPa)

2-in. Hg (5.08-cm Hg)

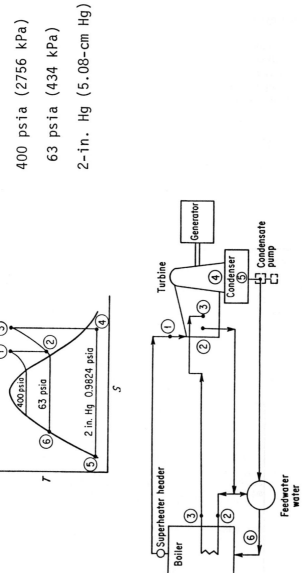

FIGURE 20 Cycle layout and TS chart of steam conditions.

5. Compute the ideal thermal efficiency
(*d*) Use the relation, ideal thermal efficiency = (heat converted to work)/(heat supplied) = 467.55/1299.13 = 0.3598, or 35.98 percent.

6. Show other ways to heat feedwater while increasing the turbogenerator output

For years, central stations and large industrial steam-turbine power plants shut off feedwater heaters to get additional kilowatts out of a turbogenerator during periods of overloaded electricity demand. When more steam flows through the turbine, the electrical power output increases. While there was a concurrent loss in efficiency, this was ignored because the greater output was desperately needed.

Today steam turbines are built with more heavily loaded exhaust ends so that the additional capacity is not available. Further, turbine manufacturers place restrictions on the removal of feedwater heaters from service. However, if the steam output of the boiler is less than the design capacity of the steam turbine, because of a conversion to coal firing, additional turbogenerator capacity is available and can be regained at a far lower cost than by adding new generator capacity.

Compensation for the colder feedwater can be made, and the lost efficiency regained, by using a supplementary fuel source to heat feedwater. This can be done in one of two ways: (1) increase heat input to the existing boiler economizer, or (2) add a separately fired external economizer.

Additional heat input to a boiler's existing economizer can be supplied by in-duct burners, Fig. 21, from slagging coal combustors, Fig. 22, or from the furnace itself. Since the economizer in a coal-fired boiler is of sturdier construction than a heat-recovery steam generator (HRSG) with finned tubing, in-duct burners can be placed closer to the economizer, Fig. 21. Burner firing may be by coal or oil.

Slagging coal combustors are under intense development. A low-NO_x, low-ash combustor, Fig. 22, supplying combustion gases at 3000°F (1648.9°C) may soon be commercially available.

To accommodate any of the changes shown in Fig. 21, a space from 12 (3.66 m) to 15 ft (4.57 m) is needed between the bottom of the primary superheater and the top of the economizer. This space is required for the installation of the in-duct

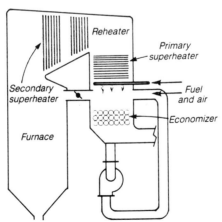

FIGURE 21 Heat input to the economizer may be increased by the addition of induct burners, by bypassing hot furnace gases into the gas path ahead of the economizer, or by recirculation (*Power*).

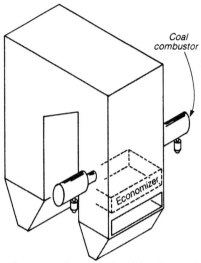

FIGURE 22 Slagging combustors can be arranged to inject hot combustion gases into gas passages ahead of economizer (*Power*).

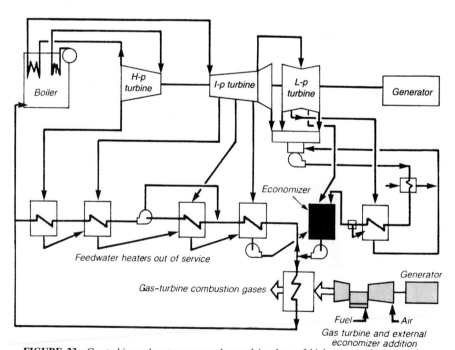

FIGURE 23 Gas-turbine exhaust gases can be used in place of high-pressure heaters, using a compact heat exchanger (*Power*).

burners or for the adequate mixing of gas streams if the furnace or an external combustor is used to supply the additional heat.

Another approach is to install a separately fired external economizer in series or parallel with the existing economizer, which could be fired by a variety of fuels. The most attractive possibility is to use waste heat from a gas-turbine exhaust, Fig. 23. The output of this simple combined-cycle arrangement would actually be higher than the combined capabilities of the derated plant and the gas turbine.

The steam-cycle arrangement for the combined plant is shown in Fig. 23. Feedwater is bypassed around the high-pressure regenerative heaters to an external low-cost, finned-tube heat exchanger, Fig. 24, where waste heat from the gas turbine is recovered.

When high-pressure feedwater heaters are shut off, steam flow through the intermediate- and low-pressure turbine sections increases and becomes closer to the full-load design flow. The reheat expansion line moves left on the Mollier chart from its derated position, Fig. 25. With steam flow closer to the design value, the exhaust losses per pound (kg) of steam, Fig. 26, are lower than at the derated load.

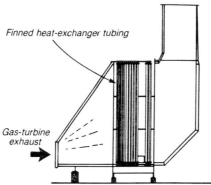

FIGURE 24 Gas-turbine heat-recovery finned-tube heat exchanger is simple and needs no elaborate controls (*Power*).

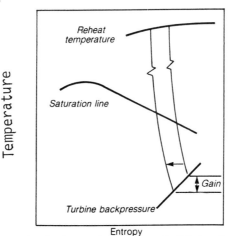

FIGURE 25 Reheat expansion line is moved to the left on the T-S chart, increasing power output (*Power*).

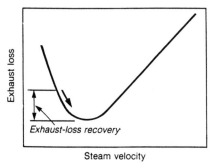

Steam velocity

FIGURE 26 Heat loss in steam-turbine exhaust is reduced when operating at rated flow (*Power*).

The data and illustrations in this step 6 are based on the work of E. S. Miliares and P. J. Kelleher, Energotechnology Corp., as reported in *Power* magazine.

TURBINE-EXHAUST STEAM ENTHALPY AND MOISTURE CONTENT

What is the enthalpy and percent moisture of the steam entering a surface condenser from the steam turbine whose Mollier chart is shown in Fig. 27? The turbine is delivering 20,000 kW and is supplied steam at 850 lb/in² (abs) (5856.5 kPa) and 900°F (482.2°C); the exhaust pressure is 1.5 in (3.81 cm) Hg absolute. The steam rate, when operating straight condensing, is 7.70 lb/delivered kWh (3.495 kg/kWh) and the generator efficiency is 98 percent.

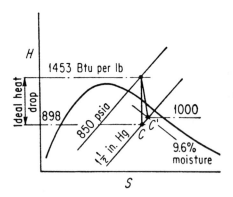

1453 Btu/lb (3385 kJ/kg)	898 Btu/lb (2092 kJ/kg)
850 psia (5856 kPa)	1000 Btu/lb (2330 kJ/kg)
1.5 in. Hg (3.8 cm Hg)	

FIGURE 27 Mollier chart for turbine exhaust conditions.

Calculation Procedure:

1. Compute the engine efficiency of the turbine
Use the relation $E_e = 3413/(w_s)(H_1 - H_c)$, where E_e = engine efficiency; w_s = steam rate of the turbine when operating straight condensing in the units given above; enthalpies H_1 and H_c are as shown in the Mollier chart. Substituting, $E_e = 3413/(7.7)(1453 - 898) = 0.7986$ for ideal conditions.

2. Find the Rankine engine efficiency for the actual turbine
The Rankine engine efficiency for this turbine is: $(0.7986/0.98) = 0.814 = (H_1 - H_{c'})/(H_1 - H_c)$. Solving, $(H_1 - H_{c'}) = 0.814(555) = 452.3$ Btu/lb (1053.8 kJ/kg).
   At the end of the actual expansion of the steam in the turbine, $H_{c'} = 1453 - 452.3 = 1000.7$ Btu/lb (2331.6 kJ/kg) enthalpy.

3. Determine the moisture of the steam
Referring to the Mollier chart where $H_{c'}$ crosses the pressure line of 1.5 in (3.81 cm) Hg, the moisture percent is found to be 9.6 percent.
   Related Calculations. The Mollier chart can be a powerful and quick reference for solving steam expansion problems in plants of all types—utility, industrial, commercial, and marine.

STEAM TURBINE NO-LOAD AND PARTIAL-LOAD STEAM FLOW RATES

A 40,000-kW straight-flow condensing industrial steam turbogenerator unit is supplied steam at 800 lb/in² (abs) (5512 kPa) and 800°F (426.7°C) and is to exhaust at 3 in (76 cm) Hg absolute. The half-load and full-load throttle steam flows are estimated to be 194,000 lb/h (88,076 kg/h) and 356,000 lb/h (161,624 kg/h), respectively. The mechanical efficiency of the turbine is 99 percent and the generator efficiency is 98 percent. Find (a) the no-load throttle steam flow; (b) the heat rate of the unit expressed as a function of the kW output; (c) the internal steam rate of the turbine at 30 percent of full load.

Calculation Procedure:

1. Find the difference between full-load and half-load steam rates and the no-load rate
(a) Assume a straight-line rating characteristic and plot Fig. 28a. This assumption is a safe one for steam turbines in this capacity range. Then, the difference between full-load and half-load steam rates is $356,000 - 194,000 = 162,000$ lb/h (73,548 kg/h). The no-load steam rate will then be = (half-load rate) − (difference between full-load and half-load rates) = $194,000 - 162,000 = 32,000$ lb/h (14,528 kg/h).

2. Determine the steam rate and heat rate at quarter-load points
(b) Using Fig. 28b, we see that the actual turbine efficiency, $E_t = 3413/(w_s)(H_1 - H_f)$, w_s = steam flow, lb/kWh (kg/kWh); H_1 = enthalpy of entering steam, Btu/lb (kJ/kg); H_f = enthalpy of condensate at the exhaust pressure, Btu/lb (kJ/kg).
   Further, turbine heat rate = $3413/E_t$ Btu/kWh (kJ/kWh) = $w_k(H_1 - H_f)$, where

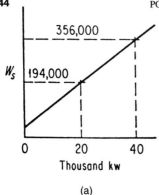

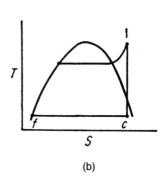

(a)   (b)

356,000 lb/hr (161,624 kg/hr)   194,000 lb/hr (88,076 kg/hr)

FIGURE 28 (a) Straight-line rating characteristic. (b) T-S diagram.

w_k = the actual steam rate, lb/kWh (kg/kWh) = w = w_s/kW output, where the symbols are as defined earlier.

Substituting w_s = 32,000 no-load throttle flow + (difference between full-load and half-load throttle flow rate/kW output at half load)(kW output) = 32,000 + (162,000/20,000)(kW) = 32,000 + 8.1 kW for this turbine-generator set. Also w_k = (32,000/kW) + 8.1.

Using the steam tables, we find H_1 = 1398 Btu/lb (3257.3 kJ/kg); H_f = 83 Btu/lb (193.4 kJ/kg). Then, $H_1 - H_f$ = 1315 Btu/lb (3063.9 kJ/kg). Substituting, heat rate = [(1315)(32,000)/(kW)] + (1316)(8.1) = 10.651.5 + (42,080,000/kW).

Computing the steam rate and heat rate for the quarter-load points for this turbine-generator we find:

At full load, w_s = 8.9 lb/kWh (4.04 kg/kWh)

At ¾ load, w_s = 9.17 lb/kWh (4.16 kg/kWh)

At ½ load, w_s = 9.7 lb/kWh (4.4 kg/kWh)

At ¼ load, w_s = 11.3 lb/kWh (5.13 kg/kWh)

At full load, heat rate = 11,700 Btu/kWh (27,261 kJ/kWh)

At ¾ load, heat rate = 12,080 Btu/kWh (28,146 kJ/kWh)

At ½ load, heat rate = 12,770 Btu/kWh (29,754 kJ/kWh)

At ¼ load, heat rate = 14,870 Btu/kWh (34,647 kJ/kWh)

3. Determine the internal steam rate of the turbine

(c) For the turbine and generator combined, E_e = 3413/$(w_k)(H_1 - H_c)$, where E_e = turbine engine efficiency; H_c = enthalpy of the steam at the condenser; other symbols as given earlier. Since, from the steam tables, H_1 = 1398 Btu/lb (3257.3 kJ/kg); H_c = 912 Btu/lb (2124.9 kJ/kg); then $(H_1 - H_c)$ = 486 Btu/lb (1132.4 kJ/kg).

From earlier steps, w_s = 356,000 lb/h (161,624 kg/h) at full-load; w_s = 32,000 lb/h (14,528 kg/h) at no-load. For the full-load range the total change is 356,000 − 32,000 = 324,000 lb/h (147,096 kg/h). Then, w_s at 30 percent load = [(32,000) + 0.30(324,000)]/0.30(40,000) = 10.77 lb/kWh (4.88 kg/kWh). Then, E_e = 3413/(10.77)(486) = 0.652 for combined turbine and generator.

If the internal efficiency of the turbine (not including the friction loss) E_i, then $E_i = 2545/(w_a)(H_1 - H_c)$. Thus $E_i = E_e/$(turbine mechanical efficiency)(generator efficiency). Or $E_i = 0.652/(0.99)(0.98) = 0.672$. Then, the actual steam rate per horsepower (kW) is $w_a = 2545/(E_i)(H_1 - H_c) = 2545/(0.672)(486) = 7.79$ lb/hp (4.74 kg/kW).

   Related Calculations.   Use this approach to analyze any steam turbine—utility, industrial, commercial, marine, *etc.*—to determine the throttle steam flow and heat rate.

POWER PLANT PERFORMANCE BASED ON TEST DATA

A test on an industrial turbogenerator gave these data: 29,760 kW delivered with a throttle flow of 307,590 lb/h (139,646 kg/h) of steam at 245 lb/in² (abs) (1688 kPa) with superheat at the throttle of 252°F (454°C); exhaust pressure 0.964 in (2.45 cm) Hg (abs); pressure at the one bleed point, Fig. 29a, 28.73 in (72.97 cm) Hg (abs); temperature of feedwater leaving bleed heater 163°F (72.8°C). For the corresponding ideal unit, find: (*a*) percent throttle steam bled, (*b*) net work for each pound of throttle steam, (*c*) ideal steam rate, and (*d*) cycle efficiency. For the actual unit find, (*e*) the combined steam rate, (*f*) combined thermal efficiency, and (*g*) combined engine efficiency.

Calculation Procedure:

1.   Determine the steam properties at key points in the cycle
Using a Mollier chart and the steam tables, plot the cycle as in Fig. 29b. Then, S_1 = 1.676; H_1 = 1366 Btu/lb (3183 kJ/kg); H_2 = 1160 Btu/lb (2577 kJ/kg); P_2 = 14.11 lb/in² (abs) (97.2 kPa); H_3 = 130.85 Btu/lb (304.9 kJ/kg); P_3 = 5.089 lb/in² (abs) (35.1 kPa); H_4 = 46.92 Btu/lb (109.3 kJ/kg); P_4 = 0.4735 lb/in² (abs) (3.3 kPa); H_5 = 177.9 Btu/lb (414.5 kJ/kg).

   (*a*) The percent throttle steam bled is found from: $100 \times (H_5 - H_4)/(H_2 - H_4)$ = $100 \times (177.9 - 46.92)/(1106 - 46.92) = 12.41$ percent.

2.   Find the amount of heat converted to work
(*b*) Use the relation, heat converted to work, $h_w = H_1 - H_2 + (1 - m_2)(H_2 - H_7)$, where m_2 = percent throttle steam bled, H_7 = enthalpy of exhaust steam in the condenser. Substituting, heat converted to work, h_w = (1366 − 1106) + (1 − 0.1241)(1106 − 924.36) = 419.1 Btu/lb (976.5 kJ/kg).

3.   Compute the ideal steam rate
(*c*) Use the relation, ideal steam rate, $l_r = 3413$ Btu/kWhr/h_w. Or, $l_r = 3413/419.1$ = 8.14 lb/kWh (3.69 kg/kWh).

4.   Find the cycle efficiency of the ideal cycle
(*d*) Cycle efficiency, C_e = (heat converted into work/heat supplied). Or $h_w/(H_1 - H_3)$; substituting, C_e = 419.1/(1366 − 130.85) = 0.3393, or 33.9 percent.

5.   Determine the combined steam rate
(*e*) The combined steam rate for the actual unit is R_c = lb steam consumed/kWh generated. Or R_c = 307,590/29,760 = 10.34 lb/kWh (4.69 kg/kWh).

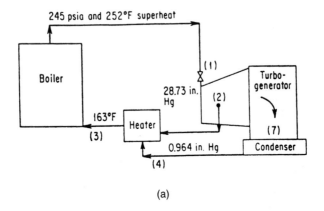

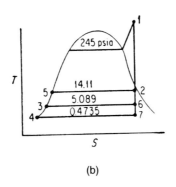

(b)

245 psia (1688 kPa)	28.73 in. Hg (72.97 cm Hg)
252°F (454°C)	163°F (72.8°C)
0.963 in. Hg (2.45 cm Hg)	

FIGURE 29   (*a*) Cycle diagram with test conditions. (*b*) *T-S* diagram for cycle.

6.   Find the combined thermal efficiency of the actual unit
(*f*) The combined thermal efficiency, TE_c = 3413/heat supplied. Or TE_c = 3413/
10.34(H_1 − H_3) = 35413/10.34(1366 − 130.85) = 0.267, or 26.7 percent.

7.   Compute the combined engine efficiency
(*g*) The combined engine efficiency TE_c/C_e, or 26.7/33.9 = 0.7876, or 78.76 percent.
    Related Calculations.   Use this general procedure to determine the percent
bleed steam, net work of each pound of throttle steam, ideal steam rate, cycle
efficiency, combined thermal efficiency and combined engine efficiency for steam-
turbine installations in central stations, industrial, municipal and marine installations. Any standard set of steam tables and a Mollier chart are sufficiently accurate
for usual design purposes.

DETERMINING TURBOGENERATOR STEAM RATE
AT VARIOUS LOADS

A 100-MW turbogenerator is supplied steam at 1250 lb/in^2 (abs) (8612.5 kPa) and 1000°F (537.8°C) with a condenser pressure of 2 in (5.08 cm) Hg (abs). At rated load, the turbine uses 1,000,000 lb (454,000 kg) of steam per hour; at zero load, steam flow is 50,000 lb/h (22,700 kg/h). What is the steam rate in pounds (kg) per kWh at $4/4$, $3/4$, $2/4$, and $1/4$ load?

Calculation Procedure:

1. Write the steam-flow equation for this turbogenerator
The curve of steam consumption, called the Willian's line, is practically a straight line for steam turbines operating without overloads. Hence, we can assume a straight line for this turbogenerator. If the Willian's line is extended to intercept the Y (vertical) axis for total steam flow per hour, this intercept represents the steam required to operate the turbine when delivering no power. This no-load steam flow—50,000 lb/h (22,700 kg/h) for this turbine—is the flow rate required to overcome the friction of the turbine and the windage, governor and oil-pump drive power, *etc.*, and for meeting the losses caused by turbulence, leakage, and radiation under no-load conditions.

Using the data provided, the steam rate equation can be written as $[(50/L) + 9.5] = (F/L) = [50 + (1000 - 50)/100(L)]/(L)$, where F = full-load steam flow, lb/h (kg/h); L = load percent.

2. Compute the steam flow at various loads
Use the equation above thus:

Load fraction	Load, MW	Steam rate lb/kWh (kg/kWh)
$1/4$	$100 \times 1/4 = 25$	$50/25 + 9.5 = 11.5$ (5.22)
$2/4$	$100 \times 2/4 = 50$	$50/50 + 9.5 = 10.5$ (4.77)
$3/4$	$100 \times 3/4 = 75$	$50/75 + 9.5 = 10.17$ (4.62)
$4/4$	$100 \times 4/4 = 100$	$50/100 + 9.5 = 10.00$ (4.54)

Related Calculations.   The Willian's line is a useful tool for analyzing steam-turbine steam requirements. As a check on its validity, compare actual turbine performance steam conditions with those computed using this procedure. The agreement is startlingly accurate.

ANALYSIS OF REHEATING-REGENERATIVE
TURBINE CYCLE

An industrial turbogenerator operates on the reheating-regenerative cycle with one reheat and one regenerative feedwater heater. Throttle steam at 400 lb/in^2 (abs) (2756 kPa) and 700°F (371°C) is used. Exhaust is at 2 in (5.1 cm) Hg (abs). Steam is taken from the turbine at a pressure of 63 lb/in^2 (abs) (434 kPa) for both reheating and feedwater heating. Reheat is to 700°F (371°C). For the ideal turbine working under these conditions find: (a) percentage of throttle steam bled for feedwater heating, (b) heat converted to work per pound (kg) of throttle steam, (c) heat supplied per pound (kg) of throttle steam, (d) ideal thermal efficiency, (e) T-S, temperature-entropy, diagram and layout of cycle.

Calculation Procedure:

1. Determine the cycle enthalpies, pressures, and entropies
Using standard steam tables and a Mollier chart, draw the cycle and T-S plot, Fig. 30a and b. Then, P_1 = 400 lb/in² (abs) (2756 kPa); t_1 = 700°F (371°C); H_1 = 1362.3 Btu/lb (3174 kJ/kg); S_1 = 1.6396; H_2 = 1178 Btu/lb (2745 kJ/kg); H_g = 1380.1 Btu/lb (3216 kJ/kg); S_g = 1.8543.
(a) Percent throttle steam bled = $(H_6 - H_5)/(H_2 - H_5)$ = (196.15/1107.9) = 0.1771, or 17.71 percent.

2. Find the amount of heat converted to work per pound (kg) of throttle steam
(b) The amount of heat converted to work per pound (kg) of throttle steam = $(H_1 - H_2)$ + $(1 - 0.1771)(H_g - H_4)$ = 467.3 Btu/lb (1088.8 kJ/kg).

3. Compute the heat supplied per pound (kg) of throttle steam
(c) The heat supplied per pound (kg) of throttle steam = $(H_1 - H_6)$ + $(H_g - H_2)$ = 1299.1 Btu/lb (3026.9 kJ/kg).

4. Determine the ideal thermal efficiency
(d) The ideal thermal efficiency = (heat recovered per pound [kg] of throttle steam)/(heat supplied per pound [kg] of throttle steam) = 467.3/1299.13 = 0.3597, or 35.97 percent. The T-S diagram and cycle layout can be drawn as shown in Fig. 30a and b.

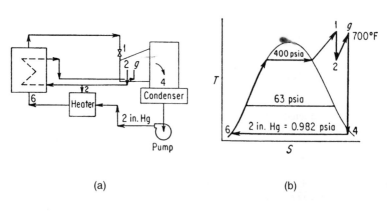

(a)   (b)

400 psia (2756 kPa)   700°F (371°C)   63 psia (434 kPa)
2 in. Hg (5.1 cm Hg)

FIGURE 30   (a) Cycle diagram. (b) T-S diagram of cycle.

Related Calculations. This general procedure can be used for any turbine cycle where reheating and feedwater heating are part of the design. Note that the enthalpy and entropy values read from the Mollier chart, or interpolated from the steam tables, may differ slightly from those given here. This is to be expected where judgement comes into play. The slight differences are unimportant in the analysis of the cycle.

The procedure outlined here is valid for industrial, utility, commercial, and marine turbines used to produce power.

STEAM RATE FOR REHEAT-REGENERATIVE CYCLE

Steam is supplied at 600 lb/in² (abs) (4134 kPa) and 740°F (393°C) to a steam turbine operating on the reheat-regenerative cycle. After expanding to 100 lb/in² (abs) (689 kPa), the steam is reheated to 700°F (371°C). Expansion then continues to 10 lb/in² (abs) (68.9 kPa) but at 30 lb/in² (abs) (207 kPa) some steam is extracted for feedwater heating in a direct-contact heater. Assuming ideal operation with no losses, find: (*a*) steam extracted as a percentage of steam supplied to the throttle, (*b*) steam rate in pounds (kg) per kWh; (*c*) thermal efficiency of the turbine, (*d*) quality or superheat of the exhaust if in the actual turbine combined efficiency is 72 percent, generator efficiency is 94 percent, and actual extraction is the same as the ideal.

Calculation Procedure:

1. Assemble the key enthalpies, entropies, and pressures for the cycle
Using the steam tables and a Mollier chart, list the following pressures, temperatures, enthalpies, and entropies for the cycle, Fig. 31*a*: P_1 = 600 lb/in² (abs) (4134 kPa); t_1 = 740°F (393°C); P_2 = 100 lb/in² (abs) (689 kPa); t_3 = 700°F (371°C); P_x = 30 lb/in² (abs) (207 kPa); p_c = 1 lb/in² (abs) (6.89 kPa); H_1 = 1372 Btu/lb (3197 kJ/kg); S_1 = entropy = 1.605; H_2 = 1188 Btu/lb (2768 kJ/kg); H_3 = 1377 Btu/lb (3208 kJ/kg); S_3 = 1.802; H_x = 1245 Btu/lb (2901 kJ/kg); H_c = 1007 Btu/lb (2346 kJ/kg); H_f = 70 Btu/lb (163 kJ/kg); H_{fx} = 219 Btu/lb (510 kJ/kg). Plot Fig. 31*b* as a skeleton Mollier chart to show the cycle processes.

2. Compute the percent steam extracted for the feedwater heater
(*a*) The steam extracted for the feedwater heater, x, = $(H_{fx} - H_f)(H_x - H_f)$ = (219 − 70)/(1245 − 70) = 0.1268, or 12.68 percent.

3. Find the turbine steam rate
(*b*) For the Rankine-cycle steam rate, w_s = 3413/($H_1 - H_c$). For this cycle, w_s = 3413/[($H_1 - H_2$) + $x(H_3 - H_x)$ + (1 − x)($H_3 - H_c$)]. Or, w_s = 3413/[1372 − 1188) + 0.1268(1377 − 1245) + (1 − 0.1268)(1377 − 1007)] = 6.52 lb/kWh (2.96 kg/kWh).

4. Calculate the turbine thermal efficiency
(*c*) The thermal efficiency, E_t = [($H_1 - H_2$) + $x(H_3 - H_x)$ + (1 − x)($H_3 - H_c$)]/[$H_3 - H_2$) + ($H_1 - H_{fx}$)]. Or, E_t = [(1372 − 1188) + (0.1268)(1377 − 1245) +

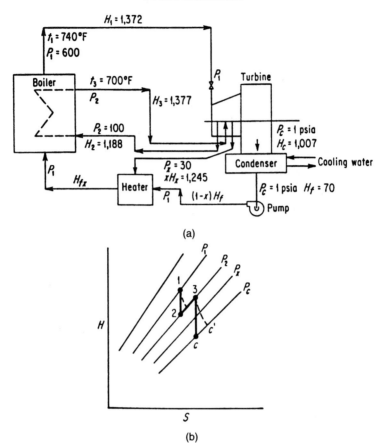

(a)

(b)

FIGURE 31 (*a*) Cycle diagram. (*b*) H-S chart for cycle in (*a*).

600 psia (4134 kPa)	740°F (393°C)	700°F (371°C)	100 psia (68.9 kPa)
1188 Btu/lb (2768 kJ/kg)	1372 Btu/lb (3197 kJ/kg)	1245 Btu/lb (2901 kJ/kg)	
1 psia (6.89 kPa)	30 psia (207 kPa)	1007 Btu/lb (2346 kJ/kg)	70 Btu/lb (163 kJ/kg)

$(1 - 0.1268)(1377 - 1007)]/[(1377 - 1188) + (1372 - 219) = 0.3903$, or 39 percent. It is interesting to note that in an ideal cycle the thermal efficiency of the turbine is the same as that of the cycle.

5. Determine the condition of the exhaust
(*d*) The engine efficiency of the turbine alone = (actual turbine combined efficiency)/actual generator efficiency). Or, using the given data, engine efficiency of the turbine alone = 0.72/0.94 = 0.765.

Using the computed engine efficiency of the turbine alone and the Mollier chart, $(H_3 - H_{c'}) = 0.765(H_3 - H_C) = 283$. Solving, $H_{c'} = H_3 - 283 = 1094$ Btu/lb (2549 kJ/kg). From the Mollier chart, the condition at $H_{c'}$ is 1.1 percent moisture. The exhaust steam quality is therefore 100 − 1.1 = 98.9 percent.

Related Calculations. This procedure is valid for a variety of cycle arrangements for industrial, central-station, commercial and marine plants. By using a combination of the steam tables, Mollier chart and cycle diagram, a full analysis of the plant can be quickly made.

BINARY CYCLE PLANT EFFICIENCY ANALYSIS

A binary cycle steam and mercury plant is being considered by a public utility. Steam and mercury temperature will be 1000°F (538°C). The mercury is condensed in the steam boiler, Fig. 32a at 10 lb/in² (abs) (68.9 kPa) and the steam pressure is 1200 lb/in² (abs) (8268 kPa). Condenser pressure is 1 lb/in² (abs) (6.89 kPa). Expansions in both turbines are assumed to be at constant entropy. The steam cycle has superheat but no reheat. Find the efficiency of the proposed binary cycle. Find the cycle efficiency without mercury.

Calculation Procedure:

1. *Tabulate the key enthalpies and entropies for the cycle*
Set up two columns, thus:

Mercury cycle	Steam cycle
H_{m1} = 151.1 Btu/lb (352 kJ/kg)	H_{s1} = 1499.2 Btu/lb (3493 kJ/kg)
S_{m1} = 0.1194	S_{s1} = 1.6293
S_{me} = 0.0094	H_{sf} = 69.7 Btu/lb (162.4 kJ/kg)
H_{mf} = 22.6 Btu/lb (52.7 kJ/kg)	

2. *Compute the quality of the exhaust for each vapor*
Since expansion in each turbine is at constant entropy, Fig. 32b, the quality for the mercury exhaust, x_m is: 0.1194 = 0.0299 + x_m(0.1121); x_m = 0.798.
    For the steam cycle, the quality, x_s is: 1.6293 = 0.1326 + x_s(1.8456); x_s = 0.81.

3. *Find the exhaust enthalpy for each vapor*
Using the properties of mercury from a set of tables, the enthalpy of the mercury exhaust, H_{me} = 22.6 + 0.798(123) = 120.7 Btu/lb (281.2 kJ/kg). The enthalpy of the condensed mercury, H_{mf} = 22.6 Btu/lb (52.7 kJ/kg).
    For the exhaust steam, the enthalpy H_{se} = 69.7 + 0.81(1036.3) = 909.1 Btu/lb (2118 kJ/kg), using steam-table data. The enthalpy of the condensed steam, H_{sf} = 69.7 Btu/lb (162.4 kJ/kg).
    Assuming 98 percent quality steam leaving the mercury condenser, then the enthalpy of the wet steam leaving the mercury condenser, H_{sw} = 571.7 + 0.98(611.7) = 1171.2 Btu/lb (2728 kJ/kg).

4. *Write the heat balance around the mercury condenser*
The steam heat gain = $H_{sw} - H_{sf}$ = 1171.2 − 69.7 = 1101.5 Btu/lb (2566.5 kJ/kg). Now, the mercury heat loss = $H_{me} - H_{mf}$ = 120.7 − 98.1 = 98.1 Btu/lb

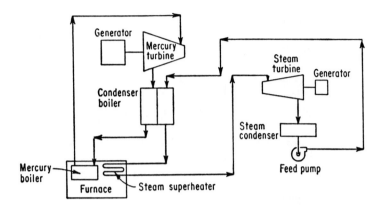

(a)

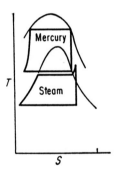

(b)

FIGURE 32   (a) Binary cycle. (b) T-S diagram for binary cycle.

(228.6 kJ/kg). The weight of mercury per pound (kg) of steam = steam heat gain/mercury heat loss = 1101.5/98.1 = 11.23.

5.   Determine the heat input and work done per pound (kg) of steam
The heat input per pound of steam is: For mercury = (lb Hg/lb steam)(H_{m1} − H_{mf}) = 11.23(151.1 − 22.6) = 1443.05 Btu (1522.5 J). For steam = (H_{s1} − H_{sw}) = 1499.2 − 1171.7 = 327.5 Btu (345.5 J). Summing these two results gives 1443.05 + 327.5 = 1770.55 Btu (1867.9 J) as the heat input per pound (kg) of steam.

The work done per pound (kg) of steam is: For mercury = (lb Hg/lb steam)(h_{m1} − H_{me}) = 11.23(151.1 − 120.7) = 341.4 Btu (360.2 J). For steam = H_{c1} − H_{se} = 1499.2 − 909.1 = 590.1 Btu (622.6 J). Summing, as before, the total work done per pound (kg) of steam = 931.5 Btu (982.7 J).

6. Compute the binary cycle efficiency
The binary cycle efficiency = (work done per pound (kg) of steam)/(heat input per pound (kg) of steam). Or binary cycle efficiency = 931.5/1770.55 = 0.526, or 52.6 percent.

7. Calculate the steam cycle efficiency without the mercury topping turbine
The steam cycle efficiency without the mercury topping turbine = (work done per pound (kg) of steam)/(H_{s1} − H_{sf}) = 590.1/(1499.2 − 69.7) = 0.4128, or 41.3 percent.

Related Calculations. Any binary cycle being considered for an installation depends on the effects of the difference in thermodynamic properties of the two pure fluids involved. For example, steam works under relatively high pressures with an attendant relatively low temperature. Mercury, by comparison, has the vapor characteristic of operating under low pressures with attendant high temperature.

In a mercury-vapor binary cycle, the pressures are selected so the mercury vapor condenses at a temperature higher than that at which steam evaporates. The processes of mercury vapor condensation and steam evaporation take place in a common vessel called the condenser-boiler, which is the heart of the cycle.

In the steam portion of this cycle, condenser water carries away the heat of steam condensation; in the mercury portion of the cycle it is the steam which picks up the heat of condensation of the mercury vapor. Hence, there is a great saving in heat and the economies effected reflect the consequent improvement in cycle efficiency.

The same furnace serves the mercury boiler and the steam superheater. Mercury vapor is only condensed, not superheated. And if the condenser-boiler is physically high enough above the mercury boiler, the head of mercury is great enough to return the liquid mercury to the boiler by gravity, making the use of a mercury feed pump unnecessary.

To avoid the high cost entailed with using mercury, a number of man-made solutions have been developed for binary vapor cycles. Their use, however, has been limited because the conventional steam cycle is usually lower in cost. And with the advent of the aero-derivative gas turbine, which is relatively low cost and can be installed quickly in conjunction with heat-recovery steam generators, binary cycles have lost popularity. But it is useful for engineers to have a comprehension of such cycles. Why? Because they may return to favor in the future.

Conventional Steam Cycles

FINDING COGENERATION SYSTEM EFFICIENCY VS. A CONVENTIONAL STEAM CYCLE

An industrial plant has 60,000 lb/h (27,240 kg/h) of superheated steam at 1000 lb/in^2 (abs) (6890 kPa) and 900°F (482.2°C) available. Two options are being considered for use of this steam: (1) expanding the steam in a steam turbine having a 70 percent efficiency to 1 lb/in^2 (abs) (6.89 kPa), and (2) expand the steam in a turbine to 200 lb/in^2 (abs) (1378 kPa) generating electricity and utilizing the low-pressure exhaust steam for process heating. Evaluate the two schemes for energy efficiency when the boiler has an 82 percent efficiency on a HHV basis.

Calculation Procedure:

1. Determine the enthalpies of the steam at the turbine inlet and after isentropic expansion

Cogeneration systems generate power and process steam from the same fuel source. Process plants generating electricity from steam produced in a boiler and using the same steam after expansion in a steam turbine for process heating of some kind are examples of cogeneration systems.

Conventional steam-turbine power plants have a maximum efficiency of about 40 percent as most of the energy is wasted in the condensing-system cooling water. In a typical cogeneration system the exhaust steam from the turbine is used for process purposes after expansion through the steam turbine; hence, its enthalpy is fully utilized. Thus, cogeneration schemes are more efficient.

At 1000 lb/in^2 (abs) (6890 kPa) and 900°F (482.2°C), the enthalpy, $h_1 = 1448$ Btu/lb (3368 kJ/kg) from the steam tables. The entropy of steam at this condition, from the steam tables, is 1.6121 Btu/lb °F (6.748 kJ/kg K). At 1 lb/in^2 (abs) (6.89 kPa), the entropy of the saturated liquid, s_f, is 0.1326 Btu/lb °F (0.555 kJ/kg K), and the entropy of the saturated vapor, s_g is 1.9782 Btu/lb °F (8.28 kJ/kg K), again from the steam tables.

Now we must determine the quality of the steam, X, at the exhaust of the steam turbine at 1 lb/in^2 (abs) (6.89 kPa) from (entropy at turbine inlet condition) = (entropy at outlet condition)(X) + (1 − X)(entropy of the saturated fluid at the outlet condition); or $X = 0.80$. The enthalpy of steam corresponding to this quality condition is h_{2s} = (enthalpy of the saturated steam at 1 lb/in^2 (abs))(X) + (enthalpy of the saturated liquid at 1 lb/in^2 (abs))(1 − X) = (1106)(0.80) + (1 − 0.80)(70) = 900 Btu/lb (2124 kJ/kg).

2. Compute the power output of the turbine

Use the equation $P = (W_s)(e_t)(h_1 - h_{2s})/3413$, where P = electrical power generated, kW; W_s = steam flow through the turbine, lb/h (kg/h); e_t = turbine efficiency expressed as a decimal; h_1 = enthalpy of the steam at the turbine inlet, Btu/lb (kJ/kg); h_{2s} = enthalpy of the steam after isentropic expansion through the turbine, Btu/lb (kJ/kg). Substituting, $P = (60,000)(0.70)(1448 - 900)/3413 = 6743$ kW = (6743)(3413) = 23 MM Btu/h (24.26 MM kJ).

3. Find the steam enthalpy after expansion in the cogeneration scheme

The steam is utilized for process heating after expansion to 200 lb/in^2 (abs) (1378 kPa) in the backpressure turbine. We must compute the enthalpy of the steam after expansion in order to find the energy available.

At 200 lb/in^2 (abs) (1378 kPa), using the same procedure as in step 1 above, $h_{2s} = 1257.7$ Btu/lb (2925.4 kJ/kg). Since we know the turbine efficiency we can use the equation, $(e_t)(h_1 - h_{2s}) = (h_1 - h_2)$; or $(0.70)(1448 - 1257.7) = (1448 - h_2)$; $h_2 = 1315$ Btu/lb (3058.7 kJ/kg), h_2 = actual enthalpy after expansion, Btu/lb (kJ/kg).

4. Determine the electrical output of the cogeneration plant

Since the efficiency of the turbine is already factored into the exhaust enthalpy of the cogeneration turbine, use the relation, $P = W_s(h_1 - h_2)/3413$, where the symbols are as defined earlier. Or, $P = 60,000(1448 - 1315)/3413 = 2338$ kW.

5. Compute the total energy output of the cogeneration plant

Assuming that the latent heat of the steam at 200 lb/in^2 (abs) (1378 kPa) is available for industrial process heating, the total energy output of the cogeneration scheme = electrical output + (steam flow, lb/h)(latent heat of the exhaust steam, Btu/lb). Since, from the steam tables, the latent heat of steam at 200 lb/in^2 (abs) (1378 kPa) = 834 Btu/lb (1939.9 kJ/kg), total energy output of the cogeneration cycle = (2338 kW)(3413) + (60,000)(834) = 58 MM Btu/h (61.2 MM kJ/h).

Since the total energy output of the conventional cycle was 23 MM Btu/h (24.3 MM kJ/h), the ratio of the cogeneration output vs. the conventional output = 58/23 = 2.52. Thus, about 2.5 times as much energy is derived from the cogeneration cycle as from the conventional cycle.

6. Find the comparative efficiencies of the two cycles

The boiler input = (weight of steam generated, lb/h)(enthalpy of superheated steam at boiler outlet, Btu/lb − enthalpy of feedwater entering the boiler, Btu/lb)/(boiler efficiency, expressed as a decimal). Or, boiler input = (60,000)(1448 − 200)/0.82 = 91.3 MM Btu/h (96.3 MM kJ/h). The efficiency of the conventional cycle is therefore (23/91.3)(100) = 25 percent. For the cogeneration cycle, the efficiency (58/91.3)(10) = 63.5 percent.

Related Calculations. This real-life example shows why cogeneration is such a popular alternative in today's world of power generation. In this study the cogeneration scheme is more than twice as efficient as the conventional cycle—63.5 percent vs. 25 percent. Higher efficiencies could be obtained if the boiler outlet steam pressure were higher than 1000 lb/in^2 (abs) (6890 kPa). However, the pressure used here is typical of today's industrial installations using cogeneration to save energy and conserve the environment.

This procedure is the work of V. Ganapathy, Heat Transfer Specialist, ABCO Industries, Inc.

BLEED-STEAM REGENERATIVE CYCLE LAYOUT AND T-S PLOT

Sketch the cycle layout, *T-S* diagram, and energy-flow chart for a regenerative bleed-steam turbine plant having three feedwater heaters and four feed pumps. Write the equations for the work-output available energy and the energy rejected to the condenser.

Calculation Procedure:

1. Sketch the cycle layout

Figure 33 shows a typical practical regenerative cycle having three feedwater heaters and four feedwater pumps. Number each point where steam enters and leaves the turbine and where steam enters or leaves the condenser and boiler. Also number the points in the feedwater cycle where feedwater enters and leaves a heater. Indicate the heater steam flow by m with a subscript corresponding to the heater number. Use W_p and a suitable subscript to indicate the pump work for each feed pump, except the last, which is labeled W_{pF}. The heat input to the steam generator is Q_a; the work output of the steam turbine is W_e; the heat rejected by the condenser is Q_r.

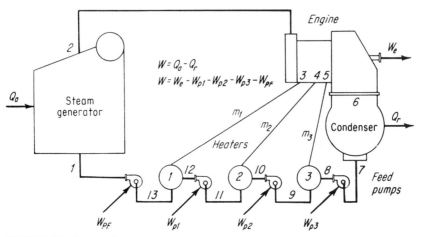

FIGURE 33 Regenerative steam cycle uses bleed steam.

2. Sketch the T-S diagram for the cycle

To analyze any steam cycle, trace the flow of 1 lb (0.5 kg) of steam through the system. Thus, in this cycle, 1 lb (0.5 kg) of steam leaves the steam generator at point 2 and flows to the turbine. From state 2 to 3, 1 lb (0.5 kg) of steam expands at constant entropy (assumed) through the turbine, producing work output $W_1 = H_2 - H_3$, represented by area 1-a-2-3 on the T-S diagram, Fig. 34a. At point 3, some steam is bled from the turbine to heat the feedwater passing through heater 1. The quantity of steam bled, m_1 lb is less than the 1 lb (0.5 kg) flowing between points 2 and 3. Plot stages 2 and 3 on the T-S diagram, Fig. 34a.

From point 3 to 4, the quantity of steam flowing through the turbine is $1 - m_1$ lb. This steam produces work output $W_2 = H_3 - H_4$. Plot point 4 on the T-S diagram. Then, area 1-3-4-12 represents the work output W_2, Fig. 34a.

At point 4, steam is bled to heater 2. The weight of this steam is m_2 lb. From point 4, the steam continues to flow through the turbine to point 5, Fig. 34a. The weight of the steam flowing between points 4 and 5 is $1 - m_1 - m_2$ lb. Plot point 5 on the T-S diagram, Fig. 34a. The work output between points 4 and 5, $W_3 = H_4 - H_5$, is represented by area 4-5-10-11 on the T-S diagram.

At point 5, steam is bled to heater 3. The weight of this bleed steam is m_3 lb. From point 5, steam continues to flow through the turbine to exhaust at point 6, Fig. 34a. The weight of steam flowing between points 5 and 6 is $1 - m_1 - m_2 - m_3$ lb. Plot point 6 on the T-S diagram, Fig. 34a.

The work output between points 5 and 6 is $W_4 = H_5 - H_6$, represented by area 5-6-7-9 on the T-S diagram, Fig. 34a. Area Q_r represents the heat given up by 1 lb (0.5 kg) of exhaust steam. Similarly, the area marked Q_a represents the heat absorbed by 1 lb (0.5 kg) of water in the steam generator.

3. Alter the T-S diagram to show actual cycle conditions

As plotted in Fig. 34a, Q_a is true for this cycle since 1 lb (0.5 kg) of water flows through the steam generator and the first section of the turbine. But Q_r is much too large; only $1 - m_1 - m_2 - m_3$ lb of steam flows through the condenser. Likewise, the net areas for W_2, W_3, and W_4, Fig. 34a, are all too large, because less than 1

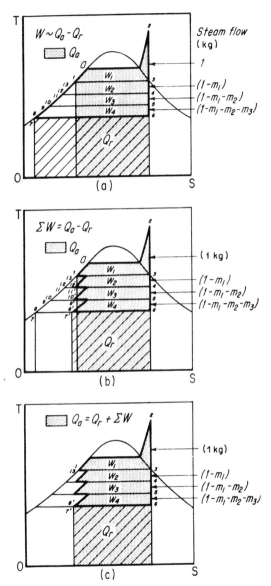

FIGURE 34 (a) T-S chart for the bleed-steam regenerative cycle in Fig. 10; (b) actual fluid flow in the cycle; (c) alternative plot of (b).

lb (0.5 kg) of steam flows through the respective turbine sections. The area for W_1, however, is true.

A true *proportionate-area* diagram can be plotted by applying the factors for actual flow, as in Fig. 34b. Here W_2, outlined by the heavy lines, equals the similarly labeled area in Fig. 34a, multiplied by $1 - m_1$. The states marked 11′ and 12′, Fig. 34b, are not true state points because of the ratioing factor applied to the area for W_2. The true state points 11 and 12 of the liquid before and after heater pump 3 stay as shown in Fig. 34a.

Apply $1 - m_1 - m_2$ to W_3 of Fig. 34a to obtain the proportionate area of Fig. 34b; to obtain W_4, multiply by $1 - m_1 - m_2 - m_3$. Multiplying by this factor also gives Q_r. Then all the areas in Fig. 34b will be in proper proportion for 1 lb (0.5 kg) of steam entering the turbine throttle but less in other parts of the cycle.

In Fig. 34b, the work can be measured by the difference of the area Q_a and the area Q_r. There is no simple net area left, because the areas coincide on only two sides. But area enclosed by the heavy lines *is* the total net work W for the cycle, equal to the sum of the work produced in the various sections of the turbine, Fig. 34b. Then Q_a is the alternate area $Q_r + W_1 + W_2 + W_3 + W_4$, as shaded in Fig. 34c.

The sawtooth approach of the liquid-heating line shows that as the number of heaters in the cycle increases, the heating line approaches a line of constant entropy. The best number of heaters for a given cycle depends on the steam state of the turbine inlet. Many medium-pressure and medium-temperature cycles use five to six heaters. High-pressure and high-temperature cycles use as many as nine heaters.

4. Draw the energy-flow chart

Choose a suitable scale for the heat content of 1 lb (0.5 kg) of steam leaving the steam generator. A typical scale is 0.375 in per 1000 Btu/lb (0.41 cm per 1000 kJ/kg). Plot the heat content of 1 lb (0.5 kg) of steam vertically on line 2-2, Fig. 35. Using the same scale, plot the heat content in energy streams m_1, m_2, m_3, W_e, W, W_p, W_{pF}, and so forth. In some cases, as W_{p1}, W_{p2}, and so forth, the energy stream may be so small that it is impossible to plot it to scale. In these instances, a single thin line is used. The completed diagram, Fig. 35, provides a useful concept of the distribution of the energy in the cycle.

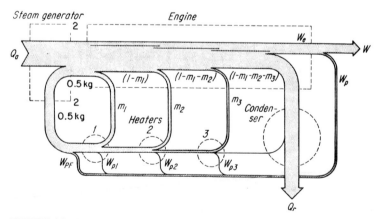

FIGURE 35   Energy-flow chart of cycle in Fig. 33.

Related Calculations. The procedure given here can be used for all regenerative cycles, provided that the equations are altered to allow for more, or fewer, heaters and pumps. The following calculation procedure shows the application of this method to an actual regenerative cycle.

BLEED REGENERATIVE STEAM CYCLE ANALYSIS

Analyze the bleed regenerative cycle shown in Fig. 36, determining the heat balance for each heater, plant thermal efficiency, turbine or engine thermal efficiency, plant heat rate, turbine or engine heat rate, and turbine or engine steam rate. Throttle steam pressure is 2000 lb/in^2 (abs) (13,790.0 kPa) at 1000°F (537.8°C); steam-generator efficiency = 0.88; station auxiliary steam consumption (excluding pump work) = 6 percent of the turbine or engine output; engine efficiency of each turbine or engine section = 0.80; turbine or engine cycle has three feedwater heaters and bleed-steam pressures as shown in Fig. 36; exhaust pressure to condenser is 1 inHg (3.4 kPa) absolute.

Calculation Procedure:

1. Determine the enthalpy of the steam at the inlet of each heater and the condenser
From a superheated-steam table, find the throttle enthalpy H_2 = 1474.5 Btu/lb (3429.7 kJ/kg) at 2000 lb/in^2 (abs) (13,790.0 kPa) and 1000°F (537.8°C). Next find the throttle entropy S_2 = 1.5603 Btu/(lb · °F) [6.5 kJ/(kg · °C)], at the same conditions in the superheated-steam table.

Plot the throttle steam conditions on a Mollier chart, Fig. 37. Assume that the steam expands from the throttle conditions at constant entropy = constant S to the inlet of the first feedwater heater, 1, Fig. 36. Plot this constant S expansion by drawing the straight vertical line 2-3 on the Mollier chart, Fig. 37, between the throttle condition and the heater inlet pressure of 750 lb/in^2 (abs) (5171.3 kPa).

Read on the Mollier chart H_3 = 1346.7 Btu/lb (3132.4 kJ/kg). Since the engine or turbine efficiency $e_e = H_2 - H_3/(H_2 - H_3)$ = 0.8 = 1474.5 − H_3/(1474.5 − 1346.7); H_3 = actual enthalpy of the steam at the inlet to heater 1 = 1474.5 − 0.8(1474.5 − 1346.7) = 1372.2 Btu/lb (3191.7 kJ/kg). Plot this enthalpy point on

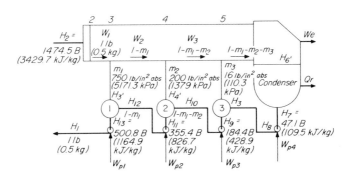

FIGURE 36 Bleed-regenerative steam cycle.

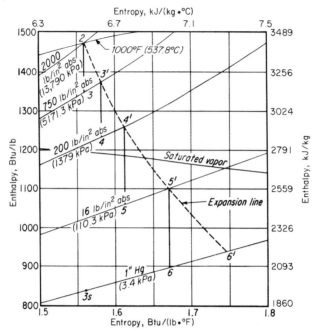

FIGURE 37 Mollier-chart plot of the cycle in Fig. 36.

the 750-lb/in² (abs) (5171.3-kPa) pressure line of the Mollier chart, Fig. 37. Read the entropy at the heater inlet from the Mollier chart as $S_{3'}$ = 1.5819 Btu/(lb·°F) [6.6 kJ/(kg·°C)] at 750 lb/in² (abs) (5171.3 kPa) and 1372.2 Btu/lb (3191.7 kJ/kg).

Assume constant-S expansion from $H_{3'}$ to H_4 at 200 lb/in² (abs) (1379.0 kPa), the inlet pressure for feedwater heater 2. Draw the vertical straight line 3'-4 on the Mollier chart, Fig. 37. By using a procedure similar to that for heater 1, $H_{4'}$ = $H_{3'} - e_e(H_{3'} - H_4)$ = 1372.2 − 0.8(1372.2 − 1230.0) = 1258.4 Btu/lb (2927.0 kJ/kg). This is the actual enthalpy of the steam at the inlet to heater 2. Plot this enthalpy on the 200-lb/in² (abs) (1379.0-kPa) pressure line of the Mollier chart, and find $S_{4'}$ = 1.613 Btu/(lb·°F) [6.8 kJ/(kg·°C)], Fig. 37.

Using the same procedure with constant-S expansion from $H_{4'}$, we find H_5 = 1059.5 Btu/lb (2464.4 kJ/kg) at 16 lb/in² (abs) (110.3 kPa), the inlet pressure to heater 3. Next find $H_{5'} = H_{4'} - e_e(H_{4'} - H_5)$ = 1258.4 − 0.8(1258.4 − 1059.5) = 1099.2 Btu/lb (2556.7 kJ/kg). From the Mollier chart find $S_{5'}$ = 1.671 Btu/(lb·°F) [7.0 kJ/(kg·°C)], Fig. 37.

Using the same procedure with constant-S expansion from $H_{5'}$ to H_6, find H_6 = 898.2 Btu/lb (2089.2 kJ/kg) at 1 inHg absolute (3.4 kPa), the condenser inlet pressure. Then $H_{6'} = H_{5'} - e_e(H_{5'} - H_6)$ = 1099.2 − 0.8(1099.2 − 898.2) = 938.4 Btu/lb (2182.7 kJ/kg), the actual enthalpy of the steam at the condenser inlet. Find, on the Mollier chart, the moisture in the turbine exhaust = 15.1 percent.

2. Determine the overall engine efficiency

Overall engine efficiency e_e is higher than the engine-section efficiency because there is partial available-energy recovery between sections. Constant-S expansion from the throttle to the 1-inHg absolute (3.4-kPa) exhaust gives H_{3s}, Fig. 37, as 838.3 Btu/lb (1949.4 kJ/kg), assuming that all the steam flows to the condenser. Then, overall $e_e = H_2 - H_{6'}/(H_2 - H_{3S}) = 1474.5 - 938.4/1474.5 - 838.3 = 0.8425$, or 84.25 percent, compared with 0.8 or 80 percent, for individual engine sections.

3. Compute the bleed-steam flow to each feedwater heater

For each heater, energy in = energy out. Also, the heated condensate leaving each heater is a saturated liquid at the heater bleed-steam pressure. To simplify this calculation, assume negligible steam pressure drop between the turbine bleed point and the heater inlet. This assumption is permissible when the distance between the heater and bleed point is small. Determine the pump work by using the chart accompanying the compressed-liquid table in Keenan and Keyes—*Thermodynamic Properties of Steam,* or the ASME—*Steam Tables.*

For heater 1, energy in = energy out, or $H_{3'}m_1 + H_{12}(1 - m_1) = H_{13}$, where m = bleed-steam flow to the feedwater heater, lb/lb of throttle steam flow. (The subscript refers to the heater under consideration.) Then $H_{3'}m_1 + (H_{11} + W_{p2})(1 - m_1) = H_{13}$, where W_{p2} = work done by pump 2, Fig. 36, in Btu/lb per pound of throttle flow. Then $1372.2m_1 + (355.4 + 1.7)(1 - m_1) = 500.8$; $m_1 = 0.1416$ lb/lb (0.064 kg/kg) throttle flow; $H_1 = H_{13} + W_{p1} = 500.8 + 4.7 = 505.5$ Btu/lb (1175.8 kJ/kg), where W_{p1} = work done by pump 1, Fig. 36. For each pump, find the work from the chart accompanying the compressed-liquid table in Keenan and Keyes—*Steam Tables* by entering the chart at the heater inlet pressure and projecting vertically at constant entropy to the heater outlet pressure, which equals the next heater inlet pressure. Read the enthalpy values at the respective pressures, and subtract the smaller from the larger to obtain the pump work during passage of the feedwater through the pump from the lower to the higher pressure. Thus, $W_{p2} = 1.7 - 0.0 = 1.7$ Btu/lb (4.0 kJ/kg), from enthalpy values for 200 lb/in² (abs) (1379.0 kPa) and 750 lb/in² (abs) (5171.3 kPa), the heater inlet and discharge pressures, respectively.

For heater 2, energy in = energy out, or $H_4m_2 + H_{10}(1 - m_1 - m_2) = H_{11}(1 - m_1)H_4m_2 + (H_9 + W_{p3})(1 - m_1 - m_2) = H_{11}(1 - m_1)1258.4m_2 + (184.4 + 0.5)(0.8584 - m_2) = 355.4(0.8584)m_2 = 0.1365$ lb/lb (0.0619 kg/kg) throttle flow.

For heater 3, energy in = energy out, or $H_5m_3 + H_8(1 - m_1 - m_2 - m_3) = H_9(1 - m_1 - m_2)H_5m_3 + (H_7 + W_{p4})(1 - m_1 - m_2 - m_3) = H_9(1 - m_1 - m_2)1099.2m_3 + (47.1 + 0.1)(0.7210 - m_3) = 184.4(0.7219)m_3 = 0.0942$ lb/lb (0.0427 kg/kg) throttle flow.

4. Compute the turbine work output

The work output per section W Btu is $W_1 = H_2 - H_{3'} = 1474.5 - 1372.1 = 102.3$ Btu (107.9 kJ), from the previously computed enthalpy values. Also $W_2 = (H_{3'} - H_{4'})(1 - m_1) = (1372.2 - 1258.4)(1 - 0.1416) = 97.7$ Btu (103.1 kJ); $W_3 = (H_{4'} - H_{5'})(1 - m_1 - m_2) = (1258.4 - 1099.2)(1 - 0.1416 - 0.1365) = 115.0$ Btu (121.3 kJ); $W_4 = (H_{5'} - H_{6'})(1 - m_1 - m_2 - m_3) = (1099.2 - 938.4)(1 - 0.1416 - 0.1365 - 0.0942) = 100.9$ Btu (106.5 kJ). The total work output of the turbine = $W_e = \Sigma W = 102.3 + 97.7 + 115.0 + 100.9 = 415.9$ Btu (438.8 kJ). The total $W_p = \Sigma W_p = W_{p1} + W_{p2} + W_{p3} + W_{p4} = 4.7 + 1.7 + 0.5 + 0.1 = 7.0$ Btu (7.4 kJ).

Since the station auxiliaries consume 6 percent of W_e, the auxiliary consumption = $0.6(415.9) = 25.0$ Btu (26.4 kJ). Then, net station work $w = 415.9 - 7.0 - 25.0 = 383.9$ Btu (405.0 kJ).

5. Check the turbine work output

The heat added to the cycle Q_a Btu/lb $= H_2 - H_1 = 1474.5 - 505.5 = 969.0$ Btu (1022.3 kJ). The heat rejected from the cycle Q_r Btu/lb $= (H_{6'} - H_7)(1 - m_1 - m_2 - m_3) = (938.4 - 47.1)(0.6277) = 559.5$ Btu (590.3 kJ). Then $W_e - W_p = Q_a - Q_r = 969.0 - 559.5 = 409.5$ Btu (432.0 kJ).

Compare this with $W_e - W_p$ computed earlier, or $415.9 - 7.0 = 408.9$ Btu (431.4 kJ), or a difference of $409.5 - 408.9 = 0.6$ Btu (0.63 kJ). This is an accurate check; the difference of 0.6 Btu (0.63 kJ) comes from errors in Mollier chart and calculator readings. Assume 408.9 Btu (431.4 kJ) is correct because it is the lower of the two values.

6. Compute the plant and turbine efficiencies

Plant energy input $= Q_a/e_b$, where $e_b =$ boiler efficiency. Then plant energy input $= 969.0/0.88 = 1101.0$ Btu (1161.6 kJ). Plant thermal efficiency $= W/(Q_a/e_b) = 383.9 = 1101.0 = 0.3486$. Turbine thermal efficiency $= W_e/Q_a = 415.9/969.0 = 0.4292$. Plant heat rate $= 3413/0.3486 = 9970$ Btu/kWh (10,329.0 kJ/kWh), where $3413 =$ Btu/kWh. Turbine heat rate $= 3413/0.4292 = 7950$ Btu/kWh (8387.7 kJ/kWh). Turbine throttle steam rate $=$ (turbine heat rate)/($H_2 - H_1$) $= 7950/(1474.5 - 505.5) = 8.21$ lb/kWh (3.7 kg/kWh).

Related Calculations. By using the procedures given, the following values can be computed for any actual steam cycle: engine or turbine efficiency e_e; steam enthalpy at the main-condenser inlet; bleed-steam flow to a feedwater heater; turbine or engine work output per section; total turbine or engine work output; station auxiliary power consumption; net station work output; plant energy input; plant thermal efficiency; turbine or engine thermal efficiency; plant heat rate; turbine or engine heat rate; turbine throttle heat rate. To compute any of these values, use the equations given and insert the applicable variables.

REHEAT-STEAM CYCLE PERFORMANCE

A reheat-steam cycle has a 2000 lb/in² (abs) (13,790-kPa) throttle pressure at the turbine inlet and a 400-lb/in² (abs) (2758-kPa) reheat pressure. The throttle and reheat temperature of the steam is 1000°F (537.8°C); condenser pressure is 1 inHg absolute (3.4 kPa); engine efficiency of the high-pressure and low-pressure turbines is 80 percent. Find the cycle thermal efficiency.

Calculation Procedure:

1. Sketch the cycle layout and cycle T-S diagram

Figures 38 and 39 show the cycle layout and T-S diagram with each important point numbered. Use a cycle layout and T-S diagram for every calculation of this type because it reduces the possibility of errors.

2. Determine the throttle-steam properties from the steam tables

Use the superheated steam tables, entering at 2000 lb/in² (abs) (13,790 kPa) and 1000°F (537.8°C) to find throttle-steam properties. Applying the symbols of the T-S diagram in Fig. 39, we get $H_2 = 1474.5$ Btu/lb (3429.7 kJ/kg); $S_2 = 1.5603$ Btu/(lb·°F) [6.5 kJ/(kg·°C)].

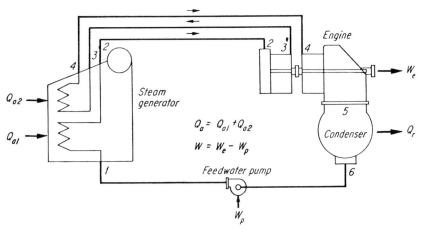

FIGURE 38 Typical steam reheat cycle.

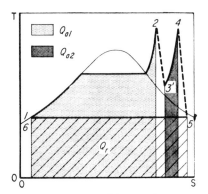

FIGURE 39 Irreversible expansion in reheat cycle.

3. Find the reheat-steam enthalpy

Assume a constant-entropy expansion of the steam from 2000 to 400 lb/in² (13,790 to 2758 kPa). Trace this expansion on a Mollier (H-S) chart, where a constant-entropy process is a vertical line between the initial [2000 lb/in² (abs) or 13,790 kPa] and reheat [400 lb/in² (abs) or 2758 kPa] pressures. Read on the Mollier chart $H_3 = 1276.8$ Btu/lb (2969.8 kJ/kg) at 400 lb/in² (abs) (2758 kPa).

4. Compute the actual reheat properties

The ideal enthalpy drop, throttle to reheat $= H_2 - H_3 = 1474.5 - 1276.8 = 197.7$ Btu/lb (459.9 kJ/kg). The actual enthalpy drop = (ideal drop)(turbine efficiency) = $H_2 - H_{3'} = 197.5(0.8) = 158.2$ Btu/lb (368.0 kJ/kg) $= W_{e1} =$ work output in the high-pressure section of the turbine.

Once W_{e1} is known, $H_{3'}$ can be computed from $H_{3'} = H_2 - W_{e1} = 1474.5 - 158.2 = 1316.3$ Btu/lb (3061.7 kJ/kg).

The steam now returns to the boiler and leaves at condition 4, where $P_4 = 400$ lb/in² (abs) (2758 kPa); $T_4 = 1000°F$ (537.8°C); $S_4 = 1.7623$ Btu/(lb·°F) [7.4 kJ/(kg·°C)]; $H_4 = 1522.4$ Btu/lb (3541.1 kJ/kg) from the superheated-steam table.

5. Compute the exhaust-steam properties

Use the Mollier chart and an assumed constant-entropy expansion to 1 inHg (3.4 kPa) absolute to determine the ideal exhaust enthalpy, or $H_5 = 947.4$ Btu/lb (2203.7 kJ/kg). The ideal work of the low-pressure section of the turbine is then $H_4 - H_5 = 1522.4 - 947.4 = 575.0$ Btu/lb (1338 kJ/kg). The actual work output of the low-pressure section of the turbine is $W_{e2} = H_4 - H_{5'} = 575.0(0.8) = 460.8$ Btu/lb (1071.1 kJ/kg).

Once W_{e2} is known, $H_{5'}$ can be computed from $H_{5'} = H_4 - W_{e2} = 1522.4 - 460.0 = 1062.4$ Btu/lb (2471.1 kJ/kg).

The enthalpy of the saturated liquid at the condenser pressure is found in the saturation-pressure steam table at 1 inHg absolute (3.4 kPa) = $H_6 = 47.1$ Btu/lb (109.5 kJ/kg).

The pump work W_p from the compressed-liquid table diagram in the steam tables is $W_p = 5.5$ Btu/lb (12.8 kJ/kg). Then the enthalpy of the water entering the boiler $H_1 = H_6 + W_p = 47.1 + 5.5 = 52.6$ Btu/lb (122.3 kJ/kg).

6. Compute the cycle thermal efficiency

For any reheat cycle,

$$e = \text{cycle thermal efficiency}$$

$$= \frac{(H_2 - H_{3'}) + (H_4 - H_{5'}) - W_p}{(H_2 - H_1) + (H_4 - H_{3'})}$$

$$= \frac{(1474.5 - 1316.3) + (1522.4 - 1062.4) - 5.5}{(1474.5 - 52.6) + (1522.4 - 1316.3)}$$

$$= 0.3766, \text{ or } 37.66 \text{ percent}$$

Figure 40 is an energy-flow diagram for the reheat cycle analyzed here. This diagram shows that the fuel burned in the steam generator to produce energy flow Q_{a1} is the largest part of the total energy input. The cold-reheat line carries the major share of energy leaving the high-pressure turbine.

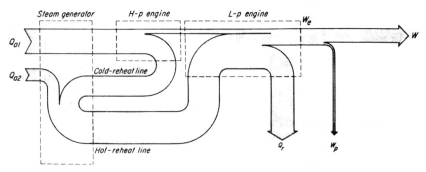

FIGURE 40 Energy-flow diagram for reheat cycle in Fig. 38.

Related Calculations.   Reheat-regenerative cycles are used in some large power plants. Figure 41 shows a typical layout for such a cycle having three stages of feedwater heating and one stage of reheating. The heat balance for this cycle is computed as shown above, with the bleed-flow terms m computed by setting up an energy balance around each heater, as in earlier calculation procedures.

By using a *T-S* diagram, Fig. 42, the cycle thermal efficiency is

$$e = \frac{W}{Q_a} = \frac{Q_a - Q_r}{Q_a} = 1 - \frac{Q_r}{Q_{a1} + Q_{a2}}$$

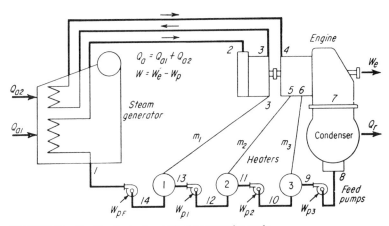

FIGURE 41   Combined reheat and bleed-regenerative cycle.

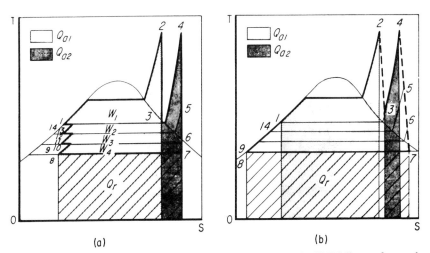

FIGURE 42   (a) *T-S* diagram for ideal reheat-regenerative-bleed cycle; (b) *T-S* diagram for actual cycle.

Based on 1 lb (0.5 kg) of working fluid entering the steam generator and turbine throttle,

$$Q_r = (1 - m_1 - m_2 - m_3)(H_7 - H_8)$$

$$Q_{a1} = (H_2 - H_1)$$

$$Q_{a2} = (1 - m_1)(H_4 - H_3)$$

Figure 43 shows the energy-flow chart for this cycle.

Some high-pressure plants use two stages of reheating, Fig. 44, to raise the cycle efficiency. With two stages of reheating, the maximum number generally used, and values from Fig. 44.

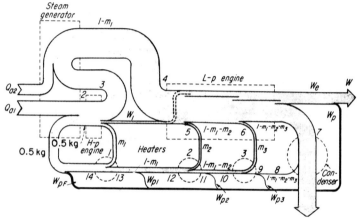

FIGURE 43   Energy flow of cycle in Fig. 41.

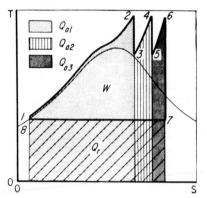

FIGURE 44   *T-S* diagram for multiple re-heat stages.

$$e = \frac{(H_2 - H_3) + (H_4 - H_5) + (H_6 - H_7) - W_p}{(H_2 - H_1) + (H_4 - H_3) + (H_6 - H_5)}$$

MECHANICAL-DRIVE STEAM-TURBINE POWER-OUTPUT ANALYSIS

Show the effect of turbine engine efficiency on the condition lines of a turbine having engine efficiencies of 100 (isentropic expansion), 75, 50, 25, and 0 percent. How much of the available energy is converted to useful work for each engine efficiency? Sketch the effect of different steam inlet pressures on the condition line of a single-nozzle turbine at various loads. What is the available energy, Btu/lb of steam, in a noncondensing steam turbine having an inlet pressure of 1000 lb/in² (abs) (6895 kPa) and an exhaust pressure of 100 lb/in² (gage) (689.5 kPa)? How much work will this turbine perform if the steam flow rate to it is 1000 lb/s (453.6 kg/s) and the engine efficiency is 40 percent?

Calculation Procedure:

1. Sketch the condition lines on the Mollier chart
Draw on the Mollier chart for steam initial- and exhaust-pressure lines, Fig. 45, and the initial-temperature line. For an isentropic expansion, the entropy is constant during the expansion, and the engine efficiency = 100 percent. The expansion or condition line is a vertical trace from h_1 on the initial-pressure line to h_2, on the exhaust-pressure line. Draw this line as shown in Fig. 45.

For zero percent engine efficiency, the other extreme in the efficiency range, $h_1 = h_2$ and the condition line is a horizontal line. Draw this line as shown in Fig. 45.

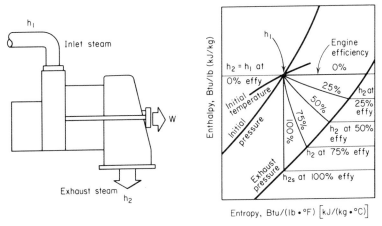

FIGURE 45 Mollier chart of turbine condition lines.

Between 0 and 100 percent efficiency, the condition lines become more nearly vertical as the engine efficiency approaches 100 percent, or an isentropic expansion. Draw the condition lines for 25, 50, and 75 percent efficiency, as shown in Fig. 45.

For the isentropic expansion, the available energy $= h_1 - h_{2s}$, Btu/lb of steam. This is the energy that an ideal turbine would make available.

For actual turbines, the enthalpy at the exhaust pressure $h_2 = h_1 -$ (available energy)(engine efficiency)/100, where available energy $= h_1 - h_{2s}$ for an ideal turbine working between the same initial and exhaust pressures. Thus, the available energy converted to useful work for any engine efficiency = (ideal available energy, Btu/lb)(engine efficiency, percent)/100. Using this relation, the available energy at each of the given engine efficiencies is found by substituting the ideal available energy and the actual engine efficiency.

2. Sketch the condition lines for various throttle pressures
Draw the throttle- and exhaust-pressure lines on the Mollier chart, Fig. 46. Since the inlet control valve throttles the steam flow as the load on the turbine decreases, the pressure of the steam entering the turbine nozzle is lower at reduced loads. Show this throttling effect by indicating the lower inlet pressure lins, Fig. 46, for the reduced loads. Note that the lowest inlet pressure occurs at the minimum plotted load—25 percent of full load—and the maximum inlet pressure at 125 percent of full load. As the turbine inlet steam pressure decreases, so does the available energy, because the exhaust enthalpy rises with decreasing load.

3. Compute the turbine available energy and power output
Use a noncondensing-turbine performance chart, Fig. 47, to determine the available energy. Enter the bottom of the chart at 1000 lb/in² (abs) (6895 kPa) and project vertically upward until the 100-lb/in² (gage) (689.5-kPa) exhaust-pressure curve is intersected. At the left, read the available energy as 205 Btu/lb (476.8 kJ/kg) of steam.

With the available energy, flow rate, and engine efficiency known, the work output = (available energy, Btu/lb)(flow rate, lb/s)(engine efficiency/100)/[550

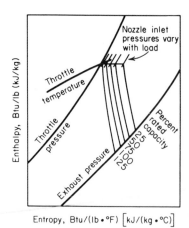

FIGURE 46 Turbine condition line shifts as the inlet steam pressure varies.

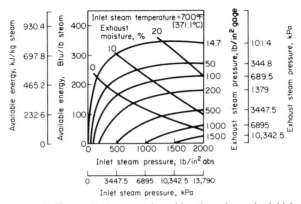

FIGURE 47 Available energy in turbine depends on the initial steam state and the exhaust pressure.

ft · lb/(s · hp)]. [*Note:* 550 ft · lb/(s · hp) = 1 N · m/(W · s).] For this turbine, work output = (205 Btu/lb)(1000 lb/s)(40/100)/550 = 149 hp (111.1 kW).

Related Calculations. Use the steps given here to analyze single-stage non-condensing mechanical-drive turbines for stationary, portable, or marine applications. Performance curves such as Fig. 47 are available from turbine manufacturers. Single-stage noncondensing turbines are for feed-pump, draft-fan, and auxiliary-generator drive.

CONDENSING STEAM-TURBINE POWER-OUTPUT ANALYSIS

What is the available energy in steam supplied to a 5000-kW turbine if the inlet steam conditions are 1000 lb/in² (abs) (6895 kPa) and 800°F (426.7°C) and the turbine exhausts at 1 inHg absolute (3.4 kPa)? Determine the theoretical and actual heat rate of this turbine if its engine efficiency is 74 percent. What are the full-load output and steam rate of the turbine?

Calculated Procedure:

1. *Determine the available energy in the steam*
Enter Fig. 48 at the bottom at 1000-lb/in² (abs) (6895.0-kPa) inlet pressure, and project vertically upward to the 800°F (426.7°C) 1-in (3.4-kPa) exhaust-pressure curve. At the left, read the available energy as 545 Btu/lb (1267.7 kJ/kg) of steam.

2. *Determine the heat rate of the turbine*
Enter Fig. 49 at an initial steam temperature of 800°F (426.7°C), and project vertically upward to the 1000-lb/in² (abs) (6895.0-kPa) 1-in (3.4-kPa) curve. At the left, read the theoretical heat rate as 8400 Btu/kWh (8862.5 kJ/kWh).

When the theoretical heat rate is known, the actual heat rate is found from: actual heat rate HR, Btu/kWh = (theoretical heat rate, Btu/kWh)/(engine efficiency). Or, actual HR = 8400/0.74 = 11,350 Btu/kWh (11,974.9 kJ/kWh).

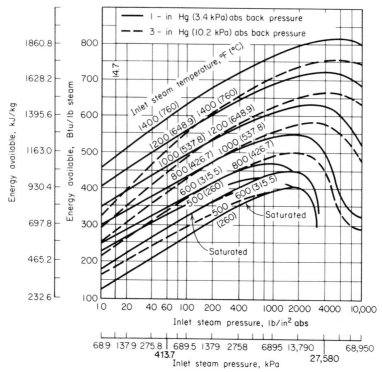

FIGURE 48 Available energy for typical condensing turbines.

3. Compute the full-load and steam rate

The energy converted to work, Btu/lb of steam = (available energy, Btu/lb of steam)(engine efficiency) = (545)(0.74) = 403 Btu/lb of steam (937.4 kJ/kg).

For any prime mover driving a generator, the full-load output, Btu = (generator kW rating)(3413 Btu/kWh) = (5000)(3413) = 17,060,000 Btu/h (4999.8 kJ/s).

The steam flow = (full-load output, Btu/h)/(work output, Btu/lb) = 17,060,000/403 = 42,300 lb/h (19,035 kg/h) of steam. Then the full-load steam rate of the turbine, lb/kWh = (steam flow, lb/h)/(kW output at full load) = 42,300/5000 = 8.46 lb/kWh (3.8 kg/kWh).

Related Calculations. Use this general procedure to determine the available energy, theoretical and actual heat rates, and full-load output and steam rate for any stationary, marine, or portable condensing steam turbine operating within the ranges of Figs. 48 and 49. If the actual performance curves are available, use them instead of Figs. 48 and 49. The curves given here are suitable for all preliminary estimates for condensing turbines operating with exhaust pressures of 1 or 3 inHg absolute (3.4 or 10.2 kPa). Many modern turbines operate under these conditions.

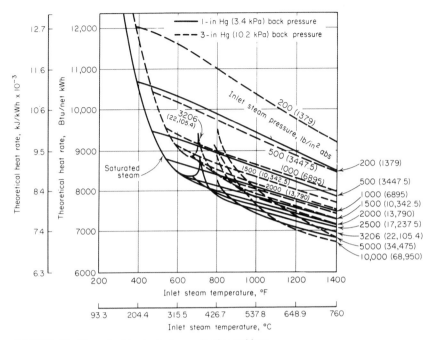

FIGURE 49 Theoretical heat rate for condensing turbines.

STEAM-TURBINE REGENERATIVE-CYCLE PERFORMANCE

When throttle steam is at 1000 lb/in² (abs) (6895 kPa) and 800°F (426.7°C) and the exhaust pressure is 1 inHg (3.4 kPa) absolute, a 5000-kW condensing turbine has an actual heat rate of 11,350 Btu/kWh (11,974.9 kJ/kWh). Three feedwater heaters are added to the cycle, Fig. 50, to heat the feedwater to 70 percent of the maximum possible enthalpy rise. What is the actual heat rate of the turbine? If 10 heaters instead of 3 were used and the water enthalpy were raised to 90 percent of the maximum possible rise in these 10 heaters, would the reduction in the actual heat rate be appreciable?

Calculation Procedure:

1. Determine the actual enthalpy rise of the feedwater
Enter Fig. 51 at the throttle pressure of 1000 lb/in² (abs) (6895 kPa), and project vertically upward to the 1-inHg (3.4-kPa) absolute backpressure curve. At the left, read the maximum possible feedwater enthalpy rise as 495 Btu/lb (1151.4 kJ/kg). Since the actual rise is limited to 70 percent of the maximum possible rise by the conditions of the design, the actual enthalpy rise = (495)(0.70) = 346.5 Btu/lb (805.9 kJ/kg).

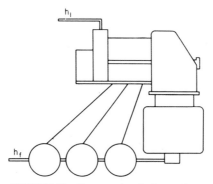

FIGURE 50   Regenerative feedwater heating.

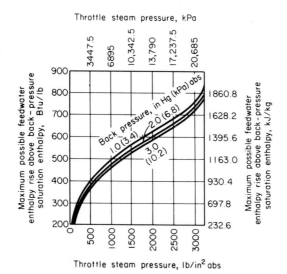

FIGURE 51   Feedwater enthalpy rise.

2. Determine the heat-rate and heater-number correction factors

Find the theoretical reduction in straight-condensing (no regenerative heaters) heat rates from Fig. 52. Enter the bottom of Fig. 52 at the inlet steam temperature, 800°F (426.7°C), and project vertically upward to the 1000-lb/in² (abs) (6895-kPa) 1-inHg (3.4-kPa) back-pressure curve. At the left, read the reduction in straight-condensing heat rate as 14.8 percent.

Next, enter Fig. 52 at the bottom of 70 percent of maximum possible rise in feedwater enthalpy, and project vertically to the three-heater curve. At the left, read the reduction in straight-condensing heat rate for the number of heaters and actual enthalpy rise as 0.71.

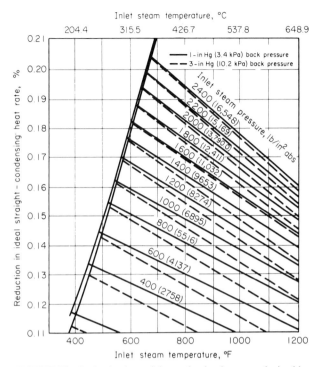

FIGURE 52 Reduction in straight-condensing heat rate obtained by regenerative heating.

3. Apply the heat-rate and heater-number correction factors
Full-load regenerative-cycle heat rate, Btu/kWh = (straight-condensing heat rate, Btu/kWh) [1 − (heat-rate correction factor)(heater-number correction factor)] = (13,350)[1 − (0.148)(0.71)] = 10,160 Btu/kWh (10,719.4 kJ/kWh).

4. Find and apply the correction factors for the larger number of heaters
Enter Fig. 53 at 90 percent of the maximum possible enthalpy rise, and project vertically to the 10-heater curve. At the left, read the heat-rate reduction for the number of heaters and actual enthalpy rise as 0.89.

Using the heat-rate correction factor from step 2 and 0.89, found above, we see that the full-load 10-heater regenerative-cycle heat rate = (11,350)[1 − (0.148)(0.89)] = 9850 Btu/kWh (10,392.3 kJ/kWh), by using the same procedure as in step 3. Thus, adding 10 − 3 = 7 heaters reduces the heat rate by 10,160 − 9850 = 310 Btu/kWh (327.1 kJ/kWh). This is a reduction of 3.05 percent.

To determine whether this reduction in heat rate is appreciable, the carrying charges on the extra heaters, piping, and pumps must be compared with the reduction in annual fuel costs resulting from the lower heat rate. If the fuel saving is greater than the carrying charges, the larger number of heaters can usually be justified. In this case, tripling the number of heaters would probably increase the

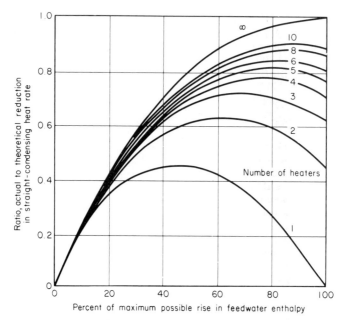

FIGURE 53 Maximum possible rise in feedwater enthalpy varies with the number of heaters used.

carrying charges to a level exceeding the fuel savings. Therefore, the reduction in heat rate is probably not appreciable.

   Related Calculations.   Use the procedure given here to compute the actual heat rate of steam-turbine regenerative cycles for stationary, marine, and portable installations. Where necessary, use the steps of the previous procedure to compute the actual heat rate of a straight-condensing cycle before applying the present procedure. The performance curves given here are suitable for first approximations in situations where actual performance curves are unavailable.

REHEAT-REGENERATIVE STEAM-TURBINE HEAT RATES

What are the net and gross heat rates of a 300-kW reheat turbine having an initial steam pressure of 3500 lb/in^2 (gage) (24,132.5 kPa) with initial and reheat steam temperatures of 1000°F (537.8°C) with 1.5 inHg (5.1 kPa) absolute back pressure and six stages of regenerative feedwater heating? Compare this heat rate with that of 3500 lb/in^2 (gage) (24,132.5 kPa) 600-mW cross-compound four-flow turbine with 3600/1800 r/min shafts at a 300-mW load.

Calculation Procedure:

1. Determine the reheat-regenerative heat rate

Enter Fig. 54 at 3500-lb/in² (gage) (24,132.5-kPa) initial steam pressure, and project vertically to the 300-mW capacity net-heat-rate curve. At the left, read the net heat rate as 7680 Btu/kWh (8102.6 kJ/kWh). On the same vertical line, read the gross heat rate as 7350 Btu/kWh (7754.7 kJ/kWh). The gross heat rate is computed by using the generator-terminal output; the net heat rate is computed after the feedwater-pump energy input is deducted from the generator output.

2. Determine the cross-compound turbine heat rate

Enter Fig. 55 at 350 mW at the bottom, and project vertically upward to 1.5-inHg (5.1-kPa) exhaust pressure midway between the 1- and 2-inHg (3.4- and 6.8-kPa) curves. At the left, read the net heat rate as 7880 Btu/kWh (8313.8 kJ/kWh). Thus, the reheat-regenerative unit has a lower net heat rate. Even at full rated load of the cross-compound turbine, its heat rate is higher than the reheat unit.

Related Calculations.   Use this general procedure for comparing stationary and marine high-pressure steam turbines. The curves given here are typical of those supplied by turbine manufacturers for their turbines.

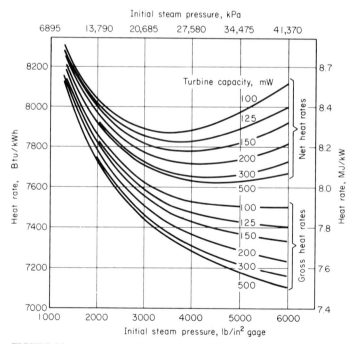

FIGURE 54 Full-load heat rates for steam turbines with six feedwater heaters, 1000°F/1000°F (538°C/538°C) steam, 1.5-in (38.1-mm) Hg (abs) exhaust pressure.

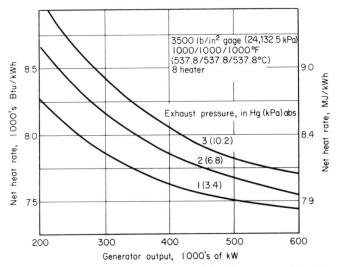

FIGURE 55   Heat rate of a cross-compound four-flow steam turbine with 3600/1800-r/min shafts.

STEAM TURBINE–GAS TURBINE CYCLE ANALYSIS

Sketch the cycle layout, *T-S* diagram, and energy-flow chart for a combined steam turbine–gas turbine cycle having one stage of regenerative feedwater heating and one stage of economizer feedwater heating. Compute the thermal efficiency and heat rate of the combined cycle.

Calculation Procedure:

1. Sketch the cycle layout
Figure 56 shows the cycle. Since the gas-turbine exhaust-gas temperature is usually higher than the bleed-steam temperature, the economizer is placed after the regenerative feedwater heater. The feedwater will be progressively heated to a higher temperature during passage through the regenerative heater and the gas-turbine economizer. The cycle shown here is only one of many possible combinations of a steam plant and a gas turbine.

2. Sketch the T-S diagram
Figure 57 shows the *T-S* diagram for the combined gas turbine–steam turbine cycle. There is irreversible heat transfer Q_T from the gas-turbine exhaust to the feedwater in the economizer, which helps reduce the required energy input Q_{a2}.

3. Sketch the energy-flow chart
Choose a suitable scale for the energy input, and proportion the energy flow to each of the other portions of the cycle. Use a single line when the flow is too small to plot to scale. Figure 58 shows the energy-flow chart.

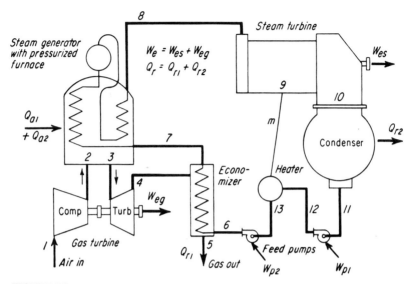

FIGURE 56 Combined gas turbine–steam turbine cycle.

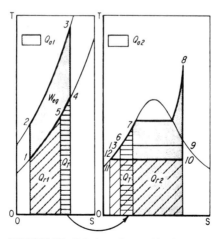

FIGURE 57 *T-S* charts for combined gas turbine–steam turbine cycle have irreversible heat transfer Q from gas-turbine exhaust to the feedwater.

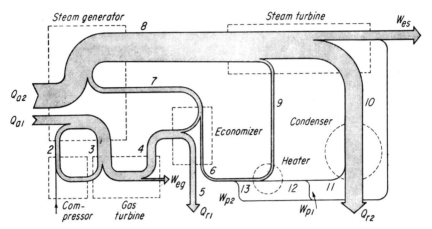

FIGURE 58   Energy-flow chart of the gas turbine–steam turbine cycle in Fig. 56.

4. Determine the thermal efficiency of the cycle

Since $e = W/Q_a$, $e = Q_a - Q_r/Q_a = 1 - [Q_{r1} + Q_{r2}/(Q_{a1} + Q_{a2})]$, given the notation in Figs. 56, 57, and 58.

The relative weight of the gas w_g to 1 lb (0.5 kg) of water must be computed by taking an energy balance about the economizer. Or, $H_7 - H_6 = w_g(H_4 - H_5)$. Using the actual values for the enthalpies, solve this equation for w_g.

With w_g known, the other factors in the efficiency computation are

$$Q_{r1} = w_g(H_5 - H_1)$$

$$Q_{r2} = (1 - m)(H_{10} - H_{11})$$

$$Q_{a1} = w_g(H_3 - H_2)$$

$$Q_{a2} = H_8 - H_7$$

The bleed-steam flow m is calculated from an energy balance about the feedwater heater. Note that the units for the above equations can be any of those normally used in steam- and gas-turbine analyses.

Calculation Procedure:

1. Find the amount of oxygen required for complete combustion of the fuel

Eight atoms of carbon in C_8 combine with 8 molecules of oxygen, O_2, and produce 8 molecules of carbon dioxide, $8CO_2$. similarly, 9 molecules of hydrogen, H_2, in H_{18} combine with 9 atoms of oxygen, O, or 4.5 molecules of oxygen, to form 9 molecules of water, $9H_2O$. Thus, 100 percent, or the stoichiometric, air quantity required for complete combustion of a mole of fuel, C_8H_{18}, is proportional to $8 + 12.5$ moles of oxygen, O_2.

2. Establish the chemical equation for complete combustion with 100 percent air

With 100 percent air: $C_8H_{18} + 12.5\ O_2 + (3.784 \times 12.5)N_2 \rightarrow 8CO_2 + 9H_2O + 47.3N_2$, where 3.784 is a derived volumetric ratio of atmospheric nitrogen, (N_2), to oxygen, O_2, in dry air. The (N_2) includes small amounts of inert and inactive gases. See Related Calculations of this procedure.

3. Establish the chemical equation for complete combustion with 400 percent of the stoichiometric air quantity, or 300 percent excess air
With 400 percent air: $C_8H_{18} + 50\ O_2 + (4 \times 47.3)N_2 \rightarrow 8CO_2 + 9H_2O + 189.2N_2 + (3 \times 12.5)O_2$.

4. Compute the molecular weights of the components in the combustion process
Molecular weight of $C_8H_{18} = [(12 \times 8) + (1 \times 18)] = 114$; $O_2 = 16 \times 2 = 32$; $N_2 = 14 \times 2 = 28$; $CO_2 = [(12 \times 1) + (16 \times 2)] = 44$; $H_2O = [(1 \times 2) + (16 \times 1)] = 18$.

5. Compute the relative weights of the reactants and products of the combustion process
Relative weight = moles × molecular weight. Coefficients of the chemical equation in step 3 represent the number of moles of each component. Hence, for the reactants, the relative weights are: $C_8H_{18} = 1 \times 114 = 114$; $O_2 = 50 \times 32 = 1600$; $N_2 = 189.2 \times 28 = 5298$. Total relative weight of the reactants is 7012. For the products, the relative weights are: $CO_2 = 8 \times 44 = 352$; $H_2O = 9 \times 18 = 162$; $N_2 = 189.2 \times 28 = 5298$; $O_2 = 37.5 \times 32 = 1200$. Total relative weight of the products is 7012, also.

6. Compute the enthalpy of the products of the combustion process
Enthalpy of the products of combustion, $h_p = m_p(h_{1600} - h_{77})$, where m_p = number of moles of the products; h_{1600} = enthalpy of the products at 1600°F (871°C); h_{77} = enthalpy of the products at 77°F (25°C). Thus, $h_p = (8 + 9 + 189.2 + 37.5)(15,400 - 3750) = 2,839,100$ Btu [6,259,100 Btu (SI)].

7. Compute the air supply temperature at the combustion chamber inlet
Since the combustion process is adiabatic, the enthalpy of the reactants $h_r = h_p$, where h_r = (relative weight of the fuel × its heating value) + [relative weight of the air × its specific heat × (air supply temperature − air source temperature)]. Therefore, $h_r = (114 \times 19,100) + [(1600 + 5298) \times 0.24 \times (t_a - 77)] + 2,839,100$ Btu [6,259,100 Btu (SI)]. Solving for the air supply temperature, $t_a = [(2,839,100 - 2,177,400/1655.5] + 77 = 477$°F (247°C).

Related Calculations. This procedure, appropriately modified, may be used to deal with similar questions involving such things as other fuels, different amounts of excess air, and variations in the condition(s) being sought under certain given circumstances.

The coefficient, (?) = 3.784 in step 2, is used to indicate that for each unit of volume of oxygen, O_2, 12.5 in this case, there will be 3.784 units of nitrogen, N_2. This equates to an approximate composition of air as 20.9 percent oxygen and 79.1 percent "atmospheric nitrogen," (N_2). In turn, this creates a paradox, because page 200 of *Principles of Engineering Thermodynamics,* by Kiefer, et al., John Wiley & Sons, Inc., states air to be 20.99 percent oxygen and 79.01 percent atmospheric nitrogen, where the ratio $(N_2)/O_2 = (?) = 79.01/20.99 = 3.764$.

Also, page 35 of *Applied Energy Conversion,* by Skrotski and Vopat, McGraw-Hill, Inc., indicates an assumed air analysis of 79 percent nitrogen and 21 percent oxygen, where (?) = 3.762. On that basis, a formula is presented for the amount of dry air chemically necessary for complete combustion of a fuel consisting of atoms of carbon, hydrogen, and sulfur, or C, H, and S, respectively. That formula is: $W_a = 11.5C + 34.5[H - (0/8)] + 4.32S$, lb air/lb fuel (kg air/kg fuel).

The following derivation for the value of (?) should clear up the paradox and show that either 3.784 or 3.78 is a sound assumption which seems to be wrong, but in reality is not. In the above equation for W_a, the carbon hydrogen, or sulfur coefficient, $C_x = (MO_2/DO_2)M_x$, where MO_2 is the molecular weight of oxygen, O_2; DO_2 is the decimal fraction for the percent, by weight, of oxygen, O_2, in dry air containing "atmospheric nitrogen," (N_2), and small amounts of inert and inactive gases: M_x is the formula weight of the combustible element in the fuel, as indicated by its relative amount as a reactant in the combustion equation. The alternate evaluation of C_x is obtained from stoichiometric chemical equations for burning the combustible elements of the fuel, i.e., $C + O_2 + (?)N_2 \rightarrow CO_2 + (?)N_2$; $2H_2 + O_2 + (?)N_2 \rightarrow 2H_2O + (?)N_2$; $S + O_2 + (?)N_2 \rightarrow SO_2 + (?)N_2$. Evidently, $C_x = [MO_2 + (?) \times MN_2)]/M_x$, where MN_2 is the molecular weight of nitrogen, N_2, and the other items are as before.

Equating the two expressions, $C_x = [MO_2 + (? \times MN_2)]/M_x = (MO_2/DO_2)M_x$, reveals that the M_x terms cancel out, indicating that the formula weight(s) of combustible components are irrelevant in solving for (?). Then, (?) $= (1 - DO_2)[M O_2/(MN_2 \times DO_2)]$. From the above-mentioned book by Kiefer, et al., $DO_2 = 0.23188$. From *Marks' Standard Handbook for Mechanical Engineers*, McGraw-Hill, Inc., $MO_2 = 31.9988$ and $MN_2 = 28.0134$. Thus, (?) $= (1 - 0.23188)[31.9988/(28.0134 \times 0.23188)] = 3.7838$. This demonstrates that the use of (?) $= 3.784$, or 3.78, is justified for combustion equations.

By using either of the two evaluation equations for C_x, and with accurate values for M_x, i.e., $M_C = 12.0111$; $M_H = 2 \times 2 \times 1.00797 = 4.0319$; $M_S = 32.064$, from *Marks' M.E. Handbook*, the more precise values for C_C, C_H, and C_S are found out to be 11.489, 34.227, and 4.304, respectively. However, the actual C_x values, 11.5, 34.5, and 4.32, used in the formula for W_a are both brief for simplicity and rounded up to be on the safe side.

GAS TURBINE COMBUSTION CHAMBER INLET AIR TEMPERATURE

A gas turbine combustion chamber is well insulated so that heat losses to the atmosphere are negligible. Octane, C_8H_{18}, is to be used as the fuel and 400 percent of the stoichiometric air quantity is to be supplied. The air first passes through a regenerative heater and the air supply temperature at the combustion chamber inlet is to be set so that the exit temperature of the combustion gases is 1600°F (871°C). (See Fig. 59.) Fuel supply temperature is 77°F (25°C) and its heating value is to be taken as 19,000 Btu/lb$_m$ (44,190 kJ/kg) relative to a base of 77°F (15°C).

The air may be treated in calculations as a perfect gas with a constant-pressure specific heat of 0.24 Btu/(lb·°F) [1.005 kJ/(kg·°C)]. The products of combustion have an enthalpy of 15,400 Btu/lb·mol) [33,950 Btu/(kg·mol)] at 1600°F (871°C) and an enthalpy of 3750 Btu/(lb·mol) [8270 Btu/(kg·mol)] at 77°F (24°C). Determine, assuming complete combustion and neglecting dissociation, the required air temperature at the inlet of the combustion chamber. (See Note on p. 1.110.)

REGENERATIVE-CYCLE GAS-TURBINE ANALYSIS

What is the cycle air rate, lb/kWh, for a regenerative gas turbine having a pressure ratio of 5, an air inlet temperature of 60°F (15.6°C), a compressor discharge tem-

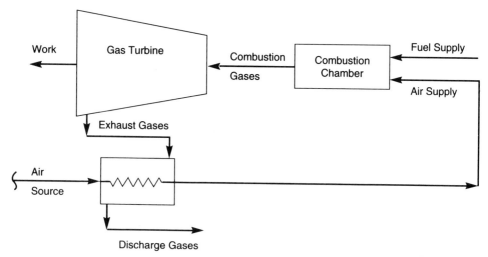

FIGURE 59   Gas turbine flow diagram.

perature of 1500°F (815.6°C), and performance in accordance with Fig. 60? Determine the cycle thermal efficiency and work ratio. What is the power output of a regenerative gas turbine if the work input to the compressor is 4400 hp (3281.1 kW)?

Calculation Procedure:

1. Determine the cycle rate
Use Fig. 60, entering at the pressure ratio of 5 in Fig. 60c and projecting to the 1500°F (815.6°C) curve. At the left, read the cycle air rate as 52 lb/kWh (23.6 kg/kWh).

2. Find the cycle thermal efficiency
Enter Fig. 60b at the pressure ratio of 5 and project vertically to the 1500°F (815.6°C) curve. At left, read the cycle thermal efficiency as 35 percent. Note that this point corresponds to the maximum efficiency obtainable from this cycle.

3. Find the cycle work ratio
Enter Fig. 60d at the pressure ratio of 5 and project vertically to the 1500°F (815.6°C) curve. At the left, read the work ratio as 44 percent.

4. Compute the turbine power output
For any gas turbine, the work ratio, percent = $100w_c/w_t$, where w_c = work input to the turbine, hp; w_t = work output of the turbine, hp. Substituting gives 44 = $100(4400)/w_t$; $w_t = 100(4400)/44 = 10,000$ hp (7457.0 kW).
    Related Calculations.   Use this general procedure to analyze gas turbines for power-plant, marine, and portable applications. Where the operating conditions are different from those given here, use the manufacturer's engineering data for the turbine under consideration.

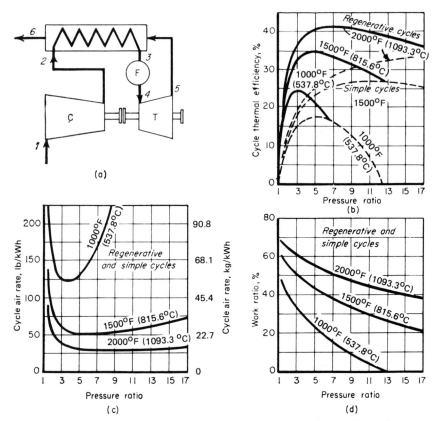

FIGURE 60   (*a*) Schematic of regenerative gas turbine; (*b*), (*c*), and (*d*) gas-turbine performance based on a regenerator effectiveness of 70 percent, compressor and turbine efficiency of 85 percent; air inlet = 60°F (15.6°C); no pressure losses.

Figure 61 shows the effect of turbine-inlet temperature, regenerator effectiveness, and compressor-inlet-air temperature on the performance of a modern gas turbine. Use these curves to analyze the cycles of gas turbines being considered for a particular application if the operating conditions are close to those plotted.

EXTRACTION TURBINE kW OUTPUT

An automatic extraction turbine operates with steam at 400 lb/in² absolute (2760 kPa), 700°F (371°C) at the throttle; its extraction pressure is 200 lb/in² (1380 kPa) and it exhausts at 110 lb/in² absolute (760 kPa). At full load 80,000 lb/h (600 kg/s) is supplied to the throttle and 20,000 lb/h (150 kg/s) is extracted at the bleed point. What is the kW output?

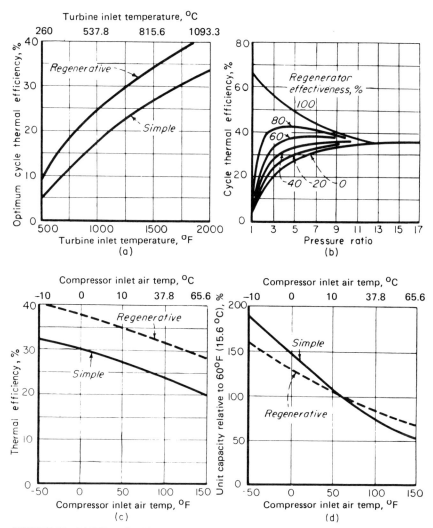

FIGURE 61 (*a*) Effect of turbine-inlet on cycle performance; (*b*) effect of regenerator effectiveness; (*c*) effect of compressor inlet-air temperature; (*d*) effect of inlet-air temperature on turbine-cycle capacity. These curves are based on a turbine and compressor efficiency of 85 percent, a regenerator effectiveness of 70 percent, and a 1500°F (815.6°C) inlet-gas temperature.

Calculation Procedure:

1. Determine steam conditions at the throttle, bleed point, and exhaust

Steam flow through the turbine is indicated by "enter" at the throttle, "extract" at the bleed point, and "exit" at the exhaust, as shown in Fig. 62*a*. The steam process is considered to be at constant entropy, as shown by the vertical isentropic line in Fig. 62*b*. At the throttle, where the steam enters at the given pressure, $p_1 = 400$

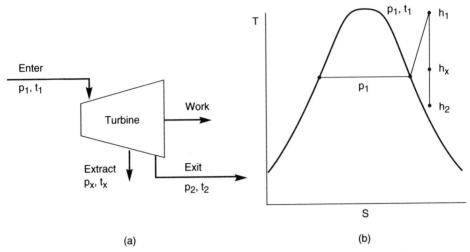

FIGURE 62 (a) Turbine steam flow diagram. (b) Temperature-entropy schematic for steam flow.

lb/in² absolute (2760 kPa) and temperature, t_1 = 700°F (371°C), steam enthalpy, h_1 = 1362.7 Btu/lb (3167.6 kJ/kg) and its entropy, s_1 = 1.6398, as indicated by Table 3, Vapor of the Steam Tables mentioned under Related Calculations of this procedure. From the Mollier chart, a supplement to the Steam Tables, the following conditions are found along the vertical isentropic line where $s_1 = s_x = s_2$ = 1.6398 Btu/(lb·°F) (6.8655 kJ/kg·°C):

At the bleed point, where the given extraction pressure, p_x = 200 lb/in² (1380 kPa) and the entropy, s_x, is as mentioned above, the enthalpy, h_x = 1284 Btu/lb (2986 kJ/kg) and the temperature t_x = 528°F (276°C). At the exit, where the given exhaust pressure, p_2 = 110 lb/in² (760 kPa) and the entropy, s_2, is as mentioned above, the enthalpy, h_2 = 1225 Btu/lb (2849 kJ/kg) and the temperature, t_2 = 400°F (204°C).

2. Compute the total available energy to the turbine
Between the throttle and the bleed point the available energy to the turbine, AE_1 = $Q_1(h_1 - h_x)$, where the full load rate of steam flow, Q_1 = 80,000 lb/h (600 kg/s); other values are as before. Hence, AE_1 = 80,000 × (1362.7 − 1284) = 6.296 × 10⁶ Btu/h (1845 kJ/s). Between the bleed point and the exhaust the available energy to the turbine, AE_2 = $(Q_1 - Q_2)(h_x - h_2)$, where the extraction flow rate, Q_x = 20,000 lb$_m$/h (150 kg/s); other values as before. Then, AE_2 = (80,000 − 20,000)(1284 − 1225) = 3.54 × 10⁶ Btu/h (1037 kJ/s). Total available energy to the turbine, $AE = AE_1 + AE_2$ = 6.296 × 10⁶ + (3.54 × 10⁶) = 9.836 × 10⁶ Btu/h (172.8 × 10³ kJ/s).

3. Compute the turbine's kW output
The power available to the turbine to develop power at the shaft, in kilowatts, kW = $AE/(Btu/kW·h)$ = 9.836 × 10⁶/3412.7 = 2880 kW. However, the actual power

developed at the shaft, $kW_a = kW \times e$, where e is the mechanical efficiency of the turbine. Thus, for an efficiency, $e = 0.90$, then $kW_a = 2880 \times 0.90 = 2590$ kW (2590 kJ/s).

Related Calculations.   The Steam Tables appear in *Thermodynamic Properties of Water Including Vapor, Liquid, and Solid Phases,* 1969, Keenan, et al., John Wiley & Sons, Inc. Use later versions of such tables whenever available, as necessary.

Steam Properties and Processes

STEAM MOLLIER DIAGRAM AND STEAM TABLE USE

(1) Determine from the Mollier diagram for steam (a) the enthalpy of 100 lb/in² (abs) (689.5-kPa) saturated steam, (b) the enthalpy of 10-lb/in² (abs) (68.9-kPa) steam containing 40 percent moisture, (c) the enthalpy of 100-lb/in² (abs) (689.5-kPa) steam at 600°F (315.6°C). (2) Determine from the steam tables (a) the enthalpy, specific volume, and entropy of steam at 145.3 lb/in² (gage) (1001.8 kPa); (b) the enthalpy and specific volume of superheated steam at 1100 lb/in² (abs) (7584.2 kPa) and 600°F (315.6°C); (c) the enthalpy and specific volume of high-pressure steam at 7500 lb/in² (abs) (51,710.7 kPa) and 1200°F (648.9°C); (d) the enthalpy, specific volume, and entropy of 10-lb/in² (abs) (68.9-kPa) steam containing 40 percent moisture.

Calculation Procedure:

1.   *Use the pressure and saturation (or moisture) lines to find enthalpy*
(a) Enter the Mollier diagram by finding the 100-lb/in² (abs) (689.5-kPa) pressure line, Fig. 63. In the Mollier diagram for steam, the pressure lines slope upward to the right from the lower left-hand corner. For saturated steam, the enthalpy is read at the intersection of the pressure line with the saturation curve *cef*, Fig. 63.

Thus, project along the 100-lb/in² (abs) (689.5-kPa) pressure curve, Fig. 63, until it intersects the saturation curve, point *g*. From here project horizontally to the left-hand scale of Fig. 63 and read the enthalpy of 100-lb/in² (abs) (689.5-kPa) saturated steam as 1187 Btu/lb (2761.0 kJ/kg). (The Mollier diagram in Fig. 63 has fewer grid divisions than large-scale diagrams to permit easier location of the major elements of the diagram.)

(b) On a Mollier diagram, the enthalpy of wet steam is found at the intersection of the saturation pressure line with the percentage-of-moisture curve corresponding to the amount of moisture in the steam. In a Mollier diagram for steam, the moisture curves slope downward to the right from the saturated liquid line *cd*, Fig. 63.

To find the enthalpy of 10-lb/in² (abs) (68.9-kPa) steam containing 40 percent moisture, project along the 1-lb/in² (abs) (68.9-kPa) saturation pressure line until the 40 percent moisture curve is intersected, Fig. 63. From here project horizontally to the left-hand scale and read the enthalpy of 10-lb/in² (68.9-kPa) wet steam containing 40 percent moisture as 750 Btu/lb (1744.5 kJ/kg).

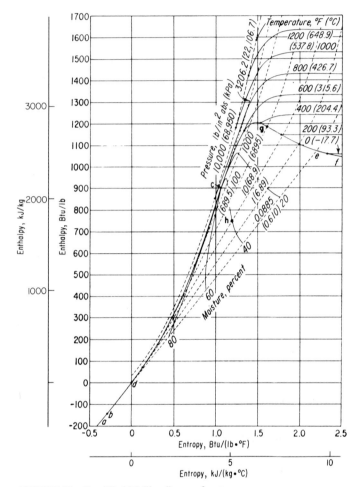

FIGURE 63 Simplified Mollier diagram for steam.

2. *Find the steam properties from the steam tables*

(*a*) Steam tables normally list absolute pressures or temperature in degrees Fahrenheit as one of their arguments. Therefore, when the steam pressure is given in terms of a gage reading, it must be converted to an absolute pressure before the table can be entered. To convert gage pressure to absolute pressure, add 14.7 to the gage pressure, or $p_a = p_g + 14.7$. In this instance, $p_a = 145.3 + 14.7 = 160.0$ lb/in² (abs) (1103.2 kPa). Once the absolute pressure is known, enter the saturation pressure table of the steam table at this value, and project horizontally to the desired values. For 160-lb/in² (abs) (1103.2-kPa) steam, using the ASME or Keenan and Keyes—*Thermodynamic Properties of Steam,* we see that the enthalpy of evaporation $h_{fg} = 859.2$ Btu/lb (1998.5 kJ/kg), and the enthalpy of saturated vapor $h_g = 1195.1$ Btu/lb (2779.8 kJ/kg), read from the respective columns of the steam tables.

The specific volume v_g of the saturated vapor of 160-lb/in^2 (abs) (1103.2-kPa) steam is, from the tables, 2.834 ft^3/lb (0.18 m^3/kg), and the entropy s_g is 1.5640 Btu/(lb·°F) [6.55 kJ/(kg·°C)].

(*c*) Every steam table contains a separate tabulation of properties of superheated steam. To enter the superheated steam table, two arguments are needed—the absolute pressure and the temperature of the steam. To determine the properties of 1100-lb/in^2 (abs) (7584.5-kPa) 600°F (315.6°C) steam, enter the superheated steam table at the given absolute pressure and project horizontally from this absolute pressure [1100 lb/in^2 (abs) or 7584.5 kPa] to the column corresponding to the superheated temperature (600°F or 315.6°C) to read the enthalpy of the superheated vapor as h = 1236.7 Btu/lb (2876.6 kJ/kg) and the specific volume of the super-heated vapor v = 0.4532 ft^3/lb (0.03 m^3/kg).

(*c*) For high-pressure steam use the ASME—*Steam Table,* entering it in the same manner as the superheated steam table. Thus, for 7500-lb/in^2 (abs) (51,712.5 kPa) 1200°F (648.9°C) steam, the enthalpy of the superheated vapor is 1474.9 Btu/lb (3430.6 kJ/kg), and the specific volume of the superheated vapor is 0.1060 ft^3/lb (0.0066 m^3/kg).

(*d*) To determine the enthalpy, specific volume, and the entropy of wet steam having y percent moisture by using steam tables instead of the Mollier diagram, apply these relations: $h = h_g - yh_{fg}/100$; $v = v_g - yv_{fg}/100$; $s = s_g - ys_{fg}/100$, where y = percentage of moisture expressed as a whole number. For 10-lb/in^2 (abs) (68.9-kPa) steam containing 40 percent moisture, obtain the needed values—h_g, h_{fg}, v_g, v_{fg}, s_g, and s_{fg}—from the saturation-pressure steam table and substitute in the above relations. Thus,

$$h = 1143.3 - \frac{40(982.1)}{100} = 750.5 \text{ Btu/lb (1745.7 kJ/kg)}$$

$$v = 38.42 - \frac{40(38.40)}{100} = 23.06 \text{ ft}^3/\text{lb (1.44 m}^3/\text{kg)}$$

$$s = 1.7876 - \frac{40(1.5041)}{100} = 1.1860 \text{ Btu/(lb·°F) [4.97 kJ/(kg·°C)]}$$

Note that Keenan and Keyes, in *Thermodynamic Properties of Steam,* do not tabulate v_{fg}. Therefore, this value must be obtained by subtraction of the tabulated values, or $v_{fg} = v_g - v_f$. The value v_{fg} thus obtained is used in the relation for the volume of the wet steam. For 10-lb/in^2 (abs) (68.9-kPa) steam containing 40 percent moisture, v_g = 38.42 ft^3/lb (2.398 m^3/kg) and v_f = 0.017 ft^3/lb (0.0011 m^3/kg). Then $v_{fg} = 38.42 - 0.017 = 28.403$ ft^3/lb (1.773 m^3/kg).

In some instances, the quality of steam may be given instead of its moisture content in percentage. The quality of steam is the percentage of vapor in the mixture. In the above calculation, the quality of the steam is 60 percent because 40 percent is moisture. Thus, quality = 1 − m, where m = percentage of moisture, expressed as a decimal.

INTERPOLATION OF STEAM TABLE VALUES

(1) Determine the enthalpy, specific volume, entropy, and temperature of saturated steam at 151 lb/in² (abs) (1041.1 kPa). (2) Determine the enthalpy, specific volume, entropy, and pressure of saturated steam at 261°F (127.2°C). (3) Find the pressure of steam at 1000°F (537.8°C) if its specific volume is 2.6150 ft³/lb (0.16 m³/kg). (4) Calculate the enthalpy, specific volume, and entropy of 300-lb/in² (abs) (2068.5-kPa) steam at 567.22°F (297.3°C).

Calculation Procedure:

1. Use the saturation-pressure table

Study of the saturation-pressure table shows that there is no pressure value for 151 lb/in² (abs) (1041.1 kPa) listed. So it will be necessary to interpolate between the next higher and next lower tabulated pressure values. In this instance, these values are 152 and 150 lb/in² (abs) (1048.0 and 1034.3 kPa), respectively. The pressure for which properties are being found [151 lb/in² (abs) or 1041.1 kPa] is called the *intermediate pressure*. At 152 lb/in² (abs) (1048.0 kPa), h_g = 1194.3 Btu/lb (2777.5 kJ/kg); v_g = 2.977 ft³/lb (0.19 m³/kg); s_g = 1.5683 Btu/(lb·°F) [6.67 kJ/(kg·°C)/ t = 359.46°F (181.9°C). At 150 lb/in² (abs) (1034.3 kPa), h_g = 1194.1 Btu/lb (2777.5 kJ/kg); v_g = 3.015 ft³/lb (0.19 m³/kg); s_g = 1.5694 Btu/(lb·°F) [6.57 kJ/(kg·°C); t = 358.42°F (181.3°C).

For the enthalpy, note that as the pressure increases, so does h_g. Therefore, the enthalpy at 151 lb/in² (abs) (1041.1 kPa), the intermediate pressure, will equal the enthalpy at 150 lb/in² (abs) (1034.3 kPa) (the lower pressure used in the interpolation) plus the proportional change (difference between the intermediate pressure and the lower pressure) for a 1-lb/in² (abs) (6.9-kPa) pressure increase. Or, at any higher pressure, $h_{gi} = h_{gl} + [(p_i - p_l)/(p_h - p_l)](h_h - h_l)$, where h_{gi} = enthalpy at the intermediate pressure; h_{gl} = enthalpy at the lower pressure used in the interpolation; h_h = enthalpy at the higher pressure used in the interpolation; p_i = intermediate pressure; p_h and p_l = higher and lower pressures, respectively, used in the interpolation. Thus, from the enthalpy values obtained from the steam table for 150 and 152 lb/in² (abs) (1034.3 and 1048.0 kPa), h_{gi} = 1194.1 + [(151 − 150)/(152 − 150)](1194.3 − 1194.1) = 1194.2 Btu/lb (2777.7 kJ/kg) at 151 lb/in² (abs) (1041.1 kPa) saturated.

Next study the steam table to determine the direction of change of specific volume between the lower and higher pressures. This study shows that the specific volume decreases as the pressure increases. Therefore, the specific volume at 151 lb/in² (abs) (1041.1 kPa) (the intermediate pressure) will equal the specific volume at 150 lb/in² (abs) (1034.3 kPa) (the lower pressure used in the interpolation) minus the proportional change (difference between the intermediate pressure and the lower interpolating pressure) for a 1-lb/in² (abs) pressure increase. Or, at any pressure, $v_{gi} = v_{gl} - [(p_i - p_l)/(p_h - p_l)](v_l - v_h)$, where the subscripts are the same as above and v = specific volume at the respective pressure. With the volume values obtained from steam tables for 150 and 152 lb/in² (abs) (1034.3 and 1048.0 kPa), v_{gi} = 3.015 − [(151 − 150)/(152 − 150)](3.015 − 2.977) = 2.996 ft³/lb (0.19 m³/kg) and 151 lb/in² (abs) (1041.1 kPa) saturated.

Study of the steam table for the direction of entropy change shows that entropy, like specific volume, decreases as the pressure increases. Therefore, the entropy at

151 lb/in² (abs) (1041.1 kPa) (the intermediate pressure) will equal the entropy at 150 lb/in² (abs) (1034.3 kPa) (the lower pressure used in the interpolation) minus the proportional change (difference between the intermediate pressure and the lower interpolating pressure) for a 1-lb/in² (abs) (6.9-kPa) pressure increase. Or, at any higher pressure, $s_{gi} = s_{gl} - [(p_i - p_l)/(p_h - p_l)](s_l - s_h) = 1.5164 - [(151 - 150)/(152 - 150)](1.5694 - 1.5683) = 1.56885$ Btu/(lb·°F) [6.6 kJ/(kg·°C)] at 151 lb/in² (abs) (1041.1 kPa) saturated.

Study of the steam table for the direction of temperature change shows that the saturation temperature, like enthalpy, increases as the pressure increases. Therefore, the temperature at 151 lb/in² (abs) (1041.1 kPa) (the intermediate pressure) will equal the temperature at 150 lb/in² (abs) (1034.3 kPa) (the lower pressure used in the interpolation) plus the proportional change (difference between the intermediate pressure and the lower interpolating pressure) for a 1-lb/in² (abs) (6.9-kPa) increase. Or, at any higher pressure, $t_{gi} = t_{gl} + [(p_i - p_l)/(p_h - p_l)](t_h - t_l) = 358.42 + [(151 - 150)/(152 - 150)](359.46 - 358.42) = 358.94$°F (181.6°C) at 151 lb/in² (abs) (1041.1 kPa) saturated.

2. Use the saturation-temperature steam table
Study of the saturation-temperature table shows that there is no temperature value of 261°F (127.2°C) listed. Therefore, it will be necessary to interpolate between the next higher and next lower tabulated values. In this instance these values are 262 and 260°F (127.8 and 126.7°C), respectively. The temperature for which properties are being found (261°F or 127.2°C) is called the intermediate temperature.

Temperature		h_g		v_g		s_g		p_g	
°F	°C	Btu/lb	kJ/kg	ft³/lb	m³/kg	Btu/(lb·°F)	kJ/(kg·°C)	lb/in² (abs)	kPa
262	127.8	1168.0	2716.8	11.396	0.71	1.6833	7.05	36.646	252.7
260	126.7	1167.3	2715.1	11.763	0.73	1.6860	7.06	35.429	244.3

For enthalpy, note that as the temperature increases, so does h_g. Therefore, the enthalpy at 261°F (127.2°C) (the intermediate temperature) will equal the enthalpy at 260°F (126.7°C) (the lower temperature used in the interpolation) plus the proportional change (difference between the intermediate temperature and the lower temperature) for a 1°F (0.6°C) temperature increase. Or, at any higher temperature, $h_{gi} = h_{gl} + [(t_i - t_l)/(t_h - t_l)](h_h - h_l)$, where h_{gl} = enthalpy at the lower temperature used in the interpolation; h_h = enthalpy at the higher temperature used in the interpolation; t_i = intermediate temperature; t_h and t_l = higher and lower temperatures, respectively, used in the interpolation. Thus, from the enthalpy values obtained from the steam table for 260 and 262°F (126.7 and 127.8°C), $h_{gi} = 1167.3 + [(261 - 260)/(262 - 260)](1168.0 - 1167.3) = 1167.65$ Btu/lb (2716.0 kJ/kg) at 260°F (127.2°C) saturated.

Next, study the steam table to determine the direction of change of specific volume between the lower and higher temperatures. This study shows that the specific volume decreases as the pressure increases. Therefore, the specific volume at 261°F (127.2°C) (the intermediate temperature) will equal the specific volume at 260°F (126.7°C) (the lower temperature used in the interpolation) minus the proportional change (difference between the intermediate temperature and the lower interpolating temperature) for a 1°F (0.6°C) temperature increase. Or, at any higher temperature, $v_{gi} = v_{gl} - [(t_i - t_l)/(t_h - t_l)](v_l - v_h) = 11.763 - [(261 - 260)/$ $(262 - 260)](11.763 - 11.396) = 11.5795$ ft^3/lb (0.7 m^3/kg) at 261°F (127.2°C) saturated.

Study of the steam table for the direction of entropy change shows that entropy, like specific volume, decreases as the temperature increases. Therefore, the entropy at 261°F (127.2°C) (the intermediate temperature) will equal the entropy at 260°F (126.7°C) (the lower temperature used in the interpolation) minus the proportional change (difference between the intermediate temperature and the lower temperature) for a 1°F (0.6°C) temperature increase. Or, at any higher temperature, $s_{gi} = s_{gl} - [(t_i - t_l)/(h_h - t_l)](s_l - s_h) = 1.6860 - [(261 - 260)/(262 - 260)](1.6860 - 1.6833) = 1.68465$ Btu/(lb·°F) [7.1 kJ/(kg·°C)] at 261°F (127.2°C).

Study of the steam table for the direction of pressure change shows that the saturation pressure, like enthalpy, increases as the temperature increases. Therefore, the pressure at 261°F (127.2°C) (the intermediate temperature) will equal the pressure at 260°F (126.7°C) (the lower temperature used in the interpolation) plus the proportional change (difference between the intermediate temperature and the lower interpolating temperature) for a 1°F (0.6°C) temperature increase. Or, at any higher temperature, $p_{gi} = p_{gl} + [(t_i - t_l)/(t_h - t_l)](p_h - p_l) = 35.429 + [(261 - 260)(262 - 260)](36.646 - 35.429) = 36.0375$ lb/in^2 (abs) (248.5 kPa) at 261°F (127.2°C) saturated.

3. Use the superheated steam table

Choose the superheated steam table for steam at 1000°F (537.9°C) and 2.6150 ft^3/lb (0.16 m^3/kg) because the highest temperature at which saturated steam can exist is 705.4°F (374.1°C). This is also the highest temperature tabulated in some saturated-temperature tables. Therefore, the steam is superheated when at a temperature of 1000°F (537.9°C).

Look down the 1000°F (537.9°C) columns in the superheated steam table until a specific volume value of 2.6150 (0.16) is found. This occurs between 325 lb/in^2 (abs) (2240.9 kPa, $v = 2.636$ or 0.16) and 330 lb/in^2 (abs) (2275.4 kPa, $v = 2.596$ or 0.16). Since there is no volume value exactly equal to 2.6150 tabulated, it will be necessary to interpolate. List the values from the steam table thus:

p		t		v	
lb/in^2 (abs)	kPa	°F	°C	ft^3/lb	m^3/kg
325	2240.9	1000	537.9	2.636	0.16
330	2275.4	1000	537.9	2.596	0.16

Note that as the pressure rises, at constant temperature, the volume decreases. Therefore, the intermediate (or unknown) pressure is found by subtracting from the higher interpolating pressure [330 lb/in^2 (abs) or 2275.4 kPa in this instance] the product of the proportional change in the specific volume and the difference in the pressures used for the interpolation. Or, $p_{gi} = p_h - [(v_i - v_h)/(v_l - v_h)](p_h - p_l)$, where the subscripts h, l, and i refer to the high, low, and intermediate (or unknown) pressures, respectively. In this instance, $p_{gi} = 330 - [(2.615 - 2.596)/(2.636 - 2.596)](330 - 325) = 327.62$ lb/in^2 (abs) (2258.9 kPa) at 1000°F (537.9 kPa) and a specific volume of 2.6150 ft^3/lb (0.16 m^3/kg).

4. *Use the superheated steam table*

When a steam pressure and temperature are given, determine, before performing any interpolation, the state of the steam. Do this by entering the saturation-pressure table at the given pressure and noting the saturation temperature. If the given temperature exceeds the saturation temperature, the steam is superheated. In this instance, the saturation-pressure table shows that at 300 lb/in² (abs) (2068.5 kPa) the saturation temperature is 417.33°F (214.1°C). Since the given temperature of the steam is 567.22°F (297.3°C), the steam is superheated because its actual temperature is greater than the saturation temperature.

Enter the superheated steam table at 300 lb/in² (abs) (2068.5 kPa), and find the next temperature lower than 567.22°F (297.3°C); this is 560°F (293.3°C). Also find the next higher temperature; this is 580°F (304.4°C). Tabulate the enthalpy, specific volume, and entropy for each temperature thus:

t		h		v		s	
°F	°C	Btu/lb	kJ/kg	ft³/lb	m³/kg	Btu/(lb·°F)	kJ/(kg·°C)
560	293.3	1292.5	3006.4	1.9218	0.12	1.6054	6.72
580	304.4	1303.7	3032.4	1.9594	0.12	1.6163	6.77

Use the same procedures for each property—enthalpy, specific volume, and entropy—as given in step 2 above; but change the sign between the lower volume and entropy and the proportional factor (temperature in this instance), because for superheated steam the volume and entropy increase as the steam temperature increases. Thus

$$h_{gi} = 1292.5 + \frac{567.22 - 560}{580 - 560}(1303.7 - 1292.5) = 1269.6 \text{ Btu/lb } (3015.9 \text{ kJ/kg})$$

$$v_{gi} = 1.9128 + \frac{567.22 - 560}{580 - 560}(1.9594 - 1.9128) = 1.9296 \text{ ft}^3/\text{lb } (0.12 \text{ m}^3/\text{kg})$$

$$s_{gi} = 1.6054 + \frac{567.22 - 560}{580 - 560}(1.6163 - 1.6054) = 1.6093 \text{ Btu/(lb·°F) } [6.7 \text{ kJ/(kg·°C)}]$$

Note: Also observe the direction of change of a property *before* interpolating. Use a *plus* or *minus* sign between the higher interpolating value and the proportional change depending on whether the tabulated value increases (+) or decreases (−).

CONSTANT-PRESSURE STEAM PROCESS

Three pounds of wet steam, containing 15 percent moisture and initially at a pressure of 400 lb/in² (abs) (2758.0 kPa), expands at constant pressure ($P = C$) to 600°F (315.6°C). Determine the initial temperature T_1, enthalpy H_1, internal energy E_1, volume V_1, entropy S_1, final entropy H_2, internal energy E_2, volume V_2, entropy S_2, heat added to the steam Q_1, work output W_2, change in initial energy ΔE, change in specific volume ΔV, change in entropy ΔS.

Calculation Procedure:

1. *Determine the initial steam temperature from the steam tables*
Enter the saturation-pressure table at 400 lb/in² (abs) (2758.0 kPa), and read the saturation temperature as 444.59°F (229.2°C).

2. *Correct the saturation values for the moisture of the steam in the initial state*
Sketch the process on a pressure-volume (P-V), Mollier (H-S), or temperature entropy (T-S) diagram, Fig. 64. In state 1, y = moisture content = 15 percent. Using the appropriate values from the saturation-pressure steam table for 40 lb/in² (abs) (2758.0 kPa), correct them for a moisture content of 15 percent:

$$H_1 = h_g - yh_{fg} = 1204.5 - 0.15(780.5) = 1087.4 \text{ Btu/lb (2529.3 kJ/kg)}$$

$$E_1 = u_g - yu_{fg} = 1118.5 - 0.15(695.9) = 1015.1 \text{ Btu/lb (2361.1 kJ/kg)}$$

$$V_1 = v_g - yv_{fg} = 1.1613 - 0.15(1.1420) = 0.990 \text{ ft}^3/\text{lb (0.06 m}^3/\text{kg)}$$

$$S_1 = s_g - ys_{fg} = 1.4844 - 0.15(0.8630) = 1.2945 \text{ Btu/(lb·°F) [5.4 kJ/(kg·°C)]}$$

3. *Determine the steam properties in the final state*
Since this is a constant-pressure process, the pressure in state 2 is 400 lb/in² (abs) (2758.0 kPa), the same as state 1. The final temperature is given as 600°F (315.6°C). This is greater than the saturation temperature of 444.59°F (229.2°C). Hence, the steam is superheated when in state 2. Use the superheated steam tables, entering at 400 lb/in² (abs) (2758.8 kPa) and 600°F (315.6°C). At this condition, $H_2 = 1306.9$ Btu/lb (3039.8 kJ/kg); $V_2 = 1.477$ ft³/lb (0.09 m³/kg). Then $E_2 = h_{2g} - P_2V_2/J = 1306.9 - 400(144)(1.477)/778 = 1197.5$ Btu/lb (2785.4 kJ/kg). In this equation, the constant 144 converts pounds per square inch to pounds per square foot, absolute, and J = mechanical equivalent of heat = 778 ft·lb/Btu (1 N·m/J). From the steam tables, $S_2 = 1.5894$ Btu/(lb·°F) [6.7 kJ/(kg·°C)].

4. *Compute the process inputs, outputs, and changes*
$W_2 = (P_1/J)(V_2 - V_1)m = [400(144)/778](1.4770 - 0.9900)(3) = 108.1$ Btu (114.1 kJ). In this equation, m = weight of steam used in the process = 3 lb (1.4 kg). Then

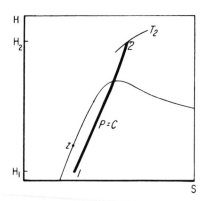
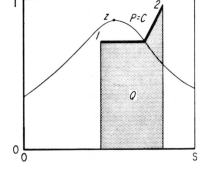

FIGURE 64 Constant-pressure process.

$$Q_1 = (H_2 - H_1)m = (1306.9 - 1087.4)(3) = 658.5 \text{ Btu } (694.4 \text{ kJ})$$

$$\Delta E = (E_2 - E_1)m = (1197.5 - 1014.1)(3) = 550.2 \text{ Btu } (580.2 \text{ kJ})$$

$$\Delta V = (V_2 - V_1)m = (1.4770 - 0.9900)(3) = 1.461 \text{ ft}^3 \ (0.041 \text{ m}^3)$$

$$\Delta S = (S_2 - S_1)m = (1.5894 - 1.2945)(3) = 0.8847 \text{ Btu}/{}^\circ\text{F } (1.680 \text{ kJ}/{}^\circ\text{C})$$

5. Check the computations

The work output W_2 should equal the change in internal energy plus the heat input, or $W_2 = E_1 - E_2 + Q_1 = -550.2 + 658.5 = 108.3$ Btu (114.3 kJ). This value very nearly equals the computed value of $W_2 = 108.1$ Btu (114.1 kJ) and is close enough for all normal engineering computations. The difference can be traced to calculator input errors. In computing the work output, the internal-energy change has a negative sign because there is a decrease in E during the process.

Related Calculations. Use this procedure for all constant-pressure steam processes.

CONSTANT-VOLUME STEAM PROCESS

Five pounds (2.3 kg) of wet steam initially at 120 lb/in² (abs) (827.4 kPa) with 30 percent moisture is heated at constant volume ($V = C$) to a final temperature of 1000°F (537.8°C). Determine the initial temperature T_1, enthalpy H_1, internal energy E_1, volume V_1, final pressure P_2, enthalpy H_2, internal energy E_2, volume V_2, heat added Q_1, work output W, change in internal energy ΔE, volume ΔV, and entropy ΔS.

Calculation Procedure:

1. Determine the initial steam temperature from the steam tables
Enter the saturation-pressure table at 120 lb/in² (abs) (827.4 kPa), the initial pressure, and read the saturation temperature $T_1 = 341.25°F$ (171.8°C).

2. Correct the saturation values for the moisture in the steam in the initial state
Sketch the process on P-V, H-S, or T-S diagrams, Fig. 65. Using the appropriate values from the saturation-pressure table for 120 lb/in² (abs) (827.4 kPa), correct them for a moisture content of 30 percent:

$$H_1 = h_g - yh_{fg} = 1190.4 - 0.3(877.9) = 927.0 \text{ Btu/lb } (2156.2 \text{ kJ/kg})$$

$$E_1 = u_g - yu_{fg} = 1107.6 - 0.3(795.6) = 868.9 \text{ Btu/lb } (2021.1 \text{ kJ/kg})$$

$$V_1 = v_g - yv_{fg} = 3.7280 - 0.3(3.7101) = 2.6150 \text{ ft}^3/\text{lb } (0.16 \text{ m}^3/\text{kg})$$

$$S_1 = s_g - ys_{fg} = 1.5878 - 0.3(1.0962) = 1.2589 \text{ Btu}/(\text{lb}\cdot{}^\circ\text{F}) \ [5.3 \text{ kJ}/(\text{kg}\cdot{}^\circ\text{C})]$$

3. Determine the steam volume in the final state
We are given $T_2 = 1000°F$ (537.8°C). Since this is a constant-volume process, $V_2 = V_1 = 2.6150$ ft³/lb (0.16 m³/kg). The total volume of the vapor equals the

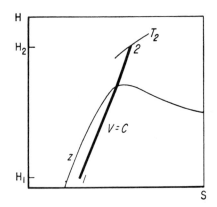
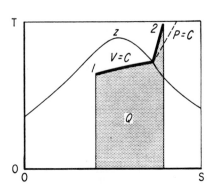

FIGURE 65 Constant-volume process.

product of the specific volume and the number of pounds of vapor used in the process, or total volume = 2.6150(5) = 13.075 ft³ (0.37 m³).

4. Determine the final steam pressure
The final steam temperature (1000°F or 537.8°C) and the final steam volume (2.6150 ft³/lb or 0.16 m³/kg) are known. To determine the final steam pressure, find in the steam tables the state corresponding to the above temperature and specific volume. Since a temperature of 1000°F (537.8°C) is higher than any saturation temperature (705.4°F or 374.1°C is the highest saturation temperature for saturated steam), the steam in state 2 must be superheated. Therefore, the superheated steam tables must be used to determine P_2.

Enter the 1000°F (537.8°C) column in the steam table, and look for a superheated-vapor specific volume of 2.6150 ft³/lb (0.16 m³/kg). At a pressure of 325 lb/in² (abs) (2240.9 kPa),

$$v = 2.636 \text{ ft}^3/\text{lb} \ (0.16 \text{ m}^3/\text{kg})$$

$$h = 1542.5 \text{ Btu/lb} \ (3587.9 \text{ kJ/kg})$$

$$s = 1.7863 \text{ Btu/(lb} \cdot \text{°F)} \ [7.48 \text{ kJ/(kg} \cdot \text{°C)}]$$

and at a pressure of 330 lb/in² (abs) (227.4 kPa)

$$v = 2.596 \text{ ft}^3/\text{lb} \ (0.16 \text{ m}^3/\text{kg})$$

$$h = 1524.4 \text{ Btu/lb} \ (3545.8 \text{ kJ/kg})$$

$$s = 1.7845 \text{ Btu/(lb} \cdot \text{°F)} \ [7.47 \text{ kJ/(kg} \cdot \text{°C)}]$$

Thus, 2.6150 lies between 325 and 330 lb/in² (abs) (2240.9 and 2275.4 kPa). To determine the pressure corresponding to the final volume, it is necessary to interpolate between the specific-volume values, or P_2 = 330 − [(2.615 − 2.596)/(2.636 − 2.596)](330 − 325) = 327.62 lb/in² (abs) (2258.9 kPa). In this equation, the volume values correspond to the upper [330 lb/in² (abs) or 2275.4 kPa], lower [325 lb/in² (abs) or 2240.9 kPa], and unknown pressures.

5. Determine the final enthalpy, entropy, and internal energy

The final enthalpy can be interpolated in the same manner, using the enthalpy at each volume instead of the pressure. Thus $H_2 = 1524.5 - [(2.615 - 2.596)/(2.636 - 2.596)](1524.5 - 1524.4) = 1524.45$ Btu/lb (3545.8 kJ/kg). Since the difference in enthalpy between the two pressures is only 0.1 Btu/lb (0.23 kJ/kg) ($= 1524.5 - 1524.4$), the enthalpy at 327.62 lb/in² (abs) could have been assumed equal to the enthalpy at the lower pressure [325 lb/in² (abs) cr 2240.9 kPa], or 1524.4 Btu/lb (3545.8 kJ/kg), and the error would have been only 0.05 Btu/lb (0.12 kJ/kg), which is negligible. However, where the enthalpy values vary by more than 1.0 Btu/lb (2.3 kJ/kg), interpolate as shown, if accurate results are desired.

Find S_2 by interpolating between pressures, or

$$S = 1.7863 - \frac{327.62 - 325}{330 - 325}(1.7863 - 1.7845) = 1.7854 \text{ Btu/(lb·°F)} [7.5 \text{ kJ/(kg·°C)}]$$

$$E_2 = H_2 - \frac{P_2 V_2}{J} = 1524.4 - \frac{327.62(144)(2.615)}{778} = 1365.9 \text{ Btu/lb (3177.1 kJ/kg)}$$

6. Compute the changes resulting from the process

Here $Q_1 = (E_2 - E_1)m = (1365.9 - 868.9)(5) = 2485$ Btu (2621.8 kJ); $\Delta S = (S_2 - S_1)m = (1.7854 - 1.2589)(5) = 2.6325$ Btu/°F (5.0 kJ/°C).

By definition, $W = 0$; $\Delta V = 0$; $\Delta E = Q_1$. Note that the curvatures of the constant-volume line on the T-S chart, Fig. 65, are different from the constant-pressure line, Fig. 64. Adding heat Q_1 to a constant-volume process affects only the internal energy. The total entropy change must take into account the total steam mass $m = 5$ lb (2.3 kg).

Related Calculations. Use this general procedure for all constant-volume steam processes.

CONSTANT-TEMPERATURE STEAM PROCESS

Six pounds (2.7 kg) of wet steam initially at 1200 lb/in² (abs) (8274.0 kPa) and 50 percent moisture expands at constant temperature ($T = C$) to 300 lb/in² (abs) (2068.5 kPa). Determine the initial temperature T_1, enthalpy H_1, internal energy E_1, specific volume V_1, entropy S_1, final temperature T_2, enthalpy H_2, internal energy E_2, volume V_2, entropy S_2, heat added Q_1, work output W_2, change in internal energy ΔE, volume ΔV, and entropy ΔS.

Calculation Procedure:

1. Determine the initial steam temperature from the steam tables

Enter the saturation-pressure table at 1200 lb/in² (abs) (8274.0 kPa), and read the saturation temperature $T_1 = 567.22°F$ (297.3°C).

2. Correct the saturation values for the moisture in the steam in the initial state

Sketch the process on P-V, H-S, or T-S diagrams, Fig. 66. Using the appropriate values from the saturation-pressure table for 1200 lb/in² (abs) (8274.0 kPa), correct them for the moisture content of 50 percent:

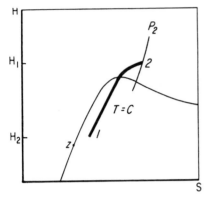 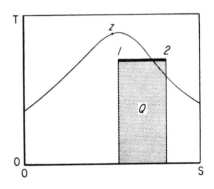

FIGURE 66   Constant-temperature process.

$$H_1 = h_g - y_1 h_{fg} = 1183.4 - 0.5(611.7) = 877.5 \text{ Btu/lb (2041.1 kJ/kg)}$$

$$E_1 = u_g - y_1 u_{fg} = 1103.0 - 0.5(536.3) = 834.8 \text{ Btu/lb (1941.7 kJ/kg)}$$

$$V_1 = v_g - y_1 v_{fg} = 0.3619 - 0.5(0.3396) = 0.19 \text{ ft}^3/\text{lb (0.012 m}^3/\text{kg)}$$

$$S_1 = s_g - y_1 s_{fg} = 1.3667 - 0.5(0.5956) = 1.0689 \text{ Btu/(lb·°F) [4.5 kJ/(kg·°C)]}$$

3. Determine the steam properties in the final state

Since this is a constant-temperature process, $T_2 = T_1 = 567.22°F$ (297.3°C); $P_2 = 300 \text{ lb/in}^2$ (abs) (2068.5 kPa), given. The saturation temperature of 300 lb/in² (abs) (2068.5 kPa) is 417.33°F (214.1°C). Therefore, the steam is superheated in the final state because 567.22°F (297.3°C) > 417.33°F (214.1°C), the saturation temperature. To determine the final enthalpy, entropy, and specific volume, it is necessary to interpolate between the known final temperature and the nearest tabulated temperatures greater and less than the final temperature.

	v		h		s	
	ft³/lb	m³/kg	Btu/lb	kJ/kg	Btu/(lb·°F)	kJ/(kg·°C)
At $T = 560°F$ (293.3°C)	1.9128	0.12	1292.5	3006.4	1.6054	6.72
At $T = 580°F$ (304.4°C)	1.9594	0.12	1303.7	3032.4	1.6163	6.76

Then

$$H_2 = 1292.5 + \frac{567.22 - 560}{580 - 560}(1303.7 - 1292.5) = 1296.5 \text{ Btu/lb (3015.7 kJ/kg)}$$

$$S_2 = 1.6054 + \frac{567.22 - 560}{580 - 560}(1.6163 - 1.6054) = 1.6093 \text{ Btu/(lb·°F) [6.7 kJ/(kg·°C)]}$$

$$V_2 = 1.9128 + \frac{567.22 - 560}{580 - 560}(1.9594 - 1.9128) = 1.9296 \text{ ft}^3/\text{lb (0.12 m}^3/\text{kg)}$$

$$E_2 = H_2 - \frac{P_2 V_2}{J} = 1296.5 - \frac{300(144)(1.9296)}{778} = 1109.3 \text{ Btu/lb (2580.2 kJ/kg)}$$

4. Compute the process changes
Here $Q_1 = T(S_2 - S_1)m$, where T_1 = absolute initial temperature, °R. So Q_1 = $(567.22 + 460)(1.6093 - 1.0689)(6) = 3330$ Btu (3513.3 kJ). Then

$$\Delta E = E_2 - E_1 = 1109.3 - 834.8 = 274.5 \text{ Btu/lb (638.5 kJ/kg)}$$

$$\Delta H = H_2 - H_1 = 1296.5 - 877.5 = 419.0 \text{ Btu/lb (974.6 kJ/kg)}$$

$$W_2 = (Q_1 - \Delta E)m = (555 - 274.5)(6) = 1.683 \text{ Btu (1.8 kJ)}$$

$$\Delta S = S_2 - S_1 = 1.6093 - 1.0689 = 0.5404 \text{ Btu/(lb} \cdot °\text{F) [2.3 kJ/(kg} \cdot °\text{C)]}$$

$$\Delta V = V_2 - V_1 = 1.9296 - 0.1921 = 1.7375 \text{ ft}^3/\text{lb (0.11 m}^3/\text{kg)}$$

Related Calculations. Use this procedure for any constant-temperature steam process.

CONSTANT-ENTROPY STEAM PROCESS

Ten pounds (4.5 kg) of steam expands under two conditions—nonflow and steady flow—at constant entropy ($S = C$) from an initial pressure of 2000 lb/in^2 (abs) (13,790.0 kPa) and a temperature of 800°F (426.7°C) to a final pressure of 2 lb/in^2 (abs) (13.8 kPa). In the steady-flow process, assume that the initial kinetic energy E_{k1} = the final kinetic energy E_{k2}. Determine the initial enthalpy H_1, internal energy E_1, volume V_1, entropy S_1, final temperature T_2, percentage of moisture y, enthalpy H_2, internal energy E_2, volume V_2, entropy S_2, change in internal energy ΔE, enthalpy ΔH, entropy ΔS, volume ΔV, heat added Q_1, and work output W_2.

Calculation Procedure:

1. Determine the initial enthalpy, volume, and entropy from the steam tables
Enter the superheated-vapor table at 2000 lb/in^2 (abs) (13,790.0 kPa) and 800°F (427.6°C), and read H_1 = 1335.5 Btu/lb (3106.4 kJ/kg); V_1 = 0.3074 ft^3/lb (0.019 m^3/kg); S_1 = 1.4576 Btu/(lb · °F) [6.1 kJ/(kg · °C)].

2. Compute the initial energy

$$E_1 = H_1 - \frac{P_1 V_1}{J} = 1335.5 - \frac{2000(144)(0.3074)}{778} = 1221.6 \text{ Btu/lb (2841.1 kJ/kg)}$$

3. Determine the vapor properties on the final state
Sketch the process on P-V, H-S, or T-S diagrams, Fig. 67. Note that the expanded steam is wet in the final state because the 2-lb/in^2 (abs) (13.8-kPa) pressure line is under the saturation curve on the H-S and T-S diagrams. Therefore, the vapor properties in the final state must be corrected for the moisture content. Read, from the saturation-pressure steam table, the liquid and vapor properties at 2 lb/in^2 (abs) (13.8 kPa). Tabulate these properties thus:

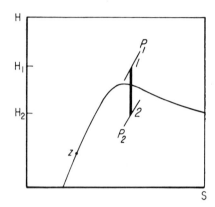 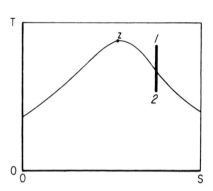

FIGURE 67   Constant-entropy process.

s_f = 0.1749 Btu/(lb·°F) [0.73 kJ/(kg·°C)] s_{fg} = 1.7451 Btu/(lb·°F) [7.31 kJ/(kg·C)]

h_f = 93.99 Btu/lb (218.6 kJ/kg) h_{fg} = 1022.2 Btu/lb (2377.6 kJ/kg)

u_f = 93.98 Btu/lb (218.6 kJ/kg) u_{fg} = 957.9 Btu/lb (2228.1 kJ/kg)

v_f = 0.016 ft³/lb (0.0010 m³/kg) v_{fg} = 173.71 ft³/lb (10.8 m³/kg)

s_g = 1.9200 Btu/(lb·°F) [8.04 kJ/(kg·C)] h_g = 1116.3 Btu/lb (2596.5 kJ/kg)

u_g = 1051.9 Btu/lb (2446.7 kJ/kg) v_g = 173.73 ft³/lb (10.8 m³/kg)

Since this is a constant-entropy process, $S_2 = S_1 = s_g - y_2 s_{fg}$. Solve for y_2, the percentage of moisture in the final state. Or, $y_2 = (s_g - S_1)/s_{fg}$ = (1.9200 − 1.4576)/1.7451 = 0.265, or, 26.5 percent. Then

$$H_2 = h_g - y_2 h_{fg} = 1116.2 - 0.265(1022.2) = 845.3 \text{ Btu/lb (1966.2 kJ/kg)}$$

$$E_2 = u_g - y_2 u_{fg} = 1051.9 - 0.265(957.9) = 798.0 \text{ Btu/lb (1856.1 kJ/kg)}$$

$$V_2 = v_g - y_2 v_{fg} = 173.73 - 0.265(173.71) = 127.7 \text{ ft}^3\text{/lb (8.0 m}^3\text{/kg)}$$

4.   Compute the changes resulting from the process
The total change in properties is for 10 lb (4.5 kg) of steam, the quantity used in this process. Thus,

$$\Delta E = (E_1 - E_2)m = (1221.6 - 798.0)(10) = 4236 \text{ Btu (4469.2 kJ)}$$

$$\Delta H = (H_1 - H_2)m = (1335.5 - 845.3)(10) = 4902 \text{ Btu (5171.9 kJ)}$$

$$\Delta S = (S_1 - S_2)m = (1.4576 - 1.4576)(10) = 0 \text{ Btu/°F (0 kJ/°C)}$$

$$\Delta V = (V_1 - V_2)m = (0.3074 - 127.7)(10) = -1274 \text{ ft}^3 \text{ (−36.1 m}^3)$$

So Q_1 = 0 Btu. (By definition, there is no transfer of heat in a constant-entropy process.) Nonflow $W_2 = \Delta E$ = 4236 Btu (4469.2 kJ). Steady flow $W_2 = \Delta H$ = 4902 Btu (5171.9 kJ).

Note: In a constant-entropy process, the nonflow work depends on the change in internal energy. The steady-flow work depends on the change in enthalpy and is larger than the nonflow work by the amount of the change in the flow work.

IRREVERSIBLE ADIABATIC EXPANSION OF STEAM

Ten pounds (4.5 kg) of steam undergoes a steady-flow expansion from an initial pressure of 2000 lb/in² (abs) (13,790.0 kPa) and a temperature 800°F (426.7°C) to a final pressure of 2 lb/in² (abs) (13.9 kPa) at an expansion efficiency of 75 percent. In this steady flow, assume $E_{k1} = E_{k2}$. Determine ΔE, ΔH, ΔS, ΔV, Q, and W_2.

Calculation Procedure:

1. Determine the initial vapor properties from the steam tables
Enter the superheated-vapor tables at 2000 lb/in² (abs) (13,790.0 kPa) and 800°F (426.7°C), and read $H_1 = 1335.5$ Btu/lb (3106.4 kJ/kg); $V_1 = 0.3074$ ft³/lb (0.019 m³/kg); $E_1 = 1221.6$ Btu/lb (2840.7 kJ/kg); $S_1 = 1.4576$ Btu/(lb·°F) [6.1 kJ/(kg·°C)].

2. Determine the vapor properties in the final state
Sketch the process on *P-V*, *H-S*, or *T-S* diagram, Fig. 68. Note that the expanded steam is wet in the final state because the 2-lb/in² (abs) (13.9-kPa) pressure line is under the saturation curve on the *H-S* and *T-S* diagram. Therefore, the vapor properties in the final state must be corrected for the moisture content. However, the actual final enthalpy cannot be determined until after the expansion efficiency $[H_1 - H_2(H_1 - H_{2s})]$ is evaluated.

To determine the final enthalpy H_2, another enthalpy H_{2s} must be computed by assuming a constant-entropy expansion to 2 lb/in² (abs) (13.8 kPa) and a temperature of 126.08°F (52.3°C). Enthalpy H_{2s} will then correspond to a constant-entropy

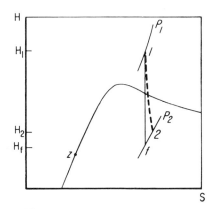
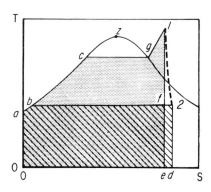

FIGURE 68 Irreversible adiabatic process.

expansion into the wet region, and the percentage of moisture will correspond to the final state. This percentage is determined by finding the ratio of $s_g - S_1$ to s_{fg}, or $y_{2s} = s_g - S_1/s_{fg} = 1.9200 - 1.4576/1.7451 = 0.265$, where s_g and s_{fg} are entropies at 2 lb/in^2 (abs) (13.8 kPa). Then $H_{2s} = h_g - y_{2s}h_{fg} = 1116.2 - 0.265(1022.2) = 845.3$ Btu/lb (1966.2 kJ/kg). In this relation, h_g and h_{fg} are enthalpies at 2 lb/in^2 (abs) (13.8 kPa).

The expansion efficiency, given as 0.75, is $H_1 - H_2/(H_1 - H_{2s}) =$ actual work/ideal work $= 0.75 = 1335.5 - H_2/(1335.5 - 845.3)$. Solve for $H_2 = 967.9$ Btu/lb (2251.3 kJ/kg).

Next, read from the saturation-pressure steam table the liquid and vapor properties at 2 lb/in^2 (abs) (13.8 kPa). Tabulate these properties thus:

$$h_f = 93.99 \text{ Btu/lb (218.6 kJ/kg)}$$

$$h_{fg} = 1022.2 \text{ Btu/lb (2377.6 kJ/kg)}$$

$$h_g = 1116.2 \text{ Btu/lb (2596.3 kJ/kg)}$$

$$s_f = 0.1749 \text{ Btu/(lb·°F) [0.73 kJ/(kg·°C)]}$$

$$s_{fg} = 1.7451 \text{ Btu/(lb·°F) [7.31 kJ/(kg·°C)]}$$

$$s_g = 1.9200 \text{ Btu/(lb·°F) [8.04 kJ/(kg·°C)]}$$

$$u_f = 93.98 \text{ Btu/lb (218.60 kJ/kg)}$$

$$u_{fg} = 957.9 \text{ Btu/lb (2228.1 kJ/kg)}$$

$$u_g = 1051.9 \text{ Btu/lb (2446.7 kJ/kg)}$$

$$v_f = 0.016 \text{ ft}^3\text{/lb (0.0010 m}^3\text{/kg)}$$

$$v_{fg} = 173.71 \text{ ft}^3\text{/lb (10.84 m}^3\text{/kg)}$$

$$v_g = 173.73 \text{ ft}^3\text{/lb (10.85 m}^3\text{/kg)}$$

Since the actual final enthalpy H_2 is different from H_{2s}, the final actual moisture y_2 must be computed by using H_2. Or, $y_2 = h_g - H_2/h_{fg} = 1116.1 - 967.9/1022.2 = 0.1451$. Then

$$E_2 = u_g - y_2 u_{fg} = 1051.9 - 0.1451(957.9) = 912.9 \text{ Btu/lb (2123.4 kJ/kg)}$$
$$V_2 = v_g - y_2 v_{fg} = 173.73 - 0.1451(173.71) = 148.5 \text{ ft}^3\text{/lb (9.3 m}^3\text{/kg)}$$
$$S_2 = s_g - y_2 s_{fg} = 1.9200 - 0.1451(1.7451) = 1.6668 \text{ Btu/(lb·°F) [7.0 kJ/kg·°C)]}$$

3. Compute the changes resulting from the process

The total change in properties is for 10 lb (4.5 kg) of steam, the quantity used in this process. Thus

$$\Delta E = (E_1 - E_2)m = (1221.6 - 912.9)(10) = 3087 \text{ Btu (3257.0 kJ)}$$
$$\Delta H = (H_1 - H_2)m = (1335.5 - 967.9)(10) = 3676 \text{ Btu (3878.4 kJ)}$$
$$\Delta S = (S_2 - S_1)m = (1.6668 - 1.4576)(10) = 2.092 \text{ Btu/°F (4.0 kJ/°C)}$$
$$\Delta V = (V_2 - V_1)m = (148.5 - 0.3074)(10) = 1482 \text{ ft}^3 \text{ (42.0 m}^3\text{)}$$

So $Q = 0$; by definition, $W_2 = \Delta H = 3676$ Btu (3878.4 kJ) for the steady-flow process.

IRREVERSIBLE ADIABATIC STEAM COMPRESSION

Two pounds (0.9 kg) of saturated steam at 120 lb/in² (abs) (827.4 kPa) with 80 percent quality undergoes nonflow adiabatic compression to a final pressure of 1700 lb/in² (abs) (11,721.5 kPa) at 75 percent compression efficiency. Determine the final steam temperature T_2, change in internal energy ΔE, change in entropy ΔS, work input W, and heat input Q.

Calculation Procedure:

1. Determine the vapor properties in the initial state
From the saturation-pressure steam tables, $T_1 = 341.25°F$ (171.8°C) at a pressure of 120 lb/in² (abs) (827.4 kPa) saturated. With $x_1 = 0.8$, $E_1 = u_f + x_1 u_{fg} = 312.05 + 0.8(795.6) = 948.5$ Btu/lb (2206.5 kJ/kg), from internal-energy values from the steam tables. The initial entropy is $S_1 = s_f + x_1 s_{fg} = 0.4916 + 0.8(1.0962) = 1.3686$ Btu/(lb·°F) [5.73 kJ/(kg·°C)].

2. Determine the vapor properties in the final state
Sketch a T-S diagram of the process, Fig. 69. Assume a constant-entropy compression from the initial to the final state. Then $S_{2s} = S_1 = 1.3686$ Btu/(lb·°F) [5.7 kJ/(kg·°C)].

The final pressure, 1700 lb/in² (abs) (11,721.5 kPa), is known, as is the final entropy, 1.3686 Btu/(lb·°F) [5.7 kJ/(kg·°C)] with constant-entropy expansion. The T-S diagram (Fig. 69) shows that the steam is superheated in the final state. Enter the superheated steam table at 1700 lb/in² (abs) (11,721.5 kPa), project across to an entropy of 1.3686, and read the final steam temperature as 650°F (343.3°C). (In most cases, the final entropy would not exactly equal a tabulated value, and it would be necessary to interpolate between tabulated entropy values to determine the intermediate pressure value.)

From the same table, at 1700 lb/in² (abs) (11,721.5 kPa) and 650°F (343.3°C), $H_{2s} = 1214.4$ Btu/lb (2827.4 kJ/kg); $V_{2s} = 0.2755$ ft³/lb (0.017 m³/lb). Then

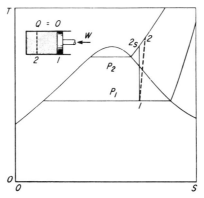

FIGURE 69 Irreversible adiabatic compression process.

$E_{2s} = H_{2s} - P_2V_{2s}/J = 1214.4 - 1700(144)(0.2755)/788 = 1127.8$ Btu/lb (2623.3 kJ/kg). Since E_1 and E_{2s} are known, the ideal work W can be computed. Or, $W = E_{2s} - E_1 = 1127.8 - 948.5 = 179.3$ Btu/lb (417.1 kJ/kg).

3. Compute the vapor properties of the actual compression

Since the compression efficiency is known, the actual final internal energy can be found from compression efficiency = ideal W/actual $W = E_{2s} - E_1/(E_2 - E_1)$, or $0.75 = 1127.8 - 948.5/(E_2 - 948.5)$; $E_2 = 1187.6$ Btu/lb (2762.4 kJ/kg). Then $E = (E_2 - E_2)m = (1187.6 - 948.5)(2) = 478.2$ Btu (504.5 kJ) for 2 lb (0.9 kg) of steam. The actual work input $W = \Delta E = 478.2$ Btu (504.5 kJ). By definition, $Q = 0$.

Last, the actual final temperature and entropy must be computed. The final actual internal energy $E_2 = (1187.6$ Btu/lb (2762.4 kJ/kg) is known. Also, the T-S diagram shows that the steam is superheated. However, the superheated steam tables do not list the internal energy of the steam. Therefore, it is necessary to assume a final temperature for the steam and then compute its internal energy. The computed value is compared with the known internal energy, and the next assumption is adjusted as necessary. Thus, assume a final temperature of 720°F (382.2°C). This assumption is higher than the ideal final temperature of 650°F (343.3°C) because the T-S diagram shows that the actual final temperature is higher than the ideal final temperature. Using values from the superheated steam table for 1700 lb/in² (abs) (11,721.5 kPa) and 720°F (382.2°C), we find

$$E = H - \frac{PV}{J} = 1288.4 - \frac{1700(144)(0.3283)}{778} = 1185.1 \text{ Btu/lb (2756.5 kJ/kg)}$$

This value is less than the actual internal energy of 1187.6 Btu/lb (2762.4 kJ/kg). Therefore, the actual temperature must be higher than 720°F (382.2°C), since the internal energy increases with temperature. To obtain a higher value for the internal energy to permit interpolation between the lower, actual, and higher values, assume a higher final temperature—in this case, the next temperature listed in the steam table, or 740°F (393.3°C). Then, for 1700 lb/in² (abs) (11,721.5 kPa) and 740°F (393.3°C),

$$E = 1305.8 - \frac{1700(144)(0.3410)}{778} = 1198.5 \text{ Btu/lb (2757.7 kJ/kg)}$$

This value is greater than the actual internal energy of 1187.6 Btu/lb (2762.4 kJ/kg). Therefore, the actual final temperature of the steam lies somewhere between 720 and 740°F (382.2 and 393.3°C). Interpolate between the known internal energies to determine the final steam temperature and final entropy. Or,

$$T_2 = 720 + \frac{1178.6 - 1185.1}{1198.5 - 1185.1}(740 - 720) = 723.7°F \text{ (384.3°C)}$$

$$S_1 = 1.4333 + \frac{1187.6 - 1185.1}{1198.5 - 1185.1}(1.4480 - 1.4333)$$

$$= 1.4360 \text{ Btu/(lb} \cdot °F) \text{ [6.0 kJ/(kg} \cdot °C)]}$$

$$\Delta S = (S_2 - S_1)m = (1.4360 - 1.3686)(2) = 0.1348 \text{ Btu/°F (0.26 kJ/°C)}$$

Note that the final actual steam temperature is 73.7°F (40.9°C) higher than that (650°F or 343.3°C) for the ideal compression.

Related Calculations. Use this procedure for any irreversible adiabatic steam process.

THROTTLING PROCESSES FOR STEAM AND WATER

A throttling process begins at 500 lb/in² (abs) (3447.5 kPa) and ends at 14.7 lb/in² (abs) (101.4 kPa) with (1) steam at 500 lb/in² (abs) (3447.5 kPa) and 500°F (260.0°C); (2) steam at 500 lb/in² (abs) (3447.5 kPa) and 4 percent moisture; (3) steam at 500 lb/in² (abs) (3447.5 kPa) with 50 percent moisture; and (4) saturated water at 500 lb/in² (abs) (3447.5 kPa). Determine the final enthalpy H_2, temperature T_2, and moisture content y_2 for each process.

Calculation Procedure:

1. Compute the final-state conditions of the superheated steam
From the superheated steam table for 500 lb/in² (abs) (3447.5 kPa) and 500°F (260.0°C), $H_1 = 1231.3$ Btu/lb (2864.0 kJ/kg). By definition of a throttling process, $H_1 = H_2 = 1231.3$ Btu/lb (2864.0 kJ/kg). Sketch the *T-S* diagram for a throttling process, Fig. 70.

To determine the final temperature, enter the superheated steam table at 14.7 lb/in² (abs) (101.4 kPa), the final pressure, and project across to an enthalpy value equal to or less than known enthalpy, 1231.3 Btu/lb (2864.0 kJ/kg). (The superheated steam table is used because the *T-S* diagram, Fig. 70, shows that the steam is superheated in the final state.) At 14.7 lb/in² (abs) (101.4 kPa) there is no tabulated enthalpy value that exactly equals 1231.3 Btu/lb (2864.0 kJ/kg). The next lower value is 1230 Btu/lb (2861.0 kJ/kg) at $T = 380°F$ (193.3°C). The next higher value at 14.7 lb/in² (abs)(101.4 kPa) is 1239.9 Btu/lb (2884.0 kJ/kg) at $T = 400°F$ (204.4°C). Interpolate between these enthalpy values to find the final steam temperature:

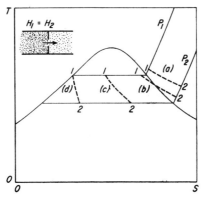

FIGURE 70 Throttling process for steam.

$$T_2 = 380 + \frac{1231.3 - 1230.5}{1239.9 - 1230.5}(400 - 380) = 381.7°F \ (194.3°C)$$

The steam does not contain any moisture in the final state because it is superheated.

2. Compute the final-state conditions of the slightly wet steam
Determine the enthalpy of 500-lb/in² (abs) (3447.5-kPa) saturated steam from the saturation-pressure steam table:

$$h_g = 1204.4 \ Btu/lb \ (2801.4 \ kJ/kg) \qquad h_{fg} = 755.0 \ Btu/lb \ (1756.1 \ kJ/kg)$$

Correct the enthalpy for moisture:

$$H_1 = h_g - y_1 h_{fg} = 1204.4 - 0.04(755.0) = 1174.2 \ Btu/lb \ (2731.2 \ kJ/kg)$$

Then, by definition, $H_2 = H_1 = 1174.2$ Btu/lb (2731.2 kJ/kg).
  Determine the final condition of the throttled steam (wet, saturated, or super-heated) by studying the T-S diagram. If a diagram were not drawn, you would enter the saturation-pressure steam table at 14.7 lb/in² (abs) (101.4 kPa), the final pressure, and check the tabulated h_g. If the tabulated h_g were greater than H_1, the throttled steam would be superheated. If the tabulated h_g were less than H_1, the throttled steam would be saturated. Examination of the saturation-pressure steam table shows that the throttled steam is superheated because $H_1 > h_g$.
  Next, enter the superheated steam table to find an enthalpy value of H_1 at 14.7 lb/in² (abs) (101.4 kPa). There is no value equal to 1174.2 Btu/lb (2731.2 kJ/kg). The next lower value is 1173.8 Btu/lb (2730.3 kJ/kg) at $T = 260°F$ (126.7°C). The next higher value at 14.7 lb/in² (abs) (101.4 kPa) is 1183.3 Btu/lb (2752.4 kJ/kg) at $T = 280°F$ (137.8°C). Interpolate between these enthalpy values to find the final steam temperature:

$$T_2 = 260 + \frac{1174.2 - 1173.8}{1183.3 - 1173.8}(280 - 260) = 260.8°F \ (127.1°C)$$

This is higher than the temperature of saturated steam at 14.7 lb/in² (abs) (101.4 kPa)—212°F (100°C)— giving further proof that the throttled steam is superheated. The throttled steam, therefore, does not contain any moisture.

3. Compute the final-state conditions of the very wet steam
Determine the enthalpy of 500-lb/in² (abs) (3447.5-kPa) saturated steam from the saturation-pressure steam table. Or, $h_g = 1204.4$ Btu/lb (2801.4 kJ/kg); $h_{fg} = 755.0$ Btu/lb (1756.1 kJ/kg). Correct the enthalpy for moisture:

$$H_1 = H_2 = h_g - y_1 h_{fg} = 1204.4 - 0.5(755.0) = 826.9 \ Btu/lb \ (1923.4 \ kJ/kg)$$

Then, by definition, $H_2 = H_1 = 826.9$ Btu/lb (1923.4 kJ/kg).
  Compare the final enthalpy, $H_2 = 826.9$ Btu/lb (1923.4 kJ/kg), with the enthalpy of saturated steam at 14.7 lb/in² (abs) (101.4 kPa), or 1150.4 Btu/lb (2675.8 kJ/kg). Since the final enthalpy is less than the enthalpy of saturated steam at the same pressure, the throttled steam is wet. Since $H_1 = h_g - y_2 h_{fg}$, $y_2 = (h_g - H_1)/h_{fg}$. With a final pressure of 14.7 lb/in² (abs) (101.4 kPa), use h_g and h_{fg} values at this pressure. Or,

$$y_2 = \frac{1150.4 - 826.9}{970.3} = 0.3335, \ or \ 33.35\%$$

The final temperature of the steam T_2 is the same as the saturation temperature at the final pressure of 14.7 lb/in² (abs) (101.4 kPa), or T_2 = 212°F (100°C).

4. Compute the final-state conditions of saturated water
Determine the enthalpy of 500-lb/in² (abs) (3447.5-kPa) saturated water from the saturation-pressure steam table at 500 lb/in² (abs) (3447.5 kPa); H_1 = h_f = 449.4 Btu/lb (1045.3 kJ/kg) = H_2, by definition. The T-S diagram, Fig. 70, shows that the throttled water contains some steam vapor. Or, comparing the final enthalpy of 449.4 Btu/lb (1045.3 kJ/kg) with the enthalpy of saturated liquid at the final pressure, 14.7 lb/in² (abs) (101.4 kPa), 180.07 Btu/lb (418.8 kJ/kg), shows that the liquid contains some vapor in the final state because its enthalpy is greater.

Since H_1 = H_2 = h_g − $y_2 h_{fg}$, y_2 = $(h_g - H_1)/h_{fg}$. Using enthalpies at 14.7 lb/in² (abs) (101.4 kPa) of h_g = 1150.4 Btu/lb (2675.8 kJ/kg) and h_{fg} = 970.3 Btu/lb (2256.9 kJ/kg) from the saturation-pressure steam table, we get y_2 = 1150.4 − 449.4/970.3 = 0.723. The final temperature of the steam is the same as the saturation temperature at the final pressure of 14.7 lb/in² (abs) (101.4 kPa), or T_2 = 212°F (100°C).

Note: Calculation 2 shows that when you start with slightly wet steam, it can be throttled (expanded) through a large enough pressure range to produce superheated steam. This procedure is often used in a throttling calorimeter to determine the initial quality of the steam in a pipe. When very wet steam is throttled, calculation 3, the net effect may be to produce drier steam at a lower pressure. Throttling saturated water, calculation 4, can produce partial or complete flashing of the water to steam. All these processes find many applications in power-generation and process-steam plants.

REVERSIBLE HEATING PROCESS FOR STEAM

Subcooled water at 1500 lb/in² (abs) (10,342.5 kPa) and 140°F (60.0°C), state 1, Fig. 71, is heated at constant pressure to state 4, superheated steam at 1500 lb/in² (abs) (10,342.5 kPa) and 1000°F (537.8°C). Find the heat added (1) to raise the

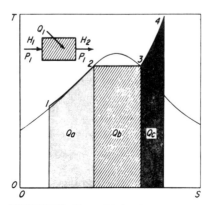

FIGURE 71 Reversible heating process.

compressed liquid to saturation temperature, (2) to vaporize the saturated liquid to saturated steam, (3) to superheat the steam to 1000°F (537.8°C), and (4) Q_1, ΔV, and ΔS from state 1 to state 4.

Calculation Procedure:

1. Sketch the T-S diagram for this process
Figure 71 is typical of a steam boiler and superheater. Feedwater fed to a boiler is usually subcooled liquid. If the feedwater pressure is relatively high, subcooling must be taken into account, if accurate results are desired. Some authorities recommend that at pressures below 400 lb/in² (abs) (2758.0 kPa) subcooling be ignored and values from the saturated-steam table be used. This means that the enthalpies and other properties listed in the steam table corresponding to the actual water temperature are sufficiently accurate. But above 400 lb/in² (abs) 2758.0 kPa), the compressed-liquid table should be used.

2. Determine the initial properties of the liquid
In the saturation-temperature steam table read, at 140°F (60.0°C), $h_f = 107.89$ Btu/lb (251.0 kJ/kg); $p_f = 2.889$ lb/in² (19.9 kPa); $v_f = 0.01629$ ft³/lb (0.0010 m³/kg); $s_f = 0.1984$ Btu/(lb · °F) [0.83 kJ/(kg · °C)].

Next, the enthalpy, volume, and entropy of the water at 1500 lb/in² (abs) (10,342.5 kPa) and 140°F (60.0°C) must be found. Since the water is at a much higher pressure than that corresponding to its temperature [1500 versus 2.889 lb/in² (abs)], the compressed-liquid portion of the steam table must be used. This table shows that three desired properties are plotted for 32, 100, and 200°F (0.0, 37.8, and 93.3°C) and higher temperatures. However, 140°F (60.0°C) is not included. Therefore, it is necessary to interpolate between 100 and 200°F (37.8 and 93.3°C). Thus, at 1500 lb/in² (abs) (10,342.5 kPa) in the compressed-liquid table:

	Temperature		
Property	100°F (37.8°C)	200°F (93.3°C)	Interpolation
$h - h_f$, Btu/lb (kJ/kg)	+3.99 (+9.28)	+3.36 (+7.82)	+3.74 (+8.70)
$(v - v_f)10^5$, ft³/lb (m³/kg)	−7.5 (−0.47)	−8.1 (−0.51)	−7.7 (−0.48)
$(s - s_f)10^3$, Btu/(lb·°F) [kJ/ (kg·°C)]	−0.86 (−3.60)	−1.79 (−7.49)	−1.23 (−5.15)

Each property is interpolated in the following way:

$$h - h_f = 3.99 - \frac{3.99 - 3.36}{200 - 100}(140 - 100) = 3.99 - 0.25$$

$$= 3.74 \text{ Btu/lb } (8.70 \text{ kJ/kg})$$

$$(v - v_f)10^5 = -7.5 - \frac{8.1 - 7.5}{200 - 100}(140 - 100) = -7.5 - 0.24$$

$$= -7.74 \text{ ft}^3/\text{lb } (-0.48 \text{ m}^3/\text{kg})$$

$$(s - s_f)10^3 = -0.86 - \frac{1.79 - 0.86}{200 - 100}(140 - 100) = -0.86 - 0.37$$

$$= -1.23 \text{ Btu/(lb·°F) } [-5.15 \text{ kJ/(kg·°C)}]$$

These interpolated values must now be used to correct the saturation data at 140°F (60.0°C) to the actual subcooled state 1 properties. Thus, at 1500 lb/in² (abs) (10,342.5 kPa) and 140°F (60.0°C).

DETERMINING STEAM ENTHALPY AND QUALITY USING THE STEAM TABLES

What is the enthalpy of 200-lb/in² absolute (1378 kPa) wet steam having a 70 percent quality? Determine the quality of 160-lb/in² absolute (1102.4 kPa) wet steam if its enthalpy is 900 Btu/lb (2093.4 kJ/kg). Use the steam tables to determine the needed values.

Calculation Procedure:

1. Compute the enthalpy of the wet steam
The enthalpy of wet steam is a function of its quality, or dryness fraction. Thus, 70 percent quality steam will be 70 percent dry and 30 percent wet. In equation form, the enthalpy of wet steam, Btu/lb (kJ/kg), $h = Xh_g + (1 - X)h_f$, where X = steam quality or dryness fraction expressed as a decimal; h_g = enthalpy of saturated steam vapor, Btu/lb (kJ/kg); h_f = enthalpy of saturated water fluid, Btu/lb (kJ/kg).

Substituting in this equation using steam-table values, $h = 0.70(1187.2) + (1 - 0.70)298.4 = 920.56$ Btu/lb (2141.22 kJ/kg). Note that the enthalpy of wet steam is not a simple product of the dryness factor (*i.e.* quality) and the enthalpy of the saturated steam. Instead, the enthalpy of the saturated liquid at the saturation pressure must also be included, adjusted for the quality of the steam.

2. Determine the quality of the steam
Knowing the absolute pressure and enthalpy of steam we can determine its quality from, $X = (h - h_f)/(h_g - h_f)$, where the symbols are defined as given above. Substituting, using steam-table values, $X = (900 - 335.93)/(1195.1 - 335.93) = 0.6565$; say 65.7 percent.

Related Calculations.   Wet steam is a fact of life in industrial plants of every type. When steam is wet it means that a larger number of pounds (kg) of steam are needed to perform a needed task. Thus, if a process requires 1000 lb (454 kg) of saturated dry steam, it will need 1050 lb (476.7 kg) of steam having a 95 percent quality, *i.e.* 5 percent more. With the 70 percent quality steam mentioned earlier, the same process would require 1300 lb (590.2 kg) of saturated dry steam. So it is easy to see why it is important to deliver dry saturated steam to a process because the overall steam consumption is reduced.

In today's energy-conscious engineering world, wet steam is an undesirable commodity unless the wetness results from doing useful work. Where steam wetness occurs because of poor pipe or equipment insulation, shoddy piping layout, or other engineering or installation errors, energy is being wasted. As the above examples show, steam consumption, and generation cost, can rise as much as 30 percent because of wet steam.

Every plant designer should keep wetness in mind when designing industrial steam plants of any type serving chemical, steel, textile, marine, automotive, aircraft, *etc.* industries.

The procedure given here can be used in any application where steam is a process fluid, and this procedure can also be used when measuring steam quality using a calorimeter of any type.

MAXIMIZING COGENERATION ELECTRIC-POWER AND PROCESS-STEAM OUTPUT

An industrial cogeneration plant is being designed for a process requiring electric power and steam for operating the manufacturing equipment in the plant. In the design analysis a steam pressure of 650 lb/in² (gage) (4478.5 kPa) saturated is being considered for a steam turbine to generate the needed electricity, with the exhaust steam being used in the process sections of the plant at 150 lb/in² (gage) (1033.5 kPa). If greater process efficiencies can be obtained in manufacturing with a higher exhaust temperature, compare the effect of greater turbine inlet pressure and temperature on the resulting exhaust steam temperature and heat content. Use typical pressure and temperature levels met in industrial steam-turbine applications.

Calculation Procedure:

1. Assemble data on typical industrial steam-turbine inlet conditions
Using data obtained from steam-turbine manufacturers, list the typical inlet pressures and temperatures used today in a tabulation such as that in Fig. 72. This listing shows typical pressure ranges from 650 lb/in² (gage) (4478.5 kPa) saturated to 1200 lb/in² (gage) (8268 kPa) at temperatures from saturated, 498°F (259°C), to 950°F (510°C), at 1200 lb/in² (gage).

2. Plot the turbine expansion process for each pressure being considered
Using an H-s diagram, Fig. 72, plot the expansion from the turbine inlet to exhaust at 150 lb/in² (gage) (1033.5 kPa). Read the temperature and the enthalpy at the exhaust and tabulate each value as in the figure.

3. Compute the percent increase in the exhaust enthalpy
Using data from the H-s chart, we see that with an inlet pressure of 650 lb/in² (gage) saturated (4487.5 kPa), the exhaust steam has 8 percent moisture and an enthalpy change of 70 Btu/lb (163.1 kJ/kg) with an exhaust temperature of 360°F (182°C).

Raising the turbine inlet steam temperature to 750°F (398.9°C) increases the enthalpy change during expansion to $1380 - 1264 = 116$ Btu/lb (270.28 kJ/kg). This is an increase of $116 - 70 = 46$ Btu/lb (107.2 kJ/kg) over the saturated-steam inlet enthalpy change during expansion from inlet to exhaust. Then, the percent increase in enthalpy change for these two inlet conditions is $100(116 - 70)/70 = 65.7$ percent; say 66 percent, as tabulated.

Continuing these calculations and tabulating the results shows that—for the pressures and temperatures considered—an increase of up to 148 percent in the enthalpy change can be obtained. Likewise, an exhaust superheat up to 320°F (160°C) can be obtained. Depending on the process served, a suitable steam pressure and temperature can be chosen for the turbine inlet conditions to maximize the efficiency of the process(es) served by this cogeneration plant.

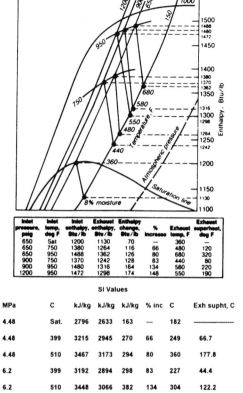

Inlet pressure, psig	Inlet temp, deg F	Inlet enthalpy, Btu/lb	Exhaust enthalpy, Btu/lb	Enthalpy change, Btu/lb	% increase	Exhaust temp, F	Exhaust superheat, deg F
650	Sat	1200	1130	70	—	360	—
650	750	1380	1264	116	66	480	120
650	950	1488	1362	126	80	680	320
900	750	1370	1242	128	83	440	80
900	950	1480	1316	164	134	580	220
1200	950	1472	1298	174	148	550	190

SI Values

MPa	C	kJ/kg	kJ/kg	kJ/kg	% inc	C	Exh supht, C
4.48	Sat.	2796	2633	163	—	182	——————
4.48	399	3215	2945	270	66	249	66.7
4.48	510	3467	3173	294	80	360	177.8
6.2	399	3192	2894	298	83	227	44.4
6.2	510	3448	3066	382	134	304	122.2
8.27	510	3430	3024	405	148	288	105.5

FIGURE 72   H-s plot of expansion in a steam turbine.
(*Power.*)

Related Calculations.   Use this general approach for steam cogeneration plant design for industries in any of these fields—chemical, petroleum, textile, food, tobacco, agriculture, manufacturing, automobile, etc. While the cogeneration plant considered here exhausts directly to the process mains, condensing turbines can also be used. With such machines, process steam is extracted at a suitable point in the cycle to provide the needed pressure and temperature.

Condensing steam turbines with steam extraction for process needs offer a significant increase in electric-power production. Further, they reduce the effects of seasonal price fluctuations for power sales, and potentially lower life-cycle costs. Capital cost of such machines, however, is higher.

Turbine suppliers are now also offering efficient, high-speed geared turbo-generators for cogeneration. One such supplier states that the incremental payback for a geared unit can be extremely high in cogeneration applications.

Much of the data in this procedure came from *Power* magazine. SI values were added by the editor.

NOTE:

Calculation Procedure:

1. Find the amount of oxygen required for complete combustion of the fuel
Eight atoms of carbon in C_8 combine with 8 molecules of oxygen, O_2, and produce 8 molecules of carbon dioxide, $8CO_2$. Similarly, 9 molecules of hydrogen, H_2, in H_{18} combine with 9 atoms of oxygen, O, or 4.5 molecules of oxygen, to form 9 molecules of water, $9H_2O$. Thus, 100 percent, or the stoichiometric, air quantity required for complete combustion of a mole of fuel, C_8H_{18}, is proportional to $8 + 12.5$ moles of oxygen, O_2.

2. Establish the chemical equation for complete combustion with 100 percent air
With 100 percent air: $C_8H_{18} + 12.5\ O_2 + (3.784 \times 12.5)N_2 \rightarrow 8CO_2 + 9H_2O + 47.3N_2$, where 3.784 is a derived volumetric ratio of atmospheric nitrogen, (N_2), to oxygen O_2, in dry air. The (N_2) includes small amounts of inert and inactive gases. See Related Calculations of this procedure.

3. Establish the chemical equation for complete combustion with 400 percent of the stoichiometric air quantity, or 300 percent excess air
With 400 percent air: $C_8H_{18} + 50\ O_2 + (4 \times 47.3)N_2 \rightarrow 8CO_2 + 9H_2O + 189.2N_2 + (3 \times 12.5)O_2$.

4. Compute the molecular weights of the components in the combustion process
Molecular weight of $C_8H_{18} = [(12 \times 8) + (1 \times 18)] = 114$; $O_2 = 16 \times 2 = 32$; $N_2 = 14 \times 2 = 28$; $CO_2 = [(12 \times 1) + (16 \times 2)] = 44$; $H_2O = [(1 \times 2) + (16 \times 1)] = 18$.

5. Compute the relative weights of the reactants and products of the combustion process
Relative weight = moles × molecular weight. Coefficients of the chemical equation in step 3 represent the number of moles of each component. Hence, for the reactants, the relative weights are: $C_8H_{18} = 1 \times 114 = 114$; $O_2 = 50 \times 32 = 1600$; $N_2 = 189.2 \times 28 = 5298$. Total relative weight of the reactants is 7012. For the products, the relative weights are: $CO_2 = 8 \times 44 = 352$; $H_2O = 9 \times 18 = 162$; $N_2 = 189.2 \times 28 = 5298$; $O_2 = 37.5 \times 32 = 1200$. Total relative weight of the products is 7012, also.

6. Compute the enthalpy of the products of the combustion process
Enthalpy of the products of combustion, $h_p = m_p(h_{1600} - h_{77})$, where m_p = number of moles of the products; h_{1600} = enthalpy of the products at 1600°F (871°C); h_{77} = enthalpy of the products at 77°F (25°C). Thus, $h_p = (8 + 9 + 189.2 + 37.5)(15,400 - 3750) = 2,839,100$ Btu [6,259,100 Btu (SI)].

7. Compute the air supply temperature at the combustion chamber inlet
Since the combustion process is adiabatic, the enthalpy of the reactants $h_r = h_p$, where h_r = (relative weight of the fuel × its heating value) + [relative weight of the air × its specific heat × (air supply temperature − air source temperature)]. Therefore, $h_r = (114 \times 19,100) + [(1600 + 5298) \times 0.24 \times (t_a - 77)] + 2,839,100$ Btu [6,259,100 Btu (SI)]. Solving for the air supply temperature, $t_a = [(2,839,100 - 2,177,400/1655.5] + 77 = 477°F$ (247°C).

Related Calculations: This procedure, appropriately modified, may be used to deal with similar questions involving such things as other fuels, different amounts of excess air, and variations in the condition(s) being sought under certain given circumstances.

The coefficient, (?) = 3.784 in step 2, in used to indicate that for each unit of volume of oxygen, O_2, 12.5 in this case, there will be 3.784 units of nitrogen, N_2. This equates to an approximate composition of air as 20.9 percent oxygen and 79.1 percent "atmospheric nitrogen," (N_2). In turn, this creates a paradox, because page 200 of *Principles of Engineering Thermodynamics*, by Kiefer, et al., John Wiley & Sons, Inc., states air to be 20.99 percent oxygen and 79.01 percent atmospheric nitrogen, where the ratio (N_2)/O_2 = (?) = 79.01/20.99 = 3.764.

Also, page 35 of *Applied Energy Conversion,* by Skrotski and Vopat, McGraw-Hill, Inc., indicates an assumed air analysis of 79 percent nitrogen and 21 percent oxygen, where (?) = 3.762. On that basis, a formula is presented for the amount of dry air chemically necessary for complete combustion of a fuel consisting of atoms of carbon, hydrogen, and sulfur, or C, H, and S, respectively. That formula is: $W_a = 11.5C + 34.5[H - (O/8)] + 4.32S$, lb air/lb fuel (kg air/kg fuel).

The following derivation for the value of (?) should clear up the paradox and show that either 3.784 or 3.78 is a sound assumption which seems to be wrong, but in reality is not. In the above equation for W_a, the carbon, hydrogen, or sulfur coefficient, $C_x = (MO_2/DO_2)M_x$, where MO_2 is the molecular weight of oxygen, O_2; DO_2 is the decimal fraction for the percent, by weight, of oxygen, O_2, in dry air containing "atmospheric nitrogen," (N_2), and small amounts of inert and inactive gases: M_x is the formula weight of the combustible element in the fuel, as indicated by its relative amount as a reactant in the combustion equation. The alternate evaluation of C_x is obtained from stoichiometric chemical equations for burning the combustible elements of the fuel, i.e., $C + O_2 + (?)N_2 \rightarrow CO_2 + (?)N_2$; $2H_2 + O_2 + (?)N_2 \rightarrow 2H_2O + (?)N_2$; $S + O_2 + (?)N_2 \rightarrow SO_2 + (?)N_2$. Evidently, $C_x = [MO_2 + (? \times MN_2)]/M_x$, where MN_2 is the molecular weight of nitrogen, N_2, and the other items are as before.

Equating the two expressions, $C_x = [MO_2 + (? \times MN_2)]/M_x = (MO_2/DO_2)M_x$, reveals that the M_x terms cancel out, indicating that the formula weight(s) of combustible components are irrelevant in solving for (?). Then, (?) = $(1 - DO_2)[MO_2/(MN_2 \times DO_2)]$. From the above-mentioned book by Kiefer, et al., $DO_2 = 0.23188$. From *Marks' Standard Handbook for Mechanical Engineers*, McGraw-Hill, Inc., $MO_2 = 31.9988$ and $MN_2 = 28.0134$. Thus, (?) = $(1 - 0.23188)[31.9988/(28.0134 \times 0.23188)] = 3.7838$. This demonstrates that the use of (?) = 3.784, or 3.78, is justified for combustion equations.

By using either of the two evaluation equations for C_x, and with accurate values of M_x, i.e., $M_C = 12.0111$; $M_H = 2 \times 2 \times 1.00797 = 4.0319$; $M_S = 32.064$, from *Marks' M.E. Handbook*, the more precise values for C_C, C_H, and C_S are found out to be 11.489, 34.227, and 4.304, respectively. However, the actual C_x values, 11.5, 34.5, and 4.32, used in the formula for W_a are both brief for simplicity and rounded up to be on the safe side.

SECTION 2
STEAM CONDENSING SYSTEMS AND AUXILIARIES

DESIGN OF CONDENSER CIRCULATING-WATER SYSTEMS FOR POWER PLANTS

Design a condenser circulating-water system for a turbine-generator steam station located on a river bank. Show how to choose a suitable piping system and cooling arrangement. Determine the number of circulating-water pumps and their capacities to use. Plot an operating-point diagram for the various load conditions in the plant. Choose a suitable intake screen arrangement for the installations.

Calculation Procedure:

1. Choose the type of circulating-water system to use
There are two basic types of circulating-water systems used in steam power plants today—the once-through systems, Fig. 1a, and the recirculating-water system, Fig. 1b. Each has advantages and disadvantages.

In the once-through system, the condenser circulating water is drawn from a nearby river or sea, pumped by circulating-water pumps at the intake structure through a pipeline to the condenser. Exiting the condenser, the water returns to the river or sea. Advantages of a once-through system include: (a) simple piping arrangement; (b) lower cost where the piping runs are short; (c) simplicity of operation—the cooling water enters, then leaves the system. Disadvantages of once-through systems include: (a) possibility of thermal pollution—*i.e.*, temperature increase of the river or sea into which the warm cooling water is discharged; (b) loss of cooling capacity in the event of river or sea level decrease during droughts; (c) trash accumulation at the inlet, reducing water flow, during periods of river or sea pollution by external sources.

2.1

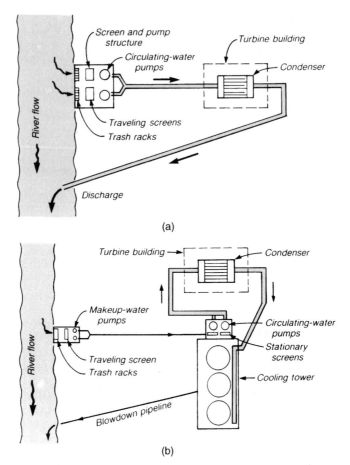

FIGURE 1 *a.* Once-through circulating-water system discharges warm water from the condenser directly to river or sea. Fig. 1*b*. Recirculating-water system reuses water after it passes through cooling tower and stationary screen. (*Power.*)

Recirculating systems use small amounts of water from the river or sea, once the system has been charged with water. Condenser circulating water is reused in this system after passing through one or more cooling towers. Thus, the only water taken from the river or sea is that needed for makeup of evaporation and splash losses in the cooling tower. The only water discharged to the river or sea is the cooling-tower blowdown. Advantages of the recirculating-water system include: (a) low water usage from the river or sea; (b) little or no thermal pollution of the supply water source because the cooling-tower blowdown is minimal; (c) remote chance of the need for service reductions during drought seasons. Disadvantages of recirculating systems include: (a) possible higher cost of the cooling tower(s) compared to the discharge piping in the once-through system; (b) greater operating complexity of the cooling tower(s), their fans, motors, pumps, etc.; (c) increased maintenance requirements of the cooling towers and their auxiliaries.

The final choice of the type of cooling system to use is based on an economic study which factors in the reliability of the system along with its cost. For the purposes of this procedure, we will assume that a once-through system with an intake length of 4500 ft (1372 m) and a discharge length of 4800 ft (1463 m) is chosen. The supply water level (a river in this case) can vary between +5 ft (1.5 m) and +45 ft (13.7 m).

2. Plot the operating-point diagram for the pumping system
The maximum cooling-water flow rate required, based on full-load steam flow through the turbine-generator, is 314,000 gpm (19,813 L/s). Intermediate flow rates of 283,000 gpm (17,857 L/s) and 206,000 gpm (12,999 L/s) for partial loads are also required.

To provide for safe 24-hour, 7-day-per-week operation of a circulating-water system, plant designers choose a minimum of two water pumps. As further safety step, a third pump is usually also chosen. That will be done for this plant.

Obtaining the pump characteristic curve from the pump manufacturer, we plot the operating-point diagram, Fig. 2, for one-pump, two-pump, and three-pump operation against the system characteristic curve for river (weir) levels of +5 ft (1.5 m) and +45 ft (13.7 m). We also plot on the operating-point diagram the seal-well weir curve.

The operating-point diagram is a valuable tool for both plant designers and operators because it shows the correct operating range of the circulating-water pumps. Proper use of the diagram can extend pump reliability and operating life.

3. Construct the energy-gradient curves for the circulating-water system
Using the head and flow data already calculated and assembled, plot the energy-gradient curve, Fig. 3, for several heads and flow rates. The energy-gradient curve, like the operating-point diagram, is valuable to both design engineers and plant operators. Practical experience with a number of actual circulating-water installations shows that early, and excessive, circulating-pump wear can be traced to the absence of an operating-point diagram and an energy-gradient curve, or to the lack of use of both these important plots by plant operating personnel.

In the once-through circulating-water system being considered here, the total conduit (pipe) length is 4500 + 4800 = 9300 ft (2835 m), or 1.76 mi (2.9 km). This conduit length is not unusual—some plants may have double this length of run. Such lengths, however, are much longer than those met in routine interior plant design where 100 ft (30.5 m) are the norm for "long" pipe runs. Because of the extremely long piping runs that might be met in circulating-water system design, the engineer must exercise extreme caution during system design—checking and double-checking all design assumptions and calculations.

4. Analyze the pump operating points
Using the operating-point diagram and the energy-gradient curves, plot the intersection of the system curves for each intake water level vs. the characteristic curves for the number of pumps operating, Fig. 3. Thus, we see that with one pump operating, the circulating-water flow is 120,000 gpm (7572 L/s) at 48.2 ft (14.7 m) total dynamic head.

With a weir level of +5 ft (1.5 m), and two pumps operating, the flow is 206,000 gpm (12,999 L/s) at 79 ft (24.1 m) total dynamic head. When three pumps are

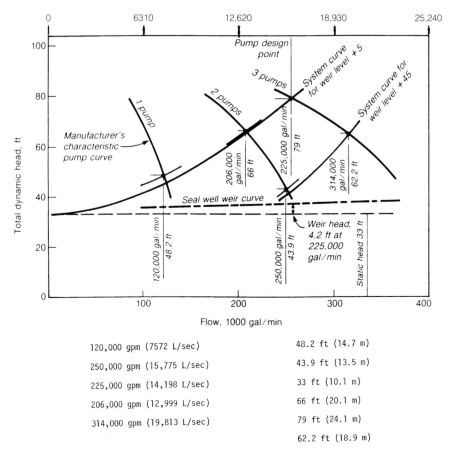

FIGURE 2 Operating-point diagram shows the correct operating range of the circulating-water pumps. (*Power.*)

used at the 5-ft (1.5 m) level, the flow is 225,000 gpm (14,198 L/s) at 79 ft (24.1 m) total dynamic head.

Using the sets of curves mentioned here you can easily get a complete picture of the design and operating challenges faced in this, and similar, plants. The various aspects of this are discussed under Related Calculations, below.

5. *Choose the type of intake structure and trash rack to use*

Every intake structure must provide room for the following components: (a) cir-culating-water or makeup-water pumps; (b) trash racks; (c) trash-removal screens—either fixed or traveling; (c) crane for handling pump removal or instal-lation; (d) screen wash pump; (e) access ladders and platforms.

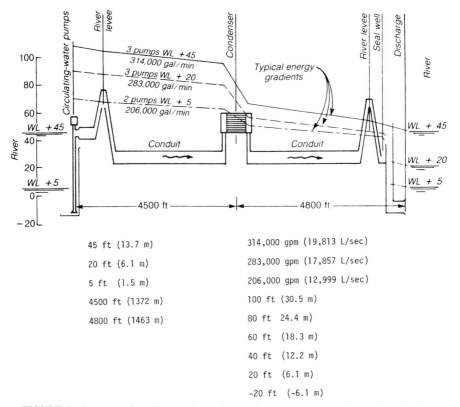

45 ft (13.7 m)   314,000 gpm (19,813 L/sec)

20 ft (6.1 m)   283,000 gpm (17,857 L/sec)

5 ft (1.5 m)   206,000 gpm (12,999 L/sec)

4500 ft (1372 m)   100 ft (30.5 m)

4800 ft (1463 m)   80 ft 24.4 m)

60 ft (18.3 m)

40 ft (12.2 m)

20 ft (6.1 m)

-20 ft (-6.1 m)

FIGURE 3 Energy-gradient diagram shows the actual system pressure values and is valuable in system design and operation. (*Power.*)

A typical intake structure having these components is shown in Fig. 4. This structure will be chosen for this installation because it meets the requirements of the design.

Trash-rack problems are among the most common in circulating-water systems and often involve unmanageable weed entanglements, rather than general debris. The type of trash rack and rack-cleaning facilities used almost exclusively in the United States and many international plants, is shown in Fig. 4. Usually, the trash rack is inclined and bars are spaced at about 3-in (76.2-mm). The trash rake may be mechanical or manual.

The two usual rake designs are the unguided rake, which rides on the trash bars, and the guided rake, which runs in guides on the two sides of the water channel. If the trash bars are vertical, the guided rake is almost a necessity to keep the rake on the bars. But neither solves all the problems.

If seaweed or grass loads are particularly severe, alternative trash rakes, such as the catenary or other moving-belt rakes, should be considered. These are rarely put into original domestic installations. There are many other alternative types of trash racks and rakes in use throughout the world that are successful in handling heavy

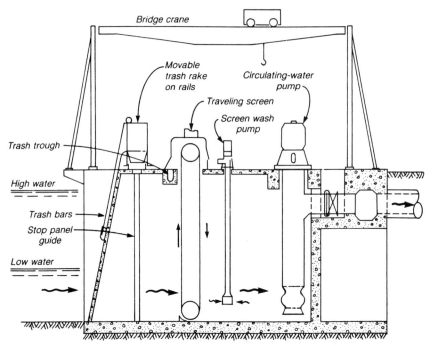

FIGURE 4   Intake structure has trash rack, traveling screen, pumps, and crane for dependable operation of the circulating-water system. (*Power.*)

loads. Log booms, skimmer walls, channel modifications, and specialized raking equipment can sometimes alleviate raking problems.

Traveling screens follow the trash racks. These usually are of the vertical flow-through type. European practice uses alternative screens, such as center-flow, dual-flow, and drum screens. Traveling screens may be one- or two-speed. Most two-speed screens operate in the range of 3 to 12 fpm (0.9 to 3.7 m/min) but speeds as high as 30 fpm (9.1 m/min) have been used. Wear is much greater at higher speeds.

Depending on the type of piping used in the circulating-water system—concrete or steel—some form of cathodic protection may be needed, in addition to the trash racks and rakes. Cathodic protection is needed primarily when steel pipe is used for the circulating water system. Concrete pipe does not, in general, require such protection. Since the piping in once-through systems can be 10 to 12 ft (3 to 3.7 m) in diameter, use of the cathodic protection is an important step in protecting an expensive investment. Cathodic protection methods are discussed elsewhere in this handbook.

Related Calculations.   Designing a condenser circulating-water system can be a complex task when the water supply is undependable. With a fixed-level supply, the design procedure is simpler. The above procedure covers the main steps in such designs. Head loss, pipe size, and other considerations are covered in detail in separate procedures given elsewhere in this handbook.

Construction of the operating-point diagram and the energy-gradient chart are important steps in the system design. Further, these two plots are valuable to operating personnel because they give the design assumptions for the system. When pressures or flow rates change, the operator will know that the system requires inspection to pinpoint the cause of the change.

The design procedure given here can be used for other circulating-water applications, such as those for refrigeration condensers, air-conditioning systems, internal-combustion-engine plants, *etc.*

Data given here are the work of R. T. Richards, Burns & Roe Inc., as reported in *Power* magazine. SI values were added by the handbook editor.

DESIGNING CATHODIC-PROTECTION SYSTEMS FOR POWER-PLANT CONDENSERS

Design a cathodic-protection system for an uncoated 10,000-tube steam condenser having an exposed waterbox/tubesheet surface area of 1000 ft^2 (92.9 m^2). Determine the protective current needed for this condenser if the design current density is 0.2 amp/ft^2 (2.15 amp/m^2) and 95 percent effective surface coverage will be maintained. How many anodes of magnesium, zinc, and aluminum would be needed in seawater to supply 50 amp for protection? Compare the number of anodes that would be needed in fresh water to supply 50 amp for protection.

Calculation Procedure:

1. *Determine the required protective current needed*
Cathodic protection of steam condensers is most often used to reduce galvanic corrosion of ferrous waterboxes coupled to copper-alloy tubesheets and tubes. Systems are also used to mitigate attack of both iron-based waterboxes and copper-alloy tubesheets in condensers tubed with titanium or stainless steel.

Cathodic protection is achieved by forcing an electrolytic direct current to flow to the structure to be protected. The name is derived from the fact that the protected structure is forced to be the cathode in a controlled electrolytic circuit.

There are two ways this current may be generated: (1) Either an external direct-current power source can be used, as in an impressed-current system, Fig. 5a, or (2) a piece of a more eletronegative metal can be electrically coupled to the structure, as in a sacrificial anode system, Fig. 5b.

The first step in the design of a cathodic-protection system is to estimate the current requirement. The usual procedure is to calculate the exposed waterbox and tubesheet area, and then compute the total current needed by assuming a current density. In practice, current needs are often estimated by applying a test current to the structure and measuring the change in structure potential.

Table 1 lists actual current densities used by utilities to protect condensers made of several different combinations of metals. The values given were taken from a survey prepared for the Electric Power Research Institute "Current Cathodic Protection Practice in Steam Surface Condensers," CS-2961, Project 1689-3, on which this procedure and its source are based.

With a design current density of 0.2 amp/ft^2 (2.15 amp/m^2), the total protective current need = 0.2 (1000) = 200 amp. With the 95 percent effective surface cov-

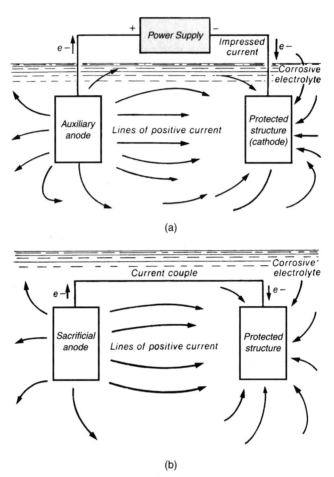

FIGURE 5 *a*. Impressed-current cathodic protection system uses external source to provide protective current. Fig. 5*b*. Sacrificial-anode cathodic protection uses piece of metal more electronegative than the structure for protection. (*Power.*)

erage, 5 percent of the surface will be exposed through coating faults. Hence, the required protective current will be 0.05(200) = 10 amp. Clearly, gross miscalculations are possible if the effectiveness of the coating is incorrectly estimated. The value of 0.2 amp/ft^2 (2.15 amp/m^2) is taken from the table mentioned above.

Another problem in estimating protective-current requirements occurs when condensers are tubed with noble alloy tubing such as stainless steel or titanium. In this case, a significant length of tubing (up to 20 ft—6.1 m) may be involved in the galvanic action, depending on the water salinity, temperature, and the tube material. This length dictates the anode/cathode area ratio and, thus, the rate of galvanic corrosion. Protective-current needs for this type of condenser can be unusually high.

TABLE 1 Current Densities Used for Various Condenser Materials*

Condenser materials			Design current density		Average water salinity
Waterbox	Tubesheet	Tubes	amp/ft^2	amp/m^2	ppm
Carbon steel	Aluminum bronze	90-10 Cu Ni	0.05	0.54	1000
Cast iron	Muntz	AL-6X stainless steel	0.1	1.08	35,000
Epoxy-coated carbon steel	Epoxy-coated copper-nickel	Titanium	0.07	0.75	35,000
Carbon steel	Muntz	Aluminum brass	0.06	0.65	1000
Carbon steel	Muntz	90-10 Cu Ni	0.06	0.65	1000
Carbon steel	Muntz	Aluminum brass	0.2	2.2	30,000

*Power

2. Select the type of protective system to use

Protective-current needs generally determine whether an impressed-current or sacrificial-anode system should be used. For a surface condenser, the sacrificial-anode system generally become impractical at current levels over 50 amp.

For a sacrificial-type system, the current output can be estimated by determining the effective voltage and the resistance between anode and structure. The effective voltage between anode and structure is defined as the anode-to-structure open-circuit voltage minus the back-emf associated with polarization at both anode and structure. This voltage depends primarily on the choice of materials, as shown in Table 2.

Resistance of the metallic path is usually negligible for an uncoated structure and the electrolytic resistance is dominant. For a coated structure, this resistance may become significant. The maximum achievable current output can be estimated by considering the case of an uncoated structure.

3. Determine the number of anodes needed for various sacrificial materials

Table 2 gives a range of current outputs estimated for different sacrificial materials with an anode of the dimensions shown in Fig. 6. Thus, for any sacrificial material, number of anodes needed = (required protective-current output, amp)/(current output for the specific sacrificial material, amp).

Since the condenser being considered here is cooled by seawater, we will use the values in the first column in Table 1. For magnesium, number of anodes required

TABLE 2 Current Output that can be Expected from Typical Sacrificial Anodes Materials*

	Current range seawater, amp	Current range fresh water, amp
Magnesium	1.4–2.3	0.014–0.023
Zinc	0.5–0.8	0.005–0.008
Aluminum	0.5–0.8	0.005–0.008

*Power

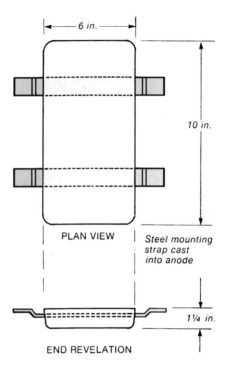

├──── 6 in. ────┤

10 in.

PLAN VIEW

Steel mounting
strap cast
into anode

1¼ in.

END REVELATION

6 IN.   (152 CM)

10 IN.   (254 CM)

1 1/4 IN.   (32 CM)

FIGURE 6 Typical sacrificial anode consists
of a flat slab of the consumable metal into which
fastening straps are cast. (*Power.*)

= 50/2.3 = 21.739; say 22 anodes. For zinc, number of anodes required = 50/
0.8 = 62.5; say 63 anodes. For aluminum, number of anodes required = 50/0.8 =
62.5; say 63. From a practical standpoint, 63 sacrificial anodes is an excessive
number to install in most condenser waterboxes.

The respective service of these anodes at 50 amp are about three months for
magnesium, six months for zinc and aluminum. This short service further reduces
the practicality of sacrificial anodes at high protective current levels.

However, in fresh water, the current output is lower and is limited by the higher
resistance of the water. Corresponding service lives are 5 to 10 years for magnesium, and 40 to 60 years for zinc and aluminum. Protective coating further reduces
the effective wetted surface area and lowers the required protective current at the
same time as it limits the current output of the anodes.

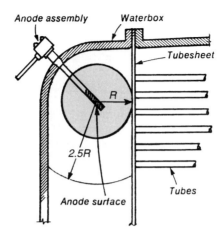

FIGURE 7 Bayonet-type impressed-current anode is located for optimum current throw onto the condenser tubesheet. (*Power.*)

4. *Choose the type of anode material to use*

Several different factors affect the choice of anode material in both sacrificial and impressed-current systems. Choice of a sacrificial-anode material is largely determined by the current density needed, but the efficiency of the material is also important. In an anode material that is 50 percent efficient, half the material deteriorates without providing any useful current. Typical electrochemical efficiencies are: magnesium, 40–50 percent; zinc, 90 percent; aluminum, 80 percent.

Here are brief features of several important anode materials: *Magnesium anodes* provide a high driving voltage, but are not as efficient as zinc or aluminum. *Zinc anodes* are excellent as sacrificial material; at temperatures above 140°F (60°C), zinc may passivate, providing almost no protective current. *Aluminum anodes* are not widely used to protect surface condensers because of performance problems. *Steel anodes* are used in a few power plants to protect copper-alloy tubesheets, but they are less efficient than traditional materials.

Impressed-current systems, Fig. 5*a*, use anodes of platinized alloy, lead alloy, or iron alloy. Platinized- and lead-alloy anodes are favored in seawater, while iron-alloy anodes are favored in low-salinity water. Platinized- and lead-alloy anodes can be operated at higher current density than those of iron alloy, so fewer anodes are needed in the waterbox.

Platinized-titanium anodes can be operated at current densities up to 1000 amp/ft² (10,764 amp/m²) and voltages up to about 8 V in seawater. Such anodes should have a service life of 10 to 20 years, depending on the current density and the platinum plating thickness.

Lead-alloy anodes are widely used in seawater applications. These anodes can be operated at current densities as high as 10 to 20 amp/ft² (107.6 to 215.3 amp/m²) with a life expectancy of more than 10 years.

Related Calculations. This procedure outlines the essentials of sizing anodes for protecting steam surface condensers. For more detailed information, refer to the report mentioned in step 1 of this procedure. Data for this procedure were compiled by John Reason and reported in *Power* magazine, using the report mentioned earlier. SI values were added by the handbook editor.

STEAM-CONDENSER PERFORMANCE ANALYSIS

(*a*) Find the required tube surface area for a shell-and-tube type of condenser serving a steam turbine when the quantity of steam condensed S is 25,000 lb/h (3.1 kg/s); condenser back pressure = 2 inHg absolute (6.8 kPa); steam temperature t_s = 101.1°F (38.4°C); inlet water temperature t_1 = 80°F (26.7°C); tube length per pass L = 14 ft (4.3 m); water velocity V = 6.5 ft/s (2.0 m/s); number of passes = 2; tube size and gage: ¾-in (1.9 cm), no. 18 BWG; cleanliness factor = 0.80. (*b*) Compute the required area and cooling-water flow rate for the same conditions as (*a*) except that cooling water enters at 85°F (29.4°C). (*c*) If the steam flow through the condenser in (*a*) decreases to 15,000 lb/h (1.9 kg/s), what will be the absolute steam pressure in the condenser shell?

Calculation Procedure:

1. *Sketch the condenser, showing flow conditions*
(*a*) Figure 8 shows the condenser and the flow conditions prevailing.

2. *Determine the condenser heat-transfer coefficient*
Use standard condenser-tube engineering data available from the manufacturer or Heat Exchange Institute. Table 3 and Fig. 9 show typical condenser-tube data used in condenser selection. These data are based on a minimum water velocity of 3 ft /s (0.9 m/s) through the condenser tubes, a minimum absolute pressure of 0.7 inHg (2.4 kPa) in the condenser shell, and a minimum Δt terminal temperature difference $t_s - t_2$ of 5°F (2.8°C). These conditions are typical for power-plant surface condensers.

Enter Fig. 9 at the bottom at the given water velocity, 6.5 ft/s (2.0 m/s), and project vertically upward until the ¾-in (1.9-cm) OD tube curve is intersected. From this point, project horizontally to the left to read the heat-transfer coefficient U =

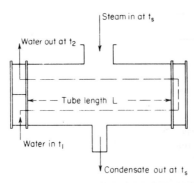

FIGURE 8 Temperatures governing condenser performance.

TABLE 3 Standard Condenser Tube Data

Tube OD. in (cm)	Tube gage BWG	Tube ID, in (cm)	Surface area, ft²/ft (m²/m)		Velocity, ft/s for 1 gal/min (m/s for 1 L/ min)	Value of k for number of tube passes		
			Outside	Inside		One	Two	Three
¾ (1.9)	18	0.652 (1.656)	0.1963 (0.0598)	0.1706 (0.0520)	0.9611 (0.0774)	0.188	0.377	0.565
	16	0.620 (1.575)	0.1963 (0.0598)	0.1613 (0.0492)	1.063 (0.0856)	0.208	0.417	0.625
	14	0.584 (1.483)	0.1963 (0.0598)	0.1528 (0.0466)	1.198 (0.0965)	0.235	0.470	0.705

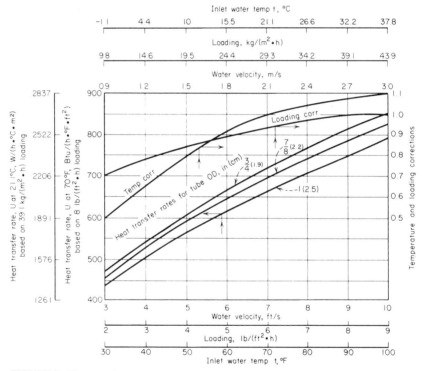

FIGURE 9 Heat-transfer and correction curves for calculating surface-condenser performances.

690 Btu/(ft² · °F) [14,104.8 kJ/(m² · °C)] LMTD (log mean temperature difference). Also read from Fig. 9 the temperature correction factor for an inlet-water temperature of 80°F (26.7°C) by entering at the bottom at 80°F (26.7°C) and projecting vertically upward to the temperature-correction curve. From the intersection with this curve, project to the right to read the correction as 1.04. Correct U for temperature and cleanliness by multiplying the value obtained from the chart by the

correction factors, or $U = 690(1.04)(0.80) = 574$ Btu/(ft$^2 \cdot$h$\cdot$°F) [11,733.6 kJ/ (m$^2 \cdot$h$\cdot$°C)] LMTD.

3. Compute the tube constant
Read from Table 3, for two passes through ¾-in (1.9-cm) OD 18 BWG tubes, $k = $ a constant $= 0.377$. Then $kL/V = 0.377(14)/6.5 = 0.812$.

4. Compute the outlet-water temperature
The equation for outlet-water temperature is $t_2 = t_s - (t_s - t_1/e^x)$, where $x = (kL /V)(U/500)$, or $x = 0.812(574/500) = 0.932$. Then $e^x = 2.7183^{0.932} = 2.54$. With this value known, $t_2 = 101.1 - (101.1 - 80/2.54) = 92.8$°F (33.8°C). Check to see that $\Delta t(t_s - t_2)$ is less than the minimum 5°F (2.8°C) terminal difference. Or, $101.1 - 92.8 = 8.3$°F (4.6°C), which is greater than 5°F (2.8°C).

5. Compute the required tube surface area
The required cooling-water flow, gal/min $= 950S/[500(t_2 - t_1)] = 950(25,000)/ [500(92.8 - 80)] = 3700$ gal/min (233.4 L/s). This equation assumes that 950 Btu is to be removed from each pound (2209.7 kJ/kg) of steam condensed. When a different quantity of heat must be removed, use the actual quantity in place of the 950 in this equation.

With the tube constant kL/V and cooling-water flow rate known, the required area is computed from $A = (kL/V)(gpm) = (0.812)(3700) = 3000$ ft^2 (278.7 m^2).

Since the value of U was not corrected for condenser loading, it is necessary to check whether such a correction is needed. Condenser loading $= S/A = 25,000/ 3000 = 8.33$ lb/ft^2 (40.7 kg/m^2). Figure 9 shows that no correction (correction factor $= 1.0$) is necessary for loadings greater than 8.0 lb/ft^2 (39.1 kg/m^2). Therefore, the loading for this condenser is satisfactory without correction.

This step concludes the general calculation procedure for a surface condenser serving any steam turbine. The next procedure shows the method to follow when a higher cooling-water inlet temperature prevails.

6. Compute the cooling-water outlet temperature
(b) Higher cooling water temperature. From Fig. 9 for 85°F (29.4°C) cooling-water inlet temperature and a 0.80 cleanliness factor, $U = 690(1.06)(0.80) = 585$ Btu/ (ft$^2 \cdot$h$\cdot$°F) [3.3 kJ/(m$^2 \cdot$°C$\cdot$s)] LMTD.

Given data from Table 3, the tube constant $kL/V = 0.377(14)/6.5 = 0.812$. Then $x = (kL/V)(U/500) = 0.812(585/500) = 0.950$. Using this exponent, we get $e^x = 2.8183^{0.950} = 2.586$. The cooling-water outlet temperature is then $t_2 = t_s - (t_s - t_1/e^x) = 101.1 - (101.1 - 85)/2.586 = 94.9$°F (34.9°C). Check to see that $\Delta t(t_s - t_2)$ is greater than the minimum 5°F (2.8°C) terminal temperature difference. Or, $101.1 - 94.9 = 6.5$°F (3.6°C), which is greater than 5°F (2.8°C).

7. Compute the water flow rate, required area, and loading
The required cooling-water flow, gal/min $= 950S/[500(t_2 - t_1)] = 950(25,000)/ [500(94.9 - 85)] = 4800$ gal/min (302.8 L/s).

With the tube constant kL/V and cooling-water flow rate known, the required area is computed from $A = (kL/V)(gpm) = 0.812(4800) = 3900$ ft^2 (362.3 m^2). Then loading $= S/A = 25,000/3900 = 6.4$ lb/ft^2 (31.2 kg/m^2).

Since the loading is less than 8 lb/ft^2 (39.1 kg/m^2), refer to Fig. 9 to obtain the loading correction factor. Enter at the bottom at 6.4 lb/ft^2 (31.2 kg/m^2), and project vertically to the loading curve. At the right, read the loading correction factor as

0.95. Now the value of U already computed must be corrected, and all dependent quantities recalculated.

8. Recalculate the condenser proportions
First, correct U for loading. Or, $U = 585(0.95) = 555$. Then $x = 0.812(555/500) = 0.90$; $e^x = 2.7183^{0.90} = 2.46$; $t_2 = 101.1 - (101.1 - 85/2.46) = 94.6°F$ (34.8°C). Check $\Delta t = t_s - t_2 = 101.1 - 94.6 = 6.5°F$ (3.6°C), which is greater than 5°F (2.8°C). The cooling-water flow rate, gal/min $= 950(25,000)/[500(94.6 - 85)] = 4950$ gal/min (312.3 L/s). Then $A = 0.812(4950) = 4020$ ft² (373.5 m²), and loading $= 25,000/4020 = 6.23$ lb/ft² (30.4 kg/m²).

Check the correction factor for this loading in Fig. 9. The correction factor is 0.94, compared with 0.95 for the first calculation. Since the value of U would be changed only about 1 percent by using the lower factor, the calculations need not be revised further. Where U would change by a larger amount—say 5 percent or more—it would be necessary to repeat the procedure just detailed, applying the new correction factor.

Note that the 5°F (2.8°C) increase in cooling-water temperature (from 80 to 85°F or 26.7 to 29.4°C) requires an additional 1020 ft² (94.8 m²) of condenser surface and 125 gal/min (7.9 L/s) of cooling-water flow to maintain the same back pressure. These increments will vary, depending on the temperature level at which the increase occurs. The effect of reduced steam flow on the steam pressure in the condenser shell will not be computed because the recalculation above is the last step in part (b) of this procedure.

(c) Reduced steam flow to condenser.

9. Determine the condenser loading
From procedure (a) above, the cooling-water flow $= 3700$ gal/min (233.4 L/s); condenser surface $A = 3000$ ft² (278.7 m²). Then, with a 15,000-lb/h (1.9-kg/s) steam flow, loading $= S/A = 15,000/3000 = 5$ lb/ft² (24.4 kg/m²).

10. Compute the heat-transfer coefficient
Correct the previous heat-transfer rate $U = 690$ Btu/(ft²·h·°F) [3.9 kJ/(m²·°C·s)] LMTD for temperature, cleanliness, and loading. Or, $U = 690(1.04)(0.80)(0.89) = 511$ Btu/(ft²·h·°F) [2.9 kJ/(m²·°C·s)] LMTD, given the correction factors from Fig. 9.

11. Compute the final steam temperature
As before, $x = (kL/V)(U/500) = (0.377)(14/6.5)(511/500) = 0.830$. Then $\Delta t = t_2 - t_1 = 950S/(500gpm) = 950(15,000)/[500(3700)] = 7.7°F$ (4.3°C). With $t_1 = 80°F$ (26.7°C), $t_2 = \Delta t + t_1 = 7.7 + 80 = 87.7°F$ (30.9°C). Since $t_2 = t_s - t_1)/e^x$, $e^x = t_s - t_1/(t_s - t_2)$, or $2.7183^{0.830} = t_s - 80/(t_s - 87.7)$. Solve for t_s; or, $t_s = 201.1 - 80/1.294 = 93.6°F$ (34.2°C).

At a saturation temperature of 93.6°F (34.2°C), the steam table (saturation temperature) shows that the steam pressure in the condenser shell is 1.59 inHg (5.4 kPa).

Check the Δt terminal temperature difference. Or, $\Delta t = t_s - t_2 = 93.6 - 87.7 = 5.9°F$ (3.3°C). Since the terminal temperature difference is greater than 5°F (2.8°C), the calculated performance can be realized.

Related Calculations. The procedures and data given here can be used to compute the required cooling-water flow, cooling-water temperature rise, quantity of steam condensed by a given cooling-water flow rate and temperature rise, required condenser surface area, tube length per pass, water velocity, steam temperature in condenser, cleanliness factor, and heat-transfer rate. Whereas Fig. 9 is suitable for all usual condenser calculations for the ranges given, check the Heat Exchange Institute for any new curves that might have been made available before you make the final selection of very large condensers (more than 100,000 lb/h or 12.6 kg/s of steam flow).

Note: The design water temperature used for condensers is either the average summer water temperature or the average annual water temperature, depending on which is higher. The design steam load is the maximum steam flow expected at the full-load rating of the turbine or engine. Usual shell-and-tube condensers have tubes that vary in length from about 8 ft (2.4 m) in the smallest sizes to about 40 ft (12.2 m) or more in the largest sizes. Each square foot of tube surface will condense 7 to 20 lb/h (0.88 to 2.5 g/s) of steam with a cooling-water circulating rate of 0.1 to 0.25 gal/(lb·min) [0.014 to 0.035 L/(kg·s)] of steam condensed. The method presented here is the work of Glenn C. Boyer.

STEAM-CONDENSER AIR LEAKAGE

The air leakage into a condenser is estimated to be 12 ft^3/min (0.34 m^3/min) of 70°F (21°C) air at 14.7 lb/in^2 (101 kPa). At the air outlet connection on the condenser, the temperature is 84°F (29°C) and the total (mixture) pressure is 1.80 inHg absolute (6.1 kPa). Determine the quantity of steam, lb$_m$/h (kg/h), lost from the condenser.

Calculation Procedure:

1. Compute the mass rate of flow per hour of the air leakage
The mass rate of flow per hour of the estimated dry air leakage into the condenser, $w_a = pV/R_aT$, where the air pressure, $p = 14.7 \times 144$ lb$_f$/ft^2 (101 kPa); volumetric flow rate, $V = 12 \times 60 = 720$ ft^3/h (20.4 m^3/h); gas constant for air, $R_a = 53.34$ ft·lb/(lb·°R) [287(m·N/kg·K)]; air temperature, $T = 70 + 460 = 530$°R (294 K). Then, $w_a = (14.7 \times 144)(720)/(53.34 \times 530) = 53.9$ lb/h (24.4 kg/h).

2. Determine the partial pressure of the air in the mixture
The partial pressure of the air in the mixture of air and steam, $p_a = p_m - p_v$, where the mixture pressure, $p_m = 1.80 \times 0.491 = 0.884$ lb/in^2 (6.09 kPa); partial vapor pressure, $p_v = 0.577$ lb/in^2 (3.98 kPa), as found in the Steam Tables mentioned under Related Calculations of this procedure. Then, $p_a = 0.884 - 0.577 = 0.307$ lb/in^2 (2.1 kPa).

3. Compute the humidity ratio of the mixture
The humidity ratio of the mixture, $w_v = R_a p_v/(R_v p_a)$, where the gas constant for steam vapor, $R_v = 85.8$ ft·lb/(lb$_m$·°R) [462(J/kg·K)], as found in a reference mentioned under Related Calculations of this procedure. Then, $w_v = 53.34 \times 0.577/(85.8 \times 0.307) = 1.17$ lb vapor/lb dry air (0.53 kg/kg).

4. Compute the rate of steam lost from the condenser
Steam is lost from the condenser at the rate of $w_h = w_v \times w_a = 1.17 \times 53.9 = 63.1$ lb/h (28.6 kg/h).

Related Calculations. The partial vapor pressure in step 2 was found at 84°F (29°C) under Table 1, Saturation: Temperatures of *Thermodynamic Properties of Water Including Vapor, Liquid, and Solid Phases,* 1969, Keenan, et al., John Wiley & Sons, Inc. Use the later versions of such tables whenever available, as necessary. The gas constant for water vapor in step 3 was obtained from *Principals of Engineering Thermodynamics,* 2d edition, by Kiefer, et al., John Wiley & Sons, Inc.

STEAM-CONDENSER SELECTION

Select a condenser for a steam turbine exhausting 150,000 lb/h (18.9 kg/s) of steam at 2 inHg absolute (6.8 kPa) with a cooling-water inlet temperature of 75°F (23.9°C). Assume a 0.85 condition factor, ⅞-in (2.2-cm) no. 18 BWG tubes, and an 8-ft/s (2.4-m/s) water velocity. The water supply is restricted. Obtain condenser constants from the Heat Exchange Institute, *Steam Surface Condenser Standards.*

Calculation Procedure:

1. Select the $t_s - t_1$ temperature difference

Table 4 shows customary design conditions for steam condensers. With an inlet-water temperature at 75°F (23.9°C) and an exhaust steam pressure of 2.0 inHg absolute (6.8 kPa), the customary temperature difference $t_s - t_1 = 26.1°F$ (14.5°C). With a sufficient water supply and a siphonic circuitry, $(t_2 - t_1)/(t_s - t_1)$ is usually between 0.5 and 0.55. For a restricted water supply or high frictional resistance and static head, the value of this factor ranges from 0.55 to 0.75.

2. Compute the LMTD across the condenser

With 75°F (23.9°C) inlet water, $t_s - t_1 = 101.14 - 75 = 26.14°F$ (14.5°C), given the steam temperature in the saturation-pressure table. Once $t_s - t_1$ is known, it is necessary to assume a value for the ratio $(t_2 - t_1)/(t_s - t_1)$. As a trial, assume 0.60, since the water supply is restricted. Then $(t_2 - t_1)/(t_s - t_1)$, $= 0.60 = (t_2 - t_1)/26.14$. Solving, we get $t_2 - t_1 = 15.68°F$ (8.7°C). The difference between the steam temperature t_s and the outlet temperature t_2 is then $t_s - t_2 = 26.14 - 15.68 = 10.46°F$ (5.8°C). Checking, we find $t_2 = t_1 + (t_2 - t_1) = 75 + 15.68 = 90.68°F$ (50.38°C); $t_s - t_2 = 101.14 - 90.68 = 10.46°F$ (5.8°C). This value is greater than the required minimum value of 5°F (2.8°C) for $t_s - t_2$. The assumed ratio 0.60 is therefore satisfactory.

TABLE 4 Typical Design Conditions for Steam Condensers

Cooling water temperature		Steam pressure		Temperature difference $t_s - t_1$	
°F	°C	inHg	kPa	°F	°C
70	21.1	1.5–2.0	5.1–6.8	21.7–31.1	12.1–17.3
75	23.9	2.0–2.5	6.8–8.5	26.1–33.7	14.5–18.7
80	26.7	2.0–4.0	6.8–13.5	21.1–45.4	11.7–25.2

Were $t_s - t_2$ less than 5°F (2.8°C), another ratio value would be assumed and the difference computed again. You would continue doing this until a value of $t_s - t_2$ greater than 5°F (2.8°C) were obtained. Then LMTD $= (t_2 - t_1)/\ln[(t_s - t_1)/(t_s - t_2)]$; LMTD $= 15.68/\ln (26.1/10.46) = 17.18°F$ (9.5°C).

3. Determine the heat-transfer coefficient

From the Heat Exchange Institute or manufacturer's data U is 740 Btu/(ft² · h · °F) [4.2 kJ/(m² · °C · s)] LMTD for a water velocity of 8 ft/s (2.4 m/s). If these data are not available, Fig. 9 can be used with complete safety for all preliminary selections.

Now U must be corrected for the inlet-water temperature, 75°F (23.9°C), and the condition factor, 0.85, which is a term used in place of the correction factor by some authorities. From Fig. 9, the correction for 75°F (23.9°C) inlet water = 1.04. Then actual $U = 740(1.04)(0.85) = 655$ Btu/(ft² · h · °F) [3.7 kJ/(m² · °C · s)] LMTD.

4. Compute the steam condensation rate

The heat-transfer rate per square foot of condenser surface with a 17.18°F (9.5°C) LMTD is U(LMTD) $= 655(17.18) = 11,252.9$ Btu/(ft² · h) [35.5 kJ/(m² · s)].

Condensers serving steam turbines are assumed, for design purposes, to remove 950 Btu/lb (2209.7 kJ/kg) of steam condensed. Therefore, the steam condensation rate for any condenser is [Btu/(ft² · h)]/950, or 1252.9/950 = 11.25 lb/(ft² · h) [15.3 g/(m² · s)].

5. Compute the required surface area and water flow

The required surface area = steam flow (lb/h)/[condensation rate, lb/(ft² · h)], or with a 150,000-lb/h (18.9-kg/s) flow, 150,000/11.25 = 13,320 ft² (1237.4 m²).

The water flow rate, gal/min $= 950S/[500(t_2 - t_1)] = 950(150,000)/[500(15.68)] = 18,200$ gal/min (1148.1 L/s).

Related Calculations. See the previous calculation procedure for steps in determining the water-pressure loss through a surface condenser.

To choose a surface condenser for a steam engine, use the same procedures as given above, except that the heat removed from the exhaust steam is 1000 Btu/lb (2326.9 kJ/kg). Use a condition (cleanliness) factor of 0.65 for steam engines because the oil in the exhaust steam fouls the condenser tubes, reducing the rate of heat transfer. The condition (cleanliness) factor for steam turbines is usually assumed to be 0.8 to 0.9 for relatively clean, oil-free cooling water.

At loads greater than 50 percent of the design load, $t_s - t_1$ follows a straight-line relationship. Thus, in the above condenser, $t_s - t_1 = 26.14°F$ (14.5°C) at the full load of 150,000 lb/h (18.9 kg/s). If the load falls to 60 percent (90,000 lb/h or 11.3 kg/s), then $t_s - t_1 = 26.14(0.60) = 15.7°F$ (8.7°C). At 120 percent load (180,00 lb/h or 22.7 kg/s), $t_s - t_1 = 26.14(1.20) = 31.4°F$ (17.4°C). This straight-line law is valid with constant inlet-water temperature and cooling-water flow rate. It is useful in analyzing condenser operating conditions at other than full load.

Single- or multiple-pass surface condensers may be used in power services. When a liberal supply of water is available, the single-pass condenser is often chosen. With a limited water supply, a two-pass condenser is often chosen.

AIR-EJECTOR ANALYSIS AND SELECTION

Choose a steam-jet air ejector for a condenser serving a 250,000-lb/h (31.5-kg/s) steam turbine exhausting at 2 inHg absolute (6.8 kPa). Determine the number of stages to use, the approximate steam consumption and the quantity of air and vapor mixture the ejector will handle.

Calculation Procedure:

1. Select the number of stages for the ejector
Use Fig. 10 as a preliminary guide to the number of stages required in the ejector. Enter at 2-inHg absolute (6.8-kPa) condenser pressure, and project horizontally to the stage area. This shows that a two-stage ejector will probably be satisfactory.

Check the number of stages above against the probable overload range of the prime mover by using Fig. 11. Enter at 2-inHg absolute (6.8-kPa) condenser pressure, and project to the two-stage curve. This curve shows that a two-stage ejector can readily handle a 25-percent overload of the prime mover. Also, the two-stage curve shows that this ejector could handle up to 50 percent overload with an increase in the condenser absolute pressure of only 0.4 inHg (1.4 kPa). This is shown by the pressure, 2.4 inHg absolute (8.1 kPa), at which the two-stage curve crosses the 150 percent overload ordinate (Fig. 11).

2. Determine the ejector operating conditions
Use the Heat Exchange Institute or manufacturer's data. Table 5 excerpts data from the Heat Exchange Institute for condensers in the range considered in this procedure.

Study of Table 5 shows that a two-stage condensing ejector unit serving a 250,000-lb/h (31.5-kg/s) steam turbine will require 450 lb/h (56.7 g/s) of 300-lb/in^2 (gage) (2068.5-kPa) steam. Also, the ejector will handle 7.5 ft^3/min (0.2 m^3/min) of free, dry air, or 33.75 lb/h (4.5 g/s) of air. It will remove up to 112.5 lb/h (14.2 g/s) of an air-vapor mixture.

The actual air leakage into a condenser varies with the absolute pressure in the condenser, the tightness of the joints, and the conditions of the tubes. Some authorities cite a maximum leakage of about 250-lb/h (31.5-g/s) steam flow. At 400,000 lb/h (50.4 kg/s), the leakage is 160 lb/h (20.2 g/s); at 250,000 lb/h (31.5 kg/s), it is 130 lb/h (16.4 g/s) of air-vapor mixture. A condenser in good condition will usually have less leakage.

For an installation in which the manufacturer supplies data on the probable air leakage, use a psychrometric chart to determine the weight of water vapor contained in the air. Thus, at 2 inHg absolute (6.8 kPa) and 80°F (26.7°C), each pound of air will carry with it 0.68 lb (0.68 kg/kg) of water vapor. In a surface condenser into which 20 lb. (9.1 kg) of air leaks, the ejector must handle 20 + 20(0.68) = 33.6 lb/h (4.2 g/s) of air-vapor mixture. Table 5 shows that this ejector can readily handle this quantity of air-vapor mixture.

Related Calculations.   When you choose an air ejector for steam-engine service, double the Heat Exchange Institute steam-consumption estimates. For most low-pressure power-plant service, a two-stage ejector with inter- and after condensers is satisfactory, although some steam engines operating at higher absolute exhaust pressures require only a single-stage ejector. Twin-element ejectors have two sets of stages; one set serves as a spare and may also be used for capacity regulation in stationary and marine service. The capacity of an ejector is constant for a given steam pressure and suction pressure. Raising the steam pressure will not increase the ejector capacity.

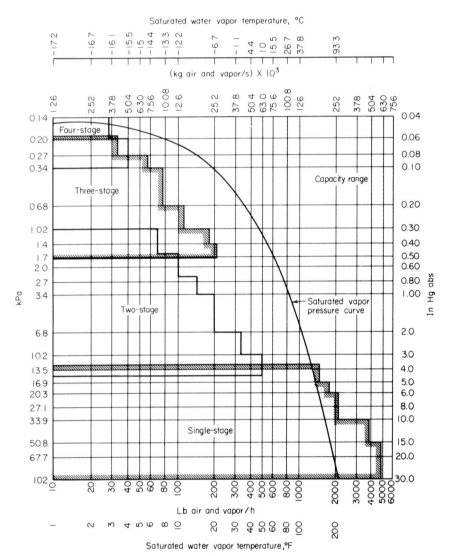

FIGURE 10   Steam-ejector capacity-range chart.

SURFACE-CONDENSER CIRCULATING-WATER PRESSURE LOSS

Determine the circulating-water pressure loss in a two-pass condenser having 12,000 ft² (1114.8 m²) of condensing surface, a circulating-water flow rate of 10,000 gal/min (630.8 L/s), ¾-in (1.9-cm) no. 16 BWG tubes, a water flow rate of 7 ft/s (2.1 m/s), external friction of 20 ft of water (59.8 kPa), and a 10-ft-of-water (29.9-kPa) siphonic effect on the circulating-water discharge.

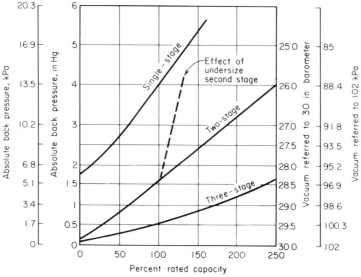

FIGURE 11   Steam-jet ejector characteristics.

TABLE 5   Air-Ejector Capacities for Surface Condensers for Steam Turbines*

Steam load, lb/h (kg/s)	Free, dry air at 70°F (21.1°C), ft³/min (cm³/s)	Air, lb/ h (g/s)	Air-vapor mixture at 30 percent dry air, lb/h (g/s)	Steam consumption at 300 lb/in² (gage) (2068.5 kPa), lb/h (g/s)
100,001– 250,000 (12.6– 31.5)	7.5 (3539.6)	33.75 (4.3)	112.5 (14.2)	450 (56.7)

*Two-stage condensing ejector unit.

Calculation Procedure:

1. Determine the water flow rate per tube

Use a tabulation of condenser-tube engineering data available from the manufacturer or the Heat Exchange Institute, or complete the water flow rate from the physical dimensions of the tube thus: ¾-in (1.9 cm) no. 16 BWG tube ID = 0.620 in (1.6 cm) from a tabulation of condenser-tube data, such as Table 3. Assume a water velocity of 1 ft/s (0.3 m/s). Then a 1-ft (0.3-m) length of the tube will contain $(12)(0.620)^2\pi/4 = 3.62$ in³ (59.3 cm³) of water. This quantity of water will flow through the tube for each foot of length per second of water velocity [194.6 cm³/ (m·s)]. The flow per minute will be 3.62 (60 s/min) = 217.2 in³/min (3559.3 cm³/min). Since 1 U.S. gal = 231 in³ (3.8 L), the gal/min flow at a 1 ft/s (0.3 m/s) velocity = 217.2/231 = 0.94 gal/min (0.059 L/s).

With an actual velocity of 7 ft/s (2.1 m/s), the water flow rate per tube is 7(0.94) = 6.58 gal/min (0.42 L/s).

2. Determine the number of tubes and length of water travel

Since the water flow rate through the condenser is 10,000 gal/min (630.8 L/s) and each tube conveys 6.58 gal/min (0.42 L/s), the number of tubes = 10,000/6.58 = 1520 tubes per pass.

Next, the total length of water travel for a condenser having A ft^2 of condensing surface is computed from A(number of tubes)(outside area per linear foot, ft^2). The outside area of each tube can be obtained from a table of tube properties, such as Table 3; or computed from (OD, in)$(\pi)(12)/144$, or $(0.75)(\pi)(12)/144 = 0.196$ ft^2/lin ft (0.06 m^2/m). Then, total length of travel = $12,000/[(1520)(0.196)] = 40.2$ ft (12.3 m). Since the condenser has two passes, the length of tube per pass = $40.2/2 = 20.1$ ft (6.1 m). Since each pass has an equal number of tubes and there are two passes, the total number of tubes in the condenser = 2 passes (1520 tubes per pass) = 3040 tubes.

3. Compute the friction loss in the system

Use the Heat Exchange Institute or manufacturer's curves to find the friction loss per foot of condenser tube. At 7 ft/s (2.1 m/s), the Heat Exchange Institute curve shows the head loss is 0.4 ft of head per foot (3.9 kPa/m) of travel for $\frac{3}{4}$-in (1.9-cm) no. 16 BWG tubes. With a total length of 40.2 ft (12.3 m), the tube head loss is $0.4(40.2) = 16.1$ ft (48.1 kPa).

Use the Heat Exchange Institute or manufacturer's curves to find the head loss through the condenser waterboxes. From the first reference, for a velocity of 7 ft/s (2.1 m/s), head loss = 1.4 ft (4.2 kPa) of water for a single-pass condenser. Since this is a two-pass condenser, the total waterbox head loss = $2(1.4) = 2.8$ ft (8.4 kPa).

The total condenser friction loss is then the sum of the tube and waterbox losses, or $16.1 + 2.8 = 18.9$ ft (56.5 kPa) of water. With an external friction loss of 20 ft (59.8 kPa) in the circulating-water piping, the total loss in the system, without siphonic assistance, is $18.9 + 20 = 38.9$ ft (116.3 kPa). Since there is 10 ft (29.9 kPa) of siphonic assistance, the total friction loss in the system with siphonic assistance is $38.9 - 10 = 28.9$ ft (86.3 kPa). In choosing a pump to serve this system, the frictional resistance of 28.9 ft (86.3 kPa) would be rounded to 30 ft (89.7 kPa), and any factor of safety added to this value of head loss.

Note: The most economical cooling-water velocity in condenser tubes is 6 to 7 ft/s (1.8 to 2.1 m/s); a velocity greater than 8 ft/s (2.4 m/s) should not be used, unless warranted by special conditions.

SURFACE-CONDENSER WEIGHT ANALYSIS

A turbine exhaust nozzle can support a weight of 100,000 lb (444,822.2 N). Determine what portion of the total weight of a surface condenser must be supported by the foundation if the weight of the condenser is 275,000 lb (1,223,261.1 N), the tubes and waterboxes have a capacity of 8000 gal (30,280.0 L), and the steam space has a capacity of 30,000 gal (113,550.0 L) of water.

Calculation Procedure:

1. Compute the maximum weight of the condenser

The maximum weight on a condenser foundation occurs when the shell, tubes, and waterboxes are full of water. This condition could prevail during accidental flooding of the steam space or during tests for tube leaks when the steam space is purposefully flooded. In either circumstance, the condenser foundation and spring supports, if used, must be able to carry the load imposed on them. To compute this load, find the sum of the individual weights:

Condenser weight, dry	275,000 lb (1,223,261.1 N)
Water in tubes and boxes = (8.33 lb/gal)(8000 gal)	66,640 lb (296,429.5 N)
Water in steam space = (8.33)(30,000)	249,900 lb (1,111,610.7 N)
Maximum weight when full of water	592,540 lb (2,631,301.2 N)

2. Compute the foundation load

The turbine nozzle can support 100,000 lb (444,822.2 N). Therefore, the foundation must support 591,540 − 100,000 = 491,540 lb (2,186,479.0 N). For foundation design purposes this would be rounded to 495,000 lb (2,201,869.9 N).

Related Calculations. When you design a condenser foundation, do the following: (1) Leave enough room at one end to permit withdrawal of faulty tubes and insertion of new tubes. Since some tubes may exceed 40 ft (12.3 m) in length, careful planning is needed to provide sufficient installation space. During the design of a power plant, a template representing the tube length is useful for checking the tube clearance on a scale plan and side view of the condenser installation. When there is insufficient room for tube removal with one shape of condenser, try another with shorter tubes.

(2) Provide enough headroom under the condenser to produce the required submergence on the condensate-pump impeller. Most condensate pumps require at least 3-ft (0.9 m) submergence. If necessary, the condensate pump can be installed in a pit under the condenser, but this should be avoided if possible.

WEIGHT OF AIR IN STEAM-PLANT SURFACE CONDENSER

The vacuum in a surface condenser is 28-in (71.12-cm) Hg referred to a 30-in (76.2-cm) barometer. The temperature in the condenser is 80°F (26.7°C). What is the percent by weight of the air in the condenser?

Calculation Procedure:

1. Find the absolute pressure in the condenser

From the steam tables at 80°F (26.7°C), 1 in (2.54 cm) Hg exerts a pressure of 0.4875 psi (3.36 kPa).

2. Determine the weight of water per lb (kg) of dry air in the condenser

In a condenser, the steam (water vapor) is condensing in contact with the tubes and may be taken as saturated. At 80°F (26.7°C), the absolute pressure of saturated steam is 0.5067 psia (3.39 kPa), from the steam tables. In the condensing condition, there are 0.622 lb (0.28 kg) of water per pound (kg) of dry air. Since the water content of the air is a function of the partial pressures, $(0.622) (0.5067)/[(2 \times 0.5067)] = 0.673$ lb of water per lb of dry air (0.305 kg).

3. Compute the percent of air by weight
Use the relation, percent of air by weight $= (100)(1)/(1 + 0.672) = (100)(1)/(1 + 0.672) = 59.8$ percent by weight of air.
   Related Calculations.   Use this general procedure for analyzing the air in surface condensers serving steam turbines of all types.

BAROMETRIC-CONDENSER ANALYSIS AND SELECTION

Select a countercurrent barometric condenser to serve a steam turbine exhausting 25,000 lb/h (3.1 kg/s) of steam at 5 inHg absolute (16.9 kPa). Determine the quantity of cooling water required if the water inlet temperature is 50°F (10.0°C). What is the required dry-air capacity of the ejector? What is the required pump head if the static head is 40 ft (119.6 kPa) and the pipe friction is 15 ft of water (44.8 kPa)?

Calculation Procedure:

1. Find the steam properties from the steam tables
At 5 inHg absolute (16.9 kPa), $h_g = 1119.4$ Btu/lb (2603.7 kJ/kg), from the saturation-pressure table. If the condensing water were to condense the steam without subcooling the condensate, the final temperature of the condensate, from the steam tables, would be 133.76°F (56.5°C), corresponding to the saturation temperature. However, subcooling almost always occurs, and the usual practice in selecting a countercurrent barometric condenser is to assume the final condensate temperature t_c will be 5°F (2.8°C) below the saturation temperature corresponding to the absolute pressure in the condenser. Given a 5°F (2.8°C) difference, $t_c = 133.76 - 5 = 128.76$°F (53.7°C). Interpolating in the saturation-temperature steam table, we find the enthalpy of the condensate h_f at 128.76°F (53.7°C) is 96.6 Btu/lb (224.8 kJ/kg).

2. Compute the quantity of condensing water required
In any countercurrent barometric condenser, the quantity of cooling water Q lb/h required is $Q = W(h_g - h_t)/(t_c - t_1)$, where $W =$ weight of steam condensed, lb/h; $t^1 =$ cooling-water inlet temperature, °F. Then $Q = 25,000(1119.4 - 96.66)/(128.76 - 50) = 325,000$ lb/h (40.9 kg/s). By converting to gallons per minute, $Q = 325,000/500 = 650$ gal/min (41.0 L/s).

3. Determine the required ejector dry-air capacity
Use the Heat Exchanger Institute or a manufacturer's tabulation of free, dry-air leakage and the allowance for air in the cooling water to determine the required dry-air capacity. Thus, from Table 6, the free, dry-air leakage for a barometric condenser serving a turbine is 3.0 ft³/min (0.08 m³/min) of air and vapor. The

TABLE 6 Free, Dry-Air Leakage

[ft³/min (m³/s) at 70°F or 21.1°C air and vapor mixture, 7½° below vacuum temperature or 4.2° for Celsius]

Maximum steam condensation		Barometric and low-level jet condensers							
		Serving turbines				Serving engines			
lb/h	kg/s	ft³/min	m³/s	lb/h	g/s	ft³/min	m³/s	lb/h	g/s
75,000–150,000	9.4–18.9	6.5	0.0031	97.5	12.3	13.0	0.0061	195.0	24.6
150,001–250,000	18.9–31.5	8.5	0.0040	127.5	16.1				
250,001–350,000	31.5–44.1	10.0	0.0047	150.0	18.9				

allowance for air in the 50°F (10.0°C) cooling water is 3.3 ft³/min (0.09 m³/min) of air at 70°F (21.1°C) per 1000 gal/min (63.1 L/s) of cooling water, Fig 12. The total dry-air leakage is the sum, or 3.0 + 3.3 = 6.3 ft³/min (0.18 m³/min). Thus, the ejector must be capable of handling at least 6.3 ft³/min (0.18 m³/min) of dry air to serve this barometric condenser at its rated load of 25,000 lb/h (3.1 kg/s) of steam.

Where the condenser will operate at a lower vacuum (i.e., a higher absolute pressure), overloads up to 50 percent may be met. To provide adequate dry-air handling capacity at this overload with the same cooling-water inlet temperature, find the free, dry-air leakage at the higher condensing rate from Table 6 and add

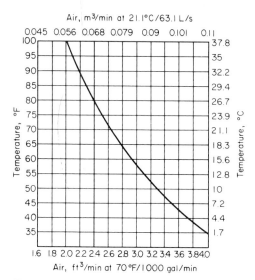

FIGURE 12 Allowance for air in condenser injection water.

this to the previously found allowance for air in the cooling water. Or, 4.5 + 3.3 = 7.8 ft³/min (0.22 m³/min). An ejector capable of handling up to 10 ft³/min (0.32 m³/min) would be a wise choice for this countercurrent barometric condenser.

4. Determine the pump head required
Since a countercurrent barometric condenser operates at pressures below atmospheric, it assists the cooling-water pump by "sucking" the water into the condenser. The maximum assist that can be assumed is $0.75V$, where V = design vacuum, inHg.

In this condenser with a 26-in (88.0-kPa) vacuum, the maximum assist is $0.75(26)$ = 19.5 inHg (66.0 kPa). Converting to feet of water, using 1.0 inHg = 1.134 ft (3.4 kPa) of water, we find 19.5(1.134) = 22.1 ft (66.1 kPa) of water. The total head on the pump is then the sum of the static and friction heads less $0.75V$, expressed in feet of water. Or, the total head on the pump = 40 + 15 − 22.1 = 32.9 ft (98.4 kPa). A pump with a total head of at least 35 ft (104.6 kPa) of water would be chosen for this condenser. Where corrosion or partial clogging of the piping is expected, a pump with a total head of 50 ft (149.4 kPa) would probably be chosen to ensure sufficient head even though the piping is partially clogged.

Related Calculations. (1) When a condenser serving a steam engine is being chosen, use the appropriate dry-air leakage value from Table 6. (2) For ejector-jet barometric condensers, assume the final condensate temperature t_c as 10 to 20°F (5.6 to 11.1°C) below the saturation temperature corresponding to the absolute pressure in the condenser. This type of condenser does not use an ejector, but it requires 25 to 50 percent more cooling water than the countercurrent barometric condenser for the same vacuum. (3) The total pump head for an ejector-jet barometric condenser is the sum of the static and friction heads plus 10 ft (29.9 kPa). The additional positive head is required to overcome the pressure loss in spray nozzles.

COOLING-POND SIZE FOR A KNOWN HEAT LOAD

How many spray nozzles and what surface area are needed to cool 10,000 gal/min (630.8 L/s) of water from 120 to 90°F (48.9 to 32.2°C) in a spray-type cooling pond if the average wet-bulb temperature is 650°F (15.6°C)? What would be the approximate dimensions of the cooling pond be? Determine the total pumping head if the static head is 10 ft (29.9 kPa), the pipe friction is 35 ft of water (104.6 kPa), and the nozzle pressure is 8 lb/in² (55.2 kPa).

Calculation Procedure:

1. Compute the number of nozzles required
Assume a water flow of 50 gal/min (3.2 L/s) per nozzle; this is a typical flow rate for usual cooling-pond nozzles. Then the number of nozzles required = (10,000 gal/min)/(50 gal/min per nozzle) = 200 nozzles. If 6 nozzles are used in each spray group, a series of crossed arms, with each arm containing one or more nozzles, then 200 nozzles/6 nozzles per spray group = 33⅓ spray groups will be needed. Since a partial spray group is seldom used, 34 spray groups would be chosen.

2. Determine the surface area required

Usual design practice is to provide 1 ft^2 (0.09 m^2) of pond area per 250 lb (113.4 kg) of water cooled for water quantities exceeding 1000 gal/min (63.1 L/s). Thus, in this pond, the weight of water cooled = (10,000 gal/min)(8.33 lb/gal)(60 min/h) = 4,998,000, say 5,000,000 lb/h (630.0 kg/s). Then, the area required, given 1 ft^2 of pond area per 250 lb of water (0.82 m^2 per 1000 kg) cooled = 5,000,000/250 = 20,000 ft^2 (1858.0 m^2).

As a cross-check, use another commonly accepted area value: 125 Btu/(ft$^2 \cdot$ °F) [2555.2 kJ/(m$^2 \cdot$ °C)] is the difference between the air wet-bulb temperature and the warm entering-water temperature. This is the equivalent of (120 − 60)(125) = 7500 Btu/ft^2 (85,174 kJ/m^2) in this spray pond, because the air wet-bulb temperature is 60°F (15.6°C) and the warm-water temperature is 120°F (48.9°C). The heat removed from the water is (lb/h of water)(temperature decrease, °F)(specific heat of water) = (5,000,000)(120 − 90)(1.0) = 150,000,000 Btu/h (43,960.7 kW). Then, area required = (heat removed, Btu/h)/(heat removal, Btu/ft^2) = 150,000,000/7500 = 20,000 ft^2 (1858.0 m^2). This checks the previously obtained area value.

3. Determine the spray-pond dimensions

Spray groups on the same header or pipe main are usually arranged on about 12-ft (3.7-m) centers with the headers or pipe mains spaced on about 25-ft (7.6-m) centers, Fig. 13. Assume that 34 spray groups are used, instead of the required 33⅓, to provide an equal number of groups in two headers and a small extra capacity.

Sketch the spray pond and headers, Fig. 13. This shows that the length of each header will be about 204 ft (62.2 m) because there are seventeen 12-ft (3.7-m) spaces between spray groups in each header. Allowing 3 ft (0.9 m) at each end of a header for fittings and clean-outs gives an overall header length of 210 ft (64.0 m). The distance between headers is 25 ft (7.6 m). Allow 25 ft (7.6 m) between the outer sprays and the edge of the pond. This gives an overall width of 85 ft (25.9 m) for the pond, if we assume the width of each arm in a spray group is 10 ft (3.0 m). The overall length will then be 210 + 25 + 25 = 260 ft (79.2 m). A

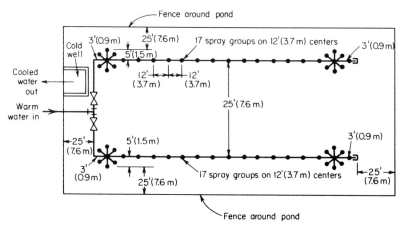

FIGURE 13   Spray-pond nozzle and piping layout.

cold well for the pump suction and suitable valving for control of the incoming water must be provided, as shown in Fig 13. The water depth in the pond should be 2 to 3 ft (0.6 to 0.9 m).

4. Compute the total pumping head

The total head, ft of water = static head + friction head + required nozzle head = 10 + 35 + 80(0.434) = 48.5 ft (145.0 kPa) of water. A pump having a total head of at least 50 ft (15.2 m) of water would be chosen for this spray pond. If future expansion of the pond is anticipated, compute the probable total head required at a future date, and choose a pump to deliver that head. Until the pond is expanded, the pump would operate with a throttled discharge. Normal nozzle inlet pressures range from about 6 to 10 lb/in² (41.4 to 69.0 kPa). Higher pressures should not be used, because there will be excessive spray loss and rapid wear of the nozzles.

Related Calculations.   Unsprayed cooling ponds cool 4 to 6 lb (1.8 to 2.7 kg) of water from 100 to 70°F/ft² (598.0 to 418.6°C/m²) of water surface. An alternative design rule is to assume that the pond will dissipate 3.5 Btu/ft²·h) (11.0 W/m²) water surface per degree difference between the wet-bulb temperature of the air and the entering warm water.

SECTION 3
COMBUSTION

COMBUSTION CALCULATIONS USING THE MILLION BTU (1.055MJ) METHOD

The energy absorbed by a steam boiler fired by natural gas is 100-million Btu/hr (29.3 MW). Boiler efficiency on a higher heating value (HHV) basis is 83 percent. If 15 percent excess air is used, determine the total air and flue-gas quantities produced. The approximate HHV of the natural gas is 23,000 Btu/lb (53,590 kJ/kg). Ambient air temperature is 80°F (26.7°C) and relative humidity is 65 percent. How can quick estimates be made of air and flue-gas quantities in boiler operations when the fuel analysis is not known?

Calculation Procedure:

1. *Determine the energy input to the boiler*
The million Btu (1.055MJ) method combustion calculations is a quick way of estimating air and flue-gas quantities generated in boiler and heater operations when the ultimate fuel analysis is not available and all the engineer is interested in is good estimates. Air and flue-gas quantities determined may be used to calculate the size of fans, ducts, stacks, etc.

It can be shown through comprehensive calculations that each fuel such as coal, oil, natural gas, bagasse, blast-furnace gas, etc. requires a certain amount of dry stochiometric air per million Btu (1.055MJ) fired on an HHV basis and that this quantity does not vary much with the fuel analysis. The listing below gives the dry air required per million Btu (1.055MJ) of fuel fired on an HHV basis for various fuels.

Combustion Constants for Fuels

Fuel	Constant, lb dry air per million Btu (kg/MW)
Blast furnace gas	575 (890.95)
Bagasse	650 (1007.2)
Carbon monoxide gas	670 (1038.2)
Refinery and oil gas	720 (1115.6)
Natural gas	730 (1131.1)
Furnace oil and lignite	745–750 (1154.4–1162.1)
Bituminous coals	760 (1177.6)
Anthracite coal	780 (1208.6)
Coke	800 (1239.5)

To determine the energy input to the boiler, use the relation $Q_f = (Q_s)/E_h$, where energy input by the fuel, Btu/hr (W); Q_s = energy absorbed by the steam in the boiler, Btu/Hr (W); Q_s = energy absorbed by the steam, Btu/hr (W); E_h = efficiency of the boiler on an HHV basis. Substituting for this boiler, $Q_f = 100/0.83 = 120.48$ million Btu/hr on an HHV basis (35.16 MW).

2. Estimate the quantity of dry air required by this boiler
The total air required $T_a = (Q_f)$(Fuel constant from list above). For natural gas, $T_a = (120.48)(730) = 87,950$ lb/hr (39,929 kg/hr). With 15 percent excess air, total air required = $(1.15)(87,950) = 101,142.5$ lb/hr (45,918.7 kg/hr).

3. Compute the quantity of wet air required
Air has some moisture because of its relative humidity. Estimate the amount of moisture in dry air in M lb/lb (kg/kg) from, $M = 0.622 (p_w)/(14.7 - p_w)$, where 0.622 is the ratio of the molecular weights of water vapor and dry air; p_w = partial pressure of water vapor in the air, psia (kPa) = saturated vapor pressure (SVP) × relative humidity expressed as a decimal; 14.7 = atmospheric pressure of air at sea level (101.3 kPa). From the steam tables, at 80 F (26.7 C), SVP = 0.5069 psia (3.49 kPa). Substituting, $M = 0.622 (0.5069 \times 0.65)/(14.7 - [0.5069 \times 0.65]) = 0.01425$ lb of moisture/lb of dry air (0.01425 kg/kg).
The total flow rate of the wet air then = 1.0142 (101,142.5) = 102,578.7 lb/hr (46,570.7 kg/hr). To convert to a volumetric-flow basis, recall that the density of air at 80°F (26.7°C) and 14.7 psia (101.3 kPa) = 39/(480 + 80) = 0.0722 lb/cu ft (1.155 kg/cu m). In this relation, 39 = a constant and the temperature of the air is converted to degrees Rankine. Hence, the volumetric flow = 102,578.7/(60 min/hr)(0.0722) = 23,679.3 actual cfm (670.1 cm m/min).

4. Estimate the rate of fuel firing and flue-gas produced
The rate of fuel firing = Q_f/HHV = (120.48 × 10^6)/23,000 = 5238 lb/hr (2378 kg/hr). Hence, the total flue gas produced = 5238 + 102,578 = 107,816 lb/hr (48,948 kg/hr).
If the temperature of the flue gas is 400°F (204.4°C) (a typical value for a natural-gas fired boiler), then the density, as in Step 3 is: 39/(400 + 460) = 0.04535 lb/cu ft (0.7256 kg/cu m). Hence, the volumetric flow = (107,816)/(60 min/hr × 0.04535) = 39,623.7 actual cfm (1121.3 cu m/min).
Related Calculations.   Detailed combustion calculations based on actual fuel gas analysis can be performed to verify the constants given in the list above. For example, let us say that the natural-gas analysis was: Methane = 83.4 percent; Ethane = 15.8 percent; Nitrogen = 0.8 percent by volume. First convert the analysis to a percent weight basis:

Fuel	Percent volume	MW	Col. 2 × Col. 3	Percent weight
Methane	83.4	16	1334.4	72.89
Ethane	15.8	30	474	25.89
Nitrogen	0.8	28	22.4	1.22

Note that the percent weight in the above list is calculated after obtaining the sum under Column 2 × Column 3. Thus, the percent methane = (1334.4)/(1334.4 + 474 + 22.4) = 72.89 percent.

From a standard reference, such as Ganapathy, *Steam Plant Calculations Manual*, Marcel Dekker, Inc., find the combustion constants, K, for various fuels and use them thus: For the air required for combustion, A_c = (K for methane)(percent by weight methane from above list) + (K for ethane)(percent by weight ethane); or A_c = (17.265)(0.7289) + (16.119)(0.2589) = 16.76 lb/lb (16.76 kg/kg).

Next, compute the higher heating value of the fuel (HHV) using the air constants from the same reference mentioned above. Or HHV = (heat of combustion for methane)(percent by weight methane) + (heat of combustion of ethane)(percent by weight ethane) = (23,879)(0.7289) + (22,320)(0.2589) = 23,184 Btu/lb (54,018.7 kJ/kg). Then, the amount of fuel equivalent to 1,000,000 Btu (1,055,000 kJ) = (1,000,000)/23,184 = 43.1 lb (19.56 kg), which requires, as computed above, (43.1)(16.76) = 722.3 lb dry air (327.9 kg), which agrees closely with the value given in Step 1, above.

Similarly, if the fuel were 100 percent methane, using the steps given above, and suitable constants from the same reference work, the air required for combustion is 17.265 lb/lb (7.838 kg/kg) of fuel. HHV = 23,879 Btu/lb (55,638 kJ/kg). Hence, the fuel in 1,000,000 Btu (1,055,000 kJ) = (1,000,000)/(23,879) = 41.88 lb (19.01 kg). Then, the dry air per million Btu (1.055 kg) fired = (17.265)(41.88) = 723 lb (328.3 kg).

Likewise, for propane, using the same procedure, 1 lb (0.454 kg) requires 15.703 lb (7.129 kg) air and 1 million Btu (1,055,000 kJ) has (1,000,000)/21,661 = 46.17 lb (20.95 kg) fuel. Then, 1 million Btu (1,055,000 kJ) requires (15.703)(46.17) = 725 lb (329.2 kg) air. This general approach can be used for various fuel oils and solid fuels—coal, coke, *etc.*

Good estimates of excess air used in combustion processes may be obtained if the oxygen and nitrogen in dry flue gases are measured. Knowledge of excess air amounts helps in performing detailed combustion and boiler efficiency calculations. Percent excess air, EA = $100(O_2-CO2)/[0.264 \times N_2-(O_2-CO/2)]$, where O_2 = oxygen in the dry flue gas, percent volume; CO = percent volume carbon monoxide; N_2 = percent volume nitrogen.

You can also estimate excess air from oxygen readings. Use the relation, EA = (constant from list below)$((O_2)/(21-O_2))$.

Constants for Excess Air Calculations

Fuel	Constant
Carbon	100
Hydrogen	80
Carbon monoxide	121
Sulfur	100
Methane	90
Oil	94.5
Coal	97
Blast furnace gas	223
Coke oven gas	89.3

If the percent volume of oxygen measured is 3 on a dry basis in a natural-gas (methane) fired boiler, the excess air, EA = (90)[3/(21–3)] = 15 percent.

This procedure is the work of V. Ganapathy, Heat Transfer Specialist, ABCO Industries.

SAVINGS PRODUCED BY PREHEATING COMBUSTION AIR

A 20,000 sq ft (1858 sq m) building has a calculated total seasonal heating load of 2,534,440 MBH (thousand Btu) (2674 MJ). The stack temperature is 600°F (316°C) and the boiler efficiency is calculated to be 75 percent. Fuel oil burned has a higher heating value of 140,000 Btu/gal (39,018 MJ/L). A preheater can be purchased and installed to reduce the breeching discharge combustion air temperature by 250°F (139°C) to 350°F (177°C) and provide the burner with preheated air. How much fuel oil will be saved? What will be the monetary saving if fuel oil is priced at 80 cents per gallon?

Calculation Procedure:

1. Compute the total combustion air required by this boiler
A general rule used by design engineers is that 1 cu ft (0.0283 cu m) of combustion air is required for each 100 Btu (105.5 J) released during combustion. To compute the combustion air required, use the relation CA = H/100 × Boiler efficiency, expressed as a decimal, where CA = annual volume of combustion air, cu ft (cu m); H = total seasonal heating load, Btu/yr (kJ/yr). Substituting for this boiler, CA = (2,534,400)(1000)/100 × 0.75 = 33,792,533 cu ft/yr (956.329 cu m).

2. Calculate the annual energy savings
The energy savings, ES = (stack temperature reduction, deg F)(cu ft air per yr)(0.018), where the constant 0.018 is the specific heat of air. Substituting, ES = (250)(33,792,533)(0.018) = 152,066,399 Btu/yr (160,430 kJ/yr).

With a boiler efficiency of 75 percent, each gallon of oil releases 0.75 × 140,000 Btu/gal = 105,000 Btu (110.8 jk). Hence, the fuel saved, FS = ES/usuable heat in fuel, Btu/gal. Or, FS = 152,066,399/105,000 = 1448.3 gal/yr (5.48 cu m/yr).

With fuel oil at $1.10 per gallon, the monetary savings will be $1.10 (1448.3) = $1593.13. If the preheater cost $6000, the simple payoff time would be $6000/1593.13 = 3.77 years.

Related Calculations. Use this procedure to determine the potential savings for burning any type of fuel—coal, oil, natural gas, landfill gas, catalytic cracker offgas, hydrogen purge gas, bagesse, sugar cane, etc. Other rules of thumb used by designers to estimate the amount of combustion air required for various fuels are: 10 cu ft of air (0.283 cu m) per 1 cu ft (0.0283 cu m) of natural gas; 1300 cu ft of air (36.8 cu m) per gal (0.003785 cu m) of No. 2 fuel oil; 1450 cu ft of air (41 cu m) per gal of No. 5 fuel oil; 1500 cu ft of air (42.5 cu m) per gal of No. 6 fuel oil. These values agree with that used in the above computation—i.e. 100 cu ft per 100 Btu of 140,000 Btu per gal oil = 140,000/100 = 1400 cu ft per gal (39.6 cu m/0.003785 cu m).

This procedure is the work of Jerome F. Mueller, P.E. of Mueller Engineering Corp.

COMBUSTION OF COAL FUEL IN A FURNACE

A coal has the following ultimate analysis (or percent by weight): C = 0.8339; H_2 = 0.0456; O_2 = 0.0505; N_2 = 0.0103; S = 0.0064; ash = 0.0533; total = 1.000 lb (0.45 kg). This coal is burned in a steam-boiler furnace. Determine the weight of air required for theoretically perfect combustion, the weight of gas formed per pound (kilogram) of coal burned, and the volume of flue gas, at the boiler exit temperature of 600°F (316°C) per pound (kilogram) of coal burned; air required with 20 percent excess air, and the volume of gas formed with this excess; the CO_2 percentage in the flue gas on a dry and wet basis.

Calculation Procedure:

1. Compute the weight of oxygen required per pound of coal

To find the weight of oxygen required for theoretically perfect combustion of coal, set up the following tabulation, based on the ultimate analysis of the coal:

Element	×	Molecular-weight ratio	=	lb (kg) O_2 required
C; 0.8339	×	32/12	=	2.2237 (1.009)
H_2; 0.0456	×	16/2	=	0.3648 (0.165)
O_2; 0.0505; decreases external O_2 required			=	−0.0505 (0.023)
N_2; 0.0103 is inert in combustion and is ignored				
S; 0.0064	×	32/32	=	0.0064 (0.003)
Ash 0.0533 is inert in combustion and is ignored				
Total 1.0000				
lb (kg) external O_2 per lb (kg) fuel			=	2.5444 (1.154)

Note that of the total oxygen needed for combustion, 0.0505 lb (0.023 kg), is furnished by the fuel itself and is assumed to reduce the total external oxygen required by the amount of oxygen present in the fuel. The molecular-weight ratio is obtained from the equation for the chemical reaction of the element with oxygen in combustion. Thus, for carbon $C + O_2 \rightarrow CO_2$, or 12 + 32 = 44, where 12 and 32 are the molecular weights of C and O_2, respectively.

2. Compute the weight of air required for perfect combustion

Air at sea level is a mechanical mixture of various gases, principally 23.2 percent oxygen and 76.8 percent nitrogen by weight. The nitrogen associated with the 2.5444 lb (1.154 kg) of oxygen required per pound (kilogram) of coal burned in this furnace is the product of the ratio of the nitrogen and oxygen weights in the air and 2.5444, or (2.5444)(0.768/0.232) = 8.4228 lb (3.820 kg). Then the weight of air required for perfect combustion of 1 lb (0.45 kg) of coal = sum of nitrogen and oxygen required = 8.4228 + 2.5444 = 10.9672 lb (4.975 kg) of air per pound (kilogram) of coal burned.

3. Compute the weight of the products of combustion

Find the products of combustion by addition:

Fuel constituents	+	Oxygen	→	Products of combustion	
				lb	kg
C; 0.8339	+	2.2237	→ CO_2	= 3.0576	1.387
H; 0.0456	+	0.3648	→ H_2O	= 0.4104	0.186
O_2; 0.0505; this is *not* a product of combustion					
N_2; 0.0103; inert but passes through furnace				= 0.0103	0.005
S; 0.0064	+	0.0064	→ SO_2	= 0.0128	0.006
Outside nitrogen from step 2			= N_2	= 8.4228	3.820
lb (kg) of flue gas per lb (kg) of coal burned				= 11.9139	5.404

4. Convert the flue-gas weight to volume

Use Avogadro's law, which states that under the same conditions of pressure and temperature, 1 mol (the molecular weight of a gas expressed in lb) of any gas will occupy the same volume.

At 14.7 lb/in^2 (abs) (101.3 kPa) and 32 °F (0°C), 1 mol of any gas occupies 359 ft^3 (10.2 m^3). The volume per pound of any gas at these conditions can be found by dividing 359 by the molecular weight of the gas and correcting for the gas temperature by multiplying the volume by the ratio of the absolute flue-gas temperature and the atmospheric temperature. To change the weight analysis (step 3) of the products of combustion to volumetric analysis, set up the calculation thus:

Products	Weight		Molecular weight	Temperature correction		Volume at	
	lb	kg				600°F, ft^3	316°C, m^3
CO_2	3.0576	1.3869	44	(359/44)(3.0576)(2.15)	=	53.6	1.518
H_2O	0.4104	0.1862	18	(359/18)(0.4104)(2.15)	=	17.6	0.498
Total N_2	8.4331	3.8252	28	(359/28)(8.4331)(2.15)	=	232.5	6.584
SO_2	0.0128	0.0058	64	(359/64)(0.0128)(2.15)	=	0.15	0.004
ft^3 (m^3) of flue gas per lb (kg) of coal burned					=	303.85	8.604

In this calculation, the temperature correction factor 2.15 = absolute flue-gas temperature, °R/absolute atmospheric temperature, °R = (600 + 460)/(32 + 460). The total weight of N_2 in the flue gas is the sum of the N_2 in the combustion air and the fuel, or 8.4228 + 0.0103 = 8.4331 lb (3.8252 kg). The value is used in computing the flue-gas volume.

5. Compute the CO$_2$ content of the flue gas

The volume of CO_2 in the products of combustion at 600°F (316°C) is 53.6 ft^3 (1.158 m^3), as computed in step 4; and the total volume of the combustion products is 303.85 ft^3 (8.604 m^3). Therefore, the percent CO_2 on a wet basis (i.e., including the moisture in the combustion products) = ft^3 CO_2/total ft^3 = 53.6/303.85 = 0.1764, or 17.64 percent.

The percent CO_2 on a dry, or Orsat, basis is found in the same manner, except that the weight of H_2O in the products of combustion, 17.6 lb (7.83 kg) from step 4, is subtracted from the total gas weight. Or, percent CO_2, dry, or Orsat basis = (53.6)/(303.85 − 17.6) = 0.1872, or 18.72 percent.

6. *Compute the air required with the stated excess flow*
With 20 percent excess air, the air flow required = (0.20 + 1.00)(air flow with no excess) = 1.20 (10.9672) = 13.1606 lb (5.970 kg) of air per pound (kilogram) of coal burned. The air flow with no excess is obtained from step 2.

7. *Compute the weight of the products of combustion*
The excess air passes through the furnace without taking part in the combustion and increases the weight of the products of combustion per pound (kilogram) of coal burned. Therefore, the weight of the products of combustion is the sum of the weight of the combustion products without the excess air and the product of (percent excess air)(air for perfect combustion, lb); or, given the weights from steps 3 and 2, respectively, = 11.9139 + (0.20)(10.9672) = 14.1073 lb (6.399 kg) of gas per pound (kilogram) of coal burned with 20 percent excess air.

8. *Compute the volume of the combustion products and the percent CO_2*
The volume of the excess air in the products of combustion is obtained by converting from the weight analysis to the volumetric analysis and correcting for temperature as in step 4, using the air weight from step 2 for perfect combustion and the excess-air percentage, or (10.9672)(0.20)(359/28.95)(2.15) = 58.5 ft³ (1.656 m³). In this calculation the value 28.95 is the molecular weight of air. The total volume of the products of combustion is the sum of the column for perfect combustion, step 4, and the excess-air volume, above, or 303.85 + 58.5 = 362.35 ft³ (10.261 m³).

By using the procedure in step 5, the percent CO_2, wet basis = 53.6/362.35 = 14.8 percent. The percent CO_2, dry basis = 53.8/(362.35 − 17.6) = 15.6 percent.

Related Calculations. Use the method given here when making combustion calculations for any type of coal—bituminous, semibituminous, lignite, anthracite, cannel, or cooking—from any coal field in the world used in any type of furnace—boiler, heater, process, or waste-heat. When the air used for combustion contains moisture, as is usually true, this moisture is added to the combustion-formed moisture appearing in the products of combustion. Thus, for 80°F (26.7°C) air of 60 percent relative humidity, the moisture content is 0.013 lb/lb (0.006 kg/kg) of dry air. This amount appears in the products of combustion for each pound of air used and is a commonly assumed standard in combustion calculations.

Fossil-fuel-fired power plants release sulfur emissions to the atmosphere. In turn, this produces sulfates, which are the key ingredient in acid rain. The federal Clean Air Act regulates sulfur dioxide emissions from power plants. Electric utilities which burn high-sulfur coal are thought to produce some 35 percent of atmospheric emissions of sulfur dioxide in the United States.

Sulfur dioxide emissions by power plants have declined some 30 percent since passage of the Clean Air Act in 1970, and a notable decline in acid rain has been noted at a number of test sites. In 1990 the Acid Rain Control Program was created by amendments to the Clean Air Act. This program further reduces the allowable sulfur dioxide emissions from power plants, steel mills, and other industrial facilities.

The same act requires reduction in nitrogen oxide emissions from power plants and industrial facilities, so designers must keep this requirement in mind when designing new and replacement facilities of all types which use fossil fuels.

Coal usage in steam plants is increasing throughout the world. An excellent example of this is the New England Electric System (NEES). This utility has been converting boiler units from oil to coal firing. Their conversions have saved cus-

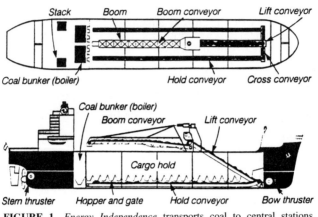

FIGURE 1 *Energy Independence* transports coal to central stations. (*Power.*)

tomers more than $60-million annually by displacing about 14-million bbl (2.2 million cu m) of oil per year.

To reduce costs, the company built the first coal-fired collier, Fig. 1, in more than 50 years in the United States, and assumed responsibility for coal transportation to its stations, cutting operating costs by more than $2-million per year. The collier makes economic sense because the utility stations in the system are not accessible by rail. This ship, the *Energy Independence*, has been an economic success for the utility. Measuring 665 ft (203 m) long by 95 ft (29 m) wide by 56 ft (17 m) deep with a 34-ft (10-m) draft, the vessel discharges a typical 40,000-ton load in 12 hours.

Data in these two paragraphs and Fig. 1 are from *Power* magazine.

PERCENT EXCESS AIR WHILE BURNING COAL

A certain coal has the following composition by weight percentages: carbon 75.09, nitrogen 1.56, ash 3.38, hydrogen, 5.72, oxygen 13.82, sulfur 0.43. When burned in an actual furnace, measurements showed that there was 8.93 percent combustible in the ash pit refuse and the following Orsat analysis in percentages was obtained: carbon dioxide 14.2, oxygen 4.7, carbon monoxide 0.3. If it can be assumed that there was no combustible in the flue gas other that the carbon monoxide reported, calculate the percentage of excess air used.

Calculation Procedure:

1. Compute the amount of theoretical air required per lb_m (kg) of coal
Theoretical air required per pound (kilogram) of coal, $w_{ta} = 11.5C' + 34.5[H'_2 - O'_2/8)] + 4.32S'$, where C', H'_2, O'_2, and S' represent the percentages by weight, expressed as decimal fractions, of carbon, hydrogen, oxygen, and sulfur, respec-

tively. Thus, w_{ta} = 11.5(0.7509) + 34.5[0.0572 − (0.1382/8)] + 4.32(0.0043) = 10.03 lb (4.55 kg) of air per lb (kg) of coal. The ash and nitrogen are inert and do not burn.

2. Compute the correction factor for combustible in the ash

The correction factor for combustible in the ash, $C_1 = (w_f C_f − w_r C_r)/(w_f \times 100)$, where the amount of fuel, w_f = 1 lb (0.45 kg) of coal; percent by weight, expressed as a decimal fraction, of carbon in the coal, C_f = 75.09; percent by weight of the ash and refuse in the coal, w_r = 0.0338; percent by weight of combustible in the ash, C_r = 8.93. Hence, C_1 = [(1 × 75.09) − (0.0338 × 8.93)]/(1 × 100) = 0.748.

3. Compute the amount of dry flue gas produced per lb (kg) of coal

The lb (kg) of dry flue gas per lb (kg) of coal, $w_{dg} = C_1(4CO_2 + O_2 + 704)/[3(CO_2 + CO)]$, where the Orsat analysis percentages are for carbon dioxide, CO_2 = 14.2; oxygen, O_2 = 4.7; carbon monoxide, CO = 0.3. Hence, w_{dg} = 0.748 × [(4 × 14.2) + 4.7 + 704)]/[3(14.2 + 0.3)] = 13.16 lb/lb (5.97 kg/kg).

4. Compute the amount of dry air supplied per lb (kg) of coal

The lb (kg) of dry air supplied per lb (kg) of coal, $w_{da} = w_{dg} − C_1 + 8[H'_2 − (O'_2/8)] − (N'_2/N)$, where the percentage by weight of nitrogen in the fuel, N'_2 = 1.56, and "atmospheric nitrogen" in the supply air, N_2 = 0.768; other values are as given or calculated. Then, w_{da} = 13.16 − 0.748 + 8[0.0572 − (0.1382/8)] − (0.0156/0.768) = 12.65 lb/lb (5.74 kg/kg).

5. Compute the percent of excess air used

Percent excess air = $(w_{da} − w_{ta})/w_{ta}$ = (12.65 − 10.03)/10.03 = 0.261, or 26.1 percent.

Related Calculations. The percentage by weight of nitrogen in "atmospheric air" in step 4 appears in *Principles of Engineering Thermodynamics*, 2nd edition, by Kiefer et al., John Wiley & Sons, Inc.

COMBUSTION OF FUEL OIL IN A FURNACE

A fuel oil has the following ultimate analysis: C = 0.8543; H_2 = 0.1131; O_2 = 0.0270; N_2 = 0.0022; S = 0.0034; total = 1.0000. This fuel oil is burned in a steam-boiler furnace. Determine the weight of air required for theoretically perfect combustion, the weight of gas formed per pound (kilogram) of oil burned, and the volume of flue gas, at the boiler exit temperature of 600°F (316°C), per pound (kilogram) of oil burned; the air required with 20 percent excess air, and the volume of gas formed with this excess; the CO_2 percentage in the flue gas on a dry and wet basis.

Calculation Procedure:

1. Compute the weight of oxygen required per pound (kilogram) of oil

The same general steps as given in the previous calculation procedure will be followed. Consult that procedure for a complete explanation of each step.

Using the molecular weight of each element, we find

Element	×	Molecular-weight ratio	=	lb (kg) O_2 required	
C; 0.8543	×	32/12	=	2.2781	(1.025)
H_2; 0.1131	×	16/2	=	0.9048	(0.407)
O_2; 0.0270; decreases external O_2 required			=	−0.0270	(−0.012)
N_2; 0.0022 is inert in combustion and is ignored					
S; 0.0034	×	32/32	=	0.0034	(0.002)
Total 1.0000					
lb (kg) external O_2 per lb (kg) fuel			=	3.1593	(1.422)

2. Compute the weight of air required for perfect combustion
The weight of nitrogen associated with the required oxygen = (3.1593)(0.768/0.232) = 10.458 lb (4.706 kg). The weight of air required = 10.4583 + 3.1593 = 13.6176 lb/lb (6.128 kg/kg) of oil burned.

3. Compute the weight of the products of combustion
As before,

Fuel constituents	+	Oxygen	=	Products of combustion
C; 0.8543 + 2.2781	=	3.1324	=	CO_2
H_2; 0.1131 + 0.9148	=	1.0179	=	H_2O
O_2; 0.270; *not* a product of combustion				
N_2; 0.0022; inert but passes through furnace	=	0.0022	=	N_2
S; 0.0034 + 0.0034	=	0.0068	=	SO_2
Outside N_2 from Step 2	=	10.458	=	N_2
lb (kg) of flue gas per lb (kg) of oil burned	=	14.6173 (6.578)		

4. Convert the flue-gas weight to volume
As before,

Products	Weight lb	Weight kg	Molecular weight	Temperature correction		Volume at 600°F, ft³	Volume at 316°C, m³
CO_2	3.1324	1.4238	44	(359/44)(3.1324)(2.15)	=	55.0	1.557
H_2O	1.0179	0.4626	18	(359/18)(1.0179)(2.15)	=	43.5	1.231
N_2 (total)	10.460	4.7545	28	(359/28)(10.460)(2.15)	=	288.5	8.167
SO_2	0.0068	0.0031	64	(359/64)(0.0068)(2.15)	=	0.82	0.023
ft³ (m³) of flue gas per lb (kg) of oil burned					=	387.82	10.978

In this calculation, the temperature correction factor 2.15 = absolute flue-gas temperature, °R/absolute atmospheric temperature, °R = (600 + 460)/(32 + 460).

The total weight of N_2 in the flue gas is the sum of the N_2 in the combustion air and the fuel, or $10.4580 + 0.0022 = 10.4602$ lb (4.707 kg).

5. Compute the CO_2 content of the flue gas
CO_2, wet basis, $= 55.0/387.82 = 0.142$, or 14.2 percent. CO_2, dry basis, $= 55.0/(387.2 - 43.5) = 0.160$, or 16.0 percent.

6. Compute the air required with stated excess flow
The pounds (kilograms) of air per pound (kilogram) of oil with 20 percent excess air $= (1.20)(13.6176) = 16.3411$ lb (7.353 kg) of air per pound (kilogram) of oil burned.

7. Compute the weight of the products of combustion
The weight of the products of combustion = product weight for perfect combustion, lb + (percent excess air)(air for perfect combustion, lb) $= 14.6173 + (0.20)(13.6176) = 17.3408$ lb (7.803 kilogram) of flue gas per pound (kilogram) of oil burned with 20 percent excess air.

8. Compute the volume of the combustion products and the percent CO_2
The volume of excess air in the products of combustion is found by converting from the weight to the volumetric analysis and correcting for temperature as in step 4, using the air weight from step 2 for perfect combustion and the excess-air percentage, or $(13.6176)(0.20)(359/28.95)(2.15) = 72.7$ ft³ (2.058 m³). Add this to the volume of the products of combustion found in step 4, or $387.82 + 72.70 = 460.52$ ft³ (13.037 m³).

By using the procedure in step 5, the percent CO_2, wet basis $= 55.0/460.52 = 0.1192$, or 11.92 percent. The percent CO_2, dry basis $= 55.0/(460.52 - 43.5) = 0.1318$, or 13.18 percent.

Related Calculations. Use the method given here when making combustion calculations for any type of fuel oil—paraffin-base, asphalt-base, Bunker C, no. 2, 3, 4, or 5—from any source, domestic or foreign, in any type of furnace—boiler, heater, process, or waste-heat. When the air used for combustion contains moisture, as is usually true, this moisture is added to the combustion-formed moisture appearing in the products of combustion. Thus, for 80°F (26.7°C) air of 60 percent relative humidity, the moisture content is 0.013 lb/lb (0.006 kg/kg) of dry air. This amount appears in the products of combustion for each pound (kilogram) of air used and is a commonly assumed standard in combustion calculations.

COMBUSTION OF NATURAL GAS IN A FURNACE

A natural gas has the following volumetric analysis at 60°F (15.5°C): $CO_2 = 0.004$; $CH_4 = 0.921$; $C_2H_6 = 0.041$; $N_2 = 0.034$; total $= 1.000$. This natural gas is burned in a steam-boiler furnace. Determine the weight of air required for theoretically perfect combustion, the weight of gas formed per pound of natural gas burned, and the volume of the flue gas, at the boiler exit temperature of 650°F (343°C), per pound (kilogram) of natural gas burned; air required with 20 percent excess air, and the volume of gas formed with this excess; CO_2 percentage in the flue gas on a dry and wet basis.

Calculation Procedure:

1. Compute the weight of oxygen required per pound of gas

The same general steps as given in the previous calculation procedures will be followed, except that they will be altered to make allowances for the differences between natural gas and coal.

The composition of the gas is given on a volumetric basis, which is the usual way of expressing a fuel-gas analysis. To use the volumetric-analysis data in combustion calculations, they must be converted to a weight basis. This is done by dividing the weight of each component by the total weight of the gas. A volume of 1 ft^3 (1 m^3) of the gas is used for this computation. Find the weight of each component and the total weight of 1 ft^3 (1 m^3) as follows, using the properties of the combustion elements and compounds given in Table 1:

		Density		Component weight = column 1 × column 2	
Component	Percent by volume	lb/ft^3	kg/m^3	lb/ft^3	kg/m^3
CO_2	0.004	0.1161	1.859	0.0004644	0.007
CH_4	0.921	0.0423	0.677	0.0389583	0.624
C_2H_6	0.041	0.0792	1.268	0.0032472	0.052
N_2	0.034	0.0739	0.094	0.0025026	0.040
Total	1.000			0.0451725	0.723

$$\text{Percent } CO_2 = 0.0004644/0.0451725 = 0.01026, \text{ or } 1.03 \text{ percent}$$
$$\text{Percent } CH_4 \text{ by weight} = 0.0389583/0.0451725 = 0.8625 \text{ or } 86.25 \text{ percent}$$
$$\text{Percent } C_2H_6 \text{ by weight} = 0.0032472/0.0451725 = 0.0718, \text{ or } 7.18 \text{ percent}$$
$$\text{Percent } N_2 \text{ by weight} = 0.0025026/0.0451725 = 0.0554, \text{ or } 5.54 \text{ percent}$$

The sum of the weight percentages = 1.03 + 86.25 + 7.18 + 5.54 = 100.00. This sum checks the accuracy of the weight calculation, because the sum of the weights of the component parts should equal 100 percent.

Next, find the oxygen required for combustion. Since both the CO_2 and N_2 are inert, they do not take part in the combustion; they pass through the furnace unchanged. Using the molecular weights of the remaining components in the gas and the weight percentages, we have

Compound	×	Molecular-weight ratio	=	lb (kg) O_2 required
CH_4; 0.8625	×	64/16	=	3.4500 (1.553)
C_2H_6; 0.0718	×	112/30	=	0.2920 (0.131)
lb (kg) external O_2 required per lb (kg) fuel			=	3.7420 (1.684)

In this calculation, the molecular-weight ratio is obtained from the equation for the combustion chemical reaction, or $CH_4 + 2O_2 = CO_2 + 2H_2O$, that is, $16 + 64 = 44 + 36$, and $C_2H_6 + \frac{7}{2}O_2 = 2CO_2 + 3H_2O$, that is $30 + 112 = 88 + 54$. See Table 2 from these and other useful chemical reactions in combustion.

TABLE 1 Properties of Combustion Elements*

Element or compound	Formula	Molecular weight	At 14.7 lb/in² (abs) (101.3 kPa), 60°F (15.6°C)		Nature		Heat value, Btu (kJ)		
			Weight, lb/ft³ (kg/m³)	Volume, ft³/lb (m³/kg)	Gas or solid	Combustible	Per lb (kg)	Per ft³ (m³) at 14.7 lb/in² (abs) (101.3 kPa), 60°F (15.6°C)	Per mole
Carbon	C	12	...	...	S	Yes	14,540 (33,820)	...	174,500
Hydrogen	H_2	2.02†	0.0053 (0.0849)	188 (11.74)	G	Yes	61,000 (141,886)	325 (12,109)	123,100
Sulfur	S	32	...	...	S	Yes	4,050 (9,420)	...	129,600
Carbon monoxide	CO	28	0.0739 (1.183)	13.54 (0.85)	G	Yes	4,380 (10,187)	323 (12,035)	122,400
Methane	CH_4	16	0.0423 (0.677)	23.69 (1.48)	G	Yes	24,000 (55,824)	1,012 (37,706)	384,000
Acetylene	C_2H_2	26	0.0686 (1.098)	14.58 (0.91)	G	Yes	21,500 (50,009)	1,483 (55,255)	562,000
Ethylene	C_2H_4	28	0.0739 (1.183)	13.54 (0.85)	G	Yes	22,200 (51,637)	1,641 (61,141)	622,400
Ethane	C_2H_6	30	0.0792 (1.268)	12.63 (0.79)	G	Yes	22,300 (51,870)	1,762 (65,650)	668,300
Oxygen	O_2	32	0.0844 (1.351)	11.84 (0.74)	G				
Nitrogen	N_2	28	0.0739 (1.183)	13.52 (0.84)	G				
Air‡	...	29	0.0765 (1.225)	13.07 (0.82)	G				
Carbon dioxide	CO_2	44	0.1161 (1.859)	8.61 (0.54)	G				
Water	H_2O	18	0.0475 (0.760)	21.06 (1.31)	G				

*P. W. Swain and L. N. Rowley, "Library of Practical Power Engineering" (collection of articles published in *Power*).
†For most practical purposes, the value of 2 is sufficient.
‡The molecular weight of 29 is merely the weighted average of the molecular weight of the constituents.

3.13

TABLE 2 Chemical Reactions

Combustible substance	Reaction	Mols	lb (kg)*
Carbon to carbon monoxide	$C + \tfrac{1}{2}O_2 = CO$	$1 + \tfrac{1}{2} = 1$	$12 + 16 = 28$
Carbon to carbon dioxide	$C + O_2 = CO_2$	$1 + 1 = 1$	$12 + 32 = 44$
Carbon monoxide to carbon dioxide	$CO + \tfrac{1}{2}O_2 = CO_2$	$1 + \tfrac{1}{2} = 1$	$28 + 16 = 44$
Hydrogen	$H_2 + \tfrac{1}{2}O_2 = H_2O$	$1 + \tfrac{1}{2} = 1$	$2 + 16 = 18$
Sulfur to sulfur dioxide	$S + O_2 = SO_2$	$1 + 1 = 1$	$32 + 32 = 64$
Sulfur to sulfur trioxide	$S + \tfrac{3}{2}O_2 = SO_3$	$1 + \tfrac{3}{2} = 1$	$32 + 48 = 80$
Methane	$CH_4 + 2O_2 = CO_2 + 2H_2O$	$1 + 2 = 1 + 2$	$16 + 64 = 44 + 36$
Ethane	$C_2H_6 + \tfrac{7}{2}O_2 = 2CO_2 + 3H_2O$	$1 + \tfrac{7}{2} = 2 + 3$	$30 + 112 = 88 + 54$
Propane	$C_3H_8 + 5O_2 = 3CO_2 + 4H_2O$	$1 + 5 = 3 + 4$	$44 + 160 = 132 + 72$
Butane	$C_4H_{10} + \tfrac{13}{2}O_2 = 4CO_2 + 5H_2O$	$1 + \tfrac{13}{2} = 4 + 5$	$58 + 208 = 176 + 90$
Acetylene	$C_2H_2 + \tfrac{5}{2}O_2 = 2CO_2 + H_2O$	$1 + \tfrac{5}{2} = 2 + 1$	$26 + 80 = 88 + 18$
Ethylene	$C_2H_4 + 3O_2 = 2CO_2 + 2H_2O$	$1 + 3 = 2 + 2$	$28 + 96 = 88 + 36$

*Substitute the molecular weights in the reaction equation to secure lb (kg). The lb (kg) on each side of the equation must balance.

2. Compute the weight of air required for perfect combustion

The weight of nitrogen associated with the required oxygen = (3.742)(0.768/0.232) = 12.39 lb (5.576 kg). The weight of air required = 12.39 + 3.742 = 16.132 lb/lb (7.259 kg/kg) of gas burned.

3. Compute the weight or the products of combustion

Fuel constituents	+	Oxygen	=	Products of combustion	
				lb	kg
CO_2; 0.0103; inert but passes through the furnace			=	0.010300	0.005
CH_4; 0.8625	+	3.45	=	4.312500	1.941
C_2H_6; 0.003247	+	0.2920	=	0.032447	0.015
N_2; 0.0554; inert but passes through the furnace			=	0.055400	0.025
Outside N_2 from step 2			=	12.390000	5.576
lb (kg) of flue gas per lb (kg) of natural gas burned			=	16.800347	7.562

4. Convert the flue-gas weight to volume

The products of complete combustion of any fuel that does not contain sulfur are CO_2, H_2O, and N_2. Using the combustion equation in step 1, compute the products of combustion thus: $CH_4 + 2O_2 = CO_2 + H_2O$; 16 + 64 = 44 + 36; or the CH_4 burns to CO_2 in the ratio of 1 part CH_4 to 44/16 parts CO_2. Since, from step 1, there is 0.03896 lb CH_4 per ft^3 (0.624 kg/m^3) of natural gas, this forms (0.03896)(44/16) = 0.1069 lb (0.048 kg) of CO_2. Likewise, for C_2H_6, (0.003247)(88/30) = 0.00952 lb (0.004 kg). The total CO_2 in the combustion products = 0.00464 + 0.1069 + 0.00952 = 0.11688 lb (0.053 kg), where the first quantity is the CO_2 in the fuel.

Using a similar procedure for the H_2O formed in the products of combustion by CH_4, we find (0.03896)(36/16) = 0.0875 lb (0.039 kg). For C_2H_6, (0.003247)(54/30) = 0.005816 lb (0.003 kg). The total H_2O in the combustion products = 0.0875 + 0.005816 = 0.093316 lb (0.042 kg).

Step 2 shows that 12.39 lb (5.58 kg) of N_2 is required per lb (kg) of fuel. Since 1 ft^3 (0.028 m^3) of the fuel weighs 0.04517 lb (0.02 kg), the volume of gas which weighs 1 lb (2.2 kg) is 1/0.04517 = 22.1 ft^3 (0.626 m^3). Therefore, the weight of N_2 per ft^3 of fuel burned = 12.39/22.1 = 0.560 lb (0.252 kg). This, plus the weight of N_2 in the fuel, step 1, is 0.560 + 0.0025 = 0.5625 lb (0.253 kg) of N_2 in the products of combustion.

Next, find the total weight of the products of combustion by taking the sum of the CO_2, H_2O, and N_2 weights, or 0.11688 + 0.09332 + 0.5625 = 0.7727 lb (0.35 kg). Now convert each weight to ft^3 at 650°F (343°C), the temperature of the combustion products, or:

Products	Weight		Molecular weight	Temperature correction		Volume at	
	lb	kg				650°F, ft³	343°C, m³
CO_2	0.11688	0.05302	44	(379/44)(0.11688)(2.255)	=	2.265	0.0641
H_2O	0.09332	0.04233	18	(379/18)(0.09332)(2.255)	=	4.425	0.1252
N_2 (total)	0.5625	0.25515	28	(379/28)(0.5625)(2.255)	=	17.190	0.4866
ft³ (m³) of flue gas per ft³ (m³) of natural-gas fuel					=	23.880	0.6759

In this calculation, the value of 379 is used in the molecular-weight ratio because at 60°F (15.6°C) and 14.7 lb/in² (abs) (101.3 kPa), the volume of 1 lb (0.45 kg) of any gas = 379/gas molecular weight. The fuel gas used is initially at 60°F (15.6°C) and 14.7 lb/in² (abs) (101.3 kPa). The ratio 2.255 = (650 + 460)/(32 + 460).

5. Compute the CO_2 content of the flue gas
CO_2, wet basis = 2.265/23.88 = 0.947, or 9.47 percent. CO_2 dry basis = 2.265/(23.88 − 4.425) = 0.1164, or 11.64 percent.

6. Compute the air required with the stated excess flow
With 20 percent excess air, (1.20)(16.132) = 19.3584 lb of air per lb (8.71 kg/kg) of natural gas, or 19.3584/22.1 = 0.875 lb of air per ft³ (13.9 kg/m³) of natural gas. See step 4 for an explanation of the value 22.1.

7. Compute the weight of the products of combustion
Weight of the products of combustion = product weight for perfect combustion, lb + (percent excess air) (air for perfect combustion, lb) = 16.80 + (0.20)(16.132) = 20.03 lb (9.01 kg).

8. Compute the volume of the combustion products and the percent CO_2
The volume of excess air in the products of combustion is found by converting from the weight to the volumetric analysis and correcting for temperature as in step 4, using the air weight from step 2 for perfect combustion and the excess-air percentage, or (16.132/22.1)(0.20)(379/28.95)(2.255) = 4.31 ft³ (0.122 m³). Add this to the volume of the products of combustion found in step 4, or 23.88 + 4.31 = 28.19 ft³ (0.798 m³).

By the procedure in step 5, the percent CO_2, wet basis = 2.265/28.19 = 0.0804, or 8.04 percent. The percent CO_2, dry basis = 2.265/(28.19 − 4.425) = 0.0953, or 9.53 percent.

Related Calculations. Use the method given here when making combustion calculations for any type of gas used as a fuel—natural gas, blast-furnace gas, coke-oven gas, producer gas, water gas, sewer gas—from any source, domestic or foreign, in any type of furnace—boiler, heater, process, or waste-heat. When the air used for combustion contains moisture, as is usually true, this moisture is added to the combustion-formed moisture appearing in the products of combustion. Thus, for 80°F (26.7°C) air of 60 percent relative humidity, the moisture content is 0.013 lb/lb (0.006 kg/kg) of dry air. This amount appears in the products of combustion for each pound of air used and is a commonly assumed standard in combustion calculations.

COMBUSTION OF WOOD FUEL IN A FURNACE

The weight analysis of a yellow-pine wood fuel is: $C = 0.490$; $H_2 = 0.074$; $O_2 = 0.406$; $N_2 = 0.030$. Determine the weight of oxygen and air required with perfect combustion and with 20 percent excess air. Find the weight and volume of the products of combustion under the same conditions, and the wet and dry CO_2. The flue-gas temperature is 600°F (316°C). The air supplied for combustion has a moisture content of 0.013 lb/lb (0.006 kg/kg) of dry air.

Calculation Procedure:

1. Compute the weight of oxygen required per pound of wood
The same general steps as given in earlier calculation procedures will be followed; consult them for a complete explanation of each step. Using the molecular weight of each element, we have

Element	$\times$	Molecular-weight ratio	$=$	lb (kg) O_2 required
C; 0.490	$\times$	32/12	$=$	1.307 (0.588)
H_2; 0.074	$\times$	16/2	$=$	0.592 (0.266)
O_2; 0.406; decreases external O_2 required			$=$	-0.406 (-0.183)
N_2; 0.030 inert in combustion				
Total 1.000				
lb (kg) external O_2 per lb (kg) fuel			$=$	1.493 (0.671)

2. Compute the weight of air required for complete combustion
The weight of nitrogen associated with the required oxygen $= (1.493)(0.768/0.232) = 4.95$ lb (2.228 kg). The weight of air required $= 4.95 + 1.493 = 6.443$ lb/lb (2.899 kg/kg) of wood burned, if the air is dry. But the air contains 0.013 lb of moisture per lb (0.006 kg/kg) of air. Hence, the total weight of the air $= 6.443 + (0.013)(6.443) = 6.527$ lb (2.937 kg).

3. Compute the weight of the products of combustion
Use the following relation:

Fuel constituents	$+$	Oxygen	$=$	Products of combustion, lb (kg)
C; 0.490	$+$	1.307	$=$	1.797 (0.809) = CO_2
H_2; 0.074	$+$	0.592	$=$	0.666 (0.300) = H_2O
O_2; not a product of combustion				
N_2; inert but passes through the furnace			$=$	0.030 (0.014) = N_2
Outside N_2 from step 2			$=$	4.950 (2.228) = N_2
Outside moisture from step 2			$=$	0.237 (0.107)
lb (kg) of flue gas per lb (kg) of wood burned			$=$	7.680 (3.458)

4. Convert the flue-gas weight to volume
Use, as before, the following tabulation:

Products	Weight		Molecular weight	Temperature correction	Volume at	
	lb	kg			600°F, ft³	316°C, m³
CO_2	1.797	0.809	44	$(359/44)(1.797)(2.15) =$	31.5	0.892
H_2O (fuel)	0.666	0.300	18	$(359/18)(0.666)(2.15) =$	28.6	0.810
N_2 (total)	4.980	2.241	28	$(359/28)(4.980)(2.15) =$	137.2	3.884
H_2O (outside air)	0.837	0.377	18	$(359/18)(0.837)(2.15) =$	35.9	10.16
Cu ft (m³) of flue gas per lb (kg) of oil					233.2	6.602

In this calculation the temperature correction factor 2.15 = (absolute flue-gas temperature, °R)/(absolute atmospheric temperature, °R) = (600 + 460)/(32 + 460). The total weight of N_2 is the sum of the N_2 in the combustion air and the fuel.

5. Compute the CO_2 content of the flue gas
The CO_2, wet basis = 31.5/233.2 = 0.135, or 13.5 percent. The CO_2, dry basis = 31.5/(233.2 − 28.6 − 35.9) = 0.187, or 18.7 percent.

6. Compute the air required with the stated excess flow
With 20 percent excess air, (1.20)(6.527) = 7.832 lb (3.524 kg) of air per lb (kg) of wood burned.

7. Compute the weight of the products of combustion
The weight of the products of combustion = product weight for perfect combustion, lb + (percent excess air)(air for perfect combustion, lb) = 8.280 + (0.20)(6.527) = 9.585 lb (4.313 kg) of flue gas per lb (kg) of wood burned with 20 percent excess air.

8. Compute the volume of the combustion products and the percent CO_2
The volume of the excess air in the products of combustion is found by converting from the weight to the volumetric analysis and correcting for temperature as in step 4, using the air weight from step 2 for perfect combustion and the excess-air percentage, or (6.527)(0.20)(359/28.95)(2.15) = 34.8 ft³ (0.985 m³). Add this to the volume of the products of combustion found in step 4, or 233.2 + 34.8 = 268.0 ft³ (7.587 m³).

By using the procedure in step 5, the percent CO_2, wet basis = 31.5/268 = 0.1174, or 11.74 percent. The percent CO_2, dry basis = 31.5/(268 − 28.6 − 35.9 − 0.20 × 0.837) = 0.155, or 15.5 percent. In the dry-basis calculation, the factor (0.20)(0.837) is the outside moisture in the excess air.

Related Calculations.   Use the method given here when making combustion calculations for any type of wood or woodlike fuel—spruce, cypress, maple, oak, sawdust, wood shavings, tanbark, bagesse, peat, charcoal, redwood, hemlock, fir, ash, birch, cottonwood, elm, hickory, walnut, chopped trimmings, hogged fuel, straw, corn, cottonseed hulls, city refuse—in any type of furnace—boiler, heating, process, or waste-heat. Most of these fuels contain a small amount of ash—usually less than 1 percent. This was ignored in this calculation procedure because it does not take part in the combustion.

Industry is making greater use of discarded process waste to generate electricity and steam by burning the waste in a steam boiler. An excellent example is that of Agrilectric Power Partners Ltd., Lake Charles, LA. This plant burns rice hulls from its own process and buys other producers' surplus rice hulls for continuous operation. Their plant is reported as the first small-power-production facility to operate on rice hulls.

By burning the waste rice hulls, Agrilectric is confronting, and solving, an environmental nuisance often associated with rice processing. When rice hulls are disposed of by being spread on land adjacent to the mill, they often smolder, creating continuous, uncontrolled burning. Installation of its rice-hull burning, electric-generating plant has helped Agrilectric avoid the costs associated with landfilling and disposal, as well as potential environmental problems.

The boiler supplies steam for a turbine-generator with an output ranging from 11.2 to 11.8 MW. Excess power that cannot be used in the plant is sold to the local utility at a negotiated price. Thus, the combustion of an industrial waste is producing useful power while eliminating the environmental impact of the waste. The advent of PURPA (Public Utility Regulatory Policies Act) requiring local utilities to purchase power from such plants has been a major factor in the design, development, and construction of many plants by food processors to utilize waste materials for combustion and power production.

MOLAL METHOD OF COMBUSTION ANALYSIS

A coal fuel has this ultimate analysis: $C = 0.8339$; $H_2 = 0.0456$; $O_2 = 0.0505$; $N_2 = 0.0103$; $S = 0.0064$; ash $= 0.0533$; total $= 1.000$. This coal is completely burned in a boiler furnace. Using the molal method, determine the weight of air required per lb (kg) of coal with complete combustion. How much air is needed with 25 percent excess air? What is the weight of the combustion products with 25 percent excess air? The combustion air contains 0.013 lb of moisture per lb (0.006 kg/kg) of air.

Calculation Procedure:

1. Convert the ultimate analysis to moles
A mole of any substance is an amount of the substance having a weight equal to the molecular weight of the substance. Thus, 1 mol of carbon is 12 lb (5.4 kg) of carbon, because the molecular weight of carbon is 12. To convert an ultimate analysis of a fuel to moles, assume that 100 lb (45 kg) of the fuel is being considered. Set up a tabulation thus:

Ultimate analysis, %	Weight		Molecular weight	Moles per 100-lb (45-kg) fuel
	lb	kg		
C = 0.8339	83.39	37.526	12	6.940
H₂ = 0.0456	4.56	2.052	2	2.280
O₂ = 0.0505	5.05	2.678	32	0.158
N₂ = 0.0103	1.03	0.464	28	0.037
S = 0.0064	0.64	0.288	32	
Ash = 0.0533	5.33	2.399	Inert	
Total	100.00	45.407	. . .	9.435

2. Compute the mols of oxygen for complete combustion

From Table 2, the burning of carbon to carbon dioxide requires 1 mol of carbon and 1 mol of oxygen, yielding 1 mol of CO_2. Using the molal equations in Table 2 for the other elements in the fuel, set up a tabulation thus, entering the product of columns 2 and 3 in column 4:

(1) Element	(2) Moles per 100-lb (45-kg) fuel	(3) Moles O_2 per 100- lb (45-kg) fuel	(4) Total moles O_2
C	6.940	1.00	6.940
H_2	2.280	0.5	1.140
O_2	0.158	Reduces O_2 required	−0.158
N_2	0.037	Inert in combustion	
S	0.020	1.00	0.020
Total moles of O_2 required	. . .	. . .	7.942

3. Compute the moles of air for complete combustion

Set up a similar tabulation for air, thus:

(1) Element	(2) Moles per 100-lb (45-kg) fuel	(3) Moles air per 100- lb (45-kg) fuel	(4) Total moles air
C	6.940	4.76	33.050
H_2	2.280	2.38	5.430
O_2	0.158	Reduces O_2 required	−0.752
N_2	0.037	Inert in combustion	
S	0.020	4.76	0.095
Total moles of air required		. . .	37.823

In this tabulation, the factors in column 3 are constants used for computing the total moles of air required for complete combustion of each of the fuel elements listed. These factors are given in the Babcock & Wilcox Company—*Steam: Its Generation and Use* and similar treatises on fuels and their combustion. A tabulation of these factors is given in Table 3.

An alternative, and simpler, way of computing the moles of air required is to convert the required O_2 to the corresponding N_2 and find the sum of the O_2 and N_2. Or, $376O_2 = N_2$; $N_2 + O_2$ = moles of air required. The factor 3.76 converts the required O_2 to the corresponding N_2. These two relations were used to convert the 0.158 mol of O_2 in the above tabulation to moles of air.

Using the same relations and the moles of O_2 required from step 2, we get $(3.76)(7.942) = 29.861$ mol of N_2. Then $29.861 + 7.942 = 37.803$ mol of air, which agrees closely with the 37.823 mol computed in the tabulation. The difference of 0.02 mol is traceable to roundings.

TABLE 3 Molal Conversion Factors

Element or compound	Mol/mol of combustible for complete combustion; no excess air					
	For combustion			Combustion products		
	O_2	N_2	Air	CO_2	H_2O	N_2
Carbon,° C	1.0	3.76	4.76	1.0	. . .	3.76
Hydrogen, H_2	0.5	0.188	2.38	. . .	1.0	1.88
Oxygen, O_2						
Nitrogen, N_2						
Carbon monoxide, CO	0.5	1.88	2.38	1.0	. . .	1.88
Carbon dioxide, CO_2						
Sulfur,° S	1.0	3.76	4.76	1.0	. . .	3.76
Methane, CH_4	2.0	7.53	0.53	1.0	2.0	7.53
Ethane, C_2H_6	3.5	13.18	16.68	2.0	3.0	13.18

°In molal calculations, carbon and sulfur are considered as gases.

4. Compute the air required with the stated excess air
With 25 percent excess, the air required for combustion = $(125/100)(37.823)$ = 47.24 mol.

5. Compute the mols of combustion products
Using data from Table 3, and recalling that the products of combustion of a sulfur-containing fuel are CO_2, H_2O, and SO_2, and that N_2 and excess O_2 pass through the furnace, set up a tabulation thus:

(1) Moles per 100-lb (45-kg) fuel	(2) Mol/mol of combustible	(3) Moles of combustion products per 100-lb (45-kg) of fuel
CO_2; 6.940	1	6.940
H_2O; 2.280 + (47.24)(0.021 + 0.158)	. . .	3.430
SO_2; 0.020	1	0.202
N_2; (47.24)(0.79)	. . .	37.320
Excess O_2; (1.25)(7.942) − 7.942	. . .	1.986

Total moles, wet combustion products = 49.878
Total moles, dry combustion products = 49.878 −3.232
= 46.646

In this calculation, the total moles of CO_2 is obtained from step 2. The moles of H_2 in 100 lb (45 kg) of the fuel, 2.280, is assumed to form H_2O. In addition, the air from step 4, 47.24 mol, contains 0.013 lb of moisture per lb (0.006 kg/kg) of air. This moisture is converted to moles by dividing the molecular weight of air, 28.95, by the molecular weight of water, 18, and multiplying the result by the moisture content of the air, or $(28.95/18)(0.013) = 0.0209$, say 0.021 mol of water per mol of air. The product of this and the moles of air gives the total moles of moisture (water) in the combustion products per 100 lb (45 kg) of fuel fired. To this is added the moles of O_2, 0.158, per 100 lb (45 kg) of fuel, because this oxygen is assumed to unite with hydrogen in the air to form water. The nitrogen in the products of combustion is that portion of the moles of air required, 47.24 mol from step 4, times the proportion of N_2 in the air, or 0.79. The excess O_2 passes through the furnace and adds to the combustion products and is computed as shown in the tabulation. Subtracting the total moisture, 3430 mol, from the total (or wet) combustion products gives the moles of dry combustion products.

Related Calculations. Use this method for molal combustion calculations for all types of fuels—solid, liquid, and gaseous—burned in any type of furnance—boiler, heater, process, or waste-heat. Select the correct factors from Table 3.

FINAL COMBUSTION PRODUCTS TEMPERATURE ESTIMATE

Pure carbon is burned to carbon dioxide at constant pressure in an insulated chamber. An excess air quantity of 20 percent is used and the carbon and the air are both initially at 77°F (25°C). Assume that the reaction goes to completion and that there is no dissociation. Calculate the final product's temperature using the following constants: Heating value of carbon, 14,087 Btu/lb (32.74×10^3 kJ/kg); constant-pressure specific heat of oxygen, nitrogen, and carbon dioxide are 0.240 Btu /lb_m (0.558 kJ/kg), 0.285 Btu/lb_m (0.662 kJ/kg), and 0.300 Btu/lb (0.697 kJ/kg), respectively.

Calculation Procedure:

1. Establish the chemical equation for complete combustion with 100 percent air
With 100 percent air: $C + O_2 + 3.78N_2 \rightarrow CO_2 + 3.78N_2$, where approximate molecular weights are: for carbon, $MC = 12$; oxygen, $MO_2 = 32$; nitrogen, $MN_2 = 28$; carbon dioxide, $MCO_2 = 44$. See the Related Calculations of this procedure for a general description of the 3.78 coefficient for N_2.

2. Establish the chemical equation for complete combustion with 20 percent excess air
With 20 percent excess air: $C + 1.2\ O_2 + (1.2 \times 3.78)N_2 \rightarrow CO_2 + 0.2\ O_2 + (1.2 \times 3.78)N_2$.

3. Compute the relative weights of the reactants and products of the combustion process
Relative weight = moles × molecular weight. Coefficients of the chemical equation in step 2 represent the number of moles of each component. Hence, for the reactants, the relative weights are: for $C = 1 \times MC = 1 \times 12 = 12$; $O_2 = 1.2 \times MO_2 = 1.2 \times 32 = 38.4$; $N_2 = (1.2 \times 3.78)MN_2 = (1.2 \times 3.78 \times 28) = 127$. For the products, relative weights are: for $CO_2 = 1 \times MCO_2 = 1 \times 44 = 44$; $O_2 = 0.2 \times MO_2 = 0.2 \times 32 = 6.4$; $N_2 = 127$, unchanged. It should be noted that the total relative weight of the reactants equal that of the products at 177.4.

4. *Compute the relative weights of the products of combustion on the basis of a per unit relative weight of carbon*

Since the relative weight of carbon, $C = 12$ in step 3; hence, on the basis of a per unit relative weight of carbon, the corresponding relative weights of the products are: for carbon dioxide, $wCO_2 = MCO_2/12 = 44/12 = 3.667$; oxygen, $wO_2 = MO_2/12 = 6.4/12 = 0.533$; nitrogen, $wN_2 = MN_2/12 = 127/12 = 10.58$.

5. *Compute the final product's temperature*

Since the combustion chamber is insulated, the combustion process is considered adiabatic. Hence, on the basis of a per unit mass of carbon, the heating value (HV) of the carbon = the corresponding heat content of the products. Thus, relative to a temperature base of $77°F$ ($25°C$), $1 \times HVC = [(wCO_2 \times c_pCO_2) + (wO_2 \times c_pO_2) + (wH_2 + c_pN_2)](t_2 - 77)$, where the heating value of carbon, $HVC = 14,087$ Btu/lb_m (32.74×10^3 kJ/kg); the constant-pressure specific heat of carbon dioxide, oxygen, and nitrogen are $c_pCO_2 = 0.300$ Btu/lb (0.697 kJ/kg), $c_pO_2 = 0.240$ Btu/lb (0.558 kJ/kg), and $c_pN_2 = 0.285$ Btu/lb (0.662 kJ/kg), respectively; final product temperature is t_2; other values as before. Then, $1 \times 14,087 = [(3.667 \times 0.30) + (0.533 \times 0.24) + (10.58 \times 0.285)(t_2 - 77)]$. Solving, $t_2 = 3320 + 77 = 3397°F$ ($1869°C$).

Related Calculations.   In the above procedure it is assumed that the carbon is burned in dry air. Also, the nitrogen coefficient of 3.78 used in the chemical equation in step 1 is based on a theoretical composition of dry air as 79.1 percent nitrogen and 20.9 percent oxygen by volume, so that $79.1/20.9 = 3.78$. For a more detailed description of this coefficient see the Related Calculations under the procedure for "Gas Turbine Combustion Chamber Inlet Air Temperature" elsewhere in this handbook.

SECTION 4

STEAM GENERATION EQUIPMENT AND AUXILIARIES

DETERMINING EQUIPMENT LOADING FOR GENERATING STEAM EFFICIENTLY

A plant has a steam generator capable of delivering up to 1000,000 lb/h (45,400 kg/h) of saturated steam at 400 lb/in^2 (gage) (2756 kPa). The plant also has an HRSG capable of generating up to 1000,000 lb/h (45,400 kg/h) of steam in the fired mode at the same pressure. How should each steam generator be loaded to generate a given quantity of steam most efficiently?

Calculation Procedure:

1. *Develop the HRSG characteristics*
In cogeneration and combined-cycle steam plants (gas turbine plus other prime movers), the main objective of supervising engineers is to generate a needed quantity of steam efficiently. Since there may be both HRSGs and steam boilers in the plant, the key to efficient operation is an understanding of the performance characteristics of each piece of equipment as a function of load.

In this plant, the HRSG generates saturated steam at 400 lb/in^2 (gage) (2756 kPa) from the exhaust of a gas turbine. It can be supplementary-fired to generate additional steam. Using the HRSG simulation approach given in another calculation procedure in this handbook, the HRSG performance at different steam flow rates should be developed. This may be done manually or by using the HRSG software developed by the author.

2. *Select the gas/steam temperature profile in the design mode*
Using a pinch point of 15°F (8.33°C) and approach point of 17°F (9.44°C), a temperature profile is developed as discussed in the procedure for HRSG simulation. The HRSG exit gas temperature is 319°F (159.4°C) while generating 25,000 lb/h (11,350 kg/h) of steam at 400 lb/in^2 (gage) (2756 kPa) using 230°F (110°C) feedwater.

3. *Prepare the gas/steam temperature profile in the fired mode*
A simple approach is to use the fact that supplementary firing is 100 percent efficient, as discussed in the procedure on HRSG simulation. All the fuel energy goes into generating steam in single-pressure HRSGs.

Compute the duty of the HRSG—*i.e.*, the energy absorbed by the steam—in the unfired mode, which is 25.4 MM Btu/h (7.44 MW). The energy required to generate 50,000 lb/h (22,700 kg/h) of steam is 50.8 MM Btu/h (14.88 MW). Hence, the additional fuel required = 50.8 − 25.4 = 25.4 MM Btu/h (7.44 MW). If a manual or computer simulation is done on the HRSG, fuel consumption will be seen to be 24.5 MM Btu/h (7.18 MW) on a Lower Heating Value (LHV) basis. Similarly, the performance at other steam flows is also computed and summarized in Table 1. Note that the exit gas temperature decreases as the steam flow increases. This aspect of an HRSG is discussed in the simulation procedure elsewhere in this handbook.

4. *Develop the steam-generator characteristics*
Develop the performance of the steam generator at various loads. Steam-generator suppliers will gladly provide this information in great detail, including plots and tabulations of the boiler's performance. As shown in Table 2, the exit-gas temper-

TABLE 1 Performance of HRSG

Load, %	25	50	75	100
Steam generation, lb/h (kg/h)	25,000 (11,350)	50,000 (22,700)	75,000 (34,050)	100,000 (45,400)
Duty, MM Btu/h (MW)	25.4 (7.4)	50.8 (14.9)	76.3 (22.4)	101.6 (29.8)
Exhaust gas flow, lb/h (kg/h)	152,000 (69,008)	153,140 (69,526)	154,330 (70,066)	155,570 (70,629)
Exit gas temperature, °F (°C)	319 (159)	285 (141)	273 (134)	269 (132)
Fuel fired, MM Btu/h LHV basis (MW)	0 (0)	24.50 (7.2)	50.00 (14.7)	76.50 (22.4)
ASME PTC 4.4 efficiency, %	70.80	83.79	88.0	89.53

Boiler pressure = 400 lb/in^2 (gage) (2756 kPa); feedwater temperature = 230°F (110°C); blowdown = 5 percent. Fuel used: natural gas; percent volume C_1 = 97; C_2 = 2; C_3 = 1; HHV = 1044 Btu/ft^3 (38.9 MJ/m^3); LHV = 942 Btu/ft^3 (35.1 MJ/m^3).

TABLE 2 Performance of Steam Generator

Load, %	25	50	75	100
Steam generation, lb/h (kg/h)	25,000 (11,350)	50,000 (22,700)	75,000 (34,050)	100,000 (45,400)
Duty, MM Btu/h (MW)	25.4 (7.4)	50.8 (14.9)	76.3 (22.4)	101.6 (29.8)
Excess air, %	30	10	10	10
Flue gas, lb/h (kg/h)	30,140 (13,684)	50,600 (22,972)	76,150 (34,572)	101,750 (46,195)
Exit gas temperature, °F (°C)	265 (129)	280 (138)	300 (149)	320 (160)
Heat losses, %				
—Dry gas loss	3.93	3.56	3.91	4.27
—Air moisture loss	0.10	0.09	0.10	0.11
—Fuel moisture loss	10.43	10.49	10.58	10.66
—Radiation loss	2.00	1.00	0.70	0.50
Efficiency, %				
—Higher Heating Value basis	83.54	84.86	84.70	84.46
—Lower Heating Value basis	92.58	94.05	93.87	93.60
Fuel fired, MM Btu/h LHV basis (MW)	27.50 (8.06)	54.00 (15.8)	81.30 (23.8)	108.60 (31.8)

Boiler pressure = 400 lb/in² (gage) (2756 kPa); feedwater temperature = 230°F (110°C); blowdown = 5 percent. Fuel used: natural gas; percent volume $C_1 = 97$; $C_2 = 2$; $C_3 = 1$; HHV = 1044 Btu/ft³ (38.9 MJ/m³); LHV = 942 Btu/ft³ (35.1 MJ/m³).

ature decreases as the load on the steam generator declines. This is because the ratio of gas/steam is maintained at nearly unity, unlike in an HRSG where the gas flow remains constant and steam flow alone is varied. Further, the radiation losses in a steam boiler increase at lower duty, while the exit-gas losses decrease. However, the boiler's efficiency falls within a narrow range. Table 3 also shows the steam generator's fuel consumption at various loads.

5. *Calculate steam vs. fuel data for combined operation of the equipment*
The next step is to develop, for combined operation of the HRSG and steam generator, a steam flow vs. fuel table such as that in Table 3. For example, 150,000

TABLE 3 Fuel Consumption at Various Steam Loads

Total steam	HRSG steam	SG steam	HRSG fuel	Sg fuel	Total fuel
lb/h	lb/h	lb/h	MM Btu/h	MM Btu/h	MM Btu/h
150,000	50,000	100,000	24.50	108.60	133.10
150,000	75,000	75,000	50.00	81.30	131.30
150,000	100,000	50,000	76.50	54.00	130.50
100,000	0	100,000	0	108.60	108.60
100,000	25,000	75,000	0	81.30	81.30
100,000	50,000	50,000	24.50	54.00	78.50
100,000	75,000	25,000	50.00	27.50	77.50
100,000	100,000	0	75.60	0	76.50
50,000	0	50,000	0	54.00	54.00
50,000	25,000	25,000	0	27.50	27.50
50,000	50,000	0	24.50	0	24.50
kg/h	kg/h	SI Units kg/h	MW	MW	MW
68,100	22,700	45,400	7.2	31.8	38.9
68,100	34,050	34,050	14.7	23.8	38.5
68,100	45,400	45,400	22.4	15.8	38.2
45,400	0	45,400	0	31.8	31.8
45,400	11,350	34,050	0	23.8	23.8
45,400	22,700	22,700	7.2	15.8	23.0
45,400	34,050	11,350	14.7	8.1	22.7
45,400	45,400	0	22.4	0	22.4
22,700	0	22,700	0	15.8	15.8
22,700	11,350	11,350	0	8.1	8.1
22,700	11,350	0	7.2	0	7.2

lb/h (68,100 kg/h) of steam could be generated in several ways—50,000 lb/h (22,700 kg/h) in the HRSG and 100,000 lb/h (45,400 kg/h) in the steam generator. Or each could generate 75,000 lb/h (34,050 kg/h); or 100,000 lb/h (45,400 kg/h) in the HRSG and the remainder in the steam generator. The table shows that maximizing the HRSG output first is the most efficient way of generating steam because no fuel is required to generate up to 25,000 lb/h (11,350 kg/h) of steam. However, this may not always be possible because of the plant operating mode, equipment availability, steam temperature requirements, *etc.*

Note also that the gas pressure drop in an HRSG does not vary significantly with load as the gas mass flow remains nearly constant. The gas pressure drop increases slightly as the firing temperature increases. On the other hand, the steam generator fan power consumption vs. load increases more in proportion to load.

It is also seen that at higher steam capacities the difference in fuel consumption between the various modes of operation is small. At 150,000 lb/h (68,100 kg/h), the difference is about 3 MM Btu/h (0.88 MW), while at 100,000 lb/h (45,400 kg/h), the difference is 30 MM Btu/h (8.79 MW). This difference should also be kept in mind while developing an operational strategy.

If a superheater is used, the performance of the superheater would have to be analyzed. Steam generators can generally maintain the steam temperature from 40 to 100 percent load, while in HRSGs the range is much larger as the steam temperature increases with firing temperature and can be controlled.

Related Calculations. Developing the performance characteristics of each piece of equipment as a function of load is the key to determining the mode of operation and loading of each type of steam producer. For best results, develop a performance curve for the steam generator, including all operating costs such as fan power consumption, pump power consumption, and gas-turbine power output as a function of load. This gives more insight into the total costs in addition to fuel cost, which is the major cost.

This procedure is the work of V. Ganapathy, Heat Transfer Specialist, ABCO Industries, Inc. The HRSG software mentioned in this procedure is available from Mr. Ganapathy.

STEAM CONDITIONS WITH TWO BOILERS SUPPLYING THE SAME STEAM LINE

Two closely adjacent steam boilers discharge equal amounts of steam into the same short steam main. Steam from boiler No. 1 is at 200 lb/in² (1378 kPa) and 420°F (215.6°C) while steam from boiler No. 2 is at 200 lb/in² (1378 kPa) and 95 percent quality. (*a*) What is the equilibrium condition after missing of the steam? (*b*) What is the loss of entropy by the higher temperature steam? Assume negligible pressure drop in the short steam main connecting the boilers.

Calculation Procedure:

1. *Determine the enthalpy of the mixed steam*
Use the T-S diagram, Fig. 1, to plot the condition of the mixed steam. Then, since equal amounts of steam are mixed, the final enthalpy, $H_3 = (H_1 + H_2)/2$. Substi-

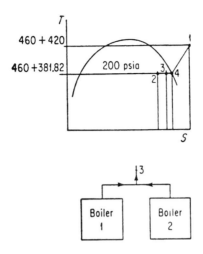

800°F (471°C)   841.8°F (449.8°C)   200 psia (1378 kPa)

FIGURE 1   *T-S* plot of conditions with two boilers on line.

tuting, using date from the steam tables and Mollier chart, $H_3 = (1225 + 1164)/2 = 1194.5$ Btu/lb (2783.2 kJ/kg).

2. Find the quality of the mixed steam

Entering the steam tables at 200 lb/in² (1378 kPa), find the enthalpy of the liquid as 355.4 Btu/lb (828.1 kJ/kg) and the enthalpy of vaporization as 843.3 Btu/lb (1964.9 kJ/kg). Then, using the equation for wet steam with the known enthalpy of the mixture from Step 1, $1194.5 = H_f + x_3 (H_{fg}) = 355.4 + x_3 (843.3)$; $x_3 = 0.995$, or 99.5 percent quality.

3. Find the entropy loss by the higher pressure steam

The entropy loss by the higher-temperature steam, referring to the Mollier chart plot, is $S_1 - S_2 = 1.575 - 1.541 = 0.034$ entropy units. The lower-temperature steam gains $S_3 - S_2 = 1.541 - 1.506 = 0.035$ units of entropy.

Related Calculations.   Use this general approach for any mixing of steam flows. Where different quantities of steam are being mixed, use the proportion of each quantity to the total in computing the enthalpy, quality, and entropy of the mixture.

GENERATING SATURATED STEAM BY DESUPERHEATING SUPERHEATED STEAM

Superheated steam generated at 1350 lb/in² (abs) (9301.5 kPa) and 950°F (510°C) is to be used in a process as saturated steam at 1000 lb/in² (abs) (6890 kPa). If the superheated steam is desuperheated continuously by injecting water at 500°F (260°C), how many pounds (kg) of saturated steam will be produced per pound (kg) of superheated steam?

Calculation Procedure:

1. Using the steam tables, determine the steam and water properties
Rounding off the enthalpy and temperature values we find that: Enthalpy of the superheated steam at 1350 lb/in² (abs) (9301.5 kPa) and 950°F (510°C) = H_1 = 1465 Btu/lb (3413.5 kJ/kg); Enthalpy of saturated steam at 1000 lb/in² (abs) (6890 kPa) = H_2 = 1191 Btu/lb (2775 kJ/kg); Enthalpy of water at 500°F (260°C) = (500 − 32) = H_3 = 488 Btu/lb (1137 kJ/kg).

2. Set up a heat-balance equation and solve it
$L X$ = lb (kg) of water at 500°F (260°C) required to desuperheat the superheated steam. Then, using the symbols given above, $H_1 + X(H_3) = (1 + X)H_2$. Solving for $X = (H_1 − H_2)/(H_2 − H_3) = (1465 − 1191)/(1191 − 488) = 0.39$. Then, $1.0 + 0.39 = 1.39$ lb (0.63 kg) of saturated steam produced per lb (kg) of superheated steam. Thus, if the process used 1000 lb (454 kg) of saturated steam at 1000 lb/in² (abs) (689 kPa), the amount of superheated steam needed to produce this saturated steam would be $1000/1.39 = 719.4$ lb (326.6 kg).

Related Calculations.   Desuperheating superheated steam for process and other use is popular because it can save purchase and installation of a separate steam generator for the lower pressure steam. While there is a small loss of energy in desuperheating (from heat losses in the piping and desuperheater), this loss is small compared to the savings made. That's why you'll find desuperheating being used in central stations, industrial, commercial and marine plants throughout the world.

DETERMINING FURNACE-WALL HEAT LOSS

A furnace wall consists of 9-in (22.9-cm) thick fire brick, 4.5-in (11.4-cm) Sil-O-Cel brick, 4-in (10.2-cm) red brick, and 0.25-in (0.64-cm) transite board. The thermal conductivity, k, values, Btu/(ft²)(°F)(ft) [kJ/(m²)(°C)(m)] are as follows: 0.82 at 1800°F (982°C) for fire brick; 0.125 at 1800°F (982°C) for Sil-O-Cel; 0.52 at 500°F (260°C) for transite. A temperature of 1800°F (982°C) exists on the inside wall of the furnace and 200°F (93.3°C) on the outside wall. Determine the heat loss per hour through each 10 ft² (0.929 m²) of furnace wall. What is the temperature of the wall at the joint between the fire brick and Sil-O-Cel?

Calculation Procedure:

1. Find the heat loss through a unit area of the furnace wall
Use the relation $Q = \Delta t/R$, where Q = heat transferred, Btu/h (W); Δt = temperature difference between the inside of the furnace wall and the outside, °F (°C); R = resistance of the wall to heat flow = $L/(kXA)$, where L = length of path

through which the heat flow, ft (m); k = thermal conductivity, as defined above; A = area of path of heat flow, ft^2 (m^2). Where there is more than one resistance to heat flow, add them to get the total resistance.

Substituting, the above values for this furnace wall, remembering that there are three resistances in series and solving for the heat flow through one square ft (0.0.0929 m^2), Q = (1800 − 200)/{[(1/0.82)(9/12)] + [(1/0.125)(4.5/12)] + [(1/052)(4/12)] + [(1/0.23)(0.25/12)]} = 344 Btu/h ft^2 (1083.6 W/m^2), or 10 (344) = 34400 Btu/h for 10 ft^2 (10,836 W/10 m^2).

2. Compute the temperature within the wall at the stated joint

Use the relation, $(\Delta t)/(\Delta t_1) = (R/R_1)$, where Δt = temperature difference across the wall, °F (°C); Δt_1 = temperature at the joint being considered, °F (°C); R = total resistance of the wall; R_1 = resistance of the first portion of the wall between the inside and the joint in question.

Substituting, $(1800 − 200)/(\Delta t_1)$ = (4.646/0.915); Δt_1 = 315°F (157.2°C). Then the interface temperature at the between the fire brick and the Sil-O-Cel is 1800 − 315 = 1485°F (807.2°C)

Related Calculations. The coefficient of thermal conductivity given here, Btu/((ft^2)(°F)(ft) is sometimes expressed in terms of per inch of thickness, instead of per foot. Either way, the conversion is simple. In SI units, this coefficient is expressed in kJ/(m^2)(°C)(m), or cm^2 and cm.

The exterior temperature of a furnace wall is an important considered in boiler and process unit design from a human safety standpoint. Excess exterior temperatures can cause injury to plant workers. Further, the higher the exterior temperature of a furnace wall, the larger the heat loss from the fired vessel. Therefore, both safety and energy conservation considerations are important in furnace design.

Typical interior furnace temperatures encountered in modern steam boilers range from 2400°F (1316°C) near the fuel burners to 1600°F (871°C) in the superheater interior. With today's emphasis on congeneration and energy conservation, many different fuels are being burned in boilers. Thus, a plant in Louisiana burns rice to generate electricity while disposing of a process waste material.

Rice hulls, which comprise 20 percent of harvested rice, are normally processed in a hammermill to increase their bulk from about 11 lb/ft^3 (176 kg/m^3) to 20 lb/ft^3 (320 kg/m^3). Then they are spread or piled on land adjacent to the rice mill. The hulls often smolder in the fields, like mine tailings from coal production. Continuous, uncontrolled burning may result, creating an environmental hazard and problem.

Burning rice hulls in a boiler furnace may create unexpected temperatures both inside and outside the furnace. Hence, it is important that the designer be able to analyze both the interior and exterior furnace temperatures using the procedure given here.

Another modern application of waste usage for power generation is the burning of sludge in a heat-recovery boiler to generate electricity. Sludge from a wastewater plant is burned in a combustor to generate steam for a turbogenerator. Not only are fuel requirements for the boiler reduced, there is also significant savings of fuel used to incinerate the sludge in earlier plants. Again, the furnace temperature is an important element in designing such plants.

The data present in these comments on new fuels for boilers is from *Power* magazine.

CONVERTING POWER-GENERATION POLLUTANTS FROM MASS TO VOLUMETRIC UNITS

In the power-generation industry, emission levels of pollutants such as CO and NO_x are often specified in mass units such as pounds per million Btu (kg per 1.055 MJ) and volumetric units such as ppm (parts per million) volume. Show how to relate these two measures for a gaseous fuel having this analysis: Methane = 97 percent; Ethane = 2 percent; Propane = 1 percent by volume, and excess air = 10 percent. Ambient air temperature during combustion = 80°F (26.7°C) and relative humidity = 60 percent; fuel higher heating value, HHV = 23,759 Btu/lb (55,358 kJ/kg); 100 moles of fuel gas is the basis of the flue gas analysis.

Calculation Procedure:

1. Find the theoretical dry air required, and the moisture in the actual air
The theoretical dry air requirements, in M moles, can be computed from the sum of (ft^3 of air per ft^3 of combustible gas)(percent of combustible in the fuel) using data from Ganapathy, *Steam Plant Calculations Manual*, Marcel Dekker, Inc. thus: M = $(9.528 \times 97) + (16.675 \times 2) + (23.821 \times 1) = 981.4$ moles. Then, with 10 percent excess air, excess air, EA = $0.1(981.4) = 98.1$ moles.

The excess oxygen, O_2 = (98.1 moles)(0.21) = 20.6 moles, where 0.21 = moles of oxygen in 1 mole of air. The nitrogen, N_2, produced by combustion = (1.1 for excess air)(981.4 moles)(0.79 moles of nitrogen in 1 mole of air) = 852.8 moles; round to 853 moles for additional calculations.

The moisture in the air = (981.4 + 98.1)(29 × 0.0142/18) = 24.69, say 24.7 moles. In this computation the values 29 and 18 are the molecular weights of dry air and water vapor, respectively, while 0.0142 is the lb (0.0064 kg) moisture per lb of dry air as shown in the previous procedure.

2. Compute the flue gas analysis for the combustion
Using the given data, CO_2 = (1 × 97) + (2 × 2) + (3 × 1) = 104 moles. For H_2O = (2 × 97) + (3 × 2) + (4 × 1) + 24.7 = 228.7 moles. From step 1, N_2 = 853 moles; O_2 = 20.6 moles.

Now, the total moles = 104 + 228.7 + 853 + 20.6 = 1206.3 moles. The percent volume of CO_2 = (104/1206.3)(100) = 8.6; the percent H_2O = (228.7/1206.1)(100) = 18.96; the percent N_2 = (853/1206.3)(100) = 70.7; the percent O_2 = (20.6/1206.3)(100) = 1.71.

3. Find the amount of flue gas produced per million Btu (1.055 MJ)
To relate the pounds per million Btu (1.055 MJ) of NO_x or CO produced to ppmv, we must know the amount of flue gas produced per million Btu (1.055 MJ). From step 2, the molecular weight of the flue gases = [(8.68 × 44) + (18.96 × 18) + (70.7 × 28) + (1.71 × 32)]/100 = 27.57.

The molecular weight of the fuel = [(97 × 16) + (2 × 30) + (1 × 44)]/100 = 16.56. Now the ratio of flue gases/fuel = (1206.3 × 27.57)/(100 × 16.56) = 20.08 lb flue gas/lb fuel (9.12 kg/kg). Hence, 1 million Btu fired produces (1,000,000)/23,789 = 42 lb (19.1 kg) fuel = (42)(20.08) = 844 lb (383 kg) wet flue gases.

4. Calculate ppm values for the gases

Let 1 million Btu fired generate N lb (kg) of NO_x. For emission calculations, NO_x is considered to have a molecular weight of 46. Also, the reference for NO_x or CO regulations is 3 percent dry oxygen by volume for steam generators. Hence, we have the relation, $N_N = 10^6(Y_x)(N/46)(MW_{fg})[(21 - 3)/(21 - O_2XY)]$, where $V_N =$ ppm dry NO_x; $Y = 100/(100 - \text{percent } H_2O)$, where percent H_2O is the percent volume water vapor in the flue gases; $N = $ lb (kg) of NO_x per million Btu (1.055 MJ) fired on an HHV basis; $MW_{fg} = $ molecular weight of wet flue gases; $W_{gm} = $ amount of wet flue gas produced per million Btu (1.055 MJ) fired.

Substituting in the above relation, $V_N = 10^6(N_x)[100/(100 - 18.96)]$ (27.57/ 844)(21 - 3)/[(21 - 1.71)(100/(100 - 18.96)] = 832N.

Similarly, $V_c = $ ppmv CO_2 generated per million Btu (1.055 MJ) fired = 1367°C, where $C = $ lb (kg) of CO generated per million Btu (1.055 MJ) and $V_c = $ amount in ppmvd (dry). The effect of excess air on these calculations is not at all significant. One may perform these calculations at 30 percent excess air and still show that $V_N = 832N$ and $V_c = 1367$ for natural gas.

Related Calculations.   These calculations for oil fuels also to show that $V_N = 783N$ and $V_c = 1286C$.

This procedure is the work of V. Ganapathy, Heat Transfer Specialist, ABCO Industries, Inc.

STEAM BOILER HEAT BALANCE DETERMINATION

A steam generator having a maximum rated capacity of 60,000 lb/h (27,000 kg/ h) is operating at 45,340 lb/h (20,403 kg/h), delivering 125-lb/in² (gage) 400°F (862-kPa, 204°C) steam with a feedwater temperature of 181°F (82.8°C). At this generating rate, the boiler requires 4370 lb/h (1967 kg/h) of West Virginia bituminous coal having a heating value of 13,850 Btu/lb (32,215 kJ/kg) on a dry basis. The ultimate fuel analysis is: C = 0.7757; H_2 = 0.0507; O_2 = 0.0519; N_2 = 0.0120; S = 0.0270; ash = 0.0827; total = 1.0000. The coal contains 1.61 percent moisture. The boiler-room intake air and the fuel temperature = 79°F (26.1°C) dry bulb, 71°F (21.7°C) wet bulb. The flue-gas temperature is 500°F (260°C), and the analysis of the flue gas shows these percentages: CO_2 = 12.8; CO = 0.4; O_2 = 6.1; N_2 = 80.7; total = 100.0. Measured ash and refuse = 9.42 percent of dry coal; combustible in ash and refuse = 32.3 percent. Compute a heat balance for this boiler based on these test data. The boiler has four water-cooled furnace walls.

Calculation Procedure:

1. Determine the heat input to the boiler

In a boiler heat balance the input is usually stated in Btu per pound of fuel as fired. Therefore, input = heating value of fuel = 13,850 Btu/lb (32,215 kJ/kg).

2. Compute the output of the boiler

The output of any boiler = Btu/lb (kJ/kg) of fuel + the losses. In this step the first portion of the output, Btu/lb (kJ/kg) of fuel will be computed. The losses will be computed in step 3.

First find W_s, lb of steam produced per lb of fuel fired. Since 45,340 lb/h (20,403 kg/h) of steam is produced when 4370 lb/h (1967 kg/h) of fuel is fired, W_s = 45,340/4370 = 10.34 lb of steam per lb (4.65 kg/kg) of fuel.

Once W_s is known, the output h_1 Btu/lb of fuel can be found from $h_1 = W_s(h_s - h_w)$, where h_s = enthalpy of steam leaving the superheater, or boiler if a superheater is not used; h_w = enthalpy of feedwater, Btu/lb. For this boiler with steam at 125 lb/in^2 (gage) [= 139.7 lb/in^2 (abs)] and 400°F (930 kPa, 204°C), h_s = 1221.2 Btu/lb (2841 kJ/kg), and h_w = 180.92 Btu/lb (420.8 kJ/kg), from the steam tables. Then h_1 = 10.34(1221.2 − 180.92) = 10,766.5 Btu/lb (25,043 kJ/kg) of coal.

3. Compute the dry flue-gas loss

For any boiler, the dry flue-gas loss h_2 Btu/lb (kJ/kg) of fuel is given by h_2 = 0.24W_g × ($T_g − T_a$), where W_g = lb of dry flue gas per lb of fuel; T_g = flue-gas exit temperature,°F; T_a = intake-air temperature,°F.

Before W_g can be found, however, it must be determined whether any excess air is passing through the boiler. Compute the excess air, if any, from excess air, percent = 100 (O_2 − ½CO)/[0.264N_2 − (O_2 − ½CO)], where the symbols refer to the elements in the flue-gas analysis. Substituting values from the flue-gas analysis gives excess air = 100(6.1 − 0.2)/[0.264 × 80.7 − (6.1 − 0.2)] = 38.4 percent.

Using the method given in earlier calculation procedures, find the air required for complete combustion as 10.557 lb/lb (4.571 kg/kg) of coal. With 38.4 percent excess air, the additional air required = (10.557)(0.384) = 4.053 lb/lb (1.82 kg/kg) of fuel.

From the same computation in which the air required for complete combustion was determined, the lb of *dry* flue gas per lb of fuel = 11.018 (4.958 kg/kg). Then, the total flue gas at 38.4 percent excess air = 11.018 + 4.053 = 15.071 lb/lb (6.782 kg/kg) of fuel.

With a flue-gas temperature of 500°F (260°C), and an intake-air temperature of 79°F (26.1°C), h_2 = 0.24(15.071)(500 − 70) = 1524 Btu/lb (3545 kJ/kg) of fuel.

4. Compute the loss due to evaporation of hydrogen-formed water

Hydrogen in the fuel is burned in forming H_2O. This water is evaporated by heat in the fuel, and less heat is available for producing steam. This loss is h_3 Btu/lb of fuel = 9H(1089 − T_f + 0.46T_g), where H = percent H_2 in the fuel ÷ 100; T_f = temperature of fuel *before* combustion,°F; other symbols as before. For this fuel with 5.07 percent H_2, h_3 = 9(5.07/100)(1089 − 79 + 0.46 × 500) = 565.8 Btu/lb (1316 kJ/kg) of fuel.

5. Compute the loss from evaporation of fuel moisture

This loss is h_4 Btu/lb of fuel = W_{mf}(1089 − T_f + 0.46T_g), where W_{mf} = lb of moisture per lb of fuel; other symbols as before. Since the fuel contains 1.61 percent moisture, in terms of *dry* coal this is (1.61)/(100 − 1.61) = 0.0164, or 1.64 percent. Then h_4 = (1.64/100)(1089 − 79 + 0.46 × 500) = 20.34 Btu/lb (47.3 kJ/kg) of fuel.

6. Compute the loss from moisture in the air

This loss is h_5 Btu/lb of fuel = 0.46W_{ma}($T_g − T_a$), W_{ma} = (lb of water per lb of dry air)(lb air supplied per lb fuel). From a psychrometric chart, the weight of moisture per lb of air at a 79°F (26.1°C) dry-bulb and 71°F (21.7°C) wet-bulb temperature is 0.014 (0.006 kg). The combustion calculation, step 3, shows that the total air required with 38.4 percent excess air = $10.557 + 4.053 = 14.61$ lb of air per lb (6.575 kg/kg) of fuel. Then, $W_{ma} = (0.014)(14.61) = 0.2045$ lb of moisture per lb (0.092 kg/kg) of air. And $h_5 = (0.46)(0.2045)(500 - 79) = 39.6$ Btu/lb (92.1 kJ/kg) of fuel.

7. Compute the loss from incomplete combustion of C to CO_2 in the stack
This loss is h_6 Btu/lb of fuel = $[CO/CO + CO_2)](C)(10.190)$, where CO and CO_2 are the percent by volume of these compounds in the flue gas by Orsat analysis; C = lb carbon per lb of coal. With the given flue-gas analysis and the coal ultimate analysis, $h_6 = 0.4/(0.4 + 12.8)[(77.57)/(100)](10.190) = 239.5$ Btu/lb (557 kJ/kg) of fuel.

8. Compute the loss due to unconsumed carbon in the refuse
This loss is h_7 Btu/lb of fuel = $W_c(14,150)$, where W_c = lb of unconsumed carbon in refuse per lb of fuel fired. With an ash and refuse of 9.42 percent of the dry coal and combustible in the ash and refuse of 32.3 percent, $h_7 = (9.42/100)(32.3/100)(14,150) = 430.2$ Btu/lb (1006 kJ/kg) of fuel.

9. Find the radiation loss in the boiler furnace
Use the American Boiler and Affiliated Industries (ABAI) chart, or the manufacturer's engineering data to approximate the radiation loss in the boiler. Either source will show that the radiation loss is 1.09 percent of the gross heat input. Since the gross heat input is 13,850 Btu/lb (32,215 kJ/kg) of fuel, the radiation loss = $(13,850)(1.09/100) = 151.0$ Btu/lb (351.2 kJ/kg) of fuel.

10. Summarize the losses; find the unaccounted-for loss
Set up a tabulation thus, entering the various losses computed earlier.

Item	Btu/lb fuel	kJ/kg fuel	Percent
1. Input	13,850.0	32,215.4	100.0
2. Output	10,770.0	25,051	77.75
Losses:			
3. Flue gas	1,524.0	3,545	11.00
4. Hydrogen	565.8	1,315	4.09
5. Water-fuel	20.3	47.2	0.15
6. Water-air	39.6	92.1	0.29
7. CO	239.5	557	1.73
8. Carbon-ash	430.2	1,001	3.11
9. Radiation	151.0	351.2	1.09
10. Unaccounted	109.6	254.9	0.79
Total	13,850.0	32,214.4	100.00

The unaccounted-for loss is found by summing all the other losses, 3 through 9, and subtracting from 100.00.

Related Calculations. Use this method to compute the heat balance for any type of boiler—watertube or firetube—in any kind of service—power, process, or heating—using any kind of fuel—coal, oil, gas, wood, or refuse. Note that step 3 shows how to compute excess air from an Orsat flue-gas analysis.

More stringent environmental laws are requiring larger investments in steam-boiler pollution-control equipment throughout the world. To control sulfur emissions, expensive scrubbers are required on large boilers. Without such scrubbers the sulfur emissions can lead to acid rain, smog, and reduced visibility in the area of the plant and downwind from it.

With the increased number of free-trade agreements between adjacent countries, cross-border pollution is receiving greater attention. The reason for this increased attention is because not all countries have the same environmental control requirements. When a country with less stringent requirements pollutes an adjacent country having more stringent pollution regulations, both political and regulatory problems can arise.

For example, two adjacent countries are currently discussing pollution problems of a cross-border type. One country's standard for particulate emissions is 10 times weaker than the adjacent country's, while its sulfur dioxide limit is 8 times weaker. With such a wide divergence in pollution requirements, cross-border flows of pollutants can be especially vexing.

All boiler-plant designers must keep up to date on the latest pollution regulations. Today there are some 90,000 environmental regulations at the federal, state, and local levels, and more than 40 percent of these regulations will change during the next 12 months. To stay in compliance with such a large number of regulations requires constant attention to those regulations applicable to boiler plants.

STEAM BOILER, ECONOMIZER, AND AIR-HEATER EFFICIENCY

Determine the overall efficiency of a steam boiler generating 56,00 lb/h (7.1 kg/s) of 600 lb/in² (abs) (4137.0 kPa) 800°F (426.7°C) steam. The boiler is continuously blown down at the rate of 2500 lb/h (0.31 kg/s). Feedwater enters the economizer at 300°F (148.9°C). The furnace burns 5958 lb/h (0.75 kg/s) of 13,100-Btu/lb (30,470.6-kJ/kg), HHV (higher heating value) coal having an ultimate analysis of 68.5 percent C, 5 percent H_2, 8.9 percent O_2, 1.2 percent N_2, 3.2 percent S, 8.7 percent ash, and 4.5 percent moisture. Air enters the boiler at 63°F (17.2°C) dry-bulb and 56°F (13.3°C) wet-bulb temperature, with 56 gr of vapor per lb (123.5 gr/kg) of dry air. Carbon in the fuel refuse is 7 percent, refuse is 0.093 lb/lb (0.2 kg/kg) of fuel. Feedwater leaves the economizer at 370°F (187.8°C). Flue gas enters the economizer at 850°F (454.4°C) and has an analysis of 15.8 percent CO_2, 2.8 percent O_2, and 81.4 percent N_2. Air enters the air heater at 63°F (17.2°C) with 56 gr/lb (123.5 gr/kg) of dry air; air leaves the heater at 480°F (248.9°C). Gas enters the air heater at 570°F (298.9°C), and 14 percent of the air to the furnace comes from the mill fan. Determine the steam generator overall efficiency, economizer efficiency, and air-heater efficiency. Figure 2 shows the steam generator and the flow factors that must be considered.

Calculation Procedure:

1. Determine the boiler output
The boiler output = $S(h_g - h_{f1}) + S_r(h_{g3} - h_{g2}) + B(h_{f3} - h_{f1})$, where S = steam generated, lb/h; h_g = enthalpy of the generated steam, Btu/lb; h_{f1} = enthalpy of inlet feedwater; S_r = reheated steam flow, lb/h (if any); h_{g3} = outlet enthalpy of reheated steam; h_{g2} = inlet enthalpy of reheated steam; B = blowoff, lb/h; h_{f3} = blowoff enthalpy, where all enthalpies are in Btu/lb. Using the appropriate steam

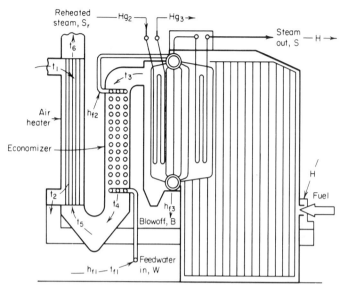

FIGURE 2 Points in a steam generator where temperatures and enthalpies are measured in determining the boiler efficiency.

table and deleting the reheat factor because there is no reheat, we get boiler output = 56,000(1407.7 − 269.6) + 2500(471.6 − 269.6) = 64,238,600 Btu/h (18,826.5 kW).

2. Compute the heat input to the boiler
The boiler input = FH, where F = fuel input, lb/h (as fired); H = higher heating value, Btu/lb (as fired). Or, boiler input = 5958(13,100) = 78,049,800 Btu/h (22,874.1 kW).

3. Compute the boiler efficiency
The boiler efficiency = (output, Btu/h)/(input, Btu/h) = 64,238,600/78,049,800 = 0.822, or 82.2 percent.

4. Determine the heat absorbed by the economizer
The heat absorbed by the economizer, Btu/h = $w_w(h_{f2} − h_{f1})$, where w_w = feed-water flow, lb/h; h_{f1} and h_{f2} = enthalpies of feedwater leaving and entering the economizer, respectively, Btu/lb. For this economizer, with the feedwater leaving the economizer at 370°F (187.8°C) and entering at 300°F (148.9°C), heat absorbed = (56,000 + 2500)(342.79 − 269.59) = 4.283,000 Btu/h (1255.2 kW). Note that the total feedwater flow w_w is the sum of the steam generated and the continuous blowdown rate.

5. Compute the heat available to the economizer
The heat available to the economizer, Btu/h = $H_g F$, where H_g = heat available in flue gas, Btu/lb of fuel = heat available in dry gas + heat available in flue-gas vapor, Btu/lb of fuel = $(t_{;3} − t_{f1})(0.24G) + (t_3 − t_{f1})(0.46)\{M_f + 8.94\text{H}_2 +$

$M_a[G - C_b - N_2 - 7.94(H_2 - O_2/8)]\}$, where $G = \{[11CO_2 + 8O_2 + 7(N_2 + CO)]/[3(CO_2 + CO)]\}(C_b + S/2.67) + S/1.60$; $M_f =$ lb of moisture per lb fuel burned; $M_a =$ lb of moisture per lb of dry air to furnace; $C_b =$ lb of carbon burned per lb of fuel burned $= C = RC_r$; $C_r =$ lb of combustible per lb of refuse; $R =$ lb of refuse per lb of fuel; H_2, N_2, C, O_2, S $=$ lb of each element per lb of fuel, as fired; CO_2, CO, O_2, $N_2 =$ percentage parts of volumetric analysis of dry combustion gas entering the economizer. Substituting gives $C_b = 0.685 - (0.093)(0.07) = 0.678$ lb/lb (0.678 kg/kg) fuel; $G = [11(0.158) + 8(0.028) + 7(0.814)]/[3(0.158)] \times (0.678 + 0.032/2.67) + 0.032/1.60$; $G = 11.18$ lb/lb (11.18 kg/kg) fuel. $H_g = (800 - 300)(0.24) \times (11.18) + (800 - 300)(0.46)\{0.045 + (8.9)(0.05) + 56/7000[11.18 - 0.678 - 0.012 - 7.94 \times (0.05 - 0.089/8)]\}$; $H_g = 1473$ Btu/lb (3426.2 kJ/kg) fuel. Heat available $= H_gF = (1473)(5958) = 8,770,000$ Btu/h (2570.2 kW).

6. Compute the economizer efficiency
The economizer efficiency $=$ (heat absorbed, Btu/h)/(heat available, Btu/h) $=$ 4,283,000/8,770,000 $= 0.488$, or 48.8 percent.

7. Compute the heat absorbed by air heater
The heat absorbed by the air heater, Btu/lb of fuel, $= A_h(t_2 - t_1)(0.24 + 0.46M_a)$, where $A_h =$ air flow through heater, lb/lb fuel $= A - A_m$; $A =$ total air to furnace, lb/lb fuel $= G - C_b - N_2 - 7.94(H_2 - O_2/8)$; $G =$ similar to economizer but based on gas at the furnace exit; $A_m =$ external air supplied by the mill fan or other source, lb/lb of fuel. Substituting shows $G = [11(0.16) + 8(0.26) + 7(0.184)]/[3(0.16)](0.678 + 0.032/2.67) + 0.032/1.60$; $G = 11.03$ lb/lb (11.03 kg/kg) fuel; $A = 11.03 - 0.69 - 0.012 - 7.94(0.05 - 0.089/8)$; $A = 10.02$ lb/lb (10.02 kg/kg) fuel. Heat absorbed $= (1 - 0.15)(10.02)(480 - 63)(0.24 + 56/7000) = 865.5$ Btu/lb (2013.2 kJ/kg fuel.

8. Compute the heat available to the air heater
The heat available to the air heater, Btu/h $= (t_5 - t_1)0.24G + (t_5 - t_1)0.46(M_f + 8.94H_2 + M_aA)$. In this relation, all symbols are the same as for the economizer except that G and A are based on the gas entering the heater. Substituting gives $G = [11(0.15) + 8(0.036) + 7(0.814)]/[3(0.15)](0.678 + 0.032/2.67) + 0.032/1.60$; $G = 11.72$ lb/lb (11.72 kg/kg) fuel. And $A = 11.72 - 0.69 - 0.012 - 7.94(0.05 - 0.089/8) = 10.71$ lb/lb (10.71 kg/kg) fuel. Heat available $= (570 - 3)(0.24)(11.72) + (570 - 63)(0.46)[0.045 + 8.94(0.05) + 56/7000(10.71)] = 1561$ Btu/lb (3630.9 kJ/kg).

9. Compute the air-heater efficiency
The air-heater efficiency $=$ (heat absorbed, Btu/lb fuel)/(heat available, Btu/lb fuel) $= 865.5/1561 = 0.554$, or 55.4 percent.
   Related Calculations. The above procedure is valid for all types of steam generators, regardless of the kind of fuel used. Where oil or gas is the fuel, alter the combustion calculations to reflect the differences between the fuels. Further, this procedure is also valid for marine and portable boilers.

FIRE-TUBE BOILER ANALYSIS AND SELECTION

Determine the heating surface in an 84-in (213.4-cm) diameter fire-tube boiler 18 ft (5.5 m) long having 84 tubes of 4-in (10.2-cm) ID if 25 percent of the upper shell ends are heat-insulated. How much steam is generated if the boiler evaporates 34.5 lb/h of water per 12 ft² [3.9 g/(m²·s)] of heating surface? How much heat is added by the boiler if it operates at 200 lb/in² (abs) (1379.0 kPa) with 200°F (93.3°C) feedwater? What is the factor of evaporation for this boiler? How much hp is developed by the boiler if 7,000,000 Btu/h (2051.4 kW) is delivered to the water?

Calculation Procedure:

1. Compute the shell area exposed to furnace gas
Shell area = $\pi DL(1 - 0.25)$, where D = boiler diameter, ft; L = shell length, ft; $1 - 0.25$ is the portion of the shell in contact with the furnace gas. Then shell area = $\pi(84/12)(18)(0.75)$ = 297 ft² (27.0 m²).

2. Compute the tube area exposed to furnace gas
Tube area = πdLN, where = tube ID, ft; L = tube length, ft; N = number of tubes in boiler. Substituting gives tube area = $\pi(4/12)(18)(84)$ = 1583 ft² (147.1 m²).

3. Compute the head area exposed to furnace gas
The area exposed to furnace gas is twice (since there are *two* heads) the exposed head area minus twice the area occupied by the tubes. The exposed head area is (total area)(1 − portion covered by insulation, expressed as a decimal). Substituting, we get $2\pi D^2/4 - (2)(84)\pi d^2/4 = 2\pi/4(84/12)^2(0.75) - (2)(84)\pi(4/12)^2/4$ = head area = 43.1 ft² (4.0 m²).

4. Find the total heating surface
The total heating surface of any fire-tube boiler is the sum of the shell, tube, and head areas, or 297.0 + 1583 + 43.1 = 1923 ft² (178.7 m²), total heating surface.

5. Compute the quantity of steam generated
Since the boiler evaporates 34.5 lb/h of water per 12 ft² [3.9 g/(m²·s)] of heating surface, the quantity of steam generated = 34.5 (total heating surface, ft²)/12 = 34.5(1923.1)/12 = 5200 lb/h (0.66 kg/s).

   Note: Evaporation of 34.5 lb/h (0.0043 kg/s) from and at 212°F (100.0°C) is the definition of the now-discarded term *boiler hp*. However, this term is still met in some engineering examinations and is used by some manufacturers when comparing the performance of boilers. A term used in lieu of boiler horsepower, with the same definition, is *equivalent evaporation*. Both terms are falling into disuse, but they are included here because they still find some use today.

6. Determine the heat added by the boiler
Heat added, Btu/lb of steam = $h_g - h_{f1}$; from steam table values 1198.4 − 167.99 = 1030.41 Btu/lb (2396.7 kJ/kg). An alternative way of computing heat added is h_g − (feedwater temperature,°F, − 32), where 32 is the freezing temperature of water on the Fahrenheit scale. By this method, heat added = 1198.4 − (200 − 32) = 1030.4 Btu/lb (2396.7 kJ/kg). Thus, both methods give the same results in this case. In general, however, use of steam table values is preferred.

7. Compute the factor of evaporation
The factor of evaporation is used to convert from the actual to the equivalent evaporation, defined earlier. Or, factor of evaporation = (heat added by boiler, Btu/lb)/970.3, where 970.3 Btu/lb (2256.9 kJ/kg) is the heat added to develop 1

boiler hp (bhp) (0.75 kW). Thus, the factor of evaporation for this boiler = 1030.4/970.3 = 1.066.

8. Compute the boiler hp output

Boiler hp = (actual evaporation, lb/h) (factor of evaporation)/34.5. In this relation, the actual evaporation must be computed first. Since the furnace delivers 7,000,000 Btu/h (2051.5 kW) to the boiler water and the water absorbs 1030.4 Btu/lb (2396.7 kJ/kg) to produce 200-lb/in² (abs) (1379.0-kPa) steam with 200°F (93.3°C) feedwater, the steam generated, lb/h = (total heat delivered, Btu/h)/(heat absorbed, Btu/lb) = 7,000,000/1030.4 = 6670 lb/h (0.85 kg/s). Then boiler hp = (6760)(1.066)/34.5 = 209 hp (155.9 kW).

The rated hp output of horizontal fire-tube boilers with separate supporting walls is based on 12 ft² (1.1 m²) of heating surface per boiler hp. Thus, the rated hp of the boiler = 1923.1/12 = 160 hp (119.3 kW). When producing 209 hp (155.9 kW), the boiler is operating at 209/160, or 1,305 times its normal rating, or (100)(1.305) = 130.5 percent of normal rating.

Note: Today most boiler manufacturers rate their boilers in terms of pounds per hour of steam generated at a stated pressure. Use this measure of boiler output whenever possible. Inclusion of the term *boiler hp* in this handbook does not indicate that the editor favors or recommends its use. Instead, the term was included to make the handbook as helpful as possible to users who might encounter the term in their work.

SAFETY-VALVE STEAM-FLOW CAPACITY

How much saturated steam at 150 lb/in² (abs) (1034.3 kPa) can a 2.5-in (6.4-cm) diameter safety valve having a 0.25-in (0.6-cm) lift pass if the discharge coefficient of the valve c_d is 0.75? What is the capacity of the same valve if the steam is superheated 100°F (55.6°C) above its saturation temperature?

Calculation Procedure:

1. Determine the area of the valve annulus

Annulus area, in² = $A = \pi DL$, where D = valve diameter, in; L = valve lift, in. Annulus area = $\pi(2.5)(0.25)$ = 1.966 in² (12.7 cm²).

2. Compute the ideal flow for this safety valve

Ideal flow F_i lb/s for any safety valve handling saturated steam is $F_i = p_s^{0.97} A/60$, where p_s = saturated-steam pressure, lb/in² (abs). For this valve, $F_i = (150)^{0.97}$ (1.966)/60 = 4.24 lb/s (1.9 kg/s).

3. Compute the actual flow through the valve

Actual flow $F_a = F_i c_d$ = (4.24)(0.75) = 3.18 lb/s (1.4 kg/s) = (3.18)(3600 s/h) = 11,448 lb/h (1.44 kg/s).

4. Determine the superheated-steam flow rate

The ideal superheated-steam flow F_{is} lb/s is $F_{is} = p_s^{0.97} A/[60(1 + 0.0065t_s)]$, where t_s = superheated temperature, above saturation temperature, °F. The F_{is} = $(150)^{0.97}(1.966)/[60(1 + 0.0065 \times 100)]$ = 3.96 lb/s (1.8 kg/s). The actual flow is $F_{as} = F_{is}c_d$ = (3.96)(0.75) = 2.97 lb/s (1.4 kg/s) = (2.97)(3600) = 10,700 lb/h (1.4 kg/s).

Related Calculations. Use this procedure for safety valves serving any type of stationary or marine boiler.

SAFETY-VALVE SELECTION FOR A WATERTUBE STEAM BOILER

Select a safety valve for a watertube steam boiler having a maximum rating of 100,000 lb/h (12.6 kg/s) at 800 lb/in² (abs) (5516.0 kPa) and 900°F (482.2°C). Determine the valve diameter, size of boiler connection for the valve, opening pressure, closing pressure, type of connection, and valve material. The boiler is oil-fired and has a total heating surface of 9200 ft² (854.7 m²) of which 1000 ft² (92.9 m²) is in waterwall surface. Use the ASME *Boiler and Pressure Vessel Code* rules when selecting the valve. Sketch the escape-pipe arrangement for the safety valve.

Calculation Procedure:

1. Determine the minimum valve relieving capacity

Refer to the latest edition of the *Code* for the relieving-capacity rules. Recent editions of the *Code* require that the safety valve have a *minimum* relieving capacity based on the pounds of steam generated per hour per square foot of boiler heating surface and waterwall heating surface. In the edition of the *Code* used in preparing this handbook, the relieving requirement for oil-fired boilers was 10 lb/(ft²·h) of steam [13.6 g/(m²·s)] of boiler heating surface, and 16 lb/(ft²·h) of steam [21.9 g/(m²·s)] of waterwall surface. Thus, the minimum safety-valve relieving capacity for this boiler, based on total heating surface, would be (8200)(10) + (1000)(16) = 92,000 lb/h (11.6 kg/s). In this equation, 1000 ft² (92.9 m²) of waterwall surface is deducted from the total heating surface of 9200 ft² (854.7 m²) to obtain the boiler heating surface of 8200 ft² (761.8 m²).

The minimum relieving capacity based on total heating surface is 92,000 lb/h (11.6 kg/s); the maximum rated capacity of the boiler is 100,000 lb/h (12.6 kg/s). Since the *Code* also requires that "the safety valve or valves will discharge all the steam that can be generated by the boiler," the minimum relieving capacity must be 100,000 lb/h (12.6 kg/s), because this is the maximum capacity of the boiler and it exceeds the valve capacity based on the heating-surface calculation. If the valve capacity based on the heating-surface steam generation were larger than the stated maximum capacity of the boiler, the *Code* heating-surface valve capacity would be used in safety-valve selection.

2. Determine the number of safety valves needed

Study the latest edition of the *Code* to determine the requirements for the number of safety valves. The edition of the *Code* used here requires that "each boiler shall have at least one safety valve and if it [the boiler] has more than 500 ft² (46.5 m²) of water heating surface, it shall have two or more safety valves." Thus, at least two safety valves are needed for this boiler. The *Code* further specifies, in the edition used, that "when two or more safety valves are used on a boiler, they may be mounted either separately or as twin valves made by placing individual valves on Y bases or duplex valves having two valves in the same body casing. Twin valves made by placing individual valves on Y bases, or duplex valves having two valves in the same body, shall be of equal sizes." Also, "when not more than two valves of different sizes are mounted singly, the relieving capacity of the smaller valve shall not be less than 50 percent of that of the larger valve."

Assume that two equal-size valves mounted on a Y base will be used on the steam drum of this boiler. Two or more equal-size valves are usually chosen for the steam drum of a watertube boiler.

Since this boiler handles superheated steam, check the *Code* requirements regarding superheaters. The *Code* states that "every attached superheater shall have one or more safety valves near the outlet." Also, "the discharge capacity of the safety valve, or valves, on an attached superheater may be included in determining the number and size of the safety valves for the boiler, provided there are no intervening valves between the superheater safety valve and the boiler, and provided the discharge capacity of the safety valve, or valves, on the boiler, as distinct from the superheater, is at least 75 percent of the aggregate valve capacity required."

Since the safety valves used must handle 100,000 lb/h (12.6 kg/s), and one or more superheater safety valves are required by the *Code*, assume that the two steam-drum valves will handle, in accordance with the above requirement, 80,000 lb/h (10.1 kg/s). Assume that one superheater safety valve will be used. Its capacity must then be at least 100,000 − 80,000 = 20,000 lb/h (2.5 kg/s). (Use a few superheater safety valves as possible, because this simplifies the installation and reduces cost.) With this arrangement, each steam-drum valve must handle 80,000/2 = 40,000 lb/h (5.0 kg/s) of steam, since there are two safety valves on the steam drum.

3. *Determine the valve pressure settings*

Consult the *Code*. It requires that "one or more safety valves on the boiler proper shall be set at or below the maximum allowable working pressure." For modern boilers, the maximum allowable working pressure is usually 1.5, or more, times the rated operating pressure in the lower [under 1000 lb/in^2 (abs) or 6895.0 kPa] pressure ranges. To prevent unnecessary operation of the safety valve and to reduce steam losses, the lowest safety-valve setting is usually about 5 percent higher than the boiler operating pressure. For this boiler, the lowest pressure setting would be 800 + 800(0.05) = 840 lb/in^2 (abs) (5791.8 kPa). Round this to 850 lb/in^2 (abs) (5860.8 kPa, or 6.25 percent) for ease of selection from the usual safety-valve rating tables. The usual safety-valve pressure setting is between 5 and 10 percent higher than the rated operating pressure of the boiler.

Boilers fitted with superheaters usually have the superheater safety valve set at a lower pressure than the steam-drum safety valve. This arrangement ensures that the superheater safety valve opens first when overpressure occurs. This provides steam flow through the superheater tubes at all times, preventing tube burnout. Therefore, the superheater safety valve in this boiler will be set to open at 850 lb/in^2 (abs) (5860.8 kPa), the lowest opening pressure for the safety valves chosen. The steam-drum safety valves will be set to open at a higher pressure. As decided earlier, the superheater safety valve will have a capacity of 20,000 lb/h (2.5 kg/s).

Between the steam drum and the superheater safety valve, there is a pressure loss that varies from one boiler to another. The boiler manufacturer supplies a performance chart showing the drum outlet pressure for various percentages of the maximum continuous steaming capacity of the boiler. This chart also shows the superheater outlet pressure for the same capacities. The difference between the drum and superheater outlet pressure for any given load is the superheater pressure loss. Obtain this pressure loss from the performance chart.

Assume, for this boiler, that the superheater pressure loss, plus any pressure losses in the nonreturn valve and dry pipe, at maximum rating, is 60 lb/in² (abs) (413.7 kPa). The steam-drum operating pressure will then be superheater outlet pressure + superheater pressure loss = 800 + 60 = 860 lb/in² (abs) (5929.7 kPa). As with the superheater safety valve, the steam-drum safety valve is usually set to open at about 5 percent above the drum operating pressure at maximum steam output. For this boiler then, the drum safety-valve set pressure = 860 + 860(0.05) = 903 lb/in² (abs) (6226.2 kPa). Round this to 900 lb/in² (abs) (6205.5 kPa) to simplify valve selection.

Some designers add the drum safety-valve blowdown or blowback pressure (difference between the valve opening and closing pressures, lb/in²) to the total obtained above to find the drum operating pressure. However, the 5 percent allowance used above is sufficient to allow for the blowdown in boilers operating at less than 1000 lb/in² (abs) (6895.0 kPa). At pressures of 1000 lb/in² (abs) (6895.0 kPa) and higher, add the drum safety-valve blowdown *and* the 5 percent allowance to the superheater outlet pressure and pressure loss to find the drum pressure.

4. *Determine the required valve orifice discharge area*
Refer to a safety-valve manufacturer's engineering data listing valve capacities at various working pressures. For the two steam-drum valves, enter the table at 900 lb/in² (abs) (6205.5 kPa), and project horizontally until a capacity of 40,000 lb/h (5.0 kg/s), or more, is intersected. Here is an excerpt from a typical manufacturer's capacity table for safety valves handling *saturated steam:*

Set pressure		Orifice area					
lb/in² (abs)	kPa	0.994 in²	6.41 cm²	1.431 in²	9.23 cm²	2.545 in²	16.42 cm²
890	6,136.6	41,750	5.26	60,000	7.56	107,200	13.5
900	6,205.5	42,200	5.32	60,900	7.67	108,000	13.6
910	6,274.5	42,700	5.38	61,600	7.76	109,300	13.8

Thus, at 900 lb/in² (abs) (6205.5 kPa) a valve with an orifice area of 0.944 in² (6.4 cm²) will have a capacity of 42,200 lb/h (5.3 kg/s) of saturated steam. This is 5.5 percent greater than the required capacity of 40,000 lb/h (5.0 kg/s) for each steam-drum valve. However, the usual selection cannot be made at exactly the desired capacity. Provided that the valve chosen has a greater steam relieving capacity than required, there is no danger of overpressure in the steam drum. Be careful to note that safety valves for saturated steam are chosen for the steam drum because superheating of the steam does not occur in the steam drum.

The superheater safety valve must handle 20,000 lb/h (2.5 kg/s) of 850 lb/in² (abs) (5860.8-kPa) steam at 900°F (482.2°C). Safety valves handling superheated steam have a smaller capacity than when handling saturated steam. To obtain the capacity of a safety valve handling superheated steam, the saturated steam capacity is multiplied by a correction factor that is less than 1.00. An alternative procedure is to divide the required superheater-steam capacity by the same correction factor to obtain the saturated-steam capacity of the valve. The latter procedure will be used here because it is more direct.

Obtain the correction factor from the safety-valve manufacturer's engineering data by entering at the steam pressure and projecting to the steam temperature, as show below.

Set pressure		Steam temperature	
lb/in² (abs)	kPa	880°F (471.1°C)	900°F (482.2°C)
800	5516.0	0.80	0.80
850	6205.5	0.81	0.80
900	5860.8	0.81	0.80

Thus, at 850 lb/in² (abs) (5860.8 kPa) and 900°F (482.2°C), the correction factor is 0.80. The required saturated steam capacity then is 20,000/0.80 = 25,000 lb/h (3.1 kg/s).

Refer to the manufacturer's saturated-steam capacity table as before, and at 850 lb/in² (abs) (5860.8 kPa) find the closest capacity as 31,500 lb/h (4.0 kg/s) for a 0.785-in² (5.1-cm²) orifice. As with the steam-drum valves, the actual capacity of the safety valve is somewhat greater than the required capacity. In general, it is difficult to find a valve with exactly the required steam relieving capacity.

5. Determine the valve nominal size and construction details

Turn to the data section of the safety-valve engineering manual to find the valve construction features. For the steam-drum valves having 0.994-in² (6.4-cm²) orifice areas, the engineering data show, for 900-lb/in² (abs) (6205.5-kPa) service, each valve is 1½-in (3.8-cm) unit rated for temperatures up to 1050°F (565.6°C). The inlet is 900-lb/in² (6205.5-kPa) 1½-in (3.8-cm) flanged connection, and the outlet is a 150-lb/in² (1034.3-kPa) 3-in (7.6-cm) flanged connection. Materials used in the valve include: body, cast carbon steel; disk seat, stainless steel AISI 321. The overall height is 27⅞ in (70.8 cm); dismantled height is 32¾ in (83.2 cm).

Similar data for the superheated steam valve show, for a maximum pressure of 900 lb/in² (abs) (6205.5 kPa), that it is a 1½-in (3.8-cm) unit rated for temperatures up to 1000°F (537.8°C). The inlet is a 900-lb/in² (6205.5-kPa) 1½-in (3.8-cm) flanged connection, and the outlet is a 150-lb/in² (1034.3-kPa) 3-in (7.6-cm) flanged connection. Materials used in the valve include: body, cast alloy steel, ASTM 217-WC6; spindle, stainless steel; spring, alloy steel; disk seat, stainless steel. Overall height is 21⅜ in (54.3 cm); dismantled height is 25¼ in (64.1 cm). Checking the *Code* shows that "every safety valve used on a superheater discharging superheated steam at a temperature over 450°F (232.2°C) shall have a casing, including the base, body, bonnet and spindle, of steel, steel alloy, or equivalent heat-resisting material. The valve shall have a flanged inlet connection."

Thus, the superheater valve selected is satisfactory.

6. Compute the steam-drum connection size

The *Code* requires that "when a boiler is fitted with two or more safety valves on one connection, this connection to the boiler shall have a cross-sectional area not less than the combined areas of inlet connections of all safety valves with which it connects."

The inlet area for each valve = $\pi D^2/4 = \pi(1.5)^2/4 = 1.77$ in² (11.4 cm²). For two valves, the total inlet area = 2(1.77) = 3.54 in² (22.8 cm²). The required minimum diameter of the boiler connection is $d = 2(A/\pi)^{0.5}$, where A = inlet area. Or, $d = 2(3.54/\pi)^{0.5} = 2.12$ in (5.4 cm). Select a 2½ × 1½ × 1½ in (6.4 × 3.8 × 3.8 cm). Y for the two steam-drum valves and a 2½-in (6.4-cm) steam-drum outlet connection.

7. *Compute the safety-valve closing pressure*

The *Code* requires safety valves to "close after blowing down not more than 4 percent of the set pressure." For the steam-drum valves the closing pressure will be $900 - (900)(0.04) = 865$ lb/in^2 (abs) (5964.2 kPa). The superheater safety valve will close at $850 - (850)(0.04) = 816$ lb/in^2 (abs) (5626.3 kPa).

8. *Sketch the discharge elbow and drip pan*

Figure 3 shows a typical discharge elbow and drip-pan connection. Fit all boiler safety valves with escape pipes to carry the steam out of the building and away from personnel. Extend the escape pipe to at least 6 ft (1.8 m) above the roof of the building. Use an escape pipe having a diameter equal to the valve outlet size. When the escape pipe is more than 12 ft (3.7 m) long, some authorities recommend increasing the escape-pipe diameter by ½ in (1.3 cm) for each additional 12-ft (3.7-m) length. Excessive escape-pipe length without an increase in diameter can cause a backpressure on the safety valve because of flow friction. The safety valve may then chatter excessively.

Support the escape pipe independently of the safety valve. Fit a drain to the valve body and rip pan as shown in Fig. 3. This prevents freezing of the condensate

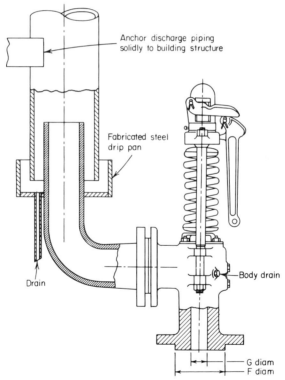

FIGURE 3 Typical boiler safety-valve discharge elbow and drip-pan connection. (*Industrial Valve and Instrument Division of Dresser Industries Inc.*)

and also eliminates the possibility of condensate in the escape pipe raising the valve opening pressure. When a muffler is fitted to the escape pipe, the inlet diameter of the muffler should be the same as, or larger than, the escape-pipe diameter. The outlet area should be greater than the inlet area of the muffler.

   Related Calculations.   Compute the safety-valve size for fire-tube boilers in the same way as described above, except that the *Code* gives a tabulation of the required area for safety-valve boiler connections based on boiler operating pressure and heating surface. Thus, with an operating pressure of 200 lb/in² (gage) (1379.0 kPa) and 1800 ft² (167.2 m²) of heating surface, the *Code* table shows that the safety-valve connection should have an area of at least 9.148 in² (59.0 cm²). A 3½-in (8.9-cm) connection would provide this area; or two smaller connections could be used provided that the sum of their areas exceeded 9.148 in² (59.0 cm²)

   Note: Be sure to select safety valves approved for use under the *Code* or local low governing boilers in the area in which the boiler will be used. Choice of an unapproved valve can lead to its rejection by the bureau or other agency controlling boiler installation and operation.

STEAM-QUALITY DETERMINATION WITH A THROTTLING CALORIMETER

Steam leaves an industrial boiler at 120 lb/in² (abs) (827.4 kPa) and 341.25°F (171.8°C). A portion of the steam is passed through a throttling calorimeter and is exhausted to the atmosphere when the barometric pressure is 14.7 lb/in² (abs) (101.4 kPa). How much moisture does the steam leaving the boiler contain if the temperature of the steam at the calorimeter is 240°F (115.6°C)?

Calculation Procedure:

1. *Plot the throttling process on the Mollier diagram*
Begin with the endpoint, 14.7 lb/in² (abs) (101.4 kPa) and 240°F (115.6°C). Plot this point on the Mollier diagram as point *A*, Fig 4. Note that this point is in the superheat region of the Mollier diagram, because steam at 14.7 lb/in² (abs) (101.4 kPa) has a temperature of 212°F (100.0°C), whereas the steam in this calorimeter has a temperature of 240°F (115.6°C). The enthalpy of the calorimeter steam is, from the Mollier diagram, 1164 Btu/lb (2707.5 kJ/kg).

2. *Trace the throttling process on the Mollier diagram*
In a throttling process, the steam expands at constant enthalpy. Draw a straight, horizontal line from point *A* to the left on the Mollier diagram until the 120-lb/in² (abs) (827.4-kPa) pressure curve is intersected, point *B*, Fig. 4. Read the moisture content of the steam as 3 percent where the 1164-Btu/lb (2707.5-kJ/kg) horizontal trace *AB*, the 120-lb/in² (abs) (827.4-kPa) pressure line, and the 3 percent moisture line intersect.

   Related Calculations.   A throttling calorimeter *must* produce superheated steam at the existing atmospheric pressure if the moisture content of the supply steam is to be found. Where the throttling calorimeter cannot produce superheated steam at atmospheric pressure, connect the calorimeter outlet to an area at a pressure less than atmospheric. Expand the steam from the source, and read the temperature at the calorimeter. If the steam temperature is greater than that corresponding to the

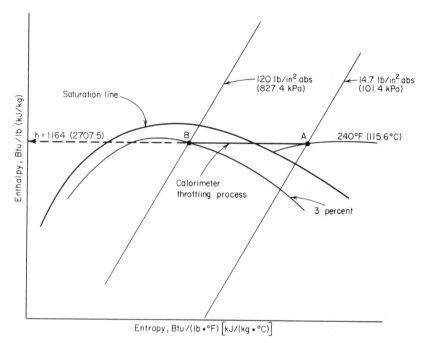

FIGURE 4   Mollier-diagram plot of a throttling-calorimeter process.

absolute pressure of the vacuum area—for example, a temperature greater than 133.76°F (56.5°C) in an area of 5 inHg (16.9 kPa) absolute pressure—follow the same procedure as given above. Point A would then be in the below-atmospheric area of the Mollier diagram. Trace to the left to the origin pressure, and read the moisture content as before.

STEAM PRESSURE DROP IN A BOILER SUPERHEATER

What is the pressure loss in a boiler superheater handling $w_s = 200,000$ lb/h (25.2 kg/s) of saturated steam at 500 lb/in² (abs) (3447.5 kPa) if the desired outlet temperature is 750°F (398.9°C)? The steam free-flow area through the superheater tubes A_s ft² is 0.500, friction factor f is 0.025, tube ID is 2.125 in (5.4 cm), developed length l of a tube in one circuit is 150 in (381.0 cm), and the tube bend factor B_f is 12.0.

Calculation Procedure:

1. Determine the initial conditions of the steam
To compute the pressure loss in a superheater, the initial specific volume of the steam v_g and the mass-flow ratio w_s/A_s must be known. From the steam table, $v_g =$

0.9278 ft^3/lb (0.058 m^3/kg) at 500 lb/in^2 (abs) (3447.5 kPa) saturated. The mass-flow ratio w_s/A_s = 200,000/0.500 = 400,000.

2. Compute the superheater entrance and exit pressure loss
Entrance and exit pressure loss p_E lb/in^2 = $v_f/8(0.00001w_s/A_s)$ = 0.9278/8[(0.00001) × (400,000)]^2 = 1.856 lb/in^2 (12.8 kPa).

3. Compute the pressure loss in the straight tubes
Straight-tube pressure loss p_s lb/in^2 = $v_f f$/ID(0.00001w_s/A_s)^2 = 0.9278(150) × (0.025)/2.125[(0.00001)(400,000)]^2 = 26.2 lb/in^2 (abs) (180.6 kPa).

4. Compute the pressure loss in the superheater bends
Bend pressure loss p_b = 0.0833B_f(0.00001w_s/A_s)^2 = 0.0833(12.0)[(0.00001) × (400,000)]^2 = 16.0 lb/in^2 (110.3 kPa).

5. Compute the total pressure loss
The total pressure loss in any superheater is the sum of the entrance, straight-tube, bend, and exit-pressure losses. These losses were computed in steps 2, 3, and 4 above. Therefore, total pressure loss p_t = 1.856 + 26.2 + 16.0 = 44.056 lb/in^2 (303.8 kPa).

Note: Data for superheater pressure-loss calculations are best obtained from the boiler manufacturer. Several manufacturers have useful publications discussing superheater pressure losses. These are listed in the references at the beginning of this section.

SELECTION OF A STEAM BOILER FOR A GIVEN LOAD

Choose a steam boiler, or boilers, to deliver up to 250,000 lb/h (31.5 kg/s) of superheated steam at 800 lb/in^2 (abs) (5516 kPa) and 900°F (482.2°C). Determine the type or types of boilers to use, the capacity, type of firing, feedwater-quality requirements, and best fuel if coal, oil, and gas are all available. The normal continuous steam requirement is 200,000 lb/h (25.2 kg/s).

Calculation Procedure:

1. Select type of steam generator
Use Fig. 5 as a guide to the usual types of steam generators chosen for various capacities and different pressure and temperature conditions. Enter Fig. 5 at the left at 800 lb/in^2 (abs) (5516 kPa), and project horizontally to the right, along AB, until the 250,000-lb/h (31.5-kg/s) capacity ordinate BC is intersected. At B, the operating point of this boiler, Fig. 5 shows that a watertube boiler should be used.

Boiler units presently available can deliver steam at the desired temperature of 900°F (482.2°C). The required capacity of 250,000 lb/h (31.5 kg/s) is beyond the range of *packaged watertube boilers*—defined by the American Boiler Manufacturer Association as "a boiler equipped and shipped complete with fuel-burning equipment, mechanical-draft equipment, automatic controls, and accessories."

Shop-assembled boilers are larger units, where all assembly is handled in the builder's plant but with some leeway in the selection of controls and auxiliaries.

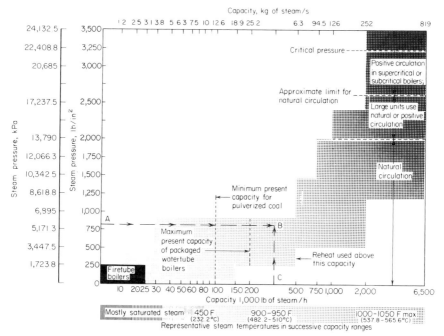

FIGURE 5 Typical pressure and capacity relationships from steam generators. (*Power.*)

The current maximum capacity of shop-assembled boilers is about 100,000 lb/h (12.6 kg/s). Thus, a standard-design, larger-capacity boiler is required.

Study manufacturers' engineering data to determine which types of watertube boilers are available for the required capacity, pressure, and temperature. This study reveals that, for this installation, a standard, field-assembled, welded-steel-cased, bent-tube, single-steam-drum boiler with a completely water-cooled furnace would be suitable. This type of boiler is usually fitted with an air heater, and an economizer might also be used. The induced- and forced-draft fans are not integral with the boiler. Capacities of this type of boiler usually available range from 50,000 to 350,000 lb/h (6.3 to 44.1 kg/s); pressure from 160 to 1050 lb/in^2 (1103.2 to 7239.8 kPa); steam temperature from saturation to 950°F (510.0°C); fuels—pulverized coal, oil, gas, or a combination; controls—manual to completely automatic; efficiency—to 90 percent.

2. Determine the number of boilers required

The normal continuous steam requirement is 200,000 lb/h (25.2 kg/s). If a 250,000-lb/h (31.5-kg/s) boiler were chosen to meet the maximum required output, the boiler would normally operate at 2000,000/250,000, or 80 percent capacity. Obtain the performance chart, Fig. 6, from the manufacturer and study it. This chart shows that at 80 percent load, the boiler efficiency is about equal to that at 100 percent load. Thus, there will not be any significant efficiency loss when the unit is operated at its normal continuous output. The total losses in the boiler are lower at 80 percent load than at full (100 percent) load.

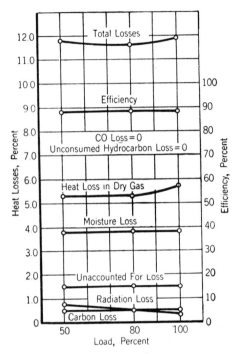

FIGURE 6 Typical watertube steam-generator losses and efficiency.

Since there is not a large efficiency decrease at the normal continuous load, and since there are not other factors that require or make more than one boiler desirable, a single boiler unit would be most suitable for this installation. One boiler is more desirable than two or more because installation of a single unit is simpler and maintenance costs are lower. However, where the load fluctuates widely and two or more boilers could best serve the steam demand, the savings in installation and maintenance costs would be insignificant compared with the extra cost of operating a relatively large boiler installed in place of two or more smaller boilers. Therefore, each installation must be carefully analyzed and a decision made on the basis of the existing conditions.

3. Determine the required boiler capacity
The stated steam load is 250,000 lb/h (31.5 kg/s) at maximum demand. Study the installation to determine whether the steam demand will increase in the future. Try to determine the rate of increase in the steam demand; for example, installation of several steam-using process units each year during the next few years will increase the steam demand by a predictable amount every year. By using these data, the rate of growth and total steam demand can be estimated for each year. Where the growth will exceed the allowable overload capacity of the boiler—which can vary from 0 to 50 percent of the full-load rating, depending on the type of unit chosen—consider installing a larger-capacity boiler now to meet future load growth. Where the future load is unpredictable or where no load growth is anticipated, a unit sized to meet today's load would be satisfactory. If this situation existed in this plant, a 250,000-

lb/h unit (31.5-kg/s) would be chosen for the load. Any small temporary overloads could be handled by operating the boiler at a higher output for short periods.

Alternatively, assume that a load of 25,000 lb/h (3.1 kg/s) will be added to the maximum demand on this boiler each year for the next 5 years. This means that in 5 years the maximum demand will be 250,000 + 25,000(5) = 375,000 lb/h (47.2 kg/s). This is an overload of (375,000 − 250,000)/250,000 = 0.50, or 50 percent. It is unlikely that the boiler could carry a continuous overload of 50 percent. Therefore it might be wise to install a 375,000-lb/h (47.2-kg/s) boiler to meet present and future demands. Base this decision on the accuracy of the future-demand prediction and the economic advantages or disadvantages of investing more money now for a demand that will not occur until some future date. Refer to the section on engineering economics for procedures to follow in economics calculations of this type.

Thus, with no increase in the future load, a 250,000-lb/h (31.5-kg/s) unit would be chosen. With the load increase specified, a 375,000-lb/h (47.2-kg/s) unit would be the choice, if there were no major economic disadvantages.

4. *Choose the type of fuel to use*
Watertube boilers of the type being considered will economically burn the three fuels available—coal, oil, or gas—either singly or in combination. In the design considered here, the furnace watercooled surfaces and boiler surfaces are integral parts of each other. For this reason the boiler is well suited for pulverized-coal firing in the 50,000- to 300,000-lb/h (6.3- to 37.8-kg/s) capacity range. Thus, if a 250,000-lb/h (31.5-kg/s) unit were chosen, it could be fired by pulverized coal. With a larger unit of 375,000 lb/h (47.2 kg/s), pulverized coal, oil, or gas firing might be used. Use an economic comparison to determine which fuel would give the lowest overall operating coast for the life of the boiler.

5. *Determine the feedwater-quality requirements*
Watertube boilers of all types require careful control of feedwater quality to prevent scale and sludge deposits in tubes and drums. Corrosion of the interior boiler surfaces must be controlled. Where all condensate is returned to the boiler, the makeup water must be treated to prevent the conditions just cited. Therefore, a comprehensive water-treating system must be planned for, particularly if the raw-water supply is poor.

6. *Estimate the boiler space requirements*
The space occupied by steam-generating units is an important consideration in plants in municipal areas and where power-plant buildings are presently crowded by existing equipment. The manufacturer's engineering data for this boiler show that for pulverized-coal firing, the hopper-type furnace bottom is best. The data also show that the smallest boiler with a hopper bottom occupies a space 21 ft (6.4 m) wide, 31 ft (9.4 m) high, and 14 ft (4.3 m) front to rear. The largest boiler occupies a space 21 ft (6.4 m) wide, 55 ft (16.8 m) high, and 36 ft (11.0 m) front to rear. Check these dimensions against the available space to determine whether the chosen boiler can be installed without major structural changes. The steel walls permit outdoor or indoor installation with top or bottom support of the boiler optional in either method of installation.

Related Calculations. Use this general procedure to select boilers for industrial, central-station, process, and marine applications.

Where a boiler is to burn hazardous industrial waste as a fuel, the designer must carefully observe two waste laws: the 1980 Superfund law and the 1976 Resource

Conservation and Recovery Act. These laws regulate the firing of hazardous wastes in boilers to control air pollution and explosion dangers.

Since hazardous wastes from industrial operations can vary in composition, it is important that the designer know what variables might be met during actual firing. Without correct analysis of the wastes, air pollution can become a severe problem in the plant locale.

The Environmental Protection Agency (EPA) and state regulatory agencies should be carefully consulted before any final design decisions are made for a new and expanded boiler plants. While the firing of hazardous wastes can be a convenient way to dispose of them, the potential impact on the environment must be considered before any design is finalized.

SELECTING BOILER FORCED- AND INDUCED-DRAFT FANS

Combustion calculations show that an oil-fired watertube boiler requires 200,000 lb/h (25.2 kg/s) of air for combustion at maximum load. Select forced- and induced-draft fans for this boiler if the average temperature of the inlet is 75°F (23.9°C) and the average temperature of the combustion gas leaving the air heater is 350°F (176.7°C) with an ambient barometric pressure of 29.9 inHg (101.0 kPa). Pressure losses on the air-inlet side are as follows, in inH$_2$O: air heater, 1.5 (0.37 kPa); air-supply ducts, 0.75 (0.19 kPa); boiler windbox, 1.75 (0.44 kPa); burners, 1.25 (0.31 kPa). Draft losses in the boiler and related equipment are as follows, in inH$_2$O: furnace pressure, 0.20 (0.05 kPa); boiler, 3.0 (0.75 kPa); superheater 1.0 (0.25 kPa); economizer, 1.50 (0.37 kPa); air heater, 2.00 (0.50 kPa); uptake ducts and dampers, 1.25 (0.31 kPa). Determine the fan discharge pressure and hp input. The boiler burns 18,000 lb/h (2.3 kg/s) of oil at full load.

Calculation Procedure:

1. Compute the quantity of air required for combustion
The combustion calculations show that 200,000 lb/h (25.2 kg/s) of air is theoretically required for combustion in this boiler. To this theoretical requirement must be added allowances for excess air at the burner and leakage out of the air heater and furnace. Allow 25 percent excess air for this boiler. The exact allowance for a given installation depends on the type of fuel burned. However, a 25 percent excess-air allowance is an average used by power-plant designers for coal, oil, and gas firing. With this allowance, the required excess air = 200,000(0.25) = 50,000 lb/h (6.3 kg/s).

Air-heater air leakage varies from about 1 to 2 percent of the theoretically required airflow. Using 2 percent, we see the air-heater leakage allowance = 200,000(0.02) = 4,000 lb/h (0.5 kg/s).

Furnace air leakage ranges from 5 to 10 percent of the theoretically required airflow. With 7.5 percent, the furnace leakage allowance = 200,000(0.075) = 15,000 lb/h (1.9 kg/s).

The total airflow required is the sum of the theoretical requirement, excess air, and leakage. Or, 200,000 + 50,000 + 4000 + 15,000 = 269,000 lb/h (33.9 kg/s). The forced-draft fan must supply at least this quantity of air to the boiler. Usual practice is to allow a 10 to 20 percent safety factor for fan capacity to ensure an adequate air supply at all operating conditions. This factor of safety is applied to the total airflow required. Using a 10 percent factor of safety, we see that fan capacity = 269,000 + 269,000(0.1) = 295,900 lb/h (37.3 kg/s). Round this to 269,000-lb/h (37.3 kg/s) fan capacity.

2. Express the required airflow in cubic feet per minute
Convert the required flow in pounds per hour to cubic feet per minute. To do this, apply a factor of safety to the ambient air temperature to ensure an adequate air supply during times of high ambient temperature. At such times, the density of the air is lower, and the fan discharges less air to the boiler. The usual practice is to apply a factor of safety of 20 to 25 percent to the known ambient air temperature. Using 20 percent, we see the ambient temperature for fan selection = 75 + 75(0.20) = 90°F (32.2°C). The density of air at 90°F (32.2°C) is 0.0717 lb/ft³ (1.15 kg/m³), found in Baumeister and Marks—*Standard Handbook for Mechanical Engineers.* Converting gives ft³/min + (lb/h)/(60 lb/ft³) = 296,000/60(0.0717) = 69,400 ft³/min (32.8 m³/s). This is the minimum capacity the forced-draft fan may have.

3. Determine the forced-draft discharge pressure
The total resistance between the forced-draft fan outlet and furnace is the sum of the losses in the air heater, air-supply ducts, boiler windbox, and burners. For this boiler, the total resistance, inH_2O = 1.5 + 0.75 + 1.75 + 1.25 = 5.25 inH_2O (1.3 kPa). Apply a 15 to 30 percent factor of safety to the required discharge pressure to ensure adequate airflow at all times. Or, fan discharge pressure, with a 20 percent factor of safety = 5.25 + 5.25(0.20) = 6.30 inH_2O (1.6 kPa). The fan must therefore deliver at least 69,400 ft³/min (32.8 m³/s) at 6.30 inH_2O (1.6 kPa).

4. Compute the power required to drive the forced-draft fan
The air hp for any fan = $0.0001753H_f$ = total head developed by fan, inH_2O; C = airflow, ft³/min. For this fan, air hp = 0.0001753(6.3)(69,400) = 76.5 hp (57.0 kW). Assume or obtain the fan and fan-driver efficiencies at the rated capacity (69,400 ft³/min, or 32.8 m³/s) and pressure (6.30 inH_2O, or 1.6 kPa). With a fan efficiency of 75 percent and assuming the fan is driven by an electric motor having an efficiency of 90 percent, we find the overall efficiency of the fan-motor combination is (0.75)(0.90) = 0.675, or 67.5 percent. Then the motor horsepower required = air hp/overall efficiency = 76.5/0.675 = 113.2 hp (84.4 kW). A 125-hp (93.2-kW) motor would be chosen because it is the nearest, next larger unit readily available. Usual practice is to choose a *larger* driver capacity when the computed capacity is lower than a standard capacity. The next larger standard capacity is generally chosen, except for extremely large fans where a special motor may be ordered.

5. Compute the quantity of flue gas handled
The quantity of gas reaching the induced-draft fan is the sum of the actual air required for combustion from step 1, air leakage in the boiler and furnace, and the weight of fuel burned. With an air leakage of 10 percent in the boiler and furnace (this is a typical leakage factor applied in practice), the gas flow is as follows:

	lb/h	kg s
Actual airflow required	296,000	37.3
Air leakage in boiler and furnace	29,600	3 7
Weight of oil burned	18,000	2.3
Total	343,600	43.3

Determine from combustion calculations for the boiler the density of the flue gas. Assume that the combustion calculations for this boiler show that the flue-gas density is 0.045 lb/ft³ (0.72 kg/m³) at the exit-gas temperature. To determine the exit-gas temperature, apply a 10 percent factor of safety to the given exit temperature, 350°F (176.6°C). Hence, exit-gas temperature = 350 + 350(0.10) = 385°F (196.1°C). Then flue-gas flow, ft³/min = (flue-gas flow, lb/h)/(60)(flue-gas density, lb/ft³) = 343,600/[(60)(0.045)] = 127,000 ft³/min (59.9 m³/s). Apply a 10 to 25 percent factor of safety to the flue-gas quantity to allow for increased gas flow. With a 20 percent factor of safety, the actual flue-gas flow the fan must handle = 127,000 + 127,000(0.20) = 152,400 ft³/min (71.8 m³/s), say 152,500 ft³/min (71.9 m³/s) for fan-selection purposes.

6. Compute the induced-draft fan discharge pressure
Find the sum of the draft losses from the burner outlet to the induced-draft fan inlet. These losses are as follows for this boiler:

	inH₂O	kPa
Furnace draft loss	0.20	0.05
Boiler draft loss	3.00	0.75
Superheater draft loss	1.00	0.25
Economizer draft loss	1.50	0.37
Air heater draft loss	2.00	0.50
Uptake ducts and damper draft loss	1.25	0.31
Total draft loss	8.95	2.23

Allow a 10 to 25 percent factor of safety to ensure adequate pressure during all boiler loads and furnace conditions. With a 20 percent factor of safety for this fan, the total actual pressure loss = 8.95 + 8.95(0.20) = 10.74 inH₂O (2.7 kPa). Round this to 11.0 inH₂O (2.7 kPa) for fan-selection purposes.

7. Compute the power required to drive the induced-draft fan
As with the forced-draft fan, air hp = $0.0001753H_fC$ = 0.0001753(11.0) × (127,000) = 245 hp (182.7 kW). If the combined efficiency of the fan and its driver, assumed to be an electric motor, is 68 percent, the motor hp required = 245/0.68 = 360.5 hp (268.8 kW). A 375-hp (279.6-kW) motor would be chosen for the fan driver.

8. Choose the fans from a manufacturer's engineering data
Use the next calculation procedure to select the fans from the engineering data of an acceptable manufacturer. For larger boiler units, the forced-draft fan is usually a backward-curved blade centrifugal-type unit. Where two fans are chosen to operate in parallel, the pressure curve of each fan should decrease at the same rate near shutoff so that the fans divide the load equally. Be certain that forced-draft fans are heavy-duty units designed for continuous operations with well-balanced rotors. Choose high-efficiency units with self-limiting power characteristics to prevent overloading the driving motor. Airflow is usually controlled by dampers on the fan discharge.

Induced-draft fans handle hot, dusty combustion products. For this reason, extreme care must be taken to choose units specifically designed for induced-draft service. The usual choice for large boilers is a centrifugal-type unit with forward-

or backward-curved, or flat blades, depending on the type of gas handled. Flat blades are popular when the flue gas contains large quantities of dust. Fan bearings are generally water-cooled.

   Related Calculations.   Use the procedure given above for the selection of draft fans for all types of boilers—fire-tube, packaged, portable, marine, and stationary. Obtain draft losses from the boiler manufacturer. Compute duct pressure losses by using the methods given in later procedures in this handbook.

POWER-PLANT FAN SELECTION FROM CAPACITY TABLES

Choose a forced-draft fan to handle 69,400 ft³/min (32.8 m³/s) of 90°F (32.2°C) air at 6.30-inH₂O (1.6-kPa) static pressure and an induced-draft fan to handle 152,500 ft³/min (72.0 m³/s) of 385°F (196.1°C) gas at 11.0-inH₂O (2.7-kPa) static pressure. The boiler that these fans serve is installed at an elevation of 5000 ft. (1524 m) above sea level. Use commercially available capacity tables for making the fan choice. The flue-gas density is 0.045 lb/ft³ (0.72 kg/m³) at 385°F (196.1°C).

Calculation Procedure:

1. Compute the correction factors for the forced-draft fan
Commercial fan-capacity tables are based on fans handling standard air at 70°F (21.1°C) at a barometric pressure of 29.92 inHg (101.0 kPa) and having a density of 0.075 lb/ft³ (1.2 kg/m³). Where different conditions exist, the fan flow rate must be corrected for temperature and altitude.

   Obtain the engineering data for commercially available forced-draft fans, and turn to the temperature and altitude correction-factor tables. Pick the appropriate correction factors from these tables for the prevailing temperature and altitude of the installation. Thus, in Table 4, select the correction factors for 90°F (32.2°C) air and 5000-ft (1524.0-m) altitude. These correction factors are $C_T = 1.018$ for 90°F (32.2°C) air and $C_A = 1.095$ for 5000-ft (1524.0-m) altitude.

   Find the composite correction factor (CCF) by taking the product of the temperature and altitude correction factors. Or, CCF = (1.018)(1.095) = 1.1147. Now divide the given cubic feet per minute (cfm) by the correction factor to find the

TABLE 4   Fan Correction Factors

Temperature		Correction factor	Altitude		Correction factor
°F	°C		ft	m	
80	26.7	1.009	4500	1371.6	1.086
90	32.2	1.018	5000	1524.0	1.095
100	37.8	1.028	5500	1676.4	1.106
375	190.6	1.255			
400	204.4	1.273			
450	232.2	1.310			

capacity-table ft³/min. Or, capacity-table ft³/min = 69,400/1.147 = 62,250 ft³/min (29.4 m³/s).

2. Choose the fan size from the capacity table

Turn to the fan-capacity table in the engineering data, and look for a fan delivering 62,250 ft³/min (29.4 m³/s) at 6.3-inH₂O (1.6-kPa) static pressure. Inspection of the table shows that the capacities are tabulated for 6.0- and 6.5-inH₂O (1.5- and 1.6-kPa) static pressure. There is no tabulation for 6.3-inH₂O (1.57-kPa) static pressure.

Enter the table at the nearest capacity to that required, 62,250 ft³/min (29.4 m³/s), as shown in Table 5. This table, excerpted with permission from the American Standard Inc. engineering data, shows that the nearest capacity of this particular type of fan is 62,595 ft³/min (29.5 m³/s). The difference, or 62,595 − 62,250 = 345 ft³/min (0.16 m³/s), is only 345/62,250 = 0.0055, or 0.55 percent. This is a negligible difference, and the 62,595-ft³/min (29.5-m³/s) fan is well suited for its intended use. The extra static pressure of 6.5 − 6.3 = 0.2 inH₂O (0.05 kPa) is desirable in a forced-draft fan because furnace or duct resistance may increase during the life of the boiler. Also, the extra static pressure is so small that it will not markedly increase the fan power consumption.

3. Compute the fan speed and power input

Multiply the capacity-table rpm and brake hp (bhp) by the composite factor to determine the actual rpm and bhp. Thus, with data from Table 5, the actual rpm = (1096)(1.1147) = 1221.7 r/min. Actual bhp = (99.08)(1.1147) = 110.5 bhp (82.4 kW). This is the hp input required to drive the fan and is close to the 113.2 hp (84.4 kW) computed in the previous calculation procedure. The actual motor hp would be the same in each case because a standard-size motor would be chosen. The difference of 113.2 − 110.5 = 2.7 hp (2.0 kW) results from the assumed efficiencies that depart from the actual values. Also, a sea-level attitude was assumed in the previous calculation procedure. However, the two methods used show how accurately fan capacity and hp input can be estimated by judicious evaluation of variables.

4. Compute the correction factors for the induced-draft fan

The flue-gas density is 0.045 lb/ft³ (0.72 kg/m³) at 385°F (196.1°C). Interpolate in the temperature correction-factor table because a value of 385°F (196.1°C) is not tabulated. Find the correction factor for 285°F (196.1°C) thus: [(Actual temperature − lower temperature)/(higher temperature − lower temperature)] × (higher temperature correction factor − lower temperature correction factor) + lower temperature correction factor. Or, [(385 − 375)/(400 − 375)](1.273 − 1.255) + 1.255 = 1.262.

TABLE 5 Typical Fan Capacities

Capacity		Outlet velocity		Outlet velocity pressure		Ratings at 6.5-inH₂O (1.6-kPa) static pressure		
ft³/min	m³/s	ft/min	m/s	inH₂O	kPa	r/min	bhp	kW
61,204	28.9	4400	22.4	1.210	0.3011	1083	95.45	71.2
62,595	29.5	4500	22.9	1.266	0.3150	1096	99.08	73.9
63,975	30.2	4600	23.4	1.323	0.3212	1109	103.0	76.8

The altitude correction factor is 1.095 for an elevation of 5000 ft (1524.0 m), as shown in Table 4.

As for the forced-draft fan, CCF = $C_T C_A$ = (1.262)(1.095) = 1.3819. Use the CCF to find the capacity-table ft^3/min in the same manner as for the forced-draft fan. Or, capacity-table ft^3/min = (given ft^3/min)/CCF = 152,500/1.3819 = 110,355 ft^3/min (52.1 m^3/s).

5. Choose the fan size from the capacity table
Check the capacity table to be sure that it lists fans suitable for induced-draft (elevated-temperature) service. Turn to the 11-inH$_2$O (2.7-kPa) static-pressure capacity table, and find a capacity equal to 110,355 ft^3/min (52.1 m^3/s). In the engineering data used for this fan, the nearest capacity at 11-inH$_2$O (2.7-kPa) static pressure is 110,467 ft^3/min (52.1 m^3/s), with an outlet velocity of 4400 ft/min (22.4 m/s), an outlet velocity pressure of 1.210 inH$_2$O (0.30 kPa), a speed of 1222 r/min, and an input hp of 255.5 bhp (190.5 kW). The tabulation of these quantities is of the same form as that given for the forced-draft fan, step 2. The selected capacity of 110,467 ft^3/min (52.1 m^3/s) is entirely satisfactory because it is only 110,467 − 110,355/110,355 = 0.00101, to 0.1 percent, higher than the desired capacity.

6. Compute the fan speed and power input
Multiply the capacity-table rpm and brake hp by the CCF to determine the actual rpm and brake hp. Thus, the actual rpm = (1222)(1.3819) = 1690 r/min. Actual brake hp = (255.5)(1.3819) = 353.5 bhp (263.6 kW). This is the hp input required to drive the fan and is close to the 360.5 hp (268.8 kW) computed in the previous calculation procedure. The actual motor horsepower would be the same in each case because a standard-size motor would be chosen. The difference in hp of 360.5 − 353.5 = 7.0 hp (5.2 kW) results from the same factors discussed in step 3.

Note: The static pressure is normally used in most fan-selection procedures because this pressure value is used in computing pressure and draft losses in boilers, economizers, air heaters, and ducts. In any fan system, the total air pressure = static pressure + velocity pressure. However, the velocity pressure at the fan discharge is not considered in draft calculations unless there are factors requiring its evaluation. These requirements are generally related to pressure losses in the fan-control devices.

Related Calculations. Use the fan-capacity table to obtain these additional details of the fan: outlet inside dimensions (length and width), fan-wheel diameter and circumference, fan maximum bhp, inlet area, fan-wheel peripheral velocity, NAFM fan class, and fan arrangement. Use the engineering data containing the fan-capacity table to find the fan dimensions, rotation and discharge designations, shipping weight, and, for some manufacturers, prices.

FAN ANALYSIS AT VARYING RPM, PRESSURE, AND AIR OR GAS CAPACITY

A fan delivers 12,000 ft^3/min (339.6 m^3/min) at a static pressure of 1 in (0.39 cm) WG at 70°F (21.1°C) when operating at 400 r/min; required power input is 4 hp (2.98 kW). (*a*) If in the same installation, 15,000 ft^3/min (424.5 m^3/min) are required, what will be the new fan speed, static pressure, and power input? (*b*) If the air temperature is increased to 200°F (93.3°C) and the fan speed remains at 400 r /min, what will be the new static pressure and power input with a flow rate of 12,000 ft³/min (339.5 m³/min)? (*c*) If the speed of the fan is increased to deliver 1 in (0.39 cm) WG at 200°F (93.3°C), what will be the new speed, capacity, and power input? (*d*) If the speed of the fan is increased so as to deliver the same weight of air at 200°F (93.3°C) as at 70°F (21.1°C), what will be the new speed, capacity, static pressure, and power input?

Calculation Procedure:

1. Determine the fan speed, static pressure, and power input at the higher flow rate

(*a*) Use the fan laws to determine the required unknowns. The first fan law states: *Air or gas capacity varies directly as the fan speed.* Thus, the new speed with higher capacity = 400(15,000/12,000) = 500 r/min. Hence, the fan speed must be increased by 25 percent, *i.e.*, 100 r/min—to have the fan handle 25 percent more air. This verifies the first fan law that capacity varies directly as fan speed.

Use the second fan law to determine the new static pressure. This law states: *Fan pressure (static, velocity, and total) varies as the square of the fan speed.* Thus, the new static pressure with the larger flow rate = 1(500/400)² = 1.5625 in (3.97 cm).

Find the new required power input at the higher flow rate and higher discharge pressure by using the third fan law, which states: *Power demand of a fan varies as the cube of the fan speed.* Hence, the new power = 4(500/400)³ = 7.8125 hp (5.82 kW).

2. Determine the new static pressure and power

(*b*) When the density of air or gas handled by a fan changes, three other fan laws apply. The first of these laws is: *At constant fan speed; i.e., rpm, and capacity; i.e., cfm (m³/min), the pressure developed and required power input vary directly as the air or gas density.* For the conditions given here the air density at 70°F (21.1°C) is 0.075 lb/ft³ (1.2 kg/m³); at 200°F (93.3°C) the air density is 0.06018 lb/ft³ (0.963 kg/m³). Then, new static pressure = 1.0(0.06018/0.075) = 0.80 in (2.04 cm). The new power is found from 4(0.06018/0.075) = 3.21 hp (2.39 kW).

3. Find speed, capacity, and power input at the new pressure

(*c*) We now have a constant-pressure output. Under these conditions, with a varying air or gas density, the fan law states that: *At constant pressure the speed, capacity, and power vary inversely as the square root of the fluid density.* Thus, new speed = 400 (0.075/0.06018)⁰·⁵ = 446.5 r/min.

The new capacity at the 1-in (0.39-cm) static pressure = 12,000(0.075/0.06018)⁰·⁵ = 13,396 r/min at 200°F (93.3°C). The new power = 4 (0.075/0.06018)⁰·⁵ = 4.46 hp (3.33 kW).

4. Compute the new speed, capacity, static pressure, and power at the increased speed

(*d*) The final fan law states: *For a constant weight of air or gas, the speed, capacity, and pressure vary inversely as the density, while the hp varies inversely as the square of the density.* Using this law, the new speed = 400(0.075/0.06018) = 498.5 r/min.

The new capacity = 12,000(0.075/0.06018) = 14,955 ft³/min (423.5 m³/min). Likewise, the new static pressure = 1.0(0.075/0.06018) = 1.246 in (3.17 cm). Finally, the new power = 4 (0.075/0.06018)² = 6.21 hp (4.63 kW).

Related Calculations. The fan laws, as given here, are powerful in the analysis of the speed, capacity, and pressure of any fan handling air or gases. These laws can be used for fans employed in air conditioning, ventilation, forced and induced draft, kitchen and hood exhausts, *etc.* The fan can be used in stationary, mobile, marine, aircraft, and similar applications. The fan laws apply equally well.

BOILER FORCED-DRAFT FAN HORSEPOWER DETERMINATION

Find the motor turbine hp needed to provide forced-draft service to a boiler that burns coal at a rate of 10 tons (9080 kg)/h. The boiler requires 59,000 ft³/min (5481 cum/min) of air under 6 in (15.2 cm) water gage (WG) from the fan which has a mechanical efficiency of 60 percent. The air is delivered at a total pressure of 6 in (15.2-cm) WG by the fan. What would be the effect on the required power to this fan if the total pressure were doubled to 12 in (30.5 cm) WG? If the required air delivery was increased to 75,000 ft³/min (2123 m³/min), at 6 in (15.2 cm) WG, what input hp would be required?

Calculation Procedure:

1. Find the required power input to the fan
Use the relation, fan hp = ft³/min(total pressure developed, lb/ft²)/33,000(fan efficiency). To apply this equation we must convert the water gage pressure to lb/ft² by (in WG/12)(62.4 lb/ft³ water density). Or (6/12)(62.4) 31.2 lb/ft² (1.49 kPa). Substituting, hp = 59,000(31.2)/33,000(0.60) = 92.96 hp (69.4 kW). Use a 100-hp (75 kW) motor or turbine to drive this induced-draft fan.

2. Determine the required power input at the higher delivery pressure
Use the same relation as in Step 1 to find, hp = 59,000(62.4)/33,000(0.60) = 185.9 hp (138.7 kW). Thus, the required power input doubles as the developed pressure doubles.

The sharp increase in the power input is a graphic example of why the pressure requirements for any type of fan must be carefully analyzed before the final choice is made. Since the cost of a fan does not rise in direct proportion to its delivery pressure, the engineer should apply a factor of safety to the computed power input to take care of possible future overloads.

3. Find the required power input at the higher flow rate
Using the same relation, hp = 118.2 (88.2 kW). Again, the required power input rises as the output from the fan is increased. This further illustrates the strong need to explore the maximum output requirements before making a final equipment choice.

Related Calculations. This approach can be used for any fan used in power-plant, HVAC, and similar applications. The key point to observe is the rise in power requirements as the fan pressure of air volume delivered increases.

EFFECT OF BOILER RELOCATION ON DRAFT
FAN PERFORMANCE

An acceptance test of a boiler shows that its induced-draft fan handles 600,000
(272,400 kg) lb per hour of flue gas at 290°F (143.3°C) against total friction of 10
in (25.4 cm) water gage (WG). The boiler is relocated to a higher elevation where
the barometric pressure is 24 inHg (60.96 cmHg), as compared to the original 30
inHg (76.2 cmHg) at sea level. If no changes are made to the equipment except
adjustments to the fan and with gas weights and temperatures as before, what are
the new volume and suction conditions for the fan design and relocation?

Calculation Procedure:

1.   Determine the new inlet condition for the fan
When a fan is required to handle air or gas at conditions other than standard, a
correction must be made in the static pressure and hp (kW). Since a fan is essentially a constant-volume machine, the ft^3/min (m^3/min) delivered will not change
materially if the speed and system configuration do not change, regardless of the
air or gas density.

The static pressure, however, changes directly with density. Hence, the static
pressure must be carefully calculated for specified conditions. For the situation
described here, assume a gas molecular weight of 28, a typical value. The density
correction factor can be computed from the ratio of the new-location barometric
pressure to the first-location barometric pressure, both expressed inHg (cmHg). Or,
density correction factor = 24 in/30 in = 0.80.

There is no temperature correction factor because the air temperature remains
the same. Therefore, at the new elevated location of the boiler the intake condition
for the fan will be (10 in WG)(0.8 correction factor) = 8 in (20.3 cm) WG.

2.   Compute the new volume condition
Use the relation Volume flow = (lb/h)(molecular weight of gas)(cfm at inlet conditions)[(absolute temperature of flue gas)/("standard" air temperature of 60°F in
absolute terms)][(reduced barometric pressure)/(reduced barometric pressure − new
suction condition)]. Substituting, new volume condition = (600,000/28)(379)[(460
+ 290)/(60 + 460)][(24)/(24 − 8)] = 17.57 × 10^6 ft^3/h (0.497 m^3/h × 10^6).

Related Calculations.   With used power-plant equipment becoming more popular throughout the world (see the classified section of any major engineering magazine) it is important that the engineer be able to determine the performance of reused equipment at different locations.

ANALYSIS OF BOILER AIR DUCTS AND
GAS UPTAKES

Three oil-fired boilers are supplied air through the breeching shown in Fig. 7a.
Each boiler will burn 13,600 lb/h (1.71 kg/s) of fuel oil at full load. The draft loss

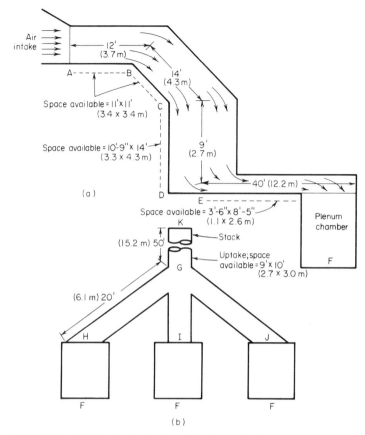

FIGURE 7   (*a*) Boiler intake-air duct; (*b*) boiler uptake ducts.

through each boiler is 8 inH$_2$O (2.0 kPa). Uptakes from the three boilers are connected as shown in Fig. 7*b*. Determine the draft loss through the entire system if a 50-ft (15.2-m) high metal stack is used and the gas temperature at the stack inlet is 400°F (204.4°C).

Calculation Procedure:

1. *Determine the airflow through the breeching*
Compute the airflow required, cubic feet per pound of oil burned, using the methods given in earlier calculation procedures. For this installation, assume that the combustion calculation shows that 250 ft^3/lb (15.6 m^3/kg) of oil burned is required. Then the total airflow required = (number of boilers)(lb/h oil burned per boiler)(ft^3/lb oil)/(60 min/h) = (3)(13,600)(250)/60 = 170,000 ft^3/min (80.2 m^3/s).

2. Select the dimensions for each length of breeching duct

With the airflow rate of 170,000 ft^3/min (80.2 m^3/s) known, the duct area can be determined by assuming an air velocity and computing the duct area. A_d ft^2 from A_d = (airflow rate, ft^3/min)/(air velocity, ft/min). Once the area is known, the duct can be sized to give this area. Thus, if 9 ft^2 (0.8 m^2) is the required duct area, a duct 3 × 3 ft (0.9 × 0.9 m) or 2 × 4.5 ft (0.6 × 1.4 m) would provide the required area.

In the usual power plant, the room available for ducts limits the maximum allowable duct size. So the designer must try to fit a duct of the required area into the available space. This is done by changing the duct height and width until a duct of suitable area fitting the available space is found. If the duct area is reduced below that required, compute the actual air velocity to determine whether it exceeds recommended limits.

In this power plant, the space available in the open area between A and C, Fig. 7, is a square 11 × 11 ft (3.4 × 3.4 m). By allowing a 3-in (7.6-cm) clearance around the outside of the duct and using a square duct, its dimensions would be 10.5 × 10.5 ft (3.2 × 3.2 m), or a cross-sectional area of (10.5)(10.5) = 110 ft^2 (10.2 m^2), closely. With 170,000 ft^3/min (80.2 m^3/s) flowing through the duct, the air velocity v ft/min = ft^3/min/A_d = 170,000/110 = 1545 ft/min (7.8 m/s). This is a satisfactory air velocity because the usual plant air system velocity is 1200 to 3600 ft/min (6.1 to 18.3 m/s).

Between C and D the open area in this power plant is 10 ft 9 in (3.3 m) by 14 ft (4.3 m). Using the same 3-in (7.6-cm) clearance all around the duct, we find the dimensions of the vertical duct CD are 10.25 × 13 ft (3.1 × 4.0 m), or a cross-sectional area of 10.25 × 13 = 133 ft^2 (12.5 m^2), closely. The air velocity in this section of the duct is v = 170,000/133 = 1275 ft/min (6.5 m/s). Since it is desirable to maintain, if possible, a constant velocity in all sections of the duct where space permits, the size of this duct might be changed so it equals that of AB, 10.5 × 10.5 ft (3.2 × 3.2 m). However, the installation costs would probably be high because the limited space available would require alteration of the power-plant structure. Also, the velocity is section CD is above the usual minimum value of 1200 ft/min (6.1 m/s). For these reasons, the duct will be installed in the 10.25 × 13 ft (3.1 × 4.0 m) size.

Between E and F the vertical distance available for installation of the duct is 3.5 ft (1.1 m), and the horizontal distance is 8.5 ft (2.6 m). Using the same 3-in (7.6-cm) clearance as before gives a 3 × 8 ft (0.9 × 2.4 m) duct size, or a cross-sectional area of (3)(8) = 24 ft^2 (2.2 m^2). At E the duct divides into three equal-size branches, one for each boiler, and the same area, 24 ft^2 (2.2 m^2), is available for each branch duct. The flow in any branch duct is then 170,000/3 = 56,700 ft^3/min (26.8 m^3/s). The velocity in any of the three equal branches is v = 56,700/24 = 2360 ft/min (12.0 m/s). When a duct system has two or more equal-size branches, compute the pressure loss in one branch only because the losses in the other branches will be the same. The velocity in branch EF is acceptable because it is within the limits normally used in power-plant practice. At F the air enters a large plenum chamber, and its velocity becomes negligible because of the large flow area. The boiler forced-draft fan intakes are connected to the plenum chamber. Each of the three ducts feeds into the plenum chamber.

3. Compute the pressure loss in each duct section

Begin the pressure-loss calculations at the system inlet, point A, and work through each section to the stack outlet. This procedure reduces the possibility of error and permits easy review of the calculations for detection of errors. Assign letters to each point of the duct where a change in section dimensions or directions, or both, occurs. Use these letters:

Point *A:* Assume that 70°F (21.1°C) air having a density of 0.075 lb/ft³ (1.2 kg/m³) enters the system when the ambient barometric pressure is 29.92 inHg (101.3 kPa). Compute the velocity pressure at point *A*, in inH₂O, from $p_v = v_2/[3.06(10^4)(460 + t)]$, where t = air temperature,°F. Since the velocity of the air at *A* is 1545 ft/min (11.7 m/s), $p_v = (1545)^2/[3.06(10^4)(530)] = 0.147$ inH₂O (0.037 kPa) at 70°F (21.1°C).

The entrance loss at *A*, where there is a sharp-edged duct, is $0.5p_v$, or 0.5(0.147) = 0.0735 inH₂O (0.018 kPa). With rounded inlet, the loss in velocity pressure would be negligible.

Section *AB:* There is a pressure loss due to duct friction between *A* and *B*, and *B* and *C*. Also, there is a bend loss at points *B* and *C*. Compute the duct friction first.

For any circular duct, the static pressure loss due to friction p_s inH₂O = $(0.03L/d^{1.24})(v/1000)^{1.84}$, where L = duct length, ft; d = duct diameter, in. To convert any rectangular or square duct with sides a and b high and wide, respectively, to an equivalent round duct of *D*-ft diameter, use the relation $D = 2ab/(a + b)$. For this duct, $d = 2(10.5)(10.5)/(10.5 + 10.5) = 10.5$ ft (3.2 m) = 126 in (320 cm) = d. Since this duct is 12 ft (3.7 m) long between *A* and *B*, $p_s = [0.03(12)/126^{1.24}](1.545/1000)^{1.84} = 0.002$ inH₂O (0.50 kPa).

Point *B:* The 45° bend at *B* has, from Baumeister and Marks—*Standards Handbook for Mechanical Engineers,* a pressure drop of 60 percent of the velocity head in the duct, for (0.60)(0.147) = 0.088 inH₂O (20.5 Pa) loss.

Section *BC:* Duct friction in the 14-ft (4.3-m) long downcomer *BC* is $p_s = [0.03(14)/126^{1.24}](1545/1000)^{1.84} = 0.0023$ inH₂O (0.56 Pa). Point *C:* The 45° bend at *C* has a velocity head loss of 60 percent of the velocity pressure. Determine the velocity pressure in this duct in the same manner as for point *A*, or $p_v = (1545)^2/[3.06(10^4)(530)] = 0.147$ inH₂O (36.1 Pa), since the velocity at points *B* and *C* is the same. Then the velocity head loss = (0.60)(0.147) = 0.088 inH₂O (21.9 Pa).

Section *CD:* The equivalent round-duct diameter is $D = (2)(10.25)(13)/(10.25 + 13) = 11.45$ ft (3.5 m) = 137.3 in (348.7 cm). Duct friction is then $p_s = [0.03(9)/137.3^{1.24}](1275/1000)^{1.84} = 0.000934$ inH₂O (0.23 Pa). Velocity pressure in the duct is $p_v = (1275)^2/[3.06(10^4)(530)] = 0.100$ inH₂O (24.9 Pa). Since there is no room for a transition piece—that is, a duct providing a gradual change in flow area between points *C* and *D*–the decrease in velocity pressure from 0.147 to 0.100 in (36.6 to 24.9 Pa), or 0.147 − 0.10 = 0.047 inH₂O (11.7 Pa), is not converted to static pressure and is lost.

Point *E:* The pressure loss in the right-angle bend at *E* is, from Baumeister and Marks—*Standard Handbook for Mechanical Engineers,* 1.2 times the velocity head, or (1.2)(0.1) = 0.12 inH₂O (29.9 Pa). Also, since this is a sharp-edged elbow, there is an additional loss of 50 percent of the velocity head, or (0.5)(0.10) = 0.05 inH₂O (12.4 Pa).

The velocity pressure at point *E* is $p_v = (2360)^2/[3.06(10^4)(530)] = 0.343$ inH₂O (85.4 Pa).

Section *EF:* The equivalent round-duct diameter is $D = (2)(3)(8)/(3 + 8) = 4.36$ ft (1.3 m) $= 52.4$ in (133.1 cm). Duct friction $p_s = [0.03(40)/52.4^{1.24}](2360/1000)^{1.84} = 0.0247$ inH$_2$O (6.2 Pa).

Air entering the large plenum chamber at *F* loses all its velocity. There is no static-pressure regain; therefore, the velocity-head loss $= 0.348 - 0.0 = 0.348$ inH$_2$O (86.6 Pa).

4. Compute the losses in the uptake and stack

Convert the airflow of 250 ft^3/lb (15.6 m^3/kg) of fuel oil to pounds of air per pound of fuel oil by multiplying by the density, or $250(0.075) = 18.75$ lb of air per pound of oil. The flue gas will contain 18.75 lb of air + 1 lb of oil per pound of fuel burned, or $(18.75 + 1)/18.75 = 1.052$ times as much gas leaves the boiler as air enters; this can be termed the *flue-gas factor.*

Point *G:* The quantity of flue gas entering the stack from each boiler (corrected to a 400°F or 204.4°C outlet temperature) is, in °R (*cfm* air to furnace)(stack, °R/air,°R)(flue-gas factor). Or stack flue-gas flow $= (56,700)[(400 + 460)/(70 + 460)](1.052) = 97,000$ ft^3/min (45.8 m^3/s) per boiler.

The total duct area available for the uptake leading to the stack is 9×10 ft (2.7 $\times$ 3.0 m) $= 90$ ft^2 (8.4 m^2), based on the clearance above the boilers. The flue-gas velocity for three boilers is $v = (3)(97,000)/90 = 3235$ ft/min (16.4 m/s). The velocity pressure in the uptake is $p = (3235)^2/[3.06(10^4)(460 + 400)] = 0.397$ inH$_2$O (98.8 Pa).

Point *H:* The flue-gas flow from all the boilers is divided equally between three ducts. *HG, IG, JG,* Fig. 7. It is desirable to maintain the same gas velocity in each duct and have this velocity equal to that in the uptake. The same velocity can be obtained in each duct by making each duct one-third the area of the uptake, or $90/3 = 30$ ft^2 (2.8 m^2). Then $v = 97,000/30 = 3235$ ft/min (16.4 m/s) in each duct. Since the velocity in each duct equals the velocity in the uptake, the velocity pressure in each duct equals that in the uptake, or 0.397 inH$_2$O (98.8 Pa).

Ducts *HG* and *JG* have two 45° bends in them, or the equivalent of one 90° bend. The velocity-pressure loss in a 90° bend is 1.20 times the velocity head in the duct; or, for either *HG* or *JG*, $(1.20)(0.397) = 0.476$ inH$_2$O (118.5 Pa).

Section *HG:* The equivalent duct diameter for a 30-ft^2 (2.8-m^2) duct is $D = 2(30/\pi)^{0.5} = 6.19$ ft (1.9 m) $= 74.2$ in (188.4 cm). The duct friction in *HG*, which equals that in *JG*, is $p_s = [0.03(20)/74.2^{1.24}](530/860)(3235/1000)^{1.84} = 0.01536$ inH$_2$O (3.8 Pa), if we correct for the flue-gas temperature with the ratio $(70 + 460)/(400 + 460) = 530/860$.

Section *GK:* The stack joins the uptake at point *G*. Assume that this installation is designed for a stack-gas area of 500 lb of oil per square foot (2441.2 kg/m^2) of stack; for three boilers, stack area $= (3)(13,600$ lb/h oil$)/500 = 81.5$ ft^2 (7.6 m^2). The stack diameter will then be $D = 2(8.15/\pi)^{0.5} = 10.18$ ft (3.1 m) $= 122$ in (309.9 cm).

The gas velocity in the stack is $v = (3)(97,000)/81.5 = 3570$ ft/min (18.1 m/s). The friction in the stack is $p_s = [0.03(50)122^{1.24}](3570/1000)^{1.84}(503/860) = 0.0194$ inH$_2$O (4.8 Pa).

5. Compute the total losses in the system

Tabulate the individual losses and find the sum as follows:

	inH$_2$O	kPa
Point A; entrance loss	0.0735	0.0183
Section AB; duct friction	0.0020	0.0005
Point B; bend loss	0.0880	0.0219
Section BC; duct friction	0.0023	0.0006
Point C; bend loss	0.0880	0.0219
Section CD; duct friction	0.0009	0.0002
Section CD; velocity-pressure loss	0.0470	0.0117
Point E; bend loss	0.1200	0.0299
Point E; sharp-edge loss	0.0500	0.0124
Section EF; duct friction	0.0247	0.0061
Section EF; plenum velocity-head loss	0.3480	0.0866
Boiler friction loss	8.0000	1.9907
Section HG; duct friction	0.0154	0.0038
Points H and G total bend loss	0.4760	0.1184
Section GK; stack friction	0.0194	0.0048
Total loss	9.3552	2.3279

The total loss computed here is the minimum static pressure that must be developed by the draft fans or blowers. This total static pressure can be divided between the forced- and induced-draft fans or confined solely to the forced-draft fans in plants not equipped with an induced-draft fan. If only a forced-draft fan is used, its static discharge pressure should be at least 20 percent greater than the losses, or $(1.2)(9.3552) = 11.21$ inH$_2$O (2.8 kPa) at a total airflow of 97,000 ft^3/min (45.8 m^3/s). If more than one forced-draft fan were used for each boiler, each fan would have a total static pressure of at least 11.21 inH$_2$O (2.8 kPa) and a capacity of less than 97,000 ft^3/min (45.8 m^3/s). In making the final selection of the fan, the static pressure would be rounded to 12 inH$_2$O (3.0 kPa).

Where dampers are used for combustion-air control, include the wide-open resistance of the dampers in computing the total losses in the system at full load on the boilers. Damper resistance values can be obtained from the damper manufacturer. Note that as the damper is closed to reduce the airflow at lower boiler loads, the resistance through the damper is increased. Check the fan head-capacity curve to determine whether the head developed by the fan at lower capacities is sufficient to overcome the greater damper resistance. Since the other losses in the system will decrease with smaller airflow, the fan static pressure is usually adequate.

Note: (1) Follow the notational system used here to avoid errors from plus and minus signs applied to atmospheric pressures and draft. Use of the plus and minus signs does not simplify the calculation and can be confusing.

(2) A few designers, reasoning that the pressure developed by a fan varies as the square of the air velocity, square the percentage safety-factor increase before multiplying by the static pressure. Thus, in the above forced-draft fan, the static discharge pressure with a 20 percent increase in pressure would be $(1.2)^2(9.3552)$ = 13.5 inH$_2$O (3.4 kPa). This procedure provides a wider margin of safety, but is not widely used.

(3) Large steam-generating units, some ship propulsion plants, and some packaged boilers use only forced-draft fans. Induced-draft fans are eliminated because there is a saving in the total fan hp required, there is no air infiltration into the boiler setting, and a slightly higher boiler efficiency can be obtained.

(4) The duct system analyzed here is typical of a study-type design where no refinements are used in bends, downcomers, and other parts of the system. This

type of system was chosen for the analysis because it shows more clearly the various losses met in a typical duct installation. The system could be improved by using a bellmouthed intake at A, dividing vanes or splitters in the elbows, a transition in the downcomer, and a transition at F. None of these improvements would be expensive, and they would all reduce the static pressure required at the fan discharge.

(5) Do not subtract the stack draft from the static pressure the forced- or induced-draft fan must produce. Stack draft can vary considerably, depending on ambient temperature, wind velocity, and wind direction. Therefore, the usual procedure is to ignore any stack draft in fan-selection calculations because this is the safest procedure.

Related Calculations. The procedure given here can be used for all types of boilers fitted with air-supply ducts and uptake breechings—heating, power, process, marine, portable, and packaged.

DETERMINATION OF THE MOST ECONOMICAL FAN CONTROL

Determine the most economical fan control for a forced- or induced-draft fan designed to deliver 140,000 ft^3/min (66.1 m^3/s) at 14 inH$_2$O (3.5 kPa) at full load. Plot the power-consumption curve for each type of control device considered.

Calculation Procedure:

1 Determine the types of controls to consider
There are five types of controls used for forced- and induced-draft fans: (*a*) a damper in the duct with constant-speed fan drive; (*b*) two-speed fan drive; (*c*) inlet vanes or inlet louvres with a constant speed fan drive; (*d*) multiple-step variable-speed fan drive; and (*e*) hydraulic or electric coupling with constant-speed drive giving wide control over fan speed.

2. Evaluate each type of fan control
Tabulate the selection factors influencing the control decision as follows, using the control letters in step 1:

Control type	Control cost	Required power input	Advantages (A), and disadvantages (D)
a	Low	High	(A) Simplicity; (D) high power input
b	Moderate	Moderate	(A) Lower input power; (D) higher cost
c	Low	Moderate	(A) Simplicity; (D) ID fan erosion
d	Moderate	Moderate	(D) Complex; also needs dampers
e	High	Low	(A) Simple; no dampers needed

3. Plot the control characteristics for the fans
Draw the fan head-capacity curve for the airflow or gasflow range considered, Fig. 8. This plot shows the maximum capacity of 140,000 ft^3/min (66.1 m^3/s) and required static head of 14 inH$_2$O (3.5 kPa), point *P*.

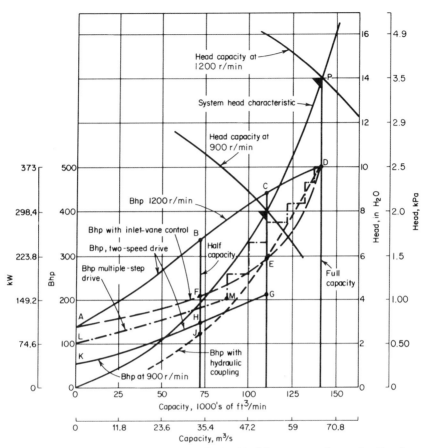

FIGURE 8 Power requirements for a fan fitted with different types of controls. (*American Standard Inc.*)

Plot the power-input curve *ABCD* for a constant-speed motor or turbine drive with damper control—type *a,* listed above—after obtaining from the fan manufacturer, or damper builder, the input power required at various static pressures and capacities. Plotting these values gives curve *ABCD.* Fan speed is 1200 r/min.

Plot the power-input curve *GHK* for a two-speed drive, type *b.* This drive might be a motor with additional winding, or it might be a second motor for use at reduced boiler capacities. With either arrangement, the fan speed at lower boiler capacities is 900 r/min.

Plot the power-input curve *AFED* for inlet-vane control on the forced-draft fan or inlet-louvre control on induced-draft fans. The data for plotting this curve can be obtained from the fan manufacturer.

Multiple-step variable-speed fan control, type *d,* is best applied with steamturbine drives. In a plant with ac auxiliary motor drives, slip-ring motors with damper integration must be used between steps, making the installation expensive. Although dc motor drives would be less costly, few power plants other than marine

TABLE 6 Fan Control Comparison

	Type of control used				
	a	b	c	d	e
Total cost, $	30,000	50,000	75,000	89,500	98,000
Extra cost, $	. . . . .	20,000	25,000	14,500	8,500
Total power saving, $	. . . . .	8,000	6,500	3,000	6,300
Return on extra investment, %	. . . . .	40	26	20.7	74.2

propulsion plants have direct current available. And since marine units normally operate at full load 90 percent of the time or more, part-load operating economics are unimportant. If steam-turbine drive will be used for the fans, plot the power-input curve *LMD*, using data from the fan manufacturer.

A hydraulic coupling or electric magnetic coupling, type *e*, with a constant-speed motor drive would have the power-input curve *DEJ*.

Study of the power-input curves shows that the hydraulic and electric couplings have the smallest power input. Their first cost, however, is usually greater than any other types of power-saving devices. To determine the return on any extra investment in power-saving devices, an economic study including a load-duration analysis of the boiler load must be made.

4. Compare the return on the extra investment

Compute and tabulate the total cost of each type of control system. Then determine the extra investment for each of the more costly control systems by subtracting the cost of type *a* from the cost of each of the other types. With the extra investment known, compute the lifetime savings in power input for each of the more efficient control methods. With the extra investment and savings resulting from it known, compute the percentage return on the extra investment. Tabulate the findings as in Table 6.

In Table 6, considering control type *c*, the extra cost of type *c* over type *b* = $75,000 − 50,000 = $25,000. The total power saving of $6500 is computed on the basis of the cost of energy in the plant for the life of the control. The return on the extra investment then = $6500/$25,000 = 0.26, or 26 percent. Type *e* control provides the highest percentage return on the extra investment. It would probably be chosen if the only measure of investment desirability is the return on the extra investment. However, if other criteria are used, such as a minimum rate of return on the extra investment, one of the other control types might be chose. This is easily determined by studying the tabulation in conjunction with the investment requirement.

Related Calculations. The procedure used here can be applied to heating, power, marine and portable boilers of all types. Follow the same steps given above, changing the values to suit the existing conditions. Work closely with the fan and drive manufacturer when analyzing drive power input and costs.

SMOKESTACK HEIGHT AND DIAMETER DETERMINATION

Determine the required height and diameter of a smokestack to produce 1.0-inH_2O (0.25-kPa) draft at sea level if the average air temperature is 60°F (15.6°C); barometric pressure is 29.92 inHg (101.3 kPa); the boiler flue gas enters the stack at 500°F (260.0°C); the flue-gas flow rate is 100 lb/s (45.4 kg/s); The flue-gas density is 0.045 lb/ft³ (0.72 kg/m³); and the flue-gas velocity is 30 ft/s (9.1 m/s). What diameter and height would be required for this stack if it were located 5000 ft (1524.0 m) above sea level?

Calculation Procedure:

1 Compute the required stack height
The required stack height S_h ft = $d_s/0.256pK$, where d_s = stack draft, inH$_2$O; p = barometric pressure, inHg; K = $1/T_a - 1/T_g$, where T_a = air temperature,°R; T_g = average temperature of stack gas,°R. In applying this equation, the temperature of the gas at the stack outlet must be known to determine the average temperature of the gas in the stack. Since the outlet temperature cannot be measured until after the stack is in use, an assumed outlet temperature must be used for design calculations. The outlet temperature depends on the inlet temperature, ambient air temperature, and materials used in the stack construction. For usual smokestacks, the gas temperature will decrease 100 to 200°F (55.6 to 111.1°C) between the stack inlet and outlet. Using a 100°F (55.6°C) gas-temperature decrease for this stack, we get S_h = (1.0) + 0.256(29.92)(1/520 − 1/910) = 159 ft (48.5 m). Apply a 10 percent factor of safety. Then the stack height = (159)(1.10) = 175 ft (53.3 m).

2. Compute the required stack diameter
Stack diameter d_s ft is found from d_s = $0.278(W_gT_g/Vd_gp)^{0.5}$, where W_g = flue-gas flow rate in stack, lb/s; V = flue-gas velocity in stack, ft/s; d_g = flue-gas density, lb/ft³. For this stack, d_s = 0.278{(100)(910)/[(30)(0.045)(29.92)]}$^{0.5}$ = 13.2 ft (4.0 m), or 13 ft 3 in (4 m 4 cm), rounding to the nearest inch diameter.
  Note: Use this calculation procedure for any stack material—masonary, steel, brick, or plastic. Most boiler and stack manufacturers use charts based on the equations above to determine the economical height and diameter of a stack. Thus, the Babcock & Wilcox Company, New York, Inc., also presents four charts for stack sizing, in *Steam: Its Generation and Use.* Combustion Engineering, Inc., also presents four charts for stack sizing, in *Combustion Engineering.* The equations used in the present calculation procedure are adequate for a quick, first approximation of stack height and diameter.

3. Compute the required stack height and diameter at 5000-ft (1524.0-m) elevation
Fuels require the same amount of oxygen for combustion regardless of the altitude at which they are burned. Therefore, this stack must provide the same draft as at sea level. But as the altitude above sea level increases, more air must be supplied to the fuel to sustain the same combustion rate, because air above sea level contains less oxygen per cubic foot than at sea level. To accommodate the larger air and flue-gas flow rate without an increase in the stack friction loss, the stack diameter must be increased.
  To determine the required stack height S_e ft at an elevation above sea level, multiply the sea-level height S_h by the ratio of the sea-level and elevated height barometric pressures inHg. Since the barometric pressure at 5000 ft (1524.0 m) is 24.89 inHg (84.3 kPa) and the sea-level barometric pressure is 29.92 inHg (101.3 kPa), S_e = (175)(29.92/24.89) = 210.2 ft (64.1 m).

The stack diameter d_e ft at an elevation above sea level will vary as the 0.40 power of the ratio of the sea-level and altitude barometric pressures, or $d_e\, d_s(p_e/p)^{0.4}$, where p_e = barometric pressure of altitude, inHg. For this stack, d_e = $(13.2)(29.92/24.89)^{0.4}$ = 14.2 ft (4.3 m), or 14 ft 3 in (4 m 34 cm).

Related Calculations. The procedure given here can be used for heating, power, marine, industrial, and residential smokestacks or chimneys, regardless of the materials used for construction. When designing smokestacks for use at altitudes above sea level, use step 3, or substitute the actual barometric pressure at the elevated location in the height and diameter equations of steps 1 and 2.

POWER-PLANT COAL-DRYER ANALYSIS

A power-plant coal dryer receives 180 tons/h (163.3 t/h) of wet coal containing 15 percent free moisture. The dryer is arranged to drain 6 percent of the moisture from the coal, and a moisture content of 1 percent is acceptable in the coal delivered to the power plant. Determine the volume and temperature of the drying gas required for the dryer, the total heat, grate area, and combustion-space volume needed. Ambient air temperature during drying is 70°F (21.1°C).

Calculation Procedure:

1. Compute the quantity of moisture to be removed
The total moisture in the coal = 15 percent. Of this, 6 percent is drained and 1 percent can remain in the coal. The amount of moisture to be removed is therefore $15 - 6 - 1 = 8$ percent. Since 180 tons (163.3 t) of coal are received per hour, the quantity of moisture to be removed per minute is $[180/(60 \text{ min/h})](2000 \text{ lb/ton})(0.08) = 480$ lb/min (3.6 kg/s).

2. Compute the airflow required through the dryer
Air enters the dryer at 70°F (21.1°C). Assume that evaporation of the moisture on the coal takes place at 125°F (51.7°C)—this is about midway in the usual evaporation temperature range of 110 to 145°F (43.3 to 62.8°C). Determine the moisture content of saturated air at each temperature, using the psychrometric chart for air. Thus, for saturated air at 70°F (21.1°C) dry-bulb temperature, the weight of the moisture it contains is w_m lb (kg) of water per pound (kilogram) of dry air = 0.0159 (0.00721), whereas at 125°F (51.7°C), w_m = 0.09537 lb of water per pound (0.04326 kg/kg) of dry air. The weight of water removed per pound of air passing through the dryer is the difference between the moisture content at the leaving temperature, 125°F (51.7°C), and the entering temperature, 70°F (21.1°C), or $0.09537 - 0.01590 = 0.07947$ lb of water per pound (0.03605 kg/kg) of dry air.

Since air at 70°F (21.1°C) has a density of 0.075 lb/ft³ (1.2 kg/m³), $1/0.075 = 13.3$ ft³ (0.4 m³) of air at 70°F (21.1°C) must be supplied to absorb 0.07947 lb of water per pound (0.03605 kg/kg) of dry air. With 480 lb/min (3.6 kg/s) of water to be evaporated in the dryer, each cubic foot of air will absorb $0.7947/13.3 = 0.005945$ lb (0.095 kg/m³), of moisture, and the total airflow must be (480 lb/min)/(0.005945) = 80,800 ft³/min (38.1 m³/s), given a dryer efficiency of 100 percent. However, the usual dryer efficiency is about 75 percent, not 100 percent. Therefore, the total actual airflow through the dryer should be $80,800/0.75 = 107,700$ ft³/min (50.8 m³/s).

Note: If desired, a table of moist air properties can be used instead of a psychrometric chart to determine the moisture content of the air at the dryer inlet and outlet conditions. The moisture content is read in the humidity ratio W_s column. See the ASHRAE—*Guide and Data Book* for such a tabulation of moist-air properties.

3. Compute the required air temperature
Assume that the heating air enters at a temperature t greater than 125°F (51.7°C). Set up a heat balance such that the heat given up by the air cooling from t to 125°F (51.7°C) = the heat required to evaporate the water on the coal + the heat required to raise the temperature of the coal and water from ambient to the evaporation temperature + radiation losses.

The heat given up by the air, Btu = (cfm)(density of air, lb/ft^3)[specific heat of air, Btu/lb·°F](t − evaporation temperature,°F). The heat required to evaporate the water, Btu = (weight of water, lb/min)(h_{fg} at evaporation temperature). The heat required to raise the temperature of the coal and water from ambient to the evaporation temperature, Btu = (weight of coal, lb/min)(evaporation temperature − ambient temperature)[specific heat of coal, Btu/(lb·°F)] + (weight of water, lb/min)(evaporation temperature − ambient temperature)[specific heat of water, Btu/(lb·°F)]. The heat required to make up for radiation losses, Btu = {(area of dryer insulated surfaces, ft^2)[heat-transfer coefficient, Btu/(ft^2·°F·h)](t − ambient temperature) + (area of dryer uninsulated surfaces, ft^2)[heat-transfer coefficient, Btu/(ft^2·°F·h)](t − ambient temperature)}/60.

Compute the heat given up by the air, Btu, as $(107,700)(0.075)(0.24)(t - 70)$, where 0.075 is the air density and 0.24 is the specific heat of air.

Compute the heat required to evaporate the water, Btu, as $(480)(1022.9)$, where $1022.9 = h_{fg}$ at 125°F (51.7°C) from the steam tables.

Compute the heat required to raise the temperature of the coal and water from ambient to the evaporation temperature, Btu, as $(6000)(t - 70)(0.30) + (480)(t - 70)(1.0)$, where 0.30 is the specific heat of the coal and 1.0 is the specific heat of water.

Compute the heat required to make up the radiation losses, assuming 3000 ft^2 (278.7 m^2) of insulated and 1500 ft^2 (139.4 m^2) of uninsulated surface in the dryer, with coefficients of heat transfer of 0.35 and 3.0 for the insulated and uninsulated surfaces, respectively. Then radiation heat loss, Btu = $(3000)(0.35)(t - 70) + (1500)(3.0)(t - 70)$.

Set up the heat balance thus and solve for t: $(107,7000)(0.075)(0.24)(t - 70) = (480)(1022.9) + (6000)(125 - 70)(0.30) + (480)(125 - 70)(1.0) + [(3000)(0.35)(t - 70) + (1500)(3.0)(t - 70)]/60$; so $t = 406°F$ (207.8°C). In this heat balance, the factor 60 is divided into the radiation heat loss to convert flow in Btu/h to Btu/min because all the other expressions are in Btu/min.

4. Determine the total heat required by the dryer
Using the equation of step 3 with $t = 406°F$ (207.8°C), we find the total heat = $(107,770)(0.075)(0.24)(406 - 70) = 651,000$ Btu/min, or $60(651,000) = 39,060,000$ Btu/h (11,439.7 kW)

5. Compute the dryer-furnace grate area
Assume that heat for the dryer is produced from coal having a lower heating value of 13,000 Btu/lb (30,238 kJ/kg) and that 40 lb/h of coal is burned per square foot [0.05 kg/(m^2·s)] of grate area with a combustion efficiency of 70 percent.

The rate of coal firing = (Btu/min to dryer)/(coal heating value, Btu/lb)(combustion efficiency) = $651,000/(13,000)(0.70) = 71.5$ lb/min, or $60(71.5)$

= 4990 lb/h (0.63 kg/s). Grate area = 4990/40 = 124.75 ft², say 125 ft² (11.6 m²).

6. Compute the dryer-furnace volume
The usual heat-release rates for dryer furnaces are about 50,000 Btu/(h·ft³) (517.5 kW/m³) of furnace volume. For this furnace, which burns 4900 lb/h (0.63 kg/s) of 13,000-Btu/lb (30,238-kJ/kg) coal, the total heat released is 4990(13,000) = 64,870,000 Btu/h (18,998.8 kW). With an allowable heat release of 50,000 Btu/(h·ft³) (517.1 kW/m³), the required furnace volume = 64,870,000/50,000 = 1297.4 ft³, say 1300 ft³ (36.8 m³).

Related Calculations. The general procedure given here can be used for any air-heated dryer used to dry moist materials. Thus, the procedure is applicable to chemical, soil, and fertilizer drying, as well as coal drying. In each case, the specific heat of the material dried must be used in place of the specific heat of coal given above.

COAL STORAGE CAPACITY OF PILES AND BUNKERS

Bituminous coal is stored in a 25-ft (7.5-m) high, 68.8-ft (21.0-m) diameter, circular-base conical pile. How many tons of coal does the pile contain if its base angle is 36°? How much bituminous coal is contained in a 25-ft (7.5-m) high rectangular pile 100 ft (30.5 m) long if the pile cross section is a triangle having a 36° base angle?

Calculation Procedure:

1. Sketch the coal pile
Figure 9a and b shows the two coal piles. Indicate the pertinent dimesions—height, the diameter, length, and base angle—on each sketch.

2. Compute the volume of the coal pile
Volume of a right circular cone, ft³ = $\pi r^2 h/3$, where r = radius, ft; h = cone height, ft. Volume of a triangular pile = $bal/2$, where b = base length, ft; a = altitude, ft; l = length of pile, ft.

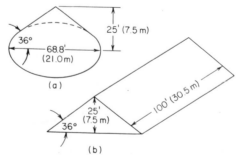

FIGURE 9 (*a*) Conical coal pile; (*b*) triangular coal pile.

For this conical pile, volume $= \pi(3.4)^2(25)/3 = 31{,}000$ ft^3 (877.8 m^3). Since 50 lb of bituminous coal occupies about 1 ft^3 of volume (800.9 kg/m^3), the weight of coal in the conical pile $= (31{,}000$ ft$^3)(50$ lb/ft$^3) = 1{,}550{,}000$ lb, or $(1{,}550{,}000$ lb)/(2000 lb/ton) $= 775$ tons (703.1 t).

For the triangular pile, base length $= 2h/\tan 36° = (2)(25)/0.727 = 68.8$ ft (21.0 m). Then volume $= (68.8)(25)(100)/2 = 86{,}000$ ft^3 (2435.2 m^3). The weight of bituminous coal in the pile is, as for the conical pile, $(86{,}000)(50) = 4{,}300{,}000$ lb, or $(4{,}300{,}000$ lb)/(2000 lb/ton) $= 2150$ tons (1950.4 t).

Related Calculations. Use this general procedure to compute the weight of coal in piles of all shapes, and in bunkers, silos, bins, and similar storage compartments. The procedure can be used for other materials also—grain, sand, gravel, coke, *etc.* Be sure to use the correct density when converting the total storage volume to total weight. Refer to Baumeister and Marks—*Standard Handbook for Mechanical Engineers* for a comprehensive tabulation of the densities of various materials.

PROPERTIES OF A MIXTURE OF GASES

A 10-ft^3 (0.3-m^3) tank holds 1 lb (0.5 kg) of hydrogen (H$_2$), 2 lb (0.9 kg) of nitrogen (N$_2$), and 3 lb (1.4 kg) of carbon dioxide (CO$_2$) at 70°F (21.1°C). Find the specific volume, pressure, specific enthalpy, internal energy, and specific entropy of the individual gases and of the mixture and the mixture density. Use Avogadro's and Dalton's laws and Keenan and Kaye—*Thermodynamic Properties of Air, Products of Combustion and Component Gases,* Krieger, commonly termed the *Gas Tables.*

Calculation Procedure:

1. Compute the specific volume of each gas
Using H, N, and C as subscripts for the respective gases, we see that the specific volume of any gas v ft^3/lb $=$ total volume of tank, ft^3 weight of gas in tank, lb. Thus, $v_H = 10/1 = 10$ ft^3/lb (0.6 m^3/kg); $v_N = 10/2 = 5$ ft^3/lb (0.3 m^3/kg); $v_C = 10/3 = 3.33$ ft^3/lb (0.2 m^3/dg). Then the specific volume of the mixture of gases is v, ft^3/lb $=$ total volume of gas in tank, ft^3/sum of weight of individual gases, lb $= 10/(1 + 2 + 3) = 1.667$ ft^3/lb (0.1 m^3/kg).

2. Determine the absolute pressure of each gas
Using $P = RTw/v_tM$, where $P =$ absolute pressure of the gas, lb/ft^2 (abs); $R =$ universal gas constant $= 1545$; $T =$ absolute temperature of the gas,°R $=$ °F $+ 459.9$, usually taken as 460; $w =$ weight of gas in the tank, lb; $v_t =$ total volume of the gas in the tank, ft^3; $M =$ molecular weight of gas. Thus, $P_H = (1545)(70 + 460)(1.0)/[(10)(2.0)] = 40{,}530$ lb/ft^2 (abs) (1940.6 kPa); $P_N = (1545)(70 + 460)(2.0)/[(10)(28)] = 5850$ lb/ft^2 (abs) (280.1 kPa); $P_C = (1545)(70 + 460)(3.0)/[(10)(44)] = 5583$ lb/ft^2 (abs) (267.3 kPa); $P_t = \Sigma P_H$, P_N, $P_C = 40{,}530 + 5850 + 5583 = 51{,}963$ lb/ft^2 (abs) (2488.0 kPa).

3. Determine the specific enthalpy of each gas
Refer to the *Gas Tables,* entering the left-hand column of the table at the absolute temperature, 530°F (294 K), for the gas being considered. Opposite the temperature, read the specific enthalpy in the h column. Thus, $h_H = 1796.1$ Btu/lb (4177.7 kJ/kg); $h_N = 131.4$ Btu/lb (305.6 kJ/kg); $h_C = 90.17$ Btu/lb (209.7 kJ/kg). The total

enthalpy of the mixture of the gases is the sum of the products of the weight of each gas and its specific enthalpy, or $(1)(1796.1) + (2)(131.4) + (3)(90.17) + 2329.4$ Btu (2457.6 kJ) for the 6 lb (2.7 kg) or 10 ft^3 (0.28 m^3) of gas. The specific enthalpy of the mixture is the total enthalpy/gas weight, lb, or $2329.4/(1 + 2 + 3) = 388.2$ Btu/lb (903.0 kJ/kg) of gas mixture.

4. Determine the internal energy of each gas
Using the *Gas Tables* as in step 3, we find $E_H = 1260.0$ Btu/lb (2930.8 kJ/kg); $E_N = 93.8$ Btu/lb (218.2 kJ/kg); $E_C = 66.3$ Btu/lb (154.2 kJ/kg). The total energy $= (1)(1260.0) + (2)(93.8) + (3)(66.3) = 1646.5$ Btu (1737.2 kJ). The specific enthalpy of the mixture $= 1646.5/(1 + 2 + 3) = 274.4$ Btu/lb (638.3 kJ/kg) of gas mixture.

5. Determine the specific entropy of each gas
Using the *Gas Tables* as in step 3, we get $S_H = 15.52$ Btu/(lb·°F) [65.0 kJ/(kg·°C)]; $S_N = 1.558$ Btu/(lb·°F) [4.7 kJ/(kg·°C)]. The entropy of the mixture $= (1)(12.52) + (2)(1.558) + (3)(1.114) = 18.978$ Btu/°F (34.2 kJ/°C). The specific entropy of the mixture $= 18.978/(1 + 2 + 3) = 3.163$ Btu/(lb·°F) [13.2 kJ/(kg·°C) of the gas mixture.

6. Compute the density of the mixture
For any gas, the total density d_t = sum of the densities of he individual gases. And since density of a gas = 1/specific volume, $d_t = 1/v_t = 1/v_H + 1/v_N + 1/v_C = 1/10 + 1/5 + 1/3.33 = 0.6$ lb/ft^3 (9.6 kg/m^3) of mixture. This checks with step 1, where $v_t = 1.667$ ft^3/lb (0.1 m^3/kg), and is based on the principle that all gases occupy the same volume.

    Related Calculations. Use this method for any gases stored in any type of container—steel, plastic, rubber, canvas, *etc.*—under any pressure from less than atmospheric to greater than atmospheric at any temperature.

STEAM INJECTION IN AIR SUPPLY

In a certain manufacturing process, a mixture of air and steam at a total mixture pressure of 300 lb/in^2 absolute (2068 kPa) and 400°F (204°C) is desired. The relative humidity of the mixture is to be 60 percent. For a required mixture flow rate of 500 lb/h (3.78 kg/s) determine (*a*) the volume flow rate of dry air in ft^3/min (m^3/s) of free air, where air is understood to be air at 14.7 lb/in^2 (101 kPa) and 70°F (21°C); and (*b*) the required rate of steam injection in lb/h (kg/s).

Calculation Procedure:

1. Determine the partial pressure of the vapor and that of the air
From Table 1, Saturation: Temperatures of the Steam Tables mentioned under Related Calculations of this procedure, at 400°F (204°C) the steam saturation pressure, $P_{vs} = 247.31$ lb/in^2 (1705 kPa), by interpolation. Since the vapor pressure is approximately proportional to the grains of moisture in the mixture, the partial pressure of vapor in the mixture, $P_{vp} = \phi P_{vs} = 0.6 \times 247.31 = 148.4$ lb/in^2 absolute (1023 kPa), where ϕ is the relative humidity as a decimal. Then, the partial pressure

of the air in the mixture, $P_a = P_m - P_{vs} = 300 - 148.4 = 151.6$ lb/in² absolute (1045 kPa), where P_m is the total mixture pressure.

2. Compute the density of air in the mixture

The air density, $\rho_a = P_a/(R_a T_a)$, where $P_a = 151.6 \times 144 = 21.83 \times 10^3$ lb/ft² (1045 kPa); the gas constant for air, $R_a = 53.3$ ft·lb/(lb·°R) [287 J/(kg·K)]; absolute temperature of the air $T_a = 400 + 460 = 860°R$ (478 K), then, $\rho_a = 21.83 \times 10^3/(53.3 \times 860) = 0.4762$ lb/ft³ (7.63 kg/m³).

3. Find the specific volume of the vapor in the mixture

From Table 3, Vapor of the Steam Tables, at 148.4 lb/in² absolute (1023 kPa) and 400°F (204°C), the specific volume of the vapor, $v_v = 3.261$ ft³/lb (0.2036 m³/kg), by interpolation.

4. Compute the density of the vapor and that of the mixture

The density of the vapor, $\rho_v = 1/v_v = 1/3.261 = 0.3066$ lb/ft³ (4.91 kg/m³). The density of the mixture, $\rho_m = \rho_a + \rho_v = 0.4762 + 0.3066 = 0.7828$ lb/ft³ (12.54 kg/m³).

5. Compute the amount of air in 500 lb/h (3.78 kg/s) of mixture

In 500 lb/h (3.78 kg/s) of mixture, w_m, the amount of air, $w_a = \rho_a \times w_m/\rho_m = 0.4762 \times 500/0.7828 = 304$ lb/h (2.30 kg/s).

6. Compute the flow rate of dry air

(a) The flow rate of dry air at 14.7 lb/in² (101 kPa) and 70°F (21°C), $V_a = w_a \times R_a \times T/P$, where the free air temperature, $T = 70 + 460 = 530°R$ (294 K); free air pressure, $P = 14.7 \times 144 = 2.117 \times 10^3$ lb/ft² (101 kPa); other values as before. Hence, $V_a = 304 \times 53.3 \times 530/(2.117 \times 10^3) = 4060$ ft³/h = 67.67 ft³/min (1.92 × 10⁻³ m³/s).

7. Compute the rate of steam injection

(b) The rate of steam injection, $w_s = w_s - w_a = 500 - 304 = 196$ lb/h (1.48 kg/s).

Related Calculations. The Steam Tables appear in *Thermodynamic Properties of Water Including Vapor, Liquid, and Solid Phases*, 1969, Keenan, et al., John Wiley & Sons, Inc. This procedure considers the air and steam as ideal gases which behave in accordance with the Gibbs-Dalton law of gas mixtures having complete homogeneous molecular dispersion and additive pressures. Also, calculations in steps 2 and 6 are based on Boyle's law and Charles' law, which relate pressure, volume, and temperature of a gas, or gas mixture. Clear and concise presentations of these and other significant definitions appear in *Thermodynamics and Heat Power*, 4th edition, by Irving Granet, Regents/Prentice-Hall, Englewood Cliffs, NJ 07632.

BOILER AIR-HEATER ANALYSIS AND SELECTION

A boiler manufacturer proposes these two alternatives to a prospective purchaser: (a) A steam-generating unit equipped with a small air heater which results in an overall steam-generating efficiency of 83 percent; (b) A similar steam-generating unit equipped with a larger air heater which results in an overall steam-generating unit efficiency of 87 percent. It is anticipated that coal delivered to the furnace will cost $60.00/ton (0.907 metric ton). The boiler is intended to operate 8000 hours per year and is to deliver one million pounds of steam per hour (126 kg/s) with an enthalpy rise of 1200 Btu per pound mass (2.791×10^6 J/kg). If the total investment charges are 20 percent, what additional cost can be paid for the larger heater? Indicate the Btu content of coal upon which the selection is based.

Calculation Procedure:

1. Compute boiler heat output during one year of intended operation
Boiler heat output per year of operation, $Q = H \times S \times \Delta h$, where the time used per year, $H = 8000$ h (2.88×10^6 s); rate of steam delivery, $S = 10^6$ lb$_m$/h (126 kg/s); enthalpy rise, $\Delta h = 1200$ Btu/lb$_m$ (2.791×10^6 J/kg). Then, $Q = 8000 \times 1200 = 9.6 \times 10^{12}$ Btu/year (22.31×10^{15} J/year).

2. Compute the annual mass of coal input for each Proposal
Proposal (a): Annual coal input, $C_a = Q/(B \times E_a)$, where the assumed Btu content of coal, $B = 13,500$ Btu/lb$_m$ (31.4×10^6 J/kg); efficiency of Proposal (a), $E_a = 0.83$. Then, $C_a = 9.6 \times 10^{12}/(13,500 \times 0.83) = 711 \times 10^6/0.83 = 857 \times 10^6$ lb$_m$/year (389×10^6 kg/year).

Proposal (b): Annual coal input, $C_b = Q/(B \times E_b)$, where efficiency of Proposal (b), $E_b = 0.87$. Then, $C_b = 711 \times 10^6/0.87 = 817 \times 10^6$ lb$_m$/year (371×10^6 kg/year).

3. Compute the annual cost of coal for each Proposal
Proposal (a): Annual cost of coal, $A_a = (C_a/2000) \times C_t$, where the cost per ton (0.907 metric ton) of coal, $C_t = $60.00. Then, $A_a = (857 \times 10^6/2000) \times 60 = $25,710,000/year.

Proposal (b): Annual cost of coal, $A_b = (C_b/2000)C_t = (817 \times 10^6/2000) \times 60 = $24,510,000/year.

4. Compute the additional investment that can be made for the larger heater
Additional investment, L, for the larger heater can be found by setting proposed annual costs for Proposal (a) equal to those for Proposal (b) in a "Break Even" equation. Thus, $(d_c \times I_a) + O_m + A_a = [d_c \times (I_a + L)] + O_m + A_b$, where the decimal fraction for total investment charges, $d_c = 0.20$; total investment charges for Proposal (a) $= I_a$, and for Proposal (b) $= (I_a + L)$; operating and maintenance charges for either Proposal $= O_m$; other items as before. Then, $(0.20 \times I_a) + O_m + $25,710,000 = [0.20 \times (I_a + L)] + O_m + $24,510,000. This reduces to, $L = ($25,710,000 - $24,510,000)/0.20 = $6,000,000$, the additional investment.

Note: The $60-per-ton price for coal used here was for example purposes only. Since coal prices vary widely with source region, transport distance, and quality, a high price was used to highlight the importance of investment decisions while reflecting the effects of inflation and the possible demand for specialty coal. The procedural steps remain the same, regardless of the dollar, franc, yen, pound, or other monetary unit price of the coal.

EVALUATION OF BOILER BLOWDOWN, DEAERATION, STEAM AND WATER QUALITY

A boiler generates 50,000 lb/h (22,700 kg/h) of saturated steam at 300 lb/in^2 (abs) (2067 kPa), out of which 10,000 lb/h (4540 kg/h) is taken for process and is returned to the deaerator, Fig. 10, as condensate at 180°F (82.2°C); the remainder is consumed. Makeup water enters the deaerator at 70°F (21.1°C) and steam is available at 300 lb/in^2 (abs) (2067 kPa) for deaeration. The deaerator operates at 25 lb/in^2 (abs) (172.3 kPa). The blowdown has total dissolved solids (TDS) of 1500 ppm (parts per million by weight) and makeup has a TDS of 100 ppm. Evaluate the blowdown and deaeration steam quantities.

Calculation Procedure:

1. Understand steam quality and steam purity

Steam purity refers to the impurities in steam in ppm. A typical value in low-pressure boilers is 1 ppm. Steam quality, by contrast, refers to the moisture in steam.

For example, the operator of a boiler plant will maintain a certain concentration of solids in the boiler drum, depending on either the American Boiler Manufacturers Association or the American Society of Mechanical Engineers recommendations, which can be found in publications of these organizations or in Ganapathy—*Steam Plant Calculations Manual,* Marcel Dekker, Inc. At 500 lb/in^2 (abs) (3445 kPa), for instance, if the boiler water concentration is 2500 ppm and steam purity is 0.5 ppm solids, the steam quality is obtained from: Percent moisture in steam = (steam purity, ppm)/(boiler water concentration, ppm)(100) = (0.5/2500)(100) = 0.02 percent; steam quality = 100.0 − 0.02 = 99.98 percent.

2. Set up the deaerator mass and energy balance

From the Fig. 10, the mass balance around the deaerator gives, using the data provided: $10,000 + D + M = F = 50,000 + B$ [Eq. (1)], where D = deaeration steam flow, lb/h (kg/h); B = blowdown, lb/h (kg/h); M = makeup flow, lb/h (kg/h); F = feedwater flow, lb/h (kg/h).

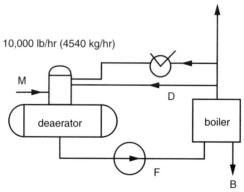

FIGURE 10 Steam plant containing deaerator and boiler blowdown.

From the energy balance around the deaerator, $100,000(148) + 1202.8(D) + M_x(38) = 209(F) = 209(50,000 + B)$, Eq. (2), where 148 Btu/lb 344.8 kJ/kg) = enthalpy of the condensate return; 1202.8 Btu/lb (2802.5 kK/kg) = enthalpy of saturated steam going into the drum for deaeration; 38 Btu/lb (88.5 kJ/kg) and 209 Btu/lb (486.9 kJ/kg) are the enthalpies of the makeup and feedwater. Enthalpy of the feedwater is computed at the deaerator operating pressure of 25 lb/in² (abs) (172.3 kPa), corresponding to a saturation temperature of 240°F (115.6°C).

A balance of the solids in the makeup and blowdown gives $100(M) = 1500(B)$, Eq. (3). In this relation we neglect the solids in the steam because they are extremely small in comparison to the solids in the makeup and the blowdown.

3. Find the blowdown, makeup, and feedwater flows for the plant
Solving the above three equations—(1), (2), and (3), we have, from (1): $D + M = 40,000 + B$, Eq. 4. Substituting Eq. (3) in Eq. (4), we have $D + 15(B) = 40,000 + B$; or $D + 14(B) = 40,000$, Eq. 5.

Substituting Eq. (5) and (3) in Eq. (2) and solving for B, we have $B = 2375$ lb/h (1078.3 kg/h); $D = 6750$ lb/h (3064.5 kg/h); $M = 35,625$ lb/h (16,173.8 kg/h); $F = 52,375$ lb/h (23,778.3 kg/h). If venting losses are considered, the engineer can add 1 percent to 3 percent to deaeration steam D.

Related Calculations. In any steam plant, when performing energy balance calculations, it is important to evaluate deaeration steam and blowdown water quantities. Interestingly, these are related to feedwater and makeup water quality and steam purity. These variables must be evaluated together and not in isolation.

Blowdown water can be flashed in a flash tank and the flash stream returned to the deaerator. This reduced the steam quantity required for deaeration. Another way to improve the performance of a deaerator is to preheat the deaerator makeup water before it enters the deaerator by using the blowdown water. Using methods similar to those in this procedure, you can study the effect of varying the amount of condensate returned on the amount of deaeration steam required.

This procedure is the work of V. Ganapathy, Heat Transfer Specialist, ABCO Industries, Inc.

HEAT-RATE IMPROVEMENT USING TURBINE-DRIVEN BOILER FANS

What is the net heat-rate improvement and net kilowatt gain in a steam power plant having a main generating unit rated at 870,000 kW at 2.5 in (6.35 cm) HgA, 0 percent makeup with motor-driven fans if turbine-driven fans are substituted? Plant data are as follows: (a) tandem-compound turbine, four-flow, 3600-r/min 33.5 in (85.1 cm) last-stage buckets with 264-ft² (24.5-m²) total last-stage annulus area; (b) steam conditions 3500 lb/in² (gage) (24,133 kPa), 1000°F/1000°F (537.8°C/537.8°C); (c) with main-unit valves wide open, overpressure with motor-driven fans, generator output = 952,000 kW at 2.5 in (6.35 cm) HgA and 0 percent makeup; net heat rate = 7770 Btu/kWh (8197.4 kJ/kWh); (d) actual fan hp = 14,000(10,444 W) at valves wide open, overpressure with no flow or head margins; (e) motor efficiency = 93 percent; transmission efficiency = 98 percent; inlet-valve efficiency = 88 percent; total drive efficiency = 80 percent; difference between the example drive efficiency and base drive efficiency = $80 - 76.7 = 3.3$ percent.

Calculation Procedure:

1. *Determine the percentage increase in net kilowatt output when turbine-driven fans are used*
Enter Fig. 11 at 264-ft² (24.5-m²) annulus area end and 14,000 required fan horsepower, and read the increase as 3.6 percent. Hence, the net plant output increase = 34,272 kW (= 0.036 × 952,000).

2. *Compute the net heat improvement*
From Fig. 12, the net heat rate improvement = 0.31 percent. Or, 0.0031(7770) = 24 Btu (25.3 J).

3. *Determine the increase in the throttle and reheater steam flow*
From Fig. 13, the increase in the throttle and reheater flow is 3.1 percent. This is the additional boiler steam flow required for the turbine-driven fan cycle.

4. *Compute the net kilowatt gain and the net heat-rate improvement*
From Fig. 14 the multipliers for the 2.5 in (6.35 cm) HgA backpressure are 0.98 for net kilowatt gain and 0.91 for net heat rate. Hence, net kW gain = 34,272(0.98) = 33,587 kW, and net heat rate improvement = 24 × 0.91 = 22.0 Btu (23.0 J).

5. *Determine the overall cycle benefits*
From Fig. 15 the correction for a drive efficiency of 80 percent compared to the base of 76.7 percent is obtained. Enter the curve with 3.3 percent (= 80 − 76.7)

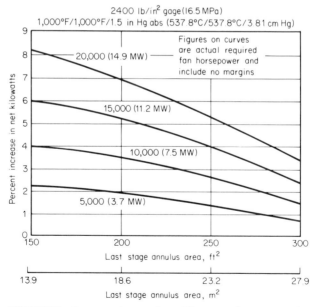

FIGURE 11   Percentage increase in net kilowatts vs. last stage annulus area for 2400-lb/in² (gage) when turbine-driven fans are used as compared to motors. (*Combustion.*)

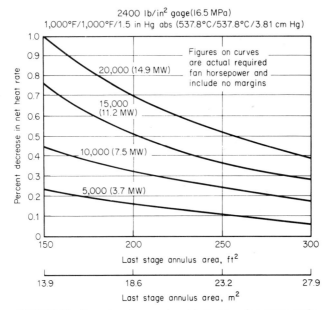

FIGURE 12 Percentage decrease in net heat rate vs. last-stage annulus area for 2400-lb/in² (gage) when turbine-driven fans are used as compared to motors. (*Combustion.*)

and read −6.6-Btu (−6.96-J) correction on the net heat rate −0.08 percent of generated kilowatts.

To determine the overall cycle benefits, add algebraically to the values obtained from step 4, or net kW gain = 33,587 + (−0.0008 × 952,000) = 32,825 kW; net heat-rate improvement = 22.1 + (−6.6) = 15.5 Btu (16.4 J).

Related Calculations. This calculation procedure can be used for any maximum-loaded main turbine in utility stations serving electric loads in metropolitan or rural areas. A maximum-loaded main turbine is one designed and sized for the maximum allowable steam flow through its last-stage annulus area.

Turbine-driven fans have been in operation in some plants for more than 10 years. Next to feed pumps, the boiler fans are the second largest consumer of auxiliary power in utility stations.

Current studies indicate that turbine-driven fans can be economic at 700 MW and above, and possibly as low as 500 MW. Although the turbine-driven fan system will have a higher initial capital cost when compared to a motor-driven fan system, the additional cost will be more than offset by the additional net output in kilowatts. In certain cases, economic studies may show that turbine drives for fans may be advantageous in constant-throttle-flow evaluations.

As power plants for utility use get larger, fan power required for boilers is increasing. Environmental factors such as use of SO_2 removal equipment are also increasing the required fan power. With these increased fan-power requirements, turbine drive will be the more economic arrangement for many large fossil plants. Further, these drives enable the plant designer to obtain a greater output from each unit of fuel input.

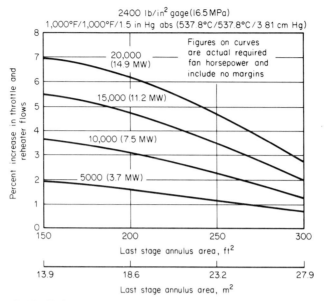

FIGURE 13   Percentage increase in throttle and reheater flows vs. last-stage annulus area for 2400-lb/in² (gage) when turbine-driven fans are used as compared to motors. (*Combustion.*)

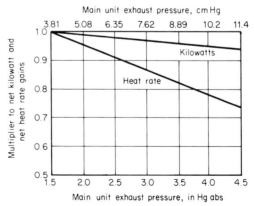

FIGURE 14   Multiplier to net kilowatt and net heat-rate gains to correct for main-unit exhaust pressure higher than 1.5 inHg (38.1 mmHg). (*Combustion.*)

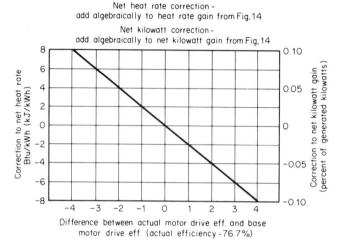

FIGURE 15   Corrections for differences in motor-drive system efficiency. (*Combustion.*)

This calculation procedure is based on the work of E. L. Williamson, J. C. Black, A. F. Destribats, and W. N. Iuliano, all of Southern Services, Inc., and F. A. Reed, General Electric Company, as reported in *Combustion* magazine and in a paper presented before the American Power Conference, Chicago.

BOILER FUEL CONVERSION FROM OIL OR GAS TO COAL

An industrial plant uses three 400,000-lb (50.4-kg/s) boilers fired by oil, a 600-MW generating unit, and two 400-MW units fired by oil. The high cost of oil, and the predictions that its cost will continue to rise in future years, led the plant owners to seek conversion of the boilers to coal firing. Outline the numerical and engineering design factors which must be considered in any such conversion.

Calculation Procedure:

1. *Evaluate the furnace size considerations*
The most important design consideration for a steam-generating unit is the fuel to be burned. Furnace size, fuel-burning and preparation equipment, heating-surface quantity and placement, heat-recovery equipment, and air-quality control devices are all fuel-dependent. Further, these items vary considerably among units, depending on the kind of fuel being used.

Figure 16 shows the difference in furnace size required between a coal-fired design boiler and an oil- or gas-fired design for the same steaming capacity in lb/h (kg/s). The major differences between coal firing and oil or natural-gas firing result from the solid form of coal prior to burning and the ash in the products of

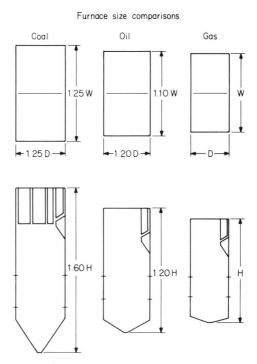

FIGURE 16   Furnace size comparisons. (*Combustion.*)

combustion. Oil produces only small amounts of ash; natural gas produces no ash. Coal must be stored, conveyed, and pulverized before being introduced into a furnace. Oil and gas require little preparation. For these reasons, a boiler designed to burn oil as its primary fuel makes a poor conversion candidate for coal firing.

2.   Evaluate the coal properties from various sources
Table 7 shows coal properties from many parts of the United States. Note that the heating values range from 12,000 Btu/lb (27,960 kJ/kg) to 6800 Btu/lb (15,844 kJ/kg). For a 600-MW unit, the coal firing rates [450,000 to 794,000 lb/h (56.7 or 99.9 kg/s)] to yield comparable heat inputs provide an appreciation of the coal storage yard and handling requirements for the various coals. On an hourly usage ratio alone, the lower-heating-value coal required 1.76 times more fuel to be handled.

Pulverizer requirements are shown in Table 8 while furnace sizes needed for the various coals are shown in Fig. 17.

3.   Evaluate conversion to coal fuel
Most gas-fired boilers can readily be converted to oil at reasonable cost. Little or no derating (reduction of steam or electricity output) is normally required.

From an industrial or utility view, conversion of oil- or gas-fired boilers not initially designed to fire coal is totally impractical from an economic viewpoint. Further, the output of the boiler would be severely reduced.

TABLE 7   Coal Properties—Nominal 600-MW Unit*

Type of coal	Eastern bituminous	Midwestern bituminous	Subbituminous C	Texas lignite	Northern plains lignite
HHV, Btu/lb (kJ/kg)	12,000 (27,912)	10,000 (23,260)	8,400 (19,538)	7,300 (16,980)	6,800 (15,817)
Moisture, %	6	12	27	32	37
lb $H_2O/10^6$ Btu (kg $H_2O/10^6$ kJ)	5 (0.00002)	12 (0.00005)	32 (0.000014)	44 (0.000019)	54 (0.000023)
Fuel fired, lb/h (kJ/h)	450,000 (202,500)	540,000 (243,000)	643,000 (289,350)	740,000 (333,000)	794,000 (357,300)

Combustion magazine.

TABLE 8   Pulverizer Requirements—Nominal 600-MW Unit*

	Eastern bituminous	Midwestern bituminous	Subbituminous	Texas lignite	Northern plains lignite
Hardgrove grindability	55	56	43	48	35
No. required	6	6	6	6	7
Nominal capacity†	50 tons/h (50.8 t/h)	63 tons/h (64 t/h)	85 tons/h (86.4 t/h)	92 tons/h (93.5 t/h)	100 tons/h (101.6 t/h)
Primary air temperature for drying coal	525°F (274°C)	640°F (338°C)	725°F (385°C)	750°F (399°C)	750°F (399°C)

Combustion magazine.
†Mill selection based on one full spare with remaining mills at 0.9 × new capacity.

For example, the overall plant site requirements for a typical station having a pair of 400-MW units designed to fire natural gas could be an area of 624,000 ft² (57,970 m²). This area would be for turbine bays, steam generators, and cooling towers. (With a condenser, the area required would be less.)

To accommodate the same facilities for a coal-fired plant with two 400-MW units, the ground area required would be 20 times greater. The additional facilities required include coal storage yard, ash disposal area, gas-cleaning equipment (scrubbers and precipitators), railroad siding, *etc.*

A coal-fired furnace is nominally twice the size of a gas-fired furnace. For some units the coal-fired boiler requires 4 times the volume of a gas-fired unit. Severe deratings of 40 to 70 percent are usually required for oil- and/or gas-fired boilers not originally designed for coal firing when they are switched to coal fuel. Further, such boilers cannot be economically converted to coal unless they were originally designed to be.

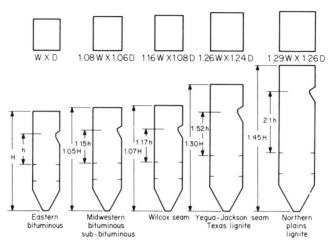

FIGURE 17   Furnace sizes needed for various coals for efficient operation. (*Combustion.*)

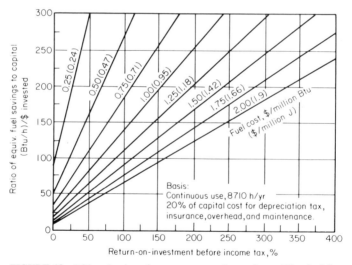

FIGURE 18   ROI evaluation of energy-conservation projects. (*Chemical Engineering.*)

As an example of the derating required, the 400,000-lb/h (50.4-kg/s) units considered here would have to be derated to 265,000 lb/h (33.4 kg/s) if converted to pulverized-coal firing. This is 66 percent of the original rating. If a spreader stoker were used to fire the boiler, the maximum capacity obtainable would be 200,000 lb/h (25.2 kg/s) of steam. Extensive physical alteration of the boiler would also be required. Thus, a spreader stoker would provide only 50 percent of the original steaming capacity.

Related Calculations.   Conversion of boilers from oil and/or gas firing to coal firing requires substantial capital investment, lengthy outage of the unit while alterations are being made, and derating of the boiler to about half the designed capacity. For these reasons, most engineers do not believe that conversion of oil-and/or gas-fired boilers to coal firing is economically feasible.

The types of boilers which are most readily convertible from oil or gas to coal are those which were originally designed to burn coal (termed *reconversion*). These are units which were mandated to convert to oil in the late 1960s because of environmental legislation.

Where the land originally used for coal storage was not sold or used for other purposes, the conversion problem is relatively minor. But if the land was sold or converted to other uses, there could be a difficult problem finding storage space for the coal.

Most of these units were designed to burn low-ash, low-moisture, high-heating-value, and high-ash-fusion coals. Fuels of this quality may no longer be available. Hence reconversion to coal firing may require significant downrating of the boiler.

Another important aspect of reconversion is the restoration of the coal storage, handling, and pulverizing equipment. This work will probably require considerable attention. Further, pulverizer capacity may not be sufficient, given the lower grade of fuel that would probably have to be burned.

This procedure is based on the work of C. L. Richards, Vice-President, Fossil Power Systems Engineering Research & Development, C-E Power Systems, Combustion Engineering, Inc., as reported in *Combustion* magazine.

To comply with environmental regulations, a number of coal-burning power plants have installed scrubbers ahead of the stack inlet to reduce sulfur dioxide emissions. Estimates show that some 22 tons/h of waste can be generated by scrubbers installed in the United States alone. This waste contains ash, limestone, and gypsum.

Research at Ohio State University is now directed at using scrubber waste to reclaim coal strip mines, fertilize farm soil by enriching it, and to create concretelike building materials.

When scrubber waste is used to treat soil from strip mines, the soil's acidity is reduced to a level where hardy grasses and alfalfa grow well. It is hoped that the barren sites of strip mines can be converted to useful fields using scrubber waste.

Grasses grown on such reclaimed sites are safe for animals to eat. Water leached from treated sites meets Environmental Protection Agency standards for agricultural use. Approval by EPA for use of scrubber waste at such sites is being sought.

Further experiments are being conducted on using the scrubber waste on acidic farmland. It will be used alone, or in combination with nutrient-rich sewer sludge. The third use for scrubber waste is as a sort of concrete for roads or the floors of feedlots.

Productive use of scrubber waste promises better control of the environment, reducing sulfur dioxide while recovering land that yielded the fuel that produced the CO_2.

ENERGY SAVINGS FROM REDUCED BOILER SCALE

A boiler generates 16,700 lb/h (2.1 kg/s) at 100 percent rating with an efficiency of 75 percent. If $\frac{1}{32}$ in (0.79 mm) of "normal" scale is allowed to form on the tubes, determine what savings can be made if 144,000-Btu/gal (40.133-MJ/m³) fuel oil costs $1 per gallon ($1 per 3.8 L) and the boiler uses 16.74 million Btu/h (4.9 MW) operating 8000 h/year.

Calculation Procedure:

1. *Determine the annual energy usage*
Compute the annual energy usage from (million Btu/h) (hours of operation annually)/efficiency. For this boiler, annual energy usage = (16.74)(8000)/0.75 = 178,560 million Btu (188,380 kJ).

2. *Find the energy loss caused by scale on the tubes*
Enter Fig. 19 at the scale thickness, ¹⁄₃₂ in (0.79 mm), and project vertically upward to the "normal" scale (salts of Ca and Mg) curve. At the left read the energy loss as 2 percent. Hence, the annual energy loss in heat units = (178,560 million Btu/year)(0.02) = 3571 million Btu/year (130.8 kW)

3. *Compute the annual savings if the scale is removed*
If the scale is removed, then the energy lost, computed in step 2, will be saved. Thus, the annual dollar savings after scale removal = (heat loss in energy units) (fuel price, $/gal)/(fuel heating value, Btu/gal). Or, savings = (3571 × 10⁶)($1.00)/144,000 = $26,049.

   Related Calculations. This approach can be used with any type of boiler—waterturbe, firetube, *etc*. The data are also applicable to tubed water heaters which are directly fired.

   Note that when the scale is high in iron and silica that the energy loss is much greater. Thus, with scale of the same thickness [¹⁄₃₂ in (0.79 mm)], the energy loss for scale high in iron and silica is 7 percent, from Fig. 19. Then the annual loss = 178,560(0.07) = 12,500 million Btu/year (3.63 MW). Removing the scale and preventing its reformation will save, assuming the same heating value and cost for the fuel oil, (12,500 × 10⁶ Btu/year) ($1.00)/144,000 = $86,805 per year.

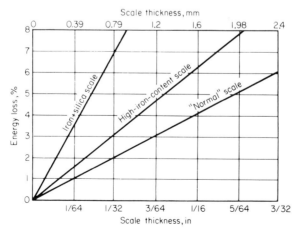

FIGURE 19   Effects of scale on boiler operation. (*Chemical Engineering.*)

While this calculation gives the energy savings from reduced boiler scale, the results also can be used to determine the amount that can be invested in a water-treatment system to prevent scale formation in a boiler, water heater, or other heat exchanger. Thus, the initial investment in treating equipment can at least equal the projected annual savings produced by the removal of scale.

This procedure is the work of Walter A. Hendrix and Guillermo H. Hoyos, Engineering Experiment Station, Georgia Institute of Technology, as reported in *Chemical Engineering* magazine.

GROUND AREA AND UNLOADING CAPACITY REQUIRED FOR COAL BURNING

An industrial plant is considering switching from oil to coal firing to reduce fuel costs. Determine the ground are required for 60 days' coal storage if the plant generates 100,000 lb/h (45,360 kg/h) of steam at a 60 percent winter load factor with a steam pressure of 150 lb/in^2 (gage) (1034 kPa), average boiler evaporation is 9.47 steam/lb coal (4.3 kg/kg), coal density = 50 lb/ft^3 (800 kg/m^3), boiler efficiency is 83 percent with an economizer, and the average storage pile height for the coal is 20 ft (6.096 m).

Calculation Procedure:

1. Determine the storage area required for the coal
The storage area, A ft^2, can be found from $A = 24WFN/EdH$, where H = steam generation rate, lb/h; F = load factor, expressed as a decimal; N = number of days storage required; E = average boiler evaporation rate, lb/h; d = density of coal, lb/ft^3; H = height of coal pile allowed, ft. Substituting yields A = 24(100,000)(0.6)(60)/[(9.47)(50)(20)] = 9123 ft^2 (847 m^2).

2. Find the maximum hourly burning rate of the boiler
The maximum hourly burning rate in tons per hour is given by $B = W/2000E$, where the symbols are as defined earlier. Substituting, we find B = 100,000/2000(9.47) = 5.28 tons/h (4.79 t/h). With 24-h use in any day, maximum daily use = 24 × 5.28 = 126.7 tons/day (115 t/day).

3. Find the required unloading rate for this plant
As a general rule, the unloading rate should be about 9 times the maximum total plant burning rate. Higher labor and demurrage costs justify higher unloading rates and less manual supervision of coal handling. Find the unloading rate in tons per hour from $U = 9W/2000E$, where the symbols are as defined earlier. Substituting gives U = 9(100,000)/2000(8.47) = 47.5 tons/h (43.1 t/h).

Related Calculations. With the price of oil, gas, wood, and waste fuels rising to ever-higher levels, coal is being given serious consideration by industrial, central-station, commercial, and marine plants. Factors which must be included in any study of conversion to (or original use of) coal include coal delivery to the plant, storage before use, and delivery to the boiler.

For land installations, coal is usually received in railroad hopper-bottom cars in net capacities ranging between 50 and 100 tons with 50- and 70-tons (45.4- and 63.5-t) capacity cars being most common.

Because cars require time for spotting and moving on the railroad siding, coal is actually delivered to storage for only a portion of the unloading time. Thawing of frozen coal and car shaking also tend to reduce the actual delivery. True unloading rate may be as low as 50 percent of the continuous-flow capacity of the handling system. Hence, the design coal-handling rate of the conveyor system serving the unloading station should be twice the desired unloading rate. So, for the installation considered in this procedure, the conveyor system should be designed to handle 2(47.5) = 95 tons/h (86.2 t/h). This will ensure that at least six rail cars of 60-ton (54.4-t) average capacity will be emptied in an 8-h shift, or about 360 tons/day (326.7 t/day).

With a maximum daily usage of 126.7 tons/day (115 t/day), as computed in step 2 above, the normal handling of coal, from rail car delivery during the day shift, will accumulate about 3 days' peak use during an 8-h shift. If larger than normal shipments arrive, the conveyor system can be operated more than 8 h/day to reduce demurrage charges.

This procedure is the work of E. R. Harris, Department Head, G. F. Connell, and F. Dengiz, all of the Environmental and Energy Systems, Argonaut Realty Division, General Motors Corporation, as reported in *Combustion* magazine.

HEAT RECOVERY FROM BOILER BLOWDOWN SYSTEMS

Determine the heat lost per day from sewering the blowdown from a 600-lb/in² (gage) (4137-kPa) boiler generating 1 million lb/day (18,939.4 kg/h) of steam at 80 percent efficiency. Compare this loss to the saving from heat recovery if the feedwater has 20 cycles of concentration (that is, 5 percent blowdown), ambient makeup water temperature is 70°F (21°C), flash tank operating pressure is 10 lb/in² (gage) (69 kPa) with 28 percent of the blowdown flashed, blowdown heat exchanger effluent temperature is 120°F (49°C), fuel cost is $2 per 10⁶ Btu [$2 per (9.5)⁶ J], and the piping is arranged as shown in Fig. 20.

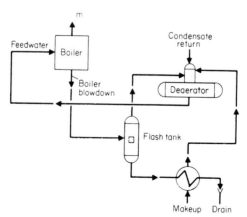

FIGURE 20 Typical blowdown heat-recovery system. (*Combustion.*)

Calculation Procedure:

1. Compute the feedwater flow rate

The feedwater flow rate, 10^6 lb/day = (steam generated, 10^6 lb/day)/(100 − blow-down percentage, or 10^6/(100 − 5) = 1.053 × 10^6 lb/day (0/48 × 10^6 kg/day).

2. Find the steam-production equivalent of the blowdown flow

The steam-production equivalent of the blowdown = feedwater flow rate − steam flow rate = 1.053 − 1.0 = 53,000 lb/day (24,090 kg/day).

3. Compute the heat loss per lb of blowdown

The heat loss per lb (kg) of blowdown = saturation temperature of boiler water − ambient temperature of makeup water. Or, heat loss = (488 − 70) = 418 Btu/lb (973.9 kJ/kg).

4. Find the total heat loss from sewering

When the blowdown is piped to a sewer (termed *sewering*), the heat in the blow-down stream is lost forever. With today's high cost of all fuels, the impact on plant economics can be significant. Thus, total heat loss from sewering = (heat loss per lb of blowdown) (blowdown rate, lb/day) = 418 Btu/lb (53,000 lb/day) = 22.2 × 10^6 Btu/day (23.4 × 10^6 J/day).

5. Determine the fuel-cost equivalent of the blowdown

The fuel-cost equivalent of the blowdown = (heat loss per day, 10^6 Btu)(fuel cost, $ per 10^6 Btu)/(boiler efficiency, %), or (22.2)(2)/0.8 = $55.50 per day.

6. Find the blowdown flow to the heat exchanger

With 28 percent of the blowdown flashed to steam, this means that 100 − 28 = 72 percent of the blowdown is available for use in the heat exchanger. Since the blowdown total flow rate is 53,000 lb/day (24,090 kg/day), the flow rate to the blowdown heat exchanger will be 0.72(53,000) = 38,160 lb/day (17,345 kg/day).

7. Determine the daily heat loss to the sewer

As Fig. 20 shows, the blowdown water which is not flashed, flows through the heat exchanger to heat the incoming makeup water and then is discharged to the sewer. It is the heat in this sewer discharge which is to be computed here.

   With a heat-exchanger effluent temperature of 120°F (49°C) and a makeup water temperature of 70°F (21°C), the heat loss to the sewer is 120 − 70 = 50 Btu/lb (116.5 kJ/kg). And since the flow rate to the sewer is 38,160 lb/day (17,345 kg/day), the total heat loss to the sewer is 50(38,160) = 1.91 × 10^6 Btu/day (2.02 × 10^6 kJ/day).

8. Compare the two systems in terms of heat recovered

The heat recovered − heat loss by sewering = heat loss with recovery = 22.2 × 10^6 Btu/day $\eth$ 1.91 × 10^6 Btu/day = 20.3 × 10^6 Btu/day (21.4 × 10^6 J/day).

9. Determine the percentage of the blowdown heat recovered and dollar savings

The percentage of heat recovered = (heat recovered, Btu/day)/(original loss, Btu/day) = 20.3/22.2)(100) = 91 percent. Since the cost of the lost heat was $55.50 per day without any heat recovery, the dollar savings will be 91 percent of this, or 0.91($55.50) = $50.51 per day, or $18,434.33 per year with 365 days of operation. And as fuel costs rise, which they are almost certain to do in future years, the annual saving will increase. Of course, the cost of the blowdown heat-recovery equipment must be offset against this saving. In general, the savings warrant the added investment for the extra equipment.

Related Calculations.   This procedure is valid for any type of steam-generating equipment for residential, commercial, industrial, central-station, or marine installations. (In the latter installation the "sewer" is the sea.) The typical range of blowdown heat recovery is in the 80 to 90 percent area. In view of the rapid rise in fuel prices, this range of heat recovery is significant. Hence, much wider use of blowdown heat recovery can be expected in all types of steam-generating plants.

To reduce scale buildup in boilers, low cycles of boiler water concentration are preferred. This means that high blowdown rates will be used. To prevent wasting expensive heat present in the blowdown, heat-recovery equipment such as that discussed above is used. In industrial plants (which are subject to many sources of condensate contamination), cycles of concentration are seldom allowed to exceed 50 (2 percent blowdown). In the above application, the cycles of concentration = 20, or 5 percent blowdown.

To prevent boiler scale buildup, good pretreatment of the makeup is recommended. Typical current selections for pretreatment equipment, by using the boiler operating pressure as the main criterion, are thus:

Boiler pressure, lb/in^2 (kPa)	Pretreatment equipment
0–600 (4137)	Sodium zeolite softening
600–900 (4137–6205)	Hot line/demineralizers
Above 900 (6205)	Demineralizers

This procedure is the work of A. A. Askew, Betz Laboratories, Inc., as reported in *Combustion* magazine.

BOILER BLOWDOWN PERCENTAGE

The allowable concentration in a certain drum is 2000 ppm. Pure condensate is fed to the drum at the rate of 85,000 gal/h (89.4 L/s). Make-up, containing 50 grains (gr)/gal (856 mg/L) of sludge-producing impurities, is also delivered to the drum at the rate of 1500 gal/h (1.58 L/s). Calculate the blowdown as a percentage of the boiler steaming capacity.

Calculation Procedure:

1. Compute the ppm of impurities per gallon of make-up water
There are 58,410 gr/gal (10^6 mg/L). See the Related Calculations of this procedure for the basis of this factor. Parts per million of impurities, $ppm_i = [(gr/gal)_i/(gr/gal)_i/(gr/gal)] \times 10^6$, where $(gr/gal)_i$ = quality of, or impurities in, the make-up water, gr/gal (mg/L); (gr/gal) = grains/gal (mg/L). Hence, $ppm_i = (50/58,410) \times 10^6 = 50 \times 17.12 = 856$.

2. Compute the blowdown rate
To maintain impurities in the drum at a certain concentration, the parts fed to the drum = parts discharged by the blowdown. Thus, $ppm_i = (gal/h)_m = ppm_a \times$

$(gal/h)_a$, where the subscripts stand for i = impurities; m = make-up; a = allowable; b = blowdown. Then, $856 \times 1500 = 2000 \times (gal/h)_b$. Solving: $(gal/h)_b = 856 \times 1500/2000 = 642$ (0.675 L/s).

3. Compute the boiler steaming capacity
The boiler steaming capacity $(gal/h)_s = (gal/h)_f + (gal/h)_m - (gal/h)_b$, where $(gal/h)_f$ = feedwater flow rate. Then, $(gal/h)_s = 85,000 + 1500 - 642 = 85,858$ (90.3 L/s).

4. Compute the blowdown percentage
Blowdown percentage = $[(gal/h)_b/(gal/h)_s] \times 100 = (642/85,858) \times 100 = 0.747$ percent.
   Related Calculations. The gr/gal factor in step 1 is based on the density of impurities being considered as equal to the maximum density of clean fresh water, 8.3443 lb/gal (1.0 kg/L). Since 1 lb = 7000 gr, then $8.3443 \times 7000 = 58,410$ gr/gal (10^6 mg/L).

SIZING FLASH TANKS TO CONSERVE ENERGY

Determine the dimensions required for a commercial flash tank if the flash tank pressure is 5 lb/in² (gage) (34.5 kPa) and 14,060 lb/h (1.77 kg/s) of flash steam is available. Would the flash tank be of the centrifugal or top-inlet type?

Calculation Procedure:

Two major types of flash tanks are in use today: top-inlet and centrifugal-inlet tanks, as shown in Fig. 21. Tank and overall height and outside diameter are also shown in Fig. 21.

1. Determine the rating and type of flash tank required
Refer to Table 9. Locate the 5-lb/in² (gage) (34.5-kPa) flash tank pressure column, and project downward to the minimum value that exceeds 14,060 lb/h (1.77 kg/s). Note that a no. 5 centrifugal flash tank with a maximum rating of 20,000 lb/h (2.5 kg/s) of flash steam is appropriate, and no standard top-inlet type has sufficient capacity at this pressure for this flow rate.

2. Determine the dimensions of the tank
In Table 9 locate tank no. 5, and read the dimensions horizontally to the right. Hence, the dimensions required for the tank are 60-in (152.4-cm) OD, 78-in (198.1-cm) tank height, 88-in (223.5-cm) overall height, inlet pipe size of 6 in (15.2 cm), steam outlet pipe of 8 in (20.3 cm), and a water outlet pipe of 6 in (15.2 cm).
   Related Calculations. Use this procedure for choosing a flash tank for a variety of applications—industrial power plants, central stations, marine steam plants, and nuclear stations. Flash tanks can conserve energy by recovering steam that might otherwise be wasted. This steam can be used for space heating, feedwater heating, industrial processes, *etc.* Condensate remaining after the flashing can be used as boiler feedwater because it is usually pure and contains valuable heat. Or the con-

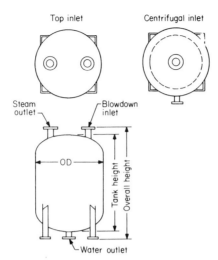

FIGURE 21 Centrifugal and top-inlet flash-tank dimensions. (*Chemical Engineering.*)

densate may be used in an industrial process requiring pure water at an elevated temperature.

Flashing steam can cause a violent eruption of the liquid from which the steam is formed. Hence, any flash tank must be large enough to act as a separator to remove entrained moisture from the steam. The dimensions given in Table 9 are for flash tanks of proven design. Hence, the values obtained from Table 9 are satisfactory for all normal design activities. The procedure given here is the work of T. R. MacMillan, as reported in *Chemical Engineering.*

FLASH TANK OUTPUT

A boiler operating with a drum pressure of 1400 lb/in² absolute (9650 kPa) delivers 200,000 lb (90,720 kg) of steam per hour and has a continuous blowdown of 2 percent of its output in order to keep the boiler water at proper dissolved solids. The water blowdown passes to a flash tank operating at slightly above atmospheric pressure in which part of the water flashes to steam, which in turn passes to an open feedwater heater. How much steam is flashed per hour?

Calculation Procedure:

1. Determine the amount of blowdown
Amount of blowdown $B = 0.02 D$, where D is the steam delivery. Hence, $B = 0.02 \times 200,000 = 4000$ lb/h (30 kg/s).

2. Find the enthalpy of the blowdown-saturated liquid
Blowdown water leaves the boiler at point d in Fig. 22a as saturated liquid, point d in Fig. 22b. Blowdown at a pressure of $p_d = 1400$ lb/in² (9650 kPa) has, from

TABLE 9 Maximum Ratings for Centrifugal and Top-Inlet Flash Tanks, 1000 lb/h (1000 kg/s)*

Tank no.	Flash-tank pressure, lb/in² (gage) (kPa)					
	1 (6.9)	5 (34.5)	10 (69.0)	20 (138.0)	50 (345.0)	100 (690.0)
	Centrifugal flash tanks					
4	6.0 (0.76)	7.1 (0.89)	8.8 (1.11)	12.0 (1.51)	21.0 (2.64)	34.0 (4.28)
5	16.0 (2.01)	20.0 (2.52)	24.0 (3.02)	32.0 (4.03)	58.0 (7.30)	100.0 (12.59)
6	27.0 (3.40)	34.0 (4.28)	42.0 (5.29)	58.0 (7.30)	105.0 (13.22)	180.0 (22.66)
	Top-inlet flash tanks					
2	1.1 (0.14)	1.3 (0.16)	1.7 (0.21)	2.2 (0.28)	4.0 (0.50)	6.90 (0.87)
3	2.2 (0.28)	2.9 (0.37)	3.5 (0.44)	4.9 (0.62)	8.7 (1.10)	14.80 (1.86)
4	4.3 (9.54)	5.2 (0.65)	6.5 (0.82)	8.7 (1.10)	15.0 (1.89)	25.0 (3.15)

Dimensions of commercial flash tanks

Tank no.	Outside diameter		Tank height		Overall height		Inlet pipe		Outlet pipe			
									Steam		Water	
	in	cm	in	cm	in	cm	in	cm	in	cm	in	cm
	Centrifugal flash tanks											
4	48	121.9	67	170.2	77	195.6	4	10.2	6	15.2	4	10.2
5	60	152.4	78	198.1	88	223.5	6	15.2	8	20.3	6	15.2
6	72	182.9	89	226.1	99	251.5	8	20.3	10	25.4	6	15.2
	Top-inlet flash tanks											
2	24	60.9	56	142.2	65.5	166.4	3	7.6	3	7.6	1.5	3.8
3	36	91.4	62	157.5	71.5	181.6	4	10.2	4	10.2	2	5.1
4	48	121.9	67	170.2	76.5	194.3	6	15.2	6	15.2	4	10.2

*Chemical Engineering magazine.

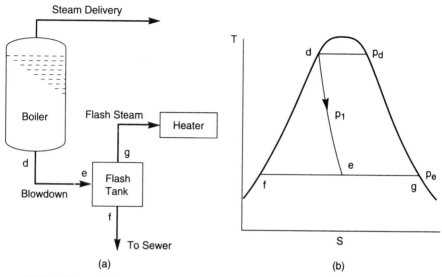

FIGURE 22 (*a*) Boiler blowdown flow diagram. (*b*) Temperature-entropy schematic for blow-down.

saturated steam tables mentioned under Related Calculations, an enthalpy $h_d = 598.7$ Btu/lb = (1392 kJ/kg).

3. Find the enthalpy of the blowdown fluid at the flash tank
The blowdown fluid is assumed to undergo an isenthalpic, or constant-enthalpy, throttling process from point *d* to the point *e* on Fig. 22*b* where, at the flash tank, $h_e = h_d$, found above.

4. Find the enthalpy of saturated liquid within the flash tank
From the saturated steam tables, at $p_e = 15$ lb/in² (103 kPa), slightly above atmospheric pressure, the enthalpy of the saturated liquid at point *f* on Fig. 22*b*, $h_f = 181.1$ Btu/lb (421 kJ/kg).

5. Find the enthalpy of evaporation within the flash tank
From the saturated steam tables, the heat required to evaporate 1 lb (0.45 kg) of water under the pressure p_e within the flash tank is $h_g - h_f = h_{fg} = 969.7$ Btu/lb (2254 kJ/kg).

6. Calculate the amount of steam flashed per hour
The tank flashes steam at the rate of $F = B[(h_e - h_f)] = 4000[598.7 - 181.1]/969.7] = 1723$ lb/h (13 kg/s).

Related Calculations. Saturation steam tables appear in *Thermodynamic Properties of Water Including Vapor, Liquid, and Solid Phases.* 1969, John Wiley & Sons, Inc.

The equation for F in step 6 stems from the presumption of an adiabatic heat balance where $F \times h_{fg} = B(h_e - h_f)$.

DETERMINING WASTE-HEAT BOILER FUEL SAVINGS

An industrial plant has 3000 standard ft³/min (1.42 m³/day) of waste gas at 1500°F (816°C) available. How much steam can be generated by this waste gas if the waste-heat boiler has an efficiency of 85 percent, the specific heat of the gas is 0.0178 Btu/(standard ft³ · °F) (1.19 kJ/cm²), the exit gas temperature is 400°F (204°C), and the enthalpy of vaporization of the steam to be generated is 970.3 Btu/lb (2256.9 kJ/kg)? What fuel savings will be obtained if the plant burns no. 6 fuel oil having a heating value of 140,000 Btu/gal (39,200 kJ/L) and a current cost of $1.00 per gallon ($1 per 3.785 L) and a future cost of $1.35 per gallon ($1.35 per 3.785 L)? The waste-heat boiler is expected to operate 24 h/day, 330 days/year. Efficiency of fuel boilers in this plant is 80 percent.

Calculation Procedure:

1. Compute the steam production rate from the waste heat
Use the relation $S = C_v V(T - t)60E/h_v$, where S = steam production rate, lb/h; C_v = specific heat of gas, Btu/(standard ft³ · °F); V = volumetric flow rate of waste gas, standard ft³/min; T = waste-gas temperature at boiler exit,°F; E = waste-heat boiler efficiency, expressed as a decimal; h_v = heat of vaporization of the steam being generated by the waste gas, Btu/lb. Substituting gives S = 0.0178(3000)(1500 − 400)60(0.85)/970.3 = 3087.7 lb/h (1403.3 kg/h).

2. Find the present and future fuel savings potential
The cost equivalent C dollars per hour of the savings produced by using the waste-heat gas can be found from $C = Sh_v K/E_b$, where the symbols are as given earlier and K = fuel cost, $ per Btu as fired ($ per 1.055 kJ), E_b = efficiency of fuel-fired boilers in the plant. Substituting for the current fuel cost of $1 per gallon, we find C = 3087.4(970.3)($1/140,000)/0.8 = $26.75. Since the waste-heat boiler will operate 24 h/day, the daily savings will be 24($26.75) = $642. With 330-days/year operation, the annual savings is (330 days)($642 per day) = $211,860. This saving could be used to finance the investment in the waste-heat boiler.

Where the exit gas temperature from the waste-heat boiler will be different from 400°F (204.4°C), adjust the steam output and dollar savings by using the difference in the equation in step 1.

Related Calculations. This procedure can be used for finding the savings possible from recovering heat from a variety of gas streams such as diesel-engine and gas-turbine exhausts, process-gas streams, refinery equipment exhausts, *etc.* To apply the procedure, several factors must be known or assumed: waste-heat boiler steam pressure, feedwater temperature, final exit gas temperature, heating value of fuel being saved, and operating efficiency of the waste-heat and fuel-fired boilers in the plant. Note that the exit gas temperature must be higher than the saturation temperature of the steam generated in the waste-heat boiler for heat transmission between the waste gas and the water in the boiler to occur.

As a guide, the exit gas temperature should be 100°F (51.1°C) above the steam temperature in the waste-heat boiler. For economic reasons, the temperature difference should be at least 150°F (76.6°C). Otherwise, the amount of heat transfer area required in the waste-heat boiler will make the investment uneconomical.

This procedure is the work of George V. Vosseller, P. E. Toltz, King, Durvall, Anderson and Associates, Inc., as reported in *Chemical Engineering* magazine.

FIGURING FLUE-GAS REYNOLDS NUMBER BY SHORTCUTS

A low-sulfur No. 2 distillate fuel oil has a chemical composition of 87.4 percent carbon and 12.6 percent hydrogen by weight, ignoring sulfur. The fuel's higher heating value (HHV), or $\Delta H_{\mathrm{gross}}$, the standard (60°F) (15.5°C) heat of combustion (based on stoichiometric air usage), is 18,993 Btu/lb (44.148 kJ/kg) of the fuel. With a volumetric proportion of 79 percent atmospheric nitrogen, including rare gases, to 21 percent oxygen, the molar ratio of N_2 to O_2 in air is 3.76:1. What is the Reynolds number for the flow of the flue gas produced by that fuel if it is completely burned in 50 percent excess air at the rate of 25.3 lb/h (11.5 kg/h) and the flue gas leaves a 1-ft (0.3-m) diameter duct at 2000°F (1093°C).

Calculation Procedure:

1. *Compute the volume flow rate of the flue gas*
Based on stoichiometric air usage, the standard (32°F, 1-atm) (0°C, 101.3-kPa) volume of flue gas (std ft³/lb) (std m³/kg) of fuel burned is $V_{\mathrm{std}} = (\Delta H_{\mathrm{gross}}/100)[1 +$ (percent excess air)/100 percent]. Then, $V_{\mathrm{std}} = (18,993/100)[1 + (50/100)] = 285$ std ft³/lb (17.79 std m³/kg).

Adjust for temperature expansion by using the ideal-gas law to get the per lb (kg) of fuel actual amount of flue gas, $V' = V_{\mathrm{std}}(460 + T)/(460 + 32)$, where T is the flue-gas temperature in°F. Thus, $V' = 285(460 + 2000)/(492) = 1425$ ft³/lb (89 m³/kg). The metric result can be verified as follows: $V_{\mathrm{std}} = 17.79[273 + (2000 - 32)/1.8]/273 = 89$ m³/kg.

The flue-gas approximate flow rate, $V = V'W/3600$, where W is the given hourly burning rate of the fuel. Hence, $V = 1425 \times 25.3/3600 = 10.0$ ft³/s (0.28 m³/s).

2. *Determine the viscosity of the flue gas*
Boiler and incinerator flue gases are composed of several gases, hence a precise calculation of the Reynolds number can be cumbersome. By assuming that the flue gas behaves like nitrogen, it is possible to obtain fast and accurate preliminary approximations for both boilers and incinerators. Then, by means of a graph in Fig. 23, read the dynamic viscosity of nitrogen as $\mu' = 0.054$ cp (54×10^{-6} Pa).

3. *Compute the Reynolds number of the flue gas*
By algebraic manipulations, as mentioned under Related Calculations of this procedure, the shortcut formula for an estimate of the Reynolds number is found to be $R_e = 73,700/[D\mu'(460 + T)]$, where the given duct diameter, $D = 1.0$ ft (0.3 m) and the other values are as previously determined. Thus, $R_e = 72,700 \times 10.0[(1)(0.054)(460 + 2000)] = 5470$.

Related Calculations.   The shortcut formula for the value of R_e is derived by algebraic manipulation of four expressions listed below. Symbols adequately defined previously are not redefined.

(1) Reynolds number $R_e = D\rho U/\mu$, where D = diameter of a circular cross-section, or equivalent diameter of some other cross-section, ft (m); ρ = density of the gas, lb/ft³ (kg/m³); U = average linear velocity of the gas, ft/s (m/s); μ = viscosity of the gas, lb/(ft · s) (Pa · s).

(2) At high temperatures, the average densities and viscosities of flue gas closely approximate those of nitrogen alone. Thus, ρ and μ for nitrogen can be used to

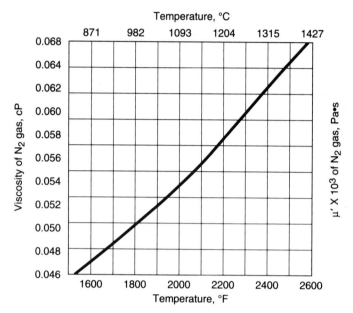

FIGURE 23   Dynamic viscosity of nitrogen gas. (*Chemical Engineering.*)

estimate R_e with no significant error. Hence, from the ideal-gas law an estimate of the density of nitrogen, $\rho = (28)(460 + 32)/[(359)(460 + T)]$, where 28 is the molecular weight of nitrogen; 359 is the volume, ft^3, of 1 lb·mol of gas at 32°F (0°C) and at atmospheric pressure, 14.7 lb/in^2 (6.89 kPa). In SI units the factor is 22.41 m^3/kg·mol.

  (3) $U = 4V/\pi D^2$

  (4) $\mu = \mu'/1488$

Turbulent flow in a boiler or incinerator assures adequate mixing and near-complete or complete combustion. Flow is turbulent when its Reynolds number is greater than 2000 or 3000, and is more turbulent when R_e is greater. This reduces the amount of excess air required for complete combustion and hence increases boiler and incinerator efficiency.

A review of the shortcut equations reveals that, for a given set of values for D, M, and T and that μ' depends on T, the degree of accuracy of the shortcut value

TABLE 10   Flue-gas components

Component	lb·mol/ lb$_m$ oil		Molar proportions		Molecular weight		Weight proportion
CO_2	0.0728/0.7763	=	0.094	×	44	=	4.14
H_2O	0.0630		0.081		18		1.46
O_2	0.0522		0.067		32		2.14
N_2	0.5883		0.758		28		21.22
Flue gas	0.7763		1.000		28.96		28.96

of R_e is reflected by the precision of the value of V_{std} found in step 1 of this procedure. This can be done by checking the stoichiometry of the combustion process per lb (kg) of the fuel with the given percentages of C and H_2, as follows: lb·mol C/lb oil $= 0.87/12 = 0.0728$; lb·mol H_2/lb $= 0.126/2 = 0.0630$.

Then, 0.0728 C $+ 0.0630$ H_2 $+ [1 +$ (percent excess air/100 percent] $(0.0728 + 0.0630/s)$ O_2 $+ [1 +$ (percent excess air/100 percent](3.76)(0.0728 $+ 0.0630/s)$ N_2 $\rightarrow 0.0728$ CO_2 $+ 0.0630$ H_2O $+$ (percent excess air/100 percent)(0.1043) O_2 $+ [1 +$ (percent excess air/100 percent)](3.76)(0.1043) N_2. Table 11 indicates that the lb·mol/lb oil is 0.7763; hence $V_{std} = 0.7763 \times 359 = 279$ std ft³/lb (17.42 std m³/kg) of oil. This shows the error in the shortcut estimate for V_{std} to be about 2 percent in this case.

Also, Table 10 shows that this flue gas has a composition of 9.4 percent CO_2; 8.1 percent H_2O; 6.7 percent O_2; 75.8 percent N_2; and has an average molecular weight of 28.96. By the method shown on page 3.279 in *Perry's Chemical Engineers' Handbook,* 6th edition, McGraw-Hill, this flue gas has a calculated mixture viscosity of $\mu' = 0.0536$ cP (53.6 $\times$ 10⁻⁶ Pa). Using this value and the average molecular weight of 28.96 instead of 28 in the gas density formula in step 2, the R_e estimate would be 5700. This indicates the shortcut estimate of 5470 to be in error by about 4 percent in this case. There are other factors that could contribute to errors in the shortcut calculations. In practice, wood and municipal sold waste contain considerable amounts of moisture, which reduces their heating values that refer to dry conditions, only. The shortcut calculations are very accurate for fossil fuels, such as coal, fuel oil, and natural gas, and wood. They are useful for wastes or waste-fuel mixtures. Errors by shortcut calculations seldom exceed ± 10 percent when excess air is less than 150 to 200 percent.

Though errors for fossil fuels with 100 percent or less excess air are generally 5 percent or less, there are factors that increase the error of the shortcut method: (1) High water content in the fuel or waste; (2) high halogen content; (3) excess air above 100 percent. However, the shortcut method can still be used to give a quick approximation even when these factors are present.

This shortcut method is based on two articles written by Irwin Frankel of The Mitrer Corp., Metrek Div., 1820 Dolley Madison Blvd., McLean, VA 22102. The articles, "Shortcut calculations for fluegas volume" and "Figure fluegas Reynolds number," appeared in the *Chemical Engineering* magazine issues of June 1, 1981, and August 24, 1981, respectively.

DETERMINING THE FEASIBILITY OF FLUE-GAS RECIRCULATION FOR NO$_x$ CONTROL IN PACKAGED BOILERS

Determine if it is feasible to recirculate flue gas in packaged boilers* to reduce NO$_x$ emissions to comply with EPA and local environmental requirements. Find the ranges of applicable parameters for packaged boilers in their typical applications.

---

*Defined as shop-assembled steam generators usually designed for oil and/or natural-gas firing in either watertube or firetube types. Packaged watertube boilers have capacities ranging up to 600,000 lb/h (75.6 kg/s) at pressures from 125 to 2000 lb/in² (gage) (860 to 13,800 kPa) with temperatures from 353 to 950°F (78 to 510°C). Higher capacities, pressures, and temperatures are possible. Firetube packaged boiler capacities can range up to some 50,000 lb/h (6.3 kg/s) at pressures up to 250 lb/in² (gage) (1720 kPa) with possible higher capacities, pressures, and temperatures.

Calculate the reduction in NO_x emissions for a 350-hp (261.1-kW) packaged boiler operating at rated capacity when flue-gas recirculation is increased from zero to 10 percent of the total flow.

Calculation Procedure:

1. Determine the suitability of flue-gas recirculation for packaged boilers
Flue-gas recirculation (FGR) for NO_x-emission control has been successfully applied on utility oil/gas-fired boilers and on industrial solid-fuel-fired units. Some engineers think that FGR is also appropriate for waste-to-energy facilities.

On oil/gas-fired steam generators, FGR acts as a flame quencher, reducing combustion temperatures by thermal dilution. In doing so, it significantly reduces excess-air requirements and flue-gas heat loss and provides a method of combustion staging. For stoker-fired units, FGR helps improve mixing of fuel and air in the fuel-bed area. Thus, it can help to reduce NO_x emissions and improve boiler efficiency. Lowering excess-air requirements minimizes the formation of thermal NO_x. Note that the technique does not affect formation of NO_x from fuel-bound nitrogen.

Recent developments in FGR have increased its range of suitable applications to include packaged firetube and watertube boilers of virtually any size. In these applications FGR functions as it does in a utility oil/gas-fired boiler—as a flame-quenching strategy, Fig. 24. The higher the recirculation rate, the greater the reduction of NO_x. Typically, a system that recirculates 10 percent of the flue gas can reduce NO_x by about 45 percent. At 20 percent recirculation, NO_x is reduced by up to 70 percent.

2. Evaluate the relationship between FGR rate and NO_x emissions
Plots of the relationship between FGR rate and NO_x emissions at various boiler capacities are available from packaged-boiler manufacturers. Figure 25 shows such plots for two typical packaged boilers of 350-hp (261.1-kW) and 200-hp (149.2-

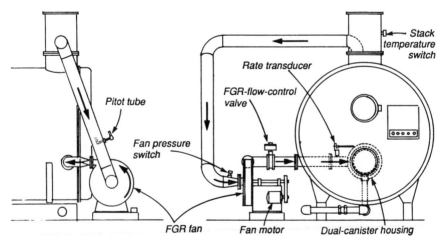

FIGURE 24 Compact system design adds little to space requirements for packaged boilers using FGR to control NO_x emissions. (*Power.*)

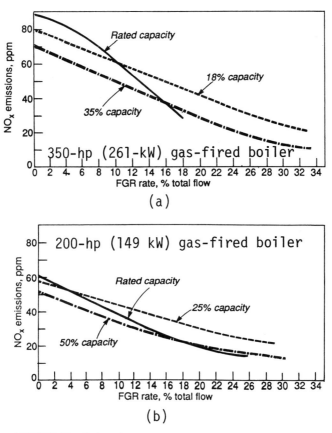

FIGURE 25 Choice of FGR rate is important for reducing NO_x emissions. Smaller units (*b*) require less recirculation to achieve desired NO_x reduction than larger units (*a*). (*Power.*)

kW). These two plots show that, in general, the NO_x emission decreases as the FGR rate, as a percentage of the total flue-gas flow, is increased. Certain limitations on the FGR rate apply, as discussed in this calculation procedure.

3. Compute the reduction in NO_x produced by FGR
Using Fig. 25 for the 350-hp (261.1-kW) packaged boiler at rated capacity, we see that the NO_x is reduced from 89 ppm to 60 ppm when the FGR rate is increased from 0 to 10 percent of the total flue-gas flow. This is a reduction of ([89 − 60]/80)(100) = 32.6 percent. A reduction of this magnitude in NO_x emission is significant.

4. Evaluate recirculation rates and burners to use
Recirculation rates can vary, depending on a particular unit's NO_x production and size. On natural gas, FGR is usually limited to 20 percent; and on oil-fired units to 10 to 12 percent at high fire. These limits are set to eliminate two common

problems with FGR: (1) Too much cooling which can quench the flame; (2) higher velocities which can push the flame away from the burner. Natural gas is far more responsive to FGR because it contains negligible amounts of fuel-bound nitrogen.

Proper introduction and recirculation of the flue gas is necessary to reduce NO_x emissions to the desired level. If the gas steam is brought into the suction side of an existing forced-draft fan, the amount of recirculated gas will be limited by the capacity of a single fan. Other factors to consider are condensation, dirt, soot collection, corrosion, and a highly variable supply of combustion air, which can reduce the capacity of the unit.

Related Calculations.   FGR is applicable to almost any type of packaged boiler used in industrial, commercial, residential, portable, marine, hotel, or other service. With ever-increasing emphasis on NO_x environmental concerns, FGR is winning more converts.

On average, uncontrolled gas-fired boilers emit 80 to 100 ppm NO_x, while average uncontrolled oil-fired boilers emit 150 to 300 ppm. FGR offers a potentially inexpensive alternative to add-on controls, such as selective catalytic and noncatalytic reduction, and more elaborate combustion modifications, including water/steam injection into the furnace, to control NO_x. Further, research into FGR reveals an additional benefit—a reduction in CO formation. This occurs as a result of the added turbulence and mixing around the flame.

Data and illustrations in this procedure are the work of Gene Tompkins, Aqua-Chem Inc., Cleaver-Brooks Div., as reported in *Power* magazine, and edited by Elizabeth A. Bretz. SI values were added by the handbook editor.

SECTION 5
FEEDWATER HEATING
METHODS

Steam-Plant Feedwater-Heating cycle
Analysis   5.1
Direct-Contact Feedwater Heater
Analysis   5.2
Closed Feedwater Heater Analysis and
Selection   5.3

Power-Plant Heater Extraction-Cycle
Analysis   5.8
Feedwater Heating with Diesel-Engine
Repowering of a Steam Plant   5.13

STEAM-PLANT FEEDWATER-HEATING CYCLE ANALYSIS

The high-pressure cylinder of a turbogenerator unit receives 1,000,000 lb per h (454,000 kg/h) of steam at initial conditions of 1800 psia (12,402 kPa) and 1050°F (565.6°C). At exit from the cylinder the steam has a pressure of 500 psia (3445 kPa) and a temperature of 740°F (393.3°C). A portion of this 500-psia (3445-kPa) steam is used in a closed feedwater heater to increase the temperature of 1,000,000 lb per h (454,000 kg/h) of 2000-psia (13,780-kPa) feedwater from 350°F (176.6°C) to 430°F (221.1°C); the remainder passes through a reheater in the steam generator and is admitted to the intermediate-pressure cylinder of the turbine at a pressure of 450 psia (3101 kPa) and a temperature of 1000°F (537.8°C). The intermediate cylinder operates nonextraction. Steam leaves this cylinder at 200 psia (1378 kPa) and 500°F (260°C). Find (a) flow rate to the feedwater heater, assuming no subcooling; (b) work done, in kW, by the high-pressure cylinder; (c) work done, in kW, by the intermediate-pressure cylinder; (d) heat added by the reheater.

Calculation Procedure:

1. Find the flow rate to the feedwater heater
(a) Construct the flow diagram, Fig. 1. Enter the pressure, temperature, and enthalpy values using the data given and the steam tables. Write an equation for flow across the feedwater heater, or $(H_2 - H_7)$ = water $(H_6 - H_5)$. Substituting using the enthalpy data from the flow diagram, flow to heater = $(1 \times 10^6)(409 - 324.4)/(1379.3 - 449.4)$ = 90.977.5 lb/h (41,303.8 kg/h).

2. Determine the work done by the high-pressure cylinder
(b) The work done = (steam flow rate, lb/h)$(H_1 - H_2)/3413$ = $(1 \times 10^6)(1511.3 - 1379.3)/3414$ = 38,675.7 kW.

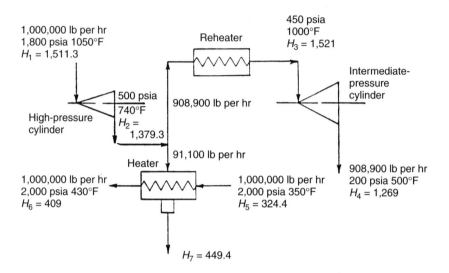

1,000,000 lb/hr (454,000 kg/hr) 1800 psia (12,402 kPa) 1050°F (565°C)
500 psia (3445 kPa) 740°F (393°C) 1379.3 Btu/lb (3214 kJ/kg) 1511.3 Btu/lb (3521 k?
2000 psia (13,780 kPa) 430°F (221°C) 409 (953 kJ/kg) 350°F (177°C) 324.4 (756 kJ/kg)
450 psia (3101 kPa) 1000°F (538°C) 1521 Btu/lb (3544 kJ/kg) 500°F (260°C)
200 psia (1378 kPa) 1269 Btu/lb (2933 kJ/kg) 324.5 Btu/lb (756 kJ/kg)
908,900 lb/hr (412,641 kg/hr) 91,100 lb/hr (41,359 kg/hr) 324.4 Btu/lb (756 kJ/kg)
449.4 Btu/lb (1047 kJ/kg)

FIGURE 1 Feedwater heating flow diagram.

3. Find the work done by the intermediate-pressure cylinder
(c) The work done = (steam flow through the cylinder)$(H_3 - H_4)/3413$ = $(1 \times 10^6 - 90.977.5 \times 10^6)(1521 - 1269)/3413$ = 67,118 kW.

4. Compute the heat added by the reheater
(d) Heat added by the reheater = (steam flow through the reheater)$(H_3 - H_2)$ = $(1 \times 10^6 - 90,977.5)(1521 - 1379.3)$ = 128.8×10^6 Btu/h (135.9 kJ/h).

   Related Calculations. Use this general procedure to determine the flow through feedwater heaters and reheaters for utility, industrial, marine, and commercial steam power plants of all sizes. The method given can also be used for combined-cycle plants using both a steam turbine and a gas turbine along with a heat-recovery steam generator (HRSG) in combination with one or more feedwater heaters and reheaters.

DIRECT-CONTACT FEEDWATER HEATER ANALYSIS

Determine the outlet temperature of water leaving a direct-contact open-type feedwater heater if 250,000 lb/h (31.5 kg/s) of water enters the heater at 100°F

(37.8°C). Exhaust steam at 10.3 lb/in^2 (gage) (71.0 kPa) saturated flows to the heater at the rate of 25,000 lb/h (31.5 kg/s). What saving is obtained by using this heater if the boiler pressure is 250 lb/in^2 (abs) (1723.8 kPa)?

Calculation Procedure:

1. Compute the water outlet temperature
Assume the heater is 90 percent efficient. Then $t_o = t_i w_w + 0.9 w_s h_g / (w_w + 0.9 w_s)$, where t_o = outlet water temperature, °F; t_i = inlet water temperature, °F; w_w = weight of water flowing through heater, lb/h; 0.9 = heater efficiency, expressed as a decimal; w_s = weight of steam flowing to the heater, lb/h; h_g = enthalpy of the steam flowing to the heater, Btu/lb.

For saturated steam at 10.3 lb/in^2 (gage) (71.0 kPa), or 10.3 + 14.7 = 25 lb/in^2 (abs) (172.4 kPa), h_g = 1160.6 Btu/lb (2599.6 kJ/kg), from the saturation pressure steam tables. Then

$$t_o = \frac{100(250,000) + 0.9(25,000)(1160.6)}{250,000 + 0.9(25,000)} = 187.5°F \ (86.4°C)$$

2. Compute the savings obtained by feed heating
The percentage of saving, expressed as a decimal, obtained by heating feedwater is $(h_o - h_i)/(h_b - h_i)$ where h_o and h_i = enthalpy of the water leaving and entering the heater, respectively, Btu/lb; h_b = enthalpy of the steam at the boiler operating pressure, Btu/lb. For this plant from the steam tables $h_o - h_i/(h_b - h_i) = 155.44 - 67.97/(1201.1 - 67.97) = 0.077$, or 7.7 percent.

A popular rule of thumb states that for every 11°F (6.1°C) rise in feedwater temperature in a heater, there is approximately a 1 percent saving in the fuel that would otherwise be used to heat the feedwater. Checking the above calculation with this rule of thumb shows reasonably good agreement.

3. Determine the heater volume
With a capacity of W lb/h of water, the volume of a direct-contact or open-type heater can be approximated from $v = W/10,000$, where v = heater internal volume, ft^3. For this heater $v = 250,000/10,000 = 25 \ ft^3$ (0.71 m^3).

Related Calculations.   Most direct-contact or open feedwater heaters store in 2-min supply of feedwater when the boiler load is constant, and the feedwater supply is all makeup. With little or no makeup, the heater volume is chosen so that there is enough capacity to store 5 to 30 min feedwater for the boiler.

CLOSED FEEDWATER HEATER ANALYSIS AND SELECTION

Analyze and select a closed feedwater heater for the third stage of a regenerative steam-turbine cycle in which the feedwater flow rate is 37,640 lb/h (4.7 kg/s), the desired temperature rise of the water during flow through the heater is 80°F (44.4°C) (from 238 to 318°F or, 114.4 to 158.9°C), bleed heating steam is at 100 lb/in^2 (abs) (689.5 kPa) and 460°F (237.8°C), drains leave the heater at the saturation temperature corresponding to the heating steam pressure [110 lb/in^2 (abs) or 689.5 kPa], and ⅝-in (1.6-cm) OD admiralty metal tubes with a maximum length of 6 ft (1.8

m) are used. Use the *Standards of the Bleeder Heater Manufacturers Association, Inc.,* when analyzing the heater.

Calculation Procedure:

1. *Determine the LMTD across heater*
When heat-transfer rates in feedwater heaters are computed, the average film temperature of the feedwater is used. In computing this the *Standards of the Bleeder Heater Manufacturers Association* specify that the *saturation temperature* of the heating steam be used. At 100 lb/in² (abs) (689.5 kPa), t_s = 327.81°F (164.3°C). Then

$$\text{LMTD} = t_m = \frac{(t_s - t_i) - (t_s - t_o)}{\ln[t_s - t_i/(t_s - t_o)]}$$

where the symbols are as defined in the previous calculation procedure. Thus,

$$t_m = \frac{(327.81 - 238) - (327.81 - 318)}{\ln[327.81 - 238/(327.81 - 318)]}$$

$$= 36.5°F\ (20.3°C)$$

The average film temperature t_f for any closed heater is then

$$t_f = t_s - 0.8t_m$$

$$= 327.81 - 29.2 = 298.6°F\ (148.1°C)$$

2. *Determine the overall heat-transfer rate*
Assume a feedwater velocity of 8 ft/s (2.4 m/s) for this heater. This velocity value is typical for smaller heaters handling less than 100,000-lb/h (12.6-kg/s) feedwater flow. Enter Fig. 2 at 8 ft/s (2.4 m/s) on the lower horizontal scale, and project vertically upward to the 250°F (121.1°C) average film temperature curve. This curve is used even though t_f = 298.6°F (148.1°C), because the standards recommend that heat-transfer rates higher than those for a 250°F (121.1°C) film temperature not be used. So, from the 8-ft/s (2.4 m/s) intersection with the 250°F (121.1°C) curve in Fig. 2, project to the left to read U = the overall heat-transfer rate = 910 Btu/(ft²·°F·h) [5.2 k]/m²·°C·s)].

Next, check Table 1 for the correction factor for U. Assume that no. 18 BWG ⅝-in (1.6-cm) OD arsenical copper tubes are used in this exchanger. Then the correction factor from Table 1 is 1.00, and U_{corr} = 910(1.00) = 910. If no. 9 BWG tubes are chosen, U_{corr} = 910(0.85) = 773.5 Btu/(ft²·°F·h) [4.4 kJ/(m²·°C·s)], given the correction factor from Table 1 for arsenical copper tubes.

3. *Compute the amount of heat transferred by the heater*
The enthalpy of the entering feedwater at 238°F (114.4°C) is, from the saturation-temperature steam table, h_{fi} = 206.32 Btu/lb (479.9 kJ/kg). The enthalpy of the leaving feedwater at 318°F (158.9°C) is, from the same table, h_{fo} = 288.20 Btu/lb (670.4 kJ/kg). Then the heater transferred H_t Btu/h is $H_t = w_w(h_{fo} - h_{fi})$, where w_w = feedwater flow rate, lb/h. Or, H_t = 37,640(288.20 − 206.32) = 3,080,000 Btu/h (902.7 kW).

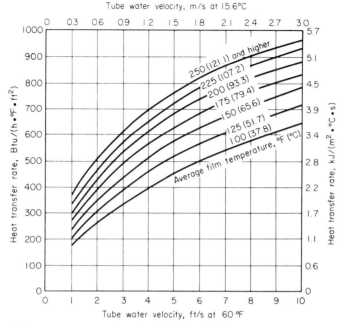

FIGURE 2 Heat-transfer rates for closed feedwater heaters. (*Standards of Bleeder Heater Manufacturers Association, Inc.*)

TABLE 1 Multipliers for Base Heat-Transfer Rates

[*For tube OD ⅝ to 1 in (1.6 to 2.5 cm) inclusive*]

BWG	As-Cu	Adm	90/10 Cu-Ni	80/20 Cu-Ni	70/30 Cu-Ni	Monel
18	1.00	1.00	0.97	0.95	0.92	0.89
17	1.00	1.00	0.94	0.91	0.87	0.85
16	1.00	1.00	0.91	0.88	0.84	0.82
15	1.00	0.99	0.89	0.86	0.82	0.79
14	1.00	0.96	0.85	0.82	0.77	0.75
13	0.98	0.93	0.81	0.78	0.73	0.70
12	0.95	0.90	0.77	0.73	0.68	0.65
11	0.92	0.87	0.74	0.70	0.65	0.62
10	0.89	0.83	0.69	0.66	0.60	0.58
9	0.85	0.80	0.65	0.62	0.56	0.54

4. Compute the surface area required in the exchanger
The surface area required A ft^2 = H_t/Ut_m. Then A = 3,080,000/[(910)(36.5)] = 92.7 ft^2 (8.6 m^2).

5. Determine the number of tubes per pass
Assume the heater has only one pass, and compute the number of tubes required. Once the number of tubes is known, a decision can be made about the number of passes required. In a closed heater, number of tubes = w_w (passes) (ft^3/s per tube)/[v(ft^2 per tube open area)], where w_w = lb/h of feedwater passing through heater; v = feedwater velocity in tubes, ft/s.

Since the feedwater enters the heater at 238°F (114.4°C) and leaves at 318°F (158.9°C), its specific volume at 278°F (136.7°C), midway between t_i and t_o, can be considered the average specific volume of the feedwater in the heater. From the saturation-pressure steam table, v_f = 0.01691 ft^3/lb (0.0011 m^3/kg) at 278°F (136.7°C). Convert this to cubic feet per second per tube by dividing this specific volume by 3600 (number of seconds in 1 h) and multiplying by the pounds per hour of feedwater per tube. Or, ft^3/s per tube = (0.01691/3600)(lb/h per tube).

Since no. 18 BWG ⅝-in (1.6-cm) OD tubes are being used, ID = 0.625 − 2(thickness) = 0.625 − 2(0.049) = 0.527 in (1.3 cm). Then, open area per tube ft^2 = (πd^2/4)/144 = 0.7854(0.527)^2/144 = 0.001525 ft^2 (0.00014 m^2) per tube. Alternatively, this area could be obtained from a table of tube properties.

With these data, compute the total number of tubes from number of tubes = [(37,640)(1)(0.01681/3600)]/[(8)(0.001525)] = 14.29 tubes.

6. Compute the required tube length
Assume that 14 tubes are used, since the number required is less than 14.5. Then, tube length l, ft = A/(number of tubes per pass)(passes)(area per ft of tube). Or, tube length for 1 pass = 92.7/[(14)(1)(0.1636)] = 40.6 ft (12.4 m). The area per ft of tube length is obtained from a table of tube properties or computed from 12π(OD)/144 = 12π(0.625)/155 = 0.1636 ft^2 (0.015 m^2).

7. Compute the actual number of passes and the actual tube length
Since the tubes in this heater cannot exceed 6 ft (1.8 m) in length, the number of passes required = (length for one pass, ft)/(maximum allowable tube length, ft) = 40.6/6 = 6.77 passes. Since a fractional number of passes cannot be used and an even number of passes permit a more convenient layout of the heater, choose eight passes.

From the same equation for tube length as in step 6, l = tube length = 92.7/[(14)(8)(0.1636)] = 5.06 ft (1.5 m).

8. Determine the feedwater pressure drop through heater
In any closed feedwater heater, the pressure loss Δp lb/in^2 is Δp = $F_1 F_2$(L + 5.5D)$N/D^{1.24}$, where Δp = pressure drop in the feedwater passing through the heater, lb/in^2; F_1 and F_2 = correction factors from Fig. 3; L = total lin ft of tubing divided by the number of tube holes in one tube sheet; D = tube ID; N = number of passes. In finding F_2, the average water temperature is taken as $t_s - t_m$.

For this heater, using correction factors from Fig. 3,

$$\Delta p = (0.136)(0.761)\left[\frac{5.06(8)(14)}{(8)(14)} + 5.5(0.527)\right]\frac{8}{0.527^{1.24}}$$

$$= 14.6 \text{ lb/in}^2 \ (100.7 \text{ kPa})$$

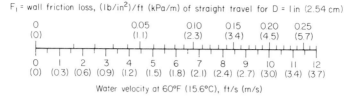

F_1 = wall friction loss, $(lb/in^2)/ft$ (kPa/m) of straight travel for D = 1 in (2.54 cm)

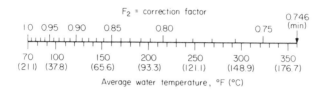

F_2 = correction factor

Average water temperature, °F (°C)

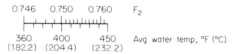

FIGURE 3 Correction factors for closed feedwater heaters. (*Standards of Bleeder Heater Manufacturers Association, Inc.*)

9. Find the heater shell outside diameter

The total number of tubes in the heater = (number of passes)(tubes per pass) = 8(14) = 112 tubes. Assume that there is $\frac{3}{8}$-in (1.0-cm) clearance between each tube and the tube alongside, above, or below it. Then the pitch or center-to-center distance between the tubes = pitch + tube OD = $\frac{3}{8}$ + $\frac{5}{8}$ = 1 in (2.5 cm).

The number of tubes per ft² of tube sheet = 166/(pitch)², or 166/1² = 166 tubes per ft² (1786.8 per m²). Since the heater has 112 tubes, the area of the tube sheet = 112/166 = 0.675 ft², or 97 in² (625.8 cm²).

The inside diameter of the heater shell = (tube sheet area, in²/0.7854)^{0.5} = (97/0.7854)^{0.5} = 11.1 in (28.2 cm). With a 0.25-in (0.6-cm) thick shell, the heater shell OD = 11.1 + 2(0.25) = 11.6 in (29.5 cm).

10. Compute the quantity of heating steam required

Steam enters the heater at 100 lb/in² (abs) (689.5 kPa) and 460°F (237.8°C). The enthalpy at this pressure and temperature is, from the superheated steam table, h_g = 1258.8 Btu/lb (2928.0 kJ/kg). The steam condenses in the heater, leaving as condensate at the saturation temperature corresponding to 100 lb/in² (abs) (689.5 kPa), or 327.81°F (164.3°C). The enthalpy of the saturated liquid at this temperature is, from the steam tables, h_f = 298.4 Btu/lb (694.1 kJ/kg).

The heater steam consumption for any closed-type feedwater heater is W, lb/h = $w_w(\Delta t)(h_g - h_f)$, where Δt = temperature rise of feedwater in heater, °F, c = specific heat of feedwater, Btu/(lb · °F). Assume c = 1.00 for the temperature range in this heater, and W = (37,640)(318 − 238)(1.00)/(1258.8 − 298.4) = 3140 lb/h (0.40 kg/s).

Related Calculations. The procedure used here can be applied to closed feedwater heaters in stationary and marine service. A similar procedure is used for selecting hot-water heaters for buildings, marine, and portable service. Various authorities recommend the following terminal difference (heater condensate temperature minus the outlet feedwater temperature) for closed feedwater heaters:

Feedwater outlet temperature		Terminal difference	
°F	°C	°F	°C
86 to 230	30.0 to 110.0	5	2.8
230 to 300	110.0 to 148.9	10	5.6
300 to 400	148.9 to 204.4	15	8.3
400 to 525	204.4 to 273.9	20	11.1

POWER-PLANT HEATER EXTRACTION-CYCLE ANALYSIS

A steam power plant operates at a boiler-drum pressure of 460 lb/in² (abs) (3171.7 kPa), a turbine throttle pressure of 415 lb/in² (abs) (2861.4 kPa) and 725°F (385.0°C), and a turbine capacity of 10,000 kW (or 13,410 hp). The Rankine-cycle efficiency ratio (including generator losses) is: full load, 75.3 percent; three-quarters load, 74.75 percent; half load, 71.75 percent. The turbine exhaust pressure is 1 inHg absolute (3.4 kPa); steam flow to the steam-jet air ejector is 1000 lb/h (0.13 kg/s). Analyze this cycle to determine the possible gains from two stages of extraction for feedwater heating, with the first stage a closed heater and the second stage a direct-contact or mixing heater. Use engineering-office methods in analyzing the cycle.

Calculation Procedure:

1. Sketch the power-plant cycle
Figure 4a shows the plant with one closed heater and one direct-contact heater. Values marked on Fig. 4a will be computed as part of this calculation procedure. Enter each value on the diagram as soon as it is computed.

2. Compute the throttle flow without feedwater heating extraction
Use the superheated steam tables to find the throttle enthalpy h_f = 1375.5 Btu/lb (3199.4 kJ/kg) at 415 lb/in² (abs) (2861.4 kPa) and 725°F (385.0°C).
Assume an irreversible adiabatic expansion between throttle conditions and the exhaust pressure of 1 inHg (3.4 kPa). Compute the final enthalpy H_{2s} by the same method used in earlier calculation procedures by finding y_{2s}, the percentage of moisture at the exhaust conditions with 1-inHg absolute (3.4-kPa) exhaust pressure. Do this by setting up the ratio $y_{2s} = (s_y - S_1)/s_{fg}$, where s_g and s_{fg} are entropies at the exhaust pressure; S_1 is entropy at throttle conditions. From the steam tables, y_{2s} = 2.0387 − 1.6468/1.9473 = 0.201. Then $H_{2s} = h_g - y_{2s}h_{fg}$, where h_g and h_{fg} are enthalpies at 1 inHg absolute (3.4 kPa). Substitute values from the steam table for 1 inHg absolute (3.4 kPa); or, H_{2s} = 1096.3 − 0.201(1049.2) = 885.3 Btu/lb (2059.2 kJ/kg).
The available energy in this irreversible adiabatic expansion is the difference between the throttle and exhaust conditions, or 1375.5 − 885.3 = 490.2 Btu/lb

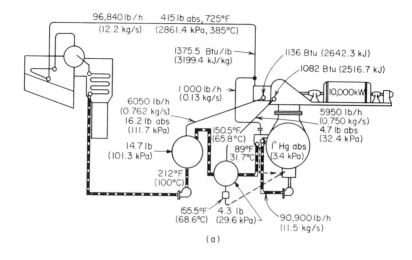

(a)

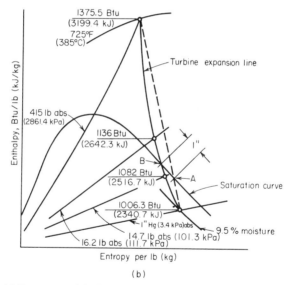

Entropy per lb (kg)

(b)

FIGURE 4 (a) Two stages of feedwater heating in a steam plant; (b) Mollier chart of the cycle in (a).

(1140.2 kJ/kg). The work at full load on the turbine is: (Rankine-cycle efficiency)(adiabatic available energy) = (0.753)(490.2) = 369.1 Btu/lb (858.5 kJ/kg). Enthalpy at the exhaust of the actual turbine = throttle enthalpy minus full-load actual work, or 1375.5 − 369.1 = 1006.4 Btu/lb (2340.9 kJ/kg). Use the Mollier chart to find, at 1.0 inHg absolute (3.4 kPa) and 1006.4 Btu/lb (2340.9 kJ/kg), that the exhaust steam contains 9.5 percent moisture.

Now the turbine steam rate SR = 3413(actual work output, Btu). Or, SR = 3413/369.1 = 9.25 lb/kWh (4.2 kg/kWh). With the steam rate known, the nonextraction throttle flow is (SR)(kW output) = 9.25(10,000) = 92,500 lb/h (11.7 kg/s).

3. Determine the heater extraction pressures

With steam extraction from the turbine for feedwater heating, the steam flow to the main condenser will be reduced, even with added throttle flow to compensate for extraction.

Assume that the final feedwater temperature will be 212°F (100.0°C) and that the heating range for each heater is equal. Both assumptions represent typical practice for a moderate-pressure cycle of the type being considered.

Feedwater leaving the condenser hotwell at 1 inHg absolute (3.4 kPa) is at 79.03°F (26.1°C). This feedwater is pumped through the air-ejector intercondensers and aftercondensers, where the condensate temperature will usually rise 5 to 15°F (2.8 to 8.3°C), depending on the turbine load. Assume that there is a 10°F (5.6°C) rise in condensate temperature from 79 to 89°F (26.1 to 31.7°C). Then the temperature range for the two heaters is 212 − 89 = 123°F (68.3°C). The temperature rise per heater is 123/2 = 61.5°F (34.2°C), since there are two heaters and each will have the same temperature rise. Since water enters the first-stage closed heater at 89°F (31.7°C), the exit temperature from this heater is 89 + 61.5 = 150.5°F (65.8°C).

The second-stage heater is a direct-contact unit operating at 14.7 lb/in² (abs) (101.4 kPa), because this is the saturation pressure at an outlet temperature of 212°F (100.0°C). Assume a 10 percent pressure drop between the turbine and heater steam inlet. This is a typical pressure loss for an extraction heater. Extraction pressure for the second-stage heater is then 1.1(14.7) = 16.2 lb/in² (abs) (111.7 kPa).

Assume a 5°F (2.8°C) terminal difference for the first-stage heater. This is a typical terminal difference, as explained in an earlier calculation procedure. The saturated steam temperature in the heater equals the condensate temperature = 150.5°F (65.8°C) exit temperature + 5°F (2.8°C) terminal difference = 155.5°F (68.6°C). From the saturation-temperature steam table, the pressure at 155.5°F (68.6°C) is 4.3 lb/in² (abs) (29.6 kPa). With a 10 percent pressure loss, the extraction pressure = 1.1(4.3) = 4.73 lb/in² (abs) (32.6 kPa).

4. Determine the extraction enthalpies

To establish the enthalpy of the extracted steam at each stage, the actual turbine-expansion line must be plotted. Two points—the throttle inlet conditions and the exhaust conditions—are known. Plot these on a Mollier chart, Fig. 4. Connect these two points by a dashed straight line, Fig. 4.

Next, measure along the saturation curve 1 in (2.5 cm) from the intersection point A back toward the enthalpy coordinate, and locate point B. Now draw a gradually sloping line from the throttle conditions to point B; from B increase the slope to the exhaust conditions. The enthalpy of the steam at each extraction point is read where the lines of constant pressure cross the expansion line. Thus, for the second-stage direct-contact heater where p = 16.2 lb/in² (abs) (111.7 kPa), h_g = 1136 Btu/lb (2642.3 kJ/kg). For the first-stage closed heater where p = 4.7 lb/in² (abs) (32.4 kPa), h_g = 1082 Btu/lb (2516.7 kJ/kg).

When the actual expansion curve is plotted, a steeper slope is used between the throttle super-heat conditions and the saturation curve of the Mollier chart, because the turbine stages using superheated steam (stages above the saturation curve) are more efficient than stages using wet steam (stages below the saturation curve).

5. *Compute the extraction steam flow*
To determine the extraction flow rates, two assumptions must be made—condenser steam flow rate and first-stage closed-heater extraction flow rate. The complete cycle will be analyzed, and the assumption checked. If the assumptions are incorrect, new values will be assumed, and the cycle analyzed again.

Assume that the condenser steam flow from the turbine is 84,000 lb/h (10.6 kg/s) when it is operating with extraction. Note that this value is less than the nonextraction flow of 92,500 lb/h (11.7 kg/s). The reason is that extraction of steam will reduce flow to the condenser because the steam is bled from the turbine after passage through the throttle but before the condenser inlet.

Then, for the first-stage closed heater, condensate flow is as follows:

From condenser	84,000 lb/h (10.6 kg/s) assumed
From steam-jet ejector	1,000 lb/h (0.13 kg/s)
From first-stage heater	5,900 lb/h (0.74 kg/s) assumed
Total	90,900 lb/h (11.5 kg/s)

The value of 5900 lb/h (0.74 kg/s) of condensate from the first-stage heater is the second assumption made. Since it will be checked later, an error in the assumption can be detected.

Assume a 2 percent heat radiation loss between the turbine and heater. This is a typical loss. Then

Steam enthalpy at heater = 1082(0.98)	= 1060.4 Btu/lb (2465.5 kJ/kg)
Enthalpy of condensate at 155.5°F (68.6°C)	= −123.4 Btu/lb (−287.0 kJ/kg)
Heat given up per lb (kg) of steam condensed	= 937.0 Btu/lb (2179.5 kJ/kg)
Enthalpy of feedwater at 150.5°F (65.8°C)	= 118.3 Btu/lb (275.2 kJ/kg)
Enthalpy of feedwater to heater at 89°F (31.7°C)	= −57.0 Btu/lb (−132.6 kJ/kg)
Heat absorbed by feedwater	= 61.3 Btu/lb (142.6 kJ/kg)

Required extraction = (total condensate flow, lb/h) [(heat absorbed by feedwater, Btu/lb)/ (heat given up per lb of steam condensed, Btu/lb)], or required extraction = (90,900)(61.3/ 937) = 5950 lb/h (0.75 kg/s)

Compare the required extraction, 5950 lb/h (0.75 kg/s), with the assumed extraction, 5900 lb/h (0.74 kg/s). The difference is only 50 lb/h (0.006 kg/s), which is less than 1 percent. Therefore, the assumed flow rate is satisfactory, because estimates within 1 percent are considered sufficiently accurate for all routine analyses.

For the second-stage direct-contact heater, condensate flow, lb/h is as follows:

From the first-stage heater	90,900 lb/h (11.5 kg/s)
Steam enthalpy at heater = 1135(0.98)	= 1112.3 Btu/lb (2587.2 kJ/kg)
Enthalpy of condensate at 212°F (100.0°C)	= −180.0 Btu/lb (−418.7 kJ/kg)
Heat given up per lb of steam condensed	= 932.3 Btu/lb (2168.5 kJ/kg)
Enthalpy of feedwater at 212°F (100.0°C)	= 180.0 Btu/lb (418.7 kJ/kg)
Enthalpy of feedwater at 150.5°F (65.8°C)	= 118.3 Btu/lb (275.2 kJ/kg)
Heat absorbed by feedwater	= 61.7 Btu/lb (143.5 kJ/kg)

The required extraction, calculated in the same way as for the first-stage heater, is (90,900)(61.7/932.2) = 6050 lb/h (0.8 kg/s).

The computed extraction flow for the second-stage heater is not compared with an assumed value because an assumption was not necessary.

6. *Compare the actual condenser steam flow*

Sketch a vertical line diagram, Fig. 5, showing the enthalpies at the throttle, heaters, and exhaust. From this diagram, the work lost by the extracted steam can be computed. As Fig. 5 shows, the total enthalpy drop from the throttle to the exhaust is 369 Btu/lb (389.3 kJ/kg). Each pound of extracted steam from the first- and second-stage bleed points causes a work loss of 75.7 Btu/lb (176.1 kJ/kg) and 129.7 Btu/lb (301.7 kJ/kg), respectively. To carry the same load, 10,000 kW, with extractions, it will be necessary to supply the following additional compensation steam to the turbine throttle: (heater flow, lb/h)(work loss, Btu/h)/(total work, Btu/h). Then

	lb/h	kg/s
First-stage closed heater:		
(5950)(75.7/369)	1220	0.15
Second-stage direct-contact heater:		
(6050)(129.7/369)	2120	0.27
Total additional throttle flow to compensate for extraction	3340	0.42

Check the assumed condenser flow using nonextraction throttle flow + additional throttle flow − heater extraction = condenser flow. Set up a tabulation of the flows as follows:

Flow	lb/h	kg/s
Throttle; nonextraction	92,500	11.65
Added flow (compensation)	3,340	0.42
Throttle; extraction	95,840	12.07
Extraction (5950 + 6050)	−12,000	−1.51
Condenser flow	83,840	10.56

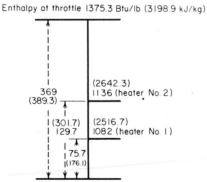

Enthalpy at throttle 1375.3 Btu/lb (3198.9 kJ/kg)

Enthalpy at exhaust 1006.3 Btu/lb (2340.7 kJ/kg)

FIGURE 5 Diagram of turbine-expansion line.

Compare this actual flow, 83,840 lb/h (10.6 kg/s), with the assumed flow, 84,000 lb/h (10.6 kg/s). The difference, 160 lb/h (0.02 kg/s), is less than 1 percent. Since an accuracy within 1 percent is sufficient for all normal power-plant calculations, it is not necessary to recompute the cycle. Had the difference been greater than 1 percent, a new condenser flow would be assumed and the cycle recomputed. Follow this procedure until a difference of less than 1 percent is obtained.

7. *Determine the economy of the extraction cycle*
For a nonextraction cycle operating in the same pressure range,

	Btu/lb	kJ/kg
Enthalpy of throttle steam	1375.3	3198.9
Enthalpy of condensate at 79°F (26.1°C)	−47.0	−109.3
Heat supplied by boiler	1328.3	3089.6

Heat chargeable to turbine = (throttle flow + air-ejector flow)(heat supplied by boiler)/(kW output of turbine) = (92,500 + 1000)(1328.3)/10,000 = 12,410 Btu/kWh (13,093.2 kJ/kWh), which is the actual heat rate HR of the nonextraction cycle.

For the extraction cycle using two heaters,

	Btu/lb	kJ/kg
Enthalpy of throttle steam	1375.3	3198.9
Enthalpy of feedwater leaving second heater	−180.0	−418.7
Heat supplied by boiler	1195.3	2780.3

As before, heat chargeable to turbine = (95,840 + 1000)(1195.3)/10,000 = 11,580 Btu/kWh (12,217.5 kJ/kWh). Therefore, the improvement = (nonextraction HR − extraction HR)/nonextraction HR = (12,410 − 11,580)/12,410 = 0.0662, or 6.62 percent.

Related Calculations. (1) To determine the percent improvement in a steam cycle resulting from additional feedwater heaters in the cycle, use the same procedure as given above for three, four, five, six, or more heaters. Plot the percent improvement vs. number of stages of extraction, Fig. 6, to observe the effect of additional heaters. A plot of this type shows the decreasing gains made by additional heaters. Eventually the gains become so small that the added expenditure for an additional heater cannot be justified.

(2) Many simple marine steam plants use only two stages of feedwater heating. To analyze such a cycle, use the procedure given, substituting the hp output for the kW output of the turbine.

(3) Where a marine plant has more than two stages of feedwater heating, follow the procedure given in (1) above.

FEEDWATER HEATING WITH DIESEL-ENGINE REPOWERING OF A STEAM PLANT

Show the economies and environmental advantages possible with Diesel-engine repowering of steam boiler/turbine plants using feedwater heating as the entree.

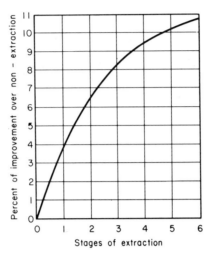

FIGURE 6 Percentage of improvement in turbine heat rate vs. stages of extraction.

Give the typical temperatures and flow rates encountered in such installations using gas and/or oil fuels.

Calculation Procedure:

1. *Determine the output ranges possible with today's diesel engines*
Medium-speed Diesel engines are available in sizes exceeding 16 MW. While this capacity may seem small when compared to gas turbines, it is appropriate for repowering of steam plants up to 600 MW via boiler feedwater heating.

Modern Diesel engines can attain simple cycle efficiencies of over 47 percent burning natural gas or heavy fuel oil (HFO). The ability to burn natural gas in Diesels is a key factor when coupled with coal-fired boilers. Since the Clean Air Act Amendments of 1990 (CAA) require these boilers to reduce both NO_x and SO_2 emissions on a lb/million Btu-fired basis (kg/MJ), a boiler feedwater heating system that can help make these reductions while simultaneously improving overall plant efficiency is attractive. Diesel engines offer these reductions when used in repowering and feedwater heating.

Today Diesel engines convert about 45 percent of mechanical energy to electricity; 30 percent becomes exhaust-gas heat; 12 percent is lost to jacket-water heat; and 6 percent is used to cool the lube oil. The remaining energy lost is generally not recoverable.

2. *Show how the diesel engine can be used in the feedwater heating cycle*
Modern steam-turbine reheat cycles, Fig. 7, use an array of feedwater heaters in a regenerative feedwater heating system. The heaters progressively increase the condensate temperature until it approaches the steam saturation temperature. Condensate then enters the final economizer and evaporator sections of the boiler.

Using the waste heat from Diesel engines to partially replace the feedwater heaters is almost completely non-intrusive to the operation of the existing system,

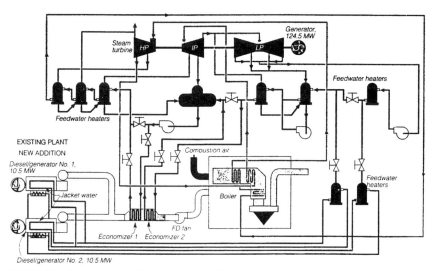

FIGURE 7 In repowering, Diesel exhaust is adjusted in temperature to the same levels expected from feedwater heaters in the existing plant. (*Power.*)

but causes several significant alterations in the cycle. Two particular cycle altera- tions are: (1) Jacket water temperature from a Diesel engine is available at about 195°F (91°C). The lube-oil cooling system produces water at about 170°F (77°C). These temperatures are appropriate for partial displacement of the boiler's low- temperature feedwater heaters.

(2) A gas/Diesel engine can operate on about 97 percent natural gas/3 percent HFO and has an exhaust temperature of 680°F (360°C). The exhaust gas can be ducted through an economizer that is equipped with selective catalytic reduction (SCR) and has heat-transfer sections that can adjust the exit temperature to match the preheated-burner-windbox air temperature. The SCR reduces NO_x emissions from the engine to about 25 ppm on leaving the economizer. This exhaust econo- mizer, Fig. 7, also elevates the temperature of the feedwater after it leaves the deaerator.

3. *Explain the environmental impact of using diesels in the feedwater heating loop*

Exhaust gas from the economizer sections, Fig. 7, is ducted to the boiler windbox. This gas serves the same function as flue-gas recirculation (FGR) in a low-NO_x burner. In the installation in Fig. 7, the two Diesel generators produce 351,600 lb/h (159,626 kg/h) of exhaust gas. Most of this gas is ducted to the boiler windbox to achieve a 17.5 percent O_2 level needed for the low NO_x burners. The balance enters the boiler as overfire air.

4. *Determine the heat-rate improvement possible*

Diesel engines are highly efficient on a simple-cycle basis. When combined with a steam turbine, as described, the cycle efficiency reaches about 56 percent on an incremental basis. In the example here, the incremental heat rate of the engine combined with the additional output from the turbine is 6060 Btu/kWh (6393 kJ/ kWh). This heat rate represents about 25 percent of the total system power and can be averaged with the heat rate of the associated plant. Total system heat rate may be improved by as much as 10 percent as a result of repowering in this fashion.

5.   Evaluate system turndown possible with this type of feedwater heating
Typically, a coal-fired boiler can be turned down to about 60 percent load while maintaining superheat and reheat temperatures. Adding Diesel feedwater heat increases system output by about 25 percent. More important, the system is almost completely non-intrusive, and can return to normal operation when the Diesel output is not required. Thus, the total turndown of the plant is increased from 40 to 52 percent, making plant operation more flexible.

6.   Compare diesels vs. gas turbines for feedwater heating
Comparing Diesels vs. gas turbines (GT) in this application, it appears that the major differences are in the temperature of the exhaust gas and the quantities of exhaust gas that must be introduced to the boiler. Most GTs have fairly high exhaust-gas rates on a per-kilowatt basis, varying from 25 to over 30 lb/kW (9 to 13.6 kg/kW). GT exhaust may contain from 14.5 to 15.5 percent O_2.

Conversely, Diesels have exhaust-gas rates of 15 to 16 lb/kW (6.8 to 7.3 kg/kW). The O_2 concentrations for Diesels vary between 11 percent for spark-ignited gas engines up to 13 percent for gas/Diesels or HFO-fired Diesels. Thus, when providing inlet gases to the boiler and adjusting the windbox concentrations to 17.5 percent O_2, the volume of gas has to be even further increased with GTs.

7.   Evaluate the cost of this type of feedwater heating
Capital cost for modifying the boiler is largely dependent on the site and boiler. Cost for a turnkey-installed Diesel facility is about \$850/kW. For a Diesel plant connected with an existing power system, net output of the existing system is increased, as noted, because of increasing flow to the steam turbine's condenser. This increased output offsets the cost of interconnection to the boiler.

Related Calculations.   The data and procedure given here represent a new approach to feedwater heating and repowering. Because three function are served—namely feedwater heating, repowering, and environmental compliance, the approach is unique. Calculation of the variables is simple because basic heat-transfer relations—covered elsewhere in this handbook—are used.

The date and methods given in this procedure are the work of F. Mack Shelor, Wartsila Diesel Inc., as reported in *Power* magazine (June 1995). SI values were added by the handbook editor.

SECTION 6

INTERNAL-COMBUSTION ENGINES

DETERMINING THE ECONOMICS OF RECIPROCATING I-C ENGINE COGENERATION

Determine if an internal-combustion (I-C) engine cogeneration facility will be economically attractive if the required electrical power and steam services can be served by a cycle such as that in Fig. 1 and the specific load requirements are those shown in Fig. 2. Frequent startups and shutdowns are anticipated for this system.

Calculation Procedure:

1. *Determine the sources of waste heat available in the typical I-C engine*
There are three primary sources of waste heat available in the usual I-C engine. These are: (1) the exhaust gases from the engine cylinders; (2) the jacket cooling water; (3) the lubricating oil. Of these three sources, the quantity of heat available is, in descending order: exhaust gases; jacket cooling water; lube oil.

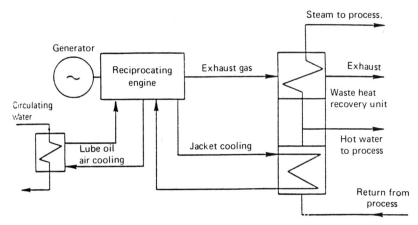

FIGURE 1 Reciprocating-engine cogeneration system waste heat from the exhaust, and jacket a oil cooling, are recovered. (*Indeck Energy Services, Inc.*)

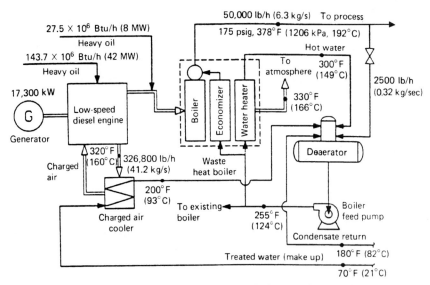

FIGURE 2 Low-speed Diesel-engine cogeneration. (*Indeck Energy Services, Inc.*)

2. Show how to compute the heat recoverable from each source

For the exhaust gases, use the relation, $H_A = W(\Delta t)(c_g)$, where W_A = rate of gas flow from the engine, lb/h (kg/h); Δt = temperature drop of the gas between the heat exchanger inlet and outlet, °F (°C); c_g = specific heat of the gas, Btu/lb °F (J/kg °C). For example, if an I-C engine exhausts 100,000 lb/h (45,400 kg/h) at 700°F (371°C) to a HRSG (heat-recovery steam generator), leaving the HRSG at 330°F (166°C), and the specific heat of the gas is 0.24 Btu/lb °F (1.0 kJ/kg °C), the heat recoverable, neglecting losses in the HRSG and connecting piping, is $H_A = 100,000(700 - 330)(0.24) = 8,880,000$ Btu/h (2602 MW).

With an average heat of vaporization of 1000 Btu/lb (2330 kJ/kg) of steam, this exhaust gas flow could generate $8,880,000/1000 = 8880$ lb/h (4032 kg/h) of steam. If oil with a heating value of 145,000 Btu/gal (40,455 kJ/L) were used to generate this steam, the quantity required would be $8,880,000/145,000 = 61.2$ gal/h (232 L/h). At a cost of 90 cents per gallon, the saving would be $0.90(61.2) = $55.08/h$. Assuming 5000 hours of operation per year, or 57 percent load, the saving in fuel cost would be $5000($55.08) = $275,400$. This is a significant saving in any plant. And even if heat losses in the ductwork and heat-recovery boiler cut the savings in half, the new would still exceed one hundred thousand dollars a year. And as the operating time increases, so too do the savings.

3. Compute the savings potential in jacket-water and lube-oil heat recovery
A similar relation can be used to compute jacket-water and lube-oil heat recovery. The flow rate can be expressed in either pounds (kg) per hour or gallons (L) per minute, depending on the designer's choice.

Since water has a specific heat of unity, the heat-recovery potential of the jacket water is $H_W = w(\Delta t_w)$, where w = weight of water flow, lb per h (kg/h); Δt_w = change in temperature of the jacket water when flowing through the heat exchanger, °F (°C). Thus, if the jacket-water flow is 25,000 lb/h (11,350 kg/h) and the temperature change during flow of the jacket water through and external heat exchanger is 190 to 70°F (88 to 21°C), the heat given up by the jacket water, neglecting losses is $H_w = 25,000(190 - 70) = 3,000,000$ Btu/h (879 MW). During 25 h the heat recovery will be $24(3,000,000) = 72,000,000$ Btu (75,960 MJ). This is a significant amount of heat which can be used in process or space heating, or to drive an air-conditioning unit.

If the jacket-water flow rate is expressed in gallons per minute instead of pounds per hour (L/min instead of kg/h), the heat-recovery potential, $H_{wg} = $ gpm$(\Delta t)(8.33)$ where $8.33 = $ lb/gal of water. With a water flow rate of 50 gpm and the same temperature range as above, $H_{wg} = 50(120)(8.33) = 49,980$ Btu/min (52,279 kJ/min).

4. Find the amount of heat recoverable from the lube oil
During I-C engine operation, lube-oil temperature can reach high levels—in the 300 to 400°F (149 to 201°C) range. And with oil having a typical specific heat of 0.5 Btu/lb °F (2.1 kJ/kg °C), the heat-recovery potential for the lube oil is $H_{w_o} = w_o(\Delta t)(c_o)$, where w_o = oil flow in lb/h (kg/h); Δt = temperature change of the oil during flow through the heat-recovery heat exchanger = oil inlet temperature − oil outlet temperature, °F or °C; c_o = specific heat of oil = 0.5 Btu/lb °F (kJ/kg °C). With an oil flow of 2000 lb/h (908 kg/h), a temperature change of 140°F (77.7°C), $H_o = 2000(140)(0.50) = 140,000$ Btu/h (41 kW). Thus, as mentioned earlier, the heat recoverable from the lube oil is usually the lowest of the three sources.

With the heat flow rates computed here, an I-C engine cogeneration facility can be easily justified, especially where frequent startups and shutdowns are anticipated. Reciprocating Diesel engines are preferred over gas and steam turbines where frequent startups and shutdowns are required. Just the fuel savings anticipated for recovery of heat in the exhaust gases of this engine could pay for it in a relatively short time.

Related Calculations. Cogeneration, in which I-C engines are finding greater use throughout the world every year, is defined by Michael P. Polsky, President, Indeck Energy Services, Inc., as "the simultaneous production of useful thermal energy and electric power from a fuel source or some variant thereof. It is more efficient to produce electric power and steam or hot water together than electric power alone, as utilities do, or thermal energy alone, which is common in industrial, commercial, and institutional plants." Figures 1 and 2 in this procedure are from the firm of which Mr. Polsky is president.

With the increased emphasis on reducing environmental pollution, conserving fuel use, and operating at lower overall cost, cogeneration—especially with Diesel engines—is finding wider acceptance throughout the world. Design engineers should consider cogeneration whenever there is a concurrent demand for electricity and heat. Such demand is probably most common in industry but is also met in commercial (hotels, apartment houses, stores) and institutional (hospital, prison, nursing-home) installations. Often, the economic decision is not over whether co-generation should be used, but what type of prime mover should be chosen.

Three types of prime movers are usually considered for cogeneration—steam turbines, gas turbines, or internal-combustion engines. Steam and/or gas turbines are usually chosen for large-scale utility and industrial plants. For smaller plants the Diesel engine is probably the most popular choice today. Where natural gas is available, reciprocating internal-combustion engines are a favorite choice, especially with frequent startups and shutdowns.

Recently, vertical modular steam engines have been introduced for use in co-generation. Modules can be grouped to increase the desired power output. These high-efficiency units promise to compete with I-C engines in the growing cogeneration market.

Guidelines used in estimating heat recovery from I-C engines, after all heat loses, include these: (1) Exhaust-gas heat recovery = 28 percent of heat in fuel; (2) Jacket-water heat recovery = 27 percent of heat in fuel; (3) Lube-oil heat recovery = 9 percent of the heat in the fuel. The Diesel Engine Manufacturers Association (DEMA) gives these values for heat disposition in a Diesel engine at three-quarters to full load: (1) Fuel consumption = 7366 Btu/bhp·h (2.89 kW/kW); (2) Useful work = 2544 Btu/bhp·h (0.999 kW/kW); (3) Loss in radiation, *etc.* = 370 Btu/bhp·h (0.145 kW/kW); (4) To cooling water = 2195 Btu/bhp·h (0.862 kW/kW); (5) To exhaust = 2258 Btu/bhp·h (0.887 kW/kW). The sum of the losses is 1 Btu/bhp·h greater than the fuel consumption because of rounding of the values.

Figure 3 shows a proposed cogeneration, desiccant-cooling, and thermal-storage integrated system for office buildings in the southern California area. While directed at the micro-climates in that area, similar advantages for other micro-climates and building types should be apparent. The data presented here for this system were prepared by The Meckler Group and are based on a thorough engineering and economic evaluation for the Southern California Gas Co. of the desiccant-cooling/thermal-energy-storage/cogeneration system, a proprietary design developed for pre- and post-Title-24 mid-rise office buildings. Title 24 is a section of the State of California Administrative Code that deals with energy-conservation standards for construction applicable to office buildings. A summary of the study was presented in *Power* magazine by Milton Meckler.

In certain climates, office buildings are inviting targets for saving energy via evaporative chilling. When waste heat is plentiful, desiccant cooling and cogeneration become attractive. In coupling the continuously available heat-rejection capacity of packaged cogeneration units, Fig. 4, with continuously operating regenerator demands, the use of integrated components for desiccant cooling, thermal-energy storage, and cogeneration increases. The combination also ensures a reasonable constant, cost-effective supply of essentially free electric power for general building use.

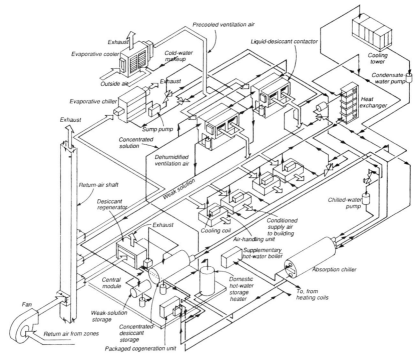

FIGURE 3 Integrated system is a proposed off-peak desiccant/evaporative-cooling configuration with cogeneration capability. (*Power* and *The Meckler Group*.)

Recoverable internal-combustion engine heat should at least match the heat requirement of the regenerator, Fig. 3. The selected engine size (see a later procedure in this section), however, should not cause the cogeneration system's Purpa (Public Utility Regulatory & Policies Act) efficiency to drop below 42.5 percent. (Purpa efficiency decreases as engine size increases.) An engine size is selected to give the most economical performance and still have a Purpa efficiency of greater than 42.5 percent.

The utility study indicated a favorable payout period and internal rate of return both for retrofits of pre-Title-24 office buildings and for new buildings in compliance with current Title-24 requirements (nominal 200 to 500 cooling tons). Although the study was limited to office-building occupancies, it is likely that other building types with high ventilation and electrical requirements would also offer attractive investment opportunities.

Based on study findings, fuel savings ranged from 3300 to 7900 therms per year. Cost savings ranged from $322,000 to $370,000 for the five-story-building case studies and from $545,000 to $656,000 for 12-story-building case studies where the synchronously powered, packaged cogeneration unit was not used for emergency power.

Where the cogeneration unit was also used for emergency power, the initial cost decreased from $257,000 to $243,000, representing a 31 percent drop in average cost for the five-story-building cases; and from $513,000 to $432,000, a 22 percent

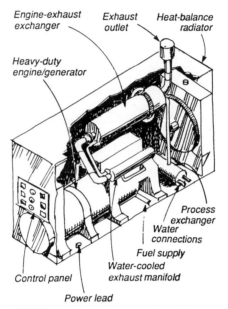

FIGURE 4 Packaged cogeneration I-C engine unit supplies waste heat to desiccant regenerator. (*Power* and *The Meckler Group*.)

dip in average cost for the 12-story-building cases. The average cost decrease shifts the discounted payback period an average of 5.6 and 5.9 years for the five- and 12-story-building cases, respectively.

Study findings were conservatively reported, since no credit was taken for potential income resulting from Purpa sales to the serving utility at off-peak hours, when actual building operating requirements fall below rated cogenerator output. This study is another example of the importance of the internal-combustion engine in cogeneration around the world today.

DIESEL GENERATING UNIT EFFICIENCY

A 3000-kW diesel generating unit performs thus: fuel rate, 1.5 bbl (238.5 L) of 25° API fuel for a 900-kWh output; mechanical efficiency, 82.0 percent; generator efficiency, 92.0 percent. Compute engine fuel rate, engine-generator fuel rate, indicated thermal efficiency, overall thermal efficiency, brake thermal efficiency.

Calculation Procedure:

1. *Compute the engine fuel rate*
The fuel rate of an engine driving a generator is the weight of fuel, lb, used to generate 1 kWh at the generator input shaft. Since this engine burns 1.5 bbl (238.5

---

*Elliott, *Standard Handbook of Power Plant Engineering*, McGraw-Hill, 1989.

L) of fuel for 900 kW at the generator terminals, the total fuel consumption is (1.5 bbl)(42 gal/bbl) = 63 gal (238.5 L), at a generator efficiency of 92.0 percent.

To determine the weight of this oil, compute its specific gravity s from $s = 141.5/(131.5 + °API)$, where °API = API gravity of the fuel. Hence, $s = 141.5(131.5 + 25) = 0.904$. Since 1 gal (3.8 L) of water weighs 8.33 lb (3.8 kg) at 60°F (15.6°C), 1 gal (3.8 L) of this oil weighs (0.904)(8.33) = 7.529 lb (3.39 kg). The total weight of fuel used when burning 63 gal is (63 gal)(7.529 lb/gal) = 474.5 lb (213.5 kg).

The generator is 92 percent efficient. Hence, the engine actually delivers enough power to generate 900/0.92 = 977 kWh at the generator terminals. Thus, the engine fuel rate = 474.5 lb fuel/977 kWh = 0.485 lb/kWh (0.218 kg/kWh).

2. Compute the engine-generator fuel rate

The engine-generator fuel rate takes these two units into consideration and is the weight of fuel required to generate 1 kWh at the generator terminals. Using the fuel-consumption data from step 1 and the given output of 900 kW, we see that engine-generator fuel rate = 474.5 lb fuel/900 kWh output = 0.527 lb/kWh (0.237 kg/kWh).

3. Compute the indicated thermal efficiency

Indicated thermal efficiency is the thermal efficiency based on the *indicated* horsepower of the engine. This is the horsepower developed in the engine cylinder. The engine fuel rate, computed in step 1, is the fuel consumed to produce the brake or shaft horsepower output, after friction losses are deducted. Since the mechanical efficiency of the engine is 82 percent, the fuel required to produce the indicated horsepower is 82 percent of that required for the brake horsepower, or (0.82)(0.485) = 0.398 lb/kWh (0.179 kg/kWh).

The indicated thermal efficiency of an internal-combustion engine driving a generator is $e_i = 3413/f_i(HHV)$, where e_i = indicated thermal efficiency, expressed as a decimal; f_i = indicated fuel consumption, lb/kWh; HHV = higher heating value of the fuel, Btu/lb.

Compute the HHV for a diesel fuel from HHV = 17,680 + 60 × °API. For this fuel, HHV = 17,680 + 60(25) = 19,180 Btu/lb (44,612.7 kJ/kg).

With the HHV known, compute the indicated thermal efficiency from $e_i = 3,413/[(0.398)(19,180)] = 0.447$ or 44.7 percent.

4. Compute the overall thermal efficiency

The overall thermal efficiency e_o is computed from $e_o = 3413/f_o(HHV)$, where f_o = overall fuel consumption, Btu/kWh; other symbols as before. Using the engine-generator fuel rate from step 2, which represents the overall fuel consumption $e_o = 3413/[(0.527)(19,180)] = 0.347$, or 34.7 percent.

5. Compute the brake thermal efficiency

The engine fuel rate, step 1, corresponds to the brake fuel rate f_b. Compute the brake thermal efficiency from $e_b = 3413/f_b(HHV)$, where f_b = brake fuel rate, Btu/kWh; other symbols as before. For this engine-generator set, $e_b = 3413/[(0.485)(19,180)] = 0.367$, or 36.7 percent.

Related Calculations. Where the fuel consumption is given or computed in terms of lb/(hp·h), substitute the value of 2545 Btu/(hp·h) (1.0 kW/kWh) in place of the value 3413 Btu/kWh (3600.7 kJ/kWh) in the numerator of the e_i, e_o, and e_b equations. Compute the indicated, overall, and brake thermal efficiencies as before. Use the same procedure for gas and gasoline engines, except that the higher heating value of the gas or gasoline should be obtained from the supplier or by test.

ENGINE DISPLACEMENT, MEAN EFFECTIVE PRESSURE, AND EFFICIENCY

A 12 × 18 in (30.5 × 44.8 cm) four-cylinder four-stroke single-acting diesel engine is rated at 200 bhp (149.2 kW) at 260 r/min. Fuel consumption at rated load is 0.42 lb/(bhp·h) (0.25 kg/kWh). The higher heating value of the fuel is 18,920 Btu/lb (44,008 kJ/kg). What are the brake mean effective pressure, engine displacement in ft^3/(min · bhp), and brake thermal efficiency?

Calculation Procedure:

1. Compute the brake mean effective pressure
Compute the brake mean effective pressure (bmep) for an internal-combustion engine from $bmep = 33,000 \, bhp_n / LAn$, where $bmep$ = brake mean effective pressure, lb/in^2; bhp_n = brake horsepower output delivered per cylinder, hp; L = piston stroke length, ft; a = piston area, in^2; n = cycles per minute per cylinder = crankshaft rpm for a two-stroke cycle engine, and 0.5 the crankshaft rpm for a four-stroke cycle engine.

For this engine at its rated hbp, the output per cylinder is 200 bhp/4 cylinders = 50 bhp (37.3 kW). Then $bmep = 33,000(50)/[(18/12)(12)^2(\pi/4)(260/2)] = 74.8$ lb/in^2 (516.1 kPa). (The factor 12 in the denominator converts the stroke length from inches to feet.)

2. Compute the engine displacement
The total engine displacement V_d ft^3 is given by $V_d = LAnN$, where A = piston area, ft^2; N = number of cylinders in the engine; other symbols as before. For this engine, $V_d = (18/12)(12/12)^2(\pi/4)(260/2)(4) = 614 \, ft^3$/min (17.4 m^3/min). The displacement is in cubic feet per minute because the crankshaft speed is in r/min. The factor of 12 in the denominators converts the stroke and area to ft and ft^2, respectively. The displacement per bhp = (total displacement, ft^3/min)/bhp output of engine = 614/200 = 3.07 ft^3/(min · bhp) (0.12 m^3/kW).

3. Compute the brake thermal efficiency
The brake thermal efficiency e_b of an internal-combustion engine is given by $e_b = 2545/(sfc)(HHV)$, where sfc = specific fuel consumption, lb/(bhp·h); HHV = higher heating value of fuel, Btu/lb. For this engine, $e_b = 2545/[(0.42)(18,920)] = 0.32$, or 32.0 percent.

Related Calculations.   Use the same procedure for gas and gasoline engines. Obtain the higher heating value of the fuel from the supplier, a tabulation of fuel properties, or by test.

ENGINE MEAN EFFECTIVE PRESSURE AND HORSEPOWER

A 500-hp (373-kW) internal-combustion engine has a brake mean effective pressure of 80 lb/in^2 (551.5 kPa) at full load. What are the indicated mean effective pressure and friction mean effective pressure if the mechanical efficiency of the engine is 85 percent? What are the indicated horsepower and friction horsepower of the engine?

Calculation Procedure:

1. Determine the indicated mean effective pressure

Indicated mean effective pressure $imep$ lb/in^2 for an internal-combustion engine is found from $imep = bmep/e_m$, where $bmep$ = brake mean effective pressure, lb/in^2; e_m = mechanical efficiency, percent, expressed as a decimal. For this engine, $imep = 80/0.85 = 94.1$ lb/in^2 (659.3 kPa).

2. Compute the friction mean effective pressure

For an internal-combustion engine, the friction mean effective pressure $fmep$ lb/in^2 is found from $fmep = imep - bmep$, or $fmep = 94.1 - 80 = 14.1$ lb/in^2 (97.3 kPa).

3. Compute the indicated horsepower of the engine

For an internal-combustion engine, the mechanical efficiency $e_m = bhp/ihp$, where ihp = indicated horsepower. Thus, $ihp = bhp/e_m$, or $ihp = 500/0.85 = 588$ ihp (438.6 kW).

4. Compute the friction hp of the engine

For an internal-combustion engine, the friction horsepower is $fhp = ihp - bhp$. In this engine, $fhp = 588 - 500 = 88$ fhp (65.6 kW).

   Related Calculations. Use a similar procedure to determine the *indicated engine efficiency* $e_{ei} = e_i/e$, where e = ideal cycle efficiency; *brake engine efficiency,* $e_{eb} = e_b e$; *combined engine efficiency* or *overall engine thermal efficiency* $e_{eo} = e_o = e_o e$. Note that each of these three efficiencies is an *engine* efficiency and corresponds to an actual thermal efficiency, e_i, e_b, and e_o.

   Engine efficiency $e_e = e_t/e$, where e_t = actual *engine* thermal efficiency. Where desired, the respective *actual* indicated brake, or overall, output can be substituted for e_i, e_b, and e_o in the numerator of the above equations if the ideal output is substituted in the denominator. The result will be the respective engine efficiency. Output can be expressed in Btu per unit time, or horsepower. Also, e_e = actual mep/ideal mep, and $e_{ei} = imep$/ideal mep; $e_{eb} = bmep$/ideal mep; e_{eo} = overall mep/ideal mep. Further, $e_b = e_m e_i$, and $bmep = e_m(imep)$. Where the actual heat supplied by the fuel, HHV Btu/lb, is known, compute $e_i e_b$ and e_o by the method given in the previous calculation procedure. The above relations apply to any reciprocating internal-combustion engine using any fuel.

SELECTION OF AN INDUSTRIAL INTERNAL-COMBUSTION ENGINE

Select an internal-combustion engine to drive a centrifugal pump handling 2000 gal/min (126.2 L/s) of water at a total head of 350 ft (106.7 m). The pump speed will be 1750 r/min, and it will run continuously. The engine and pump are located at sea level.

TABLE 1 Internal-Combustion Engine Rating Table

Diesel engines						
Continuous bhp (kW) at given rpm					No. of	
1400	1600	1750	1800	Rated bhp	cylinders	Cooling*
187 (139.5)	214 (159.6)	227 (169.3)	230 (171.6)	300 (223.8)	6	E
230 (171.6)	256 (190.0)	275 (205.2)	280 (208.9)	438 (326.7)	12	R
240 (179.0)	273 (203.7)	295 (220.0)	305 (227.5)	438 (326.7)	12	E
Gasoline engines†						
405 (302.1)	430 (320.8)	450 (335.7)	475 (354.4)	595 (438.9)	12	R

* E = heat-exchanger-cooled; R = radiator-cooled.
† Use 80 percent of tabulated power if engine is to run at continuous full load.

Calculation Procedure:

1. Compute the power input to the pump

The power required to pump water is $hp = 8.33GH/33,000e$, where G = water flow, gal/min; H = total head on the pump, ft of water; e = pump efficiency, expressed as a decimal. Typical centrifugal pumps have operating efficiencies ranging from 50 to 80 percent, depending on the pump design and condition and liquid handled. Assume that this pump has an efficiency of 70 percent. Then $hp = 8.33(2000)/(350)/[(33,000)(0.70)] = 252$ hp (187.9 kW). Thus, the internal-combustion engine must develop at least 252 hp (187.9 kW) to drive this pump.

2. Select the internal-combustion engine

Since the engine will run continuously, extreme care must be used in its selection. Refer to a tabulation of engine ratings, such as Table 1. This table shows that a diesel engine that delivers 275 continuous brake horsepower (205.2 kW) (the nearest tabulated rating equal to or greater than the required input) will be rated at 483 bhp (360.3 kW) at 1750 r/min.

The gasoline-engine rating data in Table 1 show that for continuous full load at a given speed, 80 percent of the tabulated power can be used. Thus, at 1750 r/min, the engine must be rated at $252/0.80 = 315$ bhp (234.9 kW). A 450-hp (335.7-kW) unit is the only one shown in Table 1 that would meet the needs. This is too large; refer to another builder's rating table to find an engine rated at 315 to 325 bhp (234.9 to 242.5 kW) at 1750 r/min.

The unsuitable capacity range in the gasoline-engine section of Table 1 is a typical situation met in selecting equipment. More time is often spent in finding a suitable unit at an acceptable price than is spent computing the required power output.

Related Calculations. Use this procedure to select any type of reciprocating internal-combustion engine using oil, gasoline, liquified-petroleum gas, or natural gas for fuel.

ENGINE OUTPUT AT HIGH TEMPERATURES AND HIGH ALTITUDES

An 800-hp (596.8-kW) diesel engine is operated 10,000 ft (3048 m) above sea level. What is its output at this elevation if the intake air is at 80°F (26.7°C)? What will the output at 10,000-ft (3048-m) altitude be if the intake air is at 110°F (43.4°C)? What would the output be if this engine were equipped with an exhaust turbine-driven blower?

Calculation Procedure:

1. Compute the engine output at altitude
Diesel engines are rated at sea level at atmospheric temperatures of not more than 90°F (32.3°C). The sea-level rating applies at altitudes up to 1500 ft (457.2 m). At higher altitudes, a correction factor for elevation must be applied. If the atmospheric temperature is higher than 90°F (32.2°C), a temperature correction must be applied.

Table 2 lists both altitude and temperature correction factors. For an 800-hp (596.8-kW) engine at 10,000 ft (3048 m) above sea level and 80°F (26.7°C) intake air, hp output = (sea-level hp) (altitude correction factor), or output = (800)(0.68) = 544 hp (405.8 kW).

2. Compute the engine output at the elevated temperature
When the intake air is at a temperature greater than 90°F (32.3°C), a temperature correction factor must be applied. Then output = (sea-level hp)(altitude correction factor)(intake-air-temperature correction factor), or output = (800)(0.68)(0.95) = 516 hp (384.9 kW), with 110°F (43.3°C) intake air.

3. Compute the output of a supercharged engine
A different altitude correction is used for a supercharged engine, but the same temperature correction factor is applied. Table 2 lists the altitude correction factors for supercharged diesel engines. Thus, for this supercharged engine at 10,000-ft

TABLE 2 Correction Factors for Altitude and Temperature

Engine altitude		Engine type		Intake temperature		
ft	m	Nonsuper-charged	Super-charged	°F	°C	Correction factor
7,000	2,134	0.780	0.820	90 or less	32.3 or less	1.000
8,000	2,438	0.745	0.790	95	35	0.986
9,000	2,743	0.712	0.765	100	37.8	0.974
10,000	3,048	0.680	0.740	105	40.6	0.962
12,000	3,658	0.612	0.685	110	43.3	0.950
				115	46.1	0.937
				120	48.9	0.925
				125	51.7	0.913
				130	54.4	0.900

TABLE 3 Atmospheric Pressure at Various Altitudes

Altitude		Pressure	
ft	m	inHg	mm
Sea Level		29.92	759.97
4,000	1,219	25.84	656.3
5,000	1,524	24.89	632.2
6,000	1,829	23.98	609.1
8,000	2,438	22.22	564.4
10,000	3,048	20.58	522.7
12,000	3,658	19.03	483.4

Note: A 500- to 1500-ft altitude is considered equivalent to sea level by the Diesel Engine Manufacturers Association if the atmospheric pressure is not less than 28.25 inHg (717.6 mmHg).

(3048-m) altitude with 80°F (26.7°C) intake air, output = (sea-level hp)(altitude correction factor) = (800)(0.74) = 592 hp (441.6 kW).

At 10,000-ft (3048-m) altitude with 110°F (43.3°C) inlet air, output = (sea-level hp)(altitude correction factor)(temperature correction factor) = (800)(0.74)(0.95) = 563 hp (420.1 kW).

Related Calculations.   Use the same procedure for gasoline, gas, oil, and liquefied-petroleum gas engines. Where altitude correction factors are not available for the type of engine being used, other than a diesel, multiply the engine sea-level brake horsepower by the ratio of the altitude-level atmospheric pressure to the atmospheric pressure at sea level. Table 3 lists the atmospheric pressure at various altitudes.

An engine located below sea level can theoretically develop more power than at sea level because the intake air is denser. However, the greater potential output is generally ignored in engine-selection calculations.

INDICATOR USE ON INTERNAL-COMBUSTION ENGINES

An indicator card taken on an internal-combustion engine cylinder has an area of 5.3 in² (34.2 cm²) and a length of 4.95 in (12.7 cm). What is the indicated mean effective pressure in this cylinder? What is the indicated horsepower of this four-cycle engine if it has eight 6-in (15.6-cm) diameter cylinders, an 18-in (45.7-cm) stroke, and operates at 300 r/min? The indicator spring scale is 100 lb/in (1.77 kg/mm).

Calculation Procedure:

1. *Compute the indicated mean effective pressure*
For any indicator card, *imep* = (card area, in²) (indicator spring scale, lb)/(length of indicator card, in) where *imep* = indicated mean effective pressure, lb/in². Thus, for this engine, *imep* = (5.3)(100)/4.95 = 107 lb/in² (737.7 kPa).

2. Compute the indicated horsepower

For any reciprocating internal-combustion engine, $ihp = (imep)LAn/33,000$, where ihp = indicated horsepower per cylinder; L = piston stroke length, ft; A = piston area, in^2, n = number of cycles/min. Thus, for this four-cycle engine where n = 0.5 r/min, $ihp = (107)(18/12)(6)^2(\pi/4)(300/2)/33,000 = 20.6$ ihp (15.4 kW) per cylinder. Since the engine has eight cylinders, total ihp = (8 cylinders)(20.6 ihp per cylinder) = 164.8 ihp (122.9 kW).

Related Calculations. Use this procedure for any reciprocating internal-combustion engine using diesel oil, gasoline, kerosene, natural gas, liquefied-petroleum gas, or similar fuel.

ENGINE PISTON SPEED, TORQUE, DISPLACEMENT, AND COMPRESSION RATIO

What is the piston speed of an 18-in (45.7-cm) stroke 300 = r/min engine? How much torque will this engine deliver when its output is 800 hp (596.8 kW)? What are the displacement per cylinder and the total displacement if the engine has eight 12-in (30.5-cm) diameter cylinders? Determine the engine compression ratio if the volume of the combustion chamber is 9 percent of the piston displacement.

Calculation Procedure:

1. Compute the engine piston speed

For any reciprocating internal-combustion engine, piston speed = $fpm = 2L(rpm)$, where L = piston stroke length, ft; rpm = crankshaft rotative speed, r/min. Thus, for this engine, piston speed = $2(18/12)(300)$ = 9000 ft/min (2743.2 m/min).

2. Determine the engine torque

For any reciprocating internal-combustion engine, $T = 63,000(bhp)/rpm$, where T = torque developed, in·lb; bhp = engine brake horsepower output; rpm = crankshaft rotative speed, r/min. Or $T = 63,000(800)/300 = 168,000$ in·lb (18.981 N·m).

Where a prony brake is used to measure engine torque, apply this relation: $T = (F_b - F_o)r$, where F_b = brake scale force, lb, with engine operating; F_o = brake scale force with engine stopped and brake loose on flywheel; r = brake arm, in = distance from flywheel center to brake knife edge.

3. Compute the displacement

The displacement per cylinder d_c in^3 of any reciprocating internal-combustion engine is $d_c = L_i A_i$ where L_i = piston stroke, in; A = piston head area, in^2. For this engine, $d_c = (18)(12)^2(\pi/4) = 2035$ in^3 (33,348 cm^3) per cylinder.

The total displacement of this eight-cylinder engine is therefore (8 cylinders)(2035 in^3 per cylinder) = 16,280 in^3 (266,781 cm^3).

4. Compute the compression ratio

For a reciprocating internal-combustion engine, the compression ratio $r_c = V_b/V_a$, where V_b = cylinder volume at the start of the compression stroke, in^3 or ft^3; V_a = combustion-space volume at the end of the compression stroke, in^3 or ft^3. When this relation is used, both volumes must be expressed in the same units.

In this engine, V_b = 2035 in³ (33,348 cm³); V_a = (0.09)(2035) = 183.15 in³. Then r_c = 2035/183.15 = 11.1:1.

Related Calculations. Use these procedures for any reciprocating internal-combustion engine, regardless of the fuel burned.

INTERNAL-COMBUSTION ENGINE COOLING-WATER REQUIREMENTS

A 1000-hbp (746-kW) diesel engine has a specific fuel consumption of 0.360 lb/ (bhp·h) (0.22 kg/kWh). Determine the cooling-water flow required if the higher heating value of the fuel is 10,350 Btu/lb (24,074 kJ/kg). The net heat rejection rates of various parts of the engine are, in percent: jacket water, 11.5; turbocharger, 2.0; lube oil. 3.8; aftercooling, 4.0; exhaust, 34.7; radiation, 7.5. How much 30 lb/in² (abs) (206.8 kPa) steam can be generated by the exhaust gas if this is a four-cycle engine? The engine operates at sea level.

Calculation Procedure:

1. Compute the engine heat balance

Determine the amount of heat used to generate 1 bhp·h (0.75 kWh) from: heat rate, Btu/bhp·h) = (sfc)(HHV), where sfc = specific fuel consumption, lb/(bhp· h); HHV = higher heating value of fuel, Btu/lb. Or, heat rate = (0.36)(19.350) = 6967 Btu/(bhp·h) (2737.3 W/kWh).

Compute the heat balance of the engine by taking the product of the respective heat rejection percentages and the heat rate as follows:

			Btu/(bhp·h)	W/kWh
Jacket water	(0.115)(6967)	=	800	314.3
Turbocharger	(0.020)(6967)	=	139	54.6
Lube oil	(0.038)(6967)	=	264	103.7
Aftercooling	(0.040)(6967)	=	278	109.2
Exhaust	(0.347)(6967)	=	2420	880.1
Radiation	(0.075)(6967)	=	521	204.7
Total heat loss		=	4422	1666.6

Then the power output = 6967 − 4422 = 2545 Btu/(bhp·h) (999.9 W/kWh), or 2545/6967 = 0.365, or 36.5 percent. Note that the sum of the heat losses and power generated, expressed in percent, is 100.0.

2. Compute the jacket cooling-water flow rate

The jacket water cools the jackets and the turbocharger. Hence, the heat that must be absorbed by the jacket water is 800 + 139 = 939 Btu/(bhp·h) (369 W/kWh), using the heat rejection quantities computed in step 1. When the engine is developing its full rated output of 1000 bhp (746 kW), the jacket water must absorb [939 Btu/(bhp·h)(1000 bhp) = 939,000 Btu/h (275,221 W).

Apply a safety factor to allow for scaling of the heat-transfer surfaces and other unforeseen difficulties. Most designers use a 10 percent safety factor. Applying this value of the safety factor for this engine, we see the total jacket-water heat load = 939,000 + (0.10)(939,000) = 1,032,900 Btu/h (302.5 kW).

Find the required jacket-water flow from $G = H/500\Delta t$, where G = jacket-water flow, gal/min; H = heat absorbed by jacket water, Btu/h; Δt = temperature rise of the water during passage through the jackets, °F. The usual temperature rise of the jacket water during passage through a diesel engine is 10 to 20°F (5.6 to 11.1°C). Using 10°F for this engine we find G = 1,032,900/[(500)(10)] = 206.58 gal/min (13.03 L/s), say 207 gal/min (13.06 L/s).

3. Determine the water quantity for radiator cooling
In the usual radiator cooling system for large engines, a portion of the cooling water is passed through a horizontal or vertical radiator. The remaining water is recirculated, after being tempered by the cooled water. Thus, the radiator must dissipate the jacket, turbocharger, and lube-oil cooler heat, Fig. 5.

The lube oil gives off 264 Btu/(bhp·h) (103.8 W/kWh). With a 10 percent safety factor, the total heat flow is 264 + (0.10)(264) = 290.4 Btu/(bhp·h) (114.1 W/kWh). At the rated output of 1000 bhp (746 kW), the lube-oil heat load = [290.4 Btu/(bhp·h)](1000 bhp) = 290,400 Btu/h (85.1 kW). Hence, the total heat load on the radiator = jacket + lube-oil heat load = 1,032,900 + 290,400 = 1,323,300 Btu/h (387.8 kW)

Radiators (also called fan coolers) serving large internal-combustion engines are usually rated for a 35°F (19.4°C) temperature reduction of the water. To remove 1,323,300 Btu/h (387.8 kW) with a 35°F (19.4°C) temperature decrease will require a flow of $G = H/(500\Delta t) = 1,323,300/[(500)(35)] = 76.1$ gal/min (4.8 L/s).

4. Determine the aftercooler cooling-water quantity
The aftercooler must dissipate 278 Btu/(bhp·h) (109.2 W/kWh). At an output of 1000 bhp (746 kW), the heat load = [278 Btu/(bhp·h)](1000 bhp) = 278,000 Btu/h (81.5 kW). In general, designers do not use a factor of safety for the aftercooler because there is less chance of fouling or other difficulties.

With a 5°F (2.8°C) temperature rise of the cooling water during passage through the after-cooler, the quantity of water required $G = H/(500\Delta t) = 278,000/[(500)(5)] = 111$ gal/min (7.0 L/s).

5. Compute the quantity of steam generated by the exhaust
Find the heat available in the exhaust by using $H_e = Wc\Delta t_e$, where H_e = heat available in the exhaust, Btu/h; W = exhaust-gas flow, lb/h; c = specific heat of the exhaust gas = 0.252 Btu/(lb·°F) (2.5 kJ/kg); Δt_e = exhaust-gas temperature at the boiler inlet, °F − exhaust-gas temperature at the boiler outlet, °F.

The exhaust-gas flow from a four-cycle turbocharged diesel is about 12.5 lb/(bhp·h) (7.5 kg/kWh). At full load this engine will exhaust [12.5 lb/(bhp·h)](1000 bhp) = 12,500 lb/h (5625 kg/h).

The temperature of the exhaust gas will be about 750°F (399°C) at the boiler inlet, whereas the temperature at the boiler outlet is generally held at 75°F (41.7°C) higher than the steam temperature to prevent condensation of the exhaust gas. Steam at 30 lb/in² (abs) (206.8 kPa) has a temperature of 250.33°F (121.3°C). Thus, the exhaust-gas outlet temperature from the boiler will be 250.33 + 75 = 325.33°F (162.9°C), say 325°F (162.8°C). Then H_e = (12,500)(0.252)(750 − 325) = 1,375,000 Btu/h (403.0 kW).

At 30 lb/in² (abs) (206.8 kPa), the enthalpy of vaporization of steam is 945.3 Btu/lb (2198.9 kJ/kg), found in the steam tables. Thus, the exhaust heat can gen-

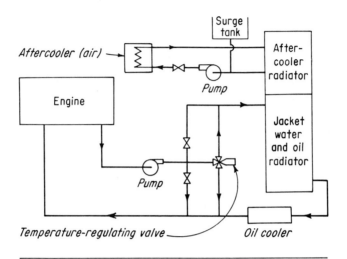

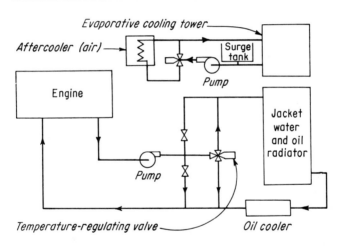

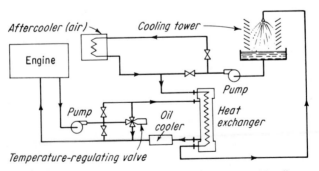

FIGURE 5   Internal-combustion engine cooling systems: (*a*) radiator type;
(*b*) evaporating cooling tower; (*c*) cooling tower. (*Power.*)

erate 1,375,000/945.3 = 1415 lb/h (636.8 kg/h) if the boiler is 100 percent efficient. With a boiler efficiency of 85 percent, the steam generated = (1415 lb/
h)(0.85) = 1220 lb/h (549.0 kg/h), or (1200 lb/h)/1000 bhp = 1.22 lb/(bhp·h)
(0.74 kg/kWh).

Related Calculations. Use this procedure for any reciprocating internal-
combustion engine burning gasoline, kerosene, natural gas, liquified-petroleum gas,
or similar fuel. Figure 1 shows typical arrangements for a number of internal-
combustion engine cooling systems.

When ethylene glycol or another antifreeze solution is used in the cooling system, alter the denominator of the flow equation to reflect the change in specific
gravity and specific heat of the antifreeze solution, a s compared with water. Thus,
with a mixture of 50 percent glycol and 50 percent water, the flow equation in step
2 becomes $G = H/(436\Delta t)$. With other solutions, the numerical factor in the denominator will change. This factor = (weight of liquid lb/gal)(60 min/h), and the
factor converts a flow rate of lb/h to gal/min when divided into the lb/h flow rate.
Slant diagrams, Fig 6, are often useful for heat-exchanger analysis.

Two-cycle engines may have a larger exhaust-gas flow than four-cycle engines
because of the scavenging air. However, the exhaust temperature will usually be 50
to 100°F (27.7 to 55.6°C) lower, reducing the quantity of steam generated.

Where a dry exhaust manifold is used on an engine, the heat rejection to the
cooling system is reduced by about 7.5 percent. Heat rejected to the aftercooler
cooling water is about 3.5 percent of the total heat input to the engine. About 2.5
percent of the total heat input to the engine is rejected by the turbocharger jacket.

The jacket cooling water absorbs 11 to 14 percent of the total heat supplied.
From 3 to 6 percent of the total heat supplied to the engine is rejected in the oil
cooler.

The total heat supplied to an engine = (engine output, bhp)[heat rate, Btu/
(bhp·h)]. A jacket-water flow rate of 0.25 to 0.60 gal/(min·bhp) (0.02 to 0.05
kg/kW) is usually recommended. The normal jacket-water temperature rise is 10°F
(5.6°C); with a jacket-water outlet temperature of 180°F (82.2°C) or higher, the
temperature rise of the jacket water is usually held to 7°F (3.9°C) or less.

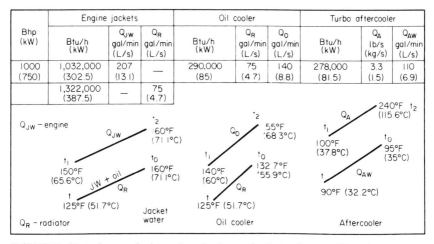

Bhp (kW)	Engine jackets			Oil cooler			Turbo aftercooler		
	Btu/h (kW)	Q_{JW} gal/min (L/s)	Q_R gal/min (L/s)	Btu/h (kW)	Q_R gal/min (L/s)	Q_O gal/min (L/s)	Btu/h (kW)	Q_A lb/s (kg/s)	Q_{AW} gal/min (L/s)
1000 (750)	1,032,000 (302.5)	207 (13.1)	—	290,000 (85)	75 (4.7)	140 (8.8)	278,000 (81.5)	3.3 (1.5)	110 (6.9)
	1,322,000 (387.5)	—	75 (4.7)						

FIGURE 6 Slant diagrams for internal-combustion engine heat exhangers. (*Power.*)

To keep the cooling-water system pressure loss within reasonable limits, some designers recommend a pipe velocity equal to the nominal pipe size used in the system, or 2ft/s for 2-in pipe (0.6 m/s for 50.8-mm); 3 ft/s for 3-in pipe (0.9 m/s for 76.2-mm); etc. The maximum recommended velocity is 10 ft/s for 10 in (3.0 m/s for 254 mm) and larger pipes. Compute the actual pipe diameter from $d = (G/2.5v)^{0.5}$, where $G =$ cooling-water flow, gal/min; $v =$ water velocity, ft/s.

Air needed for a four-cycle high-output turbocharged diesel engine is about 3.5 ft³/(min·bhp) (0.13 m³/kW); 4.5 ft³/(min·bhp)(0.17 m³/kW) for two-cycle engines. Exhaust-gas flow is about 8.4 ft³/(min·bhp) (0.32 m³/kW) for a four-cycle diesel engine; 13 ft³/(min·bhp) (0.49 m³/kW) for two-cycle engines. Air velocity in the turbocharger blower piping should not exceed 3300 ft/min (1006 m/min); gas velocity in the exhaust system should not exceed 6000 ft/min (1828 m/min). The exhaust-gas temperature should not be reduced below 275°F (135°C), to prevent condensation.

The method presented here is the work of W. M. Kauffman, reported in *Power*.

DESIGN OF A VENT SYSTEM FOR AN ENGINE ROOM

A radiator-cooled 60-kW internal-combustion engine generating set operates in an area where the maximum summer ambient temperature of the inlet air is 100°F (37.8°C). How much air does this engine need for combustion and for the radiator? What is the maximum permissible temperature rise of the room air? How much heat is radiated by the engine-alternator set if the exhaust pipe is 25 ft (7.6 m) long? What capacity exhaust fan is needed for this engine room if the engine room has two windows with an area of 30 ft² (2.8 m²) each, and the average height between the air inlet and the outlet is 5 ft (1.5 m)? Determine the rate of heat dissipation by the windows. The engine is located at sea level.

Calculation Procedure:

1. Determine engine air-volume needs

Table 4 shows typical air-volume needs for internal-combustion engines installed indoors. Thus, a 60-kW set requires 390 ft³/min (11.0 m³/min) for combustion and

TABLE 4 Total Air Volume Needs*

Set kW	ft³/min (m³/min) for combustion	ft³/min (m³/min) for radiator	Maximum ambient temperature of inlet air, °F (°C)	Room air rise, °F (°C)
20	130 (3.7)	3000 (84.9)	90 (32.2)	20 (11.1)
30	195 (5.5)	5000 (141.6)	95–105 (35–40.6)	15 (8.3)
40	260 (7.4)	5500 (155.7)	110–120 (43.3–48.9)	10 (5.6)
60	390 (1.0)	6000 (169.9)		

Power.

6000 ft³/min (169.9 m³/min) for the radiator. Note that in the smaller ratings, the combustion air needed is 6.5 ft³/(min·kW)(0.18 m³/kW), and the radiator air requirement is 150 ft³/(min·kW)(4.2 m³/kW).

2. Determine maximum permissible air temperature rise
Table 4 also shows that with an ambient temperature of 95 to 105°F (35 to 40.6°C), the maximum permissible room temperature rise is 15°C (8.3°C). When you determine this value, be certain to use the highest inlet air temperature expected in the engine locality.

3. Determine the heat radiated by the engine
Table 5 shows the heat radiated by typical internal-combustion engine generating sets. Thus, a 60-kW radiator-and fan-cooled set radiates 2625 Btu/min (12.8 W) when the engine is fitted with a 25-ft (7.6-m) long exhaust pipe and a silencer.

4. Compute the airflow produced by the windows
The two windows can be used to ventilate the engine room. One window will serve as the air inlet; the other, as the air outlet. The area of the air outlet must at least equal the air-inlet area. Airflow will be produced by the stack effect resulting from the temperature difference between the inlet and outlet air.

The airflow C ft³/min resulting from the stack effect is $C = 9.4A(h\Delta t_a)^{0.5}$, where A = free air of the air inlet, ft²; h = height from the middle of the air-inlet opening to the middle of the air-outlet opening, ft; Δt_a = difference between the average indoor air temperature at point H and the temperature of the incoming air, °F. In this plant, the maximum permissible air temperature rise is 15°F (8.3°C), from step 2. With a 100°F (37.8°C) outdoor temperature, the maximum indoor temperature would be $100 + 15 = 115$°F (46.1°C). Assume that the difference between the temperature of the incoming and outgoing air is 15°F (8.3°C). Then $C = 9.4(30)(5 \times 15)^{0.5} = 2445$ ft³/min (69.2 m³/min).

5. Compute the cooling airflow required
This 60-kW internal-combustion engine generating set radiates 2625 Btu/min (12.8 W), step 3. Compute the cooling airflow required from $C = HK/\Delta t_a$, where C = cooling airflow required, ft³/min; H = heat radiated by the engine, Btu/min; K = constant from Table 6; other symbols as before. Thus, for this engine with a fan

TABLE 5 Heat Radiated from Typical Internal-Combustion Units, Btu/min (W)*

	Cooling by radiator and fan		Cooling by radiator, fan, and city water	
Alternator, kW	40	60	40	60
Engine-alternator set, silencer, and 25 ft (7.6 m) of exhaust pipe, Btu/min (W)	1830 (8.94)	2625 (12.8)	1701 (8.3)	2500 (12.2)
Exhaust pipe beyond silencer:				
Length 5 ft (1.5 m)	24 (0.12)	35 (0.17)	20 (0.10)	22 (0.11)
Length 10 ft (3.0 m)	45 (0.22)	65 (0.32)	39 (0.19)	40 (0.20)
Length 15 ft (4.6 m)	65 (0.32)	89 (0.44)	57 (0.38)	55 (0.27)

Power.

TABLE 6   Range of Discharge Temperature*

Room fan discharge temperature range			Wind to water gage			
			Wind velocity		Inlet pressure water gage	
°F	°C	K	mph	km/h	in	mm
80–89	26.7–31.7	57	60	96.5	1.75	44.5
90–99	32.3–37.2	58	30	48.3	0.43	10.9
100–110	37.8–43.3	59				
111–120	43.9–48.9	60				
121–130	49.4–54.4	61				

* *Power.*

discharge temperature of 111 to 120°F (43.9 to 48.9°C), Table 6, K = 60; Δt_a = 15°F (8.3°C) from step 4. Then C = (2625)(60)/15 = 10,500 ft³/min (297.3 m³/min).

The windows provide 2445 ft³/min (69.2 m³/min), step 4, and the engine radiator gives 6000 ft³/min (169.9 m³/min), step 1, or a total of 2445 + 6000 = 8445 ft³/min (239.1 m³/min). Thus, 10,500 − 8445 = 2055 ft³/min (58.2 m³/min) must be removed from the room. The usual method employed to remove the air is an exhaust fan. An exhaust fan with a capacity of 2100 ft³/min (59.5 m³/min) would be suitable for this engine room.

Related Calculations.   Use this procedure for engines burning any type of fuel—diesel, gasoline, kerosene, or gas—in any type of enclosed room at sea level or elevations up to 1000 ft (304.8 m). Where windows or the fan outlet are fitted with louvers, screens, or intake filters, be certain to compute the net free area of the opening. When the radiator fan requires more air than is needed for cooling the room, an exhaust fan is unnecessary.

Be certain to select an exhaust fan with a sufficient discharge pressure to overcome the resistance of exhaust ducts and outlet louvers, if used. A propeller fan is usually chosen for exhaust service. In areas having high wind velocity, an axial-flow fan may be needed to overcome the pressure produced by the wind on the fan outlet.

Table 6 shows the pressure developed by various wind velocities. When the engine is located above sea level, use the multiplying factor in Table 7 to correct the computed air quantities for the lower air density.

An engine radiates 2 to 5 percent of its total heat input. The total heat input = (engine output, bhp) [heat rate, Btu/(bhp·h)]. Provide 12 to 20 air changes per hour for the engine room. The most effective ventilators are power-driven exhaust fans or roof ventilators. Where the heat load is high, 100 air changes per hour may be provided. Auxiliary-equipment rooms require 10 air changes per hour. Windows, louvers, or power-driven fans are used. A four-cycle engine requires 3 to 3.5 ft³/min of air per bhp (0.11 to 0.13 m³/kW); a two-cycle engine, 4 to 5 ft³/(min·bhp) (0.15 to 0.19 m³/kW).

The method presented here is the work of John P. Callaghan, reported in *Power*.

TABLE 7 Air Density at Various Elevations*

Elevation above sea level		Multiplying factor, A	Approximate air density percent compared with sea level for same temperature
ft	m		
4,000	1,219	1.158	86.4
5,000	1,524	1.202	83.2
6,000	1,829	1.247	80.2
7,000	2,134	1.296	77.2
10,000	3,048	1.454	68.8

Power.

DESIGN OF A BYPASS COOLING SYSTEM FOR AN ENGINE

The internal-combustion engine in Fig. 7 is rated at 402 hp (300 kW) at 514 r/min and dissipates 3500 Btu/(bhp·h) (1375 W/kW) at full load to the cooling water from the power cylinders and water-cooled exhaust manifold. Determine the required cooling-water flow rate if there is a 10°F (5.6°C) temperature rise during passage of the water through the engine. Size the piping for the cooling system, using the head-loss data in Fig. 8, and the pump characteristic curve, Fig. 9. Choose a surge tank of suitable capacity. Determine the net positive suction head requirements for this engine. The total length of straight piping in the cooling system is 45 ft (13.7 m). The engine is located 500 ft (152.4 m) above sea level.

Calculation Procedure:

1. Compute the cooling-water quantity required
The cooling-water quantity required is $G = H/(500\Delta t$, where G = cooling-water flow, gal/min; H = heat absorbed by the jacket water, Btu/h = (maximum engine

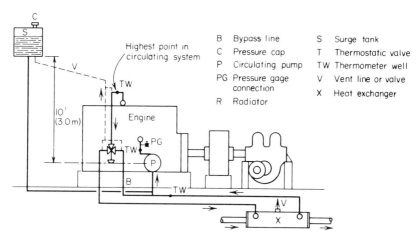

FIGURE 7 Engine cooling-system hookup. (*Mechanical Engineering.*)

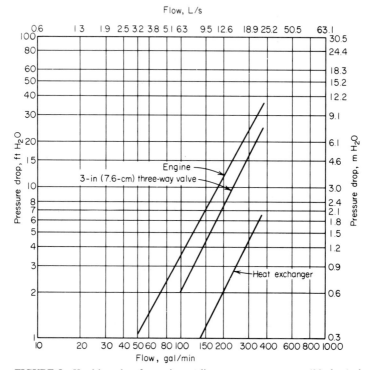

FIGURE 8 Head-loss data for engine cooling-system components. (*Mechanical Engineering.*)

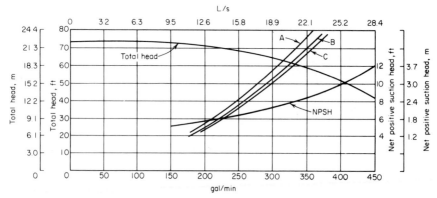

FIGURE 9 Pump and system characteristics for engine cooling system. (*Mechanical Engineering.*)

hp) [heat dissipated, Btu/(bhp·h)]; Δt = temperature rise of the water during passage through the engine, °F. Thus, for this engine, $G = (402)(3500)/[500(10)]$ = 281 gal/min (17.7 L/s)

2. Choose the cooling-system valve and pipe size

Obtain the friction head-loss data for the engine, the heat exchanger, and the three-way valve from the manufacturers of the respective items. Most manufacturers have curves or tables available for easy use. Plot the head losses, as shown in Fig. 8, for the engine and heat exchanger.

Before the three-way valve head loss can be plotted, a valve size must be chosen. Refer to a three-way valve capacity tabulation to determine a suitable valve size to handle a flow of 281 gal/min (17.7 L/s). Once such tabulation recommends a 3-in (76.2-mm) valve for a flow of 281 gal/min (17.7 L/s). Obtain the head-loss data for the valve, and plot it as shown in Fig. 8.

Next, assume a size for the cooling-water piping. Experience shows that a water velocity of 300 to 600 ft/min (91.4 to 182.9 m/min) is satisfactory for internal-combustion engine cooling systems. Using the Hydraulic Institute's *Pipe Friction Manual* or Cameron's *Hydraulic Data*, enter at 280 gal/min (17.6 L/s), the approximate flow, and choose a pipe size to give a velocity of 400 to 500 ft/min (121.9 to 152.4 m/min), i.e., midway in the recommended range.

Alternatively, compute the approximate pipe diameter from d = 4.95 [gpm/velocity, ft/min]$^{0.5}$. With a velocity of 450 ft/min (137.2 m/min), d = 4.95(281/450)$^{0.5}$ = 3.92, say 4 in (101.6 mm). The *Pipe Friction Manual* shows that the water velocity will be 7.06 ft/s (2.2 m/s), or 423.6 ft/min (129.1 m/min), in a 4-in (101.6 mm) schedule 40 pipe. This is acceptable. Using a 3½-in (88.9-mm) pipe would increase the cost because the size is not readily available from pipe suppliers. A 3-in (76.2-mm) pipe would give a velocity of 720 ft/min (219.5 m/min), which is too high.

3. Compute the piping-system head loss

Examine Fig. 7, which shows the cooling system piping layout. Three flow conditions are possible: (a) all the jacket water passes through the heat exchanger, (b) a portion of the jacket water passes through the heat exchanger, and (c) none of the jacket water passes through the heat exchanger—instead, all the water passes through the bypass circuit. The greatest head loss usually occurs when the largest amount of water passes through the longest circuit (or flow condition a). Compute the head loss for this situation first.

Using the method given in the piping section of this handbook, compute the equivalent length of the cooling-system fitting and piping, as shown in Table 8. Once the equivalent length of the pipe and fittings is known, compute the head loss in the piping system, using the method given in the piping section of this handbook with a Hazen-Williams constant of C = 130 and a rounded-off flow rate of 300 gal/min (18.9 L/s). Summarize the results as shown in Table 8.

The total head loss is produced by the water flow through the piping, fittings, engine, three-way valve, and heat exchanger. Find the head loss for the last components in Fig. 8 for a flow of 300 gal/min (18.9 L/s). List the losses in Table 8, and find the sum of all the losses. Thus, the total circuit head loss is 57.61 ft (17.6 m) of water.

Compute the head loss for 0, 0.2, 0.4, 0.6, and 0.8 load on the engine, using the same procedure as in steps 1, 2, and 3 above. Plot on the pump characteristic curve, Fig. 9, the system head loss for each load. Draw a curve A through the points obtained, Fig. 9.

TABLE 8 Sample Calculation for Full Flow through Cooling Circuit*
(*Fittings and Piping in Circuit*)

Fitting or pipe	Number in circuit	Equivalent length of straight pipe	
		ft	m
3-in (76.2-mm) elbow	1	5.5	1.7
3 × 4 (76.2 × 101.6-mm) reducer	4	7.2	2.2
4-in (101.6-mm) elbow	7	50.4	15.4
4-in (101.6-mm) tee	1	23.0	7.0
3-in (76.2-mm) pipe	· · ·	0.67	0.2
4-in (101.6-mm) pipe	· · ·	45.0	13.7
Total equivalent length of pipe:			
3-in (76.2-mm) pipe, standard weight	· · ·	13.37	4.1
4-in (101.6-mm) pipe, standard weight	· · ·	118.4	36.1

Head loss calculation: Calculation for a flow rate of 300 gal/min (18.9 L/s) through circuit:

Using the Hazen-Williams friction-loss equation with a C factor of 130 (surface roughness constant), with 300 gal/min (18.9 L/s) flowing through the pipe, the head loss per 100 ft (30.5 m) of pipe is 21.1 ft (6.4 m) and 5.64 ft (1.1 m) for the 3-in (76.2-mm) and 4-in (101.6-mm) pipes, respectively. Thus head loss in piping is[†]

$$3 \text{ in } \frac{21.1}{100} \times 13.37 = 2.83 \text{ ft } (0.86 \text{ m})$$

$$4 \text{ in } \frac{5.64}{100} \times 118.4 = 6.68 \text{ ft } (2.0 \text{ m})$$

From Fig. 5 the head loss is:	ft	m
Through engine	26.00	7.9
Through 3-in (76.2-mm) three-way valve	17.50	5.3
Through heat exchanger	4.6	1.4
Total circuit head loss	57.61	14.6

*Mechanical Engineering.
[†]Shaw and Loomis, *Cameron Hydraulic Data Book*, 12th ed., Ingersoll-Rand Company, 1951, p. 27.

Compute the system head loss for condition *b* with half the jacket water [150 gal/min (9.5 L/s)] passing through the heat exchanger and half [150 gal/min (9.5 L/s)] through the bypass circuit. Make the same calculation for 0, 0.2, 0.4, 0.6, and 0.8 load on the engine. Plot the result as curve *B*, Fig 9.

Perform a similar calculation for condition *c*—full flow through the bypass circuit. Plot the results as curve *C*, Fig. 9.

4. Compute the actual cooling-water flow rate
Find the points of intersection of the pump total-head curve and the three system head-loss curves A, B, and C, Fig. 9. These intersections occur at 314, 325, and 330 gal/min (19.8, 20.5, and 20.8 L/s), respectively.

The initial design assumed a 10°F (5.6°C) temperature rise through the engine with a water flow rate of 281 gal/min (17.7 L/s). Rearranging the equation in step 1 gives $\Delta t = H/(400G)$. Substituting the flow rate for condition *a* gives an actual temperature rise of $\Delta t = (402)(3500)/[(500)(314)] = 8.97°F (4.98°C)$. If a 180°F

(82.2°C) rated thermostatic element is used in the three-way valve, holding the outlet temperature t_o to 180°F (82.2°C), the inlet temperature t_i will be $\Delta t = t_o - t_i = 8.97$; $180 - t_i = 8.97$; $t_i = 171.03$°F (77.2°C).

5. Determine the required surge-tank capacity
The surge tank in a cooling system provides storage space for the increase in volume of the coolant caused by thermal expansion. Compute this expansion from $E = 62.4g\Delta V$, where E = expansion, gal (L); g = number of gallons required to fill the cooling system; ΔV = specific volume, ft³/lb (m³/kg) of the coolant at the operating temperature − specific volume of the coolant, ft³/lb (m³/kg) at the filling temperature.

The cooling system for this engine must have a total capacity of 281 gal (1064 L), step 1. Round this to 300 gal (1136 L) for design purposes. The system operating temperature is 180°F (82.2°C), and the filling temperature is usually 60°F (15.6°C). Using the steam tables to find the specific volume of the water at these temperatures, we get $E = 62.4(300)(0.01651 - 0.01604) = 8.8$ gal (33.3 L).

Usual design practice is to provide two to three times the required expansion volume. Thus, a 25-gal (94.6-L) tank (nearly three times the required capacity) would be chosen. The extra volume provides for excess cooling water that might be needed to make up water lost through minor leaks in the system.

Locate the surge tank so that it is the highest point in the cooling system. Some engineers recommend that the bottom of the surge tank be at least 10 ft (3 m) above the pump centerline and connected as close as possible to the pump intake. A 1½- or 2-in (38.1- or 50.8-mm) pipe is usually large enough for connecting the surge tank to the system. The line should be sized so that the head loss of the vented fluid flowing back to the pump suction will be negligible.

6. Determine the pump net positive suction head
The pump characteristic curve, Fig 9, shows the net positive suction head (NSPH) required by this pump. As the pump discharge rate increases, so does the NPSH. this is typical of a centrifugal pump.

The greatest flow, 330 gal/min (20.8 L/s), occurs in this system when all the coolant is diverted through the bypass circuit, Figs. 4 and 5. At a 330-gal/min (20.8-L/s) flow rate through the system, the required NPSH for this pump is 8 ft (2.4 m), Fig 9. This value is found at the intersection of the 330-gal/min (20.8 L /s) ordinate and the NPSH curve.

Compute the existing NPSH, ft (m), from NPSH $= H_s - H_f + 2.31(P_s - P_v)/s$, where H_s = height of minimum surge-tank liquid level above the pump centerline, ft (m); H_f = friction loss in the suction line from the surge-tank connection to the pump inlet flange, ft (m) of liquid; P_s = pressure in surge tank, or atmospheric pressure at the elevation of the installation, lb/in² (abs) (kPa); P_v = vapor pressure of the coolant at the pumping temperature, lb /in² (abs) (kPa); s = specific gravity of the coolant at the pumping temperature.

7. Determine the operating temperature with a closed surge tank
A pressure cap on the surge tank, or a radiator, will permit operation at temperatures above the atmospheric boiling point of the coolant. At a 500-ft (152.4-m) elevation, water boils at 210°F (98.9°C). Thus, without a closed surge tank fitted with a pressure cap, the maximum operating temperature of a water-cooled system would be about 200°F (93.3°C).

If a 7-lb/in² (gage) (48.3 kPa) pressure cap were used at the 500-ft (152.4-m) elevation, then the pressure in the vapor space of the surge tank could rise to $P_s =$

14.4 + 7.0 = 21.4 lb/in² (abs) (147.5 kPa). The steam tables show that water at this pressure boils at 232°F (111.1°C). Checking the NPSH at this pressure shows that NPSH = (10 − 1.02) + 2.31(21.4 − 21.4)/0.0954 = 8.98 ft (2.7 m). This is close to the required 8-ft (2.4-m) head. However, the engine could be safely operated at a slightly lower temperature, say 225°F (107.2°C).

8. Compute the pressure at the pump suction flange
The pressure at the pump suction flange P lb/in² (gage) = $0.433s(H_s − H_f)$ = $(0.433)(0.974)(10.00 − 1.02)$ = 3.79 lb/in² (gage) (26.1 kPa).

A positive pressure at the pump suction is needed to prevent the entry of air along the shaft. To further ensure against air entry, a mechanical seal can be used on the pump shaft in place of packing.

Related Calculations.   Use this general procedure in designing the cooling system for any type of reciprocating internal-combustion engine—gasoline, diesel, gas, etc. Where a coolant other than water is used, follow the same procedure but change the value of the constant in the denominator of the equation of step 1. Thus, for a mixture of 50 percent glycol and 50 percent water, the constant = 436, instead of 500.

The method presented here is the work of Duane E. Marquis, reported in *Mechanical Engineering.*

HOT-WATER HEAT-RECOVERY SYSTEM ANALYSIS

An internal-combustion engine fitted with a heat-recovery silencer and a jacket-water cooler is rated at 1000 bhp (746 kW). It exhausts 13.0 lb/(bhp · h) [5.9 kg/(bhp · h)] of exhaust gas at 700°F (371.1°C). To what temperature can hot water be heated when 500 gal/min (31.5 L/s) of jacket water is circulated through the hookup in Fig. 10 and 100 gal/min (6.3 L/s) of 60°F (15.6°C) water is heated? The jacket water enters the engine at 170°F (76.7°C) and leaves at 180°F (82.2°C).

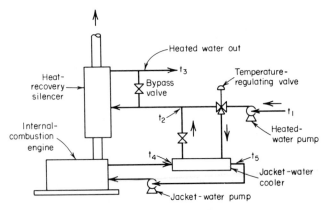

FIGURE 10   Internal-combustion engine cooling system.

Calculation Procedure:

1. Compute the exhaust heat recovered
Find the exhaust-heat recovered from $H_e = Wc\Delta t_e$, where the symbols are the same as in the previous calculation procedures. Since the final temperature of the exhaust gas is not given, a value must be assumed. Temperatures below 275°F (135°C) are undesirable because condensation of corrosive vapors in the silencer may occur. Assume that the exhaust-gas outlet temperature from the heat-recovery silencer is 300°F (148.9°C). The $H_e = (1000)(13)(0.252)(700 - 300) = 1,310,000$ Btu/h (383.9 kW).

2. Compute the heated-water outlet temperature from the cooler
Using the temperature notation in Fig. 10, we see that the heated-water outlet temperature from the jacket-water cooler is $t_z = (w_z/w_1)(t_4 - t_5) + t_1$), where $w_1 =$ heated-water flow, lb/h; $w_z =$ jacket-water flow, lb/h; the other symbols are indicated in Fig. 10. To convert gal/min of water flow to lb/h, multiply by 500. Thus, $w_1 = (100 \text{ gal/min})(500) = 50,000$ lb/h (22,500 kg/h), and $w_z = (500 \text{ gal/min})(500) = 250,000$ lb/h (112,500 kg/h). Then $t_z = (250,000/50,000)(180 - 170) + 60 = 110°F$ (43.4°C).

3. Compute the heated-water outlet temperature from the silencer
The silencer outlet temperature $t_3 = H_e/w_1 + t_z$, or $t_3 = 1,310,000/50,000 + 110 = 136.2°F$ (57.9°C).
    Related Calculations.   Use this method for any type of engine—diesel, gasoline, or gas—burning any type of fuel. Where desired, a simple heat balance can be set up between the heat-releasing and heat-absorbing sides of the system instead of using the equations given here. However, the equations are faster and more direct.

DIESEL FUEL STORAGE CAPACITY AND COST

A diesel power plant will have six 1000-hp (746-kW) engines and three 600-hp (448-kW) engines. The annual load factor is 85 percent and is nearly uniform throughout the year. What capacity day tanks should be used for these engines? If fuel is delivered every 7 days, what storage capacity is required? Two fuel supplies are available; a 24° API fuel at $0.0825 per gallon ($0.022 per liter) and a 28° API fuel at $0.0910 per gallon ($0.024 per liter). Which is the better buy?

Calculation Procedure:

1. Compute the engine fuel consumption
Assume, or obtain from the engine manufacturer, the specific fuel consumption of the engine. Typical modern diesel engines have a full-load heat rate of 6900 to 7500 Btu/(bhp·h) (2711 to 3375 W/kWh), or about 0.35 lb/(bhp·h) of fuel (0.21 kg/kWh). Using this value of fuel consumption for the nine engines in this plant, we see the hourly fuel consumption at 85 percent load factor will be (6 engines)(1000 hp)(0.35)(0.85) + (3 engines)(600 hp)(0.35)(0.85) = 2320 lb/h (1044 kg/h).
    Convert this consumption rate to gal/h by finding the specific gravity of the diesel oil. The specific gravity $s = 141.5/(131.5 + °API)$. For the 24° API oil, $s =$

141.5/(131.5 + 24) = 0.910. Since water at 60°F (15.6°C) weighs 8.33 lb/gal (3.75 kg/L), the weight of this oil is (0.910)(8.33) = 7.578 lb/gal (3.41 kg/L). For the 28° API oil, s = 141.5/(131.5 + 28) = 0.887, and the weight of this oil is (0.887)(8.33) = 7.387 lb/gal (3.32 kg/L). Using the lighter oil, since this will give a larger gal/h consumption, we get the fuel rate = (2320 lb/h)/(7.387 lb/gal) = 315 gal/h (1192 L/h).

The daily fuel consumption is then (24 h/day)(315 gal/h) = 7550 gal/day (28,577 L/day). In 7 days the engines will use (7 days)(7550 gal/day) = 52,900, say 53,000 gal (200,605 L).

2. Select the tank capacity
The actual fuel consumption is 53,000 gal (200,605 L) in 7 days. If fuel is delivered exactly on time every 7 days, a fuel-tank capacity of 53,000 gal (200,605 L) would be adequate. However, bad weather, transit failures, strikes, or other unpredictable incidents may delay delivery. Therefore, added capacity must be provided to prevent engine stoppage because of an inadequate fuel supply.

Where sufficient space is available, and local regulations do not restrict the storage capacity installed, use double the required capacity. The reason is that the additional storage capacity is relatively cheap compared with the advantages gained. Where space or storage capacity is restricted, use 1½ times the required capacity.

Assuming double capacity is used in this plant, the total storage capacity will be (2)(53,000) = 106,000 gal (401,210 L). At least two tanks should be used, to permit cleaning of one without interrupting engine operation.

Consult the National Board of Fire Underwriters bulletin *Storage Tanks for Flammable Liquids* for rules governing tank materials, location, spacing, and fire-protection devices. Refer to a tank capacity table to determine the required tank diameter and length or height depending on whether the tank is horizontal or vertical. Thus, the Buffalo Tank Corporation *Handbook* shows that a 16.5-ft (5.0-m) diameter 33.5-ft (10.2-m) long horizontal tank will hold 53,600 gal (202,876 L) when full. Two tanks of this size would provide the desired capacity. Alternatively, a 35-ft (10.7-m) diameter 7.5-ft (2.3-m) high vertical tank will hold 54,000 gal (204,390 L) when full. Two tanks of this size would provide the desired capacity.

Where a tank capacity table is not available, compute the capacity of a cylindrical tank from capacity = $5.87D^2L$, where D = tank diameter, ft; L = tank length or height, ft. Consult the NBFU or the tank manufacturer for the required tank wall thickness and vent size.

3. Select the day-tank capacity
Day tanks supply filtered fuel to an engine. The day tank is usually located in the engine room and holds enough fuel for a 4- to 8-h operation of an engine at full load. Local laws, insurance requirements, or the NBFU may limit the quantity of oil that can be stored in the engine room or a day tank. One day tank is usually used for each engine.

Assume that a 4-h supply will be suitable for each engine. Then the day tank capacity for a 1000-hp (746-kW) engine = (1000 hp) [0.35 lb/(bhp·h) fuel] (4 h) = 1400 lb (630 kg), or 1400/7.387 = 189.6 gal (717.6 L), given the lighter-weight fuel, step 1. Thus, one 200-gal (757-L) day tank would be suitable for each of the 1000-hp (746-kW) engines.

For the 600-hp (448-kW) engines, the day-tank capacity should be (600 hp)[0.35 lb/(bhp·h) fuel](4 h) = 840 lb (378 kg), or 840/7.387 = 113.8 gal (430.7 L). Thus, one 125-gal (473-L) day tank would be suitable for each of the 600-hp (448-kW) engines.

4. *Determine which is the better fuel buy*
Compute the higher heating value HHV of each fuel from HHV = 17,645 + 54(°API), or for 24° fuel, HHV = 17,645 + 54(24) = 18,941 Btu/lb (44,057 kJ/kg). For the 28° fuel, HHV = 17,645 + 54(28) = 19,157 Btu/lb (44,559 kJ/kg).

Compare the two oils on the basis of cost per 10,000 Btu (10,550 kJ), because this is the usual way of stating the cost of a fuel. The weight of each oil was computed in step 1. Thus the 24° API oil weighs 7,578 lb/gal (0.90 kg/L), while the 28° API oil weighs 7.387 lb/gal (0.878 kg/L).

Then the cost per 10,000 Btu (10,550 kJ) = (cost, \$/gal)/[HHV, Btu/lb)/10,000](oil weight, lb/gal). For the 24° API oil, cost per 10,000 Btu (10,550 kJ) = (cost, \$/gal)/[(HHV, Btu/lb)/10,000](oil weight, lb/gal). For the 24° API oil, cost per 10,000 Btu (10,550 kJ) = \$0.0825/[(18.941/10,000)(7.578)] = \$0.00574, or 0.574 cent per 10,000 Btu (10,550 kJ). For the 28° API oil, cost per 10,000 Btu = \$0.0910/[(19,157/10,000)(7387)] = \$0.00634, or 0.634 cent per 10,000 Btu (10,550 kJ). Thus, the 24° API is the better buy because it costs less per 10,000 Btu (10,550 kJ).

Related Calculations.   Use this method for engines burning any liquid fuel. Be certain to check local laws and the latest NBFU recommendations before ordering fuel storage or day tanks.

Low-sulfur diesel amendments were added to the federal Clean Air Act in 1991. These amendments required diesel engines to use low-sulfur fuel to reduce atmospheric pollution. Reduction of fuel sulfur content will not require any change in engine operating procedures. If anything, the lower sulfur content will reduce engine maintenance requirements and costs.

The usual distillate fuel specification recommends a sulfur content of not more than 1.5 percent by weight, with 2 percent by weight considered satisfactory. Refineries are currently producing diesel fuel that meets federal low-sulfur requirements. While there is a slight additional cost for such fuel at the time of this writing, when the regulations went into effect, predictions are that the price of low-sulfur fuel will decline as more is manufactured.

Automobiles produce 50 percent of the air pollution throughout the developed world. The Ozone Transport Commission, set up by Congress as part of the 1990 Clear Air Act, is enforcing emission standards for new automobiles and trucks. To date, the cost of meeting such standards has been lower than anticipated. By the year 2003, all new automobiles will be pollution-free—if they comply with the requirements of the act. Stationary diesel plants using low-sulfur fuel will emit extremely little pollution.

POWER INPUT TO COOLING-WATER AND LUBE-OIL PUMPS

What is the required power input to a 200-gal/min (12.6-L/s) jacket-water pump if the total head on the pump is 75 ft (22.9 m) of water and the pump has an efficiency of 70 percent when it handles freshwater and saltwater? What capacity lube-oil pump is needed for a four-cycle 500-hp (373-kW) turbocharged diesel engine having oil-cooled pistons? What is the required power input to this pump if the discharge pressure is 80 lb/in² (551.5 kPa) and the efficiency of the pump is 68 percent?

Calculation Procedure:

1. Determine the power input to the jacket-water pump

The power input to jacket-water and raw-water pumps serving internal-combustion engines is often computed from the relation $hp = Gh/Ce$, where hp = hp input; G = water discharged by pump, gal/min; h = total head on pump, ft of water; C = constant = 3960 for freshwater having a density of 62.4 lb/ft^3 (999.0 kg/m^3); 3855 for saltwater having a density of 64 lb/ft^3 (1024.6 kg/m^3).

For this pump handling freshwater, $hp = (200)(75)/(3960)(0.70) = 5.42$ hp (4.0 kW). A 7.5-hp (5.6-kW) motor would probably be selected to handle the rated capacity plus any overloads.

For this pump handling saltwater, $hp = (200)(75/[(3855)(0.70)]) = 5.56$ hp (4.1 kW). A 7.5-hp (5.6-kW) motor would probably be selected to handle the rated capacity plus any overloads. Thus, the same motor could drive this pump whether it handles freshwater or saltwater.

2. Compute the lube-oil pump capacity

The lube-oil pump capacity required for a diesel engine is found from $G = H/200\Delta t$, where G = pump capacity, gal/min; H = heat rejected to the lube oil, Btu/(bhp·h); Δt = lube-oil temperature rise during passage through the engine, °F. Usual practice is to limit the temperature rise of the oil to a range of 20 to 25°F (11.1 to 13.9°C), with a maximum operating temperature of 160°F (71.1°C). The heat rejection to the lube oil can be obtained from the engine heat balance, the engine manufacturer, or *Standard Practices for Stationary Diesel Engines*, published by the Diesel Engine Manufacturers Association. With a maximum heat rejection rate of 500 Btu/(bhp·h) (196.4 W/kWh) from *Standard Practices* and an oil-temperature rise of 20°F (11.1°C), $G = [500$ Btu/(bhp·h)$](1000$ hp)$/[(200)(20)] = 125$ gal/min (7.9 L/s).

By using the *lowest* temperature rise and the *highest* heat rejection rate, a safe pump capacity is obtained. Where the pump cost is a critical factor, use a higher temperature rise and a lower heat rejection rate. Thus, with a heat rejection, the above pump would have a capacity of $G = (300)(1000)/[(200)(25)] = 60$ gal/min (3.8 L/s).

3. Compute the lube-oil pump power input

The power input to a separate oil pump serving a diesel engine is given by $hp = Gp/1720e$, where G = pump discharge rate, gal/min; p = pump discharge pressure, lb/in^2, e = pump efficiency. For this pump, $hp = (125)(80)/[(1720)(0.68)] = 8.56$ hp (6.4 kW). A 10-hp (7.5-kW) motor would be chosen to drive this pump.

With a capacity of 60 gal/min (3.8 L/s), the input is $hp = (60)(80)/[(1720)(0.68)] = 4.1$ hp (3.1 kW). A 5-hp (3.7-kW) motor would be chosen to drive this pump.

Related Calculations. Use this method for any reciprocating diesel engine, two- or four-cycle. Lube-oil pump capacity is generally selected 10 to 15 percent oversize to allow for bearing wear in the engine and wear of the pump moving parts. Always check the selected capacity with the engine builder. Where a bypass-type lube-oil system is used, be sure to have a pump of sufficient capacity to handle *both* the engine and cooler oil flow.

Raw-water pumps are generally duplicates of the jacket-water pump, having the same capacity and head ratings. Then the raw-water pump can serve as a standby jacket-water pump, if necessary.

LUBE-OIL COOLER SELECTION AND OIL CONSUMPTION

A 500-hp (373-kW) internal-combustion engine rejects 300 to 600 Btu/(bhp·h) (118 to 236 W/kWh) to the lubricating oil. What capacity and type of lube-oil cooler should be used for this engine if 10 percent of the oil is bypassed? If this engine consumes 2 gal (7.6 L) of lube oil per 24 h at full load, determine its lube-oil consumption rate.

Calculation Procedure:

1. *Determine the required lube-oil cooler capacity*
Base the cooler capacity on the maximum heat rejection rate plus an allowance for overloads. The usual overload allowance is 10 percent of the full-load rating for periods of not more than 2 h in any 24 h period.

For this engine, the maximum output with a 10 percent overload is 500 + (0.10)(500) = 550 hp (410 kW). Thus, the maximum heat rejection to the lube oil would be (500 hp)[600 Btu/(bhp·h)] = 330,000 Btu/h (96.7 kW).

2. *Choose the type and capacity of lube-oil cooler*
Choose a shell-and-tube type heat exchanger to serve this engine. Long experience with many types of internal-combustion engines shows that the shell-and-tube heat exchanger is well suited for lube-oil cooling.

Select a lube-oil cooler suitable for a heat-transfer load of 330,000 Btu/h (96.7 kW) at the prevailing cooling-water temperature difference, which is usually assumed to be 10°F (5.6°C). See previous calculation procedures for the steps in selecting a liquid cooler.

3. *Determine the lube-oil consumption rate*
The lube-oil consumption rate is normally expressed in terms of bhp·h/gal. Thus, if this engine operates for 24 h and consumes 2 gal (7.6 L) of oil, its lube-oil consumption rate = (24 h)(500 bhp)/2 gal = 6000 bhp·h/gal (1183 kWh/L).

Related Calculations. Use this procedure for any type of internal-combustion engine using any fuel.

QUANTITY OF SOLIDS ENTERING AN INTERNAL-COMBUSTION ENGINE

What weight of solids annually enters the cylinders of a 1000-hp (746-kW) internal-combustion engine if the engine operates 24 h/day, 300 days/year in an area having an average dust concentration of 1.6 gr per 1000 ft³ of air (28.3 m³)? The engine air rate (displacement) is 3.5 ft³/(min·bhp) (0.13 m³/kW). What would the dust load be reduced to if an air filter fitted to the engine removed 80 percent of the dust from the air?

Calculation Procedure:

1. *Compute the quantity of air entering the engine*
Since the engine is rated at 1000 hp (746 kW) and uses 3.5 ft^3/(min·bhp) [0.133 m^3/(min·kW)], the quantity of air used by the engine each minute is (1000 hp)[3.5 ft^3/(min·hp)] = 3500 ft^3/min (99.1 m^3/min).

2. *Compute the quantity of dust entering the engine*
Each 1000 ft^3 (28.3 m^3) of air entering the engine contains 1.6 gr (103.7 mg) of dust. Thus, during every minute of engine operation, the quantity of dust entering the engine is (3500/1000)(1.6) = 5.6 gr (362.8 mg). The hourly dust intake = (60 min/h)(5.6 gr/min) = 336 gr/h (21,772 mg/h).

During the year the engine operates 24 h/day for 300 days. Hence, the annual intake of dust is (24 h/day)(300 days/year)(336 gr/h) = 2,419,200 gr (156.8 kg). Since there is 7000 gr/lb, the weight of dust entering the engine per year = 2,419,200 gr/(7000 gr/lb) = 345.6 lb/year (155.5 kg/year).

3. *Compute the filtered dust load*
With the air filter removing 80 percent of the dust, the quantity of dust reaching the engine is (1.00− 0.80)(345.6 lb/year) = 69.12 lb/year (31.1 kg/year). This shows the effectiveness of an air filter in reducing the dust and dirt load on an engine.

Related Calculations.   Use this general procedure to compute the dirt load on an engine from any external source.

INTERNAL-COMBUSTION ENGINE PERFORMANCE FACTORS

Discuss and illustrate the important factors in internal-combustion engine selection and performance. In this discussion, consider both large and small engines for a full range of usual applications.

Calculation Procedure:

1. *Plot typical engine load characteristics*
Figure 11 shows four typical load patterns for internal-combustion engines. A continuous load, Fig. 11a, is generally considered to be heavy-duty and is often met in engines driving pumps or electric generators.

Intermittent heavy-duty loads, Fig. 11b, are often met in engines driving concrete mixers, batch machines, and similar loads. Variable heavy-duty loads, Fig. 11c, are encountered in large vehicles, process machinery, and similar applications. Variable light-duty loads, Fig. 11d, are met in small vehicles like golf carts, lawn mowers, chain saws, etc.

2. *Compute the engine output torque*
Use the relation $T = 5250$ bhp/(r/min) to compute the output torque of an internal-combustion engine. In this relation, bhp = engine bhp being developed at a crankshaft speed having rotating speed of *rpm*.

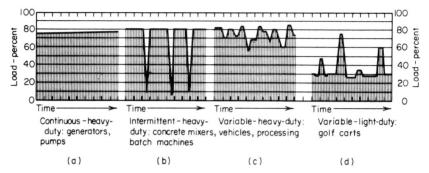

FIGURE 11 Typical internal-combustion engine load cycles: (*a*) continuous, heavy-duty; (*b*) intermittent, heavy-duty; (*c*) variable, heavy-duty; (*d*) variable, light-duty. (*Product Engineering.*)

3. Compute the hp output required

Knowing the type of load on the engine (generator, pump, mixer, saw blade, etc.), compute the power output required to drive the load at a constant speed. Where a speed variation is expected, as in variable-speed drives, compute the average power needed to accelerate the load between two desired speeds in a given time.

4. Choose the engine output speed

Internal-combustion engines are classified in three speed categories: high (1500 r/min or more), medium (750 to 1500 r/min), and low (less than 750 r/min).

Base the speed chosen on the application of the engine. A high-speed engine can be lighter and smaller for the same hp rating, and may cost less than a medium-speed or slow-speed engine serving the same load. But medium-speed and slow-speed engines, although larger, offer a higher torque output for the equivalent hp rating. Other advantages of these two speed ranges include longer service life and, in some instances, lower maintenance costs.

Usually an application will have its own requirements, such as allowable engine weight, available space, output torque, load speed, and type of service. These requirements will often indicate that a particular speed classification must be used. Where an application has no special speed requirements, the speed selection can be made on the basis of cost (initial, installation, maintenance, and operating costs), type of parts service available, and other local conditions.

5. Analyze the engine output torque required

In some installations, an engine with good lugging power is necessary, especially in tractors, harvesters, and hoists, where the load frequently increases above normal. For good lugging power, the engine should have the inherent characteristic of increasing torque with drooping speed. The engine can then resist the tendency for increased load to reduce the output speed, giving the engine good lugging qualities.

One way to increase the torque delivered to the load is to use a variable-ratio hydraulic transmission. The transmission will amplify the torque so that the engine will not be forced into the lugging range.

Other types of loads, such as generators, centrifugal pumps, air conditioners, and marine drives, may not require this lugging ability. So be certain to consult the engine power curves and torque characteristic curve to determine the speed at which the maximum torque is available.

6. Evaluate the environmental conditions

Internal-combustion engines are required to operate under a variety of environmental conditions. The usual environmental conditions critical in engine selection are altitude, ambient temperature, dust or dirt, and special or abnormal service. Each of these, except the last, is considered in previous calculation procedures.

Special or abnormal service includes such applications as fire fighting, emergency flood pumps and generators, and hospital standby service. In these applications, an engine must start and pick up a full load without warmup.

7. Compare engine fuels

Table 9 compares four types of fuels and the internal-combustion engines using them. Note that where the cost of the fuel is high, the cost of the engine is low; where the cost of the fuel is low, the cost of the engine is high. This condition prevails for both large and small engines in any service.

8. Compare the performance of small engines

Table 10 compares the principal characteristics of small gasoline and diesel engines rated at 7 hp (5 kW) or less. Note that engine life expectancy can vary from 500 to 25,000 h. With modern, mass-produced small engines it is often just as cheap to use short-life replaceable two-stroke gasoline engines instead of a single long-life diesel engine. Thus, the choice of a small engine is often based on other considerations, such as ease and convenience of replacement, instead of just hours of life. Chances are, however, that most long-life applications of small engines will still require a long-life engine. But the alternative must be considered in each case.

Related Calculations.   Use the general data presented here for selecting internal-combustion engines having ratings up to 200 hp (150 kW). For larger engines, other factors such as weight, specific fuel consumption, lube-oil consumption, etc., become important considerations. The method given here is the work of Paul F. Jacobi, as reported in *Product Engineering*.

VOLUMETRIC EFFICIENCY OF DIESEL ENGINES

A four-cycle six-cylinder Diesel engine of 4.25-in (11.4-cm) bore and 60-in (15.2-cm) stroke running at 1200 rpm has 9 percent CO_2 present in the exhaust gas. The fuel consumption is 28 lb (12.7 kg) per hour. Assuming that 13.7 percent CO_2 indicates an air-fuel ratio of 15 lb of air to 1 lb (6.6 kg to 0.45 kg), calculate the volumetric efficiency of the engine. Intake air temperature is 60°F (15.6°C) and the barometric pressure is 29.8 in (79.7 cm).

Calculation Procedure:

1. Find the percentage of N_2 in the exhaust gas

Atmospheric air contains 76.9 percent nitrogen by weight. If an analysis of the fuel oil shows zero nitrogen before combustion, all the nitrogen in the exhaust gas must come from the air. Therefore, with 13.7 percent CO_2 by volume in the dry exhaust the nitrogen content is: $N_2 = (76.9/100)(15) = 11.53$ lb (5.2 kg) N_2 per lb (0.454 kg) of fuel oil. Converting to moles, 11.53 lb (5.2 kg) $N_2/28$ lb (12.7 kg) fuel per hour = 0.412 mole N_2 per lb (0.454 kg) of fuel oil.

TABLE 9 Comparison of Fuels for Internal-Combustion Engines*

	Storage life (quantities)		Consistency, Btu/ft³	Initial cost of engine, relative	Cost of fuel	Residue	Antiknock rating	Filtering necessary	Weight		Heat content			
	Small	Large							lb/gal	kg/L	Btu/vol	mJ/vol	Btu/lb	mJ/kg
Gasoline	Good	Poor (6 months)	Good	Low	High	High	Best is costly	Medium	6.000	0.714	123,039 Btu/gal	34,291 kJ/L	20,627	47.9
Diesel: No. 1	Good	Fair (1 year)	Good	High	Low	Low if properly filtered	· · ·	High	6.850	0.815	135,800 Btu/gal	37,847 kJ/L	19,750	45.9
No. 2	Good	Fair (1 year)	Good	High	Low	Low if properly filtered	· · ·	High	7.020	0.835	139,000 Btu/gal	38,739 kJ/L	19,786	46.0
Natural gas	Not necessary	Not necessary	Poor	Medium	Medium	Low	High	Very little	· · ·	· · ·	1,000 Btu/ft³	37,250 kJ/m³		
LPG: Propane	Good	Good	Poor	Medium	Medium	Low	Good	Very little	4.235	0.504	91,740 Btu/gal	25,568 kJ/L	21,308	49.6
Butane	Good	Good	Poor	Medium	Medium	Low	Good	Very little	4.873	0.580	103,830 Btu/gal	28,937 kJ/L	20,627	47.9

*Product Engineering.

TABLE 10 Performance Table for Small Internal-Combustion Engines [Less than 7 hp (5 kW)]*

| | Variety of models available | Typical weight lb/hp (kg/kW) | Operating speeds | | Lugging ability | Torque output | Relative life expectancy, h | Relative cost | Fuel required | Shaft direction | Noise level | Starters | Integral optional Pto's | Ignition | Cost of operation | Variety of options and accessories |
			Typical maximum	Typical efficient minimum												
Lightweight: 2-stroke	Narrow	2:1 (1.2:1)	3,600 (governed) to 7,500	2,000 to 3,000	Poor to fair	Fair	500	Lowest	Gasoline oil mixed	Vertical, horizontal, or universal	High	Rope, recoil, impulse	No	Magneto	High	Standard—extremely low custom—wide
4-stroke	Wide	6:1; 10:1 (3.6:1; 6.1:1)	4,000	2,000 to 2,400	Fair to good	Good	500	1 to 2	Gasoline (LPG)	Vertical or horizontal	Moderate	Rope, recoil, impulse, electric	Several	Magneto	Moderate	Standard—wide
Heavyweight: 4-stroke	Wide	11:1; 20:1 (6.6:1; 12.1:1)	4,000	1,600 to 1,800	Good to excellent	Good	7,500	2 to 4	Gasoline (LPG)	Vertical or horizontal	Moderate	Rope, recoil, impulse, crank, electric	No	Magneto, distributor	Moderate	Standard—moderately wide
Diesel	Narrow	35:1 (21.1:1)	2,400	1,500	Excellent	Good	25,000	4	Diesel	Horizontal	Moderate to high	Electric	No	Battery, distributor, glow plugs	Low	Narrow

* Product Engineering.

6.36

2. Compute the weight of N_2 in the exhaust
Use the relation, percentage of CO_2 in the exhaust gases = $(CO_2)/(N_2 + CO_2)$ in moles. Substituting, $(13.7)/(100) = (CO_2)/(N_2 + 0.412)$. Solving for CO_2 we find $CO_2 = 0.0654$ mole.
Now, since mole percent is equal to volume percent, for 9 percent CO_2 in the exhaust gases, $0.09 = (CO_2)/(CO_2 + N_2) = 0.0654/(0.0654 + N_2)$. Solving for N_2, we find $N_2 = 0.661$ mole. The weight of N_2 therefore = $0.661 \times 28 = 18.5$ lb (8.399 kg).

3. Calculate the amount of air required for combustion
The air required for combustion is found from $(N_2) = 18.5/0.769$, where 0.769 = percent N_2 in air, expressed as a decimal. Solving, $N_2 = 24.057$ lb (10.92 kg) per lb (0.454 kg) of fuel oil.

4. Find the weight of the actual air charge drawn into the cylinder
Specific volume of the air at 60°F (15.6°C) and 29.8 in (75.7 cm) Hg is 13.03 ft³ (0.368 m³) per lb. Thus, the actual charge drawn into the cylinder = (lb of air per lb of fuel)(specific volume of the air, ft³/lb)(fuel consumption, lb/h)/3600 s/h. Or 24.1(13.02)(28)/3600 = 2.44 ft³ (0.69 m³) per second.

5. Compute the volumetric efficiency of this engine
Volumetric efficiency is defined as the ratio of the actual air charge drawn into the cylinder divided by the piston displacement. The piston displacement for one cylinder of this engine is (bore area)(stroke length)(1 cylinder)/1728 in³/ft³. Solving, piston displacement = $0.785(4.25)^2(6)(1)/1728 = 0.0492$ ft³ (0.00139 m³).
The number of suction strokes per minute = rpm/2. The volume displaced per second by the engine = (piston displacement per cylinder)(number of cylinders)(rpm/2)/60 s/min. Substituting, engine displacement = $0.0492(6)(1200/2)/60 = 2.952$ ft³/s (0.0084 m³/s).
Then, the volumetric efficiency of this engine = actual charge drawn into the cylinder/engine displacement = $2.45/2.952 = 0.8299$, or 82.99 percent.
Related Calculations. Use this general procedure to determine the volumetric efficiency of reciprocating internal-combustion engines—both gasoline and Diesel. The procedure is also used for determining the fuel consumption of such engines, using test data from actual engine runs.

SELECTING AIR-COOLED ENGINES FOR INDUSTRIAL APPLICATIONS

Choose a suitable air-cooled gasoline engine to replace a 10-hp (7.46 kW) electric motor driving a municipal service sanitary pump at an elevation of 8000 ft (2438 m) where the ambient temperature is 90°F (32.2°C). Find the expected load duty for this engine; construct a typical load curve for it.

Calculation Procedure:

1. Determine the horsepower (kW) rating required of the engine
Electric motors are rated on an entirely different basis than are internal-combustion engines. Most electric motors will deliver 25 percent more power than their rating

during a period of one or two hours. For short periods many electric motors may carry 50 percent overload.

Gasoline engines, by comparison, are rated at the maximum power that a new engine will develop on a dynamometer test conducted at an ambient temperature of 60°F (15.6°C) and a sea-level barometric pressure of 29.92 in (759.97 mm) of mercury. For every 10°F (5.56°C) rise in the intake ambient air temperature there will be a 1 percent reduction in the power output. And for every 1-in (2.5-cm) drop in barometric pressure there will be a 3.5 percent power output loss. For every 1000 ft (304.8 m) of altitude above sea level a 3.5 percent loss in power output also occurs.

Thus, for average atmospheric conditions, the actual power of a gasoline engine is about 5 to 7 percent less than the standard rating. And if altitude is a factor, the loss can be appreciable, reaching 35 percent at 10,000 ft (3048 m) altitude.

Also, in keeping with good industrial practice, a gasoline engine is not generally operated continuously at maximum output. This practice provides a factor of safety in the form of reserve power. Most engine manufacturers recommend that this factor of safety be 20 to 25 percent below rated power. This means that the engine will be normally operated at 75 to 80 percent of its standard rated output. The duty cycle, however, can vary with different applications, as Table 11 shows.

For the 10-hp (7.46 kW) electric motor we are replacing with a gasoline engine, the motor can deliver—as discussed—25 percent more than its rating, or in this instance, 12.5 hp (9.3 kW) for short periods. On the basis that the gasoline engine is to operate at not over 75 percent of its rating, the replacement engine should have a rating of 12.5/0.75 = 16.7 hp (12.4 kW).

In summary, the gasoline engine should have a rating at least 67 percent greater than the electric motor it replaces. This applies to both air- and liquid-cooled engines for sea-level operation under standard atmospheric conditions. If the engine is to operate at altitude, a further allowance must be made, resulting—in some instances—in an engine having twice the power rating of the electric motor.

2.   *Find the power required at the installed altitude and inlet-air temperature*
As noted above, altitude and inlet-air temperature both influence the required rating of a gasoline engine for a given application. Since this engine will be installed at an altitude of 8000 ft (2438 m), the power loss will be (8000/1000)(3.5) = 28 percent. Further, the increased inlet-air temperature of 90°F (32.2°C) vs. the standard of 60°F (15.6°C), or a 30° difference will reduce the power output by (30/10)(1.0) = 3.0 percent. Thus, the total power output reduction will be 28 + 3 = 31 percent. Therefore, the required rating of this gasoline engine will be at least (1.31)(16.7) = 21.87 hp (16.3 kW).

Once the power requirements of a design are known, the next consideration is engine rotative speed, which is closely related to the horsepower and service life. Larger engines with their increased bearing surfaces and lower speeds, naturally require less frequent servicing. Such engines give longer, more trouble-free life than the smaller, high-speed engines of the same horsepower (kW) rating.

The initial cost of a larger engine is greater but more frequent servicing can easily bring the cost of a smaller engine up to that of the larger one. Conversely, the smaller, higher-speed engine has advantages where lighter weight and smaller installation dimensions are important, along with a relatively low first cost.

Torque is closely associated with engine rotative speed. For most installations an engine with good lugging power is desirable, and in some installations, essential. This is especially true in tractors, harvesters, and hoists, where the load frequently increases considerably above normal.

TABLE 11 Duty Ratings for Combustion Engine Application

Key: 1—*Continuous Duty*
2—*Intermittent heavy duty*
3—*Variable load duty, heavy*
4—*Variable load duty, light*

INDUSTRIAL SERVICE

1—Standby units
2—Air compressors
3—Floor sanders
4—Shop trucks and welders

MUNICIPAL SERVICE

3—Street sweepers and flushers
3—Sanitary pumps
3—Pipe thawing rigs
4—Diesel starting units

MINING

3—Horizontal & diamond drills
3—Rocker shovels

RAILWAY MAINTENANCE

3—Tampers
3—Tie adzing machines
3—Railway maintenance cars
3—Rail grinders
3—Weed cutters
4—Rail leveling machines

HIGHWAY MAINTENANCE

1—Road rollers
1—Bituminous sprayers
2—Concrete cutters

OIL FIELD EQUIPMENT

1—Well drills and pumps
1—High pressure pumps
2—Pipe wrapping machines
3—Pipe straightening machines

AGRICULTURAL EQUIPMENT

1—Irrigation pumps
3—Combine harvesters
3—Hay balers, tractors
3—Insecticide sprayers
3—Rotary tillers
3—Potato harvesters
3—Mowers
3—Spreaders, dusters

MARINE

1—Lighthouse units
1—Water oxygenation units
3—Inboard marine engines
3—Underwater weed cutters

CONTRUCTION MACHINERY

1—Centrifugal pumps
2—Concrete mixers
3—Concrete vibrators
3—Concrete surfacing machines

CONTRUCTION MACHINERY
(Cont.)
3—Diaphragm pumps
4—Hoists and power saws

SPECIALIZED SERVICE

1—Airport service units such as air
compressors, hydraulic pumps
and generators
1—Weed burners
2—Refrigerated trucks
2—Paint sprayers
2—Portable fire fighting equipment
3—Miniature railways
3—Water purification units for
armed forces
3—Cable reelers
3—Lawn mowers and rollers
3—Post peelers
3—Portable showers for armed forces

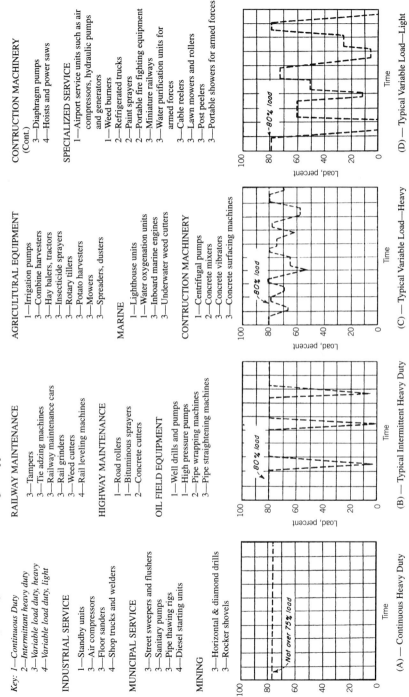

(A) — Continuous Heavy Duty

(B) — Typical Intermittent Heavy Duty

(C) — Typical Variable Load—Heavy

(D) — Typical Variable Load—Light

Not over 75% load

80% load

80% load

80% load

Load, percent

Time

If the characteristics of the engine output curve are such that the torque will increase with reducing engine speed, the tendency for the increasing load to reduce engine speed is resisted and the engine will "hang on." In short, it will have good lugging qualities, as shown in Fig. 12a. If the normal operating speed of the engine is 2000 to 2200 r/min, the maximum lugging qualities will result. Sanitary-pump drives do not—in general—require heavy lugging.

If, however, with the same curve, Fig. 12a, the normal operating speed of the engine is held at 1400 r/min or below, stalling of the engine may occur easily when the load is increased. Such an increase will cause engine speed to reduce, resulting in a decrease in torque and causing further reduction in speed until the engine finally stalls abruptly, unless the load can be quickly released.

Figure 12b shows performance curves for a typical high-speed engine with maximum power output at top speed. The torque curve for this engine is flat and the engine is not desirable for industrial or agricultural type installations.

3. Determine the duty rating; draw a load curve for the engine
Refer to Table 11 for municipal service. There you will see that sanitary pumps have a variable load, heavy duty rating. Figure 12 shows a plot of the typical load variation in such an engine when driving a sanitary pump in municipal service.

4. Select the type of drive for the engine
A variety of power takeoffs are used for air-cooled gasoline engines, Fig. 13. For a centrifugal pump driven by a gasoline engine, a flange coupling is ideal. The same is true for engines driving electric generators. Both the pump and generator run at engine speed. When a plain-flange coupling is used, the correct alignment of the gasoline engine and driven machine is extremely important. Flexible couplings and belt drives eliminate alignment problems.

In many instances a clutch is required between the engine and equipment so that the power may be engaged or disengaged at will. A manually engaged clutch is the most common type in use on agricultural and industrial equipment.

Where automatic engagement and disengagement are desired, a centrifugal clutch may be used. These clutches can be furnished to engage at any speed between 500 and 1200 r/min and the load pick-up is smooth and gradual. Typical applications for such clutches are refrigerating machines with thermostatic control for starting and stopping the engine.

Clutches also make starting of the engine easier. It is often impossible to start an internal-combustion engine rigidly connected to the load.

There are many applications where a speed reduction between the engine and machine is necessary. If the reduction is not too great, it may be accomplished by belt drive. But often a gear reduction is preferable. Gear reductions can be furnished in ratios up to 4 for larger engines, and up to 6 for smaller sizes. Many of these reductions can be furnished in either enginewise or counter-enginewise rotation, and either with or without clutches.

Related Calculations.   Table 11 shows 54 different applications and duty ratings for small air-cooled gasoline engines. With this information the engineer has a powerful way to make a sensible choice of engine, drive, speed, torque, and duty cycle.

Important factors to keep in mind when choosing small internal-combustion engines for any of the 54 applications shown are: (1) Engines should have sufficient capacity to ensure a factor of safety of 20 to 25 percent for the power output. (2) Between high- and low-speed engines, the latter have longer life, but first cost is higher. (3) In take-off couplings, the flexible types are preferred. (4) A clutch is

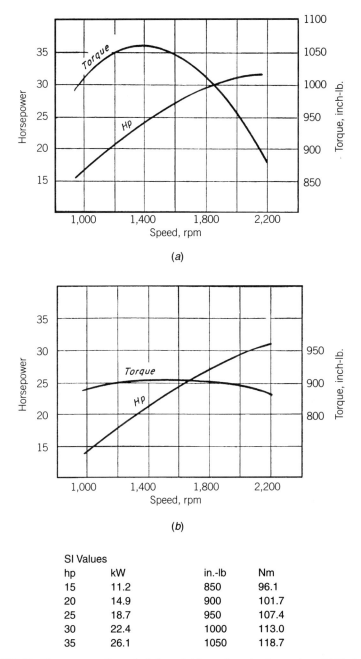

FIGURE 12 Torque curves for typical air-cooled internal-combustion engines. (*a*) Engine with good lugging quality will "hang on" as load increases. (*b*) Performance curve for a high-speed engine with maximum power output at top speed.

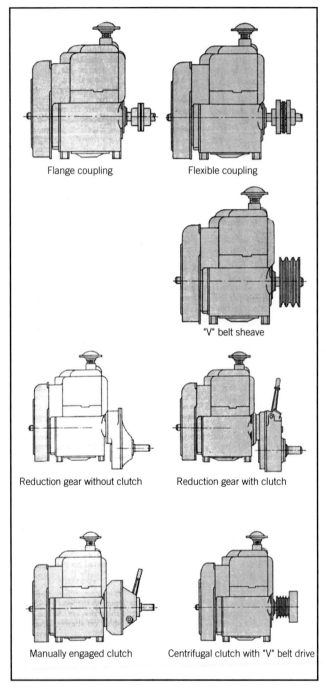

FIGURE 13 Power take-offs for air-cooled engines. Fluid couplings are also used to cushion shock loads in certain specialized applications.

desirable, especially in heavier equipment, to disconnect the load and to make engine starting easier. (5) In operations where the intake air is dusty or contains chaff, intake screens should be used. (6) An oil-bath type air cleaner should always be used ahead of the carburetor. (7) Design engine mountings carefully and locate them to avoid vibration. (8) Provide free flow of cooling air to the flywheel fan inlet and also to the hot-air outlet from the engine. Carefully avoid recirculation of the hot air by the flywheel. (9) If the engine operation is continuous and heavy, Stellite exhaust valves and valve-seat inserts should be used to ensure long life.

Valve rotators are also of considerable value in prolonging valve life, and with Stellite valves, constitute an excellent combination for heavy service.

Exclusive of aircraft, air-cooled engines are usually applied in size ranges from 1 to 30 hp (0.75 to 22.4 kW). Larger engines are being built and, depending on the inherent cooling characteristics of the system, performing satisfactorily. However, the bulk of applications are on equipment requiring about 30 hp (22.4 kW), or less. The smaller engines up to about 8 or 9 hp (5.9 to 6.7 kW) are usually single-cylinder types; from 8 to 15 hp (5.9 to 11.2 kW) two-cylinder engines are prevalent, while above 15 hp (11.2 kW), four-cylinder models are commonly used.

Within these ranges, air-cooled engines have several inherent advantages: they are light-weight, with weight varying from about 14 to 20 lb/hp (8.5 to 12.2 kg/kW) for a typical single-cylinder engine operating at 2600 r/min to about 12 to 15 lb/h (7.3 to 9.1 kg/kW) for atypical four-cylinder unit running at 1800 r/min. Auxiliary power requirements for these engines are low since there is no radiator fan or water pump; there is no danger of the engine boiling or freezing, and no maintenance of fan bearings, or water pumps; and first cost is low.

In selecting an engine of this type, the initial step is to determine the horsepower requirements of the driven load.

On equipment of entirely new design, it is often difficult to ascertain the amount of power necessary. In such instances, a rough estimate of the horsepower range (kW range) is made and one or more sample engines bracketing the range obtained for use on experimental models of the equipment. In other applications, it is possible to calculate the torque required, from which the horsepower (kW) can be determined. Or, as is not uncommon, the new piece of equipment may be another size in a line of machines. In this case, the power determination can be made on a proportional basis.

This procedure is the work of A. F. Milbrath, Vice President and Chief Engineer, Wisconsin Motor Corporation, as reported in *Product Engineering* magazine. SI values were added by the *Handbook* editor.

P · A · R · T   2

PLANT AND
FACILITIES
ENGINEERING

SECTION 7
PUMPS AND PUMPING SYSTEMS

Pump Operating Modes and Criticality

SERIES PUMP INSTALLATION ANALYSIS

A new plant addition using special convectors in the heating system requires a system pumping capability of 45 gal/min (2.84 L/s) at a 26-ft (7.9-m) head. The pump characteristic curves for the tentatively selected floor-mounted units are shown in Fig. 1; one operating pump and one standby pump, each 0.75 hp (0.56 kW) are being considered. Can energy be conserved, and how much, with some other pumping arrangement?

Calculation Procedure:

1.   Plot the characteristic curves for the pumps being considered
Figure 2 shows the characteristic curves for the proposed pumps. Point 1 in Fig. 1 is the proposed operating head and flow rate. An alternative pump choice is shown at Point 2 in Fig. 1. If two of the smaller pumps requiring only 0.25 hp (0.19 kW) each are placed in series, they can generate the required 26-ft (7.9-m) head.

2.   Analyze the proposed pumps
To analyze properly the proposal, a new set of curves, Fig. 2, is required. For the proposed series pumping application, it is necessary to establish a *seriesed pump curve.* This is a plot of the head and flow rate (capacity) which exists when both pumps are running in series. To construct this curve, double the single-pump head values at any given flow rate.

   Next, to determine accurately the flow a single pump can deliver, plot the system-head curve using the same method fully described in the previous calculation procedure. This curve is also plotted on Fig. 2.

   Plot the point of operation for each pump on the seriesed curve, Fig. 2. The point of operation of each pump is on the single-pump curve when both pumps are operating. Each pump supplies half the total required head.

   When a single pump is running, the point of operation will be at the intersection of the system-head curve and the single-pump characteristic curve, Fig. 2. At this point both the flow and the hp (kW) input of the single pump decrease. Series pumping, Fig. 2, requires the input motor hp (kW) for both pumps; this is the point of maximum power input.

3.   Compute the possible savings
If the system requires a constant flow of 45 gal/min (2.84 L/s) at 26-ft (7.9-m) head the two-pump series installation saves (0.75 hp − 2 × 0.25 hp) = 0.25 hp (0.19 kW) for every hour the pumps run. For every 1000 hours of operation, the system saves 190 kWh. Since 2000 hours are generally equal to one shift of operation per year, the saving is 380 kWh per shift per year.

   If the load is frequently less than peak, one-pump operation delivers 32.5 gal/min (2.1 L/s). This value, which is some 72 percent of full load, corresponds to doubling the saving.

   Related Calculations.   Series operation of pumps can be used in a variety of designs for industrial, commercial, residential, chemical, power, marine, and similar plants. A series connection of pumps is especially suitable when full-load demand is small; *i.e.,* just a few hours a week, month, or year. With such a demand, one pump can serve the plant's needs most of the time, thereby reducing the power bill. When full-load operation is required, the second pump is started. If there is a need for maintenance of the first pump, the second unit is available for service.

   This procedure is the work of Jerome F. Mueller, P.E., of Mueller Engineering Corp.

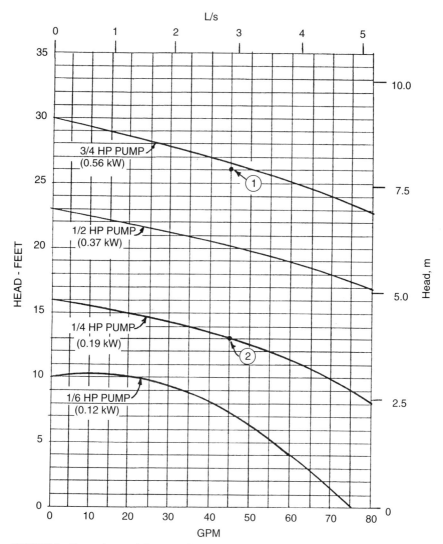

FIGURE 1 Pump characteristic curves for use in series installation.

PARALLEL PUMPING ECONOMICS

A system proposed for heating a 20,000-ft² (1858-m²) addition to an industrial plant using hot-water heating requires a flow of 80 gal/min (7.4 L/s) of 200°F (92.5°C) water at a 20°F (36°C) temperature drop and a 13-ft (3.96-m) system head. The

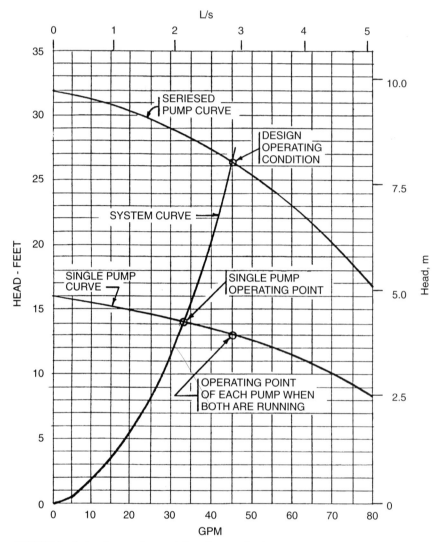

FIGURE 2   Seriesed-pump characteristic and system-head curves.

required system flow can be handled by two pumps, one an operating unit and one a spare unit. Each pump will have an 0.5-hp (0.37-kW) drive motor. Could there be any appreciable energy saving using some other arrangement? The system requires 50 hours of constant pump operation and 40 hours of partial pump operation per week.

Calculation Procedure:

1. *Plot characteristic curves for the proposed system*
Figure 3 shows the proposed hot-water heating-pump selection for this industrial building. Looking at the values of the pump head and capacity in Fig. 3, it can be seen that if the peak load of 80 gal/min (7.4 L/s) were carried by two pumps, then each would have to pump only 40 gal/min (3.7 L/s) in a parallel arrangement.

2. *Plot a characteristic curve for the pumps in parallel*
Construct the paralleled-pump curve by doubling the flow of a single pump at any given head, using data from the pump manufacturer. At 13-ft head (3.96-m) one pump produces 40 gal/min (3.7 L/s); two pumps 80 gal/min (7.4 L/s). The resulting curve is shown in Fig. 4.

The load for this system could be divided among three, four, or more pumps, if desired. To achieve the best results, the number of pumps chosen should be based on achieving the proper head and capacity requirements in the system.

3. *Construct a system-head curve*
Based on the known flow rate, 80 gal/min (7.4 L/s) at 13-ft (3.96-m) head, a system-head curve can be constructed using the fact that pumping head varies as the square of the change in flow, or $Q_2/Q_1 = H_2/H_1$, where Q_1 = known design flow, gal/min (L/s); Q_2 = selected flow, gal/min (L/s); H_1 = known design head, ft (m); H_2 = resultant head related to selected flow rate, gal/min (L/s)

Figure 5 shows the plotted system-head curve. Once the system-head curve is plotted, draw the single-pump curve from Fig. 3 on Fig. 5, and the parallelled-pump curve from Fig. 4. Connect the different pertinent points of concern with dashed lines, Fig. 5.

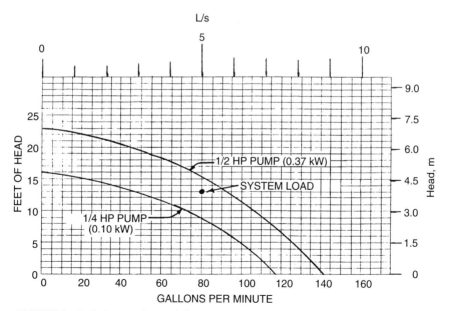

FIGURE 3 Typical pump characteristic curves.

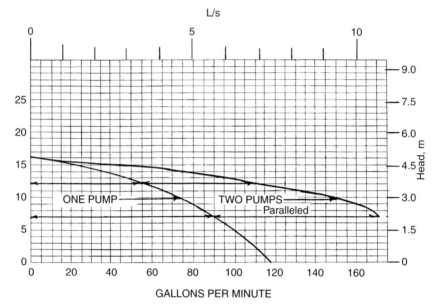

FIGURE 4   Single- and dual-parallel pump characteristic curves.

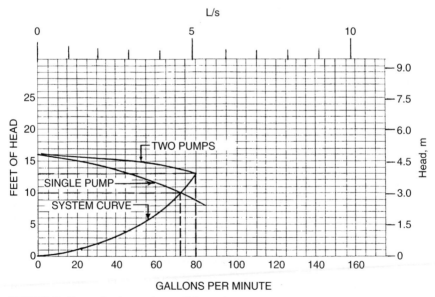

FIGURE 5   System-head curve for parallel pumping.

The point of crossing of the two-pump curve and the system-head curve is at the required value of 80 gal/min (7.4 L/s) and 13-ft (3.96-m) head because it was so planned. But the point of crossing of the system-head curve and the single-pump curve is of particular interest.

The single pump, instead of delivering 40 gal/min (7.4 L/s) at 13-ft (3.96-m) head will deliver, as shown by the intersection of the curves in Fig. 5, 72 gal/min (6.67 L/s) at 10-ft (3.05-m) head. Thus, the single pump can effectively be a standby for 90 percent of the required capacity at a power input of 0.5 hp (0.37 kW). Much of the time in heating and air conditioning, and frequently in industrial processes, the system load is 90 percent, or less.

4. *Determine the single-pump horsepower input*
In the installation here, the pumps are the inline type with non-overload motors. For larger flow rates, the pumps chosen would be floor-mounted units providing a variety of horsepower (kW) and flow curves. The horsepower (kW) for—say a 200-gal/min (18.6 L/s) flow rate would be about half of a 400-gal/min (37.2 L/s) flow rate.

If a pump were suddenly given a 300-gal/min (27.9 L/s) flow-rate demand at its crossing point on a larger system-head curve, the hp required might be excessive. Hence, the pump drive motor must be chosen carefully so that the power required does not exceed the motor's rating. The power input required by any pump can be obtained from the pump characteristic curve for the unit being considered. Such curves are available free of charge from the pump manufacturer.

The pump operating point is at the intersection of the pump characteristic curve and the system-head curve in conformance with the first law of thermodynamics, which states that the energy put into the system must exactly match the energy used by the system. The intersection of the pump characteristic curve and the system-head curve is the only point that fulfills this basic law.

There is no practical limit for pumps in parallel. Careful analysis of the system-head curve versus the pump characteristic curves provided by the pump manufacturer will frequently reveal cases where the system load point may be beyond the desired pump curve. The first cost of two or three smaller pumps is frequently no greater than for one large pump. Hence, smaller pumps in parallel may be more desirable than a single large pump, from both the economic and reliability standpoints.

One frequently overlooked design consideration in piping for pumps is shown in Fig. 6. This is the location of the check valve to prevent reverse-flow pumping. Figure 6 shows the proper location for this simple valve.

5. *Compute the energy saving possible*
Since one pump can carry the fluid flow load about 90 percent of the time, and this same percentage holds for the design conditions, the saving in energy is $0.9 \times (0.5$ kW $- .25$ kW$) \times 90$ h per week $= 20.25$ kWh/week. (In this computation we used the assumption that 1 hp = 1 kW.) The annual savings would be 52 weeks $\times 20.25$ kW/week $= 1053$ kWh/yr. If electricity costs 5 cents per kWh, the annual saving is $\$0.05 \times 1053 = \$52.65$/yr.

While a saving of some $51 per year may seem small, such a saving can become much more if: (1) larger pumps using higher horsepower (kW) motors are used; (2) several hundred pumps are used in the system; (3) the operating time is longer—168 hours per week in some systems. If any, or all, these conditions prevail, the savings can be substantial.

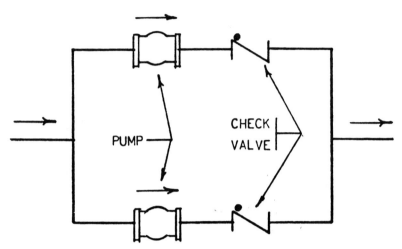

FIGURE 6   Check valve locations to prevent reverse flow.

Related Calculations.   This procedure can be used for pumps in a variety of applications: industrial, commercial, residential, medical, recreational, and similar systems. When analyzing any system the designer should be careful to consider all the available options so the best one is found.

This procedure is the work of Jerome F. Mueller, P.E., of Mueller Engineering Corp.

USING CENTRIFUGAL PUMP SPECIFIC SPEED TO SELECT DRIVER SPEED

A double-suction condenser circulator handling 20,000 gal/min (75,800 L/min) at a total head of 60 ft (18.3 m) is to have a 15-ft (4.6-m) lift. What should be the rpm of this pump to meet the capacity and head requirements?

Calculation Procedure:

1.   *Determine the specific speed of the pump*
Use the Hydraulic Institute specific-speed chart, Fig. 7, page 7.11. Entering at 60 ft (18.3 m) head, project to the 15-ft suction lift curve. At the intersection, read the specific speed of this double-suction pump as 4300.

2.   *Use the specific-speed equation to determine the pump operating rpm*
Solve the specific-speed equation for the pump rpm. Or rpm $= N_s \times H^{0.75}/Q^{0.5}$, where N_s = specific speed of the pump, rpm, from Fig. 7; H = total head on pump, ft (m); Q = pump flow rate, gal/min (L/min). Solving, rpm = $4300 \times 60^{0.75}/20,000^{0.5}$ = 655.5 r/min. The next common electric motor rpm is 660; hence, we would choose a motor or turbine driver whose rpm does not exceed 660.

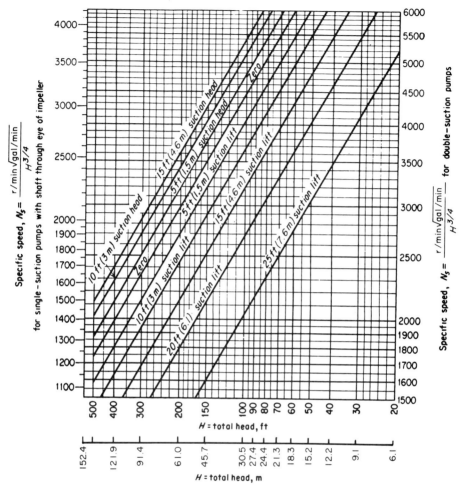

FIGURE 7 Upper limits of specific speeds of single-stage, single- and double-suction centrifugal pumps handling clear water at 85°F (29.4°C) at sea level. (*Hydraulic Institute.*)

The next lower induction-motor speed is 585 r/min. But we could buy a lower-cost pump and motor if it could be run at the next higher full-load induction motor speed of 700 r/min. The specific speed of such a pump would be: N_s = [700 $(20,000)^{0.5}$]/$60^{0.75}$ = 4592. Referring to Fig. 7, the maximum suction lift with a specific speed of 4592 is 13 ft (3.96 m) when the total head is 60 ft (18.3). If the pump setting or location could be lowered 2 ft (0.6 m), the less expensive pump and motor could be used, thereby saving on the investment cost.

Related Calculations. Use this general procedure to choose the driver and pump rpm for centrifugal pumps used in boiler feed, industrial, marine, HVAC, and similar applications. Note that the latest Hydraulic Institute curves should be used.

RANKING EQUIPMENT CRITICALITY TO COMPLY WITH SAFETY AND ENVIRONMENTAL REGULATIONS

Rank the criticality of a boiler feed pump operating at 250°F (121°C) and 100 lb/in^2 (68.9 kPa) if its Mean Time Between Failures (MTBF) is 10 months, and vibration is an important element in its safe operation. Use the National Fire Protection Association (NFPA) ratings of process chemicals for health, fire, and reactivity hazards. Show how the criticality of the unit is developed.

Calculation Procedure:

1. Determine the Hazard Criticality Rating (HCR) of the equipment
Process industries of various types—chemical, petroleum, food, *etc.*—are giving much attention to complying with new process safety regulations. These efforts center on reducing hazards to people and the environment by ensuring the mechanical and electrical integrity of equipment.

To start a program, the first step is to evaluate the most critical equipment in a plant or factory. To do so, the equipment is first ranked on some criteria, such as the relative importance of each piece of equipment to the process or plant output.

The Hazard Criticality Rating (HCR) can be determined from a listing such as that in Table 1. This tabulation contains the analysis guidelines for assessing the process chemical hazard (PCH) and the Other Hazards (O). The ratings for such a table of hazards should be based on the findings of an experienced team thoroughly familiar with the process being evaluated. A good choice for such a task is the plant's Process Hazard Analysis (PHA) Group. Since a team's familiarity with a process is highest at the end of a PHA study, the best time for rating the criticality of equipment is toward the end of such safety evaluations.

From Table 1, the NFPA rating, N, of process chemicals for Health, Fire, and Reactivity, is $N = 2$, because this is the highest of such ratings for Health. The Fire and Reactivity ratings are 0, 0, respectively, for a boiler feed pump because there are no Fire or Reactivity exposures.

The Risk Reduction Factor (RF), from Table 1, is $RF = 0$, since there is the potential for serious burns from the hot water handled by the boiler feed pump. Then, the Process Chemical Hazard, $PCH = N - RF = 2 - 0 = 2$.

The rating of Other Hazards, O, Table 1, is $O = 1$, because of the high temperature of the water. Thus, the Hazard Criticality Rating, $HCR = 2$, found from the higher numerical value of PCH and O.

2. Determine the Process Criticality Rating, PCR, of the equipment
From Table 2, prepared by the PHA Group using the results of its study of the equipment in the plant, $PCR = 3$. The reason for this is that the boiler feed pump is critical for plant operation because its failure will result in reduced capacity.

3. Find the Process and Hazard Criticality Rating, PHCR
The alphanumeric PHC value is represented first by the alphabetic character for the category. For example, Category A is the most critical, while Category D is the least critical to plant operation. The first numeric portion represents the Hazard

TABLE 1 The Hazard Criticality Rating (HCR) is Determined in Three Steps*

Hazard Criticality Rating

1. Assess the Process Chemical Hazard (PCH) by:
   - Determining the NFPA ratings (N) of process chemicals for: Health, Fire, Reactivity hazards
   - Selecting the highest value of N
   - Evaluating the potential for an emissions release (0 to 4):
     High (RF = 0): Possible serious health, safety or environmental effects
     Low (RF = 1): Minimal effects
     None (RF = 4): No effects
   - Then, $PCH = N - RF$. (Round off negative values to zero.)

2. Rate Other Hazards (O) with an arbitrary number (0 to 4) if they are:
   - Deadly (4), if:
     Temperatures $> 1000°F$
     Pressures are extreme
     Potential for release of regulated chemicals is high
     Release causes possible serious health safety or environmental effects
     Plant requires steam turbine trip mechanisms, fired-equipment shutdown systems, or toxic- or combustible-gas detectors†
     Failure of pollution control system results in environmental damage†
   - Extremely dangerous (3), if:
     Equipment rotates at >5000 r/min
     Temperatures >500°F
     Plant requires process venting devices
     Potential for release of regulated chemicals is low
     Failure of pollution control system may result in environmental damage†
   - Hazardous (2), if:
     Temperatures >300°F;
     Extended failure of pollution control system may cause damage†
   - Slightly hazardous (1), if:
     Equipment rotates at >3600 r/min
     Temperatures $> 140°F$ or pressures > 20 lb/in^2 (gage)
   - Not hazardous (0), if:
     No hazards exist

3. Select the higher value of PCH and O as the Hazard Criticality Rating

Chemical Engineering.
†Equipment with spares drop one category rating. A spare is an inline unit that can be immediately serviced or be substituted by an alternative process option during the repair period.

Criticality Rating, HCR, while the second numeric part the Process Criticality Rating, PCR. These categories and ratings are a result of the work of the PHA Group.

From Table 3, the Process and Hazard Criticality Rating, PHCR = B23. This is based on the PCR = 3 and HCR = 2, found earlier.

4. Generate a criticality list by rating equipment using its alphanumeric PHCR values

Each piece of equipment is categorized, in terms of its importance to the process, as: Highest Priority, Category A; High Priority, Category B; Medium Priority, Category C; Low Priority, Category D.

TABLE 2   Process Criticality Rating*

Process Criticality Rating	
Essential (4)	The equipment is essential if failure will result in shutdown of the unit, unacceptable product quality, or severely reduced process yield
Critical (3)	The equipment is critical if failure will result in greatly reduced capacity, poor product quality, or moderately reduced process yield
Helpful (2)	The equipment is helpful if failure will result in slightly reduced capacity, product quality or reduced process yield
Not critical (1)	The equipment is not critical if failure will have little or no process consequences

Chemical Engineering.

TABLE 3   The Process and Hazard Criticality Rating*

PHC Rankings					
Process Criticality Rating	Hazard Criticality Rating				
	4	3	2	1	0
4	A44	A34	A24	A14	A04
3	A43	B33	B23	B13	B03
2	A42	A32	C22	C12	C02
1	A41	B31	C21	CD11	D01

Note: The alphanumeric PHC value is represented first by the alphabetic character for the category (for example, category A is the most critical while D is the least critical). The first numeric portion represents the Hazard Criticality Rating, and the second numeric part the Process Criticality Rating.
Chemical Engineering.

Since the boiler feed pump is critical to the operation of the process, it is a Category B, *i.e.,* High Priority item in the process.

5.   Determine the Criticality and Repetitive Equipment, CRE, value for this equipment
This pump has an MTBF of 10 months. Therefore, from Table 4, CRE = b1. Note that the CRE value will vary with the PCHR and MTBF values for the equipment.

6.   Determine equipment inspection frequency to ensure human and environmental safety
From Table 5, this boiler feed pump requires vibration monitoring every 90 days. With such monitoring it is unlikely that an excessive number of failures might occur to this equipment.

7.   Summarize criticality findings in spreadsheet form
When preparing for a PHCR evaluation, a spreadsheet, Table 6, listing critical equipment, should be prepared. Then, as the various rankings are determined, they can be entered in the spreadsheet where they are available for easy reference.

TABLE 4 The Criticality and Repetitive Equipment Values*

	CRE Values			
	Mean time between failures, months			
PHCR	0–6	6–12	12–24	>24
A	a1	a2	a3	a4
B	a2	b1	b2	b3
C	a3	b2	c1	c2
D	a4	b3	c2	d1

Chemical Engineering.

TABLE 5 Predictive Maintenance Frequencies for Rotating Equipment Based on Their CRE Values*

	Maintenance cycles			
	Frequency, days			
CRE	7	30	90	360
a1, a2	VM	LT		
a3, a4		VM	LT	
b1, b3			VM	
c1, d1				VM

VM: Vibration monitoring.
LT: Lubrication sampling and testing.
Chemical Engineering.

TABLE 6 Typical Spreadsheet for Ranking Equipment Criticality*

		Spreadsheet for calculating equipment PHCRS								
Equipment number	Equipment description	NFPA rating				PCH	Other	HCR	PCR	PHCR
		H	F	R	RF					
TKO	Tank	4	4	0	0	4	0	4	4	A44
TKO	Tank	4	4	0	1	3	3	3	4	A34
PU1BFW	Pump	2	0	0	0	2	1	2	3	B23

Chemical Engineering.

Enter the PCH, Other, HCR, PCR, and PHCR values in the spreadsheet, as shown. These data are now available for reference by anyone needing the information.

Related Calculations. The procedure presented here can be applied to all types of equipment used in a facility—fixed, rotating, and instrumentation. Once all the equipment is ranked by criticality, priority lists can be generated. These lists can then be used to ensure the mechanical integrity of critical equipment by prioritizing predictive and preventive maintenance programs, inventories of critical spare parts, and maintenance work orders in case of plant upsets.

In any plant, the hazards posed by different operating units are first ranked and prioritized based on a PHA. These rankings are then used to determine the order in which the hazards need to be addressed. When the PHAs approach completion, team members evaluate the equipment in each operating unit using the PHCR system.

The procedure presented here can be used in any plant concerned with human and environmental safety. Today, this represents every plant, whether conventional or automated. Industries in which this procedure finds active use include chemical, petroleum, textile, food, power, automobile, aircraft, military, and general manufacturing.

This procedure is the work of V. Anthony Ciliberti, Maintenance Engineer, The Lubrizol Corp., as reported in *Chemical Engineering* magazine.

Pump Affinity Laws, Operating Speed, and Head

SIMILARITY OR AFFINITY LAWS FOR CENTRIFUGAL PUMPS

A centrifugal pump designed for a 1800-r/min operation and a head of 200 ft (60.9 m) has a capacity of 3000 gal/min (189.3 L/s) with a power input of 175 hp (130.6 kW). What effect will a speed reduction to 1200 r/min have on the head, capacity, and power input of the pump? What will be the change in these variables if the impeller diameter is reduced from 12 to 10 in (304.8 to 254 mm) while the speed is held constant at 1800 r/min?

Calculation Procedure:

1. *Compute the effect of a change in pump speed*
For any centrifugal pump in which the effects of fluid viscosity are negligible, or are neglected, the similarity or affinity laws can be used to determine the effect of a speed, power, or head change. For a *constant impeller diameter,* the laws are $Q_1/Q_2 = N_1/N_2$; $H_1/H_2 = (N_1/N_2)^2$; $P_1/P_2 = (N_1/N_2)^3$. For a *constant speed,* $Q_1/Q_2 = D_1/D_2$; $H_1/H_2 = (D_1/D_2)^2$; $P_1/P_2 = (D_1/D_2)^3$. In both sets of laws, Q = capacity, gal/min; N = impeller rpm; D = impeller diameter, in; H = total head, ft of liquid; P = bhp input. The subscripts 1 and 2 refer to the initial and changed conditions, respectively.

For this pump, with a constant impeller diameter, $Q_1/Q_2 = N_1/N_2$; $3000/Q_2 = 1800/1200$; $Q_2 = 2000$ gal/min (126.2 L/s). And, $H_1/H_2 = (N_1/N_2)^2 = 200/H_2 = (1800/1200)^2$; $H_2 = 88.9$ ft (27.1 m). Also, $P_1/P_2 = (N_1/N_2)^3 = 175/P_2 = (1800/1200)^3$; $P_2 = 51.8$ bhp (38.6 kW).

2. Compute the effect of a change in impeller diameter

With the speed constant, use the second set of laws. Or, for this pump, $Q_1/Q_2 = D_1/D_2$; $3000/Q_2 = {}^{12}\!/_{10}$; $Q_2 = 2500$ gal/min (157.7 L/s). And $H_1/H_2 = (D_1/D_2)^2$; $200/H_2 = ({}^{12}\!/_{10})^2$; $H_2 = 138.8$ ft (42.3 m). Also, $P_1/P_2 = (D_1/D_2)^3$; $175/P_2 = ({}^{12}\!/_{10})^3$; $P_2 = 101.2$ bhp (75.5 kW).

Related Calculations. Use the similarity laws to extend or change the data obtained from centrifugal pump characteristic curves. These laws are also useful in field calculations when the pump head, capacity, speed, or impeller diameter is changed.

The similarity laws are most accurate when the efficiency of the pump remains nearly constant. Results obtained when the laws are applied to a pump having a constant impeller diameter are somewhat more accurate than for a pump at constant speed with a changed impeller diameter. The latter laws are more accurate when applied to pumps having a low specific speed.

If the similarity laws are applied to a pump whose impeller diameter is increased, be certain to consider the effect of the higher velocity in the pump suction line. Use the similarity laws for any liquid whose viscosity remains constant during passage through the pump. However, the accuracy of the similarity laws decreases as the liquid viscosity increases.

SIMILARITY OR AFFINITY LAWS IN CENTRIFUGAL PUMP SELECTION

A test-model pump delivers, at its best efficiency point, 500 gal/min (31.6 L/s) at a 350-ft (106.7-m) head with a required net positive suction head (NPSH) of 10 ft (3 m) a power input of 55 hp (41 kW) at 3500 r/min, when a 10.5-in (266.7-mm) diameter impeller is used. Determine the performance of the model at 1750 r/min. What is the performance of a full-scale prototype pump with a 20-in (50.4-cm) impeller operating at 1170 r/min? What are the specific speeds and the suction specific speeds of the test-model and prototype pumps?

Calculation Procedure:

1. Compute the pump performance at the new speed

The similarity or affinity laws can be stated in general terms, with subscripts p and m for prototype and model, respectively, as $Q_p = K_d^3 N_n Q_m$; $H_p = K_d^2 K_n^2 H_m$; $\text{NPSH}_p = K_d^2 K_n^2 \text{NPSH}_m$; $P_p = K_d^5 K_n^5 P_m$, where K_d = size factor = prototype dimension/model dimension. The usual dimension used for the size factor is the impeller diameter. Both dimensions should be in the same units of measure. Also, K_n = (prototype speed, r/min)/(model speed, r/min). Other symbols are the same as in the previous calculation procedure.

When the model speed is reduced from 3500 to 1750 r/min, the pump dimensions remain the same and $K_d = 1.0$; $K_n = 1750/3500 = 0.5$. Then $Q = (1.0)(0.5)(500) = 250$ r/min; $H = (1.0)^2(0.5)^2(350) = 87.5$ ft (26.7 m); NPSH $= (1.0)^2(0.5)^2(10) = 2.5$ ft (0.76 m); $P = (1.0)^5(0.5)^3(55) = 6.9$ hp (5.2 kW). In this computation, the subscripts were omitted from the equations because the same pump, the test model, was being considered.

2. Compute performance of the prototype pump

First, K_d and K_n must be found: $K_d = 20/10.5 = 1.905$; $K_n = 1170/3500 = 0.335$. Then $Q_p = (1.905)^3(0.335)(500) = 1158$ gal/min (73.1 L/s); $H_p = (1.905)^2(0.335)^2(350) = 142.5$ ft (43.4 m); $\text{NPSH}_p = (1.905)^2(0.335)^2(10) = 4.06$ ft (1.24 m); $P_p = (1.905)^5(0.335)^3(55) = 51.8$ hp (38.6 kW).

3. Compute the specific speed and suction specific speed

The specific speed or, as Horwitz[1] says, "more correctly, discharge specific speed," is $N_s = N(Q)^{0.5}/(H)^{0.75}$, while the suction specific speed $S = N(Q)^{0.5}/(\text{NPSH})^{0.75}$, where all values are taken at the best efficiency point of the pump.

For the model, $N_s = 3500(500)^{0.5}/(350)^{0.75} = 965$; $S = 3500(500)^{0.5}/(10)^{0.75} = 13,900$. For the prototype, $N_s = 1170(1158)^{0.5}/(142.5)^{0.75} = 965$; $S = 1170(1156)^{0.5}/(4.06)^{0.75} = 13,900$. The specific speed and suction specific speed of the model and prototype are equal because these units are geometrically similar or homologous pumps and both speeds are mathematically derived from the similarity laws.

Related Calculations. Use the procedure given here for any type of centrifugal pump where the similarity laws apply. When the term *model* is used, it can apply to a production test pump or to a standard unit ready for installation. The procedure presented here is the work of R. P. Horwitz, as reported in *Power* magazine.[1]

SPECIFIC SPEED CONSIDERATIONS IN CENTRIFUGAL PUMP SELECTION

What is the upper limit of specific speed and capacity of a 1750-r/min single-stage double-suction centrifugal pump having a shaft that passes through the impeller eye if it handles clear water at 85°F (29.4°C) at sea level at a total head of 280 ft (85.3 m) with a 10-ft (3-m) suction lift? What is the efficiency of the pump and its approximate impeller shape?

Calculation Procedure:

1. Determine the upper limit of specific speed

Use the Hydraulic Institute upper specific-speed curve, Fig. 7, for centrifugal pumps or a similar curve, Fig. 8, for mixed- and axial-flow pumps. Enter Fig. 7 at the bottom at 280-ft (85.3-m) total head, and project vertically upward until the 10-ft (3-m) suction-lift curve is intersected. From here, project horizontally to the right to read the specific speed $N_S = 2000$. Figure 8 is used in a similar manner.

2. Compute the maximum pump capacity

For any centrifugal, mixed- or axial-flow pump, $N_S = (gpm)^{0.5}(rpm)/H_t^{0.75}$, where H_t = total head on the pump, ft of liquid. Solving for the maximum capacity, we get $gpm = (N_S H_t^{0.75}/rpm)^2 = (2000 \times 280^{0.75}/1750)^2 = 6040$ gal/min (381.1 L/s).

[1]R. P. Horwitz, "Affinity Laws and Specific Speed Can Simplify Centrifugal Pump Selection," *Power,* November 1964.

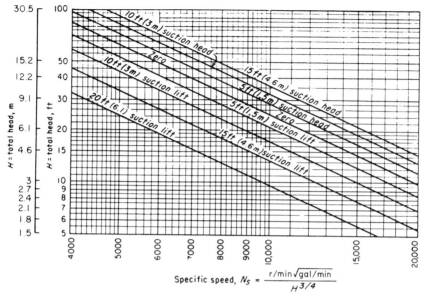

FIGURE 8 Upper limits of specific speeds of single-suction mixed-flow and axial-flow pumps. (*Hydraulic Institute.*)

3. *Determine the pump efficiency and impeller shape*

Figure 9 shows the general relation between impeller shape, specific speed, pump capacity, efficiency, and characteristic curves. At N_S = 2000, efficiency = 87 percent. The impeller, as shown in Fig. 9, is moderately short and has a relatively large discharge area. A cross section of the impeller appears directly under the N_S = 2000 ordinate.

Related Calculations.   Use the method given here for any type of pump whose variables are included in the Hydraulic Institute curves, Figs. 7 and 8, and in similar curves available from the same source. *Operating specific speed,* computed as above, is sometimes plotted on the performance curve of a centrifugal pump so that the characteristics of the unit can be better understood. *Type specific speed* is the operating specific speed giving maximum efficiency for a given pump and is a number used to identify a pump. Specific speed is important in cavitation and suction-lift studies. The Hydraulic Institute curves, Figs. 7 and 8, give upper limits of speed, head, capacity and suction lift for cavitation-free operation. When making actual pump analyses, be certain to use the curves (Figs. 7 and 8) in the latest edition of the *Standards of the Hydraulic Institute.*

SELECTING THE BEST OPERATING SPEED FOR A CENTRIFUGAL PUMP

A single-suction centrifugal pump is driven by a 60-Hz ac motor. The pump delivers 10,000 gal/min (630.9 L/s) of water at a 100-ft (30.5-m) head. The available net

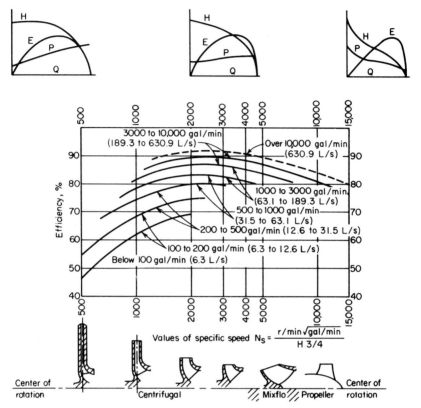

FIGURE 9 Approximate relative impeller shapes and efficiency variations for various specific speeds of centrifugal pumps. (*Worthington Corporation.*)

positive suction head = 32 ft (9.7 m) of water. What is the best operating speed for this pump if the pump operates at its best efficiency point?

Calculation Procedure:

1. Determine the specific speed and suction specific speed
Ac motors can operate at a variety of speeds, depending on the number of poles. Assume that the motor driving this pump might operate at 870, 1160, 1750, or 3500 r/min. Compute the specific speed $N_S = N(Q)^{0.5}/(H)^{0.75} = N(10,000)^{0.5}/(100)^{0.75} = 3.14N$ and the suction specific speed $S = N(Q)^{0.5}/(NPSH)^{0.75} = N(10,000)^{0.5}/(32)^{0.75} = 7.43N$ for each of the assumed speeds. Tabulate the results as follows:

Operating speed, r/min	Required specific speed	Required suction specific speed
870	2,740	6,460
1,160	3,640	8,620
1,750	5,500	13,000
3,500	11,000	26,000

2. Choose the best speed for the pump
Analyze the specific speed and suction specific speed at each of the various operating speeds, using the data in Tables 7 and 8. These tables show that at 870 and 1160 r/min, the suction specific-speed rating is poor. At 1750 r/min, the suction specific-speed rating is excellent, and a turbine or mixed-flow type pump will be suitable. Operation at 3500 r/min is unfeasible because a suction specific speed of 26,000 is beyond the range of conventional pumps.

Related Calculations. Use this procedure for any type of centrifugal pump handling water for plant services, cooling, process, fire protection, and similar requirements. This procedure is the work of R. P. Horwitz, Hydrodynamics Division, Peerless Pump, FMC Corporation, as reported in *Power* magazine.

TOTAL HEAD ON A PUMP HANDLING VAPOR-FREE LIQUID

Sketch three typical pump piping arrangements with static suction lift and submerged, free, and varying discharge head. Prepare similar sketches for the same pump with static suction head. Label the various heads. Compute the total head on each pump if the elevations are as shown in Fig. 10 and the pump discharges a maximum of 2000 gal/min (126.2 L/s) of water through 8-in (203.2-mm) schedule 40 pipe. What hp is required to drive the pump? A swing check valve is used on the pump suction line and a gate valve on the discharge line.

Calculation Procedure:

1. Sketch the possible piping arrangements
Figure 10 shows the six possible piping arrangements for the stated conditions of the installation. Label the total static head, *i.e.,* the *vertical* distance from the surface

TABLE 7 Pump Types Listed by Specific Speed*

Specific speed range	Type of pump
Below 2,000	Volute, diffuser
2,000–5,000	Turbine
4,000–10,000	Mixed-flow
9,000–15,000	Axial-flow

*Peerless Pump Division, FMC Corporation.

TABLE 8 Suction Specific-Speed Ratings*

Single-suction pump	Double-suction pump	Rating
Above 11,000	Above 14,000	Excellent
9,000–11,000	11,000–14,000	Good
7,000–9,000	9,000–11,000	Average
5,000–7,000	7,000–9,000	Poor
Below 5,000	Below 7,000	Very poor

*Peerless Pump Division, FMC Corporation.

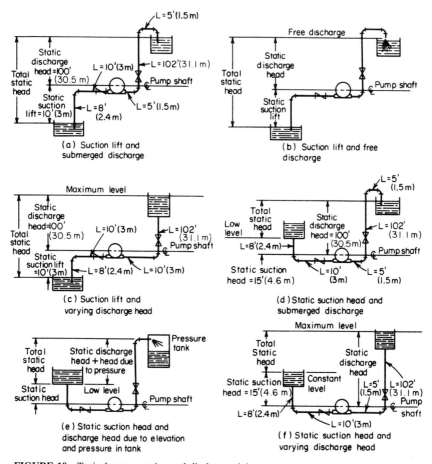

FIGURE 10   Typical pump suction and discharge piping arrangements.

of the source of the liquid supply to the free surface of the liquid in the discharge receiver, or to the point of free discharge from the discharge pipe. When both the suction and discharge surfaces are open to the atmosphere, the total static head equals the vertical difference in elevation. Use the free-surface elevations that cause the maximum suction lift and discharge head, *i.e.,* the *lowest* possible level in the supply tank and the *highest* possible level in the discharge tank or pipe. When the supply source is *below* the pump centerline, the vertical distance is called the *static suction lift;* with the supply *above* the pump centerline, the vertical distance is called *static suction head.* With variable static suction head, use the lowest liquid level in the supply tank when computing total static head. Label the diagrams as shown in Fig. 10.

2.   Compute the total static head on the pump

The total static head H_{ts} ft = static suction lift, h_{sl} ft + static discharge head h_{sd} ft, where the pump has a suction lift, s in Fig. 10*a, b,* and *c.* In these installations,

$H_{ts} = 10 + 100 = 110$ ft (33.5 m). Note that the static discharge head is computed between the pump centerline and the water level with an underwater discharge, Fig. 10a; to the pipe outlet with a free discharge, Fig. 10b; and to the maximum water level in the discharge tank, Fig. 10c. When a pump is discharging into a closed compression tank, the total discharge head equals the static discharge head plus the head equivalent, ft of liquid, of the internal pressure in the tank, or 2.31 $\times$ tank pressure, lb/in^2.

Where the pump has a static suction head, as in Fig. 10d, e, and f, the total static head H_{ts} ft $= h_{sd}$ − static suction head h_{sh} ft. In these installations, $H_t = 100 - 15 = 85$ ft (25.9 m).

The total static head, as computed above, refers to the head on the pump without liquid flow. To determine the total head on the pump, the friction losses in the piping system during liquid flow must be also determined.

3. Compute the piping friction losses

Mark the length of each piece of straight pipe on the piping drawing. Thus, in Fig. 10a, the total length of straight pipe L, ft $= 8 + 10 + 5 + 102 + 5 = 130$ ft (39.6 m), if we start at the suction tank and add each length until the discharge tank is reached. To the total length of straight pipe must be added the *equivalent* length of the pipe fittings. In Fig. 10a there are four long-radius elbows, one swing check valve, and one globe valve. In addition, there is a minor head loss at the pipe inlet and at the pipe outlet.

The equivalent length of one 8-in (203.2-mm) long-radius elbow is 14 ft (4.3 m) of pipe, from Table 9. Since the pipe contains four elbows, the total equivalent length $= 4(14) = 56$ ft (17.1 m) of straight pipe. The open gate valve has an equivalent resistance of 4.5 ft (1.4 m); and the open swing check valve has an equivalent resistance of 53 ft (16.2 m).

The entrance loss h_e ft, assuming a basket-type strainer is used at the suction-pipe inlet, is h_e ft $= Kv^2/2g$, where $K = $ a constant from Fig. 11; $v = $ liquid velocity, ft/s; $g = 32.2$ ft/s^2 (980.67 cm/s^2). The exit loss occurs when the liquid passes through a sudden enlargement, as from a pipe to a tank. Where the area of the tank is large, causing a final velocity that is zero, $h_{ex} = v^2/2g$.

The velocity v ft/s in a pipe $= gpm/2.448d^2$. For this pipe, $v = 2000/[(2.448)(7.98)^2] = 12.82$ ft/s (3.91 m/s). Then $h_e = 0.74(12.82)^2/[2(32.2)] = 1.89$ ft (0.58 m), and $h_{ex} = (12.82)^2/[(2)(32.2)] = 2.56$ ft (0.78 m). Hence, the total length of the piping system in Fig. 10a is $130 + 56 + 4.5 + 53 + 1.89 + 2.56 = 247.95$ ft (75.6 m), say 248 ft (75.6 m).

Use a suitable head-loss equation, or Table 10, to compute the head loss for the pipe and fittings. Enter Table 10 at an 8-in (203.2-mm) pipe size, and project horizontally across to 2000 gal/min (126.2 L/s) and read the head loss as 5.86 ft of water per 100 ft (1.8 m/30.5 m) of pipe.

The total length of pipe and fittings computed above is 248 ft (75.6 m). Then total friction-head loss with a 2000 gal/min (126.2-L/s) flow is H_f ft $= (5.86)(248/100) = 14.53$ ft (4.5 m).

4. Compute the total head on the pump

The total head on the pump $H_t = H_{ts} + H_f$. For the pump in Fig. 10a, $H_t = 110 + 14.53 = 124.53$ ft (37.95 m), say 125 ft (38.1 m). The total head on the pump in Fig. 10b and c would be the same. Some engineers term the total head on a pump the *total dynamic head* to distinguish between static head (no-flow vertical head) and operating head (rated flow through the pump).

The total head on the pumps in Fig. 10d, c, and f is computed in the same way as described above, except that the total static head is less because the pump has

TABLE 9 Resistance of Fittings and Valves (length of straight pipe giving equivalent resistance)

Pipe size		Standard ell		Medium-radius ell		Long-radius ell		45° Ell		Tee		Gate valve, open		Globe valve, open		Swing check, open	
in	mm	ft	m	ft	m	ft	m	ft	m	ft	m	ft	m	ft	m	ft	m
6	152.4	16	4.9	14	4.3	11	3.4	7.7	2.3	33	10.1	3.5	1.1	160	48.8	40	12.2
8	203.2	21	6.4	18	5.5	14	4.3	10	3.0	43	13.1	4.5	1.4	220	67.0	53	16.2
10	254.0	26	7.9	22	6.7	17	5.2	13	3.9	56	17.1	5.7	1.7	290	88.4	67	20.4
12	304.8	32	9.8	26	7.9	20	6.1	15	4.6	66	20.1	6.7	2.0	340	103.6	80	24.4

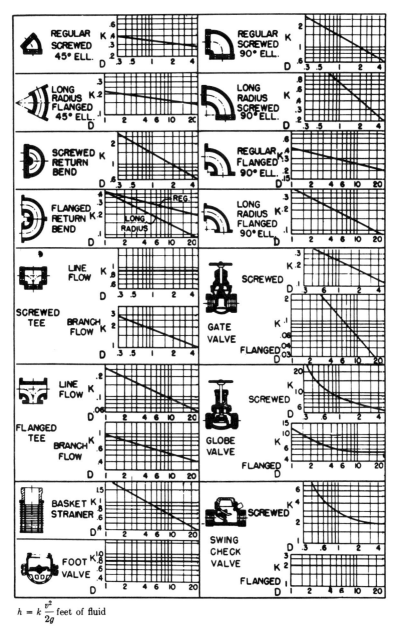

$$h = k \frac{v^2}{2g} \text{ feet of fluid}$$

FIGURE 11 Resistance coefficients of pipe fittings. To convert to SI in the equation for h, v^2 would be measured in m/s and feet would be changed to meters. The following values would also be changed from inches to millimeters: 0.3 to 7.6, 0.5 to 12.7, 1 to 25.4, 2 to 50.8, 4 to 101.6, 6 to 152.4 10 to 254, and 20 to 508. (*Hydraulic Institute.*)

TABLE 10   Pipe Friction Loss for Water (wrought-iron or steel schedule 40 pipe in good condition)

Diameter		Flow		Velocity		Velocity head		Friction loss per 100 ft (30.5 m) of pipe	
in	mm	gal/min	L/s	ft/s	m/s	ft water	m water	ft water	m water
6	152.4	1000	63.1	11.1	3.4	1.92	0.59	6.17	1.88
6	152.4	2000	126.2	22.2	6.8	7.67	2.3	23.8	7.25
6	152.4	4000	252.4	44.4	13.5	30.7	9.4	93.1	28.4
8	203.2	1000	63.1	6.41	1.9	0.639	0.195	1.56	0.475
8	203.2	2000	126.2	12.8	3.9	2.56	0.78	5.86	1.786
8	203.2	4000	252.4	25.7	7.8	10.2	3.1	22.6	6.888
10	254.0	1000	63.1	3.93	1.2	0.240	0.07	0.497	0.151
10	254.0	3000	189.3	11.8	3.6	2.16	0.658	4.00	1.219
10	254.0	5000	315.5	19.6	5.9	5.99	1.82	10.8	3.292

a static suction head. That is, the elevation of the liquid on the suction side reduces the total distance through which the pump must discharge liquid; thus the total static head is less. The static suction head is *subtracted* from the static discharge head to determine the total static head on the pump.

5.   Compute the horsepower required to drive the pump
The brake hp input to a pump $bhp_i = (gpm)(H_t)(s)/3960e$, where s = specific gravity of the liquid handled; e = hydraulic efficiency of the pump, expressed as a decimal. The usual hydraulic efficiency of a centrifugal pump is 60 to 80 percent; reciprocating pumps, 55 to 90 percent; rotary pumps, 50 to 90 percent. For each class of pump, the hydraulic efficiency decreases as the liquid viscosity increases.

Assume that the hydraulic efficiency of the pump in this system is 70 percent and the specific gravity of the liquid handled is 1.0. Then $bhp_i = (2000)(127)(1.0)/(3960)(0.70) = 91.6$ hp (68.4 kW).

The theoretical or *hydraulic horsepower* $hp_h = (gpm)(H_t)(s)/3960$, or $hp_h = (2000) = (127)(1.0)/3900 = 64.1$ hp (47.8 kW).

Related Calculations.   Use this procedure for any liquid—water, oil, chemical, sludge, *etc.*—whose specific gravity is known. When liquids other than water are being pumped, the specific gravity and viscosity of the liquid, as discussed in later calculation procedures, must be taken into consideration. The procedure given here can be used for any class of pump—centrifugal, rotary, or reciprocating.

Note that Fig. 11 can be used to determine the equivalent length of a variety of pipe fittings. To use Fig. 11, simply substitute the appropriate K value in the relation $h = Kv^2/2g$, where h = equivalent length of straight pipe; other symbols as before.

PUMP SELECTION FOR ANY PUMPING SYSTEM

Give a step-by-step procedure for choosing the class, type, capacity, drive, and materials for a pump that will be used in an industrial pumping system.

Calculation Procedure:

1. Sketch the proposed piping layout

Use a single-line diagram, Fig. 12, of the piping system. Base the sketch on the actual job conditions. Show all the piping, fittings, valves, equipment, and other units in the system. Mark the *actual* and *equivalent* pipe length (see the previous calculation procedure) on the sketch. Be certain to include all vertical lifts, sharp bends, sudden enlargements, storage tanks, and similar equipment in the proposed system.

2. Determine the required capacity of the pump

The required capacity is the flow rate that must be handled in gal/min, million gal/day, ft³/s, gal/h, bbl/day, lb/h, acre·ft/day, mil/h, or some similar measure. Obtain the required flow rate from the process conditions, for example, boiler feed

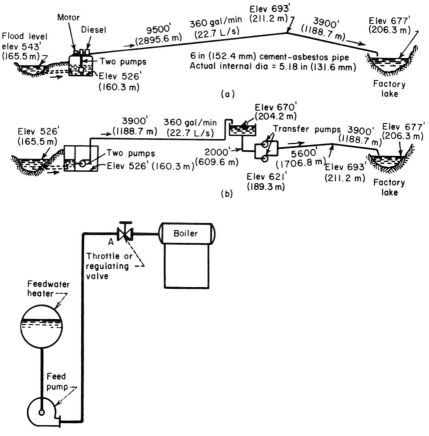

FIGURE 12 (*a*) Single-line diagrams for an industrial pipeline; (*b*) single-line diagram of a boiler-feed system. (*Worthington Corporation.*)

rate, cooling-water flow rate, chemical feed rate, *etc.* The required flow rate for any process unit is usually given by the manufacturer or can be computed by using the calculation procedures given throughout this handbook.

Once the required flow rate is determined, apply a suitable factor of safety. The value of this factor of safety can vary from a low of 5 percent of the required flow to a high of 50 percent or more, depending on the application. Typical safety factors are in the 10 percent range. With flow rates up to 1000 gal/min (63.1 L/s), and in the selection of process pumps, it is common practice to round a computed required flow rate to the next highest round-number capacity. Thus, with a required flow rate of 450 gal/min (28.4 L/s) and a 10 percent safety factor, the flow of 450 + 0.10(450) = 495 gal/min (31.2 L/s) would be rounded to 500 gal/min (31.6 L/s) *before* the pump was selected. A pump of 500-gal/min (31.6-L/s), or larger, capacity would be selected.

3. *Compute the total head on the pump*

Use the steps given in the previous calculation procedure to compute the total head on the pump. Express the result in ft (m) of water—this is the most common way of expressing the head on a pump. Be certain to use the exact specific gravity of the liquid handled when expressing the head in ft (m) of water. A specific gravity less than 1.00 *reduces* the total head when expressed in ft (m) of water; whereas a specific gravity greater than 1.00 *increases* the total head when expressed in ft (m) of water. Note that variations in the suction and discharge conditions can affect the total head on the pump.

4. *Analyze the liquid conditions*

Obtain complete data on the liquid pumped. These data should include the name and chemical formula of the liquid, maximum and minimum pumping temperature, corresponding vapor pressure at these temperatures, specific gravity, viscosity at the pumping temperature, pH, flash point, ignition temperature, unusual character-istics (such as tendency to foam, curd, crystallize, become gelatinous or tacky), solids content, type of solids and their size, and variation in the chemical analysis of the liquid.

Enter the liquid conditions on a pump selection form like that in Fig. 13. Such forms are available from many pump manufacturers or can be prepared to meet special job conditions.

5. *Select the class and type of pump*

Three *classes* of pumps are used today—centrifugal, rotary, and reciprocating, Fig. 14. Note that these terms apply only to the mechanics of moving the liquid—not to the service for which the pump was designed. Each class of pump is further subdivided into a number of *types,* Fig. 14.

Use Table 11 as a general guide to the class and type of pump to be used. For example, when a large capacity at moderate pressure is required, Table 11 shows that a centrifugal pump would probably be best. Table 11 also shows the typical characteristics of various classes and types of pumps used in industrial process work.

Consider the liquid properties when choosing the class and type of pump, be-cause exceptionally severe conditions may rule out one or another class of pump at the start. Thus, screw- and gear-type rotary pumps are suitable for handling viscous, nonabrasive liquid, Table 11. When an abrasive liquid must be handled, either another class of pump or another type of rotary pump must be used.

Summary of Essential Data Required in Selection of Centrifugal Pumps

1. Number of Units Required

2. Nature of the Liquid to Be Pumped
   Is the liquid:
   a. Fresh or salt water, acid or alkali, oil, gasoline, slurry, or paper stock?
   b. Cold or hot and if hot, at what temperature? What is the vapor pressure of the liquid at the pumping temperature?
   c. What is its specific gravity?
   d. Is it viscous or nonviscous?
   e. Clear and free from suspended foreign matter or dirty and gritty? If the latter, what is the size and nature of the solids, and are they abrasive? If the liquid is of a pulpy nature, what is the consistency expressed either in percentage or in lb per cu ft of liquid? What is the suspended material?
   f. What is the chemical analysis, pH value, etc.? What are the expected variations of this analysis? If corrosive, what has been the past experience, both with successful materials and with unsatisfactory materials?

3. Capacity
   What is the required capacity as well as the minimum and maximum amount of liquid the pump will ever be called upon to deliver?

4. Suction Conditions
   Is there:
   a. A suction lift?
   b. Or a suction head?
   c. What are the length and diameter of the suction pipe?

5. Discharge Conditions
   a. What is the static head? Is it constant or variable?
   b. What is the friction head?
   c. What is the maximum discharge pressure against which the pump must deliver the liquid?

6. Total Head
   Variations in items 4 and 5 will cause variations in the total head.

7. Is the service continuous or intermittent?

8. Is the pump to be installed in a horizontal or vertical position? If the latter,
   a. In a wet pit?
   b. In a dry pit?

9. What type of power is available to drive the pump and what are the characteristics of this power?

10. What space, weight, or transportation limitations are involved?

11. Location of installation
    a. Geographical location
    b. Elevation above sea level
    c. Indoor or outdoor installation
    d. Range of ambient temperatures

12. Are there any special requirements or marked preferences with respect to the design, construction, or performance of the pump?

FIGURE 13 Typical selection chart for centrifugal pumps. (*Worthington Corporation.*)

Also consider all the operating factors related to the particular pump. These factors include the type of service (continuous or intermittent), operating-speed preferences, future load expected and its effect on pump head and capacity, maintenance facilities available, possibility of parallel or series hookup, and other conditions peculiar to a given job.

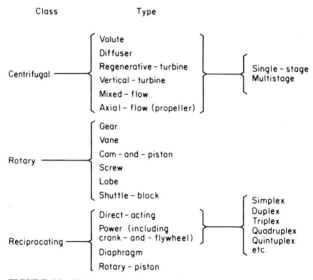

FIGURE 14   Modern pump classes and types.

Once the class and type of pump is selected, consult a rating table (Table 12) or rating chart, Fig. 15, to determine whether a suitable pump is available from the manufacturer whose unit will be used. When the hydraulic requirements fall between two standard pump models, it is usual practice to choose the next larger size of pump, unless there is some reason why an exact head and capacity are required for the unit. When one manufacturer does not have the desired unit, refer to the engineering data of other manufacturers. Also keep in mind that some pumps are custom-built for a given job when precise head and capacity requirements must be met.

Other pump data included in manufacturer's engineering information include characteristic curves for various diameter impellers in the same casing, Fig. 16, and variable-speed head-capacity curves for an impeller of given diameter, Fig. 17. Note that the required power input is given in Figs. 15 and 16 and may also be given in Fig. 17. Use of Table 12 is explained in the table.

Performance data for rotary pumps are given in several forms. Figure 18 shows a typical plot of the head and capacity ranges of different types of rotary pumps. Reciprocating-pump capacity data are often tabulated, as in Table 13.

6.   Evaluate the pump chosen for the installation
Check the specific speed of a centrifugal pump, using the method given in an earlier calculation procedure. Once the specific speed is known, the impeller type and approximate operating efficiency can be found from Fig. 9.

Check the piping system, using the method of an earlier calculation procedure, to see whether the available net positive suction head equals, or is greater than, the required net positive suction head of the pump.

Determine whether a vertical or horizontal pump is more desirable. From the standpoint of floor space occupied, required NPSH, priming, and flexibility in

TABLE 11 Characteristics of Modern Pumps

	Centrifugal		Rotary	Reciprocating		
	Volute and diffuser	Axial flow	Screw and gear	Direct acting steam	Double acting power	Triplex
Discharge flow	Steady	Steady	Steady	Pulsating	Pulsating	Pulsating
Usual maximum suction lift, ft (m)	15 (4.6)	15 (4.6)	22 (6.7)	22 (6.7)	22 (6.7)	22 (6.7)
Liquids handled	Clean, clear; dirty, abrasive; liquids with high solids content		Viscous; non-abrasive	Clean and clear		
Discharge pressure range	Low to high		Medium	Low to highest produced		
Usual capacity range	Small to largest available		Small to medium	Relatively small		
How increased head affects:						
Capacity	Decrease		None	Decrease	None	None
Power input	Depends on specific speed		Increase	Increase	Increase	Increase
How decreased head affects:						
Capacity	Increase		None	Small increase	None	None
Power input	Depends on specific speed		Decrease	Decrease	Decrease	Decrease

changing the pump use, vertical pumps may be preferable to horizontal designs in some installations. But where headroom, corrosion, abrasion, and ease of maintenance are important factors, horizontal pumps may be preferable.

As a general guide, single-suction centrifugal pumps handle up to 50 gal/min (3.2 L/s) at total heads up to 50 ft (15.2 m); either single- or double-suction pumps are used for the flow rates to 1000 gal/min (63.1 L/s) and total heads to 300 ft (91.4 m); beyond these capacities and heads, double-suction or multistage pumps are generally used.

Mechanical seals are becoming more popular for all types of centrifugal pumps in a variety of services. Although they are more costly than packing, the mechanical seal reduces pump maintenance costs.

Related Calculations. Use the procedure given here to select any class of pump—centrifugal, rotary, or reciprocating—for any type of service—power plant, atomic energy, petroleum processing, chemical manufacture, paper mills, textile mills, rubber factories, food processing, water supply, sewage and sump service, air conditioning and heating, irrigation and flood control, mining and construction, marine services, industrial hydraulics, iron and steel manufacture.

TABLE 12   Typical Centrifugal-Pump Rating Table

Size		Total head			
gal/min	L/s	20 ft, r/min—hp	6.1 m, r/min—kW	25 ft, r/min—hp	7.6 m, r/min—kW
3 CL:					
200	12.6	910—1.3	910-0.97	1010—1.6	1010—1.19
300	18.9	1000—1.9	1000—1.41	1100—2.4	1100—1.79
400	25.2	1200—3.1	1200—2.31	1230—3.7	1230—2.76
500	31.5	—	—	—	—
4 C:					
400	25.2	940—2.4	940—1.79	1040—3	1040—2.24
600	37.9	1080—4	1080—2.98	1170—4.6	1170—3.43
800	50.5	—	—	—	—

Example: 1080—4 indicates pump speed is 1080 r/min; actual input required to operate the pump is 4 hp (2.98 kW).
Source: Condensed from data of Goulds Pumps, Inc.; SI values added by handbook editor.

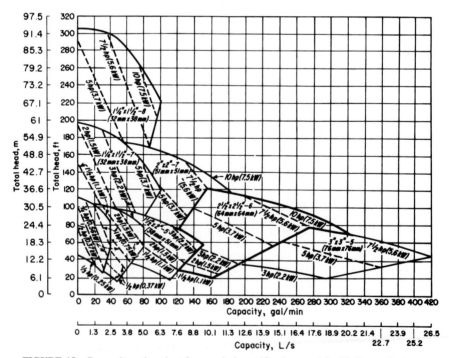

FIGURE 15   Composite rating chart for a typical centrifugal pump. (*Goulds Pumps, Inc.*)

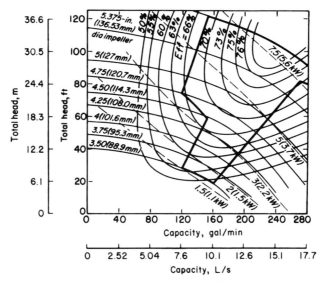

FIGURE 16 Pump characteristics when impeller diameter is varied within the same casing.

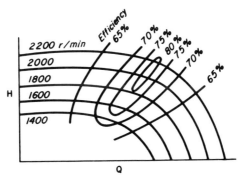

FIGURE 17 Variable-speed head-capacity curves for a centrifugal pump.

ANALYSIS OF PUMP AND SYSTEM CHARACTERISTIC CURVES

Analyze a set of pump and system characteristic curves for the following conditions: friction losses without static head; friction losses with static head; pump without lift; system with little friction, much static head; system with gravity head; system with different pipe sizes; system with two discharge heads; system with diverted flow; and effect of pump wear on characteristic curve.

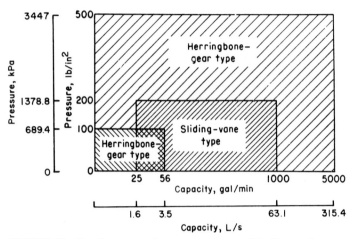

FIGURE 18 Capacity ranges of some rotary pumps. (*Worthington Corporation.*)

TABLE 13 Capacities of Typical Horizontal Duplex Plunger Pumps

Size		Cold-water pressure service			
		gal/min	L/s	Piston speed	
in	cm			ft/min	m/min
6 × 3½ × 6	15.2 × 8.9 × 15.2	60	3.8	60	18.3
7½ × 4½ × 10	19.1 × 11.4 × 25.4	124	7.8	75	22.9
9 × 5 × 10	22.9 × 12.7 × 25.4	153	9.7	75	22.9
10 × 6 × 12	25.4 × 15.2 × 30.5	235	14.8	80	24.4
12 × 7 × 12	30.5 × 17.8 × 30.5	320	20.2	80	24.4

Size		Boiler-feed service					
		gal/min	L/s	Boiler		Piston speed	
in	cm			hp	kW	ft/min	m/min
6 × 3½ × 6	15.2 × 8.9 × 15.2	36	2.3	475	354.4	36	10.9
7½ × 4½ × 10	19.1 × 11.4 × 25.4	74	4.7	975	727.4	45	13.7
9 × 5 × 10	22.9 × 12.7 × 25.4	92	5.8	1210	902.7	45	13.7
10 × 6 × 12	25.4 × 15.2 × 30.5	141	8.9	1860	1387.6	48	14.6
12 × 7 × 12	30.5 × 17.8 × 30.5	192	12.1	2530	1887.4	48	14.6

Source: Courtesy of Worthington Corporation.

Calculation Procedure:

1. Plot the system-friction curve

Without static head, the system-friction curve passes through the origin (0,0), Fig. 19, because when no head is developed by the pump, flow through the piping is zero. For most piping systems, the friction-head loss varies as the square of the liquid flow rate in the system. Hence, a system-friction curve, also called a friction-head curve, is parabolic—the friction head increases as the flow rate or capacity of the system increases. Draw the curve as shown in Fig. 19.

2. Plot the piping system and system-head curve

Figure 20a shows a typical piping system with a pump operating against a static discharge head. Indicate the total static head, Fig. 20b, by a dashed line—in this installation $H_{ts} = 110$ ft. Since static head is a physical dimension, it does not vary with flow rate and is a constant for all flow rates. Draw the dashed line parallel to the abscissa, Fig. 20b.

From the point of no flow—zero capacity—plot the friction-head loss at various flow rates—100, 200, 300 gal/min (6.3, 12.6, 18.9 L/s), etc. Determine the friction-head loss by computing it as shown in an earlier calculation procedure. Draw a curve through the points obtained. This is called the *system-head curve.*

Plot the pump head-capacity (H-Q) curve of the pump on Fig. 20b. The H-Q curve can be obtained from the pump manufacturer or from a tabulation of H and Q values for the pump being considered. The point of intersection A between the H-Q and system-head curves is the operating point of the pump.

Changing the resistance of a given piping system by partially closing a valve or making some other change in the friction alters the position of the system-head curve and pump operating point. Compute the frictional resistance as before, and plot the artificial system-head curve as shown. Where this curve intersects the H-Q curve is the new operating point of the pump. System-head curves are valuable for analyzing the suitability of a given pump for a particular application.

3. Plot the no-lift system-head curve and compute the losses

With no static head or lift, the system-head curve passes through the origin (0,0), Fig. 21. For a flow of 900 gal/min (56.8 L/s) in this system, compute the friction loss as follows, using the Hydraulic Institute *Pipe Friction Manual* tables or the method of earlier calculation procedures:

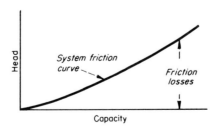

FIGURE 19 Typical system-friction curve.

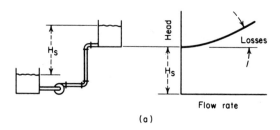

(a)

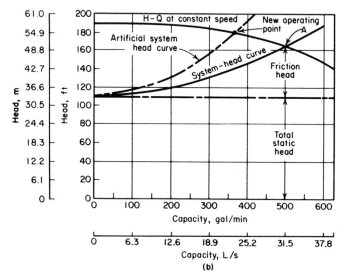

(b)

FIGURE 20 (a) Significant friction loss and lift; (b) system-head curve superimposed on pump head-capacity curve. (*Peerless Pumps.*)

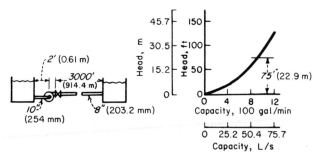

FIGURE 21 No lift; all friction head. (*Peerless Pumps.*)

	ft	m
Entrance loss from tank into 10-in (254-mm) suction pipe, $0.5v^2/2g$	0.10	0.03
Friction loss in 2 ft (0.61 m) of suction pipe	0.02	0.01
Loss in 10-in (254-mm) 90° elbow at pump	0.20	0.06
Friction loss in 3000 ft (914.4 m) of 8-in (203.2-mm) discharge pipe	74.50	22.71
Loss in fully open 8-in (203.2-mm) gate valve	0.12	0.04
Exit loss from 8-in (203.2-mm) pipe into tank, $v^2/2g$	0.52	0.16
Total friction loss	75.46	23.01

Compute the friction loss at other flow rates in a similar manner, and plot the system-head curve, Fig. 21. Note that if all losses in this system except the friction in the discharge pipe were ignored, the total head would not change appreciably. However, for the purposes of accuracy, all losses should always be computed.

4. Plot the low-friction, high-head system-head curve
The system-head curve for the vertical pump installation in Fig. 22 starts at the total static head, 15 ft (4.6 m), and zero flow. Compute the friction head for 15,000 gal/min as follows:

	ft	m
Friction in 20 ft (6.1 m) of 24-in (609.6-mm) pipe	0.40	0.12
Exit loss from 24-in (609.6-mm) pipe into tank, $v^2/2g$	1.60	0.49
Total friction loss	2.00	0.61

Hence, almost 90 percent of the total head of $15 + 2 = 17$ ft (5.2 m) at 15,000-gal/min (946.4-L/s) flow is static head. But neglect of the pipe friction and exit losses could cause appreciable error during selection of a pump for the job.

5. Plot the gravity-head system-head curve
In a system with gravity head (also called negative lift), fluid flow will continue until the system friction loss equals the available gravity head. In Fig. 23 the available gravity head is 50 ft (15.2 m). Flows up to 7200 gal/min (454.3 L/s) are obtained by gravity head alone. To obtain larger flow rates, a pump is needed to

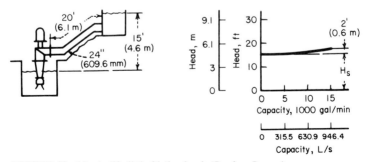

FIGURE 22 Mostly lift; little friction head. (*Peerless Pumps.*)

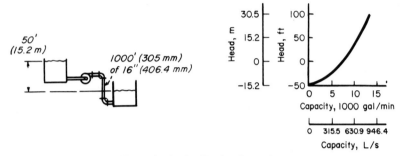

FIGURE 23 Negative lift (gravity head). (*Peerless Pumps.*)

overcome the friction in the piping between the tanks. Compute the friction loss for several flow rates as follows:

	ft	m
At 5000 gal/min (315.5 L/s) friction loss in 1000 ft (305 m) of 16-in (406.4-mm) pipe	25	7.6
At 7200 gal/min (454.3 L/s), friction loss = available gravity head	50	15.2
At 13,000 gal/min (820.2 L/s), friction loss	150	45.7

Using these three flow rates, plot the system-head curve, Fig. 23.

6. Plot the system-head curves for different pipe sizes

When different diameter pipes are used, the friction loss vs. flow rate is plotted independently for the two pipe sizes. At a given flow rate, the total friction loss for the system is the sum of the loss for the two pipes. Thus, the combined system-head curve represents the sum of the static head and the friction losses for all portions of the pipe.

Figure 24 shows a system with two different pipe sizes. Compute the friction losses as follows:

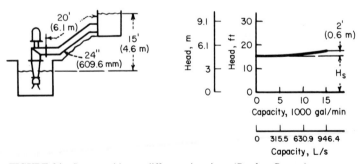

FIGURE 24 System with two different pipe sizes. (*Peerless Pumps.*)

	ft	m
At 150 gal/min (9.5 L/s), friction loss in 200 ft (60.9 m) of 4-in (102-mm) pipe	5	1.52
At 150 gal/min (9.5 L/s), friction loss in 200 ft (60.9 m) of 3-in (76.2-mm) pipe	19	5.79
Total static head for 3- (76.2-) and 4-in (102-mm) pipes	10	3.05
Total head at 150-gal/min (9.5-L/s) flow	34	10.36

Compute the total head at other flow rates, and then plot the system-head curve as shown in Fig. 24.

7. Plot the system-head curve for two discharge heads
Figure 25 shows a typical pumping system having two different discharge heads. Plot separate system-head curves when the discharge heads are different. Add the flow rates for the two pipes at the same head to find points on the combined system-head curve, Fig. 25. Thus,

	ft	m
At 550 gal/min (34.7 L/s), friction loss in 1000 ft (305 m) of 8-in (203.2-mm) pipe	10	3.05
At 1150 gal/min (72.6 L/s), friction	38	11.6
At 1150 gal/min (72.6 L/s), friction + lift in pipe 1	88	26.8
At 550 gal/min (34.7 L/s), friction + lift in pipe 2	88	26.8

The flow rate for the combined system at a head of 88 ft (26.8 m) is 1150 + 550 = 1700 gal/min (107.3 L/s). To produce a flow of 1700 gal/min (107.3 L/s) through this system, a pump capable of developing an 88-ft (26.8-m) head is required.

8. Plot the system-head curve for diverted flow
To analyze a system with diverted flow, assume that a constant quantity of liquid is tapped off at the intermediate point. Plot the friction loss vs. flow rate in the normal manner for pipe 1, Fig. 26. Move the curve for pipe 3 to the right at zero head by an amount equal to Q_2, since this represents the quantity passing through

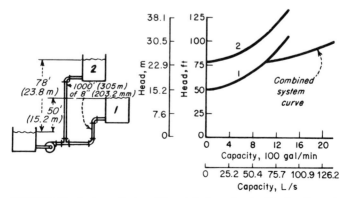

FIGURE 25 System with two different discharge heads. (*Peerless Pumps.*)

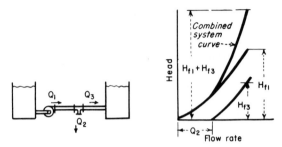

FIGURE 26   Part of the fluid flow is diverted from the main pipe. (*Peerless Pumps.*)

pipes 1 and 2 but not through pipe 3. Plot the combined system-head curve by adding, at a given flow rate, the head losses for pipes 1 and 3. With $Q = 300$ gal/min (18.9 L/s), pipe 1 = 500 ft (152.4 m) of 10-in (254-mm) pipe, and pipe 3 = 50 ft (15.2 m) of 6-in (152.4-mm) pipe.

	ft	m
At 1500 gal/min (94.6 L/s) through pipe 1, friction loss	11	3.35
Friction loss for pipe 3 (1500 − 300 = 1200 gal/min) (75.7 L/s)	8	2.44
Total friction loss at 1500-gal/min (94.6-L/s) delivery	19	5.79

9.   *Plot the effect of pump wear*
When a pump wears, there is a loss in capacity and efficiency. The amount of loss depends, however, on the shape of the system-head curve. For a centrifugal pump, Fig. 27, the capacity loss is greater for a given amount of wear if the system-head curve is flat, as compared with a steep system-head curve.

Determine the capacity loss for a worn pump by plotting its *H-Q* curve. Find this curve by testing the pump at different capacities and plotting the corresponding head. On the same chart, plot the *H-Q* curve for a new pump of the same size, Fig. 27. Plot the system-head curve, and determine the capacity loss as shown in Fig. 27.

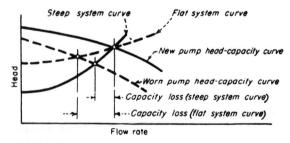

FIGURE 27   Effect of pump wear on pump capacity. (*Peerless Pumps.*)

Related Calculations. Use the techniques given here for any type of pump—centrifugal, reciprocating, or rotary—handling any type of liquid—oil, water, chemicals, etc. The methods given here are the work of Melvin Mann, as reported in *Chemical Engineering,* and Peerless Pump Division of FMC Corp.

NET POSITIVE SUCTION HEAD FOR HOT-LIQUID PUMPS

What is the maximum capacity of a double-suction condensate pump operating at 1750 r/min if it handles 100°F (37.8°C) water from a hot well in a condenser having an absolute pressure of 2.0 in (50.8 mm) Hg if the pump centerline is 10 ft (30.5 m) below the hot-well liquid level and the friction-head loss in the suction piping and fitting is 5 ft (1.52 m) of water?

Calculation Procedure:

1. Compute the net positive suction head on the pump

The net positive suction head h_n on a pump when the liquid supply is *above* the pump inlet = pressure on liquid surface + static suction head − friction-head loss in suction piping and pump inlet − vapor pressure of the liquid, all expressed in ft absolute of liquid handled. When the liquid supply is *below* the pump centerline—*i.e.,* there is a static suction lift—the vertical distance of the lift is *subtracted* from the pressure on the liquid surface instead of added as in the above relation.

The density of 100°F (37.8°C) water is 62.0 lb/ft³ (992.6 kg/m³), computed as shown in earlier calculation procedures in this handbook. The pressure on the liquid surface, in absolute ft of liquid = (2.0 inHg)(1.133)(62.4/62.0) = 2.24 ft (0.68 m). In this calculation, 1.133 = ft of 39.2°F (4°C) water = 1 inHg; 62.4 = lb/ft³ (999.0 kg/m³) of 39.2°F (4°C) water. The temperature of 39.2°F (4°C) is used because at this temperature water has its maximum density. Thus, to convert inHg to ft absolute of water, find the product of (inHg)(1.133)(water density at 39.2°F)/(water density at operating temperature). Express both density values in the same unit, usually lb/ft³.

The static suction head is a physical dimension that is measured in ft (m) of liquid at the operating temperature. In this installation, h_{sh} = 10 ft (3 m) absolute.

The friction-head loss is 5 ft (1.52 m) of water. When it is computed by using the methods of earlier calculation procedures, this head loss is in ft (m) of water at maximum density. To convert to ft absolute, multiply by the ratio of water densities at 39.2°F (4°C) and the operating temperature, or (5)(62.4/62.0) = 5.03 ft (1.53 m).

The vapor pressure of water at 100°F (37.8°C) is 0.949 lb/in² (abs) (6.5 kPa) from the steam tables. Convert any vapor pressure to ft absolute by finding the result of [vapor pressure, lb/in² (abs)] (144 in²/ft²)/liquid density at operating temperature, or (0.949)(144)/62.0 = 2.204 ft (0.67 m) absolute.

With all the heads known, the net positive suction head is h_n = 2.24 + 10 − 5.03 − 2.204 = 5.01 ft (1.53 m) absolute.

2. Determine the capacity of the condensate pump

Use the Hydraulic Institute curve, Fig. 28, to determine the maximum capacity of
the pump. Enter at the left of Fig. 28 at a net positive suction head of 5.01 ft (1.53
m), and project horizontally to the right until the 3500-r/min curve is intersected.
At the top, read the capacity as 278 gal/min (17.5 L/s).

Related Calculations: Use this procedure for any condensate or boiler-feed
pump handling water at an elevated temperature. Consult the Standards of the
Hydraulic Institute for capacity curves of pumps having different types of construc-

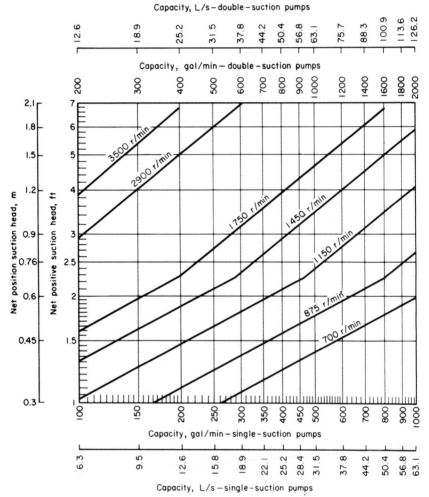

FIGURE 28 Capacity and speed limitations of condensate pumps with the shaft through the
impeller eye. (Hydraulic Institute.)

tion. In general, pump manufacturers who are members of the Hydraulic Institute rate their pumps in accordance with the *Standards,* and a pump chosen from a catalog capacity table or curve will deliver the stated capacity. A similar procedure is used for computing the capacity of pumps handling volatile petroleum liquids. When you use this procedure, be certain to refer to the latest edition of the *Standards.*

CONDENSATE PUMP SELECTION FOR A STEAM POWER PLANT

Select the capacity for a condensate pump serving a steam power plant having a 140,000 lb/h (63,000 kg/h) exhaust flow to a condenser that operates at an absolute pressure of 1.0 in (25.4 mm) Hg. The condensate pump discharges through 4-in (101.6-mm) schedule 40 pipe to an air-ejector condenser that has a frictional resistance of 8 ft (2.4 m) of water. From here, the condensate flows to and through a low-pressure heater that has a frictional resistance of 12 ft (3.7 m) of water and is vented to the atmosphere. The total equivalent length of the discharge piping, including all fittings and bends, is 400 ft (121.9 m), and the suction piping total equivalent length is 50 ft (15.2 m). The inlet of the low pressure heater is 75 ft (22.9 m) above the pump centerline, and the condenser hot-well water level is 10 ft (3 m) above the pump centerline. How much power is required to drive the pump if its efficiency is 70 percent?

Calculation Procedure:

1. Compute the static head on the pump
Sketch the piping system as shown in Fig. 29. Mark the static elevations and equivalent lengths as indicated.

The total head on the pump $H_t = H_{ts} + H_f$, where the symbols are the same as in earlier calculation procedures. The total static head $H_{ts} = h_{sd} - h_{sh}$. In this installation, $h_{sd} = 75$ ft (22.9 m). To make the calculation simpler, convert all the heads to absolute values. Since the heater is vented to the atmosphere, the pressure acting on the surface of the water in it = 14.7 lb/in² (abs) (101.3 kPa), or 34 ft (10.4 m) of water. The pressure acting on the condensate in the hot well is 1 in (25.4 mm) Hg = 1.133 ft (0.35 m) of water. [An absolute pressure of 1 in (25.4 mm) Hg = 1.133 ft (0.35 m) of water.] Thus, the absolute discharge static head = 75 + 34 = 109 ft (33.2 m), whereas the absolute suction head = 10 + 1.13 = 11.13 ft (3.39 m). Then $H_{ts} = h_{hd} - h_{sh} = 109.00 - 11.13 = 97.87$ ft (29.8 m), say 98 ft (29.9 m) of water.

2. Compute the friction head in the piping system
The total friction head $H_f =$ pipe friction + heater friction. The pipe friction loss is found first, as shown below. The heater friction loss, obtained from the manufacturer or engineering data, is then added to the pipe-friction loss. Both must be expressed in ft (m) of water.

To determine the pipe friction, use Fig. 30 of this section and Table 17 and Fig. 6 of the *Piping* section of this handbook in the following manner. Find the product of the liquid velocity, ft/s, and the pipe internal diameter, in, or *vd.* With an exhaust flow of 140,000 lb/h (63,636 kg/h) to the condenser, the condensate flow is the same, or 140,000 lb/h (63,636 kg/h) at a temperature of 79.03°F (21.6°C), corre-

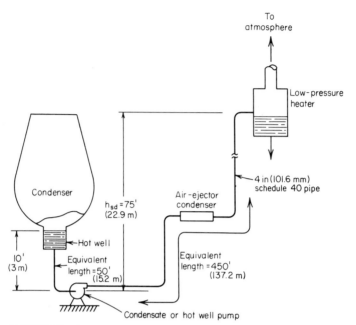

FIGURE 29   Condensate pump serving a steam power plant.

sponding to an absolute pressure in the condenser of 1 in (25.4 mm) Hg, obtained from the steam tables. The specific volume of the saturated liquid at this temperature and pressure is 0.01608 ft^3/lb (0.001 m^3/kg). Since 1 gal (0.26 L) of liquid occupies 0.13368 ft^3 (0.004 m^3), specific volume, gal/lb, is (0.01608/0.13368) = 0.1202 (1.01 L/kg). Therefore, a flow of 140,000 lb/h (63,636 kg/h) = a flow of (140,000)(0.1202) = 16,840 gal/h (63,739.4 L/h), or 16,840/60 = 281 gal/min (17.7 L/s). Then the liquid velocity $v = gpm/2.448d^2 = 281/2.448(4.026)^2 = 7.1$ ft/s (2.1 m/s), and the product $vd = (7.1)(4.026) = 28.55$.

Enter Fig. 30 at a temperature of 79°F (26.1°C), and project vertically upward to the water curve. From the intersection, project horizontally to the right to $vd = 28.55$ and then vertically upward to read $R = 250,000$. Using Table 17 and Fig. 6 of the *Piping* section and $R = 250,000$, find the friction factor $f = 0.0185$. Then the head loss due to pipe friction $H_f = (L/D)(v^2/2g) = 0.0185 (450/4.026/12)/[(7.1)^2/2(32.2)] = 19.18$ ft (5.9 m). In this computation, L = total equivalent length of the pipe, pipe fittings, and system valves, or 450 ft (137.2 m).

3.   Compute the other head losses in the system
There are two other head losses in this piping system: the entrance loss at the square-edged hot-well pipe leading to the pump and the sudden enlargement in the low-pressure heater. The velocity head $v^2/2g = (7.1)^2/2(32.2) = 0.784$ ft (0.24 m). Using k values from Fig. 11 in this section, $h_e = kv^2/2g = (0.5)(0.784) = 0.392$ ft (0.12 m); $h_{ex} = v^2/2g = 0.784$ ft (0.24 m).

4.   Find the total head on the pump
The total head on the pump $H_t = H_{ts} + H_f = 97.87 + 19.18 + 8 + 12 +$

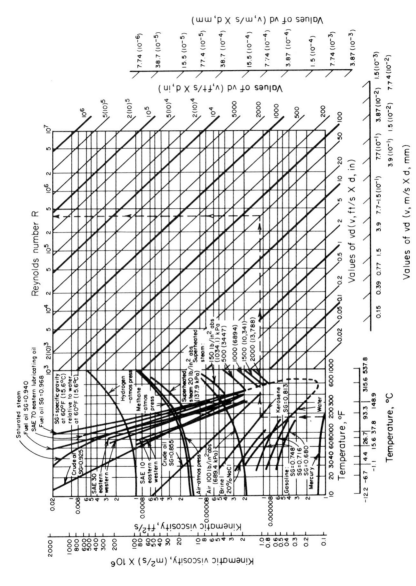

FIGURE 30 Kinematic viscosity and Reynolds number chart. *(Hydraulic Institute.)*

0.392 + 0.784 = 138.226 ft (42.1 m), say 140 ft (42.7 m) of water. In this calculation, the 8-(2.4-m) and 12-ft (3.7-m) head losses are those occurring in the heaters. With a 25 percent safety factor, total head = (1.25)(140) = 175 ft (53.3 m).

5. Compute the horsepower required to drive the pump
The brake horsepower input $bhp_i = (gpm)(H_t)(s)/3960e$, where the symbols are the same as in earlier calculation procedures. At 1 in (25.4 mm) Hg, 1 lb (0.45 kg) of the condensate has a volume of 0.01608 ft³ (0.000455 m³). Since density = 1/ specific volume, the density of the condensate = 1/0.01608 = 62.25 ft³/lb (3.89 m³/kg). Water having a specific gravity of unity weighs 62.4 lb/ft³ (999 kg/m³). Hence, the specific gravity of the condensate is 62.25/62.4 = 0.997. Then, assuming that the pump has an operating efficiency of 70 percent, we get bhp_i = (281)(175) × (0.997)/[3960(0.70)] = 17.7 bhp (13.2 kW).

6. Select the condensate pump
Condensate or hot-well pumps are usually centrifugal units having two or more stages, with the stage inlets opposed to give better axial balance and to subject the sealing glands to positive internal pressure, thereby preventing air leakage into the pump. In the head range developed by this pump, 175 ft (53.3 m), two stages are satisfactory. Refer to a pump manufacturer's engineering data for specific stage head ranges. Either a turbine or motor drive can be used.

Related Calculations. Use this procedure to choose condensate pumps for steam plants of any type—utility, industrial, marine, portable, heating, or process—and for combined steam-diesel plants.

MINIMUM SAFE FLOW FOR A CENTRIFUGAL PUMP

A centrifugal pump handles 220°F (104.4°C) water and has a shutoff head (with closed discharge valve) of 3200 ft (975.4 m). At shutoff, the pump efficiency is 17 percent and the input brake horsepower is 210 (156.7 kW). What is the minimum safe flow through this pump to prevent overheating at shutoff? Determine the minimum safe flow if the NPSH is 18.8 ft (5.7 m) of water and the liquid specific gravity is 0.995. If the pump contains 500 lb (225 kg) of water, determine the rate of the temperature rise at shutoff.

Calculation Procedure:

1. Compute the temperature rise in the pump
With the discharge valve closed, the power input to the pump is converted to heat in the casing and causes the liquid temperature to rise. The temperature rise $t = (1 - e) \times H_s/778e$, where t = temperature rise during shutoff, °F; e = pump efficiency, expressed as a decimal; H_s = shutoff head, ft. For this pump, t = (1 − 0.17)(3200)/[778(0.17)] = 20.4°F (36.7°C).

2. Compute the minimum safe liquid flow
For general-service pumps, the minimum safe flow M gal/min = 6.0(bhp input at shutoff)/t. Or, M = 6.0(210)/20.4 = 62.7 gal/min (3.96 L/s). This equation includes a 20 percent safety factor.

Centrifugal boiler-feed pumps usually have a maximum allowable temperature rise of 15°F (27°C). The minimum allowable flow through the pump to prevent the water temperature from rising more than 15°F (27°C) is 30 gal/min (1.89 L/s) for each 110-bhp (74.6-kW) input at shutoff.

3. Compute the temperature rise for the operating NPSH
An NPSH of 18.8 ft (5.73 m) is equivalent to a pressure of 18.8(0.433)(0.995) = 7.78 lb/in² (abs) (53.6 kPa) at 220°F (104.4°C), where the factor 0.433 converts ft of water to lb/in². At 220°F (104.4°C), the vapor pressure of the water is 17.19 lb/in² (abs) (118.5 kPa), from the steam tables. Thus, the total vapor pressure the water can develop before flashing occurs = NPSH pressure + vapor pressure at operating temperature = 7.78 + 17.19 = 24.97 lb/in² (abs) (172.1 kPa). Enter the steam tables at this pressure, and read the corresponding temperature as 240°F (115.6°C). The allowable temperature rise of the water is then 240 − 220 = 20°F (36.0°C). Using the safe-flow relation of step 2, we find the minimum safe flow is 62.9 gal/min (3.97 L/s).

4. Compute the rate of temperature rise
In any centrifugal pump, the rate of temperature rise t_r °F/min = 42.4(bhp input at shutoff)/wc, where w = weight of liquid in the pump, lb; c = specific heat of the liquid in the pump, Btu/(lb · °F). For this pump containing 500 lb (225 kg) of water with a specific heat, c = 1.0, t_r = 42.4(210)/[500(1.0)] = 17.8°F/min (32°C/min). This is a very rapid temperature rise and could lead to overheating in a few minutes.

    Related Calculations.   Use this procedure for any centrifugal pump handling any liquid in any service—power, process, marine, industrial, or commercial. Pump manufacturers can supply a temperature-rise curve for a given model pump if it is requested. This curve is superimposed on the pump characteristic curve and shows the temperature rise accompanying a specific flow through the pump.

SELECTING A CENTRIFUGAL PUMP TO HANDLE A VISCOUS LIQUID

Select a centrifugal pump to deliver 750 gal/min (47.3 L/s) of 1000-SSU oil at a total head of 100 ft (30.5 m). The oil has a specific gravity of 0.90 at the pumping temperature. Show how to plot the characteristic curves when the pump is handling the viscous liquid.

Calculation Procedure:

1. Determine the required correction factors
A centrifugal pump handling a viscous liquid usually must develop a greater capacity and head, and it requires a larger power input than the same pump handling water. With the water performance of the pump known—from either the pump characteristic curves or a tabulation of pump performance parameters—Fig. 31, prepared by the Hydraulic Institute, can be used to find suitable correction factors. Use this chart only within its scale limits; do not extrapolate. Do not use the chart for mixed-flow or axial-flow pumps or for pumps of special design. Use the chart only for pumps handling uniform liquids; slurries, gels, paper stock, etc., may cause

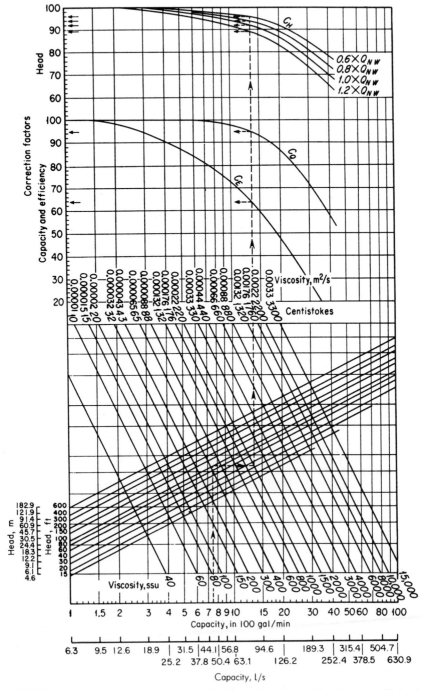

FIGURE 31 Correction factors for viscous liquids handled by centrifugal pumps. (*Hydraulic Institute.*)

incorrect results. In using the chart, the available net positive suction head is assumed adequate for the pump.

To use Fig. 31, enter at the bottom at the required capacity, 750 gal/min (47.3 L/s), and project vertically to intersect the 100-ft (30.5-m) head curve, the required head. From here project horizontally to the 1000-SSU viscosity curve, and then vertically upward to the correction-factor curves. Read $C_E = 0.635$; $C_Q = 0.95$; $C_H = 0.92$ for $1.0Q_{NW}$. The subscripts E, Q, and H refer to correction factors for efficiency, capacity, and head, respectively; and NW refers to the water capacity at a particular efficiency. At maximum efficiency, the water capacity is given as $1.0Q_{NW}$; other efficiencies, expressed by numbers equal to or less than unity, give different capacities.

2. Compute the water characteristics required
The water capacity required for the pump $Q_w = Q_v/C_Q$ where Q_v = viscous capacity, gal/min. For this pump, $Q_w = 750/0.95 = 790$ gal/min (49.8 L/s). Likewise, water head $H_w = H_v/C_H$, where H_v = viscous head. Or, $H_w = 100/0.92 = 108.8$ (33.2 m), say 109 ft (33.2 m) of water.

Choose a pump to deliver 790 gal/min (49.8 L/s) of water at 109-ft (33.2-m) head of water, and the required viscous head and capacity will be obtained. Pick the pump so that it is operating at or near its maximum efficiency on water. If the water efficiency $E_w = 81$ percent at 790 gal/min (49.8 L/s) for this pump, the efficiency when handling the viscous liquid $E_v = E_w C_E$. Or, $E_v = 0.81(0.635) = 0.515$, or 51.5 percent.

The power input to the pump when handling viscous liquids is given by $P_v = Q_v H_v s/3960E_v$, where s = specific gravity of the viscous liquid. For this pump, $P_v = (750) \times (100)(0.90)/[3960(0.515)] = 33.1$ hp (24.7 kW).

3. Plot the characteristic curves for viscous-liquid pumping
Follow these eight steps to plot the complete characteristic curves of a centrifugal pump handling a viscous liquid when the water characteristics are known: (a) Secure a complete set of characteristic curves (H, Q, P, E) for the pump to be used. (b) Locate the point of maximum efficiency for the pump when handling water. (c) Read the pump capacity, Q gal/min, at this point. (d) Compute the values of $0.6Q$, $0.8Q$, and $1.2Q$ at the maximum efficiency. (e) Using Fig. 31, determine the correction factors at the capacities in steps c and d. Where a multistage pump is being considered, use the head per stage (= total pump head, ft/number of stages), when entering Fig. 31. (f) Correct the head, capacity, and efficiency for each of the flow rates in c and d, using the correction factors from Fig. 31. (g) Plot the corrected head and efficiency against the corrected capacity, as in Fig. 32. (h) Compute the power input at each flow rate and plot. Draw smooth curves through the points obtained, Fig. 32.

Related Calculations. Use the method given here for any uniform viscous liquid—oil, gasoline, kerosene, mercury, etc—handled by a centrifugal pump. Be careful to use Fig. 31 only within its scale limits; *do not extrapolate*. The method presented here is that developed by the Hydraulic Institute. For new developments in the method, be certain to consult the latest edition of the Hydraulic Institute *Standards*.

PUMP SHAFT DEFLECTION AND CRITICAL SPEED

What are the shaft deflection and approximate first critical speed of a centrifugal pump if the total combined weight of the pump impellers is 23 lb (10.4 kg) and the pump manufacturer supplies the engineering data in Fig. 33?

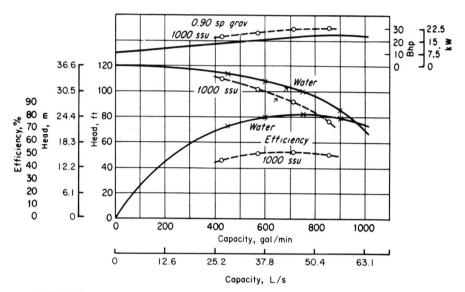

FIGURE 32   Characteristics curves for water (solid line) and oil (dashed line). (*Hydraulic Institute.*)

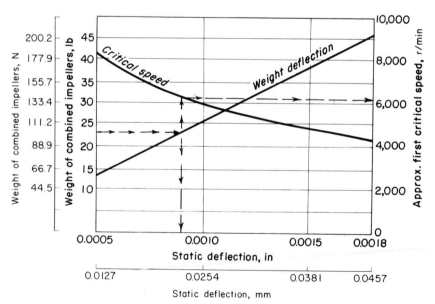

FIGURE 33   Pump shaft deflection and critical speed. (*Goulds Pumps, Inc.*)

Calculation Procedure:

1. Determine the deflection of the pump shaft
Use Fig. 33 to determine the shaft deflection. Note that this chart is valid for only
one pump or series of pumps and must be obtained from the pump builder. Such
a chart is difficult to prepare from test data without extensive test facilities.
   Enter Fig. 33 at the left at the total combined weight of the impellers, 23 lb
(10.4 kg), and project horizontally to the right until the weight-deflection curve is
intersected. From the intersection, project vertically downward to read the shaft
deflection as 0.009 in (0.23 mm) at full speed.

2. Determine the critical speed of the pump
From the intersection of the weight-deflection curve in Fig. 33 project vertically
upward to the critical-speed curve. Project horizontally right from this intersection
and read the first critical speed as 6200 r/min.
   Related Calculations.   Use this procedure for any class of pump—centrifugal,
rotary, or reciprocating—for which the shaft-deflection and critical-speed curves are
available. These pumps can be used for any purpose—process, power, marine, industrial, or commercial.

EFFECT OF LIQUID VISCOSITY ON REGENERATIVE-PUMP PERFORMANCE

A regenerative (turbine) pump has the water head-capacity and power-input characteristics shown in Fig. 34. Determine the head-capacity and power-input characteristics for four different viscosity oils to be handled by the pump—400, 600,
900, and 1000 SSU. What effect does increased viscosity have on the performance
of the pump?

Calculation Procedure:

1. Plot the water characteristics of the pump
Obtain a tabulation or plot of the water characteristics of the pump from the manufacturer or from their engineering data. With a tabulation of the characteristics,
enter the various capacity and power points given, and draw a smooth curve through
them, Fig. 34.

2. Plot the viscous-liquid characteristics of the pump
The viscous-liquid characteristics of regenerative-type pumps are obtained by test
of the actual unit. Hence, the only source of this information is the pump manufacturer. Obtain these characteristics from the pump manufacturer or their test data,
and plot them on Fig. 34, as shown, for each oil or other liquid handled.

3. Evaluate the effect of viscosity on pump performance
Study Fig. 34 to determine the effect of increased liquid viscosity on the performance of the pump. Thus at a given head, say 100 ft (30.5 m), the capacity of the
pump decreases as the liquid viscosity increases. At 100-ft (30.5-m) head, this pump
has a water capacity of 43.5 gal/min (2.74 L/s), Fig. 34. The pump capacity for
the various oils at 100-ft (30.5-m) head is 36 gal/min (2.27 L/s) for 400 SSU; 32

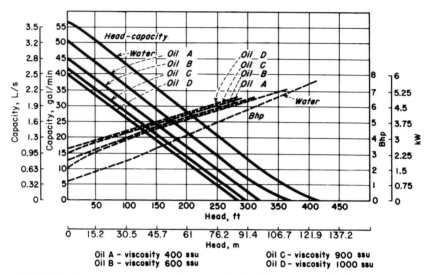

FIGURE 34 Regenerative pump performance when handling water and oil. (*Aurora Pump Division, The New York Air Brake Company.*)

gal/min (2.02 L/s) for 600 SSU; 28 gal/min (1.77 L/s) for 900 SSU; and 26 gal/min (1.64 L/s) for 1000 SSU, respectively. There is a similar reduction in capacity of the pump at the other heads plotted in Fig. 34. Thus, as a general rule, the capacity of a regenerative pump decreases with an increase in liquid viscosity at constant head. Or conversely, at constant capacity, the head developed decreases as the liquid viscosity increases.

Plots of the power input to this pump show that the input power increases as the liquid viscosity increases.

Related Calculations.   Use this procedure for a regenerative-type pump handling any liquid—water, oil, kerosene, gasoline, etc. A decrease in the viscosity of a liquid, as compared with the viscosity of water, will produce the opposite effect from that of increased viscosity.

EFFECT OF LIQUID VISCOSITY ON RECIPROCATING-PUMP PERFORMANCE

A direct-acting steam-driven reciprocating pump delivers 100 gal/min (6.31 L/s) of 70°F (21.1°C) water when operating at 50 strokes per minute. How much 2000-SSU crude oil will this pump deliver? How much 125°F (51.7°C) water will this pump deliver?

Calculation Procedure:

1. *Determine the recommended change in pump performance*
Reciprocating pumps of any type—direct-acting or power—having any number of liquid-handling cylinders—one to five or more—are usually rated for maximum

TABLE 14 Speed-Correction Factors

Liquid viscosity, SSU	Speed reduction, %	Water temperature °F	°C	Speed reduction, %
250	0	70	21.1	0
500	4	80	26.7	9
1000	11	100	37.8	18
2000	20	125	51.7	25
3000	26	150	65.6	29
4000	30	200	93.3	34
5000	35	250	121.1	38

delivery when handling 250-SSU liquids or 70°F (21.1°C) water. At higher liquid viscosities or water temperatures, the speed—strokes or rpm—is reduced. Table 14 shows typical recommended speed-correction factors for reciprocating pumps for various liquid viscosities and water temperatures. This table shows that with a liquid viscosity of 2000 SSU the pump speed should be reduced 20 percent. When 125°F (51.7°C) water is handled, the pump speed should be reduced 25 percent, as shown in Table 14.

2. Compute the delivery of the pump
The delivery capacity of any reciprocating pump is directly proportional to the number of strokes per minute it makes or to its rpm.

When 2000-SSU oil is used, the pump strokes per minute must be reduced 20 percent, or $(50)(0.20) = 10$ strokes/min. Hence, the pump speed will be $50 - 10 = 40$ strokes/min. Since the delivery is directly proportional to speed, the delivery of 2000-SSU oil $= (40/50)(100) = 80$ gal/min (5.1 L/s).

When handling 125°F (51.7°C) water, the pump strokes/min must be reduced 25 percent, or $(50)(0.5) = 12.5$ strokes/min. Hence, the pump speed will be $50.0 - 12.5 = 37.5$ strokes/min. Since the delivery is directly proportional to speed, the delivery of 125°F (51.7°C) water $= (37.5/50)(10) = 75$ gal/min (4.7 L/s).

Related Calculations. Use this procedure for any type of reciprocating pump handling liquids falling within the range of Table 14. Such liquids include oil, kerosene, gasoline, brine, water, etc.

EFFECT OF VISCOSITY AND DISSOLVED GAS ON ROTARY PUMPS

A rotary pump handles 8000-SSU liquid containing 5 percent entrained gas and 10 percent dissolved gas at a 20-in (508-mm) Hg pump inlet vacuum. The pump is rated at 1000 gal/min (63.1 L/s) when handling gas-free liquids at viscosities less than 600 SSU. What is the output of this pump without slip? With 10 percent slip?

Calculation Procedure:

1. Compute the required speed reduction of the pump

When the liquid viscosity exceeds 600 SSU, many pump manufacturers recommend that the speed of a rotary pump be reduced to permit operation without excessive noise or vibration. The speed reduction usually recommended is shown in Table 15.

With this pump handling 8000-SSU liquid, a speed reduction of 40 percent is necessary, as shown in Table 15. Since the capacity of a rotary pump varies directly with its speed, the output of this pump when handling 8000-SSU liquid = (1000 gal/min) × (1.0 − 0.40) = 600 gal/min (37.9 L/s).

2. Compute the effect of gas on the pump output

Entrained or dissolved gas reduces the output or a rotary pump, as shown in Table 16. The gas in the liquid expands when the inlet pressure of the pump is below atmospheric and the gas occupies part of the pump chamber, reducing the liquid capacity.

With a 20-in (508-mm) Hg inlet vacuum, 5 percent entrained gas, and 10 percent dissolved gas, Table 16 shows that the liquid displacement is 74 percent of the rated displacement. Thus, the output of the pump when handling this viscous, gas-containing liquid will be (600 gal/min) (0.74) = 444 gal/min (28.0 L/s) without slip.

3. Compute the effect of slip on the pump output

Slip reduces rotary-pump output in direct proportion to the slip. Thus, with 10 percent slip, the output of this pump = (444 gal/min)(1.0 − 0.10) = 369.6 gal/min (23.3 L/s).

Related Calculations. Use this procedure for any type of rotary pump—gear, lobe, screw, swinging-vane, sliding-vane, or shuttle-block, handling any clear, viscous liquid. Where the liquid is gas-free, apply only the viscosity correction. Where the liquid viscosity is less than 600 SSU but the liquid contains gas or air, apply the entrained or dissolved gas correction, or both corrections.

TABLE 15 Rotary Pump Speed Reduction for Various Liquid Viscosities

Liquid viscosity, SSU	Speed reduction, percent of rated pump speed
600	2
800	6
1,000	10
1,500	12
2,000	14
4,000	20
6,000	30
8,000	40
10,000	50
20,000	55
30,000	57
40,000	60

TABLE 16 Effect of Entrained or Dissolved Gas on the Liquid Displacement of Rotary Pumps (liquid displacement: percent of displacement)

Vacuum at pump inlet, inHg (mmHg)	Gas entrainment					Gas solubility					Gas entrainment and gas solubility combined				
	1%	2%	3%	4%	5%	2%	4%	6%	8%	10%	1% 2%	2% 4%	3% 6%	4% 8%	5% 10%
5 (127)	99	97½	96½	95	93½	99½	99	98½	97	97½	98½	96½	96	92	91
10 (254)	98½	97¼	95½	94	92	99	97½	97	95	95	97½	95	90	90	88½
15 (381)	98	96½	94½	92½	90½	97	96	94	92	90½	96	93	89½	86½	83½
20 (508)	97½	94½	92	89	86½	96	92	89	86	83	94	88	83	78	74
25 (635)	94	89	84	79	75½	90	83	76½	71	66	85½	75½	68	61	55

For example, with 5 percent gas entrainment at 15 inHg (381 mmHg) vacuum, the liquid displacement will be 90½ percent of the pump displacement, neglecting slip, or with 10 percent dissolved gas liquid displacement will be 90½ percent of the pump displacement; and with 5 percent entrained gas combined with 10 percent dissolved gas, the liquid displacement will be 83½ percent of pump replacement.

Source: Courtesy of Kinney Mfg. Div., The New York Air Brake Co.

SELECTION OF MATERIALS FOR PUMP PARTS

Select suitable materials for the principal parts of a pump handling cold ethylene chloride. Use the Hydraulic Institute recommendation for materials of construction.

Calculation Procedure:

1. Determine which materials are suitable for this pump
Refer to the data section of the Hydraulic Institute *Standards*. This section contains a tabulation of hundreds of liquids and the pump construction materials that have been successfully used to handle each liquid.

The table shows that for cold ethylene chloride having a specific gravity of 1.28, an all-bronze pump is satisfactory. In lieu of an all-bronze pump, the principal parts of the pump—casing, impeller, cylinder, and shaft—can be made of one of the following materials: austenitic steels (low-carbon 18-8; 18-8/Mo; highly alloyed stainless); nickel-base alloys containing chromium, molybdenum, and other elements, and usually less than 20 percent iron; or nickel-copper alloy (Monel metal). The order of listing in the *Standards* does not necessarily indicate relative superiority, since certain factors predominating in one instance may be sufficiently overshadowed in others to reverse the arrangement.

2. Choose the most economical pump
Use the methods of earlier calculation procedures to select the most economical pump for the installation. Where the corrosion resistance of two or more pumps is equal, the standard pump, in this instance an all-bronze unit, will be the most economical.

Related Calculations. Use this procedure to select the materials of construction for any class of pump—centrifugal, rotary, or reciprocating—in any type of service—power, process, marine, or commercial. Be certain to use the latest edition of the Hydraulic Institute *Standards,* because the recommended materials may change from one edition to the next.

SIZING A HYDROPNEUMATIC STORAGE TANK

A 200-gal/min (12.6-L/s) water pump serves a pumping system. Determine the capacity required for a hydropneumatic tank to serve this system if the allowable high pressure in the tank and system is 60 lb/in^2 (gage) (413.6 kPa) and the allowable low pressure is 30 lb/in^2 (gage) (206.8 kPa). How many starts per hour will the pump make if the system draws 3000 gal/min (189.3 L/s) from the tank?

Calculation Procedure:

1. Compute the required tank capacity
If the usual hydropneumatic system, a storage-tank capacity in gal of 10 times the pump capacity in gal/min is used, if this capacity produces a moderate run-

ning time for the pump. Thus, this system would have a tank capacity of (10)(200) = 2000 gal (7570.8 L).

2. Compute the quantity of liquid withdrawn per cycle
For any hydropneumatic tank the withdrawal, expressed as the number of gallons (liters) withdrawn per cycle, is given by $W = (v_L - v_H)/C$, where v_L = air volume in tank at the lower pressure, ft^3 (m^3); v_H = volume of air in tank at higher pressure, ft^3 (m^3); C = conversion factor to convert ft^3 (m^3) to gallons (liters), as given below.

Compute V_L and V_H using the gas law for v_H and either the gas law or the reserve percentage for v_L. Thus, for v_H, the gas law gives $v_H = p_L v_L / p_H$, where p_L = lower air pressure in tank, lb/in^2 (abs) (kPa); p_H = higher air pressure in tank lb/in^2 (abs) (kPa); other symbols as before.

In most hydropneumatic tanks a liquid reserve of 10 to 20 percent of the total tank volume is kept in the tank to prevent the tank from running dry and damaging the pump. Assuming a 10 percent reserve for this tank, $v_L = 0.1 V$, where V = tank volume in ft^3 (m^3). Since a 2000-gal (7570-L) tank is being used, the volume of the tank is 2000/7.481 ft^3/gal = 267.3 ft^3 (7.6 m^3). With the 10 percent reserve at the 44.7 lb/in^2 (abs) (308.2-kPa) lower pressure, $v_L = 0.9 (267.3) = 240.6$ ft^3 (6.3 m^3), where $0.9 = V - 0.1 V$.

At the higher pressure in the tank, 74.7 lb/in^2 (abs) (514.9 kPa), the volume of the air will be, from the gas law, $v_H = p_L v_L / p_H = 44.7 (240.6)/74.7 = 143.9$ ft^3 (4.1 m^3). Hence, during withdrawal, the volume of liquid removed from the tank will be $W_g = (240.6 - 143.9)/0.1337 = 723.3$ gal (2738 L). In this relation of the constant converts from cubic feet to gallons and is 0.1337. To convert from cubic meters to liters, use the constant 1000 in the denominator.

3. Compute the pump running time
The pump has a capacity of 200 gal/min (12.6 L/s). Therefore, it will take 723/200 = 3.6 min to replace the withdrawn liquid. To supply 3000 gal/h (11,355 L/h) to the system, the pump must start 3000/723 = 4.1, or 5 times per hour. This is acceptable because a system in which the pump starts six or fewer times per hour is generally thought satisfactory.

Where the pump capacity is insufficient to supply the system demand for short periods, use a smaller reserve. Compute the running time using the equations in steps 2 and 3. Where a larger reserve is used—say 20 percent—use the value 0.8 in the equations in step 2. For a 30 percent reserve, the value would be 0.70, and so on.

Related Calculations. Use this procedure for any liquid system having a hydropneumatic tank—well drinking water, marine, industrial, or process.

USING CENTRIFUGAL PUMPS AS HYDRAULIC TURBINES

Select a centrifugal pump to serve as a hydraulic turbine power source for a 1500-gal/min (5677.5-L/min) flow rate with 1290 ft (393.1 m) of head. The power application requires a 3600-r/min speed, the specific gravity of the liquid is 0.52, and the total available exhaust head is 20 ft (6.1 m). Analyze the cavitation potential and operating characteristics at an 80 percent flow rate.

Calculation Procedure:

1.   *Choose the number of stages for the pump*
Search of typical centrifugal-pump data shows that a head of 1290 ft (393.1 m) is too large for a single-stage pump of conventional design. Hence, a two-stage pump will be the preliminary choice for this application. The two-stage pump chosen will have a design head of 645 ft (196.6 m) per stage.

2.   *Compute the specific speed of the pump chosen*
Use the relation $N_s = $ pump $rpm(Q)^{0.5}/H^{0.75}$, where $N_s = $ specific speed of the pump; $rpm = $ r/min of pump shaft; $Q = $ pump capacity or flow rate, gal/min; $H = $ pump head per stage, ft. Substituting, we get $N_s = 3600(1500)^{0.5}/ (645)^{0.75} = 1090$. Note that the specific speed value is the same regardless of the system of units used—USCS or SI.

3.   *Convert turbine design conditions to pump design conditions*
To convert from turbine design conditions to pump design conditions, use the pump manufacturer's conversion factors that relate turbine best efficiency point (bep) performance with pump bep performance. Typically, as specific speed N_s varies from 500 to 2800, these bep factors generally vary as follows: the conversion factor for capacity (gal/min or L/min) C_Q, from 2.2 to 1.1; the conversion factor for head (ft or m) C_H, from 2.2 to 1.1; the conversion factor for efficiency C_E, from 0.92 to 0.99. Applying these conversion factors to the turbine design conditions yields the pump design conditions sought.

   At the specific speed for this pump, the values of these conversion factors are determined from the manufacturer to be $C_Q = 1.24$; $C_H = 1.42$; $C_E = 0.967$.

   Given these conversion factors, the turbine design conditions can be converted to the pump design conditions thus: $Q_p = Q_t/C_Q$, where $Q_p = $ pump capacity or flow rate, gal/min or L/min; $Q_t = $ turbine capacity or flow rate in the same units; other symbols are as given earlier. Substituting gives $Q_p = 1500/1.24 = 1210$ gal/min (4580 L/min).

   Likewise, the pump discharge head, in feet of liquid handled, is $H_p = H_t/C_H$. So $H_p = 645/1.42 = 454$ ft (138.4 m).

4.   *Select a suitable pump for the operating conditions*
Once the pump capacity, head, and rpm are known, a pump having its best bep at these conditions can be selected. Searching a set of pump characteristic curves and capacity tables shows that a two-stage 4-in (10-cm) unit with an efficiency of 77 percent would be suitable.

5.   *Estimate the turbine horsepower developed*
To predict the developed hp, convert the pump efficiency to turbine efficiency. Use the conversion factor developed above. Or, the turbine efficiency $E_t = E_p C_E = (0.77)(0.967) = 0.745$, or 74.5 percent.

   With the turbine efficiency known, the output brake horsepower can be found from bhp $= Q_t H_t E_t s/3960$, where $s = $ fluid specific gravity; other symbols as before. Substituting, we get bhp $= 1500(1290)(0.745)(0.52)/3960 = 198$ hp (141 kW).

6.   *Determine the cavitation potential of this pump*
Just as pumping requires a minimum net positive suction head, turbine duty requires a net positive exhaust head. The relation between the total required exhaust head

(TREH) and turbine head per stage is the cavitation constant $\sigma_r = \text{TREH}/H$. Figure 35 shows σ_r vs. N_s for hydraulic turbines. Although a pump used as a turbine will not have exactly the same relationship, this curve provides a god estimate of σ_r for turbine duty.

To prevent cavitation, the total available exhaust head (TAEH) must be greater than the TREH. In this installation, $N_s = 1090$ and TAEH = 20 ft (6.1 m). From Fig. 35, $\sigma_r = 0.028$ and TREH = $0.028(645) = 18.1$ ft (5.5 m). Because TAEH > TREH, there is enough exhaust head to prevent cavitation.

7. Determine the turbine performance at 80 percent flow rate

In many cases, pump manufacturers treat conversion factors as proprietary information. When this occurs, the performance of the turbine under different operating conditions can be predicted from the general curves in Figs. 36 and 37.

At the 80 percent flow rate for the turbine, or 1200 gal/min (4542 L/min), the operating point is 80 percent of bep capacity. For a specific speed of 1090, as before, the percentages of bep head and efficiency are shown in Figs. 36 and 37: 79.5 percent of bep head and percent of bep efficiency. To find the actual performance, multiply by the bep values. Or, $H_t = 0.795(1290) = 1025$ ft (393.1 m); $E_t = 0.91(74.5) = 67.8$ percent.

The bhp at the new operating condition is then bhp = 1200 $(1025)(0.678)(0.52)/3960 = 110$ hp (82.1 kW).

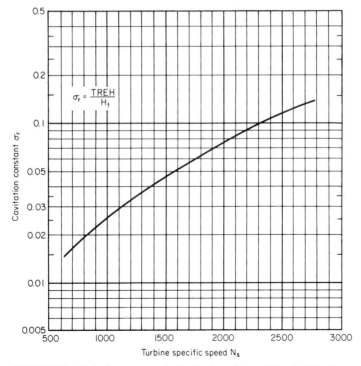

FIGURE 35 Cavitation constant for hydraulic turbines. (*Chemical Engineering.*)

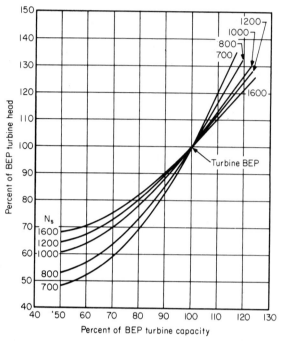

FIGURE 36   Constant-speed curves for turbine duty. (*Chemical Engineering.*)

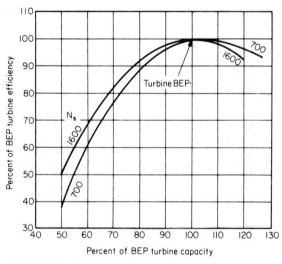

FIGURE 37   Constant-speed curves for turbine duty. (*Chemical Engineering.*)

In a similar way, the constant-head curves in Figs. 38 and 39 predict turbine performance at different speeds. For example, speed is 80 percent of bep speed at 2880 r/min. For a specific speed of 1090, the percentages of bep capacity, efficiency, and power are 107 percent of the capacity, 94 percent of the efficiency, and 108 percent of the bhp. To get the actual performance, convert as before: $Q_t = 107(1500) = 1610$ gal/min (6094 L/min); $E_t = 0.94(74.5) = 70.0$ percent; bhp $= 1.08(189) = 206$ hp (153.7 kW).

Note that the bhp in this last instance is higher than the bhp at the best efficiency point. Thus more horsepower can be obtained from a given unit by reducing the speed and increasing the flow rate. When the speed is fixed, more bhp cannot be obtained from the unit, but it may be possible to select a smaller pump for the same application.

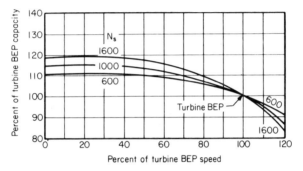

FIGURE 38   Constant-head curves for turbine duty. (*Chemical Engineering.*)

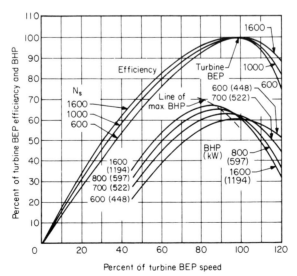

FIGURE 39   Constant-head curves for turbine only. (*Chemical Engineering.*)

Related Calculations.   Use this general procedure for choosing a centrifugal pump to drive—as a hydraulic turbine—another pump, a fan, a generator, or a compressor, where high-pressure liquid is available as a source of power. Because pumps are designed as fluid movers, they may be less efficient as hydraulic turbines than equipment designed for that purpose. Steam turbines and electric motors are more economical when steam or electricity is available.

But using a pump as a turbine can pay off in remote locations where steam or electric power would require additional wiring or piping, in hazardous locations that require nonsparking equipment, where energy may be recovered from a stream that otherwise would be throttled, and when a radial-flow centrifugal pump is immediately available but a hydraulic turbine is not.

In the most common situation, there is a liquid stream with fixed head and flow rate and an application requiring a fixed rpm; these are the turbine design conditions. The objective is to pick a pump with a turbine bep at these conditions. With performance curves such as Fig. 34, turbine design conditions can be converted to pump design conditions. Then you select from a manufacturer's catalog a model that has its pump bep at those values.

The most common error in pump selection is using the turbine design conditions in choosing a pump from a catalog. Because catalog performance curves describe pump duty, not turbine duty, the result is an oversized unit that fails to work properly.

This procedure is the work of Fred Buse, Chief Engineer, Standard Pump Aldrich Division of Ingersoll-Rand Co., as reported in *Chemical Engineering* magazine.

SIZING CENTRIFUGAL-PUMP IMPELLERS FOR SAFETY SERVICE

Determine the impeller size of a centrifugal pump that will provide a safe continuous-recirculation flow to prevent the pump from overheating at shutoff. The pump delivers 320 gal/min (20.2 L/s) at an operating head of 450 ft (137.2 m). The inlet water temperature is 220°F (104.4°C), and the system has an NPSH of 5 ft (1.5 m). Pump performance curves and the system-head characteristic curve for the discharge flow (without recirculation) are shown in Fig. 35, and the piping layout is shown in Fig. 42. The brake horsepower (bhp) of an 11-in (27.9-cm) and an 11.5-in (29.2-cm) impeller at shutoff is 53 and 60, respectively. Determine the permissible water temperature rise for this pump.

Calculation Procedure:

1. Compute the actual temperature rise of the water in the pump
Use the relation $P_0 = P_v + P_{NPSH}$, where P_0 = pressure corresponding to the actual liquid temperature in the pump during operation, lb/in² (abs) (kPa); P_v = vapor pressure in the pump at the inlet water temperature, lb/in² (abs) (kPa); P_{NPSH} = pressure created by the net positive suction head on the pumps, lb/in² (abs) (kPa). The head in feet (meters) must be converted to lb/in² (abs) (kPa) by the relation lb/in² (abs) = (NPSH, ft) (liquid density at the pumping temperature, lb/ft³)/(144 in²/ft²). Substituting yields P_0 = 17.2 lb/in² (abs) + 5(59.6)/144 = 19.3 lb/in² (abs) (133.1 kPa).

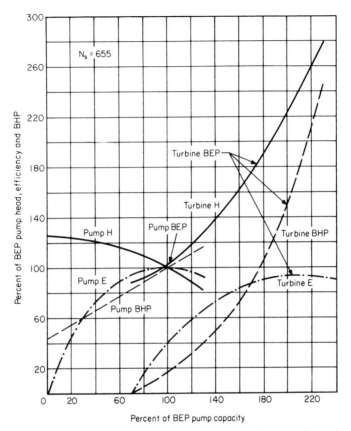

FIGURE 40 Performance of a pump at constant speed in pump duty and turbine duty. (*Chemical Engineering.*)

Using the steam tables, find the saturation temperature T_s corresponding to this absolute pressure as $T_s = 226.1°F$ (107.8°C). Then the permissible temperature rise is $T_p = T_s - T_{op}$, where T_{op} = water temperature in the pump inlet. Or, $T_p = 226.1 - 220 = 6.1°F$ (3.4°C).

2. Compute the recirculation flow rate at the shutoff head
From the pump characteristic curve with recirculation, Fig. 43, the continuous-recirculation flow Q_B for an 11.5-in (29.2-cm) impeller at an operating head of 450 ft (137.2 m) is 48.6 gal/min (177.1 L/min). Find the continuous-recirculation flow at shutoff head H_s ft (m) of 540 ft (164.6 m) from $Q_s = Q_B(H_s/H_{op})^{0.5}$, where H_{op} = operating head, ft (m). Or $Q_s = 48.6(540/450) = 53.2$ gal/min (201.4 L/min).

3. Find the minimum safe flow for this pump
The minimum safe flow, lb/h, is given by $w_{min} = 2545bhp/[C_pT_p + (1.285 \times 10^{-3})H_s]$, where C_p = specific head of the water; other symbols as before.

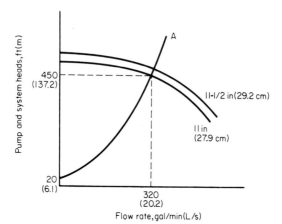

FIGURE 41   System-head curves without recirculation flow. (*Chemical Engineering.*)

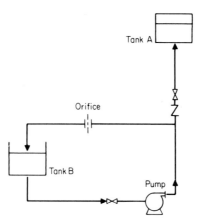

FIGURE 42   Pumping system with a continuous-recirculation line. (*Chemical Engineering.*)

Substituting, we find w_{min} = 2545(60)/[1.0(6.1) + (1.285 × 10^{-3})(540)] = 22,476 lb/h (2.83 kg/s). Converting to gal/min yields Q_{min} = w_{min}/[(ft^3/h)(gal/min)(lb/ft^3)] for the water flowing through the pump. Or, Q_{min} = 22,476/[(8.021)(59.6)] = 47.1 gal/min (178.3 L/min).

4.   Compare the shutoff recirculation flow with the safe recirculation flow
Since the shutoff recirculation flow Q_s = 53.2 gal/min (201.4 L/min) is greater than Q_{min} = 47.1 gal/min (178.3 L/min), the 11.5-in (29.2-cm) impeller is adequate to provide safe continuous recirculation. An 11.25-in (28.6-cm) impeller would not be adequate because Q_{min} = 45 gal/min (170.3 L/min) and Q_s = 25.6 gal/min (96.9 L/min).

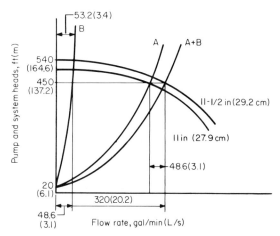

FIGURE 43 System-head curves with recirculation flow. (*Chemical Engineering.*)

Related Calculations.   Safety-service pumps are those used for standby service in a variety of industrial plants serving the chemical, petroleum, plastics, aircraft, auto, marine, manufacturing, and similar businesses. Such pumps may be used for fire protection, boiler feed, condenser cooling, and related tasks. In such systems the pump is usually oversized and has a recirculation loop piped in to prevent overheating by maintaining a minimum safe flow. Figure 41 shows a schematic of such a system. Recirculation is controlled by a properly sized orifice rather than by valves because an orifice is less expensive and highly reliable.

The general procedure for sizing centrifugal pumps for safety service, using the symbols given earlier, is this: (1) Select a pump that will deliver the desired flow Q_A, using the head-capacity characteristic curves of the pump and system. (2) Choose the next larger diameter pump impeller to maintain a discharge flow of Q_A to tank A, Fig. 41, and a recirculation flow Q_B to tank B, Fig. 41. (3) Compute the recirculation flow Q_s at the pump shutoff point from $Q_s = Q_B(H_s/H_{op})^{0.5}$. (4) Calculate the minimum safe flow Q_{min} for the pump with the larger impeller diameter. (5) Compare the recirculation flow Q_s at the pump shutoff point with the minimum safe flow Q_{min}. If $Q_s \geq Q_{min}$, the selection process has been completed. If $Q_s < Q_{min}$, choose the next larger size impeller and repeat steps 3, 4, and 5 above until the impeller size that will provide the minimum safe recirculation flow is determined.

This procedure is the work of Mileta Mikasinovic and Patrick C. Tung, design engineers, Ontario Hydro, as reported in *Chemical Engineering* magazine.

PUMP CHOICE TO REDUCE ENERGY CONSUMPTION AND LOSS

Choose an energy-efficiency pump to handle 1000 gal/min (3800 L/min) of water at 60°F (15.6°C) at a total head of 150 ft (45.5 m). A readily commercially available pump is preferred for this application.

Calculation Procedure:

1. Compute the pump horsepower required

For any pump, $bhp_i = (gpm)(H_t)(s)/3960e$, where bhp_i = input brake (motor) horsepower to the pump; H_t = total head on the pump, ft; s = specific gravity of the liquid handled; e = hydraulic efficiency of the pump. For this application where $s = 1.0$ and a hydraulic efficiency of 70 percent can be safely assumed, $bhp_i = (1000)(150)(1)/(3960)(0.70) = 54.1$ bhp (40.3 kW).

2. Choose the most energy-efficient pump

Use Fig. 44, entering at the bottom at 1000 gal/min (3800 L/min) and projecting vertically upward to a total head of 150 ft (45.5 m). The resulting intersection is within area 1, showing from Table 17 that a single-stage 3500-r/min electric-motor-driven pump would be the most energy-efficiency.

Related Calculations.   The procedure given here can be used for pumps in a variety of applications—chemical, petroleum, commercial, industrial, marine, aeronautical, air-conditioning, cooling-water, etc., where the capacity varies from 10 to 1,000,000 gal/min (38 to 3,800,000 L/min) and the head varies from 10 to 10,000 ft (3 to 3300 m). Figure 44 is based primarily on the characteristic of pump specific speed $N_s = NQ^2/H^{3/4}$, where N = pump rotating speed, r/min; Q = capacity, gal/min (L/min); H = total head, ft (m).

When N_s is less than 1000, the operating efficiency of single-stage centrifugal pumps falls off dramatically; then either multistage or higher-speed pumps offer the best efficiency.

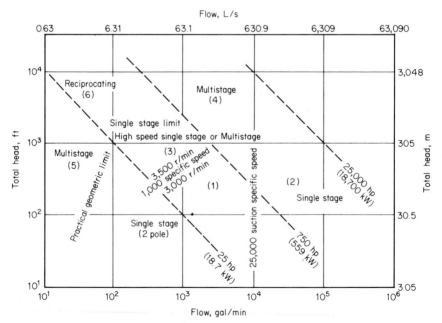

FIGURE 44   Selection guide is based mainly on specific speed, which indicates impeller geometry. (*Chemical Engineering.*)

TABLE 17 Type of Pump for Highest Energy Efficiency*

Area 1: Single-stage, 3500 r/min
Area 2: Single-stage, 1750 r/min or lower
Area 3: Single-stage, above 3500 r/min, or multistage, 3500 r/min
Area 4: Multistage
Area 5: Multistage
Area 6: Reciprocating

*Includes ANSI B73.1 standards; see area number in Fig. 38.

Area 1 of Fig. 44 is the densest, crowded both with pumps operating at 1750 and 3500 r/min, because years ago 3500-r/min pumps were not thought to be as durable as 1750-r/min ones. Since the adoption of the AVS standard in 1960 (superseded by ANSI B73.1), pumps with stiffer shafts have been proved reliable.

Also responsible for many 1750-r/min pumps in area 1 has been the impression that the higher (3500-r/min) speed causes pumps to wear out faster. However, because impeller tip speed is the same at both 3500 and 1750 r/min [as, for example, a 6-in (15-cm) impeller at 3500 r/min and a 12-in (30-cm) one at 1750 r/min], so is the fluid velocity, and so should be the erosion of metal surface. Another reason for not limiting operating speed is that improved impeller inlet design allows operation at 3500 r/min to capacities of 5000 gal/min (19,000 L/min) and higher.

Choice of operating speed also may be indirectly limited by specifications pertaining to suction performance, such as that fixing the top suction specific speed S directly or indirectly by choice of the sigma constant or by reliance on Hydraulic Institute charts.

Values of S below 8000 to 10,000 have long been accepted for avoiding cavitation. However, since the development of the inducer, S values in the range of 20,000 to 25,000 have become commonplace, and values as high as 50,000 have become practical.

The sigma constant, which relates NPSH to total head, is little used today, and Hydraulic Institute charts (which are being revised) are conservative.

In light of today's designs and materials, past restrictions resulting from suction performance limitations should be reevaluated or eliminated entirely.

Even if the most efficient pump has been selected, there are a number of circumstances in which it may not operate at peak efficiency. Today's cost of energy has made these considerations more important.

A centrifugal pump, being a hydrodynamic machine, is designed for a single peak operating-point capacity and total head. Operation at other than this best efficiency point (bep) reduces efficiency. Specifications now should account for such factors as these:

1. A need for a larger number of smaller pumps. When a process operates over a wide range of capacities, as many do, pumps will often work at less than full capacity, hence at lower efficiency. This can be avoided by installing two or three pumps in parallel, in place of a single large one, so that one of the smaller pumps can handle the flow when operations are at a low rate.

2. Allowance for present capacity. Pump systems are frequently designed for full flow at some time in the future. Before this time arrives, the pumps will operate far from their best efficiency points. Even if this interim period lasts only 2 or

3 years, it may be more economical to install a smaller pump initially and to replace it later with a full-capacity one.

3. Inefficient impeller size. Some specifications call for pump impeller diameter to be no larger than 90 or 95 percent of the size that a pump could take, so as to provide reserve head. If this reserve is used only 5 percent of the time, all such pumps will be operating at less than full efficiency most of the time.

4. Advantages of allowing operation to the right of the best efficiency point. Some specifications, the result of such thinking as that which provides reserve head, prohibit the selection of pumps that would operate to the right of the best efficiency point. This eliminates half of the pumps that might be selected and results in oversized pumps operating at lower efficiency.

This procedure is the work of John H. Doolin, Director of Product Development, Worthington Pumps, Inc., as reported in *Chemical Engineering* magazine.

Special Pump Applications

EVALUATING USE OF WATER-JET CONDENSATE PUMPS TO REPLACE POWER-PLANT VERTICAL CONDENSATE PUMPS

Evaluate the economic and application feasibility of replacing the vertical condensate pumps in a typical 1100-MW pressurized-water-reactor steam power plant having a feedwater train of two feedwater pumps, two heater drain pumps, and three vertical condensate pumps, with a water-jet pump in combination with a horizontal centrifugal pump. The flow rates, pressure heads, and related characteristics of the plant being considered are shown in Table 18.

Calculation Procedure:

1. *Develop the performance parameters for the water-jet pump*
During the past two decades, turbine generator sizes increased from about 100 MW in the 1950s to 300 MW in the 1960s, and then up to about 750 MW in the early 1970s. At this writing (1997), generator sizes for both nuclear and fossil-fuel plants are even larger than the 750 MW cited here. This drastic increase in size, plus the introduction of low pressure nuclear power cycles, brought about an increase in condensate flow to more than 10×10^6 lb/h (4.5×10^6 kg/h). Actually, in a typical 1300-MWe nuclear thermal cycle today, the condensate flow from condenser hotwell may be as high as 12×10^6 lb/h (5.5×10^6 kg/h). Current practice in condensate pumping system design is to either increase the pump capacity or increase the number of pumps operating in parallel to meet the flow requirements. However, these measures may increase:

• Initial fabrication and installation costs

• Probability of pump failures

• Routine maintenance and repair costs (and equally, if not more important, the attendant costs of plant down time).

TABLE 18A Feedwater System Pressure of a 1100-MW PWR Plant

Percent of max. design load	100	95.9	75	50	25	15
Percent of guarantee load	104.3	100	78.2	52.1	26.1	15.6
Turbine output—kW	1,210,081	1,160,596	907,560	605,040	302,519	181,512
Feedwater flow—lb/h	15,886,500	15,155,582	11,669,947	8,006,481	4,535,192	2,980,230
Feedwater flow—gal/min	36,996	35,181	26,653	17,937	9,889	6,409
Feed pump suction temp.—°F	403	398.8	377	348.1	305.7	279
Steam generator press.—lb/in² (abs)	990	990	1,036	1,073	1,091	1,093
Feed pump discharge press.—lb/in² (abs)	1,149	1,144	1,156	1,160	1,146	1,136
Feed pump suction press.—lb/in² (abs)	459	475	522	559	582	589
Feed pump TDH—lb/in²	690	669	634	601	564	547
Feed pump TDH—ft	1,856	1,794	1,673	1,555	1,420	1,360
Condensate flow—lb/h	10,500,000	10,150,000	7,820,000	5,475,000	3,125,000	2,200,000
Condensate flow—gal/min	21,212	20,505	15,798	11,061	6,313	4,444
Condensate pump discharge press—lb/in²	547.2	549.4	565.2	578.1	589.3	593.6
Condensate pump discharge press—ft	1,275	1,280	1,317	1,347	1,373	1,383
Condensate system loss—lb/in²	88.2	74.4	43.2	19.1	7.3	4.6

Notes: 1. Based on three condensate pumps and two heater drain pumps operating through the full load range.
2. The conversion factors from English Units to SI Units are tabulated below:

English Unit	SI unit
1 lb/h	1.26 × 10⁻⁴ kg/s
1 gal/min	6.309 × 10⁻⁵ m³/s
1°F	.5556 K
1 lb/in²	6.895 × 10³ Pa
1 ft	.3048 M

TABLE 18B SI Values for Feedwater System Pressure of an 1100-MW PWR Plant

Percent max. design load	100	95.9	75	50	25	15
Percent guarantee load	104.3	100	78.2	52.1	26.1	15.6
Turbine output, kW	1,210,081	1,160,596	907,560	605,040	302,519	181,512
Feedwater flow, kg/h	7,212,471	6,880,634	5,298,k56	3,634,932	2,058,977	1,353,024
Feedwater flow, L/s	2524	2220	1682	1132	624	404
Feedwater pump suction temp, °C	206	203.8	191.7	175.6	152.1	137.2
Steam generator press, kPa	6821	6821	7138	7393	7517	7531
Feedwater pump disch press, kPa	7917	7882	7965	7992	7896	7827
Feedwater pump suction press, kPa	3163	3273	3997	3852	4010	4058
Feed pump TDH, kPa	4754	4609	4368	4141	3886	3769
Feed pump TDH, m	566	547	510	474	433	415
Condensate flow, kg/h	4,767,000	4,608,100	3,550,280	2,485,650	1,418,750	998,800
Condensate flow, L/s	1338	1294	997	698	398	280
Condensate pump disch press kPa	3770	3785	3894	3983	4060	4090
Condensate pump disch press, m	389	390	401	411	418	422
Condensate system loss, kPa	608	513	298	132	50	32

The purpose of this calculation procedure is to examine the technical feasibility of using a water jet pump in combination with a horizontal centrifugal pump to replace a vertical centrifugal pump. In comparison with a typical vertical pump installation, the horizontal pump installation appears to offer a relatively higher system availability factor which should stem directly from the greater accessibility offered by the typical horizontal pump installation over the typical vertical pump installation. Marked improvement in both preventive and corrective maintenance times, even where equivalent failures or failure rates are assumed for both types of pump installations, invoke serious economic factors which cannot be overlooked in consideration of the current and inordinately high costs of plant down time. Moreover, in combination with the water jet pump, the horizontal configuration appears to offer a solution to the related NPSH problems. The combination also appears pertinent in the design of other systems, such as the heater drain systems in the feedwater cycle, where similar conditions may obtain.

A water-jet pump, Fig. 45, consists of a centrifugal pump discharging through a nozzle located at the bottom of the condenser hotwell. The operating principle of the water-jet pump is based on the transfer of momentum from one stream of fluid to another.

Water jet pumps were incorporated into the flow recirculation system of boiling water reactor design in 1965. The pumps were selected in lieu of conventional centrifugal pumps because of their basic simplicity and the economic incentives resulting from the possible reduction in the number of coolant loops and vessel nozzles by placing the water jet pump inside the pressure vessel. The reduction in the number of coolant loops permits a smaller drywell so that both primary and secondary containment structure can be designed more compactly. Concurrently, the efficiency of the water jet pump has been markedly improved through extensive development and testing programs pursued by the manufacturer. An efficiency of 41.5 percent has been obtained at a suction flow to driving flow ratio of 2.55 in the manufacturer's second generation jet pumps (1).

Figure 45 shows the proposed arrangement of a water jet pump and horizontal centrifugal pump combination to replace the conventional vertical condensate pump. The high momentum jet stream ejected from the recirculation nozzle is mixed with a low momentum stream from the condenser hotwell in the throat. This mixed flow slows down in the diffuser section where part of its momentum (kinetic energy) is converted into pressure. The flow is then led to the centrifugal pump suction through a short piping section. The pressure of the fluid is increased through the pump and a major part of this flow is then directed through the lower pressure heaters and

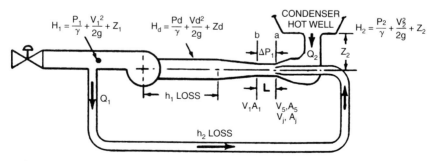

FIGURE 45 Water jet-centrifugal pump.

finally to the suction of the feedwater pump, or to the deaerator, to provide the required feedwater flow to the stream generator. The remaining flow is led through the recirculation line back to the jet pump throat to induce the suction flow from the condenser hotwell; and thereby provide for continuous recirculation. A possible turbine cycle arrangement with a water jet and horizontal centrifugal condensate pump is shown in Fig. 46.

The characteristics of a water-jet pump are defined by the following equations and nomenclature, using data from Fig. 45:

$$Q_1 = A_j V_j, \quad Q_2 = A_s V_s, \quad Q_1 + Q_2 = A_t V_t = Q_t \tag{1}$$

$$\frac{Q_2}{Q_1} = M \tag{2}$$

$$\frac{A_j}{A_t} = R \tag{3}$$

$$Q_1 = A_j \sqrt{\frac{2g(H_1 - Pa/\gamma)}{1 + K_j}} \tag{4}$$

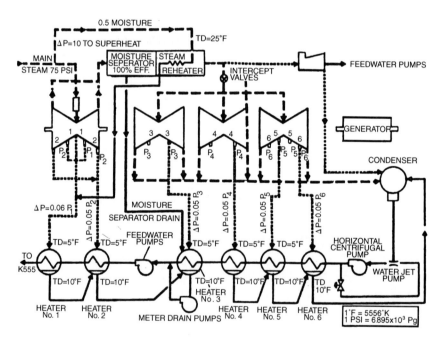

75 psi (516.6 kPa)
25°F (13.8°C)
5°F (2.8°C)
10°F (5.6°C)

FIGURE 46 Thermal cycle arrangement with water jet-centrifugal pump.

$$Q_2 = A_s \sqrt{\frac{2g(H_2 - \text{Pa}/\gamma)}{1 + K_s}} \qquad (5)$$

$$\frac{\text{Pa}}{\gamma} + \frac{V_t^2}{2g} = H_d + K_d \frac{V_t^2}{2g} \qquad (6)$$

Following Gosline and O'Brien (3), the head ratio, N, depends upon six parameters; M, R, K_s, K_j, K_t, and K_d, or:

$$N = \frac{H_d - H_2}{H_1 - H_d} = f(M, R, K_s, K_j, K_t, K_d) \qquad (7)$$

which vary with the design of the water jet pump itself and with the length of the connecting pipes. Once the design of the water jet pump is fixed, these parameters are known functions of flow. Based upon their extensive testing, the manufacturer suggests the use of the M-N curve shown in Fig. 47 for the water jet pump design evaluation. This M-N correlation is essentially a straight line and can be represented by:

$$N = N_0 - \frac{N_0}{M_0} M \qquad (8)$$

where N_0 is the value of N at $M = 0$ and M_0 is the value of M when $N = 0$. The M-N curve of the water jet pump shown in Fig. 45 may be represented by:

$$N = .246 - .04125M \qquad (9)$$

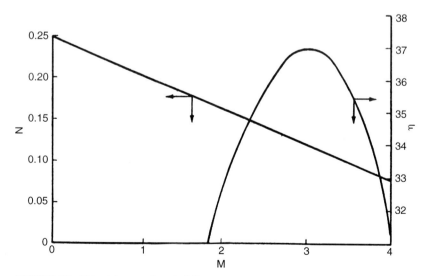

FIGURE 47 Water jet pump characteristic curve.

Nomenclature

A = cross-section area, ft² (0.0929 m²)
bhp = pump brake horsepower
D = diameter, ft (0.3048 m)
e = centrifugal pump efficiency
f = friction factor
F = brake horsepower ratio
g = acceleration of gravity, 32.2 ft/s² (9.815 m/s²)
H = hydraulic head, ft (0.3048 m)
hp = horsepower
ΔHp = pump total dynamic head, ft (0.3048 m)
K = friction parameter = fL/D
kW = kilowatt
L = length, ft (0.3048 m)
M = induced flow and driving flow ratio = Q_2/Q_1
M_0 = constant
n = pump speed, rpm
N = head ratio = $H_d - H_2/H_1 - H_d$
N_0 = constant
NHR = net heat rate, Btu/kWh (1054 J/kWh)
NPSH = net positive suction head
P = pressure, lb/ft² (47.88 Newton/m²)
q_{in} = thermal energy input, kW
Q = flow rate, ft³/s (0.02832 m³/s)
R = ratio of area of nozzle to area of throat = A_j/A_t
V = velocity, ft/s (0.3048 m/s)
Z = elevation, ft (0.3048 m)
η = jet pump efficiency
γ = specific weight of liquid, lb · ft/ft³ (16.02 kg/m³)

Subscripts

a = entrance of throat
b = end of throat
c = centrifugal pump
d = jet pump discharge
j = tip of the nozzle
s = annular area surrounding tip of nozzle
t = throat of mixing chamber
v = vertical condensate pump
$j\text{-}c$ = water jet-centrifugal pump combination

The water jet pump efficiency, η, is defined as the ratio of the total energy increase of the suction flow to the total energy decrease of the driving flow, or:

$$\eta = M \cdot N \cdot 100 \tag{10}$$

This definition of efficiency is different from the centrifugal pump efficiency, e_c, which is defined as:

$$e_c = \frac{\text{pump output}}{\text{bhp}} = \frac{Q\gamma H}{550 \times \text{bhp}} \tag{11}$$

2. Define the performance of the centrifugal pump associated with the water-jet pump

The performance of the horizontal centrifugal pump is defined in the manufacturer supplied pump characteristic curve and pump affinity laws:

$$\frac{Q}{Q_0} = \frac{n}{n_0} \quad \text{and} \quad \frac{H}{H_0} = \left(\frac{n}{n_0}\right)^2 \tag{12}$$

From Fig. 45, the pump discharge head, H_1, is:

$$H_1 = H_d - h_1 + \Delta Hp \tag{13}$$

where h_1 is the head loss from the water jet pump exit to the centrifugal pump suction.

Hydraulic Horsepower of Water Jet-Centrifugal Pump. From Fig. 45, the total flow, Q_t, through the horizontal centrifugal pump is:

$$Q_t = Q_1 + Q_2 \tag{14}$$

The hydraulic horsepower of the pump is:

$$(\text{Hydraulic hp})_{j\text{-}c} = \frac{Q_t(H_1 - H_d)}{550} \tag{15}$$

From Eqs. (7) and (8):

$$H_d = \frac{\left(N_0 - \dfrac{N_0}{M_0} M\right) H_1 + H_2}{1 + N_0 - \dfrac{N_0}{M_0} M} \tag{16}$$

Substituting Eq. (16) into Eq. (15), we have:

$$(\text{Hydraulic hp})_{j\text{-}c} = \frac{Q_t \gamma}{550} \left(\frac{H_1 - H_2}{1 + N_0 - \dfrac{N_0}{M_0} M}\right) \tag{17}$$

For the conventional vertical condensate pump, the hydraulic horsepower is:

$$(\text{Hydraulic hp})_v = \frac{Q_2 \gamma (H_1 - H_2)}{550} \tag{18}$$

From Eqs. (17) and (18):

$$\frac{(\text{Hydraulic hp})_{j\text{-}c}}{(\text{Hydraulic hp})_v} = \frac{1 + \dfrac{1}{M}}{1 + N_0 - \dfrac{N_0}{M_0} M} \tag{19}$$

The hydraulic hp ratio for different values of M is shown in Fig. 48. The man-

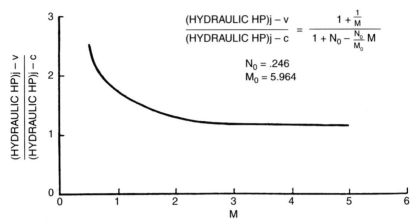

FIGURE 48   Hydraulic hp ratio vs. M.

ufacturer's suggested M-N curve, Fig. 47, is used in the calculation. From equation (11), the brake hp ratio is:

$$\frac{(\text{bhp})_{j\text{-}c}}{(\text{bhp})_v} = \frac{(\text{Hydraulic hp})_{j\text{-}c}}{(\text{Hydraulic hp})_v} \frac{e_c}{e_{j\text{-}c}} \tag{20}$$

Since the centrifugal pump efficiencies within a normal operating range do not change significantly, $e_{j\text{-}c} \approx e_c$, and:

$$\frac{(\text{bhp})_{j\text{-}c}}{(\text{bhp})_v} \approx \frac{1 + \dfrac{1}{M}}{1 + N_0 - \dfrac{N_0}{M_0} M} = F \tag{21}$$

3. Compute the effect of the water-jet pump on the net heat rate
The net heat rate is defined as:

$$\text{NHR} = \text{New Heat Rate} = \frac{q_{in}}{kW_E - kW_{AUX} - kW_{CON}} \tag{22}$$

where:

q_{in} = total thermal energy input
kW_E = generator output
kW_{AUX} = total plant auxiliary power excluding the power to condensate pumps
kW_{CON} = total power required to drive a motor driven condensate pump.

If we further define that $kW = kW_E - kW_{AUX}$,

$$(\text{Net Heat Rate})_v = \frac{q_{in}}{kW - kW_{CON}} \tag{23}$$

If a water jet-centrifugal pump combination is used to replace the vertical condensate pump while keeping q_{in} and kW unchanged, the net heat rate can be shown as:

$$\text{(Net Heat Rate)}_{j\text{-}c} = \frac{q_{in}}{kW - kW_{CON} \times F} \tag{24}$$

From the foregoing equations:

$$\frac{\text{(NHR)}_{j\text{-}c}}{\text{(NHR)}_v} = 1 + \frac{kW_{CON}(F - 1)}{kW - kW_{CON} \times F} \tag{25}$$

The effect of M on net heat ratio is calculated according to Eq. (25) and this is shown in Fig. 49. The following data of a 1100-MW PWR plant have been used in the calculation

$$kW = 1{,}160{,}596$$

$$\text{Condensate flow} = 20{,}505 \text{ gal/min (1.294 m}^3/\text{s)}$$

$$\text{Condensate pump TDH} = 1280 \text{ ft (390.14 m)}$$

$$\text{Pump efficiency} = 0.78$$

$$kW_{CON} = \frac{gpm \times H}{3960 \times e_c} \times 0.746$$

$$= 6339 \text{ kW}$$

As shown in Fig. 49, the increase in heat rate caused by the water jet pump is approximately 0.1 percent at $M = 3$ and is less than 0.2 percent at $M = 2$. However, the heat rate ratio increases very rapidly with a further decrease in M.

4. Develop the performance calculations for this installation

The performance calculations of a water jet-centrifugal pump combination in a power plant must be developed from an overall analysis of the feedwater-condensate system. To initiate the design analysis, the M-N relationship developed in Eq. (9) and shown in Fig. 47 is examined and, for obvious reasons, the peak efficiency

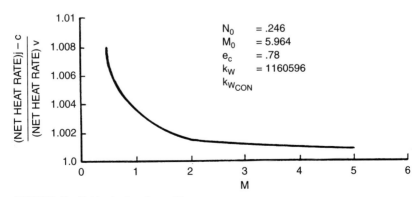

FIGURE 49 Net-heat rate ratio vs. M.

point ($M = 3$) is selected for the design of the water jet pump. At this design point, the efficiency of the water jet pump is about 37 percent.

The head and flow characteristics of the horizontal centrifugal pump are shown in Fig. 50, and the following simplifying assumptions are made in the development of the performance calculations (refer also to Fig. 45).

1. $K_2 = 0$.
2. V_s is small such that $H_2 \approx Pa/\gamma = 5$ lb/in² (34.5 kPa) or 11.65 ft (3.6 m) of water.
3. h_1 loss is 10 lb/in² (68.9 kPa) (or 23.3 ft) (7.1 m) and is a constant under all loading conditions.
4. h_2 loss is 25 lb/in² (172.2 kPa) (or 58.25 ft) (17.8 m) and is a constant under all loading conditions.

From Eqs. (7), (9), and (13), we have:

$$H_1 = H_2 + (1.246 - .04125M) \times (\Delta Hp - h_1) \qquad (26)$$

To match the vertical condensate pump at maximum design condition, the total dynamic head of the horizontal pump (ΔHp) shall be such that $H_1 = 1275$ ft (388.6

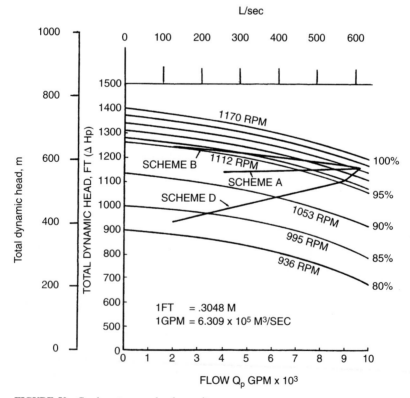

FIGURE 50   Condensate pump head-capacity curve.

m). With $h_1 = 23.3$ ft (7.1 m), $H_2 = 11.65$ ft (3.6 m), and $M = 3$, the required ΔHp from Eq. (26) is:

$$\Delta Hp = (H_1 - H_2) \div (1.246 - .04125M) + h_1 = 1149 \text{ ft (350.2 m)}$$

and H_d from Eq. (13) is:

$$H_d = H_1 + h_1 - \Delta Hp = 149.3 \text{ ft (45.5 m)}$$

From Eqs. (2) and (14):

$$Q_t = Q_2 \left(1 + \frac{1}{M}\right) \tag{27}$$

with $M = 3$, the required condensate flow of three horizontal centrifugal pump Q_t is:

$$Q_t = 10.5 \times 10^6 \times \left(1 + \frac{1}{3}\right)$$
$$= 14 \times 10^6 \text{ lb} \cdot \text{s/h or 28.283 gal/min (1.784 m}^3/\text{s)}$$

From Fig. 50, the condensate pump characteristic curve, which is generated by pump affinity law, the horizontal centrifugal pump will be running at 1147 r/min. The jet nozzle area, A_j, can be calculated from:

$$Q_1 = CA_j \sqrt{2g(H_1 - h_2 - H_2)} \tag{28}$$

which is another form of Eq. (4).
Assume that nozzle flow coefficient $C = 0.9$, we have:

$$A_j = .020948 \text{ ft}^2 \text{ (.00195 m}^2)$$

or:

$$d_j = 1.96 \text{ in (.0498 m)}$$

Substitute the value of A_j and C into Eq. (28), we have:

$$Q_1 = .1513 \sqrt{H_1 - 69.9} \tag{29}$$

Accordingly, Eqs. (2), (26), (27), and (29) define the water jet-centrifugal pump performance.

5. Determine the best drive for the water-jet centrifugal condensate pump

Pump DRIVING Schemes. Four possible schemes of driving the water jet-centrifugal condensate pump are examined. For each scheme, the water jet-centrifugal pump is designed to duplicate the head and capacity performance of the corresponding vertical condensate pump at the maximum design condition and then the schemes are examined for continuous operation at other loading conditions; namely, 100, 75, 50, 25, and 15 percent. Schematic arrangements of each scheme and sample calculations at 75 percent load are shown in Table 20.

Scheme A: Variable Speed Motor Drive and Variable M Ratio. In this scheme, variable speed electric motor is used to drive the water jet-centrifugal pump so that the condensate flow and the pressure head at the feed pump suction are identical to that of the base case which uses conventional vertical condensate pumps. To

TABLE 19   M Ratio vs. Load Variation

Load Condition	VWO	100%	75%	50%	25%	15%
M Ratio	3	2.894	2.2	1.52	.86	.6021

satisfy equation (4), the water jet pump flow ratio M changes from 3 at maximum design condition to 0.6021 at 15 percent load as shown in Table 19.

At maximum design condition, the horizontal centrifugal pump is running at 1147 r/min and 9428 gal/min (594.9 L/s). For the partial load operation, the pump will follow the curve labeled Scheme A in Fig. 50. From equations (21) and (25), the increase in net heat rate will be higher at the 25 and 15 percent partial loading operation. If long-term partial load operation is expected, this scheme should be avoided.

Scheme B: Variable Speed Motor Drive and Constant M Ratio.   In this scheme, the pump drive is identical to Scheme A except that a flow regulating control valve is installed in the water jet pump recirculation line to maintain a constant M ratio at all loading conditions. In this case, the speed of the horizontal centrifugal pump will vary according to the curve labeled Scheme B in Fig. 50. The feedwater pump operation is identical to that of Scheme A and the base case.

Scheme C: Constant Speed Motor Drive and Constant M Ratio.   In this scheme, the horizontal centrifugal pump is running at a constant speed of 1147 r/min. A control valve is used to keep the flow ratio $M = 3$. In this case, the pressure head at the water jet-centrifugal pump discharge is higher than that of the vertical condensate pump. Consequently, the feedwater pump will be running at a lower speed and lower total dynamic head to keep the steam generator pressure identical to that of the base case. The required feed pump total dynamic head and corresponding speed are shown as the curve labeled Scheme C in Fig. 51.

Scheme D: Turbine Drive Jet-Centrifugal Pump and Constant M Ratio.   In this scheme, the feedwater pump and water jet-centrifugal pump are running at a constant speed ratio and both are driven by the auxiliary turbine. A control valve is used to keep the flow ratio $M = 3$. Under these conditions, the water jet-centrifugal pump and the feedwater pump will follow the curves labeled Scheme D in Figs. 50 and 51, respectively, to produce the identical steam generator conditions in Table 18. It should be noted that the auxiliary turbine driven feedwater pump has been shown to have a better cycle efficiency than a motor driven pump in the same application for large power plants (4); intuitively, the auxiliary turbine driven jet-centrifugal pump arrangement may also provide certain gains in cycle efficiency over other water jet-centrifugal pump drive schemes.

6.   Summarize the findings for this pump application
It has been shown that a water jet-centrifugal pump can be used to replace the conventional vertical condensate pump in a steam power plant feedwater system. All four schemes discussed in the preceding section are feasible means of driving the water jet-centrifugal pump combination. While the resulting auxiliary power requirements for the jet-centrifugal pump system will be slightly higher, the increase will be insignificant if the flow rate M is kept greater than 2.

The proposed change from conventional vertical pump to a water jet-centrifugal pump may have advantages:

1. Increased feedwater system reliability and reduced plant downtime
2. Easier maintenance operations, reduced cost of maintenance

TABLE 20 Pump Driving Schemes

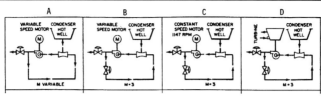

A	B	C	D
VARIABLE SPEED MOTOR / CONDENSER HOT WELL	VARIABLE SPEED MOTOR / CONDENSER HOT WELL	CONSTANT SPEED MOTOR 1147 RPM / CONDENSER HOT WELL	TURBINE / CONDENSER HOT WELL
M VARIABLE	M = 3	M = 3	M = 3

A

From Table 18 @ 75% load
Flow = 7.82×10^6 labs/h
Q_2 = 5266 gal/min
H_1 = 565.2 lb/in^2 = 1317 ft

From Eq. 29
Q_1 = .1513$\sqrt{1317 - 69.9}$
   − 5.343 ft^3/s
   = 2398 gal/min

$M = \dfrac{Q_2}{Q_1} = \dfrac{5266}{2398} = 2.196$

From Eq. 26
ΔHp =

$\dfrac{1317 - 11.65}{1.246 - .04125 \times 2.196}$
   + 23.3 = 1153 ft

$Q_t = Q_1 + Q_2$ = 7664 gal/min

From Fig. 50
Pump speed = 1123 r/min

From Eq. 13
H_d = 1317 + 23.3 − 1153
   = 187.3 ft

B

From Table 18 @ 75% load
Q_2 = 5266 gal/min
H_1 = 565.2 lb/in^2 = 1317 ft

$Q_t = Q2\left(1 + \dfrac{1}{M}\right)$
   = 7021 gal/min

From Eq. 26
ΔH = 1186.5 ft

From Eq. 6
Pump speed = 1131 RPM

From Eq. 13
H_d = 1317 + 23.3 − 1186.5
   = 153.8 ft

C

From Table 18 @ 75% load
Q_2 = 5266 gal/min
Q_t = 7021 gal/min

From Fig. 50 @ 1147 RPM
ΔHp = 1229 ft

From Eq. 26
H_1 = 11.65 + 1.12225
(1229
   −23.3) = 1364.8 ft
   = 585.7 lb/in^2

From Table 18 condensate
System head loss = 43.2 lb/in^2
Feed pump suction pressure
   = 542.3 lb/in^2
Feed Pump ΔHp
   = 1156 − 542.3
   = 613.7 lb/in^2
   = 1619 ft

From Eq. 13
H_d = 1364.8 + 23.3 − 1229
   = 159.1 ft

D

Feed Pump speed
   = 4830 RPM

Condensate pump speed
   = 1147 RPM

Gear reduction ratio = $\dfrac{4.2}{1}$

Q_t = 7021 gal/min

Iterative Procedure is Used
To find pump running
SPeeds

Try Feed Pump Speed
   = 4500 RPM

Condensate Pump RPM
   = $\dfrac{4500}{4.2}$ = 1071 rptn

From Fig. 50 ΔHp = 1061 ft

From Eq. 26
H_1 = 1176.2 ft = 504.8 lb/in^2

Feed Pump Suction Pressure
   = 504.8 − 43.3 = 461.6 lb/in^2

Feed Pump
ΔHp = 1156 − 461.6
   = 694.4 lb/in^2 = 1832.0 ft

From Fig. 511
Feed pump speed
   = 4990 rpm
Very close to assumed 4500 r/min

H_d = 1176.2 + 23.3 − 1061 = 138.5 ft

SI Values

A		A		A	
gal/min	L/s	lb/in^2	kPa	ft	m
5266	332.3	565.2	3894	1317	401.4
2398	151.3			187.3	57.1
7664	483.6				
B		**B**		**B**	
5266	332.3	565.2	3894	1317	401.4
7021	443.0			1186.5	361.6
				153.8	46.9
C		**C**		**C**	
5266	332.3	585.7	4035.5	1229	374.6
7021	443.0	43.2	297.6	1364.8	415.9
		542.3	3736.4	1619	493.5
		613.7	4228.4	159.1	48.5
D		**D**		**D**	
7021	443.0	504.8	3478.1	1061	323.4
		461.6	3180.4	1832	558.4
		694.4	4784.4	138.5	42.2

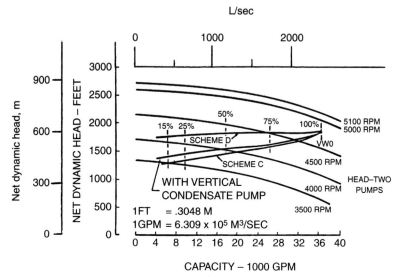

FIGURE 51   Feed pump head-capacity curve (two pumps).

3. More flexibility in plant layout which, in turn, may favorably effect on condensate system piping costs.

With the present high cost of plant outage, the improvement in system reliability alone may provide sufficient economic incentive for considering the water jet-centrifugal pump combination.

Related Calculations.   While the study here was directed at a PWR steam power plant, the approach used is valid for any steam power plant—utility, industrial, commercial, or marine—using the types of pumps considered. The water-jet pump, developed in the mid-1800s, has many inherent advantages which can be used in today's highly competitive power-generation industry. In every such installation, the condensate pump in the feedwater system of the steam electric generating power plant takes suction from the condenser hotwell and delivers the condensate through the tube side of the lower pressure feedwater heaters to the deaerator, or to the suction of the feed pump. The continuous operation of the entire plant depends upon the proper functioning of the condensate pumps. It should also be noted that the condensate pumping system consumes a significant portion of the auxiliary power, and represents a measurable portion of the plant first cost.

In power plant applications, multiple parallel pumping arrangements are employed to provide a flexible operational system. Condensate pumps are of constant speed motor-driven, vertical centrifugal type, and are located in a pit near the condenser. The difference in fluid elevations between the condenser hotwell and the first stage of the centrifugal pump is the only NPSH available to the pump because the condensate in the hotwell is always saturated.

This procedure is the work of E. N. Chu, Engineering Specialist, and F. S. Ku, Assistant Chief Mechanical Engineer, Bechtel Power Corporation, as reported in *Combustion* magazine and presented at the IEEE-ASME Joint Power Generation Conference. SI values were added by the handbook editor.

REFERENCES

1. Kudrika, A. A. and Gluntz, D. M., "Development of Jet Pumps for Boiling Water Reactor Recirculation Systems," *Journal of Engineering for Power,* Transactions of the ASME, Jan. (1974).
2. Anon., "Design and Performance of General Electric Boiling Water Reactor Jet Pumps," General Electric Company Report APED-5460, Sept. (1968).
3. Gosline, J. E. and O'Brien, M. P., "The Water Jet Pump," University of California Publication, Vol. 3, No. 3, 167 (1934).
4. Goodell, J. H. and Leung, P., "Boiler Feed Pump Drives," ASME Paper No. 64-PWR-7, IEEE-ASME National Power Conference (1964).

USE OF SOLAR-POWERED PUMPS IN IRRIGATION AND OTHER SERVICES

Devise a solar-powered alternative energy source for driving pumps for use in irrigation to handle 10,000 gal/min (37.9 m³/min) at peak output with an input of 50 hp (37.3 kW). Show the elements of such a system and how they might be interconnected to provide useful output.

Calculation Procedure:

1. Develop a suitable cycle for this application
Figure 52 shows a typical design of a closed-cycle solar-energy powered system suitable for driving turbine-powered pumps. In this system a suitable refrigerant is chosen to provide the maximum heat absorption possible from the sun's rays. Water is pumped under pressure to the solar collector, where it is heated by the sun. The water then flows to a boiler where the heat in the water turns the liquid refrigerant into a gas. This gas is used to drive a Rankine-cycle turbine connected to an irrigation pump, Fig. 52.

The rate of gas release in such a closed system is a function of (*a*) the unit enthalpy of vaporization of the refrigerant chosen, (*b*) the temperature of the water leaving the solar collector, and (*c*) the efficiency of the boiler used to transfer heat from the water to the refrigerant. While there will be some heat loss in the piping and equipment in the system, this loss is generally considered negligible in a well-designed layout.

2. Select, and size, the solar collector to use
The usual solar collector chosen for systems such as this is a parabolic tracking-type unit. The preliminary required area for the collector is found by using the rule of thumb which states: For parabolic tracking-type solar collectors the required sun-exposure area is 0.55 ft² per gal/min pumped (0.093 m² per 0.00379 m³/min) at peak output of the pump and collector. Another way of stating this rule of thumb is: Required tracking parabolic solar collector area = 110 ft² per hp delivered (13.7 m²/kW delivered).

Thus, for a solar collector designed to deliver 10,000 gal/min (37.9 m³/min) at peak output, the preliminary area chosen for this parabolic tracking solar collector

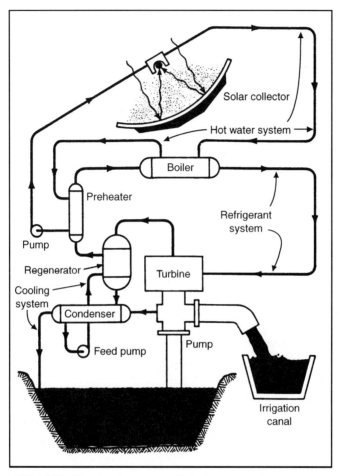

FIGURE 52   Closed-cycle system gassifies refrigerant in boiler to drive Rankine-cycle turbine for pumping water. (*Product Engineering,* Battelle Memorial Institute, and Northwestern Mutual Life Insurance Co.)

will be, A_p = (10,000 gal/min)(0.55 ft²/gal/min) = 550 ft² (511 m²). Or, using the second rule of thumb, A_p = (110)(50) = 5500 ft² (511 m²).

Final choice of the collector area will be based on data supplied by the collector manufacturer, refrigerant choice, refrigerant properties, and the actual operating efficiency of the boiler chosen.

In this solar-powered pumping system, water is drawn from a sump basin and pumped to an irrigation canal where it is channeled to the fields. The 50-hp (37.3-kW) motor was chosen because it is large enough to provide a meaningful demonstration of commercial size and it can be scaled up to 200 to 250 hp (149.2 to 186.5 kW) quickly and easily.

Sensors associated with the solar collector aim it at the sun in the morning, and, as the sun moves across the sky, track it throughout the day. These same sensing devices also rotate the collectors to a storage position at night and during storms. This is done to lessen the chance of damage to the reflective surfaces of the collectors. A backup control system is available for emergencies.

3. Predict the probable operating mode of this system

In June, during the longest day of the year, the system will deliver up to 5.6 million gallons (21,196 m³) over a 9.5-h period. Future provisions for energy storage can be made, if needed.

Related Calculations. Solar-powered pumps can have numerous applications beyond irrigation. Such applications could include domestic water pumping and storage, ornamental fountain water pumping and recirculation, laundry wash water, etc. The whole key to successful solar power for pumps is selecting a suitable application. With the information presented in this procedure the designer can check the applicability and economic justification of proposed future designs.

In today's environmentally-conscious design world, the refrigerant must be carefully chosen so it is acceptable from both an ozone-depletion and from a thermodynamic standpoint. Banned refrigerants should not, of course, be used, even if attractive from a thermodynamic standpoint.

This procedure is the work of the editorial staff of *Product Engineering* magazine reporting on the work of Battelle Memorial Institute and the Northwestern Mutual Life Insurance Co. The installation described is located at MMLI's Gila River Ranch, southwest of Phoenix AZ. SI values were added by the handbook editor.

SECTION 8
PIPING AND FLUID FLOW

Pressure Surge in Fluid Piping Systems

PRESSURE SURGE IN A PIPING SYSTEM FROM RAPID VALVE CLOSURE

Oil, with a specific weight of 52 lb/ft³ (832 kg/m³) and a bulk modulus of 250,000 lb/in² (1723 MPa), flows at the rate of 40 gal/min (2.5 L/s) through stainless steel pipe. The pipe is 40 ft (12.2 m) long, 1.5 in (38.1 mm) O.D., 1.402 in (35.6 mm) I.D., 0.049 in (1.24 mm) wall thickness, and has a modulus of elasticity, E, of 29×10^6 lb/in² (199.8 kPa $\times$ 10⁶). Normal static pressure immediately upstream of the valve in the pipe is 500 lb/in² (abs) (3445 kPa). When the flow of the oil is reduced to zero in 0.015 s by closing a valve at the end of the pipe, what is: (a) the velocity of the pressure wave; (b) the period of the pressure wave; (c) the amplitude of the pressure wave; and (d) the maximum static pressure at the valve?

Calculation Procedure:

1. Find the velocity of the pressure wave when the valve is closed
(a) Use the equation

$$a = \frac{68.094}{\sqrt{\gamma[(1/K) + (D/Et)]}}$$

where the symbols are as given in the notation below. Substituting,

$$a = \frac{68.094}{\sqrt{52\,[(1/25 \times 10^4) + (1.402/29 \times 10^6 \times 0.049]}}$$

$$= 4228 \text{ ft/s } (1288.7 \text{ m/s})$$

An alternative solution uses Fig. 1. With a D/t ratio = 1.402/0.049 = 28.6 for

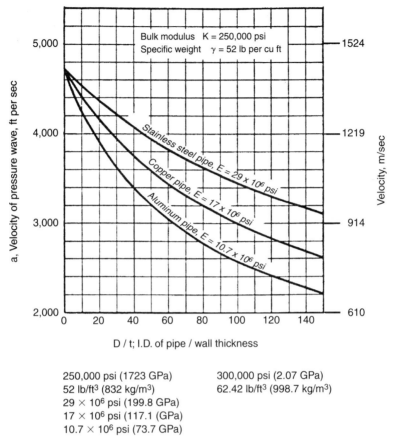

FIGURE 1 Velocity of pressure wave in oil column in pipe of different diameter-to-wall thickness ratios. (*Product Engineering.*)

stainless steel pipe, the velocity, a, of the pressure wave is 4228 ft/s (1288.7 m/s).

2. Compute the time for the pressure wave to make one round trip in the pipe
b) The time for the pressure wave to make one round trip between the pipe extremities, or one interval, is: $2L/a = 2(40)/4228 = 0.0189$ s, and the period of the pressure wave is: $2(2L/a) = 2(0.0189) = 0.0378$ s.

3. Calculate the pressure surge for rapid valve closure
c) Since the time of 01015 s for valve closure is less than the internal time $2L/a$ equal to 0.0189 s, the pressure surge can be computed from:

$$\Delta p = \gamma a V / 144g$$

for rapid valve closure.

The velocity of flow, $V = [(40)(231)(4)]/[(60)(\pi)(1.402^2)(12)]$ using the standard pipe flow relation, or $V = 8.3$ ft/s (2.53 m/s).

Then, the amplitude of the pressure wave, using the equation above is:

$$\Delta p = \frac{52 \times 4228 \times 8.3}{144 \times 32.2} = 393.5 \text{ lb/in}^2 \qquad (2711.2 \text{ kPa}).$$

4. Determine the resulting maximum static press in the pipe
d) The resulting maximum static pressure in the line, $p_{max} = p + \Delta p = 500 + 393.5 = 893.5$ lb/in^2 (abs) (6156.2 kPa).

Related Calculations.   In an industrial hydraulic system, such as that used in machine tools, hydraulic lifts, steering mechanisms, etc., when the velocity of a flowing fluid is changed by opening or closing a valve, pressure surges result. The amplitude of the pressure surge is a function of the rate of change in the velocity of the mass of fluid. This procedure shows how to compute the amplitude of the pressure surge with rapid valve closure.

The procedure is the work of Nils M. Sverdrup, Hydraulic Engineer, Aerojet-General Corporation, as reported in *Product Engineering* magazine. SI values were added by the handbook editor.

Notation

a = velocity of pressure wave, ft/s (m/s)
a_E = effective velocity of pressure wave, ft/s (m/s)
A = cross-sectional area of pipe, in^2 (mm^2)
A_o = area of throttling orifice before closure, in^2 (mm^2)
c = velocity of sound, ft/s (m/s)
C_D = coefficient of discharge
D = inside diameter of pipe, in (mm)
E = modulus of elasticity of pipe material, lb/in^2 (kPa)
F = force, lb (kg)
g = gravitational acceleration, 32.2 ft/s^2
K = bulk modulus of fluid medium, lb/in^2 (kPa)
L = length of pipe, ft (m)
m = mass, slugs
$N = T/(2L/a)$ = number of pressure wave intervals during time of valve closure
p = normal static fluid pressure immediately upstream of valve when the fluid velocity is V, lb/in^2 (absolute) (kPa)
Δp = amplitude of pressure wave, lb/in^2 (kPa)
p_{max} = maximum static pressure immediately upstream of valve, lb/in^2 (absolute) (kPa)
p_d = static pressure immediately downstream of the valve, lb/in^2 (absolute) (kPa)
Q = volume rate of flow, ft^3/s (m^3/s)
t = wall thickness of pipe, in (mm)
T = time in which valve is closed, s
v = fluid volume, in^3 (mm^3)
v_A = air volume, in^3 (mm^3)
V = normal velocity of fluid flow in pipe with valve wide open, ft/s (m/s)
V_E = equivalent fluid velocity, ft/s (m/s)
V_n = velocity of fluid flow during interval n, ft/s (m/s)

W = work, ft · lb (W)
γ = specific weight, lb/ft^3 (kg/m^3)
ϕ_n = coefficient dependent upon the rate of change in orifice area and discharge coefficient
τ = period of oscillation of air cushion in a sealed chamber, s

PIPING PRESSURE SURGE WITH DIFFERENT MATERIAL AND FLUID

(*a*) What would be the pressure rise in the previous procedure if the pipe were aluminum instead of stainless steel? (*b*) What would be the pressure rise in the system in the previous procedure if the flow medium were water having a bulk modulus, K, of 300,000 lb/in^2 (2067 MPa) and a specific weight of 62.42 lb/ft^3 (998.7 kg/m^3)?

Calculation Procedure:

1. Find the velocity of the pressure wave in the pipe
(*a*) From Fig. 2, for aluminum pipe having a D/t ratio of 28.6, the velocity of the pressure wave is 3655 ft/s (1114.0 m/s). Alternatively, the velocity could be computed as in step 1 in the previous procedure.

2. Compute the time for one interval of the pressure wave
As before, in the previous procedure, $2\ L/a = 2\ (40/3655) = 0.02188$ s.

3. Calculate the pressure rise in the pipe
Since the time of 0.015 s for the valve closure is less than the interval time of 2 L/a equal to 0.02188, the pressure rise can be computed from:

$$\Delta p = \gamma a V / 144 g$$

or,

$$\Delta p = \frac{52 \times 3655 \times 8.3}{144 \times 32.2} = 340.2 \text{ lb/in}^2 \qquad (2343.98 \text{ kPa})$$

4. Find the maximum static pressure in the line
Using the pressure-rise relation, $p_{max} = 500 + 340.2 = 840.2$ lb/in^2 (abs) (5788.97 kPa).

5. Determine the pressure rise for the different fluid
(*b*) For water, use Fig. 2 for stainless steel pipe having a D/t ratio of 28.6 to find $a = 4147$ ft/s (1264 m/s). Alternatively, the velocity could be calculated as in step 1 of the previous procedure.

6. Compute the time for one internal of the pressure wave
Using $2\ L/a = 2\ (40)/4147 = 0.012929$ s.

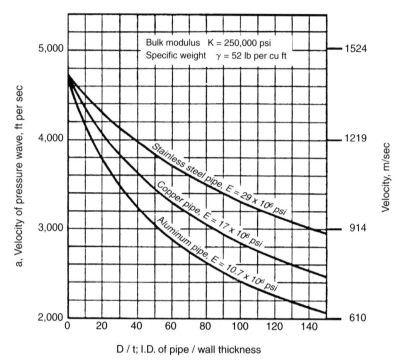

FIGURE 2   Velocity of pressure wave in water column in pipe of different diameter-to-wall thickness ratios. (*Product Engineering.*)

7. Find the pressure rise and maximum static pressure in the line

Since the time of 0.015 s for valve closure is less than the interval time $2\,L/a$ equal to 0.01929 s, the pressure rise can be computed from

$$\Delta p = \gamma a V / 144 g$$

for rapid valve closure. Therefore, the pressure rise when the flow medium is water is:

$$\Delta p = \frac{62.42 \times 4147 \times 8.3}{144 \times 32.2} = 463.4 \text{ lb/in}^2 \qquad (3192.8 \text{ kPa})$$

The maximum static pressure, $p_{max} = 500 + 463.4 = 963.4 \text{ lb/in}^2$ (abs) (6637.8 kPa).

Related Calculations.   This procedure is the work of Nils M. Sverdrup, as detailed in the previous procedure.

PRESSURE SURGE IN PIPING SYSTEM WITH COMPOUND PIPELINE

A compound pipeline consisting of several stainless-steel pipes of different diameters, Fig. 3, conveys 40 gal/min (2.5 L/s) of water. The length of each section of

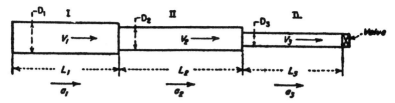

FIGURE 3 Compound pipeline consists of pipe sections having different diameters. (*Product Engineering.*)

pipe is: L_1 = 25 ft (7.6 m); L_2 = 15 ft (4.6 m); L_3 = 10 ft (3.0 m); pipe wall thickness in each section is 0.049 in (1.24 mm); inside diameter of each section of pipe is D_1 = 1.402 in (35.6 mm); D_2 = 1.152 in (29.3 mm); D_3 = 0.902 in (22.9 mm). What is the equivalent fluid velocity and the effective velocity of the pressure wave on sudden valve closure?

Calculation Procedure:

1. Determine fluid velocity and pressure-wave velocity in the first pipe
D_1/t_1 ratio of the first pipe = 1.402/0.049 = 28.6. Then, the fluid velocity in the pipe can be found from V_1 = $0.4085(G_n/(D_n)^2$, where the symbols are as shown below. Substituting, V_1 = $0.4085(40)/(1.402)^2$ = 8.31 ft/s (2.53 m/s).

Using these two computed values, enter Fig. 2 to find the velocity of the pressure wave in pipe 1 as 4147 ft/s (1264 m/s).

2. Find the fluid velocity and pressure-wave velocity in the second pipe
The D_2/t_2 ratio for the second pipe = 1.152/0.049 = 23.51. Using the same velocity equation as in step 1, above V_2 = $0.4085(40)/(1.152)^2$ = 12.31 ft/s (3.75 m/s).

Again, from Fig. 2, a_2 = 4234 ft/s (1290.5 m/s). Thus, there is an 87-ft/s (26.5-m/s) velocity increase of the pressure wave between pipes 1 and 2.

3. Compute the fluid velocity and pressure-wave velocity in the third pipe
Using a similar procedure to that in steps 1 and 2 above, V_3 = 20.1 ft/s (6.13 m/s); s_3 = 4326 ft/s (1318.6 m/s).

4. Find the equivalent fluid velocity and effective pressure-wave velocity for the compound pipe
Use the equation

$$V_E = \frac{L_1 V_1 + L_2 V_2 + \cdots + L_n V_n}{L_1 + L_2 + \cdots + L_n}$$

to find the equivalent fluid velocity in the compound pipe. Substituting,

$$V_E = \frac{25 \times 8.3 + 15 \times 12.3 + 10 \times 20.1}{25 + 15 + 10}$$

$$= 11.9 \text{ ft/s (3.63 m/s)}$$

To find the effective velocity of the pressure wave, use the equation

$$a_g = \frac{L_1 + L_2 + \cdots L_n}{(L_1/a_1) + (L_2/a_2) + \cdots; \text{pd} + (L_n/a_n)}$$

Substituting,

$$a_g = \frac{25 + 15 + 10}{(25/4147) + (15/4234) + (10/4326)}$$

$$= 4209 \text{ ft/s (1282.9 m/s)}$$

Thus, equivalent fluid velocity and effective velocity of the pressure wave in the compound pipe are both less than either velocity in the individual sections of the pipe.

Related Calculations.   Compound pipes find frequent application in industrial hydraulic systems. The procedure given here is useful in determining the velocities produced by sudden closure of a valve in the line.

$L_1, L_2, \ldots, L_n$ = length of each section of pipe of constant diameter, ft (m)
$a_1, a_2, \ldots, a_n$ = velocity of pressure wave in the respective pipe sections, ft/s (m/s)
a_g = effective velocity of the pressure wave, ft/s
$V_1, V_2, \ldots V_n$ = velocity of fluid in the respective pipe sections, ft/s (m/s)
V_E = equivalent fluid velocity, ft/s (m/s)
G_n = rate of flow in respective section, U.S. gal/min (L/s)
D_n = inside diameter of respective pipe, in (mm)

The fluid velocity in an individual pipe is

$$V_n = 0.4085G_n/D_n^2$$

This procedure is the work of Nils M. Sverdrup, as detailed earlier.

Pipe Properties, Flow Rate, and Pressure Drop

QUICK CALCULATION OF FLOW RATE AND PRESSURE DROP IN PIPING SYSTEMS

A 3-in (76-mm) Schedule 40S pipe has a 300-gal/min (18.9-L/s) water flow rate with a pressure loss of 8 lb/in² (55.1 kPa)/100 ft (30.5 m). What would be the flow rate in a 4-in (102-mm) Schedule 40S pipe with the same pressure loss? What would be the pressure loss in a 4-in (102-mm) Schedule 40S pipe with the same flow rate, 300 gal/min (18.9 L/s)? Determine the flow rate and pressure loss for a 6-in (152-mm) Schedule 40S pipe with the same pressure and flow conditions.

Calculation Procedure:

1. *Determine the flow rate in the new pipe sizes*

Flow rate in a pipe with a fixed pressure drop is proportional to the ratio of (new pipe inside diameter/known pipe inside diameter)$^{2.4}$. This ratio is defined as the *flow factor, F.* To use this ratio, the exact inside pipe diameters, known and new, must be used. Take the exact inside diameter from a table of pipe properties.

Thus, with a 3-in (76-mm) and a 4-in (102-mm) Schedule 40S pipe conveying water at a pressure drop of 8 lb/in^2 (55.1 kPa)/100 ft (30.5 m), the flow factor $F = (4.026/3.068)^{2.4} = 1.91975$. Then, the flow rate, FR, in the large 4-in (102-mm) pipe with the 8 lb/in^2 (55.1 kPa) pressure drop/100 ft (30.5 m), will be, FR $= 1.91975 \times 300 = 575.9$ gal/min (36.3 L/s).

For the 6-in (152-mm) pipe, the flow rate with the same pressure loss will be $(6.065/3.068)^{2.4} \times 300 = 1539.8$ gal/min (97.2 L/s).

2. *Compute the pressure drops in the new pipe sizes*

The pressure drop in a known pipe size can be extrapolated to a new pipe size by using a *pressure factor, P,* when the flow rate is held constant. For this condition, $P = $ (known inside diameter of the pipe/new inside diameter of the pipe)$^{4.8}$.

For the first situation given above, $P = (3.068/4.026)^{4.8} = 0.27134$. Then, the pressure drop, PD_N, in the new 4-in (102-mm) Schedule 40S pipe with a 300-gal/min (18.9-L/s) flow will be $PD_N + P(PD_K)$, where $PD_K = $ pressure drop in the known pipe size. Substituting, $PD_N = 0.27134(8) = 2.17$ lb/in^2/100 ft (14.9 kPa/30.5 m).

For the 6-in (152-mm) pipe, using the same approach, $PD_N = (3.068/6.065)^{4.8}$ (8) = 0.303 lb/in^2/100 ft (2.1 kPa/30.5 m).

Related Calculations. The flow and pressure factors are valuable timesavers in piping system design because they permit quick determination of new flow rates or pressure drops with minimum time input. When working with a series of pipe-size possibilities of the same Schedule Number, the designer can compute values for F and P in advance and apply them quickly. Here is an example of such a calculation for Schedule 40S piping of several sizes:

Nominal pipe size, new/known	Flow factor, F	Nominal pipe size, known/new	Pressure factor, P
2/1	5.092	1/2	0.0386
3/2	2.58	2/3	0.150
4/3	1.919	3/4	0.271
6/4	2.674	4/6	0.1399
8/6	1.933	6/8	0.267
10/8	1.726	8/10	0.335
12/10	1.542	10/12	0.421

When computing such a listing, the actual inside diameter of the pipe, taken from a table of pipe properties, must be used when calculating F or P.

The F and P values are useful when designing a variety of piping systems for chemical, petroleum, power, cogeneration, marine, buildings (office, commercial, residential, industrial), and other plants. Both the F and P values can be used for pipes conveying oil, water, chemicals, and other liquids. The F and P values are not applicable to steam or gases.

Note that the ratio of pipe diameters is valid for any units of measurement—inches, cm, mm—provided the same units are used consistently throughout the calculation. The results obtained using the F and P values usually agree closely with those obtained using exact flow or pressure-drop equations. Such accuracy is generally acceptable in everyday engineering calculations.

While the pressure drop in piping conveying a liquid is inversely proportional to the fifth power of the pipe diameter ratio, turbulent flow alters this to the value of 4.8, according to W. L. Nelson, Technical Editor, *The Oil and Gas Journal.*

FLUID HEAD-LOSS APPROXIMATIONS FOR ALL TYPES OF PIPING

Using the four rules for approximating head loss in pipes conveying fluid under turbulent flow conditions with a Reynolds number greater than 2100, find: (*a*) A 4-in (101.6-mm) pipe discharges 100 gal/min (6.3 L/s); how much fluid would a 2-in (50.8-mm) pipe discharge under the same conditions? (*b*) A 4-in (101.6-mm) pipe has 240 gal/min (15.1 L/s) flowing through it. What would be the friction loss in a 3-in (76.2-mm) pipe conveying the same flow? (*c*) A flow of 10 gal/min (6.3 L/s) produces 50 ft (15.2 m) of friction in a pipe. How much friction will a flow of 200 gal/min (12.6 L/s) produce? (*d*) A 12-in (304.8-mm) diameter pipe has a friction loss of 200 ft (60.9 m)/1000 ft (304.8 M). What is the capacity of this pipe?

Calculation Procedure:

1. Use the rule: At constant head, pipe capacity is proportional to $d^{2.5}$
(*a*) Applying the constant-head rule for both pipes: $4^{2.5} = 32.0$; $2^{2.5} = 5.66$. Then, the pipe capacity = (flow rate, gal/min or L/s)(new pipe size$^{2.5}$)/(previous pipe size$^{2.5}$) = $(100)(5.66)/32 = 17.69$ gal/min (1.11 L/s).

Thus, using this rule you can approximate pipe capacity for a variety of conditions where the head is constant. This approximation is valid for metal, plastic, wood, concrete, and other piping materials.

2. Use the rule: At constant capacity, head is proportional to $1/d^5$
(*b*) We have a 4-in (101.6-mm) pipe conveying 240 gal/min (15.1 L/s). If we reduce the pipe size to 3 in (76.2 mm) the friction will be greater because the flow area is smaller. The head loss = (flow rate, gal/min or L/s)(larger pipe diameter to the fifth power)/(smaller pipe diameter to the fifth power). Or, head = $(240)(4^5)/(3^5) = 1011$ ft/1000 ft of pipe (308.3 m/304.8 m of pipe).

Again, using this rule you can quickly and easily find the friction in a different size pipe when the capacity or flow rate remains constant. With the easy availability of handheld calculators in the field and computers in the design office, the fifth power of the diameter is easily found.

3. Use the rule: At constant diameter, head is proportional to gal/min $(L/s)^2$
(*c*) We know that a flow of 100 gal/min (6.3 L/s) produces 50-ft (15.2-m) friction, h, in a pipe. The friction, with a new flow will be, h = (friction, ft or m, at known

flow rate)(new flow rate, gal/min or L/s^2)/(previous flow rate, gal/min or L/s^2). Or, $h = (50)(200^2)/(100^2) = 200$ ft (60.9 m).

Knowing that friction will increase as we pump more fluid through a fixed-diameter pipe, this rule can give us a fast determination of the new friction. You can even do the square mentally and quickly determine the new friction in a matter of moments.

4. Use the rule: At constant diameter, capacity is proportional to friction, $h^{0.5}$
(d) Here the diameter is 12 in (304.8 mm) and friction is 200 ft (60.9 m)/1000 ft (304.8 m). From a pipe friction chart, the nearest friction head is 84 ft (25.6 m) for a flow rate of 5000 gal/min (315.5 L/s). The new capacity, c = (known capacity, gal/min or L/s)(known friction, ft or m$^{0.5}$)/(actual friction, ft or m$^{0.5}$). Or, $c = 5000(200^{0.5})/(84^{0.5}) = 7714$ gal/min (486.6 L/s).

As before, a simple calculation, the ratio of the square roots of the friction heads times the capacity will quickly give the new flow rates.

Related Calculations. Similar laws for fans and pumps give quick estimates of changed conditions. These laws are covered elsewhere in this handbook in the sections on fans and pumps. Referring to them now will give a quick comparison of the similarity of these sets of laws.

PIPE-WALL THICKNESS AND SCHEDULE NUMBER

Determine the minimum wall thickness t_m in (mm) and schedule number SN for a branch steam pipe operating at 900°F (482.2°C) if the internal steam pressure is 1000 lb/in^2 (abs) (6894 kPa). Use ANSA B31.1 *Code for Pressure Piping* and the ASME *Boiler and Pressure Vessel Code* valves and equations where they apply. Steam flow rate is 72,000 lb/h (32,400 kg/h).

Calculation Procedure:

1. Determine the required pipe diameter
When the length of pipe is not given or is as yet unknown, make a first approximation of the pipe diameter, using a suitable velocity for the fluid. Once the length of the pipe is known, the pressure loss can be determined. If the pressure loss exceeds a desirable value, the pipe diameter can be increased until the loss is within an acceptable range.

Compute the pipe cross-sectional area a in^2 (cm^2) from $a = 2.4Wv/V$, where W = steam flow rate, lb/h (kg/h); v = specific volume of the steam, ft^3/lb (m^3/kg); V = steam velocity, ft/min (m/min). The only unknown in this equation, other than the pipe area, is the steam velocity V. Use Table 1 to find a suitable steam velocity for this branch line.

Table 1 shows that the recommended steam velocities for branch steam pipes range from 6000 to 15,000 ft/min (1828 to 4572 m/min). Assume that a velocity of 12,000 ft/min (3657.6 m/min) is used in this branch steam line. Then, by using the steam table to find the specific volume of steam at 900°F (482.2°C) and 1000 lb/in^2 (abs) (6894 kPa), $a = 2.4(72,000)(0.7604)/12,000 = 10.98$ in^2 (70.8 cm^2). The inside diameter of the pipe is then $d = 2(a/\pi)^{0.5} = 2(10.98/\pi)^{0.5} = 3.74$ in (95.0 mm). Since pipe is not ordinarily made in this fractional internal diameter, round it to the next larger size, or 4-in (101.6-mm) inside diameter.

TABLE 1   Recommended Fluid Velocities in Piping

Service	Velocity of fluid	
	ft/min	m/s
Boiler and turbine leads	6,000–12,000	30.5–60.9
Steam headers	6,000–8,000	30.5–40.6
Branch steam lines	6,000–15,000	30.5–76.2
Feedwater lines	250–850	1.3–4.3
Exhaust and low-pressure steam lines	6,000–15,000	30.5–76.2
Pump suction lines	100–300	0.51–1.52
Bleed steam lines	4,000–6,000	20.3–30.5
Service water mains	120–300	0.61–1.52
Vacuum steam lines	20,000–40,000	101.6–203.2
Steam superheater tubes	2,000–5,000	10.2–25.4
Compressed-air lines	1,500–2,000	7.6–10.2
Natural-gas lines (large cross-country)	100–150	0.51–0.76
Economizer tubes (water)	150–300	0.76–1.52
Crude-oil lines [6 to 30 in (152.4 to 762.0 mm)]	50–350	0.25–1.78

2.   Determine the pipe schedule number
The ANSA *Code for Pressure Piping,* commonly called the *Piping Code,* defines schedule number as SN = 1000 P_i/S, where P_i = internal pipe pressure, lb/in² (gage); S = allowable stress in the pipe, lb/in², from *Piping Code.* Table 2 shows typical allowable stress values for pipe in power piping systems. For this pipe, assuming that seamless ferritic alloy steel (1% Cr, 0.55% Mo) pipe is used with the steam at 900°F (482°C), SN = (1000)(1014.7)/13,100 = 77.5. Since pipe is not ordinarily made in this schedule number, use the next *highest* readily available schedule number, or SN = 80. [Where large quantities of pipe are required, it is sometimes economically wise to order pipe of the exact SN required. This is not usually done for orders of less than 1000 ft (304.8 m) of pipe.]

3.   Determine the pipe-wall thickness
Enter a tabulation of pipe properties, such as in Crocker and King—*Piping Handbook,* and find the wall thickness for 4-in (101.6-mm) SN 80 pipe as 0.337 in (8.56 mm).

Related Calculations.   Use the method given here for any type of pipe—steam, water, oil, gas, or air—in any service—power, refinery, process, commercial, etc. Refer to the proper section of B31.1 *Code for Pressure Piping* when computing the schedule number, because the allowable stress S varies for different types of service.

The *Piping Code* contains an equation for determining the minimum required pipe-wall thickness based on the pipe internal pressure, outside diameter, allowable stress, a temperature coefficient, and an allowance for threading, mechanical strength, and corrosion. This equation is seldom used in routine piping-system design. Instead, the schedule number as given here is preferred by most designers.

PIPE-WALL THICKNESS DETERMINATION BY PIPING CODE FORMULA

Use the ANSA B31.1 *Code for Pressure Piping* wall-thickness equation to determine the required wall thickness for an 8.625-in (219.1-mm) OD ferritic steel plain-

8.13

TABLE 2 Allowable Stresses (S Values) for Alloy-Steel Pipe in Power Piping Systems*
(Abstracted from ASME Power Boiler Code and Code for Pressure Piping, ASA B31.1)

Material	ASTM specification	Grade or symbol	Minimum tensile strength		S values for metal temperatures not to exceed†					
			lb/in²	MPa	850°F	454°C	900°F	482°C	950°F	510°C
Seamless ferritic steels:										
Carbon-molybdenum	A335	P1	55,000	379.2	13,150	90.7	12,500	86.2	⋯	⋯
0.65 Cr, 0.55 Mo	A335	P2	55,000	379.2	13,150	90.7	12,500	86.2	10,000	68.9
1.00 Cr, 0.55 Mo	A335	P12	60,000	413.6	14,200	97.9	13,100	90.3	11,000	75.8

*Crocker and King—Piping Handbook.
†Where welded construction is used, consideration should be given to the possibility of graphite formation in carbon-molybdenum steel above 875°F (468°C) or in chromium-molybdenum steel containing less than 0.60 percent chromium above 975°F (523.9°C).

end pipe if the pipe is used in 900°F (482°C) 900-lb/in² (gage) (6205-kPa) steam service.

Calculation Procedure:

1. *Determine the constants for the thickness equation*
Pipe-wall thickness to meet ANSA *Code* requirements for power service is computed from $t_m = \{DP/[2(S + YP)]\} + C$, where t_m = minimum wall thickness, in; D = outside diameter of pipe, in; P = internal pressure in pipe, lb/in² (gage); S = allowable stress in pipe material, lb/in²; Y = temperature coefficient; C = end-condition factor, in.

Values of S, Y, and C are given in tables in the *Code for Pressure Piping* in the section on Power Piping. Using values from the latest edition of the *Code,* we get S = 12,500 lb/in² (86.2 MPa) for ferritic-steel pipe operating at 900°F (482°C); Y = 0.40 at the same temperature; C = 0.065 in (1.65 mm) for plain-end steel pipe.

2. *Compute the minimum wall thickness*
Substitute the given and *Code* values in the equation in step 1, or $t_m = [(8.625)(900)]/[2(12,500 + 0.4 \times 900)] + 0.065 = 0.367$ in (9.32 mm).

Since pipe mills do not fabricate to precise wall thicknesses, a tolerance above or below the computed wall thickness is required. An allowance must be made in specifying the wall thickness found with this equation by *increasing* the thickness by 12½ percent. Thus, for this pipe, wall thickness = 0.367 + 0.125(0.367) = 0.413 in (10.5 mm).

Refer to the *Code* to find the schedule number of the pipe. Schedule 60 8-in (203-mm) pipe has a wall thickness of 0.406 in (10.31 mm), and schedule 80 has a wall thickness of 0.500 in (12.7 mm). Since the required thickness of 0.413 in (10.5 mm) is greater than schedule 60 but less than schedule 80, the higher schedule number, 80, should be used.

3. *Check the selected schedule number*
From the previous calculation procedure, SN = 1000 P_i/S. From this pipe, SN = 1000(900)/12,500 = 72. Since piping is normally fabricated for schedule numbers 10, 20, 30, 40, 60, 80, 100, 120, 140, and 160, the next larger schedule number higher than 72, that is 80, will be used. This agrees with the schedule number found in step 2.

Related Calculations. Use this method in conjunction with the appropriate *Code* equation to determine the wall thickness of pipe conveying air, gas, steam, oil, water, alcohol, or any other similar fluids in any type of service. Be certain to use the correct equation, which in some cases is simpler than that used here. Thus, for lead pipe, $t_n = Pd/2S$, where P = safe working pressure of the pipe, lb/in² (gage); d = inside diameter of pipe, in; other symbols as before.

When a pipe will operate at a temperature between two tabulated *Code* values, find the allowable stress by interpolating between the tabulated temperature and stress values. Thus, for a pipe operating at 680°F (360°C), find the allowable stress at 650°F (343°C) [= 9500 lb/in² (65.5 MPa)] and 700°F (371°C) [= 9000 lb/in² (62.0 MPa)]. Interpolate thus: allowable stress at 680°F (360°C) = [(700°F − 680°F)/(700°F − 650°F)](9500 − 9000) + 9000 = 200 + 9000 = 9200 lb/in² (63.4 MPa). The same result can be obtained by interpolating downward from 9500 lb/in² (65.5 MPa), or allowable stress at 680°F (360°C) = 9500 − [(680 − 650)/ (700 − 650)](9500 − 9000) = 9200 lb/in² (63.4 MPa).

DETERMINING THE PRESSURE LOSS IN STEAM PIPING

Use a suitable pressure-loss chart to determine the pressure loss in 510 ft (155.5 m) of 4-in (101.6-mm) flanged steel pipe containing two 90° elbows and four 45° bends. The schedule 40 piping conveys 13,000 lb/h (5850 kg/h) of 20-lb/in² (gage) (275.8-kPa) 350°F (177°C) superheated steam. List other methods of determining the pressure loss in steam piping.

Calculation Procedure:

1. Determine the equivalent length of the piping
The equivalent length of a pipe L_e ft = length of straight pipe, ft + equivalent length of fittings, ft. Using data from the Hydraulic Institute, Crocker and King—*Piping Handbook*, earlier sections of this handbook, or Fig. 4, find the equivalent length of a 90° 4-in (101.6-mm) elbow as 10 ft (3 m) of straight pipe. Likewise, the equivalent length of a 45° bend is 5 ft (1.5 m) of straight pipe. Substituting in the above relation and using the straight lengths and the number of fittings of each type, we get L_e = 510 + (2)(10) + 4(5) = 550 ft (167.6 m) of straight pipe.

2. Compute the pressure loss, using a suitable chart
Figure 2 presents a typical pressure-loss chart for steam piping. Enter the chart at the top left at the superheated steam temperature of 350°F (177°C), and project vertically downward until the 40-lb/in² (gage) (275.8-kPa) superheated steam pressure curve is intersected. From here, project horizontally to the right until the outer border of the chart is intersected. Next, project through the steam flow rate, 13,000 lb/h (5900 kg/h) on scale *B*, Fig. 5, to the pivot scale *C*. From this point, project through 4-in (101.6-mm) schedule 40 pipe on scale *D*, Fig. 5. Extend this line to intersect the pressure-drop scale, and read the pressure loss as 7.25 lb/in² (50 kPa)/100 ft (30.4 m) of pipe.

   Since the equivalent length of this pipe is 550 ft (167.6 m), the total pressure loss in the pipe is (550/100)(7.25) = 39.875 lb/in² (274.9 kPa), say 40 lb/in² (275.8 kPa).

3. List the other methods of computing pressure loss
Numerous pressure-loss equations have been developed to compute the pressure drop in steam piping. Among the better known are those of Unwin, Fritzche, Spitzglass, Babcock, Guttermuth, and others. These equations are discussed in some detail in Crocker and King—*Piping Handbook* and in the engineering data published by valve and piping manufacturers.

   Most piping designers use a chart to determine the pressure loss in steam piping because a chart saves time and reduces the effort involved. Further, the accuracy obtained is sufficient for all usual design practice.

   Figure 3 is a popular flowchart for determining steam flow rate, pipe size, steam pressure, or steam velocity in a given pipe. Using this chart, the designer can

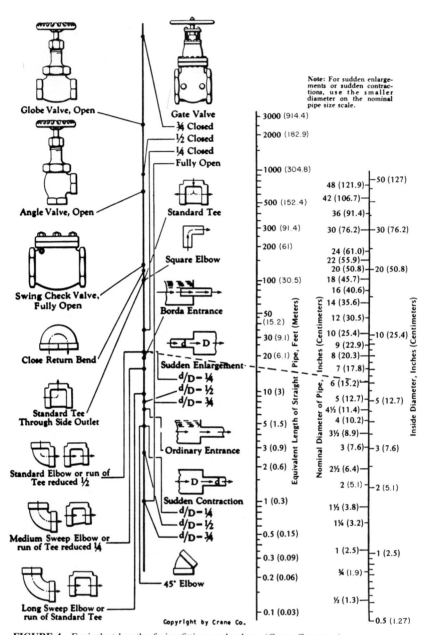

FIGURE 4   Equivalent length of pipe fittings and valves. (*Crane Company.*)

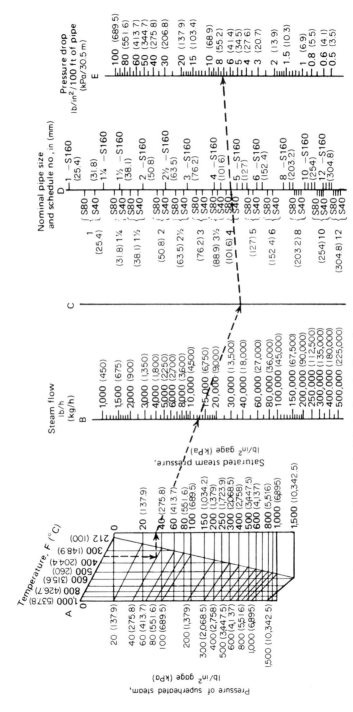

FIGURE 5 Pressure loss in steam pipes based on the Fritzche formula. (*Power.*)

8.17

determine any one of the four variables listed above when the other three are known. In solving a problem on the chart in Fig. 6, use the steam-quantity lines to intersect pipe sizes and the steam-pressure lines to intersect steam velocities. Here are two typical applications of this chart.

Example. What size schedule 40 pipe is needed to deliver 8000 lb/h (3600 kg/h) of 120-lb/in² (gage) (827.3-kPa) steam at a velocity of 5000 ft/min (1524 m/min)?

Solution. Enter Fig. 6 at the upper left at a velocity of 5000 ft/min (1524 m/min), and project along this velocity line until the 120-lb/in² (gage) (827.3-kPa) pressure line is intersected. From this intersection, project horizontally until the 8000-lb/h (3600-kg/h) vertical line is intersected. Read the *nearest* pipe size as 4 in (101.6 mm) on the *nearest* pipe-diameter curve.

Example. What is the steam velocity in a 6-in (152.4-mm) pipe delivering 20,000 lb/h (9000 kg/h) of steam at 85 lb/in² (gage) (586 kPa)?

Solution. Enter the bottom of the chart, Fig. 6, at the flow rate of 20,000 lb/h (9000 kg/h), and project vertically upward until the 6-in (152.4-mm) pipe curve is intersected. From this point, project horizontally to the 85-lb/in² (gage) (586-kPa) curve. At the intersection, read the velocity as 7350 ft/min (2240.3 m/min).

Table 3 shows typical steam velocities for various industrial and commercial applications. Use the given values as guides when sizing steam piping.

PIPING WARM-UP CONDENSATE LOAD

How much condensate is formed in 5 min during warm-up of 500 ft (152.4 m) of 6-in (152.4-mm) schedule 40 steel pipe conveying 215-lb/in² (abs) (1482.2-kPa) saturated steam if the pipe is insulated with 2 in (50.8 mm) of 85 percent magnesia and the minimum external temperature is 35°F (1.7°C)?

Calculation Procedure:

1. Compute the amount of condensate formed during pipe warm-up
For any pipe, the condensate formed during warm-up C_h lb/h $= 60(W_p)(\Delta t)(s)/h_{fg}N$, where W_p = total weight of pipe, lb; Δt = difference between final and initial temperature of the pipe, °F; s = specific heat of pipe material, Btu/(lb·°F); h_{fg} = enthalpy of vaporization of the steam, Btu/lb; N = warm-up time, min.

A table of pipe properties shows that this pipe weighs 18.974 lb/ft (28.1 kg/m). The steam table shows that the temperature of 215-lb/in² (abs) (1482.2-kPa) saturated steam is 387.89°F (197.7°C), say 388°F (197.8°C); the enthalpy h_{fg} = 837.4 Btu/lb (1947.8 kJ/kg). The specific heat of steel pipe s = 0.144 Btu/(lb·°F) [0.6 kJ/(kg·°C)]. Then C_h = 60(500 × 18.974)(388 − 35)(0.114)/[(837.4)(5)] = 5470 lb/h (2461.5 kg/h).

2. Compute the radiation-loss condensate load
Condensate is also formed by radiation of heat from the pipe during warm-up and while the pipe is operating. The warm-up condensate load decreases as the radiation load increases, the peak occurring midway (2½ min in this case) through the warm-up period. For this reason, one-half the normal radiation load is added to the warm-

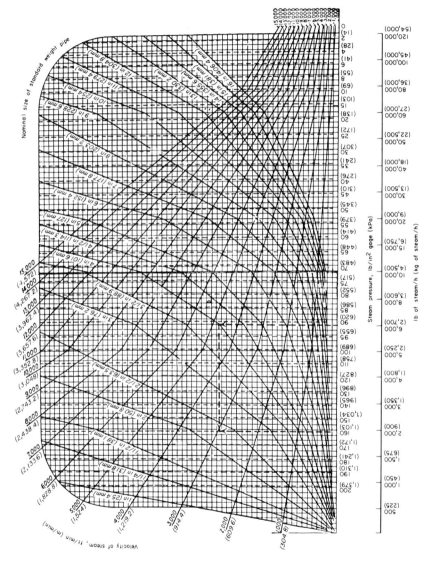

FIGURE 6 Spitzglass chart for saturated steam flowing in schedule 40 pipe.

TABLE 3 Steam Velocities Used in Pipe Design

Steam condition	Steam pressure		Steam use	Steam velocity	
	lb/in²	kPa		ft/min	m/min
Saturated	0–15	0–103.4	Heating	4,000–6,000	1,219.2–1,828.8
Saturated	50–150	344.7–1,034.1	Process	6,000–10,000	1,828.8–3,048.0
Superheated	200 and higher	1,378.8 and higher	Boiler leads	10,000–15,000	3,048.0–4,572.0

up load. Where the radiation load is small, it is often disregarded. However, the load must be computed before its magnitude can be determined.

For any pipe, $C_r = (L)(A)(\Delta t)(H)/h_{fg}$, where L = length of pipe, ft; A = external area of pipe, ft²/ft of length; H = heat loss through bare pipe or pipe insulation, Btu/(ft²·h·°F), from the piping or insulation tables. This 6-in (152.4-mm) schedule 40 pipe has an external area A = 1.73 ft²/ft (0.53 m²/m) of length. The heat loss through 2 in (50.8 mm) of 85 percent magnesia, from insulation tables, is H = 0.286　　　　Btu/(ft²·h·°F)　　　　[1.62　　　　W/(m²·°C)].　　　　Then C_r = (500) × (1.73)(388 − 35)(0.286)/837.4 = 104.2 lb/h (46.9 kg/h). Adding half the radiation load to the warm-up load gives 5470 + 52.1 = 5522.1 lb/h (2484.9 kg/h).

3. Apply a suitable safety factory to the condensate load
Trap manufacturers recommend a safety factor of 2 for traps installed between a boiler and the end of a steam main; traps at the end of a long steam main or ahead of pressure-regulating or shutoff valves usually have a safety factor of 3. With a safety factor of 3 for this pipe, the steam trap should have a capacity of at least 3(5522.1) = 16,566.3 lb/h (7454.8 kg/h), say 17,000 lb/h (7650.0 kg/h).

Related Calculations. Use this method to find the warm-up condensate load for any type of steam pipe—main or auxiliary—in power, process, heating, or vacuum service. The same method is applicable to other vapors that form condensate—Dowtherm, refinery vapors, process vapors, and others.

STEAM TRAP SELECTION FOR INDUSTRIAL APPLICATIONS

Select steam traps for the following four types of equipment: (1) the steam directly heats solid materials as in autoclaves, retorts, and sterilizers; (2) the steam indirectly heats a liquid through a metallic surface, as in heat exchangers and kettles, where the quantity of liquid heated is known and unknown; (3) the steam indirectly heats a solid through a metallic surface, as in dryers using cylinders or chambers and platen presses; and (4) the steam indirectly heats air through metallic surfaces, as in unit heaters, pipe coils, and radiators.

Calculation Procedure:

1. *Determine the condensate load*

The first step in selecting a steam trap for any type of equipment is determination of the condensate load. Use the following general procedure.

a. Solid materials in autoclaves, retorts, and sterilizers. How much condensate is formed when 2000 lb (900.0 kg) of solid material with a specific heat of 1.0 is processed in 15 min at 240°F (115.6°C) by 25-lb/in² (gage) (172.4-kPa) steam from an initial temperature of 60°F in an insulated steel retort?

For this type of equipment, use $C = WsP$, where C = condensate formed, lb/h; W = weight of material heated, lb; s = specific heat, Btu/(lb · °F); P = factor from Table 4. Thus, for this application, $C = (2000)(1.0)(0.193) = 386$ lb (173.7 kg) of condensate. Note that P is based on a temperature rise of $240 - 60 = 180$°F (100°C) and a steam pressure of 25 lb/in² (gage) (172.4 kPa). For the retort, using the specific heat of steel from Table 5, $C = (4000)(0.12)(0.193) = 92.6$ lb of condensate, say 93 lb (41.9 kg). The total weight of condensate formed in 15 min is $386 + 93 = 479$ lb (215.6 kg). In 1 h, 479(60/15) = 1916 lb (862.2 kg) of condensate is formed.

A safety factor must be applied to compensate for radiation and other losses. Typical safety factors used in selecting steam traps are as follows:

Steam mains and headers	2–3
Steam heating pipes	2–6
Purifiers and separators	2–3
Retorts for process	2–4
Unit heaters	3
Submerged pipe coils	2–4
Cylinder dryers	4–10

TABLE 4 Factors $P = (T - t)/L$ to Find Condensate Load

Pressure		Temperature		
lb/in² (abs)	kPa	160°F (71.1°C)	180°F (82.2°C)	200°F (93.3°C)
20	137.8	0.170	0.192	0.213
25	172.4	0.172	0.193	0.214
30	206.8	0.172	0.194	0.215

TABLE 5 Use These Specific Heats to Calculate Condensate Load

Solids	Btu/(lb · °F)	kJ/(kg · °C)	Liquids	Btu/(lb · °F)	kJ/(kg · °C)
Aluminum	0.23	0.96	Alcohol	0.65	2.7
Brass	0.10	0.42	Carbon tetrachloride	0.20	0.84
Copper	0.10	0.42	Gasoline	0.53	2.22
Glass	0.20	0.84	Glycerin	0.58	2.43
Iron	0.13	0.54	Kerosene	0.47	1.97
Steel	0.12	0.50	Oils	0.40–0.50	1.67–2.09

With a safety factor of 4 for this process retort, the trap capacity = (4)(1916) = 7664 lb/h (3449 kg/h), say 7700 lb/h (3465 kg/h).

b(1). *Submerged heating surface and a known quantity of liquid.* How much condensate forms in the jacket of a kettle when 500 gal (1892.5 L) of water is heated in 30 min from 72 to 212°F (22.2 to 100°C) with 50-lb/in² (gage) (344.7-kPa) steam?

For this type of equipment, $C = GwsP$, where G = gal of liquid heated; w = weight of liquid, lb/gal. Substitute the appropriate values as follows: C = (500)(8.33)(1.0) × (0.154) = 641 lb (288.5 kg), or (641)(60/3) = 1282 lb/h (621.9 kg/h). With a safety factor of 3, the trap capacity = (3)(1282) = 3846 lb/h (1731 kg/h), say 3900 lb/h (1755 kg/h).

b(2). *Submerged heating surface and an unknown quantity of liquid.* How much condensate is formed in a coil submerged in oil when the oil is heated as quickly as possible from 50 to 250°F (10 to 121°C) by 25-lb/in² (gage) (172.4-kPa) steam if the coil has an area of 50 ft² (4.66 m²) and the oil is free to circulate around the coal?

For this condition, $C = UAP$, where U = overall coefficient of heat transfer, Btu/(h · ft² · °F), from Table 6; A = area of heating surface, ft². With free convection and a condensing-vapor-to-liquid type of heat exchanger, U = 10 to 30. With an average value of $U = 20$, C = (20)(50)(0.214) = 214 lb/h (96.3 kg/h) of condensate. Choosing a safety factor 3 gives trap capacity = (3)(214) = 642 lb/h (289 kg/h), say 650 lb/h (292.5 kg/h).

b(3). *Submerged surfaces having more area than needed to heat a specified quantity of liquid in a given time with condensate withdrawn as rapidly as formed.* Use Table 7 instead of step *b*(1) or *b*(2). Find the condensate rate by multiplying the submerged area by the appropriate factor from Table 7. Use this method for heating water, chemical solutions, oils, and other liquids. Thus, with steam at 100 lb/in² (gage) (689.4 kPa) and a temperature of 338°F (170°C) and heating oil from 50 to 226°F (10 to 108°C) with a submerged surface having an area of 500 ft² (46.5 m²), the mean temperature difference (*Mtd*) = steam temperature minus the average liquid temperature = 338 − (50 + 226/2) = 200°F (93.3°C). The factor from Table 7 for 100 lb/in² (gage) (689.4 kPa) steam and a 200°F (93.3°C) *Mtd* is 56.75. Thus, the condensate rate = (56.75)(500) = 28,375 lb/h (12,769 kg/h). With a safety factor of 2, the trap capacity = (2)(28.375) = 56,750 lb/h (25,538 kg/h).

c. *Solids indirectly heated through a metallic surface.* How much condensate is formed in a chamber dryer when 1000 lb (454 kg) of cereal is dried to 750 lb (338 kg) by 10-lb/in² (gage) (68.9-kPa) steam? The initial temperature of the cereal is 60°F (15.6°C), and the final temperature equals that of the steam.

For this condition, $C = 970(W − D)/h_{fg} + WP$, where D = dry weight of the material, lb; h_{fg} = enthalpy of vaporization of the steam at the trap pressure, Btu/lb. From the steam tables and Table 4, C = 970(1000 − 750)/952 + (1000)(0.189) = 443.5 lb/h (199.6 kg/h) of condensate. With a safety factor of 4, the trap capacity = (4)(443.5) = 1774 lb/h (798.3 kg/h).

d. *Indirect heating of air through a metallic surface.* How much condensate is formed in a unit heater using 10-lb/in² (gage) (68.9-kPa) steam if the entering-air temperature is 30°F (−1.1°C) and the leaving-air temperature is 130°F (54.4°C)? Airflow is 10,000 ft³/min (281.1 m³/min).

Use Table 8, entering at a temperature difference of 100°F (37.8°C) and projecting to a steam pressure of 10 lb/in² (gage) (68.9 kPa). Read the condensate formed as 122 lb/h (54.9 kg/h) per 1000 ft³/min (28.3 m³/min). Since 10,000 ft³/min (283.1 m³/min) of air is being heated, the condensate rate = (10,000/

TABLE 6 Ordinary Ranges of Overall Coefficients of Heat Transfer

Type of heat exchanger	State of controlling resistance		Typical fluid	Typical apparatus
	Free convection, U	Forced convection, U		
Liquid to liquid	25–60 [141.9–340.7]	150–300 [851.7–1703.4]	Water	Liquid-to-liquid heat exchangers
Liquid to liquid	5–10 [28.4–56.8]	20–50 [113.6–283.9]	Oil	
Liquid to gas*	1–3 [5.7–17.0]	2–10 [11.4–56.8]	Water	Hot-water radiators
Liquid to boiling liquid	20–60 [113.6–340.7]	50–150 [283.9–851.7]	Oil	Brine coolers
Liquid to boiling liquid	5–20 [28.4–113.6]	25–60 [141.9–340.7]		
Gas* to liquid	1–3 [5.7–17.0]	2–10 [11.4–56.8]	—	Air coolers, economizers
Gas* to gas	0.6–2 [3.4–11.4]	2–6 [11.4–34.1]	—	Steam superheaters
Gas* to boiling liquid	1–3 [5.7–17.0]	2–10 [11.4–56.8]	—	Steam boilers
Condensing vapor to liquid	50–200 [283.9–1136]	150–800 [851.7–4542.4]	Steam to water	Liquid heaters and condensers
Condensing vapor to liquid	10–30 [56.8–170.3]	20–60 [113.6–340.7]	Steam to oil	
Condensing vapor to liquid	40–80 [227.1–454.2]	60–150 [340.7–851.7]	Organic vapor to water	
Condensing vapor to liquid	—	15–300 [85.2–1703.4]	Steam-gas mixture	
Condensing vapor to gas*	1–2 [5.7–11.4]	2–10 [11.4–56.8]	—	Steam pipes in air, air heaters
Condensing vapor to boiling liquid	40–100 [227.1–567.8]	—	—	Scale-forming evaporators
Condensing vapor to boiling liquid	300–800 [1703.4–4542.4]	—	Steam to water	
Condensing vapor to boiling liquid	50–150 [283.9–851.7]	—	Steam to oil	

*At atmospheric pressure.
Note: $U = $ Btu/(h·ft²·°F) [W/(m²·°C)]. Under many conditions, either higher or lower values may be realized.

TABLE 7   Condensate Formed in Submerged Steel* Heating elements, lb/(ft² · h) [kg/(m² · min)]

MTD†		Steam pressure				
°F	°C	75 lb/in² (abs) (517.1 kPa)	100 lb/in² (abs) (689.4 kPa)	150 lb/in² (abs) (1034.1 kPa)	Btu/(ft²·h)	kW/m²
175	97.2	44.3 (3.6)	45.4 (3.7)	46.7 (3.8)	40,000	126.2
200	111.1	54.8 (4.5)	56.8 (4.6)	58.3 (4.7)	50,000	157.7
250	138.9	90.0 (7.3)	93.1 (7.6)	95.7 (7.8)	82,000	258.6

*For copper, multiply table data by 2.0; for brass, by 1.6.
†Mean temperature difference, °F or °C, equals temperature of steam minus average liquid temperature. Heat-transfer data for calculating this table obtained from and used by permission of the American Radiator & Standard Sanitary Corp.

TABLE 8   Steam Condensed by Air, lb/h at 1000 ft³/min (kg/h at 28.3 m³/min)*

Temperature difference		Pressure		
°F	°C	5 lb/in² (gage) (34.5 kPa)	10 lb/in² (gage) (68.9 kPa)	50 lb/in² (gage) (344.7 kPa)
50	27.8	61 (27.5)	61 (27.5)	63 (28.4)
100	55.6	120 (54.0)	122 (54.9)	126 (56.7)
150	83.3	180 (81.0)	183 (82.4)	189 (85.1)

*Based on 0.0192 Btu (0.02 kJ) absorbed per ft³ (0.028 m³) of saturated air per °F (0.556°C) at 32°F (0°C). For 0°F (−17.8°C), multiply by 1.1.

1000)(122) = 1220 lb/h (549 kg/h). With a safety factor of 3, the trap capacity = (3)(1220) = 3660 lb/h (1647 kg/h), say 3700 lb/h (1665 kg/h).

Table 9 shows the condensate formed by radiation from bare iron and steel pipes in still air and with forced-air circulation. Thus, with a steam pressure of 100 lb/in² (gage) (689.4 kPa) and an initial air temperature of 75°F (23.9°C), 1.05 lb/h (0.47 kg/h) of condensate will be formed per ft² (0.09 m²) of heating surface in still air. With forced-air circulation, the condensate rate is (5)(1.05) = 5.25 lb/(h · ft²) [25.4 kg/(h · m²)] of heating surface.

Unit heaters have a *standard rating* based on 2-lb/in² (gage) (13.8-kPa) steam with entering air at 60°F (15.6°C). If the steam pressure or air temperature is different from these standard conditions, multiply the heater Btu/h capacity rating by the appropriate correction factor form, Table 10. Thus, a heater rated at 10,000 Btu/h (2931 W) with 2-lb/in² (gage) (13.8-kPa) steam and 60°F (15.6°C) air would have an output of (1.290)(10,000) = 12,900 Btu/h (3781 W) with 40°F (4.4°C) inlet air and 10-lb/in² (gage) (68.9-kPa) steam. Trap manufacturers usually list heater Btu ratings and recommend trap model numbers and sizes in their trap engineering data. This allows easier selection of the correct trap.

2. Select the trap size based on the load and steam pressure
Obtain a chart or tabulation of trap capacities published by the manufacturer whose trap will be used. Figure 7 is a capacity chart for one type of bucket trap manu-

TABLE 9 Condensate Formed by Radiation from Bare Iron and Steel, lb/(ft² · h) [kg/(m² · h)]

Air temperature		Steam pressure			
°F	°C	50 lb/in² (gage) (344.7 kPa)	75 lb/in² (gage) (517.1 kPa)	100 lb/in² (gage) (689.5 kPa)	150 lb/in² (gage) (1034 kPa)
65	18.3	0.82 (3.97)	1.00 (5.84)	1.08 (5.23)	1.32 (6.39)
70	21.2	0.80 (3.87)	0.98 (4.74)	1.06 (5.13)	1.21 (5.86)
75	23.9	0.77 (3.73)	0.88 (4.26)	1.05 (5.08)	1.19 (5.76)

*Based on still air; for forced-air circulation, multiply by 5.

TABLE 10 Unit-Heater Correction Factors

Steam pressure		Temperature of entering air		
lb/in² (gage)	kPa	20°F (−6.7°C)	40°F (4.4°C)	60°F (15.6°C)
5	34.5	1.370	1.206	1.050
10	68.9	1.460	1.290	1.131
15	103.4	1.525	1.335	1.194

Source: Yarway Corporation; SI values added by handbook editor.

factured by Armstrong Machine Works. Table 11 shows typical capacities of impulse traps manufactured by the Yarway Company.

To select a trap from Fig. 7, when the condensate rate is uniform and the pressure across the trap is constant, enter at the left at the condensation rate, say 8000 lb/h (3600 kg/h) (as obtained from step 1). Project horizontally to the right to the vertical ordinate representing the pressure across the trap [= Δp = steam-line pressure, lb/in² (gage) − return-line pressure with with trap valve closed, lb/in² (gage)]. Assume Δp = 20 lb/in² (gage) (138 kPa) for this trap. The intersection of the horizontal 8000-lb/h (3600-kg/h) projection and the vertical 20-lb/in² (gage) (137.9-kPa) projection is on the sawtooth capacity curve for a trap having a 9/16-in (14.3-mm) diameter orifice. If these projections intersected beneath this curve, a 9/16-in (14.3-mm) orifice would still be used if the point were between the verticals for this size orifice.

The dashed lines extending downward from the sawtooth curves show the capacity of a trap at reduced Δp. Thus, the capacity of a trap with a 3/8-in (9.53-mm) orifice at Δp = 30 lb/in² (gage) (207 kPa) is 6200 lb/h (2790 kg/h), read at the intersection of the 30-lb/in² (gage) (207-kPa) ordinate and the dashed curve extended from the 3/8-in (9.53-mm) solid curve.

To select an impulse trap from Table 11, enter the table at the trap inlet pressure, say 125 lb/in² (gage) (862 kPa), and project to the desired capacity, say 8000 lb/h (3600 kg/h), determined from step 1. Table 11 shows that a 2-in (50.8-mm) trap

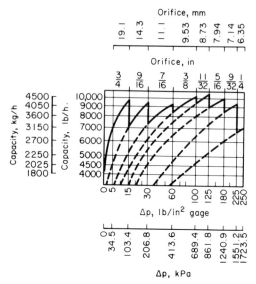

FIGURE 7   Capacities of one type of bucket steam trap.
(*Armstrong Machine Works.*)

TABLE 11   Capacities of Impulse Traps, lb/h (kg/h)
[*Maximum continuous discharge of condensate, based on
condensate at 30°F (16.7°C) below steam temperature.*]

Pressure at trap inlet		Trap nominal size	
lb/in² (gage)	kPa	1.25 in (38.1 mm)	2.0 in (50.8 mm)
125	861.8	6165 (2774)	8530 (3839)
150	1034.1	6630 (2984)	9075 (4084)
200	1378.8	7410 (3335)	9950 (4478)

Source: Yarway Corporation.

having an 8530-lb/h (3839-kg/h) capacity must be used because the next smallest size has a capacity of 5165 lb/h (2324 kg/h). This capacity is less than that required.

Some trap manufacturers publish capacity tables relating various trap models to specific types of equipment. Such tables simplify trap selection, but the condensate rate must still be computed as given here.

Related Calculations.   Use the procedure given here to determine the trap capacity required for any industrial, commercial, or domestic application including acid vats, air dryers, asphalt tanks, autoclaves, baths (dyeing), belt presses, bleach tanks, blenders, bottle washers, brewing kettles, cabinet dryers, calenders, can washers, candy kettles, chamber dryers, chambers (reaction), cheese kettles, coils (cooking, kettle, pipe, tank, tank-car), confectioners' kettles, continuous dryers, conveyor dyers, cookers (nonpressure and pressure), cooking coils, cooking kettles, cooking tanks, cooking vats, cylinder dryers, cylinders (jacketed), double-drum dryers, drum dryers, drums (dyeing), dry cans, dry kilns, dryers (cabinet, chamber, continuous, conveyor, cylinder, drum, festoon, jacketed, linoleum, milk, paper, pulp, rotary, shelf, stretch, sugar, tray, tunnel), drying rolls, drying rooms, drying tables, dye vats, dyeing baths and drums, dryers (package), embossing-press platens, evaporators, feed waterheaters, festoon dryers, fin-type heaters, fourdriniers, fuel-oil preheaters, greenhouse coils, heaters (steam), heat exchangers, heating coils and kettles, hot-break tanks, hot plates, kettle coils, kettles (brewing, candy, cheese, confectioners', cooking, heating, process), kiers, kilns (dry), liquid heaters, mains (steam), milk-bottle washers, milk-can washers, milk dryers, mixers, molding press platens, package dryers, paper dryers, percolators, phonograph-record press platens, pipe coils (still- and circulating-air), platens, plating tanks, plywood press platens, preheaters (fuel-oil), preheating tanks, press platens, pressure cookers, process kettles, pulp dryers, purifiers, reaction chambers, retorts, rotary dryers, steam mains (risers, separators), stocking boarders, storage-tank coils, storage water heaters, stretch dryers, sugar dryers, tank-car coils, tire-mold presses, tray dryers, tunnel dryers, unit heaters, vats, veneer press platens, vulcanizers, and water stills. Hospital equipment—such as autoclaves and sterilizers—can be analyzed in the same way, as can kitchen equipment—bain marie, compartment cooker, egg boiler, kettles, steam table, and urns; and laundry equipment—blanket dryers, curtain dryers, flatwork ironers, presses (dry-cleaning, laundry) sock forms, starch cookers, tumblers, etc.

When using a trap capacity diagram or table, be sure to determine the basis on which it was prepared. Apply any necessary correction factors. Thus, *cold-water capacity ratings* must be corrected for traps operating at higher condensate temperatures. Correction factors are published in trap engineering data. The capacity of a trap is greater at condensate temperatures less than 212°F (100°C) because at or above this temperature condensate forms flash steam when it flows into a pipe or vessel at atmospheric [14.7 lb/in² (abs) (101.3 kPa)] pressure. At altitudes above sea level, condensate flashes into steam at a lower temperature, depending on the altitude.

The method presented here is the work of L. C. Campbell, Yarway Corporation, as reported in *Chemical Engineering.*

SELECTING HEAT INSULATION FOR HIGH-TEMPERATURE PIPING

Select the heat insulation for a 300-ft (91.4-m) long 10-in (254-mm) turbine lead operating at 570°F (299°C) for 8000 h/year in a 70°F (21.1°C) turbine room. How much heat is saved per year by this insulation? The boiler supplying the turbine has an efficiency of 80 percent when burning fuel having a heating value of 14,000 Btu/lb (32.6 MJ/kg). Fuel costs $6 per ton ($5.44 per metric ton). How much money is saved by the insulation each year? What is the efficiency of the insulation?

Calculation Procedure:

1. Choose the type of insulation to use
Refer to an insulation manufacturer's engineering data or Crocker and King— *Piping Handbook* for recommendations about a suitable insulation for a pipe op-

erating in the 500 to 600°F (260 to 316°C) range. These references will show that calcium silicate is a popular insulation for this temperature range. Table 12 shows that a thickness of 3 in (76.2 mm) is usually recommended for 10-in (254-mm) pipe operating at 500 to 599°F (260 to 315°C).

2. Determine heat loss through the insulation
Refer to an insulation manufacturer's engineering data to find the heat loss through 3-in (76.2-mm) thick calcium silicate as 0.200 Btu/(h · ft² · °F) [1.14 W/(m² · °C)]. Since 10-in (254-mm) pipe has an area of 2.817 ft²/ft (0.86 m²/m) of length and since the temperature difference across the pipe is 570 − 70 = 500°F (260°C), the heat loss per hour = (0.200)(2.817)(50)= 281.7 Btu/(h · ft) (887.9 W/m²). The heat loss from bare 10-in (254-mm) pipe with a 500°F (260°C) temperature difference is, from an insulation manufacturer's engineering data, 4.640 Btu/(h · ft² · °F) [26.4 W/(m² · °C)], or (4.64)(2.817)(500) = 6510 Btu/(h · ft) (6.3 kW/m).

3. Determine annual heat saving
The heat saved = bare-pipe loss, Btu/h − insulated-pipe loss, Btu/h = 6510 − 281.7 = 6228.3 Btu/(h · ft) (5989 W/m) of pipe. Since the pipe is 300 ft (91.4 m) long and operates 8000 h per year, the annual heat saving = (300)(8000)(6228.3) = 14,940,000,000 Btu/year (547.4 kW).

4. Compute the money saved by the heat insulation
The heat saved in fuel as fired = (annual heat saving, Btu/year)/(boiler efficiency) = 14,940,000,000/0.80 = 18,680,000,000 Btu/year (5473 MW). Weight of fuel saved = (annual heat saving, Btu/year)/(heating value of fuel, Btu/lb)(2000 lb/ton) = 18,680,000,000/[(14,000)(2000)] = 667 tons (605 t). At $6 per ton ($5.44 per metric ton), the monetary saving is ($6)(667) = $4002 per year.

5. Determine the insulation efficiency
Insulation efficiency = (bare-pipe loss − insulated-pipe loss)/bare pipe loss, all expressed in Btu/h, or bare-pipe loss = (6510.0 − 281.7)/6510.0 = 0.957, or 95.7 percent.
    Related Calculations.   Use this method for any type of insulation—magnesia, fiber-glass, asbestos, felt, diatomaceous, mineral wool, etc.—used for piping at elevated temperatures conveying steam, water, oil, gas, or other fluids or vapors. To

TABLE 12 Recommended Insulation Thickness

Nominal pipe size		Pipe temperature			
		400–499°F	204–259°C	500–599°F	260–315°C
in	mm	in	mm	in	mm
6	152.4	2½*	63.5	2½	63.5
8	203.2	2½	63.5	3	76.2
10	254.0	2½	63.5	3	76.2
12	304.8	3	76.2	3	76.2
14 and over	355.6 and over	3	76.2	3½	88.9

*Available in single- or double-layer insulation.

coordinate and simplify calculations, become familiar with the insulation tables in a reliable engineering handbook or comprehensive insulation catalog. Such familiarity will simplify routine calculations.

ORIFICE METER SELECTION FOR A STEAM PIPE

Steam is metered with an orifice meter in a 10-in (254-mm) boiler lead having an internal diameter of $d_p = 9.760$ in (247.9 mm). Determine the maximum rate of steam flow that can be measured with a steel orifice plate having a diameter of $d_0 = 5.855$ in (148.7 mm) at 70°F (21.1°C). The upstream pressure tap is $1D$ ahead of the orifice, and the downstream tap is $0.5D$ past the orifice. Steam pressure at the orifice inlet $p_p = 250$ lb/in² (gage) (1724 kPa), temperature is 640°F (338°C). A differential gage fitted across the orifice has a maximum range of 120 in (304.8 cm) of water. What is the steam flow rate when the observed differential pressure is 40 in (101.6 cm) of water? Use the ASME Research Committee on Fluid Meters method in analyzing the meter. Atmospheric pressure is 14.696 lb/in² (abs) (101.3 kPa).

Calculation Procedure:

1. Determine the diameter ratio and steam density
For any orifice, meter, diameter ratio = β = meter orifice diameter, in/pipe internal diameter, in = 5.855/9.760 = 0.5999.

Determine the density of the steam by entering the superheated steam table at $250 + 14.696 = 264.696$ lb/in² (abs) (1824.8 kPa) and 640°F (338°C) and reading the specific volume as 2.387 ft³/lb (0.15 m³/kg). For steam, the density = 1/specific volume = d_s = 1/2.387 = 0.4193 lb/ft³ (6.7 kg/m³).

2. Determine the steam viscosity and meter flow coefficient
From the ASME publication, *Fluid Meters—Their Theory and Application,* the steam viscosity gu_1 for a steam system operating at 640°F (338°C) is $gu_1 = 0.0000141$ in · lb/(°F · s · ft²) [0.000031 N · m/(°C · s · m²)].

Find the flow coefficient K from the same ASME source by entering the 10-in (254-mm) nominal pipe diameter table at $\beta = 0.5999$ and projecting to the appropriate Reynolds number column. Assume that the Reynolds number = 10^7, approximately, for the flow conditions in this pipe. Then $K = 0.6486$. Since the Reynolds number for steam pressures above 100 lb/in² (689.4 kPa) ranges from 10^6 to 10^7, this assumption is safe because the value of K does not vary appreciably in this Reynolds number range. Also, the Reynolds number cannot be computed yet because the flow rate is unknown. Therefore, assumption of the Reynolds number is necessary. The assumption will be checked later.

3. Determine the expansion factor and the meter area factor
Since steam is a compressible fluid, the expansion factor Y_1 must be determined. For superheated steam, the ratio of the specific heat at constant pressure c_p to the specific heat at constant volume c_v is $k = c_p/c_v = 1.3$. Also, the ratio of the differential maximum pressure reading h_w, in of water, to the maximum pressure in the pipe, lb/in² (abs) = 120/246.7 = 0.454. From the expansion-factor curve in the ASME *Fluid Meters,* $Y_1 = 0.994$ for $\beta = 0.5999$ and the pressure ratio = 0.454.

And, from the same reference, the meter area factor $F_a = 1.0084$ for a steel meter operating at 640°F (338°C).

4. Compute the rate of steam flow
For square-edged orifices, the flow rate, lb/s $= w = 0.0997F_a Kd^2 Y_1 (h_w d_s)^{0.5} = (0.0997)(1.0084)(0.6486)(5.855)^2(0.994)(120 \times 0.4188)^{0.5} = 15.75$ lb/s (7.1 kg/s).

5. Compute the Reynolds number for the actual flow rate
For any steam pipe, the Reynolds number $R = 48w/(d_p g u_1) = 48(15.75)/[(3.1416)(0.760)(0.0000141)] = 1,750,000$.

6. Adjust the flow coefficient for the actual Reynolds number
In step 2, $R = 10^7$ was assumed and $K = 0.6486$. For $R = 1,750,000$, $K = 0.6489$, from ASME *Fluid Meters,* by interpolation. Then the actual flow rate $w_h =$ (computed flow rate)(ratio of flow coefficients based on assumed and actual Reynolds numbers) $= (15.75)(0.6489/0.6486)(3.600) = 56,700$ lb/h (25,515 kg/h), closely, where the value 3600 is a conversion factor for changing lb/s to lb/h.

7. Compute the flow rate for a specific differential gage deflection
For a 40-in (101.6-cm) H_2O deflection, F_a is unchanged and equals 1.0084. The expansion factor changes because $h_w/p_p = 40/264.7 = 0.151$. From the ASME *Fluid Meters,* $Y_1 = 0.998$. By assuming again that $R = 10^7$, $K = 0.6486$, as before, $w = (0.0997)(1.0084)(0.6486)(5.855)^2(0.998)(40 \times 0.4188)^{0.5} = 9.132$ lb/s (4.1 kg/s). Computing the Reynolds number as before, gives $R = (40)(0.132)/[(3.1416)(0.76)(0.0000141)] = 1,014,000$. The value of K corresponding to this value, as before, is from ASME—*Fluid Meters: K* = 0.6497. Therefore, the flow rate for a 40 in (101.6 cm) H_2O reading, in lb/h $= w_h = (0.132)(0.6497/0.6486)(3600) = 32,940$ lb/h (14,823 kg/h).

Related Calculations. Use these steps and the ASME *Fluid Meters* or comprehensive meter engineering tables giving similar data to select or check an orifice meter used in any type of steam pipe—main, auxiliary, process, industrial, marine, heating, or commercial, conveying wet, saturated, or superheated steam.

SELECTION OF A PRESSURE-REGULATING VALVE FOR STEAM SERVICE

Select a single-seat spring-loaded diaphragm-actuated pressure-reducing valve to deliver 350 lb/h (158 kg/h) of steam at 50 lb/in² (gage) (344.7 kPa) when the initial pressure is 225 lb/in² (gage) (1551 kPa). Also select an integral pilot-controlled piston-operated single-seat pressure-regulating valve to deliver 30,000 lb/h (13,500 kg/h) of steam at 40 lb/in² (gage) (275.8 kPa) with an initial pressure of 225 lb/in² (gage) (1551 kPa) saturated. What size pipe must be used on the downstream side of the valve to produce a velocity of 10,000 ft/min (3048 m/min)? How large should the pressure-regulating valve be if the steam entering the valve is at 225 lb/in² (gage) (1551 kPa) and 600°F (316°C)?

Calculation Procedure:

1. *Compute the maximum flow for the diaphragm-actuated valve*
For best results in service, pressure-reducing valves are selected so that they operate 60 to 70 percent open at normal load. To obtain a valve sized for this opening, divide the desired delivery, lb/h, by 0.7 to obtain the maximum flow expected. For this valve then, the maximum flow = 350/0.7 = 500 lb/h (225 kg/h).

2. *Select the diaphragm-actuated valve size*
Using a manufacturer's engineering data for an acceptable valve, enter the appropriate valve capacity table at the valve inlet steam pressure, 225 lb/in² (gage) (1551 kPa), and project to a capacity of 500 lb/h (225 kg/h), as in Table 13. Read the valve size as ¾ in (19.1 mm) at the top of the capacity column.

3. *Select the size of the pilot-controlled pressure-regulating valve*
Enter the capacity table in the engineering data of an acceptable pilot-controlled pressure-regulating valve, similar to Table 14, at the required capacity, 30,000 lb/h (13,500 kg/h). Project across until the correct inlet steam pressure column, 225 lb/in² (gage) (1551 kPa), is intercepted, and read the required valve size as 4 in (101.6 mm).

Note that it is not necessary to compute the maximum capacity before entering the table, as in step 1, for the pressure-reducing valve. Also note that a capacity table such as Table 14 can be used only for valves conveying saturated steam, unless the table notes state that the values listed are valid for other steam conditions.

TABLE 13 Pressure-Reducing-Valve Capacity, lb/h (kg/h)

Inlet pressure		Valve size		
lb/in² (gage)	kPa	½ in (12.7 mm)	¾ in (19.1 mm)	1 in (25.4 mm)
200	1379	420 (189)	460 (207)	560 (252)
225	1551	450 (203)	500 (225)	600 (270)
250	1724	485 (218)	560 (252)	650 (293)

Source: Clark-Reliance Corporation.

TABLE 14 Pressure-Regulating-Valve Capacity

Steam capacity		Initial steam pressure, saturated			
lb/h	kg/h	40 lb/in² (gage) (276 kPa)	175 lb/in² (gage) (1206 kPa)	225 lb/in² (gage) (1551 kPa)	300 lb/in² (gage) (2068 kPa)
20,000	9,000	6° (152.4)	4 (101.6)	4 (101.6)	3 (76.2)
30,000	13,500	8 (203.2)	5 (127.0)	4 (101.6)	4 (101.6)
40,000	18,000	—	5 (127.0)	5 (127.0)	4 (101.6)

° Valve diameter measured in inches (millimeters).
Source: Clark-Reliance Corporation.

4. Determine the size of the downstream pipe

Enter Table 14 at the required capacity, 30,000 lb/h (13,500 kg/h); project across to the valve *outlet pressure,* 40 lb/in² (gage) (275.8 kPa); and read the required pipe size as 8 in (203.2 mm) for a velocity of 10,000 ft/min (3048 m/min). Thus, the pipe immediately downstream from the valve must be enlarged from the valve size, 4 in (101.6 mm), to the required pipe size, 8 in (203.2 mm), to obtain the desired steam velocity.

5. Determine the size of the valve handling superheated steam

To determine the correct size of a pilot-controlled pressure-regulating valve handling superheated steam, a correction must be applied. Either a factor or a tabulation of corrected pressures, Table 15, may be used. to use Table 15, enter at the valve inlet pressure, 225 lb/in² (gage) (1551.2 kPa), and project across to the total temperature, 600°F (316°C), to read the corrected pressure, 165 lb/in² (gage) (1137.5 kPa). Enter Table 14 at the *next highest* saturated steam pressure, 175 lb/in² (gage) (1206.6 kPa) project down to the required capacity, 30,000 lb/h (13,500 kg/h); and read the required valve size as 5 in (127 mm).

Related Calculations.   To simplify pressure-reducing and pressure-regulating valve selection, become familiar with two or three acceptable valve manufacturers' engineering data. Use the procedures given in the engineering data or those given here to select valves for industrial, marine, utility, heating, process, laundry, kitchen, or hospital service with a saturated or superheated steam supply.

Do not oversize reducing or regulating valves. Oversizing causes chatter and excessive wear.

When an anticipated load on the downstream side will not develop for several months after installation of a valve, fit to the valve a reduced-area disk sized to handle the present load. When the load increases, install a full-size disk. Size the valve for the ultimate load, not the reduced load.

Where there is a wide variation in demand for steam at the reduced pressure, consider installing two regulators piped in parallel. Size the smaller regulator to handle light loads and the larger regulator to handle the difference between 60 percent of the light load and the maximum heavy load. Set the larger regulator to open when the minimum allowable reduced pressure is reached. Then both regulators will be open to handle the heavy load. Be certain to use the actual regulator inlet pressure and not the boiler pressure when sizing the valve if this is different

TABLE 15 Equivalent Saturated Steam Values for Superheated Steam at Various Pressures and Temperatures

Steam pressure		Steam temperature		Total temperature					
				500°F	600°F	700°F	260.0°C	315.6°C	371.1°C
lb/in² (gage)	kPa	°F	°C	Steam values, lb/in² (gage)			Steam values, kPa		
205	1413.3	389	198	171	149	133	1178.9	1027.2	916.9
225	1551.2	397	203	190	165	147	1309.9	1137.5	1013.4
265	1826.9	411	211	227	200	177	1564.9	1378.8	1220.2

Source: Clark-Reliance Corporation.

from the inlet pressure. Data in this calculation procedure are based on valves built by the Clark-Reliance Corporation, Cleveland, Ohio.

Some valve manufacturers use the valve flow coefficient C_v for valve sizing. This coefficient is defined as the flow rate, lb/h, through a valve of given size when the pressure loss across the valve is 1 lb/in^2 (6.89 kPa). Tabulations like Tables 13 and 14 incorporate this flow coefficient and are somewhat easier to use. These tables make the necessary allowances for downstream pressure less than the critical pressure (= 0.55 × absolute upstream pressure, lb/in^2, for superheated steam and hydrocarbon vapors; and 0.58 × absolute upstream pressure, lb/in^2, for saturated steam). The accuracy of these tabulations equals that of valve sizes determined by using the flow coefficient.

HYDRAULIC RADIUS AND LIQUID VELOCITY IN WATER PIPES

What is the velocity of 1000 gal/min (63.1 L/s) of water flowing through a 10-in (254-mm) inside-diameter cast-iron water main? What is the hydraulic radius of this pipe when it is full of water? When the water depth is 8 in (203.2 mm)?

Calculation Procedure:

1. Compute the water velocity in the pipe
For any pipe conveying water, the liquid velocity is v ft/s = gal/min/$(2.448d^2)$, where d = internal pipe diameter, in. For this pipe, v = 1000/[2.448(10)] = 4.08 ft/s (1.24 m/s), or (60)(4.08) = 244.8 ft/min (74.6 m/min).

2. Compute the hydraulic radius for a full pipe
For any pipe, the hydraulic radius is the ratio of the cross-sectional area of the pipe to the wetted perimeter, or $d/4$. For this pipe, when full of water, the hydraulic radius = 10/4 = 2.5.

3. Compute the hydraulic radius for a partially full pipe
Use the hydraulic radius tables in King and Brater—*Handbook of Hydraulics,* or compute the wetted perimeter by using the geometric properties of the pipe, as in step 2. From the King and Brater table, the hydraulic radius = Fd, where F = table factor for the ratio of the depth of water, in/diameter of channel, in = 8/10 = 0.8. For this ratio, F = 0.304. Then, hydraulic radius = (0.304)(10) = 3.04 in (77.2 mm).

Related Calculations.   Use this method to determine the water velocity and hydraulic radius in any pipe conveying cold water—water supply, plumbing, process, drain, or sewer.

FRICTION-HEAD LOSS IN WATER PIPING OF VARIOUS MATERIALS

Determine the friction-head loss in 2500 ft (762 m) of clean 10-in (254-mm) new tar-dipped cast-iron pipe when 2000 gal/min (126.2 L/s) of cold water is flowing.

What is the friction-head loss 20 years later? Use the Hazen-Williams and Manning formulas, and compare the results.

Calculation Procedure:

1. Compute the friction-head loss by the Hazen-Williams formula

The Hazen-Williams formula is $h_f = [v/(1.318CR_h^{0.63})]^{1.85}$, where h_f = friction-head loss per ft of pipe, ft of water; v = water velocity, ft/s; C = a constant depending on the condition and kind of pipe; R_h = hydraulic radius of pipe, ft.

For a water pipe, v = gal/min/$(2.44d^2)$; for this pipe, v = 2000/$[2.448(10)^2]$ = 8.18 ft/s (2.49 m/s). From Table 16 or Crocker and King—*Piping Handbook*, C for new pipe = 120; for 20-year-old pipe, C = 90; R_h = $d/4$ for a full-flow pipe = 10/4 = 2.5 in, or 2.5/12 = 0.208 ft (63.4 mm). Then $h_f = [8.18/(1.318 \times 120 \times 0.208^{0.63})]^{1.85}$ = 0.0263 ft (8.0 mm) of water per ft (m) of pipe. For 2500 ft (762 m) of pipe, the total friction-head loss = 2500(0.0263) = 65.9 ft (20.1 m) of water for the new pipe.

For 20-year-old pipe and the same formula, except with C = 90, h_f = 0.0451 ft (13.8 mm) of water per ft (m) of pipe. For 2500 ft (762 m) of pipe, the total friction-head loss = 2500(0.0451) = 112.9 ft (34.4 m) of water. Thus, the friction-head loss nearly doubles [from 65.9 to 112.9 ft (20.1 to 34.4 m)] in 20 years. This shows that it is wise to design for future friction losses; otherwise, pumping equipment may become overloaded.

2. Compute the friction-head loss from the Manning formula

The Manning formula is $h_f = n^2v^2/2.208R_h^{4/3}$, where n = a constant depending on the condition and kind of pipe, other symbols as before.

Using n = 0.011 for new coated cast-iron pipe from Table 17 or Crocker and King—*Piping Handbook*, we find $h_f = (0.011)^2(8.18)^2/[2.208(0.208)^{4/3}]$ = 0.0295 ft (8.9 mm) of water per ft (m) of pipe. For 2500 ft (762 m) of pipe, the total friction-head loss = 2500(0.0295) = 73.8 ft (22.5 m) of water, as compared with 65.9 ft (20.1 m) of water computed with the Hazen-Williams formula.

For coated cast-iron pipe in fair condition, n = 0.013, and h_f = 0.0411 ft (12.5 mm) of water. For 2500 ft (762 m) of pipe, the total friction-head loss = 2500(0.0411) = 102.8 ft (31.3 m) of water, as compared with 112.9 ft (34.4 m) of

TABLE 16   Values of C in Hazen-Williams Formula

Type of pipe	$C°$	Type of pipe	$C°$
Cement-asbestos	140	Cast iron or wrought iron	100
Asphalt-lined iron or steel	140	Welded or seamless steel	100
Copper or brass	130	Concrete	100
Lead, tin, or glass	130	Corrugated steel	60
Wood stave	110		

*Values of C commonly used for design. The value of C for pipes made of corrosive materials decreases as the age of the pipe increases; the values given are those that apply at an age of 15 to 20 years. For example, the value of C for cast-iron pipes 30 in (762 mm) in diameter or greater at various ages is approximately as follows: new, 130; 5 years old, 120; 10 years old, 115; 20 years old, 100; 30 years old, 90; 40 years old, 80; and 50 years old, 75. The value of C for smaller-size pipes decreases at a more rapid rate.

TABLE 17   Roughness Coefficients (Manning's n) for Closed Conduits

	Manning's n	
Type of conduit	Good construction[a]	Fair construction[a]
Concrete pipe	0.013	0.015
Corrugated metal pipe or pipe arch, 2⅔ × ½ in (67.8 × 12.7 mm) corrugation, riveted:		
Plain	0.024	

Paved invert:				
Percent of circumference paved	25	50		
Depth of flow:				
Full	0.021	0.018		
0.8D	0.021	0.016		
0.6D	0.019	0.013		

Type of conduit	Good construction[a]	Fair construction[a]
Vitrified clay pipe	0.012	0.014
Cast-iron pipe, uncoated	0.013	
Steel pipe	0.011	
Brick	0.014	0.017
Monolithic concrete:		
Wood forms, rough	0.015	0.017
Wood forms, smooth	0.012	0.014
Steel forms	0.012	0.013
Cemented-rubble masonry walls:		
Concrete floor and top	0.017	0.022
Natural floor	0.019	0.025
Laminated treated wood	0.015	0.017
Vitrified-clay liner plates	0.015	

[a] For poor-quality construction, use larger values of n.

water computed with the Hazen-Williams formula. Thus, the Manning formula gives results higher than the Hazen-Williams in one case and lower in another. However, the differences in each case are not excessive; $(73.8 - 65.9)/65.9 = 0.12$, or 12 percent higher, and $(112.9 - 102.8)/102.8 = 0.0983$, or 9.83 percent lower. Both these differences are within the normal range of accuracy expected in pipe friction-head calculations.

Related Calculations.   The Hazen-Williams and Manning formulas are popular with many piping designers for computing pressure losses in cold-water piping. To simplify calculations, most designers use the precomputed tabulated solutions available in Crocker and King—*Piping Handbook,* King and Brater—*Handbook of Hydraulics,* and similar publications. In the rush of daily work these precomputed solutions are also preferred over the more complex Darcy-Weisbach equation used in conjunction with the friction factor f, the Reynolds number R, and the roughness-diameter ratio.

Use the method given here for sewer lines, water-supply pipes for commercial, industrial, or process plants, and all similar applications where cold water at temperatures of 33 to 90°F (0.6 to 32.2°C) flows through a pipe made of cast iron, riveted steel, welded steel, galvanized iron, brass, glass, wood-stove, concrete, vitrified, common clay, corrugated metal, unlined rock, or enameled steel. Thus, either of these formulas, used in conjunction with a suitable constant, gives the friction-head loss for a variety of piping materials. Suitable constants are given in Tables 16 and 17 and in the above references. For the Hazen-Williams formula, the constant C varies from about 70 to 140, while n in the Manning formula varies from about 0.017 for $C = 70$ to 0.010 for $C = 140$. Values obtained with these formulas have been used for years with satisfactory results. At present, the Manning formula appears the more popular.

CHART AND TABULAR DETERMINATION OF FRICTION HEAD

Figure 8 shows a process piping system supplying 1000 gal/min (63.1 L/s) of 70°F (21.1°C) water. Determine the total friction head, using published charts and pipe-friction tables. All the valves and fittings are flanged, and the piping is 10-in (254-mm) steel, schedule 40.

Calculation Procedure:

1. Determine the total length of the piping
Mark the length of each piping run on the drawing after scaling it or measuring it in the field. Determine the total length by adding the individual lengths, starting at

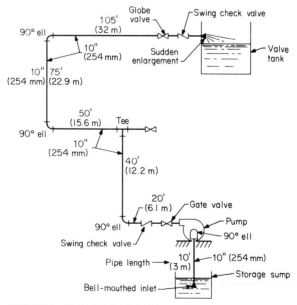

FIGURE 8 Typical industrial piping system.

the supply source of the liquid. In Fig. 8, beginning at the storage sump, the total length of piping $= 10 + 20 + 40 + 50 + 75 + 105 = 300$ ft (91.4 m). Note that the physical length of the fittings is included in the length of each run.

2. Compute the equivalent length of each fitting

The frictional resistance of pipe fittings (elbows, tees, etc.) and valves is greater than the actual length of each fitting. Therefore, the equivalent length of straight piping having a resistance equal to that of the fittings must be determined. This is done by finding the equivalent length of each fitting and taking the sum for all the fittings.

Use the equivalent length table in the pump section of this handbook or in Crocker and King—*Piping Handbook,* Baumeister and Marks—*Standard Handbook for Mechanical Engineers,* or *Standards of the Hydraulic Institute.* Equivalent length values will vary slightly from one reference to another.

Starting at the supply source, as in step 1, for 10-in (254-mm) flanged fittings throughout, we see the equivalent fitting lengths are: bell-mouth inlet, 2.9 ft (0.88 m); 90° ell at pump, 14 ft (4.3 m); gate valve, 3.2 ft (0.98 m); swing check valve, 120 ft (36.6 m); 90° ell, 14 ft (4.3 m); tee, 30 ft (9.1 m); 90° ell, 14 ft (4.3 m); 90° ell, 14 ft (4.3 m); globe valve, 310 ft (94.5 m); swing check valve, 120 ft (36.6 m); sudden enlargement = (liquid velocity, ft/s)^2/2g = $(4.07)^2$/2(32.2) = 0.257 ft (0.08 m), where the terminal velocity is zero, as in the tank. Find the liquid velocity as shown in a previous calculation procedure in this section. The sum of the fitting equivalent lengths is $2.9 + 14 + 3.2 + 120 + 14 + 30 + 14 + 14 + 310 + 120 + 0.257 = 642.4$ ft (159.8 m). Adding this to the straight length gives a total length of $642.4 + 300 = 942.4$ ft (287.3 m).

3. Compute the friction-head loss by using a chart

Figure 9 is a popular friction-loss chart for fairly rough pipe, which is any ordinary pipe after a few years' use. Enter at the left at a flow of 1000 gal/min (63.1 L/s), and project to the right until the 10-in (254-mm) diameter curve is intersected. Read the friction-head loss at the top or bottom of the chart as 0.4 lb/in^2 (2.8 kPa), closely, per 100 ft (30.5 m) of pipe. Therefore, total friction-head loss $= (0.4)(942.4/100) = 3.77$ lb/in^2 (26 kPa). Converting gives $(3.77)(2.31) = 8.71$ ft (2.7 m) of water.

4. Compute the friction-head loss from tabulated data

Using the *Standards of the Hydraulic Institute* pipe-friction table, we find that the friction head h_f of water per 100 ft (30.5 m) of pipe $= 0.500$ ft (0.15 m). Hence, the total friction head $= (0.500)(942.4/100) = 4.71$ ft (1.4 m) of water. The Institute recommends that 15 percent be added to the tabulated friction head, or $(1.15)(4.71) = 5.42$ ft (1.66 m) of water.

Using the friction-head tables in Crocker and King—*Piping Handbook,* the friction head $= 6.27$ ft (1.9 m) per 1000 ft (304.8 m) of pipe with $C = 130$ for new, very smooth pipe. For this piping system, the friction-head loss $= (942.4/1000)(6.27) = 5.91$ ft (1.8 m) of water.

5. Use the Reynolds number method to determine the friction head

In this method, the friction factor is determined by using the Reynolds number R and the relative roughness of the pipe ε/D, where ε = pipe roughness, ft, and D = pipe diameter, ft.

For any pipe, $R = Dv/v$, where v = liquid velocity, ft/s, and v = kinematic viscosity, ft^2/s. Using King and Brater—*Handbook of Hydraulics,* $v = 4.07$ ft/s

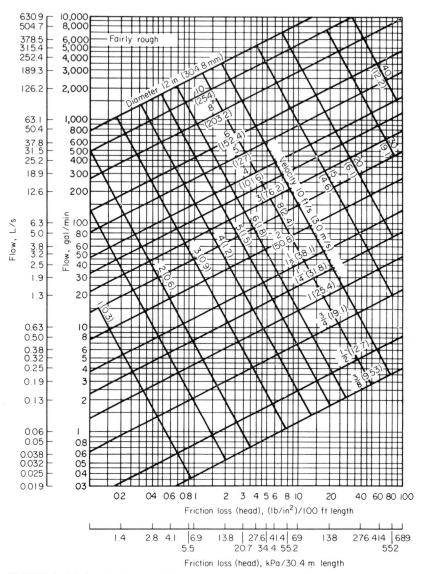

FIGURE 9 Friction loss in water piping.

(1.24 m/s), and $v = 0.00001059$ ft^2/s (0.00000098 m^2/s) for water at 70°F (21.1°C). Then $R = (10/12)(4.07)/0.00001059 = 320.500$.

From Table 18 or the above reference, $\varepsilon = 0.00015$, and $\varepsilon/D = 0.00015/(10/12) = 0.00018$. From the Reynolds-number, relative-roughness, friction-factor curve

TABLE 18 Abslute Roughness Classification of Pipe Surfaces for Selection of Friction Factor f in Fig. 10.

Commercial pipe surface (new)	Absolute roughness ϵ		Commercial pipe surface (new)	Absolute roughness ϵ	
	ft	mm		ft	mm
Glass, drawn brass, copper, lead	Smooth	Smooth	Cast iron	0.00085	0.26
Wrought iron, steel	0.00015	0.05	Wood stave	0.0006–0.003	0.18–0.91
Asphalted cast iron	0.0004	0.12	Concrete	0.001–0.01	0.30–3.05
Galvanized iron	0.0005	0.15	Riveted steel	0.003–0.03	0.91–9.14

in Fig. 10 or in Baumeister—*Standard Handbook for Mechanical Engineers,* the friction factor $f = 0.016$.

Apply the Darcy-Weisbach equation $h_f = f(l/D)(v^2/2g)$, where l = total pipe length, including the fittings' equivalent length, ft. Then $h_f = (0.016)(942.4/10/12)(4.07)^2/(2 \times 32.2) = 4.651$ ft (1.43 m) of water.

6. Compare the results obtained
Three different friction-head values were obtained: 8.71, 5.91, and 4.651 ft (2.7, 1.8, and 1.4 m) of water. The results show the variations that can be expected with the different methods. Actually, the Reynolds number method is probably the most accurate. As can be seen, the other two methods give safe results—i.e., the computed friction head is higher. The *Pipe Friction Manual,* published by the Hydraulic Institute, presents excellent simplified charts for use with the Reynolds number method.

Related Calculations. Use any of these methods to compute the friction-head loss for any type of pipe. The Reynolds number method is useful for a variety of liquids other than water—mercury, gasoline, brine, kerosene, crude oil, fuel oil, and lube oil. It can also be used for saturated and superheated steam, air, methane, and hydrogen.

RELATIVE CARRYING CAPACITY OF PIPES

What is the equivalent steam-carrying capacity of a 24-in (609.6-mm) inside-diameter pipe in terms of a 10-in (254-mm) inside-diameter pipe? What is the equivalent water-carrying capacity of a 23-in (584.2-mm) inside-diameter pipe in terms of a 13.25-in (336.6-mm) inside-diameter pipe?

Calculation Procedure:

1. Compute the relative carrying capacity of the steam pipes
For steam, air, or gas pipes, the number N of small pipes of inside diameter d_2 in equal to one pipe of larger inside diameter d_1 in is $N = (d_1^3\sqrt{d_2 + 3.6})/(d_2^3 + \sqrt{d_1} + 3.6)$. For this piping system, $N = (24^3 + \sqrt{10} + 3.6)/(10^3 + \sqrt{24} + 3.6) = 9.69$, say 9.7. Thus, a 24-in (609.6-mm) inside-diameter

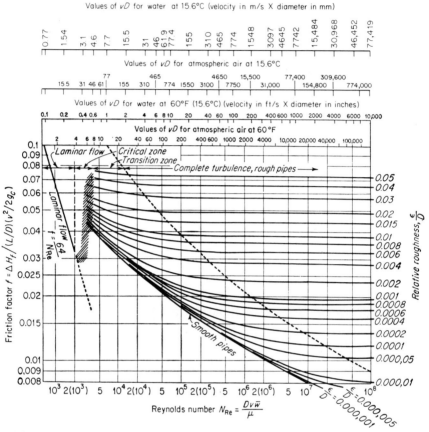

FIGURE 10   Friction factors for laminar and turbulent flow.

steam pipe has a carrying capacity equivalent to 9.7 pipes having a 10-in (254-mm) inside diameter.

2.   Compute the relative carrying capacity of the water pipes
For water, $N = (d_2/d_1)^{2.5} = (23/13.25)^{2.5} = 3.97$. Thus, one 23-in (584-cm) inside-diameter pipe can carry as much water as 3.97 pipes of 13.25-in (336.6-mm) inside diameter.

   Related Calculations.   Crocker and King—*Piping Handbook* and certain piping catalogs (Crane, Walworth, National Valve and Manufacturing Company) contain tabulations of relative carrying capacities of pipes of various sizes. Most piping designers use these tables. However, the equations given here are useful for ranges not covered by the tables and when the tables are unavailable.

PRESSURE-REDUCING VALVE SELECTION FOR
WATER PIPING

What size pressure-reducing valve should be used to deliver 1200 gal/h (1.26 L/s) of water at 40 lb/in² (275.8 kPa) if the inlet pressure is 140 lb/in² (965.2 kPa)?

Calculation Procedure:

1. Determine the valve capacity required
Pressure-reducing valves in water systems operate best when the nominal load is 60 to 70 percent of the maximum load. Using 60 percent, we see that the maximum load for this valve = 1200/0.6 = 2000 gal/h (2.1 L/s).

2. Determine the valve size required
Enter a valve capacity table in suitable valve engineering data at the valve inlet pressure, and project to the exact, or next higher, valve capacity. Thus, enter Table 19 at 140 lb/in² (965.2 kPa) and project to the next higher capacity, 2200 gal/h (2.3 L/s), since a capacity of 2000 gal/h (2.1 L/s) is not tabulated. Read at the top of the column the required valve size as 1 in (25.4 mm).

Some valve manufacturers present the capacity of their valves in graphical instead of tabular form. One popular chart, Fig. 11, is entered at the difference between the inlet and outlet pressures on the abscissa, or 140 − 40 = 100 lb/in² (689.4 kPa). Project vertically to the flow rate of 2000/60 = 33.3 gal/min (2.1 L/s). Read the valve size on the intersecting valve capacity curve, or on the next curve if there is no intersection with the curve. Figure 11 shows that a 1-in (25.4-mm) valve should be used. This agrees with the tabulated capacity.

Related Calculations. Use this method for pressure-reducing valves in any type of water piping—process, domestic, commercial—where the water temperature is 100°F (37.8°C) or less. Table 19 is from data prepared by the Clark-Reliance Corporation, Fig. 11 is from Foster Engineering Company data.

Some valve manufacturers use the valve flow coefficient C_v for valve sizing. This coefficient is defined as the flow rate, gal/min, through a valve of given size when the pressure loss across the valve is 1 lb/in² (6.9 kPa). Tabulations like Table 19 and flowcharts like Fig. 11 incorporate this flow coefficient and are somewhat easier to use. Their accuracy equals that of the flow coefficient method.

TABLE 19 Maximum Capacities of Water Pressure-Reducing Valves, gal/h (L/s)

Inlet pressure		Valve size		
lb/in² (gage)	kPa	¾ in (19.1 mm)	1 in (25.4 mm)	1¼ in (31.8 mm)
120	827.3	1550 (1.6)	2000 (2.1)	4500 (4.7)
140	965.2	1700 (1.8)	2200 (2.3)	5000 (5.3)
160	1103.0	1850 (1.9)	2400 (2.5)	5500 (5.8)

Source: Clark-Reliance Corporation.

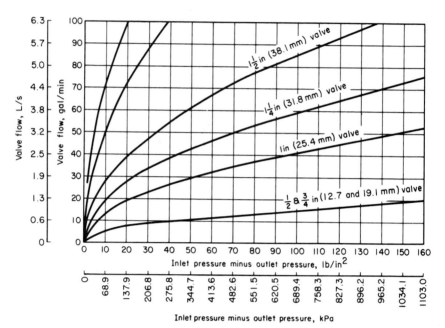

FIGURE 11 Pressure-reducing valve flow capacity. (*Foster Engineering Company.*)

SIZING A WATER METER

A 6 × 4 in (152.4 × 101.6 mm) Venturi tube is used to measure water flow rate in a piping system. The dimensions of the meter are: inside pipe diameter d_p = 6.094 in (154.8 mm); throat diameter d = 4.023 in (102.2 mm). The differential pressure is measured with a mercury manometer having water on top of the mercury. The average manometer reading for 1 h is 10.1 in (256.5 mm) of mercury. The temperature of the water in the pipe is 41°F (5.0°C), and that of the room is 77°F (25°C). Determine the water flow rate in lb/h, gal/h, and gal/min. Use the ASME Research Committee on Fluid Meters method in analyzing the meter.

Calculation Procedure:

1. Convert the pressure reading to standard conditions
The ASME meter equation constant is based on a manometer liquid temperature of 68°F (20.0°C). Therefore, the water and mercury density at room temperature, 77°F (25°C), and the water density at 68°F (20.0°C), must be used to convert the manometer reading to standard conditions by the equation $h_w = h_m(m_d - w_d)/w_s$, where h_w = equivalent manometer reading, in (mm) H_2O at 68°F (20.0°C); h_m = manometer reading at room temperature, in mercury; m_d = mercury density at room temperature, lb/ft³; w_d = water density at room temperature, lb/ft³; w_s = water density at standard conditions, 68°F (20.0°C), lb/ft³. From density values from the ASME publication *Fluid Meters: Their Theory and Application*, h_w = 10.1(844.88 − 62.244)/62.316 = 126.8 in (322.1 cm) of water at 68°F (20.0°C).

2. Determine the throat-to-pipe diameter ratio
The throat-to-pipe diameter ratio $\beta = 4.023/6.094 = 0.6602$. Then $1/(1 - \beta^4)^{0.5}$ and $1/(1 - 0.6602^4)^{0.5} = 1.1111$.

3. Assume a Reynolds number value, and compute the flow rate
The flow equation for a Venturi tube is w lb/h $= 359.0(Cd^2/\sqrt{1 - \beta^4})(w_{dp}h_w)^{0.5}$, where $C =$ meter discharge coefficient, expressed as a function of the Reynolds number; $w_{dp} =$ density of the water at the pipe temperature, lb/ft³. With a Reynolds number greater than 250,000, C is a constant. As a first trial, assume $R > 250,000$ and $C = 0.984$ from *Fluid Meters*. Then $w = 359.0(0.984)(4.023)^2(1.1111)$ $(62.426 \times 126.8)^{0.5} = 565,020$ lb/h (254,259 kg/h), or $565,020/8.33$ lb/gal $= 67,800$ gal/h (71.3 L/s), or $67,800/60$ min/h $= 1129$ gal/min (71.23 L/s).

4. Check the discharge coefficient by computing the Reynolds numbers
For a water pipe, $R = 48w_s/(\pi d_p gu)$, where $w_s =$ flow rate, lb/s $= w/3600$; $u =$ coefficient of absolute viscosity. Using *Fluid Meters* data for water at 41°F (5°C), we find $R = 48(156.95)/[(\pi \times 6.094)(0.001004)] = 391,900$. Since C is constant for $R > 250,000$, use of $C = 0.984$ is correct, and no adjustment in the computations is necessary. Had the value of C been incorrect, another value would be chosen and the Reynolds number recomputed. Continue this procedure until a satisfactory value for C is obtained.

5. Use an alternative solution to check the results
Fluid Meters gives another equation for Venturi meter flow rate, that is w lb/s $= 0.525(Cd^2/\sqrt{1 - \beta^4})[w_{dp}(p_1 - p_2)]^{0.5}$, where $p_1 - p_2$ is the manometer differential pressure in lb/in². Using the conversion factor in *Fluid Meters* for converting in of mercury under water at 77°F (25°C) to lb/in² (kPa), we get $p_1 - p_2 = (10.1)(0.4528) = 4.573$ lb/in² (31.5 kPa). Then $w = (0.525)(0.984)$ $(4.023)^2(1.1111)(62.426 \times 4.573)^{0.5} = 156.9$ lb/s (70.6 kg/s), or $(156.9)(3600$ s/ h) $= 564,900$ lb/h (254,205 kg/h), or $564,900/8.33$ lb/gal $= 67,800$ gal/h (71.3 L/s), or $67,800/60$ min/h $= 1129$ gal/min (71.2 L/s). This result agrees with that computed in step 3 within 1 part in 5600. This is much less than the probable uncertainties in the values of the discharge coefficient and the differential pressure.

Related Calculations. Use this method for any Venturi tube serving cold-water piping in process, industrial, water-supply, domestic, or commercial service.

EQUIVALENT LENGTH OF A COMPLEX SERIES PIPELINE

Figure 12 shows a complex series pipeline made up of four lengths of different size pipe. Determine the equivalent length of this pipe if each size of pipe has the same friction factor.

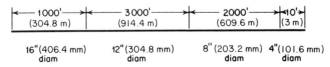

FIGURE 12 Complex series pipeline.

Calculation Procedure:

1. *Select the pipe size for expressing the equivalent length*
The usual procedure in analyzing complex pipelines is to express the equivalent length in terms of the smallest, or next to smallest, diameter pipe. Choose the 8-in (203.2-mm) size as being suitable for expressing the equivalent length.

2. *Find the equivalent length of each pipe*
For any complex series pipeline having equal friction factors in all the pipes, L_e = equivalent length, ft, of a section of constant diameter = (actual length of section, ft) (inside diameter, in, of pipe used to express the equivalent length/inside diameter, in, of section under consideration)^5.

For the 16-in (406.4-mm) pipe, $L_e = (1000)(7.981/15.000)^5 = 42.6$ ft (12.9 m). The 12-in (304.8-mm) pipe is next; for it $L_e = (3000)(7.981/12.00)^5 = 390$ ft (118.9 m). For the 8-in (203.2-mm) pipe, the equivalent length = actual length = 2000 ft (609.6 m). For the 4-in (101.6-mm) pipe, $L_e = (10)(7.981/4.026)^5 = 306$ ft (93.3 m). Then the total equivalent length of 8-in (203.2-mm) pipe = sum of the equivalent lengths = 42.6 + 390 + 2000 + 306 = 2738.6 ft (834.7 m); or, by rounding off, 2740 ft (835.2 m) of 8-in (203.2-mm) pipe will have a frictional resistance equal to the complex series pipeline shown in Fig. 12. To compute the actual frictional resistance, use the methods given in previous calculation procedures.

Related Calculations. Use this general procedure for any complex series pipeline conveying water, oil, gas, steam, etc. See Crocker and King—*Piping Handbook* for derivation of the flow equations. Use the tables in Crocker and King to simplify finding the fifth power of the inside diameter of a pipe. The method of the next calculation procedure can also be used if a given flow rate is assumed.

Choosing a flow rate of 1000 gal/min (63.1 L/s) and using the tables in the Hydraulic Institute *Pipe Friction Manual* give an equivalent length of 2770 ft (844.3 m) for the 8-in (203.2-mm) pipe. This compares favorably with the 2740 ft (835.2 m) computed above. The difference of 30 ft (9.1 m) is negligible and can be accounted for by calculator variations.

The equivalent length is found by summing the friction-head loss for 1000-gal/min (63.1-L/s) flow for each length of the four pipes—16, 12, 8, and 4 in (406, 305, 203, and 102 mm)—and dividing this by the friction-head loss for 1000 gal/min (63.1 L/s) flowing through an 8-in (203.2-mm) pipe. Be careful to observe the units in which the friction-head loss is stated, because errors are easy to make if the units are ignored.

EQUIVALENT LENGTH OF A PARALLEL PIPING SYSTEM

Figure 13 shows a parallel piping system used to supply water for industrial needs. Determine the equivalent length of a single pipe for this system. All pipes in the system are approximately horizontal.

Calculation Procedure:

1. *Assume a total head loss for the system*
To determine the equivalent length of a parallel piping system, assume a total head loss for the system. Since this head loss is assumed for computation purposes only,

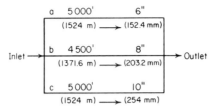

FIGURE 13 Parallel piping system.

its value need not be exact or even approximate. Assume a total head loss of 50 ft of water for each pipe in this system.

2. Compute the flow rate in each pipe in the system
Assume that the roughness coefficient C in the Hazen-Williams formula is equal for each of the pipes in the system. This is a valid assumption. Using the assumed value of C, compute the flow rate in each pipe. To allow for possible tuberculation of the pipe, assume that $C = 100$.

The Hazen-Williams formula is given in a previous calculation procedure and can be used to solve for the flow rate in each pipe. A more rapid way to make the computation is to use the friction-loss tabulations for the Hazen-Williams formula in Crocker and King—*Piping Handbook,* the Hydraulic Institute—*Pipe Friction Manual,* or a similar set of tables.

Using such a set of tables, enter at the friction-head loss equal to 50 ft (15.2 m) per 5000 ft (1524 m) of pipe for the 6-in (152.4-mm) line. Find the corresponding flow rate Q gal/min. Using the Hydraulic Institute tables, $Q_a = 270$ gal/min (17.0 L/s); $Q_b = 580$ gal/min (36.6 L/s); $Q_c = 1000$ gal/min (63.1 L/s). Hence, the total flow = 270 + 580 + 1000 = 1850 gal/min (116.7 L/s).

3. Find the equivalent size and length of the pipe
Using the Hydraulic Institute tables again, look for a pipe having a 50-ft (15.2 m) head loss with a flow of 1850 gal/min (116.7 L/s). Any pipe having a discharge equal to the sum of the discharge rates for all the pipes, at the assumed friction head, is an equivalent pipe.

Interpolating friction-head values in the 14-in (355.6-mm) outside-diameter [13.126-in (333.4-mm) inside-diameter] table shows that 5970 ft (1820 m) of this pipe is equivalent to the system in Fig. 13. This equivalent size can be used in any calculations related to this system—selection of a pump, determination of head loss with longer or shorter mains, etc. If desired, another equivalent-size pipe could be found by entering a different pipe-size table. Thus, 5310 ft (1621.5 m) of 14-in (355.6-mm) pipe [12.814-in (326.5-mm) inside diameter] is also equivalent to this system.

Related Calculations. Use this procedure for any liquid—water, oil, gasoline, brine—flowing through a parallel piping system. The pipes are assumed to be full at all times.

MAXIMUM ALLOWABLE HEIGHT FOR A LIQUID SIPHON

What is the maximum height h ft (m), Fig. 14, that can be used for a siphon in a water system if the length of the pipe from the water source to its highest point is

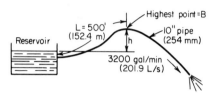

FIGURE 14   Liquid siphon piping system.

500 ft (152.4 m), the water velocity is 13.0 ft/s (3.96 m/s), the pipe diameter is 10 in (254 mm), and the water temperature is 70°F (21.1°C) if 3200 gal/min (201.9 L/s) is flowing?

Calculation Procedure:

1.   Compute the velocity of the water in the pipe
From an earlier calculation procedure, $v = gpm/(2.448d^2)$. With an internal diameter of 10.020 in (254.5 mm), $v = 3200/[(2.448)(10.02)^2] = 13.0$ ft/s (3.96 m/s).

2.   Determine the vapor pressure of the water
Using a steam table, we see that the vapor pressure of water at 70°F (21.1°C) is $p_v = 0.3631$ lb/in² (abs) (2.5 kPa), or (0.3631) $(144$ in²/ft²) $= 52.3$ lb/ft² (2.5 kPa). The specific volume of water at 70°F (21.1°C) is, from a steam table, 0.01606 ft³/lb (0.001 m³/kg). Converting this to density at 70°F (21.1°C), density $= 1/0.01606 = 62.2$ lb/ft³ (995.8 kg/m³). The vapor pressure in ft of 70°F (21.1°C) water is then $f_v = (52.3$ lb/ft²)/(62.2 lb/ft³) $= 0.84$ ft (0.26 m) of water.

3.   Compute or determine the friction-head loss and velocity head
From the reservoir to the highest point of the siphon, B, Fig. 14, the friction head in the pipe must be overcome. Use the Hazen-Williams or a similar formula to determine the friction head, as given in earlier calculation procedures or a pipe-friction table. From the Hydraulic Institute *Pipe Friction Manual*, $h_f = 4.59$ ft per 100 ft (1.4 m per 3.5 m), or $(500/100)(4.59) = 22.95$ ft (7.0 m). From the same table, velocity head $= 2.63$ ft/s (0.8 m/s).

4.   Determine the maximum height for the siphon
For a siphon handling water, the maximum allowable height h at sea level with an atmospheric pressure of 14.7 lb/in² (abs) (101.3 kPa) $= [14.7 \times (144$ in²/ft²)/ (density of water at operating temperature, lb/ft³) − (vapor pressure of water at operating temperature, ft + 1.5 × velocity head, ft + friction head, ft)]. For this pipe, $h = 14.7 \times 144/62.2 − (0.84 + 1.5 \times 2.63 + 22.95) = 11.32$ ft (3.45 m). In actual practice, the value of h is taken as 0.75 to 0.8 the computed value. Using 0.75 gives $h = (0.75)(11.32) = 8.5$ ft (2.6 m).

   Related Calculations.   Use this procedure for any type of siphon conveying a liquid—water, oil, gasoline, brine, etc. Where the liquid has a specific gravity different from that of water, i.e., less than or greater than 1.0, proceed as above, expressing all heads in ft of liquid handled. Divide the resulting siphon height by the specific gravity of the liquid. At elevations above atmospheric, use the actual atmospheric pressure instead of 14.7 lb/in² (abs) (101.3 kPa).

WATER-HAMMER EFFECTS IN LIQUID PIPELINES

What is the maximum pressure developed in a 200-lb/in^2 (1378.8-kPa) water pipeline if a valve is closed nearly instantly or pumps discharging into the line are all stopped at the same instant? The pipe is 8-in (203.2-mm) schedule 40 steel, and the water flow rate is 2800 gal/min (176.7 L/s). What maximum pressure is developed if the valve closes in 5 s and the line is 5000 ft (1524 m) long?

Calculation Procedure:

1. Determine the velocity of the pressure wave
For any pipe, the velocity of the pressure wave during water hammer is found from $v_w = 4720/(1 + Kd/Et)^{0.5}$, where v_w = velocity of the pressure wave in the pipeline, ft/s; K = bulk modulus of the liquid in the pipeline = 300,000 for water; d = internal diameter of pipe, in; E = modulus of elasticity of pipe material, lb/in^2 = 30 × 10^6 lb/in^2 (206.8 Gpa) for steel; t = pipe-wall thickness, in. For 8-in (203.2-mm) schedule 40 steel pipe and data from a table of pipe properties, $v_w = 4720/[1 + 300,000 × 7.981/(30 × 10^6 × 0.322)]^{0.5} = 4225.6$ ft/s (1287.9 m/s).

2. Compute the pressure increase caused by water hammer
The pressure increase p_1 lb/in^2 due to water hammer = $v_w v/[32.2(2.31)]$, where v = liquid velocity in the pipeline, ft/s; 32.2 = acceleration due to gravity, ft/s^2; 2.31 ft of water = 1-lb/in^2 (6.9-kPa) pressure.
   For this pipe, $v = 0.4085$ $gpm/d^2 = 0.4085(2800)/(7.981)^2 = 18.0$ ft/s (5.5 m/s). Then $p_i = (4225.6)(18)/[32.2(2.31)] = 1022.56$ lb/in^2 (7049.5 kPa). The maximum pressure developed in the pipe is then p_1 + pipe operating pressure = 1022.56 + 200 = 1222.56 lb/in^2 (8428.3 kPa).

3. Compute the hammer pressure rise caused by valve closure
The hammer pressure rise caused by valve closure p_v lb/in^2 = $2p_i L/v_w T$, where L = pipeline length, ft; T = valve closing time, s. For this pipeline, $p_v = 2(1022.56)(5000)/[(4225.6)(5)] = 484$ lb/in^2 (3336.7 kPa). Thus, the maximum pressure in the pipe will be 484 + 200 = 648 lb/in^2 (4467.3 kPa).
   Related Calculations.   Use this procedure for any type of liquid—water, oil, etc.—in a pipeline subject to sudden closure of a valve or stoppage of a pump or pumps. The effects of water hammer can be reduced by relief valves, slow-closing check valves on pump discharge pipes, air chambers, air spill valves, and air injection into the pipeline.

SPECIFIC GRAVITY AND VISCOSITY OF LIQUIDS

An oil has a specific gravity of 0.8000 and a viscosity of 200 SSU (Saybolt Seconds Universal) at 60°F (15.6°C). Determine the API gravity and Bé gravity of this oil and its weight in lb/gal (kg/L). What is the kinematic viscosity in cSt? What is the absolute viscosity in cP?

Calculation Procedure:

1. Determine the API gravity of the liquid

For any oil at 60°F (15.6°C), its specific gravity S, in relation to water at 60°F (15.6°C), is $S = 141.5/(131.5 + °API)$; or $°API = (141.5 - 131.5S)/S$. For this oil, $°API = [141.5 - 131.5(0.80)]/0.80 = 45.4 °API$.

2. Determine the Bé gravity of the liquid

For any liquid lighter than water, $S = 140/(130 + Bé)$; or $Bé = (140 - 130S)/S$. For this oil, $Bé = [140 - 130(0.80)]/0.80 = 45 Bé$.

3. Compute the weight per gal of liquid

With a specific gravity of S, the weight of 1 ft³ of oil $= (S)$[weight of 1 ft³ (1 m³) of fresh water at 60°F (15.6°C)] $= (0.80)(62.4) = 49.92$ lb/ft³ (799.2 kg/m³). Since 1 gal (3.8 L) of liquid occupies 0.13368 ft³ the weight of this oil is $(49.92)(0.13368) = 6.66$ lb/gal (0.79 kg/L).

4. Compute the kinematic viscosity of the liquid

For any liquid having an SSU viscosity greater than 100 s, the kinematic viscosity $k = 0.220 \text{(SSU)} = 135/\text{SSU}$ cSt. For this oil, $k = 0.220(200) - 135/200 = 43.325$ cSt.

5. Convert the kinematic viscosity to absolute viscosity

For any liquid, the absolute viscosity, cP = (kinematic viscosity, cSt)(density). Thus, for this oil, the absolute viscosity $= (43.325)(49.92) = 2163$ cP.

    Related Calculations. For liquids *heavier* than water, $S = 145/(145 - Bé)$. When the SSU viscosity is between 32 and 99 SSU, $k = 0.226 \text{(SSU)} - 195/\text{SSU}$ cSt. Modern terminology for absolute viscosity is dynamic viscosity. Use these relations for any liquid—brine, gasoline, crude oil, kerosene, Bunker C, diesel oil, etc. Consult the *Pipe Friction Manual* and Crocker and King—*Piping Handbook* for tabulations of typical viscosities and specific gravities of various liquids.

PRESSURE LOSS IN PIPING HAVING LAMINAR FLOW

Fuel oil at 300°F (148.9°C) and having a specific gravity of 0.850 is pumped through a 30,000-ft (9144-m) long 24-in (609.6-mm) pipe at the rate of 500 gal/min (31.6 L/s). What is the pressure loss if the viscosity of the oil is 75 cP (0.075 Pa·s)?

Calculation Procedure:

1. Determine the type of flow that exists

Flow is laminar (also termed *viscous*) if the Reynolds number R for the liquid in the pipe is less than 1200. Turbulent flow exists if the Reynolds number is greater than 2500. Between these values is a zone in which either condition may exist, depending on the roughness of the pipe wall, entrance conditions, and other factors. Avoid sizing a pipe for flow in this critical zone because excessive pressure drops result without a corresponding increase in the pipe discharge.

TABLE 20 Reynolds Number

Reynolds number R	Coefficient	Numerator			Denominator	
		First symbol	Second symbol	Third symbol	Fourth symbol	Fifth symbol
Dvp/μ	. . .	ft	ft/s	lb/ft^3	lb mass/(ft·s)	
$124dvp/z$	124	in	ft/s	lb/ft^3	cP	
$50.7G\rho/dz$	50.7	gal/min	lb/ft^3	. . .	in	cP
$6.32W/dz$	6.32	lb/h	. . .	. . .	in	cP
$35.5B\rho/dz$	35.5	bbl/h	lb/ft^3	. . .	in	cP
$7742dv/k$	7,742	in	ft/s	. . .	. . .	cP
$3162G/dk$	3,162	gal/min	. . .	. . .	in	cP
$2214B/dk$	2,214	bbl/h	. . .	. . .	in	cP
$22,735q\rho/dz$	22,735	ft^3/s	lb/ft^3	. . .	in	cP
$378.9Q\rho/dz$	378.9	ft^3/min	lb/ft^3	. . .	in	cP

Compute the Reynolds number from $R = 3.162G/kd$, where $G =$ flow rate gal/min (L/s); $k =$ kinematic viscosity of liquid, cSt $=$ viscosity z, cP/specific gravity of the liquid S; $d =$ inside diameter of pipe, in (cm). From a table of pipe properties, $d = 22.626$ in (574.7 mm). Also, $k = z/S = 75/0.85 = 88.2$ cSt. Then $R = 3162(500)/[88.2(22.626)] = 792$. Since $R < 1200$, laminar flow exists in this pipe.

2. Compute the pressure loss by using the Poiseuille formula
The Poiseuille formula gives the pressure drop p_d lb/in^2 (kPa) $= 2.73(10^{-4})luG/d^4$, where $l =$ total length of pipe, including equivalent length of fittings, ft; $u =$ absolute viscosity of liquid, cP (Pa·s); $G =$ flow rate gal/min (L/s); $d =$ inside diameter of pipe, in (cm). For this pipe, $p_d = 2.73(10^{-4})(10,000)(75)(500)/262,078 = 1.17$ lb/in^2 (8.1 kPa).

Related Calculations. Use this procedure for any pipe in which there is laminar flow of the liquid. Other liquids for which this method can be used include water, molasses, gasoline, brine, kerosene, and mercury. Table 20 gives a quick summary of various ways in which the Reynolds number can be expressed. The symbols in Table 20, in the order of their appearance, are $D =$ inside diameter of pipe, ft (m); $v =$ liquid velocity, ft/s (m/s); $p =$ liquid density, lb/ft^3 (kg/m^3); $\mu =$ absolute viscosity of liquid, lb mass/(ft·s) [kg/(m·s)]; $d =$ inside diameter of pipe, in (cm). From a table of pipe properties, $d = 22.626$ in (574.7 mm). Also, $k = z/S$ liquid flow rate, lb/h (kg/h); $B =$ liquid flow rate, bbl/h (L/s); $k =$ kinematic viscosity of the liquid, cSt; q liquid flow rate, ft^3 (m^3/s); $Q =$ liquid flow rate, ft^3/min (m^3/min). Use Table 20 to find the Reynolds number for any liquid flowing through a pipe.

DETERMINING THE PRESSURE LOSS IN OIL PIPES

What is the pressure drop in a 5000-ft (1524-m) long 6-in (152.4-mm) oil pipe conveying 500 bbl/h (22.1 L/s) of kerosene having a specific gravity of 0.813 at

65°F (18.3°C), which is the temperature of the liquid in the pipe? The pipe is schedule 40 steel.

Calculation Procedure:

1. Determine the kinematic viscosity of the oil

Use Fig. 15 and Table 21 or the Hydraulic Institute—*Pipe Friction Manual* kinematic viscosity and Reynolds number chart to determine the kinematic viscosity of the liquid. Enter Table 12 at kerosene, and find the coordinates as $X = 10.2$, $Y = 16.9$. Using these coordinates, enter Fig. 15 and find the absolute viscosity of kerosene at 65°F (18.3°C) as 2.4 cP. By the method of a previous calculation procedure, the kinematic viscosity = absolute viscosity, cP/specific gravity of the liquid = 2.4/0.813 = 2.95 cSt. This value agrees closely with that given in the *Pipe Friction Manual*.

2. Determine the Reynolds number of the liquid

The Reynolds number can be found from the *Pipe Friction Manual* chart mentioned in step 1 or computed from $R = 2214B/(dk) = 2214(500)/[(6.065)(2.95)] = 61,900$.

To use the *Pipe Friction Manual* chart, compute the velocity of the liquid in the pipe by converting the flow rate to ft³/s. Since there is 42 gal/bbl (0.16 L) and 1 gal (0.00379 L) = 0.13368 ft³ (0.00378 m³), 1 bbl = (42)(0.13368) = 5.6 ft³ (0.16 m³). With a flow rate of 500 bbl/h (79.5 m³/h) the equivalent flow = (500)(5.6) = 2800 ft³/h (79.3 m³/h), or 2800/3600 s/h = 0.778 ft³/s (0.02 m³/s). Since 6-in (152.4-mm), schedule 40 pipe has a cross-sectional area of 0.2006 ft² (0.02 m²) internally, the liquid velocity = 0.778/0.2006 = 3.88 ft/s (1.2 m/s). Then, the product (velocity, ft/s)(internal diameter, in) = (3.88)(6.065) = 23.75 ft/s. In the *Pipe Friction Manual*, project horizontally from the kerosene specific-gravity curve to the vd product of 23.75, and read the Reynolds number as 61,900, as before. In general, the Reynolds number can be found more quickly by computing it using the appropriate relation given in an earlier calculation procedure, unless the flow velocity is already known.

3. Determine the friction factor of this pipe

Enter Fig. 16 at the Reynolds number value of 61,900, and project to the curve 4 as indicated by Table 22. Read the friction factor as 0.0212 at the left. Alternatively, the *Pipe Friction Manual* friction-factor chart could be used, if desired.

4. Compute the pressure loss in the pipe

Use the Fanning formula $p_d = 1.06(10^{-4})f\rho lB^2/d^5$. In this formula, ρ = density of the liquid, lb/ft³. For kerosene, ρ = (density of water, lb/ft³)(specific gravity of the kerosene) = (62.4)(0.813) = 50.6 lb/ft³ (810.1 kg/m³). Then $p_d = 1.06(10^{-4}) \times (0.0212)(50.6)(5000)(500)^2/8206 = 17.3$ lb/in² (119.3 kPa).

Related Calculations. The Fanning formula is popular with oil-pipe designers and can be stated in various ways: (1) with velocity v ft/s, $p_d = 1.29(10^{-3})f\rho V^2 l/d$; (2) with velocity V ft/min, $p_d = 3.6(10^{-7})f\rho V^2 l/d$; (3) with flow rate in G gal/min, $p_d = 2.15(10^{-4})f\rho lG^2/d^2$; (4) with the flow rate in W lb/h, $p_d = 3.36(10^{-6})flW^2/d^5\rho$.

Use this procedure for any petroleum product—crude oil, kerosene, benzene, gasoline, naphtha, fuel oil, Bunker C, diesel oil, toluene, etc. The tables and charts presented here and in the *Pipe Friction Manual* save computation time.

VISCOSITIES

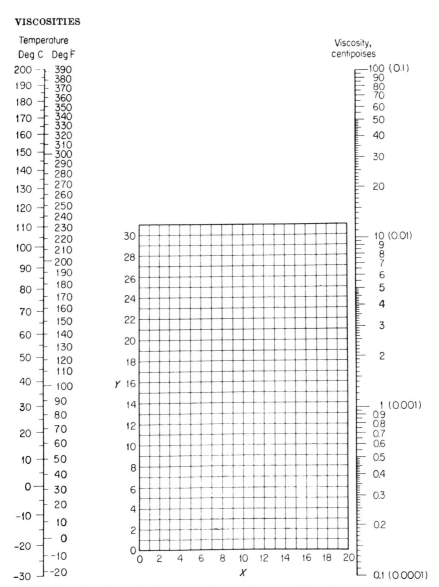

FIGURE 15 Viscosities of liquids at 1 atm. For coordinates, see Table 21.

TABLE 21 Viscosities of Liquids
Coordinates for use with Fig. 15

No.	Liquid	X	Y	No.	Liquid	X	Y
1	Acetaldehyde	15.2	4.8	56	Freon-22	17.2	4.7
	Acetic acid:			57	Freon-13	12.5	11.4
2	100%	12.1	14.2		Glycerol:		
3	70%	9.5	17.0	58	100%	2.0	30.0
4	Acetic anhydride	12.7	12.8	59	50%	6.9	19.6
	Acetone:			60	Heptene	14.1	8.4
5	100%	14.5	7.2	61	Hexane	14.7	7.0
6	35%	7.9	15.0	62	Hydrochloric acid, 31.5%	13.0	16.6
7	Allyl alcohol	10.2	14.3	63	Isobutyl alcohol	7.1	18.0
	Ammonia:			64	Isobutyric acid	12.2	14.4
8	100%	12.6	2.0	65	Isopropyl alcohol	8.2	16.0
9	26%	10.1	13.9	66	Kerosene	10.2	16.9
10	Amyl acetate	11.8	12.5	67	Linseed oil, raw	7.5	27.2
11	Amyl alcohol	7.5	18.4	68	Mercury	18.4	16.4
12	Aniline	8.1	18.7		Methanol:		
13	Anisole	12.3	13.5	69	100%	12.4	10.5
14	Arsenic trichloride	13.9	14.5	70	90%	12.3	11.8
15	Benzene	12.5	10.9	71	40%	7.8	15.5
	Brine:			72	Methyl acetate	14.2	8.2
16	CaCl₂, 25%	6.6	15.9	73	Methyl chloride	15.0	3.8
17	NaCl, 25%	10.2	16.6	74	Methyl ethyl ketone	13.9	8.6
18	Bromine	14.2	13.2	75	Naphthalene	7.9	18.1
19	Bromotoluene	20.0	15.9		Nitric acid:		
20	Butyl acetate	12.3	11.0	76	95%	12.8	13.8
21	Butyl alcohol	8.6	17.2	77	60%	10.8	17.0
22	Butyric acid	12.1	15.3	78	Nitrobenzene	10.6	16.2
23	Carbon dioxide	11.6	0.3	79	Nitrotoluene	11.0	17.0
24	Carbon disulfide	16.1	7.5	80	Octane	13.7	10.0
25	Carbon tetrachloride	12.7	13.1	81	Octyl alcohol	6.6	21.1
26	Chlorobenzene	12.3	12.4	82	Pentachloroethane	10.9	17.3
27	Chloroform	14.4	10.2	83	Pentane	14.9	5.2
28	Chlorosulfonic acid	11.2	18.1	84	Phenol	6.9	20.8
	Chlorotoluene:			85	Phosphorus tribromide	13.8	16.7
29	Ortho	13.0	13.3	86	Phosphorus trichloride	16.2	10.9
30	Meta	13.3	12.5	87	Propionic acid	12.8	13.8
31	Para	13.3	12.5	88	Propyl alcohol	9.1	16.5
32	Cresol, meta	2.5	20.8	89	Propyl bromide	14.5	9.6
33	Cyclohexanol	2.9	24.3	90	Propyl chloride	14.4	7.5
34	Dibromoethane	12.7	15.8	91	Propyl iodide	14.1	11.6
35	Dichloroethane	13.2	12.2	92	Sodium	16.4	13.9
36	Dichloromethane	14.6	8.9	93	Sodium hydroxide, 50%	3.2	25.8
37	Diethyl oxalate	11.0	16.4	94	Stannic chloride	13.5	12.8
38	Dimethyl oxalate	12.3	15.8	95	Sulfur dioxide	15.2	7.1
39	Diphenyl	12.0	18.3		Sulfuric acid:		
40	Dipropyl oxalate	10.3	17.7	96	110%	7.2	27.4
41	Ethyl acetate	13.7	9.1	97	98%	7.0	24.8
	Ethyl alcohol:			98	60%	10.2	21.3
42	100%	10.5	13.8	99	Sulfuryl chloride	15.2	12.4
43	95%	9.8	14.3	100	Tetrachloroethane	11.9	15.7
44	40%	6.5	16.6	101	Tetrachloroethylene	14.2	12.7
45	Ethyl benzene	13.2	11.5	102	Titanium tetrachloride	14.4	12.3
46	Ethyl bromide	14.5	8.1	103	Toluene	13.7	10.4
47	Ethyl chloride	14.8	6.0	104	Trichloroethylene	14.8	10.5
48	Ethyl ether	14.5	5.3	105	Turpentine	11.5	14.9
49	Ethyl formate			106	Vinyl acetate		
50	Ethyl iodide	14.7	10.3	107	Water	10.2	13.0
51	Ethylene glycol	6.0	23.6		Xylene:		
52	Formic acid	10.7	15.8	108	Ortho	13.5	12.1
53	Freon-11	14.4	9.0	109	Meta	13.9	10.6
54	Freon-12	16.8	5.6	110	Para	13.9	10.9
55	Freon-21	15.7	7.5				

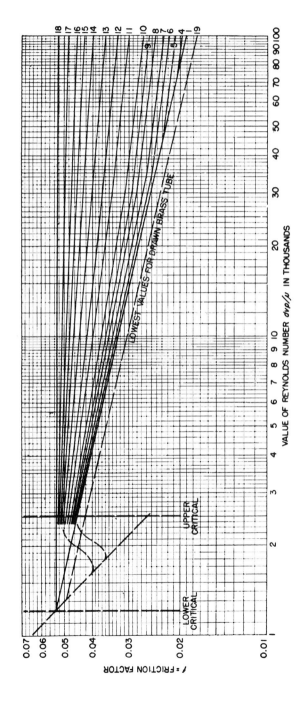

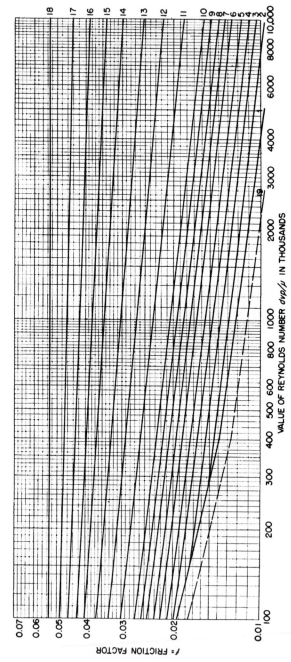

FIGURE 16 Friction-factor curves. (*Mechanical Engineering.*)

8.54

TABLE 22 Data for Fig. 16

Percentage of roughness	For value of f see curve	Drawn tubing, brass, tin, lead, glass		Clean steel, wrought iron		Clean, galvanized		Best cast iron		Average cast iron		Heavy riveted, spiral riveted	
		in	mm	in	mm	in	mm	in	mm	in	mm	in	mm
0.2	1	0.35 up	8.89 up	72	1829	—	—	—	—	—	—	—	—
1.35	4	—	—	6–12	152–305	10–24	254–610	20–48	508–1219	42–96	1067–2438	84–204	2134–5182
2.1	5	—	—	4–5	102–127	6–8	152–203	12–16	305–406	24–36	610–914	48–72	1219–1829
3.0	6	—	—	2–3	51–76	3–5	76–127	5–10	127–254	10–20	254–508	20–42	508–1067
3.8	7	—	—	1½	38	2½	64	3–4	76–102	6–8	152–203	16–18	406–457
4.8	8	—	—	1–1¼	25–32	1½–2	38–51	2–2½	51–64	4–5	102–127	10–14	254–356
6.0	9	—	—	¾	19	1¼	32	1½	38	3	76	8	203
7.2	10	—	—	½	13	1	25	1¼	32	—	—	5	127
10.5	11	—	—	⅜	9.5	¾	19	1	35	—	—	4	102
14.5	12	—	—	¼	6.4	½	13	—	—	—	—	3	76
24.0	14	0.125	3.18	—	—	⅜	9.5	—	—	—	—	—	—
31.5	16	—	—	—	—	¼	6.4	—	—	—	—	—	—
37.5	18	0.0625	1.588	—	—	⅛	3.2	—	—	—	—	—	—

Diameter (actual of drawn tubing, nominal of standard-weight pipe)

FLOW RATE AND PRESSURE LOSS IN COMPRESSED-AIR AND GAS PIPING

Dry air at 80°F (26.7°C) and 150 lb/in² (abs) (1034 kPa) flows at the rate of 500 ft³/min (14.2 m³/min) through a 4-in (101.6-mm) schedule 40 pipe from the discharge of an air compressor. What are the flow rate in lb/h and the air velocity in ft/s? Using the Fanning formula, determine the pressure loss if the total equivalent length of the pipe is 500 ft (152.4 m).

Calculation Procedure:

1. Determine the density of the air or gas in the pipe
For air or a gas, $pV = MRT$, where p = absolute pressure of the gas, lb/ft² (abs); V = volume of M lb of gas, ft³; M = weight of gas, lb; R = gas constant, ft·lb/(lb·°F); T = absolute temperature of the gas, °R. For this installation, using 1 ft³ of air, $M = pV/(RT)$, $M = (150)(144)/[(53.33)(80 + 459.7)] = 0.750$ lb/ft³ (12.0 kg/m³). The value of R in this equation was obtained from Table 23.

2. Compute the flow rate of the air or gas
For air or a gas, the flow rate W_h lb/h = (60) (density, lb/ft³)(flow rate, ft³/min); or $W_h = (60)(0.750)(500) = 22,500$ lb/h (10,206 kg/h).

3. Compute the velocity of the air or gas in the pipe
For any air or gas pipe, velocity of the moving fluid v ft/s = 183.4 $W_h/3600\ d^2\rho$, where d = internal diameter of pipe, in; ρ = density of fluid, lb/ft³. For this system, $v = (183.4)(22,500)/[(3600)(4.026)^2(0.750)] = 94.3$ ft/s (28.7 m/s).

TABLE 23   Gas Constants

Gas	R ft·lb/(lb·°F)	J/(kg·K)	C for critical-velocity equation
Air	53.33	286.9	2870
Ammonia	89.42	481.1	2080
Carbon dioxide	34.87	187.6	3330
Carbon monoxide	55.14	296.7	2820
Ethane	50.82	273.4	
Ethylene	54.70	294.3	2480
Hydrogen	767.04	4126.9	750
Hydrogen sulfide	44.79	240.9	
Isobutane	25.79	138.8	
Methane	96.18	517.5	2030
Natural gas	—	—	2070–2670
Nitrogen	55.13	296.6	2800
n-butane	25.57	137.6	
Oxygen	48.24	259.5	2990
Propane	34.13	183.6	
Propylene	36.01	193.7	
Sulfur dioxide	23.53	126.6	3870

4. *Compute the Reynolds number of the air or gas*
The viscosity of air at 80°F (26.7°C) is 0.0186 cP, obtained from Crocker and King—*Piping Handbook,* Perry et al.—*Chemical Engineers' Handbook,* or a similar reference. Then, by using the Reynolds number relation given in Table 20, $R = 6.32W/(dz) = (6.32)(22,500)/[(4.026)(0.0186)] = 1,899,000$.

5. *Compute the pressure loss in the pipe*
Using Fig. 16 or the Hydraulic Institute *Pipe Friction Manual,* we get $f = 0.0142$ to 0.0162 for a 4-in (101.6-mm) schedule 40 pipe when the Reynolds number = 3,560,000. From the Fanning formula from an earlier calculation procedure and the higher value of f, $p_d = 3.36(10^{-6})flW^2/d^5\rho$, or $p_d = 3.36(10^{-6})(0.0162)(500)(22,500)^2/[(4.026)^5(0.750)] = 17.37$ lb/in^2 (119.8 kPa).

Related Calculations. Use this procedure to compute the pressure loss, velocity, and flow rate in compressed-air and gas lines of any length. Gases for which this procedure can be used include ammonia, carbon dioxide, carbon monoxide, ethane, ethylene, hydrogen, hydrogen sulfide, isobutane, methane, nitrogen, *n*-butane, oxygen, propane, propylene, and sulfur dioxide.

Alternate relations for computing the velocity of air or gas in a pipe are $v = 144W_s/a\rho$; $v = 183.4W_s/d^2\rho$; $v = 0.0509 \, W_s v_g/d^2$, where W_s = flow rate, lb/s; a = cross-sectional area of pipe, in^2, v_g = specific volume of the air or gas at the operating pressure and temperature, ft^3/lb.

FLOW RATE AND PRESSURE LOSS IN GAS PIPELINES

Using the Weymouth formula, determine the flow rate in a 10-mi (16.1-km) long 4-in (101.6-mm) schedule 40 gas pipeline when the inlet pressure is 200 lb/in^2 (gage) (1378.8 kPa), the outlet pressure is 20 lb/in^2 (gage) (137.9 kPa), the gas has a specific gravity of 0.80, a temperature of 60°F (15.6°C), and the atmospheric pressure is 14.7 lb/in^2 (abs) (101.34 kPa).

Calculation Procedure:

1. *Compute the flow rate from the Weymouth formula*
The Weymouth formula for flow rate is $Q = 28.05[(p_i^2 - p_0^2)d^{5.33}/sL]^{0.5}$, where p_i = inlet pressure, lb/in^2 (abs); p_0 = outlet pressure, lb/in^2 (abs); d = inside diameter of pipe, in; s = specific gravity of gas; L = length of pipeline, mi. For this pipe, $Q = 28.05 \times [(214.7^2 - 34.7^2)4.026^{5.33}/0.8 \times 10]^{0.5} = 86,500$ lb/h (39,925 kg/h).

2. *Determine if the acoustic velocity limits flow*
If the outlet pressure of a pipe is less than the critical pressure p_c lb/in^2 (abs), the flow rate in the pipe cannot exceed that obtained with a velocity equal to the critical or acoustic velocity, i.e., the velocity of sound in the gas. For any gas, $p_c = Q(T_i)^{0.5}/d^2C$, where T_i = inlet temperature, °R; C = a constant for the gas being considered.

Using $C = 2070$ from Table 23, or Crocker and King—*Piping Handbook,* $p_c = (86,500)(60 + 460)^{0.5}/[(4.026)^2(2070)] = 58.8$ lb/in^2 (abs) (405.4 kPa). Since the outlet pressure $p_0 = 34.7$ lb/in² (abs) (239.2 kPa), the critical or acoustic velocity limits the flow in this pipe because $p_c > p_0$. When $p_c < p_0$, critical velocity does not limit the flow.

Related Calculations.   Where a number of gas pipeline calculations must be made, use the tabulations in Crocker and King—*Piping Handbook* and Bell—*Petroleum Transportation Handbook.* These tabulations will save much time. Other useful formulas for gas flow include the Panhandle, Unwin, Fritsche, and rational. Results obtained with these formulas agree within satisfactory limits for normal engineering practice.

Where the outlet pressure is unknown, assume a value for it and compute the flow rate that will be obtained. If the computed flow is less than desired, check to see that the outlet pressure is less than the critical. If it is, increase the diameter of the pipe. Use this procedure for natural gas from any gas field, manufactured gas, or any other similar gas.

To find the volume of gas that can be stored per mile of pipe, solve $V_m = 1.955 p_m d^2 K$, where p_m = mean pressure in pipe, lb/in² (abs) $\approx (p_i + p_0)/2$; $K = (1/Z)^{0.5}$, where Z = super compressibility factor of the gas, as given in Baumeister and Marks—*Standard Handbook for Mechanical Engineers* and Perry—*Chemical Engineer's Handbook.* For exact computation of p_m, use $p_m = (\frac{2}{3})(p_i + p_0 - p_i p_0 / p_i + p_0)$.

SELECTING HANGERS FOR PIPES AT ELEVATED TEMPERATURES

Select the number, capacity, and types of pipe hanger needed to support the 6-in (152.5-mm) schedule 80 pipe in Fig. 17 when the installation temperature is 60°F (15.6°C) and the operating temperature is 700°F (371.1°C). The pipe is insulated with 85 percent magnesia weighing 11.4 lb/ft (16.63 N/m). The pipe and unit served by the pipe have a coefficient of thermal expansion of 0.0575 in/ft (0.48

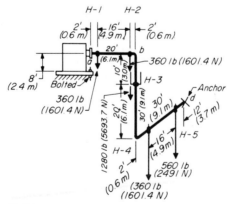

FIGURE 17   Typical complex pipe operating at high temperature.

cm/m) between the 60°F (15.6°C) installation temperature and the 700°F (371.1°C) operating temperature.

Calculation Procedure:

1. *Draw a freehand sketch of the pipe expansion*
Use Fig. 18 as a guide and sketch the expanded pipe, using a dashed line. The sketch need not be exactly to scale; if the proportions are accurate, satisfactory results will be obtained. The shapes shown in Fig. 18 cover the 11 most common situations met in practice.

2. *Tentatively locate the required hangers*
Begin by locating hangers *H*-1 and *H*-5 close to the supply and using units, Fig. 17. Keeping a hanger close to each unit (boiler, turbine, pump, engine, etc.) prevents overloading the connection on the unit.

Space intermediate hangers *H*-2, *H*-3, and *H*-4 so that the recommended distances in Table 24 or hanger engineering data (e.g., Grinnell Corporation *Pipe*

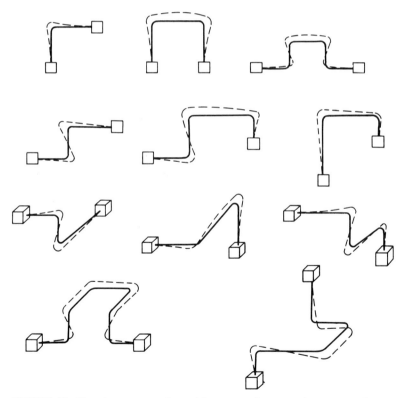

FIGURE 18 Pipe shapes commonly used in power and process plants assume the approximate forms shown by the dotted lines when the pipe temperature rises. (*Power.*)

TABLE 24 Maximum Recommended Spacing between Pipe Hangers

Nominal pipe size, in (mm)	4 (101.6)	5 (127)	6 (152.4)	8 (203.2)	10 (254)	12 (304.8)
Maximum span, ft (m)	14 (4.3)	16 (4.9)	17 (5.2)	19 (5.8)	22 (6.7)	23 (7.0)

Hanger Design and Engineering) are not exceeded. Indicate the hangers on the piping drawing as shown in Fig. 17.

3.   Adjust the hanger locations to suit structural conditions
Study the building structural steel in the vicinity of the hanger locations, and adjust these locations so that each hanger can be attached to a support having adequate strength.

4.   Compute the load each hanger must support
From a table of pipe properties, such as in Crocker and King—*Piping Handbook,* find the weight of 6-in (152.4-mm) schedule 80 pipe as 28.6 lb/ft (41.7 N/m). The insulation weighs 11.4 lb/ft (16.6 N/m), giving a total weight of insulated pipe of 28.6 + 11.4 = 40.0 lb/ft (58.4 N/m).

Compute the load on the hangers supporting horizontal pipes by taking half the length of the pipe on each side of the hanger. Thus, for hanger H-1, there is (2 ft)($\frac{1}{2}$) + (16 ft) $\times$ ($\frac{1}{2}$) = 9 ft (2.7 m) of horizontal pipe, Fig. 17, which it supports. Since this pipe weighs 40 lb/ft (58.4 N/m), the total load on hanger H-1 = (9 ft)(40 lb/ft) = 360 lb (1601.4 N). A similar analysis for hanger H-2 shows that it supports (8 + 1)(40) = 360 lb (1601.4 N).

Hanger H-3 supports the entire weight of the vertical pipe, 30 ft (9.14 m), plus 1 ft (0.3 m) at the top bend and 1 ft (0.3 m) at the bottom bend, or a total of 1 + 30 + 1 = 32 ft (9.75 m). The total load on hanger H-3 is therefore (32)(40) = 1280 lb (5693.7 N).

Hanger H-4 supports (1 + 8)(40) = 360 lb (1601.4 N), and hanger H-5 supports (8 + 6)(40) = 560 lb (2491 N).

As a check, compute the total weight of the pipe and compare it with the sum of the endpoint and hanger loads. Thus, there is 100 ft (30.5 m) of pipe weighing (80)(40) = 3200 lb (14.2 kN). The total load the hangers will support is 360 + 360 + 1280 + 360 + 560 = 2920 lb (12.9 kN). The first endpoint will support (1)(40) = 40 lb (177.9 N), and the anchor will support (6)(40) = 240 lb (1067 N). The total hanger and endpoint support = 2920 + 40 + 240 = 3200 lb (14.2 kN); therefore, the pipe weight = the hanger load.

5.   Sketch the shape of the hot pipe
Use Fig. 18 as a guide, and draw a dotted outline of the approximate shape the pipe will take when hot. Start with the first corner point nearest the unit on the left, Fig. 19. This point will move away from the unit, as in Fig. 19. Do the same for the first corner point near the other unit served by the pipe and for intermediate corner points. Use arrows to indicate the probable direction of pipe movement at each corner. When sketching the shape of the hot pipe, remember that a straight pipe expanding against a piece of pipe at right angles to itself will bend the latter. The distance that various lengths of pipe will bend while producing a tensile stress

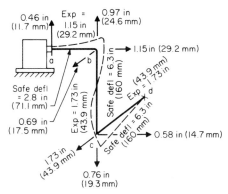

FIGURE 19 Expansion of the various parts of the pipe shown in Fig. 17. (*Power.*)

of 14,000 lb/in² (96.5 MPa) is given in Table 25. This stress is a typical allowable value for pipes in industrial systems.

6. Determine the thermal movement of units served by the pipe

If either or both fixed units (boiler, turbine, etc.) operate at a temperature above or below atmospheric, determine the amount of movement at the flange of the unit to which the piping connects, using the thermal data in Table 26. Do this by applying the thermal expansion coefficient for the metal of which the unit is made. Determine the vertical and horizontal distance of the flange face from the point of no movement of the unit. The point of no movement is the point or surface where the unit is fastened to *cold* structural steel or concrete.

The flange, point *a*, Fig. 19, is 8 ft (2.4 m) above the bolted end of the unit and directly in line with the bolt, Fig. 17. Since the bolt and flange are on a common vertical line, there will not be any *horizontal* movement of the flange because the bolt is the no-movement point of the unit.

Since the flange is 8 ft (2.4 m) away from the point of no movement, the amount that the flange will move = (distance away from the point of no movement, ft)(coefficient of thermal expansion, in/ft) = (8)(0.0575) = 0.46 in (11.7 mm) *away* (up) from the point of no movement. If the unit were operating at a temperature

TABLE 25 Deflection, in (mm), that Produces 14,000-lb/in² (96,530-kPa) Tensile Stress in Pipe Legs Acting as a Cantilever Beam, Load at Free End

Cantilever length, ft (m)	Nominal pipe size, in (mm)		
	4 (101.6)	6 (152.4)	8 (203.2)
5 (1.5)	0.26 (6.6)	0.17 (4.3)	0.13 (3.3)
10 (3.0)	1.03 (26.2)	0.70 (17.8)	0.54 (13.7)
15 (4.6)	2.32 (58.9)	1.58 (40.1)	1.21 (30.7)
20 (6.1)	4.12 (104.6)	2.80 (71.1)	2.15 (54.6)
25 (7.6)	6.44 (163.6)	4.38 (111.3)	3.35 (85.1)
30 (9.1)	9.26 (235.2)	6.30 (160.0)	4.83 (122.7)

TABLE 26 Thermal Expansion of Pipe, in/ft (mm/m) (Carbon and Carbon-Moly Steel and WI)

Operating temperature, °F (°C)	Installation temperature	
	32°F (0°C)	60°F (15.6°C)
600 (316)	0.050 (4.17)	0.0475 (3.96)
650 (343)	0.055 (4.58)	0.0525 (4.38)
700 (371)	0.060 (5.0)	0.0575 (4.79)
750 (399)	0.065 (5.42)	0.0624 (5.2)
800 (427)	0.070 (5.83)	0.0674 (5.62)

less than atmospheric, it would contract and the flange would move *toward* (down) the point of no movement. Mark the flange movement on the piping sketch, Fig. 19.

Anchor, d, Fig. 19, does not move because it is attached to either cold structural steel or concrete.

7. Compute the amount of expansion in each pipe leg
Expansion of the pipe, in = (pipe length, ft)(coefficient of linear expansion, in/ft). For length ab, Fig. 17, the expansion = (20)(0.0575) = 1.15 in (29.2 mm); for bc, (30)(0.0575) = 1.73 in (43.9 mm); for cd, (30)(0.0575) = 1.73 in (43.9 mm). Mark the amount and direction of expansion on Fig. 19.

8. Determine the allowable deflection for each pipe leg
Enter Table 25 at the nominal pipe size and find the allowable deflection for a 14,000-lb/in^2 (96.5-MPa) tensile stress for each pipe leg. Thus, for ab, the allowable deflection = 2.80 in (71.1 mm) for a 20-ft (6.1-m) long leg; for bc, 6.30 in (160 mm) for a 30-ft (9.1-m) long leg; for cd, 6.30 in (160 mm) for a 30-ft (9.1-m) long leg. Mark these allowable deflections on Fig. 19, using dashed arrows.

9. Compute the actual vertical and horizontal deflections
Sketch the vertical deflection diagram, Fig. 20a, by drawing a triangle showing the total expansion in each direction in proportion to the length of the parts at right angles to the expansion. Thus, the 0.46-in (11.7-mm) upward expansion at the flange, a, is at right angles to leg ab and is drawn as the altitude of the right triangle. Lay off 20 t (6.1 m), ab, on the base of the triangle. Since bc is parallel to the direction of the flange movement, it is shown as a point, bc, on the base of this triangle. From point bc, lay off cd on the base of the triangle, Fig. 20a, since it is at right angles to the expansion of point a. Then, by similar triangles, 50:46 = 30:x; x = 0.28 in (7.1 mm). Therefore, leg bc moves upward 0.28 in (7.1 mm) because of the flange movement at a.

Now draw the deflection diagram, Fig. 20b, showing the upward movement of leg ab and the downward movement of leg cd along the length of each leg, or 20 and 30 ft (6.1 and 9.1 m), respectively. Solve the similar triangles, or 20:x_1 = 30:(1.73 − x_1); x_1 = 0.69 in (17.5 mm). Therefore, point b moves *up* 0.69 in (17.5 mm) as a result of the expansion of leg bc. Then 1.73 − x_1 = 1.73 − 0.69 = 1.04 in (26.4 mm). Thus, point c moves *down* 1.04 in (26.4 mm) as a result of the expansion of bc. The total distance b moves up = 0.28 + 0.69 = 0.97 in (24.6

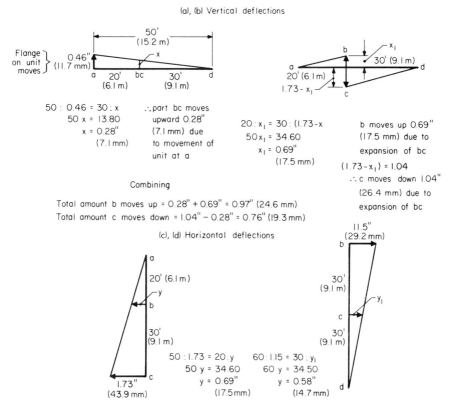

FIGURE 20 (*a*), (*b*) Vertical deflection diagrams for the pipe in Fig. 17; (*c*), (*d*) horizontal deflection diagrams for the pipe in Fig. 17. (*Power.*)

mm), whereas the total distance *c* moves down = 1.04 − 0.28 = 0.76 in (19.3 mm). Mark these actual deflections on Fig. 19.

Find the actual horizontal deflections in a similar fashion by constructing the triangle, Fig. 20*c*, formed by the vertical pipe *bc* and the horizontal pipe *ab*. Since point *a* does not move horizontally but point *b* does, lay off leg *ab* at right angles to the direction of movement, as shown. From point *b* lay off leg *bc*. Then, since leg *bc* expands 1.73 in (43.9 mm), lay this distance off perpendicular to *ac*, Fig. 20*c*. By similar triangles, 20 + 30:1.73 = 20:*y*; *y* = 0.69 in (17.5 mm). Hence, point *b* deflects 0.69 in (17.5 mm) in the direction shown in Fig. 19.

Follow the same procedure for leg *cd*, constructing the triangle in Fig. 20*d*. Beginning with point *b*, lay off legs *bc* and *cd*. The altitude of this right triangle is then the distance point *c* moves when leg *ab* expands, or 1.15 in (29.2 mm). By similar triangles, 30 + 30:1.15 = 30:y_1; y_1 = deflection of point *c* = 0.58 in (14.7 mm).

10. Select the type of pipe hanger to use

Figure 21 shows several popular types of pipe hangers, together with the movements that they are designed to absorb. For hangers *H*-1 and *H*-2, use type *E*, Fig. 21,

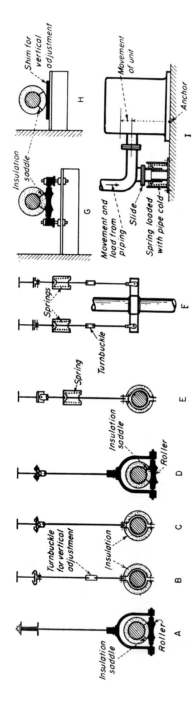

FIGURE 21 Pipe hangers chosen depend on the movement expected. Hangers *A* and *B* are suitable for pipe movement in one horizontal direction. Hangers *C* and *D* permit pipe movement in two horizontal directions. Vertical and horizontal movement requires use of hangers such as *E* for horizontal pipes and *F* for vertical pipes. (*G*) Cantilever support; (*H*) sliding movement in two horizontal directions; (*I*) base elbow support.

because the pipe moves both vertically and horizontally at these points, as Fig. 20 shows. Use type F, Fig. 21, for hanger H-3, because riser bc moves both vertically and horizontally. Hangers H-4 and H-5 should be type E, because they must absorb both horizontal and vertical movements.

Once the hangers are selected from Fig. 21, refer to hanger engineering data for the exact design details of the hangers that will be selected. During the study of the data, look for other hanger that absorb the same movement or movements but may be more adaptable to the existing structural steel conditions.

11. *Select the hanger-rod diameter for each hanger*

Use Table 27 to find the required hanger-rod diameter. Since the pipe operates at 700°F (371°C), select the maximum safe load from the 750°F (399°C) column. Tabulate the loads and diameters as follows:

Hanger	Load, lb (kN)	Rod diameter, in (mm)
H-1	360 (1.6)	⅜ (9.5)
H-2	360 (1.6)	⅜ (9.5)
H-3	1280 (5.7)	2–½ each (2–12.7)
H-4	360 (1.6)	⅜ (9.5)
H-5	560 (2.5)	½ (12.7)

Select standard springs for spring-loaded hangers from pipe-hanger engineering data. Springs are listed in the data on the basis of loading per inch of travel. For small movements [less than 1 in (25.4 mm)], it is generally desirable to select a lighter spring and precompress it at installation so that it has a light loading. Hanger movement will then load the spring to the desired value. This approach is desirable from another standpoint: any error in estimating hanger movement will not cause as large an unbalanced load on the pipe as would a heavier spring with a greater loading per inch of travel.

Related Calculations. Use this procedure for any type of pipe operating at elevated temperature—steam, oil, water, gas, etc.—serving a load in a power plant, process plant, ship, barge, aircraft, or other type of installation. In piping systems having very little or no increase in temperature during operation, the steps for computing the expansion can be eliminated. In this type of installation, the weight of the piping is the primary consideration in the choice of the hangers.

TABLE 27 Hanger-Rod Load-Carrying Capacity (Hot-Rolled Steel Rod)

Nominal diameter of rod, in (mm)	Thread root area, in² (mm²)	Maximum safe load on rod, lb (kN), at rod temperature of:	
		450°F (232°C)	750°F (399°C)
⅜ (9.5)	0.068 (43.9)	610 (2.7)	510 (2.3)
½ (12.7)	0.126 (81.3)	1130 (5.0)	940 (4.2)
⅝ (15.9)	0.202 (130.3)	1810 (8.1)	1510 (6.7)
¾ (19.1)	0.302 (194.8)	2710 (12.1)	2260 (10.1)
⅞ (22.2)	0.419 (270.3)	3770 (16.8)	3150 (14.0)
1 (25.4)	0.552 (356.1)	4960 (22.1)	4150 (18.5)

If desired, hanger loads can also be determined by taking moments about an arbitrarily selected axis on either side of the hanger. This method gives the same results as the procedure used above. The weight of bends is assumed to be concentrated at the center of gravity of each bend, whereas the weight of valves is assumed to be concentrated at the vertical centerline of the valve. Figure 22 shows typical moment arms, a and c, for valves and other fittings. The moment for W_1 about the hanger to the left of it is W_1a, and the moment for W_2 is W_2c about the hanger to the right of it. The weight of the pipe is assumed to be concentrated at a point midway between the hangers, and the moment is (weight of pipe, lb)(distance between hangers, ft/2). The method given here was developed by Frank Kamarck, Mechanical Engineer, and reported in *Power* magazine.

HANGER SPACING AND PIPE SLOPE FOR AN ALLOWABLE STRESS

An 8-in (203.2-mm) schedule 40 water pipe has an allowable bending stress of 10,000 lb/in^2 (68,950 kPa). What is the maximum allowable distance between

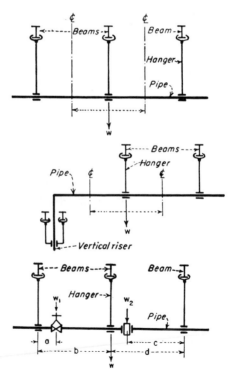

FIGURE 22 Compute hanger loads of uniformly loaded pipes as shown. Use beam relations for concentrated loads. (*Power.*)

hangers for this pipe? What slope will the allowable hanger span require to prevent pocketing of water in the pipe? Describe the method for computing hanger span and pipe slope for empty pipe. How are hanger distances computed when the pipe contains concentrated loads?

Calculation Procedure:

1. Compute the allowable span between hangers
For a pipe filled with water, $S = WL^2/8m$, where S = bending stress in pipe, lb/in^2; W = weight of pipe and water lb/lin in; L = maximum allowable distance between hangers, in; m = section modulus of pipe, in^3. By using a table of pipe properties, as in Crocker and King—*Piping Handbook*, $L = (8mS/W)^{0.5} = (8 \times 16.81 \times 10,000/4.18)^{0.5} = 568$ in, or $568/12 = 47.4$ ft (14.5 m).

2. Compute the pipe slope required by the span
To prevent pocketing of water or condensate at the low point in the pipe, the pipe must be pitched so that the outlet is lower than the lowest point in the span. When the pipe has no concentrated loads—such as valves, cross connections, or meters—the deflection of the pipe is y in = $22.5wl^4/(EI)$, where w = weight of the pipe and its contents, lb/ft; l = distance between hangers, ft; E = modulus of elasticity of pipe, lb/in^2 = 30×10^6 for steel; I = moment of inertia of the pipe, in^4. Substituting values gives $y = (22.5)(50.24)(47.4)^4/[(30 \times 10^6)(72.5)] = 2.61$ in (66.3 mm).

With the deflection y known, the pipe slope, expressed as 1 in (25.4 mm) per G ft of pipe length, is 1 in (25.4 mm) per G ft = $\frac{1}{4}y$, or $G = (47.4)/[(4)(2.61)] = 4.53$. Thus, a pipe slope of 1 in (25.4 mm) in 4.53 ft (1.38 m) is necessary to prevent pocketing of the water when the hanger span is 47.4 ft (14.5 m). With this slope, the outlet of the pipe would be $47.4/4.53 = 10.45$ in (265.4 mm) below the inlet.

3. Compute the empty-pipe hanger span and pipe slope
Use the same procedure as in steps 1 and 2, except that the empty weight of the pipe is substituted in the equations instead of the weight of the pipe when full of water. For pipes containing steam, gas, or vapor, compute the flowing-fluid weight and add it to the pipe weight. Follow the same procedure for insulated pipes, adding the insulation weight to the pipe weight.

4. Determine the hanger span and slope with concentrated loads
Hanger span and pipe slope can be computed from standard beam relations. However, most piping designers use the deflection chart and deflection factors for concentrated loads in Crocker and King—*Piping Handbook*. The chart and correction factors simplify the calculations considerably. The computation involves only simple multiplication and division.

Related Calculations. Use this procedure for piping in any type of installation—power, process, marine, industrial, or utility—for any type of liquid, vapor, or gas.

EFFECT OF COLD SPRING ON PIPE ANCHOR FORCES AND STRESSES

A carbon molybdenum pipe operates at 800°F (427°C) and has an anchor force of 5000 lb (22.2 kN) and a maximum bending stress s_b of 15,000 lb/in^2 (103.4 MPa) without cold spring. Compute the anchor force and bending stress in the hot and cold condition when the pipe is cold-sprung an amount equal to the expansion e and $0.5e$. The total expansion of the pipe is 24 in (609.6 mm).

Calculation Procedure:

1. Compute the hot-condition force and stress
The allowable cold-spring adjustment is expressed as a ratio $(e - 2S/3)/e$, where e = the total expansion of the pipe, in; S = cold-spring distance, in. This ratio is multiplied by the original anchor force and bending stress at the maximum operating temperature *without* cold spring to find the anchor force and bending stress *with* cold spring in the hot condition. If the ratio is less than $2/3$, the value of $2/3$ is used where maximum credit for cold spring is desired.

For this pipe, with maximum cold spring, the ratio $= (24 - 2 \times 24/3)/24 = 1/3$. Since this is less than $2/3$, use $2/3$. Then, the anchor force $F = (2/3)(5000) = 3333$ lb (14.8 kN), and the bending stress $x_b = (2/3)(15,000) = 10,000$ lb/in² (68.9 MPa).

With $S = 0.5e = (0.5)(24) = 12$ in (304.8 mm), the ratio $= (24 - 2 \times 12/3)/24 = 2/3$. Hence, $F = (2/3)(5000) = 3333$ lb (14.8 kN); $s_b = (2/3)(15,000) = 10,000$ lb/in² (68.9 MPa).

2. Compute the cold-condition force and stress
For the cold condition, the adjustment ratio $= -S/eM_R$, where M_R = modulus ratio for the pipe material = modulus of elasticity, lb/in², of the pipe material at the operating temperature, °F/modulus of elasticity of the pipe material, lb/in², at 70°F (21.1°C). For this pipe, $M_R = 0.865$, from a table of pipe properties. The minus sign in the ratio indicates that the anchor force and stress are reversed in the cold condition as compared with the hot condition.

For this pipe, with maximum cold spring, the ratio $= -24/[(24)(0.865)] = -1.156$. Then the anchor force in the cold condition $= (-1.156)(15,000) = -5790$ lb (25.7 kN), and the bending stress $= (-1.156)(15,000) = -17,350$ lb/in² (119.6 MPa).

With $S = 0.5e = (0.5)(24) = 12$ in (304.8 mm), the ratio $= -12/[(24)(0.865)] = -0.578$. Then the anchor force in the cold condition $(-0.578)(5000) = -2895$ lb (12.9 kN), and the bending stress $= (-0.578)(15,000) = -8670$ lb/in² (59,771 kPa).

These calculations show that cold spring reduces the anchor force and bending stress when the pipe is in the hot condition, step 1. With a cold spring of one-half the pipe expansion, the anchor force and bending stress are reduced and reversed when in the cold condition, step 2. When the cold spring equals the expansion, the anchor force and bending stress increase in the cold condition, step 2.

Related Calculations.   Use this procedure for a pipe conveying steam, oil, gas, water, and similar vapors, liquids, and gases.

REACTING FORCES AND BENDING STRESS IN SINGLE-PLANE PIPE BEND

Determine the horizontal and vertical reacting forces in the single-plane pipe bend of Fig. 20 if the pipe is 6-in (152.4-mm) schedule 40 carbon steel A106 seamless operating at 500°F (260°C). What is the maximum bending stress in the pipe and the resultant reacting or anchor force? Determine the maximum bending stress if a long-radius welded elbow is used at point C, Fig. 20. Use the tabular method of solution.

Calculation Procedure:

1. Compute the horizontal reacting or anchor force

Several methods are available for determining the reacting or anchor forces and maximum bending stress in a single-plane pipe bend. Crocker and King—*Piping Handbook* presents simplified, analytical, and graphical methods for computing forces and stresses in single- and multiplane piping systems. Another useful reference, *Design of Piping Systems,* written by members of the engineering departments of the M. W. Kellogg Company, presents both simplified and analytical methods and an excellent history and discussion of piping flexibility analysis. Probably the simplest method for routine piping flexibility analyses is that developed by the Grinnell Company, Inc., and S. W. Spielvogel. This method uses tabulated constants for specific pipe shapes in one, two, or three planes. It is satisfactory for the majority of piping problems met in normal engineering practice. To assist the practicing engineer, a number of Grinnell-Spielvogel tabulations for common pipe shapes are included here. For uncommon pipe shapes, refer to Grinnell Company—*Piping Design and Engineering* or to Spielvogel—*Piping Steam Calculations Simplified.* Both these references contain complete tabulations for a variety of pipe shapes.

To apply the Grinnell-Spielvogel solution procedure, compute the horizontal reaction force F_x lb from $F_x = k_x \, cI_p/L_2$, where k_x = a constant from Table 28 for the bend shape shown in Fig. 20; c = expansion factor = (pipe expansion, in/ 100 ft)$(EM_R/172,800)$, where E = modulus of elasticity of the pipe material being used, lb/in^2; M_R = modulus ratio = E at the operating temperature, F/E at 70°F (21.1°C) = 0.932; I_p = moment of inertia of pipe cross section, in^4; L = length of bend, ft, as shown in Fig. 23.

To enter Table 28 for the shape in Fig. 23, the values of L/a and L/h must be known, or $L/a = 40/20 = 2$; $L/h = 40/10 = 4$. Entering Table 28 at these values, read $k_x = 91$; $k_y = 21$; $k_b = 120$. From the Spielvogel c table or by computation, $c = 570$ for carbon-steel pipe operating at 500°F (260°C). From a table of pipe properties, $I_p = 28.14$ in^4 (1171.3 cm^4) for 6-in (152.4-mm) schedule 40 pipe. Then $F_x = (91)(570)(28.14)/(40)^2 = 912$ lb (4057 N).

2. Compute the vertical reacting or anchor force

Use the same procedure as in step 1, except that the vertical reacting force F_y lb = $k_y cI_p L^2$, or $F_y = (21)(570)(28.14)/(40)^2 = 211$ lb (939 N), by using the appropriate value from Table 2.

3. Compute the resultant reacting or anchor force

The resultant reacting or anchor force F lb is found by drawing and solving the force triangle in Fig. 23. From the pythagorean theorem $F = (912^2 + 211^2)^{0.5} = 936$ lb (4163 N). Draw the force triangle to scale, as shown in Fig. 23.

4. Compute the maximum bending stress in the pipe

The pipe bending stress s_b lb/in^2 is found in a similar manner from $s_b = k_b cD/L$, where k_b = bending-stress factor from Table 28; D = outside diameter of pipe, in.

TABLE 28 U Shape and Single Tangent

Reacting Force $\qquad F_x = k_x \cdot c \cdot \dfrac{I_p}{L^2}$ lb (N)

Reacting Force $\qquad F_y = k_y \, c \cdot \dfrac{I_p}{L^2}$ lb (N)

Maximum Bending Stress $\quad s_B = k_b \cdot c \cdot \dfrac{D}{L}$ lb/in^2 (Pa)

I_p in in^4 (cm^4) L in ft (m) D in in (cm)

L/h	L/a 1.5			L/a 2			L/a 3		
	k_x	k_y	k_b	k_x	k_y	k_b	k_x	k_y	k_b
1.0	2.63 (0.0261)	0.75 (0.0074)	10.5 (8,690)	2.8 (0.0278)	1.41 (0.0140)	11.3 (9,350)	3.3 (0.0327)	2.3 (0.0228)	12.5 (10,300)
2.0	14.5 (0.1439)	3.4 (0.0337)	33.6 (27,800)	16 (0.1588)	5.8 (0.0575)	38 (31,400)	20 (0.1984)	8.4 (0.0833)	42 (34,700)
3.0	39 (0.3870)	7.7 (0.0764)	67 (55,400)	45 (0.4465)	12.4 (0.1230)	75 (62,100)	53 (0.5259)	16 (0.1588)	77 (63,700)
4.0	79 (0.7838)	13.5 (0.1339)	108 (89,400)	91 (0.9029)	21 (0.2084)	120 (99,300)	108 (1.072)	26 (0.2580)	124 (103,000)
5.0	139 (1.3792)	21.8 (0.2163)	159 (131,600)	156 (1.548)	31 (0.3076)	173 (143,000)	185 (1.836)	37 (0.3671)	174 (144,000)

8.70

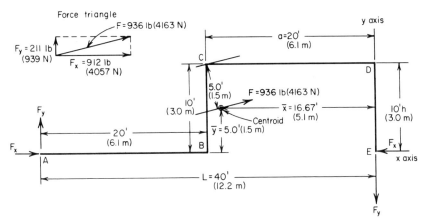

FIGURE 23   U-shaped pipe with single tangent.

For 6-in (152.4-mm) schedule 40 pipe having an outside diameter of 6.625 in (168.3 mm), $s_b = (120)(570)(6.625)/40 = 11,330$ lb/in² (78.1 MPa).

5.   Determine the bending stress in the welded elbow

The tables presented here are accurate when all the turns in the piping system analyzed are miters or rigid fittings. When all the turns are welded elbows or bends, the anchor forces derived from Table 28 are accurate for practical systems. The actual forces will be somewhat smaller than the values obtained from Table 28. Stresses in the elbows or bends may, however, exceed the values computed from Table 28 if the stress intensification factor β for these curved sections is >1. If the proportion of the straight to curved pipe is large, use the following procedure to obtain a close approximation of the stress in the curved section:

Determine the value of β from a table of pipe properties. For a 6-in (152.4-mm) schedule 40 long-radius welded elbow, $\beta = 2.22$. Therefore, the actual stress may exceed the table-computed stress, because $\beta > 1$.

Lay out the pipe bend to scale and compute the centroid of the bend by taking line moments about the x and y axes, Fig. 23.

	x axis			y axis	
AB	$(20')(0')$ =	0	AB	$(20')(30')$ =	600
BC	$(10')(5')$ =	50	BC	$(10')(20')$ =	200
CD	$(20')(10')$ =	200	CD	$(20')(10')$ =	200
DE	$(10')(5')$ =	50	DE	$(10')(0')$ =	0
	60	300		60	1000

$\bar{y} = 300/60 = 5$ ft (1.5 m) $\bar{x} = 1000/60 = 16.67$ ft (5.1 m)

In this calculation, the first value is the length of the pipe segment, and the second value is the distance of the center of gravity of the segment from the axis.

For a straight section of pipe, the center of gravity is taken as the midpoint of the pipe section. The welded elbows are ignored in stress calculations on table values.

Lay off $\bar{x}$ and $\bar{y}$ to scale, Fig. 23. Scale the distance to the tangent to the centerline of the long-radius elbow at C. This distance $d = 5.0$ ft (1.52 m). The moment at point C, $M_c = Fd$ lb·ft; or $m_c = 12Fd$ lb·in (1.36 N·m), and $m_c = 12(936)(5) = 56,160$ lb·in (6.34 kN·m), after force F is transposed from the force triangle to the centroid, Fig. 23.

The bending stress at any point in a pipe is $s_b = m\beta/S_m$, where S_m = section modulus of the pipe cross section, in³. For 6-in (152.4-mm) schedule 40 pipe, $S_m = 8.50$ in³ (139.3 cm³), from a table of pipe properties. Then $s_b = (56,160)(2.22)/8.50 = 14,700$ lb/in² (101.3 MPa). This is somewhat greater than the 11,330 lb/in² (78.1 MPa) computed in step 4 but within the allowable stress of 15,000 lb/in² (103.4 MPa) for seamless carbon steel A106 pipe at 500°F (260°C).

By inspection of the scale drawing, Fig. 23, the stress in the long-radius elbows at B and D is less than at C because the moment arm at each of these points is less than at C.

Related Calculations.   Tables 29, 30, and 31 present Grinnell-Spielvogel reaction and stress factors for three other single-plane bends—90° turn, U shape with equal tangents, and U shape with unequal legs. Use these tables and the factors in them in the same way as described above. Correct for curved elbows in the same manner. The tables can be used for piping conveying steam, water, gas, oil, and

TABLE 29   90° Turn

Reacting Force	$F_x = k_x \cdot c \cdot \dfrac{I_p}{L^2}$ lb (N)
Reacting Force	$F_y = k_y \cdot c \cdot \dfrac{I_p}{L^2}$ lb (N)
Maximum Bending Stress	$s_B = k_b \cdot c \cdot \dfrac{D}{L}$ lb/in² (Pa)

I_p in in⁴ (cm⁴) L in ft (m) D in in (cm)

L/h	k_x	k_y	k_b
1.0	12.0 (0.1191)	12.0 (0.1191)	36 (29,800)
2.0	54.0 (0.5358)	16.6 (0.1647)	102 (84,400)
3.0	150 (1.488)	23.5 (0.2332)	209 (173,000)
4.0	315 (3.125)	31.5 (0.3125)	349 (289,000)
5.0	570 (5.656)	39.5 (0.3919)	528 (437,000)

TABLE 30 U Shape with Equal Tangents

Reacting Force $\qquad F_x = k_x \cdot c \cdot \dfrac{I_P}{L^2}$ lb (N)

Maximum Bending Stress $\quad s_B = k_b \cdot c \cdot \dfrac{D}{L}$ lb/in^2 (Pa)

I_P in in^4 (cm^4) L in ft (m) D in in (cm)

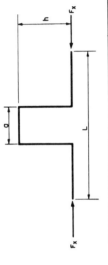

	L/a									
	2		**3**		**4**		**5**		**6**	
L/h	k_x	k_b	k_x	k_b	k_x	k_b	k_x	k_b	k_x	k_b
1.0	2.40 (0.0238)	7.20 (5,960)	2.46 (0.0244)	8.2 (6,780)	2.52 (0.0250)	8.82 (7,300)	2.58 (0.0256)	9.29 (7,690)	2.64 (0.0262)	9.69 (8,020)
2.0	12.00 (0.1191)	18.00 (14,900)	12.5 (0.1240)	21.8 (18,000)	13.24 (0.1314)	24.8 (20,500)	13.87 (0.1376)	27.1 (22,400)	14.4 (0.1429)	28.8 (23,800)
3.0	29.45 (0.2922)	29.45 (24,400)	31.2 (0.3096)	37.4 (30,900)	33.6 (0.3334)	43.7 (36,200)	35.8 (0.3552)	48.7 (40,300)	37.7 (0.3741)	52.7 (43,600)
4.0	54.9 (0.5447)	41.1 (34,000)	58.5 (0.5804)	53.6 (44,300)	64.0 (0.6350)	64.0 (53,000)	69.1 (0.6856)	72.5 (60,000)	73.6 (0.7303)	79.7 (65,900)
5.0	88.2 (0.8751)	52.9 (43,800)	95.3 (0.9456)	70.8 (58,600)	104.6 (1.0378)	85.2 (70,500)	114.7 (1.1380)	97.8 (80,900)	122.5 (1.2154)	107.5 (88,900)

TABLE 31 U Shape with Unequal Legs

Reacting Force $\qquad F_x = k_x \cdot c \cdot \dfrac{I_P}{L^2}$ lb (N)

Reacting Force $\qquad F_y = k_y \cdot c \cdot \dfrac{I_P}{L^2}$ lb (N)

Maximum Bending Stress $\quad s_B = k_b \cdot c \cdot \dfrac{D}{L}$ lb/in^2 (Pa)

I_P in in^4 (cm^4) L in ft (m) D in in (cm)

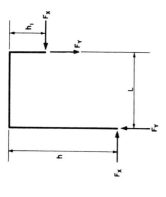

L/h	4/3			2			3		
	k_x	k_y	k_b	k_x	k_y	k_b	k_x	k_y	k_b
0.2	0.07 (0.0007)	0.6 (0.0059)	1.5 (1,240)	0.29 (0.0029)	1.8 (0.0179)	7 (5,790)	0.53 (0.0053)	3.4 (0.0337)	11 (9,100)
0.8	2.4 (0.0238)	0.9 (0.0089)	9.5 (7,860)	3.6 (0.0357)	2.5 (0.0248)	15 (12,400)	4.8 (0.0476)	4.4 (0.0436)	20 (16,500)
1.0	4.3 (0.0127)	1.2 (0.0119)	16 (13,200)	6.2 (0.0615)	3.0 (0.0298)	21 (17,400)	8 (0.0794)	4.9 (0.0486)	26 (21,500)
2.0	27 (0.2679)	2.3 (0.0228)	58 (48,000)	37 (0.3671)	6.0 (0.0595)	75 (62,100)	44 (0.4366)	8.0 (0.0883)	88 (72,800)
3.0	81 (0.8037)	3.8 (0.0377)	124 (102,600)	110 (1.0914)	10.0 (0.0992)	162 (134,000)	128 (1.2700)	15 (0.1488)	185 (153,000)

h/h_1

similar liquids, vapors, or gases. For bends of different shape, the analytical method must be used.

REACTING FORCES AND BENDING STRESS IN A TWO-PLANE PIPE BEND

Determine the horizontal reacting forces and bending and torsional stresses in the two-plane pipe bend shown in Table 32 if the dimensions of the bend are $L = 20$ ft (6.1 m); $h = 5$ ft (1.5 m); $a = 5$ ft (1.5 m); $b = 5$ ft (1.5 m). Use the tabular method of solution. The pipe is a 10-in (254-mm) carbon steel schedule 80 line operating at 750 lb/in² (gage) (5170.5 kPa) and 750°F (398.9°C). Determine the combined stress in the pipe.

Calculation Procedure:

1. Compute the tabular factors for the pipe bend
To apply the Grinnell-Spielvogel method to two-plane pipe bends, three tabular factors are required: L/a, a/b, and L/h. From the given values, $L/a = 20/5 = 4$; $a/b = 5/5 = 1$; $L/h = 20/5 = 4$.

2. Determine the force and stress factors for the pipe
From Table 32, for the factors in step 1, $k_x = 21.3$; $k_b = 24.5$; $k_t = 7.40$.

3. Compute the horizontal reacting force of the bend
The horizontal reacting force $F_x = k_x cI_p/L^2$, where the symbols are the same as in the preceding calculation procedure, except for L. Substituting the values for 10-in (254-mm) carbon-steel schedule 80 pipe operating at 750°F (399°C), we get $F_x = (21.3)(874)(244.9)/(20)^2 = 11,380$ lb (52.6 kN).

4. Compute the bending stress in the pipe
The bending stress in the pipe is found from $s_b = kcD/L$, where the symbols are the same as in the previous calculation procedure. Substituting values gives $s_b = (24.5)(874)(10.75)/(20) = 11,510$ lb/in² (79.4 MPa). Table 32 shows that the maximum combined stress in the pipe occurs at the two upper bends, D.

5. Compute the torsional stress in the pipe
The torsional stress in the pipe is found from $s_t = k_t cD/L$, where the symbols are the same as in the previous calculation procedure. Substituting values yields $s_t = (7.40)(874)(10.75)/20 = 3475$ lb/in² (24 MPa). Table 32 shows that the maximum combined stress in the pipe occurs at the two upper bends, D.

6. Determine the combined stress in the pipe
For any multiplane piping system, the combined stress s_{co} lb/in² $= 0.5\{s_1 + s_c + [4s_t^2 + (s_1 - s_c)^2]^{0.5}\}$. In this equation, $s_1 = s_b + s_p$, where s_p = pressure due to internal pressure, lb/in² (kPa); s_c = circumferential or hoop stress, lb/in² (kPa); other symbols are as given earlier. Also, $s_p = pA_i/A_m$, where p = operating pressure, lb/in² (gage) (kPa); A_i = inside area of pipe cross section, in² (cm²); A_m = metal area of pipe cross section, in² (cm²). Likewise, $s_c = p (D - t)/2t$, where D = outside diameter of pipe, in; t = pipe-wall thickness, in (cm).

TABLE 32 Two-Plane U with Tangents

Reacting Force $\quad F_x = k_x \cdot c \cdot \dfrac{I_P}{L^2}$ lb (N)

Bending Stress $\quad s_B = k_B \cdot c \cdot \dfrac{D}{L}$ lb/in^2 (Pa)

Torsional Stress $\quad s_T = k_T \cdot c \cdot \dfrac{D}{L}$ lb/in^2 (Pa)

I_P in in^4 (cm^4) $\quad L$ in ft (m) $\quad D$ in in (cm)

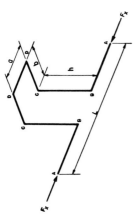

L/a = 4

L/h	a/b = 0.25			0.5			1			2		
	k_x	k_b	k_t	k_x	k_b	k_t	k_x	k_b	k_t	k_x	k_b	k_t
1	0.67 (0.0066)	D 3.20 (2,650)	...	1.22 (0.0121)	A 4.35 (3,600)	A 0.30 (250)	1.67 (0.0166)	A 5.2 (4,300)	A 0.15 (120)	2.0 (0.0198)	C 6.3 (5,200)	
2	1.35 (0.0134)	D 5.80 (6,800)	...	4.30 (0.0427)	D 9.96 (8,240)	D 2.45 (2,030)	6.96 (0.0691)	A 11.0 (9,100)	...	9.3 (0.0923)	C 15.0 (12,400)	
3	1.70 (0.0169)	D 7.00 (5,790)	...	6.23 (0.0618)	D 13.8 (11,400)	D 2.28 (1,890)	14.0 (0.1389)	D 16.5 (13,700)	D 6.55 (5,420)	21.2 (0.2103)	C 24.0 (19,900)	
4	1.88 (0.0187)	D 7.44 (6,150)	...	7.84 (0.0778)	D 16.9 (14,000)	D 2.09 (1,730)	21.3 (0.2113)	D 24.5 (20,300)	D 7.40 (6,120)	36.2 (0.3592)	C 30.0 (24,800)	
5	2.01 (0.0199)	D 7.75 (6,410)	...	8.94 (0.0887)	D 18.8 (15,600)	D 1.89 (1,560)	27.8 (0.2758)	D 31.4 (26,000)	D 7.75 (6,410)	52.6 (0.5219)	D 31.0 (25,600)	D 17.3 (14,300)

Computing stress values for this 10-in (254-mm) schedule 80 carbon-steel pipe operating at 750 lb/in² (gage) (5.2 MPa) and 750°F (399°C), and using values from a table of pipe properties, we get $s_p = (750)(71.8/18.92) = 2845$ lb/in² (19.6 MPa); $s_c = (750)(10.75 - 0.593)/2(0.593) = 6420$ lb/in² (44.3 MPa). Then $s_1 = s_b + s_p = 11,510 + 2845 = 14,355$ lb/in² (99 MPa), where s_b is from step 4. By substituting in the combined-stress equation, $s_{co} = 0.5\{14,355 + 6420 + [4 \times 3475^2 + (14,355 - 6420^2]^{0.5}\} = 16,648$ lb/in² (114.8 MPa). This is higher than the stress allowed in carbon-steel pipe by the ANSA *Piping Code,* unless the pipe conforms to the special conditions of certain paragraphs of the *Code.*

Related Calculations. Use this procedure for piping conveying steam, water, gas, oil, and similar liquids, vapors, or gases. For bends of different shape, the analytical method is generally used.

REACTING FORCES AND BENDING STRESS IN A THREE-PLANE PIPE BEND

Determine the three reacting forces and moments and bending and torsional stresses in the three-plane pipe bend shown in Table 33 if the dimensions of the bend are $L_1 = 20$ ft (6.1 m), $L_2 = 10$ ft (3.0 m), $L_3 = 5$ ft (1.5 m). The pipe is 10-in (254-mm) carbon-steel schedule 80 operating at 750 lb/in² (gage) (5.2 MPa) and 750°F (399°C).

Calculation Procedure:

1. Compute the tabular factors for the pipe bend
From the Grinnell-Spielvogel method, the two tabular factors required are $m = L_1/L_3$ and $n = L_2/L_3$; or $m = 20/5 = 4$ and $n = 10/5 = 2$.

2. Determine the force and stress factors for the pipe
From Table 33, for the factors in step 1, $k_b = 8.0$; $k_t = 3.6$; $k_x = 1.48$; $k_y = 0.13$; $k_z = 0.80$; $k_{xy} = 1.3$; $k_{xz} = 1.2$; $k_{yz} = 0.51$.

3. Compute the longitudinal reaction force of the bend
The horizontal reacting force $F_x = k_x c I_p/L_3^2$, where the symbols are the same as in the preceding calculation procedure, except for L_3. By substituting values for 10-in (254-mm) carbon-steel schedule 80 pipe operating at 750°F (398.9°C), $F_x = (1.48)(874)(244.9)/(5)^2 = 12,680$ lb (56.4 kN).

4. Compute the vertical reacting force of the bend
The vertical reacting force $F_y = k_y c I_p/L_3^2$, where the symbols are the same as in the preceding calculation procedure, except for L_3. Substituting values for this pipe, we find $F_y = (0.13)(874)(244.9)/(5)^2 = 1115$ lb (4.96 kN).

5. Compute the horizontal reacting force of the bend
The horizontal reacting force $F_2 = k_z c I_p/L_3^2$, where the symbols are the same as in the preceding calculation procedure, except for K_z and L_3. Substituting values for this pipe gives $F_z = (0.80)(874)(244.9)/(5)^2 = 6850$ lb (30.5 kN).

TABLE 33 Three-Dimensional 90° Turns

$L_1 \geq L_3$ $L_1 = m$ $\dfrac{L_2}{L_3} = n$

Bending Stress $s_B = k_b \cdot c \cdot \dfrac{D}{L}$ lb/in² (Pa)

Torsional Stress $s_T = k_t \cdot c \cdot \dfrac{D}{L_3}$ lb/in² (Pa)

Reacting Force $F_x = k_x \cdot c \cdot \dfrac{I_p}{L_3^2}$ lb (N)

Reacting Force $F_y = k_y \cdot c \cdot \dfrac{I_p}{L_3^2}$ lb (N)

Reacting Force $F_z = k_z \cdot c \cdot \dfrac{I_p}{L_3^2}$ lb (N)

Reacting Moment $M_{xy} = k_{xy} \cdot x \cdot \dfrac{I_p}{L_3}$ ft·lb (N·m)

Reacting Moment $M_{xx} = k_{xx} \cdot c \cdot \dfrac{I_p}{L_3}$ ft·lb (N·m)

Reacting Moment $M_{yz} = k_{yz} \cdot c \cdot \dfrac{I_p}{L_3}$ ft·lb (N·m)

I_p in in⁴ (cm⁴) L in ft (m) D in in (cm)

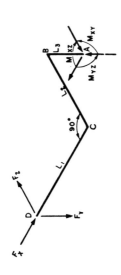

m = 3

n	k_b	k_t	k_x	k_y	k_z	k_{xy}	k_{xz}	k_{yz}
1	A 22.3 (18,500)	A 4.86 (4,020)	4.5 (0.045)	0.54 (0.0054)	1.50 (0.0148)	3.6 (0.0357)	1.62 (0.0161)	0.74 (0.0073)
2	D 9.3 (7,690)	D 0.15 (120)	1.4 (0.014)	0.22 (0.0022)	1.10 (0.0109)	1.1 (0.0109)	1.00 (0.0099)	0.71 (0.0070)
3	D 10.0 (8,270)	D 0.24 (199)	0.76 (0.007)	0.13 (0.0013)	1.08 (0.0107)	0.60 (0.0059)	0.74 (0.0073)	0.70 (0.0069)

m = 4

k_b	k_t	k_x	k_y	k_z	k_{xy}	k_{xz}	k_{yz}	n
A 24.0 (19,900)	A 6.0 (4,960)	4.80 (0.0476)	0.37 (0.0037)	1.10 (0.0110)	3.9 (0.0387)	2.0 (0.0199)	0.61 (0.0061)	1
A 8.0 (6,620)	A 3.6 (2,980)	1.48 (0.0147)	0.13 (0.0013)	0.80 (0.0079)	1.3 (0.0129)	1.2 (0.0119)	0.51 (0.0051)	2
D 7.26 (6,000)	D 0.10 (83)	0.76 (0.0075)	0.09 (0.0009)	0.65 (0.0064)	0.6 (0.0059)	0.88 (0.0088)	0.42 (0.0042)	3

6. Compute the bending and torsional stresses in the pipe
The bending stress $s_b = k_b cD/L_3$, where the symbols are the same as in the preceding calculation procedure, except for L_3. Substituting values for this pipe gives $s_b = (8.0)(874)(10.75)/5 = 15{,}020$ lb/in^2 (103.6 MPa).

The torsional stress $s_t = k_t cD/L_3$, where the symbols are the same as in the preceding calculation procedure, except for L_3. By substituting values for this pipe, $s_t = (3.6)(874)(10.75)/5 = 6760$ lb/in^2 (46.6 MPa).

7. Compute the three reacting moments at the pipe end
For each bending moment M ft·lb $= kcI_p/L_3$, where the symbols are the same as given in the previous steps in this calculation procedure, except that k is the appropriate bending-moment factor.

For the xy moment $M_{xy} = (1.3)(874)(244.9)/5 = 55{,}700$ ft·lb (75.5 kN·m). For the xz moment $M_{xz} = (1.2)(874)(244.9)/5 = 51{,}400$ ft·lb (69.7 kN·m). For the yz moment $M_{yz} = (0.51)(874)(244.9)/5 = 22{,}250$ ft·lb (30.2 kN·m).

Related Calculations. Use this procedure for piping conveying steam, water, gas, oil, and similar liquids, vapors, or gases. For bends of different shape, the analytical method must be used. Compute the combined stress in the same way as in step 6 of the previous calculation procedure. Table 33 shows that the maximum combined stress occurs at point A in this piping system.

ANCHOR FORCE, STRESS, AND DEFLECTION OF EXPANSION BENDS

Determine the deflection and anchor force in an 8-in (203.2-mm) schedule 40 double-offset expansion U bend having a radius of 64 in (1626 mm) if the bending stress is 10,000 lb/in^2 (68.9 MPa). What would the deflection and anchor force be with a bending stress of 15,000 lb/in^2 (103.4 MPa) if the bend tangents are guided and the pipe is carbon steel operating at 500°F (260°C)? With a bending stress of 8000 lb/in^2 (55.2 MPa)? Tabulate the deflection and anchor-force equations for the popular types of expansion bends when the expanding pipe is guided axially.

Calculation Procedure:

1. Compute the deflection of the pipe bend
For a double-offset expansion U bend, the deflection d in $= 0.728R^2K/D\beta$, where $R =$ bend radius, ft; $K =$ flexibility factor for curved pipe, from a table of pipe properties or from $K = (12\lambda^2 + 10)/(12\lambda^2 + 1)$, where $\lambda = 12tR/r^2$, where $t =$ pipe thickness, in, $r = (D - t)/.2$; $D =$ outside diameter of pipe, in; $\beta =$ stress coefficient for curved pipe from a table of pipe properties or from $\beta = (2K/3)[(6\lambda^2 + 5)/18]^{0.5}$ when $\lambda \leq 1.47$; $\beta = (12\lambda^2 - 2)/(12\lambda^2 + 1)$ when $\lambda > 1.47$.

For this bend, $R = 64/12 = 5.33$ ft (1.62 m); $K = 1.49$ from a table of pipe properties or by computation; $D = 8.625$ in (219.1 mm) from a table of pipe properties; $\beta = 0.86$ from a table of pipe properties, or by computation. Then $d = (0.728)(5.33)^2(1.49)/[(8.625)(0.86)] = 4.15$ in (105.4 mm).

2. Compute the anchor force of the pipe bend
For a double-offset expansion U bend, the anchor force F_x lb $= 976\, I_p/(RD\beta)$, where $I_p =$ moment of inertia of pipe cross section, in^4. For this pipe, use values

from a table of pipe properties; or computing the values, $F_x = (976)(72.5)/$ $[(5.33)(8.625)(0.86)] = 1790$ lb (7.96 kN).

3. Compute the deflection and anchor force for a larger bending stress
With a larger bending stress—15,000 lb/in² (103.4 MPa) in this instance—and a greater deflection at the higher stress d_h, solve $d_h = (d)$(allowable stress, lb/in²)/ $(10,000 M_R)$, or $d_h = (4.15)(15,000)/[10,000(0.932)] = 6.68$ in (169.7 mm). As in a preceding calculation, M_R = modulus ratio = 0.932.
   The anchor force at the larger bending stress is $F_h = F_x d_h M_R/d$, or $F_x = (1790)(6.68)(0.932)/4.15 = 2680$ lb (11,921 N).

4. Compute the deflection and anchor force for a smaller bending stress
Use the same equation as in step 3, except that the lower bending stress is substituted for the higher one. Or, $d_1 = (4.15)(8000)/[10,000(0.932)] = 3.56$ in (90.4 mm), and $F_1 = (1790)(3.56)(0.932)/4.15 = 1432$ lb (6370 N).

5. Tabulate the deflection and anchor-force equations
Do this as shown in the table on the following page.
   Related Calculations.   Use the procedures given here for piping conveying steam, water, oil, gas, air, and similar vapors, liquids, and gases. The value of E in step 5, 29×10^6 lb/in² (199.9 MPa), is satisfactory for pipes made of carbon steel, carbon moly steel, chromium moly steel, nickel steel, and chromium nickel steel. These materials are commonly used in piping systems requiring expansion bends.
   Note that the equations in step 5 apply to pipe bends having guides to direct the axial expansion of the pipe. This is the usual arrangement used today because unguided bends require too much space. For design of unrestrained bends, multiply d by 1.5 to find the deflection at the higher stress, as in step 3. This factor, 1.5, is an approximation, but it is on the safe side in almost every case. The equations given in step 5 are presented in great detail in Grinnell—*Piping Design and Engineering* and Crocker and King—*Piping Handbook.*

SLIP-TYPE EXPANSION JOINT SELECTION AND APPLICATION

Select and size slip-type expansion joints for the 20-in (508-mm) carbon-steel schedule 40 pipeline in Fig. 24 if the pipe conveys 125-lb/in² (gage) (861.6-kPa) steam having a temperature of 380°F (193°C). The minimum temperature expected in the area where the pipe is installed is 0°F (-17.8°C). Determine the anchor loads that can be expected. The steam inlet to the pipe is at A; the outlet is at F.

Calculation Procedure:

1. Determine the expansion of each section of pipe
From Fig. 25, the expansion of steel pipe at 380°F (193°C) with a 0°F (-17.8°C) minimum temperature is 3.4 in (88.9 mm) per 100 ft (30.5 m) of pipe. Expansion of each section of pipe is then e in = (3.4)(pipe length, ft/100). For AB, $e = (3.4)(140/100) = 4.76$ in (120.9 mm); for BC, $e = (3.4)(90/100) = 3.06$ in (77.7 mm); for CD, $e = (3.4)(220/100) = 7.48$ in (190 mm); for DE, $e = (3.4)(210/100) = 71.4$ in (1813.6 mm); for EF, $e = (3.4)(110/100) = 3.74$ in (95 mm).

Bend type	Deflection for 10,000 lb/in² (68.9-MPa) s_b		Anchor force for 10,000 lb/in² (68.9-MPa) s_b	
	$E = 29 \times 10^6$ lb/in²	$E = 199.9$ MPa	$E = 29 \times 10^6$ lb/in²	$E = 199.9$ MPa
Double-offset U	$d = 0.728R^2K/D\beta$	$d = 5.056 \times 10^{-3}R^2K/D\beta$	$F_x = 976\,I_p/RD\beta$	$F_x = 8070 \times I_p/RD\beta$
Expansion U bend (no tangents)	$d = 0.312R^2K/D\beta$	$d = 2.167 \times 10^{-3}R^2K/D\beta$	$F_x = 1667I_p/RD\beta$	$F_x = 13{,}780 \times I_p/RD\beta$
Expansion U bend [tangents = 2 ft (0.6 m)]	$d = [(0.312R^3 + 0.795R^2 + 0.624R)K + 0.132]/(R + 1)D\beta$	$d = [(2.167R^3 + 165R^2 + 3895R) \times 10^{-3}K + 24.7]/(R + 30)D\beta$	$F_x = \dfrac{1667I_p}{(R + 1)D\beta}$	$F_x = 13{,}780I_p/(R + 30)D\beta$
Expansion U bend (tangents = R)	$d = (0.577 + 0.011)R^2/D\beta$	$d = (4.007K + 0.076)R^2/D\beta$	$F_x = 1111I_p/RD\beta$	$F_x = 9190I_p/RD\beta$
Expansion U bend (tangents = 2R)	$d = (0.865K + 0.0662)R^2/D\beta$	$d = (6.00 \times 10^{-3}K + 0.459 \times 10^{-3})R^2/D\beta$	$F_x = 833I_p/RD\beta$	$F_x = 6890I_p/RD\beta$
Expansion U bend (tangents = 4R)	$d = (1.465K + 0.353)R^2/D\beta$	$d = (10.17K + 2.45) \times 10^{-3} \times R^2/D\beta$	$F_x = 556I_p/RD\beta$	$F_x = 4600I_p/RD\beta$
Double-offset U bend	$d = 0.260R^2K/D\beta$	$d = 1.806 \times 10^{-3}R^2K/D\beta$	$F_x = 1209I_p/RD\beta$	$F_x = 9997I_p/RD\beta$
Single-offset quarter bend	$d = 0.0366R^2K/D\beta$	$d = 2.5 \times 10^{-4}R^2K/D\beta$	$F_x = 2763I_p/RD\beta$	$F_x = 22{,}850I_p/RD\beta$
Circle bend	$d = 0.312R^2K/D\beta$	$d = 2.167 \times 10^{-3}R^2K/D\beta$	$\begin{cases} F_y = 0.066F_x \\ F_x = 1667I_p/RD\beta \end{cases}$	$\begin{cases} F_y = 0.066F_x \\ F_x = 13{,}780 \times I_p/RD\beta \end{cases}$

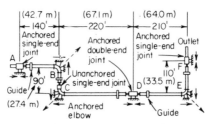

FIGURE 24 Slip-type expansion joints in a piping system. (*Yarway Corporation.*)

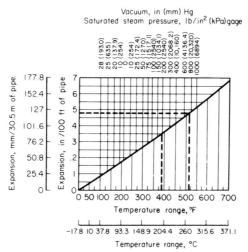

FIGURE 25 Expansion of steel pipe. (*Yarway Corporation.*)

2. Select the type and the traverse of each expansion joint

The slip-type expansion joint at *A* will absorb expansion from only one direction—the right-hand side. This expansion will occur in pipe section *AB* and is 4.76 in (120.9 mm) from step 1. Therefore, a single-end slip-type expansion joint (one that absorbs expansion on only one side) can be used. The traverse—the amount of expansion a slip joint will absorb—is usually given in multiples of 4 in (101.6 mm), that is, 4, 8, and 12 in (101.6, 203.2, and 304.8 mm). Hence, an 8-in (203.2-mm) traverse slip-type single-end joint will be suitable at *A* because the expansion is 4.76 in (120.9 mm). A 4-in (101.6-mm) traverse joint would be unsatisfactory because it could not absorb at 4.76-in (120.9-mm) expansion.

The next joint, at *C*, must absorb the expansion in the vertical pipe *BC*. Since the elbow beneath the joint is anchored, an unanchored joint can be used. With pipe expansion in only one direction—from *B* to *C*—a single-end joint can be used. Since the expansion of section *BC* is 3.06 in (77.7 mm), use a single-end 4-in (101.6-mm) traverse slip-type expansion joint, unanchored at *C*.

The expansion joint at *D* must absorb expansion from two directions—from *C* to *D* and from *E* to *D*. Therefore, a double-end joint (one that can absorb expansion on each end) must be used. The double-end joint must be anchored because the pipe expands *away* from the anchored elbow *C* in section *CD* and *away* from the anchored elbow *E* in section *DE*. In both instances the pipe expands *toward* the expansion joint at *D*.

The expansion in section *CD* is, from step 1, 7.48 in (190 mm), whereas the expansion in *DE* is 7.14 in (181.4 mm). Therefore, a double-end anchored joint with an 8-in (203.2-mm) traverse at *each* end will be suitable.

Since the pipe outlet is at *F* and there is no anchor in the pipe at *F*, the expansion joint at this point must be anchored. The pipe section between *E* and *F* will expand vertically upward into the joint for a distance of 3.74 in (95 mm), as computed in step 1. Therefore, a single-end anchored joint with a 4-in (101.6-mm) traverse will be suitable.

3. *Compute the anchor loads in the pipeline*

Use Fig. 26 to determine the anchor loads on intermediate and end anchors (those where the pipe makes a sharp change in direction). Enter Fig. 26 at the bottom at a pipe size of 20-in (508-mm) diameter, and project vertically upward to the dashed curve labeled *intermediate anchor—all pressures*. At the left read the anchor load at each intermediate anchor, *A*, *D*, and *F*, as 20,000 lb (88.9 kN). Note that the joint expansion load = joint contraction load = 20,000 lb (88.9 kN).

The end anchors, *B*, *C*, and *E*, have, from Fig. 26, a possible maximum load of 58,000 lb (258 kN), found by projecting vertically upward from the 20-in (508-mm) pipe size to 125-lb/in² (gage) (862-kPa) steam pressure, which lies midway

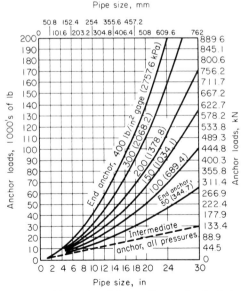

FIGURE 26 End- and intermediate-anchor loads in piping systems. (*Yarway Corporation.*)

TABLE 34   Guide and Support Spacing

Nominal pipe size, in (mm)	Distance between guide and joint, ft (m) Packing type		Distance between guides, ft (m)
	Gun	Gland	
18 (457)	24 (7.3)	11 (3.4)	100 (30.5)
20 (508)	25 (7.6)	12 (3.7)	105 (32)
24 (610)	26 (7.9)	12 (3.7)	110 (33.5)

between the 100- and 150-lb/in^2 (gage) (689.5- and 1034-kPa) curves. Indicate the possible maximum end-anchor loads by the solid arrows at each elbow, as shown in Fig. 24. The resultant R of the loads at any end anchor is found by the pythagorean theorem to be $R = (58,000^2 + 58,000^2)^{0.5} = 82,200$ lb (365.6 kN). Indicate the resultant by a dotted arrow, as shown in Fig. 24.

Contraction loads on the end anchors are in the reverse direction and consist only of friction. This friction load equals the joint expansion load, or 20,000 lb (88.9-kN). The resultant of the joint expansion loads is $(20,000^2 + 20,000^2)^{0.5} = 28,350$ lb (126.1 kN).

Locate guides within 25 or 12 ft (7.62 or 3.66 m) of the expansion joint, depending on the type of packing used, Table 34. These guides should allow free axial movement of the pipe into and out of the joint with minimum friction.

Related Calculations.   Use this procedure to choose slip-type expansion joints for pipes conveying steam, water, air, oil, gas, and similar vapors, liquids, and gases. In some instances, the gland friction and pressure thrust is used instead of Fig. 26 to determine anchor loads. With either method, the results are about the same.

CORRUGATED EXPANSION JOINT SELECTION AND APPLICATION

Select corrugated expansion joints for the 8-, 6-, and 4-in (203.2-, 152.4-, and 101.6-mm) carbon-steel pipeline in Fig. 27 if the steam pressure in the pipe is 75 lb/in^2 (gage) (517.1 kPa), the steam temperature is 340°F (171°C), and the installation temperature is 60°F (15.6°C).

Calculation Procedure:

1.   Determine the expansion of each section of pipe
From a table of thermal expansion of pipe, the expansion of carbon-steel pipe at 340°F (171°C) is 2.717 in/100 ft (2.26 mm/30.5 m) from 0 to 340°F (−17.8 to 171°C). Between 0 and 60°F (−17.8 and 15.6°C) the expansion is 0.448 in/100 ft (50 mm/30.5 m). Hence, the expansion between 60 and 340°F (15.6 and 171°C) is 2.717 − 0.448 = 2.269 in/100 ft (1.89 mm/m). This factor can now be applied to each length of pipe by finding the product of (pipe-section length, ft/100)(expansion, in/100 ft) = expansion of section, in = e.

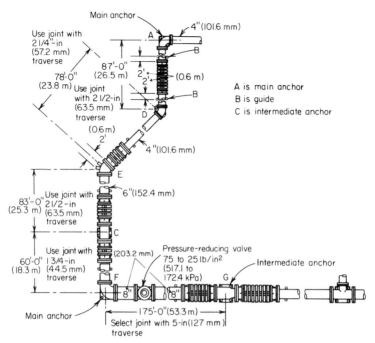

FIGURE 27 Piping system fitted with expansion joints. (*Flexonics Division, Universal Oil Products Company.*)

For section AD, $e = (87/100)(2.269) = 1.97$ in (50 mm); for DE, $e = (78/100)(2.269) = 1.77$ in (45 mm); for EC, $e = (83/100)(2.269) = 1.88$ in (47.8 mm); for CF, $e = (60/100)(2.269) = 1.36$ in (34.5 mm); for FG, $e = (175/100)(2.269) = 3.97$ in (100.8 mm).

In selecting corrugated expansion joints, the usual practice is to increase the computed expansion by a suitable safety factor to allow for any inaccuracies in temperature measurement. By applying a 25 percent safety[1] factor: for AD, $e = (1.97)(1.25) = 2.46$ in (62.5 mm); for DE, $e = (1.77)(1.25) = 2.13$ in (54.1 mm); for EC, $e = (1.88)(1.25) = 2.35$ in (59.7 mm); for CF, $e = (1.36)(1.25) = 1.70$ (43.2 mm); for FG, $e = (3.97)(1.25) = 4.96$ in (126 mm).

2. Select the traverse for, and type of, each expansion joint
Obtain corrugated-expansion joint engineering data, and select a joint with the next largest traverse for each section of pipe. Thus, traverse $AD \geq 2\frac{1}{2}$ in (63.5 mm); traverse $DE \geq 2\frac{1}{4}$ in (57.2 mm); traverse $EC \geq 2\frac{1}{2}$ in (63.5 mm); traverse $CF \geq 1\frac{3}{4}$ in (44.5 mm); traverse $FG \geq 5.0$ in (127 mm).

Two types of expansion joints are commonly used: free-flexing and controlled-flexing. Free-flexing joints are generally used where the pressures in the pipeline

---

[1]This value is for illustration purposes only. Contact the expansion-joint manufacturer for the exact value of the safety factor to use.

TABLE 35 Effective Area of Corrugated
Expansion Joints

Joint inside diameter		Joint effective area	
in	mm	in^2	cm^2
6	152.4	51.0	329.0
8	203.2	85.0	548.4
10	254.0	120.0	774.2
12	304.8	174.0	1122.6
14	355.6	215.0	1387.1
16	406.4	270.0	1741.9
18	457.2	310.0	1999.9
20	508.0	390.0	2516.1
24	609.6	540.0	3483.9

are relatively low and the required motion is relatively small. Controlled-flexing expansion joints are generally used for higher pressures and larger motions. Both types of expansion joints are available in stainless steel in both single and dual units. For precise data on a given joint being considered, consult the expansion-joint manufacturer. Corrugated expansion joints are characterized by their freedom from any maintenance needs.

3. Compute the anchor loads in the pipeline
Main anchors are used between expansion joints, as at F and A, Fig. 27, and at turns such as at F and A. The force[2] a main anchor must absorb is given by F_i lb $= F_p + F_e$, where $F_p =$ pressure thrust in the pipe, lb $= pA$, where $p =$ pressure in pipe, lb/in^2 (gage); $A =$ effective internal cross-sectional area of expansion joint, in^2 (see Table 35 for cross-sectional areas of typical corrugated joints); $F_e =$ force required to compress the expansion join, lb $= [300$ lb/in $(52.5$ N/mm)] (joint inside diameter, in) for stainless-steel self-equalizing joints, and [200 lb/in (35 N/mm)] (joint inside diameter, in) for copper nonequalizing joints. Determining the main anchor force for the 8-in (203.2-mm) pipeline gives $F_i = (75)(85) + (300)(8) = 8775$ lb (39.0 kN). In this equation, the area of 85 in^2 (548.3 cm^2) in the first term is obtained from Table 35.

The total force at a main anchor, as at A and F, Fig. 27, is the vector sum of the forces in each line leading to the anchor. Thus, at F, there is a force of 8775 lb (39.0 kN) in the 8-in (203.2-mm) line and a force of $F_i = (75)(51) + (300)(6) = 5625$ lb (25 kN) in the 6-in (152.4-mm) line connected to the elbow outlet. Since the elbow at F is a right angle, use the pythagorean theorem, or $R =$ resultant anchor force, lb $= (8775^2 + 5625^2)^{0.5} = 10,400$ lb (46.3 kN).

Where two lines containing corrugated expansion joints are connected by a bend of other than 90°, as at D and E, use a force triangle to determine the anchor force after computing F_i for each pipe. Thus, at E, F_i for the 6-in (152.4-mm) pipe $= 5625$ lb (25 kN), and $F_i = (75) \times (23.5) + (300)(4) = 2963$ lb (13.2 kN) for the 4-in

---

[2]This is an approximate method for finding the anchor force. For a specific make of expansion joint, consult the joint manufacturer.

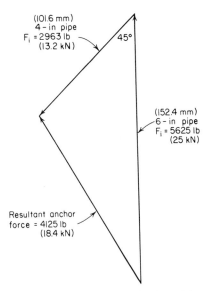

FIGURE 28   Force triangle for determining piping anchor force.

(101.6-m) pipe. Draw the force triangle in Fig. 28 with the 6-in (152.4-mm) pipe F_i and the 4-in (101.6-mm) pipe F_i as two sides and the bend angle, 45°, as the included angle. Connect the third side, or resultant, to the ends of the force vectors, and scale the resultant as 4125 lb (18.4 kN), or compute the resultant from the law of cosines. Find the resultant force at D in a similar manner as 2963 lb (13.2 kN).

Intermediate anchors, as at C and G, must withstand only one force—the unbalanced (differential) spring force. With approximate force calculations,[1] starting at C, for a 6-in (152.4-mm) expansion joint, $F_e = (300)(6) = 1800$ lb (8 kN). At G, for an 8-in (203.2-mm) expansion joint, $F_e = (300)(8) = 2400$ lb (10.7 kN). Thus, the loads the intermediate anchors must withstand are considerably less than the main-anchor loads.

Provide the pipe guides at suitable locations in accordance with the joint manufacturer's recommendations and at suitable intervals on the pipeline to prevent any lateral and buckling forces on the joint and adjacent piping. Intermediate anchors between two joints in a straight run of pipe ensure that each joint will absorb its share of the total pipe motion. Slope the pipe in the direction of fluid flow to prevent condensate accumulation. Use enough pipe hangers to prevent sagging of the pipe.

Related Calculations.   Use this procedure to choose corrugated-type expansion joints for pipes conveying steam, water, air, oil, gas, and similar vapors, liquids, and gases. When choosing a specific make of corrugated expansion joint, use the manufacturer's engineering data, where available, to determine the maximum allowable traverse. One popular make has a maximum traverse of 7.5 in (190.5 mm) or a maximum allowable lateral motion of 1.104 in (28.0 mm) in its various joint

---

[1]Consult the expansion-joint manufacturer for an exact procedure for computing the anchor forces.

sizes. The larger the lateral motion, the greater the number of corrugations required in the joint.

In some pipelines there is an appreciable pressure thrust caused by a change in direction of the pipe. This pressure or centrifugal thrust F_c is usually negligible, but the wise designer makes a practice of computing this thrust from $F_c = (2A\rho v^2/32.2) \times (\sin \theta/2)$ lb, where A = inside area of pipe, ft^2; ρ = density of fluid or vapor, lb/ft^3; v = fluid or vapor velocity, ft/s; θ = change in direction of the pipeline.

The number of corrugations required in a joint varies with the expansion and lateral motion to be absorbed. A typical free-flexing joint can absorb 6.25 in (158.8 mm) of expansion and a variable amount of lateral motion, depending on joint size and operating condition. Free-flexing joints are commonly built in diameters up to 48 in (1219 mm), while controlled-flexing joints are commonly built in diameters up to 24 in (609.6 mm). For a more precise calculation procedure, consult the Flexonics Division, Universal Oil Products Company.

DESIGN OF STEAM TRANSMISSION PIPING

Design a steam transmission pipe to supply a load that is 1700 ft (518.2 m) from the power plant. The terrain permits a horizontal run between the power plant and the load. Maximum steam flow required by the load is 300,000 lb/h (135,000 kg/h), whereas the average steam flow required is estimated as 150,000 lb/h (67,500 kg/h). The maximum steam pressure at the load must not exceed 150 lb/in^2 (abs) (1034.1 kPa) saturated. Superheated steam at 450 lb/in^2 (abs) (3102.7 kPa) and 600°F (316°C) is available at the power plant. Two schemes are proposed for the line: (1) Reduce the steam pressure to 180 lb/in^2 (abs) (1240.9 kPa) at the line inlet, thus allowing a 180 − 150 = 30-lb/in^2 (206.8-kPa) loss in the 1700-ft (518.2-m) long line. This scheme is called the *nominal pressure-loss line*. (2) Admit high-pressure steam to the line and thereby allow the steam pressure to fall to a level slightly greater than 150 lb/in^2 (abs) (1034.1 kPa). Since 600°F (316°C) steam would probably cause expansion and heat-loss difficulties in the pipe, assume that the inlet temperature of the steam is reduced to 455°F (235°C) in a desuperheater in the power plant. There is a 10-lb/in^2 (68.9-kPa) pressure loss between the power plant and the line, reducing the line inlet pressure to 440 lb/in^2 (abs) (3033.4 kPa). Since the pressure can fall about 440 − 150 = 290 lb/in^2 (1999.3 kPa), this will be called the *maximum pressure-loss line*. During design, determine which line is the most economical.

Calculation Procedure:

1. Determine the required pipe diameter for each condition
The average steam pressure in the nominal pressure-loss line is (inlet pressure + outlet pressure)/2 = (180 + 150)/2 = 165 lb/in^2 (abs) (1138 kPa). Use this average pressure to determine the pipe size, because the average pressure is more representative of actual conditions in the pipe. Assume that there will be a 5-lb/in^2 (34.5-kPa) pressure drop through any expansion bends and other fittings in the pipe. Then, the allowable friction-pressure drop = 30 − 5 = 25 lb/in^2 (172.4 kPa).

Use the Thomas saturated-steam formula to determine the required pipe diameter, or $d = (80,000W/Pv)^{0.5}$, where d = inside pipe diameter, in; W = weight of steam flowing, lb/min; P = average steam pressure, lb/in^2 (abs); v = steam velocity, ft/min. Assuming a steam velocity of 10,000 ft/min (3048 m/min), which is typical for a long steam transmission line, we get $d = [(80,000 \times 300,000/60)/ (165 \times 10,000)]^{0.5} = 15.32$ in (389.1 mm).

The inside diameter of a schedule 40 16-in (406-mm) outside-diameter pipe is, from a table of pipe properties, 15.00 in (381 mm). Assume that a 16-in (406-mm) pipe will be used if schedule 40 wall thickness is satisfactory for the nominal pressure-loss line. Note that the larger flow was used in computing the size of this line because a pipe satisfactory for the larger flow will be acceptable for the smaller flow.

The maximum pressure-loss line will have an average pressure that is a function of the inlet pressure at the pressure-reducing valve at the line outlet. Assume that there is a 10-lb/in² (68.9-kPa) drop through this reducing valve. Then steam will enter the valve at 150 + 10 = 160 lb/in² (abs) (1103 kPa), and the average line pressure = (440 + 160)/2 = 300 lb/in² (abs) (2068 kPa). Using a higher steam velocity [15,000 ft/min (4572 m/min)] for this maximum pressure-loss line than for the nominal pressure-loss line [10,000 ft/min (3048 m/min)], because there is a larger allowable pressure drop, compute the required inside diameter from the Thomas saturated-steam formula because the steam has a superheat of only 456.28 − 455.00 = 1.28°F (2.3°C). Or, $d = [(80,000 \times 300,000/60)/ (300 \times 15,000)]^{0.5} = 9.44$ in (239.8 mm). Since a 10-in (254-mm) schedule 40 pipe has an inside diameter of 10,020 in (254.5 mm), use this size for the maximum pressure-loss line.

2. Compute the required pipe-wall thickness
As shown in an earlier calculation procedure, the schedule number SN = $1000P_i/$ S. Assuming that seamless carbon-steel ASTM A53 grade A pipe is used for both lines, the *Piping Code* allows a stress of 12,000 lb/in² (82.7 MPa) for this material at 600°F (316°C). Then SN = (1000) × (435)/12,000 = 36.2; use schedule 40 pipe, the next largest schedule number for both lines. This computation verifies the assumption in step 1 of the suitability of schedule 40 for each line.

3. Check the pipeline for critical velocity
In a steam line, $p_c = W'/Cd^2$, where p_c = critical pressure in pipe, lb/in² (abs), W' = steam flow rate, lb/h; C = constant from Crocker and King—*Piping Handbook; d* = inside diameter of pipe, in.

When the pressure loss in a pipe exceeds 50 to 58 percent of the initial pressure, flow may be limited by the fluid velocity. The limiting velocity that occurs under these conditions is called the *critical velocity*, and the coexisting pipeline pressure, the *critical pressure.*

Critical velocity may limit flow in the 10-in (254-mm) maximum pressure-loss line because the terminal pressure of 150 lb/in² (abs) (1034 kPa) is less than 58 percent of 440 lb/in² (abs) (3033.4 kPa), the inlet pressure. Use the above equation to find the critical pressure. Or, $p_c = (300,000)/[(75.15)(10.02)^2] = 39.7$ lb/in² (abs) (273.7 kPa), using the constant from the *Piping Handbook* after interpolating for the initial enthalpy of 1205.4 Btu/lb (2804 kJ/kg), which is obtained from steam-table values.

Critical velocity would limit flow if the pipeline terminal pressure were equal to, or less than, 39.7 lb/in² (abs) (273.7 kPa). Since the terminal pressure of 150 lb/in² (1034.1 kPa) is greater than 39.7 lb/in² (abs) (237.7 kPa), critical velocity does not limit the steam flow. With smaller flow rates, the critical pressure will be lower because the denominator in the equation remains constant for a given pipe. Hence, the 10-in (254-mm) line will readily transmit 300,000-lb/h (135,000-kg/h) and smaller flows.

If critical pressure existed in the pipeline, the diameter of the pipe might have to be increased to transmit the desired flow. The 16-in (406.4-mm) line does not have to be checked for critical pressure because its final pressure is more than 58 percent of the initial pressure.

4. Compute the heat loss for each line
Assume that 2-in (50.8-mm) thick 85 percent magnesia insulation is used on each line and that the lines will run above the ground in an area having a minimum temperature of 40°F (4.4°C). Set up a computation form as follows:

Pipe size, in (mm)	16 (406.5)	10 (254.0)
Steam temperature, °F (°C)	373 (189)	455 (235)
Air temperature, °F (°C)	40 (4.4)	40 (4.4)
Temperature difference, °F (°C)	333 (184.6)	415 (184.6)
Insulation heat loss, Btu/(h·ft²·°F)° [W/(m²·°C)]	1.11 (6.3)	0.704 (3.99)
Heat loss, Btu/(h·1in ft) (W/m)	370 (356)	292 (281)
Heat loss, Btu/h (kW), for 1700 ft (518 m)	629,000 (184)	496,400 (145.6)
Total heat loss, Btu/h (kW), with a 25% safety factor	786,250 (230)	620,500 (182.0)
Heat loss, Btu/lb (W/kg) of steam, for 300,000-lb/h (135,000-kg/h) flow	2.62 (1.7)	2.07 (1.35)
Heat loss, Btu/lb (W/kg), for the average flow of 150,000 lb/h (67,500 kg/h)	5.24 (3.4)	4.14 (2.69)

°From table of pipe insulation, Ehret Magnesia Manufacturing Company.

In this form, the following computations were made for both pipes: heat loss, Btu/(h·lin ft) = [insulation heat loss, Btu/(h·ft·°F)] (temperature difference,°F); heat loss, Btu/h for 1700 ft (518.2 m) = [heat loss, Btu/(h·lin ft)] (1700); total heat loss, Btu/h, 25 percent safety factor = (heat loss, Btu/h) (1700 ft)(1.25); heat loss, Btu/lb steam = (total heat loss, Btu/h, with a 25 percent safety factor)/ (300,000-lb steam).

5. Compute the leaving enthalpy of the steam in each line
Acceleration of steam in each line results from an enthalpy decrease of $h_a = (v_2^2 - v_1^2)/2g(778)$, where h_a = enthalpy decrease, Btu/lb; v_2 and v_1 = final and initial velocity of the steam, respectively, ft/s; $g = 32.2$ ft/s². The velocity at any point x in the pipe is found from the continuity equation $v_x = (W'v_g)/3600 A_x$, where v_x = steam velocity, ft/s, when the steam volume is v_g ft³/lb, and A_x is the cross-sectional area of pipe, ft², at the point be considered.

For the 16-in (406.4-mm) nominal pressure-loss line with a flow of 300,000 lb/h (135,000 kg/h) at 180-lb/in² (abs) (1241-kPa) entering and 150-lb/in² (abs) (1034.1-kPa) leaving pressure, using steam and piping table values, $v_1 = (300,000)(2.53)/[(3600)(1.23)] = 171.5$ ft/s (52.3 m/s); $v_2 = 300,000(3.015)/ [(3600)(1.23)] = 205$ ft/s (62.5 m/s). Then $h_a = [(204.5)^2 - (171.5)^2]/ [(64.4)(778)] = 0.2504$ Btu/lb (0.58 kJ/kg), say 0.25 Btu/lb (0.58 kJ/kg).

By an identical calculation, h_a = 3.7 Btu/lb (8.6 kJ/kg) for the 10-in (254-mm) maximum pressure-loss line when the leaving steam is assumed to be 150 lb/in² (abs) (1034.1 kPa), saturated.

Enthalpy of the 180-lb/in² (abs) (1241-kPa) saturated steam entering the 16-in (406.4-mm) line is 1196.9 Btu/lb (2784 kJ/kg). Heat loss during 300,000-lb/h (135,000-kg/h) flow is 2.62 Btu/lb (6.1 kJ/kg), as computed in step 4. The enthalpy drop of 0.25 Btu/lb (0.58 kJ/kg) accelerates the steam. Hence, the calculated leaving enthalpy is 1196.9 − (2.62 + 0.25) = 1194.03 Btu/lb (2777.3 kJ/kg). The enthalpy of the leaving steam at 150 lb/in² (abs) (1034.1 kPa) saturated is 1194.1 Btu/lb (2777.5 kJ/kg). To have saturated steam leave the line, 1194.10 − 1194.03, or 0.07 Btu/lb (0.16 kJ/kg), must be supplied to the steam. This heat will be obtained from the enthalpy of vaporization given off by condensation of some of the steam in the line.

Make a group of identical calculations for the 10-in (254-mm) maximum pressure-loss line. The enthalpy of 440-lb/in (abs) (3033.4-kPa) 445°F (235°C) entering steam is 1205.4 Btu/lb (2803.8 kJ/kg), found by interpolation in the steam tables. Heat loss during 300,000-lb/h (135,000-kg/h) flow is 2.07 Btu/lb (4.81 kJ/kg). An enthalpy drop of 3.7 Btu/lb (8.6 kJ/kg) accelerates the steam. Hence, the calculated leaving enthalpy = 1205.4 − (2.07 + 3.7) = 1199.63 Btu/lb (2790.3 kJ/kg).

The enthalpy of the leaving steam at 150-lb/in² (abs) (1034-kPa) saturated is 1194.1 Btu/lb (2777.5 kJ/kg). As a result, under maximum flow conditions, the steam will be superheated from the entering point to the leaving point of the line. The enthalpy difference of 5.53 Btu/lb = 1199.63 − 1194.10) (12.9 kJ/kg) produces this superheat. Because the steam is superheated throughout the line length, condensation of the steam will not occur during maximum flow conditions.

For the most industrial applications, the steam leaving the line may be considered as saturated at the desired pressure. But for precise temperature regulation, some form of pressure-temperature control must be used at the end of long lines.

During average flow conditions of 150,000 lb/h (67,500 kg/h), the line heat loss is 4.14 Btu/lb (9.6 kJ/kg), as computed in step 4. The enthalpy drop to accelerate the steam is 0.925 Btu/lb (2.2 kJ/kg). As in the case of maximum flow, the steam is superheated throughout the length of the 10-in (254-mm) maximum pressure-loss line because the calculated leaving enthalpy is 1205.40 − 5.07 = 1200.33 Btu/lb (2791.9 kJ/kg).

6. Compute the quantity of condensate formed in each line

For either line, the quantity of condensate formed, lb/h = $C = W'(h_g$ at leaving pressure − calculated leaving h_g)/outlet pressure h_{fg}.

Using computed values from step 5 and steam-table values, we see the 16-in (406.4-mm) line with 300,000 lb/h (135,000 kg/h) flowing forms C = (300,000)(0.07)/863.6 = 24.35, say 24.4 lb/h (10.9 kg/h) of condensate.

Condensation during an average flow of 150,000 lb/h (67,500 kg/h) is found in the same way. The enthalpy drop to accelerate the steam is neglected for average flow in normal pressure-loss lines because the value is generally small. For the 150,000-lb/h (67,500-kg/h) flow, the calculated leaving enthalpy = 1196.90 − 5.24 = 1191.66 Btu/lb (536.3 kg/h). Hence, C = (150,000)(1194.10 − 1191.66)/863.6 = 424 lb/h (190.8 kg/h) say 425 lb/h (191.3 kg/h).

The largest amount of condensate is formed during line warm-up. Condensate-removal equipment—traps and related piping—must be sized up on the basis of the warm-up not the average steam flow. Using a warm-up time of 30 min and the method of an earlier calculation procedure, we see the condensate formed in 16-in (406.4-mm) schedule 40 pipe weighing 83 lb/ft (122.8 kg/m) is, with a 25 percent safety factor to account for radiation, $C = 1.25 \times (60)(83)(1700)(373 - 40)(0.12)/[(30)(850.8)] = 16,550$ lb/h (7448 kg/h). Thus, the trap or traps should have a capacity of about 17,000 lb/h (7650 kg/h) to remove the condensate during the 30-min warm-up period.

Condensate does not form in the 10-in (254-mm) maximum pressure-loss line during either maximum or average flow. Warm-up condensate for a 30-min warm-up period and a 25 percent safety factor is $C = 1.25(60)(40.5)(1700)(455 - 40)(0.12)/[(30)(770.0)] = 11,120$ lb/h (5004 kg/h). Thus, the trap or traps should have a capacity of about 11,500 lb/h (5175 kg/h) to remove the condensate during the 30-min warm-up period.

In general, traps sized on a warm-up basis have adequate capacity for the condensate formed during the maximum and average flows. However, the condensate formed under all three conditions must be computed to determine the maximum rate of formation for trap and drain-line sizing.

7. Determine the number of plain U bends needed

A 1700-ft (518-m) long steel steam line operating at a temperature in the 400°F (204°C) range will expand nearly 50 in (1270 mm) during operation. This expansion must be absorbed in some way without damaging the pipe. There are four popular methods for absorbing expansion in long transmission lines: plain U bends, double-offset expansion U bends, slip or corrugated expansion joints, and welded-elbow expansion bends. Each of these will be investigated to determine which is the most economical.

Assume that the governing code for piping design in the locality in which the line will be installed requires that the combined stress resulting from bending and pressure S_{bp} not exceed three-fourths the sum of the allowable stress for the piping material at atmospheric temperature S_a and the allowable stress at the operating temperature S_o of the pipe. This is a common requirement. In equation form, $S_{bp} = 0.75(S_a + S_o)$, where each stress is in lb/in².

By using allowable stress values from the *Piping Code* or the local code for 16-in (406.4-mm) seamless carbon-steel ASTM A53 grade A pipe operating at 373°F (189°C), $S_{bp} = 0.75(12,000 + 12,000) = 18,000$ lb/in² (124.1 MPa).

Determine the longitudinal pressure stress P_L by dividing the end force due to internal pressure F_e lb by the cross-sectional area of the pipe wall a_m in², or $P_L = F_e/a_m$. In this equation, $F_e = pa$, where p = pipe operating pressure, lb/in² (gage); a = cross-sectional area of the pipe, in². Since the 16-in (406.4-mm) line operates at $180 - 14.7 = 165.3$ lb/in² (gage) (1139.6 kPa) and, from a table of pipe properties, $a = 176.7$ in² (1140 cm²) and $a_m = 24.35$ in² (157.1 cm²), $P_L = (165.3)(176.7)/24.35 = 1197$, say 1200 lb/in² (8.3 MPa). The allowable bending stress at 373°F (189°C), the pipe operating temperature, is then $S_{np} - P_L = 18,000 - 1200 = 16,800$ lb/in² (115.8 MPa).

Assume that the expansion U bend will have a radius of seven times the nominal pipe diameter, or $(7)(16$ in$) = 112$ in (284.5 cm). The allowable bending stress is 16,800 lb/in² (115.8 MPa). Full *Piping Code* allowable credit will be taken for cold spring; i.e., the pipe will be cut short by 50 percent or more of the computed expansion and sprung into position.

Referring to Crocker and King—*Piping Handbook,* or a similar tabulation of allowable U-bend overall lengths for various operating temperatures, and choosing the length for 400°F (204°C), we see that the next higher tabulated temperature greater than the 373°F (189.4°C) operating temperature, an allowable length of

157.0 ft (47.9 m) is obtained for the bend. Plot a curve of the allowable bend length vs. temperature at 200, 300, 400, and 500°F (93.3, 148.9, 204.4, and 260°C). From this curve, the allowable pipe stress of 12,000 lb/in² (82.7 MPa) and no cold spring. Since the allowable stress is 16,800 lb/in² (115.8 MPa) and maximum cold spring is used, permitting a length 1.5 times the tabulated length, the total allowable length per bend = (175.0)(16,800/12,000)(1.5) = 367.5 ft (112 m). With the total length of pipe between the power plant and a load of 1700 ft (518.2 m), the number of bends required = 1700/367.5 = 4.64 bends. Since only a whole number of bends can be used, the next larger whole number, or five bends, would be satisfactory for this 16-in (406.4-mm) line. Each bend would have an overall length, Fig. 29, of 1700/5 = 340 ft (103.6 m).

Find the actual stress S_a in the pipe when five 340-ft (103.6-m) bends are used by setting up a proportion between the tabulated stress and bend length. Thus, the *Piping Handbook* chart is based on a stress of 12,000 lb/in² (82.7 MPa) without cold spring. For this stress, the maximum allowable bend length is 175 ft (53.3 m) found by graphical interpolation of the tabular values, as discussed above. When a 340-ft (103.6-m) bend with maximum cold spring is used, the pipe stress is such that the allowable bend length is 340/1.5 = 226.5 ft (69 m). The actual stress in the pipe is therefore $S_a/12,000 = 226.5/175$, or $S_a = 15,520$ lb/in² (107 MPa). This compares favorably with the allowable stress of 16,800 lb/in² (115.8 MPa). The actual stress is less because the overall bend length was reduced.

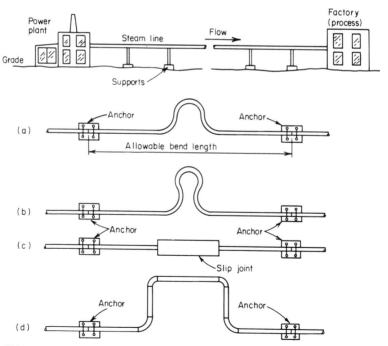

FIGURE 29 Process steam line and different schemes for absorbing pipe thermal expansion.

Use the *Piping Handbook* or the method of an earlier calculation procedure to find the anchor reaction forces for these bends. Using the *Piping Handbook* method with graphical interpolation, the anchor reacting force for a 16-in (406.4-mm) schedule 80 bend having a radius of seven times the pipe diameter is 10,550 lb (46.9 kN) at 373°F (189.4°C), based on a 12,000-lb/in² (82.7-MPa) stress in the pipe. This tabular reaction must be corrected for the actual pipe stress and for schedule 40 pipe instead of schedule 80 pipe. Thus, the actual anchor reaction, lb = (tabular reaction, lb) [(actual stress, lb/in²) (tabular stress, lb/in²)] (moment of inertia, schedule 40 pipe, in⁴/moment of inertia of schedule 80 pipe, in⁴) = (10,550)(15,520/12,000)(731.9/1156.6) = 8650 lb (38.5 kN). With a reaction of this magnitude, each anchor would be designed to withstand a force of 10,000 lb (44.5 kN). Good design would locate the bends midway between the anchor points; that is, there would be an anchor at each end of each bend. Adjustment for cold spring is not necessary, because it has negligible effect on anchor forces.

Use the same procedure for the 10-in (254-mm) maximum pressure-loss line. If 100-in (254-cm) radius bends are used, seven are required. The bending stress is 14,700 lb/in² (65.4 kN), and the anchor force is 2935 lb (13.1 kN). Anchors designed to withstand 3000 lb (13.3 kN) would be used.

8. Determine the number of double-offset U bends needed

By the same procedure and the *Piping Handbook* tabulation similar to that in step 7, the 16-in (406.4-mm) nominal pressure-loss line requires two 850-ft (259.1-m) long 112-in (284.5-cm) radius bends. Stress in the pipe is 15,610 lb/in² (107.6 MPa), and the anchor reaction is 4780 lb (21.3 kN).

The 10-in (254-mm) maximum pressure-loss line requires five 340-ft (103.6-m) long 70-in (177.8-cm) radius bends. Stress in the pipe is 12,980 lb/in² (89.5 MPa), and the anchor reaction is 2090 lb (9.3 kN).

Note that a smaller number of double-offset U bends are required—two rather than five for the 16-in (406.4-mm) pipe and five rather than seven for the 10-in (254-mm) pipe. This shows that double-offset U bends can absorb more expansion than plain U bends.

9. Determine the number of expansion joints needed

For any pipe, the total linear expansion e_t in at an elevated temperature above 32°F (0°C) is $e_t = (c_e)(\Delta t)(l)$, where c_e = coefficient of linear expansion, in/(ft · °F); Δt = operating temperature,°F − installation temperature,°F; l = length of straight pipe, ft. Using Crocker and King—*Piping Handbook* as the source for c_e for both lines, we see the expansion of the 373°F (189°C) 16-in (406.4-mm) line with a 40°F (4.4°C) installation temperature is $e_t = (12)(0.0000069)(373 − 40)(1700) = 46.8$ in (1189 mm). For the 10-in (254-mm) 455°F (235°C) line, $e_t = (12)(0.0000072)(455 − 40)(1700) = 61$ in (1549 mm). The factor 12 is used in each of these computations because Crocker and King give c_e in in/in; therefore, the pipe total length must be converted to inches by multiplying by 12.

Double-ended slip-type expansion joints that can absorb up to 24 in (609.6 mm) of expansion are available. Hence, the number of joints N needed for each line is: 16-in (406.4-mm) line, $N = 46.8/24$, or 2; 10-in (254-mm) line, $N = 61/24$, or 3.

The joints for each line would be installed midway between anchors, Fig. 29. Joints in both lines would be anchored to the ground or a supporting structure. Between the joints, the pipe must be adequately supported and free to move. Roller supports that guide and permit longitudinal movement are usually best for this service. Whereas roller-support friction varies, it is usually assumed to be about

100 lb (444.8 N) per support. At least six supports per 100 ft (30.5 m) are needed for the 16-in (406.4-mm) line and seven per 100 ft (30.5 m) for the 10-in (254-mm) line. Support friction and the number of rollers required are obtained from Crocker and King—*Piping Handbook* or piping engineering data.

The required anchor size and strength depend on the pipe diameter, steam pressure, slip-joint construction, and type of supports used. During expansion of the pipe, friction at the supports and in the joint packing sets up a force that must be absorbed by the anchor. Also, steam pressure in the joint tends to force it apart. The magnitude of these forces is easily computed. With the total force known, a satisfactory anchor can be designed. Slip-joint packing-gland friction varies with different manufacturers, type of joint, and packing used. Gland friction in one popular type of slip joint is about 2200 lb/in (385.3 N/mm) of pipe diameter. Assuming use of these joints in both lines, compute the anchor force as follows:

	lb	kN
16-in (406.4-mm) nominal pressure-loss line		
Support friction = [(1700 ft)(6 supports per 100 ft)		
(100 lb per support)]/100	= 10,200	45.3
Gland friction = (2200 lb/in diameter)(16 in)	= 35,200	156.6
Pressure force = [165.3 lb/in² (gage)](176.7-in² pipe area)	= 29,200	129.9
Total force to be absorbed by anchor	= 74,600	331.8
10-in (254-mm) maximum pressure-loss line		
Support friction = [(1700)(7)(100)]/100	= 11,900	52.8
Gland friction = (2200)(10)	= 22,000	97.9
Pressure force = (425.3)(78.9)	= 33,600	149.5
Total force to be absorbed by anchor	= 67,500	300.2

Comparing these results shows that the 10-in (254-mm) line requires smaller anchors than does the 16-in (406.4-mm) line. However, the 16-in (406.4-mm) line requires only three anchors whereas the 10-in (254-mm) line needs four anchors. The total cost of anchors for both lines will be about equal because of the difference in size of the anchors.

The advantages of slip joints become apparent when the piping layout is studied. Only a minimum of pipe is needed because the pipe runs in a straight line between the point of supply and point of use. The amount of insulation is likewise a minimum.

Corrugated expansion joints could be used in place of slip-type joints. These would reduce the required anchor size somewhat because there would be no gland friction. The selection procedure resembles that given for slip-type joints.

10.   *Select welded-elbow expansion bends*

Use the graphical analysis in Crocker and King—*Piping Handbook* or in any welding fittings engineering data. Using either method shows that three bends of the most economical shape are suitable for the 16-in (406.4-mm) line and four for the 10-in (254-mm) line. The most economical bend is obtained when the bend width, divided by the distance between the anchor points, is 0.50. With these proportions, the longitudinal stress at the top and bottom of the bend is the same. Use of such bends, although desirable, is not always feasible, because existing piping or structures interfere.

When bend dimensions other than the most economical must be used, the maximum longitudinal stress occurs at the top of the bend when the width/anchor distance < 0.5. When this ratio is > 0.5, the maximum stress occurs at the bottom of the bend. Regardless of the bend type—plain U, double-offset U, or welded—the actual stress in the pipe should not exceed 40 percent of the tensile strength of the pipe material.

11. *Determine the materials, quantities, and costs*
Set up tabulations showing the materials needed and their cost. Table 36 shows the materials required. Piping length is computed by using standard bend tables available in the cited references.

Table 37 shows the approximate material costs for each pipeline. The costs used in preparing this table were the most accurate available at the time of writing. However, the actual numerical values given in the table should not be used for similar design work because price changes may cause them to be incorrect. The important findings in such a tabulation are the differences in total cost. These differences will remain substantially constant even though prices change. Hence, if an $8000 difference exists between two sizes of pipe, this difference will not change appreciably with a moderate rise or fall in unit prices of materials.

Study of Table 37 shows that, in general, lines using double-offset U bends or welding elbows have the lowest material first cost. However, higher first costs do not rule out slip joints or plain U bends. Frequently, use of slip joints will eliminate offsets to clear existing buildings or piping because the pipe path is a straight line. Plain U bends have smaller overall heights than double-offset U bends. For this reason, the plain bend is often preferable where the pipe is run through congested areas of factories.

In some cases, past piping practice will govern line selection. For instance, in a factory that has made wide use of slip joints, the slightly higher cost of such a line might be overlooked. Preference might also be shown for plain U bends, double-offset U bends, or welded bends.

The values given in Table 37 do not include installation, annual operating costs, or depreciation. These have been omitted because accurate estimates are difficult to make unless actual conditions are known. Thus, installation costs may vary considerably according to who does the work. Annual costs are a function of the allowable depreciation, nature of process served, and location of the line. For a given transmission line of the type considered here, annual costs will usually be less for the smaller line.

The economic analysis, as made by the pipeline designer, should include all costs relative to the installation and operation of the line. The allowable cost of money and recommended depreciation period can be obtained from the accounting department.

TABLE 36 Summary of Material Requirements for Various Lines

Means used to absorb expansion	Number of anchors required		Approximate number of supports required		Approximate feet (meters) of pipe and insulation required	
	Pipe size, in (mm)					
	10 (254)	16 (406.4)	10 (254)	16 (406.4)	10 (254)	16 (406.4)
Plain U bends	9	5	127	120	2120 (646.2)	1970 (600.5)
Double-offset U bends	6	3	119	114	1985 (605.0)	1820 (554.7)
Slip joints	4	3	102	102	1700 (518.2)	1700 (518.2)
Welding elbows	5	4	106	106	1760 (536.4)	1760 (536.4)

TABLE 37 Approximate Material Costs for Various Lines

Means used to absorb expansion	Total material cost, $		Condensate removal equipment		Cost of anchors, $		Cost of supports, $		Cost of insulation, $		Cost of pipe and bends or joints, $	
	10 (254)	16 (406.4)	10 (254)	16 (406.4)	10 (254)	16 (406.4)	10 (254)	16 (406.4)	10 (254)	16 (406.4)	10 (254)	16 (406.4)
					Pipe size, in (mm)							
Plain U bends	26,500	51,350	2,000	3,000	1,000	1,800	1,800	2,400	3,700	5,650	18,000	38,500
Double-offset U bends	23,800	44,000	2,000	3,000	600	800	1,700	2,300	3,500	5,400	16,000	32,500
Slip joints	29,650	51,775	2,000	3,000	400	600	1,500	2,000	3,000	4,675	22,750	41,500
Welding elbows	23,800	43,975	2,000	3,000	400	800	1,800	2,300	3,600	5,375	16,000	32,500

12. Select the most economical pipe size

Table 37 shows that from the standpoint of first costs, the smaller line is more economical. This lower first cost is not, however, obtained without losing some large-line advantages.

Thus, steam leaves the 16-in (406.4-mm) line at 150 lb/in² (abs) (1034.1 kPa) saturated, the desired outlet condition. Special controls are unnecessary. With the 10-in (254-mm) line, the desired leaving conditions are not obtained. Slightly superheated steam leaves the line unless special controls are used. Where an exact leaving temperature is needed by the process served, a desuperheater at the end of the 10-in (254-mm) line will be needed. Neglecting this disadvantage, the 10-in (254-mm) line is more economical than is the 16-in (406.4-mm) line.

Besides lower first cost, the small line loses less heat to the atmosphere, has smaller anchor forces, and does not cause steam condensation during average flows. Lower heat losses and condensation reduce operating costs. Therefore, if special temperature controls are acceptable, the 10-in (254-mm) maximum pressure-loss line will be a more economical investment.

Such a conclusion neglects the possibility of future plant expansion. Where expansion is anticipated, installation of a small line now and another line later to handle increased steam requirements is uneconomical. Instead, installation of a large nominal pressure-loss line now that can later be operated as a maximum pressure-loss line will be found more economical. Besides the advantage of a single line in crowded spaces, there is a reduction in installation and maintenance costs.

13. Provide for condensate removal

Fit a condensate drip line for every 100 ft (30.5 m) of pipe, regardless of size. Attach a trap of suitable capacity (see step 6) to each drip line. Pitch the steam-transmission pipe toward the trap, if possible. Where the condensate must flow *against* the steam, the steam-transmission pipe *must* be sloped in the direction of condensate flow. Every vertical rise of the main line must also be dripped. Where water is scarce, return the condensate to the boiler.

Related Calculations.   Use this method to design long steam, gas, liquid, or vapor lines for factories, refineries, power plants, ships, process plants, steam heating systems, and similar installation. Follow the applicable piping code when designing the pipeline.

STEAM DESUPERHEATER ANALYSIS

A spray- or direct-contact-type desuperheater is to remove the superheat from 100,000 lb/h (45,000 kg/h) of 300-lb/in² (abs) (2068-kPa) 700°F (371°C) steam. Water at 200°F (93.3°C) is available for desuperheating. How much water must be furnished per hour to produce 30-lb/in² (abs) (206.8-kPa) saturated steam? How much steam leaves the desuperheater? If a shell-and-tube type of noncontact desuperheater is used, determine the required water flow rate if the overall coefficient of heat transfer $U = 500$ Btu/(h · ft² · °F) [2.8 kW/(m³ · °C)]. How much tube area A is required? How much steam leaves the desuperheater? Assume that the desuperheating water is not allowed to vaporize in the desuperheater.

Calculation Procedure:

1. Compute the heat absorbed by the water
Water entering the desuperheater must be heated from the entering temperature, 200°F (93.3°C), to the saturation temperature of 300-lb/in² (abs) (2068-kPa) steam, or 417.3°F (214°C). Using the steam tables, we see the sensible heat that must be absorbed by the water = h_f at 417.3°F (214°C) − h_f at 200°F (93.3°C) = 393.81 − 167.99 = 255.81 Btu/lb (525.2 kJ/kg) of water used.

Once the desuperheating water is at 417.3°F (214°C), the saturation temperature of 300°F (148.9°C) steam, the water must be vaporized if additional heat is to be absorbed. From the steam tables, the enthalpy of vaporization at 300 lb/in² (abs) (2068 kPa) is h_{fg} = 809.0 Btu/lb (1881.7 kJ/kg). This is the amount of heat the water will absorb when vaporized from 417.3°F (214°C).

Superheated steam at 300 lb/in² (abs) (2068 kPa) and 700°F (371°C) has an enthalpy of h_g = 1368.3 Btu/lb (3182.7 kJ/kg), and the enthalpy of 300-lb/in² (abs) (2068-kPa) saturated steam is h_g = 1202.8 Btu/lb (2797.7 kJ/kg). Thus 1368.3 − 1202.8 = 165.5 Btu/lb (384.9 kJ/kg) must be absorbed by the water to desuperheat the steam from 700°F (371°C) to saturation at 300 lb/in² (abs) (2068 kPa).

2. Compute the weight of water required for the spray
The weight of water evaporated by 1 lb (0.45 kg) of steam while it is being desuperheated = heat absorbed by water, Btu/lb of steam/heat required to evaporate 1 lb (0.45 kg) of water entering the desuperheater at 200°F (93.3°C), Btu = 165.5/(225.81 + 809.0) = 0.16 lb (0.07 kg) of water. Since 100,000 lb/h (45,000 kg/h) of steam is being desuperheated, the water flow rate required = (0.16)(100,000) = 16,000 lb/h (7200 kg/h). Water for direct-contact desuperheating can be taken from the feedwater piping or from the boiler.

Note that 16,000 lb/h (7200 kg/h) of additional steam will leave the desuperheater because the superheated steam is not condensed while being desuperheated. Thus, the total flow from the desuperheater = 100,000 + 16,000 = 116,000 lb/h (52,200 kg/h).

3. Compute the tube area required in the desuperheater
The total heat transferred in the desuperheater, Btu/h = UAt_m, where t_m = logarithmic mean temperature difference across the heater. Using the method for computing the logarithmic temperature difference given elsewhere in this handbook, or a graphical solution as in Perry—*Chemical Engineers' Handbook,* we find t_m = 134°F (74.4°C) with desuperheating water entering at 200°F (93.3°C) and leaving at 430°F (221.1°C), a temperature about 13°F (7°C) higher than the leaving temperature of the saturated steam, 417.3°F (214°C). Steam enters the desuperheater at 700°F (371°C). Assumption of a leaving water temperature 10 to 15°F (5.6 to 8.3°C) higher than the steam temperature is usually made to ensure an adequate temperature difference so that the desired heat-transfer rate will be obtained. If the graphical solution is used, the greatest temperature difference then becomes 700 − 200 = 500°F (278°C), and the least temperature difference = 430 − 417.3 = 12.7°F (7°C).

Then the heat transferred = (500)(*A*)(134), whereas the heat given up by the steam is, from step 1, (100,000 lb/h)(165.5 Btu/lb)[(45,000 kg/h)(384.9 kJ/kg)]. Since the heat transferred = the heat absorbed, (500)(*A*)(134) = (100,000)(165.5); *A* = 247 ft² (22.9 m²), say 250 ft² (23.2 m²).

4. Compute the required water flow

Heat transferred to the water = (500)(247)(134) Btu/h (W). The temperature rise of the water during passage through the desuperheater = outlet temperature, °F − inlet temperature = outlet temperature,°F = 430 − 200 = 230°F (127.8°C). Since the specific heat of water = 1.0, closely, the heat absorbed by the water = (flow rate, lb/h)(230)(1.0). Then the heat transferred = heat absorbed, or (500)(247)(134) = (flow rate, lb/h)(230)(1.0); flow rate = 72,000 lb/h (32,400 kg/h). Since the water and steam do *not* mix, the steam output of the desuperheater = steam input = 100,000 lb/h (45,000 kg/h).

Only about 25 percent as much water, 16,000 lb/h (7200 kg/h), is required by the direct-contact desuperheater as compared with the indirect desuperheater. The indirect type of superheater requires more cooling water because the enthalpy of vaporization, nearly 1000 Btu/lb (2326 kJ/kg) of water, is not used to absorb heat. Some indirect-type desuperheaters are designed to permit the desuperheating water to vaporize. This steam is returned to the boiler. The water-consumption determination and the calculation procedure for this type are similar to the spray-type discussed earlier. Where the water does not vaporize, it must be kept at a high enough pressure to prevent vaporization.

Related Calculations. Use this method to analyze steam desuperheaters for any type of steam system—industrial, utility, heating, process, or commercial.

STEAM ACCUMULATOR SELECTION AND SIZING

Select and size a steam accumulator to deliver 10,000 lb/h (4500 kg/h) of 25-lb/in^2 (abs) (172.4-kPa) steam for peak loads in a steam system. Charging steam is available at 75 lb/in^2 (abs) (517.1 kPa). Room is available for an accumulator not more than 30 ft (9.1 m) long, 20 ft (6.1 m) wide, and 20 ft (6.1 m) high. How much steam is required for startup?

Calculation Procedure:

1. Determine the required water capacity of the accumulator

One lb (0.45 kg) of water stored in this accumulator at 75 lb/in^2 (abs) (517.1 kPa) has a saturated liquid enthalpy h_f = 277.43 Btu/lb (645.3 kJ/kg) from the steam tables; whereas for 1 lb (0.45 kg) of water at 25 lb/in^2 (abs) (172.4 kPa), h_f = 208.42 Btu/lb (484.8 kJ/kg). In an accumulator, the stored water flushes to steam when the pressure on the outlet is reduced. For this accumulator, when the pressure on the 75-lb/in^2 (abs) (517.1-kPa) water is reduced to 25 lb/in^2 (abs) (172.4 kPa) by a demand for steam, each pound of stored 75-lb/in^2 (517.1-kPa) water flashes to steam, releasing 277.43 − 208.42 = 69.01 Btu/lb (160.5 kJ/kg).

The enthalpy of vaporization of 25 lb/in^2 (abs) (172.4-kPa) steam is h_{fg} = 952.1 Btu/lb (2215 kJ/kg). Thus, 1 lb (0.45 kg) of 75-lb/in^2 (abs) (517.1-kPa) water will form 69.01/952.1 = 0.0725 lb (0.03 kg) of steam. To supply 10,000 lb/h (4500 kg/h) of steam, the accumulator must store 10,000/0.0725 = 138,000 lb/h (62,100 kg/h) of 75-lb/in^2 (abs) (517.1-kPa) water.

Saturated water at 75 lb/in^2 (abs) (517.1 kPa) has a specific volume of 0.01753 ft^3/lb (0.001 m^3/kg) from the steam tables. Since density = 1/specific volume, the density of 75-lb/in^2 (abs) (517.1-kPa) saturated water = 1/0.01753 = 57 lb/ft^3

(912.6 kg/m³). The volume required in the accumulator to store 138,000 lb (62,100 kg) of 75-lb/in² (abs) (517.1-kPa) water = total weight, lb/density of water = 138,000/57 = 2420 ft³ (68.5 m³).

2. Select the accumulator dimensions

Many steam accumulators are cylindrical because this shape permits convenient manufacture. Other shapes—rectangular, cubic, etc.—may also be used. However, a cylindrical shape is assumed here because it is the most common.

The usual accumulator that serves as a reserve steam supply between a boiler and a load (often called a Ruths-type accumulator) can safely release steam at the rate of 0.3 [accumulator storage pressure, lb/in² (abs)] lb/ft² of water surface per hour [kg/(m²·h)]. Thus, this accumulator can release (0.3)(75) = 22.5 lb/(ft²·h) [112.5 kg/(m²·h)]. Since a release rate of 10,000 lb/h (4500 kg/h) is desired, the surface area required = 10,000/225 = 445 ft² (41.3 m²).

Space is available for a 30-ft (9.1-m) long accumulator. A cylindrical accumulator of this length would require a diameter of 445/30 = 14.82 ft (4.5 m), say 15 ft (4.6 m). When half full of water, the accumulator would have a surface area (30)(15) = 450 ft² (41.8 m²).

Once the accumulator dimensions are known, its storage capacity must be checked. The volume of a horizontal cylinder of d-ft diameter and l-ft length = $(\pi d^2/4)(l) = (\pi \times 15^2/4)(3) = 5300$ ft³ (150 m³). When half full, this accumulator could store 5300/2 = 2650 ft³ (75 m³). Since, from step 1, a capacity of 2420 ft³ (68.5 m³) is required, a 15 × 30 ft (4.6 × 9.1 m) accumulator is satisfactory. A water-level controller must be fitted to the accumulator to prevent filling beyond about the midpoint. In this accumulator, the water level could rise to about 60 percent, or (0.60)(15) = 9 ft (2.7 m), without seriously reducing the steam capacity. When an accumulator delivers steam from a more-than-half-full condition, its releasing capacity increases as the water level falls to the midpoint, where the release area is a maximum. Since most accumulators function for only short periods, say 5 or 10 min, it is more important that the vessel be capable of delivering the desired rate of flow than that it deliver the last pound of steam in its lb/h rating.

If the size of the accumulator computed as shown above is unsatisfactory from the standpoint of space, alter the dimensions and recompute the size.

3. Compute the quantity of charging steam required

To start an accumulator, it must first be partially filled with water and then charged with steam at the charting pressure. The usual procedure is to fill the accumulator from the plant feedwater system. Assume that the water used for this accumulator is at 14.7 lb/in² (abs) and 212°F (101.3 kPa and 100°C) and that the accumulator vessel is half-full at the start.

For any accumulator, the weight of charging steam required is found by solving the following heat-balance equation: (weight of starting water, lb)(h_f of starting water, Btu/lb) + (weight of charging steam, lb)(charging steam h_g, Btu/lb) = (weight of charging steam, lb + weight of starting water, lb)(h_f at charging pressure, Btu/lb). For this accumulator with a 75-lb/in² (abs) (517-kPa) charging pressure and 212°F (100°C) starting water, the first step is to compute the weight of water in the half-full accumulator. Since, from step 2, the accumulator must contain 2420 ft³ (68.5 m³) of water, this water has a total weight of (volume of water, ft³)/(specific volume of water, ft³/lb) = 2420/0.01672 = 144,600 lb (65,070 kg). However, the accumulator can actually store 2650 ft³ (75 m³) of water. Hence, the actual weight of water = 2650/0.1672 = 158,300 lb (71,235 kg). Then, with C = weight of charging steam, lb, (158,300)(180.07) + (C)(1181.9) = $(C$ + 158,300)(277.43); C = 17,080 lb (7686 kg) of steam.

Once the accumulator is started up, less steam will be required. The exact amount is computed in the same manner, by using the steam and water conditions existing in the accumulator.

Related Calculations. Use this method to size an accumulator for any type of steam service—heating, industrial, process, utility. The operating pressure of the accumulator may be greater or less than atmospheric.

SELECTING PLASTIC PIPING FOR INDUSTRIAL USE

Select the material, schedule number, and support spacing for a 1-in (25.4-mm) nominal-diameter plastic pipe conveying ethyl alcohol liquid having a temperature of 75°F (23.9°C) and a pressure of 400 lb/in² (2758 kPa). What expansion must be anticipated if a 1000-ft (304.8-m) length of the pipe is installed at a temperature of 50°F (10°C)? How does the cost of this plastic pipe compare with galvanized-steel pipe of the same size and length?

Calculation Procedure:

1. *Determine the required schedule number*
Refer to Baumeister and Marks—*Standard Handbook for Mechanical Engineers* or a plastic pipe manufacturer's engineering data for the required schedule number. Table 38 shows typical pressure ratings for various sizes and schedule number polyvinyl chloride (PVC) (plastic) piping.

Table 38 shows that schedule 40 normal-impact grade 1-in (25.4-mm) pipe is unsuitable because its maximum operating pressure with fluid at 75°F (24°C) is 310 lb/in² (2.13 MPa). Plain-end 1-in (25.4-mm) schedule 80 pipe is, however, satisfactory because it can withstand pressures up to 435 lb/in² (2.99 MPa). Note that threaded schedule 80 pipe can withstand pressures only to 225 lb/in² (1757 kPa). Therefore, plain-end normal-impact grade pipe must be used for this installation. High-impact grade pipe, in general, has lower allowable pressure ratings at 75°F

TABLE 38 Maximum Operating Pressure, PVC Pipe [normal-impact grade, fluid temperature 75°F (23.9°C) or less]

Pipe size		Schedule 40, plain end		Schedule 80			
				Plain end		Threaded	
in	mm	lb/in²	MPa	lb/in²	MPa	lb/in²	MPa
½	12.7	410	2.83	575	3.96	330	2.28
¾	19.1	335	2.31	470	3.24	285	1.97
1	25.4	310	2.14	435	2.99	255	1.76
1½	38.1	230	1.59	325	2.24	205	1.41

TABLE 39 Thermal Expansion of Plastic Pipe

Piping material	Expansion	
	in/(ft · °F)	cm/(m · °C)
Butyrate	0.00118	0.018
Kralastic	0.00067	0.010
Polyethylene	0.00108	0.016
Polyvinyl chloride	0.00054	0.008
Saran	0.00126	0.019

(24°C) because the additive used to increase the impact resistance lowers the tensile strength, temperature, and chemical resistance. Data shown in Table 38 are also presented in graphical form in some engineering data.

2. Select a suitable piping material
Refer to piping engineering data to determine the corrosion resistance of PVC to ethyl alcohol. A Grinnell Company data sheet rates PVC normal-impact and high-impact pipe as having excellent corrosion resistance to ethyl alcohol at 72 and 140°F (22.2 and 60°C). Therefore, PVC is a sizable piping material for this liquid at its operating temperature of 75°F (24°C).

3. Find the required support spacing
Use a tabulation or chart in the plastic-pipe engineering data to find the required support spacing for the pipe. Be sure to read the spacing under the correct schedule number. Thus, a Grinnell Company plastic-piping tabulation recommends a 5-ft 4-in (162.6-cm) spacing for schedule 80 1-in (25.4-mm) PVC pipe that weighs 0.382 lb/ft (0.57 kg/m) when empty. The pipe hangers should not clamp the pipe tightly; instead, free axial movement should be allowed.

4. Compute the expansion of the pipe
The temperature of the pipe rises from 50 to 75°F (10 to 24°C) when it is put in operation. This is a rise of $75 - 50 = 25$°F (14°C). Table 39 shows the thermal expansion of various types of plastic piping.

The thermal expansion of any plastic pipe is found from $E_t = LC\,\Delta t$, where E_t = total expansion, in; L = pipe length, ft; C = coefficient of thermal expansion, in/(ft · °F), from Table 39, Δt = temperature change of the pipe,°F. For this pipe, $E_t = (1000)(0.00054)(25) = 13.5$ in (342.9 mm) when the temperature rises from 50 to 75°F (10 to 24°C).

5. Determine the relative cost of the pipe
Check the prices of galvanized-steel and PVC pipe as quoted by various suppliers. These quotations will permit easy comparison. In this case, the two materials will be approximately equal in per-foot cost.

Related Calculations. Use the method given here for selecting plastic pipe for any service—process, domestic, or commercial—conveying any fluid or gas. Note that the maximum operating pressure of plastic piping is normally taken as about 20 percent of the bursting pressure. The maximum allowable operating pressure decreases with an increase in temperature. The maximum allowable operating temperature is usually 150°F (65.5°C). The pressure loss caused by pipe friction in plastic pipe is usually about one-half the pressure loss in galvanized-steel pipe of the same diameter. Pressure loss for plastic piping is computed in the same way as for steel piping.

ANALYZING PLASTIC PIPING AND LININGS FOR TANKS, PUMPS, AND OTHER COMPONENTS FOR SPECIFIC APPLICATIONS

Choose plastic piping for fire protection, process, and compressed air for an industrial application where corrosive fluids and fumes are likely to be encountered during routine operations. Show how to assemble, and evaluate key data used in choosing suitable materials for such an application. Choose which type of plastic pipe to use, A or B, when the key data you assemble show that the costs associated with each type of piping are as follows:

	Pipe A	Pipe B
First cost, $	80,000	36,000
Salvage value, $	15,000	4000
Life, years	40	15
Annual maintenance cost, $	1000	2300
Annual taxes, $/$100	1.30	1.30
Annual insurance, $/$1000	2	5

If the firm owning this industrial plant earns 6 percent on its invested capital, which type of plastic piping, A or B, is the more economical?

Calculation Procedure:

1. Assemble the data from which economic choices can be made
The world of plastic piping and equipment is comprised of a multitude of materials suitable for a variety of industrial, commercial, and power applications. Perhaps the best way to prepare oneself for the important task of materials selection is to obtain several product materials specifications and catalogs from major and specialty plastics manufacturers who make the types of products being considered, in this case, piping, pumps, and tanks.

From the data in these materials specifications and catalogs, prepare a listing, such as that in Table 40 showing the recommended applications for various plastics used for piping, tanks, and pumps. If you do not have access to specifications and catalogs of various manufacturers, Table 40 can serve as a substitute until you do obtain the needed specifications. Studying Table 40 shows that, based on the materials listed three types of plastic piping are suitable for fire protection (firewater) service, nine types are suitable for process piping, and one for compressed-air piping. This mini-survey, which is current and valid for practical applications in industry today, immediately shows the range of choices open to the designer. Thus, the widest choice is amongst process piping (nine different materials); the narrowest choice (one material) is for compressed-air piping.

Where a difference in cost exists between the choices available, as is almost always the case where more than one material is suitable, an economic study will

TABLE 40 Plastics Are Often Specified for Piping and Equipment That Must be Corrision-Resistant and Lightweight*

	Examples of Plastics Applications																		
	Epoxy	Vinyl ester	Polyester	PE	HDPE	PP	PVC	CPVC	DLPVC	PVDF	TFE	PTFE	Polyurethane	FEP	PFA	Furan	Furate	ABS	Bis-A
Firewater piping	•	•		•															
Process piping	•	•	•	•			•	•	•	•			•						
Compressed air piping																		•	
Valves	•	•					•	•	•		•	•	•						
Pumps	•	•	•				•	•			•	•	•						
Process area drains manhole & catch basins	•	•	•	•		•	•	•		•									
Tanks	•	•	•	•		•				•				•	•	•			•
Vessels & columns	•	•	•	•			•	•		•			•			•	•		
Column trays	•	•	•	•	•	•				•						•	•		•
Strippers & scrubbers	•		•	•			•	•		•			•						
Column packing	•													•					
Structural shapes & grating	•	•	•																
Fans & blowers	•	•	•			•	•	•	•	•			•						
Ducts	•	•					•	•											
Stacks & stack liners	•	•	•													•	•		•
Mist eliminators	•	•				•	•												
Fume hoods	•	•					•	•											
Filters & strainers							•	•	•										
Bearing pads											•	•							
Expansion joints										•									
Heat exchanger tubes										•									

Chemical Engineering.

show which material is the best selection for the conditions at hand, based on the investment required. Thus, plastic piping is inherently corrosion-resistant. However, other factors, Table 41, must be considered in choosing a piping material. For example, when considering plastic vs. metal, plastic may have a number of advantages, as summarized in Table 41.

Another consideration where fluids are handled is the permeation of fluids or gases through the plastic material. Table 42 lists permeation rates for selected plastic materials. Some thermoplastics have a high degree of impermeability. However, the level of impermeability may deteriorate with prolonged exposure to ultraviolet radiation, such as from the sun.

2. *Evaluate the relative corrosion resistance of the plastic material selected*

Table 43 lists the relative corrosion resistance of some plastics used for the purposes considered here: piping, tanks, and pumps. Data such as these can be assembled from manufacturer's specifications and catalogs. Or they can be used directly from Table 43 until such time as sufficient contemporary data are compiled.

To use the data in Tables 40 and 43, assume that PVC has been selected for process piping handling corrosive (acidic) liquids and fumes. Table 43 shows that

TABLE 41   Advantages of Plastic Piping vs. Metal Piping

Plastic piping:

  Has high corrosion resistance

  Does not require internal or external coating to prevent corrosion

  Does not need internal or external cathodic protection to prevent corrosion

  Can be welded at lower temperatures where ignition is a problem

  Does not usually need welding; if welding it needed, it can be done at lower temperatures

  Is nonconducting as manufactured—hence stray electric currents are not a factor in design and
    installation

  Is lighter weight than metal—hence, handling and transportation costs are lower

  Has inherently good thermal insulation in itself; in some installations additional thermal insula-
    tion may not be required

  Has inherent freeze protection and heat retention; outside supplementary protection may not be
    required

TABLE 42   Some Thermoplastics Have a High Degree of Impermeability*

| | Permeation of some themoplastic materials | | | | |
Material	Water vapor**	Oxygen†	Helium†	Nitrogen†	CO_2**
PVDF	2	20	600	30	100
PTFE	5	1500	35000	500	15000
FEP	1	2900	18000	1200	4700
PFA	8	—	17000	—	7000
PP	—	25	200	10	100
HDPE	—	30–40	20	18	200
PVC	—	3	16	1	16

*Chemical Engineering.
*$g/m^2 \cdot d$ at 1 bar and 73°F.
†Permeability through unreinforced 1-mm-thick sheet in $cm^3/m^2 \cdot d$ at 1 bar and 73°F.

PVC is resistant to caustics but not resistant to acids. Hence, you could choose PVDF (polyvinyl fluoride), which is resistant to acids. You would perform an economic study to see if PVDF was the best choice in this installation, based on the annual cost of each type of piping being considered. These same general principles apply equally well to the corrosion resistance of tanks and pumps.

3.   *Determine the annual cost of each type of piping being considered*
The first step in determining the annual cost of an asset is computation of the operating and maintenance cost. For Pipe A, the annual operating and maintenance cost, $ c = maintenance cost per year, $ + annual taxes, $ + annual insurance cost, $. Or, Pipe A, c = \$1000 + (\$1.30/\$100)(\$80,000) + (\$2/\$1000)(\$80,000) = \$2200. For Pipe B, c = \$2300 + (\$1.30/\$100)(\$36,000) + (\$5/\$1000)(\$36,000) = \$2948.

The second, and last, step in computing the annual cost of each type of pipe uses the capital-recovery equation, $A = (P - L)(CR) + Li_1 + c$, where A = annual cost, $; P = initial cost of each type of piping, $; L = salvage value of each type of piping, $; CR = capital-recovery factor from compound-interest tables for 6

TABLE 43 Relative Corrosion Resistance of Some Plastics*

	Caustics	Acids	Weak acids	Aliphatic solvent	Aromatic solvents	Alcohols	Halogenated solvents	Ketones	Deionized water
PVC	R	M	R	M	N	M	N	N	R
CPVC	R	M	R	M	N	R	N	N	R
PP	R	N	R	M	N	R	N	R	R
LDPE	R	N	R	M	N	R	N	M	R
HDPE	R	N	R	M	N	R	N	N	R
UHMWPE	R	M	R	M	N	R	N	N	R
PVDF	M	R	R	R	R	R	N	N	R
TFE	R	R	R	R	R	R	R	R	R
FEP	R	R	R	R	R	R	R	R	R
PEEK	M	M	M	—	N	—	—	M	—
Isothalic Polyester	N	N	R	R	M	M	N	N	R
Vinyl Ester	R	M	R	R	R	R	N	N	R
Epoxy Novolac Vinyl Ester	R	R	R	R	R	R	N	M	R
Bisphenol A Fumarate	R	N	R	R	M	M	N	N	R
Furan	R	M	R	R	R	R	M	R	R

R–resistant   N–not resistant   M–marginal

Chemical Engineering.

percent interest rate; i_1 = interest rate on the invested capital, 6 percent in this case; c = annual operating and maintenance cost, $, as computed earlier for each type of pipe.

Substituting for Pipe A, A = ($80,000 − $15,000)(0.06646) + $15,000(0.06) + $2200 = $7420. For Pipe B, A = ($36,000 − $4000)(0.10296) + $4000(0.06) + $2948 = $6483. Since Pipe B has a lower annual cost, it is the more economical of the two, presuming that the two piping materials have equal, or nearly equal, corrosion resistance properties.

Related Calculations. The same approach given here can be used when choosing the plastic materials for piping, tanks, pumps, ducts and other components for any industrial, commercial, or residential application. Table 44 gives suggested lining materials for piping, vessels, columns, pumps, and other structures, and gasket materials for joints. The key consideration is obtaining minimum annual cost with the desired level of corrosion or other resistance to deleterious substances. Today, plastic piping is widely used in many applications and is almost universally accepted as superior to metal where corrosion resistance is a primary requirement for the piping. Except for an unfortunate experience with plastic domestic piping installed in single-family homes and some multi-family residences, engineers have had favorable results with plastic piping.

The advantages of plastic piping listed in Table 41 apply to almost every design situation an engineer faces. And as more experience is gained by plastic piping manufacturers, the advantages cited in Table 41 are likely to increase.

TABLE 44   Which Material to Use for Piping, Vessels and Equipment*

Linings on carbon steel, concrete or FRP[1]

Thin linings
 [<0.025 in (0.64 mm)]
 [used when corrosion rate of carbon steel is ≥0.010 in (0.25 mm)/yr]

  Elastomers—sprayed
  Epoxy- and phenolic-based
    (chemically or heat cured)
  Fluoropolymer
    (sprayed and baked)

Thick linings
 [>0.025 in (0.64 mm)]
 (used when corrosion rate of carbon steel is >0.010 in (0.25 mm)/yr)

  Elastomeric sheet linings[4]
  Reinforced vinyl ester, plasticized PVC
  Fluoropolymer linings[5]

Self-supporting structures

Vessels and columns
  FRP (vinyl ester and furan)
  Rotomolded PE

Piping[2]
  FRP (vinyl ester, turan, epoxy), PE, PVC, PVDF, FEP

Valves[3]
  FRP PVC, CPVC

Dip tubes, agitators, baffles
  FRP, PTFE

Others

Gaskets
  Elastomers, fluoropolymers

Seals
  Elastomers, fluoropolymers

  Chemical Engineering.
  [1]See dual laminate constructions—Table 45.
  [2]Loose fluoropolymer linings are typically used for piping. Dual laminate and rotolined piping is also available.
  [3]Metal valves and pumps are also lined with plastics for corrosion resistance.
  [4]Also used for piping, valves, pumps, agitators and other applications.
  [5]Loose linings are used for piping, molded liners for valves and pumps. Dual laminate and roto-lined piping is also available.

For analytical purposes, it is desirable to convert the estimated costs associated with proposed alternatives to an equivalent series of uniform annual payments. The annual payment thus obtained is termed the *annual cost* of each alternative. The interest rate applied in making this conversion is the minimum investment rate considered acceptable by the organization making the investment or incurring the costs. Where alternative schemes are being evaluated on the basis of their annual cost, the usual procedure is to exclude those expenses which are identical for all schemes, since they do not affect the comparison.

Data tables in this procedure were obtained from information prepared by Benjamin S. Fultz, Engineering Supervisor, Nonmetallic Section, Materials and Quality

TABLE 45 Fluoropolymer Lining Systems*

Lining system and materials	Thickness in. (min)	Maximum size	Design limits	Fabrication[1]	Repair considerations
Adhesive bonding					
Fabric-backed		No limit	Pressure allowed. Full vacuum only at ambient temperature. Smallest nozzle is 2 in. (51 mm). Max. temp. limited by adhesive, typically 275°F (135°C)	Neoprene or epoxy adhesive, sheets welded with cap strips. Heads are thermoformed or welded	Repair is possible but testing is recommended
PVDF	0.06, 0.9 (1.5, 2.3)				
PTFE	0.08, 0.12 (2.0, 3.1)				
FEP	0.06, 0.9 (1.5, 2.3)				
ECTFE	0.06 (1.5)				
ECTFE	0.06, 0.09 (1.5, 2.3)				
PFA	0.09 (2.3)				
Rubber-backed[2]					
PVDF	0.05 (1.3)				
Dual laminate Except for rubber-backed, same as adhesive bonded		12 ft (3.7 m) dia.	No pressure allowed. Vacuum rating not determined	Liner fabricated on mandrel by hand and machine welding. FRP built up over liner	Repair is possible but testing is recommended
Sprayed dispersion		8 ft (2.4 m) dia, 40 ft (12.2 m) length	Pressure allowed. Vacuum rating not determined	Primer and multiple coats applied with conventional spray equipment. Each coat is baked	Hot patching is possible but testing is recommended
FEP	0.04 (1.0)				
PFA	0.01–0.04 (0.25–1.0)				
PFA w/mesh and carbon	0.08 (2.0)				
PVDF	0.025–0.03 (0.6–0.76)				
PVDF w/glass or carbon fabric	0.04, 0.09 (1.0, 2.3)				
Electrostatic spray-powder		8 ft (2.4 m) dia., 40 ft (12.2 m) length	Pressure allowed. Vacuum rating not determined	Primer and multiple coats applied with electrostatic spraying equipment. Each coat is baked	Hot patching is possible but testing is recommended
ETFE	up to 0.09 (2.3)				
FEP	0.01 (0.28)				
PFA	0.01 (0.28)				

TABLE

Lining system and materials	Thickness in. (min)	Maximum size	Design limits	Fabrication[1]	Repair considerations
ECTFE PVDF	0.06, 0.07 (1.5, 1.8) 0.025 (0.64)				
Rotolining ETFE PVDF ECTFE	0.1–0.2 (2.5–5.1)	8 ft (2.4 m) dia., 22 ft (6.7 m) length	Pressure allowed. Vacuum rating not determined	Rotationally molded. No seams. No primer used	Hot patching is possible but testing is recommended
Isostatic molding, paste or ram extrusion, or tape wrapping PTFE			Pressure allowed. Vacuum rating depends on lining thickness	PTFE is preformed under isostatic pressure, or is paste- or ram-extruded as tubing, and then sintered by heating. Tubing can also be built up by wrapping tape layers on a mandrel and then sintering	Hot patching with PFA possible but testing is recommended
Loose lining FEP, PFA	0.06–0.187 (1.5–4.75)	Determined by body flange	Pressure allowed. No vacuum. Gasketing required between liner and flange	Liner with nozzles hand or machine welded, then slipped inside housing	Difficult

*Chemical Engineering

[1] Nondestructive spark testing should be used, along with visual inspection for all systems except loose linings. Adhesive bonding can be done in the shop or field; other systems are shop only.

[2] Is rarely used

Services Dept., Bechtel Corp., Robert H. Rogers, Nonmetallic Engineer Specialist, Materials and Quality Services Dept., Bechtel Corp., and J. S. (Steve) Young, Engineering Specialist, Materials and Quality Services Dept., Bechtel Corp., and Pradip Khaladkar, Materials Consultant, DuPont Engineering (Table 44), as reported in *Chemical Engineering* Magazine. The economic study in Step 3 is the work of Max Kurtz, P.E.

FRICTION LOSS IN PIPES HANDLING SOLIDS IN SUSPENSION

What is the friction loss in 800 ft (243.8 m) of 6-in (152.4-mm) schedule 40 pipe when 400 gal/min (25.2 L/s) of sulfate paper stock is flowing? The consistency of the sulfate stock is 6 percent.

Calculation Procedure:

1. *Determine the friction loss in the pipe*
There are few general equations for friction loss in pipes conveying liquids having solids in suspension. Therefore, most practicing engineers use plots of friction loss available in engineering handbooks. *Cameron Hydraulic Data, Standards of the Hydraulic Institute,* and from pump engineering data. Figure 30 shows one set of

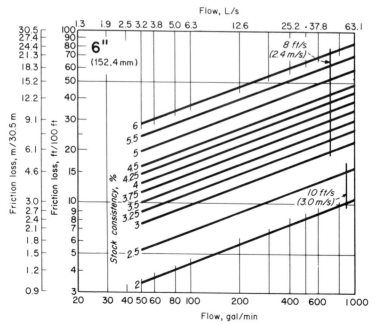

FIGURE 30 Friction loss of paper stock in 4-in (101.6-mm) steel pipe. (*Goulds Pumps, Inc.*)

typical friction-loss curves based on work done at the University of Maine on the data of Brecht and Heller of the Technical College, Darmstadt, Germany, and published by Goulds Pumps, Inc. There is a similar series of curves for commonly used pipe sizes from 2 through 36 in (50.8 through 914.4 mm).

Enter Fig. 30 at the pipe flow rate, 400 gal/min (25.2 L/s), and project vertically upward to the 6 percent consistency curve. From the intersection, project horizontally to the left to read the friction loss as 60 ft (18.3 m) of liquid per 100 ft (30.5 m) of pipe. Since this pipe is 800 ft (243.8 m) along the total friction-head loss in the pipe = (800/100)(60) = 480 ft (146.3 m) of liquid flowing.

2.   Correct the friction loss for the liquid consistency
Friction-loss factors are usually plotted for one type of liquid, and correction factors are applied to determine the loss for similar, but different, liquids. Thus, with the Goulds charts, a factor of 0.9 is used for soda, sulfate, bleached sulfate, and reclaimed paper stocks. For ground wood, the factor is 1.40.

When the stock consistency is less than 1.5 percent, water-friction values are used. Below a consistency of 3 percent, the velocity of flow should not exceed 10 ft/s (3.05 m/s). For suspensions of 3 percent and above, limit the maximum velocity in the pipe to 8 ft/s (2.4 m/s).

Since the liquid flowing in this pipe is sulfate stock, use the 0.9 correction factor, or the actual total friction head = (0.9)(480) = 432 ft (131.7 m) of sulfate liquid. Note that Fig. 30 shows that the liquid velocity is less than 8 ft/s (2.4 m/s).

Related Calculations.   Use this procedure for soda, sulfate, bleached sulfite, and reclaimed and ground-wood paper stock. The values obtained are valid for both suction and discharge piping. The same general procedure can be used for sand mixtures, sewage, slurries, trash, sludge, and foods in suspension in a liquid.

DESUPERHEATER WATER SPRAY QUANTITY

A pressure- and temperature-reducing station in a steam line is operating under the following conditions: pressure and temperature ahead of the station are 1400 lb/in^2 absolute (5650 kPa), 950°F (510°C); the reduced temperature and pressure after the station are 600°F (315°C), 200 lb/in^2 absolute (1380 kPa). If 450,000 lb/h (3400 kg/s) of steam is required at 200 lb/in^2 (1380 kPa), how much water, which is available at 200 lb/in^2 absolute (1380 kPa) and 635.8°F (335.4°C), must be sprayed in at the superheater? See Fig. 31.

Calculation Procedure:

1.   Determine the quantity of heat entering the desuperheater via the spray in terms of the amount of water
The quantity of heat entering the desuperheater via the spray, $Q = w \times h_f$, where the amount of water is w, lb/h (kg/s); from Table 2, Saturation Pressures, of the Steam Tables mentioned under Related Calculations of this procedure, heat content of water, saturated steam. At 200 lb/in^2 (1380 kPa), $h_f = 355.4$ Btu/lb$_m$ (826 kJ/kg). Thus $Q = w \times 355.4$ lb/h (kg/s). It should be noted that the Steam Tables show the saturation temperature to be 381.79°F (194.3°C) at the given pressure, shown on Fig. 31. Obviously, the 635.8°F (335.4°C) given in the problem is not

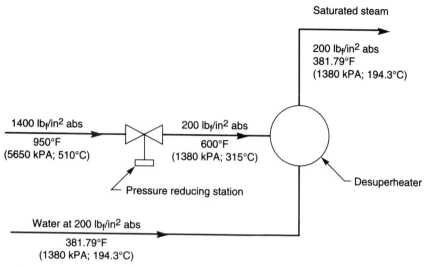

FIGURE 31 Desuperheater fluid flow diagram.

correct, because at that temperature there would either be superheated steam, vapor, or the water would have to be under a pressure of 2000 lb/in² (13.8 × 10³ kPa).

2. Find the enthalpy of the superheated steam entering the desuperheater and the enthalpy of the saturated steam leaving
From Table 3, Vapor, of the Steam Tables, superheated steam entering the desuperheater at 200 lb/in² (1380 kPa) and 600°F (315°C) has an enthalpy, $h = 1322.1$ Btu/lb (2075 kJ/kg). From Table 2, Saturation: Pressures, of the Steam Tables, saturated steam leaving the superheater at 200 lb/in² (1380 kPa) and 381.79°F (194.3°C) at saturated vapor has an enthalpy, $h_g = 1198.4$ Btu/lb$_m$ (2787 kJ/kg).

3. Compute the amount of water which must be sprayed into the desuperheater
The amount of water which must be sprayed into the desuperheater, w_w lb/h (kg/s), can be found by the use of a heat balance equation where, as an adiabatic process, the amount of heat into the superheater equals the amount of heat out. Then, $w \times h_f + (450,000 - w)h = 450,000 \times h_g$. Or, $w \times 355.4 + (450,000 - w)(1322.1) = 450,000 \times 1198.4$. Solving, $w = 450,000 \times 123.7/966.7 = 57,580$ lb/h (435.3 kg/s).

   Related Calculations.   Strictly speaking, the given pressure and temperature conditions before the pressure-reducing station were irrelevant in the Calculation Procedure. Also, the incorrect given saturation temperature of 635.8°F (335.4°C) is an example of possible distractions which should be guarded against while solving such problems. The Steam Tables appear in *Thermodynamic Properties of Water Including, Liquid, and Solid Phases,* 1969, Keenan, et al., John Wiley & Sons, Inc. Use later versions of such tables whenever available, as necessary.

SIZING CONDENSATE RETURN LINES FOR OPTIMUM FLOW CONDITIONS

An evaporator is condensing 5500 lb/h (2497 kg/h) of steam at 150-lb/in² (gage) (1033.5-kPa) supply pressure. During normal operation, a control valve maintains a pressure of 85 lb/in² (gage) (585.7 kPa) upstream of the steam trap handling the condensate from the evaporator. The condensate discharged by the steam trap is returned to an atmospheric-vented tank. What pipe line size should be used on the steam-trap outlet to provide optimum flow conditions?

Calculation Procedure:

1. Compute the percentage of flash steam in the steam-trap discharge line
The percentage of flash steam in the steam-trap discharge line is found from

$$x_{fs} = \frac{(h_{l_1} - h_{l_2})}{\Delta h_{v_2}} \times 100$$

where the symbols are as given below.

Nomenclature

A_{req} = Required cross-sectional area, ft² (m²)
D_u = Nominal pipe size, based on velocity u, in (mm)
D_{50} = Nominal pipe size, based on 50 ft/s, in (m/s, mm)
h_{l_1} = Condensate enthalpy at upstream pressure, P_1, Btu/lb (kJ/kg)
h_{l_2} = Condensate enthalpy at end-pressure, P_2, Btu/lb (kJ/kg)
Δh_{v_2} = Latent heat of vaporization at P_2, Btu/lb (kJ/kg)
Q_v = Flash-steam volumetric flow rate, ft³/h
u = New flash-steam velocity, ft/s (m/s)
v_{v_2} = Flash-steam specific volume at P_2, ft³/lb (m³/kg)
W_l = Condensate formed at P_1, lb/h (kg/h)
W_v = Flash steam formed at P_2, lb/h (kg/h)
z_{fs} = Flash steam, wt %

Essentially, what this relation does is to convert the difference of condensate enthalpies, out and in, to flash steam using the latent heat of vaporization at the outlet pressure of the steam trap as the flash-heat source. Substituting in the equation above, using steam-table data for the enthalpies, x_{fs} = [(298.4 − 180.1)/ 1150.4](100) = 10.28 percent flash steam by weight.

2. Find the weight of flash steam formed at the trap outlet
Use the relation

$$W_v = W_l \frac{x_{fs}}{100}$$

Substituting, W_v = (5500)(10.28/100) = 565.4 lb/h (256.7 kg/h).

3. Calculate the flash-steam volumetric flow rate
Use the relation

$$Q_v = W_v v_{v_2}$$

Substituting, $Q_v = (565.4)(26.8) = 15,152.7$ ft³/h (428.8 m³/h).

4. Determine the required cross-sectional area of the steam trap discharge pipe
Use the relation

$$A_{req} = \frac{Q_v}{3,600 \times 50}$$

for a flash-steam velocity of 50 ft/s (15.24 m/s), the usual value used in sizing such pipes.

Substituting, $A_{req} = (15,152.7)/(3600)(50) = 0.0842$ ft² (0.00782 m²). Converting to in², $(0.0842)(144) = 12.12$ in² (7820.5 mm²). Note that this is the required internal area of the trap discharge pipe.

5. Choose the pipe size to use
Entering a table of pipe properties, we find that a 4-in (101.6-mm) pipe is required to convey the flashing condensate from this steam trap at the chosen velocity.

Related Calculations. Undersized condensate return-lines create one of the most common problems met with process (and power-plant) steam traps. Hot condensate passing through a trap orifice loses pressure, which lowers the enthalpy of the condensate. This enthalpy change causes some of the condensate to flash into steam. The volume of the resulting two-phase mixture is usually many times that of the upstream condensate entering the trap.

The trap-outlet or downstream piping must be adequately sized to handle effectively the greater volume of the two-phase mixture. An undersize condensate return line on the trap outlet results in a high flash-steam velocity. This may cause water hammer (due to wave formation), hydrodynamic noise, premature erosion, and high backpressure. An excessively high backpressure reduces the working differential pressure across the trap and, hence, the condensate removal capability of the steam trap. In some traps excessive backpressure causes partial or full failure.

Because of the much greater volume of flash steam compared with unflashed condensate, sizing of the return line is based solely on flash steam. It is assumed that all flashing occurs across the steam trap and that the resulting vapor-liquid mixture can be evaluated at the end-pressure conditions. To ensure the condensate line does not have an appreciable pressure-drop, a low flash-steam velocity is assumed, namely, 50 ft/s (15.24 m/s).

Where there is only a small pressure drop in the discharge line, or high subcooling of the condensate, it may be necessary to size the condensate line based on the liquid velocity. Generally, a velocity of 3 ft/s (0.91 m/s). For flash-steam velocities other than the 50 ft/s (15.24 m/s) used above, the nominal pipe size can be approximated from:

$$D_u = \frac{7.07 D_{50}}{\sqrt{u}}$$

The method presented here yields a single result; thus decision-making is not required on the part of the user, as it is in other methods of steam-trap discharge-line sizing. This method can be used for new construction of all types: industrial commercial, residential, marine, HVAC, etc. It can also be used to check existing line sizes where trap performance is questionable. With minimal training, field maintenance personnel can use this method on the job.

The procedure presented here is the work of Michael V. Calogero, GESTRA, Inc., and Arthur W. Brooks, TECHMAR Engineering, Inc., as presented in *Chemical Engineering* magazine.

ESTIMATING COST OF STEAM LEAKS FROM PIPING AND PRESSURE VESSELS

Steam at 135 lb/in² (930 kPa) is leaking from a 0.05 in² (0.32 cm²) opening in a pipe and escaping to the atmosphere where it cannot be recovered. Temperature of the steam in the pipe is 450°F (232°C). Determine the cost of this leak if steam costs $2.50 per 1000 lb (454 kg) and the pipe operates 8000 h per year.

Calculation Procedure:

1. Compute the rate of steam loss through the opening
The rate of steam loss through an opening is given by $L = KA(PD)^{0.5}$, where L = steam loss, lb/h (kg/h); K = factor for steam condition; for saturated steam, $K = 1085$; for superheated steam, $K = 1138$; A = opening area, in² (cm²); P = steam pressure on the pressure side of the opening, lb/in² (abs) (kPa); D = density of the steam on the pressure side of the opening, lb/ft³ (kg/m³).

Substituting for this leak, using the superheated steam factor of 1138 because the steam temperature of 450°F (232°C) is about 100° higher than the saturation temperature of 358°F (181°C) at 150 lb/in² (abs) (= 135 lb/in² + 14.7 lb/in²) (1033.5 kPa), gives $L = 1138(0.05)(150[0.3316])^{0.5} = 401.3$ lb/h (182.2 kg/h).

2. Determine the cost of the leaking steam
The cost of a steam leak, $C\ \$ = LSH/1000$, where C = annual cost of steam leak, $, for H hours of yearly operation of the pipe; S = cost of steam, $ per 1000 lb (454 kg). Substituting, $C = 401.3$ lb/h × $2.50/1000 lb × 8000 h/yr = $8026 per yr for this leak. This is a significant cost when compared with the low cost of modern materials that can be used to stop such a leak.

Related Calculations.   Use the relation given here for steam leaks from openings in pipes, pressure vessels, traps, meters, orifices, nozzles, and other steam apparatus. The relation can also be used for leaks from the open ends of pipes. The one restriction on the use of this relation is that the pressure on the outlet side must be 0.578, or less, than the pressure on the inside of the pipe or vessel. This situation prevails in almost every case of a leaking pipe or vessel. The reason for this is that most leaks are from a pressurized source to the atmosphere. Few leaks occur from one pressurized source to another.

The value 0.578 is the critical pressure ratio for steam flow through an orifice. Velocity of the escaping steam is determined by this ratio. The higher this velocity, the larger the amount of steam escaping in a unit time.

There is greater emphasis today than ever before on preventing unnecessary steam losses through neglected leaks in piping and pressure vessels. The reason for this is that engineers and managers now recognize the chain effect of uncontrolled steam leaks. This effect is as follows: Leaking steam that does no work represents wasted fuel that was burned to generate the steam. This wasted fuel produces unnecessary pollution of the atmosphere during its combustion. Further, there is a drain or natural resources because the fuel burned produces no useful work. Thus, the total cycle is environmentally offensive when the steam is wasted by not performing any work before being exhausted to the atmosphere or a condenser. For these reasons, steam leaks are getting more attention than ever before. The same—of course—can be said about leakage of any other valuable liquid or gas—such as oil, compressed air, etc. A different equation must be used to compute such losses because the equation above applies only to steam leaks.

QUICK SIZING OF RESTRICTIVE ORIFICES IN PIPING

Choose the bore size of a restrictive orifice for a 25-gal/min (1.57 L/s) minimum bypass for a pump discharging water at 100 lb/in² (gage) (689.4 kPa) and 80°F (26.7°C) into a 50-lb/in² (gage) (344.7 kPa) drum when a 0.125-in (0.3175-cm) thick orifice plate is used. What bore size would be required for a 0.25-in (0.635-cm) thick orifice plate? Use the quick-sizing approach.

Calculation Procedure:

1. Determine the needed parameters—pressure drop, flow rate, and fluid specific gravity
The pressure drop, ΔP = 100 − 50 = 50 lb/in² (gage) (344.7 kPa). Flow rate through the orifice is given as 25 gal/min (1.57 L/s). Specific gravity of water at 80°F (26.7°C) = 1.0.

2. Compute the restrictive orifice coefficient
Use the relation $C_{vro} = Q_l/(\Delta P/S_g)^{0.5}$, where C_{vro} = restrictive orifice coefficient, dimensionless; Q_l = liquid flow rate, gal/min (L/s); ΔP = pressure drop through the orifice, lb/in² (gage); S_g specific gravity of the water, dimensionless. Substituting, $C_{vro} = 25/(50/1)^{0.5} = 3.536$.

3. Calculate the required orifice bore diameter
Use the relation $D = 0.875(C_{vro}/14.0)^{0.5}$, where D = required orifice bore (hole) diameter, in (cm); other symbols as before. Substituting, $D = 0.875(3.536/14.0)^{0.5} = 0.4397$; round to 0.440 in (1.12 cm) for ease of manufacturing.

4. Determine the orifice bore size for the thicker plate
When the plate thickness is different from 0.125 in (0.3175 cm), use the relation, $D_{corr} = D(L/0.125)^{0.2}$, D_{corr} = bore diameter in (cm) for plate thickness L, in (cm) different from 0.125 in (0.3175 cm); other symbol as before. Substituting, $D_{corr} = 0.40(0.25/0.125)^{0.2} = 0.505$ in (1.28 cm). Thus, as the orifice plate thickness increases beyond 0.125 in (0.3175 cm), the required bore diameter also increases.

Related Calculations. Restrictive orifices can be easily sized, starting with valve coefficients and making some simple assumptions, namely; (1) It is known that for a plate thickness of 0.125-in (0.3175-cm) a straight-bore orifice has a C_v of 14.0. (2) The Reynolds number is 2300 for turbulent flow. Using these two assumptions, plus the definition of C_v, one can size any restrictive orifice with a plate thickness of 0.125 in (0.3175 cm).

For plates thicker than 0.125 in (0.3175 cm), the orifice bore can be found by applying a correction factor, since I/D^5 = a constant for turbulent flow, as derived from the pressure-drop formula.

C_v = the flow in gal/min (L/s), when the medium is water with a specific gravity = 1, and the pressure drop is 1 lb/in² (6.89 kPa). When calculating C_v, use the physical variables of the flow conditions, shown above.

The C_v = 14.0 for a 0.125-in (0.3175-cm) straight-through sharp-edge orifice is taken from Scientific Apparatus Makers' Association (Washington, D.C.) data for the discharge of water for different pressures and orifice sizes. The control-valve formulas were published by Cashco Inc. (Ellsworth KS) and were modified by the author of this procedure (see below) to include the compressibility for gases and the superheat factor. For cases of two-phase flow where there is a change of cavitation, use the methods found in Fisher Controls International, Inc. (Marshalltown, IA) Catalog 10.

The control-valve formulas for gases and steam are: *For gases:* Q_g = 1360 $(C_v)(\Delta P[P_1 + P_2]/2\ S_{gTZ})^{0.5}$, where Q_g = gas flow rate, std ft³/h (std m³/h); ΔP = pressure drop across the orifice, lb/in²; P_1 = upstream pressure, lb/in²; P_2 = downstream pressure, lb/in²; T = gas temperature, R; Z = gas compressibility; other symbols as before. *For steam:* $W = 3\ C_v/K(\Delta P[P_1 + P_2]/2)^{0.5}$, where K = 1 + 0.0007 ΔT_{sh}, where W = steam flow, lb/h (kg/h); ΔT_{sh} = degree of superheat, F (C).

The restrictive orifice diameter, D, and correction for plate thickness, D_{corr}, are as given above in steps 3 and 4. If ΔP is less than or equal to 0.5 P_1, then (ΔP $[P_1 + P_2]/2$) reduces to $P_{1/2}(1.5)^{0.5}$ = 0.6124 P_1, for gases and steam only—this is sonic flow.

The orifice can be union or paddle type. Use stainless steel Type 304 or 316 for the orifice. For corrosive atmospheres, use special materials.

This procedure is the work of Herman E. Waisvisz, as reported in *Chemical Engineering* magazine.

STEAM TRACING A VESSEL BOTTOM TO KEEP THE CONTENTS FLUID

The bottom of a 4-ft (1.22-m) diameter, stainless-steel, solvent-recovery column holds a liquid that freezes at 320°F (160°C) and polymerizes at 400°F (204.4°C). The bottom head must be traced to keep the material fluid after a shutdown. Determine the required pitch of the tracing, using 150 lb/in² (gage) (1034 kPa) saturated steam. The ambient temperature is −20°F (−28.9°C), the supply steam temperature T_s = 366°F (185.6°C), thermal conductivity of stainless steel = 9.8 Btu/ (h·ft²·°F·ft) [16.95 W/(m·K)], insulation thickness = 2 in (5.1 cm), thermal conductivity of insulation = 0.3 Btu/(h·ft²·°F·ft) [0.52 W/(m·K)], and wall thickness = 0.375 in (0.95 cm). The heat-transfer coefficient between the insulation and the air is 2.0 Btu/(h·ft²·°F) [11.4 W/(m²·K)], and the inside convection coefficient is 20.

Calculation Procedure:

1. *Compute the process-fluid heat-transfer coefficient and the overall heat-transfer coefficient*
Use the relation $1/h_o = 1/h_{air} + x_{ins}/k_{inx}$, where the symbols are as defined in the previous calculation procedure. Substituting, we get $1/h_o = \frac{1}{2} + 2/0.3$; $h_o = 0.14$ Btu/(h·ft²·°F) [0.79 W/(m²·K)].
The process-fluid heat-transfer coefficient $h_t = 20$ Btu/(h·ft²·°F) [113.6 W/(m²·K)], assumed.

2. *Determine the constants A and B and the ratio B/A*
To determine the value of A, solve $A = (h_o + h_t)/Kt = (0.14 + 20)/(9.8)(0.375/12) = 65.6$, dimensionless. Also, $B = (h_i T_p + h_o T_{amb})/Kt = [(20)(320) + (0.14)(-20)]/(9.8)(0.375/12) = 20,900$. Then $B/A = 20,900/65.6 = 318$.

3. *Compute the tracer-steam outlet temperature*
The tracer steam is supplied at 50 lb/in² (gage) (344.7 kPa). Assuming a 15-lb/in² (gage) (103.4-kPa) pressure drop in the tracer system, we see the outlet pressure $= 150 - 15 = 135$ lb/in² (gage) (930.8 kPa). The corresponding saturated-steam temperature is, from the steam tables, $T_o = 358°F$ (181°C). This is the tracer outlet steam temperature.

4. *Calculate the adjusted temperature ratio*
Use the relation $(T_{mid} - B/A)/(T_o - B/A) = (320 - 318)/(358 - 318) = 0.05$.

5. *Determine the required tracing pitch*
From Fig. 32, with the adjusted temperature ratio of 0.5, $\sqrt{A}(L) = 3.7$ when $\alpha = 1$. (Here $\alpha = $ a parameter $= x/L$, where $x=$ distance along the pipe or vessel wall, ft.) Then, by solving for $L = 3.7/65.6^{1/2}$, $L = 0.46$ ft (0.14 m) $= 5.5$ in (13.97 cm).
The maximum allowable pitch for tracing the bottom of the column, Fig. 33, is 21, or 2(5.5) = 11 in (27.9 cm). A typical tracing layout is shown in Fig. 33.
Related Calculations. This procedure can be used to design steam tracing for a variety of tanks and vessels used in chemical, petroleum, food, textile, utility, and similar industries. The medium heated can be liquid, solid, vapor, etc. As with the previous calculation procedure, this procedure is the work of Carl G. Bertram, Vikram J. Desai, and Edward Interess, the Badger Company, as reported in *Chemical Engineering* magazine.

DESIGNING STEAM-TRANSMISSION LINES WITHOUT STEAM TRAPS

Design a steam line for transporting a minimum of 6.0×10^5 lb/h (2.7×10^5 kg/h) and a maximum of 8.0×10^5 lb/h (3.6×10^5 kg/h) of saturated steam at 205 lb/in² (gage) and 390°F (1413 kPa and 198.9°C). The line is 3000 ft (914.4

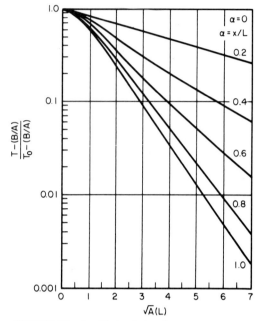

FIGURE 32 Graphical solution for steam-tracing design. (*Chemical Engineering.*)

m) long, with eight 90° elbows and one gate valve. Ambient temperatures range from −40 to 90°F (−40 to 32.2°C). The line is to be designed to operate without steam traps. Insulation 3-in (7.6-cm) thick with a thermal conductivity of 0.48 Btu · in/(h · ft^2 · °F) [0.069 W/(m · K)] will be used on the exterior of the line.

Calculation Procedure:

1. Size the pipe by using a suitable steam velocity for the maximum flow rate
The minimum acceptable steam velocity in a transmission line which is not fitted with steam traps is 110 ft/s (33.5 m/s). Assuming, for safety purposes, a steam velocity of 160 ft/s (48.8 m/s) to use in sizing this transmission line, compute the pipe diameter in inches from $d = 0.001295 f \rho \, LV^2/\Delta P$, where f = friction factor for the pipe (= 0.0105, assumed); ρ = density of the steam, lb/ft^3 (kg/m^3) [= 0.48 (7.7) for this line]; L = length of pipe, ft, including the equivalent length of fittings [= 3500 ft (1067 m) for this pipe]; V = steam velocity, ft/s [= 160 ft/s (48.8 m/s) for this line]; ΔP = pressure drop in the line between inlet and outlet, lb/in^2 [= 25 lb/in^2 (172.4 kPa) assumed for this line]. Substituting yields $d = 0.001295(0.0105)(0.48)(3500)(160)^2/25 = 22.94$ in (58.3 cm); use 24-in (61-cm) schedule 40 pipe, the nearest standard size.

2. Check the actual steam velocity in the pipe chosen
The actual velocity of the steam in the pipe can be found from $V = Q/A$, where V = steam velocity, ft/s (m/s); Q = flow rate of steam, lb/s (kg/s); A = cross-

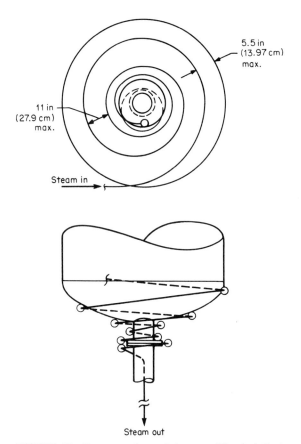

FIGURE 33 Steam-traced vessel bottom. (*Chemical Engineering.*)

sectional area of pipe, ft² (m²). Substituting gives V = (800,000 lb/h ÷ 3600 s/h)(2.08 ft³/lb for steam at the entering pressure)/2.94 ft² = 157.6 ft/s (48.0 m/s) for maximum-flow conditions; V = (600,000/3600)(2.08)/2.94 = 117.9 ft/s (35.9 m/s) for minimum-flow conditions.

3. Compute the pressure drop in the pipe for each flow condition
Use the relation ΔP = 0.001295 $f\rho LV^2/D$, where the symbols are the same as in step 1. Substituting, we find ΔP = 0.001295(0.0105)(0.48)(3500)(157.6)²/24 = 23.2 lb/in² (159.9 kPa) for maximum-flow conditions. For minimum-flow conditions by the same relation, ΔP = 0.001295 (0.0105)(0.48)(3500)(117.9)²/24 = 13.23 lb/in² (91.2 kPa). The pressure at the line outlet will be 220.0 − 23.2 = 196.5 lb/in² (1356.9 kPa) for the maximum-flow condition and 220.0 − 13.2 = 206.8 lb/in² (1425.9 kPa) for minimum-flow conditions.

4. Compute the steam velocity at the pipe outlet

Use the velocity relation in step 1. Hence, for maximum-flow conditions, $V = (800,000/3600)(2.30)/2.94 = 173.8$ ft/s (52.9 m/s). Likewise, for minimum-flow conditions, $V = (600,000/3600)(2.19)/2.94 = 124.1$ ft/s (37.8 m/s).

5. Determine the enthalpy change in the steam at maximum temperature-difference conditions

First, the heat loss from the insulated pipe must be determined for the maximum temperature-difference conditions from $Q_m = h\Delta tA$, where Q_m =heat loss at maximum flow rate, Btu/h (W); h = overall coefficient of heat transfer for the insulated pipe, Btu·in/(h·ft^2·°F) [W·cm/(m^2·°C)]; Δt = temperature difference when the minimum ambient temperature prevails,°F (°C); A = insulated area of pipe exposed to the outdoor air, ft^2 (m^2). Substituting yields $Q_m = 0.16$ (430)(3362)(6.28) = 1,452,599 Btu/lb (3378.7 MJ/kg). In this relation, 430°F = 390°F steam temperature $\pm$ (-40°F) ambient temperature; 3362 = pipe length including elbows and valves, ft; 6.28 = area of pipe per ft of pipe length, ft^2.

The enthalpy change for the maximum temperature difference will be the largest with the minimum steam flow. This change, in Btu/lb (J/kg) of steam, is $\Delta h_{max} = Q_m/F$, where F = flow rate in the line, lb/h, or $\Delta h_{max} = 1,452,599/600,000 = 2.42$ Btu/lb (5631 J/kg).

The minimum enthalpy at the pipe line outlet = inlet enthalpy − enthalpy change. For this pipe line, $h_{0min} = 1199.60 - 2.42 = 1197.18$ Btu/lb (2784.6 kJ/kg).

6. Determine the enthalpy change in the steam at the minimum temperature-difference conditions

As in step 5, $Q_{min} = h\Delta tA$, or $Q_{min} = 0.16(300)(3362)(6.28) = 1,013,441$ Btu/lb (2357.3 mJ/kg). Then $\Delta h_{min} = 1,013,441/800,000 = 1.26$ Btu/lb (2946.6 J/kg). Also $h_{2max} = 1199.60 - 1.26 = 1198.34$ Btu/lb (2787.3 kJ/kg).

7. Determine the steam conditions at the pipe outlet

From step 3, the pressure at the transmission line outlet at minimum flow and lowest ambient temperature is 206.8 lb/in^2 (1425.9 kPa), and the enthalpy is 1197.18 Btu/lb (2784.6 kJ/kg). Checking this condition on a Mollier chart for steam, we find that the steam is wet because the condition point is below the saturated-vapor line.

From steam tables, the specific volume of the steam is 2.22 ft^3/lb of total mass, while the specific volume of the condensate is 0.0000342 ft^3/lb of total mass. Thus, the percentage of condensate per volume = 100(0.0000342)/2.22 = 0.00154 percent condensate per volume. The percentage volume of dry steam therefore = 100(1.00000 − 0.00154) = 99.99846 percent dry steam per volume.

Since the velocity under these steam conditions is 124.1 ft/s (37.8 m/s), the steam will exist as a fine mist because such a status prevails when the steam velocity exceeds 110 ft/s (33.5 m/s). In the fine-mist condition, the condensate cannot be collected by a steam trap. Hence, no steam traps are required for this transmission line as long as the pressure and velocity conditions mentioned above prevail.

Related Calculations.   Some energy is lost whenever a steam trap is used to drain condensate from a steam transmission line. This energy loss continues for as long as the steam trap is draining the line. Further, a steam-trap system requires an initial investment and an ongoing cost for routine maintenance. If the energy loss and trap-system costs can be reduced or eliminated, many designers will take the opportunity to do so.

A steam transmission line carries energy from point 1 to point 2. This energy is a function of temperature, pressure, and flow rate. Along the line, energy is lost through the pipe insulation and through steam traps. A design that would reduce the energy loss and the amount of required equipment would be highly desirable.

The designer's primary concern is to ensure that steam conditions stay as close to the saturated line as possible. The steam state in the line changes according to the change in pressure due to a pressure drop and the change in enthapy due to a heat loss through insulation. These changes of condition are plotted in Fig. 34, a simplified Mollier chart for steam. Point 1 is defined by P_1 and T_1 steam conditions. Because of the variability of such parameters as flow rate and ambient temperature, the designer should consider extreme conditions. Thus, P_2 would be defined by the minimum pressure drop produced by the minimum flow rate. Similarly, h_2 would be defined by the maximum heat loss produced by the lowest ambient temperature. Point 2 on the h-s diagram is defined by the above P_2 and h_2. If point 2 is above or on the saturated-steam line, no condensate is generated and steam traps are not required.

In some cases (small pressure drop, large heat loss), point 2' is below the saturated-steam line, and some condensate is generated. The usual practice has been to provide trap stations to collect this condensate and steam traps to remove it. However, current research in two-phase flow demonstrates that the turbulent flow, produced by normal steam velocities and reasonable steam qualities, disperses any condensate into a fine mist equally distributed along the flow profile. The trap stations do not collect the condensate, and once again, the steam traps are not required. For velocities greater than 110 ft/s (33.5 m/s) and a steam fraction more than 98 percent by volume, the condensate normally generated in a transmission line exists as a fine mist that cannot be collected by steam-trap stations.

A few basic points should be followed when a steam line is operated without steam traps. All lines must be sloped. If a line is long, several low points may be required. Globe valves are used on drains for each low point and for a drain at the end of the line. Since trap stations are not required, drain valves should be located as close to the line as possible to avoid freezing. Vents are placed at all high points. All vents and drains are opened prior to warming the line. Once steam is flowing from all vent valves, they are closed. As each drain valve begins to drain steam only, it is partially closed so that it may still bleed condensate if necessary. When full flow is established, all drain valves are shut. If the flow is shut down, all valves are opened until the pipe cools and are then closed to isolate the line from the

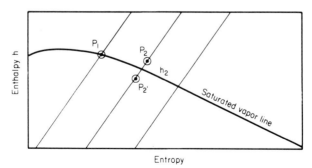

FIGURE 34 Simplified Mollier chart showing changes in steam state in a steam-transmission line.

environment. The above procedure would be the same if the steam traps were on the line.

In summary, it can be demonstrated that steam traps are not required for steam transmission lines, provided that one of the following parameters is met:

1. Steam is saturated or superheated.
2. Steam velocity is greater than 110 ft/s (33.5 m/s), and the steam fraction more than 98 percent by volume.

By using the above design, the steam energy normally lost through traps is saved, along with the construction, maintenance, and equipment costs for the traps, drip leg, strainers, etc., associated with each trap station.

This calculation procedure can be used for steam transmission lines in chemical plants, petroleum refineries, power plants, marine installations, factories, etc. The procedure is the work of Mileta Mikasinovic and David R. Dautovich, Ontario Hydro, and reported in *Chemical Engineering* magazine.

Leaks of hazardous materials from underground piping and tanks can endanger lives and facilities. To reduce leakage dangers, the EPA now requires all underground piping through which hazardous chemicals or petrochemicals flow to be designed for double containment. This means that the inner pipe conveying the hazardous material is contained within an outer pipe, giving the "double-containment" protection.

Likewise, underground tanks are governed by the new Underground Storage Tank (UST) laws. The UST laws also cover underground piping. By December 1998, all existing underground piping conveying hazardous materials will have to be retrofitted to double-containment systems to comply with EPA requirements.

Double containment of piping brings a host of new problems for the engineering designer. Expansion of the inner and outer pipes must be accommodated so that there is no interference between the two. While prefabricated double-containment piping can solve some of these problems, engineers are still faced with considerations of soil loading, pipe expansion and contraction, and fluid flow. Careful study of the EPA requirements is needed before any double-containment design is finalized. Likewise, local codes and laws must be reviewed prior to starting and before finalizing any design.

LINE SIZING FOR FLASHING STEAM CONDENSATE

REFERENCES: [1] O. Baker, *Oil & Gas J.,* July 26, 1954; [2] S. G. Bankoff, *Trans. ASME,* vol. C82, 265 (1960); [3] M. W. Benjamin and J. G. Miller, *Trans. ASME,* vol. 64, 657 (1942); [4] J. M. Chenoweth and M. W. Martin, *Pet. Ref.,* vol. 34, 151 (1955); [5] A. E. Dukler, M. Wickes, and R. G. Cleveland, *AIChE J.,* Vol. 10, 44 (1964); [6] E. C. Kordyban, *Trans. ASME,* Vol. D83, 613 (1961); [7] R. W. Lockhart and R. C. Martinelli, *Chem. Eng. Prog.,* Vol. 45, 39 (1949); [8] P. M. Paige, *Chem. Eng.,* p. 159, Aug. 14, 1967.

A reboiler in an industrial plant is condensing 1000 lb/h (0.13 kg/s) of steam of 600 lb/in^2 (gage) (4137 kPa) and returning the condensate to a nearby condensate return header nominally at 200 lb/in^2 (gage) (1379 kPa). What size condensate line will give a pressure drop of (1 lb/in^2)/100 ft (6.9 kPa/30.5 m) or less?

Calculation Procedure:

1. *Use a graphical method to determine a suitable pipe size*
Flow in condensate-return lines is usually two-phase, i.e., comprised of liquid and vapors. As such, the calculation of line size and pressure drop can be done by using a variety of methods, a number of which are listed below. Most of these methods, however, are rather difficult to apply because they require extensive physical data and lengthy computations. For these reasons, most design engineers prefer a quick graphical solution to two-phase flow computations. Figure 35 provides a rapid estimate of the pressure drop of flashing condensate, along with a determination of fluid velocity. To use Fig. 35, take these steps.

Enter Fig. 35 near the right-hand edge at the steam pressure of 600 lb/in² (gage) (4137 kPa) and project downward to the 200-lb/in² (gage) (1379-kPa) end-pressure curve.

From the intersection with the end-pressure curve, project horizontally to the left to intersect the 1000-lb/h (0.13-kg) curve. Project vertically from this intersection to one or more trial pipe sizes to find the pressure loss for each size.

Trying the 1-in (2.5-cm) pipe diameter first shows that the pressure loss—[3.0 lb/in² (gage)]/100 ft (20.7 kPa/30.5 m) exceeds the desired [1 lb/in² (gage)]/100 ft (6.9 kPa/30.5 m). Projecting to the next larger standard pipe size, 1.5 in (3.8 cm), gives a pressure drop of [2 lb/in² (gage)]/100 ft (1.9 kPa/30.5 m). This is within the desired range. The velocity in this size pipe will be 16.5 ft/s (5.0 m/s).

2. *Determine the corrected velocity in the pipe*
At the right-hand edge of Fig. 35, project upward from 600-lb/in² (gage) (4137-kPa) to 200-lb/in² (gage) (1379-kPa) end pressure to read the velocity correction factor as 0.41. Thus, the actual velocity of the flashing mixture in the pipe = 0.41 (16.5) = 6.8 ft/s (2.1 m/s).

Related Calculations. This rapid graphical method provides pressure-drop values comparable to those computed by more sophisticated techniques for two-phase flow [1–8]. Thus, for the above conditions, the Dukler [5] no-slip method gives [0.22 lb/in² (gage)]/100 ft (1.52 kPa/30.5 m), and the Dukler constant-slip method gives [0.25 lb/in² (gage)]/100 ft (1.72 kPa/30.5 m).

The chart in Fig. 35 is based on the simplifying assumption of a single homogeneous phase of fine liquid droplets dispersed in the flashed vapor. Pressure drop is computed by Darcy's equation for single-phase flow. Steam-table data were used to calculate the isenthalpic flash of liquid condensate from a saturation pressure to a lower end pressure; the average density of the resulting liquid-vapor mixture is used as the assumed homogeneous fluid density. Flows within the regime of Fig. 35 are characterized as either in complete turbulence or in the transition zones near complete turbulence.

Pressure drops for steam-condensate lines can be determined by assuming that the vapor-liquid mix throughout the lines is represented by the mix for conditions at the end pressure. This assumption conforms to conditions typical of most actual condensate systems, since condensate lines are sized for low-pressure drop, with most flashing occurring across the steam trap or control valve at the entrance.

If the condensate line is to be sized for a considerable pressure drop, so that continuous flashing occurs throughout its length, end conditions will be quite different from those immediately downstream of the trap. In such cases, an iterative calculation should be performed, involving a series of pressure-drop determinations across given incremental lengths.

This iteration is begun at the downstream end pressure and worked back to the trap, taking into account the slightly higher pressure, and thus the changing liquid-

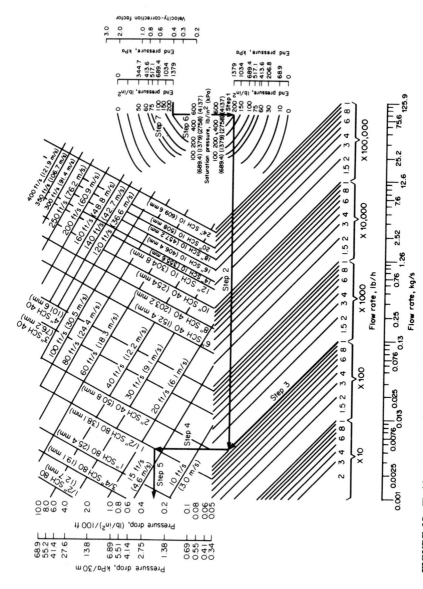

FIGURE 35 Flashing steam condensate line-sizing chart. Divide by 10⁴ to obtain numerical values for flow rate measured in kg/s. (*Chemical Engineering*.)

vapor mix, in each successive upstream incremental pipe length. The calculation is complete when the total equivalent length for the incremental lengths equals the equivalent length between the trap and the end-pressure point. This operation can be performed by using Fig. 35.

Results from Fig. 35 have also been compared to those calculated by a method suggested by a Paige [8] and based on the work of Benjamin and Miller [3]. Paige's method assumed a homogeneous liquid-vapor mixture with no liquid holdup, and thus it is similar in approach to the present method. However, Paige suggests calculation of the liquid-vapor mix based on an isentropic flash, whereas Fig. 35 is based on an isenthalpic flash; and this is believed to be more representative of steam-condensate collecting systems.

For the example, the Paige method gives $(0.26 \text{ lb/in}^2)/100 \text{ ft}$ (1.79 kPa/30.5 m) at the terminal pressure and $(0.25 \text{ lb/in}^2)/100 \text{ ft}$ (1.72 kPa/30.5 m) at a point 1000 ft (305 m) upstream of the terminal pressure, owing to the slightly higher pressure, which suppresses flashing.

The method given here is valid for sizing lines conveying flashing steam used in power plants, factories, air-conditioning systems, petroleum refineries, ships, heating systems, etc. Further, Fig. 35 is designed so that it covers the majority of steam-condensate conditions met in these applications. This calculation procedure is the work of Richard P. Ruskin, Process Engineer, Arthur G. McKee & Co., as reported in *Chemical Engineering* magazine.

DETERMINING THE FRICTION FACTOR FOR FLOW OF BINGHAM PLASTICS

REFERENCES: [1] E. Buckingham, On Plastic Flow Through Capillary Tubes, *ASTM Proc.*, Vol. 21, 1154 (1921); [2] R. W. Hanks and D. R. Pratt, On the Flow of Bingham Plastic Slurries in Pipes and between Parallel Plates, *Soc. Petrol. Eng. J.*, Vol. 1, 342 (1967); [3] R. W. Hanks and B. H. Dadis, Theoretical Analysis of the Turbulent Flow of Non-Newtonian Slurries in Pipes, *AIChE J.*, Vol. 17, 554 (1971); [4] S. W. Churchill, Friction-factor Equation Spans All Fluid-flow Regimes, *Chem. Eng.*, Nov. 7, 1977, pp. 91–92; [5] S. W. Churchill and R. A. Usagi, A General Expression for the Correlation of Rates of Transfer and Other Phenomena, *AIChE J.*, Vol. 18, No. 6, 1121–1128 (1972); [6] R. L. Whitmore, *Rheology of the Circulation*, Pergamon Press, Oxford, 1968; [7] N. Casson, A Flow Equation for Pigment-Oil Dispersions of the Printing Ink Type, Ch. 5 in *Rheology of Disperse Systems*, C. C. Mill (ed.), Pergamon Press, Oxford, 1959; [8] R. Darby and B. A. Rogers, Non-Newtonian Viscous Properties of Methacoal Suspensions, *AIChE J.*, Vol. 26, 310 (1980); [9] G. W. Govier and A. K. Azia, *The Flow of Complex Mixtures in Pipes*, Van Nostrand Reinhold, New York, 1972; [10] E. H. Steiner, The Rheology of Molten Chocolate, Ch. 9 in C. C. Mill (ed.), *op. cit.;* [11] R. B. Bird, W. I. Stewart, and E. N. Lightfoot, *Transport Phenomena,* John Wiley & Sons, New York, 1960.

A coal slurry is being pumped through a 0.4413-m (18-in) diameter schedule 20 pipeline at a flow rate of 400 m^3/h. The slurry behaves as a Bingham plastic, with the following properties (at the relevant temperature): $\tau_0 = 2 \text{ N/m}^2$ (0.0418 lbf/ ft^2); $\mu_\infty = 0.03 \text{ Pa} \cdot \text{s}$ (30 cP); $\rho = 1500 \text{ kg/m}^3$ (93.6 lbm/ft^3). What is the Fanning friction factor for this system?

Calculation Procedure:

1. *Determine the Bingham Reynolds number and the Hedstrom number*
Engineers today often must size pipe or estimate pressure drops for fluids that are
non-Newtonian in nature—coal suspensions, latex paint, or printer's ink, for example. This procedure shows how to find the friction factors needed in such calculations for the many fluids that can be described by the Bingham-plastic flow
mode. The method is convenient to use and applies to all regimes of pipe flow.
    A Bingham plastic is a fluid that exhibits a yield stress; that is, the fluid at rest
will not flow unless some minimum stress τ_0 is applied. Newtonian fluids, in contrast, exhibit no yield stress, as Fig. 36 shows.
    The Bingham-plastic flow model can be expressed in terms of either shear stress
τ versus shear rate γ, as in Fig. 36, or apparent viscosity η versus shear rate:

$$\tau = \tau_0 + \mu_\infty \dot{\gamma} \tag{1}$$

$$\eta = \frac{\tau}{\dot{\gamma}} = \frac{\tau_0}{\dot{\gamma}} + \mu_\infty \tag{2}$$

    Equation 2 means that the apparent viscosity of a Bingham plastic depends on
the shear rate. The parameter μ_∞ is sometimes called the coefficient of rigidity, but
it is really a limiting viscosity. As Eq. 2 shows, apparent viscosity approaches μ_∞
as shear rate increases indefinitely. Thus the Bingham plastic behaves almost like
a newtonian at sufficiently high shear rates, exhibiting a viscosity of μ_∞ at such
conditions. Table 46 shows values of τ_0 and μ_∞ for several actual fluids.
    For any incompressible fluid flowing through a pipe, the friction loss per unit
mass F can be expressed in terms of a Fanning friction factor f:

$$F = \frac{2fLv^2}{D} \tag{3}$$

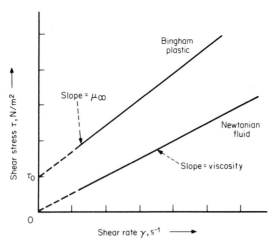

FIGURE 36 Bingham plastics exhibit a yield stress. (*Chemical Engineering.*)

TABLE 46 Values of τ_0 and μ_∞

Fluid	τ_0, N/m²	μ_∞, Pa·s	Ref.
Blood (45% hematocrit)	0.005	0.0028	[6]
Printing-ink pigment in varnish (10% by wt.)	0.4	0.25	[7]
Coal suspension in methanol (35% by vol.)	1.6	0.04	[8]
Finely divided galena in water (37% by vol.)	4.0	0.057	[9]
Molten chocolate (100°F)	20	2.0	[10]
Thorium oxide in water (50% by vol.)	300	0.403	[11]

where L is the length of the pipe section, D is its diameter, and v is the fluid velocity.

An exact description of friction loss for Bingham plastics in fully developed laminar pipe flow was first published by Buckingham [1]. His expression can be rewritten in dimensionless form as follows:

$$f_L = \frac{16}{N_{Re}}\left(1 + \frac{N_{He}}{6N_{Re}} - \frac{N_{He}^4}{3f_L^3 N_{Re}^7}\right) \qquad (4)$$

where N_{Re} is the Bingham Reynolds number (Dvp/μ_∞) and N_{He} is the Hedstrom number $(D^2 p\tau_0/\mu^2)$. Equation 4 is implicit in f_L, the laminar friction factor, but can be readily solved either by Newton's method or by iteration. Since the last term in Eq. 4 is normally small, the value of f obtained by omitting this term is usually a good starting point for iterative solution.

For this pipeline

$$N_{Re} = \frac{4Q\rho}{\pi D\mu_\infty} = \frac{4(400)(1/3)(600)(1500)}{\pi(0.4413)(0.03)} = 16,030$$

$$N_{He} = \frac{D^2 \rho\tau_0}{\mu_\infty^2} = \frac{(0.4413)^2(1500)(2)}{(0.03)^2} = 649,200$$

2. Find the friction factor f_L for the laminar-flow regime
Substituting the values for N_{Re} and N_{He} into Eq. 4, we find $f_L = 0.007138$.

3. Determine the friction factor f_T for the turbulent-flow regime
Equation 4 describes the laminar-flow sections. An empirical expression that fits the turbulent-flow regimen is

$$f_T = 10^a N_{Re}^{-0.193} \qquad (5)$$

where

$$a = -1.378\,[1 + 0.146\,\exp(-2.9 \times 10^{-5}\,N_{He})] \qquad (6)$$

We now have friction-factor expressions for both laminar and turbulent flow. Equation 6 does not apply when N_{He} is less than 1000, but this is not a practical constraint for most Bingham plastics with a measurable yield stress.

When N_{He} is above 300,000, the exponential term in Eq. 6 is essentially zero. Thus $a = -1.378$ here, and Eq. 5 becomes

$$f_T = 10^{-1.378}(16,030)^{-0.193}$$

$$= 0.006463$$

4. Find the friction factor f

Combine the f_L and f_T expressions to get a single friction factor valid for all flow regimes:

$$f = (f_L^m + f_T^m)\frac{1}{m} \tag{7}$$

where f_L and f_T are obtained from Eqs. 4 and 5, and the power m depends on the Bingham Reynolds number:

$$m = 1.7 + \frac{40,000}{N_{Re}} \tag{8}$$

The values of f predicted by Eq. 7 coincide with Hank's values in most places, and the general agreement is excellent. Relative roughness is not a parameter in any of the equations because the friction factor for non-Newtonian fluids, and particularly plastics, is not sensitive to pipe roughness.

Substituting yields $m = 1.7 + 40,000/16,030 = 4.20$, and $f = [(0.007138)^{4.20} + (0.006463)^{4.20}]^{1/4.20} = 0.00805$.

If m had been very large, the bracketed term above would have approached zero. Generally, when N_{Re} is below 4000, Eq. 8 should be solved by taking f equal to the greater of f_L and f_T.

Related Calculations.   This procedure is valid for a variety of fluids met in many different industrial and commercial applications. The procedure is the work of Ron Darby, Professor of Chemical Engineering, Texas A & M University, College of Engineering, and Jeff Melson, Undergraduate Fellow, Texas A & M, as reported in *Chemical Engineering* magazine. In their report they cite works by Hanks and Pratt [2], Hanks and Dadia [3], Churchill [4], and Churchill and Usagi [5] as important in the procedure described and presented here.

TIME NEEDED TO EMPTY A STORAGE VESSEL WITH DISHED ENDS

A tank with a 6-ft (1.8-m) diameter cylindrical section that is 16 ft (4.9 m) long has elliptical ends, each with a depth of 2 ft (0.7 m), and is half-full with ethanol, a newtonian fluid. How long will it take to empty the tank if it is set horizontally and fitted at the bottom with a drain consisting of a short tube of 2-in (5.1-cm) double extrastrong pipe? How long will it take to empty the tank if it is set vertically and fitted at the bottom with a drainpipe of 2-in (5.1-cm) double extrastrong pipe?

The drain system extends 4 ft (1.2 m) below the dished bottom and has an equivalent length of 250 ft (76.2 m).

Calculation Procedure:

1. Determine the discharge coefficient for, and orifice area of, the drain tube
Figure 37 shows the discharge coefficient is $C_d = 0.80$ for a short, flush-mounted tube. Baumeister, in *Mark's Standard Handbook for Mechanical Engineers,* indicates the internal section area of the tube is $A_n = 1.774$ in^2 (11.4 cm^2).

2. Compute the discharge time for the tank in a horizontal position
Substitute the appropriate values in the equation for t_v shown under the storage tanks in Fig. 38. Thus, $t_p = [(8)^{0.5}/[3(0.80)(1.774/144)(32.2)^{0.5}] \{16[(6)^{1.5} - (6 - 3)^{1.5}] + [2\pi(3)^{1.5}/6][6 - (3/5)(3)]\} = 2948$ s, or 49.1 min.

3. Determine the internal diameter and friction factor for the drainpipe
From Baumeister, *Mark's Standard Handbook for Mechanical Engineers,* the internal diameter of the pipe is $d = 1.503$ in (3.8 cm), or 0.125 ft (0.038 m) and the Moody friction factor is $f = 0.020$ for the equivalent length, $l = 250$ ft (76.2 m), of pipe.

4. Compute the initial and final height above the drainpipe outlet for the cylindrical section
Initial height of the liquid is $H_1 = a + b + h_o = [(16/2) + 2 + 4] = 14.0$ ft (4.3 m). Final height is $H_F = b + h_o = 2 + 4 = 6$ ft (1.8 m).

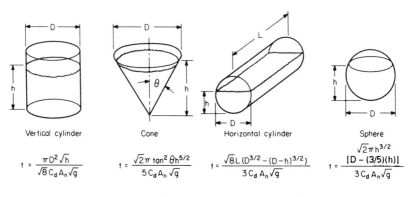

FIGURE 37 Time to empty tanks. (*Chemical Engineering.*)

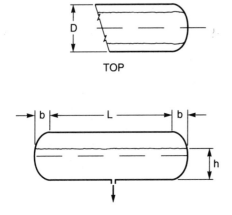

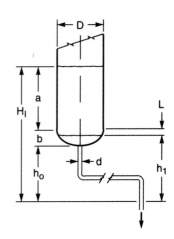

TOP

FRONT   Vertical Cylinder with dished end
Horizontal cylinder with dished ends   and drainpipe system

$$t_c = \frac{D^2}{d^2} \{[(2/g)(1 + [fl/d])]^{1/2} (H_1^{1/2} - H_F^{1/2})\}$$

$$t_e = C\{[(2 \times H_2^2/5) - (4 \times B \times H_2/3) + 2E^2)(H_2^{1/2})$$
$$- [(2 \times H_1^2/5) - (4 \times B \times H_1/3) + 2E^2)(H_1^{1/2})\}$$

$$t_p = \frac{\sqrt{8}}{3C_d A_n \sqrt{g}} \{L[D^{3/2} - (D - h)^{3/2}] + \frac{bph^{3/2}}{D}[D - (3h/5)]\}$$

FIGURE 38   Time to drain tanks. (*a*) Top and (*b*) front view of horizontal cylinder with dished ends. (*c*) Vertical cylinder with dished-end and drainpipe system. (*Chemical Engineering.*)

5.   Compute the time required to drain the cylindrical section of the tank
Substitute the appropriate values in the equation for t_c shown under the storage tanks in Fig. 38. Hence, t_c = $[(6)^2/(0.125)^2]\{(2/32.2)[1 + (0.020 \times 250/0.125)]\}^{0.5}[(14)^{0.5} - (6)^{0.5}]$ = 4751 s, or 79.2 min.

6.   Compute the initial and final liquid height above the drainpipe outlet for the elliptically dished head
Initial height of the liquid is H_1 = $b + h_o$ eq 2 + 4 = 6 ft (1.8 m). Final height is H_2 = h_o = 4 ft (1.2 m).

7.   Compute how long it will take to empty the dished bottom of the tank
In order to solve for t_e it is necessary to determine the following values: B = $h_o + b$ = 4 + 2 = 6 ft (1.8 m); E^2 = $h_o^2 + 2bh_o$ = $(4)^2 + 2(2)(4)$ = 32 ft^2 (3.0 m^2); C = $[D/(db)]^2\{[1/(2g)][1 + (f_1/d)]\}^{0.5}$ = $[6/(0.125 \times 2)]^2\{[1/(2 \times 32.2)][1 + ([0.02 \times 250]/0.125)]\}^{0.5}$ = 459.6, s/ft$^{5/2}$ (s/m$^{5/2}$).
   Then, use the values for B, E^2, C, and other relevant dimensions to find t_e from the equation shown under the storage tanks in Fig. 38. Thus, t_e = 459.6 $[(2 \times 4^2/5) - (4 \times 6 \times 4/3) + 2(32)](4)^{0.5} - [(2 \times 6^2/5) - (4 \times 6 \times 6/3) + 2(32)](6^{0.5})$ = 1073 s, or 17.9 min.

8. *Compute the time it will take to drain the half-full vertical tank*
Total time is $t_t + t_c + t_e = 4751 + 1073 = 5824$ s, or 96.1 min.

Related Calculations. Figure 38 shows the equation for computing the emptying time for a horizontal cylindrical tank with elliptically dished ends and equations fro calculating the emptying time for a vertical cylindrical tank with an elliptically dished bottom end fitted with a drain system. The symbols A_n, g, t, and C_d are defined as in the previous problem for a storage vessel without dished ends, except that A_n is now the drainpipe internal area.

The term associated with the second pair of brackets in the equation for t_p accounts for the dished ends of the horizontal tank. For hemispherical ends $b = D/2$ and for flat ends, $b = 0$.

When seeking the time required to drain a portion of the cylindrical part of the vertical tank use the formula for t_c with the appropriate values for H_l and H_F and other pertinent variables. To find the time it takes to drain a portion of the dished bottom of the vertical tank use the formula for t_e with given values of H_1 and H_2 and other applicable variables.

The relations given here are valid for storage tanks used in a variety of applications—chemical and petrochemical plants, power plants, waterworks, ships and boats, aircraft, etc. The procedure for a horizontal cylindrical tank with dished ends is the work of Jude T. Sommerfeld, and the procedure for a vertical cylindrical tank with a dished bottom end is the work of Mahnoosh Shoael and Jude T. Sommerfeld, as reported in *Chemical Engineering* magazine.

TIME NEEDED TO EMPTY A STORAGE VESSEL WITHOUT DISHED ENDS

How long will it take to empty a 10-ft (3-m) diameter spherical tank filled to a height of 8 ft (2.4 m) with ethanol, a newtonian fluid, if the drain is a short 2-in (5.1-cm) diameter tube of double extra-strong pipe?

Calculation Procedure:

1. *Determine the discharge coefficient for the drain*
Figure 37 shows that the discharge coefficient is $C_d = 0.80$ for a short, flush-mounted tube.

2. *Compute the discharge time*
Substitute the appropriate values in the equation in Fig. 37 for spherical storage tanks. Or, $t = (2)^{0.5}(\pi)(8)^{1.5}[10 - (0.6 \times 8)]/[3(0.8)(1.774/144)(32.2)^{0.5}] = 3116$ s, or 51.9 min.

Related Calculations. Figure 37 gives the equations for computing the emptying time for four common tank geometrics. The discharge coefficient C_d is constant for newtonian fluids in turbulent flow, but the coefficient depends on the shape of the orifice. Water flowing through sharp-edged orifices of 0.25-in (0.64-cm) diameter, or larger, is always turbulent. Thus, the assumption of a constant C_d is valid for most practical applications. Figure 37 lists accepted C_d values.

The relations given here are valid for storage tanks used in a variety of applications—chemical and petrochemical plants, power plants, waterworks, ships and boats, aircraft, etc. This procedure is the work of Thomas C. Foster, as reported in *Chemical Engineering* magazine.

TIME TO DRAIN A STORAGE TANK THROUGH ATTACHED PIPING

Determine the time required to drain a 50-ft (15.2-m) diameter tank to a level of 10 ft (3 m) if the tank is filled with water to a height of 25 ft (7.6 m). The drain pipe system has 474 ft (144.5 m) equivalent length of 4-in (101.6-mm) Schedule 40 pipe with a friction factor, $f = 0.0185$. The pipe outlet elevation is 0 ft (0 m). Compute the drainage time if the outlet pipe elevation is -3 ft (-1 m) instead of 0 ft (0 m).

Calculation Procedure:

1. Compute the drainage time for the first outlet elevation

The literature presents numerous equations and nomograms to determine the time to drain a vertical or horizontal cylindrical tank. However, these equations do not normally consider any associated piping through which a tank might be drained.

Using the Bernoulli equation for point 1, the tank liquid surface at any height above the bottom, and point 2, the drain pipe height, gives $t = (D^2/d^2)[2/g]\{fL/d + 1\}]^{0.5}([H_o]^{0.5} - [H_f]^{0.5})$, where t = time required to drain the tank through the attached piping, s; D = tank diameter, ft (m); d = drain pipe diameter, ft (mm); g = gravitational constant, 32.2 ft/s^2 (9.8 m/s^2); f = pipe friction factor; L = equivalent length of piping and any associated fittings, ft (m); H_o = liquid height at any time above the drain-pipe outlet, ft (m); H_f = liquid height when tank drainage is completed, ft (m).

For the first situation where the drain-pipe outlet is 0 ft (0 m), $t = (50^2/0.336^2)[(2/32.2)\{0.0185(474)/0.336\} + 1]^{0.5}([25]^{0.5} - [10]^{0.5}) = 52,496$ s, or $52,496/3600$ s/hr $= 14.58$ hr.

2. Determine the drainage time when the piping outlet is below the tank bottom

When the tank drainage-pipe outlet is below the tank bottom, H_f is a negative number. For the given situation, $H_f = -3$ ft (-0.9 m). Then, the time to drain this tank, $t = (50^2/0.336^2)[2/32.2)(0.0185(474)/0.336\} + 1]^{0.5}([28]^{0.5} - [13]^{0.5}) = 48,436$ s, or $48,436/3600 = 13.5$ h.

Related Calculations. The equation presented here can be used for any liquid—oil, water, acid, caustic, etc.—and any type of piping—steel, cast iron, plastic, wood, etc., provided the friction factor and equivalent length are adjusted to reflect the fluid and type of piping involved in the calculation.

Where frequent tank drainage is expected, this procedure can be used to determine quickly the difference in drainage time for pipes of various diameters. Then, with the cost of each diameter of piping known, an economic study will show which size piping is most attractive from an investment standpoint when each drainage time is given a relative-importance rating. Where tank drainage will be done infrequently, and drainage time is not an important factor in plant or process operation, the cheapest available drainage piping which provides safe performance is usually the best choice.

This procedure is the work of Nick J. Loiacono, P.E., Wink Engineering, as reported in *Chemical Engineering* magazine. Data on the economic aspects of drainage-pipe sizing and SI values were added by the handbook editor.

SECTION 9
AIR AND GAS COMPRESSORS AND VACUUM SYSTEMS

System Economics and Design Strategies

ESTIMATING THE COST OF AIR LEAKS IN COMPRESSED-AIR SYSTEMS

Find the cost of compressed air leaking through a 0.125-in (0.3175-cm) diameter hole in a pipe main of a typical industrial air piping system, Fig. 1, to the atmosphere at sea level when the air pressure in the pipe is 10 lb/in² (gage) (68.9 kPa), the plant, Fig. 2, operates 7500 h/yr, air temperature is 70°F (21.1°C), and the cost of compressed air is $1.25 per 1000 ft³ (28.3 m³). What is the cost of the leaking air when the pipe pressure is 50 lb/in² (gage) (344.5 kPa) and the other variables are the same as given above?

Calculation Procedure:

1. Find the volume of air discharged to the atmosphere
Air flowing through an orifice or nozzle attains a critical pressure of 0.53 times the inlet or initial pressure. This reduced pressure occurs at the throat or vena contracta, which is the point of minimum stream diameter on the outlet side of the air flow. If the outlet or back pressure exceeds the critical pressure then the vena contracta

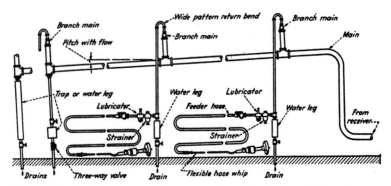

FIGURE 1 Typical compressed-air system main and branch pipes (*Factory Management and Maintenance*).

or throat pressure rises to equal the backpressure. Air flow through a hole in a pipe or tank replicates the flow through an orifice or nozzle.

When an inlet air pressure of 10 lb/in² (gage) + 14.7 = 24.7 lb/in² (abs) (170.2 kPa), the critical pressure is 0.53 × 24.7 = 13.09 lb/in² (abs) (90.2 kPa). Since 13.09 lb/in² (abs) is less than the atmospheric backpressure of 14.7 lb/in² (abs) (101.3 kPa), the throat pressure equals the backpressure, or 14.7 lb/in² (abs) (101.3 kPa). Knowing this, we can compute weight of the escaping air from $W = 1.06$ $A(P_1[P - P_1]/T)^{0.5}$, where W = leakage rate, lb/s (kg/s); A = area of leakage hole, in² (cm²); P = pipeline or initial air pressure, lb/in² (abs) (kPa); P_1 = outlet or backpressure, lb/in² (abs); T = absolute temperature of the air before leakage = °F + 460.

Substituting, using the values given above, $W = 1.06 \times 0.012272(14.7[24.7 - 14.7]/530)^{0.5} = 0.006851$ lb/s (0.0031 kg/s). Converting this air leakage rate to lb/h (kg/h), multiply by 3600 s/h, or 0.006851 × 3600 = 24.66 lb/h (11.19 kg/h). Since the cost of compressed air is expressed in $/ft³, the flow rate of the leaking air must be converted. Since air at 14.7 lb/in² (abs) (101.3 kPa) weighs 0.075 lb per ft³ (1.2 kg/m³), the rate of leakage is 24.66/0.075 = 328.8 ft³/h (9.31 m³/h).

2. Determine the annual cost of the air leakage
This compressed-air plant operates 7500 h/yr. Since the leakage rate is 328.8 ft³/h, the annual leakage through this opening is 7500 × 328.8 = 2,466,000 ft³ (58,691 m³). At a cost of $1.25 per 1000 ft³, the annual total cost of this leak is $1.25 × 2,466,000/1000 = $3,082.50. This is a sizeable charge, especially if there are several leaks of this size, or larger, in the system.

3. Find the rate of leakage at the higher line pressure
When the backpressure is less than the critical pressure, a different flow equation must be used. In the second instance, the critical pressure is 0.53 (50 lb/in² (gage) + 14.7) = 34.29 lb/in² (abs) (236.26 kPa). Since 34.29 lb/in² (abs) (236.26 kPa) is greater than the atmospheric backpressure of 14.7 lb/in² (abs) (101.3 kPa), the critical pressure is greater than the backpressure. Air leakage through the hole is now given by $W = 0.5303(ACP)/(T)^{0.5}$, where C = flow coefficient = 1.0; other symbols as before.

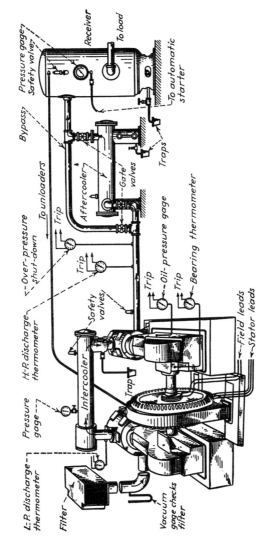

FIGURE 2 Typical compressed-air plant showing compressor and its associated piping and accessories (*Power*).

9.3

Substituting, $W = 0.5303(0.012272)(1.0)(64.7)/(530)^{0.5} = 0.01829$ lb/s (0.0083 kg/s). Converting to an hourly flow rate as earlier, $3600 \times 0.01828 = 65.84$ lb/h (29.89 kg/h).

4. Compute the annual cost of air leakage at the higher pressure
Following the same steps as earlier, annual leakage cost = 65.84 lb/h (7500 h/yr)(1.25/1000 ft³)/0.075 lb/ft³ = \$8,230.00 per year. Again, this is a significant loss of revenue. Further, the loss at the higher pressure is \$8230/3082.50 = 2.67 times as great. This points out the fact that higher pressures in a compressed-air system can cause more expensive leaks.

   Related Calculations.   Compressing air requires a power input to raise the air pressure from atmospheric to the level desired for the end use of the air. When compressed air leaks from a pipe or storage tank, the power expended in compression is wasted because the air does no useful work when it leaks into the atmosphere.

   In today's environment-conscious world, compressed-air leaks are considered to be especially wasteful because they increase pollution without producing any beneficial results. The reason for this is that the fuel burned to generate the power to compress the wasted air pollutes the atmosphere unnecessarily because the air produces only a hissing sound as it escapes through the hole in the pipe or vessel.

SELECTING AN AIR MOTOR FOR A KNOWN APPLICATION

Show how to select a suitable air motor for a reversible application requiring 2 hp (1.5 kW) at 1000 rpm for an industrial crane. Determine the probable weight of the motor, its torque output, and air consumption for this intermittent duty application. An adequate supply of air at a wide pressure range is available at the installation.

Calculation Procedure:

1. Assemble data on possible choices for the air motor
There are four basic types of air motors in use today: (1) radial-piston type; (2) axial-piston type; (3) multi-vaned type; (4) turbine type. Each type of air motor has advantages and disadvantages for various applications. Characteristics of these air motors are as follows:

   (1) Radial-piston air motors, Fig. 3, have four or five cylinders mounted around a central crankshaft similar to a radial gasoline engine. Five cylinders are preferred to supply more horsepower with evenly distributed power pulses. In such a unit there are always two cylinders having a power stroke at the same time. The radial-piston motor is usually a slow-speed unit, ranging from 85 to 1500 rpm. It is suited for heavy-duty service up to 20 hp (15 kW) where good lugging characteristics are needed. Normally they are not reversible, though reversible models are available at extra cost.

   (2) Axial-piston air motors, Fig. 4, are more compact in design and require less space than a four- or five-cylinder radial-piston motor. Air drives the pistons in translation; a diaphragm-type converter changes the translation into rotation. This arrangement supplies high horsepower per unit weight. Axial-piston motors are

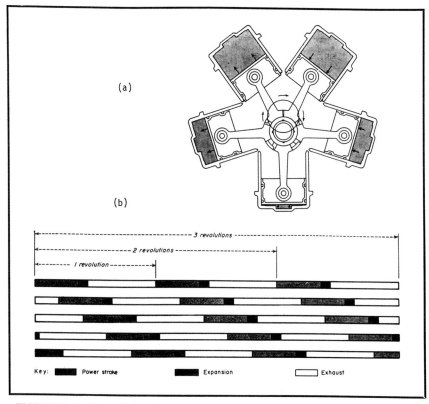

FIGURE 3 (*a*) Five-cylinder piston-type radial air motor used in sizes from about 2 hp (1.5 kW) to 22 hp (16.4 kW) and at speeds from 85 to 1500 rpm. (*b*) How five-cylinder air motor distributes power. Two cylinders are always on power stroke at any instant (*Gardner-Denver Company*).

available in sizes from 0.5 to 2.75 hp (0.37 to 2.1 kW). They run equally well in either direction. To make the motor reversible, a four-way air valve is inserted in the line.

(3) Multi-vaned motors, Fig. 5, are suitable for loads from fractional hp (kW) to 10 hp (7.5 kW). They are relatively high-speed units which must be geared down for usable speeds. The major advantages of multi-vaned motors is light weight and small size. However, if used at slow speed, the gearing may add significantly to the weight of the motor.

(4) Air-turbine motors deliver fractional horsepowers at exceptionally high speeds, from 10,000 to 150,000 rpm, and are an economical source of power. They are tiny impulse-reaction turbines in which air at 100 psi (689 kPa) impinges on buckets for the driving force. Force-feed automatic lubrication sprays a fine film of oil on to bearings continuously, minimizing maintenance.

Based on the load requirements, 2 hp (1.5 kW) at 1000 rpm, a reversible radial-piston air motor, Table 1, would be a suitable choice because it delivers up to 2.8

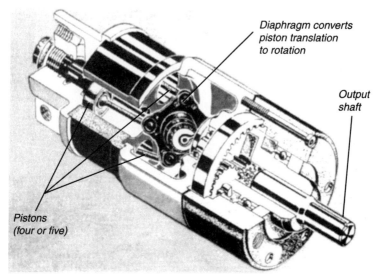

FIGURE 4   Axial-piston air motor available in various output sizes (*Keller Tool Company*).

FIGURE 5   Typical multi-vane type air motor, available in fractional hp sizes and up to some 10 hp (7.5 kW) (*Gast Manufacturing Company*).

TABLE 1 Specifications of Typical Air Motors

Rated hp* (kW)	Speed at rated hp—rpm	Free speed rpm	Weight lb (kg)	Stall torque ft—lbs	Air consumption at rated hp ft³ free air/min
Radial piston motors (non-reversible)*					
2.9 (2.2)	1,500	3,200	130 (59)	....	....
3.3 (2.5)	1,300	3.000	130 (59)	....	....
3.8 (2.8)	1,200	2,700	130 (59)	....	....
Radial piston motors (reversible)*					
2.5 (1.7)	1,200	2,200	135 (61.3)	....	....
2.8 (2.1)	1,000	1,950	135 (61.3)	....	....
3.2 (2.4)	900	1,600	135 (61.3)	....	....
5.2 (3.9)	750	1,600	200 (90.8)	....	....

*at 90 lb/in² (620 kPa).
Ingersoll-Rand.

hp (2.1 kW) at 1000 rpm with air delivered to the motor at 90 lb/in² (620 kPa). The weight of this motor, Table 1, is 135 lb (61.3 kg).

2. Compare the advantages of air motors to other types of motive power
Air motors have a number of advantages over their usual competitors—electric motors. These advantages are: (1) In explosive or gaseous environments, air motors are lower in cost than larger, heavier, explosion-proof electric motors. Air motors operate relatively trouble-free in moist, humid environments where the electric motor may suffer from a buildup of fungus and corrosion. And since the air motor requires little maintenance, it can be mounted in inaccessible locations. (2) With an air motor, the output speed can be varied from zero to free-speed no-load rotation by merely changing the volume of air supplied to the motor. Controls are simple in design and use. (3) Air motors can weigh as little as one-quarter that of electrically-powered units; their physical dimensions are about 50 percent those of electrical devices. Further, air motors do not spark; they cannot burn out from overloading; the air motor is not injured by stalling. Air motors start and stop positively; they have a consistent output torque which can be changed by varying the inlet air pressure.

Air motors do, however, have limitations. Thus: (1) Compared to electric motors, air motors are inefficient. An air motor requires about 5 hp (3.7 kW) input to the air compressor to produce one horsepower (0.7 5kW) at the motor outlet. (2) Air motors are rarely practical in sizes greater than 20 hp (15 kW). Their most efficient range is 1/20 to 20 hp (0.04 to 15 kW). (3) The initial cost of an air motor is high; in larger sizes, above 1 hp (0.75 kW), air motors cost up to five times that of equivalent electric motors.

3. Check the motor duty cycle and load against the unit's characteristics
When selecting an air motor, the first factor to be considered is the type of duty cycle, intermittent or continuous. A crane, for which this motor will be used, does have an intermittent duty cycle because it is not normally used continuously. There is a rest period while the crane load is being put on the crane and again while being off-loaded from the crane.

The great majority of air-motor applications have a low-load cycle; the air motors are used for only a few seconds continuously and have long off-duty periods.

The duty cycle will usually determine the type of motor and the size of compressor that must be used.

4. Check the horsepower and speed required
Performance curves, Fig. 6, show an air motor's torque and horsepower (kW) output at various rpm. Such curves can be varied somewhat by using governors or by modifying the air intake or exhaust ports. However, the basic shape of the performance curve depends on the fundamental design of the air motor. It is common practice to rate an air motor at its maximum output, *i.e.,* at the top of the dome-shaped performance curve. The reversible radial-piston motor chosen here has adequate horsepower and speed for the anticipated load.

5. Determine the effect of air pressure and quantity on the air motor output
Table 2 shows how the air pressure available at the motor inlet affects both the power output and rpm of typical air motors. For the motor being considered here, the output would be sufficient at the lowest air pressure listed. Thus, the motor choice is acceptable.

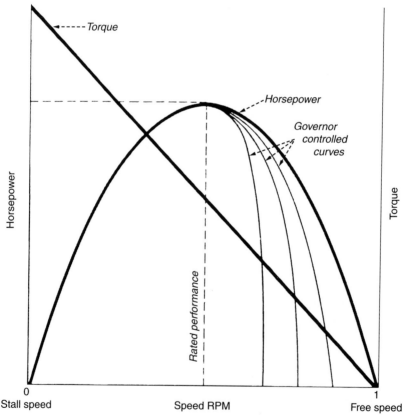

FIGURE 6   Performance curve of a typical air motor. Note how a built-in governor can change the shape of the curve by limiting the maximum speed of the air motor (*Product Engineering*).

TABLE 2 Effect of Air Pressure on Motor Performance

Motor style no.	Rated hp (kW)					rpm at rated hp					Free speed—rpm				
	1	2	3	4	5	1	2	3	4	5	1	2	3	4	5
At 60 psi lb /in²	1.9	2.5	3.2	6.4	0.4	1,200	1,100	850	800	900	2,650	2,250	2,030	1,900	1,600
(413 kPa)	(1.4)	(1.9)	(2.4)	(4.8)	(0.29)										

Gardner-Denver Company.

Related Calculations. When using tables of air-motor performance, it is important to keep in mind that the stall torque and air consumption vary for each motor. Hence, these values are not listed in the usual performance tables. There are so many variables in air-motor choice that stall torque and air consumption are unique for each application and are supplied by the motor manufacturer when the motor choice is made.

As a general rule of thumb, stall torque ranges between 2 and 2.5 times the torque developed when operating at maximum horsepower output.

In small motors, up to 2.5 hp (1.9 kW), air consumption varies from 35 to 40 ft³/min (0.99 to 1.1 m³/min) of free air per hp (0.746 kW). Larger air motors consume 20 to 25 ft³/min (0.57 to 0.70 m³/min) of free air per hp. These consumption rates apply to non-reversible motors. Reversible air motors consume 30 to 35 percent more air.

The data, tables and illustration in this procedure are from *Product Engineering* magazine.

AIR-COMPRESSOR COOLING-SYSTEM CHOICE FOR MAXIMUM COOLANT ECONOMY

Select a suitable cooling system for a two-stage 5000-hp (3730-kW) engine-driven air compressor, Fig. 7, installed in a known arid hard-water area when the rated output of the compressor is 25,000 ft³/min (708 m³/min) at 100 lb/in² (abs) (689 kPa). Water conservation is an important requirement for this compressor because of the arid nature of the area in which the unit is installed. Use standard cooling-water requirements in estimating the capacity of the cooling system.

Calculation Procedure:

1. *Assess the types of cooling systems that might be used*
Several types of cooling systems can be used for air compressors such as this. Because the air compressor is used in an arid area subject to water shortages, a recirculating system of some type is immediately indicated. Since both the engine and air-compressor cooling water require temperature reduction in such an installation, the two requirements are usually combined in one cooling system.

The first arrangement that might be chosen, Fig. 8, combines a heat exchanger for engine power-cylinder cooling and a cooling tower for raw-water cooling for the compressor. Either a natural-draft cooling tower, such as that shown, or a mechanical-draft cooling tower might be used. The cooling-tower choice depends on a number of factors. In an arid area, however, natural-draft towers are known to perform well in dry climates. Further, they require much less piping and electric wiring than mechanical-draft towers.

Another possible cooling-system arrangement uses a closed coil in the cooling tower for both the power and air cylinders, Fig. 9. This totally closed system does not allow contact between the compressor and engine cooling water with the atmosphere. This means that the compressor and engine cooling water can be treated to reduce scale formation. Raw water recirculated through the cooling tower does not contact the compressor coolant.

Where installation costs are critical, raw water can be used to cool the air-compressor cylinders, Fig. 10. The engine power cylinders, which usually operate

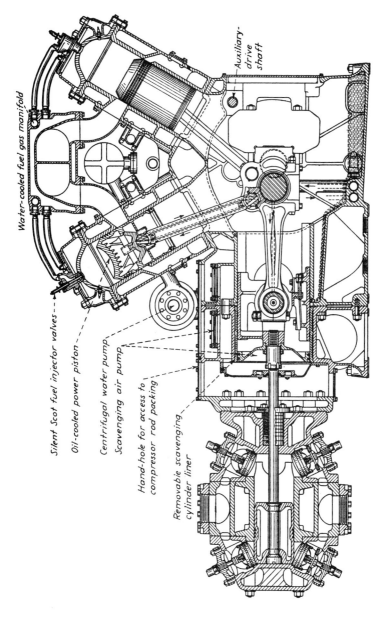

Water-cooled fuel gas manifold

Silent Scot fuel injector valves--
Oil-cooled power piston--
Centrifugal water pump
Scavenging air pump
Hand-hole for access to
compressor rod packing
Removable scavenging
cylinder liner

Auxiliary-
drive
shaft

FIGURE 7 Gas-engine driven compressor has oil-cooled power pistons. Compressor, left, uses metallic piston-rod packing (*Cooper-Bessemer Corp.*).

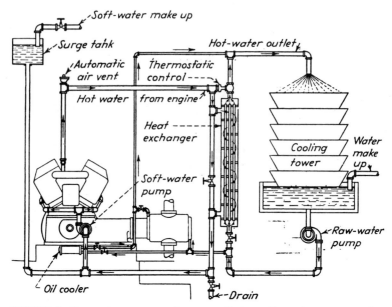

FIGURE 8   Heat exchanger, center, for power-cylinder cooling and raw-water cooling for the compressor (*Ingersoll-Rand Co.*).

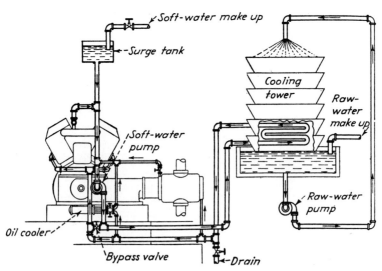

FIGURE 9   Closed cooling system for power and air cylinders utilizing pipe coil in the cooling tower (*Ingersoll-Rand Co.*).

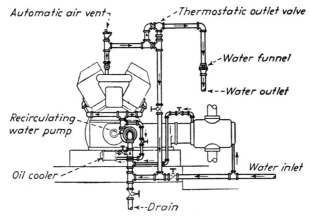

FIGURE 10 Raw water cools the air cylinders; power cylinders use closed system protected by thermostatic valve (*Ingersoll-Rand Co.*).

at a higher temperature, are cooled by a closed system protected by a thermostatic valve.

An open cooling-tower system, Fig. 11, is not recommended for installations such as this because of the possible heavy scale buildup. However, such an open cooling system might be used where the economics of the installation permit it and scale buildup is unlikely to occur.

2. *Determine the air-compressor cooling load*
Use flow rates given in Table 6, page 000 to estimate the cooling water flow rate for this air compressor. Thus, with the intercooler and jacket in series, using the

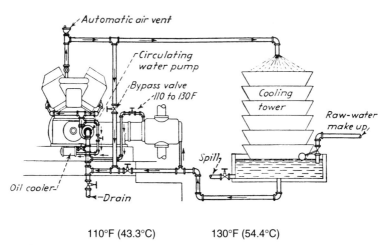

110°F (43.3°C)   130°F (54.4°C)

FIGURE 11 Cooling-tower recirculating system is not recommended because of the possibility of scale and impurities buildup (*Ingersoll-Rand Co.*).

higher flow rate of 2.8 gal/min (364.3 L/s) per 100 ft³/min (100 m³/s), the cooling-water flow required for this 25,000 ft³/min (708 m³/s) air compressor is (25,000/100)(2.8) = 700 gal/min (44.2 L/s).

3. Compute the engine jacket-water cooling load

The engine jacket water cooling load is computed separately using a cooling-water temperature rise of 10 to 20° F (5.6 to 11.1°C) during passage through the engine, as given in the Internal-Combustion Engine section of this handbook. Further, the usual jacket-water flow rate is 0.25 to 0.60 gal/(min bhp) (0.02 to 0.05 kg/kW). For a 5000-hp (3730-kW) engine, using the maximum flow rate, (5000)(0.6) = 3000 gal/min (189.3 L/s).

Additional cooling water may be used for the turbocharger, if fitted, and for aftercooling. Steps for calculating these cooling-water flows are given in the section cited above. Such cooling-water flows are usually additive to the jacket-water flow, depending on the cooling arrangement used.

Related Calculations. Cooling systems for air and gas compressors are important for reliable and safe operation of these units. Hence, great care must be exercised in choosing the most reliable and economic cooling system.

Today, both mechanical-draft and natural-draft cooling towers are popular choices. An economic study is needed to determine the best choice when the cooling effectiveness of both types of towers are about equal. Data given on cooling towers elsewhere in this handbook can be helpful to the designer in choosing the best type of tower to use for a given installation of air or gas compressors.

Straight flow-through cooling of small compressors and their drive engines is often used where adequate water supplies are available. Thus, in large cities the cooling water may be taken from the water main and discharged to the sewer after passage through the compressor and engine. Cost of the water may be small compared to the investment in a cooling tower. But with increased environmental concerns, this scheme of cooling may soon be extinct.

ECONOMICS OF AIR-COMPRESSOR INLET LOCATION

A plant designer has the option of locating an air-compressor inlet pipe either inside the compressor building or outside the structure. The prevailing average indoor temperature is 90°F (32.2°C) while the average outdoor temperature is 50°F (10°C). Air requirements for this plant from the compressor are: 1000 ft³/min (28.3 m³/min) of free air at 70°F (21.1°C) at 100 lb/in² (gage) (689 kPa) for 7500 h/yr; the 200-hp (149.1 kW) compressor drive motor operates at full load throughout the 7500 hr load year. Determine which is the best location for the compressor intake based on power savings with an electric power cost of $0.04/kWh.

Calculation Procedure:

1. Determine the power savings possible with cooler intake air

Compute the intake volume, I, required to deliver 1000 ft³/min of free air at each of the possible intake temperatures from $I = 1000$(density of air at 70 °F, lb/ft³/

density of air at inlet temperature, lb/ft^3), where I = intake volume, ft^3 (m^3), required at the air inlet temperature.

For inlet air at 50°F (21.1°C), the outside intake air temperature, using a table of air properties, I = 1000(0.07493/0.7785) = 962.49 ft^3 (27.2 m^3). With an inlet temperature of 90°F (32.2°C), using the inside-of-the-building air intake, I = 1000(0.07493/0.07219) = 1037.95 ft^3 (29.37 m^3), say 1038 ft^3 (29.37) cu m).

The power saving from using lower-temperature intake air is then hp saving = 100(intake volume required at the higher intake temperature − intake volume required at the lower intake temperature)/intake volume required at the higher intake temperature. Substituting, hp saving = 100(1038 − 962)/1038 = 7.32 percent.

2. Find the annual power saving with the lower intake temperature
The annual power saving, P kWh, can be found from: P = (hp saving/100)(motor hp)(0.746 kW/hp)(annual operating hours). Since the compressor operates at full load 7500 h/yr using 200 hp, the annual power saving is P = (7.32/100)(200 hp)(0.746)(7500) = 81,910.8 kWh.

The annual cost saving, A = (kWh/h per yr saved)(power cost, $/kWh). With a power cost of $0.04/kWh, the annual cost saving, A = (81,910.8)(0.04) = $3276.43. If an outside inlet were more expensive than an indoor inlet, this saving could be used to offset the increased cost.

Related Calculations.   As a general rule, an outside air intake, Fig. 12, is more economical than an inside air intake when the air in the building is at a higher temperature than the outside air. The only time an outside air intake might be less desirable than an indoor air intake is when the outside air is polluted with corrosive vapors, excessive dust, abrasive sand, *etc.,* which would be injurious to people or machines. Under these circumstances the designer might elect an indoor air intake. However, before choosing an indoor intake, review the efficacy of outdoor air filters of various types, Fig. 13.

Air in industrial districts may contain from 1 to 4 grains of dirt per 1000 ft^3 (28.3 m^3). If the intake air for a compressor contains only 1 gram per 1000 ft^3 (28.3 m^3), 7200 grains of dirt will pass into the air compressor in 1 week's operation. With the higher level of 4 grains per 1000 ft^3 (28.3 m^3), 28,800 grains, or over 4 lb (1.8 kg) will be carried into the compressor during 1 week's operation. Frequently, much of the dirt carried in the air is abrasive. If this dirt is allowed to get into the compressor cylinders it will mix with the lubricating oil and cause rapid wear of piston rings, cylinder walls, valves, and other parts.

Intake-air filters, Fig. 13, can reduce much of the danger of abrasive particles in the supply air. Each type has its favorable features. Viscous coated wire filters, Fig. 13a, are often used for small- and medium-size compressors. Centrifugal airflow units, Fig. 13b, and traveling-curtain oil-bath filters, Fig. 13c, are popular for larger air compressors. The final choice of an intake filter is a function of compressor capacity, intake-air quality, annual operating hours, and expected life of the compressor installation. Filter manufacturers can be most helpful to the plant designer in evaluating these factors.

The general procedure given here is valid for air compressors of all types: centrifugal, reciprocating, vane, rotary, *etc.,* and the procedure can be used for air compressors in plants of all types—chemical, petroleum, manufacturing, marine, industrial, *etc.* This procedure has universal application because it is based on the properties of air, the compressor power input, the annual operating hours, and the cost of power. These values can be found in any application of air compressors in industry today.

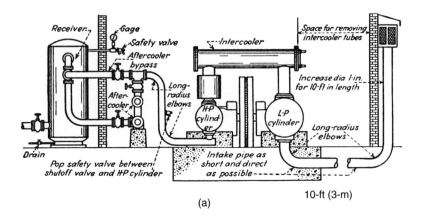

(a)

10-ft (3-m)

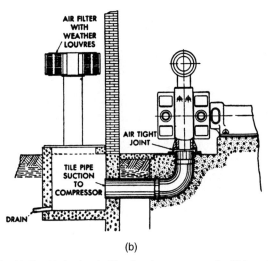

(b)

FIGURE 12   (*a*) Outside intake-air filter for air compressor should have intake pipe as short as possible and be fitted with long-radius elbows (*Ingersoll-Rand Co.*). (*b*) Glazed-tile tunnel for outdoor-air intake.

Compressed-Air and -Gas System Components and Layouts

POWER INPUT REQUIRED BY CENTRIFUGAL COMPRESSOR

A centrifugal compressor handling air draws in 12,000 ft³/min (339.6 m³/min) of air at a pressure of 14 lb/in² (abs) (96.46 kPa) and a temperature of 60°F (15.6°C).

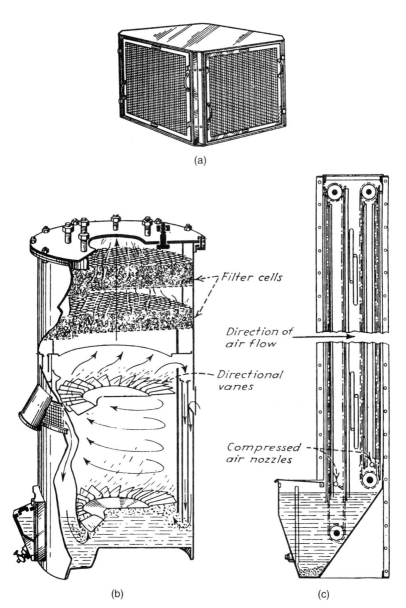

(a)

Filter cells

Direction of air flow

Directional vanes

Compressed air nozzles

(b)

(c)

FIGURE 13 (*a*) Viscous coated-wire intake-air filter (*Air-Maze Corp.*). (*b*) Centrifugal air-flow oil-bath intake-air filter also acts as a silencer. (*c*) Traveling-curtain oil-bath intake-air filter cleans itself in the oil (*American Air Filter Co., Inc.*).

The air is delivered from the compressor at a pressure of 70 lb/in² (abs) (482.4 kPa) and a temperature of 164°F (73.3°C). Suction-pipe flow area is 2.1 ft² (0.195 m²); area of discharge pipe is 0.4 ft² (0.037 m²) and the discharge pipe is located 20 ft (6.1 m) above the suction pipe. The weight of the jacket water, which enters at 60°F (15.6°C) and leaves at 110°F (43.3°C), is 677 lb/min (307.4 kg/min). What is the horsepower required to drive this compressor, assuming no loss from radiation?

Calculation Procedure:

1. Determine the variables for the compressor horsepower equation
The equation for centrifugal compressor horsepower input is,

$$\text{hp} = \frac{w}{0.707}\left[c_p(t_2 - t_1) + \frac{V_2^2 - V_1^2}{50,000} + \frac{Z_2 - Z_1}{778} \right] + \left[\frac{w_j(t_0 - t_i) + R_c}{0.707} \right]$$

In this equation we have the following variables: w = weight, lb (kg) of unit flow rate, ft³/s (m³/s) through the compressor, lb (kg), where $w = (P_1)(V_1)/R(T_1)$, where P_1 = inlet pressure, lb/in² (abs) (kPa); V_1 = inlet volume flow rate, ft³/s (m³/s); R = gas constant for air = 53.3; T_1 = inlet air temperature, degree Rankine.

The inlet flow rate of 12,000 ft³/min = 12,000/60 = 200 ft³/s (5.66 m³/s); P_1 = 14.0 lb/in² (abs) (93.46 kPa); T_1 = 60 + 460 = 520 °R. Substituting, w = 14.0(144)(200)/53.3(520) = 14.55 lb (6.6 kg).

The other variables in the equation are: c_p = specific heat of air at inlet temperature = 0.24 Btu/lb °F (1004.2 J/kg °K); t_2 = outlet temperature, R = 624 °R; t_1 = inlet temperature, R = 520°R; V_1 = air velocity at compressor entrance, ft/min (m/min); V_2 = velocity at discharge, fpm (kg/min); Z_1 = elevation of inlet pipe, ft (m); V_2 = elevation of outlet pipe, ft (m); w_j = weight of jacket water flowing through the compressor, lb/min (kg/min); t_i = jacket-water inlet temperature, °F (°C); t_0 = jacket outlet water temperature, °F (°C).

The air velocity at the compressor entrance = (flow rate, ft³/s)/(inlet area, ft²) = 200/2.1 = 95.3 ft/s (29 m/s); outlet velocity at the discharge opening = 200/0.4 = 500 ft/s (152.4 m/s).

2. Compute the input horsepower for the centrifugal compressor
Substituting in the above equation, with radiation losses, R_c = 0,

$$\text{hp} = \frac{14.55}{0.707}\left[0.24(624 - 520) + \frac{500^2 - 95.3^2}{50,000} + \frac{20}{778} \right]$$
$$+ [677/60 \times (110 - 60)]/0.707$$
$$= 20.6(24.95 + 4.8 + 0.0256) + 797 = 1,409 \text{ hp (1051 kW)}$$

Related Calculations. This equation can be used for any centrifugal compressor. Since the variables are numerous, it is a wise procedure to assemble them before attempting to solve the equation, as was done here.

COMPRESSOR SELECTION FOR COMPRESSED-AIR SYSTEMS

Determine the required capacity, discharge pressure, and type of compressor for an industrial-plant compressed-air system fitted with the tools listed in Table 3. The plant is located at sea level and operates 16 h/day.

TABLE 3 Typical Computation of Compressed-Air Requirements

Tool	(1) Air consumption ft³/min	(1) Air consumption m³/min	(2) Number of tools	(3) Air required, (1) × (2) ft³/min	(3) Air required, (1) × (2) m³/min	(4) Load factor	(5) Probable air demand, (3) × (4) ft³/min	(5) Probable air demand, (3) × (4) m³/min
Grinding wheel, 6 in (15.2 cm)	50	1.4	5	250	7.1	0.3	75	2.1
Rotary sander, 9-in (22.9-cm) pad	55	1.6	2	110	3.1	0.5	55	1.6
Chipping hammers, 13 lb (5.9 kg)	30	0.85	8	240	6.8	0.4	96	2.7
Nut setters, ⁵⁄₁₆ in (0.79 cm)	20	0.57	10	200	5.7	0.6	120	3.4
Paint spray	10	0.28	1	10	0.28	0.1	1	0.03
Plug drills	40	1.1	3	120	3.4	0.2	24	0.68
Riveters, 18 lb (8.1 kg)	35	0.99	5	175	4.9	0.4	70	1.9
Steel drill, ⅞ in (2.2 cm), 25 lb (11.3 kg)	80	2.3	5	400	11.3	0.4	160	4.5
Total							601*	16.9*

*To this sum must be added allowance for future needs and expected leakage loss, if any.

Calculation Procedure:

1. Compute the required airflow rate

List all the tools and devices in the compressed-air system that will consume air, Table 3. Then obtain from Table 4 the probable air consumption, ft³/min, of each tool. Enter this value in column 1, Table 3. Next list the number of each type of tool that will be used in the system in column 2. Find the maximum probable air consumption of each tool by taking the product, line by line, of columns 1 and 2. Enter the result in column 3, Table 3, for each tool.

The air consumption values shown in column 3 represent the airflow rate required for continuous operation of each type and number of tools listed. However, few air tools operate continually. To provide for this situation, a load factor is generally used when an air compressor is selected.

2. Select the equipment load factor

The equipment load factor = (actual air consumption of the tool or device, ft³/min)/(full-load continuous air consumption of the tool or device, ft³/min). Load factors for compressed-air operated devices are usually less than 1.0.

Two variables are involved in the equipment load factor. The first is the *time factor*, or the percentage of the total time the tool or device actually uses compressed air. The second is the *work factor*, or percentage of maximum possible work output done by the tool. The load factor is the product of these two variables.

Determine the load factor for a given tool or device by consulting the manufacturer's engineering data, or by estimating the factor value by using previous experience as a guide. Enter the load factor in column 4, Table 3. The values shown represent typical load factors encountered in industrial plants.

TABLE 4 Approximate Air Needs of Pneumatic Tools

	ft³/min	m³/min
Grinders:		
6- and 8-in (15.2- and 20.3-cm) diameter wheels	50	1.4
2- and 2½-in (5.1- and 6.4-cm) diameter wheels	14–20	0.40–0.57
File and burr machines	18	0.51
Rotary sanders, 9-in (22.9-cm) diameter pads	55	1.56
Sand rammers and tampers:		
1 × 4 in (2.5 × 10.2 cm) cylinder	25	0.71
1¼ × 5 in (3.2 × 12.7 cm) cylinder	28	0.79
1½ × 6 in (3.8 × 15.2 cm) cylinder	39	1.1
Chipping hammers:		
10 to 13 lb (4.5 to 5.9 kg)	28–30	0.79–0.85
2 to 4 lb (0.9 to 1.8 kg)	12	0.34
Nut setters:		
To ⁵⁄₁₆ in, 8 lb (0.79 cm, 3.6 kg)	20	0.57
½ to ¾ in, 18 lb (1.3 to 1.9 cm, 8.1 kg)	30	0.85
Paint spray	2–20	0.06–0.57
Plug drills	40–50	1.1–1.4
Riveters:		
³⁄₃₂- to ⅛-in (0.24- to 0.32-cm) rivets	12	0.34
Larger, weighing 18 to 22 lb (8.1 to 9.9 kg)	35	0.99
Rivet busters	35–39	0.51–0.75
Steel drills, rotary motors:		
To ¼ in (0.64 cm) weighing 1¼ to 4 lb (0.56 to 1.8 kg)	18–20	0.57–1.1
¼ to ⅜ in (0.69 to 0.95 cm) weighing 6 to 8 lb (2.7 to 3.6 kg)	20–40	1.98
½ to ¾ in (1.27 to 1.91 cm) weighing 9 to 14 lb (4.1 to 6.3 kg)	70	2.27
⅞ to 1 in (2.2 to 2.5 cm) weighing 25 lb (11.25 kg)	80	1.1
Wood borers to 1-in (2.5 cm) diameter, weighing 14 lb (6.3 kg)	40	

3. *Compute the actual air consumption*

Take the product, line by line, of columns 3 and 4, Table 3. Enter the result, *i.e.,* the probable air demand, in column 5, Table 3. Find the sum of the values in column 5, or 601 ft³/min. This is the probable air demand of the system.

4. *Apply allowances for leakage and future needs*

Most compressed-air system designs allow for 10 percent of the required air to be lost through leaks in the piping, tools, hoses, *etc.* Whereas some designers claim that allowing for leakage is a poor design procedure, observation of many installations indicates that air leakage is a fact of life and must be considered when an actual system is designed.

With a 10 percent leakage factor, the required air capacity = 1.1(601) = 661 ft³/min (18.7 m³/min).

Future requirements are best estimated by predicting what types of tools and devices will probably be used. Once this is known, prepare a tabulation similar to Table 3, listing the predicted future tools and devices and their air needs. Assume that the future air needs, column 5, are 240 ft³/min (6.8 m³/min). Then the total required air capacity = 661 + 240 = 901 ft³/min (25.5 m³/min), say 900 ft³/min (25.47 m³/min) = present requirements + leakage allowance + predicated future needs, all expressed in ft³/min.

5. *Choose the compressor discharge pressure and capacity*

In selecting the type of compressor to use, two factors are of key importance: discharge pressure required and capacity required.

Most air tools and devices are designed to operate at a pressure of 90 lb/in² (620 kPa) at the tool inlet. Hence, usual industrial compressors are rated for a discharge pressure of 100 lb/in² (689 kPa), the extra lb/in² providing for pressure loss in the piping between the compressor and the tools. Since none of the tools used in this plant are specialty items requiring higher than the normal pressure, a 100-lb/in² (689-kPa) discharge pressure will be chosen.

Where the future air demands are expected to occur fairly soon—within 2 to 3 years—the general practice is to choose a compressor having the capacity to satisfy present and future needs. Hence, in this case, a 900-ft³/min (25.5-m³/min) compressor would be chosen.

6. *Compute the power required to compress the air*

Table 5 shows the power required to compress air to various discharge pressures at different altitudes above sea level. Study of this table shows that at sea level a single-stage compressor requires 22.1 bhp/(100 ft³/min) (5.8 kW/m³) when the discharge pressure is 100 lb/in² (689 kPa). A two-stage compressor requires 19.1 bhp (14.2 kW) under the same conditions. This is a saving of 3.0 bhp/(100 ft³/min) (0.79 kW/m³). Hence, a two-stage compressor would probably be a better investment because this hp will be saved for the life of the compressor. The usual life of an air compressor is 20 years. Hence, by using a two-stage compressor, the approximate required bhp = (900/100)(19.1) = 171.9 bhp (128 kW), say 175 bhp (13.1 kW).

7. *Choose the type of compressor to use*

Reciprocating compressors find the widest use for stationary plant air supply. They may be single- or two-stage, air- or water-cooled. Here is a general guide to the types of reciprocating compressors that are satisfactory for various loads and service:

Single-stage air-cooled compressor up to 3 hp (2.2 kW), pressures to 150 lb/in² (1034 kPa), for light and intermittent running up to 1 h/day.

Two-stage air-cooled compressor up to 3 hp (2.2 kW), pressures to 150 lb/in² (1034 kPa), for 4 to 8 h/day running time.

TABLE 5 Air Compressor Brake Horsepower (kW) Input*

Altitude, ft (m)	Single-stage discharge pressure, lb/in² (gage) (kPa)			Two-stage discharge pressure, lb/in² (gage) (kPa)		
	60 (414)	80 (552)	100 (689)	60 (414)	80 (552)	100 (689)
0 (0)	16.3 (12.2)	19.5 (14.6)	22.1 (16.5)	14.7 (10.9)	17.1 (12.8)	19.1 (14.3)
2000 (610)	15.9 (11.9)	18.9 (14.1)	21.3 (15.9)	14.3 (10.7)	16.5 (12.3)	18.4 (13.7)
4000 (1212)	15.4 (11.5)	18.2 (13.6)	20.6 (15.4)	13.8 (10.3)	15.8 (11.8)	17.7 (13.2)
6000 (1820)	15.0 (11.2)	17.6 (13.1)	20.0 (14.9)	13.3 (9.9)	15.2 (11.3)	17.0 (12.7)

*Courtesy Ingersoll-Rand. Values shown are the approximate bhp input required per 100 ft³/min (2.8 m³/min) of free air actually delivered. The bhp input can vary considerably with the type and size of compressor.

Single-stage air-cooled compressor up to 15 hp (11.2 kW) for pressures to 80 lb/in² (552 kPa); above 80 lb/in² (552 kPa), use two-stage air-cooled compressor.

Single-stage horizontal double-acting water-cooled compressor for pressures to 100 lb/in² (689 kPa) hps of 10 to 100 (7.5 to 75 kW), for 24 h/day or less operating time.

Two-stage, single-acting air-cooled compressor for 10 to 100 hp (7.5 to 75 kW), 5 to 10 h/day operation.

Two-stage double-acting water-cooled compressor for 100 hp (75 kW), or more, 24 h/day, or less operating time.

Using this general guide, choose a two-stage double-acting water-cooled reciprocating compressor, because more than 100-hp (75-kW) input is required and the compressor will operate 16 h/day.

Rotary compressors are not as widely used for industrial compressed-air systems as reciprocating compressors. The reason is that usual rotary compressors discharge at pressures under 100 lb/in² (68.9 kPa), unless they are multistage units.

Centrifugal compressors are generally used for large airflows—several thousand ft³/min or more. Hence, they usually find use for services requiring large air quantities, such as steel-mill blowing, copper conversion, *etc.* As a general rule, machines discharging at pressures of 35 lb/in² (241 kPa) or less are termed *blowers*; machines discharging at pressures greater than 35 lb/in² (241 kPa) are termed *compressors.*

Using these facts as a guide enables the designer to choose, as before, a two-stage double-acting water-cooled compressor for this application. Refer to the manufacturer's engineering data for the compressor dimensions and weight.

8.   *Select the compressor drive*
Air compressors can be driven by electric motors, gasoline engines, diesel engines, gas turbines, or steam turbines. The most popular drive for reciprocating air compressors is the electric motor—either direct-connected or belt-connected. Where either dc or ac power supply is available, the usual choice is an electric-motor drive. However, special circumstances, such as the availability of low-cost fuel, may dictate another choice of drive for economic reasons. Assuming that there are no special economic reasons for choosing another type of drive, an electric motor would be chosen for this installation.

With an ac power supply, the squirrel-cage induction motor is generally chosen for belt-driven compressors. Synchronous motors are also used, particularly when power-factor correction is desired. Motor-driven air compressors generally operate at constant speed and are fitted with cylinder unloaders to vary the quantity of air delivered to the air receiver. A typical power input to a large reciprocating compressor is 22 hp (16.4 kW) per 100 ft³/min (2.8 m³/min) of free air compressed.

Air compressors are almost always rated in terms of *free air* capacity, *i.e.,* air at the compressor intake location. Since the altitude, barometric pressure, and air temperature may vary at any locality, the term *free air* does not mean air under standard or uniform conditions. The displacement of an air compressor is the volume of air displaced per unit of time, usually stated in ft³/min. In a multistage compressor, the displacement is that of the low-pressure cylinder only.

9. *Choose the type of air distribution system*

Two types of air distribution systems are in use in industrial plants: *central* and *unit*. In a central system, Fig. 14, one or more large compressors centrally located in the plant supply compressed air to the areas needing it. The supply piping often runs in the form of a loop around the areas needing air.

A unit system, Fig. 15, has smaller compressors located in the areas where air is used. In the usual plant, each compressor serves only the area in which it is located. Emergency connections between the various areas may or may not be installed.

Central systems have been used for many years in large industrial plants and give excellent service. Unit systems are used in both small and large plants but probably find more use in smaller plants today. With the large quantity of air required by this plant, a central system would probably be chosen, unless the air was needed at widely scattered locations in the plant, leading to excessive pressure losses in the distribution piping of a central system. In such a situation, a unit system with the capacity divided between compressors as necessary would be chosen.

Related Calculations. Where possible, choose a larger compressor than the calculations indicate is needed, because air use in industrial plants tends to increase. Avoid choosing a compressor having a free-air capacity less than one-third the required free-air capacity.

When choosing a water-cooled compressor instead of an air-cooled unit, remember that water cooling is more expensive than air cooling. However, the power input to water-cooled compressors is usually less than to air-cooled compressors of the same capacity. For either type of cooling, a two-stage compressor, with intercooling, is more economical when the compressor must operate 4 h or more in a 24-h period. Table 6 shows the typical cooling-water requirements of various types of water-cooled compressors.

When the inlet air temperature is above or below 60°F (15.6°C), the compressor delivery will vary. Table 7 shows the relative delivery of compressors handling air at various inlet temperatures.

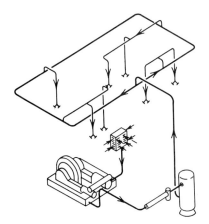

FIGURE 14 Central system for compressed-air supply.

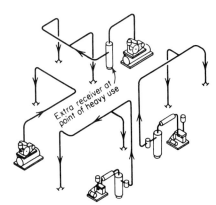

FIGURE 15 Unit system for compressed-air supply.

TABLE 6 Cooling Water Recommended for Intercoolers, Cylinder Jackets, Aftercoolers

	Actual free air, gal/min per 100 ft³/min (L/s per 100 m³/s)
Intercooler separate	2.5–2.8 (334.2–374.3)
Intercooler and jackets in series	2.5–2.8 (334.2–374.3)
Aftercoolers:	
80 to 100 lb/in² (551.6 to 689.5 kPa), two-stage	1.25 (167.1)
80 to 100 lb/in² (551.6 to 689.5 kPa), single-stage	1.8 (240.6)
Two-stage jackets alone (both)	0.8 (106.9)
Single-stage jackets:	
40 lb/in² (275.8 kPa)	0.6 (80.2)
60 lb/in² (413.7 kPa)	0.8 (106.9)
80 lb/in² (551.6 kPa)	1.1 (147.0)
100 lb/in² (689.5 kPa)	1.3 (173.8)

TABLE 7 Effect of Initial Temperature on Delivery of Air Compressors [*Based on a nominal intake temperature of 60°F (15.6°C)*]

Initial temperature				Relative delivery
°F	°R	°C	K	
40	500	4.4	277.4	1.040
50	510	10.0	283.0	1.020
60	520	15.6	288.6	1.000
70	530	21.1	294.1	0.980
80	540	26.7	299.7	0.961

TABLE 8 Air-Consumption Altitude Factors (*100-lb/in² or 689.5-kPa air supply*)

Altitude		Factor
ft	m	
6,000	1,828.8	1.224
8,000	2,438.3	1.310
10,000	3,048.0	1.404

Note: For pressure losses in compressed-air piping systems, see the index.

SIZING COMPRESSED-AIR-SYSTEM COMPONENTS

What is the minimum capacity air receiver that should be used in a compressed-air system having a compressor displacing 800 ft³/min (0.38 m³/s) when the intake pressure is 14.7 lb/in² (abs) (101.4 kPa) and the discharge pressure is 120 lb/in² (abs) (827.4 kPa)? How long will it take for this compressor to pump up a 300-ft³ (8.5-m³) receiver from 80 to 120-lb/in² (551.6 to 827.4 kPa) if the average volumetric efficiency of the compressor is 68 percent? For how long can an 80-lb/in² (abs) (551.6-kPa) tool be operated from a 120-lb/in² (abs) (827.4-kPa), 300-ft³ (8.5-m³) receiver if the tool uses 10 ft³/min (0.005 m³/s) of free air and the receiver pressure is allowed to fall to 85 lb/in² (abs) (586.1 kPa) when the atmospheric pressure is 14.7 lb/in² (abs) (101.4 kPa)? What diameter air piston is required to produce a 1000-lb (4448.2-N) force if the pressure of the air is 150 lb/in² (abs) (1034.3 kPa)?

Calculation Procedure:

1. Compute the required volume of the air receiver
Use the relation $V_m = dp_1/p_2$, where V_m = minimum receiver volume needed, ft^3; d = compressor displacement, ft^3/min (use only the first-stage displacement for two-stage compressors); p_1 = compressor intake pressure, lb/in^2 (abs); p_2 = compressor discharge pressure, lb/in^2 (abs). Thus, for this compressor, $V_m = 800(14.7 /120) = 97$ ft^3 (2.7 m^3). To provide a reserve capacity, a receiver having a volume of 150 or 200 ft^3 (4.2 or 5.7 m^3) would probably be chosen.

2. Compute the receiver pump-up time
Use the relation $t = V(p_f - p_i)/(14.7de)$, where t = receiver pump-up time, min; p_f = final pressure, lb/in^2 (abs); p_i = initial receiver pressure, lb/in^2 (abs); d = compressor piston displacement, ft^3/min; e = compressor volumetric efficiency, percent. Thus, $t = 300(120 - 80)/[14.7(800)(0.68)] = 1.5$ min. When the compressor discharge capacity is given in ft^3/min of free air instead of in terms of piston displacement, drop the volumetric efficiency term from the above relation before computing the pump-up time.

3. Compute the air supply time
Use the relation, $t_s = V(p_{max} - p_{min})/(cp_am)$, where t_s = time in minutes during which the receiver of volume V ft^3 will supply air from the receiver maximum pressure p_{max} lb/in^2 (abs) to the minimum pressure p_{min} lb/in^2 (abs); c = ft^3/min of free air required to operate the tool; p_a = atmospheric pressure, lb/in^2 (abs). Or, $t_s = 300(120 - 85)/[(10)(1.47)] = 7.15$ min.

Note that in this relation p_{min} is the minimum air pressure to operate the air tool. A higher minimum tank pressure was chosen here because this provides a safer estimate of the time duration for the supply of air. Had the tool operating pressure been chosen instead, the time available, by the same relation, would be $t_s = 81.5$ min.

This calculation shows that it is often wise to install an auxiliary receiver at a distance from the compressor but near the tools drawing large amounts of air. Use of such an auxiliary receiver, particularly near the end of a long distribution line, can often eliminate the need for purchasing another air compressor.

4. Compute the required piston diameter
Use the relation $A_p = F/p_m$, where A_p = required piston area to produce the desired force, in^2; F = force produced, lb; p_m = maximum air pressure available for the piston, lb/in^2 (abs). Or, $A_p = 1000/150 = 6.66$ in^2 (43.0 cm^2). The piston diameter d is $d = 2(A_p/\pi)^{0.5} = 2.91$ in (7.4 cm).

Related Calculations. The air consumption of power tools is normally expressed in ft^3/min of free air at sea level; the actual capacity of any type of air compressor is expressed in the same units. At locations above sea level, the quantity of free air required to operate an air tool increases because the atmospheric pressure is lower. To find the air consumption of an air tool at an altitude above sea level in terms of ft^3/min of free air at the elevation location, multiply the sea-level consumption by the appropriate factor from Table 6. Thus, a tool that consumes 10 ft^3/min (0.005 m^3/s) of free air at sea level will use 10 (1.310) = 13.1 ft^3/min (0.006 m^3/s) of 100 lb/in^2 (689.5-kPa) free air at an 8000-ft (2438.4-m) altitude.

COMPRESSED-AIR RECEIVER SIZE AND PUMP-UP TIME

What is the minimum size receiver that can be used in a compressed-air system having a compressor rated at 800 ft³/min (0.4 m³/s) of free air if the intake pressure is 14.7 lb/in² (abs) (101.4 kPa) and the discharge pressure is 120 lb/in² (abs) (827.4 kPa)? How long will it take the compressor to pump up the receiver from 60 lb/in² (abs) (413.7 kPa) to 120 lb/in² (abs) (827.4 kPa)? The compressor is a two-stage water-cooled unit. How much cooling water is required for the intercooler and jacket if they are piped in series and for the aftercooler?

Calculation Procedure:

1. Compute the required minimum receiver volume
For any air compressor, the minimum receiver volume v_m ft³ $= Dp_i/p_d$, where D = compressor displacement, ft³/min free air (use only the first-stage displacement for multistage compressors); p_i = compressor inlet pressure, lb/in² (abs); p_d = compressor discharge pressure, lb/in² (abs). For this compressor, v_m = $(800)(14.7)/(120) = 98$ ft³ (2.8 m³). To provide a reserve supply of air, a receiver having a volume of 150 or 200 ft³ (4.2 or 5.7 m³) would probably be chosen. Be certain that the receiver chosen is a standard unit; otherwise, its cost may be excessive.

2. Compute the pump-up time required
Assume that a 150-ft³ (4.2-m³) receiver is chosen. Then, for any receiver, the pump-up time t min $= v_r(p_e = p_s)/De$, where v_r = receiver volume, ft³; p_e = pressure at end of pump-up, lb/in² (abs); p_s = pressure at start of pump-up, lb/in² (abs); e = compressor volumetric efficiency, expressed as a decimal (0.50 to 0.75 for single-stage and 0.80 to 0.90 for multistage compressors). For this compressor, with a volumetric efficiency of 0.85, $t = (150)(120 - 60)/[(800)(0.85)] = 13.22$ min.

3. Determine the quantity of cooling water required   Use the Compressed Air and Gas Institute (CAGI) cooling-water recommendations given in the *Compressed Air and Gas Handbook*, or Baumeister and Marks—*Standard Handbook for Mechanical Engineers*. For 80 to 125 lb/in² (gage) (551.6 to 861.9 kPa) discharge pressure with the intercooler and jacket in series, CAGI recommends a flow of 2.5 to 2.8 gal/min per 100 ft³/min (334.2 to 374.3 L/s per 100 m³/s) of free air. Using 2.5 gal/min (334.2 L/s), we see that the cooling water required for the intercooler and jackets = $(2.5)(800/100) = 20.0$ gal/min (2673.9 L/s). CAGI recommends 1.25 gal/min per 100 ft³/min (167.1 L/s per 100 m³/s) of free air for an aftercooler serving a two-stage 80 to 125 lb/in² (gage) (551.6- to 861.9-kPa) compressor, or $(1.25)(800/100) = 10.0$ gal/min (1377.3 L/s) for this compressor. Thus, the total quantity of cooling water required for this compressor is $20 + 10 = 30$ gal/min (4010.9 L/s).

Related Calculations.   Use this procedure for any type of air compressor serving an industrial, commercial, utility, or residential load of any capacity. Follow CAGI or the manufacturer's recommendations for cooling-water flow rate. When a compressor is located above or below sea level, multiply its rated free-air capacity by the appropriate altitude correction factor obtained from the CAGI—*Compressed*

Air and Gas Handbook or Baumeister and Marks—*Standard Handbook for Mechanical Engineers.*

VACUUM-SYSTEM PUMP-DOWN TIME

An industrial vacuum system with a 200-ft³ (5.7-m³) receiver serving cleaning outlets is to operate to within 2.5 inHg (9.7 kPa) absolute of the barometer when the barometer is 29.8 inHg (115.1 kPa). How long will it take to evacuate the receiver to this pressure when a single-stage vacuum pump with a displacement of 60 ft³/min (0.03 m³/s) is used? The pump is rated to dead end at a 29.0-inHg (112.1-kPa) vacuum when the barometer is 30.0 inHg (115.9 kPa). The pump volumetric efficiency is shown in Fig. 16.

Calculation Procedure:

1. *Compute the pump operating vacuum*
The pump must operate to within 2.5 inHg (9.7 kPa) of the barometer, or a vacuum of $29.8 - 2.5 = 27.3$ inHg (105.5 kPa).

2. *Compute the quantity of free air removed from the receiver*
Select a number of absolute pressures between 29.8 inHg (115.1 kPa), the actual barometric pressure, and the final receiver pressure, 2.5 inHg (9.7 kPa); and list them in the first column of a table such as Table 9. Assume equal pressure reductions—say 3 inHg (11.6 kPa)—for each step except the last few, where smaller reductions have been assumed to ensure greater accuracy.

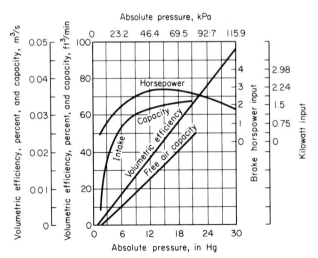

FIGURE 16 Capacity, power-input, and efficiency curves for a typical reciprocating vacuum pump.

TABLE 9 Evacuation Time Calculations

Absolute pressure in receiver, inHg (kPa)	P_r/P_a	Quantity of free air, ft^3 (m^3)		Average volumetric efficiency, Fig. 2	Free-air capacity, ft^3/min (m^3/s)	Evacuation time, min
		In receiver	Removed			
29.8 (115.1)	1.000	200.0 (61.0)	0.0 (0.0)			
26.8 (103.5)	0.899	179.8 (54.8)	20.2 (6.2)	0.91	54.6 (0.026)	0.370
23.8 (92.0)	0.798	159.6 (48.6)	20.2 (6.2)	0.81	48.6 (0.023)	0.415
20.8 (80.4)	0.698	139.6 (42.6)	20.0 (6.1)	0.72	43.2 (0.020)	0.464
17.8 (68.8)	0.597	119.4 (36.4)	20.2 (6.2)	0.62	37.2 (0.018)	0.544

Total time required 9.019

Enter in the second column of Table 9 the ratio of the absolute pressure in the receiver to the atmospheric pressure, or $P/_r/P_a$, both expressed in inHg. Thus, for the second step, P_r/P_a = 26.8/29.8 = 0.899.

The amount of air remaining in the receiver, measured at atmospheric conditions, is then the product of the receiver volume, 200 ft^3 (5.7 m^3), and the ratio of the pressures. Or, for the second pressure reduction, 200(0.899) = 179.8 ft^3 (5.1 m^3). Enter the result in the third column of Table 9. This computation is a simple application of the gas laws with the receiver temperature assumed constant. Assumption of a constant air temperature is valid because, although the air temperature varies during pumping down, the overall effect is that of a constant temperature.

Find the quantity of air removed from the receiver by successive subtraction of the values in the third column. Thus, for the second pressure step, the air removed from the receiver = 200.0 − 179.8 = 20.2 ft^3 (0.6 m^3) and so on for the remaining steps. Enter the result of each subtraction in the fourth column of Table 9.

3. Compute the actual quantity of air handled by the pump

The volumetric efficiency of a vacuum pump varies during each pressure reduction. To simplify the pump-down time calculation, an average value for the volumetric efficiency can be used for each step in the receiver pressure reduction. Find the average volumetric efficiency for this vacuum pump from Fig. 16. Thus, for the pressure reduction from 29.8 to 26.8 inHg (115.1 to 103.5 kPa), the volumetric efficiency is found at (29.8 + 26.8)/2 = 28.3 inHg (109.3 kPa) to be 91 percent. Enter this value in the fifth column of Table 9. Follow the same procedure to find the remaining values, and enter them as shown.

The actual quantity of free air this vacuum pump can handle is numerically equal to the product of the volumetric efficiency, column 5, Table 9, and the pump piston displacement. Or, for the above pressure reduction, free-air capacity = 0.91(60) = 54.6 ft^3/min (0.026 m^3/s). Enter this result in column 6, Table 9.

4. Compute the pump-down time for each pressure reduction

The second line of Table 9 shows, in column 4, that at an absolute pressure of 26.8 inHg (103.5 kPa), 20.2 ft^3 (0.6 m^3) of free air is removed from the receiver. However, the vacuum pump can handle 54.6 ft^3/min (0.03 m^3/s), column 6. Since the time required to remove air from the receiver is (ft³ removed)/(cylinder capacity, ft³/min), the time required to remove 20.2 ft³ (0.6 m³) is 20.2/54.6 ft³ = 0.370 min.

Compute the required time for each pressure step in the same manner. The total pump-down time is then the sum of the individual times, or 9.019 min, column 7, Table 9. This result is suitable for all usual design purposes because it closely approximates the actual time required, and the errors involved are so slight as to be negligible. Leakage into industrial-plant vacuum systems often equals the volume handled by the vacuum pump.

5. Use the pump-down time for compressor selection
To choose an industrial vacuum pump using the pump-down procedure described in steps 1 to 4, (a) obtain the characteristics curves for several makes and capacities of vacuum pumps; (b) compute the pump-down time for each pump, using the procedure in steps 1 to 4; (c) compute the air inflow to the system, based on the free-air capacity of each outlet and the number of outlets in the system; (d) compute how long the pump must run to handle the air inflow; and (e) choose the pump having the shortest running time and smallest required power input.

Thus, with 10 vacuum outlets each having a free-air flow of 50 ft³/h (1.4 m³/s), the total air inflow is 10(50) = 500 ft³/h (14.2 m³/s). This means that a 200-ft³ (5.7-m³) receiver would be filled 500/200 = 2.5 times per hour. Since the pump discussed in steps 1 to 4 requires approximately 9 min to reduce the receiver pressure from atmospheric to 2.5 inHg absolute (9.7 kPa), its running time to serve these outlets would be 9(2.5) = 22.5 min, approximately. The power input to this vacuum pump, Fig. 3, ranges from a minimum of about 1 hp (0.7 kW) to a maximum of about 3 hp (2.2 kW).

If another pump could evacuate this receiver in 6 min and needed only 2.5 hp (1.9 kW) as the maximum power input, it might be a better choice, provided that its first cost were not several times that of the other pump. Use the methods of engineering economics to compare the economic merits of the two pumps.

Related Calculations.   Note carefully that the procedure given here applies to industrial vacuum systems used for cleaning, maintenance, and similar purposes. The procedure should not be used for high-vacuum systems applied to production processes, experimental laboratories, *etc.* Use instead the method given in the next calculation procedure in this section.

To be certain that the correct pump-down time is obtained, many engineers include the volume of the system piping in the computation. This is done by computing the volume of all pipes in the system and adding the result to the receiver volume. This, in effect, increases the receiver volume that must be pumped down and gives a more accurate estimate of the probable pump-down time. Some engineers also add a leakage allowance of up to 100 percent of the sum of the receiver and piping volume. Thus, if the piping volume in the above system were 50 ft³ (1.4 m³), the total volume to be evacuated would be 2(200 + 50) = 500 ft³ (14.2 m³). The factor 2 in this expression was inserted to reflect the 100 percent leakage; i.e., the pump must handle the receiver and piping volume plus the leakage, or twice the sum of the receiver and piping volume.

Some industrial vacuum pumps are standard reciprocating air compressors run in the reverse of their normal direction after slight modification. The vacuum lines are connected to the receiver, from which the compressor takes it suction. After removing air from the receiver, the compressor discharges to the atmosphere.

VACUUM-PUMP SELECTION FOR HIGH-VACUUM SYSTEMS

Choose a mechanical vacuum pump for use in a laboratory fitted with a vacuum system having a total volume, including the piping, of 12,000 ft³ (339.8 m³). The operating pressure of the system is 0.10 torr (0.02 kPa), and the optimum pump-down time is 150 min. (*Note:* 1 torr = 1 mmHg = 0.2 kPa).

Calculation Procedure:

1. Make a tentative choice of pump type

Mechanical vacuum pumps of the reciprocating type are well suited for system pressures in the 0.0001- to 760-torr (2×10^{-5} to 115.6-kPa) range. Hence, this type of pump will be considered first to see whether it meets the desired pump-down time.

2. Obtain the pump characteristic curves

Many manufacturers publish pump-down factor curves such as those in Fig. 17a and b. These curves are usually published as part of the engineering data for a given line of pumps. Obtain the curves from the manufacturers whose pumps are being considered.

3. Compute the pump-down time for the pumps being considered

Three reciprocating pumps can serve this system: a single-stage pump, a compound or two-stage pump, or a combination of a mechanical booster and a single-stage backing or roughing-down pump. Figure 17 gives the pump-down factor for each type of pump.

To use the pump-down factor, apply this relation: $t = VF/d$, where t = pump-down time, min; V = system volume, ft³; F = pump-down factor for the pump; d = pump displacement, ft³/min.

Thus, for a single-stage pump, Fig. 17a shows that $F = 10.8$ for a pressure of 0.10 torr (1.5 kPa). Assuming a pump displacement of 1000 ft³/min (0.5 m³/s), $t = 12,000(10.8)/1000 = 129.6$ min, say 130 min.

For a compound pump, $F = 9.5$ from Fig. 17a. Hence, a compound pump having the same displacement, or 1000 ft³/min (0.5 m³/s), will require $t = 12,000(9.5)/1000 = 114.0$ min.

With a combination arrangement, the backing or roughing pump, a 130-ft³/min (0.06-m³/s) unit, reduces the system pressure from atmospheric, 760 torr (115.6 kPa), to the economical transition pressure, 15 torr (2.3 kPa), Fig. 17b. Then the single-stage mechanical booster pump, a 1200-ft³/min (0.6-m³/s) unit, takes over and in combination with the backing pump reduces the pressure to the desired level, or 0.10 torr (1.5 Pa). During this part of the cycle, the unit operates as a two-stage pump. Hence, the total pump-down time consists of the sum of the backing-pump and booster-pump times. The pump-down factors are, respectively, 4.2 for the backing pump at 15 torr (2.3 kPa) and 6.9 for the booster pump at 0.10 torr (1.5 Pa). Hence, the respective pump-down times are $t_1 = 12,000(4.2)/130 = 388$ min; $t_2 = 12,000(6.9)/1200 = 69$ min. The total time is thus $388 + 69 = 457$ min.

The pump-down time with the combination arrangement is greater than the optimum 150 min. Where a future lower operating pressure is anticipated, making the combination arrangement desirable, an additional large-capacity single-stage rough-

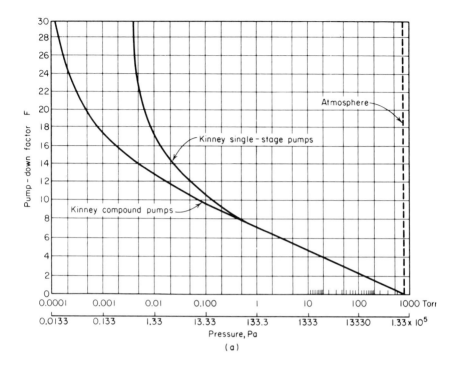

(a)

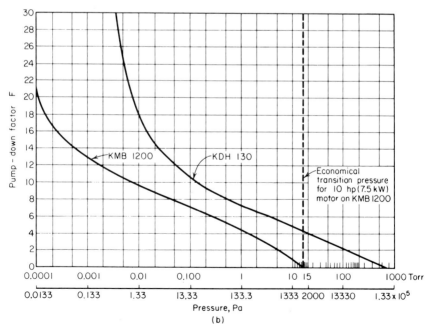

(b)

FIGURE 17 (a) Pump-down factor for single-stage and compound vacuum pumps; (b) pump-down factor for mechanical booster and backing pump. (*After Kinney Vacuum Division, The New York Air Brake Company, and Van Atta.*)

ing pump can be used to assist the 130-ft^3/min (0.06-m^3/s) unit. This large-capacity unit is operated until the transition pressure is reached and roughing down is finished. The pump is then shut off, and the balance of the pumping down is carried on by the combination unit. This keeps the power consumption at a minimum.

Thus, if a 1200-ft^3/min (0.06-m^3/s) single-stage roughing pump were used to reduce the pressure to 15 torr (2.3 kPa), its pump-down time would be $t = 12{,}000(4.0)/1200 = 40$ min. The total pump-down time for the combination would then be $40 + 69 = 109$ min, using the time computed above for the two pumps in combination.

4. Apply the respective system factors

Studies and experience show that the calculated pump-down time for a vacuum system must be corrected by an appropriate system factor. This factor makes allowance for the normal outgassing of surfaces exposed to atmospheric air. It also provides a basis for judging whether a system is pumping down normally or whether some problem exists that must be corrected. Table 10 lists typical system factors that have proved reliable in many tests. To use the system factor for any pump, apply it this way: $t_a = tS$, where t_a = actual pump-down time, min; t = computed pump-down time from step 3, min; S = system factor for the type of pump being considered.

Thus, by using the appropriate system factor for each pump, the actual pump-down time for the single-stage mechanical pump is $t_a = 130(1.5) = 195$ min. For the compound mechanical pump, $t_a = 114(1.25) = 142.5$ min. For the combination mechanical booster pump, $t_a = 190(1.35) = 147$ min.

5. Choose the pump to use

Based on the actual pump-down time, either the compound mechanical pump or the combination mechanical booster pump can be used. In the final choice of the pump, other factors should be taken into consideration—first cost, operating cost, maintenance cost, reliability, and probable future pressure requirements in the system. Where future lower pressure requirements are not expected, the compound mechanical pump would be a good choice. However, if lower operating pressures are anticipated in the future, the combination mechanical booster pump would probably be a better choice.

TABLE 10   Recommended System Factors*

| Pressure range | | System factors | | |
torr	Pa	Single-stage mechanical pump	Compound mechanical pump	Mechanical booster pump*
760–20	115.6 kPa–3000	1.0	1.0	...
20–1	3000–150	1.1	1.1	1.15
1–0.5	150–76	1.25	1.25	1.15
0.5–0.1	76–15	1.5	1.25	1.35
0.1–0.02	15–3	...	1.25	1.35
0.02–0.001	3–0.15	...	...	2.0

*Based on bypass operation until the booster pump is put into operation. Larger system factors apply if rough pumping flow must pass through the idling mechanical booster. Any time needed for operating valves and getting the mechanical booster pump up to speed must also be added.
Source: From Van Atta—*Vacuum Science and Engineering*, McGraw-Hill.

Van Atta[1] gives the following typical examples of pumps chosen for vacuum systems:

Pressure range, torr	Typical pump choice
Down to 50 (7.6 kPa)	Single-stage oil-sealed rotary; large water or vapor load may require use of refrigerated traps
0.05 to 0.01 (7.6 to 1.5 Pa)	Single-stage or compound oil-sealed pump plus refrigerated traps, particularly at the lower pressure limit
0.01 to 0.005 (1.5 to 0.76 Pa)	Compound oil-sealed plus refrigerated traps, or single-stage pumps backing diffusion pumps if a continuous large evolution of gas is expected
1 to 0.0001 (152.1 to 0.015 Pa)	Mechanical booster and backing pump combination with interstage refrigerated condenser and cooled vapor trap at the high-vacuum inlet for extreme freedom from vapor contamination
0.0005 and lower (0.076 Pa and lower)	Single-stage pumps backing diffusion pumps, with refrigerated traps on the high-vacuum side of the diffusion pumps and possibly between the single-stage and diffusion pumps if evolution of condensable vapor is expected

VACUUM-SYSTEM PUMPING SPEED AND PIPE SIZE

A laboratory vacuum system has a volume of 500 ft^3 (14.2 m^3). Leakage into the system is expected at the rate of 0.00035 ft^3/min (0.00001 m^3/min). What backing pump speed, i.e., displacement, should an oil-sealed vacuum pump serving this system have if the pump blocking pressure is 0.150 mmHg and the desired operating pressure is 0.0002 mmHg? What should the speed of the diffusion pump be? What pipe size is needed for the connecting pipe of the backing pump if it has a displacement or pumping speed of 380 ft^3/min (10.8 m^3/min) at 0.150 mmHg and a length of 15 ft (4.6 m)?

Calculation Procedure:

1. Compute the required backing pump speed
Use the relation $d_b = G/P_b$, where d_b = backing pump speed or pump displacement, ft^3/min; G = gas leakage or flow rate, mm · min/ft^3, multiply the ft^3/min by 760 mm, the standard atmospheric pressure, mmHg. Thus, $d_b = 760(0.00035)/0.150 = 1.775$ ft^3/min (0.05 m^3/min).

2. Select the actual backing pump speed
For practical purposes, since gas leakage and outgassing are impossible to calculate accurately, a backing pump speed or displacement of at least twice the computed

---

[1]C. M. Van Atta—*Vacuum Science and Engineering,* McGraw-Hill, New York, 1965.

value, or $2(1.775) = 3.550$ ft^3/min (0.1 m^3/min), say 4 ft^3/min (0.11 m^3/min), would probably be used.

If this backing pump is to be used for pumping down the system, compute the pump-down time as shown in the previous calculation procedure. Should the pump-down time be excessive, increase the pump displacement until a suitable pump-down time is obtained.

3. Compute the diffusion pump speed
The diffusion pump reduces the system pressure from the blocking point, 0.150 mmHg, to the system operating pressure of 0.0002 mmHg. (*Note:* 1 torr = 1 mmHg.) Compute the diffusion pump speed from $d_d = G/P_d$, where d_d = diffusion pump speed, ft^3/min; P_d = diffusion-pump operating pressure, mmHg. Or, $d_d = 760(0.00035)/0.0002 = 1330$ ft^3/min (37.7 m^3/min). To allow for excessive leaks, outgassing, and manifold pressure loss, a 3000- or 4000-ft^3/min (84.9- or 113.2-m^3/min) diffusion pump would be chosen. To ensure reliability of service, two diffusion pumps would be chosen so that one could operate while the other was being overhauled.

4. Compute the size of the connecting pipe
In usual vacuum-pump practice, the pressure drop in pipes serving mechanical pump is not allowed to exceed 20 percent of the inlet pressure prevailing under steady operating conditions. A correctly designed vacuum system, where this pressure loss is not exceeded, will have a pump-down time which closely approximates that obtained under ideal conditions.

Compute the pressure drop in the high-pressure region of vacuum pumps from $p_d = 1.9d_b \, L/d^4$, where p_d = pipe pressure drop, μm; d_b = backing pump displacement or speed, ft^3/min; L = pipe length, ft; d = inside diameter of pipe, in. Since the pressure drop should not exceed 20 percent of the inlet or system operating pressure, the drop for a backing pump is based on its blocking pressure, or 0.150 mmHg, or 150 μm. Hence, $p_d = 0.20(150) = 30$ μm. Then $30 = 1.9(380)(15)/d^4$, and $d = 4.35$ in (110.5 mm). Use a 5-in (127.0-mm) diameter pipe.

In the low-pressure region, the diameter of the converting pipe should equal, or be larger than, the pump inlet connection. Whenever the size of a pump is increased, the diameter of the pipe should also be increased to conform with the above guide.

Related Calculations. Use the general procedures given here for laboratory- and production-type high-vacuum systems.

DETERMINING AIR LEAKAGE IN VACUUM SYSTEMS BY CALCULATION

A 10,000-ft^3 (283 m^3) vacuum system is to be exhausted (drawn down) to an operating pressure of 90 mmHg absolute. Determine what the maximum air leakage into this vacuum system might be in both lb/h and kg/h.

Calculation Procedure:

1. Determine how the actual air leakage might be found
Methods to determine the amount of air leakage into a vacuum system can be categorized as *operational* or *empirical.* In operational methods, an actual field test

is performed to estimate the volume of air entering the vacuum system. One of the most accurate methods in this category is termed the "rate of rise" [1] method. However, performing this test is not always feasible for an existing plant, because it requires a departure from normal production and operations.

In cases where operational methods cannot be used, empirical calculations may be convenient. Although not as rigorous, empirical methods yield reasonable values for a first approach in engineering calculations. The Heat Exchange Institute (Cleveland OH) publishes a set of air leakage curves for empirical calculations, as a function of the volume of the system and the operational pressure. Such curves give maximum air leakage values for commercially tight systems.

To speed up calculation of air leakage and avoid graphical interpretation of these curves, an equation has been developed. This equation reproduces air-leakage values given by the Heat Exchange Institute's curves and it will be used here to compute the actual air leakage.

2. Compute the air leakage in the system

The air leakage is given by $MAL = AV_S^B$ for V. Coefficients A and B are given below for different ranges of operating pressure, P, in the system. Should ordinary shaft seals be used in the equipment, Myerson [2] recommends adding up to 5 lb/h (2.27 kg/h) to the calculated leakage as a correction for additional air inflow.

Coefficients for Air Leakage Calculation

| P, mmHg (abs) | MAL (lb/h) V (ft³) | | MAL (kg/h) V (m³) |
	A	B	A
90–760	0.1955	0.6630	0.9430
21–89	0.1451	0.6617	0.6966
3.1–20	0.1010	0.6579	0.4784
1–3	0.05119	0.6568	0.2415
Less than 1	0.02521	0.6639	0.1220

For this system, $MAL = 0.1995 \times 10,000^{0.6630} = 87.7$ lb/h (39.86 kg/h). Using the same data, the maximum air leakage read from the Heat Exchange Institute's curves is approximately 88 lb/h (40 kg/h).

Related Calculations. This approach can be used when designing a new vacuum system or evaluating an existing system. In such design or evaluation work, a key factor to be determined is the load of noncondensable gas. In most cases the noncondensable gas load can be taken as the mass flow of external air leaking into the system. This gas eventually passes through and is exhausted. But while in the system the gas displaces a certain volume, and alters the level of vacuum being maintained.

It is always true that if no carrier gas is being introduced into the system, and that no chemical reaction in the system is generating additional noncondensable gas, then leakage is the sole source of these noncondensables [2].

The method given here is valid for a variety of vacuum systems used in industry, including those for distillation columns, metallurgical applications: extraction, refining, degassing, melting, brazing, diffusion bonding, deposition, metallic coatings, etc. It is also applicable in other manufacturing and processing operations for pharmaceuticals, foods, mineral oils, cosmetics, epoxy resins, natural flavor extracts, etc. Likewise, the method can be used for wind tunnels and space simulation chambers, and power-plant steam condensing systems of various types.

While the pressures given here are expressed in mmHg, some designers and engineers use the torr. One torr = 1 mmHg; hence the terms are used interchangeably.

This procedure is the work of Jose Vicente Gomez, formerly Process Engineer for Maraven S.A., an affiliate of Petroleos de Venezuela, as reported in *Chemical Engineering* magazine.

CHECKING THE VACUUM RATING OF A STORAGE VESSEL

Check the vacuum rating of a cylindrical flat-ended process tank which is 12.75 ft (3.9 m) tall and 4 ft (1.22 m) in diameter. It contains fresh water at 190°F (87.8°C) and is located where $g = 32.0$ ft/s^2 (9.8 m/s^2) and the atmospheric pressure is 14.7 lb/in^2 (101.3 kPa). What is its maximum vacuum when the tank is gravity-drained? Find both the final tank vacuum and final height of liquid above the tank bottom when the tank is initially 75 percent full, first with gravity drain and then using a pumped drain, each discharging to the atmosphere. The pumped-down piping system consists of double extra-strong 2-in (5.1-cm) pipe with an equivalent length of 75 ft (22.9 m) and a pump which can discharge 40 gal/min (151 L/min) with its suction centerline 2 ft (0.61 m) below the tank bottom and has a net positive suction head of 4.5 ft (1.4 m).

Calculation Procedure:

1. Find the density of the fresh water
From a suitable source such as Baumeister, *Mark's Standard Handbook for Mechanical Engineers*, freshwater density $p = 60.33$ lb/ft^3 (966.7 kg/m^3) at the given conditions.

2. Compute the maximum vacuum rating with gravity drain
A shortcut method gives the maximum vacuum rating by use of the equation for P_s in Fig. 18 where H = overall vertical dimension of the tank; ρ = density of the fresh water; g = acceleration due to gravity at the tank's location; and g_s = 32.174 lb·ft/lb·s^2, a conversion factor. Substituting appropriate values gives $P_s = 2.036(12.75)(60.33)(32.0/[144 \times 32.174]) = 10.82$ in Hg (36.64 kPa).

3. Compute the head space volume when the tank is 75 percent full
Airspace volume $V_o = (1.00 - 0.75)(H \times 0.7854\ D^2) = 0.25(12.75 \times 0.7854 \times 4^2) = 40.1$ ft^3 (1.14 m^3).

4. Compute the final height of liquid above the tank bottom created by gravity drain
By trial and error, the final height of liquid can be computed by solving the equation for h_{fg} shown in Fig. 18 where the ambient atmospheric pressure $P_o = 14.7$ lb/in^2 (101.3 kPa); tank radius $R = 2$ ft (0.61 m); initial fluid height above tank bottom $h_o = 0.75H = 0.75(12.75) = 9.56$ ft (2.91 m); ratio of molar specific heats, C_p/C_r, is $\gamma = 1.4$ for diatomic gases. Assuming a reasonable initial value of h_{fg} on the right-hand side of the equation and substituting appropriate other values, too, gives $h_{fg} = 144(14.7)(1 - \{40.1/[40.1 + \pi(2^2)(9.56 - h_{fg})]\}1.4)/(60.33 \times 32.0/32.174)$

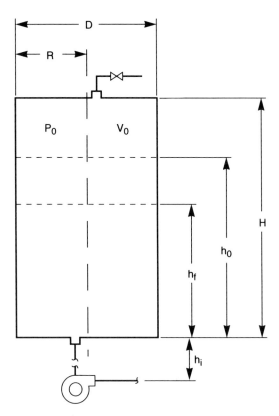

# rho	$P_s = 2.036Hd(g/[144g_c])$, in Hg$\mid$(kPa)
# rho	$h_{fg} = 144P_0(1 - \{V_0/[V_0 + pR^2(h_0 - h_{fg})]\}n)/(eg/g_c)$, ft$\mid$(m)
# pi gamma	$P_g = 29.92h_{fg}eg/(144P_0g_c)$, in Hg$\mid$(kPa)
#	$h_{fg} = NPSH + F_1 + V_p - h_i - 144P_0 \{V_0/[V_0 + pR^2(h_0 - h_{fp})]\}n/(eg/g_c)$, ft$\mid$(m)
#	$P_p = 2.036P_0\{1 - [V_0/(V_0 + pR^2h_0)]n\}$, in Hg$\mid$(kPa)

FIGURE 18   Vacuum rating of tanks.

= 7.18 ft (2.19 m), where the left-hand side value is used to repeat the interaction process until the h_{fg} values on either side of the equal sign are in close agreement.

5. Compute the final tank vacuum under gravity drain
Use the equation for P_g shown in Fig. 18 to find the final tank vacuum. Thus, P_g = 29.92 × 7.18 × 60.33 × 32.0/(144 × 14.7 × 32.174) = 6.09 in Hg (20.62 kPa).

6. Compute the water velocity in the pumped-drain system piping
Pump's flow rate q = (40 gal/min)/)[7.48 gal/ft³][60 s/min]) = 0.0891 ft³/s (0.0025 m³/s). The sectional area of the 2 in (5.1 cm) double extra-strong pipe is

$a = 0.7854d^2 = 0.7854(1.503/12)^2 = 0.0123$ ft^2 (0.00114 m^2). Thus, water mean velocity $v = q/a = 0.0891/0.0123 = 7.24$ ft/s (2.2 m/s).

7. Determine the fluid's viscosity
From Baumeister, *Mark's Standard Handbook for Mechanical Engineers*, fresh water at 190°F (87.8°C) has a dynamic viscosity of $\mu = 6.75 \times 10^{-6}$ lb·s/ft^2 (323.2 $\times 10^{-6}$ Pa·s).

8. Compute the Reynold's number for the drain system piping
Using pertinent values found previously, Reynold's number = $Re = pvd/\mu = (60.33 \times 7.24)(1.1503/12)/(6.75 \times 10^{-6}) = 8,088,690$.

9. Find the friction factor for the drain system piping
Baumeister, *Mark's Standard Handbook for Mechanical Engineers*, indicates that the relative roughness factor for the drain piping is $\epsilon/d = 150 \times 10^{-6}/0.12 = 0.0012$, hence the Moody friction factor is $f = 0.02$ for the above value of N_R.

10. Compute the friction loss of the pumped-drain piping system
Friction loss $F_1 = (fL/d)(v^2/2g)$ where F_1 is in feet (m) of fresh water and L is the equivalent length of the piping system in ft (m). Substituting, $F_1 = (0.02 \times 75/0.125)(7.24^2/[2 \times 32.174]) = 9.77$ ft (1.98 m).

11. Compute the final height of liquid above the tank bottom created by the pumped drain
In the equation for h_{fp} shown in Fig. 18, the net positive suction head (NPSH) = 4.5 ft (1.37 m); freshwater vapor pressure $V_p = 22.29$ ft (6.79 m) of water (Baumeister, *Mark's Standard Handbook for Mechanical Engineers*); height between tank bottom and centerline of pump suction $h_i = 2$ ft (0.61 m). Substituting appropriate values gives $h_{fp} = 4 + 9.77 + 22.29 - 2 - 144(14.7)\{40.1/[40.1 + \pi(2^2)(9.56 - h_{fp})]\}1.4/(60.33 \times 32.0/32.174) = 6.94$ ft (2.12 m), by trial and error, as was done for the gravity drain. From 4 to 10 trials should do it.

12. Compute the final tank vacuum for the pumped drain
Final tank vacuum is found by solving the equation for P_p shown in Fig. 32, thus $P_p = 2.036(14.7)\{1 - (40.1/[40.1 + (\pi \times 2^2)(9.56)])1.4\} = 19.44$ inHg (65.83 kPa).

   Related Calculations.   Specifying an appropriate vacuum rating could prevent the collapse of a storage vessel as the contents are being drained while the vent is inadvertently blocked. Vacuum ratings range from full vacuum to no vacuum. A full vacuum rating is advisable for tanks such as those for steam-sterilized sanitary service and those with pumped discharge. Tanks with vents that cannot be blocked require no vacuum rating.

   That the maximum gravity-drain vacuum rating occurs at 100 percent full capacity was borne out by the above calculations for P_s and P_g. However, the pumped-drain vacuum rating P_p, under more favorable conditions, still turned out to be 3.19 times greater than P_s, the maximum for gravity drain. This varies with pump capacity and the drain system piping size. These calculations assume ideal gas behavior in the head space above the fluid surface and the process is considered isothermal for drain times longer than 5 min. If the initial fluid height is set too low, it is possible for the tank to be emptied by pump drain before maximum vacuum occurs. The calculations presume a centrifugal pump will not deliver if the NPSH requirements are not met and then backflow into the tank starts. Use the equation for P_p to find the final tank vacuum if it is expected that the tank will be emptied before backflow occurs.

The procedure presented here allows the designer to choose a vacuum rating appropriate to the tank. However, it is suggested that the designer perform applicable code calculations before making a final decision of the vacuum rating for a tank. This presentation is based upon an article by Barry Wintner of Life Sciences International, and which appeared in *Chemical Engineering* magazine.

SIZING RUPTURE DISKS FOR GASES AND LIQUIDS

What diameter rupture disk is required to relieve 50,000 lb/h (6.3 kg/s) of hydrogen to the atmosphere from a pressure of 80 lb/in² (gage) (551.5 kPa)? Determine the diameter of a rupture disk required to relieve 100 gal/min (6.3 L/s) of a liquid having a specific gravity of 0.9 from 200 lb/in² (gage) (1378.8 kPa) to atmosphere.

Calculation Procedure:

1. Determine the rupture disk diameter for the gas
For a gas, use the relation $d = (W/146P)^{0.5} (1/Mw)^{0.25}$, where d = minimum rupture-disk diameter, in; W = relieving capacity, lb/h; P = relieving pressure, lb/in² (abs); Mw = molecular weight of gas being relieved. By substituting, $d = [50,000/146(94.7)]^{0.5} (1/2)^{0.25} = 1.60$ in (4.1 cm).

2. Find the rupture-disk diameter for the liquid
Use the relation $d = 0.236(Q)^{0.5} (Sp)^{0.25}/P^{0.25}$, where the symbols are the same as in step 1 except that Q = relieving capacity, gal/min; Sp = liquid specific gravity. So $d = 0.236(100)^{0.5}(0.0)^{0.25}/(214.7)^{0.25} = 0.60$ in (1.52 cm).

Related Calculations. Rupture disks are used in a variety of applications— process, chemical, power, petrochemical, and marine plants. These disks protect pressure vessels from pressure surges and are used to separate safety and relief valves from process fluids of various types.

Pressure-vessel codes give precise rules for installing rupture disks. Most manufacturers will guarantee rupture disks they size according to the capacities and operating conditions set forth in a purchase requisition or specification.

Designers, however, often must know the needed size of a rupture disk long before bids are received from a manufacturer so the designer can specify vessel nozzles, plan piping, etc.

The equations given in this procedure are based on standard disk sizing computations. They provide a quick way of making a preliminary estimate of rupture-disk diameter for any gas or liquid whose properties are known. The procedure is the work of V. Ganapathy, Bharat Heavy Electricals Ltd., as reported in *Chemical Engineering* magazine.

REFERENCES

1. *Standards for Jet Vacuum Systems, 4th Ed.,* Heat Exchange Institute, Cleveland OH 1988).
2. Myerson, E.B., "Calculate Saturated-gas Loads for Vacuum Systems," *Chemical Engineering Progress,* March 1991.

SECTION 10
MATERIALS HANDLING

CHOOSING CONVEYORS AND ELEVATORS FOR SPECIFIC MATERIALS TRANSPORTED

Determine the maximum allowable product weight between supports that can be handled by a belt conveyor at any one time when it conveys 100,000 lb/h (4540 kg/h) of pulverized aluminum oxide in an abrasive state at a belt speed of 50 ft/min (15.2 m/min) with a center-to-center distance of 32 ft (9.75 m) between belt supports. Compare this capacity with that at belt speeds of 150, 250, and 350 fpm (45.7, 76.2, and 106.7 m/min). Choose the type of conveyor and elevator to handle this material under the conditions given.

Calculation Procedure:

1. Find the maximum allowable product weight at the given belt speed
Use the relation, $P = KC/60\ S$, where P = maximum product weight on the belt at any one time, lb (kg) between belt supports; K = load, lb/h (kg/h); C = center-to-center distance between belt supports, ft (m); S = belt speed, ft/min (m/min).
  Substituting, we have, $P = (100,000)(32)/50\ (60) = 1066.7$ lb (484.3 kg).

2. Determine the maximum allowable product weight at other belt speeds
Typical conveyor belt speeds vary from a low of 150 ft/min (45.7 m/min) to a high 800 ft/min (243.8 m/min), depending on belt width, type of material conveyed, belt construction, etc.
  For this belt, using the data given earlier, $P = (100,000)(32)/150(60) = 355.6$ lb (161.4 kg) when the speed is 150 ft/min (45.7 m/min). Likewise for the two higher speeds, respectively, $P = (100,000)(32)/250(60) = 213.3$ lb (96.9 kg); $P = (100,000)(32)/(350(60) = 152.4$ lb (69.2 kg).

3. Verify the type of conveyor and elevator to use
With such a wide variety of conveyors and elevators to choose from, it is wise to verify the choice before a final decision is made. Table 1, presented by Harold V.

TABLE 1 Preferred Types of Conveyors and Elevators for Bulk and Packaged Materials

Material	Physical condition	Av wt/volume lb/ft³	Av wt/volume kg/m³	Reaction on conveyor	Preferred conveyors*	Preferred elevators*	Comment
Acid phosphate	Damp	90	1,440	Adheres	a, e	b	Sticky
Alum	Granular	60–65	960–1,040	Abrasive	a, b, c, e	g, b	
Aluminum oxide	Pulv.	60	960	Abrasive	a, e	g	
Ammonium nitrate	Pulv.	62	990	Hygroscopic	b, c, e	g, b	Explosive
Ammonium nitrate	Damp	65+	1,040+	Adheres	c, e	g, b	Sticky
Arsenic salts	Pulv.	100	1,600	Heavy	c, e	g, b	Poisonous
Ashes: dry	Granular	35–40	560–640	Abrasive	d, f	b	Dusty
wet	Sticky	45–50	720–800	Abrasive	f	b	Corrosive
Bone meal	Pulv.	55–60	880–960		a, b, c, d, e	g, b, c	
Borax	Pulv.	50–70	800–1,120	Abrasive	a, b, c, d, e	g, b	Sometimes sticky
Bran	Granular	16–20	260–320	Abrasive	a, b, c, d, e	g, b	
Brewers grains, hot	Granular	55	880	Corrosive	c, e	g, b	Fragile
Carbon black (pellets)	Granular	40	640		a, e	g, b	Packs
Cement, dry	Pulv.	90–118	1,440–1,890		a, c, d, e	g, b	Sluggish
Clays	Pulv.	35–60	560–960	Adheres	a, b, c, e	g, b	
Coal: anthracite	Lumpy	50–54	800–860		a, b, c, e	g, b	
steam sizes	Granular	50–60	800–960		a, b, c, d, e	g, b, c	
bituminous, lump	Lumpy	50–60	800–960		a, b, e	b	
bituminous, slack	Granular	50–60	800–960		a, b, c, d, e	g, b, c	
Chalk	Pulv.	70–75	1,120–1,200		a, b, c, d, e	g, b, c	Sluggish
Coffee beans	Granular	40–45	640–720		a, c, e	g, b, c	Fragile
Copra, ground	Pulv.	40	640	May be abrasive	a, c, e	g, b	Sticky
Cork, ground	Pulv.	5–15	80–240		a, b, c, d, e	g, b	Sluggish
Corn, shelled	Granular	45	720		a, c, e	g, b, c	
Cottonseed	Granular	35–40	560–640	Abrasive shell	a, b, c, d, e	g, b	Corrosive
Cullet	Granular	80–100	1,280–1,600	Sometimes sticky	a, b, e	g, b	Free-flowing
Flaxseed	Granular	45	720	Abrasive	a, b, c, d, e	g, b, c	
Flue dirt	Pulv.	100	1,600	Abrasive	b, d, e, f	g, b	
Fly ash, clean	Pulv.	35–45	560–720	Shell abrasive	a, b, c, d, e	g, b, c	Free-flowing
Glass batch	Granular	80+	1,280+	Abrasive	a, b, e	g, b	
Glue	Granular	45	720	Mild abrasive	a, c, e	g, b, c	Keep cool
Graphite (flour)	Pulv.	40	640	Abrasive	a, b, c, d, e	g, b, c	
Gravel	Granular	95–135	1,520–2,160	Lubricant	a, e, f	g, b, c	
Gypsum	Pulv.	60	960	Abrasive	a, b, c, e	g, b	
Heavy ores	Lumpy	100+	1600+		a, b, f	g, b	May be tough
Hog fuel	Stringy	15–30	240–480		a, b, d, e	g	
Lead salts	Pulv.	60–150	960–2,400		a, b, c, e		Poisonous
Lime, pebble	Granular	55–80	880–1,280	May jam	a, b, c, e		

Material	Form	lb/ft³	kg/m³	Characteristics	Symbols	Symbols	Remarks
Limestone dust	Pulv.	85–95	1,360–1,520	Sluggish	a, b, e	g, b	Sometimes difficult
Malt	Dry	45	720		a, b, c, d, e		Dusty
Manufactured products	Boxed	1–200	16–3,200	Abrasive	a, i, j		Sticky
Merchandise:							
Packaged	Boxed	15	240	May be sticky	a, b, i, j		
Garments	Hanging	5	80		i, j		
Metallic dusts	Pulv.	50–100	800–1,600		a, b, c, d, e	g, b, c	Polisher
Mica, pulverized	Pulv.	20–30	320–480		a, b, c, d, e	b	
Molybdenum conc'ts	Pulv.	110	1,760		a, b, d	g, b	
Petroleum coke	Lumpy	42	670	Abrasive	a, b, c, e	g, b, c	Difficult
Pumice	Pulv	45	720	Free-flowing	a, b, c, d, e	g, b	Corrosive if wet
Quartz (ground)	Pulv	110	1,760	Abrasive	a, b, c, d	g, b	
Rubber scrap	Stringy	50	800	Mild abrasive	a, b, e	g, b	
Salt: coarse	Granular	50	800	Mild abrasive	a, b, c, e	g, b, c	
cake	Pulv.	75–95	1,200–1,520	Very abrasive	a, b, c, d, e	g	Abrasive
Sand: dry	Granular	90–110	1,440–1,760	Sluggish	a, e, f	g	Abrasive
damp	Granular	90–110	1,440–1,760	Hygroscopic	a, e, f	g	Sticky if hot
Sawdust	Granular	15–20	240–320	Flows freely	a, b, c, d, e	g	
Sewage sludge	Pulv.	60	960	Abrasive	a, b, e, f	g	
Silica flour	Pulv.	80	1,280	Sticky	a, d, e	g	Sticky if hot
Soap flakes	Granular	10–20	160–320		a, c, e	g, c	Caustic
Soda ash: light	Pulv.	25–35	400–560	Sticky if wet	a, b, c, d, e	g, c	Caustic
heavy	Pulv.	55–65	880–1,040	Sluggish	a, b, c, d, e	g, c	
Soybean flour	Pulv.	30	480	Fragile	a, b, c, e	g, c	Explosive dust
Starch	Pulv.	30–40	480–640	Flows freely	a, b, c, e	g	Explosive dust
Sugar: raw	Granular	55–65	880–1,040	Flows freely	a, b, c, e	g	Handle gently
refined	Granular	50–55	800–880	Sticky	a, b, c, e	g, b	Explosion risk
Sulfur	Pulv.	55	880		a, b, c, d, e	g, b	Adheres to metal
Talc	Pulv.	50–60	800–960	Sticky	a, b, d, e	g	
Tobacco stems	Stringy	25	400	Corrosive if wet	a, c, d, e	g, c	Keep clean
Wheat	Granular	48	770	Mild abrasive	a, c, d, e	g, c	
Wood chips	Granular	18–20	290–320	Sluggish	a, b, c, d, e	g	Corrosive if wet
Zinc oxide	Pulv.	20–35	320–560	Free-flowing	a, b, c, d, e	g	Avoid discoloration
Zinc sulfate	Pulv.	70	1,120	May arch / May pack / May pack	a, b, c, d, e	g	

*Explanation of letter symbols:

a—belt, b—flight, c—continuous flow, d—pneumatic, e—screw, f—drag chain, g—belt and bucket, h—chain and bucket, i—overhead straight power, j—overhead power and free.

Hawkins, Manager, Product Standards and Services, Columbus McKinnon Corporation, lists preferred types of conveyors and elevators for a variety of materials in both bulk and packaged forms. Entering this table at aluminum oxide in pulverized form shows that a belt or screw conveyor is preferred, while a flight elevator is recommended for vertical lifts of this material.

While Table 1 gives general recommendations, the engineer should remember that a careful economic study is required to keep the capital investment to the minimum consistent with safe and dependable conveying of the material. Along with capital cost, the operating and maintenance costs must also be evaluated before a final choice of the conveyor and elevator is made.

Related Calculations.   Choose the belt length to accommodate the maximum expected product capacity. Belt speed should be compatible with the process equipment served and with the other materials-handling equipment associated with the conveyor belt.

Belt conveyors are suitable for bulk materials of many types. However, characteristics of the material conveyed must be considered before a final choice of the belting material is made. Thus, as outlined by K. W. Tunnell Company: (a) *Material stickiness* may prevent materials handled from discharging completely from the conveyor belt, or may interfere with the belt drive components: motors, chains, etc. (b) *Ambient temperatures* exceeding 150°F (83°C) could cause deterioration or damage to the belt materials. (c) *Chemical reactions* between the conveyed product and the belt material can cause damage. Thus, oils, chemicals, fats, and acids can damage belts. (d) *Excessively large lump size* may require an oversize belt system to handle the conveyed product safely.

One way around these problems is use of metal-belt conveyors. These are similar in design to conventional rubber and composite-material conveyor belts except that their surface is made of woven or solid metal. Popular materials include carbon steel, galvanized steel, stainless steel, and other metals and alloys.

With today's emphasis on environmental and safety aspects of engineering decisions, it is wise for the design engineer to refer to the appropriate codes and specifications governing the particular type of equipment being considered. Thus, in the materials handling field, ANSI B 20.1 "Conveyors, Cableways, and Related Equipment" should be consulted before any final design choices are made.

Likewise, state and city codes should be checked before a firm equipment selection decision. In certain instances the local code may be more restrictive than the national code. OSHA—Occupational Safety and Health Administration—regulations are important where human safety is involved. Since these regulations vary so widely with material handled, type of equipment used, and location, no generalizations about them can be made other than to recommend strongly that the regulations be studied and followed.

DETERMINING EQUIPMENT DESIGN
PARAMETERS FOR OVERHEAD CONVEYORS

Select suitable equipment for the overhead conveyor shown in Fig. 1. Determine the total chain pull and horsepower required if the conveyor is 700 ft (213 m) long, the coefficient of friction is 0.03, the total chin pull is 60 lb/ft (89.4 kg/m) comprised of the components detailed below. Use the design method presented the K. W. Tunnell Company.

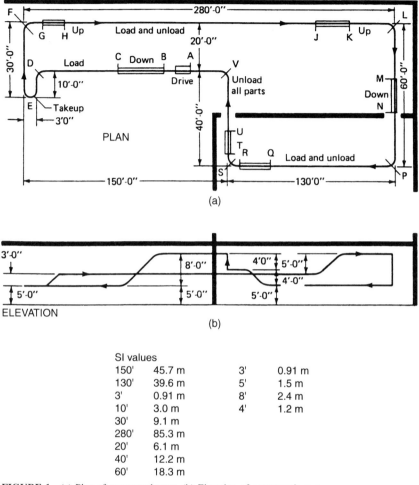

FIGURE 1 (a) Plan of conveyor layout. (b) Elevation of conveyor layout.

Calculation Procedure:

1. From the process flow charts, determine all the operations to be serviced by the conveyor

The process flow charts will be provided by the manufacturing enginner or the process engineer, depending on the type of installation the conveyor is serving. To assist in the conveyor layout and design a listing of each process served by the conveyor should be prepared.

2. Determine the path of the conveyor on a scaled plant layout

Draw a plan and elevation of the conveyor, Fig. 1, on a scaled layout of the plant. Show all obstructions the conveyor will encounter, such as columns, walls, machinery, and work aisles. Indicate the loading and unloading zones, probable drive location, and passage through walls.

3. Develop a vertical elevation to determine incline and decline dimensions
Show the inclines and declines, and their dimensions, Fig. 1b. A three-dimensional view of the installation can be prepared at this point to help people better visualize the final installation and the various routes of the conveyor.

4. Determine the material movement rate, unit load size, spacing, and carrier design for the conveyor
Information for these variables can be obtained from the flow chart and the personnel in charge of the process being served by the conveyor. It is important that the conveyor be designed for the maximum anticipated load and material size.

5. Modify turn radii to provide adequate clearances
Prepare drawings showing needed load spacing on turns, Fig. 2. Without adequate clearnaces, the conveyor may not provide the desired transportation capability needed to serve properly the process for which the conveyor is being designed.

6. Design the load spacing for clearances on inclines and declines
As inclines and declines get steeper, Fig. 2 load spacing has to be increased to provide a constant clearance or separation between loads. Table 2 gives selected clearances on inclined track for overhead conveyors for a given separation at various incline angles.

7. Redraw the conveyor path and vertical elevation views using newly determined radii and incline information
Show the new radii and incline information as determined by the redesign of the system layout, Figs. 1 and 2.

8. Compute the chain pull in the conveyor
The chain pull is the total weight of the chain, trolleys, Fig. 3, and other components, plus the weight of the carriers and load. Thus, for the given system, the tenative chain pull can be found from $C_p = L \times P_L \times f$, where C_p = tentative

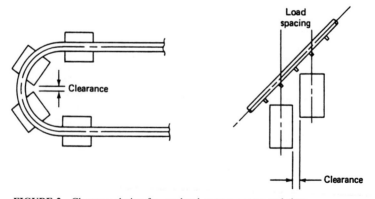

FIGURE 2   Clearance design for overhead conveyor turns and rises.

TABLE 2 Selected Load Clearance on Inclined Track for Overhead Conveyors

Load spacing, in	Incline angle, deg					
	10	20	30	40	50	60
	Horizontal centers, in					
12	11⅞	11¼	10⅜	9¼	7¾	6
16	15¾	15⅛	13⅞	12¼	10⅜	8
18	17¾	17	15⅝	13⅞	11⅜	9
24	23⅝	22⅝	20⅞	18⅜	15½	12
cm	Horizontal centers, cm					
30.5	29.9	28.6	26.4	23.5	19.7	15.2
40.6	40.0	38.7	35.2	31.1	26.4	20.3
45.7	45.1	43.2	39.7	35.2	28.9	22.9
60.9	60.0	57.5	53.0	46.7	39.4	30.5

chain pull, lb (kg); L = conveyor length, ft (m); P_L = chain load, lb/ft (kg/m); f = coefficient of friction = 0.03 for this installation.

The given chain load of 60 lb/ft (89.4 kg/m) is comprised of 10.0 lb/ft (14.9 kg/m) for the chain and trolleys, 12.5 lb/ft (18.6 kg/m) for the carriers, and 37.5 lb/ft (55.9 kg/m) for the line load. Substituting, C_p = 700(60.0)(0.03) = 1260 lb (572 kg).

For this initial calculation, inclines and declines are assumed to be level sections if the number of declines balances out the number of inclines. However, for each additional incline, the total line load rise has to be added to determine the total chain pull. If, for example, a vertical incline in this installation raises the line load 8 ft ((2.4 m), then the additional chain pull = 37.5 lb line load × 8 ft = 300 lb (136.2 kg). The total chain pull then becomes 1260 + 300 = 1560 lb (708.2 kg).

9. Select the tenative conveyor size based on the trolley load and chain pull
Use the manufacturer's data to choose the tenative conveyor size. In making your choice, try to comform to standard conveyor sizes and layouts because this will reduce the capital cost of the installation. Further, the installation will probably be made faster because there will be less customizing required.

10. Select vertical curve radii
Again, work with the standard radii available from the manufacturer, if possible. This will reduce installation costs and time.

11. Determine the conveyor power requirements and drive locations
Make point-to-point calculations of the chain pull around the complete path of the conveyor, Fig. 1. Use the following equations to compute point-to-point chain pull:
(a) Pull for each horizontal run, lb (kg), P_H = XWL, where X = 0.02 for standard ball-bearing trolleys; W = total moving weight, lb/ft (kg/m), empty or loaded as the case may be; L = length of straight run, ft (m).

(b) Pull for each traction wheel or roller turn, lb (kg), P_T = YP, where Y = 0.02 for traction wheel or roller turn; P = pull at turn, lb (kg). (c) Pull for each vertical curve, lb (kg), P_V = XWS + ZP + $HW(1 + Z)$, where X = 0.02 for standard ball-bearing trolleys; W = total moving weight, lb/ft (kg/m); S = horizontal span of

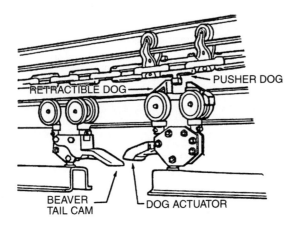

TRANSPORTATION MODE

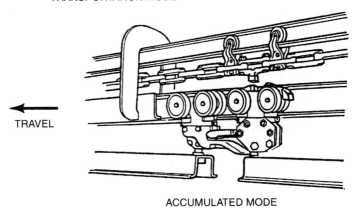

TRAVEL ⟵

ACCUMULATED MODE

FIGURE 3   Power- and free-trolley overhead conveyors.

vertical curve, ft (m); H = total change of level of conveyor, ft (m) (*plus,* when conveyor is traveling up the curve; *minus* when conveyor is traveling down the curve); Z = 0.03 for 30° incline; 0.045 for 45° incline; 0.06 for 60° incline; 0.09 for 90° incline; P = pull at start of curve, lb (kg).

Drive horsepower (kW) can be calculated from: Drive hp = (drive capacity, lb)(maximum speed, ft/min)/0.6(33,000). Thus, if the drive capacity required is 6000 lb, the maximum speed is 50 ft/min, the horsepower required = (6000)(50)/0.6(33,000) = 15.2 hp (11.3 kW).

12.   *Design the conveyor supports and superstructures*
Refer to the manufacturer's data for suitable supports and superstructures. It is best, if possible, to use standard supports and superstructures. This will save money and time for the firm owning the plant being fitted with the conveyor.

13.   *Design guards required by laws and codes*
Federal, state, and applicable codes require guards of various types under high trolley runs, particularly over aisles and work areas. Guard panels are normally

made from woven or welded wire mesh with structural angles and channels to suit the size and weight of the material being handled.

Related Calculations. The general procedure presented here is valid for overhead conveyors handling a variety of materials: manufactured goods, parts for assembly, raw materials, etc., in plants in many different industries. Since conveyor layout, sizing, and safety design are a specialized skill, the engineer should consult carefully with the conveyor manufacturer. The manufacturer's wide experience will be most helpful to the engineer in achieving an economical and safe design for the installation being considered. The steps, illustrations, and table in this procedure are the work of the K. W. Tunnell Company. SI values were added by the handbook editor.

BULK MATERIAL ELEVATOR AND CONVEYOR SELECTION

Choose a bucket elevator to handle 150 tons/h (136.1 t/h) of abrasive material weighing 50 lb/ft³ (800.5 kg/m³) through a vertical distance of 75 ft (22.9 m) at a speed of 100 ft/min (30.5 m/min). What hp input is required to drive the elevator? The bucket elevator discharges onto a horizontal conveyor which must transport the material 1400 ft (426.7 m). Choose the type of conveyor to use, and determine the required power input needed to drive it.

Calculation Procedure:

1. Select the type of elevator to use
Table 3 summarizes the various characteristics of bucket elevators used to transport bulk materials vertically. This table shows that a continuous bucket elevator would be a good choice, because it is a recommended type for abrasive materials. The second choice would be a pivoted bucket elevator. However, the continuous bucket type is popular and will be chosen for this application.

2. Compute the elevator height
To allow for satisfactory loading of the bulk material, the elevator length is usually increased by about 5 ft (1.5 m) more than the vertical lift. Hence, the elevator height = 75 + 5 = 80 ft (24.4 m).

Related Calculations. The procedure given here is valid for conveyors using rubber belts reinforced with cotton duck, open-mesh fabric, cords, or steel wires. It is also valid for stitched-canvas belts, balata belts, and flat-steel belts. The required horsepower input includes any power absorbed by idler pulleys.

Table 6 shows the minimum recommended belt widths for lumpy materials of various sizes. Maximum recommended belt speeds for various materials are shown in Table 5.

3. Compute the required power input to the elevator
Use the relation $hp = 2CH/1000$, where C = elevator capacity, tons/h; H = elevator height, ft. Thus, for this elevator, $hp = 2(150)(80)/1000 = 24.0$ hp (17.9 kW).

The power input relation given above is valid for continuous-bucket, centrifugal-discharge, perfect-discharge, and super-capacity elevators. A 25-hp (18.7-kW) motor would probably be chosen for this elevator.

TABLE 3 Bucket Elevators

	Centrifugal discharge	Perfect discharge	Continuous bucket	Gravity discharge	Pivoted bucket
Carrying paths	Vertical	Vertical to inclination 15° from vertical	Vertical to inclination 15° from vertical	Vertical and horizontal	Vertical and horizontal
Capacity range, tons/h (t/h), material weighing 50 lb/ft³ (800.5 kg/m³)	78 (70.8)	34 (30.8)	345 (312.9)	191 (173.3)	255 (231.3)
Speed range, ft/min (m/min)	306 (93.3)	120 (36.6)	100 (30.5)	100 (30.5)	80 (24.4)
Location of loading point	Boot	Boot	Boot	On lower horizontal run	On lower horizontal run
Location of discharge point	Over head wheel	Over head wheel	Over head wheel	On horizontal run	On horizontal run
Handling abrasive materials	Not preferred	Not preferred	Recommended	Not recommended	Recommended

Source: Link-Belt Div. of FMC Corp.

4. *Select the type of conveyor to use*

Since the elevator discharges onto the conveyor, the capacity of the conveyor should be the same, per unit time, as the elevator. Table 4 lists the characteristics of various types of conveyors. Study of the tabulation shows that a belt conveyor would probably be best for this application, based on the speed, capacity, and type of material it can handle. Hence, it will be chosen for this installation.

5. *Compute the required power input to the conveyor*

The power input to a conveyor is composed of two portions: the power required to move the empty belt conveyor and the power required to move the load horizontally.

Determine from Fig. 4 the power required to move the empty belt conveyor, after choosing the required belt width. Determine the belt width from Table 5.

Thus, for this conveyor, Table 5 shows that a belt width of 42 in (106.7 cm) is required to transport up to 150 tons/h (136.1 t/h) at a belt speed of 100 ft/min (30.5 m/min). [Note that the next *larger* capacity, 162 tons/h (146.9 t/h), is used when the exact capacity required is not tabulated.] Find the horsepower required to drive the empty belt by entering Fig. 4 at the belt distance between centers, 1400 ft (426.7 m), and projecting vertically upward to the belt width, 42 in (106.7 cm). At the left, read the required power input as 7.2 hp (5.4 kW).

Compute the power required to move the load horizontally from $hp = (C/100)(0.4 + 0.00345L)$, where L = distance between conveyor centers, ft; other symbols as before. For this conveyor, $hp = (150/100)(0.4 + 0.00325 \times 1400) = 6.83$ hp (5.1 kW). Hence, the total horsepower to drive this horizontal conveyor is $7.2 + 6.83 = 14.03$ hp (10.5 kW).

The total horsepower input to this conveyor installation is the sum of the elevator and conveyor belt horsepowers, or $14.03 + 24.0 = 38.03$ hp (28.4 kW).

TABLE 4 Conveyor Characteristics

	Belt conveyor	Apron conveyor	Flight conveyor	Drag chain	En masse conveyor	Screw conveyor	Vibratory conveyor
Carrying paths	Horizontal to 18°	Horizontal to 25°	Horizontal to 45°	Horizontal or slight incline, 10°	Horizontal to 90°	Horizontal to 15°; may be used up to 90° but capacity falls off rapidly	Horizontal or slight incline, 5° above or below horizontal
Capacity range, tons/h (t/h) material weighing 50 lb/ft³ (800.5 kg/m³)	2160 (1959.5)	100 (90.7)	360 (326.6)	20 (18.1)	100 (90.7)	150 (136.1)	100 (90.7)
Speed range, ft/min (m/min)	600 (182.9)	100 (30.5)	150 (45.7)	20 (6.1)	80 (24.4)	100 (30.5)	40 (12.2)
Location of loading point	Any point	Any point	Any point	Any point	On horizontal runs	Any point	Any point
Location of discharge point	Over end wheel and intermediate points by tripper or plow	Over end wheel	At end of trough and intermediate points by gates	At end of trough	Any point on horizontal runs by gate	At end of trough and intermediate points by gates	At end of trough
Handling abrasive materials	Recommended	Recommended	Not recommended	Recommended with special steels	Not recommended	Not preferred	Recommended

Source: Link-Belt Div. of FMC Corp.

TABLE 5 Capacities of Troughed Rest [tons/h (t/h) with Belt Speed of 100 ft/min (30.5 m/min)]

Belt width, in (cm)	Weight of material, lb/ft³ (kg/m³)			
	30 (480.3)	50 (800.5)	100 (1601)	150 (2402)
30 (76.2)	47 (42.6)	79 (71.7)	158 (143.3)	237 (214.9)
36 (91.4)	69 (62.6)	114 (103.4)	228 (206.8)	342 (310.2)
42 (106.7)	97 (87.9)	162 (146.9)	324 (293.9)	486 (440.9)
48 (121.9)	130 (117.9)	215 (195.0)	430 (390.1)	645 (585.1)
60 (152.4)	207 (187.8)	345 (312.9)	690 (625.9)	1035 (938.9)

Source: United States Rubber Co.

TABLE 6 Minimum Belt Width for Lumps

Belt width, in (mm)	24 (609.6)	36 (914.4)	42 (1066.8)	48 (1219.2)
Sized materials, in (mm)	4½ (114.3)	8 (203.2)	10 (254)	12 (304.9)
Unsized material, in (mm)	8 (203.2)	14 (355.6)	20 (508)	35 (889)

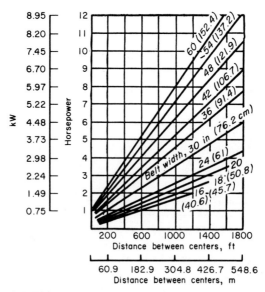

FIGURE 4 Horsepower (kilowatts) required to move an empty conveyor belt at 100 ft/min (30.5 m/min).

Related Calculations: The procedure given here is valid for conveyors using rubber belts reinforced with cotton duck, open-mesh fabric, cords, or steel wires. It is also valid for stitched-canvas belts, balata belts, and flat-steel belts. The required horsepower input includes any power adsorbed by idler pulleys.

Table 5 shows the minimum recommended belt widths for lumpy materials of various sizes. Maximum recommended belt speeds for various materials are shown in Table 6.

When a conveyor belt is equipped with a tripper, the belt must rise about 5 ft (1.5 m) above its horizontal plane of travel.

This rise must be included in the vertical-lift power input computation. When the tripper is driven by the belt, allow 1 hp (0.75 kW) for a 16-in (406.4-mm) belt, 3 hp (2.2 kW) for a 36-in (914.4-mm) belt, and 7 hp (5.2 kW) for a 60-in (1524-mm) belt. Where a rotary cleaning brush is driven by the conveyor shaft, allow about the same power input to the brush for belts of various widths.

SCREW CONVEYOR POWER INPUT AND CAPACITY

What is the required power input for a 100-ft (30.5-m) long screw conveyor handling dry coal ashes having a maximum density of 40 lb/ft³ (640.4 kg/m³) if the conveyor capacity is 30 tons/h (27.2 t/h)?

Calculation Procedure:

1. Select the conveyor diameter and speed
Refer to a manufacturer's engineering data or Table 8 for a listing of recommended screw conveyor diameters and speeds for various types of materials. Dry coal ashes are commonly rated as group 3 materials, Table 9, i.e., materials with small mixed lumps with fines.

To determine a suitable screw diameter, assume two typical values and obtain the recommended rpm from the sources listed above or Table 8. Thus, the maximum rpm recommended for a 6-in (152.4-mm) screw when handling group 3 material is

TABLE 7 Maximum Belt Speeds for Various Materials

Width of belt		Light or free-flowing materials, grains dry sand, etc.		Moderately free-flowing sand, gravel, fine stone, etc.		Lump coal, coarse stone, crushed ore		Heavy sharp lumpy materials, heavy ores, lump coke	
in	mm	ft/min	m/min	ft/min	m/min	ft/min	m/min	ft/min	m/min
12–14	305–356	400	122	250	76	—	—	—	—
16–18	406–457	500	152	300	91	250	76	—	—
20–24	508–610	600	183	400	122	350	107	250	76
30–36	762–914	750	229	500	152	400	122	300	91

TABLE 8 Screw Conveyor Capacities and Speeds

Material group	Maximum material density		Maximum r/min for diameters of:	
	lb/ft³	kg/m³	6 in (152 mm)	20 in (508 mm)
1	50	801	170	110
2	50	801	120	75
3	75	1201	90	60
4	100	1601	70	50
5	125	2001	30	25

TABLE 9 Material Factors for Screw Conveyors

Material group	Material type	Material factor
1	Lightweight:	
	Barley, beans, flour, oats, pulverized coal, etc.	0.5
2	Fines and granular:	
	Coal—slack or fines	0.9
	Sawdust, soda ash	0.7
	Flyash	0.4
3	Small lumps and fines:	
	Ashes, dry alum	4.0
	Salt	1.4
4	Semiabrasives; small lumps:	
	Phosphate, cement	1.4
	Clay, limestone	2.0
	Sugar, white lead	1.0
5	Abrasive lumps:	
	Wet ashes	5.0
	Sewage sludge	6.0
	Flue dust	4.0

90, as shown in Table 8; for a 20-in (508.0-mm) screw, 60 r/min. Assume a 6-in (152.4-mm) screw as a trial diameter.

2. Determine the material factor for the conveyor
A material factor is used in the screw conveyor power input computation to allow for the character of the substance handled. Table 9 lists the material factor for dry ashes as $F = 4.0$. Standard references show that the average weight of dry coal ashes is 35 to 40 lb/ft³ (640.4 kg/m³).

3. Determine the conveyor size factor
A size factor that is a function of the conveyor diameter is also used in the power input computation. Table 10 shows that for a 6-in (152.4-mm) diameter conveyor the size factor $A = 54$.

4. Compute the required power input to the conveyor
Use the relation $hp = 10^{-6}(ALN + CWLF)$, where hp = hp input to the screw conveyor head shaft; A = size factor from step 3; L = conveyor length, ft; N =

TABLE 10 Screw Conveyor Size Factors

Conveyor diameter, in (mm)	6 (152.4)	9 (228.6)	10 (254)	12 (304.8)	16 (406.4)	18 (457.2)	20 (508)	24 (609.6)
Size factor	54	96	114	171	336	414	510	690

conveyor rpm; C = quantity of material handled, ft^3/h; W = density of material, lb/ft^3; F = material factor from step 2. For this conveyor, given the data listed above, $hp = 10^{-6}(54 \times 100 \times 60 + 1500 \times 40 \times 100 \times 4.0) = 24.3$ hp (18.1 kW). With a 90 percent motor efficiency, the required motor rating would be $24.3/0.90 = 27$ hp (20.1 kW). A 30-hp (22.4-kW) motor would be chosen to drive this conveyor. Since this is not an excessive power input, the 6-in (152.4-mm) conveyor is suitable for this application.

If the calculation indicates that an excessively large power input, say 50 hp (37.3 kW) or more, is required, then the larger-diameter conveyor should be analyzed. In general, a higher initial investment in conveyor size that reduces the power input will be more than recovered by the savings in power costs.

Related Calculations. Use the procedure given here for screw or spiral conveyors and feeders handling any material that will flow. The usual screw or spiral conveyor is suitable for conveying materials for distances up to about 200 ft (60.9 m), although special designs can be built for greater distances. Conveyors of this type can be sloped upward to angles of 35° with the horizontal. However, the capacity of the conveyor decreases as the angle of inclination is increased. Thus the reduction in capacity at a 10° inclination is 10 percent over the horizontal capacity; at 35° the reduction is 78 percent.

The capacities of screw and spiral conveyors are generally stated in ft^3/h (m^3/h) of various classes of materials at the maximum recommended shaft rpm. As the size of the lumps in the material conveyed increases, the recommended shaft rpm decreases. The capacity of a screw or spiral conveyor at a lower speed is found from (capacity at given speed, ft^3/h) [(lower speed, r/min)/(higher speed, r/min)]. Table 8 shows typical screw conveyor capacities at usual operating speeds.

Various types of screws are used for modern conveyors. These include short-pitch, variable-pitch, cut flights, ribbon, and paddle screws. The procedure given above also applies to these screws.

DESIGN AND LAYOUT OF PNEUMATIC CONVEYING SYSTEMS

A pneumatic conveying system for handling solids in an industrial exhaust installation contains two grinding-wheel booths and one lead each for a planer, sander, and circular saw. Determine the required duct sizes, resistance, and fan capacity for this pneumatic conveying system.

Calculation Procedure:

1. Sketch the proposed exhaust system
Make a freehand sketch, Fig. 5 of the proposed system. Show the main and branch ducts and the booths and hoods. Indicate all major structural interferences, such as

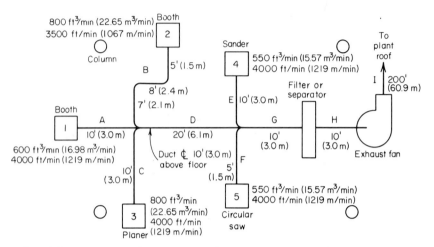

FIGURE 5   Exhaust system layout.

building columns, deep girders, beams, overhead conveyors, piping, etc. Draw the layout approximately to scale.

Mark on the sketch the length of each duct run. Avoid, if possible, vertical drops or rises in the main exhaust duct between the hoods and the fan. Do this by locating the main duct centerline 10 ft (3 m) or so above the finished floor.

Number each hood or booth, and give each duct run an identifying letter. Although it is not absolutely necessary, it is more convenient during the design process to have the hoods in numerical order and the duct runs in alphabetical order.

2. Determine the required air quantities and velocities
Prepare a listing, columns 1 and 2, Table 11, of the booths, hoods, and duct runs. Enter the required air quantities and velocities for each booth or hood and duct in Table 11, columns 3 and 4. Select the air quantities and velocities from the local code covering industrial exhaust systems, if such a code is available. If a code does not exist, use the ASHRAE *Guide* or Table 12.

Use extreme care in selecting the air quantities and velocities, because insufficient flow may cause dangerous atmospheric conditions. Harmful process wastes in the form of dust, gas, or moisture may injure plant personnel.

3. Size the main and branch ducts
Determine the required duct area by dividing the air quantity, ft³/min (m³/min), by the air velocity in the duct, or column 3/column 4, Table 11. Enter the result in column 5, Table 11.

Once the required duct area is known, find from Table 13 the nearest whole-number duct diameter corresponding to the required area. Avoid fractional diameters at this stage of the calculation, because ducts of these sizes are usually more expensive to fabricate. Later, if necessary, two or three duct sizes may be changed to fractional values. By selecting only whole-number diameters in the beginning, the cost of duct fabrication may be reduced somewhat. Enter the duct whole-number diameter in column 6, Table 11.

TABLE 11 Exhaust System Design Calculations

(1) Booth or hood	(2) Duct run	(3) ft³/min (m³/min) in duct	(4) Design velocity, ft/min (m/min)	(5) Duct area = column 3/column 4, ft² (m²)	(6) Duct diameter, in (mm)	(7) Actual velocity, ft/min (m/min)	(8) Actual velocity pressure, inH₂O (mmH₂O)	(9) Length of straight duct, ft (m)	(10) Equivalent length of elbows, ft (m)	(11) Total duct length = column 9 + column 10, ft (m)	(12) Friction per 100 ft (30 m) of duct, inH₂O (mmH₂O)	(13) Actual friction, inH₂O (mmH₂O)
1	A	600 (16.98)	4000 (1219)	0.150 (0.014)	5 (127)	4300 (1311)	1.15 (29.2)	10 (3.0)	0 (0)	10 (3.0)	5.4 (137.2)	0.54 (13.7)
2	B	800 (22.65)	3500 (1067)	0.228 (0.021)	6 (152)	4200 (1280)	1.0 (25.4)	20 (6.1)	18 (5.5)	38 (11.6)	4.0 (101.6)	1.57 (39.9)
3	C	800 (22.65)	4000 (1219)	0.200 (0.019)	6 (152)	4200 (1280)	1.0 (25.4)	10 (3.0)	6 (1.8)	16 (4.8)	4.0 (101.6)	0.64 (16.3)
	D	2200 (62.28)	4000 (1219)	0.550 (0.051)	10 (254)	4000 (1219)	1.0 (25.4)	20 (6.1)	0 (0)	20 (6.1)	2.1 (53.3)	0.42 (10.7)
4	E	550 (15.57)	4000 (1219)	0.137 (0.013)	5 (127)	4000 (1219)	1.0 (25.4)	10 (3.0)	5 (1.5)	15 (4.5)	4.6 (116.8)	0.69 (17.5)
5	F	550 (15.57)	4000 (1219)	0.137 (0.013)	5 (127)	4000 (1219)	1.0 (25.4)	5 (1.5)	5 (1.5)	10 (3.0)	4.6 (116.8)	0.46 (11.7)
	G	3300 (93.42)	4000 (1219)	0.825 (0.077)	12 (305)	4200 (1280)	1.0 (25.4)	10 (3.0)	0 (0)	10 (3.0)	1.9 (48.3)	0.19 (4.8)
	H	3300 (93.42)	3000 (1914)	1.10 (0.102)	14 (356)	3000 (914)	0.55 (13.9)	10 (3.0)	14 (4.3)	24 (7.3)	0.84 (21.3)	0.20 (5.1)
	I	3300 (93.42)	2000 (610)	1.65 (0.153)	18 (457)	2000 (610)	0.25 (6.4)	200 (60.9)	0 (0)	200 (60.9)	0.25 (6.4)	0.50 (12.7)

TABLE 11 (Continued)

	System resistance				
	Hood number				
	1	2	3	4	5
Velocity pressure in hood branch, in (mm) H_2O	1.15 (29.2) 50	1.0 (25.4) 11	1.0 (25.4) 50	1.0 (25.4) 60	1.0 (25.4) 60
Entrance loss (% of velocity pressure)	(50)	(11)	(50)	(60)	(60)
Entrance loss, in (mm) H_2O	0.58 (14.6)	0.11 (2.8)	0.50 (12.7)	0.60 (15.2)	0.60 (15.2)
Branch and main duct resistances *A*	0.54 (13.7)				
B		1.57 (39.9)			
C		.....	0.64 (16.3)		
D	0.42 (10.7)	0.42 (10.7)	0.42 (10.7)		
E	.....	.....	.....	0.69 (17.5)	
F	.....	.....	.....	.....	0.46 (11.7)
G	0.19 (4.8)	0.19 (4.8)	0.19 (4.8)	0.19 (4.8)	0.19 (4.8)
H	0.20 (5.1)	0.20 (5.1)	0.20 (5.1)	0.20 (5.1)	0.20 (5.1)
I	0.50 (12.7)	0.50 (12.7)	0.50 (12.7)	0.50 (12.7)	0.50 (12.7)
Collector or filter resistance, in (mm) H_2O	2.00 (50.8)	2.00 (50.8)	2.00 (50.8)	2.00 (50.8)	2.00 (50.8)
Total resistance in each branch, in (mm) H_2O	4.43 (112.4)	4.99 (126.8)	4.45 (113.1)	4.18 (106.1)	3.95 (100.3)

TABLE 12 Recommended Exhaust Air Quantities

Operation	ft³/min (m³/min)	Branch duct velocity, ft/min (m/min)	Branch duct diameter, in (mm)
Sanding:			
Single drum, [10-in (25.4-cm) diameter]	400 (11.32)	4000 (1219)	4 (101.6)
Disk	550 (15.57)	4000 (1219)	5 (127)
Circular saws [16- to 24-in (40.6- to 60.9-cm) diameter]	450 (12.74)	4000 (1219)	4.5 (114.3)
Shoe machinery	550 (15.57)	4000 (1219)	5 (127)
Buffing and polishing wheels [16- to 24-in (40.6- to 60.9-cm) diameter]	600 (16.98)	4500 (1372)	5 (127)
Grinding wheels [16- to 20-in (40.6- to 50.8-cm) diameter]	600 (16.98)	4500 (1372)	5 (127)
Abrasive blast rooms	. . .	3500 (1067)	
Pharmaceuticals	. . .	3000 (1067)	

Conveying velocities	
Material conveyed	Conveying velocity, ft/min (m/min)
Vapors, gases, fumes, fine dusts	1500 to 2000 (457 to 610)
Fine dry dusts	3000 (914)
Average industrial dusts	3500 (1067)
Coarse particles	3500 to 4500 (1067 to 1372)
Large particles, heavy loads, moist materials, pneumatic conveying	4500 and higher (1372 and higher)

TABLE 13 Duct Diameters and Areas

Diameter		Area	
in	mm	ft²	m²
4.0	102	0.0873	0.008
5.0	127	0.1364	0.013
6.0	152.4	0.1964	0.018
7.0	178	0.2673	0.025
8.0	203.2	0.3491	0.032
10.0	254	0.5454	0.051
12	305	0.7854	0.073
14	356	1.069	0.099
16	406.4	1.396	0.130
18	457.2	1.767	0.164
20	508	2.182	0.203
22	559	2.640	0.245
24	610	3.142	0.292

4. Compute the actual air velocity in the duct

Use Fig. 6 to determine the actual velocity in each duct. Enter the chart at the air quantity corresponding to that in the duct, and project vertically to the diameter curve representing the duct size. Read the actual velocity in the duct on the velocity scale, and enter the value in column 7 of Table 11.

The actual velocity in the duct should, in all cases, be equal to or greater than the design velocity shown in column 4, Table 11. If the actual velocity is less than the design velocity, decrease the duct diameter until the actual velocity is equal to or greater than the design velocity.

5. Compute the duct velocity pressure

With the actual velocity known, compute the corresponding velocity pressure in the duct from $h_v = (v/4005)^2$, where h_v = velocity pressure in the duct, inH_2O; v = air velocity in the duct, ft/min. Thus, for the duct run A in which the actual air velocity is 4300 ft/min (1310.6 m/min), $h_v = (4300/4005)^2 = 1.15$ in (29.2 mm) H_2O. Compute the actual velocity pressure in each duct run, and enter the result in column 8, Table 11.

6. Compute the equivalent length of each duct

Enter the total straight length of each duct, including any vertical drops, in column 9, Table 11. Use accurate lengths, because the system resistance is affected by the duct length.

Next list the equivalent length of each elbow in the duct runs in column 10, Table 11. For convenience, assume that the equivalent length of an elbow is 12 times the duct diameter in ft. Thus, an elbow in a 6-in (152.4-mm) diameter duct has an equivalent resistance of (6-in diameter/[(12 in/ft)(12)]) = 6 ft (1.83 m) of straight duct. When making this calculation, assume that all elbows have a radius equal to twice the diameter of the duct. Consider 45° bends as having the same resistance as 90° elbows. Note that branch ducts are usually arranged to enter the main duct at an angle of 45° or less. These assumptions are valid for all typical industrial exhaust systems and pneumatic conveying systems.

Find the total equivalent length of each duct by taking the sum of columns 9 and 10, Table 11, horizontally, for each duct run. Enter the result in column 11, Table 11.

7. Determine the actual friction in each duct

Using Fig. 6, determine the resistance, inH_2O (mmH_2O) per 100 ft (30.5 m) of each duct by entering with the air quantity and diameter of that duct. Enter the frictional resistance thus found in column 12, Table 11.

Compute actual friction in each duct by multiplying the friction per 100 ft (30.5 m) of duct, column 12, Table 11, by the total duct length, column 11 ÷ 100. Thus for duct run A, actual friction = 5.4(10/100) = 0.54 in (13.7 mm) H_2O. Compute the actual friction for the other duct runs in the same manner. Tabulate the results in column 13, Table 11.

8. Compute the hood entrance losses

Hoods are used in industrial exhaust systems to remove vapors, dust, fumes, and other undesirable airborne contaminants from the work area. The hood entrance loss, which depends upon the hood configuration, is usually expressed as a certain percentage of the velocity pressure in the branch duct connected to the hood, Fig. 7. Since the hood entrance loss usually accounts for a large portion of the branch resistance, the entrance loss chosen should always be on the safe side.

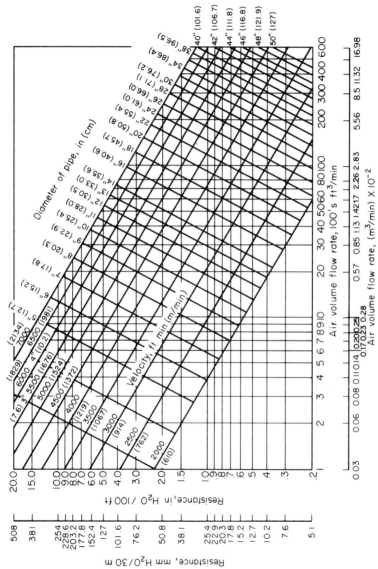

FIGURE 6 Duct resistance chart. (*American Air Filter Co.*)

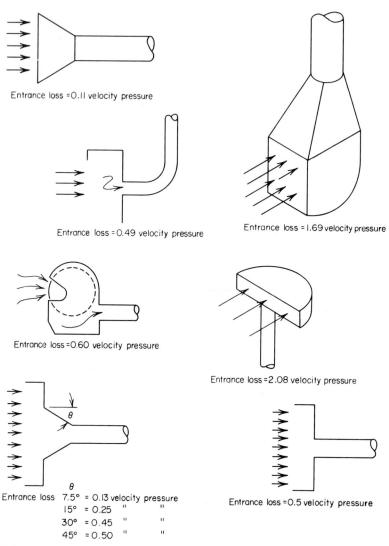

FIGURE 7   Entrance losses for various types of exhaust-system intakes.

List the hood designation number under the "System Resistance" heading, as shown in Table 11. Under each hood designation number, list the velocity pressure in the branch connected to that hood. Obtain this value from column 8, Table 11. List under the velocity pressure, the hood entrance loss form Fig. 7 for the particular type of hood used in that duct run. Take the product of these two values, and enter the result under the hood number on the "entrance loss, inH$_2$O" line. Thus, for hood 1, entrance loss = 1.15(0.50) = 0.58 in (14.7 mm) H$_2$O. Follow the same procedure for the other hoods listed.

9. Find the resistance of each branch run

List the main and branch runs, A through F, Table 9. Trace out each main and branch run in Fig. 5, and enter the actual friction listed in column 3 of Table 11. Thus for booth 1, the main and branch runs consist of A, D, G, H, and I. Insert the actual friction, in (mm) H_2O, as shown in Table 9, or A = 9.54(242.3), D = 0.42(10.7), G = 0.19(4.8), H = 0.20(5.1), I = 0.50(12.7).

Determine the filter friction loss from the manufacturer's engineering data. It is common practice to design industrial exhaust systems on the basis of dirty filters or separators; i.e., the frictional resistance used in the design calculations is the resistance of a filter or separator containing the maximum amount of dust allowable under normal operating conditions. The frictional resistance of dirty filters can vary from 0.5 to 6 in (12.7 to 152.4 mm) H_2O or more. Assume that the frictional resistance of the filter used in this industrial exhaust system is 2.0 in (50.8 mm) H_2O.

Add the filter resistance to the main and branch duct resistance as shown in Table 11. Find the sum of each column in the table, as shown. This is the total resistance in each branch, inH_2O, Table 11.

10. Balance the exhaust system

Inspection of the lower part of Table 11 shows that the computed branch resistances are unequal. This condition is usually encountered during system design. To balance the system, certain duct sizes must be changed to produce equal resistance in all ducts. Or, if possible, certain ducts can be shortened. If duct shortening is not possible, as is often the case, an exhaust fan capable of operating against the largest resistance in a branch can be chosen. If this alternative is selected, special dampers must be fitted to the air inlets of the booths or ducts. For economical system operation, choose the balancing method that permits the exhaust fan to operate against the minimum resistance.

In the system being considered here, a fairly accurate balance can be obtained by decreasing the size of ducts E and F to 4.75 in (120.7 mm) and 4.375 in (111.1 mm), respectively. Duct B would be increased to 6.5 in (165.1 mm) in diameter.

11. Choose the exhaust fan capacity and static pressure

Find the required exhaust fan capacity in ft^3/min from the sum of the airflows in the ducts, A through H, column 3, Table 11, or 3300 ft^3/min (93.5 m^3/min). Choose a static pressure equal to or greater than the total resistance in the branch duct having the greatest resistance. Since this is slightly less than 4.5 in (114.3 mm) H_2O, a fan developing 4.5 in (114.3 mm) H_2O static pressure will be chosen. A 10 percent safety factor is usually applied to these values, giving a capacity of 3600 ft^3/min (101.9 m^3/min) and a static pressure of 5.0 in (127 mm) H_2O for this system.

12. Select the duct material and thickness

Galvanized sheet steel is popular for industrial exhaust systems, except where corrosive fumes and gases rule out galvanized material. Under these conditions, plastic, tile, stainless steel, or composition ducts may be substituted for galvanized ducts. Table 14 shows the recommended metal gage for galvanized ducts of various diameters. Do not use galvanized-steel ducts for gas temperatures higher than 400°F (204°C).

Hoods should be two gages heavier than the connected branch duct. Use supports not more than 12 ft (3.7 m) apart for horizontal ducts up to 8-in (203.2-mm) diameter. Supports can be spaced up to 20 ft (6.1 m) apart for larger ducts. Fit a

TABLE 14   Exhaust-System Duct Gages

Duct diameter, in (mm)	Metal gage
Up to 8 (203.2)	22
9 to 18 (228.6 to 457.2)	20
19 to 30 (482.6 to 762)	18
31 and larger (787.4 and larger)	16

duct cleanout opening every 10 ft (3 m). Where changes of diameter are made in the main duct, fit an eccentric taper with a length of at least 5 in (127 mm) for every 1-in (25.4-mm) change in diameter. The end of the main duct is usually extended 6 in (152.4 mm) beyond the last branch and closed with a removable cap. For additional data on industrial exhaust system design, see the newest issue of the ASHRAE *Guide.*

Related Calculations.   Use this procedure for any type of industrial exhaust system, such as those serving metalworking, woodworking, plating, welding, paint spraying, barrel filling, foundry, crushing, tumbling, and similar operations. Consult the local code or ASHRAE *Guide* for specific airflow requirements for these and other industrial operations.

This design procedure is also valid, in general, for industrial pneumatic conveying systems. For several comprehensive, worked-out designs of pneumatic conveying systems, see Hudson—*Conveyors,* Wiley.

SECTION 11
HEAT TRANSFER AND HEAT EXCHANGE

SELECTING TYPE OF HEAT EXCHANGER FOR A SPECIFIC APPLICATION

Determine the type of heat exchanger to use for each of the following applications: (1) heating oil with steam; (2) cooling internal combustion engine liquid coolant; (3) evaporating a hot liquid. For each heater chosen, specify the typical pressure range for which the heater is usually built and the typical range of the overall coefficient of heat transfer U.

Calculation Procedure:

1. Determine the heat-transfer process involved
In a heat exchanger, one or more of four processes may occur: heating, cooling, boiling, or condensing. Table 1 lists each of these four processes and shows the usual heat-transfer fluids involved. Thus, the heat exchangers being considered here involve (*a*) oil heater—heating—vapor-liquid; (*b*) internal-combustion engine coolant—cooling—gas-liquid; (*c*) hot-liquid evaporation—boiling—liquid-liquid.

11.1

TABLE 1 Heat-Exchanger Selection Guide*

Heat-transfer fluids	Equipment	Action	Type†	Pressure range†	Typical range of U§
Liquid-liquid	Boiler-water blowdown exchanger	Blowdown cooled, feedwater heated	S	M, H	50-300 (0.28-1.7)
	Laundry-water heat reclaimer	Waste water cooled, feed heated	S	L	30-200 (0.17-1.1)
	Service-water heater	Waste liquid cooled, water heated	S	L, H	50-300 (0.28-1.7)
Vapor-liquid	Bleeder heater	Steam condensed, feedwater heated	S	L, H	200-800 (1.1-4/5)
	Deaerating feed heater	Steam condensed, feedwater heated	M	L, M	DC
	Jet heater	Steam condensed, water heated	M	L	DC
	Process kettle	Steam condensed, liquid heated	S	L, M	100-500 (0.57-2.8)
	Oil heater	Steam condensed, oil heated	S	L, M	20-60 (0.11-0.34)
	Service-water heater	Steam condensed, water heated	S	L, M	200-800 (1.1-4.5)
	Open flow-through heater	Steam condensed, water heated	M	L	DC
	Liquid-sodium steam superheater	Sodium cooled, steam superheated	S	M, H	50-200 (0.28-1.1)
Heating					
Gas-liquid	Waste-heat water heater	Waste gas cooled, water heated	T	L	2-10 (0.011-0.057)
	Boiler economizer	Flue gas cooled, feedwater heated	T	M, H	2-10 (0.011-0.057)
	Hot-water radiator	Water cooled, air heated	T	L	1-10 (0.0057-0.057)
Gas-gas	Boiler air heater	Flue gas cooled, combustion air heated	T, R	L	2-10 (0.011-0.057)
	Gas-turbine regenerator	Flue gas cooled, combustion air heated	T	L	2-10 (0.011-0.057)
Vapor-gas	Boiler superheater	Combustion gas cooled, steam superheated	T	M, H	2-20 (0.011-0.11)
	Steam pipe coils	Steam condensed, air heated	T	L, M	2-10 (0.011-0.057)
	Steam radiator	Steam condensed, air heated	T	L	2-10 (0.011-0.057)
Cooling					
Liquid-liquid	Oil cooler	Water heated, oil cooled	S, D	L, M	20-200 (0.11-1.1)
	Water chiller	Refrigerant boiled, water cooled	S	L, M	30-151 (0.17-0.86)
	Brine cooler	Refrigerant boiled, brine cooled	S	L, M	30-150 (0.17-0.86)
	Transformer-oil cooler	Water heated, oil cooled	S	L, M	20-50 (0.11-0.88)
Vapor-liquid	Boiler desuperheater	Boiler water heated, steam desuperheated	S, M	M, H	150-800 (0.85-4.5)

TABLE 1 (*Continued*)

Category	Subcategory	Equipment	Process	Type†	Pressure‡	Overall coefficient	U range§
Cooling	Gas-liquid	Compressor intercoolers and aftercoolers	Water heated, compressed air cooled	S	L, H	10–20	(0.057–0.11)
		Internal-combustion-engine radiator	Air heated, water cooled	T	L	2–10	(0.011–0.057)
		Generator hydrogen, air coolers	Water heated, hydrogen or air cooled	S	L	2–10	(0.011–0.057)
		Air-conditioning cooler	Water heated, air cooled	T	L	2–10	(0.011–0.057)
		Refrigeration heat exchanger	Brine heated, air cooled	T	L, M	2–10	(0.011–0.057)
	Vapor-gas	Refrigeration evaporator	Refrigerant boiled, air cooled	T	L, M	2–10	(0.011–0.057)
		Boiler desuperheater	Flue gas heated, steam desuperheated	T	M, H	2–8	(0.011–0.045)
Boiling	Liquid-liquid	Hot-liquid evaporator	Waste liquid cooled, water boiled	S	L, H	40–150	(0.23–0.85)
		Liquid-sodium steam generator	Sodium cooled, water boiled	S	M, H	500–1000	(2.8–5.7)
	Vapor-liquid	Evaporator (vacuum)	Steam condensed, water boiled	S	L	400–600	(2.3–3.4)
		Evaporator (high pressure)	Steam condensed, water boiled	S	L, M	400–600	(2.3–3.4)
		Mercury condenser-boiler	Mercury condensed, water boiled	S	M, H	500–700	(2.8–4.0)
	Gas-liquid	Waste-heat steam boiler	Flue gas cooled, water boiled	T	L, H	2–10	(0.011–0.057)
		Direct-fired steam boiler	Combustion gas cooled, water boiled	T	L, H	2–10	(0.011–0.057)
Condensing	Vapor-liquid	Refrigeration condenser	Water heated, refrigerant condensed	S, D	L, M	80–250	(0.45–1.4)
		Steam surface condenser	Water heated, steam condensed	S	L	300–800	(1.7–4.5)
		Steam mixing condenser	Water heated, steam condensed	M	L	DC	
		Intercondenser and aftercondenser	Condensate heated, steam condensed	S	L	15–300	(0.085–1.7)
	Vapor-gas	Air-cooled surface condenser	Air heated, steam condensed	T	L	2–16	(0.011–0.091)

* *Power.*

†S—shell-and-tube exchanger; M—direct contact mixing exchanger; T—tubes in path of moving fluid, or exchanger open to surrounding air; R—regenerative plate-type or simple plate-type exchanger; D—double-tube exchanger.

‡L—highest pressure ranges from 0 to 100 lb/in² (abs) (0 to 689.4 kPa); M—highest pressure from 100 to 500 lb/in² (abs) (689.4 to 3447 kPa); H—500 lb/in² (abs) (3447 kPa) up.

§Values of U represent range of overall heat-transfer coefficients that might be expected in various exchangers. Coefficients are stated in Btu/(h·°F·ft²) [W/(m²·°C)] of heating surface. Total heat transferred in exchanger, in Btu/h, is obtained by multiplying a specific value of U for that type of exchanger by the surface and the log mean temperature difference. DC indicates direct exchange of heat.

2. *Specify the heater action and the usual type selected*
Using the same identifying letters for the heaters being selected, Table 1 shows the action and usual type of heater chosen. Thus,

Action	Type
a. Steam condensed; oil heated	Shell-and-tube
b. Air heated; water cooled	Tubes in open air
c. Waste liquid cooled; water boiled	Shell-and-tube

3. *Specify the usual pressure range and typical U*
Using the same identifying letters for the heaters being selected, Table 1 shows the action and usual type of heater chosen. Thus,

		Typical U range	
Usual pressure range		Btu/(h·°F·ft²)	W/(m²·°C)
a. 0–500 lb/in² (abs) (0 to 3447 kPa)		20–60	113.6–340.7
b. 0–100 lb/in² (abs) (0 to 689.4 kPa)		2–10	11.4–56.8
c. 0–500 lb/in² (abs) (0 to 3447 kPa)		40–150	227.1–851.7

4. *Select the heater for each service*
Where the heat-transfer conditions are normal for the type of service met, the type of heater listed in step 2 can be safely used. When the heat-transfer conditions are unusual, a special type of heater may be needed. To select such a heater, study the data in Table 1 and make a tentative selection. Check the selection by using the methods given in the following calculation procedures in this section.

Related Calculations. Use Table 1 as a general guide to heat-exchanger selection in any industry—petroleum, chemical, power, marine, textile, lumber, etc. Once the general type of heater and its typical U value are known, compute the required size, using the procedure given later in this section.

SHELL-AND-TUBE HEAT EXCHANGER SIZE

What is the required heat-transfer area for a parallel-flow shell-and-tube heat exchanger used to heat oil if the entering oil temperature is 60°F (15.6°C), the leaving oil temperature is 120°F (48.9°C), and the heating medium is steam at 200 lb/in² (abs) (1378.8 kPa)? There is no subcooling of condensate in the heat exchanger. The overall coefficient of heat transfer $U = 25$ Btu/(h·°F·ft²) [141.9 W/(m²·°C)]. How much heating steam is required if the oil flow rate through the heater is 100 gal/min (6.3 L/s), the specific gravity of the oil is 0.9, and the specific heat of the oil is 0.5 Btu/(lb·°F) [2.84 W/(m²·°C)]?

Calculation Procedure:

1. *Compute the heat-transfer rate of the heater*
With a flow rate of 100 gal/min (6.3 L/s) or (100 gal/min)(60 min/h) = 6000 gal/h (22,710 L/h), the weight flow rate of the oil, using the weight of water of

specific gravity 1.0 as 8.33 lb/gal, is (6000 gal/h) (0.9 specific gravity)(8.33 lb/gal) = 45,000 lb/h (20,250 kg/h), closely.

Since the temperature of the oil rises $120 - 60 = 60°F$ (33.3°C) during passage through the heat exchanger and the oil has a specific heat of 0.50, find the heat-transfer rate of the heater from the general relation $Q = wc\ \Delta t$, where Q = heat-transfer rate, Btu/h; w = oil flow rate, lb/h; c = specific heat of the oil, Btu/(lb · °F); Δt = temperature rise of the oil during passage through the heater. Thus, $Q = (45,000)(0.5)(60) = 1,350,000$ Btu/h (0.4 MW).

2. Compute the heater logarithmic mean temperature difference

The logarithmic mean temperature difference (LMTD) is found from LMTD = $(G - L)/\ln(G/L)$, where G = greater terminal temperature difference of the heater, °F; L = lower terminal temperature difference of the heater, °F; ln = logarithm to the base e. This relation is valid for heat exchangers in which the number of shell passes equals the number of tube passes.

In general, for parallel flow of the fluid streams, $G = T_1 - t_1$ and $L = T_2 - t_2$, where T_1 = heating fluid inlet temperature, °F; T_2 = heating fluid outlet temperature, °F; t_1 = heated fluid inlet temperature, °F; t_2 = heated fluid outlet temperature, °F. Figure 1 shows the maximum and minimum terminal temperature differences for various fluid flow paths.

For this parallel-flow exchanger, $G = T_1 - t_1 = 382 - 60 = 322°F$ (179°C), where 382°F (194°C) = the temperature of 200-lb/in² (abs) (1379-kPa) saturated steam, from a table of steam properties. Also, $L = T_2 - t_2 = 382 - 120 = 262°F$ (145.6°C), where the condensate temperature = the saturated steam temperature

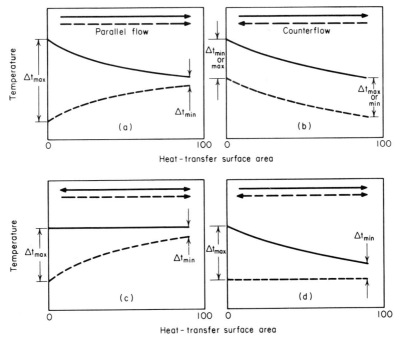

FIGURE 1   Temperature relations in typical parallel-flow and counterflow heat exchangers.

because there is no subcooling of the condensate. Then LMTD $= G - L/\ln(G/L) = (322 - 262)/\ln(322/262) = 290°F$ (161°C).

3. Compute the required heat-transfer area

Use the relation $A = Q/U \times$ LMTD, where $A =$ required heat-transfer area, ft²; $U =$ overall coefficient of heat transfer, Btu/(ft²·h·°F). Thus, $A = 1,350,000/[(25)(290)] = 186.4$ ft² (17.3 m²), say 200 ft² (18.6 m²).

4. Compute the required quantity of heating steam

The heat added to the oil $= Q = 1,350,000$ Btu/h, from step 1. The enthalpy of vaporization of 200-lb/in² (abs) (1379-kPa) saturated steam is, from the steam tables, 843.0 Btu/lb (1960.8 kJ/kg). Use the relation $W = Q/h_{fg}$, where $W =$ flow rate of heating steam, lb/h; $h_{fg} =$ enthalpy of vaporization of the heating steam, Btu/lb. Hence, $W = 1,350,000/843.0 = 1600$ lb/h (720 kg/h).

Related Calculations.   Use this general procedure to find the heat-transfer area, fluid outlet temperature, and required heating-fluid flow rate when true parallel flow or counterflow of the fluids occurs in the heat exchanger. When such a true flow does *not* exist, use a sitable correction factor, as shown in the next calculation procedure.

The procedure described here can be used for heat exchangers in power plants, heating systems, marine propulsion, air-conditioning systems, etc. Any heating or cooling fluid—steam, gas, chilled water, etc.—can be used.

To select a heat exchanger by using the results of this calculation procedure, enter the engineering data tables available from manufacturers at the computed heat-transfer area. Read the heater dimensions directly from the table. Be sure to use the next *larger* heat-transfer area when the exact required area is not available.

When there is little movement of the fluid on either side of the heat-transfer area, such as occurs during heat transmission through a building wall, the arithmetic mean (average) temperature difference can be used instead of the LMTD. Use the LMTD when there is rapid movement of the fluids on either side of the heat-transfer area and a rapid change in temperature in one, or both, fluids. When one of the two fluids is partially, but not totally, evaporated or condensed, the true mean temperature difference is different from the arithmetic mean and the LMTD. Special methods, such as those presented in Perry—*Chemical Engineers' Handbook,* must be used to compute the actual temperature difference under these conditions.

When two liquids or gases with constant specific heats are exchanging heat in a heat exchanger, the area between their temperature curves, Fig. 2, is a measure of the total heat being transferred. Figure 2 shows how the temperature curves vary with the amount of heat-transfer area for counterflow and parallel-flow exchangers when the fluid inlet temperatures are kept constant. As Fig. 2 shows, the counterflow arrangement is superior.

If enough heating surface is provided, in a counterflow exchanger, the leaving cold-fluid temperature can be raised above the leaving hot-fluid temperature. This cannot be done in a parallel-flow exchanger, where the temperatures can only approach each other regardless of how much surface is used. The counterflow arrangement transfers more heat for given conditions and usually proves more economical to use.

HEAT-EXCHANGER ACTUAL TEMPERATURE DIFFERENCE

A counterflow shell-and-tube heat exchanger has one shell pass for the heating fluid and two shell passes for the fluid being heated. What is the actual LMTD for this

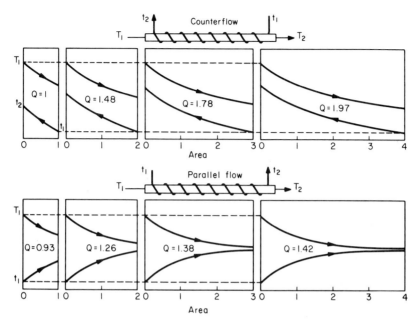

FIGURE 2 For certain conditions, the area between the temperature curves measures the amount of heat being transferred.

exchanger if $T_1 = 300°F$ (148.9°C), $T_2 = 250°F$ (121°C), $t_1 = 100°F$ (37.8°C), and $t_2 = 230°F$ (110°C)?

Calculation Procedure:

1. Determine how the LMTD should be computed
When the numbers of shell and tube passes are unequal, true counterflow does not exist in the heat exchanger. To allow for this deviation from true counterflow, a correction factor must be applied to the logarithmic mean temperature difference (LMTD). Figure 3 gives the correction factor to use.

2. Compute the variables for the correction factor
The two variables that determine the correction factor are shown in Fig. 3 as $P = (t_2 - t_1)/(T_1 - t_1)$ and $R = (T_1 - T_2)/(t_2 - t_1)$. Thus, $P = (230 - 100)/(300 - 100) = 0.65$, and $R = (300 - 250)/(230 - 100) = 0.385$. From Fig. 3, the correction factor is $F = 0.90$ for these values of P and R.

3. Compute the theoretical LMTD
Use the relation LMTD $= (G - L)/\ln(G/L)$, where the symbols for counterflow heat exchange are $G = T_2 - t_1$; $L = T_1 - t_2$; $\ln$ = logarithm to the base e. All temperatures in this equation are expressed in °F. Thus, $G = 250 - 100 = 150°F$ (83.3°C); $L = 300 - 230 = 70°F$ (38.9°C). Then LMTD $= (150 - 70)/\ln (150/70) = 105°F$ (58.3°C).

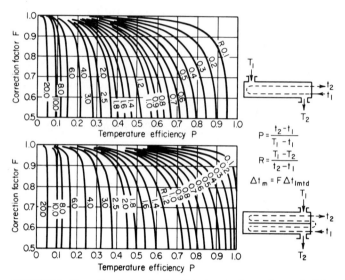

FIGURE 3 Correction factors for LMTD when the heater flow path differs from the counterflow. (*Power.*)

4. Compute the actual LMTD for this exchanger

The actual LMTD for this or any other heat exchanger is $\text{LMTD}_{\text{actual}} = F(\text{LMTD}_{\text{computed}}) = 0.9(105) = 94.5°F\ (52.5°C)$. Use the actual LMTD to compute the required exchanger heat-transfer area.

 Related Calculations. Once the corrected LMTD is known, compute the required heat-exchanger size in the manner shown in the previous calculation procedure. The method given here is valid for both two- and four-pass shell-and-tube heat exchangers. Figure 4 simplifies the computation of the uncorrected LMTD for temperature differences ranging from 1 to 1000°F (-17 to 537.8°C). It gives LMTD with sufficient accuracy for all normal industrial and commercial heat-exchanger applications. Correction-factor charts for three shell passes, six or more tube passes, four shell passes, and eight or more tube passes are published in the *Standards of the Tubular Exchanger Manufacturers Association.*

FOULING FACTORS IN HEAT-EXCHANGER SIZING AND SELECTION

A heat exchanger having an overall coefficient of heat transfer of $U = 100$ Btu/ $(\text{ft}^2 \cdot \text{h} \cdot °\text{F})$ [567.8 W/$(\text{m}^2 \cdot °\text{C})$] is used to cool lean oil. What effect will the tube fouling have on the value of U for this exchanger?

Calculation Procedure:

1. Determine the heat exchange fouling factor

Use Table 2 to determine the fouling factor for this exchanger. Thus, the fouling factor for lean oil = 0.0020.

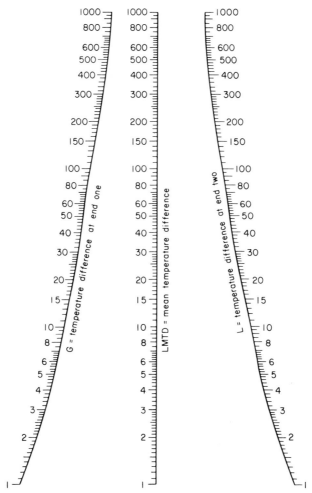

FIGURE 4 Logarithmic mean temperature for a variety of heat-transfer applications.

2. Determine the actual U for the heat exchanger

Enter Fig. 5 at the bottom with the clean heat-transfer coefficient of $U = 100$ Btu/(h · ft² · °F) [567.8 W/(m² · °C)] and project vertically upward to the 0.002 fouling-factor curve. From the intersection with this curve, project horizontally to the left to read the design or actual heat-transfer coefficient as $U_a = 78$ Btu/(h · ft² · °F) [442.9 W/(m² · °C)]. Thus, the fouling of the tubes causes a reduction of the U value of $100 - 78 = 22$ Btu/(h · ft² · °F) [124.9 W/(m² · °C)]. This means that the required heat transfer area must be increased by nearly 25 percent to compensate for the reduction in heat transfer caused by fouling.

Related Calculations. Table 2 gives fouling factors for a wide variety of service conditions in applications of many types. Use these factors as described above; or add the fouling factor to the film resistance for the heat exchanger to obtain the

TABLE 2 Heat-Exchanger Fouling Factors*

Fluid heated or cooled	Fouling factor
Fuel oil	0.0055
Lean oil	0.0020
Clean recirculated oil	0.0010
Quench oils	0.0042
Refrigerants (liquid)	0.0011
Gasoline	0.0006
Steam-clean and oil-free	0.0001
Refrigerant vapors	0.0023
Diesel exhaust	0.013
Compressed air	0.0022
Clean air	0.0011
Seawater under 130°F (54°C)	0.0006
Seawater over 130°F (54°C)	0.0011
City or well water under 130°F (54°C)	0.0011
City or well water over 130°F (54°C)	0.0021
Treated boiler feedwater under 130°F, 3 ft/s (54°C, 0.9 m/s)	0.0008
Treated boiler feedwater over 130°F, 3 ft/s (54°C, 0.9 m/s)	0.0009
Boiler blowdown	0.0022

*Condenser Service and Engineering Company, Inc.

total resistance to heat transfer. Then U = the reciprocal of the total resistance. Use the actual value U_a of the heat-transfer coefficient when sizing a heat exchanger. The method given here is that used by Condenser Service and Engineering Company, Inc.

HEAT TRANSFER IN BAROMETRIC AND JET CONDENSERS

A counterflow barometric condenser must maintain an exhaust pressure of 2 lb/in² (abs) (13.8 kPa) for an industrial process. What condensing-water flow rate is required with a cooling-water inlet temperature of 60°F (15.6°C); of 80°F (26.7°C)? How much air must be removed from this barometric condenser if the steam flow rate is 25,000 lb/h (11,250 kg/h); 250,000 lb/h (112,500 kg/h)?

Calculation Procedure:

1. Compute the required unit cooling-water flow rate
Use Fig. 6 as a quick guide to the required cooling-water flow rate for counterflow barometric condensers. Thus, entering the bottom of Fig. 6 at 2-lb/in² (abs) (13.8-kPa) exhaust pressure and projecting vertically upward to the 60°F (15.6°C) and 80°F (26.7°C) cooling-water inlet temperature curves show that the required flow rate is 52 gal/min (3.2 L/s) and 120 gal/min (7.6 L/s), respectively, per 1000 lb /h (450 kg/h) of steam condensed.

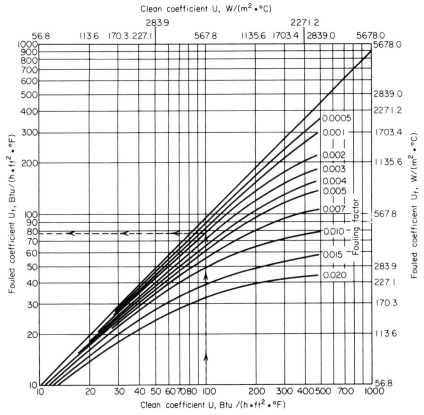

FIGURE 5   Effect of heat-exchanger fouling on the overall coefficient of heat transfer. (*Condenser Service and Engineering Co., Inc.*)

2.   Compute the total cooling-water flow rate required

Use this relation: total cooling water required, gal/min = (unit cooling-water flow rate, gal/min per 1000 lb/h of steam condensed) (steam flow, lb/h)/1000. Or, total gpm = (52)(250,000/1000) = 13,000 gal/min (820.2 L/s) of 60°F (15.6°C) cooling water. For 80°F (26.7°C) cooling water, total gal/min = (120)(250,000/1000) = 30,000 gal/min (1892.7 L/s). Thus, a 20°F (11.1°C) rise in the cooling-water temperature raises the flow rate required by 30,000 − 13,000 = 17,000 gal/min (1072.5 L/s).

3.   Compute the quantity of air that must be handled

With a steam flow of 25,000 lb/h (11,250 kg/h) to a barometric condenser, manufacturers' engineering data show that the quantity of air entering with the steam is 3 ft³/min (0.08 m³/min); with a steam flow of 250,000 lb/h (112,500 kg/h), air enters at the rate of 10 ft³/min (0.28 m³/min). Hence, the quantity of air in the steam that must be handled by this condenser is 10 ft³/min (0.28 m³/min).

Air entering with the cooling water varies from about 2 ft³/min per 1000 gal/min of 100°F (0.06 m³/min per 3785 L/min of 37.8°C) water to 4 ft³/min per 1000

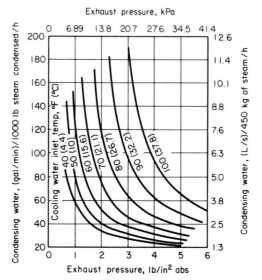

FIGURE 6 Barometric condenser condensing-water flow rate.

gal/min at 35°F (0.11 m³/min per 3785 L/min at 1.7°C). Using a value of 3 ft³/min (0.08 m³/min) for this condenser, we see the quantity of air that must be handled is (ft³/min per 1000 gal/min)(cooling-water flow rate, gal/min)(1000, or cfm of air = (3)(13,000/1000) = 39 ft³/min at 60°F (1.1 m³/min at 15.6°C). At 80°F (26.7°C) ft³/min = (3)(30,000/1000) = 90 ft³/min (2.6 m³/min).

Hence, the total air quantity that must be handled is 39 + 10 = 49 ft³/min (1.4 m³/min) with 60°F (15.6°C) cooling water, and 90 + 10 = 100 ft³/min (2.8 m³/min) with 80°F (26.7°C) cooling water. The air is usually removed from the barometric condenser by a two-stage air ejector.

Related Calculations. For help in specifying conditions for parallel-flow and counterflow barometric condensers, refer to *Standards of Heat Exchange Institute—Barometric and Low-Level Jet Condensers.* Whereas Fig. 6 can be used for a first approximation of the cooling water required for parallel-flow barometric condensers, the results obtained will not be as accurate as for counterflow condensers.

SELECTION OF A FINNED-TUBE HEAT EXCHANGER

Choose a finned-tube heat exchanger for a 1000-hp (746-kW) four-cycle turbocharged diesel engine having oil-cooled pistons and a cooled exhaust manifold. The heat exchanger will be used only for jacket-water cooling.

Calculation Procedure:

1. *Determine the heat-exchanger cooling load*

The Diesel Engine Manufacturers Association (DEMA) tabulation, Table 3, lists the heat rejection to the cooling system by various types of diesel engines. Table 3 shows that the heat rejection from the jacket water of a four-cycle turbocharged-engine having oil-cooled pistons and a cooled manifold is 1800 to 2200 Btu/(bhp·h) (0.71 to 0.86 kW/kW). Using the higher value, we see the jacket-water heat rejection by this engine is (1000 bhp)[2200 Btu/(bhp·h)] = 2,200,000 Btu/h (644.8 kW).

2. *Determine the jacket-water temperature rise*

DEMA reports that a water temperature rise of 15 to 20°F (8.3 to 11.1°C) is common during passage of the cooling water through the engine. The maximum water discharge temperature reported by DEMA ranges from 140 to 180°F (60 to 82.2°C).

TABLE 3 Approximate Rates of Heat Rejection to Cooling Systems*

	Four-cycle engines			
Engine type	Normally aspirated, dry pistons, water-jacketed exhaust manifold, Btu/(bhp·hr) (kJ/kWh)	Normally aspirated, oil-cooled pistons, water-jacketed manifold, Btu/(bhp·h) (kJ/kWh)	Turbocharged, oil-cooled pistons, dry manifold, Btu/(bhp·h) (kJ/kWh)	Turbocharged, oil-cooled pistons, cooled manifold, Btu/(bhp·h) (kJ/kWh)
Jacket water	2200–2600 (12.5–14.8)	2000–2500 (11.3–14.2)	1450–1750 (8.2–9.9)	1800–2200 (10.2–12.5)
Lubricating oil	175–350 (1.0–2.0)	300–600 (1.7–3.4)	300–500 (1.7–2.8)	300–500 (1.7–2.8)
Raw water	2375–2950 (13.5–16.7)	2300–3100 (13.1–17.6)	1750–2250 (9.9–12.8)	2100–2700 (11.9–15.3)

	Two-cycle engines		
	Loop scavenging oil-cooled pistons, Btu/(bhp·h) (kJ/kWh)	Uniflow scavenging oil-cooled pistons	
Engine type		Opposed piston, Btu/(bhp·h) (kJ/kWh)	Valve in head, Btu/(bhp·h) (kJ/kWh)
Jacket water	1300–1900 (7.4–10.8)	1200–1600 (6.8–9.1)	1700–2100 (9.6–11.9)
Lubricating oil	500–700 (2.8–4.0)	900–1100 (5.1–6.2)	400–750 (2.3–4.3)
Raw water	1800–2600 (10.2–14.8)	2100–2700 (11.9–15.3)	2100–2850 (11.9–16.2)

*Diesel Engine Manufacturers Association; SI values added by handbook editor.

Assume a 20°F (11.1°C) water temperature rise and a 160°F (71.1°C) water discharge temperature for this engine.

3. Determine the air inlet and outlet temperatures
Refer to weather data for the locality of the engine installation. Assume that the weather data for the locality of this engine show that the maximum dry-bulb temperature met in summer is 90°F (32.2°C). Use this as the air inlet temperature.
    Before the required surface area can be determined, the air outlet temperature from the radiator must be known. This outlet temperature cannot be computed directly. Hence, it must be assumed and a trial calculation made. If the area obtained is too large, a higher outlet air temperature must be assumed and the calculation redone. Assume an outlet air temperature of 150°F (65.6°C).

4. Compute the LMTD for the radiator
The largest temperature difference for this exchanger is $160 - 90 = 70°F$ (38.9°C), and the smallest temperature difference is $150 - 140 = 10°F$ (5.6°C). In the smallest temperature difference expression, 140°F (77.8°C) = water discharge temperature from the engine − cooling-water temperature rise during passage through the engine, or $160 - 20 = 140°F$ (77.8°C). Then LMTD $= (70 - 10)/[\ln(70/10)] = 30°F$ (16.7°C). (Figure 4 could also be used to compute the LMTD).

5. Compute the required exchanger surface area
Use the relation $A = Q/U \times \text{LMTD}$, where A = surface area required, ft^2; Q = rate of heat transfer, Btu/h; U = overall coefficient of heat transfer, Btu/(h · ft² · °F). To solve this equation, U must be known.
    Table 1 in the first calculation procedure in this section shows that U ranges from 2 to 10 Btu/(h · ft² · °F) [56.8 W/(m² · °C)] in the usual internal-combustion-engine finned-tube radiator. Using a value of 5 for U, we get $A = 2,200,000/[(5)(30)] = 14,650 \text{ ft}^2$ (1361.0 m²).

6. Determine the length of finned tubing required
The total area of a finned tube is the sum of the tube and fin area per unit length. The tube area is a function of the tube diameter, whereas the finned area is a function of the number of fins per inch of tube length and the tube diameter.
    Assume that 1-in (2.5-cm) tubes having 4 fins per inch (6.35 mm per fin) are used in this radiator. A tube manufacturer's engineering data show that a finned tube of these dimensions has 5.8 ft² of area per linear foot (1.8 m²/lin m) of tube.
    To compute the linear feet L of finned tubing required, use the relation $L = A/(\text{ft}^2/\text{ft})$, or $L = 14,650/5.8 = 2530$ lin ft (771.1 m) of tubing.

7. Compute the number of individual tubes required
Assume a length for the radiator tubes. Typical lengths range between 4 and 20 ft (1.2 and 6.1 m), depending on the size of the radiator. With a length of 16 ft (4.9 m) per tube, the total number of tubes required $= 2530/16 = 158$ tubes. This number is typical for finned-tube heat exchangers having large heat-transfer rates [more than 10^6 Btu/h (100 kW)].

8. Determine the fan hp required
The fan hp required can be computed by determining the quantity of air that must be moved through the heat exchanger, after assuming a resistance—say 1.0 in of water (0.025 Pa)—for the exchanger. However, the more common way of determining the fan hp is by referring to the manufacturer's engineering data.

Thus, one manufacturer recommends three 5-hp (3.7-kW) fans for this cooling load, and another recommends two 8-hp (5.9-kW) fans. Hence, about 16 hp (11.9 kW) is required for the radiator.

Related Calculations. The steps given here are suitable for the initial sizing of finned-tube heat exchangers for a variety of applications. For exact sizing, it may be necessary to apply a correction factor to the LMTD. These correction factors are published in Kern—*Process Heat Transfer,* McGraw-Hill, and McAdams—*Heat Transfer,* McGraw-Hill.

The method presented here can be used for finned-tube heat exchangers used for air heating or cooling, gas heating or cooling, and similar industrial and commercial applications.

SPIRAL-TYPE HEATING COIL SELECTION

How many feet of heating coil are required to heat 1000 gal/h (1.1 L/s) of 0.85-specific-gravity oil if the specific heat of the oil is 0.50 Btu/(lb·°F) [2.1 kJ/(kg·°C)], the heating medium is 65-lb/in² (gage) (448.2-kPa) steam, and the oil enters at 60°F (15.6°C) and leaves at 125°F (51.7°C)? There is no subcooling of the condensate.

Calculation Procedure:

1. Compute the LMTD for the heater
Steam at $65 + 14.7 = 79.7$ lb/in² (abs) (549.5 kPa) has a temperature of approximately 312°F (155.6°C), as given by the steam tables. Condensate at this pressure has the same approximate temperature. Hence, the entering and leaving temperatures of the heating fluid are approximately the same.

Oil enters the heater at 60°F (15.6°C) and leaves at 125°F (51.7°C). Therefore, the greater temperature G across the heater is $G = 312 - 60 = 252°F$ (140.0°C), and the lesser temperature difference L is $L = 312 - 125 = 187°F$ (103.9°C). Hence, the LMTD $= (G - L)/[\ln(G/L)]$, or $(252 - 187)/[\ln(252/187)] = 222°F$ (123.3°C). In this relation, $\ln$ = logarithm to the base $e = 2.7183$. (Figure 4 could also be used to determine the LMTD.)

2. Compute the heat required to raise the oil temperature
Water weighs 8.33 lb/gal (1.0 kg/L). Since this oil has a specific gravity of 0.85, it weighs $(8.33)(0.85) = 7.08$ lb/gal (0.85 kg/L). With 1000 gal/h (1.1 L/s) of oil to be heated, the weight of oil heated is $(1000 \text{ gal/h})(7.08 \text{ lb/gal}) = 7080$ lb/h (0.89 kg/s). Since the oil has a specific heat of 0.5 Btu/(lb·°F) [2.1 kJ/(kg·°C)] and this oil is heated through a temperature range of $125 - 60 = 65°F$ (36.1°C), the quantity of heat Q required to raise the temperature of the oil is $Q = (7080$ lb/h) [0.5 Btu/(lb·°F) (65°F)] $= 230,000$ Btu/h (67.4 kW).

3. Compute the heat-transfer area required
Use the relation $A = Q/(U \times$ LMTD), where Q = heat-transfer rate, Btu/h; U = overall coefficient of heat transfer, Btu/(h·ft²·°F). For heating oil to 125°F (51.7°C), the U value given in Table 1 is 20 to 60 Btu/(h·ft²·°F) [0.11 to 0.34 kW/(m²·°C)]. Using a value of $U = 30$ Btu/(h·ft²·°F) [0.17 kW/(m²·°C)] to

produce a conservatively sized heater, we find $A = 230,000/[(30)(222)] = 33.4$ ft² (3.1 m²) of heating surface.

4. Choose the coil material for the heater

Spiral-type tank heating coils are usually made of steel because this material has a good corrosion resistance in oil. Hence, this coil will be assumed to be made of steel.

5. Compute the heating steam flow required

To determine the steam flow rate required, use the relation $S = Q/h_{fg}$, where $S =$ steam flow, lb/h; h_{fg} = latent heat of vaporization of the heating steam, Btu/lb, from the steam tables; other symbols as before. Hence, $S = 230,000/901.1 = 256$ lb/h (0.03 kg/s), closely.

6. Compute the heating coil pipe diameter

Steam-heating coils submerged in the liquid being heated are usually chosen for a steam velocity of 4000 to 5000 ft/min (20.3 to 25.4 m/s). Compute the heating pipe cross-sectional area a in² from $a = 2.4Sv_g/V$, where v_g = specific volume of the steam at the coil operating pressure, ft³/lb, from the steam tables; $V =$ steam velocity in the heating coil, ft/min; other symbols as before. With a steam velocity of 4000 ft/min (20.3 m/s), $a = 2.4(256)(5.47)/4000 = 0.838$ in² (5.4 cm²).

Refer to a tabulation of pipe properties. Such a tabulation shows that the internal transverse area of a schedule 40 1-in (2.5-cm) diameter nominal steel pipe is 0.863 in² (5.6 cm²). Hence, a 1-in (2.5-cm) pipe will be suitable for this heating coil.

7. Determine the length of coil required

A pipe property tabulation shows that 2.9 lin ft (0.9 m) of 1-in (2.5-cm) schedule 40 pipe has 1.0 ft² (0.09 m²) of external area. Hence, the total length of pipe required in this heating coil = (33.1 ft²)(2.9 ft/ft²) = 96 ft (29.3 m).

Related Calculations. Use this general procedure to find the area and length of spiral heating coil required to heat water, industrial solutions, oils, etc. This procedure also can be used to find the area and length of cooling coils used to cool brine, oils, alcohol, wine, etc. In every case, be certain to substitute the correct specific heat for the liquid being heated or cooled. For typical values of U, consult Perry—*Chemical Engineers' Handbook,* McGraw-Hill; McAdams—*Heat Transmission,* McGraw-Hill; or Kern—*Process Heat Transfer,* McGraw-Hill.

SIZING ELECTRIC HEATERS FOR INDUSTRIAL USE

Choose the heating capacity of an electric heater to heat a pot containing 600 lb (272.2 kg) of lead from the charging temperature of 70°F (21.1°C) to a temperature of 750°F (398.9°C) if 600 lb (272.2 kg) of the lead is to be melted and heated per hour. The pot is 30 in (76.2 cm) in diameter and 18 in (45.7 cm) deep.

Calculation Procedure:

1. Compute the heat needed to reach the melting point

When a solid is melted, first it must be raised from its ambient or room temperature to the melting temperature. The quantity of heat required is $H =$ (weight of solid, lb)[specific heat of solid, Btu/(lb·°F)]($t_m - t_i$), where H = Btu required to raise the temperature of the solid, °F; t_1 = room, charging, or initial temperature of the solid, °F; t_m = melting temperature of the solid, °F.

For this pot with lead having a melting temperature of 620°F (326.7°C) and an average specific heat of 0.031 Btu/(lb·°F) [0.13 kJ/(kg·°C)], H = (600)(0.031)(620 − 70) = 10,240 Btu/h (3.0 kW), or (10,240 Btu/h)/(3412 Btu/kWh) = 2.98 kWh.

2. Compute the heat required to melt the solid
The heat H_m Btu required to melt a solid is H_m = (weight of solid melted, lb)(heat of fusion of the solid, Btu/lb). Since the heat of fusion of lead is 10 Btu/lb (23.2 kJ/kg), H_m = (600)(10) = 6000 Btu/h, or 6000/3412 = 1.752 kWh.

3. Compute the heat required to reach the working temperature
Use the same relation as in step 1, except that the temperature range is expressed as $t_w - t_m$, where t_w = working temperature of the melted solid. Thus, for this pot, H = (600)(0.031)(750 − 620) = 2420 Btu/h (709.3 W), or 2420/3412 = 0.709 kWh.

4. Determine the heat loss from the pot
Use Fig. 7 to determine the heat loss from the pot. Enter at the bottom of Fig. 7 at 750°F (398.9°C), and project vertically upward to the 10-in (25.4-cm) diameter pot curve. At the left, read the heat loss at 7.3 kWh/h.

5. Compute the total heating capacity required
The total heating capacity required is the sum of the individual capacities, or 2.98 + 1.752 + 0.708 + 7.30 = 12.74 kWh. A 15-kW electric heater would be chosen because this is a standard size and it provides a moderate extra capacity for overloads.

Related Calculations. Use this general procedure to compute the capacity required for an electric heater used to melt a solid of any kind—lead, tin, type metal, solder, etc. When the substance being heated is a liquid—water, dye, paint, varnish, oil, etc.—use the relation H = (weight of liquid heated, lb) [specific heat of liquid, Btu/(lb·°F)] (temperature rise desired, °F), when the liquid is heated to approximately its boiling temperature, or a lower temperature.

For space heating of commercial and residential buildings, two methods used for computing the approximate wattage required are the W/ft³ and the "35" method. These are summarized in Table 4. In many cases, the results given by these methods agree closely with more involved calculations. When the desired room temperature is different from 70°F (21.1°C), increase or decrease the required kilowatt capacity proportionately, depending on whether the desired temperature is higher than or lower than 70°F (21.1°C).

For heating pipes with electric heaters, use a heater capacity of 0.8 W/ft² (8.6 W/m²) of uninsulated exterior pipe surface per °F temperature difference between the pipe and the surrounding air. If the pipe is insulated with 1 in (2.5 cm) of insulation, use 30 percent of this value, or 0.24 (W/(ft²·°F) [4.7 W/(m²·°C)].

The types of electric heaters used today include immersion (for water, oil, plating, liquids, etc.), strip, cartridge, tubular, vane, fin, unit, and edgewound resistor heaters. These heaters are used in a wide variety of applications including liquid heating, gas and air heating, oven warming, deicing, humidifying, plastics heating, pipe heating, etc.

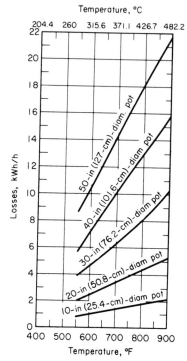

FIGURE 7   Heat losses from melting pots. (*General Electric Co.*)

TABLE 4   Two Methods for Determining Wattage for Heating Buildings Electrically*

	W/ft³ method	W/m³ method
1. Interior rooms with no or little outside exposure	0.75 to 1.25	25.6 to 44.1
2. Average rooms with moderate windows and doors	1.25 to 1.75	44.1 to 61.8
3. Rooms with severe exposure and great window and door space	1.0 to 4.0	35.3 to 141.3
4. Isolated rooms, cabins, watchhouses, and similar · buildings	3.0 to 6.0	105.9 to 211.9

	The "35" method
1. Volume in ft³ for one air change × 0.35 =	0.01 W
2. Exposed net wall, roof, or ceiling and floor in ft² × 3.5 =	0.1 W
3. Area of exposed glass and doors in ft² × 35.0 =	1 W

*General Electric Company.

For pipe heating, a tubular heating element can be fastened to the bottom of the pipe and run parallel with it. For large-wattage applications, the heater can be spiraled around the pipe. For temperatures below 165°F (73.9°C), heating cable can be used. Electric heating is often used in place of steam tracing of outdoor pipes.

The procedure presented above is the work of General Electric Company.

ECONOMIZER HEAT TRANSFER COEFFICIENT

A 4530-ft² (421-m²) heating surface counterflow economizer is used in conjunction with a 150,000-lb/h (68,040-kg/h) boiler. The inlet and outlet water temperatures are 210°F (99°C) and 310°F (154°C). The inlet and outlet gas temperatures are 640°F (338°C) and 375°F (191°C). Find the overall heat transfer coefficient in Btu/(h · ft² · °F) [W/(m² · °C)] [kJ/(h · m² · °C)].

Calculation Procedure:

1. Determine the enthalpy of water at the inlet and outlet temperatures
From Table 1, Saturation: Temperatures, of the Steam Tables mentioned under Related Calculations of this procedure, for water at inlet temperature, $t_1 = 210°F$ (99°C), the enthalpy, $h_1 = 178.14$ Btu/lb (414 kJ/kg), and at the outlet temperature, $t_2 = 310°F$ (154°C), the enthalpy, $h_2 = 279.81$ Btu/lb$_m$ (651 kJ/kg).

2. Compute the logarithmic mean temperature difference between the gas and water
As shown in Fig. 8, the temperature difference of the gas entering and the water leaving, $\Delta t_a = t_3 - t_2 = 640 - 310 = 330°F$ (166°C) and for the gas leaving and the water entering, $\Delta t_b = t_4 - t_1 = 375 - 210 = 165°F$ (74°C). Then, the logarithmic mean temperature difference, $\Delta t_m = (\Delta t_a - \Delta t_b)/[2.3 \times \log_{10}(\Delta t_a/\Delta t_b)] = (330 - 165)/[2.3 \times \log_{10}(330/165)] = 238°F$ (115°C).

3. Compute the economizer heat transfer coefficient
All the heat lost by the gas is considered to be transferred to the water, hence the heat lost by the gas, $Q = w(h_2 - h_1) = UA \Delta t_m$, where the water rate of flow, $w = 150,000$ lb/h (68,000 kg/h); U is the overall heat transfer coefficient; heating surface area, $A = 4530$ ft² (421 m²); other values as before. Then, 150,000 ×

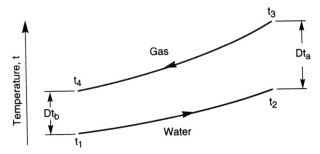

FIGURE 8   Temperature vs surface area of economizer.

(279.81 − 178.41) = $U(4530)(238)$. Solving U = [150,000 × (279.81 − 178.14)]/(4530 × 238) = 14.1 Btu/(h·ft²·°F) [80 W/(m²·°C)] [288 kJ/(h·m²·°C)].

Related Calculations.   The Steam Tables appear in *Thermodynamic Properties of Water Including Vapor, Liquid, and Solid Phases,* 1969, Keenan, et al., John Wiley & Sons, Inc. Use later versions of such tables whenever available, as necessary.

BOILER TUBE STEAM-GENERATING CAPACITY

A counterflow bank of boiler tubes has a total area of 900 ft² (83.6 m²) and its overall coefficient of heat transfer is 13 Btu/(h·ft²·°F) [73.8 W/(m²·K). The boiler tubes generate steam at a pressure of 1000 lb/in² absolute (6900 kPa). The tube bank is heated by flue gas which enters at a temperature of 2000°F (1367 K) and at a rate of 450,000 lb/h (56.7 kg/s). Assume an average specific heat of 0.25 Btu/(lb·°F) [1.05 kJ/(kg·K)] for the gas and calculate the temperature of the gas that leaves the bank of boiler tubes. Also, calculate the rate at which the steam is being generated in the tube bank.

Calculation Procedure:

1.   Find the temperature of steam at 1000 lf$_f$/in² (6900 kPa)
From Table 2, Saturation: Pressures, of the Steam Tables mentioned under Related Calculations of this procedure, the saturation temperature of steam at 1000 lb/in² (6900 kPa), t_s = 544.6°F (558 K), a constant value as indicated in Fig. 9.

2.   Determine the logarithmic mean temperature difference in terms of the flue-gas leaving temperature
The logarithmic mean temperature difference, $\Delta t_m = (\Delta t_1 - \Delta t_2)/\{2.3 \times \log_{10}[(t_1 - t_s)/(t_2 - t_s)]\}$, where Δt_1 = flue gas entering temperature = steam temperature =

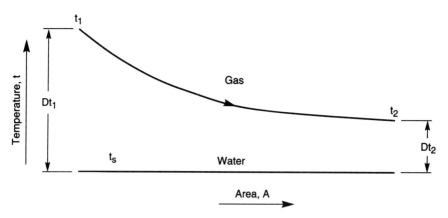

FIGURE 9   Temperature vs surface area of boiler tubes.

$(t_1 - t_s) = (2000 - 544.6)$; $\Delta t_2 =$ flue-gas leaving temperature − steam temperature $= (t_2 - t_s) = (t_2 - 544.6)$; $(\Delta t_1 - \Delta t_2) = [(2000 - 544.6) - (t_2 - 544.6)] = (2000 - t_2)$; $[(t_1 - t_s)/(t_2 - t_s)] = [(2000 - 544.6)/(t_2 - 544.6)]$. Hence, $\Delta t_m = [(2000 - t_2)/\{2.3 \times \log_{10} [(1455.4)/(t_2 - 544.6)]\}$.

3. Compute the flue-gas leaving temperature

Heat transferred to the boiler water, $Q = w_g \times c_p \times (t_1 - t_2) = UA\Delta t_m$, where the flow rate of flue gas, $w_g = 450,000$ lb/h (56.7 kg/s); flue-gas average specific heat, $c_p = 0.25$ Btu/(lb/°F) [1.05 kg/(kg·K)]; overall coefficient of heat transfer of the boiler tubes, $U = 13$ Btu/(h·ft²·°F) [73.8 W/(m²·K)]; area of the boiler tubes exposed to heat, $A = 900$ ft² (83.6 m²); other values as before.

Then, $Q = 450,000 \times 0.25 \times (2000 - t_2) = 13 \times 900 \times [(2000 - t_2)/\{2.3 \times \log_{10} [(1455.4)/(t_2 - 544.6)]\}$. Or, $\log_{10} [(1455.4/(t_2 - 544.6)] = 13 \times 900/(2.3 \times 450,000 \times 0.25) = 0.0452$. The antilog of $0.0452 = 1.11$, hence, $[(1455.4/(t_2 - 544.7)] = 1.11$, and $t_2 = (1455.4/1.11) + 544.6 = 1850°F$ (1280 K).

4. Find the heat of vaporization of the water

From the Steam Tables, the heat of vaporization of the water at 1000 lb/in² (6900 kPa), $h_{fg} = 649.5$ Btu/lb (1511 kJ/kg).

5. Compute the steam-generating rate of the boiler tube bank

Heat absorbed by the water = heat transferred by the flue gas, or $Q = w_s \times h_{fg} = w_g \times c_p \times (t_1 - t_2)$, where the mass of steam generated is w_s in lb/h (kg/s); other values as before. Then, $w_s \times 649.5 = 450,000 \times 0.25 \times (2000 - 1850) = 16.9 \times 10^6$ Btu/h (4950 kJ/s) (4953 kW). Thus, $w_s = 16.9 \times 10^6/649.5 = 26,000$ lb/h (200 kg/s).

Related Calculations. The Steam Tables appear in *Thermodynamic Properties of Water Including Vapor, Liquid, and Solid Phases,* 1969, Keenan, et al., John Wiley & Sons, Inc. Use later versions of such tables whenever available, as required.

SHELL-AND-TUBE HEAT EXCHANGER DESIGN ANALYSIS

Determine the heat transferred, shellside outlet temperature, surface area, maximum number of tubes, and tubeside pressure drop for a liquid-to-liquid shell-and-tube heat exchanger such as that in Fig. 10, when the conditions below prevail. This exchanger will be of the single tube-pass and single shell-pass design, with countercurrent flows of the tubeside and shellside fluids.

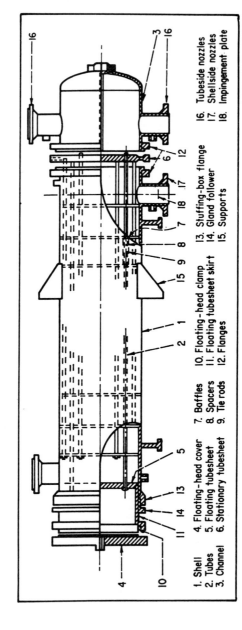

FIGURE 10 Components of shell-and-tube heat exchanger. This unit has an outside-packed stuffing box. (*Chemical Engineering.*)

1. Shell
2. Tubes
3. Channel
4. Floating-head cover
5. Floating tubesheet
6. Stationary tubesheet
7. Baffles
8. Spacers
9. Tie rods
10. Floating-head clamp
11. Floating tubesheet skirt
12. Flanges
13. Stuffing-box flange
14. Gland follower
15. Supports
16. Tubeside nozzles
17. Shellside nozzles
18. Impingement plate

Conditions	Tubeside	Shellside			
Flowrate, lb/hr	307,500	32,800	kg/h	139,605	14,891
Inlet temperature, °C	105	45	F	221	45
Outlet temperature, °C	unknown	90			
Viscosity, cp	1.7	0.3			
Specific heat, Btu/hr/°F	0.72	0.9	kJ/hr°C	3.0	3.7
Molecular weight	118	62			
Specific gravity with refercnce to water at 20°C (68°F)	0.85	0.95			
Allowable pressure drop, psi	10	10	kPa	68.9	68.9
Maximum tube length, ft	12	12	m	5.45	5.45
Minimum tube dia., in	5/8	5/8	mm	15.9	15.9
Material of construction	steel	($k = 26$)			

Calculation Procedure:

1. Determine the heat transferred in the heat exchanger

Use the relation, heat transferred, Btu/hr = (flow rate, lb/hr)(outlet temperature − inlet temperature)(liquid specific heat)(1.8 to convert from°C to°F). Substituting for this heat exchanger, we have, heat transferred = (32,800)(90 − 45)(0.9)(1.8) = 2,391,120 Btu/hr (2522.6 kJ/hr). This is the rate of heat transfer from the hot fluid to the cool fluid.

2. Find the shellside outlet temperature

The temperature decrease of the hot fluid = (rate of heat transfer)/(flow rate, lb/hr)(specific heat)(1.8 conversion factor). Or, temperature decrease = (2,391,120)/(307,500)(0.72)(1.8) = 6°C (10.8°F). Then, the shellside outlet temperature = 105 − 6 = 99°C (210.2°F). Then, the LMTD = (54 − 15)/ln (54/15) = 30.4°C (86.7°F) = ΔT_m.

3. Make a first trial calculation of the surface area of this exchanger

For a first-trial calculation, the approximate surface can be calculated using an assumed overall heat-transfer-coefficient, U, of 250 Btu/(hr) (ft^2) (°F) (44.1 W/m^2°C). The assumed value of U can be obtained from tabulations in texts and handbooks and is used only to estimate the approximate size for a first trial:

$$A = 2,391,000/(250 \times 30.4 \times 1.8) = 175 \text{ ft}^2 \ (16.3 \text{ m}^2)$$

Since the given conditions specify a maximum tube length of 12 ft and a minimum tube diameter of ⅝ in, the number of tubes required is:

$$n = 175 \ (12 \times 0.1636) = 89 \text{ tubes}$$

and the approximate shell diameter will be:

$$D_o = 1.75 \times 0.625 \times 89^{0.47} = 9 \text{ in } (228.6 \text{ cm})$$

With the exception of baffle spacing, all preliminary calculations have been made for the quantities to be substituted into the dimensional equations. For the first trial, we may start with a baffle spacing equal to about half the shell diameter. After calculating the shellside pressure drop, we may adjust the baffle spacing. Also, it is advisable to check the Reynolds number on the tubeside to confirm that the proper equations are being used.

4. Find the maximum number of tubes for this heat exchanger
To find the maximum number of tubes (n_{max}) in parallel that still permits flow in the turbulent region ($N_{Re} = 12,600$), a convenient relationship is $n_{max} = W_i/(2d_i Z_i)$. In this example, $n_{max} = 307.5/(2 \times 0.495 \times 1.7) = 183$. For any number of tubes less than 183 tubes in parallel, we are in the turbulent range and can use Eq. (1).

From Table 6, the appropriate expressions for rating are: Eq. (1) for tubeside, Eq. (11) for shellside, Eq. (18) for tube wall, and Eq. (19) for fouling. Eq. (21) and (25) respectively are used for tubeside and shellside pressure drops.

5. Compute the tubeside and shellside heat transfer
Using the equations from Table 6, Tubeside, Eq. (1):

$$\frac{\Delta T_i}{\Delta T_M} = 10.43 \left[\frac{1.7^{0.467} \times 118^{0.222}}{0.85^{0.89}} \right] \times \left[\frac{307.5^{0.2} \times 6}{30.4} \right] \left[\frac{0.495^{0.8}}{89^{0.2} \times 12} \right]$$

$$= 10.43 \times 4.27 \times 0.621 \times 0.0193 = 0.535$$

Shellside, Eq. (11):

$$\frac{\Delta T_o}{\Delta T_M} = 4.28 \left[\frac{0.3^{0.267} \times 62^{0.222}}{0.95^{0.89}} \right] \times \left[\frac{32.8^{0.4} \times 45}{30.4} \right] \left[\frac{1^{0.282} \times 5^{0.6}}{89^{0.718} \times 12} \right]$$

$$= 4.28 \times 1.89 \times 5.98 \times 0.00872 = 0.424$$

Tube Wall, Eq. (18):

$$\frac{\Delta T_w}{\Delta T_M} = 159 \left[\frac{0.72}{26} \right] \left[\frac{307.5 \times 6}{30.4} \right] \left[\frac{0.625 - 0.495}{89 \times 0.625 \times 12} \right]$$

$$= 159 \times 0.0277 \times 60.7 \times 0.000195 = 0.052$$

Fouling, Eq. (19):

$$\frac{\Delta T_s}{\Delta T_M} = 3,820 \left[\frac{0.72}{1,000} \right] \left[\frac{307.5 \times 6}{30.4} \right] \left[\frac{1}{89 \times 0.625 \times 12} \right]$$

$$= 3,820 \times 0.00072 \times 60.7 \times 0.00150 = 0.250$$

$$(\text{SOP})^* = 0.535 + 0.424 + 0.052 + 0.250 = 1.261$$

Because SOP is greater than 1, the assumed exchanger is inadequate. The surface area must be increased by adding tubes or increasing the tube length, or the performance must be improved by decreasing the baffle spacing. Since the maximum tube length is fixed by the conditions given, the alternatives are increasing the number of tubes and/or adjusting the baffle spacing. To estimate assumptions for the next trial, pressure drops are calculated.

---

*Sum of the Product—see *Related Calculations* for data.

TABLE 5 Design Features of Shell-and-Tube Heat Exchangers

Design Features	Fixed Tubesheet	Return Bend (U-Tube)	Outside-Packed Stuffing Box	Outside-Packed Lantern Ring	Pull-Through Bundle	Inside Split-Backing Ring
Is tube bundle removable?	No	Yes	Yes	Yes	Yes	Yes
Can spare bundles be used?	No	Yes	Yes	Yes	Yes	Yes
How is differential thermal expansion relieved?	Expansion joint in shell	Individual tubes free to expand	Floating head	Floating head	Floating head	Floating head
Can individual tubes be replaced?	Yes	Only those in outside rows without special designs	Yes	Yes	Yes	Yes
Can tubes be chemically cleaned, both inside and outside?	Yes	Yes	Yes	Yes	Yes	Yes
Can tubes be physically cleaned on inside?	Yes	With special tools	Yes	Yes	Yes	Yes
Can tubes be physically cleaned on outside?	No	With square or wide triangular pitch	With square or wide triangular pitch	With square or wide triangular pitch	With square or wide triangular pitch	With square or wide triangular pitch
Are internal gaskets and bolting required?	No	No	No	No	Yes	Yes
Are double tubesheets practical?	Yes	Yes	Yes	No	No	No
What number of tubeside passes are available?	Number limited by number of tubes	Number limited by number of U-tubes	Number limited by number of tubes	One or two	Number limited by number of tubes. Odd number of passes requires packed joint or expansion joint	Number limited by number of tubes. Odd number of passes requires packed joint or expansion joint
Relative cost in ascending order, least expensive = 1	2	1	4	3	5	6

TABLE 6 Empirical Heat-Transfer Relationships for Rating Heat Exchanger

Eq. No.	Mechanism or Restriction	Empirical Equation	Numerical Factor	Physical-Property Factor	Work Factor	Mechanical-Design Factor	Eq. No.
	Inside the tubes						
(1)	No phase change (liquid), $N_{Re} > 10,000$	$\dfrac{h}{cG} = 0.023\,(N_{Re})^{-0.2}\,(N_{Pr})^{-2/3}$	$\Delta T_f / \Delta T_M = 10.43$	$\times\ \dfrac{(Z_i^{0.467}M_i^{0.22})}{s_i^{0.89}}$	$\times\ \dfrac{W^{0.2}(t_H - t_L)}{\Delta T_M}$	$\times\ \dfrac{d_i^{0.8}}{n^{0.2}L}$	(1)
(2)	No phase change (gas), $N_{Re} > 10,000$	$h = 0.0144\,G^{0.8}(D_i)^{-0.2}c_p$	$\Delta T_f / \Delta T_M = 9.87$	$s_i^{0.89}$	$\times\ \dfrac{W^{0.2}(t_H - t_L)}{\Delta T_M}$	$\times\ \dfrac{d_i^{0.8}}{n^{0.2}L}$	(2)
(3)	No phase change (gas), $2,100 < N_{Re} < 10,000$	$h = 0.0059[(N_{Re})^{2/3} - 125][1 + (D/L)^{2/3}]\,(c_p/D_i)\,(\mu_f/\mu_b)^{-0.14}$	$\Delta T_f / \Delta T_M = 44,700$	$\times\ (Z_f/Z_b)^{0.14}$	$\times\ \dfrac{W(t_H - t_L)}{\Delta T_M}$	$\times\ \dfrac{1}{[(N_{Re})^{2/3} - 125][1 + (d_i N_{PT}/12L)^{2/3}]nL}$	(3)
(4)	No phase change (liquid), $2,100 < N_{Re} < 10,000$	$\dfrac{h}{cG} = 0.166\left[\dfrac{(N_{Re})^{2/3} - 125}{N_{Re}}\right][1 + (D/L)^{2/3}]\,(N_{Pr})^{-2/3}(\mu_f/\mu_b)^{-0.14}$	$\Delta T_f / \Delta T_M = 2,260$	$\times\ \left(\dfrac{M_i^{0.22}}{s_i^{0.89}\rho_b^{1/3}}\right)\left(\dfrac{Z_f}{Z_w}\right)^{0.14}$	$\times\ \dfrac{W(t_H - t_L)}{\Delta T_M}$	$\times\ \dfrac{1}{[(N_{Re})^{2/3} - 125][1 + (d_i N_{PT}/12L)^{2/3}]nL}$	(4)
(5)	No phase change (liquid), $N_{Re} < 2,100$	$\dfrac{h}{cG} = 1.86\,(N_{Re})^{-2/3}(N_{Pr})^{-2/3}(L/D_i)^{-1/3}(\mu_f/\mu_b)^{-0.14}$	$\Delta T_f / \Delta T_M = 17.5$	$\times\ \left(\dfrac{M_i^{0.22}}{\rho_b^{0.89}}\right)\left(\dfrac{Z_i}{Z_w}\right)^{0.14}$	$\times\ \dfrac{W^{2/3}(t_H - t_L)}{\Delta T_M}$	$\times\ \dfrac{1}{n^{2/3}L^{2/3}(N_{PT})^{1/3}}$	(5)
(6)	Condensing vapor, vertical, $N_{Re} < 2,100$	$h = 0.925\,k\,(g\rho_f^2/\mu\Gamma)^{1/3}$	$\Delta T_f / \Delta T_M = 4.75$	$\times\ \dfrac{(Z_L M_L)^{0.333}}{s_L^2 c_L}$	$\times\ \dfrac{W^{4/3}\lambda}{\Delta T_M}$	$\times\ \dfrac{1}{n^{4/3}d_i^{4/3}L}$	(6)
(7)	Condensing vapor, horizontal, $N_{Re} < 2,100$	$h = 0.76\,k\,(g\rho_f^2/\mu\Gamma)^{1/3}$	$\Delta T_f / \Delta T_M = 2.92$	$\times\ \dfrac{Z_L M_L^{0.333}}{s_L^2 c_L}$	$\times\ \dfrac{W^{4/3}\lambda}{\Delta T_M}$	$\times\ \dfrac{1}{n^{4/3}d_i L^{4/3}}$ (See Note 1)	(7)
(8)	Condensate subcooling, vertical	$h = 1.225\,(k/B)\,(cB\Gamma/kL_B)^{5/6}$	$\Delta T_f / \Delta T_M = 1.22$	$\times\ \left[\dfrac{(Z_L M_L)^{0.333}}{s_L^2}\right]^{1/6}$	$\times\ \dfrac{W^{0.222}(t_H - t_L)}{\Delta T_M}$	$\times\ \dfrac{1}{(n^{4/3}d_i^{4/3}L)^{1/6}}$	(8)
(9)	Nucleate boiling, vertical	$\dfrac{h}{cG} = 4.02\,(N_{Re})^{-0.3}(N_{Pr})^{-0.6}(p_L\,\sigma/P^2)^{-0.425}\,\Sigma$	$\Delta T_f / \Delta T_M = 0.352$	$\times\ \left(\dfrac{Z_L^{0.3}M_L^{0.22}\sigma_L^{0.425}}{s_L^{1.075}c_L}\right)\left(\dfrac{P_L^{0.7}}{P_i^{0.85}}\right)$	$\times\ \dfrac{W^{0.3}\gamma_i}{\Delta T_M}$	$\times\ \left(\dfrac{1}{n^{0.3}L^{0.3}}\right)\Sigma'$ (See Notes 2 and 5)	(9)

Outside the tubes

No.	Description	$\dfrac{h}{cG}$ or h		$\dfrac{\Delta T}{\Delta T_M}$		
(10)	Nucleate boiling, horizontal	$\dfrac{h}{cG} = 4.02\,(N_{Re})^{-0.3}(N_{Pr})^{0.6}(\rho_b\sigma/P^2)^{-0.425}\,\Sigma$	$\Delta T_o/\Delta T_M = 0.352$	$\times \left(\dfrac{Z_o^{0.3}M_o^{0.2}d_o^{0.425}c_o}{s_o^{1.075}c_o}\right)\left(\dfrac{p_b^{0.7}}{P_D^{0.85}}\right)$	$\times \dfrac{W_o^{0.3}\lambda_o}{\Delta T_M}$	$\times \left(\dfrac{1}{n^{0.3}L^{0.3}}\right)\Sigma$
(11)	No phase change (liquid), crossflow	$\dfrac{h}{cG} = 0.33\,(N_{Re})^{-0.4}(N_{Pr})^{-2/3}\,(0.6)$	$\Delta T_o/\Delta T_M = 4.28$	$\times \;Z_o^{0.4}M_o^{0.222}/s_o^{0.89}$	$\times \dfrac{W_o^{0.4}(T_H - T_L)}{\Delta T_M}$	$\times \dfrac{N_{PT}^{0.282}\,P_B^{0.6}}{n^{0.718}\,L}$ (See Note 3)
(12)	No phase change (gas), crossflow	$h = 0.11\,G^{0.6}\,D^{-0.4}c_p\,(0.6)$	$\Delta T_o/\Delta T_M = 7.53$		$\times \dfrac{W_o^{0.4}(T_H - T_L)}{\Delta T_M}$	$\times \dfrac{N_{PT}^{0.282}\,P_B^{0.6}}{n^{0.718}\,L}$ (See Note 4)
(13)	No phase change (gas), parallel flow	$h = 0.0144\,G^{0.8}\,D^{-0.2}c_p\,(1.3)$	$\Delta T_o/\Delta T_M = 21.7$		$\times \dfrac{W_o^{0.2}(T_H - T_L)}{\Delta T_M}$	$\times \dfrac{d_o^{1.8}\,N_{PT}^{0.685}}{n^{0.315}\,L}$
(14)	No phase change (liquid), parallel flow	$\dfrac{h}{cG} = 0.023\,(N_{Re})^{-0.2}(N_{Pr})^{-2/3}(1.3)$	$\Delta T_o/\Delta T_M = 22.9$	$\times \;Z_o^{0.467}\,M_o^{0.22}/s_o^{0.89}$	$\times \dfrac{W_o^{0.2}(T_H - T_L)}{\Delta T_M}$	$\times \dfrac{d_o^{0.8}\,N_{PT}^{0.685}}{n^{0.315}\,L}$
(15)	Condensing vapor, vertical, $N_{Re} < 2{,}100$	$h = 0.925\,k\,(g\rho_L^2/\mu\Gamma)^{1/3}$	$\Delta T_o/\Delta T_M = 4.75$	$\times \;(Z_o M_o)^{0.333}/s_o^2 c_o$	$\times \dfrac{W_o^{1/3}\lambda_o}{\Delta T_M}$	$\times \dfrac{1}{n^{4/3}d_o^{1/3}N_{PT}^{1/3}L}$
(16)	Condensing vapor, horizontal. $N_{Re} < 2{,}100$	$h = 0.76\,k\,(g\rho_L^2/\mu\Gamma)^{1/3}$	$\Delta T_o/\Delta T_M = 2.64$	$\times \;(Z_o M_o)^{0.333}/s_o^2 c_o$	$\times \dfrac{W_o^{1/3}\lambda_o}{\Delta T_M}$	$\times \dfrac{N_{PT}^{0.177}}{n^{1.156}L^{4/3}d_o}$

Tube wall

No.	Description	h				
(17)	Tube wall (sensible-heat transfer)	$h = (24\,k_w)/(d_o - d_i)$	$\Delta T_w/\Delta T_M = 159$	$\times \;c/k_w$	$\times \dfrac{W(t_H - t_L)}{\Delta T_M}$	$\times \dfrac{d_o - d_i}{nd_oL}$
(18)	Tube wall (latent-heat transfer)	$h = (24\,k_w)/(d_o - d_i)$	$\Delta T_M/\Delta T_M = 88$	$\times \;1/k_w$	$\times \dfrac{WA}{\Delta T_M}$	$\times \dfrac{d_o - d_i}{nd_oL}$

Fouling

No.	Description	h				
(19)	Fouling (sensible-heat transfer)	$h = assumed$	$\Delta T_i/\Delta T_M = 3{,}820$	$\times \;c/h$	$\times \dfrac{W(t_H - t_L)}{\Delta T_M}$	$\times \dfrac{1}{nd_oL}$
(20)	Fouling (latent-heat transfer)	$h = assumed$	$\Delta T_i/\Delta T_M = 2{,}120$	$\times \;1/h$	$\times \dfrac{WA}{\Delta T_M}$	$\times \dfrac{1}{nd_oL}$

Notes:
1. If $W_i/(ns_i d_i^{2.56}) > 0.3$, multiply $\Delta T_i/\Delta T_M$ by 1.3.
2. Surface-condition factor (Σ') for copper and steel = 1.0; for stainless steel = 1.7; for polished surfaces = 2.5.
3. For square pitch, numerical factor = 5.42.
4. For square pitch, numerical factor = 9.53.
5. $G = W_o/(A\rho_o)$.

6. Make the pressure-drop calculation for the heat exchanger
Tubeside, Eq. (21):

$$\Delta P = (17^{0.2}/0.85)(307.5/89)^{1.8}[(12/0.495) + 25]/$$
$$(5.4 \times 0.495)^{3.8} = 14.3 \text{ lb/in}^2 \text{ (98.5 kPa)}$$

Shellside, Eq. (25):

$$\Delta P = (0.326/0.95)(32.8^2)[12/(5^3 \times 9)] = 3.9 \text{ lb/in}^2 \text{ (26.9 kPa)}$$

To decrease the pressure drop on the tubeside to the acceptable limit of 10 lb/in² (68.9 kPa), the number of tubes must be increased. This will also decrease the SOP. In addition, shellside performance can be improved by decreasing the baffle spacing, since the pressure drop of 3.9 on the shellside is lower than the allowable 10 psi (68.9 kPa). Before proceeding with successive trials to balance the heat-transfer and pressure-drop restrictions, Table 8 is now set up for clarity.

7. Perform the second trial computation for heat-transfer surface and pressure drop
As a first step in adjusting the heat-transfer surface and pressure drop, calculate the number of tubes to give a pressure drop of 10 lb/in² (68.9 kPa) on the tubeside. The pressure drop varies inversely as $n^{1.8}$. Therefore, $14.3/10 = (n/89)^{1.8}$, and $n = 109$.

Each individual product of the factors is then adjusted in accordance with the applicable exponential function of the number of tubes. Since the tubeside product is inversely proportional to the 0.2 power of the number of tubes, the product from the preceding trial is multiplied by $(n_1/n_2)^{0.2}$, where n_1 is the number of tubes used in the preceding trial, and n_2 is the number to be used in the new one. The shellside product of the preceding trial is multiplied by $(n_1/n_2)^{0.718}$, and the tube-wall and fouling products by n_1/n_2. New adjusted products are then calculated as follows:

$$\text{Tubeside product} = (89/109)^{0.2} \times 0.535 = 0.514$$
$$\text{Shellside product} = (89/109)^{0.718} \times 0.424 = 0.367$$
$$\text{Tube-wall product} = (89/109) \times 0.052 = 0.042$$
$$\text{Fouling product} = (89/109) \times 0.250 = 0.204$$
$$\text{SOP} = 1.127$$

8. Make the last trial calculation
For the third trial, baffle spacing is decreased to 3.5 in (88.9 cm) from 5 in (127.0 cm). Only the shellside product must be adjusted since only it is affected by the baffle spacing. Therefore, the shellside factor of the previous trial is multiplied by the ratio of the baffle spacing to the 0.6 power:

$$(3.5/5.0)^{0.6} \times 0.367 = 0.296$$

The sum of the products (SOP) for this trial is 1.056. The shellside pressure drop $\Delta P_o = 3.9 \times (5.0/3.5)^3 = 11.4 \text{ lb/in}^2 \text{ (78.5 kPa)}$.

Because we have now reached the point where the assumed design nearly satisfies our conditions, tube-layout tables can be used to find a standard shell-size containing the next increment above 109 tubes. A 10-in-dia (254 cm) shell in a fixed-tubesheet design contains 110 tubes.

TABLE 7 Empirical Pressure-Drop Relationship for Rating Heat Exchangers

Eq. No.	Mechanism or Restriction	Empirical Equation
	Inside the tubes	
(21)	No phase change, $N_{Re} > 10,000$	$\Delta P = \dfrac{(Z_i)^{0.2}}{s_i}\left(\dfrac{W_i}{n}\right)^{1.8}\dfrac{N_{PT}[(L_o/d_i)+25]}{(5.4d_i)^{3.8}}$ (See note 1)
(22)	No phase change, $2,100 < N_{Re} < 10,000$	$\Delta P = \left(\dfrac{Z_i}{s_i}\right)\left(\dfrac{W_i}{n}\right)\dfrac{N_{PT}[(L_o/d_i)+25][(N_{Re})^{2/3}-25]}{(50.2d_i)^3}$ (See note 1)
(23)	No phase change, $N_{Re} < 2,100$	$\Delta P = \dfrac{(Z_b)^{0.326}(Z_i)^{0.14}}{s_i}\left(\dfrac{W_i}{n}\right)^{4/3}\dfrac{N_{PT}(L_o)^{2/3}}{(5.62d_i)^4}$
(24)	Condensing	$\Delta P = \dfrac{(Z_i)^{0.2}}{s_i}\left(\dfrac{W_i}{n}\right)^{1.8}\dfrac{N_{PT}[(L_o/d_i)+25]}{(5.4d_i)^{2.8}}\times 0.5$ (See note 1)
	Shellside	
(25)	No phase change, crossflow	$\Delta P = \dfrac{0.326}{s_o}(W_o)^2\dfrac{L_o}{P_B^3 D_o}$
(26)	No phase change, parallel flow	$\Delta P = \dfrac{(Z_o)^{0.2}}{s_o}\left(\dfrac{W_o}{n}\right)^{1.8}\left[\dfrac{n^{0.366}\,L_0}{(N_{PT})^{1.434}(4.912\,d_o)^{4.8}}+\dfrac{0.31n^{0.0414}\,(W_o)^{0.2}\,L_o}{d_o\,(N_{PT})^{1.76}\,(4.912d_s)^4\,Z^{0.2}\,B_o^2}\right]$ (See notes 2 and 3)
(27)	Condensing	$\Delta P = \left(\dfrac{0.081}{s_o}\right)(W_o)^2\left(\dfrac{L_o}{P_B^3\,D_o}\right)$

Notes:
1. For U-bends, use $[(L_o/d_i)+16]$ instead of $[(L_o/d_i)+25]$.
2. B_o is equal to fraction of flow area through baffle.
3. Number of baffles (N_B) = 0.48 (L_o/d_o).

TABLE 8 Results of Trial Calculations

	1st Trial	2nd Trial	3rd Trial	4th Trial
Number of tubes	89	109	109	110
Shell diameter, in	9	9	9	10
Baffle spacing, in	5	5	3½	3½
Product of factors:				
Tubeside	0.535	0.514	0.514	0.513
Shellside	0.424	0.367	0.296	0.293
Tube wall	0.052	0.042	0.042	0.042
Fouling	0.250	0.204	0.204	0.202
Total sum of products	1.261	1.127	1.056	1.050
Tubeside ΔP, psi	14.3	10	10	9.8
Shellside ΔP, psi	3.9	3.9	11.4	10.1

Again, correcting the products of the heat-transfer factors from the previous trial:

$$\text{Tubeside product} = (109/110)^{0.2} \times 0.514 = 0.513$$
$$\text{Shellside product} = (109/110)^{0.718} \times 0.296 = 0.293$$
$$\text{Tube-wall product} = (109/110) \times 0.042 = 0.042$$
$$\text{Fouling product} = (109/110) \times 0.204 = 0.202$$
$$\text{SOP} = 1.050$$

Tubeside pressure drop:

$$\Delta P_i = 10 \times (109/110)^{1.8} = 9.8 \text{ lb/in}^2 \text{ (67.5 kPa)}$$

The shellside pressure drop is now corrected for the actual shell diameter of 10 in (25.4 cm) instead of 9 in (228.6 cm).

$$\Delta P_o = 11.4 \times (9/10) = 10.1 \text{ psi (69.6 kPa)}$$

Any value of SOP between 0.95 and 1.05 is satisfactory as this gives a result within the accuracy range of the basic equations; unknowns in selecting the fouling factor do not justify further refinement. Therefore, the above is a satisfactory design for heat transfer and is within the pressure-drop restrictions specified. The surface area of the heat exchanger is $A = 110 \times 12 \times 0.1636 = 216$ ft^2 (20.1 m^2). The design overall coefficient is $U = 2,391,000/(30.4 \times 1.8 \times 216) = $ Btu/(hr) (ft^2) (°F) (35.6 W/m^2°C).

The foregoing example shows that the essence of the design procedure is selecting tube configurations and baffle spacings that will satisfy heat-transfer requirements within the pressure-drop limitations of the system.

Related Calculations. The preceding procedure was for rating a heat exchanger of single tube-pass and single shell-pass design, with countercurrent flows of tubeside and shellside fluids. Often, it will be necessary to use two or more passes for the tubeside fluid. In this case, the LMTD is corrected with the Bowman, Mueller and Nagle charts given in heat-transfer texts and the TEMA guide. If the correction factor for LMTD is less than 0.8, multiple shells should be used.

Bear in mind that n in all equations is the number of tubes in parallel through which the tubeside fluid flows; N_{PT} is the number of tubeside passes per shell (total number of tubes per shell = nN_{PT}); and L, the total-series length of path, equals shell length (L_o) (N_{PT}) × (number of shells).

The above procedure can be used for any shell-and-tube heat exchanger with sensible-heat transfer—or with no phase change of fluids—on both sides of the tubes. Also, N_{Re} on the tubeside must be greater than 10,000, and the viscosity of the fluid on the shellside must be moderate (500 cp. maximum).

As pointed out, the designer should assume as part of his job the specification of tube arrangement that will prevent the flow in the shell from taking bypass paths either around the space between the outermost tubes and the shell, or in vacant lanes of the bundle formed by channel partitions in multipass exchangers. He should insist that exchangers be fabricated in accordance with TEMA tolerances.

By using the appropriate equations from Tables 6 and 7, the technique described for rating heat exchangers with sensible-heat transfer can be used also for rating exchangers that involve boiling or condensing. The method can also be used in the design of partial condensers, or condensers handling mixtures of condensable vapors and noncondensable gases; and in the design of condensers handling vapors that form two liquid phases. However, for partial condensers and for two-phase liquid-condensate systems, a special treatment is required.

In addition to designing exchangers for specified performances, the method is also useful for evaluating the performance of existing exchangers. Here, the mechanical-design parameters are fixed, and the flow-rates and temperature conditions (work factor) are the variables that are adjusted.

The two process variables that have the greatest effect on the size (cost) of a shell-and-tube heat exchanger are the allowable pressure drops of streams, and the mean temperature difference between the two streams. Other important variables include the physical properties of the streams, the location of fluids in an exchanger, and the piping arrangement of the fluids as they enter and leave the exchanger. (See design features in Table 5.)

Selection of optimum pressure drops involves consideration of the overall process. While it is true that higher pressure drops result in smaller exchangers, investment savings are realized only at the expense of operating costs. Only by considering the relationship between operating costs and investment can the most economical pressure drop be determined.

Available pressure drops vary from a few millimeters of mercury in vacuum service to hundreds of pounds per square inch in high-pressure processes. In some cases, it is not practical to use all the available pressure drop because resultant high velocities may create erosion problems.

Reasonable pressure drops for various levels are listed below. Designs for smaller pressure drops are often uneconomical because of the large surface area (investment) required.

Pressure Level	Reasonable ΔP
Sub-atmospheric	1/10 absolute pressure
1 (6.89 kPa) to 10 lb/in² (gage) (68.9 kPa)	1/2 operating gage-pressure
10 lb/in² (gage) (68.9 kPa) and higher	5 lb/in² (34.5 kPa) or higher

In some instances, velocities of 10 to 15 ft/sec (3 to 4.6 m/sec) help to reduce fouling, but at such velocities the pressure drop may have to be from 10 (68.9 kPa) to 30 lb/in² (206.7 kPa).

Although there are no specific rules for determining the best temperature approach, the following recommendations are made regarding terminal temperature differences for various types of heat exchangers; any departure from these general limitations should be economically justified by a study of alternate system-designs:

- The greater temperature difference should be at least 20°C (36°F).
- The lesser temperature difference should be at least 5°C (9°F). When heat is being exchanged between two process streams, the lesser temperature difference should be at least 20°C (36°F).
- In cooling a process stream with water, the outlet-water temperature should not exceed the outlet process-stream temperature if a single body having one shell pass—but more than one tube pass—is used.
- When cooling or condensing a fluid, the inlet coolant temperature should not be less than 5°C (9°F) above the freezing point of the highest freezing component of the fluid.
- For cooling reactors, a 10°C (18°F) to 15°C (27°F) difference should be maintained between reaction and coolant temperatures to permit better control of the reaction.
- A 20°C (36°F) approach to the design air-temperature is the minimum for air-cooled exchangers. Economic justification of units with smaller approaches requires careful study. Trim coolers or evaporative coolers should also be considered.
- When condensing in the presence of inerts, the outlet coolant temperature should be at least 5°C (9°F) below the dewpoint of the process stream.

In an exchanger having one shell pass and one tube pass, where two fluids may transfer heat in either cocurrent or countercurrent flow, the relative direction of the fluids affects the value of the mean temperature difference. This is the log mean in either case, but there is a distinct thermal advantage to counterflow, except when one fluid is isothermal.

In concurrent flow, the hot fluid cannot be cooled below the cold-fluid outlet temperature; thus, the ability of cocurrent flow to recover heat is limited. Nevertheless, there are instances when cocurrent flow works better, as when cooling viscous fluids, because a higher heat-transfer coefficient may be obtained. Cocurrent flow may also be preferred when there is a possibility that the temperature of the warmer fluid may reach its freezing point.

These factors are important in determining the performance of a shell-and-tube exchanger:

Tube Diameter, Length.   Designs with small-diameter tubes (⅝ (15.8 cm) to 1 in (25.4 cm)) are more compact and more economical than those with larger-diameter tubes, although the latter may be necessary when the allowable tubeside pressure drop is small. The smallest tube size normally considered for a process heat exchanger is ⅝ in (15.8 cm) although there are applications where ½ (12.7 cm), ⅜ (9.5 cm), or even ¼-in (6.4 cm) tubes are the best selection. Tubes of 1 in (25.4 cm) dia are normally used when fouling is expected because smaller ones are impractical to clean mechanically. Falling-film exchangers and vaporizers generally are supplied with 1½ (38.1 cm) and 2-in (50.8 cm) tubes.

Since the investment per unit area of heat-transfer service is less for long exchangers with relatively small shell diameters, minimum restrictions on length should be observed.

Arrangement. Tubes are arranged in triangular, square, or rotated-square pitch (Fig. 11). Triangular tube-layouts result in better shellside coefficients and provide more surface area in a given shell diameter, whereas square pitch or rotated-square pitch layouts are used when mechanical cleaning of the outside of the tubes is required. Sometimes, widely spaced triangular patterns facilitate cleaning. Both types of square pitches offer lower pressure drops—but lower coefficients—than triangular pitch.

Primarily, the method given in this calculation procedure combines into one relationship the classical empirical equations for film heat-transfer coefficients with heat-balance equations and with relationships describing tube geometry, baffles and shell. The resulting overall equation is recast into three separate groups that contain factors relating to: physical properties of the fluid, performance or duty of the exchanger, and mechanical design or arrangement of the heat-transfer surface. These groups are then multiplied together with a numerical factor to obtain a product that is equal to the fraction of the total driving force—or log mean temperature-difference (LMTD or ΔT_M)—that is dissipated across each element of resistance in the heat-flow path.

When the sum of the products for the individual resistance equals one, the trial design may be assumed to be satisfactory for heat transfer. The physical significance is that the sum of the temperature drops across each resistance is equal to the total available LMTD. The pressure drop on both tubeside and shellside must be checked to assure that both are within acceptable limits. As shown in the sample calculation above, usually several trials are necessary to obtain a satisfactory balance between heat transfer and pressure drop.

Tables 6 and 7 respectively summarize the equations used with the method for heat transfer and for pressure drop. The column on the left lists the conditions to which each equation applies. The second column lists the standard form of the correlation for film coefficients that is found in texts. The remaining columns then tabulate the numerical, physical-property, work, and mechanical-design factors, all of which together form the recast dimensional equation. The product of these factors gives the fraction of total temperature drop or driving force ($\Delta T_f / \Delta T_M$) across the resistance.

As described above, the addition of $\Delta T_i / \Delta T_M$, tubeside factor; plus $\Delta T_o / \Delta T_M$, shellside factor; plus $\Delta T_s / \Delta T_M$, fouling factor; plus $\Delta T_w / \Delta T_M$, tube-wall factor, determine the heat-transfer adequacy. Any combination of $\Delta T_i / \Delta T_M$ and $\Delta T_o / \Delta T_M$ may be used, as long as a horizontal orientation on the tubeside is used with a

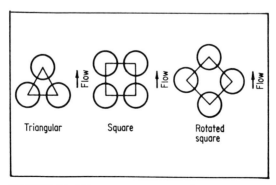

FIGURE 11 Tube arrangements used for shell-and-tube heat exchangers. (*Chemical Engineering.*)

horizontal orientation on the shellside, and a vertical tubeside orientation has a corresponding shellside orientation.

The units in the pressure-drop equations (Table 7) are consistent with those used for heat transfer. The pressure drop pin psi, is calculated directly. Because the method is a shortcut approach to design, certain assumptions pertaining to thermal conductivity, tube pitch and shell diameter are made:

For many organic liquids, thermal conductivity data are either not available or difficult to obtain. Since molecular weights (M) are known, for most design purposes the Weber equation, which follows, yields thermal conductivities with quite satisfactory accuracies:

$$k = 0.86 \ (cs^{4/3}/M^{1/3})$$

An important compound for which the Weber equation does not work well is water (the calculated thermal conductivity is less than the actual value). Figure 12 gives the physical-property factor for water (as a function of fluid temperature) that is to be substituted in the equations for sensible-heat transfer with water, or for condensing with steam.

If the thermal conductivity is known, it is best to obtain a pseudo-molecular weight by:

$$M = 0.636 \ (c/k)^3 s^4$$

This value is substituted in the applicable equation to solve for the physical-property factor.

Tube pitch for both triangular and square-pitch arrangements is assumed to be 1.25 times the tube diameter. This is a standard pitch used in the majority of shell-and-tube heat exchangers. Slight deviations do not appreciably affect results.

Shell diameter is related to the number of tubes (nN_{PT}) by the empirical equation:

$$D_o = 1.75 \ d_o (nN_{PT})^{0.47}$$

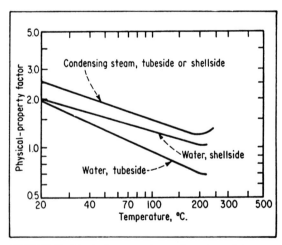

FIGURE 12 Physical-property factors for water and steam vs temperature. (*Chemical Engineering.*)

This gives the approximate shell diameter for a packed floating-head exchanger. The diameter will differ slightly for a fixed tubesheet, U-bend, or a multipass shell. For greater accuracy, tube-layout tables can be used to find shell diameters.

The following shows how the design equations are developed for a heat exchanger with sensible-heat transfer and Reynolds number 10,000 on the tube side, and with sensible-heat transfer and crossflow (flow perpendicular to the axis of tubes) on the shellside. Equations with other heat-transfer mechanisms are derived similarly.

For the film coefficient or conductance, h, and the heat-balance, these equations apply:

Value of h	Heat-Balance Eq.
$h_i = \dfrac{0.023 \, c_i G_i}{(c_i \mu_i / k_i)^{2/3}(D_i G_i / \mu_i)^{0.2}}$	$W_i c_i (t_H - t_L) = h_i A \Delta T_i$
	$W_i c_i (t_H - t_L) = h_w A \Delta T_w$
$h_w = 24 \, k_w / (d_o - d_i)$	$W_o c_o (T_H - T_L) = h_o A \Delta T_o$
$h_o = \dfrac{0.33 \, c_o G_o (0.6)}{(c_o \mu_o / k_o)^{2/3}(D_o G_o / \mu_o)^{0.4}}$	$W_i c_i (t_H - t_L) = h_s A \Delta T_s$
h_s = assumed value	

Since the resistances involved in a tube-and-shell exchanger are the tubeside film, the tube wall, the scale caused by fouling, and the shellside film, then:

$$\Delta T_i + \Delta T_w + \Delta T_s + \Delta T_o = \Delta T_M$$

therefore:

$$\Delta T_i / \Delta T_M + \Delta T_w / \Delta T_M + \Delta T_s / \Delta T_M + \Delta T_o / \Delta T_M = 1$$

or:

$$\frac{W_i c_i (t_H - t_L)}{h_i A \Delta T_M} + \frac{W_i c_i (t_H - t_L)}{h_w A \Delta T_M} + \frac{W_i c_i (t_H - t_L)}{h_o A \Delta T_M} + \frac{W_s c_o (T_H - T_L)}{h_o A \Delta T_M} = 1$$

Tubeside product	Tube-wall product	Fouling product	Shellside product

This last equation is obtained by dividing each heat-balance equation by ΔT_M and solving for $\Delta T_f / \Delta T_M$. The design equations are derived by substituting for h the appropriate correlation for the coefficient; for k, the value obtained from the Weber equation; for A, the equivalent of the surface area in terms of the number of tubes, outside diameter and length, according to the relation $A = \pi n (d_o / 12)L$; for mass velocity on the tubeside, $G_i = 183 \, W_i / (d_i^2 n)$; and for mass velocity on the shellside, $G_o = 411.4 \, W_o / (d_o n N_{PT}^{0.47} P)$.

The resulting equation is rearranged to separate the physical-property, work, and mechanical-design parameters into groups. To obtain consistent units, the numerical factor in the equation combines the constants and coefficients. The form of the equations shown in Table 6 as Eq. (1), (11), (18) and (19) omits dimensionless groups such as Reynolds or Prandtl numbers, but includes single functions of the common design parameters such as number of tubes, tube diameter, tube length, baffle pitch, etc.

The individual products calculated from the four equations are added to give the sum of the products (SOP). A valid design for heat transfer should give SOP = 1. If SOP comes out to be less or more than one, the products for each resistance are adjusted by the appropriate exponential function of the ratio of the new design parameter to that used previously.

More-sophisticated rating methods are available that make use of complex computer programs; the described method is intended only as a general, shortcut approach to shell-and-tube heat-exchanger selection. Accuracy of the technique is limited by the accuracy with which fouling factors, fluid properties and fabrication tolerances can be predicted. Nevertheless, test data obtained on hundreds of heat exchangers attest to the method's applicability.

This procedure is the work of Robert C. Lord, Project Engineer, Paul E. Minton, Project Engineer, and Robert P. Slusser, Project Engineer, Engineering Department, Union Carbide Corporation, as reported in *Chemical Engineering* magazine. SI values were added by the handbook editor.

Nomenclature

A	Outside surface area, ft²	t	Temperature on tubeside, deg C
B	Film thickness, $[0.00187Z\Gamma/g_c s^2]^{1/3}$, ft	ΔT_M	Logarithmic mean temperature difference (LMTD), deg C
c	Specific heat, Btu/(lb) (°F)	U	Overall coefficient of heat transfer, Btu/[(hr) (ft²) (°F)]
D_i	Inside tube diameter, ft	W	Flowrate, (lb/hr)/1,000
D_o	Inside shell diameter, in	Z	Viscosity, cp
d	Tube diameter, in	Γ	Tube loading, lb/(hr) (ft)
f	Fanning friction factor, dimensionless	λ	Heat of vaporization, Btu/lb
G	Mass velocity, lb/(hr) (ft² cross-sectional area)	θ	Time, hr
g_c	Gravitational constant, (4.18 × 10⁸) ft (hr)²	μ	Viscosity, lb/(hr) (ft)
h	Film coefficient of heat transfer, Btu/[(hr) (ft²) (°F)]	ρ_v	Vapor density, lb/ft³
k	Thermal conductivity, Btu/[(hr) (ft²)]	ρ_L	Liquid density, lb/ft³
L	Total series length of tubes, (L_o N_{PT} × number of shells), ft	σ	Surface tension, dynes/cm
L_A	Length of condensing zone, ft	Σ, Σ'	Surface condition factor, dimensionless
	Subscripts		
L_B	Length of subcooled zone, ft	o	Conditions on shellside or outside tubes
L_o	Length of shell, ft	i	Conditions on tubeside or inside tubes
M	Molecular weight, lb/(lb-mol)	b	Bulk fluid properties
N_{PT}	Number of tube passes per shell, dimensionless	f	Film fluid properties
n	Number of tubes per pass (or in parallel) dimensionless	H	High temperature
P	Pressure, lb/in² (abs)	L	Low temperature
P_B	Baffle spacing, in	s	Scale or fouling material
ΔP	Pressure drop, lb/in²	w	Wall or tube material
Q	Heat transferred, Btu		*Dimensionless Groups*

s	Specific gravity (referred to water at 20°C), dimensionless	N_{Re}	Reynolds Number, DG/μ	
T	Temperature on shellside, deg C	N_{Pr}	Prandtl Number, $c\mu/k$	
		N_{St}	Stanton Number, h/cG	
		N_{Nu}	Nusselt Number, hD/k	

See procedure for SI values for the variables above.

DESIGNING SPIRAL-PLATE HEAT EXCHANGERS

Design a liquid-to-liquid spiral-plate heat exchanger for liquids in laminar flow under the conditions given below.

Conditions	Hot Side	Cold Side			
Flowrate, lb/hr	6,225	5,925	kg/h	5925	2690
Inlet temperature, °C	200	60		392°F	140°F
Outlet temperature, °C	120	150.4		248°F	270.7°F
Viscosity, cp	3.35	8			
Specific heat, Btu/lb/°F	0.71	0.66	kJ/hr°C	2.95	3.75
Molecular weight	200.4	200.4			
Specific gravity	0.843	0.843			
Allowable pressure drop, psi	1	1	kPa	6.89	6.89
Material of construction	stainless steel	$(k = 10)$			
$(Z_f/Z_b)^{0.14}$	1	1			

Determine the heat transferred, the required heat-exchanger surface area, pressure-drop through the exchanger, and the final dimensions of the heat exchanger.

Calculation Procedure:

1. *Find the heat-transfer rate and log mean temperature difference*
Figure 13 shows several possible arrangements of spiral-flow heat exchangers. Using the same relation as in the previous procedure, we have, heat transfer rate, Btu/hr = (flow rate, lb/hr)(inlet temperature − outlet temperature)(specific heat, Btu/lb F)(1.8 conversion factor for temperature). Or, heat-transfer rate = (6225)(200 − 120)(0.71)(1.8) = 636,444 Btu/hr (671.4 kJ/hr). Then, the log mean temperature difference, LMTD, T_M = (60 − 49.4)/ln (60/49.4) = 54.5°C (129.9°F).

2. *Find the surface area required for this heat exchanger*
For the first trial, the approximate surface area for this exchanger can be computed using an assumed overall heat-transfer coefficient, U, of 50 Btu/hr ft²°F (8.8 W/m²°C). Then, A = 636,444/(50)(54.5)(1.8) = 129.75 ft², say 130 ft² (12 m²).

Because at 130 ft² (12 m²), this is a small heat exchanger, we will assume a plate width of 24 in (60.9 cm). Then, the plate length, L = 130/(2)(2) = 32.5 ft (9.9 m). Assume a channel spacing of ⅜ in (0.95 cm) for both fluids.

3. *Find the Reynolds number of the flow conditions*
The Reynolds number for spiral flow can be computed from N_{Re} = 10,000 (W/HZ), where W = flow rate, lb/hr/1000; H = channel width, in; Z = fluid viscosity,

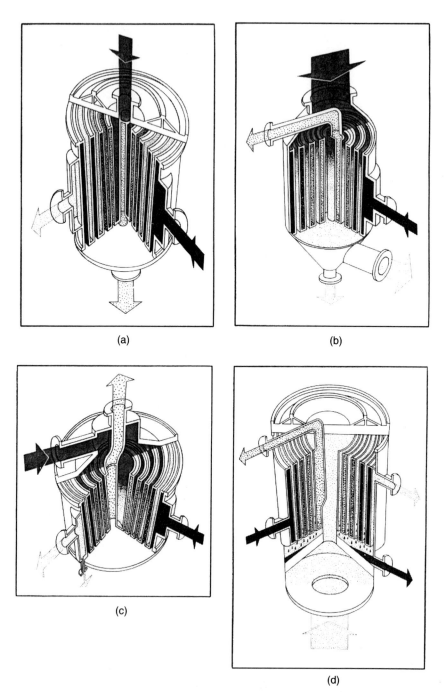

FIGURE 13 (*a*) Spiral flow in both channels is widely used. (*b*) Spiral flow in one channel, axial flow in the other; (*c*) combination flow is used to condense vapors; (*d*) modified combination flow serves on column. (*Chemical Engineering.*)

cp. Substituting, we have, for the hot side, N_{Re} = (10,000)(6225/1000)/ (24)(3.35) = 774.4. For the cold side, N_{Re} = (10,000)(5925/1000)/(24)(8) = 308.6. Because both fluids are in laminar flow, spiral flow will be chosen for this heat exchanger. From Table 9, the appropriate expressions for rating are Eq. (3) for both fluids, Eq. (10) for the plate, Eq. (12) for fouling and Eq. (15) for pressure drop.

4. Make the heat-transfer calculations for the heat exchanger
Now, substitute values:
Hot side, Eq. (3):

$$\frac{\Delta T_h}{\Delta T_M} = 32.6 \left[\frac{200.4^{0.222}}{0.843^{0.889}}\right] \times \left[\frac{6.225^{2/3} \times 80}{54.5}\right]\left[\frac{0.375}{24^{2/3} \times 32.5}\right]$$

$$= 32.6 \times 3.775 \times 4.967 \times 0.001387 = 0.848$$

Cold side, Eq. (3):

$$\frac{\Delta T_c}{\Delta T_M} = 32.6 \left[\frac{200.4^{0.222}}{0.843^{0.889}}\right]\left[\frac{5.925^{2/3} \times 90.4}{54.5}\right] \times \left[\frac{0.375}{24^{2/3} \times 32.5}\right]$$

$$32.6 \times 3.775 \times 5.431 \times 0.001387 = 0.927$$

Fouling, Eq. (12):

$$\frac{\Delta T_s}{\Delta T_M} = 6,000 \left[\frac{0.66}{1,000}\right]\left[\frac{5.925 \times 90.4}{54.5}\right]\left[\frac{1}{32.5 \times 24}\right]$$

$$= 6,000 \times 0.00066 \times 9.828 \times 0.001282 = 0.050$$

Plate, Eq. (10):

$$\frac{\Delta T_w}{\Delta T_M} = 500 \left[\frac{0.66}{10}\right]\left[\frac{5.925 \times 90.4}{54.5}\right]\left[\frac{0.125}{32.5 \times 24}\right]$$

$$= 500 \times 0.066 \times 9.828 \times 0.0001603 = 0.052$$

Sum of Products (SOP):

$$SOP = 0.848 + 0.927 + 0.050 + 0.052 = 1.877 \ (18.6 \ m)$$

Because SOP is greater than 1, the assumed heat exchanger is inadequate. The surface area must be enlarged by increasing the plate width or the plate length. Because, in all the equations, L applies directly, the following new length is adopted:

$$1.877 \times 32.5 = 61 \ ft$$

5. Compute the pressure drops in the hot and cold sides
Hot side, Eq. (15):

$$\Delta P = \left[\frac{0.001 \times 61}{0.843}\right]\left[\frac{6.225}{0.375 \times 24}\right]$$

$$\times \left[\frac{1.035 \times 3.35^{1/2} \times 1 \times 24^{1/2}}{(0.375 + 0.125) 6.225^{1/2}} + 1.5 + \frac{16}{61}\right]$$

$$\Delta P = 0.07236 \times 0.6917 \times 9.202 = 0.461 \ psi$$

Cold side, Eq. (15):

TABLE 9 Empirical Heat-Transfer and Pressure-Drop Relationships for Rating Spiral-Plate Heat Exchangers

Eq. No.	Mechanism of Restriction	Empirical Equation—Heat Transfer	Numerical Factor	Physical Property Factor	Work Factor	Mechanical Design Factor	
Spiral Flow							
(1)	No phase change (liquid), $N_{Re} > N_{Rec}$	$h = (1 + 3.54 \, D_e/D_H) \, 0.023 \, cG \, (N_{Re})^{-0.2}(Pr)^{-2/3}$	$\dfrac{\Delta T_r}{\Delta T_M} = 20.6$	$\times \dfrac{Z^{0.467}M^{0.222}}{s^{0.889}}$	$\times \dfrac{W^{0.2}(T_H - T_L)}{\Delta T_M}$	$\times \dfrac{d_e}{LH^{0.2}}$	(See Note 1)
(2)	No phase change (gas), $N_{Re} > N_{Rec}$	$h = (1 + 3.54 \, D_e/D_H) \, 0.0144 \, cG^{0.8}(D_e)^{-0.2}$	$\dfrac{\Delta T_r}{\Delta T_M} = 19.6$		$\times \dfrac{W^{0.2}(T_H - T_L)}{\Delta T_M}$	$\times \dfrac{d_e}{LH^{0.2}}$	(See Note 1)
(3)	No phase change (liquid), $N_{Re} < N_{Rec}$	$h = 1.86 \, cG \, (N_{Re})^{-2/3}(Pr)^{-1/3}(L/D_s)^{-1/3}(\mu_f/\mu_b)^{-0.14}$	$\dfrac{\Delta T_r}{\Delta T_M} = 32.6$	$\times \dfrac{M^{2/9}(Z_f)^{0.14}}{s^{8/9}(Z_b)^{0.14}}$	$\times \dfrac{W^{2/3}(T_H - T_L)}{\Delta T_M}$	$\times \dfrac{d_s}{LH^{2/3}}$	(See Note 1)
Spiral or Axial Flow							
(4)	Condensing vapor, vertical, $N_{Re} < 2,100$	$h = 0.925 \, k \, [g_c \, \rho_L^2/\mu\Gamma]^{1/3}$	$\dfrac{\Delta T_r}{\Delta T_M} = 3.8$	$\times \dfrac{M^{1/3}Z^{1/3}}{cs^2}$	$\times \dfrac{W^{4/3}}{\Delta T_M}\lambda$	$\times \dfrac{1}{L^{4/3}H}$	
(5)	Condensate subcooling, vertical, $N_{Re} < 2,100$	$h = 1.225 \, k/B \, [cB/kL_B]^{5/6}$	$\dfrac{\Delta T_r}{\Delta T_M} = 1.18$	$\times \dfrac{M^{1/18}Z^{1/18}}{s^{1/3}}$	$\times \dfrac{W^{2/9}(T_H - T_L)}{\Delta T_M}$	$\times \dfrac{1}{H^{1/6}L^{2/9}}$	
Axial Flow							
(6)	No phase change (liquid), $N_{Re} > 10,000$	$h = 0.023 \, cG \, (N_{Re})^{-0.2}(Pr)^{-2/3}$	$\dfrac{\Delta T_r}{\Delta T_M} = 167$	$\times \dfrac{Z^{0.467}M^{0.222}}{s^{0.889}}$	$\times \dfrac{W^{0.2}(T_H - T_L)}{\Delta T_M}$	$\times \dfrac{d_e}{HL^{0.2}}$	
(7)	No phase change (gas), $N_{Re} > 10,000$	$h = 0.0144 \, cG^{0.8}(D_e)^{-0.2}$	$\dfrac{\Delta T_r}{\Delta T_M} = 158$		$\times \dfrac{W^{0.2}(T_H - T_L)}{\Delta T_M}$	$\times \dfrac{d_e}{HL^{0.2}}$	
(8)	Condensing vapor, horizontal, $N_{Re} < 2,100$	$h = 0.76 \, k \, [g_c \, \rho_L^2/\mu\Gamma]^{1/3}$	$\dfrac{\Delta T_r}{\Delta T_M} = 16.1$	$\times \dfrac{Z^{1/3}M^{1/3}}{cs^2}$	$\times \dfrac{W^{4/3}}{\Delta T_M}\lambda$	$\times \dfrac{1}{L^{4/3}H^{4/3}}$	
(9)	Nucleate boiling, vertical	$h = 4.02 \, cG \, (N_{Re})^{-0.3}(Pr)^{-0.6}(\rho_L \, \sigma/P^2)^{-0.425} \, \Sigma$	$\dfrac{\Delta T_r}{\Delta T_M} = 0.619 \times \dfrac{M^{0.2}Z^{0.3}\sigma^{0.425} \, p_c^{0.7}}{cs^{1.075} \, P^{0.85}}$		$\times \dfrac{W^{0.3}\lambda}{\Delta T_M}$	$\times \dfrac{d_e^{0.35}\Sigma'}{L^{0.3}H^{0.3}}$	(See Notes 2 and 3)
Plate							
(10)	Plate, sensible heat transfer	$h = 12 \, k_w/p$	$\dfrac{\Delta T_w}{\Delta T_M} = 500 \times \dfrac{c}{k_w}$		$\times \dfrac{W(T_H - T_L)}{\Delta T_M}$	$\times \dfrac{p}{LH}$	
(11)	Plate, latent heat transfer	$h = 12 \, k_w/p$	$\dfrac{\Delta T_w}{\Delta T_M} = 278 \times \dfrac{1}{k_w}$		$\times \dfrac{W\lambda}{\Delta T_M}$	$\times \dfrac{p}{LH}$	

Eq. No.	Mechanism or Restriction	Empirical Equation—Pressure Drop		
	Fouling			
(12)	Fouling, sensible heat transfer	$h = assumed$	$\dfrac{\Delta T_s}{\Delta T_M} = 6,000 \times \dfrac{c}{h}$	$\times \dfrac{W(T_H - T_L)}{\Delta T_M} \times \dfrac{1}{LH}$
(13)	Fouling, latent heat transfer	$h = assumed$	$\dfrac{\Delta T_s}{\Delta T_M} = 3,333 \times \dfrac{1}{h}$	$\times \dfrac{W\lambda}{\Delta T_M} \times \dfrac{1}{LH}$
	Spiral Flow			
(14)	No phase change, $N_{Re} > N_{Rec}$	$\Delta P = 0.001 \dfrac{L}{s} \left[\dfrac{W}{d_s H}\right]^2 \left[\dfrac{1.3\, Z^{1/3}}{(d_s + 0.125)} \left(\dfrac{H}{W}\right)^{1/3} + 1.5 + \dfrac{16}{L}\right]$	(See Note 1)	
(15)	No phase change, $100 < N_{Re} < N_{Rec}$	$\Delta P = 0.001 \dfrac{L}{s} \left[\dfrac{W}{d_s H}\right] \left[\dfrac{1.035\, Z^{1/2}}{(d_s + 0.125)} \left(\dfrac{Z_l}{Z_b}\right)^{0.17} \left(\dfrac{H}{W}\right)^{1/2} + 1.5 + \dfrac{16}{L}\right]$	(See Note 1)	
(16)	No phase change, $N_{Re} < 100$	$\Delta P = \dfrac{L s Z}{3,385(d_s)^{.275}} \left(\dfrac{Z_l}{Z_b}\right)^{0.17} \left(\dfrac{W}{H}\right)$		
(17)	Condensing	$\Delta P = 0.005 \dfrac{L}{s} \left[\dfrac{W}{d_s H}\right]^2 \left[\dfrac{1.3\, Z^{1/3}}{(d_s + 0.125)} \left(\dfrac{H}{W}\right)^{1/3} + 1.5 + \dfrac{16}{L}\right]$		
	Axial Flow			
(18)	No phase change, $N_{Re} > 10,000$	$\Delta P = \dfrac{4 \times 10^{-5}}{s\, d_s^2} \left(\dfrac{W}{L}\right)^{1.8} 0.0115\, Z^{0.2} \dfrac{H}{d_s} + 1 + 0.03\, H$		
(19)	Condensing	$\Delta P = \dfrac{2 \times 10^{-5}}{s\, d_s^2} \left(\dfrac{W}{L}\right)^{1.8} \left[0.0115\, Z^{0.2} \dfrac{H}{d_s} + 1 + 0.03\, H\right]$		

Notes:
1. $N_{Rec} = 20,000\, (D_e / D_H)^{0.32}$
2. $G = W_o \rho_L /(A \rho_b)$
3. Surface-condition factor (Σ') for copper and steel = 1.0; for stainless steel = 1.7; for polished surfaces = 2.5.

11.41

$$\Delta P = \left[\frac{0.001 \times 61}{0.843} \right] \left[\frac{5.925}{0.375 \times 24} \right] \times \left[\frac{1.035 \times 8^{1/2} \times 1 \times 24^{1/2}}{(0.375 + 0.125) \, 5.925^{1/2}} + 1.5 + \frac{16}{61} \right]$$

$$\Delta P = 0.07236 \times 0.6583 \times 13.55 = 0.645 \; \text{lb/in}^2 \; (4.4 \; \text{kPa})$$

Because the pressure drop is less than the allowable, the spacing can be decreased. For the second trial, ¼ in (0.64 cm) spacing for both channels is adopted.

Because the heat-transfer equation for every factor except the plate varies directly with d_s, a new SOP can be calculated.

$$\Delta T_h / \Delta T_M = 0.848 \, (0.25/0.375) = 0.565$$

$$\Delta T_c / \Delta T_M = 0.927 \, (0.25/0.375) = 0.618$$

$$\Delta T_s / \Delta T_M = 0.052 \, (0.25/0.375) = 0.035$$

$$\Delta T_w / \Delta T_M = 0.050$$

$$\text{SOP} = 0.565 + 0.618 + 0.050 + 0.052 = 1.285$$

$$L = 1.285 \times 32.5 = 41.8 \; \text{ft} \; (12.7 \; \text{m})$$

$$A = 41.8 \times 2 \times 2 = 167 \; \text{ft}^2 \; (15.5 \; \text{m}^2)$$

The new pressure drop becomes:

Hot side:

$$\Delta P = \left[\frac{0.001 \times 41.8}{0.843} \right] \left[\frac{6.225}{0.25 \times 24} \right]$$

$$\times \left[\frac{1.035 \times 3.35^{1/2} \times 1 \times 24^{1/2}}{0.375 \times 6.225^{1/2}} + 1.5 + \frac{16}{41.8} \right]$$

$$\Delta P = 0.04958 \times 1.037 \times 11.80 = 0.607 \; \text{lb/in}^2 \; (4.2 \; \text{kPa})$$

Cold side;

$$\Delta P = \left[\frac{0.001 \times 41.8}{0.843} \right] \left[\frac{5.925}{0.25 \times 24} \right]$$

$$\times \left[\frac{1.035 \times 8^{1/2} \times 1 \times 24^{1/2}}{0.375 \times 5.925^{1/2}} + 1.5 + \frac{16}{41.8} \right]$$

$$\Delta P = 0.04958 \times 0.9875 \times 17.59 = 0.861 \; (5.9 \; \text{kPa})$$

The pressure drops are less than the maximum allowable. The plate spacing cannot be less than ¼ in (0.64 cm) for a 24-in (59.4 cm) plate width; decreasing the width would result in a higher than allowable pressure drop. Therefore, the design is acceptable.

6. Establish the final design of the spiral heat exchanger

The diameter of the outside spiral can now be calculated with Table 10 and the following equation:

$$D_S = [15.36 \times L \, (d_{sc} + d_{sh} + 2p) + C^2]^{1/2}$$

$$D_S = \{ 15.36 \, (41.8) \, [0.25 + 0.25 + 2 \, (0.125)] + 8^2 \}^{1/2}$$

$$D_S = 23.4 \; \text{in} \; (59.4 \; \text{cm})$$

TABLE 10   Some Spiral-Plate/Spiral Tube Exchanger Standards

Plate Widths, in	Outside Dia., Maximum, in	Core Dia., in
4	32	8
6	32	8
12	32	8
12	58	12
18	32	8
18	58	12
24	32	8
24	58	12
30	58	12
36	58	12
48	58	12
60	58	12
72	58	12

Channel spacings, in: $\frac{3}{16}$ (12 in maximum width), $\frac{1}{4}$ (48 in maximum width), $\frac{5}{16}$, $\frac{3}{8}$, $\frac{1}{2}$, $\frac{5}{8}$, $\frac{3}{4}$ and 1.
Plate thicknesses: stainless steel, 14-3 U.S. gage; carbon steel, $\frac{1}{8}$, $\frac{3}{16}$, $\frac{1}{4}$ and $\frac{5}{16}$ in.
See procedure for SI values.

Spiral-Tube Exchanger Design Standards

No. Tubes	Tube Spacing, in	Shellside Flow Area, sq in	Standard Lengths, ft		Heat-Transfer Area, sq ft
			Tubeside	Shellside	
Tube O.D., $\frac{1}{2}$ in					
30	$\frac{5}{16}$	6.30	19.14	22.2	75.0
30	$\frac{5}{16}$	6.30	27.5	30.8	108.0
30	$\frac{5}{16}$	6.30	33.41	37.2	132.15

For a spiral-plate exchanger, the best design is often that in which the outside diameter approximately equals the plate width.

Design summary:	
Plate width	24 in (60.9 cm)
Plate length	41.8 ft (12.7 m)
Channel spacing	$\frac{1}{4}$ in (both sides) (0.64 cm)
Spiral diameter	23.4 in (59.4 cm)
Heat-transfer area	167 ft^2 (15.5 m^2)
Hot-side pressure drop	0.607 lb/in^2 (4.2kPa)
Cold-side pressure drop	0.861 lb/in^2 (5.9 kPa)
U	38.8 Btu/(hr) (ft^2) (°f) (220.4 W/m^2°C)

Related Calculations.   Spiral heat exchangers have a number of advantages over conventional shell-and-tube exchangers: centrifugal forces increase heat transfer; the compact configuration results in a shorter undisturbed flow length; relatively easy cleaning; and resistance to fouling. These curved-flow units (spiral plate and spiral tube) are particularly useful for handling viscous or solids-containing fluids.

A spiral-plate exchanger is fabricated from two relatively long strips of plate, which are spaced apart and wound around an open, split center to form a pair of concentric spiral passages. Spacing is maintained uniformly along the length of the spiral by spacer studs welded to the plates.

For most services, both fluid-flow channels are closed by alternate channels welded at both sides of the spiral plate (Fig. 13). In some applications, one of the channels is left completely open (Fig. 13*d*), the other closed at both sides of the plate. These two types of construction prevent the fluids from mixing.

Spiral-plate exchangers are fabricated from any material that can be cold worked and welded, such as: carbon steel, stainless steels, Hastelloy B and C, nickel and nickel alloys, aluminum alloys, titanium, and copper alloys. Baked phenolic-resin coatings, among others, protect against corrosion from cooling water. Electrodes may also be wound into the assembly to anodically protect surfaces against corrosion.

Spiral-plate exchangers are normally designed for the full pressure of each passage. Because the turns of the spiral are of relatively large diameter, each turn must contain its design pressure, and plate thickness is somewhat restricted—for these three reasons, the maximum design pressure is 150 lb/in^2 (1033.5 kPa), although for smaller diameters the pressure may sometimes be higher. Limitations of materials of construction govern design temperature.

The shortcut rating method for spiral-plate exchangers depends on the same technique as that for shell-and-tube heat exchangers (which were discussed by Lord, Minton and Slusser in the previous procedure.)

Primarily, the method combines into one relationship the classical empirical equations for film heat-transfer coefficients with heat-balance equations and with correlations that describe the geometry of the heat exchanger. The resulting overall equation is recast into three separate groups that contain factors relating to the physical properties of the fluid, the performance or duty of the exchanger, and the mechanical design or arrangement of the heat-transfer surface. These groups are then multiplied together with a numerical factor to obtain a product that is equal to the fraction of the total driving force—or log mean temperature difference (ΔT_M or LMTD)—that is dissipated across each element of resistance in the heat-flow path.

When the sum of the products for the individual resistance equals 1, the trial design may be assumed to be satisfactory for heat transfer. The physical significance is that the sum of the temperature drops across each resistance is equal to the total available ΔT_M. The pressure drops for both fluid-flow paths must be checked to ensure that both are within acceptable limits. Usually, several trials are necessary to get a satisfactory balance between heat transfer and pressure drop.

Table 13 summarizes the equations used with the method for heat transfer and pressure drop. The columns on the left list the conditions to which each equation applies, and the second columns gives the standard forms of the correlations for film coefficients that are found in texts. The remaining columns in Table 13 tabulate the numerical, physical property, work and mechanical design factors—all of which together form the recast dimensional equation. The product of these factors gives the fraction of total temperature drop or driving force ($\Delta T_f/\Delta T_M$) across the resistance.

As stated, the sum of $\Delta T_h/\Delta T_M$ (the hot-fluid factor), $\Delta T_c/\Delta T_M$ (the cold-fluid factor), $\Delta T_s/\Delta T_M$ (the fouling factor),and $\Delta T_w/\Delta T_M$ (the plate factor) determines the adequacy of heat transfer. Any combinations of $\Delta T_f/\Delta T_M$ may be used, as long as the orientation specified by the equation matches that of the exchanger's flow-path.

The units in the pressure-drop equations are consistent with those used for heat transfer. Pressure drop is calculated directly in psi.

This procedure is the work of Paul E. Minton, Project Engineer, Engineering Department, Union Carbide Corporation, as reported in *Chemical Engineering* magazine. SI values were added by the handbook editor.

Nomenclature

A	Heat-transfer area, ft^2
B	Film thickness $(0.00187\ Z\Gamma/g_c s^2)^{1/3}$, ft
C	Core dia., in
c	Specific heat, Btu/(lb) (°F)
D_e	Equivalent dia., ft
D_H	Helix or spiral dia., ft
D_S	Exchanger outside dia., in
d_s	Channel spacing, in
f	Fanning friction factor, dimensionless
G	Mass velocity, lb/(hr) (ft^2)
g_c	Gravitational constant, ft/(hr)^2 (4.18 × 10^8)
H	Channel plate width, in
h	Film coefficient of heat transfer, Btu/(hr) (ft^2) (°F)
k	Therman conductivity, Btu/(hr) (ft^2) (°F/ft)
L	Plate length, ft
M	Molecular weight, dimensionless
P	Pressure, lb/in^2 (abs)
p	Plate thickness, in
ΔP	Pressure drop, psi
Q	Heat transferred, Btu
s	Specific gravity (referred to water at 20°C)
ΔT_M	Logarithmic mean temperature difference (LMTD), °C
U	Overall heat-transfer coefficient, Btu/(hr) (ft^2) (°F)
W	Flowrate, (lb/hr)/1,000
Γ	Condensate loading, lb/(hr) (ft)
Z	Viscosity, cp
θ	Time, hr
λ	Heat of vaporization, Btu/lb
μ	Viscosity, lb/(hr) (ft)
ρ_L	Liquid density, lb/ft^3
ρ_v	Vapor density, lb/ft^3
Σ, Σ'	Surface condition factor, dimensionless
σ	Surface tension, dynes/cm

Subscripts

b	Bulk fluid properties
c	Cold stream
f	Film fluid properties
H	High temperature
h	Hot stream
L	Low temperature
m	Median temperature (see Fig. 5)
s	Scale or fouling material
w	Wall, plate material

Dimensionless Groups

N_{Re}	Reynolds number
N_{Rec}	Critical Reynolds number
N_{Pr}	Prandtl number

SPIRAL-TUBE HEAT-EXCHANGER DESIGN

Design a liquid-condensing spiral-tube heat exchanger for the following conditions:

Conditions	Tubeside	Shellside		Shellside	Tubeside
Flowrate, lb/hr	30,000	3,422	kg/hr	13,620	1554
Inlet temperature, °C	20	121		−6.7°C	49.4°C
Outlet temperature, °C	80	121		26.7°C	49.4°C
Viscosity, liquid, cp	0.55	0.23			
Viscosity, vapor, cp	—	0.013			
Specific heat, (Btu/lb) (°F)	0.998	1.015	kJ/kg°C	4.2	4.25
Thermal conductivity, Btu/(ft²)					
(hr) (°F/ft)	0.368	0.398			
Specific gravity	0.999	0.95			
Heat of vaporization, Btu/lb	—	945		kJ/kg	2196
Vapor density, lb/ft³	—	0.0727		kg/m³	1.17
Material of construction	steel	($k = 26$)			
Allowable pressure drop, psi	15	5	kPa	103.3	34.5

Calculation Procedure:

1. Determine the rate of heat transfer and molecular weight tubeside and shellside for the heat exchanger
The rate of heat transfer, Btu/hr = (flow rate, lb/hr)(heat of vaporization of the shellside fluid) = (3422)(945) = 3,233,790 Btu/hr (3412 MJ/hr).

Use the relation for pseudomolecular weight when the thermal conductivity of one of the fluids is known. This relation is $M = 0.636 \, (c/k)^3 s^4$, where the symbols are as given in the Nomenclature list in this procedure. Substituting, we have:
Tubeside:

$$M = 0.636 \times 0.998^3 \times 0.999^4/0.368^3 = 12.63$$

Shellside:

$$M = 0.636 \times 1.015^3 \times 0.95^4/0.398^3 = 8.59 \Delta T_M (\text{or LMTD})$$

$$= (101 - 41)/\ln (101/41) = 66.5°C$$

2. Make a first trial to determine the heat-transfer surface area
Assume an overall heat-transfer coefficient, U, of 300 Btu/hr ft²°F (1.7 W/m² h °C). Then, $A = (3,233,790)/(300)(1.8)(66.5) = 90.05$ ft² (8.4 m²).

From a table of heat-exchanger design standards, such as Table 10, choose 30 0.5-in-outside-diameter tubes, 27.5-ft long, with a net free-flow area of 6.3 in²; $A =$ heat-transfer area = 108 ft²; $d_s = 0.3125$ in Then:

The maximum number of tubes for turbulent flow can be approximated by the term $W_i/2d_i z_i$:

$$n_{max} = 30/2 \times 0.402 \times 0.55 = 68$$

Because flow will be turbulent for the exchanger selected, the proper heat-transfer equations from Table 11 are: Eq. (1), tubeside; Eq. (8), shellside; Eq. (11), tube wall; and Eq. (13) fouling.

3. Perform the heat-transfer calculation
Tubeside, Eq. (1):

$$\frac{\Delta T_i}{\Delta T_M} = 9.07 \left[\frac{0.55^{0.467} \times 12.63^{0.222}}{0.999^{0.889}} \right] \times \left[\frac{30^{0.2} \times 60}{66.5} \right] \left[\frac{0.402^{0.8}}{30^{0.2} \times 27.5} \right]$$

$$= 9.07 \times 1.329 \times 1.781 \times 0.008884 = 0.191$$

Shellside, Eq. (8):

$$\frac{\Delta T_o}{\Delta T_M} = 2.92 \left[\frac{0.23^{1/3} \times 8.59^{1/3}}{0.95^2 \times 1.015} \right] \left[\frac{3.422^{4/3} \times 945}{66.5} \right] \times \left[\frac{1}{30 \times 27.5^{4/3} \times 0.5} \right]$$

$$= 2.92 \times 1.370 \times 73.28 \times 0.0008032 = 0.235$$

Tube wall, Eq. (11):

$$\frac{\Delta T_s}{\Delta T_M} = 88 \left[\frac{1}{26} \right] \left[\frac{3.422 \times 945}{66.5} \right] \left[\frac{0.098}{30 \times 27.5 \times 0.5} \right]$$

$$= 3.385 \times 48.63 \times 0.0002376 = 0.039$$

Fouling, Eq. (13):

$$\frac{\Delta T_w}{\Delta T_M} = 2{,}120 \left[\frac{1}{1{,}000} \right] \left[\frac{3.422 \times 945}{66.5} \right] \left[\frac{1}{30 \times 27.5 \times 0.5} \right]$$

$$= 2.120 \times 48.63 \times 0.002424 = 0.250$$

Sum of the Products (SOP):

$$SOP = 0.191 + 0.235 + 0.250 + 0.039 = 0.715$$

The assumed exchanger is adequate with regard to heat transfer. The next step is to check both sides for pressure drop.

4. Make the pressure-drop computation
Tubeside, Eq. (14):

$$\Delta P = 0.00268 \left[\frac{0.5^{0.15}}{0.999} \right] \left[\frac{30^{1.85}}{30^{1.85}} \right] \left[\frac{(27.5/0.402) + 16}{0.402^{3.75}} \right]$$

$$= 0.00268 \times 0.9151 \times 1.0 \times 2{,}574 = 6. \ (43.4 \ \text{kPa})$$

Shellside, Eq. (22):

$$\Delta P = 4 \times 10^{-4} \left[\frac{62.4}{0.00727} \right] \left[\frac{3.422}{27.5 \times 0.3125} \right]^2 \times 30$$

$$= 0.00004 \times 858.3 \times 0.1586 \times 30 = 0.163 \ \text{lb/in}^2 \ (1.1 \ \text{kPa})$$

Both the shellside and tubeside pressure drops are much lower than the allowable. Therefore, the next step is to try the next-smaller exchanger, which has a tube length of 19.14 ft.

TABLE 11 Empirical Heat-Transfer and Pressure-Drop Relationships for Rating Spiral-Tube Heat Exchangers

Eq. No.	Mechanism or Restriction	Empirical Equation—Heat Transfer	Numerical Factor	Physical Property Factor	Work Factor	Mechanical Design Factor	
	Tube Side						
(1)	No phase change (liquid), $N_{Re} > N_{Rec}$	$h = (1 + 3.54 D_e / D_H)\, 0.023\, cG\, (N_{Re})^{-0.2}(Pr)^{-2/3}$	$\dfrac{\Delta T_i}{\Delta T_M} = 9.07$	$\times \dfrac{Z_i^{0.467} M_i^{0.222}}{S_i^{0.889}}$	$\times \dfrac{W^{0.2}(t_H - t_L)}{\Delta T_M}$	$\times \dfrac{d_i^{0.8}}{n^{0.2} L}$	(See Note 1)
(2)	No phase change (gas), $N_{Re} > N_{Rec}$	$h = (1 + 3.54 D_e / D_H)\, 0.0144\, cG^{0.8}(D_i)^{-0.2}$	$\dfrac{\Delta T_i}{\Delta T_M} = 8.58$		$\times \dfrac{W^{0.2}(t_H - t_L)}{\Delta T_M}$	$\times \dfrac{d_i^{0.8}}{n^{0.2} L}$	(See Note 1)
(3)	No phase change (liquid), $N_{Rec} < N_{Re}$	$h = 1.86\, cG\, (N_{Re})^{-2/3}(Pr)^{-2/3}(L/D_e)^{-1/3}(\mu_f / \mu_b)^{-0.14}$	$\dfrac{\Delta T_i}{\Delta T_M} = 13.0$	$\times \dfrac{M_i^{0.222} Z_f^{0.14}}{S_i^{0.889} Z_b^{0.14}}$	$\times \dfrac{W^{2/3}(t_H - t_L)}{\Delta T_M}$	$\times \dfrac{d_i^{1/3}}{n^{2/3} L}$	(See Note 1)
(4)	Condensing vapor, horizontal, $N_{Re} < 2{,}100$	$h = 0.76\, k\, [g_c\, \rho_L^2 / \mu \Gamma]^{1/3}$	$\dfrac{\Delta T_i}{\Delta T_M} = 2.92$	$\times \dfrac{Z_i^{1/3} M_i^{1/3}}{c_p s_i^2}$	$\times \dfrac{W^{4/3}\lambda}{\Delta T_M}$	$\times \dfrac{1}{(nL)^{4/3} d_i}$	(See Note 1)
	Shell Side						
(5)	No phase change (liquid), $N_{Re} > N_{Rec}$	$h = (1 + 3.54 D_e / D_H)\, 0.023\, cG\, (N_{Re})^{-0.2}(Pr)^{-2/3}$	$\dfrac{\Delta T_o}{\Delta T_M} = 12.3$	$\times \dfrac{Z_o^{0.467} M_o^{0.222}}{S_o^{0.889}}$	$\times \dfrac{W_o^{0.2}(T_H - T_L)}{\Delta \mathbf{T}_M}$	$\times \dfrac{a}{(nd_o)^{1/2} L}$	(See Note 1)
(6)	No phase change (gas), $N_{Re} > N_{Rec}$	$h = (1 + 3.54 D_e / D_H)\, 0.0144\, cG^{0.8}(D_o)^{-0.2}$	$\dfrac{\Delta T_o}{\Delta T_M} = 11.5$		$\times \dfrac{W_o^{0.2}(T_H - T_L)}{\Delta T_M}$	$\times \dfrac{a}{(nd_o)^{1/2} L}$	(See Note 1)
(7)	No phase change (liquid), $N_{Re} < N_{Rec}$	$h = 1.86\, cG\, (N_{Re})^{-2/3}(Pr)^{-2/3}(L/D_e)^{-1/3}(\mu_f / \mu_b)^{-0.14}$	$\dfrac{\Delta T_o}{\Delta T_M} = 15.4$	$\times \dfrac{M_o^{0.222} Z_f^{0.14}}{S_o^{0.889} Z_b^{0.14}}$	$\times \dfrac{W^{2/3}(T_H - T_L)}{\Delta T_M}$	$\times \dfrac{a}{(nd_o)^{5/3} L}$	(See Note 1)
(8)	Condensing vapor, horizontal, $N_{Re} < 2{,}100$	$h = 0.76\, k\, [g_c \rho_L^2 / \mu \Gamma]^{1/3}$	$\dfrac{\Delta T_o}{\Delta T_M} = 2.92$	$\times \dfrac{Z_o^{1/3} M_o^{1/3}}{c_o s_o^2}$	$\times \dfrac{W^{4/3}\lambda}{\Delta T_M}$	$\times \dfrac{1}{nL^{4/3} d_o}$	(See Note 1)
(9)	Nucleate boiling, horizontal	$h = 4.02\, cG\, (N_{Re})^{-0.3}(Pr)^{-0.6}(\rho_v\, \sigma / P^2)^{-0.425}\, \Sigma$	$\dfrac{\Delta T_o}{\Delta T_M} = 0.352 \times \left[\dfrac{Z_o^{0.3} M_o^{0.2} \rho_o^{0.425}}{c_o s_o^{1.075}} \right]\left[\dfrac{\rho_o^{0.7}}{P_o^{0.85}} \right]$		$\times \left[\dfrac{W_o^{0.3}\lambda}{\Delta T_M} \right]$	$\times \left[\dfrac{1}{n^{0.3} d_o^{1.3}} \right] \Sigma'$	(See Notes 2 and 3)
	Tube Wall						
(10)	Tube wall, sensible heat	$h = (24 k_w)/(d_o - d_i)$	$\dfrac{\Delta T_w}{\Delta T_M} = 3{,}820$	$\times c/k_w$	$\times W(t_H - t_L)/\Delta T_M$	$\times (d_o - d_i)/nd_o L$	
(11)	Tube wall, latent heat	$h = (24 k_w)/(d_o - d_i)$	$\dfrac{\Delta T_w}{\Delta T_M} = 2{,}120$	$\times 1/k_w$	$\times W\lambda/\Delta T_M$	$\times (d_o - d_i)/nd_o L$	
	Fouling						
(12)	Fouling, sensible heat	$h =$ assumed	$\dfrac{\Delta T_i}{\Delta T_M} = 159$	$\times c/k_w$	$\times W(t_H - t_L)/\Delta T_M$	$\times (d_o - d_i)/nd_o L$	
(13)	Fouling, latent heat	$h =$ assumed	$\dfrac{\Delta T_i}{\Delta T_M} = 88$	$\times 1\, k_w$	$\times W\lambda/\Delta T_M$	$\times (d_o - d_i)/nd_o L$	

11.48

Eq. No.	Mechanism or Restriction	Empirical Equation—Pressure Drop
	Tube Side	
(14)	No phase change, $N_{Re} > N_{Rec}$	$$\Delta P = 0.00268\left[\frac{Z_i^{0.15}}{s_i}\right]\left[\frac{W_i}{n}\right]^{1.85}\left[\frac{L/d_i + 16}{d_i^{1.75}}\right]N_{PT}$$ (See Note 1)
(15)	No phase change, $100 < N_{Re} < N_{Rec}$	$$\Delta P = 0.00195\left[\frac{W_i^{1/3}Z_i^{2/3}Z_i^{0.14}L}{s_i Z_b^{0.14}n^{4/3}d_i^{13/3}}\right]$$ (See Note 1)
(16)	No phase change, $N_{Re} < 100$	$$\Delta P = 0.000544\left[\frac{LW_i Z_i}{ns_i d_i^4}\right]$$
(17)	Condensing	$$\Delta P = 0.00134\left[\frac{Z_i^{0.15}}{s_i}\right]\left[\frac{W_i}{n}\right]^{1.85}\left[\frac{L/d_i + 16}{d_i^{1.75}}\right]N_{PT}$$
	Shell Side	
(18)	No phase change, $N_{Re} > N_{Rec}$	$$\Delta P = 0.00152\left[\frac{Z_o^{0.15}}{s_o}\right]\left[L(nd_o)^{1.15}\right]\left[\frac{W_o^{1.85}}{a^3}\right]N_{PT}^{1.15}$$ (See Note 1)
(19)	No phase change, $100 < N_{Re} < N_{Rec}$	$$\Delta P = 0.0011\left[\frac{W_o^{1/3}Z_o^{2/3}(nd_o)^{5/3}L\,Z_o^{0.14}}{s_o\,a^3 Z_b^{0.14}}\right]N_{PT}^{5/3}$$ (See Note 1)
(20)	No phase change, $N_{Re} < 100$	$$\Delta P = 0.000308\left[\frac{LW_o Z_o}{s_o}\right]\left[\frac{(nd_o)^2}{a^3}\right]N_{PT}^2$$
(21)	Condensing (spiral flow)	$$\Delta P = 0.000757\left[\frac{W_o^{1.85}}{a^8}\right]\left[L(nd_o)^{1.15}\right]N_{PT}^{1.15}\left[\frac{Z_o^{0.15}}{s_o}\right]$$
(22)	Condensing (axial flow)	$$\Delta P = 4\times10^{-5}\times\frac{1}{s_o}\left[(W_o/Ld_o)^2 n\,N_{PT}\right]$$

Notes:
1. $N_{Rec} = 20,000\,(D_e/D_H)^{0.32}$
2. $G = W_{rL}/(A\rho_v)$
3. Surface-condition factor (Z') for copper and steel = 1.0; for stainless steel = 1.7; for polished surfaces = 2.5.

5. Correct the heat-transfer factors, if necessary
The four preceding heat-transfer factors must now be corrected as follows:

$$\text{Tubeside: } 0.191 \ (27.5/19.14)^{0.2} = 0.205$$
$$\text{Shellside: } 0.235 \ (27.5/19.14)^{4/3} = 0.381$$
$$\text{Fouling: } 0.250 \ (27.5/19.14) = 0.359$$
$$\text{Tube wall: } 0.039 \ (27.5/19.4) = 0.056$$

The new SOP becomes:

$$\text{SOP} = 0.205 + 0.381 + 0.359 + 0.056 = 1.001$$

This exchanger, which is adequate for heat transfer, will have the following pressure drops:
Tubeside:

$$\Delta P = 6.3 \left[\frac{(19.14/0.402) + 16}{(27.5/0.402) + 16} \right] = 4.8 \text{ lb/in}^2 \text{ (abs) (33.1 kPa)}$$

Shellside:

$$\Delta P = 0.163(27.5/19.14)^2 = 0.336 \text{ lb/in}^2 \text{ (2.3 kPa)}$$

The second exchanger selected is adequate with respect to heat transfer and pressure drop for both fluids. An exchanger with the same transfer area but with larger tubes would be more expensive. One with fewer or smaller tubes would have an excessive tubeside pressure drop.

6. Summarize the final design chosen

Design summary:

No. of tubeside passes	1	
Tube length, ft	19.14	5.8 m
Casing length, ft	22.2	6.8 m
Tube, outside dia., in	0.50	12.7 mm
Tube, inside dia., in	0.402	10.2 mm
Heat-transfer area, ft²	75.0	6.96 m²
Tubeside pressure drop, psi	4.8	33.1 kPa
Shellside pressure drop, psi	0.336	2.3 kPa
U, Btu/(hr) (ft²) (°F)	359	2.0 W/m² h °C

Related Calculations. Spiral-tube heat exchangers (like spiral-plate heat exchangers) offers several advantages over conventional shell-and-tube heat exchangers: secondary flow caused by centrifugal force that increases heat transfer; compact, short, undisturbed flow lengths that make spiral-tube exchangers ideal for heating and cooling viscous fluids, sludges and slurries; less fouling; relatively easy cleaning.

Spiral-tube exchangers are generally more expensive than shell-and-tube exchangers having the same heat-transfer surface. However, the spiral-tube exchanger's better heat transfer and lower maintenance cost often make it the more profitable choice.

Spiral-tube exchangers consist of one or more concentric, spirally wound coils clamped between a cover plate and a casing. Both ends of each coil are attached to a manifold fabricated from pipe or bar stock (Fig. 14).

The coils, which are stacked on top of each other, are held together by the cover plate and casing. Spacing is maintained evenly between each turn of the coil to create a uniform, spiral-flow path for the shellside fluid.

Coils can be formed from almost any material of construction, with some of the more common ones being carbon steel, copper and copper alloys, stainless steels, and nickel and nickel alloys. Tubes may have extended surfaces. Casings are made of cast iron, cast bronzes, and carbon and stainless steel.

Tubes may be attached to the manifolds by soldering, brazing, welding or, in some cases, rolling. Draining or venting can be facilitated by various manifold arrangements and casing connections. Flow through both the coil and casing may be single- or multipass (the latter by means of baffling).

Spiral-tube exchangers are available in sizes up to 325 ft^2., and pressures up to 600 psi. Tubeside pressures may be even higher.

The spiral-tube exchanger offers the following advantages over the shell-and-tube exchanger: (1) it is especially suited for low flows or small heat loads; (2) it is particularly effective for heating or cooling viscous fluids; since L/D ratios are much lower than those of straight-tube exchangers, laminar-flow heat transfer is much higher with spiral tubes; (3) its flows can be countercurrent (as with the spiral-plate exchanger, flows are not truly countercurrent; but again, the correction for this can be ignored); (4) it does not present the problems usually associated with differential thermal expansion; and (5) it is compact and easily installed.

The following are the chief limitations of the spiral-tube exchanger: (1) Its manifolds are usually small, making the repair of leaks at tube-to-manifold joints dif-

FIGURE 14 Spiral-tube heat exchanger. (*Chemical Engineering.*)

ficult (leaks, however, do not occur frequently); (2) it is limited to services that do not require mechanical cleaning of the inside of tubes (it can be cleaned mechanically on the shellside, and both sides can be cleaned chemically); (3) for some of its sizes, stainless steel coils must be provided with spacers to maintain a uniform shellside flow area—and these spacers increase pressure drop (this increase is not accounted for in the equations presented later).

The shortcut rating method given above for spiral-tube exchangers depends on the same technique as used for shell-and-tube exchangers (which is discussed by Lord, Minton and Slusser earlier in this section).

Primarily, the method combines into one relationship the classical empirical equations for film heat-transfer coefficients with heat-balance equations and with correlations that describe the geometry of the heat exchanger. The resulting overall equation is recast into three separate groups that contain factors relating to: the physical properties of the fluid, the performance or duty of the exchanger, and the mechanical design or arrangement of the heat-transfer surface. These groups are then multiplied together with a numerical factor to obtain a product that is equal to the fraction of the total driving force—or log mean temperature difference (δT_M or LMTD)—that is dissipated across each element of resistance in the heat-flow path.

When the sum of the products for the individual resistances equals 1, the trial design may be assumed satisfactory for heat transfer. The physical significance is that the sum of the temperature drops across each resistance is equal to the total available ΔT_M. The pressure drop for both fluid-flow paths must be checked to ensure that they are within acceptable limits. Usually, several trials are necessary to get a satisfactory balance between heat transfer and pressure drop.

Table 11 summarizes the equations used with spiral-tube exchangers. The column on the left presents the conditions to which each equation applies, and the second column gives the standard form of the film-coefficient correlation found in texts. The remaining columns tabulate the numerical, physical property, work, and mechanical design factors—all of which together form the recast dimensional equation. The product of these factors gives the fraction of the total temperature drop or driving force ($\Delta T_f/\Delta T_M$) across the resistance.

As stated, the sum of $\Delta T_i/\Delta T_M$ (the tubeside factor), $\Delta T_o/\Delta T_M$ (the shellside factor), $\Delta T_s/\Delta T_M$ (the fouling factor) and $\Delta T_w/\Delta T_M$ (the tube wall factor) determine the adequacy of heat transfer. Any combinations of $\Delta T_f/\Delta T_M$ may be used as long as the orientation specified by the equation matches that of the exchanger's flow-path.

The units in the pressure-drop equations are consistent with those used for heat transfer. Pressure drop is calculated directly in psi.

For many organic liquids, thermal conductivity data are either not available or difficult to obtain. Because molecular weights (M) are known, the Weber equation (which follows) yields thermal conductivities whose accuracies are satisfactory for most design purposes:

$$k = 0.86\ (cs^{4/3}/M^{1/3})$$

If, on the other hand, the thermal conductivity is known, a pseudomolecular weight may be used:

$$M = 0.636\ (c/k)^3 s^4$$

This procedure is the work of Paul E. Minton, Project Engineer, Engineering Department, Union Carbide Corporation, as reported in *Chemical Engineering* magazine. In his credits in his article, Mr. Minton thanks Graham Manufacturing Company "for permission to use certain design standards." Also, "he is grateful to the Union Carbide Corp. for permission to publish this article." Further, he notes that "The design method presented in this article is that used by Union Carbide Corp. for the thermal and hydraulic design of spiral-tube exchangers, and is somewhat different from that used by the fabricator." SI values were added by the handbook editor.

Nomenclature

A	Heat-transfer area, ft^2
a	Net free-flow area, in^2
B	Film thickness $(0.00187\ Z\Gamma/g_o s^2)^{1/3}$, ft
c	Specific heat, Btu/(lb) (°F)
D_e	Equivalent dia., ft
D_H	Helix or spiral dia., ft
D_i	Inside tube dia., ft
d	Tube dia., in
f	Fanning friction factor, dimensionless
G	Mass velocity, lb/(hr) (ft^2)
g_c	Gravitational constant, ft/(hr)^3 (4.18 × 10^3)
h	Film coefficient of heat transfer, Btu/(hr) (ft^2) (°F)
k	Thermal conductivity, Btu/(hr) (ft^2) (°F/ft)
L	Tube length, ft
M	Molecular weight, dimensionless
N_{PT}	No. tube passes/shell, dimensionless
n	No. tube/pass, dimensionless
P	Pressure, lb/in^2 (abs)
ΔP	Pressure drop, psi
Q	Heat transferred, Btu
s	Specific gravity (referred to water at 20 C)
T	Temperature shellside, °C
t	Temperature tubeside, °C
ΔT_M	Logarithmic mean temperature difference (LMTD), °C
U	Overall heat-transfer coefficient, Btu/(hr) (ft^2) (°F)
W	Flowrate, (lb/hr)/1,000
Γ	Condensate loading, lb/(hr) (ft)
Z	Viscosity, cp.
θ	Time, hr
λ	Heat of vaporization, Btu/lb
μ	Viscosity, lb/(hr) (ft)
ρ_L	Liquid density, lb/ft^3
ρ_v	Vapor density, lb/ft^3
Σ, Σ'	Surface condition factor, dimensionless
σ	Surface tension, dynes/cm

Subscripts

b	Bulk fluid properties
f	Film fluid properties
H	High temperature
i	Tubeside conditions
L	Low temperature
o	Shellside conditions
s	Scale or fouling material

w	Wall, tube material
Dimensionless Groups	
N_{Re}	Reynolds number
N_{Rec}	Critical Reynolds number
N_{Pr}	Prandtl number

See procedure for SI values.

HEAT-TRANSFER DESIGN FOR INTERNAL STEAM TRACING OF PIPELINES

Size a steam-tracing system to maintain an intermittently used fuel-oil unloading line at 140°F (60°C) when the following conditions apply:

Average temperature	20°F (-6.7°C)
Wind speed	20 mi/h (23.2 km/hr)
Line length	500 ft (152.4 m)
Size	12-in. nominal bore (304.8 mm)
Max. unloading rate	500 tons/h (450 t/hr)
Steam available at 65 lb/in² gage (447.9 kPa)	
Specific gravity of oil	0.985
Viscosity (see Fig. 5)	
Note: Viscosity in cP $\times$ 2.42 = viscosity in lb/(ft)(h)	
Thermal conductivity	0.0788 Btu/(h)(ft²)(°F/ft) (0.134 W/m°C)
Specific heat	0.47 Btu/(lb)(°F) (1.97 kJ/kg°C)

The maximum internal tracer length normally used is 150 ft (45.7 m).

1. Choose the type of steam tracing to use for the pipeline
As this fuel-oil unloading pipeline will be used only intermittently (as when a tanker or tank truck delivers a load of fuel oil), use internal steam tracing and an unlagged (no insulation) pipeline.

When internal-trace pipes are installed, all the available heat-transfer surface is utilized. The disadvantages are: (*a*) reduction in the equivalent internal diameter of the pipeline, (*b*) loss of ability to clean the pipeline by pigging or by using rotary brushes, (*c*) difficulty in cleaning fouled heat-transfer surfaces.

The trace pipe can only be installed in straight lengths of pipeline that are free of valves. Trace-pipe lengths have to be short, to prevent problems in supporting the trace pipe. It must enter and leave the pipeline frequently, increasing the possibility of leaks. Stresses arising due to differential expansion of the trace and the pipeline should be considered.

One possible application of the internal type of trace is for a fuel-oil unloading line, where the operating temperature is such that it is uneconomical to insulate the line, and where the line is used only intermittently. (Insulation is essential for externally traced lines.) Internal tracing is acceptable only if leakage of steam into the product conveyed by the pipeline can be tolerated.

2. Determine the Reynolds number for the pipeline
Using the nomenclature given, apply the Reynolds number equation. Or:
Heat loss—Neglecting the effect of the internal tracer at this point:

$$\text{Reynolds No., } N_{Re} = \frac{6.31W}{\mu D_{ip}}$$

where W_1 = mass flowrate
= 500 × 2,240 lb/h (1017 kg/hr)
μ = viscosity
= 200 cP at 140°F
D_{ip} = pipe dia.
= 12 in

$$\therefore N_{Re} = \frac{6.31 \times 2,240 \times 500}{200 \times 12}$$

= 2,940
L/D = 500

3. Find the tubeside heat transfer
Using, Fig. 15, the chart for determining the tubeside heat transfer based on the Reynolds number and Colburn factor,

$$\left(\frac{h_i D_{ip}}{k_a \times 12}\right)\left(\frac{C_p \cdot \mu}{k_a}\right)^{-1/3}\left(\frac{\mu}{\mu_w}\right)^{-0.14} = 8.0$$

whence,

$$h_i = \left(\frac{12k_a}{D_{ip}}\right)\left(\frac{C_p\mu}{k_a}\right)^{1/3}\left(\frac{\mu}{\mu_w}\right)^{0.14} \times 8.0$$

Assuming for the first trial that $t_s = 90°F$:

D_{ip} = 12 in
μ = 200 × 2.42 lb/(ft)(h) [at 140°F] (60°C)
μ_w = 520× 2.42 lb/(ft)(h) [at 90°F] (32.2°C)
C_p = 0.47 Btu/(lb)(°F) (1.97 kJ/kg C)
ρ = 0.985 × 62.4 lb/ft³ (1000 kg/m³)
k_a = 0.0778 Btu/(h)(ft³)(°F/ft)

$$\therefore h_i = \left(\frac{8 \times 12 \times 0.0778}{12}\right)\left(\frac{0.47 \times 200 \times 2.42}{0.0778}\right)^{1/3}\left(\frac{200 \times 2.42}{520 \times 2.42}\right)^{0.14}$$

= 0.621 (2,920)$^{1/3}$ (0.385)$^{0.14}$
= 0.621 (14.3) (0.875)
= 7.78 Btu/(h)(ft²)(°F)

take h = 8.0 Btu/(h)(ft²)(°F) (45.4 W/m² C)

Correcting to O.D. of pipe:

$$h_i = 8.0 \times (12.7/12) = 8.5 \text{ Btu/(h)(ft²)(°F)}$$

Hence,

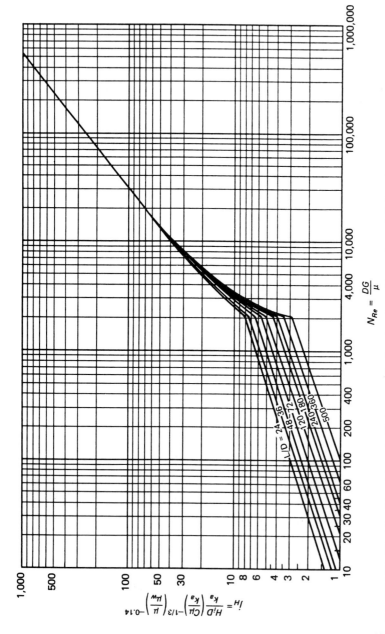

FIGURE 15 Chart for determining tubeside heat transfer, based on Reynolds number and Colburn factor. (*Chemical Engineering.*)

Axis label (x-axis): $N_{Re} = \dfrac{DG}{\mu}$

Axis label (y-axis): $j_H = \dfrac{H_i D}{k_a}\left(\dfrac{c\mu}{k_a}\right)^{-1/3}\left(\dfrac{\mu}{\mu_w}\right)^{-0.14}$

Curve labels: $L/D = 24{-}36$, $48{-}72$, $120{-}180$, $240{-}360$, 500

$$Q_{conv} = h_i A (t_p - t_s)$$
$$= 8.5A \ (140 - t_s) \ \text{Btu}/(\text{h})(\text{ft}^2)$$

and

$$Q_a = h_a A (t_s - t_a)$$

From Tables 12 and 13,

$$h_a = 1.83 \times 2.76$$
$$= 5.04 \ \text{Btu}/(\text{h})(\text{ft}^2)(°\text{F})$$
$$[\text{corrected for wind velocity}]$$
$$\therefore Q_a = 5.04A \ (t_s - 20) \ \text{Btu}/(\text{h})(\text{ft}^2)$$

4. Compute the heat loss from the pipeline
Since $Q_{conv} = Q_a$

$$8.5A \ (140 - t_s) = 5.04A \ (t_s - 20)$$

i.e.,

TABLE 12 Values of h_s for Pipes in Still Air*

Nominal pipe dia., in	$(t_s - t_a)$, °F [For an unlagged pipe $t_s = t_w$]							
	50	100	150	200	250	300	400	500
½	2.12	2.48	2.76	3.10	3.41	3.75	4.47	5.30
1	2.03	2.38	2.65	2.98	3.29	3.62	4.33	5.16
2	1.93	2.27	2.52	2.85	3.14	3.47	4.18	4.99
4	1.84	2.16	2.41	2.75	3.01	3.33	4.02	4.83
8	1.76	2.06	2.29	2.60	2.89	3.20	3.83	4.68
12	1.71	2.01	2.24	2.54	2.82	3.12	3.83	4.61
24	1.64	1.93	2.15	2.45	2.72	3.03	3.70	4.48

h_a = Btu/(h)(ft²)(°F) based on still air.
From McAdams, W. H., "Heat Transmission," 3rd ed., McGraw-Hill, New York, 1954, p. 179.

TABLE 13 Correction Factor for h_a at Different Wind Velocities*

Wind velocity mi/h	$(t_s - t_a)$, °F [For an unlagged pipe, $t_s - t_w$]				
	100	200	300	400	500
2.5	1.46	1.43	1.40	1.36	1.32
5.0	1.74	1.69	1.64	1.59	1.53
10.0	2.16	2.10	2.02	1.93	1.84
15.0	2.50	2.42	2.33	2.27	2.08
20.0	2.76	2.69	2.58	2.45	2.30
25.0	2.98	2.89	2.78	2.64	2.49
30.0	3.15	3.06	2.94	2.81	2.66
35.0	3.30	3.21	3.10	2.97	2.81

From Thermon Data Book, Premaberg (GB) Ltd.

$$1{,}190 - 8.5 \, t_s = 5.04 \, t_s - 100.8$$

$$\therefore t_s = 1{,}290/13.54 = 95°F \ (35°C)$$

There is an insignificant difference in μ_w between 90° (32.2°C) and 95°F; similarly, h_a does not change.

$$\therefore \text{heat loss} = (8.5) \, \pi (12.7/12)(140 - 95)$$
$$= 1{,}290 \ \text{Btu}/(\text{h})(\text{ft of pipe}) \ (4465 \ \text{kJ/hr m})$$

Due to the low temperature of the pipe, radiation heat loss has been neglected.

5. Calculate the tracer pipe size
For first trial, assume tracer is 1.5 in I.D. (1.9 in O.D.)

$$\text{Equivalent diameter, } D_e = \frac{144 - (1.9)^2}{12} = 11.7 \ \text{in} \ (297.2 \ \text{mm})$$

$$N_{Re} = D_e \frac{G}{\mu}$$

where:

where: G = mass velocity, lb/(h)(ft^2)

$$= \frac{500 \times 2{,}240}{\frac{\pi}{4}\left(1 - \frac{(1.9)^2}{144}\right)}$$
$$= 1.45 \times 10^6 \ \text{lb/(h)(ft}^2) \ (7.1 \times 10^6 \ \text{kg/hr m}^2)$$
$$D_e = 0.975 \ \text{ft}$$
$$\mu = 200 \ \text{cP at } 140°F$$
$$\therefore N_{Re} = \frac{11.7 \times 1.45 \times 10^6}{12 \times 200 \times 2.42} = 2.8 \times 10^3$$

From Fig. 15,

$$j_H = 9.0; \ N_{Re} = 2{,}800; \ L/D_e$$
$$= 150/0.975 = 154$$

Heat-transfer coefficient for heat transferred to the fuel oil based on I.D. of the 12 in (304.8 mm) nominal-bore pipe at the steam tracer surface where the fuel-oil viscosity is:

$$\mu_w = \frac{h_i D_e}{k_a} \left(\frac{C_p \mu}{k_a}\right)^{-1/3} \left(\frac{\mu}{\mu_w}\right)^{-0.34} = p$$

Assuming a 20-lb/in^2 (137.8 kPa) pressure drop along the steam tracer, the average steam pressure = 55 lb/in^2 gage (378.9 kPa) (steam available at 65 lb/in^2 gage) (447.9 kPa).

Average steam temperature = 303°F (150.6°C)

$$\therefore \mu_w = 1.0 \times 2.42 \text{ lb/(ft)(h)[for fuel oil]}$$

and heat-transfer coefficient at steam-tracer outside surface,

$$h_o = 9 \times \left(\frac{0.0778}{0.975}\right)\left(\frac{0.47 \times 200 \times 2.42}{0.0778}\right)^{1/3}\left(\frac{200}{1.0}\right)^{0.14}$$

$$= 21.6 \text{ Btu/(h)(ft}^2)(°F) \ (122.6 \text{ W/m}^2 \text{ C})$$

$$\frac{1}{U_o} = \frac{1}{h_o} + 0.005 = \frac{1}{21.6} + 0.005$$

$$= 0.051 \text{ Btu/(h)(ft}^2)(°F), \text{ approx.}$$

$$\therefore U_o = 19.6 \text{ Btu/(h)(ft}^2)(°F)$$

Heat transfer from steam tracer = heat lost from pipe without steam tracer:

$$Q_p = U_o A_p (t_{st} - t_p)$$

$$\therefore 1,290 = 19.6 A_p (303\text{-}140)$$

Hence,

$$A_p = \frac{1,290}{19.6 \times 163}$$

$$= 0.40 \text{ ft}^2/\text{ft run of pipe } (0.12 \text{ m}^2/\text{m})$$

Surface area of 1-in nominal-bore pipe = 0.344 ft^2/ft (0.10 m^2/m). Surface area of 1½-in nominal-bore pipe = 0.498 ft^2/ft (0.15 m^2/m). Therefore, 1½-in (38.1 mm) nominal-bore tracing will be acceptable.

6. Check the effect of fluid velocity on the outside film coefficient at the pipe wall
Since the internal tracer pipe increases the fluid velocity in the pipeline,

$$N_{Re} = \frac{D_e G}{\mu} = 1 \times \frac{1.45 \times 10^6}{2.42 \times 200} = 3,010$$

where G is based on the equivalent diameter of the pipe, and μ is at 140°F (60°C), the desired fuel-oil temperature.

$$j_H = 8.05 \text{ [from Fig. 15]}$$

There will be no significant increase in the calculated heat loss.

7. Find the pressure drop in the pipeline
Correction factor

$$= \frac{(D_i)^{2.4}(D_i + 0.5 D'_o)^{1.2}}{(D_i^2 - D'^2_o)^{1.8}}\left(\frac{\mu_w}{\mu}\right)^{0.14}$$

$$= \frac{(12)^{2.4}(12 + 0.95)^{1.2}}{(144 - 3.6)^{1.8}}\left(\frac{520}{200}\right)^{0.14} = 1.33$$

Check steam tracer pressure-drop

Maximum tracer length = 150 ft (45.7 m)

Heat load   = 150 × 1,290 Btu/h

Latent heat of steam   = 901 Btu/lb

$$\therefore \text{ Steam load} = \frac{150 \times 1,290}{901}$$

= 215 lb/h per tracer (97.6 kg/hr)

From Fig. 16, steam pressure drop at inlet conditions:

0.55 lb/(in²)(100-ft run)

$$\therefore \text{ pressure-drop/tracer} = 1.5 \times 0.275$$
$$= 0.412 \text{ lb/in}^2 \text{ (2.8 kPa)}$$

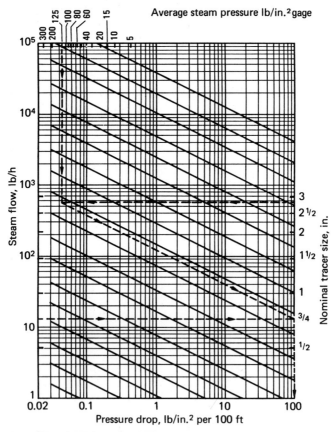

FIGURE 16 Chart for determining pressure drop in steam pipes. (*Chemical Engineering.*)

∴ the average steam temperature $= 312°F$ (65 lb/in^2 gage) (155.6°C, 447.9 kPa)

$$U_o = 19.6 \text{ Btu/(h)(ft}^2\text{)(°F)}$$

$$\Delta T = 312 - 140 = 172°F \ (77.8°C)$$

$$\therefore \text{ area of tracer required} = \frac{1,290}{(19.6 \times 172)} = 0.383 \text{ ft}^2/\text{ft} \ (0.117 \text{ m}^2/\text{m})$$

The surface area of 1½-in (38.1 mm) nominal-bore pipe is 0.498 (0.15 m^2/m) ft^2/ft.

Related Calculations. The internal heat tracer is less used than the external heat tracer. Reasons given in Step 1 above are sufficient enough to discourage designers from choosing internal tracing, except for specialized applications, such as the one considered here. Where the internal heat tracer can be used it provides maximum heat transfer because the heat source is immersed in the medium that is to be heated. Further, the internal heat tracer does not require out-of-round pipe insulation because the heating line is inside the pipe that is being heated.

This procedure is the work of I. P. Kohli, Consultant, as reported in *Chemical Engineering* magazine. SI values and numbered procedure steps were added by the handbook editor.

Nomenclature

A	Area of pipe, ft^2/ft of length
A_p	Outside area of tracer pipe, ft^2/ft length
A_1	Outside area of pipe lagging, ft^2/ft length
C_p	Specific heat, Btu/(lb)(°F)
D	Dia., in
D_e	Equivalent pipe dia., ft
D_i	Inside dia., in
D_{ip}	Pipe I.D., in
D_o	Outside dia., in
D_o'	Outside dia. of tracer pipe, in
d_i	I.D. of lagging, in
d_o	O.D. of lagging, in
G	Mass velocity, lb/(h)(ft^2)
h_a	Film coefficient to air, Btu/(h)(ft^2)
h_i	Inside film coefficient, Btu/(h)(ft^2)
h_o	Outside film coefficient, Btu/(h)(ft^2)
j_H	Colburn factor, dimensionless
k_a	Thermal conductivity, Btu/(h)(ft^2)(°F/ft)
k_1	Thermal conductivity of lagging, Btu/(h)(ft^2)(°F/ft)
L	Pipe length, ft
N_{Re}	Reynolds number, dimensionless
Q_a	Heat flux to air, Btu/(h)(ft^2)
Q_{conv}	Heat flux by convection, Btu/(h)(ft^2)
Q_p	Heat loss from pipe without tracer, Btu/(h)(ft^2)
q_2	Heat loss, Btu/(h)(ft of length)
t_a	Air temperature, °F
t_p	Temperature inside pipe, °F
t_s	Temperature, surface of lagging, °F
t_{st}	Average steam temperature, °F
t_w	Pipewall temperature, °F
T_p	Pipe temperature, °F

U_o	Heat transfer coefficient based on outside surface, Btu/(h)(ft²)(°F)
W_1	Mass flowrate, lb/h

Greek letters

μ	Viscosity, lb/(ft)(h)
μ_w	Viscosity at wall temperature, lb/(ft)(h)
ρ	Fluid density, lb/ft³

See procedure for SI values.

DESIGNING HEAT-TRANSFER SURFACES FOR EXTERNAL STEAM TRACING OF PIPELINES

Size an external steam tracing system, Fig. 17, to maintain a 3-in (76.2-mm) pipeline at 320°F (160°C). The line will be lagged with 1.5 in (38.1 mm) of insulation. Conditions at the pipeline are:

$$k_1 = 0.033 \text{ Btu/hr ft}^2\text{F ft } (0.057 \text{ W/m°C})$$

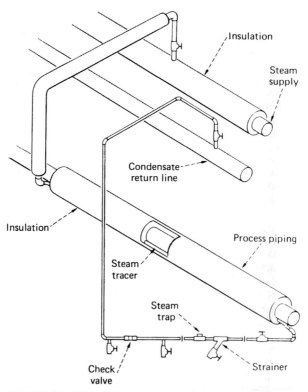

FIGURE 17   Typical steam-tracing system using external tracer pipe. (*Chemical Engineering.*)

TABLE 14 C_t, Thermal Conductance, Tracer to Pipe*

Tube size	C_t with no heat-transfer cement, Btu/(h)(°F)(ft of pipe)	C_t with heat-transfer cement Btu/(h)(°F)(ft of pipe)
⅜	0.295	3.44
½	0.393	4.58
⅝	0.490	5.73

See procedure for SI values.

$$\text{Wind speed} = 20 \text{ mph (23.2 km/h)}$$

$$\text{Minimum air temperature, } T_a = 20°F \ (-6.7°C)$$

$$\text{Line length} = 100 \text{ ft (30.5 m)}$$

$$\text{Steam for tracing is available at 150 lb/in}^2 \text{ (gage) (1033.5 kPa)}$$

$$\text{Pipe temperature, } t_p = 320°F \ (160°C)$$

Calculation Procedure:

1. Determine the heat load for the tracer pipe

Assume that one of the insulation schemes shown in Fig. 18 will be used for this pipeline. Further, assume a 5 lb/in² (34.5 kPa) pressure drop along the external steam tracer pipe. Then, the average steam temperature is, using the nomenclature in the previous procedure:

$$t_{st} = \frac{366 + 363°F}{2} = 364.5°F, \text{ say, } 364°F \ (184°C)$$

$$\therefore \text{ average temperature} = 0.5(t_{st} + t_p) \ [°F]$$

$$0.5(364 + 320) \ [°F]$$

$$= 342°F \ (172°C)$$

Allowing for tracer, I.D. of lagging = 4 in (101.6 mm)

$$\therefore q_2 = \frac{2\pi k_1(t_p - t_s)}{\log_e \frac{d_o}{d_i}} = h_a A_1(t_s - t_a)$$

$$d_o = 7 \text{ in}, d_i = 4 \text{ in (101.6 mm)}$$

Assuming for the first trial that $h_a = 4.0$ Btu/(h)(ft²)(°F) (22.7 W/m²°C):

$$q_2 = \frac{2 \times \pi \times 0.033(342 - t_s)}{2.303 \log_{10} (7.0/4.0)}$$

$$= 4.0\pi \times (7.0/12) \times 1(t_s - 20)$$

or

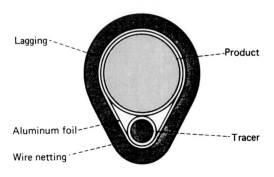

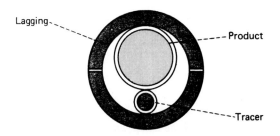

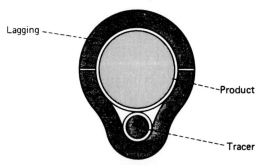

FIGURE 18 Various configurations of tracer piping and insulation lagging. (*Chemical Engineering.*)

$$\frac{0.0895(342 - t_s)}{0.245} = 7.35(t_s - 20)$$

or

$$0.368(342 - t_s) = 7.35t_s - 147$$

or

$$126 - 0.368t_s = 7.35t_s - 147$$

$$\therefore t_s = \frac{273}{7.718} = 35°F \ (1.6°C)$$

Tables 12, 13, and 14 are not sufficiently accurate in this region, but the value of 4.0 taken for h_a is on the safe side, allowing for wind speed.

$$\therefore q_2 = 7.35(t_s - 20) \ Btu/(h)(ft \ run \ of \ pipe)$$

$$= 7.35(35 - 20)$$

$$= 113 \ Btu/(h)(ft \ run \ of \ pipe) \ (391 \ kJ/hr \ m)$$

2. Compute the required size of the steam-tracer pipe
Do this by using the thermal-conductance data on tracer-to-pipe: Btu/(h)(°F)(ft of pipe). This takes into account the heat-transfer coefficient and pipe surface area. These have been found by experiment.

If one ½-in (12.7-mm) tracer without cement is used, heat transfer

$$= 0.393(t_{st} - t_p)$$

$$= 0.393(364 - 320) = 17.3 \ Btu/(h)(ft \ run) \ (59.9 \ kJ/hr \ m)$$

This does not meet the heat load requirement. Two ½-in (12.7-mm) tracers without cement do not overcome the problem. If one ½-in tracer with heat-transfer cement is installed, the heat-transfer rate is:

$$4.58(364 - 320) = 201 \ Btu/(h)(ft \ of \ run) \ (695.7 \ kJ/hr \ m)$$

A ⅜-in (9.5-mm) tracer with cement will yield a heat-transfer rate of:

$$3.44(364 - 320) = 151 \ Btu/(h)(ft \ of \ run) \ (522.7 \ kJ/hr \ m)$$

Standard practice requires the installation of ½-in (12.7-mm) tracer pipes. There is always some overdesign with steam tracers.

Related Calculations. External tracing involves the placing of steam pipes or tubes outside the pipeline. The traced line is then insulated by using preformed sectional insulation.

Heat transfer from trace to pipeline is by conduction, convection in the air space, and radiation. The contact area between the trace and the pipeline is quite small. However, when heat-sensitive liquids are being heated, or when the pipe is plastic-lined, this contact may give rise to undesirable hot spots. In such cases, asbestos packing rings are often used to eliminate any direct contact between the trace and the pipeline.

The simplest method of external tracing is to wrap copper tube around the pipeline, and then to cover the traced pipe with insulation. This is the only way to trace around valves, pipe fittings and instruments (Fig. 19). This procedure is unsuitable for horizontal runs because steam condensate collects at low points and may freeze during a shutdown.

It is essential to ensure that the trace lines are self-draining. Copper tubing of 0.5-in (12.7-mm) O.D. and 0.035-in (0.89-mm) wall thickness is usually used with straight piping. If the tubing has to be bent into a small radius (as when tracing valves), 0.375-in-O.D. (9.5-mm) tubing may be used.

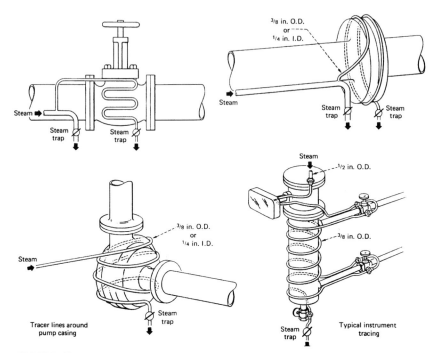

FIGURE 19   How various fittings, pumps, and instruments are steam-traced. (*Chemical Engineering.*)

The length of a single trace tube (from steam supply valve to steam trap) is limited by pressure drop in the trace. The trap should have a condensate drainage capacity to match the heating load. At a steam pressure of 100 lb/in² (gage) (689 kPa) or higher, the length of a single trace should not exceed 200 ft (60.9 m). If the steam pressure is lower, a tracer length of 100 ft (30.5 m) is recommended.

To improve contact, the trace tube may be wired to the pipeline. Even then, the conductive heat-transfer rate is quite low. It may be increased by putting a layer of heat-conducting cement (graphite mixed with sodium silicate or other binders) between the trace and the pipeline. This provides much more surface for conductive heat transfer.

When higher heat loads are desirable, straight lengths of ½-in (12.7-mm) carbon-steel pipe are clipped along the pipeline. The number of tracer pipes depends upon the heat load—large-diameter pipes carrying liquids that have a high melting point may require up to ten tracers. Steel bands, fitted by using a packing-case banding machine, help to minimize air gaps between the pipeline and tracers. At pipeline bends, the tracers are also bent.

At valves and flanges, the tracers are formed into loops that also function as expansion joints. The loops are formed in a nearly horizontal plane, to ensure self-drainage. The traced pipeline is covered with larger-sized preformed lagging.

In some cases, the traced line is wrapped with aluminum foil and then covered with shaped lagging (Fig. 4). The foil increases the radiation heat transfer. It is essential that the space between the pipeline and the tracer be kept free of particles of lagging material.

Conductive heat transfer may be enhanced by welding the trace on the pipeline. However, welding causes problems due to differential expansion of the pipeline and trace. Because horizontal pipelines are traced on the lower half, welding is a difficult operation. It is more convenient to use heat-transfer cement, as noted earlier.

This procedure is the work of I. P. Kohli, Consultant, as reported in *Chemical Engineering* magazine. SI values were added by the handbook editor.

AIR-COOLED HEAT EXCHANGER: PRELIMINARY SELECTION

Kerosene flowing at a rate of 250,000 lb/h (31.5 kg/s) is to be cooled from 160°F (71°C) to 125°F (51.6°C), for a total heat duty of 4.55 million Btu/h (1.33 MW). How large an air cooler (sometimes called a *dry* heat exchanger) is needed for this service if the design dry-bulb temperature of the air is 95°F (35°C)?

Calculation Procedure:

1. Determine the temperature rise of the air during passage through the cooler
From Table 15 estimate the overall heat-transfer coefficient for an air cooler handling kerosene at 55 Btu/(h·ft²·°F) [312.3 W/(m²·K)]. Then the air-temperature rise is $t_2 - t_1 = 0.005U\{[(T_1 + T_2)/2] - t_1\}$, where t = inlet air temperature, °F or °C; t_2 = outlet air temperature, °F or °C; U = overall heat-transfer coefficient, Btu/(h·ft²·°F) [W/(m²·K)]. T_1 = cooled fluid inlet temperature, °F or °C; T_2 = cooled fluid outlet temperature, °F or °C. Substituting yields $t_2 - t_1 = 0.005(55)\{[(160 + 125)/2] - 95\} = 13.06$°F (7.2°C).

Next, from Fig. 20, the correction factor for a process-fluid temperature rise of $160 - 125 = 35$°F (19.4°C) is 0.94. So the corrected temperature rise = $f(t_2 - t_1) = 0.94(13.06) = 12.28$°F (6.8°C). Therefore, $t_2 = 95 + 12.28 = 107.28$°F (41.8°C).

2. Find the log mean temperature difference (LMTD) for the heat exchanger
Use the relation LMTD = $(\Delta t_2 - \Delta t_1)/\ln(\Delta t_2/\Delta t_1)$. Or, LMTD = $[(160 - 107.28) - (125 - 95)]/\ln[(160 - 107.28)/(125 - 95)] = 40.30$. This value of

TABLE 15   Heat-Transfer Coefficients for Air-Cooled Heat Exchangers*

Liquid cooled	Heat-transfer coefficient	
	Btu/(h·ft²·°F)	W/(m²·K)
Diesel oil	45–55	255.5–312.3
Kerosene	55–60	312.3–340.7
Heavy naphtha	60–65	340.7–369.1

Chemical Engineering.

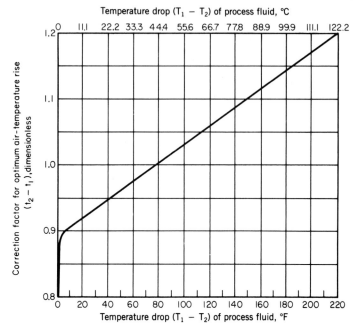

FIGURE 20   Correction factors for estimated temperature rise. (*Chemical Engineering.*)

the LMTD must be corrected by using Fig. 21 for temperature efficiency P and a correlating factor R. Thus, $P = (t_2 - t_1)/(T_1 - t_1) = (107.28 - 95)/(160 - 95) = 0.189$. Also, $R = (T_1 - T_2)/(t_2 - t_1) = (160 - 125)/(107.28 - 95) = 2.85$. Then, from Fig. 21, LMTD correction factor $= 0.95$, and the corrected LMTD $= f(\text{LMTD}) = 0.95(40.30) = 38.29°\text{F}$ (21.2°C).

3. Determine the hypothetical bare-tube area needed for the exchanger
Use the relation $A = Q/U\Delta T$, where A = hypothetical bare-tube area required, ft² (m²); Q = heat transferred, Btu/h (W); ΔT = effective temperature difference across the exchanger = corrected LMTD. Substituting gives $A = 4,550,000/55(38.29) = 2160$ ft² (200.7 m²).

4. Choose the cooler size and number of fans
Enter Table 16 with the required bare-tube area, and choose a 12-ft (3.6-m) wide cooler with either four rows of 40-ft (12-m) long tubes with two fans, for a total bare surface of 2284 ft² (205.6 m²), or five rows of 32-ft (9.6-m) long with two fans for 2288 ft² (205.5 m²) of surface. From Fig. 22, the fan hp for the cooler would be $1.56(2284/100) = 35.63$ hp (25.6 kW).

Related Calculations.   Air coolers are widely used in industrial, commercial, and some residential applications because the fluid cooled is not exposed to the atmosphere, air is almost always available for cooling, and energy is saved because there is no evaporation loss of the fluid being cooled.

Typical uses in these applications include process-fluid cooling, engine jacket-water cooling, air-conditioning condenser-water cooling, vapor cooling, etc. Today

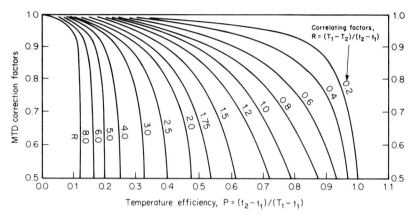

FIGURE 21 MTD correction factors for one-pass crossflow with both shell side and tube side unmixed. T represents hot-fluid characteristics, and t represents cold-fluid characteristics. Subscripts 1 and 2 represent inlet and outlet, respectively. (*Chemical Engineering.*)

TABLE 16 Typical Air-Cooled Heat-Exchanger Cooling Area, ft^2 (m^2)*

Approximate cooler width		Tube length		Fans per unit	No. of 1-in (2.5-cm) tube rows in depth on 2⅜-in (6-cm) pitch	
ft	m	ft	m		4	5
12	3.66	32	9.8	2	1827 (169.7)	2288 (212.6)
		36	10.9	2	2056 (191.0)	2574 (239.1)
		40	12.2	2	2284 (212.2)	2861 (265.8)
14	4.27	14	4.3	1	931 (86.5)	1166 (108.3)
		16	4.9	1	1064 (98.8)	1333 (123.8)

Chemical Engineering.

there are about seven leading design manufacturers of air coolers in the United States.

The procedure given here depends on three key assumptions: (1) an overall heat-transfer coefficient is assumed, depending on the fluid cooled and its temperature range; (2) the air temperature rise $t_2 - t_1$ is calculated by an empirical formula; (3) bare tubes are assumed and fan hp (kW) is estimated on this basis to avoid the peculiarities of one fin type. By using the empirical formula given in step 1 of this procedure, the size air cooler obtained will be within 25 percent of optimum. This is adjusted for greater accuracy through use of the correction factor shown in Fig. 20.

Since no existing computer program is capable of considering all variables in optimizing air coolers, the procedure given here is useful as a first trial in calculating an optimum design. The flow pattern and correction factors used for this estimating procedure are those for one-pass cross-flow with both tube fluid and air unmixed as they flow through the exchanger.

Where additional correction factors are needed for different flow patterns across the exchanger, the designer should consult the standards of the Tubular Exchanger

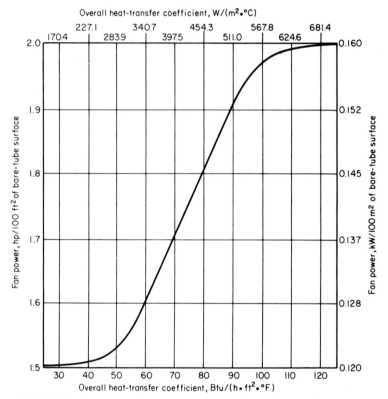

FIGURE 22 Approximate fan power requirements for air coolers. (*Chemical Engineering.*)

Manufacturers Association (TEMA). Similar data will be found in reference books on heat exchange.

The procedure given here is the work of Robert Brown, General Manager, Happy Division, Therma Technology, Inc., as reported in *Chemical Engineering* magazine. Note that the procedure given is for a preliminary selection. The final selection will usually be made in conjunction with advice and guidance from the manufacturer of the air cooler.

HEAT EXCHANGERS: QUICK DESIGN AND EVALUATION

Find the required surface area and shell-side flow rate of a cross-flow heat exchanger with four single-pass tube rows being designed to meet the following conditions: tube mass flow rate $\dot{m}$ = 22,200 lb/h (10,070 kg/h); tube specific heat capacity, c = 0.20 Btu/lb·°F (0.84 kJ/kg·°C); shell specific heat capacity, C = 0.24 Btu/lb·°F (1.00 kJ/kg·°C); tube-side inlet fluid temperature, t_1 = 500°F

(260°C); tube-side outlet fluid temperature, $t_2 = 320$°F (160°C); shell-side inlet liquid temperature, $T_1 = 86$°F (30°C); shell-side outlet liquid temperature, $T_2 = 131$°F (55°C); overall heat-transfer coefficient, $U = 9.0$ Btu/h $\cdot$ ft² $\cdot$ °F (51.1 W/m² $\cdot$ °C) (183.9 kJ/h $\cdot$ m² $\cdot$ °C).

Another unit, a 1-shell-pass and 2-tube-pass heat exchanger with a vertical shell-side baffle for divided flow, performs as follows: $\dot{m} = 33,500$ lb/h (15,200 kg/h); $c = 0.98$ Btu/lb $\cdot$ °F (4.10 kJ/kg $\cdot$ °C); shell mass flow rate $\dot{M} = 50,000$ lb/h (22,680 kg/h); $C = 0.60$ Btu/lb $\cdot$ °F (2.51 kJ/kg $\cdot$ °C); $t_1 = 270$°F (132.2°C); $T_1 = 520$°F (271.1°C); $U = 12.6$ Btu/h $\cdot$ ft² $\cdot$ °F (71.5 W/m² $\cdot$ °C) (257.4 kJ/h $\cdot$ m² $\cdot$ °C); heat exchanger surface area, $A = 2200$ ft² (204.4 m²). Evaluate the performance of this unit by finding its thermal effectiveness, its efficiency, the outlet temperature of the tube-side fluid, and the outlet temperature of the shell-side vapor.

Calculation Procedure:

1. Compute the ratio of shell-side liquid to tube-side fluid temperature differences and the thermal effectiveness, or temperature efficiency, of the cross-flow unit
The shell-to-tube ratio of temperature differences is found by $R = (T_1 - T_2)/(t_2 - t_1) = (86 - 131)/(320 - 500) = 0.25$. Thermal effectiveness, $P = (t_2 - t_1)/(T_1 - t_1) = (320 - 500)/(86 - 500) = 0.43$.

2. Compute the shell-side vapor flow rate
Shell-side vapor flow rate is $\dot{M} = \dot{m}c/RC = (22,200)(0.20)/[(0.25)(0.24)] = 74,000$ lb/h (33,570 kg/h).

3. Determine the number of transfer units and heat exchanger efficiency
The point where $R = 0.25$ and $P = 0.43$ on Fig. 31, shown amongst several figures appearing after step 7, corresponds to values for the number of transfer units, NTU = 0.62 and heat exchanger efficiency, $F = 0.99$. This value for F shows that the design has an efficiency close to that for a pure countercurrent configuration where $F = 1.00$.

4. Compute the heat exchanger surface area
To find the area use the formula $A = (\text{NTU})(\dot{m}c)/U = (0.62)(22,200)(0.20)/9.0 = 306$ ft² (28.4 m²).

5. Compute R and NTU for the shell-and-tube unit
Substitute appropriate values into the following equations, thus $R = \dot{m}c/\dot{M}C = (33,500)(0.98)/(50,000)(0.60) = 1.09$ and NTU $= UA/\dot{m}c = (12.6)(2200)/(33,500)(0.98) = 0.84$.

6. Determine the thermal effectiveness and heat exchanger efficiency
The point where curves for R and NTU intersect on Fig. 26 corresponds to a thermal effectiveness, $P = 0.42$ and an efficiency, $F = 0.89$. For the configuration of this unit, the value of F can be considered acceptable.

7. Find the exit temperatures of the tube-side fluid and the shell-side vapor
The tube-side fluid exit temperature, $t_2 = P(T_1 - t_1) + t_1 = 0.42(520 - 270) + 270 = 375$°F (190.6°C), and the shell-side vapor exit temperature, $T_2 = T_1 - R(t_2 - t_1) = 520 - 1.09(375 - 270) = 405$°F (297.2°C).

Related Calculations.   On the design and performance charts, Figs. 23 through 35, heat exchanger efficiency, F = true mean temperature difference/logarithmic temperature difference for countercurrent flow and relates the actual rate of heat transfer, Q, to the theoretical rate, $U \times A \times$ logarithmic temperature difference. True mean temperature differences has been solved analytically for each configuration. This, in conjunction with the new NTU curves, eliminates the need for trial-and-error calculations to determine the design and evaluate the performance of specific heat exchangers. By establishing desired conditions which in effect specify any two of the four parameters, the other two parameters may then be read directly from a chart and used to design a unit and/or to evaluate its performance.

The 1-shell-pass–3-tube-pass and 1-shell-pass–2-tube-pass heat exchangers represent conventional shell-and-tube units such as those for steam heating, heating of one process stream by cooling another, and condensation and cooling by a cooling-water utility. Shell-side pressure drops through divided-flow units are typically one-eighth of those through conventional shell-and-tube heat exchangers; hence divided flow units are recommended where low shell-side pressure drops are required.

Cross-flow heat exchangers differ from the conventional and divided-flow shell-and-tube units in that they have tube-bank arrangements over which another stream flows perpendicular to the tubes. Typical applications of cross-flow units include air cooling of overhead condensate streams and trim-product coolers.

This calculation procedure is based upon the work of Jeff Bowman, E.I. du Pont de Nemours Co., and Richard Turton, assistant professor of chemical engineering at West Virginia University, as reported in *Chemical Engineering* magazine. Note that final selection of a unit will usually be made in conjunction with advice and guidance from the manufacturer of the unit.

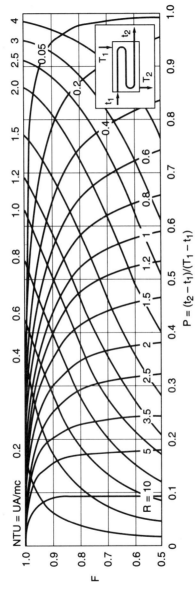

FIGURE 23 Design and performance chart for a 1-shell-pass and 3-tube-pass exchanger. (*Chemical Engineering.*)

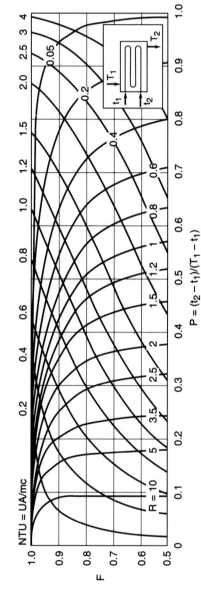

FIGURE 24 Design and performance chart for a 1-shell-pass and 4-tube-pass exchanger. (*Chemical Engineering.*)

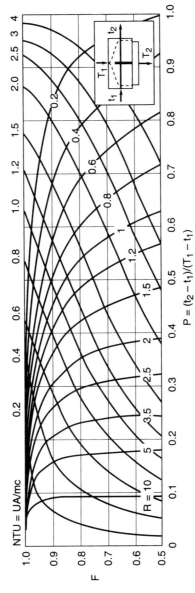

FIGURE 25 Design and performance chart for a 1-shell-pass and 1-tube-pass exchanger with a vertical shell-side baffle. (*Chemical Engineering.*)

11.75

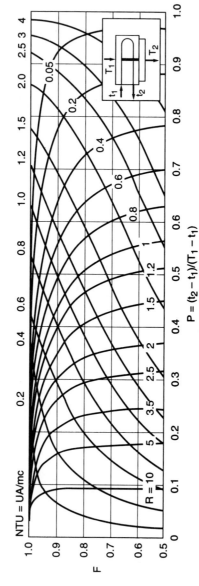

FIGURE 26 Design and performance chart for a 1-shell-pass and 2-tube-pass exchanger with a vertical shell-side baffle. (*Chemical Engineering.*)

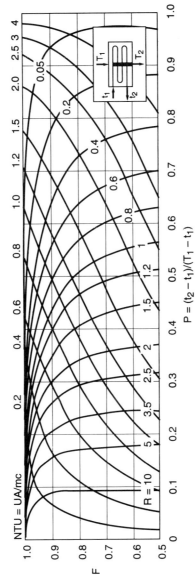

FIGURE 27 Design and performance chart for a 1-shell-pass and 4-tube-pass exchanger with a vertical shell-side baffle. (*Chemical Engineering.*)

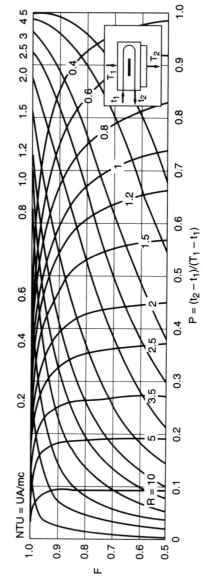

FIGURE 28 Design and performance chart for a 1-shell-pass and 2-tube pass exchanger with a horizontal shell-side baffle. (*Chemical Engineering.*)

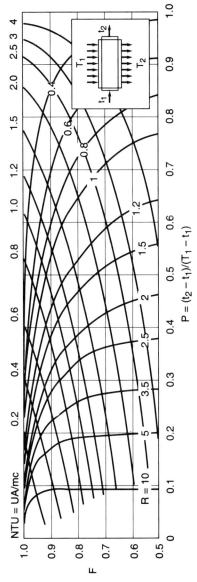

FIGURE 29 Design and performance chart for cross-flow exchanger with two single-pass row tubes. (*Chemical Engineering.*)

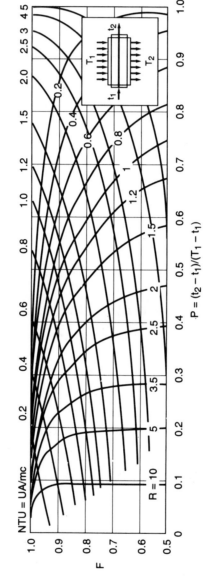

FIGURE 30 Design and performance chart for a cross-flow exchanger with three single-pass tube rows. (*Chemical Engineering.*)

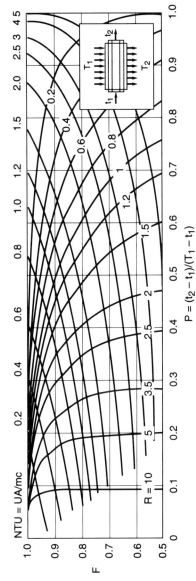

FIGURE 31 Design and performance chart for cross-flow exchanger with four single-pass tube rows. (*Chemical Engineering.*)

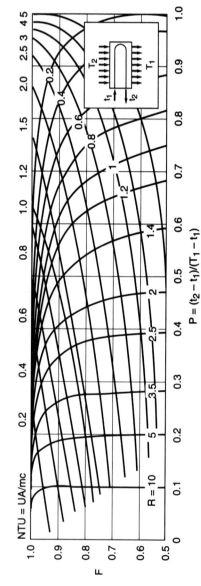

FIGURE 32 Design and performance chart for a cross-flow exchanger with a 2-tube-pass. (*Chemical Engineering.*)

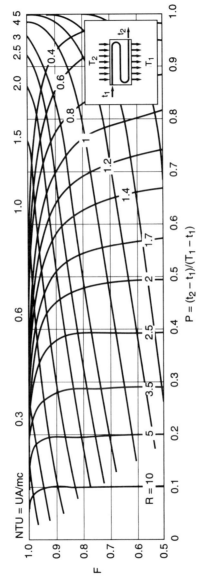

FIGURE 33 Design and performance chart for a cross-flow exchanger with a 3-tube pass. (*Chemical Engineering.*)

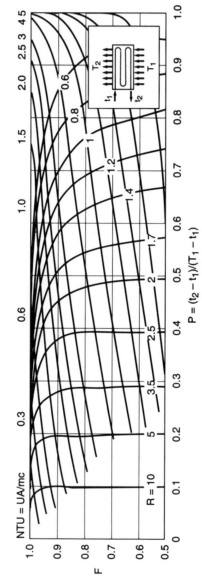

FIGURE 34 Design and performance chart for a cross-flow exchanger with a 4-tube pass. (*Chemical Engineering.*)

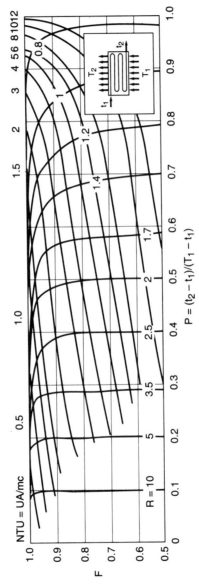

FIGURE 35 Design and performance chart for a cross-flow exchanger with a 5-tube pass. (*Chemical Engineering.*)

SECTION 12
REFRIGERATION

REFRIGERATION REQUIRED TO COOL AN OCCUPIED BUILDING

The building in Fig. 1 is to be maintained at 75°F (23.9°C) dry bulb and 64.4°F (18.0°C) wet-bulb temperatures. This building is situated between two similar units which are not cooled. There is a second-floor office above and a basement below. The south wall, containing 45 ft² (4.18 m²) of glass area, has a southern exposure. On the north side of the building there are two show windows which are ventilated to the outside and are at outside temperature conditions. Between the show windows is a doorway. This doorway is normally closed but it is frequently opened and allows an average of 600 ft³/min (16.98 m³) of outside air to be admitted. Opening of the door by customers will cause slightly more than two air changes per hour in the building. The number of persons in the building is 35; lighting is 1100 watts

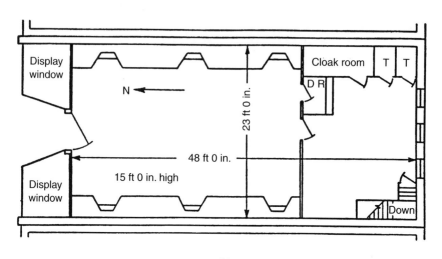

SI Values

ft	m
48	14.6
15	4.6
23	7.0

FIGURE 1 Plan of building cooled by refrigeration.

on a sunny day. Basement temperature is 80°F (26.7°C). The maximum outside conditions for design purposes are 95°F (35.0°C) dry bulb and 78°F (25.6°C) wet-bulb temperature. There is a 0.5 hp (373 W) fan motor in the interior of the building. What is the refrigeration load for cooling this building?

Calculation Procedure:

1. Assemble the overall coefficients of heat transfer for the building materials
Using the ASHRAE handbook, find the U values, Btu/ft² · h · °F as follows: East and west walls (24-in [60.96-cm] brick, plaster one side), 0.16; North partition (1.25-in [3.18-cm] tongue-and-groove wood), 0.60; Plate-glass door, 1.0; South wall (13-in [33-cm] brick, plaster one side), 0.25; Windows (single-thickness glass), 1.13; Floor (1-in [2.54-cm] wood, paper, 1-in [2.54-cm] wood over joists), 0.21; Ceiling (2-in [5.08-cm] wood on joists, lath and plaster), 0.14. To determine the SI overall coefficient, multiply the given value above by 5.68 to obtain the W/m² · °C. Thus, the values are: 0.908; 3.4; 5.68; 1.42; 6.42; 1.19; 0.79, respectively.

2. Compute the temperature differences for the walls, ceiling, and floor
For the walls and ceiling the temperature difference = outside design temperature − indoor design temperature = 95 − 75 = 20°F (36°C). For the floor, the temperature difference = 80 − 75 = 5°F (9°C).

3. Calculate the heat flow into the building
The heat leakage for any surface = U(ft² [m²] surface area)(temperature difference). For each of the surfaces in this building the heat leakage is computed thus: East and west walls = (0.16)(1440)(20) = 4610 Btu/h (1350.7 W); North partition =

(0.6)(299)(20) = 3590 Btu/h (1051.9 W); Glass door = (1.0)(35)(20) = 700 Btu/h (205.1 W); South wall = (0.25)(300)(20) = 1500 Btu/h (439.5 W); Windows = (1.13)(91)(20) = 2060 (603.5 W); Floor = (0.21)(1104)(5) = 1160 (339.9W); Ceiling = (0.14)(1104)(20) = 3090 (905.4 W). Summing these individual heat leakages gives 16710 Btu/h (4896 W).

4. Determine the sensible heat load
Using the conventional heat loads for people, lights, motor hp, the sensible heat load is: Occupants = (35)(300) = 10,500 Btu/h (3076.5 W); Lights = (1100)(3.413) = 3760 Btu/h (1101.7 W); Motor horsepower = (0.5)(2546) = 1275 Btu/h (373.6 W).

5. Find the air leakage heat load
Use the relation: Air leakage heat load, Btu/h (W) = (air change, ft^3/h)(specific heat of air)(temperature difference)/(specific volume of air, ft^3/lb). Or (36,000)(0.24)(95 − 75)/13.70 = 12,600 Btu/h (rounded off) (3691.8 W).

6. Calculate the sun effect heat load
For the glass on the south wall, the sun effect = (45 ft^2)(30 Btu/h ft^2) = 1350 Btu/h (399.6 W). The sun effect on the south wall = (300 ft^2)(0.25)(120 − 95) = 1875 Btu/h (549.4 W).

7. Find the dry tons (W) of refrigeration required
Sum the heat gains computed above thus: 16,710 + 10,500 + 3760 + 1275 + 12,600 + 1350 + 1875 = 48,070 Btu/h (14,084.5 W) = grand total heat loss. The dry tons (W) of refrigeration is then found from 48,070/12,000 Btu/ton = 4.01 tons (14.1 kW).

8. Evaluate the moisture latent heat load
List the air leakage conditions thus:

Conditions	Outside air	Inside air
Dry bulb	95°F (35°C)	75°F (23.9°C)
Wet bulb	78°F (25.6°C)	64.4°F (18.0°C)
Percent relative humidity	47	55
Dew point	71°F (21.7°C)	58°F (14.4°C)
Grains per lb	114.4 (16,327.0 mg/kg)	71.9 (10,262.1 mg/kg)
Specific volume	—	13.7 cu ft/lb (0.85 m³/kg)

First, determine the pounds (kg) of air per hour from (ft^3/h)(outside air gr/lb − inside air gr/lb)/(specific volume of air)(7000 gr/lb). Or (36,000)(114.4 − 71.9)/(13.7)(7000) = 15.95 lb/h (7.24 kg/h).

The, the air latent heat = (lb/h)(latent heat of air, Btu/lb) = (15.95)(1040) = 16,588 Btu/h (4860.3 W). Person load = (35)(100) = 3500 Btu/h (1025.5 W). Total latent heat = 16,588 + 3500 = 20,088 Btu/h (5885.8 W), or 20,088/12,000 = 1.67 tons. Then, the total refrigeration load = 4.01 + 1.67 = 5.68 tons (19.98 kW).

Related Calculations. As you can see, if you are to perform repeated calculations for buildings and rooms, a form listing both the equations and items to be computed will be helpful in saving you time. The procedure given here is useful for the occasional computation of buildings of all types: residential, commercial, industrial, etc. The same general procedure can be used for trucks, ships, aircraft and other mobile applications.

DETERMINING THE DISPLACEMENT OF A RECIPROCATING REFRIGERATION COMPRESSOR

What is the needed displacement of a reciprocating refrigeration compressor rated at 50 tons (45.4 t) when operating with a refrigerant at 0°F (-17.8°C) in the evaporator expansion coils? At this temperature, the heat absorbed by the evaporation of 1 lb (0.45 kg) of the refrigerant is 500 Btu (527.5 kJ) refrigerating effect, and the specific volume is 9 ft³/lb (0.56 m³/kg). The vapor enters the compressor in the saturated state. If the compressor speed of rotation is 180 r/min, and the stroke is 1.2 × bore, what is the bore and stroke of this single-acting compressor?

Calculation Procedure:

1. Determine heat absorbed and compressor displacement
The heat to be absorbed (tons of refrigeration)(heat equivalent of 1 ton of refrigeration, Btu/min). Or the heat absorbed = 50(200 Btu/min/ton refrigeration) = 10,000 Btu/min (10,550 kJ/min).

The compressor displacement = (heat absorbed)(specific volume of the refrigerant)/(refrigerating effect). Substituting for this refrigerant, displacement = 10,000(9)/(500) = 180 ft³ (5.09 m³).

2. Find the compressor bore
Let N = the compressor speed, rpm; D = compressor cylinder bore, ft (m); L = compressor stroke, ft (m); V = piston displacement, ft³ (m³) per stroke. Then:

$$V = 0.785D^2L \qquad N \times V = 180 \text{ ft}^3/\text{min}$$

$$= 180/N = {}^{180}\!/_{180} = 1 \text{ ft}^3$$

$$L = 1.2D \qquad V = 0.785D^2 \times 1.2D = 1 \text{ ft}^3$$

Rearranging and transposing, we see that $D = (1/1.2)(0.785)^{0.333} = 1.02$ ft (0.31 m), or 12.24 in.

3. Find the compressor piston stroke
Using the relation in step 2, $L = 1.2D = 1.2 \times 1.02 = 1.224$ ft (0.373 m), or 14.69 in. The bore and stroke as computed here are typical for a compressor of this capacity.

Related Calculations.   With the phasing out of chlorofluorcarbons (CFC) because of environmental restrictions, engineers must be able to evaluate the performance of alternative refrigerants. The procedure given above shows exactly how to perform this evaluation for any refrigerant whose thermodynamic and physical characteristics are known by the engineer, or can be obtained from standard data reference.

While many liquids boil at temperatures low enough for refrigeration, few are suitable for refrigeration purposes. Those liquids suitable for practical refrigeration applications are termed refrigerants. For any refrigerant, increased pressure on it raises its boiling point. Reducing the pressure on a refrigerant lowers its boiling point. Refrigeration occurs when a refrigerant boils at a low temperature, permitting heat flow from an item or area to be cooled to the refrigerant. Boiling of the refrigerant takes place in the evaporator which is located in the area to be cooled, Fig. 2a.

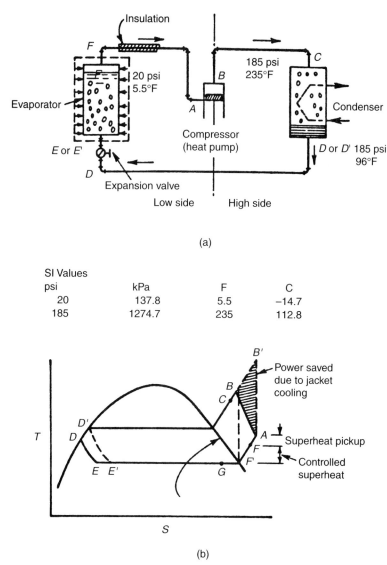

SI Values			
psi	kPa	F	C
20	137.8	5.5	−14.7
185	1274.7	235	112.8

FIGURE 2 (a) Typical reciprocating-compressor refrigeration cycle. (b) T-S plot of refrigerant cycle.

When the boiling refrigerant removes sensible heat from environment at a rate equivalent to the melting of one ton (2000 lb; 980 kg) of water ice in 24 h, the rate of heat removal is a ton (3.516 kW) of refrigeration. Since the heat of melting (sublimation) of 1 lb (0.454 kg) of water ice is 144 Btu (151.9 kJ), a ton of refrigeration is equivalent to 2000 lb (144 Btu/24h) = 12,000 Btu/h (3516 W, or 3,516 kW).

By comparison, the heat of sublimation (melting) of dry ice (CO_2) is 275 Btu/lb (640.8 kJ/kg). To say that a refrigeration machine has a capacity of 10 tons (35.16 kW) is to say that the rate of refrigeration is $10 \times 200 = 2000$ Btu/min (35.16 kW). Note that 1 ton of refrigeration equals a rate of 200 Btu/min (3.516 kW).

To determine the amount of refrigerant that must be circulated, divide the refrigerating effect of the refrigerant in But/lb (kg) into 200 Btu/min (W). Thus, with a refrigerating effect 25 Btu/lb (58.3 kJ/kg), the quantity of refrigerant to be circulated is $200/24 = 4$ lb/min (1.82 kg/min).

The work of compression is the amount of heat added to the refrigerant during compression in the cylinder or rotary compressor. It is measured by subtracting the heat content of 1 lb (0.454 kg) of refrigerant at the compressor suction conditions, point F', F, or A in Fig. 2b from the heat content of the same pound (kg) at the compressor discharge conditions, point B or B' in Fig. 2b.

The theoretical horsepower (kW) requirements of a refrigeration compressor can be found by multiplying the work compression in Btu/lb (kJ/kg) by the pounds (kg) of refrigerant circulated in one hour, and dividing this product by 2545 Btu/hp-h, hp_t = (work of compression)(refrigerant circulated)/2545. Multiply by 0.746 to obtain theoretical kW input.

A good example of the practical value of this calculation is in the recent real-life example of the upgrading of the HVAC system in a 400-unit apartment complex. Two older refrigerating machines using CFC-refrigerants were replaced by two new chillers using HCFC-123 refrigerant.

The older machines required an input of 0.81 kW per ton of cooling capacity (refrigeration) while the newer machines require only 0.55 kW per ton. Annual savings of more than 30 percent in energy costs for refrigeration are expected with the new machines and refrigerant. The new machines also reduce greenhouse gas emissions because of reduced electrical power needed to run them. Payback time for the new machines will be less than 2.5 years because the energy savings are so significant. Several examples of typical piping arrangements for reciprocating refrigeration compressors and chillers are shown in Fig. 3 through Fig. 6.

Another example of using this procedure is substitution of natural-gas fueled engine drives for refrigeration chillers using new refrigerants. In one department store installation of such a chiller the estimated annual energy cost savings are $54,875. Such drives reduce electric demand charges, are compact in size, are environmentally friendly, and are used in hospitals, nursing homes, schools, colleges, office buildings, retail, and industrial/process facilities. Some of the newest centrifugal chillers on the market report a required input of just 0.20 kW/ton when operating at 60 percent load with entering condenser water at 55°F (12.8°C).

HEAT-RECOVERY WATER-HEATING FROM REFRIGERATION UNITS

How much heat can be obtained from heating water for an apartment house having 150 apartment units served by two 200-ton (180 t) air-conditioning units if a 70°F

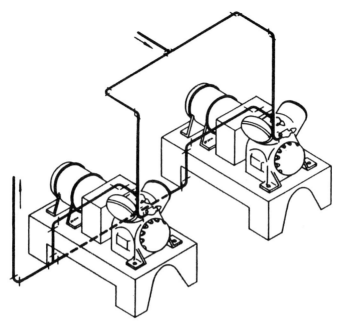

FIGURE 3 Layout of suction and hot-gas lines for multiple-compressor operation. (*Carrier Corporation.*)

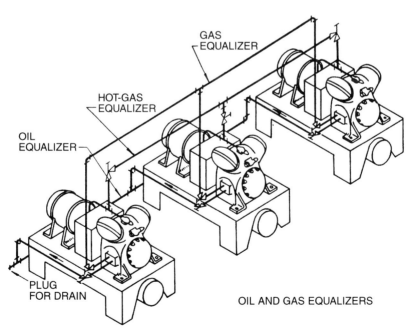

GAS
EQUALIZER

HOT-GAS
EQUALIZER

OIL
EQUALIZER

PLUG
FOR DRAIN

OIL AND GAS EQUALIZERS

FIGURE 4 Interconnecting piping for multiple condensing reciprocating refrigeration units. (*Carrier Corporation.*)

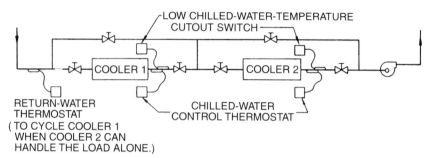

FIGURE 5 Piping arrangement for two centrifugal coolers in series arrangement. (*Carrier Corporation.*)

(38.9°C) temperature rise of the incoming cold water is required? Determine the number of gallons (L) of water that can be heated per hour and the total gallonage (L) of heated water that can be delivered with the air-conditioning units operating 8, 12, and 16 hours per day.

Calculation Procedure:

1. *Determine the quantity of heat available*
Heat is available from the high-pressure gas at the refrigeration compressor discharge. This is valid regardless of the type of compressor used: reciprocating, rotary, or centrifugal. The quantity of heat available from a specific compressor depends on the outlet-gas temperature, gas flow rate, and the efficiency of the heat exchanger used.

To recover heat from the hot gas, a heat exchanger, Fig. 7, is placed in the compressor discharge line, ahead of the regularly used condenser. Cold water from either the building's outside water supply line, or from the building's heated-water storage tank, is pumped through the heat exchanger in the compressor discharge line. Leaving the heat exchanger, the heated water returns to the hot-water storage tank.

Experience shows that a typical well-designed heat exchanger, such as a conventional water-cooled condenser, can transfer 25 to 35 percent of the Btu (kJ) rating of the refrigeration compressor, i.e., the air-conditioning unit's rating. Using the lower value in this range for these units gives, Heat Available = 0.25 (200) (12,000 Btu/h/ton) = 0.25 (200)(12,000) = 600,000 Btu/h (633,000 kJ/h). With 35 percent, Heat Available = 0.35 (200)(12,000) = 840,000 Btu/h (886,200 kJ/h). With two refrigerating units the heat available would be double the computed amount, or 1,200,000 Btu/h (1,266,000 kJ/h) and 1,680,000 Btu/h (1,772,400 kJ/h).

2. *Find the hourly water heating rate for the system*
Water weighs 8.34 lb/gal (1.02 kg/L). To raise the temperature of one pound of water (0.454 kg) 1°F (0.55°C) requires a heat input of 1 Btu (1.055 kJ). With the specified 70°F (38.9°C) water-temperature rise required in this building, the rate of water heating will be: gal/h (L/h) = (heat available)/(lb/gal)(temperature rise required).

With 25 percent heat transfer, we have, *gph* heated = (600,000)/(8.33)(70) = 1029 gal/h (3900 L/h). And with 35 percent heat transfer, *gph* = (840,000)/

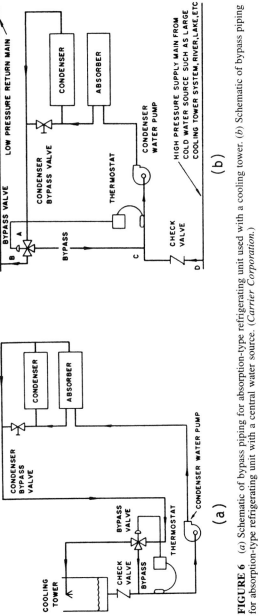

FIGURE 6 (*a*) Schematic of bypass piping for absorption-type refrigerating unit used with a cooling tower. (*b*) Schematic of bypass piping for absorption-type refrigerating unit with a central water source. (*Carrier Corporation.*)

12.9

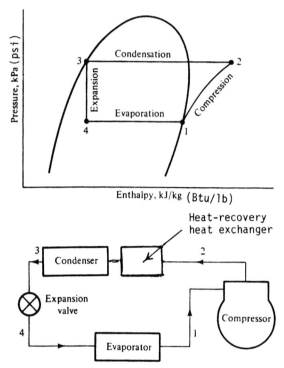

FIGURE 7 Heat-recovery heat-exchanger location in a refrigeration cycle.

$(8.33)(70) = 1440$ gal/h (5458 L/h). Again, with two units the hourly heating rate will be doubled, or 2058 gal/h (7800 L/h) and 2880 (10,915 L/h).

3. Compute the daily total gallonage of hot water produced

Since air-conditioning refrigeration units operate varying numbers of hours per day, depending on the outside weather conditions, the gallonage of hot water available from heat recovery will vary. For the range of operating hours specified, we have, per 200-ton (180-t) unit:

	Gallonage available (L) per hours of operation		
	8	12	16
25 percent	8232 (30,458)	12,348 (45,688)	16,464 (62,399)
35 percent	11,520 (42,624)	17,280 (65,491)	23,040 (87,322)

Again, we double these numbers for two units in the building.

Related Calculations. The normal discharge temperature of modern refrigerants makes heat recovery for domestic and/or process water heating an attractive option, especially in an environmentally conscious world. Further, heat recovery is such a simple design challenge that the investment is often recovered in fuel savings in less than two years.

For maximum efficiency, the designer must try to match hot-water needs with the operating time of the refrigeration unit. If there is a disparity between these two variables, a sufficiently large water storage tank can be designed into the system to store hot water during times when the refrigeration unit is not operating. Such a storage tank is not expensive and it will have a long life if properly maintained.

Heat recovery for heating incoming cold water can be used in office buildings, apartment houses, hotels, motels, factories, and commercial buildings wherever a need for hot water exists and a refrigeration unit of some kind operates in the structure. Energy savings can range from 25 to 100 percent of the cost of heating water for either domestic or process uses. A pump, Fig. 8, is used to force-circulate

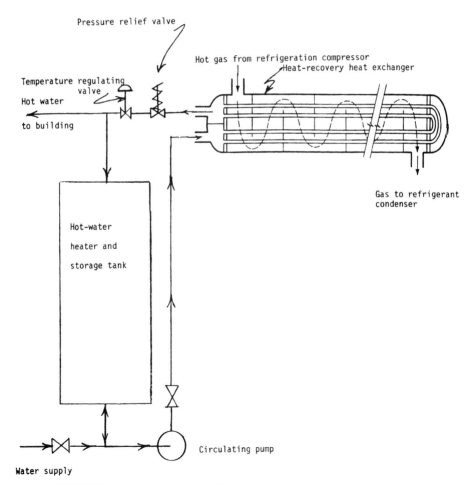

FIGURE 8   Layout of heat-recovery heat-exchanger piping for refrigeration cycle.

water through the hot-gas heat exchanger while the refrigeration unit is operating. This action recovers heat when it is available and stores the heated water for future use.

A regulating valve, Fig. 8, controls water flow and the temperature of the water leaving the heat exchanger. The valve ensures that maximum heat recovery will occur. When designing the piping for heat recovery, be certain to arrange to have the coldest water enter the heat exchanger. Then maximum benefit will be obtained from the heat recovery.

COMPUTING REFRIGERATING CAPACITY NEEDED FOR AIR-CONDITIONING LOADS

An air-conditioned space is supplied 10,000 ft^3/min (283 m^3/min) of air at 65°F (18.3°C) and 60 percent relative humidity. Air is supplied to the dehumidifier at 85°F (29.4°C); wet-bulb temperature is 70°F (21.1°C). Condensate leaves the dehumidifier at the same temperature as the outgoing air, which is at standard atmospheric pressure throughout the system. Air leaves the dehumidifier in the saturated state. Determine the refrigerating load for this air-conditioned space.

Calculation Procedure:

1. Draws sketches of the system arrangement and a skeleton psychrometric chart of the processes
Figures 9 and 10 show the system and the psychrometric chart with important state points identified by number. Show the values of the enthalpies and humidities at the various state points before attempting to make the air-conditioning design.

Thus, at point 3, v_m = 13.39 ft^3/lb of dry air (0.83 m^3/kg). Then, the weight flow rate of air, M_a = flow rate, ft^3/min/specific volume = 10,000/13.39 = 746 lb dry air per min (338.7 kg/min).

2. Compute the quantity of moisture condensed and the air dew point
Moisture condensed = (moisture content of incoming air − moisture content of leaving air)/7000 gr/lb = (86 − 55.5)/7000 = 0.00435 lb/lb dry air (0.00197 kg/kg). Figure 11 shows the psychrometric process. The dew point, from the chart, t_2 = 51°F (10.6°C); enthalpy h_{fc} = 51 − 32 = 19 Btu/lb of water (44.2 kJ/kg).

3. Set up an energy balance about the dehumidifier
The energy balance for an adiabatic mixing process is given by $M_1(h_{m1}) + M_2(h_{m2}) = M_3(h_{m3})$, where the M values are the weight flow rates in the respective air-vapor

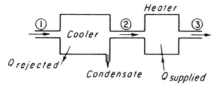

FIGURE 9 Basic dehumidifier consisting of cooler followed by heater.

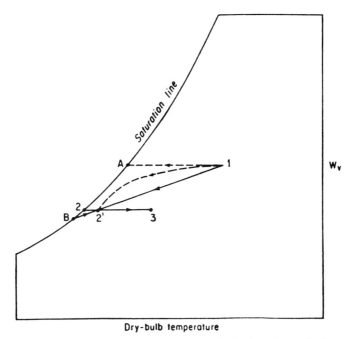

FIGURE 10 Dehumidification process on skeltonized psychrometric chart.

mixtures, and the h values are the respective enthalpies. Then, the energy relation for the cooler is $Q_R = (h_{m1} - h_{m2}) - (W_{v1} - W_{v2})h_{fc}$ where the symbols are as given above.

Substituting, $Q_R = (33.96 - 20.85) - (0.00436)(19) = 13.03$ Btu/lb (30.36 kJ/kg). Then, the total heat removed at the dehumidifier $= (13.03)(746) = 9740$ Btu/min (171.2 kW).

4. Find the refrigeration capacity required
Since 200 Btu/min = 1 ton of refrigeration, the refrigeration capacity required = 9740/200 = 48.7 tons (43.6 t).

Related Calculations. This procedure could have been performed using the pressure and humidity relations of the air streams. Approximately the same result would have been obtained. By considering the enthalpy of the condensate the estimated refrigeration load is decreased by about 0.2 ton (0.18 t), which for this installation is insignificant. The tonnage reduction will always be small when the final dew point is low, and in most cases it can be neglected. Further, the true state of which the condensate is removed is usually difficult to establish. Neglecting the energy of the condensate will give a refrigeration capacity requirement that is on the safe side, i.e., a larger value than would be arrived at with the exact solution.

Depending on the number of banks of spray nozzles, the direction of the spray, and the air velocity, the percentage of untreated air passing through an air washer, Fig. 12, may range from 5 to 35 percent for two banks and one bank of nozzles, respectively, according to ASHRAE. For a coil-type dehumidifier the percentage of untreated air may range from 2 to 39 percent for coils eight rows to one row deep operating with a face velocity of 300 ft/min (91.4 m/min). In general, the refrig-

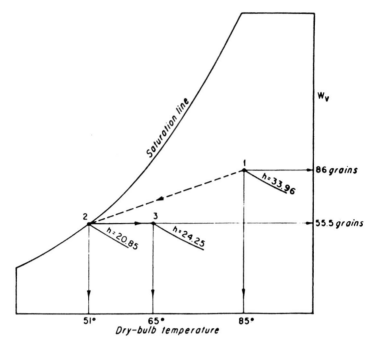

FIGURE 11   Values for the dehumidification process.

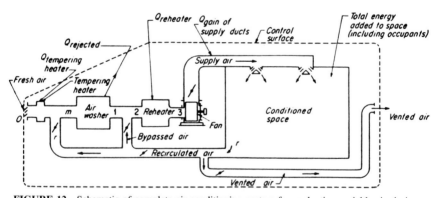

FIGURE 12   Schematic of complete air-conditioning system for evaluating variables in design.

eration power requirements of an air-conditioning system are dependent upon the heat-transfer characteristics of the dehumidifier used.

This procedure, and Figs. 9 through 12, are the work of Norman R. Sparks, Professor and Head, Department of Mechanical Engineering, The Pennsylvania State University, and Charles C. DiLio, Associate Professor of Mechanical Engineering, The Pennsylvania State University. SI values were added by the handbook editor.

WATER-VAPOR REFRIGERATION-SYSTEM ANALYSIS

A water-vapor refrigerating machine, Fig. 13, is to produce 250 gal/min (15.8 L/ s) of chilled water at 45°F (7.2°C). The makeup and recirculated water are at 60°F (15.6°C), and the quality of the vapor leaving the evaporator and entering the compressor is 0.97. Find: (a) the capacity of the machine, tons; (b) the pounds (kg) of vapor to be removed from the evaporator per minute; (c) the volume of vapor to be removed from the evaporator, ft³/min (m³/min) and ft³/(min · ton) (m³/t). Further, consider that the machine in Fig. 13 is equipped with a centrifugal compressor and mechanical vacuum pumps. The condenser pressure is 2 in Hg (5.1 cm) abs. Compression efficiency is 0.65. Mechanical efficiency of the compressor is 98 percent. The condensate leaves the condenser at 90°F (32.2°C). Power required to drive the condensate and air pumps is 6 percent of the total power input. Find: (d) the compressor hp (kW) and hp/ton (kW/t); (e) the heat rejected in the condenser; Btu/min (kJ/min); (f) the coefficient of performance.

Calculation Procedure:

1. Draw a sketch of the system showing the important state points
Figure 13 shows a schematic of the system and the state points.

2. Compute the rate of refrigerant flow
Using Fig. 13,

$h_1 = 28.06 = h_M; h_2 = 13.06; h_3 = 1049.4; v_3 = 1975$ ft³/lb (123.1 m³/kg)

$M_2 = 250(8.35) = 2087.5$ lb/min (947.7 kg/min)

3. Calculate refrigerating effect, vapor removed, and vapor volume at the compressor suction

(a) Refrigerating effect = 2087.5(15) = 31,312 Btu/min = 156.6 tons (140.9 t)

(b) Vapor removed, $M_3 = 31,312/1049.4 - 28.06 = 30.66$ lb/min (13.9 kg/min)

(c) Vapor volume at compressor suction = 30.66(1975) = 60,550 ft³/min (1714 cu³/min) or 387 ft³/min/ton (12.2 m³/t)

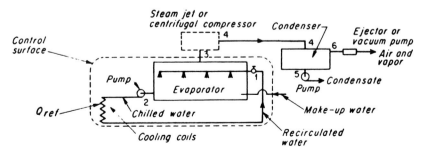

FIGURE 13   Schematic of water-vapor refrigeration system.

4. Find the actual change in enthalpy

$h_3 = 1049.4$; $s_3 = 2.0794$; $h_5 = 58$ (135.1 kJ/kg)

After isentropic compression to 2 in Hg, $h_4' = 1168$ (2721.4 kJ/kg)

Isentropic $\Delta h = 1168 - 1049.4 = 118.6$ Btu/lb $= h_4' - h_3$ (276.3 kJ/kg)

Actual $\Delta h = \dfrac{118.6}{0.65} = 182.5$ Btu/lb $= h_4 - h_3$ (425.2 kJ/kg)

$h_4 = 1049.4 + 182.5 = 1231.9$ (2 in Hg and 380°F) (2870.3 kJ/kg) (5.08 cm Hg and 193.3°C)

5. Compute the compressor and total power input

$\dfrac{W_i}{J} = 30.66(182.5) = 5595$ Btu/min (5902.7 kJ/min)

Compressor hp $= \dfrac{5595}{42.4(0.98)} = 134.6$, or 0.859 hp/ton (0.71 kW/t)

Total hp $= \dfrac{134.6}{0.94} = 143.2$ or 0.914 hp/ton (0.758 kW/t)

6. Determine the condenser heat rejection and COP

$Q = 30.66(1231.9 - 58) = 35{,}992$ Btu/min, or 230 Btu/min/ton (269.6 kJ/t)

Coefficient of performance $= \dfrac{200}{0.914(42.4)} = 5.17$ (based on total hp)

See Fig. 14 for a typical water-vapor refrigeration system using steam ejectors and a surface condenser. Figure 15 shows a water-vapor refrigeration system using a steam-jet (barometric condenser).

Related Calculations.   The rather high coefficient of performance (COP) of this machine is caused by the comparatively favorable temperature range through which the machine operates. With today's environmental concern over safer refrigerants, water has a 0.00 ozone-depletion potential. Further, water also has a 0.00 immediate global warming potential, an a 0.00 100-year global warming potential. And since such refrigeration systems can be gas-fired, they have less potential atmospheric pollution compared to coal or oil. Likewise, water has zero flammability.

Air cooling is one industrial application of refrigeration that does not ordinarily require, in the refrigerating sense, low temperatures. Hence, the water-vapor refrigerating system is ideal for such applications. Further, water is a cheap refrigerant, and it is truly a safe one, and as a result, it is finding greater use in industry today, especially in view of the new, stricter environmental regulations that seem to be imposed every year.

This procedure, and Figs. 13 through 15, are the work of Norman R. Sparks, Professor and Head, Department of Mechanical Engineering, The Pennsylvania State University, and Charles C. DiLio, Associate Professor of Mechanical Engineering, The Pennsylvania State University. SI values were added by the handbook editor.

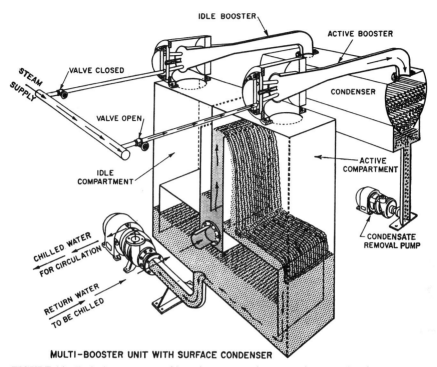

FIGURE 14 Typical water-vapor refrigeration system using steam ejectors and surface condenser. (*Ingersoll-Rand Co.*)

ANALYZING A STEAM-JET REFRIGERATION SYSTEM FOR CHILLED-WATER SERVICE

A steam-jet water-vapor refrigerating machine is to produce 250 gal/min (15.8 L/s) of chilled water at 45°F (7.2°C). Makeup and recirculated water are at 60°F (15.6°C), and the quality of the vapor leaving the evaporator and entering the ejector is 0.97. Condenser pressure is 2 in (5.08 cm) Hg abs. Condensate leaves the condenser at 90°F (32.2°C). The motive steam is supplied at 140 lb/in² (abs) (946 kPa), 370°F (187.8°C). System manufacturer supplies the following efficiencies: Nozzle efficiency = 90 percent. Entrainment efficiency = 65 percent. Diffuser efficiency = 75 percent. Steam consumption of the auxiliary ejectors is 6 percent of the total steam requirement. Neglecting the power demands of the water pumps, find: (*a*) Steam consumption of the main ejector; (*b*) the total steam consumption; (*c*) heat rejected in the primary condenser.

Calculation Procedure:

1. Determine the pertinent steam enthalpies and quality
Referring to Fig. 16 for symbols, we have

h_A = 1203.5 at 140 lb/in² (abs), 370°F (2804 kJ/kg at 964.6 kPa, 187.8°C)

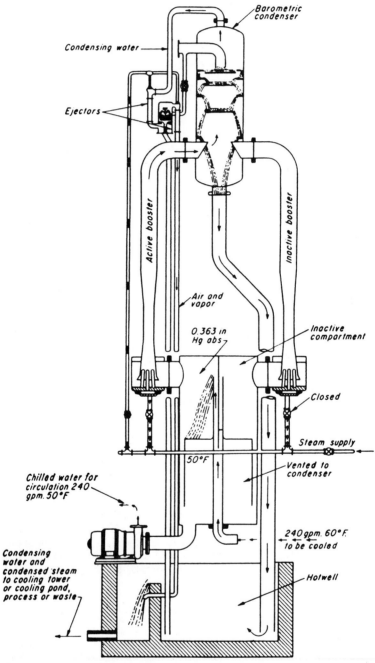

FIGURE 15 Steam-jet (barometric condenser), water-vapor refrigeration system. (*Ingersoll-Rand Co.*)

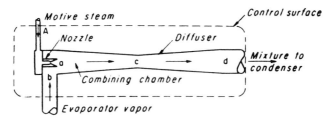

FIGURE 16 Steam-jet compressor.

h'_a = 803 (after isentropic expansion to 0.3 in Hg) (1871 kJ/kg) (0.76 cm Hg)

$h_A - h'_a$ = 400.5 Btu/lb motive steam (933.2 kJ/kg)

$h_A - h_a$ = 0.9(400.5) = 360.5 Btu/lb = KE_a (839.9 kJ/kg)

x_b = 0.97; h_b = 1049.4 (2445.1 kJ/kg)

2. Compute M_a and test its accuracy as a first trial
First trial:

Assume x_c = 0.87, for which h_c = 943 (2197.2 kJ/kg)

Then h'_d = 1043 (after isentropic compression to 2 in Hg) (2430.2 kJ/kg at 5.08 cm Hg)

$h'_d - h_c$ = 100 Btu; $h_d - h_c = \dfrac{100}{0.75}$ = 133.3 Btu/lb (310.6 kJ/kg)

h_d = 943 + 133.3 = 1076.3 Btu/lb (2507.8 kJ/kg)

From $M_A h_A + h_b = (M_A + 1)h_d$; $M_A = \dfrac{h_d - \mathbf{h}_b}{h_A - h_d}$

$$M_a = \frac{1076.3 - 1049.4}{1203.5 - 1076.3} = 0.211 \text{ lb/lb evaporator vapor (0.096 kg/kg)}$$

From $e_e M_a \dfrac{V_a^2}{2gJ} = (M_a + 1)\dfrac{V_c^2}{2gJ}$; $e_e M_a \text{KE}_a = (M_a + 1)\text{KE}_c$

$$0.65(0.211)(360.5) \neq (0.211 + 1)\,133.3$$

$$49.5 \neq 161.5$$

Examination of this last equation and the results of this first trial indicate that M_a is too low. Therefore, h_c must be increased appreciably.

3. Make a second computation of M_a
Second trial:

Assume h_c = 1000 Btu/lb (2330 kJ/kg)

h'_d = 1108; $h'_d - h_c$ = 108; $h_d - h_c = \dfrac{108}{0.75}$ = 144 (335.5 kJ/kg)

h_d = 1000 + 144 = 1144 (2665.5 kJ/kg)

$$M_a = \frac{1144 - 1049.4}{1205.5 - 1144} = 1.59 \text{ lb/lb evaporator vapor } (0.72 \text{ kg/kg})$$

$$0.65(1.59)(360.5) = (2.59)(144)$$

$$373 = 373$$

Since the entrainment energy equation shows a perfect check, the assumed h_c and the value of M_a derived from this second trial are considered correct.

4. Calculate the system variables

(*a*) Data, as necessary, have been taken from the previous procedure.

Evaporator vapor = 30.66 lb/min

Ejector steam consumption = 1.59(30.66) = 48.7 lb/in, or 2920 lb/h (1325.7 kg/h)

$$\text{Steam rate} = \frac{48.7}{156.6} = 0.311 \text{ lb/(min} \cdot \text{ton), or } 18.7 \text{ lb/(h} \cdot \text{ton) } [9.43 \text{ kg/(h} \cdot \text{t)}]$$

(*b*) Total steam consumption = 2920/0.94 = 3110 lb/h, or 19.9 lb/(h · ton) [10.04 kg/(h · t)]

(*c*) Steam to primary condenser = 30.66 + 48.7 = 79.36 lb/min (36 kg/min)

$$Q = 79.36(1144 - 58) = 86{,}200 \text{ Btu/min,}$$

$$\text{or } 550 \text{ Btu/(min} \cdot \text{ton) } [644.7 \text{ Btu/(min} \cdot \text{t)}]$$

The heat required for the generation of the motive steam will be $h_A - h_5$, or, in this instance, $1203.5 - 58 = 1145.5$ Btu/lb (2669 kJ/kg). The ratio of the refrigerating effect to the heat supplied to produce that effect is therefore 12,000/19.9(1145.5) = 0.526.

Related Calculations. Like the absorption system, the steam-jet water-vapor cycle requires but a very small portion of mechanical energy for its operation. A coefficient of performance (COP) may not, therefore, be truly expressed for this type of machine. But a pseudocoefficient can be arbitrarily devised for the purpose of comparison with other types of systems.

While there is no standard procedure for setting up a pseudocoefficient, if the motive steam were used in, say, a turbine to drive a compressor, about 50 percent of the energy available in expanding to condenser pressure might be considered as the turbine work output and the compressor input. For this procedure, the energy available per pound (kg) of motive steam between 140 lb/in^2 (abs) (964.6 kPa), 370°F (187.8°C) and 2 in (5.08 cm) Hg abs is 317.5 Btu/lb (739.8 kJ/kg). The coefficient of performance would thus be, according to this standard, 12,000/[19.9 (0.5)(317.5)] = 3.80, based on actual steam consumption. The corresponding horsepower per ton = 1.237 (1.025 kW/t).

As a practical problem, the determination of the minimum available energy of the motive steam at which the ejector will operate is not of particular importance since excessive steam consumption will render the operation unsatisfactory long before this point is reached. However, the condition of the motive steam to be supplied to an ejector for any given maximum allowable steam consumption, as dictated, for example, by cooling-water considerations, may frequently be of interest. This may be closely approximated by the application of the principles given in this, and the preceding, calculation procedures.

This procedure is the work of Norman R. Sparks, Professor and Head, Department of Mechanical Engineering, The Pennsylvania State University, and Charles C. DiLio, Associate Professor of Mechanical Engineering, The Pennsylvania State University. SI values were added by the handbook editor.

HEAT PUMP AND COGENERATION COMBINATION FOR ENERGY SAVINGS

In the heat-pump system shown in Fig. 17, the building requires 100,000 Btu/h (29.3 kW) for heating. Calculate for this system: (*a*) The capacity of the heat pump, tons (t) as a refrigerating machine; (*b*) the motor horsepower (kW); (*c*) the hourly energy input to the motor; (*d*) the overall coefficient of performance; (*e*) the heat received from the outside air; (*f*) the electrical consumption per hour for direct heating; (*g*) the coal consumption for direct heating with a 50 percent furnace efficiency; (*h*) the oil consumption for direct heating with a 65 percent furnace

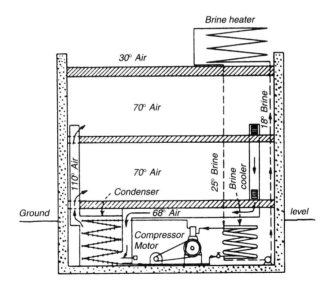

SI Values
30 F air (−1.1 C)
70 F air (21.1 C)
68 F air (20.0 C)
18 F brine (−7.8 C)
110 F air (43.3 C)
25 F brine (−3.9 C)

FIGURE 17 Schematic of heat pump used to heat a building.

efficiency; (*i*) the natural-gas consumption for direct heating with a 70 percent furnace efficiency; (*j*) the manufactured-gas consumption for direct heating with 70 percent furnace efficiency; (*k*) the energy-saving potential with a conventional diesel engine driving the compressor in a cogeneration mode.

Calculation Procedure:

1. Specify the power and fuel constants to use in the design
For this system the following power and fuel constants will be sued; such constants are readily available in standard reference works: Thus, the power consumption of motor for the unit, considered as a standard refrigerating machine, will be 1.7 hp/ ton (1.49 kW/t). Motor efficiency is 87 percent. Heat of combustion of the fuels: coal, 14,000 Btu/lb (32,620 kJ/kg); oil, 19,000 Btu/lb (44,270 kJ/kg); natural gas, 1100 Btu/ft^3 (41,007 kJ/m^3); manufactured gas, 600 Btu/ft^3 (22,367 kJ/m^3). Specific gravity of the fuel oil = 0.88.

2. Find the total heat output of the heat pump

$$1.7 \text{ hp} = 1.7(2544) = 4325 \text{ Btu/h (1.27 kW)}$$

$$\text{Motor input} = \frac{4325}{0.87} = 4970 \text{ Btu/(h · ton) (1.62 kW/t)}$$

The motor loss is therefore 4970 − 4325 = 645 Btu/(h · ton) [0.21 kW/h · t)] which may be considered as applied to direct heating.

Heat output of condenser (and cylinder jacket) per ton = 12,000 + 4325 = 16,325 Btu/h (4.8 kW)

Total heat output of machine per ton = 16,325 + 645 = 16,970 Btu/h (4.97 kW)

In this instance, this is equivalent to 12,000 + 4970.

3. Compute the system characteristics

(*a*) Capacity of unit, considered as a refrigerating machine = 100,000/16,970 = 5.89 tons (6.5 t)
(*b*) Motor hp = 5.89(1.7) = 10.0 (7.46 kW)
(*c*) Motor input = 10(2544)/0.87 = 29,240 Btu/h or 8.58 kW
(*d*) Coefficient of performance = 100,000/29,240 = 3.42
(*e*) Heat received from outside air = 5.89(12,000) = 70,680 Btu/h (20.7 kW)
(*f*) Electric power for direct heating = 100,000/3412 = 29.3 kW
(*g*) Coal for direct heating = 100,000/0.5(14,000) = 14.3 lb/h (6.5 kg/h)
(*h*) Oil for direct heating = 100,000/0.65(19,000) = 8.1 lb/h or 8.1/0.88(8.33) = 1.1 gal/h (4.2 L/h)
(*i*) Natural gas for direct heating = 100,000/0.7(1100) = 130 ft^3/h (12.1 m^3/h)
(*j*) Manufactured gas for direct heating = 100,000/0.7(600) = 238 ft^3/h (22.1 m^3/h)

4. *Determine the cogeneration potential for the installation*
With a Diesel-engine drive for the compressor, the specific fuel consumption will
typically be 0.45 lb/bhp·h (0.27 kg/kW·h). Then, the fuel required per ton (t)
refrigerating capacity will be 1.7 (0.45) = 0.765 lb/h (0.347 kg/h), representing
0.765 (19,000) = 14,535 Btu/h (4.26 kW). Of this, 4325 Btu/h (1.26 kW) leave
the engine in the form of work. The remainder, 10,210 Btu/h·ton (10,722 kJ/h·
t), is rejected in cooling water, exhaust, and by radiation. Much of this loss can be
recovered and applied to direct heating.

Assuming that 80 percent of the waste can be recovered, a typically safe assumption, the heat supplied to the building per ton (t) capacity is 16,325 + 0.80
(10,210) = 24,495 Btu/h (7.17 kW), using data from the steps above in this procedure. Then:

Capacity required = 100,000/24,495 = 4.08 tons (3.67 t)

Power required = 4.08(1.7) = 6.94 hp (5.2 kW)

Fuel required = 6.94(0.45) = 3.12 lb/h or 0.425 gal/h (1.6 L/h)

Coefficient of performance, based on shaft work = 100,000/6.94(2545) = 5.65

Related Calculations. Relative fuel and electrical energy costs can be obtained
for any given locality from the utility serving the area, and with the deregulation
taking place throughout the electrical power industry at this writing, competitive
costs may be obtained from two, or more, utilities. This means that even more
favorable rates, in general, will be obtained.

Cost calculations may indicate that, on the basis of fuel consumption alone, the
Diesel-driven heating unit would be more economical for the system above than
any other heating system, including direct heating with coal, in many localities. As
a heating unit alone, however, the initial cost of the heating cycle is greater than
that of any other system. But when a substantial proportion of the capacity of the
system may be applied to cooling in the summer, the combined heating and air-
conditioning first cost should compare favorably with that of a conventional heating
plant plus air-conditioning equipment. For this reason, air conditioning should preferably be incorporated with the heating cycle.

Recently developed rare-earth additives for Diesel engines simultaneously cut
NO_x emissions and eliminate more than 90 percent of particulates in the engine
exhaust. The system combines a cerium-based organic fluid and a conventional
ceramic particulate trap. Using such emission controls allows Diesel engines in
heavily populated areas where they formerly might not have been welcome. Thus,
Diesel-engine-drive of heat pumps in a cogeneration mode could be more popular
in the future. The additive and trap for Diesel engines results in NO_x of 2.0 grams
per brake-horsepower-hour (2.68 g/kWh and 0.013 g/kWh), compared to 0.1 g/
bhp·h (0.13g/kWh) for best traps as of this writing. EPA has set emission limits
of 2.5 g/bhp·h NO_x and 0.1 g/bhp·h, to go into effect by 2004. The rare-earth
additive will increase fuel cost by 2 to 4 percent in the U.S.

For winter operation of this heat pump, a humidifier would be used. Provision
would be made for ventilation by admission of the required amount of fresh air to
be conditioned and mixed with that which is recirculated. For summer use, the
condenser and evaporator in Fig. 17 would be so connected that they would
exchange functions, the summer evaporator acting as an air cooler and dehumidifier.
Heat removed from the inside air would be rejected, along with the heat equivalent
of the compressor work, to the atmosphere or to the water which acts as the source
of heat for cold-weather operation, and as the cooling medium in warm weather.

Unitary heat pumps (UHPs) using closed-water loops are used extensively today in office buildings because of their flexibility. As described by Gershon Meckler, P.E., in *HPAC* magazine,* UHPs are small, individual air conditioners located and controlled within each building zone, that dehumidify and cool or heat at the terminal without activating a central chiller or other UHPs. Best of all, they can be switched back and forth between heating and cooling, and some can be heating while others are cooling. Figure 18 shows a conventional unitary heat-pump system.

To retain the benefits of UHPs, while making them more responsive to the new utility rate structures and more energy efficient, Gershon Meckler developed a UHP system that reduces electric demand in several ways, amongst which are: (*a*) Shifts the dehumidification load from the UHPs to a small, central ice plant that operates off-peak (or, in an alternative version, to an efficient two-stage desiccant system); (*b*) distributes dehumidification via a small quantity of very dry ventilation air (no increase in primary air quantity over a conventional UHP system); (*c*) uses UHPs for sensible heating/cooling only, with no condensation at terminals; (*d*) operates UHPs in cooling mode at 55°F (12.8°C) refrigerant coil temperature (instead of at 40 to 45°F [4.4 to 7.2°C]), resulting in a 35 to 40 percent reduction in compressor horsepower (kW); (*e*) deactivates the UHP compressors when winter cooling and night heating are required.

As described by Gershon Meckler, P.E., Fig. 19 depicts schematically and Fig. 20 charts psychrometrically the ice dehumidification/unitary heat-pump system de-

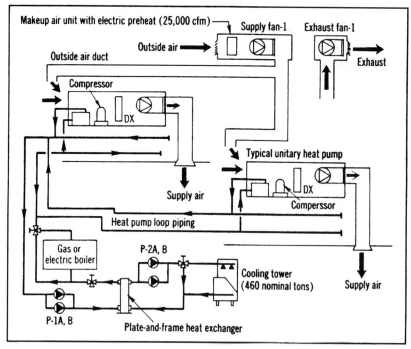

FIGURE 18 Conventional unitary heat-pump system heats, cools, and dehumidifies at terminal unitary heat pumps (UHPs). (*Gershon Meckler, P.E., and* HPAC *magazine.*). See procedure for SI values.

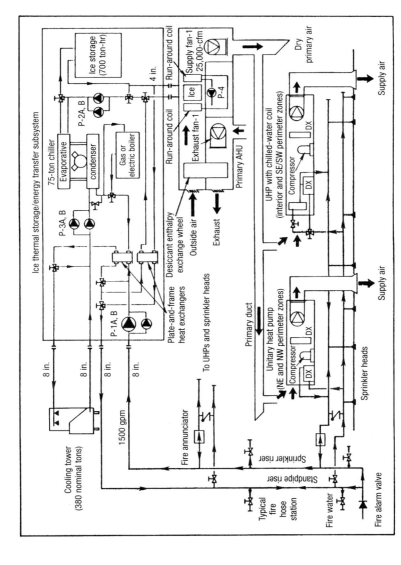

FIGURE 19 Office-building UHP system with central plant ice dehumidification and UHP/water-coil sensible heating and cooling (U.S. Patent 3918525 and other patents pending). (*Gershon Meckler, P.E. and HPAC magazine.*)

12.25

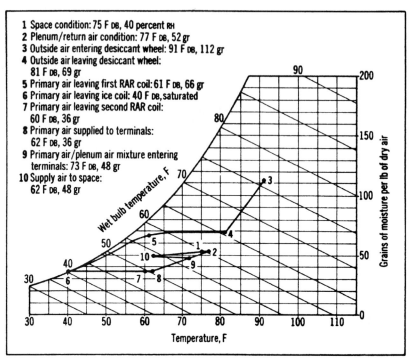

The figure contains the following legend:

1 Space condition: 75 F DB, 40 percent RH
2 Plenum/return air condition: 77 F DB, 52 gr
3 Outside air entering desiccant wheel: 91 F DB, 112 gr
4 Outside air leaving desiccant wheel:
   81 F DB, 69 gr
5 Primary air leaving first RAR coil: 61 F DB, 66 gr
6 Primary air leaving ice coil: 40 F DB, saturated
7 Primary air leaving second RAR coil:
   60 F DB, 36 gr
8 Primary air supplied to terminals:
   62 F DB, 36 gr
9 Primary air/plenum air mixture entering
   terminals: 73 F DB, 48 gr
10 Supply air to space:
   62 F DB, 48 gr

FIGURE 20 Psychrometric process for office building UHP system with central plant ice dehumidification and UHP sensible cooling. (*Gershon Meckler, P.E. and* HPAC *magazine.*) See procedure for SI values.

signed for a six-story, 159,000 ft² (14,771 m²) office building in northern New Jersey. As shown, three techniques are employed in this patented system in the central plant to limit the ice load or system demand:

(1) Incoming outside air passes through a desiccant-impregnated enthalpy exchange wheel that handles 30 to 50 percent of the building's dehumidification task without adding to the refrigeration/ice load. The rotating wheel absorbs both heat and moisture from the incoming air stream and transfers them to the drier exhaust air stream.

(2) A run-around coil further limits the load on the ice plant by precooling/reheating the air entering/leaving the ice cooling coil in the primary air handler. The very dry primary/ventilation leaves the air handler air at 60°F (15.6°C) and 36 gr/lb (79.3 gr/kg) (grains of moisture per lb or kg of dry air)—dry enough for the 0.18 ft³/min/ft² (0.05 m³/m²) distributed to handle 100 percent of the building's dehumidification load.

(3) Ice that is produced off peak melts during the day to make 34°F (1.1°C) chilled water that dehumidifies via the primary air handler cooling coil. The dehumidification or latent cooling load that is shifted off peak typically represents 20 to 40 percent of the total cooling load (24 percent of the Fig. 18 building).

Recent research and development on heat pumps has focused on absorption heat pumps and ground-coupled geothermal heat pumps.** Much of the stimulus for this R & D results from the ban of CFC refrigerants to protect the ozone layer. The year 2040 is the absolute deadline for complete banning of these refrigerants around the world. It is thought that heat-actuated absorption systems for mobile air-conditioning/heat-pump use powered by automotive waste heat could save up to 70 percent of the energy consumed.

New controls for air-, water-, or ground-source heat pumps reduce the annual energy used by reducing the number of defrosts a system must undergo during a heating season. Heat-transfer enhancements have been made on both the air and the refrigerant sides of the heat-exchanger equipment for heat pumps. Material requirements have been reduced by about 15 to 20 percent while maintaining the same overall heat transfer as earlier designs.**

Absorption heat pumps and chillers are energy-conversion machines driven by heat.** Fuel cost for an absorption machine is usually lower than for an electric machine but the first cost of an absorption machine is larger. The absorption heat-pump industry is based on lithium bromide/water technology. A significant number of large tonnage absorption chillers serve building-cooling uses. Today single-, double-, and triple-effect absorption machines are in use. The *effect* term refers to the number of times the input heat transfer is recycled internally to produce additional vapor and additional capacity. There is a tradeoff—each effect requires more equipment.

Absorption-cycle improvements are possible primarily through advanced-cycle design which incorporate higher heat input temperatures.** Examples of current developments include triple-effect cycles and generator-absorber heat-exchange (GAX) cycles. The GAX cycle is another method for increasing the performance of an ammonia-water absorption cycle. For low-temperature lift applications (such as air conditioning), the properties of ammonia-water allow the absorption cycle to be arranged so the high-temperature end of the absorber (heat-rejection device) overlaps the low-temperature end of the generator (heat-input device). Due to this temperature overlap, a portion of the absorber heat rejection can be used to provide heat to the generator. This cycle holds promise as a gas-fired residential heat pump.

Ground-coupled heat-pump systems use the ground as the thermal source or sink for the heat-pumping process.** Such systems include those in which heating/cooling coils are placed in horizontal trenches, vertical boreholes, under a building or parking lot, or in bodies of water, such as a pond. They may also use underground water directly in what is usually called a groundwater system. The primary advantage of this technology is that the earth provides a relatively constant temperature for heat transfer, improving the energy efficiency (COP) over that of conventional air systems and reducing electric-utility peak loads. Several electrical utilities have estimated that the installation of a single residential ground-coupled system reduces summer peak loads by 1 to 2 kW and winter peak loads by 4 to 8 kW. Because of the reduced peak demands, it is often more economical to encourage home owners and others to install heat-pump systems than to provide additional generating capacity.

A number of advances in heat-pump and ground heat-transfer technology have occurred: (1) System reliability has improved with new materials and system installation techniques; (2) New configurations for the in-ground heat exchanger help reduce material requirements and system cost. Today's in-ground piping systems typically consist of thermally fused polyethylene or polybutylene piping with an expected lifetime of at least 50 years. Installation techniques have also been improved, requiring less space and excavation for the in-ground piping.**

This procedure is the work of Norman R. Sparks, Professor and Head, Department of Mechanical Engineering, The Pennsylvania State University, and Charles C. DiLio, Associate Professor of Mechanical Engineering, The Pennsylvania State University. SI values were added by the handbook editor.

Additional information in this procedure comes from *Gershon Mecker, P.E., "Unitary Heat Pumps Plus Ice Storage," *HPAC* magazine, August, 1989, and **Karen Den Braven, Keith Herold, Viung Mei, Dennis O'Neal, and Steve Penoncello, "Improving Heat Pumps and Air Conditioning," *Mechanical Engineering*, September, 1993.

COMPREHENSIVE DESIGN ANALYSIS OF AN ABSORPTION REFRIGERATING SYSTEM

An ammonia absorption refrigerating system is to have a capacity of 100 tons (90 t). The cooling-water temperature is such that a condenser pressure of 170 lb/in² (abs) (1171 kPa) exists. The brine temperature requires an evaporator pressure of 30 lb/in² (abs) (206.7 kPa). Pressures throughout the system are: generator, 175 lb/in² (abs) (1206 kPa); rectifier entrance, 174 lb/in² (abs) (1199 kPa); rectifier exit, 171 lb/in² (abs) (1178 kPa); absorber, 29 lb/in² (abs) (199.8 kPa); aqua pump suction, 25 lb/in² (abs) (172.3 kPa); aqua pump discharge, 185 lb/in² (abs) (1274.7 kPa). Temperatures are: generator, 220°F (104°C); vapors and drip from rectifier, 110°F (43°C); liquid from condenser and to expansion valve, 76°F (24°C); evaporator exit, 10°F (−12°C); weak aqua from exchanger and to absorber, 100°F (38°C); strong aqua from absorber, 80°F (27°C). The aqua pump handles 570 lb (259 kg) of strong ammonia per minute. Steam is supplied to the generator coils at 20 lb/in² (abs) (137.8 kPa) dry saturated. Find: (*a*) the heat supplied to the generator, and the heat removed from the absorber, rectifier, and condenser per lb (kg) of ammonia circulated; (*b*) the pump horsepower (kW) required per ton (t), and total, if the combined hydraulic and mechanical efficiency is 65 percent; (*c*) the pump piston displacement with 20 percent slip; (*d*) pounds (kg) of steam required per minute per ton (t); (*e*) an energy balance of the system on the basis of Btu/min (kJ/minute) per ton (t). Show an alternate arrangement that might be used for this design.

Calculation Procedure:

1. Compute the absorber variables
Three equations may be written that deal with the mass flow through the absorber, Fig. 21*a*. They are

$$M_A + M_B = M_C$$

for the total quantity passing;

$$X_A M_A + M_B = X_C M_C$$

for the ammonia alone; and

$$(1 - X_A)M_A = (1 - X_C)M_C$$

for the water. The unknowns are usually either M_A and M_C or X_C depending upon

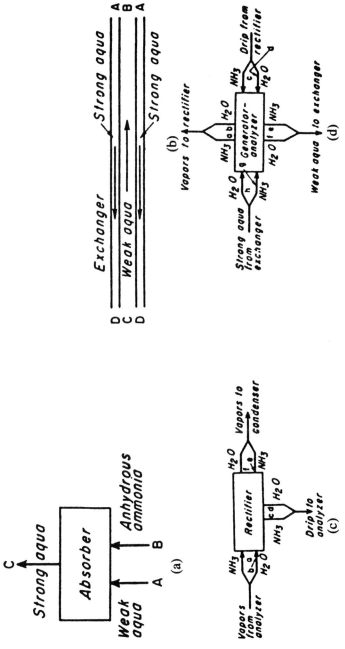

FIGURE 21 (a) Absorber flows. (b) Double-pipe heat-exchanger diagram. (c) Rectifier flows. (d) Generator-analyzer flows.

the data given. By solving two of the above equations simultaneously, either of these cases may be satisfied, or the following analysis may be used.

The weight ratio of ammonia to water in terms of the weight concentration X may be expressed for the absorber

$$\frac{\text{lb NH}_3 \text{ in strong aqua}}{\text{lb H}_2\text{O in strong aqua}} = \frac{X_C}{1 - X_C}$$

Similarly,

$$\frac{\text{lb NH}_3 \text{ in weak aqua}}{\text{lb H}_2\text{O in weak aqua}} = \frac{X_A}{1 - X_A}$$

When working with the energy quantities, it will be convenient to deal with the individual components rather than with solutions or mixtures as such. Letting the primed symbols represent ammonia in the solution or mixture and doubled-primed symbols represent the water, the energy equation for the absorber is

$$X_A M_A h_A' + (1 - X_A)M_A h_A'' + M_B h_B + M_B q = X_C M_C h_C' + (1 - X_C)M_C h_C'' + Q_R$$

or, more simply,

$$M_A' h_A' + M_A'' h_A'' + M_B h_B + M_B q = M_C' h_C' + M_C'' h_C'' + Q_R$$

and, for the heat rejected to the cooling water,

$$Q_R = M_A' h_A' + M_A'' h_A'' + M_B h_B + M_B q - M_C' h_C' - M_C'' h_C''$$

In this equation, the enthalpies at A and C are those of the particular liquids at their existing temperatures, and q is the differential heat of absorption for which the average concentration is used.

Now, for the absorber:

Enthalpy of liquid entering the expansion valve at 76°F = 127.4 (296.8 kJ/kg)

Enthalpy of vapor leaving evaporator at 30 lb/in² (abs) and 10°F = 617.8 (1439.5 kJ/kg)

Refrigerating effect per lb ammonia = 617.8 − 127.4 = 490.4 Btu (1142.6 kJ /kg)

Pound anhydrous ammonia per (min · ton) = 200/490.4 = 0.408 (0.21 kg/min · t)

Pound anhydrous ammonia per (min · ton) = 0.408(100) = 40.8 [20.6 kg/(min · t)]

For the following, refer to Fig. 21a for notation.

M_B = 40.8 lb/min, or 0.408 lb/(min · ton)

M_C = 570 lb/min, or 5.7 lb/(min · ton), or 14 lb/lb (6.4 kg/kg) anhydrous ammonia

M_A = 570 − 40.8 = 529.2 lb/min, or 5.29 lb/(min · ton), or 13 lb/lb (5.9 kg/ kg) anhydrous ammonia

t_A = 100°F; h_A' = 155.2; h_A'' = 68; t_C = 80°F; h_C' = 132; h_C'' = 48; h_B at 30 lb/ in² (abs), 10°F = 617.8

From generator conditions (175 lb/in² (abs), 220°F), (1205.8 kPa, 104°C)

$X_A = 0.325$

$X_C = \dfrac{0.325(13) + 1}{14} = 0.373$

Average concentration $= \dfrac{0.325 + 0.373}{2} = 0.3495$

$q = 345 - 250(0.3495) - 2550(0.3495^3) = 148.8$ Btu/lb (345.7 kJ/kg) anhydrous ammonia absorbed

$M'_A = 0.325(13) = 4.23$ lb/lb anhydrous ammonia (1.9 kg/kg)

$M''_A = 13 - 4.23 = 8.77$ lb/lb anhydrous ammonia (3.9 kg/kg)

$M'_C = 4.23 + 1 = 5.23$ lb/lb anhydrous ammonia (2.4 kg/kg)

$Q_R = 4.23(155.2) + 8.77(68) + 617.8 + 148.8 - 5.23 - 8.77 = 908$ Btu/lb anhydrous ammonia, or $908(0.408) = 370$ Btu/(min·ton), or $908(40.8) = 37{,}000$ Btu/min (39,035 kJ/min)

2. Determine the aqua pump power input
The power representing the useful work actually done on the fluid may be termed the hydraulic hp, and the ratio of this to the power input to the pump is the combine hydraulic and mechanical efficiency. The hydraulic work, determined by a steady-flow analysis applied to a fluid, the pressure of which is increased under reversible adiabatic action, may be stated mathematically as:

$$ W = \int VdP $$

in which V = volume of fluid pumped, ft^3
dP = infinitesimal increase of pressure, lb/ft^2
W = work, ft-lb

Since most liquids are virtually incompressible, this equation may be readily integrated between the pressure limits P_d and P_s, the discharge and suction pressure, respectively, as follows:

$$ W = V(P_d - P_s) $$

but

$$ V = M'V $$

in which M' is the weight of the fluid pumped and V is the specific volume of the fluid. If M expresses the weight rate of flow in pounds per minute the equation for the hydraulic hp is

$$ \text{Hp} = \dfrac{Mv(P_d - P_s)}{33{,}000} $$

In reciprocating pumps, the liquid delivered is theoretically equivalent in volume to the piston or plunger displacement, but actually a lesser quantity is discharged. The difference between theoretical and actual volumetric discharge expressed in terms of the piston displacement is called *slip*, which is thus the proportion of the displacement that is unproductive in so far as the output of the pump is concerned.

The weight of strong aqua to be pumped is given as 570 lb/min (258.8 kg/min)

$X = 0.373$; $P_d = 185$ lb/in^2 (abs) (1274.7 kPa); $P_a = 25$ lb/in^2 (abs) (172 kPa)

v'_f at 80° F = 0.0267; v'_f at 80°F = 0.0161

With the usual 16 percent absorption of the ammonia liquid in these systems,

v_{sol} = 0.84(0.373) (0.0267) + 0.627(0.0161) = 0.0185 ft³/lb (0.0012 m³/kg)

Hydraulic hp = $\dfrac{144(570)\ (185 - 25)\ (0.0185)}{33,000}$ = 7.37 (5.5 kW)

Hp input = $\dfrac{7.37}{0.65}$ = 11.33 or 0.1133 hp/ton (0.09 kW/t)

Piston displacement = $\dfrac{570(0.0185)}{0.8}$ = 13.2 ft³/min, or 0.132 ft³/(min · ton)

(0.004 m³/t)

The work input to the aqua pump,

W/J = 11.33(42.4) = 480. Btu/min, or 4.80 Btu/(min · ton) [5.6 kJ/min · t)] or 11.8 Btu/lb (27.4 kJ/kg) anhydrous ammonia

Of energy supplied to drive the pump, all but a fraction of that converted to heat by mechanical friction is represented in the strong aqua by an increase in enthalpy. Although this increase is small per pound of solution, it becomes noticeable when expressed in energy units per minute or per minute per ton, or when appearing in an energy balance for the system. In the exchanger calculations which follow in the next section, it will be noted that the enthalpy of the strong aqua leaving the absorber has been augmented by the pump work before entering the exchanger. This is assuming that all of the energy input to the pump is taken up by the liquid.

3. Calculate the exchanger heat transfer
The weight relationships for the exchanger are simple since there is no change in concentration of either fluid. The weights at A and at D are equal for both the total fluid and components, and the same relation holds for the weights at B and C, Fig. 21b.

The process for the weak solution is that of simple cooling. The heating of the strong aqua, however, may be accompanied by vaporization of a small portion of the constituents if the temperature is carried high enough. This will tend to repress the temperature rise of the strong aqua and will make the final temperature somewhat difficult to predict. When this temperature is desired, it may be found by trial and error, but in problems involving energy quantities alone the temperature is unimportant so long as the energy of the solution leaving the exchanger may be determined for application to the generator. This may be done, irrespective of any vaporization that may occur, by adding to the energy of the strong aqua entering the exchanger the heat removed from the weak aqua. Since in the strong aqua the relative quantities of ammonia and water have not changed during the heating, the heat of absorption need not be considered until the generator calculations are carried out.

The energy equation is

$$M'_C h'_C + M''_C h''_C + M'_A h'_A + M''_A h''_A = M'_B h'_B + M''_B h''_B + M'_D h'_D + M''_D h''_D$$

or

$$M'_D h'_D + M''_D h''_D = M'_C h'_C + M''_C h''_C + M'_A h'_A + M''_A h''_A - M'_B h'_B - M''_B h''_B$$

where $M'_D h'_D + M''_D h''_D$ may be considered as the energy of the strong solution entering the analyzer if the differential heat of absorption is subsequently taken into account. The heat exchanged through the surface is written:

$$Q = M'_C h'_C + M''_C h''_C - M'_B h'_B - M''_B h''_B = M'_{CB}(h'_C - h'_B) + M''_{CB}(h''_C - h''_B)$$

For every pound of anhydrous ammonia circulated in the system,

$$M'_A = M'_D = 5.23 \text{ lb}; \ M'_C = M'_B = 4.23 \text{ lb};$$
$$M''_A = M''_D = M''_C = M''_B = 8.77 \text{ lb (3.9 kg)}$$
$$h'_f \text{ at } 80°F = 132; \ h''_f \text{ at } 80°F = 48 \text{ (111.5 kJ/kg)}$$

$M'_A h'_f + M''_A h''_f$ at absorber exit $= 5.23(132) + 8.77(48) = 111$ Btu/lb (2581.7 kJ/kg) anhydrous ammonia

The work input to the aqua pump,

W/J (from pump computation) $= 11.8$ Btu/lb (27.4 kJ/kg) anhydrous ammonia
$M'_A h'_A + M''_A h''_A = 1111 + 11.8 = 1122.8$ Btu/lb (2609.1 kJ/kg) anhydrous ammonia
$h'_B = 155.2; \ h''_C = 68$

At generator temperature, 220°F (104°C)

$$h'_C = 313; \ h''_C = 188.1$$

$M'_D + M''_D h''_D = 4.23(313) + 8.77(188.1) + 1122.8 - 4.23(155.2) - 8.77(68)$
$= 2846$ Btu/lb anhydrous ammonia, or 1161 Btu/(min·ton), or 116,100 Btu/min (122,486 kJ/kg)

(This is determined solely for use in generator calculations.)

The heat transferred through the exchanger surface from weak to strong aqua:
$Q = 4.23(313 - 155.2) + 8.77(188.1 - 68) = 1723$ Btu/lb anhydrous ammonia, or 702 Btu/(min·ton), or 70,200 Btu/min (74,061 kJ/kg)

4. Find the analyzer vapor properties

The analyzer is a direct-contact heat exchanger that serves to further preheat the strong aqua before entrance to the generator by cooling the vapors leaving the generator and thus dehydrating them to a certain extent. The analyzer may be integral with the generator, i.e., built into the generator structure, or it may be external to or separate from the generator. Irrespective of type, the analyzer consists of a series of trays over which the strong aqua and the rectifier drip, introduced at the top, flow by gravity counter to the flow of vapors. In this manner, by exposing a large solution surface to the vapors, a good transfer of heat is effected. After passing through the analyzer, the vapors go to the rectifier for more complete dehydration while the strong aqua is introduced to the generator for further heating.

Because the weight of strong aqua is several times that of the vapors and because the specific heat of the solution is more than double that of the vapors, the cooling effect on the latter is considerably greater, measured in temperature, than is the heating effect on the solution. Depending upon the efficiency of the analyzer, a temperature difference between incoming solution and outgoing vapors of from 10 to 20° may be had or, in cases where the exchanger is functioning very effectively, the temperature difference may be less than 10°. The vapors leaving will closely approximate equilibrium conditions for the pressure and temperature since there is continuous contact with the solution at nearly the same temperature. The effect of the analyzer on the vapors is to progressively increase the ammonia content and reduce the proportion of water vapor, while the concentration of the strong aqua is, of course, also somewhat reduced at the same time.

In system design, it is convenient and satisfactory to regard the generator and analyzer as a single unit. Viewed in this way, the analyzer affects the calculations for the generator simply by reducing the quantity of energy leaving in the vapors, decreasing by an equivalent amount the heat to be supplied.

That part of this calculation procedure which concerns the analyzer alone is confined to the determination of the properties of the vapors leaving.

The temperature of the strong aqua leaving the exchanger and entering the analyzer may be estimated as 180 to 190°F (82 to 88°C). The temperature of the vapors at analyzer exit may therefore be taken as 200°F (93°C).

Properties of vapors:

Pressure $= 175$ lb/in^2 (abs) (1205.8 kPa)

Partial pressure of water vapor $= 7.2$ lb/in^2 (abs) (49.6 kPa)

Partial pressure of ammonia vapor $= 167.8$ lb/in^2 (abs) (1156.1 kPa)

Temperature of both $= 200°F$ (93.3°C)

For the water vapor, $h = 1147.8$, $v = 54.29$ (126.1 kJ/kg)

For the ammonia vapor, $h = 707.1$, $v = 2.423$

Ammonia vapor per lb water vapor $= 54.29/2.423 = 22.4$ lb (10.2 kg)

5. *Solve for the rectifier variables*

The rectifier flow diagram is shown in Fig. 21c. Three combined quantities are involved: the vapors from the analyzer, the drip to the analyzer, and the vapors to the condenser. In the diagram, these quantities are shown as divided into their respective components. Thus, in the vapors entering, a represents ammonia and b water vapor; in the drip leaving, c is ammonia and d is water; and, at the vapor exit, ammonia vapor is at e and water vapor at f. The weights at each of these points will be determined first.

From analyzer data, the weight ratio of ammonia to water vapor in the mixture entering the rectifier M_a/M_b is known. The exit pressure and temperature, assuming all of the drip to leave at this point, determine the concentration X of the solution, and the weight concentration may be expressed as $M_c/(M_c + M_d)$. From this, the weight ratio of ammonia to water, M_c/M_d, may be found by using the equation

$$\frac{M_c}{M_d} = \frac{X}{1 - X}$$

Similarly, from exit pressure and temperature, the partial pressures of ammonia and water vapor in the vapors leaving may be determined. These permit the properties to be found as well as, from the specific volumes, the weight ratio M_e/M_f.* The quantity of ammonia M_e passing to the condenser must be the same as that passing through the evaporator and returning to the absorber, and it is therefore known. This permits M_f to be immediately solved for. The weight relationship for the respective fluids are

---

*The volume of vapors to the condenser is equal to the volume of ammonia vapor at its partial pressure or the volume of water vapor at its partial pressure. Mathematically, $V = M_e v_e = M_f v_f$.

$$M_a = M_c + M_e$$

for the ammonia, and

$$M_b = M_d + M_f$$

for the water and water vapor, in which both M_e and M_f are known for any particular case. With the ratios M_a/M_b and M_c/M_d also known, either of the foregoing equations can be written in the terms of the other, and, by solving simultaneously, any one of the four unknown quantities may be found, with the remaining three then easily determined.

The energy quantities for the rectifier may be handled in the customary manner, treating the various components individually. The heat of absorption associated with the formation of the solution must in this instance be used in its entirety because the solution is entirely formed within the rectifier. The error involved in taking the average concentration over a wide range, as in this case, is relatively small and the average may be so taken for simplicity without vitally affecting the results. When aqua ammonia concentrations exceed 0.45, the ammonia absorbed above this concentration supposedly produces no heat of absorption. Therefore, in dealing with the formation of a solution from the pure constituents, the concentration cannot exceed 0.45 for that portion of the ammonia that produces heat of absorption; and for the remainder of the ammonia, with the solution above a weight concentration of 0.45, there will be no heat absorption. It thus becomes necessary, in calculating the heat of absorption for a strong solution, to apply the integrated mean heat of absorption for a concentration from zero to 0.45 to that quantity of ammonia M_x that will give a solution strength of 0.45. It is also necessary to ignore the heat of absorption for any additional ammonia that may go into solution above this concentration.

The rectifier energy equation is

$$M_a h_a + M_b h_b + m_x q = M_c h_c + M_d h_d + M_e h_e + M_f h_f + Q_R$$

or

$$Q_R = M_a h_a + M_b h_b + M_x q - M_c h_c - M_d h_d - M_e h_e - M_f h_f$$

The known data are: At rectifier exit, $p = 171$ lb/in² (abs); $t = 110°F$; X (drip) $= 0.74$, partial pressure of water vapor $= 0.18$ lb/in² (abs), $M_e = 40.8$ lb/min; and $M_a/M_b = 22.5$, or $M_a = 22.4M_b$.

$$\frac{M_c}{M_d} = \frac{0.74}{0.26} = 2.84, \text{ or } M_c = 2.84$$

Partial pressure of ammonia vapor at exit $= 170.83$ lb/in² (abs) (1176.9 kPa)

Specific volume of ammonia vapor at exit, $v_e = 1.879$

Specific volume of water vapor at exit $= \dfrac{1545(570)}{18 \times 144 \times 0.18} = 1886*$

$$\frac{M_e}{M_f} = \frac{1886}{1.879} = 1002 M_f = \frac{40.8}{1002} = 0.0407 \text{ lb/min (0.018 kg/min)}$$

---

*This may also be obtained by extrapolating the superheated vapor properties in steam tables.

Expressing in terms of M_b and M_d,

$$22.4M_b = 2.84M_d + 40.8$$

$$M_b = M_d + 0.0407$$

Solving simultaneously,

M_d = 2.04 lb/min

M_b = 2.04 + 0.0407 = 2.08 lb/min; M_a = 1.08(22.4) = 46.6 lb/min (21.2 kg/min); M_c = 46.6 − 40.8 = 5.8 lb/min (2.6 kg/min)

For the energy quantities:

h_a = 707.8; h_b = 1147.8; h_c = 167; h_d = 77.94; h_e = 648.9; h_f = 1,108.5
q = 345 − 125(0.45) − 637.5(0.45)^3 = 230.6 Btu/lb (536 kJ/kg) ammonia absorbed up to a concentration of 0.45

Ammonia absorbed to bring concentration to 0.45 = 0.45(2.04)/0.55 = 1.67 lb/min (0.76 kg/min) = M_x. Remaining ammonia has no heat of absorption. Q_R = 46.6(707.1) + 2.08(1146.8) + 1.67(230.6) − 5.8(167) − 2.04(77.9) − 40.8(648.9) − 0.0407(1109.5) = 8068 Btu/min, or 80.7 Btu/(min·ton), or 197.8 Btu/lb (460 kJ/kg) anhydrous ammonia circulated.

6. Compute the generator variables

The generator is the still, or boiler, for the system. It consists of a shell with suitable connections for the introduction and withdrawal of fluids and, when in operation, is partially filled with aqua ammonia solution. Heat is usually supplied by low-pressure steam that is passed into coils or tubes built into the generator beneath the liquid level. At the outlet, the steam coil is provided with a trap that permits the passage of condensate only.

As stated earlier, the analyzer will be considered in calculations as a part of the generator, and the two will be treated as an integral unit. The strong aqua and the drip enter the generator by way of the analyzer from the exchanger and the rectifier, respectively, and the vapors depart for the rectifier by way of the analyzer. The weak aqua is usually taken from the bottom of the generator as far as possible from the point where the strong aqua is introduced. The generator action, like that in the absorber but in the reverse direction, is simply a process for the restoration of equilibrium in the solution. The aqua ammonia from the exchanger, being too strong for the generator conditions, is rapidly reduced in ammonia content until an equilibrium concentration is established. This is as far as the process can be carried, and the solution is then removed to be subjected to the reverse action in the absorber.

From exchanger data:

$M_g h_g + M_h h_h$ = 116,100 Btu/min
$M_e h_e$ = 40.8(4.23)(313) = 54,000 Btu/min
$M_f h_f$ = 40.8(8.77)(188.1) = 67,300 Btu/min (71,001 kJ/min)

From rectifier data:

$M_a h_a$ = 46.6(707.1) = 32,950 Btu/min

$M_b h_b = 2.08(1147.8) = 2385$ Btu/min

$M_c h_c = 5.80(167) = 968$ Btu/min

$M_d h_d = 2.04(77.9) = 159$ Btu/min

$M_x q_{rect} = 385$ Btu/min (406.2 kJ/min)

Average generator concentration $X = 0.373 + 0.325/2 = 0.3495$

$q_{gen} = 345 - 250(0.3495) - 2550(0.3495)^3 = 148.8$ Btu/lb (345.7 kJ/kg) anhydrous ammonia circulated

$Mq = 385 + 40.8(148.8) = 6455$ Btu/min (6810 kJ/min)

$Qs = 32{,}950 + 2385 + 54{,}000 + 67{,}300 + 6455 - 116{,}100 - 968 - 159 = 45{,}863$ Btu/min, or 458.6 Btu/(min · ton), or 1125 Btu/lb (2614.2 kJ/kg) anhydrous ammonia

7. *Show the use of an alternate generator-energy-balance solution*

Figure 22 shows the analysis for the first law of thermodynamics applied to the several components of the absorption system. When energy quantities have been established for the rectifier, absorber, and aqua pump, the steady-flow analysis for the portion of the system bounded by the imaginary control surface of Fig. 22 materially simplifies the calculation of the heat quantity to be transferred at the generator. Referring to Fig. 22 and the values previously established, the generator energy balance is as shown in the table below. (*Note:* The water vapor into the

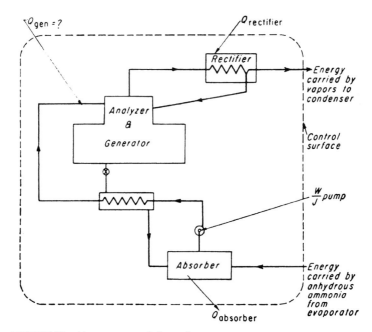

FIGURE 22 Alternate energy-balance diagram.

condenser and out of the evaporator has not been included because of its insignificance.)

Energy into system, Btu/min	Energy out of system, Btu/min
40.8 (617.8) = 25,206 (26,592 kJ/min)	40.8 (648.9) = 26,475 (27,931 kJ/min)
Pump work = 480 (506.4 kJ/min)	$Q_{rectifier}$ = 8,068 (8512 kJ/min)
$Q_{generator} = Q_{gen}$	$Q_{absorber}$ = 37,000 (39,035 kJ/min)
Sum of energy into system = 25,686 + Q_{gen}	Sum of energy out of system = 71,543

Therefore, $Q_{generator}$ = 45,857 Btu/min (48.379 kJ/min) which checks closely the answer by the detailed balance of the previous solution.

Steam consumption: Assuming the condensate to leave at 220°F; h at 20 lb/in² (abs), dry, = 1156.3; h_f at 220°F = 188.1
Steam consumption = 45,863/1156.3 − 188.1 = 47.35 lb/min, or 0.474 lb/(min·ton) [0.24 kg/min·t]

In completing this procedure, an energy balance is to be made for the system. The condenser is the only part that has not been computed. This is done in the usual way except that, the evaporation of the water vapor having been accounted for in the generator, the condensation of the water vapor should be considered in the condenser. This same quantity of water passes through the evaporator and the absorber but is neglected in those cases because the individual computations are insensitive to the small amount involved.

The energy balance then consists simply of a balance between the summation of the energies into and out of the system from or to an outside source. The balance may be made on any basis desired. In the present case, the basis is to be Btu/minute per ton capacity.

Solution. (Condenser, and system energy balance.)

Condenser:

For the ammonia entering, h (from rectifier) = 648.9
For the ammonia leaving, h_f at 76°F = 127.4
For the water vapor entering, h (from rectifier) = 1,109.5
For the water leaving, h_f at 76°F = 44

Q_R = 40.8(648.9 − 127.4) + (0.0407(1109.5 − 44) = 21,313 Btu/min, or 213.1 Btu/(min·ton) [249.8 kJ/min·t)]

Energy balance:

Part of system	Energy into system		Energy out of system	
	Btu/(min · ton)	[kJ/(min · t)]	Btu/(min · ton)	(kJ/(min · t)]
Absorber			370	433.7
Aqua pump	4.8	5.6		
Rectifier			80.7	94.6
Generator	459	538		
Condenser			213.1	249.8
Evaporator	200.0	234.4		
Total	663.8	778.1	663.8	778.1

Related Calculations. Reviewing the ammonia absorption system in the light of this calculation procedure, it may be seen that this type of machine requires approximately seven times the energy input* for a given refrigerating capacity that the compression machine needs. The quantity of energy alone, however, is hardly a reasonable criterion for comparison because the compression machine consumes high-grade energy and the absorption system may utilize almost entirely relatively low-grade energy. It would obviously be uneconomical to operate an absorption machine on an input of electrical energy, but it might prove the more economical system when the energy is obtainable from a low-cost source such as low-pressure exhaust or process steam, or cheap fuel.

The cooling-water load, measured in heat units to be disposed of, is about 2½ times as heavy in the absorption as in the compression machine. But because of the possibility of using the water through a considerably greater temperature range in the absorption system, there is not such a disparity in actual cooling-water consumption, although this will always be heavier than for a compression system under comparable conditions.

Concerning the actual operation, the absorption machine undoubtedly demands more careful attention than the compression system. This applies to the regulation of load and to the general care required in more frequent purging, in the prevention of corrosion, and in checking solution strength and vapor rectification.

Absorption refrigerating systems, once thought to be on a declining usage curve, are returning with great vigor. Applications include gas-turbine intake-air cooling, Fig. 23, and combined-cycle cogeneration, Fig. 24. In the plant shown in Fig. 24, a gas turbine provides electricity, steam, and chilled water for a health center in Canada. The 1500-ton (1350-t) absorption chiller keeps the gas-turbine intake air at the best temperature for this turbine, namely 50°F (10°C). The absorption chiller is powered by 15-lb/in² (gage) (103.4-kPa) steam extracted from the steam turbine in the cycle. This unit was reported to be the world's first gas turbine to exceed 40 percent thermal efficiency. It is part of a natural-gas fired combined-cycle cogeneration plant, a type which is becoming more popular every year.

Another cogeneration plant, Fig. 25, provides steam generated by gas-turbine exhaust gas to power an ammonia absorption refrigerating system producing 550

---

*Absorption refrigeration systems are often compared on the basis of the ratio of refrigerating effect to the heat input.

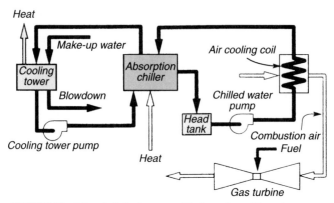

FIGURE 23 Integrated plant system with absorption-refrigerating-system chiller gas-turbine intake air. (*Power.*)

tons (495 t) of ice per day. Using this gas-turbine-exhaust generated steam allows the shutdown of electrically driven 300-ton (270 t) reciprocating refrigeration compressors which are expensive to operate during summer months when ice-making demand peaks. This cogeneration plant, as the one in Fig. 24, uses an HRSG to recover heat from the gas turbine exhaust. Electricity from both these plants is sold to local utilities. Inlet air for the cogeneration plant in Fig. 25 is cooled to within 2°F (1.1°C) of the prevailing wet-bulb temperature.

Lithium-bromide absorption refrigeration systems are popular for chilled-water service in large air-conditioning units. Developed by Carrier Corporation, these units find wide use in large office, residential, and commercial buildings. Advantages of absorption refrigerating systems ass detailed by Carrier Corporation include compactness, vibrationless operation, ease of installation anywhere in a building, from the basement to the roof, where space and a heat source are available. Such machines use the cheapest, safest, and most reliable refrigerant available, ordinary tap water.

Favorable situations outlined by Carrier Corporation for lithium-bromide absorption systems include: (1) where low-cost fuel is available; (2) where electric rates are high; (3) where steam or gas utilities desire to promote summer loads; (4) where low-pressure heating boiler capacity is largely, or wholly, unused during the cooling season; (5) where waste steam is available; (6) where there is a lack of adequate electric facilities for installing a conventional compression machine; the absorption machine uses only 2 to 9 percent of the electric power required by compression equipment. Absorption machines can be used with gas or diesel engines, gas and steam turbines.

Ammonia absorption refrigerating systems are popular for large and heavy industrial applications in a variety of industries, ice-making, oil refining, chemical manufacturing, food chilling and storage, etc. While the gas does have some negative characteristics, it has a zero ozone depletion potential. Handled in industrial situations, ammonia has proven to be relatively safe and its cost is considerably less than the conventional Freon refrigerants.

This procedure is the work of Norman R. Sparks, Professor and Head, Department of Mechanical Engineering, The Pennsylvania State University. SI values were added by the handbook editor. Illustrations of cogeneration applications came from

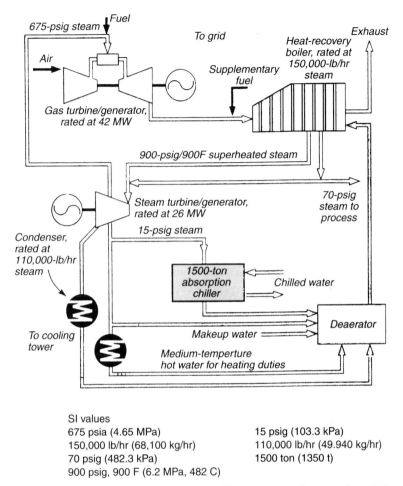

SI values
675 psia (4.65 MPa)   15 psig (103.3 kPa)
150,000 lb/hr (68,100 kg/hr)   110,000 lb/hr (49.940 kg/hr)
70 psig (482.3 kPa)   1500 ton (1350 t)
900 psig, 900 F (6.2 MPa, 482 C)

FIGURE 24 Combined-cycle cogeneration facility having over 40 percent thermal efficiency. (*Power.*)

Power and *HPAC* magazines. Data on lithium-bromide absorption systems came from Carrier Corporation.

CYCLE COMPUTATION FOR A CONVENTIONAL COMPRESSION REFRIGERATION PLANT

An ammonia-compression refrigeration plant is to be designed for a capacity of 30 tons (27 t). The cooling-water temperature requires a condenser pressure of 160 lb/in² (abs) (1102 kPa) and the brine temperature a pressure of 25 lb/in² (abs) (358 kPa) in the brine cooler. The following temperatures will exist in the system: Com-

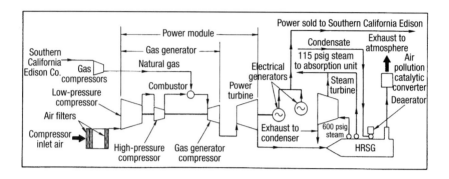

SI values
115 psig (792.4 kPa)
600 psig (4134 kPa)

FIGURE 25 Cogeneration plant which powers an ammonia absorption refrigerating plant. (*HPAC* magazine.)

pressor suction, 0°F (−17.8°C); compressor discharge, 230°F (110°C); entering condenser, 210°F (98.9°C); leaving condenser, 60°F (15.6°C); at expansion valve, 68°F (20°C); leaving evaporator, −5°F (−20.6°C). Wiredrawing through the compressor valves: Suction 5 lb/in² (34.5 kPa); discharge, 10 lb/in² (68.9 kPa). A two-cylinder, vertical, single-acting compressor is to be used at 400 ft/min (121.9 m/min) piston speed. Mechanical efficiency = 0.80; volumetric efficiency = 0.75; ratio of stroke/bore = 1.2. Using P-V and T-s diagrams, determine the properties of the refrigerant at each point in the cycle, and find: (*a*) Weight of ammonia circulated per minute; (*b*) ihp of compressor per ton and total; (*c*) power required to operate the machine; (*d*) heat rejected to the cylinder jackets per minute; (*e*) heat rejected in the condenser per minute; (*f*) compressor piston displacement per minute per ton and per minute (*g*) bore, stroke, and rpm of compressor; (*h*) coefficient of performance. Show two methods for computing the indicated horsepower (ihp) (kW) and heat rejected to the cylinder jackets for this machine.

Calculation Procedure:

1. Determine the conditions at the end of the suction and compression strokes
Draw the P-V and T-s diagrams, Fig. 26 for the cycle. Using the diagrams and table of refrigerant properties, p_1 = 25 lb/in² (abs) (172.3 kPa); t_1 = 0°F (−17.8°C); h_1 = 613.8 Btu/lb (1426.3 kJ/kg); v_1 = 11.19 ft³/lb (0.697 m³/kg).

The condition at the end of the suction stroke with a pressure reduction of 5 lb/in² (34.5 kPa) and no change in enthalpy is p_s = 20 lb/in² (abs) (137.8 kPa); t_s = −3.1°F (−19.5°C); h_s = 613.8 Btu/lb (1426.3 kJ/kg); s_s = 1.3869; v_s = 14 ft³/lb (0.87 m³/kg); p_1 = 160 lb/in² (abs) (1102.4 kPa); t_2 = 230°F (110°C); h_1 = 725.8 Btu/lb (1691.1 kJ/kg).

The condition at the end of the compression stroke with a pressure drop of 10 lb/in² (68.9 kPa) through the discharge valve and no change in enthalpy is: p_d = 170 lb/in² (abs) (1171.3 kPa); h_d = 725.8 Btu/lb (1691.1 kJ/kg); t_d = 231.5°F (110.8°C); v_d = 2.437 ft³/lb (0.152 m³/kg); s_d = 1.3439; h_3 = 713.9 Btu/lb (1663.4 kJ/kg); h_i = 109.2 (254.5 kJ/kg); h_{x8} = 118.3 Btu/lb (275.6 kJ/kg); x_8 = (118.3 − 34.3)/574.8 = 0.146; h_9 = 610.9 Btu/lb (1423.4 kJ/kg).

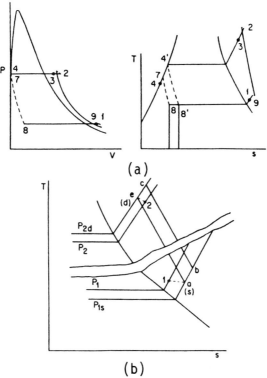

FIGURE 26 (*a*) P-V and T-s diagrams for vapor compression refrigeration system. (*b*) Magnified T-s diagram of induction, compression, and discharge processes for a reciprocating refrigeration compressor.

2. Find the refrigerating effect of this cycle

(*a*) Refrigerating effect in brine cooler = 30 × 200 = 6000 Btu/min (6330 kJ/min)

Refrigerating effect per lb of ammonia = 610.9 − 118.3 = 492.6 Btu (1144.7 kJ/kg)

Ammonia circulated per minute = 6000/492.6 = 12.2 lb or 0.407 lb/(min·ton) [2.0 kg/(min·t)]

3. Compute the horsepower (kW) input to the compressor

(*b*) First method

$$\text{Average } n: \left(\frac{v_s}{v_d}\right)^n = \frac{P_d}{P_s}; \ 5.75^n = 8.5; \ n = 1.22$$

$$\frac{W}{J} = 0.407 \left(\frac{1.22}{0.22}\right)(144)\left(\frac{20 \times 14 - 170 \times 2.437}{778.3}\right) = 55.5 \text{ Btu/(min·}$$
ton) [65.1 kJ/(min·t)]

$$\text{Ihp per ton} = \frac{55.4}{42.4} = 1.31 \ (1.08 \ \text{kW/t})$$

Total ihp $= 1.31(30) = 39.3 \ (29.3 \ \text{kW})$

Second method

$h_2 - h_1 = 725.8 - 613.8 = 112 \ \text{Btu} \ (118.2 \ \text{kJ})$

Q per lb of ammonia $= (1.3869 - 1.3439) \ (574.2) = 24.7 \ \text{Btu} \ (57.4 \ \text{kJ/kg})$

$\dfrac{W}{J}$ per (min · ton) $= 0.407(112 + 24.7) = 55.6 \ \text{Btu/(min · ton)} \ [65.1 \ \text{kJ/(min · t)}]$

$$\text{Ihp per ton} = \frac{55.6}{42.4} = 1.31 \ (1.08 \ \text{kW/t})$$

Total ihp $= 1.31(30) = 39.3 \ (29.3 \ \text{kW})$

(c) Power input to machine $= \dfrac{39.3}{0.8} = 49.1 \ \text{hp} \ (36.6 \ \text{kW})$

4. Find the heat rejected to the cylinder jacket

(d) Heat rejected to cylinder jackets:
First method

$$Q = \frac{W}{J} - M \ (h_2 - h_1) = 39.3(42.4) - 12.2(112)$$
$$= 297 \ \text{Btu/min} \ (313.3 \ \text{kJ/min})$$

Second method

From part b, $Q = 24.7 \ \text{Btu/lb}$ ammonia $(57.4 \ \text{kJ/kg})$

Q total $= 24.7(12.2) = 301 \ \text{Btu/min} \ (317.6 \ \text{kJ/min})$

5. Compute the other variables for this refrigerating machine

(e) Heat rejected in condenser $= 12.2(713.9 - 109.2) = 7360 \ \text{Btu/min} \ (7764.8 \ \text{kJ/min})$

(f) Piston displacement $= 0.407(11.19)/0.75 = 6.08 \ \text{ft}^3/(\text{min · ton})$ or $6.08(30) = 182.4 \ \text{ft}^3/\text{min} \ (5.2 \ \text{m}^3/\text{min})$

(g) Since this is a single-acting machine, the volume swept through by the two pistons during their respective suction strokes $= 182.4 \ \text{ft}^3/\text{min}$

$$\text{Piston area} = \frac{182.44(144)}{400} = 65.7 \ \text{in}^2 \ (423.9 \ \text{cm}^2) \ \text{per cylinder}$$

$$\text{Cylinder bore} = \sqrt{\frac{4(65.7)}{\pi}} = 9.15 \ \text{in} \ (23.2 \ \text{cm})$$

Stroke $= 1.2(9.15) = 11 \ \text{in} \ (27.9 \ \text{cm})$

$$\text{Rpm} = \frac{400(12)}{2(11)} = 218$$

(*h*) Coefficient of performance = 6000/49.1(42.4) = 2.88

$$\text{Hp input per ton} = \frac{49.1}{30} = 1.64 \ (1.36 \ \text{kW/t})$$

Related Calculations. Use this general method of analysis for any reciprocating refrigeration compressor using any refrigerant for which property plots and tables are available. Further, the machine can be used for any type of refrigeration load: air conditioning, cold-room cooling, ice-making, skating rink freezing, transport cooling, etc.

Data on temperatures and pressures in the cycle are available from machine manufacturers. Refrigerant property plots and tables are available form refrigerant suppliers.

This procedure is the work of Norman R. Sparks, Professor and Head, Department of Mechanical Engineering, The Pennsylvania State University, and Charles C. DiLio, Associate Professor of Mechanical Engineering, The Pennsylvania State University. SI values were added by the handbook editor.

DESIGN OF COMPOUND COMPRESSION-REFRIGERATION PLANT WITH WATER-COOLED INTERCOOLER

Design a 100-ton (90-t) compound ammonia-compression refrigeration system having a water-cooled intercooler when the following conditions apply: Condenser pressure = 200 lb/in² (abs) (1378 kPa); evaporator pressure = 22 lb/in² (abs) (151.6 kPa); intercooler pressure = 70 lb/in² (abs) (482.3 kPa). Volumetric efficiency, low-pressure end = 85 percent; high-pressure end = 75 percent. Wiredrawing through the compressor valves produces these pressure losses: Low-pressure suction = 2 lb/in² (13.8 kPa); low-pressure discharge = 5 lb/in² (34.5 kPa); high-pressure suction = 4 lb/in² (27.6 kPa); high-pressure discharge = 10 lb/in² (68.9 kPa). Ammonia may be cooled to 90°F (32.2°C) in the intercooler and subcooled as liquid to 85°F. Suction temperature = 0°F (17.8°C). Temperature leaving brine cooler = 5°F (−20.6°C). Low-pressure compression is adiabatic. High-pressure compression, *n* = 1.27. Both cylinders in this compressor are double-acting. Find: (*a*) Quantity of refrigerant to be circulated; (*b*) ihp (kW) of low-pressure cylinder; (*c*) ihp (kW) of high-pressure cylinder; (*d*) total ihp (kW); (*e*) heat rejected in the intercooler; (*f*) piston displacement of low-pressure cylinder; (*g*) piston displacement of high-pressure cylinder; (*h*) total piston displacement; (*i*) coefficient of performance based on ihp (kW).

Calculation Procedure:

1. Sketch the system layout and refrigerant state diagrams
Figure 27 shows the system layout with the water-cooled intercooler located between the high-pressure and low-pressure cylinders. Note that this layout can be

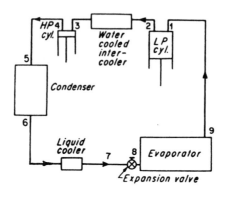

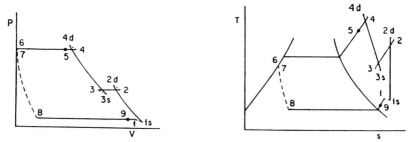

FIGURE 27 Compound vapor-compression cycle with water-cooled intercooler and its associated P-V and T-s diagrams.

used with any type of refrigerant whose thermodynamic characteristics are available in either tabular or graphic form.

2. Determine the refrigerant circulation rate and key pressures

(a) $h_s = h_7 = 137.8$; $h_9 = 612$

Ammonia circulated per min = $100(200)/612 - 137.8 = 42.2$ lb (19.2 kg/min)

(b) Suction pressure $p_s = 22 - 2 = 20$ lb/in^2 (abs); $h_1 = 614.8 = h_s$; $v_1 = 12.77$; $s_s = 1.3895 = s_d$; $t_s = -1°$

Discharge pressure $p_d = 70 + 5 = 75$ lb/in^2 (abs); $h_d = 695.5 = h_2$

Ihp, low pressure = $42.2(695.5 - 614.8)/42.4 = 80.5$ (60.1 kW)

3. Compute power input, intercooler heat rejection, and piston displacement

(c) Suction pressure $p = 70 - 4 = 66$ lb/in^2 (abs); $h_3 = h_s$, $v_3 = 4.72$, $v_s = 5.00$

Discharge pressure $p_d = 200 + 10 = 210$ lb/in^2 (abs) (1447 kPa)

$v_d = 5/(210/66)^{1/1.27} = 2.01$ ft^3/lb (0.13 m^3/kg)

Ihp, high-pressure = $1.27(144)$ (42.2) $(66 \times 5.00 - 210 \times 2.01)/(1.27 - 1)$ $(33,000) = 79.5$ (59.3 kW)

(d) Total ihp = $80.5 + 79.5 = 160$ (119.4 kW); ihp per ton = 1.6 (1.32 kW/t) (*e*) $h_2 = 695.5$, $h_3 = 654.6$

Heat rejected in intercooler = $42.2(695.5 - 654.6) = 1728$ Btu/min (1823 kJ/min)

(*f*) $v_1 = 1277$

Piston displacement, low-pressure = $42.2(12.77)/0.85 = 634$ ft³/min (17.9 m³/min)

(*g*) $v_s = 4.72$ ft³/lb

Piston displacement, high-pressure = $42.2(4.72)/0.78 = 255.3$ ft³/min (7.2 m³/min)

(*h*) Total piston displacement = $634 + 255.3 = 889.3$ ft³/min (25.1 m³/min)

4. Find the coefficient of performance

(*i*) Coefficient of performance = $\dfrac{20{,}000}{(42.4)\,(160)} = 2.95$

Related Calculations. The procedure given here can be used for a compound refrigerating compressor using any refrigerant. Computations are made as shown, using the properties of the refrigerant chosen for the cycle. The compound machine may be considered to be two simple machines in series with the proper terminal conditions for each.

With intercooling between cylinders, the overall rise in temperature of the refrigerant is reduced. With a suitable heat exchanger, the heat removed by the intercooler can be recovered for use in an industrial process of some kind, or for domestic water heating. Figure 27 shows the potential heat recovery possible with intercooling.

This procedure is the work of Norman R. Sparks, Professor and Head Department of Mechanical Engineering, The Pennsylvania State University, and Charles C. DiLio, Associate Professor of Mechanical Engineering, The Pennsylvania State University. SI values were added by the handbook editor.

ANALYSIS OF A COMPOUND COMPRESSION-REFRIGERATION PLANT WITH A WATER-COOLED INTERCOOLER AND LIQUID FLASH COOLER

Design a 100-ton (90-t) compound ammonia-compression refrigeration system having a water-cooled intercooler when the following conditions apply: Condenser pressure = 200 lb/in² (abs) (1378 kPa); evaporator pressure = 22 lb/in² (abs) (151.6 kPa); intercooler pressure = 70 lb/in² (abs) (482.3 kPa). Volumetric efficiency, low-pressure end = 85 percent; high-pressure end = 78 percent. Wiredrawing through the compressor valves produces these pressure losses: Low-pressure suction = 2 lb/in² (13.8 kPa); low-pressure discharge = 5 lb/in² (34.5 kPa); high-pressure suction = 4 lb/in² (27.6 kPa); high-pressure discharge = 10 lb/in² (68.9 kPa). Ammonia may be cooled to 90°F (32.2°C) in the intercooler and subcooled as liquid to 85°F. Suction temperature = 0°F (−17.8°C). Temperature leaving brine cooler = 5°F (−20.6°C). Low-pressure compression is adiabatic. High-pressure compression, $n = 1.27$. Both cylinders in this compressor are double-acting. Find: (*a*) Quantity of refrigerant to be circulated; (*b*) quantity of refrigerant passing through high-pressure cylinder and condenser; (*c*) ihp (kW) of low-pressure cylinder; (*d*) ihp (kW) of high-pressure cylinder; (*e*) total ihp (kW); (*f*) piston displacement of low-pressure cylinder; (*g*) piston displacement of high-pressure cylinder; (*h*) total piston displacement; (*i*) coefficient of performance based on ihp (kW).

Calculation Procedure:

1. Sketch the system layout and refrigerant state diagrams
Figure 28 shows the system layout with the water-cooled intercooler located between the high-pressure and low-pressure cylinders and the liquid flash cooler between the condenser and the evaporator. Note that this layout can be used with any type of refrigerant whose thermodynamic characteristics are available in either tabular or graphic form.

2. Determine the refrigerant circulation rate and low-pressure power input

$h_1 = 614.8 = h_{1s}$; $v_1 = 12.77$; $p_{1s} = 20$ lb/in² (abs); $t_{1s} = -1°F (-18.3°C)$;
$s_{1s} = 1.3891 = s_{2d}$; $p_{2d} = 75$ lb/in² (abs); $h_{2d} = 695.3 = h_2$; $h_3 = 654.6$ (1521.2 kJ/kg);
$h_7 = 137.8 = h_8$; $h_9 = 622.4$; $h_{10} = 84.2 = h_{11}$; $h_{12} = 612$ (1422.2 kJ/kg)

(*a*) Low-pressure ammonia per min = 100(200)/612 − 84.2 = 37.9 lb (17.2 kg/min)

(*b*) $M_{10} = 37.9$ lb/min; $M_9 = 37.9(137.8 − 84.2)/633.4 − 137.8 = 4.2$ lb/min (1.9 kg/min)
High-pressure ammonia per min = 37.9 + 4.2 = 42.1 lb (19.1 kg/min)

(*c*) Low-pressure ihp = 37.9(695.3 − 614.8)/42.4 = 72.0 (53.7 kW)

3. Compute total power input and piston displacement

(*d*) $h_4 = 37.9(654.6) + 4.2(622.4)/42.1 = 650$ Btu/lb (1510.5 kJ/kg)
$p_4 = 70$ lb/in² (abs); $t_4 = 82°F$; $v_4 = 4.635$; $h_4 = h_{4s} = 650$ Btu (1510.5 kJ/kg)
$t_{4s} = 81°F$; $p_{4s} = 66$ lb/in² (abs); $v_{4s} = 4.92$ (0.31 m³/kg)
$p_{5d} = 210$ lb/in² (abs); $v_{5d} = 4.92$ $(66/210)_{1/1.27} = 1.98$ ft³ per lb (0.12 m³/kg)

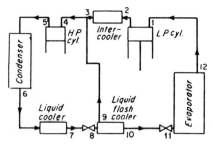

FIGURE 28 Compound vapor-compression cycle with liquid-flash intercooler.

High-pressure ihp = 1.27(144) (42.1)/0.27(33,000) × (66 × 4.92 − 210 × 1.98) = 78.6 (58.6 kW)

(e) Compressor ihp, total, = 72.0 + 78.6 = 150.6 (112.3 kW)
Ihp per ton = 1.51 (1.25 kW/t)

(f) Piston displacement, low-pressure = 37.9(12.77/0.85 = 570 ft^3/min (16.1 m^3/min)

(g) Piston displacement, high-pressure = 42.1(4.635)/0.78 = 250 ft^3/min (7.1 m^3/min)

(h) Total piston displacement = 570 + 250 = 820 ft^3/min (23.2 m^3/min)

4. Find the coefficient of performance

(i) Coefficient of performance = 20,000/42.4(150.6) = 3.13 based on ihp

Related Calculations. The procedure given here can be used for a compound refrigerating compressor using any refrigerant. Computations are made as shown, using the properties of the refrigerant chosen for the cycle. The compound machine may be considered to be two simple machines in series with the proper terminal conditions for each.

With intercooling between cylinders, the overall rise in temperature of the refrigerant is reduced. With a suitable heat exchanger, the heat removed by the intercooler can be recovered for use in an industrial process of some kind, or for domestic water heating.

Figure 28 shows the cycle layout for a compound system having an intercooler and liquid flash cooler. Two expansion valves in series are fitted. Liquid refrigerant leaving the condenser expands through the first expansion valve into the liquid flash cooler. Gas formed during this expansion is returned to be recompressed in the high-pressure cylinder. This provides a liquid temperature at the second expansion valve much below that obtainable in the normal compression cycle, i.e., the cooling-water temperature. The net effect is to produce a larger refrigerating capacity from the same size compound compression machine because the liquid refrigerant enters the evaporator at a temperature less than that of the condenser cooling water.

This compound cycle owes its higher efficiency entirely to the fact that the portion of refrigerant (evaporated in the liquid flash cooler) to cool the refrigerant liquid need be compressed through only a part of the pressure range of the entire system. The vapor used in subcooling in the liquid flash cooler will, in evaporating at the intermediate pressure, require less work for compression than will the refrigerant which vaporizes at the low pressure, i.e., at the exit of the first expansion valve in Fig. 28. Transfer of a part of the refrigerating load to the first expansion valve improves the efficiency of the cycle.

Although it is theoretically possible to fit several expansion valves and flash tanks, practical limitations restrict the number. The reason for this is that the number of compression stages are restricted, usually to two for ammonia and three in some systems with extreme pressure ratios using other refrigerants. As is frequently true for cycles of this general nature, the number of coolers or heaters, or stages of compression or expansion, are practically limited to the point where an additional number would thwart the purpose of the scheme by introducing losses or disadvantages in excess of the gain to be reasonably expected. The steam power-plant cycle (covered elsewhere in this handbook) has many points of similarity, in the thermodynamic sense, to the compound refrigeration cycle when the number devices and stages of compression are considered.

This procedure is the work of Norman R. Sparks, Professor and Head Department of Mechanical Engineering, The Pennsylvania State University, and Charles

C. DiLio, Associate Professor of Mechanical Engineering, The Pennsylvania State University. SI values were added by the handbook editor.

COMPUTATION OF KEY VARIABLES IN A COMPRESSION REFRIGERATION CYCLE WITH BOTH WATER- AND FLASH-INTERCOOLING

Design a 100-ton (90-t) compound ammonia-compression refrigeration system having a water-cooled intercooler when the following conditions apply: Condenser pressure = 200 lb/in² (abs) (1378 kPa); evaporator pressure = 22 lb/in² (abs) (151.6 kPa); intercooler pressure = 70 lb/in² (abs) (482.3 kPa). Volumetric efficiency, low-pressure end = 85 percent; high-pressure end = 78 percent. Wiredrawing through the compressor valves produces these pressure losses: Low-pressure suction = 2 lb/in² (13.8 kPa); low-pressure discharge = 5 lb/in² (34.5 kPa); high-pressure suction = 4 lb/in² (27.6 kPa); high-pressure discharge = 10 lb/in² (68.9 kPa). Ammonia may be cooled to 90°F (32.2°C) in the intercooler and subcooled as liquid to 85°F. Suction temperature = 0°F (-17.8°C). Temperature leaving brine cooler = 5°F (−230.6°C). Low-pressure compression is adiabatic. High-pressure compression, n = 1.27. Both cylinders in this compressor are double-acting. Find: (a) Quantity of refrigerant to be circulated; (b) quantity of refrigerant passing through high-pressure cylinder and condenser; (c) ihp (kW) of low-pressure cylinder; (d) ihp (kW) of high-pressure cylinder; (e) total ihp (kW); (f) piston displacement of low-pressure cylinder; (g) piston displacement of high-pressure cylinder; (h) total piston displacement; (I) coefficient of performance based on ihp (kW).

Calculation Procedure:

1. Sketch the system layout and refrigerant state diagrams
Figure 29 shows the system layout with the liquid cooler at the condenser outlet, the flash intercooler at the outlet of the first expansion valve, and the intermediate gas cooler at the outlet of the low-pressure cylinder. Note that this layout can be

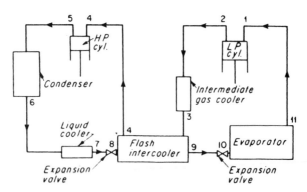

FIGURE 29 Compound vapor-compression cycle with water-cooled and flash intercoolers.

used with any type of refrigerant whose thermodynamic characteristics are available in either tabular or graphic form.

2. *Determine the refrigerant circulation rate and low-pressure power input*

$h_1 = 614.8$; $v_1 = 12.77$; $t_1 = 0°F$; $t_{1s} = -1°F$ (-18.3°C); $p_{1s} = 20$ lb/in² (abs) (137.8 kPa);

$s_{1s} = 1.3891 = s_{2d}$; $h_{1s} = 614.8$; $p_{2d} = 75$ lb/in² (abs); $h_{2d} = 695.3 = h_2$;

$h_3 = 654.6$; $h_4 = 622.4$; $v_4 = 4.15$; $t_4 = 37.7°F$; $p_{4s} = 66$ lb/in² (abs) (454.7 kPa);

$s_{4s} = 1.2719 = s_{5d}$; $p_{5d} = 210$ lb/in² (abs); $h_{5d} = 693.2 = h_5$; $h_7 = 137.8 = h_s$;

$h_9 = 84.2 = h_{10}$; $h_{11} = 612$ (1422.2 kJ/kg)

(*a*) Ammonia per min in low-pressure circuit = $100(200)/612 - 84.2 = 37.9$ lb (17.2 kg/min)

(*b*) Ammonia per min in high-pressure circuit = $37.9(654.6 - 84.2)/622.4 - 137.8 = 44.65$ lb (20.3 kg/min)

(*c*) Low-pressure ihp = $37.9(695.3 - 614.8)/42.4 = 72$ (53.7 kW)

3. *Compute total power input and piston displacement*

(*d*) High-pressure ihp = $44.65(693.2 - 622.4)/42.4 = 74.7$ (55.7 kW)

(*e*) Total ihp = $72 + 74.7 = 146.7$ (109.4 kW)
Ihp per ton = 1.47 (1.2 kW/t)

(*f*) Piston displacement, low-pressure = $37.9(12.77)/0.85 = 570$ ft³/min (16.1 m³/min)

(*g*) Piston displacement, high-pressure = $44.65(4.15)/0.78 = 237.6$ ft³/min (6.7 m³/min)

(*h*) Total piston displacement = $570 + 237.6 = 807.6$ ft³/min (22.9 m³/min)

4. *Find the coefficient of performance*

(*i*) Coefficient of performance = $20,000/42.4(146.7) = 3.22$
    Related Conditions.   The procedure given here can be used for a compound refrigerating compressor using any refrigerant. Computations are made as shown, using the properties of the refrigerant chosen for the cycle. The compound machine may be considered to be two simple machines in series with the proper terminal conditions for each.
    To further improve the compound compression cycle, the arrangement in Fig. 29 can be used. The water-cooled intercooler is retained at the condenser outlet because it is desirable to reject heat from the system wherever possible. This intercooler is followed by the first expansion valve and then a flash intercooler, or intermediate receiver, Fig. 29. The flash intercooler can be thought of as a vessel partially filled with liquid refrigerant at intermediate pressure, with gas from the low-pressure cylinder introduced below the surface of the liquid refrigerant to accelerate the cooling effect. There is provision for removing both liquid at the bottom and accumulated gas at the top of the flash intercooler, Fig. 30.
    In this intercooler, the relatively warm liquid from the high side of the system and the gas from the water-cooled intercooler are cooled by evaporation of the liquid in the flash intercooler until a condition of thermal equilibrium is established

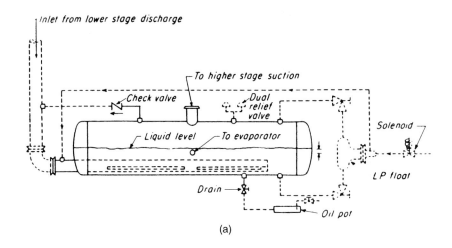

(a)

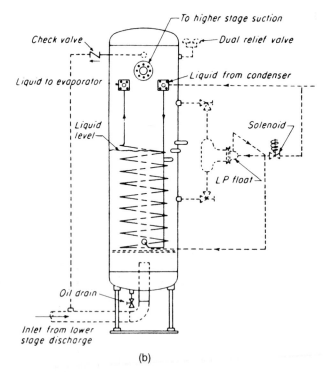

(b)

FIGURE 30   (*a*) Horizontal flash-type intercooler. (*York International Corporation.*) (*b*) Vertical coil-type intercooler. (*York International Corporation.*)

and the contents are saturated liquid and vapor. The vapor then passes to the high-pressure cylinder for compression to condenser pressure, while the liquid refrigerant is expanded through the second expansion valve to the evaporator.

Analysis will show that, in so far as quantities are concerned, there are two distinct circuits through which the working fluid travels. In the low-pressure circuit, the refrigerant passes successively through the second expansion valve, evaporator, low-pressure cylinder, intermediate gas cooler, and flash intercooler, producing in the evaporator the useful refrigeration output of the plant. The high-pressure circuit comprises the high-pressure cylinder, condenser, receiver, liquid cooler, primary expansion valve, and flash intercooler. This circuit is responsible for most of the heat rejected from the system. The quantities circulating in each part of the system are related to each other, but the relationship is dependent upon the conditions under which the machine operates, and the solution of the problem requires a knowledge of the conditions of the refrigerant entering the flash intercooler and leaving the evaporator and of the intermediate pressure. Assuming these as known and referring to Fig. 29 for notation, the solution may be carried out as follows. Obviously $M_8 = M_4$, since they are both quantities in the same circuit, and $M_3 = M_9$. The latter are directly determined by the capacity of the plant and the conditions of the refrigerant entering and leaving the evaporator. As the conditions leaving the gas cooler, point 3, and entering the expansion valve from the liquid cooler, point 7, are known, h_3 and h_7, or h_8, may be found. For thermal equilibrium within the flash intercooler, the conditions at point 4 and 9 are, respectively, that of dry saturated vapor and saturated liquid, both at intermediate pressure, and the enthalpies at these two points h_4 and h_9 may be readily determined. Then, by application of the energy equation for steady flow, balancing the ingoing against the outgoing energy quantities and neglecting any heat which may be absorbed from the outside, which in practice is extremely small,

$$M_8 h_8 + M_3 h_3 = M_4 h_4 + M_9 h_9$$

and, since $M_3 = M_9$,

$$M_8 = M_4 = \frac{M_9(h_3 - h_9)}{h_4 - f_8}$$

The weights of refrigerant circulated in each part of the system being thus determined, it only remains to use the proper quantity when calculating data for either cylinder of the compressor or for any other portion of the system. In this type of compressor, as in other compound machines, the low-pressure cylinder will not be water-jacketed, and in most cases the high-pressure cylinder is also unjacketed owing to the fact that it too receives vapor at a fairly low temperature from the flash intercooler. Data from actual machines of this type show the compression in both cylinders to be practically adiabatic. The term *adiabatic* in this case would indicate that the heat received by the gas during the first part of the compressor cycle is approximately balanced by the heat given up by the gas during the last part, rather than being truly adiabatic in the thermodynamic sense. However, the difference between the two in so far as this application is concerned is purely academic and, though worthy of note, need not affect computations.

Figure 30a shows a typical horizontal flash intercooler. The gas from the low-pressure stage enters the lowest portion of the flash tank through the slotted openings by which gas is distributed. The hot gas bubbles through the liquid refrigerant and the heat transfer involved causes some of the liquid refrigerant to vaporize. The gas above the liquid will be practically dry and saturated. Thus all the gas and liquid leaving the intercooler will be at the saturated temperature corresponding to the intercooler pressure. To make up for the liquid evaporated, i.e., to maintain a constant liquid level, a low-pressure float valve which acts as an expansion valve is provided. The resulting low-quality mixture flows into the gas supply line, at the left of the diagram.

The vertical-coil type of intercooler is shown in Fig. 30*b*. In this type of intercooler, the gas from the low-pressure stage is brought into the bottom of the shell, and gas is distributed by a multiopening plate. As this gas proceeds upward through the liquid, it is cooled, while some of the liquid refrigerant vaporizes. To make up for the liquid evaporated, some of the high-pressure liquid is admitted by the float valve, as in the case of the horizontal flash intercooler, directly into the lower portion of the vertical shell. The main body of high-pressure liquid from the condenser passes inside the coil which is submerged in the low-pressure liquid. Cooling of the liquid at the expense of further evolution of gas in the shell side results. Again, the gas and liquid leave the intercooler at the saturation temperature corresponding to the pressure of the shell side of the intercooler. Thermodynamically, the functions of the vertical and horizontal type are exactly the same. The coil type of flash intercooler may be used because of limitations in floor space.

Reciprocating vapor-compression refrigerating machines are popular in a variety of applications. For example, in the making of ice for skating, hockey, and curling rinks, compound reciprocating compressors are often chosen. The refrigeration capacity chosen ranges from 0.4 to 0.85 ton (0.36 to 0.77 t) per 100 ft^2 (9.2 m^2) of skating rink surface for indoor rinks. Outdoor rinks exposed to the sun and other heat loads can require double, or more, capacity. To provide good skating conditions, the surface temperature of the ice must be kept at a suitable temperature—usually 27°F (-2.8°C) with a variation of no more than $\pm$ 1.0°F (0.56°C).

Brine is used to cool the floor of the rink to a suitable temperature. The brine is circulated through 1- or 1.5-in (25.4 or 38.1-mm) pipe spaced 3 to 6 in (76 to 152 mm) on centers. Flow of the brine averages 10 gal/(min · ton) of refrigeration (0.63 L/s per 0.9 t). This flow corresponds to 1 gal/min per 10 to 20 ft^2 of piping surface (0.063 L/s per 0.93 to 1.86 m^2). Skating rinks range from some 400 ft^2 to 16,000 ft^2 (37 to 1486 m^2) in size.

Other applications of compound vapor-compression machines include cold storage, air conditioning, ice-making, etc. Cryogenic, ultra-low-temperature systems, also use reciprocating compressors.

This procedure is the work of Norman R. Sparks, Professor and Head, Department of Mechanical Engineering, The Pennsylvania State University, and Charles C. DiLio, Associate Professor of Mechanical Engineering, The Pennsylvania State University. SI values were added by the handbook editor. Data on skating rinks is from Fred P. Roslyn, as reported in Baumeister—*Marks' Mechanical Engineers Handbook.*

REFRIGERATION SYSTEM SELECTION

Choose a refrigeration system for a given load. Show the steps that the designer should follow in choosing a suitable refrigeration system for various types of loads.

Calculation Procedure:

1. *Determine the refrigeration load*
Use the method given in the next calculation procedure. In any refrigeration plant, the total refrigeration load = heat gain from external sources, tons + product load, tons + sensible heat load, tons.

2. *Choose the type of refrigeration system to use*
Table 1 shows the usual compressor choices for various refrigeration loads. Thus, reciprocating compressors find wide use for refrigeration loads up to 400 tons (362.9 t). Up to loads of about 5 tons (4.5 t), *unit systems* that combine the compressor, drive, evaporator, and condenser in a compact unit are popular. In some instances, larger-capacity unit systems may be available from certain manufacturers. Some large unit systems, called *central-station systems,* are built with capacities of 100 to 150 tons (90.7 to 136.1 t).

From 5- to 400-ton (4.5- to 362.9-t) capacity, *built-up central systems* are popular. In these systems, the manufacturer supplies the compressor, evaporator, and condenser as separate units. These are connected by suitable piping. The refrigeration equipment manufacturer may or may not supply the compressor driving unit. This driver may be an electric motor, steam turbine, internal-combustion engine, or some other type of prime mover.

TABLE 1   Typical Refrigeration System Choices*

System load		System type		
tons	t	Often used	Occasionally used	Rarely used
0–5	0–4.5	Unit system with reciprocating compressor	Central-station built-up system; reciprocating compressor	Central-station built-up units
5–25	4.5–22.7	Central-station built-up systems; reciprocating compressor	Central station built-up systems; reciprocating compressor	Absorption or adsorption units
25–50	22.7–45.4	Central-station built-up systems; reciprocating compressor	Central-station built-up systems; centrifugal compressor	Absorption units
50–400	45.4–362.9	Central-station built up systems; reciprocating compressor	Central-station built-up systems; steam-jet and centrifugal compressors	
400 and up	362.9 and up	Central system; centrifugal and/or absorption unit	Central-station built-up steam-jet unit	

*Adapted from ASHRAE data.
   Adapted from ASHRAE data.

For loads greater than 400 tons (362.9 t), the centrifugal refrigeration compressor is often chosen. Whereas this may be a built-up system, more and more manufacturers today supply completely fabricated systems containing all the needed components, including the controls, driver, etc.

Steam-jet refrigeration units find some application for loads of 50 tons (45.4 t) or more. The steam-jet refrigeration system is used for a large number of applications where steam is available.

Typical applications include comfort air conditioning, industrial process cooling, and similar service. In recent years, some large office buildings have used steam-jet systems mounted in the building penthouse. These units provide the cooling needed for the building air-conditioning system.

Absorption refrigeration systems were once popular for a variety of cooling tasks in industry, food storage, etc. In recent years, the absorption system has found renewed use in medium- and large-size-air-conditioning systems. The usual absorbent used today is lithium bromide; the refrigerant is ordinary tap water. Absorption refrigeration systems are popular in areas where fuel costs are low, electric rates are high, waste steam is available, low-pressure heating boilers are unused during the cooling season, or steam or gas utility companies desire to promote summer loads. Absorption refrigeration systems can be installed in almost any location in a building where the floor is of adequate strength and reasonably level. Absence of heavy moving parts practically eliminates vibration and reduces the noise level to a minimum.

Combination absorption-centrifugal refrigeration systems are well suited for many large-tonnage air-conditioning and industrial loads. These systems are extremely economical where medium- or high-pressure steam is used as the energy source.

Using the expected refrigeration load from step 1 and the data above, make a preliminary choice of the type of refrigeration systems to use. Remember that the necessity for part-load operation might change the preliminary choice of the system type.

3. Choose the system components

Manufacturers' engineering data generally list compatible components for a given capacity compressor. These components include the condenser, expansion valve, evaporator, receiver, cooling tower, etc. Later calculation procedures in this section give specific instructions for selecting these and other important components of the system. When a unit system is chosen, the important components are preselected by the manufacturer.

4. Have the system choice verified

Have the manufacturer whose equipment will be used verify the selection for the given load. This ensures a correct choice.

Related Calculations. Use this general procedure to select the type of refrigeration system serving air-conditioning, product-cooling, liquid-cooling, ice-making, and similar applications in stationary (land) and marine service.

SELECTION OF A REFRIGERATION UNIT FOR PRODUCT COOLING

What capacity and type of refrigeration system are needed for a walk-in cooler having inside dimensions of $8 \times 6 \times 10$ ft ($2.4 \times 1.8 \times 3.1$ m) if it is insulated with 4-in (10.2-cm) thick cork? The user estimates that a maximum of 400 lb (181.4 kg) of beef will be placed in the cooler daily, arriving at 70°F (21.1°C). The average hottest summer day in the cooler locality is, according to weather bureau records, 92°F (33.3°C). The meat is to be stored at 36°F (2.2°C). A ½-hp (0.12-kW) blower circulates air in the cooler. What refrigeration capacity is required for the same cooler, if the meat is stored at −10°F (−23.3°C) and the cork insulation is 8 in (20.3 cm) thick? Two ⅓-hp (0.09-kW) blowers will be used in the cooler.

Calculation Procedure:

1. Compute the outside area of the cooler
The outside dimensions of this cooler are 9 ft (2.7 m) high, 7 ft (2.1 m) wide, and 11 ft (3.4 m) long, including the cork insulation and the supporting structure. Hence, the total outside area of the cooler, including the floor and roof, is $2(9 \times 7) + 2(9 \times 11) + 2(7 \times 11) = 478$ ft^2 (44.4 m^2).

2. Compute the heat gain and service load
There is a heat gain into the cooler through the insulated surfaces caused by the difference between the inside and outside temperatures. Also, there is a service load, that is, a heat gain caused by the opening and shutting of the cooler door. Since meat will be loaded only once a day, it is safe to assume that the service load is a normal one—i.e., the door will be opened less than 5 times per hour.

For product storage, cooling, heat, and service load, Btu/h = (total outside area of cooler, ft^2)(maximum outside temperature,°F − minimum inside temperature, °F)(factor from Table 2), or $(478)(92 − 36)(0.110) = 2944$ Btu/h (0.86 kW).

3. Compute the product heat load
Use this relation: product heat load, Btu/h = (lb/h of product cooled)(temperature of product entering cooler,°F − temperature of product leaving cooler,°F)[specific heat of product, Btu/(lb·°F)]. For this cooler, given the specific heat from Table 3, the product heat load = $(400 \text{ lb}/24 \text{ h})(70 − 36)(0.8) = 453$ Btu/h (132.8 W).

4. Compute the total heat load
The total heat load = sum of heat gain and service load + product heat load + supplementary heat load, Btu/h, or $2994 + 453 + 424 = 3821$ Btu/h (1.1 kW).

5. Compute the refrigeration-system capacity required
In cooler operation, it is essential to ensure defrosting of the evaporator during the off cycle. To permit this defrosting, select a condensing unit to operate 18 h per 24-h day. With an 18-h operating time, the required condensing-unit capacity to handle the 24-h load is (24 h/operating time, h)(total heat load, Btu/h) = $(24/18)(3821) = 5082$ Btu/h (1.5 kW).

6. Select the refrigeration unit
Since the required capacity of this refrigeration system is between 0 and 5 tons (0 and 4.5 t), the previous calculation procedure indicates that a unit system with a reciprocating compressor is the most common type used. Referring to a manufacturer's engineering data shows that a 5000-Btu/h (1.46-kW) 0.5-hp (0.37-kW) air-cooled unit having a 20°F (−6.7°C) suction temperature is available. This unit will operate about 18.5 h/day to carry the actual heat load of 5082 Btu/h (1.48 kW) if

TABLE 2 Heat Leakage Factors*

	Btu and kJ per degree temperature difference per ft² (m²) of outside surface											
	Insulation thickness											
	in						cm					
	1	2	4	6	8	10	2.5	3.1	10.2	15.2	20.3	25.4
Heat leakage only	0.178	0.127	0.079	0.059	0.046	0.038	0.0010	0.0007	0.0005	0.0003	0.0003	0.0002
Heat leakage plus normal service load	0.216	0.163	0.110	0.090	0.077	0.069	0.0012	0.0009	0.0006	0.0005	0.0004	0.0004

*Brunner Manufacturing Company; SI values added by handbook editor.
Note: Light duty—multiply factor by 0.90; heavy duty—multiply factor by 1.10; single glass—multiply factor by 15; double glass—multiply factor by 6.5; triple glass—multiply factor by 5. If any wall or ceiling of a cooler is exposed to the sun, increase the temperature difference by 20°F (11.1°C) for that wall.

TABLE 3 Typical Specific and Latent Heats*

Article	Specific heat				Latent heat of freezing		Cold-storage temperature	
	Above freezing		Below freezing					
	Btu/(lb·°F)	kJ/(kg·°C)	Btu/(lb·°F)	kJ/(kg·°C)	Btu/(lb·°F)	kJ/(kg·°C)	°F	°C
Canned goods:								
Fruits	As fresh	As fresh	As fresh	As fresh	....	....	35–40	1.7–4.4
Meats	As fresh	As fresh	As fresh	As fresh	....	....	35–40	1.7–4.4
Sardines	0.760	3.18	0.410	1.72	101.0	234.9	35–40	1.7–4.4
Butter, eggs, etc.:								
Butter	0.302	1.26	0.238	1.00	18.4	42.8	18–20	−7.8–6.7
Cheese	0.480	2.01	0.305	1.28	50.5	117.5	34	1.1
Eggs	0.760	3.18	0.410	1.72	100.0	232.6	31	−0.6
Milk, ice cream	0.900	3.77	0.462	1.93	124.0	288.4	35	1.7
Flour, meal (wheat)	0.26–0.38	1.1–1.6	0.21–0.28	0.9–1.2	14.4–28.8	34–67	36–40	2.2–4.4
Vegetables:								
Asparagus	0.952	3.99	0.482	2.02	134.0	311.7	34–35	1.1–1.7
Cabbage	0.928	3.88	0.473	1.98	131.0	304.7	34–35	1.1–1.7
Carrots	0.864	3.62	0.449	1.88	119.5	278.0	34–35	1.1–1.7
Celery (edible portion)	0.952	3.99	0.482	2.02	135.0	314.0	34–35	1.1–1.7
Dried beans	0.300	1.26	0.237	0.99	18.0	41.9	32–45	0.0–7.2
Dried corn	0.284	1.19	0.231	0.97	15.1	35.1	35–45	1.7–7.2

TABLE 3 (*Continued*)

Dried peas	0.276	1.16	0.224	0.94	13.7	31.9	35–45	1.7–7.2
Onions	0.900	3.77	0.462	1.93	126.0	293.1	36	2.2
Parsnips	0.864	3.62	0.449	1.88	119.5	278.0	34–35	1.1–1.7
Potatoes	0.792	3.32	0.422	1.97	106.5	247.7	36–40	2.2–4.4
Sauerkraut	0.912	3.82	0.467	1.45	128.0	297.7	35	1.7
Miscellaneous:								
Cigars, tobacco	…	…	…	…	…	…	35–42	1.7–5.6
Furs, woolens, etc.	…	…	…	…	…	…	35	1.7
Honey	0.344	1.44	0.254	1.06	25.9	60.2	36–40	2.2–4.4
Hops	…	…	…	…	…	…	32–40	0.0–4.4
Maple syrup	0.488	2.08	0.308	1.29	51.8	120.5	40–45	4.4–7.2
Maple sugar	0.240	1.00	0.215	0.98	7.2	16.7	40–45	4.4–7.2
Poultry, dressed and iced	0.790	3.31	0.421	1.76	105.0	244.7	28–30	−2.2–1.1
Poultry, dry-packed	0.720	3.01	0.395	1.65	93.5	217.5	26–28	−2.2–1.1
Poultry, scalded	0.800	3.35	0.425	1.65	108.0	251.2	20	−6.7
Game, frozen	0.680	2.85	0.380	1.59	86.5	201.2	15–28	−9.4–2.2
Poultry, frozen	0.680	2.85	0.380	1.59	86.5	201.2	15–28	−9.4–2.2
Nuts (dried)	0.21–0.29	0.9–1.2	0.20–0.24	0.8–1.0	4.3–14.4	10–34	35–40	1.7–4.4
Water	1.000	4.20	0.500	2.09	144.0	334.9	…	…
Meats:								
Fresh (typical only)	0.800	3.35	0.404	1.69	…	…	20–40	−6.7–4.4
Fruits:								
Fresh (typical only)	0.700	2.93	0.387	1.62	…	…	32–55	0.0–12.2

*Brunner Manufacturing Company; SI values added by handbook editor.

the evaporator is chosen on a 16°F (36°F − 20°F) (8.9°C) temperature difference between the room and refrigerant.

The exact size of a condensing unit cannot be selected until a choice is made of the evaporating temperature, or suction pressure, at which the compressor is to work. In general, a difference of between 10 and 20°F (5.6 and 11.1°C) should be maintained between the product or room temperature and the evaporator temperature. Thus, the 16°F (8.9°C) temperature difference used above is within the normal working range.

A better plan for this product cooler would be to select a larger evaporator on a 10°F (5.6°C) temperature-difference basis. The running time of the condensing unit would be decreased because of the higher operating suction temperature, that is 26 instead of 20°F (−3.3 instead of −6.7°C).

A standard refrigeration unit having the same characteristics as the unit described above, except that the evaporating temperature is 25°F (−3.9°C), has a capacity of 5550 Btu/h (1.6 kW) with refrigerant 12. The evaporating pressure is 24.6 lb/in² (169.6 kPa). The compressor is a two-cylinder unit and is belt-driven by an electric motor. A finned-tube air-cooled condenser is used. The receiver is mounted below the compressor, on the same frame. Refrigerant 12 (formerly called Freon-12) is most satisfactory for low-temperature systems.

7. Compute the required capacity at below-freezing temperature

The six steps above are for product storage at temperatures above freezing, i.e., above 32°F (0°C). For temperatures below 32°F (0°C), the same procedure is followed except that the product load is computed in three steps—cooling to 32°F (0°C), freezing, and cooling to the final temperature.

Thus, heat and service load = 478[92 − (−10)](0.085) = 4140 Btu/h (1.2 kW), given the heavy-duty factor from Table 2.

For cooling to 32°F (0°C), product load = (400/24)(70 −32)(0.8) = 504 Btu/h (0.15 kW). For the freezing process, heat removal = (lb/h of product cooled)(latent heat or enthalpy of freezing, Btu/lb, from Table 3), or (400 lb/24 h)(98) = 1635 Btu/h (0.48 kW). To cool below freezing, heat removal = (lb/h of product cooled)(32°F − temperature of storage room,°F)[specific heat of product at temperature below freezing, Btu/(lb·°F), from Table 3], or (400/24)[32 − (−10)](0.404) = 282 Btu/h (0.08 kW). Then the total product load = 504 + 1635 + 282 = 2421 Btu/h (0.71 kW).

The supplementary load with two ⅓-hp (0.09-kW) blowers is (2)(⅓)(2545) = 635 Btu/h (0.19 kW).

The total load is thus 4140 + 2421 + 635 = 7196 Btu/h (2.1 kW). Assuming a 16-h/day operating time for the refrigeration unit, the condensing capacity required is 7196(24/16) = 10,800 Btu/h (3.2 kW).

Choose an evaporator for a 10°F (5.6°C) temperature difference with a capacity of 10,800 Btu/h (3.2 kW) at a suction temperature of −20°F (−28.9°C). Checking a manufacturer's engineering data shows that a 3-hp (2.2-kW) air-cooled two-cylinder unit will be suitable.

Related Calculations. Use this general procedure to choose refrigeration units for stationary, mobile (truck), and marine applications of walk-in coolers, display cases, milk and bottle coolers, ice cream freezers and hardeners, air conditioning, etc. Note that one procedure is used for applications above 32°F (0°C) and another for applications below 32°F (0°C)

In general, choose a unit for a 10 to 20°F (5.6 to 11.1°C) difference between the product and evaporator temperatures. Thus, where a room is maintained at 40°F (4.4°C), choose a condensing unit capacity corresponding to above a 25°F (−3.9°C) evaporator temperature. If brine is to be cooled to 5°F (−15.0°C), select a condensing unit for about −10°F (−23.3°C).Where a high relative humidity is desired in a cold room, select cooling coils with a large surface area, so that the minimum operating differential temperature can be maintained between the room and the coil. The procedure and data given here were published by the Brunner Manufacturing Company based on ASHRAE data.

ENERGY REQUIRED FOR STEAM-JET REFRIGERATION

A steam-jet refrigeration system operates with an evaporator temperature of 45°F (7.2°C) and a chilled-water inlet temperature of 60°F (15.6°C). The condenser operating pressure is 1.135 lb/in² (abs) (7.8 kPa), and the steam-jet ejectors use 3.1 lb of boiler steam per pound (1.4 kg/kg) of vapor removed from the evaporator. How many pounds of boiler steam are required per ton of refrigeration produced? How much steam is required per hour for a 100-ton (90.6-t) capacity steam-jet refrigeration unit?

Calculation Procedure:

1. Determine the system pressures and enthalpies
Using the steam tables, find the following values. At 45°F (7.2°C), P = 0.1475 lb/in² (abs) (1.0 kPa); h_f = 13.06 Btu/lb (30.6 kJ/kg); h_{fg} = 1068.4 Btu/lb (2485.1 kJ/kg). At 60°F (15.6°C), h_f = 28.06 Btu/lb (65.3 kJ/kg). At 1.135 lb/in² (abs) (7.8 kPa), h_f = 73.95 Btu/lb (172.0 kJ/kg), where P = absolute pressure, lb/in² (abs); h_f = enthalpy of liquid, Btu/lb; h_{fg} = enthalpy of vaporization, Btu/lb.

2. Compute the chilled-water heat pickup
The chilled-water inlet temperature is 60°F (15.6°C), and the chilled-water outlet temperature is the same as the evaporator temperature, or 45°F (7.2°C), as shown in Fig. 31. Hence, the chilled-water heat pickup = enthalpy at 60°F (15.6°C) − enthalpy at 45°F (7.2°C), both expressed in Btu/lb. Or, heat pickup = 28.06 − 13.06 = 15.0 Btu/lb (34.9 kJ/kg).

3. Compute the required chilled-water flow rate
Since a ton of refrigeration corresponds to a heat removal rate of 12,000 Btu/h (3.5 kW), the chilled-water flow rate = (12,000 Btu/h)/(chilled-water heat pickup, Btu/lb) = 12,000/15 = 8000 lb/(h·ton) [1.0 kg/(s·t)].

4. Compute the quantity of chilled water that vaporizes
Figure 31 shows the three fluid cycles involved: (*a*) chilled-water flow from the evaporator to the cooling coils and back, (*b*) chilled-water flow from the evaporator through the ejector to the condenser and back as makeup, and (*c*) boiler steam flow from the boiler to the ejector tot he condenser and back to the boiler as condensate.
    Base the calculations on 1 lb (0.5 kg) of chilled water flowing through the cooling coils. For the throttling process from 3 to 4 in the evaporator, Fig. 31, the enthalpy remains constant, but part of the chilled water vaporizes at the lower, or evaporator, pressure. Hence, $H_3 = H_4 = h_f + xh_{fg}$, where x = lb of vapor formed per lb of chilled water entering, or 28.06 = 13.06 + x(1068.4); x = 0.01405 lb of

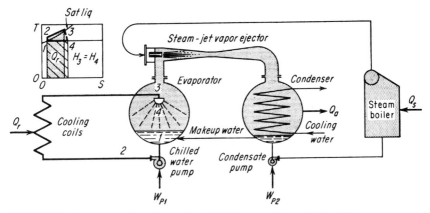

FIGURE 31 Steam-jet refrigeration unit and T-S diagram of its operating cycle.

vapor per lb (0.0063 kg/kg) of chilled water entering. The quantity of chilled water remaining at 1 in the evaporator is $1.0 - 0.01405 = 0.98595$ lb/lb (0.4436 kg/kg) of chilled water recirculating.

5. Compute the quantity of makeup vaporized
Some of the condensate in the condenser returns to the evaporator as makeup, Fig. 31. This makeup throttles into the evaporator and part of it evaporates. Hence, $H_m = h_f + x_m h_{fg}$, where H_m = enthalpy of condensate, Btu/lb; x_m = quantity of makeup vaporized, lb/lb of makeup water. Since the enthalpy of the condensate at the condenser pressure of 1.135 lb/in² (abs) (7.8 kPa) is 73.95 Btu/lb (172.0 kJ/kg), $73.95 = 13.06 + x_m(1068.4)$; $x_m = 0.057$ lb of makeup vaporized per lb (0.025 kg/kg) of makeup water entering the evaporator.

Makeup vapor simply recirculates between the evaporator and the condenser. So the total makeup water entering the evaporator must replace both the chilled-water vapor and the makeup vapor formed by the two throttling processes.

6. Compute the makeup vapor and water quantities
The lb of makeup vapor per lb of makeup water remaining in the evaporator = $x_m/(1.0 - x_m) = 0.0570/(1.0 - 0.0570) = 0.0604$.

The total makeup water to the evaporator needed to replace the vapor = $x(1 +$ lb of makeup vapor per lb of makeup water) = $0.01405(1 + 0.0604) = 0.01491$ lb/lb (0.0067 kg/kg) of chilled water circulating. This is also the vapor removed from the evaporator by the ejector.

7. Compute the total vapor removed from the evaporator
The total vapor removed from the evaporator = [lb/(h·ton) chilled water] × (makeup water per lb of chilled water circulated) = $(8000)(0.01491) = 119.3$ lb/ton (54.1 kg/t) of refrigeration.

8. Compute the boiler steam required
The boiler steam required = (vapor removed from the evaporator, lb/ton of refrigeration)(steam-jet steam, lb/lb of vapor removed from the evaporator) = $(119.3)(3.1) = 370$ lb of boiler steam per ton of refrigeration (167.8 kg/t). For a

100-ton (90.6-t) machine, the boiler steam required = (100)(370) = 37,000 lb/h (4.7 kg/s).

Related Calculations.   Use this general method for any steam-jet refrigeration system using water and steam to produce a low temperature for air conditioning, product cooling, manufacturing processes, or other applications. Note that any of the eight items computed can be found when the other variables are known.

REFRIGERATION COMPRESSOR CYCLE ANALYSIS

An ammonia refrigeration compressor takes its suction from the evaporator, Fig. 32a, at a temperature of −20°F (−28.9°C) and a quality of 95 percent. The compressor discharges at a pressure of 100 lb/in² (abs) (689.5 kPa). Liquid ammonia leaves the condenser at 50°F (10.0°C). Find the heat absorbed by the evaporator, the work input to the compressor, the heat rejected to the condenser, the coefficient of performance (COP) of the cycle, hp per ton of refrigeration, the quality of the refrigerant at state 2, quantity of refrigerant circulated per ton of refrigeration, required rate of condensing-water flow for a 100-ton (90.6-t) load, compressor displacement for a 100-ton (90.6-t) capacity. What cylinder dimensions are required for a 100-ton (90.6-t) capacity if the stroke = 1.3(cylinder bore) and the compressor makes 200 r/min?

Calculation Procedure:

1. Compute the enthalpy and entropy at cycle points
Assume a constant-entropy compression process for this cycle. This is the usual procedure in analyzing a refrigeration compressor whose actual performance is not known.

Using Fig. 32b as a guide, we see that $H_3 = h_f + xh_{fg}$, where H_3 = enthalpy at point 3, Btu/lb; h_f = enthalpy of liquid ammonia, Btu/lb from Table 4; h_{fg} = enthalpy of evaporation, Btu/lb, from the same table; x = vapor quality, expressed as a decimal. Since point 3 represents the suction conditions of the compressor, H_3 = 21.4 + 0.95(583.6) = 575.8 Btu/lb (1339.3 kJ/kg).

The entropy at point 3 is $S_3 = s_f + xs_{fg}$, where the subscripts refer to the same fluid states as above and the S and s values are the entropy. Or, S_3 = 0.0497 + 0.0497 + 0.95(1.3277) = 1.3110 Btu/(lb · °F) [5.465 kJ/(kg · °C)].

2. Compute the final cycle temperature and enthalpy
The compressor discharges at 100 lb/in² (abs) (689.5 kPa) at an entropy of S_4 = 1.3110 Btu/(lb · °F) [5.465 kJ/(kg · °C)]. Inspection of the saturated ammonia properties, Table 4, shows that at 100 lb/in² (abs) (689.5 kPa) the entropy of saturated vapor is less than that computed. Hence, the vapor discharged by the compressor must be superheated.

Enter Table 4 at S_4 = 1.3110 Btu/(lb · °F) [5.465 kJ/(kg · °C)]. Inspection shows that the final cycle temperature T_4 lies between 120 and 130°F (48.9 and 54.4°C) because the actual entropy value lies between the entropy values for these two temperatures. Interpolating gives T_4 = 130 − [(S_{130} − S_4)/(S_{130} − S_{120})] × (130 − 120), where the subscripts refer to the respective temperatures. Or, T_4 = 130 − [(1.3206 − 1.3110)/(1.3206 − 1.3104)](130 − 120) = 120.6°F (49.2°C).

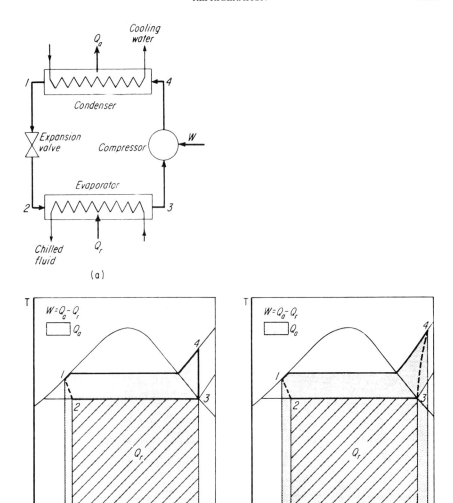

FIGURE 32 (*a*) Components of a vapor refrigeration system; (*b*) ideal refrigeration cycle *T-S* diagram; (*c*) actual refrigeration cycle *T-S* diagram.

Interpolating in a similar fashion for the final enthalpy, using the enthalpy at 130°F (54.4°C) as the base, we find $H_4 = 673.3 - [(1.3206 - 1.3110)/(1.3206 - 1.3104)](673.3 - 667.3) = 667.7$ Btu/lb (1553.1 kJ/kg).

3.   Compute the heat absorbed by the evaporator
The heat absorbed by the evaporator is $Q_r = H_3 - H_2$, where Q_r = Btu/lb of refrigerant. Or, for this system, $Q_r = 575.8 - 97.9 = 477.9$ Btu/lb (1111.6 kJ/kg).

TABLE 4 Thermodynamic Properties of Ammonia*

Saturated ammonia

Temperature t, °F (°C)	Pressure p, lb/in² (abs) (kPa)	Volume, ft³/lb (m³/kg) Liquid v_f	Volume Vapor v_g	Enthalpy, Btu/lb (kJ/kg) Liquid h_f	Enthalpy Evaporation h_{fg}	Enthalpy Vapor h_g	Entropy, Btu/(lb·°F) [kJ/(kg·°C)] Liquid s_f	Entropy Vapor s_g
0 (−17.8)	30.42 (209.7)	0.0242 (0.00151)	9.116 (0.569)	42.9 (99.8)	568.9 (1323.3)	611.8 (1423.0)	0.0975 (0.408)	1.3352 (5.590)
20 (−6.7)	48.21 (332.4)	0.0247 (0.00154)	5.910 (0.369)	64.7 (150.5)	553.1 (1286.5)	617.8 (1437.0)	0.1437 (0.564)	1.2969 (5.430)
100 (37.8)	211.9 (1461.1)	0.0272 (0.00170)	1.419 (0.089)	155.2 (361.0)	477.8 (1111.4)	633.0 (1472.4)	0.3166 (1.326)	1.1705 (4.901)
120 (48.9)	286.4 (1974.7)	0.0284 (0.00177)	1.047 (0.065)	179.0 (416.4)	455.0 (1058.3)	634.0 (1474.7)	0.3576 (1.497)	1.1427 (4.787)

Superheated ammonia

Temperature, °F (°C)	50 lb/in² (abs) (344.8 kPa) [21.67°F] (−5.7°C) saturation v	h	s	100 lb/in² (abs) (689.5 kPa) [56.05°F] (13.4°C) saturation v	h	s	150 lb/in² (abs) (1034.3 kPa) [78.81°F] (26.0°C) saturation v	h	s
100 (37.8)	6.843 (0.427)	663.7 (1543.8)	1.3816 (5.783)	3.304 (0.206)	655.2 (1524.0)	1.2891 (5.397)	2.118 (0.132)	645.9 (1502.4)	1.2289 (5.145)
120 (48.9)	7.117 (0.444)	674.7 (1569.4)	1.4009 (5.865)	3.454 (0.216)	667.3 (1552.1)	1.3104 (5.486)	2.228 (0.139)	659.4 (1533.98)	1.2526 (5.244)
140 (60.0)	7.387 (0.461)	685.7 (1594.9)	1.4195 (5.943)	3.600 (0.225)	679.2 (1579.8)	1.3305 (5.571)	2.334 (0.146)	672.3 (1563.8)	1.2745 (5.336)

4. Compute the work input to the compressor
Find the work input to the compressor from $W = H_4 - H_3$, where W = work input, Btu/lb of refrigerant. Or, $W = 667.7 - 575.8 = 91.9$ Btu/lb (213.8 kJ/kg) of refrigerant circulated.

5. Compute the heat rejected to the condenser
The heat rejected to the condenser is $Q_a = H_4 - H_1$, where Q_a = heat rejection, Btu/lb of refrigerant. Or, $Q_a = 667.7 - 97.9 = 569.8$ Btu/lb (1325.4 kJ/kg) of refrigerant circulated.

6. Compute the coefficient of performance of the machine
For any refrigerating machine, the coefficient of performance (COP) = Q_r/W, where the symbols are defined earlier. Or COP = $477.9/91.9 = 5.20$.

7. Compute the horsepower per ton for this system
For any refrigerating system, the horsepower per ton $hp_t = 4.72/\text{COP}$. Or, for this system, $hp_t = 4.72/5.20 = 0.908$ hp/ton (0.68 kW).

8. Compute the refrigerant quality at the evaporator inlet
At the evaporator inlet, or point 2, the quality of the refrigerant $x = (H_2 - h_f)/h_{fg}$, where the enthalpies are those at $-20°F$ ($-28.9°C$), the evaporator operating temperature. Or, $x = (97.9 - 21.4)/583.6 = 0.1311$, or 13.11 percent quality.

9. Compute the quantity of refrigerant circulated per ton capacity
Find the quantity of refrigerant circulated, lb/(min · ton) of refrigeration produced from $q_t = 200/Q_r$, or $q_t = 200/477.9 = 0.419$ lb/(min · ton) [(0.0035 kg/(s·t)] of refrigeration.

10. Compute the required rate of condensing-water flow
The heat rejected to the condenser Q_a must be absorbed by the condenser cooling water. The quantity of water that must be circulated is $q_w = Q_a/\Delta t$, where q_w = weight of water circulated per lb of refrigerant; Δt = temperature rise of the cooling water during passage through the condenser, °F. Assuming a 20°F (11.1°C) temperature rise of the cooling water, we find $q_w = 569.8/20 = 28.49$, say 28.5 lb of water per lb (12.8 kg/kg) of refrigerant circulated.

Since 0.419 lb/(min · ton) [0.0035 kg/(t·s)] of ammonia must be circulated, step 9, at a load of 100 tons (90.7 t), the quantity of refrigerant circulated will be $100(0.419) = 41.9$ lb/min (0.32 kg/s). The condenser cooling water required is then $(28.5)(41.9) = 1191$ lb/min (9.0 kg/s), or $1191/8.33 = 143.4$ gal/min (9.1 L/s).

11. Compute the compressor displacement
Use the relation $V_d = q_t v_g T$, where V_d = required compressor displacement, ft³/min; q_t = quantity of refrigerant circulated, lb/(ton · min); v_g = specific volume of suction gas, ft³/lb; T = refrigeration capacity, tons. For a 100-ton (90.7-t) capacity with the suction gas at $-20°F$ ($-28.9°C$), $V_d = (0.419)(14.68)(100) = 614$ ft³/min (0.29 m³/s), given the specific volume for $-20°F$ ($-28.9°C$) suction gas from Table 4.

12. Compute the compressor cylinder dimensions
For any reciprocating refrigeration compressor, V_d = (shaft rpm)(piston displacement, ft³/stroke) = v_d, or $614 = 200v_d$; $v_d = 3.07$ ft³ (0.087 m³).

Also, $D = (V_d/0.785)^{1/3}$, where D = piston diameter, ft; r = ratio of stroke length to cylinder bore. Or, $D = [3.07/(0.785 \times 1.3)]^{1/3} = 1.447$ ft (0.44 m). Then $L = 1.3D = 1.3(1.447) = 1.88$ ft (0.57 m).

Related Calculations. Employ the method given here for any reciprocating compressor using any refrigerant. Note that where the volumetric efficiency E_V of a compressor is given, the actual volume of gas drawn into the cylinder, ft³ = E_V × piston displacement, ft³. When analyzing an actual compressor, be sure you use the enthalpies which actually prevail. Thus, the gas entering the compressor suction may be superheated instead of saturated, as assumed here.

RECIPROCATING REFRIGERATION COMPRESSOR SELECTION

Choose the compressor capacity and hp, and determine the heat rejection rate for a 36-ton (32.7-t) load, a 30°F (−1.1°C) evaporator temperature, a 20°F (−6.7°C) evaporator coil superheat, a suction-line pressure drop of 2 lb/in² (13.8 kPa), a condensing temperature of 105°F (40.6°C), a compressor speed of 1750 r/min, a subcooling of the refrigerant of 5°F (2.8°C) in the water-cooled condenser, and use of refrigerant 12. Determine the required condensing-water flow rate when the entering water temperature is 70°F (21.1°C). How many gal/min of chilled water can be handled if the water temperature is reduced 10°F (5.6°C) by the evaporator chiller?

Calculation Procedure:

1. Compute the compressor suction temperature
With refrigerant 12, a pressure change of 1 lb/in² (6.9 kPa) at 0°F (−17.8°C) is equivalent to a temperature change of 2°F (1.1°C); at 50°F (10.0°C), a 1-lb/in² (6.9-kPa) pressure change is equivalent to 1°F (0.6°C) temperature change. At the evaporator temperature of 30°F (−1.1°C), the temperature change is about 1.4°F · in²/lb (0.11°C/kPa), obtained by interpolation between the ranges given above. Then, suction temperature, °F = evaporator temperature, °F − (suction-line loss, °F · in²/lb)(suction-line pressure drop, lb/in²), or $30 − 1.4 \times 2 = 27.2$, say 27°F (−2.8°C).

2. Compute the compressor equivalent capacity
To compute the compressor equivalent capacity, two correction factors must be applied: the superheat correction factor and the subcooling correction factor. Both are given in the engineering data available from compressor manufacturers.

To apply correction-factor listings, such as those in Table 5, use the following as guides: (*a*) Superheating of the suction gas can result from heat pickup by the gas outside the cooled space. Superheating increases the refrigeration compressor capacity 0.3 to 1.0 percent per 10°F (5.6°C) with refrigerant 12 or 500 if the heat absorbed represents useful refrigeration, such as coil superheat, and not superheating from a liquid suction heat exchanger. (*b*) Subcooling increases the potential refrigeration effect by reducing the percentage of liquid flashed during expansion. For each °F of subcooling, the compressor capacity is increased about 0.5 percent owing to the increased refrigeration effect per pound of refrigerant flow.

Applying guide (*a*) to a 27°F (−2.8°C) suction, 20°F (−6.7°C) superheat, interpolate in Table 6 between the 40 and 50°F (4.4 and 10.0°C) actual suction-gas

TABLE 5 Open Compressor Ratings

Suction temperature		Condensing temperature, 105°F (40.6°C)					
		Capacity		Power input		Heat rejection	
°F	°C	tons	t	bhp	kW	tons	t
10	−12.2	26.2	23.8	41.3	30.8	34.9	31.7
20	−6.7	34.0	30.8	45.3	33.8	43.6	39.6
30	−1.1	43.0	39.0	48.6	36.2	53.2	48.3

TABLE 6 Rating Basis and Capacity Multipliers—Refrigerant 12 and Refrigerant 500*

Saturated suction temperature		Actual suction gas temperature to compressor			
		30°F (−1.1°C)	40°F (4.4°C)	50°F (10.0°C)	60°F (15.6°C)
°F	°C				
20	−6.7	0.969	0.978	0.987	0.996
30	−1.1	0.970	0.979	0.987	0.996
40	4.4	. . .	0.987	0.992	0.997

*Carrier Air Conditioning Company; SI values added by handbook editor.

temperatures for a 30°F (−1.1°C) saturated suction temperature, because the actual suction temperature is 27 + 20 = 47°F (8.3°C) and the saturated suction temperature is given as 30°F (−1.1°C). Or, (0.987 − 0.979)[(47 − 40)/(50 − 40)] + 0.979 = 0.9846, say 0.985.

Applying guide (b), we see that subcooling = 5°F (2.8°C), as given. Then subcooling correction = 1 − 0.0005(15 − 5) = 0.95 , where 0.005 = 0.5 percent, expressed as a decimal; 15°F (−9.4°C) = the liquid subcooling on which the compressor capacity is based. This value is given in the compressor rating, Table 6.

With the superheat and subcooling correction factors known, compute the compressor equivalent capacity from (load, tons)/[(superheat correction factor)(subcooling correction factor)], or 36/[(0.985)(0.95)] = 38.5 tons (34.9 t).

3. Select the compressor unit
Use Table 5. Choose an eight-cylinder compressor. Interpolate for a 27°F (−2.8°C) suction and 105°F (40.6°C) condensing temperature to find compressor capacity = 40.3 tons (36.6 t); power input = 47.6 bhp (35.5 kW); heat rejection = 50.3 tons (45.6 t).

4. Compute the required condensing-water flow rate
From step 3, the condensing temperature of the compressor chosen is 105°F (40.6°C). Assume a condenser-water outlet temperature of 95°F (35.0°C), atypical value. Then the required condenser-water flow rate, gal/min = 24 × condenser load/(condensing-water outlet temperature, °F − entering condenser-water temperature, °F). Or 24(50.3)/(95 − 70) = 48.4 gal/min (3.1 L/s). This is within the normal flow for water-cooled condensing units. Thus, city-water quantities range

from 1 to 2 gal/(min.·ton) [0.07 to 0.14 L/(s·t)]; cooling-tower quantities are usually chosen for 3 gal/(min·ton) [0.21 L/s·t)].

5. *Compute the quantity of chilled water that can be handled*

Use this relation: chilled water, gal/min = 24 × capacity, tons/chilled-water temperature range, or inlet − outlet temperature, °F. Since, from step 3, the compressor capacity is 40.3 tons (36.6 t) and the chilled-water temperature range is 10°F (5.6°C), *gpm* = 24(40.3/10 = 96.7 gal/min (6.1 L/s).

The temperature of the chilled water leaving the evaporator chiller is selected so that it equals the inlet temperature required at the heat-load source. The required inlet temperature is a function of the type of heat exchanger, type of load, and similar factors.

Related Calculations. The standard operating conditions for an air-conditioning refrigeration system, as usually published by the manufacturer, are based on an entering saturated refrigerant vapor temperature of 40°F (4.4°C), an actual entering refrigerant vapor temperature of 55°F (12.8°C), a leaving saturated refrigerant vapor temperature of 105°F (40.6°C), and an ambient of 90°F (32.2°C) and no liquid subcooling.

The Air Conditioning and Refrigeration Institute (ARI) standards for a reciprocating compressor liquid-chilling package establish a standard rating condition for a water-cooled model of a leaving chilled-water temperature of 44°F (6.7°C), a chilled-water range of 10°F (5.6°C), a 0.0005 fouling factor in the cooler and the condenser, a leaving condenser-water temperature of 95°F (35.0°C), and a condenser-water temperature rise of 10°F (5.6°C). The standard rating conditions for a condenserless model are leaving chilled-water temperatures of 44°F (6.7°C), a chilled-water temperature range of 10°F (5.6°C), a 0.0005 fouling factor in the cooler, and a condensing temperature of 105 or 120°F (40.6 or 48.9°C).

Use these standard rating conditions to make comparisons between compressors. When catalog ratings of compressors of different manufacturers are compared, the rating conditions must be known, particularly the amount of subcooling and superheating needed to produce the capacities shown.

General guides for reciprocating compressors using refrigerants 12, 22 and 500 are as follows:

1. Lowering the evaporator temperature 10°F (5.6°C) from a base of 40 and 105°F (4.4 and 40.6°C) reduces the system (evaporator) capacity about 24 percent and at the same time increases the compressor hp/ton by about 18 percent.

2. Increasing the condensing temperature 15°F (8.3°C) from a base of 40 and 105°F (4.4 and 40.6°C) reduces the capacity about 13 percent and at the same time increases the compressor hp/ton by about 27 percent.

3. In air-conditioning service at normal loads, a piping loss equivalent to approximately 2°F (1.1°C) is allowed in the suction piping and to 2°F (1.1°C) in the hot-gas discharge piping. Thus when an evaporator requires a refrigerant temperature of 42°F (5.6°C) to handle a load, the compressor must be selected for a 40°F (4.4°C) suction temperature. Correspondingly, if the condenser requires 103°F (39.4°C) to reject the proper amount of heat, the compressor must be selected for a 103 + 2 = 105°F (40.6°C) condensing temperature.

4. Compressor manufacturers generally state the operating limits for each compressor in the capacity table describing it. These limits should not be exceeded.

5. To select a condenser to match a compressor, the heat rejection of the compressor must be known. For an open-type compressor, heat rejection, tons = 0.212

(compressor power input, bhp) + tons refrigeration capacity of the compressor. For a gas-cooled hermetic-type compressor, heat rejection, tons = 0.285 (kW input to the compressor) + refrigeration capacity, tons. The selection procedure and other data given here were developed by the Carrier Air Conditioning Company.

Environmental restrictions on chlorofluorocarbons (CFC) require that they be phased out over a period of time. This will restrict the use of R-12 refrigerant, which is being replaced in automotive applications by a non-CFC refrigerant named R-134a. Some changes may be required in the automotive refrigerant system when R-134a is substituted for R-12.

In air conditioning systems for buildings, ships, and other similar installations, R-123 is being substituted for R-11 and R-12 refrigerants. Conversion of existing refrigeration systems to the new non-CFC refrigerants is considered essential. Replacement parts for existing CFC plants will gradually become scarcer, as will qualified repair personnel.

CENTRIFUGAL REFRIGERATION MACHINE LOAD ANALYSIS

Select a centrifugal refrigeration machine to cool 720 gal/min (45.4 L/s) of chilled water from an entering temperature of 60°F (15.6°C) to a leaving temperature of 45°F (7.2°C).

Calculation Procedure:

1. Compute the load on the machine
Use this relation: load, tons = $gpm \times \Delta t/24$, where gpm = quantity of chilled water cooled, gal/min; Δt = temperature reduction of the chilled water during passage through the evaporator chiller, °F. For this machine, load = 720/(50 − 45)/24 = 450 tons (408.2 t).

2. Choose the compressor to use
Table 7 shows typical hermetic centrifugal refrigeration machine ratings. In a hermetic machine, the driver is built into the housing, completely isolating the refrigerant space from he atmosphere. An open machine has a shaft that projects outside the compressor housing. The shaft must be fitted with a suitable seal to prevent refrigerant leakage. Open machines are available in capacities up to approximately 4500 tons (4085 t) at air-conditioning load temperatures. Hermetic machines are available in capacities up to approximately 2000-ton (1814-t) capacity.

Study of Table 7 shows that a 450-ton (408.5-t) unit is available with a leaving chilled-water temperature of 45°F (7.2°C) and a leaving condenser-water temperature of 85°F (29.4°C). If the condenser water were available at temperatures of 75°F (23.9°C) or lower, this machine would probably be chosen.

Related Calculations. The factors involved in the selection of a centrifugal machine are load; chilled-water, or brine quantity; temperature of the chilled water or brine; condensing medium (usually water) to be used; quantity of the condensing medium and its temperature; type and quantity of power available; fouling-factor allowance; amount of usable space available; and the nature of the load, whether

TABLE 7 Typical Hermetic Centrifugal Machine Ratings* [Refrigeration Capacity, tons (t)]

Leaving chilled-water temperature		Leaving condenser-water temperature					
°F	°C	85°F	29.4°C	90°F	32.2°C	95°F	35.0°C
44	6.7	442†	401.0†	435	394.6	424	384.6
45	7.2	450	408.2	441	400.1	430	390.1
46	7.8	457†	414.6†	447	405.5	435	394.6

*Carrier Air Conditioning Company.
†These ratings require less than 330-kW input. All ratings shown are based on a two-pass cooler using 380 to 1260 gal/min (24.0 to 79.5 L/s) and on a two-pass condenser using 430 to 1430 gal/min (27.1 to 90.2 L/s).

variable or constant. The final selection is usually based on the least expensive combination of machine and heat rejection device, as well as a reasonable machine operating cost.

Brine cooling normally requires special selection of the machine by the manufacturer. As a general rule, multiple-machine applications are seldom made on normal air-conditioning loads less than about 400 tons (362.9 t).

The optimum machine selection involves matching the correct machine and cooling tower as well as the correct entering chilled-water temperature and temperature reduction. A selection of several machines and cooling towers often results in finding one combination having a minimum first cost. In many instances, it is possible to reduce the condenser-water quantity and increase the leaving condenser-water temperature, resulting in a smaller tower.

Centrifugal refrigeration machines are used for air-conditioning, process, marine, manufacturing, and many other cooling applications throughout industry.

HEAT PUMP CYCLE ANALYSIS AND COMPARISON

Determine the quantity of water required to supply heat to a heat pump that must deliver 70,000 Btu/h (20.5 kW) to a building. Refrigerant 12 is used; the temperature of the water in the heat sink is 50°F (10.0°C). Air must be delivered to the heating system at a temperature of 118°F (47.8°C).

Calculation Procedure:

1. Determine the compressor suction temperature to use
To produce sufficient heat transfer between the water and the evaporator, a temperature difference of at least 10°F (5.6°C) must exist. With a water temperature of 50°F (10.0°C), this means that a suction temperature of 40°F (4.4°C) might be satisfactory. A suction temperature of 40°F (4.4°C) corresponds to a suction pressure of 51.68 lb/in² (abs) (356.3 kPa), as a table of thermodynamic properties of refrigerant 12 shows.

Since water entering the evaporator heat exchanger cannot be reduced to 40°F (4.4°C), the refrigerant temperature, the actual outlet temperature must be either assumed or computed. Assume that the water leaves the evaporator heat exchanger at 44°F (6.7°C). Then, each pound of water passing through the evaporator yields 50°F − 44°F = 6 Btu (6.3 kJ). Since 1 gal of water weighs 8.33 lb (1 kg/L), the quantity of heat released by the water is (6 Btu/lb)(8.33 lb/gal) = 49.98 Btu/gal, say 50 Btu/gal (13.9 kJ/L).

As an alternative solution, assume a suction temperature of 35°F (1.7°C) and an evaporator exit temperature of 39°F (3.9°C). Then each pound of water will yield 50 − 39 = 11 Btu (11.6 kJ). This is equal to (11)(8.33) = 91.6 Btu/gal (25.5 kJ/L). This comparison indicates that for every°F the cooling range of the water heat sink is extended, an additional 8.33 Btu/gal (2.3 kJ/L) of water is obtained.

2. Evaluate the effect of suction-temperature decrease

As the compressor suction temperature is reduced, the specific volume of the suction gas increases. Thus the compressor must handle more gas to evaporate the same quantity of refrigerant. However, the displacement of the usual reciprocating compressor used in a heat-pump system cannot be varied easily, if at all, in some designs. Also, at the lower suction temperature, the enthalpy of vaporization of the refrigerant increases only slightly.

Study of a table of thermodynamic properties of refrigerant 12 shows that reducing the suction temperature from 40 to 35°F (4.4 to 1.7°C) increase the specific volume from 0.792 to 0.862 ft^3/lb (0.0224 to 0.0244 m^3/kg). The enthalpy of vaporization increases from 65.71 to 66.28 Btu/lb (152.8 to 154.2 kJ/kg), but the total enthalpy decreases from 82.71 to 82.16 Btu/lb (192.4 to 191.1 kJ/kg). Hence, the advisability of reducing the suction temperature must be carefully investigated before a final decision is made.

3. Determine the required compressor discharge temperature

Air must be delivered to the heating system at 118°F (47.8°C), according to the design requirements. To produce a satisfactory transfer of heat between the condenser and the air, a 10°F (5.6°C) temperature difference is necessary. Hence, the compressor discharge temperature must be at least 118 + 10 = 128°F (53.3°C).

Checking a table of thermodynamic properties of refrigerant 12 shows that a temperature of 128°F (53.3°C) corresponds to a discharge pressure of 190.1 lb/in^2 (abs) (1310.7 kPa). The table also shows that the enthalpy of the vapor at the 118°F (47.8°C) condensing temperature is 90.01 Btu/lb (209.4 kJ/kg), whereas the enthalpy of the liquid is 35.65 Btu/lb (82.9 kJ/kg).

With a suction temperature of 40°F (4.4°C), the enthalpy of the vapor is 82.71 Btu/lb (192.4 kJ/kg). Hence, the heat supplied by the evaporator is: enthalpy of vapor at 40°F (4.4°C) − enthalpy of liquid at 118°F = 82.71 − 35.65 = 47.06 Btu/lb (109.5 kJ/kg). This heat is abstracted from the water that is drawn from the heat sink.

The gas leaving the evaporator contains 82.71 Btu/lb (192.4 kJ/kg). When this gas enters the condenser, it contains 90.64 Btu/lb (210.8 kJ/kg). The difference, or 90.64 − 82.71 = 7.93 Btu/lb (18.4 kJ/kg), is added to the gas by the compressor and represents a portion of the work input to the compressor.

4. Compute the evaporator and compressor heat contribution

The total heat delivered to the air = evaporator heat + compressor heat = 47.06 + 7.93 = 54.99 Btu/lb (127.9 kJ/kg). Then the evaporator supplies 47.06/54.99

= 0.856, or 85.6 percent of the total heat, and the compressor supplies 7.93/54.99 = 0.144, or 14.4 percent of the total heat.

5. Determine the actual evaporator and compressor heat contribution

Since this heat pump is rated at 70,000 Btu/h (20.5 kW), the evaporator contributes $0.856 \times 70,000 = 59,920$ Btu/h (17.5 kW), and the compressor supplies 0.144(70,000) = 10,080 Btu/h (3.0 kW). As a check, 59,920 + 10,080 = 70,000 Btu/h (20.5 kW).

6. Compute the sink-water flow rate required

The evaporator obtains its heat, or 59,920 Btu/h (17.5 kW), from the sink water. Since, from step 1, each gallon of water delivers 50 Btu (52.8 kJ) at a 40°F (4.4°C) suction temperature, the flow rate, required to contribute the evaporator heat is 59,920/50 = 1198.4 gal/h, or 1198.4/60 = 19.9 gal/min (1.3 L/s).

7. Evaluate the lower suction temperature

At 35°F (1.7°C), the evaporator will supply 82.16 − 35.65 = 46.51 Btu/lb (108.2 kJ/kg), by the same reasoning as in step 3. The balance, or 90.64 − 82.16 = 8.48 Btu/lb (19.7 kJ/kg), must be supplied by the compressor.

8. Compute the required refrigerant gas flow

At a 40°F (4.4°C) suction temperature, a table of thermodynamic properties of refrigerant 12 shows that the specific volume of the gas is 0.792 ft³/lb (0.050 m³/ kg). Step 4 shows that the heat pump smut deliver 54.99 Btu/lb (127.9 kJ/kg) of refrigerant to the air, or 54.99/0.792 = 69.4 Btu/ft³ (2585.8 kJ/m³) of gas. With a total heat requirement of 70,000 Btu/h (20.5 kW), the compressor must handle 70,000/69.4 = 1010 ft³/h (0.0079 m³/s).

As noted earlier, the cubic capacity of a reciprocating compressor is a fixed value at a given speed. Hence, a compressor chosen to handle this quantity of gas cannot handle a larger heat load.

At a 35°F (1.7°C) suction temperature, using the same procedure as above, we see that the required heat content of the gas is 54.99/0.862 = 63.7 Btu/ft³ (2373.4 kJ/m³). The compressor capacity must be 70,000/63.7 = 1099 ft³/h (0.0086 m³/s).

If the compressor were selected to handle 1010 ft³/h (0.0079 m³/s), then reducing the suction temperature to 35°F (1.7°C) would give a heat capacity of only (1010)(63.7) = 64,400 Btu/h (18.9 kW). This is inadequate because the system requires 70,000 Btu/h (20.5 kW).

9. Compute the water flow rate at the lower suction temperature

Step 6 shows the procedure for finding the sink-water flow rate required at a 40°F (4.4°C) suction temperature. Suppose, however, that the heat output of 64,400 Btu/h (18.9 kW) at the 35°F (1.7°C) suction temperature was acceptable. The evaporator portion of this load, by the method of step 4, is (82.16 − 35.65)/54.99 = 0.847, or 84.7 percent. Hence, the quantity of water required is (64,400)(0.847)/ [(91.6)(60)] = 9.92 gal/min (0.63 L/s). In this equation, the value of 91.6 Btu/gal is obtained from step 1. The factor 60 converts hours to minutes. Thus, reducing the suction temperature from 40 to 35°F (4.4 to 1.7°C) just about halves the water quantity—from 19.9 to 9.92 gal/min (1.26 to 0.63 L/s).

10. *Compare the pumping power requirements*
The power input to the pump is hp = 8.33(*gpm*)(head, ft)/33,000(pump efficiency)(motor efficiency). If the total head on the pump is computed as being 40
ft (12.2 m), the efficiency of the pump is 60 percent, and the efficiency of the motor
is 85 percent, then with a 40°F (4.4°C) suction temperature, the pump horsepower
is 8.33(19.9)(40)/33,000(0.60)(0.85) = 0.394, say 0.40 hp (0.30 kW).

At a 35°F (1.7°C) suction temperature with a flow rate of 9.92 gal/min (0.03
L/s) and all the other factors the same, hp = 8.33(9.92)(40)/33,000(0.60)(0.85) =
0.1965, say 0.20 hp (0.15 kW). Thus, the 35°F (1.7°C) suction temperature requires
only half the pump hp that the 40°F (4.4°C) suction temperature requires.

11. *Compute the compressor power input and power cost*
At the 40°F (4.4°C) suction temperature, the compressor delivers 7.93 Btu/lb (18.4
kJ/kg) of refrigerant gas, step 3. Since the total weight of gas delivered by the
compressor per hour is (70,000 Btu/h)/(54.99 Btu/lb) = 1272 lb/h (0.16 kg/s),
the compressor's total heat contribution is (1272 lb/h)(7.93 Btu/lb) = 10,100
Btu/h (3.0 kW).

With a compressor-driving motor having an efficiency of 85 percent, the hourly
motor input is equivalent to 10,100/0.85 = 11,880 Btu/h (3.5 kW), or 11,880/
2545 Btu/(hp·h) = 4.66 hp·h = (4.66)[746 Wh/(hp·h)] = 3480 Wh = 3.48
kWh. Also, the pump requires 0.4 hp·h, step 10, or (0.4)(746) = 299 Wh = 0.299
kWh. Hence, the total power consumption at a 40°F (4.4°C) suction temperature is
3.48 + 0.299 = 3.779 kWh. At a power cost of 5 cents per kilowatthour, the energy
cost is 3.779(5.0) = 18.9 cents per hour.

With a 35°F (1.7°C) suction temperature and using the lower heating capacity
obtained with the smaller, fixed-capacity compressor, step 9, we see the weight of
gas handled by the compressor will be (64,400 Btu/h)/(54.99 Btu/lb) = 1172 lb/
h (0.15 kg/s). From step 7, the compressor must supply 8.48 Btu/lb (19.7 kJ/kg).
Therefore, the compressor's total heat contribution is (1172 lb/h)(8.48 Btu/lb) =
9950 Btu/h (2.9 kW). With a motor efficiency of 85 percent, the hourly motor
input is equivalent to 9950/0.85 = 11,700 Btu/h (3.4 kW).

Using the same procedure as above, we see that the electric power input to the
compressor will be 3.43 kWh, while the pump electric power input is 0.150 kWh.
The total electric power input is 3.58 kWh at a cost of (3.58)(5.0) = 17.9 cents
per hour. Hence, the hourly savings with the 35°F (1.7°C) suction temperature is
18.9 − 17.9 = 1.0 cent. Note, however, that the heat output at the 35°F (1.7°C)
suction temperature, 64,400 Btu/h (18.9 kW), is 5600 Btu/h (1.6 kW) less than at
the 40°F (4.4°C) suction temperature. If the lower heat output were unacceptable,
the higher suction temperature or a larger compressor would have to be used. Either
alternative would increase the power cost.

Related Calculations.   With a water sink as the heat source, the usual water
consumption of a heat pump ranges fro 1.1 to more than 4 gal/(min·ton) [0.06 to
0.23 L/(s·t)]. A consumption range this broad requires that the actual flow rate be
computed because a guess could be considerably in error.

Either air or the earth may be used as a heat source instead of water. When the
cooling load rather than the heating load establishes the basic equipment size, an
ideal situation exists for the use of the heat pump with air as the heat source. This
occurs in localities where the minimum outdoor temperature in the winter is 20°F
(−6.7°C) or higher.

Ground coils can be bulky, costly, and troublesome. One study shows that the
temperature difference between the evaporating refrigerant in a ground coil and the

surrounding earth is about equal to the number of Btu/h that may be drawn from each linear foot of coil. Thus, with a temperature difference of 15°F (8.3°C) and a 70,000-Btu (73,853.9-kJ) load of which 85 percent is supplied by the coil, the length of coil needed is (70,000)(0.85)/15 = 3970 ft (120.1 m).

The coefficient of performance of a heat pump = heat rejected by condenser, Btu/heat equivalent of the net work of compression, Btu. The usual single-stage air source heat pump has a coefficient of performance ranging between 2.25 and 3.0. The procedure and data presented were developed by Robert Henderson Emerick, P.E., Consulting Mechanical Engineer.

CENTRAL CHILLED-WATER SYSTEM DESIGN TO MEET CHLOROFLUOROCARBON (CFC) ISSUES

Choose a suitable storage tank size and capacity for a thermally stratified water-storage system for a large-capacity thermal-energy storage system for off-peak air conditioning for these conditions: Thermal storage capacity required = 100,000 ton-h (35,169 kWh); difference between water inlet and outlet temperatures = $T = 20°F$ (36°C); allowable nominal soil bearing load in one location is 2500 lb/ft^2 (119.7 kPa), in another location 4000 lb/ft^2 (19.5 kPa). Compare thank size for the two locations.

Calculation Procedure:

1. Compute the required tank capacity in gallons (liters) to serve this system
Use the relation $C = 1800 \, S/\Delta T$, where C = required tank capacity, gal; S = system capacity, ton-h; ΔT = difference between inlet and outlet temperature, °F (°C). For this installation, $C = 1800 \, (100,000)/20 = 9,000,000$ gal (34,065 m^3).

2. Determine the tank height and diameter for the allowable soil bearing loads
Depending on the proposed location of the storage tank, either the height or diameter may be a restricted dimension. Thus, tank height may be restricted by local zoning laws or possible interference with aircraft landing or takeoff patterns. Tank diameter may be restricted by the ground area available. And the allowable nominal soil bearing load will determine if the required amount of water can be stored in one tank or if more than one tank will be required.

Starting with 2500-lb/ft^2 (119.7-kPa) bearing-load soil, assume a standard tank height of 40 ft (12.2 m). Then, the required tank volume will be $V = 0.134C$, where $0.134 = $ ft^3/gal; or $V = 0.134 \, (9,000,000) = 1,206,000$ ft^3 (34,130 m^3). The tank diameter is $d = (4V/\pi h)^{0.5}$, where d = diameter in feet (m). Or $d = [4(1,206,000)/\pi 40)]^{0.5} = 195.93$ ft; say 196 ft (59.7 m). This result is consistent with the typical sizes, heights, and capacities used in actual practice, as shown in Table 8.

Checking the soil load, the area of the base of this tank is $A = \pi d^2/4 = \pi (196)^2/4 = 30,172$ ft^2 (2803 m^2). The weight of the water in the tank is $W = 8.35C = 75,150,000$ lb (34,159 kg). This will produce a soil bearing pressure of $p = W/A$ lb/ft^2 (kPa). Or, $p = 75,150,000/30,172 = 2491$ lb/ft^2 (119.3 kPa). This bearing load is within the allowable nominal specified load of 2500 lb/ft^3.

TABLE 8 Typical Thermal Storage Tank Sizes, Heights, Capacities*

Rated thermal energy storage capacity, ton-hours			Tank shell height (and nominal soil bearing load)					
			64-ft shell height (4000 lb/ft²soil)		40-ft shell height (2500 lb/ft²)		24-ft shell height (1500 lb/ft² soil)	
ΔT, 10°F	ΔT, 15°F	ΔT, 20°F	Gross volume, gal	Tank diameter, ft	Gross volume, gal	Tank diameter, ft	Gross volume, gal	Tank diameter, ft
40,000	60,000	80,000	6,880,000	135	7,200,000	175	7,880,000	236
50,000	75,000	100,000	8,610,000	151	9,000,000	196	9,850,000	264
60,000	90,000	120,000	10,330,000	166	10,800,000	214	11,820,000	290

Source: Chicago Bridge & Iron Company.
*See calculation procedures for SI values.

Where a larger soil bearing load is permitted, tank diameter can be reduced as the tank height is increased. Thus, using a standard 64-ft (19.5-m) high tank with the same storage capacity, the required diameter would be $d = [4(1,206,000)/\pi 64)]^{0.5} = 154.9$ ft (47.2 m). Soil bearing pressure will then be $W/A = 75,140,000 /[\pi(154.9)^2/4] = 3987.8$ lb/ft² (190.0 kPa). This is within the allowable soil bearing load of 4000 lb/ft² (191.5 kPa).

By reducing the storage capacity of the tank 4 percent to 8,610,000 gal (32,589 m³), the diameter of the tank can be made 151 ft (46 m). This is a standard dimension for 64-ft (19-m) high tanks with a 4000-lb/ft² (191.5-kPa) soil bearing load.

Related Calculations. Thermal energy storage (TES) is environmentally desirable because it uses heating, ventilating, and air-conditioning (HVAC) equipment and a storage tank to store heated or cooled water during off-peak hours, allowing

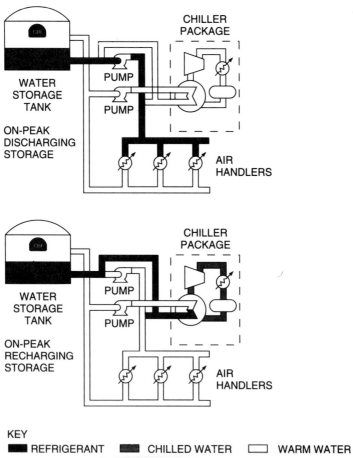

KEY

■ REFRIGERANT ▨ CHILLED WATER ▢ WARM WATER

FIGURE 33 On-peak and off-peak storage discharging and recharging of thermally stratified water-storage system. (*Chicago Bridge & Iron Company.*)

more efficient use of electric generating equipment. The stored water is used to serve HVAC or industrial process loads during on-peak hours.

To keep investment, operating, and maintenance costs low, one storage tank can be used to store both cool and warm water. Thermal stratification permits a smaller investment in the tank, piping, insulation, and controls to produce a higher-efficiency system. Lower-density warm water is thermally stratified from higher-density cool water without any mechanical separation in a full storage tank.

While systems using 200,000 gal (757 m³) of stored water are feasible, the usual minimum size storage tank is 500,000 gal (1893 m³). Tanks as large as 4.4 million gal (16,654 m³) are currently in use in TES for HVAC and process needs. TES is also used for schools, colleges, factories, and a variety of other applications. Where chlorofluorocarbon (CFC)-based refrigeration systems must be replaced with less environmentally offensive refrigerants, TES systems can easily be modified because the chiller (Fig. 33) is a simple piece of equipment.

As an added environmental advantage, the stored water in TES tanks can be used for fire protection. The full tank contents are continuously available as an emergency fire water reservoir. With such a water reserve, the capital costs for fire-protection equipment can be reduced. Likewise, fire-insurance premiums may also be reduced. Where an existing fire-protection-water tank is available, it may retrofitted for TES use.

Above-ground storage tanks (Fig. 34) are popular in TES systems. Such tanks are usually welded steel, leak-free with a concrete ring wall foundation. Insulated to prevent heat gain or loss, such tanks may have proprietary internal components for proper water distribution and stratification.

Some TES tanks may be installed partially, or fully, below grade. Before choosing partially or fully below-grade storage, the following factors should be considered: (1) system hydraulics may be complicated by a below-ground tank; (2) the

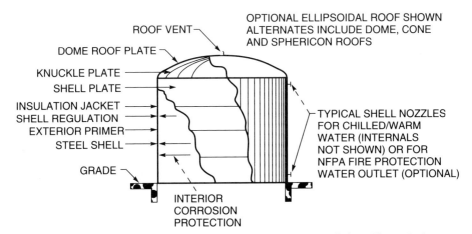

TYPICAL STRATA-THERM CHILLED WATER
THERMAL ENERGY STORAGE INSTALLATION

ROOF VENT

OPTIONAL ELLIPSOIDAL ROOF SHOWN
ALTERNATES INCLUDE DOME, CONE
AND SPHERICON ROOFS

DOME ROOF PLATE

KNUCKLE PLATE
SHELL PLATE
INSULATION JACKET
SHELL REGULATION
EXTERIOR PRIMER
STEEL SHELL

TYPICAL SHELL NOZZLES
FOR CHILLED/WARM
WATER (INTERNALS
NOT SHOWN) OR FOR
NFPA FIRE PROTECTION
WATER OUTLET (OPTIONAL)

GRADE

INTERIOR
CORROSION
PROTECTION

FIGURE 34 Typical chilled-water thermal-energy storage tank installation. (*Chicago Bridge & Iron Company.*)

tank must be designed for external pressure, particularly when the tank is empty; (3) soil and groundwater conditions may make the tank more costly; (4) local and national regulations for underground tanks may increase costs; (5) the choice of water-treatment methods may be restricted for underground tanks; (6) the total cost of an underground tank may be twice that of an above-ground tank.

The data and illustrations for this procedure were obtained from the Strata-Therm Thermal Systems Group of the Chicago Bridge & Iron Company.

ENVIRONMENTAL CONTROL

SECTION 13
WASTEWATER TREATMENT AND CONTROL

Kevin D. Wills, M.S.E., P.E.
Consulting Engineer
Stanley Consultants, Inc.

DESIGN OF A COMPLETE-MIX ACTIVATED SLUDGE REACTOR

Domestic wastewater with an average daily flow of 4.0 Mgd (15,140 m³/d) has a five day Biochemical Oxygen Demand (BOD₅) of 240 mg/L after primary settling. The effluent is to have a BOD₅ of 10 mg/L or less. Design a complete-mix activated sludge reactor to treat the wastewater including reactor volume, hydraulic retention time, quantity of sludge wasted, oxygen requirements, food to microorganism ratio, volumetric loading, and WAS and RAS requirements.

Calculation Procedure:

1. Compute the reactor volume
The volume of the reactor can be determined using the following equation derived from Monod kinetics:

$$V_r = \frac{\theta_c QY(S_o - S)}{X_a (1 + k_d\theta_c)}$$

where V_r = Reactor volume (Mgal) (m³)
   θ_c = Mean cell residence time, or the average time that the sludge remains in the reactor (sludge age). For a complete-mix activated sludge process, θ_c ranges from 5 to 15 days. The design of the reactor is based on θ_c on the assumption that substantially all the substrate (BOD) conversion occurs in the reactor, A θ_c of 8 days will be assumed.
   Q = Average daily influent flow rate (Mgd) = 4.0 Mgd (15,140 m³/d)
   Y = Maximum yield coefficient (mg VSS/mg BOD$_5$). For the activated sludge process for domestic wastewater Y ranges from 0.4 to 0.8. A Y of 0.6 mg VSS/mg BOD$_5$ will be assumed. Essentially, Y represents the maximum mg of cells produced per mg organic matter removed.
   S_O = Influent substrate (BOD$_5$) concentration (mg/L) = 240 mg/L
   S = Effluent substrate (BOD$_5$) concentration (mg/L) = 10 mg/L
   X_a = Concentration of microorganisms in reactor = Mixed Liquor Volatile Suspended Solids (MLVSS) in mg/L. It is generally accepted that the ratio MLVSS/MLSS ≈ 0.8, where MLSS is the Mixed Liquor Suspended Solids concentration in the reactor. MLSS represents the sum of volatile suspended solids (organics) and fixed suspended solids (inorganics). For a complete-mix activated sludge process, MLSS ranges from 1,000 to 6,500 mg/L. An MLSS of 4,500 mg/L will be assumed.
   => MLVSS = (0.8)(4500 mg/L) = 3600 mg/L.
   k_d = Endogenous decay coefficient (d^{-1}) which is a coefficient representing the decrease of cell mass in the MLVSS. For the activated sludge process for domestic wastewater k_d ranges from 0.025 to 0.075 d^{-1}. A value of 0.06 d^{-1} will be assumed.

Therefore:

$$V_r = \frac{(8\ d)(4.0\ \text{Mgd})(0.6\ \text{mg VSS/mg BOD}_5)(240 - 10)\text{mg/L}}{(3600\ \text{mg/L})(1 + (0.06\ d^{-1})\ (8\ d))}$$
$$= 0.83\ \text{Mgal}\ (110{,}955\ \text{ft}^3)\ (3140\ \text{m}^3)$$

2. Compute the hydraulic retention time
The hydraulic retention time (θ) in the reactor is the reactor volume divided by the influent flow rate: V_r/Q. Therefore, θ = (0.83 Mgal)/(4.0 Mgd) = 0.208 days = 5.0 hours. For a complete-mix activated sludge process, θ is generally 3–5 hours. Therefore, the hydraulic retention time is acceptable.

3. Compute the quantity of sludge wasted
The observed cell yield, Y_{obs} = $Y/1 + k_d\theta_c$ = 0.6/(1 + (0.06 d^{-1})(8 d)) = 0.41 mg/mg represents the actual cell yield that would be observed. The observed cell yield is always less than the maximum cell yield (Y).
   The increase in MLVSS is computed using the following equation:

$$P_x = Y_{obs}Q(S_O - S)(8.34\ \text{lb/Mgal/mg/L})$$

where P_x is the net waste activated sludge produced each day in (lb VSS/d).
   Using values defined above:

$$P_x = \left(0.41\ \frac{\text{mg VSS}}{\text{mg BOD}_5}\right)(4.0\ \text{Mgd})\left(240\ \frac{\text{mg}}{\text{L}} - 10\ \frac{\text{mg}}{\text{L}}\right)\left(8.34\ \frac{\text{lb/Mgal}}{\text{mg/L}}\right)$$
$$= 3146\ \text{lb VSS/d}\ (1428.3\ \text{kg VSS/d})$$

This represents the increase of volatile suspended solids (organics) in the reactor. Of course the total increase in sludge mass will include fixed suspended solids (inorganics) as well. Therefore, the increase in the total mass of mixed liquor suspended solids (MLSS) = $P_{x(ss)}$ = (3146 lb VSS/d)/(0.8) = 3933 lb SS/d (1785.6 kg SS/d). This represents the total mass of sludge that must be wasted from the system each day.

4. Compute the oxygen requirements based on ultimate carbonaceous oxygen demand (BOD_L)
The theoretical oxygen requirements are calculated using the BOD_5 of the wastewater and the amount of organisms (P_x) wasted from the system each day. If all BOD_5 were converted to end products, the total oxygen demand would be computed by converting BOD_5 to ultimate BOD (BOD_L), using an appropriate conversion factor. The "Quantity of Sludge Wasted" calculation illustrated that a portion of the incoming waste is converted to new cells which are subsequently wasted from the system. Therefore, if the BOD_L of the wasted cells is subtracted from the total, the remaining amount represents the amount of oxygen that must be supplied to the system. From stoichiometry, it is known that the BOD_L of one mole of cells is equal to 1.42 times the concentration of cells. Therefore, the theoretical oxygen requirements for the removal of the carbonaceous organic matter in wastewater for an activated-sludge system can be computed using the following equation:

lb O_2/d = (total mass of BOD_L utilized, lb/d)

− 1.42 (mass of organisms wasted, lb/d)

Using terms that have been defined previously where f = conversion factor for converting BOD_5 to BOD_L (0.68 is commonly used):

$$\text{lb } O_2/d = \frac{Q(S_O - S)\left(8.34\ \frac{\text{lb/Mgal}}{\text{mg/L}}\right)}{f} - (1.42)(P_x)$$

Using the above quantities:

$$\text{lb } O_2/d = \frac{(4.0\ \text{Mgd})(240\ \text{mg/L} - 10\ \text{mg/L})(8.34)}{0.68} - (1.42)(3146\ \text{lb/d})$$

$$= 6816\ \text{lb } O_2/d\ (3094.5\ \text{kg } O_2/d)$$

This represents the theoretical oxygen requirement for removal of the influent BOD_5. However, to meet sustained peak organic loadings, it is recommended that aeration equipment be designed with a safety factor of at least 2. Therefore, in sizing aeration equipment a value of (2)(6816 lb O_2/d) = 13,632 lb O_2/d (6188.9 kg O_2/d) is used.

5. Compute the food to microorganism ratio (F:M) and the volumetric loading (V_L)
In order to maintain control over the activated sludge process, two commonly used parameters are (1) the food to microorganism ratio (F:M) and, (2) the mean cell residence time (θ_c). The mean cell residence time was assumed in Part 1 "Compute Reactor Volume" to be 8 days.
    The food to microorganism ratio is defined as:

$$F:M = S_O/\theta X_a$$

where F:M is the food to microorganism ratio in d^{-1}.

F:M is simply a ratio of the "food" or BOD_5 of the incoming waste, to the concentration of "microorganisms" in the aeration tank or MLVSS. Therefore, using values defined previously:

$$F:M = \frac{240 \text{ mg/L}}{(0.208 \text{ d})(3600 \text{ mg/L})} = 0.321 \text{ } d^{-1}$$

Typical values for F:M reported in literature vary from 0.05 d^{-1} to 1.0 d^{-1} depending on the type of treatment process used.

A low value of F:M can result in the growth of filamentous organisms and is the most common operational problem in the activated sludge process. A proliferation of filamentous organisms in the mixed liquor results in a poorly settling sludge, commonly referred to as "bulking sludge."

One method of controlling the growth of filamentous organisms is through the use of a separate compartment as the initial contact zone of a biological reactor where primary effluent and return activated sludge are combined. This concept provides a high F:M at controlled oxygen levels which provides selective growth of floc forming organisms at the initial stage of the biological process. An F:M ratio of at least 2.27 d^{-1} in this compartment is suggested in the literature. However, initial F:M ratios ranging from 20–25 d^{-1} have also been reported.

The volumetric (organic) loading (V_L) is defined as:

$$V_L = S_O Q/V_r = S_O/\theta$$

V_L is a measure of the pounds of BOD_5 applied daily per thousand cubic feet of aeration tank volume. Using values defined previously:

$$V_L = (240 \text{ mg/L})/(0.208 \text{ d}) = 1154 \text{ mg/L} \cdot \text{d} = 72 \text{ lb}/10^3 \text{ft}^3 \cdot \text{d} \text{ } (1.15 \text{ kg/Mm}^3 \cdot \text{d})$$

Volumetric loading can vary from 20 to more than 200 $\text{lb}/10^3 \text{ft}^3 \cdot \text{d}$ (0.32 to 3.2 $\text{kg/Mm}^3 \cdot \text{d}$), and may be used as an alternate (although crude) method of sizing aeration tanks.

6. Compute the waste activated sludge (WAS) and return activated sludge (RAS) requirements

Control of the activated sludge process is important to maintain high levels of treatment performance under a wide range of operating conditions. The principle factors used in process control are (1) maintaining dissolved-oxygen levels in the aeration tanks, (2) regulating the amount of Return Activated Sludge (RAS), and (3) controlling the Waste Activated Sludge (WAS). As outlined previously in Part 5 "Compute the Food to Microorganism Ratio and the Volumetric Loading," the most commonly used parameters for controlling the activated sludge process are the F:M ratio and the mean cell residence time (θ_c). The Mixed Liquor Volatile Suspended Solids (MLVSS) concentration may also be used as a control parameter. Return Activated Sludge (RAS) is important in maintaining the MLVSS concentration and the Waste Activated Sludge (WAS) is important in controlling the mean cell residence time (θ_c).

The excess waste activated sludge produced each day (see step 3 "Compute the Quantity of Sludge Wasted") is wasted from the system to maintain a given F:M or mean cell residence time. Generally, sludge is wasted from the return sludge line because it is more concentrated than the mixed liquor in the aeration tank, hence smaller waste sludge pumps are required. The waste sludge is generally discharged to sludge thickening and digestion facilities. The alternative method of sludge wasting is to withdraw mixed liquor directly from the aeration tank where the concentration of solids is uniform. Both methods of calculating the waste sludge flow rate are illustrated below.

Use Figs. 1 and 2 when performing mass balances for the determination of RAS and WAS.

X = Mixed Liquor Suspended Solids (MLSS) - see Part 1 "Compute the Reactor Volume."

Q_r = Return activated sludge pumping rate (Mgd)

X_r = Concentration of sludge in the return line (mg/L). When lacking site specific operational data, a value commonly assumed is 8000 mg/L.

Q_e = Effluent flow rate (Mgd)

X_e = Concentration of solids in effluent (mg/L). When lacking site specific operational data, this value is commonly assumed to be zero.

Q_w = Wasted Activated Sludge (WAS) pumping rate from the reactor (Mgd)

$Q_{w'}$ = Waste Activated Sludge (WAS) pumping rate from the return line (Mgd)

Other variables are as defined previously.

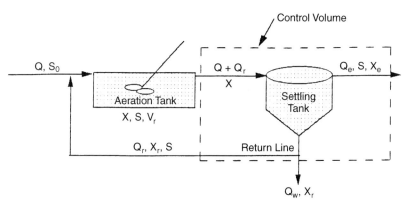

FIGURE 1 Settling tank mass balance.

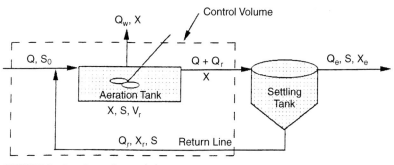

FIGURE 2 Aeration tank mass balance.

The actual amount of liquid that must be pumped to achieve process control depends on the method used and the location from which the wasting is to be accomplished. Also note that because the solids capture of the sludge processing facilities (i.e., thickeners, digesters, etc.) is not 100 percent and some solids are returned, the actual wasting rate will be higher than the theoretically determined value.

(a) *Waste Activated Sludge (WAS) pumping rate from the return line.* If the mean cell residence time is used for process control and the wasting is from the sludge return line (Fig. 1), the wasting rate is computed using the following:

$$\theta_c = \frac{V_r X}{(Q_{w'} X_r + Q_e X_e)}$$

Assuming that the concentration of solids in the effluent from the settling tank (X_e) is low, then the above equation reduces to:

$$\theta_c \approx \frac{V_r X}{Q_{w'} X_r} \Rightarrow Q_{w'} = \frac{V_r X}{\theta_c X_r}$$

Using values defined previously:

$$Q_{w'} = \frac{(0.83 \text{ Mgal})(4500 \text{ mg/L})}{(8 \text{ d}) (8000 \text{ mg/L})} = 0.0584 \text{ Mgd} = 58,400 \text{ gal/day} \ (221 \text{ m}^3/\text{d})$$

To determine the WAS pumping rate using this method, the solids concentration in both the aeration tank and the return line must be known.

If the food to microorganism ratio (F:M) method of control is used, the WAS pumping rate from the return line is determined using the following:

$$P_{x(ss)} = Q_{w'} X_r (8.34 \text{ lb/Mgal/mg/L})$$

Therefore:

$$Q_{w'} = \frac{3933 \text{ lb/d}}{(8000 \text{ mg/L})(8.34)} = 0.059 \text{ Mgd} = 59,000 \text{ gal/day} \ (223.3 \text{ m}^3/\text{d})$$

In this case, the concentration of solids in the sludge return line must be known. Note that regardless of the method used for calculation, if wasting occurs from the return line, the WAS pumping rate is approximately the same.

(b) *Waste Activated Sludge (WAS) pumping rate from the aeration tank.* If the mean cell residence time is used for process control, wasting is from the aeration tank (Fig. 2), and the solids in the plant effluent (X_e) are again neglected, then the WAS pumping rate is estimated using the following:

$$\theta_c \approx \frac{V_r}{Q_w} \Rightarrow Q_w \approx \frac{V_r}{\theta_c}$$

Using values defined previously:

$$Q_w = \frac{0.83 \text{ Mgal}}{8 \text{ d}} = 0.104 \text{ Mgd} = 104,000 \text{ gal/day} \ (393.6 \text{ m}^3/\text{d})$$

Note that in case (a) or (b) above, the weight of sludge wasted is the same (3933 lb SS/d) (1785.6 kg SS/d), and that either wasting method will achieve a θ_c of 8 days. As can be seen, wasting from the aeration tank produces a much higher waste flow rate. This is because the concentration of solids in the bottom of the settling tank (and hence the return line) is higher than in the aeration tank. Consequently, wasting a given mass of solids per day is going to require a larger WAS pumping rate (and larger WAS pumps) if done from the aeration tank as opposed to the return line. The Return Activated Sludge (RAS) pumping rate is determined by performing a mass balance analysis around either the settling tank or the aeration tank. The appropriate control volume for either mass balance analysis is illustrated in Fig. 1 and 2 respectively. Assuming that the sludge blanket level in the settling tank remains constant and that the solids in the effluent from the settling tank (X_e) are negligible, a mass balance around the settling tank (Fig. 1) yields the following equation for RAS pumping rate:

$$Q_r = \frac{XQ - X_r Q_{w'}}{X_r - X}$$

Using values defined previously, the RAS pumping rate is computed to be:

$$Q_r = \frac{(4500 \text{ mg/L})(4.0 \text{ Mgd}) - (8000 \text{ mg/L})(0.0584 \text{ Mgd})}{8000 \text{ mg/L} - 4500 \text{ mg/L}}$$

$$= 5.0 \text{ Mgd } (18{,}925 \text{ m}^3/\text{d})$$

As outlined above, the required RAS pumping rate can also be estimated by performing a mass balance around the aeration tank (Fig. 2). If new cell growth is considered negligible, then the solids entering the tank will equal the solids leaving the tank. Under conditions such as high organic loadings, this assumption may be incorrect. Solids enter the aeration tank in the return sludge and in the influent flow to the secondary process. However, because the influent solids are negligible compared to the MLSS in the return sludge, the mass balance around the aeration tank yields the following equation for RAS pumping rate:

$$Q_r = \frac{X(Q - Q_w)}{X_r - X}$$

Using values defined previously, the RAS pumping rate is computed to be:

$$Q_r = \frac{(4500 \text{ mg/L})(4.0 \text{ Mgd} - 0.104 \text{ Mgd})}{8000 \text{ mg/L} - 4500 \text{ mg/L}} = 5.0 \text{ Mgd } (18{,}925 \text{ m}^3/\text{d})$$

The ratio of RAS pumping rate to influent flow rate, or recirculation ratio (α), may now be calculated:

$$\alpha = \frac{Q_r}{Q} = \frac{5.0 \text{ Mgd}}{4.0 \text{ Mgd}} = 1.25$$

Recirculation ratio can vary from 0.25 to 1.50 depending upon the type of activated sludge process used. Common design practice is to size the RAS pumps so that they are capable of providing a recirculation ratio ranging from 0.50 to 1.50.

It should be noted that if the control volume were placed around the aeration tank in Fig. 1 and a mass balance performed, or the control volume placed around the settling tank in Fig. 2 and a mass balance performed, that a slightly higher RAS pumping rate would result. However, the difference between these RAS pumping rates and the ones calculated above is negligible.

DESIGN OF A CIRCULAR SETTLING TANK

Domestic wastewater with an average daily flow of 4.0 Mgd (15,140 m³/d) exits the aeration tank of a standard activated sludge treatment process. Design a circular settling tank to separate the sludge from the effluent. The settling tank will work in conjunction with the aeration tank. Assume a peaking factor of 2.5.

Calculation Procedure:

1. Determine the peak flow
Conventional examples of circular settling tank design utilize settling tests to size the tanks. However, it is more common that settling facilities must be designed without the benefit settling tests. When this situation develops, published values of surface loading and solids loading rates are generally used. Because of the large amount of solids that may be lost in the effluent if design criteria are exceeded, surface loading rates should be based on peak flow conditions. Using a peaking factor of 2.5, the daily peak flow (Q_p) is:

$$Q_p = 2.5 \times 4.0 \text{ Mgd} = 10.0 \text{ Mgd (37,850 m}^3\text{/d)}$$

2. Find the settling tank surface area using surface loading criteria
The recommended surface loading rates (settling tank effluent flow divided by settling tank area) vary depending upon the type of activated sludge process used. However, surface loading rates ranging from 200 to 800 gal/day/ft² (8.09 to 32.4 L/m²·d) for average flow, and a maximum of 1,000 gal/day/ft² (40.7 L/m²·d) for peak flow are accepted design values.

The recommended solids loading rate on an activated sludge settling tank also varies depending upon the type of activated sludge process used and may be computed by dividing the total solids applied by the surface area of the tank. The preferred units are lb/ft²·h (kg/m²·h). In effect, the solids loading rate represents a characteristic value for the suspension under consideration. In a settling tank of fixed area, the effluent quality will deteriorate if solids loading is increased beyond the characteristic value for the suspension. Without extensive experimental work covering all seasons and operating variables, higher rates should not be used for design. The recommended solids loading rates vary depending upon the type of activated sludge process selected. However, solids loading rates ranging from 0.8 to 1.2 lb/ft²·h (3.9 to 5.86 kg/m²·h) for average flow, and 2.0 lb/ft²·h (9.77 kg/m²·h) for peak flow are accepted design values.

For a Q_p of 10.0 Mgd and a design surface loading rate of 1,000 gal/day/ft² at peak flow, the surface area (A) of a settling tank may be calculated:

$$1000 \text{ gal/day/ft}^2 = \frac{10 \times 10^6 \text{ gal/day}}{A} \Rightarrow$$

$$A = \frac{10 \times 10^6 \text{ gal/day}}{1000 \text{ gal/day/ft}^2} = 10,000 \text{ ft}^2 \text{ (929 m}^2\text{)}$$

3. Find the settling tank surface area using solids loading criteria
The total solids load on a clarifier consists of contributions from both the influent and the return activated sludge (RAS). Assume the following (see "Design of A Complete-Mix Activated Sludge Reactor"):

Q_r = RAS flow rate = $1.25(Q)$ = $1.25(4.0$ Mgd$)$ = 5.0 Mgd $(18,925$ m^3/d)

X = MLSS in aeration tank = 4,500 mg/L

Therefore, the maximum solids loading occurs at peak flow and maximum RAS flow rate. The maximum solids entering the clarifier is calculated using:

Max. Solids (lb/d) = $(Q_p + Q_r)(X)(8.34$ lb $\cdot$ L/mg $\cdot$ Mgal$)$

Using values given above;

Max Solids (lb/d) = (10 Mgd + 5 Mgd)(4,500 mg/L)(8.34 lb $\cdot$ L/mg $\cdot$ Mgal)

= 562,950 lb/d

= 23,456 lb/h (10,649 kg/h) of Suspended Solids

Therefore, using a solids loading rate of 2.0 lb/ft$^2 \cdot$ h (9.77 kg/m$^2 \cdot$ h) at peak flow, the surface area of a settling tank may be calculated:

$$A = \frac{23,456 \text{ lb/h}}{2.0 \text{ lb/ft}^2 \cdot \text{h}} = 11,728 \text{ ft}^2 \ (1089.5 \text{ m}^2)$$

In this case, the solids loading dominates and dictates the required settling tank area.

4. Select the number of settling tanks
Generally, more than one settling thank would be constructed for operational flexibility. Two tanks will be sufficient for this example. Therefore, the surface area of each tank will be 11,728 ft^2/2 = 5,864 ft^2 (544.8 m^2). The diameter of each settling tank is then 86.41 ft. Use 87 ft (26.5 m). The total area of the two settling tanks is 11,889 ft^2 (110.4.5 m^2). For an average flow rate of 4.0 Mgd (15,140 m^3/d) and a RAS flow rate of 5.0 Mgd (18,925 m^3), the total solids entering the settling tanks at average daily flow is:

Solids (lb/d) = (4.0 Mgd + 5.0 Mgd)(4,500 mg/L) (8.34 lb $\cdot$ L/mg $\cdot$ Mgal)

= 337,770 lb/d = 14,074 lb/h (6389.6 kg/h)

Therefore, the solids loading on the settling tanks at design flow is:

$$\text{Solids Loading} = \frac{14,074 \text{ lb/h}}{11,889 \text{ ft}^2} \approx 1.18 \text{ lb/ft}^2 \cdot \text{h} \ (5.77 \text{ kg/m}^2 \cdot \text{h})$$

which is within the solids loading rate design criteria stated above.
   Related Calculations.   Liquid depth in a circular settling tank is normally measured at the sidewall. This is called the sidewater depth. The liquid depth is a factor in the effectiveness of suspended solids removal and in the concentration of the return sludge. Current design practice favors a minimum sidewater depth of 12 ft (3.66 m) for large circular settling tanks. However, depths of up to 20 ft (6.1 m)

have been used. The advantages of deeper tanks include greater flexibility in operation and a larger margin of safety when changes in the activated sludge system occur.

THICKENING OF A WASTE-ACTIVATED SLUDGE USING A GRAVITY BELT THICKENER

A wastewater treatment facility produces 58,000 gal/day (219.5 m³/d) of waste activated sludge containing 0.8 percent solids (8,000 mg/L). Design a gravity belt thickener installation to thicken sludge to 5.0 percent solids based on a normal operation of 6 h/d and 5 d/wk. Use a gravity belt thickener loading rate of 1,000 lb/h (454 kg/h) per meter of belt width. Calculate the number and size of gravity belt thickeners required, the volume of thickened sludge cake, and the solids capture in percent.

Calculation Procedure:

1. Find the dry mass of sludge that must be processed
Gravity belt thickening consists of a gravity belt that moves over rollers driven by a variable speed drive unit. The waste activated sludge is usually pumped from the bottom of a secondary settling tank, conditioned with polymer and fed into a feed/distribution box at one end. The box is used to distribute the sludge evenly across the width of the moving belt. The water drains through the belt as the sludge is carried toward the discharge end of the thickener. The sludge is ridged and furrowed by a series of plow blades placed along the travel of the belt, allowing the water released from the sludge to pass through the belt. After the thickened sludge is removed, the belt travels through a wash cycle.

The 58,000 gal/day (219.5 m³/d) of waste activated sludge contains approximately 3933 lb/d (1785.6 kg/d) of dry solids: See *Design of a Complete-Mix Activated Sludge Reactor*, step 3—"Compute the Quantity of Sludge Wasted," and step 6—"Compute the WAS and RAS Requirements."

Based on an operating schedule of 5 days per week and 6 hours per day, the dry mass of sludge that must be processed is:

Weekly Rate: (3,933 lb/d)(7 d/wk) = 27,531 lb/wk (12,499 kg/wk)

Daily Rate: (27,531 lb/wk)/(5 d/wk) = 5506 lb/d (2499.7 kg/d)

Hourly Rate: (5506 lb/d)/(6 h/d) = 918 lb/h (416.8 kg/h)

2. Size the belt thickener
Using the hourly rate of sludge calculated above, and a loading rate of 1,000 lb/h per meter of belt width, the size of the belt thickener is:

$$\text{Belt Width} = \frac{918 \text{ lb/h}}{1000 \text{ lb/h} \cdot \text{m}} = 0.918 \text{ m (3.01 ft)}$$

Use one belt thickener with a 1.0 m belt width. Note that one identical belt thickener should be provided as a spare.

The thickened sludge flow rate (S) in gal/day (m^3/d) and the filtrate flow rate (F) in gal/day (m^3/d) are computed by developing solids balance and flow balance equations:

(a) *Solids balance equation.* Solids in = solids out, which implies that: solids in sludge feed = solids in thickened sludge + solids in filtrate. Assume the following:

- Sludge feed specific gravity (s.g.) = 1.01
- Thickened sludge s.g. = 1.03
- Filtrate s.g. = 1.0
- Suspended solids in filtrate = 900 mg/L = 0.09%

Therefore, the solids balance equation on a daily basis becomes:

$$5506 \text{ lb} = (S, \text{ gal/day})(8.34 \text{ lb/gal})(1.03)(0.05)$$
$$+ (F, \text{ gal/day})(8.34 \text{ lb/gal})(1.0)(0.0009) \qquad (1)$$
$$\Rightarrow 5506 \text{ lb/d} = 0.4295(S) + 0.0075(F)$$

(b) *Flow balance equation.* Flow in = flow out, which implies that: influent sludge flow rate + washwater flow rate = thickened sludge flow rate + filtrate flow rate. Daily influent sludge flow rate = (58,000 gal/day)(7/5) = 81,200 gal/day (307.3 m^3/d).

3. Compute the thickened sludge and filtrate flow rates

Washwater flow rate is assumed to be 16 gal/min (1.0 L/s). Washwater flow rate varies from 12 gal/min (0.757 L/s) to 30 gal/min (1.89 L/s) depending on belt thickener size. Therefore, with an operating schedule of 6 h/d (360 min/d) the flow balance equation on a daily basis becomes:

$$81,200 \text{ gal/day} + (16 \text{ gal/min})(360 \text{ min/d}) = S + F$$
$$\Rightarrow 86,960 \text{ gal/day} = S + F \qquad (2)$$

Putting (EQ1) and (EQ 2) in matrix format, and solving for thickened sludge flow rate (S) and filtrate flow rate (F):

$$\begin{bmatrix} 0.4295 & 0.0075 \\ 1 & 1 \end{bmatrix} \begin{bmatrix} S \\ F \end{bmatrix} = \begin{bmatrix} 5506 \\ 86,960 \end{bmatrix}$$

S = 11,502 gal/day (43.5 m^3/d) of thickened sludge at 5.0% solids

F = 75,458 gal/day (285.6 m^3/d) of filtrate

Therefore, the volume of thickened waste activated sludge exiting the gravity belt thickener is 11,502 gal/day (43.5 m^3/d) at 5.0 percent solids.

4. Determine the solids capture

The solids capture is determined using the following:

$$\text{Solids Capture (\%)} = \frac{\text{Solids in Feed} - \text{Solids in Filtrate}}{\text{Solids in Feed}} \times (100\%)$$

Using values defined previously:

$$\text{Solids Capture } (\%) = \frac{5506 \text{ lb/d} - [(75{,}458 \text{ gal/day})(8.34 \text{ lb/gal})(1.0)(0.0009)]}{5506 \text{ lb/d}}$$

$$\times \ (100\%) = 89.7\%$$

Since only 89.7 percent of the solids entering the gravity belt thickener are captured, the thickener will actually be required to operate $(360 \text{ min/d})/(0.897) = 400 \text{ min/d}$ (6 hours 40 minutes per day) five days per week in order to waste the 3,933 lb/d (1785.6 kg/d) of dry solids required. This implies that the actual volume of thickened sludge will be $(11{,}502 \text{ gal/day})/(0.897) = 12{,}823 \text{ gal/day}$ (48.5 m³/d). The actual filtrate flow rate will be $(75{,}458 \text{ gal/day})/(0.897) = 84{,}123 \text{ gal/day}$ (318.4 m³/d).

The thickened sludge is generally pumped immediately to sludge storage tanks or sludge digestion facilities. If the thickener is operated 6.67 h/d, the thickened sludge pumps (used to pump the thickened sludge to downstream processes) will be sized based on the following thickened sludge flow rate:

$$S = \frac{12{,}823 \text{ gal/day}}{(6.67 \text{ h/d})(60 \text{ min/h})} = 32 \text{ gal/min (2.02 L/s)}$$

Hydraulic (thickened sludge) throughput for a gravity belt thickener ranges from 25 (1.6 L/s) to 100 (6.3 L/s) gal/min per meter of belt width. The filtrate flow of 84,123 gal/day is generally returned to the head of the wastewater treatment facility for reprocessing.

DESIGN OF AN AEROBIC DIGESTER

An aerobic digester is to be designed to treat the waste sludge produced by an activated sludge wastewater treatment facility. The input waste sludge will be 12,823 gal/day (48.5 L/d) (input 5 d/wk only) of thickened waste activated sludge at 5.0 percent solids—See *Thickening of a Waste Activated Sludge Using a Gravity Belt Thickener*. Assume the following apply:

1. The minimum liquid temperature in the winter is 15°C (59°F), and the maximum liquid temperature in the summer is 30°C (86°F).
2. The system must achieve a 40 percent Volatile Suspended Solids (VSS) reduction in the winter.
3. Sludge concentration in the digester is 70 percent of the incoming thickened sludge concentration.
4. The volatile fraction of digester suspended solids is 0.8.

Calculation Procedure:

1. *Find the daily volume of sludge for disposal*
Factors that must be considered in designing aerobic digesters include temperature, solids reduction, tank volume (hydraulic retention time), oxygen requirements and energy requirements for mixing.

Because the majority of aerobic digesters are open tanks, digester liquid temperatures are dependent upon weather conditions and can fluctuate extensively. As with all biological systems, lower temperatures retard the process, whereas higher temperatures accelerate it. The design of the aerobic digester should provide the necessary degree of sludge stabilization at the lowest liquid operating temperature and should supply the maximum oxygen requirements at the maximum liquid operating temperature.

A major objective of aerobic digestion is to reduce the mass of the solids for disposal. This reduction is assumed to take place only with the biodegradable content (VSS) of the sludge, although there may be some destruction of the inorganics as well. Typical reduction in VSS ranges from 40 to 50 percent. Solids destruction is primarily a direct function of both basin liquid temperature and sludge age, as indicated in Fig. 3. The plot relates VSS reduction to degree-days (temperature × sludge age).

To ensure proper operation, the contents of the aerobic digester should be well mixed. In general, because of the large amount of air that must be supplied to meet the oxygen requirement, adequate mixing is usually achieved. However, mixing power requirements should always be checked.

The aerobic digester will operate 7 days per week, unlike the thickening facilities which operate intermittently due to larger operator attention requirements. The thickened sludge is input to the digester at 12,823 gal/day (48.5 L/d), 5 days per week. However, the volume of the sludge to be disposed of daily by the digester will be lower due to its operation 7 days per week (the "bugs" do not take the weekends off). Therefore the volume of sludge to be disposed of daily (Q_i) is:

$$Q_i = (12,823 \text{ gal/day})(5/7) = 9,159 \text{ gal/day} = 1,224 \text{ ft}^3/d \ (34.6 \text{ m}^3/d)$$

2. *Determine the required VSS reduction*
The sludge age required for winter conditions is obtained from Fig. 3 using the minimum winter temperature and required VSS reduction.

To achieve a 40 percent VSS reduction in the winter, the degree-days required from Fig. 3 is 475°C·d. Therefore, the required sludge age is 475°C·d/15°C = 31.7 days. During the summer, when the liquid temperature is 30°C, the degree-days required is (30°C)(31.7 d) = 951°C·d. From Fig. 3, the VSS reduction will be 46 percent.

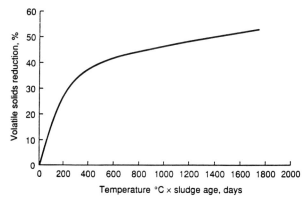

FIGURE 3   VSS reduction in aerobic digester vs. liquid temperature × sludge age. (Metcalf & Eddy, *Wastewater Engineering: Treatment, Disposal, and Reuse*, 3rd Ed., McGraw-Hill.)

The total mass of solids processed by the digester will be 3,933 lb/d (1785.6 kg/d) which is the total mass of solids wasted from the treatment facility—See *Design of a Complete-Mix Activated Sludge Reactor*, step 3. The total mass of VSS input to the digester is:

$$(0.8)(3,933 \text{ lb/d}) = 3146 \text{ lb/d} (1428.3 \text{ kg/d}).$$

Therefore, during the winter:

- VSS reduction = (3,146 lb/d)(0.40) = 1258 lb VSS reduced/d (571.1 kg/d)
- Digested (stabilized) sludge leaving the digester = 3933 lb/d − 1258 lb/d = 2675 lb/d (1214.5 kg/d).

3. *Compute the volume of digested sludge*
The volume of digested sludge is:

$$V = \frac{W_s}{(\rho)(\text{s.g.})(\% \text{ solids})}$$

where V = Sludge volume (ft³) (m³)
     W_s = Weight of sludge (lb) (kg)
     ρ = density of water (62.4 lb/ft³) (994.6 kg/m³)
   s.g. = specific gravity of digested sludge (assume s.g. = 1.03)
% solids = percent solids expressed as a decimal (incoming sludge: 5.0%)

Therefore, the volume of the digested sludge is:

$$V = \frac{2675 \text{ lb/d}}{(62.4 \text{ lb/ft}^3)(1.03)(0.05)} = 832 \text{ ft}^3/\text{d} = 6223 \text{ gal/day} (23.6 \text{ L/d})$$

During the summer:

- VSS reduction = (3,146 lb/d)(0.46) = 1447 lb VSS reduced/d (656.9 kg/d)
- Digested (stabilized) sludge leaving the digester = 3933 lb/d − 1447 lb/d = 2486 lb/d (1128.6 kg/d).
- Volume of digested sludge:

$$V = \frac{2486 \text{ lb/d}}{(62.4 \text{ lb/ft}^3)(1.03)(0.05)} = 774 \text{ ft}^3/\text{d} = 5790 \text{ gal/day} (21.9 \text{ L/d})$$

4. *Find the oxygen and air requirements*
The oxygen required to destroy the VSS is approximately 2.3 lb O_2/lb VSS (kg/kg) destroyed. Therefore, the oxygen requirements for winter conditions are:

$$(1258 \text{ lb VSS/d})(2.3 \text{ lb } O_2/\text{lb VSS}) = 2893 \text{ lb } O_2/\text{d} (1313.4 \text{ kg/d})$$

The volume of air required at standard conditions (14.7 lb/in² and 68°F) (96.5 kPa and 20°C) assuming air contains 23.2 percent oxygen by weight and the density of air is 0.075 lb/ft³ is:

$$\text{Volume of Air} = \frac{2893 \text{ lb O}_2/\text{d}}{(0.075 \text{ lb/ft}^3)(0.232)} = 166,264 \text{ ft}^3/\text{d} (4705.3 \text{ m}^3/\text{d})$$

For summer conditions:

- Oxygen required = $(1447 \text{ lb/d})(2.3 \text{ lb O}_2/\text{d}) = 3328 \text{ lb O}_2/\text{d} (1510.9 \text{ kg/d})$
- Volume of Air = 3328 lb O_2/d/(0.075 lb/ft³)(0.232) = 191,264 ft³/d (5412.8 m³/d)

Note that the oxygen transfer efficiency of the digester system must be taken into account to get the actual volume of air required. Assuming diffused aeration with an oxygen transfer efficiency of 10 percent, the actual air requirements at standard conditions are:

- Winter: volume of air = 166,264 ft³/d/(0.1)(1,400 min/d) = 1155 ft³/min (32.7 m³/min)
- Summer: volume of air = 191,264 ft³/d/(0.1)(1,440 min/d) = 1328 ft³/min (37.6 m³/min)

To summarize winter and summer conditions:

Parameter	Winter	Summer
Total Solids In, lb/d (kg/d)	3933 (1785.6)	3933 (1785.6)
VSS In, lb/d (kg/d)	3146 (1428.3)	3146 (1428.3)
VSS Reduction, (%)	40	46
VSS Reduction, lb/d (kg/d)	1258 (571.1)	1447 (656.9)
Digested Sludge Out, gal/day (L/d)	6223 (23.6)	5790 (21.9)
Digested Sludge Out, lb/d (kg/d)	2675 (1214.5)	2486 (1128.6)
Air Requirements @ S.C., ft³/min (m³/min)	1155 (32.7)	1328 (37.6)

5. Determine the aerobic digester volume
From the above analysis it is clear that the aerobic digester volume will be calculated using values obtained under the winter conditions analysis, while the aeration equipment will be sized using the 1328 ft³/min (37.6 m³/min) air requirement obtained under the summer conditions analysis.

The volume of the aerobic digester is computed using the following equation, assuming the digester is loaded with waste activated sludge only:

$$V = \frac{Q_i X_i}{X(K_d P_v + 1/\theta_c)}$$

where V = Volume of aerobic digester, ft³ (m³)
Q_i = Influent average flow rate to the digester, ft³/d (m³/d)
X_i = Influent suspended solids, mg/L (50,000 mg/L for 5.0% solids)
X = Digester total suspended solids, mg/L
K_d = Reaction rate constant, d^{-1}. May range from 0.05 d^{-1} at 15°C (59°F) to 0.14 d^{-1} at 25°C (77°F) (assume 0.06 d^{-1} at 15°C)
P_v = Volatile fraction of digester suspended solids (expressed as a decimal) = 0.8 (80%) as stated in the initial assumptions.
θ_c = Solids retention time (sludge age), d

Using values obtained above with winter conditions governing, the aerobic digester volume is:

$$V = \frac{(1{,}224 \text{ ft}^3/\text{d})(50{,}000 \text{ mg/L})}{(50{,}000 \text{ mg/L})(0.7)[(0.06\ d^{-1})(0.8) + 1/31.7\ d]} = 21{,}982 \text{ ft}^3 \ (622.2 \text{ m}^3)$$

The air requirement per 1,000 ft³ (2.8 m³) of digester volume with summer conditions governing is:

$$\text{Volume of Air} = \frac{1328 \text{ ft}^3/\text{min}}{21.982 \ 10^3\text{ft}^3} = 60.41 \text{ ft}^3/\text{min}/10^3\text{ft}^3 \ (0.97 \text{ m}^3/\text{min}/\text{Mm}^3)$$

The mixing requirements for diffused aeration range from 20 to 40 ft³/min/10³ft³ (0.32 to 0.64 m³/min/Mm³). Therefore, adequate mixing will prevail.

DESIGN OF AN AERATED GRIT CHAMBER

Domestic wastewater enters a wastewater treatment facility with an average daily flow rate of 4.0 Mgd (15,140 L/d). Assuming a peaking factor of 2.5, size an aerated grit chamber for this facility including chamber volume, chamber dimensions, air requirement, and grit quantity.

Calculation Procedure:

1. *Determine the aerated grit chamber volume*
Grit removal in a wastewater treatment facility prevents unnecessary abrasion and wear of mechanical equipment such as pumps and scrappers, and grit deposition in pipelines and channels. Grit chambers are designed to remove grit (generally characterized as nonputrescible solids) consisting of sand, gravel, or other heavy solid materials that have settling velocities greater than those of the organic putrescible solids in the wastewater.

In aerated grit chamber systems, air introduced along one side near the bottom causes a spiral roll velocity pattern perpendicular to the flow through the tank. The heavier particles with their correspondingly higher settling velocities drop to the bottom, while the rolling action suspends the lighter organic particles, which are carried out of the tank. The rolling action induced by the air diffusers is independent of the flow through the tank. Then non flow dependent rolling action allows the aerated grit chamber to operate effectively over a wide range of flows. The heavier particles that settle on the bottom of the tank are moved by the spiral flow of the water across the tank bottom and into a grit hopper. Screw augers or air lift pumps are generally utilized to remove the grit from the hopper.

The velocity of roll governs the size of the particles of a given specific gravity that will be removed. If the velocity is too great, grit will be carried out of the chamber. If the velocity is too small, organic material will be removed with the grit. The quantity of air is easily adjusted by throttling the air discharge or using adjustable speed drives on the blowers. With proper adjustment, almost 100 percent grit removal will be obtained, and the grit will be well washed. Grit that is not well washed will contain organic matter and become a nuisance through odor emission and the attraction of insects.

Wastewater will move through the aerated grit chamber in a spiral path as illustrated in Fig. 4. The rolling action will make two to three passes across the bottom of the tank at maximum flow and more at lesser flows. Wastewater is introduced in the direction of the roll.

At peak flow rate, the detention time in the aerated grit chamber should range from 2 to 5 minutes. A detention time of 3 minutes will be used for this example. Because it is necessary to drain the chamber periodically for routine maintenance, two redundant chambers will be required. Therefore, the volume of each chamber is:

$$V \text{ (ft}^3\text{)} = \frac{(\text{peak flow rate, gal/day})(\text{detention time, min})}{(7.48 \text{ gal/ft}^3)(24 \text{ h/d})(60 \text{ min/h})}$$

Using values from above, the chamber volume is:

$$V \text{ (ft}^3\text{)} = \frac{(2.5)(4 \times 10^6 \text{ gal/day})(3 \text{ min})}{(7.48 \text{ gal/ft}^3)(24 \text{ h/d})(60 \text{min/h})} = 2785 \text{ ft}^3 \text{ (78.8 m}^3\text{)}$$

2. Determine the dimensions of the grit chamber
Width-depth ratio for aerated grit chambers range from 1:1 to 5:1. Depths range from 7 to 16 feet (2.1 to 4.87 m). Using a width-depth ratio of 1.2:1 and a depth of 8 feet (2.43 m), the dimensions of the aerated grit chamber are:

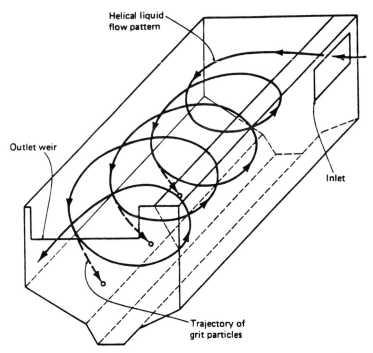

FIGURE 4   Aerated grit chamber. (Metcalf & Eddy, *Wastewater Engineering: Treatment, Disposal, and Reuse*, 3rd Ed., McGraw-Hill.)

Width = (1.2)(8 ft) = 9.6 ft (2.92 m)

$$\text{Length} = (\text{volume})/[(\text{width})(\text{depth})] = \frac{2785 \text{ ft}^3}{(8 \text{ ft})(9.6 \text{ ft})} = 36.3 \text{ ft } (11.1 \text{ m})$$

Length-width ratios range from 3:1 to 5:1. As a check, length to width ratio for the aerated grit chamber sized above is: 36.3 ft/9.6 ft = 3.78:1 which is acceptable.

3. Determine the air supply required
The air supply requirement for an aerated grit chamber ranges from 2.0 to 5.0 ft³/min/ft of chamber length (0.185 to 0.46 m³/min·m). Using 5.0 ft³/min/ft (0.46 m³/min·m) for design, the amount of air required is:

Air required (ft³/min) = (5.0 ft³/min/ft)(36.3 ft) = 182 ft³/min (5.2 m³/min)

4. Estimate the quantity of grit expected
Grit quantities must be estimated to allow sizing of grit handling equipment such as grit conveyors and grit dewatering equipment. Grit quantities from an aerated grit chamber vary from 0.5 to 27 ft³/Mgal (3.74 to 201.9 m³/L) of flow. Assume a value of 20 ft³/Mgal (149.5 m³/L). Therefore, the average quantity of grit expected is:

Volume of grit (ft³/d) = (20 ft³/Mgal)(4.0 Mgd) = 80 ft³/d (2.26 m³/d)

Some advantages and disadvantages of the aerated grit chamber are listed below:

Advantages	Disadvantages
The same efficiency of grit removal is possible over a wide flow range.	Power consumption is higher than other grit removal processes.
Head loss through the grit chamber is minimal.	Additional labor is required for maintenance and control of the aeration system.
By controlling the rate of aeration, a grit of relatively low putrescible organic content can be removed.	Significant quantities of potentially harmful volatile organics and odors may be released from wastewaters containing these constituents.
Preaeration may alleviate septic conditions in the incoming wastewater to improve performance of downstream treatment units.	Foaming problems may be created if influent wastewater has surfactants present.
Aerated grit chambers can also be used for chemical addition, mixing, preaeration, and flocculation ahead of primary treatment.	

DESIGN OF A SOLID-BOWL CENTRIFUGE FOR SLUDGE DEWATERING

A 4.0 Mgd (15,140 m³/d) municipal wastewater treatment facility produces 6,230 gal/day (23.6 m³/d) of aerobically digested sludge at 5.0 percent solids. Determine design parameters for the specification of a solid bowl centrifuge for dewatering the sludge including: number of centrifuges, solids feed rate, percent solids recovery, dewatered sludge (cake) discharge rate, centrifugal force, polymer dosage, and polymer feed rate. Assume the following apply:

- Feed sludge is aerobically digested at 5.0 percent solids.
- Dewatered sludge (cake) is to be 25 percent solids.
- Centrate assumed to be 0.3 percent solids.
- Polymer solution concentration is 25 percent.

Calculation Procedure:

1. Select the number of centrifuges

The separation of a liquid-solid sludge during centrifugal thickening is analogous to the separation process in a gravity thickener. In a centrifuge, however, the applied force is centrifugal rather than gravitational and usually exerts 1,500 to 3,500 times the force of gravity. Separation results from the centrifugal force-driven migration of the suspended solids through the suspending liquid, away from the axis of rotation. The increased settling velocity imparted by the centrifugal force as well as the short settling distance of the particles accounts for the comparatively high capacity of centrifugal equipment.

Centrifuges are commonly used for thickening or dewatering Waste Activated Sludge (WAS) and other biological sludges from secondary wastewater treatment. In the process, centrifuges reduce the volume of stabilized (digested) sludges to minimize the cost of ultimate disposal. Because centrifuge equipment is costly and sophisticated, centrifuges are most commonly found in medium to large wastewater treatment facilities.

The capacity of sludge dewatering to be installed at a given facility is a function of the size of a facility, capability to repair machinery on-site, and the availability of an alternative disposal means. Some general guidelines relating the minimal capacity requirements are listed in Table 1. This table is based on the assumption that there is no alternative mode of sludge disposal and that the capacity to store solids is limited.

Using Table 1, the number of centrifuges recommended for a 4.0 Mgd (15,410 m³/d) wastewater treatment facility is one operational + one spare for a total of two centrifuges.

TABLE 1 Facility Capacity & Number of Centrifuges

Facility size, Mgd (m³/d)	Dewatering operation, h/d	Centrifuges operating + spare @ gal/min (L/s)
2 (7,570)	7	1 + 1 @ 25 (1.58)
5 (18,930)	7.5	1 + 1 @ 50 (3.16)
20 (75,700)	15	2 + 1 @ 50 (3.16)
50 (189,250)	22	2 + 1 @ 75 (4.73)
100 (378,500)	22	3 + 2 @ 100 (6.31)
250 (946,250)	22	4 + 2 @ 200 (12.62)

(*Design Manual for Dewatering Municipal Wastewater Sludges*, U.S. EPA)

2. *Find the sludge feed rate required*
If the dewatering facility is operated 4 h/d, 7 d/wk, then the sludge feed rate is:

$$\text{Sludge Feed Rate} = (6{,}230 \text{ gal/day})/[(4 \text{ h/d})(60/\text{min/h})]$$

$$= 26 \text{ gal/min } (1.64 \text{ L/s})$$

Although a 4 h/d operation is below that recommended in Table 1, the sludge feed rate of 26 gal/min (1.64 L/s) is adequate for the size of centrifuge usually found at a treatment facility of this capacity. A longer operational day would be necessary if the dewatering facilities were operated only 5 days per week, or during extended period of peak flow and solids loading.

Assume a feed sludge specific gravity of 1.03. The sludge feed in lb/h is calculated using the following equation:

$$W_s = \frac{(V)(\rho)(\text{s.g.})(\% \text{ solids})(60 \text{ min/h})}{7.48 \text{ gal/ft}^3}$$

W_s = Weight flow rate of sludge feed, lb/h (kg/h)
V = Volume flow rate of sludge feed, gal/min (L/s)
s.g. = specific gravity of sludge
% solids = percent solids expressed as a decimal
ρ = density of water, 62.4 lb/ft^3 (994.6 kg/m^3)

Using values obtained above, the sludge feed in lb/h is:

$$W_s = \frac{(26 \text{ gal/min})(62.4 \text{ lb/ft}^3)(1.03)(0.05)(60 \text{ min/h})}{7.48 \text{ gal/ft}^3}$$

$$= 670 \text{ lb/h of dry solids } (304.2 \text{ kg/h})$$

3. *Compute the solids capture*
Since the solids exiting the centrifuge are split between the centrate and the cake, it is necessary to use a recovery formula to determine solids capture. Recovery is the mass of solids in the cake divided by the mass of solids in the feed. If the solids content of the feed, centrate and cake are measured, it is possible to calculate percent recovery without determining total mass of any of the streams. The equation for percent solids recovery is:

$$R = 100 \left(\frac{C_S}{F}\right)\left[\frac{F - C_C}{C_S - C_C}\right]$$

where R = Recovery, percent solids
C_s = Cake solids, percent solids (25%)
F = Feed solids, percent solids (5%)
C_c = Centrate solids, percent solids (0.3%)

Therefore, using values defined previously:

$$R = 100(25/5) \left[\frac{(5 - 0.3)}{25 - 0.3}\right] = 95.14\%$$

4. *Determine the Dewatered Sludge Cake Discharge Rate*
The dewatered sludge (cake) discharge rate is calculated using the following:

Cake discharge rate (lb/h) dry solids = (sludge feed rate, lb/h)(solids recovery)

$$= (670 \text{ lb/h})(0.9514) = 637.5 \text{ lb/h } (289.4 \text{ kg/h}) \text{ dry cake}$$

The wet cake discharge in lb/h is calculated using the following:

$$\text{Wet cake discharge (lb/h)} = \frac{\text{Dry cake rate, lb/h}}{\text{Cake \% solids}}$$

$$\Rightarrow \text{Wet cake discharge (lb/h)} = \frac{637.5 \text{ lb/h}}{0.25}$$

$$= 2550 \text{ lb/h wet cake } (1157.7 \text{ kg/h})$$

The volume of wet cake, assuming a cake density of 60 lb/ft³ is calculated as follows:

$$\text{Volume of wet cake (ft}^3/\text{h)} = \frac{\text{Wet cake rate, lb/h}}{\text{Cake density, lb/ft}^3} = \frac{2550 \text{ lb/h}}{60 \text{ lb/ft}^3}$$

$$= 42.5 \text{ ft}^3/\text{h } (1.2 \text{ m}^3/\text{h}) \text{ wet cake}$$

For a dewatering facility operation of 4 h/d, the volume of dewatered sludge cake to be disposed of per day is:

$$(42.5 \text{ ft}^3/\text{h})(4 \text{ h/d}) = 170 \text{ ft}^3/\text{d} = 1272 \text{ gal/day } (4.81 \text{ L/d})$$

5. *Find the percent reduction in sludge volume*
The percent reduction in sludge volume is then calculated using the following:

$$\% \text{ Volume Reduction} = \frac{\text{Sludge volume in} - \text{Sludge volume out}}{\text{Sludge volume in}}$$

$$\times 100\%$$

$$= \frac{6{,}230 \text{ gal/day} - 1{,}272 \text{ gal/day}}{6{,}230 \text{ gal/day}}$$

$$\times 100\% = 79.6\%$$

Centrifuges operate at speed ranges which develop centrifugal forces from 1,500 to 3,500 times the force of gravity. In practice, it has been found that higher rotational speeds usually provide significant improvements in terms of performance, particularly on wastewater sludges.

In most cases, a compromise is made between the process requirement and O&M considerations. Operating at higher speeds helps achieve optimum performance which is weighed against somewhat greater operating and maintenance costs. Increasing bowl speed usually increases solids recovery and cake dryness. Today most centrifuges used in wastewater applications can provide good clarity and solids concentration at G levels between 1,800 and 2,500 times the force of gravity.

6. Compute the centrifugal force in the centrifuge

The centrifugal acceleration force (G), defined as multiples of gravity, is a function of the rotational speed of the bowl and the distance of the particle from the axis of rotation. In the centrifuge, the centrifugal force, G, is calculated as follows:

$$G = \frac{(2\pi N)^2 R}{32.2 \text{ ft/s}^2}$$

where N = Rotational speed of centrifuge (rev/s)
R = Bowl radius, ft (cm)

The rotational speed and bowl diameter of the centrifuge will vary depending upon the manufacturer. However, a rotational speed of 2,450 r/min and a bowl diameter of 30 inches (72.6 cm) are common for this type of sludge dewatering operation.
Therefore, the centrifugal force is:

$$G = \frac{((2\pi)(2450 \text{ r/min}/60 \text{ s/min}))^2 (30 \text{ in}/12 \text{ in/ft})(0.5)}{32.2 \text{ ft/s}^2} = 2{,}555 \text{ } Gs$$

7. Find the polymer feed rate for the centrifuge

The major difficulty encountered in the operation of centrifuges is the disposal of the centrate, which is relatively high in suspended, non-settling solids. The return of these solids to the influent of the wastewater treatment facility can result in the passage of fine solids through the treatment system, reducing effluent quality. Two methods are used to control the fine solids discharge and increase the capture. These are: (1) increased residence time in the centrifuge, and (2) polymer addition. Longer residence time of the liquid is accomplished by reducing the feed rate or by using a centrifuge with a larger bowl volume. Better clarification of the centrate is achieved by coagulating the sludge prior to centrifugation through polymer addition. Solids capture may be increased from a range of 50 to 80 percent to a range of 80 to 95 percent by longer residence time and chemical conditioning through polymer addition.

In order to obtain a cake solids concentration of 20 to 28 percent for an aerobically digested sludge, 5 to 20 pounds of dry polymer per ton of dry sludge feed (2.27 to 9.08 kg/ton) is required. 15 lb/ton (6.81 kg/ton) will be used for this example. Usually this value is determined through pilot testing or plant operator trial and error.

The polymer feed rate in lb/h of dry polymer is calculated using the following:

$$\text{Polymer feed rate (lb/h)} = \frac{(\text{polymer dosage, lb/ton})(\text{dry sludge feed, lb/h})}{2000 \text{ lb/ton}}$$

Using values defined previously, the polymer feed rate is:

$$\text{Polymer feed rate (lb/h)} = \frac{(15 \text{ lb/ton})(670 \text{ lb/h})}{2000 \text{ lb/ton}}$$

$$= 5.0 \text{ lb/h of dry polymer } (2.27 \text{ kg/h})$$

Polymer feed rate in gal/h is calculated using the following:

$$\text{Polymer feed rate (gal/h)} = \frac{\text{polymer feed rate lb/h)}}{(8.34\ \text{lb/gal})(\text{s.g.})(\%\ \text{polymer concentration})}$$

where s.g. = specific gravity of the polymer solution
% polymer concentration expressed as a decimal

Using values defined previously:

$$\text{Polymer feed rate (gal/h)} = \frac{5.0\ \text{lb/h}}{(8.34\ \text{lb/gal})(1.0)(0.25)}$$

$$= 2.4\ \text{gal/h of 25\% polymer solution (0.009 L/h)}$$

The polymer feed rate is used to size the polymer dilution/feed equipment required for the sludge dewatering operation.

Related Calculations. Selection of units for dewatering facility design is dependent upon manufacturer's rating and performance data. Several manufacturers have portable pilot plant units, which can be used for field testing if sludge is available. Wastewater sludges from supposedly similar treatment processes but different localities can differ markedly from each other. For this reason, pilot plant tests should be run, whenever possible, before final design decisions regarding centrifuge selection are made.

SIZING OF A TRAVELING-BRIDGE FILTER

Secondary effluent from a municipal wastewater treatment facility is to receive tertiary treatment, including filtration, through the use of traveling bridge filters. The average daily flow rate is 4.0 Mgd (2778 gal/min) (15,140 m³/d) and the peaking factor is 2.5. Determine the size and number of traveling bridge filters required.

Calculation Procedure:

1. *Determine the peak flow rate for the filter system*
The traveling bridge filter is a proprietary form of a rapid sand filter. This type of filter is used mainly for filtration of effluent from secondary and advanced wastewater treatment facilities. In the traveling bridge filter, the incoming wastewater floods the filter bed, flows through the filter medium (usually sand and/or anthracite), and exits to an effluent channel via an underdrain and effluent ports located under each filtration cell. During the backwash cycle, the carriage and the attached hood (see Fig. 5) move slowly over the filter bed, consecutively isolating and backwashing each cell. The washwater pump, located in the effluent channel, draws filtered wastewater from the effluent chamber and pumps it through the effluent port of each cell, forcing water to flow up through the cell thereby backwashing the filter medium of the cell. The backwash pump located above the hood draws water with suspended matter collected under the hood and transfers it to the backwash water trough. During the backwash cycle, wastewater is filtered continuously through the cells not being backwashed.

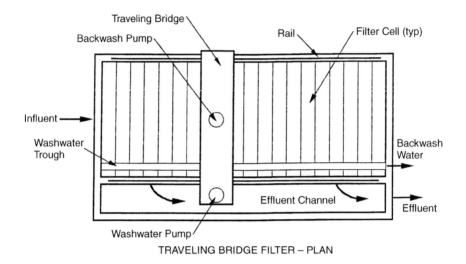

TRAVELING BRIDGE FILTER – PLAN

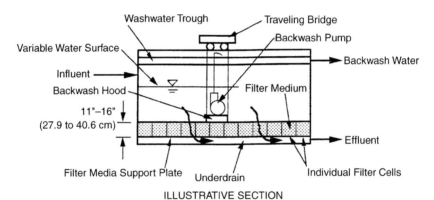

ILLUSTRATIVE SECTION

FIGURE 5   Traveling bridge filter.

Filtration in a traveling bridge filter is accomplished at a hydraulic loading typically in the range of 1.5 to 3.0 gal/min per square foot of filter surface (1.02 to 2.04 L/s·m²) at average daily flow. The maximum hydraulic loading used for design is typically 4.0 gal/min/ft² (2.72 L/s·m²) at peak flow. The peak hydraulic loading is used to size the traveling bridge filter.

The peak flow for this treatment facility is calculated as follows:

$$\text{Peak flow} = (\text{peaking factor})(\text{average daily flow}) = (2.5)(4.0 \text{ Mgd})$$

$$= 10 \text{ Mdg} = 6944 \text{ gal/min (438.2 L/s)}$$

2. Find the required filter surface area
Filter surface area required is calculated using:

$$\text{Filter surface area required (ft}^2) = \frac{\text{peak flow (gal/min)}}{\text{Hydraulic loading (gal/min} \cdot \text{ft}^2)}$$

Using values from above:

$$\text{Filter surface area required (ft}^2) = \frac{6944 \text{ gal/min}}{4.0 \text{ gal/min} \cdot \text{ft}^2} = 1736 \text{ ft}^2 \text{ (161.3 m}^2)$$

3. Determine the number of filters required
Standard filter widths available from various manufacturers are 8, 12, and 16 ft (2.44, 3.66, and 4.88 m). Using a width of 12 feet (3.66 m) and length of 50 feet (15.2 m) per filter, the area of each filter is:

$$\text{Area of each filter} = (12')(50') = 600 \text{ ft}^2 \text{ (55.7 m}^2)$$

The number of filters required is 1736 ft²/600 ft² per filter = 2.89. Use 3 filters for a total filter area of 1800 ft² (167.2 m²).

It must be kept in mind that most state and local regulations stipulate that "rapid sand filters shall be designed to provide a total filtration capacity for the maximum anticipated flow with at least one of the filters out of service." Therefore, 4 traveling bridge filters should be provided, each with filtration area dimensions of 12' wide × 50' long (3.66 m × 15.2 m).

The media depth for traveling bridge filters ranges from 11" to 16" (27.9 to 40.6 cm). Dual media may be used with 8" (20.3 cm) of sand underlying 8" (20.3 cm) of anthracite.

4. Find the hydraulic loading under various service conditions
The hydraulic loadings with all filters in operation is:

$$\text{Average flow:} \quad \frac{2778 \text{ gal/min}}{4(600 \text{ ft}^2)} = 1.16 \text{ gal/min} \cdot \text{ft}^2 \text{ (0.79 L/s} \cdot \text{m}^2)$$

$$\text{Peak flow:} \quad \frac{6944 \text{ gal/min}}{4(600 \text{ ft}^2)} = 2.89 \text{ gal/min} \cdot \text{ft}^2 \text{ (1.96 L/s} \cdot \text{m}^2)$$

Hydraulic loading with 1 filter out of service (3 active filters) is:

$$\text{Average flow:} \quad \frac{2778 \text{ gal/min}}{3(600 \text{ ft}^2)} = 1.54 \text{ gal/min} \cdot \text{ft}^2 \text{ (1.05 L/s} \cdot \text{m}^2)$$

$$\text{Peak flow:} \quad \frac{6944 \text{ gal/min}}{3(600 \text{ ft}^2)} = 3.86 \text{ gal/min} \cdot \text{ft}^2 \text{ (2.62 L/s} \cdot \text{m}^2)$$

These hydraulic loadings are acceptable and may be used in specifying the traveling bridge filter.

The amount of backwash water produced depends upon the quantity and quality of influent to the filter. The backwash pumps are usually sized to deliver approximately 25 gal/min (1.58 L/s) during the backwash cycle. Backwash water is generally returned to the head of the treatment facility for reprocessing.

DESIGN OF A RAPID-MIX BASIN AND FLOCCULATION BASIN

1.0 Mdg (3785 m³/d) of equalized secondary effluent from a municipal wastewater treatment facility is to receive tertiary treatment through a direct filtration process which includes rapid mix with a polymer coagulant, flocculation and filtration. Size the rapid mix and flocculation basins necessary for direct filtration and determine the horsepower of the required rapid mixers and flocculators.

Calculation Procedure:

1. *Determine the required volume of the rapid mix basin*
A process flow diagram for direct filtration of a secondary effluent is presented in Fig. 6. This form of tertiary wastewater treatment is used following secondary treatment when an essentially "virus-free" effluent is desired for wastewater reclamation and reuse.

The rapid mix basin is a continuous mixing process in which the principle objective is to maintain the contents of the tank in a completely mixed state. Although there are numerous ways to accomplish continuous mixing, mechanical mixing will be used here. In mechanical mixing, turbulence is induced through the input of energy by means of rotating impellers such as turbines, paddles, and propellers.

The hydraulic retention time of typical rapid mix operations in wastewater treatment range from 5 to 20 seconds. A value of 15 seconds will be used here. The required volume of the rapid mix basin is calculated as follows:

$$\text{Volume } (V) = (\text{hydraulic retention time})(\text{wastewater flow})$$

$$V = \frac{(15 \text{ s})(1 \times 10^6 \text{ gal/day})}{86,400 \text{ s/d}} = 174 \text{ gal} \cong 24 \text{ ft}^3 \ (0.68 \text{ m}^3)$$

2. *Compute the power required for mixing*
The power input per volume of liquid is generally used as a rough measure of mixing effectiveness, based on the reasoning that more input power creates greater

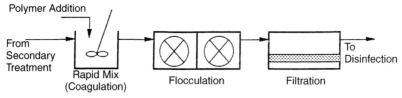

FIGURE 6   Process flow for direct filtration.

turbulence, and greater turbulence leads to better mixing. The following equation is used to calculate the required power for mixing:

$$G = \sqrt{\frac{P}{\mu V}}$$

where G = Mean velocity gradient (s^{-1})
$\quad$ P = Power requirement (ft·lb/s) (kW)
$\quad$ μ = Dynamic viscosity (lb·s/ft^2) (Pa·s)
$\quad$ V = Volume of mixing tank (ft^3) (m^3)

G is a measure of the mean velocity gradient in the fluid. G values for rapid mixing operations in wastewater treatment range from 250 to 1,500 s^{-1}. A value of 1,000 s^{-1} will be used here. For water at 60°F (15.5°C), dynamic viscosity is 2.36 × 10^{-5} lb·s/ft^2 (1.13 × 10^{-3} Pa·s). Therefore, the required power for mixing is computed as follows:

$$P = G^2\mu V = (1{,}000 \text{ s}^{-1})^2(2.36 \times 10^{-5} \text{ lb·s/ft}^2)(24 \text{ ft}^3) = 566 \text{ ft·lb/s}$$

$$= 1.03 \text{ horsepower } (0.77 \text{ kW})$$

Use the next largest motor size available = 1.5 horsepower (1.12 kW). Therefore, a 1.5 horsepower (1.12 kW) mixer should be used.

3. *Determine the required volume and power input for flocculation*
The purpose of flocculation is to form aggregates, or flocs, from finely divided matter. The larger flocs allow a greater solids removal in the subsequent filtration process. In the direct filtration process, the wastewater is completely mixed with a polymer coagulant in the rapid mix basin. Following rapid mix, the flocculation tanks gently agitate the wastewater so that large "flocs'''' form with the help of the polymer coagulant. As in the rapid mix basins, mechanical flocculators will be utilized.

For flocculation in a direct filtration process, the hydraulic retention time will range from 2 to 10 minutes. A retention time of 8 minutes will be used here. Therefore, the required volume of the flocculation basin is:

$$V = \frac{(8 \text{ min})(1 \times 10^6 \text{ gal/day})}{1{,}440 \text{ min/d}} = 5{,}556 \text{ gal} \cong 743 \text{ ft}^3 \text{ (21 m}^3)$$

G values for flocculation in a direct filtration process range from 20 to 100 s^{-1}. A value of 80 s^{-1} will be used here. Therefore, the power required for flocculation is:

$$P = G^2\mu V = (80 \text{ s}^{-1})^2(2.36 \times 10^{-5})(743 \text{ ft}^3)$$

$$= 112 \text{ ft·lb/s} = 0.2 \text{ horsepower } (0.15 \text{ kW})$$

Use the next largest motor size available = 0.5 horsepower (0.37 kW). Therefore, a 0.5 horsepower (0.37 kW) flocculator should be used.

It is common practice to taper the energy input to flocculation basins so that flocs initially formed will not be broken as they leave the flocculation facilities. In the above example, this may be accomplished by providing a second flocculation basin in series with the first. The power input to the second basin is calculated using a lower G value (such as 50 s^{-1}) and hence provides a gentler agitation.

Related Calculations. If the flows to the rapid mix and flocculation basin vary significantly, or turn down capability is desired, a variable speed drive should be provided for each mixer and flocculator. The variable speed drive should be controlled via an output signal from a flow meter immediately upstream of each respective basin.

It should be noted that the above analysis provides only approximate values for mixer and flocculator sizes. Mixing is in general a "black art," and a mixing manufacturer is usually consulted regarding the best type and size of mixer or flocculator for a particular application.

SIZING A POLYMER DILUTION / FEED SYSTEM

1.0 Mgd (3,785 m^3/d) of equalized secondary effluent from a municipal wastewater treatment facility is to undergo coagulation and flocculation in a direct filtration process. The coagulant used will be an emulsion polymer with 30 percent active ingredient. Size the polymer dilution/feed system including: the quantity of dilution water required, and the amount of neat (as supplied) polymer required.

Calculation Procedure:

1. *Determine the daily polymer requirements*
Depending on the quality of settled secondary effluent, organic polymer addition is often used to enhance the performance of tertiary effluent filters in a direct filtration process: see *Design of a Rapid Mix Basin and Flocculation Basin.* Because the chemistry of the wastewater has a significant effect on the performance of a polymer, the selection of a type of polymer for use as a filter aid generally requires experimental testing. Common test procedures for polymers involve adding an initial polymer dosage to the wastewater (usually 1 part per million, ppm) of a given polymer and observing the effects. Depending upon the effects observed, the polymer dosage should be increased or decreased by 0.5 ppm increments to obtain an operating range. A polymer dosage of 2 ppm (2 parts polymer per 1×10^6 parts wastewater) will be used here.

In general, the neat polymer is supplied with approximately 25 to 35 percent active polymer, the rest being oil and water. As stated above, a 30 percent active polymer will be used for this example. The neat polymer is first diluted to an extremely low concentration using dilution water, which consists of either potable water or treated effluent from the wastewater facility. The diluted polymer solution usually ranges from 0.005 to 0.5 percent solution. The diluted solution is injected into either a rapid mix basin or directly into a pipe. A 0.5 percent solution will be used here.

The gallons per day (gal/day) (L/d) of active polymer required is calculated using the following:

$$\text{Active polymer (gal/day)} = \text{(wastewater flow, Mgd)}$$

$$\cdot \text{(active polymer dosage, ppm)}$$

Using the values outlined above:

Active polymer = (1.0 Mgd)(2 ppm) = 2 gal/day active polymer (pure polymer)

$$= 0.083 \text{ gal/hr (gal/h) (0.31 L/h)}$$

2. Find the quantity of dilution water required
The quantity of dilution water required is calculated using the following:

$$\text{Dilution water (gal/h)} = \frac{\text{active polymer, gal/h}}{\% \text{ solution used (as a decimal)}}$$

Therefore, using values obtain above:

$$\text{Dilution water} = \frac{0.083 \text{ gal/h}}{0.005} = 16.6 \text{ gal/h (62.8 L/h)}$$

3. Find the quantity of neat polymer required
The quantity of neat polymer required is calculated as follows:

$$\text{Neat polymer (gal/h)} = \frac{\text{active polymer, gal/h}}{\% \text{ active polymer in emulsion as supplied}}$$

Using values obtained above:

$$\text{Neat polymer} = \frac{0.083 \text{ gal/h}}{0.30} = 0.277 \text{ gal/h (1.05 L/h)}$$

This quantity of neat polymer represents the amount of polymer used in its "as supplied" form. Therefore, if polymer is supplied in a 55 gallon (208.2 L) drum, the time required to use one drum of polymer (assuming polymer is used 24 h/d, 7 d/wk) is:

$$\text{Time required to use one drum of polymer} = \frac{55 \text{ gal}}{0.277 \text{ gal/h}} \cong 200 \text{ h} = 8 \text{ days}$$

DESIGN OF A TRICKLING FILTER USING THE NRC EQUATIONS

A municipal wastewater with a flow rate of 1.0 Mgd (3,785 m³/d) and a BOD_5 of 240 mg/L is to be treated by a two stage trickling filter system. The effluent wastewater is to have a BOD_5 of 20 mg/L. Both filters are to have a depth of 7 feet (2.1 m) and a recirculation ratio of 2. Filter media will consist of rock. Size both stages of the trickling filter assuming the efficiency (E) of each stage is the same.

Calculation Procedure:

1. Find the efficiency of the trickling filters
The modern trickling filter, shown in Fig. 7, consists of a bed of highly permeable medium to which microorganisms are attached and through which wastewater is

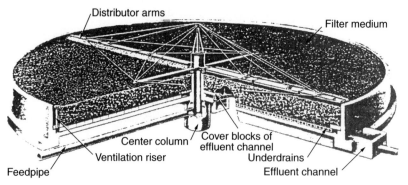

FIGURE 7   Cutaway view of a trickling filter. (Metcalf & Eddy, *Wastewater Engineering: Treatment, Disposal, and Reuse*, 3rd Ed., McGraw-Hill.)

percolated or trickled. The filter media usually consists of either rock or a variety of plastic packing materials. The depth of rock varies but usually ranges from 3 to 8 feet (0.91 to 2.44 m). Trickling filters are generally circular, and the wastewater is distributed over the top of the bed by a rotary distributor.

Filters are constructed with an underdrain system for collecting the treated wastewater and any biological solids that have become detached from the media. This underdrain system is important both as a collection unit and as a porous structure through which air can circulate. The collected liquid is passed to a settling tank where the solids are separated from the treated wastewater. In practice, a portion of the treated wastewater is recycled to dilute the strength of the incoming wastewater and to maintain the biological slime layer in a moist condition.

The organic material present in the wastewater is degraded by a population of microorganisms attached to the filter media. Organic material from the wastewater is absorbed onto the biological slime layer. As the slime layer increases in thickness, the microorganisms near the media face lose their ability to cling to the media surface. The liquid then washes the slime off the media, and a new slime layer starts to grow. The phenomenon of losing the slime layer is called "sloughing" and is primarily a function of the organic and hydraulic loading on the filter.

Two possible process flow schematics for a two-stage trickling filter system are shown in Fig. 8.

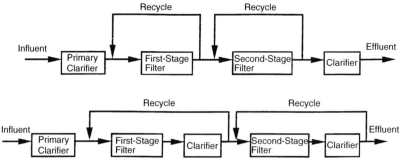

FIGURE 8   Two-stage trickling filter process flow schematics.

The NRC equations for trickling filter performance are empirical equations which are primarily applicable to single and multistage rock systems with recirculation.

The overall efficiency of the two-stage trickling filter is calculated using:

$$\text{Overall efficiency} = \frac{\text{influent BOD}_5 - \text{effluent BOD}_5}{\text{influent BOD}_5} \times 100$$

Using the influent and effluent BOD_5 values presented in the problem statement, the overall efficiency is:

$$\text{Overall efficiency} = \frac{240 \text{ mg/L} - 20 \text{ mg/L}}{240 \text{ mg/L}} \times 100 = 91.7\%$$

Also, overall efficiency $= E_1 + E_2(1 - E_1)$, and $E_1 = E_2$.

where E_1 = The efficiency of the first filter stage, including recirculation and settling (%)
E_2 = The efficiency of the second filter stage, including recirculation and settling (%)

Substituting E_1 for E_2, setting up as a quadratic equation, and solving for E_1:

$$E_1^2 - 2E_1 + 0.917 = 0 \Rightarrow E_1 = 0.712 \text{ or } 71.2\%$$

Therefore, the efficiency of each trickling filter stage is 71.2 percent.

2. Analyze the first stage filter
For a single stage or first stage rock trickling filter, the NRC equation is:

$$E_1 = \frac{100}{1 + 0.0561 \sqrt{\dfrac{W}{VF}}}$$

where W = BOD_5 loading to the filter, lb/d (kg/d)
V = Volume of the filter media, 10^3 ft^3 (m^3)
F = Recirculation factor

2a. Compute the recirculation factor of the filter
Recirculation factor represents the average number of passes of the influent organic matter through the trickling filter. The recirculation factor is calculated using:

$$F = \frac{1 + R}{(1 + (R/10))^2}$$

where R = Recirculation ratio = Q_r/Q
Q_r = Recirculation flow
Q = Wastewater flow

Using values from above, the recirculation factor is:

$$F = \frac{1 + 2}{(1 + (2/10))^2} = 2.08$$

2b. Compute the BOD_5 loading for the first stage filter
The BOD_5 loading for the first stage filter is calculated using:

$$W = (\text{Influent } BOD_5, \text{ mg/L})(\text{Wastewater flow, Mgd})(8.34 \text{ lb/Mgal/mg/L})$$

Using values from above, the BOD_5 loading for the first stage filter is:

$$W = (240 \text{ mg/L})(1.0 \text{ Mgd})(8.34) = 2,002 \text{ lb } BOD_5/d \text{ (908.9 kg/d)}$$

2c. Compute the volume and diameter of the first stage filter
Therefore, the volume of the first stage trickling filter is calculated as follows:

$$71.2 = \frac{100}{1 + 0.0561\sqrt{\dfrac{2,002 \text{ lb/d}}{V(2.08)}}} \Rightarrow V = 18.51 \ 10^3 \text{ ft}^3 \text{ (523.8 m}^3\text{)}$$

Using the given depth of 7 feet (21.m), and a circular trickling filter, the area and diameter of the first stage filter are:

$$\text{Area} = \frac{\text{volume}}{\text{depth}} \Rightarrow \frac{18.51 \times 10^3 \text{ ft}^3}{7 \text{ ft}} = 2,644 \text{ ft}^2 \Rightarrow \text{diameter} = 58.02 \text{ ft (17.7 m)}$$

3. Analyze the second stage filter
The BOD_5 loading for the second stage trickling filter is calculated using:

$$W' = (1 - E_1)W$$

where W' = BOD_5 loading to the second stage filter
$$W' = (1 - 0.712)(2,002 \text{ lb/d}) = 577 \text{ lb } BOD_5/d \text{ (261.9 kg/d)}$$

The NRC equation for a second stage trickling filter is:

$$E_2 = \frac{100}{1 + \dfrac{0.0561}{1 - E_1}\sqrt{\dfrac{W'}{VF}}}$$

Using terms defined previously and values calculated above, the volume of the second stage trickling filter is:

$$E_2 = \frac{100}{1 + \dfrac{0.0561}{1 - 0.712}\sqrt{\dfrac{577 \text{ lb/d}}{V(2.08)}}} \Rightarrow V = 64.33 \ 10^3 \text{ ft}^3 \text{ (1.82 m}^3\text{)}$$

The area and diameter of the second stage filter are:

$$\text{Area} = \frac{64.33 \times 10^3 \text{ ft}^3}{7 \text{ ft}} = 9,190 \text{ ft}^2 \Rightarrow \text{diameter} = 108.17 \text{ ft (32.97 m)}$$

4. Compute the BOD_5 loading and hydraulic loading to each filter
The BOD_5 (organic) loading to each filter is calculated by dividing the BOD_5 loading by the volume of the filter in 10^3 ft^3 (m^3):

First stage filter: BOD$_5$ loading $= \dfrac{2{,}002 \text{ lb/d}}{18.51 \ 10^3 \text{ ft}^3} = 108.2 \ \dfrac{\text{lb}}{10^3 \text{ ft}^3 \cdot \text{d}}$

$$(1.74 \text{ kg/m}^3 \cdot \text{d})$$

Second stage filter: BOD$_5$ loading $= \dfrac{577 \text{ lb/d}}{64.33 \ 10^3 \text{ ft}^3} = 8.97 \ \dfrac{\text{lb}}{10^3 \text{ ft}^3 \cdot \text{d}}$

$$(0.14 \text{ kg/m}^3 \cdot \text{d})$$

BOD$_5$ loading for a first stage filter in a two stage system typically ranges from 60 to 120 lb/10^3 ft$^3 \cdot$ d (0.96 to 1.93 kg/m$^3 \cdot$ d). The second stage filter loading typically ranges from 5 to 20 lb/10^3 ft$^3 \cdot$ d (0.08 to 0.32 kg/m$^3 \cdot$ d).

The hydraulic loading to each filter is calculated as follows:

$$\text{Hydraulic loading} = \frac{(1 + R)(Q)}{(\text{Area})(1{,}440 \text{ min/d})}$$

First stage filter:

$$\text{Hydraulic loading} = \frac{(1 + 2)(1 \times 10^6 \text{ gal/day})}{(2{,}644 \text{ ft}^2)(1{,}440 \text{ min/d})}$$

$$= 0.79 \text{ gal/min} \cdot \text{ft}^2 \ (0.54 \text{ L/s} \cdot \text{m}^2)$$

Second stage filter:

$$\text{Hydraulic loading} = \frac{(1 + 2)(1 \times 10^6 \text{ gal/day})}{(9{,}190 \text{ ft}^2)(1{,}440 \text{ min/d})}$$

$$= 0.23 \text{ gal/min} \cdot \text{ft}^2 \ (0.156 \text{ L/s} \cdot \text{m}^2)$$

Hydraulic loading for two stage trickling filter systems typically ranges from 0.16 to 0.64 gal/min $\cdot$ ft^2 (0.11 to 0.43 L/s $\cdot$ m^2).

Related Calculations.   In practice, the diameter of the two filters should be rounded to the nearest 5 ft (1.52 m) to accommodate standard rotary distributor mechanisms. To reduce construction costs, the two trickling filters are often made the same size. When both filters in a two stage trickling filter system are the same size, the efficiencies will be unequal and the analysis will be an iterative one.

DESIGN OF A PLASTIC MEDIA TRICKLING FILTER

A municipal wastewater with a flow of 1.0 Mgd (694 gal/min) (3,785 m^3/d) and a BOD$_5$ of 240 mg/L is to be treated in a single stage plastic media trickling filter without recycle. The effluent wastewater is to have a BOD$_5$ of 20 mg/L. Determine the diameter of the filter, the hydraulic loading, the organic loading, the dosing rate, and the required rotational speed of the distributor arm. Assume a filter depth of 25 feet (7.6 m). Also assume that a treatability constant ($k_{20/20}$) of 0.075 (gal/min)$^{0.5}$/ft^2 was obtained in a 20 foot (6.1 m) high test filter at 20°C (68°F). The wastewater temperature is 30°C (86°F).

Calculation Procedure:

1. *Adjust the treatability constant for wastewater temperature and depth*
Due to the predictable properties of plastic media, empirical relationships are available to predict performance of trickling filters packed with plastic media. However,
the treatability constant must first be adjusted for both the temperature of the wastewater and the depth of the actual filter.

Adjustment for temperature.   The treatability constant is first adjusted from the
given standard at 20°C (68°F) to the actual wastewater temperature of 30°C (86°F)
using the following equation:

$$k_{30/20} = k_{20/20}\theta^{T-20}$$

where $k_{30/20}$ = Treatability constant at 30°C (86°F) and 20 foot (6.1 m) filter depth
  $k_{20/20}$ = Treatability constant at 20°C (68°F) and 20 foot (6.1 m) filter depth
  θ = Temperature activity coefficient (assume 1.035)
  T = Wastewater temperature

Using above values:

$$k_{30/20} = (0.075 \text{ (gal/min)}^{0.5}/\text{ft}^2)(1.035)^{30-20} = 0.106 \text{ (gal/min)}^{0.5}/\text{ft}^2$$

Adjustment for depth.   The treatability constant is then adjusted from the standard depth of 20 feet (6.1 m) to the actual filter depth of 25 feet (7.6 m) using the
following equation:

$$k_{30/25} = k_{30/20}(D_1/D_2)^x$$

where $k_{30/25}$ = Treatability constant at 30°C (86°F) and 25 foot (7.6 m) filter depth
  $k_{30/20}$ = Treatability constant at 30°C (86°F) and 20 foot (6.1 m) filter depth
  D_1 = Depth of reference filter (20 feet) (6.1 m)
  D_2 = Depth of actual filter (25 feet) (7.6 m)
  x = Empirical constant (0.3 for plastic medium filters)

Using above values:

$$k_{30/25} = (0.106 \text{ (gal/min)}^{0.5}/\text{ft}^2)(20/25)^{0.3}$$

$$= 0.099 \text{ (gal/min)}^{0.5}/\text{ft}^2 \ (0.099 \text{ (L/s)}^{0.5}/\text{m}^2)$$

2. *Size the plastic media trickling filter*
The empirical formula used for sizing plastic media trickling filters is:

$$\frac{S_e}{S_i} = \exp[-kD\,(Q_v)^{-n}]$$

where S_e = BOD_5 of settled effluent from trickling filter (mg/L)
  S_i = BOD_5 of influent wastewater to trickling filter (mg/L)
  k_{20} = Treatability constant adjusted for wastewater temperature and filter
     depth = $(k_{30/25})$
  D = Depth of filter (ft)
  Q_v = Volumetric flowrate applied per unit of filter area (gal/min · ft²) (L/s ·
     m²) = Q/A
  Q = Flowrate applied to filter without recirculation (gal/min) (L/s)

A = Area of filter (ft^2) (m^2)
n = Empirical constant (usually 0.5)

Rearranging and solving for the trickling filter area (A):

$$A = Q \left(\frac{-\ln(S_e/S_i)}{(k_{30/25})D} \right)^{1/n}$$

Using values from above, the area and diameter of the trickling filter are:

$$A = 694 \text{ gal/min} \left(\frac{-\ln(20/240)}{(0.099)25 \text{ ft}} \right)^{1/0.5} = 699.6 \text{ ft}^2 \Rightarrow \text{Diameter} = 29.9 \text{ ft (9.1 m)}$$

3. Calculate the hydraulic and organic loading on the filter
The hydraulic loading (Q/A) is then calculated:

Hydraulic loading = 694 gal/min/699.6 ft^2 = 0.99 gal/min · ft^2 (0.672 L/s · m^2)

For plastic media trickling filters, the hydraulic loading ranges from 0.2 to 1.20 gal/min · ft^2 (0.14 to 0.82 L/s · m^2).
The organic loading to the trickling filter is calculated by dividing the BOD$_5$ load to the filter by the filter volume as follows:

$$\text{Organic loading} = \frac{(1.0 \text{ Mgd})(240 \text{ mg/L})(8.34 \text{ lb} \cdot \text{L/mg} \cdot \text{Mgal})}{(699.6 \text{ ft}^2)(25 \text{ ft})(10^3 \text{ ft}^3/1000 \text{ ft}^3)}$$

$$= 114 \frac{\text{lb}}{10^3 \text{ ft}^3 \cdot \text{d}} \text{ (557 kg/m}^2 \cdot \text{d)}$$

For plastic media trickling filters, the organic loading ranges from 30 to 200 lb/10^3 ft^3 · d (146.6 to 977.4 kg/m^2 · d).

4. Determine the required dosing rate for the filter
To optimize the treatment performance of a trickling filter, there should be a continual and uniform growth of biomass and sloughing of excess biomass. To achieve uniform growth and sloughing, higher periodic dosing rates are required. The required dosing rate in inches per pass of distributor arm may be approximated using the following:

$$\text{Dosing rate} = (\text{organic loading, lb/10}^3 \text{ ft}^3 \cdot \text{d})(0.12)$$

Using the organic loading calculated above, the dosing rate is:

$$\text{Dosing rate} = (114 \text{ lb/10}^3 \text{ ft}^3 \cdot \text{d})(0.12) = 13.7 \text{ in/pass (34.8 cm/pass)}$$

Typical dosing rates for trickling filters are listed in Table 2. To achieve the typical dosing rates, the speed of the rotary distributor can be controlled by (1) reversing the location of some of the existing orifices to the front of the distributor arm, (2) adding reversed deflectors to the existing orifice discharges, and (3) by operating the rotary distributor with a variable speed drive.

5. Determine the required rotational speed of the distributor
The rotational speed of the distributor is a function of the instantaneous dosing rate and may be determined using the following:

TABLE 2 Typical Dosing Rates for Trickling
Filters

Organic loading lb $BOD_5/10^3$ $ft^3 \cdot d$ $(kg/m^2 \cdot d)$	Dosing rate, in/pass (cm/pass)
<25 (122.2)	3 (7.6)
50 (244.3)	6 (15.2)
75 (366.5)	9 (22.9)
100 (488.7)	12 (30.5)
150 (733.0)	18 (45.7)
200 (977.4)	24 (60.9)

*(Wastewater Engineering: Treatment, Disposal, and Re-
use,* Metcalf & Eddy, 3rd Ed.)

$$n = \frac{1.6(Q_T)}{(A)(DR)}$$

where n = Rotational speed of distributor (rpm)
  Q_T = Total applied hydraulic loading rate (gal/min · ft²) (L/s · m²) = $Q + Q_R$
  Q = Influent wastewater hydraulic loading rate (gal/min · ft²) (L/s · m²)
  Q_R = Recycle flow hydraulic loading rate (gal/min · ft²) (L/s · m²) Note: re-
    cycle is assumed to be zero in this example
  A = Number of arms in rotary distributor assembly
  DR = Dosing rate (in/pass of distributor arm)

Assuming two distributor arms (two or four arms are standard), and using values
from above, the required rotational speed is:

$$n = \frac{1.6(0.99 \text{ gal/min} \cdot \text{ft}^2)}{(2)(13.7 \text{ in/pass})} = 0.058 \text{ rpm}$$

This equates to one revolution every 17.2 minutes.

SIZING OF A ROTARY-LOBE SLUDGE PUMP

A municipal wastewater treatment facility produces approximately 6,230 gal/day
(23.6 m³/d) of aerobically digested sludge at 5 percent solids. The digested sludge
is pumped to a sludge dewatering facility which is operated 4 h/d, 7 d/wk; 1000
ft (304.8 m) of 3 in (76.3 mm) equivalent length pipe exists between the aerobic
digester and the dewatering facility. The equivalent length of pipe includes all
valves, fittings, discharge pipe, and suction pipe lengths in the sludge piping system.
The static head on the sludge pump is 10 feet (3.0 m). Size a rotary-lobe pump
transfer sludge from the digester to the dewatering facility, including pump discharge and head condition, rpm and motor horsepower (kW).

Calculation Procedure:

1. *Find the flow rate required for the sludge pump*
A schematic of the sludge handling system is illustrated in Fig. 9. Rotary-lobe
pumps are positive-displacement pumps in which two rotating synchronous lobes

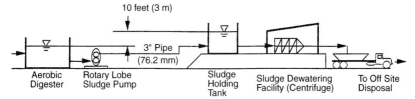

FIGURE 9   Sludge handling system.

push the fluid through the pump. Although these types of pumps are advertised as self-priming, they are generally located so as to have a suction head as shown in Fig. 9. A schematic of a rotary lobe pump is shown in Fig. 10.

The required flow rate for the sludge pump using the 4 h/d, 7 d/wk operation scheme is:

$$\text{Flow rate (gal/min)} = \frac{6{,}230 \text{ gal/day}}{(4 \text{ h/d})(60 \text{ min/h})} = 26 \text{ gal/min (1.64 L/s)}$$

2.   Compute the headloss in the piping system

The head loss through the piping system is calculated using the Hazen-Williams formula:

$$H = (0.2083) \left(\frac{100}{C}\right)^{1.85} \left(\frac{Q^{1.85}}{D^{4.8655}}\right) \left(\frac{L}{100}\right)$$

where H = Dynamic head loss for clean water, ft (m)
C = Hazen-Williams constant (use 100)
Q = Flow rate in pipe, gal/min (L/s)
D = Pipe diameter, in (mm)
L = Equivalent length of pipe, ft (m)

Note: There is an SI version of the Hazen-Williams formula; use it for SI calculations.

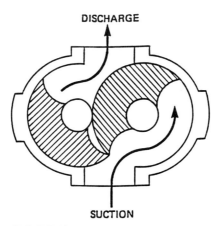

FIGURE 10   Rotary-lobe pump schematic.

Using values given above, the dynamic head loss in the piping system at the design flow rate of 26 gal/min (1.64 L/s) is:

$$H = (0.2083)\left(\frac{100}{100}\right)^{1.85}\left(\frac{26\ \text{gal/min})^{1.85}}{(3\ \text{in})^{4.8655}}\right)\left(\frac{1000\ \text{ft}}{100}\right) = 4.12\ \text{ft}\ (1.26\ \text{m})$$

This represents head loss in the sludge piping system for clean water only. To determine head loss when pumping sludge at 5 percent solids, a multiplication factor (k) is used. The value of k is obtained from empirical curves (see Fig. 11) for a given solids content and pipeline velocity. The head loss when pumping sludge is obtained by multiplying the head loss of water by the multiplication factor.

The velocity of 26 gal/min (1.64 L/s) in a 3 in (76.3 mm) diameter pipe is:

$$\text{Velocity} = \frac{26\ \text{gal/min}}{(7.48\ \text{gal/ft}^3)((\pi/4)(3\ \text{in}/12\ \text{in/ft})^2)(60\ \text{s/min})} = 1.2\ \text{ft/s}\ (0.37\ \text{m/s})$$

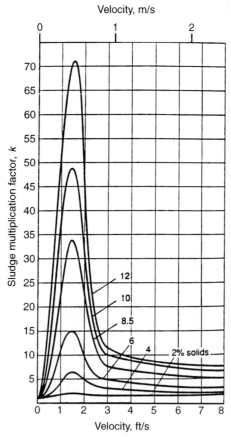

FIGURE 11   Head-loss multiplication factor for different pipe velocities vs. sludge concentration. (Metcalf & Eddy, *Wastewater Engineering: Treatment, Disposal, and Reuse*, 3rd Ed., McGraw-Hill.)

Using Fig. 11 with a velocity of 1.2 ft/s (0.37 m/s) and a solids content of 5 percent, the multiplication factor (k) is 12. Therefore, the dynamic head loss for this system when pumping 5 percent solids is:

$$\text{Dynamic Head Loss}_{5\%} = (\text{Dynamic Head Loss}_{water})(k)$$

$$=> \text{Dynamic Head Loss}_{5\%} = (4.12')(12) = 49.44 \text{ feet (15.1 m)}$$

Use 50 feet of head loss (15.2 m)

The total head loss is the sum of the static head and the dynamic head loss$_{5\%}$. Therefore, the total head loss (Total Dynamic Head or TDH) for the system is:

$$10' + 50' = 60 \text{ feet (18.3 m)}$$

This translates to a discharge pressure on the pump of:

$$\text{TDH} = (60')/(2.31 \text{ ft/psi}) = 26 \text{ lb/in}^2 \text{ (179.1 kPa)}$$

Therefore, the design condition for the rotary lobe pump is 26 gal/min (1.64 L/s) at 26 lb/in² (179.1 kPa).

3. Choose the correct pump for the application

At this point, a rotary lobe pump manufacturers catalog is required in order to choose the correct pump curve for this application. This is accomplished by choosing a pump performance curve that meets the above design condition. An example of a manufacturers curve that satisfies the design condition is shown in Fig. 12.

Plotting a horizontal line from 26 gal/min (1.64 L/s) on the left to the 26 lb/in² (179.1 kPa) pressure line and reading down gives a pump speed of approximately 175 rpm. This means that for this pump to deliver 26 gal/min (1.64 L/s) against a pressure of 26 lb/in² (179.1 kPa), it must operate at 175 rpm.

The motor horsepower required for the rotary lobe pump is calculated using the following empirical formula (taken from the catalog of Alfa Laval Pumps, Inc. of Kenosha, WI):

$$\text{Hp} = \frac{N}{7124} \left[(0.043)(q)(P) + \frac{(N)(S_f)}{1000} + \frac{(q)(N_f)}{26.42} \right]$$

Hp = Motor horsepower
N = Pump speed (rpm)
q = Pump displacement, gal/100 revolutions (L/100 revs)
P = Differential pressure or TDH, psi, (kPa)
S_f = Factor related to pump size which is calculated using: $S_f = (q)(0.757) + 3$
N_f = Factor related to the viscosity of the pumped liquid which is calculated using: $N_f = 2.2\sqrt[3]{\text{Viscosity (cp)}}$

The pump speed was found in Fig. 12 to be 175 rpm; the pump displacement is taken from the pump curve and is 25 gal/100 revs (94.6 L/100 revs); the differential pressure or TDH was calculated above to be 26 lb/in² (179.1 kPa); $S_f = (25 \text{ gal}/100 \text{ revs})(0.757) + 3 = 21.925$.

4. Determine the pump horsepower

The viscosity of a sewage sludge is dependent upon the percent solids and may be found using Table 3.

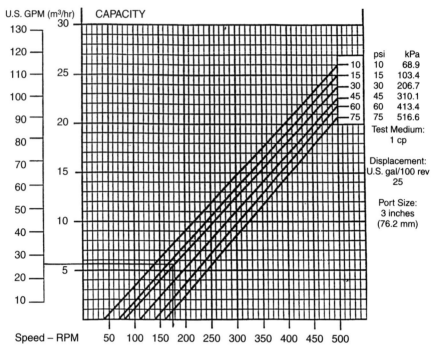

FIGURE 12 Pump performance curve. (*Alfa Laval Pumps, Inc.*)

TABLE 3 Viscosity of Sewage Sludge vs. Percent Solids

Percent solids	Viscosity in centipoise (cp)
1	10
2	80
3	250
4	560
5	1,050
6	1,760
7	2,750
8	4,000
9	5,500
10	7,500

(*Courtesy Alfa Laval Pumps, Inc.*)

Using Table 3, a 5 percent solids sludge has a viscosity of 1,050 cp. Therefore:

$$N_f = 2.2\sqrt[3]{1,050 \text{ cp}} = 22.36$$

Using the values outlined above, the motor horsepower is calculated:

$$Hp = \frac{175 \text{ rpm}}{7124} \left[(0.043)(25 \text{ gal}/100 \text{ revs})(26 \text{ psi}) + \frac{(175 \text{ rpm})(21.925)}{1000} + \frac{(25)(22.36)}{26.42} \right]$$

$$= 1.3 \text{ hp } (0.97 \text{ kW})$$

The resulting horsepower of 1.3 hp (0.97 kW) is at best an informed estimate of the horsepower which will be absorbed at the pump shaft during pumping. It includes no service factor or margin for error, neither does it allow for inefficiency in power transmission systems.

5. *Find the installed horsepower of the pump*
Actual installed horsepower will need to be greater than this calculated horsepower to allow for the random torque increases due to large solids and to provide a prudent safety factor. Therefore, the minimum installed motor horsepower is calculated as follows:

$$\text{Installed Motor hp} = \frac{(1.2)(\text{Calculated hp})}{\text{Motor Drive Efficiency (\%)}} \times 100$$

Assuming a motor efficiency of 90 percent, the installed motor horsepower is:

$$\text{Installed Motor hp} = \frac{(1.2)(1.3 \text{ hp})}{90\%} \times 100 = 1.73 \text{ hp } (1.29 \text{ kW})$$

Use the next largest motor size available, which is 2.0 hp (1.49 kW). Therefore, the rotary lobe sludge pump will have a 2.0 hp (1.49 kW) motor.

To summarize, the requirements for a rotary lobe pump for this application are: design condition of 26 gal/min (1.64 L/s) at 26 lb/in² (179.1 kPa), operating at 175 rpm with a 2.0 hp (1.49 kW) motor. Often a variable speed drive is provided so that the pump rpm (hence flow rate) may be adjusted to suit varying sludge flow rate demands at the downstream dewatering facility. Also, a second pump is generally provided as a spare.

DESIGN OF AN ANAEROBIC DIGESTOR

A high rate anaerobic digestor is to be designed to treat a mixture of primary and waste activated sludge produced by a wastewater treatment facility. The input sludge to the digester is 60,000 gal/day (227.1 m³/d) of primary and waste activated sludge with an average loading of 25,000 lb/d (11,350 kg/d) of ultimate BOD (BOD_L). Assume the yield coefficient (Y) is 0.06 lb VSS/lb BOD_L (kg/kg), and the endogenous coefficient (k_d) is 0.03 d⁻¹ at 35°C (95°F). Also assume that the efficiency of waste utilization in the digester is 60 percent. Compute the digester volume required, the volume of methane gas produced, the total volume of digester gas produced, and the percent stabilization of the sludge.

Calculation Procedure:

1. *Determine the required digester volume and loading*
Anaerobic digestion is one of the oldest processes used for the stabilization of sludge. It involves the decomposition of organic and inorganic matter in the absence

of molecular oxygen. The major applications of this process are in the stabilization of concentrated sludges produced from the treatment of wastewater.

In the anaerobic digestion process, the organic material is converted biologically, under anaerobic conditions, to a variety of end products including methane (CH_4) and carbon dioxide (CO_2). The process is carried out in an airtight reactor. Sludge, introduced continuously or intermittently, is retained in the reactor for varying periods of time. The stabilized sludge, withdrawn continuously or intermittently from the reactor, is reduced in organic and pathogen content and is nonputrescible.

In the high rate digestion process, as shown in Fig. 13, the contents of the digester are heated and completely mixed. For a complete-mix flow through digester, the mean cell residence time (θ_c) is the same as the hydraulic retention time (θ).

In the United States, the use and disposal of sewage sludge is regulated under 40 CFR Part 503 promulgated February 1993. The new regulation replaces 40 CFR Part 257—the original regulation governing the use and disposal of sewage sludge, in effect since 1979. The new regulations state that "for anaerobic digestion, the values for the mean-cell-residence time and temperature shall be between 15 days at 35°C (95°F) to 55°C (131°F) and 60 days at 20°C (68°F)."

Therefore, for an operating temperature of 35°C (95°F), a mean cell residence time of 15 days will be used. The influent sludge flow rate (Q) is 60,000 gal/day = 8,021 ft³/d (226.9 m³/d). The digester volume V required is computed using:

$$V = \theta_c Q \Rightarrow V = (15 \text{ d})(8,021 \text{ ft}^3/\text{d}) = 120,315 \text{ ft}^3 \ (3,404.9 \text{ m}^3)$$

The BOD entering the digester 25,000 lb/d (11,350 kg/d). Therefore, the volumetric loading to the digester is:

$$\text{Volumetric Loading} = \frac{25,000 \text{ lb BOD}/\text{d}}{120,315 \text{ ft}^3} = 0.21 \text{ lb/ft}^3 \cdot \text{d} \ (3.37 \text{ kg/m}^3 \cdot \text{d})$$

For high rate digesters, loadings range from 0.10 to 0.35 lb/ft³·d (1.6 to 5.61 kg /m³·d).

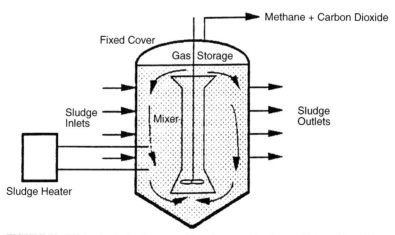

FIGURE 13 High-rate single-stage complete-mix anaerobic digester. (Adapted from Metcalf & Eddy, *Wastewater Engineering: Treatment, Disposal, and Reuse*, 3rd Ed., McGraw-Hill.)

Assuming 60 percent waste utilization, the BOD_L exiting the digester is:

$$(25,000 \text{ lb/d})(1 - 0.6) = 10,000 \text{ lb/d } (4,540 \text{ kg/d})$$

2. Compute the daily quantity of volatile solids produced
The quantity of volatile solids produced each day is computed using:

$$P_x = \frac{Y[(BOD_{in}, \text{ lb/d}) - (BOD_{out}, \text{ lb/d})]}{1 + k_d \theta_c}$$

where P_x = Volatile solids produced, lb/d (kg/d)
  Y = Yield coefficient (lb VSS/lb BOD_L)
  k_d = Endogenous coefficient (d^{-1})
  θ_c = Mean cell residence time (d)

Using values obtained above, the volatile solids produced each day are:

$$P_x = \frac{(0.06 \text{ lb VSS/lb } BOD_L)[25,000 \text{ lb/d}) - (10,000 \text{ lb/d})]}{1 + (0.03 \text{ d}^{-1})(15 \text{ d})}$$

$$= 621 \text{ lb/d } (281.9 \text{ kg/d})$$

3. Determine the volume of methane produced
The volume of methane gas produced at standard conditions (32°F and 1 atm) (0°C and 101.3 kPa) is calculated using:

$$V_{CH_4} = 5.62 \text{ ft}^3/\text{lb}[(BOD_{in}, \text{ lb/d}) - (BOD_{out}, \text{ lb/d}) - 1.42 P_x]$$

where V_{CH_4} = Volume of methane gas produced at standard conditions (ft^3/d) (m^3/d)

Using values obtained above:

$$V_{CH_4} = 5.62 \text{ ft}^3/\text{lb}[(25,000 \text{ lb/d}) - (10,000 \text{ lb/d}) - 1.42(621 \text{ lb/d})]$$

$$= 79,344 \text{ ft}^3/\text{d } (2,245 \text{ m}^3/\text{d})$$

Since digester gas is approximately 2/3 methane, the volume of digester gas produced is:

$$(79,344 \text{ ft}^3/\text{d})/0.67 = 118,424 \text{ ft}^3/\text{d } (3,351.4 \text{ m}^3/\text{d})$$

4. Calculate the percent stabilization
Percent stabilization is calculated using:

$$\% \text{ Stabilization} = \frac{[(BOD_{in}, \text{ lb/d}) - (BOD_{out}, \text{ lb/d}) - 1.42 P_x]}{BOD_{in}, \text{ lb/d}} \times 100$$

Using values obtained above:

$$\% \text{ Stabilization} = \frac{[(25,000 \text{ lb/d}) - (10,000 \text{ lb/d}) - 1.42(621 \text{ lb/d})]}{25,000 \text{ lb/d}}$$

$$\times 100 = 56.5\%$$

Related Calculations. This disadvantages and advantages of the anaerobic treatment of sludge, as compared to aerobic treatment, are related to the slow growth rate of the methanogenic (methane producing) bacteria. Slow growth rates require a relatively long retention time in the digester for adequate waste stabilization to occur. With methanogenic bacteria, most of the organic portion of the sludge is converted to methane gas, which is combustible and therefore a useful end product. If sufficient quantities of methane gas are produced, the methane gas can be used to operate duel-fuel engines to produce electricity and to provide building heat.

DESIGN OF A CHLORINATION SYSTEM FOR WASTEWATER DISINFECTION

Chlorine is to be used for disinfection of a municipal wastewater. Estimate the chlorine residual that must be maintained to achieve a coliform count equal to or less than 200/100 ml in an effluent from an activated sludge facility assuming that the effluent requiring disinfection contains a coliform count of 10^7/100 ml. The average wastewater flow requiring disinfection is 0.5 Mgd (1,892.5 m³/d) with a peaking factor of 2.8. Using the estimated residual, determine the capacity of the chlorinator. Per regulations, the chlorine contact time must not be less than 15 minutes at peak flow.

Calculation Procedure:

1. Find the required residual for the allowed residence time
The reduction of coliform organisms in treated effluent is defined by the following equation:

$$\frac{N_t}{N_0} = (1 + 0.23C_t t)^{-3}$$

where N_t = Number of coliform organisms at time t
N_0 = Number of coliform organisms at time t_0
C_t = Total chlorine residual at time t (mg/L)
t = Residence time (min)

Using values for coliform count from above:

$$\frac{2 \times 10^2}{1 \times 10^7} = (1 + 0.23C_t t)^{-3}$$

$$\Rightarrow 2.0 \times 10^{-5} = (1 + 0.23C_t t)^{-3}$$

$$\Rightarrow 5.0 \times 10^4 = (1 + 0.23C_t t)^{3}$$

$$\Rightarrow 1 + 0.23C_t t = 36.84$$

$$\Rightarrow C_t t = 155.8 \text{ mg} \cdot \text{min/L}$$

For a residence time of 15 minutes, the required residual is:

$$C_t = (155.8 \text{ mg} \cdot \text{min/L})/15 \text{ min} = 10.38 \text{ mg/L}$$

The chlorination system should be designed to provide chlorine residuals over a range of operating conditions and should include an adequate margin of safety. Therefore, the dosage required at peak flow will be set at 15 mg/L.

2. Determine the required capacity of the chlorinator
The capacity of the chlorinator at peak flow with a dosage of 15 mg/L is calculated using:

$$Cl_2 \text{ (lb/d)} = \text{(Dosage, mg/L)(Avg flow, Mgd)(P.F.)(8.34)}$$

where:
Cl_2 = Pounds of chlorine required per day (kg/d)
Dosage = Dosage used to obtain coliform reduction
Avg. Flow = Average flow
P.F. = Peaking factor for average flow
8.34 = 8.34 lb · L/Mgal · mg

Using values from above:

$$Cl_2 \text{ (lb/d)} = \text{(15 mg/L)(0.5 Mgd)(2.8)(8.34)} = 175 \text{ lb/d (79.5 kg/d)}$$

The next largest standard chlorinator size is 200 lb/d (90.8 kg/d). Therefore, two 200 lb/d (90.8 kg/d) chlorinators will be used with one serving as a spare. Although the peak capacity will not be required during most of the day, it must be available to meet chlorine requirements at peak flow.

3. Compute the daily consumption of chlorine
The average daily consumption of chlorine assuming an average dosage of 15 mg /L is:

$$Cl_2 \text{ (lb/d)} = \text{(15 mg/L)(0.5 Mgd)(8.34)} = 62.5 \text{ lb/d (28.4 kg/d)}$$

A typical chlorination flow diagram is shown in Fig. 14. This is a compound

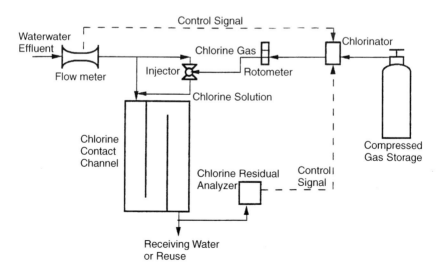

FIGURE 14   Compound-loop chlorination system flow diagram.

loop system which means the chlorine dosage is controlled through signals received from both effluent flow rate and chlorine residual.

SANITARY SEWER SYSTEM DESIGN

What size main sanitary sewer is required for a midwestern city 30-acre (1.21×10^5 m^2) residential area containing six-story apartment houses if the hydraulic gradient is 0.0035 and the pipe roughness factor $n = 0.013$? One-third of the area is served by a branch sewer. What should the size of this sewer be? If the branch and main sewers must also handle groundwater infiltration, determine the required sewer size. The sewer is below the normal groundwater level.

Calculation Procedure:

1.   *Compute the sanitary sewage flow rate*
Table 4 shows the typical population per acre for various residential areas and the flow rate used in sewer design. Using the typical population of 500 persons per acre (4046 m^2) given in Table 4, we see the total population of the area served is (30 acres)(500 persons per acres) = 15,000 persons.

Since this is a midwestern city, the sewer design basis, per capita, used for Des Moines, Iowa, 200 gal/day (8.76 mL/s), Table 4, appears to be an appropriate value. Checking with the minimum flow recommended in Table 4, 100 gal/day (4.38 mL/s), we see the value of 200 gal/day (8.76 mL/s) seems to be well justified. Hence, the sanitary sewage flow rate that the main sewer must handle is (15,000 persons)(200 gal/day) = 3,000,000 gal/day.

2.   *Convert the flow rate to cfs*
Use the relation $cfs = 1.55 (gpd/10^6)$, where cfs = flow rate, ft^3/s; gpd = flow rate, gal/24 h. so $cfs = 1.55(3,000,000/1,000,000) = 465$ ft^3/s (0.13 m^3/s).

3.   *Compute the required size of the main sewer*
Size the main sewer on the basis of its flowing full. This is the usual design procedure followed by experienced sanitary engineers.

Two methods can be used to size the sewer pipe. (*a*) Use the chart in Fig. 4, Section 14, for the Manning formula, entering with the flow rate of 4.65 ft^3/s (0.13 m^3/s) and projecting to the slope ratio or hydraulic gradient of 0.0035. Read the required pipe diameter as 18 in (457 mm).

(*b*) Use the Manning formula and the appropriate *conveyance factor* from Table 5. When the conveyance factor C_f is used, the Manning formula becomes $Q = C_f S^{1/2}$, where Q = flow rate through the pipe, ft^3/s; C_f = conveyance factor corresponding to a specific n value listed in Table 5; S = pipe slope or hydraulic gradient, ft/ft. Since Q and S are known, substitute and solve for C_f, or $C_f = Q/S^{1/2} = 4.65/(0.0035)^{1/2} = 78.5$. Enter Table 5 at $n = 0.013$ and $C_f = 78.5$, and project to the exact or next higher value of C_f. Table 5 shows that C_f is 64.70 for 15-in (381-mm) pipe and 105.1 for 18-in (457-mm) pipe. Since the actual value of C_f is 78.5, a 15-in (381-mm) pipe would be too small. Hence, an 18-in (457-mm) pipe would be used. This size agrees with that found in procedure *a*.

TABLE 4 Sanitary Sewer Design Factors

Population data		
	Typical population	
Type of area	Per acre	Per km²
Light residential	15	3,707
Closely built residential	55	13,591
Single-family residential	100	24,711
Six-story apartment district	500	123,555

Sewage-flow data		
	Sewer design basis, per capita	
City	gal/day	mL/s
Berkeley, California	92	4.03
Cranston, Rhode Island	167	7.32
Des Moines, Iowa	200	8.76
Las Vegas, Nevada	250	10.95
Little Rock, Arkansas	100	4.38
Shreveport, Louisiana	150	6.57

Typical sewer design practice		
	Design flow, per capita	
Sewer type	gal/day	mL/s
Laterals and submains	400	17.5
Main, trunks, and outfall	250	10.95
New sewers	Never 100	Never 4.38

TABLE 5 Manning Formula Conveyance Factor

Pipe diameter, in (mm)	Pipe cross-sectional area, ft² (m²)	*n*			
		0.011	0.013	0.015	0.017
6 (152)	0.196 (0.02)	6.62	5.60	4.85	4.28
8 (203)	0.349 (0.03)	14.32	12.12	10.50	9.27
10 (254)	0.545 (0.05)	25.80	21.83	18.92	16.70
12 (305)	0.785 (0.07)	42.15	35.66	30.91	27.27
15 (381)	1.227 (0.11)	76.46	64.70	56.07	49.48
18 (457)	1.767 (0.16)	124.2	105.1	91.04	80.33
21 (533)	2.405 (0.22)	187.1	158.3	137.2	121.1

4. Compute the size of the lateral sewer

The lateral sewer serves one-third of the total area. Since the total sanitary flow from the entire area is 4.65 ft³/s (0.13 m³/s), the flow from one-third of the area, given an even distribution of population and the same pipe slope, is 4.65/3 = 1.55 ft³/s (0.044 m³/s). Using either procedure in step 3, we find the required pipe size = 12 in (305 mm). Hence, three 12-in (302-mm) laterals will discharge into the main sewer, assuming that each lateral serves an equal area and has the same slope.

5. Check the suitability of the main sewer size

Compute the value of $d^{2.5}$ for each of the lateral sewer pipes discharging into the main sewer pipe. Thus, for one 12-in (305-mm) lateral line, where d = smaller pipe diameter, in, $d^{2.5} = 12^{2.5} = 496$. For three pipes of equal diameter, $3d^{2.5} = 1488 = D^{2.5}$, where D = larger pipe diameter, in. Solving gives $D^{2.5} = 1488$ and $D = 17.5$ in (445 mm). Hence, the 18-in (4570 mm) sewer main has sufficient capacity to handle the discharge of three 12-in (305-mm) sewers. Note that Fig. 4, Section 14, shows that the flow velocity in both the lateral and main sewers exceeds the minimum required velocity of 2 ft/s (0.6 m/s).

6. Compute the sewer size with infiltration

Infiltration is the groundwater that enters a sewer. The quantity and rate of infiltration depend on the character of the soil in which the sewer is laid, the relative position of the groundwater level and the sewer, the diameter and length of the sewer, and the material and care with which the sewer is constructed. With tile and other jointed sewers, infiltration depends largely on the type of joint used in the pipes. In large concrete or brick sewers, the infiltration depends on the type of waterproofing applied.

Infiltration is usually expressed in gallons per day per mile of sewer. With very careful construction, infiltration can be kept down to 5000 gal/(day · mi) [0.14 L/(km · s)] of pipe even when the groundwater level is above the pipe. With poor construction, porous soil, and high groundwater level, infiltration may amount to 100,000 gal/(day · mi) [2.7 L/(km · s)] or more. Sewers laid in dense soil where the groundwater level is below the sewer do not experience infiltration except during and immediately after a rainfall. Even then, the infiltration will be in small amounts.

Assuming an infiltration rate of 20,000 gal/(day · mi) [0.54 L/(km · s)] of sewer and a sewer length of 1.2 mi (1.9 km) for this city, we see the daily infiltration is 1.2 (20,000) = 24,000 gal (90,850 L).

Checking the pipe size by either method in step 3 shows that both the 12-in (305-mm) laterals and the 18-in (457-mm) main are of sufficient size to handle both the sanitary and infiltration flow.

Related Calculations. Where a sewer must also handle the runoff from fire-fighting apparatus, compute the quantity of fire-fighting water for cities of less than 200,000 population from $Q = 1020(P)^{0.5} [1 - 0.01(P)^{0.5}]$, where Q = fire demand, gal/min; P = city population in thousands. Add the fire demand to the sanitary sewage and infiltration flows to determine the maximum quantity of liquid the sewer must handle. For cities having a population of more than 200,000 persons, consult the fire department headquarters to determine the water flow quantities anticipated.

Some sanitary engineers apply a demand factor to the average daily water requirements per capita before computing the flow rate into the sewer. Thus, the maximum monthly water consumption is generally about 125 percent of the average annual demand but may range up to 200 percent of the average annual demand. Maximum daily demands of 150 percent of the average annual demand and maximum hourly demands of 200 to 250 percent of the annual average demand are commonly used for design by some sanitary engineers. To apply a demand factor, simply multiply the flow rate computed in step 2 by the appropriate factor. Current practice in the use of demand factors varies; sewers designed without demand factors are generally adequate. Applying a demand factor simply provides a margin of safety in the design, and the sewer is likely to give service for a longer period before becoming overloaded.

Most local laws and many sewer authorities recommend that no sewer be less than 8 in (203 mm) in diameter. The sewer should be sloped sufficiently to give a flow velocity of 2 ft/s (0.6 m/s) or more when flowing full. This velocity prevents the deposit of solids in the pipe. Manholes serving sewers should not be more than 400 ft (121.9 m) apart.

Where industrial sewage is discharged into a sanitary sewer, the industrial flow quantity must be added to the domestic sewage flow quantity before the pipe size is chosen. Swimming pools may also be drained into sanitary sewers and may cause temporary overflowing because the sewer capacity is inadequate. The sanitary sewage flow rate from an industrial area may be less than from a residential area of the same size because the industrial population is smaller.

Many localities and cities restrict the quantity of commercial and industrial sewage that may be discharged into public sewers. Thus, one city restricts commercial sewage from stores, garages, beauty salons, etc., to 135 gal/day per capita. Another city restricts industrial sewage from factories and plants to 50,000 gal/(day·acre) [0.55 mL/(m·s)]. In other cities each proposed installation must be studied separately. Still other cities prohibit any discharge of commercial or industrial sewage into sanitary sewers. For these reasons, the local authorities and sanitary codes, if any, must be consulted before the design of any sewer is begun.

Before starting a sewer design, do the following: (a) Prepare a profile diagram of the area that will be served by the sewer. Indicate on the diagram the elevation above grade of each profile. (b) Compile data on the soil, groundwater level, type of paving, number and type of foundations, underground services (gas, electric, sewage, water supply, etc.), and other characteristics of the area that will be served by the sewer. (c) Sketch the main sewer and lateral sewers on the profile diagram. Indicate the proposed direction of sewage flow by arrows. With these steps finished, start the sewer design.

To design the sewers, proceed as follows: (a) Size the sewers using the procedure given in steps 1 through 6 above. (b) Check the sewage flow rate to see that it is 2 ft/s (0.56 m/s) or more. (c) Check the plot to see that the required slope for the pipes can be obtained without expensive blasting or rock removal.

Where the outlet of a building plumbing system is below the level of the sewer serving the building, a pump must be used to deliver the sewage to the sewer. Compute the pump capacity, using the discharge from the various plumbing fixtures in the building as the source of the liquid flow to the pump. The head on the pump is the difference between the level of the sewage in the pump intake and the centerline of the sewer into which the pump discharges, plus any friction losses in the piping.

SELECTION OF SEWAGE-TREATMENT METHOD

A city of 100,000 population is considering installing a new sewage-treatment plant. Select a suitable treatment method. Local ordinances required that suspended matter in the sewage be reduced 80 percent, that bacteria be reduced 60 percent, and that the biochemical oxygen demand be reduced 90 percent. The plant will handle only domestic sanitary sewage. What are the daily oxygen demand and the daily suspended-solids content of the sewage? If an industrial plant discharges into this system sewage requiring 4500 lb (2041.2 kg) of oxygen per day, determine the population equivalent of the industrial sewage.

Calculation Procedure:

1. Compute the daily sewage flow
With an average flow per capita of 200 gal/day (8.8 mL/s), this sewage treatment plant must handle per capita (200 gal/day)(100,000 population) = 20,000 gal/day (896.2 L/s).

2. Compute the sewage oxygen demand
Usual domestic sewage shows a 5-day oxygen demand of 0.12 to 0.17 lb/day (0.054 to 0.077 kg/day) per person. With an average of 0.15 lb (0.068 kg) per person per day, the daily oxygen demand of the sewage is (0.15)(100,000) = 15,000 lb/day (78/7 g/s).

3. Compute the suspended-solids content of the sewage
Usual domestic sewage contains about 0.25 lb (0.11 kg) of suspended solids per person per day. Using this average, we see the total quantity of suspended solids that must be handled is (0.25)(100,000) = 25,000 lb/day (0.13 kg/s).

4. Select the sewage-treatment method
Table 6 shows the efficiency of various sewage-treatment methods. Since the desired reduction in suspended matter, biochemical oxygen demand (BOD), and bacteria is known, this will serve as a guide to the initial choice of the equipment.

Study of Table 6 shows that a number of treatments are available which will reduce the suspended matter by 80 percent. Hence, any one of these methods might be used. The same is true for the desired reduction in bacteria and BOD. Thus, the system choice resolves to selection of the most economical group of treatment units.

For a city of this size, four steps of sewage treatment would be advisable. The first step, *preliminary* treatment, could include screening to remove large suspended solids, grit removal, and grease removal. The next step, *primary treatment*, could include sedimentation or chemical precipitation. *Secondary treatment*, the next step, might be of a biological type such as the activated-sludge process or the trickling filter. In the final step, the sewage might be treated by cholorination. Treated sewage can then be disposed of in fields, streams, or other suitable areas.

Choose the following units for this sewage-treatment plant, using the data in Table 6 as a guide: rocks or screens to remove large suspended solids, grit chambers to remove grit, skimming tanks for grease removal, plain sedimentation, activated-sludge process, and chlorination.

Reference to Table 6 shows that screens and plain sedimentation will reduce the suspended solids by the desired amount. Likewise, the activated-sludge process reduces the BOD by up to 95 percent and the bacteria up to 95+ percent. Hence, the chosen system satisfies the design requirements.

5. Compute the population equivalent of the industrial sewage
Use the relation $P_e = R/D$, where P_e = population equivalent of the industrial sewage, persons; R = required oxygen of the sewage, lb/day; D = daily oxygen demand, lb per person per day. So $P_e = 4500/0.15 = 30,000$ persons.

TABLE 6 Typical Efficiencies of Sewage-Treatment Methods*

Treatment	Percentage reduction		
	Suspended matter	BOD	Bacteria
Fine screens	5–20	. . .	10–20
Plain sedimentation	35–65	25–40	50–60
Chemical precipitation	75–90	60–85	70–90
Low-rate trickling filter, with pre- and final sedimentation	70–90+	75–90	90+
High-rate trickling filter with pre- and final sedimentation	70–90	65–95	70–95
Conventional activated sludge with pre- and final sedimentation	80–95	80–95	90–95+
High-rate activated sludge with pre- and final sedimentation	70–90	70–95	80–95
Contact aeration with pre- and final sedimentation	80–95	80–95	90–95+
Intermittent sand filtration with presedimentation	90–95	85–95	95+
Chlorination:			
Settled sewage	. . .	†	90–95
Biologically treated sewage	. . .	†	98–99

*Steel—*Water Supply and Sewerage*, McGraw-Hill.
†Reduction is dependent on dosage.

Related Calculations. Where sewage is combined (i.e., sanitary and storm sewage mixed), the 5-day per-capita oxygen demand is about 0.25 lb/day (0.11 kg/day). Where large quantities of industrial waste are part of combined sewage, the per-capita oxygen demand is usually about 0.5 lb/day (0.23 kg/day). To convert the strength of an industrial waste to the same base used for sanitary waste, apply the population equivalent relation in step 5. Some cities use the population equivalent as a means of evaluating the load placed on the sewage-treatment works by industrial plants.

Table 7 shows the products resulting from various sewage-treatment processes per million gallons of sewage treated. The tabulated data are useful for computing the volume of product each process produces.

Environmental considerations are leading to the adoption of biogas methods to handle the organic fraction of municipal solid waste (MSW). Burning methane-rich biogas can meet up to 60 percent of the operating cost of waste-to-energy plants. Further, generating biogas avoids the high cost of disposing of this odorous by-product. A further advantage is that biogas plants are exempt from energy or carbon taxes. Newer plants also handle industrial wastes, converting them to biogas.

Biogas plants are popular in Europe. The first anaerobic digestion plant capable of treating unsorted MSW handles 55,000 mt/yr. It treats wastestreams with solids contents of 30 to 35 percent. Automated sorting first removes metals, plastics, paperboard, glass, and inerts from the MSW stream. The remaining organic fraction is mixed with recycled water from a preceding compost-drying press to form a 30 to 35 percent solids sludge which is pumped into one of the plant's three 2400-m³ (84,720 ft³) disgesters.

Residence time in the digester is about 3 weeks with a biogas yield of 99 m³/mt of MSW (3495 ft³/t), or 146 m³/mt (5154 ft³/t) of sorted organic fraction.

TABLE 7 Sludge and Other Products of Sewage-Treatment Processes per Million Gallons of Sewage Treated*

Data	Treatment process							
	Racks	Fine screens	Grit chambers	Plain sedimentation	Septic tanks	Imhoff or separate tanks	Activated sludge	Trickling filter humus tanks
Character of product	Screenings	Screenings	Grit	Raw sludge	Digested sludge	Digested sludge	Raw sludge	Raw sludge
Average amount per million gallons	4–8 ft³ (0.11–0.23 m³)	10–30 ft³ (0.28–0.83 m³)	2.5 ft³ (0.07 m³)	2500 gal (9462.5 L)	900 gal (3406.5 L)	500 gal (1892.5 L)	13,500 gal (51,098 L)	500 gal (1892.5 L)
Average moisture content, percent	80	80	15	95	90	85	99	92.5
Specific gravity				1020	1040	1040	1005	1025
Usual disposal methods	Burying, burning, or shredding and digestion with sludge	Burying, burning, or digesting with sludge	Filling land	Processing, digestion, or drying	Drying	Drying	Processing, digestion, or lagooning	Digestion and drying

*O'Rourke—*General Engineering Handbook*, McGraw-Hill.

Overflow liquid from the digester is pressed, graded, and sold as compost. Mixtures of MSW, sewage sludges, and animal slurries can also be digested in this process (Valorga) developed by Valorga SA (Vendargues, France). This is termed a *dry* process.

Wet processes handle wastestreams with only 10 to 15 percent solids content. Featuring more than one digestion stage, it is easier to control parameters such as pH and solids concentration than dry fermentation. The first plant to use wet digestion to process MSW is 20,000-mt/yr installation in Denmark. About two-thirds of the annual operating cost of $2 million is recovered through the sale of biogas. In a 14,000-mt/yr plant in Finland the biogas produced is used to fire a gas turbine. Multiple stages are said to make wet fermentation 65 percent faster than single-stage processes, with a 50 percent higher gas yield.

These developments show that sanitary engineers will be more concerned than ever with the environmental aspects of their designs. With the world population growing steadily every year and the longer lifespan of older individuals, biogas and similar recovery-conversion processes will become standard practice in every major country.

The data on biogas given above was reported in *Chemical Engineering.*

SECTION 14
WATER-SUPPLY AND STORM-WATER SYSTEM DESIGN

Water-Well Analysis

DETERMINING THE DRAWDOWN FOR GRAVITY WATER-SUPPLY WELL

Determine the depth of water in a 24-in (61-cm) gravity well, 300 ft (91-m) deep, without stopping the pumps, while the well is discharging 400 gal/min (25.2 L/s). Tests show that the drawdown in a test borehole 80 ft (24.4 m) away is 4 ft (1.2 m), and in a test borehole 20 ft (6.1 m) away, it is 18 ft (5.5 m). The distance to the static groundwater table is 54 ft (16.5 m).

Calculation Procedure:

1. Determine the key parameters of the well
Figure 1 shows a typical gravity well and the parameters associated with it. The Dupuit formula, given in step 2, below, is frequently used in analyzing gravity wells. Thus, from the given data, $Q = 400$ gal/min (25.2 L/s); $h_e = 300 - 54 = 246$ ft (74.9 m); $r_w = 1$ (0.3 m) for the well, and 20 and 80 ft (6.1 and 24.4 m), respectively, for the boreholes. For this well, h_w is unknown; in the nearest borehole it is $246 - 18 = 228$ ft (69.5 m); for the farthest borehole it is $246 - 4 = 242$ ft (73.8 m). Thus, the parameters have been assembled.

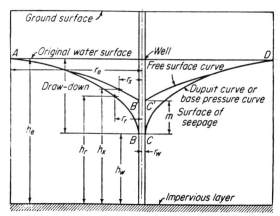

FIGURE 1   Hypothetical conditions of underground flow into a gravity well. (*Babbitt, Doland, and Cleasby.*)

2.   Solve the Dupuit formula for the well

Substituting in the Dupuit formula

$$Q = K \frac{h_e^2 - h_w^2}{\log_{10}(r_e/r_w)} = K \frac{(h_e - h_w)(h_e + h_w)}{\log_{10}(r_e/r_w)}$$

we have,

$$300 = K \frac{(246 + 228)(246 - 228)}{\log_{10}(r_e/20)} = K \frac{(246 + 242)(246 - 242)}{\log_{10}(r_e/80)}$$

Solving, $r_e = 120$ and $K = 0.027$. Then, for the well,

$$300 = 0.027 \frac{(246 + h_w)(246 - h_w)}{\log_{10}(120/1)}$$

Solving $h_w = 195$ ft (59.4 m). The drawdown in the well is $246 - 195 = 51$ ft (15.5 m).

   Related Calculations.   The graph resulting from plotting the Dupuit formula produces the "base-pressure curve," line ABCD in Fig. 1. It has been found in practice that the approximation in using the Dupuit formula gives results of practical value. The results obtained are most nearly correct when the ratio of drawdown to the depth of water in the well, when not pumping, is low.

   Figure 1 is valuable in analyzing both the main gravity well and its associated boreholes. Since gravity wells are, Fig. 2, popular sources of water supply through-out the world, an ability to analyze their flow is an important design skill. Thus, the effect of the percentage of total possible drawdown on the percentage of total possible flow from a well, Fig. 3, is an important design concept which finds wide use in industry today. Gravity wells are highly suitable for supplying typical weekly water demands, Fig. 4, of a moderate-size city. They are also suitable for most industrial plants having modest process-water demand.

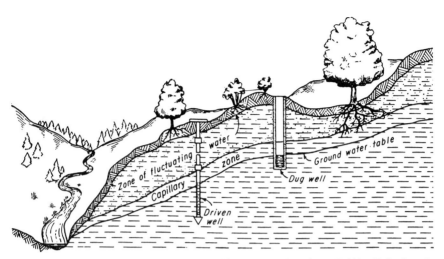

FIGURE 2 Relation between groundwater table and ground surface. (*Babbitt, Doland, and Cleasby.*)

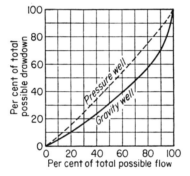

FIGURE 3 The effect of the percentage of total possible drawdown on the percentage of total possible flow from a well. (*Babbitt, Doland, and Cleasby.*)

This procedure is the work of Harold E. Babbitt, James J. Doland, and John L. Cleasby, as reported in their book, *Water Supply Engineering,* McGraw-Hill.

FINDING THE DRAWDOWN OF A DISCHARGING GRAVITY WELL

A gravity well 12 in (30.5 cm) in diameter is discharging 150 gal/min (9.5 L/s), with a drawdown of 10 ft (3 m). It discharges 500 gal/min (31.6 L/s) with a

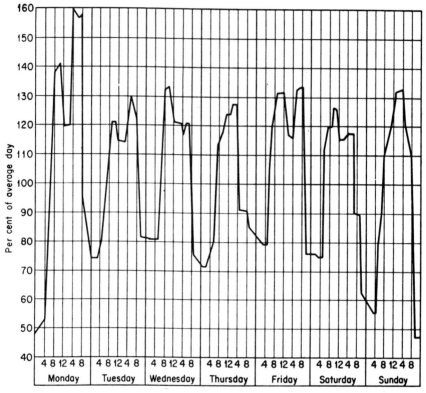

FIGURE 4 Demand curve for a typical week for a city of 100,000 population. (*Babbitt, Doland, and Cleasby.*)

drawdown of 50 ft (15 m). The static depth of the water in the well is 150 ft (45.7 m). What will be the discharge from the well with a drawdown of 20 ft (6 m)?

Calculation Procedure:

1. Apply the Dupuit formula to this well

Using the formula as given in the previous calculation procedure, we see that:

$$150 = K \frac{(10)(290)}{\log_{10}(150C/0.5)} \quad \text{and} \quad 500 = K \frac{(50)(250)}{\log_{10}(500C/0.5)}$$

Solving for C and K we have:

$$C = 0.21 \quad \text{and} \quad K = \frac{(500)(\log 210)}{12,500} = 0.093;$$

then
$$Q = 0.093 \frac{(20)(280)}{\log_{10}(0.210Q/0.5)}$$

2. Solve for the water flow by trial

Solving by successive trial using the results in step 1, we find $Q = 257$ gal/min (16.2 L/s).

Related Calculations. If it is assumed, for purposes of convenience in computations, that the radius of the circle of influence, r_e, varies directly as Q for equilibrium conditions, then $r_e = CQ$. Then the Dupuit equation can be rewritten as

$$Q = K \frac{(h_e + h_w)(h_e - h_w)}{\log_{10}(CQ/r_w)}$$

From this rewritten equation it can be seen that where the drawdown $(h_e - h_w)$ is small compared with $(h_e + h_w)$ the value of Q varies approximately as $(h_e - h_w)$. This straight-line relationship between the rate of flow and drawdown leads to the definition of the *specific capacity* of a well as the rate of flow per unit of drawdown, usually expressed in gallons per minute per foot of drawdown (liters per second per meter). Since the relationship is not the same for all drawdowns, it should be determined for one special foot (meter), often the first foot (meter) of drawdown. The relationship is shown graphically in Fig. 3 for both gravity, Fig. 1, and pressure wells, Fig. 5. Note also that since K in different aquifers is not the same, the specific capacities of wells in different aquifers are not always comparable.

It is possible, with the use of the equation for Q above, to solve some problems in gravity wells by measuring two or more rates of flow and corresponding drawdowns in the well to be studied. Observations in nearby test holes or boreholes are unnecessary. The steps are outlined in this procedure.

This procedure is the work of Harold E. Babbitt, James J. Doland, and John L. Cleasby, as reported in their book, *Water Supply Engineering,* McGraw-Hill. SI values were added by the handbook editor.

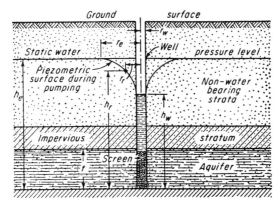

FIGURE 5 Hypothetical conditions for flow into a pressure well. (*Babbitt, Doland, and Cleasby.*)

ANALYZING DRAWDOWN AND RECOVERY FOR WELL PUMPED FOR EXTENDED PERIOD

Construct the drawdown-recovery curve for a gravity well pumped for two days at 450 gal/min (28.4 L/s). The following observations have been made during a test of the well under equilibrium conditions: diameter, 2 ft (0.61 m); h_e = 50 ft (15.2 m); when Q = 450 gal/min (28.4 L/s), drawdown = 8.5 ft (2.6 m); and when r_x = 60 ft (18.3 m), $(h_e - h_x)$ = 3 ft (0.91 m). The specific yield of the well is 0.25.

Calculation Procedure:

1. Determine the value of the constant k
Use the equation

$$Q = \frac{k(h_e - h_x)h_e}{C_x \log_{10} (r_e/0.1h_e)} \quad \text{and} \quad k = \frac{QC_x \log_{10} (r_e/0.1h_e)}{(h_e - h_x)(h_e)}$$

Determine the value of C_x when r_w is equal to the radius of the well, in this case 1.0. The value of k can be determined by trial. Further, the same value of k must be given when $r_x = r_e$ as when r_x = 60 ft (18.3 m). In this procedure, only the correct assumed value of r_e is shown—to save space.

Assume that r_e = 350 ft (106.7 m). Then, $1/350$ = 0.00286 and, from Fig. 6, C_x = 0.60. Then k = (1)(0.60)(log 350/5)/(8)(50) = (1)(0.6)(1.843)/400 = 0.00276, r_x/r_e = 60/350 = 0.172, and C_x = 0.225. Hence, checking the computed value of k, we have k = (1)(0.22)(1.843)/150 = 0.0027, which checks with the earlier computed value.

2. Compute the head values using k from step 1
Compute $h_e - (h_e^2 - 1.7\ Q/k)^{0.5}$ = 50 − (2500 − 1.7/0.0027)$^{0.5}$ = 6.8.

3. Find the values of T to develop the assumed values of r_e
For example, assume that r_e = 100. Then T = (0.184)(100)^2(0.25)(6.8)/1 = 3230 sec = 0.9 h, using the equation

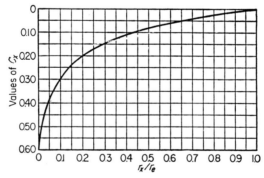

FIGURE 6 Values of C_x for use in calculations of well performance. (*Babbitt, Doland, and Cleasby.*)

$$T = \left(h_e - \sqrt{h_e^2 - 1.7\frac{Q}{k}} \right) \frac{0.184 r_e^2 f}{Q}$$

4. Calculate the radii ratio and d_0

These computations are: $r_e/r_w = 100/1 = 100$. Then, $d_0 = (6.8)(\log_{10} 100)/2.3 = 5.9$ ft (1.8 m), using the equation

$$d_0 = \frac{1}{2.3}\left(h_e - \sqrt{h_e^2 - 1.7\frac{Q}{k}} \right) \log_{10} \frac{r_0}{r_w}$$

5. Compute other points on the drawdown curve

Plot the values found in step 4 on the drawdown-recovery curve, Fig. 7. Compute additional values of d_0 and T and plot them on Fig. 7, as shown.

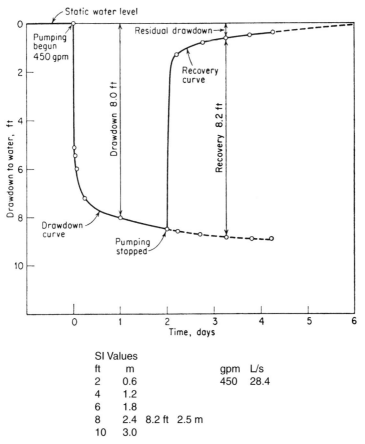

SI Values

ft	m	gpm	L/s
2	0.6	450	28.4
4	1.2		
6	1.8		
8	2.4	8.2 ft	2.5 m
10	3.0		

FIGURE 7 Drawdown-recovery curves for a gravity well. (*Babbitt, Doland, and Cleasby.*)

6. Make the recovery-curve computations

The recovery-curve, Fig. 7, computations are based the assumption that by imposing a negative discharge on the positive discharge from the well there will be in effect zero flow from the well, provided the negative discharge equals the positive discharge. Then, the sum of the drawdowns due to the two discharges at any time T after adding the negative discharge will be the drawdown to the recovery curve, Fig. 7.

Assume some time after the pump has stopped, such as 6 h, and compute r_e, with Q, f, k, and h_e as in step 3, above. Then $r_e = [(6 \times 3600 \times 1)/(0.184 \times 0.25 \times 6.8)]^{0.5} = 263$ ft (80.2 m). Then, $r_e/r_w = 263$; check.

7. Find the value of d_0 corresponding to r_e in step 6

Computing, we have $d_0 = (6.8)(\log_{10})/2.3 = 7.15$ ft (2.2 m). Tabulate the computed values as shown in Table 1 where the value 7.15 is rounded off to 7.2.

Compute the value of r_e using the total time since pumping started. In this case it is $48 + 6 = 54$ h. Then $r_e = [(54 \times 3600 \times 1)/(0.184 \times 0.25 \times 6.8)]^{0.5} = 790$ ft (240.8 m). The d_0 corresponding to the preceding value of $r_e = 790$ ft (240.8 m) is $d_0 = (6.8)(\log_{10} 790)/2.3 = 8.55$ ft (2.6 m).

8. Find the recovery value

The recovery value, $d_r = 8.55 - 7.15 = 1.4$ ft (0.43 m). Coordinates of other points on the recovery curve are computed in a similar fashion. Note that the recovery curve does not attain the original groundwater table because water has been removed from the aquifer and it has not been restored.

Related Calculations. If water is entering the area of a well at a rate q and is being pumped out at the rate Q' with Q' greater than q, then the value of Q to be used in computing the drawdown recovery is $Q' - q$. If this difference is of appreciable magnitude, a correction must be made because of the effect of the inflow from the aquifer into the cone of depression so the groundwater table will ultimately be restored, the recovery curve becoming asymptotic to the table.

This procedure is the work of Harold E. Babbitt, James J. Doland, and John L. Cleasby, as reported in their book, *Water Supply Engineering,* McGraw-Hill. SI values were added by the handbook editor.

TABLE 1 Coordinates for the Drawdown-Recovery Curve of a Gravity Well

(1) Time after pump starts, hr	(2) $\frac{r_e}{r_x} = r_e'$	(3) $2.95 \times \log_{10} \frac{r_e}{r_w} = d_0$	(4) Time after pump starts, hr	(5) $\frac{r_e}{r_x} = r_e'$	(6) $2.95 \times \log_{10} \frac{r_e}{r_w} = d_0$	(7) Time after pump stops, hr	(8) $\frac{r_e}{r_x} = r_e'$	(9) $2.95 \times \log_{10} \frac{r_e}{r_w} = d_0$	(10) Col 6 minus col 9 = d_r
0.25	54	5.10	54	784	8.5	6	263	7.2	1.3
0.50	76	5.45	66	872	8.7	18	455	7.9	0.8
1.00	107	6.0	78	950	8.8	30	587	8.2	0.6
6	263	7.2	90	1,020	8.9	42	694	8.4	0.5
24	526	8.0	102	1,085	8.9	54	784	8.5	0.4
48	745	8.5							

Conditions: $r_w = 1.0$ ft; $h_e = 50$ ft. When $Q = 1$ ft³/s and $r_x = 1.0$ ft, $(h_e - h_x) = 8.0$ ft. When $Q = 1$ ft³/s and $r_x = 60$ ft, $(h_e - h_x) = 3.0$ ft. Specific yield = 0.25; k, as determined in step 1 of example, = 0.0027; and $h_e - (h_e^2 - 1.79Q/k)^{0.5} = 6.8$.

SELECTION OF AIR-LIFT PUMP FOR WATER WELL

Select the overall features of an air-lift pump, Fig. 8, to lift 350 gal/min (22.1 L/s) into a reservoir at the ground surface. The distance to groundwater surface is 50 ft (15.2 m). It is expected that the specific gravity of the well is 14 gal/min/ft (2.89 L/s/m).

Calculation Procedure:

1. Find the well drawdown, static lift, and depth of this well
The drawdown at 350 gal/min is $d = 350/14 = 25$ ft (7.6 m). The static lift, h, is the sum of the distance from the groundwater surface plus the drawdown, or $h = 50 + 25 = 75$ ft (22.9 m).

Interpolating in Table 2 gives a submergence percentage of $s = 0.61$. Then, the depth of the well, D ft is related to the submergence percentage thus: $s = D/(D + h)$. Or, $0.61 = D/(D + 75)$; $D = 117$ ft (35.8 m). The depth of the well is, therefore, $75 + 117 = 192$ ft (58.5 m).

2. Determine the required capacity of the air compressor
The rate of water flow in cubic feet per second, Q_w is given by $Q_w = $ gal/min/(60 min/s)(7.5 ft³/gal) $= 350/(60)(7.5) = 0.78$ ft³/s (0.022 m³/s). Then the volume of free air required by the air-lift pump is given by

$$Q_a = \frac{Q_w(h + h_1)}{75E \log r}$$

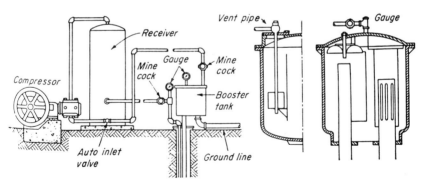

FIGURE 8   Sullivan air-lift booster. (*Babbitt, Doland, and Cleasby.*)

TABLE 2   Some Recommended Submergence Percentages for Air Lifts

Lift, ft	Up to 50	50–100	100–200	200–300	300–400	400–500
Lift, m	Up to 15	15–30	30–61	61–91	91–122	122–152
Submergence percentage	70–66	66–55	55–50	50–43	43–40	40–33

where Q_a = volume of free air required, ft³/min (m³/min); h_1 = velocity head at discharge, usually taken as 6 ft (1.8 m) for deep wells, down to 1 ft (0.3 m) for shallow wells; E = efficiency of pump, approximated from Table 3; r = ratio of compression = $(D + 34)/34$. Substituting, using 6 ft (1.8 m) since this is a deep well, we have, Q_a = $(0.779 \times 81)/(75 \times 0.35 \times 0.646)$ = 3.72 ft³/s (0.11 m³/s).

3. Size the air pipe and determine the operating pressures

The cross-sectional area of the pipe = Q'_a/V. At the bottom of the well, Q'_a = 3.72 $(34/151)$ = 0.83 ft³/s (0.023 m³/s). With a flow velocity of the air typically at 2000 ft/min (610 m/min), or 33.3 ft/s (10 m/s), the area of the air pipe is 0.83/33.3 = 0.025 ft², and the diameter is $[(0.025 \times 4)/\pi]^{0.5}$ = 0.178 ft or 2.1 in (53.3 mm); use 2-in (50.8 mm) pipe.

The pressure at the start is 142 ft (43 m); operating pressure is 117 ft (35.7 m).

4. Size the eductor pipe

At the well bottom, $A = Q/V$. $Q = Q_w + Q'_a = 0.78 + 0.83 = 1.612$ ft³/s (0.45 m³/s). The velocity at the entrance to the eductor pipe is 4.9 ft/s (1.9 m/s) from a table of eductor entrance velocities, available from air-lift pump manufacturers. Then, the pipe area, $A = Q/V = 1.61/4.9 = 0.33$. Hence, $d = [(4 \times 0.33)/\pi]^{0.5}$ = 0.646 ft, or 7.9 in Use 8-in (203 mm) pipe.

If the eductor pipe is the same size from top to bottom, then V at top = $(Q_a + Q_w)/A = (3.72 + 0.78)(4)/(\pi \times 0.667^2)$ = 13 ft/s (3.96 m/s). This is comfortably within the permissible maximum limit of 20 ft/s (6.1 m/s). Hence, 8-in pipe is suitable for this eductor pipe.

Related Calculations. In an air-lift pump serving a water well, compressed air is released through an air diffuser (also called a foot piece) at the bottom of the eductor pipe. Rising as small bubbles, a mixture of air and water is created that has a lower specific gravity than that of water alone. The rising air bubbles, if sufficiently large, create an upward water flow in the well, to deliver liquid at the ground level.

Air lifts have many unique features not possessed by other types of well pumps. They are the simplest and the most foolproof type of pump. In operation, the air-lift pump gives the least trouble because there are no remote or submerged moving parts. Air lifts can be operated successfully in holes of any practicable size. They can be used in crooked holes not suited to any other type of pump. An air-lift pump can draw more water from a well, with sufficient capacity to deliver it, than any other type of pump that can be installed in a well. A number of wells in a group can be operated from a central control station where the air compressor is located.

The principal disadvantages of air lifts are the necessity for making the well deeper than is required for other types of well pumps, the intermittent nature of the

TABLE 3 Effect of Submergence on Air Lift*

Ratio D/h	8.70	5.46	3.86	2.91	2.25
Submergence ratio, D/(D + h)	0.896	0.845	0.795	0.745	0.693
Percentage efficiency	26.5	31.0	35.0	36.6	37.7
Ratio D/h		1.86	1.45	1.19	0.96
Submergence ratio, D/(D + (h)		0.650	0.592	0.544	0.490
Percentage efficiency		36.8	34.5	31.0	26.5

*At Hattiesburg MS.

flow from the well, and the relatively low efficiencies obtained. Little is known of the efficiency of the average air-lift installation in small waterworks. Tests show efficiencies in the neighborhood of 45 percent for depths of 50 ft (15 m) down to 20 percent for depths of 600 ft (183 m). Changes in efficiencies resulting from different submergence ratios are shown in Table 3. Some submergence percentages recommended for various lifts are shown in Table 2.

This procedure is the work of Harold E. Babbitt, James J. Doland, and John L. Cleasby, as reported in their book, *Water Supply Engineering,* McGraw-Hill. SI values were added by the handbook editor.

Water-Supply and Storm-Water System Design

WATER-SUPPLY SYSTEM FLOW-RATE AND PRESSURE-LOSS ANALYSIS

A water-supply system will serve a city of 100,000 population. Two water mains arranged in a parallel configuration (Fig. 9a) will supply this city. Determine the flow rate, size, and head loss of each pipe in this system. If the configuration in Fig. 9a were replaced by the single pipe shown in Fig. 9b, what would the total head loss be if $C = 100$ and the flow rate were reduced to 2000 gal/min (126.2 L/s)? Explain how the Hardy Cross method is applied to the water-supply piping system in Fig. 11.

Calculation Procedure:

1. Compute the domestic water flow rate in the system
Use an average annual domestic water consumption of 150 gal/day (0.0066 L/s) per capita. Hence, domestic water consumption = (150 gal per capita per

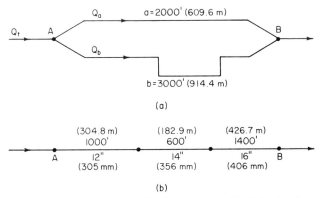

FIGURE 9 (*a*) Parallel water distribution system; (*b*) single-pipe distribution system.

day)(100,000 persons) = 15,000,000 gal/day (657.1 L/s). To this domestic flow, the flow required for fire protection must be added to determine the total flow required.

2. Compute the required flow rate for fire protection
Use the relation $Q_f = 1020(P)^{0.5} [1 - 0.01(P)^{0.5}]$, where Q_f = fire flow, gal/min; P = population in thousands. Substituting gives $Q_f = 1020(100)^{0.5} [1 - 0.01(100)^{0.5}] = 9180$, say 9200 gal/min (580.3 L/s).

3. Apply a load factor to the domestic consumption
To provide for unusual water demands, many design engineers apply a 200 to 250 percent load factor to the average hourly consumption that is determined from the average annual consumption. Thus, the average daily total consumption determined in step 1 is based on an average annual daily demand. Convert the average daily total consumption in step 1 to an average hourly consumption by dividing by 24 h or 15,000,000/24 = 625,000 gal/h (657.1 L/s). Next, apply a 200 percent load factor. Or, design hourly demand = 2.00(625,000) = 1,250,000 gal/h (1314.1 L/s), or 1,250,000/60 min/h = 20,850, say 20,900 gal/min (1318.6 L/s).

4. Compute the total water flow required
The total water flow required = domestic flow, gal/min + fire flow, gal/min = 20,900 + 9200 = 30,100 gal/min (1899.0 L/s). If this system were required to supply water to one or more industrial plants in addition to the domestic and fire flows, the quantity needed by the industrial plants would be added to the total flow computed above.

5. Select the flow rate for each pipe
The flow rate is not known for either pipe in Fig. 9a. Assume that the shorter pipe a has a flow rate Q_a of 12,100 gal/min (763.3 L/s), and the longer pipe b a flow rate Q_b of 18,000 gal/min (1135.6 L/s). Thus, $Q_a + Q_b = Q_t = 12,100 + 18,000 = 30,100$ gal/min (1899.0 L/s), where Q = flow, gal/min, in the pipe identified by the subscript a or b; Q_t = total flow in the system, gal/min.

6. Select the sizes of the pipes in the system
Since neither pipe size is known, some assumptions must be made about the system. First, assume that a friction-head loss of 10 ft of water per 1000 ft (3.0 m per 304.8 m) of pipe is suitable for this system. This is a typical allowable friction-head loss for water-supply systems.

Second, assume that the pipe is sized by using the Hazen-Williams equation with the coefficient $C = 100$. Most water-supply systems are designed with this equation and this value of C.

Enter Fig. 10 with the assumed friction-head loss of 10 ft/1000 ft (3.0 m/304.8 m) of pipe on the right-hand scale, and project through the assumed Hazen-Williams coefficient $C = 100$. Extend this straight line until it intersects the pivot axis. Next, enter Fig. 10 on the left-hand scale at the flow rate in pipe a, 12,100 gal/min (763.3 L/s), and project to the previously found intersection on the pivot axis. At the intersection with the pipe-diameter scale, read the required pipe size as 27-in (686-mm) diameter. Note that if the required pipe size falls between two plotted sizes, the next *larger* size is used.

Now in any parallel piping system, the friction-head loss through any branch connecting two common points equals the friction-head loss in any other branch connecting the same two points. Using Fig. 10 for a 27-in (686-mm) pipe, find the

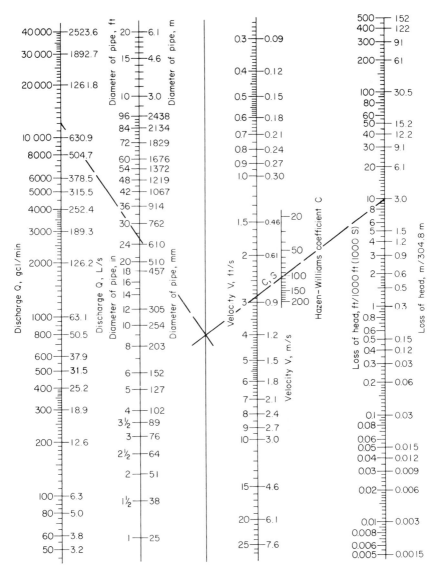

FIGURE 10   Nomogram for solution of the Hazen-Williams equation for pipes flowing full.

actual friction-head loss at 8 ft/1000 ft (2.4 m/304.8 m) of pipe. Hence, the total friction-head loss in pipe *a* is (2000 ft long)(8 ft/1000 ft) = 16 ft (4.9 m) of water. This is also the friction-head loss in pipe *b*.

Since pipe *b* is 3000 ft (914.4 m) long, the friction-head loss per 1000 ft (304.8 m) is total head loss, ft/length of pipe, thousands of ft = 16/3 = 5.33 ft/1000 ft (1.6 m/304.8 m). Enter Fig. 10 at this friction-head loss and *C* = 100. Project in the same manner as described for pipe *a*, and find the required size of pipe *b* as 33 in (838.2 mm).

If the district being supplied by either pipe required a specific flow rate, this flow would be used instead of assuming a flow rate. Then the pipe would be sized in the same manner as described above.

7. Compute the single-pipe equivalent length

When we deal with several different sizes of pipe having the same flow rate, it is often convenient to convert each pipe to an *equivalent length* of a common-size pipe. Many design engineers use 8-in (203-mm) pipe as the common size. Table 4 shows the equivalent length of 8-in (203-mm) pipe for various other sizes of pipe with C = 90, 100, and 110 in the Hazen-Williams equation.

From Table 4, for 12-in (305-mm) pipe, the equivalent length of 8-in (203-mm) pipe is 0.14 ft/ft when C = 100. Thus, total equivalent length of 8-in (203-mm) pipe = (1000 ft of 12-in pipe)(0.14 ft/ft) = 140 ft (42.7 m) of 8-in (203-mm) pipe. For the 14-in (356-mm) pipe, total equivalent length = (600)(0.066) = 39.6 ft (12.1 m), using similar data from Table 4. For the 16-in (406-mm) pipe, total equivalent length = (1400)(0.034) = 47.6 ft (14.5 m). Hence, total equivalent length of 8-in (203-mm) pipe = 140 + 39.6 + 47.6 = 227.2 ft (69.3 m).

8. Determine the friction-head loss in the pipe

Enter Fig. 10 at the flow rate of 2000 gal/min (126.2 L/s), and project through 8-in (203-mm) diameter to the pivot axis. From this intersection, project through C = 100 to read the friction-head loss as 100 ft/1000 ft (30.5 m/304.8 m), due to the friction of the water in the pipe. Since the equivalent length of the pipe is 227.2 ft (69.3 m), the friction-head loss in the compound pipe is (227.2/1000)(110) = 25 ft (7.6 m) of water.

Related Calculations.   Two pipes, two piping systems, or a single pipe and a system of pipes are said to be *equivalent* when the losses of head due to friction for equal rates of flow in the pipes are equal.

TABLE 4 Equivalent Length of 8-in (203-mm) Pipe for C = 100

Pipe diameter		C = 90	C = 100	C = 110
in	mm			
2	51	1012	851	712
4	102	34	29	24.3
6	152	4.8	4.06	3.4
8	203	1.19	1.00	0.84
10	254	0.40	0.34	0.285
12	305	0.17	0.14	0.117
14	356	0.078	0.066	0.055
16	406	0.040	0.034	0.029
18	457	0.023	0.019	0.016
20	508	0.0137	0.0115	0.0096
24	610	0.0056	0.0047	0.0039
30	762	0.0019	0.0016	0.0013
36	914	0.00078	0.00066	0.00055

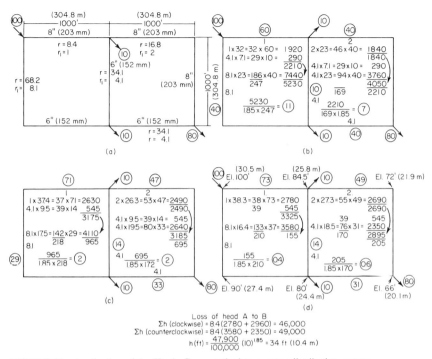

FIGURE 11 Application of the Hardy Cross method to a water distribution system.

To determine the flow rates and friction-head losses in complex waterworks distribution systems, the Hardy Cross method of network analysis is often used. This method[1] uses trial and error to obtain successively more accurate approximations of the flow rate through a piping system. To apply the Hardy Cross method: (1) Sketch the piping system layout as in Fig. 11. (2) Assume a flow quantity, in terms of percentage of total flow, for each part of the piping system. In assuming a flow quantity note that (*a*) the loss of head due to friction between any two points of a closed circuit must be the same by any path by which the water may flow, and (*b*) the rate of inflow into any section of the piping system must equal the outflow. (3) Compute the loss of head due to friction between two points in each part of the system, based on the assumed flow in (*a*) the clockwise direction and (*b*) the counterclockwise direction. A difference in the calculated friction-head losses in the two directions indicates an error in the assumed direction of flow. (4) Compute a counterflow correction by dividing the difference in head, Δh ft, by $n(Q)^{n-1}$, where $n = 1.85$ and $Q =$ flow, gal/min. Indicate the direction of this counterflow in the pipe by an arrow starting at the right side of the smaller value of h and curving toward the larger value, Fig. 11. (5) Add or subtract the counterflow to or from the assumed flow, depending on whether its direction is the same

---

[1] O'Rourke—*General Engineering Handbook,* McGraw-Hill.

or opposite. (6) Repeat this process on each circuit in the system until a satisfactory balance of flow is obtained.

To compute the loss of head due to friction, step 3 of the Hardy Cross method, use any standard formula, such as the Hazen-Williams, that can be reduced to the form $h = rQ^nL$, where h = head loss due to friction, ft of water; r = a coefficient depending on the diameter and roughness of the pipe; Q = flow rate, gal/min; n = 1.85; L = length of pipe, ft. Table 5 gives values of r for 1000-ft (304.8-m) lengths of various sizes of pipe and for different values of the Hazen-Williams coefficient C. When the percentage of total flow is used for computing Σh in Fig. 11, the loss of head due to friction in ft between any two points for any flow in gal/min is computed from $h = [\Sigma h$ (by percentage of flow)/100,000] (gal/min/100)$^{0.85}$. Figure 11 shows the details of the solution using the Hardy Cross method. The circled numbers represent the flow quantities. Table 6 lists values of numbers between 0 and 100 to the 0.85 power.

TABLE 5 Values of r for 1000 ft (304.8 m) of Pipe Based on the Hazen-Williams Formula[°]

d, in (mm)	$C = 90$	$C = 100$	$C = 110$	$C = 120$	$C = 130$	$C = 140$
4 (102)	340	246	206	176	151	135
6 (152)	47.1	34.1	28.6	24.3	21.0	18.7
8 (203)	11.1	8.4	7.0	6.0	5.2	4.6
10 (254)	3.7	2.8	2.3	2.0	1.7	1.5
12 (305)	1.6	1.2	1.0	0.85	0.74	0.65
14 (356)	0.72	0.55	0.46	0.39	0.34	0.30
16 (406)	0.38	0.29	0.24	0.21	0.18	0.15
18 (457)	0.21	0.16	0.13	0.11	0.10	0.09
20 (508)	0.13	0.10	0.08	0.07	0.06	0.05
24 (610)	0.052	0.04	0.03	0.03	0.02	0.02
30 (762)	0.017	0.013	0.011	0.009	0.008	0.007

Example: r for 12-in (305-mm) pipe 4000 ft (1219 m) long, with $C = 100$, is $1.2 \times 4.0 = 4.8$.
[°]Head loss in ft (m) $= r \times 10^{-5} \times Q^{1.85}$ per 1000 ft (304.8 m), Q representing gal/min (L/s).

TABLE 6 Value of the 0.85 Power of Numbers

N	0	1	2	3	4	5	6	7	8	9
0	0	1.0	1.8	2.5	3.2	3.9	4.6	5.2	5.9	6.5
10	7.1	7.7	8.3	8.9	9.5	10.0	10.6	11.1	11.6	12.2
20	12.8	13.3	13.8	14.4	14.9	15.4	15.9	16.4	16.9	17.5
30	18.0	18.5	19.0	19.5	20.0	20.5	21.0	21.5	22.0	22.5
40	23.0	23.4	23.9	24.3	24.8	25.3	25.8	26.3	26.8	27.3
50	27.8	28.2	28.7	29.1	29.6	30.0	30.5	31.0	31.4	31.9
60	32.4	32.9	33.3	33.8	34.2	34.7	35.1	35.6	36.0	36.5
70	37.0	37.4	37.9	38.3	38.7	39.1	39.6	40.0	40.5	41.0
80	41.5	42.0	42.4	42.8	43.3	43.7	44.1	44.5	45.0	45.4
90	45.8	46.3	46.7	47.1	47.6	48.0	48.4	48.8	49.2	49.6

WATER-SUPPLY SYSTEM SELECTION

Choose the type of water-supply system for a city having a population of 100,000 persons. Indicate which type of system would be suitable for such a city today and 20 years hence. The city is located in an area of numerous lakes.

Calculation Procedure:

1. Compute the domestic water flow rate in the system
Use an average annual domestic water consumption of 150 gal per capita day (gcd) (6.6 mL/s). Hence, domestic water consumption = (150 gal per capita day)(100,000 persons) = 15,000,000 gal/day (657.1 L/s). To this domestic flow, the flow required for fire protection must be added to determine the total flow required.

2. Compute the required flow rate for fire protection
Use the relation $Q_f = 1020(P)^{0.5} [1 - 0.01(P)^{0.5}]$, where Q_f = fire flow, gal/min; P = population in thousands. So $Q_f = 1020(100)^{0.5} [1 - 0.01 \times (100)^{0.5}] = 9180$, say 9200 gal/min (580.3 L/s).

3. Apply a load factor to the domestic consumption
To provide for unusual water demands, many design engineers apply a 200 to 250 percent load factor to the average hourly consumption that is determined from the average annual consumption. Thus, the average daily total consumption determined in step 1 is based on an average annual daily demand. Convert the average daily total consumption in step 1 to an average hourly consumption by dividing by 24 h, or 15,000,000/24 = 625,000 gal/h (657.1 L/s). Next, apply a 200 percent load factor. Or, design hourly demand = 2.00(625,000) = 1,250,000 gal/h (1314.1 L/s), or 1,250,000/(60 min/h) = 20,850, say 20,900 gal/min (1318.4 L/s).

4. Compute the total water flow required
The total water flow required = domestic flow, gal/min + fire flow, gal/min = 20,900 + 9200 = 30,100 gal/min (1899.0 L/s). If this system were required to supply water to one or more industrial plants in addition to the domestic and fire flows, the quantity needed by the industrial plants would be added to the total flow computed above.

5. Study the water supplies available
Table 7 lists the principal sources of domestic water supplies. Wells that are fed by groundwater are popular in areas having sandy or porous soils. To determine whether a well is suitable for supplying water in sufficient quantity, its specific capacity (i.e., the yield in gal/min per foot of drawdown) must be determined.

Wells for municipal water sources may be dug, driven, or drilled. Dug wells seldom exceed 60 ft (18.3 m) deep. Each such well should be protected from surface-water leakage by being lined with impervious concrete to a depth of 15 ft (4.6 m).

Driven wells seldom are more than 40 ft (12.2 m) deep or more than 2 in (51 mm) in diameter when used for small water supplies. Bigger driven wells are constructed by driving large-diameter casings into the ground.

TABLE 7 Typical Municipal Water Sources

Source	Collection method	Remarks
Groundwater	Wells (artesian, ordinary, galleries)	30 to 40 percent of an area's rainfall becomes groundwater
Surface freshwater (lakes, rivers, streams, impounding reservoirs)	Pumping or gravity flow from submerged intakes, tower intakes, or surface intakes	Surface supplies are important in many areas
Surface saltwater	Desalting	Wide-scale application under study at present

Drilled wells can be several thousand feet deep, if required. The yield of a driven well is usually greater than any other type of well because the well can be sunk to a depth where sufficient groundwater is available. Almost all wells require a pump of some kind to lift the water from its subsurface location and discharge it to the water-supply system.

Surface freshwater can be collected from lakes, rivers, streams, or reservoirs by submerged-, tower-, or crib-type intakes. The intake leads to one or more pumps that discharge the water to the distribution system or intermediate pumping stations. Locate intakes as far below the water surface as possible. Where an intake is placed less than 20 ft (6.1 m) below the surface of the water, it may become clogged by sand, mud, or ice.

Choose the source of water for this system after studying the local area to determine the most economical source today and 20 years hence. With a rapidly expanding population, the future water demand may dictate the type of water source chosen. Since this city is in an area of many lakes, a surface supply would probably be most economical, if the water table is not falling rapidly.

6. *Select the type of pipe to use*
Four types of pipes are popular for municipal water-supply systems: cast iron, asbestos cement, steel, and concrete. Wood-stave pipe was once popular, but it is now obsolete. Some communities also use copper or lead pipes. However, the use of both types is extremely small when compared with the other types. The same is true of plastic pipe, although this type is slowly gaining some acceptance.

In general, cast-iron pipe proves dependable and long-lasting in water-supply systems that are not subject to galvanic or acidic soil conditions.

Steel pipe is generally used for long, large-diameter lines. Thus, the typical steel pipe used in water-supply systems is 36 or 48 in (914 or 1219 mm) in diameter. Use steel pipe for river crossings, on bridges, and for similar installations where light weight and high strength are required. Steel pipe may last 50 years or more under favorable soil conditions. Where unfavorable soil conditions exist, the lift of steel pipe may be about 20 years.

Concrete-pipe use is generally confined to large, long lines, such as aqueducts. Concrete pipe is suitable for conveying relatively pure water through neutral soil. However, corrosion may occur when the soil contains an alkali or an acid.

Asbestos-cement pipe has a number of important advantages over other types. However, it does not flex readily, it can be easily punctured, and it may corrode in acidic soils.

Select the pipe to use after a study of the local soil conditions, length of runs required, and the quantity of water that must be conveyed. Usual water velocities in municipal water systems are in the 5-ft/s (1.5-m/s) range. However, the velocities in aqueducts range from 10 to 20 ft/s (3.0 to 6.1 m/s). Earthen canals have much lower velocities—1 to 3 ft/s (0.3 to 0.9 m/s). Rock- and concrete-lined canals have velocities of 8 to 15 ft/s (2.4 to 4.6 m/s).

In cold northern areas, keep in mind the occasional need to thaw frozen pipes during the winter. Nonmetallic pipes—concrete, plastic, etc., as well as nonconducting metals—cannot be thawed by electrical means. Since electrical thawing is probably the most practical method available today, pipes that prevent its use may put the water system at a disadvantage if subfreezing temperatures are common in the area served.

7. Select the method for pressurizing the water system

Water-supply systems can be pressurized in three different ways: by gravity or natural elevation head, by pumps that produce a pressure head, and by a combination of the first two ways.

Gravity systems are suitable where the water storage reservoir or receiver is high enough above the distribution system to produce the needed pressure at the farthest outlet. The operating cost of a gravity system is lower than that of a pumped system, but the first cost of the former is usually higher. However, the reliability of the gravity system is usually higher because there are fewer parts that may fail.

Pumping systems generally use centrifugal pumps that discharge either directly to the water main or to an elevated tank, a reservoir, or a standpipe. The water then flows from the storage chamber to the distribution system. In general, most sanitary engineers prefer to use a reservoir or storage tank between the pumps and distribution mains because this arrangement provides greater reliability and fewer pressure surges.

Surface reservoirs should store at least a 1-day water supply. Most surface reservoirs are designed to store a supply for 30 days or longer. Elevated tanks should have a capacity of at least 25 gal (94.6 L) of water per person served, *plus* a reserve for fire protection. The capacity of typical elevated tanks ranges from a low of 40,000 gal (151 kL) for a 20-ft (6.1-m) diameter tank to a high of 2,000,000 gal (7.5 ML) for an 80-ft (24.4-m) diameter tank.

Choose the type of distribution system after studying the topography, water demand, and area served. In general, a pumped system is preferred today. To ensure continuity of service, duplicate pumps are generally used.

8. Choose the system operating pressure

In domestic water supply, the minimum pressure required at the highest fixture in a building is usually assumed to be 15 lb/in² (103.4 kPa). The maximum pressure allowed at a fixture in a domestic water system is usually 65 lb/in² (448.2 kPa). High-rise buildings (i.e., those above six stories) are generally required to furnish the pressure increase needed to supply water to the upper stories. A pump and overhead storage tank are usually installed in such buildings to provide the needed pressure.

Commercial and industrial buildings require a minimum water pressure of 75 lb/in² (517.1 kPa) at the street level for fire hydrant service. This hydrant should deliver at least 250 gal/min (15.8 L/s) of water for fire-fighting purposes.

Most water-supply systems served by centrifugal pumps in a central pumping station operate in the 100-lb/in² (689.5-kPa) pressure range. In areas of one- and two-story structures, a lower pressure, say 65 lb/in² (448.2 kPa), is permissible.

Where the pressure in a system falls too low, auxiliary or booster pumps may be used. These pumps increase the pressure in the main to the desired level.

Choose the system pressure based on the terrain served, quantity of water required, allowable pressure loss, and size of pipe used in the system. Usual pressures required will be in the ranges cited above, although small systems serving one-story residences may operate at pressures as low as 30 lb/in^2 (206.8 kPa). Pressures over 100 lb/in^2 (689.5 kPa) are seldom used because heavier piping is required. As a rule, distribution pressures of 50 to 75 lb/in^2 (344.7 to 517.1 kPa) are acceptable.

9. Determine the number of hydrants for fire protection

Table 8 shows the required fire flow, number of standard hose streams of 250 gal/min (15.8 L/s) discharged through a $1\frac{1}{8}$-in (28.6-mm) diameter smooth nozzle, and the average area served by a hydrant in a high-value district. A standard hydrant may have two or three outlets.

Table 8 indicates that a city of 100,000 persons requires 36 standard hose streams. This means that 36 single-outlet or 18 dual-outlet hydrants are required. More, of course, could be used if better protection were desired in the area. Note that the required fire flow listed in Table 8 agrees closely with that computed in step 2 above.

Related Calculations. Use this general method for any water-supply system, municipal or industrial. Note, however, that the required fire-protection quantities vary from one type of municipal area to another and among different industrial exposures. Refer to *NFPA Handbook of Fire Protection,* available from NFPA, 60 Batterymarch Street, Boston, Massachusetts 02110, for specific fire-protection requirements for a variety of industries. In choosing a water-supply system, the wise designer looks ahead for at least 10 years when the water demand will usually exceed the present demand. Hence, the system may be designed so it is oversized for the present population but just adequate for the future population. The American Society for Testing and Materials (ASTM) publishes comprehensive data giving the usual water requirements for a variety of industries. Table 9 shows a few typical water needs for selected industries.

TABLE 8 Required Fire Flow and Hydrant Spacing°

Population	Required fire flow, gal/min (L/s)	Number of standard hose streams	Average area served per hydrant, ft² (m²)†	
			Direct streams	Engine streams
22,000	4,500 (284)	18	55,000 (5,110)	90,000 (8,361)
28,000	5,000 (315)	20	40,000 (3,716)	85,000 (7,897)
40,000	6,000 (379)	24	40,000 (3,716)	80,000 (7,432)
60,000	7,000 (442)	28	40,000 (3,716)	70,000 (6,503)
80,000	8,000 (505)	32	40,000 (3,716)	60,000 (5,574)
100,000	9,000 (568)	36	40,000 (3,716)	55,000 (5,110)
125,000	10,000 (631)	40	40,000 (3,716)	48,000 (4,459)
150,000	11,000 (694)	44	40,000 (3,716)	43,000 (3,995)
200,000	12,000 (757)	48	40,000 (3,716)	40,000 (3,716)

°National Board of Fire Underwriters.
†High-value districts.

TABLE 9 Selected Industrial Water and Steam Requirements°

	Water	Steam
Air conditioning	6000 to 15,000 gal (22,700 to 57,000 L) per person per season	. . .
Aluminum	1,920,000 gal/ton (8.0 ML/t)	. . .
Cement, portland	750 gal/ton cement (3129 L/t)	. . .
Coal, by-product coke	1430 to 2800 gal/ton coke (5967 to 11,683 L/t)	570 to 860 lb/ton (382 to 427 kg/t)
Rubber (automotive tire)	. . .	120 lb (54 kg) per tire
Electricity	80 gal/kW (302 L/kW) of electricity	. . .

°Courtesy of American Society for Testing and Materials.

To determine the storage capacity required at present, proceed as follows: (1) Compute the flow needed to meet 50 percent of the present domestic daily (that is, 24-h) demand. (2) Compute the 4-h fire demand. (3) Find the sum of (1) and (2).

For this city, procedure (1) = (20,900 gal/min)(60 min/h)(24 h/day)(0.5) = 15,048,000 gal (57.2 ML) with the data computed in step 3. Also procedure (2) = (4 h)(60 min/h)(9200 gal/min) = 2,208,000 gal (8.4 ML), using the data computed in step 2, above. Then, total storage capacity required = 15,048,000 + 2,208,000 = 17,256,000 gal (65.3 ML). Where one or more reliable wells will produce a significant flow for 4 h or longer, the storage capacity can be reduced by the 4-h productive capacity of the wells.

SELECTION OF TREATMENT METHOD FOR WATER-SUPPLY SYSTEM

Choose a treatment method for a water-supply system for a city having a population of 100,000 persons. The water must be filtered, disinfected, and softened to make it suitable for domestic use.

Calculation Procedure:

1. Compute the domestic water flow rate in the system
When water is treated for domestic consumption, only the drinking water passes through the filtration plant. Fire-protection water is seldom treated unless it is so turbid that it will clog fire pumps or hoses. Assuming that the fire-protection water is acceptable for use without treatment, we consider only the drinking water here.

Use the same method as in steps 1 and 3 of the previous calculation procedure to determine the required domestic water flow of 20,900 gal/min (1318.6 L/s) for this city.

2. Select the type of water-treatment system to use
Water supplies are treated by a number of methods including sedimentation, coagulation, filtration, softening, and disinfection. Other treatments include disinfection, taste and odor control, and miscellaneous methods.

Since the water must be filtered, disinfected, and softened, each of these steps must be considered separately.

3. Choose the type of filtration to use

Slow sand filters operate at an average rate of 3 million gal/(acre·day) [2806.2 L/(m²·day)]. This type of filter removes about 99 percent of the bacterial content of the water and most tastes and odors.

Rapid sand filters operate at an average rate of 150 million gal/(acre·day) [1.6 L/(m²·s)]. But the raw water must be treated before it enters the rapid sand filter. This preliminary treatment often includes chemical coagulation and sedimentation. A high percentage of bacterial content—up to 99.98 percent—is removed by the preliminary treatment and the filtration. But color and turbidity removal is not as dependable as with slow sand filters. Table 10 lists the typical limits for certain impurities in water supplies.

The daily water flow rate for this city is, from step 1, (20,900 gal/min)(24 h/day)(60 min/h) = 30,096,000 gal/day (1318.6 L/s). If a slow sand filter were used, the required area would be (30.096 million gal/day)/[3 million gal/(acre·day)] = 10+ acres (40,460 m²).

A rapid sand filter would require 30.096/150 = 0.2 acre (809.4 m²). Hence, if space were scarce in this city—and it usually is—a rapid sand filter would be used. With this choice of filtration, chemical coagulation and sedimentation are almost a necessity. Hence, these two additional steps would be included in the treatment process.

Table 11 gives pertinent data on both slow and rapid sand filters. These data are useful in filter selection.

4. Select the softening process to use

The principal water-softening processes use: (*a*) lime and sodium carbonate followed by sedimentation or filtration, or both, to remove the precipitates and (*b*) zeolites of the sodium type in a pressure filter. Zeolite softening is popular and is widely used in municipal water-supply systems today. Based on its proven usefulness and economy, zeolite softening will be chosen for this installation.

5. Select the disinfection method to use

Chlorination by the addition of chlorine to the water is the principal method of disinfection used today. To reduce the unpleasant effects that may result from using chlorine alone, a mixture of chlorine and ammonia, known as chloramine, may be used. The ammonia dosage is generally 0.25 ppm or less. Assume that the chloramine method is chosen for this installation.

TABLE 10 Typical Limits for Impurities in Water Supplies

Impurity	Limit, ppm	Impurity	Limit, ppm
Turbidity	10	Iron plus manganese	0.3
Color	20	Magnesium	125
Lead	0.1	Total solids	500
Fluoride	1.0	Total hardness	100
Copper	3.0	Ca + Mg salts	

TABLE 11 Typical Sand-Filter Characteristics

Slow sand filters	
Usual filtration rate	2.5 to 6.0 $\times$ 10^6 gal/(acre·day) [2339 to 5613 L/(m²·day)]
Sand depth	30 to 36 in (76 to 91 cm)
Sand size	35 mm
Sand uniformity coefficient	1.75
Water depth	3 to 5 ft (0.9 to 1.5 m)
Water velocity in underdrains	2 ft/s (0.6 m/s)
Cleaning frequency required	2 to 11 times per year
Units required	At least two to permit alternate cleaning
Fast sand filters	
Usual filtration rate	100 to 200 $\times$ 10^6 gal/(acre·day) [24.7 to 49.4 kL/(m²·day)]
Sand depth	30 in (76 cm)
Gravel depth	18 in (46 cm)
Sand size	0.4 to 0.5 mm
Sand uniformity coefficient	1.7 or less
Units required	At least three to permit cleaning one unit while the other two are operating

6. Select the method of taste and odor control

The methods used for taste and odor control are: (a) aeration, (b) activated carbon, (c) prechlorination, and (d) chloramine. Aeration is popular for groundwaters containing hydrogen sulfide and odors caused by microscopic organisms.

Activated carbon absorbs impurities that cause tastes, odors, or color. Generally, 10 to 20 lb (4.5 to 9.1 kg) of activated carbon per million gallons of water is used, but larger quantities—from 50 to 60 lb (22.7 to 27.2 kg)—may be specified. In recent years, some 2000 municipal water systems have installed activated carbon devices for taste and odor control.

Prechlorination and chloramine are also used in some installations for taste and odor control. Of the two methods, chloramine appears more popular at present.

Based on the data given for this water-supply system, method b, c, or d would probably be suitable. Because method b has proven highly effective, it will be chosen tentatively, pending later investigation of the economic factors.

Related Calculations. Use this general procedure to choose the treatment method for all types of water-supply systems where the water will be used for human consumption. Thus, the procedure is suitable for municipal, commercial, and industrial systems.

Hazardous wastes of many types endanger groundwater supplies. One of the most common hazardous wastes is gasoline which comes from the estimated 120,000 leaking underground gasoline-storage tanks. Major oil companies are replacing leaking tanks with new noncorrosive tanks. But the soil and groundwater must still be cleaned to prevent pollution of drinking-water supplies.

Other contaminants include oily sludges, organic (such as pesticides and dioxins), and nonvolatile organic materials. These present especially challenging removal and disposal problems for engineers, particularly in view of the stringent environmental requirements of almost every community.

A variety of treatment and disposal methods are in the process of development and application. For oily waste handling, one process combines water evaporation and solvent extraction to break down a wide variety of hazardous waste and sludge from industrial, petroleum-refinery, and municipal-sewage-treatment operations. This process typically produces dry solids with less than 0.5 percent residual hydrocarbon content. This meets EPA regulations for nonhazardous wastes with low heavy-metal contents.

Certain organics, such as pesticides and dioxins, are hydrophobic. Liquified propane and butane are effective at separating hydrophobic organics from solid particles in tainted sludges and soils. The second treatment method uses liquified propane to remove organics from contaminated soil. Removal efficiencies reported are: polychlorinated biphenyls (PCBs) 99.9 percent; polyaromatic hydrocarbons (PAHs) 99.5 percent; dioxins 97.4 percent; total petroleum hydrocarbons 99.9 percent. Such treated solids meet EPA land-ban regulations for solids disposal.

Nonvolatile organic materials at small sites can be removed by a mobile treatment system using up to 14 solvents. Both hydrophobic and hydrophilic solvents are used; all are nontoxic; several have Food and Drug Administration (FDA) approval as food additives. Used at three different sites (at this writing) the process reduced PCB concentration from 500 to 1500 ppm to less than 100 ppm; at another site PCB concentration was reduced from an average of 30 to 300 ppm to less than 5 ppm; at the third site PCBs were reduced from 40 ppm to less than 3 ppm.

STORM-WATER RUNOFF RATE AND RAINFALL INTENSITY

What is the storm-water runoff rate from a 40-acre (1.6-km^2) industrial site having an imperviousness of 50 percent if the time of concentration is 15 min? What would be the effect of planting a lawn over 75 percent of the site?

Calculation Procedure:

1. Compute the hourly rate of rainfall
Two common relations, called the *Talbot formulas,* used to compute the hourly rate of rainfall R in/h are $R = 360/(t + 30)$ for the heaviest storms and $R = 105/(t + 15)$ for ordinary storms, where $t =$ time of concentration, min. Using the equation for the heaviest storms because this relation gives a larger flow rate and produces a more conservative design, we see $R = 360/(15 + 30) = 8$ in/h (0.05 mm/s).

2. Compute the storm-water runoff rate
Apply the *rational method* to compute the runoff rate. This method uses the relation $Q = AIR$, where $Q =$ storm-water runoff rate, ft^3/s; $A =$ area served by sewer, acres; $I =$ coefficient of runoff or percentage of imperviousness of the area; other symbols as before. So $Q = (40)(0.50)(8) = 160$ ft^3/s (4.5 m^3/s).

3. Compute the effect of changed imperviousness
Planting a lawn on a large part of the site will increase the imperviousness of the soil. This means that less rainwater will reach the sewer because the coefficient of

TABLE 12 Coefficient of Runoff for Various Surfaces

Surface	Coefficient
Parks, gardens, lawns, meadows	0.05–0.25
Gravel roads and walks	0.15–0.30
Macadamized roadways	0.25–0.60
Inferior block pavements with uncemented joints	0.40–0.50
Stone, brick, and wood-block pavements with tightly cemented joints	0.75–0.85
Same with uncemented joints	0.50–0.70
Asphaltic pavements in good condition	0.85–0.90
Watertight roof surfaces	0.70–0.95

imperviousness of a lawn is lower. Table 12 lists typical coefficients of imperviousness for various surfaces. This tabulation shows that the coefficient for lawns varies from 0.05 to 0.25. Using a value of $I = 0.10$ for the $40(0.75) = 30$ acres of lawn, we have $Q = (30)(0.10)(8) = 24$ ft^3/s (0.68 m^3/s).

The runoff for the remaining 10 acres (40,460 m^2) is, as in step 2, $Q = (10)(0.5)(8) = 40$ ft^3/s (1.1 m^3/s). Hence, the total runoff is $24 + 40 = 64$ ft^3/s (1.8 m^3/s). This is $160 - 64 = 96$ ft^3/s (2.7 m^3/s) less than when the lawn was not used.

Related Calculations.   The time of concentration for any area being drained by a sewer is the time required for the maximum runoff rate to develop. It is also defined as the time for a drop of water to drain from the farthest point of the watershed to the sewer.

When rainfall continues for an extended period T min, the coefficient of imperviousness changes. For impervious surfaces such as watertight roofs, $I = T/(8 + T)$. For improved pervious surfaces, $I = 0.3T/(20 + T)$. These relations can be used to compute the coefficient in areas of heavy rainfall.

Equations for R for various areas of the United States are available in Steel—*Water Supply and Sewerage,* McGraw-Hill. The Talbot formulas, however, are widely used and have proved reliable.

The time of concentration for a given area can be approximated from $t = I(L/Si^2)^{1/3}$, where $L =$ distance of overland flow of the rainfall from the most remote part of the site, ft; $S =$ slope of the land, ft/ft; $i =$ rainfall intensity, in/h; other symbols as before. For portions of the flow carried in ditches, the time of flow to the inlet can be computed by using the Manning formula.

Table 13 lists the coefficient of runoff for specific types of built-up and industrial areas. Use these coefficients in the same way as shown above. Tables 12 and 13 present data developed by Kuichling and ASCE.

SIZING SEWER PIPES FOR VARIOUS FLOW RATES

Determine the size, flow rate, and depth of flow from a 1000-ft (304.8-m) long sewer which slopes 5 ft (1.5 m) between inlet and outlet and which must carry a

TABLE 13 Coefficient of Runoff for
Various Areas

Area	Coefficient
Business:	
Downtown	0.70–0.95
Neighborhood	0.50–0.70
Residential:	
Single-family	0.30–0.50
Multiunits, detached	0.40–0.60
Multiunits, attached	0.60–0.75
Residential (suburban)	0.25–0.40
Apartment dwelling	0.50–0.70
Industrial:	
Light industry	0.50–0.80
Heavy industry	0.60–0.90
Playgrounds	0.20–0.35
Railroad yards	0.20–0.40
Unimproved	0.10–0.30

flow of 5 million gal/day (219.1 L/s). The sewer will flow about half full. Will this sewer provide the desired flow rate?

Calculation Procedure:

1. Compute the flow rate in the half-full sewer
A flow of 1 million gal/day = 1.55 ft³/s (0.04 m³/s). Hence, a flow of 5 million gal/day = 5(1.55) = 7.75 ft³/s (219.1 L/s) in a *half-full* sewer.

2. Compute the full-sewer flow rate
In a *full sewer*, the flow rate is twice that in a half-full sewer, or 2(7.75) = 15.50 ft³/s (0.44 m³/s) for this sewer. This is equivalent to 15.50/1.55 = 10 million gal/day (438.1 L/s). Full-sewer flow rates are used because pipes are sized on the basis of being full of liquid.

3. Compute the sewer-pipe slope
The pipe slope S ft/ft = $(E_i - E_0)/L$, where E_i = inlet elevation, ft above the site datum; E_0 = outlet elevation, ft above site datum; L = pipe length between inlet and outlet, ft. Substituting gives S = 5/1000 = 0.005 ft/ft (0.005 m/m).

4. Determine the pipe size to use
The Manning formula $v = (1.486/n)R^{2/3}S^{1/2}$ is often used for sizing sewer pipes. In this formula, v = flow velocity, ft/s; n = a factor that is a function of the pipe roughness; R = pipe hydraulic radius = 0.25 pipe diameter, ft; S = pipe slope, ft/ft. Table 14 lists values of n for various types of sewer pipe. In sewer design, the value n = 0.013 for pipes flowing full.

Since the Manning formula is complex, numerous charts have been designed to simplify its solution. Figure 12 is one such typical chart designed specifically for sewers.

Enter Fig. 12 at 15.5 ft³/s (0.44 m³/s) on the left, and project through the slope ratio of 0.005. On the central scale between the flow rate and slope scales, read the

TABLE 14 Values of n for the Manning Formula

Type of surface of pipe	n
Ditches and rivers, rough bottoms with much vegetation	0.040
Ditches and rivers in good condition with some stones and weeds	0.030
Smooth earth or firm gravel	0.020
Rough brick; tuberculated iron pipe	0.017
Vitrified tile and concrete pipe poorly jointed and unevenly settled; average brickwork	0.015
Good concrete; riveted steel pipe; well-laid vitrified tile or brickwork	0.013*
Cast-iron pipe of ordinary roughness; unplaned timber	0.012
Smoothest pipes; neat cement	0.010
Well-planed timber evenly laid	0.009

*Probably the most frequently used value.

next larger standard sewer-pipe diameter as 24 in (610 mm). When using this chart, always read the next larger pipe size.

5. Determine the fluid flow velocity
Continue the solution line of step 4 to read the fluid flow velocity as 5 ft/s (1.5 m/s) on the extreme right-hand scale of Fig. 12. This is for a sewer flowing *full*.

6. Compute the half-full flow depth
Determine the full-flow capacity of this 24-in (610-mm) sewer by entering Fig. 12 at the slope ratio, 0.005, and projecting through the pipe diameter, 24 in (610 mm). At the left read the full-flow capacity as 16 ft³/s (0.45 m³/s).

The required half-flow capacity is 7.75 ft³/s (0.22 m³/s), from step 1. Determine the ratio of the required half-flow capacity to the full-flow capacity, both expressed in ft³/s. Or 7.75/16.0 = 0.484.

Enter Fig. 13 on the bottom at 0.484, and project vertically upward to the discharge curve. From the intersection, project horizontally to the left to read the depth-of-flow ratio as 0.49. This means that the depth of liquid in the sewer at a flow of 7.75 ft³/s (0.22 m³/s) is 0.49(24 in) = 11.75 in (29.8 cm). Hence, the sewer will be just slightly less than half full when handling the designed flow quantity.

7. Compute the half-full flow velocity
Project horizontally to the right along the previously found 0.49 depth-of-flow ratio until the velocity curve is intersected. From this intersection, project vertically downward to the bottom scale to read the ratio of hydraulic elements as 0.99. Hence, the fluid velocity when flowing half-full is 0.99(5.0 ft/s) = 4.95 ft/s (1.5 m/s).

Related Calculations. The minimum flow velocity required in sanitary sewers is 2 ft/s (0.6 m/s). At 2 ft/s (0.6 m/s), solids will not settle out of the fluid. Since the velocity in this sewer is 4.95 ft/s (1.5 m/s), as computed in step 7, the sewer meets, and exceeds, the minimum required flow velocity.

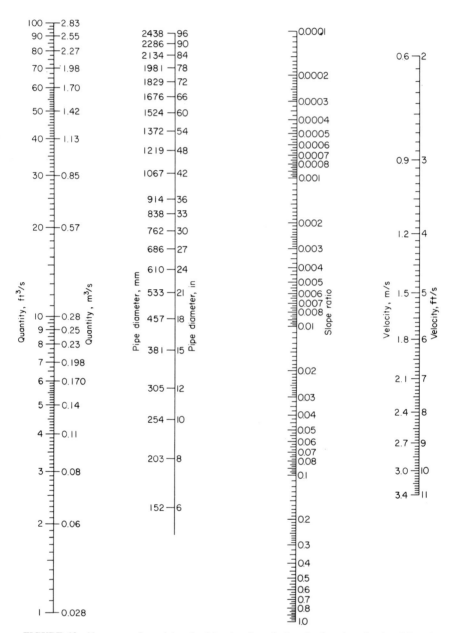

FIGURE 12 Nomogram for solving the Manning formula for circular pipes flowing full and $n = 0.013$.

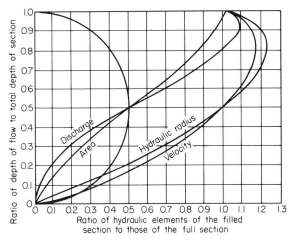

FIGURE 13 Hydraulic elements of a circular pipe.

Certain localities have minimum slope requirements for sanitary sewers. The required slope produces a minimum flow velocity of 2 ft/s (0.6 m/s) with an n value of 0.013.

Storm sewers handling rainwater and other surface drainage require a higher flow velocity than sanitary sewers because sand and grit often enter a storm sewer. The usual minimum allowable velocity for a storm sewer is 2.5 ft/s (0.76 m/s); where possible, the sewer should be designed for 3.0 ft/s (0.9 m/s). If the sewer designed above were used for storm service, it would be acceptable because the fluid velocity is 4.95 ft/s (1.5 m/s). To prevent excessive wear of the sewer, the fluid velocity should not exceed 8 ft/s (2.4 m/s).

Note that Figs. 12 and 13 can be used whenever two variables are known. When a sewer flows at 0.8, or more, full, the partial-flow diagram, Fig. 13, may not give accurate results, especially at high flow velocities.

SEWER-PIPE EARTH LOAD AND BEDDING REQUIREMENTS

A 36-in (914-mm) diameter clay sewer pipe is placed in a 15-ft (4.5-m) deep trench in damp sand. What is the earth load on this sewer pipe? What bedding should be used for the pipe? If a 5-ft (1.5-m) wide drainage trench weighing 2000 lb/ft (2976.3 kg/m) of length crosses the sewer pipe at right angles to the pipe, what load is transmitted to the pipe? The bottom of the flume is 11 ft (3.4 m) above the top of the sewer pipe.

Calculation Procedure:

1. *Compute the width of the pipe trench*
Compute the trench width from $w = 1.5d + 12$, where w = trench width, in; d = sewer-pipe diameter, in So $w = 1.5(36) + 12 = 66$ in (167.6 cm), or 5 ft 6 in (1.7 m).

2. *Compute the trench depth-to-width ratio*
To determine this ratio, subtract the pipe diameter from the depth and divide the result by the trench width. Or, $(15 - 3)/5.5 = 2.18$.

3. *Compute the load on the pipe*
Use the relation $L = kWw^2$, where L = pipe load, lb/lin ft of trench; k = a constant from Table 15; W = weight of the fill material used in the trench, lb/ft³; other symbol as before.

Enter Table 15 at the depth-to-width ratio of 2.18. Since this particular value is not tabulated, use the next higher value, 2.5. Opposite this, read $k = 1.70$ for a sand filling.

Enter Table 16 at damp sand, and read the weight as 115 lb/ft³ (1842.1 kg/m³). With these data the pipe load relation can be solved.

Substituting in $L = kWw^2$, we get $L = 1.70(115)(5.5)^2 = 5920$ lb/ft (86.4 N/ mm). Study of the properties of clay pipe (Table 17) shows that 36-in (914-mm) extra-strength clay pipe has a minimum average crushing strength of 6000 lb (26.7 kN) by the three-edge-bearing method.

TABLE 15 Values of k for Use in the Pipe Load Equation[°]

Ratio of trench depth to width	Sand and damp topsoil	Saturated topsoil	Damp clay	Saturated clay
0.5	0.46	0.46	0.47	0.47
1.0	0.85	0.86	0.88	0.90
1.5	1.18	1.21	1.24	1.28
2.0	1.46	1.50	1.56	1.62
2.5	1.70	1.76	1.84	1.92
3.0	1.90	1.98	2.08	2.20
3.5	2.08	2.17	2.30	2.44
4.0	2.22	2.33	2.49	2.66
4.5	2.34	2.47	2.65	2.87
5.0	2.45	2.59	2.80	3.03
5.5	2.54	2.69	2.93	3.19
6.0	2.61	2.78	3.04	3.33
6.5	2.68	2.86	3.14	3.46
7.0	2.73	2.93	3.22	3.57
7.5	2.78	2.98	3.30	3.67

[°]Iowa State Univ. Eng. Exp. Sta. Bull. 47.

TABLE 16 Weight of Pipe-Trench Fill

Fill	lb/ft³	kg/m³
Dry sand	100	1601
Damp sand	115	1841
Wet sand	120	1921
Damp clay	120	1921
Saturated clay	130	2081
Saturated topsoil	115	1841
Sand and damp topsoil	100	1601

TABLE 17 Clay Pipe Strength

Pipe size, in (mm)	Minimum average strength, lb/lin ft (N/mm)	
	Three-edge-bearing	Sand-bearing
4 (102)	1000 (14.6)	1500 (21.9)
6 (152)	1100 (16.1)	1650 (24.1)
8 (203)	1300 (18.9)	1950 (28.5)
10 (254)	1400 (20.4)	2100 (30.7)
12 (305)	1500 (21.9)	2250 (32.9)
15 (381)	1750 (25.6)	2625 (38.3)
18 (457)	2000 (29.2)	3000 (43.8)
21 (533)	2200 (32.1)	3300 (48.2)
24 (610)	2400 (35.0)	3600 (52.6)
27 (686)	2750 (40.2)	4125 (60.2)
30 (762)	3200 (46.7)	4800 (70.1)
33 (838)	3500 (51.1)	5250 (76.7)
36 (914)	3900 (56.9)	5850 (85.4)

4. *Apply the loading safety factor*
ASTM recommends a factor of safety of 1.5 for clay sewers. To apply this factor of safety, divide it into the tabulated three-edge-bearing strength found in step 3. Or, 6000/1.5 = 4000 lb (17.8 kN).

5. *Compute the pipe load-to-strength ratio*
Use the strength found in step 4. Or pipe load-to-strength ratio (also called the *load factor*) = 5920/4000 = 1.48.

6. *Select the bedding method for the pipe*
Figure 14 shows methods for bedding sewer pipe and the strength developed. Thus, earth embedment, type 2 bedding, develops a load factor of 1.5. Since the computed load factor, step 5, is 1.48, this type of bedding is acceptable. (In choosing a type of bedding be certain that the load factor of the actual pipe is less than, or equals, the developed load factor for the three-edge-bearing strength.)

The type 2 earth embedment, Fig. 14, is a highly satisfactory method, except that the shaping of the lower part of the trench to fit the pipe may be expensive. Type 3 granular embedment may be less expensive, particularly if the crushed stone, gravel, or shell is placed by machine.

7. *Compute the direct load transmitted to the sewer pipe*
The weight of the drainage flume is carried by the soil over the sewer pipes. Hence, a portion of this weight may reach the sewer pipe. To determine how much of the flume weight reaches the pipe, find the weight of the flume per foot of width, or 2000 lb/5 ft = 400 lb/ft (5.84 kN/mm) of width.

Since the pipe trench is 5.5 ft (1.7 m) wide, step 1, the 1-ft (0.3-m) wide section of the flume imposes a total load of 5.5(400) = 2200 lb (9.8 kN) on the soil beneath it.

To determine what portion of the flume load reaches the sewer pipe, compute the ratio of the depth of the flume bottom to the width of the sewer-pipe trench, or 11/5.5 = 2.0.

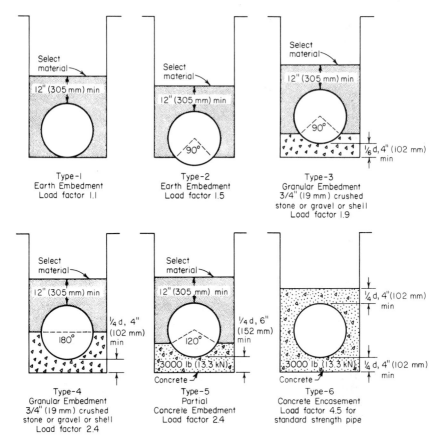

FIGURE 14 Strengths developed for various methods of bedding sewer pipes. (*W. S. Dickey Clay Manufacturing Co.*)

Enter Table 18 at a value of 2.0, and read the load proportion for sand and damp topsoil as 0.35. Hence, the load of the flume reaching each foot of sewer pipe is 0.35(2200) = 770 lb (3.4 kN).

Related Calculations. A load such as that in step 7 is termed a *short load;* i.e., it is shorter than the pipe-trench width. Typical short loads result from automobile and truck traffic, road rollers, building foundations, etc. *Long loads* are imposed by weights that are longer than the trench is wide. Typical long loads are stacks of lumber, steel, and poles, and piles of sand, coal, gravel, etc. Table 19 shows the proportion of long loads transmitted to buried pipes. Use the same procedure as in step 7 to compute the load reaching the buried pipe.

When a sewer pipe is placed on undisturbed ground and covered with fill, compute the load on the pipe from $L = kWd^2$, where d = pipe diameter, ft; other symbols as in step 3. Tables 18 and 19 are the work of Prof. Anson Marston, Iowa State University.

To find the total load on trenched or surface-level buried pipes subjected to both fill and long or short loads, add the proportion of the long or short load reaching the pipe to the load produced by the fill.

TABLE 18 Proportion of Short Loads Reaching Pipe in Trenches

Depth-to-width ratio	Sand and damp topsoil	Saturated topsoil	Damp clay	Saturated clay
0.0	1.00	1.00	1.00	1.00
0.5	0.77	0.78	0.79	0.81
1.0	0.59	0.61	0.63	0.66
1.5	0.46	0.48	0.51	0.54
2.0	0.35	0.38	0.40	0.44
2.5	0.27	0.29	0.32	0.35
3.0	0.21	0.23	0.25	0.29
4.0	0.12	0.14	0.16	0.19
5.0	0.07	0.09	0.10	0.13
6.0	0.04	0.05	0.06	0.08
8.0	0.02	0.02	0.03	0.04
10.0	0.01	0.01	0.01	0.02

TABLE 19 Proportion of Long Loads Reaching Pipe in Trenches

Depth-to-width ratio	Sand and damp topsoil	Saturated topsoil	Damp yellow clay	Saturated yellow clay
0.0	1.00	1.00	1.00	1.00
0.5	0.85	0.86	0.88	0.89
1.0	0.72	0.75	0.77	0.80
1.5	0.61	0.64	0.67	0.72
2.0	0.52	0.55	0.59	0.64
2.5	0.44	0.48	0.52	0.57
3.0	0.37	0.41	0.45	0.51
4.0	0.27	0.31	0.35	0.41
5.0	0.19	0.23	0.27	0.33
6.0	0.14	0.17	0.20	0.26
8.0	0.07	0.09	0.12	0.17
10.00	0.04	0.05	0.07	0.11

Note that sewers may have several cross-sectional shapes—circular, egg, rectangular, square, etc. The circular sewer is the most common because it has a number of advantages, including economy. Egg-shaped sewers are not as popular as circular and are less often used today because of their higher costs.

Rectangular and square sewers are often used for storm service. However, their hydraulic characteristics are not as desirable as circular sewers.

STORM-SEWER INLET SIZE AND FLOW RATE

What size storm-sewer inlet is required to handle a flow of 2 ft³/s (0.057 m³/s) if the gutter is sloped ¼ in/ft (2.1 cm/m) across the inlet and 0.05 in/ft (0.4 cm/m) along the length of the inlet? The maximum depth of flow in the gutter is estimated to be 0.2 ft (0.06 m), and the gutter is depressed 4 in (102 mm) below the normal street level.

Calculation Procedure:

1. *Compute the reciprocal of the gutter transverse slope*
The *transverse slope* of the gutter across the inlet is ¼ in/ft (2.1 cm/m). Expressing the reciprocal of this slope as r, compute the value for this gutter as $r = 4 \times 12/1 = 48$.

2. *Determine the inlet capacity per foot of length*
Enter Table 20 at the flow depth of 0.2 ft (0.06 m), and project to the depth of depression of the gutter of 4 in (102 mm). Opposite this depth, read the inlet capacity per foot of length as 0.50 ft³/s (0.014 m³/s).

3. *Compute the required gutter inlet length*
The gutter must handle a maximum flow of 2 ft³/s (0.057 m³/s). Since the inlet has a capacity of 0.50 ft³/(s · ft) [0.047 m³/(m · s)] of length, the required length, ft = maximum required capacity, ft³/s/capacity per foot, ft³/s = 2.0/0.50 = 4.0 ft (1.2 m). A length of 4.0 ft (1.2 m) will be satisfactory. Were a length of 4.2 or 4.4 ft (1.28 or 1.34 m) required, a 4.5-ft (1.37-m) long inlet would be chosen. The reasoning behind the choice of a longer length is that the extra initial investment for the longer length is small compared with the extra capacity obtained.

4. *Determine how far the water will extend from the curb*
Use the relation $l = rd$, where l = distance water will extend from the curb, ft; d = depth of water in the gutter at the curb line, ft; other symbols as before. Substituting, we find $l = 48(0.2) = 9.6$ ft (2.9 m). This distance is acceptable because the water would extend out this far only during the heaviest storms.

   Related Calculations.   To compute the flow rate in a gutter, use the relation $F = 0.56(r/n)s^{0.5}d^{8/3}$, where F = flow rate in gutter, ft³/s; n = roughness coefficient, usually taken as 0.015; s = gutter slope, in/ft; other symbols as before. Where the computed inlet length is 5 ft (1.5 m) or more, some engineers assume that a portion of the water will pass the first inlet and enter the next one along the street.

TABLE 20 Storm-Sewer Inlet Capacity per Foot (Meter) of Length

Flow depth in gutter, ft (mm)	Depression depth, in (mm)	Capacity per foot length, ft³/s (m³/s)
0.2 (0.06)	0 (0)	0.062 (5.76)
	1 (25.4)	0.141 (13.10)
	2 (50.8)	0.245 (22.76)
	3 (76.2)	0.358 (33.26)
	4 (101.6)	0.500 (46.46)
0.3 (0.09)	0 (0)	0.115 (10.69)
	1 (25.4)	0.205 (19.05)
	2 (50.8)	0.320 (29.73)
	3 (76.2)	0.450 (41.81)
	4 (101.6)	0.590 (54.82)

STORM-SEWER DESIGN

Design a storm-sewer system for a 30-acre (1.21×10^5-m²) residential area in which the storm-water runoff rate is computed to be 24 ft³/s (0.7 m³/s). The total area is divided into 10 plots of equal area having similar soil and runoff conditions.

Calculation Procedure:

1. *Sketch a plan of the sewer system*
Sketch the area and the 10 plots as in Fig. 15. A scale of 1 in = 100 ft (1 cm = 12 m) is generally suitable. Indicate the terrain elevations by drawing the profile curves on the plot plan. Since the profiles (Fig. 15) show that the terrain slopes from north to south, the main sewer can probably be best run from north to south. The sewer would also slope downward from north to south, following the general slope of the terrain.

Indicate a storm-water inlet for each of the areas served by the sewer. With the terrain sloping from north to south, each inlet will probably give best service if it is located on the southern border of the plot.

Since the plots are equal in area, the main sewer can be run down the center of the plot with each inlet feeding into it. Use arrows to indicate the flow direction in the laterals and main sewer.

2. *Compute the lateral sewer size*
Each lateral sewer handles 24 ft³/s/10 plots = 2.4 ft³/s (0.07 m³/s) of storm water. Size each lateral, using the Manning formula with $n = 0.013$ and full flow in the

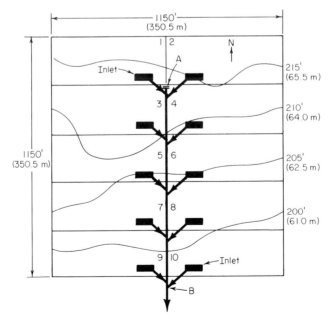

FIGURE 15   Typical storm-sewer plot plan and layout diagram.

pipe. Assume a slope ratio of 0.05 for each inlet pipe between the inlet and the main sewer. This means that the inlet pipe will slope 1 ft in 20 ft (0.3 m in 6.1 m) of length. In an installation such as this, a slope ratio of 0.05 is adequate.

By using Fig. 12 for a flow of 2.4 ft³/s (0.0679 m³/s) and a slope of 0.05, an 8-in (203-mm) pipe is required for each lateral. The fluid velocity is, from Fig. 12, 7.45 ft/s (2.27 m/s). This is a high enough velocity to prevent solids from settling out of the water. [The flow velocity should not be less than 2 ft/s (0.61 m/s).]

3. Compute the size of the main sewer

There are four sections of the main sewer (Fig. 15). The first section, section 3-4, serves the two northernmost plots. Since the flow from each plot is 2.4 ft³/s (0.0679 m³/s), the storm water that this portion of the main sewer must handle is 2(2.4) = 4.8 ft³/s (0.14 m³/s).

The main sewer begins at point A, which has an elevation of about 213 ft (64.9 m), as shown by the profile. At point B the terrain elevation is about 190 ft (57.8 m). Hence, the slope between points A and B is about 213 − 190 = 23 ft (7.0 m), and the distance between the two points is about 920 ft (280.4 m).

Assume a slope of 1 ft/100 ft (0.3 m/30.5 m) of length, or 1/100 = 0.01 for the main sewer. This is a typical slope used for main sewers, and it is within the range permitted by a pipe run along the surface of this terrain. Table 21 shows the minimum slope required to produce a flow velocity of 2 ft/s (0.61 m/s).

Using Fig. 12 for a flow of 4.8 ft³/s (0.14 m³/s) and a slope of 0.01, we see the required size for section 3-4 of the main sewer is 15 in (381 mm). The flow velocity in the pipe is 4.88 ft/s (1.49 m/s). The size of this sewer is in keeping with general design practice, which seldom uses a storm sewer less than 12 in (304.8 mm) in diameter.

Section 5-6 conveys 9.6 ft³/s (0.27 m³/s). Using Fig. 12 again, we find the required pipe size is 18 in (457.2 mm) and the flow velocity is 5.75 ft/s (1.75 m/s). Likewise, section 7-8 must handle 14.4 ft³/s (0.41 m³/s). The required pipe size is 21 in (533 mm), and the flow velocity in the pipe is 6.35 ft/s (1.94 m/s). Section 9-10 of the main sewer handles 19.2 ft³/s (0.54 m³/s), and must be 24 in (609.6 mm) in diameter. The velocity in this section of the sewer pipe will be 6.9 ft/s (2.1 m/s). The last section of the main sewer handles the total flow, or 24 ft³/s

TABLE 21 Minimum Slope of Sewers°

Sewer diameter, in (mm)	Minimum slope, ft/100 ft (m/30.5 m) of length
4 (102)	1.20 (0.366)
6 (152)	0.60 (0.183)
8 (203)	0.40 (0.122)
10 (254)	0.29 (0.088)
12 (305)	0.22 (0.067)
15 (381)	0.15 (0.046)
18 (457)	0.12 (0.037)
20 (505)	0.10 (0.030)
24 (610)	0.08 (0.024)

°Based on the Manning formula with n = 0.13 and the sewer flowing either full or half full.

(0.7 m³/s). Its size must be 27 in (686 mm), Fig. 12, although a 24-in (610.0-mm) pipe would suffice if the slope at point *B* could be increased to 0.012.

Related Calculations. Most new sewers built today are the *separate* type, i.e., one sewer for sanitary service and another sewer for storm service. Sanitary sewers are usually installed first because they are generally smaller than storm sewers and cost less. *Combined sewers* handle both sanitary and storm flows and are used where expensive excavation for underground sewers is necessary. Many older cities have combined sewers.

To size a combined sewer, compute the sum of the maximum sanitary and storm-water flow for each section of the sewer. Then use the method given in this procedure after having assumed a value for *n* in the Manning formula and for the slope of the sewer main.

Where a continuous slope cannot be provided for a sewer main, a pumping station to lift the sewage must be installed. Most cities require one or more pumping stations because the terrain does not permit an unrestricted slope for the sewer mains. Motor-driven centrifugal pumps are generally used to handle sewage. For unscreened sewage, the suction inlet of the pump should not be less than 3 in (76 mm) in diameter.

SECTION 15
PLUMBING AND DRAINAGE FOR BUILDINGS AND OTHER STRUCTURES

Facilities Planning and Layout

WATER-METER SIZING AND LAYOUT FOR PLANT AND BUILDING WATER SUPPLY

Select a suitable water meter for a building having a maximum fresh water demand of 9000 gal/h (34,110 L/h) for process and domestic use. Choose a suitable storage method for the water and for an emergency reserve for fire protection when there are no local rivers or lakes for water storage. Show how the water-supply piping would be connected to a wet-pipe sprinkler system for fire protection of the building and its occupants.

Calculation Procedure:

1. *Determine a suitable water-meter size for the installation*
Refer to a water-meter manufacturer's data for the capacity rating of a suitable water meter. The American Water Works Association (AWWA) standard for cold water meters of the displacement type is designated AWWA C700-71. It covers displacement meters known as nutating- or oscillating-piston or disk meters, which are practically positive in action.

The standard establishes maximum output or delivery classifications for each meter size as follows:

⅝-in—20 gal/min (15.9 mm—1.26 L/s)
¾-in—30 gal/min (19 mm—1.89 L/s)
1-in—50 gal/min (25.4 mm—3.1 L/s)
1.5-in—100 gal/min (38.1 mm— 6.3 L/s)
2-in—160 gal/min (50 mm—10.1 L/s)
3-in—300 gal/min (75 mm—18.9 L/s)
4-in—500 gal/min (100 mm—31.5 L/s)
6-in—100 gal/min (150 mm—63 L/s)

The standard also establishes the maximum pressure loss corresponding to the standard maximum capacities as follows:

15 lb/in² (103 kPa) for the ⅝-in (15.9-mm), ¾-in (19.0-mm) and 1-in (25.4-mm) meter sizes

20 lb/in² (138 kPa) for the 1.5-in (38.1-mm), 2-in (50-mm), 3-in (75-mm), 4-in (100-mm), and 6-in (150-mm) meter sizes

For estimating pressure loss in displacement-type cold-water meters, Fig. 1 is provided. Pressure loss in meters for flow at less than the maximum rates for any given size of meter can be estimated from Fig. 1.

Since the maximum flow through the meter will be 9000 gal/h (34,110 L/h), we can convert this to gal/min by 9000 gal/h/60 min/h = 150 gal/min (568.5 L-

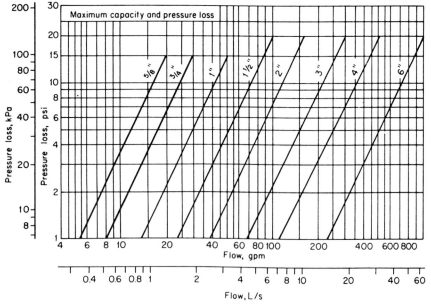

FIGURE 1 Pressure loss in displacement-type cold-water meters.

/min). Referring to the listing above, we see that a 2-in (50.8-mm) water meter will handle 160-gal/min (606.4 L/min). Since the required flow for this plant is 150 gal/min, a 2-in meter will be satisfactory.

Figure 2*a* shows how the 2-in water meter would be installed. Normal water-utility practice is to install two identical equal-size water meters with bypass piping and valves to allow cleaning or repair of one meter while the other is still in service. Where a compound meter will be installed, the piping would be laid out as shown in Fig. 2*b*.

2.   Choose the type of storage method for the system served

Fig. 3 shows three different arrangements for water storage at above-ground levels. The reservoir in Fig. 3*a* serves only the plant and domestic water needs. It does not have a provision for emergency water for fire-protection purposes.

The constant-head elevated tank in Fig. 3*b* has an emergency reserve for fire-fighting purposes. Local faire codes usually specify the reserve quantity required. The amount is usually a function of the building size, occupancy level, materials of construction, and other factors. Hence, the designer *must* consult the local applicable fire-prevention code before choosing the final capacity of the constant-head storage tank.

A vertical cylindrical standpipe is shown in Fig. 3*c*. While storing more water on the same ground area, this type of tank is sometimes thought to be visually less attractive than the elevated tanks in Fig. 3*a* and 3*b*.

The alternative to the tanks shown in Fig. 3 is an artificial lake, if space is available at the plant site. Such a solution has its own set of requirements: (1) Sufficient land area; (2) Suitable soil characteristics for water retention; (3) Fencing to prevent accidents and vandalism; (4) Approval by the local zoning board for construction of such a facility; (5) Treatment of the water prior to use to make it suitable for process and human use. A final decision on the choice of storage method is usually based on both economic factors and local zoning requirements.

3.   Show how the water supply would be connected to a wet-pipe sprinkler system

The most common types of fire-suppression systems rely on water as their extinguishing agent. Hence, it is essential that adequate supplies of water be available and be maintained available for use at all times.

The minimum recommended pipe size for fire protection is 6 in (152.4 mm). Where a pipe network is used for fire protection, a looped grid pattern is designed for the plant or building, or both. It is often cost-effective to use larger pipe sizes in a grid because the installation costs are relatively the same. Table 1 shows the relative pipe capacity for different size pipes.

The wet sprinkler system, Fig. 4, is connected to the plant water supply which can include a gravity tank, fire pump, reservoir or pressure tank and/or connection by underground piping to a city water main. As Fig. 4 shows, the sprinkler connection includes an alarm test valve, alarm shutoff and check valve, pressure gages for water and air, a fire-department connection to allow hookup of a pumper, and an air compressor.

Within the building itself, Fig. 5, the main riser is hooked into cross mains to supply each of the floors. The wet-pipe sprinkler system accounts for about 75 percent of the systems installed. Where freezing might occur in a building a dry-type sprinkler system is used.

Related Calculations.   Plumbing-system design begins at the water supply for the structure served. The most important objective in sizing the water-supply system is the satisfactory supply of potable water to all fixtures, at all times, and at proper

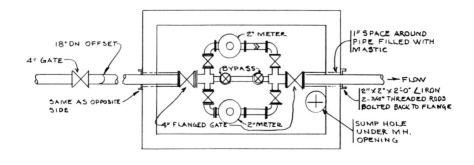

4" SERVICE – DUAL METERS

———— NOT TO SCALE ————

(a)

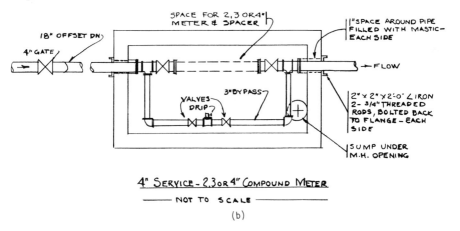

4" SERVICE – 2, 3 OR 4" COMPOUND METER

———— NOT TO SCALE ————

(b)

FIGURE 2   (*a*) Dual water-service meters installed in a pit; (*b*) Compound water-service meter installed in a pit. (*Mueller Engineering Corp.*)

pressure and flow rate for normal fixture operation. This goal is achieved only if adequate pipe sizes and fixtures are provided.

Pipe sizes chosen must be large enough to prevent negative pressures in any part of the system during peak demand. Such pipe sizes avoid the hazard of water-supply contamination caused by backflow and back siphonage from potential sources of pollution. One cause of backflow can be fire-engine pumpers connected to a water main and drawing water out of it in large quantities for fire-fighting use. Pressure in the water main can decrease quickly during such emergency uses, leading to back flow from a building's internal water system. Hence, sizing of building

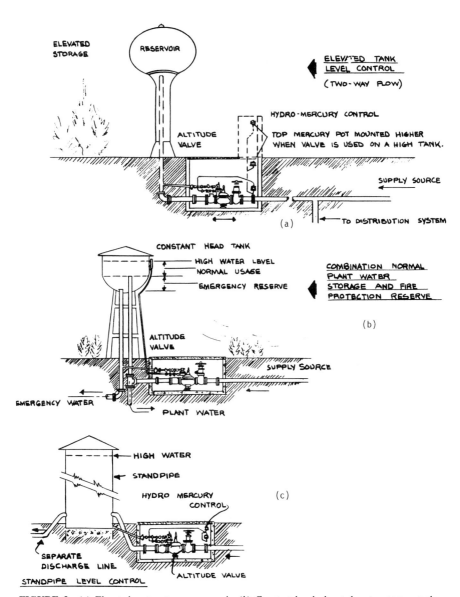

FIGURE 3 (*a*) Elevated water-storage reservoir. (*b*) Constant-head elevated water-storage tank having an emergency reserve for fire-fighting use. (*c*) Vertical standpipe for water storage. (*Mueller Engineering Corp.*)

TABLE 1   Table for Estimating Demand

Supply systems predominantly for flush tanks			Supply systems predominantly for Flushometer valves		
Load			Load		
Water supply fixture units (WSFU)	Demand		Water supply fixture units (WSFU)	Demand	
	gal/min	L/s		gal/min	L/s
1	3.0	0.19			
2	5.0	0.32			
3	6.5	0.41			
4	8.0	0.51			
5	9.4	0.59	5	15.0	0.95
6	10.7	0.68	6	17.4	1.10
7	11.8	0.74	7	19.8	1.25
8	12.8	0.81	8	22.2	1.40
9	13.7	0.86	9	24.6	1.55
10	14.6	0.92	10	27.0	1.70
12	16.9	1.01	12	28.6	1.80
14	17.0	1.07	14	30.2	1.91
16	18.0	1.14	16	31.8	2.01
18	18.8	1.19	18	33.4	2.11
20	19.6	1.24	20	35.0	2.21
25	21.5	1.36	25	38.0	2.40
30	23.3	1.47	30	42.0	2.65
35	24.9	1.57	35	44.0	2.78
40	26.3	1.66	40	46.0	2.90
45	27.7	1.76	45	48.0	3.03
50	29.1	1.84	50	50.0	3.15
60	32.0	2.02	60	54.0	3.41
70	35.0	2.21	70	58.0	3.66
80	38.0	2.40	80	61.2	3.86
90	41.0	2.59	90	64.3	4.06
100	43.5	2.74	100	67.5	4.26
120	48.0	3.03	120	73.0	4.61
140	52.5	3.31	140	77.0	4.86
160	57.0	3.60	160	81.0	5.11
180	61.0	3.85	180	85.5	5.39
200	65.0	4.10	200	90.0	5.68
250	75.0	4.73	250	101.0	6.37
300	85.0	5.36	300	108.0	6.81
400	105.0	6.62	400	127.0	8.01
500	124.0	7.82	500	143.0	9.02
750	170.0	10.73	750	177.0	11.17
1000	208.0	13.12	1000	208.0	13.12
1250	239.0	15.08	1250	239.0	15.08
1500	269.0	16.97	1500	269.0	16.97
2000	325.0	20.50	2000	325.0	20.50
2500	380.0	23.97	2500	380.0	23.97
3000	433.0	27.32	3000	433.0	27.32
4000	525.0	33.12	4000	525.0	33.12
5000	593.0	37.41	5000	593.0	37.41

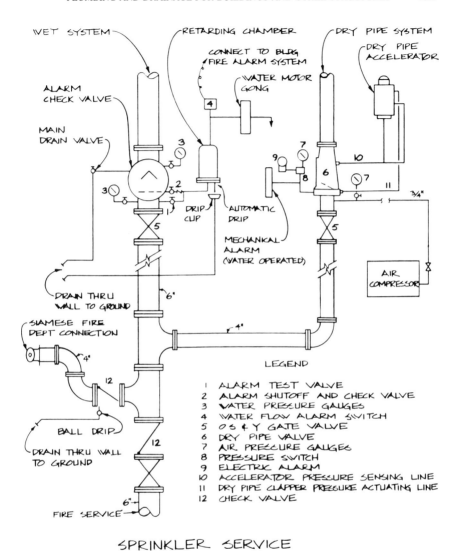

FIGURE 4 Wet-pipe sprinkler system service piping with typical fittings and devices. (*Mueller Engineering Corp.*)

water supply systems is a matter of vital concern in protecting health and is regulated by codes.

Other important objectives in the design of water-supply systems are: (1) to achieve economical sizing of piping and eliminate overdesign; (2) to provide against potential supply failure due to gradual reduction of pipe bore with the passing of time, such as may result from deposits of corrosion or hard-water scale in the

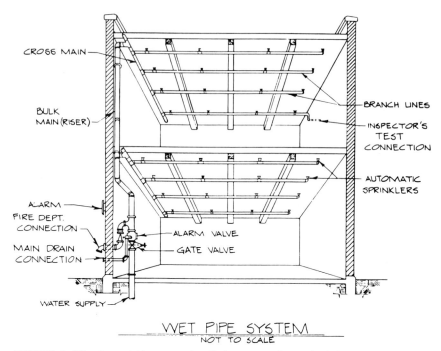

FIGURE 5 Wet-pipe sprinkler system installation on two floors of a building. (*Mueller Engineering Corp.*)

piping; (3) to avoid erosion-corrosion effects and potential pipe failure or leakage conditions owing to corrosive characteristics of the water and/or to excessive design velocities of flow; and (4) to eliminate water-hammer damage and objectional whistling noise effects in the piping due to excessive design velocities of flow.

Every designer of plumbing systems should familiarize himself/herself with the local plumbing code *before* starting to design. Then there will be fewer demands for re-design prior to final approval.

Data in this procedure come from the *National Plumbing Code,* Mueller Engineering Corporation, and L. C. Nelsen—*Standard Plumbing Engineering,* McGraw-Hill. SI values were added by the handbook editor.

PNEUMATIC WATER SUPPLY AND STORAGE SYSTEMS

Design a pneumatic water supply for use with (*a*) well-water pump, and (*b*) a municipal water supply augmented by an elevated water tank. Provide design criteria for each type of system.

Calculation Procedure:

1. *Determine the maximum water flow required for cold-, hot-, and process services*

Use the procedures given later in this section to determine the flow rate and pressure required for the building served. With a well-pump supply, Fig. 6, the pump should have a capacity to 1.5 times the maximum water flow required. Such a capacity will ensure that the pump does not operate continuously.

A booster system such as that shown in Fig. 7 is used when the city or private utility water system pressure is undependable—*i.e.*, the pressure may be consistently, or intermittently, lower than that required by various fixtures in the system. The booster pump discharge pressure is set so that it equals, or exceeds, that required by the fixtures or processes in the building. Water quantity supplied by the utility, public or private, is sufficient to meet the building demands. However, the utility pressure can vary unpredictably. As a rule of thumb, the pump must be capable of delivering a pressure at least 25 percent over that required in the plumbing supply system.

2. *Find the required air compressor discharge pressure for the system*

Well-water systems generally do not have the capacity to handle a building's peak water service demands. Hence, a storage tank of sufficient capacity to handle this demand is installed, Fig. 6, either underground or in the building itself. Once the water is in the storage tank, the well pump has served its purpose. A booster pump, Fig. 6, supplies the needed volume and pressure for the building water supply.

Since it is undesirable to have the booster pump operate continuously to supply needed water, a pressure tank and air compressor are fitted, Fig. 6. The air compressor maintains pressure on the water in the pressure tank sufficient to deliver water throughout the building at the desired pressure and in suitable quantities. Air pressure in the pressure tank is often set at 25 to 50 lb/in² (173 to 345 kPa) higher than the pressure needed in the water system. The pressure tank is provided with a pressure relief valve so excessive pressure are avoided.

Float switches in the storage and pressure tanks start the well-water or booster pump when the water level falls below a predetermined height. And when the

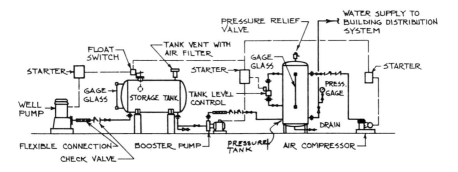

COMBINED STORAGE & HYDRO-PNEUMATIC WELL WATER SYSTEM

NO SCALE

FIGURE 6   Pneumatic well-water system for building service. (*Mueller Engineering Corp.*)

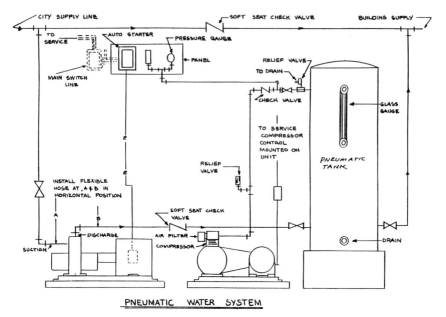

FIGURE 7 Pneumatic water system serving city-water supply. (*Mueller Engineering Corp.*)

hydraulic pressure in the pressure tank falls below a level sufficient to deliver the needed water throughout the building, the air compressor starts.

As a general rule, the minimum pressure required at ordinary faucets of plumbing fixtures is 8 lb/in² (55 kPa). At direct supply-connected flush valves (Flushometers), the minimum pressure should be 25 lb/in² (172 kPa) for blow-out-type water closets and 15 lb/in² (103 kPa) for other types of fixtures. For any type of plumbing fixture, domestic or process, the minimum pressure provided should be that recommended by the fixture manufacturer.

In a combined system, Fig. 7, there is a check valve in the bypass line around the booster system. This check valve is extremely important. The valve prevents back pressurization of the city water by the building booster system water which is at a higher pressure than the city water. Under normal operation the city water can only flow to the booster pump. Further, the booster pump cannot pull water backwards out of the pressurized building water system.

In a tall building a rooftop water storage tank can replace the booster system for the lower floors where there is sufficient head to operate the fixtures at the needed pressure. In a high-rise building the booster pump raises the water pressure sufficiently to overcome the static and friction pressure of the water-consuming fixtures on the upper floors. The booster system can also be designed to pump water into the rooftop storage tank for delivery to the lower floors.

Related Calculations. Pneumatic water systems find use in a variety of buildings: residential, commercial, industrial, etc. While they are more expensive than a simple metered system supplied at a suitable pressure and flow rate, pneumatic systems do ensure adequate water flow in buildings to which they are fitted. Where water flow is a critical concern, duplicate pumps, compressors, and tanks can be fitted.

Data in this procedure come from Mueller Engineering Corporation and L. C. Nielsen: *Standard Plumbing Engineering Design,* McGraw-Hill. SI values were added by the handbook editor.

SELECTING AND SIZING STORAGE-TANK HOT-WATER HEATERS

Size a domestic hot-water storage-tank heater for an office building with public toilets, pantry sinks, domestic-type dishwashers, and service sinks when the usable storage volume of the tank is 70 percent of the tank volume and the following numbers of fixtures are fitted: 16 lavatories; 6 sinks; 2 dishwashers; 2 service sinks. Use ASHRAE and ASPE information and representative hot-water temperatures and hot-water demand data in the computation.

Calculation Procedure:

1. Determine the hot-water consumption of the fixtures
ASHRAE publishes hot-water demand per fixture in the *ASHRAE Handbook, HVAC Applications.* Using data from that source, we have the following hot-water consumption: 16 lavatories at 2 gal/h = 32 gal/h; 6 sinks at 10 gal/h = 60 gal/h; 2 dishwashers at 15 gal/h = 30 gal/h; 2 service sinks at 20 gal/h = 40 gal/h; total possible maximum demand = 32 + 60 + 30 + 40 = 162 gal/h (614 L/h).

2. Find the probable maximum demand on the hot-water heater
ASHRAE publishes demand factors for a variety of hot-water services for apartment houses, clubs, gymnasiums, hospitals, hotels, industrial plants, office buildings, private residences, schools, YMCAs, etc. The ASHRAE demand factor for office buildings is 0.30. Hence, the probable maximum demand on the water heater = 162 × 0.30 = 48.6 gal/h (184 L/h).

3. Compute the storage capacity required for the hot-water heater
ASHRAE also publishes storage capacity factors for hot-water heaters in the reference cited above. For office buildings, the published storage capacity factor is 2.0. This is the ratio of storage-tank capacity to probable maximum demand per hour. Thus, for this heater, storage capacity without considering the usable storage volume = 48.6 × 2.0 = 97.2 gal (368 L).

Since 70 percent of the tank volume is the usable storage volume, the storage factor = 1/0.70 = 1.43. Then, storage capacity of the tank = 97.2 × 1.43 = 138.99 gal; say 139 gal (527 L).

Related Calculations.   There are a number of ways to generate hot water for commercial and institutional buildings. The most common method is to use a storage-tank type water heater, Fig. 8. Storage-type hot-water heaters generally are selected when the load profile has peaks that can be met from an adequate volume of hot water stored in the heater. Thus, the heater size and fuel/energy input are not based on the instantaneous peak load, permitting a more economical equipment selection.

Storage-tank hot-water heaters should be selected and sized based on the specific requirements for the building. Items to be considered in the selection process include: (1) type of facility served; (2) required water volume and peak loads; (3)

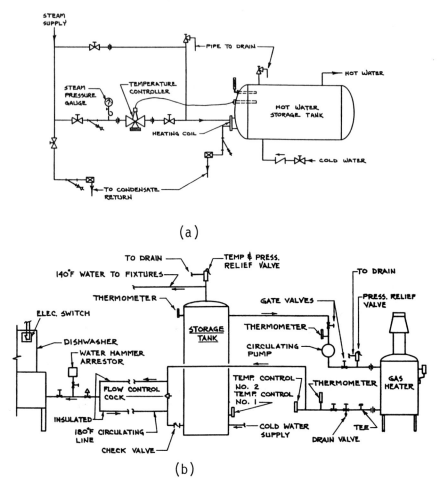

FIGURE 8 (*a*) Storage-tank hot-water heater. (*b*) Gas-fired hot-water heater. (*Mueller Engineering Corp.*)

type and number of fixtures served; (4) required water temperature(s); (5) fuel/energy sources for heating the water.

Storage-tank hot-water heaters may be heated either directly or indirectly by the fuel/energy source. Direct fuel-fired heaters may use either gas or fuel oil. In electric units the water is heated by resistance immersion heaters.

Indirect-fired storage hot-water heaters are heated by steam, hot water, or another hot fluid via a heat exchanger. This heat exchanger may be either within the water storage shell or remote from it.

Storage-tank hot-water heaters range in size from 2 to several thousand gallons (7.6 L to several thousand liters) capacity. The very small units are typically used in plumbing-code jurisdictions that prohibit the use of instantaneous hot-water heaters.

Typically, the maximum temperature for domestic hot water serving lavatories, showers, and sinks is approximately 120°F (49°C) at the fixture. The maximum desired water temperature from a fixture for personal use can be obtained by blending hot and cold water; mixing faucets are preferred over separate hot- and cold-water faucets. Or, thermostatic mixing valves may be installed near the point(s) of use. For bathing, a temperature-compensated shower valve should be used. The preferred type is a balanced-pressure model with a high-temperature limit.

ASHRAE lists hot-water utilization temperatures for various types and uses of equipment. Facilities requiring a higher water temperature than that required for personal use may have a separate hot-water heating system for the higher temperature water if there is a significant load. Otherwise, a booster heater often is used, as with a commercial dishwasher. The lowest temperature generally used is 75°F (24°C) for a chemical sanitizing glass washer, while the highest temperature is 195°F (91°C) in commercial hood or rack-type dishwashers.

Hot-water distribution temperatures may be higher than 120°F (49°C) because of the concern over *Legionella pneumophila* (Legionnaires' Disease). This bacterium, which can cause serious illness when inhaled, can grow in domestic hot-water systems at temperatures of 115°F (46°C), or less. Bacteria colonies have been found in system components, such as shower heads, faucet aerators, and in uncirculated sections of storage-type hot-water heaters.

A water temperature of approximately 140°F (60°C) is recommended to reduce the potential of growth of this bacterium. This higher temperature, however, increases the possibility of scalding during use of the water. Scalding is of particular concern for small children, the elderly and infirm, patients in health-care facilities, and occupants of nursing homes.

All storage-tank hot-water heaters are required to have temperature and pressure relief valves. Separate valves may be used, or a combination temperature/pressure-relief valve may be installed. Temperature-relief valves and combination temperature/pressure-relief valves must be installed so that the temperature-sensing element is located in the top 6-in (15.2-cm) of the storage tank.

The temperature-relief valve opens when the stored-water temperature exceeds 210°F (99°C). Its water discharge capacity should equal or exceed the heat input rating of the heater.

A thermal expansion tank, Fig. 9, should also be provided in the cold-water line adjacent to the heater whenever the system thermal expansion is restricted. Check

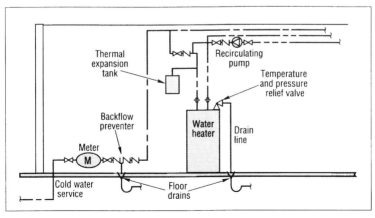

FIGURE 9 Water heater fitted with thermal expansion tank. (*Heating/Piping/Air Conditioning magazine*)

valves, pressure valves, and backflow preventers, when used on the cold-water line to the heater, restrict expansion of the water when it is heated. This results in excessive pressure buildup and can lead to tank failure. ASME construction is required on all heaters greater than 200,000 Btu/h (58.6 kW) gas input or 120 gal (455 L) storage. Additional data on sizing such hot-water heaters is available in the *ASPE Data Book,* published by the American Society of Plumbing Engineers. Use the steps in this procedure to select and size storage-tank hot-water heaters for the 10 types of applications listed in step 2 above, and for similar uses.

This procedure is the work of Joseph Ficek, Plumbing Designer, McGuire Engineers, as reported in *Heating/Piping/Air Conditioning* magazine, October, 1996. SI values were added by the handbook editor.

SIZING WATER-SUPPLY SYSTEMS FOR HIGH-RISE BUILDINGS

A 102-family multiple dwelling, seven stories and basement in height, fronts on a public street and is to be supplied by direct street pressure from an 8-in public water main located beneath the street in front of the building. The public system is of cast iron and a hydrant flow test indicates a certified minimum available pressure of 75 lb/in² (517 kPa). Top floor fixture outlets are 65 ft 8 in (20 m) above the public main and require 8 lb/in²flow pressure for satisfactory operation.

Authoritative water analysis reports show that the public water supply has a pH of 6.9, carbon dioxide content of 3 ppm, dissolved solids content of 40 ppm, and is supersaturated with air. Reports show that the public water supply has no significant corrosion effect on red brass for temperatures up to 150°F (65.6°C).

Cement-lined cast iron, class B, corporation water pipe, valves, and fittings have been selected for the water service pipe. Red brass pipe, standard pipe size, has been selected for the water distributing system inside the building.

Water supply for the building is to be metered at the point of entry by a compound meter installed in the basement. The system is to be of the upfeed riser type. A horizontal hot water storage tank is to provide hot water to the entire building, and is to be equipped with automatic tank control of water temperature set for 140°F (60°C). The tank is to have a submerged heat exchanger.

The most extreme run of piping from the public main to the highest and most remote outlet is 420 ft (128 m) in developed length, consisting of the following: 83 ft (25.3 m) of water service, 110 ft (33.5 m) of cold water piping from the water service valve to the hot water storage tank, and 227 ft (69.2 m) of hot water piping from the tank to the top floor hot water outlet at the kitchen sink. Plans of the entire water supply system are available.

The building has a basement and seven above-grade stories. The basement floor is 3 ft 8 in (1.1 m) below curb level, the first floor is 5.0 ft (1.5 m) above curb level, and the public water main is 5.0 ft (1.5 m) below curb level. Each of the above-grade stories is 9 ft 4 in in height from floor to floor. The highest fixture outlet is 3 ft above floor level.

Fixtures provided on the system for the occupancies are as follows:

1. There are 17 dwelling units on each of the second, third, fourth, fifth, sixth, and seventh floors; and each dwelling unit is provided with a sink and domestic dishwashing machine in the kitchen, and a close-coupled water closet and flush tank combination, a lavatory, and a bathtub with shower head above in a private bathroom.

2. This first floor is occupied for administrative and general purposes, and has the following provisions for such occupancy: one flush-valve supplied water closet and one lavatory in an office toilet room; one flush-valve supplied water closet, one flush-valve supplied urinal and one lavatory in a men's toilet room; two flush-valve supplied water closets and one lavatory in each of two women's toilet rooms; a sink and domestic dishwashing machine in a demonstration kitchen; one sink in an office kitchen; one sink in a craft room; and two drinking fountains in the public hall.

3. The basement is occupied for building equipment rooms, storage, utility, laundry, and general purposes and has the following provisions for such occupancy: one flush-valve supplied water closet and one lavatory in a women's toilet room; one flush-valve supplied water closet, one lavatory, and one shower stall in a men's toilet room; one service sink and six automatic laundry washing machines in a general laundry room; one faucet above a floor drain in the boiler room; and one valve-controlled primary water supply connection to the building heating system.

4. At each story and in the basement, a service sink is provided in a janitor's closet in the public hall.

5. Four outside hose bibs (only two to be used at any time) are provided for lawn watering at appropriate locations on the exterior of the building.

Fixture arrangements are typical on the six upper floors of the building, and 24 sets of risers are provided. Of these, 5 sets are for back-to-back bathrooms, 2 sets are for back-to-back kitchens, 4 sets are for back-to-back kitchen and bathroom groups, 9 sets are for separate kitchens, 3 sets are for separate bathrooms, and one set is for a service sink on each floor above the basement. Fixtures on the first floor are connected to adjacent risers. Basement fixtures are connected to overhead mains, which also supply directly the four outside hose bibs.

Design a suitable water-supply systems for this building. Choose pipe sizes for each riser, fluid velocity, pressure drop, and piping material.

Calculation Procedure:

1. *Assemble the information needed for the design*
Obtain data on the applicable plumbing code, characteristics of the water supply, location and source of the water supply, pressure available at the water entrance to the site, elevations associated with the height of the building, minimum pressure required at the highest water outlets, and any special water services required in the building. Contact local responsible authorities for any missing data over which they have control. You must have as much pertinent information as possible before the design job is started.

2. *Prepare a schematic elevation of the building water-supply system*
Figure 10 shows a schematic elevation of the building water-supply system being designed in this procedure. This drawing was developed using the building and system plans. All piping connections are shown in proper sequence for the system. The developed lengths for each section of the basic design circuit are determined from the building and system plans. Fixtures and risers are identified by combinations of letters and numbers. Those fixtures and branches having quick-closing outlets are specially identified by an asterisk. Important information for establishing a proper design basis are shown on the left side of Fig. 10.

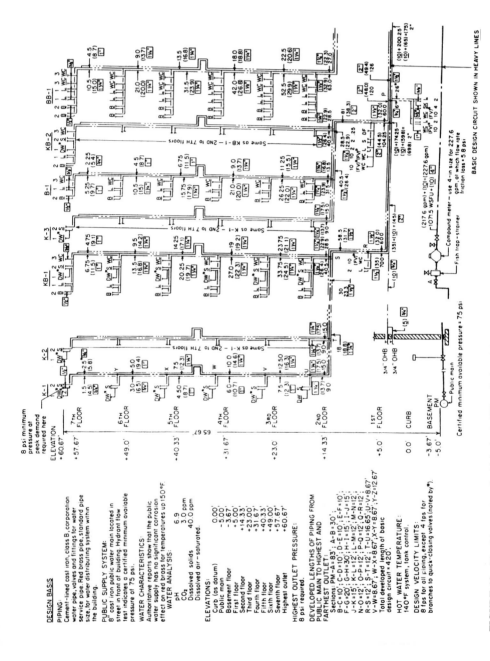

FIGURE 10 Plumbing system for high-rise building designed in the accompanying procedure.

15.16

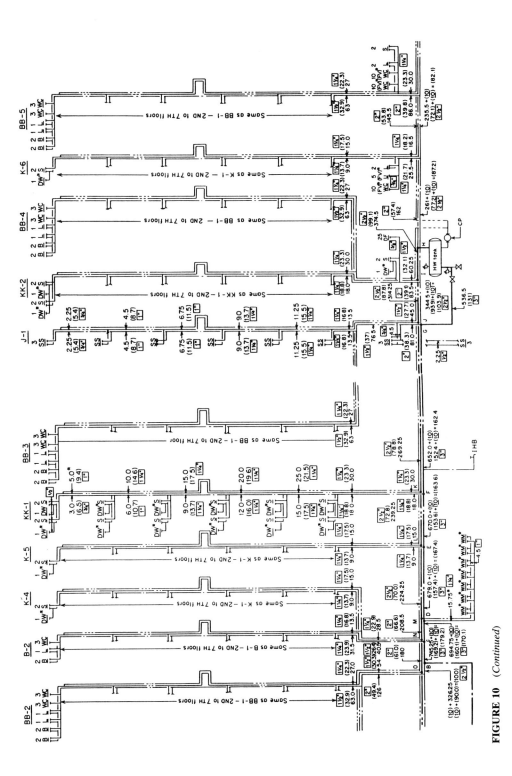

FIGURE 10 (*Continued*)

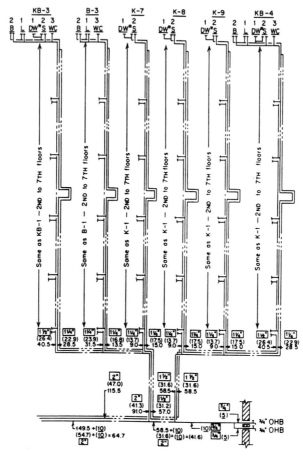

KEY FOR NOTATIONS MADE IN SIZING PROCEDURE
Load in water supply fixture units is shown unenclosed.
Load in gpm of flow is shown in ( ).
Continuous load in gpm of flow is shown in (_).
Sizes selected are shown in ☐.

FIGURE 10   (*Continued*)

3. Show hot- and cold-water loads for each section in terms of water-supply fixture units

List fixture-unit values as shown unenclosed by parentheses. Obtain the fixture-unit values from tabulations as given later in this procedure.

4. List the demand in gal/min (L/s) adjacent to the fixture-unit load

Use Table 1 to determine the demand in gal/min (L/s), applying the values shown under the heading "Supply Systems Predominantly for Flush Tanks" for all piping except for the short branch piping which supplies water to water closets and urinals equipped with flush valves on the first floor and in the basement. (This procedure uses both flush tanks and flush valves to show how to handle both in design. Remember: Flush tanks are still widely used in developing countries around the world.)

5. Determine the water demands of any special fixture
The special fixtures in this building are the four outside hose bibs, Fig. 10. Only two of these hose bibs will be used at the same time. Show this on the design sheet, along with the flow in gal/min (L/s). Obtain the normal demand for these fixtures from Table 2.

6. Size the individual fixture supply pipes to water outlets
Use Standard Code Regulations to size these pipes, as given in Table 11, later in this section of the handbook. Choose the minimum sizes recommended in Table 11.

7. Using velocity limitations established for the design, size the remainder of the system
The velocity limitations adopted for this system are 8 ft/s (2.4 m/s) for all piping, except 4 ft/s (1.2 m/s) for branches to quick-closing valves as noted by asterisks on Fig. 10. Size each line using the total fixture units of load corresponding to the total demand of each section. For those sections of the cold-water header in the basement which convey both the demand of the intermittently used fixtures and the continuous demand of hose bibs, the total demand in gal/min (L/s) was converted to equivalent water-supply fixture units of load and proper pipe sizes determined for them. Proper sizing could also have been done simply on the demand rate in gal/min (L/s).

8. Calculate the amount of pressure available at the topmost fixture
Assume conditions of no flow in the system and calculate the amount of pressure available at the topmost fixture in excess of the minimum pressure required at such

TABLE 2 Demand at Individual Water Outlets

	Demand	
Type of outlet	gal/min	L/s
Ordinary lavatory faucet	2.0	0.126
Self-closing lavatory faucet	2.5	0.158
Sink Faucet, ⅜″ (9.52 mm) or ½″ (12.7 mm)	4.5	0.284
Sink faucet, ¾″ (19 mm)	6.0	0.378
Bath faucet, ½″ (12.7 mm)	5.0	0.315
Shower head, ½″ (12.7 mm)	5.0	0.315
Laundry faucet, ½″ (12.7 mm)	5.0	0.315
Ball cock in water closet flush tank	3.0	0.189
1″ (25.4 mm) flush valve [25 lb/in² (172 kPa) flow pressure]	35.0	2.210
1″ (25.4 mm) flush valve [15 lb/in² (103 kPa) flow pressure]	27.0	1.703
¾″ (19.0 mm) flush valve [15 lb/in² (103 kPa) flow pressure]	15.0	0.946
Drinking fountain jet	0.75	0.047
Dishwashing machine (domestic)	4.0	0.252
Laundry machine [8 lb (3.6 kg) or 16 lb (7.3 kg)]	4.0	0.252
Aspirator (operating room or laboratory)	2.5	0.158
Hose bib or sill cock, ½″ (12.7 mm)	5.0	0.315

a fixture for satisfactory supply conditions. The calculated excess pressure is the limit to which friction losses may be permitted for flow during peak demand in the system. Then, excess pressure = 75 lb/in^2 − 8 lb/in^2 − (65.67 ft to highest outlet × 0.433 lb/in^2/ft of water) = 38.6 lb/in^2 (266 kPa). (*Note:* 1 ft of water column = 0.433 lb/in^2 and 1 m of water column = 9.79 kPa pressure).

9. Determine which piping circuit of the system is the basic design circuit (BDC)

The basic design circuit (BDC) is the most extreme run of piping through which water flows from the public main, or other pressure source of supply, to the highest and most distant water outlet. Heavy lines in Fig. 10 show the BDC for this structure.

There are 26 sections in the BDC in Fig. 10. For each of these sections, the developed length is computed as shown in Fig. 10, for a total of 420 ft (128 m). Then, using the BDC length and other data for the installation, the pressure loss in the BDC, is found thus, as shown in Table 3.

10. Mark on the system schematic the pressure loss through any special fixtures in the system

Obtain from the special fixture manufacturer(s) the rated pressure loss due to friction corresponding to the computed demand through any water meter, water softener, or instantaneous or tankless hot-water heating coil that may be provided in the basic design circuit.

Thus, the rated pressure loss through the compound water meter selected for this system was found from the manufacturer's meter data to be 5.8 lb/in^2 (40 kPa) for the peak demand flow rate of 227.6 gal/min (862.6 L/min). Note this on the design sheet, Fig. 10. The rated pressure loss for flow through the horizontal hot-

TABLE 3 Pressure Calculations for Basic Design Circuit

Minimum at public main	75.0 lb/in^2
Loss in rise to top outlet (65.67 ft × 0.433)	−28.4 lb/in^2
Static pressure at top outlet	46.6 lb/in^2
Minimum pressure at top outlet	− 8.0 lb/in^2
Excess static pressure at top outlet available for friction loss	38.6 lb/in^2
Friction loss through 4-in compound meter at 227 gal/min flow rate (manufacturer's charts)	− 5.8 lb/in^2
	32.8 lb/in^2
Friction loss through horizontal hot water storage tank assumed for rate flow at 8 ft/s	− 0.7 lb/in^2
Maximum pressure remaining for friction in pipe, valves, and fittings	32.1 lb/in^2
Developed length of circuit from public main to top outlet	420 ft
Equivalent length for valves and fittings in circuit (based on sizes established on velocity limitation basis)	363 ft
Total equivalent length of circuit	783 ft
Maximum uniform pressure loss for friction in basic design circuit (32.1 lb/in^2/783 ft)	0.04 lb/in^2/ft
	or 4.0 lb/in^2/100 ft

water storage tank, *i.e.*, entrance and exit losses, is assumed to be about 1.6 ft head (0.49 m), 0.7 lb/in² (4.8 kPa).

11.   *Calculate the amount of pressure remaining*
We must now calculate the amount of pressure remaining and available for dissipation as friction loss during peak demand through the piping, valves, and fittings in the basic design circuit. Deduct from the excess static pressure available at the topmost fixture (determined in step 8) the rated friction losses for any water meters, water softeners, or water heating coils provided in the basic design circuit, as determined in step 10.

The thus, the amount of pressure available for dissipation as friction loss during peak demand through the piping, valves, and fittings in the BDC is: 38.6 − 5.8 − 0.7 = 32.1 lb/in² (221 kPa).

12.   *Compute the total equivalent length of the basic design circuit*
Pipe sizes established on the basis of velocity limitations in step 7 for main lines and risers must be considered just tentative at this stage, but may be deemed appropriate for determining the corresponding equivalent lengths of fittings and valves in this step. Using the tentative sizes for the BDC, compute corresponding equivalent lengths for valves and fittings. Add the values obtained to the developed length to obtain the total equivalent length of the circuit.

The equivalent length of valves and fittings, using the methods given elsewhere in this handbook, is 363.2 ft (110.7 m). When added to the developed length, we have a total equivalent length of the BDC of 420 + 363.6 = 783.2 ft (238.7 m).

13.   *Calculate the permissible uniform pressure loss for friction in the piping of the BDC*
The amount of pressure available for dissipation as friction loss due to pipe, fittings, and valves, determined in step 11, should be divided by the total equivalent length of the circuit, determined in step 12. This establishes the pipe friction limit for the circuit in terms of pressure loss, in lb/in²/ft (Pa/m) for the total equivalent pipe length. Multiply this value by 100 to express the pipe friction in terms of lb/in² per 100 ft (Pa/100 m).

Thus, the maximum uniform pressure loss for friction in the basic design circuit is: 32.1/783.2 ft = 0.04 lb/in²/ft, or 4.0 lb/in²/100 ft (0.9 kPa/100 m). This is the pipe friction for the BDC. Apply it for sizing all the main lines and risers supplying water to fixtures on the upper floors of the building.

14.   *Set up a pipe sizing table showing the rates of flow for the system*
Set up the sizing table showing the rates of flow based on the permissible uniform pressure loss for the pipe friction calculated for the basic design circuit determined in step 13. In Table 4, the flow rates have been tabulated for various sizes of brass pipe of standard internal diameter that correspond to the velocity limit of 4 and 8 ft/s (1.2 and 2.4 m/s), and to the friction limit of 4.0 lb/in²/100 ft (0.9 kPa/100 m) of total equivalent piping length. The values shown for various velocity limitations were taken from the data cited in step 7. Values shown for friction limitations were taken directly from Fig. 11. This chart is suitable, in view of the water-supply conditions and a "fairly smooth" surface condition.

15.   *Adjust the chosen pipe sizes, as necessary*
All the main lines and risers on the design sheet have been sized in accordance with the friction limitation for the basic design circuit. Where sizes determined in

TABLE 4 Sizing Table for System
Red brass pipe, standard pipe size

| Nominal pipe size, in | Velocity limit flow rate at | | | | Friction limit flow rate at 4.0 lb/in²/100 ft, gal/min |
| | V = 4 ft/s | | V = 8 fps | | |
	WSFU (col. A)	gal/min	WSFU (col. A)	gal/min	
½	1.5	3.8	3.7	7.6	2.8
¾	3.0	6.6	8.4	13.2	5.8
1	6.3	11.1	26.4	22.0	11.7
1¼	16.8	18.3	75.0	36.6	22.5
1½	36.3	25.2	130.0	50.4	33.0
2	92.0	41.6	291.0	83.2	66.0
2½	181.0	61.2	492.0	122.4	112.0
3	335.0	92.0	842.0	184.0	288.0
4	685.0	158.0	1920.0	316.0	380.0

Note: Apply the column headed "Velocity limit, l' = 4 ft/s," to size branches to quick-closing valves. Apply the column headed "Velocity limit, l' = 8 ft/s," to all piping other than individual fixture supplies. Apply the column headed "Friction limit," just for sizing piping that conveys water to top floor outlets. Where two columns apply and two different sizes are indicated, select the larger size.

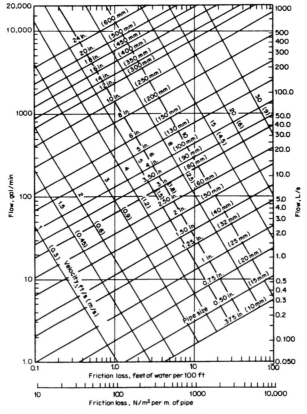

FIGURE 11 Water-piping pressure-loss chart.

this step were larger than previously determined in step 7, based on velocity limitation, the increased size was noted directly on the design sheet. Increased sizes were made in all risers and in some parts of the main lines in this system. For example, in the BDC, sections J-K, K-L, and L-M were increased from 2-in (50.8-mm) to 2.5-in (63.6-mm); sections O-P and P-Q were increased from 1.5-in (38.1-mm) to 2-in (50.8-mm); sections Q-R, R-S, and S-T were increased from 1.25-in to 1.5-in (31.8-mm to 38.1-mm); T-U, U-V, and V-W were increased from 1-in to 1.25-in (25.4-mm to 31.8-mm); section W-X was increased from 0.75-in to 1.25-in (19-mm to 31.8-mm); and section X-Y was increased from 0.75-in to 1-in (19-mm to 25.4-mm).

16. *Determine if the water supply is such that pipe sizing must be changed*
From the characteristics of the water supply given by the municipal authority, it is recognized that the water is relatively noncorrosive and nonscaling. Hence, there is no need for additional allowance in sizing in this case.

Related Calculations.   The method given here is valid for a variety of water-supply designs for apartment houses, hotels, commercial and industrial buildings, clubhouses, schools, hospitals, retirement homes, nursing homes, and residences of all sizes. As a designer, you should be certain to follow all applicable plumbing codes so the system meets every requirement of the locality.

This procedure is the work of L. C. Nielsen, as given in his *Standard Plumbing Engineering Design,* McGraw-Hill. SI values were added by the handbook editor.

Plumbing-System Design

DETERMINATION OF PLUMBING-SYSTEM PIPE SIZES

A two-story industrial plant has the following plumbing fixtures: first floor—six wall-lip urinals, three valve-operated water closets, three large-size lavatories, and six showers, each with a separate head; second floor—three wall-lip urinals, three valve-operated water closets, three large-size lavatories, and three showers, each with a separate head. Size the waste and vent stacks and the building house drain for this system. Use the *National Plumbing Code* (*NPC*) as the governing code for the plant locality. The branch piping and house drain will be pitched ¼ in (6.4 mm) per ft (m) of length.

Calculation Procedure:

1. *Select the upper-floor branch layout*
Sketch the layout of the proposed plumbing system, beginning with the upper, or second, floor. Figure 12 shows a typical plumbing-system sketch. Assume in this plant that the second-floor urinals, water closets, and lavatories are served by one branch drain and the showers by another branch. Both branch drains discharge into a vertical soil stack.

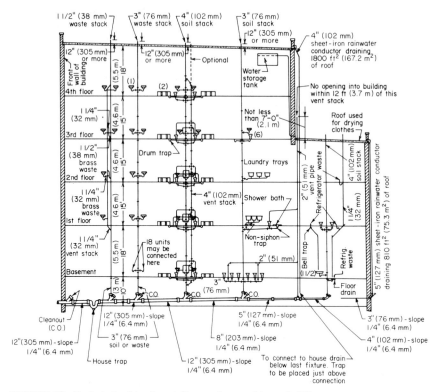

FIGURE 12   Typical plumbing layout diagram for a multistory building.

2. Compute the upper-floor branch fixture units
List each plumbing as in Table 5.

Obtain the data for each numbered column of Table 5 in the following manner.
(1) List the number of the floor being studied and number of each branch drain
from the system sketch. Since it was decided to use two branch drains, number
them accordingly. (2) List the name of each fixture that will be used. (3) List the
number of each type of fixture that will be used. (4) Obtain from the *National
Plumbing Code,* or Table 6, the number of *fixture units per fixture,* i.e., the average
discharge, during use, of an arbitrarily selected fixture, such as a lavatory or toilet.
Once this value is established in a plumbing code, the discharge rates of other types
of fixtures are stated in terms of the basic unit. Plumbing codes adopted by various
localities usually list the fixture units they recommend in a tabulation similar to
Table 6. (5) Multiply the number of fixtures, column 3, by the fixture units, column
4, to obtain the result in column 5. Thus, for the urinals, (3 urinals)(4 fixture units
per urinal fixture) = 12 fixture units. Find the sum of the fixture units for each
branch.

3. Size the upper-floor branch pipes
Refer to the *National Plumbing Code,* or Table 7, for the number of fixture units
each branch can have connected to it. Thus, Table 7 shows that a 4-in (102-mm)

TABLE 5 Floor-Fixture Analysis

(1) Floor	(2) Fixture name	(3) No. of fixtures	(4) Fixture units per fixture	(5) Total no. of fixture units
Floor 2	Urinals, wall-lip	3	4	12
Branch drain 1	Water closets, valve- operated	3	8	24
	Lavatories, large-size	3	2	6
Total	. . .	9	. . .	42
Branch drain 2	Showers	3	3	9
Total	. . .	3	. . .	9
Floor 1	Urinals, wall-lip	6	4	24
	Water closets, valve- operated	3	8	24
Branch drain 3	Lavatories, large-size	3	2	6
Total	. . .	12	. . .	54
Branch drain 4	Showers	6	3	18
Total	. . .	6	. . .	18

branch pipe must be used for branch drain 1 because no more than 20 fixture units can be connected to the next smaller, or 3-in (76-mm) pipe. Hence, branch drain 1 will use a 4-in (102-mm) pipe because it serves 42 fixture units, step 2.

Branch drain 2 serves 9 fixture units, step 2. Hence, a 2½-in (64-mm) branch pipe will be suitable because it can serve 12 fixture units or less (Table 7).

4. *Size the upper-floor stack*

The two horizontal branch drains are sloped toward a vertical *stack* pipe that conducts the waste and water from the upper floors to the sewer. Use Table 7 to size the stack, which is three stories high, including the basement. The total number of second-floor fixture units the stack must serve is 42 + 9 = 51. Hence, for a 4-in (102-mm) stack, Table 7 must be used.

5. *Size the upper-story vent pipe*

Each branch drain on the upper floor must be vented. However, the stack can be extended upward and each branch vent connected to it, if desired. Use the *NPC,* or Table 8, to determine the vent size.

TABLE 6 Fixture Units per Fixture or Group°

Fixture type	Fixture-unit value as load factors	Minimum size of trap, in (mm)
One bathroom group consisting of water closet, lavatory, and bathtub or shower stall	Tank water closet, 6; flush-valve water closet, 8	
Bathtub† (with or without overhead shower)	2	1½ (38)
Bathtub†	3	2 (51)
Bidet	3	Nominal, 1½ (38)
Combination sink and tray	3	1½ (38)
Combination sink and tray with food-disposal unit	4	Separate traps, 1½ (38)
Dental unit or cuspidor	1	1¼ (32)
Dental lavatory	1	1¼ (32)
Drinking fountain	½	1 (25)
Dishwasher, domestic	2	1½ (38)
Floor drains‡	1	2 (51)
Kitchen sink, domestic	2	1½ (38)
Kitchen sink, domestic, with food-waste grinder	3	1½ (38)
Lavatory§	1	Small PO, 1¼ (32)
Lavatory§	2	Large PO, 1½ (38)
Lavatory, barber, beauty parlor	2	1½ (38)
Lavatory, surgeon's	2	1½ (38)
Laundry tray (one or two compartments)	2	1½ (38)
Shower stall, domestic	2	2 (51)
Showers (group) per head	3	
Sinks:		
Surgeon's	3	1½ (38)
Flushing rim (with valve)	8	3 (76)
Service (trap standard)	3	3 (76)
Service (P trap)	2	2 (51)
Pot, scullery, etc.	4	1½ (38)
Urinal, pedestal, siphon jet, blowout	8	Nominal, 3 (76)
Urinal, wall lip	4	1½ (38)
Urinal, stall, washout	4	2 (51)
Urinal trough [each 2-ft (0.61-m) section]	2	1½ (38)
Wash sink (circular or multiple) each set of faucets	2	Nominal, 1½ (38)
Water closet, tank-operated	2	Nominal, 3 (76)
Water closet, valve-operated	8	3 (76)

°From *National Plumbing Code.*
†A shower head over a bathtub does not increase the fixture value.
‡Size of floor drain shall be determined by the area of surface water to be drained.
§Lavatories with 1¼- (32-mm) or 1½-in (38-mm) trap have the same load value; larger PO (plumbing orifice) plugs have greater flow rate.

TABLE 7 Horizontal Fixture Branches and Stacks°

Diameter of pipe, in (mm)	Maximum number of fixture units that may be connected to			
	Any horizontal† fixture branch	One stack of three stories in height or three intervals	More than three stories in height	
			Total for stack	Total at one story or branch interval
1¼ (32)	1	2	2	1
1½ (38)	3	4	8	2
2 (51)	6	10	24	6
2½ (64)	12	20	42	9
3 (76)	20‡	30§	60§	16‡
4 (102)	160	240	500	90
5 (127)	360	540	1100	200
6 (152)	620	960	1900	350

°From *National Plumbing Code.*
†Does not include branches of the building drain.
‡Not over two water closets.
§Not over six water closets.

TABLE 8 Sizes of Building Drains and Sewers°

Diameter of pipe, in (mm)	Maximum number of fixture units that may be connected to any portion† of the building drain or the building sewer			
	Fall per foot (meter)			
	⅟₁₆ in (1.6 mm)	⅛ in (3.2 mm)	¼ in (6.4 mm)	½ in (12.7 mm)
2 (51)	. . .	. . .	21	26
2½ (64)	. . .	. . .	24	31
3 (76)	. . .	20‡	27‡	36‡
4 (102)	. . .	180	216	250
5 (127)	. . .	390	480	575
6 (152)	. . .	700	840	1000

°From *National Plumbing Code.*
†Includes branches of the building drain.
‡Not over two water closets.

As a guide, the diameter of a branch vent or vent stack is one-half or more of the branch or stack it serves, but not less than 1¼ (32 mm). Thus branch drain 1 would have a 4/2 = 2-in (51-mm) vent, whereas branch drain 2 would have a 2½/2 = 1¼-in (32-mm) vent.

6. Select the lower-floor branch layout
Assume that the six urinals, three water closets, and three lavatories are served by one branch drain and the six showers by another. Indicate these on the system sketch. Further, arrange both branch drains so that they discharge into the vertical stack serving the second floor.

7. Compute the lower-floor branch fixture units
Use the same procedure as in step 2, listing the fixtures and their respective fixture units in the lower part of Table 5.

8. Size the lower-floor branch pipes
By Table 7, branch drain 3 must be 4 in (102 mm) because it serves a total of 54 fixture units. Branch 4 must be 3 in (76 mm) because it serves a total of 18 fixture units.

9. Size the lower-floor stack
The lower-floor stack serves both the upper- and lower-floor branch drains, or a total of $42 + 9 + 54 + 18 = 123$ fixture units. Table 7 shows that a 4-in (102-mm) stack will be satisfactory.

10. Size the lower-floor vents
By the one-half rule of step 5, the vent for branch drain 3 must be 2 in (51 mm), whereas that for branch drain 4 must be $1\frac{1}{2}$ in (38 mm).

11. Size the building drain
The building drain serves all the fixtures installed in the building and slopes down toward the city sewer. Hence, the total number of fixture units it serves $42 + 9 + 54 + 18 = 123$. This is the same as the vertical stack. Table 8 shows that a 4-in 9102-mm) drain that is sloped $\frac{1}{4}$-in/ft (21 mm/m) will serve 216 fixture units. Thus, a 4-in (102-mm) drain will be satisfactory. The house trap that is installed in the building drain should also be a 4-in (102-mm) unit.

   Related Calculations.   Where a local plumbing code exists, use it instead of the *NPC*. If no local code exists, follow the *NPC* for all classes of buildings. Use the general method given here to size the various pipes in the system. Select piping materials (cast iron, copper, clay, steel, brass, wrought iron, lead, *etc.*) in accordance with the local or *NPC* recommendations. Where the house drain is below the level of the public sewer line, it is often arranged to discharge into a suitably size *sump pit*. Sewage is discharged from the sump pit to the public sewer by a pneumatic ejector or motor-drive pump.

   With the increased emphasis on the environmental aspects of plumbing, many large cities are urging building owners to convert water closets to Ultra-Low-Flow (ULF) units. These ULF closets use 1.6 gal (6.06 L) per flush, as contrasted to 5 to 7 gal (18.95 to 26.5 L) per flush for conventional water closets. Thus, in a building having 300 water closets the water savings could range up to $[(300 \times 7) - (300 \times 1.6)] = 1620$ gal (6140 L) with just one flush per unit per day. With an average of ten flushes per day per water closet, the daily saving could be $(10 \times 1620) = 16,200$ gal (61,398 L). Using a 5-day week for an office or industrial building and a 52-week working year, the water savings could be (5 days)(52 weeks)(16,200 gal/day) = 4,212,000 gal (15,963,480 L) per year.

   When water savings of this magnitude are translated into reduced pumping power, lower electricity costs, and smaller piping sizes, the savings can be significant. This is why many large cities around the world are urging building owners to install ULF water closets and urinals, along with reduced-flow shower heads, and lavatories.

   Once disadvantage of ULF units is that the reduced water flow can cause accumulation of solids in horizontal drain piping. To remove solids, the horizontal pipes must be snaked out at regular intervals, depending on the system usage. While this does not occur in every installation of ULF units, it is being studied to determine possible remedies.

Building owners in one large city are currently receiving a bonus of $240 for each ULF water closet installed in an existing structure. This is leading to wide-scale replacement of existing water closets which use excessive amounts of water, in view of today's new environmental laws and regulations.

A further benefit of the ULF units is the smaller amount of water that must be treated for each flush. The reduced water flow allows the central sewage treatment plant to handle more buildings and their water closets, showers, sinks, and other fixtures. As cities grow, it is important that sewage-treatment plants be able to handle and process the increased waste flow. Thus, the ULF unit saves water during usage and reduces the post-usage need for waste-water treatment. It is for these two reasons that large cities are urging building owners to install ULF units.

DESIGN OF ROOF AND YARD RAINWATER DRAINAGE SYSTEMS

An industrial plant is 300 ft (91.4 m) long and 100 ft (30.5 m) wide. The roof of the building is flat except for a 50-ft (15.2-m) long, 100-ft (30.5-m) wide, 80-ft (24.4-m) high machinery room at one end of the roof. Size the leaders and horizontal drains for this roof for a maximum rainfall of 4 in/h (102 mm/h). What size storm drain is needed if the drain is sloped $\frac{1}{4}$ in/ft (2.1 cm/m) of length?

Calculation Procedure:

1. *Sketch the building roof*
Figure 13 shows the building roof and machinery room roof. Indicate on the sketch the major dimensions of the roof and machinery room.

2. *Compute the roof area to be drained*
Two roof areas must be drained, the machinery-room roof and the main roof. The respective areas are: machinery room roof area = 50 × 100 = 5000 ft² (464.5 m²); main roof area = 250 × 100 = 25,000 ft² (2322.5 m²).

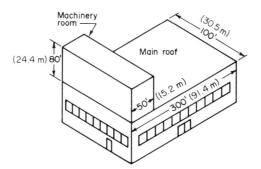

FIGURE 13 Building roof areas.

The wall of the machinery room facing the main roof will also collect rain to some extent. This must be taken into consideration when the roof leaders are sized. Do this by computing the area of the wall facing the main roof and adding one-half this area to the main roof area. Thus, wall area = 80 × 100 = 8000 ft² (743.2 m²). Adding half this area to the main roof area gives 25,000 + 8000/2 = 29,000 ft² (2694 m²).

3. Select the leader size for each roof

Decide whether the small roof area, *i.e.*, the machinery room roof, will be drained by separate leaders to the ground or to the main roof area. If the small roof area is drained separately, treat it as a building unto itself. Where the small roof drains onto the main roof, add the two roof areas to determine the leader size.

By treating the two roofs as separate units, Table 9 shows that a 5-in (127-mm) leader is needed for the 5000-ft² (464.5-m²) machinery room roof. This same table shows that an 8-in (203-mm) leader is needed for the 29,000-ft² (2694-m²) main roof, including the machinery room wall.

TABLE 9   Sizes of Vertical Leaders and Horizontal Storm Drains°

Vertical leaders	
Size of leader or conductor,† in (mm)	Maximum projected roof area, ft² (m²)
2 (51)	720 (66.9)
2½ (64)	1,300 (120.8)
3 (76)	2,200 (204.4)
4 (102)	4,600 (427.3)
5 (127)	8,650 (803.6)
6 (152)	13,500 (1,254.2)
8 (203)	29,000 (2,694)

Horizontal storm drains

Diameter of drain, in (mm)	Maximum projected roof area for drains of various slopes, ft² (m²)		
	⅛-in (3.2-mm) slope	¼-in (6.4-mm) slope	½-in (12.7-mm) slope
3 (76)	822 (76.4)	1,160 (107.8)	1,644 (152.7)
4 (102)	1,880 (174.4)	2,650 (246.2)	3,760 (349.3)
5 (127)	3,340 (310.3)	4,720 (438.5)	6,680 (620.6)
6 (152)	5,350 (497.0)	7,550 (701.4)	10,700 (994.0)
8 (203)	11,500 (1,068.4)	16,300 (1,514.3)	23,000 (2,136.7)
10 (254)	20,700 (1,923.0)	29,200 (2,712.7)	41,400 (3,846.1)
12 (305)	33,300 (3,093.6)	47,000 (4,366.3)	66,600 (6,187.1)
15 (381)	59,500 (5,527.6)	84,000 (7,803.6)	119,000 (11,055.1)

°From *National Plumbing Code*.
†The equivalent diameter of square or rectangular leader may be taken as the diameter of that circle that may be inscribed within the cross-sectional area of the leader.

4. *Size the storm drain for each roof*
The lower portion of Table 9 shows that a 6-in (152-mm) storm drain is needed for the 5000-ft² (464.5-m²) roof. A 10-in (254-mm) storm drain (Table 9) is needed for the 29,000-ft² (2694-m²) main roof.

When any storm drain is connected to a building sanitary drain or storm sewer, a trap should be used at the inlet to the sanitary drain or storm sewer. The trap prevents sewer gases entering the storm leader.

Related Calculations.   Size roof leaders in strict accordance with the *National Plumbing Code* (*NPC*) or the local applicable code. Undersized roof leaders are dangerous because they can cause water buildup on a roof, leading to excessive roof loads. Where gutters are used on a building, size them in accordance with Table 10.

When a roof leader discharges into a sanitary drain, convert the roof area to equivalent fixture units to determine the load on the sanitary drain. To convert roof area to fixture units, take the first 1000 ft² (92.9 m²) of roof area as equivalent to 256 fixture units when designing for a maximum rainfall of 4 in/h (102 mm/h). Where the total roof area exceeds 1000 ft² (92.9 m²), divide the remaining roof area by 3.9 ft² (0.36 m²) per fixture unit to determine the fixture load for the remaining area.

Thus, the machinery room roof in the above plant is equivalent to 256 + 4000/3.9 = 1281 fixture units. The main roof and machinery room wall are equivalent to 256 + 28,000/3.9 = 7436 fixture units. These roofs, if taken together, would place a total load of 1281 + 7436 = 8717 fixture units on a sanitary drain.

Where the rainfall differs from 4 in/h (102 mm/h), compute the load on the drain in the same way as described above. Choose the drain size from the appropriate table. Then multiply the drain size by actual maximum rainfall, in (mm)/4. If the drain size obtained is nonstandard, as will often be the case, use the next *larger* standard drain size. Thus, with a 6-in (152-mm) rainfall and a 5-in (127-mm) leader based on the 4-in (102-mm) rainfall tables, leader size = (5)(6/4) = 7.5 in (191 mm). Since this is not a standard size, use the next larger size, or 8 in (203 mm). Roof areas should be drained as quickly as possible to prevent excessive structural stress caused by water accumulations.

To compute the required size of drains for paved areas, yards, courts, and courtyards, use the same procedure and tables as for roofs. Where the rainfall differs

TABLE 10   Size of Gutters°

Diameter of gutter,† in (mm)	Maximum projected roof area for gutters of various slopes, ft² (m²)			
	¹⁄₁₆-in (1.6-mm) slope	⅛-in (3.2-mm) slope	¼-in (6.4-mm) slope	½-in (12.7-mm) slope
3 (76)	170 (15.8)	240 (22.3)	340 (31.6)	480 (44.6)
4 (102)	360 (33.4)	510 (47.4)	720 (66.9)	1,020 (94.8)
5 (127)	625 (58.1)	880 (81.8)	1,250 (116.1)	1,770 (164.4)
6 (152)	960 (89.2)	1,360 (126.3)	1,920 (178.4)	2,770 (257.3)
7 (178)	1,380 (128.2)	1,950 (181.2)	2,760 (256.4)	3,900 (362.3)
8 (203)	1,990 (184.9)	2,800 (260.1)	3,980 (369.7)	5,600 (520.2)
10 (254)	3,600 (334.4)	5,100 (473.8)	7,200 (668.9)	10,000 (929.0)

°From *National Plumbing Code.*
†Gutters other than semicircular may be used provided they have an equivalent cross-sectional area.

from 4 in (102 mm), apply the conversion ratio discussed in the previous paragraph. Note that the flow capacity of floor and roof drains must equal, or exceed, the flow capacity of the leader to which either unit is connected.

SIZING COLD- AND HOT-WATER-SUPPLY PIPING

An industrial building has the following plumbing fixtures: 2 showers, 200 private lavatories, 200 service sinks, 20 public lavatories, 1 dishwasher, 25 flush-valve water closets, and 20 stall urinals. Size the cold- and hot-water piping for these fixtures, using an upfeed system. The highest fixture is 50 ft (15.2 m) above the water main. The minimum water pressure available in the water main is 60 lb/in² (413.6 kPa); the pressure loss in the water meter is 8.3 lb/in² (57.2 kPa).

Calculation Procedure:

1. Sketch the proposed piping system
Draw a single-line diagram of the proposed cold- and hot-water piping. Thus, Fig. 14a shows the proposed basement layout of the water piping, and Fig. 14b shows two of the risers used in this industrial plant. Indicate on each branch line the "weight" in *fixture units* of fixtures served and the required water flow. Table 11 shows the rate of flow and required pressure during flow to different types of fixtures.

2. Compute the demand weight of the fixtures
List the fixtures as in Table 12. Next to the name and number of each fixture, list the demand weight for cold or hot water, or both, from Table 13. Note that when a fixture has both a cold-water and hot-water supply, only three-fourths of the fixture weight listed in Table 13 is used for each cold-water and each hot-water outlet. Thus, with a total demand weight of 1 for a private lavatory, the cold-water demand weight is 0.75(1) = 0.75 fixture unit, and the hot-water demand weight is 0.75(1) = 0.75 fixture unit.

Find the product of the number of each type of fixture and the demand weight per fixture for cold and hot water; enter the result in the last two columns of Table 12. The sum of the cold- and hot-water fixture demand weights, 986 and 636 fixture units, respectively, gives the total demand weight for the building, in fixture units, except for the dishwasher.

3. Compute the building water demand
Using Fig. 15a, enter at the bottom with the number of fixture units and project vertically upward to the curve. At the left read the demand—210 gal/min (13.3 L/s) of cold water and 160 gal/min (10.1 L/s) of hot water, excluding the dishwasher.

Table 13 shows that a dishwasher serving 500 people in an industrial plant requires 250 gal/h (0.26 L/s) with a demand factor of 0.40. This is equivalent to a demand of (demand, gal/h) (demand factor), or (250)(0.40) = 100 gal/h (0.11 L/s) or 100 gal/h/(60 min/h) = 1.66 gal/min (0.10 L/s), say 2.0 gal/min (0.13 L/s). Hence, the total hot-water demand is 160 + 2 = 162 gal/min (10.2 L/s). The total building water demand is therefore 210 + 162 = 372 gal/min (23.5 L/s).

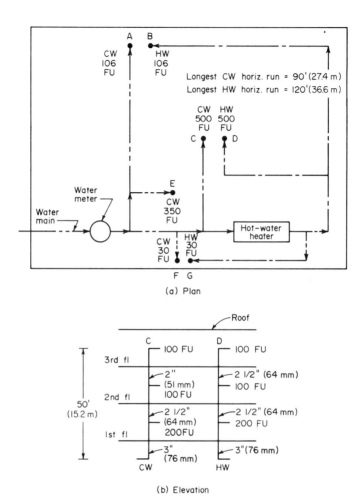

FIGURE 14 (*a*) Plan of industrial-plant water piping; (*b*) elevation of building water-supply risers.

4. Compute the allowable piping pressure drop

The minimum inlet water pressure generally recommended for a plumbing fixture is 8 lb/in² (55.2 kPa), although some authorities use a lower limit of 5 lb/in² (34.5 kPa). Flushometers normally require an inlet pressure of 15 lb/in² (103.4 kPa). Table 11 lists the usual inlet pressure and flow rates required for various plumbing fixtures.

Assume a 15-lb/in² (103.4-kPa) inlet pressure at the highest fixture. This fixture is 50 ft (15.2 m) above the water main (Fig. 14). To convert elevation in feet to pressure in pounds per square inch, multiply by 0.434, or (50 ft)(0.434) = 21.7 lb/in² (149.6 kPa). Last, the pressure loss in the water meter is 8.3 lb/in² (57.2 kPa), as given in the problem statement. Thus, the pressure loss in this or any other water-supply system, not considering piping friction loss, is fixture inlet pressure,

TABLE 11   Rate of Flow and Required Pressure during Flow for Different Fixtures°

Fixture	Flow pressure†		Flow rate	
	lb/in²	kPa	gal/min	L/s
Ordinary basin faucet	8	55.2	3.0	0.19
Self-closing basin faucet	12	82.7	2.5	0.16
Sink faucet, ⅜ in (9.5 mm)	10	68.9	4.5	0.28
Sink faucet, ½ in (12.7 mm)	5	34.5	4.5	0.28
Bathtub faucet	5	34.5	6.0	0.38
Laundry-tub cock, ½ in (12.7 mm)	5	34.5	5.0	0.32
Shower	12	82.7	5.0	0.32
Ball cock for closet	15	103.4	3.0	0.19
Flush valve for closet	10–20	68.9–137.9	15–40‡	0.95–2.5
Flush valve for urinal	15	103.4	15.0	0.95
Garden hose, 50 ft (15.2 m) and sill cock	30	206.8	5.0	0.32

°From *National Plumbing Code.*
†Flow pressure is the pressure in the pipe at the entrance to the particular fixture considered.
‡Wide range due to variation in design and type of flush-valve closets.

TABLE 12   Fixture Demand Weight

Fixture name	No. of fixtures	Demand weight per fixture in fixture units		Total fixture demand weight in fixture units	
		Cold water	Hot water	Cold water	Hot water
Shower	2	3	3	6	6
Lavatory, private	200	0.75	0.75	150	150
Lavatory, public	20	1.5	1.5	30	30
Sink, service	200	2.25	2.25	450	450
Dishwasher	1	. . . .	°	. . .	. . .
Water closet, flush-valve	25	10	. . .	250	. . .
Urinal, stall	20	5	. . .	100	. . . .
Total	. . .	. . . .	. . .	986	636

°Not given in *National Plumbing Code* tabulation.

lb/in², + vertical elevation loss, lb/in², + water-meter pressure loss, lb/in² = 15 + 21.7 + 8.3 = 45 lb/in² (310.2 kPa). Hence, the pressure available to overcome the piping frictional resistance = 60 − 45 = 15 lb/in² (103.4 kPa).

Note: The pressure loss in water meters of various sizes can be obtained from manufacturers' engineering data, or Fig. 16, for disk-type meters.

5. Compute the allowable friction loss in the piping
Figure 14a shows that the longest horizontal run of pipe is 90 + 50 = 140 ft (42.7 m). Allowing 50 percent of the straight run for the equivalent length of valves and fittings in the longest run and riser gives the total equivalent length of cold-water piping as 140 + 0.50 = 210 ft (64.0 m).

Compute the allowable friction loss per 100 ft (30.5 m) of cold water pipe from $F = 100$ (pressure available to overcome piping frictional resistance, lb/in²)/equiv-

TABLE 13 Demand Weight of Fixtures in Fixture Units°

Fixture or group†	Occupancy	Type of supply control	Weight in fixture units‡
Water closet	Public	Flush valve	10
Water closet	Public	Flush tank	5
Pedestal urinal	Public	Flush valve	10
Stall or wall urinal	Public	Flush valve	5
Stall or wall urinal	Public	Flush tank	3
Lavatory	Public	Faucet	2
Bathtub	Public	Faucet	4
Shower head	Public	Mixing valve	4
Service sink	Office, etc.	Faucet	3
Kitchen sink	Hotel or restaurant	Faucet	4
Water closet	Private	Flush valve	6
Water closet	Private	Flush tank	3
Lavatory	Private	Faucet	1
Bathtub	Private	Faucet	2
Shower head	Private	Mixing valve	2
Bathroom group	Private	Flush valve for closet	8
Bathroom group	Private	Flush tank for closet	6
Separate shower	Private	Mixing valve	2
Kitchen sink	Private	Faucet	2
Laundry trays (one to three)	Private	Faucet	3
Combination fixture	Private	Faucet	3

°From *National Plumbing Code.* For supply outlets likely to impose continuous demands, estimate continuous supply separately and add to total demand for fixtures.
†For fixtures not listed, weights may be assumed by comparing the fixture to a listed one using water in similar quantities and at similar rates.
‡The given weights are for total demand. For fixtures with both hot- and cold-water supplies, the weights for maximum separate demands may be taken as three-fourths the listed demand for supply.

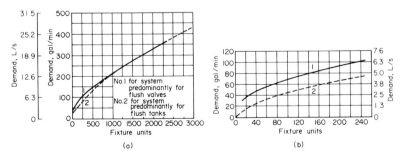

FIGURE 15 (*a*) Domestic water demand for various fixtures; (*b*) enlargement of low-demand portion of (*a*).

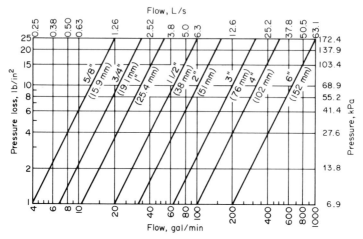

FIGURE 16   Pressure loss in disk-type water meters.

alent length of cold-water piping, ft. Or, $F = 100(15)/210 = 7.14$ lb/in^2 per 100 ft (1.62 kPa/m); use 7.0 lb/in^2 per 100 ft (1.58 kPa/m) for design purposes.

By the same procedure for the hot-water pipe, $F = 100(15)/255 = 5.88$ lb/in^2 per 100 ft (1.33 kPa/m); use 5.75 lb/in^2 per 100 ft (1.30 kPa/m). Reducing the design pressure loss for the cold- and hot-water piping design pressure loss to the next lower convenient pressure is done only to save time. If desired, the actual computed valve can be used. *Never* round off to the next higher convenient pressure loss because this can lead to undersized pipes and reduced flow from the fixture.

6.   Size the water main
Step 3 shows that the total building water demand is 372 gal/min (23.5 L/s). Using the cold-water friction loss of 7.0 lb/in^2 per 100 ft (1.58 kPa), enter Fig. 17 at the bottom at 7.0 and project vertically upward to 372 gal/min (23.5 L/s). Read the main size as 4 in (102 mm). This size would be run to the water heater (Fig. 14) unless the run were extremely long. With a long run, the main size would be reduced after each branch takeoff to the risers to reduce the cost of the piping.

7.   Compute the water flow in each riser
List the risers in Fig. 14 as shown in Table 14. Next to the letter identifying a riser, list the water it handles (hot or cold), the number of fixture units served by the riser, and the flow. Find the flow by entering Fig. 15 with the number of fixture units served by the riser and projecting up to the flush-valve curve. Read the gallons per minute (liters per second) at the left of Fig. 15.

8.   Choose the riser size
Enter the pressure loss, lb/in^2 per 100 ft (kPa per 30.5 m), found in step 5 next to each riser (Table 14). Using Fig. 15 and the appropriate pressure loss, size each riser and enter the chosen size in Table 14. Thus, riser *A* conveys 70 gal/min (4.4 L/s) with a pressure loss of 7.0 lb/in^2 per 100 ft (1.58 kPa/m). Figure 15 shows that a 2-in (50.8-mm) riser is suitable. When Fig. 15 indicate a pipe size that is between two standard pipe sizes, use the next *larger* pipe size.

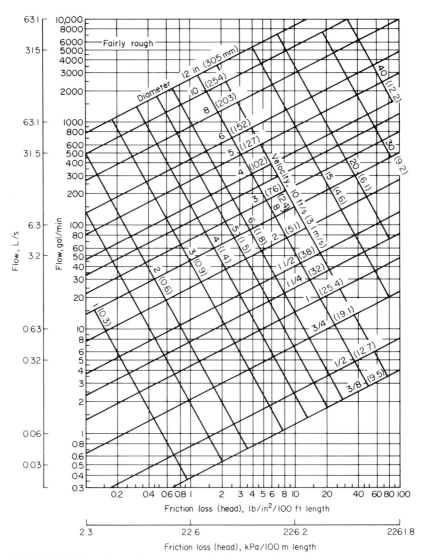

FIGURE 17   Chart for selecting water-pipe size for various flow rates.

9.   Choose the fixture supply-pipe size
Use Table 15 as a guide for choosing the fixture supply-pipe size. Note that these tabulated sizes are the minimum recommended. Where the supply-pipe run is more than 3 ft (1 m), or where more than one fixture is served, use a larger size.

10.   Select the hot-water-heater capacity
Table 16 shows that the demand factor for a hot-water heater in an industrial plant is 0.40 times the hourly hot-water demand. Step 3 shows that the total hot-water

TABLE 14 Riser Sizing Calculations

Riser	Type[*]	Fixture units	Rate gal/min	Rate L/s	Pressure loss lb/in^2	Pressure loss kPa	Riser size in	Riser size mm
A	CW	106	70	4.4	7.0	48.3	2	51
B	HW	106	70	4.4	5.75	39.6	2½	64
C	CW	500	150	9.5	7.0	48.3	3	76
D	HW	500	150	9.5	5.75	39.6	3	76
E	CW	350	130	8.2	7.0	48.3	2½	64
F	CW	30	40	2.5	7.0	48.3	2	51
G	HW	30	40	2.5	5.75	39.6	2	51

[*]CW stands for cold water; HW stands for hot water.

TABLE 15 Minimum Sizes for Fixture-Supply Pipes[°]

Type of fixture or device	Pipe size in	Pipe size mm	Type of fixture or device	Pipe size in	Pipe size mm
Bathtubs	½	12.7	Shower (single head)	½	12.7
Combination sink and tray	½	12.7	Sinks (service, slop)	½	12.7
Drinking fountain	⅜	9.5	Sinks, flushing rim	¾	19.1
Dishwasher (domestic)	½	12.7	Urinal (flush tank)	½	12.7
Kitchen sink, residential	½	12.7	Urinal (direct flush valve)	¾	19.1
Kitchen sink, commercial	¾	19.1	Water closet (tank type)	⅜	9.5
Lavatory	⅜	9.5	Water closet (flush valve type)	1	25.4
Laundry tray, one, two or three compartments	½	12.7	Hose bibs	½	12.7
			Wall hydrant	½	12.7

[°]From *National Plumbing Code.*

demand is 162 gal/min (10.2 L/s), or 162(60) = 9720 gal/h (10.2 L/s). Therefore, this hot-water heater must have a heating coil capable of heating at least 0.4(9720) = 3888 gal/h, say 3900 gal/h (4.1 L/s).

TABLE 16 Hot-Water Demand per Fixture for Various Building Types[°]

Type of fixture	Apartment house	Hospital	Hotel	Industrial plant	Office building
Basins, private lavatories	2 (7.6)	2 (7.6)	2 (7.6)	2 (7.6)	2 (7.6)
Basins, public lavatories	4 (15.1)	6 (22.7)	8 (30.3)	12 (45.4)	6 (22.7)
Showers	75 (283.9)	75 (283.9)	75 (283.9)	225 (851.6)	—
Slop sinks	20 (75.7)	20 (75.7)	30 (113.6)	20 (75.7)	15 (56.8)
Dishwashers (per 500 people)	250 (946.3)	250 (946.3)	250 (946.3)	250 (946.3)	250 (946.3)
Pantry sinks	5 (18.9)	10 (37.9)	10 (37.9)	—	—
Demand factor	0.30 (1.1)	0.35 (1.32)	0.25 (0.95)	0.40 (1.51)	0.30 (1.1)
Storage factor	1.25 (4.7)	0.60 (2.27)	0.80 (3.0)	1.00 (3.79)	2.00 (7.57)

[°]Based on average conditions for types of buildings listed, gallons of water per hour (liters per hour) per fixture at 140°F (60°C).

The storage capacity should equal the product of hourly water demand and the storage factor from Table 16. Thus, storage capacity = 9720(1.0) = 9720 gal (36,790 L). Table 17 shows the usual hot-water temperature used for various services in different types of structures.

Related Calculations.   Size the risers serving each floor, using the same procedure as in steps 6 and 7. Thus, risers *C* and *D* are each 3 in (76 mm) up to the first-floor branch. Between this and the second-floor branch, a 2½-in (64-mm) riser is needed. Between the second and third floors, a 2-in (51-mm) cold-water riser and a 2½-in (64-mm) hot-water riser are needed.

In a *downfeed* water-supply system, an elevated roof tank generally supplies cold water to the fixtures. To provide a 15-lb/in² (103.4-kPa) inlet pressure to the highest fixtures, the bottom of the tank must be (15 lb/in²)(2.31 ft · in²/lb of water) = 34.6 ft (10.6 m) above the fixture inlet. Where this height cannot be obtained because the building design prohibits it, tank-type fixtures requiring only a 3 lb/in² (20.7 kPa) or (3 /b/in²)(2.31) = 6.93-ft (2.1-m) elevation at the fixture inlet may be used on the upper floors. Valve-type fixtures are used on the lower floors where the tank elevation provides the required 15-lb/in² (103.4 kPa) inlet pressure.

To design a downfeed system: (*a*) Compute the pressure available at the highest fixture resulting from the tank elevation from lb/in² = 0.434 (tank elevation above inlet to highest fixture, ft) (9.8 kPa/m). (*b*) Subtract the required inlet pressure to the highest fixture from the pressure obtained in *a*. (*c*) Compute the pressure available to overcome the friction in 100 ft (30.5 m) of piping, using the method of step 5 of the upfeed design procedure and substituting the value found in item *b*. (*d*) Size the main from the tank so it is large enough to provide the needed flow to all the upper- and lower-floor fixtures. (*e*) Note that the pressure in each supply main increases as the distance from the tank bottom becomes greater. Thus, the hydraulic pressure increases 0.43 lb/in² · ft) (9.8 kPa/m) of distance from the tank bottom. Usual design practice allows a 15-lb/in² (103.4-kPa) drop through the fittings and valves in the main. The remaining pressure produced by the tank elevation is then available for overcoming pipe friction.

Note that both cold and hot water can be supplied from separate overhead tanks. However, hot water is usually supplied from the building basement by a pump. In exceptionally high buildings, water tanks may be located on several intermediate floors as well as the roof. Hot-water heaters may also be located on intermediate floors, although the usual location is in the basement.

In a *zoned* system, one water tank and one set of hot-water heaters serve several floors or one or more wings of a building. The piping in each zone is designed as described above, using the appropriate method for an upfeed or downfeed system.

To provide hot water as soon as possible after a fixture is opened, the water may be continuously recirculated to the fixtures (Fig. 18). Recirculation is used with both upfeed and downfeed systems. To determine the required hot-water temperature in a system, use Table 13, which shows the usual hot-water temperatures used for various services in buildings of different types. Hot-water piping is generally insulated to reduce heat loss.

TABLE 17   Hot-Water Temperatures for Various Services, °F

Cafeterias (serving areas)	130	(54)
Lavatories and showers	130	(54)
Slop sinks (floor cleaning)	150	(65.6)
Slop sinks (other cleaning)	130	(54)
Cafeteria kitchens	130 + steam	(54 + steam)

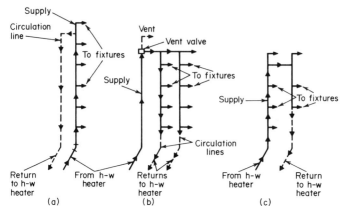

FIGURE 18   Hot-water piping systems.

SPRINKLER-SYSTEM SELECTION AND DESIGN

Select and design a sprinkler system for the warehouse building shown in Fig. 19. The materials stored in this warehouse are not flammable. The warehouse is built of fire-resistive materials.

Calculation Procedure:

1. Determine the type of occupancy of the building
The classifications of occupancy used by the National Board of Fire Underwriters (NBFU) are (1) light hazard, such as apartment houses, asylums, clubhouses, col-

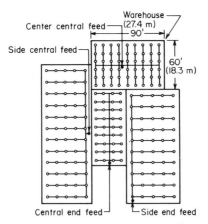

FIGURE 19   Typical arrangement of sprinkler piping.

leges, churches, dormitories, hospitals, hotels, libraries, museums, office buildings, and schools; (2) ordinary hazards, such as mercantile buildings, warehouses, manufacturing plants, and occupancies not classed as light or extra hazardous; (3) extrahazard occupancies, *i.e.*, those buildings or portions of buildings where the inspection agency having jurisdiction determines that the hazard is severe.

Since this is a warehouse used to store nonflammable materials, it can be tentatively classed as an ordinary-hazard occupancy.

2. *Compute the number of sprinkler head required*
Consult the local fire-prevention code and the fire underwriters regarding the type, size, and materials required for sprinkler systems. Typical codes recommend that each sprinkler head in an ordinary-hazard fire-resistive building protect not more than 100 ft² (9.29 m²) and that the sprinkler branch pipes, and the sprinklers themselves, be not more than 12 ft (3.7 m) apart, center to center.

The area of the warehouse floor is 90(60) = 5400 ft² (501.7 m²). With each sprinkler protecting 100 ft² (9.29 m²) of area, the number of sprinkler heads required is 5400 ft²/100 ft² (501.7 m²/9.29 m²) per sprinkler = 54 heads.

3. *Sketch the sprinkler layout*
If the warehouse has a centrally located support column or piping cluster, a center central feed pipe (Fig 19) can be used. Assuming the sprinkler branch pipes are spaced on 10-ft (3.05-m) centers, sketch the branches and heads as shown in Fig. 19. Use a small circle to indicate each sprinkler head.

Space the end sprinkler heads and branch pipes away from the walls by an amount equal to one-half the center-to-center distance between branch pipes. Thus, the end sprinkler heads and branch pipes will be 10/2 = 5 ft (1.5 m) from the walls.

4. *Size the branch and main sprinkler pipes*
Use the local code, Fig. 20, or Table 18. Table 18 shows that a 1¼-in (32-mm) branch line will be suitable for three sprinklers in an ordinary-hazard occupancy such as this warehouse. Hence, each branch line having three sprinklers will be this size.

The horizontal overhead main supplying the branches will progressively decrease in diameter as it runs farther from the vertical center central feed and serves fewer sprinklers. To the right of the vertical feed, the horizontal main serves 30 sprinklers. Table 18 shows that a 3-in (76-mm) pipe can serve up to 40 sprinklers. Hence, this size will be used because the next smaller size, 2½ in (64 mm), can serve only 20 sprinklers.

Since the first branch has six sprinklers, a 3-in (76-mm) pipe is still needed for the main because Table 18 shows that a 2½-in (64-mm) pipe can serve only 20, or fewer, sprinklers. However, beyond the second branch, the diameter of the horizontal main can be reduced to 2½ (64 mm) because the number of sprinklers serves is 30 − 12 = 18. Beyond the fourth branch, the main size can be reduced to 2 in (51 mm) because only six sprinklers are served. Size the left-hand horizontal main in the same way.

The vertical center central feed pipe serves 54 sprinklers. Hence, a 3½-in (89-mm) pipe must be used, according to Table 18.

5. *Choose the primary and secondary water supply*
Usual codes require that each sprinkler system have two water supplies. The primary supply should be automatic and must have sufficient capacity and pressure to

in	mm
1/2	12.7
5/8	15.9
3/4	19.1
1-1/4	32
1-1/2	38
2	51
2-1/2	64
3	76
3-1/2	89
4	102
5	127

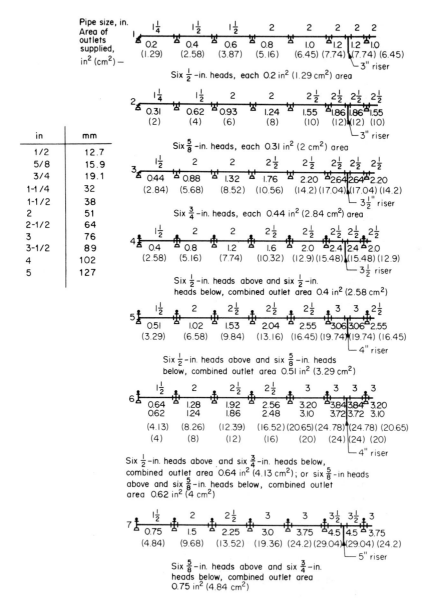

FIGURE 20 Sprinkler pipe sizes.

TABLE 18 Pipe-Size Schedule for Typical
Sprinkler Installations

Occupancy and pipe size, in (mm)	No. of sprinklers
Light hazard	
1 (25)	2
1¼ (32)	3
1½ (38)	5
2 (51)	10
2½ (64)	40
3 (76)	No limit
Ordinary hazard	
1 (25)	2
1¼ (32)	3
1½ (38)	5
2 (51)	10
2½ (64)	20
3 (76)	40
3½ (89)	65
4 (102)	100
5 (127)	160
6 (152)	250
Extra hazard	
1 (25)	1
1¼ (32)	2
1½ (38)	5
2 (51)	8
2½ (64)	15
3 (76)	27
3½ (89)	40
4 (102)	55
5 (127)	90
6 (152)	150

serve the system. Local codes usually specify the minimum pressure and capacity acceptable for sprinklers serving various occupancies.

The secondary supply is often a motor-driven, automatically controlled fire pump supplied from a water main or taking its suction under pressure from a storage system having sufficient capacity to meet the water requirements of the structure protected.

For light-hazard occupancy, the pump should have a capacity of at least 250 gal/min (15.8 L/s); when the pump supplies both sprinklers and hydrants, the capacity should be at least 500 gal/min (31.5 L/s). Where the occupancy is classed as an ordinary hazard, as this warehouse is, the capacity of the pump should be at least 500 gal/min (31.6 L/s) or 750 gal/min (47.3 L/s), depending on whether hydrants are supplied in addition to sprinklers. For extra-hazard occupancy, consult the underwriter and local fire-protection authorities.

Related Calculations.   For fire-resistive construction and light-hazard occupancy, the area protected by each sprinkler should not exceed 196 ft² (18.2 m²), and the center-to-center distance of the sprinkler pipes and sprinklers themselves should not exceed 14 ft (4.3 m). For extra-hazard occupancy, the area protected by each sprinkler should not exceed 90 ft² (8.4 m²); the distance between pipes and between sprinklers should not be more than 10 ft (3.05 m). Local fire-protection codes and underwriters' requirements cover other types of construction, including mill, semimill, open-joist, and joist-type with a sheathed or plastered ceiling.

For protection of structures against exposure to fires, outside sprinklers may be used. They can be arranged to protect cornices, windows, side walls, ridge poles, mansard roofs, etc. They are also governed by underwriters' requirements. Figure 20 shows the pipe sizes used for sprinklers protecting outside areas of buildings, including cornices, windows, side walls, etc.

Four common types of automatic sprinkler systems are in use today: wet pipe, dry pipe, preaction, and deluge. The type of system used depends on a number of factors, including occupancy classification, local code requirements, and the requirements of the building fire underwriters. Since the requirements vary from one area to another, no attempt is made here to list those of each locality or underwriter. The Standards of the National Board of Fire Underwriters, as recommended by the National Fire Protection Association, are excerpted instead because they are so widely used that they are applicable for the majority of buildings. In general, the type of sprinkler chosen does not change the design procedure given above. Figure 21 shows a typical layout of the water-supply piping for an industrial-plant sprinkler system. Figure 22 shows how sprinklers are positioned with respect to a building ceiling.

Use the same general design procedure presented here for sprinklers in other types of buildings—hotels, office buildings, schools, churches, dormitories, colleges, museums, libraries, clubhouses, hospitals, and asylums.

Note: Do not finalize a sprinkler system design until after it is approved by local fire authorities and the fire underwriters insuring the building.

SIZING GAS PIPING FOR HEATING AND COOKING

An industrial building has two 8-gal/min (0.5-L/s) water heaters and ten ranges, each of which has four top burners and one oven burner. What maximum gas

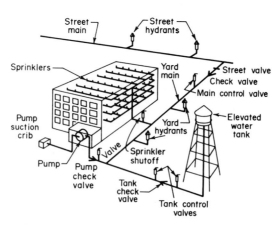

FIGURE 21   Water-supply piping for sprinklers.

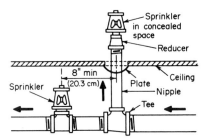

FIGURE 22 Sprinkler positioning with respect to a building ceiling

consumption must be provided for if carbureted water gas is used as the fuel? Determine the pressure in the longest run of gas pipe in this building if the total equivalent length of pipe in the longest run is 150 ft (45.7 m) and the specific gravity of the gas is 0.60 relative to air. What would the pressure loss of a 0.35-gravity gas be?

Calculation Procedure:

1. Compute the heat input to the appliances

Table 19 lists the typical heat input to various gas-burning appliances. By using the tabulated data for the 8-gal/min (0.5-L/s) water heaters and the four-burner stove, the maximum heat input = 2(300,000) + 10(62,500) = 1,225,000 Btu/h (359.0 kW). The gas-supply pipe must handle sufficient gas to supply this heat input because all burners might be operated simultaneously.

2. Compute the required gas-flow rate

Table 20 shows that the heating value of carbureted water gas is 508 Btu/ft³ (18,928 kJ/m³). Using a value of 500 Btu/ft³ (18,629 kJ/m³) to provide a modest safety factor, we find gas flow required, ft³ = maximum heat input required, Btu/h fuel heating value, Btu/ft³ = 1,225,000/500 = 2450 ft³/h (69.4 m³/h).

TABLE 19 Heat Input to Common Appliances

Unit	Approximate input, Btu/h (kW)
Water heater, side-arm or circulating type	25,000 (7.32)
Water heater, automatic instantaneous:	
4 gal/min (0.25 L/s)	150,000 (43.9)
6 gal/min (0.38 L/s)	225,000 (65.9)
8 gal/min (0.5 L/s)	300,000 (87.9)
Refrigerator	2,500 (0.73)
Ranges, domestic:	
Four top burners, one oven burner	62,500 (18.3)
Four top burners, two oven burners	82,500 (24.2)

TABLE 20 Typical Heating Values of Commercial Gases

Gas	Net heating value, Btu/ft³ (kJ/m³)
Natural gas (Los Angeles)	971 (36,178.4)
Natural gas (Pittsburgh)	1,021 (38,041.3)
Coke-oven gas	514 (19,151.1)
Carbureted water gas	508 (18,927.5)
Commercial propane	2,371 (88,340.9)
Commercial butane	2,977 (110,919.8)

3. Compute the pressure loss in the gas pipe

The longest equivalent run is 150 ft (45.7 m). Gas flows through the pipe at the rate of 2450 ft³/h (69.4 m³/h). Enter Table 21 at this flow rate, or at the next *larger* tabulated flow rate and project horizontally to the first pressure drop listed, or 3.5 in/100 ft (89 mm/30 m) in a 2-in (51-mm) pipe.

The pressure loss listed in Table 21 is for 100 ft (30.5 m) of pipe if the gas has a specific gravity of 0.6 in relation to air. To find the pressure loss in 150 ft (45.7 m) of pipe, use the relation actual pressure loss, in of water = (tabulated pressure loss in per 100 ft)(actual pipe length, ft/tabulated pipe length, ft) = 3.5(150/100) = 5.25 in (13.3 mm) of water. Since the actual flow rate is less than 3000 ft³/h (84.9 m³/h), the actual pressure drop will be less than computed.

TABLE 21 Capacities of Gas Pipes [Losses of pressure are shown in inches of water per 100 ft (millimeters per 30.4 m) of pipe, due to the flow of gas with a specific gravity of 0.6 with respect to air]°

Rate of flow, ft³/h (m³/h)	Size of pipe, in (mm)			
	1¼ (32)	1½ (38)	2 (51)	2½ (64)
1000 (28.3)	. . . . . .	1.6 (40.6)	0.80 (20.3)	0.15 (3.8)
1500 (42.5)	8.8 (223.5)	3.6 (91.4)	0.86 (21.8)	0.33 (8.4)
2000 (56.6)	. . . . . .	6.3 (160.0)	1.50 (38.1)	0.60 (15.2)
3000 (84.9)	. . . . . .	. . . . . .	3.5 (88.9)	0.94 (23.9)
5000 (141.5)	. . . . . .	. . . . . .	8.8 (223.5)	3.6 (91.4)

Rate of flow, ft³/h (m³/h)	Size of pipe, in (mm)			
	3 (76)	3½ (89)	4 (102)	5 (127)
1000 (28.3)	0.05 (1.3)	. . . . . .	. . . . . .	. . . . . .
1500 (42.5)	0.11 (2.8)	0.05 (1.3)	. . . . . .	. . . . . .
2000 (56.6)	0.18 (4.6)	0.08 (2.0)	0.05 (1.3)	. . . . . .
3000 (84.9)	0.44 (11.2)	0.19 (4.8)	0.10 (2.5)	. . . . . .
5000 (141.5)	1.2 (30.5)	0.54 (13.7)	0.34 (8.6)	0.08 (2.0)

°To determine head losses for othe rlengths of pipe, multiply the head losses in this table by thelength of the pipe and divide by 100 ft. (30.4 m). For head losses due to flow of gases with specific gravity other than 0.6, use the figures given in Table 22.

4. Compute the pressure loss of the lighter gas

To correct for a gas of different specific gravity, multiply the actual gas flow by the appropriate factor from Table 22. Thus, equivalent flow rate for this plant when a gas of 0.5 gravity is flowing is 2450(0.77) = 1882 ft³/h (53.5 m³/h); say 2000 ft³/h (56.6 m³/h).

Entering Table 21 shows that a flow of 2000 ft³/h (56.6 m³/h) will have a pressure loss of 6.3 in (160 mm) of water in 100 ft (30.5 m) of 1½-in (38-mm) pipe. In 150 ft (45.7 m) of 1½-in (38-mm) pipe, the pressure loss will be 6.3(150/100) = 9.45 in (240 mm) of water. Increasing the pipe size to 2 in (51 mm) would reduce the pressure loss to 2.25 in (57.2 mm) of water.

Related Calculations. When gas flows upward in a vertical pipe to serve upper floors, there is a *gain* in the gas pressure if the gas is lighter than air. Table 23 shows the gain in gas pressure per 100 ft (30.5 m) of rise in a vertical pipe for gases of various specific gravities. This pressure gain must be recognized when piping systems are designed.

As with piping and fixtures for plumbing systems, gas piping and fixtures are subject to code regulations in most cities and towns. Natural and manufactured gases are widely used in stoves, water heaters, and space heaters of many designs. Since gas can form explosive mixtures when mixed with air, gas piping must be absolutely tight and free of leaks at all times. Usual codes cover every phase of gas-piping size, installation, and testing. The local code governing a particular building should be carefully followed during design and installation.

For gas supply, the usual practice is for the public-service gas company to run its pipes into the building cellar, terminating with a brass shutoff valve and gas meter inside the cellar wall. From this point, the plumbing contractor or gas-pipe fitter runs lines through the building to the various fixture outlets. When the pressure of the gas supplied by the public-service company is too high for the devices in

TABLE 22 Factors by Which Flows in Table 21 Must be Multiplied for Gases of Other Specificity Gravity°

Specific gravity of gas	0.35	0.40	0.45	0.50	0.55	0.60	0.65	0.70
Factor	0.77	0.82	0.87	0.91	0.96	1.00	1.04	1.08

TABLE 23 Changes in Pressure in Gas Pipes

Gain in pressure per 100 ft (30.4 m) of rise in vertical pipe, in of water (mm of water)	0.96 (24.4)	0.89 (22.6)	0.81 (20.6)	0.74 (18.8)	0.66 (16.8)	0.59 (15.0)	0.52 (13.2)	0.44 (11.2)
Specific gravity of gas compared to air	0.35	0.40	0.45	0.50	0.55	0.60	0.65	0.70

the building, a pressure-reducing valve can be installed near the point where the line enters the building. The valve is usually supplied by the gas company.

Besides municipal codes governing the design and installation of gas piping and devices, the gas company serving the area will usually have a number of regulations that must be followed. In general, gas piping should be run in such a manner that it is unnecessary to locate the meter near a boiler, under a window or steps, or in any other area where it may be easily damaged. Where multiple-meter installations are used, the piping must be plainly marked by means of a metal tag showing which part of the building is served by the particular pipe. When two or more meters are used in a building to supply separate consumers, there should be no interconnection on the outlet side of the meters.

Materials used for gas piping include black iron, steel, and wrought iron. Copper tubing is also finding some use, and the values listed in Table 21 apply to it as well as schedule 40 (standard weight) pipe made of the materials listed above. Use the procedure given here to size gas pipes for industrial, commercial, and residential installations.

SWIMMING POOL SELECTION, SIZING, AND SERVICING

Choose a swimming pool to serve 140 bathers with facilities for diving and swimming contests. Size the pumps for the pool. Select a suitable water-treatment system and the inlet and outlet pipe sizes. Determine the size of heater required for the pool, should heating of the water be required.

Calculation Procedure:

1. Compute the swimming-pool are a required
Usual swimming pools are sized in accordance with the recommendations of the Join Committee on Swimming Pools, which uses 25 ft^2 (2.3 m^2) per bather as a desirable pool area. With 140 bathers, the recommended area = 25(140) = 3500 ft^2 (325.2 m^2).

2. Choose the pool dimensions
Use Table 24 as a guide to usual pool dimensions. This tabulation shows that a 105-ft (32-m) long by 35-ft (10.7-m) wide pool is suitable for 147 bathers. Since the next smaller pool will handle only 108 bathers, the larger pool must be used.

To provide for swimming contests, lanes at least 7 ft (2.1 m) wide are required. Thus, this pool could have 35 ft/7 ft = 5 lanes for swimming contests. If more lanes are desired, the pool width must be increased, if there is sufficient space. Also, consideration of the pool length is required if swimming meets covering a specified distance are required. Assume that the 105 × 35 ft (32 × 10.7 m) pool chosen earlier is suitable with respect to contests and space.

To provide for diving contests, a depth of more than 9 ft (2.7 m) is recommended at the deep end of the pool. Table 24 shows that this pool has an actual maximum depth of 10 ft (3.05 m), which makes it better suited for diving contests. Some swimming specialists recommend a depth of at least 10 ft (3.05 m) for diving contests. Assume, therefore, that 10 ft (3.05 m) is acceptable for this pool.

TABLE 24 Dimensions of Official Swimming Pools

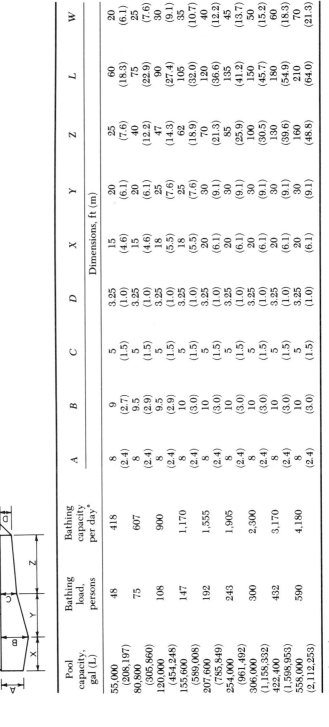

Pool capacity, gal (L)	Bathing load, persons	Bathing capacity per day[a]	A	B	C	D	X	Y	Z	L	W
			Dimensions, ft (m)								
55,000 (208,197)	48	418	8 (2.4)	9 (2.7)	5 (1.5)	3.25 (1.0)	15 (4.6)	20 (6.1)	25 (7.6)	60 (18.3)	20 (6.1)
80,800 (305,860)	75	607	8 (2.4)	9.5 (2.9)	5 (1.5)	3.25 (1.0)	15 (4.6)	20 (6.1)	40 (12.2)	75 (22.9)	25 (7.6)
120,000 (454,248)	108	900	8 (2.4)	9.5 (2.9)	5 (1.5)	3.25 (1.0)	18 (5.5)	25 (7.6)	47 (14.3)	90 (27.4)	30 (9.1)
155,600 (589,008)	147	1,170	8 (2.4)	10 (3.0)	5 (1.5)	3.25 (1.0)	18 (5.5)	25 (7.6)	62 (18.9)	105 (32.0)	35 (10.7)
207,600 (785,849)	192	1,555	8 (2.4)	10 (3.0)	5 (1.5)	3.25 (1.0)	20 (6.1)	30 (9.1)	70 (21.3)	120 (36.6)	40 (12.2)
254,000 (961,492)	243	1,905	8 (2.4)	10 (3.0)	5 (1.5)	3.25 (1.0)	20 (6.1)	30 (9.1)	85 (25.9)	135 (41.2)	45 (13.7)
306,000 (1,158,332)	300	2,300	8 (2.4)	10 (3.0)	5 (1.5)	3.25 (1.0)	20 (6.1)	30 (9.1)	100 (30.5)	150 (45.7)	50 (15.2)
422,400 (1,598,953)	432	3,170	8 (2.4)	10 (3.0)	5 (1.5)	3.25 (1.0)	20 (6.1)	30 (9.1)	130 (39.6)	180 (54.9)	60 (18.3)
558,000 (2,112,253)	590	4,180	8 (2.4)	10 (3.0)	5 (1.5)	3.25 (1.0)	20 (6.1)	30 (9.1)	160 (48.8)	210 (64.0)	70 (21.3)

[a] Based on 8-h turnover.

15.49

The pool will have a capacity of 155,600 gal (588,946 L) of water (Table 24). If installed indoors, the pool would probably be faced with tile or glazed brick. An outdoor pool of this size is usually constructed of concrete, and the walls are a smooth finish.

3. Determine the pump capacity required

To keep the pool water as pure as possible, three *turnovers* (i.e., the number of times the water in the pool is changed each day) are generally used. This means that the water will be changed once each 24 h/3 changes = 8 h. The water is changed by recirculating it through filters, a chlorinator, strainer, and heater. Thus, the pump must handle water at the rate of pool capacity, gal/8 h = 155,600/8 = 19,450 gal/h (73,626 L/h), or 19,450/60 min/h = 324.1 gal/min (1266.9 L/min); say 325 gal/min (1230.3 L/min).

4. Choose the pump discharge head

Motor-driven centrifugal pumps find almost universal application for swimming pools. Reciprocating pumps are seldom suitable because they produce pulsations in the delivery pipe and pool filters. Either single- or double-suction single-stage centrifugal pumps can be used. The double-suction design is usually preferred because the balance impeller causes less wear.

The discharge head that a swimming pool circulating pump must develop is a function of the resistance of the piping, fittings, heater, and filters. Of these four, the heater and filters produce the largest head loss.

The usual swimming pool heater causes a head loss of up to 10 ft (3.05 m) of water. Sand filters cause a head loss of about 50 ft (15.2 m) of water, whereas diatomaceous earth filters cause a head loss of about 90 ft (27.4 m) of water. To choose the pump discharge head, find the sum of the pump suction lift, piping and fitting head loss, and heater and filter head loss. Add a 10 percent allowance for overload. The result is the required pump discharge head in feet of water.

Most pools are equipped with two identical circulating pumps. The spare pump ensures constant operation of the pool should one pump fail. Also, the spare pump permits regular maintenance of the other pump.

5. Compute the quantity of makeup water required

Swimmers splash water over the gutter line of the pool. This water is drained away to the sewer in some pools; in others the water is treated and returned to the pool for reuse. Gutter drains are usually spaced at 15-ft (4.6-m) intervals.

Since the pool waterline is level with the gutter, every swimmer who enters the pool displaces some water, which enters the gutter and is drained away. This drainage must be made up by the pool recirculating system.

The water displaced by a swimmer is approximately equal to his or her weight. Assuming each swimmer weighs 160 lb (72.7 kg) this weight of water will be displaced into the gutter. Since 1 gal (3.79 L) of water weighs 8.33 lb (3.75 kg), each swimmer will displace 160 lb/(8.33 lb/gal) = 19.2 gal (72.7 L). With a maximum of 140 swimmers in the pool, the total quantity of water displaced is (140)(19.2) = 2695 gal/h (10,201.7 L/h), or 2695/(60 min/h) = 44.9 gal/min (169.9 L/min), say 45 gal/min (170.3 L/min).

Thus, to keep this pool operating, the water-supply system must be capable of delivering at least 45 gal/min (170.3 L/min). This quantity of water can come from a city water system, a well, or recirculation of the gutter water after purification.

6. Compute the required filter-bed area

Two types of filters are used in swimming pools: sand and diatomaceous earth. In either type, a flow rate of 2 to 4 gal/(min·ft²) [81.5 to 162.5 L/(min·m²)] is generally used. The lower flow rates, 2 to 2.5 gal/(min·ft²) [81.5 to 101 L/(min·m²)] are usually preferred. Assuming that a flow rate of 2.5 gal/(min·ft²) [101 L/(min·m²)] is used and that 325 gal/min (1230.3 L/min) flows through the filters, as computed in step 3, the filter-bed area required is 325 gal/min/[2.5 gal/(min·ft²)] = 130 ft² (12.1 m²).

Two filters are generally used in swimming pools to ensure continuity of service and back-washing of one filter while the other is in use. The required area of 130 ft² (12.1 m²) could then be divided between the two filters. Some pools use three or more filters. Regardless of how many filters are used, the required area can be evenly divided among them.

7. Choose the number of water inlets and outlets for the pool

The pool inlets supply the recirculation water required. Usual practice rates each inlet for a flow of 10 to 20 gal/min (37.9 to 75.7 L/min). Given a 10-gal/min (37.9-L/min) flow for this pool, the number of inlets required is 325 gal/min/10 gal/min per inlet = 32.5, say 32.

Locate the inlets around the periphery of the pool and on each end. Space the inlets so that they provide an even distribution of the water. In general, inlets should not be located more than 30 ft (10 m) apart.

Size the pool drain to release the water in the pool within the desired time interval, usually 4 to 12 h. Since there is no harm in emptying a pool quickly—if the sewer into which the pool discharges has sufficient capacity—size the discharge line liberally. Thus, a 12-in (305-mm) discharge line can handle about 2000 gal/min (7570.8 L/min) when a swimming pool is drained.

8. Compute the quantity of disinfectant required

Chlorine, bromine, and ozone are some of the disinfectants used in swimming pools. Chlorine is probably the most popular. It is used in quantities sufficient to maintain 0.5 ppm chlorine in the water. Since this pool contains 155,600 gal (589,008 L) of water that is recirculated three times per day, the quantity of chlorine disinfectant that must be added each day is (155,600 gal)(3 changes per day)(8.33 lb/gal) [0.5 lb of chlorine per 10^6 lb (454,545.5 kg) of water] = 1.95 lb (0.89 kg) of chlorine per day. The required chlorine can be pumped into the pool inlet water or fed from cylinders.

9. Size the water heater for the pool

The usual swimming pool heater has a heating capacity, in gallons per hour, which is 10 times the gallons-per-minute rating of the circulating pump. Since the circulating pump for this pool is rated at 325 gal/min (12230.6 L/min), the heater should have a capacity of 10(325) = 3250 gal/h (12,306.6 L/h).

Since the entering water temperature may be as low as 40°F (4.4°C) in the winter, the heater should be chosen for this entering temperature. The outlet temperature of the water should be at least 80°F (26.7°C). To heat the entire contents of the pool from 40°F (4.4°C) to about 70°F (21°C), at least 48 h is generally allowed. Instantaneous hot-water heaters are usually chosen for swimming pool service.

10. Select the backwash sump pump

When the filter backwash flow cannot be discharged directly to a sewer, the usual practice is to pipe the backwash to a sump in the pool machinery room. The accumulated backwash is then pumped to the sewer by a sump pump mounted in the sump.

The sump should be large enough to store sufficient backwash to prevent overflowing. Assuming that either of the 130-ft^2 (12.1-m^2) filter beds is backwashed with a flow of 12.5 gal/(min · ft^2) [509.3 L/(min · m^2)] of filter-bed area, the quantity of water entering the sump will be (12.5)(130) = 1725 gal/min. If there is room for a 5-ft deep, 8-ft wide, and 5-ft long (1.5-m deep, 2.4-m wide, and 1.5-m long) sump, its capacity will be 5 × 8 × 5 × 7.5 gal/ft^3 = 1500 gal (5678 L). The difference, of 1625 − 1500 = 125 gal/min (473.2 L/min), must be discharged by the pump to prevent overflow of the sump. A 150-gal/min (567.8-L/min) sump pump should probably be chosen to provide a margin of safety. Further, it is usual practice to install duplicate sump pumps to ensure pool operation in the event one pump fails. Where water is collected from other drains and discharged to the sump, the pump capacity may have to be increased accordingly.

Related Calculations. Use the general procedure given here to choose swimming pools and their related equipment for schools, recreation centers, hotels, motels, cities, towns, etc. Wherever possible, follow the recommendations of local codes and of the Joint Committee on Swimming Pools.

SELECTING AND SIZING BUILDING SEWAGE EJECTION PUMPS

Choose a suitable ejection pump or pumps for a building have 30 flush-valve water closets, 12 flush-valve urinals, and 14 lavatories. The pump discharge line will contain one check valve, one gate valve, and one 90-degree elbow. Static head on the pump, obtained from building plans, will be 17 ft (5.2 m).

Calculation Procedures:

1. *Determine the sewage flow rate the pump must handle*
Use the Fixture Unit (FU) method to determine the flow rate of sewage into the pump. This method is fully explained in this section of this handbook in the procedure titled "Determination of Plumbing-System Pipe Sizes." Using Table 6 in that procedure, we find the flow rate as: (30 flush-valve water closets × 8 FU) + (12 flush-valve urinals × 8 FU) + (14 lavatories × 1 FU) = 350 FU. Convert this fixture-unit load to gal/min by using Fig. 15 of this section to find the flow rate as 110 gpm (6.9 L/s).

2. *Size the contaminant-tank basin size*
Before the basin can be sized you must know the depth of the lowest incoming piping. The reason for this is that the incoming piping should never be flooded by the fluid in the basin. Hence, the depth determined by flooding prevention is the starting point for finding basin size. A rule of thumb is that the inlet should never be less than 2 ft (0.6 m) from the top of the basin, Fig. 23.

The depth of the inlet will vary depending on the distance from the farthest fixture to the basin. Consult the local plumbing code for the minimum required slopes for sewers.

A properly sized sump basin should include the following features: (1) The basin must be large enough to accommodate the pump. Consult the pump manufacturer

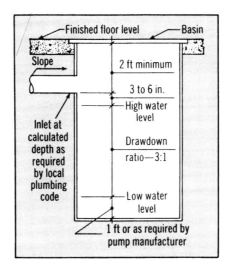

SI Values
1 ft   (0.3 m)
2 ft   (0.6 m)
3 in.   (7.6 cm)
6 in.   (15.2 cm)

FIGURE 23 Typical basin installation. (*HPAC magazine*)

to ensure this; however, basin diameters are generally as shown in Table 25. (2) A drawdown ratio of 3:1, or a storage capacity of three times the pump capacity or incoming flow rate, between the high-water level and the low-water level should be maintained. The reason for this is that with lower amounts, the pump will "short cycle" and will be operating too often for too short a time. This could cause pump damage or shorten the life of the pump.

For example, with a sewage flow of 110 gal/min (6.9 L/s), choose duplex pumps rate at 67 percent of total flow, or $0.67 \times 110 = 74$ gal/min (4.7 L/s) per pump. Total flow $= 2 \times 74 = 148$ gal/min (9.3 L/s). Then, the drawdown $= 3 \times 148 =$

TABLE 25 Pump Capacity and Basin Diameter

Pump capacity, gal/min	One pump (simplex) basin diameter, in	Two pumps (duplex) basin diameter, in	SI Values		
			L/s	cm	cm
50 to 125	30	36	3.2 to 7.9	76.2	
125 to 200	36 to 42	42 to 48	7.9 to 12.6	91.4 to 106.7	91.4 to 121.9
200 to 300	36 to 42	48 to 60	12.6 to 18.9	91.4 to 106.7	121.9 to 152.4
350 to 500	36 to 48	48 to 60	22.1 to 31.6	91.4 to 121.9	121.9 to 152.4
Larger	Consult manufacturer			Consult manufacturer	

444 gal (1683 L). From Table 25, 444 gal would require a 48 to 60 in (122 to 152 cm) diameter basin for duplex pumps.

A minimum depth at the bottom of the basin should be maintained as required by the pump manufacturer. However, a rule of thumb is 1 ft (0.3 m) minimum depth. This ensures that the pumps will maintain their prime, constantly sealed by water. A dry pump that has lost its prime will fail to operate or run out of control.

A 3- to 6-in. (7.6- to 15.2-cm) difference between elevations of the inlet and the high-water level should be maintained for alarm purposes. If the pumps fail to operate or become overloaded and cannot keep up with the incoming flow, a simple float-operated switch will active an alarm.

Now that the basin diameter parameters, the inlet depth, and the drawdown are known, various combinations of basin depths can be used to finish sizing the basin. Using the basin capacities shown in Table 26, divide the determined drawdown by the capacities listed. For example, 48-in (122-cm) diameter = 444 gal drawdown/ 95 gal = 4 ft 8 in (1.4 m), and 60-in diameter = 444 gal drawdown/150 gal = 3 ft (0.9 m).

Either tank size will work for this building. The design engineer or contractor must choose the basin that best fits the installation. Both sizes will be economically the same. However, the 48-in (122-cm) size, which requires less floor space, will be deeper. Location of the sewage ejector in the building may dictate which size should be used, depending on the constraints of the building configuration or structure.

Figure 24 shows the basin that could be selected for this installation. The basin could be 48-in (122-cm) in diameter by 8 ft deep (2.4 m), or 60-in (152-cm) in diameter by 6.5 ft (2 m) deep. Note that 4 in (10 cm) were added to the 48-in basin and 6 in (15 cm) to the 60-in basin to round them off to standard manufactured depths, which are in 6-in (15-cm) increments. The difference will be used in setting the controls and is where the 3- and 6-in (7.6- to 15-cm) difference occurs between the inlet and high-water level.

3. Find the total dynamic head on the ejector

Determining the total dynamic head (TDH) is a means for solving for how much force the pump must produce to send the incoming sewage flow to its destination. The TDH is a summation of the vertical lift (static head), the resistance of pipe and fittings to flow (friction loss), and the backpressure found in a sewer that is flowing.

Static head is determined by measuring the vertical distance between the low-water level in the basin and the highest point of discharge, Fig. 24.

TABLE 26   Circular Basin Capacities

Diameter, in	Capacity per ft depth, gal	SI values	
		cm	L/m
18	14	45.7	174
24	24	60.9	298
30	38	76.2	473
36	53	91.4	659
42	77	106.7	957
48	95	121.9	1181
60	150	152.4	1865
72	212	182.9	2636

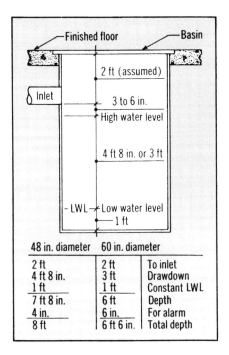

FIGURE 24   Basin parameters as calculated in procedure. (*HPAC magazine*)

To determine the friction losses in the discharge line, one must know the discharge pipe size. Generally, a velocity of 2 fps (0.6 m/s) is acceptable. This velocity allows sufficient flow to ensure carrying solids. Table 27 is based on that flow velocity. Accepted solid sizes are generally 2 in (5 cm), and piping smaller than 2 in (5 cm) diameter should be verified with local plumbing code. Note that this table is for both plastic and steel piping. Since the acceptable materials can vary from one locality to another, the local plumbing code must be consulted before making a final materials choice.

For this installation, with a pipe length of 10 ft (3 m), a 2-in pipe is beyond the listed friction heads for 148 gal/min (9.3 L/s) in Table 27. Hence, the next larger size, 2.5 in (6.4-cm) will have to be used. Then the friction loss for a flow rate of

TABLE 27 Friction Head in feet per 100 ft of Schedule 40 Pipe.*

Flow, gal/min	Pipe size, in									
	1¼		1½		2		2½		3	
	Plastic	Steel	Plastic	Steel	Plastic	Steel	Plastic	Steel	Plastic	Steel
4	0.34	0.35								
6	0.71	0.72	0.33	0.34						
8	1.19	1.20	0.56	0.57						
10	1.78	1.74	0.83	0.85						
12	2.48	2.45	1.16	1.18	0.34	0.35				
14	3.29	3.24	1.54	1.51	0.45	0.46				
16	4.21	4.15	1.97	1.93	0.58	0.59				
18	5.25	5.17	2.41	2.40	0.72	0.73				
20	6.42	6.31	2.96	2.92	0.88	0.88				
25	10.39	9.61	4.80	4.80	1.38	1.39				
30	13.60	13.00	6.27	1.81	1.82	0.75	0.77			
35	19.20	18.20	8.82	8.82	2.40	2.40	1.01	0.99		
40			10.70	10.80	3.12	3.10	1.28	1.30		
45			14.00	14.00	3.80	3.80	1.50	1.60	0.55	0.56
50			16.50	16.50	4.70	4.70	1.90	1.90	0.66	0.68
60					6.50	6.60	2.70	2.70	0.94	0.91
70					8.60	8.80	3.70	3.60	1.20	1.20
80					11.10	11.40	4.70	4.60	1.60	1.60
90					13.80	14.30	5.80	5.80	2.00	2.00
100					16.80	17.50	7.10	7.10	2.40	2.40
125							10.90	10.90	3.70	3.60
150							15.90	15.90	5.20	5.10
175									6.90	6.90

*See text of procedure for SI values.

148 gpm using the loss for 150 gal/min from Table 27 is for 10 ft of 2.5-in pipe = (10 ft × 15.9 ft)/100 ft = 1.59 ft (0.48 m) of head.

Using Table 28 for determining the friction losses in the discharge piping and fittings, we have: check valve = 20.6 ft; gate valve = 1.7 ft; 90-degree elbow = 6.2 ft; total = 20.6 + 1.7 + 6.2 = 28.5 ft (8.7 m). The head loss through these fittings will be, as computed for the straight pipe: (28.5 × 15.9 ft)/100 = 4.53 ft (1.38 m) of head.

TABLE 28 Friction Factors for Pipe Fittings in Terms of Equivalent Feet of Straight Pipe.*

Nominal pipe size, in	90 deg elbow	45 deg elbow	Tee (through-flow)	Tee (branch flow)	Swing check valve	Gate valve
1¼	3.5	1.8	2.3	6.9	11.5	0.9
1½	4.0	2.2	2.7	8.1	13.4	1.1
2	5.2	2.8	3.5	10.3	17.2	1.4
2½	6.2	3.3	4.1	12.3	20.6	1.7
3	7.7	4.1	5.1	15.3	25.5	2.0

*See text of procedure for SI values.

The backpressure from sewers is usually 2 to 3 ft (0.6 to 0.91 m) of head loss caused by the pump operating against a sewer in operation. (We will use 3-ft head in this calculation). Adding this to the computed friction losses and static head will give the TDH. For this procedure, TDH = 17 ft static head + 1.59 ft pipe loss + 4.5 ft fittings loss + 3 ft backpressure = 26 ft (7.9 m) TDH. This total dynamic head will be used to enter the pump manufacturer's pump curves, along with the gpm required, to select the correct pump.

4. Select the solids-handling capability
As stated earlier, solids handling is generally acceptable since most modern fixtures will not pass larger solids. Some sewage ejectors may not require 2-in (5-cm) solids handling capability if they serve a sanitary sewer that contains no raw sewage, such as effluent pumps in drainage fields. In all designs, consult the local plumbing code to determine the specific requirements in the locality. In this design, water closets were used, and 2-in size (5-cm) would be a minimum discharge size.

5. Evaluate pump type to use
The style of pump chosen depends largely on the type of building served. What this really means is that the larger the expected flows, the larger the handling capacity of the pumps. A submersible pump, found in the basement of many homes, is adequate to handle the flow and frequency of operation associated with the average family.

A commercial building, such as an office structure, may use larger submersible pumps, Fig. 25a, with flow rates in the range of 1000 gal/min (63.1 L/s). Where floor space is at a premium, a submersible pump is beneficial because the pump is located inside the basin. Since the pump is immersed directly in the sewage, removal, replacement, and standard servicing are not simple tasks.

Vertical lift pumps, Fig. 25b, are used in many commercial buildings with adequate floor space, when flow rates are in the range of 50 to 1500 gal/min (3 to 95 L/s). The advantage of a vertical lift pump is that the driver is mounted at floor level with a drive shaft extending into the sewage, where the impeller is located. Service and maintenance are simplified by not having personnel enter the containment tank. Because the impeller is located in the sewage where the work is actually performed, sizes are relatively limited. Impeller clogging can still require complete removal.

Self-priming pumps, Fig. 25c, are found in both commercial and industrial buildings. Flow rates range from 50 to 2000 gal/min (3 to 126 L/s). The advantage of a self-priming pump is that the driver and impeller are mounted at the floor level with a suction pipe extending into the sewage. Since both impeller and driver are located above the floor, caution should be used in selecting self-primers for buildings with inadequate floor space.

6. Choose the pump to use
Using the pump manufacturer's pump curves, look for a pump that has its highest efficiency close to, or at, the intersection of lines drawn vertically from the gal/min (L/s) required and horizontally from the TDH. Figure 26 shows typical manufacturer's pump curve.

Basically, the higher the rpm of the pump, the higher the TDH at lower gal/min with the same horsepower (kW). As you can see from Fig. 26, (a) At 1750 r/min = 1 hp (0.75 kW) with a 5.75-in (14.6-cm) diameter impeller the efficiency is close to peak at 52 percent; (b) with 1150 r/min = 1 hp (0.75 kW) with an 8-in (20.3-cm) diameter impeller, efficiency is not at peak at 52 percent.

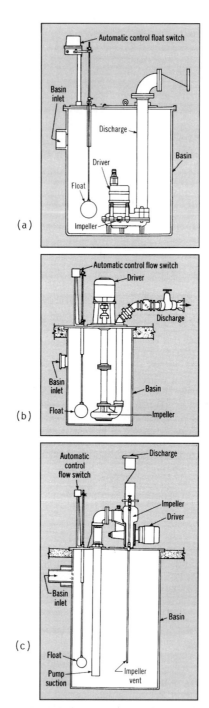

FIGURE 25 Three types of sewage-ejection pumps. (*a*) Submersible pump; (*b*) Vertical-lift pump; (*c*) Self-priming pump. (*HPAC magazine*)

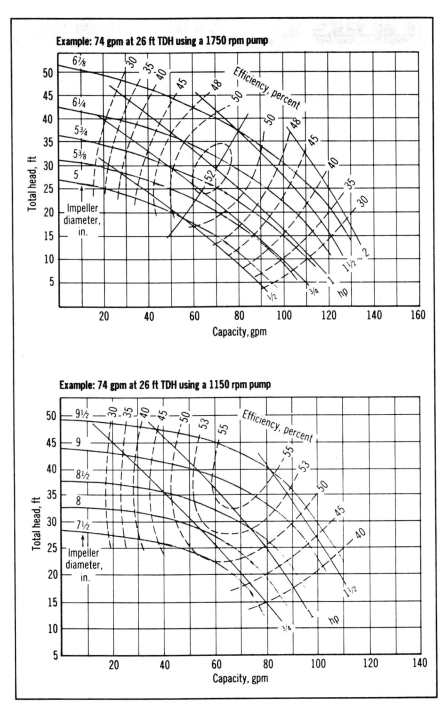

FIGURE 26 Typical manufacturer's curves for sewage-ejection pumps. See text for SI values. (*HPAC magazine*)

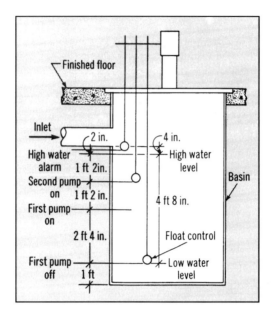

SI Values

2 in.	(5 cm)
1' 2"	(0.36 m)
2' 4"	(0.71 m)
1'	(0.3 m)
4 in.	(10 cm)
4' 8"	(1.4 m)

FIGURE 27 Locations of controls for 4.67-ft (1.4-m) diameter deep basin. (*HPAC magazine*)

All pumps have one feature in common: they are most efficient when operating at their peak performance load. A pump that is oversized or undersized and is not operating at is peak performance for the greatest amount of its running time will either consume more energy than is necessary or burn itself out prematurely. Try to select a pump that has its highest efficiency at the TDH and gal/min (L/s) the job requires. Following this guideline, choose the 1750-rpm pump, which is usually the choice with smaller flows and shorter running times.

7. *Determine which controls to use*
Controls are manufactured in many different forms. There are simple float controls like those found in tank-type water closets. Then there are pressure-sensitive electronic devices with no moving parts.

All controls turn the pump on and off. Depending on the relative importance of controlling the pump, deciding on the right controls can be a matter of economics. The more sophisticated the controls, in general, the more expensive they are. In the case of duplex pumps, the controls should also alternate the lead pump so that the pumps wear evenly.

Setting the locations of the controls is a matter of defining predetermined locations from the water levels (step 2, above), Fig. 27. The first or lead pump turns on when about two-thirds of the total draindown depth is filled. Next, the second pump turns on when the level continues to rise to half the depth left to the high-water level. The high-water alarm activates when half the depth from the high-water level to the inlet remains.

Related Calculations.   Sewage ejectors are highly reliable. When selected, installed, and engineered correctly, they can survive the life of the building. Inexpensive steps that can minimize problems with sewage ejectors are: (1) Use a precast or concrete slab under the basin when installing it. Set the slab on undisturbed earth, sand, or compacted backfill. This will prevent the basin from settling. (2). When using a glass fiber basin in poor soil conditions, set the basin inside a larger corrugated metal pipe or concrete bell preset in concrete and fill the voids between them with concrete or mortar to set up a concrete envelope. This will keep the glass fiber basin from being lifted by surrounding groundwater. (3) Consider connecting the sewage ejector to a second source of power—either an emergency generator or a dual-feed from the power source. (4) Always vent the basin to atmosphere through the roof to prevent odors. Most local plumbing codes require such ventilation anyway.

Keep in mind the following considerations when designing any sewage-ejection system. (1) Sewage ejectors are pumping systems designed specifically to lift emulsified solids in liquid from a building's interior or clear liquid from drainage fields to a preferred destination. (2) Such a pumping system usually consists of a driver (either a motor or engine), an impeller, a containment tank, and control devices to operate the driver automatically. (3) Be certain to consult equipment manufacturers when designing systems for special applications or uses not commonly found in handling raw, untreated sewage.

This procedure is the work of Larry Robertson, CIPE, Perkins and Will, as reported in *Heating/Piping/Air Conditioning* magazine, August, 1990. SI values were added by the handbook editor.

SECTION 16
HEATING, VENTILATING, AND AIR CONDITIONING

Economics of Interior Climate Control

DETERMINING COOLING-TOWER FAN HORSEPOWER REQUIREMENTS

A cooling tower serving an air-conditioning installation is designed for these conditions: Water flowrate, L = 75,000 gal/min (4733 L/s); inlet water temperature, T_i = 110°F (43.3°C); outlet water temperature, T_o = 90°F (32.2°C); atmospheric wet-bulb temperature, T_w = 82°F (27.8°C); total fan efficiency as given by tower manufacturer, E_T = 75%; recirculation of air in tower, given by tower manufacturer, R_c = 8.5%; total air pressure drop through the tower, as given by manufacturer, ΔP = 0.477 in (1.21 cm) H_2O. What is the required fan horsepower input under these conditions? If the weather changes and the air outlet temperature becomes 102°F (38.9°C) with a wet-bulb temperature of 84°F (28.9°C)?

Calculation Procedure:

1. Determine the fan horsepower for the given atmospheric and flow conditions

The fan brake horsepower input for a cooling tower is given by the relation: $BHP = L \times \Delta T \times V_{sp} \times R_c \times \Delta P / \Delta H \times 6356 \times E_T$, where $\Delta T = T_o - T_i$; V_{sp} = specific volume of outlet air, ft³/lb (m³/kg); ΔH = difference between enthalpy of outlet air and inlet air, Btu/lb (kJ/kg); other symbols as given earlier. Determine the enthalpy difference between the outlet air and inlet air by referring to a psychrometric chart where you will find that the enthalpy of the outlet air for the first case above, H_o = 72 Btu/lb (167.5 kJ/kg); from the same source the enthalpy of the inlet air, H_i = 46 Btu/lb (107.0 kJ/kg); likewise, V_{sp} = 15.1 ft³/lb (93.7 m³/ kg) from the chart. Substituting in the equation above, BHP = (75,000 × 8.337 × 20 × 1.085 × 15.1 × 0.477)/(72 − 24) × 6356 × 0.75 = 788.51; say 789 hp (588.3 kW). In the above equation the constants 8.337 and 1.085 are used to convert gal/min to lb/min and air flow to ft³/lb, respectively.

2. Determine the power input required for the second set of conditions

For the second set of conditions the air outlet temperature, T_o = 102°F (38.9°C) and the wet-bulb temperature is 84°F (28.9°C). Using the psychometric chart again, ΔH = 75 − 48 = 27 Btu/lb (62.8 kJ/kg), and the specific volume of the air at this temperature—from the chart—15.2 ft³/lb (0.95 m³/kg). Substituting as before, BHP = 764 hp (569.6 kW).

    Related Calculations. This procedure can be used with any type of cooling tower employed in air conditioning, steam power plants, internal combustion engines, or gas turbines. The method is based on knowing the tower's air outlet temperature. Use the psychometric chart to determine volumes and temperatures for various air states. As presented here, this method is the work of Ashfaq Noor, Dawood Hercules Chemicals Ltd., as reported in *Chemical Engineering* magazine.

    With the greater environmental interest in reducing stream pollution of all types, including thermal, cooling towers are receiving more attention as a viable way to eliminate thermal problems in streams and shore waters. The cooling tower is a nonpolluting device whose only environmental impact is the residue left in its bottom pans. Such residue is minor in amount and easily disposed of in an environmentally acceptable manner.

CHOOSING AN ICE STORAGE SYSTEM FOR FACILITY COOLING

Select an ice storage cooling system for a 100-ton (350-kW) peak cooling load, 10-h cooling day, 75 percent diversity factor, \$8.00/month kW demand charge, 12-month ratchet—i.e., the utility term for a monthly electrical bill surcharge based on a previous month's higher peak demand. Analyze the costs for a partial-storage and for a full-storage system.

Calculation Procedure:

1. Analyze partial-storage and full-storage alternatives
Stored cooling systems use the term *ton-h* instead of *tons of refrigeration,* which is the popular usage for air-conditioning loads. Figure 1 shows a theoretical cooling load of 100 tons (350 kW) maintained for 100 h, or a 1000 ton-h (3500 kWh) cooling load. Each of the squares in the diagram represents 10 ton-h (35 kWh).

No building air-conditioning system operates at 100 percent capacity for the entire daily cooling cycle. Air-conditioning loads peak in the afternoon, generally from 2:00 to 4:00 pm, when ambient temperatures are highest. Figure 2 shows a typical building air-conditioning load profile during a design day.

As Fig. 2 shows, the full 100-ton chiller capacity (350 kW) is needed for only 2 of the 10 h in the cooling cycle. For the other 8 h, less than the total chiller capacity is required. Counting the tinted squares shows only 75, each representing 10 ton-h (35 kWh). The building, therefore, has a true cooling load of 750 ton-h (2625 kWh). A 100-ton (350 kW) chiller must however, be specified to handle the peak 100-ton (250 kW) cooling load.

The *diversity factor,* defined as the ratio of the actual cooling load to the total potential chiller capacity, or diversity factor, percent = 100 (Actual ton-hours)/total

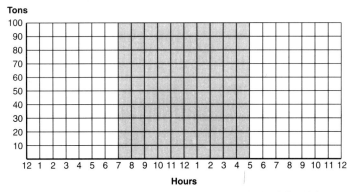

FIGURE 1 Cooling load of 100 tons (351.7 kW) maintained for 10 h, or a 1000 ton-h cooling load. (*Calmac Manufacturing Corporation.*)

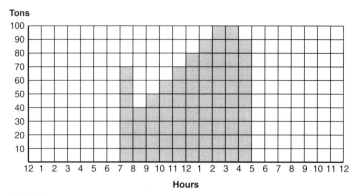

FIGURE 2 Typical building air-conditioning load profile during a design day. (*Calmac Manufacturing Corporation.*)

potential ton-hours. For this installation, diversity factor = 100(750)/1000 = 75 percent. If a system's diversity factor is low, its cost efficiency is also low.

Dividing the total ton-hours of the building by the number of hours the chiller is in operation gives the building's average load throughout the cooling period. If the air-conditioning load can be shifted to off-peak hours or leveled to the average load, 100 percent diversity can be achieved, and better cost efficiency obtained.

When electrical rates call for complete load shifting, i.e., are excessively high, a conventionally sized chiller can be used with enough energy storage to shift the entire load into off-peak hours. This is called a *full-storage system* and is used most often in retrofit applications using existing chiller capacity. Figure 3 shows the same building air-conditioning load profile but with the cooling load completely shifted into 14 off-peak hours. The chiller is used to build and store ice during the night. The 32°F (0°C) energy stored in the ice then provides the required 750 ton-h (2625 kWh) of cooling during the day. The average load is lowered to (750 ton-h)/14 h = 53.6 tons (187.6 kW), which results in significantly reduced demand charges.

In new construction, a partial-storage system is the most practical and cost effective load-management strategy. In this load-leveling method, the chiller runs

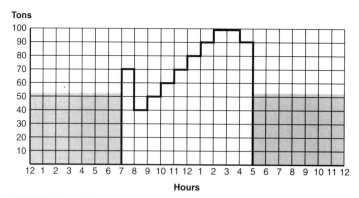

FIGURE 3 Building air-conditioning load profile of Fig. 2 with the cooling load shifted into 14 off-peak hours. (*Calmac Manufacturing Corporation.*)

continuously. It charges the ice storage at night and cools the load directly during the day with help from stored cooling. Extending the hours of operation from 14 to 24 h results in the lowest possible Average Load, (750 ton-h)/24 hours = 31.25 tons (109.4 kW, as shown by the plot in Fig. 4). Demand charges are greatly reduced and chiller capacity can often be decreased by 50 to 60 percent, or more.

2. Compute partial-storage demand savings

Cost estimates for a conventional chilled-water air-conditioning system comprised of a 100-ton (350 kW) chiller with all accessories such as cooling tower, fan coils, pumps, blowers, piping, controls, etc., show a price of $600/ton, or 100 tons × $600/ton = $60,000. The distribution system for this 100-ton (350-kW) plant will cost about the same, or $60,000. Total cost therefore = $60,000 + $60,000 = $120,000.

With partial-storage using a 40 percent size chiller with ice storage at a 75 percent diversity factor, the true cooling load translates into 750 ton-h (2626 kWh) with the chiller providing 400 ton-h (1400-kWh) and stored ice the balance, or 750 − 400 = 350 ton-h (1225 kWh). Hence, cost of 40-ton chiller at $600/ton = $24,000. From the manufacturer of the stored cooling unit, the installed cost is estimated to be $60/ton-h, or $60 × 350 ton-h = $21,000. The distribution system, as before, costs $60,000. Hence, the total cost of the partial-storage system will be $24,000 + $21,000 + $60,000 = $105,000. Therefore, the purchase savings of the partial-storage system over the conventional chilled-water air-conditioning system = $120,000 − $105,000 = $15,000.

The electrical demand savings, which continue for the life of the installation, are: (100 tons − 40 tons chiller capacity)(1.5-kW/ton at peak summer demand conditions, including all accessories)($8.00/mo/kW demand charge)(12 mo/yr) = $8640.

3. Determine full-storage savings

With full-storage, 100-ton (350-kW) peak cooling load, 10-h cooling day, 75 percent diversity factor, 1000-h cooling season, $8.00/mo/kW demand charge, 12-month ratchet, $0.03/kWh off-peak differential, the chiller cost will be (10 h)(100 tons)(75 percent)($60/ton-h, installed) = $45,000. The demand savings will be, as

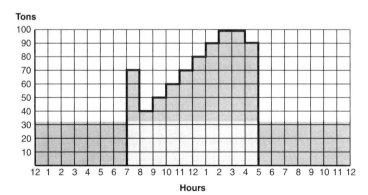

FIGURE 4 Extending the hours of operation for 14 to 24 results in the lowest possible average load = 750 ton-h/24 = 31.25. Demand charges are greatly reduced and chiller capacity can often be decreased 50 to 60 percent, or more. (*Calmac Manufacturing Corporation.*)

before, (100 tons)(1.5 kW/ton)(12 mo)($8.00/mo/kW demand charge)($8.00) = $14,400/year. Energy savings are computed using the electric company's off-peak kWh off-peak differential, or ($0.03/kWh)(1000 h)(100 tons)(1.2 average kW/ton) = $3,600/year. The simple payback time for this project = (equipment cost, $)/ (demand savings, $ + energy savings, $) = $45,000/$18,000 = 2.5 yr. After the end of the payback time there is an annual energy savings of $18,000/year. And as rates increase, which they usually do, the annual savings will probably increase above this amount.

Related Calculations. Ice storage systems are becoming more popular for a variety of structures: office buildings, computer data centers, churches, nursing homes, police stations, public libraries, theaters, banks, medical centers, hospitals, hotels, convention centers, schools, colleges, universities, industrial training centers, cathedrals, medical clinics, manufacturing plants, warehouses, museums, country clubs, stock exchanges, government buildings, and courthouses.

There are several reasons for this growing popularity: (1) utility power costs can be reduced by shifting electric power demand to off hours by avoiding peak-demand charges; (2) lower overall electric rates can be obtained for the facility if the kilowatt demand is reduced, thereby eliminating the need for the local utility to build new generating facilities; (3) ice storage can provide uninterrupted cooling in times of loss of outside, or inside, electric generating capability during natural disasters, storms, or line failures—the ice storage system acts like a battery, giving the cooling required until the regular coolant supply can be reactivated; (4) environmental regulations are more readily met because less power input is required, reducing the total energy usage; (5) by making ice at night, the chiller operates when the facilities' electrical demands are lowest and when a utility's generating capacity is underutilized; (6) provision can be made to use more environmentally friendly HCFC-123, thereby complying with current regulations of federal and state agencies; (7) facility design can be planned to include better control of indoor air quality, another environmentally challenging task faced by designers today; (8) new regulations, specifically The Energy Policy Act of 1992 (EPACT), curtails the use of and eliminates certain fluorescent and incandescent lamps (40-W T12, cool white, warm white, daylight white, and warm white deluxe), which will change both electrical demand and replacement bulb costs in facilities, making cooling costs more important in total operating charges.

Designers now talk of "greening" a building or facility, i.e., making it more environmentally acceptable to regulators and owners. An ice storage system is one positive step to greening a facility while reducing the investment required for cooling equipment. The procedure given here clearly shows the savings possible with a typical well-designed ice storage system.

There are three common designs used for ice storage systems today: (1) *direct-expansion ice storage* where ice is frozen directly on metal refrigerant tubes submerged in a water tank; cooled water in the tank is pumped to the cooling load when needed; (2) *ice harvester system* where a thin coat of ice is frozen on refrigerated metal plates and periodically harvested into a bin or water tank by melting the bond of the ice to the metal plates; the chilled water surrounding the ice is pumped to the cooling load when needed; (3) *patented ice bank system* uses a modular, insulated polyethylene tank containing a spiral-wound plastic tube heat exchanger surrounded with water; at night a 26°F (−3.33°C) 75 percent water/25 percent glycol solution from a standard packaged air-conditioning chiller circulates through the heat exchanger, freezing solid all the water in the tank; during the day the ice cools the solution to 44°F (6.66°C) for use in the air-cooling coils where it cools the air from 75°F (23.9°C) to 55°F (12.8°C).

The patented system has several advantages over the first two, namely: (1) ice is the storage medium, rather than water. One pound (0.45 kg) of ice can store 144 Btu (152 J) of energy, while one pound of water in a stratified tank stores only 12 to 15 Btu (12.7 J to 15.8 J). Hence, such an ice storage system needs only about one-tenth the space for energy storage. This small space requirement is important in retrofit applications where space is often scarce.

(2) Patented systems are closed; there is no need for water treatment or filtration; pumping power requirements are small; (3) power requirements are minimal; (4) installation of the insulated modular tanks is fast and inexpensive since there are no moving parts; the tanks can be installed indoors or outdoors, stacked or buried to save space. Currently these tanks are available in three sizes: 115, 190, and 570 ton-h (402.5, 665, and 1995 kWh). (5) A low-temperature duct system can be used with 45°F (7.2°C) air instead of the conventional 55°F (12.8°C) air in the air-conditioning system. This can permit further large savings in initial and operating costs. The 45°F (7.2°C) primary air requires much lower air flow [ft³/min (m³/m)] than 55°F (12.8°C) air. This reduces the needed size of both the air handler and duct system; both may be halved. Energy savings from the smaller air-handler motors may total 20 percent, even after figuring the additional energy required for the small mixing-box motors.

This procedure is based on data provided by the Calmac Manufacturing Corporation, Englewood, NJ. The economic analysis was provided by Calmac, as were the illustrations in this procedure. Calmac manufactures the *Levload* modular insulated storage tanks mentioned above that are used in their *Ice Bank Stored Cooling System.* Their system, when designed with a low-temperature heat-recovery loop, can also make the chiller into a water-source heat pump for winter heating. Thus, office and similar buildings often require heating warm-up in the morning on winter days, but these same buildings may likewise require cooling in the afternoon because of lights, people, computers, etc. Ice made in the morning to provide heating supplies free afternoon cooling and melts to be ready for the next day's warm-up. Even on coldest days, low-temperature waste heat (such as cooling water or exhaust air), or off-peak electric heat can be used to melt the ice. Oil or gas connections to the building can thus be eliminated.

Nontoxic eutectic salts are available to lower the freezing point of the water in Calmac Ice Banks to either 28°F (−2.2°C) or 12°F (−11.1°C) and, consequently, the temperature of the resulting ice. Twenty-eight-degree ice, for example, can provide cold, dry primary air for many uses, including extra-low temperature airside applications. Twelve-degree ice can be used for on-ground aircraft cooling, off-peak freezing of ice rinks, and for industrial process applications requiring colder liquids. Other temperatures can be provided for specialized applications, such as refrigerated warehouses.

Figure 5 shows the charge cycle using a partial storage system for an air-conditioning installation. At night a water-glycol solution is circulated through a standard packaged air-conditioning chiller and the Ice Bank heat exchanger, by-passing the air-handler coil. The cooling fluid is at 26°F (−3.3°C) and freezes the water surrounding the heat exchanger.

During the day, Fig. 6, the water-glycol solution is cooled by the Ice Bank from 52°F (11.1°C) to 34°F (1.1°C). A temperature-modulating valve, set at 44°F (6.7°C) in a bypass loop around the Ice Bank, allows a sufficient quantity of 52°F (11.1°C) fluid to bypass the Ice Bank, mix with 34°F (1.1°C) fluid, and achieve the desired 44°F (6.7°C) temperature. The 44°F (6.7°C) fluid enters the coil, where it cools the air passing over the coil from 75°F (23.9°C) to 55°F (12.8°C). Fluid leaves the coil at 60°F (15.6°C), enters the chiller and is cooled to 52°F (11.1°C).

FIGURE 5 Counterflow heat-exchanger tubes used in the Ice Bank. (*Calmac Manufacturing Corporation.*)

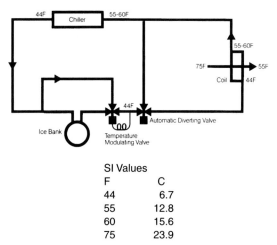

SI Values

F	C
44	6.7
55	12.8
60	15.6
75	23.9

FIGURE 6 Charge cycle. (*Calmac Manufacturing Corporation.*)

Note that, while making ice at night, the chiller must cool the water-glycol to 26°F (−3.3°C), rather than produce 44°F (6.7°C) or 45°F (7.2°C) water temperatures required for conventional air-conditioning systems. This has the effect of "derating" the nominal chiller capacity by about 30 percent. Compressor efficiency, however, is only slightly reduced because lower nighttime temperatures result in cooler condenser water from the cooling tower (if used) and help keep the unit operating efficiently. Similarly, air-cooled chillers benefit from cooler condenser entering air-temperatures at night.

The temperature-modulating valve in the bypass loop has the added advantage of providing unlimited capacity control. During many mild-temperature days in the spring and fall, the chiller will be capable of providing all the cooling needed for the building without assistance from stored cooling. When the building's actual cooling load is equal to or lower than the chiller capacity, all of the system coolant flows through the bypass loop, as in Fig. 8.

Using 45°F (7.2°C) rather than 55°F (12.8°C) system air in the air-conditioning system permits further large savings in initial and operating costs. The 45°F (7.2°C)

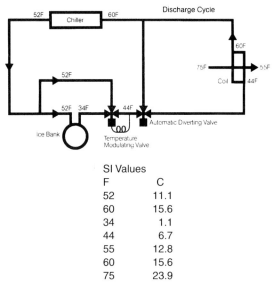

SI Values	
F	C
52	11.1
60	15.6
34	1.1
44	6.7
55	12.8
60	15.6
75	23.9

FIGURE 7 Discharge cycle. (*Calmac Manufacturing Corporation.*)

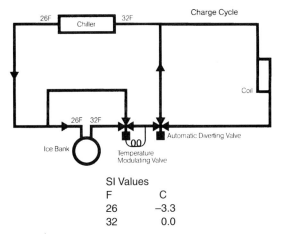

SI Values	
F	C
26	−3.3
32	0.0

FIGURE 8 When cooling load equals, or is lower than chiller capacity, all coolant flows through bypass loop. (*Calmac Manufacturing Corporation.*)

low-temperature air is achieved by piping 38°F (3.3°C) water-glycol solution from the stored cooling Ice Bank to the air handler coil instead of mixing it with bypassed solution, as in Fig. 6. The 45°F (7.2°C) air is used as primary air and is distributed to motorized fan-powered mixing boxes where it is blended with room air to obtain the desired room temperature. Primary 45°F (7.2°C) air requires much lower ft^3/min (m^3/min) than 55°F (12.8°C) air. Consequently, the size and cost of the air handler and duct system may be cut in about half. Energy savings of the smaller air handler motors total 20 percent, even counting the additional energy required for the small mixing-box motors.

The recommended coolant solution for these installations is an ethylene glycol-based industrial coolant such as Union Carbide Corporation's UCARTHERM® or Dow Chemical Company's DOWTHERM® SR-1. Both are specially formulated for low viscosity and superior heat-transfer properties, and both contain a multi-component corrosion inhibitor system effective with most materials of construction, including aluminum, copper, solder, and plastics. Standard system pumps, seals, and air-handler coils can be sued with these coolants. However, because of the slight difference in the heat-transfer coefficient between water-glycol and plain water, air-handler coil capacity should be increased by about 5 percent. Further, the water and glycol must be thoroughly mixed before the solution enters the system.

Another advantage of ice storage systems for cooling and heating is provision of an uninterrupted power supply (UPS) in the event of the loss of a building's cooling or heating facilities. Such an UPS can be important in data centers, hospitals, research laboratories, and other installations where cooling or heating are critical.

Figure 9a shows the conventional "ice builder" and Fig. 9b the LEVLOAD Ice Bank. When ice is stored remote from the refrigerating system evaporator, as in Fig. 9b, the evaporator is left free to aid the cooling during the occupied hours of a building or other structure. Figure 10 compares the chiller performance of a Partial Storage Ice Bank, Fig. 9b (upper curve), with an ice-builder system, Fig. 9a (lower

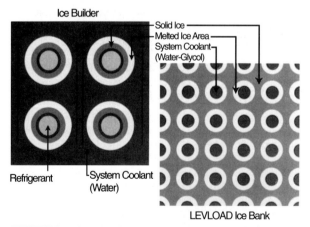

FIGURE 9 (*a*) Typical ice-builder arrangement. (*b*) LEVLOAD Ice Bank method of ice burn-off (ice melting). (*Calmac Manufacturing Corporation.*)

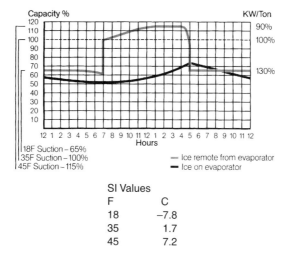

18F Suction – 65%
35F Suction – 100%
45F Suction – 115%

— Ice remote from evaporator
— Ice on evaporator

SI Values

F	C
18	−7.8
35	1.7
45	7.2

FIGURE 10 Comparison of chiller performance of a Partial Storage Ice Bank (upper curve) and conventional ice-builder system (lower curve). (*Calmac Manufacturing Corporation.*)

curve), on a typical design day. Note that when compressor cooling is done through ice on the evaporator, suction temperatures are low and kW/ton is increased.

With discussions still taking place about chlorofluorocarbon refrigerants suitable for environmental compliance, designers have to seek the best choice for the system chiller. One approach finding popularity today as an interim solution is to choose a chiller which can use an energy-efficient refrigerant today and the most environmentally friendly refrigerant in the future. Thus, for some chillers, CFC-11 is the most energy-efficient refrigerant today. The future most environmentally friendly refrigerant is currently thought to be HCFC-123. By choosing and sizing a chiller that can run on HCFC-123 in the future, energy savings can be obtained today, and, if environmental requirements deem a switch in the future, the same chiller can be used for CFC-11 today and HCFC-123 in the future. It is also possible that the same chiller can be retrofitted to use a non-CFC refrigerant in the future. A number of ice storage systems have adopted this design strategy. Many firms that installed ice storage systems in recent years are so pleased with the cost savings (energy, equipment, ducting, UPS, etc.) that they plan to expand such systems in the future.

Chillers using CFC-11 normally produce ice at 0.64 to 0.75 kW/ton, depending on the amount of ice produced. Power consumption is lower when larger quantities of ice are produced. When HCHC-123 is used, the power input ranges between 0.7 and 0.8 kW/ton, again depending on the number of tons produced. As before, power consumption is lower when larger tonnages are produced. New centrifugal chillers produce cooling at power input ranges close to 0.5 kW/ton.

In all air-conditioning systems, designers must recognize that there are three courses of action open to them when CFC refrigerants are no longer available: (1) continue to use existing CFC-based equipment, taking every precaution possible to stop leaks and conserve available CFC supplies; (2) retrofit existing chilling equipment to use non-CFC refrigerants; this step requires added investment and changed operating procedures; (3) replace existing chillers with new chillers specifically designed for non-CFC refrigerants; again, added investment and changed operating procedures will be necessary.

To avoid CFC problems, new high-efficiency chlorine-free screw chillers are being used. And there are packaged ammonia screw chiller available also. Likewise, a variety of alternative refrigerants are now being produced for new and retrofit refrigeration and air-conditioning uses.

Table 1 shows an economic analysis of typical partial-storage and full-storage installations. A conventional chilled-water air-conditioning system is compared with a partial-storage 40 percent size chiller with Ice Banks in the partial-storage analysis. Full-storage produces the simple payback time of 2.5 years for the investment. Data in this analyze are from Calmac Manufacturing Corporation. When using a similar analysis, be certain to obtain current prices for components, demand charges, and electricity. Values given here are for illustration purposes only.

ANNUAL HEATING AND COOLING ENERGY LOADS AND COSTS

A 2000 ft² (185.8 m²) building has a 100,000 Btu/h (29.3 kW) heat loss in an area where the heating season is 264 days' duration. Average winter outdoor temperature is 42°F (5.6°C); design conditions are 70°F (21.1°C) indoors and 0°F (−17.8°C) outdoors. The building also has a summer cooling load of 7.5 tons (26.4 kW) with an estimated full-load cooling time of 800 operating hours. What are the total winter and summer estimated loads in Btu/h (kW)? If oil is 90 cents/gallon and electricity is 7 cents/kWh, what are the winter and summer energy costs? Use a 24-h heating day for winter loads and a boiler efficiency of 75 percent when burning oil with a higher heating value (HHV) of 140,000 Btu/gal (39,018 MJ/L).

Calculation Procedure:

1. Compute the winter operating costs
The winter seasonal heating load, WL = N × 24 h/day × Btu/h heat loss × (average indoor temperature, °F − average outdoor temperature, °F)/(average indoor temperature, °F − outside design temperature, °F), where N = number of days in the heating season. For this building, WL = (264 × 24 × 100,000)(70 − 42)/ (70 − 0) = 253,440,000 Btu (267.4 MJ).

The cost of the heating oil, CO = Btu seasonal heating load × oil cost $/gal/ boiler efficiency × HHV. Or, for this building, CO = (253,440,000)(0.90)/0.75 × 140,000 = $2,172.34 for the winter heating season.

2. Calculate the summer cooling cost
The summer seasonal electric consumption in kWh, SC = tons of air conditioning × kW/ton of air conditioning × number of operating hours of the system. The kW/design ton factor is based on both judgment and experience. General consensus amongst engineers is that the kW/ton varies from 1.8 for small window-type systems to 1.0 for large central-plant systems. The average value of 1.4 kW/design ton is frequently used and will be used here. Thus, the summer electric consumption, SC = 7.5 × 1.4 × 800 = 8400 kWh. This energy will cost 8400 kWh × $0.07 = $588.00.

TABLE 1 Economic Analysis of Typical Partial Storage and Full Storage Installations*

Partial storage
Assume: 100-ton peak cooling load, 10-h cooling day, 75 percent diversity factor, $8.00/mo/kW demand charge, 12-mo ratchet.*
Conventional Chilled Water Air Conditioning System:

100-ton chiller at $600/ton, installed**	$ 60,000
Distribution system	60,000
Total	$120,000

Partial storage (40% size chiller with Ice Banks):
At 75 percent diversity factor, the true cooling load translates into 750 ton-h with the chiller providing 400 ton-h and stored cooling the balance, or 350 ton-h. Therefore:

40-ton chiller at $600/ton, installed	$ 24,000
Stored cooling at $60/ton-h, installed	21,000
Distribution system	60,000†
Total	$105,000

Purchase savings: $ 15,000
Demand savings:

$60 \text{ tons} \times 1.5 \text{ kW/ton}‡ \times 12 \text{ mo} \times \$8.00 = \$8,640/\text{yr}$

*Utility term for a monthly electrical bill surcharge based on a previous month's higher peak demand.

Full storage
Assume: 100-ton peak cooling load, 10-hour cooling day, 75% diversity factor, 1000-hour cooling season, $8.00/mo/kW demand charge, 12-mo ratchet, $0.03/kWh off-peak differential.
Full storage:

10 h × 100 tons × 75% × $60/ton-h, installed	$45,000

Demand savings:

100 tons × 1.5 kW/ton × 12 mo × $8.00	$14,400/yr

Energy savings:

$0.03/kWh × 1000 h × 100 tons × 1.2 Avg. kW/ton	$ 3600/yr
	$18,000/yr

Total savings:
Simple payback: $45,000 ÷ $18,000 = 2.5 yr.

**The $600/ton includes all accessories, such as cooling tower, fan coils, pumps, blowers, piping, controls, etc.
†Figure shown is for conventional temperature system. This cost could be reduced by 50 percent by using a low temperature duct system.
‡The 1.5 kW/ton is figured at the peak summer demand conditions and also includes all accessories.

Specifications	Model 1098	Model 1190	Model 1500
Total ton-hour capacity	115	190	570
Tube surface/ton-h, ft²	12.0	12.0	12.0
Nominal discharge time, h	6-12	6-12	6-12
Latent storage cap, ton-h	98	162	486
Sensible storage cap., ton-h	17	28	84
Maximum operating temp., °F	100	100	100
Maximum operating press., lb/in²	90	90	90
Outside diameter, in	89	89	—
Length × width, in	—	—	268×96
Height, in	68	101	102
Weight, unfilled, lb	1,060	1,550	4,850
Weight, filled, lb	9,940	16,750	50,450

Specifications	Model 1098	Model 1190	Model 1500
Volume of water/ice, gals.	980	1620	4860
Volume of solution in HX, gals.	90	148	555
Press. drop (25% glycol, 28°F), PSI			
20 gal/min	4.0	—	—
40 gal/min	9.7	5.0	—
60 gal/min	18.8	9.0	—
80 gal/min	—	13.0	2.8
160 gal/min	—	—	7.0
240 gal/min	—	—	13.0

The outlet temperatures from the tanks vary with the rate at which the tanks are discharged. See LEVLOAD Performance Manual for details.
LEVLOAD and CALMAC are registered trademarks of Calmac Manufacturing Corporation. The described product and its applications are protected by United States Patents 4,294,078; 4,403,645; 4,565,069; 4,608,836; 4,616,390; 4,671,347 and 4,687,588.
*Calmac Manufacturing Corporation
SI values in procedure text.

The total annual energy cost is the sum of the winter and summer energy costs, $2172.34 + $588.00 = $2760.34.

Related Calculations.   Use this procedure to compute the energy costs for any type of structure, industrial, office, residential, medical, educational, etc., having heating or cooling loads, or both. Any type of fuel, oil, gas, coal, etc., can be used for the structure.

This procedure is the work of Jerome F. Mueller, P.E. of Mueller Engineering Corp.

HEAT RECOVERY USING A RUN-AROUND SYSTEM OF ENERGY TRANSFER

A hospital operating room suite requires 6000 ft^3/min (169.8 m^3/s) of air in the supply system with 100 percent exhaust and 100 percent compensating makeup air. Winter outdoor design temperature is 0°F (-17.8°C); operating-room temperature is 80°F (26.7°C) with 50 percent relative humidity year-round. How much energy can be saved by installing coils in both the supply and exhaust air ducts with a pump circulating a non-freeze liquid between the two coils, absorbing heat from the exhaust air and transferring this heat to the makeup air being introduced?

Calculation Procedure:

1.   Choose the coils to use
In the winter the exhaust air is at 80°F (26.7°C) while the supply air is at 0°F (-17.8°C). Hence, the coil in the exhaust air duct will transfer heat to the nonfreeze liquid. When this liquid is pumped through the coil in the intake-air duct it will release heat to the incoming air. This transfer of otherwise wasted heat will reduce the energy requirements of the system in the winter.

As a first choice, select a coil area of 12 ft^2 (1.1 m^2) with a flow of 6000 ft^3/min (169.8 m^3/s). While a number of coil arrangements are possible, the listing below shows typical coil conditions at face velocities of 500 ft/min (152.4 m/min) and 600 ft/min (182.8 m/min) with a coil having 8-fins/in coil. Entering coolant temperature = 45°F (7.2°C); entering air temperature = 80°F (26.7°C) dry bulb, 67°F (19.4°C) wet bulb. Various manufacturers' values may vary slightly from these values.

Ft/min (m/min) face velocity	Temp rise, °F (°C)	No. of rows	Total MBtu/h (kWh)	Leaving dry bulb, °F (°C)	Leaving wet bulb, °F (°C)
500 (152.4)	12 (21.6)	6	16.6 (4.86)	57.3 (14.1)	56.5 (13.6)
500 (152.4)	12 (21.6)	8	20.1 (5.89)	54.2 (12.3)	53.9 (12.2)
600 (182.9)	10 (18)	4	14.3 (4.19)	62.1 (16.7)	59.7 (15.4)

The middle coil listed above, if placed in the exhaust duct, would produce 20.1 MBtu/h (5.89 kWh) with a 12°F (21.6°C) temperature rise with a leaving air temperature of 53.9°F (12.2°C) when the liquid coolant enters the coil at 45°F (7.2°C) and leaves 57°F (13.9°C).

2. *Determine the coil heating capacity*

The heating capacity of a coil is the product of (coil face area, ft²)[heat release, Btu/(h·ft²) of coil face area]. For this coil, heating capacity = 12 × 20,100 = 241,200 Btu/h (70.7 kWh). The incoming makeup air can be heated to a temperature of: (heating capacity, Btu)/(makeup air flow, ft³/min)(1.08) = 241,200/6000 × 1.08 = 37.2°F (2.9°C). Hence, the makeup air is heated from 0°F (−17.8°C) to 37.2°F (2.9°C).

The energy saved is—assuming 1000 Btu/lb of steam (2330 kJ/kg)—241,200 /1000 = 241.2 lb/h (109.5 kg/h). With a 200-day heating season and 10-h operation/day, the saving will be 200 × 10 × 241.2 = 482,400 lb/yr (219,010 kg/yr). And if steam costs $20/thousand pounds ($20/454 kg), the saving will be (482,400/1000)($20) = $9,648,00/year. Such a saving could easily pay for the heating coil in one year.

Related Calculations.   Use this general approach to choose heating coils for any air-heating application where waste heat can be utilized to increase the temperature of incoming air, thereby reducing the amount of another heating medium that might be required. Most engineers use the 1000 Btu/lb (2300 kJ/kg) latent heat of steam as a safe number to convert quickly from hourly heat savings in Btu (kg) to pounds of steam. This procedure can be used for industrial, commercial, residential, and marine applications.

The procedure is the work of Jerome F. Mueller, P.E., of Mueller Engineering Corp.

Figure 11 is a typical run-around coil detail that is very commonly used in energy recovery systems in which the purpose is to exract heat from air that must be exhausted. Normally about 40 to 60 percent of the heat being wasted can be recovered. This seemingly simple detail has two points that should be carefully noted. The difference in fluid temperatures in this closed system creates small expansion and contraction problems and an expansion tank is required. Most impor-

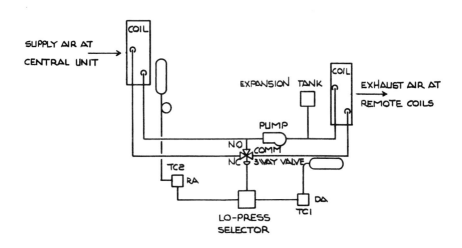

FIGURE 11   Heat-recovery-loop schematic. (*Jerome F. Mueller.*)

tantly the temperature of the incoming supply air can create a coil temperature so
low that the coil in the exhaust air stream begins to ice up. Normally this betins at
some 35°F (1.67°C). This is when the three-way bypass valve comes into play and
the glycol is not circulated through the outside air coil. Obviously if the outside air
temperature is low enough, the system will go into full bypass and the ciruclating
pump should be stopped.

ROTARY HEAT EXCHANGER ENERGY SAVINGS

A hospital heating, ventilating and air-conditioning installation has a one-pass system supplying and exhausting 10,000 ft³/min (2260 m³/min) with these operating
conditions: Summer outdoor design temperature 95°F dry bulb (35°C), 78°F wet
bulb (25.6°C); summer inside exhaust temperature 75°F dry bulb (35°C), 62.5°F wet
bulb (16.9°C); winter outdoor design temperature 0°F (−17.8°C); winter outdoor
exhaust temperature 75°F (35°C). How much energy can be saved if a rotary heat
exchanger (heat or thermal wheel) is used as an energy-saving device?

Calculation Procedure:

1. Determine the cooling-load savings
A rotary heat exchanger generally consists of an all-metallic rotor wheel with radial
partitions containing removable heat-transfer media sections made of aluminum,
stainless steel, or Monel, Fig. 12. A purge section permits cleaning of the heat-
transfer media using water, steam, solvent spray, or compressed air to eliminate
bacteria growth, especially in hospitals and laboratories. Data supplied by rotary
heat exchanger manufacturers give an average sensible heat transfer efficiency as
80 percent, and an enthalpy efficiency of 65 percent.

From a psychrometric chart the enthalpy of air at 95°F dry bulb (35°C) and 78°F
wet bulb (25.6°C) is 41.6 Btu/lb (96.9 kJ/kg); at 75°F dry bulb (35°C) and 62.5°F
wet bulb (16.9°C) it is 28.3 Btu/lb (65.9 kJ/kg). The specific volume of the air,
from the chart, is 13.6 ft³/lb (0.85 m³/kg). Then the cooling load heat saving,
Btu/h (W) = (heat-wheel efficiency)(air flow, ft³/min)(60 min/h)(enthalpy difference, Btu/lb)/(specific volume of the air, ft³/lb). Substituting, heat saving =
0.65(10,000)(60)(41.6 − 28.3)/13.6 = 381,397 Btu/h (111.7 kW), or 31.8 tons of
refrigeration.

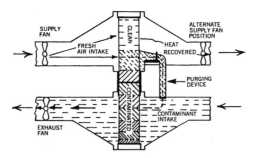

FIGURE 12 Heat or thermal wheel rotates at 1 to 3
r/min and permits recovery of heat from exhaust air.
Wheel can also be used to cool incoming air.

2. Find the heat-load saving
The heat-load saving, Btu/h (W) = (heat-wheel efficiency)(1.08)(air flow, cfm)(indoor temperature − winter outdoor design temperature). Substituting, heat saving = 0.80(1.08)(10,000)(75 − 0) = 648,000 Btu/h (189.9 kWh). Using 1000 Btu/lb (2330 kJ/kg) as the latent heat of steam, the saving will be 648,000/1000 = 648 lb/h (294.2 kg/h).

3. Find the leaving air temperature for each condition
For the summer air cooling load, the temperature of the supply air leaving the rotary heat exchanger = (summer outdoor design temperature) − (sensible heat-transfer efficiency)(summer outdoor design temperature − summer indoor exhaust temperature) = 95 − 0.80 (97 − 75) = 79°F dry bulb (26.1°C).

To determine the summer wet-bulb temperature, use the relation: Summer wet-bulb temperature = temperature at the enthalpy found from (enthalpy at summer outdoor design condition) − (sensible heat-transfer efficiency)(enthalpy at summer outdoor design condition − enthalpy at summer indoor exhaust temperature) = 41.6 − 0.65(41.6 − 28.3) = 32.96 Btu/lb (76.8 kJ/kg). Entering the psychrometric chart, read the wet-bulb temperature as 68.8°F (20.4°C).

For the winter air condition, the supply air leaving the rotary heat exchanger has a temperature of (outdoor design temperature) + (rotary heat exchanger sensible-heat efficiency)(indoor air temperature − outdoor design temperature). Or temperature of supply air leaving the rotary heat exchanger = 0 + 0.80(75 − 0) = 60°F dry bulb (15.6°C).

The winter wet-bulb temperature is found from (indoor air enthalpy) − (efficiency)(indoor air enthalpy − outdoor air enthalpy); once the enthalpy is known, the wet-bulb temperature can be found from the psychrometric chart. For this rotary heat exchanger, 28.3 − 0.65(28.3 − 1.0) = 10.56 Btu/lb (kJ/kg). Entering the psychrometric chart, find the wet-bulb temperature as 32°F (0°C). In this calculation the enthalpy of the 0°F (−17.8°C) air is taken as 1.0 Btu/lb (2.33 kJ/kg) because that is the value of the enthalpy at the 0°F (−17.8°C) temperature.

Related Calculations.   Rotary heat exchangers find use in many applications: industrial, commercial, residential, etc. The key to using any rotary heat exchanger is the trade-off between heater cost vs. the savings anticipated. Thus, if a rotary heat exchanger can be paid for in either two years, or less, most firms will find the heater acceptable.

Rotary heat exchangers are also called *thermal wheels* and they are popular for energy conservation. Current standard designs can handle clean filtered air from ambient temperature to 500°F (260°C). Wheels designed to handle high-temperature air are rated at air temperatures to 1500°F (816°C). Normal rotative speed is 1 to 3 r/min.

Properly designed heat-recovery thermal wheels can recover 60 to 80 percent of the sensible heat from exhaust air and transmit this heat to incoming outside air. Where both sensible- and latent-heat recovery are desired, specially designed thermal wheels will also recover 60 to 80 percent of the heat in the exhaust air stream and transmit it to the incoming air.

With the increased emphasis on indoor air quality (IAQ), some regulatory groups prohibit use of thermal wheels where there is the possibility of leakage from the exhaust stream contaminating the incoming air. Hence, the designer must carefully check local code requirements before specifying use of a thermal wheel. Both the exhaust stream and the incoming air stream should be filtered to prevent contamination. Usual choice is 2-in-thick (50.4-mm) "roughing" filters.

The thermal wheel can cool incoming air when the exhaust air is at a lower temperature. Thus, with 75°F (24°C) incoming air and 60°F (16°C) exhaust air, the incoming air temperatures can be reduced to, possibly, 65°F (18°C). Hence, the thermal wheel can produce savings in both directions, i.e., heating or cooling incoming supply air.

The procedure given here is the work of Jerome F. Mueller, P.E., Mueller Engineering Corp. Supplementary data on heat wheels is from Grimm and Rosaler—*Handbook of HVAC Design*, McGraw-Hill.

SAVINGS FROM "HOT DECK" TEMPERATURE RESET

An office building has a dual-duct heating and cooling system rated at 30,000 ft³/min (849 m³/min). The winter heating season is 37 weeks and the summer cooling season 15 weeks. Following Federal guideline suggestions, a decision has been made to reduce (reset) the hot deck by 6°F (10.8°C) in the summer and by 4°F (7.2°C) in the winter. How much energy will be saved if the building occupied cycle is 60 h/wk?

Calculation Procedure:

1. Compute the summer energy saving
The energy saved in the summer, S, can be found from S = ft³/min(0.5)(1.08)(°F by which hot deck temperature is lowered)(weeks of cooling)(occupied cycle, h/wk). Or, S = 30,000(0.5)(1.08)(6)(15)(60) = 87,480,000 (92,291 MW). In this equation the factor 0.5 is used because only one of the dual ducts is the "hot deck."

2. Determine the winter energy saving
Use the same equation, substituting the winter temperature reduction and the duration of the heating season. Or winter saving, W = 30,000(0.5)(1.08)(4)(37)(60) = 143,856,000 Btu (151,768 MJ).

3. Compute the annual energy saving
The annual energy saving, SA, is the sum of the summer and winter savings, or SA = 87,480,000 + 143,856,000 = 231,336,000 Btu (244,059 MJ).

Related Calculations.   Use this procedure for any type of building having a dual-duct heating and cooling system: industrial, office, commercial, residential, medical, health-care, etc. Be certain to use the actual reset temperature reduction when analyzing the potential savings.

This procedure is the work of Jerome F. Mueller, P.E., Mueller Engineering Corp.

AIR-TO-AIR HEAT-EXCHANGER PERFORMANCE

A laboratory heating, ventilating and air-conditioning system requires 100 percent exhaust of its 5000-ft³/min (141.5-m³/min) air supply. The operating conditions are: Summer outdoor design temperature 95°F dry bulb (35°C); 78°F wet bulb (25.6°C); Summer laboratory room exhaust temperature 75°F dry bulb (23.9°C); 62.5 wet bulb (16.9°C); Winter outdoor design temperature 0°F (−18°C); Winter laboratory room exhaust temperature 75°F dry bulb (23.9°C); 62.5°F wet bulb (16.9°C). Because the high moisture content of the laboratory room air must be maintained year round, the design requires that an energy-saving device include a moisture-saving feature needing no energy to operate. What is a suitable system? How much savings can be obtained?

Calculation Procedure:

1. Evaluate potential systems for this installation

The design requirements dictate some form of direct heat-transmission interchange which can be achieved by a run-around coil system or a heat wheel. But the further requirement of no energy used in the recovery system dictates a slightly different approach.

There are insulated air-to-air exchangers available using cross-flow cartridges, Fig. 13, which are non-clogging and bacteriostatic. Efficiencies are generally 75 percent for sensible heat and 60 percent enthalpic. The cartridges in such exchangers are constructed of alternating layers of corrugated and flat sheets separating the exhaust and supply air streams. Cross-contamination and leakage are less than 0.3 percent.

During summer and winter operation, moisture is entirely in the vapor phase. Permeation of moisture from the humid air stream to the dry air stream is effected without chemical impregnation. In most applications, there is little risk of ice formation even at low winter design temperatures.

To determine if a given application presents a possibility of icing, plot the outside air condition and design air exhaust condition on a psychrometric chart and draw a straight line through these two points. If this straight line between the two points does not intersect the saturation curve on the psychrometric chart, no danger of icing exists.

In the event the saturation curve is intersected, draw a line from the exhaust-air temperature condition tangent to the saturation curve and extend this tangent to the horizontal dry-bulb temperature line. The difference between the temperature value of the horizontal line at this intersection and the actual outside air temperature defines the number of degrees the outside air temperature must be raised by preheating.

2. Determine the properties of the air

From a psychrometric chart, the enthalpy at 95°F dry bulb (35°C) and 78°F wet bulb (25.6°C) is 41.6 Btu/lb (96.9 kJ/kg). At 75°F dry bulb (23.9°C) and 62.5°F wet bulb (16.9°C), the enthalpy is 28.3 Btu/lb (65.9 kJ/kg). At 0°F (−17.8°C) the air is very dry and the enthalpy is generally about 1.0 Btu/lb (2.33 kJ/kg).

3. Compute the cooling-load saving

The cooling-load saving, C Btu/h = (efficiency)(ft^3/min)(enthalpy difference, Btu /lb)(60 min/h)/specific volume of the air, ft^3/lb. Substituting, using the enthalpic efficiency, $C = 0.60(5000)(41.6 − 28.3)(60)/13.6 = 176,029$ Btu/h (51.6 kW), or 14.7 tons of refrigeration.

4. Find the heating-load savings

The heating-load saving, H Btu/h (kW) = (sensible-heat efficiency)(1.08)(ft^3/ min)(inside room temperature − outside air temperature), or $H = 0.75(1.08)(5000)(75 − 0) = 303,750$ Btu/h (88.9 kN), or 303.8 lb (139.8 kg) of

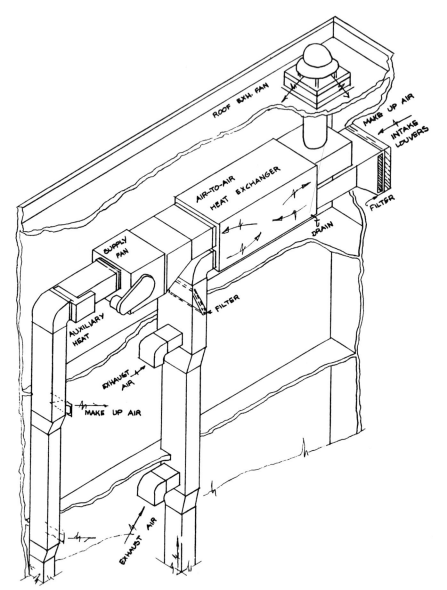

FIGURE 13 Typical air-to-air heat exchanger. System shown is used where there are no contaminants in the exhaust and intake air and a portion of the return air stream is being exhausted. (*Jerome F. Mueller.*)

steam, using an enthalpy of vaporization of 1000 Btu/lb (2330 kJ/kg) of steam, which is a safe assumption.

5. Determine summer and winter enthalpies and temperatures
In the summer, the supply air leaving the heat exchanger has a dry-bulb temperature of: (summer outdoor dry-bulb temperature) − (sensible-heat efficiency)(outdoor dry-bulb − indoor dry-bulb) = 95 − 0.75(95 − 75) = 80°F (26.7°C).

The summer wet-bulb temperature is found at the enthalpy for: (enthalpy at summer design outdoor wet-bulb) − (enthalpic efficiency)(summer outdoor wet-bulb enthalpy − summer indoor wet-bulb enthalpy); or summer wet-bulb = 41.6 − 0.60(41.6 − 28.3) = 33.62 Btu/lb (78.3 kJ/kg). On the psychrometric chart, read the wet-bulb temperature at 33.62 Btu/lb (78.3 kJ/kg) as 69.5°F (20.8°C).

In the winter, the supply air dry-bulb temperature leaving the heat exchanger will be at: (outdoor air design temperature) + (sensible-heat efficiency)(winter exhaust temperature − outdoor design temperature); or 0 + 0.75(75 − 0) = 56.25°F (13.5°C).

The winter wet-bulb temperature is found at the enthalpy for: (indoor wet-bulb enthalpy) − (enthalpic efficiency)(indoor wet-bulb enthalpy − outdoor design temperature enthalpy); or 28.3 − 0.60(28.3 − 1) = 11.92 Btu/lb (27.8 kJ/kg). On the psychrometric chart, for 11.92 Btu/lb (27.8 kJ/kg), read the wet-bulb temperature as 32°F (0°C).

Related Calculations.   Use this general procedure to evaluate the performance of any air-to-air heat exchanger used in an energy-recovery application. Buildings in which such a heat exchanger would be useful include office, factory, commercial, residential, medical, hospitals, etc.

This procedure is the work of Jerome F. Mueller, P.E., Mueller Engineering Corp.

STEAM AND HOT-WATER HEATING CAPACITY REQUIREMENTS FOR BUILDINGS

A building with a volume of 500,000 ft³ (14,150 m³) is to be heated in 0°-weather (−17.8°C) to 70°F (21.1°C). The wall and roof surfaces aggregate 28,000 ft² (2601.2 m²) and the glass area aggregates 7000 ft² (650.3 m²). Air in the building is changed three times every hour (a 20-min air change). Allowing transmission coefficients of 0.25 Btu/(ft²·h·°F) of 0.25 for the wall and roof surfaces(1.42 m²), and 1.13 Btu/(ft²·h·°F) (6.42 W/m²) for the single-paned glass windows, determine the square feet (m²) of steam and hot-water radiation required if each square foot emits 240 Btu/h [2725.5 kJ/(m²·h)] for steam and 150 Btu/h [1703.4 kJ/(m²·h)] for hot water.

Calculation Procedure:

1. Find the wall, roof, and glass heat losses
The wall and roof losses = (heat-transmission coefficient)(area)(temperature difference). Or, for this building, wall and roof losses = (0.25)(28,000)(70 − 0) = 490,000 Btu/h (143.6 kW). For the glass, using the same relation, heat loss = (1.13)(7000)(70 − 0) = 555,000 Btu/h (162.2 kW).

2. Compute the ventilation heat load
The ventilation heat load = (volume of air inflow)(number of air changes/ hour)(density of air)(specific heat of air)(temperature rise of entering air). For this building, ventilation heat load = $(500,000)(3)(0.075)(0.24)(70 - 0) = 1.89 \times 10^6$ Btu/h (553.8 kW).

3. Calculate the amount of steam and hot-water radiation required
First sum the various heat loads, namely walls, roof, glass, and ventilation, and divide by the heat emitted by each square foot (m^2) of radiation. Or $(490,000 + 555,000 + 1.89 \times 10^6)/240 = 12,229.2$ ft^2 of equivalent direct radiation (EDR), or [138,878.4 kJ/($m^2 \cdot$h)]. For hot-water heating the required area is 19,567 ft^2 EDR [222,208.7 kJ/($m^2 \cdot$h)].

    Related Calculations. To find the fuel consumption for a building such as this, divide the total heat load by the efficiency of the heating system times the heating value of the fuel as fired. In such calculations, remember that 4 ft^2 (0.37 m^2) of steam radiation are equivalent to a condensation rate of 1 lb (0.454 kg) of steam/ hour for low-pressure heating systems.

HEATING STEAM REQUIRED FOR SPECIALIZED ROOMS

A control room for an oil refinery unit is to be heated and ventilated by a central duct system. Ventilation is to be at the rate of 3 ft^3/min (0.085 m^3/min) of outside air/square foot (m^2) of floor area. The room to be ventilated is 40×60 ft (12.2 $\times$ 18.3 m), and is to be pressurized to keep out hazardous gases. Outside design temperature is $-10°$F ($-23.3°$C). Determine the steam consumption rate for maximum design conditions with the use of 5-lb/in^2 (gage) (34.5-kPa) saturated steam for heating.

Calculation Procedure:

1. Compute the amount of outside air needed
The rate of outside air to be handled by the ventilating system is (floor are)(ft^3/ min/ft^2 of floor area) = $(40 \times 60)(3) = 7200$ ft^3/min (203.8 m^3/min).

2. Find the heating load for this system
Use the relation, heating load = (ft^3/min)(1.08)(temperature rise of the air). In this relation the constant is 1.08 = (specific heat of air)(minutes/hour)/(specific volume of air). For air at normal atmospheric conditions, $(0.24)(60)/(13.3) = 1.0827$; this value is normally rounded to 1.08, as given above. Substituting, heating load, $Q = (7200)(1.08)(75 - [-10])660,960$ Btu/h (193.7 kW).

3. Calculate the rate of steam consumption
Saturated steam at 5 lb/in^2 (gage) has a heat of condensation of 960 Btu/lb (2236.8 kJ/kg). The steam rate is then $660,960/960 = 688.6$ lb/h (312.6 kg/h).

    Related Calculations. Heating systems generally use lower-pressure steam as their heating medium because (1) piping, valve, and fitting costs are lower; (2) the heat of condensation (or latent heat) of lower pressure steam is higher (larger), meaning that more heat is absorbed by the condensation of each pound (kg) of steam. While high-pressure steam may be used under specialized circumstances, the majority of steam-heating systems use low-pressure steam.

DETERMINING CARBON DIOXIDE BUILDUP IN OCCUPIED SPACES

An office space has a total volume of 75,000 ft^3 (2122.5 m^3). Equipment occupies 25,000 ft^3 (707.5 m^3). The space is occupied by 100 employees. If all outside air supply is cut off, how long will it take to render the space uninhabitable?

Calculation Procedure:

1. *Determine the cubage of the space*
For carbon dioxide buildup measurements, the *net volume* or (*cubage*) of the space is used. The net volume of a space = total volume − volume of equipment, files, machinery, etc. For this space, net volume, NV = total volume − machinery and equipment volume = 75,000 − 25,000 = 50,000 ft^3 (1415 m^3).

2. *Compute the time to vitiate the inside air*
Use the relation, $T = 0.04V/P$, where T = time to vitiate the inside air, h; V = net volume, ft^3; P = number of people occupying the space. Substituting, $T = 0.04(50,000)/100 = 20$ h. During this time the oxygen content of the air will be reduced from a nominal 21 percent by volume to 17 percent.

It is a general rule to consider that after 5 h, or one-quarter of the calculated time of 20 h that the air would become stale and affect worker efficiency. Atmospheres containing less than 12 percent oxygen or more than 5 percent carbon dioxide are considered dangerous to occupants. The formula used above is popular for determining the time for carbon dioxide to build up to 3 percent with a safety factor.

Related Calculations. In today's environmentally conscious world, smoking indoors is prohibited in most office and industrial structures throughout the United States. Much of the Western world appears to be considering adoption of the same prohibition, albeit slowly. Part of the reason for prohibiting smoking inside occupied structures is the oxygen depletion of the air caused by smokers.

Today, indoor air quality (IAQ) is one of the most important design considerations faced by engineers. A variety of environmental rules and regulations control the design of occupied spaces. These requirements cannot be overlooked if a building or space is to be acceptable to regulatory agencies.

For years, occupied spaces which were not air conditioned were designed using *general ventilation* rules. In most buildings, exhaust fans located high in the side walls, or on the roof, were used to draw outside air into the building through windows or louvers. The air movement produced an air flow throughout the space to remove smoke, fumes, gases, excess moisture, heat, odors, or dust. A constant inflow of fresh, outside air was relied on for the removal of foul, stale air.

Today, with the increase in external air pollution, combined with the outgassing of construction and furnishing materials, general ventilation is a much more complex design problem. No longer can the engineer rely on clean, unpolluted outside air. Instead, careful choice of the location of outside-air intakes must be made. Other calculation procedures in this handbook deal with this design challenge.

COMPUTING BYPASS-AIR QUANTITY AND DEHUMIDIFIER EXIT CONDITIONS

A space to be conditioned has a sensible heat load of 10,000 Btu/min (10,550 kJ /min) and a moisture load of 26,400 grains/min (1,710,667 mg/min); 2300 lb (1044.2 kg) of air are to be introduced each minute to this space for its conditioning to 80°F (26.7°C) dry bulb and 50 percent relative humidity. How much air should be bypassed in this system, Fig. 14, and what is the amount and temperature of the air leaving the dehumidifier?

Calculation Procedure:

1. Set up a listing of the air conditions for this system
Using a psychrometric chart and a table of air properties, set up the list thus:

	Room conditions	Air leaving dehumidifier
Dry bulb	80°F (26.7°C)	54°F (12.2°C)
Wet bulb	67°F (19.4°C)	54°F (12.2°C)
Dew point	60°F (15.6°C)	54°F (12.2°C)
Relative humidity	50 percent	100 percent
Total heat	31.15 Btu/lb (72.58 kJ/kg)	22.54 Btu/lb (52.52 kJ/kg)
Grains/lb	77.3 (11,033 mg/kg)	62.1 (8863.5 mg/kg)
Specific volume, ft³/lb	13.84 (0.863 m³/kg)	13.13 (0.818 m³/kg)
Lb/min air flow	2300 (1044.2 kg/min)	1610 (730.9 kg/min)

2. Find the moisture load in the system
The moisture load = (grains/min)/(7000 grains/lb)(latent heat of air, Btu/lb); or (26,400/7000)(1040) = 3920 Btu/min (4135.6 kJ/min), rounded from 3922.3 Btu/min. The sensible heat load = 10,000 Btu/min (10,550 kJ/min). Total load = 3920 + 10,000 = 13,920 Btu/min (14,685.6 kJ/min).

3. Using trial and error, find the air quantities in the system
Solve by trial and error, assuming 53°F (11.7°C) air leaving the dehumidifier. Then, the total heat at 80°F (26.7°C) and 50 percent relative humidity = 31.15 Btu (32.86 kJ); total heat at 53°F (11.7°C) and 100 percent relative humidity = 21.87 Btu (23.07 kJ). The difference in total heat content is the pickup in the dehumidifier. Or, 31.15 − 21.87 = 9.28 Btu (9.79 kJ).

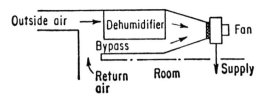

FIGURE 14 Dehumidifier fitted with bypass-air control.

On the basis of our first trial, the weight of air circulated = (total load, Btu/min)/(heat pickup, Btu); or 13,920/9.28 = 1500 lb/min (681 kg/min). Check this result using (lb/min computed)(specific heat of air)(temperature difference, dry bulb − assumed temperature of air leaving the dehumidifier, or 53°F [11.7°C] in this case). Solving, (1500)(0.24)(80 − 53) = 9720 Btu/min (10,254.6 kJ/min). This value is not enough because the sensible heat load is larger, i.e., 10,000 Btu/min (10,550 kJ/min).

Using trial and error again, assume 54°F (12.2°C) air leaving the dehumidifier. Then, as before: Total heat at 80°F (26.7°C) and 50 percent relative humidity = 31.15 Btu (32.86 kJ); total heat at 54°F (12.2°C) and 100 percent relative humidity = 22.54 Btu (23.78 kJ). The difference = 31.15 − 22.54 = 8.61 Btu (9.08 kJ).

The air circulated is now 13,920/8.61 = 1620.2 lb/min (735.6 kdg/min). Checking as before, (1620.2)(0.24)(80 − 54) = 10,110 Btu/min (10,666.1 kJ/min). The 10,110 Btu/min is slightly higher than the 10,000 Btu/min required, actually 1.1 percent. This is acceptable for usual design purposes.

4. Find the amount of air leaving the dehumidifier
Using the assumed 54°F (12.2°C) leaving temperature, the amount of air leaving the dehumidifier is, from step 3, 1620.2 lb/min (735.6 kg/min).

5. Compute the quantity of air bypassed
The quantity of air bypassed = (lb of air introduced/minute − quantity of air leaving the dehumidifier); or air bypassed = 2300 − 1620.2 = 679.8 lb/min (308.6 kg/min). The temperature of the air leaving the dehumidifier is the assumed value of 54°F (12.2°C).

Related Calculations.   Strict standards have been introduced governing use of outside air in bypass air-conditioning systems. The reason for this is the increased air pollution in urban areas. In some instances, outside-air intakes have been found close to truck and bus driveways, leading to polluted air being drawn into the air-conditioning unit.

The Environmental Protection Agency publishes guidelines for allowable contaminants in outside air used for air-conditioning units. These guidelines must be used if a design is to be accepted by governing authorities. The guidelines are discussed in other calculation procedures presented in this handbook.

DETERMINATION OF EXCESSIVE VIBRATION POTENTIAL IN MOTOR-DRIVEN FAN

Determine if the motor-driven fan in Fig. 15 will have excessive vibration. The motor is 110-V, 60 Hz, 2400 r/min; armature weight 40 lb (18.2 kg); radius of gyration = 5 in (12.7 cm). This motor drives a 3-bladed fan weighing 10 lb (4.54 kg) with a radius of gyration of 9 in (22.86 cm); the drive shaft is steel. Is this design acceptable? If not, what changes should be made in the design?

Calculation Procedure:

1. Find the moment of inertia for each part of the assembly
The motor and fan are arranged as in Fig. 15. Find the moment of inertia of the fan from I_f = (fan weight/32.2 ft/s · s)(1/12 in/ft)(radius of gyration^2); or I_f = (10

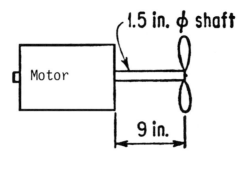

SI Value	
in.	cm
1.5	3.8
9	22.9

FIGURE 15   Motor-driven fan.

lb/32.2)($\frac{1}{12}$)(9 × 9) = 2.094 in⁴ (87.15 cm⁴). For the motor, I_m = (40 lb/ 32.2)($\frac{1}{12}$)(5 × 5) = 2.587 in⁴ (107.68 cm⁴).

The torsional constant for the steel shaft, k = (modulus of elasticity of the steel)/(radius of gyration × 0.495). Or, k = (11.5 × 10⁶)/(9)(0.495) = 6.3 × 10⁵.

2.   Find the frequency of the assembly
Use the relation, frequency, $f = (\frac{1}{2}\pi)[I_f + I_m)(k)/(I_f \times I_m)]^{0.5}$ where the symbols are as defined earlier. Substituting, $f = (\frac{1}{2}\pi)[(2.094 + 2.587)(6.3 \times 10^5)/(2.094 \times 2.587)]^{0.5}$ = 117.47 cycles/s.

The motor frequency = 2400 r/min/60 cycles/s = 40 cycles/s. There may be excessive vibration when the system starts or is shut down if the rate of increase or decrease in speed is small—say 40 cycles/s. Recognizing this, the system should be redesigned.

To redesign this system to eliminate the danger of excessive vibration, increase the radius of gyration of both the fan and the motor armature. Also, increase the shaft diameter, or decrease its length until the frequency of the entire system is below 40 cycles/s.

Related Calculations.   Use this general approach to design connected system so there is no danger of excessive vibration. While vibration may be tolerated during the starting and stopping of connected units, it is best to design the assembly so there is no vibration during any of the normal speeds encountered in the design.

POWER INPUT REQUIRED BY CENTRIFUGAL COMPRESSOR

A centrifugal compressor handling air draws in 12,000 ft³/min (339.6 m³/min) of air at a pressure of 14 lb/in² (abs) (96.46 kPa) and a temperature of 60° F (15.6°C). The air is delivered from the compressor at a pressure of 70 lb/in² (abs) (482.4 kPa) and a temperature of 164°F (73.3°C). Suction-pipe flow area is 2.1 ft² (0.195 m²); area of discharge pipe is 0.4 ft² (0.037 m²) and the discharge pipe is located

20 ft (6.1 m) above the suction pipe. The weight of the jacket water, which enters at 60°F (15.6°C) and leaves at 110°F (43.3°C) is 677 lb/min (307.4 kg/min). What is the horsepower required to drive this compressor, assuming no loss from radiation?

Calculation Procedure:

1. Determine the variables for the compressor horsepower equation
The equation for centrifugal compressor horsepower input is,

$$ \text{hp} = \frac{w}{0.707}\left[c_p(t_2 - t_1) + \frac{V_2^2 - V_1^2}{50,000} + \frac{Z_2 - Z_1}{778} \right] + \left[\frac{w_j(t_o - t_i) + R_c}{0.707} \right] $$

In this equation, we have the following variables: w = weight, lb (kg) of unit flow rate, ft^3/s (m^3/s) through the compressor, lb (kg), where $w = (P_1)(V_1)/R(T_1)$, where P_1 = inlet pressure, lb/in^2 (abs) (kPa); V_1 = inlet volume flow rate, ft^3/s (m^3/s); R = gas constant for air = 53.3; T_1 = inlet air temperature, °R.
   The inlet flow rate of 12,000 ft^3/min = 12,000/60 = 200 ft^3/s (5.66 m^3/s); P_1 = 14.0 lb/in^2 (abs) (93.46 kPa); T_1 = 60 + 460 = 520 R. Substituting, w = 14.0(144)(200)/53.3(520) = 14.55 lb (6.6 kg).
   The other variables in the equation are: c_p = specific heat of air at inlet temperature = 0.24 Btu/(lb · °F) [1004.2 J/(kg · K)]; t_2 = outlet temperature, °R = 624 R; t_1 = inlet temperature, °R = 520 R; V_1 = air velocity at compressor entrance, ft/min (m/min); V_2 = velocity at discharge, ft/min (kg/min); Z_1 = elevation of inlet pipe, ft (m); V_2 = elevation of outlet pipe, ft (m); w_j = weight of jacket water flowing through the compressor, lb/min (kg/min); t_i = jacket-water inlet temperature,°F (°C); t_o = jacket outlet water temperature,°F (°C).
   The air velocity at the compressor entrance = (flow rate, ft^3/s)/(inlet area, ft^2) = 200/2.1 = 95.3 ft/s (29 m/s); outlet velocity at the discharge opening = 200/0.4 = 500 ft/s (152.4 m/s).

2. Compute the input horsepower for the centrifugal compressor
Substituting in the above equations, with radiation losses, $R_c = 0$,

$$ \text{hp} = \frac{14.55}{0.707}\left[0.24(624 - 520) + \frac{500^2 - 95.3^2}{50,000} + \frac{20}{778} \right] $$

$$ + [677/60 \times (110 - 60)]/0.707 $$

$$ = 20.6(24.95 + 4.8 + 0.0256) + 797 = 1409 \text{ hp } (1051 \text{ kW}). $$

   Related Calculations.   This equation can be used for any centrifugal compressor. Since the variables are numerous, it is a wise procedure to assemble them before attempting to solve the equation, as was done here.

EVAPORATION OF MOISTURE FROM OPEN TANKS

A paper-mill machine room produces 50 tons (45.5 t) of finished paper/day. Studies show that 1.5 lb (0.68 kg) of water must be evaporated for every pound (kg) of finished paper as the paper goes over the dryer rolls. What capacity exhaust fan is needed if the room conditions are 100°F (37.8°C) and 40 percent relative humidity and tempered air enters the room at 70°F (21.1°C) and 50 percent relative humidity? Determine the air flow and exhaust fan capacity required if the room temperature remains at 100°F (37.8°C) but the exhaust relative humidity of the exhaust air could be raised to 60 percent.

Calculation Procedure:

1. Determine the amount of water evaporated into the room atmosphere/unit time
With 50 tons (45.5 t) of paper being produced/24 h, the weight of paper = 50 (2000) = 100,000 lb (45,400 kg). Then, the amount of paper produced/minute = (100,000 lb)/(24 h)(60 min/h) = 69.4 lb (31.5 kg)/min. Since water is evaporated at the rate of 1.5 lb/lb of paper (1.5 kg/kg), the total evaporation rate = 1.5 (69.4) 104.1 lb (47.3 kg)/minute.

2. Find the amount of moisture removed/unit of air flow
From the psychrometric chart or a table of air properties, find the moisture content of the entering and leaving air for this room. Thus, at an entering air temperature of 70°F (21.1°C) and 50 percent relative humidity, each 100 ft^3/min (2.83 m^3/min) contains 0.059 lb (0.0268 kg) of moisture. The leaving exhaust air at 100°F (37.8°C) and 40 percent relative humidity contains 0.117 lb (0.053 kg) of moisture/100 ft^3 /min (2.83 m^3/min). The moisture absorbed by the air during passage through the room therefore is 0.117 − 0.059 = 0.058 lb/100 ft^3/min (0.026 kg/2.83 m^3/min).

3. Compute the air flow required to remove the moisture generated
The air flow required = (quantity of moisture to be removed, lb or kg/unit time)/ (moisture absorbed/unit time, lb or kg). For this plant, with a total evaporation rate of 104.1 lb (47.3 kg)/minute, air flow required = (104.1/0.058)(100) = 179,482.8 ft^3/min (5079.3 m^3/min). An exhaust fan with a capacity of 180,000 ft^3/min (5094 m^3/min) would be chosen for this application. If one fan of this capacity was too large for the space available, two fans of 90,000 ft^3/min (2547 m^3/min) could be chosen instead. Any other combination of capacities that would give the desired flow could also be chosen.

4. Calculate the air flow required with a higher exhaust relative humidity
With the room temperature remaining at 100°F (37.8°C) but the exhaust air at 60 percent relative humidity, the moisture content would be 0.175 lb (0.079 kg)/100 ft^3/min (2.83 m^3/min). Then, the new absorption, as before, = 0.0175 − 0.058 = 0.117 lb/100 ft^3/min (0.053 kg/2.83 m^3/min). Then, air flow required = (104.1/ 0.117)(100) = 88,974 ft^3/min (2517.9 m^3/min). Thus, we see that the required air flow is reduced by (179,483 − 88,974) = 90,509 ft^3/min (2561 m^3/min), or 50.4 percent. An exhaust fan capacity of 90,000 ft^3/min (2547 m^3/min) would be chose for this installation.
    Related Calculations.   This procedure can be used for any installation where air borne moisture must be removed from a closed space, such as a factory, meeting room, ballroom, restaurant, indoor swimming pool, etc. In every such installation, the required air flow to remove a given quantity of moisture will decrease as the relative humidity of the exhaust air is increased. Reducing the required air flow will save money in several ways: on the initial cost, installation cost, operating cost,

and maintenance cost of the exhaust fan(s) chosen. Hence, it is important that the engineer carefully analyze the entering and leaving air relative humidity or moisture content.

Selecting an "environmentally gentle" fan of lower capacity requiring a smaller power input is a design objective of many businesses and institutions today. Hence, careful analysis of conditions in the installation will be rewarded with reduced overall and life-cycle costs while meeting environmental goals.

Where steam or moisture is released to working spaces from open tanks or similar vessels, with resultant high humidity conditions, the engineer is faced with the problem of estimating the rate of evaporation and providing for a reduction in the moisture content of the room air. The procedure above shows one popular way to control the room moisture content.

For steam escaping into a closed space, the escape rate is easily calculated. However, moisture given off by industrial processes is not as easily computed. Further, there isn't much information in the technical literature covering moisture generation by industrial processes. High humidity affects worker comfort and can produce mild to severe condensation problems with moisture dripping from walls and ceilings. Typical industries with this "wet" heat problem and the requirements to reduce it are:

Textile industry: Elimination of fog and condensation in dye houses, bleacheries, and finishing departments. *Paper and pulp mills:* Elimination of high vapor generation in machine, heater, and grinder rooms. *Steel and metal goods:* Elimination of excessive vapor generation in pickling rooms. *Food industries:* Control of large vapor generation in kettle, canning, blanching, bottle washing, and bottle-filling areas. *Process industries:* Control of vapor generation in electroplating, coating, and chemical processing.

CHECKING FAN AND PUMP PERFORMANCE FROM MOTOR DATA

Determine the air flow from a fan driven by a 460-V motor when the current draw is 7 a, the fan delivers air at a pressure of 4-in (10.2-cm) water gage, if the fan and motor efficiencies are 65 and 90 percent respectively and the power factor = 0.80. Likewise, determine the efficiency of a pump delivering 90 gal/min (5.7 L/ s) at a 1000 lb/in² (6890 kPa) pressure differential when the motor current is 100 a, voltage is 460 V, power factor = 0.85, and motor efficiency = 90 percent.

Calculation Procedure:

1. Compute the rate of air delivery by the fan

It can be shown that the relationship between fan horsepower and motor power consumption gives the equation $q = (14.757)$(motor efficiency)(EI)(cos ϕ)(fan efficiency)/(fan pressure developed, in water), where q = air delivery rate, ft³/min (m³/min); E = motor voltage; I = motor current flow, A; cos ϕ = power factor. Substituting, $q = (14.757)(0.90)(460)(7)(0.8)(0.65)/4 = 5559.6$ ft³/min (2624 L/ s).

With the computed delivery rate known, we can now easily compare the actual fan efficiency against manufacturer's-guarantee data. If the results do not agree with the guarantee, suitable action can be taken.

2. Calculate the pump efficiency

It can likewise be shown that pump efficiency, P_e = (flow rate, gal/min)(pressure differential across the pump, lb/in^2)/4(EI)(motor efficiency)(power factor). Or, motor efficiency = (90)(1000)/(4)(460)(100)(0.90)(0.85) = 0.6393; say 64 percent.

Using this computed efficiency, we can refer to pump efficiency curves to see if the plotted efficiency at the flow and head agree. If the computed efficiency is significantly different from the plotted value, then further investigation of the pump performance is warranted.

Related Calculations.   For fans used for forced-draft boiler applications, it can be shown, by using the MM Btu (kJ) method of combustion analysis, that for a fixed boiler output the air flow required in lb/h (kg/h) is a constant at a given excess air requirement. Hence, irrespective of density conditions, the same mass flow of air must be delivered to the burner. (Computation of air flow from a fan at different densities and altitudes is discussed elsewhere in this handbook.)

Further, a boiler's back pressure in inches (cm) of water is a function of mass flow of air and its density. Hence, as the density of air decreases (as at higher temperatures or altitudes), the pressure head to be delivered to a boiler increases, while the mass flow and head that fan can deliver decreases. Thus, fan performance must be checked at the lowest density conditions. This is an important point that should not be overlooked in practical performance calculations.

A pump delivers the same flow in gal/min (L/s) and head in feet (m) of liquid at any temperature. However, because of changes in fluid density, the flow in lb/h (kg/h), pressure in lb/in^2 (kPa), and brake horsepower (kW) input will change. Thus, a boiler feed pump which must deliver a stated flow in lb/h (kg/h) at a given lb/in^2 (kPa) will require a larger power input as the fluid density decreases. Likewise, if a pump must deliver a given flow in gal/min (L/s) at a given head in feet (m), then as the density of the fluid decreases the brake horsepower input also decreases. However, this is not the situation in boiler applications.

This procedure is the work of V. Ganapathy, Heat Transfer Specialist, ABCO Industries, Inc.

CHOICE OF AIR-BUBBLE ENCLOSURE FOR KNOWN USAGE

Choose a suitable air-bubble enclosure for a maintenance shop requiring 100,000 ft^2 (9290 m^2) of covered area. The enclosure will house 75 maintenance workers in a southwestern area of the United States where daytime temperatures can reach 110°F (43.3°C) and nighttime temperatures can fall to 0°F (-17.8°C). Determine the size, and number, of blowers required for this enclosure, probable power consumption, enclosure cost, and precautions in construction and use of an air-bubble enclosure.

Calculation Procedure:

1. Determine the general requirements of the air-bubble enclosure

Most air-bubble enclosures are rectangular in shape, though there are no restrictions on square and round shapes. Typical plastic-reinforced fabric structures range in size from 80 to 300 ft (24.4 to 91.4 m) wide, 80 to 450 ft (24.4 to 137.2 m) long, and 30 to 75 ft (9.1 to 22.9 m) high. Width constrains the size of the structure, and the limits of material strength determines the structure's width.

The cost of air-bubble structures varies from \$6.00 to \$11.00/ft² (\$65 to \$118/m²), compared to \$36 to \$80/ft² (\$388 to \$861/m²) for a prefabricated light-duty steel building. With 100,000 ft² (9290 m²) enclosed space, and the dimension constraints given above, this enclosure could be 250 ft wide by 400 ft long (76.2 by 122 m), depending on the size of plot available and its access roads. The enclosure height will depend on the clearance required for the machinery needed for the maintenance work performed in the enclosure. For this enclosure, we'll assume that the height of the enclosure is 50 ft (15.2 m).

2. Find the capacity, and number, of blowers required by the enclosure
Air leakage from an air-bubble enclosure is typically 6000 to 8000 ft³/min (2832 to 3776 L/s). This air leakage occurs mainly at doors, air locks, and perimeter anchor points. Electrical operating cost to drive the blower to replace air leakage and keep the enclosure inflated depends on how often the air is changed in the enclosure. The frequency of air changes should be suited to the requirements of the work done in the enclosure, for example, if painting is being done in the enclosure, additional ventilation is necessary.

A positive interior pressure must be maintained to keep the fabric rigid. The enclosure inflation system must automatically respond to wind gusts, equipment failures, power losses, fabric ruptures, and other unpredictable events. For a long-term air-bubble enclosure, such as a maintenance shop, the inflation system should consist of a primary, a secondary, and an emergency blower and a vent system.

When specifying air-change frequency, the design engineer should take into account such considerations as expected air leakage, the inside temperature during the summer, the diluent effect of the inside air volume, and the vehicular traffic, if any, inside the enclosure. Three to four air changes/hour have been shown to be adequate for a 100,000 ft² (9290 m²) air-bubble enclosure.

The primary inflation system should replace normal air losses and high winds. For a typical 100,000 ft² (9290 m²) air-bubble enclosure the internal air pressure is 0.85 to 0.95 in water (21.6 to 24.1 mm). A typical inflation blower for 3 air changes/hour for this enclosure would be a 36,000 ft³/min (16,992 L/s) unit with a 15-hp (11.2 kW) drive motor. The backup blower would have a capacity equal to the primary blower. An emergency blower of 9000 ft³/min (4248 L/s) would be chosen to handle somewhat more than the expected air losses.

3. Determine what types of controls are suitable for the blowers
Simple, flexible controls are normally specified for air-bubble enclosure blowers. Typically, primary blowers are controlled by an on-off switch, are run continuously, with louvre vents located at a distance from the blowers on the perimeter of the enclosure. Vent cycling open and closed is regulated by a pressure switch.

Design all vents so they are regulated by a pressure switch. Further, all vents should be spring-to-close upon power failure and have grease fittings for easy lubrication. In the event of a sudden large pressure drop within the enclosure, the vents will stay closed to maintain the internal pressure in the enclosure. If the pressure loss is catastrophic, i.e., larger than the primary blowers can handle—the secondary blowers should have switches to turn them on automatically.

Secondary blowers are sometimes activated manually, for example when trucks are operating inside the enclosure, but are usually left on automatic control. Where there is more than one secondary blower, that start automatically in sequence as needed to maintain the needed internal pressure for the enclosure. Each blower has a vent located elsewhere on the perimeter that cycles open and closed as regulated by a pressure switch. Normal controls include a high-pressure shutdown switch and a backup high-pressure shutdown. All blowers should have gravity-operated back-draft dampers to prevent air loss when they are not operating.

4.   Choose a suitable emergency power supply and controls
The emergency inflation system for an air-bubble enclosure should be completely automatic to ensure the structure's stability in the event of power loss. There has been disagreement among designers and engineers over whether to rely on gener-ators or directly driven internal-combustion-engine blowers, or both. To simplify maintenance, use the least number of units possible in an emergency system. At the same time, the emergency unit must start reliably.

For these reasons, the emergency system often consists of two propane-fueled internal-combustion engines that directly drive the blowers. As noted above, the typical emergency blower has a capacity of 9000 ft³/min (4248 L/s). The control system includes automatically charged batteries for reliable starting of the engine driving the blower.

To prevent overpressurization, a critical design consideration, a high-pressure switch for shutting down the main and emergency blower motor can be coupled with a low-pressure switch for restarting it. Such a system should be backed up with a completely independent high-pressure switch system having separate air-sensing tubing. Relief vents can present resealing problems and can be accidentally opened during storms or if the structure moves.

5.   Select air locks, fabric, drainage, inside temperature, and lighting
Size air locks to handle the largest equipment that will enter the enclosure. Thus, some enclosures have air locks permitting the entrance of large trucks and even earth-moving equipment.

For vehicles the air lock usually consists of outer and inner doors, electrically interlocked to prevent them from being opened simultaneously. Each air-lock door should be equipped with a manual chain-operated opener. Specify a heavy roll-up type door. Roll-up flaps allow doors and air locks to be removed. Personnel doors are usually the revolving type. Avoid swinging personnel doors because they add stress to the structure and could leak air.

For a 100,000 ft² (9290 m²) structure, the fabric may have to be three or four sections. The structural design of an air-bubble enclosure should meet local building codes. Such codes may include fabric specifications.

The typical air-bubble enclosure is designed to withstand winds of 80 mi/h (128.8 km/h), with gusts up to 100 mi/h (161 km/h). Snow loads can be up to 10 lb/ft² (48.9 kg/m²) which will be produced by 12 to 14 in (30.5 to 35.5 cm) of snow or 2 in (5.1 cm) of ice. The contour of the structure and its flexibility cause snow and ice to slide off. Also, the internal pressure may be boosted to offset snow or wind loading.

Air-bubble structures can be made from many types of fabrics. If a structure life of more than 2 years is intended, the options become limited because of ultraviolet deterioration. This can be prevented by bonding a polyvinylfluoride film to the exterior of the fabric. While this coating is expensive, it can be justified by the increased service life of the fabric.

Selecting fabrics is not easy because of the variety of reinforcement methods for distributing the structural loading to the anchors. The fabric should be chosen so it resists dry rot, mildew, and weathering. Fabrics for structures over 200-ft (61-

m) wide must be more carefully selected than usual because of the higher stresses. Of proven quality is a 28-oz/yd (0.86 kg/m) coated vinyl-polyester fabric.

Sewn seams should be avoided to limit stretching, and air and water leakage. Fusion-welded seams are generally preferred. A double-layered fabric should be considered if the structure will be air conditioned. Manufacturers of air-bubble enclosures and fabrics provide repair kits with complete instructions for plant personnel.

A reinforcement system over the fabric consisting of a metallic net or a web of high-strength plastic material can be used to help distribute the structural loading to the enclosure anchors. The same considerations that apply to fabric selection apply to the reinforcement system. Further, the reinforcement system should not expose the fabric to wear or tearing. For example, a cable system, which is preferred for structures over 200 ft (61 m) wide, should be coated with a thermoplastic. The engineer designer should ensure that if one or two cables or webs break, a domino effect will not be created that will lead to a structural failure. Design of the structure should relieve stresses in all directions as much as possible.

Condensate can form on the inside surface of an air-bubble enclosure if it is not double-walled or if the inside is not air conditioned. Specify double walling if moisture drippage cannot be tolerated. For a typical 100,000 ft² (9290 m²) air-bubble enclosure in the southern United States, about 1000 gal/day (3790 L/day) of condensate can form during spring or autumn. Interior gutters can be installed to collect condensate and route it outside.

Specify allowable water and air leakage. Air-bubble enclosures usually allow little leakage. For this 100,000 ft² (9290 m²) enclosure, water leakage has been estimated at less than 10 gal (37.9 L) during a 6-in (15.2-cm) rain storm. Water leaks are usually confined to areas around air locks, doors, and anchors. Such leaks can be minimized by close inspections during construction and startup.

The inside temperature of an uninsulated and unventilated air-bubble enclosure can be 10°F to 20°F (5.6°C to 11.1°C) higher than outside. Three or four air changes an hour might hold the inside temperature to 10°F (5.6°C) above ambient.

The typical air-bubble fabric is very translucent. On a clear day, the fabric can let in an average of 300–400 footcandles (1028 to 1370 cd/m²), almost nine times the light required inside a typical warehouse. More light can be let in by placing clear panels In the fabric. However, translucency must be limited as necessary to reduce solar heat gain. Because the enclosure fabric is highly reflective, a few lights are adequate at night.

6. *Choose a suitable anchoring system*

Geological soil data, including soil bearing pressure, at the site are needed for designing the anchoring. Good anchoring helps ensure the structure's reliability and extends its life. Among the anchors used are steel helicoils, concrete piers, and the continuous-grade beam. Helicoils and piers are adequate for short-term structures. The continuous-grade beam is recommend for long-term ones.

Although the continuous-grade beam is the most expensive type of anchor, it minimizes the likelihood of the domino type of simultaneous reinforcement failures. Further, such beams divert rainwater effectively, limiting its entry. This is accomplished by sloping the top of the grade beam at least 1.5 to 2 in (3.8 to 5.1 cm). The beam also permits continuous clamping of the enclosure fabric.

A continuous rope-bead edge on the fabric ensures a structurally sound clamping of the fabric to concrete. The gap between the fabric and concrete is sealed by a gasketed steel angle. Nothing, such as signs, fences, gates, etc., should be closer than 2.5 ft (0.76 m) from the fabric to keep it from being torn during deflations for maintenance work. The surrounding area, outside of the grade beam, should slope away to run off rainwater.

Related Calculations. The general principles presented in this procedure are valid for air bubble enclosures used for industrial purposes: shops, warehouses, construction shelters, materials-storage areas, conveyor housings, etc. In such applications, they can have important environmental uses because of the Resource Conservation and Recovery Act which regulates pollution resulting from rainfall. Long-term air-bubble enclosures can be insulated and the enclosed space can be heated and air conditioned. Such a structure can last more than 20 years.

Other applications of air-bubble enclosures include sports facilities for tennis, indoor golf, running tracks, swimming pools, etc. The same general principles apply for their design and application. Agricultural uses include greenhouses, farm markets, and farm animal housing.

Air-bubble enclosure manufacturers can be extremely helpful in the planning and layout of these structures. However, since typical enclosures require the usual building services, electric, water, sanitation, heating, and possibly air conditioning, a project engineer should be assigned to supervise the installation. It is the task of the project engineer to coordinate the work of civil, mechanical, electrical, and other engineers assigned to the successful installation of the air-bubble enclosure.

This procedure is the work of Charles W. Hawk, Jr., P.E., Project Manager, Olin Corporation's Southeast Engineering Group, as reported in *Chemical Engineering* magazine. SI values were added by the handbook editor.

SIZING HYDRONIC-SYSTEM EXPANSION TANKS

Determine the needed expansion-tank volume for a hydronic heating system containing 10,00 gal (37,900 L) of water, having a fill pressure of 15 lb/in^2 (gage) (104.4 kPa) at the tank, with the maximum pressure limit set at 25 lb/in^2 (gage) (172.3 kPa), designed to operate over a temperature range of 70°F (21.1°C) at fill to 220°F (104.4°C) during operation. Show the various equations which can be used and the results obtained with each.

Calculation Procedure:

1. *Select the equations which can be used for expansion-tank sizing*
Derivation of equations for sizing expansion tanks in hydronic systems is fundamental if it is assumed that the air cushion in the tank behaves as a perfect gas. For such equations, all the necessary values that are not established as design parameters are readily available from the steam tables.

The only complications that may be met with such equations are those relative to how the tank is used in the system. For example, if it is assumed that the water in the expansion tank always remains at its initial temperature, that compression and expansion of the air in the tank is isothermal, and that the air in the tank was initially compressed from atmospheric pressure in the tank, Eq. 1, from Table 2, is readily derived and applies. Thus, if the designer uses Eq. 1 and anticipates system performance to be in accordance with the design, all possible steps must be taken to assure that the actual design satisfies the assumptions. This might include leaving the tank uninsulated in such a way that thermal circulation between the tank and piping will be minimal.

To see the results obtainable with Eq. 1, substitute as follows: $V_T = 10,000$ $\{[(0.01677/0.01606)] - 1 - 3\ (0.000006)(220 - 70)\}/[(14.7/29/7) - (14.7/39/7)] = 3314.6$ gal (12,562 L).

2. Find the expansion-tank volume with different design assumptions

If the design assumptions are that (a) the initial charge of water in the expansion tank changes temperature with the main volume of water, (b) that the air in the tank is at its initial charge temperature and compresses and expands isothermally, and (c) that the air in the expansion tank was initially compressed from atmospheric pressure, Eq. 2 results (see Table 2).

TABLE 2 Equations for expansion tank sizing.[1]

Equation no.	Equation	Assumptions	Example tank size, gal (L)
1	$V = \dfrac{V_w[(v_2/v_1 - 1) - 3\alpha\Delta t]}{(p_a/p_1 - p_a/p_2)}$	• Air compresses isothermally (t_1) °F (°C)	
		• Water in tank is at temperature t_1 °F (°C)	3314.6 gal (12,562 L)
		• Initial air charge is atmospheric	
2	$V_T =$ $\dfrac{V_w[(v_2/v_1 - 1) - 3\alpha\Delta t]}{(p_a/p_1 - p_a/p_2) - (v_2/v_1 - 1)(1 - p_a/p_1)}$	• Air compresses isothermally (t_1) °F (°C)	
		• Water in tank is at temperature t_2 °F (°C)	4033.9 gal (15,288.8 L)
		• Initial air charge is atmospheric	
3	$V_T = \dfrac{V_w[(v_2/v_1 - 1) - 3\alpha\Delta t]}{(1 - p_1/p_2)}$	• Air compresses isothermally (t_1) °F (°C)	1639.6 gal (6214.1 L)
		• Initial air charge is at pressure p_1	

where, with volumes in consistent units:

V_w = volume of water in system (piping, heat exchangers, etc.)* gal (L)

V_T = volume of expansion tank gal (L)

p_a = atmospheric pressure, lb/in² (abs) (kPa)

p_1 = pressure at lower temperature, lb/in² (abs) (kPa)

p_2 = pressure at higher temperature, lb/in² (abs) (kPa)

t_1 = lower temperature, °F (°C)

t_2 = higher temperature, °F (°C)

v_1 = specific volume of water at temperature t_1 °F (°C)

v_2 = specific volume of water at temperature t_2 °F (°C)

α = linear coefficient of thermal expansion, 1/°F (1/1.8 °C)

Δt = higher temperature − lower temperature, °F (°C)

*At t, and not including water in the tank.
[1]HPAC Magazine.

Equation 2 becomes a bit more complex if the air in the tank is assumed to increase in temperature with the liquid (note that the total pressure of the gas in the tank is the sum of the partial pressures of the air and water vapor and that a saturated condition always exists). Although not totally accurate, Eq. 2 would be a fair approximation of the condition where a portion of a thermal storage tank is used to provide the expansion cushion.

Substituting in Eq. 2, using data from Eq. 1, we find that $V_T = 4033.9$ gal (15,289 L). Note that the expansion-tank volume given by Eq. 2 is 100 (4033.9 − 3314.6)/4033.9 = 17.8 percent smaller than that computed by Eq. 1. This difference is the result of the change in the assumptions made between the two designs.

3. Determine expansion-tank volume with a precharged diaphragm

If the initial charge in the tank is not compressed from atmospheric pressure in the tank itself, but rather is forced into the tank at a design operating pressure (either from a compressed-air system or as a precharged diaphragm type) and the air is assumed to compress and expand isothermally, Eq. 3 results (see Table 2).

Substituting in Eq. 3, we find $V_T = 1639.6$ gal (6214.1 L). Again, this differs from the two earlier computed values because of the changes in the design assumptions made. Further, the first cost of a precharged tank will be higher than that of a simple storage tank with no special attachments or fittings.

Related Calculations. The calculated expansion-tank sizes for the three different conditions show how much the computed sizes can vary. These significant variations indicate that in the selection of an expansion tank for a hydronic system the designer must: (1) determine what operating assumptions are to be used in selecting the tank sizing equations; (2) design the system to achieve the assumed conditions as closely as possible.

While a number of other equations for sizing expansion tanks have been developed by various authorities, the equations, in general, resemble those presented here. The reason for this is that the variables—water volume, pressure, and temperature, enter all the equations developed for the analysis. Further, the results closely approximate those found here.

From the standpoints of thermodynamics and hydraulics, the equations given in this procedure can be used to size expansion tanks with an equal degree of accuracy for hot-water heating systems, chilled-water systems, and dual-temperature water systems. When a tank with a liquid-gas interface is used in a chilled-water system, water logging may occur. Extreme precaution must be taken in design because in such systems there is a continual pumping effect that removes air from the tank by absorption in the water. This causes the small tanks designed by the equations in this procedure to water log frequently.

One option to avoid water logging is to provide an oversize thank to minimize the frequency of needed air charging. Another option is to design such systems to prevent the absorption phenomenon; but this can have numerous other detrimental effects on the system.

The fundamental components of any hydronic system are: (a) the heat source; (b) the load; (c) the circulator; (d) piping; and (e) the expansion tank. Strangely enough, the most complex device of the five, the expansion tank, may appear to be the least complicated.

The expansion tank serves a dual purpose in the hydronic system. It allows for the volumetric changes in the fluid, resulting from fluid temperature changes, to occur between designed pressure limits. Further, the expansion tank establishes the point of constant or known pressure in the hydronic system. And, in many hydronic systems, the tank serves the additional purpose of being an integral part of the air control subsystem.

The correct sizing of expansion tanks is becoming even more critical as larger volume hydronic systems are being designed and built. Not only are large volumes a result of higher capacity systems, but they are also an integral part of solar systems and other power-conserving installations that utilize thermal storage through liquid-phase temperature changes. The sizing of an expansion tank, as noted by Lockhart and Carlson,* relates not only to the volume of fluid in the hydronic system, the temperature limits, and the pressure limits, but also to how the tank is designed into the system.

This procedure is based on the work of William J. Coad, Vice-President, Charles J. R. McClure & Associates, and Affiliate Professor of Mechanical Engineering, Washington University, as reported in *HPAC* Magazine. SI values were added by the handbook editor.

Additional design pointers given by Jerome F. Mueller, P.E., are as follows: In any hydraulic system there is a basic requirement to keep system pressure at a desired normal operational level. The primary objectives of the expansion tank system are to limit the pressure of all the equipment in the system to the allowable working pressure, to maintain minimum pressure for all normal operating temperatures, to vent air, to prevent cavitation at the pump suction and the boiling of system water, and to accomplish of all this with a minimum addition of water to the system.

In this book the most common basic types of expansion tank systems are depicted. The first type is a system sized to accommodate the volume created by the water expansion, with sufficient gas space to keep the pressure range within the design limits of the system. Air is normally used and the system must remain as tight as possible because every recharging cycle introduces additional oxygen with the air and thus promotes corrosion. In the initial start-up, the oxygen reacts with the system components, leaving basically a nitrogen atmosphere in the expansion tank. The initial fill pressure set a minimum level of water in the tank. The compression of the gas space because of water expansion determines the maximum system pressure.

An alternative to this arrangement is the diaphragm type expansion tank, which is precharged to the system fill pressure and sized to accept the normally expected expansion of water. The tank's air charge and the system water are permanently separated by a flexible elastomer diaphragm. For a high temperature hot-water system the pressurization may be either by steam or inert gas, commonly nitrogen. Other details will depict nitrogen pressurization systems for high-temperature hot-water applications.

For all expansion tank systems the location of the system pump in relation to the expansion tank connection determines whether the pump pressure is added to or subtracted from the system static pressure. This is due tot he fact that the junction of the tank with the system is the point of no pressure change whether the pump is in operation or not.

In our details, which depict common expansion-tank situations, we show the expansion tank on the suction side of the pump. The reason for this location is that when the pump is discharging away from the boiler and the expansion tank, the full pump pressure will appear as an increase at the pump discharge. All points downstream will show a pressure equal to the pump pressure minus the friction loss from the pump at that point. The fill pressure need to be only slightly higher than the system static pressure. If, on the other hand, the pump discharges into the

---

*Lockhart, H.A. and Carlson, G. F., "Compression Tank Selection for Hot Water Heating Systems," *Heating/Piping/Air Conditioning,* April, 1953.

boiler and the expansion tank, which is common in small residential and small commercial systems, full pump pressure is reduced and system fill pressure is increased. Therefore, until the fill pressure is higher than the pump pressure, a vacuum can be created in the system. Normally, a pump discharging into the boiler and a pressure tank system is used only in low-rise buildings, small systems, or single-family residences in which the pumps need to have only a low total head capability.

Sizing of expansion tanks is generally based on a system determined from the standard ASME formula that is shown in the *ASHRAE Systems Handbook* and other publications. The size of the expansion tank is determined by the volume of water in the system; the range of water temperatures normal to the operation of the system; the pressure of air in the expansion tank when the fill water first enters the tank; the relation of the height of the boiler to the high point of the system, which is usually but not always the item in the system with the lowest working pressure; the characteristics of the expansion tank; the high point of the system; the pressure developed by the circulating pump; and the location of the circulating pump with respect to the expansion tank and the boiler. Finally, it should be noted that any time there is a change in water temperature in a hot- or chilled-water system, an expansion tank should be used. Expansion tanks are therefore needed not only in hot water systems and run-around coil heat-recovery systems used for energy conservation designs but also in chilled-water systems.

Factory Pressurized Tanks. Figures 16 through 18 illustrate the application of a factory pressurized tank with an expandable diaphragm. Figure 16 shows a typical expansion tank installation in a system that supplies either hot or chilled water to the system. The boiler pump shown is seemingly inconsistent with the arrangement described in our previous discussion of expansion tank location, but in this application the pump is performing a specific task. It is used to provide circulation in the boiler under certain special conditions.

A system pump by definition supplies the system. Note that on the chiller side we again have a special circulating pump for the chiller. The object of these two separate pumps is not merely to serve the system but also to balance and maintain the flow from within the chiller and the boiler under varying overall flow and temperature conditions. Note that in this detail, as in many others, the items related to the expansion tank are described. For example, in the supply of makeup water there is a filling control unit tied into the return side of the system. The system

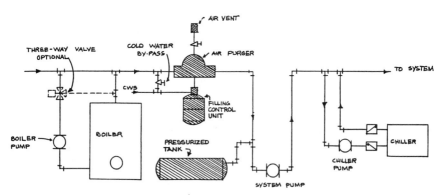

FIGURE 16 Factory-pressurized expansion tank with dual temperatures and separate boiler and chiller pumps. (*Jerome F. Mueller.*)

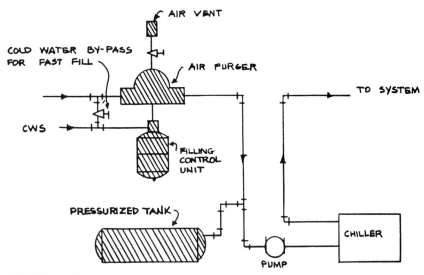

FIGURE 17   Factory-pressurized tank system for chilled-water supply. (*Jerome F. Mueller.*)

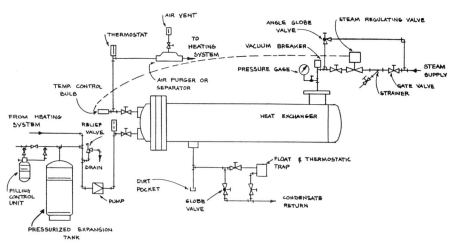

FIGURE 18   Converter/factory-pressurized expansion-tank installation. (*Jerome F. Mueller.*)

passes through an automatic air purger with a manual or automatic vent. Finally, note that the pressurized tank is tied into the suction side of the system pump.

Chilled Water Expansion Tank.   An expansion tank is needed in a chilled water system, as well as in a hot water system. Figure 17 shows the application of an expansion tank to a chilled-water system only. This application is common when the chilled water and hot water are in separate piping systems or are otherwise separated. Again, the return fluid is passed through an air purger and vent to remove air. There is an automatic control of makeup water in the filling control unit. The pressurized tank, shown in previous details, is illustrated again in this detail.

In Fig. 18 we show the application of an expansion tank in the usual steam or hot-water converter installation. The converter is a steam converter. We show the basic steam piping connections to the converter. At this point, whether the device is a converter or a boiler, the basic premise of the system is still as shown. System water is pumped into the heat exchanger. Frequently the way to make the heat exchanger perform mist efficiently is to create flow under pressure. The vent on the system is at the high point on top of the heat exchanger. It might also be noted that the pumping system for the converter is often on the floor of the equipment room, and the heat exchanger, sometimes separated from the converter by a considerable distance, is at the ceiling.

Air Source Expansion Tank. In Figs. 19 and 20 we illustrate the air source expansion tank. Here there is no diaphragm separating air from water. This arrangement has been common for expansion tank installation in many applications for a very long time. The boiler pumps and other devices are not clearly shown but are implied with the notation to pump suction and with the obvious notation that the air strainer is on the suction line of the pump. As can be seen, all the connections rise vertically and pitch up to the expansion tank and to the cold-water fill line. The makeup water and expansion tank lines are tied into the air separator, which is designed for this particular type of system. Both the air separator and the expansion tank have drain valves to drain excess water from the expansion tank. The cold water fill line shown has a pressure-reducing valve to supply makeup water to the system at an acceptable pressure.

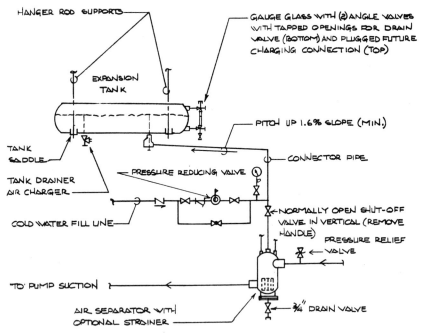

FIGURE 19   Air control and piping connections for water-system expansion tank. (*Jerome F. Mueller.*)

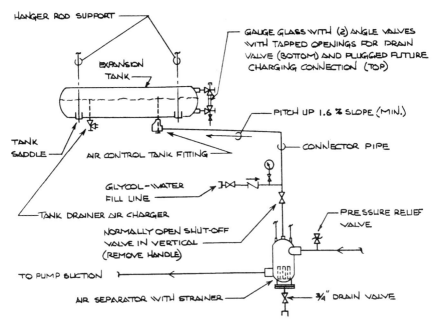

FIGURE 20   Air control and piping connections for glycol-water-system expansion tank. (*Jerome F. Mueller.*)

Figure 20 is a nearly identical installation that is used for a special system. When snow melting or run-around energy conservation coils are used, the system fluid is a combination of glycol and water. The air type of system has the advantage of not having a diaphragm that can be affected by the corrosive nature of this commonly used nonfreeze solution. The detail is similar to the one in Fig. 19 except that there is no cold water makeup. Instead there is a glycol makeup which comes from a special pumped glycol-water solution thank that is not shown in this detail but depicted in a snow melting detail in Fig. 25, pg. 16.66.

Nitrogen Pressurization.   In Fig. 21 we show the complete piping detail around a nitrogen-fed high-temperature hot-water expansion tank system. This type of system is special to the high-temperature hot-water installation. The piping shown on this system has some unusual points. First, as can be seen, the expansion tank is still on the suction side of the high-temperature hot-water circulating pump. Second, in a high-temperature hot-water system there usually is a soft-water service. We note in our detail where the connection of this treated water goes into our expansion-tank system. Controlling the amount of water in the high-temperature hot-water expansion system is extremely important. In the lower right side of the detail the makeup pump with its low level switches and controls carefully adds water to the expansion tank. As a further safety precaution there are high and low pressure cutoffs and alarm switches. Knowing the system's limits is also very important.

Finally, high-pressure water will, if the pressure is released, turn instantly into high-pressure steam. Thus, the pressure-relief valve used is similar to that used in a high-pressure boiler. The connection of the pressure-relief valve drip pan and its exhaust pipe through the roof is nearly identical tot hat of a 125 lb/in^2 (gage) (861.3 kPa) steam system. If there is a loss of pressure in the expansion tank, the

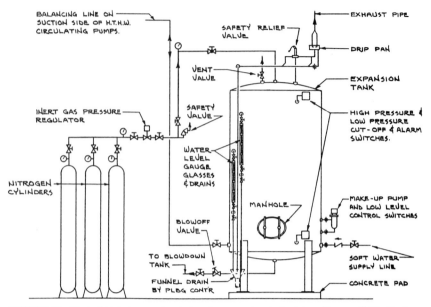

FIGURE 21   Typical high-temperature hot-water expansion-tank piping. (*Jerome F. Mueller.*)

result will cause the water to flash into steam. Note the line that goes to a blowdown tank; it provides protection against this occurrence. The pressurization for this system is normally provided by specially piped dry nitrogen cylinders. These cylinders and their output are controlled by a pressure regulator that has a safety valve to insure that the pressurization to the system is maintained at the proper value at all times. Finally, there are times when excess nitrogen does have to be vented. This is accomplished through a small vent valve in the top of the tank.

The system designer wants to reduce the area of contact between a gas and water, thereby reducing the absorption of gas in the water. This is why the tank is shown installed vertically, which is the generally preferred arrangement.

The ratings of fittings, valves, piping, and equipment generally are based on a minimum pressure, which is about 25 to 50 lb/in^2 (gage) (172.3 to 344.5 kPa) above the maximum saturation pressure. An imposed additional pressure head above the vapor pressure must be sufficient to prevent steaming in the high-temperature hot-water generator at all times, even under unusual flow conditions, such as firing rates at which the created flow of two or more generators is not evenly matched. This is a critical condition since a gas-pressurized system does not have separate safety valves. The pressure varies with changes in water level in the expansion vessel. When the system water volume increases because of a temperature rise, the expansion of the system water into the vessel compresses the inert gas, raising the system pressure. The reverse condition takes place on a drop in system water temperature. The pressure is permitted to vary from a minimum point above saturation to a maximum that is determined by the materials used in the system. The expansion tank itself can be sized for the sum of the volumes required for pressurization plus the volume required for expansion and the volume required for sludge and reserve.

The data and illustrations in this portion of this procedure are the work of Jerome F. Mueller, P.E., Mueller Engineering Corporation.

System Analysis and Equipment Selection

BUILDING OR STRUCTURE HEAT-LOSS DETERMINATION

An industrial building has 8-in (203.2 mm) thick uninsulated brick walls, a 2-in (60.8-mm) thick concrete uninsulated roof, and a concrete floor. The building is 150 ft (45.7 m) long, 75 ft (22.9 m) wide, and 15 ft (4.6 m) high. Each long wall contains eight 5 × 10 ft (1.5 × 3 m) double-glass windows, and each short wall contains two 4 × 8 ft (1.2 × 2.4 m) double-glass doors. What is the heat loss of this building per hour if the required indoor temperature is 70°F (21.1°C) and the design outside temperature is 0°F (−17.8°C)? How much will the heat loss increase if infiltration causes two air changes per hour?

Calculation Procedure:

1. *Compute the heat loss through the glass*
The usual heat-loss computation begins with the glass areas of a building. Hence, this procedure is followed here. However, a heat-loss computation can be started with any part of the building, provided each part of the structure is eventually considered.

To compute the heat loss through a building surface, use the general relation $H_L = UA \, \Delta t$, where H_L = heat loss, Btu/h, through the surface; U = overall coefficient of heat transmission for the material, Btu/(h · °F · ft²); A = area of heat-transmission surface, ft²; ΔT = temperature difference, $F = t_i - t_o$, where t_i = inside temperature, °F; t_o = outside temperature, °F. Find U from Table 3 for the material in question.

This building has sixteen 5 × 10 ft (1.5 × 3 m) double-glass windows and four 4 × 8 ft (1.2 × 2.4 m) double-glass doors. Hence, the total glass area = 16 × 5 × 10 + 4 × 4 × 8 = 928 ft² (86.2 m²). The value of U for double glass is, from Table 3, 0.45. Thus, $H_L = UA \, \Delta t = 0.45(928)(70 - 0) = 29,200$ Btu/h (8639.8 W).

TABLE 3 Typical Overall Coefficients of Heat Transmission, Btu/(ft² · h · °F) [W/(m² · K)]

Building surface	Type	Type of insulation
Walls	8-in (203.2-mm) brick	0.50 (2.84)
Roof	2-in (50.8-mm) concrete	0.82 (4.66)
Windows	Single glass	1.13 (6.42)
	Double glass	0.45 (2.56)

2. Compute the heat loss through the building walls
Use the same relation as in step 1, substituting the wall heat-transfer coefficient and wall area. Thus, $U = 0.50$, and $A = 2 \times 150 \times 15 + 2 \times 75 \times 15 - 928 = 5822$ ft^2 (540.9 m^2). Then $H_L = UA\,\Delta t = 0.50(5822)(70-0) = 204,000$ Btu/h (59,746.5 W).

3. Compute the heat loss through the building roof
Use the same relation as in step 1, substituting the roof heat-transfer coefficient and roof area. Thus, $U = 0.82$ and $A = 150 \times 75 = 11,250$ ft^2 (1045.1 m^2). Then $H_L = UA\,\Delta t = 0.82 \times (11,250)(70-0) = 646,000$ Btu/h (189,197.3 W).

4. Compute the total heat loss of the building
The total heat loss of a building is the sum of the individual heat losses of the walls, glass areas, roofs, and floor. In large buildings the heat loss through concrete floors is usually negligible and can be ignored. Hence, the total heat loss of this building caused by transmission through the building surfaces is $H_T = 29,200 + 204,000 + 646,000 = 879,200$ Btu/h (257,495.7 W).

5. Compute the infiltration heat loss
The building volume is $150 \times 75 \times 15 = 168,500$ ft^3 (4773.1 m^3). With two air changes per hour, the volume of infiltration air that must be heated is $2 \times 168,500 = 337,000$ ft^3/h (9546.1 m^3/h). The heat that must be supplied to raise the temperature of this air is $H_i = $ (ft^3/h)(Δt)/55 $= (337,000)(70-0)/55 = 429,000$ Btu/h (125,643.4 W). Thus, the total heat loss of this building, including infiltration, is $H_T = 879,200 + 429,000 = 1,308,200$ Btu/h (383.1 kW).

   Related Calculations. Determine the design outdoor temperature for a given locality from Baumeister—*Standard Handbook for Mechanical Engineers* or the ASHRAE *Guide and Data Book,* published by the American Society of Heating, Refrigerating and Air-Conditioning Engineers. Both these works are also suitable sources of comprehensive listings of U values for various materials and types of building constructions. Since the winter-design outdoor temperature is usually for nighttime conditions, no credit is taken for heat given off by machinery, lights, people, etc., unless the structure will always operate on a 24-h basis. The safest design ignores these heat sources because the machinery in the building can be removed, the operating cycle changed, or the heat sources eliminated in some other way. However, where an internal heat source of any kind will be a permanent part of a building, simply subtract the hourly heat release from the total building heat loss. The result is the net heat loss of the building and is used in choosing the heating equipment for the building.

   Most heat-loss calculations for large structures are made on a form available from heat equipment manufacturers. Such a form helps organize the calculations. The steps followed, however are identical to those given above. Another advantage of the calculation form is that it helps the designer remember the various items—walls, roof, glass, infiltration, etc.—to consider.

   When some areas in a structure will be kept at a lower temperature than others, compute the heat loss from one area to another in the manner shown above. Substitute the lower indoor temperature for the outdoor design temperature. For areas exposed to prevailing winds, some manufacturers recommend increasing the computed heat loss by 10 percent. Thus, if the north wall of a building were exposed to the prevailing winds and its heat loss is 50,000 Btu/h (14,655.0 W), this heat loss would be increased to 1.1(50,000) = 55,000 Btu/h (16,120.5 W).

HEATING-SYSTEM SELECTION AND ANALYSIS

Choose a heating system suitable for an industrial plant consisting of a production area 150 ft (45.7 m) long and 75 ft (22.9 m) wide and an office area 75 ft (22.9 m) wide and 60 ft (18.3 m) long. The heat loss from the production area is 1,396,000 Btu/h (409.2 kW); the heat loss from the office area is 560,00 Btu/h (164.1 kW). Indoor design temperature for both areas is 70°F (21.1°C); outdoor design temperature is 0°F (−17.8°C). What will the fuel consumption of the chosen heating system be if the annual degree-days for the area in which the plant is located is 6000? Compare the annual fuel consumption of gas, oil, and coal.

Calculation Procedure:

1. Choose the type of heating system to use
Table 4 lists the various types of heating systems used today and typical applications. Study of this tabulation shows that steam unit heaters would probably be best for the production area because it is relatively large and open. Either a forced-warm-air or a two-pipe steam heating system could be used for the office area. Since the production area will use steam unit heaters, a two-pipe steam heating system would probably be best for the office area. Since steam unit heaters are almost universally two-pipe, the same method of supply and return is best chosen for the office system.

Note that Table 4 lists six different types of two-pipe steam heating systems. Choice of a particular type of two-pipe system depends on a number of factors,

TABLE 4 Typical Applications of Heating Systems

System type	Fuel°	Typical applications
Gravity warm air	G, O, C	Small residences, wooden or masonry
Forced warm air	G, O, C	Small and large residences, wooden or masonry; small and medium-sized industrial plants, offices
Steam heating:		
One-pipe	G, O, C	Small residences, wooden or masonry
Two-pipe†	G, O, C	Small and large residences, wooden or masonry; small and large industrial plants, offices. High-pressure systems [30 to 150 lb/in² (gage) (206.8 to 1034.1 kPa)] may be used in large industrial buildings having unit heaters or fan units. Unit heaters are used for large, open areas
Hot water:		
Gravity	G, O, C	Small residences, wooden or masonry
Forced‡	G, O, C	Small and large residences; small and large industrial plants
Radiant	G, O, C	Small and large residences and plants
Electric	Electricity	Small residences and plants

°G—gas; O—oil; C—coal.
†May be low-pressure, two-pipe vapor; two-pipe vacuum; two-pipe subatmospheric; two-pipe orifice; high-pressure.
‡May be one-pipe; two-pipe.

including economics, steam pressure required for nonheating (i.e., process) services in the building, type and pressure rating of boiler used, etc. Where high-pressure steam—30 to 150 lb/in² (gage) (206.8 to 1034.1 kPa), or higher—is used for process, the two-pipe system fitted with pressure-reducing valves between the process and heating mains is often an economical choice.

Hot-water heating is unsuitable for this building because unit heaters are required for the production area. An inlet air temperature has less than 30°F (−1.1°C) is generally not recommended for hot-water unit heaters. Since the inlet-air temperature can be as low as 0°F (−17.8°C) in this plant, hot-water unit heaters could not be used.

2. Compute the annual fuel consumption of the system

Use the degree-day method to compute the annual fuel consumption. To apply the degree-day method, substitute the appropriate values in $F_G = DU_gRC$, for gas heating; $F_o = DU_cH_TC/1000$, for oil; $F_c = DU_cH_TC/1000$, for coal, where F = fuel consumption, the type of fuel being identified by the subscript, g for gas, o for oil, c for coal, with the unit of consumption being the therm for gas, gal for oil, and lb for coal; D = degree-days during the heating season; U = unit fuel consumption, the unit again being identified by the subscript; R = ft² EDR (equivalent direct radiation) in the heating system; C = a correction factor for the outdoor design temperatures. Values of U and C are given in Table 5. Note the fuel heating values

TABLE 5   Factors for Estimating Fuel Consumption*

Fuel	Consumption per degree-day	Heating conditions	
		Steam	Hot water
Gas	Therms [100,000 Btu (105,500 kJ)]	0.00127 [300-ft² (28-m²) EDR] 0.00121 [300–700 ft² (28–65 m²) EDR] 0.00116 [700-ft² (65-m²) EDR]	0.000743 [500-ft² (46-m²) EDR] 0.000709 [500–1000 ft² (46–92 m²) EDR] 0.000675 [1200-ft² (111-m²) EDR]
		Heating-plant efficiency	
Oil [heating value = 141,000 Btu/gal (39,297 MJ/L)]	Gal per 1000-Btu/h heat loss [L/(MJ·h)]	70% 0.00437 (0.01568)	80% 0.00383 (0.01374)
Coal [heating value = 12,000 Btu/lb (27,912 kJ/kg)]	Lb per 1000-Btu/h heat loss [kg/(MJ·h)]	70% 0.0507 (0.02163)	80% 0.0444 (0.01893)

Outside-design temperature correction factor					
Outside design temperature, °F (°C)	−20 (−28.9)	−10 (−23.3)	0 (−17.8)	+10 (−12.2)	+20 (−6.7)
Correction factor	0.778	0.875	1.000	1.167	1.400

*ASHRAE *Guide and Data Book* with permission.

on which this table is based. For other heating values, see Related Calculations, below.

For gas heating, $F_g = DU_gRC$, therms, where 1 therm = 100,000 Btu (105,500 kJ). To select the correct value for U_g, compute the ft^2 EDR (see the next calculation procedure) from $H_T/240$, or ft^2 (m^2) EDR = 1,956,000/240 = 8150(757.1). Hence, $U_g = 0.00116$ from Table 5. From the same table, $C = 1.00$ for an outdoor design temperature of 0°F (-17.8°C). Hence, $F_g = (6000)(0.00116)(8150)(1.00) = 56,800$ therms. Assuming gas having a heating value h_v of 1000 Btu/ft^3 (37,266 kJ/m^3) is burned in this heating system, the annual gas consumption is $F_g \times 10^5/h_v$. Or, $56,800 \times 10^5/1000 = 5,680,000$ ft^3 (160,801 m^3).

Using Table 5 for oil heating and assuming a heating plant efficiency of 70 percent, $U_o = 0.00437$ gal per 1000 Btu/h [0.01568 L/(MJ·h)] heat loss, and $C = 1.00$ for design conditions. Then $F_o = DU_oH_TC/1000 = (6000)(0.00437)(1,956,000)(1.00)/1000 = 51,400$ gal (194,570 L).

Using Table 5 for coal heating and assuming a heating-plant efficiency of 70 percent, we get $U_c = 0.0507$ lb coal per 1000 Btu/h [0.2163 kg/(MJ·h)] heat loss, and $C = 1.00$ for the design conditions. Then, $F_c = DU_cH_TC/1000 = (6000)(0.0507)(1,956,000)(1.00)/1000 = 595,000$ lb, or 297 tons of 2000 lb (302 t) each.

Related Calculations. When the outdoor-design temperature is above or below 0°F (-18°C), a correction factor must be applied as shown in the equations given in step 2. The appropriate correction factor is given in Table 5. Fuel-consumption values listed in Table 5 are based on an indoor-design temperature of 70°F (21°C) and an outdoor-design temperature of 0°F (-18°C).

The oil consumption values are based on oil having a heating value of 141,000 Btu/gal (39,296.7 MJ/L). Where oil having a different heating value is burned, multiply the fuel consumption value selected by 141,000/(heating value, Btu/gal, of oil burned).

The coal consumption values are based on coal having a heating value of 12,000 Btu/lb (27,912 kJ/kg). Where coal having a different heating value is burned, multiply the fuel consumption value selected by 12,000/(heating value, Btu/lb, of coal burned).

For example, if the oil burned in the heating system in step 2 has a heating value of 138,000 Btu/gal (38,460.6 MJ/L), $U_o = 0.00437(141,000/138,000) = 0.00446$. And if the coal has a heating value of 13,500 Btu/lb (31,401 kJ/kg), $U_c = 0.0507(12,000/13,500) = 0.0451$.

Steam consumption of a heating system can also be computed by the degree-day method. Thus, the weight of steam W required for the degree-day period—from a day to an entire heating season is $W = 24H_TD/1000$, where all the symbols are as given earlier. Thus, for the industrial building analyzed above, $W = 24(1,956,000) \times (6000)/1000 = 281,500,000$ lb (127,954,546 kg) of steam per heating season. The denominator in this equation is based on low-pressure steam having an enthalpy of vaporization of approximately 1000 Btu/lb (2326 kJ/kg). Where high-pressure steam is used and the enthalpy of vaporization is lower, substitute the actual enthalpy in the denominator to obtain more accurate results.

Steam consumption for building heating purposes can also be given in pounds (kilograms) of steam per 1000 ft^3 (28.3 m^3) of building space per degree-day and of steam per 1000 ft^2 (92.9 m^2) of EDR per degree-day. On the building-volume basis, steam consumption in the United States can range from a low of 0.130 (0.06) to a high of 2.07 lb (0.94 kg) of steam per 1000 ft^3/degree-day (28.3 m^3/degree-day), depending on the building type (apartment house, bank, church, department store, garage, hotel or office building) and the building location (southwest or far north). Steam consumption can range from a low of 21 lb (10.9 kg) per 1000 ft² (92.9 m²) EDR per degree-day to a high of 120 lb (54.6 kg) per 1000 ft² (92.9 m²) EDR per degree-day, depending on the building type and location. The ASHRAE *Guide and Date Book* lists typical steam consumption values for buildings of various types in different locations.

REQUIRED CAPACITY OF A UNIT HEATER

An industrial building is 150 ft (45.7 m) long, 75 ft (22.9 m) wide, and 30 ft (9.1 m) high. The heat loss from the building is 350,000 Btu/h (102.6 kW). Choose suitable unit heaters for this building if 5-lb/in² (gage) (30.5-kPa) steam and 200°F (93.3°C) hot water are available to supply heat. Air enters the unit heater at 0°F (−18°C); an indoor temperature of 70°F (21°C) is desired in the building. What capacity unit heaters are needed if 20,000 ft³/min (566.2 m³/min) is exhausted from the building?

Calculation Procedure:

1. Compute the total heat loss of the building
The heat loss through the building walls and roof is given as 350,000 Btu/h (102.6 kW). However, there is an additional heat loss caused by infiltration of outside air into the building. Compute this loss as follows:

Find the cubic content of the building from volume, ft³ = LWH, where L = building length, ft; W = building width, ft; H = building height, ft. Thus, volume = (150)(75)(30) = 337,500 ft³ (9554.6 m³).

Determine the heat loss caused by infiltration by estimating the number of air changes per hour caused by leakage of air into and out of the building. For the usual industrial building, one to two air changes per hour are produced by infiltration. At one air change per hour, the quantity of infiltration air that must be heated from the outside to the inside temperature = building volume = 337,500 ft³/h (9554.6 m³/h). Had two air changes per hour been assumed, the quantity of infiltration air that must be heated = 2 × building volume.

Compute the heat required to raise the temperature of the infiltration air from the outside to the inside temperature from H_i = (ft³/h)(Δt)/55, where H_i = heat required to raise the temperature of the air, Btu/h, through Δt, where $\Delta t = t_i - t_o$, and t_i = inside temperature, °F; t_o = outside temperature, °F. For this building, H_i = 337,500(70 − 0)/55 = 429,000 Btu/h (125.7 kW). Hence the total heat loss of this building H_t = 350,000 + 429,000 Btu/h (228.3 kW), without exhaust ventilation.

2. Determine the extra heat load caused by exhausting air
When air is exhausted form a building, an equivalent amount of air must be supplied by infiltration or ventilation. In either case, an amount of air equal to that exhausted must be heated from the outside temperature to room temperature.

With an exhaust rate of 10,000 ft³/min = 60(10,000) = 600,000 ft³/h (19,986 m³/h), the heat required is H_e = (ft³/h)(Δt)/55, where H_e = heat required to raise the temperature of the air that replaces the exhaust air, Btu/h. Or, H_e = 600,000(70 − 0)/55 = 764,000 Btu/h (223.9 kW).

To determine the total heat loss from a building when both infiltration and exhaust occur, add the larger of the two heat requirements—infiltration or exhaust—to the heat loss caused by transmission through the building walls and roof. Since, in this building, $H_e > H_i$, $H_t = 350,000 + 764,000 = 1,114,000$ Btu/h (326.5 kW).

3. Choose the location and number of unit heaters
This building is narrow; i.e., it is half as wide as it is long. In such a building, three vertical-discharge unit heaters (Fig. 22) will provide good distribution of the heated air. With three heaters, the capacity of each should be $764,000/3 = 254,667$ Btu/h (74.6 kW) without ventilation and $1,114,000/3 = 371,333$ Btu/h (108.8 kW) with ventilation. Once the capacity of each unit heater is chosen, the spread diameter of the heated air discharge by the heater can be checked to determine whether it is sufficient to provide the desired comfort.

4. Select the capacity of the unit heaters
Use the engineering data published by a unit-heater manufacturer, such as Table 6, to determine the final air temperature, Btu delivered per hour, cubic feet of air handled, and the quantity of condensate formed. Thus, Table 6 shows that vertical-discharge model D unit heater delivers 277,900 Btu/h (81.5 kW) when the entering air is at 0°F (-17.8°C) and the heating steam is at a pressure of 5 lb/in^2 (gage) (34.5 kPa). This heater discharges 3400 ft^3/min (96.3 m^3/min) of heated air at 76°F (24.4°C) and forms 290 lb/h (131.8 kg/h) of condensate. The capacity table for a

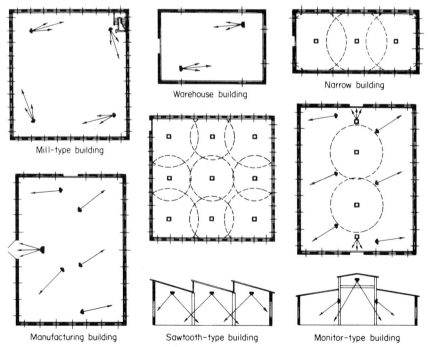

Mill-type building
Warehouse building
Narrow building
Manufacturing building
Sawtooth-type building
Monitor-type building

FIGURE 22   (*a*) Recommended arrangements of unit heaters in various types of buildings. (*Modine Manufacturing Company.*)

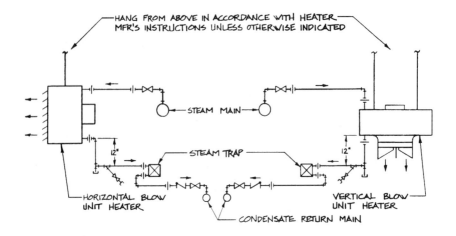

SI Values
12 in.   30.5 cm

FIGURE 22 (*b*) Horizontal and vertical blow steam unit-heater piping. (*Jerome F. Mueller.*)

TABLE 6   Typical Vertical-Delivery Unit-Heater Capacities*

Model	Mtg. ht., ft (m)	Spread diam., ft (m)	Motor speed, r/min	ft³/min† (m³/ min)	0°F (−17.8°C) entering air			50°F (10°C) entering air		
					Btu/h (kW)	Final temp., °F (°C)	Cond., lb/h (kg/h)	Btu/h (kW)	Final temp., °F (°C)	Condensate, lb/ h (kg/h)
D	14 (4.3)	48, 54 (14.8, 16.5)	1,135	3,400 (96)	277,900 (81.5)	76 (24.4)	290 (161)	208,900 (61.2)	111 (43.9)	218 (99)
E	15 (4.6)	56, 62 (17.3, 18.9)	1,135	4,920 (139.3)	388,400 (113.8)	73 (22.8)	404 (183)	292,000 (85.6)	109 (42.8)	304 (151)

Mounting height correction factors				
Steam press., lb/in² (gage) (kPa)	Water temp., °F (°C)	Normal room temp., °F (°C)		
		60 (15.6)	70 (21.1)	80 (26.7)
. . . . . .	210 (98.9)	1.05	1.10	1.20
0–5 (0–34.4)	220 (104.4)	1.00	1.07	1.14
6–15 (41.4–103.4)	. . . . . .	0.88	0.94	1.00
16–30 (110.3–206.8)		0.77	0.81	0.86
31–50 (213.7–344.7)	. . . . . .	0.70	0.73	0.77

*5-lb/in² (gage) (34. 5-kPa) steam supply.
†ft³/min capacity at *final* air temperature. For horizontal discharge unit heaters, the ft³/min capacity is usually stated at the *entering* air temperature.

horizontal-discharge unit heater is similar to Table 6. When the building is ventilated, a model E unit heater delivering 388,400 Btu/h (113.8 kW) could be used with entering air at 0°F (−17.8°C). This heater, as Table 6 shows, delivers 4920 ft³/min (139.3 m³/min) of air at 73°F (22.8°C) and forms 404 lb (183.6 kg) of condensate.

5. Check the spread diameter produced by the heater
Table 6 shows the different spread diameters, i.e., diameter of the heated-air blast at the floor level for different mounting heights of the unit heater. Thus, at a 14-ft (4.3-m) mounting level above the floor, model D will produce a spread diameter of 48 ft (14.6 m) or 54 ft (16.5 m), depending on the type of outlet cone used. These spread diameters are based on 2-lb/in² (gage) (13.8-kPa) steam and 60°F (15.6°C) room temperature. For 5-lb/in² (gage) (34.5-kPa) steam and 70°F (21.1°C) room temperature, multiply the tabulated spread diameter and mounting height by the correction factor shown at the bottom of Table 6. Thus, for model D, spread diameter = 1.07(48) = 51.4 ft (15.7 m) and 1.07(54) = 57.8 ft (17.6 m), whereas the mounting height could be 1.07(14) = 14.98 ft (4.6 m).

Find the spread diameter for model E in the same way, or, 56 ft (17.1 m) and 62 ft (18.9 m) at a 15-ft (4.6-m) height with 2-lb/in² (gage) (13.8-kPa) steam. With 5-lb/in² (gage) (34.5-kPa) steam, the spread diameters are 1.07(56) = 59.9 ft (18.3 m) and 1.07(62) = 66.4 ft (20.2 m), whereas the mounting height could be 1.07(15) = 16 ft (4.9 m).

6. Compute the hot-water-heater capacity required
Study several manufacturers' engineering data to determine the capacity of a suitable hot-water unit heater. This study will show that hot-water unit heaters are generally not available to inlet-air temperatures less than 30°F (−1.1°C). Since the inlet-air temperature in this building is 0°F (−17.8°C), a hot-water unit heater would be unsuitable; hence, it cannot be used. The *minimum* outlet-air temperature often recommended for unit heaters is 95°F (35.0°C).

Related Calculations.   Use the same general method to choose horizontal-delivery unit heaters. The heated air delivered by these units travels horizontally or can be deflected down toward the floor. Tables in manufacturers' engineering data list the heat-throw distance for horizontal-delivery unit heaters.

Standard ratings of steam unit heaters are given for 2-lb/in² (gage) (13.8-kPa) steam and 60°F (15.6°C) entering air; hot-water unit heaters for 180°F (82.2°C) entering water and 60°F (15.6°C) entering air. For other steam pressures, water temperatures, or entering air temperatures, *divide* the Btu per hour at stated conditions by the appropriate correction factor from Table 7 to obtain the required rating at *standard* conditions. Thus, a steam unit heater rated at 18,700 Btu/h (5.5 kW) at 20 lb/in² (gage) (137.9 kPa) and 70°F (21.1°C) entering air has a standard rating of 18,700/1/178 = 15,900 Btu/h (4.7 kW), closely, at 2 lb/in² (gage) (13.8 kPa) and 60°F (15.6°C) entering air temperature, using the correction factor from Table 7. Conversely, a steam unit heater rated at 228,000 Btu/h (66.8 kW) at 2 lb/in² (gage) (13.8 kPa) and 60°F (15.6°C) will deliver 228,000(1.421) = 324,000 Btu/h (94.9 kW) at 50 lb/in² (gage) (344.7 kPa) and 70°F (21.1°C) entering air. Electric and gas-fired unit heaters are also rated on the basis of 60°F (15.6°C) entering air.

When a unit heater is supplied both outside and recalculating air from within the building, use the temperature of the combined airstreams as the inlet temperature. Thus, with 1000 ft³/min (28.3 m³/min) of 10°F (−12.2°C) outside air and 4000 ft³/min (113.2 m³/min) of 70°F (21.1°C) recirculated air, the temperature of

TABLE 7   Unit-Heater Conversion Factors

	Entering air temperature, °F (°C)			
	40 (4.4)	50 (10.0)	60 (15.6)	70 (21.1)
Steam pressure, lb/in² (gage) (kPa)	Horizontal-delivery steam heaters			
2 (13.8)	1.153	1.076	1.000	0.927
5 (34.5)	1.209	1.131	1.055	0.981
10 (68.9)	1.288	1.209	1.132	1.057
20 (137.9)	1.413	1.333	1.254	1.178
40 (275.8)	1.593	1.510	1.430	1.351
50 (344.7)	1.664	1.582	1.500	1.421
Entering water temperature, °F (°C)	Horizontal-delivery hot-water heaters			
150 (65.6)	0.911	0.790	0.676	0.568
160 (71.1)	1.027	0.900	0.783	0.670
170 (76.7)	1.142	1.012	0.890	0.773
180 (82.2)	1.262	1.127	1.000	0.880
190 (87.8)	1.384	1.245	1.115	0.990
200 (93.3)	1.503	1.365	1.231	1.102

the air entering the heater is $t_e = (t_{oa}cfm_{oa} + t_r cfm_r)/cfm_t$, where t_e = temperature of air entering heater, °F; t_{oa} = temperature of outside air, °F; cfm_{oa} = quantity (cubic feet per minute) of outside air entering the heater, ft³/min; t_r = temperature of room air entering heater, °F; cfm_r = quantity of room air entering heater, ft³/min; cfm_t = total cfm entering heater = cfm_{oa} + cfm_r. Thus, $t_e = (10 \times 1000 + 70 \times 4000)/5000 = 58$°F (14.4°C). This relation is valid for both steam and hot-water unit heaters.

To find the approximate outlet air temperature of a unit heater when the capacity, cfm, and entering air temperature are known, use the relation $t_o = t_i + (460 + t_i)/$ [575 ft³ · h/(min · Btu)] − 1, where t_o = unit-heater outlet-air temperature, °F; t_i = temperature of air entering the heater, °F; cfm = quantity of air passing through the heater, ft³/min; Btu/h = rated capacity of the unit heater. Thus, for a heater rated at 73,500 Btu/h (21.5 kW), 1530 ft³/min (43.4 m³/min), with 50°F (10.0°C) entering air temperature, $t_o = 50 + (460 + 50)/(575 \times 1530/73,500) - 1 = 94.1$°F (34.5°C). The unit-heater capacity used in this equation should be the capacity at standard conditions: 2 lb/in² (gage) (13.8 kPa) steam or 180°F (82.2°C) water and 60°F (15.6°C) entering air temperature. Results obtained with this equation are only approximate. In any event, the outlet air temperature should never be less than the room temperature. For the actual outlet temperature of a specific unit heater, refer to the manufacturers' engineering data.

The air discharged by a unit heater should be at a temperature greater than the room temperature because air in motion tends to chill the occupants of a room. Choose the outlet temperature by referring to the heated-air velocity. Thus, with a velocity of 20 ft/min (6.1 m/min), the air temperature should be at least 76°F (24.4°C) at the heater outlet. As the air velocity increases, higher air temperatures are required.

The outlet air velocity and distance of blow of typical unit heaters are shown in Table 8.

When a unit heater of any type discharges against an external resistance, such as a duct or grille, its heating capacity and air capacity, as compared to standard conditions, are reduced. However, the final air temperature usually increases a few degrees over that at standard conditions.

To convert the rated output of any steam unit heater to ft^2 (m^2) of equivalent direct radiation [abbreviated ft^2 EDR (m^2 EDR)], divide the unit-heater rated capacity in Btu/h (W) by 240 Btu/(h·ft^2 EDR) (70.3 W/m^2 EDR). Thus, a unit heater rated at 240,000 Btu/h (70.34 W) has a heat output of 240,000/240 = 1000 ft^2 EDR (1000 m^2 EDR). For hot-water unit heaters, use the conversion factor of 150 Btu/(h·ft^2 EDR) (43.9 W/m^2 EDR).

To determine the rate of condensate formation in a steam unit heater, divide the rated output in Btu/h (W) by an enthalpy of 930 Btu/lb (2163.2 kJ/kg) of steam. Most unit-heater rating tables list the rate of condensate formation for each heater. Table 9 shows typical pipe sizes recommended for various condensate loads of steam unit heaters. thus, with 1000 lb/h (454.6 kg/h) of condensate and 30-lb/in^2 (gage) (206.8-kPa) steam supply to the unit heater, the supply main should be 2½-in (64-mm) nominal diameter and the return main should be 1¼-in (32-mm) nominal diameter.

Figure 22a shows how unit heaters of any type should be located in buildings of various types. The diagrams are also useful in determining the approximate number of heaters needed once the heat loss is known. Locate unit heaters so that the following general conditions prevail, if possible.

TABLE 8   Unit-Heater Outlet Velocity and Blow Distance

Unit-heater type	Outlet velocity, ft/min (m/min)	Blow distance, ft (m)
Centrifugal fan	1500–2500 (457–762)	20–200 (6–61)
Horizontal propeller fan	400–1000 (122–305)	30–100 (9–30)
Vertical propeller fan	1200–2200 (366–671)	70 (21)

TABLE 9   Typical Steam Unit-Heater Pipe Diameters,* in (mm)

Condensate, lb/h (kg/h)	Steam-supply pressure, lb/in^2 (gage) (kPa)			
	5 (34.5)		30 (206.8)	
	Supply	Return†	Supply	Return
100 (45)	2 (51)	1 (25)	1¼ (32)	¾ (19)
400 (180)	3 (76)	2 (51)	2 (51)	1 (25)
800 (360)	4 (102)	2½ (64)	2½ (64)	1¼ (32)
1000 (450)	5 (127)	2½ (64)	2½ (64)	1¼ (32)

*Modine Manufacturing Company.
†Gravity return.

Unit heater type	Desirable conditions
Horizontal delivery	Discharge should wipe exposed walls at an angle of about 30°. With multiple units, the airstreams should support each other.
Vertical delivery	With only vertical units, the airstream should blanket exposed walls with warm air.

The unit-heater arrangements shown in Fig. 22a illustrate a number of important principles.[1] The basic principle of unit-heater location is shown in the *mill-type* building. Here the heated-air flow from each unit heater supports the air flow from the other unit heaters and tends to set up a general circulation of air in the space heated. In the *warehouse* building arrangement, maximum area coverage is obtained with a minimum number of units. The *narrow* building uses vertical-discharge unit heaters that blanket the building walls with warmed air.

In the *manufacturing* building, circular air movement is sacrificed to offset a large roof heat loss and to permit short runouts from a single steam main. Note how a long-throw unit heater blankets a frequently used doorway.

Vertical-discharge unit heaters are used in the medium-height *sawtooth-type* building shown in Fig. 22a. The *monitor-type*-building installation combines both horizontal and vertical unit heaters. Horizontal-discharge unit heaters are located in the low-ceiling areas and vertical-discharge units in the high-ceiling areas above the craneway. Much of the data in this procedure were supplied by Modine Manufacturing Company. Figure 22b shows typical horizontal and vertical-blow unit-heater piping.

STEAM CONSUMPTION OF HEATING APPARATUS

Determine the probable steam consumption of the non-space-heating equipment in a building equipped with the following: three bain-maries, each 100 ft² (9.3 m²), two 50-gal (189.3-L) coffee urns, one jet-type dishwasher, one plate warmer having a 60-ft³ (1.7-m³) volume, two steam tables, each having an area of 50 ft² (4.7 m²) and one water still having a capacity of 75 gal/h (283.9 L/h). The available steam pressure is 40 lb/in² (gage) (275.8 kPa); the kitchen equipment will operate at 20 lb/in² (gage) (137.9 kPa) and the water still at 40 lb/in² (gage) (275.8 kPa).

Calculation Procedure:

1. *Determine steam consumption of the equipment*
The general procedure in determining heating-equipment steam consumption is to obtain engineering data from the manufacturer of the unit. When these data are unavailable, Table 10 will provide enough information for a reasonably accurate first approximation. Hence, this tabulation is used here to show how the data are applied.

---

[1]Modine Manufacturing Company.

TABLE 10   Typical Steam Consumption of Heating Equipment

Equipment	Steam, lb/h at 20–40 lb/in² (gage) (kg/h at 138–276 kPa)	
Bain-marie, per ft² (m²) of surface	3.0	(14.5)
Coffee urns, per gal (L)	2.5–3.0	(0.30–0.36)
Dishwashers, jet-type	60	(27.0)
Plate warmer, per 20 ft³ (0.57 m³)	30	(13.5)
Soup or stock kettle, 60 gal (0.23 L); 40 gal (0.15 L)	60; 45	(27.0; 20.3)
Steam table, per ft² (m²) of surface	1.5	(7.3)
Vegetable steamer, per compartment 5-lb (2.3-kg) press	30	(13.5)
Water still, per gal (L) capacity per h	9	(1.1)

Since the supply steam pressure is 40 lb/in² (gage) (275.8 kPa), a pressure-reducing valve will have to be used between the steam main and the kitchen supply main. The capacity of this valve depends on the steam consumption of the equipment. Hence, the valve capacity cannot be determined until the equipment steam consumption is known.

During equipment operation the steam consumption is different from the consumption during startup. Since the operating consumption must be known before the starting consumption can be computed, the former is determined first.

Using data from Table 10, we see the three 100-ft² (9.3-m²) bain-maries will require (3 lb/h)(3 units)(100 ft² each) = 900 lb/h (409.1 kg/h) of steam. The two 50-gal (189.3-L) coffee urns require (2.75 lb/h)(2 units)(50 gal per unit) = 275 lb/h (125 kdg/h), using the average steam consumption. One jet-type dishwasher will require 60 lb/h (27.3 kg/h) of steam. A 60-ft³ (1.7-m³) plate warmer will require (60 ft³)/(1 unit) [(30 lb/h)/20-ft³ unit] = 90 lb/h (40.9 kg/h). Two 50-ft² (4.7-m²) steam tables require (1.5 lb/h)(2 units) (50 ft² per unit) = 150 lb/h. The 75-gal/h (283.9 L/h) water still will consume (9 lb/h)(1 unit)(75 gal/h) = 675 lb/h (306.8 kg/h). Hence, the total operating consumption of 20-lb/in² (gage) (137.9 kPa) steam is 900 + 275 + 60 + 90 + 150 = 1475 lb/h (670.5 kg/h). The water still will consume 675 lb/h (306.8 kg/h) of 40-lb/in² (gage) (275.8-kPa) steam.

Since the 20-lb/in² (gage) (137.9-kPa) steam must pass through the pressure-reducing valve before entering the 20-lb/in² (gage) (137.9-kPa) main, the required *operating* capacity of this valve is 1475 lb/h (670.5 kg/h). However, the total steam consumption during operation, without an allowance for condensation in the pipelines, is 1475 + 675 = 2150 lb/h (977.3 kg/h).

2.   Compute the system condensation losses
Condensation losses can range from 25 to 50 percent of the steam supplied, depending on the type of insulation used on the piping, the ambient temperature in the locality of the pipe, and the degree of superheat, if any, in the steam. Since the majority of the steam used in this building is 20-lb/in² (gage) (137.9-kPa) steam reduced in pressure from 40 lb/in² (gage) (275.8 kPa) and there will be a small amount of superheating during pressure reduction, a 25 percent allowance for pipe condensation is probably adequate. Hence, the total operating steam consumption = (1.25)(2150) = 2688 lb/h (1221.8 kg/h).

3. Compute the startup consumption

During equipment startup there is additional condensation caused by the cold metal and, possibly, some cold products in the equipment. Therefore, the startup steam consumption is different from the operating consumption.

One rule of thumb estimates the startup steam consumption as two times the operating consumption. Thus, by this rule of thumb, startup steam consumption = 2(2688) = 5376 lb/h (2443.6 kg/h). Note that this consumption rate is of relatively short duration because the metal parts are warmed rapidly. However, the pressure-reducing valve must be sized for this flow rate unless slower warming is acceptable.

The actual rate of condense formation can be computed if the weight of the equipment, the specific heat of the materials of construction, and initial and final temperatures of the equipment are known. Use the relation steam condensation, lb/h = $60 \, Ws(\Delta t)/h_{fg}T$, where W = weight of equipment and piping being heated, lb; s = specific heat of the equipment and piping, Btu/(lb $\cdot$ °F); Δt = temperature rise of the equipment from the cold to the hot state, °F; h_{fg} = latent heat of vaporization of the heating steam, Btu/lb; T = heating period, min. In SI units, condensation is found from the same relationship, except W is in kg; s is in kJ/(kg $\cdot$ °C); t = temperature change, °C; h_{fg} = latent heat of vaporization, kJ/kg; other variables the same.

This relation assumes that the final temperature of the equipment approximately equals the temperature of the heating steam. Where the specific heat of the equipment is different from that of the piping, solve for the steam condensation rate of each unit and sum the results. Where products in the equipment must be heated, use the same relation but substitute the produce weight and specific heat.

Related Calculations.   Use the general method given here to compute the steam consumption of any type of industrial equipment for which the unit steam consumption is known or can be determined.

SELECTION OF AIR HEATING COILS

Select a steam heating coil to heat 80,000 ft³/m (2264.8 m³/min) of outside air from 10°F (-12.2°C) to 150° (65.6°C) for steam at 15 lb/in² (abs) (103.4 kPa). The heated air will be used for factory space heating. Illustrate how a steam coil is piped. Show the steps for choosing a hot-water heating coil.

Calculation Procedure:

1. Compute the required face area of the coil

If the coil-face air velocity is not given, a suitable air velocity must be chosen. In usual air-conditioning and heating practice, the air velocity across the face of the coil can range from 300 to 1000 ft/min (91.4 to 305 m/min) with 500, 800, and 1000 ft/min (152.4, 243.8, and 305 m/min) being common choices. The higher velocities—up to 1000 ft/min (305 m/min)—are used for industrial installations where noise is not a critical factor. Assume a coil face velocity of 800 ft/min (243.8 m/min) for this installation.

Compute the required face area from $A_c = cfm/V_a$, where A_c = required coil face area, ft² : cfm = quantity of air to be heated by the coil, ft³/min, at 70°F (21.1°C) and 29.92 in (759.97 mm) Hg; V_a = air velocity through the coil, ft/min.

To correct the air quantity to standard conditions, when the air is being delivered at nonstandard conditions, multiply the flow in at the other temperature by the appropriate factor from Table 11. Thus, with the incoming air at 10°F (−12.2°C), ft³/m at 70°F (21.1°C) and 29.92 in (759.97 mm) Hg = (1.128)(80,000) = 90,400 ft³/m (2559.2 m³/min). Hence, A_c = 90,400/800 = 112.8 ft² (10.5 m²).

2. Compute the coil outlet temperature

The capacity, final temperature, and condensate formation rate for steam heating coils for air-conditioning and heating systems are usually based on steam supplied at 5 lb/in² (abs) (34.5 kPa) and inlet air at 0°F (−17.8°C). At other steam pressures a correction factor must be applied to the tabulated outlet temperature for 5-lb/in² (abs) (34.5-kPa) coils with 0°F (−17.8°C) inlet air.

Table 12 shows an excerpt from a typical coil-rating table and excerpts from coil correction-factor tables. To use such a tabulation for a coil supplied steam at 5 lb/in² (abs) (34.5 kPa), enter at the air inlet temperature and coil face velocity. Find the final air temperature equal to, or higher than, the required final air temperature. Opposite this read the number of rows of tubes required. Thus, in a 5-lb /in² (abs) (34.5-kPa) coil with 0°F (−17.8°C) inlet air, 800-ft/min (243.8-m/min) face velocity, and a 165°F (73.9°C) final air temperature, Table 12 shows that five rows of tubes would be required. This table shows that the coil forms condensate at the rate of 149.8 lb/(h · ft²) [732.9 kg/(h · m²)] of net fin area when the final air temperature is 166°F (74.4°C). The coil thus chosen is the first-trail coil, which must be checked against the actual steam conditions as described below.

When a coil is supplied steam at a pressure different from 5 lb/in² (abs) (34.5 kPa), multiply the final air temperature given in the 5-lb/in² (abs) (34.5-kPa) table for 0°F (−17.8°C), at the face velocity being used, by the correction factor given in Table 12 for the actual steam pressure and actual inlet air temperature. Thus, for this coil, which is supplied steam at 15 lb/in² (abs) (103.4 kPa) and has an inlet air temperature of 10°F (−12.2°C), the temperature correction factor from Table 12 is 1.056. Add the product of the correction factor and the tabulated final air temperature to the inlet air temperature to obtain the actual final air temperature. Several trials may be necessary before the desired outlet temperature is obtained.

The desired final air temperature for this coil is 150°F (65.6°C). Using the 124°F (51.1°C) final air temperature from Table 12 as the first-trial valve, we get the actual final air temperature = (1.056)(124) + 10 = 141°F (60.6°C). This is too low. Trying the next higher final air temperature gives the actual final air temperature = (1.056)(148) + 10 = 166.5°F (74.7°C). This is higher than required, but the steam

TABLE 11 Air-Volume Conversion Factors*†

Air temp., °F (°C)	Factor	Air temp., °F (°C)	Factor
0 (−17.18)	1.152	90 (32.2)	0.964
10 (−12.2)	1.128	100 (37.8)	0.946
20 (−6.7)	1.104	110 (43.3)	0.930
30 (−1.1)	1.082	120 (48.9)	0.914
40 (4.4)	1.060	130 (54.4)	0.898
50 (10.0)	1.039	140 (60.0)	0.883
60 (15.6)	1.019	150 (65.6)	0.869
70 (21.1)	1.000	160 (71.1)	0.855
80 (26.7)	0.981	170 (76.7)	0.841

TABLE 12 Steam-Heating-Coil Final Temperatures and Condensate-Formation Rate

Inlet air temp., °F (°C)	Rows of tubes	Face velocity, 800 ft/min (243.8 m/min)	
		Final air temp., °F, (°C)	Condensate, lb/h (kg/h)
0 (−17.8)	2	51 (10.6)	83.7 (37.7)
0 (−17.8)	3	124 (51.1)	111.7 (50.3)
0 (−17.8)	4	148 (64.4)	132.9 (59.8)
0 (−17.8)	5	166 (74.4)	149.8 (67.4)

Temperature-rise correction factor

Actual inlet air temp., °F (°C)	Steam pressure, lb/in² (abs) (kPa)		
	10 (68.9)	15 (103.4)	20 (137.9)
0 (−17.8)	1.054	1.100	1.139
10 (−12.2)	1.010	1.056	1.095
20 (−6.7)	0.966	1.011	1.051

Condensate correction factors

0 (−17.8)	1.063	1.117	1.165
10 (−12.2)	1.019	1.072	1.120
20 (−6.7)	0.974	1.027	1.075

supply can be reduced to produce the desired final air temperature. Thus, the coil will be four rows of tubes deep, as Table 12 shows. Hence, the five rows of coils originally indicated will not be needed. Instead, four rows will suffice.

3. Compute the quantity of condensate produced by the coil
Use the same general procedure as in step 2, or actual condensate formed, lb/(h · ft²) = [lb/(h · ft²)] [condensate from 5-lb/in² (abs), 0°F (34.5-kPa, −17.8°C) table] (correction factor from Table 12); or for this coil, (132.9)(1.072) = 142.47 lb/(h · ft²) [690.1 kg/(h · m²)]. Since the coil has a net fin face area of 112.8 ft² (10.5 m²), the total actual condensate formed = (112.8)ft²(142.47) = 16,100 lb/h (7318.2 kg/h).

4. Determine the coil friction loss
Most manufacturers publish a chart or table of coil friction losses in coils having various face velocities and tube rows. Thus, Fig. 23a shows that a coil having a face velocity of 800 ft/min (243.8 m/min) and four rows of tube has a friction loss of 0.45 in (11.4 mm) of water. Figure 23a is a typical friction-loss chart and can be safely used for all routine preliminary coil selections. However, when the final choice of heating coil is made, use the friction chart or table prepared by the manufacturer of the coil chosen.

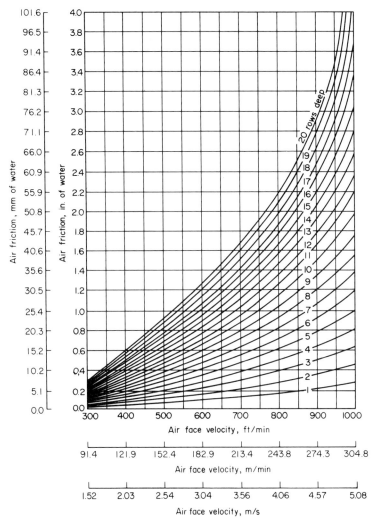

FIGURE 23 (*a*) Heating-coil air-friction chart for air at standard conditions of 70°F (21.1°C) and 29.92 in (76 cm) (*McQuay, Inc.*)

5. Determine the coil dimensions

Refer to the manufacturer's engineering data for the dimensions of the coil chosen. Each manufacturer has certain special construction features. Hence, there will be some variation in dimensions from one manufacturer to another.

6. Indicate how the coil will be piped

The ASHRAE *Guide and Data Book* shows piping arrangements for low- and high-pressure steam heating coils as recommended by various coil manufacturers. Follow

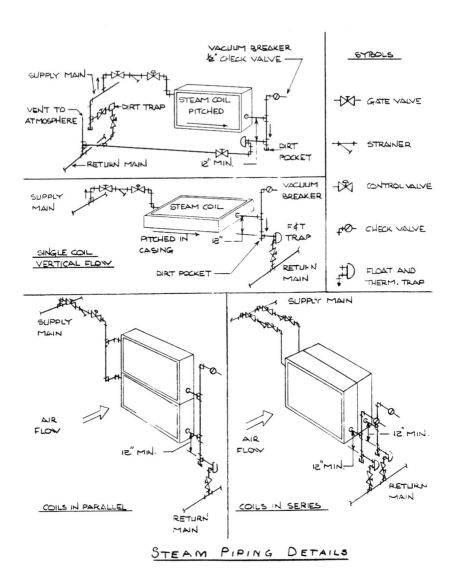

SI Values
1/2 in.   12.7 mm
12 in.   30.5 cm

FIGURE 23   (*b*) Several typical piping arrangements for steam-heated air-heating coils. (*Jerome F. Mueller.*)

the recommendations of the manufacturer whose coil is actually used when the final selection is made. Figure 23*b* shows several typical piping arrangements for steam-heated air-heating coils.

   Related Calculations.   Typical variables met in heating-coil selection are the *face velocity*, which varies from 300 to 1000 ft/min (91.4 to 304.8 m/min), with the higher velocities being used for industrial applications, the lower velocities for nonindustrial applications; the *final* air temperature, which ranges between 50 and 300°F (10 and 148.9°C), the lower temperatures being used for ventilation, the higher ones for heating; *steam pressures,* which vary from 2 to 150 lb/in² (gage) (13.8 to 1034.1 kPa), with the lower pressures—2 to 15 lb/in² (gage) (13.8 to 103.4 kPa)—being the most popular for heating.

   Hot-water heating coils are also used for air heating. The general selection procedure is: (*a*) Compute the heating capacity, Btu/h, required from (1.08)(temperature rise of air, °F)(cfm heated). (*b*) Compute the coil face area required, ft², from cfm/face velocity, ft/min. Assume a suitable face velocity using the guide given above. (*c*) Compute the logarithmic mean-effective-temperature difference across the coil, using the method given elsewhere in this handbook. (*d*) Compute the required hot-water flow rate, gallons per minute (gal/min), from Btu/h heating capacity/(500)(temperature drop of water, °F). The usual temperature drop of the hot water during passage through the heating coil is 20°F (11.1°C), with water supplied at 150 to 225°F (65.6 to 107.2°C). (*e*) Determine the tube water velocity, ft/s, from (8.33)(gal/min)/(384)(number of tubes in heating coil). The number of tubes in the coil is obtained by making a preliminary selection of the coil, using heating-capacity tables similar to Table 10. The usual hot-water heating coil has a water velocity between 2 and 6 ft/s (0.6 and 1.8 m/s). (*f*) Compute the number of tube rows required from Btu/h heating capacity/(face area ft²)(logarithmic mean-effective-temperature difference from step 3)(*K* factor from the manufacturer's engineering data). (*g*) Compute the coil air resistance or friction loss, using the manufacturer's chart or table. The usual friction loss ranges from 0.375 to 0.675 in (9.5 to 17.2 mm) of water for commercial applications to about 1.0 in (25 mm) of water for industrial installations.

RADIANT-HEATING-PANEL CHOICE AND SIZING

One room of a building has a heat loss of 13,900 Btu/h (4074.1 W). Choose and size a radiant heating panel suitable for this room. Illustrate the trial method of panel choice. The floor of the room is made of wooden blocks.

Calculation Procedure:

1.   *Choose the type and location of the heating coil*
Compute the heat loss for a given room or building, using the method given earlier in this section. Once the heat loss is known, choose the type of heating panel to use—ceiling or floor. In some rooms or buildings, a combination of floor and ceiling panels may prove more effective than either type used alone. Wall panels are also used but not as extensively as floor and ceiling panels.

   In general, ceiling panels are embedded in concrete or plaster, as are wall panels. Floor panels are almost always embedded in concrete. Hence, use of another type of floor—block, tile, wood, or metal—may rule out the use of floor panels. Since this room has a wooden-block floor, a ceiling panel will be chosen.

2. Size the heating panel

Table 13 shows the maximum Btu/h (W) heat output of ⅜-in (9.5-mm) copper-tube ceiling panels. The ⅜-in (9.5-mm) size is popular; however, other sizes—½-, ⅝-, and ⅞-in (12.7-, 15.9-, and 22.2-mm) diameter—are also used, depending on the heat load served. Using ⅜-in (9.5-mm) tubing on 6-in (152.4-mm) centers embedded in a plaster ceiling will provide a heat output of 60 Btu/(h · ft^2) (189.1 W/m^2) of tubing , as shown in Table 13. Figure 24 shows a typical piping diagram for radiant ceiling heating and cooling coils.

To obtain the area of the heated panel, A ft^2, use the relation $A = H_L/P$, where H_L = room or building heat loss, Btu/h; P = panel maximum heat output, Btu/(h · ft^2). For this room, $A = 13,900/60 = 232$ ft^2 (21.6 m^2).

3. Determine the total length of tubing required

Use the appropriate tube-length factor from Table 13, or $(2.0)(232) = 464$ lin ft (141.4 m) of ⅜-in (9.5-mm) copper tubing.

4. Find the maximum panel tube length

To stay within the commercial limits of smaller hot-water circulating pumps, the maximum tube lengths per panel circuit given in Table 13 are generally used. This tabulation shows that the maximum panel unit tube length for ⅜-in (9.5-mm) tubing on 6-in (152.4-mm) centers is 165 ft (50.3 m). Such a length will not require a pump head of more than 4 ft (1.22 m), excluding the head loss in the mains.

5. Determine the number of panels required

Find the number of panels required by dividing the linear tubing length needed by the maximum unit tube length, or $464/165 = 2.81$. Use three panels, the next larger whole number, because partial panels cannot be used. To conserve tubing and reduce the first cost of the installation, three panels, each having 155 lin ft (47.2 m) of piping, would be used. Note that the tubing length chosen for the actual panel must be *less than* the maximum length listed in Table 13.

6. Find the required piping main size required

Determine the number of panels required in the remainder of the building. Use Table 13 to select the proper main size for a pressure loss of about 0.5 ft/100 ft (150 mm/30 m) of the main, including the supply and return lines. Thus, if 12 ceiling panels of maximum length were used in the building, a 1½-in (38.1-mm) main would be used.

7. Use the trial method to choose the main size

If the size of the main for the panels required cannot be found in Table 13, compute the total Btu required for the panels from Table 13. Then, by trial, find the size of the main from Table 13 that will deliver approximately, and preferably somewhat more, than this total Btu/h requirement.

For instance, suppose an industrial building requires the following panel circuits:

TABLE 13   Heating Panel Characteristics

Maximum Btu/(h·ft²) (W/m²) of ⅜-in (9.5-mm) copper tube embedded in ceiling plaster		Maximum Btu/(h·ft²) (W/m²) of copper tube embedded in concrete floor	
	Approx.		*Approx.*
4½ in (114.3 mm) center to center	75 (236.4)	½ in (12.7 mm) 9 in (228.6 mm) center to center	50 (157.6)
6 in (152.4 mm) center to center	60 (189.1)	¾ in (19.1 mm) 9 in (228.6 mm) center to center	50 (157.6)
9 in (228.6 mm) center to center	45 (141.8)	¾ in (19.1 mm) 12 in (304.8 mm) center to center	50 (157.6)
		1 in (25.4 mm) 12 in (304.8 mm) center to center	50 (157.6)

Total length of tube required, ft (m)†

2.7 (0.82)	Where 4½-in (114.3-mm) centers are required
2 (0.61)	Where 6-in (152.4-mm) centers are required
1.3 (0.40)	Where 9-in (228.6-mm) centers are required
1 (0.30)	Where 12-in (304.8-mm) centers are required

Maximum panel unit tube length‡

Ceilings				Floors			
Nominal size, in (mm)	Centers, in (mm)	Btu/(h·ft) (W/m) of tube	Ft (m)	Nominal size, in (mm)	Centers, in (mm)	Btu/(h·ft) (W/m) of tube	Ft (m)
⅜ (9.5)	4½ (114.3)	27 (25.9)	175 (47.9)	½ (12.7)	9 (228.6)	38 (36.5)	220 (67.1)
⅜ (9.5)	6 (152.4)	30 (28.9)	165 (50.3)	¾ (19.1)	9 (228.6)	38 (36.5)	400 (121.9)
⅜ (9.5)	9 (228.6)	34 (32.7)	150 (45.7)	¾ (19.1)	12 (304.8)	50 (48.1)	350 (106.7)
				1 (25.4)	12 (304.8)	50 (48.1)	550 (167.7)

Number of panel circuits of maximum length§

Mains	Ceiling		Floor	
Diam., in (mm)	⅜ in—4½, 6, 9 center-to-center (9.53 mm—114.3, 152.4, 228.6 center-to-center)	½ in—9 center-to-center (12.7 mm—228.6 center-to-center)	¾ in— 9, 12 center-to-center (19.1 mm—228.6, 304.8 center-to-center)	1 in—12 center-to-center (25.4 mm—304.8 center-to-center)
2 (50.8)	27	16	8	5
1½ (38.1)	12	7	4	2
1¼ (31.6)	8	5	2	1
1 (25.4)	4	3	1	1
¾ (19.1)	2	1		

†To arrive at the required lin ft of tube per panel, multiply the ft² of heated panel by the factors given.
‡To keep within the commercial limits of the smaller pumps, as an example the above maximum tube lengths per panel circuit are suggested. These lengths alone will require not more than about a 4-ft (1.2-m) head. This does not include loss in mains.
§Use the information given in the section on maximum panel unit tube length, as given above, which can be supplied, allowing about 0.5 ft (150 mm) head required per 100 ft (30 m) of main (supply and return).

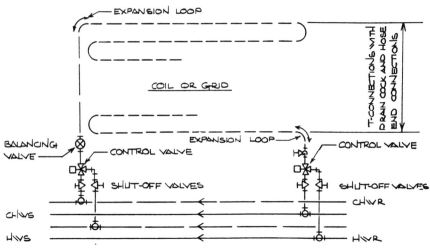

FIGURE 24 Typical piping diagram for radiant ceiling heating and cooling coils. (*Jerome F. Mueller.*)

No. of panel circuits	Tube size and location	Btu (kJ) required from Table 13 (number of circuits × Btu/ft of tube × maximum panel tube length)
4	⅜-in (9.5-mm) tubes on 4½-in (114.3-mm) centers, ceiling	4 × 27 × 175 = 18,900 (19,940)
1	½-in (12.7-mm) tubes on 9-in (228.6-mm) centers, floor	1 × 38 × 220 = 8,360 (8,820)
1	1-in (25.4-mm) tubes on 12-in (304.8-mm) centers, floor	1 × 50 × 550 = 27,500 (29,013)
3	¾-in (19.1-mm) tubes on 9-in (228.6-mm) centers, floor	3 × 38 × 400 = 45,600 (48,108)
9		Total Btu/h (kJ/h) required = 100,360 (105,881)

Trial 1. Assume 100 ft of 1½-in (30 m of 38-mm) main is used for seven floor circuits that are each 220 ft (67.1 m) long and made of ½-in (12.7-mm) tubing on 9-in (228.6-mm) centers. From Table 13 the output of these seven floor circuits is 7 circuits × 38 Btu/(h·ft) of tube (36.5 W/m) × 220 ft (67.1 m) of tubing = 58,520 Btu/h (17.2 kW). Since this output is considerably less than the required output of 100,360 Btu/h (29.4 kW), a 1½-in (38.1-mm) main is not large enough.

Trial 2. Assume 100 ft of 2-in (30 m of 50.8-mm) main for 16 circuits that are each 220 ft (67.1 mm) long and are made of ½-in (12.7-mm) tubing on 9-in (228.6-mm) centers. Then, as in trial 1, 16 × 38 × 220 = 133,760 Btu/h (39.2 kW) delivered. Hence, a 2-in (50.8-mm) main is suitable for the nine panels listed above.

Related Calculations. Use the same general method given here for heating panels embedded in the concrete floor of a building. The liquid used in most panel heating systems is water at about 130°F (54.4°C). This warm water produces a panel temperature of about 85°F (29.5°C) in floors and about 115°F (46.1°C) in ceilings. The maximum water temperature at the boiler is seldom allowed to exceed 150°F (65.6°C).

When the first-floor ceiling of a multistory building is not insulated, the floor above a ceiling panel develops about 17 Btu/(ft$^2 \cdot$ h) (53.6 W/m^2) from the heated panel below. If this type of construction is used, the radiation into the room above can be deducted from the heat loss computed for that room. It is essential, however, to calculate only the heat output in the floor area directly above the heated panel.

Standard references, such as ASHRAE *Guide and Data Book,* present heat-release data for heating panels in both graphical and tabular form. Data obtained from charts are used in the same way as described above for the tabular data. Tubing data for radiant heating are available from Anaconda American Brass Company.

SNOW-MELTING HEATING-PANEL CHOICE AND SIZING

Choose and size a snow-melting panel to melt a maximum snowfall of 3 in/h (76.2 mm/h) in a parking lot that has an area of 1000 ft^2 (92.9 m^2). Heat losses downward, at the edges, and back of the slab are about 25 percent of the heat supplied; also, there is an atmospheric evaporation loss of 15 percent of the heat supplied. The usual temperature during snowfalls in the locality of the parking lot is 32°F (0°C).

Calculation Procedure:

1. *Compute the hourly snowfall weight rate*
The density of the snow varies from about 3 lb/ft^3 at 5°F (48 kg/m^3 at -15°C) to about 7.8 lb/ft^3 at 34°F (124.9 kg/m^3 at 1.1°C). Given a density of 7.3 lb/ft^3 (116.9 kg/m^3) for this installation, the hourly snowfall weight rate per ft^2 is (area, ft)(depth, ft)(density) = $(1.0)(3/12)(7.3) = 1.83$ lb/(h$\cdot$ ft^2) [8.9 kg/(h$\cdot$m^2)]. In this computation, the rate of fall of 3 in/h (76.2 mm/h) is converted to a depth in ft by dividing by 12 in/ft.

2. *Compute the heat required for snow melting*
The heat of fusion of melting snow is 144 Btu/lb (334.9 kJ/kg). Since the snow accumulates at the rate of 1.83 lb/(h$\cdot$ft^2) [8.9 kg/(h$\cdot$m^2)], the amount of heat that must be supplied to melt the snow is $(1.83)(144) = 264$ Btu/(ft$^2 \cdot$h) (832.1 W/m^2).

Of the heat supplied, the percent lost is $25 + 15 = 40$ percent as given. Of this total loss, 25 percent is lost downward and 15 percent is lost to the atmosphere. Hence, the total heat that must be supplied is $(1.0 + 0.40)(264) = 370$ Btu/(h$\cdot$ft^2) (1166 W/m^2). With an area of 1000 ft ^{2}(92.9 m^2) to be heated, the panel system must supply $(1000)(370) = 370,000$ Btu/h (108.6 kW).

3. *Determine the length of pipe or tubing required*
Consult the ASHRAE *Guide and Data Book* or manufacturer's engineering data to find the heat output per ft of tubing or pipe length. Suppose the heat output is 50 Btu/(h$\cdot$ft) (48.1 W/m) of tube. Then the length of tubing required is 370,000/50 = 7400 ft (2256 m).

Some manufacturers rate their pipe or tubing on the basis of rainfall equivalent of the snowfall and the wind velocity across the heated surface. Where this method is used, compute the heat required to melt the snow as the equivalent amount of heat to vaporize the water. This is $Q_e = 1074(0.002V + 0.055)(0.185 - v_a)$, where Q_e = heat required to vaporize the water, Btu/(h·ft²); V = wind velocity over the heated surface, mi/h; v_a = vapor pressure of the atmospheric air, in Hg. Figure 25 shows the piping arrangement for a typical snow-melting system.

4. Determine the quantity of heating liquid required
Use the relation gal/min = 0.125 H_t/dc Δt, for ethylene glycol, the most commonly used heating liquid. In this relation, H_t = total heat required for snow melting, Btu/h; d = density of the heating liquid, lb/ft³; c = specific heat of the heating liquid, Btu/(lb·°F); Δt = temperature loss of the heating liquid during passage through the heating coil, usually taken as 15 to 20°F (−9.4 to −6.7°C).

Assuming that a 60 percent ethylene glycol solution is used for heating, we have d = 68.6 lb/ft³ (1098.3 kg/m³); c = 0.75 Btu/(lb·°F) [3140 J/(kg·K)]. Since the piping must supply 370,000 Btu/h (108.5 kW), gal/min = (0.125)(370,000)/(68.6)(0.75)(20) = 45 gal/min (170.3 L/min) when the temperature loss of the heating liquid is 20°F (−6.7°C).

5. Size the heater for the system
The heater must provide at least 370,000 Btu/h (108.5 kW) to the ethylene glycol. If the heater has an overall efficiency of 60 percent, then the required heat input to deliver 370,000 Btu/h (108.5 kW) is 370,000/0.60 = 617,000 Btu/h (180.6 kW).

To avoid a long warmup time at the start of a snowfall, the usual practice is to operate the system for several hours prior to an expected snowfall. The heating liquid temperature during warmup is kept at about 100°F (37.8°C) and the pump is operated at half-speed.

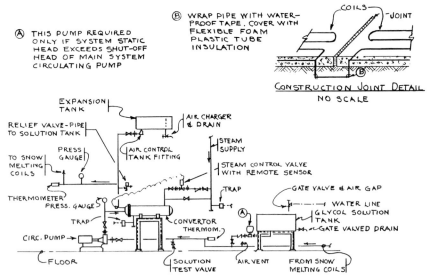

FIGURE 25   Piping arrangement for typical snow-melting system. (*Jerome F. Mueller.*)

Related Calculations. Use this general method to size snow-melting systems for sidewalks, driveways, loading docks, parking lots, storage yards, roads, and similar areas. To prevent an excessive warmup load on the system, provide for prestorm operation. Without prestorm operation, the load on the heater can be twice the normal hourly load.

Cooper tubing and steel pipe are the most commonly used heating elements. For properties of tubing and piping important in snow-melting calculations, see Baumeiser—*Standard Handbook for Mechanical Engineers.*

HEAT RECOVERY FROM LIGHTING SYSTEMS FOR SPACE HEATING

Determine the quantity of heat obtainable from 30 water-cooled fluorescent luminaires rated at 200 W each if the entering water temperature is 70°F (21.1°C) and the water flow rate is 1.0 gal/min (3.8 L/min). How much heat can be recovered from 10 air-cooled fluorescent luminaires rated at 100 W each?

Calculation Procedure:

1. Compute the wear-cooled luminaire heat recovery
Luminaire manufacturers publish heat-recovery data in chart form (Fig. 26). This chart shows that with a flow rate of 0.5 gal/min (1.9 L/min) and a 70°F (21.1°C) entering water temperature, the heat recovery from a luminaire is 74 percent of the total input to the fixture.

For a group of lighting fixtures, total input, W = (number of fixtures)(rating per fixture, W). Or, for this installation, input, W = (30)(200) = 6000 W. Since 74 percent of this input is recoverable by the cooling water, recovered input = 0.74(6000) = 4400 W.

To convert incandescent lighting watts to Btu/h, multiply by 3.4. Where fluorescent lights are used, apply a factor of 1.25 to include the heat gain in the lamp

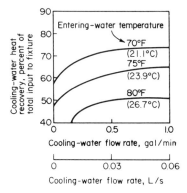

FIGURE 26 Heat recovery in water-cooled lighting fixtures.

ballast. Thus, the heat available for recovery from these fluorescent lamps is $(3.4)(1.25)(4440) = 18,900$ Btu/h (5.5 kW).

2. Compute the temperature of the water

Find the temperature rise of the water from $\Delta t =$ (Btu/h)/500 gal/min, where Δt = temperature rise of the water, F; Btu/h = heat available; gal/min = water flow rate through the luminaire. Or, $\Delta t = 18,9000/500(1.0) = 37.8°F$ (21.0°C). A temperature rise of this magnitude is seldom used in practice. However, this calculation shows the large amount of heat recoverable with water-cooled lighting fixtures.

3. Compute the heat recoverable with air cooling

In the usual air-cooled luminaire, 50 to 70 percent of the input energy is recoverable. Assuming a 60 percent recovery, with a total input of $(10)(100) = 1000$ W, we see that the energy recoverable is $0.60(1000) = 600$ W. Converting to the heat recoverable gives $(3.4)(1.25)(600) = 2545$ Btu/h (745.9 W).

Related Calculations. Heat recovery from lighting fixtures is receiving increasing attention for many different structures because substantial fuel savings are possible. The water or air heated by the lighting is used to heat the supply or return air supplied to the conditioned space. Where the heat recovered must be rejected, as in the summer, either a cooling tower (for water) or an air-cooled condenser (for air) may be used.

Other popular sources of heat are refrigeration condensers and electric motors. In many installations the heat is absorbed from the condenser or motors, or both, by air.

AIR-CONDITIONING-SYSTEM HEAT-LOAD DETERMINATION—GENERAL METHOD

Show how to compute the total heat load for an air-conditioned industrial building fitted with windows having shades, internal heat loads from people and machines, and heat transmission gains through the walls, roof, and floor. Use the ASHRAE *Guide and Data Book* as a data source.

Calculation Procedure:[1]

1. Determine the design outdoor and indoor conditions

Refer to the ASHRAE *Guide and Data Book* (called *Guide*) for the state and city in which the building is located. Read from the *Guide* table for the appropriate city the design outdoor dry-bulb and wet-bulb temperatures. At the same time, determine from the *Guide* the indoor design condtions—temperature and relative humidity for the type of application being considered. The *Guide* lists a variety of typical applications such as apartment houses, motels, hotels, industrial plants, etc. It also lists the average summer wind velocity for a variety of locations. Where the exact location of a plant is not tabulated in the *Guide,* consult the nearest local branch of the weather bureau for information on the usual summer outdoor high and low dry- and wet-bulb temperatures, relative humidity, and velocity.

---

[1]SI units are given in later numerical procedures.

2. Compute the sunlight heat gain

The sunlight heat gain results from the solar radiation through the glass in the building's windows and the materials of construction in certain of the building's walls. If the glass or wall of a building is shaded by an adjacent solid structure, the sunlight heat gain for that glass and wall is usually neglected. The same is true for the glass and wall of the building facing the north.

Compute the glass sunlight heat gain, Btu/h, from (glass area, ft^2)(equivalent temperature difference from the appropriate *Guide* table)(factor for shades, if any are used). The equivalent temperature difference is based on the time of day and orientation of the glass with respect to the points of the compass. A latitude correction factor may have to be applied if the building is located in a tropical area. Use the equivalent temperature difference for the time of day on which the heat-load estimate is based. Several times may be chosen to determine at which time the greatest heat gain occurs. Where shades are used in the building, choose a suitable shade factor from the appropriate *Guide* table and insert it in the equation above.

Compute the sunlight heat gain, Btu/h, for the appropriate walls and the roof from (wall area, ft^2)(equivalent temperature difference from *Guide* for walls) [coefficient of heat transmission for the wall, Btu/(h · ft^2 · °F)]. For the roof, find the heat gain, Btu/h, from (roof area, ft^2)(equivalent temperature difference from the appropriate *Guide* table for roofs) [coefficient of heat transmission for the roof, Btu/(h · ft^2 · °F)].

3. Compute the transmission heat gain

All the glass in the building windows is subject to transmission of heat from the outside to the inside as a result of the temperature difference between the outdoor and indoor dry-bulb temperatures. This transmission gain is commonly called the *all-glass gain*. Find the all-glass transmission heat gain, Btu/h, from (total window-glass area, ft^2)(outdoor design dry-bulb temperature, °F—indoor design dry-bulb temperature, °F)[coefficient of heat transmission of the glass Btu/(h · ft^2 · °F), from the appropriate *Guide* table].

Compute the heat transmission, Btu/h, through the shaded walls, if any, from (total shaded wall area, ft^2)(equivalent temperature difference for shaded wall from the appropriate *Guide* table, °F) [coefficient of heat transmission of the wall material, Btu/(h · ft^2 · °F), from the appropriate *Guide* table)].

Where the building has a machinery room or utility room that is not air-conditioned and is next to a conditioned space and the temperature in the utility room is higher than in the conditioned space, find the heat gain, Btu/h, from (area of utility or machine room partition, ft^2)(utility or machine room dry-bulb temperature, °F − conditioned-space dry-bulb temperature, °F)[coefficient of heat transmission of the utility or machine room partition, Btu/(h · ft^2 · °F), from the appropriate *Guide* table].

For buildings having a floor contacting the earth, or over unventilated and unheated basements, there is generally *no* heat gain through the floor because the ground is usually at a lower temperature than the floor. Where the floor is above the ground and in contact with the outside air, find the heat gain, Btu/h, through the floor from (floor area, ft^2)(design outside dry-bulb temperature, °F—design inside dry-bulb temperature, °F) [coefficient of heat transmission of the floor material, Btu/(h · ft^2 · °F), from the appropriate *Guide* table]. When a machine room or utility room is below the floor, use the same relation but substitute the machine room or utility room dry-bulb temperature for the design outside dry-bulb temperature.

For floors above the ground, some designers reduce the difference between the design outdoor and indoor dry-bulb temperatures by 5°F (−15°C); other designers use the shaded-wall equivalent temperature difference from the *Guide*. Either method, or that given above, will provide safe results.

4. Compute the infiltration heat gain

Use the relation infiltration heat gain, Btu/h = (window crack length, ft) [window infiltration, ft^3/(ft · min) from the appropriate *Guide* table] (design outside dry-bulb temperature, °F)(1.08). Three aspects of this computation require explanation.

The window crack length used is usually one-half the total crack length in all the windows. Infiltration through cracks is caused by the wind acting on the building. Since the wind cannot act on all sides of the building at once, one-half the total crack length is generally used (but never less than one-half) in computing the infiltration heat gain. Note that the crack length varies with different types of windows. Thus the *Guide* gives, for metal sash, crack length = total perimeter of the movable section. For double-hung windows, the crack length = three times the width plus twice the height.

The window infiltration rate, ft^3/(min · ft) of crack, is given in the *Guide* for various wind velocities. Some designers use the infiltration rate for a wind velocity of 10 mi/h (16.1 km/h); others use 5 mi/h (8.1 km/h). The factor 1.08 converts the computed infiltration to Btu/h ($\times$ 0.32 = W).

5. Compute the outside-air bypass heat load

Some outside air may be needed in the conditioned space to ventilate fumes, odors, and other undesirables in the conditioned space. This ventilation air imposes a cooling or dehumidifying load on the air conditioner because the heat or moisture, or both must be removed from the ventilation air. Mot air conditioners are arranged to permit some outside air to bypass the cooling coils. The bypassed outdoor air becomes a load within the conditioned space similar to infiltration air.

Determine heat load, Btu/h, of the outside air bypassing the air conditioner from (cfm of ventilation air)(design outdoor dry-bulb-temperature, °F − design indoor dry-bulb temperature, °F)(air-conditioner bypass factor)(1.08).

Find the ventilation-air quantity by multiplying the number of people in the conditioned space by the ft^3/min per person recommended by the *Guide*. The ft^3/min (m^2/min) per person can range from a minimum of 5 (0.14) to a high of 50 (1.42) where heaving smoking is anticipated. If industrial processes within the conditioned space require ventilation, the air may be supplied by increasing the outside air flow, by a local exhaust system at the process, or by a combination of both. Regardless of the method used, outside air must be introduced to make up for the air exhausted from the conditioned space. The sum of the air required for people and processes is the total ventilation-air quantity.

Until the air conditioner is chosen, its bypass factor is unknown. However, to solve for the outside-air bypass heat load, a bypass factor must be applied. Table 14 shows typical bypass factors for various applications.

6. Compute the heat load from internal heat sources

Within an air-conditioned space, heat is given off by people, lights, appliances, machines, pipes, etc. Find the sensible heat, Btu/h, given off by people by taking the product (number of people in the air-conditioned space)(sensible heat release per person, Btu/h, from the appropriate *Guide* table). The sensible heat release per person varies with the activity of each person (seated, at rest, doing heavy work) and the dry-bulb temperature. Thus, at 80°F (26.7°C), a person doing heavy work in a factory will give off 465 Btu/h (136.3 W) sensible heat; seated at rest in a theater at 80°F (26.7°C), a person will give off 195 Btu/h (57.2 W), sensible heat.

TABLE 14   Typical Bypass Factors*

a. For various applications

Coil bypass factor†	Type of application	Example
0.30–0.50	A *small* total load or a load that is somewhat larger with a low sensible heat factor (high latent load)	Residence
0.20–0.30	Typical comfort application with a *relatively small* total load or a low sensible heat factor with a somewhat larger load	Residence; small retail shop; factory
0.10–0.20	Typical comfort application	Department store; bank; factory
0.05–0.10	Applications with high internal sensible loads or requiring a large amount of outdoor air for ventilation	Department store; restaurant; factory
0–0.10	All outdoor air applications	Hospital; operating room; factory

b. For finned coils

	Without sprays		With sprays†	
	8 fins/in (25.4 mm)	14 fins/in (25.4 mm)	8 fins/in (25.4 mm)	14 fins/in (25.4 mm)
Depth of coils, rows	Velocity, ft/min (m/min)			
	300–700 (91.4–213.4)	300–700 (91.4–213.4)	300–700 (91.4–213.4)	300–700 (91.4–213.4)
2	0.42–0.55	0.22–0.38		
3	0.27–0.40	0.10–0.23		
4	0.19–0.30	0.50–0.14	0.12–0.22	0.03–0.10
5	0.12–0.23	0.02–0.09	0.08–0.14	0.01–0.08
6	0.08–0.18	0.01–0.06	0.06–0.11	0.01–0.05
8	0.03–0.08	. . .	0.02–0.05	

*Carrier Air Conditioning Company.
†The bypass factor with spray coils is decreased because spray provides more surface for contacting the air.

Find the heat, Btu/h, given off by electric lights from (wattage rating of all installed lights) (3.4). Where the installed lighting capacity is expressed in kilowatts, use the factor 3413 instead of 3.4.

For electric motors, find the heat, Btu/h, given off from (total installed motor hp)(2546)/motor efficiency expressed as a decimal. The usual efficiency assumed for electric motors is 85 percent.

Many other sensible-heat-generating devices may be used in an air-conditioned space. These devices include restaurant, beauty shop, hospital, gas-burning, and kitchen appliances. The *Guide* lists the heat given off by a variety of devices, as well as pipes, tanks, pumps, etc.

7. Compute the room sensible heat
Find the sum of the sensible heat gains computed in steps 2 (sunlight heat gain), 3 (transmission heat gain), 4 (infiltration heat gain, 5 (outside air heat gain), 6 (internal heat sources). This sum is the room sensible-heat subtotal.

A further sensible heat gain may result from supply-duct heat gain, supply-duct leakage loss, and air-conditioning-fan horsepower. To the sum of these losses, a safety factor is usually added in the form of a percentage, since all the losses are also generally expressed as a percentage. The *Guide* provides means to estimate each loss and the safety factor. Assuming the sum of the losses and safety factor is x percent, the room sensible heat load, Btu/h, is $(1 + 0.01 \times x)$(room sensible heat subtotal, Btu/h).

8. Compute the room latent heat load
The room latent heat load results from the moisture entering the room with the infiltration air and bypass ventilation air, the moisture given off by room occupants, and any other moisture source such as open steam kettles, sterilizers, etc.

Find the infiltration-air latent heat load, Btu/h from (cfm infiltration)(moisture content of outside air design at design outdoor conditions, g/lb—moisture content of the conditioned air at the design indoor conditions, g/lb)(0.68). Use a similar relation for the bypass ventilation air, or Btu/h latent heat = (cfm ventilation)(moisture content of the outside air at design outdoor conditions, g/lb − moisture content of conditioned air at design indoor conditions, g/lb)(bypass factor)(0.68).

Find the latent heat gain from the room occupants, Btu/h, from (number of occupants in the conditioned space)(latent heat gain, Btu/h per person, from the appropriate *Guide* table). Be sure to choose the latent heat gain that applies to the activity *and* conditioned-room dry-bulb temperature.

Nonhooded restaurant, hospital, laboratory, and similar equipment produces both a sensible and latent heat load in the conditioned space. Consult the *Guide* for the latent heat load for each type of unit in the space. Find the latent heat load of these units, Btu/h, from (number of units of type)(latent heat load, Btu/h, per unit).

Take the sum of the latent heat loads for infiltration, ventilation bypass air, people, and devices. This sum is the room latent-heat subtotal, if water-vapor transmission through the building surfaces is neglected.

Water vapor flows through building structures, resulting in a latent heat load whenever a vapor-pressure difference exists across a structure. The latent heat load from this source is usually insignificant in comfort applications and need to be considered only in low or high dew-point applications. Compute the latent heat gain from this source, using the appropriate *Guide* table, and add it to the room latent-heat subtotal.

Factors for the supply-duct leakage loss and for a safety margin are usually applied to the above sum. When all the latent heat subtotals are summed, the result is the room latent-heat total.

9. Compute the outside-air heat
Air brought in for space ventilation imposes a sensible and latent heat load on the air-conditioning apparatus. Compute the sensible heat load, Btu/h, from (cfm outside air)(design outdoor dry-bulb temperature − design indoor dry-bulb temperature)(1 − bypass factor)(1.08). Compute the latent heat load, Btu/h, from (cfm outside air) (design outdoor moisture content, g/lb − design indoor moisture content, g/lb)(1 − bypass factor)(0.68). Apply percentage factors for return duct heat

and leakage gain, pump horsepower, dehumidifier, and piping loss, see the *Guide* for typical values.

10. Compute the grand-total heat and refrigeration tonnage
Take the sum of the room total heat and outside-air heat. The result is the grand-total heat load of the space, Btu/h.

Compute the refrigeration load, tons, from (grand-total heat, Btu/h)[12,000 Btu/(h·ton) (3577 W) of refrigeration]. A refrigeration system having the next higher standard rating is generally chosen.

11. Compute the sensible heat factor and the apparatus dew point
For any air-conditioning system, sensible heat factor = (room sensible heat, Btu/h)/(room total heat, Btu/h). Using a psychrometric chart and the known room conditions, we find the apparatus dew point. An alternate, and quicker, way to find the apparatus dew point is to use the Carrier *Handbook of Air Conditioning System Design* tables.

12. Compute the quantity of dehumidified air required
Determine the dehumidified air temperature rise, °F, from (1 − bypass factor)(design indoor temperature, °F − apparatus dew point, °F). Compute the dehumidified air quantity, ft³/min, from (room sensible heat, Btu/h)/1.08 (dehumidified air temperature rise, °F)

Related Calculations.   The general procedure given above is valid for all types of air-conditioned spaces—offices, industrial plants, residences, hotels, apartment houses, motels, etc. Use the ASHRAE *Guide and Data Book* or the Carrier *Handbook of Air Conditioning System Design* as a source of data for the various calculations. Application of this method to an actual building is shown in the next calculation procedure.

In actual design work, a calculations form incorporating the calculations shown above is generally used. Such forms are obtainable from equipment manufacturers. Since the usual form does not provide any explanation of the calculations, the present calculation procedure is a useful guide to using the form. Refer to the Carrier *Handbook of Air Conditioning System Design* for one such form.

In using SI units in air-conditioning design calculations, the same general steps are followed as given above. Numerical usage of SI is shown in the next calculation procedure.

AIR-CONDITIONING-SYSTEM HEAT-LOAD DETERMINATION—NUMERICAL COMPUTATION

Determine the required capacity of an air-conditioning system to serve the industrial building shown in Fig. 27. The outside walls are 8-in (203.2-mm) brick with an interior finish of ⅜-in (9.50-mm) gypsum lath plastered and furred. A 6-in (152.4-mm) plain poured-concrete partition separates the machinery room from the conditioned space. The roof is 6-in (152.4-mm) concrete covered with ½-in (12.7-mm) thick insulating board, and the floor is 2-in (50.8-mm) concrete. The windows are double-hung, metal-frame locked units with light-colored shades three-quarters drawn. Internal heat loads are: 100 people doing light assembly work; twenty-five

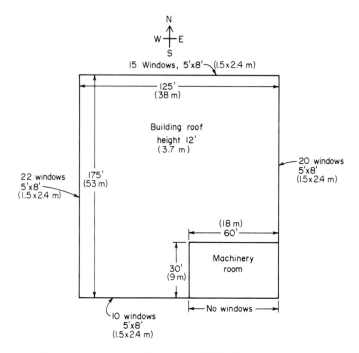

N wall area = 125 (12) – 15 (40) = 900 ft²(83.6 m²)

S wall area = 125 (12) – 5 x 8 x 10 = 1100 ft²(102.2 m²)

E wall area = (175)(12) – 5 x 8 x 20 = 1300 ft²(120.8 m²)

W wall area = (175)(12) – 5 x 8 x 22 = 1220 ft²(113.3 m²)

Roof area = (175)(125) = 21,875 ft²(2032.3 m²)

Glass area = (20 x 40) + (10 x 40) + 22(40) + 15(40) = 2680 ft²(248.9 m²)

Partition area = 90 x 12 = 1080 ft²(100.3 m²)

FIGURE 27   Industrial building layout.

1-hp (746-W) motors running continuously at full load; 20,000 W of light kept on at all times. The building is located in Port Arthur, Texas, at about 30° north latitude. The desired indoor design conditions are 80°F (26.7°C) dry bulb, 67°F (19.4°C) wet bulb, and 51 percent relative humidity. Air-conditioning equipment will be located in the machinery room. Use the general method given in the previous calculation procedure.

Calculation Procedure:

1. *Determine the design outdoor and indoor conditions*
The ASHRAE *Guide and Data Book* lists the design dry-bulb temperature in common use for Port Arthur, Texas, as 95°F (35°C) and the design wet-bulb temperature

as 79°F (26.1°C). Design indoor temperature and humidity conditions are given; if they were not given, the recommended conditions given in the *Guide* for an industrial building housing light assembly work would be used.

2. Compute the sunlight heat gain

The east, south, and west windows and walls of the building are subject to sunlight heat gains. North-facing walls are neglected because the sunlight heat gain is usually less than the transmission heat gain. Reference to the *Guide* table for sunlight radiation through glass shows that the largest amount of heat radiation occurs through the east and west walls. The maximum radiation is 181 Btu/(h · ft²) (570.5 W/m²) of glass area at 8 a.m. for the east wall and the same for the west wall at 4 p.m. Radiation through the glass in the south wall never reaches this magnitude. Hence, only the east or west wall need be considered. Since the west wall has 22 windows compared with 20 in the east wall, the west-wall sunlight heat gain will be used because it has a *larger* heat gain. (If both walls had an equal number of windows, either wall could be used.) When the window shades are normally three-quarters drawn, a value 0.6 times the tabulated sunlight radiation can be used. Hence, the west-glass sunlight heat gain = (22 windows)(5 × 8 ft each)(181)(0.6) = 95,600 Btu/h (28,020.4 W).

For the same time of day, 4 p.m., the *Guide* table shows that the east-glass radiation is 0 Btu/(h · ft²) and the south glass is 2 Btu/(h · ft²) (6.3 W/m²). Hence, the south-glass sunlight heat gain = (10 windows) (5 × 8 ft each)(2)(0.6) = 480 Btu/h (140.7 W).

The same three walls, and the roof, are also subject to sunlight heat gains. Reference to the *Guide* shows that with 8-in (203-mm) walls the temperature difference resulting from sunlight heat gains is 15°F (8.3°C) for south walls and 20°F (−6.7°C) for east and west walls. At 4 p.m. the roof temperature difference is given as 40°F (4.4°C). Hence, sunlight gain, south wall = (wall area, ft²)(temperature difference, °F)[wall coefficient of heat transfer, Btu/(h · ft² · °F)], or (1100)(15)(0.30) = 4950 Btu/h (1450.9 W). Likewise, the east-wall sunlight heat gain = (1300)(20)(0.30) = 7800 Btu/h (2286.2 W); the west-wall sunlight heat gain = (1220)(20)(0.30) = 7,320 Btu/h (2145.5 W); the roof sunlight heat gain = (21,875)(40)(0.33) = 289,000 Btu/h (84.7 kW). Note that the wall and roof coefficients of heat transfer are obtained from the appropriate *Guide* table.

The sum of the sunlight heat gains gives the total sunlight gain, or 405,150 Btu/h (118.7 kW).

3. Compute the glass transmission heat gain

All the glass in the building is subject to a transmission heat gain. Find the all-glass transmission heat gain, Btu/h, from (total glass area, ft²)(outdoor design dry-bulb temperature, °F − indoor design dry-bulb temperature, °F)[coefficient of heat transmission of glass, Btu/(h · ft² · °F), from *Guide*], or (2680)(95 − 80)(1.13) = 45,400 Btu/h (13.3 kW).

The transmission heat gain of the south, east, and west walls can be neglected because the sunlight heat gain is greater. Hence, only the north-wall transmission heat gain need be computed. For unshaded walls, the transmission heat gain, Btu/h, is (wall area, ft²)(design outdoor dry-bulb temperature, °F − design indoor dry-bulb temperature, °F)[coefficient of heat transmission, Btu/(h · ft² · °F)] = (900)(95 − 80)(0.30) = 4050 Btu/h (1187.1 W).

The heat gain from the ground can be neglected because the ground is usually at a lower temperature than the floor. Thus, the total transmission heat gain is the sum of the individual gains, or 66,530 Btu/h (19.5 kW).

4. Compute the infiltration heat gain
The total crack length for double-hung windows is 3 × width + 2 × height, or
(67 windows)[(3 × 5) + (2 × 8)] = 2077 ft (633.1 m). By using one-half the total
length, or 2077/2 = 1039 ft (316.7 m), and a wind velocity of 10 mi/h (16.1
km/h), the leakage ft³/min is (crack length, ft)(leakage per ft of crack) =
(1039)(0.75) = 770 ft³/min, or 60(779) = 46,740 ft³/h (1323 m³/h).

The heat gain due to infiltration through the window cracks is (leakage, ft³/min)
(design outdoor dry-bulb temperature, °F − design indoor dry-bulb temperature,
°F)(1.08), or (779)(95 − 80)(1.08) = 12,610 Btu/h (3.7 kW).

5. Compute the outside-air bypass heat load
For factories, the *Guide* recommends a ventilation air quantity of 10 ft³/min (0.28
m³/min) per person. Local codes may require a larger quantity; hence, the codes
should be checked before a final choice is made of the ventilation-air quantity used
per person. Since there are 100 people in this factory, the required ventilation
quantity is 100(10) = 100 ft³/min (28 m³/min). Next, the bypass factor for the air-
conditioning equipment must be chosen.

Table 14 shows that the usual factory air-conditioning equipment has a bypass
factor ranging between 0.10 and 0.20. Assume a value of 0.10 for this installation.

The heat load, Btu/h, of the outside air bypassing the air conditioner is (cfm of
ventilation air)(design outdoor dry-bulb temperature, °F − design indoor dry-bulb
temperature, °F)(air-conditioner bypass factor)(1.08). Hence, (1000)(95 −
80)(0.10)(1.08) = 1620 Btu/h (474.8 W).

6. Compute the heat load from internal heat sources
The internal heat sources in this building are people, lights, and motors. Compute
the sensible heat load of the people, from Btu/h = (number of people in the air-
conditioned space)(sensible heat release per person, Btu/h, from the appropriate
Guide table). Thus, for this building with an 80°F (26.7°C) indoor dry-bulb temperature and 100 occupants doing light assembly work, the heat load produced by
people = (100)(210) = 21,000 Btu/h (6.2 kW).

The motor heat load, Btu/h, is (motor hp)(2546)/motor efficiency. Given an 85
percent motor efficiency, the motor heat load = (25)(2546)/0.85 = 75,000 Btu/h
(21.9 kW). Thus, the total internal heat load = 21,000 + 68,000 + 75,000 =
164,000 Btu/h (48.1 kW).

7. Compute the room sensible heat
Find the sum of the sensible heat gains computed in steps 2, 3, 4, 5, and 6. Thus,
sensible heat load = 405,150 + 66,530 + 12,610 + 1620 + 164,000 = 649,910
Btu/h (190.5 kW), say 650,000 Btu/h (190.5 kW). Using an assumed safety factor
of 5 percent to cover the various losses that may be encountered in the system, we
find room sensible heat = (1.05)(650,000) = 682,500 Btu/h (200.0 kW).

8. Compute the room latent-heat load
The room latent load results from the moisture entering the air-conditioned space
with the infiltration and bypass air, moisture given off by room occupants, and any
other moisture sources.

Find the infiltration heat load, Btu/h, from (cfm infiltration)(moisture content of
outside air at design outdoor conditions, g/lb − moisture content of the conditioned
air at the design indoor conditions, g/lb)(0.68). Using a psychrometric chart or the
Guide thermodynamic tables, we get the infiltration latent-heat load = (779)(124
− 78)(0.68) = 24,400 Btu/h (7.2 kW).

Using a similar relation for the ventilation air gives Btu/h latent heat = (cfm ventilation air)(moisture content of outside air at design outdoor conditions, g/lb − moisture content of the conditioned air at the design indoor conditions, g/lb)(bypass factor)(0.68). Or, (1000)(124 − 78)(0.10)(0.68) = 3130 Btu/h (917.4 W).

The latent heat gain from room occupants is Btu/h = (number of occupants in the conditioned space)(latent heat gain, Btu/h per person, from the appropriate *Guide* table). Or, (100)(450) = 45,000 Btu/h (917.4 W).

Find the latent heat gain subtotal by taking the sum of the above heat gains, or 24,400 + 3130 + 45,000 = 72,530 Btu/h (21.2 kW). Using an allowance of 5 percent for supply-duct leakage loss and a safety margin gives the latent heat gain = (1.05)(72,530) = 76,157 Btu/h (22.3 kW).

9. Compute the outside heat

Compute the sensible heat load of the outside ventilation air from Btu/h = (cfm outside ventilation air)(design outdoor dry-bulb temperature − design indoor dry-bulb temperature)(1 − bypass factor)(1.08). For this system, with 1000 ft^3/min (28.3 m^3/min) outside air ventilation, sensible heat = (1000)(95 − 80)(1 − 0.010)(1.08) = 14,600 Btu/h (4.3 kW).

Compute the latent-heat load of the outside ventilation air from (cfm outside air) × (design outdoor moisture content, g/lb − design indoor moisture content, g/lb)(1 − bypass factor)(0.68). Using the moisture content from step 8 gives (1000)(124 − 78)(1 − 0.1) × (0.68) = 28,200 Btu/h (8.3 kW).

10. Compute the grand-total heat and refrigeration tonnage

Take the sum of the room total heat and the outside-air sensible and latent heat. The result is the grand-total heat load of the space, Btu/h W or kW).

The room total heat = room sensible-heat total + room latent-heat total = 682,500 + 76,157 = 768,657 Btu/h (225.3 kW). Then the grand-total heat = 768,657 + 14,600 + 28,200 = 811,457 Btu/h, say 811,500 Btu/h (237.9 kW).

Compute the refrigeration load, tons, from (grand-total heat, Btu/h)/(12,000 Btu/h per ton of refrigeration), or 811,500/12,000 = 67.6 tons (237.7 kW), say 70 tons (246.2 kW).

The quantity of cooling water required for the refrigeration-system condenser is Q gal/min = 30 × tons of refrigeration/condenser water-temperature rise, °F. Assuming a 75°F (23.9°C) entering water temperature and 95°F (35°C) leaving water temperature, which are typical values for air-conditioning practice Q = 30(70)/95 − 75 = 105 gal/min (397.4 L/min).

11. Compute the sensible heat factor and apparatus dew point

For any air-conditioning system, sensible heat factor = (room sensible heat, Btu/h)/(room total heat, Btu/h) = 682,500/768,657 = 0.888.

The *Guide* or the Carrier *Handbook of Air Conditioning Design* gives an apparatus dew point of 58°F (14.4°C), closely.

12. Compute the quantity of dehumidified air required

Determine the dehumidified air temperature rise first from F = (1 − bypass factor)(design indoor temperature, °F − apparatus dew point, °F), or F = (1 − 0.1)(80 − 58) = 19.8° F (−6.8°C).

Next, compute the dehumidified air quantity, ft^3/min from (room sensible heat, Btu/h)/1.08(dehumidified air temperature rise, °F), or (682,500)/1.08(10.8) = 34,400 ft^3/min (973.9 m^3/min).

Related Calculations. Use this general procedure for any type of air-conditioned building or space—industrial, office, hotel, motel, apartment house, residence, laboratories, school, etc. Use the ASHRAE *Guide and Data Book* or the Carrier *Handbook of Air Conditioning System Design* as a source of data for the various calculations. In comparing the various values from the *Guide* used in this procedure, note that there may be slight changes in certain tabulated values from one edition of the *Guide* to the next. Hence, the values shown may differ slightly from those in the current edition. This should not cause concern, because the procedure is the same regardless of the values used.

AIR-CONDITIONING-SYSTEM COOLING-COIL SELECTION

Select an air-conditioning cooling coil to cool 15,000 ft³/min (424.7 m³/min) of air from 85°F (29.4°C) dry bulb, 67°F (19.4°C) wet bulb, 57°F (13.9°C) dew point, 38 percent relative humidity, to a dry-bulb temperature of 65°F (18.3°C) with cooling water at 50°F (10°C). Suppose the air were cooled below the dew point of the entering air. How would the calculation procedure differ?

Calculation Procedure:

1. Compute the weight of air to be cooled
From a psychrometric chart (Fig. 28) find the specific volume of the entering air as 13.75 ft³/lb (0.86 m³/kg). To convert the air flow in ft³/min to lb/h, use the relation lb/h = 60 cfm/v_s, where cfm = air flow, ft³/min; v_s = specific volume of the entering air, ft³/lb. Hence for this cooling coil, lb/h = 60(15,000)/13.75 = 65,500 lb/h (29,773 kg/h).

Where a cooling coil is rated by the manufacturer for air at 70°F (21.1°C) dry bulb and 50 percent relative humidity, as is often done, use the relation lb/h = 4.45*cfm*. Thus if this coil were rated for air at 70°F (21.1°C) dry bulb, the weight of air to be cooled would be lb/h = 4.45(15,000) = 66,800 (30,364 kg/h).

Since this air quantity is somewhat greater than when the entering air specific volume is used, and since cooling coils are often rated on the basis of 70°F (21.1°C) dry-bulb air, this quantity, 66,800 lb/h (30,364 kg/h), will be used. The procedure is the same in either case. [*Note:* 4.45 = 60/135 ft³/lb, the specific volume of air at 70°F (21.1°C) and 50 percent relative humidity.]

2. Compute the quantity of heat to be removed
Use the relation $H_r = ws\Delta t$, where H_r = heat to be removed from the air, Btu/h; w = weight of air cooled, lb/h; s = specific heat of air = 0.24 Btu/(lb · °F); Δt = temperature drop of the air = entering dry-bulb temperature, °F − leaving dry-bulb temperature, °F. For this coil, H_r = (66,800)(0.24)(85 − 65) = 321,000 Btu/h (94.1 kW).

3. Compute the quantity of cooling water required
The quantity of cooling water required in gal/min = $H_r/500 \Delta t_w$ = leaving water temperature°F − entering water temperature, °F. Since the leaving water temperature is not known, a value must be assumed. The usual temperature rise of water during passage through an air-conditioning cooling coil is 4 to 12°F (2.2 to 6.7°C). As-

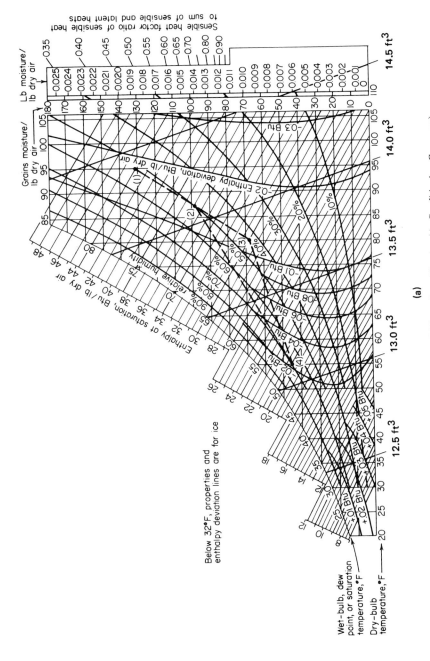

FIGURE 28 (a) Psychrometric charts for normal temperatures—USCS version. (*Carrier Air Conditioning Company.*)

16.79

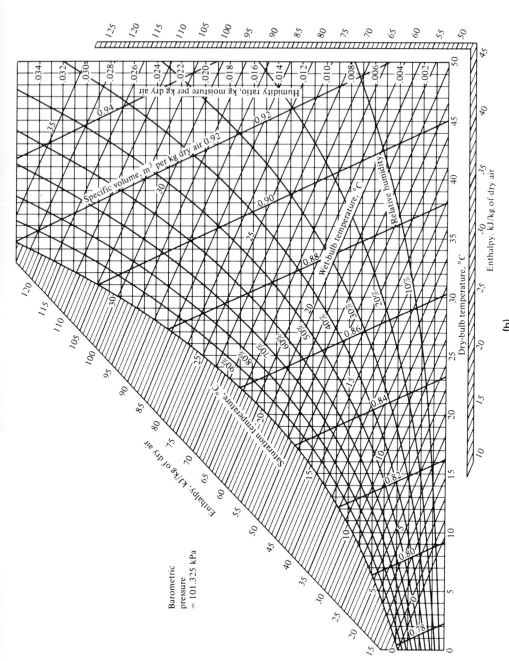

FIGURE 28 (b) SI version (Stoecker and Jones, *Refrigeration and Air Conditioning*, McGraw-Hill.)

Barometric pressure = 101.325 kPa

suming a 10°F (5.6°C) rise, which is a typical value, gal/min required = 321,000 /500(10) = 64.2 gal/min (4.1 L/s).

4. Determine the logarithmic mean temperature difference
Use Fig. 4 in the heat transfer section of this handbook to determine the logarithmic mean temperature difference for the cooling coil. In this chart, greatest terminal difference = entering air temperature, °F − leaving water temperature, °F = 85 − 60 = 25°F (13.9°C), and least terminal temperature difference = leaving air temperature, °F − entering water temperature, °F = 65 − 50 = 15°F (8.3°C). Entering Fig. 4 at these two temperature values gives a logarithmic mean temperature difference (LMTD) of 19.5°F (10.8°C).

5. Compute the coil core face area
The coil core face area is the area exposed to the air flow; it does not include the area of the mounting flanges. Compute the coil core face area from $A_c = cfm/V_a$, where V_a = air velocity through the coil, ft/min. The usual air velocity through the coil, often termed face velocity, ranges from 300 to 800 ft/min (91.4 to 243.8 m/min) although special designs may use velocities down to 200 ft/min (60.9 m/min) or up to 1200 ft/min (365.8 m/min). Assuming a face velocity of 500 ft/min (152.4 m/min) gives A_c = 15,000/500 = 30 ft^2 (2.79 m^2).

6. Select the cooling coil for the load
Using the engineering data provided by the manufacturer whose coil is to be used, choose the coil. Table 13 summarizes typical engineering data provided by a coil manufacturer. This table shows that two 15.4-ft^2 (1.43-m^2) coils placed side by side will provide a total face area of 2 × 15.4 = 30.8 ft^2. Hence, the actual air velocity through the coil is V_a = cfm/A_c = 15,000/30.8 = 487 ft/min (148.4 m/min).

7. Compute the water velocity in the coil
Table 15 shows that the coil water velocity, ft/min = 3.59*gpm*. Since the required flow is, from step 3, 64.2 gal/min (243 L/min), the water flow for each unit will be half this, or 64.2/2 = 32.1 gal/min (121.5 L/min). Hence the water velocity = 3.59(32.1) = 115.2 ft/min (35.1 m/min).

8. Determine the coil heat-transfer factors
Table 15 lists typical heat-transfer factors for various water and air velocities. Interpolating between 400- and 500-ft/min (121.9- and 152.4-m/min) air velocities at a water velocity of 115 ft/min (35.1 m/min) gives a heat transfer factor of k = 165 Btu/(h·ft^2·°F) [936.9 W/(m^2·K)] LMTD row. The increase in velocity from 15 to 15.2 ft/min (4.63 m/min) is so small that it can be ignored. If the actual velocity is midway, or more, between the two tabulated velocities, interpolate vertically also.

9. Compute the number of tube rows required
Use the relation number of tube rows = H_r/(LMTD)(A_c)(k). Thus, number of rows = 321,000/(19.5)(30.8)(165) = 3.24, or four rows, the next larger *even* number.

Water cooling coils for air-conditioning service are usually built in units having two, four, six or eight rows of coils. If the above calculation indicates that an odd number of coils should be used (i.e., the result was 3.24 rows), use the next smaller or larger *even* number of rows after increasing or decreasing the air and water velocity. Thus, to decrease the air velocity, use a coil having a larger face area. Recompute the air velocity and water velocity; find the new heat-transfer factor and

TABLE 15 Typical Cooling-Coil Characteristics

Face area, ft^2 (m^2)	14.0 (1.3)	15.4 (1.43)	17.9 (1.66)
Tube length, ft · in (m)	5–6 (1.68)	6–0 (1.83)	7–0 (2.13)
Water velocity, ft/min (m/ min)	3.59 gal/min (4.14 L/min)	3.59 gal/min (4.14 L/min)	3.59 gal/min (4.14 L/min)

Coil heat-transfer factors, k = Btu/(h·ft^2·°F) LMTD row [W/(m^2·K)]

Water velocity, ft/min (m/min)	Air velocity, ft/min (m/min)		
	400 (121.9)	500 (152.4)	550 (167.6)
113 (34.4)	154 (46.9)	162 (49.4)	170 (51.8)
115 (35.1)	158 (48.2)	167 (50.9)	175 (53.3)
117 (35.7)	161 (49.1)	172 (52.4)	176 (53.6)

Coil water pressure drop, ft (m) of water per row

Tube length, ft · in (m)	Water velocity, ft/min (m/min)		
	90 (27.4)	120 (36.6)	150 (45.7)
5-6 (1.68)	0.16 (0.04)	0.26 (0.08)	0.38 (0.12)
6-0 (1.83)	0.18 (0.05)	0.29 (0.09)	0.42 (0.13)
7-0 (2.13)	0.21 (0.06)	0.32 (0.10)	0.46 (0.14)

Header water pressure drop, ft (m) of water

Coil type	Water velocity, ft/min (m/min)		
	90 (27.4)	120 (36.6)	150 (45.7)
A	0.26 (0.08)	0.48 (0.15)	0.72 (0.22)
B	0.34 (0.10)	0.62 (0.19)	0.92 (0.28)

the required number of rows. Continue doing this until a suitable number of rows is obtained. Usually only one recalculation is necessary.

10.   *Determine the coil water-pressure drop*
Table 15 shows the water-pressure drop, ft (m) of water, for various tube lengths and water velocities. Interpolating between 90 and 120 ft/min (27.4 and 36.6 m/ min) for a 6-ft (1.83-m) long tube gives a pressure drop of 0.27 ft (0.08 m) of water per row at a water velocity of 115.2 ft/min (35.1 m/min). Since the coil has four rows, total tube pressure drop = 4(0.27) = 1.08 ft (0.33 m) of water.

There is also a water pressure drop in the coil headers. Table 15 lists typical values. Interpolate between 90 and 120 ft/min (27.4 and 36.6 m/min) for a B-type coil gives a header pressure loss of 0.57 ft (0.17 m) of water at a water velocity of 115.2 ft/min (35.1 m/min). Hence, the total pressure loss in the coil is 1.08 + 0.57 = 1.65 ft (0.50 m) of water = coil loss + header loss.

11. Determine the coil resistance to air flow
Table 16 lists the resistance of coils having two to six rows of tubes and various air velocities. Interpolating for four tube rows gives a resistance of 0.225 in H_2O for an air velocity of 487 ft/min (148.4 m/min). The increase in resistance with a wet tube surface is, from Table 16, 28 percent at a 500-ft/min (152.4-m/min) air velocity. This occurs when the air is cooled below the entering air dew point and is discussed in step 13.

12. Check the coil selection in a coil-rating table
Many manufacturers publish precomputed coil-rating tables as part of their engineering data. Table 16 shows a portion of one such table. This tabulation shows that with an air velocity of 500 ft/min (152.4 m/min), four tube rows, an entering-air temperature of 85°F (29.4°C), and an entering water temperature of 50°F (10.0°C), the cooling coil has a cooling capacity of 10,300 Btu/(h·ft²) (32,497 W/m²), and a final air temperature of 65°F (18.3°C). Since the actual air velocity of 487 ft/min (148.4 m/min) is close to 500 ft/min (152.4 m/min), the tabulated cooling capacity closely approximates the actual cooling capacity. Hence the required heat-transfer area is $A_c = H_r/10,300 = 321,000/10,300 = 31.1$ ft² (2.98 m²). This agrees closely with the area of 30.8 ft² (2.86 m²) found in step 6.

In actual practice, designers use a coil cooling capacity table whenever it is available. However, the procedure given in steps 1 through 11 is also used when an exact analysis of a coil is desired or when a capacity table is not available.

13. Compute the heat removal for cooling below the dew point
When the temperature of the air leaving the cooling coil is lower than the dew point of ht entering air, H_r = (weight of air cooled, lb)(total heat of entering air at its wet-bulb temperature, Btu/lb − total heat of the leaving air at its wet-bulb temperature, Btu/lb). Once H_r is known, follow all the steps given above except

TABLE 16 Typical Cooling-Coil Resistance Characteristics

No. of tube rows	Air-flow resistance, in H_2O for 70°F air (mm H_2O for 21.1°C air)		
	Air-face velocity, ft/min (m/min)		
	400 (121.9)	500 (152.4)	600 (182.9)
2	0.081 (2.06)	0.122 (3.10)	0.164 (4.17)
4	0.162 (4.11)	0.234 (5.94)	0.318 (8.08)
6	0.234 (5.94)	0.344 (8.74)	0.472 (11.99)
8	0.312 (7.92)	0.454 (11.53)	0.622 (15.80)

Resistance increase due to wet tube surface, percent		
32	28	24

Air velocity, ft/min (m/min)	No. of tube rows	Coil cooling capacity, Btu/(h·ft²) (W/m²) face area and final air temperature, °F (°C)	Entering water temperature, °F (°C)	
		Entering air temperature, °F (°C)	45 (7.2)	50 (10)
500 (152.4)	4	85 (29.4)	11,900 (36,545) 63 (17.2)	10,300 (32,497) 65 (18.3)

that (*a*) a correction must be applied in step 11 for a wet tube surface. Obtain the appropriate correction factor from the manufacturer's engineering data, and apply it to the air-flow-resistance data, for the coil selected. (*b*) Also, the usual coil-rating table presents only the sensible-heat capacity of the coil. Where the ratio of sensible heat removed to latent heat removed is more than 2:1, the usual coil-rating table can be used. If the ratio is less than 2:1, use the procedure in steps 1 through 13.

Related Calculations.   Use the method given here in steps 1 through 11 for any finned-type cooling coil mounted perpendicular to the air flow and having water as the cooling medium where the final air temperature leaving the cooling coil is *higher than* the dew point of the entering air. Follow step 13 for cooling below the dew point of the entering air. Figure 29 shows a typical air-handling unit for an air-conditioning system and the location of the coil module in it. Two piping arrangements for air cooling coils are shown in Fig. 30*a* and 30*b*.

Cooling and dehumidifying coils used in air-conditioning systems generally serve the following ranges of variables: (1) dry-bulb temperature of the entering air is 60 to 100°F (15.6 to 37.8°C); wet-bulb temperature of entering air is 50 to 80°F (10 to 26.7°C); (2) coil core face velocity can range from 200 to 1200 ft/min (60.9 to 365.8 m/min) with 500 to 800 ft/min (152.4 to 243.8 m/min) being the most common velocity for comfort cooling applications; (3) entering water-temperature ranges from 40 to 65°F (4.4 to 18.3°C); (4) the water-temperature rise ranges from 4 to 12°F (2.2 to 6.7°C) during passage through the coil; (5) the water velocity ranges from 2 to 6 ft/s (0.61 to 1.83 m/s).

To choose an air-cooling coil using a direct-expansion refrigerant, follow the manufacturer's engineering data. Since most of the procedures are empirical, it is

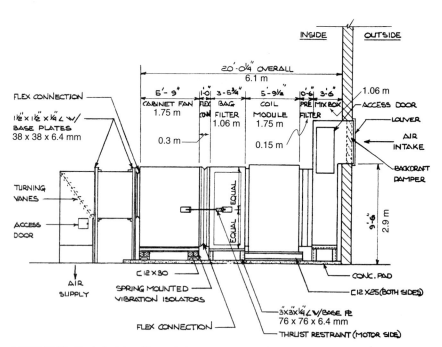

FIGURE 29 Typical air-handling unit for an air-conditioning system and the location of the coil module in it. (*Jerome F. Mueller.*)

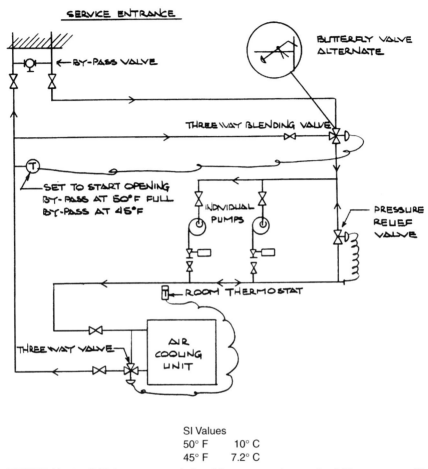

SI Values
50° F   10° C
45° F   7.2° C

FIGURE 30 (*a*) Chilled-water system designed for constant volume and variable temperature. (*b*) Chilled water system designed for constant temperature and variable volume. (*Jerome F. Mueller.*)

difficult to generalize about which procedure to use. However, the usual range of the volatile refrigerant temperature at the coil suction outlet is 25 to 55°F (−3.9 to 12.8°C). Where chilled water is circulated through the coil, the usual quantity range is 2 to 6 gal/(m · ton) (8.4 to 25 L/t).

MIXING OF TWO AIRSTREAMS

An air-conditioning system is designed to deliver 100,000 ft³/min (2831 m³/min) of air to a conditioned space. Of this total, 90,000 ft³/min (2548 m³/min) is recirculated indoor air at 72°F (22.2°C) and 40 percent relative humidity; 10,000 ft³/min (283.1 m³/min) is outdoor air at 0°F (−17.8°C). What are the enthalpy, temperature, moisture content, and relative humidity of the resulting air mixture? If air

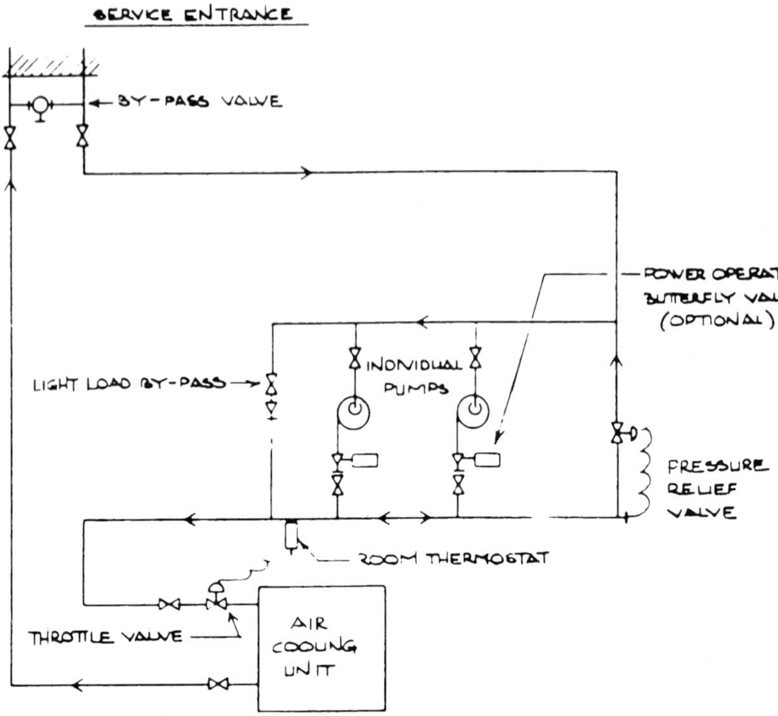

FIGURE 30   *Continued.*

enters the room from the outlet grille at 60°F (15.6°C) after leaving the apparatus at a 50°F (10°C) dew point and the return air is at 75°F (23.9°C), what proportion of conditioned air and bypassed return air must be used to produce the desired outlet temperature at the grille?

Calculation Procedure:

1. Determine the proportions of each airstream
Use the relations $p_r = r/t$ and $p_0 = o/t$, where p_r = percent recirculated room air, expressed as a decimal; r = recirculated air quantity, ft^3/min; t = total air quantity, ft^3/min; p_0 = percent outside air, expressed as a decimal; o = outside air quantity, ft^3/min. For this system, p_r = 90,000/100,000 = 0.90, or 90 percent; p_0 = 10,000/100,000 = 0.10, or 10 percent. (The computation is SI units is identical.)

2. Determine the enthalpy of each airstream
Use a psychrometric chart or table to find the enthalpy of the recirculated indoor air as 24.6 Btu/lb (57.2 kJ/kg).

The enthalpy of the outdoor air is 0.0 Btu/lb (0.0 kJ/kg) because in considering heating or humidifying processes in winter it is always safest to assume that the outdoor air is completely dry. This condition represents the greatest heating and humidifying load because the enthalpy and the water-vapor content of the air are at a minimum when the air is considered dry at the outdoor temperature.

3. Determine the moisture content of each airstream

The moisture content of the indoor air at a 72°F (22.2°C) dry-bulb temperature and 40 percent relative humidity is, from the psychrometric chart (Fig. 28), 47.2 gr/lb (6796.8 mg/kg). From a psychrometric table the moisture content of the 0°F (−17.8°C) outdoor air, which is assumed to be completely dry, is 0.0 gr/lb (0.0 mg/kg).

4. Compute the enthalpy of the air mixture

Use the relation $h_m = (oh_0 + rh_r)/t$, where h_m = enthalpy of mixture, Btu/lb; h_0 = enthalpy of the outside air, Btu/lb; h_r = enthalpy of the recirculated room air, Btu/lb; other symbols as before. Hence, $h_m = (10,000 \times 0 + 90,000 \times 24.6)/100,000 = 22.15$ Btu/lb (51.5 kJ/kg).

5. Compute the temperature of the air mixture

Use a similar relation to that in step 4, substituting the air temperature for the enthalpy. Or, $t_m = (ot_0 + rt_r)/t$, where t_m = mixture temperature, °F; t_0 = temperature of outdoor air, °F; t_r = temperature of recirculated room air, °F; other symbols as before. Hence, $t_m = (10,000 \times 0 + 90,000 \times 72)/100,000 = 64.9$°F (18.3°C).

6. Compute the moisture content of the air mixture

Use a similar relation to that in step 4, substituting the moisture content for the enthalpy. Or, $g_m = (og_0 + rg_r)/t$, where g_m = gr of moisture per lb of mixture; g_0 = gr of moisture per lb of outdoor air; g_r = gr of moisture per lb of recirculated room air; other symbols as before. Thus, $g_m = (10,000 \times 0 + 90,000 \times 47.2)/100,000 = 42.5$ gr/lb (6120 mg/kg).

7. Determine the relative humidity of the mixture

Enter the psychrometric chart at the temperature of the mixture, 64.9°F (18.3°C), and the moisture content, 42.5 gr/lb (6120 mg/kg). At the intersection of the two lines, find the relative humidity of the mixture as 47 percent relative humidity.

8. Determine the required air proportions

Set up an equation in which x = proportion of conditioned air required to produce the desired outlet temperature at the grille and y = the proportion of bypassed air required. The air quantities will also be proportional to the dry-bulb temperatures of each airstream. Since the dew point of the air leaving an air-conditioning apparatus = dry-bulb temperature of the air, $50x + 75y = 60(x + y)$, or $15y = 10x$. Also, the sum of the two airstreams $x + y = 1$. Substituting and solving for x and y, we get $x = 60$ percent; $y = 40$ percent. Multiplying the actual air quantity supplied to the room by the percentage representing the proportion of each airstream will give the actual ft³/min required for supply and bypass air.

Related Calculations. Use this general procedure to determine the properties of any air mixture in which two airstreams are mixed without compression, expansion, or other processes involving a marked change in the pressure or volume of either or both airstreams.

SELECTION OF AN AIR-CONDITIONING SYSTEM FOR A KNOWN LOAD

Choose the type of air-conditioning system to use for comfort conditioning of a factory having a heat gain that varies from 500,000 to 750,000 Btu/h (146.6 to

219.8 kW) depending on the outdoor temperature and the conditions inside the building. Indicate why the chosen system is preferred.

Calculation Procedure:

1. Review the types of air-conditioning systems available
Table 17 summarizes the various types of air-conditioning systems *commonly used* for different applications. Economics and special designs objectives dictate the final choice and modifications of the systems listed. Where higher-quality air conditioning is desired (often at a higher cost), certain other systems may be considered. These are dual-duct, dual-conduit, three-pipe induction and fan-coil, four-pipe induction and fan-coil, and panel-air systems.

Study of Table 17 shows that four main types of air-conditioning systems are popular: direct-expansion (termed DX), all-water, all-air, and air-water. These classifications indicate the methods used to obtain the final within-the-space cooling and heating. The air surrounding the occupant is the end medium that is conditioned.

2. Select the type of air-conditioning system to use
Table 17 indicates that direct-expansion and all-air air-conditioning systems are *commonly used* for factory comfort conditioning. The load in the factory being considered may range from 500,000 to 750,000 Btu/h (146.6 to 219.8 kW). This is the equivalent of a maximum cooling load of $750,000/12,000$ Btu/(h·ton) of refrigeration = 62.5 tons (219.8 kW) of refrigeration.

Where a building has a varying heat load, bypass control wherein neutral air is recirculated from the conditioned space while the amount of cooling air is reduced is often used. With this arrangement, the full quantity of supply air is introduced to the cooled area at all times during system operation.

Self-contained direct-expansion systems can serve large factory spaces. Their choice over an all-air bypass system is largely a matter of economics and design objectives.

Where reheat is required, this may be provided by a reheater in a zone duct. Reheat control maintains the desired dry-bulb temperature within a space by replacing any decrease in sensible loads by an artificial heat load. Bypass control maintains the desired dry-bulb temperature within the space by modulating the amount of air to be cooled. Since the bypass all-air system is probably less costly for this building, it will be the first choice. A complete economic analysis would be necessary before this conclusion could be accepted as fully valid.

Related Calculations. Use this general method to make a preliminary choice of the 11 different types of air-conditioning systems for the 31 applications listed in Table 17. Where additional analytical data for comparison of systems are required, consult Carrier Air Conditioning Company's *Handbook of Air Conditioning System Design* or ASHRAE *Guide and Data Book.*

With the greater emphasis on the environmental aspects of air-conditioning systems, more attention is being paid to natural-gas cooling. There are three types of natural-gas cooling systems used today: (1) engine, (2) desiccant, and (3) absorption. In each type of system, natural gas is used to provide the energy required for the cooling.

In the engine-type natural-gas cooling system, gas fuels an internal-combustion engine which drives a conventional chiller. The advantage of engine-driven chillers

TABLE 17   Systems and Applications*

Applications	Systems†	Applications	Systems†
Single-purpose occupancies		Multipurpose occupancies	
Residential:		Office buildings	2e 2i 2j
Medium	1c	Hotels, dormitories	1e 1f 2j 2k
Large	1d 1e 2i	Motels	1e
Restaurants:		Apartment buildings	1f 2j 2k
Medium	1d 1h	Hospitals	1f 2h 2j
Large	1d 2f 2g 2h 2i	Schools and colleges	1f 2e 2f 2g 2h
Variety and specialty shops	1d	Museums	2h 2i
Bowling alleys	1d 2f	Libraries:	
Radio and TV studios:		Standard	2h 2f 2h 2i
Small	1d 2f 2h 2i	Rare books	2h
Large	1d 2f 2h 2i	Department stores	1d 2f
Country clubs	1d 2f 2h 2i	Shopping centers	1d 2f 2i
Funeral homes	1d 2i	Laboratories:	
Beauty salons	1c 1d	Small	1d 2e 2h 2i
Barber shops	1c 1d	Large building	2e 2g 2j
Churches	1d 2f 2i	Marine	2g 2j
Theaters	2f		
Auditoriums	2f		
Dance and roller skating			
pavillions	1d 2e 2f		
Factories (comfort)	1d 2f 2h		

*Carrier Air Conditioning—*Handbook of Air Conditioning Systems Design*, McGraw-Hill.
†The systems in the table are:

1. Individual room or zone unit systems
   a. DX self-contained
   b. All-water
   c. Room DX self-contained 0.5 to 2 tons (1.76 to 7.0 kW)
   d. Zone DX self-contained 2 tons and over (7.0 kW and over)
   e. All-water room fan-coil recirculating air
   f. All-water room fan-coil with outdoor air
2. Central station apparatus systems
   a. All-air
   b. Air-water
   c. All-air, single airstream
   d. Air-water, primary air systems
   e. All-air, single airstream, variable volume
   f. All-air, single airstream, bypass
   g. All-air, single airstream, reheat at terminal
   h. All-air, single airstream, reheat zone in duct
   i. All-air, single airstream, multizone single duct
   j. Air-water primary air systems, secondary water H-V H-P induction
   k. Air-water primary air systems, room fan-coil with outside air

    Systems listed for a particular application are the systems most commonly used. Economics and design objectives dictate the choice and deviations of systems listed above, other systems as listed in note 2, and some entirely new systems.

    Several systems are used in many of these applications when higher-quality air conditioning is desired (often at higher expense). They are dual-duct, dual-conduit, three-pipe induction and fan-coil, four-pipe induction and fan-coil, and panel-air.

is that their fuel consumption of natural gas is low. Further, such engines are clean-burning with minimum atmospheric pollution and easily comply with the Clean Air Act requirements when properly maintained. Such natural-gas fueled engine-driven chiller systems are popular in large office buildings, factories, and similar installations. As a further economy and environmental plus, engine jacket water can be used to heat domestic hot water for the building being cooled. In one installation the engine jacket water is also used to heat water in a swimming pool. This combines natural-gas cooling with cogeneration.

Desiccant systems may use one or more heat wheels to capture heat from the building's exhaust air stream. Where more than one heat wheel is used, the first wheel may be a total-energy recovery unit with a desiccant section, cooling coil, and electric reheat coil. The second wheel may be a polymer-coated sensible-heat-only wheel with a conventional chilled-water or direct-expansion refrigerant coil to cool and dehumidify or heat and humidify the outdoor air, depending on ambient conditions.

With the enormous emphasis on indoor quality (IAQ) today, desiccant cooling and preconditioning of air is gaining importance. Thus, in a school system originally designed for 5 ft^3/min (0.14 m^3/min) per student, and later redesigned for 15 ft^3/min (0.42 m/min) per student, desiccant preconditioning maintains the indoor relative humidity at 50 to 52 percent in conjunction with packaged HVAC units in a very energy-efficient manner.* Further, the indoor air quality produced removes the risk of mold and mildew resulting from a lack of humidity control in the spaces being served.

Absorption natural-gas cooling can use direct-fired chillers to provide the needed temperature reduction. A variety of direct-fired chillers are available from various manufacturers to serve the needs of such installations. Such chillers can range in capacity from 30 tons (105.5 kW) to more than 600 tons (2.1 MW). Using natural gas as the heat source provides low-cost cooling and heating for a variety of buildings: office, factory, commercial, etc. Natural gas is a low-pollution fuel and absorption cooling installations using it usually find it easy to comply with the provisions of the Clean Air Act.

SIZING LOW-VELOCITY AIR-CONDITIONING-SYSTEMS DUCTS—EQUAL-FRICTION METHOD

An industrial air-conditioning system requires 36,000 ft^3/min (1019.2 m^3/min) of air. This low-velocity system will be fitted with enough air outlets to distribute the air uniformly throughout the conditioned space. The required operating pressure for each duct outlet is 0.20 in (5.1 mm) wg. Determine the duct sizes required for this system by using the equal-friction method of design. What is the required fan static discharge pressure?

Calculation Procedure:

1. Sketch the duct system
The required air quantity, 36,000 ft^3/min (1019.2 m^3/min), must be distributed in approximately equal quantities to the various areas in the building. Sketch the pro-

---

*Smith, James C., "Schools Resolve IAQ/Humidity Problems with Desiccant Preconditioning," *Heating/Piping/Air Conditioning*, April, 1996.

posed duct layout as shown in Fig. 31. Locate air outlets as shown to provide air to each area in the building.

Determine the required capacity of each air outlet from air quantity required, ft³/min per number of outlets, or outlet capacity = 36,000/18 = 2000 ft³/min (56.6 m³/min) per outlet. This is within the usual range of many commercially available outlets. Where the required capacity per outlet is extremely large, say 10,000 ft³/min (283.1 m³/min), or extremely small, say 5 ft³/min (0.14 m³/min), change the number of outlets shown on the duct sketch to obtain an air quantity within the usual capacity range of commercially available outlets. Relocate each outlet so it serves approximately the same amount of floor area as each of the other outlets in the system. Thus, the duct sketch serves as a trial-and-error analysis of the outlet location and capacity.

Where a building area requires a specific amount of air, select one or more outlets to supply this air. Size the remaining outlets by the method described above, after subtracting the quantity of air supplied through the outlets already chosen.

2. Determine the required outlet operating pressure

Consult the manufacturer's engineering data for the required operating pressure of each outlet. Where possible, try to use the same type of outlets throughout the system. This will reduce the initial investment. Assume that the required outlet operating pressure is 0.20-in (5.1-mm) wg for each outlet in this system.

3. Choose the air velocity for the main duct

Use Table 18 to determine a suitable air velocity for the main duct of this system. Table 18 shows that an air velocity up to 2500 ft/min (762 m/min) can be used for main ducts where noise is the controlling factor; 3000 ft/min (914.4 m/min)

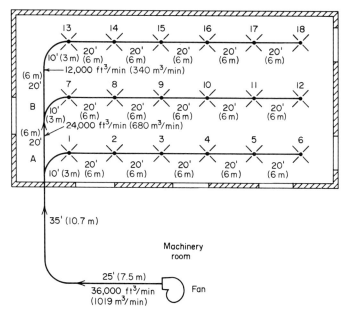

FIGURE 31 Duct-system layout.

TABLE 18 Recommended Maximum Duct Velocities for Low-Velocity Systems, ft/min (m/min)*

| Application | Controlling factor— noise generation, main ducts | Controlling factor—duct friction | | | |
| | | Main ducts | | Branch ducts | |
		Supply	Return	Supply	Return
Residences	600 (183)	1000 (300)	800 (244)	600 (183)	600 (183)
Apartments, hotel bedrooms, hospital bedrooms	1000 (300)	1500 (457)	1300 (396)	1200 (366)	1000 (300)
Private offices, directors rooms, libraries	1200 (366)	2000 (610)	1500 (457)	1600 (488)	1200 (366)
Theaters, auditoriums	800 (244)	1300 (396)	1100 (335)	1000 (300)	800 (244)
General offices, high-class restaurants, high-class stores, banks	1500 (457)	2000 (610)	1500 (457)	1600 (488)	1200 (366)
Average stores, cafeterias	1800 (549)	2000 (610)	1500 (459)	1600 (488)	1200 (366)
Industrial	2500 (762)	3000 (914)	1800 (549)	2200 (671)	1500 (457)

*Carrier Air Conditioning Company.

where duct friction is the controlling factor. A velocity of 2500 ft/min (762 m/min) will be used for the main duct in this installation.

4. Determine the dimensions of the main duct
The required duct area A ft^2 = (ft^3/min)/(ft/min) = 36,000/2500 = 14.4 ft^2 (1.34 m^2). A nearly square duct, i.e., a duct 46 × 45 in (117 × 114 cm), has an area of 14.38 ft^2 (1.34 m^2) and is a good first choice for this system because it closely approximates the outlet size of a standard centrifugal fan. Where possible, use a square main duct to simplify fan connections. Thus a 46 × 46 in (117 × 117 cm) duct might be the final choice for this system.

5. Determine the main-duct friction loss
Convert the duct area to the equivalent diameter in inches d, using $d = 2(144A/\pi)^{0.5} = 2(144 \times 144/\pi)^{0.5} = 51.5$ in (130.8 cm).
   Enter Fig. 32 at 36,000 ft^3/min (1019.2 m^3/min) and project horizontally to a round-duct diameter of 51.5 in (130.8 cm). At the top of Fig. 32 read the friction loss as 0.13 in (3.3 mm) wg per 100 ft (30 m) of equivalent duct length.

6. Size the branch ducts
For many common air-conditioning systems the equal-friction method is used to size the ducts. In this method the supply, exhaust, and return-air ducts are sized so they have the same friction loss per foot of length for the entire system. The equal-friction method is superior to the velocity-reduction method of duct sizing because the former requires the less balancing for symmetrical layouts.
   The usual procedure in the equal-friction method is to select an initial air velocity in the main duct near the fan, using the sound level as the limiting factor. With this initial velocity and the design air flow rate, the required duct diameter is found, as in steps 4 and 5, above. Once the duct diameter is known, the friction loss is found from Fig. 32, as in step 5. This same friction loss is then maintained

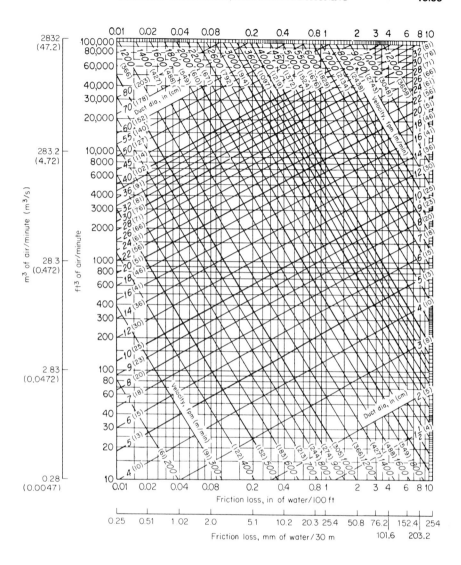

Friction loss for usual air conditions. This chart applies to smooth round galvanized iron ducts. See table below for corrections to apply when using other pipe.

Type of pipe	Degree of roughness	Velocity		Roughness factor (use as multiplier)
		ft/min	m/min	
Concrete	Medium rough	1000–2000	300–610	1.4
Riveted steel	Very rough	1000–2000	300–610	1.9
Tubing	Very smooth	1000–2000	300–610	0.9

FIGURE 32 Friction loss in round ducts.

throughout the system, and the equivalent round-duct diameter is chosen from Fig. 32.

To expedite equal-friction calculations, Table 19 is often used instead of the friction chart. (It is valid for SI units also.) This provides the same duct sizes. Duct areas are determined from Table 19, and the area found is converted to a round-, rectangular-, or square-duct size suitable for the installation. This procedure of duct sizing automatically reduces the air velocity in the direction of air flow. Hence, the equal-friction method will be used for this system.

TABLE 19 Percentage of Section Area in Branches for Maintaining Equal Friction*

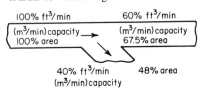

ft³/min (m³/min) capacity, %	Duct area,† %	ft³/min (m³/min) capacity, %	Duct area, %	ft³/min (m³/min) capacity, %	Duct area, %	ft³/min (m³/min) capacity, %	Duct area, %
1	2.0	26	33.5	51	59.0	76	81.0
2	3.5	27	34.5	52	60.0	77	82.0
3	5.5	28	35.5	53	61.0	78	83.0
4	7.0	29	36.5	54	62.0	79	84.0
5	9.0	30	37.5	55	63.0	80	84.5
6	10.5	31	39.0	56	64.0	81	85.5
7	11.5	32	40.0	57	65.0	82	86.0
8	13.0	33	41.0	58	65.5	83	87.0
9	14.5	34	42.0	59	66.5	84	87.5
10	16.5	35	43.0	60	67.5	85	88.5
11	17.5	36	44.0	61	68.0	86	89.5
12	18.5	37	45.0	62	69.0	87	90.0
13	19.5	38	46.0	63	70.0	88	90.5
14	20.5	39	47.0	64	71.0	89	91.5
15	21.5	40	48.0	65	71.5	90	92.0
16	23.0	41	49.0	66	72.5	91	93.0
17	24.0	42	50.0	67	73.5	92	94.0
18	25.0	43	51.0	68	74.5	93	94.5
19	26.0	44	52.0	69	75.5	94	95.0
20	27.0	45	53.0	70	76.5	95	96.0
21	28.0	46	54.0	71	77.0	96	96.5
22	29.5	47	55.0	72	78.0	97	97.5
23	30.5	48	56.0	73	79.0	98	98.0
24	31.5	59	57.0	74	80.0	99	99.0
25	32.5	50	58.0	75	80.5	100	100.0

*Carrier Air Conditioning Company.
†The same duct area percentage applies when flow is measured in m³/min or m³/s.

Compute the duct areas, using Table 19. Tabulate the results, using the duct run having the highest resistance. The friction loss through all elbows and fittings in the section must be included. The total friction loss in the duct having the highest resistance is the loss the fan must overcome.

Inspection of the duct layout (Fig. 31) shows that the duct run from the fan to outlet 18 probably has the highest resistance because it is the longest run. Tabulate the results as shown.

(1)	(2)	(3)	(4)	(5)	(6)
	Air quantity,	ft³/min (m³/min) capacity,	Duct,	Area, ft²	
Duct section	ft³/min (m³/ min)	percent	percent	(m²)	Duct size, in (cm)
Fan to A	36,000 (1,019.2)	100	100	14.4 (1.34)	46 × 45 (117 × 114)
A–B	24,000 (679.4)	67	73.5	10.6 (0.98)	39 × 39 (99 × 99)
B–13	12,000 (339.7)	33	41.0	5.9 (0.55)	30 × 29 (76 × 74)
13–14	10,000 (283.1)	28	35.5	5.1 (0.47)	27 × 27 (69 × 69)
14–15	8,000 (226.5)	22	29.5	4.3 (0.40)	25 × 25 (64 × 64)
15–16	6,000 (169.9)	17	24.0	3.5 (0.33)	23 × 22 (58 × 56)
16–17	4,000 (113.2)	11	17.5	2.5 (0.23)	20 × 18 (51 × 46)
17–18	2,000 (56.6)	6	10.5	1.5 (0.14)	15 × 15 (38 × 38)

The values in this tabulation are found as follows. Column 1 lists the longest duct run in the system. In column 2, the air leaving the outlets in branch A, or (6 outlets) (2000 ft³/min per outlet) = 12,000 ft³/min (339.7 m³/min), is subtracted from the quantity of air, 36,000 ft³/min (1019.2 m³/min), discharged by the fan to give the air quantity flowing from A-B. A similar procedure is followed for each successive duct and air quantity.

Column 3 is found by dividing the air quantity in each branch listed in columns 1 and 2 by 36,000, the total air flow, and multiplying the result by 100. Thus, for run B-13, column 3 = 12,000 (100)/36,000 = 33 percent.

Column 4 values are found from Table 19. Enter that table with the ft³/min capacity from column 3 and read the duct area, percent. Thus, for branch 13-14 with 28 percent ft³/min capacity, the duct area from Table 19 is 35.5 percent. Determine the duct area, column 6, by taking the product, line by line, of column 4 and the main duct area. Thus, for branch 13-14, duct area = (0.355)(14.4) = 5.1 ft² (0.47 m²). Convert the duct area to a nearly square, or a square, duct by finding two dimensions that will produce the desired area.

Duct sections A through 6 and B through 12 have the same dimensions as the corresponding duct sections B through 18.

7. Find the total duct friction loss

Examination of the duct sketch (Fig. 31) indicates that the duct run from the fan to outlet 18 has the highest resistance. Compute the total duct run length and the equivalent length of the two elbows in the run thus as shown.

(1)	(2)	(3)		(4) Elbow equivalent length	
		Length			
Duct section	System part	ft	m	ft	m
Fan to A	Duct	60	18.3		
	Elbow	. . .		30	9.1
A–B	Duct	20	6.1		
B–13	Duct	30	9.1		
	Elbow	. . .		15	4.6
13–14	Duct	20	6.1		
14–15	Duct	20	6.1		
15–16	Duct	20	6.1		
16–17	Duct	20	6.1		
17–18	Duct	20	6.1		
Total		210	64.0	45	13.7

Note several factors about this calculation. The duct length, column 3, are determined from the system sketch, Fig. 31. The equivalent length of the duct elbows, column 4, is determined from the *Guide* or Carrier *Handbook of Air Conditioning Design*. The total equivalent duct length = column 3 + column 4 = 210 + 45 = 255 ft (77.7 m).

8. Compute the duct friction loss

Use the general relation $h_T = Lf$, where h_T = total friction loss induct, in wg; L = total equivalent duct length, ft; f = friction loss for the system, in wg per 100 ft (30 m). With the friction loss of 0.13 in (3.3 mm) wg per 100 ft (30 m), as determined in step 5, $h_r = (229/100)(0.13) = 0.2977$ in (7.6 mm) wg; say 0.30 in (7.6 mm) wg.

9. Determine the required fan static discharge pressure

The total static pressure required at the fan discharge = outlet operating pressure + duct loss − velocity regain between first and last sections of the duct, all expressed in wg. The first two variables in this relation are already known. Hence, only the velocity regain need be computed.

The velocity v ft/min of air in any duct is v = ft³/min/duct area. For duct section A, $v = 36,000/14.4 = 2500$ ft/min (762 m/min); for the last duct section, 17-18, $v = 2000/1.5 = 1333$ ft/min (406.3 m/min).

When the fan discharge velocity is higher than the duct velocity in an air-conditioning system, use this relation to compute the static pressure regain $R = 0.75[(v_f/4000)^2 - (v_d/4000)^2]$, where R = regain, in wg; v_f = fan outlet velocity, ft/min; v_d = duct velocity, ft/min. Thus, for this system, $R = 0.75[(2500/4000)^2 - (1333/4000)^2] = 0.21$ in (5.3 mm) wg.

With the regain known, compute the total static pressure required as 0.20 + 0.30 − 0.21 = 0.29 in (7.4 mm) wg. A fan having a static discharge pressure of at least 0.30 in (7.6 mm) wg would probably be chosen for this system.

If the fan outlet velocity exceeded the air velocity in duct section A, the air velocity in this section would be used instead of the air velocity in the last duct section. Thus, in this circumstance, the last section becomes the duct connected to the fan outlet.

Figure 33 shows details of duct hangers for ducts of various dimensions. Shown in Fig. 34 are details of rectangular duct takeoffs for air supply to specific rooms or areas.

Related Calculations. Where the velocity in the fan outlet duct is *higher* than the fan outlet velocity, use the relation $l = 1.1 [(v_d/4000)^2 - (v_f/4000)^2]$, where l = loss, in wg. This loss is the additional static pressure required of the fan. Hence, this loss must be *added* to the outlet operating pressure and the duct loss to determine the total static pressure required at the fan discharge.

The equal-friction method does not satisfy the design criteria of uniform static pressure at all branches and air terminals. To obtain the proper air quantity at the beginning of each branch it is necessary to include a splitter damper to regulate the flow to the branch. It may also be necessary to have a control device (vanes, volume damper, or adjustable-terminal volume control) to regulate the flow at each terminal for proper air distribution.

The *velocity-reduction method* of duct design is not too popular because it requires a broad background of duct-design experience and knowledge to be within

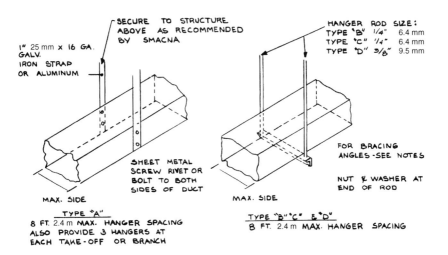

DUCT SCHEDULE			
DUCT DIMENSIONS			TYPE
INCHES		cm	HANGER
UP THRU 12		31	A
13	18	33 46	A
19	30	48 76	A/B
31	42	79 107	B
43	54	109 137	B
55	60	140 152	B
61	84	155 213	C
85	96	216 244	C
OVER	96	Over 244	D

NOTES:

1. FOR SEVERAL DUCTS ON ONE HANGER TYPE "B"-"C" OR "D" MAY BE USED. SIZE OF HANGER WILL BE SELECTED ON THE SUM OF DUCT WIDTHS EQUAL TO MAX. WIDTH OF DUCT SCHEDULE.

2. SCHEDULE FOR ANGLES FOR BRACING: TYPE "B" 1½" X 1½" x ⅛" 38 x 38 x 3 mm ANGLE, MAX. SPACING 8'-0" CENTERS; TYPE "C" 1½" x 1½" x 3/16" 38 x 38 x 5 mm ANGLE MAX SPACING 8'-0" CENTERS; TYPE "D" 2" X 2"x ¼" MAX SPACING 51 x 51 x 6 mm 4'-0" 1.2 m CENTERS.

FIGURE 33 Duct hangers for ducts of various dimensions. (*Jerome F. Mueller.*)

RECTANGULAR DUCT TAKE-OFF. SEE FIGURE A,B,C,D FOR ORDER OF PREFERENCE. PROVIDE A SPLITTER DAMPER FOR EACH TAKE OFF.

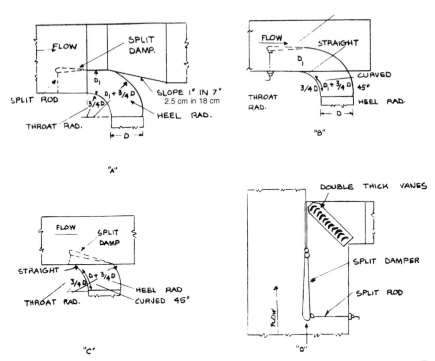

FIGURE 34 Rectangular-duct takeoffs for air supply to specific rooms or areas. (*Jerome F. Mueller.*)

reasonable accuracy. It should be used for only the simplest layouts. Splitters and dampers should be included for balancing purposes.

To apply the velocity-reduction method: (1) Select a starting velocity at the fan discharge. (2) Make arbitrary reductions in velocity down the duct run. The starting velocity should not exceed the values in Table 18. Obtain the equivalent round-duct diameter from Fig. 32. Compute the required duct area from the round-duct diameter, and from this the duct dimensions, as shown in steps 4 and 5 above. (3) Determine the required fan static discharge pressure for the supply by using the longest run of duct, including all elbows and fittings. Note, however, that the longest run is not necessarily the run with the greatest friction loss, as shorter runs may have more elbows, fittings, and restrictions.

The equal-friction and velocity-reduction methods of air-conditioning system duct design are applicable only to low-velocity systems, i.e., systems in which the maximum air velocity is 3000 ft/min (914.4 m/min), or less. The methods presented in this calculation procedure are those used by the Carrier Air Conditioning Company at the time of this writing.

SIZING LOW-VELOCITY AIR-CONDITIONING DUCTS—STATIC-REGAIN METHOD

Using the same data as in the previous calculation procedure, an air velocity of 2500 ft/min (762.0 m/min) in the main duct section, an unvaned elbow radius of $R/D = 1.25$, and an operating pressure of 0.20 in (2.5 mm) wg for each outlet, size the system ducts, using the static-regain method of design for low-velocity systems.

Calculation Procedure:

1. Compute the fan outlet duct size

The fan outlet duct, also called the main duct section, will have an air velocity of 2500 ft/min (762.0 m/min). Hence, the required duct area is $A = 36,000/2500 = 14.4$ ft^2 (1.34 m^2). This corresponds to a round-duct diameter of $d = 2(144A/\pi)^{0.5} = 2(144 \times 14.4/\pi)^{0.5} = 51.5$ in (130.8 cm). A nearly square duct, i.e., a duct 46 $\times$ 45 in (116.8 $\times$ 114.3 cm), has an area of 14.38 ft^2 (1.34 m^2) and is a good first choice for this system because it closely approximates the outlet size of a standard centrifugal fan.

Where possible, use a square main duct to simplify fan connections. Thus, a 46 $\times$ 46 in (116.8 $\times$ 116.8 cm) duct might be the final choice of this system.

2. Compute the main-duct friction loss

Using Fig. 32 find the main-duct friction loss as 0.13 in (3.3 mm) wg per 100 ft (30 m) of equivalent duct length for a flow of 36,000 ft^3/min (1019.2 m^3/min) and a diameter of 51.5 in (130.8 cm).

3. Determine the friction loss up to the first branch duct

The length of the main duct between the fan and the first branch is $25 + 35 = 60$ ft (18.3 m). The equivalent length of the elbow is, from the *Guide* or Carrier *Handbook of Air Conditioning Design,* 26 ft (7.9 m). Hence, the total equivalent length $= 60 + 30 = 90$ ft (27.4 m). The friction loss is then $h_T = L_f = (90/100)(0.13) = 0.117$ in (2.97 mm) wg.

4. Size the longest duct run

The longest duct run is from A to outlet 18 (Fig. 31). Size the duct using the following tabulation, preparing it as described on the next page.

List in column 1 the various duct sections in the longest duct run, as shown in Fig. 31. In column 2 list the air quantity flowing through each duct section. Tabulate in column 3 the equivalent length of each duct. Where a fitting is in the duct section, as in B-13, assume a duct size and compute the equivalent length using the *Guide* or Carrier fitting table. When the duct section does not have a fitting, as with section 13-14, the equivalent length equals the distance between the centerlines of two adjacent outlets.

Next, determine the L/Q ratio for each duct section, using Fig. 35. Enter Fig. 35 at the air quantity in the duct and project vertically upward to the curve representing the equivalent length of the duct. At the left read the L/Q ratio for this

(1) Section number	(2) Air flow, ft³/min (m³/min)	(3) Equivalent length, ft (m)	(4) L/Q ratio	(5) Velocity, ft/ min (m/min)	(6) Duct area, ft² (m²) (2)/(5)	(7) Duct size, in (cm)
Fan to A	36,000 (1,019.2)	86 (26.2)	. . .	2,500 (762.0)	14.4 (1.34)	46 × 45 (116.8 × 114.3)
A–B	24,000 (679.4)	20 (6.1)	0.034	2,410 (734.6)	9.95 (0.92)	38 × 38 (96.5 × 96.5)
B–13	12,000 (339.7)	26° (7.9)°	0.088	2,200 (670.6)	5.45 (0.51)	28 × 28 (71.1 × 71.1)
13–14	10,000 (283.1)	20 (6.1)	0.072	2,040 (621.8)	4.90 (0.46)	27 × 27 (68.6 × 68.6)
14–15	8,000 (226.5)	20 (6.1)	0.083	1,850 (563.9)	4.33 (0.40)	25 × 25 (63.5 × 63.5)
15–16	6,000 (169.9)	20 (6.1)	0.098	1,700 (518.2)	3.53 (0.33)	24 × 23 (60.9 × 58.4)
16–17	4,000 (113.2)	20 (6.1)	0.130	1,520 (463.3)	2.63 (0.24)	20 × 19 (50.8 × 48.3)
17–18	2,000 (56.6)	20 (6.1)	0.195	1,300 (396.2)	1.54 (0.14)	15 × 15 (38.1 × 38.1)
B–7	12,000 (339.7)	25° (7.6)°	. . .	. . .	. . .	28 × 28 (71.1 × 71.1)
7–8	10,000 (283.1)	20 (6.1)	. . .	. . .	. . .	27 × 27 (68.6 × 68.6)
8–9	8,000 (226.5)	20 (6.1)	. . .	. . .	. . .	25 × 25 (63.5 × 63.5)
9–10	6,000 (169.9)	20 (6.1)	. . .	. . .	. . .	25 × 23 (60.9 × 58.4)
10–11	4,000 (113.2)	20 (6.1)	. . .	. . .	. . .	20 × 19 (50.8 × 48.3)
11–12	2,000 (56.1)	20 (6.1)	. . .	. . .	. . .	15 × 15 (38.1 × 38.1)
A–1	12,000 (339.7)	25° (7.6)	. . .	. . .	. . .	28 × 28 (71.1 × 71.1)
1–2	10,000 (283.1)	20 (6.1)	. . .	. . .	. . .	27 × 27 (68.6 × 68.6)
2–3	8,000 (226.5)	20 (6.1)	. . .	. . .	. . .	25 × 25 (63.5 × 64.5)
3–4	6,000 (169.9)	20 (6.1)	. . .	. . .	. . .	24 × 23 (60.9 × 58.4)
4–5	4,000 (113.2)	20 (6.1)	. . .	. . .	. . .	20 × 19 (50.8 × 48.3)
5–6	2,000 (56.1)	20 (6.1)	. . .	. . .	. . .	15 × 15 (38.1 × 38.1)

°See text.

section of the duct. Thus, for duct section 13-14, Q = 10,000 ft³/min (283.1 m³/min), and L = 20 ft (6.1 m). Entering the chart as detailed above shows that L/Q = 0.72. Proceed in this manner, determining the L/Q ratio for each section of the duct in the longest duct run.

Determine the velocity of the air in the duct by using Fig. 36. Enter Fig. 36 at the L/Q ratio for the duct section, say 0.072 for section 13-14. Find the intersection of the L/Q curve with the velocity curve for the preceding duct section: 2200 ft/min (670.6 m/min) for section 13-14. At the bottom of Fig. 36 read the velocity in the duct section, i.e., after the previous outlet and in the duct section under consideration. Enter this velocity in column 5. Proceed in this manner, determining the velocity in each section of the duct in the longest duct run.

Determine the required duct area from column 2/column 5, and insert the result in column 6. Find the duct size, column 7, by converting the required duct area to a square- or rectangular-duct dimension. Thus a 27 × 17 in (68.6 × 43.2 cm) square duct has a cross-sectional area slightly greater than 4.90 ft² (0.46 m²).

5. Determine the sizes of the other ducts in the system
Since the ducts in runs A and B are symmetric with the duct containing the outlets in the longest run, they can be given the same size when the same quantity of air flows through them. Thus, duct section 7-8 is sized the same as section 13-14 because the same quantity of air, 10,000 ft³/min (4.72 m³/s), is flowing through both sections.

Where the duct section contains a fitting, as B-7 and A-1, assume a duct size and find the equivalent length, using the *Guide* or Carrier fitting table. These sections are marked with an asterisk.

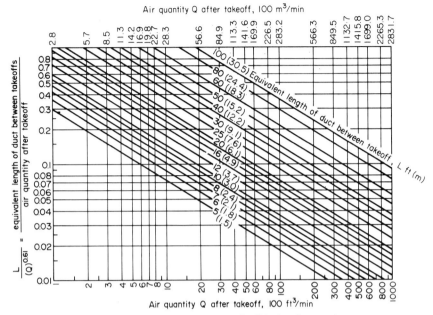

FIGURE 35 *L/Q* ratio for air ducts. (*Carrier Air Conditioning Company.*)

6. *Determine the required fan discharge pressure*

The total pressure required at the fan discharge equals the sum of the friction loss in the main duct plus the terminal operating pressure. Hence the required fan static discharge pressure = 0.117 + 0.20 = 0.317 in (8.1 mm) wg.

Related Calculations.   The basic principle of the static-regain method is to size a duct run so that the increase in static pressure (regain due to the reduction in velocity) at each branch or air terminal just offsets the friction loss in the succeeding section of duct. The static pressure is then the same before each terminal and at each branch.

As a *general* guide to the results obtained with the static-regain and equal-friction duct-design methods, the following should be helpful:

	Static regain	Equal friction
Main-duct sizes	Same	Same
Branch-duct sizes	Larger	Smaller
Sheet-metal weight	Greater	Less
Fan horsepower	Less	Greater
Balancing time	Less	Greater
Operating costs	Less	Greater

Note that these tabulated results are *general* and may not apply to every system. The method presented in this calculation procedure is that used by the Carrier Air Conditioning Company at the time of this writing.

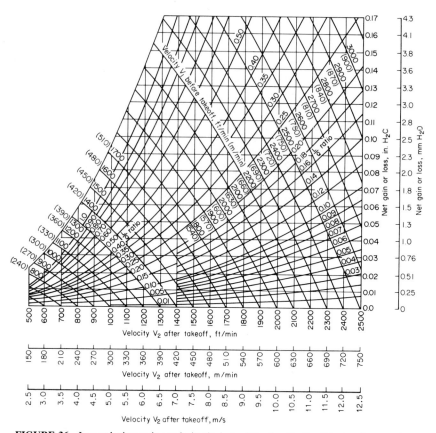

FIGURE 36　Low-velocity static-regain in air ducts. (*Carrier Air Conditioning Company.*)

HUMIDIFIER SELECTION FOR DESIRED ATMOSPHERIC CONDITIONS

A paper mill has a storeroom with a volume of 500,000 ft³ (14,155 m³). The lowest recorded outdoor temperature in the mill locality is 0°F (−17.8°C). What capacity humidifier is required for this storeroom if a 70°F (21°C) dry-bulb temperature and a 65 percent relative humidity are required in it? Moisture absorption by the paper products in the room is estimated to be 450 lb/h (204.6 kg/h). The storeroom ventilating system produces three air changes per hour. What capacity humidifier is required if the room temperature is maintained at 60°F (15.5°C) and 65 percent relative humidity? The products release 400 lb/h (181.8 kg/h) of moisture. Steam at 25 lb/in² (gage) (172.4 kPa) is available for humidification. The outdoor air has a relative humidity of 50 percent and a minimum temperature of 5°F (−15°C).

Calculation Procedure:

1. *Determine the outdoor design temperature*
In choosing a humidifier, the usual procedure is to add 10°F (5.6°C) to the minimum outdoor recorded temperature because this temperature level seldom lasts more than a few hours. The result is the design outdoor temperature. Thus, for this mill, design outdoor temperature = 0 + 10 = 10°F (−12°C).

2. *Compute the weight of moisture required for humidification*
Enter Table 20 at an outdoor temperature of 10°F (−12.2°C), and project across to the desired relative humidity, 65 percent. Read the quantity of steam required as 1.330 lb/h (0.6 kg/h) per 1000 ft³ (28.3 m³) of room volume for two air changes per hour. Since this room has three air changes per hour, the quantity of moisture required is (3/2)(1330) = 1.995 lb/h (0.91 kg/h) per 1000 ft³ (28.3 m³) of volume.

The amount of moisture in the form of steam required for this storeroom = (room volume, ft³/1000)(lb/h of steam per 1000 ft³) = (500,000/1000)(1.995) = 997.5 lb (453.4 kg) for humidification of the air. However, the products in the storeroom absorb 450 lb/h (202.5 kg/h) of moisture. Hence, the total moisture quantity required = moisture for air humidification + moisture absorbed by products = 997.5 + 450.0 = 1447.5 lb/h (657.9 kg/h), say 1450 lb/h (659.1 kg/h) for humidifier sizing purposes.

3. *Select a suitable humidifier*
Table 21 lists typical capacities for humidifiers having orifices of various sizes and different steam pressures. Study of Table 21 shows that one 1¼-in (32-mm) orifice humidifier and two ⅜-in (9.5-cm) orifice humidifiers will discharge 1130 + (2)(174) = 1478 lb/h (671.8 kg/h) of steam when the steam supply pressure is 25 lb/in² (gage) (172.4 kPa). Since the required capacity is 1450 lb/h (659.1 kg/h), these humidifiers may be acceptable.

Large-capacity steam humidifiers usually must depend on existing ducts or large floor-type unit heaters for distribution of the moisture. When such means of distri-

TABLE 20 Steam Required for Humidification at 70°F (21°C)*†

Outdoor temp		Relative humidity desired indoors, percent									
		40		50		60		65		70	
°F	°C	lb	kg	lb	kg	lb	kg	lb	kg	lb	kg
50	10.0	0.045	0.02	0.271	0.12	0.501	0.23	0.616	0.28	0.731	0.33
40	4.4	0.307	0.14	0.537	0.24	0.767	0.35	0.882	0.40	1.000	0.45
30	1.1	0.503	0.23	0.734	0.33	0.964	0.43	1.079	0.49	1.194	0.54
20	−6.7	0.654	0.29	0.883	0.40	1.115	0.50	1.230	0.55	1.345	0.61
10	−12.2	0.754	0.34	0.985	0.44	1.215	0.55	1.330	0.60	1.445	0.65
0	−17.8	0.819	0.37	1.049	0.47	1.279	0.58	1.394	0.63	1.509	0.68
−10	−23.3	0.860	0.39	1.090	0.49	1.320	0.59	1.435	0.65	1.550	0.70
−20	−28.9	0.885	0.30	1.115	0.50	1.345	0.61	1.460	0.66	1.575	0.71

* Armstrong Machine Works.
† Pounds (kilograms) of steam per hour required per 1000 ft³ (28.3 m³) of space to secure desired indoor relative humidity at 70°F (21°C), with various outdoor temperatures. Assuming two air changes per hour and outdoor relative humidity of 75 percent.

TABLE 21 Humidifier Capacities*

Steam pressure, lb/in² (gage) (kPa)	Orifice size, in (mm)			
	⁷⁄₁₆ (1.11)	⅜ (9.5)	1¼ (31.8)	1¾₄ (28.2)
5 (34.5)	100	. . .	340	42
		. . .		
		. . .		
		. . .		
10 (68.9)	140	. . .	610	
		. . .		
15 (103.4)	170	. . .	810	74
		138		
20 (137.9)	. . .	158	980	80
25 (172.4)	. . .	174	1130	90
30 (206.8)	. . .	190	1280	100

*Armstrong Machine Works.
†Continuous discharge capacity with steam pressures as indicated. No allowance for pressure drop after solenoid valve opens.

bution are not available, choose a larger number of smaller-capacity humidifiers and arrange them as shown in Fig. 37c. Thus, if ⅜-in (9.5-mm) orifice humidifiers were selected, the number required would be (moisture needed, lb/h)/(humidifier capacity, lb/h) = 1450/174 = 8.33, or 9 humidifiers.

4. Choose a humidifier for the other operating conditions
Where the desired room temperature is different from 70°F (231°C), use Table 20 instead of Table 21. Enter Table 22 at the desired room temperature, 60°F (15.6°C), and read the moisture content of saturated air at this temperature, as 5.795 gr/ft³ (13.26 mL/dm³). The outdoor air at 5 + 10 = 15°F (9.4°C) contains, as Table 22 shows, 0.984 gr/ft³ (2.25 mL/dm³) of moisture when fully saturated.

Find the moisture content of the air at the room and the outdoor conditions from moisture content, gr/ft³ = (relative humidity of the air, expressed as a decimal) (moisture content of saturate air, gr/ft³). For the 60°F (15.6°C), 65 percent relative humidity room air, moisture content = (0.65)(5.795) = 3.77 gr/ft³ (8.63 mL/dm³). For the 15°F (−9.4°C) 50 percent relative humidity outdoor air, moisture content = (0.50)(0.984) = 0.492 gr/ft³ (1.13 mL/dm³). Thus, the humidifier must add the difference or 3.77 = 0.492 = 3.278 gr/ft³ (7.5 mL/dm³).

This storeroom has a volume of 500,000 ft³ (14,155 m³) and three air changes per hour. Thus, the weight of moisture that must be added per hour is (number of air changes per hour)(volume, ft³)(gr/ft³ of air)/7000 gr/lb or, for this storeroom, (3)(500,000)(3.278)/7000 = 701 lb/h (0.09 kg/s) excluding the product load. Since the product load is 400 lb/h (0.05 kg/s), the total humdification load is 701 + 400 = 1101 lb/h (0.14 kg/s). Choose the humidifiers for these conditions in the same way as described in step 4.

Related Calculations. Use the method given here to choose a humidifier for any normal industrial or comfort application. Table 23 summarizes typical recommended humidities and temperatures for a variety of industrial operations. The

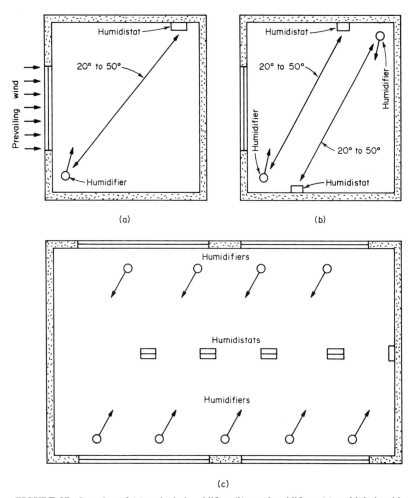

FIGURE 37   Location of (*a*) a single humidifier, (*b*) two humidifiers, (*c*) multiple humidifiers.

relative humidity maintained in industrial plants is extremely important because it can control the moisture content of hygroscopic materials.

Where the number of hourly air changes is not specified, assume two air changes, except in cotton mills where three or four may be necessary. If the plant ventilating system provides more than two air changes per hour, use the actual number of changes in computing the required humidifier capacity.

Many types of manufactured goods and raw materials absorb or release moisture during processing and storage. Since product quality usually depends directly on the moisture content, carefully controlled humidity will often reduce the number of rejects. The room humidifier must supply sufficient moisture for humidification of the air, plus any moisture absorbed by the products or materials in the room. Where these products or materials continuously release moisture to the atmosphere in the

TABLE 22   Moisture Content of Saturated Air

°F	°C	Grains of water Per ft³	Grains of water Per m³
15	−9.4	0.984	34.8
20	−6.7	1.242	43.9
40	4.4	2.863	101.1
50	10.0	4.106	145.0
60	15.6	5.795	204.7
70	21.1	8.055	284.5

TABLE 23   Recommended Industrial Humidities and Temperatures

Industry	Degrees °F	Degrees °C	Relative humidity, %
Ceramics:			
Drying refractory shapes	110–150	43–65	50–60
Molding room	80	26	60
Confectionery:			
Chocolate covering	62–65	17–18	50–55
Hard-candy making	70–80	21–27	30–50
Electrical:			
Manufacture of cotton-covered wire, storage, general	60–80	21–27	60–70
Food storage:			
Apple	31–34	−0.5–1.1	75–85
Citrus fruit	32	0	80
Grain	60	16	30–45
Meat ripening	40	4	80
Paper products:			
Binding	70	21	45
Folding	77	25	65
Printing	75	24	60–78
Storage	75–80	24–27	40–60
Textile:			
Cotton carding	75–80	24–27	50–55
Cotton spinning	60–80	16–27	50–70
Rayon spinning, throwing	70	21	85
Silk processing	75–80	24–27	60–70
Wool spinning, weaving	75–80	24–27	55–60
Miscellaneous:			
Laboratory, analytical	60–70	16–21	60–70
Munitions, fuse loading	70	21	55
Cigar and cigarette making	70–75	21–24	55–65

room, the quantity released can be subtracted from the moisture required for humidification. However, this condition can seldom be relied on. The usual procedure then is to select the humidifier on the basis of the moisture required for humidification of the air. The humidistat controls the operation of the humidifier, shutting it off when the products release enough moisture to supply the room requirements.

Correct locations for one or more humidifiers are shown in Fig. 32. Proper location of humidifiers is necessary if the design is to take advantage of the prevailing wind in the plant locality. Also, correct location provides a uniform, continuous circulation of air throughout the humidified area.

When only one humidifier is used, it is placed near the prevailing wind wall and arranged to discharge parallel to the wall exposed to the prevailing wind, Fig. 37a. Two humidifiers, Fig. 37b, are generally located in opposite corners of the manufacturing space and their discharges are used to produce a rotary air motion. Installations using more than two humidifiers generally have a slightly greater number of humidifiers on the windward wall to take advantage of the natural air drift form one side of the room to the other.

Pipe spray humidifiers are as shown in Fig. 38 unless the manufacturer advises otherwise. Size the return lines as shown in Table 24.

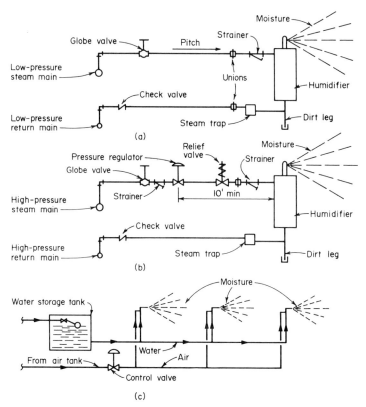

FIGURE 38 Piping for spray-type humidifiers. (*a*) Low-pressure steam; (*b*) high-pressure steam; (*c*) water spray.

TABLE 24   Steam- and Return-Pipe Sizes, in (mm)

Steam or condensate flow, lb/h (kg/h)	Steam pressure, lb/in² (gage) (kPa)				Length of return pipe, ft (m)	
	5 (34.5)	10 (68.9)	50 (344.7)	100 (689.4)	100 (30)	200 (60)
100 (45)	1½ (38.1)	1¼ (32)	1 (25)	1 (25)	1 (25)	1 (25)
200 (90)	2 (51)	2 (51)	1¼ (32)	1¼ (32)	1¼ (32)	1¼ (32)
400 (180)	3 (76)	2½ (64)	2 (51)	1½ (38.1)	1½ (38)	2 (51)
500 (225)	3 (76)	2½ (64)	2 (51)	2 (51)	2 (51)	2 (51)
1000 (450)	3½ (88.9)	3 (76)	2½ (64)	2½ (64)	2½ (64)	2½ (64)
2000 (900)	5 (127)	4 (102)	3 (76)	3 (76)	3 (76)	3 (76)
4000 (1800)	6 (152)	5 (127)	4 (102)	4 (102)	4 (102)	4 (102)

Humidistats to start and stop the flow of moisture into the room may be either electrically or air (hygrostat) operated, according to the type of activities in the space. Where electric switches and circuits might cause a fire hazard, use an air-operated hygrostat instead of a humidistat. Locate either type of control to one side of the humidifying moisture stream, 20 to 50 ft (6.1 to 15.2 m) away.

USE OF THE PSYCHROMETRIC CHART IN AIR-CONDITIONING CALCULATIONS

Determine the properties of air at 80°F (26.7°C) dry-bulb (db) temperature and 65°F (18.3°C) wet-bulb (wb) temperature, using the psychrometric chart. Determine the same properties of air if the wet-bulb temperature is 75°F (23.9°C) and the dew-point temperature is 67°F (19.4°C). Show on the psychrometric chart an air-conditioning process in which outside air at 95°F (35°C) db and 80°F (26.7°C) wb is mixed with return air from the room at 80°F (26.7°C) db and 65°F (18.3°C) wb. Air leaves the conditioning apparatus at 55°F (12.8°C) db and 50°F (10°C) wb.

Calculation Procedure:

1. Determine the relative humidity of the air
Using Fig. 28, enter the bottom of the chart at the first dry-bulb temperature, 80°F (26.7°C),and project vertically upward until the slanting 65°F (18.3°C) wet-bulb temperature line is intersected. At the intersection, or *state point,* read the relative humidity as 45 percent on the sloping curve. Note that the number representing the wet-bulb temperature appears on the saturation, or 100 percent relative humidity, curve and that the wet-bulb temperature line is a straight line sloping downward from left to right. The relative humidity curves slope upward from left to right and have the percentage of relative humidity marked on them.

When the wet-bulb and dew-point temperatures are given, enter the psychro-metric chart at the wet-bulb temperature, 75°F (23.9°C) on the saturated curve. From here project downward along the wet-bulb temperature line until the horizontal line representing the dew-point temperature, 67°F (19.4°C), is intersected. At the intersection, or state point, read the dry-bulb temperature as 94.7°F (34.8°C) on the bottom scale of the chart. Read the relative humidity at the intersection as 40.05 percent because the intersection is very close to the 40 percent relative humidity curve.

2. Determine the moisture content of the air
Read the moisture content of the air in grains on the right-hand scale by projecting horizontally from the intersection, or state point. Thus, for the first condition of 80°F (26.7°C) dry bulb and 65°F (18.3°C) wet bulb, projection to the right-hand scale gives a moisture content of 68.5 gr/lb (9.9 gr/kg) of dry air.

For the second condition, 75°F wet bulb and 67°F (19.4°C) dew point, projection to the right-hand scale gives a moisture content of 99.2 gr/lb (142.9 gr/kg).

3. Determine the dew point of the air
This applies to the first condition only because the dew point is known for the second condition. From the intersection of the dry-bulb temperature, 80°F (26.7°C), and the wet-bulb temperature, 65°F (18.3°C), that is, the state point, project horizontally to the left to read the dew point on the horizontal intersection with the saturation curve as 56.8°F (13.8°C). Note that the temperatures plotted along the saturation curve correspond to both the wet-bulb and dew-point temperatures.

4. Determine the enthalpy of the air
Find the enthalpy (also called *total heat*) by reading the value on the sloping line on the central scale above the saturation curve at the state point for the air. Thus, for the first condition, 80°F (26.7°C) dry bulb and 65°F (18.3°C) wet bulb, the enthalpy is 30 Btu/lb (69.8 kJ/kg). The enthalpy value on the psychrometric chart includes the heat of 1 lb (0.45 kg) of dry air and the heat of the moisture in the air, in this case, 68.5 gr (98.6 gr) of water vapor.

For the second condition, 75°F (23.9°C) wet bulb and 67°F (19.4°C) dew point, read the enthalpy as 38.5 Btu/lb (89.6 kJ/kg) at the state point.

5. Determine the specific volume of the air
The specific volume lines slope downward from left to right from the saturation curve to the horizontal dry-bulb temperature. Values of specific volume increase by 0.5 ft³/lb (0.03 m³/kg) between each line.

For the first condition, 80°F (26.7°C) dry-bulb and 65°F (18.3°C) wet-bulb temperature, the stage point lies just to the right of the 13.8 line, giving a specific volume of 13.81 ft³/lb (0.86 m³/kg). For the second condition, 75°F (23.9°C) wet-bulb and 67°F (19.4°C) dew-point temperatures, the specific volume, read in the same way, is 14.28 ft³/lb (0.89 m³/kg).

The weight of the air-vapor mixture can be found from 1.000 + 68.5 gr/lb of air/(7000 gr/lb) = 1.0098 lb (0.46 kg) for the first condition and 1.000 + 99.2/7000 = 1.0142 lb (0.46 kg). In both these calculations the 1000 lb (454.6 kg) represents the weight of the *dry* air and 68.5 gr (4.4 gr) and 99.2 gr (6.43 gr) represents the weight of the moisture for each condition.

6. Determine the vapor pressure of the moisture in the air
Read the vapor pressure by projecting horizontally from the state point to the extreme left-hand scale. Thus, for the first condition the pressure of the water vapor is 0.228 lb/in² (1.57 kPa). For the second condition the pressure of the water is 0.328 lb/in² (2.26 kPa).

7. Plot the air-conditioning process on the psychrometric chart

Air-conditioning processes are conveniently represented on the psychrometric chart. To represent any process, locate the various state points on the chart and convert the points by means of lines representing the process.

Thus, for the air-conditioning process being considered here, start with the outside air at 95°F (35°C) db and 80°F (26.7°C) wb, and plot point (Fig. 28) at the intersection of the two temperature lines. Next, plot point 3, the return air from the room at 80°F (26.7°C) db and 65°F (18.3°C) wb. Point 2 is obtained by computing the final temperature of two airstreams that are mixed, using the method of the calculation procedure given earlier in this section. Plot point 4, using the given leaving temperatures for the apparatus, 55°F (12.8°C) db and 50°F (10°C) wb.

The process in this system is as follows: Air is supplied to the conditioned space along line 4-3. During passage along this line on the chart, the air absorbs heat and moisture from the room. While passing from point 3 to 2, the air absorbs additional heat and moisture while mixing with the warmer outside air. From point 1 to 2, the outside air is cooled while it is mixed with the indoor air. At point 2, the air enters the conditioning apparatus, is cooled, and has its moisture content reduced.

Related Calculations. Use the psychrometric chart for all applied air-conditioning problems where graphic representation of the state of the air or a process will save time. At any given state point of air, the relative humidity in percent can be computed from [partial pressure of the water vapor at the dew-point temperature, lb/in^2 (abs) + partial pressure of the water vapor at saturation corresponding to the dry-bulb temperature of the air, lb/in^2 (abs)](100). Determine the partial pressures from a table of air properties or from the steam tables.

In an *air washer* the temperature of the entering air is reduced. Well-designed air washers produce a leaving-air dry-bulb temperature that equals the wet-bulb and dew-point temperatures of the leaving air. The humidifier portion of an air-conditioning apparatus adds moisture to the air while the dehumidifier removes moisture from the air. In an ideal air washer, adiabatic cooling is assumed to occur.

By using the methods of step 7, any basic air-conditioning process can be plotted on the psychrometric chart. Once a process is plotted, the state points for the air are easily determined from the psychrometric chart.

When you make air-conditioning computations, keep these facts in mind: (1) The total enthalpy, sometimes termed *total heat,* varies with the wet-bulb temperature of the air. (2) The sensible heat of air depends on the wet-bulb temperature of the air; the enthalpy of vaporization, also called the *latent heat,* depends on the dew-point temperature of the air; the dry-bulb, wet-bulb, and dew-point temperatures of air are the same for a saturated mixture. (3) The dew-point temperature of air is fixed by the amount of moisture present in the air.

DESIGNING HIGH-VELOCITY AIR-CONDITIONING DUCTS

Design a high-velocity air-distribution system for the duct arrangements shown in Fig. 39 if the required total air flow is 5000 ft³/min (2.36 m³/s).

Calculation Procedure:

1. Determine the main-duct friction loss

Many high-velocity air-conditioning systems are designed for a main-header velocity of 4000 ft/min (1219.2 m/min) and a friction loss of 1.0 in H_2O per 100 ft

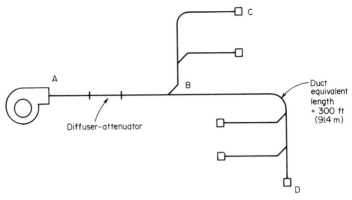

FIGURE 39   High-velocity air-duct-system layout.

(0.08 cm/m) of equivalent duct length. The fan usually discharges into a combined air-diffuser noise-attenuator in which the static pressure of the air increases. This pressure increase must be considered in the choice of the fan-outlet static pressure, but the duct friction loss must be calculated first, as shown below.

Determine the main-duct friction loss by assuming a 1 in/100 ft (0.08 cm/m) static pressure loss for the main duct and a fan-outlet and main-duct velocity of 4000 ft/min (1219.2 m/min). Size the duct by using the equal-friction method. Thus, for the 300-ft (91.4-m) equivalent-length main duct in Fig. 39, the friction pressure loss will be (300 ft)(1.0 in/100 ft) = 3.0 in (76 mm) H_2O.

2. Compute the required fan-outlet pressure

The total friction loss in the duct = duct friction, in H_2O + diffuser-attenuator static pressure in H_2O. In typical installations the diffuser-attenuator static pressure varies from 0.3 to 0.5 in (7.6 to 9.7 mm) H_2O. This is the inlet pressure required to force air through the diffuser-attenuator with all outlets open. Using a value of 0.5 in (12.7 mm) give the total friction loss in the duct = 3.0 + 0.5 = 3.5 in (8.9 mm).

At the fan outlet the required static pressure is less than the total friction loss in the main duct because there is static regain at each branch takeoff to the outlets. This static regain is produced by the reduction in velocity that occurs at each takeoff from the main duct. There is a recovery of static pressure (velocity regain) at the takeoff that offsets the friction loss in the succeeding duct section.

Assume that the velocity in branch C (Fig. 39) is 2000 ft/min (609.6 m/min). This is the usual maximum velocity in takeoffs to terminals. Then, the maximum static regain that could occur $R = (v_i/4005)^2 - (v_f/4000)^2$, where R = static regain, in of H_2O; v_i = initial velocity of the air, ft/min; v_f = final velocity of the air, ft/min. For this system with an initial velocity of 4000 ft/min (1219.2 m/min) and a final velocity of 2000 ft/min (609.6 m/min), $R = (4000/4005)^2 - (2000/4005)^2 = 0.75$.

The maximum static regain is seldom achieved. Actual static regains range from 0.5 to 0.8 of the maximum. With a value of 0.8, the actual static regain = 0.8(0.75) = 0.60 in (15.2 mm) H_2O. This static regain occurs at point B, the takeoff, and reduces the required fan discharge pressure to total friction loss in the duct − static region at first takeoff = 3.5 − 0.60 = 2.9 in (73.7 mm) H_2O. Thus, a fan developing a static discharge pressure of 3.0 in (76 mm) H_2O would probably be chosen for this system.

3. Find the branch-duct pressure loss

To find the branch-duct pressure loss, find the pressure in the main duct at the takeoff point. Use the standard duct-friction chart (Fig. 32) to determine the pressure loss from the fan to the takeoff point. Subtract the sum of this loss and the diffuser-attentuator static pressure from the fan static discharge pressure. The result is the pressure available to force air through the branch duct. Size the branch duct by using the equal-friction method.

Related Calculations. Note that the design of a high-velocity duct system (i.e., a system design in which the air velocities and static pressures are higher than in conventional systems) is basically the same as for a low-velocity duct system designed for static regain. The air velocity is reduced at each takeoff to the riser and air terminals. Design of any high-velocity duct system involves a compromise between the reduced duct sizes (with a saving in materials, labor, and space costs) and higher fan horsepower.

Class II centrifugal fans (Table 25) are generally required for the higher static pressures used in high-velocity air-conditioning systems. Extra care must be taken in duct layout and construction. The high-velocity ducts are usually sealed to prevent air leakage that may cause objectionable noise. Round ducts are preferred to rectangular ones because of the greater rigidity of the round duct.

Use as many symmetric duct runs as possible in designing high-velocity duct systems. The greater the system symmetry, the less time required for duct design, layout, balancing, construction, and installation.

The initial starting velocity used in the supply header depends on the number of hours of operation. To achieve an economic balance between first cost and operating cost, lower air velocities in the header are recommended for 24-h operation, where space permits. Table 26 shows typical air velocities used in high-velocity air-conditioning systems. Use this tabulation to select suitable velocities for the main and branch ducts in high-velocity systems.

Carrier Air Conditioning Company recommends that the following factors be considered in laying out header ductwork for high-velocity air-conditioning systems.

1. The design friction losses from the fan discharge to a point immediately upstream of the first riser takeoff from each branch header should be as nearly equal as possible.
2. To satisfy principle 1 above as applied to multiple headers leaving the fan, and to take maximum advantage of the allowable high velocity, adhere to the fol-

TABLE 25 Classes of Construction for Centrifugal Fans

Class	Maximum total pressure, in H_2O (mm H_2O)
I	3¾ (95)—standard
II	6¾ (172)—standard
III	12¾ (324)—standard
IV	More than 12¾ (324)—recommended

TABLE 26 Typical High-Velocity-System Air Velocities*

	Velocity, ft/min (m/min)
Header or main duct:	
12-h operation	3000–4000 (914.4–1219.2)
24-h operation	2000–3500 (609.6–1066.8)
Branch ducts:†	
90° conical tee	4000–5000 (1219.1–1524.0)
90° tee	3500–4000 (1066.8–1219.2)

*Carrier Air Conditioning Company.
†Branches are defined as a branch header or riser having four to five, or more, takeoffs to terminals.

lowing basic rule whenever possible: Make as nearly equal as possible the ratio of the total equivalent length of each header run (fan discharge to the first riser takeoff) to the initial header diameter (L/D ratio). Thus, the longest header run should preferably have the highest air quantity so that the highest velocities can be used throughout.

3. Unless space conditions dictate otherwise, use a 90° tee or 90° conical tee for the takeoff from the header rather than a 45° tee. Fittings of 90° provide more uniform pressure drops to the branches throughout the system. Also, the first cost is lower.

AIR-CONDITIONING-SYSTEM OUTLET- AND RETURN-GRILLE SELECTION

Choose an air grille to deliver 425 ft³/min (0.20 m³/s) of air to a broadcast studio having a 12-ft (3.7-m) ceiling height. The room is 10 ft (3.0 m) long and 10 ft (3.0 m) wide. Specify the temperature difference to use, the air velocity, grille static resistance, size and face area.

Calculation Procedure:

1. Choose the outlet-grille velocity
The air velocity specified for an outlet grille is a function of the type of room in which the grille is used. Table 27 lists typical maximum outlet air velocities used in grilles serving various types of rooms. Assuming a velocity of 350 ft/min (106.7 m/min) for the outlet grille in this broadcast studio, compute the grille area required from $A = cfm/v$, where A = grille area, ft²; cfm = air flow through the grille, ft³/min; v = air velocity, ft/min. Hence, $A = 425/350 = 1.214$ ft² (0.11 m²).

2. Select the outlet-grille size
Use the selected manufacturer's engineering data, such as that in Table 28. Examination of this table shows that there is no grille rated at 425 ft³/min (0.20 m³/s). Hence, the next larger capacity, 459 ft³/min (0.22 m³/s) must be used. This grille, as the third column from the right of Table 28 shows, is 24 in wide and 8 in high (60.9 cm wide and 20.3 cm high).

TABLE 27 Typical Air Outlet Velocities

Type of room	Maximum velocity, ft/min (m/min)	
Broadcast studio	300–500	(91.4–152.4)
Apartments, private residences, churches, hotel	500–750	(152.4–228.6)
bedrooms, legitimate theatres, private offices	500–750	(152.4–228.6)
	500–750	(152.4–228.6)
	500–750	(152.4–228.6)
	500–750	(152.4–228.6)
	500–750	(152.4–228.6)
Movie theaters	1000	(304.8)
General offices	1200–1500	(365.8–457.2)
Stores—upper floors	1500	(365.8)
Stores—main floors	2000	(609.6)

3. Choose the grille throw distance
Throw is the horizontal distance the air will travel after leaving the grille. With a *fan spread,* the throw of this grille is 10 ft (3.0 m), Table 28. This throw is sufficient if the duct containing the grille is located at any point in the room, i.e., along one wall, in the center, etc. If desired, the grille can be adjusted to reduce the throw, but the throw cannot be increased beyond the distance tabulated. Hence, a fan-spread grille will be used.

4. Select the grille-mounting height
The grille-mounting height is a function of several factors: the difference between the temperature of the entering air and the room air, the room-ceiling height, and the air *drop* (i.e., the distance the air falls from the time it passes through the outlet until it reaches the end of the throw).

By assuming a temperature difference of 20°F (11.1°C) between the entering air and the room air, Table 28 shows that the minimum ceiling height for this grille is 10 ft (3.0 m). Since the room is 12 ft (3.67 m) high, the grille can be mounted at any distance above the floor of 10 ft (3.0 m) or higher.

5. Determine the actual air velocity in the grille
Table 28 shows that the actual air velocity in the grille is 375 ft/min (114.3 m/min). Table 27 shows that an air velocity of 300 to 500 ft/min (91.4 to 152.4 m/min) is suitable for broadcast studios. Hence, this grille is acceptable. If the actual velocity at the grille outlet were higher than that recommended in Table 27, a larger grille giving a velocity within the recommended range would have to be chosen.

6. Determine the grille static resistance
Table 28 shows that the grille static resistance is 0.01 in (0.25 mm) H_2O. This is within the usual static resistance range of outlet grilles.

7. Determine the outlet-grille area
Table 28 shows that the outlet grille has an area of 1.224 ft^2 (0.11 m^2), or 176 in^2 (1135.5 cm^2). This agrees well with the area computed in step 1, or 1.214 ft^2 (0.11 m^2).

TABLE 28 Air-Grille-Selection Table*

Air flow, ft³/min (m³/s)	Wall area per outlet, ft² (m²)		Throw, ft (m)†	Min. ceiling height, ft (m) for temp. difference of			Air velocity, ft/min (m/min)	Grille static resist., in H₂O (cm H₂O)	Outlet size, in (cm)	Grille face area	
	Max.	Min.		15°F (8.3°C)	20°F (11.1°C)	25°F (13.9°C)				ft² (m²)	in² (cm²)
306 (0.14)	25 (2.32)	8 (0.74)	S:11 (3.35)	13 (3.96)	14 (4.27)	15 (4.57)	250 (76.2)	0.005 (0.01)	24 × 8 (60.9 × 20.3)	1.224 (0.11)	176 (1135.5)
			F:6 (1.83)	10 (3.05)	10 (3.05)	10 (3.05)					
459 (0.22)	57 (5.30)	17 (1.58)	S:20 (6.1)	16 (4.88)	17 (5.18)	18 (5.49)	375 (114.3)	0.01 (0.03)	24 × 8 (60.9 × 20.3)	1.224 (0.11)	176 (1135.5)
			F:10 (3.0)	10 (3.05)	10 (3.05)	11 (3.35)					

°Waterloo Register.
†S—straight; F—fan spread.

8. Select the air-return grille

Table 29 shows typical air velocities used for return grilles in various locations. By assuming that the air is returned through a wall louvre, a velocity of 500 ft/min (152.4 m/min) might be used. Hence, by using the equation of step 1, grille area $A = \text{cfm}/v = 425/500 = 0.85\ \text{ft}^2\ (0.08\ \text{m}^2)$.

If a lattice-type return intake having a free area of 60 percent is used, Table 30 shows that the pressure drop during passage of the air through the grille is 0.04 in (1.02 mm) H_2O. Locate the return grille away from the supply grille to prevent short circuiting of the air and excessive noise. The pressure losses in Table 29 are typical for return grilles. Choice of the pressure drop to use is generally left with the system designer. Figure 40 shows details of grille takeoffs and installations.

Related Calculations.　Use this general method to choose outlet and return grilles for industrial, commercial, and domestic applications. Be certain not to exceed the tabulated velocities where noise is a factor in an installation. Excessive noise can lead to complaints from the room occupants.

The outlet-table excerpt presented here is typical of the table arrangements used by many manufacturers. Hence, the general procedure given for selecting an outlet is similar to that for any other manufacturer's outlet.

Many modern-design ceiling outlets are built so that the leaving air entrains some of the room air. The air being discharged by the outlet is termed *primary air,*

TABLE 29　Lattice-Type Return-Grille Pressure Drop, in H_2O (mm H_2O)

Free area of grille, percent	Face velocity, ft/min (m/min) 400 (121.9)	600 (182.9)	Return-intake-air velocities° Intake location	Velocity over gross area, ft/min (m/min)
50	0.04 (1.02)	0.09 (2.29)	Above occupied zone	800 and up (243.8 and up)
60	0.03 (0.76)	0.06 (1.52)	In occupied zone:	
70	0.02 (0.51)	0.05 (1.27)	Not near seats	600–800　(182.9–243.8)
80	0.01 (0.25)	0.03 (0.76)	Near seats	400–600　(121.9–182.9)
			Door or wall louvers	500–700　(152.4–213.4)
			Undercut door (through undercut area)	600　　(182.9)

° ASHRAE *Guide.*

TABLE 30　Approximate Pressure Drop for Lattice Return Intakes, in (mm) H_2O*

Percentage of free area	Face velocity, ft/min (m/min) 400 (121.9)	600 (182.9)	800 (243.8)	1000 (304.8)
50	0.06 (1.52)	0.13 (3.30)	0.22 (5.59)	0.35 (8.89)
60	0.04 (1.02)	0.09 (2.29)	0.16 (4.06)	0.24 (6.10)
70	0.03 (0.76)	0.07 (1.78)	0.12 (3.05)	0.18 (4.57)
80	0.02 (0.51)	0.05 (1.27)	0.09 (2.29)	0.14 (3.56)

° ASHRAE *Guide and Data Book.*

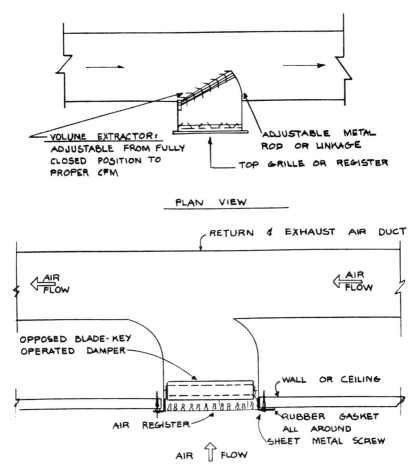

FIGURE 40 Details of grille takeoffs and installations. (*Jerome F. Mueller.*)

and the room air is termed *secondary air.* The induction ratio R_i = (total air, ft³/min)/(primary air, ft³/min). Typical induction ratios run in the range of 30 percent.

For a given room, the total air in circulation, ft³/min = (outlet ft³/min)(induction ratio). Also, average room air velocity, ft/min = 1.4 (total ft³/min in circulation)/area of wall, ft², opposite the outlet or outlets. The wall area in the last equation is the *clear* wall area. Any obstructions must be deducted. The multiplier 1.4 allows for blocking caused by the airstream. Where the room circulation factor K must be computed, use the relation K = (average room air velocity, ft/min)/1.4(induction ratio). The ideal room-air velocity for most applications is 25 ft/min (7.62 m/min). However, velocities up to 300 ft/min (91.4 m/min) are used in some factory air-conditioning applications.

The types of outlets commonly used today are grille (perforated, fixed-bar, adjustable-bar), slotted, ejector, internal induction, pan, diffuser, and perforated ceiling. Choice of a given type depends on the room ceiling height, desired air-temperature difference blow, drop, and spread, as well as other factors that are a function of the room, air quantity, and the activities in the room.

As a general guide to outlet selection, use the following pointers: (1) Choose the number of outlets for each room after considering the quantity of air required, throw or diffusion distance available, ceiling height, obstructions, etc. (2) Try to arrange the outlets symmetrically in the space available as shown by the room floor plan.

SELECTING ROOF VENTILATORS FOR BUILDINGS

A 10-bay building is 200 ft (60.9 m) long, 100 ft (30.5 m) wide, 50 ft (15.2 m) high to the top of the pitched roof, and 35 ft (01.7 m) high to the eaves. The building houses 15 turbine-driven generators and is classed as an engine room. Choose enough roof ventilators to produce a suitable number of air changes in the building. During reduced-load operating periods between 12 midnight and 7 a.m. on weekdays, and on weekends, only half the full-load air changes are required. The prevailing summer-wind velocity against the long side of the building is 10 mi/h (16.1 km/h). The total available open-window area on each long side is 300 ft² (27.9 m²). The minimum difference between the outdoor and indoor temperatures will be 40°F (22.2°C).

Calculation Procedure:

1. Determine the cubic volume of the building
To compute the cubic volume of a pitched-roof building, the usual procedure is to assume an average height from the eaves to the ridge. Since this building has a 15-ft (4.6-m) high ridge from the eaves, the average height $= 15/2 = 7.5$ ft (4.6 m). Since the height from the ground to the eaves is 35 ft (10.7 m), the building height to be used in the volume computation is $35 + 7.5 = 42.5$ ft (13.9 m). Hence, the volume of the building, $ft^3 = V =$ length $\times$ width $\times$ average height, all measured in ft $= 200 \times 100 \times 42.5 = 850,000$ ft³ (24,064 m³).

2. Determine the number of air changes required
Table 29 shows that four to six air changes per hour are normally recommended for engine rooms. Using five air changes per hour will probably be satisfactory, and the roof ventilators will be chose on this basis. During the early morning, and on weekends, 2.5 air changes will be satisfactory, since only half the normal number of air changes is needed during these periods.

3. Compute the required hourly air flow
The required hourly air flow, $ft^3/h = Q =$ (number of air changes per hour)(building volume, ft^3) $= (5)(850,000) = 4,250,000$ ft³/h (120,318 m³/h). During the early morning hours and on weekends when 2.5 air changes are used, $Q = 2.5(850,000) = 2,125,000$ ft³/h (60,159 m³/h).

4. Compute the air flow produced by natural ventilation
The ASHRAE *Guide and Data Book* lists the prevailing winter and summer wind velocities for a variety of locations. Usual practice, in designing natural-ventilation

systems, is to use one-half of tabulated wind velocity for the season being considered. Since summer ventilation is usually of greater importance than winter ventilation, one-half the prevailing summer wind velocity is generally used in natural-ventilation calculations. As the prevailing summer wind velocity in this locality is 8 mi/h (12.9 kg/h), a velocity of 8/2 = 4 mi/h (6.4 km/h) will be used to compute the air flow produced by the wind.

Use the relation $Q = VAE$ to find the air flow produced by the wind. In this relation, Q = air flow produced by the wind, ft^3/min; V = design wind velocity, ft/min = 88 × mi/h; A = free area of the air-inlet openings, ft^2; E = effectiveness of the air inlet openings—use 0.50 to 0.60 for openings perpendicular to the wind and 0.25 to 0.35 for diagonal winds.

Assuming $E = 0.50$, we get $Q = VAE = (4 \times 88)(300)(0.50) = 52,800$ ft^3/min (1494.8 m^3/min) or 60(52,800) = 3,168,000 ft^3/h (89,686.1 m^3/h). Step 3 shows that the required air flow is 4,250,000 ft^3/h (120,318 m^3/h) when all turbines are operating. Hence, the air flow produced by natural ventilation is inadequate for full-load operation. However, since the required flow of 2,125,000 ft^3/h (60,159 m^3/h) for the early morning hours and weekends is less than the natural ventilation flow of 3,168,000 ft^3/h (89,686.1 m^3/h), natural ventilation may be acceptable during these periods.

5. *Determine the number of stationary-type ventilators needed*

A stationary-type roof ventilator (i.e., one that depends on the wind and air-temperature difference to produce the desired air movement) may be suitable for this application. If the stationary-type is not suitable, a power-fan type of roof ventilator will be investigated and must be used. The Breidert-type ventilator is investigated here because the procedure is similar to that used for other stationary-type roof ventilators.

Stationary ventilators produce air flow out of a building by two means: suction caused by wind action across the ventilators and the stack effect caused by the temperature difference between the inside and outside air.

Figure 41 shows the air velocity produced in a stationary Breidert ventilator by winds of various velocities. Thus, with the average 5-mi/h (8.1-km/h) wind assumed earlier for this building, Fig. 41 shows that the air velocity through the ventilator produced by this wind velocity is 220 ft/min (67.1 m/min), closely.

Table 32 shows that a 1.0-ft^2 (0.093-m^2) ventilator installed on a 50-ft (15.2-m) high building having an air-temperature difference of 40°F (22.2°C) will produce, owing to the stack effect, an air-flow velocity of 482 ft/min (146.9 m/min). Hence, the total velocity through this ventilator resulting from the wind and stack action is 220 + 482 = 702 ft/min (213.9 m/min).

Since air flow, ft^3/min = (air velocity, ft/min)(area of ventilator opening, ft^2), an air flow of (702)(1.0) = 702 ft^3/min (19.9 m^3/min) will be produced by each square foot of ventilator-neck or inlet-duct area. Thus, to produce a flow of (4,250,000 ft^3/h)/(60 min/h) = 70,700 ft^3/min (2002 m^3/min) will require a ventilator area of 70,700/702 = 101 ft^2 (9.38 m^2). A 48-in (121.9-cm) Breidert ventilator has a neck area of 12.55 ft^2 (1.17m^2). Hence, a 101/12.55 = 8.05, or eight ventilators will be required. Alternatively, the Breidert capacity table in the engineering data prepared by the manufacturer shows that a 48-in (121.9-cm) ventilator has a ventilating capacity of 8835 ft^3/min (250.1 m^3/mm) when it is used for a 5-mi/h (8.1-km/h), 50-ft (15.2-m) high, 40°F (22.2°C) temperature-difference application. With this capacity, the number of ventilators required = 70,700/8835 = 8.02, say eight ventilators. These ventilators will be suitable for both full- and part-load operation.

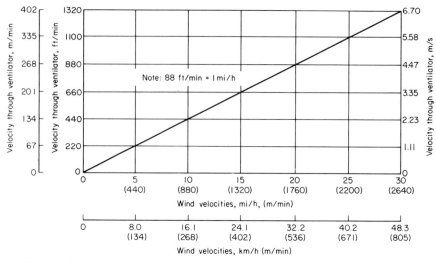

FIGURE 41   Roof-ventilator air-exhaust capacity for various wind velocities. Add the extra velocity for temperature difference given in Table 32. (*G. C. Breidert Co.*)

TABLE 31   Number of Air Changes Required per Hour*

Auditoriums and assembly rooms	10–15	Libraries	3
Boiler rooms	10–15	Machine shops	6
Churches	10–15	Paint shops	10–15
Engine rooms	4–6	Paper mills	15–20
Factory buildings (general)	4	Pump rooms	8–10
Factory buildings (where excessive	15–20	Railroad shops	4
conditions of fumes, moisture, etc., are		Schools	10–12
present)		Textile mills (general)	4
Foundries	12	Textile mill dye houses	15–20
Garages	10–15	Theaters	5–8
General offices	3	Waiting rooms	4
Hotel dining rooms	4	Warehouses	4
Hotel kitchens	10–20	Wood-working shops	8
Laundries	15–25		

*DeBothezat Fans Division, AMETEK Inc.

6.   *Determine the number of powered ventilators needed*

Powered ventilators are equipped with single- or two-speed fans to produce a positive air flow independent of wind velocity and stack effect. For this reason, some
engineers prefer powered ventilators where it is essential that air movement out of
the building be maintained at all times.

Two-speed powered ventilators are usually designed so that the reduced-speed
rpm is approximately one-half the full-speed rpm. The air flow at half-speed is
about one-half that at full speed.

Checking the capacity table of a typical powered-ventilator manufacturer shows
that ventilator capacities range from about 2100 ft³/min (59.5 m³/min) for a 21-in
(53.3-cm) diameter unit at a ⅛-in (3.18-mm) static pressure difference to about

24,000 ft³/min (679 m³/min) for a 36-in (91.4-cm) diameter ventilator at the same pressure difference. With a 27-in (68.6-cm) diameter powered ventilator which has a capacity of 14,9000 ft³/min (421.8 m³/min), the number required is [70,700 ft³/min (2002 m³/m)]/[14,900 ft³/min (422 m³/min) per unit] = 4.76 or 5.

7. Choose the type of ventilator to use
Either a stationary or powered ventilator might be chosen for this application. Since a large amount of heat is generated in an engine room, the powered ventilator would probably be a better choice because there would be less chance of overheating during periods of little or no wind.

Related Calculations.   Use the general method given here to choose stationary or powered ventilators for any of the 25 applications listed in Table 31. Usual practice is to locate one ventilator in each bay or sawtooth of a building.

With the greater interdiction of smoking in public places (factories, offices, hotels, restaurants, schools, etc.), special exhaust fans—often termed "smoke eaters"—are being installed. These high-velocity fans draw smoke-laden air from a designated smoking area and exhaust it to the atmosphere or to treatment devices.

Local building codes govern smoking in structures of various types, so the engineer must consult the local code before choosing the type of exhaust fan to sue for a specific building. Many cities now prohibit all smoking inside a building. In such cities special exhaust fans are not needed to handle cigarette, cigar, or pipe smoke.

Restaurants, bowling alleys, billiard rooms, taverns, and similar gathering places still permit some indoor smoking. Most cities have not prohibited smoking in such establishments because it is a part of the ambiance of these places. Some cities, however, recommend—or require—designated smoking areas, particularly in restaurants. It is here that the engineer must select and specify a suitable exhaust fan to rid the area of tobacco smoke.

Follow the same procedure given above to choose the fan or fans. Be certain to use the required number of air changes specified by any local building code. While an excessive number of changes will increase the winter heating load, many engineers overdesign to be certain they meet clean air requirements. Tobacco smoke must be handled decisively so that all patrons of an establishment are comfortable.

VIBRATION-ISOLATOR SELECTION FOR AN AIR CONDITIONER

Choose a vibration isolator for a packaged air conditioner operating at 1800 r/min. What minimum mounting deflection is required if the air conditioner is mounted on a basement floor? On an upper-story floor made of light concrete?

Calculation Procedure:

1. Determine the suggested isolation efficiency
Table 33 lists the suggested isolation efficiency for various components used in air-conditioning and refrigeration systems. This tabulation shows that the suggested isolation efficiency for a packaged air conditioner is 90 percent. This means that the vibration isolator or mounting should absorb 90 percent, or more, of the vibration caused by the machine. At this efficiency only 10 percent of the machine vibration would be transmitted to the supporting structure.

TABLE 32 Flow of Air in Natural-Draft Flues, ft³/(min · ft²) [m³/(min · m²)]*

Difference in temperature, °F (°C)	Height of flue, ft (m), same as height of room or building			
	30 (9.1)	40 (12.2)	50 (15.2)	60 (18.3)
10 (5.6)	188 (57.3)	217 (66.1)	242 (73.7)	264 (80.4)
20 (11.1)	265 (80.6)	306 (93.2)	342 (104.2)	373 (113.7)
30 (16.7)	325 (99.0)	375 (114.3)	419 (127.7)	461 (140.5)
40 (22.2)	374 (113.9)	431 (131.3)	482 (140.2)	529 (161.2)
50 (27.8)	419 (127.7)	484 (147.5)	541 (164.9)	594 (181.0)
60 (33.3)	460 (140.2)	532 (162.1)	595 (181.4)	650 (198.1)

*G. C. Breidert Co.

TABLE 33 Suggested Isolation Efficiencies

Equipment	Installed efficiency, %
Absorption units	95
Steam generators	95
Centrifugal compressors	98
Reciprocating compressors:	
Up to 15 hp (11.2 kW)	85
20–60 hp (14.9–44.8 kW)	90
75–150 hp (56.0–111.9 kW)	95
Packaged air conditioners	90
Centrifugal fans:	
80 r/min and above; all diameters	90–95
350–800 r/min; all diameters	70–90
200–350 r/min; 48-in (121.9-cm) diameter or smaller	°
200–350 r/min; 54-in (137.2-cm) diameter or larger	70–80
Centrifugal pumps	95
Cooling towers	85
Condensers	80
Fan coil units	80
Piping	95

°Installed for noise isolation only.

2. Determine the static deflection caused by the vibration
Use Fig. 42 to find the static deflection caused by the vibration. Enter at the bottom of Fig. 42 at 1800 r/min the disturbing frequency, and project vertically upward to the 90 percent efficiency curve. At the left read the static deflection as 0.11 in (2.79 mm).

3. Select the type of vibration isolator to use
Project to the right from the intersection with the efficiency curve (Fig. 42) to read the type of isolator to use. Thus, neoprene pads or neoprene-in-shear mounts will safely absorb up to 0.25-in (6.35-mm) static deflection. Hence, either type of isolator mounting could be used.

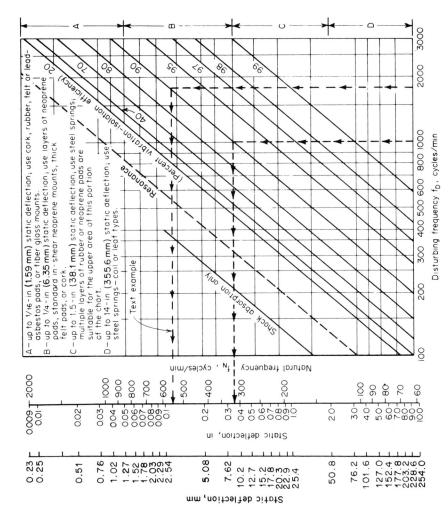

FIGURE 42 Vibration-isolator deflection for various disturbing frequencies. (*Power.*)

16.123

4. Check the isolator selection

Use Table 34 to check the theoretical isolation efficiency of the mounting chosen. Enter at the top at the rpm of the machine, and project vertically downward until an efficiency equal to, or greater than, that desired is intersected.

For this machine operating at 1800 r/min, single-deflection rubber mountings have an efficiency of 94 percent. Since neoprene is also called synthetic rubber, the isolator choice is acceptable because it yields a higher efficiency than required.

5. Determine the minimum mounting deflection required

Table 35 lists the minimum mounting deflection required at various operating speeds for machines installed on various types of floors. Thus, at 1800 r/min, machines mounted on a basement floor must have isolator mountings that will absorb deflections up to 0.10 in (2.54 mm). Since the neoprene mountings chosen in step 3 will absorb up to 0.25 in (6.35 mm) of deflection, they will be acceptable for use on a basement-mounted machine.

For mounting on a light-concrete upper-story floor, Table 35 shows that the mounting must be able to absorb a deflection of 0.80 in (20.3 mm) for machines operating at 1800 r/min. Since the neoprene isolators can absorb only 0.25 in (6.35 mm), another type of mounting is needed if the machine is installed on an upper floor. Figure 42 shows that steel springs will absorb up to 1.5 in (38.1 mm) of static deflection. Hence, this type of mounting would be used for machines installed on upper floors of the building.

Related Calculations. Use this general procedure for engines, compressors, turbines, pumps, fans, and similar rotating and reciprocating equipment. Note that the suggested isolation efficiencies in Table 33 are for air-conditioning equipment located in critical areas of buildings, such as office, hospitals, etc. In noncritical areas, such as basements or warehouses, an isolation efficiency of 70 percent may be acceptable. Note that the efficiencies given in Table 33 are useful as general guides for all types of rotating machinery.

Although much emphasis is placed on atmospheric, soil, and water pollution, greater attention is being placed today on audio pollution than in the past. Audio pollution is the discomfort in human beings produced by excessive or high-pitch noise. One good example is the sound produced by jet aircraft during takeoff and landing.

Audio pollution can be injurious when it is part of the regular workplace environment. At home, audio pollution can interfere with one's life-style, making both indoor and outdoor activities unpleasant. For these reasons, regulatory agencies are taking stronger steps to curb audio pollution.

Control of audio pollution almost always reverts to engineering design. For this reason, engineers will be more concerned with the noise their designs produce because it is they who have more control of it than others in the design, manufacture, and use of a product.

SELECTION OF NOISE-REDUCTION MATERIALS

A concrete-walled test laboratory is 25 ft (7.62 m) long, 20 ft (6.1 m) wide, and 10 ft (3.05 m) high. The laboratory is used for testing chipping hammers. What noise reduction can be achieved in this laboratory by lining it with acoustic materials?

TABLE 34 Theoretical Vibration-Isolation Efficiencies*

Isolation material	Average static deflection, in (mm)		Average natural frequency	Efficiencies, percent									
				350 r/min	500 r/min	600 r/min	800 r/min	1000 r/min	1200 r/min	1500 r/min	1800 r/min	3000 r/min	3600 r/min
2-in (50.8-mm) thick standard-density cork	0.08	(2.03)	By test 1420	...	...	...	...	...	...	...	...	72	82
Type W waffle pad	Curvature corrected, 0.035	(0.89)	1000	...	...	...	...	...	...	20	55	87	92
Two layers of W waffle pad	Curvature corrected, 0.070	(1.78)	710	...	...	...	...	...	46	71	82	93	96
Single-deflection rubber mountings	0.20	(5.08)	420	...	...	...	62	79	86	91	94	98	99
Double-deflection rubber mountings	0.40	(10.16)	300	...	44	67	84	90	93	96	97	99	Almost perfect
Standard spring mountings	1.00	(25.4)	188	70	85	89	94	96	97	98	99	Almost perfect	Almost perfect
Double-deflection rubber and spring mountings	1.40	(35.6)	160	75	89	93	96	97	98	99	Almost perfect	Almost perfect	Almost perfect

*Power magazine.

TABLE 35 Minimum Mounting Deflections

Operating speed, r/min	Basement—negligible floor deflection, in (mm)	Rigid concrete floor, in (mm)	Upper story—light-concrete floor, in (mm)	Wood floor, in (mm)
300	1.50 (38.1)	3.00 (76.2)	3.50 (88.9)	4.00 (101.6)
500	0.63 (16.0)	1.25 (31.8)	1.65 (41.9)	1.95 (49.5)
800	0.25 (6.35)	0.60 (15.2)	1.00 (25.4)	1.25 (31.6)
1200	0.20 (5.08)	0.45 (11.4)	0.80 (20.3)	1.00 (25.4)
1800	0.10 (2.54)	0.35 (8.9)	0.80 (20.3)	1.00 (25.4)
3600	0.03 (0.76)	0.20 (5.08)	0.80 (20.3)	1.00 (25.4)
7200	0.03 (0.76)	0.20 (5.08)	0.80 (20.3)	1.00 (25.4)

Calculation Procedure:

1. Determine the noise level of devices in the room
Table 36 shows that a chipping hammer produces noise in the 130-dB range. Hence, the noise level of this room can be assumed to be 130 dB. This is rated as deafening by various authorities. Therefore, some kind of sound-absorption material is needed in this room if the uninsulated walls do not absorb enough sound.

2. Compute the total sound absorption of the room
The *sound-absorption coefficient* of bare concrete is 0.1. This means that 10 percent of the sound produced in the room is absorbed by the bare concrete walls.

To find the total sound absorption by the walls and ceiling, find the product of the total area exposed to the sound and the sound absorption coefficient of the material. Thus, concrete area, excluding the floor but including the ceiling = two walls (25 ft long × 10 ft high) + two walls (20 ft wide × 10 ft high) + one ceiling (25 ft long × 20 ft wide) = 1400 ft² (130.1 m²). Then the total sound absorption = (1400)(0.1) = 140.

3. Compute the total sound absorption with acoustical materials
Table 37 lists the sound- or noise-reduction coefficients for various acoustic materials. Assume that the four walls and ceiling are insulated with membrane-faced mineral-fiber tile having a sound absorption coefficient of 0.90, from Table 37.

Then, by the procedure of step 2, total noise reduction = (1400 ft²)(0.90) = 1260 ft² (117.1 m²).

4. Compute the noise reduction resulting from insulation use
Use the relation noise reduction, dB = 10 log (total absorption *after* treatment/total absorption *before* treatment) = 10 log (1260/140) = 9.54 dB. Thus, the sound level in the room would be reduced to 130.00 − 9.54 = 120.46 dB. This is a reduction of 9.54/130 = 0.0733, or 7.33 percent. To obtain a further reduction of the noise in this room, the floor could be insulated or, preferably, the noise-producing device could be redesigned to give off less noise.

Related Calculations.   Use this general procedure to determine the effectiveness of acoustic materials used in any room in a building, on a ship, in an airplane, etc.

TABLE 36 Power and Intensity of Noise Sources

Sound source	Power range, W	Decibel range $(10^{-13}$ W)
Ram jet	100,000.0	180
Turbojet with 7000-lb (31,136-N) thrust	10,000.0	170
	1,000.0	160
Four-propeller airliner	100.0	150
75-piece orchestra, pipe organ; small aircraft engine	10.0	140
Chipping hammer	1.0	130
Piano, blaring radio	0.1	120
Centrifugal ventilating fan at 13,000 ft^3/min (6.14 m^3/s)	0.01	110
Automobile on roadway; vane-axial ventilating fan	0.001	100
Subway car, air drill	0.0001	90
Conversational voice; traffic on street corner	0.000 01	80
Street noise, average radio	0.000 001	70
Typical office	0.000 000 1	60
	0.000 000 01	50
Very soft whisper	0.000 000 001	40

Reference values relate decibel scales		
db scale	Definition	Reference quantity
Sound-power level	$\text{PWL} = 10 \log \dfrac{W}{W \text{ re}}$	$W \text{ re} = 10^{-13}$ W
Sound-intensity level	$\text{IL} = 10 \log \dfrac{I}{I \text{ re}}$	$I \text{ re} = 10^{12} \text{ W/m}^2$ $= 10^{-16} \text{ W/cm}^2$
Sound-pressure level	$\text{SPL} = 10 \log \dfrac{P^2}{P^2 \text{ re}} = 20 \log \dfrac{P}{P \text{ re}}$	$P \text{ re} = 0.000,02 \text{ N/m}^2$ $= 0.0002 \ \mu\text{bar}$ $= 0.0002 \text{ dyn/cm}^2$

*$Power$ magazine.

CHOOSING DOOR AND WINDOW AIR CURTAINS FOR VARIOUS APPLICATIONS

Show how to choose environmental air curtains for shopping malls, supermarkets, hospitals, schools, restaurants, warehouses, service take-out windows, and manufacturing facilities. Detail air-flow requirements to provide (1) thermal barriers to reduce energy consumption, and (2) insect, dust, and odor control.

Calculation Procedure:

1. *Determine the door and window dimensions*
Commercially available air-curtain units, Fig. 43, are designed for specified door and window widths and heights. Such units consist of an electric-motor-driven fan

TABLE 37 Noise-Reduction Coefficients*

Type	Material	Noise-reduction coefficient range [¾ in (19.1 mm) thick]
1	Regularly perforated cellulose-fiber tile	0.65–0.85
2	Randomly perforated cellulose-fiber tile	0.60–0.75
3	Textured, perforated, fissured, cellulose tile	0.50–0.70
4	Cellulose-fiber lay-in panels	0.50–0.60†
5	Perforated mineral-fiber tile	0.65–0.85
6	Fissured mineral-fiber tile	0.65–0.80
7	Textured, perforated or smooth mineral-fiber tile	0.65–0.85
8	Membrane-faced mineral-fiber tile	0.30–0.90
9	Mineral-fiber lay-in panels	0.20–0.90
10	Perforated-metal pans with mineral-fiber pads	0.60–0.80†
11	Perforated-metal lay-in panels with mineral-fiber pads	0.75–0.85
12	Mineral-fiber tile—fire-resistive assemblies	0.55–0.90
13	Mineral-fiber lay-in panels—fire-resistive units	0.65–0.75‡
14	Perforated-asbestos panels with mineral-fiber pads	0.65–0.75§
15	Sound-absorbent duct lining	0.65–0.75†
16	Special acoustic panels and materials	

°Acoustical and Insulating Materials Association.
†Noise-reduction coefficient 1 in (25.4 mm) thick.
‡Noise-reduction coefficient ⅝ in (15.9 mm) thick.
§Noise-reduction coefficient ¹⁵⁄₁₆ in (23.8 mm) thick.

fitted with suitable deflecting vanes mounted in a metal or plastic casing that is easily installed over the door or window opening. An air heater is usually fitted to units used in colder climates.

Popular standard air-curtain units are available for door widths up to 20 ft (6.1 m), or more, and door heights up to 22 ft (6.7 m). Special high-velocity units can be designed for greater door widths and heights. Both heated and unheated air-curtain units are available. Heated units can use steam, hot water, gas, or electricity as their heat source.

Find the door or window width and height from the drawings of the installation. Where a door or window is excessively wide—say more than 14 ft (4.3 m), two or more air-curtain units may be required.

2. Choose the type of air-curtain unit to use

Heated air-curtain units are used where temperature control is required in the installation—termed *climate control* by air-curtain unit manufacturers. Typical climate-control unit applications include customer entrances to shopping areas, receiving/service doors, service take-out windows in fast-food establishments and banks, refrigerated rooms, warehouse doors, hospital entrances, etc.

Where climate control is required—i.e. warmer air is being projected from the ceiling of an opening to the floor or counter of the opening—a heater of some type is required. The type of heater chosen depends on the heating medium available—steam, hot water, gas, or electric. No matter what type of heating medium is chosen, an electric-motor-driven fan fitted with directional vanes at the air outlet nozzles is required. The directional control vanes allow adjustment of the air flow for cold weather, versus the vane setting for control of dust, insects, odors, and

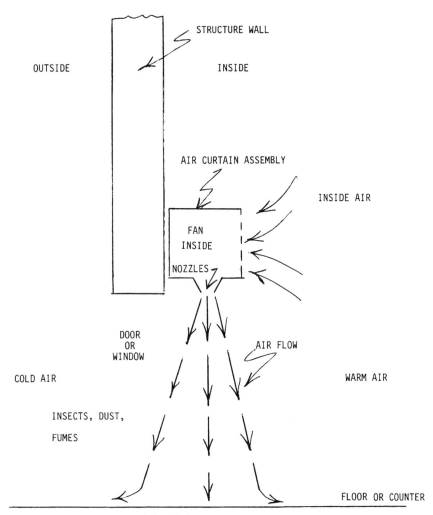

FIGURE 43 Typical air curtain installation.

fumes. Extra-high-velocity fans are often used for wider doors—those exceeding 12 ft (3.7 m) in height.

   To choose the specific air curtain to use, refer to a manufacturer's catalog showing the types of units available. Such catalogs present tabulations of both unheated and heated air-curtain units for various applications. The selection tables also specify which units are best for insect, fume, and dust barriers. Diagrams in the catalogs also suggest installation methods for a variety of door types—track, roll-up, vertical-lift, and sliding. Air-nozzle length and projection angle are also detailed.

   An air curtain effectively separates cold (or warm) outside air from the conditioned air inside a structure. Air curtains can also be used inside buildings to separate manufacturing areas from office areas, guard shacks from interior or exterior spaces, etc.

Typical air-flow rates for heavy-duty air curtains separating inside air from outside air can range from about 12,000 cfm (5664 L/s) for a 10-ft to 14-ft (3- to 4.3-m) high door to 33,000 cfm (15,567 L/s) for a 22-ft (6.7-m) high door. Air velocities for such doors range from about 3500 fpm (1067 m/min) to 6000 fpm (1829 m/min).

Motor horsepower for air curtains ranges from a low of about 0.5 hp (0.37 kW) for the smallest doors to 5 hp (3.7 kW) for the smaller door in the above paragraph to 20 hp (14.9 kW) for the larger door. Weight of air-curtain units varies with the materials of construction of the box enclosure. Typical weights range from 600 lb (272 kg) to 1200 lb (545 kg). Tables in manufacturers catalogs give detailed information on the door height served, air flow at the nozzle, average and maximum air flow in cfm (L/s), motor horsepower (kW), unit weight, and heating capacity with various heating mediums—hot water, steam, gas, and electric. Using these data, it is an easy task to select a suitable air curtain for specific conditions.

Heating capacity of air-curtain units range from less than 100,000 Btu/hr (29.3 kW) to more than 1.5 million Btu/hr (439.5 kW). Heated air curtains reduce thermal stratification in buildings by recirculating cold air near the floor to higher levels in the structure where it can mix with warm air in ceiling pockets. Outside winds in the 30 mph (48.3 km/hr) range can be kept from entering a building when a properly designed air curtain is chosen. In areas of high sustained wind velocity, higher powered air-curtain fans are usually chosen. Many air-curtain package assemblies are installed inside the building door when heat loss and prevention of cold air ingress are desired.

Dual air-curtain assemblies can be mounted side by side for excessively wide doors. Either metal or plastic assembly boxes can be used. Control of flies and other insects is another popular application of air curtains. The air curtain effectively keeps such flying insects out of the conditioned space, where they might cause human infection or other problems.

Related Calculations. While air curtains can be designed from scratch, the best solution to selecting a suitable unit usually is to pick from a premanufactured assembly. Air-curtain manufacturers spend large sums on researching and testing their units so they perform most efficiently. Further, the manufacturers are extremely willing to cooperate on any special design needs you might have for the air-curtain installation you are designing.

By choosing a pre-engineered air curtain, the designer can save time and money for the installation being considered. Off-the-shelf air curtains can be delivered quickly to the site where they are needed, again saving time and energy.

Since the use of air curtains continues to grow, the designer is urged to consult an experienced manufacturer for the latest developments. Data in this procedure were excerpted from information presented in several air-curtain manufacturer's catalogs. Two suppliers of air curtains are Leading Edge, Inc. Miami FL, and Mars Sales Co., Inc., Gardena CA, both of whom are long-established and highly experienced air-curtain manufacturers.

SECTION 17
SOLAR ENERGY

Economics and Applications

ANALYSIS OF SOLAR ELECTRIC GENERATING SYSTEM LOADS AND COSTS

Analyze the feasibility of a solar electric generating system (SEGS) for a power system located in a sub-tropical climate. Compare generating loads and costs with conventional fossil-fuel and nuclear generating plants.

Calculation Procedure:

1. *Determine when a solar electric generating system can compete with conventional power*
Solar electric generation, by definition, requires abundant sunshine. Without such sunshine, any proposed solar electric generating plant could not meet load demands. Hence, such a plant could not compete with conventional fossil-fuel or nuclear plants. Therefore, solar electric generation, is at this time, restricted to areas having high concentrations of sunshine. Such areas are in both the subtropical and tropical regions of the world.

One successful solar electric generating system is located in the Mojave Desert in southern California. At this writing, it has operated successfully for some 12 years with a turbine cycle efficiency of 37.5 percent for a solar field of more than 2-million ft² (1,805,802 m²). A natural-gas backup system has a 39.5 percent efficiency. Both these levels of efficiency are amongst the highest attainable today with any type of energy source.

2. Sketch a typical cycle arrangement
Technology developed by Luz International Ltd. uses a moderate-pressure state-of-the-art Rankine-cycle steam-generating system using solar radiation as its primary energy source, Fig. 1. In the Mojave Desert plant mentioned above, a solar field comprised of parabolic-trough solar collectors which individually track the sun using sun sensors and microprocessors provides heat for the steam cycle.

Collection troughs in the Mojave Desert plant are rear surface mirrors bent into the correct parbolic shape. These specially designed mirrors focus sunlight onto heat-collection elements (HCE). Each mirror is washed every two weeks with de-mineralized water to remove normal dust blown off the desert. The mirrors must be clean to focus the optimal amount of the sun's heat on the HCE.

3. Detail the sun collector arrangement and orientation
With the parabolic mirrors described above, sun sensors begin tracking the sun before dawn. Microprocessors prompt the troughs to follow the sun, rotating 180° each day. A central computer facility at the Mojave Desert plant monitors and controls each of the hundreds of individual solar collectors in the field and all of the power plant equipment and systems.

During summer months when solar radiation is strongest, some mirrors must be turned away from the sun because there is too much heat for the turbine capacity.

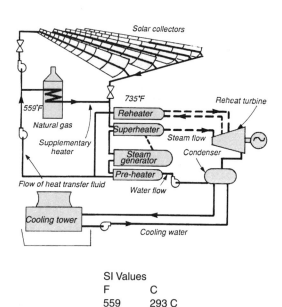

SI Values	
F	C
559	293 C
735	391 C

FIGURE 1 Solar-generating-method schematic traces flow of heat-transfer fluid. (*Luz International Ltd. and Power.*)

When this occurs, almost every other row of mirrors must be turned away. However, in the winter, when solar radiation is the weakest, every mirror must be employed to produce the required power.

In the Mojave Desert plant, the mirrors focus the collected heat on the HCEs—coated steel pipes mounted inside vacuum-insulated glass tubes. The HCEs contain a synthetic-oil heat-transfer fluid, which is heated by the focused energy to approximately 735°F (390.6°C) and pumped through a series of conventional heat exchangers to generate superheated steam for the turbine-generator.

In the Mojave Desert plant, several collectors are assembled into units called solar collector assemblies (SCA); generally, each 330-ft (100.6-m) row of collectors comprises one SCA. The SCAs are mounted on pylons and interconnected with flexible hoses. An 80-MW field consists of 852 SCAs arranged in 142 loops. Each SCA has its own sun sensor, drive motor, and local controller, and is comprised of 224 collector segments, or almost 5867 ft² (545 m²) of mirrored surface and 24 HCEs. From this can be inferred that some (5867/80) = 73.3 ft² (6.8 m²) per MW is required at this installation.

4. Plan for an uninterrupted power supply

To ensure uninterrupted power during peak demand periods, an auxiliary natural-gas fired boiler is available at the Mojave Desert plant as a supplemental source of steam. However, use of this boiler is limited to 25 percent of the time by federal regulations. This boiler serves as a backup in the event of rain, for night production when called for, or if "clean sun" is unavailable. According to Luz International, clean sun refers to solar radiation untainted by smog, clouds, or rain. Figure 2 shows the firing modes for typical summer (left) and winter (right) days. Correlation of solar generation to peaking-power requirements is evident.

As shown in the cycle diagram, the balance-of-plant equipment consists of the turbine-generator, steam generator, solar superheater, two-cell cooling tower, and an intertie with the local utility company, Southern California Edison Co. The Mojave Desert installation represents some 90 percent of the world's solar power production. Since installing it first solar electric generating system in 1984, a 13.8-MW facility, Luz has built six more SEGS of 30 MW each. Units 6 and 7 use third-generation mirror technology.

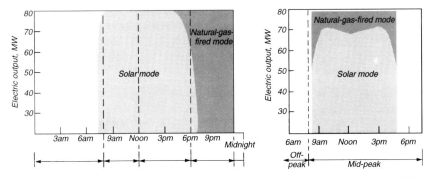

FIGURE 2 Firing modes are shown for typical summer day, left, and typical winter day, right. Correlation of solar generation to peaking power requirements is evident. (*Luz International Ltd. and Power.*)

5. Determine the costs of solar power

SEGS are suited to utility peaking service because they provide up to 80 percent of their output during those hours of a utility's greatest demand, with minimal production during low-demand hours.

Cost of Luz's solar-generated power is less than that of many nuclear plants—$0.08/kWh, down from $0.24/kWh for the first SEGS, according to company officials. Should the price of oil go up beyond $20/barrel, solar will become even more competitive with conventional power.

But the advantages over conventional power sources include more than cost-competitiveness. Emissions levels are much lower—10 ppm—because the sun is essentially non-polluting. SEGS are equipped with the best available technology for emissions cleanup during the hours they burn natural gas, the only time they produce emission.

Related Calculations. Luz International Ltd. has installed more capacity at the Mojave Desert plant mentioned here, proving the acceptance and success of its approach to this important technical challenge. That data in this procedure can be useful to engineers studying the feasibility of solar electric generation for other sites around the world. Luz received an Energy Conservation Award from *Power* magazine, from which the data and illustrations in this procedure were obtained. There are estimates showing that the sunshine impinging the southwestern United States is more than enough to generate the entire electrical needs of the country—when efficient conversion apparatus is developed. It may be that the equipment described here will provide the efficiency needed for large-scale pollution-free power generation. Results to date have been outstanding and promise greater efficiency in the future.

ECONOMICS OF INVESTMENT IN AN INDUSTRIAL SOLAR-ENERGY SYSTEM

Determine the rate of return and after tax present value of a new industrial solar energy system. The solar installation replaces all fuel utilized by an existing fossil-fueled boiler when optimum weather conditions exist. The existing boiler will be retained as an auxiliary unit. Assume a system energy output (E_s) of 3×10^9 Btu /yr ($3.17 \text{ kJ} \times 10^9$/yr) an initial cost for the total system of $503,000 based on a collector area (A_c) of 10,060 ft^2 (934.6 m^2), a depreciation life (DP) of 12 yr, a tax rate (τ) of 0.4840, a tax credit factor (TC) of 0.25, a system life of 20 yr, an operating cost fraction (OMPI) of 0.0250, an initial fuel cost (P_{f0}) of $3.11/MBtu ($3.11/947.9 MJ) and a fuel price escalation rate (e) of 0.1450.

Calculation Procedure:

1. Compute unit capacity cost (K_s) in $/million Btu per year

$$K_s = \frac{\text{initial cost of system}}{E_s} = \frac{\$503,000}{3 \times 10^9 \text{ Btu/yr}}$$

$$= \frac{\$167.67}{1 \text{ million Btu/yr}} \quad (\$167.67/947.9 \text{ MJ/yr}).$$

2. Compute levelized coefficient of initial costs (M) over the life of the system

$$M = \text{OMPI} + \frac{\text{CRF}_{R,N}}{1 - \tau}\left[1 - \left(\frac{\text{TC}}{1 + R}\right) - (\tau \times \text{DEP})\right]$$

$\text{CRF}_{R,N}$ is the capital recovery factor which is a function of the market discount rate (R)* over the expected lifetime of the system (20 yr) and is determined as follows:

$$\text{CRF}_{R,20} = \frac{R}{1 - (1 + R)^{-20}}$$

DEP is the depreciation which will be calculated by an accelerated method, the sum of the years digits (SOYD), in accordance with the following formula:

$$\text{DEP} = \frac{2}{\text{DP}(\text{DP} + 1)R}\left(\text{DP} - \frac{1}{\text{CRF}_{R,\text{DP}}}\right)$$

where DP is an allowed depreciation period, or tax life, of 12 yr.
    Prepare a tabulation (see below) of M values for various market discount rates (R).

3. Compute the levelized cost of solar energy (S), for the life cycle of the system in $/million Btu ($/MJ)

Use the relation, $S = (K_s)(M)$. Since M varies with R, refer to the tabulation of S for various market discount rates.

4. Compute the levelized cost of fuel (F) in $/million Btu ($/MJ) and compare to S

$$F = \frac{P_{f0}}{\eta}\left[\text{CRF}_{R,N}\left(\frac{1 + e}{R - e}\right)\left(1 - \left[\frac{1 + e}{1 + R}\right]^N\right)\right]$$

where η is the boiler efficiency for a fossil fuel system which supplies equivalent heat. Referring to the tabulation, the value of F is tabulated at various market discount rates for η values of 70, 80, and 100 percent. The rate of return for the solar installation is that value of R at which $F = S$. For $\eta = 70$ percent, R is between 7.5 and 8.0 percent. For $\eta = 80$ percent, R is between 6.5 and 7.0 percent. These rates of return should exceed current interest (discount) rates to attain economic feasibility.

5. Compute the after tax present value (PV) of the solar investment if the existing boiler installation has an efficiency of 70 percent

$$\text{PV} = E_s\left(\frac{1 - \tau}{\text{CRF}_{R,N}}\right)(F - S)$$

In order to have a positive value of PV, F must exceed S. Therefore, select a market discount rate (R) from the tabulation which satisfies this criteria. For example, at a 5 percent discount rate,

---

*See tabulation on page 17.6.

$$PV = \frac{3 \times 10^9}{10^6} \left(\frac{1 - 0.484}{0.08024} \right) (20.00 - 13.91) = \$117,489.02$$

Note that at 8 percent or higher PV will be negative and the investment proves uneconomical against other investment options.

$R*$	$CRF_{R,N}$	$CRF_{R,DP}$	$DEP_{(SOYD)}$	M	S	$\eta = 100\%$	$\eta = 80\%$	$\eta = 70\%$
							F	
4.5	0.07688	0.10967	0.8210	0.07915	13.27	14.29	17.86	20.41
5.0	0.08024	0.11283	0.8044	0.08295	13.91	14.00	17.50	20.00
5.5	0.08368	0.11603	0.7882	0.08686	14.56	13.71	17.14	19.59
6.0	0.08718	0.11928	0.7727	0.09093	15.25	13.43	16.79	19.19
6.5	0.09076	0.12257	0.7577	0.09511	15.95	13.16	16.45	18.80
7.0	0.09439	0.12590	0.7431	0.09940	16.67	12.89	16.11	18.41
7.5	0.09809	0.12928	0.7290	0.10382	17.41	12.63	15.79	18.04
8.0	0.10185	0.13270	0.7154	0.10833	18.16	12.38	15.47	17.69

*As used in engineering economics, R, discount rate and interest rate refer to the same percentage. The only difference is that interest refers to a progression in time, and discount to a regression in time. See "Engineering Economics for P.E. Examinations," Max Kurtz, McGraw-Hill.

REFERENCE

Brown, Kenneth C., "How to Determine the Cost-Effectiveness of Solar Energy Projects," *Power* magazine, March, 1981.

DESIGNING A FLAT-PLATE SOLAR-ENERGY HEATING AND COOLING SYSTEM

Give general design guidelines for the planning of a solar-energy heating and cooling system for an industrial building in the Jacksonville, Florida, area to use solar energy for space heating and cooling and water heating. Outline the key factors considered in the design so they may be applied to solar-energy heating and cooling systems in other situations. Give sources of pertinent design data, where applicable.

Calculation Procedure:

1. Determine the average annual amount of solar energy available at the site
Figure 3 shows the average amount of solar energy available, in Btu/(day · ft²) (W/m²) of panel area, in various parts of the United States. How much energy is collected depends on the solar panel efficiency and the characteristics of the storage and end-use systems.

Tables available from the National Weather Service and the American Society of Heating, Refrigerating and Air Conditioning Engineers (ASHRAE) chart the monthly solar-radiation impact for different locations and solar insolation [total radiation form the sun received by a surface, measured in Btu/(h · ft²) (W/m²);

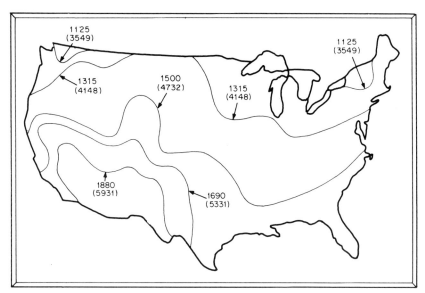

FIGURE 3 Average amount of solar energy available, in Btu/(day · ft²) (W/m²), for different parts of the United States. (*Power.*)

insolation is the sum of the direct, diffuse, and reflected radiation] for key hours of a day each month.

Estimate from these data the amount of solar radiation likely to reach the surface of a solar collector over 1 yr. Thus, for this industrial building in Jacksonville, Florida, Fig. 3 shows that the average amount of solar energy available is 1500 Btu/(day · ft²) (4.732 W/m²).

When you make this estimate, keep in mind that on a clear, sunny day direct radiation accounts for 90 percent of the insolation. On a hazy day only diffuse radiation may be available for collection, and it may not be enough to power the solar heating and cooling system. As a guide, the water temperatures required for solar heating and cooling systems are:

Space heating	Up to 170°F (76.7°C)
Space cooling with absorption air conditioning	From 200 to 240°F (93.3 to 114.6°C)
Domestic hot water	140°F (60°C)

2. *Choose collector type for the system*
There are two basic types of solar collectors: flat-plate and concentrating types. At present the concentrating type of collector is not generally cost-competitive with the flat-plate collector for normal space heating and cooling applications. It will probably find its greatest use for high-temperature heating of process liquids, space cooling, and generation of electricity. Since process heating applications are not the subject of this calculation procedure, concentrating collectors are discussed separately in another calculation procedure.

Flat-plate collectors find their widest use for building heating, domestic water heating, and similar applications. Since space heating and cooling are the objective of the system being considered here, a flat-plate collector system will be a tentative choice until it is proved suitable or unsuitable for the system. Figure 4 shows the construction details of typical flat-plate collectors.

3. Determine the collector orientation

Flat-plate collectors should face south for maximum exposure and should be tilted so the sun's rays are normal to the plane of the plate cover. Figure 5 shows the optimum tilt angle for the plate for various insolation requirements at different latitudes.

Since Jacksonville, Florida, is approximately at latitude 30°, the tilt of the plate for maximum year-round insolation should be 25° from Fig. 5. As a general rule for heating with maximum winter insolation, the tilt angle should be 15° plus the angle of latitude at the site; for cooling, the tilt angle equals the latitude (in the south, this should be the latitude minus 10° for cooling); for hot water, the angle of tilt equals the latitude plus 5°. For combined systems, such as heating, cooling, and hot water, the tilt for the dominant service should prevail. Alternatively, the tilt for maximum year-round insolation can be sued, as was done above.

When collector banks are set in back of one another in a sawtooth arrangement, low winter sun can cause shading of one collector by another. This can cause a

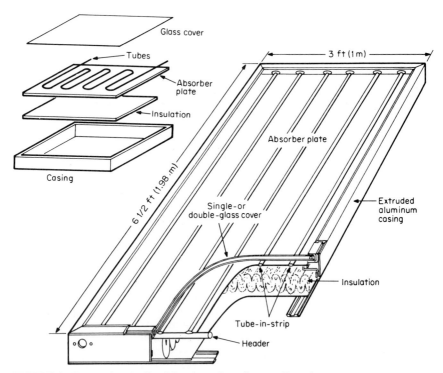

FIGURE 4   Construction details of flat-plate solar collectors. (*Power.*)

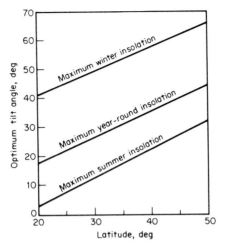

FIGURE 5 Spacing of solar flat-plate collectors to avoid shadowing. (*Power.*)

loss in capacity unless the units are carefully spaced. Table 1 shows the minimum spacing to use between collector rows, based on the latitude of the installation and collector tilt.

4. *Sketch the system layout*

Figure 6 shows the key components of a solar system using flat-plate collectors to capture solar radiation. The arrangement provides for heating, cooling, and hot-water production in this industrial building with sunlight supplying about 60 percent of the energy needed to meet these loads—a typical percentage for solar systems.

For this layout, water circulating in the rooftop collector modules is heated to 160°F (71.1°C) to 215°F (101.7°C). The total collector area is 10,000 ft² (920 m²). Excess heated hot water not need for space heating or cooling or for domestic water is directed to four 6000-gal (22,740-L) tanks for short-term energy storage. Conventional heating equipment provides the hot water needed for heating and cooling during excessive periods of cloudy weather. During a period of 3 h around noon on a clear day, the heat output of the collectors is about 2 million Btu/h (586 kW), with an efficiency of about 50 percent at these conditions.

TABLE 1 Spacing to Avoid Shadowing, ft (m)°

Collector angle, deg	Latitude of installation, deg			
	30	35	40	45
30	9.9 (3.0)	11.1 (3.4)	12.6 (3.8)	13.7 (4.2)
45	10.6 (3.2)	12.2 (3.7)	14.4 (4.4)	16.0 (4.9)
60	10.7 (3.3)	12.6 (3.8)	15.4 (4.7)	17.2 (5.2)

° *Power* magazine.

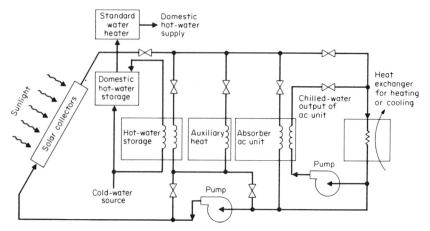

FIGURE 6   Key components of a solar-energy system using flat-plate collectors. (*Power.*)

For this industrial building solar-energy system, a lithium-bromide absorption air-conditioning unit (a frequent choice for solar-heated systems) develops 100 tons (351.7 kW) of refrigeration for cooling with a coefficient of performance of 0.71 by using heated water from the solar collectors. Maximum heat input required by this absorption unit is 1.7 million Btu/h (491.8 kW) with a hot-water flow of 240 gal/min (909.6 L/min). Variable-speed pumps and servo-actuated valves control the water flow rates and route the hot-water flow from the solar collectors along several paths—to the best exchanger for heating or cooling of the building, to the absorption unit for cooling of the building, to the storage tanks for use as domestic hot water, or to short-term storage before other usage. The storage tanks hold enough hot water to power the absorption unit for several hours or to provide heating for up to 2 days.

Another—and more usual—type of solar-energy system is shown in Fig. 7. In it a flat-plate collector absorbs heat in a water/antifreeze solution that is pumped to a pair of heat exchangers.

From unit no. 1 hot water is pumped to a space-heating coil located in the duct work of the hot-air heating system. Solar-heated antifreeze solution pumped to unit no. 2 heats the hot water for domestic service. Excess heated water is diverted to fill an 8000-gal (30,320-L) storage tank. This heated water is used during periods of heavy cloud cover when the solar heating system cannot operate as effectively.

5. *Give details of other techniques for solar heating*
Wet collectors having water running down the surface of a tilted absorber plate and collected in a gutter at the bottom are possible. While these "trickle-down" collectors are cheap, their efficiency is impaired by heat losses from evaporation and condensation.

Air systems using rocks or gravel to store heat instead of a liquid find use in residential and commercial applications. The air to be heated is circulated via ducts to the solar collector consisting of rocks, gravel, or a flat-plate collector. From here other ducts deliver the heated air to the area to be heated.

In an air system using rocks or gravel, more space is needed for storage of the solid media, compared to a liquid. Further, the ductwork is more cumbersome and

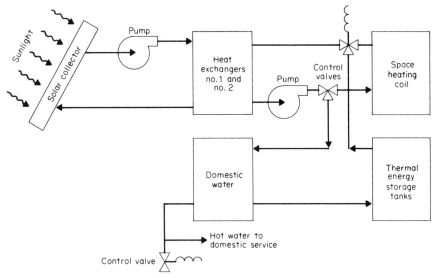

FIGURE 7   Solar-energy system using flat-plate collectors and an antifreeze solution in a pair of heat exchangers. (*Power.*)

occupies more space than the piping for liquid heat-transfer media. And air systems are generally not suitable for comfort cooling or liquid heating, such as domestic hot water.

Eutectic salts can be used to increase the storage capacity of air systems while reducing the volume required for storage space. But these salts are expensive, corrosive, and toxic, and they become less effective with repeated use. Where it is desired to store thermal energy at temperatures above 200°F (93.3°C), pressurized storage tanks are attractive.

Solar "heat wheels" can be used in the basic solar heating and cooling system in the intake and return passages of the solar system. The wheels permit the transfer of thermal energy from the return to the intake side of the system and offer a means of controlling humidity.

For solar cooling, high-performance flat-plate collectors or concentrators are needed to generate the 200 to 240°F (93.3 to 115.6°C) temperatures necessary for an absorption-chiller input. These chillers use either lithium bromide or ammonia with hot water to for an absorbent/refrigerant solution. Chiller operation is conventional.

Solar collectors can be used as a heat source for heat-pump systems in which the pump transfers heat to a storage tank. The hot water in the tank can then be used for heating, while the heat pump supplies cooling.

In summary, solar energy is a particularly valuable source of heat to augment conventional space-heating and cooling systems and for heating liquids. The practical aspects of system operation can be troublesome—corrosion, deterioration, freezing, condensation, leaks—but these problems can be surmounted. Solar energy is not "free" because a relatively high initial investment for equipment must be paid off over a long period. And the equipment requires some fossil-fuel energy to fabricate.

But even with these slight disadvantages, the more solar energy that can be put to work, the longer the supply of fossil fuels will last. And recent studies show that solar energy will become more cost-competitive as the price of fossil fuels continues to rise.

6. Give design guides for typical solar systems

To ensure the best performance from any solar system, keep these pointers in mind:

a. For space heating, size the solar collector to have an area of 25 to 50 percent of the building's floor area, depending on geographic location, amount of insulation, and ratio of wall to glass area in the building design.

b. For space cooling, allow 250 to 330 ft^2 (23.3 to 30.7 m^2) of collector surface for every ton of absorption air conditioning, depending on unit efficiency and solar intensity in the area. Insulate piping and vessels adequately to provide fluid temperatures of 200 to 240°F (93.3 to 115.6°C).

c. Size water storage tanks to hold between 1 and 2 gal/ft^2 (3.8 to 7.6 L/m^2) of collector surface area.

d. In larger collector installations, gang collectors in series rather than parallel. Use the lowest fluid temperature suitable for the heating or cooling requirements.

e. Insulate piping and collector surfaces to reduce heat losses. Use an overall heat-transfer coefficient of less than 0.04 Btu/(h · ft^2 · °F) [0.23 W/(m^2 · K)] for piping and collectors.

f. Avoid water velocities of greater than 4 ft/s (1.2 m/s) in the collector tubes, or else efficiency may suffer.

g. Size pumps handling antifreeze solutions to carry the additional load caused by the higher viscosity of the solution.

Related Calculations. The general guidelines given here are valid for solar heating and cooling systems for a variety of applications (domestic, commercial, and industrial), for space heating and cooling, and for process heating and cooling, as either the primary or supplemental heat source. Further, note that solar energy is not limited to semitropical areas. There are numerous successful applications of solar heating in northern areas which are often considered to be "cold." And with the growing energy consciousness in all field, there will be greater utilization of solar energy to conserve fossil-fuel use.

Energy experts in many different fields believe that solar-energy use is here to stay. Since there seems to be little chance of fossil-fuel price reductions (only increases), more and more energy users will be looking to solar heat sources to provide some of or all their energy needs. For example, Wagner College in Staten Island, New York, installed, at this writing, 11,100 ft^2 (1032.3 m^2) of evacuated-tube solar panels on the roof of their single-level parking structure. These panels provide heating, cooling, and domestic hot water for two of the buildings on the campus. Energy output of these evacuated-tube collectors is some 3 billion Btu (3.2 × 10^9 kJ), producing a fuel-cost savings of $25,000 during the first year of installation. The use of evacuated-tube collectors is planned in much the same way as detailed above. Other applications of such collectors include soft-drink bottling plants, nursing homes, schools, etc. More applications will be found as fossil-fuel price increases make solar energy more competitive in the years to come. Table 2 gives a summary of solar-energy collector choices for quick preliminary use.

Data in this procedure are drawn from an article in *Power* magazine prepared by members of the magazine's editorial staff and from Owens-Illinois, Inc.

TABLE 2 Solar-Energy Design Selection Summary

Energy collection device	Typical heat-transfer applications	Typical uses of collected heat
Flat-plate tubed collector	Liquid heating	Space heating or cooling; water heating
Wet collectors	Liquid heating	Water or liquid heating
Rock or gravel collectors	Air heating	Space heating
Flat-plate air heaters	Air heating	Space heating or cooling
Eutectic salts	Air heating	Space heating
Evacuated-tube collectors	Liquid heating to higher temperature levels	Space cooling and heating; process heating
Concentrating collectors	Steam generation	Electric-power generation
Pressurized storage tank	Storage of thermal energy above 200°F (93.3°C)	Industrial processes
Heat wheels	Heat transfer	Humidity control

DETERMINATION OF SOLAR INSOLATION ON SOLAR COLLECTORS UNDER DIFFERING CONDITIONS

A south-facing solar collector will be installed on a building in Glasgow, Montana, at latitude 48°13′N. What is the clear-day solar insolation on this panel at 10 a.m. on January 21 if the collector tilt angle is 48°? What is the daily surface total insolation for January 21, at this angle of collector tilt? Compute the solar insolation at 10:30 a.m. on January 21. What is the actual daily solar insolation for this collector? Calculate the effect on the clear-day daily solar insolation if the collector tilt angle is changed to 74°.

Calculation Procedure:

1. Determine the insolation for the collector at the specified location
The latitude of Glasgow, Montana, is 48°13′N. Since the minutes are less than 30, or one-half of a degree, the ASHRAE clear-day insolation table for 48° north latitude can be used. Entering Table 3 (which is an excerpt of the ASHRAE table) for 10 a.m. on January 21, we find the clear-day solar insolation on a south-facing collector with a 48° tilt is 206 Btu/(h · ft^2) (649.7 W/m^2). The daily clear-day surface total for January 21 is, from the same table, 1478 Btu/(day · ft^2) (4661.6 W/m^2) for a 48° collector tilt angle.

2. Find the insolation for the time between tabulated values
The ASHRAE tables plot the clear-day insolation at hourly intervals between 8 a.m. and 4 p.m. For other times, use a linear interpolation. Thus, for 10:30 a.m., interpolate in Table 3 between 10:00 and 11:00 a.m. values. Or, (249 − 206)/2 + 206 = 227.5 Btu/(h · ft^2) (717.5 W/m^2), where the 249 and 206 are the insolation values at 11 and 10 a.m. respectively. Note that the difference can be either added to or subtracted from the lower, or higher, clear-day insolation value, respectively.

TABLE 3 Solar Position and Insolation Values for 48°N Latitude[°]

Date	Solar time a.m.	Solar time p.m.	Solar position Alt.	Solar position Azm.	Btu·h/ft² (W/m²) total insolation on surfaces Normal	Horiz.	South-facing surface angle with horizontal 38	48	58	68	90
Jan 21	8	4	3.5	54.6	37 (116.6)	4 (12.6)	17 (53.6)	19 (59.9)	21 (66.2)	22 (69.4)	22 (69.4)
	9	3	11.0	42.6	185 (583.2)	46 (145.0)	120 (378.3)	132 (416.1)	140 (441.4)	145 (457.1)	139 (438.2)
	10	2	16.9	29.4	239 (753.4)	83 (261.7)	190 (598.9)	206 (649.4)	216 (680.9)	220 (693.6)	206 (649.4)
	11	1	20.7	15.1	261 (822.8)	107 (337.3)	231 (728.2)	249 (784.9)	260 (819.7)	263 (829.1)	243 (766.1)
	12		22.0	0.0	267 (841.7)	115 (362.5)	245 (772.4)	264 (832.3)	275 (866.9)	278 (876.4)	255 (803.9)
	Surface daily totals				1710 (5390.7)	596 (1878.9)	1360 (4287.4)	1478 (4659.4)	1550 (4886.4)	1578 (4974.6)	1478 (4659.4)

[°]From ASHRAE, excerpted; used with permission from ASHRAE. Metrication supplied by handbook editor.

3. *Find the actual solar insolation for the collector*
ASHRAE tables plot the clear-day solar insolation for particular latitudes. Dust, clouds, and water vapor will usually reduce the clear-day solar insolation to a value less than that listed.

To find the actual solar insolation at any location, use the relation $i_A = pi_T$, where i_A = actual solar insolation, Btu/(h·ft²) (W/m²); p = percentage of clear-day insolation at the location, expressed as a decimal; i_T = ASHRAE-tabulated clear-day solar insolation, Btu/(h·ft²) (W/m²). The value of $p = 0.3 + 0.65(S/100)$, where S = average sunshine for the locality, percent, from an ASHRAE or government map of the sunshine for each month of the year. For January, in Glasgow, Montana, the average sunshine is 50 percent. Hence, $p = 0.30 + 0.65(50/100) = 0.625$. Then $i_A = 0.625(1478) = 923.75$, say 923.5 Btu/(day·ft²) (2913.7 W/m²), by using the value found in step 1 of this procedure for the daily clear-day solar insolation for January 21.

4. *Determine the effect of a changed tilt angle for the collector*
Most south-facing solar collectors are tilted at an angle approximately that of the latitude of the location plus 15°. But if construction or other characteristics of the site prevent this tilt angle, the effect can be computed by using ASHRAE tables and a linear interpolation.

Thus, for this 48°N location, with an actual tilt angle of 48°, a collector tilt angle of 74° will produce a clear-day solar insolation of $i_T = 1578[(74 - 68)/(90 - 68)](1578 - 1478) = 1551.0$ Btu/(day·ft²) (4894.4 W/m²), by the ASHRAE tables. In the above relation, the insolation values are for solar collector tilt angles of 68° and 90°, respectively, with the higher insolation value for the smaller angle. Note that the insolation (heat absorbed) is greater at 74° than at 48° tilt angle.

Related Calculations. This procedure demonstrates the flexibility and utility of the ASHRAE clear-day solar insolation tables. Using straight-line interpolation, the designer can obtain a number of intermediate clear-day values, including solar insolation at times other than those listed, insolation at collector tilt angles different from those listed, insolation on both normal (vertical) and horizontal planes, and surface daily total insolation. The calculations are simple, provided the designer carefully observes the direction of change in the tabulated values and uses the latitude table for the collector location. Where an exact-latitude table is not available, the designer can interpolate in a linear fashion between latitude values less than and greater than the location latitude.

Remember that the ASHRAE tables give clear-day insolation values. To determine the actual solar insolation, the clear-day values must be corrected for dust, water vapor, and clouds, as shown above. This correction usually reduces the amount of insolation, requiring a larger collector area to produce the required heating or cooling. ASHRAE also publishes tables of the average percentage of sunshine for use in the relation for determining the actual solar insolation for a given location.

SIZING COLLECTORS FOR SOLAR-ENERGY HEATING SYSTEMS

Select the required collector area for a solar-energy heating system which is to supply 70 percent of the heat for a commercial building situated in Grand Forks,

Minnesota, if the computed heat loss 100,000 Btu/h (29.3 kW), the design indoor temperature is 70°F (21.1°C), the collector efficiency is given as 38 percent by the manufacturer, and collector tilt and orientation are adjustable for maximum solar-energy receipt.

Calculation Procedure:

1. Determine the heating load for the structure
The first step in sizing a solar collector is to compute the heating load for the structure. This is done by using the methods given for other procedures in this handbook in Sec. 16 under Heating, Ventilating and Air Conditioning, and in Sec. 16 under Electric Comfort Heating. Use of these procedures would give the hourly heating load—in this instance, it is 100,000 Btu/h (29.3 kW).

2. Compute the energy insolation for the solar collector
To determine the insolation received by the collector, the orientation and tilt angle of the collector must be known. Since the collector can be oriented and tilted for maximum results, the collector will be oriented directly south for maximum insolation. Further, the tilt will be that of the latitude of Grand Forks, Minnesota, or 48°, since this produces the maximum performance for any solar collector.

Next, use tabulations of mean percentage of possible sunshine and solar position and insolation for the latitude of the installation. Such tabulations are available in ASHRAE publications and in similar reference works. List, for each month of the year, the mean percentage of possible sunshine and the insolation in Btu/(day · ft²) (W/m²), as in Table 4.

TABLE 4   Solar Energy Available for Heating

Month	Mean sunshine, percent	Total insolation Btu/ (ft²·day)	Total insolation W/ (m²·day)	Efficiency of collector, percent	Energy available from collector* Btu/ (ft²·day)	Energy available from collector* W/ (m²·day)	Heating-season energy available Btu/ (month·ft²)	Heating-season energy available W/ (month·m²)
Jan.	49	1,478	4,663.1	38	275.2	868.3	8,531.2	26,915.9
Feb.	54	1,972	6,221.7	38	404.7	1,276.8	11,331.6	35,751.2
Mar.	55	2,228	7,029.3	38	465.7	1,469.3	14,436.7	45,547.8
Apr.	57	2,266	7,149.2	38	490.8	1,548.5	14,724.0	46,454.2
May	60	2,234	7,048.3	38	509.4	1,607.2	15,791.4	49,821.9
June	64	2,204	6,953.6	38	NA†	NA	NA	NA
July	72	2,200	6,941.0	38	NA†	NA	NA	NA
Aug.	69	2,200	6,941.0	38	NA†	NA	NA	NA
Sept.	60	2,118	6,682.3	38	482.9	1,523.6	14,487.0	45,706.5
Oct.	54	1,860	5,868.3	38	381.7	1,204.3	11,832.7	37,332.2
Nov.	40	1,448	4,568.4	38	220.1	694.4	6,603.0	20,832.5
Dec.	40	1,250	3,943.8	38	190.0	599.5	5,890.0	18,582.9
						Total =	103,627.6 [326.9 kW/ (month·m²)]	

*Values in this column = (mean sunshine, %)(total insolation, Btu/(ft²·day))(efficiency of collector, %).
†Not applicable because not part of the heating season; June, July, and August are ignored in the calculation.

Using a heating season of September through May, we find total solar energy available from the collector for these months is 103,627.6 Btu/ft^2 (326.9 kW/m^2), found by taking the heat energy per month (= mean sunshine, percent)[total insolation, Btu/(ft^2·day)](collector efficiency, percent) and summing each month's total. Heat available during the off season can be used for heating water for use in the building hot-water system.

3. Find the annual heating season heat load
Since the heat loss is 100,000 Btu/h (29.3 kW), the total heat load during the 9-month heating season from September through May, or 273 days, is H_a = (24 h)(273 days)(100,000 = 655,200,000 Btu (687.9 MJ).

4. Determine the collector area required
The calculation in step 2 shows that the total solar energy available during the heating season is 103,627.6 Btu/ft^2 (326.9 kW)/m^2). Then the collector area required is A ft^2 (m^2) = H_a/S_a, where S_a = total solar energy available during the heating season, Btu/ft^2. Or A = 655,200,000/103,627.6 = 6322.64 ft^2 (587.4 m^2) if the solar panel is to supply all the heat for the building. However, only 70 percent of the heat required by the building is to be supplied by solar energy. Hence, the required solar panel area = 0.7(6322.6) = 4425.8 ft^2 (411.2 m^2).

With the above data, a collector of 4500 ft^2 (418 m^2) would be chosen for this installation. This choice agrees well with the precomputed collector sizes published by the U.S. Department of Energy for various parts of the United States.

Related Calculations. The procedure shown here is valid for any type of solar collector—flat-plate, concentrating, or nonconcentrating. The two variables which must be determined for any installation are the annual heat loss for the structure and the annual heat flow available from the solar collector. Once these are known, the collector area is easily determined.

The major difficulty in sizing solar collectors for either comfort heating or water heating lies in determining the heat output of the collector. Factors such as collector tilt angle, orientation, and efficiency must be carefully evaluated before the collector final choice is made. And of these three factors, collector efficiency is probably the most important in the final choice of a collector.

F CHART METHOD FOR DETERMINING USEFUL ENERGY DELIVERY IN SOLAR HEATING

Determine the annual heating energy delivery of a solar space-heating system using a double-glazed flat-plate collector if the building is located in Bismarck, North Dakota, and the following specifications apply:

Building
Location: 47°N latitude
Space-heating load: 15,000 Btu/(°F·day) [8.5 kW/(m^2·K·day)]

Solar System
Collector loss coefficient: $F_R U_C$ = 0.80 Btu/(h·ft^2·°F) [4.5 W/(m^2·K)]
Collector optical efficiency (average): $F_R(\overline{\tau\alpha})$ = 0.70
Collector tilt: $\beta = L + 15° = 62°$

Collector area: $A_c = 600$ ft² (55.7 m²)
Collector fluid flow rate: $\dot{m}_c/A_c = 11.4$ lb/(h·ft²)(water) [0.0155 kg/(s·m²)]
Collector fluid heat capacity—specific gravity product: $c_{pc} = 0.9$ Btu/(lb·°F) [3.8 kJ/(kg·K)] (antifreeze)
Storage capacity: 1.85 gal/ft²(water) (75.4 L/m²)
Storage fluid flow rate: $\dot{m}_s/A_c = 20$ lb/(h·ft²)(water) [0.027 kg/(s·m²)]
Storage fluid heat capacity: $c_{ps} = 1$ Btu/(lb·°F)(water) [4.2 kJ/(kg·K)]
Heat-exchanger effectiveness: 0.75

Climatic Data
Climatic data from the NWS are tabulated in Table 5.

Calculation Procedure:

1. Determine the solar parameter P_s
The F chart is a common calculation procedure used in the United States to ascertain the useful energy delivery of active solar heating systems. The F chart applies only to the specific system designs of the type shown in Fig. 8 for liquid systems and Fig. 9 for air systems. Both systems find wide use today.

The F chart method consists of several empirical equations expressing the monthly solar heating fraction f_s as a function of dimensionless groups which relate system properties and weather data for a month to the monthly heating requirement. The several dimensionless parameters are grouped into two dimensionless groups call the *solar parameter* P_s and the *loss parameter* P_L.

The solar parameter P_s is the ratio of monthly solar energy absorbed by the collector divided by the monthly heating load, or

$$P_s = \frac{K_{ldhx}F_{hx}(F_R\overline{\tau\alpha})\bar{I}_cN}{L}$$

TABLE 5 Climatic and Solar Data for Bismarck, North Dakota

Month	Average ambient temperature		Heating, degree-days		Horizontal solar radiation, langleys/day
	°F	°C	°F	°C	
Jan.	8.2	−13.2	1761	978	157
Feb.	13.5	−10.3	1442	801	250
Mar.	25.1	−3.8	1237	687	356
Apr.	43.0	6.1	660	367	447
May	54.4	12.4	339	188	550
June	63.8	17.7	122	68	590
July	70.8	21.5	18	10	617
Aug.	69.2	20.7	35	19	516
Sept.	57.5	14.2	252	140	390
Oct.	46.8	8.2	564	313	272
Nov.	28.9	−1.7	1083	602	161
Dec.	15.6	−9.1	1531	850	124

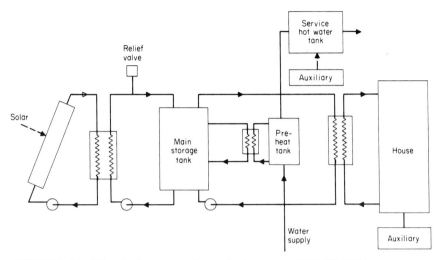

FIGURE 8   Liquid-based solar space- and water-heating system. (*DOE/CS-0011.*)

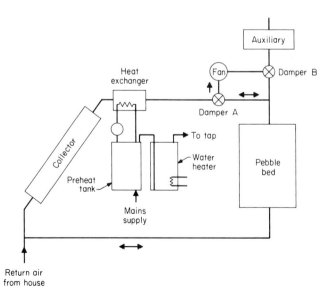

FIGURE 9   Air-based solar space- and water-heating system. (*DOE/CS-0011.*)

where F_{hx} = heat-exchanger penalty factor (see Fig. 10)

$F_R\tau\alpha$ = average collector optical efficiency = 0.95 × collector efficiency curve intercept $F_R(\tau\alpha)_n$

$\bar{I}_c$ = monthly average insolation on collector surface from a listing of monthly solar and climatic data

N = number of days in a month

L = monthly heating load, *net of any passive system delivery* as calculated by the P chart, solar load ratio (SLR), or any other suitable method, Btu/month

K_{ldhx} = load-heat-exchanger correction factor for liquid systems, Table 6

(The P chart method and the SLR method are both explained in Related Calculations below.)

The value of P_s is found for each month of the year by substituting appropriate unit values in the above equation and tabulating the results (Table 7).

2. *Determine the loss parameter P_L*

The loss parameter P_L is related to the long-term energy losses from the collector divided by the monthly heating load:

$$P_L = (K_{stor}K_{flow}K_{DHW})\frac{F_{hx}(F_R U_c)(T_r - \bar{T}_a)\,\Delta t}{L}$$

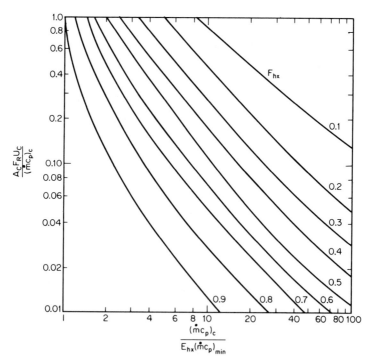

FIGURE 10 Heat-exchanger penalty factor F_{hx}. When no exchanger is present, $F = 1$. (*Kreider*—The Solar Heating Design Process, *McGraw-Hill.*)

TABLE 6 F Chart K Factors

Correction factor	Air or liquid system	Correction factor	Validity range for factor
K_{flow}	A°	$\{[2\ \text{ft}^3/(\text{min}\cdot\text{ft}_c^2)]/\text{actual flow}\}^{-0.28}$ $\{[0.61\ \text{m}^3/(\text{min}\cdot\text{m}^2)]/\text{actual flow}\}^{-0.28}$	1–$4\ \text{ft}^3/(\text{min}\cdot\text{ft}_c^2)$ $[0.03$ to $0.11\ \text{m}^3/(\text{min}\cdot\text{m}^2)]$
	L	Small effect included in F_R and F_{hx} only	
K_{stor}	A†	$[(0.82\ \text{ft}^3/\text{ft}_c^2)/\text{actual volume}]^{0.30}$ $[(0.25\ \text{m}^3/\text{m}^2)/\text{actual volume}]^{0.30}$	0.4–$3.3\ \text{ft}^3/\text{ft}_c^2$ $(0.012$ to $0.10\ \text{m}^3/\text{m}^2)$
	L	$[(1.85\ \text{gal}/\text{ft}_c^2)/\text{actual volume}]^{0.25}$ $[(0.0754\ \text{L}/\text{m}^2)/\text{actual volume}]^{0.25}$	0.9–$7.4\ \text{gal}/\text{ft}_c^2$ $(36.7$ to $301.5\ \text{L}/\text{m}^2)$
K_{DHW}‡	L	$\dfrac{(1.18T_{w,o} + 3.86T_{w,i} - 2.32\overline{T}_a - 66.2)}{212 - \overline{T}_a}$	
K_{ldhx}§	L	$0.39 + 0.65\exp[-0.139UA/E_L(\dot{m}c_p)_{air}]$	$0.5 < \dfrac{E_L(\dot{m}c_p)_{air}}{UA} < 50$
	A	NA	

°User must also include the effect of flow rate in F_R, i.e., in $F_R(\overline{\tau\alpha})$ and F_RU_c. Refer to manufacturer's data for this.

†For air systems using latent-heat storage, see J. J. Jurinak and S. I. Abdel-Khalik, *Energy*, vol. 4, p. 503 (1979) for the expression for K_{stor}.

‡Only applies for liquid storage DHW systems (air collectors can be used, however); $T_{w,o}$ = hot-water supply temperature, $T_{w,i}$ = cold-water supply temperature to water heater, both °F. Applies only to a specific water use schedule; predictions of performance for other schedules will have reduced accuracy.

§UA = unit building heat load, Btu/(h·°F); E_L = load-heat-exchanger effectiveness; $(\dot{m}c_p)_{air}$ = load-heat-exchanger air capacitance rate, Btu/(h·°F) = density × 60 × cubic feet per minute × 0.24.

Note: Metrication supplied by handbook editor.

where F_RU_c = magnitude of collector efficiency curve slope (can be modified to include piping and duct losses)

$\overline{T}_a$ = monthly average ambient temperature, °F (°C)[1]

Δt = number of hours per month = $24N$

T_r = reference temperature = 212°F (100°C)

K_{stor} = storage volume correction factor, Table 6

K_{flow} = collector flow rate correction factor, Table 6

K_{DHW} = conversion factor for parameter P_L when *only* a water heating system is to be studied, Table 6

3. Determine the monthly solar fraction

The monthly solar fraction f_s depends only on these two parameters, P_s and P_L. For liquid heating systems using solar energy as their heat source, the monthly solar fraction is given by

$$f_s = 1.029P_s - 0.065P_L - 0.245P_s^2 + 0.0018P_L^2 + 0.0215P_s^3$$

if $P_s > P_L/12$ (if not, $f_s = 0$).

---

[1]Note that $T_a \neq 65° -$ (degree heating days/N), contrary to statements of many U.S. goverment contractors and in many government reports. The equality is only valid for those months when $T_a < 65°F$ every day, i.e., only 3 or 4 months of the year at the most. The errors propagated through the solar industry by assuming the equality to be true are too many to count.

TABLE 7 *F* Chart Summary

Month	Collector-plane radiation Btu/(day·ft²)	W/m²	Monthly energy demand L million Btu	MJ/month	P_L^a	P_s^a	f_s	Monthly delivery million Btu	MJ/month
Jan.	1506	4749.9	26.41	27.86	2.68	0.72	0.46	12.15	12.82
Feb.	1784	5626.7	21.63	22.82	2.88	0.95	0.60	12.98	13.69
Mar.	1795	5661.4	18.55	19.57	3.50	1.22	0.73	13.54	14.28
Apr.	1616	5096.9	9.90	10.44	5.75	2.00	0.94	9.31	9.82
May	1606	5065.3	5.08	5.36	10.78	>3.00	1.00	5.08	5.36
June	1571	4954.9	1.83	1.93	>20.00	>3.00	1.00	1.83	1.93
July	1710	5393.3	0.27	0.28	>20.00	>3.00	1.00	0.27	0.28
Aug.	1712	5399.6	0.52	0.55	>20.00	>3.00	1.00	0.52	0.55
Sept.	1721	5428.0	3.78	3.99	13.76	>3.00	1.00	3.78	3.99
Oct.	1722	5431.2	8.46	8.93	6.79	2.58	1.00	8.46	8.93
Nov.	1379	4349.4	16.24	17.13	3.79	1.04	0.61	9.91	10.46
Dec.	1270	4005.6	22.96	24.22	2.98	0.70	0.43	9.87	10.41
Annual			135.66	143.1			0.65	87.70	92.52

$^a P_s > 3.0$ or $P_L > 20.0$ implies $P_s = 3.0$ and $P_L = 20$, i.e., $f_s = 1.0$. The annual solar fraction $\bar{f}_s$ is 87.70/135.66, or 6 percent.

For air-based solar heating systems, the monthly solar heating fraction is given by

$$f_s = 1.040P_s - 0.065P_L - 0.159P_s^2 + 0.00187P_L^2 + 0.0095P_s^3$$

if $P_s > 0.07P_L$ (if not, $f_s = 0$).

Flow rate, storage, load heat-exchanger, and domestic hot-water correction factors for use in the equations for P_s and P_L, in steps 1 and 2 above, are given in Table 11 and Fig. 11.

When you use the *F* chart method of calculation for any system, follow this order: collector insolation, collector properties, monthly heat loads, monthly ambient temperatures, and monthly values of P_s and P_L. Once the parameter values are known, the monthly solar fraction and monthly energy delivery are readily calculated, as shown in Table 7.

The total of all monthly energy deliveries is the total annual useful energy produced by the solar system. And the total annual useful energy delivered divided by the total annual load is the annual solar load fraction.

4. Compute the monthly energy delivery

Set up a tabulation such as that in Table 7. Using weather data for Bismarck, North Dakota, list the collector-plane radiation, Btu/(day·ft²), monthly energy demand [= space-heating load, Btu/(°F·day)](degree days for the month, from weather data), P_L computed from the relation given, P_s computed from the relation given, f_s computed from the appropriate relation (water or air) given earlier, and the monthly delivery found from f_s (monthly energy demand).

Related Calculations.   In applying the *F* chart method, it is important to use a consistent area basis for calculating the efficiency curve information and the solar and loss parameters, P_s and P_L. The early National Bureau of Standards (NBS) test

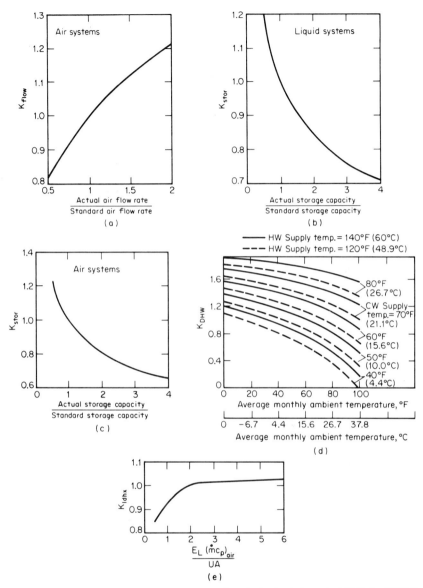

FIGURE 11　*F* chart correction factors. (*a*) K_{flow} {standard value is 2 standard ft^2/(min · ft^2)[0.01 m^3/(s · m^2)]} (*b*), (*c*) K_{stor} [standard liquid value is 1.85 gal/ft^2 (75.4 L/m^2); standard rock value is 0.82 ft^3/ft^2 (0.025 m^3/m^2)]. (*d*) K_{DHW}. (*e*) K_{ldhx} (standard value of abscissa is 2.0). (*U.S. Dept. of Housing and Urban Development and Kreider*—The Solar Heating Design Process, *McGraw-Hill.*)

procedures based the collector efficiency on net glazing area. A more recent and more widely used test procedure developed by ASHRAE (93.77) uses the gross-area basis. The gross area is the area of the glazing plus the area of opaque weatherstripping, seals, and supports. Hence, when ASHRAE test data are used, the solar and loss parameters must be based on gross area. The efficiency curve basis and F chart basis must be consistent for proper results.

The F chart method can be used for a number of other solar-heating calculations, including performance of an associated heat-pump backup system, collectors connected in series, etc. For specific steps in these specialized calculation procedures, see Kreider—*The Solar Heating Design Process,* McGraw-Hill. The calculation procedure given here is based on the Kreider book, with numbers and SI units being added to the steps in the calculation by the editor of this handbook.

The P chart mentioned as part of the P_s calculation is a trademark of the Solar Energy Design Corporation, POB 67, Fort Collins, Colorado 80521. Developed by Arney, Seward, and Kreider for passive predictions of solar performance, the P chart uses only the building heat load in Btu/(°F·day). The P chart will specify the solar fraction and optimum size of three passive systems.

The monthly solar load ratio is an empirical method of estimating monthly solar and auxiliary energy requirements for passive solar systems. For more data on both the P chart and SLR methods, see the Kreider work mentioned above.

DOMESTIC HOT-WATER-HEATER COLLECTOR SELECTION

Select the area for a solar collector to provide hot water for a family of six people in a residential building in Northport, New York, when the desired water outlet temperature is 140°F (60°C) and the water inlet temperature is 50°F (10°C). A pumped-liquid type of domestic hot-water (DHW) system (Fig. 12) is used. Compare the collector area required for 60, 80, and 90 percent of the DHW heading load.

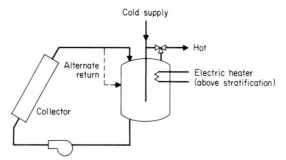

FIGURE 12 Direct-heating pump circulation solar water heater. (*DOE/CS-0011.*)

Calculation Procedure:

1. Find the daily DHW heating load
A typical family in the United States uses about 20 gal (9.1 L) of hot water per person per day. Hence, a family of six will use a total of 6(20) = 120 gal (54.6 L) per day. Since water has a specific heat of unity (1.0) and weighs 8.34 lb/gal (1.0 kg/dm³), the daily DHW heating load is L = (120 gal/day)(8.34 lb/gal) [1.0 Btu /(lb·°F)](140 − 50) = 90,072 Btu/day (95,026 kJ/day). This is the 100 percent heating load.

2. Determine the average solar insolation for the collector
Use the month of January because this usually gives the minimum solar insolation during the year, providing the maximum collector area. Using Fig. 13 for eastern Long Island, where Northport is located, we find the solar insolation H = 580 Btu/(ft²·day) [1829.3 W/(m²·day)] on a horizontal surface. (The horizontal surface insolation is often used in DHW design because it provides conservative results.)

3. Find the HA/L ratio for this installation
Use Fig. 14 to find the HA/L ratio, where A = collector area, ft² (m²). Enter Fig. 14 on the left at the fraction F of the annual load supplied by solar energy; project to the right to the tinted area and then vertically downward to the HA/L ratio. Thus, for F = 60 percent, HA/L = 1.0; F = 80 percent, HA/L = 1.5; for F = 90 percent, HA/L = 2.0.

4. Compute the solar collector area required
Use the relation HA/L = 1 for the 60 percent fraction. Or, HA/L = 1, A = L/H = 0.6(90,072)/580 = 93.18 ft² (8.7 m²). For HA/L = 1.5 for the 80 percent fraction, A = 0.8(90,072)/580 = 124.24 ft² (11.5 m²). And for the 90 percent fraction, A = 0.9(90.072)/580 = 139.77 ft² (12.98 m²).

Comparing these areas shows that the 80 percent factor area is 33 percent larger than the 60 percent factor area, while the 90 percent factor area is 50 percent larger. To evaluate the impact of the increased area, the added cost of the larger collector must be compared with the fuel that will be saved by reducing the heat input needed for DHW heating.

Related Calculations. Figures 13 and 14 are based on computer calculations for 11 different locations for the month of January in the United States ranging from Boulder, Colorado, to Boston; New York; Manhattan, Kansas; Gainesville, Florida; Santa Maria, California; St. Cloud, Minnesota; Washington; Albuquerque, New Mexico; Madison, Wisconsin; Oak Ridge, Tennessee. The separate curve above the shaded band in Fig. 14 is the result for Seattle, which is distinctly different from other parts of the country. Hot-water loads used in the computer computations range from 50 gal/day (189.2 L/day) to 2000 gal/day (7500 L/day). The sizing curves in Fig. 14 are approximate and should not be expected to yield results closer than 10 percent of the actual value.

Remember that the service hot-water load is nearly constant throughout the year while the solar energy collected varies from season to season. A hot-water system sized for January, such as that in Fig. 14 with collectors tilted at the latitude angle, will deliver high-temperature water and may even cause boiling in the summer. But a system sized to meet the load in July will not provide all the heat needed in

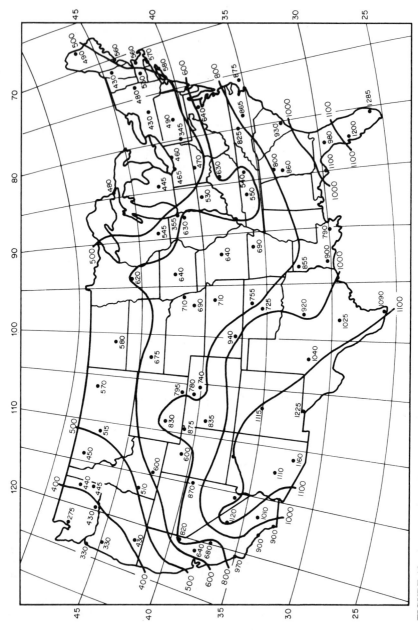

FIGURE 13 Average solar radiation, Btu/ft² (× 3.155 = W/m²), horizontal surface in the month of January. (*DOE/CS-0011.*)

17.26

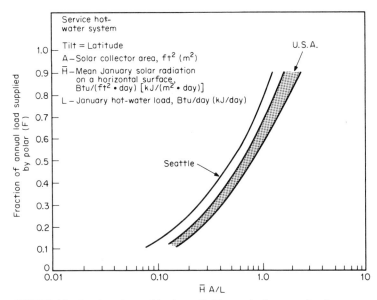

FIGURE 14 Fraction of annual load supplied by a solar hot-water heating system as a function of January conditions. (*DOE/CS-0011.*)

winter. Orientation of the collector can partially overcome the month-to-month fluctuations in radiation and temperature.

Solar-energy water heaters cost from $300 for a roof-mounted collector to over $2000 for a collector mounted on a stand adjacent to the house. The latter are nonfreeze type collectors fitted with a draindown valve, 50 ft² (4.7 m²) of collector surface area, an 80-gal (302.8-L) water tank, and the needed pumps and controls. Estimates of the time to recover the investment in such a system can range from as little as 3 to as long as 8 yr, depending on the cost of the fuel saved. The charts used in this procedure were originally published in DOE/CS-0011, *Introduction to Solar Heating and Cooling—Design and Sizing,* available from the National Technical Information Service, Springfield, Virginia 22161.

DOE/CS-0011 notes that a typical family of four persons requires, in the United States, about 80 gal (302.8 L) of hot water per day. At a customary supply temperature of 140°F (60°C), the amount of heat required if the cold inlet water is at 60°F (15.5°C) is about 50,000 Btu/day (52,750 kJ/day). Further, there is a wide variation in the solar availability from region to region and from season to season in a particular location. There are also the short-term radiation fluctuations owning to cloudiness and the day-night cycle.

Seasonal variations in solar availability result in a 200 to 400 percent difference in the solar heat supply to a hot-water system. In the winter, for example, an average recovery of 40 percent of 1200 Btu/ft² (3785 W/m²) of solar energy of sloping surface would require approximately 100 ft² (9.3 m²) of collector for the 50,000-Btu (52,750-kJ) average daily requirement. Such a design would provide essentially all the hot-water needs on an average winter day, but would fall short on days of less than average sunshine. By contrast, a 50 percent recovery of an average summer radiant supply of 2000 Btu/ft² (6308 W/m²) would involve the need for only 50 ft² (4.6 m²) of collector to satisfy the average hot-water requirements.

If a 50-ft² (4.6-m²) solar collector were installed, it could supply the major part of, or perhaps nearly all, the summer hot-water requirements, but it could supply less than half the winter needs. And if a 100-ft² (9.3-m²) solar collector were used so that winter needs could be more nearly met, the system would be oversized for summer operation and excess solar heat would be wasted. In such circumstances, if an aqueous collection medium were used, boiling in the system would occur and collector or storage venting of steam would have to be provided.

The more important disadvantage of the oversized solar collector (for summer operation) is the economic penalty associated with investment in a collector that is not fully utilized. Although the cost of the 100-ft² (9.3-m²) solar collector system would not be double that of the 50-ft² (4.6-m²) unit, its annual useful heat delivery would be considerably less than double. It would, of course, deliver about twice as much heat in the winter season, when nearly all the heat could be used. But, in the other seasons, particularly in summer, heat overflow would occur. The net effect of these factors is a lower economic return, per unit of investment, by the larger system. Stated another way, more Btu (kJ) per dollar of investment (hence cheaper solar heat) can be delivered by the smaller system.

If it is sized on average daily radiation in the sunniest months, the solar collector will be slightly oversized and a small amount of heat will be wasted on days of maximum solar input. On partly cloudy days during the warm season, some auxiliary heat must be provided. In the month of lowest average solar energy delivery, typically one-half to one-third as much solar-heated water can be supplied as during the warm season. Thus, fuel requirements for increasing the temperature of solar-heated water to the desired (thermostated) level could involve one-half to two-thirds of the total energy needed for hot-water heating in a midwinter month.

One disadvantage of solar DHW heating systems is the possibility of the water in the collector and associated pipe freezing during unexpectedly cold weather. Since they were introduced on a wide scale, thousands of solar DHW systems have suffered freeze damage, even relatively warm areas of the world. Such damage is both costly and wasteful of energy.

Three ways are used to prevent freeze damage in solar DHW systems:

1. Pump circulation of warm water through the collector and piping during the night hours reduces the savings produced by the solar DHW heating system because the energy required to run the pump must be deducted from the fuel savings resulting from use of the solar panels.

2. Use an automatic drain-down valve or mechanism to empty the system of water during freezing weather. Since the onset of a freeze can be sudden, such systems must be automatic if they are to protect the collector and piping while the occupants of the building are away. Unfortunately, there is no 100 percent reliable drain-down valve or mechanism. A number of "fail-safe" systems have frozen during unusually sharp or sudden cold spells. Research is still being conducted to find the completely reliable drain-down device.

3. Indirect solar DHW systems use a nonfreeze fluid in the collector and piping to prevent freeze damage. The nonfreeze fluid passes through a heat exchanger wherein it gives up most of its heat to the potable water for the DHW system. To date, the indirect system gives the greatest protection against freezing. Although there is a higher initial cost for an indirect system, the positive freeze protection is felt to justify this additional investment.

There are various sizing rules for solar DHW heating systems. Summarized below are those given by Kreider and Keith—*Solar Heating and Cooling,* Hemisphere and McGraw-Hill:

Collector area: 1 ft^2/(gal · day) [0.025 m^2/(L · day)]; DHW storage tank capacity: 1.5 to 2 gal/ft^2 (61.1 to 81.5 L/m^2) of collector area; collector water flow rate: 0.025 gal/(min · ft^2) [0.000017 m^3/(s · m^2); indirect system storage flow rate: 0.03 to 0.04 gal/(min · ft^2) [0.0002 to 0.00027 m^3/(s · m^2)] of collector area; indirect system heat-exchanger area of 0.05 to 0.1 ft^2/collector ft^2 (0.005 to 0.009 m^2/collector m^2); collector tilt; latitude $\pm 5°$; indirect system expansion-tank volume: 12 percent of collector fluid loop; controller turnon ΔT: 15 to 20°F (27 to 36°C); controller turnoff: 3 to 5°F (5.4 to 9°C); system operating pressure: provide 3 lb/in^2 (20.7 kPa) at topmost collector manifold; storage-tank insulation: R-25 to R-30; mixing-valve set point: 120 to 140°F (48.8 to 59.9°C); pipe diameter: to maintain fluid velocity below 6 ft/s (1.83 m/s) and above 2 ft/s (0.61 m/s).

Most domestic solar hot-water heaters are installed to reduce fuel cost. Typically, domestic hot water is heated in an oil-burning boiler or heater. A solar collector reduces the amount of oil needed to heat water, thereby reducing fuel cost. Simple economic studies will show how long it will take to recover the cost of the collector, given the estimated fuel saving.

A welcome added benefit obtained when using a solar collector to heat domestic water is the reduced atmospheric pollution because less fuel is burned to heat the water. All combustion produces carbon dioxide, which is believed to contribute to atmospheric pollution and the possibility of global warming. Reducing the amount of fuel burned to heat domestic water cuts the amount of carbon dioxide emitted to the atmosphere.

Although reduced carbon dioxide emission is difficult to evaluate on an economic basis, it is a positive factor to be considered in choosing a hot-water heating system. With greater emphasis on environmentally desirable design, solar heating of domestic hot water will receive more attention in the future.

PASSIVE SOLAR HEATING SYSTEM DESIGN

A south-facing passive solar collector will be designed for a one-story residence in Denver, Colorado. Determine the area of collector required to maintain an average inside temperature of 70°F (21°C) on a normal clear winter day for a corner room 15 ft (4.6 m) wide, 14 ft (4.3 m) deep, and 8 ft (2.4 m) high. The collector is located on the 15-ft (4.6-m) wide wall facing south, and the 14-ft (4.3-m) sidewall contains a 12-ft^2 (1.11 m^2) window. The remaining two walls adjoin heated space and so do not transfer heat. Find the volume and surface area of thermal storage material needed to prevent an unsuitable daytime temperature increase and to store the solar gain for nighttime heating. Estimate the passive solar-heating contribution for an average heating season.

Calculation Procedure:

1. *Compute the heat loss*
The surface areas and the coefficients of heat transmission of collector, windows, doors, walls, and roofs must be known to calculate the conductive heat losses of a

space. The collector area can be estimated for purposes of heat-loss calculations from Table 8.

Table 8 lists ranges of the estimated ratio of collector area to floor area, g, of a space for latitudes 36°N or 48°N based on 4°F (2.2°C) intervals of average January temperature and on various types of passive solar collectors. Average January temperatures can be selected from government weather data. Denver has an average January temperature of 32°F (0°C). Choosing a direct-gain system for this installation, read down to the horizontal line for t_o = 32°F (0°C), and then read right to the column for a direct-gain system. To find the estimated ratio of collector area to floor area, use a linear interpolation. Thus for Denver, which is located at approximately 40°N, interpolate between 48°N and 36°N values. Or, $(0.24 - 0.20)/12 \times (40 - 36) + 0.20 = 0.21$, where 0.24 and 0.20 are the ratios at 48°N and 36°N, respectively; 12 is a constant derived from $48 - 36$; and 40 is the latitude for which a ratio is sought.

Next, find the collector area by using the relation $A_C = (g)(A_F)$, where A_C = collector area, ft^2 (m^2); g = ratio of collector area to floor area, expressed as a decimal; and A_F = floor area, ft^2 (m^2). Therefore, $A_C = (0.21)(2.10) = 44$ ft^2 (4.1 m^2).

To compute the conductive heat loss through a surface, use the general relation $H_C = UA \, \Delta t$, where H_C = conductive heat loss, Btu/h (W); U = overall coefficient of heat transmission of the surface, Btu/(h · ft^2 · °F) [W/(m^2 · K)]; A = area of heat transmission surface, ft^2 (m^2); and Δt = temperature difference,°F = $65 - t_o$ (°C = $18.33 - t_o$), where t_o = average monthly temperature,°F (°C). The U values of materials can be found in ASHRAE and architectural handbooks.

Since a direct-gain system was selected, the total area of glazing is the sum of the collector and noncollector glazing 44 ft^2 (4.1 m^2) + 12 ft^2 (1.1 m^2) = 56 ft^2 (5.2 m^2). Double glazing is recommended in all passive solar designs and is found to have a U value of 0.42 Btu/(h · ft^2 · °F) [2.38 W/(m^2 · K)] in winter. Thus, the conductive heat loss through the glazing is $H_C = UA \, \Delta t = (0.42)(56)(65 - 32) = 776$ Btu/h (227.4 W).

The area of opaque wall surface subject to heat loss can be estimated by multiplying the wall height by the total wall length and then subtracting the estimated glazed areas from the total exterior wall area. Thus, the opaque wall area of this

TABLE 8 Estimated Ratio of Collector Area to Floor Area, $g = h_L(65 - t_o)/i_r$ for 36 to 48° North Latitude°

Average January temperature T_o, °F (°C)	Direct gain g	Water wall g	Masonry wall g
20 (−6.7)	0.27–0.32	0.54–0.64	0.69–0.81
24 (−4.4)	0.25–0.29	0.49–0.58	0.63–0.74
28 (−2.2)	0.22–0.27	0.44–0.52	0.56–0.67
32 (0)	0.20–0.24	0.39–0.47	0.50–0.60
36 (+2.2)	0.17–0.21	0.35–0.41	0.44–0.53
40 (+4.4)	0.15–0.18	0.30–0.35	0.38–0.45
44 (+6.7)	0.13–0.15	0.25–0.30	0.32–0.38

For SI temperatures, use the relation $g = h_L(18.33 - t_o)/i_T$.
°Based on a heat loss of 8 Btu/(day · ft^2 · °F) [0.58 W/(m^2 · K)].

space is $(8)(15 + 14) - 56 = 176$ ft^2 (16.3 m^2). Use the same general relation as above, substituting the U value and area of the wall. Thus, $U = 0.045$ Btu/(h·ft^2 · °F) [0.26 W/(m^2·K)], and $A = 176$ ft^2 (16.3 m^2). Then $H_C = UA \Delta t = (0.045)(176)(65 - 32) = 261$ Btu/h (76.5 W).

To determine the conductive heat loss of the roof, use the same general relation as above, substituting the U value and area of the roof. Thus, $U = 0.029$ Btu/(h· ft^2·°F) [0.16 W/(m^2·K)] and $A = 210$ ft^2 (19.5 m^2). Then $H_C = UA \Delta t = (0.029)(210)(65 - 32) = 201$ Btu/h (58.9 W).

To calculate infiltration heat loss, use the relation $H_i = Vn \Delta t/55$, where $V = $ volume of heated space, ft^3 (m^3); $n = $ number of air changes per hour, selected from Table 9. The volume for this space is $V = (15)(14)(8) = 1680$ ft^3 (47.6 m^3). Entering Table 9 at the left for the physical description of the space, read to the right for n, the number of air changes per hour. This space has windows on two walls, so $n = 1$. Thus, $H_i = (1680)(1)(65 - 32)/55 = 1008$ Btu/h (295.4 W).

The total heat loss of the space is the sum of the individual heat losses of glass, wall, roof, and infiltration. Therefore, the total heat loss for this space is $H_T = 776 + 261 + 201 + 1008 = 2246$ Btu/h (658.3 W). Convert the total hourly heat loss to daily heat loss, using the relation $H_D = 24H_T$, where $H_D = $ total heat loss per day, Btu/day (W). Thus, $H_D = 24(2246) = 53{,}904$ Btu/day (658.3 W).

2. Determine the daily insolation transmitted through the collector
Use government data or ASHRAE clear-day insolation tables. The latitude of Denver is 39°50′N. Since the minutes are greater than 30, or one-half of a degree, the ASHRAE table for 40°N is used. The collector is oriented due south. Hence, the average daily insolation transmitted through vertical south-facing single glazing for a clear day in January is $i_T = 1626$ Btu/ft^2 (5132 W/m^2), or double the half-day total given in the ASHRAE table. Since double glazing is used, correct the insolation transmitted through single glazing by a factor of 0.875. Thus, $i_T = (1626)(0.875) = 1423$ Btu/(day·ft^2) (4490 W/m^2) of collector.

3. Compute the area of unshaded collector required
Determine the area of unshaded collector needed to heat this space on an average clear day in January. An average clear day is chosen because sizing the collector for extreme or cloudy conditions would cause space overheating on clear days. January is used because it generally has the highest heating load of all the months.

To compute the collector area, use the relation $A_C = H_D/(E)(i_T)$, where $E = $ a rule of thumb for energy absorptance efficiency of the passive solar-heating system

TABLE 9 Air Changes per Hour for Well-Insulated Spaces°

Description of space	Number of air changes per hour n
No windows or exterior doors	0.33
Windows or exterior doors on one side	0.67
Windows or exterior doors on two sides	1.0
Windows or exterior doors on three sides	1.33

°These figures are based on spaces with weatherstripped doors and windows or spaces with storm windows or doors. If the space does not have these features, increase the value listed for n by 50%.

used, expressed as a decimal. Enter Table 10 for a direct gain system to find $E = 0.91$. Therefore, $A_C = 53,904/(0.91)(1423) = 42$ ft^2 (3.9 m^2).

If the area of unshaded collector computed in this step varies by more than 10 percent from the area of the collector estimated for heat-loss calculations in step 1, the heat loss should be recomputed with the new areas of collector and opaque wall. In this example, the computed and estimated collector areas are within 10 percent of each other, making a second computation of the collector area unnecessary.

4. Compute the insolation stored for nighttime heating

To compute the insolation to be stored for nighttime heating, the total daily insolation must be determined. Use the relation $i_D = (A_C)(i_T)(E)$, where i_D = total daily insolation collected, Btu (J). Therefore, $i_D = (42)(1423)(0.91) = 54,387$ Btu (57.4 kJ).

Typically 35 percent of the total space heat gain is used to offset daytime heat losses, requiring 65 percent to be stored for nighttime heating. Therefore, $i_S = (0.65)i_D$, where i_S = insolation stored, Btu (J). Thus, $i_S = (0.65)(54,387) = 35,352$ Btu (37.3 kJ). This step is not required for the design of thermal storage wall systems since the storage system is integrated within the collector.

5. Compute the volume of thermal storage material required

For a direct-gain system, use the formula $V_M = i_S/(d)(c_p)(\Delta t_s)(C_S)$, where V_M = volume of thermal storage material, ft^3 (m^3); d = density of storage material, lb/ft^3 (kg/m^3); c_p = specific heat of the material, Btu/(lb·°F) [kJ/(kg·K)]; Δt_s = temperature increase of the material, °F (°C); and, C_S = fraction of insolation absorbed by the material due to color, expressed as a decimal.

Select concrete as the thermal storage material. Entering Table 11, we find the density and specific heat of concrete to be 144 lb/ft^3 (2306.7 kg/m^3) and 0.22 Btu/(lb·°F) [0.921 kJ/(kg·K)], respectively.

A suitable temperature increase of the storage material in a direct-gain systems is $\Delta t_s = +15$°F ($+8.3$°C). A range of $+10$ to $+20$°F ($+5.6$ to 11.1°C) can be used with smaller increases being more suitable. Select from Table 12. In this space, thermal energy will be stored in floors and walls, resulting in a weighted average of $C_S = 0.60$. Thus, $V_M = 35,352/(144)(0.22)(15)(0.60) = 124$ ft^3 (3.5 m^3).

As a rule of thumb for thermal storage wall systems, provide a minimum of 1 ft^3 (0.30 m^3) of dark-colored thermal storage material per square foot (meter) of collector for masonry walls or 0.5 ft^3 (0.15 m^3) of water per square foot (meter) of collector for a water wall. This will provide enough thermal storage material to maintain the inside space temperature fluctuation within 15°F (8.33°C).

TABLE 10   Energy Absorptance Efficiency of Passive Solar-Heating Systems

System	Efficiency E
Direct gain	0.91
Water thermal storage wall or roof pond	0.46
Masonry thermal storage wall	0.36
Attached greenhouse	0.18

TABLE 11 Properties of Thermal Storage Materials

Material	Density d		Specific heat c_p		Heat capacity	
	lb/ft^3	kg/m^3	Btu/(lb·°F)	kJ/(kg·K)	Btu/(ft^3·°F)	kJ/(m^3·K)
Water	62.4	999.0	1.00	4184.0	62.40	4180.8
Rock	153	2449.5	0.22	920.5	33.66	2255.2
Concrete	144	2305.4	0.22	920.5	31.68	2122.6
Brick	123	1969.2	0.22	920.5	27.06	1813.0
Adobe	108	1729.1	0.24	1004.2	25.92	1736.6
Oak	48	768.5	0.57	2384.9	27.36	1833.1
Pine	31	496.3	0.67	2803.3	20.77	1391.6

TABLE 12 Insolation Absorption Factors for Thermal Storage Material Based on Color

Color/Material	Factor C_S
Black, matte	0.95
Dark blue	0.91
Slate, dark gray	0.89
Dark green	0.88
Brown	0.79
Gray	0.75
Quarry tile	0.69
Red brick	0.68
Red clay tile	0.64
Concrete	0.60
Wood	0.60
Dark red	0.57
Limestone, dark	0.50
Limestone, light	0.35
Yellow	0.33
White	0.18

6. Determine the surface area of storage material for a direct-gain space

In a direct-gain system, the insolation must be spread over the surface area of the storage material to prevent overheating. Generally, the larger the surface area of material, the lower the inside temperature fluctuation, and thus the space is more comfortable. To determine this area, enter Fig. 15 at the lower axis to select an acceptable space temperature fluctuation. Project vertically to the curve, and read left to the A_S/A_C ratio. This is the ratio of thermal storage material surface to collector area, where A_S = surface of storage material receiving direct, diffused or reflected insolation, ft^2 (m^2). In this example, 15°F (8.33°C) is selected, requiring A_S/A_C = 6.8. Thus, A_S = (6.8)(42) = 286 ft^2 (26.5 m^2).

This step is not required for the design of thermal storage wall systems in which $A_S = A_C$.

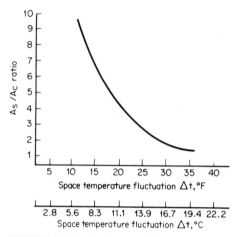

FIGURE 15   Ratio of mass surface area to collector area. (*Based on data in DOE/CS-0127-2* Passive
Solar Design Handbook, *vol. 2.*)

7. Determine the average daily inside temperatures

To verify that the collector and thermal storage material are correctly sized, the
average inside temperature must be determined. Use $t_I = t_o + 5 + (i_D)(65 - t_o)/H_D$, where t_I = average daily inside temperature, °F (°C), and 5°F (2.8°C) is an
assumed inside temperature increase owing to internal heat generation such as
lights, equipment, and people. Thus, $t_I = 32 + 5 + (54,387)(65 - 32)/53,904 =$
70.3°F (21.27°C).

To determine the average daily low and high temperatures, use $t_L = t_I - \Delta t/2.5$, and $t_H = t_I + \Delta t/1.67$, where t_L = minimum average space temperature,°F
(°C); t_H = maximum average space temperature, °F (°C); and Δt = inside space
temperature fluctuation used in step 6. Thus, $t_L = 70.3 - 15/2.5 = 64.3$°F (17.9°C),
and $t_H = 70.3 + 15/1.67 = 79.3$°F (26.27°C).

8. Estimate the passive solar-heating contribution

To estimate the passive solar-heating contribution (SHC) for an average month, use
$\text{SHC}_M = 100(i_D)(p)/H_D$, where SHC_M = solar heating contribution of the total
monthly space-heating needs, percent, and p = an insolation factor based on the
percentage of clear days, expressed as a decimal. The value of $p = 0.30 +$
$0.65(S/100)$, where S = average sunshine for the month, percent, from an ASHRAE
or government map of sunshine for each month. The average January sunshine for
Denver is 67 percent. Hence, $p = 0.30 + 0.65(67/100) = 0.74$. Thus for this room
in January, $\text{SHC}_M = 100(54,387)(0.74)/53,904 = 74.7$ percent of the total average
space-heating needs are provided by the passive solar-heating system.

To estimate the average annual solar-heating contribution for a building, repeat
steps 1, 2, and 7 for each space for each month of the heating season. Use the
collector area computed in step 3 for an average clear day in January to determine
i_D for each month unless part of the collector is shaded (in which case, determine
the unshaded area and use that figure). Use $\text{SHC}_A = 100\Sigma (i_D)(p)(D)/\Sigma(H_D)(D)$,
where SHC_A = annual passive solar-heating contribution, percent, and D = number
of days of the month. The summation of the heat gains for each space for each

month of the heating season is divided by the summation of the heat losses for each space for each month.

Related Calculations.   These design procedures are suitable for buildings with skin-dominated heat loads such as heat losses through walls, roofs, perimeters, and infiltration. They are not applicable to buildings which have internal heat loads or buildings which are so deep that it is difficult to collect solar heat. Therefore, these procedures generally should be limited to small and medium-size buildings with good solar access.

These procedures use an average clear-day method as a basis for sizing a passive solar-heating system. Average monthly and yearly data also are used. If the actual weather conditions vary substantially from the average, the performance of the system will vary. For instance, if a winter day is unseasonably warm, the passive solar-heating system will collect more heat than is required to offset the heat loss on that day, possibly causing space overheating. Since passive solar-heating systems rely on natural phenomena, temperature fluctuation and variability in performance are inherent in the system. Adjustable shading, reflectors, movable insulation, venting mechanisms, and backup heating systems are often used to stabilize system performance.

Since passive systems collect, store, and distribute heat through natural physical means, the system must be integrated with the architectural design. The actual efficiency of the system is highly variable and dependent on this integration within the architectural design. Efficiency ratings given in this procedure are rules of thumb. Detailed analyses of many variables and how they affect system performance can be found in DOE/CS-0127-2 and 3, *Passive Solar Design Handbook,* volumes 2 and 3, available from the National Technical Information Service, Springfield, Virginia, 22161. *The Passive Solar Energy Book,* by Edward Mazria, available from Rodale Press, Emmaus, PA, examines various architectural concepts and how they can be utilized to maximize system performance.

If thermal collection and storage to provide heating on cloudy days is desired, the collector area can be oversized by 10 percent. This necessitates the oversizing of the thermal storage material to store 75 percent of the total daily heat gain rather than 65 percent, as used in step 4. Oversizing the system will increase the average inside temperature. Step 7 should be used to verify that this higher average temperature is acceptable. Oversizing the system for cloudy-day storage is not recommended for excessively hazy or cloudy climates. Cloudy climates do not have enough clear days in a row to accumulate reserve heat for cloudy-day heating. This increased collector area may increase heat load in these climates. Cloudy-day storage should be considered only for climates with a ratio of several clear days to each cloudy day.

Passive solar-heating systems may overheat buildings if insolation reaches the collector during seasons when heating loads are low or nonexistent. Shading devices are recommended in passive solar-heated buildings to control unwanted heat. Shading devices should allow low-angle winter insolation to penetrate the collector but block higher-angle summer insolation. The shading device should allow enough insolation to penetrate the collector to heat the building during the lower-heating-load seasons of autumn and spring without overheating spaces. If shading devices are used, the area of unshaded collector must be calculated for each month to determine i_D. Methods to calculate the area of unshaded collector can be found in *The Passive Solar Energy Book* and in *Solar Control and Shading Devices,* by V. and A. Olgyay, available from Princeton University Press.

Passive solar-heating systems should be considered only for tightly constructed, well-insulated buildings. The cost of a passive system is generally higher than that of insulating and weather-stripping a building. A building that has a relatively small heat load will require a smaller collection and storage system and so will have a lower construction cost. The cost effectiveness of a passive solar-heating system is inversely related to the heat losses of the building. Systems which have a smaller ratio of collector area to floor area are generally more efficient.

Significant decreases in the size of the collector can be achieved by placing movable insulation over the collector at night. This is especially recommended for extremely cold climates in more northern latitudes. If night insulation is used, calculate heat loss for the uninsulated collector for 8 h with the daytime average temperature and for the insulated collector for 16 h with the nighttime average temperature.

Table 8 is based on a heat loss of 8 Btu/(day·ft²) of floor area per °F [W/(m² ·K)]. Total building heat loss will increase with the increase in the ratio of collector to floor area because of the larger areas of glazing. However, it is assumed that this increase in heat loss will be offset by providing higher insulation values in noncollector surfaces. The tabulated values correspond to a residence with a compact plan, 8-ft-high ceilings, R-30 roof insulation, R-19 wall insulation, R-10 perimeter insulation, double glazing, and one air change per hour. It is provided for estimating purposes only. If the structure under consideration differs, the ratio of collector area to floor area, g, can be estimated for heat-loss calculations by using $g = h(65 - t_o)/i_T$, where h_L = estimated heat loss, Btu/(day·ft²·°F) [W/(m²·K)],

Passive solar heating is nonpolluting and is environmentally attractive. Other than the pollution (air, stream, and soil) possibly created in manufacturing the components of a passive solar heating system, this method of space heating is highly desirable from an environmental standpoint.

Solar heating does not provide carbon dioxide, as does the combustion of coal, gas, oil, and wood. Thus, there is no accumulation of carbon dioxide in the atmosphere from solar heating. It is the accumulated carbon dioxide in the earth's atmosphere that traps heat from the sun's rays and earth reradiation that leads to global warming.

Computer models of the earth's atmosphere and the warming that might be caused by excessive accumulation of carbon dioxide show that steps must be taken to control pollution. Although there is some disagreement about the true effect of carbon dioxide on global warming, most scientists believe that efforts to reduce carbon dioxide emissions are worthwhile. Both a United Nations scientific panel and research groups associated with the National Academy of Sciences recommend careful study and tracking of the possibility of global warming.

For these reasons solar heating will receive more attention from designers. With more attention being paid to the environment, solar heating offers a nonpolluting alternative that can easily be incorporated in the design of most buildings.

DETERMINING IF A SOLAR WATER HEATER WILL SAVE ENERGY

An engineer for a small office building of 2000 ft² (186 m²) area housing 12 people 250 days a year believes the building can more economically heat its required domestic hot water with a 6-ft² (0.56-m²) solar collector water heater. Is this true?

Calculation Procedure:

1. Assemble the data needed to analyze the water heating needs
Data needed for this procedure: (*a*) daily hot-water consumption in gallons (L) per
person; (*b*) total solar insolation incident on a vertical surface in the area of the
building in Btu/(yr· ft²) [kJ/(yr·m²)]; (*c*) percent annual sunshine for the building
location; (*d*) difference between incoming city water temperature and the hot-water
service temperature *e* (= 100°F [180°C]) for this building. For the typical office
building, you can use data from the ASHRAE *Guide*. Assembling the data gives:
(*a*) = 1 gal (3.79 L) per person per day; (*b*) 732,000 Btu/(yr·ft²) [8345 MJ/(yr·
m²)]; (*c*) 59 percent; (*d*) 100°F (180°C).

2. Determine if a solar water heater will save energy
The annual water consumption for this building is 1 gal (3.79 L) per person per
day (12 people)(250 days per year) = 3000 gal/yr (11,370 L/yr). The energy
required to heat this water, *H* Btu/yr = (gal consumed/yr)(required temperature
rise, °F) 8.3, where the constant 9.3 converts gallons to pounds. Or, *H* =
3000(100)(8.3) = 2,490,000 Btu/yr (2627 MJ/yr).

The usable energy from the solar water heater, *S* = (percent annual sunshine)(solar insolation on collector surface, Btu/yr·ft²)(solar collect heating absorbing area, ft²); or *S* = 0.59 (732,000)(6) = 2,591,280 Btu/yr (2734 MJ/yr).

Comparing the heat required to warm the domestic water for this building,
2,490,000 Btu/yr, versus the heat available from the solar water heater, 2,591,280
Btu/yr, shows that the solar water heater could be substituted for the existing water
heater in this building. The energy costs saved in heating the water can be used to
pay for the solar collector. Figure 16 shows a possible arrangement for a solar water
heater used as a booster heater to reduce water heating costs in an occupied building.

Related Calculations. Use this procedure to analyze the suitability of a substitute water-heating method in any type of building: office, commercial, residential,
medical, health-care, hospital, etc. When evaluating a solar water heater, be certain
to use accurate data on the availability of sunshine in the area of the structure, and
the solar insolation for the area.

This procedure is the work of Jerome F. Mueller, P.E., Mueller Engineering
Corp.

SIZING A PHOTOVOLTAIC SYSTEM FOR ELECTRICAL SERVICE

A small home has the electrical load shown in the tabulation below. Select suitable
photovoltaic modules to serve the electrical load for two locations, Albuquerque,
NM and Pittsburgh, PA. Show how this load would be serviced by the modules.

Calculation Procedures:

1. Detail the electrical load of the building
Construct a tabulation such as that below for the building being considered.

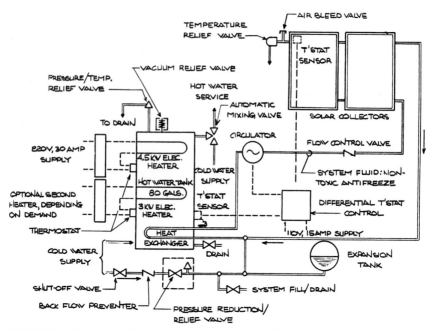

FIGURE 16   Booster solar heater uses nonfreeze glycol solution that is pumped through the heat exchanger in the hot-water tank. (*Jerome F. Mueller.*)

Load	Daily usage, h		Wattage		Energy consumption, Watt-hour (W-h)
Radio	2	×	25	=	50
Television	6	×	60	=	360
Lamps, fluorescent	3	×	27	=	81
VCR	0.5	×	30	=	15
	Total daily energy consumption = 506 W-h				

Thus, for this building, a photovoltaic system producing an average daily energy output of 506 W-h is required. Since sunlight (solar energy) is the source of power for photovoltaic systems, we must determine the daily amount of sunlight in the location of the structure. Because photovolaic systems are rated by peak Watt output, the electricity produced at noon on a clear day is of critical importance in system design.

2. *Show how the system will be hooked up for each location*
Figure 17 shows the typical hookup for a photovoltaic system serving a building. The module contains solar cells which convert the sun's rays to electricity. An inverter and storage batteries are used to change the direct current from the cells to alternating current and to store the direct current.

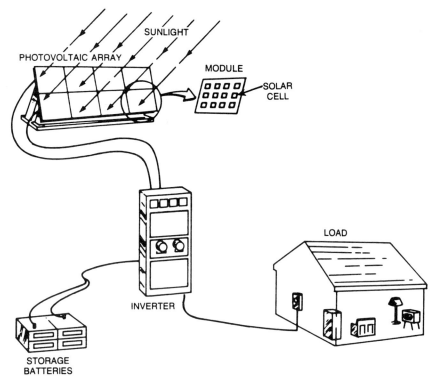

FIGURE 17   Photovoltaic array for providing electricity to a residence. (*U.S. Department of Energy.*)

   When the sunlight strikes a typical photovoltaic or solar cell, Fig. 18, photons are absorbed and electrons are freed to flow as shown. Each cell consists of a thin layer of phosphorus-doped silicon in close contact with a layer of boron-doped silicon. By chemically treating silicon in this manner, a permanent electric field is created. The electrons freed when sunlight hits the cell flow through metal contacts to generate electricity.
   Photovoltaics, defined as electricity produced solely from the energy of the sun, is modular in design. It can grow with the electrical demands placed on it. An individual cell will produce only a small amount of electricity, but placing several cells together increases the amount of electricity produced. Groups of cells are mounted on a rigid plate and are electrically interconnected to form a photovoltaic *module* which produces a nominal 12 V. Then, groups of modules are mounted together on a permanently attached frame to form a *panel*. Panels are interconnected to form a photovoltaic *array* for differing power levels, Fig. 19.
   Photovoltaic systems are a useful form of renewable energy. Thus, the modules have no moving parts, are east to install, require little maintenance, contain no fluids, consume no fuel, produce no pollution, and have a long life span—more than 20 yr, and they are equally or more reliable than competing power sources,

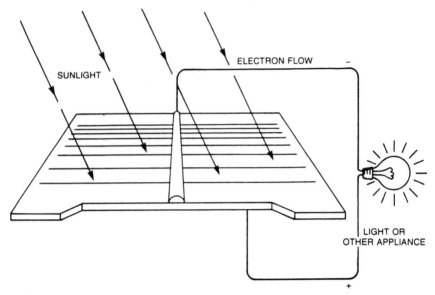

FIGURE 18   Electron flow from the array generates electricity. (*U.S. Department of Energy.*)

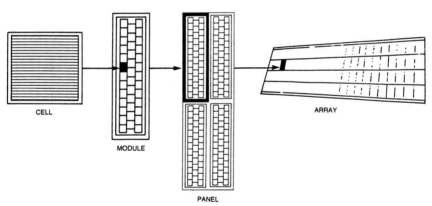

FIGURE 19   Steps in the interconnection of a cell to a module to a panel to an array. (*U.S. Department of Energy.*)

such as batteries and generators. However, components, such as the load and controls, also affect the reliability of the entire system.

3. *Determine the number of modules required*
Reference data show that the winter average peak wattage for a module in Albuquerque, NM is 6.1 W-h; for Pittsburgh, PA, a cloudier area, the same module will produce 2.4 W-h. To determine the peak Watts of the system, divide the total energy consumption by the W-h for the location. Then divide the result by 0.8 to account

for inefficiency; multiply the result by 1.2 (a 20 percent factor of safety) to account for any items you might have overlooked.

Thus, for Albuquerque, we have (506 W-h/6.1/0.8)(1.2) = 124.4 peak Watts. If we're using 50-W modules, a standard and convenient size, we would need 124.4/50 =2.488 modules. Since modules are made only in whole sizes, three modules would be used for this application.

For Pittsburgh, PA, following the same procedure, we have: (506 W-h/2.4/ 0.8)(1.2) = 316.3 peak Watts. Using 50-W modules, we have 316.3/50 = 6.32 modules; seven would be chosen.

For precise accuracy in module selection, designers suggest using month-by-month averages for average W-h for a location. Then the lowest monthly average is actually used in the calculations.

Related Calculations. Today, the most common applications for photovoltaics are remote or stand-alone systems. A stand-alone system is not connected to the local utility and is used in remote areas of the world where there are no nearby power lines. Thus, some 50,000 homes, worldwide, are powered by photovoltaics. Connecting such remote homes to a utility costs much more than generating electricity from the sun.

Developing nations use photovoltaics to refrigerate medicines, pump water, or light villages. Developed countries use photovoltaics to light roads, billboards, and road signs. Recreational vehicles and boats also use solar energy to operate various devices.

Other uses of solar power include railway cabooses, airport emergency lighting systems, marine buoys and coastal markers, and electric carts for transporting sight-seers.

Hybrid photovoltaic systems use two or more power-generating sources connected together. The secondary source for the solar modules can range from gas or diesel generators to wind generators or hydroelectric generators. However, the most common secondary source is the gas or Diesel generator. In the hybrid system, the secondary generator automatically starts when the battery voltage of the photovoltaic array drops below a safe level. By using a hybrid system, the solar energy array can be reduced in size. This allows the secondary source to supply the remaining needed energy during times of minimal sunshine caused by seasonal or weather factors.

Currently, the price of a peak Watt is in the $3 to $5 range. As greater use is made of photovoltaics, the price per peak Watt is expected to decline. Since this source of power is more expensive than conventional sources, photovoltaics is still confined to specialized uses.

Data in this procedure are from the U.S. Department of Energy Conservation and Renewable Energy Inquiry and Referral Service.

SECTION 18
ENVIRONMENTAL CONTROL AND ENERGY CONSERVATION

Environmental Contamination Analysis and Prevention

RECYCLE PROFIT POTENTIALS IN MUNICIPAL WASTES

Analyze the profit potential in typical municipal wastes listed in Table 1. Use data on price increases of suitable municipal waste to compute the profit potential for a typical city, town, or state.

TABLE 1 Examples of Price Changes in Municipal Wastes*

	Price per ton, $	
	Last year	Current year
Newspapers	60	150
Corrugated cardboard	18	150
Plastic jugs, bottles	125	600
Copper wire and pipe	9060	1200

*Based on typical city wastes.

Calculation Procedure:

1. Compute the percentage price increase for the waste shown

Municipal waste may be classed in several categories: (1) newspapers, magazines, and other newsprint; (2) corrugated cardboard; (3) plastic jugs and bottles—clear or colored; (4) copper wire and pipe. Other wastes, such as steel pipe, discarded internal combustion engines, electric motors, refrigerators, air conditioners, etc., require specialized handling and are not generated in quantities as large as the four numbered categories. For this reason, they are not normally included in estimates of municipal wastes for a given locality.

For the four categories of wastes listed above, the percentage price increases in one year for an Eastern city in the United States were as follows: Category 1—newspaper: Percentage price increase = 100(current price, $ − last year's price, $)/last year's price, $. Or 100(150 − 60)/60 = 150 percent. Category 2: Percentage price increase = 100(150 − 18)/18 = 733 percent. Category 3: Percentage price increase = 100(600 − 125)/125 = 380 percent. Category 4: Percentage price increase = 100(1200 − 960)/960 = 25 percent.

2. Determine the profit potential of the wastes considered

Profit potential is a function of collection costs and landfill savings. When collection of several wastes can be combined to use a single truck or other transport means, the profit potential can be much higher than when more than one collection method must be used. Let's assume that a city can collect Category 1, newspapers, and Category 3, plastic, in one vehicle. The profit potential, P, will be: P = (sales price of the materials to be recycled, $ per ton − cost per ton to collect the materials for recycling, $). With a cost of $80 per ton for collection, the profit for collecting 75 tons of Category 1 wastes would be P = 75($150 − $80) = $5250. For collecting 90 tons of Category 3 wastes, the profit would be P = 90($600 − 80) = $46,800.

Where landfill space is saved by recycling waste, the dollar saving can be added to the profit. Thus, assume that landfill space and handling costs are valued at $30 per ton. The profit on Category 1 waste would rise by 75($30) = $2250, while the profit on Category 3 wastes would rise by 90($30) = $2700. When collection is included in the price paid for municipal wastes, the savings can be larger because the city or town does not have to use its equipment or personnel to collect the wastes. Hence, if collection can be included in a waste recycling contract the profits to the municipality can be significant. However, even when the municipality performs the collection chore, the profit from selling waste for recycling can still be high. In some cities the price of used newspapers is so high that gangs steal the bundles of papers from sidewalks before they are collected by the city trucks.

Related Calculations. Recyclers are working on ways to reuse almost all the ordinary waste generated by residents of urban areas. Thus, telephone books, magazines, color-printed advertisements, waxed milk jars, etc. are now being recycled and converted into useful products. The environmental impact of these activities is positive throughout. Thus, landfill space is saved because the recycled products do not enter landfill; instead they are remanufactured into other useful products. Indeed, in many cases, the energy required to reuse waste is less than the energy needed to produce another product for use in place of the waste.

Some products are better recycled in other ways. Thus, the United States discards, according to industry records, over 12 million computers a year. These computes, weighing an estimated 600 million pounds (272 million kg) contribute toxic waste to landfills. Better that these computers be contributed to schools, colleges, and universities where they can be put to use in student training. Such computers may be slower and less modern than today's models, but their value in training programs has little to do with their speed or software. Instead, they will enable students to learn, at minimal cost to the school, the fundamentals of computer use in their personal and business lives.

Recycling waste products has further benefits for municipalities. The U.S. Clean Air Act's Title V consolidates all existing air pollution regulations into one massive operating permit program. Landfills that burn pollute the atmosphere. And most of the waste we're considering in this procedure burns when deposited in a landfill. By recycling this waste the hazardous air pollutants they may have produced while burning in a landfill are eliminated from the atmosphere. This results in one less worry and problem for the municipality and its officials. In a recent year, the U.S. Environmental Protection Agency took 2247 enforcement actions and levied some $165-million in civil penalties and criminal fines against violators.

Any recycling situation can be reduced to numbers because you basically have the cost of collection balanced against the revenue generated by sale of the waste. Beyond this are nonfinancial considerations related to landfill availability and expected life-span. If waste has to be carted to another location for disposal, the cost of carting can be factored into the economic study of recycling.

Municipalities using waste collection programs state that their streets and sidewalks are cleaner. They attribute the increased cleanliness to the organization of people's thinking by the waste collection program. While stiff fines may have to be imposed on noncomplying individuals, most cities report a high level of compliance from the first day of the program. The concept of the "green city" is catching on and people are willing to separate their trash and insert it in specific containers to comply with the law.

"Green products, i.e., those that produce less pollution, are also strongly favored by the general population of the United States today. Manufacturing companies are finding a greater sales acceptance for their "green" products. Even automobile manufacturers are stating the percentage of each which is recyclable, appealing to the "green" thinking permeating the population.

Recent studies show that every ton of paper not landfilled saves 3 yd^3 (2.3 m^3) of landfill space. Further, it takes 95 percent less energy to manufacture new products from recycled materials. Both these findings are strong motivators for recycling of waste materials by all municipalities and industrial firms.

Decorative holiday trees are being recycled by many communities. The trees are chipped into mulch which are given to residents and used by the community in parks, recreation areas, hiking trails, and landfill cover. Seaside communities sometimes plant discarded holiday trees on beaches to protect sand dunes from being carried away by the sea.

CHOICE OF CLEANUP TECHNOLOGY FOR CONTAMINATED WASTE SITES

A contaminated waste site contains polluted water, solid wastes, dangerous metals, and organic contaminants. Evaluate the various treatment technologies available for such a site and the relative cost of each. Estimate the landfill volume required if the rate of solid-waste generation for the site is 1,500,000 lb (681,818 kg) per year. What land area will be required for this waste generation rate if the landfill is designed for the minimum recommended depth of fill? Determine the engineer's role in site cleanup and in the economic studies needed for evaluation of available alternatives.

Calculation Procedure:

1. *Analyze the available treatment technologies for cleaning contaminated waste sites*

Table 2 lists 13 available treatment technologies for cleaning contaminated waste sites, along with the type of contamination for which each is applicable, and the relative cost of the technology. This tabulation gives a bird's eye view of technologies the engineer can consider for any waste site cleanup.

When approaching any cleanup task, the first step is to make a health-risk assessment to determine if any organisms are exposed to compounds on, or migrating from, a site. If there is such an exposure, determine whether the organisms could suffer any adverse health effects. The results of a health-risk assessment can be used to determine whether there is sufficient risk at a site to require remediation.

This same assessment of risks to human health and the environment can also be used to determine a target for the remediation effort that reduces health and environmental risks to acceptable levels. It is often possible to negotiate with regulatory agencies a remediation level for a site based on the risk of exposure to both a maximum concentration of materials and a weighted average. The data in Table 2 are useful for starting a site cleanup having the overall goals of protecting human health and the environment.

2. *Make a health-risk assessment of the site to determine cleanup goals*[1]

Divide the health-risk assessment into these four steps: (1) *Hazard Identification*—Asks "Does the facility or site pose sufficient risk to require further investigation?" If the answer is Yes, then: (*a*) Select compounds to include in the assessment; (*b*) Identify exposed populations; (*c*) Identify exposure pathways.

(2) *Exposure Assessment*—Asks "To how much of a compound are people and the environment exposed?" For exposure to occur, four events must happen: (*a*) release; (*b*) contact; (*c*) transport; (*d*) absorption. Taken together, these four events form an *exposure pathway*. There are many possible exposure pathways for a facility or site.

(3) *Toxicity Assessment*—Asks "What adverse health effects in humans are potentially caused by the compounds in question?" This assessment reviews the threshold and nonthreshold effects potentially caused by the compounds at the environmental concentration levels.

---

[1] Hopper, David R., "Cleaning Up Contaminated Waste Sites," *Chemical Engineering*, Aug., 1989.

TABLE 2 Various Treatment Technologies Available to Clean Up a Contaminated Waste Site*

Technology	Description	Applicable contamination	Relative cost
Soil vapor extraction	Air flow is induced through the soil by pulling a vacuum on holes drilled into the soil, and carries out volatilized contaminants	Volatile and some semivolatile organics	Low
Soil washing or soil flushing	Excavated soil is flushed with water or other solvent to leach out contaminants	Organic wastes and certain (soluble) inorganic wastes	Low
Stabilization and solidification	Waste is mixed with agents that physically immobilize or chemically precipitate constituents	Applies primarily to metals; mixed results when used to treat organics	Medium
Thermal desorption	Solid waste is heated to 200–800°F to drive off volatile contaminants, which are separated from the waste and further treated	Volatile and semivolatile organics; volatile metals such as elemental mercury	Medium to high
Incineration	Waste is burned at very high temperatures to destroy organics	Organic wastes; metals do not burn, but concentrate in ash	High
Thermal pyrolysis	Heat volatilizes contaminants into an oxygen-starved air system at temperatures sufficient to pyrolzye the organic contaminants. Frequently, the heat is delivered by infrared radiation	Organic wastes	Medium to high
Chemical precipitation	Solubilized metals are separated from water by precipitating them as insoluble salts	Metals	Low

Method	Description	Applicable wastes	Cost
Aeration or air stripping	Contaminated water is pumped through a column where it is contacted with a countercurrent air flow, which strips out certain pollutants	Mostly volatile organics	Low
Steam stripping	Similar to air stripping except steam is used as the stripping fluid	Mostly volatile organics	Low
Carbon adsorption	Organic contaminants are removed from a water or air stream by passing the stream through a bed of activated carbon that absorbs the organics	Most organics, though normally restricted to those with sufficient volatility to allow carbon regeneration	Low to medium when regeneration is possible
Bioremediation	Bacterial degradation of organic compounds is enhanced	Organic wastes	Low
Landfilling	Covering solid wastes with soil in a facility designed to minimize leachate formation	Solid, nonhazardous wastes	Low but rising fast
In situ vitrification	Electric current is passed through soil or waste, which increases the temperature and melts the waste or soil. The mass fuses upon cooling	Inorganic wastes, possibly organic wastes; not applicable to very large volumes	Medium

*Chemical Engineering.

18.7

(4) *Risk Characterization*—Asks "At the exposures estimated in the Exposure Assessment, is there potential for adverse health effects to occur; if so, what kind and to what extent?" The Risk Characterization develops a hazard index for threshold effects and estimates the excess lifetime cancer-risk for carcinogens.

3. Select suitable treatment methods and estimate the relative costs

The site contains polluted water, solid wastes, dangerous metals, and organic contaminants. Of these four components, the polluted water is the simplest to treat. Hence, we will look at the other contaminants to see how they might best be treated. As Table 2 shows, thermal desorption treats volatile and semivolatile organics and volatile metals; cost is medium to high. Alternatively, incineration handles organic wastes and metals with an ash residue; cost is high. Nonhazardous solid wastes can be landfilled at low cost. But the future cost may be much higher because landfill costs are rising as available land becomes scarcer.

Polluted water can be treated with chemicals, aeration, or air stripping—all at low cost. None of these methods can be combined with the earlier tentative choices. Hence, the polluted water will have to be treated separately.

4. Determine the landfill dimensions and other parameters

Annual landfill space requirements can be determined from $V_A = W/1100$, where V_A = landfill volume required, per year, yd^3 (m^3); W = annual weight, lb (kg) of waste generated for the landfill; 1100 lb/yd^3 (650 kg/m^3) = solid waste compaction per yd^3 or m^3. Substituting for this site, $V_A = 1,500,000/1100 = 1363.6$ yd^3 (1043.2 m^3).

The minimum recommended depth for landfills is 20 ft (6 m); minimum recommended life is 10 years. If this landfill were designed for the minimum depth of 20 ft (6 m), it would have an annual required area of 1363.6×27 ft^3/yd^3 = 36,817.2 ft^3/20 ft high = 1840.8 ft^2 (171.0 m^2), or 1840.9 ft^2/43,560 ft^2/acre = 0.042 acre (169.9 m^2; 0.017 ha) per year. With a 10-year life the landfill area required to handle solid wastes generated for this site would be $10 \times 0.042 = 0.42$ acre (1699.7 m^2, 0.17 ha); with a 20-year life the area required would be $20 \times 0.042 = 0.84$ acre (3399.3 m^2; 0.34 ha).

As these calculations show, the area required for this landfill is relatively modest—less than an acre with a 20-year life. However, in heavily populated areas the waste generation could be significantly larger. Thus, when planning a sanitary landfill, the usual assumption is that each person generates 5 lb (2.26 kg) per day of solid waste. This number is based on an assumption of half the waste (2.5 lb; 1.13 kg) being from residential sources and the other half being from commercial and industrial sources. Hence, in a city having a population of 1-million people, the annual solid-waste generation would be 1,000,000 people $\times$ 5 lb/day per person $\times$ 365 days per year = 1,825,000,000 lb (828,550,000 kg).

Following the same method of calculation as above, the annual landfill space requirement would be $V_A = 1,825,000,000/1100 = 1,659,091$ yd^3 (1,269,205 m^3). With a 20-ft (6-m) height for the landfill, the annual area required would be $1,659,091 \times 27/20 \times 43,560 = 51.4$ acres (208,002 m^2; 20.8 ha). Increasing the landfill height to 40 ft (12 m) would reduce the required area to 25.7 acres (104,037 m^2; 10.4 ha). A 60-ft high landfill would reduce the required area to 17.1 acres (69,334 m^2; 6.9 ha). In densely populated areas, landfills sometimes reach heights of 100 ft (30.5 m) to conserve horizontal space.

This example graphically shows why landfills are becoming so much more expensive. Further, with the possibility of air and stream pollution from a landfill, there is greater regulation of landfills every year. This example also shows why incineration of solid waste to reduce its volume while generating useful heat is so attractive to communities and industries. Further advantages of incineration include reduction of the possibility of groundwater pollution from the landfill and the chance to recover valuable minerals which can be sold or reused. Residue from incineration can be used in road and highway construction or for fill in areas needing it.

Related Calculations. Use this general procedure for tentative choices of treatment technologies for cleaning up contaminated waste sites. The greatest risks faced by industry are where human life is at stake. Penalties are severe where human health is endangered by contaminated wastes. Hence, any expenditures for treatment equipment can usually be justified by the savings obtained by eliminating lawsuits, judgments, and years of protracted legal wrangling. A good example is the asbestos lawsuits which have been in the courts for years.

To show what industry has done to reduce harmful wastes, here are results published in the *Wall Street Journal* for the years 1974 and 1993: Lead emissions declined from 223,686 tons in 1973 to 4885 tons in 1993 or to 2.2 percent of the original emissions; carbon monoxide emissions for the same period fell from 124.8 million tons to 97.2 million tons, or 77.9 percent of the original; rivers with fecal coliform above the federal standard were 31 percent in 1974 and 26 percent in 1994; municipal waste recovered for recycling was 7.9 percent in 1974 and 22.0 percent in 1994.

The simplest way to dispose of solid wastes is to put them in landfills. This practice was followed for years, but recent studies show that rain falling on land-filled wastes seeps through and into the wastes, and can become contaminated if the wastes are harmful. Eventually, unless geological conditions are ideal, the contaminated rainwater seeps into the groundwater under the landfill. Once in the groundwater, the contaminants must be treated before the water can be used for drinking or other household purposes.

Most landfills will have a leachate seepage area, Fig. 1. There may also be a contaminant plume, as shown, which reaches, and pollutes, the groundwater. This is why more and more communities are restricting, or prohibiting, landfills. Engineers are therefore more pressed than ever to find better, and safer, ways to dispose of contaminated wastes. And with greater environmental oversight by both Federal and State governments, the pressure on engineers to find safe, economical treatment methods is growing. The suggested treatments in Table 2 are a good starting point for choosing suitable and safe ways to handle contaminated wastes of all types.

Landfills must be covered daily. A 6-in (15-cm) thick cover of the compacted refuse is required by most regulatory agencies and local authorities. The volume of landfill cover, ft^3, required each day can be computed from: (Landfill working face length, ft)(landfill working width, ft)(0.5). Multiply by 0.0283 to convert to m^3. Since the daily cover, usually soil, must be moved by machinery operated by humans, the cost can be significant when the landfill becomes high—more than 30 ft (9.1 m). The greater the height of a landfill, the more optimal, in general, is the site and its utilization. For this reason, landfills have grown in height in recent years in many urban areas.

Table 2 is the work of David R. Hopper, Chemical Process Engineering Program Manager, ENSR Consulting and Engineering, as reported in *Chemical Engineering* magazine.

CLEANING UP A CONTAMINATED WASTE SITE VIA BIOREMEDIATION

Evaluate the economics of cleaning up a 40-acre (161,872 m^2) site contaminated with petroleum hydrocarbons, gasoline, and sludge. Estimates show that some

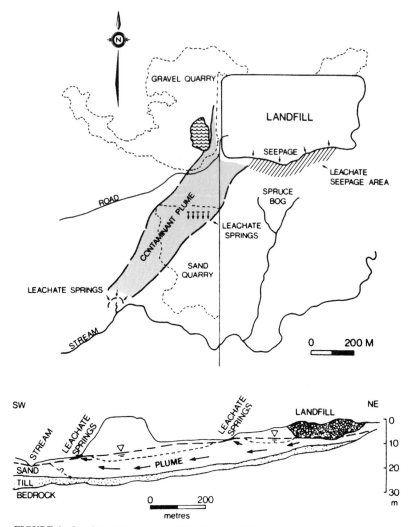

FIGURE 1   Leachate seepage in landfill. (*McGraw-Hill*).

100,000 yd³ (76,500 m³) must be remediated to meet federal and local environmental requirements. The site has three impoundments containing weathered crude
oils, tars, and drilling muds ranging in concentration from 3800 to 40,000 ppm, as
measured by the Environmental Protection Agency (EPA) Method 8015M. While
hydrocarbon concentrations in the soil are high, tests for flash point, pH, 96-h fish
bioassay, show that the soil could be classified as nonhazardous. Total petroleum
hydrocarbons are less than 500 ppm. Speed of treatment is not needed by the owner
of the project. Show how to compute the net present value for the investment in

alternative treatment methods for which the parameters are given in step 4 of this procedure.

Calculation Procedure:

1. Compare the treatment technologies available

A number of treatment technologies are available to remediate such a site. Where total petroleum hydrocarbons are less than 500 ppm, as at this site, biological land treatment is usually sufficient to meet regulatory and human safety needs. Further, hazardous and nonhazardous waste cleanup via bioremediation is gaining popularity. One reason is the high degree of public acceptance of bioremediation vs. alternatives such as incineration. The Resource Conservation and Recovery Act (RCRA) defines hazardous waste as specifically listed wastes or as wastes that are characteristically toxic, corrosive, flammable, or reactive. Wastes at this site fit certain of these categories.

Table 3 compares three biological treatment technologies currently in use. The type of treatment, and approximate cost, $/ft^3 ($/m^3), are also given. Since petroleum hydrocarbons are less than 500 ppm at this site, biological land treatment will be chosen as the treatment method.

Looking at the range of costs in Table 3 shows a minimum of $30/yd^3 ($39/m^3) for land treatment and a maximum of $250/yd^3 ($327/m^3) for bioreactor treatment. This is a ratio of $250/$30 = 8.3:1. Thus, where acceptable results will be obtained, the lowest cost treatment technology would probably be the most suitable choice.

2. Determine the cost ranges that might be encountered in this application

The cost ranges that might be encountered in this—or any other application—depend on the treatment technology which is applicable and chosen. Thus, with some 100,000 yd^3 (76,500 m^3) of soil to be treated, the cost ranges from Table 1 = 100,000 yd^3 × $/yd^3. For *biological land treatment,* cost ranges = 100,000 × $30 = $3,000,000; 100,000 × $90 = $9,000,000. For *bioventing,* cost ranges = 100,000 × $50 = $5,000,000; 100,000 × $120 = $12,000,000. For *biorector treatment,* cost ranges = 100,000 × $150 = $15,000,000; 100,000 × $250 = $250,000,000. Thus, a significant overall cost range exists—from $3,000,000 to $25,000,000, depending on the treatment technology chosen.

The wide cost range computed above shows why it is so important that the engineer choose the most cost-effective system which accomplishes the desired cleanup in accordance with federal and state requirements. With an estimated 2000 hazardous waste sites currently known in the United States, and possibly several times that number in the rest of the world, the potential financial impact on companies and their insurers, is enormous. The actual waste site discussed in this procedure highlights the financial decisions engineers face when choosing a method of cleanup.

Once a cleanup (or remediation) method is tentatively chosen—after the site investigation and feasibility study by the engineer—the controlling regulatory agencies must be consulted for approval of the method selected. The planned method of remediation is usually negotiated with the regulatory agency before final approval is given. Once such approval is obtained, it is difficult to change the remediation method chosen. Hence, the engineer, and the organization involved, should find the chosen remediation method acceptable in every way possible.

TABLE 3   Comparison of Biological Treatment Technologies*

Type/cost ($/yd³)	Advantages	Disadvantages
Land treatment $30–$90	• Can be used for in situ or ex situ treatment depending upon contaminant and soil type • Little or no residual waste streams generated • Long history of effective treatment for many petroleum compounds (gasoline, diesel) • Can be used as polishing treatment following soil washing or bioslurry treatment	• Moderate destruction efficiency depending upon contaminants • Long treatment time relative to other methods • In situ treatment only practical when contamination is within two feet of the surface • Requires relatively large, dedicated area for treatment cell
Bioventing $50–$120	• Excellent removal of volatile compounds from soil matrix • Depending upon vapor treatment method, little or no residual waste streams to dispose • Moderate treatment time • Can be used for in situ or ex situ treatment depending upon contaminant and soil type	• Treatment of vapor using activated carbon can be expensive at high concentrations of contaminants • System typically requires an air permit for operation
Bioreactor $150–$250	• Enhanced separation of many contaminants from soil • Excellent destruction efficiency of contaminants • Fast treatment time	• High mobilization and demobilization costs for small projects • Materials handling requirements increase costs • Treated solids must be dewatered • Fullscale application has only become common in recent years

Chemical Engineering magazine.

3.   *Evaluate the time requirements of each biological treatment technology*

Biological land treatment has been used for many years for treating petroleum residues. Also known as land-farming, this is the simplest and least expensive biological treatment technology. However, this method requires large amounts of land that can be dedicated to the treatment process for a period of several months to several years. Typically, land treatment involves the control of oxygen, nutrients, and moisture (to optimize microbial activity) while the soil is tilled or otherwise aerated.

Bioventing systems, Fig. 2, are somewhat more complex than land treatment, at a moderate increase in cost. They are used on soils with both volatile and nonvolatile hydrocarbons. Conventional vapor extraction technology (air stripping) of the volatile components is combined with soil conditioning (such as nutrient addition) to enhance microbial degradation. This treatment method can be used both in situ and ex situ. Relative to land treatment, space requirements are reduced. Treatment time is on the order of weeks to months.

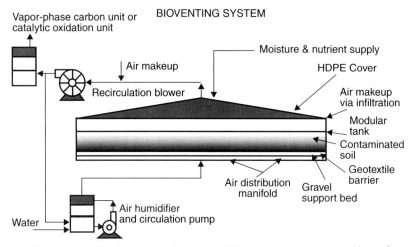

FIGURE 2   Pipes blowing air from the bottom of this enclosure separate contaminants from the soil. (*OHM Corp., Carla Magazino* and *Chemical Engineering.*)

Bioreactors are the most complex and expensive biological alternative. They can clean up contaminated water alone, or solids mixed with water (slurry bioreactors). The reactor can be configured from existing impoundments, aboveground tanks, or enclosed tanks (if emissions controls are required). Batch, semicontinuous, or continuous modes of operation can be maintained. The higher cost is often justified by the faster treatment time (on the order of hours to days) and the ability to degrade contaminants on difficult-to-treat soil matrices.

Since time is not a controlling factor in this application, biological land treatment, the least expensive method, will be chosen and applied.

4.   Compute the net present value for alternative treatment methods

Where alternative treatment methods can be used for a hazardous waste site, the method chosen can be analyzed on the basis of the present net worth of the "cash flows" produced by each method. Such "cash flows" can be estimated by converting savings in compliance, legal, labor, management, and other costs to "cash flows" for each treatment method. Determining the net present worth of each treatment method will then provide a comparative evaluation which will be an additional input in the final treatment choice decision.

The table below shows the estimated annual "cash flows" for two suitable treatment methods: Method A and Method B

Year	Method A	Method B
0	−$180,000	−$180,000
1	60,000	180,000
2	60,000	30,000
3	60,000	18,000
4	60,000	12,000

Interest rate charged on the investment is 12 percent.

Using the Net Present Value (NPV), or Discounted Cash Flow (DCF), equation for each treatment method gives, NPV, Treatment Method = Investment, first year + each year's cash flow × capital recovery factor for the interest rate on the investment. For the first treatment method, using a table of compound interest factors for an interest rate of 12 percent, NPV, treatment A = −$180,000 + $60,000/ 0.27741 = $36,286. In this relation, the cash flow for years 1, 2, and 3 repays the investment of $180,000 in the equipment. Hence, the cash flow for the fourth year is the only one used in the NPV calculation.

For the second treatment method, B, NPV = −$180,000 + $180,000/0.8929 + $30,000/0.7972 + $18,000/0.7118 + $12,000/0.6355 = $103,392. Since Treatment Method B is so superior to Treatment Method A, B would be chosen. The ratio of NPV is 2.84 in favor of Method B over Method A.

Use the conventional methods of engineering economics to compare alternative treatment methods. The prime consideration is that the methods compared provide equivalent results for the remediation process.

5. *Develop costs for combined remediation systems*

Remediation of sites always involves evaluation of a diverse set of technologies. While biological treatment alone can be used for the treatment of many waste streams, combining bioremediation with other treatment technologies may provide a more cost-effective remedial alternative.

Figure 3 shows the costs of a full-scale groundwater treatment system treating 120 gal/min (7.6 L/s) developed for a site contaminated with pentachlorophenol (PCP), creosote, and other wood-treating chemicals at a forest-products manufacturing plant. The site contained contaminated groundwater, soil, and sludges. Capital cost, prorated for the life of the project, for the biological unit is twice that of an activated carbon system. However, the lower operating cost of the biological system results in a total treatment cost half the price of its nearest competitor. Carbon polishing adds 13 percent to the base cost.

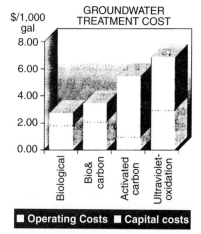

FIGURE 3   Under the right circumstances, biological treatment can be the lowest-cost option for groundwater cleanup. (*Carla Magazino* and *Chemical Engineering.*)

For the systems discussed in the paragraph above, the choice of alternative treatment technologies was based on two factors: (1) Biological treatment followed by activated carbon polishing may be required to meet governmental discharge requirements. (2) Liquid-phase activated carbon, and UV-oxidation are well established treatment methods for contaminated groundwater.

Soils and sludges in the forest-products plant discussed above are treated using a bioslurry reactor. The contaminated material is slurried with water and placed into a mixed, aerated biotreatment unit where suspended bacteria degrade the contaminants. Observation of the short-term degradation of PCP in initial tests suggested that the majority of the degradation occurred in the first 10 to 30 days of treatment. These results suggested that treatment costs could be minimized by initial processing of soils in the slurry bioreactor followed by final treatment in an engineered land-farm.

Treatment costs for a bioslurry reactor system using a 30-day batch time, followed by land treatment, are shown in Fig. 4. The minimum cost, $62/ton, occurs with a 5-year remediation lifetime, Fig. 4. An equivalent system using only the bioreactor would require an 80+-day cycle time to reach the cleanup criteria. The treatment cost can be reduced by over $45/ton using the hybrid system.

Note that the costs given above are for a specific installation. While they are not applicable to all plants, the cost charts show how comparisons can be made and how treatment costs vary with various cleanup methods. You can assemble, and compare, costs for various treatment methods using this same approach.

Related Calculations.   Bioremediation works because it uses naturally occurring microorganisms or consortia of microorganisms that degrade specific pollutants and, more importantly, classes of pollutants. Biological studies reveal degradation pathways essential to assure detoxification and mineralization. These studies also show how to enhance microbial activity, such as by the addition of supplementary oxygen and nutrients, and the adjustment of pH, temperature, and moisture.

Bioremediation can be effective as a pre- or post-treatment step for other cleanup techniques. Degradation of pollutants by microorganisms requires a carbon source, electron hydrocarbons (PAHs) found in coal tar, creosote, and some petroleum-

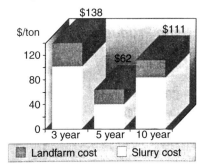

**TREATMENT COST
FOR COMBINED SLURRY
BIOREACTOR-LANDFARM SYSTEM**

FIGURE 4   The treatment cost for this system reaches a minimum value after 5 years, then rises again. (*Carla Magazino* and *Chemical Engineering.*)

compounds acceptor, nutrients, and appropriate pH, moisture, and temperature. The waste can be the carbon source or primary substrate for the organisms. Certain waste streams may also require use of a cosubstrate to trigger the production of enzymes necessary to degrade the primary substrate. Some wastes can be cometabolized directly along with the primary substrate.

Regulatory constraints are perhaps the most important factor in selecting bioremediation as a treatment process. Regulations that define specific cleanup criteria, such as land disposal restrictions under the U.S. Resource Conservation and Recovery Act (RCRA), also restrict the types of treatment technologies to be used. Other technologies, such as incineration, have been used to define the "best demonstrated available technology" (BDAT) for hazardous waste treatment of listed wastes.

The schedule for a site cleanup can also be driven by regulatory issues. A consent decree may fix the timetable for a site remediation, which may eliminate the use of bioremediation, or limit the application to a specific biological treatment technology.

Specific cleanup tasks for which biological treatment is suitable include remediation of petroleum compounds (gasoline, diesel, bunker oil); polynuclear aromatic hydrocarbons (PAHs) found in coal tar, creosote, and some petroleum compounds; soils with volatile and nonvolatile hydrocarbons; contaminated water; drilling muds; polychlorinated biphenyls (PCBs). The general approach given here can be used for the named pollutants, plus others amenable to bioremediation.

Data in this procedure are the work of Chris Jespersen, P.E., Project Manager, OHM Remediation Services Corp., Douglas E. Jerger, Technical Director, Bioremediation, OHM Remediation Services Corp., and Jurgen H. Exner, Principal and President, JHE Technology Systems, Inc., as reported in *Chemical Engineering* magazine. The data in step 4 of this procedure were prepared by the *Handbook* editor.

David R. Hopper, Chemical Process Engineering Program Manager, ENSR Consulting and Engineering, writing in *Chemical Engineering* magazine, notes that:

> Many of today's contaminated sites are the result of accepted and lawful waste-disposal practices of years ago. While the methods of disposal have improved and the regulations preventing disposal techniques that might result in future contamination are in place, there is no guarantee that today's landfilled wastes will not end up being remediated in coming years. In addition, the new regulations and technologies have come at a time of increased disposal cost and ever-diminishing landfill capacity.
>
> Waste minimization, or pollution prevention, is one way of avoiding the whole disposal problem, and its associated long-term liability. By reducing the creation of waste by the manufacturing process, or recovering and recycling potential wastes between processes, the amount of waste to be disposed of is reduced. . . .
>
> Pollution prevention programs are gaining momentum at both the federal and state levels. Several states (e.g. Texas and New Jersey) have introduced legislation aimed at promoting waste reduction. Federal agencies (e.g. EPA, Department of Defense, Department of Energy, and Department of Interior) are actively supporting research and development of waste-minimization methods. However, the major driving force remains the economic benefits of reducing the amount of waste produced. Savings in raw materials and avoidance of the disposal costs result in attractive returns on investment for waste-minimizing process improvements. Between the potential savings and the future regulatory focus, waste minimization is likely to be an active, and beneficial, aspect of future waste-management programs.

PROCESS AND EFFLUENT TREATMENT PLANT
COST ESTIMATES BY SCALE-UP METHODS

Estimate the cost of a new effluent treatment plant using the R-factor when the proposed plant will have a capacity of 800,000 tons per year; an earlier effluent treatment plant of similar design that treats 250,000 tons per year has a cost of $8,600,000. Determine the cost of this plant using the same method when the appropriate construction-cost index has risen from 325 to 387 between the time of construction of the first plant and today.

Calculation Procedure:

1. Estimate the cost of the new plant using the R-factor method
The R-factor method permits quick estimates of the cost of proposed new plants of many types. The method uses an exponent, R, which is applied to the ratio of the capacities of the proposed and known facilities, with the result being multiplied by the cost of the earlier plant. This method is also termed the 0.6-power-factor model. It was first applied to equipment cost estimates in 1947 and has since found expanded use in a number of important industrial applications. In 1950, it was expanded to include plant cost estimates. The range of R factors for plants is much wider than for equipment.

This R-factor predesign cost-estimating approach is especially useful for performing sensitivity analyses, for which a high degree of accuracy is not required. Equipment and plant costs are still being estimated by this method, and operating costs can be estimated by a variation of it.

Values for R for several hundred different types of plants and processes are available in a comprehensive compilation.[2] To permit immediate use by handbook owners, a summary of average R values is presented in Table 4. For specific processes or plants not detailed in this table, the reader should refer to Remer and Chai.[2]

TABLE 4 Average R Values Classified by Type of Process Industry

Industry	Table reference	Average R	Standard deviation
All values	2–7	0.67	0.13
Chemical plants and processes	2	0.67	0.13
Gases	3	0.65	0.10
Polymers	4	0.72	0.10
Biotechnology	5	0.67	0.13
Power plants, effluent treatment, drinking water, refrigeration and utilities	6	0.75	0.10
Miscellaneous	7	0.70	0.05

---

[2]Remer and Chai, "Estimate Costs of Scaled-Up Process Plants," *Chemical Engineering,* April, 1990, pp. 138–175.

The relationship between cost and capacity is given by $(C_2)/(C_1) = (S_2)/(S_1)^R$, where C_1 = cost of the original plant or facility, \$; C_2 = cost of the new plant or facility, \$; S_1 = size of known plant or facility, expressed in suitable units; S_2 = size of new plant or facility, expressed in the same units.

For the first effluent plant being considered, using $R = 0.75$ from Table 4, C_2 = \$8,600,000$(800,000/250,000)^{0.75}$ = \$20,575,999; say \$21,000,000 for estimating purposes. This agrees nicely with the cost ratios for size multiples in Table 5. Thus, the ratio for $(800,000/250,000)^{0.75}$ = 2.39. From the table, interpolating between $R = 0.7$ and $R = 0.8$, gives a ratio of 2.285 for a plant three times as large as the original plant. The plant here is 3.2 times ($= 800,000/250,000$). Again, this is well within the accuracies met in early cost estimates.

A rule of thumb says that with $R = 0.6$, doubling the size of a plant increases the cost by 50 percent; tripling its size increases its cost by 100 percent. In the chemical process industries, average values of R fall between 0.6 and 0.7. Other industries—power plants, effluent treatment, drinking water, refrigeration, and utilities—generally have a higher value, typically 0.75.

You must be careful when assuming a value for R when there are no references giving historical values for the situation you face. Table 5 shows potential errors that might occur if an incorrect R value is assumed. Thus, if R is assumed to be 0.7 when it actually is 0.9, the table shows that the final cost could be 28 percent in error for a five-fold scaleup.

2. Determine the plant cost using construction-cost indexes

When using a construction-cost index to update an earlier cost to today's cost you use the same equation as in step 1 and multiply it by the ratio of today's cost to the earlier cost. Or, $(C_2)/(C_1) = (S_2)/(S_1)^R(i_t/i_e)$, where i_t = today's cost index; i_e = cost index at earlier date; other symbols as before. Substituting, C_2 = \$8,600,000$(800,000/250,000)^{0.75}(387/325)$ = \$24,501,267; say \$24,500,000 for discussion and comparison purposes.

Thus, the engineer making estimates of plant or facility costs can bring these right up to date by using a suitable cost index. Popular indexes used today include: Chemical Engineering (CE) Plant Cost Index; Marshall and Swift (M&S) Equipment Cost Index; Nelson Refinery Index; Engineering News Record (ENR) Index.

Related Calculations. Scaled-up cost estimates give the engineer a fast, easy, and reliable way to compute costs of various plants and facilities at the redesign stage. Such estimates are helpful for looking at the effect of plant size on profitability when doing discounted-cash-flow-rate-of-return and payback-period calculations. These estimates are also useful for making an economic sensitivity analysis involving a large number of variables. Table 6 shows cost ratios for increasing the size of a plant as a function of the exponent R.

The exponent R tends to be higher if the process uses equipment designed for high pressure or is constructed of expensive alloys. As R approaches 1, cost becomes a linear function of capacity—that is, doubling the capacity doubles the cost. The value of R may also approach 1 if product lines will be duplicated, rather than enlarged.

Large capacity extrapolations must be done carefully. Costs must also be scaled down carefully from very large to very small plants because much of the equipment (such as computers and instruments) cost about the same regardless of price.

This procedure is the work of Donald S. Remer, Oliver C. Field Professor of Engineering, Harvey Mudd College of Engineering and Science, and Lawrence Chai, John F. Hurst Consulting Engineering, as reported in *Chemical Engineering* magazine.

TABLE 5 Potential Errors from Using the 0.6 or 0.7 as Cost-Capacity Factors

	Actual R value								
Scaleup	0.2	0.3	0.4	0.5	0.6	0.7	0.8	0.9	1.0
					Error in using 0.6, %				
2 times	+38	+23	+15	+7	0	−7	−13	−19	−24
5 times	+90	+62	+38	+17	0	−15	−28	−38	−47
10 times	+151	+100	+58	+26	0	−21	−37	−50	−60
					Error in using 0.7, %				
2 times	+41	+38	+23	+15	+7	0	−7	−13	−19
5 times	+124	+90	+62	+38	+17	0	−15	−28	−38
10 times	+216	+151	+100	+58	+26	0	−21	−37	−50

TABLE 6 Cost Ratios for Increasing the Size of a Plant as a Function of Exponent R

	Cost ratios for size multiples			
R value	2 times	3 times	5 times	10 times
0.2	1.15	1.25	1.38	1.58
0.3	1.23	1.39	1.62	2.00
0.4	1.32	1.55	1.90	2.51
0.5	1.41	1.73	2.24	3.16
0.6	1.52	1.93	2.45	3.98
0.7	1.62	2.16	3.09	5.01
0.8	1.74	2.41	3.62	6.31
0.9	1.86	2.69	4.26	7.94
1.0	2.00	3.00	5.00	10.00
1.1	2.14	3.35	5.87	12.59

DETERMINATION OF GROUND-LEVEL POLLUTANT CONCENTRATION

A pollutant is emitted at a rate of 0.13 lb/min (1 g/s) from a 164-ft (50-m) high stack on a highly turbulent day. Wind speed at a height of 32.8 ft (10 m) is 4.47 mi/h (2 m/s). What will be the maximum ground-level concentration of the pollutant under these conditions? Neglect any plume rise.

Calculation Procedure:

1. Compute the wind speed at the stack outlet
Since the emission is taking place at an elevated point source, we must first compute the wind speed at the stack outlet. Use the relation: $(V_H)/(V_R) = (H/Z)^p$, where V_H = wind speed at the elevated-source height, mi/h (m/s); V_R = wind speed at reference height R, mi/h (m/s); H = stack outlet height, ft (m); Z = reference height, ft (m); p = a dimensionless wind profile power-law exponent. The value of p for urban areas is approximately equal to 0.15, 0.25, and 0.3 for stabilities A, D, and F, respectively. Stability A is the most turbulent; stability D is neutral; stability F is the most stable.

Since this analysis is for a highly turbulent day, $p = 0.15$. Then, $V_H/2 = (50/10)^{0.15} = 2.55$ m/s (5.7 mi/h) at the stack outlet 164 ft (50 m) above the ground.

2. Compute the maximum ground-level concentration of the pollutant
Use the relation $c_{max} = 0.23\ Q/V_H \times H^2$, where c_{max} = maximum pollutant concentration at ground level, g/m^3 (lb/ft^3); Q = emission rate, g/s (lb/ft^3); other symbols as before. Substituting, $c_{max} = (0.23 \times 1.0)/[2.55(50)^2] = 4 \times 10^{-5}$ g/m^3 (2.297×10^{-9} lb/ft^3).

While this may seem like a small concentration of the pollutant, remember that the Clean Air Act (CAA) requires NO_x standards of 30 ppm, or less. For failure to comply with CAA standards there are penalties of $250,000 per day in fines, five years in prison, or both. Any designer or operator would be foolish to accept such penalties without first taking steps to determine the pollutant concentration and using any legitimate means to reduce the concentration to, or below, the legally acceptable minimum.

3. Show the procedure for a non-elevated point source
Where a non-elevated point source emits a pollutant at a rate of 1 g/s (0.13 lb/min) on a day with calm and stable atmospheric conditions, the ground-level ambient concentration at a downwind distance of 400 m (1312 ft) from the source is given by the relation $c = 36Q/Ux^2$. With F stability, which applies to this condition, as is shown below, the wind speed is assumed as 1.0 m/s (0.348 ft/s). Substituting in the above equation, $c = (36 \times 1)/[1(400)^2] = 2.25 \times 10^{-4}$ g/m³ (1.11×10^{-9} lb/ft³).

4. Give the equations for general pollutant determination
Simplified dispersion equations can provide rough and quick approximations of pollutant levels from non-elevated and elevated point sources. Two of these equations were used in the calculations above. With the severe fines and imprisonment threats now in the Clean Air Act, rapid determination of pollutant concentration is a skill all engineers should possess.

Briggs' formulas can be used to relate dispersion coefficients and downwind distance of the pollutant concentration in urban areas. Three stability conditions of the surrounding atmosphere are used for computing the maximum concentration of the pollutant in the plume. These stability conditions are A—the most turbulent atmospheric conditions, i.e., wind speed of 2 m/s (6.6 ft/s); D—neutral stability, i.e., 5 m/s (16.4 ft/s); F—the most stable atmospheric condition, i.e., 1.0 m/s (3.3 ft/s). Then, the plume's centerline concentration, c, is given by: For A stability, $c = 4Q/Ux^2$; for D stability, $c = 14.2Q/Ux^2$; for F stability, $c = 35Q/Ux^2$.

Related Calculations. The three equations presented can be used to estimate quickly ambient ground-level concentrations of pollutants from non-elevated and elevated point sources. These equations can be used for any type of pollutant emitted to the atmosphere from power, chemical, manufacturing, or waste-disposal plants. Both critical and hazardous pollutant concentration can be evaluated using these equations.

This calculation procedure is the work of Ajay Kumar, P.E., Senior Engineer, Air Group, EA Engineering, Science and Technology, Inc., as reported in *Chemical Engineering* magazine. USCS values were added by the *Handbook* editor.

ESTIMATING HAZARDOUS-GAS RELEASE CONCENTRATIONS INSIDE AND OUTSIDE BUILDINGS

At a water-treatment plant, chlorine is supplied from a 1-ton (0.91-tonne) cylinder located inside a building. The dimensions of the building are 49 ft × 33 ft × 16.4 ft (15 m × 10 m × 5 m); ambient wind speed in the area is 3.4 mi/h (1.5 m/s); molecular weight of chlorine is 70.8. If a malfunctioning valve associated with the cylinder allows its contents to leak continuously at a rate of 100 lb/day (45.4 kg/day; 5.26 g/s), estimate the chlorine concentration: (*a*) inside the building; (*b*) in the building cavity; (*c*) 66 ft (20 m) downwind of the building.

Calculation Procedure:

1. Determine the chlorine concentration inside the building

To estimate the chlorine concentration inside the building, assume the wind speed, U, in the building is 2.2 mi/h (1.0 m/s); the height, H, of a hypothetical box inside the building in which the released gas has mixed homogeneously, is 5.9 ft (1.8 m); crosswind width of the building is 33 ft (10 m). Then $C = Q/UHW$. Using SI units because they are more conventional in these calculations, C = gas concentration, g/m^3; Q = release rate of the gas, g/m^3; U = wind speed, m/s; H = hypothetical box height, m. Solving for this building, C = 0.526 g/s/(1 m/s)(10 m)(1.8 m) = 0.0292 g/m^3. To convert to parts per million, ppm, use the relation, C_{ppm} = (0.0245 × 10^6)($C_{g/m}$)/M, where M = molecular weight of leaking gas. Substituting, C_{ppm} = (0.0245 × 10^6)(0.0292)/70.8 = 10.1 ppm.

2. Compute the gas concentration in the building cavity

A building cavity is the region near the downwind side of a building. In this region, pollutants emitted from an elevated source are mixed rapidly toward the ground due to the aerodynamic turbulence induced by the building. Equations presented in this procedure are based on the assumption that the released gas dispersion pattern is assumed to be equivalent to that of a passive gas (i.e., density of a passive gas is equal to that of air).

When computing the concentration of the released gas within the building cavity, it is assumed that the hazardous gas from the building releases to the atmosphere through various building openings (such as doors and windows). Once released into ambient air, the hazardous gas will be trapped within the building cavity due to the aerodynamic effects around the building. In the building cavity it can be assumed that the released gas is vigorously mixed, resulting in a homogeneous gas concentration within the cavity.

To estimate the ambient concentration of hazardous gas inside the building cavity, use $C = Q/(1.5U_c)(A_p)$, where U_c = the critical wind speed, m/s (ft/min); A_p = cross-sectional area of the building perpendicular to the wind direction, m^2. For this building, assuming a critical wind speed of 1 m/s (3.28 ft/s), and a cross-sectional area of 10 m × 5 m = 50 m^2 (538.2 ft^2), C = 0.526 g/s/[(1.5)(1 m/s)(50 m^2)] = 0.007 g/m^3. Or, using the equation in step 1 for ppm, C_{ppm} = 0.0245 × 10^6 × 0.007/70.8 = 2.42 ppm.

3. Estimate the gas concentration outside a building cavity

Outside a building cavity the ambient concentration of a hazardous gas release can be estimated using the Gaussian equation, $C = Q/(\pi)(l_d v_d)(U)$, where l_d = lateral dispersion coefficient, m (ft); v_d = vertical dispersion coefficient, m (ft).

Up to a distance ten times the building height ($10h_b$), these dispersion coefficients are a function of building dimensions. For example, for a squat building (width greater than height), these dispersion coefficients can be estimated using: l_d = 0.35h_w + 0.067(x − 3h_b), and v_d = 0.7h_b + 0.067(x − 3h_b), where h_w = building width, m (ft), which is approximated by the diameter of a circle with an area equal to the horizontal area of the building, i.e., h_w = 0.866(L^2 + W^2)$^{0.5}$, where L = building length, m (ft); W = building width, m (ft). In the first two equations for the lateral and vertical dispersion coefficients, x is the distance from the source to the receptor at which the hazardous gas concentration is measured. Note that x is measured from the center of the building (i.e., the source) the receptor in question.

To estimate the chlorine concentration at a receptor located 20 m (66 ft) away from the leeside of the building, estimate the lateral and vertical dispersion coefficients first thus: $h_w = 0.866(15^2 + 10^2)^{0.5} = 15.61$ m (51.2 ft). Note, $x = 20$ m + 15 m/2 = 27.5 m (90.2 ft). Using the dispersion equations, $l_d = 0.35(15.61$ m) + 0.067[27.5 m − 3(5 m)] = 6.3 m (20.7 ft); $v_d = 0.7(5.0$ m) + 0.067[27.5 m − 3(5 m)] = 4.34 m. Then, the ambient concentration for chlorine can be found from $C = 0.526$ g/s/[$(\pi)(6.3$ m)(4.34 m)(1.5 m/s)] = 4.08 × 10^{-3} g/m^3. Computing ppm, $C_{ppm} = 0.0245 × 10^6(4.08 × 10^{-3})/70.8 = 1.41$.

Related Calculations. Equations presented here yield conservative estimates. They are useful for engineers who need to estimate quickly the concentration of gases resulting from a hazardous gas release on-site.

Hazardous gas leaks are a major concern to operators of municipal facilities and to those throughout many segments of the chemical process industries. For example, a water-treatment plant generally has chlorine-containing cylinders delivered to the facility and stored on-site, mostly within the confines of a storage building. From time to time, the hazardous gas may escape from a cylinder, creating a serious problem not only for the facility workers inside the building, but for those working outside the building, or nearby, as well.

In the event of a hazardous gas release, a quick estimation of the concentration within the building, and in the area surrounding the building and beyond may help to direct the response actions of all facility personnel.

The equations and procedure presented here are the work of Ajay Kumar, P.E., Senior Chemical Engineer, EA Engineering, Science and Technology, Inc., Sandi Wiedenbaum, Senior Project Manager, EA Engineering, Science and Technology, Inc., and Michael Woodman, Air Quality Scientist, EA Engineering, Science and Technology, Inc., as reported in *Environmental Engineering World* magazine (Sept–Oct '95 issue).

DETERMINING CARBON DIOXIDE BUILDUP IN OCCUPIED SPACES

An office space has a total volume of 75,000 ft^3 (2122.5 m^3). Equipment occupies 25,000 ft^3 (707.5 m^3). The space is occupied by 100 employees. If all outside air supply is cut off, how long will it take to render the space uninhabitable?

Calculation Procedure:

1. Determine the cubage of the space
For carbon dioxide buildup measurements, the *net volume* or (*cubage*) of the space is used. The net volume of a space = total volume − volume of equipment, files, machinery, etc. For this space, net volume, NV = total volume − machinery and equipment volume = 75,000 − 25,000 = 50,000 ft^3 (1415 m^3).

2. Compute the time to vitiate the inside air
Use the relation, $T = 0.04V/P$, where T = time to vitiate the inside air, hours; V = net volume, ft^3; P = number of people occupying the space. Substituting, $T = 0.04(50,000)/100 = 20$ h. During this time the oxygen content of the air will be reduced from a nominal 21 percent by volume to 17 percent.

It is a general rule to consider that after 5 h, or one-quarter of the calculated time of 20 h, the air would become stale and affect worker efficiency. Atmospheres containing less than 12 percent oxygen or more than 5 percent carbon dioxide are

considered dangerous to occupants. The formula used above is popular for determining the time for carbon dioxide to build up to 3 percent with a safety factor.

   Related Calculations.   In today's environmentally conscious world, smoking
indoors is prohibited in most office and industrial structures throughout the United
States. And much of the Western world appears to be considering adoption of the
same prohibition, albeit slowly. Part of the reason for prohibiting smoking inside
occupied structures is the oxygen depletion of the air caused by smokers.

   Today, indoor air quality (IAQ) is one of the most important design considerations faced by engineers. A variety of environmental rules and regulations control
the design of occupied spaces. These requirements cannot be overlooked if a building or space is to be acceptable to regulatory agencies.

   For years, occupied spaces which were not air-conditioned were designed using
general ventilation rules. In most buildings, exhaust fans located high in the side
walls, or on the roof, were used to draw outside air into the building through
windows or louvers. The air movement produced an air flow throughout the space
to remove smoke, fumes, gases, excess moisture, heat, odors, or dust. A constant
inflow of fresh, outside air was relied on for the removal of foul, stale air.

   Today, with the increase in external air pollution, combined with the outgassing
of construction and furnishing materials, general ventilation is a much more complex design problem. No longer can the engineer rely on clean, unpolluted outside
air. Instead, careful choice of the location of outside-air intakes must be made.
Other calculation procedures in the *Handbook* deal with this design challenge.

ENVIRONMENTAL EVALUATION OF INDUSTRIAL COOLING SYSTEMS

Evaluate the various cooling systems available and determine the most desirable
for the following types of installations: (1) steam power plant; (2) cogeneration
plant; (3) combined-cycle plant; (4) nuclear power station. For each type of plant
the following factors are important: (*a*) environmental; (*b*) permitting; (*c*) cost.

Calculation Procedure:

1.   *Review the types of environmentally acceptable cooling systems*
There are three types of cooling systems in use today for industrial and power
plants of the types mentioned above. These cooling systems are: (1) dry cooling;
(2) wet cooling; (3) wet/dry hybrid cooling systems. Each has two versions, depending on the application. Tables 7–9 list the characteristics of each type of cooling system.

   With the types and characteristics of the cooling systems known, tentative
choices can be made for the specific plants under consideration. Numeric values
can be assigned to the relative cost of each system type to help in the evaluation.

2.   *Choose a cooling system for the steam power plant*
Obtain, from equipment manufacturers, cost estimates for the various types of cooling which might be used. If actual costs cannot be obtained, try to obtain from
manufacturers relative costs for the various types of cooling that might be used for
this power plant. To show how such relative costs might be used, the handbook
editor lists, in Table 4, *assumed relative costs* for the various types of cooling

TABLE 7 Characteristics and Types of Dry Cooling Systems

Advantages of the system

Permits more flexible site selection since water supply is not a critical issue;
Shortens permitting process because local water supplies are not used;
Satisfies environmental agencies because there is neither demand nor effluent;
Plant can be closer to fuel supply—mine-mouth plants ave fuel transport costs and pollution associated with the transportation'
Plant can be closer to its expected loads, saving transmission costs and investment;
Eliminates the plume hazards of wet cooling towers which can hamper airport and highway operations.

Types of dry cooling systems

Direct dry cooling—turbine or other exhaust steam goes directly to air-cooled finned-tube heat exchangers fitted with forced-draft fans. Condensate is collected and returned to feedwater system for re-use. No external cooling water is needed or used.

Indirect dry cooling (called the Heller System) uses surface or "jet" condensers with the cooling water's temperature reduced in finned-tube "deltas" in either natural or mechanical-draft tower. The system is closed throughout—there is no contact between the water and the air and makeup water is not required. To increase the overall heat-rejection capacity, the deltas can be "deluged"—i.e., a small amount of makeup water is sprayed over the finned-tube deltas during warm weather, reducing the turbine exhaust steam backpressure.

TABLE 8 Characteristics and Types of Wet Cooling Systems

Advantages and types of wet cooling systems

Once-through wet cooling systems dispose of waste heat by discharging directly into the sea, a river, or a lake. Such systems are simpler, cost less, and are more efficient than evaporative systems. But (1) environmental regulations often prohibit the thermal and contaminant pollution inherent in these systems, and (2) powerplants are now often built where water is scare or costly.

Evaporative cooling systems use either mechanical- or natural-draft cooling towers to reduce the temperature of warm water. Natural-draft towers, which rely on atmospheric conditions, must be larger than forced-draft, and may be higher in cost. Further, in excessively warm weather they may not provide the needed cooling conditions.

TABLE 9 Characteristics of Wet/Dry Hybrid Cooling Systems

Usually used in special situations where neither a wet nor dry system could do the job alone. Two examples include sites with moderate—but not severe—water shortages and evaporatively cooled plants in urban areas that must be protected from plumes—such as near airports and highways.

Using wet and dry sections, either of which can be operated separately, or in series, water first flows through finned-tube heat exchangers before entering the cooling-tower distribution system. Mixing dry air from the heat exchangers with moist air from the cooling tower prevents plume formation. Maximum cooling efficiency is achieved by adjusting cooling water flow through the wet section, and optimizing the division of air flow through the two sections, using air-inlet switches, shutters, and fan-speed controls.

systems considered here. The editor wishes to emphasize that these assumed costs may not be accurate. Hence, the *Handbook* user *must* consult manufacturers to obtain actual relative costs. The costs given in Table 10 are used for illustrative purposes only.

Steam power plants usually have large heat rejection requirements. Since water is a more efficient coolant than air, a cooling tower appears to be a good choice as a preliminary selection. And since mechanical-draft towers can provide more cooling capacity in all weathers, a mechanical-draft tower will be the first choice here. Further, on the relative-cost basis assumed here, it appears the more economical choice.

Actual applications today make wide use of mechanical-draft towers for steam-power-plant reject heat. Newer towers use precast concrete construction, thereby reducing the maintenance cost associated with older wooden towers. Large fans, some 36-ft (10.97-m) diameter, provide adequate cooling at acceptable cost. Cooling towers are gaining favor over once-through direct cooling because of the environmental and scarcity restrictions mentioned earlier.

3. Select a cooling system for a cogeneration plant
Cogeneration plants are often found in water-scarce, stringent-permitting areas, where plume emission is either forbidden or dangerous to surrounding industry and residents. For these reasons, cogeneration plant designers, either of the original plant, or expanded facilities, often choose dry cooling with an air-cooled condenser, option 3 in Table 10.

With direct dry cooling there is no water consumption, no plume emission, and low noise generation. Hence, this choice of cooling is popular for cogeneration plants throughout the world, and with water supplies drying up worldwide, it is likely that direct dry cooling will continue growing in both popularity and number of installations.

4. Choose a cooling system for the combined-cycle plant
Combined-cycle plants use two or more prime movers, such as steam turbines and gas turbines or either type of turbine and internal-combustion engines. Steam turbines often use steam from the plant's heat-recovery steam generators (HRSG) fired by the gas turbine exhaust gases. Such combined-cycle plants can have thermal efficiencies approaching 50 percent.

A popular cooling choice for combined-cycle plants is direct dry cooling. The reasons for this choice include zero water use, no plume emission, and low noise level. Since combined-cycle plants offer such high thermal efficiencies, it is likely that they will continue to find favor worldwide in coming years.

TABLE 10 Assumed Relative Costs of Cooling Equipment*

Lowest cost	1 Once-through direct cooling
	2 Mechanical-draft cooling towers
	3 Direct dry cooling
	4 Indirect dry cooling
	5 Natural-draft cooling towers
Highest cost	6 Wet/dry hybrid cooling system

*Presented for illustrative purposes, only. Actual relative costs may vary considerably.

Where space limitations prohibit installation of direct dry-cooling units, wet/dry hybrid cooling is often used for cogeneration plants. Wet/dry cooling is popular where once-through cooling can no longer be used because of low water conditions or environmental regulations.

5. Choose a cooling system for the nuclear power station

Nuclear power stations continue to grow in popularity, worldwide. With extensive operating hours, often 8000 or more hours per year, a frequent choice for cooling is the natural-draft cooling tower. The reasons for this include: (*a*) simplicity of operation; (*b*) ease of maintenance; (*c*) minimum power requirements.

Natural-draft cooling towers continue to be a popular choice for nuclear power stations worldwide. As power demand continues to rise in developing nations, it is likely that natural-draft cooling towers will be the first choice for nuclear-plant cooling.

Related Calculations. Cooling-system choices given in this calculation procedure are "first cuts"—i.e., general indications of current practice. Before a final choice is made, the designer must: (*a*) obtain actual cost estimates for the various options available; (*b*) evaluate applicable environmental regulations; (*c*) review local zoning requirements; (*d*) study site restrictions that may dictate one choice over another. Only then can an intelligent choice be made of a cooling system.

Much of the data in this procedure was obtained from publications of the GEA Power Cooling Systems, Inc., San Diego, CA, except for the relative cost information in Table 10 which was prepared by the *Handbook* editor *solely for illustrative purposes.* Besides the cooling systems mentioned above, GEA, a worldwide organization, designs and manufactures heat exchangers for flue-gas cooling. A recent release from GEA states, in part:

A growing number of central stations use heat exchangers to cool flue gases to the scrubber's optimum operating temperature. In this application the heat is recovered, stored in the heat-exchangers circulating-water loop, and then reintroduced into the clean-gas output of the scrubber just upstream of the stack. At waste-to-energy plants, slightly different heat exchangers perform a similar function, but without water; they dry the scrubber's flue-gas output with heat recovered from its input gas stream. In both applications, heat recovery saves energy and reduces cost.

Successful use of heat exchangers for flue-gas heat recovery began about the mid 80s, when all-plastic finned tubes solved an application problem that plastic-coated steel and graphite heat exchangers could not. When flue gas is cooled through its dew point, free dust particles in the gas react with flyash, creating a potentially corrosive mixture in the acidic flue-gas environment. Where such mixtures had been adhering to and corroding the surfaces of plastic-coated steel and graphite heat exchangers, they simply slid down the plastic tubes.

Today, the tubes of heat exchangers designed for flue-gas applications are made of fluorine plastics such as polyvinylidenfluoride (PVDF) and perfluoralkoxy (PFO).

As a further example of the importance of industrial cooling, Fig. 5, shows an example of a zero-discharge steam power plant. As reported in *Power,* many power plants recycle waste water to avoid the need to dispose of it and to minimize raw-water requirements. As waste water is recycled, however, the quality often degrades to a point at which it becomes unsuitable for further use in plant systems without expensive treatment.

Power plants that are not permitted to discharge to receiving bodies of water have often relied on solar evaporation ponds to dispose of this "unsuitable" water, especially in arid and semiarid regions.

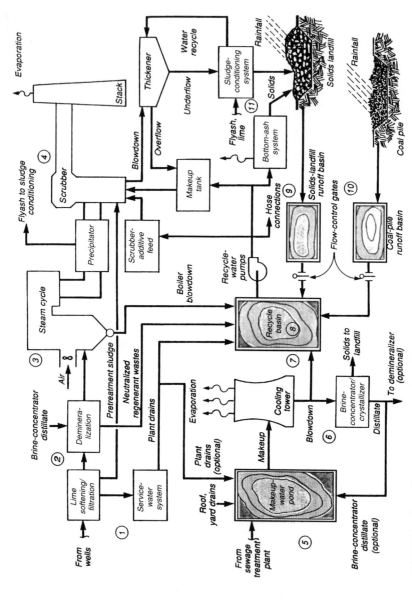

FIGURE 5 Zero-discharge steam-power plant. (*Power*).

The plant shown in Fig. 5 is in the Orlando, FL area which has a particularly wet season from June through September. Hence, evaporation ponds are not feasible. Further, the plant site is not located near any major river or lake, so discharge to surface waters is impractical. Moreover, groundwater quality has undergone extensive regulation in Florida, as the state attempts to preserve water quality to serve a growing population.

Because of these circumstances, it was imperative that this plant be designed and operated as a true zero-discharge power plant. All waters imported to the plant are ultimately evaporated or retained on site. Key elements in the detailed plan drawn up for the management of water and wastewater include: (1) Selection of the water source most appropriate for the service. At this plant, municipal sewage-treatment plant effluent is the water source. Figure 5 shows the principal water/wastewater routings, with cooling water the high-volume consumer. (2) Conservative design of water/wastewater systems to allow for changing operating conditions. For example, the pond system, with basins for makeup water, recycling water, coal-pile runoff, and solids-landfill runoff, allows equalization of flows and water quality, enhancing economical and reliable operation in a zero-discharge situation. (3) Compatibility of plant system design with the designated water source.

The report on which the above is based was edited by Thomas Elliott, of the magazine.

Strategies to Conserve Energy and Reduce Environmental Pollution

Environmental engineering is probably the fastest growing branch of engineering today. Impacting every facet of industry and society, environmental engineering is the answer to a cleaner, safer world. Regardless of where pollution control is exercised—before the pollution occurs, or afterwards—environmental engineering *is* the answer to creating a better environment for everyone, everywhere.

Environmental engineering uses the skills and technologies of almost every other branch of the profession. Thus, the environmental engineer will use methods and solutions from engineering disciplines including mechanical, civil, electrical, chemical, industrial, architectural, sanitary, nuclear, and control engineering. Today a number of engineering schools are offering a major in environmental engineering. Graduates have studied portions of the disciplines just mentioned.

This section of the *Handbook* concentrates on procedures for solving environmental problems of many types. Where procedures in related disciplines are needed, for example pipe sizing, the reader should refer to that discipline in this handbook. By combining the methods given in related sections with those in this section, an engineer should be able to develop solutions to a variety of practical, everyday environmental problems.

GENERALIZED COST-BENEFIT ANALYSIS

An engineering atmospheric control to protect the public against environmental pollution will have an incremental operating cost of $100,000. If the pollution were uncontrolled, the damage to the public would have an estimated incremental cost of $125,000. Would this atmospheric control be a beneficial investment?

Calculation Procedure:

1. Write the cost-benefit ratio for this investment

The generalized dimensionless cost-benefit equation is $0 \le C/B \le 1$, where $C =$ incremental operating cost of the proposed atmospheric control, $, or other consistent monetary units; $B =$ benefit to the public of having the pollution controlled, $, or other consistent monetary units.

2. Compute the cost-benefit ratio for this situation

Using the values given, $0 \le \$100,000/\$125,000 \le 1$. Or, $0 \le 0.80 \le 1$. This result means that 80¢ spent on environmental control will yield $1.00 in public benefits. Investing in the control would be a wise decision because a return greater than the cost of the control is obtained.

Related Calculations. In the general cost-benefit equation, $0 \le C/B \le 1$, the upper limit of unity means that $1.00 spent on the incremental operating cost of the atmospheric control will deliver $1.00 in public benefits. A cost-benefit ratio of more than unity is uneconomic. Thus, $1.25 spent to obtain $1.00 in benefits would not, in general, be acceptable in a rational analysis. The decision would be to accept the environmental pollution until a satisfactory cost-benefit solution could be found.

A negative result in the generalized equation means that money invested to improve the environment actually degrades the condition. Hence, the environmental condition becomes worse. Therefore, the technology being applied cannot be justified on an economic basis.

In applying cost-benefit analyses, a number of assumptions of the benefits to the public may have to be made. Such assumptions, particularly when expressed in numeric form, can be open to change by others. Fortunately, by assigning a number of assumed values to one or more benefits, the cost-benefit ratios can easily be evaluated, especially when the analysis is done on a computer.

SELECTION OF MOST DESIRABLE PROJECT USING COST-BENEFIT ANALYSIS

Five alternative projects for control of environmental pollution are under consideration. Each project is of equal time duration. The projects have the cost-benefit data shown in Table 11. Determine which project, if any, should be constructed.

Calculation Procedure:

1. Evaluate the cost-benefit (C/B) ratios of the projects

Setting up the C/B ratios for the five projects by the cost by the estimated benefit shows—in Table 11—that all C/B ratios are less than unity. Thus, each of the five projects passes the basic screening test of $0 \le C/B \le 1$. This being the case, the optimal project must be determined.

TABLE 11   Project Costs and Benefits

Project	Equivalent uniform net annual benefits, $	Equivalent uniform net annual costs, $	C/B ratio
A	200,000	135,000	0.68
B	250,000	190,000	0.76
C	180,000	125,000	0.69
D	150,000	90,000	0.60
E	220,000	150,000	0.68

2. *Analyze the projects in terms of incremental cost and benefit*
Alternative projects cannot be evaluated in relation to one another merely by comparing their C/B ratios, because these ratios apply to unequal bases. The proper approach to analyzing such a situation is: Each project corresponds to a specific *level* of cost. To be justified, *every* sum of money expended must generate at least an equal amount in benefits; the step from one level of benefits to the next should be undertaken only if the incremental benefits are at least equal to the incremental costs.

Rank the projects in ascending order of costs. Thus, Project D costs $90,000; Project C costs $125,000; and so on. Ranking the projects in ascending order of costs gives the sequence D-C-A-E-B.

Next, compute the incremental costs and benefits associated with each step from one level to the next. Thus, the incremental cost going from Project D to Project C is $125,000 − $90,000 = $35,000. And the benefit from going from Project D to Project C is $180,000 − $150,000 = $30,000, using the data from Table 11. Summarize the incremental costs and benefits in a tabulation like that in Table 12. Then compute the C/B ratio for each situation and list it in Table 12. This computation shows that Project E is the best of these five projects because it has the lowest cost—75¢ per $1.00 of benefit. Hence, this project would be chosen for control of environmental pollution in this instance.

Related Calculations.   There are some situations in which the minimum acceptable C/B ratio should be set at some value close to 1.00. For example, with reference to the above projects, assume that the government has a fixed sum of money that is to be divided between a project listed in Table 11 and some unrelated project. Assume that the latter has a C/B ratio of 0.91, irrespective of the sum expended. In this situation, the step from one level to a higher one is warranted only if the C/B ratio corresponding to this increment is at least 0.91.

Closely related to cost-benefit analysis and an outgrowth of it is *cost-effectiveness analysis,* which is used mainly in the evaluation of military and space

TABLE 12   Cost-Benefit Comparison

Step	Incremental benefit, $	Incremental cost, $	C/B ratio	Conclusion
D to C	30,000	35,000	1.17	Unsatisfactory
D to A	50,000	45,000	0.90	Satisfactory
A to E	20,000	15,000	0.75	Satisfactory
E to B	30,000	40,000	1.33	Unsatisfactory

programs. To apply this method of analysis, assume that some required task can be accomplished by alternative projects that differ in both cost and degree of performance. The effectiveness of each project is expressed in some standard unit, and the projects are then compared by a procedure analogous to that for cost-benefit analysis.

Note that cost-benefit analysis can be used in any comparison of environmental alternatives. Thus, cost-benefit analyses can be used for air-pollution controls, industrial thermal discharge studies, transportation alternatives, power-generation choices (windmills vs. fossil-fuel or nuclear plants), cogeneration, recycling waste for power generation, solar power, use of recycled sewer sludge as a fertilizer, and similar studies. The major objective in each comparison is to find the most desirable alternative based on the benefits derived from various options open to the designer.

For example, electric utilities using steam generating stations burning coal or oil may release large amounts of carbon dioxide into the atmosphere. This carbon dioxide, produced when a fuel is burned, is thought to be causing a global greenhouse effect. To counteract this greenhouse effect, some electric utilities have purchased tropical rain forests to preserve the trees in the forest. These trees absorb carbon dioxide from the atmosphere, counteracting that released by the utility.

Other utilities pay lumber companies to fell trees more selectively. For example, in felling the 10 percent of the marketable trees in a typical forest, as much as 40 to 50 percent of a forest may be destroyed. By felling trees more selectively, the destruction can be reduced to less than 20 percent of the forest. The remaining trees absorb atmospheric carbon dioxide, turning it into environmentally desirable wood. This conversion would not occur in these trees if they were felled in the usual foresting operation. The payment to the lumber company to do selective felling is considered a cost-benefit arrangement because the unfelled trees remove carbon dioxide from the air. The same is true of the tropical rain forests purchased by utilities and preserved to remove carbon dioxide which the owner-utility emits to the atmosphere.

Recently, a market has developed in the sale of "pollution rights" in which a utility that emits less carbon dioxide because it has installed pollution-control equipment can sell its "rights" to another utility that has less effective control equipment. The objective is to control, and reduce, the undesirable emissions by utilities.

With a potential "carbon tax" in the future, utilities and industrial plants that produce carbon dioxide as a by-product of their operations are seeking cost-benefit solutions. The analyses given here will help in evaluating potential solutions.

ECONOMICS OF ENERGY-FROM-WASTES ALTERNATIVES

A municipality requires the handling of 1500 tons/day (1524 mt/day) of typical municipal solid waste. Determine if a waste-to-energy alternative is feasible. If not, analyze the other means by which this solid-waste stream might be handled. Two waste-to-energy alternatives are being considered—mass burn and processed fuel. The expected costs are shown in Table 13. If earnings of 6 percent on invested capital are required, which alternative is more economical?

Calculation Procedure:

1. Plot the options available for handling typical municipal waste
Figure 6 shows the options available for handling solid municipal wastes or refuse. The refuse enters the energy-from-waste cycle and undergoes primary shredding.

TABLE 13 Estimated Costs of Municipal Solid Waste
Disposal Facilities

	Mass burn	Processed fuel
First cost, $	15,000,000	22,500,000
Salvage value, $	1,500,000	2,500,000
Life, years	15	25
Annual maintenance cost, $	750,000	400,000
Annual taxes, $	15,000	15,000
Annual insurance, $	100,000	150,000

Then the shredded material is separated according to its density. Heavy materials—such as metal and glass—are removed for recovery and recycling. Experience and studies show that recycling will recover no more than 35 percent of the solid wastes entering a waste-to-energy facility. And most such facilities today are able to recycle only about 20 percent of municipal refuse. Assuming this 20 percent applies to the plant facility being considered here, the amount of waste that would be recycled would be 0.20(1500) = 300 tons/day (305 mt/day).

Numerous studies show that complete recycling of municipal waste is uneconomic. Therefore, the usual solution to municipal waste handling today features four primary components: (1) source reduction, (2) recycling, (3) waste to energy, and (4) landfilling. The Environmental Protection Agency (EPA) recently proposed broad policies encouraging recycling and reduction of pollutants at their source.

Using waste-to-energy facilities reduces the volume of wastes requiring disposal while producing a valuable commodity—steam and/or electric power. Combustion control is needed in every waste-to-energy facility to limit the products of incomplete combustion which escape in the flue gas and cause atmospheric pollution. Likewise, limiting the quantities of metal entering the combustor reduces their emission in the ash or flue gas. This, in turn, reduces pollution.

2. Determine the energy available in the municipal waste

Usual municipalities generate 1 ton (0.91 Mg) of solid waste per year per capita. About 35 percent of this waste is from residences; 65 percent is from industrial and commercial establishments. The usual heating value of municipal waste is 5500 Btu/lb (12×10^6 J/kg). Table 14 shows typical industrial wastes and their average heating values. Municipalities typically spend $25 or more per ton (0.91 Mg) to dispose of solid wastes.

Because municipal wastes have a variety of ingredients, many plants burn the solid waste as a supplement to coal. The heat in the waste is recovered for useful purposes, such as generating steam or electricity. When burned with high-sulfur coal, the solid waste reduces the sulfur content discharged in the stack gases. The solid waste also increases the retention of sulfur compounds in the ash. The result is reduced corrosion of the boiler tubes by HCl. Further, acid-rain complaints are fewer because of the reduced sulfur content in the stack gases.

Where an existing or future plant burns, or will burn, oil, another approach may be taken to the use of solid municipal waste as fuel. The solid waste is first shredded; then it is partially burned in a rotary kiln in an oxygen-deficient atmosphere at 1652° (900°C). The gas produced is then burned in a conventional boiler to supplement the normal oil fuel.

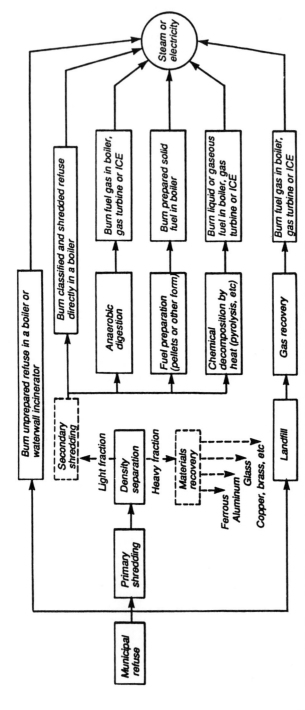

FIGURE 6 Several options for energy extraction from municipal waste. Selection should be based on local variables and economics. (*Power.*)

18.34

TABLE 14 Typical Industrial Wastes with Significant Fuel Value

	Average heating value (as fired)	
	Btu/lb	kJ/kg
Waste gases:		
Coke-oven	19,700	45,900
Blast-furnace	1,139	2,654
Carbon monoxide	579	1,349
Liquids:		
Refinery	21,800	50,794
Industrial sludge	3,700–4,200	8,621–9,786
Black liquor	4,400	10,252
Sulfite liquor	4,200	9,786
Dirty solvents	10,000–16,000	23,300–37,280
Spent lubricants	10,000–14,000	23,300–32,620
Paints and resins	6,000–10,000	13,980–23,300
Oily waste and residue	18,000	41,940
Solids:		
Bagasse	3,600–6,500	8,388–15,145
Bark	4,500–5,200	10,485–12,116
General wood wastes	4,500–6,500	10,485–15,145
Sawdust and shavings	4,500–7,500	10,485–17,475
Coffee grounds	4,900–6,500	11,417–15,145
Nut hulls	7,700	17,941
Rice hulls	5,200–6,500	12,116–15,145
Corn cobs	8,000–8,300	18,640–19,339

Source: Power; SI units added by editor.

Estimates show that about 5 percent of the energy needs of the United States could be produced by the efficient burning of solid municipal wastes in steam plants. Such plants must be located within about 100 mi (160 km) of the waste source to prevent excessive collection and transportation costs.

Combustion of, and heat recovery from, solid municipal wastes reduces waste volume considerably. But there is still ash from the combustion that must be disposed of in some manner. If landfill disposal is used, the high alkali content of the typical municipal ash must be considered. This alkali content often presents leaching and groundwater contamination problems. So, while the solid-waste disposal problem may have been solved, there are still environmental considerations that must be faced. Further, the large noncombustible items often removed from solid municipal waste before combustion—items like refrigerators and auto engine blocks—must still be disposed of in an environmentally acceptable manner. Table 15 shows a number of ash reuse and disposal options available for use today.

3. *Choose between available alternatives*

The two alternatives being considered—*mass burn* and *processed fuel*—have separate and distinct costs. These costs must be compared to determine the most desirable alternative.

In a mass-burn facility the trash is burned as received, after hand removal of large noncombustible items—sinks, bathtubs, engine blocks, etc. The remaining trash is rough-mixed by a clamshell bucket and delivered to the boiler's moving

TABLE 15 Ash Reuse and Disposal Options

	Treatment required	Use
Bottom ash	Particle-size screening	Coarse highway aggregate, concrete products
Bottom ash		Asphalt paving
Combined ash	Particle-size screening	Artificial reefs
Combined ash	Particle-size screening	Aggregate for paving
Flyash	Particle-size screening	Aggregate for paving
Flyash	Particle-size screening	Fine cement aggregate

Source: Power.

grate. Some 30 to 50 percent by weight and 5 to 15 percent by volume of the waste burned in a mass-burn facility leaves in the form of bottom ash and flyash.

In a processed-fuel facility [also called a refuse-derived-fuel (RDF) facility], the solid waste is processed in two steps. First, noncombustibles are separated from combustibles. The remaining combustible waste is reduced to uniform-sized pieces in a hammermill-type shredder. The shredded pieces are then delivered to a boiler for combustion.

Using the annual cost of each alternative (see Section 12) as the "first cut" in the choice: Operating and maintenance cost = C = maintenance cost per year, $ + annual taxes, $ + annual insurance cost, $. For mass burn, C = $750,000 + $15,000 + 0.002($15,000,000) = $795,000. For processed fuel, C = $400,000 + $15,000 + 0.002($22,500,000) = $460,000.

Next, using the capital-recovery equation from Section 12, for mass burn, the annual cost, A = ($15,000,000 − $1,500,000)(0.06646) + $750,000 + $1,500,000(0.06) + $795,000 = $1,737,210. For processed fuel, A = ($22,500,000 − $2,500,000)(0.06646) + $400,000 + $2,500,000(0.06) = $1,879,200. Therefore, mass burn is the more attractive alternative from an annual-cost basis because it is $1,879,000 − $1,737,210 = $141,990 per year less expensive than the processed-fuel alternative.

Several more analyses would be made before this tentative conclusion was accepted. However, this calculation procedure does reveal an acceptable first-cut approach to choosing between different available alternatives for evaluating an environmental proposal.

Related Calculations. Another source of usable energy from solid municipal waste is landfill methane gas. This methane gas is produced by decomposition of organic materials in the solid waste. The gas has a heating value of about 500 Btu/ft^3 (1.1 × 10^6 J/kg) and can be burned in a conventional boiler, gas turbine, or internal-combustion engine. Using landfill gas to generate steam or electricity can reduce landfill odors. But such burning does *not* reduce the space and ground-water problems produced by landfills. The cost of landfill gas can range from $0.45 to $5/million Btu ($0.45 to $5/1055 kJ). Much depends on the cost of recovering the gas from the landfill.

Methane gas is recovered from landfills by drilling wells into the field. Plastic pipes are then inserted into the wells and the gas is collected by gas compressors. East coast landfills in the United States have a lifespan of 5 to 7 years. West coast landfills have a lifespan of 15 to 18 years.

Data in this procedure were drawn from *Power* magazine and Hicks, *Power Plant Evaluation and Design Reference Guide,* McGraw-Hill.

FLUE-GAS HEAT RECOVERY AND
EMISSIONS REDUCTION

A steam boiler rated at 32,000,000 Btu/h (9376 MW) fired with natural gas is to heat incoming feedwater with its flue gas in a heat exchanger from 60°F (15.6°C) to an 80°F (26.7°C) outlet temperature. The flue gas will enter the boiler stack and heat exchanger at 450°F (232°C) and exit at 100°F (37.8°C). Determine the efficiency improvement that might be obtained from the heat recovery. Likewise, determine the efficiency improvement for an oil-fired boiler having a flue-gas inlet temperature of 300°F (148.9°C) and a similar heat exchanger.

Calculation Procedure:

1. Sketch a typical heat-recovery system hookup
Figure 7 shows a typical hookup for stack-gas heat recovery. The flue gas from the boiler enters the condensing heat exchanger at an elevated temperature. Water sprayed into the heat exchanger absorbs heat from the flue gas and is passed through a secondary external heat exchanger. Boiler feedwater flowing through the secondary heat exchanger is heated by the hot water from the condensing heat exchanger. Note that the fluid heated can be used for a variety of purposes other than boiler feedwater—process, space heating, unit heaters, domestic hot water, etc.

Flue gas from the boiler can enter the condensing heat exchanger at temperatures of 300°F (148.9°C), or higher, and exit at 100 to 120°F (37.8 to 48.9°C). The sensible and latent heat given up is transferred to the spray water. Since this sprayed

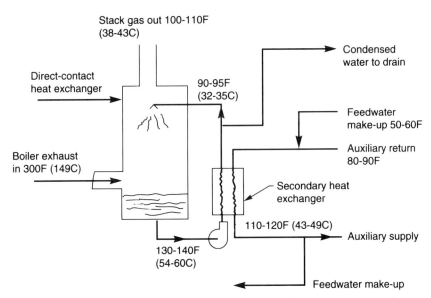

FIGURE 7 Effective condensation heat recovery depends on direct contact between flue gas and cooling medium and low gas-side pressure drop. (*Power.*)

cooling water may be contaminated by the flue gas, a secondary heat exchanger (Fig. 7), may be used. Where the boiler fuel is clean-burning natural gas, the spray water may be used directly, without a secondary heat exchanger. Since there may be acid contamination from SO_2 in the flue gas, careful analysis is needed to determine if the contamination level is acceptable in the process for which the heated water or other fluid will be used.

2. *Determine the efficiency gain from the condensation heat recovery*
Efficiency gain is a function of fuel hydrogen content, boiler flue-gas exit temperature, spray (process) water temperature, amount of low-level heat needed, fuel moisture content, and combustion-air humidity. The first four items are of maximum significance for gas-, oil-, and coal-fired boilers. Installations firing lignite or high-moisture-content biomass fuels may show additional savings over those computed here. If combustion-air humidity is high, the efficiency improvement from the condensation heat recovery may be 1 percent higher than predicted here.

The inlet water temperature is normally 20°F (36°C) lower than the flue-gas outlet temperature. And for the usual preliminary evaluation of the efficiency of condensation heat recovery, the flue-gas exit temperature from the heat exchanger is taken as 100°F (37.8°C).

For the natural-gas-burning boiler, flue-gas inlet temperature = 450°F (232°C); cold-water inlet temperature = 60°F (15.6°C); water outlet temperature = 60 + 20 = 80°F (26.7°C); flue-gas outlet temperature = 80°F (26.7°C). Find the *basic* efficiency improvement, ΔEi, from Fig. 8 as ΔEi = 14.5 percent by entering at the bottom at the flue-gas temperature of 450°F (232°C), projecting to the gas-fired curve, and reading ΔEi on the left-hand axis.

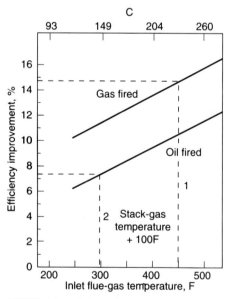

FIGURE 8 Efficiency increase depends on fuel and on temperature of flue gas. (*Power.*)

Next, find the *actual* efficiency improvement, ΔE, from $\Delta E = F(\Delta Ei)$, where F is a factor depending on the flue-gas outlet temperature. Values of F are shown in Fig. 9 for various outlet gas temperatures. With an outlet gas temperature of 80°F (26.7°C), Fig. 9a shows $F = 1.19$. Then, $\Delta E = 14.5 \times 1.19 = 17.3$ percent. Table 16 details system temperatures.

For the oil-fired boiler, flue-gas inlet temperature = 300°F (148.9°C); cold-water inlet temperature = 70°F (21.1°C); water outlet temperature = 70 + 20 = 90°F (32.2°C). Find the *basic* efficiency improvement from Fig. 8 as $Ei = 7.2$ percent. Next, find F from Fig. 9b as 1.18. Then, the *actual* efficiency = $\Delta E = 1.18(7.2) = 8.5$ percent.

The lower efficiency improvement for oil-fired boilers is generally due to the lower hydrogen content of the fuel. Note, however, that where the cost of oil is higher than natural gas, the dollar saving may be greater.

The efficiency-improvement charts given here assume that all of the low-level heat generated can be used. A plant engineer familiar with a plant's energy balance

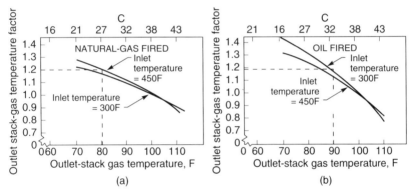

(a)   (b)

FIGURE 9 Use these curves to allow for the effect of variations in the exit temperature from the recovery unit on efficiency increase possible with heat-recovery unit. (*Power.*)

TABLE 16 Use of Figs. 8 and 9

1. Natural gas		
Inlet temperature	450°F	232°C
Inlet cold water	60°F	16°C
Outlet temperature = 60 + 20 =	80°F	27°C
ΔEb (from Fig. 8)	14.5%	14.5%
F (from Fig. 9)	1.19	1.19
$\Delta E = (14.5 \times 1.19)$	17.3%	17.3%
2. Fuel oil		
Inlet temperature	300°F	149°C
Inlet cold water	70°F	21°C
Outlet temperature = 70 + 20 =	90°F	32°C
ΔEb (from Fig. 8)	7.2%	7.2%
F (from Fig. 9)	1.18	1.18
ΔE (7.2 $\times$ 1.18)	8.5%	8.5%

is in the best position to choose the optimum level of heat recovery. Typical applications are: makeup-water preheat, low-temperature process load, space heating, and domestic hot water.

Makeup-water-preheat needs depend largely on the amount of condensate that is returned to the boiler. Generally, there is more heat available in the flue gas than can be used to preheat feedwater. If the boiler is operating at 100 percent makeup, only about 60 percent of the available heat can be transferred to the incoming feedwater. One reason for this is the low temperature of the hot water. This limitation can be handled in two ways: (1) Design the heat-recovery unit to take a slip-stream from the flue gas and only recover as much heat as can be used to heat feedwater; or (2) in multiple-boiler plants, install a heat-recovery system on one boiler only and use it to preheat feedwater for all the boilers.

Process hot water, if it is needed, can provide extremely short payback for a condensation heat-recovery unit. In food and textile processes, the hot-water needs account for 15 percent or more of the total boiler load. Any facility with hot-water requirements between 10 and 15 percent of boiler capacity and an operating schedule greater than 4000 h/yr should seriously investigate condensation heat recovery.

Space-heating economics are generally less favorable than makeup or process hot water because of the load variation, limited heating season, and the difficulty of matching demand and supply schedules. The difficulty of retrofitting heat exchangers to an existing heating system also limits the number of useful applications. However, paybacks between 2.5 and 3 years are possible in colder regions and certainly warrant preliminary investigation. Sometimes space heating can be combined with feedwater or process-water heating.

3. *Estimate the cost of the condensation heat-recovery equipment*

Figure 10 shows an approximate range of costs for equipment and installation. Note that the installed cost may be three times the equipment cost because of retrofit difficulties involved with an existing installation.

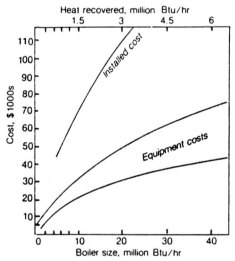

FIGURE 10 Installation cost of condensation heat-recovery unit may run as high as three times equipment cost. (*Power.*)

Operating costs are primarily fan and pump power consumption. These generally range from 5 to 10 percent of the value of the recovered heat. The lower figure applies to limited distribution of the hot water, while the higher figure applies to systems where hot water is distributed 100 ft (30.5 m) or more from the boiler or where high-pressure-drop heat-recovery units are used.

Figure 11 shows how a heat-recovery unit can be used to heat the feedwater for one or more boilers. The heat recovered, as noted above, may be more than needed to heat the feedwater for just one boiler.

Corrosion in a condensing heat-recovery unit can usually be prevented by using Type 304 or 316 stainless steel or fiberglass-reinforced plastic for the tower pump and secondary exchanger. If the flue gas is unusually corrosive, it may be advisable to do a chemical analysis before planning the recovery unit.

A unique feature of condensation heat recovery is that it recovers energy while also reducing emissions. In addition, when natural gas is burned, a small percentage of the NO_x emissions are reduced by condensation of oxides of nitrogen. SO_2

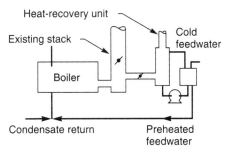

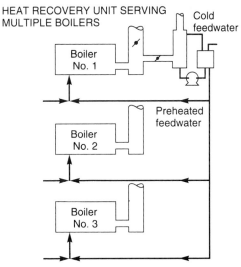

FIGURE 11 Heat recovery from the unit is more than enough to heat feedwater to one boiler. (*Power.*)

emissions can be reduced significantly by using an alkaline water spray in a pH range of 6 to 8. Natural gas depletes the ozone layer less than other fossil fuels.

The potential emission reduction can have a significant effect in nonattainment areas and could increase allowable plant capacity. But it should be pointed out that the SO_2-emission reduction from the scrubbing and condensation have not been substantiated by independent tests. Such tests should be provided for any installation that depends on emission reduction for its justification.

At the time of the preparation of the revision of this handbook, the Tennessee Valley Authority (TVA) is testing at its Shawnee plant a lime treatment to reduce boiler stack gas SO_2 emissions. Lime, in fine particle form, is suspended in the flue gas before release to the plant smokestack. Sulfur in the flue gas binds to the lime, thereby reducing the potential for acid rain. A cyclone and electrostatic precipitator separate the lime particles from the exiting flue gas. At this time the lime system is believed to have lower equipment and operating costs than other competitive systems.

This procedure is based on the work of R. E. Thompson, KVB, Inc., as reported in *Power* magazine.

ESTIMATING TOTAL COSTS OF COGENERATION-SYSTEM ALTERNATIVES

Compare the capital and operating costs of two cogenerational coal-fired industrial steam plants—Option 1: a stoker-fired (SF) plant, Fig. 12; Option 2: a pulverized-coal-fired (PF) plant, Fig. 13. Both plants operate at 600 lb/in² (gage) (4134 kPa) /750°F (399°C). The SF plant meets a demand of 200,000 lb/h (90,800 kg/h) of 150-lb/in² gage (1034-kPa) steam and 10 MW of electric power with the SF boiler and purchased power from a local utility. For the PF boiler the same demands are met using a nonextraction backpressure turbine.

Calculation Procedure:

1. Obtain, or develop, the capital costs for the boiler alternatives
Contact manufacturers of suitable boilers, asking for estimated costs based on the proposed operating capacity, pressure, and temperature. Figure 14 shows a typical plot of the data supplied by manufacturers for the boilers considered here. For coal firing, field-erected boilers are to be used in this plant. (*Note:* The costs given here are for example purposes *only*. Do *not* use the given costs for actual estimating purposes. Instead, obtain current costs from the selected manufacturers.)

From Fig. 14, the SF unit costs $61/lb ($27.70/kg) of steam generated; and the PF unit $72/lb ($32.70/kg) of steam generated per hour. Assuming that SO_2 reduction is not required by environmental considerations for either unit, a dry-scrubber/fabric-filter combination can be used to remove the total suspended particulates (TSP). To cover the cost of this combination to remove TSP, add $7/lb ($3.20/kg) of steam generated. This brings the cost to $68/lb ($30.90/kg) and $79/lb ($35.90/kg) of steam generated.

2. Compute the capital cost for the turbine installation
Before determining the capital cost of the turbine installation—often called a *turbine island*—estimate the potential electric-power generation from the process-

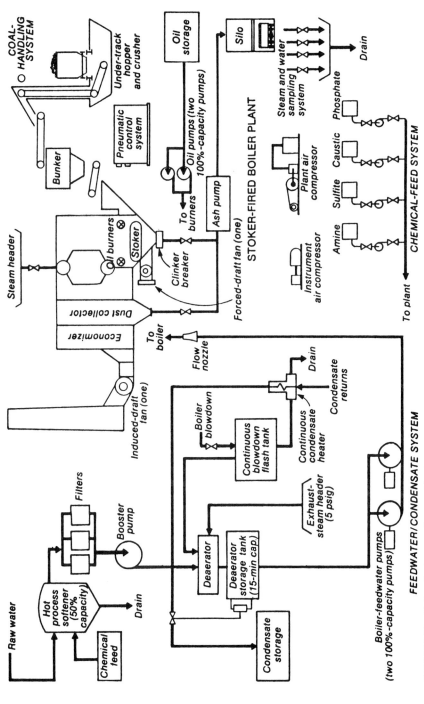

FIGURE 12 Stoker-fired steam plant used in the cost analysis includes all the components shown. (*Power.*)

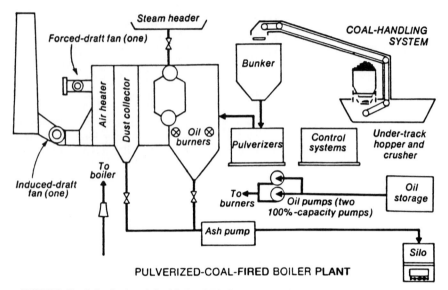

FIGURE 13 Pulverized-coal-fired industrial boilers produce from 200,000 lb/h (90,800 kg/h) to 1 million lb/h (454,000 kg/h) of steam. Unit here does not have an economizer. (*Power.*)

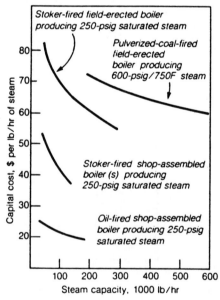

FIGURE 14 Capital costs for three boiler alternatives. (*Power.*)

steam flow based on the ASME data in Table 17. At 600 lb/in² (gage) (4134 kPa)/750°F (399°C) throttle conditions, and 150-lb/in² (gage) (1034 kPa) backpressure, 6127 kW is available, determined as follows: Theoretical steam rate from Table 17 = 23.83 lb/kWh (10.83 kg/kWh); turbine efficiency = 73 percent from Table 17. Then, actual steam rate = theoretical steam rate/efficiency. Or, actual steam rate = 23.83/0.73 = 32.64 lb/kWh (14.84 kg/kWh). Then kW available = steam flow rate, lb/h/steam rate, lb/kWh. Or, kW available = 200,000/32.64 = 6127 kW.

Referring to Fig. 15 for a 6-MW nonextraction backpressure turbine shows a capital cost of $380/kW. Summarize the capital costs in tabular form, Table 18. Thus, the SF boiler cost, Option 1 in Table 18, is (200,000 lb/h)($68/lb of steam generated) = $13,600,000. The PC-fired boiler, Option 2, will have a cost of (200,000 lb/h)($79/lb of steam generated) = $15,800,000. Since a turbine is used with the PC-fired unit, its costs must also be included. Or, 6100 kW ($380/kW) = $2,318,000. Computing the total cost for each option shows, in Table 18, that Option 2 costs $18,118,000 − $13,600,000 = $4,518,000 more than Option 1.

3. Compare operating costs of each option
Obtain from the plant owner and equipment suppliers the key data needed to compare operating costs, namely: Operating time, h/yr; boiler efficiency, percent; fuel cost, $/ton ($/tonne); electric-power use, kWh/yr; electric power cost, ¢/kWh; maintenance cost as a percent of the capital investment per year; personnel required; personnel cost; ash removal, tons/yr (tonnes/yr); ash removal cost, $/yr. Using these data, compute the operating cost for each option and tabulate the results as shown in Table 19.

With Option 1, all electric power is purchased; with Option 2, 3900 kW must be purchased. The difference in annual operating cost is $6,288,350 − $4,895,000

TABLE 17   ASME Values for Estimating Turbine Steam Rates*

	Theoretical steam rate, lb/kWh (kg/kWh)					
Exhaust pressure	600 lb/in² (gage) (4134 kPa)		900 lb/in² (gage) (6201 kPa)		1500 lb/in² (gage) (10,335 kPa)	
	750°F	(399°C)	900°F	(482°C)	900°F	(482°C)
4.0 inHg (10.2 cmHg)	7.64	3.47	6.69	3.04	6.48	2.94
5 lb/in² (gage) (34.5 kPa)	11.05	5.02	9.21	4.18	8.53	3.87
15 lb/in² (gage) (103.4 kPa)	12.16	5.52	9.98	4.53	9.06	4.11
150 lb/in² (gage) (1034 kPa)	23.83	10.82	16.91	7.68	14.30	6.49
	Turbine efficiency, %					
4.0 in Hg	77		78		78	
5 lb/in² (gage)	73		76		76	
15 lb/in² (gage)	74		76		76	
150 lb/in² (gage)	73		74		74	

*Actual turbine steam rate equals theoretical steam rate divided by efficiency.

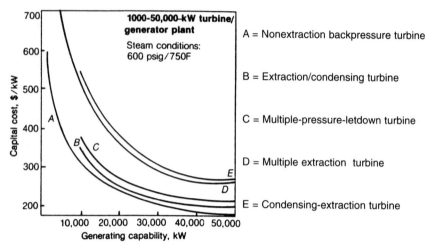

FIGURE 15   Turbine capital cost is estimated from this type of graph. (*Power.*)

TABLE 18   Summary of Capital Costs

	Option 1	Option 2
Boiler island		
Stoker-fired unit ($68/lb steam)	$13,600,000	—
($30.90/kg steam)		
PC-fired unit ($79/lb steam)	—	$15,800,000
($35.90/kg steam)		
Turbine island (6100 kW, $380/kW)	—	2,318,000
Total	$13,600,000	$18,118,000

= $1,393,350. Since Option 2 costs $4,518,000 more than Option 1, but has a $1,393,350-per-year lower operating cost, the simple payback time for the cogeneration option (2), ignoring the cost of money, is: Payback time = larger capital cost, $/annual savings, $, of the higher cost option. Or, $4,518,000/$1,393,350 = 3.24 years. Thus, the cogeneration option (2) is attractive because the payback time is relatively short.

Other economic analyses should also be conducted. For example, higher cycle efficiencies can be obtained with higher throttle conditions, but boiler capital and operating costs will be higher. Plants in the 40- to 60-MW range might benefit from an extraction–condensing-turbine arrangement generating all plant electric-power demand. Although not as efficient as a straight extraction machine, the cost, in $/kW, will be less than purchased power.

Related Calculations.   The EPA and state environmental bodies favor cogeneration of electricity because it reduces atmospheric pollution while conserving fuel. While the options considered here use steam-powered prime movers, cogeneration installations can use diesel-, gasoline-, or natural-gas-fueled prime movers of many different types—reciprocating, gas-turbine, etc. The principal objective of cogeneration is to wrest more heat from available energy streams by the simultaneous

TABLE 19 Comparison of Operating Costs for Coal-Fired Plants

Operating requirements	Option 1 Stoker-fired	Option 2, PC-fired
Equivalent full-power operating time, h/yr	5,400	5,700
Boiler efficiency, %	82	87
Fuel consumption, tons/yr (tonne/yr)	52,700 (47,798)	62,000 (56,234)
Fuel cost, $/ton ($/tonne)	50 (45.35)	42 (38.09)
Total fuel cost, $/yr	2,640,000	2,600,000
Electric-power use, kWh/yr	3,240,000	8,550,000
Electric-power cost, $/yr @ 5¢/kWh	162,000	427,000
Maintenance cost, $/yr (based on 2.5% of capital investment per year)	340,000	395,000
Personnel per shift	2.5	2.75
Personnel cost, $/yr (based on $30,000/yr for each person)	270,000	330,000
Ash removal, tons/yr (tonne/yr)	5,270 (4780)	6,200 (5,623)
Ash-removal cost, $/yr	26,350	31,000
Total operating cost, $/yr	3,438,350	3,783,500
Total operating costs, steam + purchased power		
Steam	3,438,350	3,783,500
Purchased power @ 5¢/kWh	10MW 2,850,000	3.9 MW 1,111,500
Total operating cost	$6,288,350	$4,895,000

generation of electricity and steam (or some other heated medium), thereby saving fuel while reducing atmospheric pollution.

While the possible choices for boiler fuel are oil, gas, and coal, practical choices are limited to coal for most industrial cogeneration projects. Gas firing in industrial plants is restricted to units 80,000 lb/h (36,364 kg/h) or less by the Industrial Fuel Use Act. Oil firing is allowed in field-erected boilers, but usually gives way to coal on an economic basis. Packaged oil-fired boilers top out at 200,000 lb/h (90,909 kg/h) to permit shipping.

Coal can be burned in either a stoker-fired unit or a pulverized-coal-fired unit, the essential cost differences being in coal preparation and ash handling. PC-fired units produce about 75 percent flyash and 25 percent bottom ash, and stoker-fired units the reverse. Coal supply specifications also vary between the two. Grindability is important to PC-fired units; top size and fines content is critical to stoker-fired units. Coal for stoker-fired units averages $5/ton ($5.51/tonne) more for coals of similar heating values.

PC-fired boilers always include an air heater for drying coal upstream of the pulverizer, and usually an economizer for flue-gas heat recovery. The reverse is true of stoker-fired units, but, concerning the air heater, more care is needed to ensure that the grate is not overheated during normal operation.

The greatest overlap in choosing between the two exists in the 200,000 to 300,000 lb/h (90,909 to 136,364 kg/h) size range. Above this range, stoker-fired units are limited by grate size. Below this range, PC-fired units usually do not compete economically. Shop-assembled chain-grate stoker-fired units have been shipped up to a capacity of 45,000 lb/h (20,455 kg/h).

No matter what type of firing is used, industrial power plants must meet EPA emission limits for total suspended particulates (TSP), SO_2, and NO_x. If an on-site coal pile is contemplated, water runoff control must meet National Pollution Discharge Elimination System (NPDES) standards.

NO_x formation is typically limited in the combustion process through careful choice of burners. SO_2 and TSP are usually removed from the flue gas. For TSP reduction, an electrostatic precipitator or a fabric filter is used. These will sometimes be preceded by cyclone collectors for stoker-fired boilers to reduce the total load on the final stage of the ash-collection system.

For boilers under 250-million Btu/h (73.3 MW) heat input, SO_2 formation can be limited by burning low-sulfur coal. Where SO_2 emissions reduction is required to meet the National Ambient Air Quality Standards (NAAQS), a dry-scrubber–fabric-filter combination is a satisfactory strategy. Wet scrubbers, though highly effective, create an additional sludge-disposal problem.

Operating costs for stoker- and pulverized-coal-fired plants designed to produce 200,000 lb/h (90,800 kg/h) of steam from one boiler are shown in Table 19.

To understand more about how operating costs were calculated, look at the entries in Table 19 line by line. First, equivalent full-power operating hours are determined by subtracting 336 h (2 weeks) for maintenance from the total number of hours in a year (8760), and by multiplying the result by both unit availability (85 percent for stoker, 90 percent for pulverized coal) and the assumed plant load factor—in this case 75 percent.

Fuel consumption is based on typical operating efficiencies for similar plants and a fuel heating value of 12,500 Btu/lb (29,125 kJ/kg) for coal. Note that a premium is paid for stoker coal because a relatively clean fuel of suitable size is needed to maintain efficient operation.

The general procedure given here is applicable to a variety of options because today's emphasis on industrial cogeneration calls for a method of reasonably estimating costs of the many system alternatives. The approach differs from utility cost-estimating mainly because steam capacity and power-generation capability are separate design objectives. Either the industrial power plant meets the process-steam demand and then generates whatever power that creates, or else it meets the electric-power demand and generates the required steam.

Choice of approach depends on the steam and electric-power requirements of the facility. Ideally, they balance exactly. In practice, most steam requirements will not generate enough electric power to meet the plant load. Conversely, the steam flow can rarely generate more electric power than the plant needs. Variations in steam conditions and turbine-exhaust pressure lead to many ways of matching the loads. More important, regulated buyback of excess electric power by public utilities now eases the problem of load balancing.

A profile of steam and electric-power consumption is necessary to begin evaluating alternatives. Daily, weekly, monthly, and seasonal variations are all important. During initial evaluation of the balance, use an average of 25 lb (11.4 kg) of steam/kWh as the steam rate of a small steam turbine. It is a conservative number, and the actual value will probably be lower—meaning more electric power for the steam flow—but it will give a rough idea of how close the two demands will match.

In the above procedure, capital costs are separated into costs for the boiler island and costs for the turbine island. Absolute accuracy is to within ±25 percent, not of appropriation quality but good enough to compare different plant designs. In fact, relative accuracy is closer to ±10 percent.

This procedure is the work of B. Dwight Coffin, H. K. Ferguson Co., and was reported in *Power* magazine.

CHOOSING STEAM COMPRESSOR FOR
COGENERATION SYSTEM

Select a suitable steam compressor to deliver an 80-lb/in^2 (gage) (551-kPa) discharge pressure for a cogeneration system using two 1500-kW diesel-engine-generator sets operating at 1200 rpm with 1200°F (649°C) exhaust temperature. Each engine exhaust is vented through a waste-heat (heat-recovery) boiler which generates 3800 lb/h (1725 kg/h) of steam at 110-lb/in^2 (gage) (758-kPa) saturated. The cooling system of each engine generates 5000 lb/h (2270 kg/h) of 15-lb/in^2 (gage) (103-kPa) steam. Choose the compressor to boost the 15-lb/in^2 (gage) (103-kPa) steam pressure to 80-lb/in^2 (gage) (551-kPa) to be used in a distribution system. Steam at 110 lb/in^2 (gage) (758 kPa) is first used in laundry and heat-exchange equipment before being reduced in 80 lb/in^2 (gage) (551 kPa) and combined with the compressor discharge flow. About 16,500 lb/h (7491 kg/h) of 80-lb/in^2 (gage) (551-kPa) steam satisfies the distribution system requirements, except in severe weather when the existing boilers are fired to supplement the steaming requirements.

Calculation Procedure:

1. Determine the amount of steam that can be generated by each waste-heat boiler
Exhaust gas from each diesel engine enters the waste-heat boiler at 1200°F (649°C). Using the rule of thumb that a diesel-engine exhaust heat boiler can produce 1.9 lb/h (0.86 kg/h) of 100-lb/in^2 (gage) (689-kPa) saturated steam at full load per rated horsepower, find the amount of steam generated as 1.9 (1500 kW/0.746 hp/kW) = 3820.4 lb/h (1734.5 kg/h). Since the 110-lb/in^2 (gage) (758 kPa) steam required by the cogeneration system needs slightly more heat input, round off the quantity of steam generated to 3800 lb/h (1725 kg/h).

2. Select the type of steam compressor to use
The compression ratios used in cogeneration—higher than for many process applications—dictate use of mechanical compressors. Several different thermodynamic paths may be followed during the compression process (Fig. 16). Of the three processes shown, the highest compressor coefficient of performance (COP) is exhibited by direct two-phase compression (Fig. 17). The task becomes one of selecting a suitable unit to follow this path.

Centrifugal and axial compressors, both of the general category of dynamic compressors, work on aerodynamic principles. Though capable of handling large flow rates, they are sensitive to water droplets that may cause blade erosion. Thus, two-phase flow is undesirable. These compressors are also limited to operation within a narrow range because of surging or low efficiency when conditions deviate from design.

Positive-displacement compressors—such as the screw, lobe, or reciprocating variety—are more suited to cogeneration applications. Both the screw compressor and, to a lesser extent, the reciprocating compressor, achieve high pressure ratios at high COP. Further, the units approach the isothermal condition because work which normally goes into producing sensible heat during compression simply causes additional liquid to evaporate. Intercooling is avoided, and input power requirements are acceptable (Fig. 18).

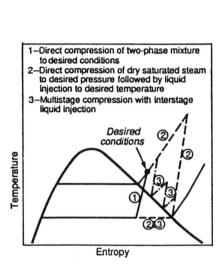

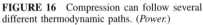

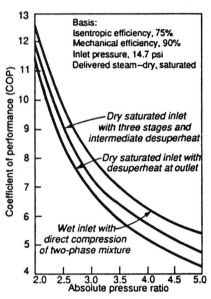

FIGURE 16 Compression can follow several different thermodynamic paths. (*Power.*)

FIGURE 17 Each compression path exhibits a different coefficient of performance. (*Power.*)

Cost considerations tend to make reciprocating units the second choice after screw and lobe units, but they are comparable on a technical basis. Note that reciprocating units require large foundations and must be driven at slow speeds. Some manufacturers offer carbon-ring units, requiring no lubrication. Even with the more common units requiring lubrication, use of synthetic lubricants keeps the amount of oil small, making steam contamination a negligible concern.

Based on the above information, two screw compressors will be chosen to boost the 15-lb/in^2 (gage) (103-kPa) steam to 80-lb/in^2 (gage) (551 kPa). Use Fig. 18 to approximate the required horsepower (kW) input to each screw compressor. With 16,500 lb/h (7491 kg/h) process steam required, Fig. 18 shows that the required horsepower input for each screw compressor will be 1250 hp (933 kW).

Note: Although Fig. 18 applies to an inlet pressure of 15 lb/in^2 (gage) (103 kPa) and an outlet pressure of 100 lb/in^2 (gage) (689 kPa), the results are accurate enough for an 80-lb/in^2 (gage) (551-kPa) outlet pressure. The required horsepower will be slightly less than that shown.

The two screw compressors will be clutch-connected at the generator end of the two engine-generator sets, as shown in Fig. 19. This diagram also shows the piping layout for the three steam systems—110 (758), 80 (551), and 15-lb/in^2 (gage) (103 kPa). Low-pressure steam separators are used to remove water from the low-pressure cogenerated steam.

In an actual system similar to that shown here, 110-lb/in^2 (gage) (758-kPa) steam is first used in a laundry and in heat-exchange equipment before being reduced to 80 lb/in^2 (gage) (551 kPa) and combined with the compressor discharge flow. In the summer, 15-lb/in^2 (gage) (103-kPa) steam is used to drive a large absorption chiller supplying 600 tons (2112 kW) of refrigeration.

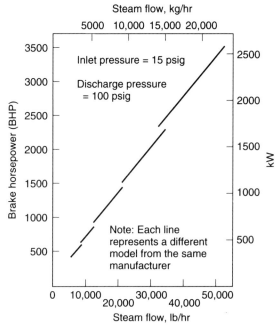

FIGURE 18   Energy input requirements for a screw compressor. (*Power.*)

Related Calculations.   The recent popularity of reciprocating-engine-based co-generation has caused a new factor to enter the economic evaluation of these systems: the value of the 15-lb/in^2 (gage) (103-kPa) steam typically recovered from the engine's cooling system. Because uses for 15-lb/in^2 (gage) (103-kPa) steam are limited, boosting the pressure to about 100 lb/in^2 (gage) (689 kPa) multiplies the practical uses of the recovered heat. This concept of pressure boosting is compatible with recent trends in cogeneration to maximize the value of the thermal output through closer coupling of the power-process interface.

Steam recompression has long been an accepted practice in the process industries where large quantities of low-pressure steam can be economically upgraded. A pound (0.45 kg) of steam vented to the atmosphere or condensed represents a loss of about 1000 Btu (1055 kJ) of heat energy. Thus, in many applications, it is less expensive (and more environmentally wise) to boost steam pressure than to produce the equivalent amount in a boiler.

Pressure ratios used to satisfy process requirements are relatively low—about 1.5 to 2. Thermocompressors most economically satisfy these ratios. They use high-pressure steam to boost low-pressure steam to a point in between the two.

Practical limitations on thermocompressors for satisfying higher ratios are two: (1) a large quantity of high-pressure steam is needed, and (2) the heat balance must be such that the steam need not be vented or condensed. For example, if 600-lb/in^2 (gage) (4134-kPa) boiler steam is available to boost 15-lb/in^2 (gage) (103-kPa) steam to 150 lb/in^2 (gage) (1034 kPa), about 12 times the quantity of high-pressure

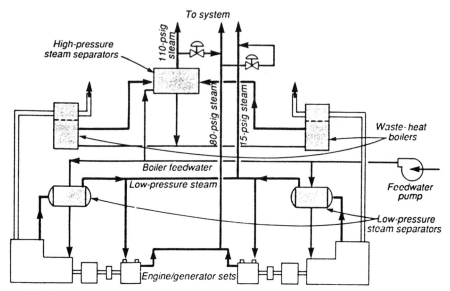

FIGURE 19   System uses common high-pressure steam separator. (*Power.*)

steam is required. So if an engine produces 5000 lb/h (2270 kg/h) of 15-lb/in^2 (gage) (103-kPa) steam, 60,000 lb/h (27,240 kg/h) of high-pressure steam is required to meet a 65,000 lb/h (29,510 kg/h) 150-lb/in^2 (1034 kPa) steam demand.

The typical reciprocating engine rejects 65 to 70 percent of its heat input to exhaust, engine cooling, lube-oil coolers, intercoolers, and radiation. About 20 percent of this heat is recoverable from the exhaust as steam at pressures up to 150 lb/in^2 (gage) (1034 kPa) and beyond, representing about 70 percent of the available heat in the exhaust. Another 28 percent represents engine cooling that can be completely recovered as hot water or as steam at a maximum pressure of 15 lb/in^2 (gage) (103 kPa). Lube-oil heat may also be recoverable, but usually not the intercooler heat.

Although hot water and 15-lb/in^2 (gage) (103-kPa) steam can be used for space heating, as a heat source for absorption refrigeration, or for domestic water heating, there usually is more demand for higher-pressure steam—such as that produced by the engine's exhaust.

As cogeneration systems maximize heat recovery from an internal-combustion engine, it is important to note that the engine exhaust no longer is a simple pipe protruding through the roof of a building. Exhaust explosions, not uncommon to engine operation, thus can be destructive to the often large and complex exhaust systems of cogeneration installations.

For this reason, it is prudent to tighten engine specifications. Partial failure of the ignition system, for example, should not be able to cause a potentially catastrophic exhaust explosion. Further, cross-limiting should be provided when fuel and air are measured at different locations.

Engine manufacturers may require that the exhaust system resist any explosion. A preferred alternative to this requirement is to insist that the engine manufacturer design to minimize exhaust explosions so that the cogeneration plant designer can confidently specify lightweight preformed exhaust ducting.

The data presented in this procedure were drawn from the work of Paul N. Garay, FMC Associates, a division of Parsons Brinckerhoff Quade & Douglas Inc., as reported in *Power* magazine.

USING PLANT HEAT NEED PLOTS FOR COGENERATION DECISIONS

An industrial process plant's heat needs are dominated by distillation. Its heat needs are represented by the fired-heat composite curve (FHCC) shown in Fig. 20. Five distinct heat sources are used in this process plant: two furnaces and steam supplied at three different pressure levels, as shown in Table 20. A gas turbine with the

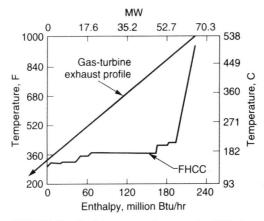

FIGURE 20   Fired-heat composite curve for BTX plant matches well with exhaust profile of gas turbine. (*Power.*)

TABLE 20   Heat Loads

Heat source	Heat load, million Btu/h (MW)
280-lb/in² (gage) (1.93 kPa) steam	25.1 (7.35)
140-lb/in² (gage) (0.96 kPa) steam	88.1 (25.8)
70-lb/in² (gage) (0.48 kPa) steam	$23.9 (7.0)
Furnace duty, 242–954°F (117–512°C) nonlinear	51.8 (15.2)
Furnace duty, 356–360°F (180–182°C)	19.4 (5.7)
Total	208.3 (61.05)

exhaust profile shown in Fig. 20 can supply all the heat needs of the process. Determine the annual fuel savings and payback time.

Calculation Procedure:

1. *Analyze the fired-heat composite curve*

Heat obtained from a cogeneration system generally displaces heat from other sources that can be traced back to direct fuel firing. Even though the fuel may not be fired at the point of use, it is almost always fired somewhere, such as in a boiler.

All these heating needs can be represented by a single FHCC. This curve of heat quantity (H) vs. temperature (T) represents the overall heating duties that, as far as possible, must be satisfied by the cogeneration system. The exhaust-heat profile of the cogeneration system can also be represented by a curve of heat quantity vs. temperature.

To see how such curves are developed, consider the three heat-acceptance profiles in Fig. 21. Profile (a) is a simple constant-heat-capacity profile, typical of heating duties in which no phase change occurs. The process stream is heated from its supply temperature to its target temperature and the heat load varies linearly between these points.

Profile (b) represents low-pressure steam raising. The first linear part of this curve corresponds to preheat, the horizontal plateau to vaporization, and the final linear section to super-heating. Profile (c) represents high-pressure steam raising.

Heat loads, unlike temperature, are additive. Thus it is possible to add the three profiles of Fig. 21 to obtain a combined heat-acceptance profile (Fig. 22). This is the FHCC and it shows total heating needs in terms of the quantity of heat required and the temperature at which it is needed.

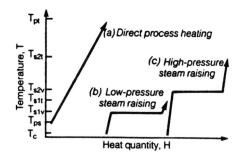

T_c = condensate to boiler
T_{ps} = process supply
T_{s1v} = steam level 1 vaporization
T_{s1t} = steam level 1 target
T_{s2v} = steam level 2 vaporization
T_{s2t} = steam level 2 target
T_{pt} = process target

FIGURE 21   Every heating duty has characteristic heat-acceptance profile. (*Power.*)

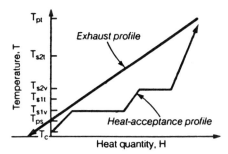

FIGURE 22   Total process heat-acceptance profile is matched with prospective exhaust profile.
(*Power.*)

The exhaust profile of the proposed cogeneration plant is also shown in Fig. 22. In this case it represents heat in the gas-turbine exhaust and is a straight line, neglecting the effect of condensation. Note that the exhaust profile lies above the heat-acceptance curve, implying that heat can be transferred from the exhaust stream to the process. The vertical separation between the two profiles is a measure of the available thermal driving force for heat transfer. Residual heat in the exhaust system, after the process duties have been satisfied, overhangs the heat-acceptance curve (at the left-hand end) and is lost up the stack.

Composite curves and profile matching provide a convenient way of representing the thermodynamics of heat recovery in cogeneration systems. Implicit within the construction of Fig. 22 are the requirements of the first law of thermodynamics, which demand a heat balance, and those of the second law, which lead to a relationship between the temperatures at which heat is required and the efficiency of the cogeneration system.

Analysis of the FHCC in Fig. 20 shows that all the needed process heat can be supplied by the gas turbine exhaust. Hence, a further evaluation of the proposed cogeneration installation is justified.

2.   Determine the annual fuel saving and payback period
Assemble the financial data in Table 21 from information available in plant records and estimates. These data show, for *this* proposed cogeneration installation, that the savings that can be obtained are: (a) boiler fuel savings, $4.1 million per year; (b) credit for cogenerated power, $13.1 million per year; total savings = $4.1 million + $13.1 million = $17.2 million per year. The additional cost is that for the cogeneration gas which is burned in the gas turbine, or $12.2 million. Thus, the net savings will be $17.2 million − $12.2 million = $5.0 million per year.

The payback time = installed cost, $/annual savings, $. Or, payback time = $15.8 million/$5.0 = 3.16, say 3.2 years. This is a relatively short payback time that would be acceptable in most industries.

Related Calculations.   Reciprocating internal-combustion engines are also often considered where gas turbines appear to be a possible choice. The reason for this is that about 20 percent of the heat content of fuel fired in a reciprocating engine is rejected in the exhaust gases and the heat-rejection profile is similar to that of a gas turbine. And even more heat, about 30 percent, is removed in cooling water at a temperature of 160°F (71°C) to 240°F (1116°C). A further 5 percent is

TABLE 21 Parameters Used to Evaluate Cogeneration
Process

Displaced furnace fuel cost, $/million Btu	2
Furnace efficiency, %	85
Boiler fuel savings, $ million/yr	4.1
Displaced or exported power, $/kWh	0.045
Gas for cogeneration system, $/million Btu	3.50
Cogeneration gas cost, $ million/yr	12.2
Operating hours per year	8000
Power output, MW	36.3
Credit for cogenerated power, $ million/yr	13.1
Cogeneration efficiency, %	78.1
Installed cost, $ million	15.8
Total cash benefit, $ million/yr	5
Estimated payback, years	3

available in the lubricating oil, usually below 180°F (82°C). The heat-rejection profile of a reciprocating engine that closely matches the composite curve of the plant's
process is also shown in Fig. 23.

A reciprocating engine has a higher overall efficiency than a gas turbine and
therefore generates a greater cash benefit for the plant owner. For the scale of
operation we are considering here, it would be necessary to use several engines
and the capital cost would be substantially greater than that of a single gas turbine.
As a result, payback periods for the two systems are about the same.

Gas turbines are often mated with steam turbines in combined-cycle cogeneration plants. In its basic form the combined-cycle power plant has the gas turbine
exhausting into a heat-recovery steam generator (HRSG) that supplies a steam-
turbine cycle. This cycle is the most efficient system for generating steam and/or
electric power commercially available today. The cycle also has significantly lower

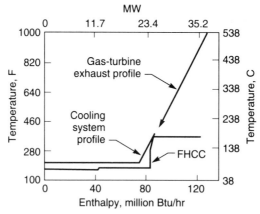

FIGURE 23 Exhaust-heat profile of reciprocating engine
is good fit with fired-heat composite curve of textile mill.
(*Power.*)

capital costs than competing nuclear and conventional fossil-fuel-fired steam/electric stations. Other advantages of the combined-cycle plant are low air emissions, low water consumption, reduced space requirements, and modular units which allow phased-in construction. And from an efficiency standpoint, even in a simple-cycle configuration, gas turbines now exhibit efficiencies of between 30 and 35 percent, comparable to state-of-the-art fossil-fuel-fired power stations.

Cogeneration, which is the simultaneous production of useful thermal energy and electric power from a fuel source, or some variant thereof, is a good match for combined cycles. Experience with cogeneration and combined-cycle power plants has been most favorable. Figure 24 shows a variety of combined-cycle cogeneration plants using reheat in an HRSG to provide steam for a steam-turbine generator. Flexibility is extended as gas turbines, steam turbines, and HRSGs are added to a system. Reheat can improve thermal efficiency and performance by several percentage points, depending on how it is integrated into the combined cycle.

Aeroderivative gas turbines, as part of a combined cycle, increasingly are finding application in cogeneration in the under 100-MW capacity range. Cogeneration has the airline and defense industries to thank for the rapid development of high-efficiency, long-running gas turbines at extremely low research cost.

And the new large gas turbines have exhaust temperatures high enough to justify reheat in the steam cycle without supplementary firing in a boiler. Depending on how the reheat cycle is configured, thermal performance at rated conditions can vary by up to three percentage points.

The Public Utilities Regulatory Policies Act (PURPA) passed by Congress to help manage energy includes incentives for efficient cogeneration systems. Cogeneration plants are allowed to sell power to local electric utilities to increase the return on investment earned from cogeneration.

A whole new energy-saving industry—termed nonutility generation (NUG)—has developed. At this writing NUG plants in the 200- to 300-MW range are common. And the pipeline industry which supplies natural-gas fuel for gas turbines is being restructured under the Federal Energy Regulatory Commission (FERC). Lower fuel costs are almost certain to result.

While lower electricity and energy costs are in the offing, these must be balanced against increased environmental requirements. The Clean Air Act Amendments of 1990 require better cleaning of stack emissions to provide a cleaner atmosphere. Yet this same 1990 act allows utilities to meet the required sulfur standard by installing suitable scrubber cleaning equipment, or by switching to a low-sulfur fuel.

A utility may buy—from another utility which exceeds the required sulfur standard—allowances to exhaust sulfur to the atmosphere. Each allowance permits a utility to emit 1 ton (tonne) of sulfur to the atmosphere. Public auctions of these allowances are now being held periodically by the Chicago Board of Trade.

Active discussions are underway at present over the suitability of selling sulfur allowances. Some opponents to sulfur pollution allowances believe that their use will delay the cleanup that ultimately must take place. Further, these opponents say, the pollution allowances delay the installation of sulfur-removal equipment. Meanwhile, sulfuric acid rain (also called acid rain) continues to plague communities in the path of a utility's sulfur effluent.

Challenging the above view is the Environmental Defense Fund. Its view is that there are too few allowances available to prevent the ultimate cleanup required by law.

The calculation data in this procedure are the work of A. P. Rossiter and S. H. Chang, ICI/Tensa Services as reported in *Power* magazine, along with John Mak-

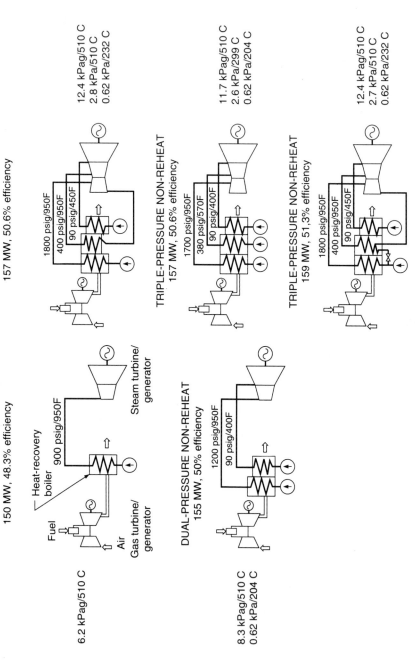

SINGLE-PRESSURE NON-REHEAT
150 MW, 48.3% efficiency

900 psig/950F

Heat-recovery boiler

Steam turbine/generator

Fuel

Air

Gas turbine/generator

6.2 kPag/510 C

DUAL-PRESSURE NON-REHEAT
155 MW, 50% efficiency

1200 psig/950F

90 psig/400F

8.3 kPag/510 C
0.62 kPa/204 C

DUAL-PRESSURE NON-REHEAT
157 MW, 50.6% efficiency

1800 psig/950F
400 psig/950F
90 psig/450F

12.4 kPag/510 C
2.8 kPa/510 C
0.62 kPa/232 C

TRIPLE-PRESSURE NON-REHEAT
157 MW, 50.6% efficiency

1700 psig/950F
380 psig/570F
90 psig/400F

11.7 kPag/510 C
2.6 kPa/299 C
0.62 kPa/204 C

TRIPLE-PRESSURE NON-REHEAT
159 MW, 51,3% efficiency

1800 psig/950F
400 psig/950F
90 psig/450F

12.4 kPag/510 C
2.7 kPa/510 C
0.62 kPa/232 C

FIGURE 24 Combined-cycle gas-turbine cogeneration plants using reheat in an HRSG to provide steam for a steam-turbine generator. (*Power.*)

ansi, executive editor, reporting in the same publication. Data on environmental laws are from the cited regulatory agency or act.

GEOTHERMAL AND BIOMASS POWER-GENERATION ANALYSES

Compare the costs—installation and operating—of a 50-MW geothermal plant with that of a conventional fossil-fuel-fired installation of the same rating. Likewise, compare plant availability for each type. Brine available to the geothermal plant free-flows at 4.3 million lb/h (1.95 million kg/h) at 450 lb/in^2 (gage) at 450°F (3100 kPa at 232°C).

Calculation Procedure:

1. Estimate the cost of each type of plant
The cost of constructing a geothermal plant (i.e., an electric-generating station that uses steam or brine from the ground produced by nature) is in the $1500 to $2000 per installed kW range. This cost includes all associated equipment and the development of the well field from which the steam or brine is obtained.

Using this cost range, the cost of a 50-MW geothermal station would be in the range of: 50 MW × ($1500/kW) × 1000 = $75 million to 50 MW × ($2000/kW) × 1000 = $100 million. Fossil-fuel-fired installations cost about the same—i.e., $1500 to $2000 per installed kW. Therefore, the two types of plants will have approximately the same installed cost.

Department of Energy (DOE) estimates give the average cost of geothermal power at 5.7¢/kWh. This compares with the average cost of 2.4¢/kWh for fossil-fuel-based plants. Advances in geothermal technology are expected to reduce the 5.7¢ cost significantly over the next 40 years.

Because of the simplicity of geothermal plant design, maintenance requirements are relatively low. Some modular plants even run unattended; and because maintenance is limited, plant availability is high. In recent years geothermal-plant availability averaged 97 percent. Thus, the maintenance cost of the usual geothermal plant is lower than a conventional fossil-fuel plant. Further, geothermal plants can meet new emission regulations with little or no pollution-abatement equipment.

2. Choose the type of cycle to use
Tapping geothermal energy from liquid resources poses a number of technical challenges—from drilling wells in a high-temperature environment to excessive scaling and corrosion in plant equipment. But DOE-sponsored and private-sector R&D programs have effectively overcome most of these problems. Currently, there are more than 35 commercial plants exploiting liquid-dominated resources. Of the 800 MW of power generated by these plants, 620 MW is produced by flash-type plants and 180 MW by binary-cycle units (Fig. 25).

The flashed-steam plant is best suited for liquid-dominated resources above 350°F (177°C). For lower-temperature sources, binary systems are usually more economical.

In flash-type plants, steam is produced by dropping the pressure of hot brine, causing it to "flash." The flashed steam is then expanded through a conventional steam turbine to produce power. In binary-cycle plants, the hot brine is directed

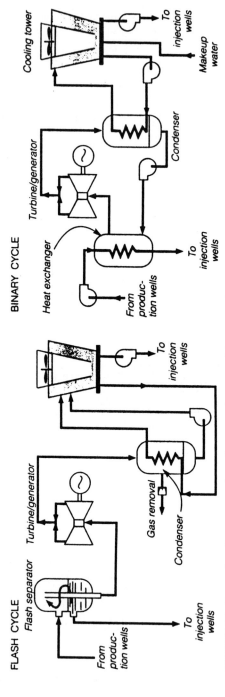

FIGURE 25 Energy from hot-water geothermal resources is converted by either a flash-type or binary-cycle plant. (*Power.*)

through a heat exchanger to vaporize a secondary fluid which has a relatively low boiling point. This working fluid is then used to generate power in a closed-loop Rankine-cycle system. Because they use lower-temperature brines than flash-type plants, binary units (Fig. 25), are inherently more complex, less efficient, and have higher capital equipment costs.

In both types of plants the spent brine is pumped down a well and reinjected into the resource field. This is done for two reasons: (1) to dispose of the brine—which can be mineral-laden and deemed hazardous by environmental regulatory authorities, and (2) to recharge the geothermal resource.

One recent trend in the industry is to collect noncondensable gases (NCGs) purged from the condenser and reinject them along with the brine. Older plants use pollution-abatement devices to treat NCGs, then release them to the atmosphere. Reinjection of NCGs with brine lowers operating costs and reduces gaseous emissions to near zero.

Major improvements in flashed-steam plants over the past decade centered around: (1) improving efficiency through a dual-flash process and (2) developing improved water treatment processes to control scaling caused by brines. The pressure of the liquid brine stream remaining after the first flash is further reduced in a secondary chamber to generate more steam. This two-stage process can generate 20 to 30 percent more power than single-flash systems.

Most of the recent improvements in binary-cycle plants have been made by applying new working fluids. The thermodynamic and transport properties of these fluids can improve cycle efficiency and reduce the size and cost of heat-transfer equipment.

To illustrate: By using ammonia rather than the more common isobutane or isopentane, capital cost can be reduced by 20 to 30 percent. It is also possible to improve the conversion efficiency by using mixtures of working fluids, which in turn reduces the required brine flow rate for a given power output.

A flashed-steam cycle will be tentatively chosen for this installation because the brine free-flows at 450°F (232°C), which is higher than the cutoff temperature of 350°F (177°C) for binary systems. An actual plant (Fig. 26), operating with these parameters uses two flashes. The first flash produces 623,000 lb/h (283,182 kg/h) of steam at 100 lb/in² (gage) (689 kPa). In the second flash an additional 262,000 lb/h (117,900 kg/h) of steam at 10 lb/in² (gage) (68.9 kPa) is produced.

Steam is cleaned in two trains of scrubbers, then expanded through a 54-MW, 3600-rpm, dual-flow, dual-pressure, five-stage turbine-generator to produce 48.9 MW. Of this total, 47.5 MW is sold to Southern California Edison Co. because of transmission losses.

The turbine exhausts into a surface condenser, coupled to a seven-cell cooling tower. About 40,000 lb/h (18,000 kg/h) of the high-pressure steam is required by the plant's air ejectors to remove NCGs from the main condenser at a rate of 6500 lb/h (2925 kg/h).

Because the liquid brine from the flash process is supersaturated, various solid compounds precipitate out of solution and must be removed to avoid scaling and fouling of the pumps, pipelines, and injection wells. This is accomplished as the brine flows to the crystallizer and clarifier tanks where, respectively, solid crystals grow and then are separated. The solids are dewatered and used in construction-grade soil cement. The clarified brine is disposed of by pumping it into three injection wells.

Related Calculations. Geothermal generating plants are environmentally friendly because there are no stack emissions from a boiler. Further, such plants do not consume fossil fuel, so they are not depleting the world's supply of such fuels.

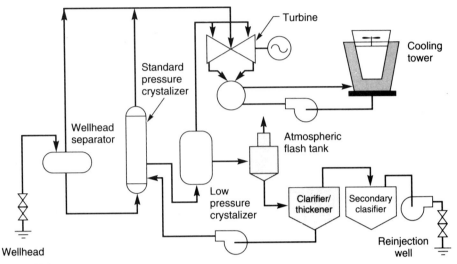

FIGURE 26   Dual-flash process extracts up to 30 percent more power than older, single-flash units. (*Power.*)

And by using the seemingly unlimited supply of heat from the earth, such plants are contributing to an environmentally cleaner and safer world while using a renewable fuel.

Another renewable fuel available naturally that is receiving—like geothermal power—greater attention today is *biomass*. The most common biomass fuels used today are waste products and residue left over from various industries, including farming, logging, pulp, paper, and lumber production, and wood-products manufacturing. Wooden and fibrous materials separated from the municipal waste stream also represent a major source of biomass.

Although biomass-fueled power plants currently account only for about 1 percent of the installed generating capacity in the United States, or 8000 MW, they play an important role in solving energy and environmental problems. Since the fuels burned in these facilities are considered waste in many cases, combustion yields the double benefits of reducing or eliminating disposal costs for the seller and providing a low-emissions fuel source for the buyer. On a global scale, biomass firing could present even more advantages, such as: (1) there is no net buildup of atmospheric CO_2 and air emissions are lower compared to many coal- or oil-fired plants. (2) Vast areas of deforested or degraded lands in tropical and subtropical regions can be converted to practical use. Because much of the available land is in the developing regions of Latin America and Africa, the fuels produced on these plantations could help improve a country's balance of payments by reducing dependence on imported oil. (3) Industrialized nations could potentially phase out agricultural subsidies by encouraging farmers to grow energy crops on idle land.

The current cost of growing, harvesting, transporting, and processing high-grade biomass fuels is prohibitive in most areas. However, proponents are counting on the successful development of advanced biomass-gasification technologies. They contend that biomass may be a more desirable feedstock for gasification than coal because it is easier to gasify and has a very low sulfur content, eliminating the need for expensive O_2 production and sulfur-removal processes.

One report indicates that integrated biomass-gasification–gas-turbine-based power systems with efficiencies topping 40 percent should be commercially available by year 2000. By 2025, efficiencies may reach 57 percent if advanced biomass-gasification–fuel-cell combinations become viable. Proponents are optimistic because this technology is currently being developed for coal gasification and can be readily transformed to biomass.

Data in this procedure are the work of M. D. Forsha and K. E. Nichols, Barber-Nichols Inc., for the geothermal portion, and Steven Collins, assistant editor, *Power,* for the biomass portion. Data on both these topics was published in *Power* magazine.

ESTIMATING CAPITAL COST OF COGENERATION HEAT-RECOVERY BOILERS

Use the Foster-Pegg* method to estimate the cost of the gas-turbine heat-recovery boiler system shown in Fig. 27 based on these data: The boiler is sized for a Canadian Westinghouse 251 gas turbine; the boiler is supplementary fired and has a single gas path; natural gas is the fuel for both the gas turbine and the boiler; superheated steam generated in the boiler at 1200 lb/in² (gage) (8268 kPa) and 950°F (510°C) is supplied to an adjacent chemical process facility; 230-lb/in² (gage) (1585-kPa) saturated steam is generated for reducing NO_x in the gas turbine; steam is also generated at 25 lb/in² (gage) (172 kPa) saturated for deaeration of boiler feedwater; a low-temperature economizer preheats underaerated feedwater obtained from the process plant before it enters the deaerator. Estimate boiler costs for two gas-side pressure drops: 14.4 in (36.6 cm) and 10 in (25.4 cm), and without, and with, a gas bypass stack. Table 22 gives other application data. *Note:* Since cogeneration will account for a large portion of future power generation, this procedure is important from an environmental standpoint. Many of the new cogeneration facilities planned today consist of gas turbines with heat-recovery boilers, as does the plant analyzed in this procedure.

Calculation Procedure:

1. Determine the average LMTD of the boiler
The average log mean temperature difference (LMTD) of a boiler is indicative of the relative heat-transfer area, as developed by R. W. Foster-Pegg, and reported in *Chemical Engineering* magazine. Thus, $LMTD_{avg} = Q_t/C_t$, where Q_t = total heat exchange rate of the boiler, Btu/s (W); C_t = conductance, Btu/s · °F (W). Substituting, using data from Table 22, $LMTD_{avg} = 81,837/1027 = 79.7$°F (26.5°C).

2. Compute the gas pressure drop through the boiler
The gas pressure drop, ΔP inH_2O (cmH_2O) = $5C_t/G$, where G = gas flow rate, lb/s (kg/s). Substituting, $\Delta P = 5(1027/355.8)$ with a gas flow of 355.8 lb/s (161.5 kg/s), as given in Fig. 17; then $\Delta P = 14.4$ inH_2O (36.6 cmH_2O). With a stack and inlet pressure drop of 3 inH_2O (7.6 cmH_2O) and a supplementary-firing pressure drop of 3 inH_2O (7.6 cmH_2O) given by the manufacturer, or determined from previous experience with similar designs, the total pressure drop = 14.4 + 3.0 + 3.0 = 20.4 inH_2O (51.8 cmH_2O).

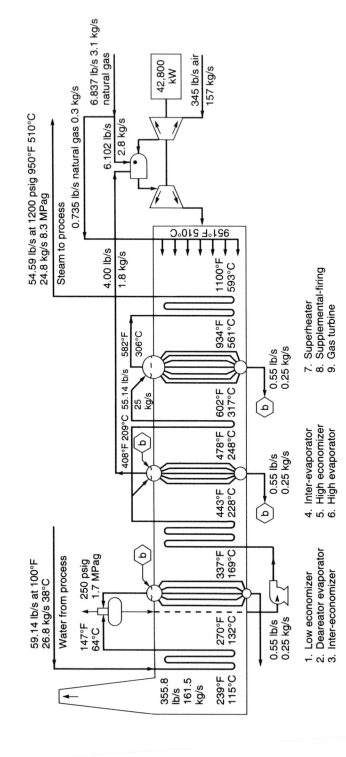

FIGURE 27 Gas-turbine and heat-recovery-boiler system. (*Chemical Engineering*.)

1. Low economizer
2. Deareator evaporator
3. Inter-economizer
4. Inter-evaporator
5. High economizer
6. High evaporator
7. Superheater
8. Supplemental-firing
9. Gas turbine

18.64

TABLE 22 Data for Heat Recovery Boiler*

	LMTD, °F	Q, Btu/s	C, Btu/s · °F	$C^{0.8}$ Btu/s · °F
Superheater	237	16,098	67.92	29.22
High evaporator	116	32,310	278.53	90.34
High economizer	40	11,583	290.3	93.39
Inter-evaporator	50.5	3,277	64.89	28.17
Inter-economizer	57	9,697	169.82	60.81
Deareator evaporator	46	6,130	134.43	50.44
Low economizer	131	2,742	20.93	11.39
Additional for superheater material				29.22
Additional for low-economizer material				11.39
Total		81,837		1,027	404.37

*See procedure for SI values in this table.
Source: Chemical Engineering.

3. Compute the system costs

The conductance cost component, Cost_{ts}, is given by Cost_{ts}, in thousands of $ = 5.65[(C_{sh}^{0.8} + C_1^{0.8} + \cdots + (C_n^{0.8}) + 2(C_n^{0.8})]$, where C = conductance, Btu/s · F (W), and the subscripts represent the boiler elements listed in Table 22. Substituting, $\text{Cost}_{ts} = 5.65(404.37) = \$2,285,000$ in 1985 dollars. To update to present-day dollars, use the ratio of the 1985 *Chemical Engineering* plant cost index (310) to the current year's cost index thus: Current cost = (today's plant cost index/310)(cost computed above).

The steam-flow cost component, Cost_w, in thousands of $ = 4.97(W_1 + W_2 + \cdots + W_n)$, where Cost_w = cost of feedwater, $; W = feedwater flow rate, lb/s (kg/s); the subscripts 1, 2, and n denote different steam outputs. Substituting, $\text{Cost}_w = 4.97(59.14) = \$294,000$ in 1985 dollars, with a total feedwater flow of 59.14 lb/s (26.9 kg/s).

The cost for gas flow includes connecting ducts, casing, stack, etc. It is proportional to the sum of the separate gas flows, each raised to the power of 1.2. Or, cost of gas flow, Cost_g, in thousands of $ = 0.236(G_1^{1.2} + G_2^{1.2} + \cdots + G_n^{1.2})$. Substituting, $\text{Cost}_g = 0.236(355.8)^{1.2} = \$272,000$ with a gas flow of 355.8 lb/s (161.5 kg/s) and no bypass stack.

The cost of a supplementary-firing system for the heat-recovery boiler in 1985 dollars is additional to the boiler cost. Typical fuels for supplementary firing are natural gas or No. 2 fuel oil, or both. The supplementary-firing system cost, Cost_f, in thousands of $ = B/1390 + 30N + 20$, where B = boiler firing capacity in Btu (kJ) high heating value; N = number of fuels burned. For this installation with *one* fuel, $\text{Cost}_f = 16,980/1390 + 30 + 20 = \$62,000$, rounded off. In this equation the 16,980 Btu/s (17,914 kJ/s) is the high heating value of the fuel and $N = 1$ since only *one* fuel is used.

The total boiler cost (with base gas ΔP and no gas bypass stack) = total material cost + erection cost, or $\$2,285,000 + 294,000 + 272,000 + 62,000 = \$2,913,000$ for the materials. A *budget estimate* for the cost of erection = 25 percent of the total material cost, or $0.25 \times \$2,913,000 = \$728,250$. Thus, the budget estimate for the erected cost = $\$2,913,000 + \$728,250 = \$3,641,250$.

The estimated cost of the entire system—which includes the peripheral equipment, connections, startup, engineering services, and related erection—can be approximated at 100 percent of the cost of the major equipment delivered to the site, but not erected. Thus, the total cost of the boiler ready for operation is approximately twice the cost of the major equipment material, or 2(boiler material cost) = 2($2,913,000) = $5,826,000.

4. Determine the costs with the reduced pressure drop

The second part of this analysis reduces the gas pressure drop through the boiler to 10 inH_2O (25.4 cmH_2O). This reduction will increase the capital cost of the plant because much of the equipment will be larger.

Proceeding as earlier, the total pressure drop, $\Delta P = 10 + 3 + 3 = 16$ inH_2O (40.6 cmH_2O). The pressure drop for normal solidity (i.e., normal tube and fin spacing in the boiler) is $\Delta P_1 = 14.4$ inH_2O (36.6 cmH_2O). For a different pressure drop, ΔP_2, the surface cost, $C_s(\$)$, is at ΔP_2, $C_s = [1.67(\Delta P_1/\Delta P_2)^{0.28} - 0.67](C_s$ at $P_1)$. Substituting, $C_s = 1.67(14.4/10)^{0.28} - 0.67 = 1.18 \times$ base cost from above. Hence, the surface cost for a pressure drop of 10 inH_2O (25.4 cmH_2O) = $1.18 \times$ ($2,285,000) = $2,696,300.

The total material cost will then be $2,696,300 + $272,000 + $62,000, using the data from above, or $3,324,300. Budget estimate for erection, as before = 1.25($3,324,300) = $4,155,375. And the estimated system cost, ready to operate = 2($3,324,300) = $6,648,600.

Adding for a gas bypass stack, the gas-flow component is the same as before, $272,000. Then the budget estimate of the installed cost of the gas bypass stack = 1.25($272,000) = $340,000. And the total cost of the boiler ready for operation at a gas-pressure drop of 10 inH_2O (25.4 cmH_2O) with a gas bypass stack = 2($3,324,000 + $272,000) = $7,192,600.

Related Calculations. To convert the costs found in this procedure to current-day costs, assume that the *Chemical Engineering* plant cost index today is 435, compared to the 1985 index of 310. Then, today's cost, $ = (today's cost index/1985 cost index)(1985 plant or equipment cost, $). Thus, for the first installation, today's cost = (435/310)($5,826,000) = $8,175,194. And for the second installation, today's cost = (435/310)($7,192,600) = $10,092,842.

Boilers for recovering exhaust heat from gas turbines are very different from conventional boilers, and their cost is determined by different parameters. Because engineers are becoming more involved with cogeneration, the differences are important to them when making design and cost estimates and decisions.

In a conventional boiler, combustion air is controlled at about 110 percent of the stoichiometric requirement, and combustion is completed at about 3000°F (1649°C). The maximum temperature of the water (i.e., steam) is 1000°F (538°C), and the temperature difference between the gas and water is about 2000°F (1093°C). The temperature drop of the gas to the stack is about 2500°F (1371°C), and the gas/water ratio is consistent at about 1.1.

By contrast, the exhaust from a gas turbine is at a temperature of about 1000°F (538°C), and the difference between the gas and water temperatures averages 100°F (56°C). The temperature drop of the gas to the stack is a few hundred degrees, and the gas/water ratio ranges between 5 and 10. Because the airflow to a heat-recovery boiler is fixed by the gas turbine, the air varies from 400 percent of the stoichiometric requirement of the fuel to the turbine (unfired boiler) to 200 percent if the boiler is supplementary fired.

In heat-recovery boilers, the tubes are finned on the outside to increase heat capture. Fins in conventional boilers would cause excessive heat flux and overheating of the tubes. Although the lower gas temperatures in heat-recovery boilers allow gas enclosures to be uncooled internally insulated walls, the enclosures in conventional boilers are water-cooled and refractory-lined.

Because the exhaust from a gas turbine is free of particles and contaminants, gas velocities past tubes can be high, and fin and tube spacings can be close, without erosion or deposition. Because the products of combustion in a conventional boiler may contain sticky residues, carbon, and ash particles, tube spacing must be wider and gas velocities lower. Because of its configuration and absence of refractories, the heat-recovery boiler used with gas turbines can be shop-fabricated to a greater extent than conventional boilers.

These differences between conventional and heat-recovery boilers result in different cost relationships. With both operating on similar clean fuels, a heat-recovery boiler will cost more per pound of steam and less per square foot (m²) of surface area than a conventional boiler. The cost of a heat-recovery boiler can be estimated as the sum of three major parameters, plus other optional parameters. Major parameters are: (1) the capacity to transfer heat ("conductance"), (2) steam flow rate, and (3) gas flow rate. Optional parameters are related to the optional components of supplementary firing and a gas bypass stack.

This procedure is the work of R. W. Foster-Pegg, Consultant, as reported in *Chemical Engineering* magazine. Note that the costs computed by the given equations are in 1985 dollars. Therefore, they *must* be updated to current costs using the *Chemical Engineering* plant cost index.

"CLEAN" ENERGY FROM SMALL-SCALE HYDRO SITES

A newly discovered hydro site provides a potential head of 65 ft (20 m). An output of 10,000 kW (10 MW) is required to justify use of the site. Select suitable equipment for this installation based on the available head and the required power output.

Calculation Procedure:

1. Determine the type of hydraulic turbine suitable for this site
Enter Fig. 28 on the left at the available head, 65 ft (20 m), and project to the right to intersect the vertical projection from the required turbine output of 10,000 kW (10 MW). These two lines intersect in the *standardized tubular unit* region. Hence, such a hydroturbine will be tentatively chosen for this site.

2. Check the suitability of the chosen unit
Enter Table 23 at the top at the operating head range of 65 ft (20 m) and project across to the left to find that a tubular-type hydraulic turbine with fixed blades and adjustable gates will produce 0.25 to 15 MW of power at 55 to 150 percent of rated head. These ranges are within the requirements of this installation. Hence, the type of unit indicated by Fig. 28 is suitable for this hydro site.

Related Calculations. Passage of legislation requiring utilities to buy electric power from qualified site developers is leading to strong growth of both site development and equipment suitable for small-scale hydro plants. Environmental concerns over fossil-fuel-fired and nuclear generating plants make hydro power more attractive. Hydro plants, in general, do not pollute the air, do not take part in the

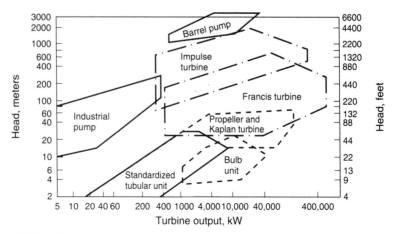

FIGURE 28 Traditional operating regimes of hydraulic turbines. New designs allow some turbines to cross traditional boundaries. (*Power.*)

acid-rain cycle, are usually remote from populated areas, and run for up to 50 years with low maintenance and repair costs. Environmentalists rate hydro power as "clean" energy available with little, or no, pollution of the environment.

To reduce capital cost, most site developers choose standard-design hydroturbines. With essentially every high-head site developed, low-head sites become more attractive to developers. Table 23 shows the typical performance characteristics of hydroturbines being used today. Where there is a region of overlap in Table 23 or Fig. 28, site-specific parameters dictate choice and whether to install large units or a greater number of small units.

Delivery time and ease of maintenance are other factors important in unit choice. Further, the combination of power-generation and irrigation services in some installations make hydroturbines more attractive from an environmental view because two objectives are obtained: (1) "clean" power, and (2) crop watering.

Maintenance considerations are paramount with any selection; each day of downtime is lost revenue for the plant owner. For example, bulb-type units for heads between 10 and 60 ft (3 and 18 m) have performance characteristics similar to those of Francis and tubular units, and are often 1 to 2 percent more efficient. Also, their compact and, in some cases, standard design makes for smaller installations and reduced structural costs, but they suffer from poor accessibility. Sometimes the savings arising from the unit's compactness are offset by increased costs for the watertight requirements. Any leakage can cause severe damage to the machine.

To reduce the costs of hydroturbines, suppliers are using off-the-shelf equipment. One way this is done is to use centrifugal pumps operated in reverse and coupled to an induction motor. Although this is not a novel concept, pump manufacturers have documented the capability of many readily available commercial pumps to run as hydroturbines. The peak efficiency as a turbine is at least equivalent to the peak efficiency as a pump. These units can generate up to 1 MW of power. Pumps also benefit from a longer history of cost reductions in manufacturing, a wider range of commercial designs, faster delivery, and easier servicing—all of which add up to more rapid and inexpensive installations.

TABLE 23 Performance Characteristics of Common Hydroturbines

Type	Operating head range		Capacity range	
	Rated head, ft (m)	% of rated head	MW	% of design capacity
Vertical fixed-blade propeller	7–120 (3–54) and over	55–125	0.25–15	30–115
Vertical Kaplan (adjustable blades and guide vanes)	7–66 (3–30) and over	45–150	1–15	10–115
Vertical Francis	25–300 (11–136) and over	50–150 and over	0.25–15	35–115
Horizontal Francis	25–500 (11–227) and over	50–125	0.25–10	35–115
Tubular (adjustable blades, fixed gates)	7–59 (3–27)	65–140	0.25–15	45–115
Tubular (fixed blades, adjustable gates)	7–120 (3–54)	55–150	0.25–15	35–115
Bulb	7–66 (3–30)	45–140	1–15	10–115
Rim generator	7–30 (3–14)	45–140	1–8	10–115
Right-angle-drive propeller	7–59 (3–27)	55–140	0.25–2	45–115
Cross flow	20–300 (9–136) and over	80–120	0.25–2	10–115

Source: Power.

Though a reversed pump may begin generating power ahead of a turbine installation, it will not generate electricity more efficiently. Pumps operated in reverse are nominally 5 to 10 percent less efficient than a standard turbine for the same head and flow conditions. This is because pumps operate at fixed flow and head conditions; otherwise efficiency falls off rapidly. Thus, pumps do not follow the available water load as well unless multiple units are used.

With multiple units, the objective is to provide more than one operating point at sites with significant flow variations. Then the units can be sequenced to provide the maximum power output for any given flow rate. However, as the number of reverse pump units increases, equipment costs approach those for a standard turbine. Further, the complexity of the site increases with the number of reverse pump units, requiring more instrumentation and automation, especially if the site is isolated.

Energy-conversion-efficiency improvements are constantly being sought. In low-head applications, pumps may require specially designed draft tubes to minimize remaining energy after the water exists from the runner blades. Other improvements being sought for pumps are: (1) modifying the runner-blade profiles or using a turbine runner in a pump casing, (2) adding flow-control devices such as wicket gates to a standard pump design or stay vanes to adjust turbine output.

Many components of hydroturbines are being improved to reduce space requirements and civil costs, and to simplify design, operation, and maintenance. Cast parts used in older turbines have largely been replaced by fabricated components. Stainless steel is commonly recommended for guide vanes, runners, and draft-tube inlets because of better resistance to cavitation, erosion, and corrosion. In special cases, there are economic tradeoffs between using carbon steel with a suitable coating material and using stainless steel.

Some engineers are experimenting with plastics, but much more long-term experience is needed before most designers will feel comfortable with plastics. Further, stainless steel material costs are relatively low compared to labor costs. And stainless steel has proven most cost-effective for hydroturbine applications.

While hydro power does provide pollution-free energy, it can be subject to the vagaries of the weather and climatic conditions. Thus, at the time of this writing, some 30 hydroelectric stations in the northwestern part of the United States had to cut their electrical output because the combination of a severe drought and prolonged cold weather forced a reduction in water flow to the stations. Purchase of replacement power—usually from fossil-fuel-fired plants—may be necessary when such cutbacks occur. Thus, the choice of hydro power must be carefully considered before a final decision is made.

This procedure is based on the work of Jason Makansi, associate editor, *Power* magazine, and reported in that publication.

CENTRAL CHILLED-WATER SYSTEM DESIGN TO MEET CHLOROFLUOROCARBON (CFC) ISSUES

Choose a suitable storage tank size and capacity for a thermally stratified water-storage system for a large-capacity thermal-energy storage system for off-peak air conditioning for these contitions: Thermal storage capacity required = 100,000 ton-h (35,169 kWh); difference between water inlet and outlet temperatures = T = 20°F (36°C); allowable nominal soil bearing load in one location is 2500 lb/ft^2

(119.7 kPa); in another location 4000 lb/ft^2 (191.5 kPa). Compare tank size for the two locations.

Calculation Procedure:

1. Compute the required tank capacity in gallons (liters) to serve this system
Use the relation $C = 1800S/\Delta T$, where C = required tank capacity, gal; S = system capacity, ton-h; ΔT = difference between inlet and outlet temperature, °F (°C). For this installation, $C = 1800(100,000)/20 = 9,000,000$ gal (34,065 m^3).

2. Determine the tank height and diameter for the allowable soil bearing loads
Depending on the proposed location of the storage tank, either the height or diameter may be a restricted dimension. Thus, tank height may be restricted by local zoning laws or possible interference with aircraft landing or takeoff patterns. Tank diameter may be restricted by the ground area available. And the allowable nominal soil bearing load will determine if the required amount of water can be stored in one tank or if more than one tank will be required.

Starting with 2500-lb/ft^2 (119.7-kPa) bearing-load soil, assume a standard tank height of 40 ft (12.2 m). Then, the required tank volume will be $V = 0.134C$, where $0.134 = $ ft^3/gal; or $V = 0.134(9,000,000) = 1,206,000$ ft^3 (34,130 m^3). The tank diameter is $d = (4V/\pi h)^{0.5}$, where $d = $ diameter in feet (m). Or $d = [4(1,206,000)/\pi 40)]^{0.5} = 195.93$ ft; say 196 ft (59.7 m). This result is consistent with the typical sizes, heights, and capacities used in actual practice, as shown in Table 24.

Checking the soil load, the area of the base of this tank is $A = \pi d^2/4 = \pi (196)^2/4 = 30,172$ ft^2 (2803 m^2). The weight of the water in the tank is $W = 8.35C = 75,150,000$ lb (34,159 kg). This will produce a soil bearing pressure of $p = W/A$ lb/ft^2 (kPa). Or, $p = 75,150,000/30.172 = 2491$ lb/ft^2 (119.3 kPa). This bearing load is within the allowable nominal specified load of 2500 lb/ft^2.

Where a larger soil bearing load is permitted, tank diameter can be reduced as the tank height is increased. Thus, using a standard 64-ft (19.5-m) high tank with the same storage capacity, the required diameter would be $d = [4(1,206,000)/\pi 64)]^{0.5} = 154.9$ ft (47.2 m). Soil bearing pressure will then be $W/A = 75,140,000/[\pi(154.9)^2/4] = 3987.8$ lb/ft^2 (190.9 kPa). This is within the allowable soil bearing load of 4000 lb/ft^2 (191.5 kPa).

By reducing the storage capacity of the tank 4 percent to 8,610,000 gal (32,589 m^3), the diameter of the tank can be made 151 ft (46 m). This is a standard dimension for 64-ft (19-m) high tanks with a 4000-lb/ft^2 (191.5-kPa) soil bearing load.

Related Calculations. Thermal energy storage (TES) is environmentally desirable because it uses heating, ventilating, and air-conditioning (HVAC) equipment and a storage tank to store heated or cooled water during off-peak hours, allowing more efficient use of electric generating equipment. The stored water is used to serve HVAC or industrial process loads during on-peak hours.

To keep investment, operating, and maintenance costs low, one storage tank can be used to store both cool and warm water. Thermal stratification permits a smaller investment in the tank, piping, insulation, and controls to produce a higher-efficiency system. Lower-density warm water is thermally stratified from higher-density cool water without any mechanical separation in a full storage tank.

TABLE 24 Typical Thermal Storage Tank Sizes, Heights, Capacities*

Rated thermal energy storage capacity, ton-hours			Tank shell height (and nominal soil bearing load)					
			64-ft shell height (4000 lb/ft² soil)		40-ft shell height (2500 lb/ft²)		24-ft shell height (1500 lb/ft² soil)	
ΔT, 10°F	ΔT, 15°F	ΔT, 20°F	Gross volume, gal	Tank diameter, ft	Gross volume, gal	Tank diameter, ft	Gross volume, gal	Tank diameter, ft
40,000	60,000	80,000	6,880,000	135	7,200,000	175	7,880,000	236
50,000	75,000	100,000	8,610,000	151	9,000,000	196	9,850,000	264
60,000	90,000	120,000	10,330,000	166	10,800,000	214	11,820,000	290

Source: Chicago Bridge & Iron Company.
*See calculation procedures for SI values.

18.72

While systems using 200,000 gal (757 m³) of stored water are feasible, the usual minimum size storage tank is 500,000 gal (1893 m³). Tanks as large as 4.4 million gal (16,654 m³) are currently in use in TES for HVAC and process needs. TES is also used for schools, colleges, factories, and a variety of other applications. Where chlorofluorocarbon (CFC)-based refrigeration systems must be replaced with less environmentally offensive refrigerants, TES systems can easily be modified because the chiller (Fig. 29), is a simple piece of equipment.

As an added environmental advantage, the stored water in TES tanks can be used for fire protection. The full tank contents are continuously available as an emergency fire water reservoir. With such a water reserve, the capital costs for fire-protection equipment can be reduced. Likewise, fire-insurance premiums may also be reduced. Where an existing fire-protection-water tank is available, it may be retrofitted for TES use.

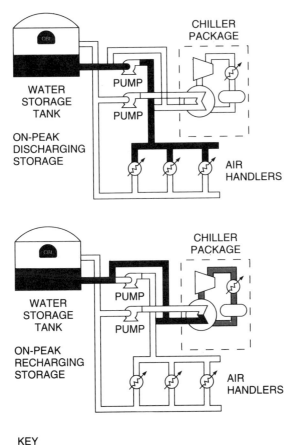

FIGURE 29 On-peak and off-peak storage discharging and recharging of thermally stratified water-storage system. (*Chicago Bridge & Iron Company.*)

TYPICAL STRATA-THERM CHILLED WATER
THERMAL ENERGY STORAGE INSTALLATION

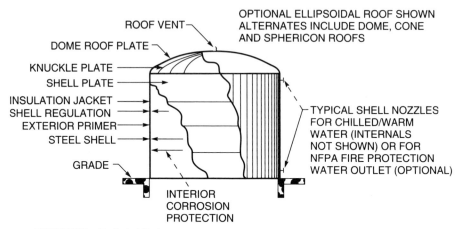

FIGURE 30   Typical chilled-water thermal-energy storage tank installation. (*Chicago Bridge & Iron Company.*)

Above-ground storage tanks (Fig. 30) are popular in TES systems. Such tanks are usually welded steel, leak-free with a concrete ringwall foundation. Insulated to prevent heat gain or loss, such tanks may have proprietary internal components for proper water distribution and stratification.

Some TES tanks may be installed partially, or fully, below grade. Before choosing partially or fully below-grade storage, the following factors should be considered: (1) system hydraulics may be complicated by a below-ground tank; (2) the tank must be designed for external pressure, particularly when the tank is empty; (3) soil and groundwater conditions may make the tank more costly; (4) local and national regulations for underground tanks may increase costs; (5) the choice of water-treatment methods may be restricted for underground tanks; (6) the total cost of an underground tank may be twice that of an above-ground tank.

The data and illustrations for this procedure were obtained from the Strata-Therm Thermal Systems Group of the Chicago Bridge & Iron Company.

WORK REQUIRED TO CLEAN OIL-POLLUTED BEACHES

How much relative work is required to clean a 300-yd (274-m) long beach coated with heavy oil, if the width of the beach is 40 yd (36.6 m), the depth of oil penetration is 20 in (50.8 cm), the beach terrain is gravel and pebbles, the oil coverage is 60 percent of the beach, and the beach contains heavy debris?

Calculation Procedure:

1. *Establish a work-measurement equation from a beach model*
After the *Exxon Valdez* ran aground on Bligh Reef in Prince William Sound, a study was made to develop a model and an equation that would give the relative amount of work needed to rid a beach of spilled oil. The relative amount of work remaining, expressed in clydes, is defined as the amount of work required to clean 100 yd (91.4 m) of lightly polluted beach. As the actual cleanup progressed, the actual work required was found to agree closely with the formula-predicted relative work indicated by the model and equation that were developed.

The work-measurement equation, developed by on-the-scene Commander Peter C. Olsen, U.S. Coast Guard Reserve, and Commander Wayne R. Hamilton, U.S. Coast Guard, is $S = (L/100)(EWPTCD)$, where S = standardized equivalent beach work units, expressed in clydes; L = beach-segment length in yards or meters (considered equivalent because of the rough precision of the model); E = degree of contamination of the beach expressed as: light oil = 1; moderate oil = 1.5; heavy oil = 2; random tar balls and very light oil = 0.1; W = width of beach expressed as: less than 30 m = 1; 30 to 45 m = 1.5; more than 45 m = 2; P = depth of penetration of the oil expressed as: less than 10 cm = 1; 10 to 20 cm = 2; more than 30 cm = 3; T = terrain of the beach expressed as: boulders, cobbles, sand, mud, solid rock without vertical faces = 1; gravel/pebbles = 2; solid rock faces = 0.1; C = percent of oil coverage of the beach expressed as: more than 67 percent coverage = 1; 50 to 67 percent = 0.8; less than 50 percent = 0.5; D = debris factor expressed as: heavy debris = 1.2; all others = 1.

2. *Determine the relative work required*
Using the given conditions, $S = (300/100)(2 \times 1.5 \times 1 \times 1 \times 0.8 \times 1.2) = 8.64$ clydes. This shows that the work required to clean this beach would be some 8.6 times that of cleaning 100 yd of lightly oiled beach. Knowing the required time input to clean the "standard" beach (100 yd, lightly oiled), the approximate time to clean the beach being considered can be obtained by simple multiplication. Thus, if the cleaning time for the standard lightly oiled beach is 50 h, the cleaning time for the beach considered here would be 50(8.64) = 432 h.

Related Calculations.   The model presented here outlines—in general—the procedure to follow to set up an equation for estimating the working time to clean any type of beach of oil pollution. The geographic location of the beach will not in general be a factor in the model unless the beach is in cold polar regions. In cold climates more time will be required to clean a beach because the oil will congeal and be difficult to remove.

A beach cleanup in Prince William Sound was defined as eliminating all gross amounts of oil, all migratory oil, and all oil-contaminated debris. This definition is valid for any other polluted beach be it in Europe, the Far East, the United States, etc.

Floating oil in the marine environment can be skimmed, boomed, absorbed, or otherwise removed. But oil on a beach must either be released by (1) scrubbing or (2) steaming and floated to the nearby water where it can be recovered using surface techniques mentioned above.

Where light oil—gasoline, naphtha, kerosene, etc.—is spilled in an accident on the water, it will usually evaporate with little damage to te environment. But heavy oil—No. 6, Bunker C, unrefined products, etc.—will often congeal and stick to rocks, cobbles, structures, and sand. Washing such oil products off a beach requires the use of steam and hot high-pressure water. Once the oil is freed from the surfaces to which it is adhering, it must be quickly washed away with seawater so that it flows to the nearby water where it can be recovered. Several washings may be required to thoroughly cleanse a badly polluted beach.

The most difficult beaches to clean are those comprised of gravel, pebbles, or small boulders. Two reasons for this are: (1) the surface areas to which the oil can adhere are much greater, and (2) extensive washing of these surface areas is required. This washing action can carry away the sand and the underlying earth, destroying the beach. When setting up an equation for such a beach, this characteristic should be kept in mind.

Beaches with larger boulders having a moderate slope toward the water are easiest to clean. Next in ease of cleaning are sand and mud beaches because thick oil does not penetrate deeply in most instances.

Use this equation as is, and check its results against actual cleanup times. Then alter the equation to suit the actual conditions and personnel met in the cleanup.

The model and equation described here are the work of Commander Peter C. Olsen, U.S. Coast Guard Reserve and Commander Wayne R. Hamilton, U.S. Coast Guard, as reported in government publications.

SIZING EXPLOSION VENTS FOR INDUSTRIAL STRUCTURES

Choose the size of explosion vents to relieve safely the maximum allowable overpressure of 0.75 lb/in² (5.2 kPa) in the building shown in Fig. 31 for an ethane/air explosion. Specify how the vents will be distributed in the structure.

Calculation Procedure:

1. Determine the total internal surface area of Part A of the building
Using normal length and width area formulas for Part A, we have: Building floor area = 100 × 25 = 2500 ft² (232.3 m²); front wall area = 12 × 100 = 1200 ft² − 12 × 20 = 960 ft² (89.2 m²); rear wall area = 12 × 100 = 1200 ft² (111.5 m²); end wall area = 2 × 25 × 12 + 2 × 25 × 3/2 = 675 ft² (62.7 m²); roof area = 2 × 3 × 100 = 600 ft² (55.7 m²). Thus, the total internal surface area of Part A of the building is 2500 + 960 + 1220 + 600 = 5935 ft² (551.4 m²).

2. Determine the total internal surface area of Part B of the building
Using area formulas, as before: Floor area = 50 × 20 = 1000 ft² (92.9 m²); side wall area = 2 × 50 × 12 = 1200 ft² (111.5 m²) front wall area = 20 × 12 = 240 ft² (22.3 m²); roof area = 50 × 20 = 1000 ft² (92.9 m²); total internal surface area of Part B is 1000 + 1200 + 240 + 1000 = 3440 ft² (319.6 m²).

3. Compute the vent area required
Using the relation $A_v = CA_s/(P_{red})^{0.5}$, where A_v = required vent area, m²; C = deflagration characteristic of the material in the building, (kPa)$^{0.5}$, from Table 25; A_s = internal surface area of the structure to be protected, m². For this industrial structure, $A_v = 0147(551.4 + 319.6)/(5.17)^{0.5} = 180.1$ m² (1939 ft²) total vent area.

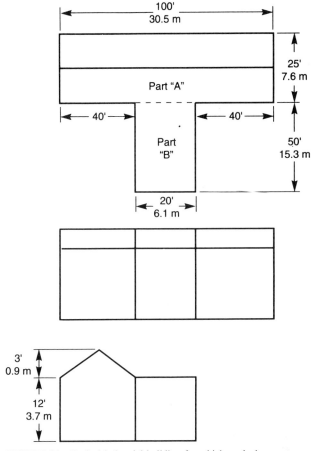

FIGURE 31   Typical industrial building for which explosion vents are sized.

The required vent area should be divided proportionately between Part A and Part B of the building, or Part A vent area = 180.1(551.4/871.0) = 114 m² (1227 ft²); Part B vent area = 180.1(319.6/871.0) = 66.1 m² (712 ft²).

The required vent area should be distributed equally over the external wall and roof areas in each portion of the building. Before final choice of the vent areas to be used, the designer should consult local and national fire codes. Such codes may require different vent areas, depending on a variety of factors such as structure location, allowable overpressure, and gas mixture.

Related Calculations.   This procedure is the work of Tom Swift, a consultant, reported in *Chemical Engineering.* In his explanation of his procedure he points out that the word *explosion* is an imprecise term. The method outlined above is intended for those explosions known as deflagrations—exothermic reactions that propagate from burning gases to unreacted materials by conduction, convection,

TABLE 25 Parameters for Vent Area Equation*

Material	$\dfrac{S_u \rho_u}{G'}$	$\dfrac{P_{max}}{P_o}$
Methane	1.1×10^{-3}	8.33
Ethane	1.2×10^{-3}	9.36
Propane	1.2×10^{-3}	9.50
Pentane	1.3×10^{-3}	9.42
Ethylene	1.9×10^{-3}	9.39

Material	C, $(kPa)^{1/2}$
Methane	0.41
Ethane	0.47
Propane	0.48
Pentane	0.51
Ethylene	0.75
ST 1 dusts	0.26
ST 2 dusts	0.30

and radiation. The great majority of structural explosions at chemical plants are deflagrations.

The equation used in this procedure is especially applicable to "low-strength" structures widely used to house chemical processes and other manufacturing operations. This equation is useful for both gas and dust deflagrations. It applies to the entire subsonic venting range. Nomenclature for Table 25 is given as follows:

A_s Internal surface area of structure to be protected, m²

A_v Vent area, m²

B Dimensionless constant

C Deflagration characteristic, $(kPa)^{1/2}$

C_D Discharge coefficient

G' Maximum subsonic mass flux through vent, kg/m²·s

P_f Overpressure, kPa

P_{max} Maximum deflagration pressure in a sealed spherical vessel, kPa

P_0 Initial (ambient) pressure, kPa

P_{red} Maximum reduced explosion pressure that a structure can withstand, kPa

S_u Laminar burning velocity, m/s

γ_b Ratio of specific heats of the combustion gases

ρ_u Density of the unburnt gases, kg/m³

λ Turbulence enhancement factor

With increased interest in the environment by regulatory authorities, greater attention is being paid to proper control and management of industrial overpressures. Explosion vents that are properly sized will protect both the occupants of the building and surrounding structures. Therefore, careful choice of explosion vents is a prime requirement of sensible environmental protection.

INDUSTRIAL BUILDING VENTILATION FOR ENVIRONMENTAL SAFETY

Determine the ventilation requirements to maintain interior environmental safety of a pump and compressor room in an oil refinery in a cool-temperate climate. Floor area of the pump and compressor room is 2000 ft² (185.8 m²) and room height is 15 ft (4.6 m); gross volume = 30,000 ft³ (849 m³). The room houses two pumps—one of 150 hp (111.8 kW) with a pumping temperature of 350°F (177°C), and one of 75 hp (55.9 kW) with a pumping temperature of 150°F (66°C). Also housed in the room is a 1000-hp (745.6-kW) compressor and a 50-hp (37.3-kW) compressor.

Calculation Procedure:

1. Determine the hp-deg for the pumps

The hp-deg = pump horsepower × pumping temperature. For these pumps, the total hp-deg = (150 × 350) + (75 × 150) = 63,750 hp-deg (19.789 kW-deg). Enter Fig. 32 on the left axis at 63,750 and project to the diagonal line representing the ventilation requirements for pump rooms in cool-temperate climates. Then extend a line vertically downward to the bottom axis to read the air requirement as 7200 ft³/min (203.8 m³/min).

The compressors require a total of 1050 hp (782.9 kW). Enter Fig. 32 on the right-hand axis at 1050 and project horizontally to cool-temperate climates for compressor and machinery rooms. From the intersection with this diagonal project vertically to the top axis to read 2200 ft³/min (62.3 m³/min) as the ventilation requirement.

Since the ventilation requirements of pumps and compressors are additive, the total ventilation-air requirement for this room is 7200 + 2200 = 9400 ft³/min (266 m³/min).

2. Check to see if the computed ventilation flow meets the air-change requirements

Use the relation $N = 60F/V$, where N = number of air changes per hour; F = ventilating-air flow rate, ft³/min (m³/min); V = room volume, ft³ (m³). Using the data for this room, $N = 60(9400)/30,000 = 18.8$ air changes per hour.

Figure 32 is based on a minimum of 10 air changes per hour for summer and 5 air changes per hour for winter. Since the 18.8 air changes per hour computed exceeds the minimum of 10 changes per hour on which the chart is based, the computed air flow is acceptable.

In preparing the chart in Fig. 32 the climate lines are based on ASHRAE degree-day listings, namely: *Cool, temperate climates,* 5000 degree-days and up; *average climates,* 2000 to 5000 degree-days; *warm climates,* 2000 degree-days maximum.

3. Select the total exhaust-fan capacity

An exhaust fan or fans must remove the minimum computed ventilation flow, or 9400 ft³/min (266 m³/min) for this room. To allow for possible errors in room size, machinery rating, or temperature, choose an exhaust fan 10 percent larger than the computed ventilation flow. For this room the exhaust fan would therefore have a capacity of 1.1 × 9400 = 10,340 ft³/min (292.6 m³/min). A fan rated at 10,500

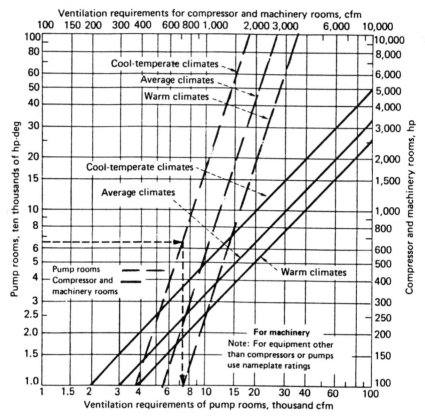

FIGURE 32   Chart for determining building ventilation requirements. (*Chemical Engineering.*)

or 11,000 ft³/min (297.2 or 311.3 m³/min), depending on the ratings available from the supplier, would be chosen.

Related Calculations.   Ventilation is environmentally important and must accomplish two goals: (1) Removal of excess heat generated by machinery or derived from hot piping and other objects; (2) removal of objectionable, toxic, or flammable gases from process pumps, compressors, and piping.

The usual specifications for achieving these goals commonly call for an arbitrary number of hourly air changes for a building or room. However, these specifications vary widely in the number of air changes required, and use inconsistent design methods for ventilation. The method given in this procedure will achieve proper results, based on actual applications.

Because of health and explosion hazards, workers exposed to toxic or hazardous vapors and gases should be protected against dangerous levels [threshold limit values (TLV)] and explosion hazards [lower explosive limit (LEL)] by diluting workspace air with outside air at adequate ventilation rates. If a workspace is protected by adequate ventilation rates for health (i.e., below TLV) purposes, the explosion hazard (LEL) will not exist. The reason for this is that the health air changes far exceed those required for explosion prevention.

To render a workspace safe in terms of TLV, the number of ft³/min (m³/min) of dilution air, A_d, required can be found from: $A_d = [1540 \times S \times T/(M \times \text{TLV})]K$, or in SI, $A_{dm} = \text{m}^3/\text{min} = 0.0283A_d$, where S = gas or vapor expelled over an 8-h period, lb (kg); M = molecular weight of vapor or gas; TLV = threshold limit value, ppm; T = room temperature, absolute °R (K); K = air-mixing factor for nonideal conditions, which can vary from 3 to 10, depending on actual space conditions and the efficiency of the ventilation-air distribution system.

If the space temperature is assumed to be 100°F (37.8°C) (good average summer conditions), the above equation becomes $A_d = [862,400 \times S/(M \times \text{TLV})]K$.

For every pound (kg) of gas or vapor expelled of an 8-h period, when $S = 1$, the second equation becomes $A_d = [862,400/(M \times \text{TLV})]K$. For values of S less or greater than unity, simple multiplication can be used.

It is only for ideal mixing that $K = 1$. Hence, K must be adjusted upward, depending on ventilation efficiency, operation, and the particular system application.

If mixing is perfect and continuous, then each air change reduces the contaminant concentration to about 35 percent of that before the air change. Perfect mixing is seldom attainable, however, so a room mixing factor, K, ranging from 3 to 10 is recommended in actual practice.

The practical mixing factor for a particular workspace is at best an estimate. Therefore, some flexibility should be built into the ventilation system in anticipation of actual operations. For small enclosures, such as ovens and fumigation booths, K-values range from 3 to 5. If you are not familiar with efficient mixing within enclosures, use a K-factor equal to 10. Then your results will be on the safe side. Figure 32 is based on a K-value equal to 8 to 10. Table 26 gives K-factors for ventilation-air distribution systems as indicated.

In some installations, heat generated by rotating equipment (pumps, compressors, blowers) process piping, and other equipment can be calculated, and the outside-air requirements for dilution ventilation determined. In most cases, however, the calculation is either too cumbersome and time consuming or impossible.

Figure 32 was developed from actual practice in the chemical-plant and oil-refinery businesses. The chart is based on a closed processing system. Hence, air quantities found from the chart are not recommended if (1) the system is not closed or (2) if abnormal operating conditions prevail that permit the escape of excessive amounts of toxic and explosive materials into the workplace atmosphere.

For these situations, special ventilation measures, such as local exhaust through hoods, are required. Vent the exhaust to pollution-control equipment or, where permitted, directly outdoors.

Figure 32 and the procedure for determining dilution-air ventilation requirements were developed from actual tests of workspace atmospheres within processing buildings. Design and operating show that by supplying outside air into a building

TABLE 26 *K*-values for Various Ventilation-Air Distribution Systems

K-values	Distribution system
1.2–1.5	Perforated ceiling
1.5–2.0	Air diffusers
2.0–3.0	Duct headers along ceiling with branch headers pointing downward
3.0 and up	Window fans, wall fans, and the like

Chemical Engineering

near the floor, and exhausting it high (through the roof or upper outside walls), safe and comfortable conditions can be attained. Use of chevron-type stormproof louvers permits outside air to enter low in the room.

The chevron feature causes the air to sweep the floor, picking up heat and diluting gases and vapors on the way up to the exhaust fan (Fig. 33).

In the system shown in Fig. 33 there are a number of features worth noting. With low-level distribution and adequate high exhaust, only the internal plant heat load (piping, equipment) is of importance in maintaining desirable workspace conditions. Wall and transmission heat loads are swept out of the building and do not reach the work areas. Even the temperature rise caused by the plant load occurs above the work level. Hence, low-level distribution of the supply air maintains the work area close to supply-air temperatures.

For any installation, it is good practice to check the ratio of hp-deg/ft^2 (kW-deg/m^2) of floor area. When this ratio exceeds 100, consider installing a totally enclosed ventilation system for cooling. This should be complete with ventilating fans taking outside air, preferably from a high stack, and discharging through ductwork into a sheetmetal motor housing.

The result is a the greater use of outside air for cooling through a confined system at a lower ventilation rate. Ventilation air flow needs may be obtained from the equipment manufacturer or directly from the chart (Fig. 32). The remainder of the building may be ventilated as usual, based either on the absence of equipment or on any equipment outside the ventilation enclosure. When designing the duct system, take care to prevent moisture entrainment with the incoming airstream.

This procedure is the work of John A. Constance, P.E., consultant, as reported in *Chemical Engineering*.

ESTIMATING POWER-PLANT THERMAL POLLUTION

A steam power plant has a 1000-MW output rating. Find the cooling-water thermal pollution by this power plant when using a once-through cooling system for the condensers if the plant thermal efficiency is 30 percent.

Calculation Procedure:

1. Determine the amount of heat added during plant operation
The general equation for power-plant efficiency is $E = W/Q_A$, where E = plant net thermal efficiency, percent; W = plant net output, MW; Q_A = heat added, MW. For this plant, $Q_A = W/E = 1000/0.30 = 3333$ MW.

2. Compute the heat rejected by this plant
The general equation for heat rejected is $Q_R = (W/E - W)$, Q_R = heat rejected, MW. For this power plant, $Q_R = (1000/0.3 - 1000) = 2333$ MW. Thus, this plant will reject 2333 MW to the condenser cooling water.

The heat rejected to the cooling water will be absorbed by the river, lake, or ocean providing the water pumped through the condenser. Depending on the thermal efficiency of the plant, the required cooling-water flow for the condenser will range from 250×10^6 lb/h, or 65,000 ft^3/min (30 m^3/s) to 400×10^6 lb/h or 100,000 ft^3/min (50 m^3/s). The discharged water in a once-through cooling system will be 20 to 25°F (11 to 14°C) higher in temperature than the entering water.

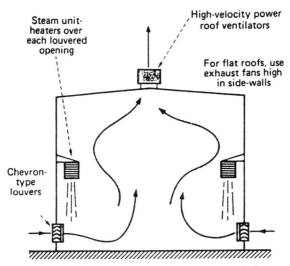

FIGURE 33 Ventilation system for effective removal of plant heat loads. (*Chemical Engineering.*)

3. *Assess the effects of this thermal pollution*

Warm water discharged in large volume to a restricted water mass may affect the ecosystem in a deleterious way. Fish and plant life, larvae, plankton, and other organisms can be damaged or have a high mortality rate. If chlorine is used to control condenser scaling, the effect on the ecosystem can be more damaging.

If the warm condenser cooling water is discharged into a large body of water, such as a major river or ocean, the effect on the ecosystem can be more beneficial than deleterious. Thus, well-planned cooling-water outlets can be used to increase fish production in hatcheries. In agriculture, the warm-water discharge can be used to markedly increase the output of greenhouses and open fields in cold climates. Thus, the overall effect of thermal pollution can be positive, if the pollution energy is used in an antipolluting manner.

Related Calculations. Since thermal and atmospheric pollution are associated with the generation of electricity, environmental engineers are seeking ways to reduce electricity use. Personal computers (PCs) are big users of electricity today. At the time of this wiring, PCs consume some 5 percent of commercial energy used in the United States.

Typical PCs use 150 to 200 W of power when in use, or just on but not in use. Some 30 to 40 percent of PCs are left on overnight and during weekends. The extra electricity which must be generated to carry this PC load leads to more thermal and air pollution.

New PCs have "sleep" circuitry which reduces the electrical load to 30 W when the computer is not being used. Such microprocessors will reduce the electrical load caused by PCs. This, in turn, will reduce thermal and air pollution produced by power-generating plants.

Internal-combustion engines—diesel, gas, and gas turbines—produce both thermal and air pollution. To curb this pollution and wrest more work from the fuel burned, cogeneration is being widely applied. Heat is extracted from the internal-combustion engine's cooling water for use in process or space heating. In addition, exhaust gases are directed through heat exchangers to extract more heat from the internal-combustion engine exhaust. Thus, environmental considerations are met while conserving fuel. This is one reason why cogeneration is so popular today.

Compute the heat recovery from cogeneration using the many concepts given earlier in this section. Try to combine both heat recovery and pollution reduction; then the required investment will be easier to justify from an economic standpoint.

DETERMINING HEAT RECOVERY OBTAINABLE BY USING FLASH STEAM

Fifty steam traps of various sizes in an industrial plant discharge a total of 95,000 lb/h (11.96 kg/s) of condensate from equipment operating at 150 lb/in² (gage) (1034.3 kPa) to a flash tank maintaining a pressure of 5 lb/in² (gage) (34.5 kPa) at a temperature close to the steam temperature. The remaining condensate is discharged. Determine the quantity, available heat, and temperature of the flash steam formed. What quantity of water would be heated by this steam in a hot-water heater having an overall efficiency of 85 percent if the temperature is raised from 40°F (4.4°C) to 140°F (55.6°C)? Determine the effect on flash steam and condensate outlet temperature for the flash tank if the terminal temperature difference (flash-down) is 25°F (45°C). What would the effect on flash steam be if the condensate in the steam traps is subcooled 13°F (23.4°C)?

Calculation Procedure:

1. Sketch the complete condensate and flash steam recovery system
Refer to Fig. 34 for a typical installation.

2. Determine the percent of flash steam formed
In Table 27, locate an initial steam pressure of 150 lb/in² (1034.3 kPa). Cross to the right to the 5-lb/in² (gage) (34.5-kPa) flash tank pressure column and read 14.8 percent of the condensate forms flash steam.

3. Compute the quantity of flash steam formed
This equals the percent of flash steam formed multiplied by the condensate discharge from the steam traps, or $(0.148)(95,000) = 14,060$ lb/h (1.8 kg/s).

4. Compute the available heat in the flash steam formed
This equals the latent heat of evaporation for a flash tank pressure of 5 lb/in² (gage) (34.5 kPa) multiplied by the quantity of flash steam formed. From Table 27, the latent head of evaporation is 960 Btu/lb (2232.9 kJ/kg) at 5 lb/in² (gage) (34.5) kPa). Hence, the available heat is $(960)(14,060) = 13,500,000$ Btu/h (3955.5 kW).

5. Determine the flash steam temperature
This equals the saturated water temperature corresponding to the saturated flash tank pressure of 5 lb/in² (gage) (34.5 kPa). From Table 27 this value is shown as 228°F (108.9°C).

6. Compute the quantity of water heated in the hot water heater
If water were heated with the energy from the flash steam (assuming all of the flash steam could be used), this quantity would be equivalent to the (flash-steam available

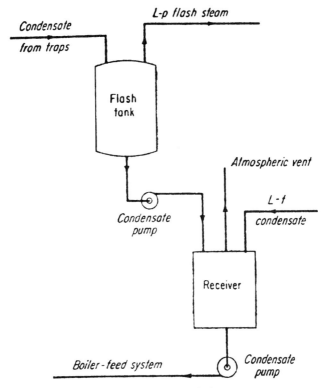

FIGURE 34 Complete condensate and flash-steam recovery system (*Chemical Engineering.*)

heat)(efficiency of the hot water heater)(temperature increase) or (13,500,000) (0.85)(100) = 114,750 lb/h (14.4 kg/s). (*Note:* Additional heat is available in the condensed flash steam.)

7. Compute the effect on flash steam and remaining condensate temperature for the flash tank with a terminal temperature difference (flashdown) of 25°F (45°C)

The temperature of the remaining condensate at the flash tank outlet equals the saturated water temperature plus flashdown, or 228 + 25 = 253°F (122.8°C). (*Note:* The flashdown represents a loss to the system, and may be necessary due to sizing considerations. The flashdown process will continue across the flash tank outlet, and until the temperature of the remaining condensate is 228°F (108.9°C), the quantity of flash steam formed as a consequence of flashdown will be reduced. Refer to Fig. 35 for a schematic of the flash tank energy balance. Note that the values shown for enthalpy (*H*) are determined from steam-table data. Hence, since energy input equals energy output, 95,000(338.7) = (*m*)(221.6) + (95,000 − *m*)(1156.3). Solving for *m*, the quantity of remaining condensate is 83,100 lb/h (10.5 kg/s). Therefore, the quantity of flash steam is 95,000 − 83,100 or 11,900 lb/h (1.5 kg/

TABLE 27 Percent Flash Steam Formed

Initial steam pressure lb/in² (gage) (kPa)	Sat. temp °F	°C	Flash-tank pressure, lb/in² (gage) (kPa)						
			0 (0)	5 (34.5)	10 (68.9)	50 (344.5)	100 (689)	125 (861.3)	150 (1033.5)
125 (861.1)	353	577.8	14.8	13.4	12.2	6.3	1.7	0	0
150 (1034.3)	366	601.2	16.8	14.8	13.7	7.8	2.3	1.6	0
175 (1206.5)	377	621.0	17.4	16.0	15.0	9.0	4.6	3.0	1.5
200 (1378.8)	388	640.8	18.7	17.5	16.2	10.4	6.0	4.4	2.8
Total heat of flash steam Btu/lb (kJ/kg)			1500 (2674.9)	1156 (2693.5)	1160 (2702.8)	1179 (2747.1)	1185 (2770.4)	1193 (2779.7)	1195 (2784.4)
Latent heat of evaporation, Btu/lb (kJ/kg)			970 (2260.1)	960 (2232.9)	952 (2218.2)	912 (2125)	881 (2052.7)	868 (2022.4)	857 (1996.8)
Heat of liquid, Btu/lb (kJ/kg)			180 (419.4)	196 (456.7)	208 (484.6)	267 (622.1)	309 (719.9)	324 (754.9)	338 (787.5)
Saturated water temperature, °F (°C)			212 (100)	228 (108.9)	240 (115.5)	298 (147.8)	338 (170)	353 (178.3)	366 (185.6)
Volume of flash steam, ft³/lb (m³/kg)			26.8 (1.67)	20.0 (1.25)	16.3 (1.02)	6.6 (0.41)	3.5 (0.24)	3.2 (0.20)	2.7 (0.17)

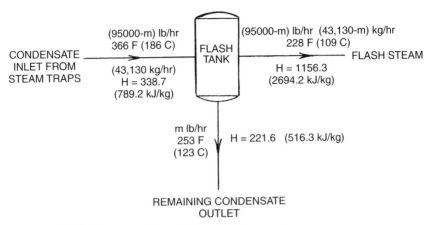

FIGURE 35   Flash-tank heat balance. (*Chemical Engineering.*)

s). And the flashdown reduces the flash steam quantity by the following: $(14,060 - 11,900)/14,060 = 0.1536$ or about 15.4 percent.

8. Compute the effect on flash steam quantity if the condensate in the steam traps is subcooled by 13°F (23.4°C)

From Table 27, read a saturated water temperature at 150 lb/in^2 (gage) as 366°F (185.6°C). Therefore, subcooling by 13°F (23.4°C) will reduce the condensate temperature to $366 - 13 = 353$°F (178.3°C). Cross to 353°F (178.3°C) in the 5-lb/in^2 (gage) (34.5 kPa) flash-tank-pressure saturation-temperature column and read 13.4 percent of the condensate forms flash steam. The quantity of flash steam may then be computed as $(95,000)(0.134) = 12,730$ lb/h (1.6 kg/s). Hence, the subcooling reduces the flash steam quantity by $14,060 - 12,730/14,060 = 0.095$ or 9.5 percent. Note that interpolation may be used for intermediate temeprature values.

Related Calculations.   This general procedure can be used for analyzing steam flow in flash tanks used in commercial, industrial, marine, and similar applications. With the great emphasis on conserving energy to reduce fuel costs, flash tanks are receiving greater attention than every before. Since flash steam contains valuable heat, every effort possible is being made to recover this heat, consistent with the investment required for the recovery.

The method given here is the work of T. R. MacMillan, as reported in *Chemical Engineering* magazine.

ENERGY CONSERVATION AND COST REDUCTION DESIGN FOR FLASH STEAM

A plant has the steam layout shown in Fig. 36. Determine the dollar value of the flashed steam and what can be done about reducing the energy loss, if any. In this plant, steam from the boiler is condensed in the heat exchanger at 100 lb/in^2 (gage) (689 kPa) and 338°F (170°C). Process water is heated from 50°F (10°C) to 150°F (65.6°C), with heat transferred at the rate of 1-million Btu/h (293 kW). Condensate

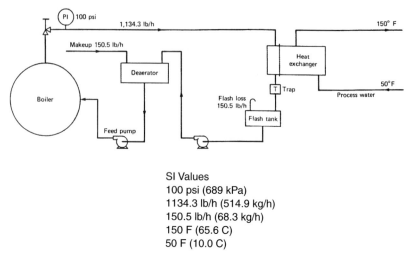

SI Values
100 psi (689 kPa)
1134.3 lb/h (514.9 kg/h)
150.5 lb/h (68.3 kg/h)
150 F (65.6 C)
50 F (10.0 C)

FIGURE 36 With steam valued at $8 per million Btu, venting flash steam results in an annual loss of almost $12,000. (*Chemical Engineering.*)

drains through a trap to a flash tank, where the flash steam is vented to the atmosphere. Analyze the benefits of reducing the supply steam pressure, and the financial benefits of recovering the flashed steam.

Calculation Procedure:

1. Determine the amount of heat lost in the flashed steam

From the steam tables, each pound of condensate at 100 lb/in² (gage) (69 kPa) contains 309.0 Btu (718.7 kJ/kg). At atmospheric pressure, each pound of condensate holds 180.2 Btu (419.2 kJ/kg) as sensible heat. The surplus, 309.0 − 180.2 = 128.8 Btu/lb (299.6 kJ/kg) flashes off 128.8 Btu/(970.6 Btu/lb) = 0.1327 pounds of steam per pound of condensate (0.06 kg/kg), or 13.27 percent. In this relation the value 970.6 is the latent heat of the condensate at 14.7 lb/in² (gage) (101.3 kPa). Since the flash steam carries its total heat with it, 13.27 percent of the 1150.8 Btu (1214.1 J) is vented per pound of condensate, or 152.7 Btu/lb (355.2 kJ/kg).

Because 1-million Btu/h (1055 kJ) is transferred in the heat exchanger, and the latent heat of 100-lb/in² (gage) steam is 881.6 Btu/lb (2050.6 kJ/kg), the steam flow from the boiler is (1,000,000 Btu/h)/(881.6 Btu/lb) = 1134.3 lb/h (514.9 kg/h). The heat vented from the flash tank is (152.7 Btu/lb)(1134.3 lb/h) = 173,207 Btu/h (50.7 kW). Makeup water at 50°F (10°C) brings in 18 Btu (18.9 J) with each 13.27 percent of 1134.3 lb (514.9 kg), or (18 Btu/lb)(150.5 lb/h) = 2709.0 Btu/h (851.8 W). Thus, the heat loss is 173,207 Btu/h − 2709 Btu/h = 170,498 Btu/h (49.9 kW). This is more than 17 percent of the useful heat transferred to the exchanger, i.e., 170,498/1,000,000 = 0.170498.

2. Compute the annual dollar cost of the lost heat

To determine the dollar cost multiply the cost of steam, per million Btu, by the hour loss, Btu, and the number of operating hours per year. Assuming continuous

24-hour operation of this plant, the annual dollar cost with steam priced at $8/ million Btu is: (170,498 Btu/h)(8760 h/yr)($8/1,000,000 Btu steam cost) = $11,949 per year. This is nearly $1000 per month lost from the flash steam.

3. *Analyze an alternative plant layout to reduce the cost of the lost heat*
To reduce the annual loss, an arrangement such as that in Fig. 37 is often proposed. A pressure-reducing valve (PT) in the diagram, in the steam-supply line lowers the boiler steam pressure to, say, 10 lb/in² (gage) (68.9 kPa) instead of the 100 lb/in² (gage) (689 kPa) used in step 1, above. The latent heat at this pressure increases to 952.9 Btu/lb (2216.5 kJ/kg), as shown in the steam tables.

With 1,000,000 Btu/h (1055 kJ) transferred in the heat exchanger, as earlier, the steam flow rate will be (1,000,000)/(952.9 Btu/lb) = 1049.4 lb/h (476.4 kg/h). At 10 lb/in² (gage) (68.9 kPa) the sensible heat is 207.9 Btu/lb (483.6 kJ/kg) from the steam tables. At 0 lb/in², i.e., atmospheric pressure to which the flash tank exhausts, the sensible heat is 180.2 Btu/lb (419.2 kJ/kg). Then, the flash steam percentage will be (207.9 Btu/lb − 180.2 Btu/lb)/(970.6 Btu/lb) = 0.02854, or 2.854.

The rate of flow of flash steam will be 0.02854 × 1049.4 = 29.95 lb/h (13.6 kg/h). The heat loss of this flash steam will be 29.95 lb/h × 1150.8 Btu/lb = 34,466 Btu/h (10.1 kW). Makeup water containing 18 Btu/lb (41.9 kJ/kg) enters at a rate of 29.95 lb/h (13.6 kg/h) providing 18 × 29.95 = 539.1 Btu/h (158 W). Then, the net heat loss = 34,466 − 539 = 33,927 Btu/h (9.94 kW), or 3.39 percent of the heat transferred in the exchanger.

Such a reduction in heat loss, amounting to about 136,571 Btu/h (40 kW), an annual saving of about $9570 (using the same steam cost as earlier), represents a substantial saving.

But at least one additional factor must be considered before installing the pressure-reducing valve. The heat exchanger that was large enough when supplied with

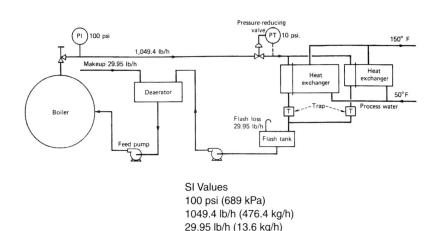

SI Values
100 psi (689 kPa)
1049.4 lb/h (476.4 kg/h)
29.95 lb/h (13.6 kg/h)
10 psi (68.9 kPa)
150° F (65.6°C)
50° F (10.0° C)

FIGURE 37   Lowering the steam pressure by using a pressure-reducing valve cuts operating cost but increases capital cost. (*Chemical Engineering.*)

18.90

100-lb/in² (gage) (689 kPa) 338°F (170°C) steam would be too small when supplied with 10-lb/in² (gage) (68.9 kPa) 240°F (115.6°C) steam.

Temperatures at the exchanger for these two cases (assuming for simplicity that arithmetic mean temperature differences are sufficiently accurate) are shown in Fig. 38. The Δt across the system has dropped from 238°F (114°C) to 140°F (60°C). If the U value remains the same, the surface area of the exchanger would have to be increased by 238°F/140°F = 1.7 times.

If $U = 150$ Btu/(ft²·°F·h), from $Q = UA \, \Delta t$; $A = 1,000,000/(150)(238) = 28$ ft² (2.6 m²). Hence, at the lower steam pressure, the exchanger surface area would have to be increased by 0.7 × 28 = 19.6 ft² (1.82 m²). Such an exchanger would cost about $550, 41,200 when installed. This would be in addition to the cost of about $600 for buying and installing the pressure-reducing valve. If such alterations must be made to produce the annual saving of $9570, other alternatives should also be considered.

4. Determine if recovering the flash steam is economically worthwhile

The minimum heat-transfer area is needed if as much as possible of the heat flow is to be from the 100-lb/in² (gage) (689-kPa) 338°-F(170°-C) steam. Suppose, however, that the high-pressure condensate were not discharged to a flash venting tank but to a flash-steam recovery vessel. (For maximum economy, let the recovery system operate at atmospheric pressure). If the flash steam were passed to a supplementary condenser fitted at the inlet side of the main exchanger (thus serving as a preheater), the atmospheric-pressure condensate would flow to the return pump without further heat loss, and the installation would be as shown in Fig. 39, with the temperature diagram shown in Fig. 40.

In Fig. 39, the latent heat of the 100-lb/in² (689-kPa) steam is 881.6 Btu/lb (2051 kJ/kg); flash steam at 0 lb/in² (gage) (0 kPa) has a latent heat of 970.6 Btu/lb (2258 kJ/kg). Each pound (0.45 kg) of condensate at 100 lb/in² (689 kPa) contains 309.0 Btu (718.7 kJ/kg); at atmospheric pressure the sensible heat of the condensate is 180.2 Btu/lb (419.1 kJ/kg). The surplus heat, 309.0 − 180.2 = 128.8

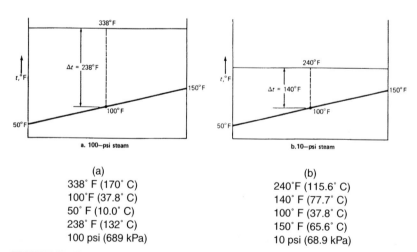

(a)	(b)
338° F (170° C)	240°F (115.6° C)
100°F (37.8° C)	140° F (77.7° C)
50° F (10.0° C)	100° F (37.8° C)
238° F (132° C)	150° F (65.6° C)
100 psi (689 kPa)	10 psi (68.9 kPa)

FIGURE 38 Temperatures around the heat exchanger are indicated for before and after the reduction in steam pressure. (*Chemical Engineering.*)

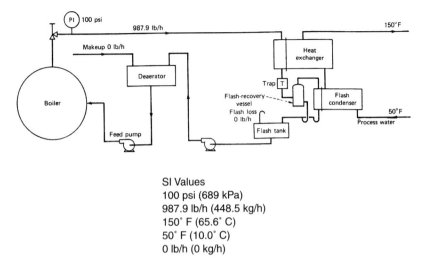

SI Values
100 psi (689 kPa)
987.9 lb/h (448.5 kg/h)
150° F (65.6° C)
50° F (10.0° C)
0 lb/h (0 kg/h)

FIGURE 39 Recovering the flash steam eliminates the venting loss and reduces the additional capital investment. (*Chemical Engineering.*)

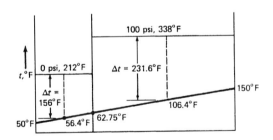

SI values

338° F	(170° C)	156° F	(86.6° C)
150° F	(65.6° C)	231.6° F	(128.5° C)
106.4° F	(41.3° C)	0 psi	(0 kPa)
62.75° F	(17.1° C)	100 psi	(689 kPa)
56.4° F	(13.6° C)		
50° F	(10.0° C)		
212° F	(100° C)		

FIGURE 40 Temperatures of exchanger and preheater with flash-steam recovery. (*Chemical Engineering.*)

Btu/lb (299.6 kJ/kg) flashes off $128.8/970.6 = 0.1327$ lb of steam per lb of condensate (0.06 kg/kg), or 13.2 percent of the condensate. The latent heat transferred $= 881.6 + 128.8 = 1010.4$ Btu (1065.9 J). The total steam flow with 1,000,000 Btu/h (293 kW) heat transfer is $1,000,000/1010.4 = 989.7$ lb (449.3 kg). The proportion of latent heat in the flash steam is $128.8/1010.4 = 0.1275$, or 12.75 percent. Temperature rise in the preheater $= (0.1275 \times 100°F) = 12.75°F$ (22.95°C). From this, the other temperatures in Fig. 40 can be derived.

Assuming again that the U value of 150 is maintained, the preheater surface area $= (127,000 \text{ Btu/h})/[150 \text{ Btu}/(\text{ft}^2 \cdot °F \cdot h)](155.6°F) = 5.5 \text{ ft}^2$ (0.51 m²). Hence, the exchanger is slightly oversized.

In retrofitting, the 28 ft² (2.6 m²) exchanger would allow the steam pressure to be less than 100 lb/in² (689 kPa). In a new installation, an exchanger having a surface area of $(0.1275 \times 1,000,000)/(1,000,000 \times 28) = 25 \text{ ft}^2$ (2.32 m²) could be installed.

The cost of the installed preheater would be about $800, to which must be added the cost of the flash-recovery vessel at about $600. The following summarizes the choices:

Installation	Extra capital cost, $	Operating cost, $/yr
Existing	0	11,964
Add pressure-reducing valve and extra 20 ft² (1.85 m²) of heat exchanger	1,800	2,394
Add flash-recovery vessel and extra 5.5 ft² (0.5 m²) of exchanger	1,400	0

In an actual installation, some of the simplifications made here would be reassessed. However, it will generally remain the case that, with the recovery of flash steam, heat exchangers can operate at the greater efficiency afforded by higher-pressure, higher-temperature steam. Flash steam, recovered at the lowest practical pressure, can be used in a preheater, or in a separate, unconnected load.

Related Calculations.   With greater emphasis on lowering environmental air pollution, reducing flash steam costs takes on more importance in plant analyses. During initial design studies, a choice must often be made between high- and low-pressure operating steam at both the boiler and at any heat exchangers in the system. For new installations, the economics of generating moderately high-pressure, rather than low-pressure, steam are usually fairly obvious.

Low-pressure boilers are physically larger, and more costly than boilers producing the same quantity of steam at higher pressure. Also, steam produced at low pressure is often wetter than high-pressure steam, leading to lower heat-transfer rates in exchangers, even if water hammer is avoided.

The choice between high- and low-pressure steam may not always be so clear. Using high-pressure steam to get the benefits of lower capital costs of smaller heat-transfer areas can mean higher operating costs. Steam losses from flash-tank vents are greater with high-pressure steam. The value of the steam lost can quickly exceed the capital-cost savings. Such considerations often lead to choosing lower operating steam pressures.

Since flash-steam losses represent fuel used to generate this steam, every effort possible should be made to control flash steam. When flash losses are reduced, fuel consumption is cut. With lower fuel consumption, there is less atmospheric pollution. Reduced pollution lowers the cost of handling flash, sulfur compounds, and other boiler effluents. So there are many good reasons for limiting flash-steam losses in any plant.

The procedure presented here can be used for any plant using steam for processes, heating, power generation, or other heat-transfer purposes. Such plants include chemical, food, textile, manufacturing, marine, cogeneration, and central-station. All can benefit from reducing flash-steam losses, as described above.

This procedure is the work of Albert Armer, Technical Adviser and Sales Specialist, Spirax Sarco, Inc., as reported in *Chemical Engineering* magazine. SI values were added by the handbook editor to the calculations and illustrations.

COST SEPARATION OF STEAM AND ELECTRICITY IN A COGENERATION POWER PLANT USING THE ENERGY EQUIVALENCE METHOD

Allocate—using the energy equivalence method—the steam and electricity costs in a power plant having a double automatic-extraction, noncondensing steam turbine for process steam and electric generation. Turbine throttle steam flow is 800,000 lb/h (100.7 kg/s) at 865 lb/in² (abs) (5964.1 kPa). Process steam is extracted from the turbine in the amounts of 100,000 lb/h (12.6 kg/s) at 335 lb/in² (abs) (2309.8 kPa) and 200,000 lb/h (25.2 kg/s) at 150 lb/in² (abs) (1034.3 kPa) and is delivered to process plants. A total of 500,000 lb/h (62.9 kg/s) is exhausted at 35 lb/in² (abs) (241.3 kPa) with 100,000 lb/h (12.6 kg/s) of this exhaust steam for deaerator heating in the cycle and 400,000 lb/h (50.4 kg/s) sent to process plants. The turbine has a gross electric output of 51,743 kW, and the heat balance for the dual-purpose turbine cycle is shown in Fig. 41. Efficiency of the steam boiler is 85.4 percent, while the fuel is priced at $0.50 per 10^6 Btu ($0.47 per MJ). If a condensing turbine is used, an attainable backpressure is 1.75 inHg (abs) [43.75 mmHg (abs)], while the assumed turbine efficiency is 82 percent and the exhaust enthalpy is h_f, = 1032 Btu/lb (2400.4 MJ/kg). Figure 42 shows the expansion-state curve of the turbine on a Mollier diagram. Final feedwater enthalpy is 228 Btu/lb (530.3 mJ/kg). Allocate the fuel cost to each energy use by using the energy equivalence method.

Calculation Procedure:

1. Compute the hourly total fuel cost
The total fuel cost for this plant per hour is C_f = (1/boiler efficiency)(0.50/10^6)(throttle steam flow rate m_t, lb/h)($h_t - h_{fw}$), where h_t = throttle enthalpy, Btu/lb, and h_{fw} = feedwater enthalpy, Btu/lb. Substituting gives C_f = (1/0.854)(0.50/10^6)(800,000)(1482 − 228) = $587.35 per hour.

2. Compute the nonextraction ultimate electric output
The ultimate electric output is $E_u = m_t(h_t - h_f)/3413$, where m_h = total turbine inlet steam flow, lb/h; h_i = turbine initial enthalpy, Btu/lb; other symbols as before. Substituting, we find E_u = (800,000)(1482 − 1032)/3413 = 105,480 kW.

3. Determine the actual electric output of the dual-purpose turbine
Use the relation E_a = W(actual)/3413 = the work done by the extraction steam between the throttle inlet and the extraction point, plus the work done by the non-

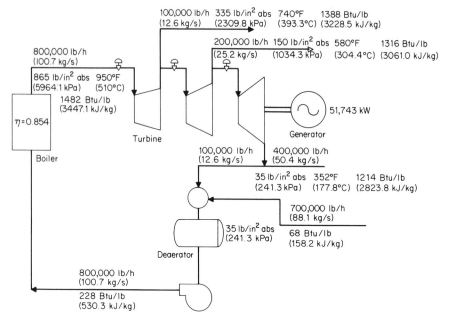

FIGURE 41   Dual-purpose turbine heat balance. (*Combustion.*)

extraction steam between the throttle and the exhaust. Or, from the turbine expansion curve in Fig. 42, $E_a = [100,000(1482 - 1388) + 200,000(1482 - 1316) + 500,000(1482 - 1214)]/3413 = 51,743$ kW.

4. Compute the extraction steam kilowatt equivalence

Again from Fig. 42, $E_{x1} = (h_{x1} - h_f)/3413 = 100,000(1388 - 1032)/3413 = 10,432$ kW; $E_{x2} = 200,000(1316 - 1032)/3413 = 16,642$ kW; $E_{x3} = 500,000(1214 - 1032)/3413 = 26,663$ kW. Hence, the nonextraction turbine ultimate electric output $= E_a + E_{x1} + E_{x2} + E_{x3} = 51,743 + 10,432 + 16,642 + 26,663 = 105,480$ kW.

5. Determine the base fuel cost of electricity and steam

The base fuel cost of electricity $= C_f/E_u$, or $587.35/105,480 = \$0.005568$ per kilowatthour, or 5.568 mil/kWh.

Now the base fuel cost of the steam at the different pressures can be found from (kW equivalence)(base cost of electricity, mil)/(rate of steam use, lb/h). Thus, for the 335-lb/in² (abs) (2309.8-kPa) extraction steam used at the rate of 100,000 lb/h (12.6 kg/s), base fuel cost $= 10,432(5.568)/100,000 = \$0.5808$ per 1000 lb ($\$0.2640$ per 1000 kg). For the 150-lb/in² (abs) (1034.3-kPa) steam, base fuel cost $= 16,642(5.568)/200,000 = \$0.4633$ per 1000 lb ($\$0.21059$ per 1000 kg). And for the 35-lb/in² (abs) (241.3-kPa) steam, base fuel cost $= 26,663(5.568)/500,000 = \$0.2969$ per 1000 lb ($\$0.13495$ per 1000 kg). since 100,000 lb/h (12.6 kg/s) of the 500,000-lb/h (62.9 kg/s) is used for deaerator heating, the cost of this heating steam $= (100,000/1000)(\$0.2969) = \$29.69$ per hour.

Since 100,000 lb (45,000 kg) of steam utilizes its energy for deaerator heating within the cycle, its equivalent electric output of $26,663/5 = 5333$ kW should be

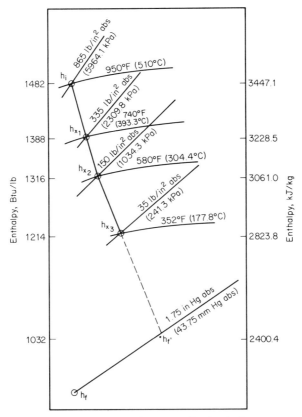

FIGURE 42   Turbine expansion curve. (*Combustion.*)

deducted from the 26,663-kW electric energy equivalency of the 35-lb/in² (abs) (241.3-kPa) steam. The remaining equivalent energy of 21,330 kW (= 26,663 − 5333) represents 35-lb/in² (abs) (241.3-kPa) extraction steam to be delivered to process plants.

6. *Determine the added unit fuel cost*
The deaerator-steam fuel cost of $29.69 per hour would be shared by both process steam and electricity in terms of energy equivalency as 105,480 − 5333 = 100,147 kW. Using this output as the denominator, we see that the added unit fuel cost for electricity based on sharing the cost of this heat energy input to the deaerator is $26.69/100,147 = $0.000296 per kilowatthour, or 0.296 mil/kWh.

Likewise, added fuel cost of 335-lb/in² (abs) (2309-kPa) steam = 10,432(100,000)/0.296 = $0.031 per 1000 lb (450 kg); added fuel cost of 150-lb /in² (abs) (1034-kPa) steam = $0.0246 per 1000 lb (450 kg); added fuel cost of 35-lb/in² (abs) (241-kPa) steam = $0.0158 per 1000 lb (450 kg). The fuel-cost allocation of steam and electricity is summarized in Table 28.

Related Calculations.   The energy equivalence method is based on the fact that the basic energy source for process steam and electricity is the heat from fuel (combustion or fission). The cost of the fuel must be charged to the process steam

TABLE 28 Energy Equivalence Method of Fuel Cost Allocation*

Utility	Base	+	Unit cost heating steam	=	Total	Total fuel cost	Percent of total
Electricity 51,743 kW†	5.568		0.296		5.864 mi/kWh	$303.40	51.65
Steam @ 335 lb/in² (abs) (2309 kPa) 100,000 lb/h (45,000 kg/h)	0.5808		0.031		$0.6118 per 1000 lb (450 kg)	$ 61.18	10.41
Steam @ 150 lb/in² (abs) (1034 kPa) 200,000 lb/h (90,000 kg/h)	0.4633		0.0246		$0.4879 per 1000 lb (450 kg)	$ 97.57	16.62
Steam @ 35 lb/in² (abs) (241 kPa) 400,000 lb/h (180,000 kg/h)	0.2969		0.0158		0.3127 per 1000 lb (450 kg)	$125.20	21.32
					Total:	$587.35	100.00

*_Combustion_ magazine.
†Net kW delivered to process plants should be delivered after deducting fixed mechanical and electrical losses of the alternator. Electricity unit cost charged to production would be slightly higher after this adjustment.

and electricity. Since the analysis does not distinguish between types of fuels or methods of heat release, this procedure can be used for coal, oil, gas, wood, peat, bagasse, etc. Also, the procedure can be used for steam generated by nuclear fission.

Cogeneration is suitable for a multitude of industries such as steel, textile, shipbuilding, air-craft, food, chemical, petrochemical, city and town district heating, etc. With the increasing cost of all types of fuel, cogeneration will become more popular than in the past. This calculation procedure is the work of Paul Leung of Bechtel Corporation, as reported at the 34th Annual Meeting of the American Power Conference and published in _Combustion_ magazine. Since the procedure is based on thermodynamic and economic principles, it has wide applicability in a variety of industries. For a complete view of the allocation of costs in cogeneration plants, the reader should carefully study the Related Calculations in the next calculation procedure.

COGENERATION FUEL COST ALLOCATION BASED ON AN ESTABLISHED ELECTRICITY COST

A turbine of the single-purpose type, operating at initial steam conditions identical to those in the previous calculation procedure, and a condenser backpressure of 1.75 in (43.75 mm) Hg (abs), would have a turbine heat rate of 9000 Btu/kWh (9495 kJ/kWh). Compute the fuel cost allocation to that of steam by using the established-electricity-cost method.

Calculation Procedure:

1. Compute the unit cost of the electricity

The unit cost of the electricity is F_e = (fuel price, $)(turbine heat rate, Btu/kWh)/ (boiler efficiency). For this plant, F_e = $(0.5/10^6)(9000)/(0.854)$ = \$0.00527 per kilowatthour, or 5.27 mil/kWh.

2. Determine where the deaerator heating steam should be charged

The turbine heat rate of 9000 Btu/kWh is a reasonable and economically justifiable heat rate of a regenerative cycle with a certain degree of feedwater heating. Hence, in this case, the deaerator heating steam should not be charged to the electricity. Instead, this portion of the deaerator-heating-steam cost should be charged to the process steam.

3. Allocate the fuel cost to steam

The total fuel cost from the previous calculation procedure is \$587.35 per hour. The electricity cost allocation = (kW generated)(cost \$/kWh) = (51,743)(0.00527) = \$273. Hence, the fuel cost to the steam is \$587.35 − \$273.00 = \$314.35.

4. Compute the power equivalence of the steam

From the previous calculation procedure, $E_x = E_{x1} + E_{x2} + E_{x3}$, where E_x = equivalent electric output of the extraction steam, kW; $E_{x1}, \ldots$ = equivalent electric output of the various extraction steam flows, kW. Hence, ΣE_x = 10,432 + 16,642 + 26,663 = 53,737 kW.

5. Determine the ratio of each extraction steam flow to the total extraction steam flow

The ratio for any flow is $E_x/\Sigma E_x$. Thus, $E_{x1}/\Sigma E_x$ = 10,432/53,737 = 0.194; $E_{x2}/\Sigma E_x$ = 16,663/53,737 = 0.310; and $E_{x3}/\Sigma E_x$ = 26,663/53,737 = 0.496.

6. Compute the base unit fuel cost of steam

Use the relation $(E_x/\Sigma E_x)$(fuel cost to steam)/m, where m = (steam flow rate, lb/ h)/1000. Hence, for 335-lb/in^2 (abs) (2309.8-kPa) steam, base unit fuel cost = (0.194)(\$314.35)/100 = \$0.610 per 1000 lb (\$0.277 per 1000 kg); for 150-lb/in^2 (abs) (1034.3-kPa) steam, base unit fuel cost = (0.310)(\$314.35)/200 = \$0.487 per 1000 lb (\$0.2213 per 1000 kg); for 35-lb/in^2 (abs) (241.3-kPa) steam, base unit fuel cost = (0.496)(\$314.35)/500 = \$0.312 per 1000 lb (\$0.1418 per 1000 kg). Since the deaeration steam is at 35 lb/in^2 (abs) (241 kPa), the cost of this steam = (100,000/1000)(\$0.312) = \$31.20 per hour.

7. Determine the unit fuel cost from sharing the cost of the deaerator heating steam

If the 5333-kW power equivalence of the deaerator heating steam is deducted from the electric power equivalence of the extraction steam, the kilowatt equivalence of all steam to production centers becomes 53,737 − 5333 = 48,404 kW. The unit fuel cost from sharing the cost of the deaerator heating steam is then (\$31.20/h)/ (48,404) = \$0.000644 per kilowatthour, or 0.644 mil/kWh.

8. Compute the added fuel cost of steam at each pressure

The added fuel cost at each pressure is (kW output at that pressure/steam flow rate, lb/h)(0.644). Thus, added fuel cost of 335-lb/in^2 (abs) (2309.8-kPa) steam = (10,431/100,000)(0.644) = \$0.067 per 1000 lb (\$0.03045 per 1000 kg); added fuel cost for 150-lb/in^2 (abs) (1034.3-kPa) steam = \$0.053 per 1000 lb (\$0.02409 per 1000 kg); added fuel cost for 35-lb/in^2 (abs) (241.3-kPa) steam = \$0.034 per 1000 lb (\$0.01545 per 1000 kg). Table 29 summarizes the fuel-cost allocation of steam and electricity by using this approach.

Related Calculations. The established-electricity-cost method is based on the assumption (or existence) of a reasonable and economically justifiable heat rate of the cycle being considered or used. The cost of the fuel must be charged to the process steam and electricity. Since the analysis does not distinguish between types of fuels or methods of heat release, this procedure can be used for coal, oil, gas, wood, peat, bagasse, etc. Also, the procedure can be used for steam generated by nuclear fission.

Cogeneration is suitable for a multitude of industries such as steel, textile, shipbuilding, aircraft, food, chemical, petrochemical, city and town district heating, etc. With the increasing cost of all types of fuels, cogeneration will become more popular than in the past.

Other approaches to cost allocations for cogeneration include: (1) capital cost segregation, (2) capital cost allocation by cost separation of major functions, (3) cost separation of joint components, (4) capital cost allocation based on single-purpose electric generating plant capital cost, (5) unit cost based on fixed annual capacity factor, and (6) unit cost based on fixed peak demand. Each method has its advantages, depending on the particular design situation.

In the two examples given here (the present and previous calculation procedures), water return to the dual-purpose turbine cycle is assumed to be of condensate quality. Hence, no capital and operating costs of water have been included. In actual cases, a cost account should be set up based on the quantity of the returned condensate. Special charges would be necessary for the unreturned portion of the water. Although the examples presented are for a fossil-fueled cycle, the methods are

TABLE 29 Established-Electricity-Cost Method of Fuel-Cost Allocation*

Utility	Base	+	Unit cost heating steam	=	Total	Total fuel cost	Percent of total
Electricity 51,743 kW†	5.27		0		5.27 mil/kWh	\$273.00	46.48
Steam @ 335 lb/in^2 (abs) (2309 kPa) 100,000 lb/h (45,000 kg/h)	0.610		0.067		\$0.677 per 1000 lb (450 kg)	\$ 67.70	11.53
Steam @ 150 lb/in^2 (abs) (1034 kPa) 200,000 lb/h (90,000 kg/h)	0.487		0.053		\$0.541 per 1000 lb (450 kg)	\$108.25	18.43
Steam @ 35 lb/in^2 (abs) (241 kPa) 400,000 lb/h (180,000 kg/h)	0.312		0.034		\$0.346 per 1000 lb (450 kg)	\$138.40	23.56
					Total:	\$587.35	100.00

Combustion magazine.
†Net kw delivered to process plants should be delivered after deducting fixed mechanical loss and electrical loss of the alternator. Electricity unit cost charged to production would be slightly higher after this adjustment.

equally valid for a nuclear steam-turbine cycle. For a contrasting approach and for more data on where this procedure can be used, review the Related Calculations portion of the previous calculation procedure.

This calculation procedure is the work of Paul Leung of Bechtel Corporation, as reported at the 34th Annual Meeting of the American Power Conference and published in *Combustion* magazine. Since the procedure is based on thermodynamic and economic principles, it has wide applicability in a variety of industries.

With utility power plants—some 3500—reaching their 30th birthday within the next few years, designers are evaluating ways of repowering. When a plant is re-powered, emissions are reduced, efficiency rises, as do reliability, output, and service life. So repowering has many attractions, including environmental benefits. More than 20 GW of capacity are estimated candidates for repowering.

Repowering replaces older facilities with new or different equipment. Several types of repowering are used today: (1) *Partial repowering*—which combines an existing plant system, infrastructure, and new equipment to provide increased output. *Example:* Combined-cycle repowering using a heat-recovery steam generator (HRSG) that recovers waste heat from the exhaust of a new gas-turbine/generator. *Example:* New gas-turbine/generator exhausts into existing boilers eliminating combustion-air-forced-draft needs while increasing the efficiency of the steam-generation cycle. Capital requirements are smaller than for an HRSG. This form of repowering is popular in Europe.

(2) *Station repowering*—reuses existing buildings, water-treatment systems, and fuel-handling system—but *not* the original steam cycle. New generating capability is installed to replace the existing steam plant—usually in the form of one or more gas turbines.

(3) *Site repowering*—uses an existing site but none of the equipment, such as boilers or turbines. Reusing an existing site eases permitting requirements, compared to developing a new site. To reduce overall project costs, it may be possible to reuse the infrastructure supporting the plant—such as power line and water- and fuel-delivery systems.

Specific methods for repowering include: (1) Combined-cycle repowering uses a new gas turbine and an HRSG to repower an existing facility by replacing or augmenting an existing boiler. (2) Gas turbines serving multiple-pressure HRSG provide power output and steam to existing steam-turbine generators. The gas-turbine power output goes directly to the utility's power lines. Natural gas fuels the gas turbine. (3) Pressurized fluidized-bed boilers are installed in place of existing boilers. Hot gases from the new boiler are used to drive a gas-turbine generator to increase the overall plant output. (4) Hot windbox repowering (also called the turbocharged boiler) adds a gas turbine/generator to an existing plant. The high-temperature exhaust from the gas turbine is used as combustion air in the existing boiler. This eliminates—in most cases—the need for a forced-draft fan while the gas turbine is operating. Plants using this method of repowering, which is prevalent in Europe, boost efficiency by 10 to 15 percent and output by 20 to 33 percent.

Data presented here on repowering were reported by Steven Collins, assistant editor, in *Power* magazine.

FUEL SAVINGS PRODUCED BY DIRECT DIGITAL CONTROL OF THE POWER-GENERATION PROCESS

A 200-MW steam-turbine generating unit supplied steam at 1000°F (538°C) and 2400 lb/in² (gage) (16,548 kPa) has an existing variability of 20°F (6.7°C) in the steam-temperature control. It is desired to reduce the variability of the steam temperature and thus allow a closer approach to the turbine design warrantee limits of 1050°F (566°C). A digital-control system will allow a 30°F (16.7°C) higher operating temperature at the turbine throttle. What will be the effect of this more precise temperature control on the efficiency and fuel cost of this unit? Fuel cost is $2.50 per 10^6 Btu ($2.38 per 10^6 J), and the plant heat rate is 9061 Btu/kWh (9559 kJ/kWh), with a turbine backpressure of 1 inHg (2.5 cmHg).

Calculation Procedure:

1. Sketch the unit flow diagram; write the overall efficiency equation
Figure 43 shows the flow diagram for this unit with the input, output, and losses indicated. The overall efficiency e of the system is determined by dividing the power output H_w, Btu/h (W), by the fuel input F, Btu h (W).

2. Express the efficiency equation with the system losses shown
The losses in a typical steam-turbine generating unit are the stack loss L_s, Btu/h (W); mechanical loss in turbine L_m, Btu/h (W); condenser loss L_c, Btu/h (W). The power output can now be expressed as $H_w = F - L_s - L_m - L_c$. Hence, $e = (F - L_s - L_m - L_c)/F$.

3. Write the loss relations for the unit
The boiler efficiency and turbine efficiency can each be assumed to be about 90 percent. This is a safe assumption for such an installation. Then $L_s = 0.1F$; $L_m = 0.1H_c$; $H_i = 0.9F$; $H_o = 0.9H_i$, by using the symbols shown in Fig. 43.

4. Write the plant efficiency equation at the higher temperature
With the temperature at the turbine inlet increased by 30°F (16.7°C), the condenser inlet enthalpy h_i' will change to 1480.9 Btu/lb (3444.6 kJ/kg), based on steam-table values. Setting up a ratio between the condenser inlet enthalpy after the throttle-temperature increase and before yields $1480.9/1023 = 1.0135h_i$. This ratio can be used for the other values, if we allow the prime symbol to indicate the values at the higher temperatures. Or, $H_i' = 1.0135 H_i$; $L_m' = 1.0135H_i$; $L_c' = 1.0058L_c$. Then $e' = (F' - L_s' - L_m' - L_c')/F'$.

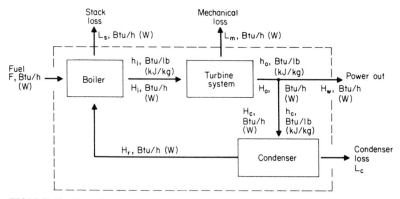

FIGURE 43 Flow diagram showing input, output, and losses. (*Combustion.*)

5. Compute the efficiency and heat-rate improvement
The improvement in plant efficiency $\Delta e = (e' - e)/e = e'/e - 1$. Substituting values, we find $\Delta e = [0.82089 - (L_c/F)(1.0058)]/\{1.0135[0.81 - (L_c/F)] - 1\}$.

With a heat rate of 9061 Btu/kWh (9559 kJ/kWh), the overall efficiency $e = (3412 \text{ Btu/kWh}/9061)100 = 37.65$ percent. Substituting this value of e in the general efficiency relation given in step 2 above shows that $L_c/F = 0.4335$. Then, substituting this value in the Δe equation above, we find $\Delta e = 0.3849/0.3816 - 1 = 0.0086$, or 0.86 percent.

6. Convert the efficiency improvement into annual fuel-cost savings
The annual fuel cost C can be computed from $C = 3.412(\text{fuel cost, \$ per } 10^6 \text{ Btu})$ (hours of operation per year) (plant MW capacity)$/e$. Or, $C = 3.412$ (2.5) (8760)(200)/0.3765 = \$39,693,386 per year. Then, with an efficiency improvement of 0.86 percent, the annual fuel-cost saving $S = 0.0086(\$39,693,386) = \$341,362$. In 10 years, with no increase in fuel costs (a highly unlikely condition), the fuel-cost savings with more precise steam-temperature control would be nearly \$3.5 million.

Related Calculation. Although this procedure is based on the use of digital-control systems, it is equally applicable for all forms of advanced control systems which can improve operator effectiveness and thereby save money in operating costs. Table 30 shows the basic reasons for using automatic control in both central-station and industrial power plants. With the ever-increasing energy costs forecast for the future, automatic control of power equipment will become of greater importance.

Table 31 shows the functions of direct digital controls (DDCs) for a variety of power-plant types. Performance improvements of up to 5:1 or more have been reported with DDC. With the expected life of today's plants at 40 years, the annual savings produced by DDC can have a significant impact on life-cycle costs. Table 32 shows the evolution of boiler controls—six phases of evolution. Central-station and large industrial plants today are in a distributed digital "revolution." Changes will occur in measurement, control, information, systems, and actuators. Clear benefits will be measured in terms of installed cost, ease of startup, reliability, control performance, and flexibility. This procedure is based on the work of M. A. Keys, Vice-President, Engineering, M. P. Lukas, Manager, Application Engineering, Bailey Controls Company, as reported in *Combustion* magazine.

SMALL HYDRO POWER CONSIDERATIONS AND ANALYSIS

A city is considering a small hydro power installation to save fossil fuel. To obtain the savings, the following steps will be taken: refurbish an existing dam, install new turbines, operate the generating plant. Outline the considerations a designer must weigh before undertaking the actual construction of such a plant.

Calculation Procedure:

1. Analyze the available head
Most small hydro power sites today will have a head of less than 50 ft (15.2 m) between the high-water level and tail-water level, Fig. 44. The power-generating capacity will usually be 25 MW or less.

TABLE 30 Basic Reasons for Using Automatic Control*

1. Increase in quantity or number of products (generation for fixed investment)
2. Improved product quality
3. Improved product uniformity (steam-temperature variability)
4. Savings in energy (improved efficiency or heat rate)
5. Raw-material savings (fixed savings)
6. Savings in plant equipment (more capacity from fixed investment)
7. Decrease in human drudgery (increased operator effectiveness)

* *Combusion* magazine.

TABLE 31 Functions of Direct Digital Controls*

1. Feedwater
2. Air flow and furnace draft
3. Fuel flow
4. RH and SH steam temperature
5. Primary air pressure and temperature
6. Minor loop control
    a. Cold-end metal temperature
    b. Cold-end metal temperature
    c. Turbine lube oil temperature
    d. Generator stator coolant to secondary coolant pressure differential
    e. Hydrogen temperature controls

* *Combustion* magazine.

TABLE 32 Evolution of Boiler Controls*

1. 1905–1920 Hand control with regulator assistance
2. 1920–1940 Analog boiler control systems acceptance
3. 1940–1950 Pneumatic direct connected analog systems
4. 1950–1960 Pneumatic transmitted analog systems
5. 1960–1970 Discrete component solid-state electric analog systems, burner control, and digital computers
6. 1970–1980 Integrated-circuit digital and analog systems

* *Combustion* magazine.

2. Relate absolute head to water flow rate

Because heads across the turbine in small hydro installations are often low in magnitude, the tail-water level is important in assessing the possibilities of a given site. At high-water flows, tail-water levels are often high enough to reduce turbine output, Fig. 45a. At some sites, the available head at high flow is extremely low, Fig. 45b.

The actual power output from a hydro station is $P = HQwe/550$, where $P =$ horsepower output; $H =$ head across turbine, ft; $Q =$ water flow rate, ft^3/s; $w =$ weight of water, lb/ft^3; $e =$ turbine efficiency. Substituting in this equation for the plant shown in Fig. 45b, for flow rates of 500 and 1500 m^3/s, we see that a tripling

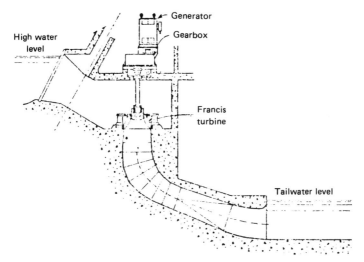

FIGURE 44   Vertical Francis turbine in open pit was adapted to 8-m head in an existing Norwegian dam. (*Power.*)

of the water flow rate increases the power output by only 38.7 percent, while the absolute head drops 53.8 percent (from 3.9 to 1.8 m). This is why the tail-water level is so important in small hydro installations.

Figure 45*c* shows how station costs can rise as head decreases. These costs were estimated by the Department of Energy (DOE) for a number of small hydro power installations. Figure 28*D* shows that station cost is more sensitive to head than to power capacity, according to DOE estimates. And the prohibitive costs for developing a completely new small hydro site mean that nearly all work will be at existing dams. Hence, any water exploitation for power must not encroach seriously on present customs, rights, and usages of the water. This holds for both upstream and downstream conditions.

3.   Outline machinery choice considerations

Small-turbine manufacturers, heeding the new needs, are producing a good range of semistandard designs that will match any site needs in regard to head, capacity, and excavation restrictions.

The Francis turbine, Fig. 44, is a good example of such designs. A horizontal-shaft Francis turbine may be a better choice for some small projects because of lower civil-engineering costs and compatibility with standard generators.

Efficiency of small turbines is a big factor in station design. The problem of full-load versus part-load efficiency, Fig. 46, must be considered. If several turbines can fit the site needs, then good part-load efficiency is possible by load sharing.

Fitting new machinery to an existing site requires ingenuity. If enough of the old powerhouse is left, the same setup for number and type of turbines might be used. In other installations the powerhouse may be absent, badly deteriorated, or totally unsuitable. Then river-flow studies should be made to determine which of the new semistandard machines will best fit the conditions.

Personnel costs are extremely important in small hydro projects. Probably very few small hydro projects centered on redevelopment of old sites can carry the

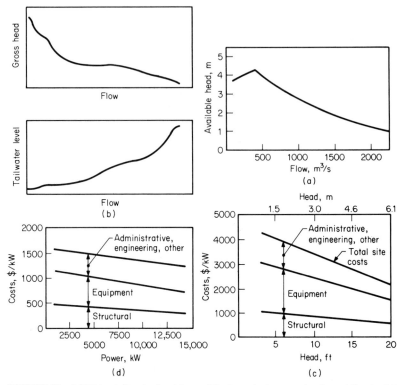

FIGURE 45 (*a*) Rising tail-water level in small hydroprojects can seriously curtail potential. (*b*) Anderson-Cottonwood dam head dwindles after a peak at low flow. (*c*) Low heads drive DOE estimates up. (*d*) Linear regression curves represent DOE estimates of costs of small sites. (*Power.*).

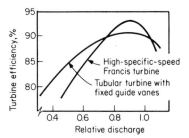

FIGURE 46 Steep Francis-turbine efficiency falloff frequently make multiple units advisable.

burden of workers in constant attendance. Hence, personnel costs should be given close attention.

Tube and bulb turbines, with horizontal or nearly horizontal shafts, are one way to solve the problem of fitting turbines into a site without heavy excavation or civil engineering works. Several standard and semistandard models are available.

In low head work, the turbine is usually low-speed, far below the speed of small generators. A speed-increasing gear box is therefore required. A simple helical-gear unit is satisfactory for vertical-shaft and horizontal-shaft turbines. Where a vertical turbine drives a horizontal generator, a right-angle box makes the turn in the power flow.

Governing and control equipment is not a serious problem for small hydro plants.

Related Calculations. Most small hydro projects are justified on the basis of continuing inflation which will make the savings they produce more valuable as time passes. Although this practice is questioned by some people, the recent history of inflation seems to justify the approach.

As fossil-fuel prices increase, small hydro installations will become more feasible. However, the considerations mentioned in this procedure should be given full weight before proceeding with the final design of any plant. The data in this procedure were drawn from an ASME meeting on the subject with information from papers, panels, and discussion summarized by William O'Keefe, Senior Editor, *Power* magazine, in an article in that publication.

RANKING EQUIPMENT CRITICALITY TO COMPLY WITH SAFETY AND ENVIRONMENTAL REGULATIONS

Rank the criticality of a boiler feed pump operating at 250°F (121°C) and 100 lb/in² (68.9 kPa) if its Mean Time Between Failures (MTBF) is 10 months, and vibration is an important element in its safe operation. Use the National Fire Protection Association (NFPA) ratings of process chemicals for health, fire, and reactivity hazards. Show how the criticality of the unit is developed.

Calculation Procedure

1. *Determine the Hazard Criticality Rating (HCR) of the equipment*
Process industries of various types—chemical, petroleum, food, etc.—are giving much attention to complying with new process safety regulations. These efforts center on reducing hazards to people and the environment by ensuring the mechanical and electrical integrity of equipment.

To start a program, the first step is to evaluate the most critical equipment in a plant or factory. To do so, the equipment is first ranked on the basis of some criteria, such as the relative importance of each piece of equipment to the process or plant output.

The Hazard Criticality Rating (HCR) can be determined from a listing such as that in Table 33. This tabulation contains the analysis guidelines for assessing the process chemical hazard (PCH) and the Other Hazards (O). The rankings for such a table of hazards should be based on the findings of an experienced team throughly

TABLE 33 The Hazard Criticality Rating (HCR) is Determined in Three Steps*

Hazard criticality rating

1. Assess the process chemical hazard (PCH) by:
   - Determining the NFPA ratings (N) of process chemicals for:
     Health, fire, reactivity hazards
   - Selecting the highest value of N
   - Evaluating the potential for an emissions release (0 to 4):
     High (RF = 0): Possible serious health, safety or environmental effects
     Low (RF = 1): Minimal effects
     None (RF = 4): No effects
   - *Then, PCH = N − RF.* (Round off negative values to zero.)

2. Rate other hazards (0) with an arbitrary number (0 to 4) if they are:
   - Deadly (4), if:
     Temperatures > 1,000°F
     Pressures are extreme
     Potential for release of regulated chemicals is high
     Release causes possible serious health safety or environmental effects
     Plant requires steam turbine trip mechanisms, fired-equipment shutdown systems, or toxic- or combustible-gas detectors*
     Failure of pollution control system results in environmental damage*
   - Extremely dangerous (3), if:
     Equipment rotates at >5,000 rpm
     Temperatures >500°F
     Plant requires process venting devices
     Potential for release of regulated chemicals is low
     Failure of pollution control system may result in environmental damage*
   - Hazardous (2), if:
     Temperatures >300°F;
     Extended failure of pollution control system may cause damage*
   - Equipment rotates at >3,600 rpm
   - Temperatures >140°F or pressures >20 psig
   - Not hazardous (0), if:
     No hazards exist

3. Select the higher value of PCH and O as the hazard criticality rating

Equipment with spares drop one category rating. A spare is an inline unit that can be immediately serviced or be substituted by an alternative process option during the repair period.
Chemical Engineering.

familiar with the process being evaluated. A good choice for such a task is the plant's Process Hazard Analysis (PHA) Group. Since a team's familiarity with a process is highest at the end of a PHA study, the best time for ranking the criticality of equipment is toward the end of such safety evaluations.

From Table 33, the NFPA rating, N, of process chemicals for Health, Fire, and Reactivity, is $N = 2$, because this is the highest of such ratings for Health. The Fire and Reactivity ratings are 0, 0, respectively, for a boiler feed pump because there are no Fire or Reactivity exposures.

The Risk Reduction Factor (RF), from Table 33, is RF = 0, since there is the potential for serious burns from the hot water handled by the boiler feed pump. Then, the Process Chemical Hazard, PCH = N − RF = 2 − 0 = 2.

The rating of Other Hazards, O, Table 33, is O = 1, because of the high temperature of the water. Thus, the Hazard Criticality Rating, HCR = 2, found from the higher numerical value of PCH and O.

2. Determine the process criticality rating, PCR, of the equipment
From Table 34, prepared by the PHA Group using the results of its study of the equipment in the plant, PCR = 3. The reason for this is that the boiler feed pump is critical for plant operation because its failure will result in reduced capacity.

3. Find the process and hazard criticality ranking, PHCR
The alphanumeric PHC value is represented first by the alphabetic character for the category. For example, Category A is the most critical, while Category D is the least critical to plant operation. The first numeric portion represents the Hazard Criticality Rating, HCR, while the second numeric part represents the Process Criticality Rating, PCR. These categories and ratings are a result of the work of the PHA Group.
From Table 35, the Process and Hazard Criticality Ranking, PHCR = B23. This is based on the PCR = 3 and HCR = 2, found earlier.

TABLE 34 The Process Criticality Rating (PCR)*

Process criticality rating	
Essential (4)	The equipment is essential if failure will result in shutdown of the unit, unacceptable product quality, or severely reduced process yield
Critical (3)	The equipment is critical if failure will result in greatly reduced capacity, poor product quality, or moderately reduced process yield
Helpful (2)	The equipment is helpful if failure will result in slightly reduced capacity, product quality or reduce process yield
Not critical (1)	The equipment is not critical if failure will have little or no process consequences

Chemical Engineering.

TABLE 35 The Process and Hazard Criticality Rating*

PHC rankings					
Process criticality rating	Hazard criticality rating				
	4	3	2	1	0
4	A44	A34	A24	A14	A04
3	A43	B33	B23	B13	B03
2	A42	A32	C22	C12	C02
1	A41	B31	C21	CD11	D01

Note: The alphanumeric PHC value is represented first by the alphabetic character for the category (for example, category A is the most critical while D is the least critical). The first numeric portion represents the Hazard Criticality Rating, and the second numeric part the Process Criticality Rating.
Chemical Engineering.

4. *Generate a criticality list by ranking equipment using its alphanumeric PHCR values*

Each piece of equipment is categorized, in terms of its importance to the process, as: Highest Priority, Category A; High Priority, Category B; Medium Priority, Category C; Low Priority, Category D.

Since the boiler feed pump is critical to the operation of this process, it is a Category B, i.e., High Priority item in the process.

5. *Determine the Criticality and Repetitive Equipment, CRE, value for this equipment*

This pump has an MTBF of 10 months. Therefore, Table 36, CRE = b1. Note that the CRE value will vary with the PCHR and MTBF values for the equipment.

6. *Determine equipment inspection frequency to ensure human and environmental safety*

From Table 37, this boiler feed pump requires vibration monitoring every 90 days. With such monitoring it is unlikely that an excessive number of failures might occur to this equipment.

TABLE 36   The Criticality and Repetitive Equipment Values*

	CRE values			
	Mean time between failures, months			
PHCR	0–6	6–12	12–24	>24
A	a1	a2	a3	a4
B	a2	b1	b2	b3
C	a3	b2	c1	c2
D	a4	b3	c2	d1

Chemical Engineering.

TABLE 37   Predictive Maintenance Frequencies for Rotating Equipment Based on Their CRE Values*

	Maintenance cycles Frequency, days			
CRE	7	30	90	360
a1, a2	VM	LT		
a3, a4		VM	LT	
b1, b3			VM	
c1, d1				VM

VM: Vibration monitoring
LT: Lubrication sampling and testing
Chemical Engineering.

7. Summarize criticality findings in spreadsheet form
When preparing for a PHCR evaluation, as spreadsheet, Table 38, listing critical equipment, should be prepared. Then, as the various rankings are determined, they can be entered in the spreadsheet where they are available for easy reference.

Enter the PCH, Other, HCR, PCR, and PHCR values in the spreadsheet, as shown. These data are now available for reference by anyone needing the information.

Related Calculations. The procedure presented here can be applied to all types of equipment used in a facility: fixed, rotating, and instrumentation. Once all the equipment is ranked by criticality, priority lists can be generated. These lists can then be used to ensure the mechanical integrity of critical equipment by prioritizing predictive and preventive maintenance programs, inventories of critical spare parts, and maintenance work orders in case of plant upsets.

In any plant, the hazards posed by different operating units are first ranked and prioritized based on a PHA. These rankings are then used to determine the order in which the hazards need to be addressed. When the PHAs approach completion, team members evaluate the equipment in each operating unit using the PHCR system.

The procedure presented here can be used in any plant concerned with human and environmental safety. Today, this represents every plant, whether conventional or automated. Industries in which this procedure finds active use include chemical, petroleum, textile, food, power, automobile, aircraft, military, and general manufacturing.

This procedure is the work of V. Anthony Ciliberti, Maintenance Engineer, The Lubrizol Corp., as reported in *Chemical Engineering* magazine.

FUEL SAVINGS PRODUCED BY HEAT RECOVERY

Determine the primary-fuel saving which can be produced by heat recovery if 150 M Btu/h (158.3 MJ/h) in the form of 650-lb/in^2 (gage) (4481.1-kPa) steam superheated to 750°F (198.9°C) is recovered. The projected average primary-fuel cost (such as coal, gas, oil, etc.) over a 12-year evaluation period for this proposed heat recovery scheme is $0.75 per 10^6 Btu ($0.71 per million joules) lower heating value (LHV). Expected thermal efficiency of a conventional power boiler to produce steam at the equivalent pressure and temperature is 86 percent, based on the LHV of the fuel.

TABLE 38   Typical Spreadsheet for Ranking Equipment Criticality*

		Spreadsheet for calculating equipment PHCRS								
Equipment number	Equipment description	NFPA rating				PCH	Other	HCR	PCR	PHCR
		H	F	R	RF					
TKO	Tank	4	4	0	0	4	0	4	4	A44
TKO	Tank	4	4	0	1	3	3	3	4	A34
PU1BFW	Pump	2	0	0	0	2	1	2	3	B23

Chemical Engineering.

Calculation Procedure:

1. Determine the value of the heat recovered during 1 year

Entering Fig. 47 at the bottom at 1 year and project vertically to the curve marked $0.75 per 10^6 LHV. From the intersection with the curve, project to the left to read the value of the heat recovered as $5400 per year per MBtu/h recovered ($5094 per MJ).

2. Find the total value of the recovered heat

The total value of the recovered heat = (hourly value of the heat recovered, $/$10^6$ Btu)(heat recovered, 10^6 Btu/h)(life of scheme, years). For this scheme, total value of recovered heat = ($5400)(150 × 10^6 Btu/h)(12 years) = $9,720,000.

3. Compute the total value of the recovered heat, taking the boiler efficiency into consideration

Since the power boiler has an efficiency of 86 percent, the equivalent cost of the primary fuel would be $0.75/0.86 = $0.872 per 10^6 Btu ($0.823 per million joules). The total value of the recovered heat if bought as primary fuel would be $9,720,000($0.872/$0.75) = $11,301,119. This is nearly $1 million a year for the 12-year evaluation period—a significant amount of money in almost any business. Thus, for a plant producing 1000 tons/day (900t/day) of a product, the heat recovery noted above will reduce the cost of the product by about $3.14 per ton, based on 258 working days per year.

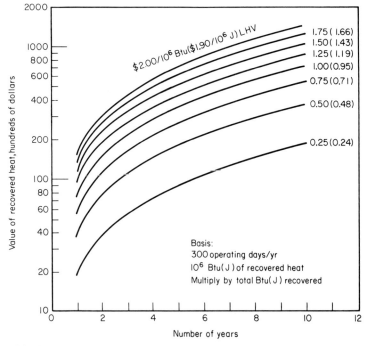

FIGURE 47 Chart yields value of 1 million Btu of recovered heat. This value is based on the projected average costs for primary fuel. (*Chemical Engineering.*)

Related Calculations. This general procedure can be used for any engineered installation where heat is available for recovery, such as power-generating plants, chemical-process plants, petroleum refineries, marine steam-propulsion plants, nuclear generating facilities, air-conditioning and refrigeration plants, building heating systems, etc. Further, the procedure can be used for these and any other heat-recovery projects where the cost of the primary fuel can be determined. Offsetting the value of any heat saving will be the cost of the equipment needed to effect this saving. Typical equipment used for heat savings include waste-heat boilers, insulation, heat pipes, incinerators, etc.

With the almost certain continuing rise in fuel costs, designers are seeking new and proven ways to recover heat. Ways which are both popular and effective include the following:

1. Converting recovered heat to high-pressure steam in the 600- to 1500-lb/in² (gage) (4137- to 10,343-kPa) range where the economic value of the steam is significantly higher than at lower pressures.

2. Superheating steam using elevated-temperature streams to both recover heat and add to the economic value of the steam.

3. Using waste heat to raise the temperature of incoming streams of water, air, raw materials, etc.

4. Recovering heat from circulating streams of liquids which might otherwise be wasted.

In evaluating any heat-recovery system, the following facts should be included in the calculation of the potential savings:

1. The economic value of the recovered heat should exceed the value of the primary energy required to produce the equivalent heat at the same temperature and/or pressure level. An efficiency factor must be applied to the primary fuel in determining its value compared to that obtained from heat recovery. This was done in the above calculation.

2. An economic evaluation of a heat-recovery system must be based on a projection of fuel costs over the average life of the heat-recovery equipment.

3. Environmental pollution restrictions must be kept in mind at all time because they may force the use of a more costly fuel.

4. Many elevated-temperature process streams require cooling over a long temperature range. In such instances, the economic analysis should credit the heat-recovery installation with the savings that result from eliminating non-heat-recovery equipment that normally would have been provided. Also, if the heat-recovery equipment permits faster cooling of a stream and this time saving has an economic value, this value must be included in the study.

5. Where heat-recovery equipment reduces primary-fuel consumption, it is possible that plant operations can be continued with the use of such equipment whereas without the equipment the continued operation of a plant might not be possible.

The above calculations and comments on heat recovery are the work of J. P. Fanaritis and H. J. Streich, both of Struthers Wells Corp., as reported in *Chemical Engineering* magazine.

Where the primary-fuel cost exceeds or is different from the values plotted in Fig. 47, use the value of $1.00 per 10^6 Btu (J) (LHV) and multiply the result by the ratio of (actual cost, dollars per 10^6 Btu/$1)($/J/$1). Thus, if the actual cost is $3 per 10^6 Btu (J), solve for $1 per 10^6 Btu (J) and multiply the result by 3. And if the actual cost were $0.80, the result would be multiplied by 0.8.

Controls in Environmental and Energy-Conservation Design

SELECTION OF A PROCESS-CONTROL SYSTEM

A continuous industrial process contains four process centers, each of which has two variables that must be controlled. If a fast process-reaction rate is required with only small to moderate dead time, select a suitable mode of control. The system contains more than two resistance-capacity pairs. What type of transmission system would be suitable for this process?

Calculation Procedure:

1. Compute the number of process capacities
The number of process capacities = (number of process centers)(number of variables per center), or, for this system, $4 \times 2 = 8$ process capacities. This is defined as a *multiple* number of process capacities because the number controlled is greater than unity.

2. Analyze the process-time lags
A small to moderate dead time is allowed in this process control system. With such a dead-time allowance and with two or more resistance-capacity pairs in the system, a mode of control that provides for any number of process-time lags is desirable.

3. Select a suitable mode of control
Table 39 summarizes the forms of control suited to processes having various characteristics. This table is a *general guide*—it provides, at best, an *approximate* aid in selecting control modes. Hence it is suitable for tentative selection of the mode of control. Final selection must be based on actual experience with similar systems.
Inspection of Table 39 shows that for a multiple number of processes with small to moderate dead time and any number of resistance-capacity pairs, a proportional plus reset mode of control is probably suitable. Further, this mode of control provides for any (i.e., fast or slow) reaction rate. Since a fast reaction rate is desired, the proportional plus reset method of control is suitable because it can handle any process-reaction rate.

4. Select the type of transmission system to use Four types of transmission systems are used for process control today: pneumatic, electric, electronic, and hydraulic. The first three types are by far the most common.
Pneumatic transmission systems use air at 3 to 20 lb/in^2 (gage) (20.7 to 137.9 kPa) to convey the control signal through small-bore metal tubing at distances ranging to several thousand feet. The air used in pneumatic systems must be clean and dry. To prevent a process from getting out of control, a constant supply of air is required. Pneumatic controllers, receivers, and valve positioners usually have small air-space volumes of 5 to 10 in^3 (81.0 to 163.9 cm^3). Air motors of the diaphragm or piston type have relatively large volumes: 100 to 5000 in^3 (1639 to 81,935 cm^3).
Pneumatic control systems are generally considered to be spark-free. Hence, they find wide use in hazardous process areas. Also, control air is readily available, and

TABLE 39 Process Characteristics versus Mode of Control*

| Number of process capacities | Process reaction rate | Process time lags | | Load changes | | Suitable mode of control |
		Resistance capacity (RC)	Dead time (transportation)	Size	Speed	
Single	Slow	Moderate to large	Small	Any	Any	Two-position; two-position with differential gap
				Moderate	Slow	Multiposition; proportional input
Single (self-regulating)	Fast	Small	Small	Any	Slow	Floating modes: Single speed, multispeed
					Moderate	Proportional-speed floating
Multiple	Slow to moderate	Moderate	Small	Small	Moderate	Proportional position
Multiple	Moderate	Any	Small	Small	Any	Proportional plus rate
Multiple	Any	Any	Small to moderate	Large	Slow to moderate	Proportional plus reset
Multiple	Any	Any	Small	Large	Fast	Proportional plus reset plus rate
Any	Faster than that of the control system	Small or nearly zero	Small to moderate	Any	Any	Wideband proportional plus fast reset

*Considine—*Process Instruments and Controls Handbook*, McGraw-Hill.

it can be "dumped" to the atmosphere safely. The response time of pneumatic control systems may be slower than that of electric or hydraulic systems.

Electric and electronic control systems are fast-response with the signal conveyed by a wire from the sensing point to the controller. In hazardous atmospheres the wire must be protected against abrasion and breakage.

Hydraulic control systems are also rapid-response. These systems are capable of high power actuation. Slower-acting hydraulic systems use fluid pressures in the 50 to 100 lb/in² (344.8 to 689.5 kPa) range; fast-acting systems use fluid pressures to 5000 lb/in² (34,475 kPa).

Dirt and fluid flammability are two factors that may be disadvantages in certain hydraulic-control-system applications. However, new manufacturing techniques and nonflammable fluids are overcoming these disadvantages.

Since a fast response is desired in this process-control system, electric, electronic, or hydraulic transmission of the signals would be considered first. With long distances between the sensing points [say 1000 ft (305 m) or more], an electric or electronic system would probably be best.

Next, determine whether the systems being considered can provide the mode of control (step 2) required. If a system cannot provide the necessary mode of control, eliminate the system from consideration.

Before a final choice of a system is made, other factors must be considered. Thus, the relative cost of each type of system must be determined. Should an electric system prove too costly, the slightly slower response time of the pneumatic system might be accepted to reduce the initial investment.

Other factors influencing the choice of the type of a control system include type of controls, if any, currently used in the installation, skill and experience of the operating and maintenance personnel, type of atmosphere in which, and type of process for which, the controls will be used. Any of these factors may alter the initial choice.

Related Calculations.   Use this general method to make a preliminary choice of controls for continuous processes, intermittent processes, air-conditioning systems, combustion-control systems, etc. Before making a final choice of any control system, be certain to weigh the cost, safety, operating, and maintenance factors listed above. Last, the system chosen *must* be able to provide the mode of control required.

PROCESS-TEMPERATURE-CONTROL ANALYSIS

A water storage tank (Fig. 48) contains 500 lb (226.8 kg) of water at 150°F (65.6°C) when full. Water is supplied to the tank at 50°F (10.0°C) and is withdrawn at the rate of 25 lb/min (0.19 kg/s). Determine the process-time constant and the zero-frequency process gain if the thermal sensing pipe contains 15 lb (6.8 kg) of water between the tank and thermal bulb and the maximum steam flow to the tank is 8 lb/min (0.060 kg/s). The steam flow to the tank is controlled by a standard linear regulating valve whose flow range is 0 to 10 lb/min (0 to 0.076 kg/s) when the valve operator pressure changes from 5 to 30 lb/in² (34.5 to 206.9 kPa).

Calculation Procedure:

1.   *Compute the distance-velocity lag*
The time in minutes needed for the thermal element to detect a change in temperature in the storage tank is the *distance-velocity lag,* which is also called the *trans-*

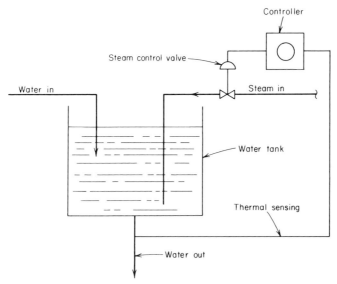

FIGURE 48   Temperature control of a simple process.

portation lag, or *dead time.* For this process, the distance-velocity lag d is the ratio of the quantity of water in the pipe between the tank and the thermal bulb—that is, 15 gal (57.01 L)—and the rate of flow of water out of the tank—that is, 25 lb/min (0.114 kg/s)—or $d = 15/25 = 0.667$ min.

2. Compute the energy input to the tank

This is a *transient-control process;* i.e., the conditions in the process are undergoing constant change instead of remaining fixed, as in *steady-state conditions.* For transient-process conditions the heat balance is $H_{in} = h_{out} + h_{stor}$, where H_{in} = heat input, Btu/min; H_{out} = heat output, Btu/min; H_{stor} = heat stored, Btu/min.

The heat input to this process is the enthalpy of vaporization h_{fg} Btu/(lb·min) of the steam supplied to the process. Since the regulating valve is linear, its sensitivity s is (flow-rate change, lb/min)/(pressure change, lb/in²). Or, by using the known valve characteristics, $s = (10 - 0)/(30 - 5) = 0.4$ (lb/min)/(lb/in²) [0.00044 kg/(kPa·s)].

With a change in steam pressure of p lb/in² (p' kPa) in the valve operator, the change in the rate of energy supply to the process is $H_{in} = 0.4$ (lb/min)/(lb/in²) $\times p \times h_{fg}$. Taking h_{fg} as 938 Btu/lb (2181 kJ/kg) gives $h_{in} = 375p$ Btu/min (6.6p' kW).

3. Compute the energy output from the system

The energy output H_{out} = lb/min of liquid outflow $\times$ liquid specific heat, Btu/(lb·°F) $\times$ (T_a − 150°F), where T_a = tank temperature,°F, at any time. When the system is in a state of equilibrium, the temperature of the liquid in the tank is the same as that leaving the tank or, in this instance, 150°F (65.6°C). But when steam is supplied to the tank under equilibrium conditions, the liquid temperature will rise to $150 + T_r$, where T_r = temperature rise,°F (T_r,°C), produced by introducing steam into the water. Thus, the above equation becomes H_{out} = 25 lb/min × 1.0 Btu/(lb · °F) × T_r = $25T_r$ Btu/min ($0.44T_r$ kW).

4. Compute the energy stored in the system
With the rapid mixing of the steam and water, H_{stor} = liquid storage, lb × liquid specific heat, Btu/(lb · °F) × $T_r q$ = 500 × 1.0 × $t_r q$, where q = derivative of the tank outlet temperature with respect to time.

5. Determine the time constant and process gain
Write the process heat balance, substituting the computed values in $H_{in} = H_{out} + H_{stor}$, or $375p = 25T_r + 500T_r q$. Solving gives $T_r/p = 375/(25 + 500q) = 15/(1 + 20q)$.

The denominator of this linear first-order differential equation gives the process-system time constant of 20 min in the expression $1 + 20q$. Likewise, the numerator gives the zero-frequency process gain of 15°F/(lb/in²) (1.2°C/kPa).

Related Calculations.   This general procedure is valid for any liquid using any gaseous heating medium for temperature control with a single linear lag. Likewise, this general procedure is also valid for temperature control with a double linear lag and pressure control with a single linear lag.

COMPUTER SELECTION FOR INDUSTRIAL PROCESS-CONTROL SYSTEMS

Select the type of computer and its speed of operation for use in an industrial control application. The computer will be used to monitor and control two continuous-flow process operations. Budget limitations restrict the investment in the computer to abut $100,000 with a typical execution time of 10 ms or better.

Calculation Procedure:

1. Analyze the computers available; select a suitable computer
Four types of computers are available for consideration in any control problem: analog; digital; hybrid—consisting of analog and digital; and special-purpose—analog or digital computers for industrial control.

Digital computers (Fig. 49) find wide use for controlling continuous-flow processes. When used to control continuous-flow processes, the digital computer is connected *online;* i.e., information reflecting the activity in the process being controlled is introduced to the data-processing system as soon as it occurs, and action is immediately initiated by the system to make any needed adjustments. Since digital computers are proven machines for process control, this type will be tentatively chosen for this process and its suitability investigated further.

The usual digital computer used in process control is a general-purpose one that can receive and transmit analog signals. A magnetic-drum-type stored memory is often used in control applications.

2. Determine the computer operating time
The speed of a computer depends on the actions needed to perform a calculation. Thus, a drum or disk memory may have a 150-μs add time with a memory access

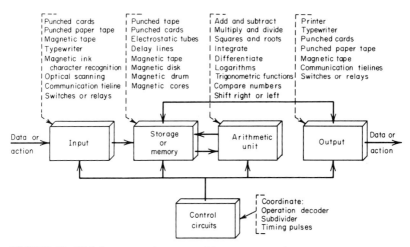

FIGURE 49 Digital-computer elements used in process control.

time of 16,000 μs. In such a computer the memory-access time is the controlling speed factor.

Figure 50 shows the execution times for a variety of digital computers of different makes in solving the same *bench-mark* or *test* problems. A typical benchmark problem consists of a pair of simultaneous equations having two unknowns.

Study of Fig. 50 shows that three machines—*A*, *B*, and *D*—meet the general requirements for this process with respect to speed and cost. Since each of these computers is produced by a different manufacturer, a fairly wide choice of units is available. When a chart such as Fig. 50 is not available, prepare one after computing the costs of machines available from various manufacturers.

3. Check the computer performance

Computer performance is rated according to three factors: speed margin, i.e., how the computer copes with the worst-case time combination of events in the process; memory-storage margin, that is, provisions for worst-case data storage, retrieval, and working space; and reliability or consistency of performance.

Word length affects computer speed. Although a computer handling 24-bit words may seem to have an operation rate identical to a 12-bit-word computer, there may

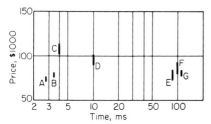

FIGURE 50 Execution time for an arithmetic benchmark problem shows the solution time required by different makes of computers.

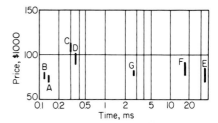

FIGURE 51 Execution time for a bench-mark problem shows no clear relationships between solution time and computer cost.

be as much as a 2:1 variation in the speed to the same degree of precision. To compare two computers, use their relative *problem-solving speeds* as a guide.

Manufacturers publish the relative problem-solving speeds of each model of computer. List these speeds for each of the suitable makes—A, B, and D, step 2. Select the computer giving the fastest speed for the smallest investment.

4. Investigate the computer logic function

In some control applications, the logical-problem-solving ability of the computer may be more important than the arithmetic ability. In the logic function, the computer uses information transfers, manipulations, and comparisons, all of which take time. Using manufacturer's data, plot the logic speed against cost for each computer for a specific bench-mark or test problem. Usually the execution time for a logical bench-mark problem shows no clear relation to price (Fig. 51). However, the plot does indicate the price range for various operating speeds.

5. Evaluate the computer-memory size

An online-control computer cannot deal with a real-time problem unless all instructions defining the action are stored and available when needed. Distribution of the storage between high-speed random-access memory sections and lower-cost cyclic-access sections, such as drums and disks, influences computer speed. Good practice stores an image of the high-speed working memory in one of the slower-speed backup memories.

In machines with 12- to 15-bit words, more than one memory word may be needed to store an instruction or data item. Compare bench-mark problems to determine the real-time need of memory addresses between different computers. In general, more process-control applications are better served by machines having expandable memories. The tendency of many control-system designers is to underestimate the memory capacity needed. So choose a machine having the largest memory possible within the prevailing financial constraints.

6. Evaluate the computer reliability

The mean time between failure (MTBF) is a good measure of computer reliability. For a typical control computer, the MTBF should be in the 1000+ -h category.

Compare the MTBF for each of the computers in step 2. Choose the machine having the highest MTBF at the desired speed. Further, the computer should be capable of operating in the 50 to 120°F (10 to 48.9°C) ambient temperature range.

Related Calculations.   Use this general method to choose a control computer for any process-type application. Where high-speed operations are involved, such as in missile-guidance applications, computers operating at nanosecond speeds are generally required. The selection procedure for such machines is different from that for process-control computers where millisecond speeds are usually satisfactory. Note that the computer prices cited here are relative; for actual prices consult the manufacturers concerned.

The program or software prepared for automatic process control can cost as much as the computer hardware. The first process applications of computers in an industry often show that the programs needed cost more than anticipated. Thus, computer programming to perform startup and shutdown sequence monitoring in a process system may require 4000 instructions. For automatic sequence control, as many as 20,000 instructions may be needed. Further, each plant is unique, and little of the programming work done for one process can be applied to another.

Programming still uses the most time of any phase of total computer operation for problem solving. So even if the computer operating time were cut to nearly zero, the saving in machine time would not be significant when compared with the programming time. On the other hand, saving in computer machine time is important in process-control applications because processes can be held more closely to optimum levels with reduced machine time.

Direct digital control (DDC) is replacing analog control in some process applications. The major advantage of DDC is that it removes the need for digital-to-analog converters and gives the computer more direct control over plant equipment while removing a source of spurious signals.

In process control, certain aspects of programming warrant special consideration. These aspects include real-time operation, memory capability, and operator misuse. Because a process computer functions in a real-time environment, a certain amount of "free" time must be available to allow for emergency reactions and special operator requests. A good rule of thumb is to allow 40 percent of free time within the computer; any less may cause the computer program to fall out of step with the process situation.

Most process-control computers use one of the variants of IBM's original Fortran. This language finds its principal applications in relatively infrequent procedures, such as plant startup and shutdown. Minute-by-minute scanning is usually handled by a standard "scan, monitor, and alarm" program prepared manually for the particular scanning sequence dictated by operating requirements.

A self-checking program is an important feature of a process-control computer system. Unlike a scientific program in which the results obtained are printed out for perusal by a scientist or engineer, a closed-loop control system utilizes the computer's calculations to act directly on the process plant itself. Should either the input to the computer or the data handling within it be in error, the resulting calculated control points and output signals will be incorrect and could result in hazardous operation. Double and triple checks may be necessary to ensure that the operating data are valid. As more and more input signals are utilized, the programming necessary to provide such validity checks becomes increasingly more complicated. Continuous calculation of a process heat balance can, for example, be useful in determining the validity of the input information, since an extensive range of input data figures in the calculation.

CONTROL-VALVE SELECTION FOR PROCESS CONTROL

Select a steam control valve for a heat exchanger requiring a flow of 1500 lb/h (0.19 kg/s) of saturated steam at 80 lb/in^2 (gage) (551.6 kPa) at full load and 300 lb/h (0.038 kg/s) at 40 lb/in^2 (gage) (275.8 kPa) at minimum load. Steam at 100 lb/in^2 (gage) (689.5 kPa) is available for heating.

Calculation Procedure:

1. Compute the valve flow coefficient
The valve flow coefficient C_v is a function of the maximum steam flow rate through the valve and the pressure drop that occurs at this flow rate. In choosing a control valve for a process-control system, the usual procedure is to assume a maximum flow rate for the valve based on a considered judgment of the overload the system may carry. Usual overloads to not exceed 25 percent of the maximum rated capacity

of the system. Using this overload range as a guide, assume that the valve must handle a 20 percent overload, or 0.20 (1500) = 300 lb/h (0.038 kg/s). Hence, the rated capacity of this valve should be 1500 + 300 = 1800 lb/h (0.23 kg/s).

The pressure drop across a steam control valve is a function of the valve design, size, and flow rate. The most accurate pressure-drop estimate usually available is that given in the valve manufacturer's engineering data for a specific valve size, type, and steam-flow rate. Without such data, assume a pressure drop of 5 to 15 percent across the valve as a first approximation. This means that the pressure loss across this valve, assuming a 10 percent drop at the maximum steam-flow rate, would be $0.10 \times 80 = 8$ lb/in^2 (gage) (55.2 kPa).

With these data available, compute the valve flow coefficient from $C_v = WK/3(\Delta p \ P_2)^{0.5}$, where W = steam flow rate, lb/h; $K = 1 + (0.007 \times °F$ superheat of the steam); p = pressure drop across the valve at the maximum steam flow rate, lb/in^2; P_2 = control-valve outlet pressure at maximum steam flow rate, lb/in^2 (abs). Since the steam is saturated, it is not superheated and $K = 1$. Then $C_v = 1500/3(8 \times 94.7)^{0.5} = 18.1$.

2. Compute the low-load steam flow rate

Use the relation $W = 3(C_v \ \Delta p \ P_2)^{0.5}/K$, where all the symbols are as before. Thus, with a 40-lb/in^2 (gage) (275.8-kPa) low-load heater inlet pressure, the valve pressure drop is $80 - 40 = 40$ lb/in^2 (gage) (275.8 kPa). The flow rate through the valve is then $W = 3(18.1 \times 40 \times 54.7)^{0.5}/1 = 598$ lb/h (0.75 kg/s).

Since the heater requires 300 lb/h (0.038 kg/s) of steam at the minimum load, the valve is suitable. Had the flow rate of the valve been insufficient for the minimum flow rate, a different pressure drop, i.e., a larger valve, would have to be assumed and the calculation repeated until a flow rate of at least 300 lb/h (0.038 kg/s) was obtained.

Related Calculations. The flow coefficient C_v of the usual 1-in (2.5-cm) diameter double-seated control valve is 10. For any other size valve, the approximate C_v valve can be found from the product $10 \times d^2$, where d = nominal body diameter of the control valve. Thus, for a 2-in (5.1-cm) diameter valve, $C_v = 10 \times 2^2 = 40$. By using this relation and solving for d, the nominal diameter of the valve analyzed in steps 1 and 2 is $d = (d_v/10)^{0.5} = (18.1/10)^{0.5} = 1.35$ in (3.4 cm); use a 1.5-in (3.8-cm) valve because the next smaller standard control valve size, 1.25 in (3.2 cm), is too small. Standard double-seated control-valve sizes are ¾, 1, 1¼, 1½, 2, 2½, 3, 4, 6, 8, 10, and 12 in (1.9, 2.5, 3.2, 3.8, 5.1, 6.4, 7.6, 10.2, 15.2, 20.3, 25.4, 30.5 cm). Figure 5 shows typical flow-lift characteristics of popular types of control valves.

To size control valves for liquids, use a similar procedure and the relation $C_v = V(G/\Delta p)$, where V = flow rate through the valve, gal/min; Δp = pressure drop across the valve at maximum flow rate, lb/in^2; G = specific gravity of the liquid. When a liquid has a specific gravity of 100 SSU or less, the effect of viscosity on the control action is negligible.

To size control valves for gases, use the relation $C_\bullet = Q(GT_a)^{0.5}/1360(\Delta p \ P_2)^{0.5}$, where Q = gas flow rate, ft^3/h at 14.7 lb/in^2 (abs) (101.4 kPa) and 60°F (15.6°C); T_a = temperature of the flowing gas,°F abs = 460 +°F; other symbols as before. When the valve outlet pressure P_2 is less than $0.5P_1$, where P_1 = valve inlet pressure, use the value of $P_1/2$ in place of $(\Delta p \ P_2)^{0.5}$ in the denominator of the above relation.

To size control valves for vapors other than steam, use the relation $C_v = W(v_2/\Delta p)^{0.5}/63.4$, where W = vapor flow rate, lb/h; v_2 = specific volume of the vapor at the outlet pressure P_2, ft^3/lb; other symbols as before. When P_2 is less than

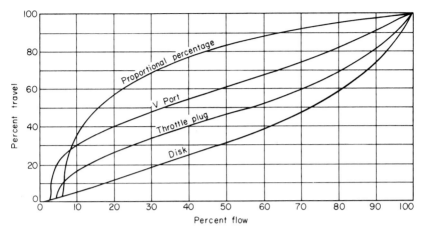

FIGURE 52 Flow-lift characteristics of control valves. (*Taylor Instrument Process Control Division of Sybron Corporation.*)

$0.5P_1$, use the value of $P_1/2$ in place of Δp and use the corresponding value of v_2 at $P_1/2$.

When the control valve handles a flashing mixture of water and steam, compute C_v by using the relation for liquids given above after determining which pressure drop to use in the equation. Use the *actual* pressure drop or the *allowable* pressure drop, whichever is smaller. Find the allowable pressure drop by taking the product of the supply pressure and the correction factor R, where R is obtained from Fig. 53. For a further discussion of control-valve sizing, see Considine—*Process Instruments and Controls Handbook*, McGraw-Hill, and G. F. Brockett and C. F. King—"Sizing Control Valves Handling Flashing Liquids," Texas A & M Symposium.

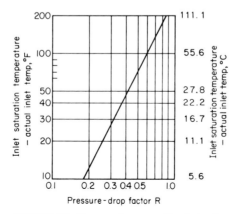

FIGURE 53 Pressure-drop correction factor for water in the liquid state. (*International Engineering Associates.*)

CONTROLLED-VOLUME-PUMP SELECTION FOR A CONTROL-SYSTEM

Select a controlled-volume pump to deliver 80 gal/h (0.084 L/s) of 100°F (37.8°C) distilled water to a chemical-feed system operating at 2000 lb/in^2 (abs) (13,790 kPa). What is the net positive suction head (NPSH) at the beginning of the pump suction stroke if the supply tank produces a 2-lb/in^2 (gage) (13.8-kPa) suction head at the pump centerline? Compute the minimum allowable NPSH for this pump if the length of the 1.5-in (3.8-cm) pipe between the pump and suction tank is 30 ft (9.1 m).

Calculation Procedure:

1. Choose the general type of pump to use
Controlled-volume pumps serve two functions when used as the final control elements in a control loop: to deliver liquid at the required pressure and to deliver liquid in the required quantities. In its second role, the pump also serves as a meter.

Two types of controlled-volume pumps are popular: plunger and diaphragm. The plunger pump is of somewhat simpler construction and is often used where contact of the plunger and liquid handled is not objectionable. Since distilled water is a relatively bland liquid, a plunger pump will be the tentative first choice for this control application.

2. Determine the pump dimensions and speed
The capacity, dimensions, speed, and efficiency of plunger-type controlled-volume pumps are given by $Q = D^2LNE/K$, where Q = pump capacity, gal/h (L/s); D = plunger diameter, in (cm); L = plunger stroke length, in (cm); N = number of strokes per minute, i.e., the pump speed; E = volumetric efficiency, %; K = dimensional constant = 4.92 for Q gal/h (0.0052 L/s), 295 for pump capacity in gal/min (18.6 L/s), and 0.0013 for pump capacity in mL/h.

Assume a pump speed of 50 strokes/min. This is a typical speed for plunger-type controlled-volume pumps. The usual efficiency of such a pump is 90 percent. With a 3-in (7.6-cm) stroke, the plunger diameter is $D = (QK/LNE)^{0.5} = [(80 \times 4.92)/(3 \times 50 \times 0.9)]^{0.5} = 1.71$ in, say 1.75 in (4.4 cm). This is a standard pump-plunger diameter.

3. Compute the pump NPSH
Use the relation NPSH $= P_a \pm P_h - P_v$, where NPSH = pump net positive suction head, lb/in^2 (abs) (kPa); P_a = atmospheric pressure at pump location = 14.7 lb/in^2 (abs) (101.4 kPa) at sea level; P_h = pressure head of liquid column above (+) or below (−) the centerline of the pump suction, lb/in^2 (gage) (kPa); P_v = vapor pressure of the liquid at the pumping temperature, lb/in^2 (abs) (kPa).

From the steam tables, $P_v = 0.949$ lb/in^2 (abs) (6.54 kPa) for water at 100°F (37.8°C). With an atmospheric pressure of 14.7 lb/in^2 (abs) (101.4 kPa), NPSH = 14.7 + 2.0 − 0.949 = 15.751 lb/in^2 (abs) (108.6 kPa).

4. Compute the pump minimum NPSH
Use the relation NPSH$_{min} = sL_pLN^2D^2/120,000D_p^2$, where NPSH$_{min}$ = minimum net positive suction head with which the pump can operate, lb/in^2 (abs) (kPa); s =

specific gravity of liquid handled; L_p = length of suction pipe, ft (m); D_p = suction pipe inside diameter, in (cm); other symbols as given before. Assuming a specific gravity of 1.0 $(NPSH_{min})$ = $(1.0 \times 30 \times 3 \times 50 \times 50 \times 1.5 \times 1.5)/(120,000 \times 1.5 \times 1.5)$ = 1.87 lb/in^2 (abs) (12.9 kPa).

Since the available NPSH [15.751 lb/in^2 (abs) (108.6 kPa), step 3] is greater than the minimum NPSH required [1.87 lb/in^2 (abs) (12.9 kPa), step 4], the pump will operate satisfactorily and without cavitation.

Related Calculations.   Use this general procedure to choose controlled-volume metering pumps for control systems requiring flows ranging from 1 mL/h to 20 gal/min (1 mL/h to 1.3 L/s) or more, at pressures ranging to 50,000 lb/in^2 (344,750 kPa). Typical applications for which this procedure is valid include chemical feed, ratioing, proportioning, and control of process variables.

STEAM-BOILER-CONTROL SELECTION AND APPLICATION

Choose a suitable feedwater regultor and combustion control for an industrial boiler serving the following loads: heating, 18,000 lb/h (2.3 kg/s); process 100,000 lb/h (12.6 kg/s); miscellaneous uses, 12,000 lb/h (1.5 kg/s). The boiler will have a maximum overload of 20 percent, and wide load fluctuations are expected at frequent intervals during operation. Pulverized-coal fuel is used to fire the boiler.

Calculation Procedure:

1. Determine the required boiler rating
Find the sum of the individual loads on the boiler, or 18,000 + 100,000 + 12,000 = 130,000 lb/h (16.4 kg/s). With a 20 percent overload, the boiler rating must be 1.2(130,000) = 156,000 lb/h (19.7 kg/s). With a 10 percent additional reserve capacity to provide for unusual loads, the rated boiler capacity should be 1.1(156,000) × 171,500 lb/h, say 175,000 lb/h (22.0 kg/s) for selection purposes.

2. Choose the type of feedwater regulator to use
Table 2 summarizes typical feedwater regulators used for boilers of various capacities. Study of Table 2 shows that a boiler in the 75,000 to 200,000 lb/h (9.4 to 25.2 kg/s) capacity range can use a relay-operated regulator with one or two elements when the load fluctuations are reasonable. With wide load swings, the relay-operated three-element regulator is a better choice. Since this boiler will encounter wide load swings, a three-element regulator is a wise and safe choice.

3. Choose the type of combustion-control system
Table 41 summarizes the important selection features of four types of combustion-control systems. Study of Table 41 shows that a stream flow-air flow type of combustion-control system would probably be best for the fuel and load conditions in this plant. Hence, this type of control system will be chosen.

Related Calculations.   Any control system selected for a boiler by using this procedure should be checked out by studying the engineering data available from the control-system manufacturer. The procedure given here is valid for heating, industrial, power, marine, and similar boilers.

TABLE 40   Boiler-Feedwater-Regulator Selector Chart*†

Boiler capacity	Type of feedwater regulator		
	Self-operated single-element	Relay-operated single- or two-element	Relay-operated three-element
Below 75,000 lb/h (9.4 kg/s)	For steady loads (building heating or continuous processes)	For irregular loads (batch processes, hoists, rolling mills, etc.)	
75,000–200,000 lb/h (9.4–25.2 kg/s)	Use only in special cases	For all steady and fluctuating loads	For extreme load and water conditions and boilers with steaming economizers
Above 200,000 lb/h (25.2 kg/s)	—	Use only on steady loads	For all types of loads

*From Kallen—*Handbook of Instrumentation and Controls*, McGraw-Hill.
†Excess pressure ahead of feedwater regulator should be at least 50 lb/in² (344.8 kPa) and should be controlled by regulation of the feed pump. Use excess-pressure valves only when excess pressure varies more than plus or minus 30 percent. Where drum level is unsteady owing to high solids concentration or boiler feed or other causes, use next-higher-class feed regulator.

CONTROL-VALVE CHARACTERISTICS AND RANGEABILITY

A flow control valve will be installed in a process system in which the flow may vary from 100 to 20 percent while the pressure drop in the system rises from 5 to 80 percent. What is the required rangeability of the control valve? What type of control-valve characteristic should be used? Show how the effective characteristic is related to the pressure drop that the valve should handle.

Calculation Procedure:

1. Compute the required valve rangeability
Use the relation $R = (Q_1/Q_2)(\Delta P_2/\Delta P_1)^{0.5}$, where R = valve rangeability; Q_1 = valve initial flow, percentage of total flow; Q_2 = valve final flow, percentage of total flow; P_1 = initial pressure drop across the valve, percentage of total pressure drop; P_2 = percentage of final pressure drop across the valve.
   Substituting gives $R = (100/20)(80/5)^{0.5} = 20$.

2. Select the type of valve characteristic to use
Table 42 lists the typical characteristics of various control valves. Study of Table 42 shows that as equal-percentage valve must be used if a rangeability of 20 is required. Such a valve has equal stem movements for equal-percentage changes in flow at a constant pressure drop based on the flow occurring just before the change

TABLE 41 Classification of Combustion-Control Systems*

	A, series-fuel	B, series-air	C, parallel	D, calorimeter or steam flow–air flow
Action	Temperature- or pressure-actuated master adjusts fuel rate; fuel meter adjusts air flow	Temperature- or pressure-actuated master adjusts air flow; air-flow meter adjusts fuel flow	Temperature- or pressure-actuated master adjusts fuel flow and air flow simultaneously	Pressure-actuated master adjusts fuel flow; steam flow adjusts air flow
Relative speed of control	Master adjusted for fast response because fuel-rate fluctuations caused by fluctuating pressure or temperature on master do not have correspondingly fast effect on that controlled variable	Master adjusted for slow response, because air-flow fluctuations following fast fluctuating master signal have a rapid effect on controlled variable and may cause hunting action if air-flow response is too fast	Master adjusted for slow response for same reason as in series-air	Master adjusted for fast response for same reason as in series-fuel. Steam flow–air flow control can be relatively rapid since steam-flow fluctuations are not so rapid as pressure variations
Used on fuels	Easily metered fuels such as oil and gas	All fuels. Oil, gas, and coal, either solid or burned in suspension	Primarily on solid fuels (grate firing)	Fuels hard to meter or fuels burned simultaneously. Commonly used on pulverized-coal-fired boilers
Advantages	When fuel may be in short supply, eliminates possibility of carrying high excess air for long period	Eliminates possibility of explosive mixture in combustion space when air fails. Eliminates need of fuel cutback for this purpose	Relatively inexpensive control system. No metering necessary	Ensures proper air-fuel ratio, even though fuel cannot be accurately metered or is of varying heat content. Ensures this condition even when burning a mixture of different fuels at the same time

*From Kallen—*Handbook of Instrumentation and Controls*, McGraw-Hill.

TABLE 42   Control-Valve Characteristics

Valve type	Typical flow rangeability	Stem movement
Linear	12-1	Equal stem movement for equal flow change
Equal-percentage	30-1 to 50-1	Equal stem movement for equal-percentage flow change°
On-off	Linear for first 25% of travel; on-off thereafter	Same as linear up to on-off range

° At constant pressure drop.

is made.[1] The equal-percentage valve finds use where large rangeability is desired and where equal-percentage characteristics are necessary to match the process characteristics.

3.   *Show how the valve effective characteristic is related to pressure drop*

Figure 54 shows the inherent and effective characteristics of typical linear, equal-percentage, and on-off control valves. The inherent characteristic is the theoretical performance of the valve.[1] If a valve is to operate at a constant load without changes in the flow rate, the characteristic of the valve is important, since only one operating point of the valve is used.

Figure 54*b* and *c* gives definite criteria for the amount of pressure drop the control valve should handle in the system. This pressure drop is not an arbitrary value such as 5 lb/in² (34.4 kPa) but rather a percentage of the total dynamic drop. The control valve should take at least 33 percent of the total dynamic system pressure drop[1] if an equal-percentage valve is used and is to retain its inherent characteristics. A linear valve should not take less than a 50 percent pressure drop if its linear properties are desired.

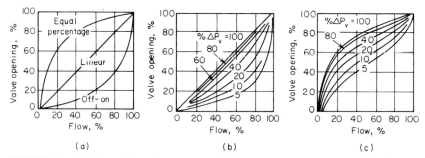

FIGURE 54   (*a*) Inherent flow characteristics of valves at constant pressure drop; (*b*) effective characteristics of a linear valve; (*c*) effective characteristics of a 50:1 equal-percentage valve.

[1]E. Ross Forman, "Fundamentals of Process Control," *Chemical Engineering,* June 21, 1965.

There is an economic compromise in the selection of every control valve. Where possible, the valve pressure drop should be as high as needed to give good control. If experience or an economic study dictates that the requirement of additional horsepower to provide the needed pressure is not worth the investment in additional pumping or compressor capacity, the valve should take less pressure drop with the resulting poorer control.

FLUID-AMPLIFIER SELECTION AND APPLICATION

Select a fluid amplifier to amplify the output of a fluidic sensor in a control system having a sensor output of 1 lb/in² (6.9 kPa) and requiring a control-valve operating pressure for modulating purposes of at least 10 lb/in² (68.9 kPa).

Calculation Procedure:

1. Compute the required amplification ratio
The amplification ratio μ = the gain of the amplifier = amplifier output, lb/in² (kPa)/amplifier input, lb/in² (kPa). For this amplifier, μ = 10/1 = 10. Hence, this fluid amplifier must increase a 1-lb/in² (6.9-kPa) input signal from the sensor to at least a 10-lb/in² (68.9-kPa) signal at the amplifier output.

2. Select a suitable fluid amplifier
Refer to a manufacturer's engineering data for the characteristic of a suitable fluid amplifier. Figure 55 shows the characteristics of a typical commercially available fluid amplifier. This amplifier is available for two gains: 9:1 and 18:1. Since the first gain is lower than required, the second or 18:1, gain would be used if this amplifier were chosen. With this gain, a 1-lb/in² (6.9-kPa) input signal would be amplified in the device to an output of an 18(1) = 18-lb/in² (124.4-kPa) signal.

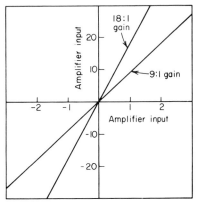

FIGURE 55 Operational-amplifier performance for two gains.

Since a 10-lb/in² (68.9-kPa) or larger signal is required to operate the control valve, this amplifier is acceptable.

Related Calculations. Many fluid amplifiers of the proportional type are packaged into fully operational modules, just as transistors and linear integrated circuits are. When so packaged, the device is termed an *operational amplifier.* An operational amplifier[1] is a module that contains within itself all the elements of a high-gain, accurate, and repeatable analog amplifier. Called *op-amps* for short, these devices are useful because they can amplify low-level signals, such as the outputs of fluidic sensors, proportionally to levels high enough to modulate control valves.

A fluid amplifier can produce a gain of pressure as, in this instance, power, or flow. Compute the gain for any of these outputs by setting up the ratio of output to input, as in step 1. Be certain to use consistent units for the output and input variables.

Where an amplifier is *not* linear (Fig. 56), the gain is different for each output. Hence, the operating points (i.e., input and output variables) must be stated for the gain being considered. Table 43 lists the selection characteristics for fluid amplifiers, control valves, and integrated circuits. Data presented in this tabulation are useful in choosing the most suitable control for a given application.

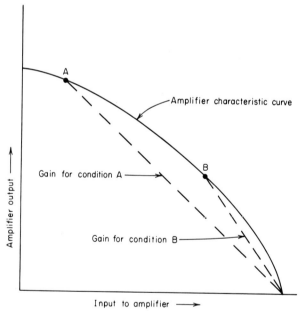

FIGURE 56 Fluid-amplifier operating characteristic curve and gain for two operating conditions.

---

[1]Frank Yeaple, "Analog Fluidic Amplifiers Are Waiting in the Wings," *Product Engineering,* Oct. 23, 1967.

TABLE 43 Control Selection Guide*†

	Fluid amplifiers	Spools, poppets	Solid-state integrated circuits
Characteristics of device needed			
Is environment-proof:			
Can be designed to operate at extremely high temperature	✓		
Can be made tolerant of any atmosphere	✓	✓	✓
Is unhurt by nuclear radiation	✓	✓	
Is unhurt by heavy vibration or shock	✓	. . .	✓
Has no moving parts:			
No stiction, hysteresis, dead zone, or jamming; also, no mechanical blocking, so no fluid hammer	✓	. . .	✓
Is tiny, stackable, monolithic:			
Whole circuits can be made in one integrated block, all permanently sealed, and with no moving parts	✓	. . .	✓
Can be supplied from any fluid source:			
Air, gas, water, oil, or process fluids will work; even the water rushing by the hull of a boat, or the air slipping past an airfoil, can be exploited	✓		
Needs no electricity:			
Is not affected in performance by radio or electrical interference	✓	✓	
Is responsive to extremely small inputs:			
Breaths of air; proximity of anything, such as tiny threads, specks, liquid surfaces, and air bubbles; motion of housing (fluid inertia effect); shock waves; fluid disturbances; spark discharges; sound waves; controlled vibration; localized heat	✓	. . .	✓
Is fast (millisecond or better)	✓	. . .	✓
Characteristics of device sought			
Has high energy output:			
Easily transduced to mechanical movement	Some	✓	
Is widely available from many sources:			
For proportional control	Some	Some	✓
For on-off control	✓	✓	✓
Is ultrafast (μsecond)	. . .	. . .	✓
Can be shut off individually when not in use:			
Power requirement drops drastically when device is switched off	. . .	✓	✓

*Checks show which control to use.
†*Product Engineering.*

CAVITATION, SUBCRITICAL-, AND CRITICAL-FLOW CONSIDERATIONS IN CONTROLLER SELECTION

Given the sizing formulas of the Fluid Controls Institute (FCI), size control valves for the cavitation, subcritical-, and critical-flow situations described below. Show how accurate the FCI formulas are.

Cavitation: Select a control valve for a situation where cavitation may occur. The fluid is steam condensate; inlet pressure P_1 is 167 lb/in² (abs) (1151.5 kPa); $\Delta P = 105$ lb/in² (724.0 kPa); inlet temperature T_1 is 180°F (82.2°C); vapor pressure P_v is 7.5 lb/in² (abs) (51.7 kPa).

Subcritical gas flow: Determine the valve capacity required at these conditions: fluid is air; flow Q_g is 160,000 standard ft³/h (1.3 standard m³/s); inlet pressure P_1 is 275 lb/in² (abs) (1896.1 kPa); $\Delta P = 90$ lb/in² (620.4 kPa); gas temperature T_1 is 60°F (15.6°C).

Critical vapor flow: A heavy-duty angle valve is suggested for a steam pressure-reducing application. Determine the capacity required, and compare an alternative valve type. The fluid is saturated steam; flow W is 78,000 lb/h (9.8 kg/s); inlet pressure P_1 is 1260 lb/in² (abs) (8686.4 kPa); outlet pressure P_2 is 300 lb/in² (abs) (2068.5 kPa).

Calculation Procedure:

1. Choose the valve type and determine its critical flow factor for the cavitation situation
If otherwise suitable (i.e., with respect to size, materials, and space considerations), a butterfly control valve is acceptable on a steam-condensate application. Find, from Table 44, the value of the critical flow factor $C_f = 0.68$ for a butterfly valve with 60° operation.

2. Compute the maximum allowable pressure differential for the valve
Use the relation $\Delta P_m = C_f^2(P_1 - P_v)$, where ΔP_m = maximum allowable pressure differential, lb/in² (kPa); P_1 = inlet pressure, lb/in² (abs) (kPa); P_v = vapor pressure, lb/in² (abs) (kPa). Substituting gives $\Delta P_m = (0.68)^2(167 - 7.5) = 74$ lb/in² (510.2 kPa). Since the actual pressure drop, 105 lb/in² (724.0 kPa), exceeds the allowable drop, 74 lb/in² (510.2 kPa), cavitation *will* occur.

3. Select another valve and repeat the cavitation calculation
For a single-port top-guided valve with flow to open plug, find $C_f = 0.90$ from Table 44. Then $\Delta P_m = (0.90)^2(167 - 7.5) = 129$ lb/in² (889.5 kPa).

In the case of the single-port top-guided valve, the allowable pressure drop, 129 lb/in² (889.5 kPa) exceeds the actual pressure drop, 105 lb/in² (724.0 kPa) by a comfortable margin. This valve is a better selection because cavitation will be avoided. A double-port valve also might be used, but the single-port valve offers lower seat leakage. However, the double-port valve offers the possibility of a more economical actuator, especially in larger valve sizes. This concludes the steps for choosing the valve where cavitation conditions apply.

4. Apply the FCI formula for subcritical flow
The FCI formula for subcritical gas flow is $C_v = Q_g/1360/(\Delta P/GT)^{0.5}[(P_1 + P_2)/2]^{0.5}$, where C_v = valve flow coefficient; Q_g = gas flow, standard ft³/h (standard m³/s); ΔP = pressure differential, lb/in² (kPa); G = specific gravity of gas at 14.7 lb/in² (abs) (101.4 kPa) and 60°F (15.6°C); T = absolute temperature of the gas, R; other symbols as given earlier. Substituting yields $C_v = 160,000/1360(90/520)^{0.5}[(275 + 185)/2]^{0.5} = 18.6$. Note that $G = 1.00$ for air.

TABLE 44 Critical Flow Factors for Control Valves at 100 Percent Lift*

Split body

A	Flow to close plug Flow to open plug Parabolic plug only	0.80 0.75
B	Flow to close plug Flow to open plug Parabolic plug only	0.50 0.90

Single-port, globe body

A	Flow to close plug Flow to open plug Parabolic plug only	0.85 0.90
B	Flow to close plug Flow to open plug Parabolic plug only	0.50 0.90

Angle body

A	Flow to close plug Flow to open plug Parabolic plug only	0.40 0.90
B	Flow to close plug Flow to open plug Parabolic plug only	0.55 0.95

Double-port, globe body

A	Parabolic plug V-port plug	0.90 1.00
B	Parabolic plug V-port plug	0.62 0.95

Butterfly

$D/d = 1$	$\alpha = 60°$ 0.68 $\alpha = 90°$ 0.58
$D/d = 2$	$\alpha = 60°$ 0.62 $\alpha = 90°$ 0.50

(A) Full-capacity trim, orifice diameter ~ 0.8 valve diameter

(B) Reduced capacity trim, 50% of (A) and less.

NOTE: The listed values apply for equal port-area valves only and do not include corrections for pipe friction.

*Henry W. Boger and *Chemical Engineering*.

18.131

5. Compute C_v, using the unified gas-sizing formula

For greater accuracy, many engineers use the unified gas-sizing formula. Assuming a single-port top-guided valve installed open to flow. Table 44 shows $C_f = 0.90$. Then $Y = (1.63/C_f)(\Delta P/P_1)^{0.5}$, where Y is defined by the equation and the other symbols are as given earlier. Substituting gives $Y = (1.63/0.90)(90/275)^{0.5} = 1.04$. Figure 57 shows the flow correlation established from actual test data for many valve configurations at maximum valve opening, and relates Y and the fraction of the critical flow rate.

Find from Fig. 58 the value of $Y - 0.148Y^3 = 0.87$. Compute $C_v = Q_g(GT)^{0.5}/834C_f(Y - 0.148Y^3)$, where all the symbols are as given earlier. Or, $C_v = 160,000(520)^{0.5}/[834(0.90)(275)(0.87)] = 20.4$. This value represents an error of approximately 10 percent in the use of the FCI formula.

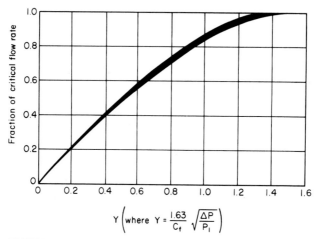

FIGURE 57 Flow correction established from actual data for many valve configurations at maximum valve opening.

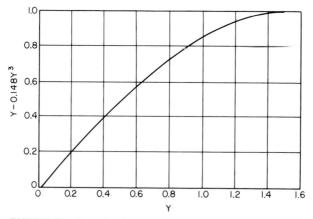

FIGURE 58 Correction-factor values.

6. Determine C_f for critical vapor flow
Assuming reduced valve trim for a heavy-duty angle valve, we find $C_f = 0.55$ from Table 44.

7. Compute the critical pressure drop in the valve
Use $\Delta P_c = 0.5(C_f)^2 P_1$, where P_c = critical pressure drop, lb/in^2 (kPa); other symbols as given earlier. So $\Delta P_c = 0.5(0.55)^2(1260) = 191$ lb/in^2 (1316.9 kPa).

8. Determine the value of C_v
Use the relation $C_v = W/1.83 C_f P_1$, where the symbols are as given earlier. Substituting yields $C_v = 78,000/[1.83(0.55)(1260)] = 61.5$. A lower C_v could be attained by using the valve flow to open, but a more economical choice is a single-port top-guided valve installed open to flow.

For a single-port top-guided valve flow to open, $C_f = 0.90$ from Table 44. Hence $C_v = 78,000/[1.83(0.90)(1260)] = 37.6$.

A lower capacity is required at critical flow for a valve with less pressure recovery. Although this may not lead to a smaller body size because of velocity and stability considerations, the choice of a more economical body type and a smaller actuator requirement is attractive. The heavy-duty angle valve finds its application generally on flashing-hydrocarbon liquid service with a coking tendency.

This calculation procedure is the work of Henry W. Boger, Engineering Technical Group Manager, Worthington Controls Co.

SERVO SYSTEM STABILITY DETERMINATION

The servo system shown in Fig. 59 is to be used in an industrial application. Determine the range of K for the system to be stable.

Calculation Procedure:

1. Solve, using the Routh criterion
Using the Routh criterion gives

$$G(s) = G_1(s)G_2(s) = \frac{K}{(s+2)(s^2+4s+20)} = \frac{K}{D}$$

We set $1 + G(s)H(s) = 0$:

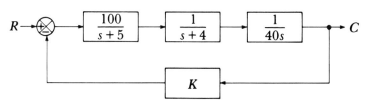

FIGURE 59 Servo system block diagram.

$$1 + \frac{K}{(s + 2)(s^2 + 4s + 20)} \frac{10}{s} = 0$$

$$s(s + 2)(s^2 + 4s + 20) + 10K = 0$$

$$s^4 + 6s^3 + 28s^2 + 40s + 10K = 0$$

2. **Set up the Routh array**

s^4	1	28	10K
s^3	6	40	0
s^2	$\dfrac{(6)(28) - (1)(40)}{6} = 21.33$	$\dfrac{60K - 1(0)}{6} = 10K$	0
s^1	$\dfrac{(21.33)40 - 60K}{21.33} = 40 - 2.813K$	0	
s^0	10K		

3. **Solve for the range of K**

$$K = 0 \qquad \text{(for the } s^0 \text{ term)}$$

$$40 - 2.813K = 0 \qquad \text{(for the } s^1 \text{ term)}$$

$$2.813K = 40$$

$$K \geq 14.22$$

Therefore K must lie between 0 and 14.22. This procedure is the work of Charles R. Hafer, P.E.

ANGULAR-POSITION SERVO SYSTEM ANALYSIS

An angular-position servo system uses potentiometer feedback and has these gain-transfer functions:

Amplifier: $G_1(s) = \dfrac{100}{(s + 5)} \qquad \dfrac{\text{V}}{\text{V}}$

Motor mechanical transfer function: $G_2(s) = \dfrac{1}{40s} \qquad \dfrac{\text{rad}}{\text{A}}$

Motor electrical transfer function: $G_3(s) = \dfrac{1}{s + 4} \qquad \dfrac{\text{A}}{\text{V}}$

Potentiometer gain constant: $G_4(s) = K \qquad \dfrac{\text{V}}{\text{rad}}$

Draw a block diagram for this system, and determine the open-loop transfer function and the closed-loop transfer function. If $K = 100$, is the system stable?

Calculation Procedure:

1. Write the open-loop transfer function

The block diagram is drawn as shown in Fig. 60. The open-loop transfer function is

$$G(s) = G_1(s)G_2(s)(G_3(s) = \frac{100}{40s(s+5)(s+4)} = \frac{2.5}{s(s+5)(s+4)}$$

2. Write the closed-loop transfer function

The closed-loop transfer function requirements suggest that the system be put in the canonical form as

$$\frac{C}{R} = \frac{G(s)}{1 + G(s)H(s)} = \frac{2.5}{(s)(s+5)(s+4) + 2.5K}$$

3. Determine whether the system is stable

The characteristic equation for this system is

$$(s)(s+5)(s+4) + 2.5K = 0$$

$$s^3 + 9s^2 + 20s + 2.5K = 0$$

Set up a Routh array to find the gain range of K for a stable system:

s^3	1	20 0
s^2	9	2.5K 0
s^1	$\dfrac{180 - 2.5K}{9}$	0
s^0	2.5K	0

Setting the s^1 and s^0 terms in the first column ≥ 0, solve for the range of K that makes the system stable. Thus,

$$\text{For } 180 - 2.5K \geq 0: \qquad K \leq 72$$

$$\text{For } 2.5K \geq 0: \qquad K \geq 0$$

So the range of K for a stable system is $0 \leq K \leq 72$ V/rad. So the system is not stable for $K = 100$. This procedure is the work of Charles R. Hafer, P.E.

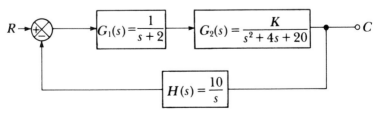

FIGURE 60   Angular-position servo system.

LOOP-GAIN FUNCTION ANALYSIS

The loop-gain function $GH(s) = K/s(s + \omega_1)(s + \omega_2)$ is represented by the straight-line approximation Bode plot in Fig. 61. The solid line represents the gain, and the dashed line represents phase. Find ω_1 and ω_2, K, phase margin ϕ_m, and the value of the K adjustment for a phase margin of 45°.

Calculation Procedure:

1. Determine the first pole and the other two breakpoints

The first pole determined by $1/s$ is at zero frequency. The other two breakpoints can be determined from Fig. 61 to be

$$\omega_1 = 10 \text{ rad/s} \qquad T_1 = \frac{1}{\omega_1} = 0.1 \text{ s}$$

$$\omega_2 = 100 \text{ rad/s} \qquad T_2 = \frac{1}{\omega_2} = 0.01 \text{ s}$$

2. Write the transfer function

The transfer function is now

$$GH(s) = \frac{K'}{s(0.1s + 1)(0.01s + 1)} = \frac{K}{s(s + 10)(s + 100)}$$

$$GH(j\omega) = \frac{K}{j\omega(j\omega + 10)(j\omega + 100)}$$

Since the approximation is a straight line, the maximum gain error occurs at the breakpoints. To solve for K with minimum error, move as far from a break as possible. Choose $\omega = 0.1$ rad/s. Then

$$|GH(j\omega)| = 60 \text{ dB} = 1000 = \left| \frac{K}{(0.1j)(0.1j + 10)(0.1j + 100)} \right|$$

$$\left| \frac{K}{(0.1/90°)(10/0.6°)(100/0.06°)} \right| = 1000$$

$$\frac{K}{100} = 1000$$

$$K = 10^5$$

3. Find the phase margin

From Fig. 61, the phase is 175° at the point where the gain is 0 dB. Hence, the phase margin is

$$\phi_m = 180° - 175° = 5°$$

This would be very marginal for stability, since 45° would be preferred.

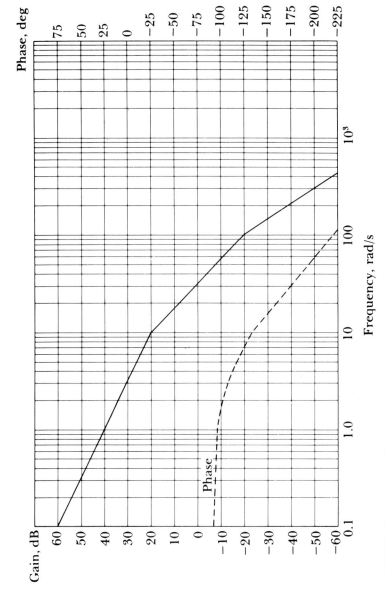

FIGURE 61 Straight-line Bode plot approximation.

18.137

4. *Determine the phase plot for a phase margin of 45°*

For a phase margin of 45°, the phase plot has a value of

$$\phi = -180° + \phi_m = -180° + 45° = -135°$$

At this value of ω on Fig. 61, the gain is 20 dB at a frequency of 10 rad/s. It is desired that the gain be 0 dB at this magnitude of phase and frequency. Therefore, it would be necessary to reduce the gain by 20 dB (a factor of 10). Now K becomes 10^4. This procedure is the work of Charles R. Hafer, P.E.

SERVO SYSTEM OVERSHOOT AND SETTLING TIME

A servo system has the block diagram shown in Fig. 62a. Determine the values of A and a from the expression $A/s(s + a)$ for a 15 percent overshoot and a 10-s settling time.

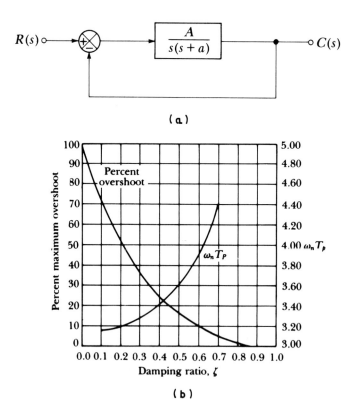

(a)

(b)

FIGURE 62 (a) Block diagram of servo system; (b) percentage of overshoot and peak time versus damping ratio for a second-order system.

Calculation Procedure:

1. Write the transfer function for the closed-loop system
The transfer function is

$$F(s) = \frac{C(s)}{R(s)} = \frac{G(s)}{1 + G(s)H(s)} = \frac{A}{s^2 + as + A}$$

This is the classic form of

$$F(s) = \frac{\omega_n^2}{s^2 + 2\xi\omega_n s + \omega_n^2}$$

2. Solve for the settling-time criteria
Assume that settling time means less than 5 percent, which it normally implies. The equation for a 5 percent settling time is

$$t_s = \frac{3}{\xi\omega_n}$$

$$PO = \frac{100 \exp(-\xi\pi)}{(1 - \omega^2)^{1/2}}$$

Solving for the settling-time criteria, we find

$$\xi\omega_n = \frac{3}{10} = 0.3$$

3. Find the damping ratio which satisfies the requirements
From the curve of Fig. 62, the value of the damping ratio which satisfies the requirements is approximately 0.53 for 15 percent of overshoot. Therefore,

$$\omega_n = \frac{0.3}{\xi} = 0.57$$

$$\omega_n^2 = 0.32$$

4. Write the closed-loop expression and solve it

$$F(s) = \frac{0.32}{s^2 + 0.6s + 0.32}$$

$$A = 0.32$$

$$a = 0.6$$

This procedure is the work of Charles R. Hafer, P.E.

ANALYSIS OF A SERVO SYSTEM WITH A LOOP DELAY

A servo system with a loop delay is represented by the block diagram in Fig. 63 and the function e^{-sT}. Find the maximum value of T for a stable system.

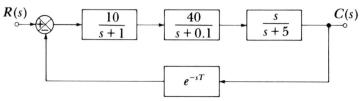

$R(s)$　　　　　　　　　　　　　　　　　　　　　　　$C(s)$

FIGURE 63　Block diagram of servo system.

Calculation Procedure:

1. Find the loop gain first

The loop gain is

$$GH(s) = G(s)H(s) = G_1(s)G_2(s)G_3(s)e^{-sT}$$

Neglecting the phase shift, we obtain $GH'(s)$:

$$GH'(s) = \frac{10}{s+1}\frac{40}{s+0.1}\frac{s}{s+5} = \frac{400s}{(s+0.1)(s+1)(s+5)}$$

$$GH'(j\omega) = \frac{400j\omega}{(0.1+j\omega)(1+j\omega)(5+j\omega)}$$

2. Plot the function

For a starting point, find $G(j\omega)$ far from a breakpoint (for accuracy purposes), and plot the function. Pick $\omega = 0.01$ rad/s, and solve for $GH(j\omega)$:

$$GH(j0.01) = \frac{4j}{(0.1)(1)(5)} = 8j$$

$$|GH(j0.01)| = 20\log_{10}8 = 18.6 \text{ dB}$$

Now plot the straight-line approximation as shown in Fig. 64.

3. Determine the phase at the crossover point

It appears as though the plot crosses 0 dB at 20 rad/s. The phase at this crossover point is

$$GH(j20) = \frac{400(j20)}{(0.1+j20)(1+j20)(5+j20)} \approx \frac{400}{(20\underline{/87.1°})(20.6\underline{/76°})}$$

$$GH(j20) = 0.97\underline{/-163.1°}$$

The gain checks since it is close to 1. The phase margin is only 16.9° (= 180° − 163.1°) and is already potentially unstable. It will surely be unstable if another 16.9° is added. Therefore, at $\omega = 20$ rad/s

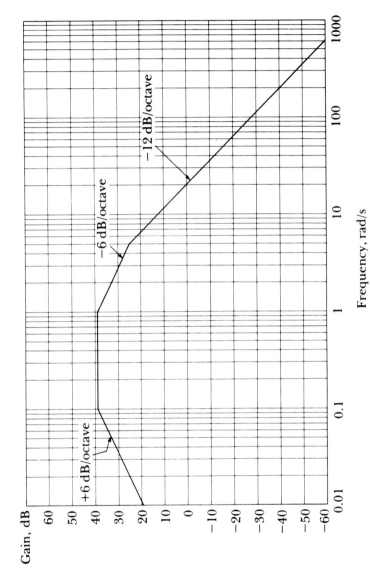

FIGURE 64 Bode plot straight-line approximation for *GH(s)*.

$$\omega T = 16.9° \times \frac{2\pi \text{ rad}}{360°}$$

$$\omega T = 0.294 \text{ rad}$$

$$T = \frac{0.294}{20} = 14.75 \text{ ms}$$

This procedure is the work of Charles R. Hafer, P.E.

SERVO SYSTEM CLOSED-LOOP TRANSFER FUNCTION CHARACTERISTICS

The servo system in Fig. 65 is to have the closed-loop transfer function $G(s) = C(s)/R(s)$ to possess the following characteristics: $\xi = 0.707$, $\omega_n = 15$. Find the transfer function of a compensating network to accomplish this, using cancellation compensation. Determine the network that will provide this function.

Calculation Procedure:

1. Describe how to obtain the desired function
Cancellation compensation is accomplished by removing a pole or zero and replacing it with another pole or zero. For example, if we have a function of the form $G(s) = 1/(s + a)$ and we desire $G'(s) = 1/(s + b)$, we can multiply our original function by $G_c(s) = (s + a)/(s + b)$, the compensating network. Thus, we can get our desired function $G'(s)$ as follows:

$$G'(s) = G(s)G_c(s) = \frac{[1/(s + a)](s + a)}{s + b} = \frac{1}{s + b}$$

2. Find the closed-loop transfer function
The closed-loop transfer function can be found from

$$G(s) = \frac{C(s)}{R(s)} = \frac{G_1(s)}{1 + G_1(s)H(s)}$$

Since $H(s) = 1$ and $G_1(s) = 1/s(s + 2)$.

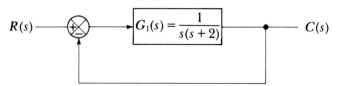

FIGURE 65 Block diagram of servo system.

$$G(s) = \frac{1}{s^2 + 2s + 1}$$

This is the general form of a second-order system described by

$$G(s) = \frac{\omega_n^2}{s^2 + 2\xi\omega_n^s + \omega_n^2}$$

For the circuit of Fig. 65 we have

$$\omega_n = 1 \qquad \xi = 1$$

For the given requirements of $\xi = 0.707$ and $\omega_n = 15$ we have $\omega_n^2 = 225$ and $2\xi\omega_n = 21.21$, and our desired closed-loop function becomes

$$G'(s) = \frac{225}{s^2 + 21.21s + 225}$$

3. Find the compensation network
Now take this expression back to open-loop form and obtain $G_1'(s)$ from

$$G'(s) = \frac{G_1'(s)}{1 + G_1'(s)H(s)} = \frac{N_1'(s)}{H(s)N_1'(s) + D_1'(s)}$$

where $$G_1'(s) = \frac{N_1'(s)}{D_1'(s)}$$

$$= \frac{225}{s^2 + 21.21s} = \frac{225}{s(s + 21.21)}$$

If we use cancellation techniques, we should cancel out our original pole with a zero and insert a new pole as described by $G_1'(s)$. Since our gain is 225, our compensation network now becomes

$$G_c(s) = \frac{225(s + 2)}{s + 21.21}$$

The new forward-loop transfer function is now

$$G_1'(s) = G_1(s)G_c(s)$$

$$= \frac{1}{s(s + 2)} \frac{225(s + 2)}{s + 21.21} = \frac{225}{s(s + 21.21)}$$

4. Use networks for an active network solution
Use an active solution because good isolation (low output impedance) is obtained. The gain of the compensating network will be negative because of the inverting amplifier. Or,

$$G_c(s) = \frac{-225(s + 2)}{s + 21.21} = \frac{-21.21(0.5s + 1)}{0.047s + 1}$$

These requirements can be satisfied by circuits from a table of networks. The transfer-impedance function is

$$G_C(s) = \frac{A(1 + sT)}{(1 + s\theta T)}$$

where
$$A = 21.22$$
$$T = 0.5$$
$$\theta T = 0.047$$
$$\theta = 0.094$$

From a table of transfer functions,

$$R_1' = \frac{A}{2} = \frac{21.22}{2} = 10.61$$

$$R_2' = \frac{A\theta}{4(1 - \theta)} = 0.55$$

$$C = \frac{4T(1 - \theta)}{A} = 0.0854$$

Choose $C = 100\ \mu F$ and scale the other values accordingly. Then

$$\alpha = \frac{0.0854}{10^{-4}} = 854$$

$$R_1 = \alpha R_1' = 9.06\ k\Omega$$

$$R_2 = \alpha R_2' = 470\ \Omega$$

Pick the nearest standard value of R_1 of 9.09 kΩ.

5. Find the dc gain function
The dc gain for the compensator is

$$G_C(s) = -21.22$$
$$\scriptstyle s \to 0$$

Therefore, since C is an open circuit at dc, our dc gain function is

$$|A| = \frac{Z_f}{Z_i}$$

$$\frac{2R_1}{R_3} = \frac{18.12 \times 10^3}{R_3}$$

$$R_3 = 854\ \Omega$$

Choose the nearest standard value of 845 Ω from a table of standard component values. Note that 180° of phase reversal has been introduced when an inverting amplifier is used. The sign of the summer must therefore be changed at the input error circuit. The circuit implementation is shown in Fig. 66 and the system block diagram in Fig. 67. This procedure is the work of Charles R. Hafer, P.E.

DEVELOPING A TRANSFER FUNCTION FOR A GIVEN PHASE MARGIN

Using a lead compensator, find the transfer function for the system shown in Fig. 68 when the gain at dc remains the same as for the uncompensated case. Provide a circuit arrangement that will satisfy the transfer function.

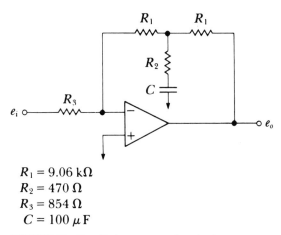

$$R_1 = 9.06 \text{ k}\Omega$$
$$R_2 = 470 \ \Omega$$
$$R_3 = 854 \ \Omega$$
$$C = 100 \ \mu\text{F}$$

FIGURE 66 Cancellation compensation circuit.

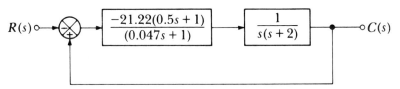

FIGURE 67 Block diagram of servo-system solution.

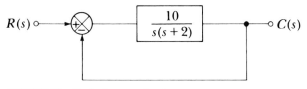

FIGURE 68 Block diagram of system.

Calculation Procedure:

1. Plot the magnitude and phase in Bode form

Start with $\omega = 0.1$ rad/s because it is far from a breakpoint and will not introduce much error. From this plot (Fig. 69), the phase ϕ_x at 0-dB gain is approximately $-146°$. This shows that the phase margin is $+34(180° - 146°)$.

2. Determine the phase shift needed

For a phase margin $+60°$, a phase shift of ϕ_m = phase margin $- 180° - \phi_x = 60 - 180 + 146 = 26°$. The normal procedure is to add a few degrees because the crossover shifts on the axis in a way to introduce more lag. Use 30°. Then

$$\sin \phi_m = \frac{\alpha - 1}{\alpha + 1}$$

$$\frac{\alpha - 1}{\alpha + 1} = 0.5$$

$$\alpha = 3$$

$$GH(\omega_m) = -10 \log \alpha = -10 \log 3 = -4.8 \text{ dB}$$

3. Find T_1 for the system

Refer to Fig. 69 and find $GH(\omega_n) = -4.8$ dB. Then ω_m is approximately 4 rad/s, which will be the new crossover. Now T_1 and $G_C(s)$ can be calculated.

$$T_1 = \frac{\alpha^{-1/2}}{\omega_m} = \frac{1}{(1.73)(4)} = 0.145s$$

and

$$G_C(s) = \frac{1 + \alpha T_1 s}{\alpha(1 + T_1 s)} = \frac{1}{3}\frac{1 + 0.43s}{1 + 0.145s}$$

$$= \frac{s + 2.3}{s + 6.9}$$

4. Choose the network that will provide the above function

Review of a table of Bode plots shows that the circuit of Fig. 70 followed by a buffer amplifier to provide isolation and gain will be suitable. The transfer function desired is

$$G_C(s) = \frac{1 + 0.43s}{3(1 + 0.145s)} = \frac{A(T_1 s + 1)}{T_3 s + 1}$$

Pick $C_1 = 100$ μF and $R_3 = 10$ kΩ and solve for our other values:

$$R_1 = \frac{T_1}{C_1} = \frac{0.435}{10^{-4}} = 4.35 \text{ k}\Omega$$

$$A = \frac{R_2}{R_1 + R_2}$$

$$R_2 = 2.15 \text{ k}\Omega$$

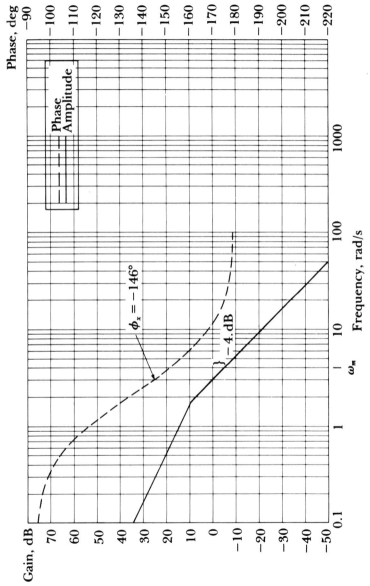

FIGURE 69 Bode plot for system having 60° phase margin.

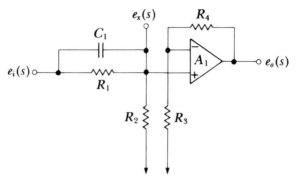

FIGURE 70　Lead-compensation circuit.

In order that the gain at dc remain the same, the gain of the amplifier must equal α. So

$$\frac{R_4 + R_3}{R_3} = \alpha = 3$$

$$R_4 = 20 \text{ k}\Omega$$

This procedure is the work of Charles R. Hafer, P.E.

FEEDBACK CONTROL SYSTEM PHASE-LAG COMPENSATOR DESIGN

For the feedback control system shown in Fig. 71, find a phase-lag compensator that will fulfill the requirements of $K_v = 20$ s^{-1}, the velocity = error constant, and $\phi_m = 45°$, phase margin. Design a circuit that will provide the compensating network.

Calculation Procedure:

1. Plot the Bode diagram for GH(s) for this system
Place $GH(s)$ in the proper format and plot, as in Fig. 72. Note that $H(s) = 1$. Then

$$GH(s) = \frac{K}{s(1 + 0.05s)(1 + 0.5s)}$$

$$= \frac{40K}{s(s + 20)(s + 2)}$$

2. Develop the transfer function for the velocity-error constant
To meet the first requirement (that the velocity-error constant be 20 s^{-1}), solve for K_v thus:

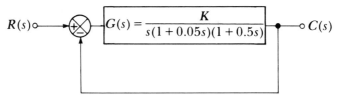

FIGURE 71 Feedback control system.

$$K_v = \lim_{s \to 0} sG(s) = \frac{40K}{(20)(2)} = 20$$

Therefore, $K = 20$ and the transfer function becomes

$$GH(s) = \frac{800}{s(s + 2)(s + 20)}$$

$$GH(j\omega) = \frac{800}{j\omega(2 + j\omega)(20 + j\omega)}$$

3. Compute the phase from the ω_0 value

The Bode plot is shown in Fig. 72 for the function in step 2. From this plot, determie that $\omega_0 = 6.3$ rad/s. Calculate the phase from this frequency. This is more accurate than obtaining the phase from the plot. Thus

$$GH(j6.3) = \frac{800}{j6.3(2 + j6.3)(20 + j6.3)}$$

$$= \frac{800}{(6.3\underline{/90°})(6.61\underline{/72.4°})(20.97\underline{/17.5°})} = 0.916\underline{/-179.9°}$$

The gain is close to 0 dB (-0.8 dB), and the phase checks. We have $0°$ phase margin and need to obtain a phase of at least $-135°$ to obtain a phase margin of $45°$. Add an additional $5°$ for a safety margin, and ϕ_c is now $-130°$. This occurs at a frequency ω_c of approximately 1.5 rad/s. The gain G_c at this frequency is 22 dB. To cross 0 dB at 1.5 rad/s, the lag compensator must reduce the gain by this amount. The transfer function of a lag compensator is represented by

$$G_c(s) = \frac{1 + \alpha T_1 s}{1 + T_1 s}$$

The constants can be determined as follows:

$$\alpha = 10^{-G_c/20} = 10^{-22/20} = 0.0794$$

$$\alpha T_1 = \frac{10}{\omega_c} = \frac{10}{1.5} = 6.67 \text{ s}$$

$$T_1 = 84 \text{ s}$$

The transfer function becomes

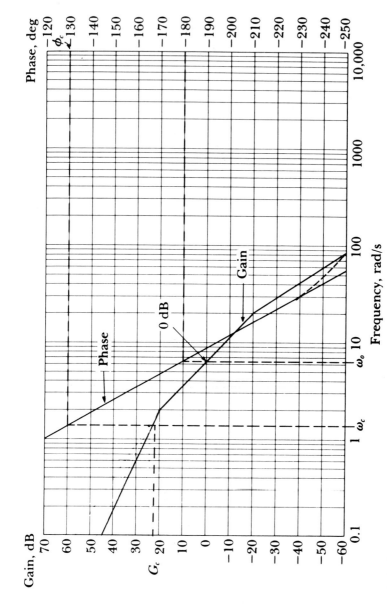

FIGURE 72 Bode plot of $GH(s)$ for feedback control system.

18.150

$$G_c(s) = \frac{1 + 6.67s}{1 + 84s}$$

The total compensated loop transfer function is now:

$$GH_c(s) = GH(s)G_c(s) = \frac{63.5(s + 0.015)}{s(s + 0.012)(s + 2)(s + 20)}$$

$$GH_c(j\omega) = \frac{63.5(0.15 + j\omega)}{j\omega(0.012 + j\omega)(2 + j\omega)(20 + j\omega)}$$

This function is plotted in Fig. 73.

4. Select a circuit to meet the $G_c(s)$ transfer function
Study circuits and their related transfer functions in a standard reference; this study will show that the circuit of Fig. 74 meets the $G_c(s)$ transfer function requirements. Follow this circuit with a buffer amplifier to prevent loading. Then

$$G_c(s) = \frac{1 + 6.67s}{1 + 84s} = \frac{T_1 s + 1}{T_3 s + 1}$$

$$T_1 = R_1 C_1 = 6.67$$

$$T_3 = (R_1 + R_2)C_1 = 84$$

Pick $C_1 = 150 \ \mu F$ and solve for the remaining component values:

$$R_1 = 44.5 \ k\Omega \qquad R_2 = 516 \ k\Omega$$

Pick the nearest standard values from a table of component characteristics. The circuit is shown in Fig. 74. This procedure is the work of Charles R. Hafer, P.E.

ANALYSIS OF TEMPERATURE-MEASURING AMPLIFIER

The operational amplifier shown in Fig. 75 is used to measure temperature. Here R_T is a temperature-sensitive resistor that varies with temperature as follows: $R_T = 1000e^{-T/25°C}$. Determine R_2 if E_o is to be 0 V at $-55°C$. What value must R_5 be if full-scale deflection at 125°C is required? The meter resistance R_M is 1 kΩ and the meter has a full-scale deflection of 1 mA.

Calculation Procedure:

1. Find the values of R_T at minimum and maximum temperatures
At $-55°C$, the minimum temperature is $R_T = 1000e^{-55/(-25)} = 9.025$ kΩ. At the maximum temperature of 125°C, $R_T = 1000e^{-125/25} = 6.74 \ \Omega$.

Gain, dB

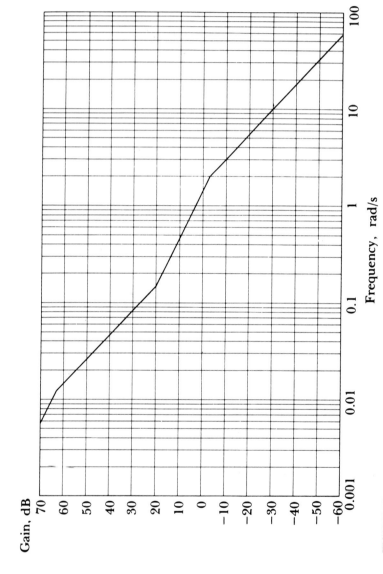

Frequency, rad/s

FIGURE 73 Bode plot of $GH_c(s)$.

18.152

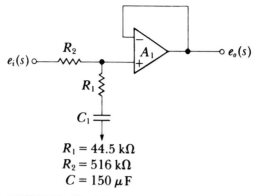

$$R_1 = 44.5 \text{ k}\Omega$$
$$R_2 = 516 \text{ k}\Omega$$
$$C = 150 \text{ } \mu\text{F}$$

FIGURE 74 Circuit lag compensator.

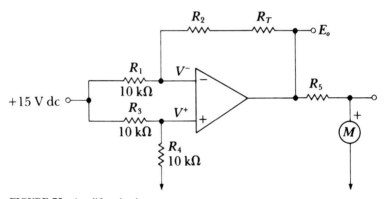

FIGURE 75 Amplifier circuit.

2. *Find R_2 for this amplifier*

Using the generalized equation for the analysis of dc amplifiers, we get

$$E_o = \frac{(V_1 - V_2)(R_3 + R_4)R_2}{(R_1 + R_2)R_3} + \frac{(R_3 + R_4)V_2}{R_3} - \frac{V_3 R_4}{R_3}$$

We substitute the terms for this amplifier, allowing $V = 15$ V dc:

$$E_o = \frac{V(R_1 + R_2 + R_T)R_4}{(R_3 + R_4)R_1} - \frac{V(R_2 + R_T)}{R_1}$$

Solve for R_2 with $E_o = 0$ V and $R_T = 9.025$ kΩ:

$$0 = \frac{15(10 \times 10^3 + R_2 + 9.025 \times 10^3)10 \times 10^3}{(20 \times 10^3)10 \times 10^3} - \frac{15(R_2 + 9.025 \times 10^3)}{10 \times 10^3}$$

So

$$0.5R_2 + 9.5125 \times 10^3 - R_2 - 9.025 \times 10^3 = 0$$

$$R_2 = 975 \ \Omega$$

From a table of standard metal-film resistor values, select $R_2 = 976 \ \Omega$.

3. Determie the values of E_o and R_s

At maximum temperature of 125°C, full-scale deflection is required. Calculate the value of E_o at 125°C ($R_T = 6.74 \ \Omega$ from step 1):

$$E_o = \frac{15(10 \times 10^3 + 975 + 6.74)10 \times 10^3}{(20 \times 10^3)(10 \times 10^3)} - \frac{15(975 + 6.74)}{10 \times 10^3}$$

$$= 6.764 \text{ V dc}$$

With this value of E_o, 1 mA is to flow into the meter. Therefore,

$$R_s + R_M = \frac{E_o}{1 \text{ mA}} = \frac{6.764}{1 \text{ mA}} = 6.764 \text{ k}\Omega$$

$$R_s = 6.764 \text{ k}\Omega$$

Choose the nearest 1 percent value; let $R_s = 5.76 \text{ k}\Omega$.
    This procedure is the work of Charles R. Hafer, P.E.

ANALYSIS OF TEMPERATURE-MEASURING INSTRUMENT

Figure 76 shows the schematic of a measuring system. In this system the motor drives a recording stylus which indicates temperature and simultaneously moves the wiper of the potentiometer that indicates the position of the stylus. Here R_1 and R_2 are internal to the recorder; R_T is the transducer and has a resistance value of 330 Ω at 25°C and a temperature coefficient of +3 Ω/°C. Find the values of R_1 and R_4 if we want to measure a temperature range of −25 to 125°C and the wiper

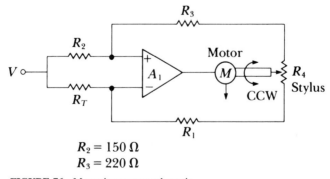

$R_2 = 150 \ \Omega$
$R_3 = 220 \ \Omega$

FIGURE 76 Measuring system schematic.

of the potentiometer is returned to ground potential. At $-25°C$, the stylus is fully counterclockwise (CCW).

Calculation Procedure:

1. Determine the value of R_T at -25 and $+125°C$

$$\text{At } -25°C: \qquad R'_T = 330 + (3\ \Omega/°C)(-50°C) = 180\ \Omega$$

$$\text{At } 125°C: \qquad R''_T = 330 + (3\ \Omega/°C)(100°C) = 630\ \Omega$$

2. Draw the equivalent circuit, and write the equations for it at $-25°C$

Draw the equivalent circuit for $-25°C$ as shown in Fig. 77. The potentiometer is full CCW. Since the voltage at the negative terminal must equal the voltage at the positive terminal, the following equations apply:

$$V^+ = V^-$$

$$\frac{(R_3 + R_4)V}{R_2 + R_3 + R_4} = \frac{R_1 V}{R'_T + R_1}$$

$$\frac{220 + R_4}{150 + 220 + R_4} = \frac{R_1}{180 + R_1}$$

Solving gives

$$R_4 = 0.833R_1 - 220$$

3. Draw the equivalent circuit, and write the equations for it at $125°C$

For the $125°C$ condition, the potentiometer is fully clockwise, and the circuit of Fig. 78 is applicable. Write these equations for $125°C$:

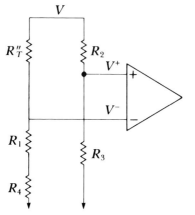

FIGURE 77 Fully counterclockwise equivalent circuit.

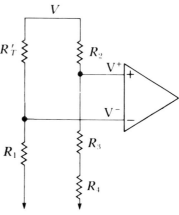

FIGURE 78 Fully clockwise equivalent circuit.

$$V^+ = V^-$$

$$\frac{(R_1 + R_4)V}{R_T'' + R_1 + R_4} = \frac{R_3 V}{R_2 + R_3}$$

$$\frac{R_1 + R_4}{630 + R_1 + R_4} = \frac{220}{370}$$

Thus

$$R_1 + R_4 = 925 \ \Omega$$

And substituting from our previous solution for R_4 gives

$$R_1 + 0.833R_1 - 220 = 925 \ \Omega$$

$$R_1 = 625 \ \Omega$$

$$R_4 = 300 \ \Omega$$

Choose the nearest standard value for R_1 from a table of component values. So $R_1 = 619$; R_4 would be a potentiometer of 300 Ω.

This procedure is the work of Charles R. Hafer, P.E.

MEASURING RMS VALUE OF A WAVEFORM

A dc ammeter is used to measure the rms values of the waveform shown in Fig. 79. Determine the value of R_1 for the circuit in Fig. 80 if the internal resistance of the meter is 1 kΩ and the meter has a full-scale deflection of 1 mA. The peak input signal level is 10 V.

Calculation Procedure:

1. Calculate the average value of the waveform voltage
A dc ammeter reads the average value of a waveform. Since we need rms values, we must find the rms and average values for the waveform and design the resistor

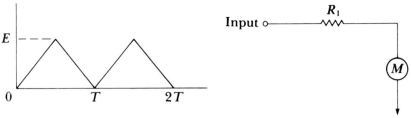

FIGURE 79 Input waveform. **FIGURE 80** Ammeter circuit.

accordingly. We must make the assumption that the natural time constant of the meter movement is much larger than the period of the waveform, or the meter movement will follow the profile of the waveform.

Calculate the average value for the waveform first. The equation for the waveform from the origin to $T/2$ is $E_1(t) = Et/(T/2) = 2Et/T$. The equation for the waveform from $T/2$ to T is $E_2(t) = 2E - 2Et/T$. The average value can be calculated as follows:

$$
\begin{aligned}
E_{av} &= \frac{1}{T}\int_0^T e \, dt = \frac{1}{T}\left[\int_0^{T/2} \frac{2Et}{T} \, dt + \int_{T/2}^T \left(2E - \frac{2Et}{T}\right) dt\right] \\
&= \frac{1}{T}\left(\frac{Et^2}{T}\Big|_0^{T/2} + 2Et - \frac{Et^2}{T}\Big|_{T/2}^T\right) \\
&= \frac{1}{T}\left(\frac{ET}{4} + 2ET - ET - ET + \frac{ET}{4}\right) = \frac{E}{2} = 5 \text{ V}
\end{aligned}
$$

2. Compute the rms value of the waveform

The rms value of the triangular waveform can be calculated as follows:

$$
\begin{aligned}
E_{rms} &= \left(\frac{1}{T}\int_0^T e^2 \, dt\right)^{1/2} \\
&= \left[\frac{1}{T}\int_0^{T/2}\left(\frac{2Et}{T}\right)^2 dt + \frac{1}{T}\int_{T/2}^T\left(2E - \frac{2Et}{T}\right)^2 dt\right]^{1/2} \\
&= \left[\frac{1}{T}\left(\frac{4E^2t^3}{3T^2}\right)_0^{T/2} + \frac{1}{T}\left(4E^2t - \frac{4E^2t^2}{T} + \frac{4E^2t^3}{3T^2}\right)_{T/2}^T\right]^{1/2} \\
&= \left[\frac{1}{T}\left(\frac{E^2T}{6}\right) + \frac{1}{T}\left(4E^2T - 2E^2T - 4E^2T + E^2T + \frac{4E^2T}{3} - \frac{E^2T}{6}\right)\right]^{1/2} \\
&= \left(\frac{E^2}{6} + 4E^2 - 2E^2 - 4E^2 + E^2 + \frac{4E^2}{3} - \frac{E^2}{6}\right)^{1/2} = \left(\frac{E^2}{3}\right)^{1/2} \\
&= 5.774 \text{ V}
\end{aligned}
$$

3. Determine the value of R_1

For a full-scale deflection of 10 V, we would want the meter resistance plus R_1 to equal 10 kΩ for a 1-mA full-scale deflection. Thus the meter would read 5 V for the average value. But we want an rms reading which is 5.774 V, so we need to increase our current by 5.774/5, which yields a current increase to 1.155 mA. Therefore,

$$
R_M + R_1 = \frac{10 \text{ V}}{1.155 \text{ mA}} = 8.66 \text{ k}\Omega
$$

$$
R_1 = 7.66 \text{ k}\Omega \qquad \text{or the nearest standard value}
$$

This procedure is the work of Charles R. Hafer, P.E.

NAND GATE CIRCUIT IMPLEMENTATION*

A 4-bit binary-coded decimal (BCD) code is shown in Table 45, appearing as inputs
A, B, C, and D, with A being the least significant bit (LSB). Find a NAND gate
implementation to convert Table 45.

Calculation Procedure:

1. Set up the standard basis for the designation numbers of inputs and outputs
Set up the standard basis which gives the designation numbers of the inputs and
then those of the outputs (W, X, Y, Z) desired. This is done as shown in Table 46.
The table is set up as follows: For the BCD portion of the table, column 1, we
have a zero code; in the bottom part of the table we also want a zero code. This
procedure is continued for all columns and corresponds to the desired results shown
in Table 45. Now, since W, X, Y, and Z are functions of A, B, C, and D, we can

TABLE 45 BCD Input, Gray Code Output

Input				Output			
A	B	C	D	W	X	Y	Z
0	0	0	0	0	0	0	0
1	0	0	0	1	0	0	0
0	1	0	0	1	1	0	0
1	1	0	0	1	1	1	0
0	0	1	0	1	1	1	1
1	0	1	0	0	1	1	1
0	1	1	0	0	0	1	1
1	1	1	0	0	0	0	1
0	0	0	1	1	0	0	1
1	0	0	1	1	1	0	1

TABLE 46 Standard Basis

A	0101	0101	01	(LSB)		
B	0011	0011	00		BCD	(Input)
C	0000	1111	00		code	
D	0000	0000	11			
W	0111	1000	11	(LSB)		
X	0011	1100	01		Gray	(Output)
Y	0001	1110	00		code	
Z	0000	1111	11			

*See pg 7.48ff for a discussion of boolean algebra with a number of pertinent calculation procedures.

use a Karnaugh map to simplify each of our expressions W, X, Y, and Z. The numbers in each square correspond to the miniterm representation in the designation number. For example,

$$0 \rightarrow \overline{ABCD} \qquad 2 \rightarrow \overline{A}B\overline{CD}$$

2. Use the Karnaugh map to simplify each expression

Go through the complete procedure for determining W, and the solutions for X, Y, and Z can be done similarly. The designation number for W is determined from Table 46.

$$W = 0111 \qquad 1000 \qquad 11XX \qquad XXXX$$

The X's denote constrained states or "don't care" conditions. The Karnaugh map of W is shown in Fig. 81. In all boxes where we want a 1 we put a diagonal line. In all squares where we don't care, we put an X. The X means that the square can be occupied by either a 1 or a 0 at the convenience of the designer. Now minimization can be achieved. All X's are taken to be 1 for this variable. Grouping as shown yields

Squares	Representation
3, 11, 9, 1	$A\overline{C}$
2, 3, 10, 11	$B\overline{C}$
8, 9, 10, 11, 12, 13, 14, 15	D
4, 12	$\overline{A}\overline{B}C$

Similar solutions for X, Y, and Z are shown in Figs. 82 through 84, respectively. The circuit implementation is shown in Fig. 85. If $\overline{A}$, $\overline{B}$, $\overline{C}$, and $\overline{D}$ are not available, inverters can be placed on A, B, C, and D.

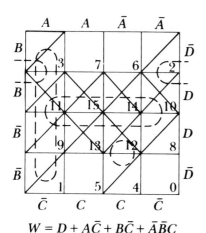

$$W = D + A\overline{C} + B\overline{C} + \overline{A}\overline{B}C$$

FIGURE 81 Karnaugh map of W.

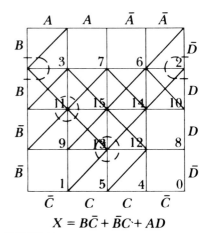

$$X = B\overline{C} + \overline{B}C + AD$$

FIGURE 82 Karnaugh map of X.

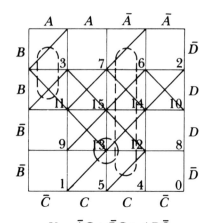

$$Y = \bar{A}C + \bar{B}C + AB\bar{C}$$

FIGURE 83 Karnaugh map of Y.

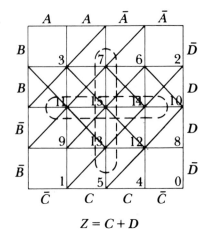

$$Z = C + D$$

FIGURE 84 Karnaugh map of Z.

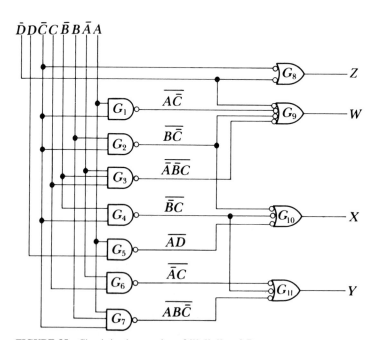

FIGURE 85 Circuit implementation of W, X, Y, and Z.

This procedure is the work of Charles R. Hafer, P.E.

USING NAND GATES TO IMPLEMENT FUNCTIONS WITH TWO LEVELS OF LOGIC

The boolean expression

$$F_1 = AB + ABC + BC$$

$$F_2 = (A\overline{B} + C)(A + \overline{B})C$$

$$F_3 = AB + (\overline{B} + \overline{C}) + \overline{A}C$$

is to be simplified by using boolean algebra and to be implemented by using a maximum of two levels of logic. Use NAND gates to implement these functions, and show the diagrams.

Calculation Procedure:

1. Simplify the given expression, using boolean algebra

$$F_1 = AB + ABC + BC$$

$$F_1 = AB(1 + C) + BC = AB + BC = B(A + C)$$

$$F_2 = (A\overline{B} + C)(A + \overline{B})C = (A\overline{B}C)(A + \overline{B})$$

$$F_2 = C(A\overline{B} + 1)(A + \overline{B}) = C(A + \overline{B}) = AC + \overline{B}C$$

$$F_3 = AB + (\overline{B} + \overline{C}) + \overline{A}C = AB + A\overline{B} + \overline{A}C + \overline{A}C$$

$$F_3 = A + \overline{A} = 1$$

(*Note:* $\overline{B}$ implies $A\overline{B}$ and $\overline{C}$ implies $\overline{A}C$.)

2. Implement, using NAND gates; show the diagram

The method for NAND gate implementation is to use AND gates and OR gates (no inversions) to implement the function and to place an inversion "ball" at the output of all AND gates and an inversion "ball" at the input of all AND gates used as OR functions. NAND gates may be added where required to provide an inverter function. Figure 86 shows how F_1 is obtained by using NAND logic, while Fig. 87 demonstrates the technique for F_2. Here F_3 is a trivial solution since it is a logic 1 level only.

This procedure is the work of Charles R. Hafer, P.E.

SIZING STEAM-CONTROL AND PRESSURE-REDUCING VALVES

Dry saturated steam at 30 lb/in² (abs) (206.9 kPa) will flow at the rate of 1000 lb/h (0.13 kg/s) through a single-seat pressure-reducing throttling valve. The de-

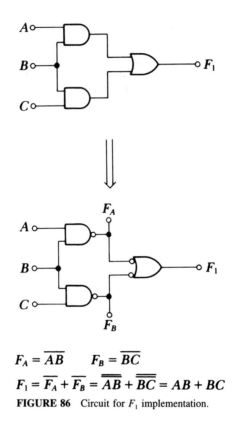

$$F_A = \overline{AB} \qquad F_B = \overline{BC}$$

$$F_1 = \overline{F_A} + \overline{F_B} = \overline{\overline{AB}} + \overline{\overline{BC}} = AB + BC$$

FIGURE 86   Circuit for F_1 implementation.

sired exit pressure is 20 lb/in² (abs) (137.9 kPa) at the valve outlet. Select a valve of suitable size.

Calculation Procedure:

1. Determine the critical pressure for the valve
Critical pressure exists in a valve and piping system when the pressure at the valve outlet is 58 percent, or less, of the absolute inlet pressure for saturated steam (55 percent for hydrocarbon vapors and superheated steam). Thus, for this system the critical outlet pressure is $P_c = 0.58 P_i$, where P_c = critical pressure for the system, lb/in² (abs) (kPa); P_i = inlet pressure, lb/in² (abs) (kPa). Or, $P_c = 0.58(30) = 17.4$ lb/in² (abs) (119.9 kPa).

Since the outlet pressure, 20 lb/in² (abs) (137.9 kPa), is greater than the critical pressure, the flow through the valve is noncritical.

2. Find the density of the outlet steam
Assume adiabatic expansion of the steam from 30 lb/in² (abs) (206.9 kPa) to 20 lb/in² (abs) (137.9 kPa). (This is a valid assumption for a throttling process such

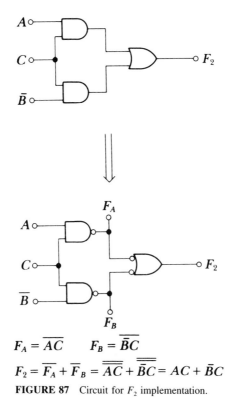

$$F_A = \overline{AC} \qquad F_B = \overline{\bar{B}C}$$

$$F_2 = \overline{F_A} + \overline{F_B} = \overline{\overline{AC}} + \overline{\overline{\bar{B}C}} = AC + \bar{B}C$$

FIGURE 87   Circuit for F_2 implementation.

as that which takes place in a pressure-reducing valve.) Using the steam tables, we find the density of the steam at the outlet pressure of 20 lb/in^2 (abs) (137.9 kPa) is 0.05 lb/ft^3 (0.8 kg/m^3).

3.   Compute the valve flow coefficient c_v
Use the relation $c_v = W/63.5\sqrt{(P_i - P_2)\rho}$, where c_v = valve flow coefficient, dimensionless; W = steam (or vapor or gas) flow rate, lb/h (kg/s); P_i = valve inlet pressure, lb/in^2 (abs); ρ = density of the vapor or gas flowing through the valve, lb/ft^3 (kg/m^3). For this valve, $c_v = 1000/63.5\sqrt{(30 - 20)0.05} = 22.3$, say 22.0 because c_v valves are usually stated in even numbers for larger-size valves.

4.   Select the control valve to use
At normal operating conditions, most engineers recommend that the flow through the valve not exceed 80 percent of the maximum flow possible. Thus the valve selected should have a c_v equal to or greater than the computed $c_v/0.80$. Thus, for this valve, choose a unit having a c_v equal to or greater than $22/0.80 = 27.5$. From Table 47 choose a 2-in (5.08-cm) single-seat valve having a c_v of 36.

The operating c_v of any valve is $c_{vo} = c_{vf}/c_{vs}$, where c_{vf} = c_v value computed by the formula in step 3 and c_{vs} = c_v of actual valve selected. Or, for this valve, $c_{vo} = 22/36 = 0.61$.

TABLE 47 Flow Coefficients for Steam-Control Valves*

Size		Straight-through throttling		Straight-through on-off	Straight-through regulators	
		Single seat	Double seat	Single seat	Single seat	Double seat
in	cm					
⅛	0.32	0.23				
¼	0.64	0.78				
⅜	0.95	1.7				
½	1.27	3.2				
¾	1.91	5.4	7.2	7	3.6	4.3
1	2.54	9	12	12	6	7.2
1¼	3.18	14	18	18	9	10.8
1½	3.81	21	28	27	14	16.8
2	5.08	36	48	42	24	28.8
2½	6.35	54	72	65	36	43.2
3	7.62	75	100	93	50	60
4	10.2	124	165	170	83	99
6	15.2	270	360	380	180	216
8	20.3	480	640	660	320	384
10	25.4	750	1000	1100	500	600
12	30.5	1080	1440	1550	720	864

Chemical Engineering.

To avoid wire drawing which occurs when the valve plug operates too close to the valve seat, c_{vo} values of less than 0.10 should not be used. Since $c_{vo} = 0.61$ for this valve, wire drawing will not occur.

Related Calculations.   To speed up the determination of c_v, Fig. 88 can be used instead of the formula in step 3. This is a performance-tested chart valid for steam control valves for blast-heating coils, tank heaters, pressure-reducing stations, and any other installations—stationary, mobile, or marine—where steam flow and pressure are to be regulated. The approach can also be used for valves handling gases other than steam.

The valve coefficient c_v is conventional; it equals the gallons per minute (liters per second) of clear cold water at 60°F (15.6°C) that will pass through the flow restriction (valve or orifice) while undergoing a pressure drop of 1 lb/in² (7.0 kPa). The c_v value is the same for liquids, gases, and steam. Tables listing c_v values versus valve size and type are published by the various valve manufacturers. General c_v values not limited to any manufacturer are given in Table 47 for a variety of valve types and sizes.

Note in Fig. 88 the relations for the density of the steam of various valve outlet pressures. In these relations P_c = critical pressure, lb/in² (abs) (kPa), as defined earlier. The solution given in Fig. 88 is for a flow rate of 200 lb/h (0.025 kg/s) of steam having a density of 0.08 lb/ft³ (1.25 kg/m³) at the valve outlet with a 10-lb/in² (abs) (68.9-kPa) pressure drop through the valve, giving a c_v of 3.8.

This procedure is the work of John D. Constance, P.E., as reported in *Chemical Engineering* magazine.

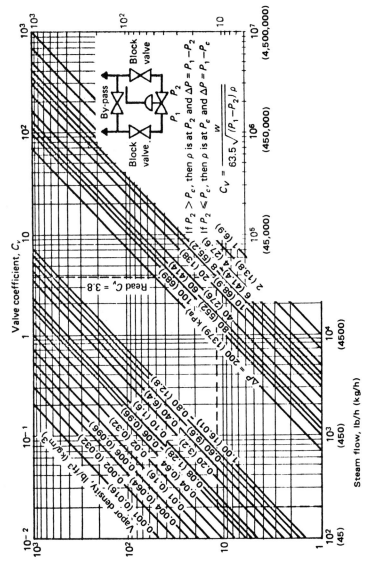

FIGURE 88 c_v valves for steam-control and pressure-reducing valves. (*Chemical Engineering.*)

BOOLEAN ALGEBRA FOR CONTROL SYSTEMS

We shall develop the basic concepts and laws of boolean algebra by relating them directly to a set of tangible objects. In this system of algebra, we are concerned with the *state* of an object, and each object has two possible states: open or closed, horizontal or vertical, red or green, etc. These objects, which are termed *components,* are arranged to form a *system,* and the system itself has two possible states.

The states of the components and of the system are expressed by use of a code consisting of the numerals 0 and 1. For example, 0 can denote that a component is open and 1 that it is closed. The expression $A = 0$ states that component A is at the state represented by 0. For brevity, we say that A has the *value* 0, although we are expressing a state of being rather than a numerical value. When referring to the system, the numeral 1 represents the required state of the system; when referring to a component, 1 represents the state of the component that enables the system to attain its required state.

Assume that a system consists of two components: A and B. The state of the system is determined by the states of A and B and by the manner in which these components are arranged. Two arrangements are possible. Under the first arrangement, the system is 1 if *either A or B* is 1 (or if both are 1); under the second arrangement, the system is 1 if and only if *both A and B* are 1. These arrangements of the components are referred to as the OR and AND relationships, respectively.

As an illustration, refer to Fig. 89a, where components A and B are arranged in parallel. Assume that we are currently at point m and wish to reach point n. To do this, we must pass through either A or B. Now assume that each component is either passable or impassable. The system itself is passable if it allows movement from m to n, and it is impassable if such is not the case. Manifestly, the system is passable if either A or B is passable (or if both are passable). Now refer to Fig. 89b, where

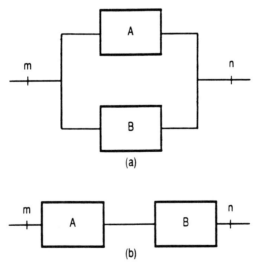

FIGURE 89   Passage through system. (*a*) Components in parallel; (*b*) components in series.

components A and B are arranged in series. To move from m to n, we must pass through both A and B. Under this arrangement, the system is passable only if both A and B are passable. In summary, the OR relationship corresponds to the parallel arrangement of the components, and the AND relationship corresponds to the series arrangement. This concept of movement across the system provides a simple means of arriving at the laws of boolean algebra.

The OR and AND relationships are denoted in this manner: The expression $A + B$ (read "A or B") signifies that the system is 1 if either A or B is 1 (or if both are 1); the expression AB (read "A and B") signifies that the system is 1 if both A and B are 1. However, it is to be emphasized that these expressions have nothing to do with addition and multiplication in ordinary algebra, despite the similarity of notation.

A system consisting of three or more components may be considered to contain *subsystems,* and the composition of a subsystem can be described by the use of parentheses. As an illustration, refer to Fig. 90. This system contains two subsystems. One consists of A and B in parallel, and the other consists of C and D in parallel. The two subsystems are arranged in series. Thus, to move from m to n, we must first pass through either A or B, and then through either C or D. The expression for this system is $(A + B)(C + D)$.

A component is described as a *variable* or a *constant* according to whether its state is alterable or unalterable, respectively. The expression $A0$ describes a system in which A is in series with a component that is always at state 0. Similarly, the expression $A + 1$ describes a system in which A is in parallel with a component that is always at state 1. Two or more components are said to be *identical* if they are always at the same state, and they are represented by an identical symbol. For example, the expression AA describes a system in which two identical components are arranged in series.

Equivalence:

Two systems are *equivalent* to each other if they are always at the same state, and equivalence is expressed by use of an equals sign. Thus, the statement $AA = A$ means that a system consisting of two components A in series is equivalent to a system having a single component A. A statement of equivalence is called an *equation.* The following equations are self-evident:

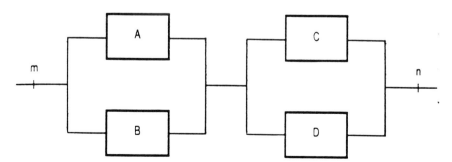

FIGURE 90   System consisting of subsystems arranged in series.

$$A + 0 = A \tag{1}$$

$$A + 1 = 1 \tag{2}$$

$$A + A = A \tag{3}$$

$$A0 = 0 \tag{4}$$

$$A1 = A \tag{5}$$

$$AA = A \tag{6}$$

$$A + B = B + A \tag{7}$$

$$AB = BA \tag{8}$$

Equations 7 and 8 state that components A and B are *commutative*.

If three components are arranged in parallel or in series, any two may be considered to form a subsystem. Therefore, the components are *associative*. Expressed symbolically,

$$(A + B) + C = A + (B + C) \tag{9}$$

$$(AB)C = A(BC) \tag{10}$$

Consider the expression $A(B + C)$, which describes the system in Fig. 91a. In moving from m to n, we have two alternative paths: A and B, and A and C. Therefore, the system is passable if both A and B are passable or if both A and C are passable. It follows that the system in Fig. 91a is equivalent to that in Fig. 91b, and we have

$$A(B + C) = AB + AC \tag{11}$$

Therefore, in the expression at the left, A is *distributive*. Reversing the procedure, we see that it is possible to *factor* the repeating variable A in the expression at the right, thereby obtaining the expression at the left.

Equation 11 is extendible. Consider the expression $(A + B)(C + D)$, which describes the system in Fig. 90. In moving from m to n, we have found alternative paths: A and C, A and D, B and C, and B and D. Therefore,

$$(A + B)(C + D) = AC + AD + BC + BD$$

Generalized Terminology:

Having developed the basic laws of boolean algebra by use of a specific application, we can now modify our terminology to make it more general. In our new terminology, a component becomes a constant or a variable, a system becomes a function, and a subsystem becomes a term in an expression. Thus, Eq. 11 now acquires the following meaning: The expression $A(B + C)$ is equivalent to the expression $AB + AC$.

A function is denoted by the letter f. The statement $f = AB + CD$ means that f assumes the value 1 if either of these conditions exists: A and B are both 1; C and D are both 1. Thus, the expression at the right is a set of specifications; it lists the requirements that must be satisfied for f to become 1. In the following material

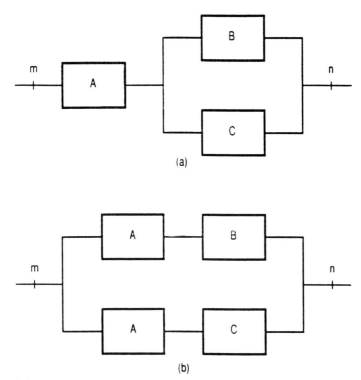

FIGURE 91 Equivalent systems. (*a*) System for $A(B + C)$; (*b*) system for $AB + AC$.

the word *expression* refers to the expression for a function, and the value of the expression is the value of the function itself.

Complementary Variables and Expressions:

The symbol $\overline{A}$ (read "not A") denotes a variable that is always at the state different from that of A. Thus, if $A = 0$, $\overline{A} = 1$, and vice versa. (The symbol A' is sometimes used to denote "not A".) It follows at once that

$$A + \overline{A} = 1 \tag{12}$$

For this reason, A and $\overline{A}$ are said to be *complementary* to each other. Similarly, we have

$$A\overline{A} = 0 \tag{13}$$

The symbol $\overline{\overline{A}}$ denotes the complement of $\overline{A}$. Then

$$\overline{\overline{A}} = A \tag{14}$$

We shall consider A to be the independent variable and $\overline{A}$ the dependent variable.

The NOT notation also applies to expressions as well as single variables. For example, the expression $\overline{AB}$ is always at the state different from that of AB. Thus, if $AB = 0$, $\overline{AB} = 1$, and vice versa. The expressions AB and $\overline{AB}$ are also said to be complementary to each other.

The conditions that make a given expression 1 are those that make its complementary expression 0. By establishing the conditions that make $AB = 0$, we arrive at this result:

$$\overline{AB} = \overline{A} + \overline{B} \tag{15}$$

By extension,

$$\overline{ABC} = \overline{A} + \overline{B} + \overline{C}$$

Similarly,

$$\overline{A + B} = \overline{A}\,\overline{B} \tag{16}$$

By extension,

$$\overline{A + B + C} = \overline{A}\,\overline{B}\,\overline{C}$$

Equations 15 and 16 are referred to as *De Morgan's laws.* We also have the following:

$$\overline{AB + AC + BC} = \overline{A}\,\overline{B} + \overline{A}\,\overline{C} + \overline{B}\,\overline{C} \tag{17}$$

$$\overline{AB + AC + BC} = (\overline{A} + \overline{B})(\overline{A} + \overline{C})(\overline{B} + \overline{C})$$

and $\tag{18}$

Equations 17 and 18 stem from the fact that $AB + AC + BC = 0$ if any two of the three variables are 0.

An expression may contain both a given variable and its complementary variable. The expression $A + \overline{A}B$ is an illustration. This expression is 1 if either of the following conditions exists: $A = 1$; $A = 0$ and $B = 1$. Thus, it is simply necessary that either A or B be 1, and

$$A + \overline{A}B = A + B \tag{19}$$

The following equations can be derived by simple logic or by applying the preceding equations:

$$A + AB = A \tag{20}$$

$$A(A + B) = A \tag{21}$$

$$(A + B)(A + C) = A + BC \tag{22}$$

$$A(A + B) = AB \tag{23}$$

$$AB + A\overline{B} = A \tag{24}$$

The terms NOR and NAND are contractions for "Not OR" and "Not AND," respectively. Thus, $\overline{A + B + C}$ is a NOR term, and $\overline{ABC}$ is a NAND term.

The EXCLUSIVE OR Relationship:

The expression $A + B$ is 1 if either A or B is 1, or if both are 1. However, in some applications of boolean algebra it is necessary to impose a more stringent requirement: The expression is to be 1 if *one and only one* variable is 1. This requirement is referred to as the EXCLUSIVE OR relationship, and it is denoted by the symbol $A \oplus B$.

Since the expression $A \oplus B$ is 1 if one variable is 1 and the other is 0, it follows that

$$A \oplus B = (A + B)(\overline{A} + \overline{B}) \tag{25}$$

$$A \oplus B = A\overline{B} + \overline{A}B$$

and $\tag{26}$

We also have the following:

$$A \oplus 0 = A \tag{27}$$

$$A \oplus 1 = \overline{A} \tag{28}$$

$$A \oplus A = 0 \tag{29}$$

Consider the expression $\overline{A \oplus B}$. The complementary expression $A \oplus B$ is 0 if both variables are 0, or if both variables are 1. Therefore,

$$\overline{A \oplus B} = AB + \overline{A}\,\overline{B} \tag{30}$$

and $\overline{A \oplus B} = (A + \overline{B})(\overline{A} + B)$ $\tag{31}$

Inclusion:

A given AND term is said to *include* a longer AND term if the variables that appear in the former also appear in the latter. For example, the term $A\overline{C}D$ includes the term $AB\overline{C}D$ because the latter contains A, $\overline{C}$, and D. The significance of inclusion is this: The term $A\overline{C}D$ requires that $A = 1$, $C = 0$, $D = 1$, and B can be 0 or 1. The term $AB\overline{C}D$ requires that $A = 1$, $C = 0$, $D = 1$, and $B = 0$. Thus, the second requirement is encompassed in the first. It is convenient to extend the definition of inclusion by saying that a given AND term includes itself.

Compatible Terms:

Two AND terms are said to be *compatible* with each other if one term can be obtained from the other by replacing one and only one variable with its complement. For example, the terms $A\overline{B}DG$ and $ABDG$ are compatible because the second term can be obtained from the first by replacing $\overline{B}$ with B.

Compatible terms that are connected by the OR relationship can be combined in accordance with Eq. 24 or an extension of it. For example, let

$$f = \overline{A}BC + \overline{A}\,\overline{B}C$$

The first term states that $f = 1$ if $A = 0$, $B = 1$, $C = 1$; the second term states that $f = 1$ if $A = 0$, $B = 0$, $C = 1$. These terms can be combined to form $f = \overline{A}C$. This condensed equation states that $f = 1$ if $A = 0$ and $C = 1$, and the value of B is irrelevant.

If a given term is incompatible with any other term in the expression, it is called a *prime implicant*.

Standard Expressions:

Assume that f is a function of n independent variables. In the expression for f, a term is said to be in *standard form* if it is either an OR or AND term and each variable (or its complement) appears in the term once and only once. The following are standard terms for a four-variable function: $\overline{A}\overline{B}CD$, $\overline{A}BCD$, $A + \overline{B} + C + \overline{D}$, $\overline{A} + B + C + D$. A standard AND term is also called a *minterm,* and a standard OR term is also called a *maxterm.*

The expression for f is said to be in *standard, elemental,* or *canonical form* if it's composed of standard terms. The following are standard four-variable expressions:

$$\overline{A}\overline{B}C\overline{D} + \overline{A}B\overline{C}D + A\overline{B}\,\overline{C}D \qquad ABCD + A\overline{B}\,\overline{C}D$$

$$(A + \overline{B} + C + D)(\overline{A} + B + C + \overline{D})$$

An AND-to-OR expression for f is one where each term is an AND expression, and these terms are linked by the OR relationship. Thus, $ABC + \overline{A}\overline{B}C$ is an AND-to-OR expression. Similarly, an OR-to-AND expression for f is one where each term is an OR expression, and these terms are linked by the AND relationship. Thus, $(A + \overline{B} + C)(\overline{A} + B + C)$ is an OR-to-AND expression. Standard AND-to-OR expressions are also called *minterm forms* or *disjunctive expressions,* and standard OR-to-AND expressions are also called *maxterm forms* or *conjunctive expressions.*

Let n denote the number of independent variables that are present. Since each standard term contains either a given independent variable or its complement, the number of standard terms that can be formed is 2^n. If we expand our definition of a standard expression to include the degenerate cases where f is a constant, we may say that a standard expression is a combination of m standard terms, where $0 < m \le 2^n$. It follows that the number of possible standard expressions is 2 raised to the 2^n power.

As an illustration, let $n = 2$. The number of possible standard expressions is $2^4 = 16$, and the AND-to-OR forms are recorded in Table 48. This table is constructed by starting with the basic forms $\overline{A}\overline{B}$, $\overline{A}B$, $A\overline{B}$, and AB and then combining them two at a time, three at a time, and four at a time. Since the last expression encompasses all possible terms, it corresponds to the case $f = 1$.

Reduction of Expressions:

A letter that represents a variable is termed a *literal*. In counting literals, we include duplicates. For example, the expression $AB + \overline{A}BCD + CD\overline{E}$ contains nine literals.

TABLE 48   Standard AND-to-OR Expressions (Minterm Forms) for Two-Variable Functions

0	$\overline{A}B + A\overline{B}$
$\overline{A}\,\overline{B}$	$\overline{A}B + AB$
$\overline{A}B$	$A\overline{B} + AB$
$A\overline{B}$	$\overline{A}\,\overline{B} + \overline{A}B + A\overline{B}$
AB	$\overline{A}\,\overline{B} + \overline{A}B + AB$
$\overline{A}\,\overline{B} + \overline{A}B$	$\overline{A}\,\overline{B} + A\overline{B} + AB$
$\overline{A}\,\overline{B} + A\overline{B}$	$\overline{A}B + A\overline{B} + AB$
$\overline{A}\,\overline{B} + AB$	$\overline{A}\,\overline{B} + \overline{A}B + A\overline{B} + AB$

A given expression is said to be *reduced to simpler form* when it is transformed to an equivalent expression with fewer literals. There are three basic methods of reducing an expression to simpler form. They are as follows:

1. *Consolidation of terms:* As previously stated, two compatible terms that are connected by the OR relationship can be combined. For example,

$$A\overline{B}D\overline{G} + ABD\overline{G} = AD\overline{G}(\overline{B} + B) = AD\overline{G}1 = AD\overline{G}$$

Thus, the combined term contains solely the variables that are common to the compatible terms.

2. *Elimination of redundancies:* For example, Eq. 19 reduces the expression $A + \overline{A}B$, which contains three literals, to $A + B$, which contains only two. The variable $\overline{A}$ in the original expression was redundant.

3. *Factoring:* For example, the equation $AB + AC = A(B + C)$ reduces an expression with four literals to an expression with three.

A given term in an AND-to-OR expression may be compatible with several terms in the expression. It is permissible to combine each compatible pair of terms. The justification is that $A + A = A$, and therefore this duplication of terms does not inject any error.

When a given expression has been reduced to the fullest extent possible by consolidating terms and by eliminating redundancies, it is said to be in *optimal form.* If the resulting expression is then reduced by factoring, the original expression is said to be in its *minimal form.* In many applications of boolean algebra, it is necessary to reduce a given expression to its minimal form, and we shall discuss systematic methods of accomplishing this objective.

Laws of Duality:

Boolean algebra is characterized by a duality that imparts an image to each expression and each equation. This duality can be exploited to considerable advantage. There are two laws of duality, as follows:

   Theorem 1. A given expression is transformed to its complementary expression if the OR and AND relationships are interchanged and each variable is replaced with its complementary variable.

This principle stems from Eqs. 15 and 16. As an illustration, let

$$f = A\bar{B}(C + \bar{D}E)$$

Then
$$\bar{f} = \bar{A} + B + \bar{C}(D + \bar{E})$$

The expression for $\bar{f}$ is clearly valid because $f = 0$ if any of these conditions exist: $A = 0$; $B = 1$; $C = 0$, and either $D = 1$ or $E = 0$.

Theorem 2. A given equation is transformed to a corresponding equation if the OR and AND relationships are interchanged and the constants 0 and 1 are interchanged.

The corresponding equation is called the *dual* of the first. Thus, Eq. 5 is the dual of Eq. 1.

ALGEBRAIC PROOF OF AN EQUATION

Applying solely the equations of boolean algebra, prove the following:

$$\bar{A} + A\bar{B}C + \overline{D(\bar{A} + E)} = \bar{A} + \bar{B}C + \bar{D} + \bar{E} \qquad (32)$$

Calculation Procedure:

1. Apply Eq. 19 to the first two terms and Eq. 15 to the third term
Let f denote the expression at the left. The specified equations transform f to the following:

$$f = \bar{A} + \bar{B}C + \bar{D} + \overline{\bar{A} + E}$$

2. Apply Eqs. 16 and 14 in turn
The result is

$$f = \bar{A} + \bar{B}C + \bar{D} + A\bar{E}$$

3. Apply Eq. 19 to the first and fourth terms
The result is

$$f = A + \bar{B}C + \bar{D} + \bar{E}$$

Equation (32) is thus proved.

PROVING AN EQUATION BY IDENTIFYING THE SATISFACTORY CONDITIONS

With reference to the preceding Calculation Procedure, prove Eq. 32 by identifying the conditions that make the expression on each side of the equation equal to 1.

Calculation Procedure:

1. *Identify and record these conditions*
Let $f1$ and $f2$ denote, respectively, the expression on the left side and on the right side of Eq. 32. In Table 49, record the alternative conditions that make $f1 = 1$ and $f2 = 1$. Number the conditions in the manner shown.

2. *Eliminate any redundancies that may exist*
Conditions 2 and 3 are relevant solely if condition 1 does not exist; i.e., if $A = 1$. Therefore, in conditions 2 and 3, the requirement $A = 1$ is redundant and should be eliminated.

3. *Compare the two sets of conditions as they now exist*
With the redundancy eliminated, the conditions that make $f1 = 1$ coincide with those that make $f2 = 1$. Equation 32 is thus proved.

USE OF TRUTH TABLES

Construct a truth table to confirm the following equation, which stems from the first law of duality:

$$\overline{A + \overline{B}C} = \overline{A}(B + \overline{C})$$

Calculation Procedure:

1. *Number the rows and assign values to the variables in the conventional manner*
A *truth table* is used to confirm an equation. In this table, we assign a column to each independent variable and to each term that appears in the equation. We then record every possible combination of values of the independent variables and the corresponding value of a given term and expression. If the expression contains n independent variables, the number of combinations of values is 2^n. In the present case, $n = 3$ and the number of combinations is $2^3 = 8$.

TABLE 49

Conditions for $f1 = 1$		Conditions for $f2 = 1$	
1	$A = 0$	4	$A = 0$
2	$A = 1$ and $B = 0$ and $C = 1$	5	$B = 0$ and $C = 1$
		6	$D = 0$
3 or	$D = 0$ $A = 1$ and $E = 0$	7	$E = 0$

Refer to Table 50. By convention, the rows are numbered consecutively, starting with 0. Values are assigned to the variables in such manner that the digits form the row number in the binary system. For example, since 5 is 101 in binary form, row 5 contains these values: $A = 1$, $B = 0$, $C = 1$.

2. Establish the values of terms $\overline{B}C$, $B + \overline{C}$, $A + \overline{B}C$, and $\overline{A}(B + \overline{C})$
Consider the term $\overline{B}C$. This is 1 if $B = 0$ and $C = 1$. Therefore, record the value 1 in rows 1 and 5 and the value 0 in all other rows. Now consider the term $B + \overline{C}$. This is 1 if $B = 1$ or $C = 0$; conversely, it is 0 if $B = 0$ and $C = 1$. Therefore, record the value 0 in rows 1 and 5 and the value 1 in all other rows. Continue in this manner to obtain the results recorded in Table 50.

3. Compare the values of $A + \overline{B}C$ and of $\overline{A}(B + \overline{C})$
In all instances, these two terms differ in value. Therefore, the given equation is confirmed.

EXPANSION OF NONSTANDARD EXPRESSION TO STANDARD FORM

The expression for f is

$$f = \overline{A}BC + A\overline{B} + B$$

Expand this expression to standard form.

Calculation Procedure:

1. Examine each term of the expression to determine which variables must be added to place the term in standard form
The first term is standard, the second term requires the addition of C (or $\overline{C}$), and the third term requires the addition of A and C (or $\overline{A}$ and $\overline{C}$).

2. Fill in the gaps by applying Eqs. 5 and 12
The result is

$$f = \overline{A}BC + A\overline{B}(C + \overline{C}) + (A + \overline{A})B(C + \overline{C})$$

TABLE 50 Truth Table

Row	A	B	C	$\overline{B}C$	$B + \overline{C}$	$A + \overline{B}C$	$\overline{A}(B + \overline{C})$
0	0	0	0	0	1	0	1
1	0	0	1	1	0	1	0
2	0	1	0	0	1	0	1
3	0	1	1	0	1	0	1
4	1	0	0	0	1	1	0
5	1	0	1	1	0	1	0
6	1	1	0	0	1	1	0
7	1	1	1	0	1	1	0

3. *Remove the parentheses by applying Eq. 11*
The result is

$$f = \overline{A}BC + A\overline{B}C + A\overline{BC} + ABC + AB\overline{C} + \overline{A}BC + \overline{A}B\overline{C}$$

4. *Eliminate duplications*
Since the first and sixth terms in the last expression are identical and $A + A = A$, delete the sixth term. The final expression is

$$f = \overline{A}BC + A\overline{B}C + A\overline{BC} + ABC + AB\overline{C} + \overline{A}B\overline{C}$$

As the previous Calculation Procedure illustrates, every nonstandard AND-to-OR expression can be expanded to standard form. From the principle of duality, it follows that every nonstandard OR-to-AND expression can also be expanded to standard form.

ALGEBRAIC METHOD OF REDUCTION

Applying purely algebraic operations, reduce the following expression to its minimal form:

$$f = A\overline{BC}D + \overline{AB}CD + \overline{ABC}D + AB\overline{C}D + \overline{ABCD} + A\overline{BCD} + \overline{A}BCD \qquad (a)$$

Calculation Procedure:

1. *Perform cycle 1 in the consolidation of terms*
We shall follow the formalized algebraic procedure for reducing a standard AND-to-OR expression to optimal form that was developed by W. V. Quine. The given expression is reduced by combining compatible terms, proceeding in cycles until all possibilities have been exhausted.

Number the terms in (a) from left to right, and record them in columnar form in Table 51. Starting with term 1, compare it with each subsequent term for compatibility. Term 1 is compatible with both term 3 and term 6. Identify each compatible pair of terms in Table 51 by placing the mark X in a unique column and on the same row as the term. Thus, the first column has an X on rows 1 and 3, and the second column has an X on rows 1 and 6. Now take term 2; it is compatible with both term 3 and term 7. Thus, the third column has an X on rows 2 and 3,

TABLE 51 Identity of Compatible Pairs of Terms

		Cycle 1								Cycle 2		
1	$A\overline{BC}D$	X	X						1	$\overline{BC}D$	X	
2	$\overline{AB}CD$			X	X				2	$\overline{AB}C$		X
3	$\overline{ABC}D$	X		X		X			3	$\overline{AC}D$		
4	$AB\overline{C}D$								4	$\overline{AB}D$		
5	$\overline{ABCD}$					X	X		5	$\overline{AB}C$		X
6	$A\overline{BCD}$		X				X		6	$\overline{BC}D$	X	
7	$\overline{A}BCD$				X							

and the fourth column has an X on rows 2 and 7. Similarly, term 3 is compatible with term 5, and term 5 is compatible with term 6. Term 4 is a prime implicant, and it is carried into the subsequent expression for f.

In Table 52, record the numbers of the terms that are compatible and the combined term they form. For example, terms 1 and 3 form $\overline{B}CD$, and terms 1 and 6 form $A\overline{B}C$. At the end of cycle 1, the expression for f has been reduced to

$$f = ABC\overline{D} + \overline{B}CD + A\overline{B}C + \overline{A}CD + \overline{A}BD + \overline{A}BC + \overline{B}CD \qquad (b)$$

However, the expression in (b) also contains compatible terms; therefore, a second cycle of consolidation is needed.

2. Perform cycle 2 in the consolidation of terms
Number all terms in (b) beyond the first and record them in Table 51. Again investigate for compatibility. Thus, term 1 is compatible with term 6, term 2 is compatible with term 5, and terms 3 and 4 are prime implicants. Complete Table 52. This table shows that both compatible pairs form $\overline{B}C$. Since $A + A = A$, take $\overline{B}C$ only once. At the end of cycle 2, the expression for f has been reduced to the following:

$$f = ABC\overline{D} + \overline{A}CD + \overline{A}BD + \overline{B}C \qquad (c)$$

The terms in (c) are incompatible with one another, and the consolidation process is now complete.

3. Examine the expression in (c) for redundancies and eliminate any that may exist
Ascertain how the terms in Eq. A are included in those in Eq. c. In Table 53, again record the terms in Eq. a in columnar form at the left, and record the terms in Eq. c in a horizontal row across the top. Take each term in Eq. c and compare it with each term in Eq. a with respect to inclusion. Thus, $ABC\overline{D}$ includes only term 4. Indicate this condition by placing the mark X at the indicated location. Also, $\overline{A}CD$ includes terms 2 and 3, $\overline{A}BD$ includes terms 2 and 7, etc.

Now proceed to construct the optimal expression for f. Each term in the original expression must be included at least once in the optimal expression. Therefore, the optimal expression must contain $ABC\overline{D}$ because that is the only term in Eq. c that includes term 4 in Eq. a. Similarly, the optimal expression must contain $\overline{B}C$ because

TABLE 52

Cycle	Compatible terms	Combined term
1	1 and 3	$\overline{B}CD$
	1 and 6	$A\overline{B}C$
	2 and 3	$\overline{A}CD$
	2 and 7	$\overline{A}BD$
	3 and 5	$\overline{A}BC$
	5 and 6	$\overline{B}CD$
2	1 and 6	$\overline{B}C$
	2 and 5	$\overline{B}C$

TABLE 53

Terms in Eq. a	Terms in Eq. c			
	$ABC\overline{D}$	$\overline{A}C D$	$\overline{A}BD$	$\overline{BC}$
1 $A\overline{B}C D$				X
2 $\overline{A}B C D$		X	X	
3 $\overline{A}BCD$		X		X
4 $ABC\overline{D}$	X			
5 $A\overline{BC}D$				X
6 $A\overline{B}CD$				X
7 $\overline{A}BC D$			X	

that is the only term in Eq. c that includes terms 1, 5, and 6 in Eq. a. Finally, the optimal expression must contain $\overline{A}BD$ because that is the only term in Eq. c that contains term 7 in Eq. a. Term $\overline{A}CD$ is redundant because terms 2 and 3 are already embodied in the optimal expression. Therefore, the optimal expression is

$$f = ABC\overline{D} + \overline{BC} + \overline{A}BD$$

(This expression will be obtained in a subsequent Calculation Procedure by a graphical method.)

4. Factor the optimal expression to obtain the minimal expression
The result is

$$f = B(AC\overline{D} + \overline{A}D) + \overline{BC}$$

CONSTRUCTION OF KARNAUGH MAP FOR THREE-VARIABLE EXPRESSION

Construct the Karnaugh map for the expression

$$f = ABC + A\overline{BC} + \overline{A}BC$$

Calculation Procedure:

1. Establish the size of the array and assign a cell to each possible term
A *Karnaugh map* (or *truth map*) affords a graphical method of reducing a given boolean AND-to-OR expression to optimal form through the combination of compatible terms. The map is a rectangular array of squares or *cells*. Each cell corresponds to a specific standard AND term, and a cell is provided for each possible term. Therefore, if n denotes the number of independent variables, the number of cells requires is 2^n. In the present case, $n = 3$, and the number of cells is 8.

The basic requirement of a Karnaugh map is this: As we move from one row or column to an adjacent row or column, only one variable can change. In the Karnaugh map, the horizontal rows are numbered from the top down, and the vertical columns are numbered from left to right, starting with the number 1 in both cases. A cell is identified by specifying its row and column, in that order. For example, cell 25 (read "two five") lies in row 2 and column 5.

Refer to Fig. 92, which is an array consisting of 2 rows and 4 columns. Each variable, both independent and dependent, must be represented by at least 1 row or column. Make the following assignments: row 1 to A; row 2 to $\bar{A}$; columns 1 and 2 to B; columns 3 and 4 to $\bar{B}$; columns 2 and 3 to C; columns 1 and 4 to $\bar{C}$. Record these assignments by placing the appropriate labels along the periphery of the array, as shown. Visualize that the map is inscribed on a torus rather than a plane, the result being that it is circular in both directions. Thus, columns 1 and 4 are adjacent to each other.

The term that corresponds to a given cell is found by taking the row and column designations of that cell. For example, in Fig. 92, cell 11 corresponds to ABC, cell 21 corresponds to $\bar{A}BC$, and cell 23 corresponds to $\bar{A}B\bar{C}$. Moreover, the manner in which rows and columns have been assigned satisfies the basic requirement of a Karnaugh map.

In a Karnaugh map, adjacent cells correspond to compatible terms. For example, cells 12 and 22 correspond, respectively, to ABC and $\bar{A}BC$, and these terms are compatible. Similarly, cells 21 and 24 (which are adjacent) correspond, respectively, to $\bar{A}BC$ and $\bar{A}B\bar{C}$, and these terms are compatible.

2. Complete the Karnaugh map by inserting 1's
The terms in the expression $ABC + A\bar{B}\bar{C} + \bar{A}BC$ are represented by cells 12, 14, and 22 in Fig. 92. Therefore, to record the expression in the map, place the numeral 1 in each of these cells. The cells marked 1 are referred to as the *p cells.*

CONSTRUCTION OF KARNAUGH MAP FOR FOUR-VARIABLE EXPRESSION

Construct the Karnaugh map for the expression

$$f = \overline{ABCD} + \overline{AB}\overline{CD} + ABCD + A\overline{BCD} + \overline{ABCD}$$

Calculation Procedure:

1. Establish the size of the array and assign a cell to each possible term
Since there are now 4 independent variables, the number of cells is $2^4 = 16$. Refer to Fig. 93. Again, visualize the map to be circular in both directions. As a result,

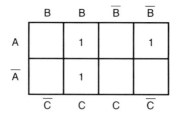

FIGURE 92 Karnaugh map for three-variable expression.

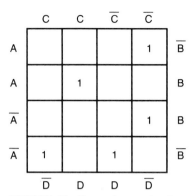

FIGURE 93 Karnaugh map for four-variable expression.

rows 1 and 4 are adjacent, and columns 1 and 4 are adjacent. Assign the rows and columns in the manner shown.

2. Complete the Karnaugh map by inserting 1's
The terms in the given expression are represented by cells 41, 34, 22, 14, and 43. Therefore, to record the given expression in the map, place the numeral 1 in each of these cells.

CONSTRUCTION OF KARNAUGH MAP FOR FIVE-VARIABLE EXPRESSION

Construct the Karnaugh map for the expression

$$f = \overline{AB}CD\overline{E} + AB\overline{C}DE + A\overline{BCDE}$$

Calculation Procedure:

1. Establish the size of the array and assign a cell to each possible term
Since there are 5 independent variables, the number of cells is $2^5 = 32$. Refer to Fig. 94, where the array consists of 8 rows and 4 columns. Since the map is considered to be circular in both directions, rows 1 and 8 are adjacent, and columns 1 and 4 are adjacent. Assign the rows and columns in the manner shown, and visualize that the subarray for $\overline{E}$ is obtained by revolving the subarray for E about line a. The basic requirement is satisfied. For example, when we move from row 4 to row 5, the sole change is from E to $\overline{E}$. Similarly, when we move from row 8 to row 1, the sole change is from $\overline{E}$ to E.

2. Complete the Karnaugh map by inserting 1's
The terms in the given expression are represented by cells 52, 23, and 84. Therefore, place the numeral 1 in each of these cells.

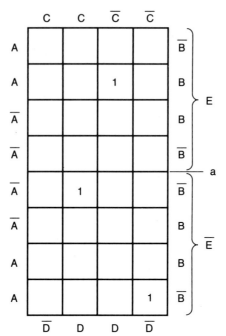

FIGURE 94 Karnaugh map for five-variable expression.

REDUCTION OF EXPRESSION TO OPTIMAL FORM BY USE OF KARNAUGH MAP

Reduce the following expression to optimal form by constructing a Karnaugh map:

$$f = ABC\overline{D} + ABCD + \overline{A}BC\overline{D} + \overline{A}BCD$$

Then verify the result by reducing the given expression algebraically.

Calculation Procedure:

1. Construct the Karnaugh map

Assume that a Karnaugh map contains 2^m adjacent p cells, where m is a positive integer. These cells are said to form an *mth-order block,* and the extent of the block is shown by enclosing it with light lines. As we shall demonstrate, the terms that correspond to the cells in the block can be combined into a single term that contains solely the variables within which the block is confined.

In general, if the given expression contains n independent variables and the block is of the mth order, the combined term corresponding to this block contains $n - m$ variables. Therefore, in forming blocks, the guiding principle is to make the blocks as large as possible. It is convenient to view an isolated p cell as a zero-order block, the justification being that $2^0 = 1$.

The given expression is mapped in Fig. 95.

2.   *Establish the optimal expression*
In Fig. 95, the p cells form a second-order block, and this block is confined within the rows for B and the columns for C. Therefore, the given expression reduces to $f = BC$. This condensed equation signifies that the values assumed by A and D are irrelevant.

3.   *Verify the result*
To reduce the given expression algebraically, factor recurrently and then apply Eq. 12, in this manner: Since each term in the given expression contains B and C, factor these variables in the first step. Then

$$f = BC(A\overline{D} + AD + \overline{A}D + \overline{A}D)$$
$$= BC[A(\overline{D} + D) + \overline{A}(\overline{D} + D)]$$
$$= BC(A + \overline{A}) = BC$$

USE OF KARNAUGH MAP WITH DISTINCTIVE BLOCKS

Reduce the following expression to optimal form by constructing a Karnaugh map:

$$f = A\overline{B}\overline{C}D + \overline{A}\overline{B}CD + \overline{A}BC\overline{D} + ABC\overline{D} + \overline{A}\overline{B}CD + A\overline{B}C\overline{D} + \overline{A}BCD$$

(This expression was reduced algebraically in an earlier Calculation Procedure.)

Calculation Procedure:

1.   *Construct the Karnaugh map*
The given expression is mapped in Fig. 96.

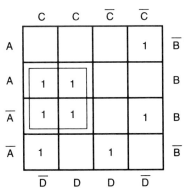

FIGURE 95   Karnaugh map with second-order block.

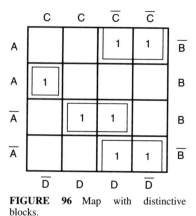

FIGURE 96 Map with distinctive blocks.

2. Establish the optimal expression

Since rows 1 and 4 are adjacent, cells 13, 14, 43, and 44 form a second-order block, and their terms combine to form $\overline{B}C$. Cells 32 and 33 form a first-order block, and their terms combine to form $\overline{A}BD$. Cell 21 is a zero-order block, and its term is $ABCD$. At this point, all p cells have been taken into account, and the optimal expression is complete. That cells 33 and 43 constitute a block is not relevant because both these cells have already been taken into account. Therefore, carrying the term $\overline{A}CD$ into the reduced expression would be redundant. Thus, the optimal expression is

$$f = \overline{B}C + \overline{A}BD + ABC\overline{D}$$

This result agrees with that obtained algebraically in the earlier Calculation Procedure.

USE OF KARNAUGH MAP WITH OVERLAPPING BLOCKS

A function f has the expression that is mapped in Fig. 97. Establish the optimal form of the expression.

Calculation Procedure:

1. Form blocks of the largest possible size

In forming blocks, it is permissible to include a given cell in more than one block. As an illustration, consider the expression

$$f = \overline{A}BC + \overline{A}\overline{B}C + A\overline{B}C$$

The first and second terms combine to form $\overline{A}B$, and the second and third terms combine to form $\overline{B}C$. Then

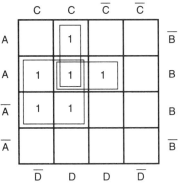

FIGURE 97 Map with overlapping blocks.

$$f = \overline{AB} + \overline{B}C$$

Thus, $f = 1$ if either of these conditions exist: $A = 0$, $B = 0$, and C can have either value; $B = 0$, $C = 1$, and A can have either value.

In the present case, form the second-order block and the two first-order blocks shown.

2. Formulate the optimal expression
The second-order block corresponds to BC, and the first-order blocks correspond to ACD and ABD. Therefore, the optimal expression is

$$f = BC + ACD + ABD$$

USE OF LARGE BLOCKS IN A KARNAUGH MAP

A function f has the expression that is mapped in Fig. 98. Establish the minimal form of the expression.

Calculation Procedure:

1. Form blocks in the most effective manner
It is possible to combine the p cells to form one second-order block and two first-order blocks. The resulting expression would have $2 + 3 + 3 = 8$ literals. On the other hand, by using overlapping blocks, it is possible to form second-order blocks exclusively, in the manner shown. The resulting expression would have $2 \times 3 = 6$ literals. Therefore, choose the second method.

2. Formulate the optimal expression
The result is

$$f = AD + BC + BD$$

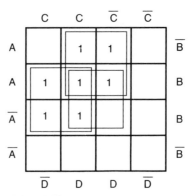

FIGURE 98 Map with overlapping second-order blocks.

3. Formulate the minimal expression by factoring

We can factor either B or D, and the following alternative forms result:

$$f = AD + B(C + D) \qquad f = D(A + B) + BC$$

USE OF KARNAUGH MAP WITH AN INCOMPLETE BLOCK

A function f has the expression that is mapped in Fig. 99. Establish the minimal form of the expression.

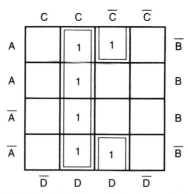

FIGURE 99 Map with an incomplete third-order block.

Calculation Procedure:

1. Form the second-order and first-order blocks shown, and obtain the optimal expression

The expression is

$$f = CD + \overline{BC}D$$

2. Obtain the minimal expression

Factor the foregoing expression and then apply Eq. 19. The result is

$$f = D(C + \overline{BC}) = D(\overline{B} + C)$$

The last expression is the minimal one.

3. Alternatively, obtain the minimal expression by using an incomplete block

Tentatively transform cells 23 and 33 to p cells, thereby forming a third-order block having D as its expression. Cells 23 and 33 form a first-order block having BCD as its term. Therefore, impose the additional requirement that BCD be 0. Then

$$f = D(\overline{BCD})$$

Now apply Eqs. 15 and 13 to obtain

$$f = D(\overline{B} + C + \overline{D}) = D(\overline{B} + C)$$

Thus, use of an incomplete block yields the minimal expression in a more direct manner.

USE OF IRRELEVANT (DON'T CARE) TERMS

The variables A, B, and C determine a function f. Table 54 exhibits the alternative sets of values that make $f = 1$ and those that make $f = 0$. The two sets of values not shown in the table have no effect on the value of f. Formulate the minimal expression for f (a) without using irrelevant terms and (b) by using these terms.

TABLE 54

Condition	A	B	C
$f = 1$	1	1	0
	1	0	1
	0	1	0
$f = 0$	1	1	1
	1	0	0
	0	1	1

Calculation Procedure:

1. Write the expression for f as given by Table 54
The table states that $f = 1$ if any of these conditions exist: $AB\overline{C} = 1$, $A\overline{B}C = 1$, $\overline{A}B\overline{C} = 1$. Therefore,

$$f = AB\overline{C} + A\overline{B}C + \overline{A}B\overline{C}$$

2. Construct the Karnaugh map and place the appropriate mark in each cell
Figure 100 is the Karnaugh map. As before, place the numeral 1 in the cells corresponding to the terms that make $f = 1$. In addition, place the numeral 0 in the cells corresponding to the terms that make $f = 0$. Now place an X in the two remaining cells. Where certain combinations of values cannot occur in practice or have no effect on the value of the function, the corresponding terms are called *irrelevant* or *don't-care terms*. Thus, the X's in the Karnaugh map identify the irrelevant terms.

3. Formulate the optimal expression by combining solely the p cells
Combine the p cells in the manner shown to obtain

$$f = A\overline{B}C + B\overline{C}$$

4. Now formulate the optimal expression by incorporating the irrelevant terms
Where irrelevant terms are present, it is permissible to enlarge the original expression for the function to include these terms if doing so yields a briefer final expression for the function. Since irrelevant terms cannot become 1 or their values are of no consequence, their inclusion in the expression does not inject any error. In Fig. 100, combine cell 23 with cell 13. We now have two first-order cells, and the resulting expression is

$$f = \overline{B}C + B\overline{C}$$

This expression is preferable to that in step 3 because it contains four literals instead of five.

ALTERNATIVE OPTIMAL EXPRESSIONS RESULTING FROM USE OF IRRELEVANT TERMS

A function f has the Karnaugh map in Fig. 101. Formulate the optimal expression for f.

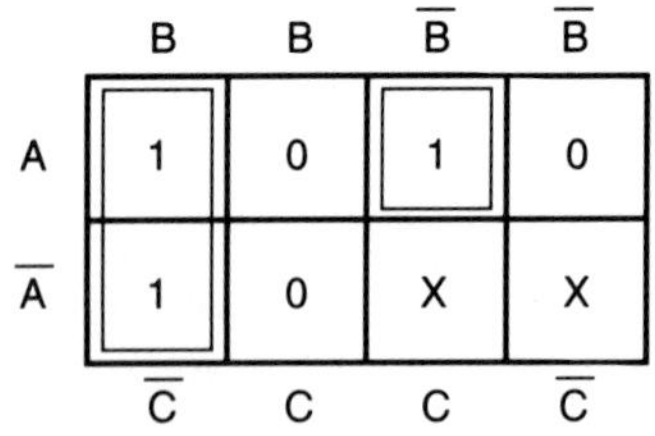

FIGURE 100 Method of identifying irrelevant terms on a map.

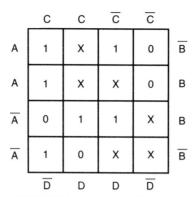

FIGURE 101   Formation of blocks containing irrelevant terms.

Calculation Procedure:

1. *Combine the terms in a suitable manner and write the corresponding expression*
Use the irrelevant terms to obtain the blocks shown in Table 55. The corresponding expression is

$$f = AC + AD + BD + \overline{B}\,C\overline{D}$$

2. *Combine the terms in an alternative manner and write the corresponding expression*
Now combine cell 41 with cell 44 rather than cell 11. The expression now becomes

$$f = AC + AD + BD + \overline{A}B\overline{D}$$

As the foregoing Calculation Procedure demonstrates, there may be alternative ways of forming blocks in the Karnaugh map where irrelevant terms are present. If such is the case, the problem of formulating the optimal expression for the function is ambiguous.

EVALUATING REPOWERING OPTIONS AS POWER-PLANT CAPACITY-ADDITION STRATEGIES

Evaluate repowering strategies for an old (20 years +), but still serviceable, steam-cycle electric-generating station to improve significantly its efficiency and/or expand capacity, while maintaining a more favorable environmental profile. Consider the various options available for existing coal-fired and gas-fired stations. Choose a suitable system for preventing damage to a gas turbine from ice ingestion.

TABLE 55

Cells in block	Term
11, 12, 21, 22	AC
12, 13, 22, 23	AD
22, 23, 32, 33	$\overline{BD}$
11, 41	$\overline{BCD}$

Calculation Procedure:

1. Outline the types of repowering options available

The types of repowering options considered include (a) *partial repowering*—replacing a boiler with a heat-recovery (HRSG) coupled to a gas turbine; (b) *station repowering*—use of existing infrastructure but not the original steam cycle; (c) *site repowering*—reusing an existing site but none of the original equipment.

While repowering with natural gas and gas turbines gets attention today, other options are worth considering. New schemes use coal gasification, Diesel engines, district heating, ultra-high-temperature steam-turbine modules, and adding a steam bottoming cycle to an existing gas turbine.

Integrated resource plans (IRP) and the creation of exempt wholesale generators (EWG) by the National Energy Policy Act of 1992 are often seen as repowering-friendly. Using existing assets and an existing site generally evaluates more favorably, based on raw economics, than a new site with all-new equipment, transmission or utility tie-in, fuel supply, etc.

An EWG operating an efficient, repowered site in an open electric-supply market may have inherent cost advantages over a new-plant EWG. Converting a site into an EWG could involve (1) a utility and non-utility generator (NUG) working together as, say, developer and operator of the asset; (2) the outright sale of the generating asset to an NUG for a quick cash infusion to the utility; or (3) the utility challenging the NUGs for the least-cost option in an IRP.

3. Give useful rules of thumb for repowering

The first rule of thumb for repowering is: *The more an existing steam cycle is relied on, the more difficult the repowering will be.* Reasons why this rule of thumb exists are given below.

The basic challenge in repowering is to match the old steam-cycle parameters to the new (repowered) part of the plant. In the case of gas-turbine/HRSG repowerings, the thermal output of the HRSG must be matched to the existing steam-turbine generator. Steam turbines are usually custom-designed for the desired output. Gas turbines, by comparison, come in discrete sizes, each characterized by a specific exhaust energy flow available to generate steam. How well this steam flow, or combination of steam flows, matches the present-day characteristics of the steam turbine determines the efficiency and output of the repowered unit.

Figure 102 shows two different approaches for repowering steam-turbine plants with gas turbines—(a) use of a single-pressure HRSG, and (b) use of a triple-pressure HRSG.

Ways of getting around the discrete size limitation of gas turbines are: (1) use of supplementary firing of the HRSG; (2) inject steam into the gas turbine if excess steam is available; (3) use evaporative coolers and/or inlet-air chillers to augment

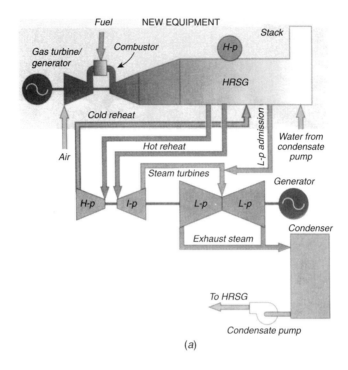

(a)

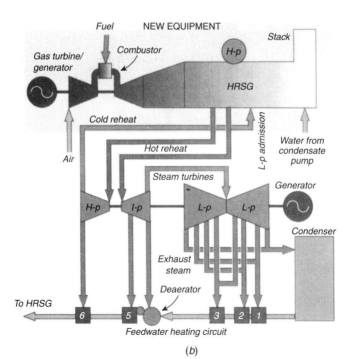

(b)

FIGURE 102   (a) Single-pressure HRSG in repowering. (b) Triple-pressure HRSG in repowering. (*Power*).

the power available from the gas turbine, depending on ambient temperatures. All add to the cost and complexity of the retrofit and influence the ultime unit heat rate. Steam or water injection into the gas turbine, for example, imposes higher maintenance costs on the gas turbine and greater water-treatment needs. So the second rule of thumb is: *Avoiding discrete gas-turbine-size constraints can lead to higher capital and maintenance costs.*

The third rule of thumb for repowering is: *The heat-rejection needs of the new cycle must match the capacity of the existing system.* If the heat-rejection needs do not meet, then extensive upgrades of the condenser and cooling towers may be required.

Two other options for repowering are: (a) the hot windbox, Fig. 103a and (b) feedwater heating repowering, Fig. 103a. In the hot-windbox option, Fig. 103b a high-temperature duct must take gas-turbine exhaust to the boiler. Addition of the high-temperature O_2-rich stream poses substantial changes to the temperature and flow profiles in the boiler. Virtually the entire boiler must be reevaluated for its ability to handle the changes. These changes dovetail into the steam cycle, also. Extensive duct, air heater, and economizer changes may be needed to make this option viable.

The feedwater-heating option, Fig. 103b, directs the gas-turbine exhaust energy to the existing feedwater heating circuit—displacing steam available for expansion through the full length of the steam turbine. The more feedwater heating steam displaced, the more efficient the repowered cycle. This cycle is often constrained by technical or environmental limitations in the heat-rejection circuit.

Additional rules of thumb growing out of these findings are: *Matching the old steam cycle parameters to the new, repowered part of the plant, is the most basic challenge in repowering designs.* And, *Discrete-size gas-turbine matching of steam flows with the existing steam turbine(s) determines the efficiency and output of the repowered unit.* Further, *Old steam-turbine-generators post greater design challenges and costs for repowering.*

3. Compute the theoretical gains from repowering
Although highly site-specific, the potential gains from repowering are enormous. The gas-turbine/HRSG repowering options, by one analysis, can provide new ca-

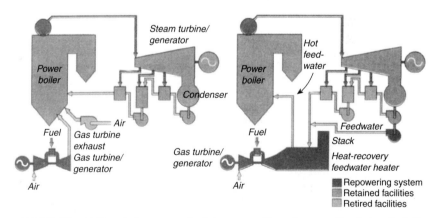

FIGURE 103 (a) Hot-windbox repowering injects gas-turbine exhaust gas directly into the boiler, replacing the forced-draft system. (b) Feedwater-heating repowering directs gas-turbine exhaust into the feedwater circuit, replacing all, or part of, the feedwater heat exchangers. (*Power*).

pacity at an incremental capital cost of about \$600/kW, a 155 percent increase in net plant output, and a net plant heat-rate improvement of 35 percent. For the feedwater-heating option, the figures are estimated to be \$700/kW, 55 percent greater plant output, and 6 percent heat-rate improvement. With the hot windbox option, the estimated figures are \$800/kW, 42 percent plant output gain, and 8 percent net improvement in heat rate.

Reductions in emissions are also impressive. When a gas-fired gas-turbine/HRSG replaces the existing boiler, emissions decrease markedly. Most of today's advanced gas turbines achieve NO_x emissions in the 9- to 25-ppm range with no downstream cleanup or steam/water injection into the gas-turbine combustor. And if the original boiler was firing oil with appreciable sulfur content, SO_2 emissions decline substantially, too. Higher efficiency means lower CO_2 emissions for equivalent output. Even in the hot-windbox repowering option, NO_x emissions can be significantly reduced. Aggregate emissions may not be as favorable, depending on output gain and capacity factor for the repowered plant.

4. *Evaluate other options for repowering*
Other forms of repowering include: integrating a coal gasifier or pressurized fluidized-bed (PFB) combustor into the gas-turbine/HRSG-repowered steam turbine; replacing a pulverized-coal fired boiler with a fluidized-bed boiler; replacing a fossil-fired boiler with a municipal solid-waste (MSW)-fired steam generator; adding district-heating capability; combining Diesel engines exhausting into a fossil-fired boiler; and even adding an ultra high-temperature steam turbine module to the existing steam-turbine train. Each of these approaches requires careful analysis of the economics of the proposed scheme to see that it earns the required rate of return for the organization sponsoring the project.

Thus, MSW may be possible when an electric utility teams with a waste-to-energy firm to repower. Besides repowering, such a plan might help solve specific solid-waste disposal problems faced by the local community.

Where very high overall thermal efficiency is desired, repowering can include modifying the steam cycle for district heating. Many large electric generating plants installed in Europe in the last decade feed extensive district-heating networks. While including district heating in a repowering scheme does not lower emissions in an absolute sense, it avoids a separate emissions source—and fuel and hardware expenses—to separately generate energy for heating or cooling.

Repowering with Diesel engines has attracted the interest of some. Here, the idea is to combine the high efficiency and great fuel flexibility of engines with the economics of existing coal-fired generation, Fig. 104. In one, scheme, oil-gas-fired engines exhaust into a coal-fired boiler modified to burn micronized coal. Heat rate in the range of 9000 Btu/kWh (8550 kJ/kWh) has been projected for the cycle.

Such a scheme requires back-end (exhaust) cleanup to achieve respectable emissions levels. Assuming a back-end cleanup system is installed, it can do double-duty—removing pollutants from both the Diesel and the Rankine-cycle portions of the plant—and perhaps allow lower-quality, less expensive fuels to be fired in the engine. Other potential advantages noted relative to gas turbines are: greater flexibility matching prime mover to existing steam cycle; less impact on performance from ambient temperature or ambient-air conditions, especially at coastal sites; greater fuel flexibility; less arduous operation and maintenance problems because of the ruggedness of engines compared to gas turbines; and better capability for meeting radically changing thermal and electric loads.

An ultra-high temperature steam-turbine/generator module, still in the development stage, may be an upcoming option for repowering without changing the fuel basis for the plant. Such a steam turbine, designed for 1300 F (704 C) steam,

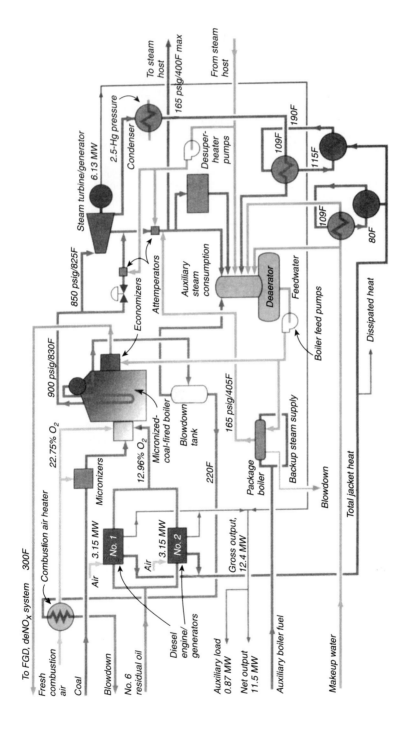

FIGURE 104 Diesel engines are used here to repower a conventional small boiler plant. The same concept can be applied to large steam turbines. (*Power*).

makes use of alloys and design techniques used in gas turbines. At that temperature, it could increase unit output by 22 percent and lower heat rate by 12 percent. Similar materials of construction are envisioned for making the boiler tubes that would have to be added to the existing steam generator.

5. Design a suitable system to prevent gas-turbine damage from ice ingestion
Ice ingestion at a gas-turbine (GT) inlet is not uncommon in northern climates. For many GTs a typical ice-ingestion event can cause primary- and secondary-compressor damage and lead to total repair bills exceeding $250,000, not including plant downtime.

Icing conditions can occur at a GT inlet at ambient temperatures well above freezing. Some GTs, for example, encounter icing at roughly 40 F (4.4 C) and 70 percent relative humidity—primarily caused by the temperature depression that ensues as the filtered air accelerates in the compressor inlet. Thus, potential icing conditions are determined by temperature, humidity, and compressor intake velocity. Figure 105 shows for one particular gas turbine a typical psychrometric chart, which is used to determine the moisture content of the ambient air, and to define warning zones where icing can occur.

Four inlet-heating design solutions are available for preventing icing at GT inlets—(1) water/glycol; (2) compressor bleed air; (3) low-pressure steam; (4) exhaust-heat systems. Each has advantages and disadvantages the designer must evaluate before making a final choice. Here are pertinent facts to help in the design choice.

Water/glycol systems often see service in plants with heat-recovery systems, such as cogeneration applications and where inlet-air chilling may be desired. The system allows a single coil, Fig. 106 in the air filter to serve for both inlet heating and cooling. Mixture ratio for most water/glycol systems is 50/50. Because GT power output is maximized at a particular inlet air temperature, it is important to control the heating system accurately, even while the ambient temperature varies.

Compressor bleed air, Fig. 107 is a popular choice for simple-cycle installations and peaking units. It is a well-proven technology and usually quite reliable and controllable. The biggest drawback with bleed-air systems is the associated performance penalty which, depending on ambient temperature and application, can

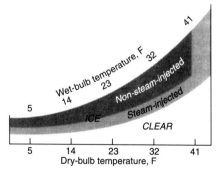

FIGURE 105 Psychrometric chart plots inlet icing conditions. The steam-injected and non-steam-injected areas show where potential icing conditions exist. (*Power*).

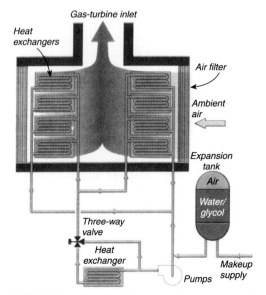

FIGURE 106   Water/glycol system is frequently used in plants with heat-recovery systems. (*Power*).

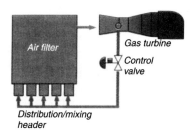

FIGURE 107   Compressor bleed-air system is a popular choice for simple-cycle and peaking units. (*Power*).

be 2–5 percent of the total power output. Plants with GTs that operate less than about 200 hr/yr under potentially icing conditions should consider this system.

Low-pressure steam is used in many cogeneration plants to provide heat for the anti-icing system. Steam-coil heat exchangers, Fig. 108 are located ahead of the first filter stage and regulated by an upstream control valve. Because the plant is typically recirculating steam otherwise wasted in this application, the heating cost is minimal.

A key drawback with steam-coil systems is this: Some cogeneration plants encounter difficulties in controlling the modulated temperature for power augmentation. This is attributable, in part, to the demands of accurately modulating the low-pressure steam supply and return pressures. To avoid these difficulties with power augmentation, GT manufacturers are likely to recommend a different inlet-air heating system.

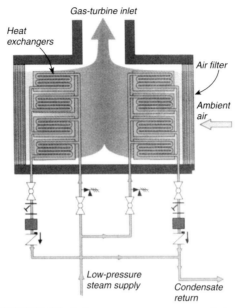

FIGURE 108   In low-pressure system, steam coils located ahead of the first filter stage are regulated by an upstream control valve. (*Power*).

Exhaust-heat systems, Fig. 109, are becoming increasingly attractive because of the desire to lower the performance penalty. Those plants requiring extended anti-icing operation, such as in locations in Canada, will probably find this option more cost-effective than a bleed-air system, provided operation exceeds 200 hr/yr.

Exhaust-heat systems either (1) directly bleed the heat from the exhaust stream back into the intake, (2) use heat exchangers and forced-air ducts. The latter option is desirable because accelerated filter blockage is not encountered and the risk of unburned fuels being ingested into the GT is eliminated.

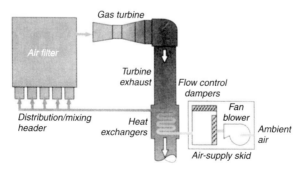

FIGURE 109   Exhaust-heat systems are attractive for reducing the performance penalty. (*Power*).

A performance penalty of less than 0.1 percent—because of added backpressure of heat exchangers in the exhaust duct—will be experienced year-round. The fan-motor parasitic load only adds to operating cost while the heating system is in operation.

Related Calculations. Repowering offers plant designers some of the greatest opportunities for creative solutions to environmentally sound expansion of existing power facilities. When the designer evaluates the many options presented here—using traditional thermodynamic and economic analyses discussed in this part of this handbook—it will be seen that the opportunities for making major savings are enormous. This procedure gives a comprehensive review of the typical options that can be implemented today.

Data in this procedure are the work of Jason Makansi, Executive Editor, *Power* magazine, through Step 4, and, for Step 5, William Calvert, Stewart & Stevenson Services Inc., as reported in *Power* magazine. SI values were added by the handbook editor.

COOLING-TOWER CHOICE FOR GIVEN HUMIDITY AND SPACE REQUIREMENTS

Select the type of cooling tower to use to cool condenser circulating water for a 450-MW steam central station located where the relative humidity rarely falls below 35 percent and it is desired to keep piping, electrical wiring, and controls to the minimum. The area in which the plant is located has relatively short summers.

Calculation Procedure:

1. Compare the installed costs of the available cooling towers

For central stations the two usual choices for cooling towers are (1) induced-draft, and (2) natural-draft. To make the cost comparison, designers base the tower capacity requirements on a constant annual heat rate. This eliminates unnecessary variables in the comparison. Where fuel costs do not change seasonally, average-annual-heat-rate evaluation is valid. We will use this approach in this procedure because it has been found suitable in a variety of installations.

Obtain cost estimates from tower manufacturers for a plant of 450-MW capacity. Here are typical costs for such an installation:

	Induced-draft	Natural-draft
First cost of towers, including fans and drives	$2,000,000	$3,400,000
Electrical controls, wiring	460,000	Nil
Incremental piping	700,000	Base
Capitalized operating costs	1,040,000	Nil
Total cost	$4,200,000	$3,400,000

Plot the results of such a study on a chart as that in Fig. 110, which gives actual data for a 225-MW unit and illustrates a standoff. Varying local conditions for this unit might move the standoff point down to about 200 MW.

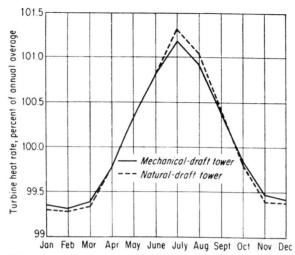

FIGURE 110 Average annual heat rate is normal evaluation basis for comparing cooling-tower types. Winter gains in heat rate balance summer dryness. (*Power*).

Based on the data in the above cost study for the 450-MW unit, a natural-draft hyperbolic cooling tower appears to be the most economic choice. However, other variables must also be considered.

2. *Evaluate other advantages of the hyperbolic cooling tower tentative choice*

Hyperbolic cooling towers are natural-draft. As such there is no fan-horsepower (kW) operating cost. In induced-draft mechanical cooling towers fan horsepower can amount to 0.5 percent of installed generating capacity. Thus, a 200-MW unit might need mechanical towers having some 14 cells, each with a 100-hp (75-kW) fan motor. (Normal induced-draft cooling-tower cells handle 15 to 20 MW each). The fan power becomes available for sale in a natural-draft cooling-tower installation.

As this brief evaluation indicates, several tangible factors can offset the higher first cost of hyperbolic natural-draft towers. There are also several intangibles, difficult to assign exact dollar values but nevertheless important.

Natural-draft hyperbolic towers, Fig. 111, need far less space than comparable induced-draft arrangements. Instead of having to be widely separated to prevent recirculation of heated water vapor, natural-draft towers can be placed on centers 1.5 times their base diameter. They can be located right beside the power station since their operation is not affected by prevailing winds. A recent space-requirement comparison for a 700-MW station made up of several units showed that an induced-draft installation needs about 72 acres (29.2 ha), while a natural-draft installation needs only 15 acres (6 ha)—an 80 percent reduction. Figure 112 shows the relative cost of mechanical-draft and natural-draft cooling towers for generating plants of various capacities.

Where land is expensive, or impossible to obtain, the reduced space requirements of natural-draft towers alone might tip the choice. For example, there are many "downtown" power plants operating a multiplicity of old, small generating units on induced-draft cooling towers. Replacement with large modern units might be

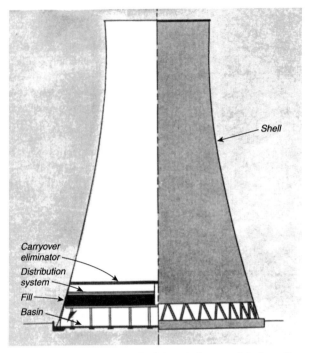

FIGURE 111 Concrete shell of hyperbolic natural-draft cooling tower is curved to agree with the vena contracta formed by air flow through the tower shell. (*Power*).

impossible if induced-draft towers were again used. But existing land might well be adequate for a natural-draft installation. In a big-city location, ground fogging might also be a distinct problem with induced-draft towers. The height of natural-draft towers lifts vapor discharge, so there's no ground fogging.

3. Compare the relative humidity constraints for each type of tower

Hyperbolic natural-draft towers operate satisfactorily under low-humidity (under 50 percent) conditions. Thus, there are many such towers at work in India. But there is an economic lower limit of application—probably about 35 percent relative humidity for design conditions. Below this value of relative humidity, cooling-tower size and cost increase rapidly.

Further, if a natural-draft tower is forced to operate at very low relative humidities (5 to 6 percent with the dry-bulb temperature higher than the incoming hot water), air inversion can occur. This will reverse the draft direction in the tower. Such inversion is not a steady-state condition; hunting takes place, reducing the tower's efficiency.

Natural-draft towers perform best when the difference between the cold-water (outlet-water) and air wet-bulb temperatures is equal to, or greater than, the difference between the hot-water (tower inlet water) and cold-water (tower outlet water) temperatures. In cooling-tower language, when the *approach* is equal to, or greater than, the *range*.

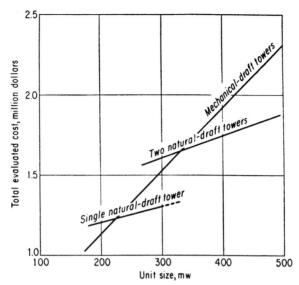

FIGURE 112   Evaluation of natural-draft vs mechanical-draft cooling towers varies with unit size and number of towers needed. (*Power*).

Operation of a natural-draft tower varies with seasonal changes in dry-bulb temperatures and relative humidity. In winter, air flow increases, producing cooler outlet water, but lower wet-bulb temperatures cut the heat-transfer force and reduce the effect to some extent. Figure 113 shows the higher capacity of natural-draft cooling towers at higher humidities.

Air-density differences produce the air flow in natural-draft towers. Thus, factors such as dry-bulb temperatures and relative humidity play an important part in tower performance. In operation, heavier outside air displaces the lighter saturated air in the tower, forcing it up and out the top. This is much like chimney operation except that water saturation rather than heat causes the change in air density. Draft losses in typical natural-draft towers range between 0.15 and 0.25 in (3.8 and 6.4 mm) of water.

Unlike the mechanical-draft tower, whose fan produces a fixed air-volume flow regardless of density, a natural-draft tower's air flow varies with changing atmospheric conditions. Studies show that natural-draft towers operate better with high relative humidity at a given wet-bulb temperature.

4.   *Evaluate tower construction alternatives*
Natural-draft cooling tower shells are circular in plan, hyperbolic in profile. Thus, the structure is a double curvature. From a thermal point of view a tower could be cylindrical. But the momentum of the entering air forms a vena contracta whose dimensions vary with the ratio of tower diameter to height of air inlet at the base. Considerable saving in materials and costs can be achieved by tapering the shell to the diameter of the vena contracta. The hyperbolic shape of the tower stiffens the concrete shell against wind forces—an added safety factor.

C

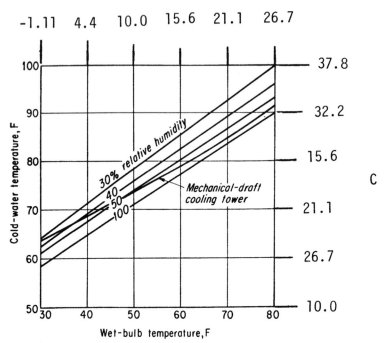

FIGURE 113 Natural-draft vs mechanical-draft cooling-tower operating curves show higher capacity of natural-draft towers at higher humidities. (*Power*).

Today's natural-draft towers have a steel-reinforced concrete shell about 18-in. (46-cm) thick at the base. The concrete thickness tapers to 4 to 5 in (10 to 13 cm) at about one-fifth the tower height and continues at this thickness to the top. Vertical ribs are sometimes poured integrally on the outside of the tower shell to break up wind forces by creating turbulence. This permits up to 40 percent theoretical increased stress.

Common tower shell design is based on 72-mph (116-km/h) wind (base reaction), the equivalent of 135 mph (217-km/h) (projected-area basis) for normal chimney design. Towers of 260-ft diameter and 340-ft height (79 to 104 m) are not unusual today.

The entire weight of the concrete shell and the wind-reaction load is supported by reinforced-concrete columns. They are inclined in the same plane as the bottom of the shell and provide a support between the base ring of the shell and the foundation at grade. Open spaces between inclined columns are air-inlet ports to the packing.

A cold-water basin beneath the tower forms the foundation in some designs; in others the basin may be separate. The circular basin can be provided with a conical bottom to collect silt from the cooling water. Blowdown valves remove concentrated silt.

Splash fill for natural-draft cooling towers uses wooden bars spaced at frequent intervals in staggered rows, Fig. 114. Falling water falls from row to row and breaks into droplets, exposing maximum surface to the upflowing air. Typical spacing puts the wooden bars 6-in. (15-cm) apart horizontally and 9-in. (23-cm) vertically with a resultant depth fill of about 15 ft (4.6 m). Stop logs can be used to cut out half the water-distribution flume (and associated radial pipes) for cleaning with the other half of the tower in operation.

Usual splash fill is California redwood or Scandinavian softwoods. Recent studies show that structurally stronger Douglas fir is the equal of redwood when properly treated. Species chosen depends on lumber-market activity affecting costs at the time of construction.

Water-film fill uses large sheets of 1/8-in. (3-mm) thick compressed asbestos cement hung from a series of concrete supports, Fig. 114. Teakwood or plastic spacers separate the sheets evenly. A thin film of water slides down the surface of the asbestos-cement sheets, exposing maximum water surface for evaporative cooling.

Both types of fill (wood and compressed asbestos cement) are specifically designed to cut draft losses and promote high air volume in the tower. Typical fill

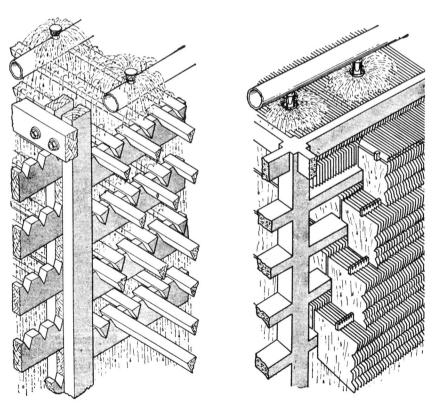

FIGURE 114 Treated-wood fill breaks up falling water droplets continually, *left*. Water film is formed on thin asbestos-cement fill sheets to give surface for cooling, *right*. (*Power*).

water loading runs 2 to 4 gal/min per ft² (1.4 to 2.7 L/s/m²). This is slightly less than current mechanical-draft tower practice.

Related Calculations. Hyperbolic natural-draft cooling towers have a considerably higher capital cost than mechanical-draft cooling towers. However, the operating cost of natural-draft towers is less than for mechanical-draft towers, as is the cost of piping, electrical wiring and controls. The inherent advantages of natural-draft towers is leading to their wider adoption throughout the world in both central-station and large industrial power plants.

A recent example of the usefulness of the hyperbolic natural-draft tower is a central station on the Gulf coast of Florida. Two hyperbolic "helper" natural-draft towers are used in conjunction with a bank of mechanical-draft towers, Fig. 115. These helper natural-draft towers reduce the thermal impact that the existing once-through cooling system has on water discharged to the Gulf of Mexico. The complex—when it became operational—was the largest concrete, saltwater hyperbolic helper-tower installation in the United States.

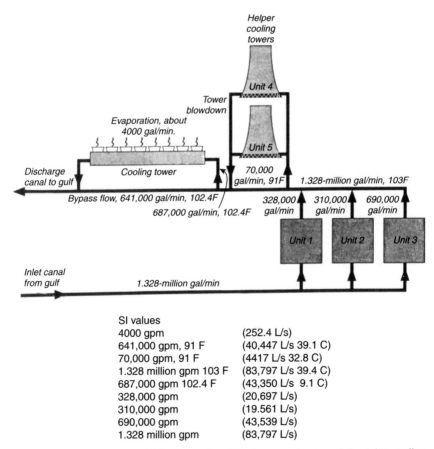

SI values	
4000 gpm	(252.4 L/s)
641,000 gpm, 91 F	(40,447 L/s 39.1 C)
70,000 gpm, 91 F	(4417 L/s 32.8 C)
1.328 million gpm 103 F	(83,797 L/s 39.4 C)
687,000 gpm 102.4 F	(43,350 L/s 9.1 C)
328,000 gpm	(20,697 L/s)
310,000 gpm	(19.561 L/s)
690,000 gpm	(43,539 L/s)
1.328 million gpm	(83,797 L/s)

FIGURE 115 Helper system withdraws a portion of discharge-canal water and directs it to cooling towers. Cooled water from towers is returned to canal, where the two streams are mixed. (*Power*).

To ensure the highest reliability, unusual cooling-tower features include the use of high-quality precast concrete for the structure, a non-clog splash fill, and a CRT-based digital system to monitor and control the towers and all auxiliary systems.

The primary current use of hyperbolic natural-draft cooling towers is in condenser cooling water temperature reduction. In this application these towers find worldwide use in both warm and temperate climates. European nuclear plants use hyperbolic natural-draft towers almost exclusively. And the use of these towers will increase as more power capacity is installed throughout the world.

In summary, as a guide to tower choice, the following guidelines are offered here: Hyperbolic natural-draft towers are most often selected over the mechanical-draft type when: (1) operating conditions couple low wet-bulb temperature and high relative humidity; (2) a combination of low wet-bulb and high inlet- and exit-water temperatures exists—that is, a broad cooling range and long cooling approach; (3) a heavy winter load is possible, and (4) a long amortization period can be arranged. Economics also tends to favor hyperbolics over mechanical-draft units as power-plant size grows and the possibility of erecting fewer (but larger) hyperbolic towers exists.

Data in this procedure are the work of B.G.A. Skrotzki, Associate Editor, *Power* magazine, Joe Lander, Florida Power Corp., and Gray Christensen, Black & Veatch, (Florida power plant details) as reported in *Power*. SI values were provided by the handbook editor.

DESIGN ENGINEERING

SECTION 19

SHAFTS, FLYWHEELS, PULLEYS, AND BELTS FOR POWER TRANSMISSION

Stresses in Solid and Hollow Shafts and Their Components

SHAFT TORQUE AND SHEARING STRESS DETERMINATION

A hydraulic turbine in a hydropower plant is rated at 12,000 hp (8952 kW). The steel vertical shaft connecting the turbine and generator is 24 in (60.96 cm) in diameter and rotates at 60 r/min. What is the maximum torque and shearing stress in the shaft at full load?

Calculation Procedure:

1. Compute the torque in the shaft at full load
Use the relation, hp $= 2\pi NT/33,000$, where N = rpm of the shaft; T = torque, lb/ft (kg/m). Solving for torque, T = hp(33,000)/$2\pi(N)$. Or, t = 12,000 × 33,000/6.28 × 60 = 1.05 × 10^6 lb·ft (1423 kN·m).

2. Find the maximum shearing stress in the shaft
Maximum shear stress occurs in the shaft at full load, i.e., 12,000 hp (8952 kW). Use the relation $S = RT/I$, where S = maximum shear stress, lb/in² (kPa); R = shaft radius, in (cm); T = torque in shaft at maximum load, lb·ft (N·m); I = polar moment of inertia of the shaft, in⁴ (cm⁴).

For a circular section, the polar moment of inertia $I = [\pi(d)^4]/32$. For this 24-in (60.96-cm) shaft, I = [3.14 × 24⁴]/32 = 32,556 in⁴ (82691 cm⁴). Substituting in the stress equation, S = (24/2)(1,050,000)(12 in/ft)/32556 = 4644 lb/in² (31.99 MPa).

Related Calculations. Use the relations here to compute the torque and shear stress in shafts of any material: steel, iron, aluminum, copper, Monel, stainless steel, plastic, etc. Obtain the polar moment of inertia by computation using the standard equations for various shapes available in any mechanical engineering handbook. The shear stress in this shaft is relatively low, a characteristic of hydraulic turbines. This low shear stress partly explains why hydraulic turbines have some of the longest lives of machines, in some cases more than 100 years.

CHOICE OF SHAFT DIAMETER TO LIMIT TORSIONAL DEFLECTION

A solid cast-iron circular shaft 60 in (152.4 cm) long carries a solid circular head 60 in in diameter at one end, Fig. 1. The bar is subjected to a torsional moment of 60,000 lb in (6780 Nm) which is applied at one end. It is desired to keep the torsional deflection of the circular head below 1/32 in (0.079 cm) when the shaft is transmitting power over its entire length in order to prevent chattering of the assembly. What should the diameter of the shaft be if the working stress is taken as 3000 lb/in² (20.7 MPa) and the transverse modulus of elasticity is 6 million lb/in² (41,340 MPa)?

Calculation Procedure:

1. Determine the shaft diameter based on the torque at full load
The torque in the shaft = (S_w)(polar moment of inertia of the circular shaft, I)/(working stress, C). For this shaft, torque = 60,000 = $3000\pi(d^3/16)$. Solving for (d_3) = (60,000 × 16)/(3000 × 3.14) = 4.66 in (11.84 cm).

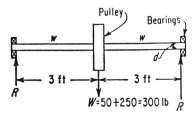

SI Values
3 ft (0.91 m)   300 lb (136.2 kg)

FIGURE 1   Shaft carrying solid circular head.

2.   Find the shaft diameter based on the allowable torsional deflection

For torsional stiffness, $\theta = (1/32)/r$, where r = radius of the head, in (cm). This equals $(1/32)(30) = 1/960$ radian. Since arc length along the head is $\theta(r)$, $1/32 = \theta(r) = \theta(30)$.

Find the shaft diameter based on the torsional deflection from $(d^4) = 32(T)(S)/\pi(E)(\theta)$, where T = given shaft torsion; S = given wheel diameter; E = given transverse modulus of elasticity; $\theta = 1/960$. Substituting, $d^4 = 32(60,000)(60)/3.14(6,000,000)(1/960) = 5870$; then $d = 8.75$ in (22.2 cm). Since 8.75 in (22.2 cm) is greater than 4.66 in (11.84 cm), the shaft must be designed for torsional stiffness, i.e., its diameter must be increased to at least 8.75 in (22.2 cm).

Related Calculations.   Use this general approach to size shafts to resist torsional deflection beyond a certain desired level. Increasing torsionals stiffness can reduce shaft chatter. With increased emphasis on noise reduction in manufacturing plants by EPA, torsional stiffness of shafts is receiving greater attention today.

SHAFT DIAMETER NEEDED TO TRANSMIT GIVEN LOAD AT STATED STRESS

What diameter steel shaft is required to transmit 2200 hp (1641 kW) at 2000 r/min with a maximum fiber stress in the shaft of 15,000 lb/in² (103.4 MPa)?

Calculation Procedure:

1.   Determine the torque in the shaft

Use the relation, $T = \text{hp}(33,000)/2\pi(\text{rpm})$, or $T = 2200(33,000)/6.28(2000) = 5780$ lb·ft (7831 N·m). Note that as the power transmitted rises, torque will increase if the rpm is constant, but if the rpm increases along with the power transmitted, the torque can remain fairly constant, depending on the relative increase of each.

2. Compute the shaft diameter required

Use the relation, $T = SZ_p$, where S = stress in shaft in given units; Z_p = polar section modulus of the shaft, in^3 (cm^3), and $Z_p = \pi(d^3)/16$. Solving these two relations for $d^3 = T(12$ in per ft)(16)$/S$ (π). Or $d^3 = 5780(12)(16)/15,000(3.14) = 2356$; $d = 2.86$ in (7.26 cm). A 3-in (7.62-cm) shaft would be chosen, unless space restrictions prevented using a shaft of this diameter.

Related Calculations. Use this procedure to determine a suitable diameter for shafts of any material: cast iron, Monel, stainless steel, plastic, etc.

MAXIMUM STRESS IN A SHAFT PRODUCED BY BENDING AND TORSION

A $1^{13}/_{16}$-in (4.6-cm) diameter steel shaft is supported on bearings 6 ft (1.82 m) apart. A 24-in (60.96-cm) pulley weighing 50 lb (22.7 kg) is attached to the center of the span. The pulley runs at 40 r/min and delivers 15 hp (11.2 kW) to the shaft, which weighs 8.77 lb/ft (13.1 kg/m). A belt exerts a 250-lb (113.5-kg) force in a vertically downward direction on the pulley. Determine the maximum stress in the shaft produced by the combination of bending and torsional stresses.

Calculation Procedure:

1. Determine the maximum bending load at the bearing at each end of the shaft

Consider the shaft to be a beam with fixed ends, Fig. 1. The maximum bending moment due to loads occurring at the bearings is given by the beam equation $BM = [(wl^2)/12 + PI/8]$, where w = shaft weight, lb/ft (kg/m); l = distance between bearings, ft (m); P = weight of load, lb (kg). Solving, $BM = [8.77(6^2)(12$ in/ft)/$12 + (300 \times 6 \times 12)/8] = 3015.7$ lb $\cdot$ in (340.8 N $\cdot$ m).

2. Find the torque delivered by the power input to the shaft

The torque, T, delivered by the power input to the shaft is given by T = hp $\times$ $33,000/2\pi \times$ rpm. Or, $T = 15 \times 33,000/6.28 \times 40 = 1970.5$ lb $\cdot$ in (222.7 N $\cdot$ m).

3. Compute the maximum shearing stress due to combined loads

Use the relation, maximum shearing stress produced by combined loads, $S_s = [1/l_p][(BM^2 + T^2]^{0.5}$. Or, $S_s = [16/\pi(1/13/16)^3] + [(3015.7)^2 + (1970.5)^2]^{0.5} = 3602.4$ lb/in^2 (24.86 MPa).

4. Determine the maximum normal stress due to combined loads

The maximum normal stress due to combined loads, $S_n = [1/l_p][3015.7 + \{(3015.7)^2 + (1970.5)^2\}]^{0.5} = 5663.8$ lb/in^2 (39 MPa).

Related Calculations. Use this procedure to find the maximum shearing stress and maximum normal stress in shafts made of any materials for which stress data are available. Thus, the relations given here are valid for cast iron, stainless steel, Monel, aluminum, plastic, etc.

COMPARISON OF SOLID AND HOLLOW SHAFT DIAMETERS

A solid circular shaft is used to transmit 200 hp (149 kW) at 1000 r/min. (*a*) What diameter shaft is required if the allowable maximum shearing stress is 20,000 lb/in^2 (137.8 MPa)? (*b*) If a hollow shaft is used having an inside diameter equal to the outside diameter of the solid shaft determined in part (*a*), what must be the outside diameter of this shaft if the angular twist of the two shafts is to be equal?

Calculation Procedure:

1. Determine the torque required to transmit the power
Use the relation $T = 33,000(\text{hp})/2\pi N$, where T = torque, lb·in (N·m); hp = horsepower (kW) transmitted; N = shaft rpm. Substituting, $T = 33,000(200)(12 \text{ in/ft})/2(3.14)(1000) = 12,611$ lb·in (1425 N·m).

2. Find the required diameter of the solid
Use two relations to find the required shaft diameter, namely: (1) $S_s = T(d)/2(I_p)$, where S_s = maximum shear stress, lb/in^2 (N·m); d = shaft diameter, in (cm); I_p = polar moment of inertia of the solid shaft, in^4 (cm^4). (2) The polar moment of inertia of the solid shaft, $I_p = \pi(d^4)/32$, where the symbols are as given earlier. Combining the two equations gives $S_s = 16T/\pi(d^3)$. Substituting $d^3 = (16)(12,611)/\pi(20,000) = 3.21$; $d = 1.475$ in (3.75 cm).

3. Compute the outside diameter of the hollow shaft
A hollow shaft having an inside diameter of 1.475 in (3.75 cm), that is the same as the outside diameter of the solid shaft, is desired. The outside diameter of the hollow shaft is to be such that its angular twist shall equal that of the solid shaft. Or, in equation form, $\theta_{ds} = T_{ds}(L_{ds})/(G_e)(I_{ps})$, where θ_{ds} = angular twist of the solid shaft, degrees; T = torque in solid shaft, lb·in (N·m); G_e = modulus of elasticity of the shaft material in shear, lb/in^2 (kPa); other symbols as before. For the circular shaft, $\theta_{dh} = T_{dh}(L_{dh})/G_e(I_{pdh})$. Symbols are the same as earlier, except that the subscript h refers to the hollow shaft.

Since the torque on the shaft and the shaft length are identical for both shafts which are made of the same material, by equating the angular twist equations to each other, $I_{ps} = I_{ph}$, or $\pi(d^4)/32 = [\pi(D_{dh}^4) - (d_{dh}^4)]/32$, where D = outside diameter of the hollow shaft, in (cm). Substituting, and solving for the outside diameter of the hollow shaft gives $D = 1.754$ in (4.45 cm). The hollow-shaft thickness will be $(1.754 - 1.475)/2 = 0.139$ in (0.35 cm).

Related Calculations. Use this general approach to determine the outside diameter of a hollow shaft, compared to that of a solid shaft. While the same angular twist was specified for these two shafts, different angular twists can be handled using the same general procedure. Any materials can be analyzed with this procedure: steel, cast iron, plastic, etc.

Hollow shafts often find favor today in an environmentally conscious design world. Thus, Richard M. Phelan, Professor of Mechanical Engineering, Cornell University, writes:

"Most shafts are solid. But in situations where weight and reliability are of great importance, hollow shafts with a ratio of $d_{inner}/D_{outer} = 0.6$ are often used. The (shaft) weight decreases more rapidly than the strength because the material near the center is not highly stressed and carries only a relatively small part of the total bending and torque loads. The reliability of the material is increased by using hollow shafts. Very large shafts are usually machined from a forged billet, and boring out is specified to remove inclusions, holes, etc., left in the center of the billet, the last region to solidify upon cooling. A hollow shaft also permits more uniform heat treatment and simplifies inspection of the finished part.

The shaft may be made hollow by boring, forging, or using cold-drawn seamless tubing. Unless the seamless tubing can be purchased so close to the required final dimensions that very little machining needs to be done, the high material cost may make it less practical than boring a hole in a solid piece."

SHAFT KEY DIMENSIONS, STRESSES, AND FACTOR OF SAFETY

A solid steel machine shaft with a safe shear stress of 7000 lb/in² (48.2 MPa) transmits a torque of 10,500 lb·in (1186.5 N·m). (a) Find the shaft diameter for these conditions. (b) A square key is used whose width is equal to one-fourth the shaft diameter and whose length equals 1.5 times the shaft diameter. Find the key dimensions and check the key for its induced shear and compressive stresses. (c) Obtain the factors of safety of the key in shear and in crushing, allowing the ultimate shearing stress of 50,000 lb/in² (344.5 MPa) and a compression stress of 60,000 lb/in² (413.4 MPa).

Calculation Procedure:

1. Find the shaft diameter for the given conditions
Use the relation $d^3 = 16T\pi(S_s)$, where d = shaft diameter, in (cm); T = torque on shaft, lb·in (N·m); S_s = shear stress, lb/in (kPa). Substituting, $d^3 = 16(10,500)/\pi(7000) = 7.643$; $d = 1.969$ in (5.00 cm). A 2-in (5.08-cm) diameter shaft would be used.

2. Determine the key dimensions
The width of the key is to be one-fourth of the shaft diameter, or 2.0 in/4 = 0.5 in (1.27 cm). Length of the key is to equal 1.5 times the shaft diameter, or 1.5 × 2 = 3.0 in (7.62 cm).

Now that we know the key dimensions, we can check it for induced shear and compressive stresses. The tangential force set up at the outside of the P_t, lb (kg), is P_t = torque, lb·in/radius of shaft, in. Or P_t = 10,500/1 = 10,500 lb (4540 kg).

The shear stress of the key is given by $S_s = (P_t)/bL$, where b = key width, in (cm); L = key length, in (cm). Substituting, $P_t = 10,500/0.5(3) = 7000$ lb/in² (48.2 MPa).

Find the crushing stress from $S_c = (2P_t)bL$, where S_c = crushing stress, lb/in (kPa). Substituting, $S_c = 2(10,500)/(0.5 × 3) = 14,000$ lb/in² (96.5 MPa).

3. Compute the factor of safety for both types of stresses
The factor of safety = allowable stress/actual stress. For shear, factor of safety, F_s = 50,000/7000 = 7.14. For crushing, the factor of safety, F_c = 60,000/14,000 = 4.29.

Related Calculations.   Use this general procedure to size keys for rotating shafts made of any of the popular materials: steel, cast iron, aluminum, plastic, etc. Recommended dimensions of various types of shaft keys can be obtained from handbooks listed in the References section of this book.

SHAFT KEY MINIMUM LENGTH FOR KNOWN TORSIONAL STRESS

What is the minimum length for a 0.875-in (2.22-cm) wide key for a 3.4375-in (8.73-cm) diameter gear-driving shaft designed to operate at a torsional working stress of 11,350 lb/in^2 (78.2 MPa)? The allowable shear stress in the key is 12,000 lb/in^2 (1.74 MPa).

Calculation Procedure:

1. *Determine the torque on the shaft*
Use the relation, $T = (d^3)S_t/5.1$, where T = shaft torque, lb $\cdot$ in (N $\cdot$ m); d = shaft diameter, in (cm); S_t = torsional working stress of the shaft, lb/in^2 (kPa). Substituting, $T = (3.4375^3)(11,350)/5.1 = 90,397$ lb $\cdot$ in (10.2 kN $\cdot$ m).

2. *Compute the tangential force on the key*
Use the relation, $P = T/r$, where P = tangential force on the key, lb (kg); r = shaft radius, in (cm). Substituting, $T = 90,397/1.71875 = 52,595$ lb (23878 kg).

3. *Find the length of the key to satisfy the given conditions*
Use the relation, $L = P/b(S_s)$, where L = key length, in (cm); P = tangential force computed in step 2; S_s = allowable shear stress in key, lb/in^2 (kPa). Substituting, $L = 52,596/0.875(12,000) = 5.00$ in (12.7 cm).
   Using the rule that $L = 1.5d$, then $L = 1.5 \times 3.4375 = 5.16$ in (13.1 cm). Therefore, the minimum computed length of 5.0 in (12.7 cm) because it closely approximates the length based on a ratio to the shaft diameter. The difference is negligible.
   Related Calculations.   Use this general procedure to size keys for any type of shaft having a known torsional stress.

SHAFT STRESS RESULTING FROM INSTANTANEOUS STOPPING

A flywheel weighing 200 lb (90.8 kg) whose radius of gyration is 15 in (38.1 cm) is secured to one end of a 6-in (15.2-cm) diameter shaft; the other end of the shaft is connected through a chain and sprocket to a motor rotating at 1800 r/min. The motor sprocket is 6 in (15.2 cm) in diameter and the shaft sprocket is 36 in (91.4 cm) in diameter. Total shaft length between flywheel and sprocket is 72 in (182.9 cm). Determine that maximum stress in the shaft resulting from instantaneous stopping of the motor drive, assuming that the sprocket and chain have no ability to absorb impact loading. Assume a shear modulus of 12,000,000 lb/in² (82,680 MPa). Neglect the effect of shaft kinetic energy.

Calculation Procedure:

1. Determine the torsional impact caused by the sudden stop
When the flywheel or any other rotating mass is stopped short, the stored kinetic energy in the rotating mass is converted to torsional impact. The magnitude of this energy is given by $E = W(\rho^2)(\Omega^2)/2g$, where W = flywheel weight, lb (kg); ρ = radius of gyration of the flywheel, ft (cm); Ω = angular velocity, radians per second; g = acceleration of gravity, 32.2 ft/s² (9.8 m/s²). Substituting, $E = 200 (1.25^2)(2\pi300/60(1/12) = 57{,}394$ lb·in (6486 N·m).

2. Find the modulus of resilience for the shaft
The shaft offers resilience to torsional twist, as detailed in Marks' "Mechanical Engineers' Handbook." Resilience, U in lb·in (N·m) is the potential energy stored in the deformed body, the shaft. The amount of resilience equals the work required to deform the shaft from zero stress to stress S. So the modulus of resilience, U_p, in lb·in/in³ (N·m/cm³), or unit resilience, is the elastic energy stored in an in³ (cm³) of the shaft material at the elastic limit. The unit of resilience for a solid shaft is $U_p = (S_s)^2/4G$, where S_s = maximum shear stress developed on instantaneous stopping, lb/in² (kPa); G = modulus of elasticity of the shaft material, lb/in² (kPa).

Find the full volume of the shaft from $V = 0.785 \times 6^2(72) = 2035$ in³ (33,348 m³). Then, substituting in the unit resilience equation for the entire shaft, U_p total $= 0.25(S_t^2)(2035)(1/12{,}000{,}000) = 57{,}394$ lb/in. Solving for $S_t = (4 \times 12{,}000{,}000 \times 57394/2035)^{0.5} = 36{,}794$ lb/in² (4158 N·m). Thus, the maximum stress in the shaft at instantaneous stopping will be 36,794 lb/in² (4159 N·m).

Related Calculations. Sudden stopping of a rotating member can cause excessive stress that may lead to failure. Therefore, it is important that the stress caused by sudden stopping be analyzed for every design where the possibility of such stopping exists. The time needed to compute the stress that might occur is small compared to the damage that might result if sudden stopping does occur. Further, having the calculations on file proves that he engineer took time to look ahead to see what might happen in the event of sudden stopping.

Shaft Applications in Power Transmission

ENERGY STORED IN A ROTATING FLYWHEEL

A 48-in (121.9-cm) diameter spoked steel flywheel having a 12-in wide × 10-in (30.5-cm × 25.4-cm) deep rim rotates at 200 r/min. How long a cut can be stamped in a 1-in (2.5-cm) thick aluminum plate if the stamping energy is obtained from this flywheel? The ultimate shearing strength of the aluminum is 40,000 lb/in² (275,789.9 kPa).

Calculation Procedure:

1. Determine the kinetic energy of the flywheel

In routine design calculations, the weight of a spoked or disk flywheel is assumed to be concentrated in the rim of the flywheel. The weight of the spokes or disk is neglected. In computing the kinetic energy of the flywheel the weight of a rectangular, square or circular rim is assumed to be concentrated at the horizontal centerline. Thus, for this rectangular rim, the weight is concentrated at a radius of $48/2 - 10/2 = 19$ in (48.3 cm) from the centerline of the shaft to which the flywheel is attached.

Then the kinetic energy $K = Wv^2/(2g)$, where K = kinetic energy of the rotating shaft, ft·lb; W = flywheel weight of flywheel rim, lb; v = velocity of flywheel at the horizontal centerline of the rim, ft/s. The velocity of a rotating rim is $v = 2\pi RD/60$, where $\pi = 3.1416$; R = rotational speed, r/min; D = distance of the rim horizontal centerline from the center of rotation, ft. For this flywheel, $v = 2\pi (200)(19/12)/60 = 33.2$ ft/s (10.1 m/s).

The rim of the flywheel has a volume of (rim height, in)(rim width, in)(rim circumference measured at the horizontal centerline, in), or $(10)(12)(2\pi)(19) = 14{,}350$ in^3 (235,154.4 cm^3). Since machine steel weighs 0.28 lb/in^3 (7.75 g/cm^3), the weight of the flywheel rim is $(14{,}350)(0.28) = 4010$ lb (1818.9 kg). Then $K = (4010)(33.2)^2/[2(32.2)] = 68{,}700$ ft·lb (93,144.7 N·m).

2. Compute the dimensions of the hole that can be stamped

A stamping operation is a shearing process. The area sheared is the product of the plate thickness and the length of the cut. Each square inch of the sheared area offers a resistance equal to the ultimate shearing strength of the material punched.

During stamping, the force exerted by the stamp varies from a maximum F lb at the point of contact to 0 lb when the stamp emerges from the metal. Thus, the average force during stamping is $(F + 0)/2 = F/2$. The work done is the product of $F/2$ and the distance through which this force moves, or the plate thickness t in. Therefore, the maximum length that can be stamped is that which occurs when the full kinetic energy of the flywheel is converted to stamping work.

With a 1-in (2.5-cm) thick aluminum plate, the work done is W ft·lb = (force, lb)(distance, ft). The work done when all the flywheel kinetic energy is used is $W = K$. Substituting the kinetic energy from step 1 gives $W = K = 68{,}700$ ft·lb (93,144.7 N·m) = $(F/2)(1/12)$; and solving for the force yields $F = 1{,}650{,}000$ lb (7,339,566.3 N).

The force F also equals the product of the plate area sheared and the ultimate shearing strength of the material stamped. Thus, $F = lts_u$, where l = length of cut, in; t = plate thickness, in; s_u = ultimate shearing strength of the material. Substituting the known values and solving for l, we get $l = 1{,}650{,}000/[(1)(40{,}000)] = 41.25$ in (104.8 cm).

Related Calculations. The length of cut computed above can be distributed in any form—square, rectangular, circular, or irregular. This method is suitable for computing the energy stored in a flywheel used for any purpose. Use the general procedure in step 2 for computing the principal dimension in blanking, punching, piercing, trimming, bending, forming, drawing, or coining.

SHAFT TORQUE, HORSEPOWER, AND DRIVER EFFICIENCY

A 4-in (10.2-cm) diameter shaft is driven at 3600 r/min by a 400-hp (298.3-kW) motor. The shaft drives a 48-in (121.9-cm) diameter chain sprocket having an output efficiency of 85 percent. Determine the torque in the shaft, the output force on the sprocket, and the power delivered by the sprocket.

Calculation Procedure:

1. Compute the torque developed in the shaft
For any shaft driven by any driver, the torque developed is T lb·in $= 63,000hp/R$, where hp = horsepower delivered to, or by, the shaft; R = shaft rotative speed, r/min. Thus, the torque developed by this shaft is $T = (63,000)(400)/3600 = 7000$ lb·in (790.9 N·m).

2. Compute the sprocket output force
The force developed at the output surface, tooth, or other part of a rotating member is given by $F = T/r$, where F = force developed, lb; r = radius arm of the force, in. In this drive the radius is $48/2 = 24$ in (61 cm). Hence, $F = 7000/24 = 291$ lb (1294.4 N).

3. Compute the power delivered by the sprocket
The work input to this shaft is 400 hp (298.3 kW). But the work output is less than the input because the efficiency is less than 100 percent. Since efficiency = work output, hp/work input, hp, the work output, hp = (work input, hp)(efficiency), or output hp = $(400)(0.85) = 340$ hp (253.5 kW).

   Related Calculations.   Use this procedure for any shaft driven by any driver—electric motor, steam turbine, internal-combustion engine, gas turbine, belt, chain, sprocket, etc. When computing the radius of toothed or geared members, use the pitch-circle or pitch-line radius.

PULLEY AND GEAR LOADS ON SHAFTS

A 500-r/min shaft is fitted with a 30-in (76.2-cm) diameter pulley weighing 250 lb (113.4 kg). This pulley delivers 35 hp (26.1 kW) to a load. The shaft is also fitted with a 24-in (61.0-cm) pitch-diameter gear weighing 200 lb (90.7 kg). This gear delivers 25 hp (18.6 kW) to a load. Determine the concentrated loads produced on the shaft by the pulley and the gear.

Calculation Procedure:

1. Determine the pulley concentrated load
The largest concentrated load caused by the pulley occurs when the belt load acts vertically downward. Then the total pulley concentrated load is the sum of the belt load and pulley weight.

   For a pulley in which the tension of the tight side of the belt is twice the tension in the slack side of the belt, the maximum belt load is $F_p = 3T/r$, where F_p = tension force, lb, produced by the belt load; T = torque acting on the pulley, lb·in; r = pulley radius, in. The torque acting on a pulley is found from $T = 63,000hp/R$, where hp = horsepower delivered by pulley; R = revolutions per minute (rpm) of shaft.

   For this pulley, $T = 63,000(35)/500 = 4410$ lb·in (498.3 N·m). Hence, the total pulley concentrated load $= 882 + 250 = 1132$ lb (5035.1 N).

2. Determine the gear concentrated load
With a gear, the turning force acts only on the teeth engaged with the meshing gear. Hence, there is no slack force as in a belt. Therefore, $F_g = T/r$, where $F_g =$ gear tooth-thrust force, lb; $r =$ gear pitch radius, in; other symbols as before. The torque acting on the gear is found in the same way as for the pulley.

Thus, $T = 63,000(25)/500 = 3145$ lb $\cdot$ in (355.3 N $\cdot$ m). Then $F_g = 3145/12 = 263$ lb (1169.9 N). Hence, the total gear concentrated load is $263 + 200 = 463$ lb (2059.5 N).

Related Calculations. Use this procedure to determine the concentrated load produced by any type of gear (spur, herringbone, worm, etc.), pulley (flat, V, or chain belt), sprocket, or their driving member. When the power transmission belt or chain leaves the belt or sprocket at an angle other than the vertical, take the vertical component of the pulley force and add it to the pulley weight to determine the concentrated load.

SHAFT REACTIONS AND BENDING MOMENTS

A 30-ft (9.1-m) long steel shaft weighing 150 lb/ft (223.2 kg/m) of length has a 500-lb (2224.1-N) concentrated gear load 10 ft (3.0 m) from the left end of the shaft and a 2000-lb (8896.4-N) concentrated pulley load 15 ft (4.6 m) from the right end of the shaft. Determine the end reactions and the maximum bending moment in this shaft.

Calculation Procedure:

1. Draw a sketch of the shaft
Figure 2a shows a sketch of the shaft. Label the left-and right-hand reactions L_R and R_R, respectively.

2. Compute the shaft end reactions
Take moments about R_R to determine the magnitude of L_R. Since the shaft has a uniform weight per foot of length, assume that the total weight of the shaft is concentrated at its midpoint. Then $30L_R - 500(20) - 150(30)(15) - 2000(15) = 0$; $L_R = 3583.33$ lb (15,939.4 N). Take moments about L_R to determine R_R. Or, $30R_R - 500(10) - 150(30)(15) - 2000(15) = 0$; $R_R = 3416.67$ lb (15,198.1 N). Alternatively, the first reaction found could be subtracted from the sum of the vertical loads, or $500 + 30 \times 150 + 2000 - 3583.33 = 3416.67$ lb (15,198.1 N). However, taking moments about each support permits checking the results, because the sum of the reactions should be equal the sum of the vertical loads, including the weight of the shaft.

3. Compute the maximum bending moment
The maximum bending moment in a shaft occurs where the shear is zero. Find the vertical shear at each point of applied load or reaction by taking the algebraic sum of the vertical forces to the left and right of the load. Use a plus sign for upward forces and a minus sign for downward forces. Designate each shear force by V with a subscript number showing its location, in feet (meters) along the shaft from the left end. Use L and R to indicate whether the shear is to the left or right of the load. The shear at the left-hand reaction is $V_{LR} = + 3583.33$ lb (+ 15,939.5 N); $V_{10L} = 3583.33 - 10 \times 150 = 2083.33$ lb (9267.1 N), where the product $10 \times$

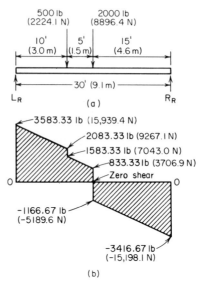

FIGURE 2 Shaft bending-moment diagram.

150 = the weight of the shaft from the point V_{LR} to the 500-lb (2224.1-N) load. At this load, V_{10R} = 2083.33 − 500 = 1583.33 lb (7043.0 N). To the right of the 500-lb (2224.1-N) load, at the 2000-lb (8896.4-N) load, V_{20L} = 1583.33 − 5 × 150 = 833.33 lb (3706.9 N). To the right of the 2000-lb (8896.4-N) load, V_{20R} = 833.33 − 2000 = −1166.67 lb (−5189.6 N). At the left of V_R, V_{30L} = −1166.67 − 15 × 150 = −3416.67 lb (−15,198.1 N). At the right hand end of the shaft V_{30R} = −3416.67 + 3416.67 = 0.

Draw the shear diagram (Fig. 2b). This diagram shows that zero shear occurs at a point 15 ft (4.6 m) from the left-hand reaction, Hence, the maximum bending moment M_m on this shaft is M_m = 3583.33(15) − 500(5) − 150(15)(7.5) = 34.340 lb · ft (46,558.8 N · m).

Related Calculations. Use this procedure for shafts of any metal—steel, bronze, aluminum, plastic, etc.—if the shaft is of uniform cross section. For non-uniform shafts, use the procedures discussed later in this section.

SOLID AND HOLLOW SHAFTS IN TORSION

A solid steel shaft will transmit 500 hp (372.8 kW) at 3600 r/min. What diameter shaft is required if the allowable stress in the shaft is 12,500 lb/in² (86,187.5 kPa)? What diameter hollow shaft is needed to transmit the same power if the inside diameter of the shaft is 1.0 in (2.5 cm)?

Calculation Procedure:

1. Compute the torque in the solid shaft
For any solid shaft, the torque T, lb · in = 63,000hp/R, where R = shaft rpm. Thus, T = 63,000(500)/3600 = 8750 lb · in (988.6 N · m).

2. Compute the required shaft diameter
For any solid shaft, the required diameter d, in $= 1.72(T/s)^{1/3}$, where $s =$ allowable stress in shaft, lb/in². Thus, for this shaft, $d = 1.72(8750/12,500)^{1/3} = 1.526$ in (3.9 cm).

3. Analyze the hollow shaft
The usual practice is to size hollow shafts such that the ratio q of the inside diameter d_i into the outside diameter d_o in is 1:2 to 1:3 or some intermediate value. With a q in this range the shaft will have sufficient thickness to prevent failure in service.
    Assume $q = d_i/d_o = 1/2$. Then with $d_i = 1.0$ in (2.5 cm), $d_o = d_i/q$, or $d_o = 1.0/0.5 = 2.0$ in (5.1 cm). With $q = 1/3$, $d_o = 1.0/0.33 = 3.0$ in (7.6 cm).

4. Compute the stress in each hollow shaft
For the hollow shaft $s = 5.1T/d_o^3(1 - q^4)$, where the symbols are defined above. Thus, for the 2-in (5.1-cm) outside-diameter shaft, $s = 5.1(8750)/[8(1 - 0.0625)] = 5950$ lb/in² (41,023.8 kPa).
    By inspection, the stress in the 3-in (7.6-cm) outside-diameter shaft will be lower because the torque is constant. Thus, $s = 5.1(8750)/[27(1 - 0.0123)] = 1672$ lb/in² (11,528.0 kPa).

5. Choose the outside diameter of the hollow shaft
Use a trial-and-error procedure to choose the hollow shaft's outside diameter. Since the stress in the 2-in (5.1-cm) outside-diameter shaft, 5950 lb/in² (41,023.8 kPa), is less than half the allowable stress of 12,500 lb/in² (86,187.5 kPa), select a smaller outside diameter and compute the stress while holding the inside diameter constant.
    Thus, with a 1.5-in (3.8-cm) shaft and the same inside diameter, $s = 5.1(8750)/[3.38(1 - 0.197)] = 16,430$ lb/in² (113,284.9 kPa). This exceeds the allowable stress.
    Try the larger outside diameter, 1.75 in (4.4 cm), to find the effect on the stress. Or $s = 5.1(8750)/[5.35(1 - 0.107)] = 9350$ lb/in² (64,468.3 kPa). This is lower than the allowable stress.
    Since a 1.5-in (3.8-cm) shaft has a 16,430-lb/in² (113,284.9-kPa) stress and a 1.75-in (4.4-cm) shaft has a 9350-lb/in² (64,468.3-kPa) stress, a shaft of intermediate size will have a stress approaching 12,500 lb/in² (86,187.5 kPa). Trying 1.625 in (4.1 cm) gives $s = 5.1(8750)/[4.4(1 - 0.143)] = 11,820$ lb/in² (81,489.9 kPa). This is within 680 lb/in² (4688.6 kPa) of the allowable stress and is close enough for usual design calculations.
    Related Calculations.   Use this procedure to find the diameter of any solid or hollow shaft acted on only by torsional stress. Where bending and torsion occur, use the next calculation procedure. Find the allowable torsional stress for various materials in Baumeister and Marks—*Standard Handbook for Mechanical Engineers.*

SOLID SHAFTS IN BENDING AND TORSION

A 30-ft (9.1-m) long solid shaft weighing 150 lb/ft (223.2 kg/m) is fitted with a pulley and a gear as shown in Fig. 3. The gear delivers 100 hp (74.6 kW) to the shaft while driving the shaft at 500 r/min. Determine the required diameter of the shaft if the allowable stress is 10,000 lb/in² (68,947.6 kPa).

Calculation Procedure:

1. Compute the pulley and gear concentrated loads

Using the method of the previous calculation procedure, we get $T = 63,000hp/R$ $= 63,000(100)/500 = 12,600$ lb $\cdot$ in (1423.6 N $\cdot$ m). Assuming that the maximum tension of the tight side of the belt is twice the tension of the slack side, we see the maximum belt load is $R_p = 3T/r = 3(12,600)/24 = 1575$ lb (7005.9 N). Hence, the total pulley concentrated load = belt load + pulley weight = 1575 + 750 = 2325 lb (10,342.1 N).

The gear concentrated load is found from $F_g = T/r$, where the torque is the same as computed for the pulley, or $F_g = 12,600/9 = 1400$ lb (6227.5 N). Hence, the total gear concentrated load is 1400 + 75 = 1475 lb (6561.1 N).

Draw a sketch of the shaft showing the two concentrated loads in position (Fig. 3).

2. Compute the end reactions of the shaft

Take moments about R_R to determine L_R, using the method of the previous calculation procedures. Thus, $L_R(30) - 2325(25) - 1475(8) - 150(30)(15) = 0$; $L_R = 4580$ lb (20,372.9 N). Taking moments about L_R to determine R_R yields $R_R(30) - 1475(22) - 2325(5) - 150(3)(15) = 0$; $R_R = 3720$ lb (16,547.4 N). Check by taking the sum of the upward forces: $4580 + 3720 = 8300$ lb (36,920.2 N) = sum of the downward forces or $2325 + 1475 + 4500 = 8300$ lb (36,920.2 N).

3. Compute the vertical shear acting on the shaft

Using the method of the previous calculation procedures, we find $V_{LR} = 4580$ lb (20,372.9 N); $V_{5L} = 4580 - 5(150) = 3830$ lb (17,036.7 N); $V_{5R} = 3830 - 2325 = 1505$ lb (6694.6 N); $V_{22L} = 1505 - 17(150) = -1045$ lb (−4648.4 N); $V_{22R} = -1045 - 1475 = -2520$ lb (−11,209.5 N); $V_{30L} = -2520 - 8(150) = -3720$ lb (−16,547.4 N); $V_{30R} = -3720 + 3720 = 0$.

4. Find the maximum bending moment on the shaft

Draw the shear diagram shown in Fig. 3. Determine the point of zero shear by scaling it from the shear diagram or setting up an equation thus: positive shear − $x(150$ lb/ft) = 0, where the positive shear is the last recorded plus value, V_{5R} in this shaft, and x = distance from V_{5R} where the shear is zero. Substituting values gives $1505 - 150x = 0$; $x = 10.03$ ft (3.1 m). Then $M_m = 4580(15.03) - 2325(10.03) - (150)(5 + 10.03)[(5 + 10.03)/2] = 28,575$ lb (127,108.3 N).

5. Determine the required shaft diameter

Use the method of maximum shear theory to size the shaft. Determine the equivalent torque T_e from $T_e = (M_m^2 + t^2)^{0.5}$, where M_m is the maximum bending moment, lb $\cdot$ ft, acting on the shaft and T is the maximum torque acting on the shaft. For this shaft, $T_e = [28,575^2 + (12,600/12)^2]^{0.5} = 28,600$ lb $\cdot$ ft (38,776.4 N $\cdot$ m), where the torque in pound-inches is divided by 12 to convert it to pound-feet. To convert T_e to $T_{e'}$ lb $\cdot$ in, multiply by 12.

Once the equivalent torque is known, the shaft diameter d in is computed from $d = 1.72(T_{e'}/s)^{1/3}$, where s = allowable stress in the shaft. For this shaft, $d = 1.72(28,500)(12)/(10,000)^{1/3} = 5.59$ in (14.2 cm). Use a 6.0-in (15.2-cm) diameter shaft.

Related Calculations. Use this procedure for any solid shaft of uniform cross section made of metal—steel, aluminum, bronze, brass, etc. The equation used in step 4 to determine the location of zero shear is based on a strength-of-materials

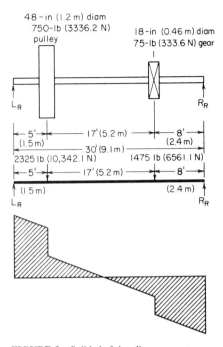

FIGURE 3   Solid-shaft bending moments.

principle: When zero shear occurs between two concentrated loads, find its location by dividing the last *positive* shear by the uniform load. If desired, the maximum principal stress theory can be used to combine the bending and torsional stresses in a shaft. The results obtained approximate those of the maximum shear theory.

EQUIVALENT BENDING MOMENT AND IDEAL TORQUE FOR A SHAFT

A 2-in (5.1-cm) diameter solid steel shaft has a maximum bending moment of 6000 lb·in (677.9 N·m) and an applied torque of 3000 lb·in (339.0 N·m). Is this shaft safe if the maximum allowable bending stress is 10,000 lb/in^2 (68,947.6 kPa)? What is the ideal torque for this shaft?

Calculation Procedure:

1.   Compute the equivalent bending moment
The equivalent bending moment M_e lb·in for a solid shaft is $M_e = 0.5[M + (M^2 + T^2)^{0.5}]$, where M = maximum bending moment acting on the shaft, lb·in; T = maximum torque acting on the shaft, lb·in. For this shaft, $M_e = 0.5[6000 + (6000^2 + 3000^2)^{0.5}] = 6355$ lb·in (718.0 N·m).

2. Compute the stress in the shaft

Use the flexure relation $s = Mc/I$, where s = stress developed in the shaft, lb/in²; $M = M_e$ for a shaft; I = section moment of inertia of the shaft about the neutral axis; in⁴; c = distance from shaft neutral axis to outside fibers, in. For a circular shaft, $I = \pi(d)^4/64 = \pi(2)^4/64 = 0.785$ in⁴ (32.7 cm⁴); $c = d/2 = 2/2 = 1.0$. Then $s = Mc/I = (6355(1.0)/0.785 = 8100$ lb/in² (55,849.5 kPa). Thus, the actual bending stress is 1900 lb/in² (13,100.5 kPa) less than the maximum allowable bending stress. Therefore the shaft is safe. Alternatively, compute the maximum equivalent bending moment from $M_e = sI/c = (10,000)(0.785)/1.0 = 7850$ lb·in (886.9 N·m). This is $7850 - 6355 = 1495$ lb·in (168.9 N·m) greater than the actual equivalent bending moment. Hence, the shaft is safe.

3. Compute the ideal torque for the shaft

The ideal torque T_i lb·in for a shaft is $T_i = M + (M^2 + T^2)^{0.5}$, where M and T are the bending and torsional moments, respectively, acting on the shaft, lb·in. For this shaft, $T_i = 6000 + (6000^2 + 3000^2)^{0.5} = 12,710$ lb·in (1436.0 N·m).

Related Calculations. Use this procedure for any shaft of uniform cross section made of metal—steel, aluminum, bronze, brass, etc.

TORSIONAL DEFLECTION OF SOLID AND HOLLOW SHAFTS

What diameter solid steel shaft should be used for a 500-hp (372.8-kW) 250-r/min application if the allowable torsional deflection is 1°, the maximum allowable stress is 10,000 lb/in² (68,947.6 kPa), and the modulus of rigidity is 13×10^6 lb/in² (89.6×10^6 kPa)? What diameter hollow steel shaft should be used if the ratio of the inside diameter to the outside diameter is 1:3, the allowable deflection is 1°, the allowable stress is 10,000 lb/in² (68,947.6 kPa), and the modulus of rigidity is 13×10^6 lb/in² (89.6×10^6 kPa)? What shaft has the greatest weight?

Calculation Procedure:

1. Determine the torque acting on the shaft

For any shaft, $T = 63,000hp/R$; or for this shaft, $T = 63,000(500)/250 = 126,000$ lb·in (14,236.1 N·m).

2. Compute the required diameter of the solid shaft

For a solid metal shaft, $d = (584Tl/G\alpha)^{1/3}$, where l = shaft length expressed as a number of shaft diameters, in; G = modulus of rigidity, lb/in²; α = angle of torsion deflection, degree.

Usual specifications for noncritical applications of shafts require that the torsional deflection not exceed 1° in a shaft having a length of equal to 20 diameters. Using this length gives $d = [584 \times 126,000 \times 20/(13 \times 10^6 \times 1.0)]^{1/3} = 4.84$ in (12.3 cm). Use a 5-in (12.7-cm) diameter shaft.

3. Compute the outside diameter of the hollow shaft

Assume that the shaft has a length equal to 20 diameters. Then for a hollow shaft $d = [584Tl/G\alpha(1 - q^4)]^{1/3}$, where $q = d_i/d_o$; d_i = inside diameter of the shaft, in; d_o = outside diameter of the shaft, in. For this shaft, $d = \{584 \times 126,000 \times 20/$

$(13 \times 10^6 \times 1.0)[1 - (1/3)^4]\}^{1/3} = 4.86$ in (12.3 cm). Use a 5-in (12.7-cm) outside-diameter shaft. The inside diameter would be $5.0/3 = 1.667$ in (4.2 cm).

4. Compare the weight of the shafts

Steel weighs approximately 480 lb/ft³ (7688.9 kg/m³). To find the weight of each shaft, compute its volume in cubic feet and multiply it by 480. Thus, for the 5-in (12.7-cm) diameter solid shaft, weight $= (\pi 5^2/4)(5 \times 20)(480)/1728 = 540$ lb (244.9 kg). The 5-in (12.7-cm) outside-diameter hollow shaft weighs $(\pi 5^2/4 - \pi 1.667^2/4)(5 \times 20)(480)/1728 = 242$ lb (109.8 kg). Thus, the hollow shaft weighs less than half the solid shaft. However, it would probably be more expensive to manufacture because drilling the central hole could be costly.

Related Calculations. Use this procedure to determine the steady-load torsional deflection of any shaft of uniform cross section made of any metal—steel, bronze, brass, aluminum, Monel, etc. The assumed torsional deflection of 1° for a shaft that is 20 times as long as the shaft diameter is typical for routine applications. Special shafts may be designed for considerably less torsional deflection.

DEFLECTION OF A SHAFT CARRYING CONCENTRATED AND UNIFORM LOADS

A 2-in (5.1-cm) diameter steel shaft is 6 ft (1.8 m) long between bearing centers and turns at 500 r/min. The shaft carries a 600-lb (2668.9-N) concentrated gear load 3 ft (0.9 m) from the left-hand center. Determine the deflection of the shaft if the modulus of elasticity E of the steel is 30×10^6 lb/in² (206.8×10^9 Pa). What would the shaft deflection be if the load were 2 ft (0.6 m) for the left-hand bearing? The shaft weighs 10 lb/ft (14.9 kg/m).

Calculation Procedure:

1. Compute the deflection caused by the concentrated load

When a beam carries both a concentrated and a uniformly distributed load, compute the deflection for each load separately and find the sum. This sum is the total deflection caused by the two loads.

For a beam carrying a concentrated load, the deflection Δ in $= Wl^3/48EI$, where $W =$ concentrated load, lb; $l =$ length of bean, in; $E =$ modulus of elasticity, lb/in²; $I =$ moment of inertia of shaft cross section, in⁴. For a circular shaft, $I = \pi d^4/64 = \pi(2)^4/64 = 0.7854$ in⁴ (32.7 cm⁴). Then $\Delta = 600(72)^3/[48(30)(10^6)(0.7854)] = 0.198$ in (5.03 mm). The deflection per foot of shaft length is $\Delta_f = 0.198/6 = 0.033$ in/ft (2.75 mm/m) for the concentrated load.

2. Compute the deflection due to shaft weight

For a shaft of uniform weight, $\Delta = 5wl^3/384EI$, where $w =$ total distributed load $=$ weight of shaft, lb. Thus, $\Delta = 5(60)(72)^3/[384(30 \times 10^6)(0.7854)] = 0.0129$ in (0.328 mm). The deflection per foot of shaft length is $\Delta_f = 0.0129/6 = 0.00214$ in/ft (0.178 mm/m).

3. Determine the total deflection of the shaft

The total deflection of the shaft is the sum of the deflections caused by the concentrated and uniform loads, or $\Delta_t = 0.198 + 0.0129 = 0.2109$ in (5.36 mm). The total deflection per foot of length is $0.033 + 0.00214 = 0.03514$ in/ft (2.93 mm/m).

Usual design practice limits the transverse deflection of a shaft of any diameter to 0.01 in/ft (0.83 mm/m) of shaft length. The deflection of this shaft is $3\frac{1}{2}$ times this limit. Therefore, the shaft diameter must be increased if this limit is not to be exceeded.

Using a 3-in (7.6-cm) diameter shaft weighing 25 lb/ft (37.2 kg/m) and computing the deflection in the same way, we find the total transverse deflection is 0.0453 in (1.15 mm), and the total deflection per foot of shaft length is 0.00755 in/ft (0.629 mm/m). This is within the desired limits. By reducing the assumed shaft diameter in $\frac{1}{8}$-in (0.32-cm) increments and computing the deflection per foot of length, a deflection closer to the limit can be obtained.

4. Compute the total deflection for the noncentral load
For a noncentral load, $\Delta = (Wc'/3EIl)[(cl/3 + cc'/3)^3]^{0.5}$, where c = distance of concentrated load from left-hand bearing, in; c' = distance of concentrated load from right-hand bearing, in. Thus $c + c' = 1$, and for this shaft $c = 24$ in (61.0 cm) and $c' = 48$ in (121.9 cm). Then $\Delta = [600 \times 48/(3 \times 30 \times 10^6 \times 0.7854 \times 72)] [(24 \times 72/x + 24 \times 48/3)^3]^{0.5} = 0.169$ in (4.29 mm).

The deflection caused by the weight of the shaft is the same as computed in step 2, or 0.0129 in (0.328 mm). Hence, the total shaft deflection is $0.169 + 0.0129 = 0.1819$ in (4.62 mm). The deflection per foot of shaft length is $0.1819/6 = 0.0303$ in (2.53 mm/m). Again, this exceeds 0.01 in/ft (0.833 mm/m).

Using a 3-in (7.6-cm) diameter shaft as in step 3 shows that the deflection can be reduced to within the desired limits.

Related Calculations. Use this procedure for any metal shaft—aluminum, brass, bronze, etc.—that is uniformly loaded or carries a concentrated load.

SELECTION OF KEYS FOR MACHINE SHAFTS

Select a key for a 4-in (10.2-cm) diameter shaft transmitting 1000 hp (745.7 kW) at 1000 r/min. The allowable shear stress in the key is 15,000 lb/in² (103,425.0 kPa), and the allowable compressive stress is 30,000 lb/in² (206,850.0 kPa). What type of key should be used if the allowable shear stress is 5000 lb/in² (34,475.0 kPa) and the allowable compressive stress is 20,000 lb/in² (137,900.0 kPa)?

Calculation Procedure:

1. Compute the torque acting on the shaft
The torque acting on the shaft is $T = 63,000hp/R$, or $T = 63,000(1000/1000) = 63,000$ lb·in (7118.0 N·m).

2. Determine the shear force acting on the key
The shear force F_s lb acting on a key is $F_s = T/r$, where T = torque acting on shaft, lb·in; r = radius of shaft, in. Thus, $T = 63,000/2 = 31,500$ lb (140,118.9 N).

3. Select the type of key to use

When a key is designed so that its allowable shear stress is approximately one-half its allowable compressive stress, a square key (i.e., a key having its height equal to its width) is generally chosen. For other values of the stress ratio, a flat key is generally used.

Determine the dimensions of the key from Baumeister and Marks—*Standard Handbook for Mechanical Engineers.* This handbook shows that a 4-in (10.2-cm) diameter shaft should have a square key 1 in wide × 1 in (2.5 cm × 2.5 cm) high.

4. Determine the required length of the key

The length of a 1-in (2.5-cm) key based on the allowable shear stress is $l = 2F_s/(w_k s_s)$, where w_k = width of key, in. Thus, $l = 31,500/[(1)(15,000)] = 2.1$ in (5.3 cm), say $2\frac{1}{8}$ in (5.4 cm).

5. Check key length for the compressive load

The length of a 1-in (2.5-cm) key based on the allowable compressive stress is $l = 2F_s/(ts_c)$, where t = key thickness, in; s_c = allowable compressive stress, lb/in². Thus, $l = 2(13,500)/[(1)(30,000)] = 2.1$ in (5.3 cm). This agrees with the key length based on the allowable shear stress. The key length found in steps 4 and 5 should agree if the key is square in cross section.

6. Determine the key size for other stress values

When the allowable shear stress does not equal one-half the allowable compressive stress for a shaft key, a flat key is generally used. A flat key has a width greater than its height.

Find the recommended dimensions for a flat key from Baumeister and Marks—*Standard Handbook for Mechanical Engineers.* This handbook shows that a 4-in (10.2-cm) diameter shaft will use a 1-in (2.5-cm) wide by $\frac{3}{4}$-in (1.9-cm) thick flat key.

The length of the key based on the allowable shear stress is $l = F_s/(w_k s_s) = 31,500/[(l)(5000)] = 6.31$ in (16.0 cm). Use a $6\frac{5}{16}$-in (16.0-cm) long key.

Checking the key length based on the allowable compressive stress yields $l = 2F_s/(ts_c) = 2(31,500)/[(0.75)(20,000)] = 4.2$ in (10.7 cm). Use the longer length, $6\frac{5}{16}$ in (16.0 cm), because the shorter key would be overloaded in compression.

Related Calculations. Use this procedure for shafts and keys made of any metal (steel, bronze, brass, stainless steel, etc.). The dimensions of shaft keys can also be found in ANSI Standard B17f, Woodruff Keys, Keyslots and Cutters. Woodruff keys are used only for light-torque applications.

SELECTING A LEATHER BELT FOR POWER TRANSMISSION

Choose a leather belt to transmit 50 hp (37.3 kW) from a 1750-r/min squirrel-cage compensator-starting motor through a 12-in (30.5-cm) diameter pulley in an oily atmosphere. What belt width is needed with a 50-hp (37.3-kW) internal-combustion engine fitted with a 17500-r/min 12-in (30.5-cm) diameter pulley operating in an oily atmosphere?

Calculation Procedure:

1. *Determine the belt speed*
The speed of a belt S is found from $S = \pi RD$, where R = rpm of driving or driven pulley; D = diameter, ft, of driving or driven pulley. Thus, for this belt, $S = \pi(1750)(12/12) = 5500$ ft/min (27.9 m/s).

2. *Determine the belt thickness needed*
Use the National Industrial Leather Association recommendation. Enter Table 1 at the bottom at a belt speed of 5500 ft/min (27.9 m/s), i.e., between 4000 and 6000 ft/min (20.3 and 30.5 m/s); and project horizontally to the next smaller pulley diameter than that actually used. Thus, by entering at the line marked 4000–6000 ft/min (20.3–30.5 m/s) and projecting to the 10-in (25.4-cm) minimum diameter pulley, since a 12-in (30.5-cm) pulley is used, we see that a 23/64-in (0.91-cm) thick double-ply heavy belt should be used. Read the belt thickness and type at the top of the column in which the next smaller pulley diameter appears.

3. *Determine the belt capacity factors*
Enter the body of Table 1 at a belt speed of 5500 ft/min (27.9 m/s), i.e., between 4000 and 6000 ft/min (20.3 and 30.5 m/s); then project to the double-ply heavy column. Interpolating by eye gives a belt capacity factor of $K_c = 14.8$.

4. *Determine the belt correction factors*
Table 2 lists motor, pulley diameter, and operating correction factors, respectively. Thus, from Table 2, the motor correction factor $M = 1.5$ for a squirrel-cage compensator-starting motor. Also from Table 2, the smaller pulley diameter correction factor $P = 0.7$; and $F = 1.35$ for an oily atmosphere.

5. *Compute the required belt width*
The required belt width, in, is $W = hpMF/(K_cP)$, where hp = horsepower transmitted by the belt; the other factors are as given above. For this belt, then, $W =$

TABLE 1   Leather-Belt Capacity Factors

Belt speed		Double ply	
		20/64 in (7.9 mm)	23/64 in (9.1 mm)
ft/min	m/s	Medium	Heavy
4000	20.3	10.9	12.6
5000	25.4	12.5	14.3
6000	30.5	13.2	15.2

Minimum pulley diameters					
Up to 2500	Up to 12.7	5 in°	12.7 cm°	8 in°	20.3 cm°
2500–4000	12.7–20.3	6°	15.2°	9°	22.9°
4000–6000	20.3–30.5	7°	17.8°	10°	25.4°

° For belts 8 in (20.3 cm) and over, add 2 in (5.1 cm) to pulley diameter.

TABLE 2 Leather-Belt Correction Factors

	Correction factor
Characteristics or condition of motor and starter:	
Squirrel-cage, compensator-starting motor	$M = 1.5$
Squirrel-cage, line-starting	$M = 2.0$
Slip-ring, high starting torque	$M = 2.5$
Diameter of small pulley, in (cm):	
4 and under (10.2 and under)	$P = 0.5$
4.5 to 8 (11.4 to 20.3)	$P = 0.6$
9 to 12 (22.9 to 30.5)	$P = 0.7$
13 to 16 (33.0 to 40.6)	$P = 0.8$
17 to 30 (43.2 to 76.2)	$P = 0.9$
Over 30 (over 76.2)	$P = 1.0$
Operating conditions:	
Oily, wet, or dusty atmosphere	$F = 1.35$
Vertical drives	$F = 1.2$
Jerky loads	$F = 1.2$
Shock and reversing loads	$F = 1.4$

$(50)(1.5)(1.35)/[14.8(0.7)] = 9.7$ in (24.6 cm). Thus, a 10-in (25.4-cm) wide belt would be used because belts are commercially available in 1-in (2.5-cm) increments.

6. Determine the belt width for the engine drive
For a double-ply belt driven by a driver other than an electric motor, $W = 2750hp/dR$, where $d =$ driving pulley diameter, in; $R =$ driving pulley, r/min. Thus, $W = 2750(50)/[(12)(1750)] = 6.54$ in (16.6 cm). Hence, a 7-in (17.8-cm) wide belt would be used.

For a single-ply belt the above equation becomes $W = 1925\ hp/dR$.

Related Calculations.   Note that the relations in steps 1, 5, and 6 can be solved for any unknown variable when the other factors in the equations are known. Where the hp rating of a belt material is available from the manufacturer's catalog or other published data, find the required width from $W = hp_bF/K_cP$, where $hp_b =$ hp rating of the belt material, as stated by the manufacturer; other symbols as before. To find the tension T_b, lb in a belt, solve $T_b = 33,000hp/S$ where $S =$ belt speed, ft/min. The tension per inch of belt width is $T_{bi} = T_b/W$. Where the belt speed exceeds 6000 ft/min (30.5 m/s), consult the manufacturer.

SELECTING A RUBBER BELT FOR POWER TRANSMISSION

Choose a rubber belt to transmit 15 hp (11.2 kW) from a 7-in (17.8-cm) diameter pulley driven by a shunt-wound dc motor. The pulley speed is 1300 r/min, and the belt drives an electric generator. The arrangement of the drive is such that the arc of contact of the belt on the pulley is 220°.

Calculation Procedure:

1. *Determine the belt service factor*
The belt *service factor* allows for the typical conditions met in the use of a belt with a given driver and driven machine or device. Table 3 lists typical service factors S_f used by the B. F. Goodrich Company. Entering Table 3 at the type of driver, a shunt-wound dc motor, and projecting downward to the driven machine, an electric generator, shows that $S_f = 1.2$.

2. *Determine the arc-of-contact factor*
A rubber belt can contact a pulley in a range from about 140 to 220°. Since the hp capacity ratings for belts are based on an arc of contact of 180°, a correction factor must be applied for other arcs of contact.

Table 4 lists the arc-of-contact correction factor C_c. Thus, for an arc of contact of 220°, $C_c = 1.12$.

3. *Compute the belt speed*
The belt speed is $S = \pi RD$, where S = belt speed, ft/min; R = pulley rpm; D = pulley diameter, ft. For this pulley, $S = \pi(1300)(7/12) = 2380$ ft/min (12.1 m/s).

TABLE 3 Service Factor S

Application	Squirrel-cage ac motor		Wound rotor ac motor (slip ring)	Single-phase capacitor motor	Dc shunt-wound motor	Diesel engine, four or more cylinders, above 700 r/min
	Normal torque, line start	High torque				
Agitators	1.0–1.2	1.2–1.4	1.2			
Compressors	1.2–1.4	. . . . . .	1.4	1.2	1.2	1.2
Belt conveyors (ore, coal, sand)	. . . . . .	1.4	. . . . . .	. .	1.2	
Screw conveyors	. . . . . .	1.8	. . . . . .	. .	1.6	
Crushing machinery	. . . . . .	1.6	1.6	. .	. . . . . .	1.4–1.6
Fans, centrifugal	1.2	. . . . . .	1.4	. .	1.4	1.4
Fans, propeller	1.4	2.0	1.6	. .	1.6	1.6
Generators and exciters	1.2		. . . . . .	. .	1.2	2.0
Line shafts	1.4	. . . . . .	1.4	1.4	1.4	1.6
Machine tools	1.0–1.2	. . . . . .	1.2–1.4	1.0	1.0–1.2	
Pumps, centrifugal	1.2	1.4	1.4	1.2	1.2	
Pumps, reciprocating	1.2–1.4	. . . . . .	1.4–1.6	. .	. . . . . .	1.8–2.0

TABLE 4 Arc of Contact Factor K—Rubber Belts

Arc of contact, °	140	160	180	200	220
Factor K	0.82	0.93	1.00	1.06	1.12

4. *Choose the minimum pulley diameter and belt ply*
Table 5 lists minimum recommended pulley diameters, belt material, and number of plies for various belt speeds. Choose the pulley diameter and number of plies for the next higher belt speed when the computed belt speed falls between two tabulated values. Thus, for a belt speed of 2380 ft/min (12.1 m/s), use a 7-in (17.8-cm) diameter pulley as listed under 2500 ft/min (12.7 m/s). The corresponding material specifications are found in the left-hand column and are four plies, 32-oz (0.9-kg) fabric.

5. *Determine the belt power rating*
Enter Table 6 at 32 oz (0.9 kg) four-ply material specifications, and project horizontally to the belt speed. This occurs between the tabulated speeds of 2000 and 2500 ft/min (10.2 and 12.7 m/s). Interpolating, we find [(2500 − 2380)/(2500 − 2000)](4.4 − 3.6) = 0.192. Hence, the power rating of the belt hp_{bi} is 4.400 − 0.192 = 4.208 hp/in (1.2 kW/cm) of width.

TABLE 5 Minimum Pulley Diameters, in (cm)—Rubber Belts of 32-oz (0.9-kg) Hard Fabric

Ply	Belt speed, ft/min (m/s)			
	2000 (149.4)	2500 (186.4)	3000 (223.7)	4000 (298.3)
3	4 (10.2)	4 (10.2)	4 (10.2)	4 (10.2)
4	5 (12.7)	6 (15.2)	6 (15.2)	7 (17.8)
5	8 (20.3)	8 (20.3)	9 (22.9)	10 (25.4)
6	11 (27.9)	11 (27.9)	12 (30.5)	13 (33.0)
7	15 (38.1)	15 (38.1)	16 (40.6)	17 (43.2)
8	18 (45.7)	19 (48.3)	20 (50.8)	21 (53.3)
9	22 (55.9)	23 (58.4)	24·(61.0)	25 (63.5)
10	26 (66.0)	27 (68.6)	28 (71.1)	29 (73.7)

TABLE 6 Power Ratings of Rubber Belts [32-oz (0.9 kg) Hard Fabric]

(*Hp = hp/in of belt width for 180° wrap*)
(*Power = kW/cm of belt width for 180° wrap*)

Ply	Belt speed, ft/min (m/s)			
	2000 (10.2)	2500 (12.7)	3000 (15.2)	4000 (20.3)
3	2.9 (0.85)	3.5 (1.03)	4.1 (1.20)	5.1 (1.50)
4	3.9 (1.14)	4.7 (1.38)	5.5 (1.61)	6.8 (2.00)
5	4.9 (1.44)	5.9 (1.73)	6.9 (2.03)	8.5 (2.50)
6	5.9 (1.73)	7.1 (2.08)	8.3 (2.44)	10.2 (2.99)
7	6.9 (2.03)	8.3 (2.44)	9.7 (2.85)	11.9 (3.49)
8	7.9 (2.32)	9.5 (2.79)	11.1 (3.26)	13.6 (3.99)
9	8.9 (2.61)	10.6 (3.11)	12.4 (3.64)	15.3 (4.49)
10	9.8 (2.88)	11.7 (3.43)	13.7 (4.02)	17.0 (4.99)

6. *Determine the required belt width*
The required belt width $W = hpS_f/(hp_{bi}C_c)$, or $W = (15)(1.2)/[(4.208)(1.12)] = 3.82$ in (9.7 cm). Use a 4-in (10.2-cm) wide belt.
   Related Calculations.   Use this procedure for rubber-belt drives of all types. For additional service factors, consult the engineering data published by B. F. Goodrich Company, The Goodyear Tire and Rubber Company, United States Rubber Company, etc.

SELECTING A V BELT FOR POWER TRANSMISSION

Choose a V belt to drive a 0.75-hp (559.3-W) stoker at about 900 r/min from a 1750-r/min motor. The stoker is fitted with a 3-in (7.6-cm) diameter sheave and the motor with a 6-in (15.2-cm) diameter sheave. The distance between the sheave shaft centerlines is 18 in (45.7 cm). The stoker handles soft coal free of hard lumps.

Calculation Procedure:

1. *Determine the design hp for the belt*
V-belt manufacturers publish service factors for belts used in various applications. Table 7 shows that a stoker is classed as heavy service and has a service factor of 1.4 to 1.6. By using the lower value, because the stoker handles soft coal free of hard lumps, the design horsepower for the belt is found by taking the product of the rated horsepower of the device driven by the belt and the service factor, or (0.75 hp)(1.4 service factor) = 1.05 hp (783.0 W). The belt must be capable of transmitting this, or a greater, horsepower.

2. *Determine the belt speed and arc of contact*
The belt speed $S = \pi RD$, where R = sheave rpm; D = sheave pitch diameter, ft = (sheave outside diameter, in $- 2X)/12$, where $2X$ = sheave dimension from Table 8. Before solving this equation, an assumption about the cross-sectional width of the belt must be made because $2X$ varies from 0.10 to 0.30 in (2.5 to 7.6 mm),

TABLE 7   Service Factors for V-Belt Drives

Typical machines	Type of service	Service factors
Domestic washing machines, domestic ironers, advertising display fixtures, small fans and blowers	Light	1.0–1.2
Fans and blowers (heavy rotors), centrifugal pumps, oil burners, home workshop machines	Medium	1.2–1.4
Stokers, reciprocating pumps and compressors, refrigerators, drill presses, grinders, lathes, meat slicers, machines for industrial use	Heavy	1.4–1.6

TABLE 8 Sheave Dimensions—Light-Duty V Belt

Belt cross section	Sheave effective OD		Groove angle, deg	W		D		2X	
	in	cm		in	mm	in	mm	in	mm
2L	Under 1.5	Under 3.8	32	0.240	6.10	0.250	6.4	0.10	2.5
	1.5–1.99	3.8–5.05	34	0.243	6.17				
	2.0–2.5	5.08–6.4	36	0.246	6.25				
	Over 2.5	Over 6.4	38	0.250	6.35				
3L	Under 2.2	Under 5.6	32	0.360	9.14	0.406	10.3	0.15	3.8
	2.2–3.19	5.6–8.10	34	0.364	9.25				
	3.2–4.2	8.13–10.7	36	0.368	9.35				
	Over 4.20	Over 10.7	38	0.372	9.45				
4L	Under 2.65	Under 6.7	30	0.485	12.32	0.490	12.4	0.20	5.1
	2.65–3.24	6.7–8.23	32	0.490	12.45				
	3.25–5.65	8.26–14.4	34	0.494	12.55				
	Over 5.65	Over 14.4	38	0.504	12.80				

and the exact cross section of the belt that will be used is not yet known. A value of $X = 0.15$ in (3.8 mm) is usually a safe assumption. It corresponds to a 3L belt cross section. Using $X = 0.15$ and the diameter and speed of the larger sheave, we see $S = \pi(1750)(6.0 - 0.15)/12 = 2675$ ft/min (13.6 m/s).

Compute the belt arc of contact from arc of contact, degrees = $180 - [60(d_1 - d_s)/l]$, where d_1 = large sheave nominal diameter, in; d_s = small sheave nominal diameter, in; l = distance between shaft centers, in. For this drive, arc = $180 - [60(6 - 3)/18] = 170°$. An arc-of-contact correction factor must be applied in computing the belt power capacity. Read this correction factor from Table 9 as $C_c = 0.98$ for a V-sheave to V-sheave drive and a 170° arc of contact. *Note:* If desired, the pitch diameters can be used in the above relation in place of the nominal diameters.

TABLE 9 Correction Factors for Arc of Contact—V-Belt Drives

Arc of contact, deg	Correction factor		Arc of contact, deg	Correction factor	
	V to V	V to flat°		V to V	V to flat°
180	1.00	0.75	130	0.86	0.86
170	0.98	0.77	120	0.82	0.82
160	0.95	0.80	110	0.78	0.78
150	0.92	0.82	100	0.74	0.74
140	0.89	0.84	90	0.69	0.69

°A V-to-flat drive has a small sheave and a larger-diameter flat pulley.

3. Select the belt to be used

The $2X$ value used in step 3 corresponds to a $3L$ cross section belt. Check the power capacity of this belt by entering Table 10 at a belt speed of 2800 ft/min (14.2 m/s), the next larger tabulated speed, and projecting across to the appropriate small-sheave diameter—3 in (7.6 cm) or larger. Read the belt horsepower rating as 0.87 hp (648.8 W). This is considerably less than the required capacity of 1.05 hp (783.0 W) computed in step 1. Therefore, the $3L$ belt is unsatisfactory.

Try a $4L$ belt, Table 11, following the same procedure. A $4L$ belt with a 3-in (7.6-cm) diameter small sheave has a rating of 1.16 hp (865.0 W). Correct this for the actual arc of contact by multiplying by C_c, or $(1.16)(0.98) = 1.137$ (847.9 W). Thus, the belt is suitable for the design hp value of 1.05 hp (783.0 W).

As a final check, compute the actual belt speed using the actual $2X$ value from Table 8. Thus, for a $4L$ belt on a 6-in (15.2-cm) sheave, $2X = 0.20$, and $S = \pi$ $(1750)(6 - 0.20)/12 = 2660$ ft/min (13.5 m/s). Hence, use of 2800 ft/min (14.2 m/s) in selecting the belt was a safe assumption. Note that the difference between the belt speed based on the assumed value of $2X$, 2675 ft/min (13.6 m/s), and the actual belt speed, 2660 ft/min (13.5 m/s), is about 0.5 percent. This is negligible.

Related Calculations. Use this procedure when choosing a single V belt for a drive. Where multiple belts are used, follow the steps given in the next calculation procedure. The data presented for single V belts is abstracted from *Standards for*

TABLE 10 Power Ratings of 3L Cross Section V Belts

(Based on 180° arc of contact on small sheave)

Belt speed		\multicolumn{8}{c}{Effective OD of small sheave}							
		1½ in (3.8 cm)		2 in (5.1 cm)		2½ in (6.4 cm)		3 in (7.6 cm) or more	
ft/min	m/s	hp	W	hp	W	hp	W	hp	W
2200	11.2	0.12	89.5	0.44	328.1	0.64	477.2	0.77	574.2
2400	12.2	0.10	74.6	0.45	335.6	0.66	492.2	0.81	604.0
2600	13.2	0.07	52.2	0.46	343.0	0.69	514.5	0.84	626.4
2800	14.2	0.04	29.8	0.46	343.0	0.70	522.0	0.87	648.8
3000	15.2	0.01	7.5	0.45	335.6	0.72	536.9	0.89	663.7

TABLE 11 Power Ratings of 4L Cross Section V Belts

(Based on 180° arc of contact on small sheave)

Belt speed		\multicolumn{8}{c}{Effective OD of small sheave}							
		2½ in (6.4 cm)		3 in (7.6 cm)		3½ in (8.9 cm)		4 in (10.2 cm) or more	
ft/min	m/s	hp	W	hp	W	hp	W	hp	W
2200	11.2	0.67	499.6	1.08	805.4	1.37	1021.6	1.58	1178.2
2400	12.2	0.68	507.1	1.12	835.2	1.43	1066.4	1.66	1237.9
2600	13.2	0.66	492.2	1.16	865.0	1.50	1118.5	1.75	1305.0
2800	14.2	0.65	484.7	1.18	879.9	1.54	1148.4	1.81	1349.7
3000	15.2	0.63	469.8	1.19	887.4	1.58	1178.2	1.87	1394.5

Light-duty or Fractional-horsepower V-Belts, published by the Rubber Manufacturers Association.

SELECTING MULTIPLE V BELTS FOR POWER TRANSMISSION

Choose the type and number of V belts needed to drive an air compressor from a 5-hp (3.7-kW) wound-rotor ac motor when the motor speed is 1800 r/min and the compressor speed is 600 r/min. The pitch diameter of the large sheave is 20 in (50.8 cm); and the distance between shaft centers is 36.0 in (91.4 cm).

Calculation Procedure:

1. Choose the V-belt section
Determine the design horsepower of the drive by finding the product of the service factor and the rated horsepower. Use Table 3 to find the service factor. The value of this factor is 1.4 for a compressor driven by a wound-rotor ac motor. Thus, the design horsepower = (5.0 hp)(1.4 service factor) = 7.0 hp (5.2 kW).

Enter Fig. 4 at 7.0 hp (5.2 kW), and project up to the small sheave speed, 600 r/min. Read the belt cross section as type B.

2. Determine the small-sheave pitch diameter
Use the speed ration of the shafts to determine the diameter of the small sheave. The speed ratio of the shafts is the ratio of the speed of the high-speed shaft to that of the low-speed shaft, or 1800/600 = 3.0. The sheave pitch diameters have the same ratio, or $20/PD_s = 3$; $PD_s = 20/3 = 6.67$ in (16.9 cm).

3. Compute the belt speed
The belt speed is $S = \pi RD$, where R = small-sheave rpm; D = small-sheave pitch diameter, ft. Thus, $S = \pi(1800)(6.67/12) = 3140$ ft/min (16.0 m/s).

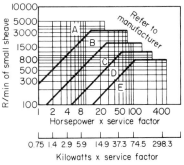

FIGURE 4 V-belt cross sectin for required hp rating.

4. Determine the belt horsepower rating

A tabulation of allowable belt horsepower ratings is used to determine the rating of a specific belt. To enter this table, the belt speed and the small-sheave equivalent diameter must be known.

Find the equivalent diameter d_e of the small sheave by taking the product of the small-sheave pitch diameter and the diameter factor, Table 12. Thus, for a speed range of 3.0 the small-diameter factor = 1.14, from Table 12. Hence, d_e = $(6.67)(1.14)$ = 7.6 in (19.3 cm).

Enter Table 12A at a belt speed of 3200 ft/min (16.3 m/s) and d_e = 7.6 in (19.3 cm). In the last column read the belt hp rating as 6.5 hp (4.85 kW). This rating must be corrected for the arc of contact and the belt length.

The arc of contact = $180 - [60(d_l - d_s)/l]$, where d_l and d_s = large- and small-sheave pitch diameters, respectively, in; l = distance between sheave shaft centers, in. Thus, arc of contact = $180 - [60(20 - 6.67)/36]$ = 157.8°. Using Table 4 and interpolating, we find the arc-of-contact correction factor C_c = 0.94.

TABLE 12 Small-Diameter Factors—Multiple V Belts

Speed ratio range	Small-diameter factor
1.000–1.019	1.00
1.020–1.032	1.01
1.033–1.055	1.02
1.056–1.081	1.03
1.082–1.109	1.04
1.110–1.142	1.05
1.143–1.178	1.06
1.179–1.222	1.07
1.223–1.274	1.08
1.275–1.340	1.09
1.341–1.429	1.10
1.430–1.562	1.11
1.563–1.814	1.12
1.815–2.948	1.13
2.949 and over	1.14

TABLE 12A Power Ratings for Premium-Quality B-Section V Belts

Belt speed		Equivalent diameter d_e			
		6.6 in (16.8 cm)		7.0+ in (17.8+ cm)	
ft/min	m/s	hp	kW	hp	kW
3000	15.2	5.90	4.40	6.26	4.67
3200	16.3	6.12	4.56	6.50	4.85
3400	17.3	6.31	4.71	6.73	5.02

Compute the belt pitch length from $L = 2l + 1.57(d_l + d_s) + (d_l - d_s)^2/(4l)$, where all the symbols are as given earlier. Thus, $L = 2(36) + 1.57(20 + 6.67) + (20 - 6.67)^2/[4(36)] = 115.1$ in (292.4 cm).

Enter Table 13 by interpolating between the standard belt lengths of 105 and 120 in (266.7 and 304.8 cm), and find the length correction factor for a B cross section belt as 1.06.

Find the product of the rated horsepower of the belt and the two correction factors—arc of contact and belt length, or $(6.5)(0.94)(1.06) = 6.47$ hp (4.82 kW).

5. *Choose the number of belts required*

The design horsepower, step 1, is 7.0 hp (5.2 kW). Thus, 7.0 design hp/6.47 rated belt hp = 1.08 belts; use two belts. Choose the next *larger* number of belts whenever a fractional number is indicated.

Related Calculations.   The tables and data used here are based on engineering information which is available from and updated by the Mechanical Power Transmission Association and the Rubber Manufacturers Association. Similar engineering data are published by the various V-belt manufacturers. Data presented here may be used when manufacturer's engineering data are not available.

SELECTION OF A WIRE-ROPE DRIVE

Choose a wire-rope drive for a 3000-lb (1360.8-kg) traction type freight elevator designed to lift freight or passengers totaling 4000 lb (1814.4 kg). The vertical lift of the elevator is 500 ft (152.4 m), and the rope velocity is 750 ft/min (3.8 m/s). The traction-type elevator sheaves are designed to accelerate the car to full speed in 60 ft (18.3 m) when it starts from a stopped position. A 48-in (1.2-m) diameter sheave is used for the elevator.

Calculation Procedure:

1. *Select the number of hoisting ropes to use*

The number of ropes required for an elevator is usually fixed by state or city laws. Check the local ordinances before choosing the number of ropers. Usual laws re-

TABLE 13 Length Correction
Factors—Multiple V Belts

Standard length designation		Belt cross section			
in	cm	A	B	C	D
96	243.8	1.08	. . .	0.92	
105	266.7	1.10	1.04	0.94	
120	304.8	1.13	1.07	0.97	0.86
136	345.4	. . .	1.09	0.99	

quire at least four ropes for a freight elevator. Assume four ropes are used for this elevator.

2. Select the rope size and strength

Standard "blue-center" steel hoisting rope is a popular choice, as is "plow-steel" and "mild plow-steel" rope. Assume that four $\%_{16}$-in (14.3-mm) six-strand 19-wires-per-strand blue-center steel ropes will be suitable for this car. The 6×19 rope is commonly used for freight and passenger elevators. Once the rope size is chosen, its strength can be checked against the actual load. The breaking strength of $\%_{16}$ in (14.3 mm), 6×19 blue-center steel rope is 13.5 tons (12.2 t), and its weight is 0.51 lb/ft (0.76 kg/m). These values are tabulated in Baumeister and Marks—*Standard Handbook for Mechanical Engineers* and in rope manufacturers' engineering data.

3. Compute the total load on each rope

The weight of the car and its contents is $3000 + 4000 = 7000$ lb (3175.1 kg). With four ropes, the load per rope is $7000/[4(2000$ lb $\cdot$ ton$)] = 0.875$ ton (0.794 t).

With a 500-ft (152.4-m) lift, the length of each rope would be equal to the lift height. Hence, with a rope weight of 0.51 lb/ft (0.76 kg/m), the total weight of the rope $= (0.51)(500)/2000 = 0.127$ ton (0.115 t).

Acceleration of the car from the stopped condition places an extra load on the rope. The rate of acceleration of the car is found from $a = v^2/(2d)$, where $a =$ car acceleration, ft/s^2; $v =$ final velocity of the car, ft/s; $d =$ distance through which the acceleration occurs, ft. For this car, $a = (750/60)^2/[2(60)] = 1.3$ ft/s^2 (39.6 cm/s^2). The value 60 in the numerator of the above relation converts from feet per minute to feet per second.

The rope load caused by acceleration of the car is $L_r = Wa/$(number of ropes)(2000 lb/ton) [$g = 32.2$ ft/s^2 (9.8 m/s^2)], where $L_r =$ rope load, tons; $W =$ weight of car and load, lb. Thus, $L_r = (7000)(1.3)/[(4)(2000)(32.2)] = 0.03351$ ton (0.03040 t) per rope.

The rope load caused by acceleration of the rope is $L_r = Wa/32.2$, where $W =$ weight of rope, tons. Or, $L_r = (0.127)(1.3)/32.2 = 0.0512$ ton (0.0464 t).

When the rope bends around the sheave, another load is produced. This bending load is, in pounds, $F_b = AE_r d_w/d_s$, where $A =$ rope area, in ^{2}; $E_r =$ modulus of elasticity of the whole rope $= 12 \times 10^6$ lb/in^2 (82.7 $\times 10^6$ kPa) for steel rope; $d_w =$ rope diameter, in; $d_s =$ sheave diameter, in. Thus, for this rope, $F_b = (0.0338)(12 \times 10^6)(0.120/48) = 1014$ lb, or 0.507 ton (0.406 t).

The total load on the rope is the sum of the individual loads, or $0.875 + 0.127 + 0.0351 + 0.507 + 0.051 = 1.545$ tons (1.4 t). Since the rope has a breaking strength of 13.5 tons (12.2 t), the factor of safety FS $=$ breaking strength, tons/ rope load, tons $= 13.5/1.545 = 8.74$. The usual minimum acceptable FS for elevator ropes is 8.0. Hence, this rope is satisfactory.

Related Calculations. Use this general procedure when choosing wire-rope drivers for mine hoists, inclined-shaft hoists, cranes, derricks, car pullers, dredges, well drilling, etc. When standard hoisting rope is chosen, which is the type most commonly used, the sheave diameter should not be less than $30d_w$; the recommended diameter is $45d_w$. For *haulage rope* use $42d_w$ and $72d_w$, respectively; for special flexible *hoisting rope*, use $18d_w$ and $27d_w$ sheaves.

DESIGN METHODS FOR NONCIRCULAR SHAFTS

Find the maximum shear stress and the angular twist per unit length produced by a torque of 20,000 lb $\cdot$ in (2258.0 N $\cdot$ m) imposed on a double-milled steel shaft with a four-splined hollow core, Fig. 5. The shaft outer diameter is 2 in (5.1 cm); the inner diameter is 1 in (2.5 cm). Modulus of rigidity of the shaft is $G = 12 \times 10^6$ lb/in² (83.3×10^6 kPa). The flat surfaces of the outer, milled shaft are cut down 0.1 in (0.25 cm), while the inner contour has a radius $r_i = 0.5$ in (1.27 cm) with four splines of half-width $A = 0.1$ in (0.25 cm) and height $B = 0.1$ in (0.25 cm).

Calculation Procedure:

1. *Find the torsional-stiffness and shear-stress factors for the outer shaft*
The proportionate mill height H/R for the outer shaft is, from Fig. 6 and the given data, $H/R = 0.1/1 = 0.1$. Entering Fig. 6 at $H/R = 0.1$, project to the torsional-stiffness factor (V) curve for shaft type 1, and read $V = 0.71$. Likewise, from Fig. 6, f = shear-stress factor = 0.82.

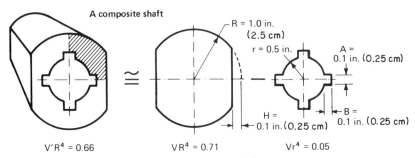

FIGURE 5 Typical comosit shaft. (*Design Engineering.*)

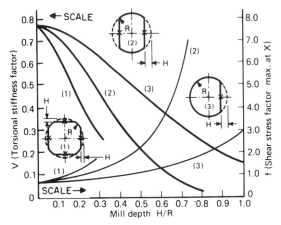

FIGURE 6 Milled shafts. (*Design Engineering.*)

2. Find the torsional-stiffness factor for the inner contour

The spline dimensions, from Fig. 7 are $A/B = 0.1/0.1 = 1.0$, and $B/R = 0.1/0.5 = 0.2$. Entering Fig. 7 at $B/R = 0.2$, project upward to the $A/B = 1.0$ curve, and read the torsional-stiffness factor $V = 0.85$.

3. Compute the angular twist of the shaft

The composite torsional stiffness is defined as $V_t R^4$, where V_t = composite-shaft stiffness factor and R = the internal diameter. For a composite shaft, $V_t R^4 = \Sigma V_i r_i^4 = (0.71)(1^4) - (0.85)(0.5)^4 = 0.66$ for this shaft. (The first term in this summation is the outer shaft; the second is the inner cross section.)

To find the angular twist θ of this composite shaft, substitute in the relation $T = 2G\theta V_t R^4$, where T = torque acting on the shaft, lb · in; other symbols as defined earlier. Substituting, we find $20{,}000 = 2(12 \times 10^6)\theta(0.66)$; $\theta = 0.0012$ rad/in (0.000472 rad/cm).

4. Find the maximum shear stress in the shaft

For a *solid* double-milled shaft, the maximum shear stress, S_s, lb/in^2 $= Tf/R^3$, where the symbols are as given earlier. Substituting for this shaft, assuming for now that the shaft is solid, we get $S_s = Tf/R^2 = (20{,}000)(0.82)/(1^2) = 16{,}400$ lb /in^2 (113,816 kPa), where the value of f is from step 1.

Because the shaft is hollow, however, the solid-shaft maximum shear stress must be multiplied by the ratio $V/V_t = 0.71/0.66 = 1.076$, where the value of V is from step 1 and the value of V_t is from step 3.

The hollow-shaft maximum shear stress is $S_s' = S_s(V/V_t) = 16{,}400(1.076) = 17{,}646$ lb/in^2 (122,993 kPa).

Related Calculations. By using charts (Figs. 6 through 15) and equations presented here, quantitative and performance factors can be calculated to well within 5 percent of a variety of widely used shafts. Thus, designers and engineers now have a solid analytical basis for choosing shafts, instead of having to rely on rules of thumb, which can lead to application problems.

Although design engineers are familiar with torsion and shear stress analyses of uniform circular shafts, usable solutions for even the most common noncircular shafts are often not only unfamiliar, but also unavailable. As a cirucular bar is twisted, each infinitesimal cross section rotates about the bar's longitudinal axis: plane cross sections remain plane, and the radii within each cross section remain straight. If the shaft cross section deviates even slightly from a circle, however, the situation changes radically and calculations bog down in complicated mathematics.

The solution for the circular cross section is straightforward: The shear stress at any point is proportional to the point's distance from the bar's axis; at each point, there are two equal stress vectors perpendicular to the radius through the point, one stress vector lying in the plane of the cross section and the other parallel to the bar's axis. The maximum stress is tangent to the shaft's outer surface. At the same time, the shaft's torsional stiffness is a function of its material, angle of twist, and the polar moment of inertia of the cross section.

The stress and torque relations can be summarized as $\theta = T/(J/G)$, or $T = G\theta J$, and $S_s = TR/J$ or $S_s = G\theta R$, where J = polar moment of inertia of a circular cross section ($= \pi R^4/2$); other symbols are as defined earlier.

If the shaft is splined, keyed, milled, or pinned, then its cross sections do not remain plane in torsion, but warp into three-dimensional surfaces. Radii do not remain straight, the distribution of shear stress is no longer linear, and the direction of shear stress are not longer perpendicular to the radius.

These changes are described by partial differential equations drawn from Saint-Venant's theory. The equations are unwieldy, so unwieldy that most common shaft

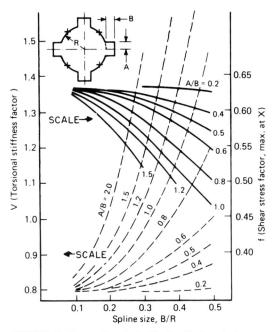

FIGURE 7   Four-spline shaft. (*Design Engineering.*)

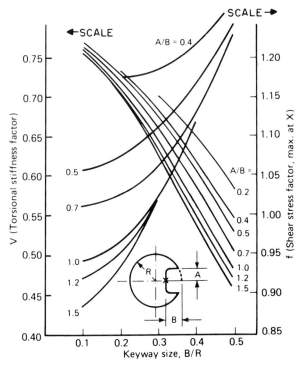

FIGURE 8   Single-keyway shaft. (*Design Engineering.*)

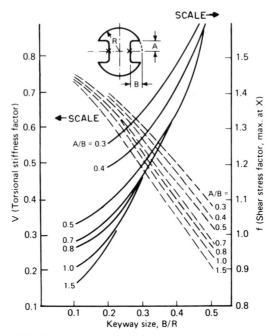

FIGURE 9 Two-keyway shaft. (*Desgin Engineering.*)

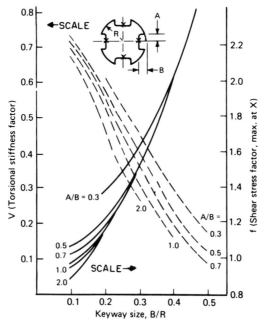

FIGURE 10 Four-keyway shaft. (*Design Engineering.*)

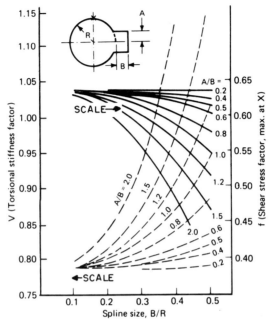

FIGURE 11   Single-spline shaft. (*Design Engineering.*)

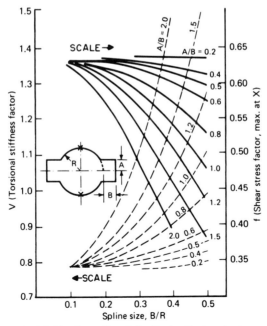

FIGURE 12   Two-spline shaft. (*Design Engineering.*)

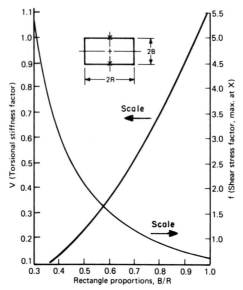

FIGURE 13   Rectangular shafts. (*Design Engineering.*)

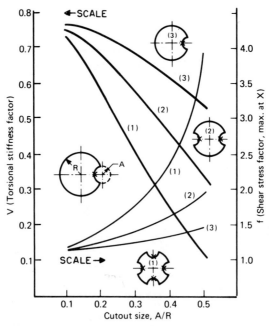

FIGURE 14   Pinned shaft. (*Design Engineering.*)

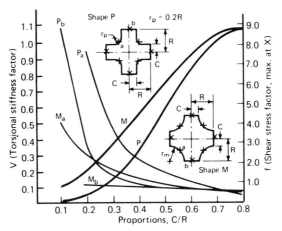

FIGURE 15 Cross-shaped shaft. (*Design Engineering.*)

problems cannot be solved in closed form, but demand numerical approximations and educated intuition.

Of the methods of solving for Saint-Venant's torsion stress functions Φ, one of the most effective is the technique of finite differences. A finite-difference computer program (called SHAFT) was developed for this purpose by the Scientific and Engineering Computer Applications Division of U.S. Army AARADCOM (Dover, New Jersey).

SHAFT analyzed 10 fairly common transmission-shaft cross sections and (in the course of some 50 computer runs for each cross section) generated dimensionless torsional-stiffness and shear-stress factors for shafts with a wide range of proportions. Since the factors were calculated for unit-radius and unit-side cross sections, they may be applied to cross sections of any dimensions. These computer-generated factors, labeled V, f, and $d\phi/ds$, are derived from Prandtl's "membrane analogy" of the Φ function.

Because of the torsional-stiffness factor V may be summed for parallel shafts, V values for various shaft cross sections may be adjusted for differing radii and then added or subtracted to give valid results for composite shaft shapes. Thus, the torsional stiffness of a 2-in (5.1-cm) diameter eight-splined shaft may be calculated (to within 1 percent accuracy) by adding the V factors of two four-splined shafts and then subtracting the value for one 2-in (5.1-cm) diameter circle (to compensate for the overlapping of the central portion of the two splined shafts).

Or, a hollow shaft (like that analyzed above) can be approximated by taking the value of VR^4 for the cross section of the hollow and subtracting it from the VR^4 value of the outer contour.

In general, any composite shaft will have its own characteristic torsional-stiffness factor V, such that $V_t R^4 = \Sigma V_i r_i^4 t$, or $V_t = \Sigma V_i (r_i/R)^4$, where R is the radius of the outermost cross section and V_i and r_i are the torsional-stiffness factors and radii for each of the cross sections combined to form the composite shaft.

By the method given here, a total of 10 different shaft configurations can be analyzed: single, two, and four keyways; single, two, and four splines; milled; rectangular; pinned; and cross-shaped. It is probably the most versatile method of shaft analysis to be developed in recent years. It was published by Robert I. Isakower, Chief, Scientific and Engineering Computer Applications Division, U.S. Army AARADCOM, in *Design Engineering.*

CALCULATING EXTERNAL INERTIA, WK², FOR ROTATING AND LINEAR MOTION

Determine the external inertia, or WK^2, for a hollow steel cylinder rotating about its own axis if the cylinder is 3 ft (0.914 m) long, 1 ft (0.305 m) in outside diameter, 0.75 ft (0.229 m) inside diameter, if it rotates at 1800 rpm. What is the kinetic energy of the rotating cylinder?

Calculation Procedure:

1. Find the radius of gyration, K, of the part
Use Table 14 to find the equation for the radius of gyration of the part. Or, for this hollow cylinder, $K = 0.354 \, (D_2^2 + D_1^2)^{0.5}$, where the symbols are as defined in the table. Or $K = 0.354 \, (1^2 + 0.75^2)^{0.5} = 0.4425$ ft (0.135 m).

2. Determine the weight of the part
Using Table 14 again, along with Table 15 for the weight of the cylinder, we have, $W = 1360 \, wL \, (D_2^2 - D_1^2) = 1360 \, (0.282)(3)(1^2 - 0.75^2) = 503.37$ lb (228.5 kg).

3. Compute WK² for the part
Use the relation for external inertia $= WK^2 = (503.37)(0.4425)^2 = 98.56$ lb-ft² (4.1457 kg m²).

4. Find the kinetic energy of the rotating part
The kinetic energy, E, of a rotating part driven by a motor, turbine, engine, compressor, or other rotary device is given by $E = (WK^2N^2)/5872$ where $N^2 =$ rotative speed, rpm. Thus, for this machine, $E = (98.56)(1880)^2/5872 = 54.383$ ft-lb (73.69 kNm).

    Related Calculations. Whenever an electric motor is to be used in a mechanical application where it must start or reverse many times per minute the engineer will be asked "What is the external inertia, or WK^2, referred to the motor shaft?" While it is strictly a mechanical function, the importance and usefulness of external inertia seem to be little known, or if known, disregarded.

    Many mechanical engineers and designers have difficulty calculating WK^2, and also seem to have confused ideas concerning the power required for acceleration of a stationary or rotating mass. In reality, the problem is comparatively simple, being one of kinetic energy, as far as the mechanical engineer is concerned.

    For any body of mass M moving linearly at a velocity of V ft/sec (m/sec), the kinetic energy of this body expressed in ft lb (Nm) is:

$$E = \tfrac{1}{2}MV^2, \tag{1}$$

For a rotating body, such as that analyzed above, the entire mass, insofar as kinetic energy is concerned, is considered to be concentrated at a certain radius, K, called the radius of gyration, from the axis of rotation. The linear velocity of the concentrated mass at radius K ft (m) will then be $2\pi KN$ ft/min (m/min), where N is the

TABLE 14 WK^2 of Parts

Part	Radius of gyration, K Ft	Weight, W w = weight per cu in. of material, lb	WK^2 lb − Ft^2
Circular cylinder — about its own axis	(See text for SI 0.354 D	conversion) 1,360 wLD^2	170.4 wLD^4
Circular cylinder — about axis at one end	$0.144 \sqrt{3D^2 + 16L^2}$	1,360 wLD^2	196 wLD^2 [3D^2 + 16L^2]
Circular cylinder — about outside axis	$0.144 \sqrt{3D^2 + 16L^2 + 48dL + 48d^2}$	1,360 wLD^2	196 wLD^2 [3D^2 + 16L^2 + 48dL + 48d^2]
Hollow circular cylinder — about its own axis	$0.354 \sqrt{D_2{}^2 + D_1{}^2}$	1,360 wL [D$_2{}^2$ − D$_1{}^2$]	170.4 wL [D$_2{}^4$ − D$_1{}^4$]
Right elliptical cylinder — about its own axis	$\sqrt{\dfrac{a^2 + b^2}{2}}$	5,426 abwL	$5{,}426 \ abwL \left[\dfrac{a^2 + b^2}{4}\right]$
Rectangle — about its own axis	$0.289 \sqrt{a^2 + c^2}$	1,728 abcw	144.5 abcw [a^2 + c^2]
Rectangle — about axis at one end	$0.289 \sqrt{a^2 + 4c^2}$	1,728 abcw	144.5 abcw [a^2 + 4c^2]
Rectangle — about outside axis	$0.289 \sqrt{a^2 + 4c^2 + 12cd + 12d^2}$	1,728 abcw	144.5 abcw [a^2 + 4c^2 + 12cd + 12d^2]

TABLE 15 Approximations for Calculating Moments of Inertia

NAME OF PART	MOMENT OF INERTIA
Fly Wheels (not applicable to belt pulleys)	Moment of inertia equal to 1.08 to 1.15 times that of rim alone
Fly Wheel (based on total weight and outside diameter)	Moment of inertia equal to 2/3 of that of total weight concentrated at the outer circumference
Spur or helical gears (teeth alone)	Moment of inertia of teeth equal to 40 percent of that of a hollow cylinder of the limiting dimensions
Spur or helical gears (rim alone)	Figured as a hollow cylinder of same limiting dimensions
Spur or helical gears (total moment of inertia)	Equal to 1.25 times the sum of that of teeth plus rim
Spur or helical gears (with only weight and pitch diameter known)	Moment of inertia considered equal to 0.60 times the moment of inertia of the total weight concentrated at the pitch circle
Motor Armature (based on total weight and outside diameter)	Multiply outer radius of armature by following factors to obtain radius of gyration: Large slow speed motor......0.75 to 0.85 Medium Speed d-c or induction motor...............0.70 to 0.80 Mill type motor............0.60' to 0.65

rpm. Dividing this by 60 to reduce to ft/sec (m/sec) and using $W/32.2$ for mass M, Equation (1) becomes Equation 2, namely

$$E = WK^2N^2/5872 \text{ ft lb (Nm)}. \qquad (2)$$

This is the first appearance of the external inertia, WK^2. It is merely the weight of the body times the radius of gyration squared. Since the kinetic or "stored" energy in the body is directly proportional to WK^2, this expression is commonly known as the "moment of inertia," although strictly speaking the moment of inertia is $WK^2/32.2$. However, in developing the most commonly used formulas, the factor 32.2 disappears or is absorbed in the constants. Hence WK^2 itself remains as a convenient medium of calculation and is generally known as "moment of inertia" even though incorrectly.

The energy calculated from Eq. 2 represents the input to the rotating body itself, and bears no definite relation to the input to the driving unit during acceleration. It is, therefore, less confusing to determine the actual torque at the driver shaft necessary to accelerate the load. The input from the line will then correspond to this value of torque.

This torque can be readily determined by considering the angular velocity of the driven body. From $F = MA$, the following equation is derived for T, the acceleration torque, $T = I\theta$, where $I =$ the true moment of inertia and it equals $WK^2/32.2$, $\theta =$ the angular acceleration of the body expressed in radians/s/s. Assuming constant torque, θ equals the final angular velocity, ω, divided by the time, t, required to accelerate from rest. Thus, where $N =$ final angular velocity in rpm,

$$\theta = \frac{\omega}{t} \text{ or } \theta = \frac{2\pi N}{60t}$$

Then, substituting in the fundamental equation,

$$T = I\theta = \frac{WK^2}{32.2} \times \frac{2\pi N}{60t}$$

or

$$T = \frac{WK^2 N}{308t}, \text{ lb ft} \tag{3}$$

Note here that Eq. 3 assumes constant angular acceleration up to the final velocity of the machine. Therefore, if the acceleration is not constant but increases or decreases as the machine gathers speed, this equation will not give accurate results. Nevertheless, even though the acceleration does vary while the machine is coming up to speed, it is possible in some cases to make use of this equation.

Equation 3 shows that the torque needed to accelerate a rotating body depends directly on the value of the term WK^2—hence on the body's moment of inertia. Thus, once the value of WK^2 is determined, the horsepower (kW) required for acceleration, Fig. 16, can be found from

$$HP = \frac{\text{Torque} \times N}{5250} \tag{4}$$

In the above equations the factors WK^2 and N pertain to the rotating body or machine element. Normally in the power train of a machine the rotating parts—such as gears and drums—rotate at a speed different from the driver. To calculate the torque required to accelerate such a system, the WK^2 of each moving part is reduced to an equivalent WK_e^2 about a common shaft, usually the driver shaft, and then the WK_e^2 of the respective parts are added together.

The WK_e^2 of a rotating part referred to the driver shaft is obtained by means of Eq. 2. Thus, where N_m is the rpm of the driver shaft, the kinetic energy in the rotating part is

$$E = \frac{WK^2 N^2}{5872} = \frac{WK_e^2 N_m^2}{5872}$$

$$WK_e^2 = WK^2(N/N_m)^2 \tag{5}$$

For example, in a motor-driven pulley or gear train, Fig. 17, running at constant speed, disregarding frictional losses, the horsepower (kW) delivered by the motor equals that of the output shaft. But the torque imposed on the motor shaft may be greater or less than the torque on the output shaft of the train. Which torque is greater depends upon the ratio of the speeds of the shafts. Where

H_m = horsepower delivered by motor shaft (kW)

H_o = horsepower delivered by output shaft (kW)

T_m = torque on motor shaft, lb ft (Nm)

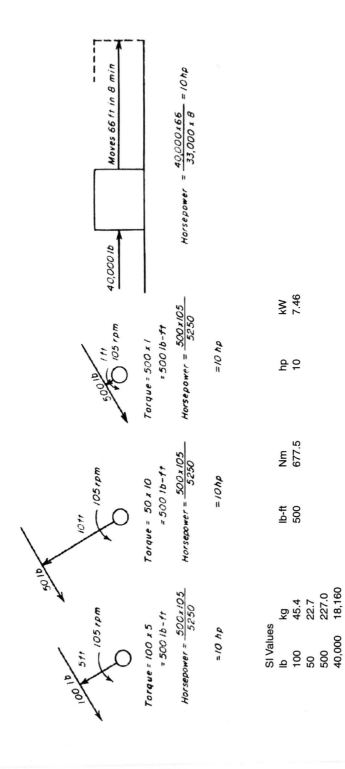

FIGURE 16 Horsepower (kW) values are found by knowing the torque and speed of a rotating body.

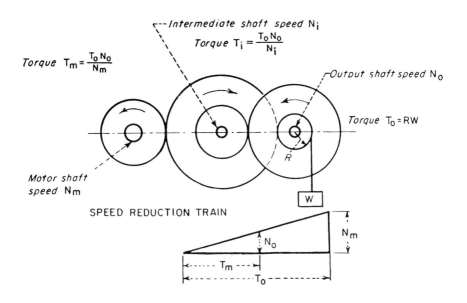

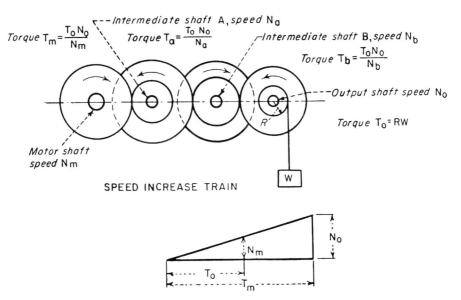

FIGURE 17 Gear trains can increase or decrease the speed and torque ratio between driver shaft and the output shaft.

T_o = torque on output shaft, lb ft (Nm)

N_m = speed of motor shaft, rpm

N_o = speed of output shaft, rpm

$$H_m = \frac{T_m N_m}{5250} = \frac{T_o N_o}{5250} = H_o$$

Therefore, the torque T_m at the driver shaft that results from torque T_o at the output shaft is

$$T_m = \frac{T_o N_o}{N_m}. \tag{6}$$

The relations between horsepower (kW) and the effects of introducing speed changes in a drive, such as the use of speed reducers, are important considerations in selecting a motor that will produce a desired torque.

The discussion thus far has dealt with rotating parts. In many applications there is a combination of rotating parts and linearly moving parts. To determine the accelerating torque for such combinations, it is necessary to reduce the mathematical consideration of the linearly moving parts to a basis equivalent to that used for the rotating parts—consideration of the latter generally being based on the speed at the driver shaft.

The linear speed of a concentrated mass at radius K ft(m) will be, as stated earlier, $2\pi K N$ ft/min (m/min), where N represents the rpm. Then the energy is 0.5 $M (2\pi KN)^2$ for the rotating body and 0.5 MV^2 for the linearly moving body. Equating these two expressions and substituting $W/32.2$ for M to obtain pounds (kg), the equation is:

$$E = \tfrac{1}{2} \frac{W}{32.2} (2\pi)^2 K^2 N^2 = \tfrac{1}{2} \frac{W}{32.2} V^2$$

or

$$WK^2(2\pi)^2 N^2 = WV^2 \tag{7}$$

Then, referred to the driver shaft, the equivalent WK_e^2 of a part moving in a straight line is similarly

$$WK_e^2(2\pi)^2 N_m^2 = WV^2$$

or

$$WK_e^2 = \frac{W}{39.48} \left(\frac{V}{N_m}\right)^2 \tag{8}$$

Equation 8 is valid only when the velocity V bears a constant relation to the driver speed N, as in the case of a rack engaged with a pinion on the driver shaft. And Eq. 8 cannot be applied to reciprocating parts because of their varying velocity, ranging from zero to a maximum.

Nevertheless, by using Eqs. 5 and 8, the calculation on all parts of a machine, that either rotate or move linearly, can be reduced to the mathematical basis corresponding to the speed of the driver or to any other reference speed. After the summation of the WK_e^2 of all parts has been made, the torque required to accelerate can be calculated from Eq. 3, and horsepower from Eq. 4. The factor N in Eq. 3 and 4 is the speed of the driver, or the reference speed, since all parts have been reduced to that speed by the summation of the part inertias.

For example, to compute the total WK_e^2 at the drive shaft for the power train in Fig. 18 where

V = lineal velocity of rack and load, ft/min (m/min)

A = total WK^2 of rotating parts on the drive shaft

B = total WK^2 of rotating parts on intermediate shaft

C = total WK^2 or rotating parts on final shaft

D = WV^2 of lineally moving rack and load

N_a = rpm of drive shaft

N_b = rpm of intermediate shaft

N_e = rpm of final shaft

WK_e^2 = total WK_e^2 referred to drive shaft for complete power train

$WK_e^2 = A + B(N_b/N_a)^2 + C(N_e/N_a)^2 + D/39.48\ N_a^2$

Torque required at the drive shaft to accelerate this WK_e^2 load from rest to a drive shaft speed of N_a rpm in t sec from Eq (3) is

$$T_a = \frac{WK_e^2 N_a}{308t}, \text{ lb ft (Nm)}$$

In a machine where the motor is coupled directly to the drive shaft, the T_m required at the motor shaft to accelerate the WK_e^2 of the load and the WK_m^2 of the motor is

$$T_m = \frac{(WK_e^2 + WK_m^2)N_a}{308t}$$

When the motor is connected to the drive shaft of the machine by a belt, the motor having a speed of N_m rpm and the drive shaft a speed of N_a rpm, the torque T_m required at the motor shaft to accelerate the WK_e^2 of the load referred to the drive shaft and the WK_m^2 of the motor referred to its shaft is

$$T_m = \frac{WK_m^2 N_m + WK_e^2(N_a/N_m)^2}{308t}$$

In addition to the torque required for acceleration, additional torque is required to overcome load torque, friction torques and break-away torques.

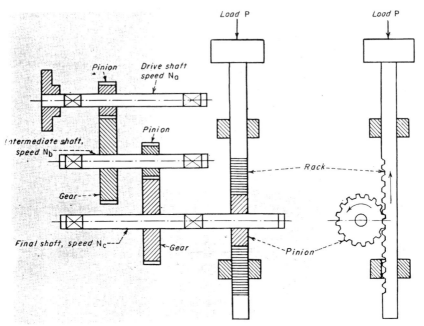

FIGURE 18   The equivalent WK^2 of the gear train and rack and pinion must be found before the input torque or the output torque can be calculated.

Torque at the drive shaft caused by load P from Eq (6) is

$$T_a' = \frac{T_c N_c}{N_a}, \text{ lb ft (Nm)}$$

or at the motor shaft

$$T_m' = T_a' N_a / N_m$$

Torque that results from the friction on parts such as bearings and gears windage, and break-away torque caused by the static friction of oil and grease in bearings, are often large enough to be considered. These torques can be expressed in terms of equivalent torque T_m'' at the motor shaft by using the same procedure as with other torques.

To start and accelerate a machine and to overcome load and friction torques, therefore, the motor must produce a torque T on its shaft such that

$$T = T_m + T_m' + T_m''$$

It will be seen from the previous discussion that WK^2 has a habit of popping up almost anywhere and that it is not only convenient but is an almost indispensable tool in the technical kit of those interested in determining accelerating torques. In fact, before any of the foregoing equations can be solved, the value of this quantity must be determined. The calculation of the weight W is simple. The determination of K, however, is relatively complicated. To simplify such calculations, Table 14 includes the radii of gyration, the weight and the WK^2, for various forms of bodies and for various axes of rotation of these bodies, as shown by the illustrations.

In most practical problems the actual rotating body will seldom correspond exactly to any one of the forms shown here. In every instance, however, it may be resolved into parts, each of which agrees absolutely or very closely with one of the forms in Table 14. For example, a spoked flywheel can usually be separated into a rim, spokes, hub, and web, each corresponding to one of the forms shown in Table 14. Then, the moment of inertia WK^2 of the wheel will be equal to the direct sum of the moments of inertia of all the parts as calculated separately, each about the same axis of rotation, usually the center of the shaft. Each separate calculation will simply be the weight of the particular part being considered multiplied by its radius of gyration squared. In ordinary calculations the radius is always taken in feet and the weight taken in pounds (meters and kilograms in SI units), unless otherwise specified.

In many instances it is not necessary to calculate each individual part exactly. If the flywheel has a heavy rim and is relatively large in diameter, it is usually sufficient to calculate the rim and then add a percentage for the spokes and hub, since they are lighter and are nearer the center and hence have a lower K. In the case of motor armatures, calculations without actual design data are usually based only on the weight and outside diameter. Table 15 gives values for K which are suitable for approximate calculations. Table 16 and Fig. 19 can also be used.

Now, having found out what a highly important part the moment of inertia plays in problems involving acceleration and having discussed ways for determining this value, there are certain practical points to be considered. The two factors making up this quantity are weight and radius. Since the latter enters into the problem as a squared quantity, increasing the radius has a greater effect on inertia than increasing the weight. Doubling the weight, for example, merely doubles the inertia, provided the radius remains the same, but doubling the radius, while the weight remains constant, increases the inertia four times. Hence, in designing high-speed parts to be started or reversed frequently, it is of first importance to keep their diameters small; next, to make them as light in weight as possible.

Even though high speed may not be involved, should the number of reversals or starts exceed, say, eight or ten per min, these same factors should be carefully considered.

To calculate torque and horsepower, another factor—speed—is encountered. It can be seen from Eqs. (3) and (4) that in determining the horsepower to accelerate a mechanism the inertia factor enters into the calculation only once, but the speed is there twice, or as a squared quantity. Thus, though the value of WK^2 should be kept low, it does not usually cause as much trouble as high speed. Often the im-

TABLE 16 Weight of Metals

Pound per Cubic Inch		kg/cu cm
Magnesium	0.0628	0.000002
Aluminum	0.0924	0.000003
Cast iron	0.260	0.000007
Steel	0.282	0.000008
Copper	0.318	0.000009
Bronze	0.320	0.000009
Lead	0.411	0.000011

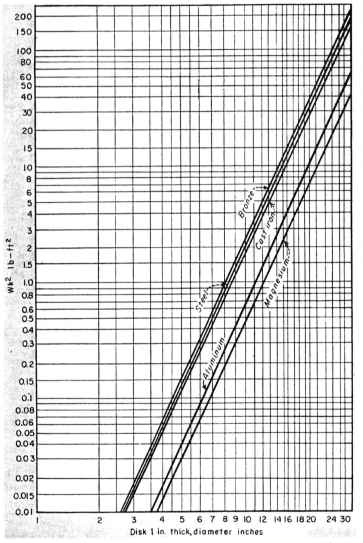

SI Values	
in.	cm
1	2.54
2	5.08
3	7.62
4	10.16
5	12.7
6	15.24
6	15.2
7	17.78
8	20.32
9	22.86
10	25.4
12	30.48
14	35.56
16	40.64
18	45.72
20	50.8
24	60.96
30	76.2

Disk 1 in. thick, diameter inches

SI Values

$lb\text{-}ft^2$	$m\text{-}kg^2$						
0.01	0.00042	0.15	0.00633	2.0	0.084	30	1.265
0.015	0.00063	0.20	0.00844	3	0.126	40	1.687
0.02	0.00084	0.3	0.0126	4	0.168	50	2.10
0.03	0.00127	0.4	0.0168	5	0.21	60	2.53
0.04	0.00169	0.5	0.021	6	0.253	80	3.37
0.05	0.00211	0.6	0.0253	8	0.337	100	4.217
0.06	0.00253	0.8	0.0337	10	0.421	150	6.326
0.08	0.00337	1.0	0.0421	15	0.632	200	8.453
0.10	0.00422	1.5	0.063	20	0.843		

FIGURE 19 WK^2 for 1 in (2.54 cm) thick disk of various diameters and different materials of construction.

portance of this point is overlooked by designers in laying out a drive which starts or reverses frequently. To have a mechanism reverse with the least expenditure of energy, the speed of its parts must be kept at the minimum.

Frequently motor manufacturers receive requests to supply reversing motors that are to run at 1800 rpm. It is usually more desirable to use motors running at 900 rpm or not more than 1200 rpm. The horsepower (kW) necessary to accelerate the 900 rpm motor is only ¼ of that required for the 1800 rpm motor plus what may be required to take care of the somewhat greater WK^2 of the slower-speed motor. Usually the gain from the reduced speed is much greater than the loss from increased WK^2. However, with the use of the slower speed, the external inertia, assuming no change, becomes a bigger percentage of the total. Actual calculations are frequently necessary to arrive at the proper combination of motor speed, and the speed of the external load.

In connection with this, reference to Eq. (5) will bring out another point in regard to speed. When finding the equivalent inertia of rotating parts, the square of the speed also enters into the calculations. Hence, if there is to be a gear train between the motor and the driven parts, it is advisable to make the reduction next to the motor as great as possible. In this way it is frequently possible to disregard the inertia of all parts beyond perhaps the first or second reducction. For example, with a 4:1 reduction adjacent to the motor and with no further reductions, anything beyond the motor shaft itself has only one-sixteenth the effect on the total inertia that it would have were it running at motor speed.

Of course, on rapid reversing jobs, all parts of the motor shaft, such as couplings or pinions, should be kept as small as possible. Too, the use of aluminum or aluminum alloys is often desirable for high-speed pulleys and sheaves.

This procedure is the work of John W. Harper, General Electric Company, as reported in *Product Engineering* magazine. SI values were added by the handbook editor.

SECTION 20
GEAR DESIGN AND APPLICATION

ANALYZING GEARS FOR DYNAMIC LOADS

A two-stage, step-up gearbox drives a compressor and has a lubrication pump mounted on one of the gear shafts, Fig. 1. Tables 1, 2, and 3 show the spur-gear data, tolerances for tooth errors, and polar moments of inertia for the masses in the compressor drive. All gears in the drive have 20° pressure angles. The gears are made of steel; the compressor is made of cast iron. A 50-hp (37.3-kW) motor drives the gearbox at 3550 rpm. What are the dynamic loads on this gearbox?

Calculation Procedure:

1. *Determine the pitchline velocity, V, applied load, W, at each mesh, and shaft speed, n*

In the equations that follow, subscripts identify the shaft, gear, or mesh under consideration. For example, the subscript A denotes Shafts A_1 and A_2; B, Shaft B; and C, Shaft C. Also, subscripts 1, 2, 3, 4 refer respectively to Gears 1, 2, 3, and 4 in the gearbox. The subscripts a and b denote the mesh of Gears 1 and 2 and the mesh of Gears 3 and 4, respectively. Finally, subscripts r and n refer to the driver gear and driven gear. Full nomenclature appears at the end of this procedure.

For Shaft A_1 and A_2,

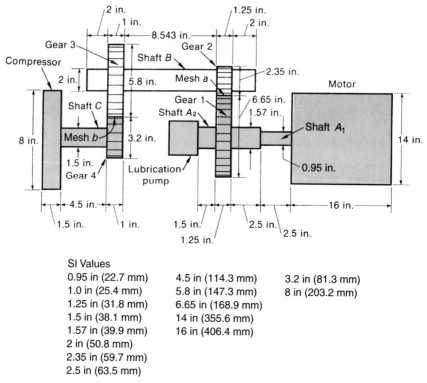

SI Values

0.95 in (22.7 mm)	4.5 in (114.3 mm)	3.2 in (81.3 mm)
1.0 in (25.4 mm)	5.8 in (147.3 mm)	8 in (203.2 mm)
1.25 in (31.8 mm)	6.65 in (168.9 mm)	
1.5 in (38.1 mm)	14 in (355.6 mm)	
1.57 in (39.9 mm)	16 in (406.4 mm)	
2 in (50.8 mm)		
2.35 in (59.7 mm)		
2.5 in (63.5 mm)		

FIGURE 1   Two-stage step-up gearbox driving a compressor (*Machine Design.*)

TABLE 1   Spur Gear Data

	Number of teeth, N	Pitch radius		Face width		Lewis form factor, y
		R (in)	mm	F (in)	mm	
Gear 1	133	3.325	84.5	1.25	31.8	0.276
Gear 2	47	1.175	29.8	1.25	31.8	0.256
Gear 3	116	2.90	73.7	1.00	25.4	0.275
Gear 4	64	1.60	40.6	1.00	25.4	0.264

$$n_A = \text{input speed}$$

$$= 3550 \text{ r/min}$$

$$P_A = \text{input power}$$

$$= 50 \text{ hp (37.3 kW)}$$

TABLE 2 Tolerances for Tooth Errors

	Tolerance for tooth profile error e_p		Tolerance for tooth spacing error e_s	
	in	mm	in	mm
Gear 1	0.0010	0.0254	0.0007	0.0178
Gear 2	0.0009	0.0229	0.00065	0.0165
Gear 3	0.0010	0.0254	0.0007	0.0178
Gear 4	0.0009	0.0229	0.00065	0.0165

TABLE 3 Polar Moment of Inertia of Cylindrical Masses

Element	Diameter, D		Length, L		Polar moment of inertia, I_o	
	in	mm	in	mm	$in^2 \cdot lb \cdot s^2/ft$	$cm^2 \cdot kg \cdot s^2/m$
Motor	14	355.6	16	406.4	534.751	5138.78
Shaft A_1	0.95	24.1	2.5	63.5	0.002	0.0192
Shaft A_2	1.57	39.88	4.0	101.6	0.021	0.2018
Gear 1	6.65	168.9	1.25	31.75	2.127	20.439
			$I_{oA} = \Sigma I_o =$		536.901	5159.44
Gear 2	2.35	59.69	1.25	31.75	0.033	0.3171
Shaft B	2.0	50.8	12.543	318.59	0.175	1.6816
Gear 3	5.8	147.3	1.0	25.4	0.985	9.465
			$I_{oB} = \Sigma I_o =$		1.193	11.463
Gear 4	3.2	81.28	1.0	25.4	0.091	0.8744
Shaft C	1.5	38.1	4.5	114.3	0.020	0.1921
Compressor	8.0	203.2	1.5	38.1	4.915	47.231
			$I_{oC} = \Sigma I_o =$		5.026	48.297

Tabulated values do not include the polar moment of inertia for the lubrication pump.
A material mass factor $B = 0.00087$ lb $\cdot$ s^2/in^3 $\cdot$ ft was assumed for steel; $B = 0.00080$ lb $\cdot$ sec^2/in^3 $\cdot$ ft (0.000073 kg $\cdot$ s^2/cm^3 $\cdot$ m) cast iron.
The polar moment of inertia for cylindrical masses was calcualted as, $I_o = BD^4L$.

$$T_A = 63{,}025 \, P_A/n_A$$

$$= 63{,}025 \,(50)/3550$$

$$= 888 \text{ lb} \cdot \text{in } (100.3 \text{ N} \cdot \text{m})$$

$$W_a = T_A/R_1$$

$$= 888/3.325$$

$$= 267 \text{ lb } (121.2 \text{ kg})$$

$$V_a = 0.5236 \, R_1 n_A$$

$$= 0.5236 \,(3.325)\,(3550)$$

$$= 6180 \text{ ft/min } (1883.7 \text{ m/min})$$

Therefore, for Shafts B and C, the following values for speed, torque, load and pitchline velocity are obtained.

$$n_B = n_A(N_1/N_2)$$

$$= 3550\,(133/47)$$

$$= 10{,}046 \text{ r/min}$$

$$T_B = T_A(N_2/N_1)$$

$$= 888\,(47/133)$$

$$= 314 \text{ lb} \cdot \text{in } (35.5 \text{ N} \cdot \text{m})$$

$$n_C = n_B(N_3/N_4)$$

$$= 10{,}046\,(116/64)$$

$$= 18{,}208 \text{ r/min}$$

$$T_C = T_B(N_4/N_3)$$

$$= 314\,(64/116)$$

$$= 173 \text{ lb} \cdot \text{in } (19.5 \text{ N} \cdot \text{m})$$

$$W_b = T_B/R_3$$

$$= 314/2.90$$

$$= 108 \text{ lb } (49 \text{ kg})$$

$$V_b = 0.5236\, R_3 n_B$$

$$= 0.5236\,(2.9)\,(10{,}046)$$

$$= 15{,}254 \text{ ft/min } (4649.4 \text{ m/min})$$

From these values, dynamic loads can be calculated.

Find the Total Effective Mass at Mesh a and Mesh b

Total effective mass m_t must be determined at each gear mesh. The parameter m_t can be calculated from

$$1/m_t = 1/m_r + 1/m_n$$

where m_r is the effective mass at the driver gear and m_n is the effective mass at the driven gear.

For rotating components such as gears, an effective mass m at the pitch radius R can be calculated from

$$m = \Sigma I_o / R^2$$

The gears in the compressor drive mesh at two points, for which the total effective mass m_t must be determined.

At Mesh a and Mesh b, m_t is

$$1/m_{ta} = 1/m_{ra} + 1/m_{na}$$

$$1/m_{tb} = 1/m_{rb} + 1/m_{nb}$$

However, a lubrication pump is directly coupled to Shaft A_2. For Mesh a, the effective mass at the driver gear m_{ra} is infinite because of the incompressible fluid in the pump that prevents instantaneous accelerations. Also at Mesh b, the effective mass at the driver gear m_{rb} is infinite because the reflected inertia of the pump is also infinite.

Consequently, the equations for m_t become

$$1/m_{ta} = 1/\infty + 1/m_{na}$$

$$m_{ta} = m_{na}$$

and

$$1/m_{tb} = 1/\infty + 1/m_{nb}$$

$$m_{tb} = m_{nb}$$

Finally, the total effective mass m_t at Mesh a and Mesh b can be calculated as

$$m_{ta} = m_{na} = \frac{I_{oB} + I_{oC}(R_3/R_4)^2}{R_2^{\,2}}$$

$$= \frac{1.193 + 5.026(2.90/160)^2}{1.175^2}$$

$$= 12.82 \ \text{lb} \cdot \text{s}^2/\text{ft} \ (19.1 \ \text{kg} \cdot \text{s}^2/\text{m})$$

$$m_{tb} = m_{nb} = I_{oC}/R_4^2$$

$$= 5.026/1.6^2$$

$$= 1.963 \ \text{lb} \cdot \text{s}^2/\text{ft} \ (2.924 \ \text{kg} \cdot \text{s}^2/\text{m})$$

3. Compute the acceleration force for the spur gears

For spur gears, the acceleration force f_1 is

$$f_1 = H m_t V^2$$

where $H = A_1 (1/R_r + 1/R_n)$. The calculation factor $A_1 = 0.00086$ for 14.5° teeth, 0.00120 for 20° teeth, and 0.00153 for 25° teeth. The parameters R_r and R_n are the pitch radii of driver and driven gears, respectively.

Therefore at Mesh a,

$$H_a = 0.00120(1R_r + 1/R_n)$$

$$= 0.00120(1/3.325 + 1/1.175)$$

$$= 0.00138$$

$$V_a = 6180 \text{ ft/min } (1883.7 \text{ m/min})$$

$$m_{ta} = 12.82 \text{ lb} \cdot \text{s}^2/\text{ft } (19.1 \text{ kg} \cdot \text{s}^2/\text{m})$$

$$f_{1a} = H_a m_{ta} V_a^2$$

$$= (0.00138)(12.82)(6180)^2$$

$$= 675,685 \text{ lb } (306,761 \text{ kg})$$

And at Mesh *b*,

$$H_b = 0.00120(1/R_r + 1/R_n)$$

$$= 0.00120(1/2.90 + 1/1.60)$$

$$= 0.00116$$

$$V_b = 15,254 \text{ ft} \cdot \text{min } (4649.4 \text{ m/min})$$

$$m_{tb} = 1.963 \text{ lb} \cdot \text{s}^2/\text{ft } (2.924 \text{ kg} \cdot \text{s}^2/\text{m})$$

$$f_{1b} = H_b m_{tb} V_b^2$$

$$= (0.00116)(1.963)(15,254)^2$$

$$= 529,841 \text{ lb } (240,548 \text{ kg})$$

4. Calculate the deflection force
The deflection force f_2 is

$$1/f_2 = 1/C + (1/C_1 + 1/C_2 + \cdots 1/C_x)$$

where C accounts for deflection from bending and compressive loads in the gear teeth, and C_x accounts for deflection from torsional loads in shafts, flexible couplings, and other components. C is the load required to deflect gear teeth by an amount equal to the error in action. C_x is the load required at the pitch radius to deflect a shaft or coupling by an amount equal to the error in action. Previously, calculations of f_2 only considered the parameter C. With the addition of C_x to the equation for f_2, the value of f_2 will always be less than the smallest value of C, $C_1, C_2 \ldots C_x$.

The parameter C is

$$C = W + 1000 \ eFA$$

where A is the load required to deflect teeth by 0.001 in (0.0254 mm).

If the Lewis form factor y is known, A can be calculated from

$$A = \frac{E_r Z_r E_n Z_n}{1000(E_r Z_r + E_n Z_n)}$$

where for the driver and driven gears, E is the modulus of elasticity, and Z is the calculation factor for A. Thus calculation factor Z is

$$Z = y/(0.242 + 7.25y)$$

The table below summarizes the results of calculations for A. For both Mesh a and Mesh b, $A = 1837$ lb (834 kg).

Summary of Calculations for Deflection Load

	Lewis form factor, y	Calculation factor, Z	Deflection load, A (lb)
Gear 1	0.276	0.123	At Mesh a,
Gear 2	0.256	0.122	$A = 1837$ (834 kg)
Gear 3	0.275	0.123	At Mesh b,
Gear 4	0.264	0.122	$A = 1837$ (834 kg)

Previously, the error in action e had to be assumed for a given class of gears, based on recommendations tabulated in handbooks. However, this error can be approximated as

$$e = \sqrt{(e_{pr} + e_{sr})^2 + (e_{pn} + e_{sn})^2}$$

where for the driver and driven gears, e_p is the tooth profile error and e_s is the tooth spacing error. If actual measurements of e_p and e_s are not available, then the tolerances for allowable e_p and e_s can be used for calculations instead. Therefore, for both Mesh a and Mesh b, the approximate error in action e is

$$e = \sqrt{(0.0010+0.0007)^2 + (0.0009 + 0.00065)^2}$$

$$= \sqrt{0.0000053}$$

$$= 0.0023 \text{ in } (0.058 \text{ mm})$$

Finally, C can be calculated for both meshes. At Mesh a,

$$C_a = W_a + 1000 e_a F_a A_a$$

$$= 267 + 1000(0.0023)(1.25)(1837)$$

$$= 5548 \text{ lb } (2519 \text{ kg})$$

and at Mesh b,

$$C_b = W_b + 1000 e_b F_b A_b$$

$$= 108 + 1000(0.0023)(1.00)(1837)$$

$$= 4333 \text{ lb } (1967 \text{ kg})$$

For steel shafts, the parameter C_x is calculated from

$$C_x = 1,080,000eD_s^4/R^2L_s$$

where D_s is the shaft diameter, and L_s is the shaft length. Generally, for other components such as flexible couplings, C_x has to be determined experimentally or from information supplied by the component manufacturer.

In the gear box, the portion of Shaft A_2 between the motor and Gear 1 deflects interdependently with Shaft A_1. On the other hand, the portion of Shaft A_2 between the lubrication pump and Gear 1 deflects independently of the shaft elements to the right of Gear 1. The table below summarizes the calculations of C_x for all shaft elements affecting each mesh.

Summary of Calculations for Shaft Deflection Load

Shaft elements for Mesh a	Shaft diameter, D_s (in)	mm	Shaft length, L_s (in)	mm	Pitch radius, R (in)	mm	Shaft deflection load, C_x (lb)	kg
Shaft A_1	0.95	24.1	2.5	63.5	3.325	84.5	73	33.1
Shaft A_2	1.57	39.9	2.5	63.5	3.325	84.5	546	247.9
Shaft A_2	1.57	39.9	1.5	38.1	3.325	84.5	910	413.1
Shaft B	2.00	50.8	8.543	216.9	1.175	29.8	3370	1529.9
Shaft elements for Mesh b								
Shaft B	2.00	50.8	8.543	216.9	2.90	73.7	553	251
Shaft C	1.5	38.1	4.5	114.3	1.60	40.6	1091	495.3

When shafts have steps of different diameter, the loads required to deflect each step torsionally by the error in action must be determined. Then the loads for each step must be combined as an inverse of the sum of the reciprocals of the loads. For example, Shaft A_2 has a different diameter than Shaft A_1 has. Consequently, the effective value for C_x between the motor and pump is

$$C_x = C_4 = C_3 + 1/(1C_2 + 1/C_1)$$

$$= 910 + 1/(1/546 + 1/73)$$

$$= 974 \text{ lb } (442.2 \text{ kg})$$

Note that values of C_x for each shaft element in the table are differentiated from each other by the use of a subscript x, enumerated as $x = 1, 2, 3, 4 \ldots$ This subscript should not be confused with numerical subscripts used to denote Gears 1, 2, 3, 4.

For Mesh a, f_2 is calculated as

$$1/f_{2a} = 1/C_a + 1/C_4 + 1/C_5$$

$$= 1/5548 + 1/974 + 1/3370$$

$$F_{2a} = 665 \text{ lb } (301.9 \text{ kg})$$

and for Mesh b, f_2 is

$$1/f_{2b} = 1/C_b + 1/C_6 + 1/C_7$$

$$= 1/4333 + 1/553 + 1/1091$$

$$f_{2b} = 338 \text{ lb } (153.5 \text{ kg})$$

5. Find the resultant force and dynamic load at each mesh
At Mesh a,

$$1/f_{aa} = 1/f_{1a} + 1/f_{2a}$$

$$= 1/675,685 + 1/665$$

$$f_{aa} = 664 \text{ lb}$$

$$W_{da} = W_a + \sqrt{f_{aa}(2f_{2a} - f_{aa})}$$

$$= 267 + \sqrt{664[2(665) - 664]}$$

$$= 932 \text{ lb } (423.1 \text{ kg})$$

and at Mesh b,

$$1/f_{ab} = 1/f_{1b} + 1/f_{2b}$$

$$= 1/529,841 + 1/338$$

$$f_{ab} = 338 \text{ lb}$$

$$W_{db} = W_b + \sqrt{f_{ab}(2f_{2b} - f_{ab})}$$

$$= 108 + \sqrt{338[2(338) - 338]}$$

$$= 446 \text{ lb } (202.5 \text{ kg})$$

At Mesh a and Mesh b, $W_d/W = 3.49$ and 4.13 respectively. Because the W_d/W ratio for both meshes is greater than two, there will be free impact between the gear teeth. Also, the drive will be noisy and may wear rapidly. The dynamic load W_d can de decreased by placing a flexible coupling between the motor and Shaft A_1. This will effectively isolate the motor mass. Also, the lubrication pump can be isolated by driving it with a quill shaft. Finally, the diameter of Shaft B can be minimized to meet the required torque capacity while increasing shaft resilience to lower the magnitude of W_d.

Related Calculations. Gears in mesh never operate under a smooth, continuous load. Factors such as manufacturing errors in tooth profile and spacing, tooth deflections under load, and imbalance all interact to create a dynamic load on gear teeth. The resulting action is similar to that of a variable load superimposed on a steady load.

Consider, for example, how tooth loads can fluctuate from manufacturing errors. The maximum, instantaneous tooth load occurs at the maximum error in action. The average load is the applied load on the teeth at the pitchline of the gears.

As each pair of teeth moves through its duration of contact, errors in action create periods of sudden acceleration that momentarily separate mating gear teeth.

Impact occurs as each pair of teeth returns to mesh to complete its contact duration. The impact loads can be significantly greater than the applied load at the pitchline. The maximum magnitude of the dynamic load depends on gear and pinion masses, connected component masses, operating speed, and material elasticity. Elastic deflections in gear teeth and drive components, such as shafts and flexible couplings, help reduce dynamic loads.

Generally, gears should be designed so that their bending and wear capacities are equal to or greater than the maximum, instantaneous dynamic load. However, the exact magnitude of the maximum dynamic load is seldom known. Although many gear studies have been conducted, there is no full agreement on the single best method for determining dynamic loads.

One of the most widely accepted methods for calculating dynamic loads considers the maximum dynamic load W_d, resulting from elastic impact, to be

$$W_d = W + W_i$$

The term W is the applied load at the pitchline of a gear. For gears, the incremental load W_i is

$$W_i = \sqrt{f_a(2f_2 - f_a)}$$

In the above equation, f_a is the resultant force required to accelerate masses in a system of elastic bodies. The resultant acceleration force f_a is

$$1/f_a = 1/f_1 + 1/f_2$$

The term f_1 is the force required to accelerate masses in a system of rigid bodies. The term f_2 is the force required to deflect the system elastically by the amount of error in action. As mentioned previously, free impact loads occur in gears when the teeth separate and return to mesh suddenly during the contact interval. For free impact, the value of the ratio W_d/W will always be greater than two.

If forces f_1, f_2, and f_a are plotted as functions of pitchline velocity, force f_2 is seen to be an asymptote of the incremental load W_i. The equations that define f_1 and f_2 depend on the type of gears being analyzed for dynamic loads. The example presented in this procedure provides equations defining f_1 and f_2 for spur gear applications.

Sometimes this method for calculating W_d gives values for dynamic load that are conservative. These high dynamic load estimates can lead to overdesigned gears. Conservative calculations of W_d can be minimized if f_2 can be calcualted more accurately. A less conservative calculation of f_2 effectively reduces W_d because f_2 is an asymptote for the incremental load W_i.

The method presented here has been refined to give more accurate values for f_2. Previously, only the elastic deflection in the gear teeth was considered in the calculation of f_2. Now, elastic deflections in other mechanical components, such as flexible couplings and shafts, are also considered for their effects on f_2 through inclusion of a parameter C_x.

Also, the calculation of f_2 is further refined by a more accurate approximation of the error in action. Previously, the error in action was assumed for a given class of gears, based on recommendations commonly tabulated in gear design handbooks. Now the error in action can be calculated from tooth profile errors and tooth-to-tooth spacing errors. The equation for approximating error in action results from tests where gears were measured for error in action, profile error, spacing error, and runout error.

This procedure shows how the refinements in the dynamic load analysis apply in spur gear applications. However, the principles can be applied to other forms of gearing as well. Furthermore, the example describes how resilience in mechanical components, such as shafts, can be applied to reduce dynamic loads on gears. Reduced dynamic loads result in lower operating noise and longer component life.

In all new applications where there is no design experience, gear drives should be analyzed for the possibility of dynamic loading. Also, all gear drives operating at peripheral speeds of about 1000 ft/min (304.8 m/min) and higher should be checked.

At low operating speeds, usually less than 1000 ft/min (304.8 m/min), gear drives follow the dynamics of rigid bodies. Consequently, elastic deflections in components have little effect on dynamic loads. Inertial forces generated by momentum variations appear to be directly proportional to mass magnitude and the square of mass velocity. At intermediate operating speeds, elastic deflections help reduce instantaneous dynamic loads. In this range, inertial forces appear to be directly proportional to mass velocity and the square root of mass magnitude. At higher operating speeds, elastic deflections have a pronounced influence on dynamic loads. The deflections tend to limit inertial forces to an asymptotic value. Consequently, inertial forces appear to be independent of mass magnitude and mass velocity.

Some applications must be carefully analyzed for dynamic loading. Examples include:

- Drives where large masses are directly connected, or connected with short shafts, to gear drives;
- Gear drives connected with solid couplings;
- Drives with fluid pumps connected directly to gears or gear shafts;
- Long gear trains, such as those in printing presses, paper machines, and process machinery;
- Auxiliary drives with small gears connected to high power systems;
- Drives with multiple power inputs;
- High-speed drives that include servodrives;
- Any gear drive where gears shown signs of distress or operate with high noise for no apparent reason.

This procedure is the work of Eliot K. Buckingham, President, Buckingham Associates, Inc., as reported in *Machine Design* magazine.

Nomenclature

A = Load which deflects gear teeth by 0.001 in, lb (0.0254 mm, kg)
A_1 = Calculation factor for H
B = Material mass factor, $\text{lb} \cdot \text{s}^2/\text{in}^3 \cdot \text{ft}$ $(\text{kg} \cdot \text{s}^2/\text{cm}^3\text{-m})$
C = Load which deflects gear teeth by an amount equal to the error in action, lb (kg)
$C_{1,2,3\cdots r}$ = Load at the pitch radius which deflects a shaft by an amount equal to the error in action, lb (kg)
D = Cylinder diameter, in (mm)
D_s = Shaft diameter, in (mm)
E = Modulus of elasticity, $\text{lb}/\text{in}^2(\text{kPa})$
e = Error in action, in (mm)
e_p = Tooth profile error, in (mm)

e_s = Tooth spacing error, in (mm)
f = Face width, in (mm)
f_a = Resultant force required to accelerate masses in a system of elastic bodies, lb (kg)
f_1 = Force required to accelerate masses in a system of rigid bodies, lb (kg)
f_2 = Force required to deflect elastically the system by the amount of error in action, lb (kg)
H = Calculation factor for f_1
I_a = Polar moment of inertia, $in^2 \cdot lb \cdot s^2/ft$ $(cm^2 \cdot kg \cdot s^2/m)$
L = Cylinder length, in (mm)
L_s = Shaft length, in (mm)
m = Effective mass, $lb \cdot s^2/ft$ $(kg \cdot s^2/m)$
m_n = Effective mass at the driven gear, $lb \cdot s^2/ft$ $(kg \cdot s^2/m)$
m_r = Effective mass at the driver gear, $lb \cdot s^2/ft$ $(kg \cdot s^2/m)$
m_t = Total effective mass at the mesh, $lb \cdot s^2/ft$ $(kg \cdot s^2/m)$
N = Number of gear teeth
n = Shaft speed, rpm
P = Power, hp (kW)
R = Pitch radius, in (mm)
T = Torque, $lb \cdot in$ $(N \cdot m)$
V = Pitchline velocity, ft/min (m/min)
W = Applied load at the pitch radius, lb (kg)
W_d = Dynamic load, lb (kg)
W_i = Incremental load, lb (kg)
y = Lewis form factor
Z = Calculation factor for C

Subscripts

A, B, C = Shaft A, B, and C
$1, 2, 3, 4$ = Gear 1, 2, 3, and 4
a, b = Mesh a and b
n, r = Driven gear, driver gear

HELICAL-GEAR LAYOUT ANALYSIS

Helical gears are to be designed for shafts 4-in (10.16-cm) apart and at right angles, Fig. 2. The normal dimeteral pitch, P_n = 20; the number of teeth, N, in the gear or worm wheel = 64; the number of teeth, n, in the pinion or threads in the worm = 32. Determine the suitable helix angle, or angles, α, of the gear or worm.

Calculation Procedure:

1. *Determine if the selected variables for this gearset are suitable*
In the preliminary layout of gear drives, the designer must determine if the diameteral pitch, gear ratio, and center distance are suitable for the proposed layout. With this settled, the designer can proceed to calculate the specific helix angle.

To check a proposed layout, use the relation,

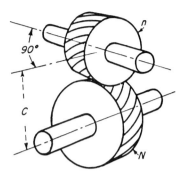

FIGURE 2 Helical-gear layout. (*Product Engineering.*)

$$\frac{2CP_n}{N} = \frac{R}{\sin a} + \frac{1}{\cos a}$$

where

C = center distance, in (mm)
P_n = normal diameteral pitch
N = number of teeth in gear or worm wheel
n = number of teeth in pinion or threads in worm
R = ratio, n/N
a = helix angle of gear or worm wheel, deg.

Find the value of the lefthand portion of this equation for this gear layout from: $2(4)(20)/64 = 2.5$. Then, $R = 32/64 = 0.5$.

2. Find the suitable helix angle or angles
Knowing the value of the lefthand side of the equation and the value of R we can assume suitable values for alpha and solve the equation. If the chosen helix angle is correct, the two sides of the equation will be equal.

As a first choice, assume a helix angle of 20°. Substituting, using values from standard trigonometric tables, we have $2.5 = (0.5/0.3584) + (1/0.9336) = 2.466$. This helix angle is not suitable because the two sides of the equation are not equal.

Try 20°30'. Or, $2.5 = (0.5/0.3486) + (1/0.9373) = 2.4968$. This is much closer than the first try. Another trial calculation shows that 20°25' is the correct helix angle for this layout.

Another angle, namely 58°30' is also suitable, based on the equation above. For greater efficiency and less wear, the gear and pinion helix angles should be nearly equal. Thus, the gear angle chosen could be 58°30' and the pinion angle becomes 31° and 30'.

Related Calculations. Use this equation and general approach for 90 deg shaft angles for helical gears in any service. Note that the trial helix angles chosen should be in the general range for the type of drive being considered. This will save time in the computation.

This procedure is the work of Wayne A. Ring, Barber-Colman Company, as reported in *Product Engineering* magazine.

ANALYZING SHAFT SPEED IN EPICYCLIC GEAR TRAINS

Analyze the input and output shaft rpm for the six epicyclic gear trains shown in Figs. 3 through 8. The gear train in Fig. 3 has arm A integral with the righthand shaft. Gears C and D are keyed to a short length of shaft which is mounted in a bearing in arm A. Gear C meshes with internal gear E which is keyed to the lefthand shaft.

Calculation Procedure:

1. *Find the ratio of the shaft speeds*
To find the ratio of the speed of shaft E to the speed of shaft A, proceed thus: Let N_b be the number of teeth in gear B, N_c the number in gear C, and so on. Let arm A, which was originally in a vertical position, be given an angular displacement, θ.

In giving arm A the angular displacement, gear C will traverse through arc *ab* on gear B. Since angles are inversely proportion to radii, or to the number of teeth, gears C and D with turn through angle $(\theta)N_b/N_c$.

While the foregoing was taking place, gears D and E were rotating on each other through the equal arcs *ed* and *ef*. Gear E will have been turned in the reverse direction through angle $(\theta)(N_b/N_c)(N_d/N_e)$.

The net effect of these two operations is to move the point of gear E, which was originally vertical at *g*, over to location *f*. Gear E has thus rotated through

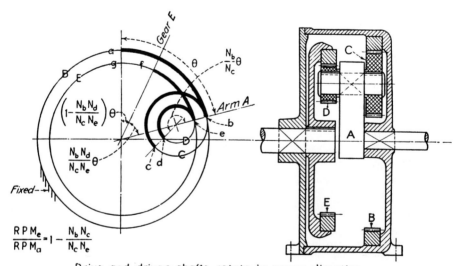

Drive and driven shafts rotate in **same** direction

FIGURE 3 Epicyclic gear with drive and driven shafts rotating in same direction. (*Product Engineering.*)

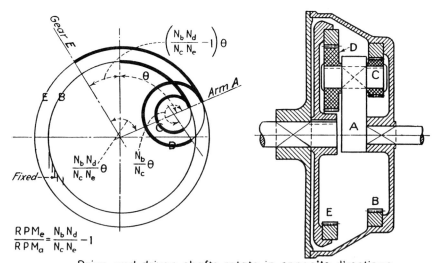

FIGURE 4 Drive and driven shafts rotate in opposite directions. (*Product Engineering.*)

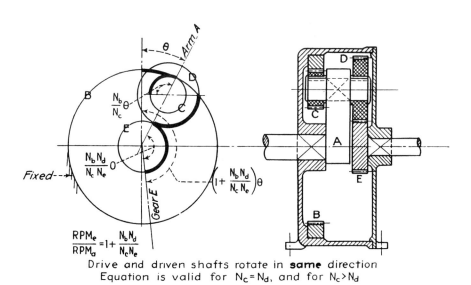

FIGURE 5 Drive and driven shafts rotate in same direction. (*Product Engineering.*)

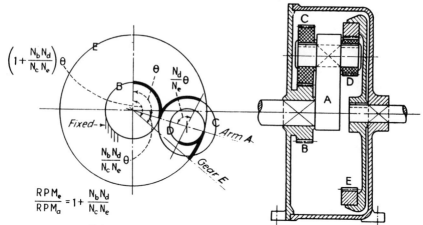

$$\left(1+\frac{N_b N_d}{N_c N_e}\right)\theta$$

$$\frac{RPM_e}{RPM_a} = 1 + \frac{N_b N_d}{N_c N_e}$$

Drive and driven shafts rotate in **same** direction
Equation is valid for $N_c = N_d$, and for $N_c > N_d$

FIGURE 6   Drive and driven shafts rotate in same direction. (*Product Engineering.*)

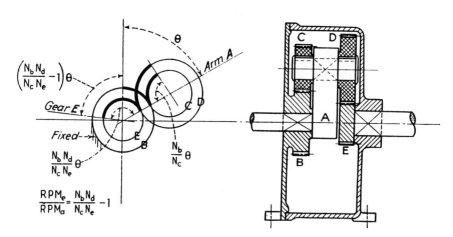

$$\left(\frac{N_b N_d}{N_c N_e} - 1\right)\theta$$

$$\frac{RPM_e}{RPM_a} = \frac{N_b N_d}{N_c N_e} - 1$$

Drive and driven shafts rotate in **opposite** directions

FIGURE 7   Drive and driven shafts rotate in opposite directions. (*Product Engineering.*)

angle $(1 - [N_b/N_d]/[N_c/N_e])(\theta)$. This latter value, when divided by θ, the angular movement of shaft A, gives the ratio of the rotations of shafts E and A, respectively.

2.   *Extend the analysis to other gear arrangements*
The above analysis can be applied to the gear arrangements shown in Fig. 4 through 8. When this is done the speed ratios for each arrangement is as given in the illustrations.

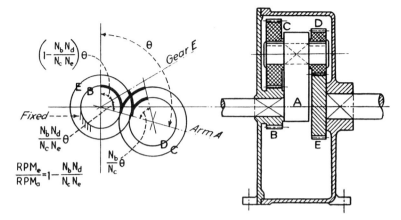

Drive and driven shafts rotate in **same** direction

FIGURE 8   Drive and driven shafts rotate in same direction. (*Product Engineering.*)

Related Calculations.   This procedure is a valuable analysis for epicyclic gear trains of all types. The resulting speed equations can be used to determine both input and output speeds of the gearset.

E. F. Spotts, Northwestern Technological Institute, prepared this procedure which was reported in *Product Engineering* magazine.

SPEEDS OF GEARS AND GEAR TRAINS

A gear having 60 teeth is driven by a 12-tooth gear turning at 800 r/min. What is the speed of the driven gear? What would be the speed of the driven gear if a 24-tooth idler gear were placed between the driving and driven gear? What would be the speed of the driven gear if two 24-tooth idlers where used? What is the direction of rotation of the driven gear when one and two idlers are used? A 24-tooth driving gear turning at 600 r/min meshes with a 48-tooth compound gear. The second gear of the compound gear has 72 teeth and drives a 96-tooth gear. What are the speed and direction of rotation of the 96-tooth gear?

Calculation Procedure:

1.   Compute the speed of the driven gear
For any two meshing gears, the speed ratio $R_D/R_d = N_d/N_D$, where R_D = rpm of driving gear; R_d = rpm of driven gear; N_d = number of teeth in driven gear; N_D = number of teeth in driving gear. By substituting the given values, $R_D/R_d = N_d/N_D$, or $800/R_d = 60/12$; $R_d = 160$ r/min.

2.   Determine the effect of one idler gear
An idler gear has *no* effect on the speed of the driving or driven gear. Thus, the speed of each gear would remain the same, regardless of the number of teeth in

the idler gear. An idler gear is generally used to reduce the required diameter of the driving and driven gears on two widely separated shafts.

3. Determine the effect of two idler gears
The effect of more than one idler is the same as that of a single idler—i.e., the speed of the driving and driven gears remains the same, regardless of the number of idlers used.

4. Determine the direction of rotation of the gears
Where an odd number of gears are used in a gear train, the first and last gears turn in the *same* direction. Thus, with one idler, one driver, and one driven gear, the driver and driven gear turn in the *same* direction because there are three gears (i.e., an odd number) in the gear train.

Where an even number of gears is used in a gear train, the first and last gears turn in the *opposite* direction. Thus, with two idlers, one driver, and one driven gear, the driver and driven gear turn in the *opposite* direction because there are four gears (i.e., an even number) in the gear train.

5. Determine the compound-gear output speed
A compound gear has two gears keyed to the same shaft. One of the gears is driven by another gear; the second gear of the compound set drives another gear. In a compound gear train, the product of the number of teeth of the driving gears and the rpm of the first driver equals the product of the number of teeth of the driven gears and the rpm of the last driven gear.

In this gearset, the first driver has 24 teeth and the second driver has 72 teeth. The rpm of the first driver is 600. The driven gears have 48 and 96 teeth, respectively. Speed of the final gear is unknown. Applying the above rule gives $(24)(72)(600)2(48)(96)(R_d)$; $R_d = 215$ r/min.

Apply the rule in step 4 to determine the direction of rotation of the final gear. Since the gearset has an even number of gears, four, the final gear revolves in the opposite direction from the first driving gear.

Related Calculations.   Use the general procedure given here for gears and gear trains having spur, bevel, helical, spiral, worm, or hypoid gears. Be certain to determine the correct number of teeth and the gear rpm before substituting values in the given equations.

SELECTION OF GEAR SIZE AND TYPE

Select the type and size of gears to use for a 100-ft³/min (0.047-m³/s) reciprocating air compressor driven by a 50-hp (37.3-kW) electric motor. The compressor and motor shafts are on parallel axes 21 in (53.3 cm) apart. The motor shaft turns at 1800 r/min while the compressor shaft turns at 300 r/min. Is the distance between the shafts sufficient for the gears chosen?

Calculation Procedure:

1. Choose the type of gears to use
Table 4 lists the kinds of gears in common use for shafts having parallel, intersecting, and nonintersecting axes. Thus, Table 14 shows that for shafts having parallel axes, spur or helical, external or internal, gears are commonly chosen. Since

TABLE 4 Types of Gears in Common Use*

Parallel axes	Intersecting axes	Nonintersecting parallel axes
Spur, external	Straight bevel	Crossed helical
Spur, internal	Zerol† bevel	Single-enveloping worm
Helical, external	Spiral bevel	Double-enveloping worm
Helical, internal	Face gear	Hypoid

°From Darle W. Dudley—*Practical Gear Design*, McGraw-Hill, 1954.
†Registered trademark of the Gleason Works.

external gears are simpler to apply than internal gears, the external type is chosen wherever possible. Internal gears are the planetary type and are popular for applications where limited space is available. Space is not a consideration in this application; hence, an external spur gearset will be used.

Table 5 lists factors to consider in selecting gears by the characteristics of the application. As with Table 4, the data in Table 5 indicate that spur gears are suitable for this drive. Table 6, based on the convenience of the user, also indicates that spur gears are suitable.

2. Compute the pitch diameter of each gear
The distance between the driving and driven shafts is 21 in (53.3 cm). This distance is approximately equal to the sum of the driving gear pitch radius r_D in and the driven gear pitch radius r_d in. Or $d_D + r_d = 21$ in (53.3 cm).

In this installation the driving gear is mounted on the motor shaft and turns at 1800 r/min. The driven gear is mounted on the compressor shaft and turns at 300 r/min. Thus, the speed ratio of the gears (R_D, driver rpm/R_d, driven rpm) = $1800/300 = 6$. For a spur gear, $R_D/R_d = r_d/r_D$, or $6 = r_d/r_D$, and $r_d = 6r_D$. Hence, substituting in $r_D + r_d = 21$, $r_D + 6r_D = 21$; $r_D = 3$ in (7.6 cm). Then $3 + r_d = 21$, $r_d = 18$ in (45.7 cm). The respective pitch diameters of the gears are $d_D = 2 \times 3 = 6.0$ in (15.2 cm); $d_d = 2 \times 18 = 36.0$ in (91.4 cm).

3. Determine the number of teeth in each gear
The number of teeth in a spur gearset, N_D and N_d, can be approximated from the ratio $R_D/R_d = N_d/N_D$, or $1800/300 = N_d/N_D$; $N_d = 6N_D$. Hence, the driven gear will have approximately six times as many teeth as the driving gear.

As a trial, assume that $N_d = 72$ teeth; then $N_D = N_d/6 = 72/6 = 12$ teeth. This assumption must now be checked to determine whether the gears will give the desired output speed. Since $R_D/R_d = N_d/N_D$, or $1800/300 = 72/12$; $6 = 6$. Thus, the gears will provide the desired speed change.

The distance between the shafts is 21 in (53.3 cm) = $r_D + r_d$. This means that there is no clearance when the gears are meshed. Since all gears require some clearance, the shafts will have to be moved apart slightly to provide this clearance. If the shafts cannot be moved apart, the gear diameter must be reduced. In this installation, however, the electric-motor driver can probably be moved a fraction of an inch to provide the desired clearance.

4. Choose the final gear size
Refer to a catalog of stock gears. From this catalog choose a driving and a driven gear having the required number of teeth and the required pitch diameter. If gears of the exact size required are not available, pick the nearest suitable stock sizes.

TABLE 5 Gear Drive Selection by Application Characteristics*

Characteristic	Type of gearbox	Kind of teeth	Range of use
High power	Simple, branched, or epicyclic	Helical	Up to 40,000 hp (29,828 kW) per single mesh; over 60,000 hp (44,742 kW) in MDT designs; up to 40,000 hp (29,282 kW) in epicyclic units
	Simple, branched, or epicyclic	Spur	Up to 4000 hp (2983 kW) per single mesh; up to 10,000 hp (7457 kW) in an epicyclic
	Simple	Spiral bevel	Up to 15,000 hp (11,186 kW) per single mesh
		Zerol bevel	Up to 1000 hp (745.7 kW) per single mesh
High efficiency	Simple	Spur, helical or bevel	Over 99 percent efficiency in the most favorable cases—98 percent efficiency is typical
Light weight	Epicyclic	Spur or helical	Outstanding in airplane and helicopter drives
	Branched-MDT	Helical	Very good in marine main reductions
	Differential	Spur or helical	Outstanding in high-torque-actuating devices
		Bevel	Automobiles, trucks, and instruments
Compact	Epicyclic	Spur or helical	Good in aircraft nacelles
	Simple	Worm-gear	Good in high-ratio industrial speed reducers
	Simple	Spiroid	Good in tools and other applications
	Simple	Hypoid	Good in auto and truck rear ends plus other applications
	Simple	Worm-gear	Widely used in machine-tool index drives
	Simple	Hypoid	Used in certain index drives for machine tools
Precision	Simple or branched	Helical	A favorite for high-speed, high-accuracy power gears
	Simple	Spur	Widely used in radar pedestal gearing, gun control drives, navigation instruments, and many other applications
	Simple	Spiroid	Used where precision and adjustable backlash are needed

*Mechanical Engineering, November 1965.

Check the speed ratio, using the procedure in step 3. As a general rule, stock gears having a slightly different number of teeth or a somewhat smaller or larger pitch diameter will provide nearly the desired speed ratio. When suitable stock gears are not available in once catalog, refer to one or more other catalogs. If suitable stock gears are still not available, and if the speed ratio is a critical factor in the selection of the gear, custom-sized gears may have to be manufactured.

Related Calculations.   Use this general procedure to choose gear drives employing any of the 12 types of gears listed in Table 4. Table 7 lists typical gear selections based on the arrangement of the driving and driven equipment. These tables are the work of Darle W. Dudley.

TABLE 6 Gear Drive Selection for the Convenience of the User*

Consideration	Kind of teeth	Typical applications	Comments
Cost	Spur	Toys, clocks, instruments, industrial drives, machine tools, transmissions, military equipment, household applications, rocket boosters	Very widely used in all manner of applications where power and speed requirements are not too great—parts are often mass-produced at very low cost per part
Ease of use	Spur or helical	Change gears in machine tools, vehicle transmissions where gear shifting occurs	Ease of changing gears to change ratio is important
	Worm-gear	Speed reducers	High ratio drive obtained with only two gear parts
Simplicity	Crossed helical	Light power drives	No critical positioning required in a right-angle drive
	Face gear	Small power drives	Simple and easy to position for a right-angle drive
	Helical	Marine main drive units for ships, generator drives in power plants	Helical teeth with good accuracy and a design that provides good axial overlap mesh very smoothly
	Spiral bevel	Main drive units for aircraft, ships, and many other applications	Helical type of tooth action in a right-angle power drive
Noise	Hypoid	Automotive rear axle	Helical type of tooth provides high overlap
	Worm-gear	Small power drives in marine, industrial, and household appliance applications	Overlapping, multiple tooth contacts
	Spiroid-Gear	Portable tools, home appliances	Overlapping, multiple tooth contacts

Mechanical Engineering, November 1965.

GEAR SELECTION FOR LIGHT LOADS

Detail a generalized gear-selection procedure useful for spur, rack, spiral miter, miter, bevel, helical, and worm gears. Assume that the drive horsepower and speed ratio are known.

Calculation Procedure:

1. *Choose the type of gear to use*
Use Table 4 of the previous calculation procedure as a general guide to the type of gear to use. Make a tentative choice of the gear type.

2. *Select the pitch diameter of the pinion and gear*
Compute the pitch diameter of the pinion from $d_p = 2c/(R + 2)$, where d_p = pitch diameter, in, of the pinion, which is the *smaller* of the two gears in mesh; c = center distance between the gear shafts, in; R = gear ratio = larger rpm, number

TABLE 7 Gear Drive Selection by Arrangement of Driving and Driven Equipment[*]

Kind of teeth	Axes	Gearbox type	Type of tooth contact	Generic family
Spur[†]	Parallel	Simple (pinion and gear), epicyclic (planetary, star, solar), branched systems, idler for reverse	Line	Coplanar
Helical[†] (single or double helical, herringbone)	Parallel	Simple, epicyclic, branched	Overlapping line	Coplanar
Bevel	Right-angle or angular, but intersecting	Simple, epicyclic, branched	(Straight) line, (Zerol)[‡] line, (spiral) overlapping	Coplanar
Worm	Right-angle, nonintersecting	Simple	(Cylindrical) overlapping line, (double-enveloping)[§] overlapping line	Nonplanar
Crossed helical	Right-angle or skew, nonintersecting	Simple	Point	Nonplanar
Face gear	Right-angle, intersecting	Simple	Line (overlapping if helical)	Coplanar
Hypoid	Right-angle or angular, nonintersecting	Simple	Overlapping line	Nonplanar
Spiroid, helicon, planoid	Right-angle, nonintersecting	Simple	Overlapping line	Nonplanar

[*]*Mechanical Engineering*, November 1965.
[†]These kinds of teeth are often used to change rotary motion to linear motion by use of a pinion and rack.
[‡]Zerol is a registered trademark of the Gleason Works, Rochester, New York.
[§]The most widely used double-enveloping worm gear is the cone-drive type.

of teeth, or pitch diameter + smaller rpm, smaller number of teeth, or smaller pitch diameter.

Compute the pitch diameter of the gear, which is the *larger* of the two gears in mesh, from $d_g = d_p R$.

3. Determine the diametral pitch of the drive
Tables 8 to 11 show typical diametral pitches for various horsepower ratings and gear materials. Enter the appropriate table at the horsepower that will be transmitted, and select the diametral pitch of the pinion.

4. Choose the gears to use
Enter a manufacturer's engineering tabulation of gear properties, and select the pinion and gear for the horsepower and rpm of the drive. Note that the rated horsepower of the pinion and the gear must equal, or exceed, the rated horsepower of the drive at this specified input and output rpm.

5. Compute the actual center distance
Find half the sum of the pitch diameter of the pinion and the pitch diameter of the gear. This is the actual center-to-center distance of the drive. Compare this value with the available space. If the actual center distance exceeds the allowable distance, try to rearrange the drive or select another type of gear and pinion.

6. Check the drive speed ratio
Find the actual speed ratio by dividing the number of teeth in the gear by the number of teeth in the pinion. Compare the actual ratio with the desired ratio. If there is a major difference, change the number of teeth in the pinion or gear or both.

Related Calculations.   Use this general procedure to select gear drives for loads up to the ratings shown in the accompanying tables. For larger loads, use the procedures given elsewhere in this section.

TABLE 8 Spur-Gear Pitch Selection Guide*

(20° pressure angle)

Gear diametral pitch		Pinion		Gear	
in	cm	hp	W	hp	W
20	50.8	0.04–1.69	29.8–1,260	0.13–0.96	96.9–715.9
16	40.6	0.09–2.46	67.1–1,834	0.22–1.61	164.1–1,200
12	30.5	0.24–5.04	179.0–3,758	0.43–3.16	320.8–2,356
10	25.4	0.46–6.92	343.0–5,160	0.70–5.12	522.0–3,818
8	20.3	0.88–10.69	656.2–7,972	1.11–7.87	827.7–5,869
6	15.2	1.84–16.63	1,372–12,401	2.28–12.39	1,700–9,239
5	12.7	3.04–24.15	2,267–18,009	3.75–17.19	2,796–12,819
4	10.2	5.29–34.83	3,945–25,973	6.36–25.17	4,743–18,769
3	7.6	13.57–70.46	10,119–52,542	15.86–51.91	11,831–38,709

*Morse Chain Company.

TABLE 9 Miter and Bevel-Gear Pitch Selection Guide*

(20° pressure angle)

Gear diametral pitch		Hardened gear		Unhardened gear	
in	cm	hp	kW	hp	kW
			Steel spiral miter		
18	45.7	0.07–0.70	0.053–0.522	0.04–0.42	0.030–0.313
12	30.5	0.15–1.96	0.112–1.462	0.09–1.17	0.067–0.873
10	25.4	0.50–4.53	0.373–3.378	0.30–2.70	0.224–2.013
8	20.3	1.56–7.15	1.163–5.331	0.93–4.26	0.694–3.177
7	17.8	1.93–9.30	1.439–6.935	1.15–5.54	0.858–4.131
			Steel miter		
20	50.8	. . . .	. . . .	0.01–0.12	0.008–0.090
16	40.6	0.07–0.73	0.053–0.544	0.02–0.72	0.015–0.537
14	35.6	. . . .	. . . .	0.04–0.37	0.030–0.276
12	30.5	0.14–2.96	0.104–2.207	0.07–1.77	0.052–1.320
10	25.4	0.39–3.47	0.291–2.588	0.23–2.07	0.172–1.544

Steel and cast-iron bevel gears

Ratio	hp	W
1.5:1	0.04–2.34	29.8–1744.9
2:1	0.01–12.09	7.5–9015.5
3:1	0.04–8.32	29.8–6204.2
4:1	0.05–10.60	37.3–7904.4
6:1	0.07–2.16	52.2–1610.7

*Morse Chain Company.

TABLE 10 Helical-Gear Pitch Selection Guide*

Gear diametral pitch		Hardened-steel gear	
in	cm	hp	W
20	50.8	0.04–1.80	29.8–1,342.3
16	40.6	0.08–2.97	59.7–2,214.7
12	30.5	0.22–5.87	164.1–4,377.3
10	25.4	0.37–8.29	275.9–6,181.9
8	20.3	†0.66–11.71	492.2–8,732.1
		‡0.49–9.07	365.4–6,763.5
6	15.2	§1.44–19.15	1,073.8–14,280.2
		†1.15–15.91	857.6–11,864.1

*Morse Chain Company.
†1-in (2.5-cm) face.
‡¾-in (1.9-cm) face.
§1½-in (3.8-cm) face.

TABLE 11   Worm-Gear Pitch Selection Guide*

Gear diametral pitch		Bronze gears					
		Single		Double		Quadruple	
in	cm	hp	W	hp	W	hp	W
12	30.5	0.04–0.64	29.8–477.2	0.05–1.21	37.3–902.3	0.05–3.11	37.3–2319
10	25.4	0.06–0.97	44.7–723.3	0.08–2.49	59.7–1856	0.13–4.73	96.9–3527
8	20.3	0.11–1.51	82.0–1126	0.15–3.95	111.9–2946	0.08–7.69	59.7–5734
				Triple			
5	12.7	0.51–4.61	380.3–3437	1.10–10.53	820.3–7852		
4	10.2	0.66–6.74	492.2–5026				

*Morse Chain Company.

SELECTION OF GEAR DIMENSIONS

A mild-steel 20-tooth 20° full-depth-type spur-gear pinion turning at 900 r/min must transmit 50 hp (37.3 kW) to a 300-r/min mild-steel gear. Select the number of gear teeth, diametral pitch of the gear, width of the gear face, the distance between the shaft centers, and the dimensions of the gear teeth. The allowable stress in the gear teeth is 800 lb/in² (55,160 kPa).

Calculation Procedure:

1.   Compute the number of teeth on the gear
For any gearset, $R_D/R_d, = N_d/N_D$, where R_D = rpm of driver; R_d = rpm of driven gear; N_d = number of teeth on the driven gear; N_D = number of teeth on driving gear. Thus, $900/300 = N_d/20$; $N_d = 60$ teeth.

2.   Compute the diametral pitch of the gear
The diametral pitch of the gear must be the same as the diametral pitch of the pinion if the gears are to run together. If the diametral pitch of the pinion is known, assume that the diametral pitch of the gear equals that of the pinion.

When the diametral pitch of the pinion is not known, use a modification of the Lewis formula, shown in the next calculation procedure, to compute the diametral pitch. Thus, $P = (\pi \, SaYv/33,000 \text{ hp})^{0.5}$, where all symbols are as in the next calculation procedure, except that $a = 4$ for machined gears. Obtain $Y = 0.421$ for 60 teeth in a 20° full-depth gear from Baumeister and Marks—*Standard Handbook for Mechanical Engineers*. Assume that v = pitch-line velocity = 1200 ft/min (6.1 m/s). This is a typical reasonable value for v. The $P = [\pi \times 8000 \times 4 \times 0.421 \times 1200/(33,000 \times 50)]^{0.5} = 5.56$, say 6, because diametral pitch is expressed as a whole number whenever possible.

3.   Compute the gear face width
Spur gears often have a face width equal to about four times the circular pitch of the gear. Circular pitch $p_c = \pi/P = \pi/6 = 0.524$. Hence, the face width of the gear $= 4 \times 0.524 = 2.095$ in, say 2⅛ in (5.4 cm).

4. Determine the distance between the shaft centers

Find the exact shaft centerline distance from $d_c = (N_p + N_g)/2P)$, where N_p = number of teeth on pinion gear; N_g = number of teeth on gear. Thus, $d_c = (20 + 60)/[2(6)] = 6.66$ in (16.9 cm).

5. Compute the dimensions of the gear teeth

Use AGMA *Standards,* Dudley—*Gear Handbook,* or the engineering tables published by gear manufacturers. Each of these sources provides a list of factors by which either the circular or diametral pitch can be multiplied to obtain the various dimensions of the teeth in a gear or pinion. Thus, for a 20° full-depth spur gear, using the circular pitch of 0.524 computed in step 3, we have the following:

	Factor		Circular pitch		Dimension, in (mm)
Addendum	=	0.3183	×	0.524 =	0.1668 (4.2)
Dedendum	=	0.3683	×	0.524 =	0.1930 (4.9)
Working depth	=	0.6366	×	0.524 =	0.3336 (8.5)
Whole depth	=	0.6866	×	0.524 =	0.3598 (9.1)
Clearance	=	0.05	×	0.524 =	0.0262 (0.67)
Tooth thickness	=	0.50	×	0.524 =	0.262 (6.7)
Width of space	=	0.52	×	0.524 =	0.2725 (6.9)
Backlash = width of space − tooth thickness	=			0.2725 − 0.262 =	0.0105 (0.27)

The dimensions of the pinion teeth are the same as those of the gear teeth.

Related Calculations. Use this general procedure to select the dimensions of helical, herringbone, spiral, and worm gears. Refer to the AGMA *Standards* for suitable factors and typical allowable working stresses for each type of gear and gear material.

HORSEPOWER RATING OF GEARS

What are the strength hp rating, durability horsepower rating, and service horsepower rating of a 600-r/min 36-tooth 1.75 in (4.4-cm) face-width 14.5° full-depth 6-in (15.2-cm) pitch-diameter pinion driving a 150-tooth 1.75-in (4.4-cm) face width 14.5° full-depth 25-in (63.5-cm) pitch-diameter gear if the pinion is made of SAE 1040 steel 245 BHN and the gear is made of cast steel 0.35/0.45 carbon 210 BHN when the gearset operates under intermittent heavy shock loads for 3 h/day under fair lubrication conditions? The pinion is driven by an electric motor.

Calculation Procedure:

1. Compute the strength horsepower, using the Lewis formula

The widely used Lewis formula gives the strength horsepower, $hp_s = SYFK_v v/(33,000P)$, where S = allowable working stress of gear material, lb/in²; Y = tooth form factor (also called the Lewis factor); f = face width, in; K_v = dynamic load factor = $600/(600 + v)$ for metal gears, $0.25 + 150/(200 + v)$ for nonmetallic gears; v = pitchline velocity, ft/min = (pinion pitch diameter, in)(pinion rpm)(0.262); P = diametral pitch, in = number of teeth/pitch diameter, in. Obtain values of S and Y from tables in Baumeister and Marks—*Standard Handbook for Mechanical Engineers,* or AGMA *Standards Books,* or gear manufacturers' engineering data. Compute the strength hp for the pinion and gear separately.

Using one of the above references for the pinion, we find S = 25,000 lb/in² (172,368.9 kPa) and Y = 0.298. The pitchline velocity for the metal pinion is v = (6.0)(600)(0.262) = 944 ft/min (4.8 m/s). Then K_v = 600/(600 + 944) = 0.388. The diametral pitch of the pinion is $P = N_p/d_p$, where N_p = number of teeth on pinion; d_p = diametral pitch of pinion, in. Or P = 36/6 = 6.

Substituting the above values in the Lewis formula gives hp_s = (25,000) (0.298)(1.75)(0.388)(944)/[(33,000)(6)] = 24.117 hp (17.98 kW) for the pinion.

Using the Lewis formula and the same procedure for the 150-tooth gear, hp_s = (20,000)(0.374)(1.75)(0.388)(944)/[33,000)(6)] = 24.2 hp (18.05 kW). Thus, the strength hp of the gear is greater than that of the pinion.

2. Compute the durability horsepower

The durability horsepower of spur gears is found from $hp_d = F_i K_r D_o C_r$ for 20° pressure-angle full-depth or stub teeth. For 14.5° full-depth teeth, multiply hp_d by 0.75. In this relation, F_i = face-width and built-in factor from AGMA *Standards;* K_r = factor for tooth form, materials, and ration of gear to pinion from AGMA *Standards;* $D_o = (d_p^2 R_p/158,000)(1 - v^{0.5}/84)$, where d_p = pinion pitch diameter, in; R_p = pinion rpm; v = pinion pitchline velocity, ft/min, as computed in step 1; C_r = factor to correct for increased stress at the start of single-tooth contact as given by AGMA *Standards.*

Using appropriate values from these standards for low-speed gears of double speed reductions yields hp_d = (0.75)(1.46)(387)(0.0865)(1.0) = 36.6 hp (27.3 kW).

3. Compute the gearset service rating

Determine, by inspection, which is the lowest computed value for the gearset—the strength or durability horsepower. Thus, step 1 shows that the strength horsepower hp_s = 24.12 hp (18.0 kW) of the pinion is the lowest computed value. Use this lowest value in computing the gear-train service rating.

Using the AGMA *Standards,* determine the service factors for this installation. The load service factor for heavy shock loads and 3 h/day intermittent operation with an electric-motor drive is 1.5 from the *Standards.* The lubrication factor for a drive operating under fair conditions is, from the *Standards,* 1.25. To find the service rating, divide the lowest computed hp by the product of the load and lubrication factors; or, service rating = 24.12/(1.5)(1.25) = 7.35 hp (9.6 kW).

Were this gearset operated only occasionally (0.5 h or less per day), the service rating could be determined by using the lower of the two computed strength horsepowers, in this case 24.12 hp (18.0 kW). Apply only the load service factor, or 1.25 for occasional heavy shock loads. Thus, the service rating for these conditions = 24.12/1.25 = 19.30 hp (14.4 kW).

Related Calculations. Similar AGMA gear construction-material, tooth-form, face-width tooth-stress, service, and lubrication tables are available for rating helical, double-helical, herringbone, worm, straight-bevel, spiral-bevel, and Zerol gears. Follow the general procedure given here. Be certain, however, to use the applicable values from the appropriate AGMA tables. In general, choose suitable stock gears first; then check the horsepower rating as detailed above.

MOMENT OF INERTIA OF A GEAR DRIVE

A 12-in (30.5-cm) outside-diameter 36-tooth steel pinion gear having a 3-in (7.6-cm) face width is mounted on a 2-in (5.1-cm) diameter 36-in (91.4-cm) long steel shaft turning at 600 r/min. The pinion drives a 200-r/min 36-in (91.4-cm) outside-diameter 108-tooth steel gear mounted on a 12-in (30.5-cm) long 2-in (5.1-cm) diameter steel shaft that is solidly connected to a 24-in (61.0-cm) long 4-in (10.2-cm) diameter shaft. What is the moment of inertia of the high-speed and low-speed assemblies of this gearset?

Calculation Procedure:

1. *Compute the moment of inertia of each gear*
The moment of inertia of a cylindrical body about its longitudinal axis is $I_i = WR^2$, where I_i = moment of inertia of a cylindrical body, in⁴/in of length; W = weight of cylindrical material, lb/in³; R = radius of cylinder to its outside surface, in. For a steel shaft or gear, this relation can be simplified to $I_i = D^4/35.997$, where D = shaft or gear diameter, in. When you are computing I for a gear, treat it as a solid blank of material. This is a safe assumption.

Thus, for the 12-in (30.5-cm) diameter pinion, $I = 12^4/35.997 = 576.05$ in⁴/in (9439.8 cm⁴/cm) of length. Since the gear has a 3-in (7.6-cm) face width, the moment of inertia for the total length is $I_t = (3.0)(576.05) = 1728.15$ in⁴ (71,931.0 cm⁴).

For the 36-in (91.4-cm) gear, $I_i = 36^4/35.997 = 46,659.7$ in⁴/in (764,615.5 cm⁴/cm) of length. With a 3-in (7.6-cm) face width, $I_t = (3.0)(46,659.7) = 139,979.1$ in⁴ (5,826,370.0 cm⁴).

2. *Compute the moment of inertia of each shaft*
Follow the same procedure as in step 1. Thus for the 36-in (91.4-cm) long 2-in (5.1-cm) diameter pinion shaft, $I_t = (2^4/35.997)(36) = 16.0$ in⁴ (666.0 cm⁴).

For the 12-in (30.5-cm) long 2-in (5.1-cm) diameter portion of the gear shaft, $I_t = (2^4/35.997)(12) = 5.33$ in⁴ (221.9 cm⁴). For the 24-in (61.0-cm) long 4-in (10.2-cm) diameter portion of the gear shaft, $I_t = (4^4/35.997)(24) = 170.69$ in⁴ (7104.7 cm⁴). The total moment of inertia of the gear shaft equals the sum of the individual moments, or $I_t = 5.33 + 170.69 = 176.02$ in⁴ (7326.5 cm⁴).

3. *Compute the high-speed-assembly moment of inertia*
The effective moment of inertia at the high-speed assembly input $= I_{thi} = I_{th} + I_{tl}/(R_h/R_l)^2$, where I_{th} = moment of inertia of high-speed assembly, in⁴; I_{tl} = moment of inertia of low-speed assembly, in⁴; R_h = high speed, r/min; R_l = low speed, r/min. To find I_{th} and I_{tl}, take the sum of the shaft and gear moments of inertia for the high- and low-speed assemblies, respectively. Or, $I_{th} = 16.0 + 1728.5 = 1744.15$ in⁴ (72,597.0 cm⁴); $I_{tl} = 176.02 + 139,979.1 = 140,155.1$ in⁴ (5,833,695.7 cm⁴).

Then $I_{thi} = 1744.15 + 140,155.1/(600/200)^2 = 17,324.2$ in⁴ (721,087.6 cm⁴).

4. *Compute the low-speed-assembly moment of inertia*
The effective moment of inertia at the low-speed assembly output is $I_{tlo} = I_{tl} + I_{th}(R_h/R_l)^2 = 140,155.1 + (1744.15)(600/200)^2 = 155,852.5$ in⁴ (6,487,070.8 cm⁴).

Note that $I_{thi} \neq I_{tlo}$. One value is approximately nine times that of the other. Thus, in stating the moment of inertia of a gear drive, be certain to specify whether the given value applies to the high- or low-speed assembly.

Related Calculations.   Use this procedure for shafts and gears made of any metal—aluminum, brass, bronze, chromium, copper, cast iron, magnesium, nickel, tungsten, etc. Compute WR^2 for steel, and multiply the result by the weight of shaft material, $lb/in^3/0.283$.

BEARING LOADS IN GEARED DRIVES

A geared drive transmits a torque of 48,000 lb·in (5423.3 N·m). Determine the resulting bearing load in the drive shaft if a 12-in (30.5-cm) pitch-radius spur gear having a 20° pressure angle is used. A helical gear having a 20° pressure angle and a 14.5° spiral angle transmits a torque of 48,000 lb·in (5423.2 N·m). Determine the bearing load it produces if the pitch radius is 12 in (30.5 cm). Determine the bearing load in a straight bevel gear having the same proportions as the helical gear above, except that the pitch cone angle is 14.5°. A worm having an efficiency of 70 percent and a 30° helix angle drives a gear having a 20° normal pressure angle. Determine the bearing load when the torque is 48,000 lb·in (5423.3 N·m) and the worm pitch radius is 12 in (30.5 cm).

Calculation Procedure:

1. Compute the spur-gear bearing load

The tangential force acting on a spur-gear tooth is $F_t = T/r$, where F_t = tangential force, lb; T = torque, lb·in; r = pitch radius, in. For this gear, $F_t = T/r = 48,000/12 = 4000$ lb (17,792.9 N). This force is tangent to the pitch-diameter circle of the gear.

The separating force acting on a spur-gear tooth perpendicular to the tangential force is $F_s = F_t \tan \alpha$, where α = pressure angle, degrees. For this gear, $F_s = (4000)(0.364) = 1456$ lb (6476.6 N).

Find the resultant force R_F lb from $R_F = (F_t^2 + F_s^2)^{0.5} = (4000^2 + 1456^2)^{0.5} = 4260$ lb (18,949.4 N). This is the bearing load produced by the gear.

2. Compute the helical-gear load

The tangential force acting on a helical gear is $F_t = T/r = 48,000/12 = 4000$ lb (17,792.9 N). The separating force, acting perpendicular to the tangential force, is $F_s = F_t \tan \alpha / \cos \beta$, where β = the spiral angle. For this gear, $F_s = (4000)(0.364)/0.986 = 1503$ lb (6685.7 N). The resultant bearing load, which is a side thrust, is $R_F = (4000^2 + 1503^2)^{0.5} = 4380$ lb (19,483.2 N).

Helical gears produce an end thrust as well as the side thrust just computed. This end thrust is given by $F_e = F_t \tan \beta$ or $F_e = (4000)(0.259) = 1036$ lb (4608.4 N). The end thrust of the driving helical gear is equal and opposite to the end thrust of the driven helical gear when the teeth are of the opposite hand in each gear.

3. Compute the bevel-gear load

The tangential force acting on a bevel gear is $F_t = T/r = 48,000/12 = 4000$ lb (17,792.9 N). The separating force is $F_s = F_t \tan \alpha \cos \theta$, where θ = pitch cone angle. For this gear, $F_s = (4000)(0.364)(0.968) = 1410$ lb (6272.0 N).

Bevel gears produce an end thrust similar to helical gears. This end thrust is $F_e = F_t \tan \alpha \sin \theta$, or $F_e = (4000)(0.364)(0.25) = 364$ lb (1619.2 N). The side thrust in a bevel gear is $F_t = 4000$ lb (17,792.9 N) and acts tangent to the pitch-diameter circle. The resultant is an end thrust produced by F_s and F_e, or $R_F = (F_s^2 + F_e^2)^{0.5} = (1410^2 + 364^2)^{0.5} = 1458$ lb (6485.5 N). In a bevel-gear drive, F_t is common to both gears, F_s becomes F_e on the mating gear, and F_e becomes F_s on the mating gear.

4. Compute the worm-gear bearing load

The worm tangential force $F_t = T/r = 48{,}000/12 = 4000$ lb (17,792.9 N). The separating force is $F_s = F_t E \tan \alpha / \sin \phi$, where $E =$ worm efficiency expressed as a decimal; $\phi =$ worm helix or lead angle. Thus, $F_s = (4999)(0.70)(0.364)/0.50 = 2040$ lb (9074.4 N).

The worm end thrust force is $F_e = F_t E \cot \phi = (4000)(0.70)(1.732) = 4850$ lb (21,573.9 N). This end thrust acts perpendicular to the separating force. Thus the resultant bearing load $R_F = (F_s^2 + F_e^2)^{0.5} = (2040^2 + 4850^2)^{0.5} = 5260$ lb (23,397.6 N).

Forces developed by the gear are equal and opposite to those developed by the worm tangential force if canceled by the gear tangential force.

Related Calculations. Use these procedures to compute the bearing loads in any type of geared drive—open, closed, or semiclosed—serving any type of load. Computation of the bearing load is necessary step in bearing selection.

FORCE RATIO OF GEARED DRIVES

A geared hoist will lift a maximum load of 1000 lb (4448.2 N). The hoist is estimated to have friction and mechanical losses of 5 percent of the maximum load. How much force is required to lift the maximum load if the drum on which the lifting cable reels is 10 in (25.4 cm) in diameter and the driving gear is 50 in (127.0 cm) in diameter? If the load is raised at a velocity of 100 ft/min (0.5 m/s), what is the hp output? What is the driving-gear tooth load if the gear turns at 191 r/min. A 15-in (38.1-cm) triple-reduction hoist has three driving gears with 48-, 42-, and 36-in (121.9-, 106.7-, and 91.4-cm) diameters, respectively, and two pinions of 12- and 10-in (30.5- and 25.4-cm) diameter. What force is required to lift a 1000-lb (4448.2-N) load if friction and mechanical losses are 10 percent?

Calculation Procedure:

1. Compute the total load on the hoist

The friction and mechanical losses *increase* the maximum load on the drum. Thus, the total load on the drum = maximum lifting load, lb + friction and mechanical losses, lb = $1000 + 1000(0.05) = 1050$ lb (4670.6 N).

2. Compute the required lifting force

Find the lifting force from $L/D_g = F/d_d$, where $L =$ total load on hoist, lb; $D_g =$ diameter of driving gear, in; $f =$ lifting force required, lb; $d_d =$ diameter of lifting drum, in. For this hoist, $1050/50 = f/10$; $f = 210$ lb (934.1 N).

3. Compute the hp input
Find the hp input from $hp = Lv/33,000$, where $v = $ load velocity, ft/min. Thus, $hp = (1050)(100)/33,000 = 3.19$ hp (2.4 kW).

Where the mechanical losses are not added to the load before the hp is computed, use the equation $hp = Lv/(1.00 - \text{losses})(33,000)$. Thus, $hp = (1000)(100)/(1 - 0.05)(33,000) = 3.19$ hp (2.4 kW), as before.

4. Compute the driving-gear tooth load
Assume that the entire load is carried by one tooth. Then the tooth load L_t, lb = 33,000 hp/v_g, where $v_g = $ peripheral velocity of the driving gear, ft/min. With a diameter of 50 in (127.0 cm) and a speed of 191 r/min, $v_g = \pi D_g R/12$, where $R = $ gear rpm. Or, $v_g = \pi(50)(191)/12 = 2500$ ft/min (12.7 m/s). Then, $L_t = (33,000)(3.19)/2500 = 42.1$ lb (187.3 N). This is a nominal tooth-load value.

5. Compute the triple-reduction hoisting force
Use the equation from step 2, but substitute the product of the three driving-gear diameters for D_g and the three driven-gear diameters for d_d. The total load = $1000 + 0.10(1000) = 1100$ lb (4893.0 N). Then $L/D_g = F/d_d$, or $1100/(48 \times 42 \times 36) = F/(15 \times 12 \times 10)$; $F = 27.2$ lb (121.0 N). Thus, the triple-reduction hoist reduces the required lifting force to about one-tenth that required by a double-reduction hoist (step 2).

Related Calculations.   Use this procedure for geared hoists of all types. Where desired, the number of gear teeth can be substituted for the driving- and driven-gear diameters in the force equation in step 2.

DETERMINATION OF GEAR BORE DIAMETER

Two helical gears transmit 500 hp (372.9 kW) at 3600 r/min. What should the bore diameter of each gear be if the allowable stress in the gear shafts is 12,500 lb/in² (86,187.5 kPa)? How should the gears be fastened to the shafts? The shafts are solid in cross section.

Calculation Procedure:

1. Compute the required hub bore diameter
The hub bore diameter must at least equal the outside diameter of the shaft, unless the gear is press- or shrink-fitted on the shaft. Regardless of how the gear is attached to the shaft, the shaft must be large enough to transmit the rated torque at the allowable stress.

Use the method of step 2 of "Solid and Hollow Shafts in Torsion" in this section to compute the required shaft diameter, after finding the torque by using the method of step 1 in the same procedure. Thus, $T = 63,000 hp/R = (63,000)(500)/3600 = 8750$ lb · in (988.6 N · m). Then $d = 1.72(T/s)^{1/3} = 1.72(8750/12,500)^{1/3}$ 1.526 in (3.9 cm).

2. Determine how the gear should be fastened to the shaft
First decide whether the gears are to be permanently fastened or removable. This decision is usually based on the need for gear removal for maintenance or replace-

ment. Removable gears can be fastened by a key, setscrew, spline, pin clamp, or a taper and screw. Large gears transmitting 100 hp (74.6 kW) or more are usually fitted with a key for easy removal. See "Selection of Keys for Machine Shafts" in this handbook for the steps in choosing a key.

Permanently fastened gears can be shrunk, pressed, cemented, or riveted to the shaft. Shrink-fit gears generally transmit more torque before slippage occurs than do press-fit gears. With either type of fastening, interference is necessary; i.e., the gear bore is made smaller than the shaft outside diameter.

Baumeister and Marks—*Standard Handbook for Mechanical Engineers* shows that press- or shrink-fit gears on shafts of 1.19- to 1.58-in (3.0- to 4.0-cm) diameter should have an interference ranging from 0.3 to 4.0 thousandths of an inch (0.8 to 10.2 thousandths of a centimeter) on the diameter, depending on the class of fit desired).

Related Calculations. Use this general procedure for any type of gear—spur, helical, herringbone, worm, etc. Never reduce the shaft diameter below that required by the stress equation, step 1. Thus, if interference is provided by the shaft diameter, *increase* the diameter; do not reduce it.

TRANSMISSION GEAR RATIO FOR A GEARED DRIVE

A four-wheel vehicle must develop a drawbar pull of 17,500 lb (77,843.9 N). The engine, which develops 500 hp (372.8 kW) and drives through a gear transmission a 34-tooth spiral bevel pinion gear which meshes with a spiral bevel gear having 51 teeth. This gear is keyed to the drive shaft of the 48-in (121.9-cm) diameter rear wheels of the vehicle. What transmission gear ratio should be used if the engine develops maximum torque at 1500 r/min? Select the axle diameter for an allowable torsional stress of 12,500 lb/in² (86,187.5 kPa). The efficiency of the bevel-gear differential is 80 percent.

Calculation Procedure:

1. Compute the torque developed at the wheel
The wheel torque = (drawbar pull, lb)(moment arm, ft), where the moment arm = wheel radius, ft. For this vehicle having a wheel radius of 24 in (61.0 cm), or $24/12 = 2$ ft (0.6 m), the wheel torque = $(17,500)(2) = 35,000$ lb · ft (47,453.6 N · m).

2. Compute the torque developed by the engine
The engine torque $T = 5250/hp/R$, or $T = (5250)(500)/1500 = 1750$ lb · ft (2372.7 N · m), where R = rpm.

3. Compute the differential speed ratio
The differential speed ratio = $N_g/N_p = 51/34 = 1.5$, where N_g = number of gear teeth; N_p = number of pinion teeth.

4. Compute the transmission gear ratio
For any transmission gear, its ratio = (output torque, lb · ft).[(input torque, lb · ft)(differential speed ratio)(differential efficiency)], or transmission gear ratio =

35,000/[(1750)(1.5)(0.80)] = 16.67. Thus, a transmission with a 16.67 ratio will give the desired output torque at the rated engine speed.

5. Determine the required shaft diameter

Use the relation $d = 1.72(T/s)^{1/3}$ from the previous calculation procedure to determine the axle diameter. Since the axle is transmitting a total torque of 35,000 lb · ft (47,453.6 N · m), each of the two rear wheels develops a torque of 35,000/2 = 17,500 lb · ft (23,726.8 N · m), and $d = 1.72(17,500.12,500)^{1/3} = 1.92$ in (4.9 cm).

Related Calculations. Use this general procedure for any type of differential—worm gar, herringbone gear, helical gear, or spiral gear—connected to any type of differential. The output torque can be developed through a wheel, propeller, impeller, or any other device. Note that although this vehicle has two rear wheels, the total drawbar pull is developed by *both* wheels. Either wheel delivers *half* the drawbar pull. If the total output torque were developed by only one wheel, its shaft diameter would be $d = 1.72(35,000/12,500)^{1/3} = 2.42$ in (6.1 cm).

EPICYCLIC GEAR TRAIN SPEEDS

Figure 9 shows several typical arrangements of epicyclic gear trains. The number of teeth and the rpm of the driving arm are indicated in each diagram. Determine the driven-member rpm for each set of gears.

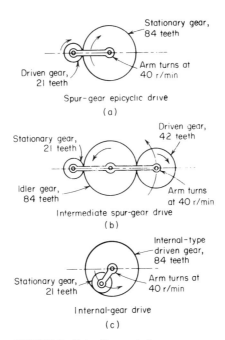

FIGURE 9 Epicyclic gear trains.

Calculation Procedure:

1. *Compute the spur-gear speed*
For a gear arranged as in Fig. 9a, $R_d = R_D(1 + N_s/N_d)$, where R_d = driven-member rpm; R_D = driving-member rpm; N_s = number of teeth on the stationary gear; N_d = number of teeth on the driven gear. Given the values given for this gear and since the arm is the driving member, $R_d = 40(1 + 84/21)$; $R_d = 200$ r/min.

Note how the driven-gear speed is attained. During one planetary rotation around the stationary gear, the driven gear will rotate axially on its shaft. The number of times the driven gear rotates on its shaft $= N_s/N_d = 84/21 = 4$ times per planetary rotation about the stationary gear. While rotating on its shaft, the driven gear makes a planetary rotation around the fixed gear. So while rotating axially on its shaft four times, the driven gear makes one additional planetary rotation about the stationary gear. Its total axial and planetary rotation is $4 + 1 = 5$ r/min per rpm of the arm. Thus, the gear ration $G_r = R_D/R_d = 40/200 = 1:5$.

2. *Compute the idler-gear train speed*
The idler gear, Fig. 9b, turns on its shaft while the arm rotates. Movement of the idler gear causes rotation of the driven gear. For an epicyclic gear train of this type, $R_d = R_D(1 - N_s/N_d)$, where the symbols are as defined in step 1. Thus, $R_d = 40(1 - 21/42) = 20$ r/min.

3. *Compute the internal gear drive speed*
The arm of the internal gear drive, Fig. 9c, turns and carries the stationary gear with it. For a gear train of this type, $R_d = R_D(1 - N_s/N_d)$, or $R_d = 40(1 - 21/84) = 30$ r/min.

Where the internal gear is the driving gear that turns the arm, making the arm the driven member, the velocity equation becomes $r_d = R_D N_D/(N_D + N_s)$, where R_D = driving-member rpm; N_D = number of teeth on the driving member.

Related Calculations. The arm was the driving member for each of the gear trains considered here. However, any gear can be made the driving member if desired. Use the same relations as given above, but substitute the gear rpm for R_D. Thus, a variety of epicyclic gear problems can be solved by using these relations. Where unusual epicyclic gear configurations are encountered, refer to Dudley— *Gear Handbook* for a tabular procedure for determining the gear ratio.

PLANETARY-GEAR SYSTEM SPEED RATIO

Figure 10 shows several arrangements of important planetary-gear systems using internal ring gears, planet gears, sun gears, and one or more carrier arms. Determine the output rpm for each set of gears.

Calculation Procedures:

1. *Determine the planetary-gear output speed*
For the planetary-gear drive, Fig 10a, the gear ratio $G_r = (1 + N_4 N_2/N_3 N_1)/(1 - N_4 N_2/N_5 N_1)$, where $N_1, N_2, \ldots, N_5$ = number of teeth, respectively on each of gears 1, 2, ..., 5. Also, for any gearset, the gear ration G_r = input rpm/output rpm, or G_r = driver rpm/driven rpm.

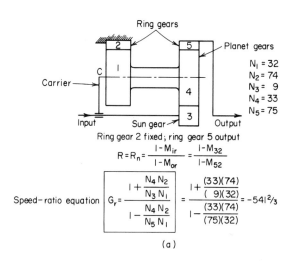

Ring gear 2 fixed; ring gear 5 output

$$R = R_n = \frac{1 - M_{ir}}{1 - M_{or}} = \frac{1 - M_{32}}{1 - M_{52}}$$

Speed-ratio equation
$$G_r = \frac{1 + \dfrac{N_4 N_2}{N_3 N_1}}{1 - \dfrac{N_4 N_2}{N_5 N_1}} = \frac{1 + \dfrac{(33)(74)}{(9)(32)}}{1 - \dfrac{(33)(74)}{(75)(32)}} = -541\,^2/_3$$

(a)

Coupled planetary drives

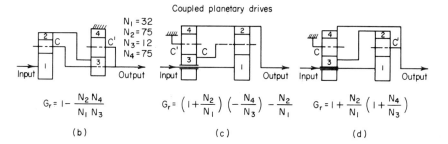

$$G_r = 1 - \frac{N_2 N_4}{N_1 N_3}$$

(b)

$$G_r = \left(1 + \frac{N_2}{N_1}\right)\left(-\frac{N_4}{N_3}\right) - \frac{N_2}{N_1}$$

(c)

$$G_r = 1 + \frac{N_2}{N_1}\left(1 + \frac{N_4}{N_3}\right)$$

(d)

Fixed-differential drives

Output is difference between speeds of two parts leading to high reduction ratios

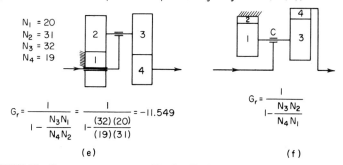

$$G_r = \frac{1}{1 - \dfrac{N_3 N_1}{N_4 N_2}} = \frac{1}{1 - \dfrac{(32)(20)}{(19)(31)}} = -11.549$$

(e)

$$G_r = \frac{1}{1 - \dfrac{N_3 N_2}{N_4 N_1}}$$

(f)

FIGURE 10 Planetary gear systems. (*Product Engineering.*)

Triple planetary drives

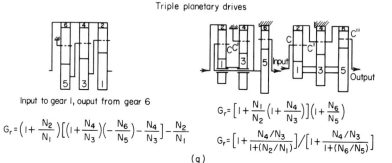

Input to gear 1, ouput from gear 6

$$G_r = \left(1 + \frac{N_2}{N_1}\right)\left[\left(1 + \frac{N_4}{N_3}\right)\left(-\frac{N_6}{N_5}\right) - \frac{N_4}{N_3}\right] - \frac{N_2}{N_1}$$

$$G_r = \left[1 + \frac{N_1}{N_2}\left(1 + \frac{N_4}{N_3}\right)\right]\left(1 + \frac{N_6}{N_5}\right)$$

$$G_r = \left[1 + \frac{N_4/N_3}{1+(N_2/N_1)}\right] / \left[1 + \frac{N_4/N_3}{1+(N_6/N_5)}\right]$$

(g)

Compound spur-bevel gear drive   Planocentric drive   Wobble-gear drive

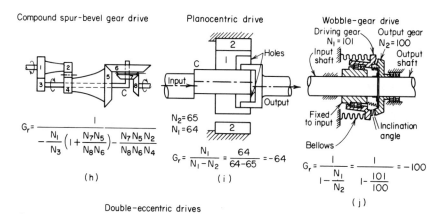

$$G_r = \frac{1}{-\frac{N_1}{N_3}\left(1 + \frac{N_7 N_5}{N_8 N_6}\right) - \frac{N_7 N_5 N_2}{N_8 N_6 N_4}}$$

(h)

$N_2 = 65$
$N_1 = 64$

$$G_r = \frac{N_1}{N_1 - N_2} = \frac{64}{64-65} = -64$$

(i)

$$G_r = \frac{1}{1 - \frac{N_1}{N_2}} = \frac{1}{1 - \frac{101}{100}} = -100$$

(j)

Double-eccentric drives
Two arrangements. Input is through double-throw crank (carrier).
Gear 1 fixed to frame.

Humpage's bevel gears

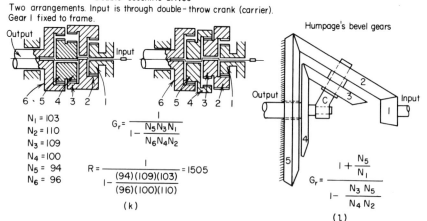

$N_1 = 103$
$N_2 = 110$
$N_3 = 109$
$N_4 = 100$
$N_5 = 94$
$N_6 = 96$

$$G_r = \frac{1}{1 - \frac{N_5 N_3 N_1}{N_6 N_4 N_2}}$$

$$R = \frac{1}{1 - \frac{(94)(109)(103)}{(96)(100)(110)}} = 1505$$

(k)

$$G_r = \frac{1 + \frac{N_5}{N_1}}{1 - \frac{N_3 N_5}{N_4 N_2}}$$

(l)

FIGURE 10 Continued.

With ring gear 2 fixed and ring gear 5 the output gear, Fig 10*a*, and the number of teeth shown, $G_r \{1 + (33)(74)/[(9)(32)]\}/\{1 - (33)(74)/[(175)(32)]\} = -541.667$. The minus sign indicates that the output shaft revolves in a direction *opposite* to the input shaft. Thus, with an input speed of 5000 r/min, G_r = input rpm/output rpm; output rpm = input rpm/G_r, or output rpm = 5000/541.667 = 9.24 r/min.

2. Determine the coupled planetary drive output speed

The drive, Fig. 10*b*, has the coupled ring gear 2, the sun gear 3, the coupled planet carriers C and C', and the fixed ring gear 4. The gear ratio is $G_r = (1 - N_2N_4/N_1N_3)$, where the symbols are the same as before. Find the output speed for any given number of teeth by first solving for G_r and then solving G_r = input rpm/output rpm.

With the number of teeth shown, $G_r = 1 - (75)(75)/[(32)(12)]; = -13.65$. Then output rpm = input rpm/G_r = 1200/13.65 = 87.9 r/min.

Two other arrangements of coupled planetary drives are shown in Fig. 10*c* and *d*. Compute the output speed in the same manner as described above.

3. Determine the fixed-differential output speed

Figure 10*e* and *f* shows two typical fixed-differential planetary drives. Compute the output speed in the same manner as step 2.

4. Determine the triple planetary output speed

Figure 10*g* shows three typical triple planetary drives. Compute the output speed in the same manner as step 2.

5. Determine the output speed of other drives

Figure 10*h*, *i*, *j*, *k*, and *l* shows the gear ratio and arrangement for the following drives: compound spur-bevel gear, plancentric, wobble gear, double eccentric, and Humpage's bevel gears. Compute the output speed for each in the same manner as step 2.

Related Calculations. Planetary and sun-gear calculations are simple once the gear ratio is determined. The gears illustrated here[1] comprise an important group in the planetary and sun-gear field. For other gear arrangements, consult Dudley—*Gear Handbook*.

---

[1]John H. Glover, "Planetary Gear Systems," *Product Engineering,* Jan. 6, 1964.

SECTION 21

TRANSMISSIONS, CLUTCHES, ROLLER-SCREW ACTUATORS, COUPLINGS, AND SPEED CONTROL

CONSTRUCTING MATHEMATICAL MODELS FOR ANALYZING HYDROSTATIC TRANSMISSIONS

Construct a mathematical model of vehicle performance for a construction vehicle powered by a hydrostatic transmission when the vehicle is driven by a 45-hp (33.6-kW) engine at 2400 rpm, with a high idle-speed of 2600 rpm. The vehicle has a supercharge pump rated at 2 hp (1.5 kW). Other vehicle data are: loaded radius, r_L = 14.5 in (36.8 cm); gross vehicle weight, W_g = 8500 lb (3825 kg); weight on drive wheels, W_w = 5150 lb (2338 kg); final drive ratio, R_{fd} = 40:1; coefficient of slip, C_s = 0.8; and coefficient of rolling resistance, C_r = 60 lb/1000 lb (27.2 kg/454 kg) of gross vehicle weight. The vehicle is powered by a hydrostatic transmission with a 2.5-in³/rev (41-mL/rev) displacement pump, rated at 5000 lb/in² (34.5-MPa). Compare the performance produced by using a 2.5-in³/rev (41-mL/rev) displacement fixed-displacement motor and a 2.5-in³/rev (41-mL/rev) displacement variable-displacement motor with an 11-degree displacement stop. Other pump and motor data are given on performance curves available from the pump manufacturer.

Calculation Procedure:

1. *Determine the vehicle speed at maximum tractive effort*
The theoretical pump displacement required to absorb the input horsepower from
the engine, using the nomenclature at the end of this procedure, is:

$$D_{pt} = \frac{396,000 \; H_p}{P_p N_p}$$

Substituting,

$$D_{pt} = \frac{396,000 \; (45 - 2)}{5,000 \; (2,400)}$$

$$= 1.42 \; \text{in}^3/\text{rev} \; (23.3 \; \text{mL/rev})$$

Next, find the horsepower-limited displacement from

$$D_p = D_{pt} E_{tp}$$

Substituting,

$$D_p = 1.42 \; (0.92)$$

$$= 1.31 \; \text{in}^3/\text{rev} \; (21.5 \; \text{mL/rev})$$

Now we must find the pump flow from

$$Q_p = \frac{D_p N_p E_{vp}}{231}$$

Substituting,

$$Q_p = \frac{1.31 \; (2,400) \; (0.88)}{231}$$

$$= 12 \; \text{gal/min} \; (0.76 \; \text{L/s})$$

Using the motor torque curve from the manufacturer for the pump being considered, similar to Fig. 1, these data give a motor torque, $T_m = 1800$ lb/in (203.3
Nm) at a motor speed of 960 rpm.
    Maximum tractive effort is given by

$$T_{emax} = \frac{T_m R_{fd} E_{fd} n_m}{r_L}$$

Substituting,

$$T_{emax} = \frac{1,800 \; (40) \; (0.9)}{14.5}$$

$$= 4,469 \; \text{lb} \; (2029 \; \text{kg})$$

The vehicle speed at this tractive effort is given by

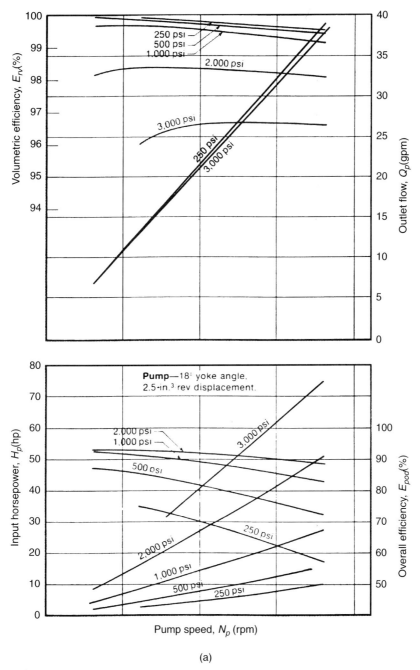

(a)

FIGURE 1 (a) Typical pump performance curves relate input horsepower, fluid outlet flow rate, speed, and pressure to volumetric and overall efficiency. (b) Motor performance curves relate output horsepower, fluid inlet flow rate, speed, and pressure to volumetric and overall efficiency. (*Machine Design.*)

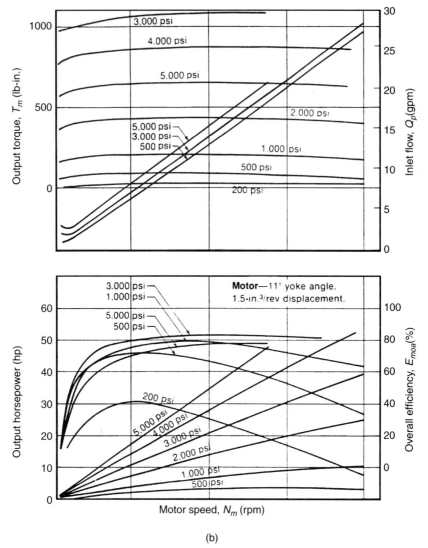

(b)

FIGURE 1 *Continued.*

SI values for Fig. 1a and 1b:

gpm	L/sec	lb-in.	Nm
0	0	0	0
5	0.32	500	56.4
10	0.63	1000	112.9
15	0.95		

gpm	L/sec	psi	MPa	hp	kW
20	1.26	200	1.4	0	0
25	1.58	250	1.72	10	7.46
30	1.89	500	3.4	20	14.9
35	2.2	1000	6.89	30	22.4
40	2.5	2000	13.8	40	29.8
		3000	20.7	50	37.3
		4000	27.6	60	44.8
		5000	34.5	70	52.2
				80	59.7

1.5 cu in/rev (24.6 mL/revf)   2.5 cu in/rev (41 mL/rev)

$$N_v = \frac{N_m r_L}{168 \, R_{fd} E_{fd}}$$

Substituting,

$$N_v = \frac{960 \, (14.5)}{168 \, (40)}$$

$$= 2.1 \text{ mi/h } (0.939 \text{ m/s})$$

Plot these computed values as point A on the tractive-effort vs. vehicle-speed curve, Fig. 2.

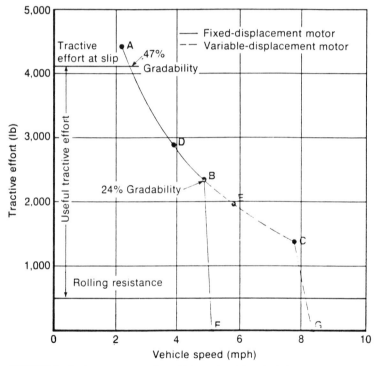

FIGURE 2 Performance curve for vehicle analyzed in calculation procedure. (*Machine Design.*)

Fig. 2 SI

lb	kg	mph	m/sec
		0	0
1000	454	2	0.89
2000	908	4	1.78
3000	1362	6	2.68
4000	1816	8	3.58
5000	2270	10	4.47

2. Determine the tractive effort at the maximum vehicle speed

From the pump performance curve obtained from the manufacturer, the maximum pump flow is 25.2 gal/min (1.59 L/s) at 2700 lb/in² 18.6 (MPa) and 2.5 in³/rev (41 mL/rev). From these data, the motor torque curves for the fixed-displacement motor give N_m = 2240 rpm and T_m = 1000 lb/in (112.9 Nm).

The maximum vehicle speed produced by the fixed-displacement motor is

$$N_v = \frac{N_m r_L}{168 \ R_{fd} E_{fd}}$$

Substituting,

$$N_v = \frac{2{,}240 \ (14.5)}{168 \ (40)}$$

$$= \ 4.8 \ \text{mi/h} \ (2.15 \ \text{m/s})$$

The tractive effort at this speed is

$$T_{emax} = \frac{T_m R_{fd} E_{fd} n_m}{r_L}$$

Substituting,

$$T_e = \frac{1{,}000 \ (40) \ (0.9)}{14.5}$$

$$= 2{,}482 \ \text{lb} \ (1127 \ \text{kg})$$

Plot these values as point B on Fig. 2.

From the curves for the variable-displacement motor, N_m = 3580 rpm and T_m = 560 lb/in (63.2 Nm). Therefore, as before, maximum vehicle speed produced by the variable-displacement motor is

$$N_v = \frac{3{,}580 \ (14.5)}{168 \ (40)}$$

$$= 7.7 \ \text{mi/h} \ (3.44 \ \text{m/s})$$

And the tractive effort, as before, is:

$$T_e = \frac{560 \ (40) \ (0.9)}{14.5}$$

$$= 1{,}390 \ \text{lb} \ (631 \ \text{kg})$$

Plot these values as point C on Fig. 2.

3. Find intermediate points on the tractive-effort vs. vehicle-speed curve

To plot an intermediate point on the curve, Fig. 2, a pump flow of 21 gal/min (1.325 L/s) is chosen arbitrarily. For the fixed-displacement motor, this flow gives N_m = 1800 rpm and T_m = 1200 lb/in (135.5 Nm). Therefore, vehicle speed and tractive effort are

$$N_v = \frac{1{,}800 \ (14.5)}{168 \ (40)}$$

$$= 3.8 \text{ mi/h } (1.698 \text{ m/s})$$

$$T_e = \frac{800 \ (40) \ (0.9)}{14.5}$$

$$= 2{,}979 \text{ lb } (1352 \text{ kg})$$

which are plotted as point D on Fig. 2.

For the variable-speed motor, $N_m = 2600$ rpm and $T_m = 800$ lb/in (90.3 Nm). Therefore, the vehicle speed and tractive effort are:

$$N_v = \frac{2{,}600 \ (14.5)}{168 \ (40)}$$

$$= 5.6 \text{ mi/h } (2.5 \text{ m/s})$$

$$T_e = \frac{800 \ (40) \ (0.9)}{14.5}$$

$$= 1{,}986 \text{ lb } (901.6 \text{ kg})$$

Plot these values as point E on Fig. 2.

4. Find the maximum theoretical speed for each motor type

The final point needed to construct the performance curve is the maximum theoretical speed of the vehicle for each type of motor. For the fixed-displacement motor, maximum motor speed is given by

$$N_{mmax} = \frac{N_{pmax} D_p E_{vp} E_{vm}}{n_m D_m}$$

Substituting,

$$N_{mmax} = \frac{2{,}600 \ (2.5) \ (0.95) \ (0.95)}{2.5}$$

$$= 2{,}346 \text{ rpm}$$

The maximum theoretical vehicle speed is found from

$$N_{vmax} = \frac{N_{mmax} \ r_L}{168 \ R_{jd}}$$

Substituting,

$$N_{vmax} = \frac{2{,}346 \ (14.5)}{168 \ (40)}$$

$$= 5.1 \text{ mi/h } (2.5 \text{ m/s})$$

which is plotted as point F on Fig. 2.

For the variable-displacement motor, maximum motor speed is

$$N_{mmax} = \frac{2,600 \ (2.5) \ (0.95) \ (0.95)}{1.5}$$

$$= 3,911 \ \text{rpm}$$

and the maximum theoretical vehicle speed is

$$N_{vmax} = \frac{3,911 \ (14.5)}{169 \ (40)}$$

$$= 8.4 \ \text{mi/h} \ (3.75 \ \text{m/s})$$

which is plotted as point G on Fig. 2.

5.   Refine the curve with rolling resistance and tractive effort at wheel slip
To refine the curve, rolling resistance, tractive effort at wheel slip, and gradability must be determined. Rolling resistance is found from

$$R_r = W_{gv} C_r$$

Substituting,

$$R_r = \frac{8,900 \ (60)}{1,000}$$

$$= 510 \ \text{lb} \ (231.5 \ \text{kg})$$

Tractive effort at wheel slip is given by

$$T_e = W_w C_s$$

Single-path

$$T_e = 0.6 \ W_w C_s$$

Dual-path

Substituting,

$$T_e = 5,150 \ (0.8)$$

$$= 4,120 \ \text{lb} \ (1870 \ \text{kg})$$

The gradability at slip is given by

$$G = \tan \left[\sin^{-1} \left(\frac{T_e - R_r}{W_{gv}} \right) \right] 100$$

Substituting,

$$G = \tan\left[\sin^{-1}\left(\frac{4{,}120 - 510}{8{,}500}\right)\right]100$$

$$= 47\%$$

for the fixed-displacement motor.

The maximum gradability for the variable-displacement motor, limited by 5000 lb/in² (34.45 MPa), is

$$G = \tan\left[\sin^{-1}\left(\frac{2{,}482 - 510}{8{,}500}\right)\right]100$$

$$= 24\%$$

These data are also shown on the tractive-effort curve, Fig. 2, which now gives a complete picture of vehicle performance.

Related Calculations. The analytical technique presented here allows the hydrostatic transmission to be evaluated on paper, and necessary changes made before the unit is actually built. The procedure uses a series of calculations that gradually define transmission and vehicle data. With these data, a curve can be constructed, Fig. 2, so that vehicle performance can be predicted for the entire operating range.

The first step in the analysis is to compare the "application values" of the vehicle and the available transmissions. This comparison provides a simple way to match vehicle requirements to transmission capabilities.

The vehicle application value expresses vehicle requirements and depends on required vehicle speed and maximum tractive effort. For single-path applications only one transmission is used. For dual-path applications, where two transmissions are used, the transmission on each side of the vehicle must be treated as if it were a single-path system. In such a case, a normal assumption is that 60 percent of the total weight on the drive wheels transfers to one side of the vehicle when it negotiates a slope or turn. The transmission application value expresses transmission capabilities and depends on motor torque and speed.

If the transmission application value is greater than that for the vehicle, the proposed transmission is viable, and calculations to size properly the transmission can be made. If the vehicle application value is greater, consideration must be given to increasing pump speed, using variable-displacement motors, increasing pump displacement, lowering maximum vehicle speed, or accepting a lower vehicle tractive effort.

If a comparison of application values indicates that a proposed transmission is adequate, a more refined procedure must be used to size the transmission. This ensures that the transmission is applied within its horsepower rating, and that it meets vehicle power requirements.

The input power available to the transmission is the net engine flywheel horsepower (kW) at full-load governed speed, less the horsepower (kW) required for supercharge and auxiliary no-load losses. For a transmission to operate satisfactorily, input horsepower (kW) must not exceed its rated horsepower (kW). In dual-path applications, power is split between two pumps and 80 percent of the total input horsepower (kW) should be used in this calculation. Thus, up to 80 percent of the total input horsepower (kW), is assumed to be directed to one pump.

Next, the transmission's ability to produce the required tractive effort and propelling speed must be checked. For these calculations, the pump and motor performance curve for the proposed transmission must be available, usually from the manufacturer.

The maximum tractive effort is limited by the pump relief-valve setting or wheel slip. For multiple-path systems, maximum tractive effort must be divided by the number of motors and multiplied by 0.6 before the comparison is made.

The final calculation required to determine whether a transmission meets vehicle requirements is to check maximum vehicle speed. If the transmission can produce the required tractive effort and speed, it is sized properly. However, if speed is too low and tractive effort acceptable, consideration should be given to increasing pump speed, using a variable-displacement motor, or decreasing the ratio of the final drive. If speed is acceptable but tractive effort too low, give consideration to increasing the final drive ratio. The resultant loss in maximum speed can be recovered by increasing pump speed or by using a variable-displacement motor.

Once the transmission is sized to meet vehicle requirements, a mathematical model can be generated to predict system performance. The calculations necessary to produce the model take into account such factors as pump speed, pump and motor displacement, and pump and motor efficiency. The expected vehicle performance is represented by a tractive-effort vs. speed curve.

The first two steps in generating the math model are to define the upper and lower limits on the curve. The upper limit is the vehicle speed produced at the maximum tractive effort; the lower limit is the tractive effort produced at maximum vehicle speed.

Typically, four intermediate points on the performance curve are sufficient to provide a rough approximation of vehicle performance. Six to eight points may be required for a complete analysis. These points are calculated as shown here, except that motor torque and speed are determined for pump displacements between maximum displacement and displacement at maximum tractive effort.

To complete the analysis, a number of factors must be calculated to determine how they affect vehicle performance. One factor that must be considered is rolling resistance, the portion of tractive effort required to overcome friction and move the vehicle. For the vehicle to move, available tractive effort must be greater than the rolling resistance. If the actual coefficient of rolling resistance is not known, the values in Table 1 can be used.

Few vehicles operate only on level ground; so the slope or grade it can climb must be determined. This factor, called gradability, is calculated as shown above. Gradability can be determined for any point along the tractive-effort vs. speed curve, up to the slip-limited effort. Gradability at wheel slip is usually the upper limit.

TABLE 1 Coefficients of Rolling Resistance

Surface	Rubber tire	kg/454 kg	Crawler	kg/454 kg
Concrete	10–20	4.5–9.1	40	18.2
Asphalt	12–22	5.5–9.9	40	18.2
Packed gravel	15–40	6.8–18.2	40	18.2
Soil	25–40	11.4–18.2	40	18.2
Mud	37–170	16.8–77.2	—	—
Loose sand	60–150	27.2–68.1	100	45.4
Snow	25–50	9.1–22.7	—	—

Units are lb/1,000 lb (kg/454 kg) gross vehicle weight.

The final factor to be considered is the tractive effort required to slip the wheels. If the actual coefficient of slip is not known, the values in Table 2 can be used.

This procedure is valid for a variety of off-the-road vehicles: tractors, draggers, bulldozers, rippers, scrapers, excavators, loaders, trenchers, hauling units, etc. It is the work of Charles Griesel, Sperry Vickers, as reported in *Machine Design* Magazine. SI values were added by the handbook editor.

Nomenclature

A_t = Transmission application value
A_r = Vehicle application value
C_r = Coefficient of rolling resistance, lb/1,000 lb (kg/454 kg)
C_s = Coefficient of slip
D_m = Motor displacement, in^3/rev (mL/rev)
D_p = Pump displacement in^3/rev (mL/rev)
D_{pt} = Theoretical pump displacement, in^3/rev (mL/rev)
E_{fd} = Final drive efficiency, %
E_{poa} = Overall pump efficiency, %
E_{tm}, E_{tp} = Pump or motor torque efficiency, %
E_{vm}, E_{vp} = Pump or motor volumetric efficiency, %
G = Gradability, %
H_p = Pump input horsepower, hp (kW)
H_r = Pump rated horsepower, hp (kW)
N_m, N_p = Motor or pump speed, rpm
N_{mmax} = Maximum motor speed, rpm
N_{pmax} = Maximum pump speed, rpm
N_{pr} = Rated pump speed, rpm
N_v = Vehicle speed, mi/h (m/s)
N_{vmax} = Maximum vehicle speed, mi/h (m/s)
n_m = Number of motors
P_p = Pump pressure, lb/in^2 (kPa)
Q_p = Pump output flow, gal/min (L/s)
R_{fd} = Final drive output ratio
R_r = Rolling resistance, lb (kg)
r_L = Loaded radius, in (cm)
T_e = Tractive effort, lb (kg)
T_{emax} = Maximum tractive effort, lb (kg)
T_m = Motor torque, lb/in (Nm)
W_{gv} = Gross vehicle weight, lb (kg)
W_w = Weight on drive wheels, lb (kg)

TABLE 2 Coefficients of Slip

Surface	Rubber tire	Crawler
Concrete or asphalt	0.8–1.0	0.5
Dry clay	0.5–0.7	0.9
Sand & Gravel	0.3–0.6	0.4
Firm soil	0.5–0.6	0.9
Loose soil	0.4–0.5	0.6

SELECTING A CLUTCH FOR A GIVEN LOAD

Choose a clutch for a lathe designed for automatic operation. There will be no gear shifting in the headstock. All speed changes will be made using hydraulically operated clutches to connect the proper gear train to the output shaft. Determine the number of plates and the operating force required for the clutch if it is to transmit a torque of 300 lb/in (33.9 Nm) under normal operating conditions. Design the clutch to slip under 300 percent of rated torque to protect the gears and other parts of the drive. Space limitations dictate an upper limit of 4-in (10.2 cm) and a lower limit of 2.5-in (6.35 cm) for the diameters of the friction surfaces. The clutch will operate in an oil atmosphere.

Calculation Procedure:

1. *Choose the type of clutch to use*
Based on the proposed application, choose a wet clutch with hardened-steel plates. (Since the clutch is operating in an oil atmosphere, use of a dry clutch could lead to operational problems.)

2. *Compute the number of friction plates needed for this clutch*
Use the general clutch relation

$$T = N\mu P\frac{D + d}{4}$$

where T = torque transmitted by clutch, lb/in (Nm); N = number of friction plates in the clutch; μ = coefficient of friction for the clutch; P = total operating force on the clutch, lb (kg); D = maximum space limitation, in (cm); d = minimum space limitation, in (cm).

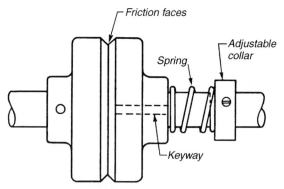

FIGURE 3   Basic friction clutch. Adjustable spring tension holds the two friction surfaces together and sets the overload limit. As soon as the overload is removed the clutch reengages. (*Product Engineering.*)

Using a pressure valve, p, of 100 lb/in² (689 kPa) for long wear of this clutch and its plates, the total operating force for this clutch will be $P = pA = 100 \times \pi \times [(D^2 - d^2)/4] = 100 \times \pi \times [(4^2 - 2.5^2)/4)] = 766$ lb (347.8 kg).

Substituting in the general clutch relation, with the torque at 300 percent of normal operating conditions, or 3×300 lb/in = 900 lb/in (101.7 Nm), and a coefficient of friction of 0.1, $900 = N \times 0.1 \times 766 \times [(4 + 2.5)/4]$. Solving for N, we find $N = 7.23$. The next larger even whole number of friction plates is 8. Therefore, eight friction planes and nine plates will be specified.

3. Determine if the chosen number of plates is optimum for this clutch

Once we've chosen the number of plates we have the option of either reducing the operating force, P, and thus the pressure on the plates, by the ratio 7.23/8 or keeping the pressure between the plates at 100 lb/in² (689 kPa) and reducing the outer diameter of the plates. Since space is important in the design of this clutch, we will determine the outer diameter required when $p = 100$ lb/in² (689 kPa) and $N = 8$.

Substituting $pA = p(\pi)[D^2 - d^2)/4]$ for P in the general clutch relation gives

$$T = N \, \mu p \pi \, \frac{D^2 - d^2}{4} \frac{D + d}{4}$$

$$900 = 8 \times 0.1 \times 100 \, \frac{D^2 - 2.5^2}{4} \frac{D + 2.5}{4}$$

Solving the resulting cubic equation for D, we find $D = 3.90$ in (9.906 cm). Solving for P, we find $P = 704$ lb (319.6 kg).

Our specifications for this clutch will be: Plates 9 (eight friction planes), hardened steel, outer diameter of friction surface = 3.90 in (9.906 cm); inner diameter friction surface = 2.50 in (6.35 cm); operating force = 704 lb (319.6 kg).

Related Calculations. Use this general procedure to choose either wet or dry clutches. The relations given here can be applied to either type of clutch. A wet clutch is chosen wherever the atmosphere in the clutch operating area is such that oil or moisture are present and cannot be conveniently removed. Using a wet clutch saves the cost of seals and other devices needed to seal the clutch from the atmospheric moisture.

Dry-plate clutches are used where there is no danger of oil or moisture getting on the plates. Most such clutches use either natural or forced convection for cooling. The drive material is a manmade composition in contact with cast iron, bronze, or steel plates.

This procedure is the work of Richard M. Phelan, Associate Professor of Mechanical Engineering, Cornell University. SI values were added by the handbook editor.

CLUTCH SELECTION FOR SHAFT DRIVE

Choose a clutch to connect a 50-hp (37.30 kW) internal-combustion engine to a 300-r/min single-acting reciprocating pump. Determine the general dimensions of the clutch.

Calculation Procedure:

1. *Choose the type of clutch for the load*

Table 3 shows a typical applications for the major types of clutches. Where economy is the prime consideration, a positive-engagement or a cone-type friction clutch would be chosen. Since a reciprocating pump runs at a slightly varying speed, a centrifugal clutch is not suitable. For greater dependability, a disk or plate friction clutch is more desirable than a cone clutch. Assume that dependability is more important than economy, and choose a disk-type friction clutch.

2. *Determine the required clutch torque at starting capacity*

A clutch must start its load from a stopped condition. Under these circumstances the instantaneous torque may be two, three, or four times the running torque. Therefore, the usual clutch is chosen so it has a torque capacity of at least twice the running torque. For internal-combustion engine drives, a starting torque of three to four times the running torque is generally used. Assume 3.5 time is used for this engine and pump combination. This is termed the *clutch starting factor.*

Since $T = 63,000 hp/R$, where T = torque, lb·in; hp = horsepower transmitted; R = shaft rpm; $T = 63,000(50)/300 = 10,500$ lb·in (1186.3 Nm). This is the required starting torque capacity of the clutch.

TABLE 3 Clutch Characteristics

Type of clutch	Typical applications*
Friction:	
Cone	Varying loads; 0 to 200 hp (0 to 149.1 kW); losing popularity for many applications, particularly in the higher hp ranges
Disk or plate	Varying loads; 0 to 500 hp (0 to 372.9 kW); widely used; more popular than the cone clutch
Rim:	
Band	Varying loads; 0 to 100 hp (0 to 74.6 kW); not too widely used
Overrunning	Constant or moderately varying loads; 0 to 200 hp (0 to 149.1 kW); engages in one direction; freewheels in the opposite direction
Centrifugal	Constant loads; 0 to 50 hp (0 to 37.3 kW)
Inflatable	Varying loads; 0 to 5000 hp (0 to 3728.5 kW); compressed air inflates clutch; have 360° friction surface
Magnetic	Varying loads; 0 to 10,000 hp (0 to 7457.0 kW); high speeds; also used where disk clutch would be overloaded
Positive-engagement	Nonslip operation; low-speed (10 to 150 r/min) engagement; has sudden starting action
Fluid	Large, varying loads; 0 to 10,000 hp (0 to 7457.0 kW); variable-speed output; can produce a desired slip
Electromagnetic	Large, varying loads; 0 to 10,000 hp (0 to 7457.0 kW); variable-speed output; characteristics similar to fluid clutches

*Clutch capacity depends on the design, materials of constructions, type of load, shaft speed, and operating conditions. The applications and capacity ranges given here are typical but should not be taken as the only uses for which the listed clutches are suitable.

3. *Determine the total required clutch torque capacity*

In addition to the clutch starting factor, a service factor is also usually applied. Table 4 lists typical clutch service factors. This tabulation shows that the service factor for a single-reciprocating pump is 2.0. Hence, the total required clutch torque capacity = required starting torque capacity × service factor = 10,500 × 2.0 = 21,000-lb · in (2372.7 Nm) torque capacity.

4. *Choose a suitable clutch for the load*

Consult a manufacturer's engineering data sheet listing clutch torque capacities for clutches of the type chosen in step 1 of this procedure. Choose a clutch having a rated torque equal to or greater than that computed in step 3. Table 5 shows a portion of a typical engineering data sheet. A size 6 clutch would be chosen for this drive.

Related Calculations. Use the general method given here to select clutches for industrial, commercial, marine, automotive, tractor, and similar applications. Note that engineering data sheets often list the clutch rating in terms of torque, lb · in, and hp/(100 r/min).

Friction clutches depend, for their load-carrying ability, on the friction and pressure between two mating surfaces. Usual coefficients of friction for friction clutches

TABLE 4 Clutch Service Factors

Type of service	Service factor
Driver:	
Electric motor:	
Steady load	1.0
Fluctuating load	1.5
Gas engine:	
Single cylinder	1.5
Multiple cylinder	1.0
Diesel engine:	
High-speed	1.5
Large, slow-speed	2.0
Driven machine:	
Generator:	
Steady load	1.0
Fluctuating load	1.5
Blower	1.0
Compressor, depending on number of cylinders	2.0–2.5
Pumps:	
Centrifugal	1.0
Reciprocating, single-acting	2.0
Reciprocating, double-acting	1.5
Lineshaft	1.5
Woodworking machinery	1.75
Hoists, elevators, cranes, and shovels	2.0
Hammer mills, ball mills, and crushers	2.0
Brick machinery	3.0
Rock crushers	3.0

TABLE 5 Clutch Ratings

Clutch number	Torque rating		Power (100 r/min)	
	lb·in	N·m	hp	kW
1	2,040	230.5	3	2.2
2	4,290	484.7	6	4.5
3	8,150	920.8	12	8.9
4	13,300	1,502.7	21	15.7
5	19,700	2,225.8	31	23.1
6	35,200	3,977.1	55	41.0
7	44,000	4,971.3	69	51.5

range between 0.15 and 0.50 for dry surfaces, 0.05 and 0.30 for greasy surfaces, and 0.05 and 0.25 for lubricated surfaces. The allowable pressure between the surfaces ranges from a low of 8 lb/in² (55.2 kPa) to a high of 300 lb/in² (2068.5 kPa).

SIZING PLANETARY LINEAR-ACTUATOR ROLLER SCREWS

A high-speed industrial robot requires a linear actuator with a 1.2-m (3.94-ft) stroke to advance a load averaging 5700 N (1281 lb) at 20 m/min (65.6 ft/min). The load to reposition the arm is 1000 N (225 lb). Positioning should be within 1 mm (0.03937 in). Find the mean load, expected life, life in million revolutions, and maximum speed of a roller screw for this application. Suggest a type of lubrication, estimate screw efficiency, and calculate the power required to drive the roller screw, Fig. 4.

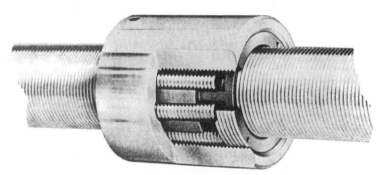

FIGURE 4 Recirculating roller screw which has high positioning accuracy is well-suited for precise work, such as refocusing lenses for laser beams. (*Machine Design.*)

Calculation Procedure:

1. *Choose the type of screw to use*
Roller screws are best suited for loads exceeding 100,000 N (22,482 lb) or speeds over 6 m/min (19.7 ft/min). A roller screw without preload has a backlash of about 0.02 to 0.04 mm (0.00079 to 0.00157 in). This backlash adds to the thread inaccuracy of the roller screw. Depending on the accuracy class, positioning is usually within 0.1 to 0.25 mm (0.00384 to 0.0098 in) over a travel of 4000 mm (157.5 in). Preloading the screw eliminates backlash, and overall positioning accuracy is not significantly affected by a variable external load. Because more than 18 mm/min (0.709 in) velocity is required, a roller screw will be used for this application, with single support bearings at each end. The machine served by this actuator will make 300 load cycles/h, operate 16 h/day, 240 days/yr, and function for 5 years.

2. *Determine the life expectancy of this roller screw actuator*
Because length and load are not excessive, a medium-duty roller screw will be chosen. To accommodate linear speed, shaft speed, and life, an initial selection of a roller screw has a 44-mm (1.73-in) diameter, a 12-mm (0.472-in) lead, and a 108,200 N (24,326-lb) dynamic nut capacity, based on manufacturer's catalog data. Manufacturers often list the dynamic load of a nut or screw for an L_{10} life of 1 million revolutions. Total life of the screw is:

$$L_{rev} = \left(\frac{l}{s}\right) l_h\, t_h\, t_d\, t_y$$

where L = life in 10^6 revolutions; I = stroke, mm (in); s = screw lead, mm/rev (in/rev); I_h = strokes/h; t_h = operating hours/day; t_y = years of service. Another factor may be included to account for variations in load alignment, acceleration, and lubrication.

Substituting for this roller screw,

$$L_{rev} = \frac{1,200\times 2}{12}\,(300)\,16\,(240)\,5$$

$$= 1,152 \times 10^6\ rev$$

3. *Find the mean load on the screw*
For the advance and retract loads, the mean load on the screw is found from:

$$F_m = \sqrt[3]{\frac{F_a^{\,3} + F_r^{\,3}}{2}}$$

where F_a = load during advance, N (lb); F_r = load during retraction, N (lb). Substituting,

$$F_m = \sqrt[3]{\frac{5,700^3 + 1,000^3}{2}}$$

$$= 4,532\ N\ (1019\ lb)$$

4. Determine the L_{10} life of this roller-screw actuator
The L_{10} value for a roller screw is found from

$$L_{10} = \left(\frac{C}{F_m}\right)^3 \times 10^6 \text{ rev}$$

where L_{10} = life corresponding to a 10 percent probability of failure, and C = dynamic nut capacity, N (lb). Substituting,

$$L_{10} = \left(\frac{108,000}{4,532}\right)^3 \times 10^6$$

$$= 13.5 \times 10^9 \text{ rev}$$

5. Find the rotational speed and type of lubrication needed
With a linear speed of 20 m/mm (65.6 ft/min), the rotational speed in rpm will be l/s, where I = stroke length, mm (in); s = screw lead, mm/rev (in/rev). Substituting, rpm = n = (20 m × 1000 mm/m)/12) = 1667 rpm is required.

Knowing the rotational speed, we can compute the nD value, where D = nominal screw diameter, mm. Substituting, nD = 1667(44) = 73,348.

Lubrication type defines the roller screw speed limit. This limit is given by the nD value computed above, or vD/s, where v = linear nut speed, mm/s (in/s); other symbols as before. Oil lubrication allows nD values as high as 140,000, while grease permits nD values to 93,000. For rolled-thread ball screws, nD values are about 64 percent of these. Since this roller screw has an nD value of 73,348, grease lubrication is acceptable.

If lubrication of the roller screw is not regular or old lubricant is used, the life figure can be modified by a factor of 0.5 to 0.66. Further, if the lubricant is likely to be contaminated, an adjustment factor of 0.33 to 0.5 can be used.

6. Compute the maximum speed of the screw shaft
The maximum permissible speed of the screw shaft is 80 percent of the first critical speed and is given by:

$$\omega_{max} = \frac{0.8(392)(10^5)\,ad_o}{l^2}$$

where ω_{max} = maximum permissible speed, rpm; a = screw support factor from Table 3; d_o = screw shaft root diameter, mm (in); l = distance between centers of screw shaft support bearings, mm (in). Substituting,

$$\omega_{max} = \frac{0.8(392\times10^5)\,2.47(42)}{1,200^2}$$

$$= 2,259 \text{ rpm}$$

7. Calculate the theoretical efficiency of this roller screw
Efficiency of converting rotary motion to linear motion is estimated with

$$e = \frac{s}{s+Kd_o}$$

where e = efficiency; K = friction angle factor, 0.0375 for heavy duty and 0.0325 for medium-duty designs. At low loads, less than 10 percent of dynamic capacity, e is within 10 percent of the calculated value. As load increases to the dynamic capacity, efficiency estimates are less certain, usually within 25 percent of the calculated value. Substituting,

$$e = \frac{12}{12+0.0325(42)}$$

$$= 0.90$$

8. Find the input power to the roller screw
Input power to drive the load at a given speed is found from

$$P = \frac{F_m S \, \omega}{60,000e}$$

where P = power, W (hp); other symbols as before. Substituting,

$$P = \frac{4,532(12) \, 1,667}{60,000 \, (0.9)}$$

$$= 1,679 \text{ W (2.25 hp)}$$

9. Calculate the screw root diameter required to avoid buckling
Buckling of roller screws becomes a problem when the screw shaft is loaded in compression. To avoid buckling, the screw root diameter must exceed the critical or minimum screw diameter for the load, or

$$d_0 > \sqrt[4]{\frac{F_m l}{34,000b}}$$

where b = screw support factor from Table 6. Substituting,

$$d_0 > \sqrt[4]{\frac{4,532(1,200)}{34,000(1)}}$$

$$d_0 > 3.56 \text{ mm (0.14 in.)}$$

The root diameter of the selected shaft is $d_o = 44$ mm (1.74 in). Since this is greater than the 3.56 mm (0.14 in), the shaft is unlikely to buckle.

The limiting factor in this application appears to be the maximum shaft speed. If the bearings are encased at one end of the shaft, allowing $a = 3.85$ and $b = 2$, a 36-mm (1.41-in) diameter screw with 66,000 N (14,838 lb) dynamic capacity and the same lead will also perform adequately.

Related Calculations. Roller screws are cost-effective alternatives to ball screws in applications requiring high speed, long life, and high load capacity. The load-bearing advantage of roller screws lies in their contact area. Ball screws transfer load through point contact on the balls. Consequently, load is carried through a discrete number of points. Roller screws, by contrast, transmit load through contacts on each roller. Unlike a ball, contact on a roller can be ground precisely for a duty requirement.

TABLE 6 Screw Speed and Compression Load Support Factors***

Support bearings on shaft	Speed factor, a	Compression factor, b
**	0.88	0.25
*	2.47	1
	3.85	2
	5.6	4

*= simply supported shaft end
**= encased end
***Machine Design

Roller screws have other advantages. Because the roller and nut threads have the same helix angles, the rollers do not move axially as they roll inside the nut. Hence, no recirculation is required. Further, a planetary gear in each housing end turns the rollers during movement so that if something hinders rolling motion, they are driven past the problem area. Rolling also minimizes the friction penalty of area contact.

The absence of recirculation has the added benefit of allowing greater speeds than ball screws. Because the rollers are in constant contact with the screw, as opposed to being lifted and repositioned, roller screws can be driven at higher rotational linear speeds than ball screws. Nonrecirculation also means the rollers are not subjected to cyclic stressing. This further improves fatigue life.

Load is a good parameter for starting the sizing process for a roller screw. While the load often fluctuates and reverses with each cycle, once a mean load is calculated, a unit can be selected by using the nut's dynamic capacity. Constant mean load is given by

$$F_m = \sqrt[3]{\frac{F_1^3 L_1 + F_2^3 L_2 + \cdots + F_n^3 L_n}{L_1 + L_2 + \cdots L_n}}$$

where F_m = constant mean load, N (lb); F_1 through F_n = constant loads encountered during operation, N (lb). These loads correspond to L_1 through L_n, which are the number of revolutions under a particular constant load.

If the loads are fairly constant as the screw advances or retracts, mean load is found from

$$F_m = \sqrt[3]{\frac{F_a{}^3 + F_r{}^3}{2}}$$

where F_a = load during advance, N (lb); F_r = load during retraction N (lb).

Lubrication of a roller screw is similar to that for a roller bearing. A circulating lubrication system is preferred, especially in hot or dirty environments. Here, cool oil reduces frictional heat buildup during operation and flushes debris from the threads. Oil should have a viscosity of 100 ISO or about SAE 30 during operation. With heavy load or at low speed, an EP additive is recommended to improve film strength.

Grease lubrication is acceptable when oil is impractical. Wipers should be used on the nut to prevent dirty grease from entering the threads. The bearing greases most often used are NLGI 2. Lithium-base greases are generally suitable from -30 to 110°C (-20°F to 230°F).

Alignment and shock can severely alter roller-screw bearing life. If the load is not completely axial, the constant mean load, F_m, may be adjusted by a factor of 1.05 to 1.1. If traverse loads are applied to the screw, F_m is adjusted by a factor of 1.1 to 1.5.

If accelerations are low and speed variations in the application are smooth, no adjustment to the screw capacity factor is necessary. However, if the speed varies rapidly, or vibrations or high-frequency oscillations are present, an adjustment factor of 1.05 to 1.2 should be applied to F_m.

Roller screws find many applications including catapult reset after aircraft launch from aircraft carriers, automated arms for industrial presses, rudder control on large aircraft, robot actuators, etc.

The data and calculation in this procedure are the work of Pierre C. Lemor, Manager, Linear Components, SKF Component Systems, as reported in *Machine Design* magazine. SI units were added by the handbook editor.

DESIGNING A ROLLING-CONTACT TRANSLATION SCREW

A rolling-contact screw-operated translation device is being designed to raise a 10,000-lb (44,480-N) load a distance of 15 in (38.1 cm) as rapidly as possible. The proposed design is shown schematically in Fig. 5. Modified square threads have been proposed for this power screw. The nut will be made of bronze; the screw of AISI 3140 steel oil-quenched and tempered at 1000°F (537.8°C). All thrust and guide bearings will be rolling-contact types with negligible friction. Determine (*a*) the dimensions of the screw and nut for a factor of safety of 2; (*b*) the time required to raise the load; (*c*) the required horsepower (kW) of the electric drive motor.

Calculation Procedure:

1. Determine the required screw diameter
Consider the screw as a column. Then, the J. B. Johnson formula applies. Checking thread-property tables, it is found that a 1-in (2.54-cm) four-threads-per-inch (2.54 cm) is the smallest modified square thread that will be satisfactory for this design.

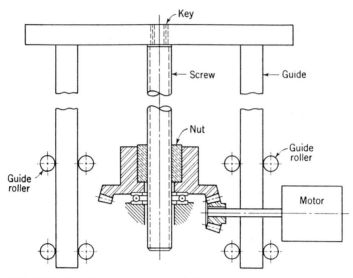

FIGURE 5   Translation device to raise load.

However, the unsupported length of this screw must also carry the screw torque, and screw strength must be checked under the combined stress of the column and torque loads.

2.   Find the screw torque

$$T = \frac{Qd}{2} \frac{\cos \alpha \tan \lambda + \mu}{\cos \alpha - \mu \tan \lambda} \text{ lb-in (Nm)}$$

where T = screw torque, lb-in (Nm); Q = screw load, lb (kg); d = screw pitch diameter, in (mm); trigonometric functions are defined in Figs. 6 and 7; $\cos \alpha$ = 0.9962; α = 5 deg; $\tan \lambda = l/\pi d = (1/4)/(\pi)(0.875) = 0.0909$ for single-thread screw, μ = 0.14 from Table 7, which gives the starting friction for high-grade materials, superior-quality workmanship, and best running conditions. Thus,

$$T = \frac{10,000 \times 0.875}{2} \frac{0.9962 \times 0.0909 + 0.14}{0.9962 - 0.14 \times 0.0909} = 1,163 \text{ lb-in (131.4 Nm)}$$

3.   Check screw strength under combined stress
Make the check by comparing the factor of safety under combined stress with the specified factor of safety of 2. Or,

$$\text{f.s.} = \frac{s_y/2}{\sqrt{(s/2)^2 + (s_s)^2}}$$

where s_y = yield point, lb/in² (kPa) from an AISI-3140 steel properties plot = 132,000 lb/in² (909.5 MPa); s = equivalent normal stress for the column load, lb/in² (kPa); s_s = shear stress produced by the torque, lb/in² (kPa).

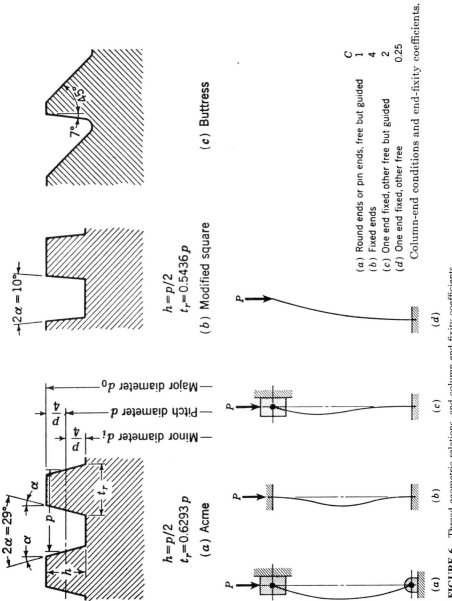

FIGURE 6 Thread geometric relations, and column end-fixity coefficients.

$h = p/2$
$t_r = 0.6293 p$
(a) Acme

$h = p/2$
$t_r = 0.5436 p$
(b) Modified square

(c) Buttress

Major diameter d_0
Pitch diameter d
Minor diameter d_i

		C
(a)	Round ends or pin ends, free but guided	1
(b)	Fixed ends	4
(c)	One end fixed, other free but guided	2
(d)	One end fixed, other free	0.25

Column-end conditions and end-fixity coefficients.

(a)   (b)   (c)   (d)

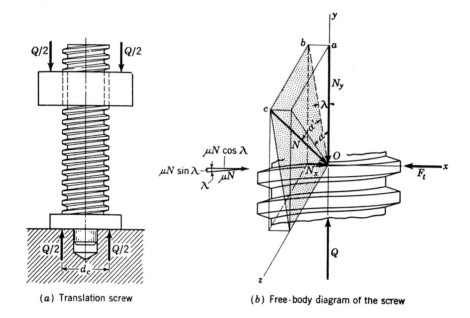

(a) Translation screw

(b) Free-body diagram of the screw

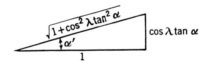

FIGURE 7   Translation screw geometric relations.

To compute an equivalent normal stress for the column load we will use

$$S_{equiv} = \frac{Q}{A_i} \frac{1}{[s_s(L/k)^2/4C\pi^2 E]}$$

with

$Q = 10,000$ lb (44,480 N)
$A_i = \pi d_i^2/4 = \pi(0.750)^2/4 = 0.442$ in² (2.85 cm²)
$s_y = 132,000$ lb/in² (909.5 MPa) from a table of steel properties
$L = 15$ in (38.1 cm)
$k = d_i/4 = 0.750/4 = 0.1875$ in (0.48 cm)
$C = 3$, from Fig. 6 on the basis that the screw is essentially a column with fixed ends but that the normal clearance between the threads at the nut will give a slight decrease in the rigidity at that end
$E = 30,000$ lb/in² (206,700 MPa)

Thus,

TABLE 7 Coefficients of Friction for Screw Threads and Thrust Collars*

Steel screw and bronze or cast-iron nut			Thrust-collar friction		
	Average coefficient of friction, μ			Average coefficient of friction, μ_c	
Conditions	Starting	Running	Materials	Starting	Running
High-grade materials and workmanship and best running coditions	0.14	0.10	Soft steel on cast iron . .	0.17	0.12
Average quality of materials and workmanship and average running conditions	0.18	0.13	Hardened steel on cast iron	0.15	0.09
Poor workmanship or very slow and infrequent motion with indifferent lubrication or newly machined surfaces . . .	0.21	0.15	Soft steel on bronze . . . Hardened steel on bronze	0.10 0.08	0.08 0.06

*After C. W. Ham and D. G. Ryan, An Experimental Investigation of the Friction of Screw Threads, *Univ. Illinois Eng. Expt. Sta. Bull.* 247.

$$S_{equiv} = \frac{10,000}{0.442} \frac{1}{1 - \{[132,000 \times (15/0.1875^2]/(4 \times 3 \times \pi^2 \times 30,000,000)}$$

$$= 29,700 \text{ lb/in}^2 \text{ (204.6 MPa)}$$

To calculate the shear stress due to torque, we will use:

$$s_s = \frac{Tc}{J} = \frac{t}{J/c}$$

and $T = 1,163$ lb-in.
$J/c = \pi d_i^3/16 = [\pi \times (0.750)^3]/16 = 0.0828$

Thus,

$$s_s = \frac{1,163}{0.0828} = 14,000 \text{ lb/in}^2 \text{ (96.5 MPa)}$$

Solving for the factor of safety we find:

$$\text{f.s.} = \frac{132,000/2}{\sqrt{(29,700/2)^2 + (14,000)^2}} = 3.23$$

Therefore, since 3.23 is greater than 2, the screw has more than adequate strength as a column.

4. *Determine the length of thread engagement*

The specified length of thread engagement will be the greatest of the lengths required for adequate shear strength of the screw threads, adequate shear strength of the nut threads, and satisfactory wear life. In this design, we are interested in raising the load as rapidly as possible, and the wear rate will probably be the most critical factor. The procedure we will follow will be to calculate the lengths of engagement required for adequate strength, then select a length at least equal to the larger of the two, and to determine the maximum operating speed of the nut for a reasonable wear rate.

The shear of the screw threads with a factor of safety of 2 and the shear stress determined earlier gives a length of engagement of

$$L_{e,screw} = \text{f.s.} \ \frac{pQ}{\pi d_i t_i s_{s, \ screw}}$$

with

f.s. = 2
p = ¼ = 0.250 in (0.635 cm)
Q = 10,000 lb (44,480 N)
d_i = 0.750 in (1.91 cm)
$t_i = t_r$ = 0.1359 in (0.345 cm)
$s_s = s_{sy} = s_y/2$ = 132,000/2 = 66,000 lb/in² (454.7 MPa)

Thus,

$$L_{e,screw} = 2 \ \frac{0.250 \times 10,000}{\pi \times 0.750 \times 0.1359 \times 66,000} = 0.237 \text{ in } (0.60 \text{ cm})$$

The shear of the nut threads is found from

$$L_{e,nut} = \text{f.s.} \ \frac{pQ}{\pi d_p t_o s_{s,nut}}$$

with all the values the same as for the screw except that the factor of safety = 4 because the design will be based upon the ultimate shearing strength of the bronze nut. In the above relation, d_o = 1.00 in (2.54 cm); $s_s = s_{su}$ = 35,000 lb/in² (241 MPa) from AISI 3140 steel property data. Thus,

$$L_{e,nut} = 4 \ \frac{0.250 \times 10,000}{\pi \times 1.000 \times 0.1359 \times 35,000} = 0.669 \text{ in } (1.699 \text{ cm})$$

Therefore, any length of thread engagement equal to or greater than 0.669 in (1.699 cm) will result in adequate shear strength of the threads. Since wear is a function of the product of the bearing pressure and sliding velocity, it is evident that the maximum speed of operation will be possible when the bearing pressure is a minimum. Hence, the length of engagement should be as large as practical. The load will not be uniformly distributed over the several threads, being carried mostly by the first thread or two. Consequently, there is little to be gained by using a length of engagement more than 1.0 or 1.5 times the screw diameter. Here, we shall use L_e = 1.5 × 1 in (2.54 cm), calculate the bearing pressure, and see what sliding speed will be permissible on the basis of values in Table 8, using

TABLE 8 Allowable Bearing Pressures for Screws

| Application | Material | | Allowable bearing pressure, psi | MPa | Sliding speed at thread pitch diameter | m/min |
	Screw	Nut				
Hand press	Steel	Bronze	2,500–3,500	17.2–24.1	Low speed, well lubricated	
Jack screw	Steel	Cast iron	1,800–2,500	12.4–17.2	Low speed, 8 fpm	2.43
Jack screw	Steel	Bronze	1,600–2,500	11.0–17.2	Low speed, 10 fpm	3.04
Hoisting screw	Steel	Cast iron	600–1,000	4.1–6.89	Medium speed, 20–40 fpm	6.09–12.2
Hoisting screw	Steel	Bronze	800–1,400	5.5–9.65	Medium speed, 20–40 fpm	6.06–12.2
Lead screw	Steel	Bronze	150–240	1.03–1.65	High speed, 50 fpm	15.2

$$L_e = \frac{4pQ}{\pi(d_o^2 - d_i^2)s_b} \text{ in}$$

Thus,

$$1.5 = \frac{4 \times \frac{1}{4} \times 10,000}{\pi \times (1.000^2 - 0.750^2)s_b}$$

Solving for s_b, we find

$$s_b = 4,850 \text{ lb/in}^2 \text{ (33.4 MPa)}$$

which is higher than recommended for even low-speed, well-lubricated operation. Therefore, a new approach to the design is required. Nevertheless, our efforts have not been wasted entirely because we now know that any screw selected for this application on the basis of wear will be more than adequately strong. The difference is so great that, except for the increased resistance of the hardened surface to wear, there is little to be gained in using the heat-treated alloy steel. If conditions warrant, the material may be charged to cold-drawn low-carbon bar stock, 1020 or equivalent.

5. Perform any needed redesign

The design will now be based entirely on wear considerations. A reasonable combination of allowable bearing pressure and sliding speed is 800 lb/in² (5.5 MPa) at 40 ft/min (12.2 m/min) from Table 8. The size of screw that will give a bearing pressure of 800 lb/in² (5.5 MPa) or less, with a length of engagement of 1.5 times the nominal diameter, d_o, must be determined by trial and error as in the table below.

d_o, in	cm	d_i, in*	cm	p*, in	cm	L_e, in	cm	s_b, psi	MPa
2.000	5.08	1.5556	3.952	1/2.25	1.1288	3.00	7.62	1194	8.226
2.500	6.35	2.000	5.08	1/2	1.27	3.75	9.525	755	5.2

*From a table of thread dimensions for translation screws

Hence, a 2.5-in (63.5-cm) two-threads-per-inch (25.4 mm) modified square thread will be satisfactory.

6. Determine the time required to raise the load

The pitch-line sliding velocity is:

$$V = \frac{\pi dn}{12} \text{ ft/min (m/min)}$$

with these values: $V = 40$ ft/min (12.2 m/min); $d = 2.50 - h = 2.50 - 0.250 = 2.250$ in (5.72 cm); $n = $ rpm of nut. Solving for n, we find $n = 67.9$ rpm. Each revolution will raise the load a distance equal to the lead. If a single-thread screw is used, the lead will be 0.500 in (1.27 cm), and the lifting speed will be

$$V_{load} = nl \ 67.9 \times 0.500 = 33.8 \text{ in/min (85.9 cm/min)}$$

and the time to raise the load 15 in (38.1 cm) will be:

$$t = \frac{S}{V} = \frac{15}{33.8} = 0.444 \text{ min or } 26.6 \text{ sec}$$

If 26.6 s is too long, a double-thread screw will require only half the time, or 13.3 sec, a triple-thread screw will required only 8.9 sec, etc.

7. Find the motor horsepower (kW) requirements
If a lift time of 13.3 s for the double-thread screw is considered reasonable, the drive horsepower can be determined in terms of torque and rotative speed from

$$\text{hp} = \frac{Tn}{63,000}$$

where T = torque, lb/in (Nm), and n = rpm.

The torque calculated in the first part of the solution is no longer valid, because both the lead and the diameter have been changed, and another calculation must be made. Since the motor will probably have a starting torque of 200 percent of rated torque, there is considerable justification for determining the power requirements under running conditions. However, there is an appreciable difference between the starting and running coefficients of friction, 0.14 and 0.10, respectively, from Table 7. A slight change in lubrication conditions could increase the coefficient of friction to the point where overheating of the motor might become a problem. Thus, the power requirement will be based on the starting coefficient of friction. The excess of the motor will produce a more rapid acceleration of the load. The required torque is found from

$$T = \frac{Qd}{2} \frac{\cos \alpha \tan \lambda + \mu}{\cos \alpha - \mu \tan \lambda} \text{ lb/in (Nm)}$$

where Q = 10,000 lb (44,480 N); D = 2.25 in (6.35 cm); $\cos \alpha$ = 0.9962; $\tan \lambda$ = $l/\pi(d)$ = $(2 \times 1/2)/(\pi \times 2.25)$ = 0.1415; μ = 0.14. Substituting,

$$T = \frac{10,000 \times 2.250}{2} \frac{0.9962 \times 0.1415 + 0.14}{0.9962 - 0.14 \times 0.1415} = 3,960 \text{ lb/in (447.4 Nm)}$$

Therefore, if the relatively small power losses in the thrust bearings, the guide bearing, and the gear train between the motor and the nut are neglected, the power required is

$$\text{hp} = \frac{3,960 \times 67.9}{63,000} = 4.27 \text{ (3.19 kW)}$$

Thus, a 5-hp (3.2-kW) motor would be specified. Note that for the double-thread screw, μ = 0.14 and $\cos \alpha \times \tan \lambda$ = 0.1409. Hence, the screws is not self-locking. The guide bearings, gears, etc., might possibly provide enough additional friction to hold the load at any point under starting or static friction. However, if for any reason, such as pronounced vibration, the friction coefficient decreases to the running value, the load would rapidly lower itself. The solution is to use either the self-locking single-thread screw with its slower motion (and smaller power requirement) or to use a brake of some type, such as that in Fig. 8. In this design we will

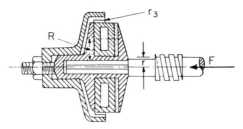

FIGURE 8   Brake for translation screw.

assume that the saving in lift time will justify the additional complication of adding a brake.

Our specifications for this actuator will therefore be: Screw and nut: double-thread, modified-square-thread, 2.5-in (6.35-cm)-diameter, four-threads-per-inch (2.54-cm), low-carbon-steel screw and bronze nut with length of engagement (nut height) of 3.75 in (9.52 cm). Operation time: 13.3 s (neglecting acceleration and deceleration). Motor: 5 hp (3.19 kW).

Related Calculations.   Use this general procedure to design ball and roller bearing translation screws. Note that various adjustments must be made to initial assumptions as the design work proceeds.

This procedure is the work of Richard M. Phelan, Associate Professor of Mechanical Engineering, Cornell University. SI vales were added by the handbook editor.

SELECTION OF A RIGID FLANGE-TYPE SHAFT COUPLING

Choose a steel flange-type coupling to transmit a torque of 15,000 lb · in (1694.4 Nm) between two 2½-in (6.4-cm) diameter steel shafts. The load is uniform and free of shocks. Determine how many bolts are needed in the coupling if the allowable bolt shear stress is 3000 lb/in^2 (20,685.0 kPa). How thick must the coupling flange be, and how long should the coupling hub be if the allowable stress bearing for the hub is 20,000 lb/in^2 (137,900.0 kPa) and in shear 6000 lb/in^2 (41,370.0 kPa)? The allowable shear stress in the key is 12,000 lb/in^2 (82,740.0 kPa). There is no thrust force acting on the coupling.

Calculation Procedure:

1.   *Choose the diameter of the coupling bolt circle*
Assume a bolt-circle diameter for the coupling. As a first choice, assume the bolt-circle diameter is three times the shaft diameter, or 3 × 2.5 = 7.5 in (19.1 cm). This is a reasonable first assumption for most commercially available couplings.

[1]John H. Glover, "Planetary Gear Systems," *Product Engineering,* Jan. 6, 1964.

2. Compute the shear force acting at the bolt circle

The shear force F_s lb acting at the bolt-circle radius r_b in is $F_s = T/r_b$, where T = torque on shaft, lb · in. Or, $F_s = 15,000/(7.5/2) = 4000$ lb (17,792.9 N).

3. Determine the number of coupling bolts needed

When the allowable shear stress in the bolts is known, compute the number of bolts N required from $N = 8F_s/(\pi d^2 s_s)$, where d = diameter of each coupling bolt, in; s_s = allowable shear stress in coupling bolts, lb/in^2.

The usual bolt diameter in flanged, rigid couplings ranges from $\frac{1}{4}$ to 2 in (0.6 to 5.1 cm), depending on the torque transmitted. Assuming that $\frac{1}{2}$-in (1.3-cm) diameter bolts are used in this coupling, we see that N = $8(4000)/[\pi(0.5)^2(3000)]$ = 13.58, say 14 bolts.

Most flanged, rigid couplings have two to eight bolts, depending on the torque transmitted. A coupling having 14 bolts would be a poor design. To reduce the number of bolts, assume a larger diameter, say 0.75 in (1.9 cm). Then N = $8(4000)/[\pi(0.75)^2(3000)]$ = 6.03, say eight bolts, because an odd number of bolts are seldom used in flanged couplings.

Determine the shear stress in the bolts by solving the above equation for s_s = $8F_s/(\pi d^2 N) = 8(4000)/[\pi(0.75)^2(8)]$ = 2265 lb/in^2 (15,617.2 kPa). Thus, the bolts are not overstressed, because the allowable stress is 3000 lb/in^2 (20,685.0 kPa).

4. Compute the coupling flange thickness required

The flange thickness t in for an allowable bearing stress s_b lb/in^2 is $t = 2F_s/(Nds_b)$ = $2(4000)/[(8)(0.75)(20,000)]$ = 0.0666 in (0.169 cm). This thickness is much less than the usual thickness used for flanged couplings manufactured for off-the-shelf use.

5. Determine the hub length required

The hub length is a function of the key length required. Assuming a $\frac{3}{4}$-in (1.9-cm) square key, compute the hub length l in from $l = 2F_{ss}/(t_k s_t)$, where F_{ss} = force acting at shaft outer surface, lb; t_k = thickness, in. The force $F_{ss} = T/r_h$, where r_h = inside radius of hub, in = shaft radius = $2.5/2$ = 1.25 in (3.2 cm) for this shaft. Then $F_{ss} = 15,000/1.25 = 12,000$ lb · in (1355.8 Nm). Then $l = 2(12,000)/[(0.75)(20,000)]$ = 1.6 in (4.1 cm).

When the allowable design stress for bearing, 20,000 lb/in^2 (137,895.1 kPa) here, is less than half the allowable design stress for shear, 12,000 lb/in^2 (82,740.0 kPa) here, the longest key length is obtained when the bearing stress is used. Thus, it is not necessary to compute the thickness needed to resist the shear stress for this coupling. If it is necessary to compute this thickness, find the force acting at the surface of the coupling hub from $F_h = T/r_h$, where r_h = hub radius, in. Then $t_s = F_h/\pi d_h s_s$, where d_h = hub diameter, in; s_s = allowable hub shear stress, lb/in^2.

Related Calculations. Couplings offered as standard parts by manufacturers are usually of sufficient thickness to prevent fatigue failure.

Since each half of the coupling transmits the total torque acting, the length of the key must be the same in each coupling half. The hub diameter of the coupling is usually 2 to 2.5 times the shaft diameter, and the coupling is generally made the same thickness as the coupling flange. The procedure given here can be used for couplings made of any metallic material.

SELECTION OF FLEXIBLE COUPLING FOR A SHAFT

Choose a stock flexible coupling to transmit 15 hp (11.2 kW) from a 1000-r/min four-cylinder gasoline engine to a dewatering pump turning at the same rpm. The pump runs 8 h/day and is an uneven load because debris may enter the pump. The pump and motor shafts are each 1.0 in (2.5 cm) in diameter. Maximum misalignment of the shafts will not exceed 0.5°. There is no thrust force acting on the coupling, but the end float or play may reach $\frac{1}{16}$ in (0.2 cm).

Calculation Procedure:

1. Choose the type of coupling to use
Consult Table 9 or the engineering data published by several coupling manufacturers. Make a tentative choice from Table 9 of the type of coupling to use, based on the maximum misalignment expected and the tabulated end-float capacity of the coupling. Thus, a roller-chain-type coupling (one in which the two flanges are connected by a double roller chain) will be chosen from Table 9 for this drive because it can accommodate 0.5° of misalignment and an end float of up to $\frac{1}{16}$ in (0.2 cm).

2. Choose a suitable service factor
Table 10 lists typical service factors for roller-chain-type flexible couplings. Thus, for a four-cylinder gasoline engine driving an uneven load, the service factor SF = 2.5.

3. Apply the service factor chosen
Multiply the horsepower or torque to be transmitted by the service factor to obtain the coupling design horsepower or torque. Or, coupling design $hp = (15)(2.5) = 37.5$ hp (28.0 kW).

4. Select the coupling to use
Refer to the coupling design horsepower rating table in the manufacturer's engineering data. Enter the table at the shaft rpm, and project to a design horsepower slightly greater than the value computed in step 3. Thus, in Table 11 a typical rating tabulation shows that a coupling design horsepower rating of 38.3 hp (28.6 kW) is the next higher value above 37.5 hp (28.0 kW).

TABLE 9 Allowable Flexible Coupling Misalignment

Coupling type	Angular misalignment	Parallel misalignment		End float	
		USCS	SI	USCS	SI
Plastic chain	Up to 1.0°	0.005 in	0.1 mm	$\frac{1}{16}$ in	2 mm
Roller chain	Up to 0.5°	2% of chain pitch	2% of chain pitch	$\frac{1}{16}$ in	2 mm
Silent chain	Up to 0.5°	2% of chain pitch	2% of chain pitch	$\frac{1}{4}$ to $\frac{3}{4}$ in	0.6 to 1.9 cm
Neoprene biscuit	Up to 5.0°	0.01 to 0.05 in	0.3 to 1.3 mm	Up to $\frac{1}{2}$ in	Up to 1.3 cm
Radial	Up to 0.5°	0.01 to 0.02 in	0.3 to 0.5 mm	Up to $\frac{1}{16}$ in	Up to 2 mm

TABLE 10 Flexible Coupling Service Factors*

Type of drive			
Engine,† less than six cylinders	Engine, six cylinders or more	Electric motor; steam turbine	Type of load
2.0	1.5	1.0	Even load, 8 h/day; nonreversing, low starting torque
2.5	2.0	1.5	Uneven load, 8 h/day; moderate shock or torque, nonreversing
3.0	2.5	2.0	Heavy shock load, 8 h/day; reversing under full load, high starting torque

*Morse Chain Company.
†Gasoline or diesel.

TABLE 11 Flexible Coupling hp Ratings*

r/min			Bore diameter, in (cm)	
800	1000	1200	Maximum	Minimum
16.7	19.9	23.2	1.25 (3.18)	0.5 (1.27)
32.0	38.3	44.5	1.75 (4.44)	0.625 (1.59)
75.9	90.7	105.0	2.25 (5.72)	0.75 (1.91)

*Morse Chain Company.

5. Determine whether the coupling bore is suitable
Table 11 shows that a coupling suitable for 38.3 hp (28.6 kW) will have a maximum bore diameter up to 1.75 in (4.4 cm) and a minimum bore diameter of 0.625 in (1.6 cm). Since the engine and pump shafts are each 1.0 in (2.5 cm) in diameter, the coupling is suitable.

The usual engineering data available from manufactures include the stock keyway sizes, coupling weight, and principal dimensions of the coupling. Check the overall dimensions of the coupling to determine whether the coupling will fit the available space. Where the coupling bore diameter is too small to fit the shaft, choose the next larger coupling. If the dimensions of the coupling make it unsuitable for the available space, choose a different type or make a coupling.

Related Calculations. Use the general procedure given here to select any type of flexible coupling using flanges, springs, roller chain, preloaded biscuits, etc., to transmit torque. Be certain to apply the service factor recommended by the manufacturer. Note that biscuit-type couplings are rated in hp/100 r/min. Thus, a biscuit-type coupling rated at 1.60 hp/100 r/min (1.2 kW/100 r/min) and a maximum allowable speed of 4800 r/min could transmit a maximum of (1.80 hp)(4800/100) = 76.8 hp (57.3 kW).

SELECTION OF A SHAFT COUPLING FOR TORQUE AND THRUST LOADS

Select a shaft coupling to transmit 500 hp (372.9 kW) and a thrust of 12,500 lb (55,602.8 N) at 100 r/min from a six-cylinder diesel engine. The load is an even one, free of shock.

Calculation Procedure:

1. Compute the torque acting on the coupling
Use the relation $T = 5252hp/R$ to determine the torque, where T = torque acting on coupling, lb·ft; hp = horsepower transmitted by the coupling; R = shaft rotative speed, r/min. For this coupling, $T = (5252)(500)/100 = 26,260$ lb·ft (35,603.8 Nm).

2. Find the service torque
Multiply the torque T by the appropriate service factor from Table 10. This table shows that a service factor of 1.5 is suitable for an even load, free of shock. Thus, the service torque = $(26,260$ lb·ft$)(1.5) = 39,390$ lb·ft, say 39,500 lb·ft (53,554.8 Nm).

3. Choose a suitable coupling
Enter Fig. 9 at the torque on the left, and project horizontally to the right. Using the known thrust, 12,500 lb (55,602.8 N), enter Fig. 9 at the bottom and project vertically upward until the torque line is intersected. Choose the coupling model represented by the next higher curve. This shows that a type A coupling having a maximum allowable speed of 300 r/min will be suitable. If the plotted maximum rpm is lower than the actual rpm of the coupling, use the next plotted coupling type rated for the actual, or higher rpm.

In choosing a specific coupling, use the manufacturer's engineering data. This will resemble Fig. 9 or will be tabulation of the ranges plotted.

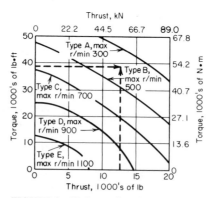

FIGURE 9 Shaft-coupling characteristics.

Related Calculations. Use this procedure to select couplings for industrial and marine drives where both torque and thrust must be accommodated. See the Marine Engineering section of this handbook for an accurate way to compute the thrust produced on a coupling by a marine propeller. Always check to see that the coupling bore is large enough to accommodate the connected shafts. Where the bore is too small, use the next larger coupling.

HIGH-SPEED POWER-COUPLING CHARACTERISTICS

Select the type of power coupling to transmit 50 hp (37.3 kW) at 200 r/min if the angular misalignment varies from a minimum of 0 to a maximum of 45°. Determine the effect of angular misalignment on the shaft position, speed, and acceleration at angular misalignments of 30 and 45°.

Calculation Procedure:

1. Determine the type of coupling to use
Table 12, developed by N. B. Rothfuss, lists the operating characteristics of eight types of high-speed couplings. Study of this table shows that a universal joint is the only type of coupling among those listed that can handle an angular misalignment of 45°. Further study shows that a universal coupling has a suitable speed and hp range for the load being considered. The other items tabulated are not factors in this application. Therefore, a universal coupling will be suitable. Table 13 compares the functional characteristics of the couplings. Data shown support the choice of the universal joint.

2. Determine the shaft position error
Table 14, developed by David A. Lee, shows the output variations caused by misalignment between the shafts. Thus, at 30° angular misalignment, the position error is 4°06′42″. This means that the output shaft position shifts from −4°06′42″ to +4°06′42″ twice each revolution. At a 45° misalignment the position error, Table 14, is 9°52′26″. The shift in position is similar to that occurring at 30° angular misalignment.

3. Compute the output-shaft speed variation
Table 14 shows that at 30° angular misalignment the output-shaft speed variation is ±15.47 percent. Thus, the output-shaft speed varies between 200(1.00 ± 0.1547) = 169.06 and 230.94 r/min. This speed variation also occurs *twice* per revolution.

For a 45° angular misalignment the speed variation, determined in the same way, is 117.16 to 282.84 r/min. This speed variation also occurs twice per revolution.

4. Determine output-shaft acceleration
Table 14 lists the ratio of maximum output-shaft acceleration A to the square of the input speed, ω^2, expressed in radians. To convert r/min to rad/s, use $rps = 0.1047$ r/min $= 0.0147(200) = 20.94$ rad/s.

For 30° angular misalignment, from Table 14 $A/\omega^2 = 0.294571$. Thus, $A = \omega^2(0.294571) = (20.94)^2(0.294571) = 129.6$ rad/s^2. This means that a constant

TABLE 12 Operating Characteristics of Couplings*

	Contoured diaphragm	Axial spring	Laminated disk	Universal joint	Ball-race	Gear	Chain	Elastomeric
Speed range, r/min	0–60,000	0–8,000	0–20,000	0–8,000	0–8,000	0–25,000	0–6,300	0–6,000
Power range, kW/100 r/min	1–500	1–9,000	1–100	1–100	1–100	1–2,000	1–200	0–400
Angular misalignment, degrees	0–8.0	0–2.0	0–1.5	0–45	0–40	0–3	0–2	0–4
Parallel misalignment, mm	0–2.5	0–2.5	0–2.5	None	None	0–2.5	0–2.5	0–2.5
Axial movement, cm	0–0.5	0–2.5	0–0.5	None	None	0–5.1	0–2.5	0–0.8
Ambient temperature, °C	900	Varies	900	Varies	Varies	Varies	Varies	Varies
Ambient pressure, kPa	Sea level to zero	Varies	Sea level to zero	Varies	Varies	Varies	Varies	Varies

Product Engineering.

TABLE 13 Functional Characteristics of Couplings*

	Contoured diaphragm	Axial spring	Laminated disk	Universal joint	Ball-race	Gear	Chain	Elastomeric
No lubrication	√	:	√	:	:	:	:	√
No backlash	√	√	√	†	†	:	:	√
Constant velocity ratio	√	:	:	†	√†	†	†	√
Containment	√	:	†	:	√†	√	√	
Angular only	√	:	√	√	√	√	√	√
Axial and angular	√	:	√	:	√	√	√	√
Axial and parallel	√	√	√	:	:	√	√	√
Axial, angular, and parallel	√	√	√	:	:	√	√	
High temperature	√	:	√					
High altitude	√	:	√	√	√	√	√	
High torsional spring rate	√	:	√	√	√	√	√	
Low bending moment	√	√	√	√	√	√	√	√
No relative movement		:	:	:	:	:	:	√

*Product Engineering.
†Zero backlash and containment can be obtained by special design.
‡Constant velocity ratio at small angles can be closely approximated.

21.37

TABLE 14   Universal Joint Output Variations*

Misalignment angle, deg	Maximum position error	Maximum speed error, percent	Ratio A/ω^2
5	0°06′34″	0.382	0.011747
10	0°26′18″	1.543	0.030626
15	0°59′36″	3.526	0.069409
20	1°46′54″	6.418	0.124966
25	2°48′42″	10.338	0.198965
30	4°06′42″	15.470	0.294571
35	5°42′20″	22.077	0.417232
40	7°36′43″	30.541	0.576215
45	9°52′26″	41.421	0.787200

*Caused by misalignment of the shaft. Table from *Machine Design*.

input speed of 200 r/min produces an output acceleration ranging from $-$ 129.6 to $+$ 129.6 rad/s², and back, at a frequency of 2(200 r/min) = 400 cycles/min.

At a 45° angular misalignment, the acceleration range of the output shaft, determined in the same way, is -346 to $+346$ rad/s². Thus, the acceleration range at the larger shaft angle misalignment is 2.67 times that at the smaller, 30°, misalignment.

Related Calculations.   Table 12 is useful for choosing any of seven other types of high-speed couplings. The eight couplings listed in this table are popular for high-horsepower applications. All are classed as rigid types, as distinguished from entirely flexible connectors such as flexible cables.

Values listed in Table 12 are nominal ones that may be exceeded by special designs. These values are guideposts rather than fixed; in borderline cases, consult the manufacturer's engineering data. Table 13 compares the functional characteristics of the couplings and is useful to the designer who is seeking a unit with specific operating characteristics. Note that the values in Table 12 are maximum and not additive. In other words, a coupling *cannot* be operated at the maximum angular and parallel misalignment and at the maximum horsepower and speed simultaneously—although in some cases the combination of maximum angular misalignment, maximum horsepower, and maximum speed would be acceptable. Where shock loads are anticipated, apply a suitable correction factor, as given in earlier calculation procedures, to the horsepower to be transmitted before entering Table 12.

SELECTION OF ROLLER AND INVERTED-TOOTH (SILENT) CHAIN DRIVES

Choose a roller chain and the sprockets to transmit 6 hp (4.5 kW) from an electric motor to a propeller fan. The speed of the motor shaft is 1800 r/min and of the driven shaft 900 r/min. How long will the chain be if the centerline distance between the shafts is 30 in (76.2 cm)?

Calculation Procedure:

1. Determine, and apply, the load service factor

Consult the manufacturer's engineering data for the appropriate load service factor. Table 15 shows several typical load ratings (smooth, moderate shock, heavy shock) for various types of driven devices. Use the load rating and the type of drive to determine the service factor. Thus, a propeller fan is rated as a heavy shock load. For this type of load and an electric-motor drive, the load service factor is 1.5, from Table 15.

Apply the load service factor by taking the product of it and the horsepower transmitted, or (1.5)(6 hp) = 9.0 hp (6.7 kW). The roller chain and sprockets must have enough strength to transmit this horsepower.

2. Choose the chain and number of teeth in the small sprocket

Using the manufacturer's engineering data, enter the horsepower rating table at the small-sprocket rpm and project to a horsepower value equal to, or slightly greater than, the required rating. At this horsepower rating, read the number of teeth in the small sprocket, which is also listed in the table. Thus, in Table 16, which is an excerpt from a typical horsepower rating tabulation, 0.9 hp (6.7 kW) is not listed at a speed of 1800 r/min. However, the next higher horsepower rating, 9.79 hp

TABLE 15 Roller Chain Loads and Service Factors*

Load rating	
Driven device	Type of load
Agitators (paddle or propeller)	Smooth
Brick and clay machinery	Heavy shock
Compressors (centrifugal and rotary)	Moderate shock
Conveyors (belt)	Smooth
Crushing machinery	Heavy shock
Fans (centrifugal)	Moderate shock
Fans (propeller)	Heavy shock
Generators and exciters	Moderate shock
Laundry machinery	Moderate shock
Mills	Heavy shock
Pumps (centrifugal, rotary)	Moderate shock
Textile machinery	Smooth

Service factor			
	Internal-combustion engine		
Type of load	Hydraulic drive	Mechanical drive	Electric motor or turbine
Smooth	1.0	1.2	1.0
Moderate shock	1.2	1.4	1.3
Heavy shock	1.4	1.7	1.5

*Excerpted from Morse Chain Company data.

TABLE 16 Roller Chain Power Rating*

[Single-strand, ⅝-in (1.6-cm) pitch roller chain]

No. of teeth in small sprocket	Small sprocket rpm					
	1500		1800		2100	
	hp	kW	hp	kW	hp	kW
14	10.7	7.98	8.01	5.97	6.34	4.73
15	11.9	8.87	8.89	6.63	7.03	5.24
16	13.1	9.77	9.79	7.30	7.74	5.77
17	14.3	10.7	10.7	7.98		

*Excerpted from Morse Chain Company data.

(7.3 kW), will be satisfactory. The table shows that at this power rating, 16 teeth are used in the small sprocket.

This sprocket is a good choice because most manufacturers recommend that at least 16 teeth be used in the smaller sprocket, except at low speeds (100 to 500 r/min).

3. Determine the chain pitch and number of strands
Each horsepower rating table is prepared for a given chain pitch, number of chain strands, and various types of lubrication. Thus, Table 16 is for standard single-strand ⅝-in (1.6-cm) pitch roller chain. The 9.79-hp (7.3-kW) rating at 1800 r/min for this chain is with type III lubrication—oil bath or oil slinger—with the oil level maintained in the chain casing at a predetermined height. See the manufacturer's engineering data for the other types of lubrication (manual, drip, and oil stream) requirements.

4. Compute the drive speed ratio
For a roller chain drive, the speed ration $S_r = R_h/R_l$, where R_h = rpm of high-speed shaft; R_l = rpm of low-speed shaft. For this drive, $S_r = 1800/900 = 2$.

5. Determine the number of teeth in the large sprocket
To find the number of teeth in the large sprocket, multiply the number of teeth in the small sprocket, found in step 2, by the speed ratio, found in step 4. Thus, the number of teeth in the large sprocket = (16)(2) = 32.

6. Select the sprockets
Refer to the manufacturer's engineering data for the dimensions of the available sprockets. Thus, one manufacturer supplies the following sprockets for ⅝-in (1.6-cm) pitch single-strand roller chain: 16 teeth, OD = 3.517 in (8.9 cm), bore = ⅝ in (1.6 cm); 32 teeth, OD = 6.721 in (17.1 cm), bore = ⅝ or ¾ in (1.6 or 1.9 cm). When choosing a sprocket, be certain to refer to data for the size and type of chain selected in step 3, because each sprocket is made for a specific type of chain. Choose the type of hub—setscrew, keyed, or taper-lock bushing—based on the torque that must be transmitted by the drive. See earlier calculation procedures in this section for data on key selection.

7. *Determine the length of the chain*

Compute the chain length in pitches L_p from $L_p = 2C + (S/2) + K/C$, where C = shaft center distance, in/chain pitch, in; S = sum of the number of teeth in the small and large sprocket; K = a constant from Table 17, obtained by entering this table with the value D = number of teeth in large sprocket − number of teeth in small sprocket. For this drive, $C = 30/0.625 = 48$; $S = 16 + 32 = 48$; $D = 32$ − $16 = 16$; $K = 6.48$ from Table 17. Then, $L_p = 2(48) + 48/2 + 6.48/48 = 120.135$ pitches. However, a chain cannot contain a fractional pitch; therefore, use the next higher number of pitches, or $L_p = 121$ pitches.

Convert the length in pitches to length in inches, L_i, by taking the product of the chain pitch p in and L_p. Or $L_i = L_p p = (121)(0.625) = 75.625$ in (192.1 cm).

Related Calculations. At low-speed ratios, large-diameter sprockets can be used to reduce the roller-chain pull and bearing loads. At high-speed ratios, the number of teeth in the high-speed sprocket may have to be kept as small as possible to reduce the chain pull and bearing loads. The Morse Chain Company states: Ratios over 7:1 are generally not recommended for single-width roller chain drives. Very slow-speed drives (10 to 100 r/min) are often practical with as few as 9 or 10 teeth in the small sprocket, allowing ratios up to 12:1. In all cases where ratios exceed 5:1, the designer should consider the possibility of using compound drive to obtain maximum service life.

When you select standard inverted-tooth (silent) chain and high-velocity inverted-tooth silent-chain drives, follow the same general procedures as given above, except for the following changes.

Standard inverted-tooth silent chain: (*a*) Use a minimum of 17 teeth, and an odd number of teeth on one sprocket, where possible. This increases the chain life. (*b*) To achieve minimum noise, select sprockets having 23 or more teeth. (*c*) Use the proper service factor for the load, as given in the manufacturer's engineering data. (*d*) Where a long or fixed-center drive is necessary, use a sprocket or shoe idler where the largest amount of slack occurs. (*e*) Do not use an idler to reduce the chain wrap on small-diameter sprockets. (*f*) Check to see that the small-diameter sprocket bore will fit the high-speed shaft. Where the high-speed shaft diameter exceeds the maximum bore available for the chosen smaller sprocket, increase the number of teeth in the sprocket or choose the next larger chain pitch. (This general procedure also applies to roller chain sprockets.) (*g*)Compute the

TABLE 17 Roller Chain Length Factors*

D	K	D	K
1	0.03	11	3.06
2	0.10	12	3.65
3	0.23	13	4.28
4	0.41	14	4.96
5	0.63	15	5.70
6	0.91	16	6.48
7	1.24	17	7.32
8	1.62	18	8.21
9	2.05	19	9.14
10	2.53	20	10.13

*Excerpted from Morse Chain Company data.

chain design horsepower from (drive hp)(chain service factor). (*h*) Select the chain pitch, number of teeth in the small sprocket, and chain *width* from the manufacturer's rating table. Thus, if the chain design horsepower = 36 hp (26.8 kW) and the chain is rated at 4 hp/in (1.2 kW/cm) of width, the required chain width = 36 hp/(4 hp/in) = 9 in (22.9 cm).

High-velocity inverted-tooth silent chain: (*a*) Use a minimum of 25 teeth and an odd number of teeth on one sprocket, where possible. This increases the chain life. (*b*) To achieve minimum noise, select sprockets with 27 or more teeth. (*c*) Use a larger service factor than the manufacturer's engineering data recommends, if trouble-free drives are desired. (*d*) Use a wider chain than needed, if an increased chain life is wanted. Note that the chain width is computed in the same way as described in item (*h*) above. (*e*) If a longer center distance between the drive shafts is desired, select a larger chain pitch [usual pitches are ¾, 1, 1½, or 2 in (1.9, 2.5, 3.8, or 5.1 cm)]. (*f*) Provide a means to adjust the centerline distance between the shafts. Such an adjustment *must* be provided in vertical drives. (*g*) Try to use an even number of pitches in the chain to avoid an offset link.

CAM CLUTCH SELECTION AND ANALYSIS

Choose a cam-type clutch to drive a centrifugal pump. The clutch must transmit 125 hp (93.2 kW) at 1800 r/min to the pump, which starts and stops 40 times per hour throughout its 12-h/day, 360-day/year operating period. The life of the pump will be 10 years.

Calculation Procedure:

1. Compute the maximum torque acting on the clutch
Compute the torque acting on the clutch from $T = 5252hp/R$, where the symbols are the same as in the previous calculation procedure. Thus, for this clutch, $T = 5252 \times 125/1800 = 365$ lb·ft (494.9 Nm).

2. Analyze the torque acting on the clutch
For installations free of shock loads during starting and stopping, the running torque is the maximum torque that acts on the clutch. But if there is a shock load during starting and stopping, or at other times, the shock torque must be added to the running torque to determine the total torque acting. Compute the shock torque using the relation in step 1 and the actual hp and speed developed by the shock load.

3. Compute the total number of load applications
With 40 starts and stops (cycles) per hour, a 12-h day, and 360 operating days per year, the number of cycles per year is (40 cycles/h)(12 h/day)(360 days/year) = 172,800. In 10 years, the clutch will undergo (172,800 cycles/year)(10 years) = 1,728,000 cycles.

4. Choose the clutch size
Enter Fig. 10 at the maximum torque, 365-lb·ft (494.9 Nm), on the left, and the number of load cycles, 1,728,000, on the bottom. Project horizontally and vertically until the point of intersection is reached. Select the clutch represented by the next higher curve. Thus a type A clutch would be used for this load. (Note that the

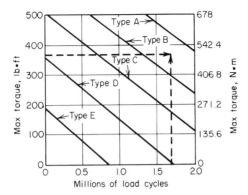

FIGURE 10   Cam-type-clutch selection chart.

clutch capacity could be tabulated instead of plotted, but the results would be the same.)

5.   Check the clutch dimensions
Determine whether the clutch bore will accommodate the shafts. If the clutch bore is too small, choose the next larger clutch size. Also check to see whether the clutch will fit into the available space.

   Related Calculations.   Use this general procedure to select cam-type clutches for business machines, compressors, conveyors, cranes, food processing, helicopters, fans, aircraft, printing machinery, pumps, punch presses, speed reducers, looms, grinders, etc. When choosing a specific clutch, use the manufacturer's engineering data to select the clutch size.

TIMING-BELT DRIVE SELECTION AND ANALYSIS

Choose a toothed timing belt to transmit 20 hp (14.9 kW) from an electric motor to a rotary mixer for liquids. The motor shaft turns at 1750 r/min and the mixer shaft is to turn at 600 ± 20 r/min. This drive will operate 12 h/day, 7 days/week. Determine the type of timing belt to use and the driving and driven pulley diameters if the shaft centerline distance is about 27 in (68.6 cm).

Calculation Procedure:

1.   Choose the service factor for the drive
Timing-belt manufacturers publish service factors in their engineering data based on the type of prime mover, the type of driven machine (compressor, mixer, pump, etc.), type of drive (speedup), and drive conditions (continuous operation, use of an idler, etc.).

   Usual service factors for any type of driver range from 1.3 to 2.5 for various types of driven machines. Correction factors for speed-up drives range from 0 to 0.40; the specific value chosen is *added* to the machine-drive correction factor.

Drive conditions, such as 24-h continuous operation or the use of an idler pulley on the drive, cause an additional 0.2 to be added to the correction factor. Seasonal or intermittent operation *reduces* the machine-drive factor by 0.2.

Look up the service factor in Table 18, if the manufacturer's engineering data are not readily available. Table 18 gives safe data for usual timing-belt applications and is suitable for preliminary selection of belts. Where a final choice is being made, use the manufacturer's engineering data.

For a liquid mixer shock-free load, use a service factor of 2.0 from Table 18, since there are no other features which would required a larger value.

2. Compute the design horsepower for the belt

The design horsepower $hp_d = hp_l \times SF$, where hp_l = load horsepower; SF = service factor. Thus, for this drive, $hp_d = (20)(2) = 40$ hp (29.8 kW).

3. Compute the drive speed ratio

The drive speed ration $S_r = R_h/R_l$, where R_h = rpm of high-speed shaft; R_l = rpm of low-speed shaft. For this drive $S_r = 1750/600 = 2.92:1$, the rated rpm. If the driven-pulley speed falls 10 r/min, $S_r = 1750/580 = 3.02:1$. Thus, the speed ratio may vary between 2.92 and 3.02.

4. Choose the timing-belt pitch

Enter Table 19, or the manufacturer's engineering data, at the design horsepower and project to the driver rpm. Where the exact value of the design horsepower is not tabulated, use the next higher tabulated value. Thus, for this 1750-r/min drive

TABLE 18   Typical Timing-Belt Service Factors*

Type of drive	Type of load	Service factor
Electric motors, hydraulic motors, internal-combustion engines, line shafts	Shock-free	2.0
	Shocks	2.5
	Continuous operation or idler use	2.7
	Speed-up	3.0

*Use only for preliminary selection of belt. From Morse Chain Company data.

TABLE 19   Typical Timing-Belt Pitch*

Design power		Speed of high-speed shaft, r/min					
		3500		1750		1160	
hp	kW	in	cm	in	cm	in	cm
25	18.6	½, ⅞	1.3, 2.2	½, ⅞	1.3, 2.2	⅞	2.2
50	37.3	½, ⅞	1.3, 2.2	⅞	2.2	⅞, 1¼	2.2, 3.2
60 and up	44.7 and up	⅞	2.2	⅞	2.2	⅞, 1¼	2.2, 3.2

*Morse Chain Company.

having a design horsepower of 40 (29.8 kW), Table 19 shows that a ⅞-in (2.2-cm) pitch belt is required. This value is found by entering Table 19 at the next higher design hp, 50 (37.3 kW), and projecting to the 1750-r/min column. If 40 horsepower (29.8 kW) were tabulated, the table would be entered at this value.

5. *Choose the number of teeth for the high-speed sprocket*
Enter Table 20, or the manufacturer's engineering data, at the timing-belt pitch and project across to the rpm of the high-speed shaft. Opposite this value read the minimum number of sprocket teeth. Thus, for a 1750-r/min ⅞-in (2.2-cm) pitch timing belt, Table 19 shows that the high-speed sprocket should have no less than 24 teeth nor a pitch diameter less than 6.685 in (17.0 cm). (If a smaller diameter sprocket were used, the belt service life would be reduced.)

6. *Select a suitable timing belt*
Enter Table 21, or the manufacturer's engineering data, at either the exact speed ratio, if tabulated, or the nearest value to the speed-ratio range. For this drive, having a ratio of 2.92:3.02, the nearest value in Table 21 is 3.00. This table shows that

TABLE 20 Minimum Number of Sprocket Teeth*

Belt pitch		High-speed shaft, r/min	Minimum sprocket pitch distance		No. of teeth
in	cm		in	cm	
½	1.3	3500	3.501	8.9	20
		1750	3.183	8.1	18
		1160	2.865	7.3	16
⅞	2.2	3500	7.241	18.4	26
		1750	6.685	17.0	24
		1160	6.127	15.6	22
1¼	3.2	3500	10.345	26.3	26
		1750	9.549	24.3	24
		1160	8.753	22.2	22

*Morse Chain Company.

TABLE 21 Timing-Belt Center Distances*

Speed ratio	No. of sprocket teeth		Center distance					
			XH		XH		XH	
	Driver	Driven	in	cm	in	cm	in	cm
2.80	30	84	22.81	57.94	30.11	76.48	37.30	94.74
3.00	24	72	27.17	69.01	34.34	87.22	41.46	105.3
3.20	30	96	19.19	48.74	26.84	68.17	34.19	86.84

*Morse Chain Company.

with a 24-tooth driver and a 72-tooth driven sprocket, a center distance of 27.17 in (69.0 cm) is obtainable. Since a center distance of about 27 in (68.6 cm) is desired, this belt is acceptable.

Where an exact center distance is specified, several different sprocket combinations may have to be tried before a belt having a suitable center distance is obtained.

7. Determine the required belt width
Each center distance listed in Table 21 corresponds to a specific pitch and type of belt construction. The belt construction is often termed XL, L, H, XH, and XXH. Thus, the belt chosen in step 6 is an XH construction.

Refer now to Table 22 or the manufacturer's engineering data. Table 22 shows that a 2-in (5.1-cm) wide belt will transmit 38 hp (28.3 kW) at 1750 r/min. This is too low, because the design horsepower rating of the belt is 40 hp (29.8 kW). A 3-in (7.6-cm) wide belt will transmit 60 hp (44.7 kW). Therefore, a 3-in (7.6-cm) belt should be used because it can safely transmit the required horsepower.

If five, or less, teeth are in mesh when a timing belt is installed, the width of the belt must be increased to ensure sufficient load-carrying ability. To determine the required belt width to carry the load, divide the belt width by the appropriate factor given below.

Teeth in mesh	5	4	3	2
Factor	0.80	0.60	0.40	0.20

Thus, a 3-in (7.6-cm) belt with four teeth in mesh would have to be widened to $3/0.60 = 5.0$ in (12.7 cm) to carry the desired load.

Related Calculations. Use this procedure to select timing belts for any of these drives: agitators, mixers, centrifuges, compressors, conveyors, fans, blowers, generators (electric), exciters, hammer mills, hoists, elevators, laundry machinery, line shafts, machine tools, paper-manufacturing machinery, printing machinery, pumps, sawmills, textile machinery, woodworking tools, etc. For exact selection of a specific make of belt, consult the manufacturer's tabulated or plotted engineering data.

TABLE 22 Belt Power Rating*

[⅞ in (2.2 cm) pitch XH]

No. of teeth in high-speed sprocket	Belt width		Sprocket rpm					
			1700		1750		2000	
	in	cm	hp	kW	hp	kW	hp	kW
24	2	5.1	37	27.6	38	28.3	43	32.1
	3	7.6	59	44.0	60	44.7	67	50.0
	4	10.2	83	61.9	85	63.4	95	70.8

*Morse Chain Company.

GEARED SPEED REDUCER SELECTION AND APPLICATION

Select a speed reducer to lift a sluice gate weighing 200 lb (889.6 N) through a distance of 6 ft (1.8 m) in 5 s or less. The door must be opened and closed 12 times per hour. The drive for the door lifter is a 1150-r/min electric motor that operates 10 h/day.

Calculation Procedure:

1. Choose the type of speed reducer to use
There are many types of speed reducers available for industrial drives. Thus, a roller chain with different size sprockets, a V-belt drive, or a timing-belt drive might be considered for a speed-reduction application because all will reduce the speed of a driven shaft. Where a load is to be raised, often geared speed reducers are selected because they provide a positive drive without slippage. Also, modern geared drives are compact, efficient units that are easily connected to an electric motor. For these reasons, a right-angle worm-gear speed reducer will be tentatively chosen for this drive. If upon investigation this type of drive proves unsuitable, another type will be chosen.

2. Determine the torque that the speed reducer must develop
A convenient way to lift a sluice door is by means of a roller chain attached to a bracket on the door and driven by a sprocket keyed to the speed reducer output shaft. As a trial, assume that a 12-in (30.5-cm) diameter sprocket is used.

The torque T lb·in developed by sprocket $= T = Wr$, where $W =$ weight lifted, lb; $r =$ sprocket radius, in. For this sprocket, by assuming that the starting friction in the sluice-door guides produces an additional load of 50 lb (222.4 N), $T = (200 + 50)(6) = 1500$ lb·in (169.5 Nm).

3. Compute the required rpm of the output shaft
The door must be lifted 6 ft (1.8 m) in 5 s. This is a speed of (6 ft × 60 s/min)/ 5 s = 72 ft/min (0.4 m/s). The circumference of the sprocket is $\pi d = \pi(1.0) = 3.142$ ft (1.0 m). To lift the door at a speed of 72 ft/min (0.4 m/s), the output shaft must turn at a speed of (ft/min)/(ft/r) = 72/3.142 = 22.9 r/min. Since a slight increase in the speed of the door is not objectionable, assume that the output shaft turns at 23 r/min.

4. Apply the drive service factor
The AGMA *Standard Practice for Single and Double Reduction Cylindrical Worm and Helical Worm Speed Reducers* lists service factors for geared speed reducers driven by electric motors and internal-combustion engines. These factors range from a low of 0.80 for an electric motor driving a machine producing a uniform load for occasional 0.5-h service to a high of 2.25 for a single-cylinder internal-combustion engine driving a heavy shock load 24 h/day. The service factor for this drive, assuming a heavy shock load during opening and closing of the sluice gate, would be 1.50 for 10-h/day operation. Thus, the drive must develop a torque of at least (load torque, lb·in)(service factor) = (1500)(1.5) = 2250 lb·in (254.2 Nm).

TABLE 23 Speed Reducer Torque Ratings*

(Single-reduction worm gear)

Input power at 1150 r/min		Drive output				
			Torque		OHL†	
hp	kW	r/min	lb·in	N·m	lb	N
1.54	1.15	28.7	2416	273.0	1367	6080.7
1.24	0.92	23.0	2300	259.9	1367	6080.7
0.93	0.69	19.2	1970	222.6	1367	6080.7

*Extracted from Morse Chain Company data.
†Allowable overhung load on drive.

5. Choose the speed reducer

Refer to Table 23 or the manufacturer's engineering data. Table 23 shows that a single-reduction worm-gear speed reducer having an input of 1.24 hp (924.7 W) will develop 2300 lb·in (254.2 N·n) of torque at 23 r/min. This is an acceptable speed reducer because the required output torque is 2250 lb·in (254.2 Nm) at 23 r/min. Also the allowable overhung load, 1367 lb (6080.7 N), is adequate for the sluice-gate weight. A 1.5-hp (1118.5-W) motor would be chosen for this drive.

Related Calculations. Use this general procedure to select geared speed reducers (single- or double-reduction worm gears, single-reduction helical gears, gear motors, and miter boxes) for machinery drives of all types, including pumps, loaders, stokers, welding positioners, fans, blowers, and machine tools. The starting friction load, applied to the drive considered in this procedure, is typical for applications where a heavy friction load is likely to occur. In rotating machinery of many types, the starting friction load is usually nil, except where the drive is connected to a loaded member, such as a conveyor belt. Where a clutch disconnects the driver from the load, there is negligible starting friction.

Well-designed geared speed reducers generally will not run at temperatures higher than 100°F (55.6°C) *above* the prevailing ambient temperature, measured in the lubricant sump. At higher operating temperatures the lubricant may break down, leading to excessive wear. Fan-cooled speed reducers can carry heavier loads than noncooled reducers without overheating.

POWER TRANSMISSION FOR A VARIABLE-SPEED DRIVE

Choose the power-transmission system for a three-wheeled contractor's vehicle designed to carry a load of 1000 lb (4448.2 N) at a speed of 8 mi/h (3.6 m/s) over rough terrain. The vehicle tires will be 16 in (40.6 cm) in diameter, and the engine driving the vehicle will operate continuously. The empty vehicle weighs 600 lb (2668.9 N), and the engine being considered has a maximum speed of 4200 r/min.

Calculation Procedure:

1. Compute the horsepower required to drive the vehicle

Compute the required driving horsepower from $hp = 1.25\ Wmph/1750$, where W = total weight of *loaded* vehicle, lb (N); mph = maximum loaded vehicle speed, mi/h (km/h). Thus, for this vehicle, $hp = 1.25(1000 + 600)(8)/1750 = 9.15$ hp (6.8 kW).

2. Determine the maximum vehicle wheel speed

Compute the maximum wheel rpm from $rpm_w = $ (maximum vehicle speed, mi/h) $\times$ (5280 ft/mi)/15.72 (tire rolling diameter, in). Or, $rpm_w = $ (8)(5280)/[(15.72)(16)] = 167.8 r/min.

3. Select the power transmission for the vehicle

Refer to engineering data published by drive manufacturers. Choose a drive suitable for the anticipated load. The load on a typical contractor's vehicle is one of sudden starts and stops. Also, the drive must be capable of transmitting the required horse-power. A 10-hp (7.5-kW) drive would be chosen for this vehicle.

Small vehicles are often belt-driven by means of an infinitely variable transmission. Such a drive, having an overdrive or speed-increase ration of 1:1.5 or 1:1, would be suitable for this vehicle. From the manufacturer's engineering data, a drive having an input rating of 10 hp (7.5 kW) will be suitable for momentary overloads of up to 25 percent. The operating temperature of any part of the drive should never exceed 250°F (121.1°C). For best results, the drive should be operated at temperatures well below this limit.

4. Compute the required output-shaft speed reduction

To obtain the maximum power output from the engine, the engine should operate at its maximum rpm when the vehicle is traveling at its highest speed. This prevents lugging of the engine at lower speeds.

The transmission transmits power from the engine to the driving axle. Usually, however, the transmission cannot provide the needed speed reduction between the engine and the axle. Therefore, a speed-reduction gear is needed between the transmission and the axle. The transmission chosen for this drive could provide a 1:1 or a 1:1.5 speed ratio. Assume that the 1:1.5 speed ratio is chosen to provide higher speeds at the maximum vehicle load. Then the speed reduction required = (maximum engine speed, r/min)(transmission ratio)/(maximum wheel rpm) = (4200)(1.5)/167.8 = 37.6.

Check the manufacturer's engineering data for the ratios of available geared speed reducers. Thus, a study of one manufacturer's data shows that a speed-reduction ratio of 38 is available by using a single-reduction worm-gear drive. This drive would be suitable if it were rated at 10 hp (7.5 kW) or higher. Check to see that the gear has a suitable horsepower rating before making the final selection.

Related Calculations. Use the general procedure given here to choose power transmissions for small-vehicle compressors, hoists, lawn mowers, machine tools, conveyors, pumps, snow sleds, and similar equipment. For nonvehicle drives, substitute the maximum rpm of the driven machine for the maximum wheel velocity in steps 2, 3, and 4.

SECTION 22
BEARING DESIGN AND SELECTION

DETERMINING STRESSES, LOADING, BENDING MOMENTS, AND SPRING RATE IN SPOKED BEARING SUPPORTS

Spoked bearing supports are used in gas turbines, large air-cooling fans, electric-motor casings slotted for air circulation, and a variety of other applications. For the shaft and 6-spoked bearing system in Fig. 1 having three rotor masses and these parameters and symbols,

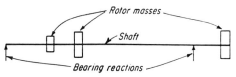

FIGURE 1   Shaft and 6-spoked bearing system having three rotor masses. (*Product Engineering.*)

	SI values
$P = 10,000$ lb (at either bearing)	(44,480 N)
$I_S = 25$ in^4	(1040.6 cm^4)
$I_R = 0.3$ in^4	(12.5 cm^4)
$A = 7$ in^2 (for ring also)	(45.2 cm^2)
$E_S = E_R = 10 \times 10^6$ psi	(68,900 MPa)
$L = 10$ in	(25.4 cm)
$R = 12$ in	(30.5 cm)
$C_R = 0.40$ in	(8.9 cm)
	(1.02 cm)

Symbols	SI values
A = spoke cross-section area, in^2	(cm^2)
C_S = distance, neutral axis to extreme fiber (of spoke), in	(cm)
C_R = distance, neutral axis to extreme fiber (of ring), in	(cm)
E_R, E_S = elasticity moduli (ring and spoke), psi	(kPa)
R_{R1} = axial loading in inclined spokes, lb; (+ for upper two, − for lower two)	(N)
F_{R2} = axial loading in vertical spokes, lb; (+ for top, − for bottom)	(N)
F_T = tangential load at OD of inclined spokes, lb; (clockwise on left side, counterclockwise on right side)	(N)
I_R = outer-ring moment of inertia about neutral axis pependicular to plane of support, in^4	(cm^4)
I_S = spoke moment of inertia about neutral axis perpendicular to plane of support, in^4	(cm^4)
k = spring rate with respect to outer shell, lb/in	(N/cm)
L = spoke length, in	(cm)
M = max bending moment (6-spoked support) in outer ring at OD of vertical spokes, in-lb; (+ at inner-fiber upper point and outer-fiber lower point)	(Nm)
= max bending moment at OD of all spokes in 4-spoked support	
P = bearing radial-load, lb	(N)
R = ring radius, in	(cm)
T = axial loading in outer ring at OD of vertical spokes in 6-spoked support; all spokes in 4-spoked support (− at upper points, + at lower points)	lb (N)
+ denotes tension	
− denotes compression	

find (*a*) the maximum bending moment in the outer ring of the support, (*b*) the axial loading in the outer ring, (*c*) the total stress in the outer ring at the top vertical spoke, Fig. 2, (*d*) the total axial loading in one of the inclined spokes, (e) the bending moment in the spoke at the hub, and (f) the spring rate of the spoked bearing support. Use the free-body diagram, Fig. 3, to analyze this bearing support.

Calculation Procedure:

1. Find the maximum bending moment in the outer ring of the 6-spoke bearing support
Use the relation

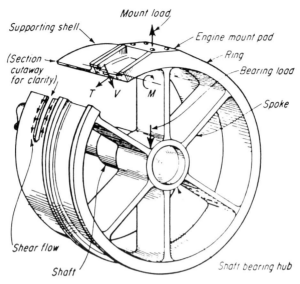

FIGURE 2 Typical 6-spoke bearing support having the mount load at the top for an aircraft gas turbine; in a stationary plant, mount load would be at the bottom of the support. (*Product Engineering.*)

$$\left(\frac{I_S}{I_R}\right)\left(\frac{R}{L}\right)^3\left(\frac{E_S}{E_R}\right) = \text{abscissae parameter, Fig. 4}$$

where the symbols are as shown above. Substituting, we find the parameter = 144, from:

$$\left(\frac{25}{0.3}\right)\left(\frac{12}{10}\right)^3\left(\frac{1}{1}\right)$$

Using the M curve in Fig. 4 for a parameter value of 144 gives

$$\frac{100M}{PR} = 1.45$$

Solving for M, we have

$$M = \frac{1.45\ PR}{100} = \frac{1.45(10,000)(12)}{100} = 1740 \text{ in/lb (196.6 Nm)}$$

2. Determine the axial loading in the outer ring at the outside diameter (OD)
Find T, the axial loading from Fig. 4 for the 144 parameter as

$$\frac{10T}{P} = 1.55$$

Substituting,

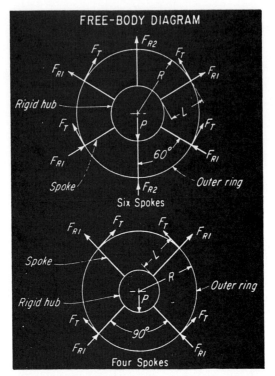

FIGURE 3 Free-body diagram for 6- and 4-spoke bearing supports. (*Product Engineering.*)

$$T = \frac{1.55P}{10} = \frac{(1.55)(10,000)}{10} = 1550 \text{ lb (6894 N)}$$

3. Compute the total stress in the outer ring at the top vertical spoke
Use the relation

$$M(C_R/I_R) + \frac{T}{A} = 1740\frac{(0.4)}{(0.3)} + \frac{1550}{7} = 2320 + 220 = 2540 \text{ lb/in}^2 \text{ (17501 kPa)}$$

4. Find the total axial loading in one of the inclined spokes
Using the F_{R1} curve in Fig. 4 for the same parameter, 144,

$$\frac{10F_{R1}}{P} = 0.82 \ F_{R1} = \frac{0.82 \ P}{10} = \frac{(0.82)(10,000)}{10} = 820 \text{ lb (3647 N)}$$

Also,

$$\frac{10F_T}{P} = 1.47 \ F_T = \frac{(1.47)(P)}{10} = \frac{(1.47)(10,000)}{10} = 1470 \text{ lb (6539 N)}$$

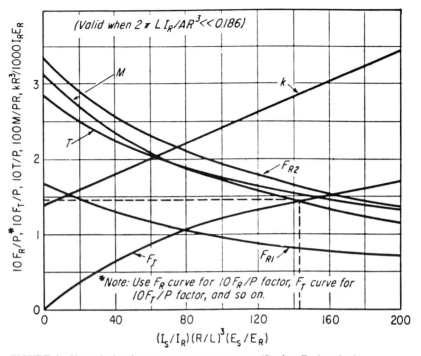

FIGURE 4 Six-spoke bearing-support parameter curves. (*Product Engineering.*)

5. Determine the bending moment in the spoke at the hub
The bending moment in the spoke at the hub is $(F_T)(L) = (1470)(10) = 14,700$ in-lb (1661 Nm) $= M_S$.
Then, the total stress in this section is

$$M_S \left(\frac{C_S}{I_S}\right) + \frac{F_{R1}}{A} = (14,700)\frac{(3.5)}{25} + \frac{820}{7}$$
$$= 2060 + 120 = 2180 \ \text{lb/in}^2 \ (15,020 \ \text{kPa})$$

6. Find the spring rate of the spoked bearing support
For the 144 parameter,

$$\frac{kR^3}{I_R E_R} = 2850, \ \text{whence} \ k = \frac{(2850) \ I_R E_R}{R^3} \ k = \frac{(2850)(0.3)(10)(10^6)}{(12)^3}$$

$$= 4.93 \times 10^6 \ \text{lb/in} \ (8633 \ \text{kN/cm})$$

Related Calculations. Figure 5 gives values for 4-spoke bearing supports. Use it in the same way that Fig. 4 was used in this procedure.

With more jet aircraft being built, an increase in the use of aero-derivative gas turbines in central-station and industrial power plants, wider use of air conditioning throughout the world, and construction of larger and larger electric motors, the

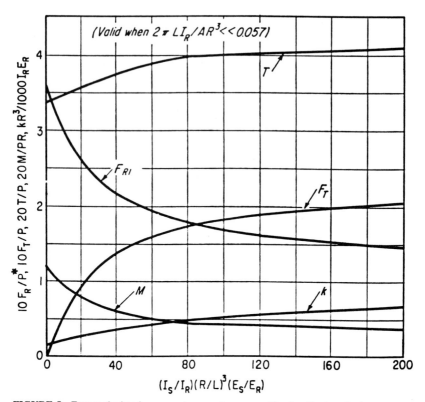

FIGURE 5 Four-spoke bearing-support parameter curves. (*Product Engineering.*)

spoked bearing support is gaining greater attention. The procedure presented here can be used for any of these applications, plus many related ones.

In calculations for spoked bearings the rings are assumed supported by sinusoidally varying tangential skin-shear reactions. Spokes are also assumed pinned to the ring but rigidly attached to the hub.

This procedure is the work of Lawrence Berko, Supervising Design Engineer, Walter Kide & Co., as reported in *Product Engineering* magazine. SI values were added by the handbook editor.

GRAPHIC COMPUTATION OF BEARING LOADS ON GEARED SHAFTS

Geared shafts having loads and reactions as shown in Fig. 6, are arranged as shown in Fig. 7. The physical characteristics of the gears are:

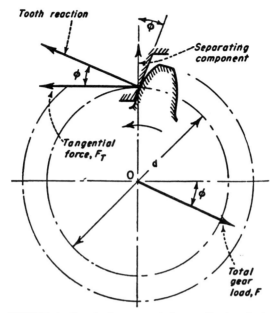

FIGURE 6 Gear loads on a typical gear. (*Product Engineering.*)

Pitch dia of gears	Moment arm
A = 2 in (5.08 cm)	a = 1.50 in (3.81 cm)
B = 1.50 in (3.81 cm)	b = 3.50 in (8.89 cm)
C = 4.00 in (10.16 cm)	c = 5.00 in (12.7 cm)
Driver = 1.75 in (4.45 cm)	d = 7.00 in (17.78 cm)
Angle, deg	Torque on driver = 100 lb/in (11.3 Nm)
alpha = 55	Torque delivered by A = 40 percent of torque on center shaft
beta = 48	Torque delivered by B = 60 percent on center shaft
tau = 45	Pressure angle of all gears, ϕ = 20 deg

Determine the bearing loads resulting from gear action using the total force directly. Use a combined numerical and graphical solution.

Calculation Procedure:

1. Find the load on each gear in the set
The tangential forces, F_T on the driver, Fig. 6, is $F_T = 2T/D$, where T = torque transmitted by the gear, lb/in (Nm); D = gear pitch diameter, in (cm). Substituting with the given torque on the driver of 100 lb/in (11.3 Nm), we have, $F_T = 2(100)/1.75 = 114$ lb (508 N). Then, the torque on the center shaft = $D(F_T) = 2(114) = 228$ lb/in (25.8 Nm).

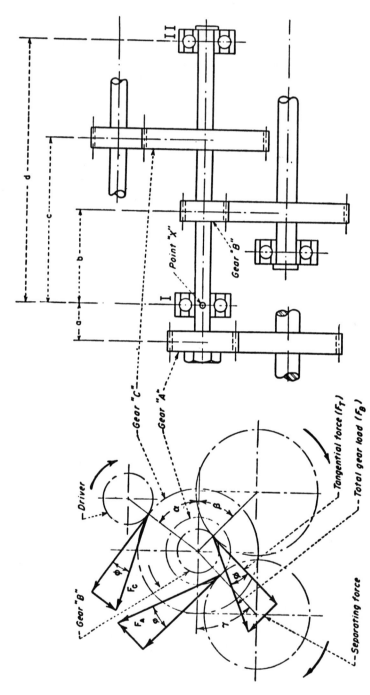

FIGURE 7 Typical gear set for which the bearing loads are computed. (*Product Engineering.*)

The gear loads are computed from F_A = (percent torque on shaft)(2)(torque on center shaft, lb/in)(sec of pressure angle on the gear)/D, where F_A = gear load, lb (N), on gear A, Fig. 7. Substituting using the given and computed values, F_A = 0.4 (2)(228)(sec 20°)/2 = 97 lb (431.5 N). Likewise, for gear B using a similar relation, F_B = 0.6 (2)(228)(sec 20°)/1.5 = 195 lb (867.4 N). Further, F_C, for gear C = 114 (sec 20°) = 121.5 lb (540 N).

2. Collect the needed data to prepare the graphical solution
Assemble the data as shown in Table 1.

3. Prepare the couple diagram, Fig. 8
When all the data are collected, draw the couple diagram, Fig. 8. When drawing this couple diagram it is important to note that: (a) Vectors representing negative couples are drawn in the same direction but in opposite sense to the forces causing them; (b) The direction of the closing should be such as to make the sum of all couples equal to zero. Thus, the direction of P_{II} is the direction of bearing reaction. The bearing load has the same direction but is of opposite sense.

In Fig. 8, the vector P_{II} measures 1149 lb/in (129.7 Nm) to scale. Therefore, the reaction on the bearing *II* is P_{II} = 1148/7 = 164 lb (729.7 N) at 11.5°.

TABLE 1 Bearing Loads on Geared Shafts

Gear or bearing	Distance from X, in (cm)	Force lb (N)	Couple lb/in° (Nm°)	Angular position
A	−1.5 (−3.81)	97 (431.5)	−145.5 (−16.4)	115
I	0	P_I	0	0
B	3.5	195 (867.4 N)	621 (70.2)	202
C	5.00	121.5 (68.7)	608 (68.7)	165
II	7.00	P_{II}	7 P_{II}	Δ

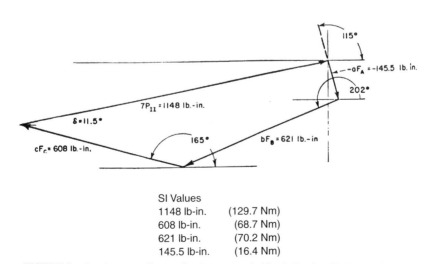

SI Values

1148 lb-in.	(129.7 Nm)
608 lb-in.	(68.7 Nm)
621 lb-in.	(70.2 Nm)
145.5 lb-in.	(16.4 Nm)

FIGURE 8 Couple vector diagram for the gear set in Fig. 7 (*Product Engineering.*)

4. *Construct the force vector diagram*

The value of P_{II} found in step 3 is now used to construct the force vector diagram of forces acting at point X, Fig. 9. Drawing the closing line gives the value and direction of the reaction on bearing I. The force vectors are drawn in the usual way in their respective directions and sense. Then, the loading on bearing I is $P_I = 184$ lb (818.4 N) at 27.5°.

    Related Calculations.   The principles used in this procedure to obtain loads on bearings supporting the center shaft in Fig. 7 can be applied to obtain loads on bearings carrying shafts with any number of gears. It is limited, however, to those cases which are statistically determinate.

    Bending-moment and shear diagrams can now be constructed since the magnitude and direction of all forces acting on the shaft are known. To calculate the bearing loads resulting from gear action, both the magnitude and direction of the tooth reaction must be known. This reaction is the force at the pitch circle exerted by the tooth in the direction perpendicular to, and away from the tooth surface. Thus, the tooth reaction of a gear is always in the same general direction as its motion.

    Most techniques for evaluating bearing loads separate the total force acting on the gear into tangential and separating components. This tends to complicate the solution. The method given in this procedure uses the total force directly.

    Since a force can be replaced by an equal force acting at a different point, plus a couple, the total gear force can be considered as acting at the intersection of the shaft centerline and a line passing through the mid-face of the gear, if the appropriate couple is included. For example, in Fig. 7, the total force on gear B is equivalent to a force F_B applied at point X plus the couple $b \times F_B$. In establishing the couples for the other gears, a sign convention must be adopted to distinguish clockwise and counterclockwise moments.

    If a vector diagram is now drawn for all couples acting on the shaft, the closing line will be equal (to scale) to the couple resulting from the reaction at bearing II.

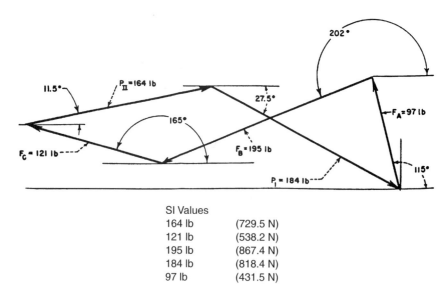

SI Values	
164 lb	(729.5 N)
121 lb	(538.2 N)
195 lb	(867.4 N)
184 lb	(818.4 N)
97 lb	(431.5 N)

FIGURE 9   Force vector diagram for the gear set in Fig. 7. (*Product Engineering.*)

Knowing the distance between the two bearings, the load on bearing II can be found, the direction being the same as that of the couple caused by it.

The load on bearing I is found in the same manner by drawing force vector diagram for all the forces acting at X, including the load on bearing II found from the couple diagram.

This procedure is the work of Zbigniew Jania, Project Engineer, Ford Motor Company, as reported in *Product Engineering*. SI values were added by the handbook editor.

SHAFT BEARING LOAD ANALYSIS USING POLAR DIAGRAMS

Determine the allowable directional side loads in the sheave drive in Fig. 10 if the maximum permissible radial load in bearing A is 2100 lb (9341 N) and 3750 lb (16,680 N) in bearing B when the maximum torque transmitted by the pinion is 1600 lb/in (181 Nm). The tangential and separating forces at the point of tooth contact are 1970 lb (8763 N) and 718 lb (3194 N) respectively. Use numerical and graphical analysis methods.

Calculation Procedure:

1. *Construct a vector diagram to determine the side forces on the bearings*
Draw the vector diagram in Fig. 10 to combine the tangential and separating forces of 1970 lb (8763 N) and 718 lb (3195 N) vectorially to give a resultant R of 2100 lb (9341 N). This resultant, R, can be replaced by a parallel force, R', of equal magnitude at the shaft centerline and a moment of 1970×4 in $= 7880$ lb/in (890 Nm) in a plane normal to the shaft centerline.

Construct a load diagram, of the shaft, Fig. 11. Taking moments, we find that the resultant loads at the bearings are: Bearing A $= 1575$ lb (7006 N); Bearing B $= 525$ lb (2335 N). Next, each bearing load condition must be treated separately.

2. *Use the fictitious-force approach to find the bearing load conditions*
The key to the solution is to replace R' with fictitious forces in the plane of the output sheave, Fig. 10, of such magnitudes that the bearing loads are unchanged.

Figures 12*a* and 12*b* show the procedure and the necessary forces with their directions of application. These fictitious forces, if plotted to some convenient scale on polar coordinate paper with the leading edge at the origin, will represent the bearing loads resulting from gear action.

If the side load is applied in the same direction as fictitious force R'_A, it will be limited by the allowable bearing capacity of bearing A. Since the maximum permissible radial load is given as 2100 lb (9341 N) by the bearing manufacturer, the resultant radial force, R'_A, acting in the plane of the output sheave must not exceed 2400 lb (10,675 N), since from Fig. 12*a*: $(1575 + 525)4 = (1800 + 600)3.5$. Of this 2400 lb (10,675 N), the fictitious force of 1800 lb (8006 N) must be considered a definite component. The actual applied load would then be limited to 600 lb (2669 N) ($= 2400 - 1800$), in the direction of R'_A.

If applied in the opposite direction, the limiting side load becomes 4200 lb (18,682 N) ($= 2400 + 1800$), since it acts to nullify rather than to complement the fictitious force in producing bearing load. Recalling the principles of polar diagrams, it is now apparent that a circle of 2400 lb (10,675 N) equivalent radius,

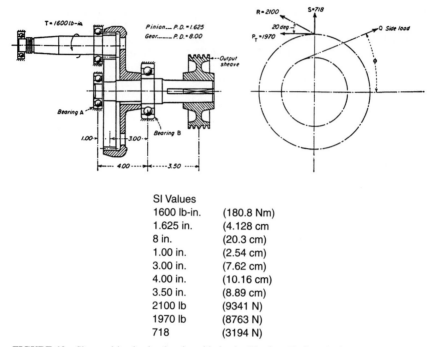

SI Values
1600 lb-in.	(180.8 Nm)
1.625 in.	(4.128 cm
8 in.	(20.3 cm)
1.00 in.	(2.54 cm)
3.00 in.	(7.62 cm)
4.00 in.	(10.16 cm)
3.50 in.	(8.89 cm)
2100 lb	(9341 N)
1970 lb	(8763 N)
718	(3194 N)

FIGURE 10   Sheave drive having bearing side loads. (*Product Engineering.*)

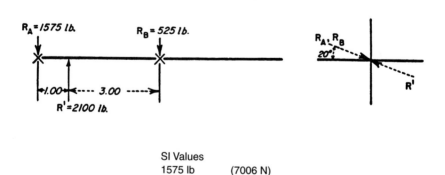

SI Values
1575 lb	(7006 N)
525 lb	(2335 N)
1 in.	(2.54 cm)
3 in.	(7.62 cm)
2100 lb	(9341 N)

FIGURE 11   Load diagram for sheave drive shaft. (*Product Engineering.*)

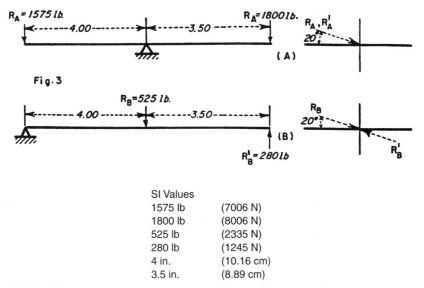

Fig·3

SI Values

1575 lb	(7006 N)
1800 lb	(8006 N)
525 lb	(2335 N)
280 lb	(1245 N)
4 in.	(10.16 cm)
3.5 in.	(8.89 cm)

FIGURE 12   Load diagram for sheave dive shaft using fictitious forces. (*Product Engineering.*)

R'_{AL}, would result in a polar diagram, Fig. 13, of permissible side loading for the design life of Bearing A. The validity of this solution and operation is understood by noting that the equivalent radius is always the resultant of the fictitious force, R'_A and the rotating vector which represents the applied side load.

3.   Analyze the other bearing in the system using a similar procedure
We will repeat the same procedure for Bearing B, Fig. 10, which has a design life (either assumed or given by the manufacturer) compatible with the 3750 lb (16,680 N) maximum rated load. When superimposed on one another, these diagrams give a confined realm of loading (*k-l-m-n*), Fig. 13.

Using the chosen scale, it is now possible to conveniently measure load limitations in any direction. When dividing the torque transmitted at the sheave, namely 7880 lb/in (890 Nm), Fig. 10, by any maximum indicated side load, the corresponding minimum sheave diameter is obtained.

4.   Analyze the shaft stresses based on desired bearing life
Now that the side-load limitations, based on desired bearing life, have been established, the imposed shaft stresses can be analyzed. Consider the stress to be critical under Bearing B. For safe loading limit, the maximum equivalent bending will be taken as 7200 lb/in (814 Nm). The equation for combined bending and torsion is:

$$M_c = \tfrac{1}{2} M + \tfrac{1}{2} \sqrt{M^2 + T^2}$$

Substituting known quantities,

$$7{,}200 = \tfrac{1}{2}M + \tfrac{1}{2} \sqrt{M^2 + 7{,}880^2}$$

The maximum allowable bending moment is then 5000 lb/in (565 Nm). At the

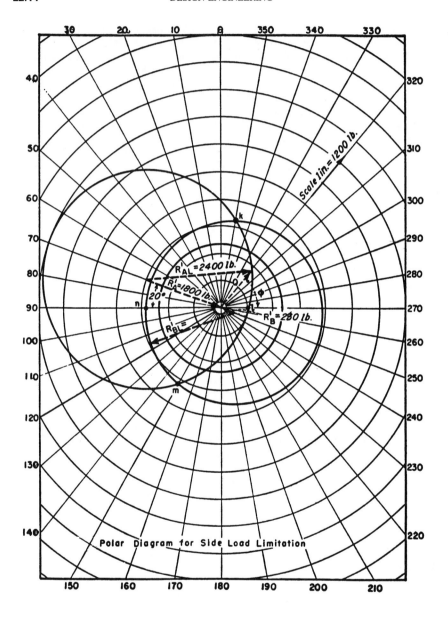

FIGURE 13 Polar diagram of sheave drive showing side forces. (*Product Engineering.*)

SI Values
2400 lb   (10,675 N)
280 lb   (1245 N)
1800 lb   (8006 N)
(2.54 cm = 5338 N)

specified distance of overhang, the side load, Q, caused by chain or belt pull is therefore limited by the shaft strength of 1430 lb (6361 N) in any direction. This limitation is shown on the polar diagram, Fig. 13, by a circle with its center at the origin and a radius equivalent to 1430 lb (6361 N). Study of the polar diagram, Fig. 10, shows that Bearing B is no longer a limiting factor since it completely encloses the now smaller region of safe loading.

Related Calculations. When applying belt and chain drives to geared power transmitting devices, excessive side loads are a frequent source of trouble in the bearings. Improper selection of sheaves and arrangement of connections can result in poor bearing life and early shaft breakage. A thorough analysis of bearing loads and shaft stresses is necessary if these troubles are to be avoided.

Use of polar diagrams provides a unique method of determining the maximum permissible side loading and the minimum size sheaves on an overhung shaft where gear loading is present. This type of analysis gives a graphical representation of limiting side loads with relation to their corresponding directions of pull. In essence it becomes a permanent data sheet for an individually designed unit to which future reference is readily available. The approach given here is valid for geared drives in factories, waterworks installations, marine applications, air conditioning and ventilation, etc., wherever an overhung shaft is used.

This procedure is the work of Richard J. Derks, Assistant Professor of Mechanical Engineering, University of Notre Dame, formerly Design Engineer, Twin Disc Clutch Company, as reported in *Product Engineering.*

JOURNAL BEARING FRICTIONAL HORSEPOWER LOSS DURING OPERATION

An 8-in (20.3-cm) diameter journal bearing is designed for 140-degree optimum oil-film pressure distribution, Fig. 14, when the journal length is 9 in (22.9 cm), shaft rotative speed is 1800 rpm, and the total load is 20,000 lb (9080 kg). What is the frictional horsepower loss when this journal bearing operates under stated conditions with oil of optimum viscosity?

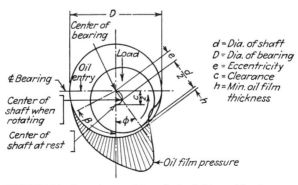

FIGURE 14 Operating condition of a loaded journal bearing.

Calculation Procedure:

1. Find the journal rubbing speed

The rubbing speed of a journal bearing is given by $V_r = \pi(D)(\text{rpm})$, where $D =$ journal diameter, ft (m). Substituting, $V_r = 3.1416(8.12)(1800) = 3770$ ft/min (1149 m/min).

2. Compute the rubbing surface area

The rubbing surface area of a journal bearing, $A_r = (d/2)(\alpha)(\pi)(L)$, where $d =$ bearing diameter, in; $\alpha =$ optimum bearing angle, degrees; $L =$ bearing length, in. Substituting, $A_r = (8/2)(140/180)(\pi)(9) = 87.96$ in^2 (567.5 cm^2).

3. Calculate the bearing pressure

With a total load of 20,000 lb (9080 kg), the bearing pressure = total load/bearing rubbing surface area, or $20,000/87.96 = 227.4$ lb/in^2 (1567 kPa).

4. Find the frictional horsepower (kW) loss

Use the relation $f_{hl} = \mu(P)(V_r)/33,000$, where $f_{hl} =$ frictional hp (kW) loss; $\mu =$ coefficient of friction for the journal bearing; $P =$ total load on the bearing, lb (kg); $V_r =$ rubbing speed, ft/min.

Using standard engineering handbooks, the coefficient of friction can be found to be 0.002 for a journal bearing having the computed rubbing speed and the given bearing pressure. Substituting,

$$f_{hl} = (0.002)(20,000)(377)/33,000 = 4.569 \text{ hp (3.4 kW)}.$$

Related Calculations. Use this general procedure to find the frictional power loss in journal bearings serving motors, engines, pumps, compressors, and similar equipment. Data on the coefficient of friction is available in standard handbooks. Figure 15 shows a typical plot of the coefficient of friction for journal bearings. The coefficient of friction chose for a well-designed journal bearing is that for full fluid film lubrication.

JOURNAL BEARING OPERATION ANALYSIS

A 2.5-in (6.36-cm) diameter journal bearing is subjected to a load of 1000 lb (454 kg) while the shaft it supports is rotating at 200 rpm. If the coefficient of friction is 0.02 and $L/D = 3.0$, find (a) bearing projected area, (b) pressure on bearing, (c) total work, (d) work of friction, (e) total heat generated, and (f) heat generated per minute per unit area.

Calculation Procedure:

1. Determine the bearing dimensions and projected area

The length/diameter ratio, $L/D = 3.0$; hence $L = 3D = 3 \times 2.25 = 6.75$ in (17.15 cm). (a) Then the projected area = $L \times D = 6.75 \times 2.5 = 15.19$ in^2 (97.99 cm^2).

2. Find the pressure on the bearing and the total work

(b) The pressure on the bearing = total load on bearing, P lb/(projected area, in^2) = $1000/15.19 = 65.8$ lb/in^2 (453.6 kPa). (c) The total friction work transmitted

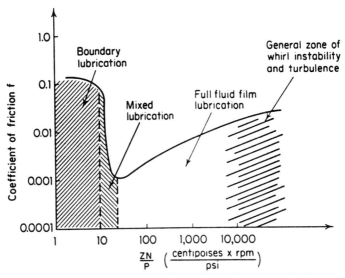

FIGURE 15 Various zones of possible lubrication for a journal bearing.

by the bearing, W = friction factor $(P)(\pi)$(bearing diameter, in/12 in/ft)(rpm). Substituting, $W = 0.02(1000)(\pi)(2.25/12)(200) = 2356$ ft-lb/min (53.2 W).

3. *Compute the work of friction and total heat generated*
(d) The work of friction, w = W/LD, where the symbols are as defined earlier. Substituting, $w = 2356/15.19 = 155.1$ ft-lb/min (3.5 W). (e) The total heat generated = $Q = W/778 = 155.1/778 = 3.03$ Btu/min (3.19 kJ/min).

4. *Find the heat generated per unit area*
The heat generated per unit area, $q = w/778 = 155.1/778 = 0.199$ Btu/in^2 min (0.0325 kJ/cm^2 min).
    Related Calculations.   This procedure shows that the analysis of journal bearings is a simple task requiring knowledge only of the physical dimensions or ratios of the bearing, the coefficient of friction and the shaft rpm. Using these data, any journal bearing can be designed for the mechanical requirements of the machine or structure.

BEARING-TYPE SELECTION FOR A KNOWN LOAD

Choose a suitable bearing for a 3-in (7.6-cm) diameter 100-r/min shaft carrying a total radial load of 12,000 lb (53,379 N). A reasonable degree of shaft misalignment must be allowed by the bearing. Quiet operation of the shaft is desired. Lubrication will be intermittent.

Calculation Procedure:

1. Analyze the desired characteristics of the bearing

Two major types of bearings are available to the designer, *rolling* and *sliding*. Rolling bearings are of two types, *ball* and *roller*. Sliding bearings are also of two types, *journal* for radial loads and *thrust* for axial loads only or for combined axial and radial loads. Table 2 shows the principal characteristics of rolling and sliding bearings. Based on the data in Table 2, a sliding bearing would be suitable for this application because it has a *fair* misalignment tolerance and a *quiet* noise level. Both factors are key considerations in the bearing choice.

2. Choose the bearing materials

Table 3 shows that a porous-bronze bearing, suitable for intermittent lubrication, can carry a maximum pressure load of 4000 lb/in² (27,580.0 kPa) at a maximum shaft speed of 1500 ft/min (7.62 m/s). By using the relation $l = L/(Pd)$, where l = bearing length, in, L = load, lb, d = shaft diameter, in, the required length of this sleeve bearing is $l = L/(Pd) = 12,000/[(4000)(3)] = 1$ in (2.5 cm).

Compute the shaft surface speed V ft/min from $V = \pi dR/12$, where d = shaft diameter, in; R = shaft rpm. Thus, $V = \pi(3)(100)/12 = 78.4$ ft/min (0.4 m/s).

With the shaft speed known, the PV, or pressure-velocity, value of the bearing can be computed. For this bearing, with an operating pressure of 4000 lb/in² (27,580.0 kPa), PV = 4000 × 78.4 = 313,600 (lb/in²) (ft/min) (10,984.3 kPa · m /s). This is considerably in excess of the PV limit of 50,000 (lb/in²) (ft/min) (1751.3 kPa · m/s) listed in Table 3. To come within the recommended PV limit, the operating pressure of the bearing must be reduced.

Assume an operating pressure of 600 lb/in² (4137.0 kPa). Then $l = L/(Pd) = 12,000/[(600)(3)] = 6.67$ in (16.9 cm), say 7 in (17.8 cm). The PV value of the bearing then is (600)(78.4) = 47,000 (lb/in²) (ft/min) (1646.3 kPa · m/s). This is a satisfactory value for a porous-bronze bearing because the recommend limit is 50,000 (lb/in²) (ft/min) (1751.3 kPa · m/s).

3. Check the selected bearing size

The sliding bearing chosen will have a diameter somewhat in excess of 3 in (7.6 cm) and a length of 7 in (17.8 cm). If this length is too great to fit in the allowable space, another bearing material will have to be studied, by using the same procedure. Figure 16 shows the space occupied by rolling and sliding bearings of various types.

Table 4 shows the load-carrying capacity and maximum operating temperature for oil-film journal sliding bearings that are regularly lubricated. These bearings are termed *full film* because they receive a supply of lubricant at regular intervals. Surface speeds of 20,000 to 25,000 ft/min (101.6 to 127.0 m/s) are common for industrial machines fitted with these bearings. This corresponds closely to the surface speed for ball and roller bearings.

4. Evaluate oil-film bearings

Oil-film sliding bearings are chosen by the method of the next calculation procedure. The bearing size is made large enough that the maximum operating temperature listed in Table 4 is not exceeded. Table 5 lists typical design load limits for oil-film bearings in various services. Figure 17 shows the typical temperature limits for rolling and sliding bearings made of various materials.

TABLE 2 Key Characteristics of Rolling and Sliding Bearings*

	Rolling	Sliding
Life	Limited by fatigue properties of bearing metal	Unlimited, except for cyclic loading
Load:		
Unidirectional	Excellent	Good
Cyclic	Good	Good
Starting	Excellent	Poor
Unbalance	Excellent	Good
Shock	Good	Fair
Emergency	Fair	Fair
Speed limited by:	Centrifugal loading and material surface speeds	Turbulence and temperature rise
Starting friction	Good	Poor
Cost	Intermediate, but standardized, varying little with quantity	Very low in simple types or in mass production
Space requirements (radial bearing):		
Radial dimension	Large	Small
Axial dimension	⅛ to ½ shaft diameter	¼ to 2 times shaft diameter
Misalignment tolerance	Poor in ball bearings except where designed for at sacrifice of load capacity; good in spherical roller bearings; poor in cylindrical roller bearings	Fair
Noise	May be noisy, depending on quality and resonance of mounting	Quiet
Damping	Poor	Good
Low-temperature starting	Good	Poor
High-temperature operation	Limited by lubricant	Limited by lubricant
Type of lubricant	Oil or grease	Oil, water, other liquids, grease, dry lubricants, air, or gas
Lubrication, quantity required	Very small, except where large amounts of heat must be removed	Large, except in low-speed boundary-lubrication types
Type of failure	Limited operation may continue after fatigue failure but not after lubricant failure	Often permits limited emergency operation after failure
Ease of replacement	Function of type of installation; usually shaft need not be replaced	Function of design and installation; split bearings used in large machines

Product Engineering.

5. Evaluate rolling bearings

Rolling bearings have lower starting friction (coefficient of friction $f = 0.002$ to 0.005) than sliding bearings ($f = 0.15$ to 0.30). Thus, the rolling bearings is preferred for applications requiring low staring torque [integral-horsepower electric motors up to 500 hp (372.9 kW), jet engines, etc.]. By pumping oil into a sliding bearing, its starting coefficient of friction can be reduced to nearly zero. This arrangement is used in large electric generators and certain mill machines.

TABLE 3 Materials for Sleeve Bearings*

[Cost figures are for a 1-in (2.54-cm) sleeve bearing ordered in quantity]

	Maximum load		Maximum speed		PV limit		Maximum operating temperature		Cost, $
	lb/in²	kPa	ft/min	m/s	(lb/in²)(ft/min)	kPa·m/s	°F	°C	
Porous bronze	4,000	27,579.0	1,500	7.6	50,000	1,751.3	150	65.6	0.11
Porous iron	8,000	55,158.1	800	4.1	50,000	1,751.3	150	65.6	0.09
Teflon fabric	60,000	413,685.4	50	0.3	25,000	875.6	500	260.0	0.04
Phenolic	6,000	41,368.5	2,500	12.7	15,000	525.4	200	93.3	0.05
Wood	6,000	41,368.5	2,000	10.2	15,000	525.4	150	65.6	0.40
Carbon-graphite	600	4,136.9	2,500	12.7	10,000	350.3	750	398.9	0.39
Reinforced Teflon	2,500	17,236.9	2,500	12.7	10,000	350.3	500	260.0	0.45
Nylon	1,000	6,894.8	1,000	5.1	3,000	105.1	200	93.3	0.04
Delrin	1,000	6,894.8	1,000	5.1	3,000	105.1	180	82.2	0.03
Lexan	1,000	6,894.8	1,000	5.1	3,000	105.1	220	104.4	0.05
Teflon	500	3,447.4	100	0.5	1,000	35.0	500	260.0	1.00

*Product Engineering.

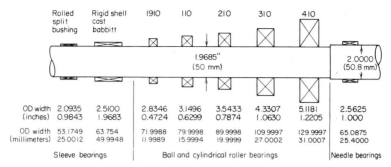

FIGURE 16 Relative space requirements of sleeve and rolling-element bearings to carry the same diameter shaft. (*Product Engineering.*)

TABLE 4 Oil-Film Jorunal Bearing Characteristics*

Bearing material	Load-carrying capacity		Maximum operating temperature	
	lb/in²	kPa	°F	°C
Tin-based babbitt	800–1,500	5,516.0–10,342.5	300	148.9
Lead-based babbitt	800–1,200	5,516.0–8,274.0	300	148.9
Alkali-hardened steel	1,200–1,500	8,274.0–10,342.5	500	260.0
Cadmium base	1,500–2,000	10,342.5–13,789.5	500	260.0
Copper-lead	1,500–2,500	10,342.5–17,236.9	350	176.7
Tin bronze	4,000	27,580.0	500+	260.0+
Lead bronze	3,000–4,000	20,685.0–27,580.0	450	232.2
Aluminum alloy	4,000	27,580.0	250	121.1
Silver (overplated)	4,000	27,580.0	500	260.0
Three-component bearings babbitt-surfaced	2,000–4,000	13,790.0–27,580.0	225–300	107.2–148.9

Product Engineering.

The running friction of rolling bearings is in the range of f = 0.001 to 0.002. For oil-film sliding bearings, f = 0.002 to 0.005.

Rolling bearings are more susceptible to dirt than are sliding bearings. Also, rolling bearings are inherently noisy. Oil-film bearings are relatively quiet, but they may allow higher amplitudes of shaft vibration.

Table 6 compares the size, load capacity, and cost of rolling bearings of various types. Briefly, ball bearings and roller bearings may be compared thus: ball bearings (*a*) run at higher speeds without undue heating, (*b*) cost less per pound of load-carrying capacity for light loads, (*c*) have friction torque at light loads, (*d*) are available in a wider variety of sizes, (*e*) can be made in smaller sizes, and (*f*) have seals and shields for easy lubrication. Roller bearings (*a*) can carry heavier loads, (*b*) are less expensive for larger sizes and heavier loads, (*c*) are more satisfactory under shock and impact loading, and (*d*) may have lower friction at heavy loads. Table 7 shows the speed limit, termed the *dR limit* (equals bearing shaft bore d in mm multiplied by the shaft rpm R), for ball and roller bearings. Speeds higher than those shown in Table 7 may lead to early bearing failures. Since the *dR* limit is

TABLE 5 Typical Design Load Limits for
Oil-Film Bearings*

Bearing	Maximum load on projected area	
	lb/in²	kPa
Electric motors	200	1,379.0
Steam turbines	300	2,068.4
Automotive engines:		
Main bearings	3,500	24,131.6
Connecting rods	5,000	34,473.8
Diesel engines:		
Main bearings	3,000	20,684.7
Connecting rods	4,500	31,026.4
Railroad car axles	350	2,413.2
Steel mill roll necks:		
Steady	2,000	13,789.5
Peak	5,000	24,473.8

Product Engineering.

proportional to the shaft surface speed, the dR value gives an approximate measure of the bearing power loss and temperature rise.

  Related Calculations.   Use this general procedure to select shaft bearings for any type of regular service conditions. For unusual service (i.e., excessively high or low operating temperatures, large loads, etc.) consult the specific selection procedures given elsewhere in this section.

  Note that the PV value of a sliding bearing can also be expressed as PV = $L/(dl) \times \pi dR/12 = \pi LR/(12l)$. The bearing load and shaft speed are usually fixed by other requirements of a design. When the PV equation is solved for the bearing length l and the bearing is too long to fit the available space, select a bearing material having a higher allowable PV value.

SHAFT BEARING LENGTH AND HEAT GENERATION

How long should a sleeve-type bearing be if the combined weight of the shaft and gear tooth load acting on the bearing is 2000 lb (8896 N)? The shaft is 1 in (2.5 cm) in diameter and is oil-lubricated. What is the rate of heat generation in the bearing when the shaft turns at 60 r/min? How much above an ambient room temperature of 70°F (21.1°C) will the temperature of the bearing rise during operation in still air? In moving air?

Calculation Procedure:

1. Compute the required length of the bearing
The required length l of a sleeve bearing carrying a load of L lb is $l = L/(Pd)$, where l = bearing length, in; L = bearing load, lb = bearing reaction force, lb; P

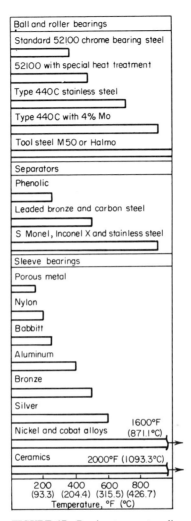

FIGURE 17   Bearing temperature limits. (*Product Engineering.*)

= allowable mean bearing pressure, lb/in² [ranges from 25 to 2500 lb/in² (172.4 to 17,237.5 kPa) for normal service and up to 8000 lb/in² (55,160.0 kPa) for severe service], on the projected bearing area, in² = ld; d = shaft diameter, in. Thus, for this bearing, assuming an allowable mean bearing pressure of 400 lb/in² (2758.0 kPa), $l = L/(Pd) = 2000/[(400)(1)] = 5$ in (12.7 cm).

2. Compute the rate of bearing heat generation

The rate of heat generation in a plain sleeve bearing is given by $h = fLdR/3000$, where h = rate of heat generation in the bearing, Btu/min; f = bearing coefficient of friction for the lubricant used; R = shaft rpm; other symbols as in step 1.

TABLE 6 Relative Load Capacity, Cost, and Size of Rolling Bearings*

Bearing type (for 50-mm bore)	Radial capacity	Axial capacity	Cost	Outer diameter	Width
Ball bearings:					
Deep groove (Conrad)	1.0	1.0	1.0	1.0	1.0
Filling notch	1.2	Low	1.2	1.0	1.0
Double row	1.5	1.1	2.2	1.0	1.6
Angular contact	1.1	1.9	1.6	1.0	1.0
Duplex	1.8	1.9	2.0	1.0	2.0
Self-aligning	0.7	0.2	1.3	1.0	1.0
Ball thrust	0	0.9	0.8	0.7	0.8
Roller bearings:					
Cylindrical	1.6	. . .	1.9	1.0	1.0
Tapered	1.3	0.8	0.9	1.0	1.0
Spherical	3.0	1.1	5.0	1.0	1.5
Needle	1.0	0	0.3	0.5	1.6
Flat thrust	0	4.0	3.8	0.8	0.9

Product Engineering.

TABLE 7 Speed Limits for Ball and Roller Bearings*

Lubrication	DN limit, mm $\times$ r/min
Oil:	
Conventional bearing designs	300,000–350,000
Special finishes and separators	1,000,000–1,500,000
Grease:	
Conventional bearing designs	250,000–300,000
Silicone grease	150,000–200,000
Special finishes and separators high-speed greases	500,000–600,000

Product Engineering.

The coefficient of friction for oil-lubricated bearings ranges from 0.005 to 0.030, depending on the lubricant viscosity, shaft rpm, and mean bearing pressure. Given a value of $f = 0.020$, $h = (0.020)(2000)(1)(60)/3000 = 0.8$ Btu/min, or $H = 0.8(60$ min/h$) = 48.0$ Btu/h (14.1 kW).

3. Compute the bearing wall area

The wall area A of a small sleeve-type bearing, such as a pillow block fitted with a bushing, is $A = (10$ to $15)dl/144$, where A = bearing wall area, ft^2; other symbols as before. For larger bearing pedestals fitted with a cast-iron or steel bearing shell, the factor in this equation varies from 18 to 25.

Since this is a small bearing having a 1-in (2.5-cm) diameter shaft, the first equation with a factor of 15 to give a larger wall area can be used. The value of 15 was chosen to ensure adequate radiating surface. Where space or weight is a factor, the value of 10 might be chosen. Intermediate values might be chosen for other conditions. Substituting yields $A = (15)(1)(5)/144 = 0.521$ ft^2 (0.048 m^2).

4. Determine the bearing temperature rise

In *still air*, a bearing will dissipate $H = 2.2A (t_w - t_a)$ Btu/h, where t_w = bearing wall temperature, °F; t_a ambient air temperature, °F; other symbols as before. Since H and A are known, the temperature rise can be found by solving for $t_w - t_s = H/(2.2A) = 48.0/[(2.2)(0.521)] = 41.9$°F (23.3°C), and $t_w = 41.9 + 70 = 111.9$°F (44.4°C). This is a low enough temperature for safe operation of the bearing. The maximum allowable bearing operating temperature for sleeve bearings using normal lubricants is usually assumed to be 200°F (93.3°C). To reduce the operating temperature of a sleeve bearing, the bearing wall area must be increased, the shaft speed decreased, or the bearing load reduced.

In *moving air*, the heat dissipation from a sleeve-type bearing is $H = 6.5A(t_w - t_s)$. Solving for the temperature rise as before, we get $t_w - t_a = H/(6.5A) = 48.0/[(6.5)(0.521)] = 14.2$°F (7.9°C), and $t_w = 14.2 + 70 = 84.2$°F (29.0°C). This is a moderate operating temperature that could be safely tolerated by any of the popular bearing materials.

Related Calculations. Use this procedure to analyze sleeve-type bearings used for industrial line shafts, marine propeller shafts, conveyor shafts, etc. Where the ambient temperature varies during bearing operation, use the highest ambient temperature expected, in computing the bearing operating temperature.

ROLLER-BEARING OPERATING-LIFE ANALYSIS

A machine must have a shaft of about 5.5 in (14.0 cm) in diameter. Choose a roller bearing for this 5.5-in (14.0-cm) diameter shaft that turns at 1000 r/min while carrying a radial load of 20,000 lb (88,964 N). What is the expected life of this bearing?

Calculation Procedure:

1. Determine the bearing life in revolutions

The operating life of rolling-type bearings is often stated in millions of revolutions. Find this life from $R_L = (C/L)^{10/3}$, where R_L = bearing operating life, millions of revolutions; C = dynamic capacity of the bearing, lb; L = applied radial load on bearing, lb.

Obtain the dynamic capacity of the bearing being considered by consulting the manufacturer's engineering data. Usual values of dynamic capacity range between 2500 lb (11,120.6 N) and 750,000 lb (3,338,166.5 N) depending on the bearing design, type, and bore. For a typical 5.5118-in (14.0-cm) bore roller bearing, $C = 92,400$ lb (411,015.7 N).

With C known, compute $R_L = (C/L)^{10/3} = (92,400/20,000)^{10/3} = 162 \times 10^6$.

2. Determine the bearing life

The minimum life of a bearing in millions of revolutions, R_L, is related to its life in hours, h, by the expression $R_L = 60Rh/10^6$, where R = shaft speed, r/min. Solving gives $h = 10^6 R_L/(60R) = (10^6)(162)/[(60)(1000)] = 2700$-h minimum life.

Related Calculations. This procedure is useful for those situations where a bearing must fit a previously determined shaft diameter or fit in a restricted space. In these circumstances, the bearing size cannot be varied appreciably, and the machine designer is interested in knowing the minimum probable life that a given size of bearing will have. Use this procedure whenever the bearing size is approximately predetermined by the installation conditions in motors, pumps, engines, portable tools, etc. Obtain the dynamic capacity of any bearing under consideration from the manufacturer's engineering data.

ROLLER-BEARING CAPACITY REQUIREMENTS

A machine must be fitted with a roller bearing that will operate at least 30,000 h without failure. Select a suitable bearing for this machine in which the shaft operates at 3600 r/min and carries a radial load of 5000 lb (22,241 N).

Calculation Procedure.

1. Determine the bearing life in revolutions
Use the relation $R_L = 60Rh/10^6$, where the symbols are the same as in the previous calculation procedure. Thus, $R_L = 60(3600)(30,000)/10^6$; $R_L = 6480$ million revolutions (Mr).

2. Determine the required dynamic capacity of the bearing
Use the relation $R_L = (C/L)^{10/3}$, where the symbols are the same as in the previous calculation procedure. So $C = L(R_L)^{3/10} = (5000)(6480)^{3/10} = 69,200$ lb (307,187.0 N).

Choose a bearing of suitable bore having a dynamic capacity of 69,200 lb (307,187.0 N) or more. Thus, a typical 5,9055-in (15.0-cm) bore roller bearing has a dynamic capacity of 72,400 lb (322,051.3 N). It is common practice to undercut the shaft to suit the bearing bore, if such a reduction in the shaft does not weaken the shaft. Use the manufacturer's engineering data in choosing the actual bearing to be used.

Related Calculations. This procedure shows a situation in which the life of the bearing is of greater importance than its size. Such a situation is common when the reliability of a machine is a key factor in its design. A dynamic rating of a given amount, say 72,400 lb (322,051.3 N), means that if in a large group of bearings of this size each bearing has a 72,400-lb (322,051.3-N) load applied to it, 90 percent of the bearings in the group will complete, or exceed, 10^6 r before the first evidence of fatigue occurs. This average life of the bearing is the number of revolutions that 50 percent of the bearings will complete, or exceed, before the first evidence of fatigue develops. The average life is about 3.5 times the minimum life.

Use this procedure to choose bearings for motors, engines, turbines, portable tools, etc. Where extreme reliability is required, some designers choose a bearing having a much larger dynamic capacity than calculations show is required.

RADIAL LOAD RATING FOR ROLLING BEARINGS

A mounted rolling bearing is fitted to a shaft driven by a 4-in (10.2-cm) wide double-ply leather belt. The shaft is subjected to moderate shock loads about one-third of the time while operating at 300 r/min. An operating life of 40,000 h is required of the bearing. What is the required radial capacity of the bearing? The bearing has a normal rated life of 15,000 h at 500 r/min. The weight of the pulley and shaft is 145 lb (644.9 N).

Calculation Procedure:

1. Determine the bearing operating factors

To determine the required radial capacity of a rolling bearing, a series of operating factors must be applied to the radial load acting on the shaft: life factor f_L, operating factor f_O, belt tension factor f_B, and speed factor f_S. Obtain each of the four factors from the manufacturer'ss engineering data because there may be a slight variation in the factor value between different bearing makers. Where a given factor does not apply to the bearing being considered, omit it from the calculation.

2. Determine the bearing life factor

Rolling bearings are normally rated for a certain life, expressed in hours. If a different life for the bearing is required, a life factor must be applied. The bearing being considered here has a normal rated life of 15,000 h. The manufacturer's engineering data show that for a mounted bearing which must have a life of 40,000 h, a life factor $f_L = 1.340$ should be sued. For this particular make of bearing, f_L varies from 0.360 at a 500-h to 1.700 at a 100,000-h life for mounted units. At 15,000 h, $f_L = 1.000$.

3. Determine the bearing operating factor

A rolling-bearing operating factor is used to show the effect of peak and shock loads on the bearing. Usual operating factors vary from 1.00 for steady loads with any amount of overload to 2.00 for bearings with heavy shock loads throughout their operating period. For this bearing with moderate shock loads about one-third of the time, $f_O = 1.32$.

    A combined operating factor, obtained by taking the product of two applicable factors, is used in some circumstances. Thus, when the load is an oscillating type, an additional factor of 1.25 must be applied. This type of load occurs in certain linkages and pumps. When the outer race of the bearing revolves, as in sheaves, truck wheels, or gyrating loads, an additional factor of 1.2 is used. To find the combined operating factor, first find the normal operating factor, as described earlier. Then take the product of the normal and the additional operating factors. The result is the combined operating factor.

4. Determine the bearing belt-tension factor

When the bearing is used on a belt-driven shaft, a belt-tension factor must be applied. Usual values of this factor range from 1.0 for a chain drive to 2.30 for a single-ply leather belt. For a double-ply leather belt, $f_B = 2.0$.

5. Determine the bearing speed factor

Rolling bearings are rated at various speeds. When the shaft operates at a speed different from the rated speed, a speed factor f_S must be applied. Since this shaft operates at 300 r/min while the rated speed of the bearing is 500 r/min, $f_S = 0.860$, from the manufacturer's engineering data. For a 500-r/min bearing, f_S varies from 0.245 at 5 r/min to 1.87 at 4000 r/min.

6. Determine the radial load on the bearing

The radial load produced by a leather belt can vary from 130 lb/in (227.7 N/cm) of width for normal-tension belts to 450 lb/in (788/1 N/cm) of width for very tight belts. Assuming normal tension, the radial load for a double-ply leather belt is, from engineering data, 180 lb/in (315.2 N/cm) of width. Since this belt is 4 in (10.2 cm) wide, the radial belt load = 4(180) = 720 lb (3202.7 N). The total radial load R_T is the sum of the belt, shaft, and pulley loads, or $R_T = 720 + 145 = 865$ lb (3847.7 N).

7. Compute the required radial capacity of the bearing
The required radial capacity of a bearing $R_C = R_T f_L f_O f_B f_S = (865)(1.340)(1.32)(2.0)(0.86) = 2630$ lb (11,698.8 N).

8. Select a suitable bearing
Enter the manufacturer's engineering data at the shaft rpm (300 r/min for this shaft) and project to a bearing radial capacity equal to, or slightly greater than, the computed required radial capacity. Thus, one make of bearing, suitable for 2- and 2³⁄₁₆-in (5.1- and 5.6-cm) diameter shafts, has a radial capacity of 2710 at 300 r/min. This is close enough for general selection purposes.

  Related Calculations. Use this general procedure for any type of rolling bearing. When comparing different makes of rolling bearings, be sure to convert them to the same life expectancies before making the comparison. Use the life-factor table presented in engineering handbooks or a manufacturer's engineering data for each bearing to convert the bearings being considered to equal lives.

ROLLING-BEARING CAPACITY AND RELIABILITY

What is the required basic load rating of a ball bearing having an equivalent radial load of 3000 lb (13,344.7 N) if the bearing must have a life of 400×10^6 r at a reliability of 0.92? The ratio of the average life to the rating life of the bearing is 5.0. Show how the required basic load rating is determined for a roller bearing.

Calculation Procedure.

1. Compute the required basic load rating
Use the Weibull two-parameter equation, which for a life ratio of 5 is $L_e/L_B = (1.898/R_L^{0.333})(\ln 1/R_e)^{0.285}$, where L_e = equivalent radial load on the bearing, lb; L_B = the required basic load rating of the bearing, lb, to give the desired reliability at the stated life; R_L = bearing operating life, Mr; ln = natural or Naperian logarithm to the base e; R_e = required reliability, expressed as a decimal.
  Substituting gives $3000/L_B = (1.898/400^{0.333})(\ln 1/0.92)^{0.285}$; $L_B = 23,425$ lb (104,199.6 N). Thus, a bearing having a basic load rating of at least 23,425 lb (104,199.6 N) would provide the desired reliability. Select a bearing having a load rating equal to, or slightly in excess of, this value. Use the manufacturer's engineering data as the source of load-rating data.

2. Compute the roller-bearing basic load rating
Use the following form of the Weibull equation for roller bearings: $L_e/L_B = (1.780/R_L^{0.30})(\ln 1/R_e)^{0.s.257}$. Substitute values and solve for L_B, as in step 1.
  Related Calculations. Use the Weibull equation as given here when computing bearing life in the range of 0.9 and higher. The ratio of average life/rating life = 5 is usual for commercially available bearings.[1]

---

[1]C. Mischke, "Bearing Relaibility and Capacity," *Machine Design*, Sept. 30, 1965.

POROUS-METAL BEARING CAPACITY AND FRICTION

Determine the load capacity ψ and coefficient of friction of a porous-metal bearing for a 1-in (2.5-cm) diameter shaft, 1-in (2.5-cm) bearing length, 0.2-in (0.5-cm) thick bearing, 0.001-in (0.003-cm) radial clearance, metal permeability $\phi = 5 \times 10^{-10}$ in (1.3×10^{-9} cm), shaft speed = 1500 r/min, eccentricity ratio $\epsilon = 0.8$, and an SAE-30 mineral-oil lubricant with a viscosity of 6×10^{-6} lb·s/in² (4.1×10^{-6} N·s/cm²).

Calculation Procedure:

1. Sketch the bearing and shaft
Figure 18 shows the bearing and shaft with the various known dimensions indicated by the identifying symbols given above.

2. Compute the load capacity factor
The load capacity factor $\psi = \phi H / C^3$, where ϕ = metal permeability, in; H = bearing thickness, in; C = radial clearance of bearing, $R_b - r = 0.001$ in (0.003 cm) for this bearing. Hence, $\psi = (5 \times 10^{-10})(0.2)/(0.001)^3 = 0.10$.

3. Compute the bearing thickness-length ratio
The thickness-length ratio = $H/b = 0.2/1.0 = 0.2$.

4. Determine the $S(d/b)^2$ value for the bearing
In the $S(d/b)^2$ value for the bearing, S = the Summerfeld number for the bearing; the other values are as shown in Fig. 18.
   Using the ψ, ϵ, and H/b values, enter Fig. 19a, and read $S(b/d)^2 = 1.4$ for an eccentricity ratio of $\epsilon = e/C = 0.8$. Substitute in this equation the known values for b and d and solve for S, or $S(1.0/1.0)^2 = 1.4$; $S = 1.4$.

5. Compute the bearing load capacity
Find the bearing load capacity from $S = (L/R_i \eta b)(C/r)^2$, where η = lubricant viscosity, lb·s/in²; r = shaft radius, in; R_i = shaft velocity, in·s. Solving gives $L = (SR_i \eta b)/(C/r)^2 = (1.4)(78.5)(6 \times 10^{-6})(1.0)/(0.001/0.5)^2 = 164.7$ lb (732.6 N). (The shaft rotative velocity must be expressed in in/s in this equation because the lubricant viscosity is given in lb·s/in²).

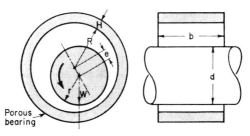

FIGURE 18 Typical porous-metal bearing. (*Product Engineering.*)

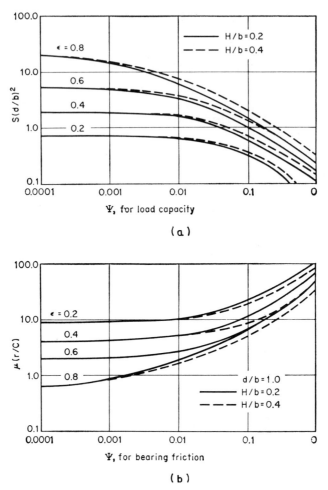

FIGURE 19   (a) Bearing load capacity factors; (b) bearing friction factors. (*Products Engineering.*)

6. *Determine the bearing coefficient of friction*

Enter Fig. 19b with the known values of ψ, ϵ, and H/b and read $u(r/C) = 7.4$. Substitute in this equation the known values for r and C, and solve for the bearing coefficient of friction μ, or $\mu = 7.4C/r = (7.4)(0.001)/0.5 = 0.0148$.

Related Calculations.   Porous-metal bearings are similar to conventional sliding-journal bearings except that the pores contain an additional supply of lubricant to replace that which may be lost during operation. The porous-metal bearing is useful in assemblies where there is not enough room for a conventional lubrication system or where there is a need for improved lubrication during the starting and stopping of a machine. The permeability of the finished porous metal greatly influences the ability of a lubricant to work its way through the pores. Porous-metal bearings are used in railroad axle supports, water pumps, generators, machine tools, and other equipment. Use the procedure given here when choosing porous-metal bearings for any of these applications.

The method given here was developed by Professor W. T. Rouleau, Carnegie Institute of Technology, and C. A. Rhodes, Senior Research Engineer, Jet Propulsion Laboratory, California Institute of Technology.

HYDROSTATIC THRUST BEARING ANALYSIS

An oil-lubricated hydrostatic thrust bearing must support a load of 107,700 lb (479,073.5 N). This vertical bearing has an outside diameter of 16 in (40.6 cm) and a recess diameter of 10 in (25.4 cm). What oil pressure and flow rate are required to maintain a 0.006-in (0.15-mm) lubricant film thickness with an SAE-20 oil having an absolute viscosity of $\eta = 42.4 \times 10^{-7}$ lb/(s·in²) [2.9 × 10⁻⁶ N /(s·cm²)] if the shaft turns at 750 r/min? What are the pumping loss and the viscous friction loss? What is the optimum lubricant film thickness?

Calculation Procedure:

1. Determine the required lubricant-supply pressure
The design equations and methods developed at Franklin Institute by Dudley Fuller, Professor of Mechanical Engineering, Columbia University, are applicable to vertical hydrostatic bearings, Fig. 20, using oil, grease, or gas lubrication. By substituting the appropriate value for the lubricant viscosity, the same set of design equations can be used for any of the lubricants listed above. These equations are accurate, simple, and reliable; they are therefore used here.

Solve Fuller's applied load equation, $L = (p_i \pi / 2)\{r^2 - r_i^2 / [\ln (r/r_i)]\}$, for the lubricant-supply inlet pressure. In this equation, L = applied load on the bearing, lb; p_i = lubricant-supply inlet pressure, lb/in²; r = shaft radius, in; r_i = recess or step radius, in; ln = natural or Naperian logarithm to the base e. Solving gives $p_i = 2L/\pi\{r^2 - r_i^2/[\ln (r/r_i)]\} = 2(107,700)/\pi[8^2 - 5^2/(\ln 8.5)]$, or $p_i = 825$ lb/ in² (5688.4 kPa).

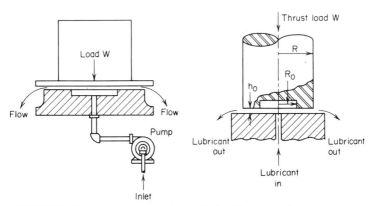

FIGURE 20 Hydrostatic thrust bearings. (*Product Engineering.*)

2. Compute the required lubricant flow rate

By Fuller's flow-rate equation, $Q = p_i \pi h^3 / [6\eta \ln (r/r_i)]$, where Q = lubricant flow rate, in^3/s; h = lubricant-film thickness, in; η = lubricant absolute viscosity, $\text{lb} \cdot \text{s} / \text{in}^2$; other symbols as in step 1. Thus, with $h = 0.006$ in, $Q = (825 \, \pi)(0.006)^3 / [6(42.4 \times 10^{-6})(0.470)] = 46.85 \, \text{in}^3/\text{s} \ (767.7 \, \text{cm}^3/\text{s})$.

3. Compute the pumping loss

The pumping loss results from the work necessary to force the lubricant radially outward through the film space, or $H_p = Q(p_i - p_o)$, where H_p = power required to pump the lubricant = pumping loss, in $\cdot$ lb/s; p_o = lubricant outlet pressure, lb $/\text{in}^2$; other symbols as in step 1. For circular thrust bearings it can be assumed that the lubricant outlet pressure p_o is negligible, or $p_o = 0$. Then $H_p = 467.85(825 - 0) = 38,680 \, \text{in} \cdot \text{lb/s} \ (4370.3 \, \text{Nm/s}) = 38,680/[550 \, \text{ft} \cdot \text{lb}/(\text{min} \cdot \text{hp})](12 \, \text{in/ft}) = 5.86 \, \text{hp} \ (4.4 \, \text{kW})$.

4. Compute the viscous friction loss

The viscous-friction-loss equation developed by Fuller is $H_f = [(R^2\eta)/(58.05h)](r^4 - r_0^4)$, where H_f = viscous friction loss, in $\cdot$ lb/s; R = shaft rpm; other symbols as in step 1. Thus, $H_f = \{(750)^2(42.4 \times 10^{-7})/[58.05)(0.006)]\}(8^4 - 5^4)$, or $H_f = 23,770 \, \text{in} \cdot \text{lb/s} \ (2685.6 \, \text{Nm/s}) = 23,770/[(550)(12)] = 3.60 \, \text{hp} \ (2.7 \, \text{kW})$.

5. Compute the optimum lubricant-film thickness

The film thickness that will produce a minimum combination of pumping loss and friction loss can be evaluated by determining the minimum point of the curve representing the sum of the respective energy losses (pumping and viscous friction) when plotted against film thickness.

With the shaft speed, lubricant viscosity, and bearing dimensions constant at the values given in the problem statement, the viscous-friction loss becomes $H_f = 0.0216/h$ for this bearing. Substitute various values for h ranging between 0.001 and 0.010 in (0.0254 and 0.254 mm) (the usual film thickness range), and solve for H_f. Plot the results as shown in Fig. 21.

Combine the lubricant-flow and pumping-loss equations to express H_p in terms of the lubricant-film thickness, or $H_p = (1000h)^3/36.85$, for this bearing with a pump having an efficiency of 100 percent. For a pump with a 50 percent efficiency, this equation becomes $H_p = (1000h)^3/18.42$. Substitute various values of h ranging between 0.001 and 0.010, and plot the results as in Fig. 21 for pumps with 100 and 50 percent efficiencies, respectively. Figure 21 shows that for a 100 percent efficient pump, the minimum total energy loss occurs at a film thickness of 0.004 in (0.102 mm). For 50 percent efficiency, the minimum total energy loss occurs at 0.0035-in (0.09-mm) film thickness, Fig. 21.

Related Calculations.   Similar equations developed by Fuller can be used to analyze hydrostatic thrust bearings of other configurations. Figures 22 and 23 show the equations for modified square bearings and circular-sector bearings. To apply these equations, use the same general procedures shown above. Note, however, that each equation uses a factor K obtained from the respective design chart.

Also note that a hydrostatic bearing uses an externally fed pressurized fluid to keep two bearing surfaces *completely* separated. Compared with hydrodynamic bearings, in which the pressure is self-induced by the rotation of the shaft, hydrostatic bearings have (1) lower friction, (2) higher load-carrying capacity, (3) a lubricant-film thickness insensitive to shaft speed, (4) a higher spring constant, which leads to a self-centering effect, and (5) a relatively thick lubricant film permitting cooler operation at high shaft speeds. Hydrostatic bearings are used in rolling mills, instruments, machine tools, radar, telescopes, and other applications.

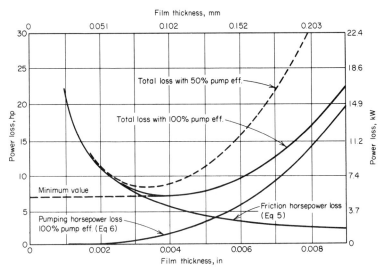

FIGURE 21   Oil-film thickness for minimum power loss in a hydrostatic thrust bearing. (*Product Engineering.*)

HYDROSTATIC JOURNAL BEARING ANALYSIS

A 4.000-in (10.160-cm) metal shaft rests in a journal bearing having an internal diameter of 4.012 in (10.190 cm). The lubricant is SAE-30 oil at 100°F (37.8°C) having a viscosity of 152×10^{-7} reyn. This lubricant is supplied under pressure through a groove at the lowest point in the bearing. The length of the bearing is 6 in (15.2 cm), the length of the groove is 3 in (7.5 cm), and the load on the bearing is 3600 lb (16,013.6 N). What lubricant-inlet pressure and flow rate are required to raise the shaft 0.002 in (0.051 mm) and 0.004 in (0.102 mm)?

Calculation Procedure:

1. Determine the radial clearance and clearance modulus
The design equations and methods developed at Franklin Institute by Dudley Fuller, Professor of Mechanical Engineering, Columbia University, are applicable to hydrostatic journal bearings using oil, grease, or gas lubrication. These equations are accurate, simple, and reliable; therefore they are used here.

By Fuller's method, the radial clearance c, in $= r_b - r_s$, Fig. 16, where $r_b =$ bearing internal radius, in; $r_s =$ shaft radius, in. Or, $c = (4.012/2) - (4.000/2) = 0.006$ in (0.152 mm).

Next, compute the clearance modulus m from $m = c/r_s = 0.006/2 = 0.003$ in /in (0.003 cm/cm). Typical values of m range from 0.005 to 0.003 in/in (0.005 to 0.003 cm/cm) for hydrostatic journal bearings.

2. Compute the shaft eccentricity in the clearance space
The numerical parameter used to describe the eccentricity of the shaft in the bearing clearance space is the ratio $\epsilon = 1 - h/(mr)$, where $h =$ shaft clearance, in, during

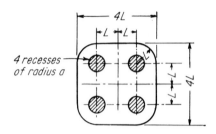

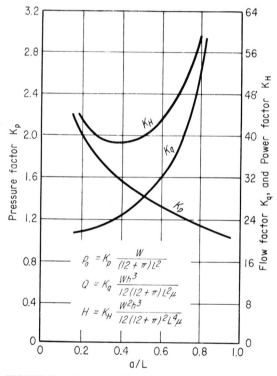

FIGURE 22 Constants and equations for modified square hydrostatic bearings. (*Product Engineering.*)

operation. With a clearance of $h = 0.002$ in (0.051 mm), $\epsilon = 1 - [0.002/(0.003 \times 2)] = 0.667$. With a clearance of $h = 0.004$ in (0.102 mm), $\epsilon = 1 - [0.004/(0.003 \times 2)] = 0.333$.

3. Compute the eccentricity constants

The eccentricity constant $A_k = 12[2 - \epsilon/(1 - \epsilon)^2] = 12[2 - 0.667/(1 - 0.667)^2] = 144.6$. A second eccentricity constant B is given by $B_k = 12\{\epsilon(4 - \epsilon^2)/[2(1 - \epsilon^2)^2] + 2 + \epsilon^2/(1 - \epsilon^2)^{2.5} \times \arctan[1 + \epsilon/(1 - \epsilon^2)^{0.5}]\}$. Since this relation is awkward to handle, Fig. 25 was developed by Fuller. From Fig. 25, $B_k = 183$.

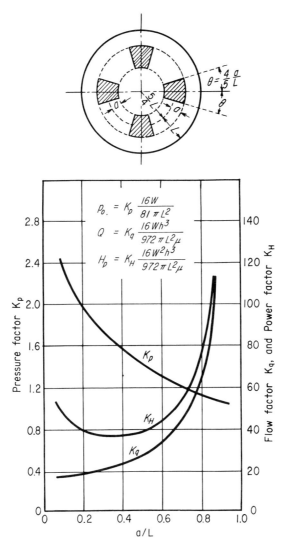

FIGURE 23 Constants and equations for circular-sector hydrostatic bearings. (*Product Engineering.*)

4. *Compute the required lubricant flow rate*

The lubricant flow rate is found from Q in^3/s $= 2Lm^3r_s/(\eta A_k)$, where L = load acting on shaft, lb; η = lubricant viscosity, reyns. For the bearing with $h = 0.002$ in (0.051 mm), $Q = 2(3600)(0.003)^3(2)/[152 \times 10^{-7})(144.6)] = 0.177$ in^3/s (2.9 cm^3/s), or 0.0465 gal/min (2.9 mL/s).

5. *Compute the required lubricant-inlet pressure*

The lubricant-inlet pressure is found from $p_i = \eta QB/(2bm^3r_s^2)$, where p_i = lubricant

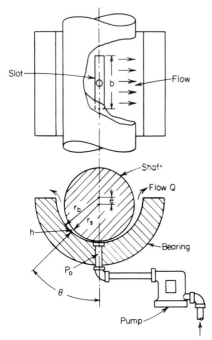

FIGURE 24 Hydrostatic journal bearing. (*Product Engineering.*)

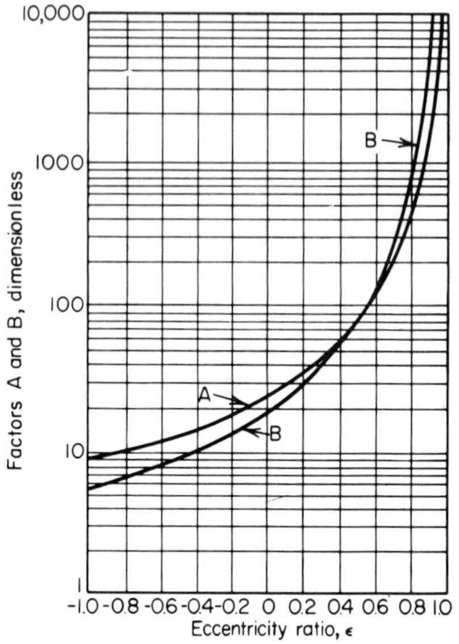

FIGURE 25 Constants for hydrostatic journal bearing oil flow and load capacity. (*Product Engineering.*)

inlet pressure, lb/in²; other symbols as before. Thus, $p_i = (152 \times 10^{-7})(0.177)(183)/[(2)(3)(0.003)^3(2)^2] = 759$ lb/in² (5233.3 kPa).

6. Analyze the larger-clearance bearing
Use the same procedure as in steps 1 through 5. Then $\epsilon = 0.333$; $A = 45.0$; $B = 42.0$; $Q = 0.560$ in³/s (9.2 cm³/s) = 0.1472 gal/min (9.2 mL/s); $p_i = 551$ lb/in² (3799.1 kPa).

   Related Calculations.   Note that the closer the shaft is to the center of the bearing, the smaller the lubricant pressure required and the larger the oil flow. If the larger flow requirements can be met, the design with the thicker oil film is usually preferred, because it has a greater ability to absorb shock loads and tolerate thermal change.

   Use the general design procedure given here for any applications where a hydrostatic journal bearing is applicable and there is no thrust load.

HYDROSTATIC MULTIDIRECTION BEARING ANALYSIS

Determine the lubricant pressure and flow requirements for the multidirection hydrostatic bearing shown in Fig. 26 if the vertical coplanar forces acting on the plate are 164,000 lb (729,508.4 N) upward and downward, respectively. The lubricant

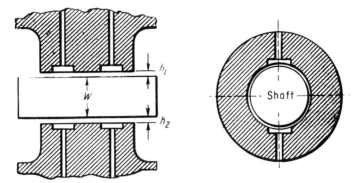

FIGURE 26 Double-acting hydrostatic thrust bearing. (*Product Engineering.*)

viscosity $\eta = 393 \times 10^{-7}$ reyn film thickness $= 0.005$ in (0.127 mm), $L = 7$ in (17.8 cm), $a = 3.5$ in (8.9 cm). What would be the effect of decreasing the film thickness h on one side of the plate by 0.001-in (0.025-mm) increments from 0.005 to 0.002 in (0.127 to 0.051 mm)? What is the bearing stiffness?

Calculation Procedure:

1. Compute the required lubricant-inlet pressure
By Fuller's method, Fig. 27 shows that the bearing has four pressure pads to support the plate loads. Figure 27 also shows the required pressure, flow, and power equations, and the appropriate constants for these equations. The inlet-pressure equation is $p_i = K_p L_s / (16 L^2)$, where p_i = required lubricant inlet pressure, lb/in²; K_p = pressure constant from Fig. 27; L_s = plate load, lb; L = bearing length, in.

Find K_p from Fig. 27 after setting up the ratio $a/L = 3.5/7.0 = 0.5$, where a = one-half the pad length, in. Then $K_p = 1.4$. Hence $p_i = (1.4)(164,000)/[(16)(7)^2] = 293$ lb/in² (2020.2 kPa).

2. Compute the required lubricant flow rate
From Fig. 27, the required lubricant flow rate $Q = K_q L_s h^3 / (192 L^2 \eta)$, where K_q = flow constant; h = lubricant-film thickness, in; other symbols as before. For $a/L = 0.5$, $K_q = 36$. Then $Q = (36)(164,000)(0.005)^3 / [(192)(7)^2(393 \times 10^{-7})] = 1.99$ in³/s (32.6 cm³/s). This can be rounded off to 2.0 in³/s (32.8 cm³/s) for usual design calculations.

3. Compute the pressure and load for other plate clearance
The sum of the plate lubricant-film thicknesses, $h_1 + h_2$, Fig. 26, is a constant. For this bearing, $h_1 + h_2 = 0.005 + 0.005 = 0.010$ in (0.254 mm). With no load on the plate, if oil is pumped into both bearing faces at the rate of 2.0 in³/s (32.8 cm³/s), the maximum recess pressure will be 293 lb/in² (2020.2 kPa). The force developed on each face will be 164,000 lb (729,508.4 N). Since the lower face is pushed up with this force and the top face is pushed down with the same force, the net result is zero.

With a downward external load imposed on the plate such that the lower film thickness h_2 is reduced to 0.004 in (0.102 mm), the upper film thickness will become 0.006 in (0.152 mm), since $h_1 + h_2 = 0.010$ in (0.254 mm) = a constant for

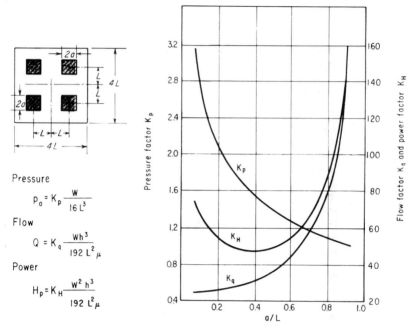

Pressure

$$p_o = K_p \frac{W}{16\,L^3}$$

Flow

$$Q = K_q \frac{Wh^3}{192\,L^2\,\mu}$$

Power

$$H_p = K_H \frac{W^2\,h^3}{192\,L^2\,\mu}$$

FIGURE 27   Dimensions and equations for thrust-bearing design. (*Product Engineering.*)

this bearing. If the lubricant flow rate is held constant at 2.0 in³/s (32.8 cm³/s), then $K_q = 36$ from Fig. 27, since $a/L = 0.5$. With these constants, the load equation becomes $L_s = 0.0205/h^3$, and the inlet-pressure equation becomes $p_i = L_s/560$.

Using these equations, compute the upper and lower loads and inlet pressures for h_2 and h_1 ranging from 0.004 to 0.002 and 0.006 to 0.008 in (0.102 to 0.051 and 0.152 to 0.203 mm), respectively. Tabulate the results as shown in Table 8. Note that the allowable load is computed by using the respective film thickness for

TABLE 8   Load Capacity of Dual-Direction Bearing

Film thickness, in (cm)		Inlet pressure, lb/in² (kPa)		Load, lb (N)		Load capacity, lb (N)
h_2	h_1	p_{i2}	p_{i1}	L_{s2}	L_{s1}	$L_{s2} - L_{s1}$
0.005	0.005	293	293	164,000	164,000	0
(0.127)	(0.127)	(2,020.2)	(2,020.0)	(729,508.4)	(729,508.4)	(0.0)
0.004	0.006	571	170	320,000	95,000	225,000
(0.102)	(0.152)	(3,937.0)	(1,172.2)	(1,423,431.0)	(422,581.1)	(1,000,850.0)
0.003	0.007	1,360	106	760,000	59,700	700,300
(0.076)	(0.178)	(9,377.2)	(730.9)	(3,380,648.7)	(265,558.9)	(3,115,089.9)
0.002	0.008	4,570	71	2,560,000	40,000	2,520,000
(0.051)	(0.203)	(31,510.2)	(489.5)	(11,387,448.3)	(177,928.9)	(11,209,519.4)

the lower and upper parts of the plate. The same is true of the lubricant-inlet pressure, except that the corresponding load is used instead.

Thus, for $h_2 = 0.004$ in (0.102 mm), $L_{s2} = 0.0205/(0.004)^3 = 320,000$ lb (1,423,431.0 N). Then $p_{i2} = 320,000/560 = 571$ lb/in^2 (3937.0 kPa). For $h_k = 0.006$ in (0.152 mm), $L_{s1} = 0.0205/(0.006)^3 = 95,000$ lb (422,581.1 N). Then $p_{i1} = 95,000/560 = 169.6$, say 170, lb/in^2 (1172.2 kPa).

The load difference $L_{s2} - L_{i1}$ = the bearing load capacity. For the film thicknesses considered above, $L_{s2} - L_{s1} = 320,000 - 95,000 = 225,000$ lb (1,000,850.0 N). Load capacities for various other film thicknesses are also shown in Table 8.

4. Determine the bearing stiffness

Plot the net load capacity of this bearing vs. the lower film thickness, Fig. 28. A tangent to the curve at any point indicates the stiffness of this bearing. Draw a tangent through the origin where $h_2 = h_1 = 0.005$ in (0.127 mm). The slope of this tangent = vertical value/horizontal value = $(725,000 - 0)/(0.005 - 0.002) = 241,000,000$ lb/in (42,205,571 N/m) = the bearing stiffness. This means that a load of 241,000,000 lb (1,072,021,502 N) would be required to displace the plate

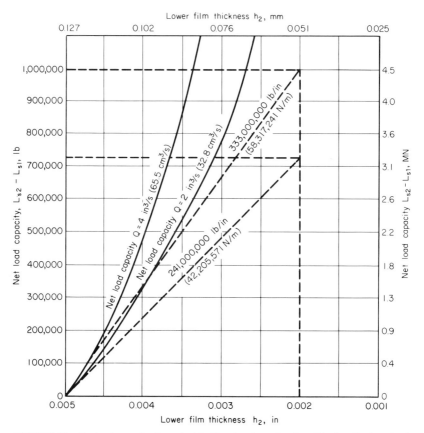

FIGURE 28 Net load capacity for a double-direction thrust bearing. (*Product Engineering.*)

1.0 in (2.5 cm). Since the plate cannot move this far, a load of 241,000 lb (1,072,021.5 N) would move the plate 0.001 in (0.025 mm).

If the lubricant flow rate to each face of the bearing were doubled, to $Q = 4.0$ in^3/s (65.5 cm^3/s), the stiffness of the bearing would increase to 333,000,000 lb/in (58,317,241 N/m), as shown in Fig. 28. This means that an additional load of 333,000 lb (1,481,257.9 N) would displace the plate 0.001 in (0.025 mm). The stiffness of a hydrostatic bearing can be controlled by suitable design, and a wide range of stiffness values can be designed into the bearing system.

Related Calculations.   Hydrostatic bearings of the design shown here are useful for a variety of applications. Journal bearings for multidirectional loads are analyzed in a manner similar to that described here.

LOAD CAPACITY OF GAS GEARINGS

Determine the load capacity and bearing stiffness of a hydrostatic air bearing, using 70°F (21.1°C) air if the bearing orifice radius = 0.0087 in (0.0221 cm), the radial clearance h = 0.0015 in (0.038 mm), the bearing diameter d = 3.00 in (7.6 cm), the bearing length L = 3 in (7.6 cm), the total number of air orifices N = 8, the ambient air pressure p_a = 14.7 lb/in^2 (abs) (101.4 kPa), the air supply pressure p_s = 15 lb/in^2 (gage) = 29.7 lb/in^2 (abs) (204.8 kPa), the (gas constant)(total temperature) = RT = 1.322 × 10^8 in^2/s^2 (8.5 × 10^8 cm^2/s^2), ϵ = the eccentricity ratio = 0.30, air viscosity η = 2.82 × 10^{-9} lb·s/in^2 (1.94 × 10^{-9} N·s/cm^2), the orifice coefficient α = 0.63, and the shaft speed ω = 2100 rad/s = 20,000 r/min.

Calculation Procedure:

1.   Compute the bearing factors Λ and Λ_r
For a hydrostatic gas bearing, $\Lambda = (6\eta\omega/p_a)(d/zh)^2 = [(6)(2.82 × 10^{-9})(2100)/14.7] [3/(2 × 0.0015)]^2 = 2.41$.

Also, $\Lambda_T = 6\eta N a^2 \alpha (RT)^{0.5}/p_a h^3 = (6)(2.82 × 10^{-9})(8)(0.0087)^2(0.63)(1.322 × 10^8)^2/[(14.7)(0.0015)^3] = 1.5$.

2.   Determine the dimensionless load
Since $L/d = 3/3 = 1$, and $p_s/p_a = 29.7/14.7 \approx 2$, use the first chart[1] in Fig. 29. Before entering the chart, compute $1/\Lambda = 1/2.41 = 0.415$. Then, from the chart, the dimensionless load = $0.92 = L_d$.

3.   Compute the bearing load capacity
The bearing load capacity $L_s = L_d p_a LD\epsilon = (0.92)(14.7)(3.00)(3.00)(0.3) = 36.5$ lb (162.4 N). If there were no shaft rotation, $\Lambda = 0$, $\Lambda_T = 1.5$, and $L_d = 0.65$, from the same chart. Then $L_s = L_d p_a LD\epsilon = (0.65)(14.7)(3.00)(3.00)(0.3) = 25.8$ lb (114.8 N). Thus, rotation of the shaft increases the load-carrying ability of the bearing by $[(36.5 - 25.8)/25.8](100) = 41.5$ percent.

4.   Compute the bearing stiffness
For hydrostatic gas bearing, the bearing stiffness $B_s = L_s/(h\epsilon) = 36.5/[0.0015)(0.3)] = 81,200$ lb/in (142,203.0 N/cm).

SECTION 23
SPRING SELECTION AND ANALYSIS

PROPORTIONING HELICAL SPRINGS BY MINIMUM WEIGHT

The detent helical spring in Fig. 1 is to be designed so that force P_1 for the extended condition is to be 20 lb (88.9 N). After an additional compression of 0.625 in (1.59 cm) the shearing stress in the spring is to be 75,000 lb/in^2 (517 MPa). The spring index, c, is approximately 8, based on past experience, and the modulus of elasticity in shear is 11,500,000 lb/in^2 (79.2 GPa). Find the diameter of the wire, helix radius, and number of active coils for the smallest amount of material in the spring.

Calculation Procedure:

1. Find the diameter of the spring wire
A spring which will contain the smallest possible amount of material will have design parameters selected so that the maximum force, stress, and deflection are

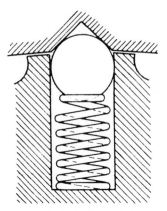

FIGURE 1 Spring-loaded detent. (*Product Engineering.*)

exactly twice the minimum force, stress, and deflection, respectively. With such parameters, the spring wire diameter,

$$d = \sqrt{\frac{16P_1 c}{\pi (S_s)_{\max}}}$$

where d = spring wire diameter, in (cm); P_1 = force in the extended condition, lb (N); c = spring index, dimensionless; π = 3.1416; $(S_s)_{\max}$ = shear stress in the spring wire, lb/in² (kPa). Substituting, $d = (16 \times 20 \times 8/75{,}000\pi)^{0.5} = 0.1042$ in (0.265 cm). Referring to a spring wire table, choose No. 12 wire with $d = 0.1055$ in (0.268 cm) as the nearest commercially available wire size.

2. Compute the mean radius of the spring helix
The helix mean radius,

$$R = \frac{\pi d^3 (S_s)_{\max}}{32 P_1}$$

where R = mean radius of the helix, in (cm); other symbols as defined earlier. Substituting, $R = (\pi \times 0.1055^3 \times 75{,}000)/32 \times 20 = 0.4323$ in (1.09 cm).

3. Determine the number of active coils in the spring
Use the relation, N = number of active coils in the spring,

$$N = \frac{(\delta_2 - \delta_1)\, dG}{2\pi R^2 (S_s)_{\max}}$$

where, δ_2 = maximum compression of spring, in (cm); δ_1 = minimum compression of the spring, in (cm); G = modulus of elasticity in shear; other symbols as before. Substituting, $N = (0.625 \times 0.1055 \times 11{,}500{,}000)/(2\pi \times 0.4323^2 \times 75{,}000 = 8.61$ active coils.

4. Find the volume of the spring
Volume of a helical spring is given by $V = 0.5 \times \pi^2 \times d^2 \times RN$, where the symbols are as given earlier. Substituting, $V = 0.5 \times 3.14^2 \times 0.1055^2 \times 0.4323 \times 8.61 = 2.044$ in^3 (30.0335 dm^3).

Related Calculations. Because of the many variables involved, the proportioning of a helical spring is usually done by trial and error. An unlimited number of springs can be obtained that will satisfy a given set of equations for stress and deflection. However, all springs found this way will contain more material than necessary unless the conditions for the mathematical minimum are fulfilled.

The functioning of a helical spring usually occurs when the spring is extended, and is least able to exert a force. For example, in the indexing mechanism shown in Fig. 1, when the ball is in the detent, the holding power of the mechanism depends on the force exerted by the spring. After relative motion between the parts occurs, the ball is out of the detent and the additional force from the increased compression serves no useful purpose and may be a disadvantage. In this condition the shearing stress in the wire is usually the governing factor.

Many similar applications arise in which the design is controlled by (*a*) the spring force in the extended condition and (*b*) the shearing stress for the most compressed condition. It has been proved for this instance, that the spring will contain the smallest possible amount of material when the design parameters are selected so that the maximum force, stress, and deflection are exactly twice the minimum force, stress, and deflection, respectively.

Although not directly specified, the designer can usually estimate a suitable value for the spring index, *c*. Taking the maximum load as twice the minimum load, the wire-diameter, *d*, is computed using the equation given in step 1, above. Next the helix radius, *R*, is computed using the *d* value from step 1. Then the number of active coils is computed, step 3. Lastly, the minimum volume of the spring is determined, as in step 4.

This procedure is the work of M. F. Spotts, Northwestern Technological Institute, as reported in *Product Engineering* magazine.

DETERMINING SAFE TORSIONAL STRESS FOR A HELICAL SPRING

The load on a helical spring is 1600 lb (726.4 kg) and the corresponding deflection is to be 4 in (10.2 cm). Rigidity modulus is 11×10^6 lb/in^2 (75.8 GPa) and the maximum intensity of safe torsional stress is 60,000 lb/in^2 (413.4 MPa). Design the spring for the total number of turns if the wire is circular in cross section with a diameter of 0.625 in (1.5875 cm) and a centerline radius of 1.5 in (3.81 cm).

Calculation Procedure:

1. Find the number of active coils in the spring
Use the relation $y = N(64)(P)(r^3)/G(d^4)$, where y = total deflection of the spring, in (cm); N = number of active coils in the spring; P = load on spring, lb (N); r = radius as axis to centerline of wires, in (cm); G = modulus of elasticity of spring material in shear, lb/in^2 (kPa); d = spring wire diameter, in (cm). Solving for the number of active coils, $N = y(G)(d^4)/64(P)(r^3)$. Or $N = 4(11 \times 11^6)(0.625^4)/64(1600)(1.5^3) = 19.4$ active coils.

2. Check for the safe limit of torsional stress in the spring
Use the relation $S_t = 16(P)(r)/\pi(d^3)$. Solving, $S = 16(1600)(1.5)/\pi(0.625^3) = 50,066$ lb/in^2 (344.95 MPa). This torsional stress is less than the safe limit of 60,000 lb/in^2 (413.4 MPa) chosen for this spring. Hence, the spring is acceptable.

The total number spring turns for two inactive coils (one at each end) is $N + 2 = 19.4 + 2 = 21.4$, or 22 coils.

Related Calculations.   Use this general approach for any helical spring you wish to analyze for safe torsional stress.

DESIGNING A SPRING WITH A SPECIFIC NUMBER OF COILS

A coiled spring with an outside diameter of 1.75 in (4.45 cm) is required to work under a load of 140 lb (622.7 N) and have seven active coils with the ends closed and round. Determine unit deflection, total number of coils, and length of spring when under load when $G = 12 \times 10^6$ lb/in^2 (82.7 GPa) and mean spring diameter (outside diameter − wire diameter) is to be 0.779 in (1.98 cm).

Calculation Procedure:

1. Find the safe shearing stress in the spring
Use the relation, $S_t = 8PD/\pi d^3$, where S = fiber stress in shear, lb/in^2 (kPa); P = axial load on spring, lb (N); D = mean diameter of spring as defined above, in (cm); d = diameter of spring wire, in (cm). Substituting, $S = 9(140)(0.779)(2)/\pi (0.192^3) = 78,475$ lb/in^2 (540.7 MPa).

2. Determine the deflection of the spring
Use the relation, $y = 4(\pi)(N)(r^2)(S)/Gd$, where y = total deflection, in (cm); N = number of active coils; G = rigidity modulus of the spring material in shear, lb/in^2 (kPa); other symbols as defined earlier. Substituting, $y = 4(\pi)(7)(0.779^2)(78,475)/12 \times 10^6(0.192) = 1.818$ in (4.6 cm), or about $1^{13}/_{16}$ in.

3. Compute the total number of coils in the spring
For the ends of a spring to be closed and ground smooth, 1.5 coils should be taken as inactive. In compression springs the number of active coils depends on the style of ends, as follows: *open ends, not ground*—all coils are active; *open ends, ground*—0.5 coil inactive; *closed ends, not ground*—1 coil inactive; *closed ends, ground*—1.5 coils inactive; *squared ends, ground*—2 coils inactive. Since this spring is to have ends that are closed and round; i.e., closed and ground, use 1.5 inactive coils. Hence, the total number of coils will be $7 + 1.5 = 8.5$, say 9 coils.

4. Find the length of the spring when under load
The solid height of the spring when it is entirely compressed is 9 coils × 0.192 = 1.728 in (4.39 cm), say $1^3/_4$ in.

The total deflection under load is $1^{13}/_{16}$ in (4.6 cm). Using a total free space between coils of 1 in (2.54 cm), the free length of the spring will be $1^3/_4 + 1^{13}/_{16} + 1 = 4^9/_{16}$ in (11.59 cm). Hence, the length of the spring when under load = $4^9/_{16} - 1^{13}/_{16} = 2^3/_4$ in (6.98 cm).

Related Calculations.   The procedure given here is valid for sizing any spring when it must have a specified number of coils.

DETERMINING SPRING DIMENSIONS BEFORE AND AFTER LOAD APPLICATION

A cylindrical helical spring of circular cross-section wire is to be designed to safely carry an axial compressive load of 1200 lb (5338 N) at a maximum stress of 110,000 lb/in^2 (757.9 MPa). The spring is to have a deflection scale of approximately 150 lb/in (262 N/cm). Spring proportions are to be: Mean diameter of coil/diameter of wire = 6 to 8; spring length when closed/mean diameter of coil = 1.7 to 2.3. Determine (*a*) mean diameter of coil; (*b*) diameter of wire; (*c*) length of coil when closed, and (*d*) length of coil before load application. Use *G* = rigidity modulus of steel in shear = 11.5 × 10^6 (79,235 MPa).

Calculation Procedure:

1. Find the trial wire size
In this trial design we can assume the ratio of D/d to be 6:8, as given. Then $d = D/7$, using the midpoint ratio, where D = mean diameter of spring coil/diameter of spring wire. Use the relation, $D = S(\pi)(d^3)/8(P)$, where S = spring stress, lb/in^2 (kPa); D = mean diameter of the spring = outside diameter − wire diameter, in (cm); d = diameter of spring wire, in (cm). Substituting, $D = (110,000)(3.1416)[(D/7)^3]/8(1200)$; $D = 4.45$ in (2.02 cm).

The wire diameter is $d = 4.45/7 = 0.635$ in (1.6 cm). The nearest wire standard wire size is 0.6666 in (1.69 cm). Solving for $D/d = 4.45/0.6666 = 6.68$. This lies within the limits of 6 to 8 set for this spring for D/d.

2. Determine the spring scale
Use the relation $P/y = G(d^4)/N(64)(r^3)$, where y = total deflection of the spring, in (cm); N = number of coils; other symbols as before. For this spring, $P/y = 150$. Substituting, $150 = (11.5 \times 10^6)(0.625^4)/N(64)(2.38^3)$. Solving, $N = 13.6$ active coils.

Total turns for this spring, assuming closed ends ground = 13.6 + 1.5 = 15.1.

3. Evaluate the length closed/mean diameter ratio
This ratio must be between 1.7 and 2.3 if the spring is to meet its design requirements. Using the relation, (*d*)(total turns)/*D*, we have (15.5)(0.6666)/4.45 = 2.26, which is within the required limits. Then, the length of the closed coil = 15.1 × 0.6666 = 10.065 in (25.56 cm).

4. Find the coil length before load application
To find the length of the spring coil before the load is applied we add the closed length of the coil to the total deflection of the spring. To find the total deflection, use the relation, $y = N(64)(P)(2.38^3)/11.5 \times 10^6(0.6666^4)$; or $y = 13.6(64)(1200)(2.38^3)/11.5 \times 10^6(0.6666^4) = 6.2$ in (15.75 cm). Then, the length of the coil before load application = 10 + 6.2 = 16.2 in (41.1 cm).

Related Calculations. The steps in this procedure are useful when designing a spring to meet certain preset requirements. Using the given steps you can develop a spring meeting any of several dimensional or stress parameters.

ANALYSIS AND DESIGN OF FLAT METAL SPRINGS

Determine the width and thickness of the leaves of a six-leaf steel cantilever spring 13-in (33-cm) long to carry a load of 375 lb (1668 N) with a deflection of 1.25 in (3.2 cm). The maximum stress allowed in this spring is 50,000 lb/in² (344.5 MPa); the modulus of elasticity of the steel spring material is 30×10^6 lb/in² (206,700 MPa).

Type of spring	W, safe load =	F, deflection =
Flat parallel spring W	$\dfrac{Sbt^2}{6l}$	$\dfrac{4Wl^3}{Ebt^3} = \dfrac{2}{3}\dfrac{Sl^2}{Et}$
Flat triangular spring W	$\dfrac{Sbt^2}{6l}$	$\dfrac{6Wl^3}{Ebt^3} = \dfrac{Sl^2}{Et}$
Leaf spring W	$\dfrac{SNbt^2}{6l}$ Where N = No. of leaves	$\dfrac{6Wl^3}{ENbt^3} = \dfrac{Sl^2}{Et}$ Where N = No. of leaves

Calculation Procedure:

1. Find the leaf thickness for this steel spring
See the list of equations below. Knowing the deflection, we can use the equation, $F = S(l^2)/E(t)$, where F = deflection, in (cm); S = safe tensile stress in the spring metal, lb/in^2 (kPa); l = spring length, in (cm); E = modulus of elasticity as given above; t = spring leaf thickness in (cm). substituting, $1.25 = 50,000(13^2)/30 \times 10^6(t)$; $t = 0.225$ in (0.57 cm).

W = save load or pull, lb (N)
F = deflection at point of application, in (cm)
S = safe tensile stress of material, lb/in^2 (kPa)
E = modulus of elasticity, 30×10^6 for steel (kPa)

Other symbols as shown above.

2. Determine the width of each leaf in the spring
Use the relation, $W = S(N)(b)(t^2)/6(l)$, where the symbols are as defined above and W = safe load or pull, lb (N). Substituting, $375 = 50,000(6)(b)(.225^2)/6(13)$; $b = 1.93$ in (4.9 cm).

   Related Calculations.   Use the spring layout and equations above to design any of the three types of metal springs shown. Flat metal springs find wide use in a variety of mechanical applications. The three types shown above—flat parallel, flat triangular, and flat leaf—are the most popular today. Using the data given here, engineers can design a multitude of flat springs for many different uses.

SIZING TORSIONAL LEAF SPRINGS

Size a torsional leaf spring, Fig. 2, for a 38-in $\times$ 38-in (96.5-cm $\times$ 96.5-cm) work platform weighing 145 lb (65.8 kg) which requires an assist spring to reduce the lifting force that must be exerted by a human being. A summation of the moments about a line through both platform pivots, Fig. 3, shows that a lifting force of only 23 lb (102.3 N) is needed if a torsional spring provides an assist moment of $M_t = 1880$ lb $\cdot$ in (212.4 N $\cdot$ m). An assist moment is required to raise the platform from either the stowed or the operating position. As the platform is rotated from one position to the other, the assist spring undergoes a load reversal. The required spring wind-up angle, $\theta = \pi/2$ radians; the spring length = 36 in (91.4 cm). AISI 4130 alloy steel heat-treated to 31 Rc minimum with an allowable shear stress of 68,000 lb/in^2 (468.5 MPa) and a shear modulus of $G = 11.5 \times 10^6$ lb/in^2 (79,235 MPa) is to be used for this spring. Size the spring, determining the required dimensions for blade thickness, number of blades, aspect ratio, and blade width.

Calculation Procedure:

1. Estimate the blade thickness
Use the equation

$$\beta = \frac{b}{l} = \frac{\tau}{2G\theta},$$

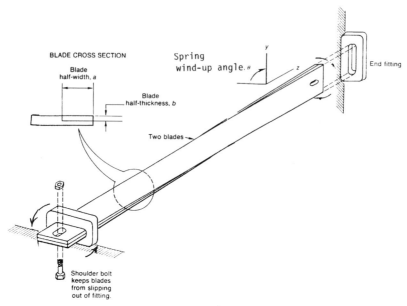

FIGURE 2 Torsional leaf springs normally are twisted to a specific wind-up angle. This two-blade spring is wound at right angles. (*Machine Design.*)

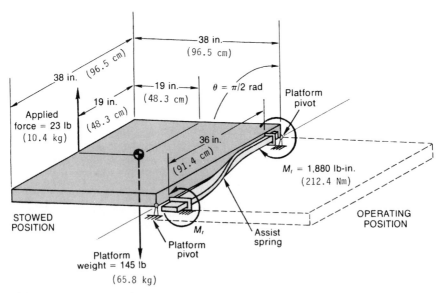

FIGURE 3 Work platform requires an assist spring to reduce required lifting force. (*Machine Design.*)

where the symbols are as given below. Solving for the blade thickness using the given data, $b = (68.000)(36)/2(11.5 \times 10^6)(\pi/2) = 0.068$ in (0.172 cm). Note that this is the blade *half-thickness.*

To keep manufacturing costs low, a stock plate thickness would be chosen. The nearest stock plate thickness, $2b_f$, is 0.125 in (0.3175 cm). So $b_f = 0.125/2 = 0.0625$ in (0.159 cm), which is less than $b = 0.068$ in (0.172 cm) the computed required blade half-thickness.

Nomenclature

$a =$ blade half-width, in (cm)
$a_f =$ selected blade half-width, in (cm)
$a_r =$ required blade half-width, in (cm)
$b =$ blade half-thickness, in (cm)
$b_f =$ selected blade half-thickness, in (cm)
$F =$ torsional stress factor, $1/\text{in}^3$ $(1/\text{cm}^3)$
$f =$ normalized torsional stress factor

$G =$ shear modulus, lb/in^2 (kPa)
$K =$ torsional stiffness factor, in^4 (cm^4)
$k =$ normalized torsional stiffness factor
$l =$ spring length, in (cm)
$M_t =$ applied moment, $\text{lb} \cdot \text{in}$ $(\text{N} \cdot \text{m})$
$m_t =$ normalized applied moment
$n =$ number of blades
$n_f =$ selected number of blades
$R =$ aspect ratio

$R_0 =$ estimated aspect ratio
$R_r =$ required aspect ratio
$s =$ normalized shear stress
$\alpha =$ normalized half-width
$\beta =$ normalized half-thickness
$\theta =$ spring wind-up angle, rad
$\tau =$ torsional stress, lb/in^2 (kPa)
$\tau_a =$ allowable torsional stress, lb/in^2 (kPa)

2. Select the number of blades for this torsion leaf spring
Use the equation,

$$nR = \frac{3l(M_t/\theta)}{16Gb_f^4}$$

where the symbols are as given above. Substituting, $nR = 3(36)[1880/(\pi/2)]/16(11.5 \times 10^6)(0.0626^4) = 46.$

Next, use the equation,

$$4 < R_0 = \frac{nR}{n_f} < 15,$$

letting $n_f = 5$; then $R_0 = 46/5 = 9.2$.

3. Choose the aspect ratio
The required aspect ratio, R_r, is established by employing the computed value for R_0 in step 2 as an initial estimate in the equation

$$R_{i+1} = \frac{\left(\dfrac{l}{G}\right)\left(\dfrac{M_t}{\theta}\right)}{n_f b_f^4 \left[\dfrac{16}{3} - \dfrac{3.36}{R_i}\left(1 - \dfrac{1}{12R_i^4}\right)\right]}$$

where $i = 0, 1, 2, \ldots$

Usually, after three or so iterations, computed values R_{i+1} converge to a single value which is the required aspect ratio, R_r. Substituting, using $R_0 = 9.2$, as computed in step 2, for R_1,

$$R_1 = \frac{\left(\dfrac{36}{11.5 \times 10^6}\right)\left(\dfrac{1880}{\pi/2}\right)}{5(0.0625^4)\left[\dfrac{16}{3} - \dfrac{3.36}{9.2}\left(1 - \dfrac{1}{12(9.2^4)}\right)\right]}$$

$$= \frac{49.108}{\dfrac{16}{3} - \dfrac{3.36}{9.2}\left(1 - \dfrac{1}{12(9.2^4)}\right)}$$

$$= 9.88$$

Using the R_1 value of 9.88 found above, substitute it in the equation again,

$$R_2 = \frac{49.108}{\dfrac{16}{3} - \dfrac{3.36}{9.88}\left(1 - \dfrac{1}{12(9.88^4)}\right)}$$

$$= 9.83 = R_r$$

Further iterations show that the aspect ratio converges to a value of 9.83. Therefore, let $R_r = 9.83$.

4. Determine the blade width

Given $R_r = 9.83$ and $2b_f = 0.125$ in (0.3175 cm), the required blade width, $2a$, is found from $2a_r = R_r(2b_f)$. Or, $2a_r = (9.83)(0.125) = 0.123$ in (0.31 cm). For manufacturing simplicity, the blades will be cut to a 1.25-in (3.18-cm) width from stock 0.125-in (3.18-cm) thick plates. Therefore, let $2a_f = 1.25$ in (3.18 cm).

5. Check the assist moment and stress

The assist spring has been sized to have five 36-in (91.4-cm)-long blades with a 0.125-in (3.18-cm) wide by 0.125-in (3.18-cm) thick cross section. The assist moment provided by such a spring stack is determined from

$$M_t = \frac{KG\theta}{l}$$

$$= \frac{n_f G}{l} a_f b_f^3 \left[\frac{16}{3} - 3.36\frac{b_f}{a_f}\left(1 - \frac{b_f^4}{12a_f^4}\right)\right]\theta$$

$$\approx M_{t\,required}$$

Substituting,

$$M_t = \frac{5(11.5 \times 10^6)}{36} (0.625)(0.0625^3)$$

$$\times \left[\frac{16}{3} - 3.36 \left(\frac{0.0625}{0.625} \right) \left(1 - \frac{0.0625^4}{12(0.625^4)} \right) \right] \frac{\pi}{2}$$

$$= 1913 \text{ lb} \cdot \text{in}$$

$$\approx M_{t\,\text{required}} = 1880 \text{ lb} \cdot \text{in} \qquad (212.4 \text{ N} \cdot \text{m})$$

Because the actual assist moment of 1913 lb·in (216.2-N·m) is slightly greater than the required moment of 1880 lb·in (212.4 N·m), the lifting force needed will be less than 23 lb (102.3 N).

6. Find the shear stress in each blade
Use the equation,

$$\tau = FM_t$$

$$= \frac{1}{n_f} \left(\frac{3a_f + 1.8b_f}{8a_f^2 b_f^2} \right) M_t$$

$$< \tau_a$$

to determine the shear stress in each blade. Substituting,

$$\tau = \frac{1}{5} \left[\frac{3(0.625) + 1.8(0.0625)}{8(0.625^2)(0.0625^2)} \right] (1913)$$

$$= 62,300 \text{ lb/in}^2 < \tau_a = 68,000 \text{ lb/in}^2$$

Since the blade shear stress of 62,300 lb/in² (429.2 MPa) is less than the allowable shear stress of 68,000 lb/in² (468.5 MPa), this leaf spring can be used as sized here.

   Related Calculations.   Leaf springs, in most applications where they transmit forces, are subjected primarily to bending loads. Design of such springs normally is based upon cantilever beam theory, which has been well-documented in the literature on structures and machine elements.

   It is not widely known that leaf springs also can be designed to twist. When so used, leaf springs can generate high torques while occupying a small space. Such torsional leaf springs are most commonly used as assist devices on hatches, doors, and folding platforms or walkways. Typical applications are on a variety of vehicles in which space is restricted. The procedure presented here applies to blades having identical dimensions, the usual configuration.

   A manual available from the Society of Automotive Engineers provides design techniques for leaf springs for vehicular applications. The SAE manual describes a design procedure for leaf springs consisting of blades with either the same or different widths and thicknesses. The procedure given here is said, by its developer (see below), to be a more straightforward method.

   Stresses in torsional leaf springs can be reduced significantly by increasing either the number of blades, the aspect ratio, or both. Single torsional springs made of square, round, or rectangular bar materials have higher stresses than do torsional leaf springs, given the same assist moment and angular deflection.

According to Timoshenko and Goddier, a local irregularity in the stress distribution occurs near the attachment points when the aspect ratio of the spring is large. Because no specific information is generally available, experience and good judgment must be used to determine how large the aspect ratio can be before local attachment stresses become significant. Generally, the aspect ratio should be greater than 4 but less than 15. Also, the number of blades should be limited so that the spring stack height does not exceed the spring width.

This procedure is the work of Russel Lilliston, Staff Engineer, Martin Marietta Aerospace, as reported in *Machine Design* magazine.

DESIGNING A SPRING TO FIRE A PROJECTILE

A 22-coil squared and ground spring is designed to fire a 10-lb (4.54-kg) projectile into the air. The spring has a 6-in (15.24-cm) diameter coil with 0.75-in (1.9-cm) diameter wire. Free length of the spring, Fig. 4, is 26 in (66 cm); it is compressed to 8 in when loaded or set. The shear elastic limit for the spring material is 85,000 lb/in^2 (585.65 MPa). The spring constant Wahl factor is 1.18. Shear modulus of elasticity of the spring material is 12×10^6 lb/in^2(82,680 MPa). Determine (*a*) the height to which the projectile will be fired; (*b*) the safety factor for this spring.

Calculation Procedure:

1. *Find the actual stress in the spring*
Use the relation, actual stress, $S_a = (L_c)(G)(d)(W)/(4\pi)(n)(r^2)$, where S_c = actual stress, lb/in^2 (kPa); L = compressed length of spring, in (cm); G = shear modulus of elasticity, lb/in^2 (kPa); d = spring-wire diameter, in (cm); W = Wahl factor; n = number of coils; r = spring radius = diameter/2, in (cm). Substituting, $S_c = (8)(12 \times 10^6)(0.75)(1.18)/(4\pi)(22)(3^2) = 34{,}146$ lb/in^2 (235.2 MPa).

2. *Compute the force to which the spring is stressed when loaded*
Use the relation, $P = 0.1963(d^3)(S_c)/(r)(W)$, where P = loaded force, lb (N); other symbols as given earlier. Substituting, $P = (0.1963)(0.75^3)(34{,}146)/(3)(1.18) = 798.8$ lb (3553 N).

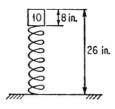

SI Values
8 in.	(20.3 cm)
26 in.	(66 cm)

FIGURE   4 Spring for firing a projectile.

3. Calculate the distance the projectile will be fired
Use the relation (weight, lb)(distance fired, s) = (P)(deflection)/2. Or s = (798.8)(8)/(2)(10) = 319.5 in (811.6 cm).

4. Determine the factor of safety for the spring
The factor of safety = (stress at elastic limit)/(actual stress) = 85,000/34,146 = 2.487.

  Related Calculations. Use this general procedure to determine the height to which a projectile can be fired for toys, military devices, naval applications, etc.

SPRING SUPPORT OF MACHINERY TO CONTROL VIBRATION FORCES

A machine weighing 2 tons (1816 kg) creates a disturbing force of 1200 lb (5338 N) at a frequency of 2000 cpm. The machine is to be supported on six springs, each taking an equal share of the load in such a way that the force transmitted to the building housing the machine is not to exceed 20 lb (89 N). Guides limit the machine motion to just the vertical direction. What is the required scale, lb/in (kg/cm), of each spring?

Calculation Procedure:

1. Set up the frequency and amplitude equations for this machine
The natural frequency of free vibration of this system is given by

$$f = \frac{1}{2\pi} \sqrt{\frac{kg}{W}}$$

where k = spring modulus, lb/in (kg/cm); g = gravitational constant, in/s^2 (cm/s^2); W = weight of each spring, lb (kg).
  The amplitude of the forced vibration of this machine, A, is given by

$$A = x_s \frac{1}{1 - (f_1/f)^2}$$

where x_s = displacement of the spring, in (cm), caused by steady-state vibration of the machine; f_1 = $W/2\pi$, or frequency of the exciting force, cps; f = natural frequency of the system, cps.

2. Set up the magnification factor
The ratio of $f_{1/f}$ is called the frequency ratio, and the ratio of $1/(1 - [f_{1/f}]^2)$ may be interpreted as the *magnification factor*. Now, because of proportionality factors, we can say

$$\frac{20}{1200} = \frac{1}{1 - \left[\dfrac{(2000/60)^2}{f^2}\right]}$$

For the reduction in the disturbing force which we seek, the ratio A/x_s is negative. Solving for f in the above equation gives $f = 4.339$ cps.

3. Find the spring modulus
Substitute the value found for f in the first equation above for natural frequency and we find $f = 4.339 = (1/6.28)(k \times 386.4/[4000/6]^{0.5})$; $k = 128.1$ lb/in (2243 N/cm).

    Related Calculations. Isolating the vibration caused by machinery is an important design challenge, especially in heavily populated structures. Springs are widely used to isolate the vibrations produced by reciprocating engines and compressors, pumps, presses, and similar machinery. This general procedure gives a useful approach to analyzing springs for such applications.

SPRING SELECTION FOR A KNOWN LOAD AND DEFLECTION

Give the steps in choosing a spring for a known load and an allowable deflection. Show how the type and size of spring are determined.

Calculation Procedure:

1. Determine the load that must be handled
A spring may be required to absorb the force produced by a falling load or the recoil of a mass, to mitigate a mechanical shock load, to apply a force or torque, to isolate vibration, to support moving masses, or to indicate or control a load or torque. Analyze the load to determine the magnitude of the force that is acting and the distance through which it acts.

    Once the magnitude of the force is known, determine how it might be absorbed—by compression or extension (tension) of a spring. In some applications, either compression or extension of the spring is acceptable.

2. Determine the distance through which the load acts
The load member usually moves when it applies a force to the spring. This movement can be in a vertical, horizontal, or angular direction, or it may be a rotation. With the first type of movement, a *compression,* or *tension,* spring is generally chosen. With a torsional movement, a *torsion-type* spring is usually selected. Note that the movement in either case may be negligible (i.e., the spring applies a large restraining force), or the movement may be large, with the spring exerting only a nominal force compared with the load.

3. Make a tentative choice of spring type
Refer to Table 1, entering at the type of load. Based on the information known about the load, make a tentative choice of the type of spring to use.

4. Compute the spring size and stress
Use the methods given in the following calculation procedures to determine the spring dimensions, stress, and deflection.

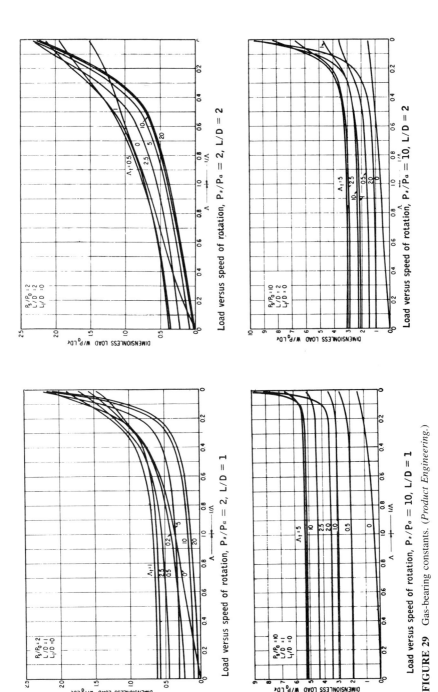

FIGURE 29 Gas-bearing constants. (*Product Engineering.*)

Related Calculations. Use this procedure for the selection of gas bearings where the four charts presented here are applicable. The data summarized in these charts result from computer solutions of the complex equations for "hybrid" gas bearings. The work was done by Mechanical Technology Inc., headed by Beno Sternlicht.

TABLE 1 Metal Spring Selection Guide

Type of load	Suitable spring type	Relative magnitude of load on spring	Deflection absorbed
Compression	Helical	Small to large	Small to large
	Leaf	Large	Moderate
	Flat	Small to large	Small to large
	Belleville	Small to large	Moderate
	Ring	Large	Small
Tension	Helical	Small to large	Small to large
	Leaf	Large	Moderate
	Flat	Small to large	Small to large
Torsion	Helical torsion	Small to large	Small to large
	Spiral	Moderate	Moderate
	Torsion bar	Large	Small

5. Check the suitability of the spring
Determine (a) whether the spring will fit in the allowable space, (b) the probable spring life, (c) the spring cost, and (d) the spring reliability. Based on these findings, use the spring chosen, if it is satisfactory. If the spring is unsatisfactory, choose another type of spring from Table 1 and repeat the study.

SPRING WIRE LENGTH AND WEIGHT

How long a wire is needed to make a helical spring having a mean coil diameter of 0.820 in (20.8 mm) if there are five coils in the spring? What will this spring weigh if it is made of oil-tempered spring steel 0.055 in (1.40 mm) in diameter?

Calculation Procedure:

1. Compute the spring wire length
Find the spring length from $l = \pi n d_m$, where l = wire length, in; n = number of coils in the spring, in. Thus, for this spring, $l = \pi(5)(0.820) = 12.9$ in (32.8 cm).

2. Compute the weight of the spring
Find the spring weight from $w = 0.224 l d^2$, where w = spring weight, lb; d = spring wire diameter, in. For this spring, $w = 0.224(12.9)(0.055)^2 = 0.0087$ lb (0.0387 kg).

   Related Calculations.   The weight equation in step 2 is valid for springs made of oil-tempered steel, chrome vanadium steel, silica-manganese steel, and silicon-chromium steel. For stainless steels, use a constant of 0.228, in place of 0.224, in the equation. The relation given in this procedure is valid for any spring having a continuous coil—helical, spiral, etc. Where a number of springs are to be made, simply multiply the length and weight of each by the number to be made to determine the total wire length required and the weight of the wire.

HELICAL COMPRESSION AND TENSION SPRING ANALYSIS

Determine the dimensions of a helical compression spring to carry a 5000-lb (22,241.1-N) load if it is made of hard-drawn steel wire having an allowable shear stress of 65,000 lb/in² (448,175.0 kPa). The spring must fit in a 2-in (5.1-cm) diameter hole. What is the deflection of the spring? The spring operates at atmospheric temperature, and the shear modulus of elasticity is 5×10^6 lb/in² (34.5×10^9 Pa).

Calculation Procedure:

1. *Choose the tentative dimensions of the spring*
Since the spring must fit inside a 2-in (5.1-cm) diameter hole, the mean diameter of the coil should not exceed about 1.75 in (4.5 cm). Use this as a trial mean diameter, and compute the wire diameter from $d = [8Ld_m k/(\pi s_s)]^{1/3}$, where $d =$ spring wire diameter, in; $L =$ load on spring, lb; $D_m =$ mean diameter of coil, in; $k =$ spring curvature correction factor $= (4c - 1)/(4c - 4) + 0.615/c$ for heavily coiled springs, where $c = 2r_m/d = d_m/d$; $s_s =$ allowable shear stress material, lb/in². For lightly coiled springs, $k = 1.0$. Thus, $d = [8 \times 5000 \times 1.75 \times 1.0/(\pi \times 65,000)]^{1/3} = 0.70$ in (1.8 cm). So the outside diameter d_o of the spring will be $d_o = d_m + 2(d/2) = d_m + d = 1.75 + 0.70 = 2.45$ in (6.2 cm). But the spring must fit a 2-in (5.1-cm) diameter hole. Hence, a smaller value of d_m must be tried.

Using $d_m = 1.5$ in (3.8 cm) and following the same procedure, we find $d = [8 \times 5000 \times 1.50 \times 1.0/(\pi \times 65,000)]^{1/3} = 0.665$ in (1.7 cm). Then $d_o = 1.5 + 0.665 = 2.165$ in (5.5 cm), which is still too large.

Using $d_m = 1.25$ in (3.2 cm), we get $d = [8 \times 1.25 \times 1.0/(\pi \times 65,000)]^{1/3} = 0.625$ in (1.6 cm). Then $d_o = 1.25 + 0.625 = 1.875$ in (4.8 cm). Since this is nearly 2 in (5.1 cm), the spring probably will be acceptable. However, the value of k should be checked.

Thus, $c = 2r_m/d = 2(1.25/2)/0.625 = 2.0$. Note that $r_m = d_m/2 = 1.25/2$ in this calculation for the value of c. Then $k = [(4 \times 2) - 1]/[(4 \times 2) - 4] + 0.615/2 = 2.0575$. Hence, the assumed value of $k = 1.0$ was inaccurate for this spring. Recalculating, $d = [8 \times 5000 \times 1.25 \times 2.0575/(\pi \times 65,000)]^{1/3} = 0.796$ in (2.0 cm). Now, $1.25 + 0.796 = 2.046$ in (5.2 cm), which is still too large.

Using $d_m = 1.20$ in (3.1 cm) and assuming $k = 2.0575$, then $d = [8 \times 5000 \times 1.20 \times 2.0575/(\pi \times 65,000)]^{1/3} = 0.785$ in (2.0 cm) and $d_o = 1.20 + 0.785 = 1.985$ in (5.0 cm). Checking the value of k gives $c = 1.20/0.785 = 1.529$ and $k = [(4 \times 1.529) - 1]/[(4 \times 1.529) - 4] + 0.615/1.529 = 2.820$. Recalculating again, $d = [8 \times 5000 \times 1.20 \times 2.820/(\pi \times 65,000)]^{1/3} = 0.872$ in (2.2 cm). Then $d_o = 1.20 + 0.872 = 2.072$ in (5.3 cm), which is worse than when $d_m = 1.25$ in (3.2 cm) was used.

It is now obvious that a practical trade-off must be utilized so that a spring can be designed to carry a 5000-lb (22,241.1-N) load and fit in a 2-in (5.1-cm) diameter hole, d_h, with suitable clearance. Such a trade-off is to use hard-drawn steel wire of greater strength at higher cost. This trade-off would not be necessary if d_h could be increased to, say, 2.13 in (5.4 cm).

Using $d_m = 1.25$ in (3.2 cm) and a clearance of, say, 0.08 in (0.20 cm), then $d = d_h - d_m - 0.08 = 2.00 - 1.25 - 0.08 = 0.67$ in (1.7 cm). Also, $c = d_m/d = 1.25/0.67 = 1.866$ and $k = [(4 \times 1.866) - 1]/[(4 \times 1.866) - 4] + 0.615/1.866$

= 2.196. The new allowable shear stress can now be found from $s_s = 8Ld_m k / \pi d^3$ = 8 × 5000 × 1.25 × 2.196/(π × 0.67³) = 116,206 lb/in² (801,212.1 kPa). Hard-drawn spring wire (ASTM A-227-47) is available with s_s = 117,000 lb/in² (806,686.6 kPa) and G = 11.5 × 10⁶ lb/in² (79.29 × 10⁹ Pa).

Hence, d = [8 × 5000 × 1.25 × 2.196/(π × 117,000)]$^{1/3}$ = 0.668 in (1.7 cm). Thus, d_o = 1.25 + 0.668 = 1.918 in (4.9 cm). Now, the spring is acceptable because further recalculations would show that d_o will remain less than 2 in (5.1 cm), regardless.

2. Compute the spring deflection
The deflection of a helical compression spring is given by $f = 64nr_m^3 L/(d^4 G) = 4\pi n r_m^2 s_s/(dGk)$, where f = spring deflection, in; n = number of coils in this spring; r_m = mean radius of spring coil, in; L, d, G, s_s, and k are as determined for the acceptable spring.

Assuming n = 10 coils, we find f = 64 × 10 × (1.25/2)³ × 5000/(0.668⁴ × 11.5 × 10⁶) = 0.341 in (0.9 cm). Or, f = 4π × 10 × (1.25/2)² × 117,000/(0.668 × 11.5 × 10⁶ × 2.196) = 0.340 (0.9 cm), a close agreement.

The number of coils n assumed for this spring is based on past experience with similar springs. However, where past experience does not exist, several trial values of n can be used until a spring of suitable deflection and length is obtained.

Related Calculations. Use this general procedure to analyze helical coil compression or tension springs. As a general guide, the outside diameter of a spring of this type is taken as (0.96)(hole diameter). The active solid height of a compression-type spring, i.e., the height of the spring when fully closed by the load, usually is nd, or (0.9) (final height when compressed by the design load).

SELECTION OF HELICAL COMPRESSION AND TENSION SPRINGS

Choose a helical compression spring to carry a 90-lb (400.3-N) load with a stress of 50,000 lb/in² (344,750.0 kPa) and a deflection of about 2.0 in (5.1 cm). The spring should fit in a 3.375-in (8.6-cm) diameter hole. The spring operates at about 70°F (21.1°C). How many coils will the spring have? What will the free length of the spring be?

Calculation Procedure:

1. Determine the spring outside diameter
Using the usual relation between spring outside diameter and hole diameter, we get d_o = 0.96d_h, where d_h = hole diameter, in. Thus, d_o = 0.96(3.375) = 3.24 in, say 3.25 in (8.3 cm).

2. Determine the required wire diameter
he equations in the previous calculation procedure can be used to determine the required wire diameter, if desired. However, the usual practice is to select the wire diameter by using precomputed tabulations of spring properties, charts of spring properties, or a special slide rule available from some spring manufacturers. The tabular solution will be used here because it is one of the most popular methods.

Table 2 shows typical loads and spring rates for springs of various outside diameters and wire diameters based on a corrected shear stress of 100,000 lb/in² (689,500.0 kPa) and a shear modulus of $G = 11.5 \times 10^6$ lb/in² (79.3×10^9 Pa).

Before Table 2 can be used, the actual load must be corrected for the tabulated stress. Do this by taking the product of (actual load, lb)(table stress, lb/in²)/(allowable spring stress, lb/in²). For this spring, tabular load, lb = (90)(100,000/50,000) = 180 lb (800.7 N). This means that a 90-lb (400.3-N) load at a 50,000-lb/in² (344,750.0-kPa) stress corresponds to a 180-lb (800.7-N) load at 100,000-lb/in² (689,500-kPa) stress.

Enter Table 2 at the spring outside diameter, 3.25 in (8.3 cm), and project vertically downward in this column until a load of approximately 180 lb (800.7 N) is intersected. At the left read the wire diameter. Thus, with a 3.25-in (8.3-cm) outside diameter and 183-lb (814.0-N) load, the required wire diameter is 0.250 in (0.635 cm).

3. Determine the number of coils required
The allowable spring deflection is 2.0 in (5.1 cm), and the spring rate per single coil, Table 2, is 208 lb/in (364.3 N/cm) at a tabular stress of 100,000 lb/in² (689,500 kPa). We use the relation, deflection f, in = load, lb/desired spring rate, lb/in, S_R; or, $2.0 = 90/S_R$; $S_R = 90/20 = 45$ lb/in (78.8 N/cm).

4. Compute the number of coils in the spring
The number of active coils in a spring is n = (tabular spring rate, lb/in)/(desired spring rate, lb/in). For this spring, $n = 208/45 = 4.62$, say 5 coils.

5. Determine the spring free length
Find the approximate length of the spring in its free, expanded condition from l in = $(n + i)d + f$, where l = approximate free length of spring, in; i = number of inactive coils in the spring; other symbols as before. Assuming two inactive coils for this spring, we get $l = (5 + 2)(0.25) + 2 = 3.75$ in (9.5 cm).

Related Calculations. Similar design tables are available for torsion springs, spiral springs, coned-disk (Belleville) springs, ring springs, and rubber springs. These design tables can be found in engineering handbooks and in spring manu-

TABLE 2 Load and Spring Rates for Helical Compression and Tension Springs*

Spring wire diameter		Outside diameter of spring coil					
		in	cm	in	cm	in	cm
in	cm	3	7.6	3.25	8.3	3.5	8.9
0.207	0.5258	113†	502.6	104	462.6	97.2	432.4
		121	211.9	93.6	163.9	74.1	129.8
0.250	0.6350	198	880.7	183	814.0	170	756.2
		270	472.8	208	364.3	163	285.5
0.283	0.7188	285	1267.7	263	1169.9	247	1098.7
		460	805.6	352	616.4	276	483.4

*After H. F. Ross, "Application of Tables for Helical Compression and Extension Spring Design," *Transactions ASME*, vol. 69, p. 727.
†First figure given is loads in lb at 100,000-lb/in² (in N at 689,500-kPa) stress. Second figure is spring rate in lb/in (N/cm) per coil, $G = 11.5 \times 10^6$ lb/in² (79.3×10^9 Pa).

facturers' engineering data. Likewise, spring design charts are available from many of these same sources. Spring design slide rules are generally available free of charge to design engineers from spring manufacturers.

SIZING HELICAL SPRINGS FOR OPTIMUM DIMENSIONS AND WEIGHT

Determine the dimensions of a helical spring having the minimum material volume if the initial, suddenly applied, load on the spring is 15 lb (66.7 N), the mean coil diameter is 1.02 in (2.6 cm), the spring stroke is 1.16 in (2.9 cm), the final spring stress is 100,000 lb/in^2 (689,500 kPa), and the spring modulus of torsion is 11.5 $\times$ 10^6 lb/in^2 (79.3 $\times$ 10^9 Pa).

Calculation Procedure:

1. Compute the minimum spring volume
Use the relation $v_m = 8fLG/s_f^2$, where v_m = minimum volume of spring, in^3; f = spring stroke, in^3; L = initial load on spring, lb; G = modulus of torsion of spring material, lb/in^2; s_f = final stress in spring, lb/in^2. For this spring, v_m = 8(1.16)(15)(11.5 $\times$ 10^6)/(100,000)^2 = 0.16 in^3 (2.6 cm^3). Note: $s_f = 2s_s$, where s_s = shear stress due to a static, or gradually applied, load.

2. Compute the required spring wire diameter
Find the wire diameter from $d = [16Ld_m/(\pi s_f)]^{1/3}$, where d = wire diameter, in; d_m = mean diameter of spring, in; other symbols as before. For this spring, $d = [16 \times 15 \times 1.02/(\pi \times 100,000)]^{1/3}$ = 0.092 in (2.3 mm).

3. Find the number of active coils in the spring
se the relation $n = 4v_m/(\pi^2 d^2 d_m)$, where n = number of active coils; other symbols as before. Thus, $n = 4(0.16)/[\pi^2(0.092)^2(1.02)]$ = 7.5 coils.

4. Determine the active solid height of the spring
The solid height $H_s = (n + 1)d$, in, or $H_s = (7.5 + 1)(0.092)$ = 0.782 in (2.0 cm). For a practical design, allow 10 percent clearance between the solid height and the minimum compressed height H_c. Thus, $H_c = 1.1H_s = 1.1(0.782)$ = 0.860 in (2.2 cm). The assembled height $H_a = H_c + f$ = 0.860 + 1.16 = 2.020 in (5.13 cm).

5. Compute the spring load-deflection rate
The load-deflection rate $R = Gd^4/(8d_m^3 n)$, where R = load-deflection rate, lb/in; other symbols as before. Thus, $R = (11.5 \times 10^6)(0.092)^4/[8(1.02)^3(7.5)]$ = 12.9 lb/in (2259.1 N/m).

The initial deflection of the spring is $f_i = L/R$ in, or f_i = 15/12.9 = 1.163 in (3.0 cm). Since the free height of a spring $H_f = H_a + f_i$, the free height of this spring is H_f = 2.020 + 1.163 = 3.183 in (8.1 cm).

Related Calculations. The above procedure for determining the minimum spring volume can be used to find the minimum spring weight by relating the spring weight W lb to the density of the spring material ρ lb/in^3 in the following manner: For the required initial load L_1 lb, $W_{min} = \rho(8fL_1G/s_f^2)$. For the required energy capacity E in · lb, $W_{min} = \rho(4ED/s_f^2)$. For the required final load L_2 lb, $W_{min} = \rho(2f_2L_2G/s_f^2)$.

The above procedure assumes the spring ends are open and are not ground. For other types of end conditions, the minimum spring volume will be greater by the following amount: For squared (closed) ends, $v_m = 0.5\pi^2 d^2 d_m$. For ground ends, $v_m = 0.25\pi^2 d^2 d_m$. The methods presented here were developed by Henry Swieskowski and reported in *Producted Engineering*.

SELECTION OF SQUARE- AND RECTANGULAR-WIRE HELICAL SPRINGS

Choose a square-wire spring to support a load of 500 lb (2224.1 N) with a deflection of not more than 1.0 in (2.5 cm). The spring must fit in a 4.25-in (10.8-cm) diameter hole. The modulus of rigidity for the spring material is $G = 11.5 \times 10^6$ lb/in² (79.3 $\times 10^9$ Pa). What is the shear stress in the spring? Determine the corrected shear stress for this spring.

Calculation Procedure:

1. Determine the spring dimensions
Assume that a 4-in (10.2-cm) diameter square-bar spring is used. Such a spring will fit the 4.25-in (10.8-cm) hole with a small amount of room to spare.

As a trial, assume that the width of the spring wire = 0.5 in (1.3 cm) = a. Since the spring is square, the height of the spring wire = 0.5 in (1.3 cm) = b.

With a 4-in (10.2-cm) outside diameter and a spring wire width of 0.5 in (1.3 cm), the mean radius of the spring coil $r_m = 1.75$ in (4.4 cm). This is the radius from the center of the spring to the center of the spring wire coil.

2. Compute the spring deflection
The deflection of a square-wire tension spring is $f = 45Lr_m^3 n/(Ga^4)$, where f = spring deflection, in; L = load on spring, lb; n = number of coils in spring; other symbols as before. To solve this equation, the number of coils must be known. Assume, as a trial value, five coils. Then $f = 45(500)(1.75)^3(5)/[(11.5 \times 10^6)(0.5)^4] = 0.838$ in (2.1 cm). Since a deflection of not more than 1.0 in (2.5 cm) is permitted, this spring is probably acceptable.

3. Compute the shear stress in the spring
Find the shear stress in a square-bar spring from $S_s = 4.8Lr_m/a^3$, where S_s = spring shear stress, lb/in²; other symbols as before. For this spring, $S_s = (4.8)(500)(1.75)/(0.5)^3 = 33,600$ lb/in² (231,663.8 kPa). This is within the allowable limits for usual spring steel.

4. Determine the corrected shear stress
Find the shear stress in a square-bar spring from $S_s = 4.8Lr_m/a^3$, where s_s = spring correction factor $k = 1 + 1.2/c + 0.56/c^2 + 0.5/c^3$, where $c = 2r_m/a$. For this spring, $c = (2 \times 1.75)/0.5 = 7.0$. Then $k = 1 + 1.2/7 + 0.56/7^2 + 0.5/7^3 = 1.184$. Hence, the corrected shear stress is $S_s' = ks_s$, or $S_s' = (1.184)(33,600) = 39,800$ lb/in² (274,411.3 kPa). This is still within the limits for usual spring steel.

Related Calculations. Use a similar procedure to select rectangular-wire springs. Once the dimensions are selected, compute the spring deflection from $f = 19.6Lr_m^3 n/[Gb^2(a - 0.566)]$, where all the symbols are as given earlier in this calculation procedure. Compute the uncorrected shear stress from $S_s = Lr_m(3a + 1.8b)/(a^2b^2)$. To correct the stress, use the Liesecke correction factor given in Wahl—*Mechanical Springs*. For most selection purposes, the uncorrected stress is satisfactory.

CURVED SPRING DESIGN ANALYSIS

Find the maximum load P, maximum deflection F, and spring constant C for the curved rectangular wire spring shown in Fig. 5 if the spring variables expressed in metric units are $E = 14,500$ kg/mm^2, $S_b = 55$ kg/mm^2, $b = 1.20$ mm, $h = 0.30$ mm, $r_1 = 0.65$ mm, $r_2 = 1.75$ mm, $L = 9.7$ mm, $u_1 = 1.7$ mm, and $u_2 = 5.6$ mm.

Calculation Procedure:

1. Divide the spring into analyzable components
Using Fig. 6, developed by J. Palm and K. Thomas of West Germany, as a guide, divide the spring to be analyzed into two or more analyzable components, Fig. 5. Thus, the given spring can be divided into two springs—a type D (Fig. 6), called system I, and a type A (Fig. 6), called system II.

2. Compute the spring force
The spring force $P = P_1 = P_{II}$. Since $(u_2 + r_2) > (u_1 + r_1)$, the spring in system II exerts a larger force. From Fig. 6 for $\beta = 90°$, $P = S\sigma_{max}/(u_2 + r_2)$, where $S =$ section modulus, mm^3, of the spring wire. Since $S = bh^2/6$ for a rectangle, $P = bh^2\sigma_{max}/[6(u_2 + r_2)]$ where $b =$ spring wire width, mm; $h =$ spring wire height, mm; $\sigma_{max} =$ maximum bending stress in the spring, kg/mm^3; other symbols as given in Fig. 5. Then $P = (1.20)(0.30)^2 \times (55)/[6(5.6 + 1.75)] = 0.135$ kg.

3. Compute the spring deflection
The total deflection of the springs is $F = 2F_1 + F_{II}$, where $F =$ spring deflection, mm, and the subscripts refer to each spring system. Taking the sum of the deflections as given in Fig. 6, we get $F = [2P/(3EI)][2K_1r_1^3(m_1 + \beta_1/2)^2 + (v_1 - u_1)^3 + K_2r_2^3(m_2 + \beta_2)^3]$, where $E =$ Young's modulus, kg/mm^2; $I =$ spring wire moment of inertia, mm^4; $K =$ correction factor for the spring from Fig. 7, where the subscripts refer to the radius being considered in the relation u/r; $m = u/r$; $\beta =$ angle

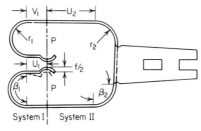

FIGURE 5 Typical curved spring. (*Product Engineering.*)

Spring type	Spring deflection	Spring force and bending stresses
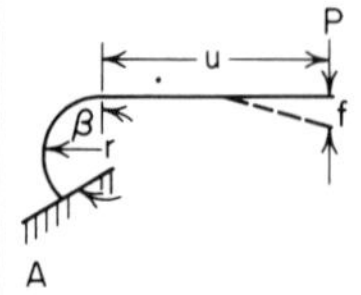 **A**	$F_1 = \dfrac{KPr^3}{3EI}(m+\beta)^3$ where $\alpha = \beta$ for finding K	When $\alpha = 0°$ to $90°$ When $\alpha = 90°$ to $180°$ $P = \dfrac{S\sigma}{u+\sin\beta}$ $P = \dfrac{S\sigma}{u+r}$ $\sigma = \dfrac{Pr(m+\sin\beta)}{S}$ $\sigma = \dfrac{Pr(m+1)}{S}$
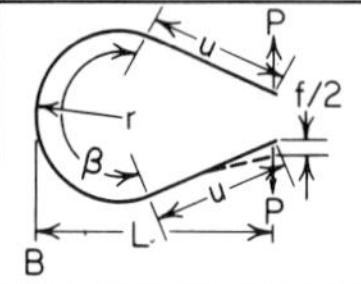 **B**	$F_2 = \dfrac{2KPr^3}{3EI}\left(m+\dfrac{\beta}{2}\right)^3$ where $\alpha = \dfrac{\beta}{2}$ for finding K	$P = \dfrac{S\sigma}{L}$
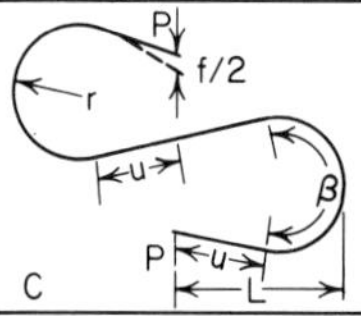 **C**	$F_3 = 2F_2 = \dfrac{4KPr^3}{3EI}\left(m+\dfrac{\beta}{2}\right)^3$ where $\alpha = \dfrac{\beta}{2}$ for finding K	$\sigma = \dfrac{PL}{S}$
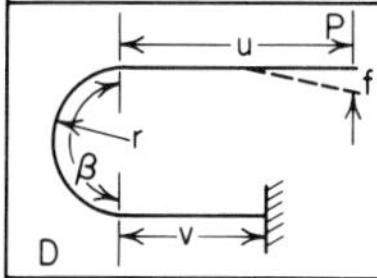 **D** **E**	$F_4 = F_5 = \dfrac{P}{3EI}\left[2Kr^3\left(m+\dfrac{\beta}{2}\right)^3 + (v-u)^3\right]$ where $\alpha = \dfrac{\beta}{2}$ for finding K	$P = \dfrac{S\sigma}{\lambda} = \dfrac{P\lambda}{S}$ <table><tr><td>First condition</td><td>Second condition</td><td>λ</td></tr><tr><td>$u \geq v$</td><td>- - -</td><td>$u+r$</td></tr><tr><td>$u < v$</td><td>$(u-v) < (u+r)$</td><td>$u+r$</td></tr><tr><td>$u < v$</td><td>$(v-u) > (u+r)$</td><td>$v-u$</td></tr><tr><td>$u = 0$</td><td>$v \leq r$</td><td>r</td></tr><tr><td>$u = 0$</td><td>$v > r$</td><td>v</td></tr></table>

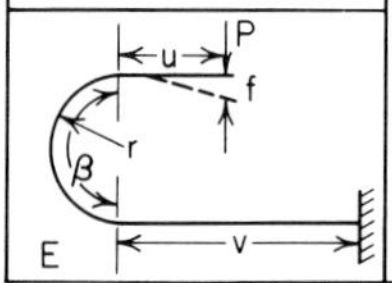

FIGURE 6   Deflection, force, and stress relations for curved springs. (*Product Engineering.*)

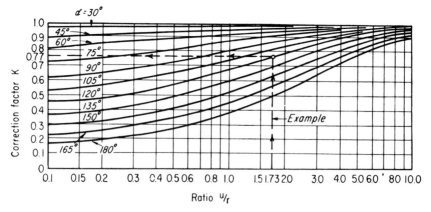

FIGURE 7   Correction factors for curved springs. (*Product Engineering.*)

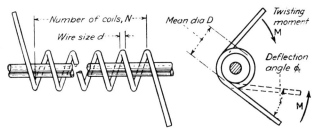

FIGURE 8   Typical torsion spring. (*Product Engineering.*)

of spring curvature, rad. Where the subscripts 1 and 2 are used in this equation, they refer to the respective radius identified by this subscript. Since $I = bh^3/12$ for a rectangle, or $I = (1.20)(0.30)^3/12 = 0.0027$ mm⁴, $F = \{[2(0.135)/[(3)(14,500)(0.0027)]\}[2(0.92)(0.65)(2.62 + 1.57)^3 + 0 + 0.94(1.75)^3(3.2 + 1.57)^3] = 1.34$ mm.

4.   Compute the spring constant
The spring constant $C = P/F = 0.135 = 0.135/1.34 = 0.101$ kg/mm.

   Related Calculations.   The relations given here can also be used for round-wire springs. For accurate results, h/r for flat springs and d_o/r for round-wire springs should be less than 0.6. The various symbols used in this calculation procedure are defined in the text and illustrations. Since the equations given here analyze the springs and do not contain any empirical constants, the equations can be used, as presented, for both metric and English units. Where a round spring is analyzed, $h = b = d_o$, where d_o = spring outside diameter, mm or in.

ROUND- AND SQUARE-WIRE HELICAL TORSION-SPRING SELECTION

Choose a round-music-wire torsion spring to handle a moment load of 15.0 lb·in (1.7 N·m) through a deflection angle of 250°. The mean diameter of the spring should be about 1.0 in (2.5 cm) to satisfy the space requirements of the design. Determine the required diameter of the spring wire, the stress in the wire, and the number of turns required in the spring. What is the maximum moment and angular deflection the spring can handle? What is the maximum moment and deflection without permanent set?

Calculation Procedure:

1.   Select a suitable wire diameter
To reduce the manufacturing cost of a spring, a wire of standard diameter should be used, whenever possible, for the spring, Fig. 8. Usual torsion-spring wire diameters and the side of square-wire springs range from 0.02 to 0.60 in (0.05 to 1.52 cm), depending on the moment the spring must carry and the angular deflection.

   Assume a wire diameter of 0.10 in (0.25 cm) and a bending stress of 150,000 lb/in² (1.03×10^9 Pa) as trial values for this spring. [Typical round-wire and square-wire torsion-spring bending stresses range from 100,000 to 200,000 lb/in² (689.5 × 10⁶ to 1.38 × 10⁹ Pa), depending on the material used in the spring.]

Compute the twisting moment corresponding to the assumed stress from $M_i = \pi d^3 S_b / 32$, where M_i = twisting moment load, lb·in; d = spring wire diameter, in; S_b = bending stress in spring, lb/in². Thus, $M_i = \pi(0.10)^3(150,000)/32 = 14.7$ lb·in (1.66 N·m). This is very close to the actual moment load of 15.0 lb·in (1.7 N·m). Therefore, the assumed spring diameter and bending stress are acceptable, thus far.

2. Compute the actual spring stress
Use the following relation to find the actual bending stress S_b lb/in² in the spring: S_b = (actual spring moment lb · in/computed spring moment, lb · in)(assumed stress, lb/in²); S_b = (15.0/14.7)(150,000) = 153,000 lb/in² (1.05 × 10⁹ Pa).

3. Check the actual vs. recommended spring stress
Enter Fig. 9 at the wire diameter of 0.10 in (0.25 cm), and project vertically upward to the music-wire curve to read the recommended bending stress for music wire as 159,000 lb/in² (1.10 × 10⁹ Pa). Since the actual stress, 153,000 lb/in² (1.05 × 10⁹ Pa), is less than but reasonably close to the recommended stress, the selected wire diameter is acceptable for the planned load on the spring. This chart and calculation procedure were developed by H. F. Ross and reported in *Product Engineering*.

4. Determine the angular deflection per spring coil
Compute the angular deflection per coil from $\phi = 360 S_b d_m / (Ed)$, where ϕ = [fj angular deflection per spring coil, degrees; d_m = mean diameter of spring, in; E = Young's modulus for spring material = 30 × 10⁶ lb/in² (206.9 × 10⁹ Pa) for spring steel; other symbols as before. Thus, by using the *assumed* bending stress in the spring, $\phi = 360(150,000)(1.0)/[(30 \times 10^6)(0.1)] = 18°$. This value is the maximum safe deflection per coil for the spring.

5. Compute the number of coils required
The number of coils n required in a helical torsion spring is $n = \phi_t$(assumed stress, lb/in²)/ϕ (actual stress, lb/in²), where ϕ_t = total angular deflection of spring, degrees; ϕ = maximum safe deflection per coil, degrees. Thus, $n = 250(150,000)/[(18)(153,000)] = 13.6$ coils; use 14 coils.

6. Determine the maximum moment the spring can handle
On the basis of the maximum recommended stress, the moment can be increased to M_i = [(maximum recommended stress, lb/in²)/(assumed stress, lb/in²)](actual moment, lb · in). Read the maximum recommended stress from Fig. 9 as 159,000 lb/in² (1.10 × 10⁹ Pa) for 0.1-in (0.25-cm) diameter music wire, as in step 3. Thus, $M_i = (159,000/150,000)(14.7) = 15.6$ lb · in (1.8 N · m).

7. Compute the maximum angular deflection
The maximum angular deflection per coil is ϕ = [(maximum recommended stress, lb/in²)/(assumed stress, lb/in²)](computed angular deflection per coil, degrees) = 159,000/150,000 × 18 = 19.1° per coil.

8. Determine the special-case moment and deflection
The maximum moment M_{max} and deflection ϕ_{max}, without permanent set, can be one-third greater than in steps 6 and 7, or M_{max} = 15.6(1.33) = 20.8 lb · in (2.4 N · m), and ϕ_{max} = 19.1(1.33) = 25.5° per coil. These maximum values allow for overloads on the spring.

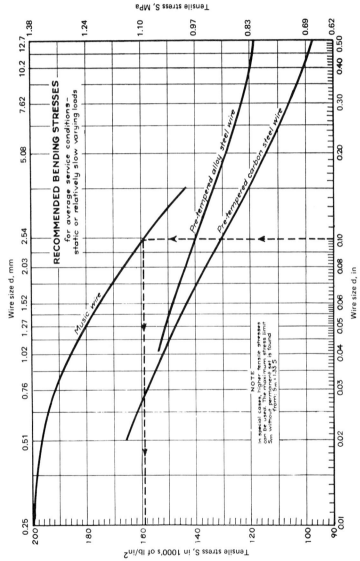

FIGURE 9 Recommended bending stresses for torsion springs. (*Product Engineering.*)

23.25

Related Calculations. Use the same procedure for square-wire helical torsion springs, but substitute the length of the side of the square for d in each equation where d appears.

TORSION-BAR SPRING ANALYSIS

What must the diameter of a torsion bar be if it is to have a spring rate of 2400 lb·in/rad (271.2 N·m/rad) and a total angle of twist of 0.20 rad? The bar is made of 302 stainless steel, which has a proportional limit in tension of 35,000 lb/in^2 (241.3 × 10^6 Pa), and G = torsional modulus of elasticity = 10^7 lb/in^2 (68.95 × 10^9 Pa). The length of the torsion bar is 26.0 in (66.0 cm), and it is solid throughout. What size square torsion bar would be required? What size equilateral triangular section would be required? What is the energy storage of each bar form?

Calculation Procedure:

1. Determine the proportional limit in shear
For stainless steel, the proportional limit S_s lb/in^2 in shear is 0.55 times that in tension, or S_s = 0.55(35,000) = 19,250 lb/in^2 (132.7 × 10^6 Pa).

2. Compute the required diameter of the bar
Use the relation $d = 2S_s l/(G\theta)$, where d = torsion-bar diameter, in; l = torsion-bar length, in; θ = total angle of twist of torsion bar, rad; other symbols as before. Thus, d = 2(19,250)(26.0)/[10^7(0.20)] = 0.50 in (1.3 cm).

3. Compute the square-bar size
Use the relation $d = 1.482S_s l/(G\theta)$, where d = side of the square bar, in. Thus d = 1.482(19,250)(26.0)/[10^7(0.2)] = 0.371 in (0.9 cm).

4. Compute the triangular-bar size
Use the relation $d = 2.31S_s l/(G\theta)$, where d = side of the triangular bar, in. Thus, d = 2.31(19,250)(26.0)/[10^7(0.2)] = 0.578 in (1.5 cm).

5. Compute the energy storage of each bar
For a solid circular torsion spring, the energy storage $e = S_s^2/(4G)$, where e = energy storage in the bar, in·lb/in^3. Thus, e = (19.250)^2/[4(10^7)] = 9.25 in·lb/in^3 (6.4 N·cm/cm^3).

For a square bar, $e = S_s^2/(6.48G)$, where the symbols are the same as before, or, e = (19,250)^2/[6.48(10^7)] = 5.71 in·lb/in^3 (3.9 N·cm/cm^3).

For a triangular bar, $e = S_s^2/(7.5G)$, where the symbols are the same as before. Or, e = (19,250)^2/[7.5(10^7)] = 4.94 in·lb/in^3 (3.4 N·cm/cm^3).

Related Calculations. Use this procedure for torsion-bar springs made of any metal. The energy-storage capacity of various springs in terms of the spring weight is as follows:

| | Energy storage of spring | |
Type of spring	in·lb/lb	N·m/kg
Leaf	300–450	74.7–112.1
Helical round-		
wire coil	700–1100	174.4–274.0
Torsion-bar	1000–1500	249.1–373.6
Volute	500–1000	124.5–249.1
Rubber in shear	2000–4000	498.2–996.4

The analyses in this calculation procedure are based·on the work of Donald Bastow and D. A. Derse and are reported in *Product Engineering.*

MULTIRATE HELICAL SPRING ANALYSIS

Determine the required spring rates, number of coils, coil clearances, and free length of two helical coil springs if spring 1 has preload of 1.2 lb (5.3 N) and spring 2 has a preload of 19.1 lb (85.0 N) in a double preload mechanism. The rod is to deflect 0.46 in (1.2 cm) before building up to the preload of 19.1 lb (85.0 N). Total deflection is to be 3.0 in (7.6 cm) with a load of 78 lb (347.0 N). The mean spring diameter d_m = 1.29 in (3.28 cm) for both springs; the wire diameter is d = 0.148 in (3.76 mm) for spring 1; d = 0.156 in (3.96 mm) for spring 2; G = 11.5 × 10⁶ lb/in² (79.3 × 10⁹ Pa) for both springs.

Calculation Procedure:

1. Determine the spring rate for each spring
The spring rate, lb/in, is R_s = (preload spring 2, lb − preload spring 1, lb)/ deflection, in, before full preload. Thus, for spring 1, R_{s1} = (19.1 − 1.2)/0.46 = 38.9 lb/in (68.1 N/cm).

For the combination of the two springs R_{st} = (78 − 19.1)/(3.0 − 0.46) = 23.1 lb/in (40.5 N/cm).

For spring 2, R_{s2} = $R_{s1} R_{st}/(R_{s1} − R_{st})$, where the symbols are the same as before. Or, R_{s2} = (38.9)(23.1)/(38.0 − 23.1) = 56.9 lb/in (99.6 N/cm).

2. Check the spring rate against the spring deflection
The deflection, in, is $f = L/R$, where L = load on the spring, lb; R = spring rate, lb/in. Thus for spring 1, f_1 = (78 − 1.2)/38.9 = 1.97 in (5.0 cm). For spring 2, $f_2 = L_2/R_{s2}$ = (78 − 19.1)/56.9 = 1.03 in (2.6 cm). For the two springs, $F_t = f_1 + f_2$ = 1.97 + 1.03 = 3.00 in (7.6 cm). This agrees with the allowable deflection of 3 in (7.6 cm) at the full load of 78 lb (347.0 N). Therefore, the computed spring rates and preloads are acceptable.

3. Compute the number of coils for each spring
The number of coils $n = Gd^4/(8d_m^3 R)$, where the symbols are as defined before. Thus, n_1 = (11.5 × 10⁶)(0.148)⁴/[8*(1.29)³(38.9)] = 8.25 coils. And n_2 = (11.5 × 10⁶)(0.156)⁴/[8(1.29)³(56.9)] = 7 coils.

4. *Compute the solid height of each spring*

Allowing one inactive coil for each end of each spring, so that the ends may be squared and ground, we find the solid height $h_s = d$(number of coils + 2). Or h_{s1} = (0.148)(8.25 + 2) = 1.517 in (3.85 cm). And h_{s2} = (0.1567)(7 + 2) + 1.404 in (3.57 cm).

5. *Determine the coil clearances*

Assume a coil clearance of 3 times the spring wire diameter. Then the coil clearance c, in, for each spring is c_1 = (3)(0.148) = 0.444 in (1.128 cm) and c_2 = (3)(0.156) = 0.468 in (1.189 cm).

6. *Compute the free length of each spring*

The free length of a helical spring = l_f = solid height + coil clearance + deflection + [preload, lb/(spring rate, lb/in)]. For spring 1, l_{f1} = 1.517 + 0.444 + 1.970 + (1.2/38.9) = 3.962 in (10.06 cm). For spring 2, l_{f2} = 1.404 + 0.468 + 1.030 + (19.1/56.9) = 3.235 in (8.22 cm).

   Related Calculations. Use this procedure for springs made of any metal. This analysis is based on the work of K. A. Flesher, as reported in *Product Engineering*.

BELLEVILLE SPRING ANALYSIS FOR SMALLEST DIAMETER

What are the minimum outside radius r_o and thickness t for a steel Belleville spring that carries a load of 1000 lb (4448.2 N) at a maximum compressive stress of 200,000 lb/in^2 (1.38 × 10^9 Pa) when compressed flat?

Calculation Procedure:

1. *Determine the spring radius ratio and the height-thickness ratio*

The radius ratio $r_r = r_o/r_i$ for a Belleville spring, where r_o = outside radius of spring, in; r_i = inside radius of spring, in = radius of hole in spring, in. Table 3 summarizes recommended values for the radius ratio for various values of the height-thickness ratio to produce the smallest diameter spring. In general, an r_r value of 1.75 usually produces a spring of suitably small size. When r_i = 1.75, Table 3 shows that the height-thickness ratio h/t with both values expressed in

TABLE 3 Design Constants for Belleville Springs*

h/t	r_o/r_i
1.00	1.25
1.25	1.50
1.50	1.75
1.75	2.00
2.00	2.50

Product Engineering.

inches is 1.5. Assume that these two values are valid, and proceed with the calculation.

2. Determine the spring outside radius

Table 4 shows the stress constant $r_o s_c / L^{0.5}$, where s_c = maximum compressive stress on the top surface at the inner edge, lb/in², Fig. 10; L = total axial load on spring, lb. For $h/t = 1.5$ and $r_r = 1.75$, the stress constant $r_o s_c / L^{0.5} = -19,050$. Solving gives $r_o = -19,050 L^{0.5} / s_c$. By substituting the given values, $r_o = 19,050(1000)^{0.5} / -200,000 = 3.01$ in (7.65 cm). The negative sign is used for the spring stress because it is a compressive stress.

3. Determine the radius of the hole in the spring

For this Belleville spring, $r_i = r_o / r_r = 3.01/1.75 = 1.72$ in (4.37 cm).

4. Compute the spring thickness

The thickness of a Belleville spring is given by $t = [s_c r_o^2 / (KE)]^{0.5}$, where K = a stress constant from Table 4; E = modulus of elasticity of the spring material, lb/in²; other symbols as before. Thus, with $E = 30 \times 10^6$ (206.9 × 10⁹ Pa), $t = [-200,000 \times 3.01^2 / (-5.6279 \times 30 \times 10^6)]^{0.5} = 0.1037$ in (2.63 mm).

5. Compute the spring height

Since $h/t = 1.5$ for this spring, $h = 1.5(0.1037) = 0.156$ in (3.96 mm).

   Related Calculations. Professor M. F. Spotts developed the analytical procedure and data presented here. His studies show that space is usually the limiting factor in spring selection, and the designer generally must determine the minimum

TABLE 4 Stress Constants for Belleville Springs*

	$r_o / r_i = 1.75$	
h/t	K	$r_r s_c / L^{0.5}$
1.00	− 3.2455	− 13,460
1.25	− 4.3734	− 16,220
1.50	− 5.6279	− 19,050
1.75	− 7.0090	− 21,970

Product Engineering.

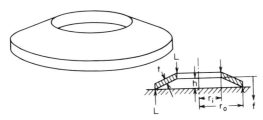

FIGURE 10 Belleville spring: appearance and dimensions. (*Product Engineering.*)

permissible outside diameter of the spring to carry a given load at a specified stress. Further, the ratio of the outside to the inside diameter for the smallest spring is about 1.75, assuming that the load spring is compressed nearly flat, which is the usual design assumption. A value of h/t of 1.5 is recommended for most spring applications. Belleville springs are used in disk brakes, the preloading of bolted assemblies, ball bearings, etc. The analysis presented here is useful for all usual applications of Belleville springs.

BELLEVILLE SPRING COMPUTATIONS FOR DISK DEFLECTION, LOAD, AND NUMBER

Compute for a Belleville spring the deflection, f, and maximum stress, S, for a single disk when $D = 3$ in (7.62 cm); $D/d = 2$; $h/t = 1.6$; $t = 0.06$ in (0.152 cm); $P = 440$ lb (1957 N).

Calculation Procedure:

1. Determine the deflection for a single-disk Belleville spring
Use the general equation for Belleville springs,

$$P = \frac{4Et^4}{(1 - \mu^2)MD^2} \frac{f}{t} \times \left[\left(\frac{h}{t} - \frac{f}{t} \right) \left(\frac{h}{t} - \frac{f}{2t} \right) + 1 \right]$$

$$4E/(1 - \mu^2) = 1.3 \times 10^8$$

where the nomenclature is as given below. In this relation, the constants M, C_1 and C_2 are dependent on the ratio D/d; values are given Table 5 and Fig. 11.

Nomenclature

$P =$ spring load, lb (N); loading and support uniformly distributed around the circumference
$S =$ maximum stress, lb/in² (kPa)
$D =$ outer diameter of disk, in (cm)
$d =$ inner diameter of disk, in (cm)
$t =$ spring thickness, in (cm)

TABLE 5 Load and Stress Constants for Disk (Belleville) Springs

D/d	M	C_1	C_2
1.2	0.295	1.015	1.043
1.4	0.455	1.07	1.13
1.6	0.57	1.123	1.22
1.8	0.64	1.17	1.30
2.0	0.69	1.225	1.38
3.0	0.785	1.424	1.738
4.0	0.80	1.60	2.06
5.0	0.79	1.775	2.374

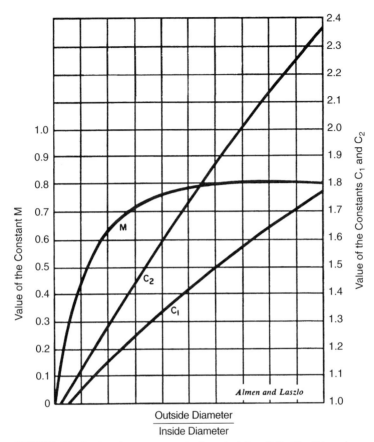

FIGURE 11 Load and stress constants for calculating Belleville disk springs. (*Product Engineering* and *Transactions of the American Society of Mechanical Engineers.*)

$$h = \text{spring free height, in (cm)}$$
$$f = \text{axial deflection, in (cm)}$$
$$E = \text{modulus of elasticity of spring material, lb/in}^2 \text{ (kPa)}$$
$$\mu = \text{Poisson's ratio; for steel } \mu = 0.3$$
$$D/d, \, h/t, \, f/t = \text{dimensionless ratios}$$
$$M, \, C_1, \, C_2 = \text{constants that depend on the ratio } D/d$$

Substituting in the general equation for Belleville springs with $E = 30 \times 10^6$ lb/in² (206.7 GPa), we find $f = 0.057$ in (0.145 cm).

2. Find the tensile stress in this single-disk Belleville spring
To find the tensile stress in a Belleville spring use the general relation,

$$S = \frac{4Et^2}{(1 - \mu^2)MD^2} \frac{f}{t} \times \left[C_1 \left(\frac{h}{t} - \frac{f}{2t} \right) + C_2 \right]$$

where the symbols are as defined above. Substituting, we find, $S = 195,000$ lb/in^2 (1.34 GPa).

Related Calculations.   The two equations above, plus the data in the table and chart allow analysis of Belleville springs used as single disks, in series, parallel, or series and parallel arrangements.

In a series arrangement the Belleville disks are stacked alternately in a bellows-like arrangement. The spring load capacity is limited by that required to produce the maximum permissible deflection in a single disk. Load capacity is not influenced by the number of disks stacked in series. The total deflection of the stack is the product of the number of disks and the deflection of an individual disk. Thus, if 6 disks were used in the above situation, the total deflection would be $6 \times 0.057 = 0.342$ in (0.868 cm). Springs loaded in series are subject to friction only at the areas of loading and of support contacts.

In the parallel arrangement, the disks are nested. In practice, it is seldom that more than three Belleville springs are placed or arranged in a nest. Load capacity is proportional to the number of disks. Friction between the surfaces of the disks will reduce the load capacity. The total deflection of two or three Belleville springs arranged in parallel equals the deflection of any individual spring in the group.

In a series and parallel arrangement combined, the deflection range of single springs, the load capacity of the assembly, and the amount of friction damping can be varied by combining series and parallel arrangements in an assembly. Using the data given for the three possible arrangements listed here, the number of disks required to accommodate a given load or deflection can be computed, along with the stress in individual disks.

This procedure is the work of Herbert Steuer, Engineer, VDI, as reported in *Product Engineering* and is based on the formulas given by J. O. Almen and A. Laszlo in "The Uniform-Section Disk Spring," reported in the *Transactions of the American Society of Mechanical Engineers.*

RING-SPRING DESIGN ANALYSIS

Determine the major dimensions of a ring spring made of material having an allowable stress of 175,000 lb/in^2 (1.21 $\times$ 10^9 Pa), $E = 29 \times 10^6$ lb/in^2 (199.9 $\times$ 10^9 Pa), a coefficient of friction of 0.12, an inside diameter of 7.0 in (17.8 cm), an outside diameter of 9.0 in (22.9 cm) or less, a taper angle of 14°, an axial load of 56 tons (50.8 t), and a deflection of not more than 8.0 in (20.3 cm).

Calculation Procedure:

1. Determine the inner-ring dimensions
For the usual ring spring, the ring height h is 15 percent of the allowable outside diameter, or $(0.15)(9.0) = 1.35$ in (3.4 cm), Fig. 12. The axial gap between the rings g is usually 25 percent of the ring height.

Compute the area of the internal ring from $A_i = L/(\pi K_c s_i)$, where A_i = area of internal ring, in^2; L = axial load on spring, lb; K_c = spring constant from Fig. 13; s_i = allowable stress in the inner ring of the spring, lb/in^2. With a coefficient of friction $\mu = 0.12$ and a taper angle of 14°, $K_c = 0.38$. Then $A_i = 56 \times 2000/[\pi(0.38)(175,000)] = 0.537$ in^2 (3.47 cm^2).

The width w_i of the inner ring is $w_i = [A_i - (h_i^2 \tan \theta)/4]/h_i$, where $\tan \theta =$

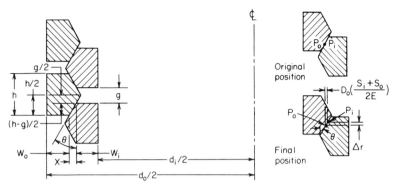

FIGURE 12   Ring-spring positions and dimensions. (*Product Engineering.*)

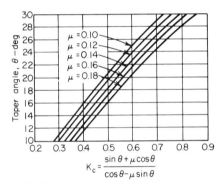

FIGURE 13   Ring-spring compression constant in terms of the taper angle for various values of the coefficient of friction. (*Product Engineering.*)

tangent of taper angle; h_i = height of inner ring, in. Thus, w_i = [0.537 − (1.35² tan 14°)/4]/1.35 = 0.314 (7.98 mm).

Use a trial-and-error process to determine the dimensions of the outer ring. Do this by assuming a cross-sectional area for the outer ring; then compute whether the outside diameter and stress meet the specifications for the spring.

Assume that A_o = 0.609 in² (3.93 cm²). Then s_o = $L/(\pi A_o K_c)$, where s_o = stress in outer ring, lb/in²; other symbols as before. So s_o = 56 × 2000/[π(0.609)(0.38)] = 154,200 lb/in² (1.06 × 10⁹ Pa). This stress is within the allowable limits.

In the usual ring spring, h_o = h_i = 1.35 in (3.4 cm) for this spring. Then, by using a relation similar to that for the inner ring, w_o = [A_o − (h_o^2 tan θ)/4]/h_o = [0.609 − (1.35² tan 14°)/4]/1.35 = 0.366 in (9.3 mm).

Find the outside diameter of the ring from d_o = d_i + 2w_i + 2w_o + (h − g) tan θ, where d_o = outside diameter of outer ring, in; d_i = inside diameter of inner ring, in; g = axial gap of rings, in = 25 percent of ring height for this spring, or 0.25(1.35) = 0.3375 in (8.57 mm). Hence, d_o = 7.0 + 2(0.314) + 2(0.366) + 1.35 − 0.3375) tan 14° = 8.613 in (21.9 cm). This is close enough to the maximum allowable outside diameter of 9 in (22.9 cm) to be acceptable. Were the value of d_o unacceptable, another value of A_o would be assumed and the calculation repeated until the stress and d_o values were acceptable.

2. Compute the number of rings required

Find the axial deflection per ring f, in, from $f = d_a[(s_i + s_o)/(2E)]$ cot θ, where d_a = mean diameter of the spring, in; E = modulus of elasticity of the spring material, lb/in². Compute $d_a = [(d_o - 2w_o) + (d_i + 2w_i)]/2 = [(8.613 - 2 \times 0.366) + (7.0 + 2 \times 0.314)]/2 = 7.755$ in (19.7 cm). Then $f = 7.755[(175,000 + 154,200)/(2 \times 29 \times 10^6)]$ cot $14° = 0.176$ in (4.47 mm). Since the axial deflection must not exceed 8 in (20.3 cm), the number of rings required = axial deflection, in/deflection per ring, in = 8.0/0.176 = 45.5, or 46 rings. Figure 14 shows the spring dimensions.

Related Calculations. Ring springs are suitable for pipe-vibration isolation, shock absorbers, plows, trench diggers, railroad couplers, etc. The recommended approximate proportions of ring springs are as follows: (1) Compressed height should be at least 4 times the deflection of the spring. (2) Ring height should be 15 to 20 percent of the ring outside diameter. (3) Spring outside diameter and height are usually as large as space permits. (4) Thin ring sections are preferred to thick ones. (5) Ring taper should be 1:4. (6) Coefficient of friction for ring springs varies from 0.10 to 0.18. (7) Allowable spring stresses are 160,000 lb/in² (1.10 × 10⁹ Pa) for nonmachined steel, 200,000 lb/in² (1.38 × 10⁹ Pa) for machined steel. For vibratory loads, the allowable stress is about one-half these values. (8) Load capacities of ring springs vary between 2 and 150 tons (1.8 and 136.1 t). (9) Spring deflections vary between 1 in (2.5 cm) and 1 ft (0.3 m). (10) The equations given above can be used for spring design or for analysis of an existing spring.

The design method given here was developed by Tyler G. Hicks and reported in *Product Engineering.*

LIQUID-SPRING SELECTION

Select a liquid spring to absorb a 50,000-lb (222,411.1-N) load with a 5-in (12.7-cm) stroke. The rod diameter is 1 in (2.5 cm). What is the probable temperature rise per stroke? Compare this spring with metal-coil, Belleville, and ring springs.

Calculation Procedure:

1. Compute the liquid volume required

Assume that the final pressure of the compressed liquid is 50,000 lb/in² (344,750 kPa) and that the liquid is compressed 18 percent on application of full load on the spring. This means that 82 percent (100 − 18) of the original volume remains after application of the load.

Compute the liquid volume required from $v = \pi S d^2/(4c)$, where v = liquid volume required, in³; S = stroke length, in; d = rod diameter, in; c = liquid compressibility, expressed as a decimal. Thus $v = \pi(5)(1)^2/[4(0.18)] = 21.8$ in³ (357.2 cm³).

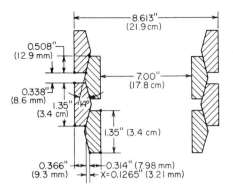

FIGURE 14 Dimensions of a typical ring spring. (*Product Engineering.*)

2. Determine the cylinder length

In a liquid spring, the cylinder inside diameter d_i is usually greater than that of the rod. Assuming an inside diameter of 1.8 in (4.6 cm) for the cylinder, we find length $= 4v/(\pi d_i^2)$, where d_i = cylinder inside diameter, in; other symbols as before. For this cylinder, length $= 4(21.8)/[\pi(1.8)^2] = 8.56$, say 8.6, in (21.8 cm).

3. Determine the cylinder dimensions

With a 1.8-in (4.6-cm) inside diameter, a 3-in (7.6-cm) outside diameter will be required, based on the usual cylinder proportions. Allowing 3.4 in (8.6 cm) for the cylinder ends and seals and 5 in (12.7 cm) for the stroke, we find that the total length of the cylinder will be $8.6 + 3.4 + 5.0 = 17.0$ in (43.2 cm).

4. Compute the cylinder temperature rise

Assume that the average friction load is 10 percent of the load on the spring, or $0.1 \times 50,000 = 5000$ lb (22,241.1 N). A friction load of 10 percent is typical for liquid springs.

The energy absorbed per stroke of the spring is $e = Fl$, where e = energy absorbed, ft · lb; F = friction force, lb; l = stroke length, ft. For this spring, $e = 5000(\frac{5}{12}) = 2085$ ft · lb (2826.9 N · m). Since 778.2 ft · lb = 1 Btu = 1.1 kJ, $e = 2085/778.2 = 2.68$ Btu (2827.6 J).

An assembly of the dimensions counted in step 3 will weigh about 35 lb (15.9 kg) and will have an average overall specific heat of 0.15 Btu/(lb ·°F) [628.0 J/(kg ·°C)]. Hence, the temperature rise per stroke will be: Btu of heat generated per stroke/[(specific heat)(cylinder weight, lb)] = 0.51°F (0.28°C) per stroke. A temperature rise of this magnitude is easily dissipated by the external surfaces of the cylinder. But a smaller liquid spring under rapidly fluctuating loads may have an excessive temperature rise. Each spring must be analyzed separately.

5. Compare the various types of springs

By using previously presented calculation procedures, Table 6 can be constructed. This table and the spring analysis given above are based on the work of Lloyd M. Polentz, Consulting Engineer. The tabulation shows that the liquid spring is the shortest and has the smallest diameter for the load in question. Figure 15 shows

TABLE 6 Performance of Four Typical Spring Types*

		Nested		
	Coil	Belleville washers	Tapered rings	Liquid
Useful range:				
Low load	1 oz (28.3 g)	20 lb (9.1 kg)	2 tons (1.8 t)	100 lb (45.4 kg)
High load	10 ton (9.1 t)	100 ton (90.7 t)	150 ton (136.1 t)	200 ton (181.4 t)
Force vs. deflection	Low to high	High	High	Medium to high
Stroke	Short to long	Short	Short to medium	Short to long
Damping ability	Low	Low	Low	Low to high
Relative cost	Low	Low	Medium	High

Product Engineering.
Note: An example: For 50,000-lb (222,411.1-N) load, 5-in (12.7-cm) stroke:

Size	Length, in (cm)	68 (172.7)	37 (94.0)	24 (61.0)	17 (43.2)
	Diameter, in (cm)	11.5 (29.5)	8 (20.3)	5 (12.7)	3 (7.6)

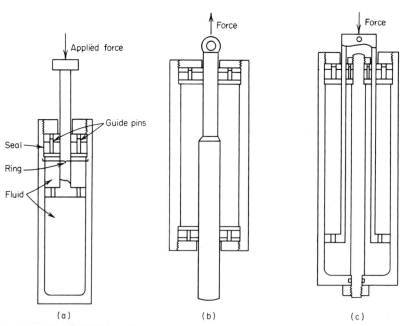

FIGURE 15 Typical liquid springs. (*a*) General design; (*b*) tension type; (*c*) long-stroke type. (*Product Engineering.*)

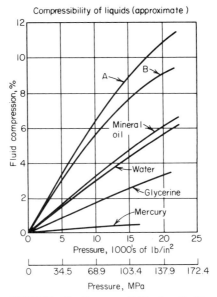

FIGURE 16 Common fluids for liquid springs are Dow-Corning type F-4029, curve A, and type 200, curve B. (*Product Engineering.*)

typical liquid springs; Fig. 16 shows the compressibility of liquids used in various liquid springs.

Related Calculations. Use the method given here to select liquid springs for applications in any of a variety of services where a large load must be absorbed. The seals at the cylinder ends must be absolutely tight. Liquid springs are best applied in atmospheres where the temperature variation is minimal.

SELECTION OF AIR-SNUBBER DASHPOT DIMENSIONS

Determine the required orifice area, peak actuator pressure, peak negative acceleration, and the time required for the stroke of a 3-in³ (49.2-cm³) capacity air snubber if the total load mass $M = 0.1$ lbs · s²/in (17.9 g · s²/cm); the snubber pressure $P_i = 100$ lb/in² (689.5 kPa); piston area $A_p = 3$ in² (19.4 cm²); initial snubber active length $S = 1.0$ in (2.5 cm); initial piston velocity $v_i = 100$ in/s (254 cm/s); piston velocity at the end of travel $v = 29$ in/s (73.7 cm/s); constant external force on snubber $F = 150$ lb (667.2 N); initial gas temperature $T_i = 530$ R (294.1 K); gas constant $R = [639.6$ in · lb/(lb ·°R)] (air); $C_D =$ orifice discharge coefficient = 0.9 dimensionless.

Calculation Procedure:

1. *Compute the snubber dimensionless parameters*
The first dimensionless parameter K_E = stored energy/kinetic energy = $P_i V_i (M v_i^2)$ = $(100)(3)/[(0.1)(100)^2]$ = 0.3. The next parameter K_F = constant external force/ initial pressure force = $F/(P_i A_p)$ = $150/[(100)(3)]$ = 0.5. The third parameter K_v = piston velocity at end of stroke/initial piston velocity = v/v_i = $20/100$ = 0.20.

2. *Determine the actual value of the orifice parameter*
The parameter K_w = initial orifice flow/initial displacement flow = $w_i/(\rho A_p v_i)$, where ρ = gas density, lb/in^3. Figure 17 gives values of K_w for K_F = 0 and K_F = 1.0. However, K_F for this snubber = 0.5. Therefore, it is necessary to interpolate between the charts for K_F = 0 and K_F = 1.0.
Interpolate by constructing a chart, Fig. 18, using values of K_w read from each chart in Fig. 17. Thus, when K_F = 1, K_v = 0.2, K_E = 0.3, K_w = 0.295. After the curve is constructed, read K_w = 0.375 for K_f = 0.5.

3. *Compute the true flow through the orifice*
The true initial flow rate w_i, lb/s = $K_w[P_i/(RT_i)]A_p v_i$, where all the symbols are as defined earlier. Thus, w_i = $(0.375)\{100/[(639.6)(530)]\}(3)(100)$ = 0.0332 lb/s (15.1 g/s).

4. *Compute the required orifice area*
Use the equation $A_o = w_i/P_i C_D\{(kg/RT_i)[2/(k+1)](k+1)/(k-1)\}^{0.5}$, where k = 1.4; g = 32.2 ft/s^2 (9.8 m/s^2); other symbols as before. Thus, A_o = 0.0332/ $[100(0.9)]\{(1.4 \times 32.2/639.6 \times 530)[2/(1.4+1)](1.4+1)/(1.4-1)\}^{0.5}$ = 0.016 in^2 (0.103 cm^2).

5. *Determine the maximum pressure at the end of the stroke*
Read, from Fig. 19, $K_{p,max}$ = 10.2 for K_F = 0.5, K_w = 0.375. Then the true P_{max} at end of stroke = $K_p P_i$ = 10.2(100) = 1020 lb/in^2 (7032.9 kPa).

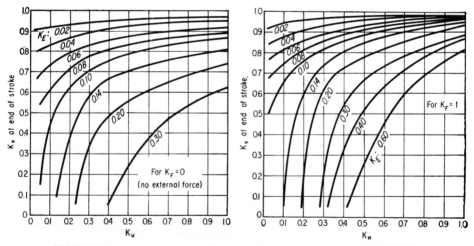

FIGURE 17 Impact velocity vs. orifice flow (dimensionless) for air snubber. (*Product Engineering.*)

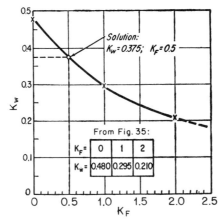

FIGURE 18 Cross plot for an air snubber. (*Product Engineering.*)

6. Determine the maximum acceleration of the piston

For an air snubber, $K_{a,\text{max}} = K_F - K_p = 0.5 - 10.2 = -9.7$. Also, the maximum acceleration $a_{\text{max}} = K_a P_i A_p / M = (-9.7)(100)(3.0)/0.1 = -29,100$ in/s^2 (-739.1 m/s^2).

7. Determine the approximate travel time of the piston

The travel time for the piston is $t = K_t S / v_i$, where t = travel time, s. Or, $t = 0.95 \times (1.0)/100 = 0.0095$ s, assuming $K_t = 0.95$.

    Related Calculations. The equations in this procedure were developed by Tom Carey and T. T. Hadeler, and are based on these assumptions: (1) They apply only to a piston-orifice-type dashpot; (2) the piston is firmly stopped at the end of the stroke and does not rebound, oscillate, or bounce; (3) friction is zero; (4) the external force is constant; (5) $k = 1.4$, which means that the equations are valid for air, hydrogen, nitrogen, oxygen, and any other gas having a specific-heat ratio of about 1.4; (6) the contained air or gas is ideal; (7) compression is adiabatic; (8) flow through the bleed orifice is critical (a valid assumption except when the dashpot initial pressure is atmospheric, as in screendoor snubbers). When actual friction exists, there is a slight increase in the value of K_t. Figure 20 shows a simplified design of a typical air snubber.

DESIGN ANALYSIS OF FLAT REINFORCED-PLASTIC SPRINGS

A large shaker unit in a vibrating screen system is supported on a series of six steel leaf springs, each a cantilever 6 in (15.2 cm) wide by 0.125 in (3.18 mm) thick. How thick should a single epoxy-glass leaf spring of the same width be if it is to replace the composite steel spring? The cantilever is 30 in (76.2 cm) long with a 24-in (60.9-cm) free length; maximum deflection = 0.375 in (9.53 mm); axial load per spring = 2500 lb (11,120.6 N); safety factor = 8; $E = 4.5 \times 10^6$

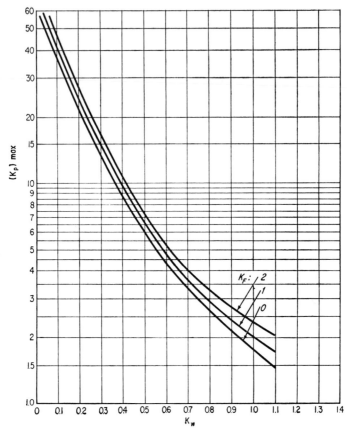

FIGURE 19   Maximum pressure at impact (dimensionless). (*Product Engineering.*)

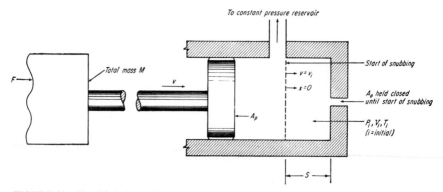

FIGURE 20   Simplified air-snubber design. (*Product Engineering.*)

lb/in^2 (31.0×10^9 Pa) for the plastic spring; ultimate flexure strength = 100,000 lb/in^2 (689,475.7 kPa).

Calculation Procedure:

1. Compute the spring thickness for minimum bending stress

The equation for the thickness giving the minimum bending stress is $t = [4Ll^2/(wE)]^{1/3}$, where t = spring thickness, in; L = axial load on spring, lb; l = spring free length, in; w = spring width, in; E = modulus of elasticity of spring material, lb/in^2. For this spring, $t = [4 \times 2500 \times 24^2/(6 \times 4.5 \times 10^6)]^{1/3} = 0.598$, say 0.6 in (1.5 cm).

2. Determine the total combined stress in the beam

The maximum combined stress $s_B = (3Et/l^2 + 6L/wt^2)f$, where f = spring deflection, in; other symbols as before. Thus, $s_B = \{e \times 4.5 \times 10^6 \times 0.6/[24^2 + 6 \times 2500/(6 \times 0.6^2)]\} \times 0.375 = 7875$ lb/in^2 (54,290 kPa).

The total stress in the spring $s_T = s_B + L/A$, where A = spring cross-sectional area. Thus, $s_T = 7875 + 2500/(6 \times 0.6) = 8570$ lb/in^2 (59,082 kPa). The allowable stress = ultimate flexure strength, lb/in^2/factor of safety = 100,000/8 = 12,500 lb/in^2 (86,175.5 kPa). Since $s_T < 12,500$ lb/in^2 (86,175.5 kPa), the dimensions of the spring are satisfactory.

3. Check the critical buckling stress of the spring

To prevent buckling of the spring, the following must hold: $L/A \leq \pi^2 Et^2/36^2 = s_{CR}$, or $2500/(6 \times 0.6) \leq \pi^2 \times 4.5 \times 10^6 \times 0.6^2/(3 \times 24^2) = 694$ lb/in^2 (4784 kPa) < 9240 lb/in^2 (63,701 kPa). Hence, the spring dimensions are satisfactory.

4. Determine whether the computed thickness gives adequate stiffness

For a plastic spring to have a stiffness equal to a steel spring having n leaves, the plastic spring should have a thickness of $t = t_s(nE_s/E)^{1/3}$, where t_s and E_s refer to the thickness, in, and modulus of elasticity, lb/in^2, respectively, of the steel spring; other symbols as before. Thus, $t = 0.125[6 \times 30 \times 10^6/(4.5 \times 10^6)]^{1/3} = 0.43$ in (1.1 cm). Since $t = 0.60$ in (1.5 cm), as computed in step 1, the plastic spring is slightly too stiff.

5. Check the thickness required for equivalent thickness

Using the equations for s_B and s_T from step 2, and the equation for s_{CR} from step 3, compute the respective stresses for values of t less than, and greater than, 0.43. Thus

t		s_B		Q/A		s_T		s_{CR}	
in	cm	lb/in^2	kPa	lb/in^2	kPa	lb/in^2	kPa	lb/in^2	kPa
0.500	1.270	8,143	56,144.0	833	5,743.5	8,976	61,889.5	6,400	44,128.0
0.600	1.524	7,875	54,296.2	695	4,792.0	8,570	59,090.2	9,240	63,709.8
0.625	1.588	7,900	54,468.6	666	4,592.1	8,566	59,062.6	10,000	68,947.6

Plot the results as in Fig. 21. This plot clearly shows that $t = 0.43$ in (1.09 cm) gives $s_T < 12,500$ lb/in^2 (86,175 kPa), and $Q/A < s_{CR}$. Hence, this thickness is satisfactory.

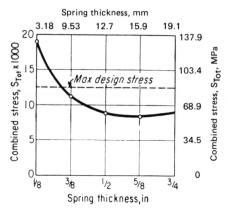

FIGURE 21 Combined stress in a plastic spring. (*Product Engineering.*)

6.   *Determine whether a thinner spring can be used*
A thinner spring will save money. From Fig. 21, $t = 0.375$ in (9.53 mm) gives an s_T value well below the maximum design stress, and the actual spring stress is one-third the critical buckling stress. If tests on a 0.375-in (9.53-mm) thick spring show no serious disruption of harmonic operation, then specify the thinner material to lower the cost. Otherwise, use the thicker 0.43-in (1.09-cm) spring.

   Related Calculations.   Use this procedure for unidirectional, cross-plied, or iso-tropic-ply plastic springs. Obtain the allowable stress for the spring from the plastic manufacturer. The method given here is the work of L. A. Heggernes, reported in *Product Engineering.*

LIFE OF CYCLICALLY LOADED MECHANICAL SPRINGS

What is the probable life in cycles of a Belleville spring under a bending load if it is made of carbon steel having a Rockwell hardness of C48?

Calculation Procedure:

1.   *Determine the spring material tensile strength*
Enter Fig. 22 at the Rockwell hardness C48, and project vertically upward to read the tensile strength of the carbon-steel spring material as 235,00 lb/in² (1620 MPa).

2.   *Compute the actual stress in the spring*
Using the spring dimensions and the equations presented in the Belleville spring calculation procedure, compute the actual stress in the spring. For the spring in question, the actual stress is found to be 150,000 lb/in² (1034 MPa). This is $150,000(100)/235,000 = 63.8$, say 64, percent of the spring material tensile strength.

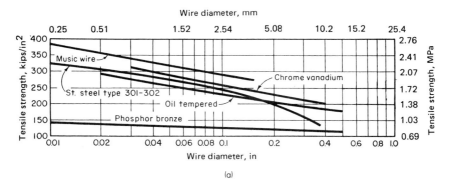

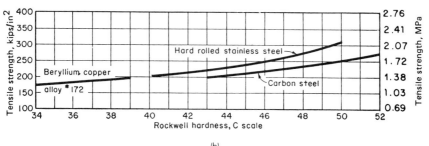

FIGURE 22 (*a*) Tensile strength of spring wire; (*b*) tensile strength of spring strip. (*Product Engineering.*)

3. Estimate the spring cycle life

Enter the upper part of Table 7 for springs in bending. This tabulation shows that at a stress of 65 percent of the tensile strength, the spring will have a life between 10,000 and 100,000 stress cycles. Actual test of the spring caused failure at about 100,000 cycles.

Related Calculations. Use this procedure for helical torsion springs, cantilever springs, wave washers, flat springs, motor springs, helical compression and extension springs, torsion bars, and Belleville springs. Be sure to enter the proper portion of Table 7 when finding the approximate number of repetitive stress cycles. The method presented here is the work of George W. Kuasz and William R. Johnson and is reported in *Product Engineering.*

SHOCK-MOUNT DEFLECTION AND SPRING RATE

Determine the maximum probable acceleration, the shock isolator deflection, and the isolator spring rate for a 25-lb (11.3-kg) piece of electronic equipment which drops from a 24-in (61.0-cm) high tailgate of a truck onto a concrete road. The product lands on one corner point and should be considered as rigid steel for analysis purposes. In its carton, the load will be supported by 16 shock isolators.

TABLE 7   Design Stresses for Springs*

No. of repetitive stress cycles	Maximum design stress (percent of the tensile strength shown in charts)
Design stress for springs in bending†	
10,000	80
	65‡
100,000	53
1,000,000	50
10,000,000	48
Design stress for springs in torsion§	
10,000	45
	35‡
100,000	35
1,000,000	33
10,000,000	30

*Product Engineering.
†For example, helical torsion springs, Bellevilles, cantilever springs, wave washers, flat springs, and motor springs.
‡For stainless-steel and phosphor-bronze materials. Tests show that such materials have low yield points.
§For example, helical compression springs, helical extension springs, and torsion bars.

Calculation Procedure:

1. Compute the acceleration of the load

Use the relation $g = (72/t)(h)^{0.5}$, where g = load acceleration in g[1 g = 32.2 ft/ s^2 (9.8 m/s²) at sea level]; t = shock-rise time, ms, from Table 8; h = drop height, in. From Table 8, t = 2 ms for rigid steel making pint contact with concrete. Then $g = 72/2 \times (24)^{0.5} = 176.5g$.

TABLE 8   Typical Value for Shock-Time Rise

Condition	Shock-time rise, ms	
	Flat face	Point
Rigid steel against concrete	1	2
Rigid steel against wood or mastic	2–3	5–6
Steel or aluminum against compact earth	2–4	6–8
Steel or aluminum against sand	5–6	15
Product case against mud	15	20
Product case against 1-in (2.5-cm) felt	20	30

Note: Mass of struck surface is assumed to be at least 10 times the striking mass. Point contact with spherical radius of 1 in (2.5 cm).

2. *Compute the isolator deflection*

Use the relation $d = 2h/(g - 1)$, where d = isolator deflection in; other symbols as before. For this load, $d = 2 \times 24/(176.5 - 1) = 0.273$ in (6.93 mm).

3. *Compute the required specific spring rate for the isolator*

Use the relation $K = g/d$, where D = isolator specific spring rate, lb/(in · lb). Thus $K = 176.5/0.273 = 646$ lb/(in · lb) [254.3 N/(N · cm)].

4. *Determine the required spring rate per isolator*

With n shock isolators, the required spring rate, lb/in per isolator is $k = KW/n$, where W = weight of part, lb; n = number of isolators used in the carton. Thus, $k = (646)(25)/16 = 1020$ lb/in (182.2 kg/cm).

Related Calculations.   Some of the largest stock loads encountered by equipment occur during transportation. Thus, vertical accelerations on the body of a 2-ton (1.8-t) truck traveling at 30 mi/h (13.4 m/s) on good pavement range from 1 to 2 g, with a rise time of 10 to 15 ms. Higher speeds, rougher roads, stiffer truck springs, and careless driving all decrease the rise time and thus double or triple the acceleration loads.

The highest acceleration forces in railroad freight cars occur during humping, when the impact loads on a product container may range from 4.5 to 28 g.

For most components that are sensitive to shock, suppliers include maximum safe acceleration loads in engineering data. Maximum allowable loads on vacuum tubes are 2 to 5 g; relays may withstand higher accelerations, depending on the type and direction of the acceleration. Transistors have low mass and good rigidity and are highly resistant to shock when properly supported. Ball-bearing races may be indented by the balls; sleeve bearings are usually much more resistant to shock.

The function of a shock mount is to provide enough protection to avoid damage under expected conditions. But overdesign can be costly, both in the design of the product and in the shock-mount components. Underdesign can lead to failures of the shock mount in service and possible damage to the product. Therefore, careful design of shock mounts is important. The method presented here is the work of Raymond T. Magner, reported in *Product Engineering.*

MECHANICAL AND ELECTRICAL BRAKES

BRAKE SELECTION FOR A KNOWN LOAD

Choose a suitable brake to stop a 50-hp (37.3-kW) motor automatically when power is cut off. The motor must be brought to rest within 40 s after power is shut off. The load inertia, including the brake rotating member, will be about 200 lb·ft² (82.7 N·m²); the shaft being braked turns at 1800 r/min. How many revolutions will the shaft turn before stopping? How much heat must the brake dissipate? The brake operates once per minute.

Calculation Procedure:

1. Choose the type of brake to use
Table 1 shows that a shoe-type electric brake is probably the best choice for stopping a load when the braking force must be applied automatically. The only other possible choice—the eddy-current brake—is generally used for larger loads than this brake will handle.

2. Compute the average brake torque required to stop the load
Use the relation $T_a = Wk^2 n/(308t)$, where T_a = average torque required to stop the load, lb·ft; Wk^2 = load inertia, including brake rotating member, lb·ft², n = shaft speed prior to braking, r/min; t = required or desired stopping time, s. For this brake, $T_a = (200)(1800)/[308(40)] = 29.2$ lb·ft, or 351 lb·in (39.7 N·m).

3. Apply a service factor to the average torque
A service factor varying from 1.0 to 4.0 is usually applied to the average torque to ensure that the brake is of sufficient size for the load. Applying a service factor of 1.5 for this brake yields the required capacity = 1.5(351) = 526 in·lb (59.4 N·m).

TABLE 1 Mechanical and Electrical Brake Characteristics

Type of brake	Typical characteristics
Block	Wooden or cast-iron shoe bearing on iron or steel wheel; double blocks prevent bending of shaft; used where economy is prime consideration; leverage 5:1
Band	Asbestos fabric bearing on metal wheels; fabric may be reinforced with copper wire and impregnated with asphalt; bands are faced with wooden blocks; used where economy is a major consideration; leverage 10:1
Cone	Friction surface attached to metal cone; popular for cranes; coefficient of friction = 0.08 to 0.10; useful for intermittent braking applications
Disk	Have one or more flat braking surfaces; effective for large loads; continuous application
Internal-shoe	Popular for vehicles where shaft rotation occurs in both directions; self-energizing, i.e., friction makes shoe follow rotating brake drum; capable of large braking power
Eddy-current	Used for flywheels requiring quick braking and where large kinetic energy of rotating masses precludes use of block brakes because of excessive heating
Electric, shoe-type	Used where automatic application of brake is required as soon as power is turned off; spring-activated brake shoes apply the braking action
Electric, friction-disk type	Best for duty cycles requiring a number of stops and starts per minute; may have one or multiple disks

4. Choose the brake size
Use an engineering data sheet from the selected manufacturer to choose the brake size. Thus, one manufacturer's data show that a 16-in (40.6-cm) diameter brake will adequately handle the load.

5. Compute the revolutions prior to stopping
Use the relation $R_s = tn/120$, where R = number of revolutions prior to stopping; other symbols as before. Thus, $R_s = (40)(1800)/120 = 600$ r.

6. Compute the heat the brake must dissipate
Use the relation $H = 1.7FWk^2(n/100)^2$, where H = heat generated at friction surfaces, ft $\cdot$ lb/min; F = number of duty cycles per minute; other symbols as before. Thus, $H = 1.7(1)(200)(1800/100)^2 = 110,200$ ft $\cdot$ lb/min (2490.2 N $\cdot$ m/s).

7. Determine whether the brake temperature will rise
From the manufacturer's data sheet, find the heat dissipation capacity of the brake while operating and while at rest. For a 16-in (40.6-cm) shoe-type brake, one manufacturer gives an operating heat dissipation $H_o = 150,000$ ft $\cdot$ lb/min (3389 5 N $\cdot$ m/s) and an at-rest heat dissipation of $H_v = 35,000$ ft $\cdot$ lb/min (790.9 N $\cdot$ m/s).

Apply the cycle time for the event; i.e., the brake operates for 400 s, or 40/60 of the time, and is at rest for 20 s, or 20/60 of the time. Hence, the heat dissipation of the brake is $(150,000)(40/60) + (35,000)(20/60) = 111,680$ ft $\cdot$ lb/min (2523.6 N $\cdot$ m/s). Since the heat dissipation, 111,680 ft $\cdot$ lb/min (2523.6 N $\cdot$ m/s), exceeds the heat generated. 110,200 ft $\cdot$ lb/min (2490.2 N $\cdot$ m/s), the temperature of the brake will remain constant. If the heat generated exceeded the heat dissipated, the brake temperature would rise constantly during the operation.

Brake temperatures high than 250°F (121.1°C) can reduce brake life. In the 250 to 300°F (121.1 to 148.9°C) range, periodic replacement of the brake friction surfaces may be necessary. Above 300°F (148.9°C), forced-air cooling of the brake is usually necessary.

Related Calculations. Because electric brakes are finding wider industrial use, Tables 2 and 3, summarizing their performance characteristics and ratings, are presented here for easy reference.

The coefficient of friction for brakes must be carefully chosen; otherwise, the brake may "grab," i.e., attempt to stop the load instantly instead of slowly. Usual values for the coefficient of friction range between 0.08 and 0.50.

The methods given above can be used to analyze brakes applied to hoists, elevators, vehicles, etc. Where Wk^2 is not given, estimate it, using the moving parts of the brake and load as a guide to the relative magnitude of load inertia. The method presented is the work of Joseph F. Peck, reported in *Product Engineering.*

MECHANICAL BRAKE SURFACE AREA AND COOLING TIME

How much radiating surface must a brake drum have if it absorbs 20 hp (14.9 kW), operates for half the use cycle, and cannot have a temperature rise greater than 300°F (166.7°C)? How long will it take this brake to cool to a room temperature of 75°F (23.9°C) if the brake drum is made of cast iron and weighs 100 lb (45.4 kg)?

Calculation Procedure:

1. Compute the required radiating area of the brake
Use the relation $A = 42.4 hpF/K$, where A = required brake radiating area, in^2; hp = power absorbed by the brakes; F = brake load factor = operating portion of use cycle; K = constant = Ct_r, where C = radiating factor from Table 4, t_r = brake temperature rise, °F. For this brake, assuming a full 300°F (166.7°C) temperature rise and using data from Table 4, we get $A = 42.4(20)(0.5)/[(0.00083)(300)] =$ 1702 in^2 (10,980.6 cm^2).

2. Compute the brake cooling time
Use the relation $t = (cW \ln t_r)/(K_c A)$, where t = brake cooling time, min; c = specific heat of brake-drum material, Btu/(lb·°F); W = weight of brake drum, lb; t_r = drum temperature rise, °F; ln = log to base e = 2.71828; K_c = a constant varying from 0.4 to 0.8; other symbols as before. Using $K_c = 0.4$, $c = 0.13$, $t =$ (0.13 × 100 ln 300)/[(0.4)(1702)] = 0.1088 min.

Related Calculations. Use this procedure for friction brakes used to stop loads that are lifted or lowered, as in cranes, moving vehicles, rotating cylinders, and similar loads.

TABLE 2 Performance Characteristics of Electric Brakes

Brake type	Operational mode		Design characteristics					Brake functions performed					On-off duty-cycling capability
	On-off	Continuous	Torque adjustment	Torque-control range	Wear adjustment	Residual drag	Heat dissipation	Instant stop	Cushioned stop	Retard (drag)	Hold	Failsafe brake	
Magnetic particle	Yes	Yes	Electrical	Wide	Nonwearing	High	Limited	No	Yes	Yes	Yes	No	Limited by heat-dissipation capability to low-inertia loads
Eddy-current, air-cooled	Yes	Yes	Electrical	Wide	Nonwearing	Moderate to low	Good	No	Yes	Yes	No	No	Limited to long time cycles
Eddy-current, water-cooled	Yes	Yes	Electrical	Wide	Nonwearing	High to moderate	Excellent	No	Yes	Yes	No	No	Limited to long time cycles
Single-disk friction, electrically actuated	Yes	Yes	Electrical	Wide	Self-compensating	None	Excellent	Yes	Yes	Yes	Yes	No	Excellent—up to several hundred stops per minute
Multidisk friction, electrically actuated, direct-acting	Yes	No	Electrical	Moderate		Low	Limited	Yes	Yes	No	Yes	No	Same as comparable size electric motor: 12 stops per minute (maximum)
Multidisk friction, electrically actuated, indirect-acting	Yes	No	Mechanical	Limited	Mechanical			Yes	Semisoft	No	Yes	No	Same as comparable size electric motor: 12 stops per minute (maximum)
Multidisk friction, spring-actuated	Yes	No	Mechanical	Limited	Mechanical	Low	Limited	Yes	Semisoft	No	Yes	Yes	Same as comparable size electric motor: 12 stops per minute (maximum)
Shoe brake, spring-actuated	Yes	No	Mechanical	Limited	Mechanical	None	Good	Yes	Semisoft	No.	Yes	Yes	Generally not over 3 stops per minute without derating

TABLE 3 Representative Range of Ratings and Dimensions for Electric Brakes

Brake type	hp (W)	Torque, maximum, lb·ft (N·m)	Shaft speed, maximum, r/min	Diameter, in (cm)	Length, in (cm)	Inertia of rotating member, lb·ft² (N·m²)
Magnetic particle brakes	$\frac{1}{60}$-25 (14.9–18,643)	0.6–150 (0.8–203.4)	1,000–2,000	2–10 (5.1–25.4)	2–6 (5.1–15.2)	1.5×10^{-4}–0.27 (6.2×10^{-5}–0.11)
Eddy-current brakes:						
Air cooled	$\frac{3}{4}$-75 (559.3–55,928)	5–1,740 (6.8–2,359)	2,000–900	6½–24¾ (16.5–62.9)	9½–43½ (24.1–110.5)	0.12–100 (0.05–41.3)
Water-cooled	40–800 (29,828–596,560)	130–4,600 (176.3–6,237)	1,800–1,200	14¾–36½ (37.5–92.7)	18½–43 (47.0–109.2)	8.5–725 (3.5–299.6)
Friction disk brakes:						
Single-disk, electrically actuated	$\frac{1}{60}$-200 (14.9–149,140)	0.17–700 (0.23–949.1)	10,000–1,800	1½–15¼ (3.8–38.7)	1½–4½ (3.8–11.4)	0.000125–3 (0.000052–1.2)
Multiple-disk, electrically actuated	¼–2,000 (186.4–1,491,400)	3–15,000 (4.1–20,337)	5,000–750	2¼–21 (5.7–53.3)	2–8 (5.1–20.3)	Up to 90 (Up to 37.2)
Multiple-disk, spring actuated	¼–2,000 (186.4–1,491,400)	4–7,500 (5.4–10,169)	5,000–1,200	4–29 (10.2–73.7)	2½–16½ (6.4–41.9)	
Shoe brakes, spring-actuated	1–2,500 (745.7–1,864,250)	3–10,000 (4.1–13,558)	10,000–1,200	2–28 (5.1–71.1)	4½–12 (11.4–30.5)	0.023–485 (0.010–200.4)

24.5

TABLE 4 Brake Radiating Factors

Temperature rise of brake		Radiating factor C
°F	°C	
100	55.6	0.00060
200	111.1	0.00075
300	166.7	0.00083
400	222.2	0.00090

BAND BRAKE HEAT GENERATION, TEMPERATURE RISE, AND REQUIRED AREA

A construction-industry hoisting engine is to be designed to lower a maximum load of 6000 lb (2724 kg) using a band brake, Fig. 1, having a 48-in (121.9-cm) drum diameter and a drum width of 8-in (20.3-cm). The brake band width is 6-in (15.2-cm) with an arc of contact of 300°. Maximum distance for lowering a load is 200 ft (61 m). The cycle of the engine is 1.5 min hoisting and 0.75 min lowering, with 0.3 min for loading and unloading. If a temperature rise of 300°F (166.5°C) is permitted, determine the heat generated, radiating surface required, and the actual temperature rise of the drum if the brake is made of cast iron and weighs 600 lb (272.4 kg).

Calculation Procedure:

1. *Determine the heat generated, equivalent to the power developed*
Use the relation, $H_g = (wd/T_L)(33,000)$, where H_g = heat generated during lowering of the load of w, lb (kg), hp (kW); d = distance the load is lowered, ft (m); T_L = time for lowering, minutes. Substituting for this hoisting engine, we have, $H_g = (6000)(200)/(0.75)(33,000) = 48.48$ hp (36.2 kW).

2. *Find the load factor for the brake*
The load factor, q, for a brake = $T_L/(T_H + T_L + T_{LU})$, where T_H = hoisting time, minutes; T_{LU} = time to load and unload, minutes; other symbols as before. Substituting, $q = (0.75)/(1.5 + 0.75 + 0.3) = 0.294$.

3. *Compute the required radiating area for the brake*
Use the relation, $A_R = 42.4(q)(H_g)/Ct_r$, where A_R = required radiating area, in² (cm²); Ct_r = brake radiating factor from Table 4. Assuming a temperature rise of 300°F (166.7°C), and substituting, $A_R = (42.4)(0.294)(48.5)/0.25 = 2418.3$ in² (15,601.9 cm²).

It is necessary to assume a temperature rise when analyzing a brake because the rise is a factor in the computation. Without such an assumption the required radiating area cannot be determined.

Using the given dimensions for this brake, we can find the area of the drum from $A_d = 2(\pi)(48 \times 8) - (\pi)(48 \times 6)(300°/360°) = 1657.9$ in² (10,696.1 cm²). Since the required radiating area is 2418 in²., as computed above, the excess area required will be $2418 - 1658 = 760$ in² (4903.2 cm²). This area can be provided

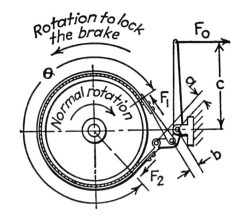

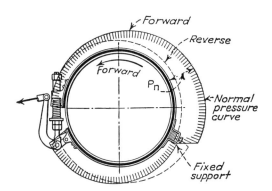

FIGURE 1 (*a*) Self-locking band brake. (*b*) Pressure variation along the surface of a band brake.

by the brake flanges and web. To be certain that sufficient area is available for heat radiation, check the physical dimensions of the brake flanges and webs to see if the needed surface is present.

4. Find the temperature rise during brake operation

The actual temperature of the brake drum will vary slightly above and below the assumed 300°F (166.7°C) temperature rise during operation. The reason for this is because heat is radiated during the whole cycle of operation but is generated only during the lowering cycle, which is 29.4 percent of the total cycle.

The temperature change of the drum during the braking operation will be $T_r = [1/(778)(W_r)(c)][Wh - Ct_r A_r m(778)]$, where c = specific heat of the drum material = 0.13 for cast iron; m = lowering time in minutes; other symbols as given earlier. Substituting, $T_r = [1/(778)(600)(0.13)][(6000 \times 200) - (0.25)(2418)(0.75)(778)] = 14°F$ (7.78°C). Thus, the drum temperature can range about 14°, or about 7° above and 7° below the average operating temperature of 300°F (148.9°C).

Related Calculations.   The actual temperature attained by a brake drum, and the time required for it to cool, cannot be accurately calculated. But the method given here is suitable for preliminary calculations. In the final design of a new brake, heating should be checked by a proportional comparison with a brake already known to give good performance in actual service.

An approximation of the time required for a brake to cool can be made using the relation given in step 2 of the previous procedure. Note that the value of K selected in that relation will directly influence brake cooling time. Thus, a lower value chosen for K will increase the estimated cooling time while a higher value will decrease the time. For safety reasons, engineers will often select lower K values so the brake will be given more time to cool, or will be provided with a larger capacity cooling mechanism.

This procedure is the work of Alex. Vallance, Chief Designer, Reed Roller Bit Co. and Venton L. Doughtie, Professor of Mechanical Engineering, University of Texas.

DESIGNING A BRAKE AND ITS ASSOCIATED MECHANISMS

Design a hoist for the building industry to lift a cubic yard (0.76 m³) of concrete at the rate of 200 ft/min (61 m/min). A cubic yard of concrete weighs approximately 400 lb (1816 kg) and the bucket weighs 1250 lb (568 kg). Since the hoist may be used for other construction purposes, the design capacity will be 6000 lb (2725 kg). There will be no counterweights in this hoist and the cable drum will be connected to a 1750-r/min electric motor through a reduction gear train. Lowering will be controlled by manual operation of a brake. This brake must automatically hold the load at any position when the motor is not driving the hoist. Figure 2 shows the proposed arrangement of parts and the limiting dimensions of this hoist control. It has been proposed that a spring-loaded band brake be used that will be manually released by the operator during lowering of the load. An overrunning clutch is to be provided to disengage the brake automatically when the torque from the motor through the gear train is sufficient to raise the load. Determine (*a*) the dimensions *q* and *n*, Fig. 2, for maximum self-energization without self-locking if the coefficient of friction, μ, varies between 0.20 and 0.50—the wide range is selected to cover possible changes in service due to unavoidable entrance of small amounts of water, oil, or dirt into the brake; (*b*) the spring force required to ensure that the brake will not slip when μ is between 0.20 and 0.50; (*c*) the force exerted by the operator when the rated load first starts to lower under normal conditions; i.e., when $\mu = 0.35$; (*d*) the minimum width of brake lining; and (*e*) the maximum lowering speed for a reasonable wear life.

Calculation Procedure:

1. Determine the actuating arm dimensions
The force F_1 and F_2 must act as shown, Fig. 2, for the braking torque to oppose the load torque. This brake will be self-energizing when the F_1 moment tends to

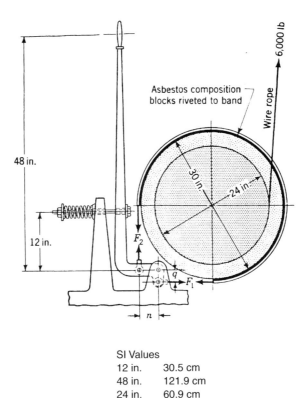

FIGURE 2 Band brake used in hoisting application.

SI Values

12 in.	30.5 cm
48 in.	121.9 cm
24 in.	60.9 cm
6000 lb	2724 kg

apply the brake, that is q is less than n, and becomes self-locking when $F_1 q \geq F_2 n$. The appropriate equation is

$$\frac{F_1}{F_2} = e^{\mu\theta}$$

and the critical design condition will be when F_1/F_2 is a maximum, $\mu = 0.50$. Thus,

$$\frac{F_1}{F_2} = \frac{n}{q} = e^{\mu\theta} = e^{0.50\times(3/2)\pi} = 10.55$$

This unit will be most compact when q is as small as possible. The strengths of the pin and the lever will be the major factors in determining the minimum dimension for q. However, if q is estimated to be 1.5 in (3.8 cm), it can be seen that there is not enough space in which to place the lever, as shown, when $n = 10.55q$. Therefore, under the specified conditions, it is not practical to use self-energization, and the proposed design will be modified to make $q = 0$, as shown in Fig. 3. Choose the dimension n as 2.0 in (5.08 cm).

2. *Find the required spring force*

Maximum spring force will be required when the coefficient of friction is a minimum, that is when $\mu = 0.20$. For this case,

$$\frac{F_1}{F_2} = e^{\mu\theta} = e^{0.20\times(3/2)\pi} = 2.57$$

and

$$F_1 = 2.57F_2$$

$$T = (F_1 - F_2)\frac{D}{2}$$

and

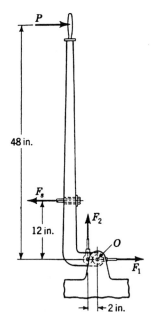

SI Values

48 in.	121.9 cm
12 in.	30.5 cm
2 in.	5.1 cm

FIGURE 3 Lever for band brake actuation.

$$T = F_{cable} \frac{D_{cable\ drum}}{2} = 6000 \times 24/2 = 72{,}000 \text{ lb} \cdot \text{in.} \quad (8136 \text{ N} \cdot \text{m})$$

$$\frac{D}{2} = 30/2 = 15 \text{ in} \quad (38.1 \text{ cm})$$

Solving for F_1, we find

$$F_1 = F_2 + 4800 \text{ lb}$$

Solving the two equations above simultaneously, we find

$$F_2 = 3060 \text{ lb} \quad (13{,}611 \text{ N})$$

and

$$F_1 = 7860 \text{ lb} \quad (34{,}961 \text{ N})$$

$$\Sigma M_O = 0$$

$$F_s \times 12 - F_2 \times 2 = F_s \times 12 - 3060 \times 2 = 0$$

Solving for the spring force we find it to be 510 lb (2268 N).

3. *Compute the operating force for rated load under normal conditions*
The operator must push the lever to the right to lower the load. When $\mu = 0.35$,

$$\frac{F_1}{F_2} = e^{\mu\theta} = e^{0.35 \times (3/2)\pi} = 5.20$$

and

$$F_1 = 5.20F_2$$

Solving the equations for force simultaneously, we find

$$F_1 = 5940 \text{ lb} \quad (26421 \text{ N}) \qquad F_2 = 1140 \text{ lb} \quad (5071 \text{ N})$$

Then,

$$\Sigma M_O = 0$$

$$-P \times 48 + F_s \times 12 - F_2 \times 2 = 0$$

$$-P \times 48 + 510 \times 12 - 1140 \times 2 = 0$$

Solving for P, we find $P = 80$ lb (356 N). This force is too large for an operator to exert and the situation becomes even worse when the hoist is lightly loaded.

For example, if the load is considered to be negligible, the force required to release the brake will be $510/4 = 127.5$ lb (567 N). A reasonable design solution would be to use a compound lever, Fig. 4, in place of the single lever we have been analyzing. A force of 30 lb (133 N) will be satisfactory for releasing the brake when the load is negligible.

Based on the analysis above, we shall consider 2 in (5.08 cm) as the value for the distance from the point of spring attachment to the pin B on lever 2, Fig. 4. It

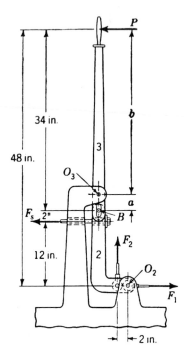

SI Values
48 in.	121.9 cm
34 in.	86.4 cm
2 in.	5.1 cm
12 in.	30.5 cm

FIGURE 4 Compound lever for band brake.

should be noted that the operator must now pull the lever to the left to release the brake. Taking moments we have:

$$\Sigma M_{O_2} = 0$$

$$-F_B \times 14 + F_s \times 12 = -F_B \times 14 + 510 \times 12 = 0$$

from which
$$F_B = 437 \text{ lb}$$

$$\Sigma M_{O_3} = 0$$

$$Pb - F_B a = 30 \times (34 - a) - 437a = 0$$

from which
$$a = 2.18 \text{ in}$$

and
$$b = 34 - 2.18 = 31.82 \text{ in}$$

The operating force for rated load with the compound lever under normal conditions is:

$$\Sigma M_{O_2} = 0$$

$$-F_B \times 14 + F_s \times 12 - F_2 \times 2 = -F_B \times 14 + 510 \times 12 - 1{,}140 \times 2 = 0$$

from which

$$F_B = 274 \text{ lb}$$

$$\Sigma M_{O_3} = 0$$

$$P \times 31.82 - F_B \times 2.18 = P \times 31.82 - 274 \times 2.18 = 0$$

Solving for P, we find it to be 18.8 lb (83.6 N), which is satisfactory.

4. Determine the required width of the brake lining

Design data specifically applicable to band brakes are not available. Hence, we must use the information available for shoe brakes given in Table 5. The main difficulty is that the pressure distribution for band brakes is much less uniform than for shoe brakes. Hence, data based on projected area must be used with caution. A conservative procedure will be to calculate the band width by limiting the maximum value of the pressure, p, and the product of the pressure and the velocity, pV, to those given for shoe brakes, based on the projected area of the brake. The maximum pressure will be limited to 100 lb/in^2 (689 kPa).

The basic relation for band width, b, is

$$b = \frac{F}{pR} = \frac{F}{pD/2}$$

When the hoist is used to raise concrete, the lowering load will be essentially the weight of the bucket and adhering concrete. But, since the hoist will be sold for general use, the brake should have a reasonable wear life with the rated load of 6000 lb (2724 kg) under normal conditions, i.e., with $\mu = 0.35$. The maximum pressure will be at the F_1 end of the lining, Fig. 2. Since F_1 was determined for these conditions, in the calculations for operating force, to be 5940 lb (26,421 N),

TABLE 5 Coefficients of Friction and Allowable Pressures

Material	μ	p, lb/in^2	kPa
Asbestos in rubber compound, on metal	0.3–0.4	75–100	516.8–689.0
Asbestos in resin binder, on metal:			
Dry	0.3–0.4	75–100	516.8–689.0
In oil	0.10	600	
Sintered metal on cast iron:			4134.0
Dry	0.20–0.40	400	
In oil	0.05–0.08		2756.0

$pV \leq 30{,}000$ for continuous application of load and poor dissipation of heat.
$pV \leq 60{,}000$ for intermittent application of load, comparatively long periods of rest, and poor dissipation of heat.
$pV \leq 84{,}000$ for continuous application of load and good dissipation of heat, as in an oil bath.

$$b = \frac{5940}{100 \times {}^{30}\!/\!_2} = 3.96 \text{ or } 4 \text{ in } (10.2 \text{ cm})$$

5. Compute lowering velocity of the brake

The brake will be in almost continuous use for relatively long periods of time; therefore, pV will be limited to 30,000. Thus

$$V = \frac{pV}{p} = \frac{30,000}{100} = 300 \text{ ft/min } (91.4 \text{ m/min})$$

The load velocity corresponding to the brake sliding velocity of 300 ft/min (91.4 m/min) will be

$$V_{\text{load}} = 300 \times {}^{24}\!/\!_{30} = 240 \text{ ft/min } (73.2 \text{ m/min})$$

Summarizing this design we have the following: Brake lever—compound lever, tight side of band to pivot—see Fig. 4; Spring force = 510 lb (2268 N); release force = 18.8 lb (80.1 N) with rated load of 6000 lb (2724 kg) and μ = 0.35; lining width = 4 in (10.2 cm); lowering velocity = 240 ft/min (91.4 m/min) with rated load of 6000 lb (2724 kg).

Related Calculations. The primary function of a brake is to slow, and stop, the rotation of a shaft. No matter where a brake is used, it will have the stopping of the rotation of a shaft as its primary function.

Thus, brakes used in hoists, such as this one, elevators, motor vehicles, aircraft landing gears, etc., all stop, or slow, a rotating shaft. Energy absorbed by the brake during its stopping or slowing action is dissipated as heat. In some applications, such as motor vehicles and aircraft, the brake is usually outdoors where there is an infinite heat sink to absorb the dissipated heat. But in other applications, such as passenger elevators, the brake may be indoors where heat dissipation is not as certain because the heat sink may be restricted by enclosures, heating systems, etc. Hence, careful analysis of the brake operating temperature is necessary.

Three factors governing brake performance are (1) the pressure between the brake shoe and drum; (2) the coefficient of friction of the brake-shoe lining material; (3) the heat dissipating capacity of the brake. Each of these must be checked carefully before accepting a final brake design.

Brakes may also be used to position a part at rest or prevent an unwanted reversal of the direction of rotation of a shaft. With the greater attention to environmental aspects of machine design today, asbestos brake linings are receiving intense study because of the nature of this material. Asbestos is used in several brakes types —shoe, band, and disk—it is not used in hydrodynamic brakes. The major disadvantage of the hydrodynamic brake is that it cannot stop motion entirely and a shoe or band brake is required to stop the motion and hold the member in position. A hydrodynamic brake is essentially a fluid coupling with the output rotor stationary so the coupling operates with 100 percent slip at all times. Water is generally used as the fluid.

Disk brakes are becoming more popular every year. They are used on automobiles, airplanes, bicycles, trains, and trucks. Almost all the newer designs have the anti-locking feature which prevents accidents from locked brakes. Advantages cited for disk brakes are increased braking surface and better heat dissipation.

This procedure is the work of Richard M. Phelan, Associate Professor of Mechanical Engineering, Cornell University.

INTERNAL SHOE BRAKE FORCES AND TORQUE CAPACITY

An internal shoe brake of the type shown in Fig. 5 has a diameter of 12 in (30.5 cm). The actuating forces F are equal and the shoes have a face width of 1.5 in

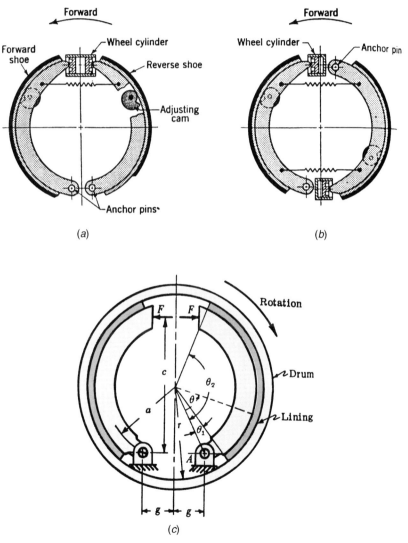

FIGURE 5 (*a*) Internal shoe brake with single actuating cylinder. (*b*) Dual actuating cylinders. (*c*) Mathematical relations for an internal shoe brake.

(3.8 cm). For a coefficient of friction of 0.3 and a maximum permissible pressure of 150 lb/in^2 (1033.5 kPa), with $\theta_1 = 0$, $\theta_2 = 130°$, $\theta_m = 90°$, $a = 5$ in (12.7 cm), and $c = 9$ in (22.9 cm), determine the value of the actuating forces F and the brake torque capacity.

Calculation Procedure:

1. Find the moment of the frictional forces about the brake-shoe pivot

The moment of the frictional forces, M_f, about the right-hand pivot of the brake is given by

$$M_f = \frac{fp_m wr}{\sin \theta_m} \int_0^{\theta_2} (\sin \theta)(r - a \cos \theta) \, d\theta = \frac{fp_m wr}{\sin \theta_m} \left[r - r \cos \theta_2 - \frac{1}{2} a \sin^2 \theta_2 \right]$$

$$= 3400 \text{ in} \cdot \text{lb} \quad (384.2 \text{ N} \cdot \text{m})(0.03)(150)(1.5)(6)[6 + 6(0.643) - 2.5(0.766)^2]$$

where f = coefficient of friction; p_m = maximum permissible pressure on the shoe, lb/in^2 (kPa); w = band width, in (cm); r = brake radius, in (cm); other symbols as given in Fig. 5c.

2. Compute the moment of the normal forces about the brake-shoe pivot

The moment, M_n, of the normal forces about the right-hand pivot is given by

$$M_n = \frac{p_m wra}{\sin \theta_m} \int_0^{\theta_2} \sin^2 \theta \, d\theta = \frac{p_m wra}{\sin \theta_m} \left[\frac{1}{2} \theta_2 - \frac{1}{4} \sin 2\theta_2 \right]$$

$$= 9300 \text{ in} \cdot \text{lb} \ (1050.9 \text{ N} \cdot \text{m})$$

$$F = (M_n - M_f)/c = (9300 - 3400)/9 = 656 \text{ lb} \quad (2917.9 \text{ N})$$

where the symbols are as given earlier and in the procedure statement.

3. Determine the brake torque capacity

The brake torque capacity of the right shoe is

$$T = fp_m wr^2 \left(\frac{\cos \theta_1 - \cos \theta_2}{\sin \theta_m} \right) = 4000 \text{ lb} \quad (17,792 \text{ N})$$

For the left shoe, $T = 1860$ in $\cdot$ lb (210.2 N $\cdot$ m) based on $p'_m = 69.7$ lb/in^2 (480.2 kPa) from

$$p'_m = \frac{Fcp_m}{M_n + M_f}$$

4. Find the brake torque capacity

The total torque, or brake torque, capacity = $4000 + 180 = 5860$ in $\cdot$ lb (662.2 N $\cdot$ m).

Related Calculations. Internal brake shoes are popular in a variety of applications including vehicles of various types, hoisting machinery, etc. Figures 5a and 5b show vehicle application with hydraulic cylinders for actuation of the internal shoes. Disk brakes are more popular today in automobile applications because they have a number of advantages over internal shoe brakes.

This procedure is the work of Allen S. Hall, Jr., Professor of Mechanical Engineering, Purdue University; Alfred R. Holowenko, Professor of Mechanical Engineering, Purdue University; and Herman G. Laughlin, Associate Professor of Mechanical Engineering, Purdue University.

ANALYZING FAILSAFE BRAKES FOR MACHINERY

Determine the torque that a brake must handle when the power to be dissipated by the brake is 5 hp (3.7 kW) at 3600 r/min. The duty cycle (number of times the brake is used) is frequent. Show how to compute the brake torque, energy absorbed per stop, heat produced by the brake, temperature rise from brake-generated heat, average brake-disk temperature, peak temperature of the brake, and the service life of the brake.

Calculation Procedure:

1. Find the torque the brake must handle

The torque a brake must handle is given by $T = 5250PK/s$, where T = torque brake handles, lb·ft (N·m); P = hp absorbed (kW); K = a safety factor which varies between 1.5 and 5 as the duty cycle of the brake increases; s = brake rotation speed, rpm. Substituting for this brake using a value of $K = 5$ because this brake is used frequently, $T = 5250(5)(5)/3600 = 36.46$ lb·ft (49.4 N·m).

This equation may be the only means available for brake selection, especially if the actual load torques are difficult to calculate and the inertia of the system cannot be determined.

Torque can also be computed from $T = WR^2 \, \Delta s/308t$, where W = weight of body stopped, lb (kg); R = radius of gyration, ft (m); Δs = change in speed of the brake, rpm. Note that this expression takes into account the torque necessary to overcome the inertia of the system. However, it is important to note that frictional torque is not considered in this relationship.

2. Show how to compute the energy absorbed per brake stop

In general, the major factor influencing brake service life is the energy absorbed per stop and the resulting heat produced in the friction material. For a caliper brake, the energy absorbed by the brake per stop is: $E_s = WR^2N^2/5872$, where E_s = energy absorbed per stop, ft·lb (N·m); R = radius of gyration, ft (m); N = number of turns to stop. This expression, as the second one in step 1, uses the inertia of the system. Hence, the designer faces the necessity of either determining, by computation, the inertia of the system, or having it supplied by the machine builder.

3. Find the heat produced in the brake

The heat produced in the brake by the energy absorbed is found from $H = E_s/778.3$, where H = heat absorbed per stop, Btu (kJ).

4. Compute the temperature rise of the brake

The temperature rise resulting from the heat produced in the brake is $T_r = H/0.12W$, where T_r = temperature rise per stop, °F (°C).

5. Calculate the average disk temperature per stop

Use the relation $T_d = HT_a/2.25A$, T_d = disk temperature per stop, °F (°C); T_a = average disk temperature, °F (°C); A = disk surface area, ft² (m²). In this expression the factor 2.25 is the cooling index used when the disk is stationary during cooling—the usual condition following an emergency stop. However, if the disk rotates during cooling, a factor of 4.5 should be used instead.

6. Compute the peak temperature of the brake

The peak temperature of the brake is $T_p = T_d + 0.5T_r$, where all the temperatures are either in °F or (°C).

7. Find the brake service life

To find the service life, L, in number of stops for a brake, use the relation, $L = (1.98 \times 10^6)ZY/E_s$, where the service factor, Z, is found from standard curves available from brake manufacturers and friction-material suppliers; Y = total friction material volume, in³ (cm³).

Related Calculations.   Failsafe brakes are "opposite-mode" devices that activate when a machine is off and disengage when the machine is on. These brakes store energy that is released to apply the brake when the machine's power supply is either turned off intentionally or lost through an equipment malfunction. In this manner failsafe brakes provide a reliable method for automatically stopping a potentially dangerous machine. Figure 6 compares brake characteristics for five different types of failsafe brakes.

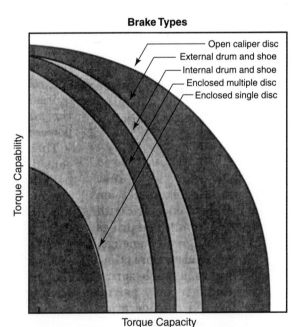

Brake Types

- Open caliper disc
- External drum and shoe
- Internal drum and shoe
- Enclosed multiple disc
- Enclosed single disc

Torque Capability

Torque Capacity

FIGURE 6 Comparison of the operating characteristics of five different types of failsafe brakes. (*Machine Design.*)

Failsafe brake characteristics are difficult to compare quantitatively because they depend on many operating variables such as load, weight, speed, and environment. The dimensionless curves in Fig. 6 are useful, however, in positioning the various failsafe brakes available. These curves indicate general trends for important operating characteristics, engagement, and disengagement methods.

High torque capability is generally associated with a proportionately low energy capacity for each brake. These torque and energy limits are primarily due to the rate at which the brake dissipates heat and the ability of the brake material to withstand high temperatures.

The most common type of failsafe brake is the energy-absorbing or dynamic brake that decelerates a rotating shaft or other machine part until it comes safely to rest. A second type, called a parking or static brake, holds the position of a moving part after it has been stopped. Both types apply braking force in the absence of normal equipment power.

Failsafe brakes are used in almost any type of equipment that contains moving parts. Applications include shears, punch presses, machine tools, and other manufacturing equipment. Public conveyances using these brakes include trains and elevators. Large pieces of farm and construction machinery use failsafe brakes along with missile launchers, antenna drives, and other aerospace equipment.

The brakes used in these heavy-duty applications typically are large devices capable of dissipating high levels of kinetic energy to stop machines quickly. However, not all failsafe brakes are used on large equipment. Because of OSHA restrictions, failsafe brakes are being used increasingly on consumer-operated power equipment such as garden tractors, lawn mowers, and golf carts. Failsafe brakes are also being used on the mechanical portions of electronic equipment, such as computers and medical equipment.

In any failsafe brake, the release element that overcomes the force of brake engagement may consist of a hydraulic, pneumatic, or electrical system. Friction surfaces are molded phenolics, copper, brass, ceramics in a sintered matrix, sintered iron, or sintered brass. Maple blocks have also been used.

Electric, hydraulic, and pneumatic release systems have high torque capability and response speed. Pneumatic release systems are often used because they are relatively inexpensive, safe, clean, and simple to install and maintain. Mechanical release systems provide the lowest torque and speed capabilities and are used only in a few limited applications.

The equations and data in this procedure are the work of Herbert S. Peterson, President, Simplatrol Products; Jack W. Moss, Chief Engineer, Wichita Clutch; and Roger W. Eisbrener, Marketing Manager, Formsprag-Gerbing, as reported in *Machine Design* magazine. SI values were added by the handbook editor.

SECTION 25

HYDRAULIC AND PNEUMATIC SYSTEMS DESIGN

DETERMINING RESPONSE TIME OF PILOT-OPERATED SOLENOID-ENERGIZED SPOOL VALVES IN HYDRAULIC SYSTEMS

A pilot-operated solenoid-energized spool control valve in a hydraulic system has the dimensions, operating pressures, and performance given in Table 1. Pilot supply pressure is 100 lb/in² (689 kPa); main supply pressure is 500 lb/in² (3445 kPa). Find the maximum velocity of this valve, its acceleration time, and the total response time. Next, using the same dimensions and main operating pressure, find the same unknowns when the pilot pressure is made equal to the main operating pressure i.e., 500 lb/in² (3445 kPa). As a further modification, a small actuating piston is placed at each end of the main spool, Fig. 3, to increase the longitudinal velocity for a given pilot-fluid flow rate. Trial and error would normally be used to calculate the most effective diameter for the actuating piston. In this procedure we will use a diameter $d_x = 1.4$ in (3.56 cm) for this small actuating piston. If the dimensions and operating pressures are unchanged, analyze the valve for the same unknowns as above.

TABLE 1   Dimensions and Operating Conditions*

	Pilot Spool	Main spool
Diameter, in	$d = 0.25$	$D = 2.5$
Mass, lb-sec²/in	$m = 0.0002$	$M = 0.05$
Stroke, in	$s = 0.375$	$S = 1.5$
Land length, in	—	$L = 6.0$
Radial clearance, in	—	$C = 0.0003$
Coefficient of friction	—	$F_R = 0.04$
Solenoid force, lb (initial; final)	$F_{SOL} = 1; 8.5$	—
Back pressure, lb/in²	—	$p_B = 20$
Supply pressure, lb/in²	$p = 100$	$P = 500$
Differential pressure, lb/in²	$\Delta p = 70$ (approx)	$\Delta P = 450$ (approx)
Port area, in²	$a_0 = 0.05$	$A_M = 1.2$
Flow coefficient	$f = 0.6$	$f = 0.6$
Viscosity, CP	$\mu = 80$	$\mu = 80$
Density, lb-sec²/in⁴	0.000,085	0.000,085

*SI values given in calculation procedure.

Calculation Procedure:

1. Compute the axial force on the main spool of this valve
The forces acting on the main spool at maximum velocity are: Pilot backpressure, p_B; viscous damping force, D_V; and radial jet force P_{rad}, Fig. 2. From the equation,

$$P_{as} = 2F_r f A_M \Delta P$$

where the symbols are as given, Table 2. Then, $P_{ax} = 2(0.04)(0.6)(1.2)(450) = 26$ lb (115.6 N). converting to pressure by dividing by the area of the main spool valve end, we have 26/4.9 = 5.3 lb/in² (36.5 kPa).

2. Compute the combined hydrodynamic resistance of the valve
Provisionally, estimate that D_V is equivalent to 3.2 lb/in² (22 kPa) and P_B = pilot-valve backpressure = 20 lb/in² (138.8 kPa). The combined hydrodynamic resistance is then the sum of: Radial pressure, lb/in² (kPa) + Viscous drag, lb/in² (kPa) + Pilot-valve backpressure, lb/in² (kPa). Or combined hydrodynamic resistance = 5.3 + 3.2 + 20 = 28.5 lb/in² (196.4 kPa).

3. Calculate the pilot-valve flow rate
The pilot-valve pressure differential, delta P = 100 − 28.5 = 71.5 lb/in² (492.6 kPa). Hence, the valve flow rate is, using the equation below

$$q = f a_o \sqrt{\frac{2\Delta p}{\rho}}$$

where q = flow, in³/s (mL/s); f = flow factor, dimensionless, ranging from 0.55 to 0.70 depending on valve type; a_o = cross-sectional area, in² (cm²); of the min-

TABLE 2 Valve Symbology*

		Pilot	Actuating piston	Main valve
Spool Dimensions and Mass	Diameter, in	d	d_x	D
	Cross-sectional area, in²	—	a_p	A_s
	Mass, lb-sec²/in	m	—	M
	Stroke: Intermediate	x	—	—
	Full	s	S^a	S
	Engagement (length in contact), in	—	l_p	—
	Land length, in (total)	—	—	L
	Spool-to-bore radial clearance, in	C	C	C
Solenoid Forces	Initial, lb	A	—	—
	Gradient, lb/in	B	—	—
	Final, lb	F_{SOL}	—	—
	Ratio, A/B	r	—	—
Drag Forces	Back pressure, psi	p_B	p_B	p_B
	Viscous drag, lb (or psi)	—	d_V	D_V
	Radial jet, lb	—	P_{rad}	P_{rad}
	Coefficient of friction	—	F_R	F_R
	Axial jet, lb	—	P_{ax}	P_{zx}
	Acceleration force, lb	$F = ma$	—	$F = Ma$
Oil pressure, flow, and port size	Pressure: Supply, psi	p	p	P
	Pilot downstream, psi	p_1	p_1	p_1
	Differential, psi	Δp	Δp	ΔP
	Port area, in²(effective orifice)	a_0	—	A_M
	Flow coefficient (0.55 to 0.70)	f	f	f
	Viscosity, centipoise	μ	μ	μ
	Oil density, lb-sec²/in⁴	ρ	ρ	ρ
	Flow rate, in³/sec	q	q	—
	Oil velocity, in/sec (through port)	—	—	V_0
	Oil mass flow, lb-sec/in	—	—	M_f
	Oil-jet deflection angle, deg	—	—	a
Valve response	Acceleration time, sec	t_a	T_a	T_a
	Shifting velocity, in/sec	v_p	v	V
	Shifting time, sec (after energization)	t	T	T

*SI values given in calculation procedure.

imum port opening—usually the drilled port hole; $\Delta p = p - p_1$ = differential pressure, lb/in² (kPa) measured across the pilot inlet and outlet ports; p = fluid mass density, lb-s²/in⁴, normally 0.000085 for oil. Substituting, q = 0.6 $(0.5)[(2)(71.5)/0.000085)]^{0.5}$ = 40 in³/s (656 mL/s), using a value of f = 0.6 for this valve.

4. Determine the maximum velocity of the main spool and the viscous damping force
The maximum velocity of the main spool, Fig. 1, V = (flow rate, in³/s/(area of spool end, in²) = 40/4.9 = 8.2 in/s (20.8 cm/s). Knowing the velocity, we can find the damping force, D_V, from

$$D_V = \frac{D\pi LV\mu}{C \times 6.9 \times 10^6}$$

where D = spool diameter, in (cm); L = length of spool lands, in (cm); V = main spool velocity, in/s (cm/s); mu = absolute viscosity, centipoise; C = spool-to-bore radial clearance, in (cm). If the temperature varies more than 30 to 50 degrees, it is nearly impossible to compute the viscous resistance. Substituting, D_V = $2.5\pi(6)(8.2)(80)/(0.0003)(6.9 \times 10^6)$ = 3.05 lb/in² (21 kPa). Thus, the provisional estimate of D_V = 3.2 was close enough (within 4.9 percent) and recalculation is not necessary.

5. Find the accelerating pressure and acceleration time of the spool
The forces acting upon the spool during acceleration are: p_R, P_{ax}, D_V, and F, where $F = Ma$. Assuming a mean value for initial port opening A_M = 0.4 in² (2.58 cm²), then from

$$P_{as} = 2fA_M\Delta P \cos \alpha$$

where α normally varies from 70 degrees at initial opening to 90 degrees at full opening. In calculations, use the axial jet pressure during initial opening, and the axial component of radial pressure during the remainder of travel. Substituting, P_{ax} = 2(0.6)(0.4)(450)(0.26) = 56 lb (248.1 N). Then 56/4.9 = 11.4 lb/in² (78.5 kPa), α = 75 deg; cos α = 0.26.

Viscous drag will be the average: D_V = 3.2/2 = 1.6 lb/in² (11 kPa). Backpressure is still p_B = 20 lb/in² (137.8 kPa). So the total is 11.4 + 1.6 + 20 = 33 lb/in² (227.4 kPa).

Therefore, accelerating pressure = 100 − 33 = 67 lb/in² (461.6 kPa). Converting to force, we have 67 (4.9) = 328 lb (1441.2 N). The acceleration time, t_a s = MV/F = 0.05 (8.2)/328 = 0.0013 s.

6. Determine the main spool displacement and the energization time interval
The displacement of the main spool during the acceleration period is negligible, being less than 1 percent of the total stroke. Time for the total stroke of 1.5 in (3.8 cm) is 1.5/8.2 = 0.182 s, and the time interval from energization of the solenoid to completion of the main valve stroke, T = 0.190 s.

7. Analyze the valve with the higher pilot pressure
Much larger flow will pass through the pilot valve because of the higher pressure. Maximum velocity period: P_{ax} = 5.3 lb/in² (36.5 kPa), the same as before; D_V = 7.7 lb/in² (53.1 kPa)—a higher estimate, proportional to the anticipated velocity;

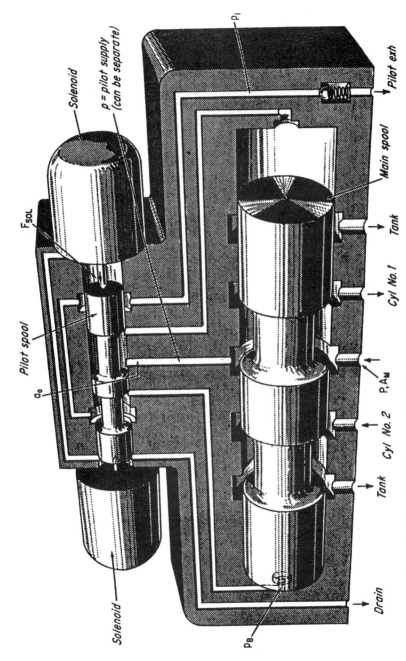

FIGURE 1 Typical solenoid-energized pilot-operated spool valve.

25.5

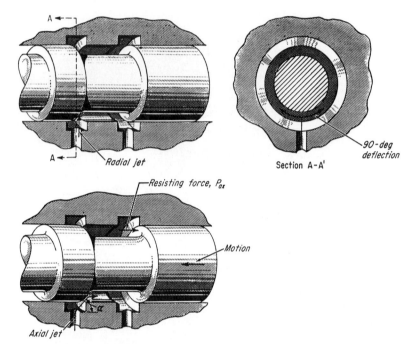

FIGURE 2   Jet-force drag in pilot-operated spool valves.

$p_B = 20.0$ lb/in^2 (137.8 kPa), the same as before. The total is 33.0 lb/in^2 (227.5 kPa).

The new $\Delta P = 467$ lb/in^2 (3217.6 kPa), and $Q = 4.7 (467)^{0.5} = 102$ in^3/s (1671.5 mL/s); $V = 102/4.9 = 20.8$ in/s (52.1 cm/s); $D_V = 1.82 (20.8) = 37.8$ lb (168.1 N) = 7.75 lb/in^2 (53.4 kPa), which proves out the assumption of 7.7 lb /in^2 (53.1 kPa).

Accelerating time, $t_a = (0.05)(20.8)/(2280) = 0.0005$ s. The 1.5-in (3.81-cm) stroke takes $1.5/20.8 = 0.072$ s. Total time = 0.081 s.

The flow rate of the pilot oil is more important than pressure intensity in obtaining a fast-acting valve. A slightly larger pilot valve and enlarged porting have a marked effect on the operational speed of the main valve.

Note that increasing the pilot pressure fivefold, from 100 lb/in^2 to 500 lb/in^2 (689 kPa to 3445 kPa) only doubles the speed of response, from 0.19 s to 0.08 s. Increasing the port area can result in an nearly proportional gain in speed, and no additional pressure is necessary, saving on pumping costs. Costs of producing a 0.375-in (9.52-mm) pilot spool are not much greater than those for a 0.25-in (6.35-mm) spool. The increase in capacity is 50 percent without the additional heat losses entailed by an increase in pressure.

8.   *Analyze the valve fitted with actuating pistons*

For the valve, Fig. 3, with the small actuating pistons, taking the summation of the viscous drag, ΣD_V, and inserting the known optimum values in parentheses after the computed values, we have:

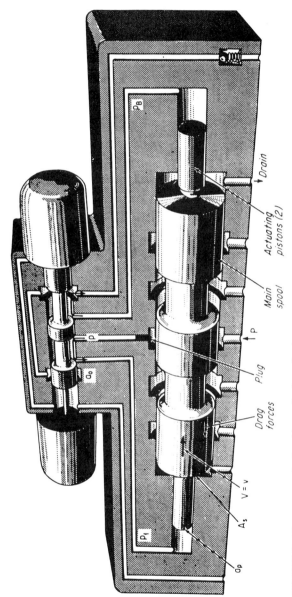

FIGURE 3 Piston-operated solenoid-energized spool valve.

$$\Sigma D_V = \frac{8.2 \times 4.9 \times 4 \times 80}{0.0003 \times 619 \times 10^6 \times 1.4}$$

$$\times \left(\frac{2.5 \times 6}{1.4} + 2 \times 1.5 \right)$$

$$= \frac{12800}{2075 \times 1.4} \left(\frac{15}{1.4} + 3 \right)$$

$$= 60.5 \text{ lb (269.1 N) [59.0 lb optimum; 262.4 N]}$$

Introducing the value of ΣD_V in the equation,

$$a_p = \frac{3k_3}{2k_2} = \frac{3(P_{ax} + \Sigma D_V)}{2(p - p_B)}$$

we have

$$a_p = \frac{3(26 + 60.5)}{2(100 - 20)} = 1.62 \text{ in (10.45 sq cm) [1.59 in}^2\text{; 10.26 cm}^2\text{]}$$

$d = 1.44$ in (3.66 cm) [1.425 in, 3.62 cm].

With optimum $a_p = 1.59$ cu in (26.06 mL), piston velocity using

$$q = fa_o \sqrt{\frac{2\Delta p}{\rho}}$$

$$v = k_1 \sqrt{\frac{k_2}{a_p^2} - \frac{k_3}{a_p^3}}$$

$$= \sqrt{\frac{k_1^2 k_2}{a_p^2} - \frac{k_1^2 k_3}{a_p^3}}$$

is

$$v_p = 4.6 \sqrt{\frac{80}{2.53} - \frac{(26 + 59)}{4.0}}$$

$$= 15.0 \text{ in/s} \quad (38.1 \text{ cm/s})$$

The total time, $T = 0.100 + 0.009 = 0.109$ s. Using a pilot pressure of 500 pst (3445 kPa), $V = 20.8$ in/s (52.8 cm/s) and $d = 1.06$ in (2.69 cm). Then:

$$\Sigma DV = \frac{20.8 \times 4.9 \times 4.80}{0.0003 \times 6.9 \times 10^6 \times 1.06}$$

$$\times \left(\frac{15}{1.06} + 3 \right) = 258 \text{ lb}$$

$$a_p = \frac{3(26 + 258)}{2(500 - 20)} = 0.886 \text{ in}^2$$

$$d_z = 1.06 \text{ in}$$

$$v_p = 4.6 \sqrt{\frac{480}{0.784} - \frac{284}{0.61}}$$

$$= 55 \text{ in/s}$$

$$T = 0.036 \text{ s}$$

Table 3 and Fig. 4 depict the effect of actuating-piston area upon the spool shifting velocity and shifting time.

Related Calculations. Pilot-operated flow-control valves are probably the most common valves used in industrial hydraulic systems. Speed of response of these valves is important during the design and operation of any hydraulic system. The procedure given here analyzes the speed of response of a typical valve in terms of the fluid flow rate, characteristic force-vs-airgap curve of the solenoid; shape, size, clearance, and displacement of each spool; and the fluid viscosity.

The method given in this procedure relates the above parameters for the valve in Fig. 1. and can be applied to any other pilot-operated spool valve. And the procedure includes a special technique for a large spool valve, Fig. 3, actuated by a small auxiliary piston.

In the sequence of operation of solenoid-energized pilot-operated spool valve, here is what happens. The solenoid is energized, the pilot spool moves quickly to the full open position, Fig. 5, and the main spool is shifted at a rate determined by the amount of fluid that can move through the pilot ports against these five resisting forces: (1) pilot system backpressure, lb/in^2 (kPa); (2) viscous damping force, lb (N); (3) radial jet force, lb (N); (4) axial jet force, lb (N); (5) acceleration force, lb (N).

Figure 6 shows a time-displacement chart for the spool as it shifts to a new position. This chart is for a 0.25-in (0.635-cm) diameter closed-center pilot spool controlling flow to a 2.5-in (6.35-cm) main spool.

Pilot pressure chosen for this procedure is 100 lb/in^2 (689 kPa), which is less than the main supply pressure. A separate supply for the pilot is required. There are no hard and fast rules for establishing pilot pressures, but if possible keep the pressure in the range of a few hundred lb/in^2 (kPa) if this range will do the job, to avoid possible distortion or leakage in the pilot system.

In analyzing any pilot-operated valves, some simplifying assumptions must be made; otherwise there is no practical mathematical solution. For one, assume that the backpressure of the pilot system, set by the pilot exhaust valve, is constant. Ignore line resistance because the connecting lines are short. Neglect viscous damp-

TABLE 3 Effect of Adding Actuating Pistons*

Pilot pressure p, psi	Main valve diameter, in	Maximum valve velocity, without piston in/sec	Total shift time T, sec	Maximum valve velocity, with piston in sec	Piston diameter, in	Total shift time T, sec
100	2.50	8.2	0.190	15.0	1.425	0.109
500	2.50	20.8	0.081	55.0	1.06	0.036

*SI values given in calculation procedure.

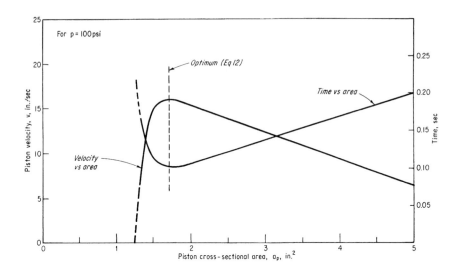

SI Values			
in./sec	cm/sec	in.^{2}	cm^2
5	12.7	1	6.45
10	25.4	2	12.9
15	38.1	3	19.4
20	50.8	4	25.8
25	63.5	5	32.3

FIGURE 4   Effect of varying piston diameter.

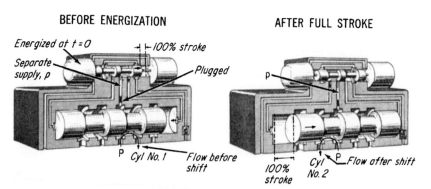

FIGURE 5   Before energization and after full stroke of a solenoid-energized spool valve.

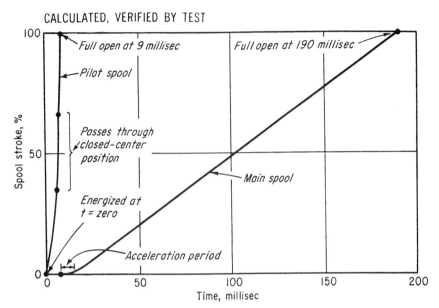

FIGURE 6 Time-displacement graph for solenoid-energized pilot-operated spool valve.

ing except at the full-velocity portion of the stroke. Then the five dynamic resistances can be handled with the simple equations presented in this procedure.

The pilot-system backpressure is usually 5 to 7 percent of the pilot pressure, p. Select the higher value if the operating pressures are over 200 lb/in² (1378 kPa), because it adds a margin of safety that compensates for spool rubbing friction. The friction is from metal-to-metal contact at points where the oil film is partially destroyed.

Above 400-lb/in² (2756 kPa) operating pressure, a separate pilot supply usually is provided. Pilot pressure in these instances ought to be at least 7 percent of the main operating pressure to ensure adequate force to move the main spool.

This procedure is the work of Louis Dodge, Hydraulics Consultant, as reported in *Product Engineering* magazine.

HYDRAULIC-SYSTEM RESERVOIR AND HEAT EXCHANGER SELECTION AND SIZING

(1) Determine if a "first-pass" reservoir choice, Fig. 7, can dissipate enough heat to keep oil temperature below 120°F (48.9°C) in a 70°-F (21.1°-C) ambient (50°-F [27.8°-C]) temperature difference, T_D. The source of heat is a 20-gal/min (1.26-L /s) constant-delivery pump operating in a 60-percent overall efficiency system. The hydraulic working unit, including the piping, is not helping to dissipate heat. Reservoir tank size is small with a cooling surface of 21 ft² (1.95 m²), based on a reservoir volume of twice pump flow—or $2 \times 20 = 40$ gal (75.8 L). System pressure = 750 lb/in² (5167.5 kPa); overall heat transfer coefficient, $k = 5$ Btu/ft²

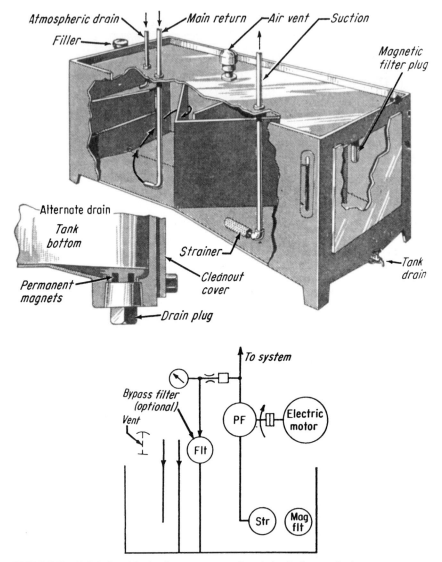

FIGURE 7　Well-designed hydraulic-system reservoir and circuit diagram for it.

hr °F (28.4 W/m² °C), a conservative value. (2) The return flow of the hydraulic fluid in another industrial hydraulic system must be cooled continuously to 125°F (51.7°C). The hottest uncooled drain temperature of the fluid is 140°F (60.0°C). Flow of the hydraulic fluid through the system is 12 gal/min (0.76 L/s); the cooling-water inlet temperature = 65°F (18.3°C); outlet temperature = 85°F (29.4°C); k = 90 Btu/hr-ft² °F (511.2 W/m² °C). Use a counterflow, single-pass heat exchanger, Fig. 8, in this analysis. (3) Lastly, calculate the temperature of a standard 60-gal (227.4 L) reservoir after 5 hr of operation. Pump discharge is 20 gal/min (1.26 L/s) at 750 lb/in² (5.17 MPa). Cooling surface is 28 ft² (0.792 sq m). An attached heat-dissipating working unit weighs W = 800 lb (362.2 kg) and its effective surface area A = 5.5 ft² (0.156 m²). With an initial system oil temperature, T_{oil} = 70°F (21.1°C), and an ambient temperature, T_{air} = 50°F (10.0°C), the initial temperature-over-ambient, T_p = 70 − 50 = 20°F(11.1°C). The estimated median value of k = 4 Btu/ft² hr °F(22.7 W/sq m °C). The 60 gallons of oil weigh 444 lb (201.6 kg).

Calculation Procedure:

1. Find the total heat generated in the system

Find the total heat generated in the system using the equation

$$E_L = 1.48 \times Q \times P(1 - \mu)$$

where the symbols are as given below. Substituting

$$E_L = 1.48(20 \text{ gal/min})(750 \text{ lb/in}^2)(1 - 0.60) = 8880 \text{ Btu/hr (2601.1 W)}.$$

2. Compute the required reservoir cooling area

Use the equation

$$_{max}T_D = E_L/\Sigma kA$$

and solve for the required area, A. Or, A = 8880/(50 × 5) = 35.5 ft² (3.3 m²). Since this reservoir has only 21 ft² (1.95 m²) of cooling surface, the tank area is not large enough to dissipate the heat generated. Hence, a larger reservoir cooling area must be provided for this installation.

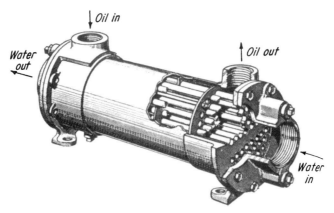

FIGURE 8 Single-pass shell-and-tube heat exchanger for industrial hydraulic system.

3. Determine the heat-transfer area and cooling-water flow rate required
Use the equation

$$E_{exch} = \Delta T_{oil} \times Q_{oil} \times 210$$

$$= \Delta T_{water} \times Q_{water} \times 500$$

to find the heat exchanger heat load. Substituting, we have $E_{exch} = (140 - 125)$ (12)(210) = 37,800 Btu/hr (11.1 kW). The maximum temperature difference, $\Delta T_{max} = 125 - 65 = 60°F$ (51.7°C). Minimum temperature difference $= \Delta T_{min} = 140 - 85 = 55°F$ (30.6°C). Log-mean temperature difference, computed as shown elsewhere in this handbook (see index) is $\Delta T_{mean} = 57.5°$ (31.9°C). Figure 9 shows the oil and water temperature changes in a generalized manner.

Find the required heat-transfer area from

$$E_{exch} = kA\Delta T_{mean} \text{Btu/hr}$$

solving for A. Or A = 37,800/(90 × 57.5) = 7.3 ft² (0.68 m²).

To find the required cooling-water flow rate, use $Q_{water} = E_{exch}/(500 \times \text{gal/min})$, where 500 = a constant to convert gal/min to gph. Substituting, $Q_{water} = 38,800/(500 \times 20) = 3.78$ gal/min (0.2385 L/s).

Refer to manufacturer's catalogs for the size of a heat exchanger with the proper surface area. Be certain to check the heat exchanger pressure rating if the system pressure exceeds 150 lb/in² (1033.5 kPa).

4. Find the heat exchanger heat load
Use the same relation as in step 1, above, to find $E_L = 8880$ Btu/hr (2601.1 W).

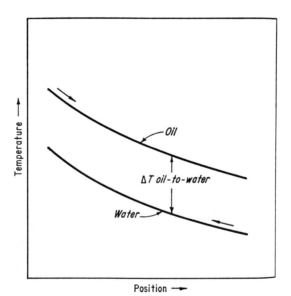

FIGURE 9 Hydraulic oil and cooling water temperature plot for heat exchanger in Fig. 8.

5. *Compute the heating variables for the fluid and reservoir*
The heat balance for the oil and attached heat dissipating working unit are given by $cW = 0.4(444) + 0.1(800) = 257.6$ Btu/°F (489.4 kJ/°C). For the tank and working unit, the $kA = 4(28 + 5.5) = 134$ Btu/hr°F (254.6 kJ/hr°C).

6. *Determine the temperature above ambient for the hydraulic fluid*
Use the equation

$$T_D = \frac{E_L}{\Sigma kA} (1 - e^{-\Sigma kA/\Sigma cW^t}) + (_{\text{init}}T_D)e^{-\Sigma kA/\Sigma cW^t}$$

$$T_D = \frac{8880}{134} (1 - e^{-134/257 \times 5}) + 20\ e^{-134/257 \times 5}$$

Substituting, we have $T_D = 62.8$°F (17.1°C).

The maximum operating temperature over ambient, $_{\max}T_D = E_L/kA = 8880/134 = 66.3$°F (19°C). Then, the oil temperature $= 50 + 66.3 = 116.3$°F (46.8°C). Based on these results, no heat exchanger is required.

Note that the result of this calculation depends on the correct evaluation of k, which depends on air circulation around the reservoir and attached heat-dissipating unit. The influence of the initial temperature difference is minor. Practical experience with system design is most important.

Related Calculations. The highest recommended temperature for oil in a conventional hydraulic systems reservoir is 120°F (48.9°C). The procedure presented above shows ways to prevent the oil temperature from exceeding that level. There are certain exceptions to the rule given above. Some conventional hydraulic systems are designed to operate at 150°F (65.6°C). So-called super-systems with special fluids and seals can operate at 500°F (260°C), and higher. But for any level of operating temperature, the same heat-transfer principles apply.

In designing a fluid system's heat-transfer, after you've established the basic system and reservoir design, follow the simple heat-balance method given above. If the reservoir's peak temperature calculated this way is less than 120°F (48.9°C) (or some other desired temperature), no further work is necessary. That's why a heat exchanger was not required in step 6, above. If the calculated reservoir temperature is higher than desired, you have two alternatives: (1) improve heat dissipation by modifying the reservoir tank, components, or piping; (2) add a heat exchanger, using the rating method given in this procedure.

Most of the heat in industrial hydraulic systems comes from in-the-system components. Exceptions are systems in hot environments or adjacent to heat-producing equipment, but the same heat-balance principles apply.

Heat is generated whenever hydraulic oil is throttled or otherwise restricted. Examples of heat-producing devices include pressure regulators, relief valves, undersize piping, dirty or undersize filters, leakage points, and areas of turbulence anywhere in the system.

A good measure of internally generated heat loss is the difference between pump input power, Btu/hr (W), and system useful work, Btu/hr (W). The energy loss, $E_L = E_{\text{in}}(1 - \mu)$ where $\mu =$ system efficiency.

Figure 10 shows a typical duty cycle in an industrial hydraulic system. The utilization pressure, measured at the inlet to the working device (fluid motor or cylinder) will always be lower than the source pump pressure, depending on the amount of throttling or other regulation required in the system.

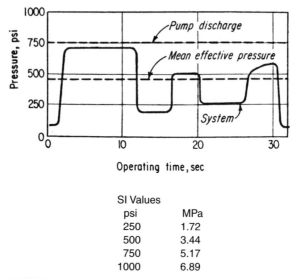

SI Values	
psi	MPa
250	1.72
500	3.44
750	5.17
1000	6.89

FIGURE 10   Typical duty cycle for industrial hydraulic system showing the mean effective pressure, system pressure, and pump discharge pressure.

The difference in energies—pumped vs utilized—must be absorbed during transients and eventually dissipated by the system. An approximate measure of overall system efficiency, μ, is the ratio of the mean effective pressure, Fig. 10, to pump pressure, where the mean effective pressure is calculated from the area under the curve, Fig. 10, divided by the time base.

Figure 11 compares reservoir tank temperature for two different pumps: (1) a fixed-delivery pump, with constant flow and pressure and a full-flow relief valve for bypass flow during idling of the workload, and (2) a variable-delivery pump, with pressure and flow automatically varied to match load requirements.

Note that in the constant-pressure system, Fig. 11, the greatest rate of heat generation is during idling of the workload; all the flow is throttled back to the reservoir and does no useful work. By comparison, the variable-delivery pump does not waste energy at idle because the flow is automatically reduced to nearly zero.

Both Fig. 10 and 11 indicate that savings in energy are possible if only the needed oil is pumped. Excess capacity is forced back to the reservoir through the relief valve, and the energy is converted to waste heat. Auxiliary pumps are great offenders if they are operated when not needed. Additional useful guides for reservoir selection and sizing are given below.

Plan for a reservoir capacity that can feed the pump system for two or three minutes, neglecting any return flow. With a tank that big, several related requirements usually are met automatically: (1) There will be enough reserve fluid to fill the hydraulic system at startup without exposing the filter and strainer; (2) A fairly stable oil level will be maintained despite normal fluctuations in flow; (3) Enough hydraulic fluid will be available to sustain the system while the rotating parts coast to a stop during emergency shutdown if a return line breaks; (4) Thermal capacity will be available to absorb unexpected heat for short periods or to store heat during

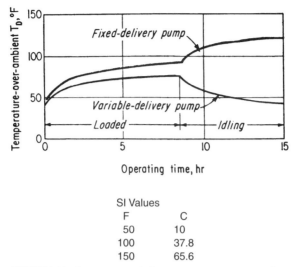

SI Values	
F	C
50	10
100	37.8
150	65.6

FIGURE 11 Proper pump choice can conserve energy and reduce system temperature.

idle periods in a cold environment; (5) Enough surface area (reservoir tank walls) is available for natural cooling during normal operation.

If the reservoir tank volume in gallons (L) is less than twice pump flow in gallons per min (L/min)—that is, if the tank can be pumped dry in less than 2 minutes—add a heat exchanger to the system circuit to avoid excessive temperature fluctuations. For any size reservoir tank, specify an oil-level indicator or sight glass, in addition to whatever automatic level controls are provided.

When designing a reservoir tank, include each device shown in Fig. 7 to provide reliable service for the system. The suction-line filter should be ½ to ¾ in (1.3 to 1.9 cm) above the tank bottom. Strainer oil flow capacity should be two to four times the pump capacity. A vacuum gage on the pump suction will show if the strainer is clogged. A permanent-magnet filter can be specified as a drain plug or mounted on the baffle plate in a region of concentrated return oil flow.

The main return oil flow should discharge below the reservoir oil level about one inch (2.5 cm) above the tank bottom. Backpressure in the return line will be 5 to 10 lb/in^2 (34.5 to 68.9 kPa), or higher. Atmospheric return lines, including seal-leakage lines, are at zero pressure and should be discharged above the hydraulic oil level.

If the atmospheric lines have high flow and a high air content, they should be discharged above the oil level into a chute sloping gradually (5 to 10 degrees) into the tank fluid. The chute slows and fans out the flow, enabling the oil to free itself of air. This is important, because oil saturated with air and operating at high pressures will run 25 percent hotter than air-free oil. This is caused partly by the heat of compression of the air and partly by its low thermal conductivity.

Internal baffles between the return pipe and pump inlet will slow the fluid circulation, help settle out dirt particles, give air the chance to escape, and allow dissipation of heat. The top of the baffle should be submerged about 30 percent below the surface of the fluid.

Keep the hydraulic oil temperature between 120 and 150°F (54 and 66°C)—preferably at the lower value for oil viscosities from 100 to 300 SSU based on 100°F (38°C). Temperatures up to 160°F (71°C) are permissible if the hydraulic fluid viscosity is from 300 to 750 SSU, based on 100°F (38°C). Higher operating temperatures require special design.

Tank walls should be thin to permit good thermal conductivity. Make them approximately 1/16 in (0.16 cm) for tank capacities up to 25 gal (95 L); 1/8 in (0.32 cm) for capacities up to 100 gal (379 L); 1/4 in (0.64 cm) for 100 gal (379 L) or more. Use slightly heavier plate for the bottom. Give the top plate four times wall thickness to assure vibration-free operation and to hold alignment of pump and motor shafts. Specify a thermometer to be mounted on the tank top where the operator can see it.

Avoid designing industrial hydraulic system machines with integral tanks. It is better to have a separate reservoir, accessible from all sides. Small reservoir tanks can even be mounted on castors. Tanks within the machine frame are troublesome to maintain; be sure to work out maintenance details of such a tank before committing yourself to the design.

Equip the reservoir tank with cleanout doors and slope the bottom toward the doors. Provide a drain cock or discharge valve at the low point of the bottom and at other low points if needed for complete drainage. Put a manhole cover on the tank top for removing filters and strainers. Design a connection for hooking to a portable filtration unit.

If the reservoir tank is made of cast iron, don't paint the interior surface. Be sure that all grit and core sand are removed before putting the tank into service. Surfaces must be sandblasted.

This procedure is the work of Louis dodge, Hydraulics Consultant, as reported in *Product Engineering* magazine. SI values were added by the handbook editor.

Heat-transfer terminology and symbols

Heat loss and efficiency
μ = System efficiency, E_{used}/E_{in}, %
E_{used} = Energy utilized in system, Btu/hr (W)
E_{in} = Pump input power, Btu/hr (W)
E_L = Heat loss generated in system, Btu/hr (W)
E_A = Heat absorbed by oil, tank and components, Btu/hr (W)
E_D = Heat dissipated to atmosphere or coolant, Btu/hr (W)
E_{exch} = Heat exchanger load, Btu/hr (W)

Fluid conditions and flow

t = Operating time, hr
Q = Flow, gal/min (L/s)
P = Pump gage pressure, lb/in^2 (kPa)
T_D = Temperature-over-ambient for oil, °F: $T_D = T_{oil} - T_{air}$ (values are mean) (C)
ΔT = Heat exchanger only: $\Delta T_{water} = T_{out} - T_{in}$; $\Delta T_{oil} = T_{in} - T_{out}$; ΔT_{mean} = log-mean ΔT, oil-to-water (C)

Equation constants

c = Specific heat (mean value), Btu/lb-°F (kJ/kg °C)
k = Overall heat-transfer coefficient, Btu/sq ft-hr-°F (W/m^2 °C)
W = Combined weight of oil and system components, lb (kg)
A = Surface area for dissipating heat, sq ft (sq m)
e = Base for natural logs = 2.718
Σ = Summation sign. ΣcW = effective cW for all of system components

Typical values for c and k

c, Btu/lb-°F: (kJ/kg °C) Oil, 0.40; aluminum, 0.18; iron, 0.11; copper, 0.09
k, Btu/sq ft-hr-°F: (W mm/m² °C)
2 to 5—Tank inside machine or with inhibited air circulation
5 to 10—Steel tank in normal air
10 to 13—Tank with good air circulation (guided air current)
25 to 60—Forced air cooling or oil-to-air heat exchanger
80 to 100—Oil-to-water heat exchanger (k values increase slightly with temperature)

CHOOSING GASKETS FOR INDUSTRIAL HYDRAULIC PIPING SYSTEMS

Choose a suitable gasket to seal industrial hydraulic fluid at 1200 lb/in² (8.27 MPa) and 180°F (82.2°C). Flanges are 1.5-in (3.8-cm) raised-face, 600-lb (2668.8-N) weld-neck type made from Type 304 stainless steel. There are four bolts, 0.75-in (1.9-cm) 10 NC, made from ASTM A193 grade B7 alloy steel. Hydrotest pressure is specified as 2.5 times operating pressure.

Calculation Procedure:

1. Determine the total bolt force and torque for these flanges
Assuming that torque wrenches will be used to check this installation, as is almost universally done today, select the bolt-stress method to calculate the total bolt force. This method uses the equation,

$$F_b = N_b S_b A_b$$

where the symbols are as given below.
From Table 5, the stress area for 3/3-10 NC bolt is 0.3340 sq in. (2.15 sq cm). The bolt material specified can easily take a stress of 30,000 lb/in² (206.7 MPa) without yielding. This can be verified from

$$F_b = 16,000 D_b/A_b$$

which gives $F_b = (16,000)(0.75/0.3340) = 36,000$ lb (160.1 kN).

TABLE 4 Pressure-Temperature Value for Gasket Materials*

Material	Max $P_i \times T$	Max Temp, °F
Rubber	15,000	300
Vegetable fiber	40,000	250
Rubberized cloth	125,000	400
Compressed asbestos**	250,000	850
Metal types	>250,000	depends on metal

*SI values given in procedure.
**Or acceptable substitute.

TABLE 5 Stress Areas for Flange Bolts*

	Coarse Threads		Fine Threads	
Nominal dia, in	Threads per in	Stress area, sq in	Threads per in	Stress area, sq in
0.125 (No. 5)	40	0.0079	44	0.0082
0.138 (No. 6)	32	0.0090	40	0.0101
0.164 (No. 8)	32	0.0139	36	0.0146
0.190 (No. 10)	24	0.0174	32	0.0199
0.216 (No. 12)	24	0.0240	28	0.0257
1/4	20	0.0317	28	0.0362
5/16	18	0.0522	24	0.0579
3/8	16	0.0773	24	0.0876
7/16	14	0.1060	20	0.1185
1/2	13	0.1416	20	0.1597
1/2	12	0.1374		
9/16	12	0.1816	18	0.2026
5/8	11	0.2256	18	0.2555
3/4	10	0.3340	16	0.3724
7/8	9	0.4612	14	0.5088
1	8	0.6051	12	0.6624
1 1/8	7	0.7627	12	0.8549
1 1/4	7	0.9684	12	0.0721
1 3/8	6	1.1538	12	1.3137
1 1/2	6	1.4041	12	1.5799

*SI values given in procedure.

From the bolt-stress equation, $F_b = 4(30{,}000)(0.3340) = 40{,}080$ lb (178.3 kN). The torque required to produce this stress level at installation is given by,

$$T = 0.2D_bS_bA_b$$

Or, $T = 0.2(0.75)(30{,}000)(0.3340)/12 = 125$ ft-lb (169.4 Nm). This torque will be specified on the system assembly drawings so it is used during construction.

2. ***Choose a suitable gasket material***
The pressure-temperature relation for this installation is $1200 \times 180 = 216{,}000$ in USCS units and 6798 in SI units. This, from Table 4, suggests choosing a compressed-asbestos (or acceptable substitute) type gasket. This would be compatible with industrial hydraulic fluid.

The gasket area is 4.73 sq in. (30.5 sq cm), calculated from an outside diameter of 2 7/8 in (7.07 cm), the same as the OD of the raised flange, per ASA-B16.5, and an ID of 1.5 in (3.8 cm). The seating stress is computed from

$$S_g = F_b/A_g$$

Or, $S_g = 40{,}080/4.73 = 8470$ lb/in^2 (58.4 MPa).

Table 6 shows this stress will easily seat the selected compressed-asbestos (or acceptable substitute) gasket. Tentatively choose asbestos (or acceptable substitute) with CR (neoprene) binder for oil resistance; thickness 1/32 in (0.079 cm).

MECHANICS OF A GASKETED JOINT

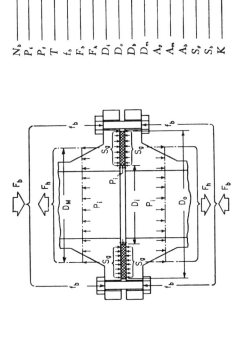

LIST OF SYMBOLS

N_b	Number of bolts
P_i	Operating pressure, psi (kPa)
P_t	Test pressure, psi (kPa)
T	Bolt torque, in.-lb (Nm)
f_b	Force per bolt, lb (N)
F_b	Total bolt force, lb (N)
F_h	Hydrostatic end force, lb (N)
D_i	Gasket ID, in. (cm)
D_o	Gasket OD, in. (cm)
D_b	Nominal bolt diameter, in. (cm)
D_m	Gasket mid-diameter, in. (cm)
A_t	Gasket contact area, sq in. (sq cm)
A_m	Area subject to pressure P_i, (sq cm)
A_b	Stress area per bolt, sq in.
S_g	Gasket seating stress, psi (kPa)
S_b	Bolt stress, psi (kPa)
K	Safety factor, dimensionless

TABLE 6 Minimum Seating Stresses for Typical Gasket Materials*

Material	Type (see Table IV)	Thickness, in	Minimum seating stress, lb/in²	Flange-surface finish 1st choice	2nd choice
Rubber sheet					
SBR (75 Durometer)	Flat	$\frac{1}{32}$ and up	200		
CR (60 Durometer)	Flat	$\frac{1}{32}$ and up	175		
Compressed asbestos**					
SBR binder	Flat	$\frac{1}{64}$	3,000		
		$\frac{1}{32}$	2,000		
		$\frac{1}{16}$	1,600		
		$\frac{1}{8}$	1,200		
CR binder	Flat	$\frac{1}{64}$	3,750	Concentric-serrated	All other types
		$\frac{1}{32}$	2,500		
		$\frac{1}{16}$	2,000		
		$\frac{1}{8}$	1,500		
Rubberized cloth	Folded	2-ply	2,500		
		3-ply	2,100		
		4-ply	1,800		
Vegetable-fiber sheet	Flat	all	750		
Fluorocarbon (TFE)					
Virgin	Flat	$\frac{1}{64}$	14,000	80 rms ($\frac{1}{64}$ in only)	All other types
		$\frac{1}{32}$	6,500		
		$\frac{1}{16}$	3,700		
		$\frac{1}{8}$	1,600		
Glass-filled	Flat	$\frac{1}{64}$	14,000	Concentric-serrated (all other thicknesses)	All other types
		$\frac{1}{32}$	11,000		
		$\frac{1}{16}$	6,000		
		$\frac{1}{8}$	3,000		
Asbestos cloth (impreg.)**	Flat	$\frac{3}{32}$	1,600		

NONMETAL

	Gasket type	Facing	Thickness, in	Seating stress	Surface finish	Surface finish
M E T A L	Flat metals Aluminum Copper Carbon steel Monel Stainless	Flat	1/32 and 1/16	20,000 45,000 68,750 81,250 93,750		80 rms or less
	Stamped metals Lead Aluminum Copper Carbon steel Monel Stainless	Corrugated jacket-metal filler	%64 only	500 1,000 2,500 3,500 4,500 6,000		
	Machined metals Aluminum Copper Carbon steel Monel Stainless	Profile (1/8-in pitch)	All thicknesses	25,000 35,000 55,000 65,000 75,000	80 rms or less	150 to 200 rms
M E T A L S	Metal (asbestos-filler) Aluminum Copper Carbon steel Stainless Monel	Corrugated and corded	1/8 only	2,000 2,500 3,000 3,500 4,000	150 to 200 rms	Concentric-serrated
	Lead Aluminum Copper Carbon steel Nickel Monel Stainless	Plain metal jacket	1/8 only	500 2,500 4,000 6,000 6,000 7,500 10,000	80 rms or less	
	Same as stamped metals	Corrugated jacket				
	Stainless	Spiral-wound	0.125 and 0.175	3,000 to 30,000		150 to 200 rms

*SI values given in procedure.
**Or acceptable substitute.

25.23

3. Determine the hydrostatic end force for the chosen gasket
The mean area acted upon by the pressure in the hydraulic line is defined by a diameter of $(2.875 + 1.5)/2 = 3.74$ sq in (24.1 sq cm). Selecting a safety factor of 1.5 from Table 8, the end force is calculated and balanced against the total bolt force by the equation,

$$F_b \geqq KP_t A_m$$

Or,
Thus, there is no end-force balance problem with bolts stressed to 30,000 lb/in^2 (206.7 MPa).

4. Select a suitable surface finish for the flange
Table 6 shows that a concentric-serrated surface finish on the flange is best. Economy may dictate a conventional spiral-serrated surface, which Table 6 shows will work in this case.

5. Prepare the final specification for the gasket
Include in the specifications the material type, dimensions, and bolt-torque data computed in step 1, above. For greater torque-wrench accuracy, specify uniform fit on all bolts and lubrication before installation.

   Related Calculations.   While the procedure given here is directed at industrial hydraulic systems, the steps and data are valid for choosing gaskets for any piping system: steam, condensate, oil, fuel, etc. Just be certain that the pressures and temperatures are within the ranges in the tables and equations presented here.

   Three main design factors govern the selection of a gasket material—whether sheet packing, metal, or a combination of materials. These factors are:

   Fluid compatability at the pressure-temperature condition being designed for must be checked first. Refer to data available from gasket manufacturers—there is much of it available free to designers.

   The *pressure-temperature combination* determines whether the gasket material is inherently strong enough to resist blow-out. One rule-of-thumb criterion is the product of operating pressure, P_i, and operating temperature, T. Table 4 lists values of this product for several basic types of gasket material. These figures are based on experience, test data, and analysis of current technical literature.

   The total bolt force at installation must be sufficient to: (1) flow the gasket surface into the flange surface to make an effective seal; (2) prevent the internal pressure from opening the flanges. This demands careful matching of gasket material, seating area, bolt selection, and flange-surface finish. The procedure presented here gives a logical way to achieve the right balance among these factors for the majority of gasket joint applications.

   Where asbestos is the recommended gasket material in this procedure, the designer must review the environmental aspects of the design. Acceptable substitute materials may be required by local environmental regulations. Hence, these regulations must be carefully studied before a final design choice is made.

   This procedure is the work of J. J. Whalen, Staff Engineer, Johns-Manville Corporation, as reported in *Product Engineering* magazine.

COMPUTING FRICTION LOSS IN INDUSTRIAL HYDRAULIC SYSTEM PIPING

(1) Determine the friction loss for hydraulic-system fluid having a viscosity of 110 centistokes at 120°F (48.9°C) when flowing through a 1-in ID (25.4 mm) pipe 50

TABLE 7 Typical Gasket Cross-Sections

Type (see Table 4)	Cross-section	Description
Flat		Simplest gasket form; available in wide variety of materials for different conditions of fluid media, temperature and pressure; most easily manufactured for nonsymmetrical shapes.
Folded		Fabric-based gaskets especially designed for extremely rough and wavy flanges with low bolting loads.
Plain metal jacket		Asbestos filler enclosed partially or totally by metal jacket; combines easy compression with resistance to high pressure. Temperature resistance depends on jacket material chosen.
Corrugated jacket: metal-filled asbestos-filled*		Corrugated metal or asbestos core enclosed by a corrugated-metal jacket and top washer. Better than flat gaskets for rough flanges—corrugations give greater resilience.
Corrugated and corded*		Deeply corrugated metal with twisted asbestos cord cemented into the corrugations on both faces; aluminized for nonsticking. Well-suited to rough or warped flanges.
Profile		Heavy solid metal with concentric V-shape contact ridges that provide multiple seating surfaces. Each ridge is 0.010 in wide. For N ridges, effective arera is then: $\pi \times$ mean contact dia $\times 0.010 \times N$. Seating stress increases with decrease in pitch.
Spiral-wound		Interlocked plies of preformed metal strip are spiral-wound with an interleaving cushion of asbestos or fluorocarbon plastic strip. Ratio of metal to filler can be closely controlled to vary seating stress over wide range. Well suited to fluctuating-temperature applications.

*Or acceptable substitute.

TABLE 8 Safety Factors for Gasketed Joints

K-factor	When to Apply
1.2 to 1.4	For minimum-weight applications where all installation factors (bolt lubrication, tension, parallel seating, etc.) are carefully controlled; low-temperature applications; where adequate proof pressure is applied.
1.5 to 2.5	For most normal designs where weight is not a major factor, vibration is moderate, and temperatures do not exceed 700 to 800 F. Use high end of range where bolts are not lubricated.
2.6 to 4.0	For cases of extreme fluctuations in pressure, temperature or vibration; where no test pressure is applied, or where uniform bolt tension is difficult to ensure.

ft (15.2 m) long at a velocity of 20 ft/s (6.1 m/s); specific gravity = 0.88. (2) Find the pressure loss for a "light" hydraulic oil having a viscosity of 32 centistokes at 100°F (37.8°C) when flowing through a 100-ft (30.5-m) long 2-in (50.8-mm) ID commercial steel pipe at a velocity of 30 ft/s (9.1 m/s); specific gravity = 0.88.

Calculation Procedure:

1. Find the Reynolds number for the hydraulic fluid
Use the relation

$$R = \frac{92{,}900}{12}\frac{VD}{v} = 7{,}740\,\frac{VD}{v}$$

where R = Reynolds number, dimensionless; V = hydraulic fluid velocity in the piping, ft/s (m/s); D = pipe diameter, in (mm); v = kinematic viscosity of hydraulic fluid, centistokes. Substituting, $R = (7740)(20)(1)/110 = 1407.3$.

2. Determine the relative roughness of the piping
Since the Reynolds number for this piping is less than 2000, roughness of the pipe does not enter into the calculation. See Fig. 12.

3. Find the friction factor for the piping
Use Fig. 12 and the Reynolds number. Since R is less than 2000, f = friction factor = 64/R = 64/1407.3 = 0.045.

4. Compute the pressure loss in the piping
Use the relation $p_f = 0.0808\, f\, (L/D)\, V^2(s)$, where p_f = pressure loss in piping, lb/in² (kPa); L = pipe length, ft (m); s = specific gravity of the hydraulic fluid; other symbols as before. Substituting, $p_f = 0.0808\,(0.045)(50)(400)(0.88) = 63.99$ lb/in²; say 64 lb/in² (440.9 kPa).

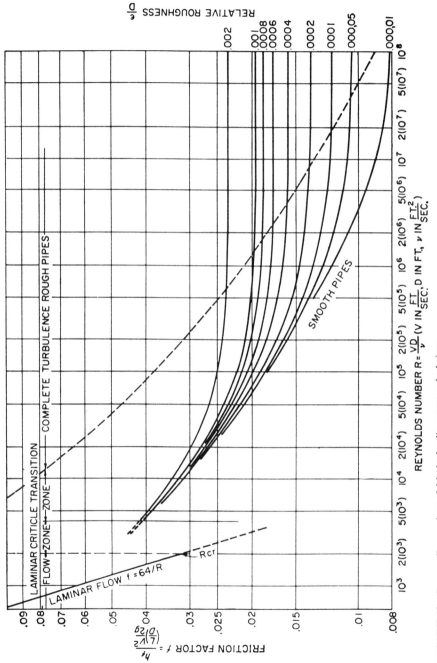

FIGURE 12 Stanton diagram is useful in hydraulic-system calculations

25.27

5. Find the Reynolds number for the second hydraulic fluid
Use the same equation as in step 1, $R = (7740)(30)(2)/32 = 14,512.5$.

6. Determine the relative roughness of the pipe
Use Fig. 13, entering at the pipe diameter at the bottom and projecting vertically upwards to the commercial steel pipe curve to read the relative roughness, e/D as 0.0009.

7. Find the friction factor, f
Enter Fig. 12 at Reynolds number and relative roughness to read $f = 0.03$.

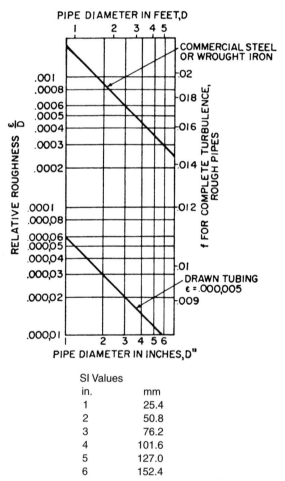

SI Values

in.	mm
1	25.4
2	50.8
3	76.2
4	101.6
5	127.0
6	152.4

FIGURE 13 Relative roughness as a function of pipe diameter for various types of piping.

8. *Compute the friction loss in the piping*

Using the same equation as in step 4, we have p_f = (0.0808) (0.03)(100)(900)(0.88)/2 = 95.99 lb/in^2 (661.4 kPa).

Related Calculations. Viscosity-temperature curves are shown for typical hydraulic oils in Fig. 14. The Herschel relationship, which expresses the viscosity-temperature function between two viscosities, μ, and, μ_0, existing at temperatures T and T_0, is

$$\mu = \mu_0 \left(\frac{T_0}{T}\right)^K$$

The above relationship holds for only a relatively restricted range of temperatures and should not be extrapolated beyond the range of validity. The table, which is part of Fig. 14, gives exponents for the hydraulic oils shown in the chart, covering a temperature range of 70 to 130°F (21 to 54°C). As a fair approximation, the viscosity of oils most commonly used in hydraulic work changes with the third power of the temperature gradient within the normal operating range. With a commonly encountered temperature gradient of about 2:1 between cold start and maximum operating temperature, viscosities vary as 8:1.

A graph representing the value of the friction factor, f, as a function of the Reynolds number, R, is often called the Stanton chart, after its developer, who was the first to employ this representation of the friction factor.

A chart taking advantage of the functional relationships established by research was drawn up by Lewis F. Moody, and is reproduced in Fig. 12 in a form convenient for the user of this handbook. In Fig. 12, the friction factor, f, is shown as a function of the Reynolds number, R, and the relative roughness, e/D, e being a linear quantity in feet or meters representing the absolute roughness. An auxiliary chart is given in Fig. 13, from which e/D can be taken for any size and type of pipe.

The procedure given here is valid for industrial hydraulic systems used in hydraulic presses; drilling, boring, and honing machines; planers; grinders; milling, transfer, and broaching machines; die-casting and plastic molding machines; hydraulic steering mechanisms in ships, aircraft, trucks, etc. In each instance, the basic approach given here is valid.

This procedure is the work of Walter Ernst, Hydraulic Consulting Engineer. SI values were added by the handbook editor.

HYDRAULIC-CYLINDER CLEARANCE FOR DAMPING END-OF-STROKE FORCES

An undamped hydraulic cylinder, Fig. 15, is fitted with an annular clearance in the cavity of the cylinder cap, Fig. 16, as a flow restriction. The cylinder is then provided with a piston having a velocity of 1.1 ft/s (0.34 m/s) having a mass, M, of 64 lb-s^2/ft (95.3 kg-s^2/m), a length, L = 4 in (10.16 cm), a dashpot radius of R = 1 in (2.54 cm), a dashpot capacity of 12.5 cu in (204.8 cu cm), a coefficient of discharge, C_D = 0.62, and a pressure differential, ΔP, of 30 lb/in^2 (206.7 kPa). What annular clearance is needed when handling hydraulic fluid with a specific gravity of 0.85?

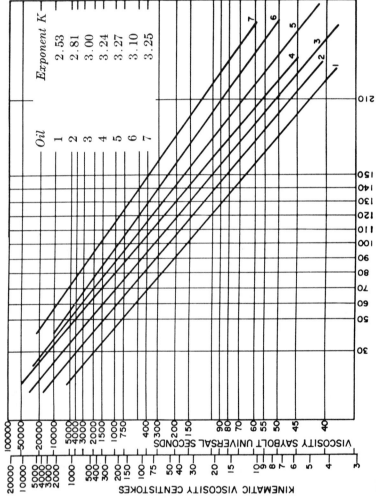

FIGURE 14 Viscosity-temperature curves for typical hydraulic oils.

Oil	Exponent K
1	2.53
2	2.81
3	3.00
4	3.24
5	3.27
6	3.10
7	3.25

VISCOSITY SAYBOLT UNIVERSAL SECONDS

KINEMATIC VISCOSITY CENTISTOKES

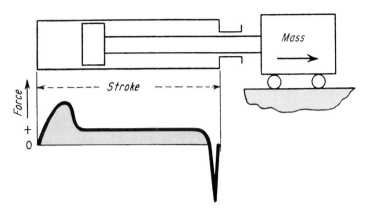

FIGURE 15 Undamped cylinder experiences shock forces at, and during, reversal of mass.

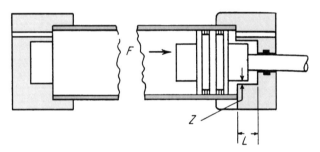

FIGURE 16 Annular clearance in cavity of cylinder serves as flow restriction for dashpot. Hydraulic fluid then escapes at outlet port.

Calculation Procedure:

1. Find the mean dashpot resistance

Use the relation, $F = MV^2/2L$, where F = mean dashpot resistance, lb (N); other symbols as given above. Substituting, $F = (64)(1.1)^2(12)/(2)(4) = 116.2$ lb (516.9 N).

2. Compute the piston acceleration and deceleration

Use the relation, $a = F/M$, where a = piston acceleration or deceleration, ft/s²; other symbols as given earlier. Substituting, $a = F/M = 116.2/64 = 1.816$ ft/s² (0.55 m/s²).

3. Determine the stopping time of the piston

Use the relation, $T = 2 V/a$, where T = stopping time, sec; other symbols as before. Substituting, $T = (2)(1.1)/1.81 = 1.22$ s.

4. Find the liquid discharge from the cylinder

Use the relation, $Q_D = 0.26 B/T$, where Q_D = cylinder liquid discharge, gal/min (L/s); B = dashpot capacity, cu in (cu cm); other symbols as before. Substituting, $Q_D = 0.26(12.5)/(1.22) = 2.66$ gal/min (0.168 L/s).

5. Calculate the annular clearance needed
Use the relation, $Z = Q_G(S)^{0.5}/[238(RC_D)(\Delta P)^{0.5}]$ where Z = annular clearance required, in (cm); Q_G = liquid discharge, gal/min (L/s); S = specific gravity of the hydraulic fluid handled; R = radius of dashpot, in (cm); other symbols as before. Substituting, $Z = (2.66)(0.92)/(238)(1)(0.62)(5.47) = 0.00303$ in (0.0077 cm).

Related Calculations. The hydraulic shock absorber, also called a dashpot, limits axial piston velocity where and when desired by trapping oil ahead of the piston, then releasing it through a restriction, slowly and with predetermined control. In this procedure we assumed that at the start of dashpot action inertia forces alone are dissipated through the ejection of dashpot oil. Hence, the kinetic energy of the moving parts equals the work done during penetration of the dashpot.

The assumed value of the coefficient of discharge, C_D, may be checked against the Reynolds number of the calculated flow and adjusted if it deviates too much from what experience shows as reasonable. Use the previous calculation procedure to check the Reynolds number. For Reynolds numbers below 100, C_D may vary from 0.1 to 0.7.

Hydraulic damping and shockless reversal can be obtained with flow-restriction or pressure-reducing devices. These include servo-controlled variable pumps, flow-control valves, cylinder modifiers (orifices, special pistons), and other power cutouts and reducers.

This procedure is the work of Louis Dodge, Hydraulics Consultant, as reported in *Product Engineering* magazine. SI values were added by the handbook editor.

HYDRAULIC SYSTEM PUMP AND DRIVER SELECTION

Choose the pump and the driver horsepower for a rubber-tired tractor bulldozer having four-wheel drive. The hydraulic system must propel the vehicle, operate the dozer, and drive the winch. Each main wheel will be driven by a hydraulic motor at a maximum wheel speed of 59.2 r/min and a maximum torque of 30,000 lb·in (3389.5 N·m). The wheel speed at maximum torque will be 29.6 r/min; maximum torque at low speed will be 74,500 lb·in (8417.4 N·m). The tractor speed must be adjustable in two ways: for overall forward and reverse motion and for turning, where the outside wheels turn at a faster rate than do the inside ones. Other operating details are given in the appropriate design steps below.

Calculation Procedure:

1. Determine the propulsion requirements of the system
Usual output requirements include speed, torque, force, and power for each function of the system, through the full capacity range.

First analyze the *propel* power requirements. For any propel condition, $hp = Tn/63,000$, where hp = horsepower required; T = torque, lb·in, at n r/min. Thus at maximum speed, $hp = (30,000)(59.2)/63,000 = 28.2$ horsepower (21.0 kW). At maximum torque, $hp = 74,500 \times 29.6/63,000 = 35.0$ (26.1 kW); at maximum speed and maximum torque, $hp = (74,500)(59.2)/63,000 = 70.0$ (52.2 kW).

The drive arrangement for a bulldozer generally uses hydraulic motors geared down to wheel speed. Choose a 3000-r/min step-variable type of motor for each wheel of the vehicle. Then each motor will operate at either of two displacements. At maximum vehicle loads, the higher displacement is used to provide maximum torque at low speed; at light loads, where a higher speed is desired, the lower displacement, producing reduced torque, is used.

Determine from a manufacturer's engineering data the motor specifications. For each of these motors the specifications might be: maximum displacement, 2.1 in^3/r (34.4 cm^3/r); rated pressure, 6000 lb/in^2 (41,370.0 kPa); rated speed, 3000 r/min; power output at rated speed and pressure, 90.5 horsepower (67.5 kW); torque at rated pressure, 1900 lb·in (214.7 N·m).

The gear reduction ratio GR between each motor and wheel = (output torque required, lb·in)/(input torque, lb·in, × gear reduction efficiency). Assuming a 92 percent gear reduction efficiency, a typical value, we find GR = 74,500/(1900 × 0.92) = 42.6:1. Hence, the maximum motor speed = wheel speed × GR = 59.2 × 42.6 + 2520 r/min. At full torque the motor speed is, by the same relation, 29.6 × 42.6 = 1260 r/min.

The required oil flow for the four motors is, at 1260 r/min, in^3/r × 4 motors × (r/min)/(231 in^3/gal) = 2.1 × 4 × 1260/231 = 45.8 gal/min (2.9 L/s). With a 10 percent leakage allowance, the required flow = 50 gal/min (3.2 L/s), closely, or 50/4 = 12.5 gal/min (0.8 L/s) per motor.

As computed above, the power output per motor is 35 horsepower (26.1 kW). Thus, the four motors will have a total output of 4(35) = 140 horsepower (104.4 kW).

2. Determine the linear auxiliary power requirements

The dozer uses a linear power output. Two hydraulic cylinders each furnish a maximum force of 10,000 lb (44,482.2 N) to the dozer at a maximum speed of 10 in/s (25.4 cm/s). Assuming that the maximum operating pressure of the system is 3500 lb/in^2 (24,132.5 kPa), we see that the piston are a required per cylinder is: force developed, lb/operating pressure, lb/in^2 = 10,000/3500 = 2.86 in^2 (18.5 cm^2), or about a 2-in (5.1-cm) cylinder bore. With a 2-in (5.1-cm) bore, the operating pressure could be reduced in the inverse ratio of the piston areas. Or, 2.86/($2^2\pi$/4) = p/3500, where p = cylinder operating pressure, lb/in^2. Hence, p = 3180 lb/in^2, say 3200 lb/in^2 (22,064.0 kPa).

By using a 2-in (5.1-cm) bore cylinder, the required oil flow, gal, to each cylinder = (cylinder volume, in^3)(stroke length, in)/(231 in^3/gal) = ($2^2\pi$/4)(10)/231 = 0.1355 gal/s, or 0.1355 gal/s, or 0.1355 × (60 s/min) = 8.15 gal/min (0.5 L/s), or 16.3 gal/min (1.0 L/s) for two cylinders. The power input to the two cylinders is hp = 16.3(3200)/1714 = 30./4 horsepower (22.7 kW).

3. Determine rotary auxiliary power requirements

The winch will be turned by one hydraulic motor. This winch must exert a maximum line pull of 20,000 lb (88,964.4 N) at a maximum linear speed of 280 ft/min (1.4 m/s) with a maximum drum torque of 200,000 lb·in (22,597.0 N·m) at a drum speed of 53.5 r/min.

Compute the drum horsepower from hp = Tn/63,000, where the symbols are the same as in step 1. Or, hp = (200,000)(53.5)/63,000 = 170 horsepower (126.8 kW).

Choose a hydraulic motor having these specifications: displacement = 6 in^3/r (98.3 cm^3/r); rated pressure = 6000 lb/in^2 (41,370.0 kPa); rated speed = 2500 r/min; output torque at rated pressure = 5500 lb·in (621.4 N·m); power output at rated speed and pressure = 218 horsepower (162.6 kW). This power output rating is somewhat greater than the computed rating, but it allows some overloading.

The gear reduction ratio GR between the hydraulic motor and winch drum, based on the maximum motor torque, is GR = (output torque required, lb·in)/(torque at rated pressure, lb·in, × reduction gear efficiency) = 20,000/(5500 × 0.92) = 39.5:

1. Hence, by using this ratio, the maximum motor speed = 53.5 × 39.5 = 2110 r/min. Oil flow rate to the motor = in³/r × (r/min)/231 = 6 × 2110/231 = 54.8 gal/min (3.5 L/s), without leakage. With 5 percent leakage, flow rate = 1.05(54.8) = 57.2 gal/min (3.6 L/s).

4. Categorize the required power outputs
List the required outputs and the type of motion required—rotary or linear. Thus: propel = rotary; dozer = linear; winch = rotary.

5. Determine the total number of simultaneous functions
There are two simultaneous functions: (a) propel motors and dozer cylinders; (b) propel motors at slow speed and drive winch.

For function a, maximum oil flow = 50 + 16.3 = 66.3 gal/min (4.2 L/s); maximum propel motor pressure = 6000 lb/in² (41,370.0 kPa); maximum dozer cylinder pressure = 3200 lb/in² (22,064.0 kPa). Data for function a came from previous steps in this calculation procedure.

For function b, the maximum oil flow need not be computed because it will be less than for function a.

6. Determine the number of series nonsimultaneous functions
These are the dozer, propel, and winch functions.

7. Determine the number of parallel simultaneous functions
These are the propel and dozer functions.

8. Establish function priority
The propel and dozer functions have priority over the winch function.

9. Size the piping and valves
Table 9 lists the normal functions required in this machine and the type of valve that would be chosen for each function. Each valve incorporates additional func-

TABLE 9   Hydraulic-System Valving and Piping

Valving	
Function	Type of valve
Step variable selector	Three-way, two-position
Propel directional	Four-way three-position, tandem-center
Winch directional	Four-way, three-position, tandem-center
Dozer directional	Four-way, four-position

Piping			
Branch of circuit	Propel motor	Dozer cylinder	Winch motor
Maximum flow, gal/min (L/s)	12.5 (0.8)	16.3 (1.0)	57.2 (3.6)
Maximum pressure, lb/in² (kPa)	6000 (41,370)	3200 (22,064)	6000 (41,370)
Tube size, in (cm)	¾ (1.9)	¾ (1.9)	1½ (3.8)
Tube material, ASTM	4130	4130	4130
Tube wall, in (mm)	0.120 (3.05)	0.109 (2.77)	0.250 (6.35)

tions: The step variable selector valve has a built-in check valve; the propel directional valve and winch directional valve have built-in relief valves and motor overload valves; the dozer directional valve has a built-in relief valve and a fourth position called *float*. In the float position, all ports are interconnected, allowing the dozer blade to move up or down as the ground contour varies.

10. Determine the simultaneous power requirements

These are: Horsepower for propel and dozer = (gal/min)(pressure, lb/in^2)/1714 for the propel and dozer functions, or $(50)(6000)/1714 + (16.3)(3200)/1714 = 205.4$ horsepower (153.2 kW). Winch horsepower, by the same relation, is $(57.2)(6000)/1714 = 200$ horsepower (149.1 kW). Since the propel-dozer functions do not operate at the same time as the winch, the prime mover power need be only 205.4 horsepower (153.2 kW).

11. Plan the specific circuit layouts

To provide independent simultaneous flow to each of the four propel motors, plus the dozer cylinders, choose two split-flow piston-type pumps having independent outlet ports. Split the discharge of each pump into three independent flows. Two pumps rated at $66.3/2 = 33.15$ gal/min (2.1 L/s) each at 6000 lb/in^2 (41,370.0 kPa) will provide the needed oil. Figure 17 shows a schematic of the piping, valves and motors for this bulldozer, while Fig. 18 shows the valving.

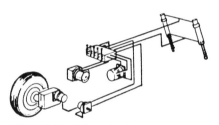

FIGURE 17 Schematic of the piping, valves, and motors for bulldozer.

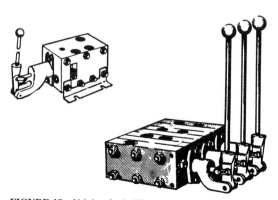

FIGURE 18 Valving for bulldozer.

When the vehicle is steered, additional flow is required by the outside wheels. Design the circuit so oil will flow from three pump pistons to each wheel motor. Four pistons of one split-flow pump are connected through check valves to all four motors. With this arrangement, oil will flow to the motors with the least resistance.

To make use of all or part of the oil from the propel-dozer circuits for the winch circuit, the outlet series ports of the propel and dozer valves are connected into the winch circuit, since the winch circuit is inoperative only when both the propel *and* the dozer are operating. When only the propel function is in operation, the winch is able to operate slowly but at full torque.

12. *Investigate adjustment of the winch gear ratio*
As computed in step 3, the winch gear ratio is based on torque. Now, because a known gal/min (gallons per minute) is available for the winch motor from the propel and dozer circuits when these are not in use, the gear ratio can be based on the motor speed resulting from the available gal/min.

Flow from the propel and dozer circuit = 66.3 gal/min (4.2 L/s); winch motor speed = 2450 r/min; required winch drum speed = 53.5 r/min. Thus, GR = 2450/53.5 = 45.8:1.

With the proposed circuit, the winch gear reduction should be increased from 39.5:1 to 45.8:1. The winch circuit pressure can be reduced to (39.5/45.8) (6000) = 5180 lb/in^2 (35,716.1 kPa). The required size of the winch oil tubing can be reduced to 0.219 in (5.6 mm).

13. *Select the prime mover horsepower*
Using a mechanical efficiency of 89 percent, we see that the prime mover for the pumps should be rated at 205.4/0.89 = 230 horsepower (171.5 kW). The prime mover chosen for vehicles of this type is usually a gasoline or diesel engine. Figure 19 shows the final tractor-dozer hydraulic circuits.

Related Calculations.   The method presented here is also valid for fixed equipment using a hydraulic system, such as presses, punches, and balers. Other applications for which the method can be used include aircraft, marine, and on-highway vehicles. Use the method presented in an earlier section of this handbook to determine the required size of the connecting tubing.

The procedure presented above is the work of Wes Master, reported in *Product Engineering.*

HYDRAULIC PISTON ACCELERATION, DECELERATION, FORCE, FLOW, AND SIZE DETERMINATION

What net acceleration force is needed by a horizontal cylinder having a 10,000-lb (4500kg) load and 500-lb (2.2-kN) friction force, if 1500 lb/in^2 (gage) (10,341 kPa) is available at the cylinder port, there is zero initial piston velocity, and a 100-ft/min (30.5-m/min) terminal velocity is reached after 3-in (76.2-mm) travel at constant acceleration with the rod extending? Determine the required piston diameter and maximum fluid flow needed.

What pressure will stop a piston and load within 2 in (50.8 mm) at constant deceleration if the cylinder is horizontal, the rod is extending, the load is 5000 lb (2250 kg), there is a 500-lb (2224-N) friction force, the driving pressure at the head

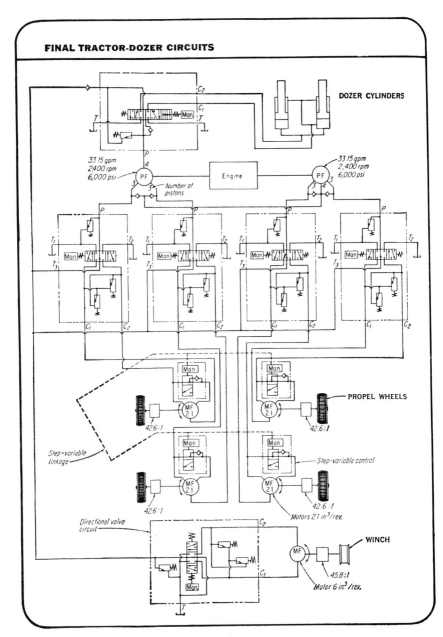

FIGURE 19 Final tractor-dozer hydraulic circuits.

end is 800 lb/in² (gage) (5515.2 kPa), and the initial velocity is 80 ft/min (24.4 m/min)? The rod diameter is 1 in (25.4 mm), and the piston diameter is 1.5 in (38.1 mm).

Calculation Procedure:

1. Find the needed accelerating force

Use the relation $F_A = Ma = M \, \Delta V/\Delta t$, where F_A = net accelerating force, lb (N); M = mass, slugs or lb·s²/ft (N·s²/m); a = linear acceleration, ft/s²(m/s²), assumed constant; ΔV = velocity change during acceleration, ft/s (m/s); Δt = time to reach terminal velocity, s. Substituting for this cylinder, we find $M = 10,000/32.17 = 310.85$ slugs.

Next $\Delta S = 3$ in/(12 in/ft) = 0.25 ft (76.2 mm). Also $\Delta V = (100$ ft/min)/(60 s/min) = 1.667 ft/s (0.51 m/s). Then $F_A = 0.5(310.85)(1.667)^2/0.25 = 1727.6$ lb (7684.4 N).

2. Determine the piston area and diameter

Add the friction force and compute the piston area and diameter thus: $\Sigma F = F_A + F_F$, where ΣF = sum of forces acting on piston, i.e., pressure, friction, inertia, load, lb; F_F = friction force, lb. Substituting gives $\Sigma F = 1727.6 + 500 = 2227.6$ lb (9908.4 N).

Find the piston area from $A = \Sigma F/P$, where P = fluid gage pressure available at the cylinder port, lb/in². Or, $A = 2227.6/1500 = 1.485$ in² (9.58 cm²). The piston diameter D, then, is $D = (4A/\pi)^{0.5} = 1.375$ in (34.93 mm).

3. Compute the maximum fluid flow required

The maximum fluid flow Q required is $Q = VA/231$, where Q = maximum flow, gal/min; V = terminal velocity of the piston, in/s; A = piston area, in². Substituting, we find $Q = (100 \times 12)(1.485)/231 = 7.7$ gal/min (0.49 L/s).

4. Determine the effective driving force for the piston with constant deceleration

The driving force from pressure at the head end is F_D = [fluid pressure, lb/in² (gage)](piston area, in²). Or, $F_D = 800(1.5)^2\pi/4 = 1413.6$ lb (6287.7 N). However, there is a friction force of 500 lb (2224 N) resisting this driving force. Therefore, the effective driving force is $F_{ED} = 1413.6 - 500 = 913.6$ lb (4063.7 N).

5. Compute the decelerating forces acting

The mass, in slugs, is $M = F_A/32.17$, from the equation in step 1. By substituting, $M = 5000/32.17 = 155.4$ slugs.

Next, the linear piston travel during deceleration is $\Delta S = 2$ in/(12 in/ft) = 0.1667 ft (50.8 mm). The velocity change is $\Delta V = 80/60 = 1.333$ ft/s (0.41 m/s) during deceleration.

The decelerating force $F_A = 0.5M(\Delta V^2)/\Delta S$ for the special case when the velocity is zero at the start of acceleration at the end of deceleration. Thus $F_A = 0.5(155.4)(1.333)^2/0.1667 = 828.2$ lb (3684 N).

The total decelerating force is $\Sigma F = F_A + F_{ED} = 827.2 + 913.6 = 1741.8$ lb (7748 N).

6. Find the cushioning pressure in the annulus

The cushioning pressure is $P_c = \Sigma F/A$, where A differential area = piston area − rod area, both expressed in in². For this piston, $A = \pi(1.5)^2/4 - \pi(1.0)^2/4 = 0.982$

in^2 (6.34 cm^2). Then $P = \Sigma F/A = 1741.8/0.982 - 1773.7$ lb/in^2 (gage) (12,227.9 kPa).

Related Calculations. Most errors in applying hydraulic cylinders to accelerate or decelerate loads are traceable to poor design or installation. In the design area, miscalculation of acceleration and/or deceleration is a common cause of problems in the field. The above procedure for determining acceleration and deceleration should eliminate one source of design errors.

Rod buckling can also result from poor design. A basic design rule is to allow a compressive stress in the rod of 10,000 to 20,000 lb/in^2 (68,940 to 137,880 kPa) as long as the effective rod length-to-diameter ratio does not exceed about 6:1 at full extension. A firmly guided rod can help prevent buckling and allow at least four times as much extension.

With rotating hydraulic actuators, the net accelerating, or decelerating torque in lb·ft (N·m) is given by $T_A = J\alpha = MK^2$ rad/s$^2 = 0.1047\ MK\ \Delta N/\Delta T = WK^2$ $\Delta N/(307)\ \Delta t$, where J = mass moment of inertia, slugs·ft^2, or lb·s^2·ft; α = angular acceleration (or deceleration), rad/s^2; K = radius of gyration, ft; ΔN = r/min change during acceleration or deceleration; other symbols as given earlier.

For the special case where the r/min is zero at the start of acceleration or end of deceleration, $T_A = 0.0008725MK^2\ (\Delta N)^2/\Delta revs$; in this case, $\Delta revs$ = total revolutions = average r/min $\times\ \Delta t/60 = 0.5\ \Delta N\Delta T/60$; $\Delta t = 120(\Delta revs/\Delta t)$. For the linear piston and cylinder where the piston velocity at the start of acceleration is zero, or at the end of deceleration is zero, $\Delta t = \Delta S/$average velocity $= \Delta S/(0.5\ \Delta V)$.

High water base fluids (HWBF) are gaining popularity in industrial fluid power cylinder applications because of lower cost, greater safety, and biodegradability. Cylinders function well on HWBF if the cylinder specifications are properly prepared for the specific application. Some builders of cylinders and pumps offer designs that will operate at pressures up to several thousand pounds per square inch, gage. Most builders, however, recommend a 1000-lb/in^2 (gage) (6894-kPa) limit for cylinders and pumps today.

Robotics is another relatively recent major application for hydraulic cylinders. There is nothing quite like hydrostatics for delivering high torque or force in cramped spaces.

This procedure is the work of Frank Yeaple, Editor, *Design Engineering,* as reported in that publication.

HYDROPNEUMATIC ACCUMULATOR DESIGN FOR HIGH FORCE LEVELS

Design a hydropneumatic spring to absorb the mechanical shock created by a 300-lb (136.4-kg) load traveling at a velocity of 20 ft/s (6.1 m/s). Space available to stop the load is limited to 4 in (10.2 cm).

Calculation Procedure:

1. Determine the kinetic energy which the spring must absorb

Figure 20 shows a typical hydropneumatic accumulator which functions as a spring. The spring is a closed system made up of a single-acting cylinder (or sometimes a rotary actuator) and a gas-filled accumulator. As the load drives the piston, fluid (usually oil) compresses the gas in the flexible rubber bladder. Once the load is

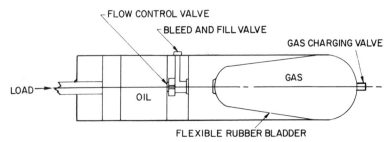

FIGURE 20   Typical hydropneumatic accumulator. (*Machine Design.*)

removed, either partially or completely, the gas pressure drives the piston back for the return cycle.

The flow-control valve limits the speed of the compression and return strokes. In custom-designed springs, flow-control valves are often combinations of check valves and fixed or variable orifices. Depending on the orientation of the check valve, the compression speed can be high with low return speed, or vice versa. Within the pressure limits of the components, speed and stroke length can be varied by changing the accumulator precharge. Higher precharge pressure gives shorter strokes, slower compression speed, and faster return speed.

The kinetic energy that must be absorbed by the spring is given by $E_k = 12WV^2/2g$, where E_k = kinetic energy that must be absorbed, in · lb; W = weight of load, lb; V = load velocity, ft/s; g = acceleration due to gravity, 32.2 ft/s². From the given data, $E_k = 12(300)(20)^2/2(32.2) = 22{,}360$ in · lb (2526.3 N · m).

2. Find the final pressure of the gas in the accumulator
To find the final pressure of the gas in the accumulator, first we must assume an accumulator size and pressure rating. Then we check the pressure developed and the piston stroke. If they are within the allowable limits for the application, the assumptions were correct. If the limits are exceeded, we must make new assumptions and check the values again until a suitable design is obtained.

For this application, based on the machine layout, assume that a 2.5-in (6.35-cm) cylinder with a 60-in³ (983.2-cm³) accumulator is chosen and that both are rated at 2000 lb/in² (13,788 kPa) with a 1000-lb/in² (abs) (6894-kPa) precharge. Check that the final loaded pressure and volume are suitable for the load.

The final load pressure p_2 lb/in² (abs) (kPa) is found from $p_2^{(n-1)/n} = p_2^{(n-1)/n}\{[E_k(n-1)/(p_1v_1)] + 1\}$, where p_1 = precharge pressure of the accumulator, lb/in² (abs) (kPa); n = the polytropic gas constant = 1.4 for nitrogen, a popular charging gas; v_1 = accumulator capacity, in³ (cm³). Substituting gives $p_2^{(1.4-1)/1.4} = 1000^{(1.4-1)/1.4}\{[22{,}360(1.4-1)/(1000 \times 60)] + 1\} = 1626$ lb/in² (abs) (11,213.1 kPa). Since this is within the 2000-lb/in² (abs) limit selected, the accumulator is acceptable from a pressure standpoint.

3. Determine the final volume of the accumulator
Use the relation $v_2 = v_2(p_1/p_2)^{1/n}$, where v_2 = final volume of the accumulator, in³; v_1 = initial volume of the accumulator, in³; other symbols as before. Substituting, we get $v_2 = 60(1000/1626)^{1/1.4} = 42.40$ in³ (694.8 cm³).

4. Compute the piston stroke under load
Use the relation $L = 4(v_1 - v_2)\pi D^2$, where L = length of stroke under load, in; D = piston diameter, in. Substituting yields $L = 4(60 - 42.40)/(\pi \times 2.5^2)$ 3.58 in

(9.1 cm). Since this is within the allowable travel of 4 in (10 cm), the system is acceptable.

Related Calculations. Hydropneumatic accumulators have long been used as shock dampers and pulsation attenuators in hydraulic lines. But only recently have they been used as mechanical shock absorbers, or springs.

Current applications include shock absorption and seat-suspension systems for earth-moving and agricultural machinery, resetting mechanisms for plows, mill-roll loading, and rock-crusher loading. Potential applications include hydraulic hammers and shake tables.

In these relatively high-force applications, hydropneumatic springs have several advantages over mechanical springs. First, they are smaller and lighter, which can help reduce system costs. Second, they are not limited by metal fatigue, as mechanical springs are. Of course, their life is not infinite, for it is limited by wear of rod and piston seals.

Finally, hydropneumatic springs offer the inherent ability to control load speeds. With an orifice check valve or flow-control valve between actuator and accumulator, cam speed can be varied as needed.

One reason why these springs are not more widely used is that they are not packaged as off-the-shelf items. In the few cases where packages exist, they are often intended for other uses. Thus, package dimensions may not be those needed for spring applications, and off-the-shelf springs may not have all the special system parameters needed. But it is not hard to select individual off-the-shelf accumulators and actuators for a custom-designed system. The procedure given here is an easy method for calculating needed accumulator pressures and volumes. It is the work of Zeke Zahid, Vice President and General Manager, Greer Olaer Products Division, Greer Hydraulics, Inc., as reported in *Machine Design.*

MEMBRANE VIBRATION IN HYDRAULIC PRESSURE-MEASURING DEVICES

A pressure-measuring device for an industrial hydraulic system is to be constructed of a 0.005-in (0.0127-cm) thick alloy steel circular membrane stretched over a chamber opening, as shown in Fig. 21. The membrane is subjected to a uniform tension of 2000 lb (8900 N) and then secured in position over a 6-in (15.24-cm) diameter opening. The steel has a modulus of elasticity of 30,000,000 lb/in² (210.3 GPa) and weights 0.3 lb/in³ (1.1 N/cm³). Vibration of the membrane due to pres-

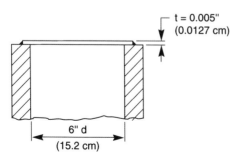

FIGURE 21 Membrane for pressure-measuring device.

sure in the chamber is to be picked up by a strain gage mechanism; in order to calibrate the device, it is required to determine the fundamental mode of vibration of the membrane.

Calculation Procedure:

1. Compute the weight of the membrane per unit area
Weight of the membrane per unit area, $w = w_u \times t$, where the weight per unit volume, $w_u = 0.3$ lb/in³ (1.1 N/cm³); membrane thickness, $t = 0.005$ in (0.0127 cm). Hence, $w = 0.3 \times 0.005 = 0.0015$ lb/in² (0.014 N/cm²).

2. Compute the uniform tension per unit length of the membrane boundary
Uniform tension per unit length of the membrane boundary, $S = F/L$, where the uniformly applied tensile force, $F = 2000$ lb (8900 N); length of the membrane boundary, $L = d = 6$ in (15.24 cm). Thus, $S = 2000/6 = 333$ lb/in (584 N/cm).

3. Compute the area of the membrane
The area of the membrane, $A = \pi d^2/4 = \pi(6)^2/4 = 28.27$ in² (182.4 cm²).

4. Compute the frequency of the fundamental mode of vibration in the membrane
From *Marks' Standard Handbook for Mechanical Engineers,* 9th edition, McGraw-Hill, Inc., the frequency of the fundamental mode of vibration of the membrane, $f = (\alpha/2\pi)[(gS)/(wA)]^{1/2}$, where the membrane shape constant for a circle, $\alpha = 4.261$; gravitational acceleration, $g = 32.17 \times 12 = 386$ in/s² (980 cm/s²); other values as before. Then, $f = (4.261/2\pi)[(386 \times 333)/(0.0015 \times 28.27)]^{1/2} = 1181$ Hz.

Related Calculations. To determine the value for S in step 2 involves a philosophy similar to that for the hoop stress formula for thin-wall cylinders, i.e., the uniform tension per unit length of the membrane boundary depends on tensile forces created by uniformly stretching the membrane in all directions. Therefore, for symmetrical shapes other than a circle, such as those presented in *Marks' M. E. Handbook,* the value for L in the equation for S as given in this procedure is the length of the longest line of symmetry of the geometric shape of the membrane. The shape constant and other variable values change accordingly.

POWER SAVINGS ACHIEVABLE IN INDUSTRIAL HYDRAULIC SYSTEMS

An industrial hydraulic system can be designed with three different types of controls. At a flow rate of 100 gal/min (6.31 L/s), the pressure drop across the controls is as follows: Control A, 500 lb/in² (3447 kPa); control B, 100 lb/in² (6894 kPa); control C, 2000 lb/in² (13,788 kPa). Determine the power loss and the cost of this loss for each control if the cost of electricity is 15 cents per kilowatthour. How much more can be spent on a control if it operates 3000 h/year?

Calculation Procedure:

1. Compute the horsepower lost in each control

The horsepower lost during pressure drop through a hydraulic control is given by $horsepower = 5.82(10^{-4})Q \, \Delta P$, where Q = flow rate through the control, gal/min; ΔP = pressure loss through the control. Substituting for each control and using the letter subscript to identify it, we find $horsepower_A = 5.82(10^{-4})(100)(500) = 29.1$ horsepower (21.7 kW); $horsepower_B = 5.82(10^{-4})(100)(1000) = 5.82$ horsepower (43.4 kW); $horsepower_c = 5.82(10^{-4})(100)(2000) = 116.4$ horsepower (86.8 kW).

2. Find the cost of the pressure loss in each control

The cost in dollars per hour wasted w = kW($/kWh) = $horsepower(0.746)($/kWh). Substituting and using a subscript to identify each control, we get $w_A = 21.7(\$0.15) = \$3.26$; $w_B = 43.4(\$0.15) = \$6.51$; $w_c = 86.8(\$0.15) = \$13.02$.

The annual loss for each control with 3000-h operation is $w_{A,an} = 3000(\$3.26) = \$9780$; $w_{B,an} = 3000(\$6.51) = \$19,530$; $w_{C,an} = 3000(\$13.02) = \$39,060$.

3. Determine the additional amount that can be spent on a control

Take one of the controls as the base or governing control, and use it as the guide to the allowable extra cost. Using control C as the base, we can see that it causes an annual loss of $39,060. Hence, we could spend up to $39,060 for a more expensive control which would provide the desired function with a smaller pressure (and hence, money) loss.

The time required to recover the extra money spent for a more efficient control can be computed easily from ($39,060 − loss with new control, $), where the losses are expressed in dollars per year.

Thus, if a new control costs $2500 and control C costs $1000, while the new control reduced the annual loss to $20,060, the time to recover the extra cost of the new control would be ($2500 − $1000)/($39,060 − $20,060) = 0.08 year, or less than 1 month. This simple application shows the importance of careful selection of energy control devices.

And once the new control is installed, it will save $39,060 − $20,060 = $19,000 per year, assuming its maintenance cost equals that of the control it replaces.

Related Calculations. This approach to hydraulic system savings can be applied to systems serving industrial plants, aircraft, ships, mobile equipment, power plants, and commercial installations. Further, the approach is valid for any type of hydraulic system using oil, water, air, or synthetic materials as the fluid.

With greater emphasis in all industries on energy conservation, more attention is being paid to reducing unnecessary pressure losses in hydraulic systems. Dual-pressure pumps are finding wider use today because they offer an economical way to provide needed pressures at lower cost. Thus, the alternative control considered above might be a dual-pressure pump, instead of a throttling valve.

Other ways that pressure (and energy) losses are reduced is by using accumulators, shutting off the pump between cycles, modular hydraulic valve assemblies, variable-displacement pumps, electronic controls, and shock absorbers. Data in this procedure are from *Product Engineering* magazine, edited by Frank Yeaple.

PNEUMATIC-CIRCUIT ANALYSIS USING VARIOUS EQUATIONS AND COEFFICIENTS

The pneumatic system in Fig. 22 has been designed for use in an industrial application; the flow rate through this system is 250 scfm (7.075 cu m/min). As part of

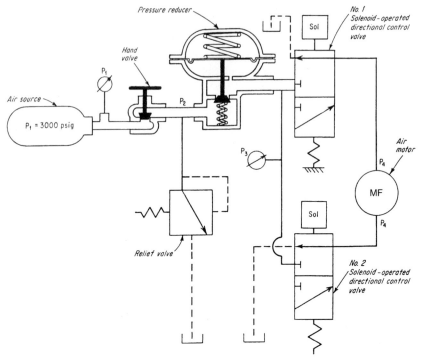

FIGURE 22   Pneumatic circuit for industrial application.

the design, the hand valve, pressure reducer, and air motor have already been specified and catalog performance data assembled for them. The only unknown is the size and capacity of the control valve. Before this valve can be specified, the minimum acceptable value for the flow factor, F, must be determined for each component in the system. Find this value for each unit in this decision.

Calculation Procedure:

1.   *Tabulate the known flow coefficients and their defining equations*
Obtain from the component manufacturers the flow coefficients and defining equations and tabulate them as shown in Table 10. Use the symbols given as part of the table. For this procedure, the National Bureau of Standards (NBS) flow factor F is used as the standard flow coefficient. However, any other standard flow coefficient the handbook user would like to use will work equally well. Equations based on the NBS flow factor F are the most accurate and there are many published design techniques to simplify their use.

The procedure given her shows how to convert to F from C_V and D_0. This will be done for each component of the pneumatic system.

2.   *Convert each known flow coefficient to the flow factor,* F
Use the conversion equations in Table 11, correlating them with the given values in Table 10. Thus:

TABLE 10 Flow Coefficients and Defining Equations

Component	Flow coefficient	Defining equation
Hand valve	$C_V = 1.26$	$C_V = \dfrac{Q \times 60}{1360} \sqrt{\dfrac{GT_U}{\Delta P \times P_U}}$
Pressure reducer	$D_0 = 0.25$	$D_0 = \sqrt{\dfrac{Q}{33P_U}} \times \dfrac{1}{\sqrt{r(r^{0.43} - r^{0.71})}}$
Control valve	$F = $ unknown	$F = \dfrac{Q}{P_U \sqrt{8/5}} \sqrt{\dfrac{1}{r(1 - r)(3 - r)}}$
Air motor	$Q = $ scfm; $P = 600$ lb/in² (abs) (given)	

Symbols

Q = air flow in standard units, scfm (14.7 lb/in² (abs), 68°F) (cu m/min)
q = air flow at actual conditions, cfm. $Q = q(P/14.7)(528/T)$ (cu m/min)
V = velocity, fps (average through valve)
P = pressure in absolute units, psia (subscript D = downstream, U = upstream) (kPa)
p = gage pressure, psi (kPa)
ΔP = pressure drop, psi (kPa)
r = pressure ratio P_D/P_U
ρ = density, lb/ft³ (kg/m³)
G = specific gravity, ρ_{gas}/ρ_{air}
T = absolute temperature, deg R = deg $F + 460$
A = Inlet pipe area, in² (cm²)
D_c = diameter of equivalent sharp-edge orifice, in (coefficient of discharge $C_D = 0.6$)
M = molecular weight, lb (M = 29 lb for air) (kg)
W = water weight flow, lb/sec (kg/sec)
C_v, K, = typical flow coefficients (also called flow constants and flow factors) in a flow
F, D_o equation

Hand valve: $F = 0.556C_V = 0.556 \times 1.26 = 0.7$
Reducer: $F = 10D_0^2 = 10 \times (0.25)^2 = 0.625$

Control valve: Calculate the control-valve flow factor F from the known flow and pressure at P_4 and the calculated pressure at P_3, Fig. 21. Do this by starting at the supply pressure $P_1 = 3014.7$ lb/in² (abs) (20.77 MPa), and knowing that the flow is 250 scfm (7.075 cu m/min), find P_2 and P_3 by substitution in the NBS flow equation—Equation 9 in Table 12—for each component in the system thus:

Hand valve:

$$F = \frac{Q}{P_U \sqrt{8/5}} \sqrt{\frac{1}{r(1 - r)(3 - r)}}$$

where $P_V = P_1 = 3014.7$ lb/in² (abs) (20.77 MPa); $Q = 250$ scfm (7.075 cu m/min); $F = 0.7$; $r = P_2/P_1$. Solve for r by trial and error. The abbreviated list of values in Table 13 for $r(1 - r)(3 - r)$ will help.

TABLE 11 Conversion Equations

	D_o	C_v	F	K — $r = 1.0$	K — $r = 0.75$	K — $r = 0.5$
D_o		$= 0.236\sqrt{C_v}$	$= 0.316\sqrt{F}$	$= 1.456\,\dfrac{\sqrt{A}}{K^{1/4}}$	$= 1.521\,\dfrac{\sqrt{A}}{K^{1/4}}$	$= 1.641\,\dfrac{\sqrt{A}}{K^{1/4}}$
C_v	$= 18.0D_0^2$		$= 1.8F$	$= 38.2\,\dfrac{A}{\sqrt{K}}$	$= 41.5\,\dfrac{A}{\sqrt{K}}$	$= 48.3\,\dfrac{A}{\sqrt{K}}$
F	$= 10D_0^2$	$= 0.556C_v$		$= 21.2\,\dfrac{A}{\sqrt{K}}$	$= 23.1\,\dfrac{A}{\sqrt{K}}$	$= 26.9\,\dfrac{A}{\sqrt{K}}$
K $\quad r = 1.0$	$= 4.5\,\dfrac{A^2}{D_0^4}$	$= 1460\,\dfrac{A^2}{C_v^2}$	$= 450\,\dfrac{A^2}{F^2}$			
$\quad\quad r = 0.75$	$= 5.36\,\dfrac{A^2}{D_0^4}$	$= 1725\,\dfrac{A^2}{C_v^2}$	$= 534\,\dfrac{A^2}{F^2}$			
$\quad\quad r = 0.5$	$= 7.29\,\dfrac{A^2}{D_0^4}$	$= 2330\,\dfrac{A^2}{C_v^2}$	$= 724\,\dfrac{A^2}{F^2}$			

NOTE: The K factor varies with r and A and you must know which values the manufacturer used to derive his published K. For example, if K was derived at $r = 0.75$ and valve inlet pipe area $A = 0.2$, then $F = 23.1 \times 0.2/\sqrt{K} = 4.62/\sqrt{K}$.

TABLE 12 Typical Air-Flow Equations (Taken from Catalogs)

Eq no	Equation for flow (sub-critical) (Q = scfm-standard ft³/min)	Flow coefficient defined
1	$Q = 33\, D_o^2 P_U \sqrt{r(r^{0.43} - r^{0.71})}$	D_o = equivalent sharp-edged orifice (coeff of discharge $C_D = 0.6$) $\quad = \sqrt{\dfrac{Q}{33\,P_U}} \times \dfrac{1}{\sqrt{r(r^{0.43} - r^{0.71})}}$
2	$Q = \dfrac{963}{60}\, C_V \sqrt{\dfrac{\Delta P(P_U + P_D)}{GT_U}}$	C_V = valve coefficient $\quad = \dfrac{Q \times 60}{963}\sqrt{\dfrac{GT_U}{\Delta P(P_U + P_D)}}$
3	$Q = \dfrac{1360}{60}\, C_V \sqrt{\dfrac{\Delta P \times P_U}{GT_U}}$	C_V = flow coefficient $\quad = \dfrac{Q \times 60}{1360}\sqrt{\dfrac{GT_U}{\Delta P \times P_U}}$
4	$Q = \dfrac{1390}{60}\, C_V \sqrt{\dfrac{\Delta P \times P_D}{GT_U}}$	C_V = capacity factor $\quad = \dfrac{Q \times 60}{1390}\sqrt{\dfrac{GT_U}{\Delta P \times P_D}}$
5	$Q = \dfrac{5180}{60}\, C_V \sqrt{\dfrac{P_U^2 - P_D^2}{MT_U}}$	C_V = valve-flow coefficient $\quad = \dfrac{Q \times 60}{5180}\sqrt{\dfrac{MT_U}{P_U^2 - P_D^2}}$
6	$Q = \dfrac{963}{60}\, C_V \sqrt{\dfrac{P_U^2 - P_D^2}{GT_U}}$	C_V = flow coefficient (Eq 2 = Eq 6) $\quad = \dfrac{Q \times 60}{963}\sqrt{\dfrac{GT_U}{P_U^2 - P_D^2}}$
7	$Q = \dfrac{2.32^{0.443}}{60}\, C_G \sqrt{\dfrac{\Delta P^{0.443} \times P_U^{0.6}}{GT_U/520}}$	C_G = gas-flow coefficient $\quad = \dfrac{Q \times 60}{(2.32)^{0.443}} \times \dfrac{\sqrt{GT_U/520}}{\Delta P^{0.443} \times P_U^{0.6}}$
8	$Q = FP_U\sqrt{4/3}\sqrt{1 - r^2}$	F = NBS flow factor (#1) $\quad = \dfrac{Q}{P_U\sqrt{4/3}}\sqrt{\dfrac{1}{1 - r^2}}$
9	$Q = FP_U\sqrt{8/5}\sqrt{r(1 - r)(3 - r)}$	F = NBS flow factor (#2 – better) $\quad = \dfrac{Q}{P_U\sqrt{8/5}}\sqrt{\dfrac{1}{r(1 - r)(3 - r)}}$
10	$Q = 38.1\, P_U A \sqrt{\dfrac{1 - r}{K}}$	K = K factor $\quad = \dfrac{2g}{\rho V^2}\,\Delta P$

TABLE 13 Computed r Values

r	$r(1 - r)(3 - r)$
0.5	0.625
0.7	0.483
0.8	0.352
0.9	0.189
0.92	0.153
0.94	0.116
0.96	0.0784
0.98	0.0396
0.99	0.0199
0.995	0.0100
1.000	0.0000

For this hand valve, $r = 0.995$. Therefore, $P_3 = 0.995 \times 3014.7 = 2999.63$ lb /in^2 (abs) ((20.67 MPa). Then, $P = 3014.7 - 2999.6 = 15.1$ lb/in^2 (abs) (105.0 kPa).

Reducer: Determine P_3 in the same way. This gives $r = 0.994$; $P_3 = 0.994 \times 3000 = 2982$ lb/in^2 (abs) (20.5 MPa); $P = 3000 - 2982 = 18$ lb/in^2 (124 kPa).

Control valve: Compute the minimum flow factor F for the control valve from

$$F = \frac{Q}{P_U\sqrt{8/5}} \sqrt{\frac{1}{r(1 - r)(3 - r)}}$$

where $Q = 250$ scfm (7.075 cu m/min); $P_V = P_3 = 2982$ lb/in^2 (abs) (20.5 MPa); $r = P_4/P_3 = 600/2982 = 0.201$.

Note: The value $r = 0.201$ is less than the critical flow ratio of $r = 0.5$; therefore, F by definition is $Q/P_V = 250/2982 = 0.0838$. A valve manufacturer will accept a flow coefficient such as F as a specification for minimum flow because the coefficient completely defines flow and pressure drop.

3. Find the flow coefficient for the relief valve
The relief valve ($A/[K]^{0.5} = 0.021$) is not part of the flow path. However, you can convert its flow coefficient to F and compute the valve's relieving capacity. The pressure ratio r is less than 0.5 because the valve discharges to atmosphere. Hence, the conversion equation is

$$F = 26.9 \, \frac{A}{\sqrt{K}} = 0.565$$

The maximum flow is then $Q = F \, P_V = 0.565 \times 3000 = 1695$ scfm (47.96 cvu m/min). The relieving capacity is $1695/250 = 6.78$ times the system design capacity.

Related Calculations. During the design of any series of pneumatic systems it is wise to standardize the flow coefficient that will be used. Then the results will be consistent for all systems designed. The NBS coefficient given here is an acceptable and valid design value for pneumatic systems used in industrial machines, aircraft, ships, etc.

This procedure is the work of Dominic Lapera, Chief Engineer, Kemp Aero Products, and Franklin D. Yeaple, Associate Editor, *Product Engineering* magazine. SI values were added by the handbook editor.

AIR FLOW THROUGH CLOSE-CLEARANCE ORIFICES IN PNEUMATIC SYSTEMS

(1) Find the mass rate of flow of air through a capillary orifice, Fig. 23, when the pressure drop is small, 0.32 in H_2O (0.81 cm)[1.67 psf], and the following conditions prevail: length, $L = 0.25$ in (0.685 cm); diameter, $D = 0.0625$ in (0.158 cm); air density $= 0.0743$ lb/cu ft (1.19 kg/cu m); and viscosity $= 4.79 \times 10^{-7}$ lb/s/ ft^2 (229.35 Pas $\times 10^{-7}$). (2) Find the air flow through a capillary with a larger pressure drop for these conditions: diameter, $D = 0.0156$ in (0.0396 cm); length, $L = 0.135$ in (0.342 cm); air temperature $= 200°F$ (93.3°C); air viscosity $=$ the same as in (1) above: $P_1 = 21$ lb/in^2 (abs) (144.7 kPa); $P_2 = 14.7$ lb/in^2 (abs) (101.3 kPa); flow is isentropic ($n = 1.4$); coefficient of specific heat at constant pressure, $c_p = 0.24$ Btu/lb°F (1004.2 J/kg°C).

Calculation Procedure:

1. Compute the air velocity through the capillary
Use the equation

$$V = K\sqrt{2g(P_1 - P_2)/\rho}$$

where the symbols are as given below. Then

$$V = K\sqrt{64.4 \times 1.67/0.0743}$$

$$V = 37.9K \text{ ft/s} \quad (11.55\ K \text{ m/s})$$

Thus, the fluid velocity, V, ft/s (m/s) is a function of K, the dimensionless orifice coefficient.

2. Determine the Reynolds number for the flow situation
Use the relations,

$$W = \rho VA$$

$$R_e = WD/A\mu g$$

$$\rho VAD/A\mu g$$

$$= \rho VD/\mu g$$

Substituting,

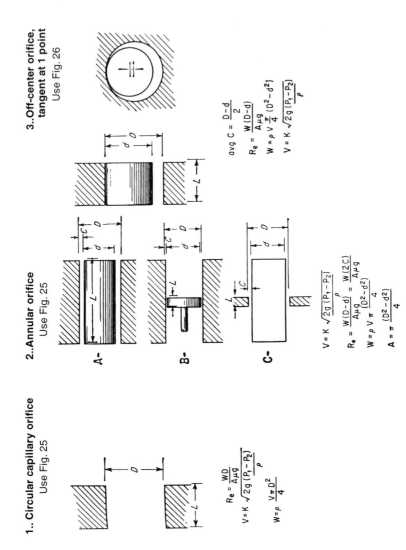

FIGURE 23 Three typical orifices and their associated analysis equations.

$$R_e = (0.0743 \times 37.9K \times 0.0625)/$$

$$(12 \times 4.79 \times 10^{-7} \times 32.2)$$

$$= 950K$$

Now we must make one or more trial solutions using the correction chart in Fig. 25 to determine the actual Reynolds number for this orifice. We will try different values of K to see what Reynolds number each will yield. Thus:
   First trial: Let $K = 0.5$; then $R = 950 \times 0.5 = 475$. Figure 25 shows that for $D/2L = 0.0625/0.54 = 0.116$, and $K = 0.5$, $R_e = 180$. This is a wrong guess. *Second trial:* Try $K = 0.7$; then $R_e = 950 \times 0.7 = 665$. Figure 25 shows $R_e = 900$. Again, this is a wrong guess. *Third trial:* Try $K = 0.65$, then $R_e = 950 \times 0.65 = 617.5$; Fig. 25 shows $R_e = 600$, which is close enough to the computed 617.5, value within 3 percent.

3.   Find the velocity and mass flow rate for the orifice
From step 1, the velocity, $V = 37.9 K = 37.9 \times 0.65 = 24.63$ ft/s (7.5 m/s). The mass rate of flow, $W = 0.0743 (24.63)( )(0.0625)^2/(4 \times 144) = 0.000039$ lb/s (0.0000018 kg/s).

4.   Find the Reynolds number for flow through the larger orifice
Use the following two relations and substitute as shown below

$$W = \frac{223.8AP_1}{R} \times$$

$$\left(\sqrt{\frac{c_p}{T_1}} \left[\left(\frac{P_2}{P_1}\right)^{2/n} - \left(\frac{P_2}{P_1}\right)^{(n+)/n} \right] \right) K \qquad (2)$$

$$R_e = \frac{W(2C)}{A\mu g} = \frac{223.8P_1}{R} \times$$

$$\left(\sqrt{\frac{c_p}{T_1}} \left[\left(\frac{P_2}{P_1}\right)^{2/n} - \left(\frac{P_2}{P_1}\right)^{(n+1)/n} \right] \right) K \frac{2C}{\mu g} \quad (2) \qquad (3)$$

$$= \frac{223.8 \times 21 \times 144}{53.3 \times 4.79 \times 10^{-7}} \times$$

$$\sqrt{\frac{.24}{660}} \left[\left(\frac{14.7}{21}\right)^{2/1.4} - \left(\frac{14.7}{21}\right)^{(1.4+1)/1.4} \right]$$

$$\times K \frac{0.0156}{12 \times 32.2} = 4900K$$

5.   Determine the mass rate of flow for the larger orifice
Proceed as in step 2, above, with $D/2L = 0.0156/0.27 = 0.05777$. Using Fig. 25, assume values for K, as before. *First trial:* $K = 0.5$; $R_e = 4900 \times 0.5 = 2450$. From Fig. 25, $R_e = 350$, which is a wrong guess. *Second trial:* Try $K = 0.6$; then

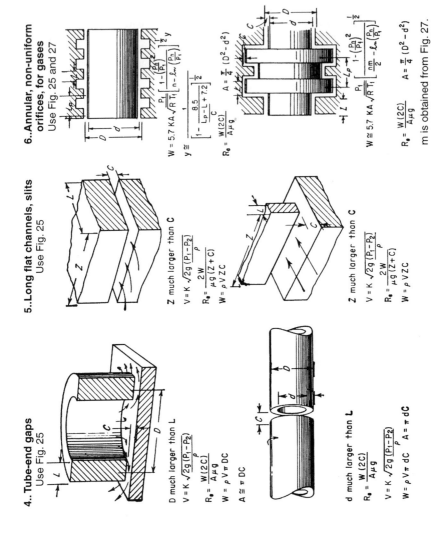

FIGURE 24 Three more orifices and their associated analysis equations.

K values for symmetrical restrictions

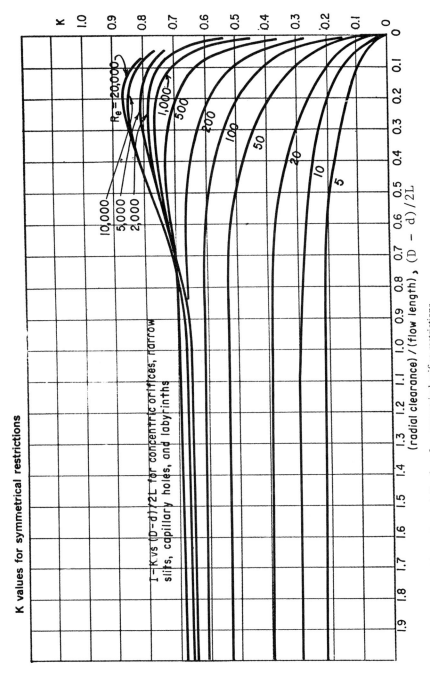

FIGURE 25 Reynolds number and *K* values for symmetrical orifice restrictions.

25.53

$R_e = 700$, from Fig. 25; wrong again. *Third trial:* Try $K = 0.715$; then $R_e = 3503.5$; Fig. 25 shows $R_e = 3550$, which is close enough, within 1.3 percent.

Substituting in the first equation above $W = 0.55 \times 10^{-4}$ lb/s (0.2497 kg $\times 10^{-4}$ kg/s).

Related Calculations. When a restriction is too long for pure orifice flow analysis, and too short for line flow analysis, as is the case for many short capillary orifices and close-clearance labyrinths, Figs. 23 and 24, considered in this procedure, empirical solutions must be used. Published test results which agree well with computed data are summarized on a mean-value basis in Figs. 25 and 26.

The basic equation selected for the analysis in this procedure is

$$V = K\sqrt{2gH}$$

$$V = K\sqrt{2g(P_1 - P_2)/\rho} \tag{1}$$

Another equation could have been selected but this is adequate. In this equation, the variables are: V = velocity; K = empirical coefficient, Figs. 24 and 25; H = head loss = $(P_1 - P_2)/\rho$, where P_1 = inlet pressure, psf (kPa); P_2 = outlet pressure, psf (kPa); ρ = fluid density in suitable units. This is basically the equation for incompressible flow through an orifice.

The devices discussed in this procedure are not pure orifices. Neither are they long tubes. The correction charts in Figs. 25 and 26, based on actual tests, take care of discrepancies. Another equation could have been chosen for tubes, and the method would still work, except that a different set of correction charts would be needed.

Flow of compressible fluids also can be calculated with this method. If the density change from inlet to outlet is slight, then assume the gas is incompressible and continue to use Equation 1, above. If there is considerable expansion, then substitute the following conventional equation for polytropic compressible flow:

$$W = \frac{KA_2 P_1}{R}$$

$$\times \sqrt{2gJ \frac{c_p}{T_1} \left[\left(\frac{P_2}{P_1}\right)^{2/n} - \left(\frac{P_2}{P_1}\right)^{(n+1)/n} \right]} \tag{2}$$

Units for Equation 2 are given below. Use Fig. 25 for the value of coefficient K. In the case of a labyrinth seal, special versions of Equation 2 have been developed. These accompany the sketches of labyrinths in Fig. 24. The seal coefficient, m, is plotted in Fig. 27 for all common values of the length-clearance ratio L/C.

the Reynolds number is the heart of this method. Here is a convenient form for round orifices and capillaries:

$$R_e \doteq WD/A\mu g \tag{3A}$$

For annular orifices, a different version is needed:

$$R_e = W(D - d)/A\mu g \tag{3B}$$

And for slots:

$$R_e = W(2C)/A\mu g \tag{3C}$$

Equations 3B and 3C were derived using the concept of hydraulic radius, R_H,

K values for eccentric restrictions

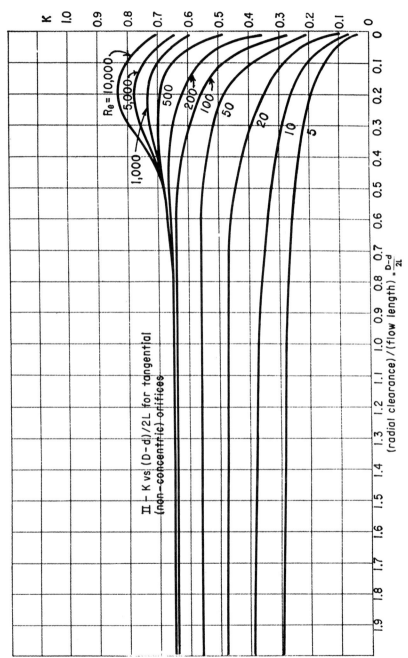

II – K vs (D–d)/2L for tangential (non-concentric) orifices

FIGURE 26 Reynolds number and K values for eccentric restrictions.

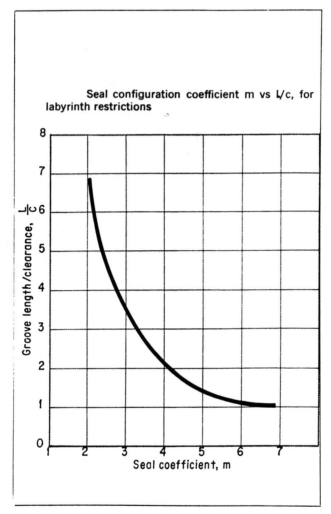

FIGURE 27   Seal configuration coefficient m vs L/C for labyrinth restrictions.

where $4R_H$ is known to be roughly equivalent to diameter in figuring the Reynold's
number. Hydraulic radius is:

$$R_H = \frac{\text{area of flowing fluid}}{\text{wetted perimeter}} \tag{4}$$

For annular orifices, $A = \pi(D^2 - d^2)/4$ and perimeter $= \pi(D + d)$. Thus, radius
$R_H = (D - d)/4$, and $4R_H = D - d$, or $2C$, where C is the radial clearance.
For slots, $A = CZ$, perimeter $= 2(C + Z) =$ approximately $2Z$. Thus, $R_H = 2C$,
where C is the clearance.

Note: There is some disagreement among experts here. For instance, Crane Technical Paper 410 (Crane Company), pp 1–4, states that the hydraulic radius for a very thin slot is equal to the narrower dimension, C. Thus, the equivalent diameter would be $4C$, instead of $2C$. Equations $3B$ and $3C$ would be affected by this change.

To use this method, solve Equation 1 or 2 first, yielding velocity V in terms of the unknown coefficient K. Convert to mass flow W with the equations provided above. Then solve for the Reynold's number using Equation 3. It will contain the unknown coefficient K.

Solve for the coefficient K by trial and error, as shown above. Then calculate the Reynold's number. Check the value against that in Figs. 25 or 26 for the given value of $(D - d)/L$. If it's wrong, try another value for K. Three or four guesses should get the desired result. Knowing K, you can calculate the desired velocity and mass flow.

Accuracy of this method is as good as that for simple orifices or pipes, but the same chances for inaccuracy exist. Remember that poor surface finish, slight rounding of the edges, inaccurate dimensions, and many other physical variations will affect the results greatly. It is better to build and test a model, where possible.

This procedure is the work of Andrew Lenkei, Project Engineer, Research Department, Worthington Corporation, as reported in *Product Engineering* magazine. SI values were added by the handbook editor.

Symbols

A = Flow area, ft² (m²)
C = Radial clearance, ft (m)
c_p = coefficient of specific heat at constant pressure, Btu/lb°F (kJ/kg °C)
d = Minor diameter of annular orifice, ft (m)
D = Major diameter of annular orifice, ft (m)
g = gravitational constant, 32.2 ft/s²
H = Head, ft (m)

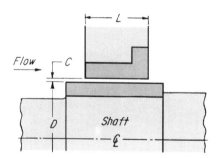

$$H_T = \frac{V^2}{2g}\left[1.5 + \frac{fL}{2C}\right]$$

$$V = \sqrt{\frac{2g\,H_T}{1.5 + \dfrac{fL}{2C}}}$$

FIGURE 28 Positive-clearance seal for centrifugal pumps and other liquid-handling applications has the lowest cost.

K = Orifice coefficient (dimensionless)
J = Mechanical equivalent of heat, 778 ft lb/btu
L = Flow length; also labyrinth groove axial distance, ft (m)
m = seal configuration factor
n = Polytropic exponent ($n = k = 1.4$ for air in isentropic process)
N = Number of sealing points
L_p = Pitch of sealing points, ft (m)
P = Pressure, psf (kPa)
R = Gas constant (53.3 for air)
R_H = Hydraulic radius (R_H = flow area/wetted perimeter)
R_e = Reynold's number, $W(D - d)/A \mu g$
T = Temperature, °R
V = Fluid velocity, ft/s (m/s)
μ = Absolute viscosity, lb s/ft^2
W = Mass rate of flow lb/s (kg/s)
ρ = Density, lb/ft^3 (kg/m^3)
In = Log$_e$ (m)
Z = Slot width, ft

Subscripts
1 First term
2 Second term
n nth term

LABYRINTH SHAFT SEAL LEAKAGE DETERMINATION

(1) Determine the fluid leakage through the sleeve seal in Fig. 28 when the known pressure drop through the seal, $H_T = 300$ lb/in^2 = 693 ft (2067 kPa; 211.2 m); shaft diameter, $D = 10$ in (25.4 cm); total axial length of shaft seal, $L = 2$ in (5.08 cm); radial clearance, $C = 0.020$ in (0.050 cm); Reynolds number $N_R = 2 \times 10^5$ CV; velocity through the seal clearance is assumed to be $V = 180$ ft/s (54.9 m/s), which is 85 percent of the free discharge velocity of $2g \times 693)^{0.5}$. (2) Find the fluid leakage through the non-interlocking labyrinth seal, Fig. 30, when the known pressure drop, and diameter are the same as in (1) above, and $L = 0.125$ in (0.32 cm); $C = 0.02$ in (0.0508 cm); $n = 8$ stages; velocity through the seal = 35 percent of free discharge.

Calculation Procedure:

1. Compute the Reynolds number for the sleeve seal
Use the relation $N_R = 2 \times 10^5 \ CV = (2 \times 10^5)(0.020/12)(180) = 6 \times 10^4$.

2. Calculate the friction factor for this seal
Use the relation, friction factor, $f = 0.316/(N_R)^{0.25}$ for turbulent flow. Substituting, $f = 0.316/(6 \times 10^4)^{0.25} = 0.02019$.

3. Find the fluid velocity through the seal
Use the velocity relation in Fig. 28. Or, $V = ([62.4 \times 693]/[1.5 + 0.02019 \times 2/2 \times 0.02])^{0.5} = 131.27$ ft/s (40.0 m/s).

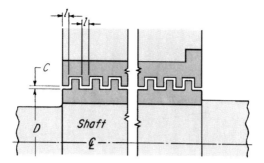

$$H_T = \frac{V^2}{2g} \left[1.5 + \frac{fL}{2C} + 2.0 \ (n\text{-}1) \right]$$

$$V = \sqrt{\frac{2g \ H_T}{1.5 + \dfrac{fL}{2C} + 2.0 \ (n\text{-}1)}}$$

FIGURE 29 Interlocking labyrinths seal liquids and gases. Requires split assembly but has low leakage.

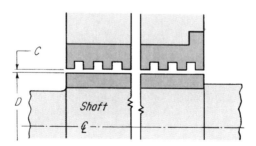

$$H_T = \frac{V^2}{2g} \left[1.5 + \frac{fL}{2C} + 1.5 \ (n\text{-}1) \right]$$

$$V = \sqrt{\frac{2g \ H_T}{1.5 + \dfrac{fL}{2C} + 1.0 \ (n\text{-}1)}}$$

FIGURE 30 Non-interlocking seal is a compromise between Figs. 28 and 29; seals liquid and gas; no split necessary.

4. Compute the fluid flow through the seal

Use the relation, fluid flow through seal, $Q = VA$, where Q = flow, ft^3/s (m^3/s); A = seal area, ft^2. The value of $A = (\pi)(10)(0.02)/144 = 0.00436$ ft^2 (0.000405 m^2). Then, fluid flow $= Q = VA = 131.27 (.00436) = 0.572$ ft^3/s (0.053 m^3/s).

5. Determine the flow through the non-interlocking seal

Follow the same steps as above. Thus, $V = 0.35 (2$ g $\times 693)^{0.5} = 72.78$ ft/s (22.18 m/s). The $N_R = 2 \times 10^5 (0.02/12 \times 72.78) = 2.266 \times 10^4$. Then, $f = 0.316/(2.266 \times 10^4)^{0.25} = 0.02575$. Further, $V = ([62.4 \times 693]/[1.5 + 0.025 \times \{0.125/2\} \times 0.02 + 1.0 \times 7])^{0.5} = 71.0$ ft/s (21.6 m/s). Finally, $Q = 71 \times 0.00436 = 0.3095$ ft^3/s (0.02876 m^3/s).

Related Calculations. Shaft seals are primarily used to reduce leakage of fluid from a hydraulic system. The greater the obstruction to fluid flow, the more efficient the seal. While ten basic seal designs are shown in these procedures, there are hundreds of variants.

The majority of labyrinth seals are machined of bronze or a similar alloy. Seals can be made accurately; they are mechanically strong and they can withstand high temperatures. However, the metal selected for the seal must not gall or melt during wear-in. There must be no residue or distortion to lessen the effectiveness of the seal.

There is a limit to the complexity of the seal. Tests show a diminishing advantage beyond a certain number of stages, particularly where the shaft does not stay centered according to calculations. In such cases, simpler labyrinths are as good as the more serpentine, and much less expensive.

The tapered teeth in Figs. 32 and 34 are considered the most efficient for sealing gas or air. Usually the teeth are rings inserted into the seal sleeve and staked. They are tapered to an edge of about 0.0120 in (0.0254 cm) at the shaft diameter, and no clearance is allowed. The tips are quickly ground off when the machine begins to operate. Water lubrication can be used to prevent excessive heating during wear-in.

In the hydraulic field, it is conventional practice to set minimum and maximum limits for the clearance, C, between the shaft and seal, Fig. 28. The clearance values depend on the strength of the rotating shaft and its deflection. Tests prove that the speed of the shaft has no effect on the performance of the seal. Typical recommended clearances are shown in Fig. 31.

The head loss and flow equations used in this procedure are approximate and cannot take the place of actual seal tests. These equations show trends accurately, however, and are workable once you apply correction factors for your own designs.

In any seal, the effective total resistance to flow, measured in feet (meters) of lost head, is the sum of three types of unit resistance: (1) turbulent conversion of initial static pressure to velocity; (2) wall friction; (3) turbulence caused by abrupt changes of section in the flow path.

The first two, H_V and H_F, are explained in the symbols and equations listed below. For the third type of resistance, which includes both the entrance loss, H_E, and the head loss for one or more stages, H_S, the explanation is: The first abrupt change in section is at the entrance, and head loss according to the conventional entrance equation is $H_E = 0.5\ V^2/2g$. The succeeding changes in section depend on the number of edges in each labyrinth. In Fig. 29, there are four edges per stage and a rough estimate of head loss $H_S = 4 \times 0.5\ V^2/2g$. For Fig. 30 and 32, there are two edges per stage, and the relationship is $H_S = 1.0\ V^2/2g$. The summation is

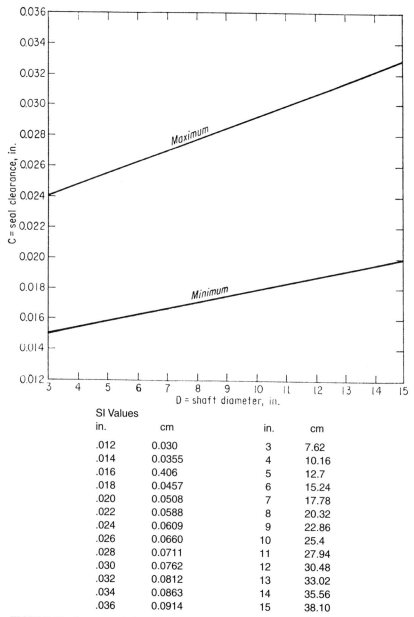

SI Values			
in.	cm	in.	cm
.012	0.030	3	7.62
.014	0.0355	4	10.16
.016	0.406	5	12.7
.018	0.0457	6	15.24
.020	0.0508	7	17.78
.022	0.0588	8	20.32
.024	0.0609	9	22.86
.026	0.0660	10	25.4
.028	0.0711	11	27.94
.030	0.0762	12	30.48
.032	0.0812	13	33.02
.034	0.0863	14	35.56
.036	0.0914	15	38.10

FIGURE 31 Recommended clearances for seals in Figs. 28, 29, and 30.

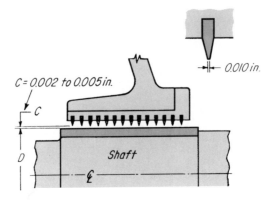

$$H_T = \frac{V^2}{2g}\left[1.5 + 1.5\,(n\text{-}1)\right]$$

$$V = \sqrt{\frac{2g\,H_T}{1.5 + 1.5\,(n\text{-}1)}}$$

FIGURE 32 Tapered-tooth seal for low-leakage applications.

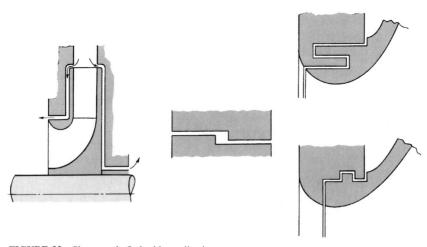

FIGURE 33 Sleeve seals find wide application.

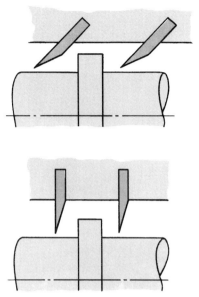

FIGURE 34 Different types of labyrinths for shaft seals.

$$H_T = H_V + H_F + H_E + \Sigma H_S \qquad (1)$$

where the value of ΣH_S is explained in the symbols and equations panel below.

To express Equation 1 in terms of velocity, insert the known expressions for head loss given in the equations and symbols panel. This useful relationship results:

$$H_T = [V^2/2g][1 + fL/2C + 0.5 + a\,(n - 1)] \qquad (2)$$

where $a = 2$ for the seal in Fig. 29, and $a = 1$ for the seals in Figs. 30 and 32.

Equation 2 applies to gases and liquids. For gases, the accuracy is problematical because velocity and density change from one point in the seal to the next point in the seal, and outside of it. To accommodate these changes you have to assume average values during flow through the seal.

Leakage flow through the seal at any given point, for gas or liquid, is simply the average velocity, V, at that point, times the cross-sectional area, A, of the seal. The relationship is $Q = VA$.

For gases, accurate predictions of leakage flow using a particular equation are not possible unless the design of the seal happens to match the conditions of the test upon which the flow equation is based. There are many more published flow equations than are discussed here. Variations in calculated leakage flow might exceed 2:1. However, for first approximations, the equations in this procedure will suffice.

One way to improve the accuracy of these calculations is to use trial and error. Assume a Reynolds number, calculate friction factor, compute leakage flow, check, out the assumed Reynolds number, correct it if necessary, and try again. Figure 35 gives viscosity values for calculating the Reynolds number for fluids and gases used in hydraulic and pneumatic systems.

This procedure is the work of Louis Dodge, Hydraulics Consultant, as reported in *Product Engineering* magazine. SI values were added by the handbook editor.

Another approach to labyrinth seal design is given by V. L. Peickii, Director of Research & Engineering, and Dan A. Christensen, Research Engineer, National Seal Division, Federal-Mogul-Bower Bearings Inc., writing in the same publication where they note:

Labyrinth or positive-clearance seals are so specialized that no standard types or designs have evolved. Design is usually controlled by the tolerable leak-rate, from which one can calculate gap clearances and number of elements required, Fig. 36.

The number of rings to limit leakage to a given flow can be found from:

$$N = (40\ P - 2600)\ (W/A)/540\ (W/A) - P$$

where N = Number of rings
  W = Permissible leakage, lb/sec (kg/sec)
  $A = C_\pi D$, cross-sectional area, in^2 (cm^2)
  C = Clearance, in (cm)
  D = Diameter, in (cm)
  P = Absolute pressure, lb/in^2 (abs) (kPa)

In this equation, to find N_{12} for a leakage from pressure level P_1 to a lower pressure level, P_2, the N for each must be found; then $N_{12} = N_1 - N_2$.

The leakage flow rate can be found from:

$$W = 25\ KA \sqrt{\frac{\left(\dfrac{P_1}{V_1}\right) - \left[1 - \left(\dfrac{P_2}{P_1}\right)^2\right]}{\left[N - \mathrm{Log}_n\left(\dfrac{P_2}{P_1}\right)\right]}}$$

where W = Flow rate, lb/hr (kg/hr)
  V_1 = Initial specific volume, ft^3/lb (m^3/kg)
  K = Experimental coefficient

For interlocking labyrinths, $K = 55$ approximately, if the velocity is effectively throttled between labyrinths. It is independent of clearance in the usual range. For non-interlocking labyrinths, K varies with the ratio of labyrinth spacing divided by radial clearance. For a ratio of 5, $K = 100$; for a ratio of 50, $K = 60$.

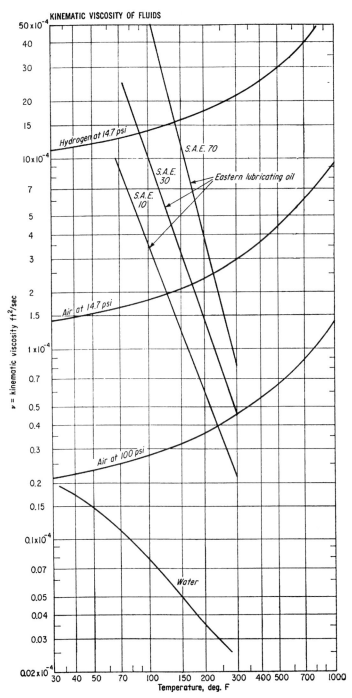

FIGURE 35a Viscosities for various hydraulic-system fluids.

SI Values			
deg F	deg C	ft²/sec	m²/sec
30	−1.1	50×10^{-4}	4.5×10^{-4}
40	4.4	40	3.7
50	10.0	30	2.8
70	21.1	20	1.9
100	37.8	15	1.4
150	65.6	10×10^{-4}	0.92×10^{-4}
200	93.3	7	0.65
300	148.9	5	0.46
400	204.4	4	0.37
500	260.0	3	0.28
700	371.1	2	0.19
1000	537.8	1.5	0.14
		1×10^{-4}	0.09×10^{-4}
		0.7	0.065
		0.5	0.046
		0.4	0.037
		0.3	0.028
		0.2	0.018
		0.15	0.014
		0.1×10^{-4}	0.009×10^{-4}
		0.07	0.006
		0.05	0.0046
		0.04	0.0037
		0.03	0.0028
		0.02×10^{-4}	0.0018×10^{-4}

FIGURE 35b   (*Continued*).

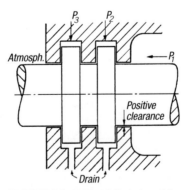

FIGURE 36   Typical labyrinth seal for which empirical formulas have been derived.

Equations and Symbols

A = area of annular clearance, ft^2 = $D\pi C$ (m^2)
C = radial clearance, in (cm)
D = diameter of shaft, in (cm)
f = friction factor (dimensionless)
    $f = 64/N_R$ for laminar flow ($N_R < 2320$)
    $f = 0.316/N_R^{0.25}$ for turbulent flow
H_S = head loss for one stage, ft (m)
    $H_S = 2V^2/2g$ for Fig 2 (4 edges)
    $H_S = V^2/2g$ for Figs 3 and 4 (2 edges)
    $\Sigma H_S = H_s (n - 1)$
H_V = velocity head, ft = $V^2/2g$ (m)
H_F = friction head loss, ft = $f \dfrac{L}{2C} \dfrac{V^2}{2g}$ (m)
H_T = total head loss of all stages, ft (m)
    $H_T = H_E + \Sigma H_S + H_V + H_F$
L = total axial length of shaft seal, in (cm)
n = number of seals or teeth
    $n - 1$ = number of stages or spaces
N_R = Reynolds number, dimensionless
    $N_R = 2CV/12\,v = 2 \times 10^5\ CV$ for 80°F water
Q = flow, ft^3/sec = VA (m^2/sec)
V = fluid velocity, ft/sec (m/sec)
v = kinematic viscosity, ft^2/sec (m^2/sec)

SECTION 26

METALWORKING AND NONMETALLIC MATERIALS PROCESSING

Economics of Machining

ESTIMATING CUTTING TIME AND COST WITH DIFFERENT TOOL MATERIALS

A 9-in (22.86-cm) diameter steel shaft is to be "heavy roughed" with either of two cutting tools—high-speed steel (HSS), or cemented carbide. The work material is AISI 1050 having a hardness of 200 BHN. Feed rate is 0.125 in/r (3.17 mm/r); depth of cut = 1.0 in (25.4 mm); tool life is based on 0.030-in (0.726-mm) flank wear. Choose the most effective tool to use if the tool signature is: −6, 10, 6, 6, 15, 15, $\frac{1}{16}$ R; the tool-changing time = 4 min for both tools; the cost of a sharp tool = $0.50 for HSS and $2.00 for cemented carbide; and M = machine labor plus overhead rate, $/min = 15 cents for each type of tool.

Calculation Procedure:

1. Determine the minimum-cost tool life for each type of tool material
Analyses of the economics metal of cutting with different types of cutting-tool materials are often plotted on two bases—Figs. 1 and 2. Figure 1 shows the machining cost, tool cost, and nonproductive cost added to show the total cost per piece. In Fig. 2, the machine time, tool-changing time, and nonproductive time are added and plotted as the total time per piece.

Studies show that the cutting speed and production rate resulting from minimum-cost tool life of approximately the same value is much higher for carbide tools than for high-speed steel tools—150 ft/min (45.7 m/min) cutting speed for carbide tools vs. 30 ft/min (9.14 m/min) for high-speed steel tools. These two values of cutting speed will be used in this procedure.

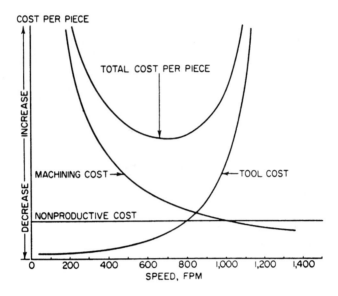

SI Values

200 fpm	60.9m/min
400	121.9
600	182.9
800	243.8
1000	304.8
1200	365.8
1400	426.7

FIGURE 1 Total cost per piece is found by adding the plots of machining costs, tool costs, and nonproductive costs. (*T. E. Hayes and American Machinist.*)

The minimum-cost tool life, T_c, is a function of the slope, n, of the tool-life curve, Fig. 3. It can be said that n is one of the controlling influences on Hi-E cutting conditions.* Thus, for high-speed steel, the expression for T_c is:

$$T_c = \left(\frac{1}{n} - 1\right)\left(\frac{t}{M} + TCT\right)$$

where T_c = minimum-cost tool life, min; n = slope of tool-life curve; M = machine labor plus overhead rate, \$/min; TCT = tool-changing time, min. Substituting,

*The Hi-E term was originally coined by Thomas E. Hayes, Service Engineer, Metallurgical Products Department, General Electric Company, and first published in his article, "How to Cut Costs with Carbides by 'Hi-E' Machining."

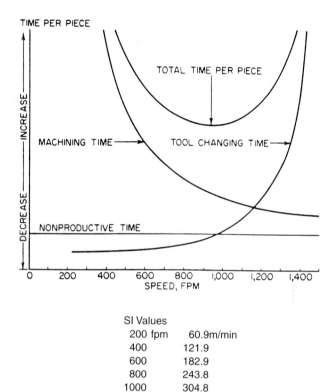

SI Values

200 fpm	60.9m/min
400	121.9
600	182.9
800	243.8
1000	304.8
1200	365.8
1400	426.7

FIGURE 2   Total time per piece is found by adding the plots of machine times, tool-changing time, and nonproductive time. (*T. E. Hayes and American Machinist.*)

$$T_c = \left(\frac{1}{125} - 1\right)\left(\frac{0.50}{0.15} + 4\right)$$

$$= 51.3 \text{ min}$$

For cemented carbide, we have

$$T_c = = \left(\frac{1}{n} - 1\right)\left(\frac{t}{M} + TCT\right)$$

$$= \left(\frac{1}{0.25} - 1\right)\left(\frac{2}{0.15} + 4\right)$$

$$= 52 \text{ min}$$

Thus, the T_c, values for both tools are approximately the same.

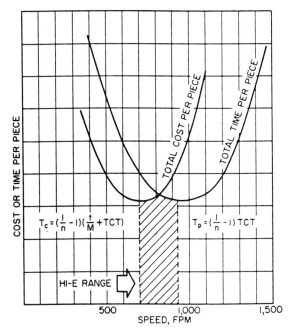

FIGURE 3   A combination of the total cost per piece and total time per piece plots on a single graph forms the Hi-E range between their respective minimum points. (*Brierley and Siekmann.*)

2.   Compute the tool life for maximum productive rate

The tool life for maximum productive rate T_p, min, is given by

$$T_p = \left(\frac{1}{n} - 1\right) TCT$$

where symbols are as before.

Substituting for high-speed steel we have

$$T_p = \frac{1}{0.125} - 1 = 28 \text{ min}$$

Entering Fig. 3 at 28 min and projecting to the HSS plot, we find that the cutting speed should be 33 ft/min (10.1 m/min).

Using the same relation for cemented carbide, we find, entering Fig. 3 at 12 minute and projecting up to the cemented-carbide plot, the cutting speed to be 220 ft/min (67.1 m/min).

3.   Tabulate the results of the calculations

List the cutting conditions for each type of tool material, as in Table 1. Studying the results in Table 1 shows that only about 20 percent as much time is required per piece with cemented-carbide tools as with HSS tools, and the total cost per

TABLE 1 Operation of the Job Illustrated in Figure 1 at Minimum Cost-Cutting Conditions Results in the Following Economic Comparison. Machining Costs are Halved and Production is Tripled*

Cutting conditions	HSS	Cemented carbide
Machine time per piece	45 min	9.1 min
Nonproductive time per piece	10 min	10 min
Labor plus overhead rate	$0.15	$0.15
Machine cost per piece	$6.75	$1.36
Nonproductive cost per piece	$1.50	$1.50
Tool cost per piece	$0.50	$2.00
Total cost per piece	$8.75	$4.86
Total time per piece	55 min	19.1 min
Pieces per hour	1.1	3.1

*Brierley and Seikmann.

piece is only about 55 percent of that of HSS. Thus, the higher tool cost results in greater productivity (3.1 pieces per hour vs. 1.1 pieces per hour).

Related Calculations. This procedure is the work of Robert G. Brierley, Tool Applications Specialist, Metallurgical Products Department, General Electric Company and H. J. Siekmann, Vice President, Marketing, Martin Metals Company, Division of Martin Marietta Corporation. If reflects the Hi-E approach used at General Electric Company, plus the basics of metalworking physics.

The Hi-E range is shown in Fig. 4, which depicts a combination of the tool cost per piece and total time per piece plotted on a single graph. The Hi-E range is between the respective minimum points.

Since tool-life plots are important in the Hi-E analyses of machining economics, the value of n is of much interest. Although n varies slightly as machining conditions are changed, Brierley and Siekmann cite the following values for practical everyday use to satisfy the calculations for the Hi-E range: For high-speed steel, $n = 0.125$ and $([1/n] - 1) = 7$; for carbide, $n = 0.25$ to 0.30 and $([1/n] - 1) = 3$ for the 0.25 value; for cemented oxide or ceramic tools, $n = 0.50$ to 0.70 and $([1/n] - 1) = 1$ for the 0.50 value. More exact values can be obtained from tabulations available from ASTME.

The procedure given here was presented by the above two authors in their book *Machining Principles and Cost Control*, McGraw-Hill.

COMPARING FINISH MACHINING TIME AND COSTS FOR DIFFERENT TOOL MATERIALS

Compare machining costs and times for cemented-carbide and cemented-oxide tools for a high-speed finishing operation using the data given in Fig. 5 and the equations in the previous procedure. Tabulate the results for comparison.

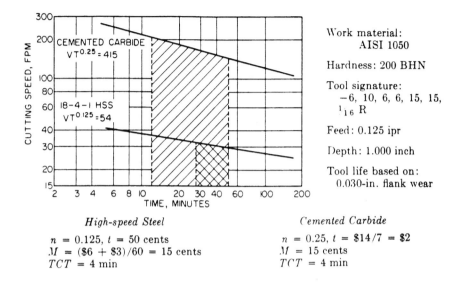

Work material:
    AISI 1050

Hardness: 200 BHN

Tool signature:
    −6, 10, 6, 6, 15, 15,
    1_{16} R

Feed: 0.125 ipr

Depth: 1.000 inch

Tool life based on:
    0.030-in. flank wear

High-speed Steel

$n = 0.125$, $t = 50$ cents
$M = (\$6 + \$3)/60 = 15$ cents
$TCT = 4$ min

Cemented Carbide

$n = 0.25$, $t = \$14/7 = \$2$
$M = 15$ cents
$TCT = 4$ min

0.125 ipr	3.175 mm
1.000 in.	25.4 mm
0.030 in.	0.762 mm

FIGURE 4 Heavy roughing of a steel shaft with carbide widens the Hi-E range compared with using high-speed steel. (*Brierley and Siekmann.*)

Calculation Procedure:

1. Find the minimum-cost tool life for each tool material
Use the T_c equation of step 1 of the previous procedure with the same symbols. Then, for cemented carbide,

$$T_c = \left(\frac{1}{n} - 1\right)\left(\frac{t}{M} + TCT\right)$$

$$= \left(\frac{1}{0.3} - 1\right)\left(\frac{0.25}{0.15} + 1\right)$$

$$= 6.22 \text{ min}$$

Likewise, using the same equation for cemented oxide,

$$T_c = \left(\frac{1}{n} - 1\right)\left(\frac{t}{M} + TCT\right)$$

$$= \left(\frac{1}{0.7} - 1\right)\left(\frac{0.375}{0.15} + 1\right)$$

$$= 1.57 \text{ min}$$

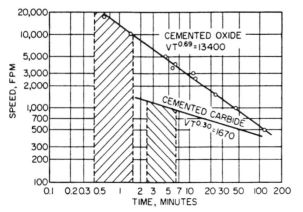

Hardness: 170–173 BHN

Tool Signature:
    Oxide: -10, -10, 10,
       10, 15, 15, $\frac{1}{16}$ R
    Carbide: -5, -5, 5,
       5, 15, 15, $\frac{1}{32}$ R

Feed: 0.010 ipr

Depth: 0.100 inch

Tool life based on:
    0.012 in. flank wear

Cemented Carbide   *Cemented Oxide*

$n = 0.3$, $t = 25$ cents, $M = 15$ cents, $n = 0.7$, $t = 37.5$ cents, $M = 15$ cents,
    $TCT = 1$ $TCT = 1$

SI Values	
0.010 ipr	0.254 mm
1.000 in.	25.4 mm
0.030 in.	0.762 mm

FIGURE 5   A high-speed finishing operation switched to cemented oxide. (*Brierley and Siekmann.*)

2. Determine the tool life for the maximum productive rate

As in step 1, above, use the equation and symbols from step 2 in the previous procedure. Thus, for cemented-carbide tools,

$$T_p = \left(\frac{1}{n} - 1\right) TCT \quad = 2.33 \text{ min}$$

Projecting from 2.33 min on the horizontal scale in Fig. 5, we find the cutting speed to be 1150 ft/min (350.5 m/min).
    For cemented-oxide tools,

$$T_p = \left(\frac{1}{n} - 1\right) TCT$$

$$= 0.45 = >20{,}000 \text{ ft/min}$$

Plotting from 0.45 min, we find that the cutting speed would exceed 20,000 ft/min (6096 m/min)

3. Summarize the calculations in tabular form

Table 2 summarizes the calculations for these two tooling materials. As you can see, there is a significant difference in the machine time per piece: 1 7.2 min vs.

TABLE 2 Minimum Cost-Cutting Conditions Using Cemented Oxide Rather Than Carbide Halve the Machining Costs of This Finishing Operation While Production Is Doubled*

Cutting conditions	Cemented carbide	Cemented oxide
Machine time per piece	17.2 min	1.63 min
Nonproductive time per piece	10 min	10 min
Labor plus overhead rate	$0.15	$0.15
Machine cost per piece	$2.50	$0.245
Nonproductive cost per piece	$1.50	$1.50
Tool cost per piece	$0.25	$0.375
Total cost per piece	$4.25	$2.120
Total time per piece	27.2 min	11.63 min
Pieces per hour	2.2	5.4

*Brierley and Seikmann.

1.63 min. Likewise, the cost is at a 10-times ratio: $0.245 vs. $2.50, and the piece output is more than double: 5.4 pieces per hour vs. 2.2 pieces per hour. As in the previous procedure, the more expensive tool significantly increases the output while reducing production costs.

Related Calculations. This procedure, like the previous one, is the work of Brierley and Siekmann. Full citation information is given in the previous procedure.

In building their approach to the economics of machining, Brierley and Siekmann give a number of key equations that lead up to the steps presented in this and the previous procedure. These equations are: (1) *Machining cost* = (machining time per piece)(labor + overhead rate); (2) *Machining time* = [(length of piece cut)(cut)]/(feed)(rpm of cutter); (3) *Tool cost* = (tool-changing cost + tool-grinding cost per edge + tool depreciation per edge + tool inventory cost)/(production per edge); (4) *Cost to change the tool* = (tool-changing time)(the machine operator's rate + overhead); (5) *Tool-grinding cost per edge* = [(grinding time)(grinder's rate + overhead)]/(edges per grind); (6) Brazed-tool depreciation cost per edge = (cost of tool)/(number of regrinds + 1); (7) For disposable-insert toolholder or milling-cutter head, *Tool depreciation cost per edge* = [(cost of disposable insert/number of cutting edges per insert) + (cost of holder or head)]/[(number of inserts in life of holder) (number of edges per insert)]; (8) For on-end insert toolholder or regindable inserted-blade milling-cutter head, *Tool depreciation cost per edge* = (cost of insert)/[(number of regrinds per insert)(number of edges per grind)] + (cost of holder or head)/[(number of in life of holder or head)(number of regrinds per insert)(number of edges per grind)]; (9) *Tool inventory cost* = (number of tools at machine + number of tools in grinding room)(cost per tool)(inventory cost rate); (10) *Nonproductive cost* = load and unload time + (other noncutting time)(operator labor + overhead rate); (11) *Total machining time* = machine time from Eq. (1) + tool changing time + nonproductive time.

Using the above eleven equations and the relations given in Figs. 3, 4, and 5, the economics of machining can be planned in a preliminary way for a given machine. Then the Hi-E approach and advances in it should be considered for in-depth analysis of the economics of a given machining application.

FINDING MINIMUM COST AND MAXIMUM PRODUCTION TOOL LIFE FOR DISPOSABLE TOOLS

Find the minimum cost and maximum production tool life for a disposable tool having the following characteristics: price of insert plus toolholder depreciation, $P = \$1.80$; total cutting edges in the life of the insert, $E = 6$; machine operator's rate, $MR = \$4.00/h$; machine overhead rate, $MO = \$8.00/h$; tool-changing time, $TCT = 1$ min; the constant for the slope of the tool-life line for carbide tools, $n = 3.5$.

Calculation Procedure:

1. Find the minimum-cost tool life
The expression for the minimum-cost tool life, T_c, is given by

$$T_c = \left(\frac{1}{n} - 1\right)\left(\frac{t}{M} + TCT\right)$$

where

$$t = \frac{\text{price of insert} + \text{toolholder depreciation}}{\text{total cutting edges in life of insert}} = \frac{1.80}{6} = 0.30$$

$$M = \frac{\text{labor per hour} + \text{overhead per hour}}{60} = \frac{4.00 + 8.00}{60} = 0.20$$

$$TCT = \text{tool-changing time (min)} = 1$$

$$\left(\frac{1}{n} - 1\right) = \text{a constant (3.5) based on the slope}$$
$$\text{of the tool-life line}$$

Substituting,

$$T_c(\text{min}) = 3.5\left(\frac{0.30}{0.20} + 1\right)$$

$$T_c(\text{min}) = 3.5 \times 2.5$$

$$T_c(\text{min}) = 8.75$$

2. Calculate the maximum production tool life for this tool
To solve for T_p, we need only the constant, n, and the tool-changing time. Or,

$$T_p = \left(\frac{1}{n} - 1\right)TCT$$

$$T_p = 3.5 \times 1$$

$$T_p = 3.5$$

3. Show an alternative approach to the calculation of T_c and T_p

Convert the known quantities to desired values. Thus, (*a*) Total tool cost per cutting edge, $P/E = t$, $1.80/6 = $0.30; (*b*) cost of labor plus overhead per hour $= MR + RO = MRO = $4.00 + $8.00 = $12.00; cost of labor plus overhead per minute $= MRO/60 = M = $12.00/60 = $0.20; (*c*) combined tool cost plus tool-changing cost per edge $= [t + (M \times TCT)] = [0.30 + (0.20 \times 1)] = $0.50.

Now we are ready to calculate the tool life for minimum part cost. Since the standard formula for carbide tools is $3.5(C/M) = TC = 3.5(0.50/0.20) = 8.75$ min.

The next step is to convert the tool-life values for T_c and T_p to cutting speeds. Further, it may be desirable to convert cutting speeds from linear dimensions to revolutions per minute for the operator's convenience.

Related Calculations.   The computations shown here can be done on a pre-printed form or by using a computer program specially prepared for this purpose. Brierley and Siekmann, in the previously cited reference, present preprinted forms for this purpose. The procedure given here is from that reference. Anyone seeking to use the Hi-E method, or its advancements, will find much help in the preprinted forms available to them. The approach given here is valuable for anyone seeking the most economical machining tools.

COMPUTING MINIMUM COST AND MAXIMUM PRODUCTION TOOL LIFE FOR REGRINDABLE TOOLS

Find the minimum cost and maximum production tool life for regrindable brazed-type tools, on-end slugs, or others which are normally resharpened by grinding. The variables for the tool are: Price of tool, $P = $3.50; tool cutting edges in life of tool, $E = 7$; tool grinder's rate, $GR = $4.00/h; toolroom overhead rate, $GO = $8.00/h; grinding time per edge, $GT = 5$ min; machine operator's rate, $MR = $4.00/h; machine overhead rate, $MO = $8.00/h; tool-changing time, $TCT = $ min.

Calculation Procedures:

1. Find the total tool cost per cutting edge

(*a*) The tool cost per cutting edge, $P/E = $3.50/7 = $0.50. (*b*) Grinding cost per minute $= (GR + GO)/60 = ($4.00 + $8.00)/60 = $0.20. (*c*) Grinding cost per cutting edge $= [(GC/m)GT]GC/E = $0.20 \times 5 = $1.00. (*d*) Total cost per edge (tool cost + grinding cost) $= [(P/E) + (GC/E) = t] = 0.50 + 1.00 = 1.50.

2. Find the total cost of labor plus overhead

Cost per hour $= MR + MO = MRO = $4.00 + $8.00 = $12.00. (*b*) Cost per minute $= M = MRO/60 = $12.00/60 = $0.20.

3. Find the combined tool cost plus tool-changing cost per edge

Combined cost $= C = [t + M(TCT)] = $1.50 + ($0.20 \times 3) = $2.10.

4. Calculate tool life for minimum part cost

Use the standard formula for carbide tools $= T_p = (3.5C/M) = 3.5(2.10/0.20) = 36.75$ min.

5. Calculate tool life for maximum production rate
The standard formula for carbide tools is $3.5(TCT) = T_p = 3.5 \times 3 = 10.5$ min.

6. Convert tool life to cutting speed

	Desired tool life
Minimum part cost, T_c	36.75 min
Maximum production rate, T_p	10.5 min

    Related Calculations. The greater cost of getting a new cutting edge on the job as reflected in the higher first cost of the tool, the added cost of grinding, and the longer time to change cutting edges results in a longer tool life for minimum part cost. This illustrates the need for tools that are low in cost per edge and easily and quickly changed if low unit costs are achieved. As in the previous procedure, the next step is to convert tool-life values to cutting speeds and then to revolutions per minute to suit the diameter or diameters being machined.
    This procedure is the work of Brierley and Siekmann, as reported in the reference cited earlier.

Machining Process Calculations

TOTAL ELEMENT TIME AND TOTAL OPERATION TIME

The observed times for a turret-lathe operation are as follows: (1) material to bar stop, 0.0012 h; (2) index turret, 0.0010 h; (3) point material, 0.0005 h; (4) index turret, 0.0012 h; (5) turn 0.300-in (0.8-cm) diameter part, 0.0075 h; (6) clear hexagonal turret, 0.0009 h; (7) advance cross-slide tool, 0.0008 h; (8) cutoff part, 0.0030 h; (9) aside with part, 0.0005 h. What is the total element time? What is the total operation time if 450 parts are processed? Pointing of the material was later found unnecessary. What effect does this have on the element and operation total time?

Calculation Procedure:

1. Compute the total element time
Compute the total element time by finding the sum of each of the observed times in the operation, or sum steps 1 through 9: $0.0012 + 0.0010 + 0.0005 + 0.0012 + 0.0075 + 0.0009 + 0.0008 + 0.0030 + 0.0005 = 0.0166$ h = 0.0166 (60 min/h) = 0.996 minute per element.

2. Compute the total operation time
The total operation time = (element time, h)(number of parts processed). Or, $(0.0166)(450) = 7.47$ h.

3. Compute the time savings on deletion of one step
When one step is deleted, two or more times are usually saved. These times are the machine preparation and machine working times. In this process, they are steps 2 and 3. Subtract the sum of these times from the total element time, or 0.0166 − (0.0010 + 0.0005) = 0.0151 h. Thus the total element time decreases by 0.0015 h. The total operation time will now be (0.0151)(450) = 6.795 h, or a reduction of (0.0015)(450) = 0.6750 h. Checking shows 7.470 − 6.795 = 0.675 h.

Related Calculations.   Use this procedure for any multiple-step metalworking operation in which one or more parts are processed. These processes may be turning, boring, facing, threading, tapping, drilling, milling, profiling, shaping, grinding, broaching, hobbing, cutting, etc. The time elements used may be from observed or historical data.

Recent introduction of international quality-control specifications by the International Organization for Standardization (ISO) will require greater accuracy in all manufacturing calculations. The best-known set of specifications at this time is ISO 9000 covering quality standards and management procedures. All engineers and designers everywhere should familiarize themselves with ISO 9000 and related requirements so that their products have the highest quality standards. Only then will their designs survive in the competitive world of international commerce and trading.

CUTTING SPEEDS FOR VARIOUS MATERIALS

What spindle rpm is needed to produce a cutting speed of 150 ft/min (0.8 m/s) on a 2-in (5.1-cm) diameter bar? What is the cutting speed of a tool passing through 2.5-in (6.4-cm) diameter material at 200 r/min? Compare the required rpm of a turret-lathe cutter with the available spindle speeds.

Calculation Procedure:

1. Compute the required spindle rpm
In a rotating tool, the spindle rpm $R = 12C/\pi d$, where C = cutting speed, ft/min; d = work diameter, in. For this machine, $R = 12(150)/\pi(2) = 286$ r/min.

2. Compute the tool cutting speed
For a rotating tool, $C = R\pi d/12$. Thus, for this tool, $C = (200)(\pi)(2.5)/12 = 131$ ft/min (0.7 m/s).

The cutting-speed equation is sometimes simplified to $C = Rd/4$. Using this equation for the above machine, we see $C = 200(2.5)/4 = 125$ ft/min (0.6 m/s). In general, it is wiser to use the exact equation.

3. Compare the required rpm with the available rpm
Consult the machine nameplate, *American Machinist's Handbook*, or a manufacturer's catalog to determine the available spindle rpm for a given machine. Thus, one Warner and Swasey turret lathe has spindle speed of 282 compared with the 286 r/min required in step 1. The part could be cut at this lower spindle speed, but the time required would be slightly greater because the available spindle speed is 286 − 282 = 4 r/min less than the computed spindle speed.

When preparing job-time estimates, be certain to use the available spindle speed, because this is frequently less than the computed spindle speed. As a result, the actual cutting time will be longer when the available spindle speed is lower.

Related Calculations. Use this procedure for a cutting tool having a rotating cutter, such as a lathe, boring mill, automatic screw machine, etc. Tables of cutting speeds for various materials (metals, plastics, etc.) are available in the *American Machinist's Handbook,* as are tables of spindle rpm and cutting speed.

DEPTH OF CUT AND CUTTING TIME FOR A KEYWAY

What depth of cut is needed for a ¾-in (1.9-cm) wide keyway in a 3-in (7.6-cm) diameter shaft? The keyway length is 2 in (5.1 cm). How long will it take to mill this keyway with a 24-tooth cutter turning at 130 r/min if the feed is 0.005 per tooth?

Calculation Procedure:

1. Sketch the shaft and keyway
Figure 6 shows the shaft and keyway. Note that the depth of cut D in $= W/2 + A$, where $W =$ keyway width, in; $A =$ distance from the key horizontal centerline to the top of the shaft, in.

2. Compute the distance from the centerline to the shaft top
For a machined keyway, $A = [d - (d^2 - W^2)^{0.5}]/2$, where $d =$ shaft diameter, in. With the given dimensions, $A = [3 - (3^2 - 0.75^2)^{0.5}]/2 = 0.045$ in (1.1 mm).

3. Compute the depth of cut for the keyway
The depth of cut $D = W/2 + A = 0.75/2 + 0.045 = 0.420$ in (1.1 cm).

4. Compute the keyway cutting time
For a single milling cutter, cutting time, min = length of cut, in/[(feed per tooth) × (number of teeth on cutter)(cutter rpm)]. Thus, for this keyway, cutting time = $2.0/[(0.005)(24)(130)] = 0.128$ min.

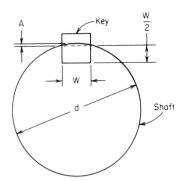

FIGURE 6   Keyway dimensions.

Related Calculations. Use this procedure for square or rectangular keyways. For Woodruff key-seat milling, use the same cutting-time equation as in step 4. A Woodruff key seat is almost a semicircle, being one-half the width of the key *less than* a semicircle. Thus, a $\%_{16}$-in (1.4-cm) deep Woodruff key seat containing a $\frac{3}{8}$-in (1.0-cm) wide key will be $(\frac{3}{8})/2 = \frac{3}{16}$ in (0.5 cm) less than a semicircle. The key seat would be cut with a cutter having a radius of $\%_{16} + \frac{3}{16} = {}^{12}\!/_{16}$, or $\frac{3}{4}$ in (1.9 cm).

MILLING MACHINE TABLE FEED AND CUTTER APPROACH

A 12-tooth milling cutter turns at 40 r/min and has a feed of 0.006 per tooth per revolution. What table feed is needed? If this cutter is 8 in (20.3 cm) in diameter and is facing a 2-in (5.1-cm) wide part, determine the cutter approach.

Calculation Procedure:

1. Compute the required table feed
For a milling machine, the table feed F_T in/min $= f_t nR$, where f_t = feed per tooth per revolution; n = number of teeth in cutter; R = cutter rpm. For this cutter, $F_T = (0.006) \times (12)(400) = 28.8$ in/min (1.2 cm/s).

2. Compute the cutter approach
The approach of a milling cutter A_c in $= 0.5D_c - 0.5(D_c^2 - w^2)^{0.5}$, where D_c = cutter diameter, in; w = width of face of cut, in. For this cutter, $A_c = 0.5(8) - 0.5(8^2 - 2^2)^{0.5} = 0.53$ in (1.3 cm).

Related Calculations. Use this procedure for any milling cutter whose dimensions and speed are known. These cutters can be used for metals, plastics, and other nonmetallic materials.

DIMENSIONS OF TAPER AND DOVETAILS

What are the taper per foot (TPF) and taper per inch (TPI) of an 18-in (45.7-cm) long part having a large diameter d_l of 3 in (7.6 cm) and a small diameter of d_s of 1.5 in (3.8 cm)? What is the length of a part with the same large and small diameters as the above part if the TPF is 3 in/ft (25 cm/m)? Determine the dimensions of the dovetail in Fig. 7 if $B = 2.15$ in (5.15 cm), $C = 0.60$ in (1.5 cm), and $a = 30°$. A $\frac{3}{8}$-in (1.0-cm) diameter plug is used to measure the dovetail.

Calculation Procedure:

1. Compute the taper of the part
For a round part TPF in/ft $= 12(d_l - d_s)/L$, where L = length of part, in; other symbols as defined above. Thus for this part, TPF $= 12(3.0 - 1.5)/18 = 1$ in/ft (8.3 cm/m). And TPI in/in $= (d_l - d_s)/L_2$, or $(3.0 - 1.5)/18 = 0.0833$ in/in (0.0833 cm/cm).

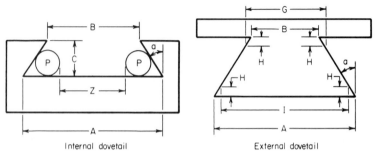

FIGURE 7 Dovetail dimensions.

The taper of round parts may also be expressed as the angle measured from the shaft centerline, that is, one-half the included angle between the tapered surfaces of the shaft.

2. Compute the length of the tapered part
Converting the first equation of step 1 gives $L = 12(d_l - d_s)/\text{TPF}$. Or, $L = 12(3.0 - 1.5)/3.0 = 6$ in (15.2 cm).

3. Compute the dimensions of the dovetail
For external and internal dovetails, Fig. 7, with all dimensions except the angles in inches, $A = B + CF = I + HF$; $B = A - CF = G - HF$; $E = P \cot (90 + a/2) + P$; $D = P \cot (90 - a/2) + P$; $F = 2 \tan a$; $Z = A - D$. Note that $P =$ diameter of plug used to measure dovetail, in.
   With the given dimensions, $A = B + CF$, or $A = 2.15 + (0.60)(2 \times 0.577) = 2.84$ in (7.2 cm). Since the plug P is $\frac{3}{8}$ in (1.0 cm) in diameter, $D = P \cot (90 - a/2) + P = 0.375 \cot (90 - {}^{30}\!/_2) + 0.375 = 1.025$ in (2.6 cm). Then $Z = A - D = 2.840 - 1.025 = 1.815$ in (4.6 cm). Also $E = P \cot (90 + a/2) + P = 0.375 \cot (90 + {}^{30}\!/_2) + 0.375 = 0.591$ in (1.5 cm).
   With flat-cornered dovetails, as at I and G, and $H = \frac{1}{8}$ in (0.3 cm), $A = I + HF$. Solving for I, we get $I = A - HF = 2.84 - (0.125)(2 \times 0.577) = 2.696$ in (6.8 cm). Then $G = B + HF = 2.15 + (0.125)(2 \times 0.577) = 2.294$ in (5.8 cm).
   Related Calculations.   Use this procedure for tapers and dovetails in any metallic and nonmetallic material. When a large number of tapers and dovetails must be computed, use the appropriate tables in the *American Machinist's Handbook.*

ANGLE AND LENGTH OF CUT FROM GIVEN DIMENSIONS

At what angle must a cutting tool be set to cut the part in Fig. 8? How long is the cut in this part?

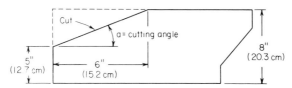

FIGURE 8 Length of cut of a part.

Calculation Procedure:

1. Compute the angle of the cut

Use trigonometry to compute the angle of the cut. Thus, tan a = opposite side/ adjacent side = $(8 - 5)/6 = 0.5$. From a table of trigonometric functions, a = cutting angle = 26°34′, closely.

2. Compute the length of the cut

Use trigonometry to compute the length of cut. Thus, sin a = opposite side/hypotenuse, or $0.4472 = (8 - 5)$/hypotenuse; length of cut = length of hypotenuse = $3/0.4472 = 6.7$ in (17.0 cm).

Related Calculations.   Use this general procedure to compute the angle and length of cut for any metallic or nonmetallic part.

TOOL FEED RATE AND CUTTING TIME

A part 3.0 in (7.6 cm) long is turned at 100 r/min. What is the feed rate if the cutting time is 1.5 min? How long will it take to cut a 7.0-in (17.8-cm) long part turning at 350 r/min if the feed is 0.020 in/r (0.51 mm/r)? How long will it take to drill a 5-in (12.7-cm) deep hole with a drill speed of 1000 r/min and a feed of 0.0025 in/r (0.06 mm/r)?

Calculation Procedure:

1. Compute the tool feed rate

For a tool cutting a rotating part, $f = L/(Rt)$, where t = cutting time, min. For this part, $f = 3.0/[(100)(1.5)] = 0.02$ in/r (0.51 mm/r).

2. Compute the cutting time for the part

Transpose the equation in step 1 to yield $t = L/(Rf)$, or $t = 7.0/[(350)(0.020)] = 1.0$ min.

3. Compute the drilling time for the part

Drilling time is computed using the equation of step 2, or $t = 5.0/[(1000)(0.0025)] = 2.0$ min.

Related Calculations.   Use this procedure to compute the tool feed, cutting time, and drilling time in any metallic or nonmetallic material. Where many computations must be made, use the feed-rate and cutting-time tables in the *American Machinist's Handbook.*

TRUE UNIT TIME, MINIMUM LOT SIZE, AND TOOL-CHANGE TIME

What is the machine unit time to work 25 parts if the setup time is 75 min and the unit standard time is 5.0 min? If one machine tool has a setup standard time of 9 min and a unit standard time of 5.0 min, how many pieces must be handled if a machine with a setup standard of 60 min and a unit standard time of 2.0 min is to be more economical? Determine the minimum lot size for an operation requiring 3 h to set up if the unit standard time is 2.0 min and the maximum increase in the unit standard may not exceed 15 percent. Find the unit time to change a lathe cutting tool if the operator takes 5 min to change the tool and the tool cuts 1.0 min/cycle and has a life of 3 h.

Calculation Procedure:

1. Compute the true unit time
The true unit time for a machine $T_u = S_u/N + U_s$, where S_u = setup time, min; N = number of pieces in lot; U_s = unit standard time, min. For this machine, $T_u =$ 75/75 + 5.0 = 6.0 m in.

2. Determine the most economical machine
Call one machine X, the other Y. Then (unit standard time of X, min)(number of pieces) + (setup time of X, min) = (unit standard time of Y, min)(number of pieces) + (setup time of Y, min). For these two machines, since the number of pieces Z is unknown, $5.0Z + 9 = 2.0Z + 60$. So $Z = 17$ pieces. Thus, machine Y will be more economical when 17 more pieces are made.

3. Compute the minimum lot size
The minimum lot size $M = S_u/(U_s K)$, where K = allowable increase in unit-standard time, percent. For this run, $M = (3 \times 60)/[(2.0)(0.15)] = 600$ pieces.

4. Compute the unit tool-changing time
The unit tool-changing time U_t to change from dull to sharp tools is $U_t = T_c C_t/l$, where T_c = total time to change tool, min; C_t = time tool is in use during cutting cycle, min; l = life of tool, min. For this lathe, $U_t = (5)(1)/[(3)(60)] = 0.0278$ min.

Related Calculations. Use these general procedures to find true unit time, the most economical machine, minimum lot size, and unit tool-changing time for any type of machine tool—drill, lathe, milling machine, hobs, shapers, thread chasers, etc.

TIME REQUIRED FOR TURNING OPERATIONS

Determine the time to turn a 3-in (7.6-cm) diameter brass bar down to a 2½-in (6.4-cm) diameter with a spindle speed of 200 r/min and a feed of 0.20 in (0.51 mm) per revolution if the length of cut is 4 in (10.2 cm). Show how the turning-time relation can be used for relief turning, pointing of bars, internal and external chamfering, hollow mill work, knurling, and forming operations.

Calculation Procedure:

1. *Compute the turning time*

For a turning operation, the time to turn T_t min $= L/(fR)$, where L = length of cut, in; f = feed, in/r; R = work rpm. For this part, $T_t = 4/[(0.02)(200)] = 1.00$ min.

2. *Develop the turning relation for other operations*

For *relief turning* use the same relation as in step 1. Length of cut is the length of the relief, Fig. 9. A small amount of time is also required to handfeed the tool to the minor diameter of the relief. This time is best obtained by observation of the operations.

The time required to *point a bar*, called *pointing*, is computed by using the relation in step 1. The length of cut is the distance from the end of the bar to the end of the tapered point, measured parallel to the axis of the bar, Fig. 9.

Use the relation in step 1 to compute the time to cut an internal or external chamfer. The length of cut of a chamfer is the horizontal distance L, Fig. 9.

A hollow mill reduces the external diameter of a part. The cutting time is computed by using the relation in step 1. The length of cut is shown in Fig. 9.

Compute the time to knurl, using the relation in step 1. The length of cut is shown in Fig. 9.

Compute the time for forming, using the relation in step 1. Length of cut is shown in Fig. 9.

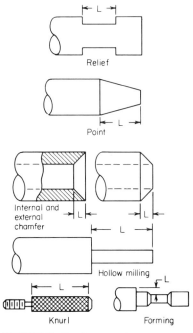

FIGURE 9 Turning operations.

TIME AND POWER TO DRILL, BORE, COUNTERSINK AND REAM

Determine the time and power required to drill a 3-in (7.6-cm) deep hole in an aluminum casting if a ¾-in (1.9-cm) diameter drill turning at 1000 r/min is used and the feed is 0.030 in (0.8 mm) per revolution. Show how the drilling relations can be used for boring, countersinking, and reaming. How long will it take to drill a hole through a 6-in (15.2-cm) thick piece of steel if the cone height of the drill is 0.5 in (1.3 cm), the feed is 0.002 in/r (0.05 mm/r), and the drill speed is 100 r/min?

Calculation Procedure:

1. Compute the time required for drilling
The time required to drill T_d min $= L/fR$, where L = depth of hole = length of cut, in. In most drilling calculations, the height of the drill cone (point) is ignored. (Where the cone height is used, follow the procedure in step 4.) For this hole, T_d $= 3/[(0.030) \times (1000)] = 0.10$ min.

2. Compute the power required to drill the hole
The power required to drill, in hp, is $hp = 1.3LfCK$, where C = cutting speed, ft/min, sometimes termed surface feet per minute $sfpm = \pi DR/12$; K = power constant from Table 3. For an aluminum casting, $K = 3$. Then $hp = (1.3)(3)(0.030)(\pi \times 0.75 \times 1000/12)(3) = 66.0$ hp (49.2 kW). The factor 1.3 is used to account for dull tools and for overcoming friction in the machine.

3. Adapt the drill relations to other operations
The time and power required for boring are found from the two relations given above. The length of the cut = length of the bore. Also use these relations for undercutting, sometimes called *internal relieving* and for counterboring. These same relations are also valid for countersinking, center drilling, start or spot drilling, and reaming. In reaming, the length of cut is the total depth of the hole reamed.

4. Compute the time for drilling a deep hole
With parts having a depth of 6 in (15.2 cm) or more, compute the drilling time from $T_d = (L + h)/(fR)$, where h = cone height, in. For this hole, $T_d = (6 + 0.5)/[(0.002)(100)] = 32.25$ min. This compares with $T_d = L/fR = 6/[(0.002)(100)] = 30$ min when the height of the drill cone is ignored.

TIME REQUIRED FOR FACING OPERATIONS

How long will it take to face a part on a lathe if the length of cut is 4 in (10.2 cm), the feed is 0.020 in/r (0.51 mm/r) and the spindle speed is 50 r/min? Determine the facing time if the same part is faced by an eight-tooth milling cutter turning at 1000 r/min and having a feed of 0.005 in (0.13 mm) per tooth per revolution. What table feed is required if the cutter is turning at 50 r/min? What is the feed per tooth with a table feed of 4.0 in/min (1.7 mm/s)? What added table

TABLE 3 Power Constants for Machining

Material	Power constant
Carbon steel C1010 to C1025	6
Manganese steel T1330 to T1350	9
Nickel steel 2015 to 2320	7
Molybdenum	9
Chromium	10
Stainless steels	11
Cast iron:	
Soft	3
Medium	3
Hard	4
Aluminum alloys:	
Castings	3
Bar	4
Copper	4
Brass (except manganese)	4
Monel metal	10
Magnesium alloys	3
Malleable iron:	
Soft	3
Medium	4
Hard	5

travel is needed when a 4-in (10.2-cm) diameter cutter is cutting a 4-in (10.2-cm) wide piece of work?

Calculation Procedure:

1. Compute the lathe facing time
For lathe facing, the time to face T_f min $= L/(fR)$, where the symbols are the same as given for previous calculation procedures in this section. For this part, $T_f = 4/[(0.02)(50)] = 4.0$ min.

2. Compute the facing time using a milling cutter
With a milling cutter, $T_f = L/(f_t nR)$, where $f_t =$ feed per tooth, in/r; $n =$ number of teeth on cutter; other symbols as before. For this part, $T_f = 4/[(0.005)(8) \times (1000)] = 0.10$ min.

3. Compute the required table feed
In a milling machine, the table feed F_t in/min $= f_t nR$. For this machine, $F_t = (0.005) \times (8)(50) = 2.0$ in/min (0.85 mm/s).

4. Compute the feed per tooth
For a milling machine, the feed per tooth, in/r, $f_t = F_t/Rn$. In this machine, $f_t = 4.0/[(50)(8)] = 0.01$ in/r (0.25 mm/r).

5. Compute the added table travel
In face milling, the added table travel A_t in $= 0.5[D_c - (D_c^2 - W^2)^{0.5}]$, where the

symbols are the same as given earlier. For this cutter and work, $A_t = 0.5[\, 4 - (4^2 - 4^2)^{0.5}] = 2.0$ in (5.1 cm).

THREADING AND TAPPING TIME

How long will it take to cut a 4-in (10.2-cm) long thread at 100 r/min if the rod will have 12 threads per inch and a button die is used? The die is backed off at 200 r/min. What would the threading time be if a self-opening die were used instead of a button die? What will the threading time be for a single-pointed threading tool if the part being threaded is aluminum and the back-off speed is twice the threading speed? The rod is 1 in (2.5 cm) in diameter. How long will it take to tap a 2-in (5.1-cm) deep hole with a 1-14 solid tap turning at 100 r/min? How long will it take to mill-thread a 1-in (2.5-cm) diameter bolt having 15 threads per inch 3 in (7.6 cm) long if a 4-in (10.2-cm) diameter 20-flute thread-milling hob turning at 80 r/min with a 0.003 in (0.08-mm) feed is used?

Calculation Procedure:

1. Compute the button-die threading time
For a multiple-pointed tool, the time to thread $T_t = Ln_t/R$, where $L =$ length of cut = length of thread measured parallel to thread longitudinal axis, in; $n_t =$ number of threads per inch. For this button die, $T_t = (4)(12)/100 = 0.48$ min. This is the time required to cut the thread.

Compute the back-off time B min from $B = Ln_t/R_B$, where $R_B =$ back-off rpm, or $B = (4)(12)/200 = 0.24$ min. Hence, the total time to cut and back-off $= T_t + B = 0.48 + 0.24 = 0.72$ min.

2. Compute the self-opening die threading time
With a self-opening die, the die opens automatically when it reaches the end of the cut thread and is withdrawn instantly. Therefore, the back-off time is negligible. Hence, the time to thread $= T_t = Ln_t/R = (4)(12)/100 = 0.48$ min. One cut is usually sufficient to make a suitable thread.

3. Compute the single-pointed tool cutting time
With a single-pointed tool, more than one cut is usually necessary. Table 4 lists the number of cuts needed with a single-pointed tool working on various materials. The maximum cutting speed for threading and tapping is also listed.

Table 4 shows that four cuts are needed for an aluminum rod when a single-pointed tool is used. Before computing the cutting time, compute the cutting speed to determine whether it is within the recommended range given in Table 4. From a previous calculation procedure, $C = R\pi d/12$, or $C = (100)(\pi)(1.0)/12 = 26.2$ ft/min (13.3 cm/s). Since this is less than the maximum recommended speed of 30 r/min, Table 4, the work speed is acceptable.

Compute the time to thread from $T_t = Ln_t c/R$, where $c =$ number of cuts to thread, from Table 4. For this part, $T_t = (4)(12)(4)/100 = 1.92$ min.

If the tool is backed off at twice the threading speed, and the back-off time $B = Ln_t c/R_B$, $B = (4)(12)(4)/200 = 0.96$ min. Hence, the total time to thread and back off $= T_t + B = 1.92 + 0.96 = 2.88$ min. In some shapes, a single-pointed tool may not be backed off; the tool may instead be repositioned. The time required for this approximates the back-off time.

TABLE 4 Number of Cuts and Cutting Speed for Dies and Taps

| | No. of cuts° | Cutting speed† | |
		ft/min	m/s
Aluminum	4	30	0.15
Brass (commercial)	3	30	0.15
Brass (naval)	4	30	0.15
Bronze (ordinary)	5	30	0.15
Bronze (hard)	7	20	0.10
Copper	5	20	0.10
Drill rod	8	10	0.05
Magnesium	4	30	0.15
Monel (bar)	8	10	0.05
Steel (mild)	5	20	0.10
Steel (medium)	7	10	0.05
Steel (hard)	8	10	0.05
Steel (stainless)	8	10	0.05

°Single-pointed threading tool; maximum spindle speed, 250 r/min.
†Maximum recommended speed for single- and multiple-pointed tools; maximum spindle speed for multiple-pointed tools = 150 r/min for dies and taps.

4. *Compute the tapping time*
The time to tap T_t min = Ln_t/R. With a solid tap, the tool is backed out at twice the tapping speed. With a collapsing tap, the tap is withdrawn almost instantly without reversing the machine or tap.

For this hole, T_t = (2)(12)/100 = 0.28 min. The back-off time $B = Ln_t/R_B$ = (2)(14)/200 = 0.14 min. Hence, the total time to tap and back off = $T_t + B$ = 0.28 + 0.14 = 0.42 min.

The maximum spindle speed for tapping should not exceed 250 r/min. Use the cutting-speed values given in Table 4 in computing the desirable speed for various materials.

5. *Compute the thread-milling time*
The time for thread milling is $T_t = L/(fnR)$, where L = length of cut, in = circumference of work, in; f = feed per flute, in; n = number of flutes on hob; R = hob rpm. For this bolt, T_t = 3.1416/[(0.003)(20)(80)] = 0.655 min.

Note that neither the length of the threaded portion nor the number of threads per inch enters into the calculation. The thread hob covers the entire length of the threaded portion and completes the threading in one revolution of the work head.

TURRET-LATHE POWER INPUT

How much power is required to drive a turret lathe making a ½-in (1.3-cm) deep cut in cast iron if the feed is 0.015 in/r (0.38 mm/r), the part is 2.0 in (5.1 cm) in diameter, and its speed is 382 r/min? How many 1.5-in (3.8-cm) long parts can be cut from a 10-ft (3.0-m) long bar if a ¼-in (6.4-mm) cutoff tool is used? Allow for end squaring.

Calculation Procedure:

1. Compute the surface speed of the part

The cutting, or surface, speed, as given in a previous calculation procedure, is $C = R\pi d/12$, or $C = (382)(\pi)(2.0)/12 = 200$ ft/min (1.0 m/s).

2. Compute the power input required

For a turret lathe, the hp input $hp = 1.33DfCK$, where D = cut depth, in; f = feed, in/r; K = material constant from Table 5. For cast iron, $K = 3.0$. Then $hp = (1.33)(0.5)(0.015)(200)(3.0) = 5.98$, say 6.0 hp (4.5 kW).

3. Compute the number of parts that can be cut

Allow 2 in (5.1 cm) on the bar end for checking and ½ in (1.3 cm) on the opposite end for squaring. With an original length of 10 ft = 120 in (304.8 cm), this leaves $120 - 2.5 = 117.5$ in (298.5 cm) for cutting.

Each part cut will be 1.5 in (3.8 cm) long + 0.25 in (6.4 mm) for the cutoff, or 1.75 in (4.4 cm) of stock. Hence, the number of pieces which can be cut = $117.5/1.75 = 67.1$, or 67 pieces.

Related Calculations. Use this procedure to find the turret-lathe power input for any of the materials, and similar materials, listed in Table 5. The parts cutoff computation can be used for any material—metallic or nonmetallic. Be sure to allow for the width of the cutoff tool.

TIME TO CUT A THREAD ON AN ENGINE LATHE

How long will it take an engine lathe to cut an acme thread having a length of 5 in (12.7 cm), a major diameter of 2 in (5.1 cm), four threads per inch (1.575 threads per centimeter), a depth of 0.1350 in (3.4 mm), a cutting speed of 70 ft/min (0.4 m/s), and a depth of cut of 0.005 in (0.1 mm) per pass if the material cut is medium steel? How many passes of the tool are required?

TABLE 5 Turret-Lathe Power Constant

Material	Constant K
Bronze	3
Cast iron	3
SAE steels:	
1020	6
1045	8
3250	9
4150	9
4615	6
X1315	6
Straight tubing	6
Steel castings and forgings	9
Heat-treated steels:	
4150	10
52100	10

TABLE 6 Thread Cutting Speeds

Material	Cutting speed	
	ft/min	m/s
Soft nonferrous metals	250	1.25
Mild steel	100	0.50
Medium steel	75	0.38
Hard steel	50	0.25

Calculation Procedure:

1. Compute the cutting time

For an acme, square, or worm thread cut on an engine lathe, the total cutting time T_t min, excluding the tool positioning time, is found from $T_t = Ld_t Dn_t/(4Cd_c)$, where L = thread length, in, measured parallel to the thread longitudinal axis; d_t = thread major diameter, in; D = depth of thread, in; n_t = number of threads per inch; C = cutting speed, ft/min; d_c = depth of cut per pass, in.

For this acme thread, $T_t = (5)(2)(0.1350)(4)/[4(70)(0.005)] = 3.85$ min. To this must be added the time required to position the tool for each pass. This equation is also valid for SI units.

2. Compute the number of tool passes required

The depth of cut per pass is 0.005 in (0.1 mm). A total depth of 0.1350 in (3.4 mm) must be cut. Therefore, the number of passes required = total depth cut, in/ depth of cut per pass, in = 0.1350/0.005 = 27 passes.

Related Calculations. Use this procedure for threads cut in ferrous and nonferrous metals. Table 6 shows typical cutting speeds.

TIME TO TAP WITH A DRILLING MACHINE

How long will it take to tap a 4-in (10.2-cm) deep hole with a 1½-in (3.8-cm) diameter tap having six threads per inch (2.36 thread per centimeter) if the tap turns at 75 r/min?

Calculation Procedure:

1. Compute the tap surface speed

By the method of previous calculation procedure, $C = R\pi d/12 = (75)(\pi)(1.5)/12 = 29.5$ ft/min (0.15 m/s).

2. Compute the time to tap the hole and withdraw the tool

For tapping with a drilling machine, $T_t = Dn_t D_c \pi/(8C)$, where D = depth of cut = depth of hole tapped, in; n_t = number of threads per inch; D_c = cutter diameter, in = tap diameter, in. For this hole, $T_t = (4)(6)(1.5)\pi/[8(29.5)] = 0.48$ min, which is the time required to tap and withdraw the tool.

Related Calculations. Use this procedure for tapping ferrous and nonferrous metals on a drill press. The recommended tap surface speed for various metals is: aluminum, soft brass, ordinary bronze, soft cast iron, and magnesium: 30 ft/min (0.15 m/s); naval brass, hard bronze, medium cast iron, copper and mild steel: 20 ft/min (0.10 m/s); hard cast iron, medium steels, and hard stainless steel: 10 ft/min (0.05 m/s).

MILLING CUTTING SPEED, TIME, FEED, TEETH NUMBER AND HORSEPOWER

What is the cutting speed of a 12-in (30.5-cm) diameter milling cutter turning at 190 r/min? How many teeth are needed in the cutter at this speed if the feed is 0.010 in (0.3 mm) per tooth, the depth of cut is 0.075 in (1.9 mm); the length of cut is 5 in (12.7 cm), the power available at the cutter is 14 hp (10.4 kW), and the mill is cutting hard malleable iron? How long will it take the mill to make this cut? What is the maximum feed rate that can be used? What is the power input to the cutter if a 20-hp (14.9-kW) machine is used?

Calculation Procedure:

1. Compute the cutter cutting speed
For a milling cutter, use the simplified relation $C = Rd/4$, where the symbols are given earlier in this section. Or, $C = (190)(12)/4 = 570$ ft/min (2.9 m/s).

2. Compute the number of cutter teeth required
For a carbide cutter, $n = K_m hp_c/(Df_t LR)$, where n = number of teeth on cutter; K_m = machinability constant or K factor from Table 7; hp_c = horsepower available

TABLE 7 Machinability Constant K_m

Aluminum	2.28
Brass, soft	2.00
Bronze, hard	1.40
Bronze, very hard	0.65
Cast iron, soft	1.35
Cast iron, hard	0.85
Cast iron, chilled	0.65
Cast magnesium	2.50
Malleable iron	0.90
Steel, soft	0.85
Steel, medium	0.65
Steel, hard	0.48
Steel:	
100 Brinell	0.80
150 Brinell	0.70
200 Brinell	0.65
250 Brinell	0.60
300 Brinell	0.55
400 Brinell	0.50

at the milling cutter; D = depth of cut, in, f_t = cutter feed, inches per tooth; L = length of cut, in; R = cutter rpm.

Table 7 shows that K_m = 0.90 for malleable iron. Then n = (0.90)(14)(0.075) (0.01)(5)(190) = 17.68, say 18 teeth. For general-purpose use, the Metal Cutting Institute recommends that n = 1.5(cutter diameter, in) for cutters having a diameter of more than 3 in (7.6 cm). For this cutter, n = 1.5(12) = 18 teeth. This agrees with the number of teeth computed with the cutter equation.

3. Compute the milling time
For a milling machine, the time to cut T_t min = $L/(f_t nR)$, where L = length of cut, in; f_t = feed per tooth, inches per tooth per revolution; n = number of teeth on the cutter; R = cutter rpm. Thus, the time to cut is T_t = 5/(0.01)(18)(190) = 0.146 min.

4. Compute the maximum feed rate
For a milling machine, the maximum feed rate f_m in/min = $K_m hp_c/(DL)$, where L = length of cut; other symbols are the same as in step 2. Thus, f_m = (0.90)(14)/ [(0.075)(5)] = 33.6 in/min (1.4 cm/s).

5. Compute the power input to the machine
The power available at the cutter is 14 hp (10.4 kW). The power required hp_c = $DLnRf_t/K_m$, where all symbols are as given above. Thus, hp_c = (0.075)(5)(18) (190)(0.01)/0.90 = 14.25 hp (10.6 kW). This is slightly more than the available horsepower.

Milling machines have overall efficiencies ranging from a low of 40 percent to a high of 80 percent, Table 8. Assume a machine efficiency of 65 percent. Then the required power input is 14.25/0.65 = 21.9 hp (16.3 kW). Therefore, a 20- or 25-hp (14.9- or 18.6-kW) machine will be satisfactory, depending on its actual operating efficiency.

Related Calculations. After selecting a feed rate, check it against the suggested feed per tooth for milling various materials given in the *American Machinist's Handbook*. Use the method of a previous calculation procedure in this section to determine the cutter approach. With the approach known, the maximum chip thick-

TABLE 8 Typical Milling-Machine Efficiencies

Rated power of machine		Overall efficiency, percent
hp	kW	
3	2.2	40
5	3.7	48
7.5	5.6	52
10	7.5	52
15	11.2	52
20	14.9	60
25	18.6	65
30	22.4	70
40	29.8	75
50	27.3	80

ness, in = (cutter approach in)(table advance per tooth, in)/(cutter radius, in). Also, the feed per tooth, in = (feed rate, in/min)/[(cutter rpm)(number of teeth on cutter)].

GANG-, MULTIPLE-, AND FORM-MILLING CUTTING TIME

How long will it take to gang mill a part if three cutters are used with a spindle speed of 70 r/min and there are 12 teeth on the smallest cutter, a feed of 0.015 in/r (0.4 mm/r) and a length of cut of 8 in (20.3 cm)? What will be the unit time to multiple mill four keyways if each of the four cutters has 20 teeth, the feed is 0.008 in (0.2 mm) per tooth, spindle speed is 150 r/min, and the keyway length is 3 in (7.6 cm)? Show how the cutting time for form milling is computed, and how the cutter diameter for straddle milling is computed.

Calculation Procedure:

1. Compute the gang-milling cutting time

For any gang-milling operation, from the dimensions of the smallest cutter, the time to cut $T_t = L/f_t nR$, where L = length of cut, in; f_t = feed per tooth, in/r; n = number of teeth on cutter; R = spindle rpm. For this part, $T_t = 8/[(0.015)(12)(70)]$ = 0.635 min.

Note that in all gang-milling cutting-time calculations, the number of teeth and feed of the *smallest* cutter are used.

2. Compute the multiple-milling cutting time

In multiple milling, the cutting time $T_t = L/(f_t nR_m)$, where n = number of milling cutter used. In multiple milling, the cutting time is termed the unit time. For this machine, $T_t = 3/[(0.008)(20)(150)(4)]$ = 0.0303 min.

3. Show how form milling time is computed

Form-milling cutters are used on surfaces that are neither flat nor square. The cutters used for form milling resemble other milling cutters. The cutting time is therefore computed from $T_t = L/(f_t nR)$, where all symbols are the same in step 1.

4. Show how the cutter diameter is computed for straddle milling

In straddle milling, the cutter diameter must be large enough to permit the work to pass under the cutter arbor. The minimum-diameter cutter to straddle mill a part = (diameter of arbor, in) + 2 (face of cut, in + 0.25). The 0.25 in (6.4 mm) is the allowance for clearance of the arbor.

Related Calculations. Use the equation of step 1 to compute the cutting time for metal slitting, screw slotting, angle milling, T-slot milling, Woodruff key-seat milling, and profiling and routing of parts. In T-slot milling, two steps are required—milling of the vertical member and milling of the horizontal member. Compute the milling time of each; the sum of the two is the total milling time.

SHAPER AND PLANER CUTTING SPEED, STROKES, CYCLE TIME, POWER

What is the cutting speed of a shaper making 54-strokes/min if the stroke length is 6 in (15.2 cm)? How many strokes per minute should the ram of a shaper make if it is shaping a 12-in (30.5-cm) long aluminum bar at a cutting speed of 200 ft/min? How long will it take to make a cut across a 12-in (30.5-cm) face of a cast-iron plate if the feed is 0.050 in (1.3 mm) per stroke and the ram makes 50 strokes/min? What is the cycle time of a planer if its return speed is 200 ft/min (1.0 m/s), the acceleration-deceleration constant is 0.05, and the cutting speed is 100 ft/min (0.5 m/s)? What is the planer power input if the depth of cut is ⅛ in (3.2 mm) and the feed is ¹⁄₁₆ in (1.6 mm) per stroke?

Calculation Procedure:

1. Compute the shaper cutting speed
For a shaper, the cutting speed, ft/min, is $C = SL/6$, where S = strokes/min; L = length of stroke, in; where the cutting-stroke time = return-stroke time. Thus, for this shaper, $C = (54)(6)/6 = 54$ ft/min (0.3 m/s).

2. Compute the shaper stroke rate
Transpose the equation of step 1 to $S = 6C/L$. Then $S = 6(200)/12 = 100$ strokes/min.

3. Compute the shaper cutting time
For a shaper the cutting time, min, is $T_t = L/(fS)$, where L = length of cut, in; f = feed, in/stroke; S = strokes/min. Thus, for this shaper, $T_t = 12/[(0.05)(50)] = 4.8$ min. Multiply T_t by the number of strokes needed; the result is the total cutting time, min.

4. Compute the planer cycle time
The cycle time for a planer, min, = $(L/C) + (L/R_c) + k$, where R_c = cutter return speed, ft/min; k = acceleration-deceleration constant. Since the cutting speed is 100 ft/min (0.5 m/s) and the return speed is 200 ft/min (1.0 m/s), the cycle time = $(12/100) + (12/200) + 0.05 = 0.23$ min.

5. Compute the power input to the planer
Table 9 lists typical power factors for planers planing cast iron and steel. To find the power required, multiply the power factor by the cutter speed, ft/min. For the planer in step 3 with a cutting speed of 100 ft/min (0.5 m/s) and a power factor of 0.0235 for a ⅛-in (3.2-mm) deep cut and a ¹⁄₁₆-in (1.6-mm) feed, $hp_{input} = (0.0235)(100) = 2.35$ hp (1.8 kW).

For steel up to 40 points carbon, multiply the above result by 2; for steel above 40 points carbon, multiply by 2.25.

Related Calculations. Where a shaper has a cutting stroke time that does not equal the return-stroke time, compute its cutting speed from $C = SL/(12)$(cutting-stroke time, min/sum of cutting- and return-stroke time, min). Thus, if the shaper in step 1 has a cutting-stroke time of 0.8 min and a return-stroke time of 0.4 min, $C = (54)(6)/[(12) \times (0.8/1.2)] = 40.5$ ft/min (0.2 m/s).

TABLE 9 Power Factors for Planers*

Depth of cut		Feed		
in	cm	$\frac{1}{32}$ in (0.8 mm) per stroke	$\frac{1}{16}$ in (1.6 mm) per stroke	$\frac{1}{8}$ in (3.2 mm) per stroke
$\frac{1}{8}$	0.3	0.0115	0.0235	0.047
$\frac{1}{4}$	0.6	0.023	0.047	0.094
$\frac{3}{8}$	1.0	0.035	0.070	0.141
$\frac{1}{2}$	1.3	0.047	0.094	0.189
$\frac{5}{8}$	1.6	0.063	0.118	0.236
$\frac{3}{4}$	1.9	0.080	0.142	0.284
$\frac{7}{8}$	2.2	0.087	0.165	0.331
1	2.5	0.094	0.189	0.378

*Excerpted from the Cincinnati Planer Company and *American Machinist's Handbook*.

GRINDING FEED AND WORK TIME

What is the feed of a centerless grinding operation if the regulating wheel is 8 in (20.3 cm) in diameter and turns at 100 r/min at an angle of inclination of 5°? How long will it take to rough grind on an external cylindrical grinder a brass shaft that is 3.0 in (7.6 cm) in diameter and 12 in (30.5 cm) long, if the feed is 0.003 in (0.076 mm), the spindle speed is 20 r/min, the grinding-wheel width is 3 in (7.6 cm) and the diameter is 8 in (20.3 cm), and the total stock on the part is 0.015 in (0.38 mm)? How long would it take to make a finishing cut on this grinder with a feed of 0.001 in (0.025 mm), stock of 0.010 in (0.25 mm), and a cutting speed of 100 ft/min (0.5 m/s)?

Calculation Procedure:

1. Compute the feed rate for centerless grinding
In centerless grinding, the feed, in/min, $f = \pi dR \sin \infty$, where $\pi = 3.1416$; $d =$ diameter of the regulating wheel, in; $R =$ regulating wheel rpm; $\infty =$ angle of inclination of the regulating wheel. For this grinder, $f = \pi(8)(100)(\sin 5°) = 219$ in/min (9.3 cm/s). Centerless grinders will grind as many as 50,000 1-in (2.5-cm) parts per hour.

2. Compute the rough-grinding time
The rough-grinding time T_t min $= Lt_s d/(2WfC)$, where $L =$ length of ground part, in; $t_s =$ total stock on part, in; $W =$ width of grinding-wheel face, in; $C =$ cutting speed, ft/min.
   Compute the cutting speed first because it is not known. By the method of previous calculation procedures, $C = \pi dR/12 = \pi(8)(20)/12 = 42$ ft/min (0.2 m/s). Then $T_t = (12)(0.015)(3)/[2(3)(0.003)(42)] = 0.714$ min.

3. Compute the finish-grinding time
For finish grinding, use the same equation as in step 2, except that the factor 2 is omitted from the denominator. Thus, $T_t = Lt_s d/(WfC)$, or $T_t = (12)(0.010)(3)/[(3)(0.001)(100)] = 1.2$ min.

Related Calculations.   Use the same equations as in steps 1 and 4 for internal cylindrical grinding. In surface grinding, about 250 in^2/min (26.9 cm^2/s) can be ground 0.001 in (0.03 mm) deep if the material is hard. For soft materials, about 1000 in^2 (107.5 cm^2) and 0.001 in (0.03 mm) deep can be ground per minute.

In honing cast iron, the average stock removal is 0.006 to 0.008 in/(ft · min) [0.008 to 0.011 mm/(m · s)]. With hard steel or chrome plate, the rate of honing averages 0.003 to 0.004 in/ft · min [0.004 to 0.006 mm/(m · s)].

BROACHING TIME AND PRODUCTION RATE

How long will it take to broach a medium-steel part if the cutting speed is 20 ft/min (0.1 m/s), the return speed is 100 ft/min (0.5 m/s), and the stroke length is 36 in (91.4 cm)? What will the production rate be if starting and stopping occupy 2 s and loading 5 s with an efficiency of 85 percent?

Calculation Procedure:

1. *Compute the broaching time*
The broaching time T_t min $= (L/C) + (L/R_c)$, where L = length of stroke, ft; C = cutting speed, ft/min; R_c = return speed, ft/min; for this work, $T_t = (3/20) + (3/100) = 0.18$ min.

2. *Compute the production rate*
In a complete cycle of the broaching machine there are three steps: broaching; starting and stopping; and loading. The cycle time, at 100 percent efficiency, is the sum of these three steps, or $0.18 \times 60 + 2 + 5 = 17.8$ s, where the factor 60 converts 0.18 min to seconds. At 85 percent efficiency, the cycle time is greater, or $17.8/0.85 = 20.9$ s. Since there are 3600 s in 1 h, production rate $= 3600/20.9 = 172$ pieces per hour.

HOBBING, SPLINING, AND SERRATING TIME

How long will it take to hob a 36-tooth 12-pitch brass spur gear having a tooth length of 1.5 in (3.8 cm) by using a 2.75-in (7.0-cm) hob? The whole depth of the gear tooth is 0.1789 in (4.5 mm). How many teeth should the hob have? Hob feed is 0.084 in/r (2.1 mm/r). What would be the cutting time for a 47° helical gear? How long will it take to spline-hob a brass shaft which is 2.0 in (5.1 cm) in diameter, has 12 splines, each 10 in (25.4 cm) long, if the hob diameter is 3.0 in (7.6 cm), cutter feed is 0.050 in (1.3 mm), cutter speed is 120 r/min, and spline depth is 0.15 in (3.8 mm)? How long will it take to hob 48 serrations on a 2-in (5.1-cm) diameter brass shaft if each serration is 2 in (5.1 cm) long, the 18-flute hob is 2.5 in (6.4 cm) in diameter, the approach is 0.3 in (7.6 mm), the feed per flute is 0.008 in (0.2 mm), and the hob speed is 250 r/min?

Calculation Procedure:

1. *Compute the hob approach*
The hob approach $A_c = \sqrt{d_g(D_c - d_g)}$, where d_g = whole depth of gear tooth, in;

TABLE 10 Gear-Hobbing Cutting Speeds

| Gear material | Spur gears | | Helical gears° | |
| | Cutting speed | | | |
	ft/min	m/s	Angle,°	Percentage of feed to use
Brass	150	0.8	0–36	100
Fiber	150	0.8	36–48	80
Cast iron (soft)	100	0.5	48–60	67
Steel (mild)	100	0.5	60–70	50
Steel (medium)	75	0.4	70–90	33
Steel (hard)	50	0.3		

°Reduce feed by percentage shown when helical gears are cut.

D_c = hob diameter, in. For this hob, $A_c = \sqrt{0.1798(2.7500 - 0.798)} = 0.68$ in (1.7 cm).

2. Determine the cutting speed of the hob
Table 10 shows that a cutting speed of $C = 150$ ft/min (0.8 m/s) is generally used for brass gears. With a 2.75-in (7.0-cm) diameter hob, this corresponds to a hob rpm of $R = 12C/(\pi D_c) = (12)(150)/[\pi(2.75)] = 208$ r/min.

3. Compute the hobbing time
The time to hob a spur gear T_t min $= N(L + A_c)/fR$, where N = number of teeth in gear to be cut; L = length of a tooth in the gear, in; A_c = hob approach, in; f = hob feed, in/r; R = hob rpm. For this spur gear, $T_t = (36)(1.5 + 0.68)/[(0.084)(208)] = 4.49$ min.

4. Compute the cutting time for a helical gear
Table 10 shows that the feed for a 47° helical gear should be 80 percent of that for a spur gear. By the relation in step 3, $T_t = (36)(1.5 + 0.68)/[(0.80)(0.084)(208)] = 5.61$ min.

5. Compute the time to spline hob
Use the same procedures as for hobbing. Thus, $A_c = \sqrt{d_g(D_c - d_g)} = \sqrt{0.15(3.0 - 0.15}} = 0.654$ in (1.7 cm). Then $T_t = N(L + A)/fR$, where N = number of splines; L = length of spline, in; other symbols as before. For this shaft, $T_t = (12)(10 + 0.654)/[(0.05)(120)] = 21.3$ min.

6. Compute the time to serrate
The time to hob serrations T_t min $= N(L + A)/(fnR)$, where N = number of serrations; L = length of serration, in; n = number of flutes on hob; other symbols as before . For this shaft, $T_t = (48)(2 + 0.30)/[(0.008)(18)(250)] = 3.07$ min.

TIME TO SAW METAL WITH POWER AND BAND SAWS

How long will it take to saw a rectangular piece of alloy-plate aluminum 6 in (15.2 cm) wide and 2 in (5.1 cm) thick if the length of cut is 6 in (15.2 cm), the power hacksaw makes 120 strokes/min, and the average feed per stroke is 0.0040 in (0.1 mm)? What would the sawing time be if a band saw with a 200-ft/min (1.0-m/s) cutting speed, 16 teeth per inch (6.3 teeth per centimeter), and a 0.0003-in (0.008-mm) feed per tooth is used?

Calculation Procedure:

1. Compute the sawing time for a power saw
For a power saw with positive feed, the time to saw T_t min $= L/(Sf)$, where $L =$ length of cut, in; $S =$ strokes/min of saw blade; $f =$ feed per stroke, in. In this saw, $T_t = (6)/[(120)(0.0040)] = 12.5$ min.

2. Compute the band-saw cutting time
For a band saw, the sawing time T_t min $= L/(12Cnf)$, where $L =$ length of cut, in; $C =$ cutting speed, ft/min; $n =$ number of saw teeth per inch; $f =$ feed, inches per tooth. With this band saw, $T_t = (6)/[(12)(200)(16)(0.0003)] = 0.521$ min.

   Related Calculations. When nested round, square, or rectangular bars are to be cut, use the greatest *width* of the nested bars as the length of cut in either of the above equations.

OXYACETYLENE CUTTING TIME AND GAS CONSUMPTION

How long will it take to make a 96-in (243.8-cm) long cut in a 1-in (2.5-cm) thick steel plate by hand and by machine? What will the oxygen and acetylene consumption be for each cutting method?

Calculation Procedure:

1. Compute the cutting time
For any flame cutting, the cutting time T_t min $= L/C$, where $L =$ length of cut, in; $C =$ cutting speed, in/min, from Table 11. With manual cutting, $T_t = 96/8 = 12$ min, using the lower manual cutting speed given in Table 11. At the higher manual cutting speed, $T_t = 96/12 = 8$ min. With machine cutting, $T_t = 96/14 = 6.86$ min, by using the lower machine cutting speed in Table 11. At the higher machine cutting speed, $T_t = 96/18 = 5.34$ min.

2. Compute the gas consumption
From Table 11 the oxygen consumption is 130 to 200 ft³/h (1023 to 1573 cm³/s). Thus, actual consumption, ft³ $=$ (cutting time, min/60)(consumption, ft³/h) $=$ (12/60)(130) $= 26$ ft³ (0.7 m³) at the minimum cutting speed and minimum oxygen consumption. For this same speed with maximum oxygen consumption, actual ft³ used $=$ (12/60)(200) $= 40$ ft³ (1.1 m³).

   Compute the acetylene consumption in the same manner, or (12/60)(13) $= 2.6$ ft³ (0.07 m³), and (12/60)(16) $= 3.2$ ft³ (0.09 m³). Use the same procedure to compute the acetylene and oxygen consumption at the higher cutting speeds.

   Related Calculations. Use the procedure given here for computing the cutting time and gas consumption when steel, wrought iron, or cast iron is cut. Thicknesses ranging up to 5 ft (1.5 m) are economically cut by an oxyacetylene torch. Alloying

TABLE 11 Oxyacetylene Cutting Speed and Gas Consumption

| Metal thickness | | Speed | | | | Gas consumption | | | |
| | | Manual | | Machine | | Oxygen | | Acetylene | |
in	cm	in/min	mm/s	in/min	mm/s	ft³/h	cm³/s	ft³/h	cm³/s
0.25	0.6	16–18	6.8–7.6	20–26	8.5–11.0	50–90	393.3–707.9	8–11	62.9–86.5
0.50	1.3	12–15	5.1–6.4	17–22	7.2–9.3	90–125	707.9–983.2	10–13	78.7–102.3
1	2.5	8–12	3.4–5.1	14–18	5.9–7.6	130–200	1023–1573	13–16	102.3–125.9
2	5.1	5–7	2.1–3.0	10–13	4.2–5.5	200–300	1573–2360	16–20	125.9–157.3
4	10.2	4–5	1.7–2.1	7–9	3.0–3.8	300–400	2360–3146	21–26	165.2–204.5
6	15.2	3–4	1.3–1.7	5–7	2.1–3.0	400–500	3146–3933	26–32	204.5–251.7
8	20.3	3–6	1.3–2.5	4–6	1.7–2.5	500–650	3933–5113	28–35	220.2–275.3
10	25.4	2–3	0.8–1.3	3–4	1.3–1.7	700–1000	5506–7860	30–38	236.0–298.9
12	30.5	2.5–3.5	1.1–1.5	3–4	1.3–1.7	720–880	5663–6922	42–52	330.4–409.0

elements in steel may require preheating of the metal to permit cutting. To compute the gas required per lineal foot, divide the actual consumption for the length cut, in inches, by 12.

COMPARISON OF OXYACETYLENE AND ELECTRIC-ARC WELDING

Determine the time required to weld a 4-ft (1.2-m) long seam in a ⅜-in (9.5-mm) plate by the oxyacetylene and electric-arc methods. How much oxygen and acetylene are required? What weight of electrode will be used? What is the electric-power consumption? Assume that one weld bead is run in the joint.

Calculation Procedure:

1. Compute the welding time
For any welding operation, the time required to weld T_t min $= L/C$, where $L =$ length of weld, in; $C =$ welding speed, in/min. When oxyacetylene welding is used, $T_t = 48/1.0 = 48$ min, when a welding speed of 1.0 in/min (0.4 mm/s) is used. With electric-arc welding, $T_t = 48/18 = 2.66$ min when the welding speed $= 18$ in/min (7.6 mm/s) per bead. For plate thicknesses under 1 in (2.5 cm), typical welding speeds are in the range of 1 to 2 in/min (0.4 to 0.8 mm/s) for oxyacetylene and 18 in/min (7.6 mm/s) for electric-arc welding. For thicker plates, consult *The Welding Handbook*, American Welding Society.

2. Compute the gas consumption
Gas consumption for oxyacetylene welding is given in cubic feet per foot of weld. Using values from *The Welding Handbook*, or a similar reference, we see that oxygen consumption $= (\text{ft}^3\ O_2$ per ft of weld)(length of weld, ft); acetylene consumption $= (\text{ft}^3$ acetylene per ft of weld)(length of weld, ft). For this weld, with only one bead, oxygen consumption $= (10.0)(4) = 40\ \text{ft}^3$ (1.1 m³); acetylene consumption $= (9.0)(4) = 36\ \text{ft}^3$ (1.0 m³).

3. Compute the weight of electrode required
The Welding Handbook tabulates the weight of electrode for various types of welds—square grooves, 90° grooves, etc., per foot of weld. Then the electrode weight required, lb $=$ (rod consumption, lb/ft)(weld length, ft).

For oxyacetylene welding, the electrode weight required, from data in *The Welding Handbook*, is $(0.597)(4) = 2.388$ lb (1.1 kg). For electric-arc welding, weight $= (0.18)(4) = 0.72$ lb (0.3 kg).

4. Compute the electric-power consumption
In electric-arc welding the power consumption is $kW = (V)(A)/(1000)(\text{efficiency})$. *The Welding Handbook* shows that for a ⅜-in (9.5-mm) thick plate, $V = 40$, $A = 450$ A, efficiency $= 60$ percent. The power consumption $= (40)(450)/[(1000)(0.60)] = 30$ kW. For this press, $F = (8)(0.5)(16.0) = 64$ tons (58.1 t).

Related Calculations.   Where more than one pass or bead is required, multiply the time for one bead by the number of beads deposited. If only 50 percent penetration is required for the bead, the welding speed will be twice that where full penetration is required.

PRESSWORK FORCE FOR SHEARING
AND BENDING

What is the press force to shear an 8-in (20.3-cm) long 0.5-in (1.3-cm) thick piece of annealed bronze having a shear strength of 16.0 tons/in² (2.24 t/cm²)? What is the stripping load? Determine the force required to produce a U bend in this piece of bronze if the unsupported length is 4 in (10.2 cm), the bend length is 6 in (15.2 cm), and the ultimate tensile strength is 32.0 tons/in² (4.50 t/cm²).

Calculation Procedure:

1. Compute the required shearing force
For any metal in which a straight cut is made, the required shearing force, tons = $F = Lts$, where L = length of cut, in; t = metal thickness, in; s = shear strength of metal being cut, tons/in². Where round, elliptical, or other shaped holes are being cut, substitute the sum of the circumferences of all the holes for L in this equation.

2. Compute the stripping load
For the typical press, the stripping load is 3.5 percent of the required shearing force, or $(0.035)(64) = 2.24$ tons (2.0 t).

3. Compute the required bending force
When U bends or channels are pressed in a metal, $F = 2Lt^2s_t/W$, where s_t = ultimate tensile strength of the metal, tons/in²; W = width of unsupported metal, in = distance between the vertical members of a channel or U bend, measured to the *outside* surfaces, in. For this U bend, $F = 2(6)(0.5)^2(32)/4 = 24$ tons (21.8 t).
  Related Calculations. Right-angle edge bends require a bending force of $F = Lt^2s_t/(2W)$, while free V bends with a centrally located load require a bending force of $F = Lt^2s_t/W$. All symbols are as given in steps 1 and 2.

MECHANICAL-PRESS MIDSTROKE CAPACITY

Determine the maximum permissible midstroke capacity of single- and twin-driven 2-in (5.1-cm) diameter crankshaft presses if the stroke of the slide is 12 in (30.5 cm) for each.

Calculation Procedure:

1. Compute the single-driven press capacity
For a single-driven crankshaft press with a heat-treated 0.35 to 0.45 percent carbon-steel crankshaft having a shear strength of 6 tons/in² (0.84 t/cm²), the maximum permissible midstroke capacity F tons = $2.4d^3/S$, where d = shaft diameter at main bearing, in; S = stroke length, in; or $F = (2.4)(2)^3/12 = 1.6$ tons (1.5 t).

2. *Compute the twin-driven press capacity*
Twin-driven presses with main (bull) gears on each end of the crankshaft have a maximum permissible midstroke capacity of $F = 3.6d^3/S$, when the shaft shearing strength is 9 tons/in^2. For this press, $F = 3.6(2)^3/12 = 2.4$ tons (2.2 t).

   Related Calculations.   Use the equation in step 2 to compute the maximum permissible midstroke capacity of all wide (right-to-left) double-crank presses. Since gear eccentric presses are built in competition with crankshaft presses, their midstroke pressure capacity is within the same limit as in crankshaft presses. The diameters of the fixed pins on which the gear eccentrics revolve are usually made the same as the crankshaft in crankshaft presses of the same rated capacity.

STRIPPING SPRINGS FOR PRESSWORKING METALS

Determine the force required to strip the work from a punch if the length of cut is 5.85 in (14.9 cm) and the stock is 0.25 in (0.6 cm) thick. How many springs are needed for the punch if the force per inch deflection of the spring is 100 lb (175.1 N/cm)?

Calculation Procedure:

1. *Compute the required stripping force*
The required stripping force F_p lb needed to strip the work from a punch is $F_p = Lt/0.00117$, where L = length of cut, in; t = thickness of stock cut, in. For this punch, $F_p = (5.85)(0.25)/0.00117 = 1250$ lb (5560.3 N).

2. *Compute the number of springs required*
Only the first ⅛-in (0.3-cm) deflection of the spring can be used in the computation of the stripping force produced by the spring. Thus, for this punch, number of springs required = stripping force, lb/force, lb, to produce ⅛-in (0.3-cm) deflection of the spring, or 1250/100 = 12.5 springs. Since a fractional number of springs cannot be used, 13 springs would be selected.

   Related Calculations.   In high-speed presses, the springs should not be deflected more than 25 percent of their free length. For heavy, slow-speed presses, the total deflection should not exceed 37.5 percent of the free length of the spring. The stripping force for aluminum alloys is generally taken as one-eighth the maximum blanking pressure.

BLANKING, DRAWING, AND NECKING METALS

What is the maximum blanking force for an aluminum part if the length of the cut is 30 in (76.2 cm), the metal is 0.125 in (0.3 cm) thick, and the yield strength is 2.5 tons/in^2 (0.35 t/cm^2)? How much force is required to draw a 12-in (30.5-cm) diameter 0.25-in (0.6-cm) thick stainless steel shell if the yield strength is 15 tons/in^2 (2.1 t/cm^2)? What force is required to neck a 0.125-in (0.3-cm) thick aluminum

TABLE 12   Metal Yield Strength

Metal	Yield strength	
	tons/in^2	t/cm^2
Aluminum, 2S annealed	2.5	0.35
Aluminum, 24S heat-treated	23.0	2.23
Low brass, ¼ hard	24.5	3.46
Yellow brass, annealed	10.0	1.41
Cold-rolled steel, ¼ hard	16.0	2.25
Stainless steel, 18-8	15.0	2.11

Note: As a general rule, the necking angle should not exceed 35°.

shell from a 3- to a 2-in (7.6- to 5.1-cm) diameter if the necking angle is 30° and the ultimate compressive strength of the material is 14 tons/in^2 (1.97 t/cm^2)?

Calculation Procedure:

1. *Compute the maximum blanking force*
The maximum blanking force for any metal is given by $F = Lts$, where $F =$ blanking force, tons; $L =$ length of cut, in (= circumference of part, in); $t =$ metal thickness, in; $s =$ yield strength of metal, tons/in^2. For this part, $F = (30)(0.125)(2.5) = 0.375$ tons (0.34 t).

2. *Compute the maximum drawing force*
Use the same equation as in step 1, substituting the drawing-edge length or perimeter (circumference of part) for L. Thus, $F = (12\pi)(0.25)(15) = 141.5$ tons (128.4 t).

3. *Compute the required necking force*
The force required to neck a shell is $F = ts_c(d_1 - d_s)/\cos$ (necking angle), where $F =$ necking force, tons; $t =$ shell thickness, in; $s_c =$ ultimate compressive strength of the material, tons/in^2; $d_1 =$ large diameter of shell, i.e., the diameter *before* necking, in; $d_s =$ small diameter of shell, i.e., the diameter *after* necking, in. For this shell, $F = (0.125)(14)(3.0 - 2.0)/\cos 30° = 2.02$ tons (1.8 t).
   Related Calculations.   Table 12 presents typical yield strengths of various metals which are blanked or drawn in metalworking operations. Use the given strength as shown above.

METAL PLATING TIME AND WEIGHT

How long will it take to electroplate a 0.004-in (0.1-mm) thick zinc coating on a metal plate if a current density of 25 A/ft^2 (269.1 A/m^2) is used at an 80 percent plating efficiency? How much zinc is required to produce a 0.001-in (0.03-mm) thick coating on an area of 60 ft^2 (5.6 m^2)?

TABLE 13   Electroplating Current and Metal Weight

Metal	Time to deposit, Ah		Metal density	
	0.001 in/ft^2 at 100% efficiency	0.01 mm/m^2 at 100% efficiency	lb/in^3	g/cm^3
Antimony, Sb	10.40	0.038	0.241	6.671
Cadmium, Cd	9.73	0.036	0.312	8.636
Chromium, Cr	51.80	0.189	0.256	7.086
Cobalt(ous), Co	19.00	0.069	0.322	8.913
Copper(ous), Cu	8.89	0.033	0.322	8.913
Copper(ic), Cu	17.80	0.065	0.322	8.913
Gold(ous), Au	6.20	0.023	0.697	19.29
Gold(ic), Au	18.60	0.068	0.697	19.29
Nickel, Ni	19.00	0.069	0.322	8.913
Platinum	27.80	0.102	0.775	21.45
Silver, Ag	6.20	0.023	0.380	10.52
Tin(ous), Sn	7.80	0.029	0.264	7.307
Tin(ic), Sn	15.60	0.057	0.264	7.307
Zinc, Zn	14.30	0.052	0.258	7.141

Calculation Procedure:

1. *Compute the metal plating time*
The plating time T_p min $= 60\, An/(A_a e)$, where A = A/ft^2 required to deposit 0.001 in (0.03 mm) of metal at 100 percent cathode efficiency; n = number of thousandths of inch actually deposited; A_a = current actually supplied, A/ft^2; e = plating efficiency, expressed as a decimal. Table 13 gives typical values of A for various metals used in electroplating. For plating zinc, from the value in Table 13, T_p = $60(14.3)(4)/[(25)(0.80)]$ = 171.5 min, or 171.5/60 = 2.86 h.

2. *Compute the weight of metal required*
The plating metal weight = (area plated, in^2)(plating thickness, in)(plating metal density, lb/in^3). For this plating job, given the density of zinc from Table 13, the plating metal weight = $(60 \times 144)(0.004)(0.258)$ = 8.91 lb (4.0 kg) of zinc. In this calculation the value 144 is used to convert 60 ft^2 to square inches.

Related Calculations.   The efficiency of finishing cathodes is high, ranging from 80 to nearly 100 percent. Where the actual efficiency is unknown, assume a value of 80 percent and the results obtained will be safe for most situations.

SHRINK- AND EXPANSION-FIT ANALYSES

To what temperature must an SAE 1010 steel ring 24 in (61.0 cm) in inside diameter be raised above a 68°F (20°C) room temperature to expand it 0.004 in (0.10 mm) if the linear coefficient of expansion of the steel is 0.0000068 in/(in · °F) [0.000012 cm/(cm · °C)]? To what temperature must a 2-in (5.08-cm) diameter SAE steel shaft be reduced to fit it into a 1.997-in (5.07-cm) diameter hole for an expansion fit? What cooling medium should be used?

TABLE 14   Metal Shrinkage with Nitrogen Cooling

Metal	Shrinkage, in/in (cm/cm) of shaft diameter
Magnesium alloys	0.0046
Aluminum alloys	0.0042
Copper alloys	0.0033
Cr-Ni alloys (18-8 to 18-12)	0.0029
Monel metals	0.0023
SAE steels	0.0022
Cr steels (5 to 27% Cr)	0.0019
Cast iron (not alloyed)	0.0017

Calculation Procedure:

1.   *Compute the required shrink-fit temperature rise*
The temperature needed to expand a metal ring a given amount before making a
shrink fit is given by $T = El/(Kd)$, where T = temperature rise *above* room temperature, °F; K = linear coefficient of expansion of the metal ring, in/(in · °F); d =
ring internal diameter, in. For this ring, $T = 0.004/[(0.0000068)(24)] = 21.5°F$
(11.9°C). With a room temperature of 68°F (20.0°C), the final temperature of the
ring must be $68 + 21.5 = 89.5°F$ (31.9°C) or higher.

2.   *Compute the temperature for an expansion fit*
Nitrogen, air, and oxygen in liquid form have a low boiling point, as does dry ice
(solid carbon dioxide). Nitrogen and dry ice are considered the safest cooling media
for expansion fits because both are relatively inert. Liquid nitrogen boils at
−320.4°F (−195.8°C) and dry ice at −109.3°F (−78.5°C). At −320°F (−195.6°C)
liquid nitrogen will reduce the diameter of metal parts by the amount shown in
Table 14. Dry ice will reduce the diameter by about one-third the values listed in
Table 14.
    With liquid nitrogen, the diameter of a 2-in (5.1-cm) round shaft will be reduced
by $(2.0)(0.0022) = 0.0044$ in (0.11 mm), given the value for SAE steels from Table
14. Thus, the diameter of the shaft at −320.4°F (−195.8°C) will be $2.000 − 0.0044$
$= 1.9956$ in (5.069 cm). Since the hole is 1.997 in (5.072 cm) in diameter, the
liquid nitrogen will reduce the shaft size sufficiently.
    If dry ice were used, the shaft diameter would be reduced $0.0044/3 = 0.00146$
in (0.037 mm), giving a final shaft diameter of $2.00000 − 0.00146 = 1.99854$ in
(5.076 cm). This is too large to fit into a 1.997-in (5.072-cm) hole. Thus, dry ice
is unsuitable as a cooling medium.

PRESS-FIT FORCE, STRESS, AND SLIPPAGE TORQUE

What force is required to press a 4-in (10.2-cm) outside-diameter cast-iron hub on
a 2-in (5.1-cm) outside-diameter steel shaft if the allowance is 0.001-in interference
per inch (0.001 cm/cm) of shaft diameter, the length of fit is 6 in (15.2 cm), and

the coefficient of friction is 0.15? What is the maximum tensile stress at the hub bore? What torque is required to produce complete slippage of the hub on the shaft?

Calculation Procedure:

1. Determine the unit press-fit pressure
Figure 10 shows that with an allowance of 0.001 in interference per inch (0.001 cm/cm) of shaft diameter and a shaft-to-hub diameter ratio of $2/4 = 0.5$, the unit press-fit pressure between the hub and the shaft is $p = 6800$ lb/in² (46,886.0 kPa).

2. Compute the press-fit force
The press-fit force F tons $= \pi f p d L / 2000$, where $f =$ coefficient of friction between hub and shaft; $p =$ unit press-fit pressure, lb/in²; $d =$ shaft diameter, in; $L =$ length of fit, in. For this press fit, $F = (\pi)(0.15)(6800)(2.0)(6)/2000 = 19.25$ tons (17.4 t).

3. Determine the hub bore stress
Use Fig. 11 to determine the hub bore stress. Enter the bottom of Fig. 11 at 0.0010 in (0.0010 cm) interference allowance per in of shaft diameter and project vertically to $d/D = 0.5$. At the left read the hub stress at 11,600 lb/in² (79,982 kPa).

4. Compute the slippage torque
The torque, in · lb, required to produce complete slippage of a press fit is $T = 0.5 \pi f p L d^2$, or $T = 0.5(3.1416)(0.15)(6800)(6)(2)^2 = 38,450$ in · lb (4344.1 N · m).
    Related Calculations.   Figure 12 shows the press-fit pressures existing with a steel hub on a steel shaft. The three charts presented in this calculation procedure are useful for many different press fits, including those using a hollow shaft having

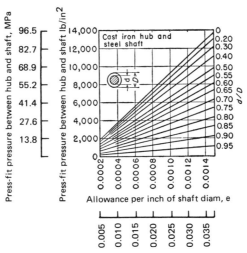

FIGURE 10   Press-fit pressures between steel hub and shaft.

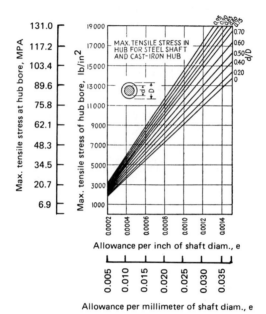

FIGURE 11 Variation in tensile stress in cast-iron hub in press-fit allowance.

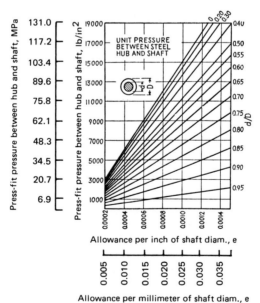

FIGURE 12 Press-fit pressures between cast-iron hub and shaft.

an internal diameter less than 25 percent of the external diameter and for all solid steel shafts.

LEARNING-CURVE ANALYSIS AND CONSTRUCTION

A short-run metalworking job requires five operators. The longest individual learning time for the new task is 3 days; 2 days are allowed for group familiarization with the task. If the normal output is 1000 units per 8-h day, determine the daily allowance per operator when the standard for 100 percent performance is 0.8 worker-hour per 100 units produced.

Calculation Procedure:

1. *Plot the learning curve*
A learning curve shows the improvement that occurs with repetition of a task. Figure 13 is a typical learning curve with the learning period, days, plotted against the percent of methods time measurement (MTM) determined normal task. The shape of the curve, once determined for a given operation, does not change. The horizontal scale division is, however, changed to suit the minimum learning period for 100 percent performance. Thus, for a 3-day learning period the horizontal scale becomes 3 days. The coordinate at each of these three points (i.e., days) becomes the minimum expected task for each day. Performance above these tasks rates a bonus. The base of 60 percent of normal performance for the first day of learning for all jobs is attainable and meets management's minimum requirements.

2. *Determine the learning period to allow*
(*a*) Find the learning time, by test, for each work station in the group. (*b*) Select the longest individual learning time—in this instance, 3 days. (*c*) Add a group familiarization allowance when the group exceeds three operators—2 days here. (*d*) Find the sum of $b + \bar{c}$, or $3 + 2 = 5$ days. This is the learning period to allow.

3. *Find the task for each day*
Divide the horizontal learning-period axis into five parts, one part for each of the 5 learning-period days allowed. Draw an ordinate for each day, and read the percentage task for that day at the intersection with the learning curve, or: 60.0; 70.5; 75.5; 80.0; 87 percent for days 1, 2, 3, 4, and 5, respectively.

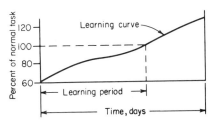

FIGURE 13 Typical learning curve for a metalworking task.

4. Compute the daily task and daily allowance

With a normal (100 percent) task of 1000 units for an 8-h day, set up a table (like Table 15) of daily tasks and time allowance during the learning period. Begin with a column listing the number of learning days. In the next column, list the percentage learning performance read from Fig. 13. Find the daily task in units, column 3 of Table 15, by taking the product of 1000 units and the percentage learning performance, column 2, expressed as a decimal. Last, compute the daily allowance per operator by finding the product of 8 h and the difference between 1.00 and the percentage of learning performance; i.e., for day 1: $(1.00 - 0.60)(8) = 3.2$ h. Tabulate the results in the fourth column of Table 15.

5. Compute the incentive pay for the group

In this plant the incentive pay is found by taking the product of the production in units and the standard set for 100 percent performance, or 0.80 worker-hour per 100 units produced. Thus, production of 600 units on day 1 will earn $(600/100)(0.80) = 4.8$-h pay for each group. Add to this the learning allowance of 3.2 h for day 1, and each group has earned $4.8 + 3.2 = 8$-h pay for 8-h work.

If the group produced 700 units during day 1, it would earn $(700/100)(0.80) = 5.6$-h pay at this standard. With the learning allowance of 3.2 h, the daily earnings would be $5.6 + 3.2 = 8.8$-h pay for 8-h work. This is exactly what is desired. The operator is rewarded for learning quickly.

Related Calculations. Select the length of the learning period for any new short-run task by conferring with representatives of the manufacturing, industrial engineering, and industrial relations departments. A simple operation that will be performed 1000 to 2000 times in an 8-h period would require a 3-day learning period. This is considered the minimum time for bringing such an operation up to normal speed. This is also true if a small group (three or less operators) perform equally simple operations. With larger groups (four or more operators), both simple and complex operations require an additional allowance for operators to adjust themselves to each other. Two days is a justified allowance for up to 15 operators learning to cooperate with one another under incentive conditions.

To prepare a plant-wide learning curve, keep records of the learning rates for a number of short-run tasks. Combine these data to prepare a typical learning curve for a particular plant. The method developed here was first described in *Factory,* now *Modern Manufacturing,* magazine.

LEARNING-CURVE EVALUATION OF MANUFACTURING TIME

A metalworking process requires 1.00 h for manufacture of the first unit of a production run. If the operator has an improvement or learning rate of 90 percent, determine the time required to manufacture the 2d, 4th, 8th, and 16th units. What is the cumulative average unit time for the 16th unit? If 100 units are manufactured, what is the cumulative average time for the 100th unit? What is the unit manufacturing time for the 100th item?

Calculation Procedure:

1. Compute the unit time for the production cycle

The learning curve relates the production time to the number of units produced. When the number of units produced doubles, the time required to produce the unit

TABLE 15 Learning-Curve Analysis

Learning days	Percentage of learning performance	Daily task, units	Daily allowance per operator, in units
1	60.0	600	40% × 8 = 3.20
2	70.5	705	29.5% × 8 = 2.36
3	75.5	755	24.5% × 8 = 1.96
4	80.0	800	20.0% × 8 = 1.60
5	87.0	870	13.0% × 8 = 1.04
6	100.0	1000	0% × 8 = 0

representing the doubled quantity is: (Learning rate, percent)(time, h or min, to produce the unit representing one-half the doubled quantity). Or, for the production line being considered here:

Unit number	Production time, h
1	1.00
2	0.90(1.00) = 0.900
4	0.90(0.90) = 0.810
8	0.90(0.81) = 0.729
16	0.90(0.729) = 0.656

2.   *Compute the cumulative average unit time*
The cumulative average unit time for any unit in a production run = (Σ unit time for each item in the run)/(number of items in the run). Thus, computing the time for items 1 through 16 as shown in step 1, and taking the sum, we get the cumulative average unit time = 12.044 h/16 units = 0.752 h.

3.   *Compute the cumulative average time for the 100th unit*
Set up a ratio of the learning factor for the 100th unit/learning factor for the 16th unit, and multiply the ratio by the cumulative average 16th unit time. Or, from the factors in Table 16, (0.497/0.656)(0.752 h) = 0.570 h.

4.   *Compute the unit time for the 100th unit*
Using the factor for the 90 percent learning curve in Table 16, the unit time for the 100th unit made = (1.00 h)(0.497) = 0.497 h.

   Related Calculations.   When using learning curves, be extremely careful to distinguish between *unit time* and *cumulative average unit time.* The unit time is the time required to make a particular unit in a production run, say the 10th, 16th, etc. Thus, a unit time of 0.5 h for the 16th unit in a production run means that the time required to make the 16th unit is 0.5 h. The 15th unit will require *more* time to make it; the 17th unit will require *less* time.

   The cumulative average unit time is the *average* time to manufacture a given number of identical items. To obtain the cumulative average unit time for any given number of items, take the sum of the time required for each item up to and including that item and divide the sum by the number of items.

   Either the *unit time or cumulative average unit time* can be used in manufacturing time or cost estimates, as long as the estimator knows which time value is

TABLE 16   Learning-Curve Factors

No. of units	Learning rate, percent		
	85	90	95
	Time or cost, percent of unit 1.	Time or cost, percent of unit 1	Time or cost, percent of unit 1
1	1.000	1.000	1.000
2	0.850	0.900	0.950
4	0.723	0.810	0.903
8	0.614	0.729	0.857
16	0.522	0.656	0.815
32	0.444	0.591	0.774
64	0.377	0.531	0.735
100	0.340	0.497	0.711

Learning-curve slopes

Learning rate, percent	Curve slope
70	−0.514
75	−0.415
80	−0.322
85	−0.234
90	−0.152
95	−0.074

being used. Failure to recognize the respective time values can result in serious errors.

A learning curve plotted on log-log coordinates is a straight line, Fig. 14. The slope of typical learning curves is listed in Table 16. Since a learning curve slopes downward—i.e., the unit manufacturing time decreases as more units are produced—the slope is expressed as a negative value.

Typical improvement or learning rates are: machining, drilling, etc., 90 to 95 percent; short-cycle bench assembly, 85 to 90 percent; equipment maintenance, 75 to 80 percent; electronics assembly and welding, 80 to 90 percent; general assembly,

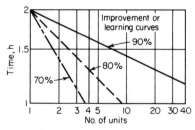

FIGURE 14   Learning curves plotted on log-log scale.

70 to 80 percent. When an operation consists of several tasks having different learning rates, compute the overall learning rate for the task by taking the sum of the product of each learning rate (LR) and the percentage of the total task it represents. Thus, with $LR_1 = 0.90$ for 60 percent of the total task; $LR_2 = 0.80$ for 20 percent of the total task; $LR_3 = 0.70$ for 10 percent of the total task, the overall learning rate $LR = 0.90(0.60) + 0.80(0.20) + 0.70(0.10) = 0.77$.

Note that in machine-paced operations—i.e., those in which the speed of the machine controls the operator's activities—there is less chance for the operator to learn. Hence, the learning rate will be higher—90 to 95 percent—than in worker-paced operations that have learning rates of 70 to 80 percent. When learning or improvement ceases, the operator has reached the level-off point, and the task cannot be performed any more rapidly. The ratio set up in step 3 can use any two items in a production run, provided that the cumulative average time for the smaller item is multiplied by the ratio.

DETERMINING BRINELL HARDNESS

A 3000-kg load is put on a 10-mm diameter ball to determine the Brinell hardness of a steel. The ball produces a 4-mm-diameter indentation in 30 s. What is the Brinell hardness of the steel?

Calculation Procedure:

1. Determine the Brinell hardness by using an exact equation
The standard equation for determining the Brinell hardness is $BHN = F/(\pi d_1/2)(d_1 - \sqrt{d_1^2 - d_s^2})$, where F = force on ball, kg; d_1 = ball diameter, mm; d_s = indentation diameter, mm. For this test, $BHN = 3000/(\pi \times 10/2)(10 - \sqrt{10^2 - 4^2}) = 229$.

2. Compute the Brinell hardness by using an approximate equation
One useful approximate equation for Brinell hardness is $BHN = (4F/\pi d_s^2) - 10$. For this test, $BHN = (4 \times 3000/\pi \times 4^2) - 10 = 228.5$. This compares favorably with the exact formula. For Brinell hardness exceeding 200, the approximate equation gives results that are less than 0.1 percent in error.

Related Calculations. Use this procedure for iron, steel, brass, bronze, and other hard or soft metals. A 500-kg test load is used for soft metals (brass, bronze, etc.). For Brinell hardness above 500, use a tungsten-carbide ball. The metal tested should be at least 10 times as thick as the indentation depth and wide enough so that no metal flows toward the edges of the specimen. The metal surface must be clean and free of defects.

ECONOMICAL CUTTING SPEEDS AND PRODUCTION RATES

A cutting tool used to cut beryllium costs $6 with its shank and can be reground for reuse five times. The average tool-changing time is 5 min. What is the most economical cutting speed if the machine labor rate is $3 per hour and the overhead is 200 percent? What is the cutting speed for the maximum production rate? The cost of regrinding the tool is 35 cents per edge.

Calculation Procedure:

1. Determine the tool cost factor
The cost factor $T_c + Y/X$ for a tool is composed of T_c = time to change tool, min; Y = tool cost per cutting edge, including prorated initial cost plus reconditioning costs, cents; X = machining rate, including labor and overhead, cents/min.

For this tool, T_c = 5 min. The tool can be reground five times after its original use, giving a total of 5 + 1 = 6 cutting edges (five regrindings + the original edge) during its life. Since the tool costs $6 new, the prorated cost per edge = $6/6 edges = $1, or 100 cents. The regrinding cost = 35 cents per edge; thus Y = 100 + 35 = 135 cents per edge.

With a machine labor rate of $3 per hour and an overhead factor of 200 percent, or 2.00($3) = $6, the value of X = machining rate = $3 + $6 = $9, or 900 cents per hour, or 900/60 = 15 cents per minute. Then, the cost factor $T_c + Y/X$ = 5 + 135/15 = 14.

2. Determine the cutting speed for minimum cost
Enter Fig. 15 at a cost factor of 14 and project vertically upward until the cutting-speed curve is intersected. At the left, read the cutting speed for minimum tool cost as 320 surface ft/min.

3. Determine the probable tool life
Project upward from the cost factor of 14 in Fig. 15 to the tool-life curve. At the right, read the tool life as 66 min.

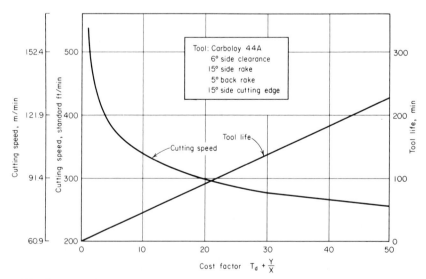

FIGURE 15 Optimum cutting-speed chart. (*American Machinist.*)

4. *Determine the speed and life for the maximum production rate*
Substitute the value of T_c for the cost factor $T_c + Y/X$ on the horizontal scale of Fig. 15. As before, read the cutting speed and tool life at the intersection with the respective curves. The plotted values apply when the chip-removal suction devices will operate efficiently at the cutting speeds indicated by the curves. Thus, with $T_c = 5$, the cutting speed is 370 surface ft/min, and the tool life is 30 min.

Related Calculations. Figure 15, and similar optimum cutting-speed charts, is plotted for a specific land wear—in this case 0.010 in (0.03 cm). For a land wear of 0.015 in (0.04 cm), multiply the cutting speeds obtained from Fig. 15 by 1.13. However, a land wear of 0.015 in (0.04 cm) is not recommended because the wear rates are accelerated. If Carboloy 883 tools are used in place of the 44A grade plotted in Fig. 15, multiply the cutting speeds obtained from this chart by 1.12 for a 0.010-in (0.03-cm) wear land or 1.26 for a 0.015-in (0.04-cm) wear land.

Charts similar to Fig. 15 for other tool materials can be obtained from tool manufacturers. Do not use Fig. 15 for any tool material other than Carboloy 44A. The method presented here is the work of D. R. Walker and J. Gubas, as reported in *American Machinist.*

OPTIMUM LOT SIZE IN MANUFACTURING

A manufacturing plant has a demand for 900 of its products per month on which the setup cost is $10. The cost of each unit is $5; the annual inventory charge is 12 percent/year of the average dollar value held in stock; the period for which the demand has occurred is 1/12 year. What is optimum manufacturing lot size? Plot a cost chart for this plant.

Calculation Procedure:

1. *Determine the optimum manufacturing lot size*
Optimum manufacturing lot size can be found from: {2(demand, units, per period)(cost per setup, $)/[(demand period, fraction of a year)($ cost per unit)(annual inventory charge, percent of average $ value held in stock)]}$^{0.5}$. For this run, optimum lot size = $\{2(900)(\$10)/(1/12)(\$5)(0.12)]\}^{0.5}$ = 600 units.

2. *Plot a cost chart for this plant*
Figure 16 shows a typical cost chart. Plot each curve using production runs of 100, 200, 300, 400, . . . , 1600 units. The values for each curve are determined from: inventory carrying charges = (number of units in run)($ cost per unit)(annual inventory charge, percent)(demand period)/2; setup and startup costs = (demand during period, units)($ cost of setup)/(number of units in run); total changing costs = inventory carrying charges + setup and startup costs.

The units for these equations are the same as given in step 1. Note that the total-changing-costs curve is a minimum at the point where the inventory-carrying-charges curve and setup-and-startup-costs curve intersect. Also, the two latter curves intersect at the optimum manufacturing lot size—600 units, as computed in step 1.

Related Calculations. Economical lot-size relations are readily adaptable to machine-shop computations. With only slight changes, the same principles can be applied to determine of optimum-quantity purchases. The procedure described here is the work of I. Heitner, as reported in *American Machinist.*

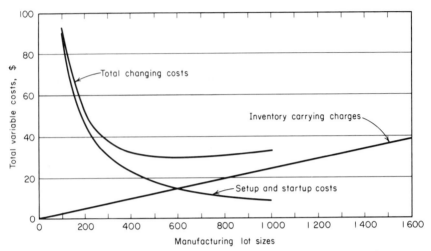

FIGURE 16 Changing costs associated with manufacturing lot sizes. (*American Machinist.*)

PRECISION DIMENSIONS AT VARIOUS TEMPERATURES

A magnesium workpiece with a dimension of 12.5000 to 12.4996 in (31.750 to 31.749 cm) is at a temperature of 85°F (29.4°C) after machining. The steel gage with which the dimensions of the workpiece will be checked is at 75°F (23.9°C). The workpiece must be gaged immediately to determine whether further grinding is necessary. Tolerance on the work is ±0.0002 in (0.005 mm). What should the dimensions of the workpiece be if there is not enough time available to allow the gage and workpiece temperatures to equalize? The standard reference temperature is 68°F (20.0°C).

Calculation Procedure:

1. Compute the actual work dimensions
The temperature of the workpiece is 85°F (29.4°C), or 85 − 68 = 17°F (9.4°C) above the standard reference of 68°F (20.0°C). Since the actual temperature of the workpiece is greater than the standard temperature, the dimensions of the workpiece will be larger than at the standard temperature because the part expands as its temperature increases.

To find the amount by which the workpiece will be oversize at the actual temperature, multiply the nominal dimension of the workpiece, 12.5 in (31.75 cm), by the coefficient of linear expansion of the material and by the difference between the actual and standard temperatures. For this magnesium workpiece, the average oversize amount at the actual temperature, 85°F (29.4°C), is $(12.5)(14.4 \times 10^{-6})(85 - 68) = 0.003060$ in (0.078 mm).

2. Compute the actual gage dimension
Compute the actual gage dimension in a similar way, using the same dimension, 12.5 in (31.75 cm), but the coefficient of linear expansion of the gage material,

steel, and the gage temperature, 75°F (23.9°C). Or, $(12.5)(6.4 \times 10^{-6})(75 - 68) = 0.000560$ in (0.014 mm).

3. Compute the workpiece dimension as a check

The workpiece dimension, corrected for tolerance, plus the difference between the oversize amounts computed in steps 1 and 2 is the dimension to which the part should be checked at the existing shop and gage temperature.

Applying the tolerance, ± 0.0002 in (± 0.005 mm), to the drawing dimension, $12.5000 - 12.4996$, gives a drawing dimension of 12.4998 ± 0.0002 in (31.7495 $\pm$ 0.0005 cm). Adding the difference between oversize dimensions, $0.003060 - 0.000560 = 0.002500$ in (0.0635 mm), to 12.4998 ± 0.0002 in (31.7495 $\pm$ 0.0005 cm) gives a checking dimension of 12.5023 ± 0.0002 in (31.7558 $\pm$ 0.0005 cm). If personnel check the workpiece at this dimension, they will have full confidence that it will be the right size.

Related Calculations. This procedure can be used for any metal—bronze, aluminum, cast iron, etc.—for which the coefficient of linear expansion is known. Obtain the coefficient from Baumeister and Marks—*Standard Handbook for Mechanical Engineers,* or a similar reference. When a workpiece is at a temperature less than the National Bureau of Standards standard of 68°F (20.0°C), the part contracts instead of expanding. The dimension change computed in step 1 is then negative. This is also true of the gage, if it is at a temperature of less than 68°F (20.0°C). Note that the tolerance is constant regardless of the actual temperature of the part.

The procedure given here is the work of H. K. Eitelman, as reported in *American Machinist.*

HORSEPOWER REQUIRED FOR METALWORKING

What is the input horsepower required for machining, on a geared-head lathe, a 4-in (10.2-cm) diameter piece of AISI 4140 steel having a hardness of 260 BHN if the depth of cut is 0.25 in (0.6 cm), the cutting speed is 300 ft/min (1.5 m/s), and the feed per revolution is 0.025 in (0.6 mm)?

Calculation Procedure:

1. Determine the metal removal rate

Compute the metal removal rate (MRR) in^3/min from MRR $= 12fDC$, where $f =$ tool feed rate, in/r; $D =$ depth of cut, in; $C =$ cutting speed, ft/min. For this workpiece, MRR $= 12(0.025)(0.25)(300) = 22.5$ in^3/min (6.1 cm^3/s).

2. Determine the unit horsepower required

Table 17 lists the average unit horsepower required for cutting various metals. The unit horsepower hp_u is the power required to remove 1 in^3 (1 cm^3) of metal per minute at 100 percent efficiency of the machine. Table 17 shows that AISI 4130 to 4345 of 250 to 300 BHN, the range into which AISI 4140 260 BHN falls, has a unit hp of 0.70 (8.5 unit kW).

The unit horsepower must be corrected for feed. From Fig. 17 and a feed of 0.025 in/r (0.6 mm/r), the correction factor is found to be 0.90. Thus, the true unit horsepower $= (0.70)(0.90) = 0.63$ hp/(in$^3 \cdot$ min)[28.7 W/(cm$^3 \cdot$ min)].

TABLE 17 Average Unit hp (kW) Factors for Ferrous Metals and Alloys*

Material classification	Brinell hardness number		
	201–250	251–300	301–350
AISI 3160–3450	0.62 (7.52)	0.75 (9.1)	0.87 (10.6)
AISI 4130–4345	0.58 (7.04)	0.70 (8.5)	0.83 (10.1)
AISI 4615–4820	0.58 (7.04)	0.70 (8.5)	0.83 (10.1)

*General Electric Company.

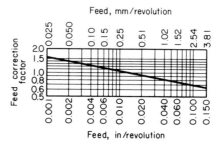

FIGURE 17 Feed correction factors based on normal tool geometries. (*General Electric Co.*)

3. Compute the horsepower required at the cutter

The horsepower required at the cutter $hp_c = (hp_u)(MRR)$, or $hp_c = (0.63)(22.5) = 14.18$ hp (10.6 kW).

4. Compute the motor horsepower required

The power required at the cutter is the input necessary after allowing for losses in gears, bearings, and other parts of the drive. Table 18 lists typical overall machine-tool efficiencies. A gear-head lathe has an efficiency of 70 percent. Thus, $hp_m = hp_c/e$, where e = machine-tool efficiency, expressed as a decimal. Or, $hp_m = 14.18/0.70 = 20.25$ hp (15.1 kW). A 20-hp (14.9-kW) motor would be satisfactory for this machine.

Related Calculations. Use this procedure for single or multiple tools. When more than one tool is working at the same time, compute hp_c for each tool and add the individual values to find the total hp_c. Divide the total hp_c by the machine efficiency to determine the required motor horsepower hp_m. This procedure makes ample allowance for dulling of the tools.

Compute metal removal rates for other operations as follows. *Face milling:* MRR $= WDF_T$, where W = width of cut, in; D = depth of cut, in; F_T = table feed, in/min. *Slot milling:* MRR $= WDF_T$, where all symbols are as before. *Planing or shaping:* MRR $= DfLS$, where D = depth of cut, in; f = feed, in per stroke or revolution; L = length of workpiece, in; S = strokes/min. *Multiple tools:* MRR $= (d_1^2 - d_s^2)\pi fR/4$, where d_1 = original diameter of workpiece, in, *before* cutting; d_s

TABLE 18 Efficiencies of Metalworking Machines*

Typical overall machine-tool efficiency values (except milling machines), percent	Typical overall efficiencies for milling machines		
	Rated power of machine		Overall efficiency, percent
	hp	kW	
Direct spindle drive, 90	3	2.2	40
	5	3.7	48
One-belt drive, 85	7.5	5.6	52
Two-belt drive, 70	10	7.5	52
	15	11.2	52
Geared head, 70	20	14.9	60
	25	18.6	65
	30	22.4	70
	40	29.8	75
	50	37.3	80

*General Electric Company.

= workpiece diameter *after* cutting, in; R = rpm of workpiece; other symbols as before.

The procedure given here is the work of Robert G. Brierley and H. J. Siekmann as reported in *Machining Principles and Cost Control.*

CUTTING SPEED FOR LOWEST-COST MACHINING

What is the optimal cutting speed for a part if the maximum feed for which an acceptable finish is obtained at 169 r/min of the workpiece is 0.011 in/r (0.3 mm/r) when the cost of labor and overhead is $0.24 per hour, the number of pieces produced per tool change is 15, the cost per tool change is $0.62, and the length of cut is 8 in (20.3 cm)?

Calculation Procedure:

1. Compute the optimization factor
When the lowest-cost machining speed for an operation is determined, one popular procedure is to choose any speed and feed at which the operation meets the finish requirements. If desired, the speed and feed at which the operation is now running might be chosen. By keeping the speed constant, the feed is increased to the maximum value for which the finish is acceptable. This is the optimal value for the feed and is called the *optimal feed.* The number of pieces produced under these conditions is measured between tool changes. Then the optimal cutting speed is

computed from: optimal cutting speed, r/min = (chosen speed, r/min)(optimization factor).

For any operation, the optimization factor = {(labor and overhead cost, $/min)(number of pieces per tool change)(length of cut, in)/[(3)(cost per tool change, $)(chosen speed, r/min)(optimal feed, in/r)]}$^{-4}$. Substitute the given values. Thus, the optimization factor = {(0.24)(15)(8)/[(3)(0.62)(169)/(0.011)]}$^{-4}$ = 1.7.

2. Compute the optimal cutting speed
From the relation given in step 1, optimal cutting speed = (chosen speed, r/min)(optimization factor) = (169)(1.7) = 287 r/min.

Related Calculations. The relation given in step 1 for the optimization factor is valid for carbide tools. It can be modified to apply to high-speed tools by changing the fourth root to an eight root and changing the 3 in the denominator to 7.

REORDER QUANTITY FOR OUT-OF-STOCK PARTS

A metalworking process uses 10 parts during the lead time. How many parts should be reordered if an out-of-stock situation can be accepted for 10 percent of the time? For 35 percent of the time?

Calculation Procedure:

1. Determine the out-of-stock factor
Table 19 lists out-of-stock factors for various times during which a part might be out of stock. Thus, the acceptable out-of-stock factor for 10 percent is 1.29, and for 35 percent it is 0.39.

TABLE 19 Out-of-Stock Factors*

Acceptable percentage of stock-outs	Out-of-stock factor
50	0.00
45	0.13
40	0.26
35	0.39
25	0.68
15	1.04
10	1.29
5	1.65
4	1.76
3.5	1.82
3	1.89
2	2.06
1	2.33
0	4.0

*Nyles V. Reinfeld in *American Machinist.*

2. Compute the reorder quantity

For any manufacturing process, reorder quantity = (out-of-stock factor)(usage during lead time)$^{0.5}$ + (usage during lead time). Thus, the reorder point for this process with an acceptable out-of-stock factor for 10 percent is $(1.29)(10)^{0.5} + (10) = 14.08$ parts, say 15 parts. With 35 percent, reorder point = $(0.29)(10)^{0.5} + 10 = 11.23$, or 12 parts.

Related Calculations. Use this procedure for any types of parts ordered from either an internal or external source. In general, reducing the allowable stock-out time will increase the time during which a process using the parts can operate.

SAVINGS WITH MORE MACHINABLE MATERIALS

What are the gross and net savings made with a more machinable material that reduces the production time by 36 s per part when 800 lb (362.9 kg) of steel is required for 1000 parts and the total machine operating cost is $6 per hour? The more machinable material costs 4 cents per pound (8.8 cents per kilogram) more than the less machinable material, and 5000 parts are produced per day.

Calculation Procedure:

1. Compute the gross savings possible

The gross saving possible in a machining operation when a more machinable material is used is: gross saving, cents/lb = (machining time saved with new material, s per piece)(total cost of operating machine, cents/h)/[(3.6)(weight of material to make 1000 pieces, lb)]. For this operation, the gross saving = $(36)(600)/[(3.6)(800)]$ = 7.5 cents per pound (16.5 cents per kilogram). With a production rate of 5000 parts per day, the gross saving is (5000 parts)(800 lb/1000 parts)(7.5 cents/lb) = 30,000 cents, or $300.

2. Compute the net savings possible

The more machinable materials cost 4 cents more per pound than the less machinable material. Hence, the net saving is $(7.5 - 4.0) = 3.5$ cents per pound (7.7 cents per kilogram), or (5000 parts)(800 lb/1000 parts)(3.5 cents/lb) = 14,000 cents, or $140.

Related Calculations. Use this general procedure for parts made of any material—steel, brass, bronze, aluminum, plastic, etc.

TIME REQUIRED FOR THREAD MILLING

How long will it take to thread-mill a $2\frac{7}{8}$-in (7.3-cm) diameter hard steel bolt with a $2\frac{1}{2}$-in (6.4-cm) diameter 18-flute hob?

Calculation Procedure:

1. Determine the cutting speed and feed of the hob

Table 20 lists typical cutting speeds and feeds for various materials. For hard steel, the usual cutting speed in thread milling is 50 ft/min (0.3 m/s), and the feed per flute is 0.002 in (0.05 mm).

TABLE 20 Thread-Milling Speeds and Feeds

Material threaded	Speed		Feed per flute	
	ft/min	m/s	in	mm
Aluminum	500	2.5	0.0015	0.038
Brass	250	1.3	0.0015	0.038
Mild steel	100	0.5	0.0020	0.051
Medium steel	75	0.4	0.0020	0.051
Hard steel	50	0.3	0.0020	0.051

2. Compute the time required for thread milling
The time required for thread milling, T, min $= \pi d/(fnR)$, where $d =$ work diameter, in; $f =$ feed, in per flute; $n =$ number of flutes on hob; $R =$ hob rpm.
From a previous calculation procedure, $R = 12C/(\pi d)$, where $C =$ hob cutting speed, ft/min; $d =$ hob diameter, in. For this hob, $R = (12)(50)/[\pi(2.5)] = 76.4$ r/min. Then $T_t = \pi(2\frac{7}{8})/[(0.002)(18)(76.4)] = 3.29$ min.
Related Calculations. Use this procedure for any metallic or nonmetallic material—aluminum, brass, mild steel, medium steel, hard steel, plastics, etc.

DRILL PENETRATION RATE AND CENTERLESS GRINDER FEED RATE

What is the drill penetration rate when a drill turns at 1000 r/min and has a feed of 0.006 in/r (0.15 mm/r)? What is the feed rate of a centerless grinder having a 12-in (30.5-cm) diameter regulating wheel running at 60 r/min if the angle of inclination between the regulating and grinding wheel is 5°?

Calculation Procedure:

1. Compute the rate of drill penetration
The rate of drill penetration P in/min $= fR$, where $f =$ drill feed, in/r; $R =$ drill rpm. For this drill, $P = (0.006)(1000) = 6.0$ in/min (2.5 mm/s).

2. Compute the grinder feed rate
The work feed f in/min in a centerless grinder is $f = \pi dR \sin a$, where $d =$ regulating-wheel diameter, in; $R =$ regulating-wheel rpm; $a =$ angle of inclination between the regulating and grinding wheel. For this grinder, $f = \pi(12)(60)(\sin 5°) = 197.6$ in/min (8.4 cm/s).

BENDING, DIMPLING, AND DRAWING METAL PARTS

What is the minimum bend radius R in for 0.02-gage Vascojet 1000 metal if it is bent transversely to an angle of 130°? What is the minimum radius R in of a bend in 0.040-gage Rene 41 metal bent longitudinally at an angle of 52° at room temperature? Determine the maximum length of dimple flange H in for AM-350 metal at 500°F (260°C) when the bend angle is 42° and the edge radius R is 0.250 in (6.4 mm). Find the maximum blank diameter and maximum cup depth for drawing Rene 41 metal at 400°F (204°C) when using a die diameter of 10 in (25.4 cm) and 0.063-gage material. Figure 18*a*, *b*, and *c* shows the anticipated manufacturing conditions.

Calculation Procedure:

1. *Compute the minimum bend radius*

Table 21 shows that the critical bend angle (i.e., maximum bend angle α without breakage) for Vascojet 1000 metal is 118°. Hence, the required bend angle is greater than the critical bend angle. Therefore, the required bend limit equals the critical bend limit, and $R/T = 1.30$, from Table 21. Hence, the minimum radius $R_m = (R/T)(T) = (1.30)(0.02) = 0.026$ in (0.66 mm).

With Rene 41 metal, bent longitudinally at room temperature, the critical bend angle is 122°, from Table 21. Since the required bend angle of 52° is less than critical, find the R/T value in the right-hand portion of Table 21. When the actual bend angle is between two tabulated angles, interpolate thus:

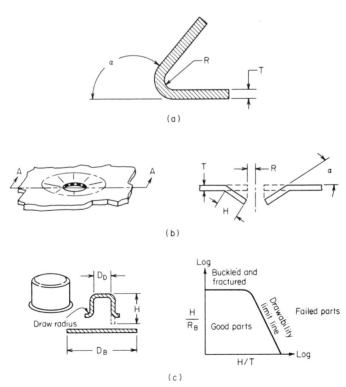

(a)

(b)

(c)

FIGURE 18   (*a*) Brake-bent part shape and parameters; (*b*) ram-coin dimpling setup; (*c*) drawing setup. (*American Machinist.*)

TABLE 21 Brake-Bend Parts Parameters*

Material	L/T	F	∞	R/T	30	45	60	75	90	105	120
										R/T for angles ∞ below critical	
Titanium (13V-11Cr-3Al)	L/T	RT	105	2.40	0.34	0.68	1.16	1.80	2.25	2.40	
Vascojet 1000	L/T	RT	118	1.30	0.18	0.38	0.64	0.92	1.13	1.26	1.30
USS 12 MoV	L/T	RT	119	1.20	0.16	0.34	0.60	0.84	1.04	1.16	1.20
17-7PH	L/T	RT	122	0.80	0.10	0.22	0.37	0.54	0.66	0.75	0.79
AM-350	L/T	RT	122	0.80	0.10	0.22	0.37	0.54	0.66	0.75	0.79
PH 15-7 Mo	L/T	RT	121	0.86	0.11	0.23	0.42	0.60	0.72	0.80	0.84
A-286	L/T	RT	124	0.66	0.07	0.15	0.29	0.43	0.54	0.62	₀.65
Hastelloy X	L/T	RT	120	1.00	0.12	0.26	0.47	0.67	0.84	0.95	1.00
Inconel X	L/T	RT	124	0.64	0.06	0.14	0.28	0.41	0.52	0.60	0.63
Rene 41	L	RT	122	0.80	0.10	0.22	0.37	0.54	0.66	0.75	0.79
Rene 41	T	RT	113	1.64	0.28	0.53	0.84	1.16	1.44	1.58	1.64
J-1570	L	RT	124	0.68	0.08	0.16	0.30	0.45	0.56	0.64	0.67

*American Machinist, LTV, Inc.; USAF.
Note: L/T = grain direction, where L = longitudinal and T = transverse; F = bending temperature; ∞ = critical bend; R/T = critical bend limits.

Angle,°	R/T value
60	0.37
52	
45	0.22

$[(52 - 45)/(60 - 45)](0.37 - 0.22) = 0.07$. Then R/T for $52° = 0.22 + 0.07 = 0.29$. With R/T known for $52°$, compute the minimum radius $R_m = (R/T)(T) = (0.29)(0.040) = 0.0116$ in $(0.2946$ mm$)$.

2. Determine the dimple-flange length

Table 22 shows a typical dimpling limits to avoid radial splitting at the edge of the hole of various modern materials. With a bend angle between the tabulated angles, interpolate thus:

Angle, °	H/R value
45	1.10
42	
40	1.43

$[(42 - 40)/(45 - 40)](1.10 - 1.43) = -0.132$, and $H/R = 1.43 + (-0.132) = 1.298$ at $42°$. Then the maximum dimple-flange length $H_m = (H/R)(R) = (1.298)(0.250) = 0.325$ in $(8.255$ mm$)$.

3. Determine the maximum blank diameter

Table 23 lists the drawing limits for flat-bottom cups made of various modern materials. For this cup, $D_D/T = 10/0.063 = 158.6$, say 159. The corresponding

TABLE 22 Dimpling Limits to Avoid Radial Splitting at Hole Edge*

Material	Temperature, °F (°C)	Dimpling limit H/R				
		Standard, for various bend angles a; above and below standard bend angle				
		30°	35°	40°	45°	50°
2024-T3	70 (21.1)	2.15	1.60	1.20	0.93	0.80
Ti-8-1-1	70 (21.1)	1.88	1.42	1.08	0.82	0.70
TZM Moly	70 (21.1)	1.98	1.50	1.12	0.87	0.73
Cb-752	70 (21.1)	2.28	1.70	1.30	0.98	0.83
PH 15-7 Mo	500 (260.0)	2.43	1.84	1.40	1.07	0.90
AM-350	500 (260.0)	2.46	1.87	1.43	1.10	0.93
Ti-8-1-1	1200 (648.9)	2.30	1.72	1.30	1.00	0.85
Ti-13-11-3	1200 (648.9)	2.58	1.95	1.48	1.15	0.95

*American Machinist, LTV, Inc.; USAF.

TABLE 23 Drawing Limits for Flat-Bottom Cups*

Material	Temperature ratio, °F (°C)	Die to blank diameter ratios D_B/D_D; cup-depth ratios H/D_D							
		For various D_D/T ratios							
		25	50	100	150	200	250	300	400
Am-350	500 (260.0) D_B/D_D	2.22	2.18	2.00	1.71	1.54	1.42	1.40	1.30
	H/D_D	0.97	0.95	0.75	0.50	0.37	0.30	0.26	0.20
A-286	1000 (537.8) D_B/D_D	2.22	2.46	2.16	1.85	1.64	1.49	1.41	1.36
	H/D_D	1.00	1.21	0.87	0.57	0.42	0.34	0.28	0.22
Rene 41	400 (204.4) D_B/D_D	2.22	2.22	1.92	1.73	1.52	1.48	1.42	1.33
	H/D_D	0.97	0.97	0.73	0.51	0.37	0.31	0.27	0.21
L-605	500 (260.0) D_B/D_D	2.22	2.29	2.00	1.68	1.54	1.45	1.44	1.38
	H/D_D	0.97	1.05	0.74	0.47	0.35	0.30	0.26	0.21
T1-13-11-3	1200 (648.9) D_B/D	2.38	2.53	2.28	1.92	1.67	1.58	1.45	1.44
	H/D_D	1.15	1.34	0.94	0.60	0.44	0.35	0.30	0.24
Tungsten	600 (315.6) D_B/D_D	2.08	2.11	1.98	1.66	1.53	1.46	1.38	1.34
	H/D_D	0.83	0.87	0.69	0.45	0.35	0.29	0.24	0.20

*American Machinist, LTV, Inc.; USAF.

D_B/T and H/D_D ratios are not tabulated. Therefore, interpolate between D_D/T values of 150 and 200. Thus, for Rene 41:

D_D/T	D_B/D_D
200	1.52
159	
150	1.73

$[(159 - 150)/(200 - 150)](1.52 - 1.73) = -0.0378$, and $D_B/D_D = 1.73 + (-0.0378) = 1.692$, when $D_D/T = 159$. Then, the *maximum* value of $D_{Bm} = (D_B/D_D)(D_D) = (1.692)(10) = 16.92$ in (43.0 cm).

Interpolating as above yields $H/D_D = 0.48$ when $D_D/T = 159$. Then the maximum height $H_m = (H/D_D)(D) = (0.48)(10) = 4.8$ in (12.2 cm).

Related Calculations. The procedures given here are typical of those used for the newer "exotic" metals developed for use in aerospace, cryogenic, and similar advanced technologies. The three tables presented here were developed by LTV, Inc., for the U.S. Air Force, and reported in *American Machinist.*

BLANK DIAMETERS FOR ROUND SHELLS

What blank diameter D in is required for the round shells in Fig. 19a and b if $d = 12$ in (30.5 cm), $d_1 = 12$ in (30.5 cm), $d_2 = 14$ in (35.6 cm), and $h = 14$ in (35.6 cm)?

Calculation Procedure:

1. Compute the plain-cup blank diameter

Figure 19a shows the plain cup. Compute the required blank diameter from $D = (d^2 + 4dh)^{0.5} = (12^2 + 4 \times 12 \times 4)^{0.5} = 18.33$ in (46.6 cm).

2. Compute the flanged-cup blank diameter

Figure 19b shows the flanged cup. Compute the required blank diameter from $D = (d_2^2 + fd_1h)^{0.5} = (14^2 + 4 \times 12 \times 4)^{0.5} = 19.7$ in (49.3 cm).

Related Calculations. Figure 19 gives the equations for computing 12 different round-shell blank diameters. Use the same general procedures as in steps 1 and 2 above. These equations were derived by Ferene Kuchta, Mechanical Engineer, J. Wiss & Sons Co., and reported in *American Machinist.*

BREAKEVEN CONSIDERATIONS IN MANUFACTURING OPERATIONS

A manufacturing plant has the net sales and fixed and variable expenses shown in Table 24. What is the breakeven point for this plant in units and sales? Plot a conventional and an alternative breakeven chart for this plant.

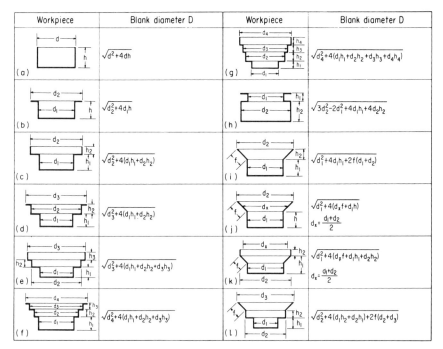

Workpiece	Blank diameter D	Workpiece	Blank diameter D
(a)	$\sqrt{d^2+4dh}$	(g)	$\sqrt{d_4^2+4(d_1h_1+d_2h_2+d_3h_3+d_4h_4)}$
(b)	$\sqrt{d_2^2+4d_1h}$	(h)	$\sqrt{3d_2^2-2d_1^2+4d_1h_1+4d_2h_2}$
(c)	$\sqrt{d_2^2+4(d_1h_1+d_2h_2)}$	(i)	$\sqrt{d_1^2+4d_1h_1+2f(d_1+d_2)}$
(d)	$\sqrt{d_3^2+4(d_1h_1+d_2h_2)}$	(j)	$\sqrt{d_1^2+4(d_xf+d_1h)}$ $d_x=\dfrac{d_1+d_2}{2}$
(e)	$\sqrt{d_3^2+4(d_1h_1+d_2h_2+d_3h_3)}$	(k)	$\sqrt{d_1^2+4(d_xf+d_1h_1+d_2h_2)}$ $d_x=\dfrac{d_1+d_2}{2}$
(f)	$\sqrt{d_4^2+4(d_1h_1+d_2h_2+d_3h_3)}$	(l)	$\sqrt{d_2^2+4(d_1h_2+d_2h_1)+2f(d_2+d_3)}$

FIGURE 19   Blank diameters for round shells. (*American Machinist.*)

TABLE 24   Manufacturing Business Income and Expenses

Condensed Income Statement
For year ending Dec. 31, 19—

	Variable	Fixed	
Net sales (60,000 units @ $20 per unit)			$1,200,000
Less costs and expenses:			
Direct material	$195,000	. . .	
Direct labor	215,000	. . .	
Manufacturing expenses	100,000	$200,000	
Selling expenses	50,000	150,000	
General and administrative expenses	160,000	50,000	
Total	720,000	400,000	1,120,000
Net profit before federal income taxes			80,000

Calculation Procedure:

1.   *Compute the breakeven units*
Use the relation BE_u = fixed expenses, $/[(sales income per unit, $ − variable costs per unit, $)], where BE_u = breakeven in units. By substituting, BE_u = $400,000/($20 − $12) = 50,000 units.

2. Compute the breakeven sales

Two methods can be used to compute the breakeven sales. With the breakeven units known from step 1, $BE_s = BE_u$(unit sales price, $), where BE_s = breakeven sales, $. By substituting, $BE_s = (50,000)(\$20) = \$1,000,000$.

Alternatively, compute the profit-volume (PV) ratio: $PV = $ (sales, $ − variable costs, $)/sales, $ = ($1,200,000 − $720,000)/$1,200,000 = 0.40$. Then $BE_s = $ fixed costs, $/PV = \$400,000/0.40 = \$1,000,000$. This is identical to the breakeven sales computed in the previous paragraph.

3. Draw the conventional and alternative breakeven charts

Figure 20a shows the conventional breakeven chart for this plant. Construct this chart by drawing the horizontal line F for the fixed expenses, the solid sloping line I for the income or sales, and the dotted sloping line C for the total costs. Note that the vertical axis is for the expenses and income and the horizontal axis is for the sales, all measured in monetary units.

The breakeven point is at the intersection of the income and total-cost curves, point B, Fig. 20a. Projecting vertically downward shows that point B corresponds to a sales volume of $1,000,000, as computed in step 2.

Alternative breakeven charts are shown in Fig. 20b and c. Both charts are constructed in a manner similar to Fig. 20a.

Related Calculations.   Breakeven computations are valuable tools for analyzing any manufacturing operation. The concepts are also applicable to other business activities. Thus, typical PV values for various types of businesses are:

Business	Typical activity	Typical PV
Consumer appliances	Fully automated; high-volume output	0.15–0.25
Standard centrifugal pumps	Batch output in large volume	0.20–0.30
Acid-handling centrifugal pumps	Batch output in small volume	0.25–0.35
Standard prototype, one-of-a-kind	Ships, machine tools	0.30–0.40
Special-design one-of-a-kind prototype	Buildings, factories	0.35–0.50

CALCULATING GEOMETRIC DIMENSIONS OF DRAWN PARTS

A round metal piece, Fig. 21, having a diameter of 25 in (63.5 cm), is drawn to a 30 percent reduction in its area, Fig. 22. Find the percent decrease in the diameter of the part, the percent increase in length, and the diameter after the drawing operation is finished. A square metal piece 30-in (76.2-cm) wide before drawing is

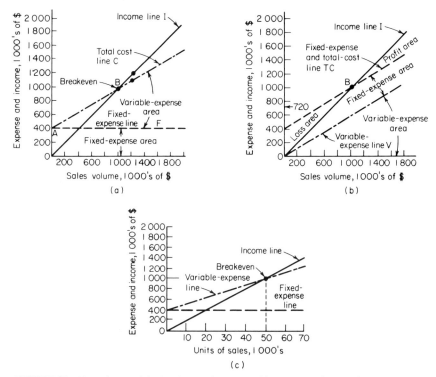

FIGURE 20   Three forms of the breakeven chart as used in metalworking activities.

reduced to 25 in (63.5 cm) after drawing; its length before drawing is 1200 in (3048 cm). Determine its change in area, change in thickness, and change in length produced by the drawing.

Calculation Procedure:

1. Give the general relation for change in dimensions produced by drawing
When a part is plastically deformed, its geometrical dimensions become interdependent. If the volume is kept constant, any changes in one dimension will produce a change in the other dimensions of the part. For any given drawing process, all changed dimensions can be computed from the factors and formulas in Table 25. The basis for this tabulation is:

$$V = A_0 \times l_0 = A_1 \times l_1 \qquad \text{(refer to Fig. 2)}$$

where V = volume
A_0 = cross-sectional area before deformation
A_1 = cross-sectional area after deformation
l_0 = length before deformation
l_1 = length after deformation

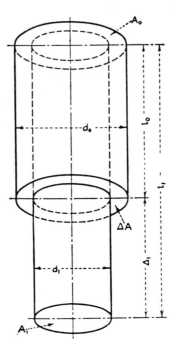

FIGURE 21 Changes in dimensions of a drawn part. (*Product Engineering.*)

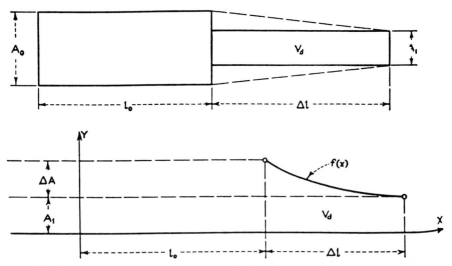

FIGURE 22 Changes in volume of a drawn part. (*Product Engineering.*)

TABLE 25 Formulas for Various Material Profiles

Symbol		Formulas for rounds and symmetrical profiles						Unit of Measurement
V	Volume before and after drawing	$V = A \times l$						cu.in. (cm^3)
A_0	Cross-section before	$d^2 \times 0.785$	$d^2 \times 0$	d^2	$d^2 \times 0.705$	$d^2 0.867$	$d^2 \times 0.828$	sq.in. (cm^2)
A_1	Cross-section after							
l_0	length before	$l = V/A$						in. (cm)
l_1	length after							
d_0	Diameter or thickness before	$\sqrt{1.273A}$	$\sqrt{1.733A}$	$\sqrt{A}$	$\sqrt{1.418A}$	$\sqrt{1.154A}$	$\sqrt{1.208A}$	in. (cm)
d_1	Diameter or thickness after							
ΔA	Percent decrease in cross-section	$\Delta A = \dfrac{A_0 - A_1}{A_0} \times 100$						percent
Δl	Percent increase in length	$\Delta l = \dfrac{l_1 - l_0}{l_0} \times 100$						percent
Δd	Percent decrease in diameter or thickness	$\Delta d = \dfrac{d_0 - d_1}{d_0} \times 100$						percent

To compute the work done in deformation, and the other factors, the displaced volume, V_0, must be determined. This value of V_d is derived from

$$V_d = \log_e \frac{A_0}{A_1} \times V = 2 \log_e \frac{d_0}{d_1} \times V \cong \frac{A_0 + A_1}{2} \times \Delta l \qquad \text{(see Fig. 22)}$$

or according to Table 26

$$V_d = V \times \text{factor}$$

2. *Determine the changes produced in drawing the round part*

We are given that the drawing produces a 30 percent decrease, ΔA, in the part's cross sectional area when the original diameter is 25 in (63.5 cm). Entering Table 26 at $\Delta A = 30$ percent, we find that the part's diameter decrease, $\Delta d = 16.33$ percent. The new diameter will be, from Table 26, $d_1 = 25 \times 0.8367 = 20.92$ in

TABLE 26 Factors for Use in Drawing Calculations

	Calculations for		Calculations for			Calculations for			Calculations for
ΔA	$A_0 =$ $A_1 \times F$	$A_1 =$ $A_0 \div F$	Δl %	$l_0 =$ $l_1 \times F$	$l_1 =$ $l_0 \div F$	Δd %	$d_1 =$ $d_0 \div F$	$d_1 =$ $d_0 \times F$	$V_d = V \times F$
				$F =$ Factor					
1	1.0101	0.99	1.01	0.99	1.0101	0.50	1.006	0.9950	0.0100
2	1.0204	0.98	2.04	0.98	1.0204	1.01	1.010	0.9899	0.0202
3	1.0309	0.97	3.09	0.97	1.0309	1.51	1.015	0.9849	0.0305
4	1.0417	0.96	4.17	0.96	1.0417	2.02	1.020	0.9798	0.0408
5	1.0526	0.95	5.26	0.95	1.0526	2.53	1.025	0.9747	0.0513
6	1.0638	0.94	6.38	0.94	1.0638	3.05	1.031	0.9695	0.0619
7	1.0753	0.93	7.53	0.93	1.0753	3.56	1.037	0.9644	0.0726
8	1.0870	0.92	8.70	0.92	1.0870	4.08	1.042	0.9592	0.0834
9	1.0989	0.91	9.89	0.91	1.0989	4.61	1.048	0.9539	0.0943
10	1.1111	0.90	11.11	0.90	1.1111	5.13	1.054	0.9487	0.1054
11	1.1236	0.89	12.36	0.89	1.1236	5.66	1.060	0.9434	0.1165
12	1.1364	0.88	13.64	0.88	1.1364	6.19	1.066	0.9381	0.1278
13	1.1494	0.87	14.94	0.87	1.1494	6.73	1.072	0.9327	0.1393
14	1.1628	0.86	16.28	0.86	1.1628	7.26	1.078	0.9274	0.1508
15	1.1765	0.85	17.65	0.85	1.1765	7.81	1.084	0.9219	0.1625
16	1.1905	0.84	19.05	0.84	1.1905	8.35	1.091	0.9165	0.1744
17	1.2048	0.83	20.48	0.83	1.2048	8.90	1.097	0.9110	0.1863
18	1.2195	0.82	21.95	0.82	1.2195	9.45	1.104	0.9055	0.1985
19	1.2346	0.81	23.46	0.81	1.2346	10.00	1.111	0.9000	0.2107
20	1.2500	0.80	25.00	0.80	1.2500	10.56	1.118	0.8944	0.2237
21	1.2658	0.79	26.58	0.79	1.2658	11.12	1.124	0.8888	0.2357
22	1.2821	0.78	28.21	0.78	1.2821	11.68	1.132	0.8832	0.2485
23	1.2987	0.77	29.87	0.77	1.2987	12.25	1.140	0.8775	0.2614
24	1.3158	0.76	31.58	0.76	1.3158	12.82	1.147	0.8718	0.2744
25	1.3333	0.75	33.33	0.75	1.3333	13.40	1.155	0.8660	0.2877
26	1.3514	0.74	35.14	0.74	1.3514	13.98	1.162	0.8602	0.3011
27	1.3699	0.73	36.99	0.73	1.3699	14.56	1.170	0.8544	0.3147
28	1.3889	0.72	38.89	0.72	1.3889	15.15	1.177	0.8485	0.3285
29	1.4085	0.71	40.85	0.71	1.4085	15.74	1.186	0.8426	0.3425
30	1.4286	0.70	42.86	0.70	1.4286	16.33	1.195	0.8367	0.3567
31	1.4493	0.69	44.93	0.69	1.4493	16.93	1.204	0.8307	0.3711
32	1.4706	0.68	47.06	0.68	1.4706	17.54	1.213	0.8246	0.3857
33	1.4925	0.67	49.25	0.67	1.4925	18.15	1.221	0.8185	0.4005
34	1.5152	0.66	51.52	0.66	1.5152	18.76	1.231	0.8124	0.4155
35	1.5385	0.65	53.85	0.65	1.5385	19.38	1.240	0.8062	0.4308
36	1.5625	0.64	56.25	0.64	1.5625	20.00	1.250	0.8000	0.4463
37	1.5873	0.63	58.73	0.63	1.5873	20.63	1.259	0.7937	0.4620
38	1.6129	0.62	61.29	0.62	1.6129	21.26	1.270	0.7874	0.4780
39	1.6393	0.61	63.93	0.61	1.6393	21.90	1.280	0.7810	0.4943
40	1.6667	0.60	66.67	0.60	1.6667	22.54	1.291	0.7746	0.5108
41	1.6949	0.59	69.49	0.59	1.6949	23.19	1.302	0.7681	0.5276
42	1.7241	0.58	72.41	0.58	1.7241	23.84	1.313	0.7616	0.5447
43	1.7544	0.57	75.44	0.57	1.7544	24.50	1.324	0.7550	0.5621
44	1.7857	0.56	78.57	0.56	1.7857	25.17	1.336	0.7483	0.5758
45	1.8182	0.55	81.82	0.55	1.8182	25.84	1.348	0.7416	0.5978
46	1.8519	0.54	85.19	0.54	1.8519	26.52	1.360	0.7348	0.6162
47	1.8868	0.53	88.68	0.53	1.8868	27.20	1.373	0.7280	0.6349
48	1.9231	0.52	92.31	0.52	1.9231	27.89	1.386	0.7211	0.6539
49	1.9608	0.51	96.08	0.51	1.9608	28.59	1.400	0.7141	0.6734
50	2.0000	0.50	100.00	0.50	2.0000	29.29	1.414	0.7071	0.6932

(53.1 cm). Likewise, from Table 26, the part's increase in length, $\Delta l = 42.86$ percent. The area of the part before drawing is, $A_0 = (0.785)(25)^2 = 490.6$ in^2 (3165.3 cm^2). After drawing, the area is, with a 30 percent reduction, $(0.7)(490.6) = 343.4$ in^2 (2215.6 cm^2).

3. Find the changes in the square piece after the drawing operation
The area before drawing, $A_0 = (30)^2$ for a square piece $= 900$ in^2 (5806.4 cm^2); after drawing, $A_0 = (25)^2 = 625$ in^2 (4032.3 cm^2). Using Table 25, the decrease in area, $\Delta A = (100)(900 - 625)/900 = 30.56$ percent. Again, from Table 25, the change in thickness after drawing is, $\Delta d = (100)(30 - 25)/30 = 16.67$ percent. This could be approximated from Table 26 by interpolating between given values.

The percent increase in length, Δl, can be found by interpolating between $\Delta A = 31$ and 30.5 in Table 26, giving $\Delta l = 42.89$ percent. Hence, the new length $= 1200(1.4289) = 1714.7$ in (4355.3 cm).

Related Calculations. Use these relations for any material that plastically deforms on drawing. These generalized equations are valid for any drawing operation in which the volume of the material drawn is constant. The equations in Table 25 and factors in Table 26 were first published in *Product Engineering* magazine. The handbook editor added the SI values.

ANALYZING STAINLESS-STEEL MOLDING METHODS

Compare the methods available for molding stainless-steel parts to eliminate cracks, distortions, shrinkage, and breaks during cooling. Compare the overall cost of the various methods available for molding stainless steel. Likewise, compare degree of complexity, tooling cost, suitability for large and small parts, life of tooling and patterns, accuracy and finish, intricacy, time for first sample, dimensional stability, and stress distribution.

Calculation Procedure:

1. Prepare a molding comparison tabulation
List, as in Table 27, the various molding methods available for the material being considered. Thus, for stainless steel we might use investment casting, baked-sand casting, precision forgings, welded assemblies, and machined parts. List these processes across the top of the table.

Next, list vertically, the various comparisons important in the decision process, starting with the overall cost, Table 27. Once the methods and comparison parameters are listed, the analysis can begin.

2. Enter comparison data obtained from competent sources
Starting with *overall cost,* Table 27, list the comparison data. Often, this information must be obtained from someone familiar with the specific process because it is difficult for one person to know everything about every manufacturing process. Thus, a manufacturing engineer might provide the data for overall cost. Such data might show that baked-sand casting and precision forgings are cheaper than shell molding for stainless steel.

Continue the comparison, line for line, using data input from qualified experts. On completion your tabulation will be like that shown in Table 27.

TABLE 27   Comparison of Shell Molding and Overlapping Processes

	Investment casting vs. shell molding	Baked-sand casting vs. shell molding	Precision forgings vs. shell molding	Welded assemblies vs. shell molding	Machined parts vs. shell molding
Overall cost...............	Shell Cheaper	Baked Sand Cheaper	Forgings Cheaper	Shell Cheaper	Shell Cheaper
Degree of complexity	Investment Better	Shell Better	Shell Better	Evenly Matched	Shell Better
Tooling cost	Shell Cheaper	Baked Sand Cheaper	Shell Cheaper	Welding Cheaper	Hard to Compare
For large parts	Shell Better	Baked Sand Better	Forgings Better	Welding Better	Machining Better
For small parts.............	Investment Better	Shell Better	Shell Better	Evenly Matched	Machining Better
Life of tooling and pattern ...	Evenly Matched	Shell Better	Shell Better	Welding Better	Machining Better
Accuracy and finish.........	Investment Better	Shell Better	Shell Better	Evenly Matched	Machining Better
Intricacy..................	Investment Better	Shell Better	Shell Better	Shell Better	Machining Better
Time for first sample........	Shell Better	Baked Sand Better	Evenly Matched	Welding Better	Machining Better
Dimensional stability........	Evenly Matched	Evenly Matched	Shell Better	Shell Better	Hard to Compare
Stress distribution...........	Shell Better	Shell Better	Shell Better	Shell Better	Cannot Compare

3.   Analyze the variables in stainless-steel molding

Stresses in stainless-steel molded parts must be considered between 1500°F (815.6°C) and 2600°F (1426.7°C) solidification temperature, Fig. 23. No cracks, distortions, or breaks should occur while parts cool through this temperature range.

The freezing characteristics of molten metal must be kept in mind when designing shell moldings. Since castings solidify from the outside in, shrinkage may cause trouble. If there are isolated areas of heavy mass in the casting, they may become disconnected from the flow of liquid metal before they are sound, resulting in voids. Avoid sudden changes of mass by generous use of tapered sections in the design of the casting. Avoid large flat areas of thin wall because they will not stay flat and oil canning may result.

Sharp reentrant angles produce areas of stress concentration. They develop a notch effect in the casting and cracking becomes difficult to prevent. Small holes tend to complicate the foundry problem if they are cast in. Usually, small holes in castings are cheaper to drill or machine. Wherever possible, the parting plane should permit pouring of the molten metal in a vertical position. From a cost standpoint, the parting plane of a casting should be restricted to a single plane.

Related Calculations.   The considerations in this design procedure can be used for stainless steel parts of any kind. Since stainless steel is used only for more expensive castings, good manufacturing procedures are extremely important.

This procedure is the work of Walter H. Dunn, Foundry Superintendent and Robert E. Day, Staff Engineer, Solar Aircraft Company, as reported in *Product Engineering*. SI values were added by the handbook editor.

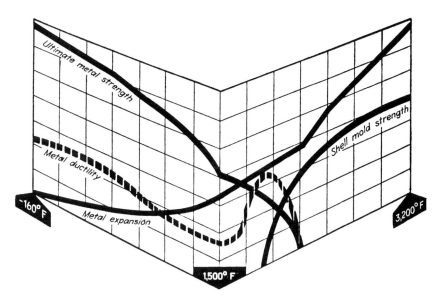

FIGURE 23   Critical temperature range in which stresses are liable to develop in a shell-molded casting is between 1500°F (815.6°C) and 2600°F (1426.7°C). Solidification of metal occurs near upper end of this range.

REDUCING MACHINING COSTS BY DESIGNING WITH SHIMS

The idler shaft of a gear transmission comprising four gears is arranged as shown in Fig. 24. To assure proper end play in the assembly, the overall dimensions of the gears must be held to 4.600 ± 0.005 in (116.84 mm ± 0.127 mm), defined as $N \pm t$. Machining tolerances on the four gears are listed in Table 28. The shims can be manufactured to a tolerance of ±0.001 in (0.0254 mm).

Calculation Procedure:

1.   *Find the thickness of the first shim*
The thickness of the first shim, in (mm) $S_1 = (2t - Z) \pm Z$, where t = tolerance range, in (mm); Z = manufacturing tolerance on the shim, in (mm). Substituting, $S_1 = [(2 \times 0.005) - 0.001] \pm 0.001 = 0.009 \pm 0.001$ in (0.2286 ± 0.0254 mm).

2.   *Determine the lower limit of the stack (assembly) dimension*
The lower limit of the stack dimension where the first shim, S_1, becomes unusable is next determined because this is the smallest dimension of the assembly stack. Use the relation, $L_1 = (N - t) - (S_1 - Z)$, where L_1 lower limit of stack dimension, where S_1 becomes unusable, in (mm); N = overall nominal dimension of the stack, in (mm), Fig. 25; other symbols as defined earlier. Substituting, $L_1 = (4.600 - 0.005) - (0.009 - 0.001) = 4.587$ in (116.509 mm).

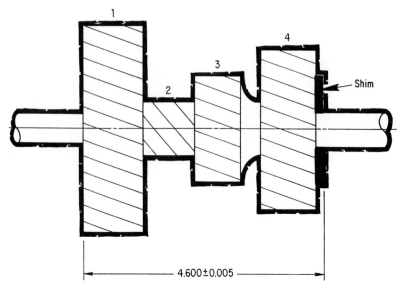

FIGURE 24 Four-gear stack assembly. Gear dimensions are shown in Table 28 (*Machine Design.*)

TABLE 28 Gear Widths for Example Problem

Gear	Nominal dimension, in (mm)				Tolerance, in (mm)	
	Original		Final			
1	1.200	(30.48)	1.200	(30.48)	±0.003	(0.0762)
2	1.000	(25.40)	1.000	(25.40)	±0.003	(0.0762)
3	1.300	(33.02)	1.292	(32.82)	±0.003	(0.0762)
4	1.100	(27.94)	1.100	(27.94)	±0.004	(0.3302)
	4.600	(116.84)	4.592	(116.636)	±0.013	
	$(=N)$		$(=N_F)$		$(=\pm T)$	

3. Compare the step 2 limit with an adjusted nominal total dimension

Test the limit in step 2 with the equation, $L_2 \leq N_F - T$, where L_2 = minimum required tolerance with shim S_2, both expressed in in. (mm); N_F = adjusted nominal dimension, in (mm). For stacks whose measured total dimension is less than L_7, a second shim, S_2, must be chosen. It is desirable to use the largest possible shim to minimize the number of shims stocked in an assembly area. The size of the largest shim is the difference between the high limit of the required tolerance zone and L_1, or $S_2 = (N + t) - L_1 - Z$, with the symbols as given earlier.

Substituting, $L_2 = 4.592 - 0.013 = 4.579$ in (116.306 mm). However, the value in step 2, 4.587, is found to be greater than the $N_F - T = 4.579$ in (116.306 mm) value; therefore a second shim must be chosen.

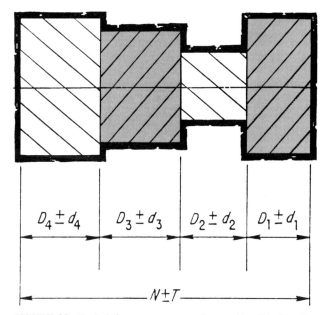

FIGURE 25 Typical four-component stack assembly. (*Machine Design.*)

4. Find the thickness of the second shim

Use the relation $S_2 = (N + t) - L_1 - Z$, where S_2 = second shim thickness, in (mm); other symbols as before. Substituting, $S_2 = (4.600 + 0.005) - 4.587 - 0.001 = 0.017$ in (0.4318 mm).

5. Determine the lower limit for the second shim

The lower limit for shim 2, $L_2 = (N - t) - (S_2 - Z) = (4.600 - 0.005) - (0.017 - 0.001) = 4.579$ in (116.306 mm). And since the value of $L_2 \leq 4.579$ in (116.306 mm), no other shims need be used. Table 29 can now be constructed to indicate the proper shim size to be used with the various ranges of stack dimensions.

Related Calculations. Cumulative tolerances in gears and other stacked assemblies are difficult to control without incurring the penalty of paying either for extremely close machining tolerances on each component or for a selective assembly operation. Designing such assemblies to include shims often saves both time and machining dollars.

TABLE 29 Shim Sizes for Example Problem

Stack dimension		Shim size	
(in)	(mm)	(in*)	(mm)
4.605 to 4.595	(116.96 to 116.71)	None	
4.595 to 4.587	(116.71 to 116.50)	$S_1 = 0.009$	(0.2286)
4.587 to 4.579	(116.60 to 116.30)	$S_2 = 0.017$	(0.4318)

The method described in this calculation procedure can be applied to gearshafts, optical equipment, bearing and clutch assemblies, electrical contactors, and similar assemblies. Areas of application include equipment of all types and sizes ranging from the smallest mechanisms in electronic typewriters to components in earth-moving vehicles.

Assuming a uniform distribution of machining tolerances (to assure 100 percent assembly) is always safe for assemblies comprising up to four or five components. For stacked assemblies made up of a greater number of parts, the statistical distribution of the total stack dimension should be considered. In such cases, the statistical high and low limits are treated the same as the arithmetic limits, and the shim dimensions are calculated accordingly. Taking advantage of the statistical distribution in this manner results in fewer shims being necessary.

When shims are used in a product, they are included in the specification; their dimensions and tolerances must be determined before the overall design is completed. Also, a table must be prepared, along with assembly instructions, listing shim dimensions and corresponding dimensions of the assembly with which each shim is used. In actual assembly, the correct matching of shims can be simplified with the use of a precalibrated dial indicator and color-marked shims.

A typical assembly, made up of four stacked components is shown in Fig. 25. The overall nominal dimension is N and the overall tolerance (the sum of the manufacturing tolerances of the parts) is $\pm T$. In most such assemblies, the difference between the maximum and minimum stack dimensions $(N + T) - (N - T)$ is too large and is unacceptable for proper function of the product. One solution is to tighten the machining tolerances on the several components; another is to evaluate the use of shims in the assembly.

The first step is to determine the acceptable tolerance range for the overall stacked assembly, $N \pm t$, where $t < T$. Then, let the high limit of the acceptable range, $(N + t) = (N_F + T)$, where N_F is an adjusted nominal total dimension and T (as before) is the total of the plus tolerances of the individual components.

For example, if the dimension to be held is 1.250 ± 0.003 in (31.75 ± 0.0762 mm), the upper limit of the stack, $N_F + T$, is $1.250 + 0.003 = 1.253$ in (31.826 mm). The nominal dimension of one or more of the individual parts is then reduced so that the total of the nominal dimensions of the several parts and their plus tolerances equals 1.253 in (31.826 mm). With this situation (all dimensions on the high limit), no shim is required, and no combination of components can exceed the acceptable high limit.

Adding a shim is necessary, of course, when the stack dimension (measured at assembly) falls below the low limit, $N - t$.

You can size the shims by choosing the size of the first shim so that its addition to a stack which is just under the low limit of the required range, $N - t$, will, in no case, put the total dimension over the high limit, $N + t$. Its maximum size, therefore, cannot exceed the range between the high and low limits, $2t$. Tolerances on shims must also be considered, so the size of S_1, the first shim is: $S_1 = (2t - Z) \pm Z$, where the symbols are as given above.

The next step is to determine the smallest dimension of the stack, L_1, where the first shim becomes unusable. This value is obtained by subtracting the smallest shim dimension, $S_1 - Z$, from the low limit of the required stack dimension: $L_1 = (N - t) - (S_1 - Z)$.

For stacks whose measured total dimension is less than L_1, a second shim, S_2, must be chosen. It is desirable to use the largest possible shim to minimize the number of shims stocked. Use the relation in step 4 above to find S.

Shim S_2 is usable until the stack dimension is so small that the addition of the smallest shim, $S_2 - Z$ fails to bring the total dimension up to the minimum required tolerance.

The same procedure is repeated until a lower limit is obtained that is equal to or less than the minimum possible measured stack dimensions, that is: $L_n \leq N_F - T$. Shim sizes thus are: $S_1 = 2t - Z$; $S_n = (N = t) - L_n - 1 - Z$, and the lower limits associated with the shims are: $L_n = (N - t) - (S_n - Z)$.

This procedure is the work of R. L. Weiss, Staff Engineer, Systems Products Division, IBM Corporation, as reported in *Machine Design* magazine.

ANALYZING TAPER FITS FOR MANUFACTURING AND DESIGN

Machine components must be assembled with a taper fit having a single $7/8$-in (2.22-cm) diameter bolt, Fig. 26c, with an end pressure plate applying force to a tapered hub. The hub dimensions are: $D_0 = 7$ in (17.78 cm); $d_0 = 5$ in (12.7 cm); and $L = 10$ in (25.4 cm). The shaft taper is characterized by four parameters: $\theta = 5.7106°$; $D = 6$ in (15.24 cm); $d = 4$ in (10.16 cm); $d_i = 3$ in (7.62 cm). Assume a friction coefficient, $\mu = 0.15$; safety factor $S_F = 3$; and the bolt torque coefficient, $K = 0.2$. For an applied bolt torque $T = 4000$ lb · in. (452.0 N · m), determine the shaft and hub stresses and the torque resistance of the assembly.

Calculation Procedure:

1. Determine the contact pressure in this tapered fit
Use Eq. 3, having the nomenclature shown below. Or

$$P = \frac{2Tn \cos \theta}{Kb(D + d)\pi L(\mu \cos \theta + \sin \theta)} \tag{3}$$

Then,

$$P = \frac{2(4000)(1) \cos 5.7106°}{0.2(7/8)(6 + 4)\pi 10(0.15 \cos 5.7106° + \sin 5.7106°)}$$

$$= 582.05 \text{ lb/in}^2 \quad (4010.3 \text{ kPa})$$

2. Compute the shaft stresses
Use Eqs. 4 and 5, below. Thus

$$S_{rs} = -P \tag{4}$$

$$S_{ts} = -P \left(\frac{d^2 + d_i^2}{d^2 - d_i^2} \right) \tag{5}$$

$$S_{rs} = -582.05 \text{ lb/in}^2 \quad (-4010.3 \text{ kPa})$$

$$S_{ts} = -582.05 \left(\frac{4^2 + 3^2}{4^2 - 3^2} \right)$$

$$= -2078.76 \text{ lb/in}^2 \quad (-14.3 \text{ MPa})$$

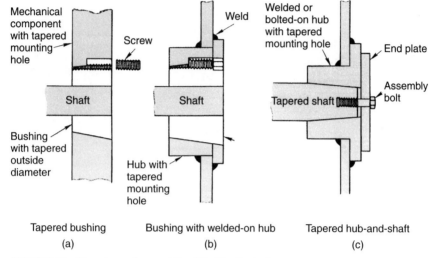

FIGURE 26   Three types of tapered fits. (*Machine Design.*)

3.   Find the hub stresses
Use Eqs. 6, 7, and 8. Thus,

$$S_{rh} = -P \tag{6}$$

$$S_{thl} = P\left(\frac{D_o^2 + D^2}{D_o^2 - D^2}\right) \tag{7}$$

$$S_{ths} = P\left(\frac{d_o^2 + d^2}{d_o^2 - d^2}\right) \tag{8}$$

$$S_{rh} = -582.05 \text{ lb/in}^2 \quad (-4010.3 \text{ kPa})$$

$$S_{thl} = 582.05\left(\frac{7^2 + 6^2}{7^2 - 6^2}\right)$$

$$= 3805.73 \text{ lb/in}^2 \quad (26.2 \text{ MPa})$$

$$S_{ths} = 582.05\left(\frac{5^2 + 4^2}{5^2 - 4^2}\right)$$

$$= 2651.57 \text{ lb/in}^2 \quad (18.27 \text{ MPa})$$

4.   Determine the torque resistance of the assembly
Use Eq. 9. Thus,

$$M = \frac{\mu P \pi L (D + d)^2}{S_F 8 \cos \theta} \tag{9}$$

$$M = \frac{0.15(582.05)\pi10(6 + 4)^2}{3(8) \cos 5.7106°}$$

$$= 11,485.57 \text{ lb} \cdot \text{in} \quad (1297.9 \text{ N} \cdot \text{m})$$

Related Calculations. One of the most effective methods of mounting mechanical components onto shafts uses taper fits. Such fits rely on a wedging of mating surfaces to lock components in place. Machine elements with taper fits usually are mounted and disassembled more easily than are those with alternative shrink fits. Also, taper fits eliminate the sometimes unwieldy heating and cooling of members required for shrink-fit assembly.

Bushings with a tapered outside diameter, Fig. 26a, probably are the most commonly used devices for obtaining taper fits. These bushings slip easily onto conventional shafts and firmly grip sheaves, pulleys, sprockets, and couplings when integral screws or bolts are tightened. To permit such assembly, the drive components require a mounting hole with a tapered diameter. If a tapered mounting hole cannot be machined into components, then a special tapered hub must be welded or bolted on.

Certain applications cannot accommodate a tapered bushing between the hub and shaft. For these cases, the hub must be mounted directly onto a shaft that has one end machined in a taper, Fig. 26b. Such a tapered hub-and-shaft mounting configuration has had little coverage in the literature, unlike mountings with taper-fit bushings. Consequently, there are few guidelines on how to analyze and size tampered hub-and-shaft mountings. Nevertheless, appropriate mathematical expressions can be developed by applying simple force balance principles from statics and by applying stress equations for interference fits.

Four parameters must be analyzed to design effective taper fits. These are: (1) contact pressure; (2) shaft stresses, (3) hub stresses; (4) torque resistance. The following equations, used above, can be used to analyze taper fits.

For most practical purposes, the friction coefficient in taper fits, μ, can be assumed to fall in the range of 0.12 to 0.20. Using Fig. 27 and the nomenclature for reference, the contact pressure can be found using Eq. 1, or

$$P = \frac{2F \cos \theta}{(D + d)\pi L(\mu \cos \theta + \sin \theta)} \tag{1}$$

For cases where the assembly force is applied by bolts threaded into the shaft, the contact pressure in the taper fit must be expressed in terms of the bolt torque, T. Then, the bolt force for n bolts is given by Eq. 2,

$$F_B = \frac{Tn}{Kb} \tag{2}$$

Values for K, which is the bolt torque coefficient, are given in Table 30. These data assume a bolt friction of 0.15. When such data are not available, it usually can be assumed that $K = 0.2$.

Assuming the bolt force F_B = the assembly force, F, Eq. 2 can be substituted into Eq. 1 to give Eq. 3, above.

From the literature on press fits, the radial and tangential stresses on a hollow shaft are defined as those given in Eqs. 4 and 5, above. Equations 4 and 5 were derived for a hollow shaft with an inside diameter d_i. For solid shafts, $d_i = 0$.

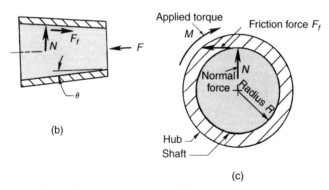

Geometry of a taper fit

FIGURE 27   (*a*) Geometry of a tapered fit. (*b*) Free-body diagram shows that an assembly force applied to the hub end produces force reactions at the inside hub surface. (*c*) Taper fits depend on frictional forces between the hub and shaft to provide resistance to applied torque. (*Machine Design.*)

Hub stresses are given by three equations, namely Eqs. 6, 7, and 8, above. Torque resistance is given by Eq. 9, above.

This procedure is the work of Thomas L. Angle, Supervisor, Equipment Engineering, WEMCO, as reported in *Machine Design* magazine. SI values were added by the handbook editor.

Nomenclature

A = Contact area, in^2 (cm^2)
b = Bolt nominal diameter, in (cm)
D = Shaft large inside diameter and hub large outside diameter, in (cm)
D_o = Hub large outside diameter, in (cm)
d = Shaft small outside diameter and hub small inside diameter, in (cm)
d_i = Shaft inside diameter, in (cm)
d_o = Hub small outside diameter, in (cm)
F = Assembly force, lb (N)
F_B = Bolt force, lb (N)
F_f = Frictional force, lb (N)
K = Bolt torque coefficient

L = Length of contact, in (cm)
M = Applied torque, lb $\cdot$ in (N $\cdot$ m)
N = Normal force, lb (N)
n = Number of bolts
P = Contact pressure, lb/in² (kPa)
R = Shaft outside radius, in (cm)
S_F = Safety factor
S_{rh} = Radial hub stress, lb/in² (kPa)
S_{rs} = Radial shaft stress, lb/in² (kPa)
S_{thl} = Tangential hub stress at large end of taper, lb/in² (kPa)
S_{ths} = Tangential hub stress at small end of taper, lb/in² (kPa)

S_{ts} = Tangential shaft stress at outer surface, lb/in² (kPa)
T = Bolt torque, lb $\cdot$ in (N $\cdot$ m)
θ = Taper angle, °
μ = Friction coefficient between tapered shaft and hub

DESIGNING PARTS FOR EXPECTED LIFE

A machined and ground rod has an ultimate strength of s_u = 90,000 lb/in² (620,350 kPa) and a yield strength s_y = 60,000 lb/in² (413,700 kPa). It is grooved by grinding and has a stress concentration factor of K_f = 1.5. The expected loading in bending

TABLE 30 Data for Computing bolt Torque*

Bolt size	Bolt torque coefficient, K	Bolt size	Bolt torque coefficient, K
¼-20 NC	0.210	⁹⁄₁₆-12 NC	0.198
¼-28 NF	0.205	⁹⁄₁₆-18 NF	0.193
⁵⁄₁₆-18 NC	0.210	⅝-11 NC	0.199
⁵⁄₁₆-24 NF	0.205	⅝-18 NF	0.193
⅜-160 NC	0.204	¾-10 NC	0.194
⅜-24 NF	0.198	¾-16 NF	0.189
⁷⁄₁₆-14 NC	0.205	⅞-9 NC	0.194
⁷⁄₁₆-20 NF	0.200	⅞-14 NF	0.189
½-13 NC	0.201	1–8 NC	0.193
½-20 NF	0.195	1–12 NF	0.188

*Shigley—*Machine Design,* McGraw-Hill.

is 10,000 to 60,000 lb/in² (68,950 to 413,700 kPa) for 0.5 percent of the time; 20,000 to 50,000 lb/in² (137,900 to 344,750 kPa) for 9.5 percent of the time; 20,000 to 45,000 lb/in² (137,900 to 310,275 kPa) for 20 percent of the time; 30,000 to 40,000 lb/in² (206,850 to 275,800 kPa) for 30 percent of the time; 30,000 to 35,000 lb/in² (206,850 to 241,325 kPa) for 40 percent of the time. What is the expected fatigue life of this part in cycles?

Calculation Procedure:

1. Determine the material endurance limit
For s_u = 90,000 lb/in² (620,550 kPa), the endurance limit of the material s_e = 40,000 lb/in² (275,800 kPa), closely, from Fig. 28.

2. Compute the equivalent completely reversed stress
The largest equivalent completely reversed stress for each load in bending is s_F = $(s_e/s_y)s_a + K_{fsa}$ lb/in²; where s_a = average or steady stress, lb/in²; other symbols as before. Since s_e/s_y = 40,000/60,000 = ⅔ and K_f = 1.5, s_v = ⅔s_a + 1.5s_a. Then

$$s_{v1} = (⅔)(35,000) + (1.5)(25,000) = 60,830 \text{ lb/in}^2 \text{ (419,422.9 kPa)}$$

$$s_{v2} = (⅔)(35,000) + (1.5)(15,000) = 45,830 \text{ lb/in}^2 \text{ (315,997.9 kPa)}$$

$$s_{v3} = (⅔)(32,500) + (1.5)(12,500) = 40,420 \text{ lb/in}^2 \text{ (278,695.9 kPa)}$$

$$s_{v4} = (⅔)(35,000) + (1.5)(5000) = 30,830 \text{ lb/in}^2 \text{ (212,572.9 kPa)}$$

$$s_{v5} = (⅔)(32,500) + (1.5)(2500) = 25,420 \text{ lb/in}^2 \text{ (175,270.9 kPa)}$$

3. Compute the fatigue life for the initial stress
The initial stress is s_{v1}; the fatigue life at this stress is, in cycles, N_1 = $1000(s_u/s_{vi})^{3/\log_{1(s_u/s_e)}}$ where s_{vi} = equivalent completely reversed stress, lb/in²; other

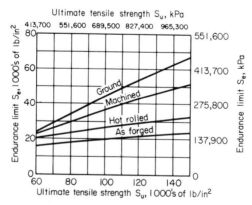

FIGURE 28 Relationship between endurance limit and ultimate tensile strength. (*Product Engineering.*)

symbols as before. By taking the first value of $s_{vi} = s_{v1} = 60,830$ lb/in^2 (419,422.9 kPa), $N_1 = 1000(90,000/60,830)^{3/\log 2.25} = 28,100$ cycles.

4. Compute the exponent for the fatigue-life equation

The exponent for the fatigue-life equation is $2.55/\log(s_u/s_e) = 2.55/\log(90,000/40,000) = 7.2406$.

5. Compute the factors for the fatigue-life equation

The factors needed for the fatigue-life equation are s_{v1}/s_{vi}, $(s_{vi}/s_{v1})^{2.55/\log(s_u/s_e)}$, and $\alpha_i(s_{vi}/s_{v1})^{2.55/\log(s_u/s_e)}$. In these factors, the value of $s_{vi} = s_{v2}$, s_{v3}, and so forth, as summarized in Table 31. The value $\alpha_i =$ percent-time duration of a stress, expressed as a decimal. The numerical values computed are summarized in Table 31.

6. Compute the part fatigue life in cycles

The part fatigue life, in cycles, is $N = N_1/\alpha_1 + \alpha_2[1/(s_{v1}/s_{v2})]^{2.55\log(s_u/s_e)} + 3[1/(s_{v1}/s_{v3})]^{2.55\log(s_u/s_e)} + \ldots$ for each bending load. In this equation, $\alpha_1, \alpha_2, \ldots$ = percent-time duration of a stress, expressed as a decimal, the subscript referring to the stress mentioned in step 2, above.

Since Table 31 summarizes the denominator of the fatigue-life equation, $N = N_1/0.03050 = 28,100/0.03050 = 922,000$ cycles.

Related Calculations.   Data on the endurance limit, yield point, and ultimate strength of ferrous materials are tabulated in Baumeister and Marks—*Standard Handbook for Mechanical Engineers.* The equations presented in this calculation procedure hold for both simple and complex loading. These equations can be used for analysis of an existing part or for design of a part to fail after a selected number of cycles. The latter procedure is sometimes used for components in an assembly in which the principal part has an accurately known life.

The method presented here is the work of Professor M. F. Spotts, reported in *Product Engineering.*

WEAR LIFE OF ROLLING SURFACES

Determine the maximum allowable bearing load in various bearing materials to avoid pronounced wear before 40×10^6 stress cycles if the roller bearing made of these materials has these dimensions: outside diameter = 4.3307 in (11.0 cm), bore

TABLE 31   Values for Cycles-to-Failure Analysis

s_{vi}		s_{v1}/s_{vi}	$(s_{vi}/s_{v1})^{7.2406}$	i	$i(s_{vi}/s_{v1})^{7.2406}$
lb/in^2	kPa				
60,830	419,422.9	1.00	1.00	0.005	0.00500
45,830	315,997.9	1.3273	0.12873	0.095	0.01223
40,420	278,695.9	1.5051	0.05179	0.200	0.01036
30,830	212,572.9	1.9730	0.007296	0.300	0.00219
25,420	175,270.9	2.3934	0.001802	0.400	0.00072
					0.03050

= 2.3622 in (6.0 cm), width = 0.866 in (2.2 cm), inner-race radius r_2 = 1.439 in (3.655 cm), roller diameter = 0.468 in (1.19 cm), roller width f = 0.468 in (1.19 cm), number of rollers n = 16. Materials being considered are gray cast iron, Meehanite; and hardened-steel rollers on cast-iron races, on heat-treated cast-iron races, on heat-treated and medium-steel races, and on carburized low-carbon steel races.

Calculation Procedure:

1. Determine the load-stress factor for each material

Table 32 lists the load-stress factor K for various materials at varying load cycles. Thus for gray cast iron, K = 1300 at 40×10^6 cycles.

TABLE 32 Load-Stress K Factors for Various Materials*

Roller 1		Roller 2	K of roller 1 at number of cycles			
Material	Hardness	Material	10^6	10^7	4×10^7	10^8
Gray cast iron	130–180 BHN		4,000	2,000	1,300	
GM Meehanite	190–240 BHN	Same as roller 1	4,000	2,500	1,950	
Nodular cast iron	207–241 BHN		10,000	5,600	....	3,400
Gray cast iron	270–290 BHN		7,500	5,300	4,200	
Gray cast iron, phosphate-coated	140–160 BHN		2,600	1,400	1,000	
	160–190 BHN		3,200	1,900	1,300	
	270–290 BHN		5,500	4,000	3,100	
SAE 1020 steel, phosphate-coated	130–150 BHN		4,500	2,700	1,700	
SAE 4150 steel, chromium-plated	270–300 BHN	Carbon tool steel	13,500	11,000	....	9,000
SAE 6150 steel	270–300 BHN		2,600	1,300		
SAE 1020 steel, induction-hardened	45–55 RC	60–62 Re	21,000	14,500	....	10,000
SAE 1340 steel, case-hardened	50–58 RC		26,000	20,000	....	15,000
Phosphor bronze	67–77 BHN		3,600	1,600	1,000	
Yellow brass	Drawn		5,600	3,000	2,000	
	Extruded		4,500	2,400	1,700	
Zinc diecasting	....		1,100	500	320	
Laminated-graphitized phenolic	....		1,700	1,300	1,000	
Cast aluminum SAE 39	60–65 BHN	Gray cast iron 340–360 BHN	1,200	500	300	

*Product Engineering. Based on data presented by W. C. Cram at a University of Michigan symposium on surface damage.

2. Compute the maximum allowable bearing load

Use the relation $F = nfK/[5(1/r_1 + 1/r_2)]$ where all symbols are defined in the problem statement above. Thus, $F = (16)(0.468)(1300)/[5(1/0.234 + 1/1.439)] = 391$ lb (1739.3 N) for gray cast iron.

For the other materials, by using the appropriate K value from Table 32 and the above procedure, the allowable load is as follows:

Bearing materials	Allowable load, lb (N)
Meehanite rollers and races	587 (2,611.1)
Hardened-steel rollers, cast-iron races	300 (1,334.5)
Hardened-steel rollers, heat-treated cast-iron races	933 (4,150.2)
Hardened-steel rollers, heat-treated medium-steel races	2,700 (12,010.2)
Hardened-steel rollers, carburized low-carbon steel races	3,912 (17,401.4)

Related Calculations. The same, or similar, procedures can be used for computing the wear life of gears, cams, bearings, clutches, chains, and other devices having rolling surfaces. Thus, in a joint composed of a pin having radius r_1 in a hole of radius r_2, $F = fK/(1/r_1 - 1/r_2)$. This relation also applies to a roller chain. For a cam, $F = fK \cos \infty/(1/\rho - 1/r)$, where ∞ = cam pressure angle; ρ = cam radius of contact, in; r = radius of contact, in.

The method presented here is the work of Professor Donald J. Myatt, reported in *Product Engineering.*

FACTOR OF SAFETY AND ALLOWABLE STRESS IN DESIGN

Determine the cross section dimension b of a uniform square bar of structural steel having a tensile yield strength $s_y = 33,000$ lb/in² (227,535.0 kPa), if the bar carries a center load of 1000 lb (4448.2 N) as a beam with a span of 24 in (61.0 cm) with simply supported ends. Use both the allowable-stress and ultimate-strength methods to determine the required dimension.

Calculation Procedure:

1. Design first on the basis of allowable stress

From Table 33, in the section on buildings, the working or allowable stress $s_w = 0.6s_y = 0.6(33,000) = 19,800$ lb/in² (136,521.0 kPa).

2. Compute the maximum bending moment

For a central load and simple supports, the maximum bending moment is, from earlier sections of this handbook, $M_m = (W/2)(L/2)$, where W = load on beam, lb; L = length of beam, in. Thus, $M_m = (1000/2)(24/2) = 6000$ in · lb (677.9 N · m).

TABLE 33 Illustrative Allowable Stresses and Factors of Safety*

Application	Materials	Allowable stress s_w	Approximate factor of safety
Buildings and other structures	Structural steel	Direct tension s_w = 0.6s_y; 0.45s_y on net section at pin holes; bending s_w = 0.6s_y; shear s_w = 0.45s_y on rivets, 0.40s_y on girder webs; bearing s_w = 1.35s_y on rivets in double shear; 1s_y in single shear	1.70 for beams; 1.85 for continuous frames
	Structural aluminum 6061-T6, 6062-T6 s_u = 38,000 lb/in² (262,010 kPa), s_y = 35,000 lb/in² (241,325 kPa)	Direct tension s_w = 17,000 lb/in² (117,215 kPa); bending structural shapes, s_w = 17,000 lb/in² (117,215 kPa); rectangular sections s_w = 23,000 lb/in² (158,585 kPa); shear s_w = 10,000 lb/in² (68,950 kPa) on cold-driven rivets, 11,000 lb/in² (75,845 kPa) on girder webs; bearing s_w = 30,000 lb/in² (206,850 kPa) on rivets	1.8 for beams; 2 for columns
	Reinforced concrete	Bending, compression in concrete, s_w = 0.45s_c'; bending, tension in steel, s_w = 0.40s_y; bending, tension in plain concrete footings, s_w = 0.03s_c'; shear, on concrete in unreinforced web, s_w = 0.03s_c'; compression in concrete column, s_w = 0.225s_c'	6 in general
	Wood	Bending, s_w = ⅚s'; long compression, s_w = ⅚s'; transverse-compression, s_w = ⅚ elastic limit (400 for Douglas fir); shear parallel to grain s_w = 120 for Douglas fir	
Bridges	Structural metals and reinforced concrete	s_w about 0.9s_w for buildings	2.05
	Wood	s_w same as for buildings	6

26.82

TABLE 33 (Continued)

Machinery	Steel (shafts, etc.)	Steady tension, compression or bending, $s_w = s_y/n$; pure shear, $s_w = s_y/2n$; tension s_t plus shear s_s; $s_t^2 + 4s_s^2 \leqq s_y/n$; alternating stress s_a plus mean stress ns_m; point representing an alternating stress $nk_f s_a$ and a mean stress ns_m must lie below the Goodman diagram, Fig. 1	*n* usually between 1.5 and 2
	Steel (SAE 1095), leaf springs, thickness = t in $t > 0 < 0.10$	Static loading $s_w = 230{,}000 - 1{,}000{,}000t$ lb/in^2 (1,585,850 − 6,895,000t kPa); variable loading, 10^7 cycles, $s_w = 200{,}000 - 800{,}000t$ lb/in^2 (1,379,000 − 5,516,000t kPa); dynamic loading, 10^7 cycles, $s_w = 155{,}000 - 600{,}000t$ lb/in^2 (1,068,725 − 4,137,000t kPa)	
	Steel (wire, ASTM-A228, helical springs)	$s_w = 100{,}000$ lb/in^2 (689,500 kPa) [for $d = 0.2$ in (0.5 cm); 10^7 cycles repeated stress]	
Pressure vessels (unfired)	Carbon steel	Membrane stress, $s_w = 0.211s_u$; membrane plus discontinuity stresses, $s_w = 0.9s_y$ or $0.6s_u$	5
	Alloy steels	Membrane stress, $s_w = 0.25s_u$; membrane plus discontinuity stresses, $s_w = 0.95s_y$ or $0.6s_u$	4
	Cast iron	Membrane stress, $s_w = 0.1s_u$; bending stress $s_w = 0.15s_u$	10, 6.67
	Nonferrous metals	Same rule as alloy steels	4
Airplanes	Aluminum alloy and steel	Ultimate strength design	1.5 against ultimate
	Wood	Ultimate strength design	1 against yield

*From Roark—*Formulas for Stress and Strain,* 4th ed., McGraw-Hill.
Note: s_w = allowable or working stress; s_u = ultimate tensile strength; s_y = tensile yield strength; s' = modulus of rupture in cross-bending of rectangular bar; s_s' = ultimate shear strength; s_c' = ultimate compressive strength; s_e = endurance limit or endurance strength for specified life; n = dividing factor applied to s_u, s_y, or s_c to obtain s_w.

3. *Compute the required cross section dimension*
For a beam of square cross section, $I/c = [b(b)^3/12]/(b/2) = b^3/6$. Also, $s_w = M_m c/I$. Substituting and solving for b gives $b^3 = (6000)(6)/19,800$; $b = 1.22$ in (3.1 cm).

4. *Design the beam on the basis of ultimate strength and a factor of safety*
The safety factor from Table 33 is 1.70 for beams. This safety factor is applied by designing the beam to fail just under a load of $1.7(1000 \text{ lb}) = 1700$ lb (7562.0 N). Thus, to generalize, the design failure load = (factor of safety)(load on part, lb) = W_u.

For structural steel and other materials capable of fully plastic behavior, a simple beam or other member will collapse when the maximum moment equals the *plastic moment* M_p, which is developed when the stress throughout the section becomes equal to s_y. Hence, $M_p = s_y z$, where Z = plastic modulus = arithmetical sum of the static moments of the upper and lower parts of the cross section about the horizontal axis that divides the area in half. For a square with its edges horizontal and vertical, as assumed here, $Z = (\frac{1}{2}b^2)(\frac{1}{4}b) + (\frac{1}{2}b^2)(\frac{1}{4}b) = \frac{1}{4}b^3$. Hence, $M_p = 33,000(\frac{1}{4}b^3)$.

Set M_p = the ultimate bending moment, or $(W_u/2)(L/2)$, where W_u = design failure load = (factor of safety)(load on part, lb) = $(1.70)(1000) = 1700$ lb (7562.0 N) for this beam. Thus, $M_p = 33,000(\frac{1}{4}b^3) = (1700/2)(24/2)$; $b = 1.07$ in (2.7 cm).

By using the allowable stress, $b = 1.22$ in (3.1 cm), as compared with 1.07 in (2.7 cm). The difference in results for the two methods (about 12 percent) can be traced to the ratio of Z to I/c, called the *shape factor.* For the case under consideration, this ratio is $(\frac{1}{4}b^3)/(\frac{1}{6}d^3) = 1.5$.

5. *Determine the beam size for a vertical diagonal*
Here $Z = 0.2357b^3$, $I/c = 0.1178b^3$, and the shape factor = $Z/(I/c) = 0.2357/0.1178 = 2.0$.

Designing by allowable stress, using steps 1, 2, and 3, gives $(0.1178b^3)(19,800) = 6000$; $b = 1.37$ in (3.5 cm). Designing by ultimate strength gives $(0.2357b^3)(33,000) = 10,200$; $b = 1.095$ in (2.8 cm).

This computation shows that a more economical design is generally obtained by ultimate-strength design, with the advantage becoming greater as the shape factor becomes larger.

Related Calculations. For conventional structural sections, such as I beams, the shape factor is not much greater than 1, and the two methods yield about the same result for statically determinate problems, such as this one. If the problem involves a statically indeterminate beam or a rigid frame, the advantage of the ultimate-strength method becomes more apparent because it takes account of the fact that collapse cannot occur, as a rule, until the plastic moment is developed at each of two or more sections.

The method given here is the work of Professor Raymond J. Roark, reported in *Product Engineering.*

RUPTURE FACTOR AND ALLOWABLE STRESS IN DESIGN

Determine the proper thickness t of a circular plate 40 in (101.6 cm) in diameter if the edge of the plate is simply supported and the plate carries a uniformly distributed pressure of 200 lb/in² (1379.0 kPa). The plate is made of cast iron having an ultimate strength of 50,000 lb/in² (344,750.0 kPa).

Calculation Procedure:

1. Design on the basis of allowable stress

From Table 33 of the previous calculation procedure, note in the section for pressure vessels that the value of the working or allowable stress s_w lb/in² for tension due to bending is 0.15 s_u, where s_u = ultimate tensile strength of the material, lb/in². Thus, s_w = 0.15(50,000) = 7500 lb/in² (51,712.5 kPa).

2. Compute the required plate thickness

The maximum stress for a simply supported plate is, from Roark—*Formulas for Stress and Strain*, s_{max} = $(3W/8\pi t^2)(3 + v)$, where W = total load on plate, lb; t = plate thickness, in; v = Poisson's ratio. For this plate, W = (200 lb/in²)(π)(20)² = 251,330 lb (1,117,971.6 N). Assuming v = 0.3 and solving for t, we find 7500 = $[(3)(251,330)/8\pi t^2](3 + 0.3)$; t = 3.63 in (9.2 cm).

3. Design on the basis of ultimate strength

For ultimate-strength design, from Table 33 of the previous calculation procedure, use the value of 6.67 for the factor of safety. Design the plate to break, theoretically, under a load. Hence, the breaking load would be (factor of safety) W, or 6.67(251,330) = 1,667,000 lb (7,415,186.1 N).

4. Apply the rupture factor to the design

With a brittle material like cast iron, the concept of the plastic moment does not apply. Use instead the *rupture factor*, which is the ratio of the calculated maximum tensile stress at rupture to the ultimate tensile strength, both expressed in lb/in². Whereas the rupture factor must be determined experimentally, a number of typical values are given in Table 34 for a variety of cases.

For case 2, which corresponds to this problem, the rupture factor for cast iron is R_i = 1.9. Using this in the same equation for s_w as in step 1, s_w = 1.9(50,000) = 95,000 lb/in² (655,025.0 kPa). Then, by using the procedure and equation in step 2 but substituting the breaking load for the plate, 95,000 = $[(3)(1,677,000)/8\pi t^2] (3 + 0.3)$; t = 2.63 in (6.7 cm).

Related Calculations.   The reliability of the solution using the ultimate-strength design technique depends on the accuracy of the rupture factor. When experimental or tabulated values of R_i are not available, the modulus of rupture may be used in place of $R_i s_u$. Typical values of Poisson's ratio v for various materials are given in Table 35.

The method presented here is the work of Professor Raymond J. Roark, reported in *Product Engineering*.

FORCE AND SHRINK FIT STRESS, INTERFERENCE, AND TORQUE

A 0.5-in (1.3-cm) thick steel band having a modulus of elasticity of E = 30 × 10⁶ lb/in² (206.8 × 10⁹ Pa) is to be forced on a 4-in (10.2-cm) diameter steel shaft. The maximum allowable stress in the band is 24,000 lb/in² (165,480.0 kPa). What interference should be used between the band and the shaft? How much torque can

TABLE 34 Values of the Rupture Factor for Brittle Materials*

Form of member and manner of loading	Rupture factor; ratio of computed maximum stress at rupture to ultimate tensile strength		Ratio of computed maximum stress at rupture to modulus of rupture in bending or torsion	
	Cast iron	Plaster	Cast iron	Plaster
1. Rectangular beam, end support, center loading, $l/d = 8$ or more	1.70	1.60	1	1
2. Solid circular plate, edge support, uniform loading, $a/t = 10$ or more	1.9	1.71	...	1.07
3. Solid circular plate edge support, uniform loading on concentric circular area	$2.4-0.5(r_o/a)^{1/6}$	$2.2-0.5(r_o/a)^{1/6}$	$1.40-0.3(r_o/a)^{1/6}$	$1.4-0.3(r_o/a)^{1/6}$

*From Roark—*Formulas for Stress and Strain*, 4th ed., McGraw-Hill.

TABLE 35 Poisson's Ratio for Various Materials

Material	Poisson's ratio
Aluminum:	
Cast	0.330
Wrought	0.330
Brass, cast, 66% Cu, 34% Zn	0.350
Bronze, cast, 85% Cu, 7.2% Zn, 6.4% Sn	0.358
Cast iron	0.260
Copper, pure	0.337
Phosphor bronze, cast, 92.5% Cu, 7.0% Sn,	
0.5% Ph	0.380
Steel:	
Soft	0.300
1% C	0.287
Cast	0.280
Tin, cast, pure	0.330
Wrought iron	0.280
Nickel	0.310
Zinc	0.210

the fit develop if the band is 3 in (7.6 cm) long and the coefficient of friction is 0.20?

Calculation Procedure:

1. Compute the required interference

Use the relation $i = sd/E$, where i = the required interference to produce the maximum allowable stress in the band, in; s = stress in band or hub, lb/in^2; d = shaft diameter, in; E = modulus of elasticity of band or hub, lb/in^2. For this fit, $i = (24,000)(4.0)/(30 \times 10^6) = 0.0032$ in (0.081 mm).

2. Compute the torque the fit will develop

Use the relation $T = Eitl\pi f$, where T = fit torque, lb·in; t = band or hub thickness, in; l = band or hub length, in; f = coefficient of friction between the materials. For this joint, $T = 30 \times 10^6 \times 0.0032 \times 0.5 \times 3.0 \times \pi \times 0.20 = 90,432$ lb·in (10,217.4 N·m).

Related Calculations.   Use this general procedure for either shrink or press fits. The axial force required for a press fit of two members made of the same material is F_a = axial force for the press fit, lb; p_c = radial pressure between the two members, lb/in^2 = $iE(d_c^2 - d_i^2)(d_o^2 - d_c^2)/zd_c^3(d_o^2 - d_i^2)$, where d_o = outside diameter of the external member, in; d_c = nominal diameter of the contact surfaces, in; d_i = inside diameter of the inner member, in.

SELECTING BOLT DIAMETER FOR BOLTED PRESSURIZED JOINT

Select a suitable bolt diameter for the typical bolted joint in Fig. 29 when the joint is used on a pressurized cylinder having a flanged head clamped to the body of the

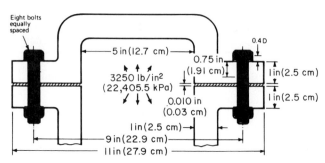

FIGURE 29 Typical bolted joint analyzed in the calculation procedure. (*Product Engineering.*)

cylinder by eight equally spaced bolts. The vessel internal pressure, which may be produced by hydraulic fluid or steam, varies from 0 to 3250 lb/in² (0 to 22,405.5 kPa). What clamping force must be applied by each bolt to ensure that no leakage will occur? Check the selected bolt size to ensure long fatigue life under static and fluctuating loading.

Calculation Procedure:

1. Determine the axially pressurized area of the cylinder
The internal diameter of the cylinder is, as shown, 5 in (12.7 cm). The axially pressurized area of the cylinder head is $A = \pi D^2/4 = 5^2/4 = 19.63$ in² (126.62 cm²).

2. Find the applied working load on each bolt
Since eight bolts are specified for this flanged joint, each bolt will carry one-eighth of the total load found from $F_A = PA/8$, where F_A = axial applied working load, lb; P = maximum pressure in vessel, lb/in²; A = axially pressurized area, in². Or, $F_A = 3250(19.63)/8 = 7947.7$ lb, say 8000 lb (3633.4 kg) for calculation purposes.

3. Compute the bolt load produced by torquing
An air-stall power wrench will be used to tighten (torque) the nuts on these bolts. Such a wrench has a torque tightening factor C_T of 2.5 maximum, as shown in Table 36. Using $C_T = 2.5$, we find the load on each bolt P_Y, lb, causing a yield stress is $P_Y = C_T F_A = 2.5(8000) = 20,000$ lb (9090.9 kg).

4. Find the ultimate tensile strength P_U for the bolt
With the yield strength typically equal to 80 percent of the ultimate tensile strength of the bolt, $P_U = P_Y/0.80 = 20,000/0.80 = 25,000$ lb (11,363.6 kg).

5. Determine the required bolt area and diameter
Select a grade 8 bolt having an ultimate strength of 150,000 lb/in² (1034.1 MPa). The nominal bolt area must then be $A_b = P_U/U_s = 25,000/150,000 = 0.1667$ in² (1.08 cm²). Then the bolt diameter is $D_b = (4A_b/\pi)^{0.5} = [4(0.1667)/\pi]^{0.5} = 0.4607$ in (1.17 cm).

The closest standard bolt size is 0.5 in (1.27 cm). Choose 0.5-13NC bolts, keeping in mind that coarse-thread bolts generally have stronger threads.

TABLE 36   Torque Tightening Factor*

Method	C_T
Electronic bolt-torquing systems	1.0 to 1.5
Torque or power wrench with direct torque control	1.6 to 1.8
Power wrench by elongation measurement of calibrated bolts with the original clamped part	1.4 to 1.6
Power wrench using air-stall principle	1.7 to 2.5

Product Engineering.

6. Find the spring rate of the bolt chosen

The general equation for the spring rate K, lb/in, for a part under tension loading is $K = AE/L$, where A = part cross-sectional area, in^2; E = Young's modulus, lb/in^2; L = length of section, in. To find the spring rate of a part with different cross-sectional areas, as a bolt has, add the reciprocal of the spring rate of each section, or $1/K_{\text{total}} = 1/K_1 + 1/K_2 + \cdots$.

For a bolt, which consists of three parts (head, unthreaded portion, and nut), the spring rate determined by G. H. Junker, a consultant with Unbrako-SPS European Division, is given by $1/K_B = (1/E)[(0.4D/A_1) + (L_1/A_1) + (L_T/A_M) + (0.4D_M/A_M)]$, where D = nominal bolt diameter, in; A_1 = cross-sectional area of unthreaded portion of bolt, in^2; A_m = cross-sectional area, in^2, of minor threaded diameter D_M; L_1 = length of unthreaded portion, 0.75 in (1.91 cm) for this bolt; L = length of threaded portion being clamped, in. Values of $0.4D$ and $0.4D_M$ pertain to the elastic deformation in the head and nut areas, respectively, and were derived in tests in Germany.

For a 0.5-in (1.27-cm) diameter bolt, the cross-sectional area A_1 of the unthreaded portion of the body is $A_1 = \pi(0.5)^2/4 = 0.1964$ in^2 (1.27 cm^2).

The minor thread diameter of the 0.5-13NC thread bolt is, from a table of thread dimensions, 0.4056 in (1.03 cm). Thus, $A_M = \pi(0.4056)^2/4 = 0.1292$ in^2 (0.83 cm^2).

Substituting the appropriate values in the spring-rate equation for the bolt gives $1/K_B = (1/30 \times 10^6)[0.4 \times 0.5/0.1964 + 0.75/0.1964 + (2.010 - 0.75)/0.1292 + 0.4 \times 0.4056/0.1292]$; $K_B = 1.893 \times 10^6$ lb/in (338,752 kg/cm).

7. Compute the spring rate of the joint

Calculations of the spring rate of the joint can be simplified by assuming that the bolt head and nut, when compressing the joint as the bolt is tightened, will cause a stress distribution in the shape of a hollow cylinder, with most of the joint compression occurring in the vicinity under the bolt head and nut.

To calculate an equivalent area for the joint A_J in^2 for use in the spring-rate equation for the joint, the designer has a choice of one of three equations:

Case 1—When most of the outside diameter of the joint is equal to or smaller than the bolt-head diameter, as when parts of a bushing are clamped, Fig. 30, then $A_J = (\pi/4)(D_0^2 - D_H^2)$, where D_0 = outside diameter of the joint or bushing, in; D_H = diameter of bolt hole, in.

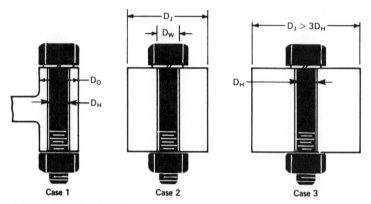

FIGURE 30 Joint size influences the general equations for spring rate of the assembly. (*Product Engineering.*)

Case 2—When the outer diameter of the joint D_J in is greater than the effective bolt-head diameter or washer D_W in, but less than $3D_H$, then $A_J = (\pi/4)(D_W^2 - D_N^2) + (\pi/8)(D_J/D_W - 1)(D_W L_J/5 + L_J^2/100)$.

Case 3—When the joint diameter D_J in is equal to or greater than $3D_H$, then $A_J = (\pi/4)(D_W + L_J/10 - D_H^2)$.

For this bolted joint, case 2 applies. Assuming that the bearing diameter of the head or nut is 0.75 in (1.9 cm), then $A_J = (\pi/4)(0.75^2 - 0.5^2) + (\pi/8)(2/0.75 - 1)[0.75(2.010)/5 + (2.010)^2/100] = 0.4692$ in²(3.03 cm²). Inserting this value of A_J in the spring-rate equation of step 6 gives $K_J = 0.4692(30 \times 10^6)/2.010 = 7.003 \times 10^6$ lb/in (1.253 × 10⁶ kg/cm).

8. Find the portion of the working load that unloads the clamped joint
The loading constant C_L considers the bolt and joint elasticity and is given by $C_L = K_B/(K_B + K_J)$. For this joint, $C_L = 1.893/(1.893 + 7.003) = 0.2128$. Then F_p, the portion of the working load F_A that unloads the clamped joint, is $F_P = F_A(1 - C_L) = 8000(1 - 0.2128) = 6298$ lb (2862.7 kg).

9. Determine the loss of clamping force due to embedding
Some embedding occurs after a bolt is tightened at its assembly. A recommended value is 10 percent. Or specific values can be obtained from tests. Thus, the loss of clamping force F_Z, lb $= 0.10F_A = 0.10(8000) = 800$ lb (363.6 kg).

10. Find the clamping force required for the joint
Since the working load in the vessel fluctuates from 0 to 8000 lb (0 to 3636.4 kg), the minimum required clamping force $F_K = 0$. Now the maximum required clamping force F_M, lb, can be found from $F_M = C_T(F_P + F_K + F_Z)$, since all the variables are known. Or, $F_M = 2.5(6298 + 0 + 800) = 17,745$ lb (8065.9 kg).

Since the bolt has a yield strength P_Y of 20,000 lb (9090.9 kg), it has sufficient strength for static loading. For dynamic loading, however, the additional loading is $F_S = F_A C_L$, where $F_S =$ portion of working load that additionally loads the bolt, lb. Or, $F_S = 8000(0.2128) = 1702$ lb (773.6 kg).

11. Check the endurance limit of the selected bolt
The endurance limit S_E, lb, should not be exceeded for long-life operation of the joint. Exact values of S_E can be obtained from the bolt manufacturer or computed from standard endurance-limit equations. For grade 8 bolts, $S_E = 4600D^{1.59}$, or $S_E = 4600(0.5)^{1.59} = 1527$ lb (694.1 kg).

The value for F_S that was computed for the 0.5-in (1.27-cm) bolt, 1702 lb (773.6 kg), should not have exceeded 1527 lb (694.1 kg). Since it did, a bolt with a larger diameter must now be selected. Checking a 0.625-in (1.59-cm) bolt (0.625-11), we find the endurance limit becomes $S_E = 4600(0.625)^{1.59} = 2179$ lb (990.5 kg).

Thus, although the 0.5-in (1.27-cm) bolt would have sufficed for static loading, it would not have provided the necessary endurance strength for long-life operation. So the 0.625-in (1.59-cm) bolt is selected, and a new clamping force $F_M = 16,970$ lb (7713.6 kg) is calculated, by using the same series of steps detailed above.

Related Calculations. The procedure given here can be used to analyze bolted joints used in pressure vessels in many different applications—power plants, hydraulic systems, aircraft, marine equipment, structures, and piping systems. New thread forms permit longer fatigue life for bolted joints. For this reason, the bolted joint is becoming more popular than ever. Further, electronic bolt-tightening equipment allows a bolt to be tightened precisely up to its yield point safely. This gets the most out of a particular bolt, resulting in product and assembly cost savings.

The key to economical bolted joints is finding the minimum bolt size and clamping force to provide the needed seal. Further, the bolt size chosen and clamping force used must be correct. But some designers avoid the computations for these factors because they involve bolt loading, elasticity of the bolt and joint, reduction in preload resulting from embedding of the bolt, and the method used to tighten the bolt. The procedure given here, developed by G. H. Junker, a consultant with Unbrako-SPS European Division, simplifies the computations. Data given here were presented in *Product Engineering* magazine, edited by Frank Yeaple, M.E.

DETERMINING REQUIRED TIGHTENING TORQUE FOR A BOLTED JOINT

Determine the bolt-tightening torque required for the 0.625-in 11NC (1.59-cm) bolt analyzed in the previous calculation procedure. The bolt must provide a 17,000-lb (7727.3-kg) preload. The coefficient of friction between threads f_T is assumed to be 0.12, while the coefficient of friction between the nut and the washer f_N is assumed to be 0.14.

Calculation Procedure:

1. Determine the dimensions of the bolt
From tables of thread dimensions, the minor thread diameter $D_M = 0.5135$ in (1.3 cm), and the pitch diameter of the bolt threads $D_P = 0.5660$ in (1.44 cm). The mean bearing diameter of the nut is $D_N = 1.25D$, where D = nominal bolt diameter, in. Or, $D_N = 1.25(0.625) = 0.781$ in (1.98 cm).

For the thread coefficient of friction, $\tan \phi = 0.12$; $\phi = 6.84°$. The helix angle of the thread, $\beta = 2.93°$, is found from $\sin \beta = 1/\pi (11)(0.5660) = 0.05112$.

2. Find the dimensionless thread angle factor C_A
The thread angle factor $C_A = \tan (\beta + \phi)/\cos \alpha = \tan (2.36 + 6.84)/\cos 30° = 0.2$.

3. Compute the tightening torque applied
The tightening torque applied to the nut or bolt head, in·lb, is $T = F_M(D_N f_N + C_A D_P)/2$. Substituting gives $T = 17,000[0.781(0.14) + 0.2(0.566)]/2 = 1886$ in·lb $= 157$ ft·lb (212.7 J).

4. Find the combined stress induced in the bolt
The combined stress S_C in a bolt in a bolted joint is $S_C = T\{0.89 + 1.66[1 + (5.2C_A D_P/D_M)^2]^{0.5}\}/D_M^2(f_n D_N + C_A D_P)$. Substituting, we find $S_C = 1886\{0.89 + 1.66[1 + (5.2 \times 0.2 \times 0.566/0.5135)^2]^{0.5}\}/\{0.5135^2[0.14(0.781) + 0.2(0.566)]\} = 109,766$ lb/in² (756,812.4 kPa).

Generally, S_C should be kept within 68 percent of the ultimate strength S_U. Thus, $S_U = 1.47S_C$. Substituting for the above bolt gives $S_U = 1.47(109,766) = 161,356$ lb/in² (1.1 MPa).

Related Calculations. This procedure can be used for determining the bolt-tightening torque for any application in which a bolted joint is used. Numerous applications are listed in the previous calculation procedure. The equations and approach given here are those of Bernie J. Cobb, Mechanical Engineer, Missile Research and Development Command, Redstone Arsenal, as reported in *Product Engineering,* edited by Frank Yeaple.

SELECTING SAFE STRESS AND MATERIALS FOR PLASTIC GEARS

Determine the safe stress, velocity, and material for a plastic spur gear to transmit 0.125 hp (0.09 W) at 350 r/min 8 h/day under a steady load. Number of teeth in the gear $= 75$; diametral pitch $= 32$; pressure angle $= 20°$; pitch diameter $= 2.34375$ in (5.95 cm); face width $= 0.375$ in (0.95 cm).

Calculation Procedure:

1. Compute the velocity at the gear pitch circle
Use the relation $V = rpm(D_p)\pi/12$, where $V =$ velocity at pitch-circle diameter, ft/min; $rpm =$ gear speed, r/min; $D_p =$ pitch diameter, in. Solving yields $V = 350(2.34375)\pi/12 = 215$ ft/min (2.15 m/s).

2. Find the safe stress for the gear
Use the relation $S_S = 55(600 + V)PC_S H_P/(FYV)$, where $S_S =$ safe stress on the gear, lb/in²; $P =$ diametral pitch, in; $C_S =$ service factor from Table 37; $H_P =$ horsepower transmitted by the gear; $F =$ face width of the gear, in; $Y =$ tooth form factor, Table 38. Substituting, we find $S_S = 55(600 + 215)(32)(1.0)(0.125)/[0.375(0.434)(215)] = 5124$ lb/in² (35,561 kPa).

3. Select the gear material
Enter Table 39 at the safe stress, 5124 lb/in² (35,561 kPa), and choose either nylon or polycarbonate gear material because the safe stress falls within the allowed range,

TABLE 37 Service Factor C_S for Horsepower Equations*

Type of load	8–10 h/day	24 h/day	Intermittent, 3 h/day	Occasional, 0.5 h/day
Steady	1.00	1.25	0.80	0.50
Light shock	1.25	1.50	1.00	0.80
Medium shock	1.50	1.75	1.25	1.00
Heavy shock	1.75	2.00	1.50	1.25

Product Engineering.

TABLE 38 Tooth-Form Factor Y for Horsepower Equations*

Number of teeth	14½° Involute or cycloidal	20° Full-depth involute	20° Stub-tooth involute	20° Internal full depth Pinion	Gear
50	0.352	0.408	0.474	0.437	0.613
75	0.364	0.434	0.496	0.452	0.581
100	0.371	0.446	0.506	0.462	0.581
150	0.377	0.459	0.518	0.468	0.565
300	0.383	0.471	0.534	0.478	0.534
Rack	0.390	0.484	0.550	. . . .	. . . .

Product Engineering.

TABLE 39 Safe Stress Values for Horsepower Equations*

Plastic	Unfilled lb/in²	kPa	Glass-reinforced lb/in²	kPa
ABS	3,000	20,682	6,000	41,364
Acetal	5,000	34,470	7,000	48,258
Nylon	6,000	41,364	12,000	82,728
Polycarbonate	6,000	41,364	9,000	62,046
Polyester	3,500	24,129	8,000	55,152
Polyurethane	2,500	17,235	. . . .	. . . .

Product Engineering.

600 lb/in² (41,364 kPa), for these two materials. If a glass-reinforced gear is to be used, any of a number of materials listed in Table 39 would be suitable.

Related Calculations. Two other v equations are used in the analysis of plastic gears. For helical gears: $H_P = S_S FYV/[423(78 + V^{0.5})P_N C_S]$. For straight bevel gears, $H_P = S_S FYV(C - F)^4/[55(600 + V)PCC_S]$, where all the symbols are as given earlier; C = pitch-cone diameter, in; P_N = normal diametral pitch.

In growing numbers of fractional-horsepower applications up to 1.5 hp (1.12 kW), gears molded of plastics are being chosen. Typical products are portable power tools, home appliances, instrumentation, and various automotive components.

The reasons for this trend are many. Besides offering the lowest initial cost, plastic gears can be molded as one piece to include other functional parts such as cams, ratchets, lugs, and other gears without need for additional assembly or finishing operations. Moreover, plastic gears are lighter and quieter than metal gears, and are self-lubricating, corrosion-resistant, and relatively free from maintenance. Also, they can be molded inexpensively in colors for coding during assembly or just for looks.

Improved molding techniques achieve high accuracy. Also new special gear-tooth forms improve strength and wear. "We now can hold tooth-to-tooth composite error to within 0.0005 in (0.00127 cm)," says Samuel Pierson, president of ABA Tool & Die Co. It is made possible by electric-discharge machining of the metal molds and computerized analysis of the effects of moisture absorption and other factors on size change of the plastic gears.

"Furthermore, we now are recommending special tooth forms we developed to utilize full-fillet root radii for increased fatigue strength and tip relief for more uniform motion when teeth flex under load. Usually, it is too expensive for designers of machined metal gears to deviate from standard AGMA tooth forms. But with molded plastic gears, deviations add little to the cost," reports Mr. Pierson in *Product Engineering* magazine.

Plastic gears, however, are weaker than metal gears, have a relatively high rate of thermal expansion, and are temperature-limited.

If performance cannot be achieved solely with change in tooth form, try fillers that stabilize the molded part, boost load-carrying abilities, and improve self-lubricating and wear characteristics. Popular fillers include glass, polytetrafluoroethylene (PTFE), molybdenum disulfide, and silicones.

In short hairlike fibers, miniscule beads, or fine-milled powder, glass can markedly increase the tensile strength of a gear and reduce thermal expansion to as little as one-third the original value. Molybdenum disulfide, PTFE, and silicones, as built-in lubricants, reduce wear. Plastic formulations containing both glass and lubricant are becoming increasingly popular to combine strength and lubricity.

Six common plastics for molded gears are nylon, acetal, ABS, polycarbonate, polyester, and polyurethane.

The most popular gear plastic is still nylon. This workhorse has good strength, high abrasion resistance, and a low coefficient of friction. Furthermore, it is self-lubricating and unaffected by most industrial chemicals. Numerous manufacturers make nylon resins.

Nylon's main drawback is its tendency to absorb moisture. This is accompanied by an increase in the gear dimensions and toughness. The effects must be predicted accurately. Also, nylon is harder to mold than acetal—one of its main competitors.

All plastics have higher coefficients of thermal expansion than metals do. The coefficient for steel between 0 and 30°C is 1.23×10^5 per degree, whereas for nylon it lies between 7 and 10×10^{-5}.

The grade usually preferred is nylon 6/6. Frequently it is filled with about 25 percent short glass fibers and a small amount of lubricant fillers. Do not hesitate to ask the injection molder to custom-blend resins and fillers.

The combination of a nylon gear running against an acetal gear results in a lower coefficient of friction than either a nylon against nylon or an acetal against acetal can. A good idea is to intersperse nylon with acetal in gear trains. Acetal is less expensive than nylon and is generally easier to mold. Strength is lower, however.

Two types of acetals are popular for gears: acetal homopolymer, which was the original acetal developed by DuPont, available under the tradename of Delrin; and acetal copolymer, manufactured by Celanese under the tradename Celcon. Other examples are glass-filled acetals (Fulton 404, from LNP Corp.).

All acetals are easily processed and have good natural lubricity, creep resistance, chemical resistance, and dimensional stability. But they have a high rate of mold shrinkage.

The main attraction of ABS plastic is its low cost, probably the lowest of the six classes of plastics used for gears. Some ABS is translucent and has a high gloss surface. It also offers ease of processibility, toughness, and rigidity. Two typical tradenames of ABS are Cycolac (Borg Warner), and Kralastic (Uniroyal).

The polycarbonates have high impact strength, high resistance to creep, a useful temperature range of −60 to 240°F (−51 to 115.4°C), low water absorption and thermal expansion rates, and ease and accuracy in molding.

Because of a rather high coefficient of friction, polycarbonate formulations are available to boost flexible strength of the gear teeth.

Recently, thermoplastic polyesters have been available for high-performance injection-molded parts. Some polyesters are filled with reinforcing glass fibers for gear applications. The adhesion between the polyester matrix and the glass fibers results in a substantial increase in strength and produces a rigid material that is creep-resistant at elevated temperature [330°F(148.7°C)].

The glossy surface of some polyesters seems to improve lubricity against other thermoplastics and metals. It withstands most organic solvents and chemicals at room temperature and has long-term resistance to gasoline, motor oil, and transmission fluids up to 140°F (60°C).

The polyurethanes are elastomeric resins sought for noise dampening or shock absorption, as in gears for bedroom clocks or sprockets for snowmobiles. Many proprietary polyurethane versions are available, including Cyanaprene (American Cyanimide), Estane (B. F. Goodrich), Texin (Mobay Chemical), and Voranol (Dow Chemical).

BIBLIOGRAPHY

Part 1 POWER GENERATION

Power Generation

REFERENCES: El-Wakil—*Powerplant Technology*, McGraw-Hill; Goss—*Factors Affecting Power Plant Waste Heat Utilization*, Pergamon Press; Polimeros—*Energy Cogeneration Handbook*, Industrial Press; Yu—*Electric Power System Dynamics*, Academic Press; Hagel—*Alternative Energy Strategies*, Praeger; Aschner—*Planning Fundamentals of Thermal Power Plants*, Israel University Press; Komanoff—*Power Plant Cost Escalation*, VNR; Seeley—*Elements of Thermal Technology*, Dekker; Hunt—*Handbook of Energy Technology*, VNR; Blair, Cassel, and Edelstein—*Geothermal Energy: Investment Decisions and Commercial Development*, Wiley; Goodman and Love—*Geothermal Energy Projects: Planning and Management*, Pergamon Press; Edgerton—*Available Energy and Environmental Economics*, Heath; Meyers—*Handbook of Energy Technology and Economics*, Wiley; Babcock & Wilcox Company—*Steam; Its Generation and Use*; Combustion Engineering Corporation—*Combustion Engineering*; Skrotzki and Vopat—*Power Station Engineering*.

Combustion

REFERENCES: Chigier—*Energy, Combustion and Environment*, McGraw-Hill; Lewis and Von Elbe—*Combustion, Flames and Explosion of Gases*, Pergamon Press; Zung—*Evaporation-Combustion of Fuel*, American Chemical Society; Johnson and Auth—*Fuels and Combustion Handbook*, McGraw-Hill; Babcock & Wilcox Company—*Steam: Its Generation and Use*; Combustion Engineering Corporation—*Combustion Engineering*; Gaffert—*Steam Power Stations*, McGraw-Hill; Skrotzki and Vopat—*Applied Energy Conversion*, McGraw-Hill; Popovich-Hering—*Fuels and Lubricants*, Wiley; ASME—*Power Test Code for Steam Boilers*; Moore—*Coal*, Wiley; Moore—*Liquid Fuels*, The Technical Press, Ltd., London; American Gas Association—*Combustion*; Dunstan—*Science of Petroleum*, Oxford, London; Trinks—*Industrial Furnaces*, Wiley; Perry—*Chemical Engineers Handbook*, McGraw-Hill.

Internal-Combustion Engines

REFERENCES: Benson—*Internal Combustion Engines*, Pergamon; Kates and Luck—*Diesel and High-Compression Gas Engines*, American Technical Society; Ranney—*Fuel Additives for Internal Combustion Engines*, Noyes; Blackman and Thomas—*Fuel Economy of the Gasoline Engine*, Halsted Press; Sitkei—*Heat Transfer and Thermal Loading in Internal Combustion Engines*, International Publications Services; Baxa—*Noise Control in Internal Combustion Engines*, Wiley; Diesel Engine Manufacturers Associations—*Standard Practices for Stationary Diesel Engines*; Lichty—*Internal-Combustion Engines*, McGraw-Hill; Allen—*The Modern Diesel*, Prentice-Hall; Maleev—*Internal-Combustion Engines*, McGraw-Hill; *Diesel Engineering Handbook*, Diesel Publications; Adams—*Elements of Diesel Engineering*, Henley; Severns and Degler—*Steam, Air and Gas Power*, Wiley; Ricardo—*The High-Speed Internal-Combustion Engine*, Blackie; Obert—*Internal Combustion Engines*, International Textbooks; Fors—*Practical Marine Diesel Engineering*, Simmons-Boardman.

Part 2 PLANT AND FACILITIES ENGINEERING

Pumps and Pumping Systems

REFERENCES: Karassik—*Pump Handbook,* McGraw-Hill; Warring—*Pumps—Selection, Systems, and Applications,* Trade and Technical Press (England); Crawford—*Marine and Off-shore Pumping and Piping Systems,* Butterworth; *Europump Terminology: Glossary of Pump Applications in English, German, Italian, and Spanish,* International Ideas; Isman—*Fire Service Pumps and Hydraulics,* Delmar; Pollak—*Pump User's Handbook,* Gulf Publishing; Anderson—*Centrifugal Pumps,* Trade and Technical Press (England); Walker—*Pump Selection,* Ann Arbor Science Press; Bartlett—*Pumping Stations for Water and Sewage,* Halsted Press; Koutitas—*Elements of Computational Hydraulics,* Chapman and Hall; Blevins—*Applied Fluid Dynamics Handbook,* VNR; Herbich—*Offshore Pipeline Design Elements,* Dekker; Zienkiew-icz—*Numerical Methods in Offshore Engineering,* Wiley; The Hydraulic Institute—*Standards of the Hydraulic Institute;* Allis-Chambers Manufacturing Company—*Pic-A-Pump;* Hicks and Edwards—*Pump Application Engineering,* McGraw-Hill; Stepanoff—*Centrifugal and Axial Flow Pumps,* Wiley; Karassik and Carter—*Centrifugal Pumps,* McGraw-Hill; Allen—*Using Centrifugal Pumps,* Oxford; Buffalo Pumps—*Centrifugal Pump Applications Manual;* Kristal and Annett—*Pumps,* McGraw-Hill; Economy Pumps, Inc.—*Pump Data;* Molloy—*Pumps and Pumping,* Chemical Publishing; Moore et al.—*The Vertical Pump,* Johnston Pump Company; Karassik—*Engineers' Guide to Centrifugal Pumps,* McGraw-Hill; Kovats and Desmur—*Pompes, Ventilateurs, Compresseurs,* Dunod, Paris; Fuchslocher and Schulz—*Die Pumpen,* Springer-Verlag, Berlin; Pfleiderer—*Die Kreiselpumpen,* Springer-Verlag, Berlin.

Piping and Fluid Flow

REFERENCES: Severud and Marr—*Elevated Temperature Piping Design,* ASME; Jeppson—*Analysis of Flow in Pipe Networks,* Butterworths/Ann Arbor Science Press; Sherwood and Whistance—*Piping Guide,* Syentek Books; Williams—*Pipelines and Permafrost,* Longmans; Watters—*Modern Analysis and Control of Unsteady Flow in Pipelines,* Butterworths/Ann Arbor Science Press; Lambert—*Pipeline Instrumentation and Controls Handbook,* Gulf Publishing; Marks—*Oceanic Pipeline Computations,* Penwell; Kentish—*Industrial Pipework,* McGraw-Hill (UK); Brebbia and Ferrante—*Computational Hydraulics,* Butterworths; King and Crocker—*Piping Handbook,* McGraw-Hill; ANSA—*Code for Pressure Piping* (commonly called the *Piping Code*); ASME—*Fluid Meters—Their Theory and Application;* King and Brater—*Handbook of Hydraulics,* McGraw-Hill; Ingersoll-Rand Company—*Cameron Hydraulic Data;* The Hydraulic Institute—*Standards of the Hydraulic Institute;* Baumeister and Marks—*Standard Handbook for Mechanical Engineers,* McGraw-Hill; Littleton—*Industrial Piping,* McGraw-Hill; The Hydraulic Institute—*Pipe Friction Manual;* Black Sivalls, and Bryson—*Valve Sizing Book* and *Cv Book;* Fluid Controls Institute—*Recommended Voluntary Standard Formulas for Sizing Control Valves;* Bell—*Petroleum Transportation Handbook,* McGraw-Hill; Perry—*Chemical Engineers' Handbook,* McGraw-Hill; Spielvogel—*Piping Stress Calculations Simplified,* Spielvogel Publishing; Grinnell Company, Inc.—*Piping Design and Engineering;* M. W. Kellogg Co.—*Design of Piping Systems,* Wiley; National Valve and Manufacturing Co.—*Piping Catalog;* McClain—*Fluid Flow in Pipes,* Industrial Press; Tube Turns Division of Chemetron Corp.—*Piping Engineering.*

Air and Gas Compressors and Vacuum Systems

REFERENCES: Hawthorne—*Aerodynamics of Turbines and Compressors,* Princeton University Press; Tramm and Dean—*Centrifugal Compressor and Pump Stability, Stall, and Surge,* ASME; *Chemical Engineering* Magazine—*Fluid Movers: Pumps, Compressors, Fans and Blowers,* McGraw-Hill; Martini—*Practical Seal Design,* Dekker; Cheremisinoff and Gupta—*Handbook of Fluids in Motion,* Butterworths; Van Atta—*Vacuum Science and Engineering,* McGraw-Hill; Dushman—*Scientific Foundations of Vacuum Technique,* Wiley; Guthrie and Wakerling—*Vacuum Equipment and Techniques,* McGraw-Hill; Yarwood—*High Vac-*

uum Techniques, Wiley; Lewin—*Vacuum Science and Technology,* McGraw-Hill; Pirani and Yarwood—*Principles of Vacuum Engineering,* Reinhold; Reimann—*Vacuum Technique,* Chapman and Hall; Steinherz—*Handbook of High Vacuum Engineering,* Reinhold; Compressed Air and Gas Institute—*Compressed Air and Gas Handbook;* Ingersoll-Rand Company—*Compressed Air Data.*

Materials Handling

REFERENCES: Apple—*Materials Handling System Design,* Wiley; Wasp—*Slurry Pipeline Transportation,* Trans Tech; Machinery Studies—*Materials Handling Equipment,* Business Trends; Bolz—*Materials Handling Handbook,* Wiley; Reisner and Eisenhart—*Bins and Bunkers for Handling Bulk Materials,* Trans Tech; Chemical Engineering Magazine—*Pneumatic Conveying of Bulk Materials,* McGraw-Hill; Hudson—*Conveyors,* Wiley; Buffalo Forge Company—*Fan Engineering;* Stanier—*Plant Engineering Handbook,* McGraw-Hill; Baumeister and Marks—*Standard Handbook for Mechanical Engineers,* McGraw-Hill.

Heat Transfer and Heat Exchangers

REFERENCES: Goldstein—*Heat Transfer in Energy Conservation,* ASME; Isachenko—*Heat Transfer,* Mir (Moscow); Karlekar and Desmond—*Engineering Heat Transfer,* West; Kays and Crawford—*Convective Heat and Mass Transfer,* McGraw-Hill; Butterworth and Hewitt—*Two Phase Flow and Heat Transfer,* Oxford University Press; French—*Heat Transfer and Fluid Flow in Nuclear Systems,* Pergamon Press; Frost—*Heat Transfer at Low Temperatures,* Plenum Press; McAdams—*Heat Transmission,* McGraw-Hill; Kern—*Process Heat Transfer,* McGraw-Hill; General Electric Company—*Electric Heaters and Heating Devices;* Jakob—*Heat Transfer,* Wiley; Bosworth—*Heat Transfer Phenomena,* Wiley; Kays and London—*Compact Heat Exchangers,* McGraw-Hill; Kraus—*Cooling Electronic Equipment,* Prentice-Hall; Fraas and Ozisik—*Heat Exchanger Design,* Wiley; Heat Transfer Research, Inc.—*Design Manual;* API Standards—*Heat Exchangers for General Refinery Service;* Giedt—*Principles of Engineering Heat Transfer,* Van Nostrand; Eckert and Drake—*Heat and Mass Transfer,* McGraw-Hill; Schnieder—*Conduction Heat Transfer,* Addison-Wesley; Kreith—*Principles of Heat Transfer,* International Textbook; Perry—*Chemical Engineers' Handbook,* McGraw-Hill; Carslaw and Jaeger—*Conduction of Heat in Solids,* Oxford; Wilkes—*Heat Insulation,* Wiley.

Refrigeration

REFERENCES: Trott—*Refrigeration and Air Conditioning,* McGraw-Hill; Hallowell—*Cold and Freezer Storage Manual,* AVI; Munton and Stott—*Refrigeration at Sea,* Applied Science (England); International Institute of Refrigeration—*Low Temperature and Electric Power,* Pergamon; Betts—*Refrigeration and Thermometry below One Kelvin,* Crane-Russak; Emerick—*Heating Design and Practice,* McGraw-Hill; Carrier Air Conditioning Company—*Handbook of Air Conditioning System Design,* McGraw-Hill; Severns and Fellows—*Air Conditioning and Refrigeration,* Wiley; ASHRAE—*Guide and Data Book: Fundamentals and Equipment;* ASHRAE—*Guide and Data Book: Applications;* Strock and Koral—*Handbook of Heating, Air Conditioning, and Ventilation,* Industrial Press; American Blower Corporation—*Air Conditioning and Engineering;* MacIntire-Hutchinson—*Refrigeration Engineering,* Wiley.

Part 3 ENVIRONMENTAL CONTROL

Wastewater Treatment, Stormwater Handling, and Water Supply

REFERENCES: McClelland and Evans—*Individual Onsite Wastewater Systems,* National Sanitation Foundation; Harbold—*Sanitary Engineering Problems and Calculations for Professional Engineers,* Ann Arbor Science; Feachem—*Water, Waste and Health in Hot Climates,*

Wiley; Nobel—*Sanitary Land Fill Design,* Technomic; Rich—*Environmental Systems Engineering,* McGraw-Hill; Kalbermatten—*Appropriate Sanitation Alternatives: A Technical and Economic Appraisal,* John Hopkins; *Sanitation Details in SI Metric,* International Ideas; Sawyer and McCarty—*Chemistry for Environmental Engineering,* McGraw-Hill; Fair—*Sewage Treatment,* Wiley; Steel—*Water Supply and Sewerage,* McGraw-Hill; Gurnham—*Principles of Industrial Waste Treatment,* Wiley; Babbitt, Dolan, and Cleasby—*Water Supply Engineering,* McGraw-Hill; Wright—*Rural Water Supply and Sanitation,* Wiley; Ehlers and Steel—*Municipal and Rural Sanitation,* McGraw-Hill; Babbitt and Baumann—*Sewerage and Sewage Treatment,* Wiley; Chow—*Handbook of Applied Hydrology,* McGraw-Hill; American Society of Civil Engineers—*Design and Construction of Sanitary and Storm Sewers;* King and Brater—*Handbook of Hydraulics,* McGraw-Hill; Woods—*Highway Engineering Handbook,* McGraw-Hill; Hicks—*Pump Application Engineering,* McGraw-Hill; Fair, Geyer, and Okun—*Water Supply and Wastewater Engineering,* Wiley; Besselievre—*Industrial Waste Treatment,* McGraw-Hill; Federation of Sewage and Industrial Wastes Association—*Chlorination of Sewage and Industrial Wastes;* American Society of Civil Engineers—*Sewage Treatment Plant Design;* Imhoff and Fair—*Sewage Treatment,* Wiley; American Society of Civil Engineers—*Filtering Materials for Sewage Treatment Plants;* Mahlie—*Manual for Sewage Plant Operators,* Texas Water & Sewage Works Association; Escritt—*Sewerage and Sewage Disposal,* Contractors Record, Ltd. (London); Escritt—*Pumping Station Equipment and Design,* C. R. Books, Ltd.

Plumbing and Drainage

REFERENCES: Church—*Practical Plumbing Design Guide,* McGraw-Hill; Ripka—*Plumbing Installation and Design,* American Technical Society; Galeno—*Plumbing Estimating Handbook,* Van Nostrand; Page and Nation—*Estimator's Piping Manhour Manual,* Gulf Publishing; Nielson—*Standard Plumbing Engineering Design,* McGraw-Hill; Blenderman—*Design of Plumbing and Drainage Systems,* Industrial Press; D'Arcangelo—*Mathematics for Plumbers and Pipefitters,* Delmar; Miller and Gallina—*Estimating and Cost Control in Plumbing Design,* Van Nostrand; Manas—*National Plumbing Code Handbook,* McGraw-Hill; Babbitt—*Plumbing,* McGraw-Hill; American Standards Institute—*National Plumbing Code;* National Bureau of Standards—*Water Distributing Systems for Buildings;* National Association of Plumbing Contractors—*Water Supply Piping for the Plumbing System;* Copper and Brass Research Association—*Brass Pipe Handbook for Plumbing Installation* and *Copper Tube Handbook on Plumbing and Heating.*

Heating, Ventilating, and Air Conditioning

REFERENCES: Edwards—*Automatic Controls for Heating and Air Conditioning,* McGraw-Hill; Bowyer—*Central Heating,* David and Charles (England); Kut—*Heating and Hot Water Services in Buildings,* Pergamon; Stoecker—*Using SI Units in Heating, Air Conditioning and Refrigeration,* Business News; Oliker—*Cogeneration District Heating Applications,* ASME; Rizzi—*Design and Estimating for Heating, Ventilating and Air Conditioning,* Van Nostrand Reinhold; Edwards—*Handbook of Geothermal Energy,* Gulf Publishing; Cheremisinoff—*Cooling Towers: Selection, Design and Practice,* Ann Arbor Science; Dubin and Long—*Energy Conservation Standards for Building Design, Construction and Operation,* McGraw-Hill; Stoecker and Jones—*Refrigeration and Air Conditioning,* McGraw-Hill; Carrier Air Conditioning Company—*Handbook of Air Conditioning System Design,* McGraw-Hill; American Society of Heating, Refrigerating, and Air Conditioning Engineers—*Guide and Data Book, Fundamentals and Equipment* and *Guide and Data Book, Applications;* Buffalo Forge Company—*Fan Engineering;* Strock and Koral—*Handbook of Heating, Air Conditioning, and Ventilation,* Industrial Press; Carrier, Cherne, Grant, and Roberts—*Modern Air Conditioning, Heating and Ventilating,* Pitman; Emerick—*Heating Design and Practice,* McGraw-Hill; Holmes—*Air Conditioning in Summer and Winter,* McGraw-Hill; The Trane Company—*Air-Conditioning Manual.*

Solar Energy

REFERENCES: ASHRAE—*Handbook of Fundamentals;* Balcomb, et al.—*Passive Solar Design Handbook, Vol. 2: Passive Solar Design Analysis,* DOE/CS-0127-2; Mazria—*The Passive Solar Energy Book,* Rodale Press; Strock and Koral—*Handbook of Air Conditioning, Heating and Ventilating,* Industrial Press; Kreider and Kreith—*Solar Heating and Cooling,* McGraw-Hill; Balcomb—*Passive Solar Design Handbook,* Solar Energy Information; Greeley—*Solar Heating and Cooling of Buildings,* Ann Arbor Science; Howell—*Thermal.*

Energy Conservation

REFERENCES: Hunt—*Windpower,* Van Nostrand Reinhold; Burberry—*Building for Energy Conservation,* Halsted Press; Chiogioji—*Industrial Energy Conservation,* Dekker; Courtney—*Energy Conservation in the Built Environment,* Longman; Culp—*Principles of Energy Conversion,* McGraw-Hill; Dorf—*The Energy Factbook,* McGraw-Hill; Dubin—*Energy Conservation Standards,* McGraw-Hill; *Energy Conservation in the International Energy Agency,* OECD; Grant—*Energy Conservation in the Chemical & Process Industries,* Institute of Chemical Engineers, England; Helcke—*The Energy Saving Guide,* Commission of the European Communities; Jarmul—*The Architect's Guide to Energy Conservation,* McGraw-Hill; Kovah—*Thermal Energy Storage,* Pergamon; Meckler—*Energy Conservation in Buildings & Industrial Plants,* McGraw-Hill; Payne—*Energy Managers' Handbook,* Butterworth; Pindyck—*The Structure of the World Energy Demands,* M.I.T. Press; Reay—*Industrial Energy Conservation,* Pergamon; Smith—*Industrial Energy Management for Cost Reduction,* Ann Arbor Science; Yaverbaum–*Energy Saving by Increasing Boiler Efficiency,* Noyes; Considine—*Energy Technology Handbook,* McGraw-Hill.

Controls in Environmental and Energy-Conservation Design

REFERENCES: Morris—*Control Engineering,* McGraw-Hill; Kuo—*Automatic Control Systems,* Prentice-Hall; Schwartz—*Multivariable Technical Control Systems,* Elsevier North-Holland; Berkovitz—*Optimal Control Theory,* Springer-Verlag; Bryson and Ho—*Applied Optimal Control,* Hemisphere; Craven—*Mathematical Programming & Control Theory,* Chapman & Hall; Fallside—*Control System Design by Pole-Zero Assignment,* Cambridge University Press; Hafer—*Electronics Engineering for Professional Engineers' Examinations,* McGraw-Hill; Considine—*Process Instruments and Controls Handbook,* McGraw-Hill; Merritt—*Hydraulic Control Systems,* Wiley; Considine and Ross—*Handbook of Applied Instrumentation,* McGraw-Hill; Eckman—*Automatic Process Control,* Wiley; Kallen—*Handbook of Instrumentation and Controls,* McGraw-Hill; Farrington—*Fundamentals of Automatic Control,* Wiley; ASME—*Fluid Meters;* Shinskey—*Process-Control Systems,* McGraw-Hill; Graham-McRuer—*Analysis of Nonlinear Control Systems,* Wiley; Harriott—*Process Control,* McGraw-Hill; Mesarovic—*The Control of Multivariable Systems,* Wiley; Savas—*Computer Control of Industrial Processes,* McGraw-Hill; Newton-Gould-Kaiser—*Analytical Design of Linear Feedback Controls,* Wiley; Burr-Brown Research Corp.—*Handbook of Operational Amplifier Active RC Networks:* Bower-Schultheiss—*Introduction to Design of Servomechanisms,* Wiley; Gibson-Tuteur—*Control System Components,* McGraw-Hill; Bode—*Network Analysis and Feedback Amplifier Design,* D. Van Nostrand; Doss—*Information Processing Equipment,* Reinhold; ASME—*Flowmeter Computation Handbook;* ASME—*Flow Measurement;* Dommasch and Laudeman—*Principles Underlying Systems Engineering,* Pitman.

Environmental Data Studies

REFERENCES: McGraw-Hill *Encyclopedia of Environmental Science and Engineering,* McGraw-Hill; Lesage and Jackson—*Groundwater Contamination and Analysis at Hazardous Waste Sites,* Marcel Dekker; Corbitt—*Standard Handbook of Environmental Engineering,* McGraw-Hill; Woodslide—*Hazardous Materials and Hazardous Waste Management,* Wiley;

Fthenakis—*Prevention and Control of Accidental Releases of Hazardous Gases,* Van Nostrand Reinhold; LaGrega, Buckingham and Evans—*Hazardous Waste Management,* McGraw-Hill; Cheremisinoff—*Air Pollution Control and Design for Industry,* Marcel Dekker; Freeman—*Standard Handbook of Hazardous Waste Treatment and Disposal,* McGraw-Hill; Office of Technology Assessment—*Green Products by Design—Choices for a Cleaner Environment,* U.S. Government Printing Office; Arbuckle—*Environmental Law Handbook,* Government Institute, Inc.; Lund—*The McGraw-Hill Recycling Handbook,* McGraw-Hill; Holmes—*Refuse Recycling and Recovery,* Wiley; Gabor—*Beyond the Age of Waste—A Report to the Club of Rome,* Pergamon; Porteous—*Refuse Derived Fuels,* Halsted; Council on Environmental Quality—*Environmental Quality,* U.S. Government Printing Office; Jain et al.—*Environmental Assessment,* McGraw-Hill; Morgan—*Renewable Resource Utilization for Development,* Pergamon; White and Plaskett—*Biomass as Fuel,* Academic; Herwan and Boyce—*Gas Turbine Engineering Handbook,* Gulf; Wetzel and Murphy—*Treating Industrial-Waste Interferences at Publicly-Owned Treatment Works,* Noyes Data; Beranek and Ver—*Noise and Vibration Control Engineering,* Wiley; Neporozhny—*Thermal Power Plants and Environmental Control,* MIR Publishing; Payne—*Cogeneration Sourcebook,* Fairmont Press; Fortuna and Lennett—*Hazardous Waste Regulation—The New Era,* McGraw-Hill; Hu—*Cogeneration,* Reston; McDermott—*Handbook of Ventilation for Contaminant Control,* Butterworths; Goldstick and Thumann—*Principles of Waste Heat Recovery,* Fairmont Press; Vutukuri and Lama—*Environmental Engineering in Mines,* Cambridge University Press; Hallenbeck and Cunningham—*Quantitative Risk Assessment for Environmental and Occupational Health,* Lewis Publishers; Nejat and Veziroglu—*Alternative Energy Sources,* Hemisphere Publishing; Spiewak—*Cogeneration and Small Power Production,* Fairmont Press; Plunkett—*Handbook of Industrial Toxicology,* Chemical Publishing; Marine Board et al.—*Dredging Coastal Ports,* National Academy Press; Thumann and Miller—*Fundamentals of Noise Control Engineering,* Prentice Hall.

Statistical Analyses and Economic Studies

REFERENCES: Kurtz—*Handbook of Engineering Economics,* McGraw-Hill; Barish and Kaplan—*Economic Analysis for Engineering and Managerial Decision Making,* McGraw-Hill; DeGarmo et al.—*Engineering Economy,* Macmillan; Grant and Leavenworth—*Principles of Engineering Economy,* Ronald Press; Kasmer—*Essentials of Engineering Economics,* McGraw-Hill; Smith—*Engineering Economy,* Iowa State University Press; Cissell—*Mathematics of Finance,* Houghton Mifflin; Clifton and Fyffe—*Project Feasibility Analysis,* Wiley; Sullvian and Claycombe—*Fundamentals of Forecasting,* Reston; Weston and Brigham—*Essentials of Managerial Finance,* Dryden Press; Lock—*Engineer's Handbook of Management Techniques,* Grove Press (London, England); Jelen—*Project and Cost Engineers' Handbook,* American Association of Cost Engineers; Kharbanda—*Process Plant and Equipment Cost Estimation,* Vivek Enterprises (Bombay, India); Johnson and Peters—*A Computer Program for Calculating Capital and Operating Costs,* Bureau of Mines Information Circular 8426, U.S. Department of Interior; Ostwald—*Cost Estimation for Engineering and Management,* Prentice-Hall; American Association of Cost Engineers—*Cost Engineers' Notebook;* Gass—*Linear Programming: Methods and Applications,* McGraw-Hill; Hadley—*Linear Programming,* Addison-Wesley; Bellman and Dreyfus—*Applied Dynamic Programming,* Princeton University Press; Hadley—*Nonlinear and Dynamic Programming;* Addison-Wesley; Allen—*Probability and Statistics, and Queuing Theory,* Academic Press; Gross and Harris—*Fundamentals of Queuing Theory,* Wiley; Beightler—*Foundations of Optimization,* Prentice-Hall; Blum and Rosenblatt—*Probability and Statistics,* W. B. Saunders; Brownlee—*Statistical Theory and Methodology in Science and Engineering,* Wiley; Quinn—*Probability and Statistics,* Harper & Row; Newnan—*Engineering Economic Analysis,* Engineering Press; Park—*Cost Engineering,* Wiley; Taylor—*Managerial and Engineering Economy,* VNR; Mishan—*Cost-Benefit Analysis,* Praeger; Jelen and Black—*Cost and Optimization Engineering,* McGraw-Hill; White et al.—*Principles of Engineering Economic Analysis,* Wiley; Riggs—*Engineering Economics,* McGraw-Hill, Guenther—*Concepts of Statistical Inference,* McGraw-Hill; Lindgren—*Statistical Theory,* Macmillan; Meyer—*Introductory Probability and Statistical Applications,*

Addison-Wesley; Renwick—*Introduction to Investments and Finance*, Macmillan; O'Brien—*CPM in Construction Management*, McGraw-Hill; Gupta and Cozzolino—*Fundamentals of Operations Research for Management*, Holden-Day.

Part 4 DESIGN ENGINEERING

Machine Design and Analysis

REFERENCES: Deutschman—*Machine Design: Theory and Practice*, Macmillan; Johnson—*Mechanical Design Synthesis*, Kreiger; Stephenson and Collander—*Engineering Design*, Wiley-Interscience; Creamer—*Machine Design*, Addison-Wesley; Artobolevskii—*Mechanisms in Modern Engineering Design*, MIR Publishers (Moscow); Sandor and Erdman—*Advanced Mechanism Design: Analysis and Synthesis*, Prentice-Hall; Dhillon—*Reliability Engineering in Systems Design and Operation*, VNR; Chakraborty and Dhande—*Kinematics and Geometry of Planar and Spatial Cam Mechanisms*, Wiley: Lynwander—*Gear Drive Systems: Design and Application*, Dekker; Schwartz—*Composite Materials Handbook*, McGraw-Hill; Taraman—*CAD/CAM: Meeting Today's Productivity Challenge*, SME; Shtipelman—*Design and Manufacture of Hypoid Gears*, Wiley: Roark—*Formulas for Stress and Strain*, McGraw-Hill; Church—*Mechanical Vibrations*, Wiley; *Machinery's Handbook*, Industrial Press; Johnson—*Optimum Design of Mechanical Elements*, Wiley; Slaymaker—*Mechanical Design Analysis*, Wiley; Chironis—*Spring Design and Application*, McGraw-Hill; Spotts—*Design of Machine Elements*, Prentice-Hall; AGMA *Standards Books*, American Gear Manufacturers Association; Doughtie and Vallance—*Design of Machine Members*, McGraw-Hill; Buckingham—*Manual of Gear Design*, Industrial Press; Fuller—*Theory and Practice of Lubrication for Engineers*, Wiley; Dudley—*Gear Handbook*, McGraw-Hill; Churchman—*Prediction and Optimal Decision*, Wiley; Crandall—*Engineering Analysis*, McGraw-Hill; Ver Planck and Téare—*Engineering Analysis*, Wiley; Wahl—*Mechanical Springs*, McGraw-Hill; Haberman—*Engineering Systems Analysis*, Merrill; Shigley—*Mechanical Engineering Design*, McGraw-Hill; Ryder—*Creative Engineering Analysis*, Prentice-Hall; Baumeister and Marks—*Standard Handbook for Mechanical Engineers*, McGraw-Hill; Church—*Kinematics of Machines*, Wiley: Carmichael—*Kent's Mechanical Engineers' Handbook*, Wiley; Faires—*Design of Machine Elements*, Macmillan; Black—*Machine Design*, McGraw-Hill; Maleev—*Machine Design*, International; Bradford and Eaton—*Machine Design*, Wiley; Dudley—*Practical Gear Design*, McGraw-Hill; Shigley—*Simulation of Mechanical Systems*, McGraw-Hill.

Metalworking

REFERENCES: Blazynski—*Metal Forming: Tool Profiles and Flow*, Halsted Press; Lippmann—*Engineering Plasticity: Theory of Metal Forming Processes*, Springer-Verlag; Ross—*Handbook of Metal Treatments and Testing*, Tavistock (England); Le Grand—*American Machinist's Handbook*, McGraw-Hill; Boston—*Metal Processing*, Wiley; Nordhoff—*Machine-Shop Estimating*, McGraw-Hill; *Machinery's Handbook*, Industrial Press; *Welding Handbook*, American Welding Society; ASTME—*Tool Engineer's Handbook*, McGraw-Hill; *Procedure Handbook of Arc Welding Design and Practice*, The Lincoln Electric Company; Black—*Theory of Metal Cutting*, McGraw-Hill; Doyle—*Manufacturing Processes and Materials for Engineers*, Prentice-Hall; Brierly and Siekmann—*Machining Principles and Cost Control*, McGraw-Hill; Reason—*The Measurement of Surface Texture*, Cleaver-Hume; Bolz—*Production Processes: Their Influence on Design*, Penton; Harris—*A Handbook of Woodcutting*, HMSO, London; *Application Data, Cemented Carbides, Cemented Oxides*, Metallurgical Products Department, General Electric Company; Wood—*Final Report on Advanced Theoretical Formability Manufacturing Technology*, LTV, Inc., and USAF; Maynard—*Handbook of Business Administration*, McGraw-Hill; Niedzwiedzki—*Manual of Machinability and Tool Evaluation*, Huebner Publications, Cleveland; Hendriksen—*Chipbreakers*, The National Machine Tool Builders Association; ASTME—*Fundamentals of Tool Design*, Prentice-

Hall; Crane—*Plastic Working of Metals and Power Press Operations,* Wiley; Jones—*Die Design and Die Making Practice,* Industrial Press; DeGarmo—*Materials and Processes in Manufacturing,* Macmillan; Jevons—*The Metallurgy of Deep Drawing and Pressing,* Wiley; Stanley—*Punches and Dies,* McGraw-Hill.

INDEX